D0753456

Biomechanics of Impact Injury and Injury Tolerances of the Head-Neck Complex

PT-43

Edited by

Stanley H. Backaitis

Published by:
Society of Automotive Engineers, Inc.
400 Commonwealth Drive
Warrendale, PA 15096-0001
U.S.A.
Phone: (412) 776-4841
Fax: (412) 776-5760

ISBN 1-56091-363-0
SAE/PT-93/43

Library of Congress Catalog Card Number 92-063368

TABLE OF CONTENTS

Section 1: Head-Neck Anthropometry, Physiology and Morphology

Section 2: Head-Neck Kinematics and Impact Response

Section 3: Head-Neck Impact Tolerance and Injury Criteria

Section 4: Brain Injury

Section 5: Face-Skull Injury

Section 6: Neck Injury

Section 7: Head-Neck Structural Properties and Modeling

PREFACE

This compendium, *Biomechanics of Impact Injury and Injury Tolerances of the Head-Neck Complex* (PT-43), is the first of four books that SAE plans to publish on the technical aspects of trauma to the human body in the automotive crash environment. PT-43 contains papers that provide up-to-date references on the fundamentals of human head-neck anatomy and the biomechanics of impact injury and injury tolerances of these two body segments.

This 43rd book in the SAE "Progress in Technology" series includes previously published technical papers that were selected by a peer review panel from the SAE Global Mobility Database and several other books and publications that contained new or additional information. Several papers were brought to the editor's attention by members of the review panel who are active and prominent in head-neck injury research.

The contents selected for this compendium underwent a two-stage evaluation process. The reviews were performed by an invited peer review panel made up of nationally and internationally distinguished scientists and researchers in the field of impact injury to the head-neck complex. The first stage of selections consisted of evaluations and nominations from a pre-selected list of 304 SAE and 29 non-SAE papers that, in the judgment of the reviewers, met the criteria of technical integrity, scholarship, uniqueness of biomechanical procedures, long-term value, and historical perspective. The results from the first review process were tabulated by typical category, and any paper that received four or more nominations became a candidate for the second review cycle. This process yielded 138 SAE papers and 21 non-SAE papers.

For the second evaluation, the reviewers were asked to reevaluate the reduced list with a particular emphasis on uniqueness and outstanding contribution to the state-of-the-art, avoiding duplication of subject matter, assuring that papers fit into topical categories and were of superior quality and usefulness to all needing to use this information. The final review yielded 68 SAE papers for full reprinting in the text and 44 SAE papers for listing by author, title and abstract in the Bibliography (Appendix 1). The remaining papers are listed by author and title in the Related Reading section (Appendix 2). The final selections were based on those nominations which received at least six outstanding nominations and had a cumulative score of 12 of outstanding and desirable quality. A paper that received an outstanding score between 3 to 5 and a cumulative score of outstanding and desirable quality of at least 8 but below 12 was selected for listing in the Bibliography.

While the editor set out to select and reprint the best published articles on head-neck impact injury biomechanics and impact tolerances in automotive type accidents in PT-43, this in no way diminishes the importance of other similar articles that were either not considered or not selected during this review process. As in most publications, the inclusion of articles is based on rather arbitrary cut-off points based on the number of votes received, their relevance to the state-of-the-art, and their fit within a particular topical category. The reader of the book may also note that papers within a particular category may address both the head and neck topics. Their placement in one or another category reflects more or less the emphasis that the original author placed in writing the paper.

The review panel consisted of the following participants:

Professor Dr. Voigt Hodgson, Director
School of Medicine, Neurosurgery
Wayne State University
550 E. Canfield, Room 116
Detroit, MI 48201

Priv. Doz. Dr. rer. nat. Dimitrios Kallieris
Institute Fuer Rechtsmedizin
Klinikum Der Universitaet Heidelberg
PostFach 10 30 69
6900 Heidelberg, Germany

Mr. Jeffrey Marcus
Manager, Protection and Survival Laboratory
Civil Aeromedical Institute
Mike Monroney Aeronautical Center, FAA
P.O. Box 25082
Oklahoma City, Oklahoma 73125

Dr. John Melvin
Sr. Staff Research Engineer
Biomedical Science Department
GMRL
GM Technical Center
Warren, MI 48090

Dr. Harold Mertz
Staff Engineer
W3-SCS Engineer
Current Product Engineering
GM Technical Center
Warren, MI 48090

Allan Nahum, M.D.
Professor Emeritus of Surgery
School of Medicine
University of California at San Diego
8494 El Paseo Grande
La Jolla, CA 92037

Dr. James Newman
President
Biokinetics and Associates, Ltd.
2470 Don Reid Drive
Ohawa, Ontario KIH BP5

The editor is greatly indebted to all members of the review panel for their excellent cooperation, timely responses, and constructive participation in this very time-consuming selection process. The reviewers' voluntary contribution in this very extensive effort reflects their professional dedication and their desire to reduce the causes of injuries that occur in automotive collisions.

It is hoped that the final selection of articles in this publication reflects also the best judgments of each reviewer in spite of some compromises that had to be made to reach consensus. The editor is grateful to Mr. Barry Felrice, Associate Administrator, Rulemaking, and Dr. Patricia Breslin, Office of Vehicle Safety Standards, National Highway Safety Administration (NHTSA), for their support and encouragement in carrying out this project.

Sincere thanks for excellent cooperation and assistance are due to a number of other people who helped to bring this compendium to reality, particularly to Miss Paula Orr, a co-op student from the Virginia Polytechnic Institute for her help in processing some of the responses, Miss Carlita Ballard for the secretarial support, both of NHTSA, and to the SAE Publications Group staff in Warrendale for their enthusiastic assistance and support in getting this publication together.

Stanley H. Backaitis
National Highway Traffic Safety Administration
Washington, D.C.

Chairman, SAE Occupant Protection Committee

Section 1:
Head-Neck Anthropometry, Physiology and Morphology

700195

Synopsis of Anatomy and Medical Terminology

Donald F. Huelke
University of Michigan

THIS PAPER by no means represents a complete presentation of gross anatomy as would be given to medical students. It is merely intended to present some very elementary anatomical principles in a simplified form for the beginner or semiprofessional interested in automobile collision research and the production of occupant injury. Anatomical details and injuries in various body areas will be described in the other chapters. This paper includes, in Appendix A, reference books to which the reader may go for further information and more detailed study. Also listed in the appendix is a series of anatomical and medical related terms, many of which must become a part of the engineers' terminology so that communication with physicians and nurses will be enhanced, and medical records more understandable.

Anatomy is the study of human body structure. The field of anatomy is conveniently subdivided into gross anatomy and microscopic anatomy. A special area of study of the nervous system is called neuroanatomy. The study of the development of the human body before birth is called embryology. In each area are subdivisions which, for our purposes, need not be detailed.

In order for us to understand various body parts, a standardized orientation of the body—the anatomical position—is used. This is a standing individual facing forward with the palms of the hands outward (Fig. 1). From this orientation then, various directions can be more clearly related (Table 1). That portion of the body which is closer to the head or above another part is said to be cranial (or superior). Conversely that which is closer to the feet would be caudal (or inferior). The front of the body is referred to as the anterior or ventral side, and the back of the body then is the posterior or dorsal side. Especially in the limbs there are two other terms used to reference the relative position of one structure to another. A structure or part that is closer to the attachment of the limb to the torso, would be proximal; any part farther away from the root of the limb than some other structure is distal. For example, the shoulder joint is proximal to the elbow joint, and conversely the wrist is distal to the elbow. Thus, in hospital descriptions you may see, for example, that the distal end of the radius is fractured. The radius, one of the bones of the forearm, has a fracture that is closer to the wrist than to the upper end of the bone which is at the elbow joint. Similarly in the lower limb, the knee joint is proximal to the foot.

Table 1 - Positional and Descriptive Terminology

Anterior - or ventral - Toward the front of the body.
Posterior - or dorsal - Toward the back of the body.
Medial - Nearer the median plane than some other part.
Lateral - Toward the side, or farther from the median plane.
Internal - Deeper to some other part.
External - More superficial than some other part.
Superior - or Cranial - Toward the head, or closer to the head than some other structure.
Inferior - or caudal - Toward the tail, or farther from the head than some other structure.
Central - In the center of the body, or nearer the center than some other part.
Peripheral - The surface of the body, or farther from the center than some other part.
Proximal - Nearer the root of the limb than some other part.
Distal - Farther from the root of the limb than some other part.

ANATOMICAL TERMINOLOGY

The terms used in anatomy and medicine in general all have specific meanings and have been derived from Latin and Greek

Planes Of Direction

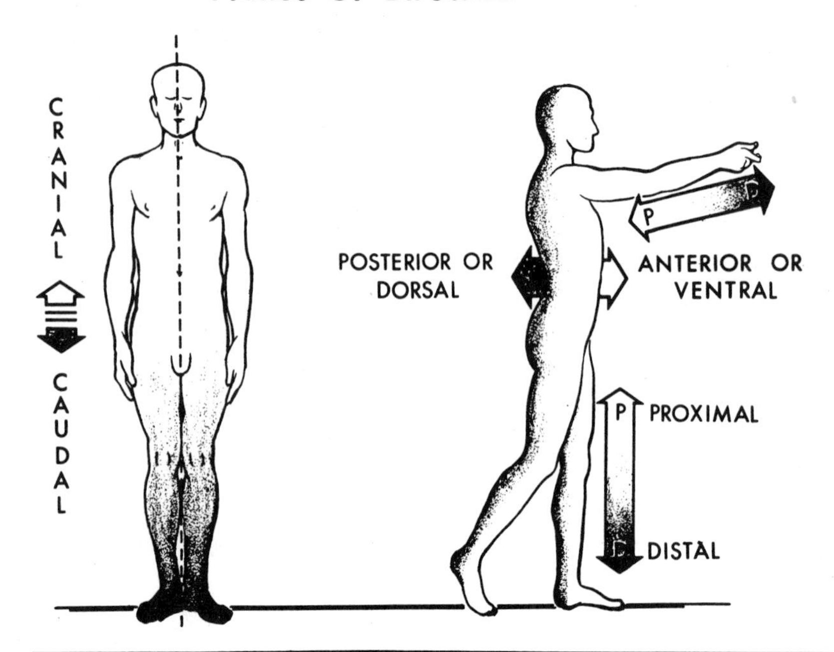

Fig. 1 - Planes of direction

origins. It is highly recommended that a medical dictionary be within arms reach of anyone learning this new terminology —the language of the healing arts. In addition, the dictionary will aid in the proper pronounciation of the word. Nothing is more annoying to a physician or health scientist than to hear medical or anatomical terms mispronounced! In Tables 1 and 2 are presented the beginnings of this new medical language.

Anatomical and medical terms are frequently formed from several subterms. The familiar "biceps brachii" muscle is in the arm—between the shoulder and elbow. Actually, its name tells one where it is located—brachii (the arm); the word biceps indicates a two (bi-) headed (-ceps) muscle. Thus, triceps brachii is the three-headed muscle in the arm. Arteries, for example, frequently take the name of the area in which they are located (brachial artery, facial artery, etc.) or their area of supply (coronary artery, gastric, splenic, femoral, etc.) Frequently, structures have even more descriptive names; the external abdominal oblique is that muscle of the abdominal wall which has fibers running obliquely and is the outermost muscle in the area. Its name also indicates that there must be an internal abdominal oblique muscle, for if there would be only one abdominal oblique muscle, that would be its name, for anatomists do not use more words than are necessary in naming body parts. One last example: the longest named muscle in the body is the levator labii superioris dilator alaequae nasi. This small muscle is found on the side of the nose. Its name tells one that it lifts (levator) the upper lip (labii superioris) and dilates the nose (dilator alaequae nasi). Thus anatomical terminology has a rational basis; but to understand and to use such terms, one must know the body regions and related structures which often form the basis for the name of a specific structure.

REGIONS OF THE BODY

The body is divisible into many parts and regions, but for ease in classification seven areas will be recognized: head, neck, thorax, abdomen, pelvis, upper and lower extremities. Specific subregions must also be studied for many of the structures in a specific area take on the name of the region in which they are found (Table 2).

The head is that area above the chin including the skull and surrounding soft tissue. Most importantly, the skull encloses the brain, and two separate subdivisions of the head are recognized—the face and cranium (Fig. 2). Many of the external environmental stimuli are received in the facial area; food and water are taken in through the mouth and air through the mouth or nose. The special sense organs for taste, smell, and vision are also found in the facial area. The cranium encloses the brain and the organs of hearing and equilibrium.

The neck region consists of the throat structures in front and the cervical portion of the vertebral column and associated muscle mass behind. On each side of the neck, covered by muscles, are the large vessels that traverse the neck to supply structures of the head and the brain. The lower extent of the neck is approximately at the level of the clavicles (collar bones).

The thorax can be considered as the area outlined by the ribs. Posteriorly, the ribs articulate with the vertebral column and, anteriorly, with the sternum (breast bone). Below, the thorax is closed by respiratory diaphragm, which forms the roof of the abdominal cavity. The thorax houses the lungs, heart, and great vessels, and the lower portion of the trachea (windpipe).

Extending from the diaphragm above, the abdominopelvic cavity is limited by the vertebral column posteriorly, the soft

Table 2 - Anatomical Terminology of Body Regions

Region	The Region of the:
Antibrachial	forearm
Axilla	armpit
Brachial	shoulder to elbow (arm)
Cervical	neck
Cubital	elbow
Epigastric	above the stomach
Femoral	thigh
Hypochondriac	under the rib cartilages
Hypogastric	below the stomach
Iliac (inguinal)	over the hips
Lumbar	small of back
Pectoral	anterior chest
Popliteal	back of knee
Thoracic	chest
Umbilical	navel

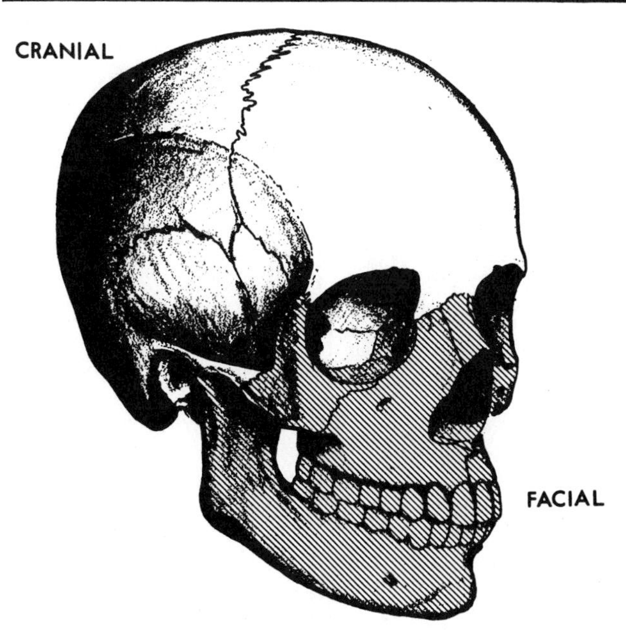

Fig. 2 - Units of the skull

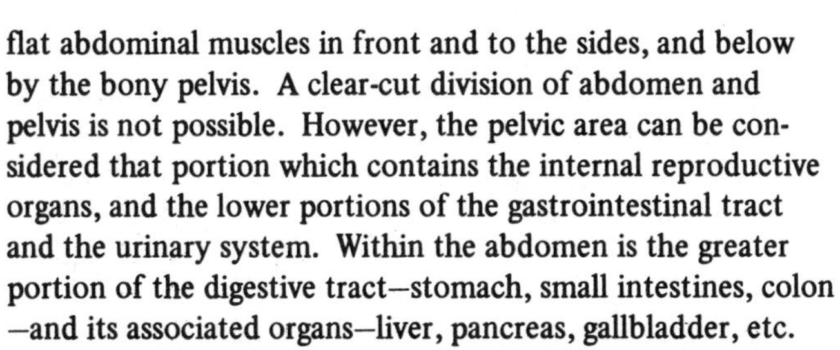

flat abdominal muscles in front and to the sides, and below by the bony pelvis. A clear-cut division of abdomen and pelvis is not possible. However, the pelvic area can be considered that portion which contains the internal reproductive organs, and the lower portions of the gastrointestinal tract and the urinary system. Within the abdomen is the greater portion of the digestive tract–stomach, small intestines, colon –and its associated organs–liver, pancreas, gallbladder, etc.

The upper extremity includes the shoulder girdle, consisting of the scapula (shoulder blade) and clavicle (collar bone). Extending from these structures are the subregions of the upper extremity: the arm, forearm, wrist, and hand. Each area is interconnected by freely movable articulations (joints). Similarly, the lower extremity is attached to the pelvic girdle and is divided into the thigh, leg, ankle, and foot.

It is very important that correct terminology be used for these various named regions. Thus, the thigh is that area between the hip and knee joints; it is not the upper leg. The upper leg is that area just below the knee. Likewise, in the upper extremity the terms are "arm and forearm," not the "upper arm" and "lower arm."

THE SKELETON

The skeleton is the structural framework of the body. It consists of an internal set of bones which in the complete adult skeleton number 206. For convenience, the skeleton is divided into the axial and the appendicular skeletons. The axial skeleton consists of the skull, the hyoid (a small bone underneath the chin), the vertebral column, the ribs, and the sternum (breast bone) (Fig. 3). The appendicular skeleton is composed of the upper and lower limbs. Many of the bones of the skeleton–especially those of the upper and lower limbs–have a length greater than their breadth. These are the long bones of the body. Even though some of these bones may be quite short, such as those in the fingers and toes, they are still considered long bones for their length is greater than their breadth. On the other hand, short bones are those which are small cuboidal or rectangular in character; they are typified by the bones in the wrist and ankle. Flat bones are those of the skull which enclose the brain. In addition, the sternum (breast bone), the ribs, and the scapulae (shoulder blades) are also flat bones. Flat bones are characterized by having thin plates of compact bone separated by thin sponge-like bony spaces.

Long bones consist of three parts, a shaft, or diaphysis, and two ends, the epiphyses. Quite a few of the long bones have a smooth and more or less spherical end which is called the "head," beneath which there is generally a constriction called the "neck." The other end of some of the bones, as the humerus and femur, have smooth and rounded eminences called "condyles," while others, as the metacarpals (hand bones) and metatarsals (foot bones) have squarish rough ends called "bases."

Long bones are not solid, but have a cavity in the shaft which is called the "marrow cavity." The epiphyses are not hollow but are porous and spongy, and also contain marrow.

The shaft of long bones is hard and dense, and is called "compact bone," while the porous epiphyses of long bones, the bodies of vertebrae, and the inside of flat bones are made up of spongy bone.

In Table 3 is presented the distribution of the various bones in the body. It is strongly recommended that the parts of the major bones of the body be well learned.

The skull, in particular, is a unit of the body which is made up of 29 bones. Almost all of these bones are well joined together through immovable joints called "sutures." The only movable bone of the skull is the mandible (lower jaw). That

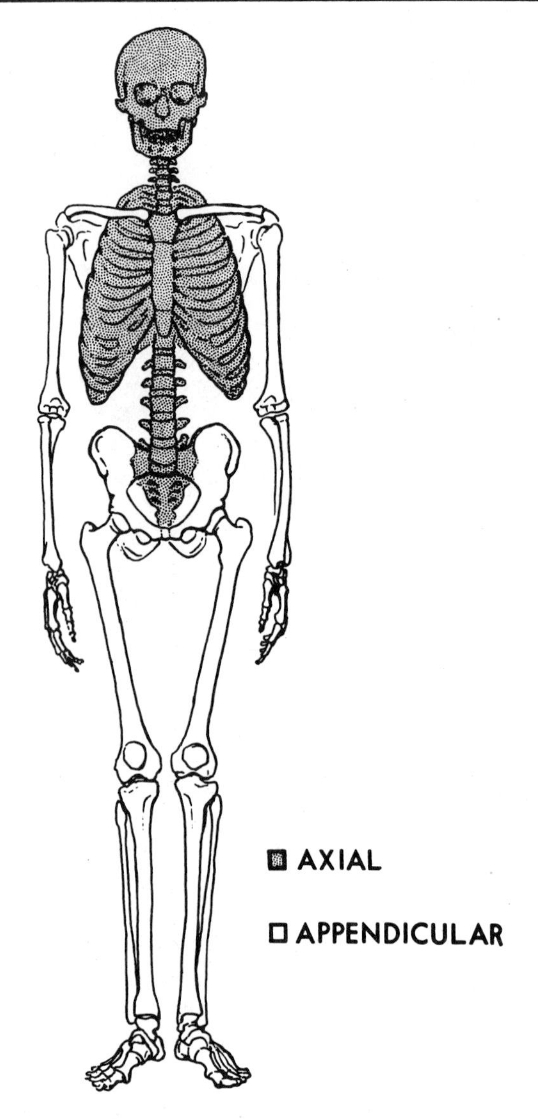

Fig. 3 - Axial and appendicular skeleton

part of the skull which encloses the brain is called the "cranium" and consists of eight individual flat bones. The face consists of 14 bones (Fig. 2). The jaw bones are those portions of the skull which support the teeth. The lower jaw is named the mandible and the upper jaw, the maxilla.

The names of the various portions of the extremities are important for knowledgeable conversation with those in the medical sciences. The shoulder girdle is that portion of the upper extremity which includes the scapula (shoulder blade) and clavicle (collar bone), (Fig. 3). These two interconnect at the shoulder joint with the bone of the arm, the humerus. The elbow joint is made up of the distal end of the humerus and the proximal ends of the radius and ulna. The area of the body between the elbow joint and wrist is the forearm. The wrist is a group of articulations (joints) between the forearm and the hand.

In the lower extremity the pelvic girdle articulates with the femur at the hip joint. The femur is the bone of the thigh; it joins with the leg at the knee joint. Within the leg are two

Table 3 - The Number of Skeletal Elements

Skull		22
Cranium	8	
Face	14	
Ear bones		6
Hyoid		1
Thorax		25
Ribs	24	
Sternum	1	
Vertebral column		26 (in adult)
Upper extremities		64
Lower extremities		62
Total		206

bones—the larger one, the tibia, and the small thin bone on the outside called the fibula. These articulate with the foot bones through the ankle joint.

THE JOINTS

Wherever one bone joins with another, there always is a joint between the two; not always are these joints movable. Typically, movable joints are surrounded by ligaments which are strong and tough, but yet pliable; as a group they have very little elasticity to them. At the ends of the bones in a movable joint there is a covering of cartilage and within the enclosed joint space a bit of lubricant called synovial fluid. This allows for the smooth movement at the joint area. The joints are generally classified according to their degree of movement.

JOINT CLASSIFICATION BASED ON THE DEGREE OF MOVEMENT

I. NONMOVABLE (Synarthroses) - joints where no movement is possible—the sutures of the skull.

II. MODERATELY MOVABLE (Amphiarthroses) - joints where only slight movement is possible.

A - Symphysis—symphysis pubis, sacroiliac joint, and the intervertebral joints.

B - Syndesmosis—the distal tibiofibular joint, articulation between the shafts of the ulna and radius, coracoclavicular joint, etc.

III. FREELY MOVABLE (Diarthroses) - joints permitting free movement in various directions.

A - Hinge—elbow (humerus-ulna), knee (femur-tibia), and the interphalangeal joints.

B - Pivot—atlanto-axial and proximal radioulnar joints.

C - Ball and socket—the shoulder and hip joints and the humeroradial articulation.

D - Gliding—the joints of the vertebral column, intercarpal joints of the wrist, intertarsal joints of the ankle, sternoclavicular joint.

E - Saddle—the joint between the greater multiangular and the first metacarpal bones.

F - Condyloid—metacarpophalangeal, metatarsophalangeal joints, etc.

G - Ellipsoidal—Temporamandibular and wrist joints.

A synarthrosis joint is one in which there is no movement possible typified by the sutures of the skull. An amphiarthrosis joint is one in which there is slight movement whereas a diarthrosis is a joint that has free movement.

The strength of a joint is determined by several factors: its bony structure, the ligaments that span the joint, the muscles and tendons that also cross over the joint, the tightness of fit between the articulating surfaces of the bones as well as the skin and connective tissue (fascia), muscles, and other soft-tissue structures.

All the joints of the body do not have the same degree of strength or stability. Some, as the hip joint, sacroiliac, and intervertebral joints, are fairly stable, while the shoulder, knee, and ankle joints are less stable and, as a result, are more easily injured. The strength and degree of movement of joints varies. In the shoulder joint, stability is sacrificed for movement, while in the hip joint or intervertebral joints, movement is sacrificed for stability.

THE MUSCLES

Myology is a study of muscles and their properties. Muscles form almost half of the body weight and, although they appear bulky and solid, they are approximately 75% water.

It is almost impossible to make a definite statement as to the exact number of muscles in any one human body. The reasons for this are: (1) Some individuals have extra muscles that do not usually occur. (2) Some muscles can be considered either as a separate muscle, or as a part of a larger muscle. (3) Some muscles may be absent on one or both sides of the body. It is safe to say, however, that there are over 600 main muscles in the average human body, 240 of which have different names. The difference in the two figures is due to the fact that many of the muscles occur in two or more pairs (Fig. 4).

Fig. 4 - Superficial muscles of the body

Muscles vary in size, shape, weight, structure, and manner of attachment. Some muscles, as the latissimus dorsi, are large, while others, as the rotatores, are very small. Muscles in the forearm or the back of the thigh are long and narrow, while others between the ribs, the intercostal muscles, are short and wide; the gluteus maximus of the buttock is thick and heavy, the eyelid muscles are thin and delicate. There are muscles, such as the quadratus femoris, that are square in shape, others, as the pronator teres, are round, while still others are triangular, rhomboid, or trapezoid in shape. Muscles also vary in the direction they run; the rectus femoris runs vertically on the front of the thigh, the external abdominal oblique extends obliquely across the abdomen, and the transverse abdominus and muscles run horizontally.

The names of muscles are quite descriptive and are derived as follows:

1. *Action Produced* - Adduction (bringing toward the median plane), extension (decreasing the dorsal angle), supination (turning the palm upward, etc.

2. *Attachment* - Omohyoid (shoulder to hyoid bone), Sternocleidomastoid (sternum-clavicle-mastoid process).

3. *Direction* - Oblique (at an angle), rectus (straight), transversus (horizontal).

4. *Location* - Brachii (arm), pectoralis (chest), spinatus (spine), tibialis (tibia).

5. *Relative Size* - Major (large), medius (intermediate), minor (small), minimus (smallest).

6. *Shape* - Serratus (saw-toothed), teres (round), trapezius (trapezoid).

7. *Structure* - Biceps (two heads), diagastric (two bellies), quadriceps (four heads), triceps (three heads).

8. *Combinations of the Above* - It is usually by the combination of two or more of the above terms that muscles are named: biceps brachii, gluteus maximus, pectoralis major, flexor carpi ulnaris.

Muscles may produce one or several of the following actions:

Abductors pull skeletal elements away from the median line, as in spreading the legs apart.

Adductors pull skeletal elements toward the median plane, as holding the arm against the body.

Extensors decrease the dorsal angle between skeletal elements, as in straightening the elbow when pointing. Extension of the arm is represented by the backward movement of the arm in bowling or pitching a softball.

Flexors increase the dorsal angle between skeletal elements (except knee) as pulling the forearm toward the arm (bending the elbow).

Rotators revolve a skeletal element around a long axis, as

Table 4 - The Nervous System

Central Nervous System	Peripheral Nervous System
Brain	Cranial nerves - 12 pairs
Cerebrum	Spinal nerves - 31 pairs
Brain stem	Autonomic nervous system
Spinal cord	Parasympathetic
	Cranial part
	Sacral part
	Sympathetic
	Thoracic part
	Lumbar part

turning the arm around an axis running the length of the humerus with the forearm extended.

Lateral rotation of the arm or thigh occurs when the thumb or great toe is turned away from the median line.

Medial rotation of the arm or thigh occurs when the thumb or great toe is turned toward the median plane.

Circumduction is the movement of the arm or thigh in the shape of a cone, as in throwing the arm around in a circle. Sometimes a pitcher's windup approaches circumduction. It is a sequential combination of the above actions.

Pronators turn the palm of the hand downward when the arm is flexed.

Supinators turn the palm of the hand upward when the arm is flexed.

THE NERVOUS SYSTEM

The muscles are powerless to produce movement without some mechanism to stimulate them into action. This control mechanism is supported by the nervous system which supplies individual fibers to all muscles, as well as to the skin, viscera, and joints.

The general function of the nervous system is to bring about an integration of the body (Fig. 5). Not only does the nervous system provide the means by which muscles are controlled and made to respond either at will or automatically, but it controls the action of all organs of the body, thus causing them to act in a coordinated manner. It is by means of nervous connections between our eyes, ears, nose, and skin and our muscles that we are aware of our surroundings and are able to respond in an intelligent manner to external stimuli. All motor nerves going to skeletal muscle are connected directly or indirectly with the motor area of the cerebral cortex, and as a result, are under its control (Table 4).

The nervous system is analogous to our lighting or telephone systems in which a predominating control system exists. The cerebrum represents the main exchange, whereas the large cables leading away from the central exchange are represented by the spinal cord. The wires going to the various parts of the city are represented by the spinal nerves, and the wires to the factories or homes or office buildings are represented by individual nerve fibers going to various muscles, organs, and other tissues.

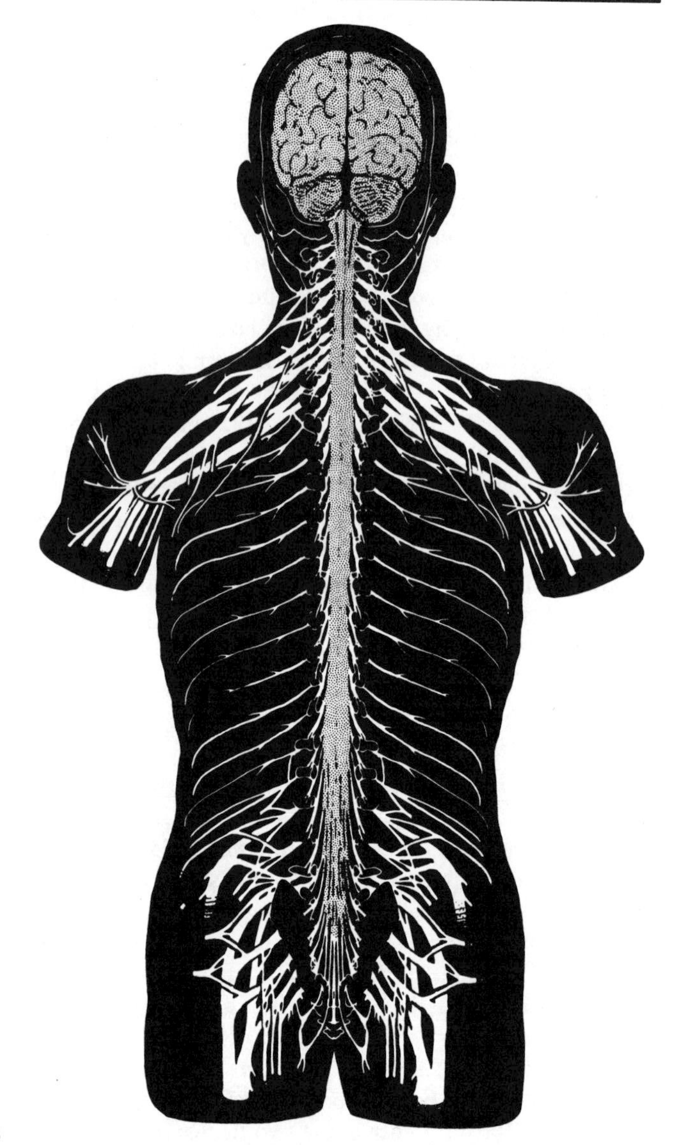

Fig. 5 - General plan of the nervous system

In the telephone system there are subexchanges that make connections independently of the main exchange, so in the nervous system there are subexchanges that make connections with other nerves without the impulse going to the cortex. These subexchanges, or connecting places, are distributed throughout the brain stem, cord, and in the body ganglia between the spinal cord and viscera, and in the walls of the viscera themselves. A connection place in the central nervous system (CNS) is called a nucleus and is defined as a group of nerve cell bodies and connections inside the central nervous system. A ganglion is defined as a group of nerve cell bodies and its connections outside the central nervous system. Impulses making connections in subexchanges without going to the cortex bring about reflex action, that is, an action produced without cerebral involvement.

There are three kinds of nerves—afferent, efferent, and

mixed. The afferent or sensory nerves are those which conduct nerve impulses from the periphery (for example, the skin, the eyes, the nose, etc.) to the central nervous system (the spinal cord or brain). Efferent nerves are those which conduct impulses from the central nervous system to the periphery (glands and muscles). The efferent nerves that conduct nerve impulses to the voluntary muscles are motor nerves. Those nerves which conduct nerve impulses to the visceral muscles, glands, and the heart belong to the autonomic nervous system. The mixed nerves are composed of both afferent and efferent fibers and hence conduct nerve impulses either to or from the central nervous system. Of the 12 pairs of cranial nerves, some are afferent, some are efferent, and some are mixed. The 31 pairs of spinal nerves are all mixed (motor and sensory) nerves.

In certain regions of the body–neck, shoulder, pelvic–the spinal nerves join together after they leave the vertebral canal to form nerve plexuses. A nerve plexus is defined as an intermingling and redistribution of nerve fibers.

THE CIRCULATORY SYSTEM

This closed transportation system consists of a pump (the heart), an aeration unit (the lungs), arteries to transport blood to all parts of the body, and veins for the return of the blood with the metabolic waste products to the heart. Interconnecting the arteries and veins are meshworks of very small vessels called capillaries. It is through the capillary walls that food and other substances pass into the tissue spaces.

The heart is a four-chambered reservoir pump; it can also be thought of as being two individual pumps joined together. Blood from all parts of the body enters the right reservoir (the right atrium; passing through valves blood enters the pump chamber (right ventricle) which upon contraction forces the blood into the artery (pulmonary artery) leading to the lungs. Returning to the left side of the heart via the pulmonary vein the blood enters the left atrium, passes into the left ventricle, and then is pumped out through the aorta (Fig. 6).

The aorta is the largest artery of the body; from it arteries branch to all structures except the lungs. Each artery of the body is named, usually according to the area that it supplies. The splenic, uterine, ovarian, and femoral arteries are typical examples.

Veins return blood to the heart. Receiving blood from the capillaries, small veins (venules) join together to form larger veins. Always at least one or two veins accompany each artery.

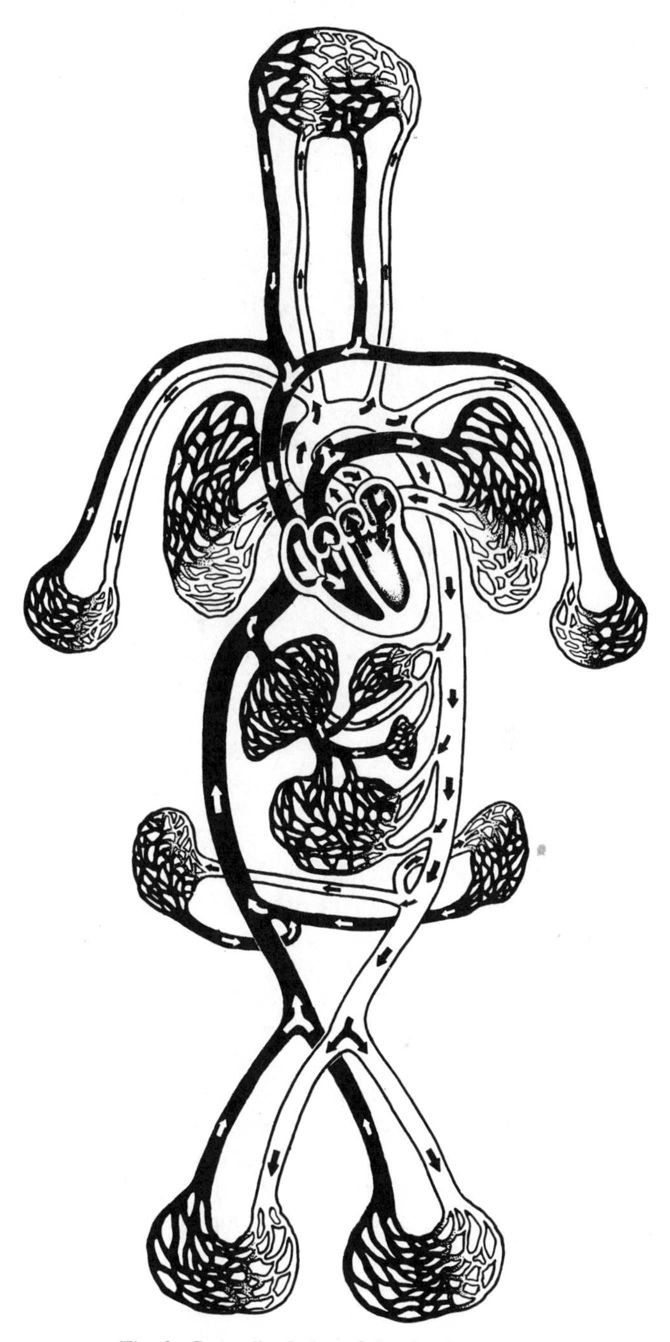

Fig. 6 - Generalized plan of the circulatory system

THE SKIN

The skin is the largest organ of the body. It has many functions, most of which are protective. It not only protects the deeper structures against mechanical injury and bacterial invasion but prevents loss of body fluids, protects the body from harmful sun rays via skin pigments, and acts as an excretory organ via perspiration.

The skin is the most frequently injured organ of the body; however, fortunately, almost all mechanical injuries to the skin typically found in motor vehicle crashes are not life-threatening.

Some of the more common skin injuries are:

1. Abrasion - a scrape
2. Amputation - cutting off of a limb or other part
3. Avulsion - tearing away of a part or structure
4. Contusion - a bruise; produced without laceration
5. Laceration - a cut
6. Rupture - forceful tearing or breaking of a part

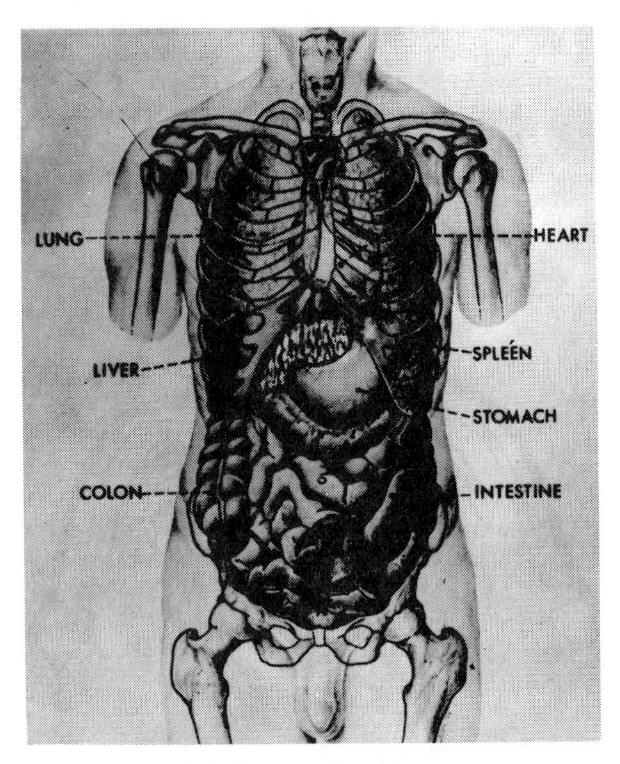

Fig. 7 - Thoracic and abdominal viscera

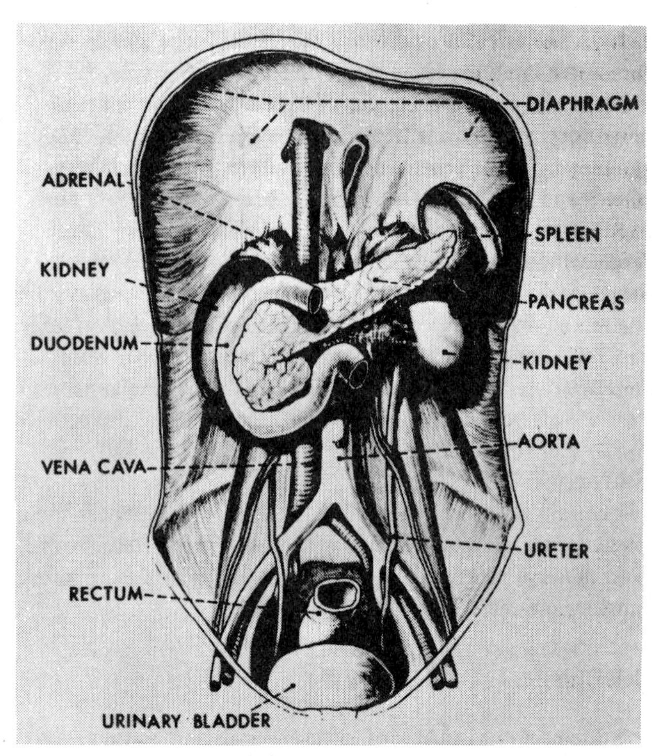

Fig. 8 - Deep abdominal viscera

THE THORAX

The thorax or chest is bounded by the thoracic portion of vertebral column posteriorly, the ribs, and in front of the sternum. Below, the respiratory diaphragm separates the thorax from the abdominal cavity. The diaphragm is dome-shaped, with its margins attached to the lower ribs. The dome of the diaphragm varies in position during respiration extending as high as the level of the nipples in forced expiration. Between the ribs are muscles and blood vessels. These intercostal arteries and veins course along with the ribs.

The thorax can be divided into three compartments–the right and left pleural (chest) cavities and the mediastinum. The mediastinum is a partition; it is a group of structures–the trachea, heart, aorta, large veins, esophagus–that together form a central partition separating the right from the left chest cavity (Fig. 7).

THE ABDOMEN AND PELVIS

For our discussion the abdomen and pelvis will be considered together. Bounded above by the diaphragm, the lumbar portion of the vertebral column posteriorly and the flat abdominal muscles on the sides and in front, and the bony pelvis below, the abdomen and pelvis contain the digestive, urinary, and reproductive systems.

The viscera found in the abdominal cavity are:

Stomach	Small intestine	Spleen
Liver	Large intestine	Adrenals
Gallbladder	Sigmoid colon	Kidneys
Pancreas	Rectum	Ureters
Urinary bladder	Vas deferens	Inferior vena cava
In the Male:	Seminal vesicles	Portal vein
Uterus	Prostate gland	Nerves
Vagina	In the Female:	Nerve plexuses
Abdominal aorta and its branches	Ovaries	
	Uterine tubes	

Because of the lack of adequate protection by bones, the abdominal contents are highly susceptible to blunt trauma. Abdominal organs are either thin-walled (stomach, intestines, etc.) or are spongelike and blood-filled (liver, spleen, kidney). In addition, numerous blood vessels are found in the abdomen for the supply of these organs (Fig. 8).

Thus, blunt impact to the abdomen can cause rupture of the stomach or intestines which release their contents into the abdominal cavity. Bacterial contamination thus can occur (peritonitis). Also, massive bleeding can occur if the blood-filled liver or spleen is ruptured, or if major arteries are damaged. Surgical repair of blunt abdominal injuries is almost always indicated.

The liver is one of the most frequently injured organs of the abdomen. It is about one-fortieth of the total body weight and occupies the upper right area of the abdomen. It is attached to the undersurface of the right side of the diaphragm and usually does not extend below the rib margins. Thus, although the liver is an abdominal organ, it is under the cover of the lower ribs on the right side. Hence, it is not unusual to have a blunt impact to the thorax which produces a liver injury.

Abdominal injuries can be produced by the unrestrained occupants of the car impacting the steering wheel, instrument panel, door panel, or the front seat back. Also, seat belts, when worn improperly, can cause injuries to viscera by direct pressure on the soft abdominal wall.

Within the bony pelvis are located the lower end of the gastrointestinal tract, the urinary bladder, and the uterus in the female. Also, numerous blood vessels and nerves, not only for the supply of pelvic structures but also for the lower extremities, are found here. The urinary bladder and uterus (especially the pregnant uterus) can be sheared off from its attachments by blunt, low abdominal impacts. Fractures of the bony pelvis can injure the organs or blood vessels within. The forces of walking are transmitted through the bony pelvis; thus, a pelvic fracture will be incapacitating and debilitating.

ANATOMICAL PROBLEMS ASSOCIATED WITH AGE

The infant and small child have specific anatomical characteristics which are unique to them. The skulls of these children are relatively thin bones not affording the impact protection as found in the adult. The sutures only incompletely join the skull bones together. Deciduous teeth (baby teeth) may not yet have erupted and the permanent teeth are yet forming in the jaws. Thus, jaw fractures can have a marked effect upon future tooth eruption and proper tooth alignment.

The soft tissue and bones of children heal rapidly, for they are in the growing stages. Not infrequently scars of soft tissue can hardly be found some years later because of tissue repair and growth.

Infants and children live in an adult world of automobile design. Thus, the child requires special and unique considerations for impact protection. Because they are smaller, they frequently will not contact areas in the car that have been designed for adult impact-force amelioration.

The bones of children and early teenagers are not completely formed. In infants the bones are small with cartilage separating the epiphysis from the diaphysis. This is the growth center of the bones and if disrupted can produce growth disturbances in terms of bone length, normal joint arrangements, and possibly in limiting the range of motion at a joint.

The bones of children are highly elastic and the greenstick fracture just above the wrist is a characteristic fracture of children. The bony thorax is no exception. Infants have highly elastic ribs. The space for the mediastinal structures is very small. Thus, chest impacts could easily deform the chest wall, and the heart and great vessels, producing fatal injuries with minimal forces.

In children the liver is massive in relation to the size of the other abdominal organs. Therefore, it is very susceptible to blunt impacts. The potbelly abdomen and the narrow space between the abdominal wall and the front of the thigh in the sitting position makes restraining children with a lap belt very difficult.

In children the head is large and sits on a thin neck which has muscles of minimal strength. Thus, neck injuries from violent head movements can be expected.

The elderly have similar and also different problems from those of children. Whereas bones of children are elastic and incompletely formed, older people have fully formed brittle bones. Thus, rib and long bone fractures are not uncommon in the elderly as well as the potential internal organ injuries associated with blunt impact. The elderly, as a group, heal much more slowly; bones take longer to unite, joints longer to repair, and soft tissue more time to heal. Also, long-term immobilization can cause problems of blood clots, pneumonia, and other complications. Many of the elderly have medical problems which when associated with impact trauma can unite to produce a serious if not fatal outcome. Problems in blood clotting, the body reflexes to prevent or minimize shock, etc., may play a significant role in the final outcome.

APPENDIX A

Selected References

Listed below are suggested references in ascending order of difficulty. Those marked with an asterisk (*) are strongly recommended for those unacquainted with anatomy and medical terminology.

1. Beginning Texts:

*"The Question and Answer Book About the Human Body," Ann McGovern, New York; Random House, 1965.

*"The How and Why Wonderbook of the Human Body," Martin Keen, New York; Wonderbooks, 1961.

*"The Wonders of the Human Body," Martin Keen, New York; Grosset and Dunlap, 1966.

"Textbook of Anatomy and Physiology," C. P. Anthony, C. V. Mosby Co., Seventh edition, 1967.

Data-Guides, Human Anatomy Reference Charts, Data-Guide, Inc., Flushing, New York.

Elementary Human Anatomy - Andrew J. Berger, John Wiley & Sons, Inc., 1964.

Introduction to Human Anatomy - Carl C. Francis, C. V. Mosby Co., Fourth edition, 1964.

Human Anatomy and Physiology - B. G. King and M. J. Showers, W. B. Saunders Co., Fifth edition, 1966.

2. Advanced Texts:

Essentials of Human Anatomy - R. T. Woodburne, Oxford University Press, Fourth edition, 1969.

Anatomy of the Human Body - H. Gray, Lea & Febiger, edited by C. M. Goss, Twenty-seventh edition, 1959.

Anatomy of the Human Body - R. D. Lockhart, et al, J. B. Lippincott Co., 1959.

Outline of Human Anatomy - Saul Wischinitzer, McGraw-Hill Book Co., Blakiston Division, 1963.

3. Gross Anatomy Atlases:

Illustrations of Regional Anatomy - E. B. Jamieson, The Williams & Wilkins Co., Eighth edition, 1959.

An Atlas of Anatomy - J. C. B. Grant, The Williams & Wilkins Co., Fifth edition, 1962.

4. Medical Dictionaries:

Dorland's Medical Dictionary - W. B. Saunders Co., Twenty-fourth edition, 1965.

Blakiston's New Gould Medical Dictionary - The Blakiston Co., First edition, 1949.

Stedman's Medical Dictionary - The Williams & Wilkins Co., Twenty-first edition, 1966.

APPENDIX B

A SELECTED GLOSSARY OF ANATOMICAL TERMINOLOGY

A- (G prefix, without).

Ab- (L prefix, away from).

Acetabulum (L, a vinegar cup). A large cup-shaped cavity in which the head of the femur fits.

Acromion (G, *akros*, top; *omos*, shoulder). A projection on the scapula forming the point of the shoulder.

Ad- (L prefix, toward, upon).

Afferent (L, *ad*, to; *fero*, bear or bring). To bring to—a sensory nerve that carries impulses to the central nervous system.

Alveolar (L, a little cavity).

Ankylosis (G, *ankylos*, bent). To make a joint immobile.

Aorta (G, *aorte*, to lift). Large blood vessel leading from the left ventricle of the heart.

Atlas (G, *tlao*, to bear). First cervical vertebra.

Atrium (L, a court). One of the chambers of the heart; an expanded place.

Atrophy (L, not; to nourish). A deficiency or reduction in the size of a structure.

Auricle (L, *auricula*, a little ear). An earlike projection of the atrium of the heart.

Axilla, axillary (L, *axilla*, a little axis). Refers to the armpit.

Biceps (L, *bi*, two; *caput*, head). A two-headed muscle.

Brachial (L, *brachium*, arm). Refers to the arm, or any arm-like process.

Bronchus (G, *bronchos*, windpipe). A division of the trachea.

Carpal (L, *carpalis*, wrist). Refers to the wrist, as the carpal bones.

Caudal (L, *cauda*, tail). Refers to the tail.

Cephalic (G, *kepnale*, head). Refers to the head.

Cerebellum (L, a little brain). A division of the brain, lying inferior to the cerebrum, and posterior to the brain stem.

Cervical (L, *cervix*, neck). Refers to the neck region.

Circulation (L, *circulo*, to form a circle). To flow in a circle. The flow of blood through its blood vessels through the body.

Clavicle (L, clavis, key). The key-like bone of the shoulder girdle.

Colon (G, *kolon*, member). The portion of the large intestine between the cecum and the rectum.

Costal (L, *costa*, rib). Refers to the ribs.

Cutaneous (L, *cutis*, skin). Refers to the skin.

Deltoid (L, *deltoides*, triangular). A triangular-shaped muscle of the shoulder.

Diaphragm (G, *dia*, between; *phragnymi*, to enclose). A membrane that closes a cavity, or separates two cavities, as the respiratory diaphragm.

Dura mater (L, hard mother). The thick outermost meningeal layer around the brain and cord.

Efferent (L, effero, to bear away from). Carrying a fluid or nerve impulse away from a certain part; as efferent nerves which carry impulses away from the cord.

Eip- (G prefix, upon).

Epidermis (G, *epi*, upon; *derma*, skin). Outermost layer of the skin.

Epiphysis (G, an outgrowth). The terminal ends of long bones.

Esophagus (G, to carry; to eat). The upper part of the alimentary tract, extending from the pharynx to the stomach.

Ex- (L prefix, out, outside).

Fibula (L, *fibula*, clasp, buckle). Lateral and smaller of the two bones of the leg.

Gastric (L, *gaster*, stomach). Refers to the stomach.

Gluteus (G, *gloutos*, rump). Refers to the buttocks.

Ileum (G, *eilo*, twist). Refers to the distal two-thirds of the small intestine.

Infra (L prefix, below).

Inguinal (L, *inguen*, groin). Refers to region of the groin.

Intestine (L, *intestinus*, inside or internal). The part of the digestive tract extending from the stomach to the anus.

Jejunum (L, dry or empty). That portion of the digestive tract extending between the duodenum and the ileum.

Jugular (L, *jugulum*, the collar bone or throat). Refers to veins in the neck, draining the head.

Manubrium (L, handle). Superior part of the sternum.

Maxilla (L, *maxilla*, jaw bone). The upper jaw bone.

Membrane (L, *membrum*, member). A thin layer of tissue which covers or lines an organ or surface, or divides a space or organ.

Mesentery (G, *mesos*, middle; *enteron*, gut). Fold of peritoneum which attaches the intestine to the posterior abdominnal wall.

Met-, meta- (L or G prefix, between, after, reversely).

Metacarpal (G, *meta*, after; *karpos*, wrist). Refers to bones between the wrist and fingers.

Metatarsal (G, *meta*, after; *tarsos*, a flat surface of the foot). Refers to bones of the foot between the tarsal bones and the toes.

Occipital (L, *occiput*, back of the head). Pertaining to the back of the head, as the occipital bone.

Omo- (G prefix, *omos*, shoulder). Shows some relation to the shoulder.

Oral (L, *os*, mouth). Refers to the mouth as the oral cavity.

Orbital (L, *orbita*, orbit). Pertains to the orbit or eye socket.

Ossification (L, os, ossis, bone; facio, to manufacture). The conversion of any substance into bone.

Pectoral (L, *pectoralis*, chest). Refers to the breast or chest.

Pericardium (L, around; heart). Membranous sac around the heart.

Peritoneum (G, *peri*, around; teino, stretch). A serous membrane that lines the abdominal walls and invests the organs contained therein.

Peritonitis (G, *peri*, around; *teino*, stretch, inflammation). Inflammation of the peritoneum.

Pharynx (G, *pharynx*, throat). Proximal end of the digestive tract into which the oral and nasal cavities empty.

Pia mater (L, tender mother). Innermost layer of the meninges around the brain.

Pleura (G, *pleura*, rib, side). The serous membrane that lines the thoracic cavity and invests the lungs.

Plexus (L, braid or interweaving). Refers to a network or interweaving of nerves and veins.

Portal (L, *porta*, gate). A hilum through which vessels and nerves enter a gland or organ. In the circulatory system refers to a vein entering the porta of the liver.

Pre- (L prefix, before).

Pro- (L or G prefix, before).

Pulmonary (L, *pulmon*, lung). Refers to the lungs.

Renal (L, *renes*, kidneys). Pertains to the kidneys.

Retro- (L prefix, back, backward).

Serratus (L, *serra*, saw). A saw-toothed muscle of the shoulder girdle.

Sternum (G, *sternon*, breast bone). The bone of the axial skeleton to which the costal cartilages of the true ribs attach.

Sub- (L prefix, under).

Supine (L, *supino*, to put on the back). The position a person is in when lying on the back.

Supra- (L prefix, above).

Tendon (L, *tendo*, to stretch). A fibrous cord coming from muscles.

Tissue (L, *texo*, to weave). An aggregation of similar cells united in the performance of a particular function. Examples: muscle tissue, lung tissue, etc.

Trachea (G, *trachys*, rough). The part of the air passage leading from the larynx to the bronchii.

Triceps,(L, *tri*, three; *caput*, head). A three-headed muscle.

Umbilical (L, *umbilicus*, navel). Refers to the umbilicus or navel.

Ventricle (L, *venter*, belly). A cavity, especially in the heart and brain.

700195

The Anatomy and Physiology of the Head and Neck

Donald S. Strachan
University of Michigan

PARTS OF THIS CHAPTER are written so that the reader must participate in looking, feeling, and sensing the anatomy of the area. A mirror is absolutely essential and a tongue blade and pen light certainly will help one to see and understand much of the oral-facial anatomy to be presented. Anatomy is "there," it is real and should be no secret to the observer.

The best way to study this chapter is to scan through the figures first and try to visualize the areas and dimensions of the areas to be covered. Many figures have information that pertains to several sections of the text. Become familiar with the figures and with your own anatomy.

The cranium and the face are the components of the head. The cranium contains the brain, and the face is comprised of all the tissues and structures found to the anterior and inferior of the cranium. Proportionally there is appreciably more growth of the face from birth than there is in the cranium.

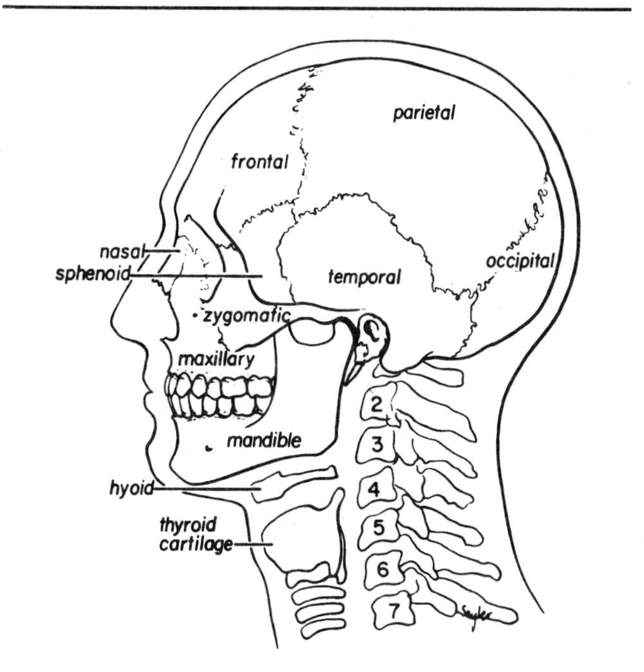

Fig. 1 - Lateral surface of skull showing bones of cranium (frontal, parietal, occipital, temporal and sphenoid) and some facial bones (nasal, zygomatic, maxiallary and mandible)

BONES OF THE CRANIUM

The bones of the cranium enclose and form a solid protective covering for the brain (Figs. 1 and 2). These bones for the most part are flat-surfaced bones with an outer and inside surface of solid compact bone and an in-between layer of less dense ("spongy") bone. There are seven bones which contribute to the cranium: 1 frontal, 2 parietal, 1 occipital, 2 temporal, and 1 sphenoid (also the superior portion of the ethmoid should be included). The frontal bone, parietal bones, and posterior surface of the occipital bone are broad convex bones which make up the anterior, superior, and posterior sections and much of the lateral surfaces of the brain case. Parts of the occipital bone form the posterior part of the base of the cranium. In the base of the occipital bone is the opening (foramen magnum) for the spinal cord and for a pair of major arteries and veins to the brain.

ARTERIES OF THE BRAIN

The common carotid artery gives off the external carotid artery in the neck region and continues in the carotid sheath as the internal carotid artery. There are no branches of this artery until it passes through the base of the skull. The artery enters the temporal bone anterior and lateral to the foramen magnum and follows a bony canal which is directed forward to reach the middle cranial fossa where the arteries (one from each side) connect with branches from the basilar artery, a common branch of the two vertebral arteries. The two vertebral arteries enter the cranium through the foramen magnum.

Therefore, there are four arteries to the brain—right and left internal carotid arteries and right and left vertebral arteries. These arteries are interconnected by means of a basilar artery and by communicating arteries. These interconnections allow alternative pathways of blood supply and can accommodate fluctuations in arterial flow from these four major vessels.

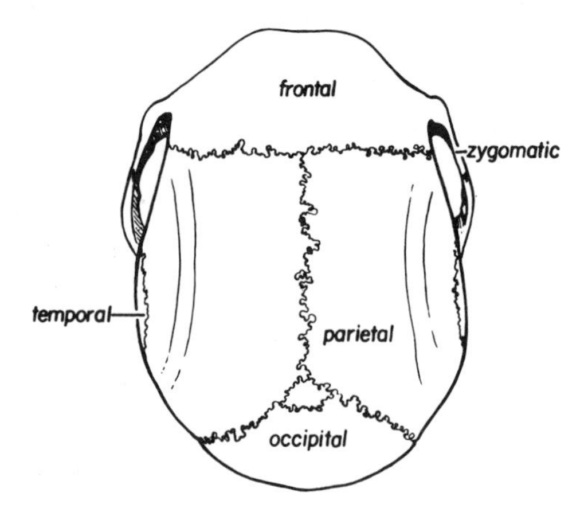

Fig. 2 - Bones of cranium viewed from above. Frontal bone is anterior and occipital bone is posterior.

BLOOD SUPPLY OF THE MENINGES

The meninges (coverings) of the brain have an arterial blood supply that is different from that of the brain itself. There are several of these meningeal arteries which originate from larger arteries on the outside of the cranium. These arteries pass through the foramina (openings) in the bones of the skull and follow the inside surfaces of the bones comprising the cranium. The path of the larger meningeal arteries is clearly marked by grooves found on these inner bony surfaces. These arteries supply the meninges (dura, arachnoid, and pia mater). The venous drainage of these connective tissue coverings passes to venous sinuses which also serve the venous drainage of the brain.

THE ORBIT AND ITS CONTENTS

The bony orbits are conical in shape and contain the eyes and their associated vessels and nerves (Fig. 3). The orbits also contain the lacrimal glands and the several muscles which are responsible for the movements of the eyes and for elevation of the upper lid.

The outer rims of the orbits are formed by portions of the frontal, zygomatic, and maxillary bones. The outer rims are made up of thickened compact bone. The upper half of the rims are prominent and bulky and are formed by parts of the frontal bone. The outside lower edges of the rim are formed by the zygomatic bones. Portions of the maxillary bone form the lower inner parts of the rim of the orbit. The inner surfaces of the orbits are formed by several facial bones which are thin and flat. Immediately above the bony roof of the orbits are the frontal lobes of the brain. Medial to the medial wall of the orbit, the ethmoid air sinuses are found superiorly and the nasal cavities inferiorly. Below the orbits are the paired maxillary sinuses.

The eyeball itself is a fluid-filled fibrous tissue capsule containing several specialized tissues such as the lens, iris, and retina. The cornea (anterior surface) is clear and covered with a smooth lubricated epithelium called the conjunctiva. Except

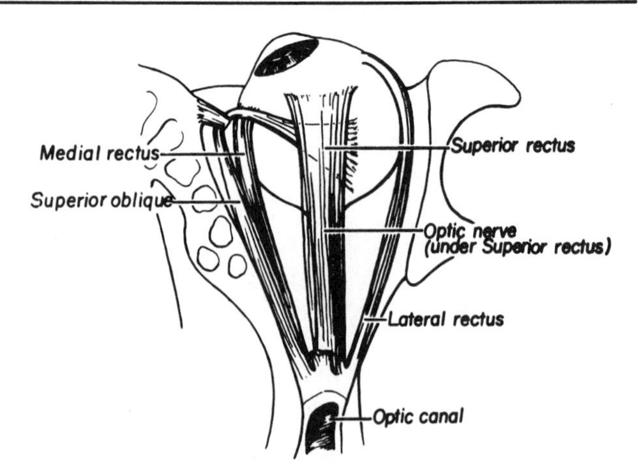

Fig. 3 - Diagram of right orbit viewed from above with roof of bony orbit removed. Not shown are nerves and vessels. Fatty tissue fills in space behind eyeball. Optic nerve is found below superior rectus. Note common origin of these muscles from outer surfaces of optic canal. Only origin of superior levator palpebrae muscle is shown. Its normal position is above superior oblique muscle

for this anterior part, the inner surface of the eyeball is lined by the retina which contains several layers of photosensitive receptors and nerve cells. Thousands of axons from these nerve cells join at the rear of the eye to form the optic nerve. (cranial nerve II).* This prominent nerve passes through fatty tissue which is found in the rear part of the orbit, exits through the optic canal, and travels to the brain. The optic canal is a thick-walled bony foramen located at the apex or posterior part of the orbit. The outer surfaces of the optic canal serve as attachment for the origin of several of the muscles which move the eye. In each orbit there are six muscles which move the eyeball, and one muscle which inserts into the upper eyelid. Four ocular muscles originate as small tendons from the area of the optic canal and insert as flattened narrow strips into the fibrous outer covering of the eye. These muscles are the superior rectus, medial rectus, inferior rectus, and lateral rectus muscles.

The superior oblique muscle travels in the superior medial aspect of the orbit. On the inside surface near the rim of the orbit this muscle passes through a connective tissue sling and reflects backward around this pulley-like structure to insert on the superior surface of the eye. There is also the inferior oblique muscle which originates on the inferior surface of the orbit and inserts on the inferior surface of the eyeball. The superior levator palpebrae muscle inserts into the upper eyelid in a broad flat attachment. Contraction of this muscle raises the upper lid.

The skin on the outside of the upper eyelid is very loosely

*Convention among anatomists dictates that cranial nerves be designated by Roman numerals. The cranial nerves exit directly from the brain and are: I, olfactory; II, optic; III, oculomotor; IV, trochlear; V, trigeminal; VI, abducens; VII, facial; VIII, staticoacoustic; IX, glossopharyngeal; X, vagus; XI, accessory; XII, hypoglossal.

attached to a dense connective tissue plate of tissue found on the inside of this eyelid. The inside of the eyelids are lined with conjunctiva which reflects onto the outer surface of the eye.

The nerves to all these muscles come directly from the brain and pass through the superior orbital fissure to reach these muscles. The superior orbital fissure is found near the apex of the orbit. The abducens nerve (cranial nerve VI) supplies the lateral rectus muscle. The trochlear nerve (cranial nerve IV) innervates the superior oblique muscle. The oculomotor nerve (cranial nerve III) supplies the remaining muscles and also supplies nerve fibers to involuntary muscles which control the size of the iris and the shape of the lens. The superior orbital fissure also transmits another cranial nerve, the opthalmic branch of the trigeminal nerve (cranial nerve V). This nerve is the sensory nerve supply to the contents of the orbit and to the nearby nasal sinuses. Branches from this nerve continue through the orbit and exit onto the surface of the face to supply sensory nerve fibers for the forehead area, nose, and eyelids.

Branches of the internal carotid arteries, the right and left ophthalmic arteries, supply the orbit and its contents. These vessels arise from the internal carotid arteries on the inside of the cranium and pass through the optic canal with the optic nerve and enter the apex (rear) of the orbit.

The lacrimal gland secretes a watery fluid which lubricates and protects the conjunctival surfaces of the eye and eyelid. This gland is located in a fossa in the anterior superior lateral aspect of the orbit. There are several ducts (openings) which are found on the superior aspect of the inner surface of the upper lid.

The lacrimal fluid (tears) flows across the eye and the flow is assisted by blinking of the eyelids. If in excess, the fluid can enter two small pores found in the inner corners of the eye. If one looks carefully, an upper and lower orifice can be seen on little papillae that project near the medial surfaces of each eyelid. These pores lead to canals which join behind the inner corner of the eye and then travel downward in an expanded duct (nasolacrimal duct) to exit into the front of the nasal cavity below the inferior conchae.

The normal functioning of the lacrimal gland and of the muscles surrounding the eye which cause blinking are necessary for the integrity and health of the conjunctiva.

THE NOSE AND NASAL CAVITIES

Except for the bridge of the nose, which has a bony support, the nose is made up of dense connective tissue and several small cartilage plates (Figs. 4 and 5). The inside surface is covered with epidermis which contains many long hairs called vibrissae. Immediately inside the nose are right and left nasal cavities divided by a nasal septum. The nasal septum is a thin flat wall in an anterior posterior direction, and has a cartilagenous framework anteriorly and a bony framework, posteriorly. The lateral walls of each nasal cavity have bony projections; the superior, middle, and inferior conchae. Practically all surfaces of the nasal cavities are covered by respiratory epithelium. This epithelium contains cells which have mobile

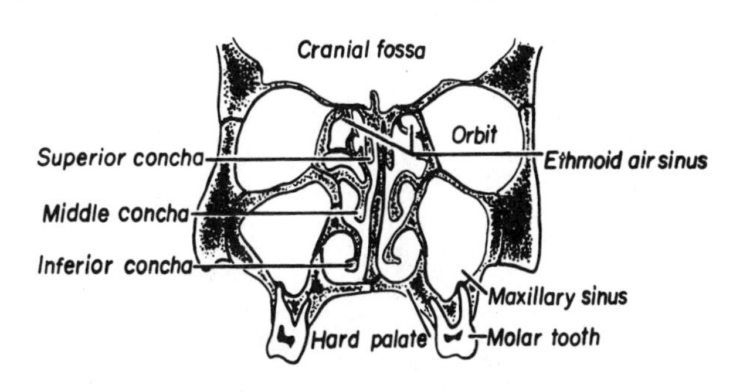

Fig. 4 - Graphic description of a frontal section showing relative positions of nasal cavities (right and left), maxillary sinuses and orbits. Note midline nasal septum and projections from lateral walls of nasal cavities (superior, middle and inferior chonchae)

cilia on the surface of the cells. The cilia beat rapidly and can move mucinous material and small entrapped debris outward toward the nasopharynx. Cells and glands in the epithelium produce the protective and moistening mucin.

The tissue underlying the covering epithelium is very vascular. Inflammatory processes or sensitivity reactions cause the veins to enlarge and swell with venous blood. This is the reason for the "stuffed up" nasal feeling during colds. Decongestants so frequently used for any number of nasal problems contain chemical vasoconstrictors which constrict the muscular walls of these vessels. This same kind of epithelium and underlying tissue lines the nasal sinuses.

The nasal sinuses all have connections or exits to the nasal chambers. These sinuses are found inside parts of the frontal, ethmoid, sphenoid, and maxillary bones. They are small at birth and attain their maximum size with adulthood. The paired frontal sinuses are found directly above the orbit, in the rim of the orbit, and near the midline. The ethmoid sinuses are a number of sinuses found directly in the roof of the nasal cavities and between the upper parts of the orbits. These sinuses are "pea"-shaped or smaller and are irregular in morphology. There are anterior, middle, and inferior ethmoid sinuses with no distinct boundaries between one another. Immediately posterior is the single sphenoid sinus. This sinus is deep in the skull, about 2-1/2 in. directly posterior from the bridge of the nose, and is indented from above by a fossa for the pituitary gland. This gland is directly attached to the brain and secretes growth hormone and regulatory hormones to the adrenal gland, ovary or testis, kidney, thyroid gland, and mammary glands. The maxillary sinuses are large irregular-shaped sinuses (total volume of each would approximate the volume of a golf ball). They are located above the maxillary bicuspid and molar teeth, lateral to most of the nasal cavities, and below the orbits.

All nasal sinuses have their openings to the nasal cavities and all are lined by epithelia similar to that of the nasal cavities.

NASOPHARYNX

At the posterior border of the nasal septum is the nasopharynx (Fig. 5). The back wall of the nasopharynx is comprised

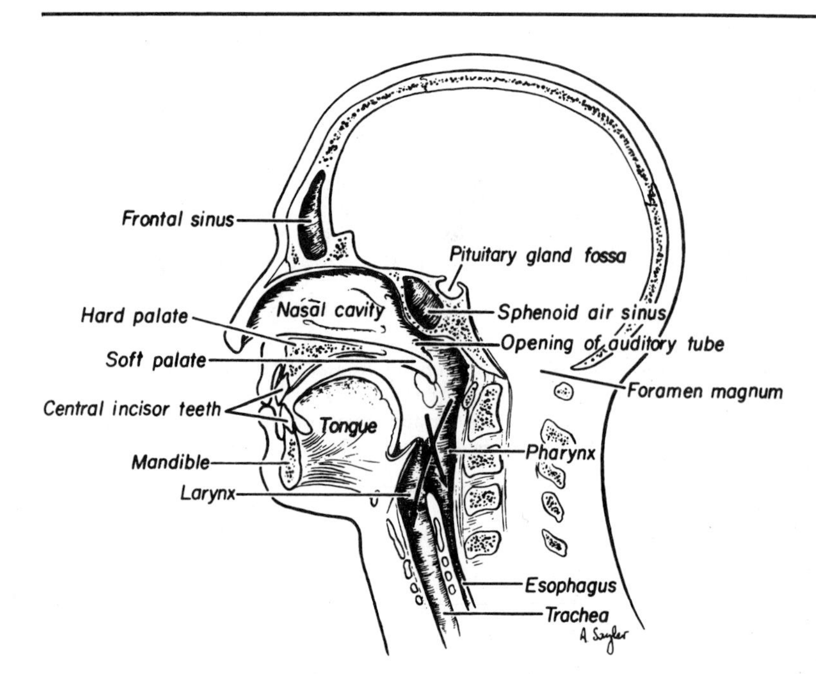

Fig. 5 - Midline depiction of head and neck area. Nasal septum has been removed. Arrows show normal passage for inspiratory air into trachea, and for food into esophagus

of an epithelial surface covering an underlying muscular wall. This muscular wall is continued throughout the pharynx. High in the posterior superior aspect of the nasopharynx is the location of the adenoid tonsil. This tissue is many times surgically removed along with the palatine tonsil (the "tonsil"). The palatine tonsil is found posterior and lateral to the tongue in the oropharynx. Tonsilar tissue has growth cycles and is largest at around 7 years and before puberty and then regresses thereafter.

In the lateral wall of the nasopharynx are two muscles which insert into the soft palate. These muscles lift and elevate the soft palate closing off the nasophamyx from below during swallowing and speech. In the lateral walls are the openings to the inner ear. These openings lead to the auditory tube which is a cartilaginous and bony canal which opens directly to the tympanic membrane (eardrum) of the ear. Pressures on the outside of the eardrum can be equalized and relieved on the inside by opening of the auditory tube (opening of this tube occurs with swallowing). Both palatal muscles mentioned above have partial origins from the auditory tube and these muscles are active during the initial phases of swallowing. Contraction of these muscles opens the tube. Swallowing and its relationship to relieving pressure on the ear is well known to air travellers.

SENSORY NERVES OF THE HEAD AND NECK

The sensory nerves of the neck and posterior aspect of the head region are individual nerves which come from several levels of the spinal cord in the cervical (neck) region. The face and the mouth are innervated by three major divisions of the fifth cranial nerve. (Fig. 6) The three divisions and their area of supply are: (1) opthalmic division to the forehead and nose; (2) maxillary division to the areas around the zygomatic bone, below the eye and to the upper lip and maxillary teeth; (3) mandibular division to the areas above the ear overlying the mandible itself and to the lower lip and mandibular teeth. The mandibular division also carries a motor component to all the muscles of mastication and to one of the palatal muscles. Sensory nerves from the tongue travel in branches of the seventh (VII) and ninth (IX) cranial nerves. The pharynx and tonsilar regions are innervated by the ninth (IX) and tenth (X) cranial nerves. The tenth (X) cranial nerve (the vagus) carries the sensory nerves from the laryngeal and esophageal areas.

The sensations of smell, sight, taste, and hearing, and the sense of equilibrium are transmitted by cranial nerves carrying sensory fibers. The olfactory nerves (Cranial nerve I) are connected to specialized receptor cells in the olfactory epithelium found in the superior part of the nasal chambers. These cranial nerves transmit impulses from this region to the brain to be translated into the sense of smell. The optic nerve (Cranial nerve II) transmits information from the retina of the eye to the brain. Cranial nerve VII (facial nerve) and Cranial nerve IX (glossopharyngeal nerve) carry sensory nerves for taste from the tongue. Cranial nerve VIII (statico-acoustic) receives impulses from the three semicircular canals which are embedded deep in the bone near the inner ear. These impulses carry information about the position of the head in relation to the force of gravity. This nerve also carries the impulses from the sound-transducing mechanisms found in the inner ear.

MUSCLES OF FACIAL EXPRESSION

The muscles of facial expression include all the superficial muscles of the facial complex except the muscles of mastication (Figs. 6 and 7). These muscles differ from the majority of skeletal muscles in that they (1) do not have an insertion from one bone to another allowing motion of one bone in relation to the other; (2) can originate as interdigitations from

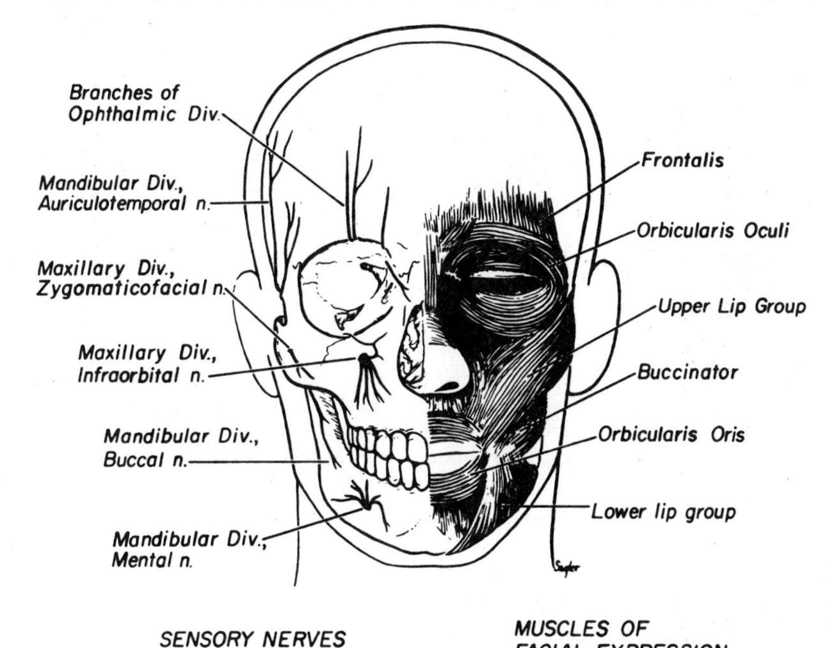

Fig. 6 - Frontal view of head showing sensory nerves on left and muscles of facial expression on right. Only larger branchings of the sensory nerves are shown

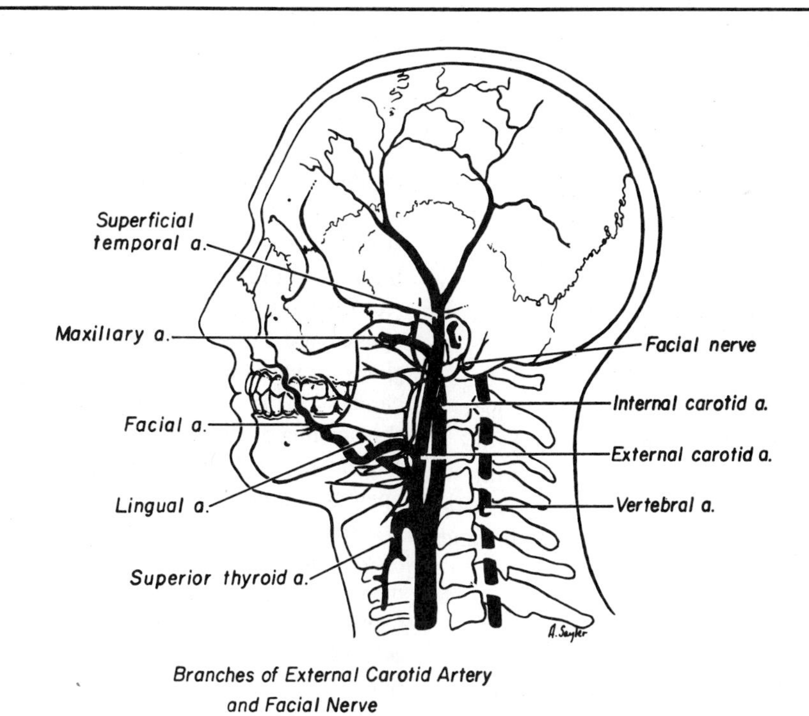

Fig. 7 - Lateral view of head and neck area showing major vessels and path of facial nerve. Internal carotid artery and vertebral arteries enter cranium deep to this lateral surface and supply brain. External carotid arteries supply anterior parts of neck and all of face. Terminal branches of maxillary artery and lingual arteries passing deep to mandible are not shown. Facial nerve exits from temporal bone, divides several times and supplies muscles of facial expression. Many fine terminal branchings of this nerve are not presented here

another muscle; (3) rarely have a bony insertion but rather have insertions into other muscles or more commonly into the connective tissue layer underlying the skin; (4) they are flat and thin muscles. With these muscles very finite and discrete actions are possible. For purposes of description, these muscles can be organized into several groups.

Ocular Group: orbicularis oculi, a flat circularly arranged muscle which constricts the tissues surrounding the eye. There are also small muscles inserting into the skin above the bridge of the nose.

Upper Lip Group: zygomaticus major and levator anguli oris insert into muscles at the angle of the mouth, and levator labii superioris inserts into muscles of the lip.

Lip and Cheek Group: orbicularis oris is the circularly arranged muscle in the upper and lower lips. All the muscles of the upper and lower lip group and the cheek muscles insert and interdigitate with orbicularis oris adding to the wide range of expression and movements possible about the mouth. The major muscle of the cheek wall is the buccinator muscle. It interdigitates in front with orbicularis oris. Deep to the mandible and behind the third molar teeth the buccinator interdigitates with the muscles forming the pharynx wall. The muscle wall of the cheek is then directly connected with the muscle wall of the pharynx.

Lower Lip Group: depressor anguli oris inserts into muscles at the corner of the mouth and the mentalis muscle inserts into the skin overlying the chin area and into the fibers of the orbicularis oris near the midline.

In addition to the muscles described above, there are other muscles of facial expression which are associated with the nose and the ear and superficial tissues of the neck. Also there are two muscles which insert into the connective tissue coverings of the scalp. These muscles are the frontalis, above the orbits, and the occipitalis found on the posterior aspect of the skull.

All the above muscles of facial expression are innervated by branches of the facial (VII) nerve. This nerve exits the cranium from a foramen between the mastoid and styloid processes of the temporal bone. Entering the substance of the parotid gland, this nerve branches into temporal, zygomatic, buccal, mandibular, and cervical groups of nerves. There are many interconnections between these groups. The several nerves exit from the parotid gland and pass into the superficial tissues of the face to go directly to the muscles of facial expression. These nerves continue to branch and become much smaller and more numerous. There are many unions forming plexuses between the numerous branches.

The normal muscle-nerve relationship and function of the muscles of facial expressions is extremely important. One's appearance and expression have social importance to the individual. Slight alterations in normal expressions are easily noted. In addition, many functions, such as speech and mastication, require that these tissues function in an integrated and complete manner.

MANDIBLE

The mandible (lower jaw) has three major components: the body and two rami (Fig. 8). The upper part of the body (alveolar process) contains the bony sockets for the mandibular teeth. The lower part (base) is more bulky, serves as attachment for several muscles, and contains an interior channel (mandibular canal) for the nerves and vessels to the mandibular teeth. A branch is given off from these components and exits from the mental foramen as the mental nerve and supplies the immediate area with blood supply and sensory nerve supply.

TEMPOROMANDIBULAR JOINT

The mandible articulates with the temporal bone of the skull at the temporomandibular joint (Fig. 8). This specialized joint has a fibrous disc interposed between the mandibular condyle and the mandibular fossa of the temporal bone. The disc is attached at its periphery to the joint capsule. Ligaments, prominent on the lateral surface, add fibrous support to this articulation. The upper and lower joint cavities are filled with a viscous lubricating fluid (high molecular weight polymers) which is secreted by cells on the inside periphery of this capsule.

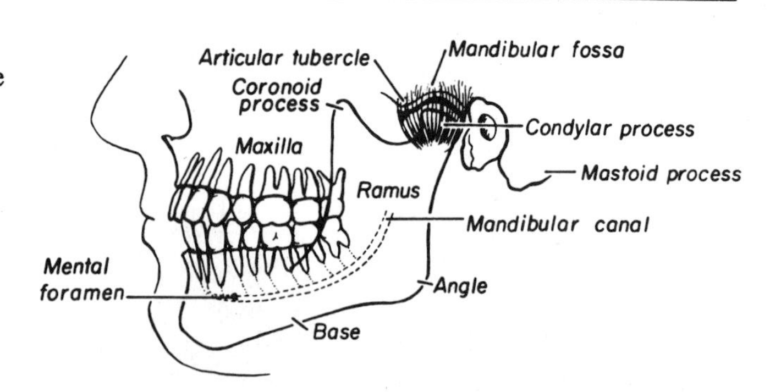

Fig. 8 - Lateral view of mandible and maxilla including temporomandibular joint. Coronoid process and condylar process are parts of ramus. Alveolar process (not labeled) contains roots of teeth. Fibrous disc of temporomandibular joint is shown encircled by fibers of joint capsule. This disc separates joint cavity into upper and lower compartments

Fibrous cartilage covers the articulating bony surfaces of the condyle and the mandibular fossa. This tissue is smooth and nonvascular with few, if any, nerve endings. The ligaments around the joint, the fibrous nature of the capsule itself, and the vectors of force generated by muscle groups limit and direct the movements possible at this joint. When the jaw is opened and closed the head of the condyle rotates and also glides forward on to the posterior surface of the articular tubercle. This is easily felt by placing both index fingers into the opening of the ears and pressing the fingers forward. The movements of the condyle can then be easily palpated when opening and closing the jaw. Place your fingers over the lateral surfaces of the joint and repeat the opening and closing movements. Again note the anterior and downward movement of the condyle. (Clicking of the joint during these movements is relatively common.) When the jaw is moved to one side, feel again over the joint surface. The condyle on the one side will be forced laterally (outward), but the condyle on the opposite side will have a large excursion downward and forward somewhat as in opening the jaw.

TEETH AND OCCLUSION

The adult dentition is composed of 32 teeth. For purposes of description the dental arches can be divided at the midline into quadrants: maxillary (upper) right and left and mandibular (lower) right and left. Starting anteriorly the names of the eight teeth in each quadrant are central incisor, lateral incisor, cuspid, first bicuspid, second bicuspid, first molar, second molar, and third molar. When the teeth are in a normal closed position the upper anterior teeth form an arch to the outside of the lower arch. The anterior maxillary teeth are larger than

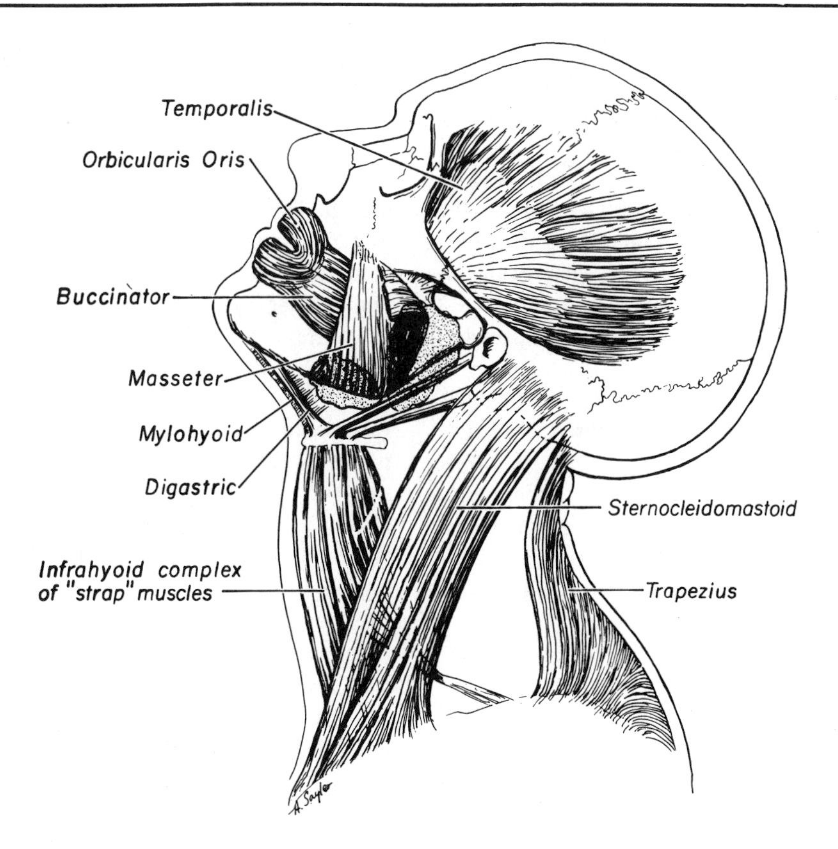

Fig. 9 - Lateral view of head and neck area showing musculature and parotid and submandibular salivary glands. Posterior origin of masseter muscle on zygomatic arch has been reflected back to show attachment of temporalis muscle on coronoid process of mandible. Parotid gland (shaded dotted area) is behind the mandible and the submandibular gland is found under mandible near the angle of jaw. Two muscles of facial expression are shown on this diagram–orbicularis oris and buccinator muscles

the mandibular anterior teeth so the lower arch fits inside the upper arch. The lateral cusps (buccal cusps) on the upper molar teeth are also found to the outside of the buccal cusps of the lower molar teeth. The medial cusps (palatal cusps) of the maxillary molars intedigitate between the buccal (lateral) and lingual (medial) cusps of the lower molars. The above description of the occlusion is only found in the usual or most common sense. Many variations of the occlusal pattern occur depending on the size and shape and jaw-to-jaw relationship of the mandible and maxilla. However, this static position has little relation to normal function. The teeth are normally apart. Seldom are the teeth in contact except when chewing (or when grinding of the teeth–a habit destructive to the dentition and to the supporting socket and gums). The muscles of mastication (to be described below), support the jaw so that there is always 2-4 mm of space found between the teeth. These muscles are responsible for mandibular movements. In this "rest position" there is a continuous weak "steady-state" contraction in these muscles which exists when the person is standing or sitting erect. These are postural type reflexes and are the result of nerve sensors in the muscle reacting to stretch in the muscle fibers. From these proprioceptive (position) receptors, nerves conduct impulses to brain centers and motor impulses from these centers then return via motor nerves to cause this continued regulated weak muscle contraction. This sensory-motor loop of muscle receptor, sensory nerve, brain motor nerve, and muscle fiber does not reach the level of consciousness (i.e., cerebral control). This is a sensitive system and imbalance can occur. Examples of imbalance include tooth loss and shifting of teeth, dental restorations which are too "high," fractures not reunited to the original occlusal pattern of the individual (which is not the same in every person).

MUSCLES OF MASTICATION

There are several muscles which move the mandible–these are the muscles of mastication (Fig. 9). They are powerful bulky muscles and are located so that such actions as opening, closing, protrusion (moving the jaw forward), and retrusion (moving the jaw backward), and all combinations are possible. These muscles are innervated by branches of the fifth (V) cranial nerve.

Place your fingers over the ramus of the mandible and clench the teeth intermittently. The muscle mass felt contracting is the masseter muscle. This muscle arises from the zygomatic arch and attaches to the outside surface of the mandible. Now place your hands high on the side of the face above a line between the eyes and ears. Again clench the teeth. The muscle felt is the temporalis, which originates on a larger area of the side of the head, and the muscle fibers come together on the inside of the zygomatic arch to insert on the coronoid process of the mandible. The morphology of the origin and insertion of the masseter and temporalis muscles show that the principal actions of these muscles is directed in closing the jaw. The medial pterygoid is a muscle similar to the masseter but located on the inside of the jaw. The medial pterygoid muscle is a thick bulky muscle which originates in the region directly behind the maxillary sinus, the infratemporal fossa, and then inserts on the posterior aspect of the inside of the mandible. The masseter, temporalis, and medial pterygoid all assist to close the jaw and are oriented primarily in an up and down direction.

The lateral pterygoid has an origin near the medial pterygoid but travels in a horizontal direction and inserts into the condyle of the mandible. This muscle helps to protrude the mandible and also initiates opening movements. Other muscles of mastication insert into the base of the mandible from below and aid in opening movements. These muscles (the digastric and mylohyoid) are situated so they can help in the opening movements of the mandible.

THE TONGUE

The tongue is a muscular mass of tissue covered by a very specialized epithelium. The epithelium is roughened by many varied projections called papillae. Taste buds are distributed throughout the surface of the tongue and concentrations of these are found in association with large rounded papillae found on the posterior part of the tongue. In addition to taste receptors, there are numerous other sensory receptors in the tongue which give very discrete perception of touch and pressure sensations. The tongue is a strong muscle and also is capable of very quick and finite motions.

In speech, mastication, and swallowing, the movements of the tongue are varied, very precise, and controlled. There are numerous muscles which make up the major mass of the tongue. The intrinsic muscles of the tongue run horizontally and vertically and longitudinally. Muscles from the mandible, hyoid bone, and styloid process also insert into the tongue and their actions contribute to the varied movements and actions of the tongue. The underside of the tongue, easily visible in a mirror by placing the tip of the tongue into the roof of the mouth, has a thin epithelium and underlying veins can easily be seen. There are also small salivary glands located in this area.

THE SALIVARY GLANDS

There are major and minor salivary glands (Fig. 9). The minor ones are small groups of glandular tissues found on the inside surfaces of the lips, cheek, and the under surface of the tongue. The salivary glands in the lips can be palpated and the orifices of these glands can be seen by everting the lip, wiping the inner surface dry with a cloth and then watching the bead-like appearance of the secretion products. The major salivary glands are the parotid glands, palatine gland, submandibular glands, and the sublingual glands. The sublingual glands are found in the floor of the mouth and the position of these glands can be seen as raised ridges when the tongue tip is placed back in the roof of the mouth. The submandibular gland is a walnut-shaped tissue located underneath and below the posterior inferior part of the mandible. The ducts from these glands and from much of the sublingual gland enter the oral cavity on the underside of the tongue in the floor of the mouth. When the tongue is lifted back a prominent vertical fold of tissue can be seen bridging from the inside of the mandible to the tongue. This is the sublingual frenulum and the two papillae-like folds of tissue approximately in the middle of the frenulum are the exits of the ducts. Dry this area and then watch saliva flow from these ducts. The parotid gland is a large gland which occupies a deep cleft in the area behind the mandible and in front of the ear. The duct for this gland passes on the outer surface of the ramus of the mandible and enters the cheek anterior to the ramus of the mandible to exit inside the mouth at a raised-tissue papillae. This is found near the upper second bicuspid or first molar tooth. Again opening the mouth wide, pull back the cheek and dry this area. Saliva should promptly flow from this duct. The palatine gland is located in the connective tissue near the junction of the hard and soft palate. The hard palate is the bony structure and overlying tissue that form the roof of the mouth and the floor of the nasal cavities. The soft palate is the muscular appendage which hangs into the pharyngeal region separating the nasopharynx from the oropharynx. The junction can be palpated and in this region glandular tissue is located. These glands have several ducts which open to the oral surface.

Saliva is necessary for the continued lubrication of the oral tissues, especially during speech and mastication, and for the moistening of food so that swallowing can be facilitated. The rate of flow is controlled by secretory motor nerves are stimulated by a number of sensations, e.g., taste, olfactory, tactile, emotional, or psychological stimuli.

MASTICATION AND SWALLOWING

Mastication is a complex function of integrated movements, involving many groups of muscles including the muscles of mastication, the muscles of facial expression, especially the orbicularis oris and buccinator muscles and the muscles of the tongue (Fig. 5). After incising food with the anterior teeth the tongue flips the food to the occlusal table of the molar teeth. This is invariably to the same side. Most everyone has a favored "chewing side" where the majority of mastication takes place. Mastication of the food is not just an up-and-down movement of the mandible but rather it is an elliptical movement. Experiment by chewing a cracker in front of a mirror and notice the actions described above. The tongue on the inside and the buccinator muscle of the cheek play an important role in keeping the food up on top of the chewing surfaces. The buccinator actually has a roll-like action during chewing. The tongue moves quickly and accurately during mastication. These actions are best noticed by using yourself as subject and observer.

When the food is to be swallowed, the tongue initiates this action by pushing the food back, and the space between the tongue and the roof of the palate is closed off. Rapidly the

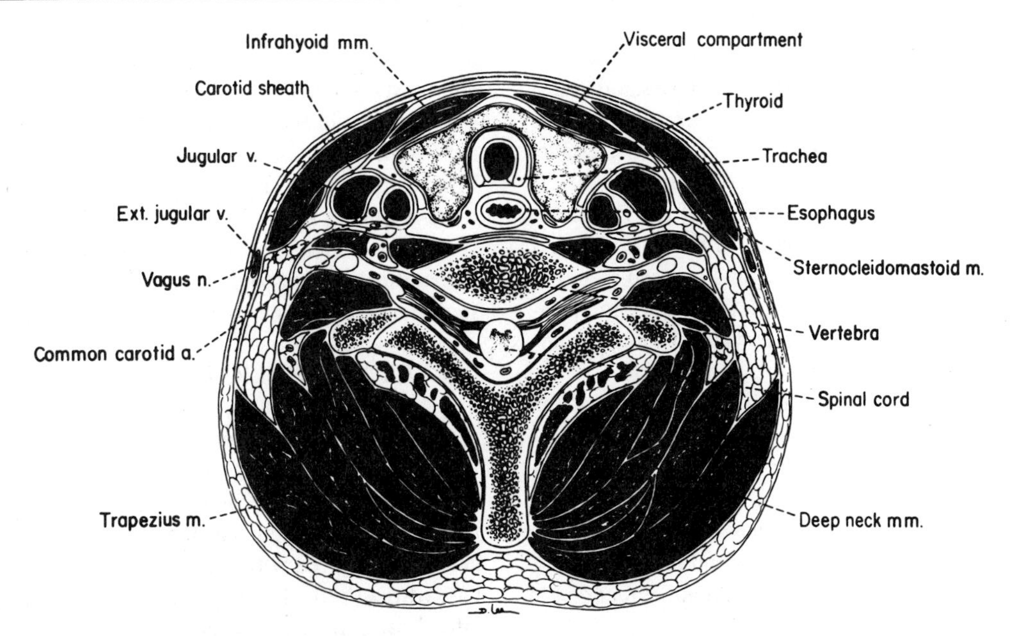

Fig. 10 - Cross section of neck at a level below thyroid cartilage. Compartments of neck are surrounded by connective tissue layers. Note relationships of trachea (anterior) to esophagus (posterior). In carotid sheath jugular vein is lateral to carotid artery

soft palate is raised and the pharynx is lifted, and the laryngeal opening is closed. The laryngeal tissues are raised and tipped forward during this stage of swallowing. These movements aid in protecting the openings to the lungs. The middle parts and inferior parts of the pharynx constrict in sequence forcing the food directly to the esophagus. By rapid contraction of the muscles in the esophageal wall from superior to inferior the food is driven into the stomach. Again, using yourself as subject, place your hand over the larynx and notice the direction of movement during swallowing.

BLOOD SUPPLY OF THE NECK AND FACIAL REGIONS

In the neck region the common carotid artery gives off the internal and external carotid arteries (Fig. 7). The internal carotid artery continues on superiorly and enters the cranium to supply the brain. The external carotid artery supplies all the tissues of the facial region and also tissues of the anterior parts of the neck. The branches of this artery are:

1. Superior thyroid to the anterior tissues of the neck and thyroid gland.
2. Lingual artery to the tongue.
3. Facial artery which crosses the exterior surface of the mandible just anterior to the insertion of the masseter muscle and continues in a tortuous path giving off branches to the lower and upper lip and nose and terminates near the inner corner of the eye.
4. Maxillary artery which passes deep to the ramus of the mandible supplying the tissues in this deep facial region and then terminating in many branches which supply the maxillary sinus, upper and lower teeth, palatal region, and lateral wall of the nasal cavities. One branch continues in a bony canal in the floor of the orbit and exits from the infraorbital foramen which is found below the lower rim of the orbit. Branches of this artery supply the lower eyelid region, nose, and parts of the upper lip.
5. Superficial temporal. After the maxillary artery is given off in the substance of the parotid gland the superficial temporal artery continues superiorly and sends branches to the lateral surfaces of the cranial region.

There are other smaller branches of this external carotid artery to the tonsilar and pharyngeal regions. Also a posterior branch supplies the region of the occipital bone.

COMPARTMENTS OF THE NECK

The neck when viewed in cross section can be described as consisting of several compartments (Fig. 10). These compartments are surrounded by connective tissue sheaths. The compartments are not absolute boundaries as there are pathways for nerves and vessels between these compartments. These connective tissue boundaries do allow for movements of one compartment in relationship to another during various movements of the head and neck.

The posterior cervical compartment includes the cervical vertebrae and the several deep muscles of the neck. These muscles surround the vertebrae and are primarily concerned with holding the head erect, with rotational movements, and with extension and flexion of the head. These same kind of muscles extend through the deep back region surrounding the vertebral column.

The carotid sheath is another functional envelope containing the common carotid artery, the internal jugular vein, and the vagus nerve. The common carotid arteries and internal jugular veins are the principal arteries and veins of the head and neck region. Vessels arising from the subclavian system of arteries and veins in the shoulder region supply lower parts of the neck and also give rise to the vertebral arteries and veins which partially supply the brain. The vagus nerve gives branches to pharyngeal musculature both to actuate muscles and also to supply sensory information from this region. The laryngeal tissues are innervated by branches of the vagus. The major components of the vagus nerve serve thoracic and abdominal viscera. Impulses from the vagus slow respiration and heart rate and increase secretion of digestive glands and peristalsis (mobility) of the smooth muscle walls of the gastrointestinal system.

There is a visceral compartment in the anterior part of the neck. Located in the anterior part of this region are the laryngeal tissues and the trachea and in the posterior part the esophagus. The thyroid gland is found here. This gland is very vascular and secretes a hormone–thyroxine–which increases the rate of general body metabolism. Also found here are the small bean-shaped parathyroid glands which are embedded in the connective tissue of the posterior parts of the thyroid gland. These glands secrete hormones which regulate the levels of calcium and phosphate ions found in body tissues and in the blood.

Visualize the closeness of the larynx and trachea to the very front of the neck. The laryngeal protuberance ("Adam's apple") can be seen and palpated. Press with your fingers below this area and several pliable rings of cartilage can be felt. Immediately inferior to the thyroid cartilage is the cricoid cartilage and again inferiorly are found the several cartilage rings of the trachea.

Tracheotomies are performed by incising the skin and connective tissues directly over the trachea permitting air to enter and exit to and from the lungs. This procedure is instituted in emergency situations in which trauma to anterior facial and neck tissues destroys a patent airway or when foreign bodies or swelling of laryngeal tissues block the airway. Emergency tracheotomies are usually done immediately below the thyroid cartilage in the cricoid cartilage, or more inferiorly in the trachea proper.

Anterior to the visceral compartment are several flat muscles that comprise the infrahyoid group of muscles. These are the "strap" muscles of the neck and are found between the sternum, thyroid cartilage, and hyoid bone. They assist in movements of the thyroid cartilage and hyoid bone and also serve to stabilize these components for the action of other muscles which originate from these same components. These other muscles insert into the muscles of the pharynx and tongue and into the mandible.

The trapezius and sternocleidomastoid muscles are located in the superficial connective tissue envelope of the neck. The trapezius is a broad flat muscle which covers most of the back of the neck region. It inserts onto the occipital bone and originates on the scapula in the thorax.

The sternocleidomastoid muscle is a large thick bundle which is easily seen in the front of the neck, especially when the the head is turned to one side. The anterior border becomes prominent during this movement. This muscle travels obliquely from its origin on the mastoid process of the temporal bone to its insertion on the medial part of the clavicle and on the sternum.

Also found in the superficial tissues of the neck are prominent veins which assist in the draining of the facial tissues. The position of these superficial veins is variable; however, there is usually an external jugular vein which arises from the veins near the angle of the mandible and passes superficial to the sternocleidomastoid and terminates in deep veins in the shoulder region.

CERVICAL VERTEBRAE

The neck contains seven cervical vertebrae (Fig. 1). These vertebrae are connected by a comprehensive series of ligaments and joints. The first cervical vertebra is referred to as the atlas. On the upper surface of the atlas there are two facets, each covered with cartilage, which articulate with the base of the skull. These articulating facets are found on the lateral parts of the atlas and form the inferior parts of the joint. The upper parts of the joint are the condyles of the occipital bone of the skull. The condyles are smooth-surfaced and are covered with hyaline cartilage. These joints are completed with a fibrous joint capsule. Additional ligaments strengthen and limit the movements between the atlas and the skull. Much of the up and down movement (extension and flexion) of the head occurs at this joint. These joints are close (in an anterior-lateral direction) to the foramen magnum so that movements about these joints do not create undue tension to the spinal cord, its coverings, and associated vessels. More limited extension and flexion can occur between the several other cervical vertebra. The articulation between the first cervical vertebrae (atlas) and the second cervical vertebrae (axis) is also unique. Greater angular rotation (moving head from side to side) is possible at this joint than is possible at the other articulations between remaining vertebrae. There are additional ligaments found supporting these joints and limiting the extent of movement found at the articulations of the axis and atlas.

PHONATION

Speech involves the combined integrated actions of the abdominal diaphram, lungs, trachea, larynx, pharynx, tongue, soft palate, and cheeks and lips. Air passing the vocal cords in the larynx causes sound vibrations to be emitted and these sounds are greatly modified by the size and shape of the pharyngeal and oral tissues. Several muscles in a complex arrangement are situated in the laryngeal area so they can reduce or increase the tension on the vocal cords to change the pitch and some of the quality of the sound. Pronounce the vowels a.e.i, o, and u and notice the wide range of openings and constrictions possible in the oral cavity and pharyngeal regions. Especially notice the positions of the tongue. In other speech sounds, such as consonants, the tongue acts as an intermittent valve mechanism against the lips, palate, and upper teeth. Speak the consonant "f" and note the valve action between lower lip and maxilliary teeth, "t" between the anterior palate and tongue.

Observe the positions of the oral and palatal tissues with other sounds such as k, g, th, l, s, r (as in rough), etc. Small differences in the positions of these tissues causes pronounced differences in the sound.

REFERENCES FOR FURTHER STUDY

For further study of the head and neck, an anatomy textbook written for nurses or dental hygienists would be advisable for study. These texts usually cover the function (physiology) as well as the morphology (anatomy) and do allow for a more comprehensive and initial treatment.

Recently many good texts have been written on human anatomy for the undergraduate college student. These would also be good to start with. The best resource would be a medical or dental school library to browse through the anatomy texts and consult with the librarians. For the more serious student, one should consider (1) self-study movies on human anatomy. These usually are of excellent quality but still do not simulate the knowledge gained from actual specimens. (2) enroll in (or audit) courses in human anatomy occasionally given for lawyers, undergraduate college students, medical illustrators, paramedical or paradental assistants. These initial experiences give the learner a good overall view and perspective to then proceed with more detailed study.

The coverage in this short chapter is extremely simplified and generalized. There are several complete texts covering only the head and neck areas. Literally every facet on every bone and every visually observable structure has been studied in detail and the anatomy of this area is well documented. Even for the casual observer, it would be enlightening to peruse a copy of the classic Grey's ANATOMY or to look at the contents of any of the several anatomical journals (Anatomical Record, Acta Anatomica, Journal of Anatomy, to name a few) to assess the "state of the art" in these human morphological disciplines.

700195

Morphology of the Nervous System as Related to Trauma

Charles L. Votaw
University of Michigan

THE HUMAN BRAIN subserves all known intellectual processes. Although the nervous system of man cannot function alone, it is essential to normal human endeavor. Therefore, the body has attempted to protect the brain in special ways, both in shielding and in internal arrangement. A general knowledge of these protective mechanisms is helpful in understanding the results of trauma.

The way in which the nervous system subserves such processes as memory, learning, and emotion are unknown. The nervous system is a complex arrangement of interrelating conducting elements. These interrelationships are precisely arranged and delicately balanced in order to subserve these complex functions.

No attempt will be made to describe the details of the nervous system in this presentation. Rather, a presentation of certain basic relationships and generalities of the nervous system and how these are affected by trauma will follow, beginning first with a discussion of the morphology and physiology of the nervous system and then progressing to descriptions of functional disturbances related to trauma.

THE COMPONENTS

The basic component of the nervous system, as with all living organ systems, is a living cell. This cell is called a neuron. Like other cells of the body, it is made up of a cell body, a nucleus, and living protoplasm (Fig. 1). In addition, the neuron is specifically specialized for the conduction of nerve impulses. These specializations are in the nature of long tentacle-like processes which project from the nerve cell body (Fig. 1). Those processes which receive incoming information are called dendrites, and a single process of each cell which carries information away from the cell body is called an axon. The metabolic processes of the nerve cell are particularly designed to subserve the transmission of the nerve impulse.

Only the axon actually transmits a nerve impulse. A potential difference is established across the axon membrane such that the inside of the axon is negative with respect to the space outside. This potential difference is created by ionic concentrations established by the living cell. During the process of the transmission of the nerve impulses, a depolarization occurs across this membrane so that in the immediate area of activity there is a breakdown of the normally maintained potential. This area of depolarization has the characteristic of being able to depolarize the membrane immediately next to it. In this way the depolarization travels along the axon and is known as the nerve impulse. The nerve impulse is thus an electrochemical phenomenon and is not similar to an electric current. The axon itself serves as a very poor cable and will not transmit an electric current in the sense that a copper wire will. Since the electrochemical process depends upon the amount of the membrane potential, which remains constant, each nerve impulse is like every other nerve impulse. This is known as the "all-or-none" law.

Certain protective sheaths surround the conducting fiber. Immediately around the axon is a structure known as the myelin sheath (Fig. 1). This sheath is actually a modified living cell membrane. The protoplasm of the cell whose membrane forms the myelin forms a second sheath around the axon. There is a full range of variation in size of the myelin sheath from not being present at all to a level in which this sheath is 3-4 times larger than the axon itself. All of the axons are covered by a cellular sheath wherever they are found in the body. These sheaths appear to be important in both the speed of conduction of the impulse, which varies from 10-400 ft/sec, and in isolating each axon to eliminate the possibility of cross-talk in conducting elements.

The axon of a nerve cell, as it reaches its termination, branches into a large number of terminals. Each terminal ends in close relationship to the dendrite, cell body, or axon of another cell or, in certain instances, comes into relationship with an effector organ such as a voluntary muscle, smooth muscle, or gland cell. The area of termination close to a second neuron is known as the synapse. The synapse consists of a bulb-

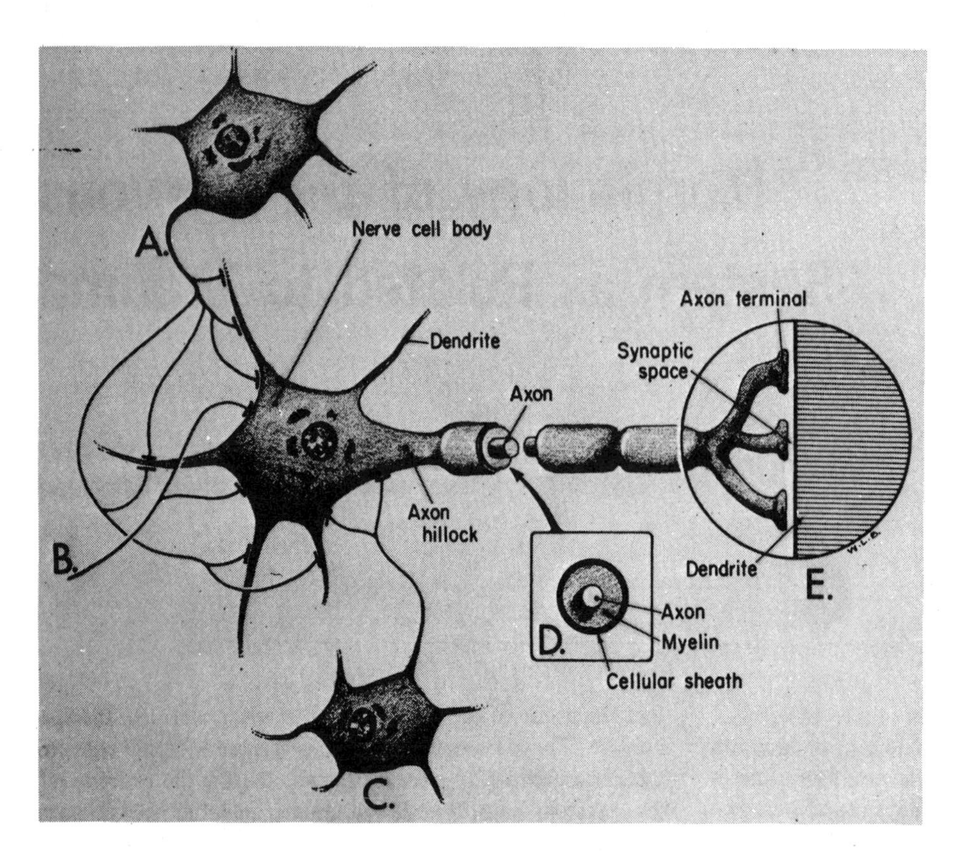

Fig. 1 - A, B, C - Schematic representation of a neuron, its parts, and interrelationships. D - Enlarged representation of an axon and its sheaths. E - Magnified representation of a synaptic ending.

like ending at the end of an axon with a space between it and the membrane of the second neuron (Fig. 1). The relative sizes are such that a single neuron can be covered by hundreds to thousands of synaptic end bulbs. When the nerve impulse reaches the end-bulb region, it effects the release of a chemical substance which crosses the synaptic space and causes a change in membrane potential of the second neuron. Any potential change within the protoplasm of the dendrites or nerve-cell body is reflected in a change of transmembrane potential in the region of the axon hillock. The nerve impulse normally is generated only at the axon hillock. The size of this change will be determined by the total number of acting synaptic end bulbs, the time duration of this action and the distance of the end bulb from the axon hillock. If the result of this synaptic activity decreases the transmembrane potential in the region of the axon hillock to a critical level, a nerve impulse will be generated and then passed down the axon of the cell. If, on the other hand, the results of the synaptic transmitter chemicals increase the membrane potential in the axon hillock region, it will be more difficult to create an impulse. The former situation is known as excitation and the second condition as inhibition. A single nerve cell receives synaptic endings from a large number of axons, some of which are excitatory and some of which are inhibitory.

Since there is only one basic component in the conducting system of the brain and since a nerve impulse follows the all-or-none law, the only alterations in information being transmitted must be created by variations in the synaptic relationships between cells. If there is a very large number of excitatory synaptic endings active at any particular instant of time on a given cell, then the second cell will respond to this activity by sending out a nerve impulse. The summation of a large number of acting end bulbs is known as spatial summation. The duration of a nerve impulse is approximately 1ms, but the duration of the activity at a particular synaptic bulb occurs over several milliseconds. Therefore, if a large number of impulses come down the same axon in a brief period of time, the results add up over a period of time to cause the cell to respond. Such a summation of activity is known as temporal summation. Both spatial and temporal summation can result in excitation or inhibition of the cell. If both excitatory and inhibitory influences are working upon a cell simultaneously, the resultant activity will be some balance between the two influences which will produce nerve impulses in the second cell which are temporarily coded. The different temporal codings which result are meaningful to the nervous system in terms of information being transmitted.

Unlike the multiple components of electronic instruments, the nervous system has but a single operating component, the neuron. There are variations of neurons in terms of size, in the number of endings on a particular axon, and the number and length of dendrites. These are morphologic variations which seem to be related to specific functions. Dendrites are usually rather short and located in the immediate vicinity of the neuron. A given nerve cell body varies from 2-150 μ in size. The length of the axon is extremely variable, ranging all the way from 15-20 μ to 4 ft.

THE CHASSIS

Each conducting element of the nervous system is surrounded by a sheath which is a cellular structure. Within the

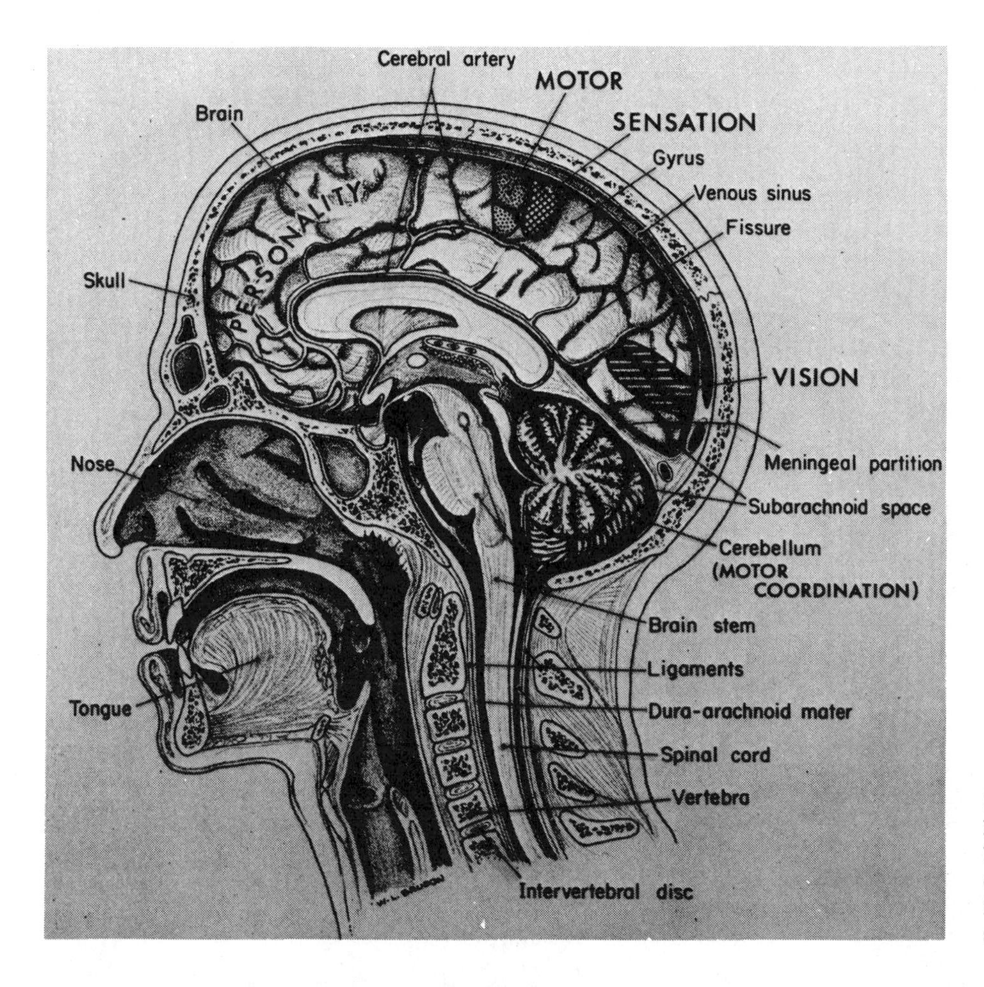

Fig. 2 - View of the brain within the head which has been sectioned midway between right and left in a front-to-back direction. Morphologic structures and function of various brain parts are shown.

substance of the brain these cells are known as glia. These glial cells fill up the entire space between neurons and actually support and give form to the brain. They also serve in supporting the metabolic processes of the nervous system, but do not, at least according to present knowledge, take part in transmission of neural information except in a supportive way.

The portion of the brain which lies within the head and backbone is known as the central nervous system. The parts of of the central nervous system are the brain, the brain stem, the cerebellum, and the spinal cord (Fig. 2). In general terminology the large portion of the nervous system within the head, which is familiar to most persons, is known as the brain. There is also a substance of material leading from the brain to the spinal cord known as the brain stem. On top of the brain stem is another structure known as the cerebellum which is important in normal motor operation of the individual. Finally, the spinal cord is a portion of the nervous system within the vertebral canal.

Fig. 2 illustrates the parts of the brain as if the human body were cut in half in a front-to-back direction. An indication of the anatomical terminology for locating brain parts is shown.

THE CIRCUITRY

The neurons are arranged within the nervous system in a constant organizational pattern. The function of nerve cells and fibers can often be related to their position. This is dependent upon the peripheral input and output of the system to be referred to later. Axons which tend to mediate information of a similar nature tend to accumulate together and are known as a track of the central nervous system. The nerve cell bodies which give origin to these axons also tend to accumulate in functionally definable morphologic areas and such are termed "nucleii." We do not have precise information on location of tracks and nucleii for many of the functions of the nervous system. However, enough information is known which makes it reasonable to consider specific functional areas.

Fig. 3 illustrates a cross section of the spinal cord. There is a central area within the spinal cord which contains nerve cell nucleii. As is indicated in the drawing those nerve cells located in the ventral portion of this "H"-shaped structure are functionally related to the voluntary muscles of the body. Similarly the nucleii of the dorsal part of this region are related to sensory input. Around the outside of the area of nucleii of the spinal cord are all of the conducting axons carrying information to or from the brain. Illustrated in Fig. 3 is the location of the nerve fibers which transmit information which is interpreted by the individual as pain and/or temperature sense. There is a further localization within this track for each region of the human body. Thus, pain fibers from the foot are located dorsally in this track and pain being relayed from the hand is located more ventrally. Also indicated within the cross section of the spinal cord is an area for discriminative senses

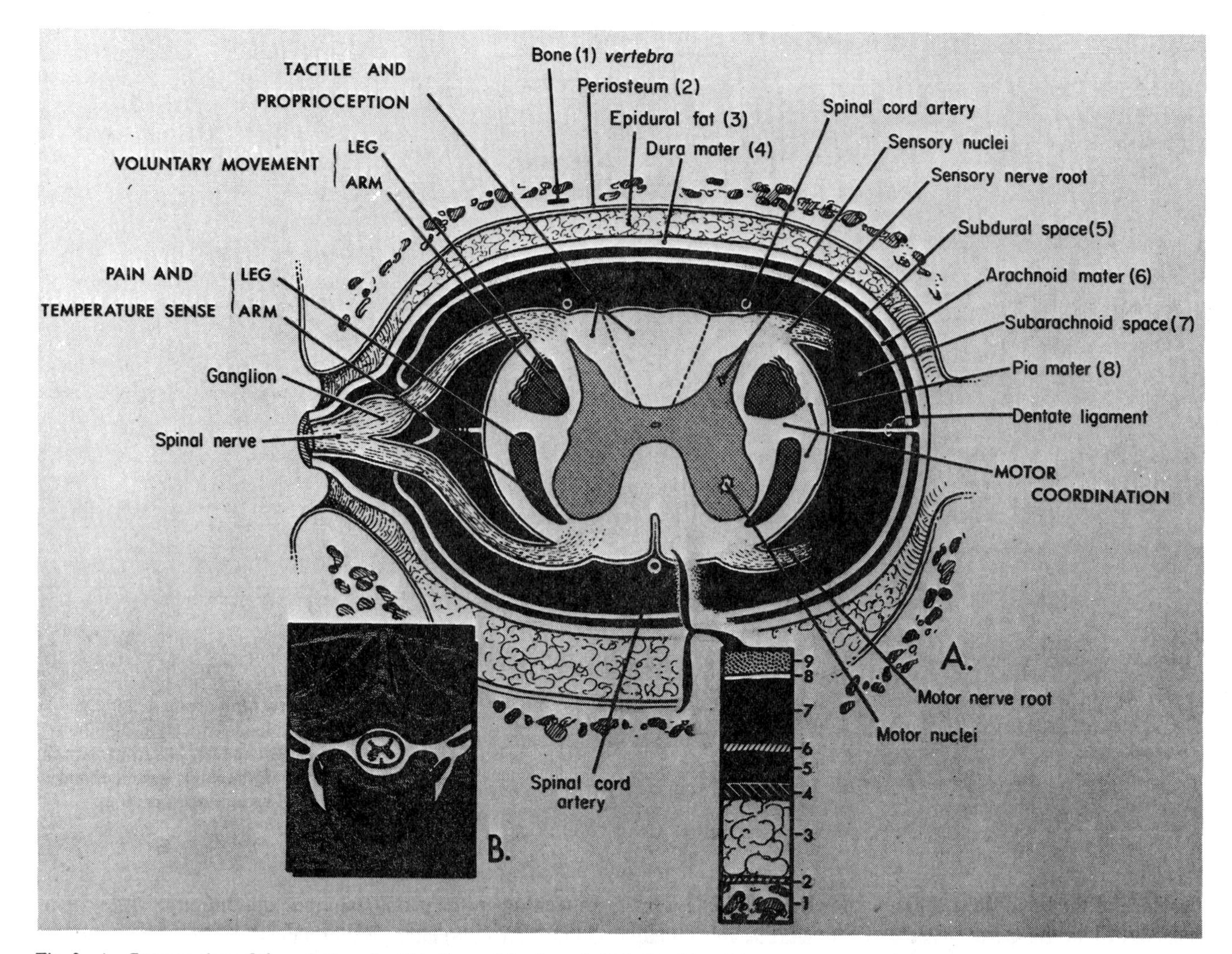

Fig. 3 - A - Cross section of the spinal cord within the vertebral canal. Relationship of the spinal cord to its coverings is shown. Numerical representation is the same as that used in Fig. 9. B - Illustration of the position of cord within the posterior body wall.

which relays information concerning the position of our body parts in space (proprioception). The position of fibers in the voluntary motor system is also indicated in Fig. 3 and consists of fibers which are going to the motor region of the area of nucleii of the spinal cord. This pathway is necessary for all voluntary movement. There is also a location within this path such that those fibers which will operate the voluntary muscles of the leg are found toward the outside of the spinal cord and those of the arm are toward the center of the spinal cord.

Voluntary motor movement alone is not sufficient since all movement needs to be regulated and coordinated. This coordination is controlled in part by the cerebellum and input by tracks other than the voluntary motor system. These fibers are scattered throughout the nonlabeled area in Fig. 3.

Fig. 4 is a representation of the lateral surface of the brain itself. In this figure and in Fig. 2 are indicated many folds and grooves. The folds are known as gyri and the grooves as fissures. These folds and grooves are constant and, therefore, useful as an anatomic localization. Certain functions can be associated with particular gyri. Indicated in the drawings are the areas of the brain that have to do with sensory input; (that is, vision, general body sensation, audition) and motor output; (that is, motor cortex, eye movement, and voluntary speech). Other areas not specified in the drawings are associated with more complex functions of the nervous system such as personality, memory, learning, etc.

A knowledge of the functional location of the parts of the nervous system is very useful in understanding defects which result from injuries to the central nervous system. In general the left side of the brain controls the right side of the body and vice versa. This crossing of functions occurs when large tracks of axons cross from one side to the other in various parts of the brainstem.

THE ENVELOPE

The central nervous system is arranged in such a way that it actually floats within a space filled with fluid. This space is lined, within the bony structure, by membranes known as meninges (Figs. 2, 3, 6). There is a membrane which applies directly to the outer side of the central nervous system known as the pia mater. Immediately outside of the pia is a space which is bounded on the outside by another membrane known as the arachnoid mater. The space which lies between the pia and

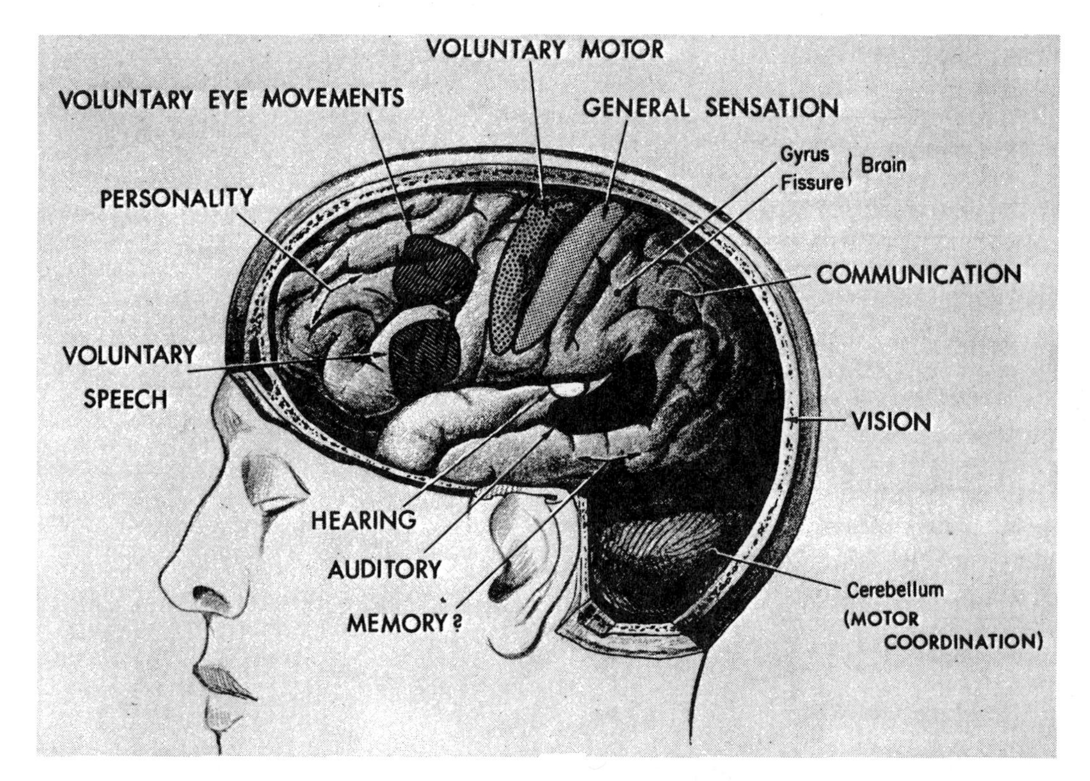

Fig. 4 - View of the brain within the head. Indicated are the functional areas and structures of the brain.

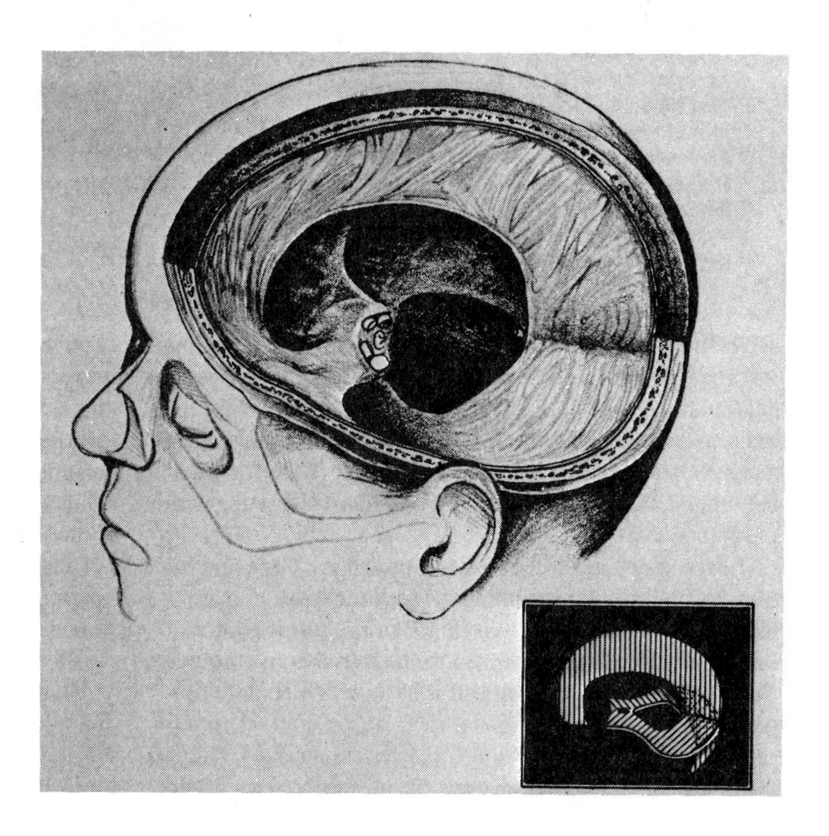

Fig. 5 - Illustration of the inside of the skull as it would appear if the brain were removed and the meninges left intact. Spaces created for various parts of the brain by the meningeal partitions are indicated. Inset is a diagrammatic representation of the configuration of the meningeal partitions

the arachnoid is known as the subarachnoid space. This space is filled with a fluid, the cerebral spinal fluid, which is produced by specialized cells found deep within the brain. Surrounding the outside of the arachnoid and attached closely but not always firmly to the bony structure is the dura mater. In normal life there is no space between the arachnoid and dura mater; however, the union of these two ligamentous membranes is not strong and can be separated rather easily.

Within the cranial cavity the arachnoid and dura form certain partitions as indicated in Fig. 5. These partitions support the brain in place and separate one neural part from another. Such a meningal partition is shown in Fig. 2, lying between the

Fig. 6 - A - Illustration of the brain within the skull sectioned in a right-to-left direction midway between front and back of the head. B - Enlarged representation demonstrating the configuration of meninges, the formation of a dural sinus, formation of the subarachnoid space, and the configuration of bridging veins. C - Magnified representation of the configuration of the meninges emphasizing the formation of subarachnoid space and the differences between blood supply of the brain and of the meninges

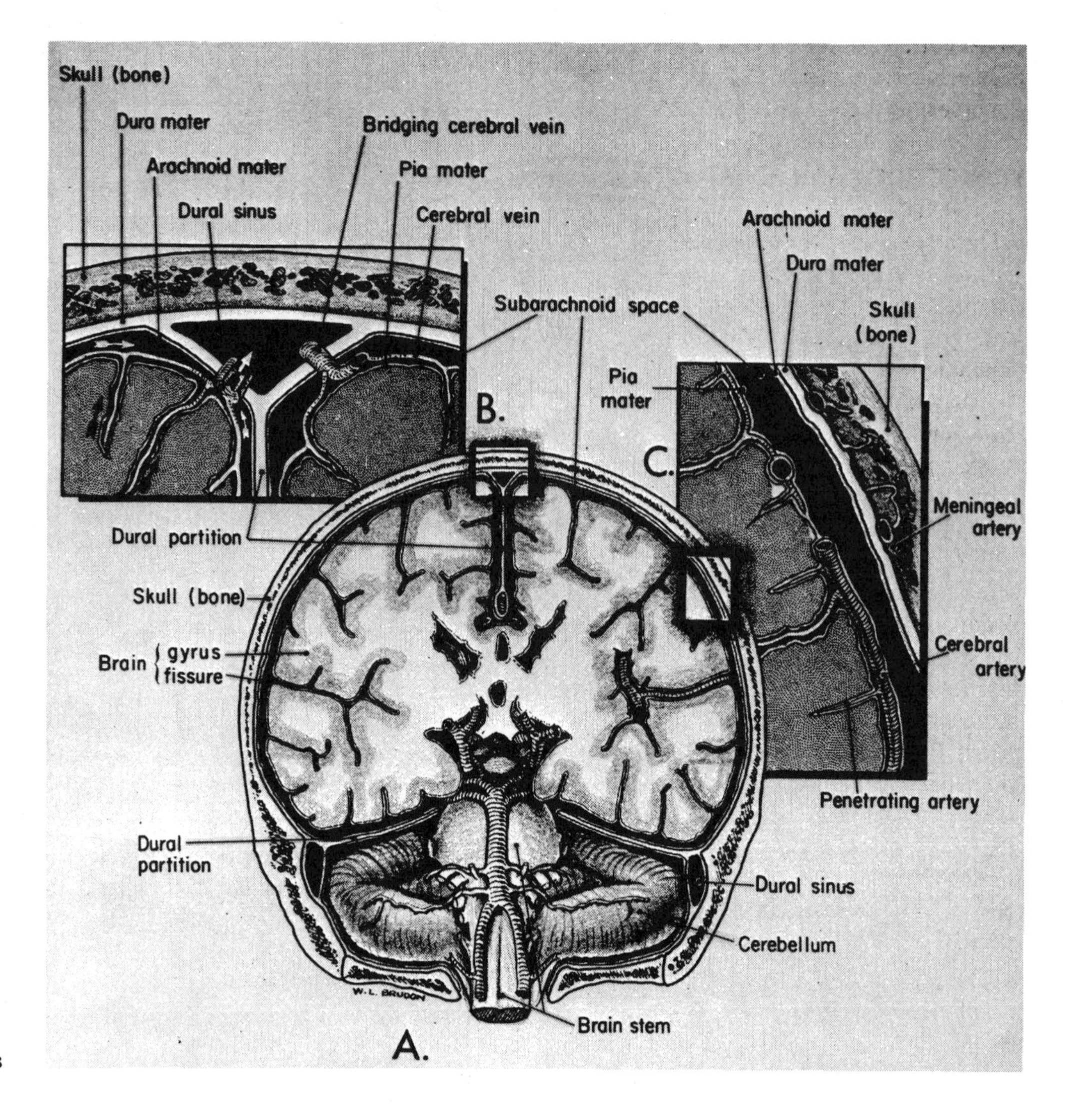

cerebellum and brain (tentorium cerebelli). These partitions can serve as sharp instruments or space-limiting structures during traumatic episodes. The only specific fixation of the brain to outside structures is through connections of the cranial nerves which arise from the brain stem. Therefore, within the limited area available the semisolid brain is floating within the cranial cavity.

The meninges and spaces created around the brain and brain stem and cerebellum is continuous with the space that surrounds the spinal cord as well (Fig. 2). The spinal cord is somewhat more firmly attached to the lateral bony structures than the brain. This attachment is by way of a dentate ligament which connects the lateral surface of the spinal cord on each side with the lateral area of the vertebral canal. This ligamentous membrane fixes the spinal cord firmly in the center of the canal.

THE SHIELD

The firm and hard protective shield of the nervous system is the bony covering plus certain ligamentous structures. The brain and brain stem lie within the cranial cavity which is formed by the cranial bones (Figs. 2, 4-6). Several separate bones make up this shield. In the adult the joints between the separate bones are firm and fixed; therefore, in effect producing a cavity with a fixed capacity. During the developmental phases of growth of the human being the separate bones of the cranium are not firmly attached to each other. This allows for enlargement to accommodate the growing brain. Since the bones are not firmly attached to one another, the dynamics of the fluid spaces within the cranial cavity differ in the young person from that of the adult. These cranial bones become fused to each other sometime during growth, usually in the early to middle teen age period.

The spinal cord lies within the vertebral canal which is made up by the vertical arrangement of the several vertebral bones (Fig. 2). This results in a vertebral canal, in which lies the spinal cord, not completely surrounded by bony structures. There are several openings into the vertebral canal between bony structures which are filled with ligamentous structures. These ligaments attach one vertebral bone to another (Figs. 2, 3). Lying between each body of the vertebrae is an intervertebral disc which is made up of a semisoft, compared to bone, material. This disc allows for some flexibility and compressibility of the vertebral column. As can be seen from this description and in Fig. 8, the spinal cord is not so totally protected in terms of a shield as is the brain. This is related to the back being designed as a semiflexible structure as compared to the inflexible head.

PROTECTION AGAINST DETERIORATION

The brain is a living organ demanding a constant supply of nutrient materials to support its metabolic processes. These materials are brought in by way of the arterial system, and waste products of metabolism are eliminated through the venous system.

The arterial supply of the brain and the spinal cord can be divided into two subsystems. First, there is an arterial supply to the nervous tissue itself and second there is a supply to the meninges which has a characteristic different position in relationship to the former system.

The arterial supply to the brain (Figs. 2, 6) reaches the structure through openings in the base of the skull which transmit major arteries into the cranial cavity. These arteries traverse the spaces from the bone to the brain and branch into cerebral arteries through which blood is distributed to the brain. These arteries are located on the surface of the brain within the pia mater. Branches of the cerebral arteries penetrate and supply the substance within the brain. In general, there are two arterial inputs; one, the internal carotid system, lying forward in the skull and supplying most of the front, top, and inside portions of the brain. A second arterial system enters through the large opening in the back of the skull, foramen magnum, through which the spinal cord joins the brain stem, and supplies the brain stem and the back portions of the brain.

The arterial supply to the spinal cord (Fig. 3) enters the vertebral canal through the openings which occur between the various vertebrae. After these arteries cross the spaces around the spinal cord, they are found within the pia mater. These arteries then distribute branches to the substance of the spinal cord.

After it has traversed through the capillaries and an exchange of metabolic materials has occurred, the blood enters the venous system. The veins of the brain, like the arteries, run on the surface of the pia mater. At some place the vein must turn to cross the subarachnoid space and enter the venous drainage system found within the dura mater. These particular venous channels (Figs. 2, 6) within the dura mater are known as sinuses. The venous sinuses empty out into the venous channels within the neck (jugular system) which return the blood to the heart. In contrast, the venous supply of the spinal cord is returned by veins which run the same course as the arteries and there are no separate venous channels found within the dura around the spinal cord.

The weakest point of the entire vascular system of the brain and spinal cord is that point at which the channels must cross the subarachnoid space. The arterial supply is under the force of arterial blood pressure which is operating normally at approximately 120 mm Hg. This pulsating blood pressure is transmitted to the inside of the skull and, since the skull is a fluid-filled space, this pressure is transmitted to all the structures within the cavity. There is, therefore, a normal intracranial pressure which is not so high as blood pressure but is above atmospheric pressure (100-150 cm of water). The portion of the artery which crosses the subarachnoid space is structured so that the blood pressure does not blow out the vessel.

The venous return from the brain also has its weakest point in that area in which the vessel leaves the surface of the brain and crosses the subarachnoid space to enter a venous sinus. Since the head normally is higher than the heart and blood is returning to the heart largely by hydrostatic pressure, the pressure in the venous side of the circulation is rather low. The strength of the walls of the veins do not appear of necessity to be as strong as the arteries crossing the space. The area of the vein as it enters the dural sinuses is particularly prone to injury.

The other arterial system is that which supplies the meninges. This arterial system also enters the cranial cavity by way of an opening in the skull. The branches of this system lie very close and actually within grooves of the cranial bones on the outside of the dura mater (Fig. 6). The venous return of this system follows the distribution of the arteries.

The fluid in which the brain is suspended is cerebral spinal fluid which is produced by gland cells deep within the brain substance. Some of the waste products of metabolism are eliminated from the brain into the fluid. This fluid is continually taken up by the dural sinuses and then carried away from the brain by means of the venous return.

INPUT AND OUTPUT

The nervous system can obviously not function without obtaining information. This information arrives at the central nervous system by means of peripheral nerves which in turn receive their information from specialized sensors distributed throughout the body. Similarly, the nervous system could not be an active instrument unless there was some mechanism for information to go from the central nervous system to a series of effectors. Therefore, peripheral nerves have fibers in them which carry information to the voluntary muscle system, the involuntary muscle system (smooth muscles) and gland cells.

The peripheral nervous system consists of all parts of the nervous system which are left after subtracting the central nervous system. Therefore, the peripheral nervous system contains all the nerves of the body. These peripheral nerves consist of nerve cell axons with various protective surrounding sheaths. The axons carry information in one of two directions. Those that carry information into the nervous system are called afferent fibers and those that carry information from the central nervous system to the effector are known as efferent fibers.

Distributed throughout the body is a number of specialized sensors. Each sensor is tuned to a particular type of incoming energy. Thus, for example, the eye responds to a particular portion of light spectral energy but not sound or other forms of electromagnetic energies. The ear responds to vibratory energy which takes place over a certain frequency range only.

Similarly, there are special types of endings responding to pain, temperature, tactile, proprioceptive (the knowledge of a position of a body part in space), visceral (sensory information from the stomach, intestinal tract, and other internal organs) stimuli. There is also a considerable amount of input to the central nervous system which we cannot categorize in a conscious sense. This input is used by the nervous system to regulate and control the life processes of the human body and much of this control lies outside of our voluntary abilities to appreciate or regulate.

It will be recalled that the nerve impulse in all axons is the same and that each nerve impulse follows the all-or-nothing law. Therefore, the ability of the nervous system to distinguish one sensory modality from another depends upon the hook up of the nerve fibers to the specific sensor and the location where this input arrives within the central nervous system. Also the temporal relationships of the nerve impulses in terms of pattern and frequency is dependent upon certain physiologic variations such as intensity or frequency of the inciting energy. The particular specific mechanism by which the sensors induce a nerve impulse in an afferent nerve fiber is not clearly understood at this time.

The afferent nerve fibers have their nerve cell bodies in accumulations of nerve cells which lie outside the central nervous system and are called ganglia. There is but a single axon which extends from the sensor to the central nervous system (making some axons as long as 4 ft).

Information which leaves the nervous system going to the effectors also goes by way of the peripheral nerves. In this instance the cell body of the axons which lie in the peripheral nerve are located within the central nervous system. In the case of the voluntary muscle system a single axon travels from the central nervous system to the effector which is a voluntary muscle. The relationship between the efferent axon and the striated muscle is remarkably similar to a synapse. The action potential traveling down the efferent fiber causes the release of a chemical at the terminations of the axon which in turn travels across the space and causes the muscle to contract. There are a large number of axons supplying a given muscle of the body. The intensity, the time course, and duration with which a muscle contracts depends upon the total number of active nerve fibers going to the muscle at any given time and the pattern of organization of impulses within the same and related axons.

The efferent system which supplies the involuntary motor system and gland cells differs in that it requires two neurons to span the distance from the central nervous system to the effector. The second neuron gives off an axon which goes directly to a smooth muscle cell or gland cell where a relationship is created not unlike a synapse. The smooth muscles of our body are used to regulate the digestive process, the caliber of blood vessels, and, in general, maintain proper body functions. This regulation is effected by the nervous system using afferent sensory information so that the body is maintained in a certain normal physiologic state. This regulation is outside our voluntary control. When demands of particular activity are created so that a change from normal static conditions is required it is also effected through this involuntary nerve system.

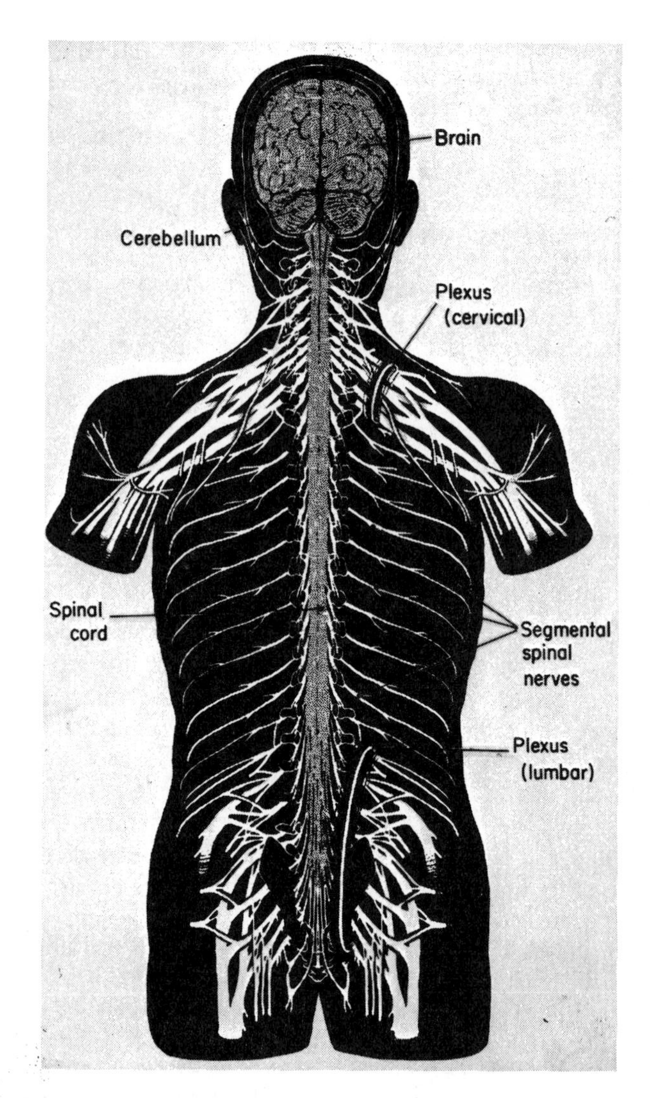

Fig. 7 Illustration of the segmental nature of the spinal nerves and formation of the nerve plexuses

Both afferent and efferent nerve fibers are found within the same peripheral nerve of the body. We, therefore, refer to most nerves as mixed in that they carry both motor and sensory information. There is no obvious morphologic difference between those fibers which carry incoming and those which carry outgoing information. There are 12 peripheral nerves which arise from the brain stem within the skull and these are called cranial nerves. All of the remaining nerves of the body arise from the spinal cord and are known as spinal nerves. The nerves are enclosed within a fibrous sheet which is usually very firm so that the peripheral nerve has a good deal of tensile strength.

The cranial nerves subserve our specific senses of smell, taste, sight, and hearing, as well as the general senses from the face, mouth, and tongue. The spinal nerves are segmental in nature in that there is one spinal nerve for each vertebra (Fig. 7). At times in the distribution of nerve fibers to the body, individual nerves will unite so that the nerve axons within each of the spinal nerves can redistribute and form new nerves made up of axons from the cell bodies over a large region of the spinal cord. These areas of redistribution are known as plexuses (Fig. 7). There is an anterior-to-posterior segmental arrangement of spinal nerves and this is roughly associated with an anterior-to-posterior arrangement of distributions within the body. Therefore, the nerves supplying the leg go into portions of the spinal cord most distant from the brain and the fibers which supply the upper extremity enter the spinal cord in the region of the neck. Other areas of the body are distributed in sequence on each side of the examples given.

TROUBLE SHOOTING

When abnormalities caused by a disease process of any sort interferes with normal functioning of the nervous system certain symptoms will result. As with any complicated communicating system determining the area of destruction from the symptoms is not a simple matter. Also, since the nervous system is a living system inside a living animal, there is a large variation from one person to another in the seriousness, extent, and duration of defects which show up after seemingly similar disease processes.

A portion of the nervous system does not have to be directly involved in a disease process or destruction to be in a nonfunctioning condition. Pressure, which can be a local anesthetic, caused by diseases elsewhere will cause a nerve to stop conducting impulses. Disease processes themselves sometimes cause a reaction in terms of tissue fluid which in turn produces a swelling similar to swellings which occur in bumps and bruises in joints and skin. This swelling also interferes with normal functioning of the nervous tissue through the pressure created. The symptoms that develop from any particular accident depend on the specific area of destruction plus the total area of nervous system which is involved either primarily or secondarily in the process.

One class of symptoms resulting from abnormal functioning of the nervous system can be classified as an absence of function. If enough of the nervous system is not functioning so that incoming information is not being processed and outgoing information cannot be initiated, the individual is unable to respond and we say that he is in a coma or is unconscious. Coma and unconsciousness then are symptoms resulting from a massive nonfunctioning of the nervous system. Certain generalized functions of the brain, such as intelligence, abstract thinking, memory, and personality, also depend more on the total amount of normal nervous tissue than they do on a specific area of the brain. Therefore, destructions of brain sometimes result in a decrease in intellectual abilities of the individual.

If particular parts of the nervous system are destroyed, then precise symptomotology occurs. Thus, if the area of the brain which has to do with vision is destroyed, the person becomes blind. If the area of the brain a person uses to hear is destroyed, the person becomes deaf. If that portion of the brain which has to do with voluntary motor movements is destroyed, the person can no longer voluntarily move the muscle and we say that he is paralyzed. If a mixed nerve to a particular area of the body is destroyed, then the input cannot reach the central nervous system and there is a sensory loss over the area which had been supplied by the nerve. Also, with respect to the same nerve, any muscle that had been supplied by that nerve would be paralyzed since no information could reach it from the central nervous system.

There are specific patterns of loss which are important in terms of symptomotology. Thus, as previously indicated, the spinal cord track for voluntary motor movement is arranged so those fibers which cause the leg to move are on the outside, the arm on the inside. Thus, if the destruction is on the side of the spinal cord, it is the leg that will be paralyzed; if it is in the center of the spinal cord, it will be the arm that is paralyzed. The pattern of loss will, of course, vary whether a spinal nerve, a plexus, or a nerve resulting from the redistribution of fibers in a plexus is involved in the destruction.

There are certain symptoms in the nervous system which result from irritation instead of destruction. This irritation may be caused by swelling or fluid in the brain, blood in the brain, or by scars which develop after destruction in the brain. It appears that the disease process is able to cause nerve cells in the vicinity of the disease to overly react. Epilepsy or a convulsion resulting from abnormally increased central nervous system functioning is an example of such an irritation. There are also irritations which may occur in the brain resulting in peculiar phenomena. If that area of the brain related to memory is irritated, the person often recalls a particular incident in his past which does not seem to be related with ongoing activity of the day. He also may report that he sees objects which are not there or hears noises which are not there which can be caused by irritation of the visual or auditory areas of the cerebral cortex. In cases where a loss of body part occurs, such as an amputated leg, the nerves which went to the leg must be cut off in the area of the amputation. There is then left within the body a portion of the nerve which supplied the amputated portion. If any process, such as that of an artificial leg, stimulates the end of this nerve, the central nervous system will interpret that information as coming from the part that had been removed since localization depends upon central connections as well as peripheral. Thus, a person with an amputation often complains that he can feel in a finger or toe which he no longer has (phantom limb).

Symptoms can also result from the release of activity. Of necessity the nervous system must be ready to respond and there are built-in mechanisms of inhibition. If this inhibition is removed through a disease process, then there is a resultant release of the activity which was being held in check. Thus, if certain areas of the nervous system are destroyed, a person will show a tremor or other types of unwanted and purposeless muscle movement. Sometimes the muscles will become very stiff. This increase in rigidity can be caused because checks are taken off the normal process of making muscles move.

REGENERATIVE POWERS

Recovery of function after a complete division of a peripheral nerve can take place only when the divided ends lie in

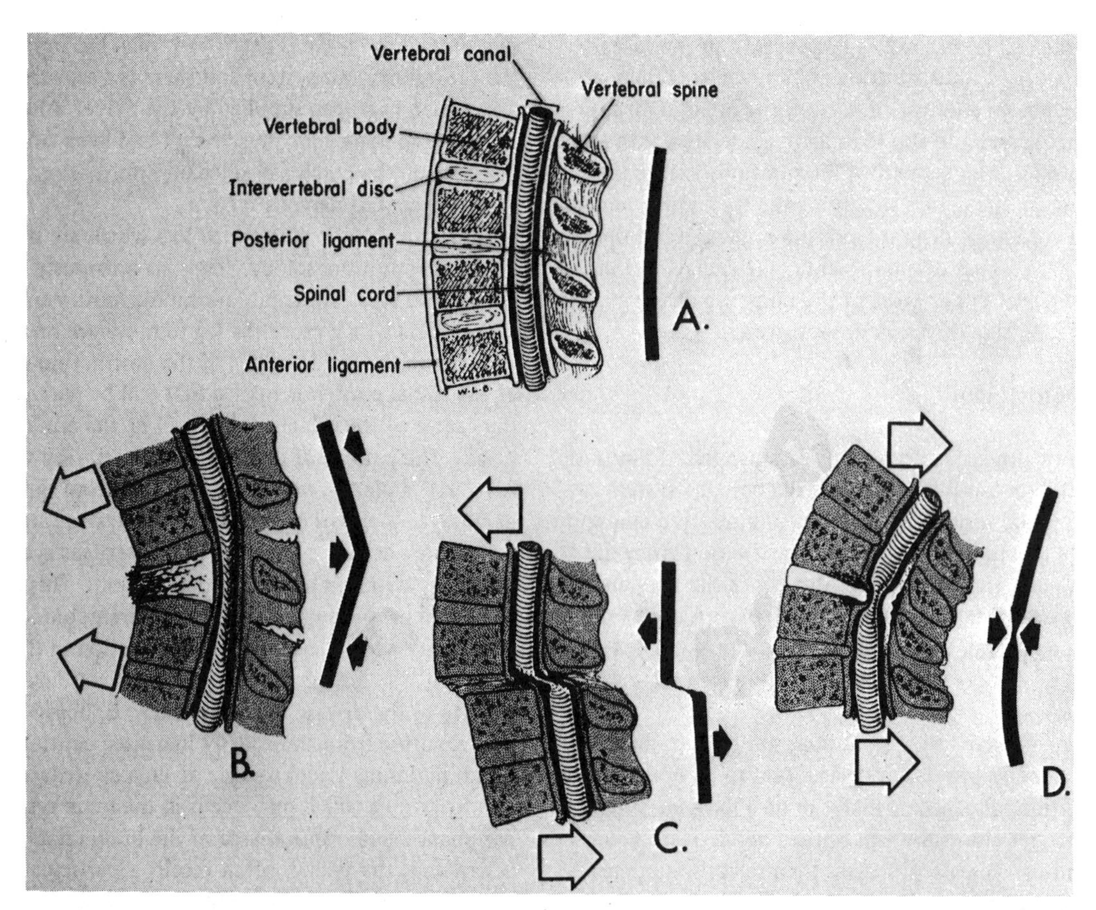

Fig. 8 - Representation of the spinal cord within the vertebral canal under various traumatic conditions. The heavy black line to the right of each illustration represents the configuration of the spinal cord. The large white arrows indicate the direction of the forces causing the injury. The black arrows indicate the direction of forces upon the spinal cord in the various conditions indicated. A - Normal condition. B - Compression fracture of a single vertebra. C - Fracture dislocation of one vertebra upon another. D - Hyperextension injuries

close position or have been sutured together. The growth of the axons of the peripheral nerve will occur at the rate of 1-2 mm per day.

Once destruction has occurred within the central nervous system, regeneration is not possible. Once a nerve cell is destroyed, the loss is permanent. Thus, any help for a person with a destruction of the central nervous system must be to maintain what he has left since it is impossible to regain any portion of the central nervous system that has been lost.

CONCUSSION AND CONTUSION

Sometimes, when a person is struck in the region of the head, he is rendered temporarily unconscious. He may recover consciousness a short time later (within minutes) and suffer no further symptoms. There has been a temporary cessation of nervous function. The neural location for this difficulty may be in the brain stem where normal functioning is necessary to keep a person awake. The mechanism of such a temporary unconsciousness without further defect is not really known.

As previously described, the brain floats in a closed, fluid-filled cavity. The fluid, as with all fluids, is not compressible while the brain is. Therefore, when a sudden force is applied to the outside of the cavity, the force must be transmitted through the vessel. It is possible, but only a suggestion, that there is a loss of fluid tensile forces of the liquid which causes sudden cavitation changes, which in turn produce sudden pressure changes in certain locations (brain stem in our example) which result in a bruising of the nervous tissue. This in turn causes an outflowing of fluid into brain substance causing a swelling similar to the response to injury we have learned to expect outside of the nervous system (for example, sprained ankle). However, the outflow of fluid is in this closed space, thus, squeezing the only compressible structure within the cavity, the brain. The condition is temporary and completely reversible.

If the spinal cord is involved in such an injury, the defect will be a loss of one or more of the functions of the spinal cord and will not involve consciousness. A local paralysis, because of involvement of the motor tract or a sensory loss because of involvement of sensory areas would result. If the head is suddenly hyperextended (thrown back) as in a whiplash injury, the spinal cord, which is essentially a tube, will be severely bent and, therefore, flattened (Fig. 8d). Since the cord is flattened and actually pulled by its lateral attachment, the dentate ligaments, it may get squeezed in the changing pattern of vertebral canal size. The ligament which goes across the back of the vertebra will have to change its thickness to allow for its sudden decrease in length. In this process the spinal cord may be temporarily squeezed and a concussion of the spinal cord result.

An increase in the forces which produce a concussion could lead to a contusion. In this case the local injury to the area of the nervous system is increased and a variable amount of bleeding (hemorrhage) will result. If the hemorrhage is large, there will be a major defect. However, if the hemorrhage is small and yet significant enough to cause a loss of blood supply to nerve tissue, cell death occurs. In such a case the symptoms will be the same as under concussion but would be more permanent. Thus, in the case of a whiplash noted above, if hemorrhage has occurred, then there will be some permanent loss. Many injuries are combinations of concussion and contusion effects.

If the pressure (from edema fluid or hemorrhage) which is acting as a local anesthetic in these injuries is relieved soon enough, a relief of symptoms results. If the pressure is long lasting enough, causing nerve tissue death, permanent defects result. There is a predictive quality in the fact that the longer the symptomogology remains after the time of injury, the less likely a total recovery will occur. The longer the unconsciousness, the worse the final outcome will be.

MAJOR DEFECTS

Peripheral Nervous System - If there is a sudden rendering apart or severing of a peripheral nerve, a motor and sensory loss will occur over the peripheral distribution of the nerve (Fig. 7). The pattern of the loss will be that of the distribution of the particular nerve which has been injured.

On occasion the injuries are closer to the area of exit of these nerves from the spinal cord. This becomes particularly important when the injury involves the cervical plexus which contains all the nerve fibers supplied to the arm or the lumbosacral plexus which contains all the nerve fibers supplying the leg (Fig. 7). In such instances, instead of segmental, there is a different organization of the loss of function which involves several segments. If the severed ends of a peripheral nerve are rejoined under favorable conditions, return of function can occur.

Fractures and Fracture Dislocations - The various vertebrae of the vertebral column are subject to fractures or dislocations which cannot occur in the absence of a fracture. A vertical force will on occasion cause a compression fracture of a body of a vertebra (Fig. 8b). When such a fracture occurs, there is a sharp angulation of the canal through which the spinal cord passes and injury to the cord occurs. Forces in a lateral direction to the vertical arrangement of the vertebral column may cause a fracture–dislocation of one vertebra upon another. In this instance the canal for the spinal cord is severely compromised and in effect the spinal cord is severed (Fig. 8c). A complete loss of all the functions of the spinal cord below the level of the destruction will result. The defect is not repairable and the loss is permanent.

On occasion the intervertebral disc which lies between two vertebral bodies is pushed out of its area because of squeezing forces applied from both directions. Should this disc move in the direction of the spinal cord (Fig. 8d) or one of the peripheral nerves, a compromise of nervous function will result. If the pressure which has been applied is reasonably minimal and the disc can be removed by surgery after the injury, a good deal of nervous function of the spinal cord may return.

Fractures of the cranium also may occur upon a sudden traumatic impact. The fracture may be very simple for the adult in that the bones are not displaced and no artery receives serious injury. Such a patient may suffer a brief concussion with unconsciousness, recover, and the fracture is found only on X-ray. In this case the forces of impact have been dissipated with the creation of the fracture, which in reality protects against any destruction from occurring to the brain and, therefore, minimal symptomotology results. It might be said that if one is to get hit on the head rather hard, that less damage will occur if a linear nondisplacing fracture that does not involve major blood vessels results.

If the object causing a fracture to the cranial cavity is blunt so that a chip of bone is broken off, this chip may penetrate into the brain material. When this occurs, brain tissue is destroyed and hemorrhage will occur within the substance of the brain. This hemorrhage, known as hematoma, will expand under arterial pressure and replace a significant portion of brain, and cause excess pressure. A large portion of brain will die because of lack of blood supply. In such an instance, the bony fragments must be removed or elevated away from the brain and the particular hemorrhage removed and bleeding stopped in a neurosurgical operation if possible. Recovery will never be total, but can be satisfactory. Since the many bones which make up a child's skull are not firmly fixed to each other, a certain amount of shifting is allowed. This shifting partially protects the brain, and injuries in children are less severe than in adults, assuming equal amounts of force.

If a linear fracture line goes across a meningeal artery, severe difficulties will result. Since the meningeal arteries lie next to or within a groove of bone, the artery may be suddenly severed in the creation of a fracture. Bleeding will occur under arterial blood pressure levels. This hemorrhage will force its way between the bone and the dura mater (Fig. 9e) and will rapidly expand. Since the hemorrhage is outside of the dura, just under the bone, it is known as an epidural hemorrhage. An illustrative case may be as follows. A young man is hit on the head during an accident. He is rendered unconscious immediately. If nothing is done to help him, his breathing will become slow and labored and he will become unreactive to all sensory input. The vital nervous system functions maintaining respiration and proper internal visceral (heart, blood pressure, etc.) regulation is compromised because of pressure to the brain stem. Death occurs rapidly under these conditions. If the patient is seen in time so that openings can be drilled through the cranial bones, the hemorrhage removed and the bleeding controlled, the chance of recovery is good. The emphasis in such a case is on speed.

Penetrating Wounds - The case above of the depressed cranial fracture is really that of a penetrating wound. On the other hand, if a fragment of any kind (for example, bullet, knife,

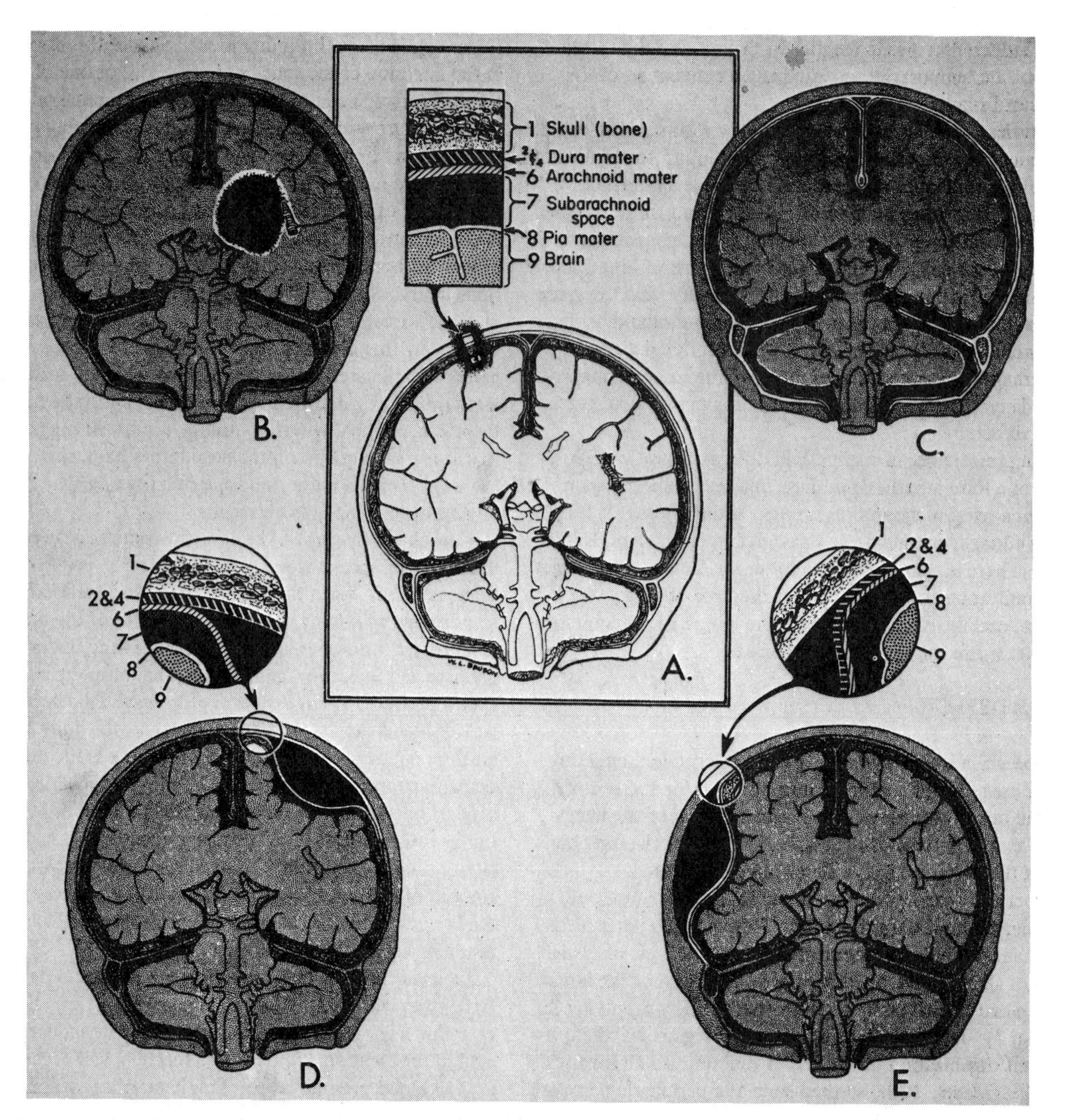

Fig. 9 - Representation of types of hemorrhage which occur in the cranial cavity. A - Normal representation of a cross section of the brain within the skull similar to Fig. 6. Inset shows detail of meninges in relation to the brain and skull. Numerical representation is the same as in Fig. 2. B - Intracerebral hemorrhage. C - Subarachnoid hemorrhage. D - Subdural hemorrhage; inset illustrates the location of hemorrhage between the dura and arachnoid mater. E - Epidural hemorrhage; inset shows location of hemorrhage between dura mater and bone

sharp metal, etc.) breaks through bone and enters the nervous system there is a penetrating wound which will cause the direct destruction of nervous tissue and a secondary destruction from the hemorrhage resulting from the severing of blood vessels in the area. Removal of the foreign body and the control of the bleeding will often limit the loss to the specific area destroyed.

Closed Vascular Difficulties - As previously indicated, the brain is floating in a fluid-filled, closed cavity. When rapid changes in acceleration occur in the head (as during an accident) the inertia of the floating brain will cause it to move in the opposite direction relative to the applied force. If this movement is sudden enough, the brain is liable to come in contact with the skull or with one of the meningeal partitions. If the striking force is strong enough, the central nervous system will be bruised. If this is the only result of injury, the picture will be that of a contusion.

On occasion when the head is struck hard in an accident, the brain shifts the maximum distance it can inside the skull. This is particularly true if a fracture does not occur. The branching veins which enter the venous sinuses (Fig. 6) may be pulled out of the sinus during this shift of brain position. If this occurs, there can be venous hemorrhage into the area between the dura mater and the arachnoid mater. Such a hemorrhage is known as a subdural hemorrhage since it is immediately underneath the dura (Fig. 9d). In this instance a patient may be hit on the head in an accident and lose consciousness temporarily. He then regains consciousness and may even be symptom-free. Following this episode by 1 week to several months the patient may again complain of headaches and begin to lose consciousness. He may lose complete consciousness or be just generally obtundated (functioning at a low level). The hemorrhage in this case is subdural and is occurring under venous pressures. The original injury renders the patient unconscious because of the mechanisms of concussion. Since the venous pressure is low, the hemorrhage is slow but relentless. As it increases in size it begins to apply pressure to the brain and shut off normal vascular supply to the area underneath. As the pressure increases slowly, the symptomotology begins to appear. These patients can be helped greatly by neurosurgery. The goal of surgery is to remove the hemorrhage and stop the bleeding vein. Recovery in many instances is almost complete.

In some instances the dynamics of the motion of the brain may cause one of the cerebral arteries on the surface to break open and the bleeding is into the subarachnoid space. Since this space (filled with spinal fluid) is open all around the brain, the blood which is released immediately surrounds the entire nervous system (Fig. 9c). This blood is a tremendous irritant to the brain and the symptoms are immediate, severe, and dramatic. The patient will complain of severe headache, will be rendered unconscious, and will remain so unless the bleeding stops on its own or the site can be found and repaired.

On occasion the injury can cause a rupture of a blood vessel within the substance of the brain. This may occur with a penetrating injury or it may occur because of a sudden rendering of the vessel in some fashion. In this case the blood is released under arterial pressure directly into the brain substance and in effect causes an expanding nonfunctioning pressure-producing area (Fig. 9b). The symptomatology resulting will depend on where the hemorrhage is located. This type of hemorrhage is usually not approachable or treatable and, therefore, treatment of the patient is supportive and measures taken to try to limit the spread of the hemorrhage.

Long-Term Sequelae - In some injuries to the nervous system a reaction of the supporting glial tissue results and a type of scar forms. This scar can be irritating to the surrounding nervous tissue. If it is irritating to an extreme extent, it may cause the nervous system cells around it to send out impulses when they should not and in a markedly rapid fashion. This condition is known as epilepsy, and, although rather rare, it is possible that epileptic seizures can result from scarring which has resulted in turn from a prior injury.

On occasion there is a condition within the nervous system in which once an area of destruction exists, it seems to cause destruction of the surrounding nervous tissue and this continually expands until a good deal of brain is destroyed. The basic mechanism of this process is unknown, and it is rather rare. However, if the continual destruction is present, certain symptomatology can result, usually that based on the general functions of the nervous system, which means a continual dropping of intellectual level of functioning and the occurrence of behavior defects such as a personality change. It is highly probable that the long-term defects occurring from trauma are not nearly so frequent as many persons believe. However, the possibility does exist.

SUMMARY

The nervous system is a complex instrument composed of the conducting elements of nerve cell fibers and a network of synapses at which a complex information transfer occurs. If this highly organized pattern is interrupted in any way, a defect in nerve function occurs and symptomatology results. The symptomatology resulting depends on the specific area of nervous tissue destroyed, the amount of nervous tissue destroyed, and the character of the lesion.

The relationship of the brain and spinal cord floating within a bony and ligamentous cavity filled with fluid is a situation which explains the characteristic results of trauma. Results of trauma vary, depending on whether a fracture of the bony enclosure has occurred and whether particular vascular structures are incorporated within the destructive forces.

Whenever a blow occurs to the nervous system, a concussion or contusion may result. The mechanism of concussion is not known, although a theory is suggested.

The regenerative powers of the nervous system are limited. In the peripheral nervous system, if severed ends of nerves are sutured together or if the destruction of the nerve has occurred without destroying the physical continuity of the fibers, then regeneration is likely. Any destruction of the central nervous system (brain, brain stem, spinal cord) is permanent.

REFERENCES

1. E. Gardner, "Fundamentals of Neurology," 5th edition. Philadelphia: W. B. Saunders, 367 pp., 1968.
2. F. E. Jackson, "The Pathophysiology of Head Injuries." Ciba Clinical Symposium, 18(3), 1966.
3. H. A. Matzke, and F. M. Foltz, "Synopsis of Neuroanatomy." Oxford: Oxford University Press, 149 pp., 1967.
4. S. Mullan, "Essentials of Neurosurgery." New York: Springer, 273 pp. (particular reference to chapters 1, 6-10), 1961.
5. D. Wooldridge, "The Machinery of the Brain." New York: McGraw-Hill, 252 pp. 1963.

Section 2: Head-Neck Kinematics and Impact Response

670919

Investigation of the Kinematics and Kinetics of Whiplash

H. J. Mertz, Jr. and L. M. Patrick
Wayne State University

THE SO-CALLED WHIPLASH SYNDROME constitutes the most prevalent trauma to occupants of automobiles struck from the rear. It is particularly insidious with subtle pathology that often does not show up with radiological or other quantitative diagnostic techniques. Acute or chronic symptoms sometimes persist for years, and in some cases where there is no immediate obvious morbidity, injuries attributed to the accident show up months later. Fortunately, whiplash is seldom disabling during the recovery period, and it is generally amenable to conservative treatment. Some orthopedic surgeons advocate operative procedures only after 2 years of other types of treatment are unsuccessful. Lesions ranging from vague aches, pains, vertigo, and dysphagia to torn muscles, ligament damage, joint injuries, and bone damage have been reported by physicians. Experiments have reproduced some of the lesions in laboratory animals and human cadavers with various types of sleds and accelerators. However, the mechanics of whiplash are not well known, and the correlation of animals, cadavers, and living human beings is not accurately established.

Many mechanisms for explaining injuries associated with acceleration hyperextension have been postulated. The most obvious is one in which the neck is treated as a beam in bending with posterior compression injuries to the cervical vertebrae and/or to the intervertebral discs. Tensile forces due to bending produce anterior tissue damage. While these injuries have been observed clinically and experimentally, tension-type lesions to the soft tissue at the back of the neck have also been found.

Tensile damage to the soft tissue at the back of the neck has led to the theory that the injury in the whiplash accident results not during the hyperextension portion of the cycle but, instead, on the flexion rebound portion. This explains the torn tissues at the back of the neck but is hard to justify due to the difficulty in reproducing this type of injury in the laboratory and also due to the fact that the chin provides a mechanical stop or limitation to the flexion as it strikes the chest. When the chin strikes the chest, the back of the neck will still be in tension but the stresses would be expected to be below those producing injury due to the large moment arm from the chin to back of the neck.

It has been hypothesized that injury occurs during the

ABSTRACT

The kinematics of rear-end collisions based on published acceleration pulses of actual car-to-car collisions (10 and 23 mph) were reproduced on a crash simulator using anthropomorphic dummies, human cadavers, and a volunteer. Comparison of the responses of subjects without head support were based on the reactions developed at the base of the skull (occipital condyles). The cadavers gave responses which were representative of persons unaware of an impending collision. The responses of both dummies used were not comparable with those of the cadavers or volunteer, or to each other.

An index based on voluntary human tolerance limits to statically applied head loads was developed and used to determine the severity of the simulations for the unsupported head cases. Results indicated that head torque rather than neck shear or axial forces is the major factor in producing neck injury.

When the head was initially supported by a flat, padded headrest, all subjects gave comparable headrest loads. Using this configuration, the volunteer withstood an equivalent 44 mph simulation with only slight discomfort.

With the head separated from the headrest by 2-3/4 in. the head load increased from 150 to 390 lb, but with additional padding, the load was increased to only 250 lb.

Controlled seat back rotation decreased the magnitudes of the head loads and neck reactions for the supported and unsupported head cases, respectively.

initial part of the acceleration cycle from shear action caused by a presumed relative motion between the head and torso prior to appreciable rotation. If this theory is correct, the head support proposed as a means of eliminating the acceleration-extension injury would have to be against the head at all times with little or no padding for comfort.

The general purpose of this research program is to establish the actual injury mechanism, while the detailed objectives are to:

1. Analyze forces and moments at the base of the skull during acceleration extension.
2. Verify the analysis experimentally.
3. Compare the experimental results of anthropomorphic dummies, cadavers, and human volunteers.
4. Study the effectiveness of head support in mitigating whiplash injury.
5. Determine the effect of seat back rotation on acceleration-extension parameters.
6. Correlate human voluntary static forces and moments with impact severity.

Using the comparison of the results from anthropomorphic dummy and cadaver impacts with those from the human volunteer as a basis, a logical extrapolation from the sub-injury to the injury impact severity is expected to be achieved in future work. The present program includes only two degrees of severity (severe and nonsevere) which were simulated from data of actual car-to-car collisions published by Severy and Mathewson (1)*.

VOLUNTARY HUMAN NECK TOLERANCES TO STATICALLY APPLIED HEAD LOADS

BASIS FOR STATIC TOLERANCE LEVEL - Analyzing the head as a free body (Fig. 1), the important parameters are the reactions at the base of the skull. These reactions must accelerate the head during whiplash and consequently cause the neck to hyperextend. Since the motion is two dimensional, these reactions can be resolved into a shear force R_{811} acting perpendicular to the vertebral column at the occipital condyles and lying in the mid-sagittal plane, an axial force R_{812} acting along the axis of the second cervical vertebra, and a resulting couple T_8 acting about an axis passing through the point of intersection of the shear and axial forces and perpendicular to the mid-sagittal plane.

Maximum tolerable static limits for these reactions were determined for various loading configurations and serve as a basis for voluntary neck tolerances to statically applied head loads.

PROCEDURE - Five different static loading configurations were investigated using the same volunteer who was also subjected to the simulated whiplash environment. In the first four configurations the subject was strapped in a seated position in a rigid chair. The loads were applied by the volunteer using a block and tackle arrangement shown in Fig. 2 and were transferred to the head through a tightly fitted plastic headband from a welding visor. An axial load cell inserted between the block and tackle and the headband sensed the applied load.

In the fifth configuration the volunteer assumed a standing position with his head supported in the occipital and mandible regions by a cradle constructed of seat belt webbing material. The cradle was fastened to a load cell which was secured to the ceiling. Load was applied to the neck by the weight of the subject with additional force being applied by the subject pulling up against a fixed horizontal bar.

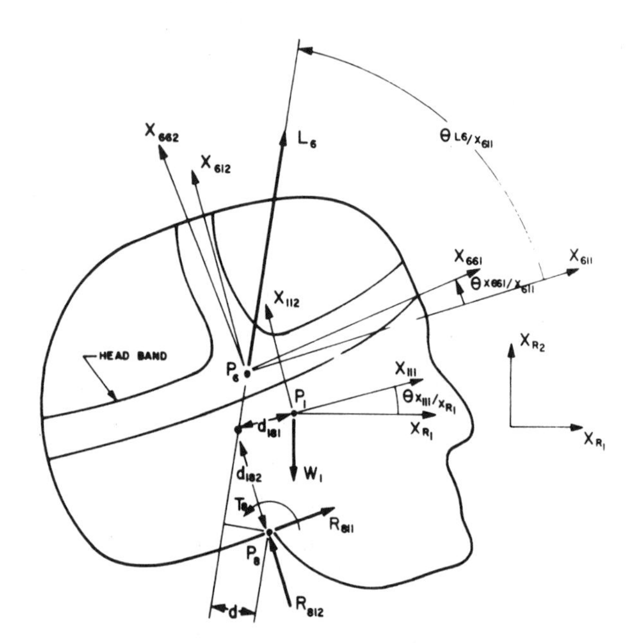

Fig. 1 - Free body diagram of statically loaded head

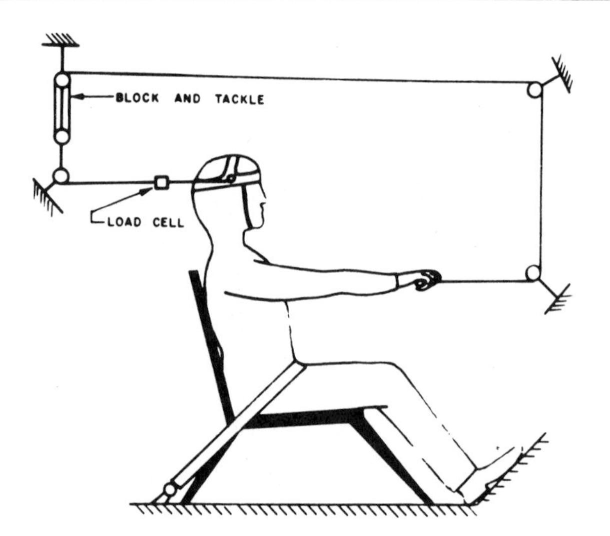

Fig. 2 - Setup for applying static head loads

*Numbers in parentheses designate References at end of paper.

The instrumentation was the same for the five configurations. The load was measured by a strain gage axial sensitive load cell whose output was recorded continuously on light-sensitive paper by a light beam galvanometer. Direction and point of application of the load were recorded incrementally using a 35 mm SLR Nikon camera equipped with a battery powered shutter firing and film advance system. The shutter pulse was recorded on the light-sensitive paper to give a time correlation between the photograph and the applied load.

METHOD OF DATA REDUCTION - The neck reactions for all configurations were computed by considering the head as a free body (Fig. 1). Four orthogonal coordinate systems were assumed. The $(X_R)_i$ set was chosen as an absolute system with X_{R1} axis being horizontal with respect to the ground. The $(X_{66})_i$ system was assumed fixed to the head with the origin P_6 at the point of application of the applied load with the X_{661} axis parallel to the superior edge of the circumferential part of the headband. The $(X_{61})_i$ system rotated with the head, with the origin also taken at point P_6. However, the X_{611} axis was directed in a posterior to anterior (P-A) direction parallel to the direction of the shear reaction at the base of the skull, and the X_{612} axis was directed in an inferior to superior (I-S) direction. The fourth system $(X_{11})_i$ was affixed to the skull at P_1, the center of gravity of the head. The X_{111} axis was oriented in the P-A directioı with the X_{112} axis superiorly directed. Point P_8 was taken at the occipital condyles and had to be approximated from surface landmarks of the side of the face. To determine the reactions at P_8, the equations of static equilibrium were applied:

$$\sum F_{611} = 0 \qquad (1a)$$

$$\sum F_{612} = 0 \qquad (1b)$$

$$\sum M_8 = 0 \qquad (1c)$$

Resolving the applied load, L_6, and the head weight, W_1, into components and substituting these values into Eq. 1, the following expressions for neck reactions are obtained:

$$R_{811} = W_1 \sin\left(\theta_{X_{111}} / X_{R_1}\right) - L_6 \cos\left(\theta_{L_6} / X_{611}\right) \qquad (2a)$$

$$R_{812} = W_1 \cos\left(\theta_{X_{111}} / X_{R_1}\right) - L_6 \sin\left(\theta_{L_6} / X_{611}\right) \qquad (2b)$$

where:

$\theta_{X_{111}} / X_{R_1}$ = Angle between X_{111} co-ordinate axial and the horizontal

θ_{L_6} / X_{611} = Angle between L_6 and X_{611} axis.

The magnitude of the resulting torque is given by

$$T_8 = L_6 (d) + W_1 \cos\left(\theta_{X_{111}} / X_{R_1}\right) d_{181} - W_1 \sin\left(\theta_{X_{111}} / X_{R_1}\right) d_{182} \qquad (3)$$

To solve these equations for the neck reactions, the magnitudes of the applied loads were obtained from the light beam oscillograph records and the geometric parameters were measured from the corresponding photographs. The positive directions of these results are shown on the free body diagram.

RESULTS AND DISCUSSION - The resulting maximum reactions for the five configurations which were evaluated are presented in Table 1.

For the first configuration (runs S-1 and S-2), the head was held upright in a normal position. The load was applied essentially in the A-P (anterior to posterior) direction as indicated by the angle θ_{L_6} / X_{611}. The maximum torque which could be resisted by the neck muscles before the head began to rotate was -10.6 ft-lb; the maximum shear load developed during this loading sequence was 39.5 lb. The axial load is of no significance, since the principal load was not in this direction.

In the second configuration (runs S-4 and S-5), the head was flexed forward from its normal position, causing the applied load to place a compressive load on the neck vertebrae. In this position the mechanical advantages of the neck muscles which limit extension of the neck have increased, resulting in a higher maximum resistive neck torque (-12.7 ft-lb) than was obtained when the head was held upright. Also, for this configuration the maximum shear load and compressive axial loads were increased to 56.6 and 37.8 lb, respectively.

The advantages of this head position compared to the normal upright position in a rear-end collision are twofold:

1. The neck torque has been increased from -10.6 to -12.7 ft-lb, which implies that a more severe impact can be withstood.

2. The angle through which the head would rotate before severe hyperextension occurs would be greater, resulting in more energy being dissipated during the rotation, which would reduce the degree of hyperextension.

In the third configuration the head was rotated rearward from a normal upright position with load being applied in an A-P, S-I direction. In this position the maximum neck torque was increased to -17.5 ft-lb. This increase was primarily due to a change in the point of rotation of the head with respect to the neck. Instead of rotating about the condyles at the base of the skull, the head rotates about the posterior portion of the first cervical vertebra. Consequently, the effective moment arm of the neck muscles, which restrict this rotation, has been increased, resulting in a larger resistive neck couple for a given resultant muscle force. Also, the anterior ligaments of the neck vertebrae are elongated, resulting in additional load-carrying capacity. The maximum measured shear and axial tension forces were 68.1 and -86.2 lb, respectively.

On a comparative basis, this configuration gives a higher resistive neck torque level than the previous two, which implies that the volunteer could withstand a more severe rear-end collision in this position.

Configuration 4 was evaluated to determine a suitable position which a person could assume if he is aware of an impending rear-end collision. In this configuration the volunteer grasped his hands together behind his head, providing support against head rotation with his arms. The load was applied in an A-P direction and his head remained in a normal upright position. The maximum applied load was 175 lb as compared to maximum applied loads of 41, 64, and 89 lb for the unsupported normal, flexed, and extended positions of the head, respectively. Consequently, this configuration offers the greatest degree of safety against hyperextension of the head. Further increase in safety could be accomplished by using this method with the head initially bent forward.

Table 1 - Maximum Reactions for Various Static Loading Configurations
(Volunteer - LMP, W_1 = 10.8 lb, d_{181} = 0.75 in., d_{182} = 2.44 in.)

					Max. Value		
Run	Configuration[a]	L_6, lb	θ_{L_6}/X_{611}, deg	$\theta_{X_{111}}/X_{R_1}$, deg	R_{811}, lb	R_{812}, lb	T_8, ft-lb
S-1	1	34	187	1	33.7	14.9	-9.1
S-1	1	34	192	-3	33.3	17.9	-8.1
S-2	1	41	196	-8	39.5	21.8	-10.1
S-2	1	40	191	-3	39.2	18.4	-10.6
S-4	2	64	205	-18	55.0	36.9	-12.7
S-4	2	61	207	-21	54.4	37.8	-10.4
S-5	2	64	201	-17	56.6	33.7	-10.8
S-6	3	68	159	25	68.1	-14.5	-17.5
S-7[b]	3	41	123	64	32.0	-29.6	-7.3
S-8[c]	3	89	75	82	-12.7	-86.2	-14.5
S-8	3	63	82	75	1.7	-62.5	-14.6
S-9	4	164	*	*	*	*	*
S-10	4	148	*	*	*	*	*
S-11	4	175	*	*	*	*	*
S-12	5	330	53	36	-192.0	-254.0	0

*Not measured.

[a] Configuration 1 - Head in normal position, load A-P.
2 - Head in flexed position, load A-P.
3 - Head in extended position, load A-P.
4 - Bracing head with hands, load A-P.
5 - Hanging position.

[b] Headband digging into forehead.

[c] Rubber inserted between headband and forehead.

To determine the strength of the neck in tension, the volunteer was subjected to a hanging type of loading in configuration 5. In this position a maximum axial tension load of -254 lb was achieved. The corresponding shear load was -192 lb.

Additional voluntary human tolerance levels applied to the reactions at the base of the skull can be obtained by modifying other researchers' published results. Since the method of modifying these results may be doubtful, an explanation of the assumptions and calculations used in each modification follows.

From a paper presented by Stapp (2) it was stated that an "upward seat ejection safe limit" was 20 g. Assuming an average weight of a head of 12 lb, a 20 g acceleration of the head would require an axial compressive force at the base of the skull of 240 lb. This value is probably on the low side, since a pilot would be wearing a helmet which would increase the weight of the head. Taking into account this additional assumption, a voluntary neck axial compressive force of 250 lb should not be injurious.

In a paper by Carroll, et al. (3), five human volunteers were subjected to static and dynamic P-A head loads with the head being initially in a normal upright position. The load was applied through a headband which was positioned 1/2 in. superiorly to the occipital prominence. Their results indicated than an average static "neck torque" of 40.7 ft-lb could be developed. Through personal communication it was learned that the point of rotation for computing these torques was taken as the mid-clavicle, resulting in a range of moment arms of 7-7-1/2 in. The average distance from the mid-clavicle to the midpoint of the external auditory canal was 5-1/8 in. Since the torques developed at the occipital condyles are needed to compare with the results obtained using volunteer LMP, the torques given by Carroll, et al. were recalculated. Based on a moment arm of 7-1/4 in., the average 40.7 ft-lb moment was produced by applying a 67.3 lb force. Assuming that the distance from the applied load to the occipital condyle is 2-3/4 in., a resulting couple and shear force at the base of the skull of 15.4 ft-lb and -67.3 lb, respectively, are needed to maintain equilibrium. The difference between this torque (15.4 ft-lb with the load applied in the P-A direction) and the maximum torque (-10.6 ft-lb with the load applied in the A-P direction) is not as great as expected, since the largest neck muscles are attached to the skull in the occipital region, which implies that a person should be able to resist much larger flexual torques than extending torques. Consequently, the average value given by Carroll, et al. is more conservative than those obtained from the volunteer LMP. In order to compare the two results, the maximum static neck torque of 49.5 ft-lb withstood by one of the five volunteers used by Carroll, et al. will be used. Relating this torque to the reactions at the occipital condyles gives values of 18.8 ft-lb and 882 lb for the neck torque and shear force, respectively.

Table 2 - Summary of Voluntary Static Human Tolerance Levels Based on Reactions Acting at the Occipital Condyles

Head Position	Shear Force, lb P-A	Shear Force, lb A-P	Axial Force, lb I-S	Axial Force, lb S-I	Couple, ft-lb (+)	Couple, ft-lb (-)
Normal	40[a]	80[a,b]	250[c]	---	19.0[b]	10.5
Extended	70[a]	190	---	255	----	17.5
Flexed	55[a]	---	40[a]	---	----	12.5

[a]Value is not a maximum tolerable load.

[b]Based on paper by Carroll, et al. (3).

[c]Based on paper by Stapp (2), a dynamic value.

NOTES: All values based on volunteer LMP except where noted. Loads and torques rounded off to nearest 5 lb and 1/2 ft-lb, respectively. Directions based on free body diagram shown in Fig. 1.

Table 3 - Pertinent Kinematic Parameters from Car-to-Car Rear-End Collisions

Struck Car	Struck Car Kinematics: Impact Velocity, mph	Change In Velocity, mph	Pulse Time, ms	Peak Accel., g	Mean Accel., g	Accel. Dist., in.
1956 Olds	10	9.1	135	5.9	3.07	10.8
1956 Olds	23	14.8	132	10.0	5.10	17.2
1955 Nash	23	15.2	135	10.3	5.13	18.1

Values based on data presented by Severy and Mathewson (1). Impacting car was 1955 Hudson.

A rough estimate of the breaking strength of the neck is given by Simmons and Herting (4). From an eye witness report of three executions (by hanging) of criminals in Japan in which two of the three died by strangulation instead of broken necks, an approximation for the hanging force based on energy relationships was made. It was concluded that the adult human neck is capable of withstanding an applied hanging force of 2000 lb.

Using the maximum resistive forces and moments from the various sources presented, preliminary voluntary static human tolerance levels for reactions at the base of the skull can be established and are summarized in Table 2. On an average these values are certainly lower bounds to an injury tolerance level.

It should be noted that none of the P-A shear loads represent true maximums since in the configurations used, the neck torques limited the application of higher shear loads. Also, because of the torque limitation, the listed values of the A-P shear load for the normally positioned head and of the I-S axial load for the flexed head are not maximums.

Since the emphasis was on obtaining tolerance levels pertinent to hyperextension of the neck, values were not obtained for all possible combinations of head position and neck reactions. An important omission is the value for the maximum positive couple with the head fully flexed. This couple is of significance in the evaluation of upper torso restraint systems for frontal impacts.

The majority of the values presented in this table are based on only one volunteer. Data from additional volunteers are needed to verify the various tolerance levels.

Taking into account these limitations, these voluntary static human tolerance levels will be used as a basis for evaluation of the severity of neck reactions produced in dummies, cadavers, and volunteers during simulated rear-end collisions.

SIMULATED REAR-END COLLISIONS

EXPERIMENTAL SETUP - The simulated rear-end collisions were conducted on the horizontal accelerator shown in Fig. 3. The sled travels on two horizontal rails and is accelerated pneumatically with a maximum speed of 40 mph. Deceleration is accomplished by a hydraulic snubber which can be regulated to subject the sled to a rectangular or triangular deceleration pulse. The maximum deceleration is structurally limited to 25 g. The stopping distance is continuously variable up to 22 in., depending on the sled velocity.

The characteristics of the sled acceleration pulses for the simulations were obtained from data presented by Severy and Mathewson (1) of actual car-to-car rear-end collisions. Table 3 lists the pertinent kinematic parameters for two severe and one nonsevere rear-end collisions. The changes in velocities of the struck cars were obtained by numerically integrating the curves of the longitudinal acceleration history of a point on the frame of the cars. Also, the pulse durations and peak accelerations were obtained from these curves. The g levels indicated in Table 3 could not be duplicated on the acceleration stroke of the horizontal simulator; consequently, the simulation was produced on the deceleration stroke. In order to duplicate Severy's data in this configuration, the sled velocity, stopping distance, and snubber deceleration wave shape had to be prescribed. Assuming a constant acceleration wave shape and letting the sled velocity and decelerating pulse time equal the change in velocity and acceleration-time duration of the struck car, the required stopping distance for the sled was calculated from

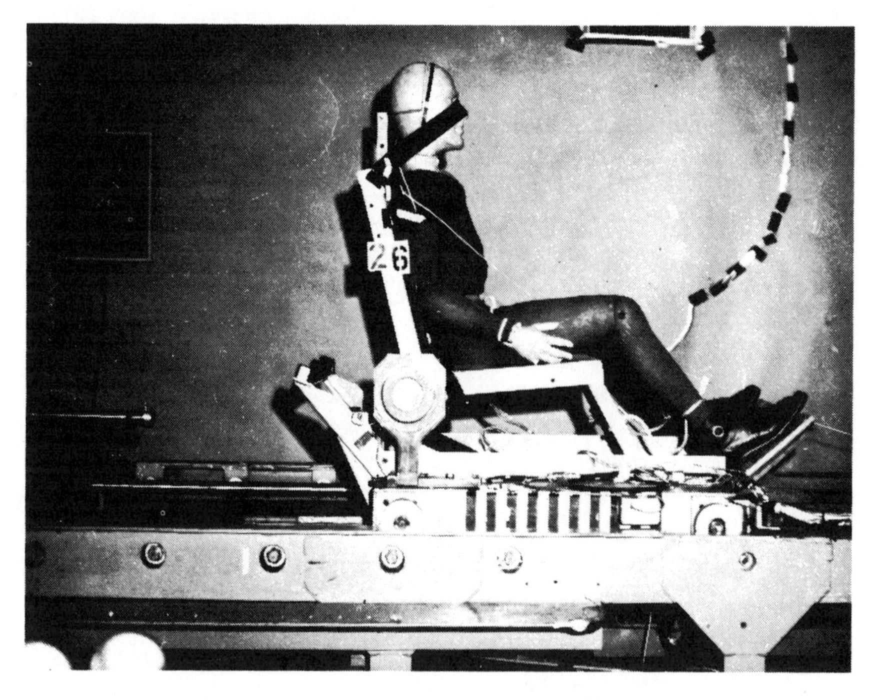

Fig. 3 - Horizontal accelerator - overall view

the kinematic equations of constant acceleration. To duplicate the 10 mph, nonsevere rear-end collision, a sled velocity of 9 mph and a stopping distance of 10 in. were used. For the equivalent 23 mph, severe rear-end collision, a sled velocity of 15 mph and a stopping distance of 17.5 in. were used.

To maintain position of the subjects prior to deceleration, restraints were placed on the head, chest, and pelvic area during the acceleration stroke and were mechanically removed at the end of the stroke when the sled was moving at the prescribed constant velocity.

The seat was rigidly constructed using steel angles for main structural components and plywood coverings for the seat back and bottom. The rigidity of the seat back was controlled by the setting of a frictional torque limiter which allowed the seat back to rotate at prescribed levels of constant torques. The seat was mounted to the sled in a rearward facing direction. An adjustable, removable headrest was attached to the seat back.

Two types of headrest configurations, a curved and flat surface, were used. Padding to prevent localized head forces consisted of one layer of 5/8 in. Rubatex in each case. To monitor the head forces, the headrest surfaces were affixed to a biaxial load cell which was rigidly mounted to the seat back.

DETERMINATION OF NECK REACTIONS - To evaluate the severity of the whiplash simulation and the effectiveness of the safety devices, the neck reactions and headrest loads were determined. The headrest loads were measured directly using a biaxial load cell. The neck reactions were obtained by applying the equations of dynamic equilibrium to the head. In this analysis the head is considered to be a rigid body undergoing plane motion as shown in the free body diagram (Fig. 4).

The relevant notation is:

P_1 = Center of gravity of the head

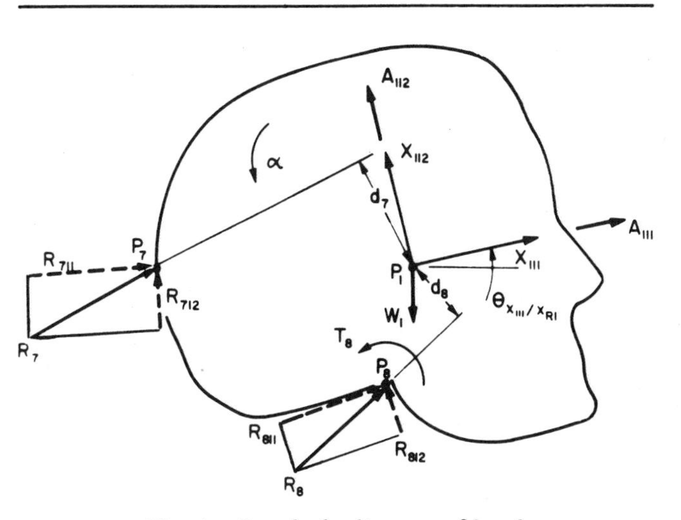

Fig. 4 - Free body diagram of head

P_7 = Point of application of the headrest load

P_8 = Occipital condyles

W_1 = Weight of head

I_1 = Mass moment of inertia of head with respect to c.g. of head

R_{811} = Shear force (described previously)

R_{812} = Axial force (described previously)

R_8 = Result of the shear and axial forces

T_8 = Torque (described previously)

R_{711} = Headrest load on head in X_{111} direction

R_{712} = Headrest load on head in X_{112} direction

R_7 = Result of the headrest loads R_{711} and R_{712}

d_7 = Moment arm for R_7 with respect to c.g.

d_8 = Moment arm for R_8 with respect to c.g.

$\theta_{X_{111}}/X_{R_1}$ = Angle between X_{111} coordinate axis and horizontal

α = Absolute angular acceleration of head

ω = Absolute angular velocity of head

A_{111} = Absolute acceleration of c.g. of head in X_{111} direction

A_{112} = Absolute acceleration of c.g. of head in X_{112} direction

The shear reaction is given by,

$$\sum F_{X_{111}} = \frac{W_1}{g} A_{111}$$

or

$$R_{811} = \frac{W_1}{g} A_{111} - R_{711} + W_1 \sin\left(\theta_{X_{111}} / X_{R_1}\right) \quad (4)$$

The axial force is calculated from

$$\sum F_{X_{112}} = \frac{W_1}{g} A_{112}$$

or

$$R_{812} = \frac{W_1}{g} A_{112} - R_{712} + W_1 \cos\left(\theta_{X_{111}} / X_{R_1}\right) \quad (5)$$

And the corresponding moment, T_8, is computed from

$$\sum M_1 = I_1 \alpha$$

or

$$T_8 = I_1 \alpha + R_7 d_7 - R_8 d_8 \qquad (6)$$

By letting $R_7 = 0$, this set of equations is valid for the case when the head is not in contact with the headrest or when the headrest is not used. The weight of the head, the mass moment of inertia, the magnitude of the headrest load, and the geometric head properties can be determined. The only parameters which must still be evaluated are the accelerations of the c.g. of the head and angular acceleration.

To obtain these parameters, the total accelerations of two points on the head were measured using biaxial accelerometers. The kinematic relationship between these two points, P_2 and P_3, is:

$$A_2 = A_3 +\!\!\!\!\rightarrow A_{2/3}$$

or since both points lie on the same rigid body

$$A_{211} +\!\!\!\!\rightarrow A_{212} = A_{311} +\!\!\!\!\rightarrow A_{312}$$

$$+\!\!\!\!\rightarrow (d_{2/3})\, \alpha +\!\!\!\!\rightarrow (d_{2/3})\, \omega^2 \qquad (7)$$

where:

$d_{2/3}$ = Relative distance between P_2 and P_3

Since the first four quantities (from left to right) are known in both direction and magnitude from the accelerometer outputs and the distance between the points is also known, as well as the directions of the relative acceleration terms, this vector equation can be solved for the two unknown magnitudes, α and ω.

Table 4 - Geometric and Inertial Properties of Dummies

Dummy Segment	Length, in.		C.G. Position, in.		Weight, lb		C.G., lb-in.-sec^2	
	1[a]	2[b]	1	2	1	2	1	2
Head - c.g. measured from head pivot; length to top of head	6.1	7.3	1.1	2.2	12.1	11.4	0.29	0.31
Neck - c.g. measured from torso pivot point; length to head pivot	5.5	5.5	2.3	2.5	2.5	4.4	0.04	0.03
Torso - c.g. measured from H-point; length to pivot point of 1st neck vertebra	19.9	21.5	9.5	11.8	35.5	66.0	3.29	8.31
Upper Arm[c] - c.g. measured from shoulder pivot; length to elbow pivot	10.6	11.4	2.9	5.1	7.7	4.5	0.35	0.17
Lower Arm and Hand - c.g. measured from elbow pivot; length to fingertips	17.1	17.5	6.6	7.4	4.0	4.6	0.24	0.26
Upper Leg - c.g. measured from H-point; length to knee point	16.4	16.9	6.4	7.3	14.4	16.5	1.26	1.37
Lower Leg, Foot and Shoe - c.g. measured from knee pivot; length to sole of shoe	19.9	20.6	13.4	11.3	12.0	11.6	1.60	1.54

[a]Dummy 1 - Alderson F5-AU, 5th percentile, vertebraed with rib cage, weight 124 lb, height 66 in.

[b]Dummy 2 - Sierra No. 292-750, 50th percentile, vertebraed with rib cage, weight 159 lb, height 71 in.

[c]Upper arm for dummy 2 does not include houlder pivot shaft.

NOTES: Weights and moments of inertia of arms and legs for right side only. Methods of measurement similar to those described by Naab (5).

Knowing α and ω, a relative acceleration equation can be written between either P_2 or P_3 and the c.g., P_1. This equation is

$$A_1 = A_2 \nrightarrow A_{1/2}$$

or in component form

$$A_{111} \nrightarrow A_{112} = A_{211} \nrightarrow A_{212}$$

$$\nrightarrow (d_{1/2}) \alpha \nrightarrow (d_{1/2}) \omega^2 \qquad (8)$$

Since the distance between P_1 and P_2 $(d_{1/2})$ can be measured, all the quantities on the right-hand side of the equation are known in both direction and magnitude. Consequently, this vector equation can be solved for the desired components of the acceleration of the c.g., A_{111} and A_{112}.

With these accelerations determined, Eqs. 4-6 can be used to obtain the neck reactions.

SUBJECTS - Two different types of anthropomorphic dummies, two cadavers, and a human volunteer were used as subjects for the simulations. The geometric and inertial properties of the Alderson Model F5-AU and the Sierra 50th percentile, vertebrae dummies are listed in Table 4, and their relative joint stiffnesses are presented in Table 5.

The neck of the Alderson dummy, which consisted of three steel segments, was modified by the insertion of thin pieces of rubber between each of the neck segments. These rubber interfaces attenuate the high frequency acceleration which occurs when two adjacent neck segments "bottom out." The head and neck segments were fastened to the torso by means of a steel cable. The tension in the cable was used to regulate the stiffness of the neck. Relative rotation of the head with respect to the torso in the sagittal plane was approximately ± 65 deg.

The neck assembly for the Sierra dummy consisted of seven steel segments, ball and socketed together. The ends of the assembly were bolted rigidly in place to the head and torso, respectively. The relative stiffness between adjacent segments is individually adjustable. A sandwich-type rubber disc consisting of a thick layer of spongy rubber between two layers of thin, hard rubber between each pair of segments

Table 5 - Relative Joint Stiffnesses of Dummies

Joint	Load, lb 1[a]	Load, lb 2[b]	Moment Arm, ft 1	Moment Arm, ft 2	Moment, ft-lb 1	Moment, ft-lb 2
Head	12.9	6.5	0.23	0.33	3.0	2.2
Hip	6.3	7.5	2.00	1.00	12.6	7.5
Knee	34.5	2.5	1.00	1.00	34.5	2.5
Shoulder	22.5	2.3	1.00	1.00	22.5	2.3
Elbow	23.0	9.0	0.78	1.00	17.9	9.0

[a]Dummy 1, Alderson, see Table 4.

[b]Dummy 2, Sierra, see Table 4.

NOTES: Inferior end of moment arm for head taken as pivot point of head with respect to neck, P8. Values for dual segments are averages of right and left sides.

Table 6 - Cadaver and Volunteer Statistics

				Head Parameters						
Subject	Age, yr	Wt., lb	Ht., in.	Wt., lb	ICG, lb-in.-sec^2	Circum., in.	Width, in.	Length (A-P), in.	C.G. to Occipital Condyles, in.	Hip to Shoulder, in.
Cadaver 1035[a]	66	134	64	6.3	0.11	20.8	5.7	7.4	1.8	21.5
Cadaver 1089[b]	69	130	67	9.1	0.19	24.0	6.5	7.9	2.4	23.5
Volunteer (LMP)	47	160	68	10.8	0.20	22.3	5.9	7.5	2.4	21.5

[a]Died 7/14/66 of natural causes.

[b]Died 10/12/66 of spinal artery thrombosis.

NOTES: Head parameters for cadavers taken from their decapitated heads. Head parameters for volunteer estimated from cadaver data using method described in Ref. 6 and the Ph.D. dissertation, "Kinematics and Kinetics of Whiplash," by H. J. Mertz, Jr., Wayne State University, 1967. All subjects were male.

prevented metal-to-metal contact between vertebrae. The allowable sagittal rotation of the head with respect to the torso without these interfaces was ± 72 deg, and was somewhat less with them in place.

Table 6 lists the pertinent information for the two cadavers and volunteer. Dummy 1 had the heaviest head, 12.1 lb, as compared to 11.4, 6.3, 9.1, and 10.8 lb for dummy 2, cadavers 1035 and 1089 and the volunteer, respectively, but was classified by the manufacturer as a 5th percentile dummy.

INSTRUMENTATION - A complete listing of the types of transducers used and the corresponding conditioning and read-out systems are presented in Table 7. The sled instrumentation consisted of a sled accelerometer and a velocity transducer. Seat back resisting torque and rotation were monitored when the back was allowed to rotate. When used, the sagittal headrest load was measured by two axes of a triaxial load cell.

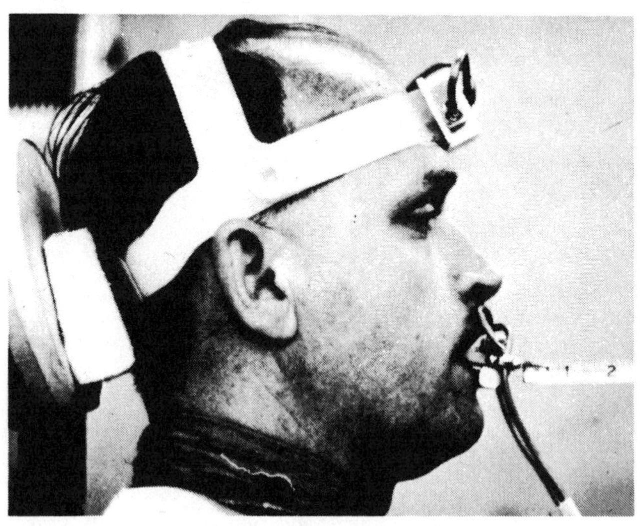

Fig. 5 - Head instrumentation of volunteer

Table 7 - Transducer, Electronic Conditioning, and Recording Details

Trace	Transducer	Amplification	Galvanometer Frequency[a]
Head acceleration	Statham accelerometers, Models A52-100-350 and A6-100-350	Heiland carrier amplifier, Model 119-B1	1650
Head load cell	2 axes of triaxial strain gage load cell (G.M. and W.S.U. design)	Heiland carrier amplifier, Model 119-B1	1650
Seat back torque	Strain gaged seat shaft	Heiland carrier amplifier, Model 119-B1	3300
Seat back rotation	Rotary potentiometer used in Wheatstone bridge	Kin Tel amplifier, Model 112A	Tape[b]
Lap belt load	Strain gaged axial load cells	Heiland carrier amplifier, Model 119-B1	Tape[b]
Sled acceleration	Statham accelerometer, Model A5-100-350	Kin Tel amplifier[c], Model 112A	1650[d] and tape
Sled velocity	Magnetic pickup	None	5000

[a]Data recorded using either a Honeywell Visicorder Model 906A with Honeywell subminiature galvanometers and/or an Electro-Medi-Dyne 8 channel 1/2 in. tape recorder (linear to 5000 cps at tape speed of 15 ips).

[b]Trace recorded using M1650 galvanometer when head load was not measured.

[c]For runs when cadaver 1089 was used, no amplification needed because a very sensitive M100-120 galvanometer was used to filter high frequency structural vibrations. Bridge output shunted by 175 ohm resistor to critically damp this galvanometer.

[d]Output recorded with both galvanometer and tape for synchronization.

All subjects' heads were instrumented at two points with two uniaxial accelerometers whose axes were orthogonal. In the case of the dummies the accelerometers were mounted within the head cavity, a pair posteriorly mounted and a pair anteriorly mounted. For the cadavers one pair of accelerometers was screwed to the superior portion of the skull and the other pair was mounted to dental acrylic which was molded to the interior of the oral cavity and protruded through the mouth. For the volunteer a pair of accelerometers was mounted to a fitted biteplate made of dental acrylic (Fig. 5). The other pair was attached to a plastic headband in the forehead region. Axial load cells were used to measure the seat belt loads.

Because of the multiplicity of transducer channels, two different read-out systems, tape and oscillograph, were employed. Synchronization between these systems was accomplished by splitting the sled acceleration trace and timing signal and recording both on each system.

High-speed cinematography of all simulations was obtained using a 16 mm, 500 f/sec, 160 deg shutter Milliken camera Model No. DSM-4B with a Kodak Cine Ekta 25 mm, f1.4 lens. The camera was placed 18 ft from the vertical plane containing the sagittal plane of the subject. The film used was Ektachrome ERB, high-speed, ASA 125. Film speed was obtained from a neon bulb flashing every 1/120 sec. Synchronization between the film and the transducer outputs was obtained by firing a flash bulb at sled contact with the snubber and recording the firing voltage with a galvanometer.

PROCEDURE - For both dummies and both cadavers the simulation sequences were with minor exceptions identical. Two series, one with the head supported followed by one with no head support, were conducted using these subjects. For the head supported series, the first run was with a rigid seat back at the nonsevere, 9 mph and 10 in. stopping distance condition. The remaining runs of the series were conducted at the severe level, 15 mph and 17.5 in stopping distance. The degree of seat back rigidity was incrementally increased with the seat back being rigid for the last run of the series. The simulation sequence for the case of no head support was identical.

Similarly, the simulation sequence for the volunteer consisted of two series, head supported and unsupported. However, for the supported head series the first run was conducted at 9 mph and 22 in. stopping distance with the severity of each succeeding run being increased. The most severe run was 14.7 mph with a 10 in. stopping distance. After this run the volunteer was still willing to undergo higher severity runs, but because of fatigue runs at increased severity were not conducted. Two runs were conducted with no head support, the first at 8.4 mph and 22 in. stopping distance and the second at 8.9 mph and 10 in. stopping distance, which correspond to the nonsevere simulation condition. After the last run the volunteer expressed the opinion that he did not care to increase the severity level at that time. In all cases for the volunteer the seat back was rigid.

Three different restraint systems, free, lap belted, and lap and diagonal chest belted, were employed with the Alderson dummy. However, preliminary runs indicated that no load was applied to the chest straps. Consequently, all other subjects were restrained only by a lap belt. The headrest used for the Alderson dummy was curved, for the Sierra dummy the curved surface and a flat surface were employed, and for the cadavers and volunteer only the flat surface headrest was used.

When cadavers were used as subjects, a preliminary set of X-rays of the cervical spine was taken before they were subjected to the simulation. Subsequent sets of X-rays were taken after any simulation in which damage to the cervical spine was suspected to have occurred. All X-ray sets consisted of three lateral shots of the cervical spine with the neck in normal, flexed, and extended positions; an anterior to posterior shot of the head, neck, and upper thoracic vertebrae with the head in a normal position; and a posterior to anterior shot of the cervical spine with the head extended so as to reveal the odontoid process.

METHOD OF DATA ANALYSIS - All the traces on the oscillograph records of each simulation were read out in various time increments. From the high-speed film the frames corresponding to these time increments were analyzed for relative head position and seat back angle. These data were used as input information for a computer program which calculated, based on Eqs. 4-8, the acceleration of the c.g. of the head, the angular acceleration and velocity of the head, and the corresponding neck reactions for each time increment. Also the computer program corrected each pair of accelerometer readings to give the acceleration at points described by the intersection of their sensitive axes.

RESULTS AND DISCUSSION

COMPARISON OF THE RESPONSES OF VARIOUS SUBJECTS WITH NO HEADREST - The first comparison is for the case where the subjects, with the exception of the volunteer, who underwent only the 10 mph simulation, were subjected to the simulated 10 and 23 mph rear-end collisions without a headrest. In each simulation the subject was lap belted, and the seat back was not allowed to rotate.

The oscillograph records for the various subjects are presented in Figs. 6-14. The head acceleration traces are identified by three letters. The first two letters, H.A., signify head acceleration, while the third letter designates the anatomical position of the accelerometers as described previously. The symbols S-I (superior toward inferior) and P-A (posterior toward anterior) indicate the approximate direction of the sensitive axis of the accelerometers relative to the head and are indicated by arrows. The vertical line labeled t = 0 represents the beginning of the simulated pulse. The small vertical hash marks above the sled acceleration trace subdivide the time after t = 0 into 50 ms intervals. The timing signal gives a continuous 10 ms time base. Calibration factors for each trace along with an equivalent 1 in. scale are given.

Because of the differences in location of accelerometers of each subject, direct comparison of traces can be made only in special cases and then only when differences in calibration factors are noted. Direct comparison of all traces between dummy 1 with dummy 2 and cadaver 1035 with cadaver 1089 can be made, but comparisons between the traces of the dummies, cadavers, and volunteer can be made only for the anteriorly located accelerometers in the P-A and S-I directions.

For the 10 mph simulation, the general shapes of the corresponding traces for the two cadavers (Figs. 8 and 9) are in good agreement with each other. The similarity between

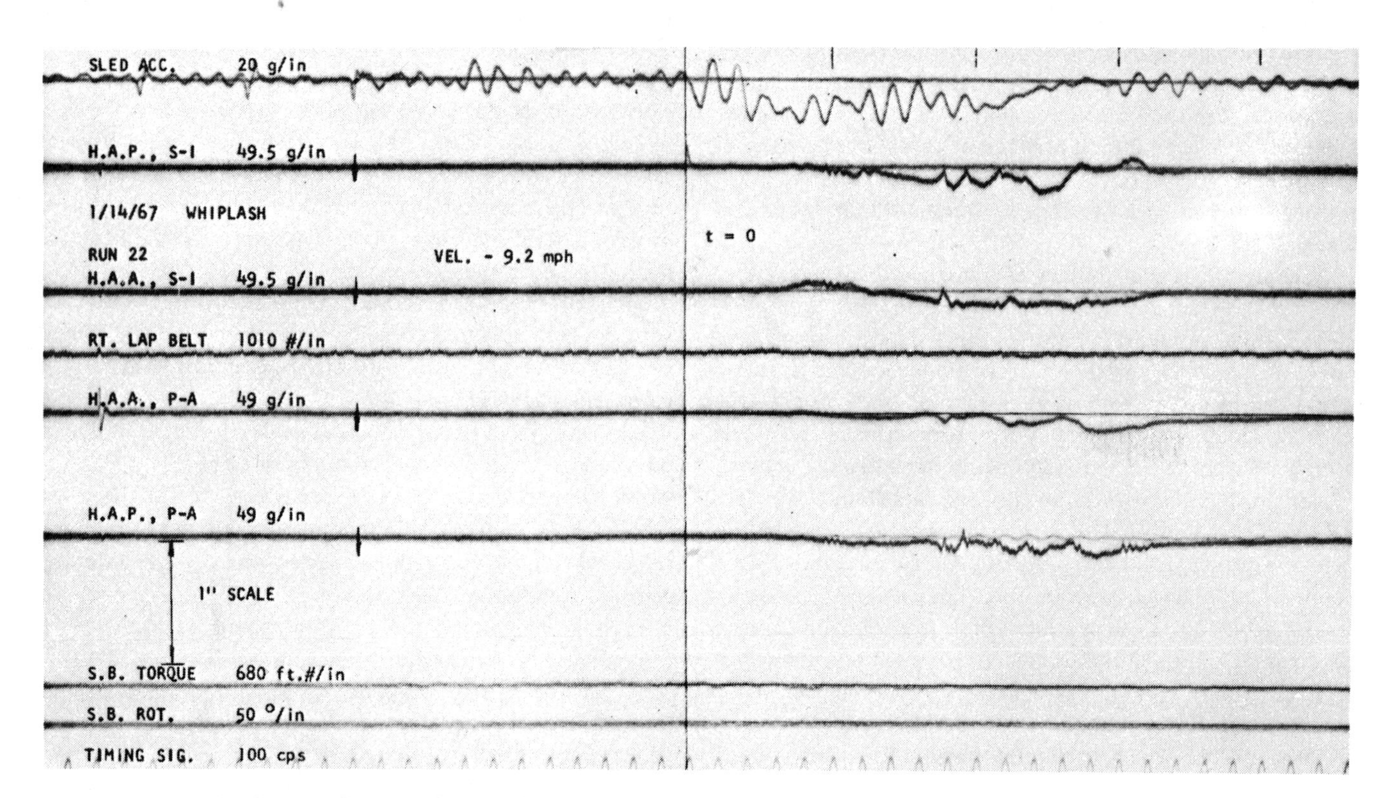

Fig. 6 - Oscillograph record of 10 mph simulation, dummy 1, rigid seat back, no headrest

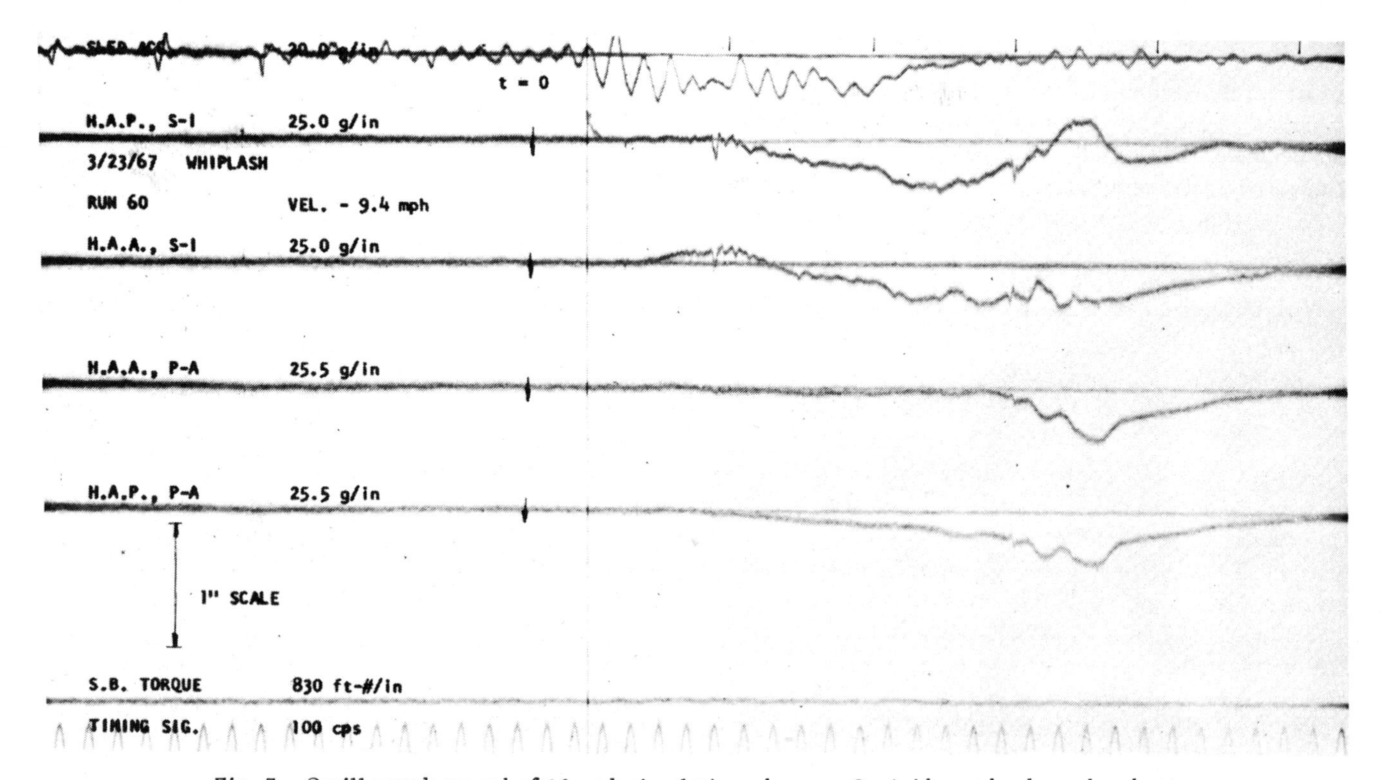

Fig. 7 - Oscillograph record of 10 mph simulation, dummy 2, rigid seat back, no headrest

accelerometer response for the dummies (Figs. 6 and 7) is not good. In particular, the corresponding H.A.P., S-I and the H.A.A., P-A traces are not similar even though the accelerometers are located in identical head positions. The H.A.A., S-I traces for both cadavers and both dummies compare in general shape to the H.A.M., S-I trace (Fig. 10) of the volunteer. The pulse times for the volunteer traces are longer than for the cadavers or dummies due to the tensing of muscles prior to the simulation.

For the 23 mph equivalent rear-end collision, the cor-

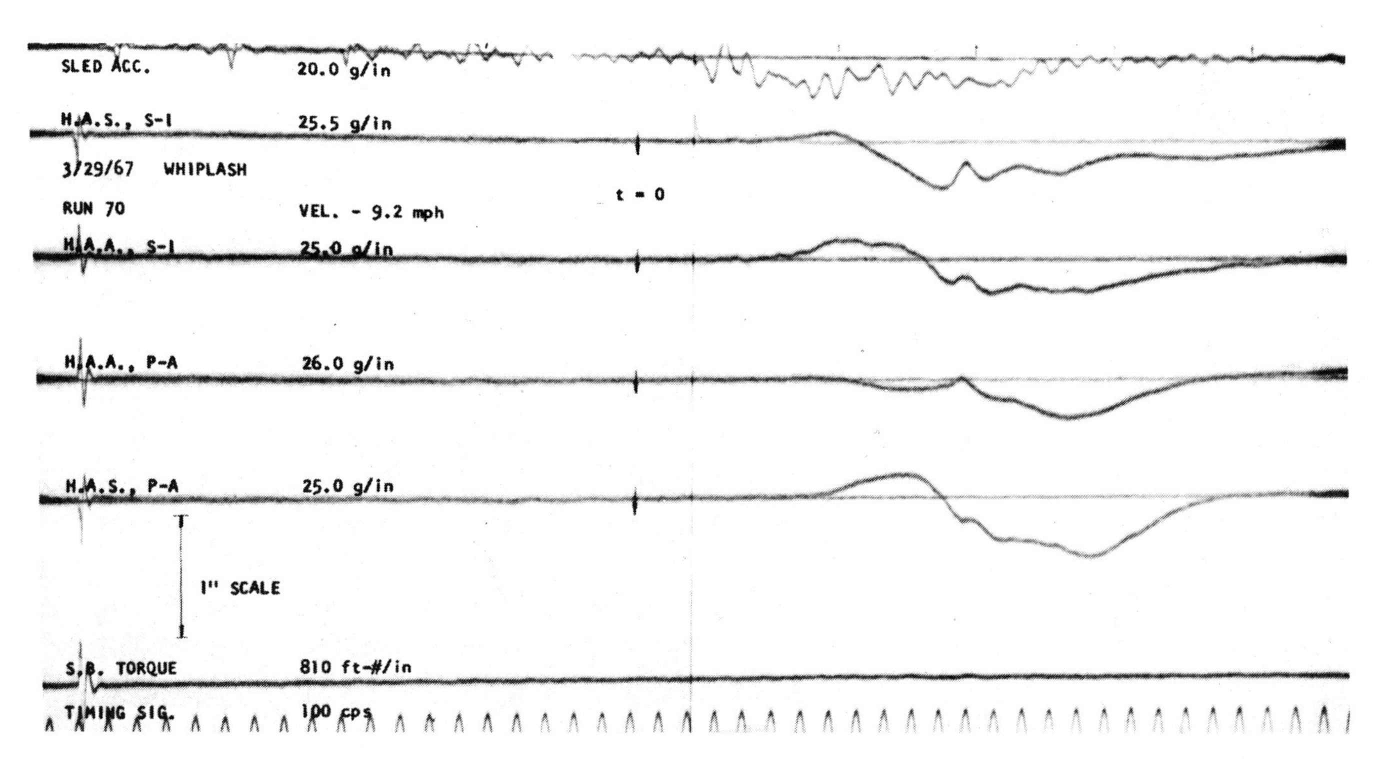

Fig. 8 - Oscillograph record of 10 mph simulation, cadaver 1035, rigid seat back, no headrest

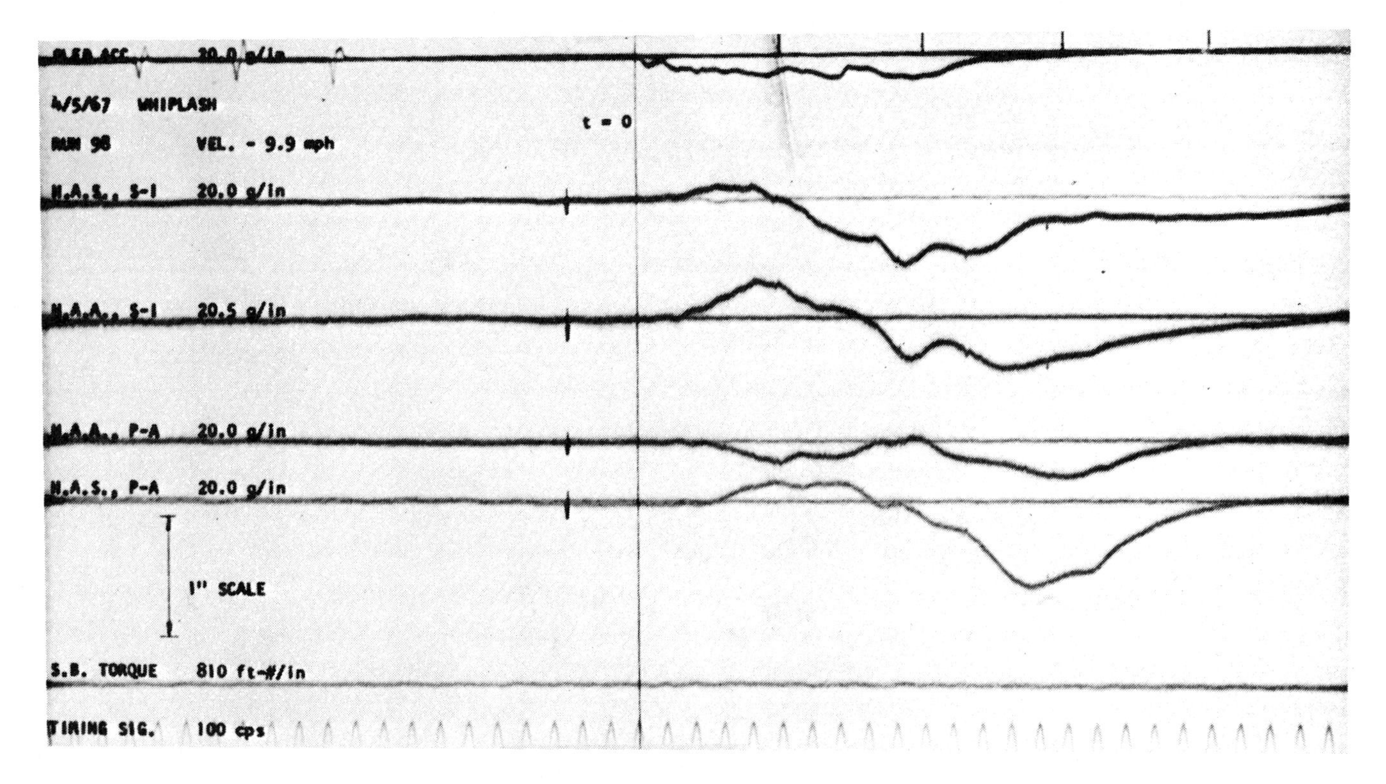

Fig. 9 - Oscillograph record of 10 mph simulation, cadaver 1089, rigid seat back, no headrest

responding traces (Figs. 13 and 14) between cadavers are again in good agreement, while the dummy traces (Figs. 11 and 12) are not. A typical high g artifact caused by metal-to-metal contact of the neck segments is shown on the H.A.A., P-A for dummy 1 at approximately 165 ms (Fig. 11). A similar artifact occurs on all head traces for dummy 2 at approximately 130 ms (Fig. 12). Pieces of rubber were inserted between the segments of the neck of each dummy to attenuate this effect; however, complete elimination would make head instrumentation and record read-out more reliable.

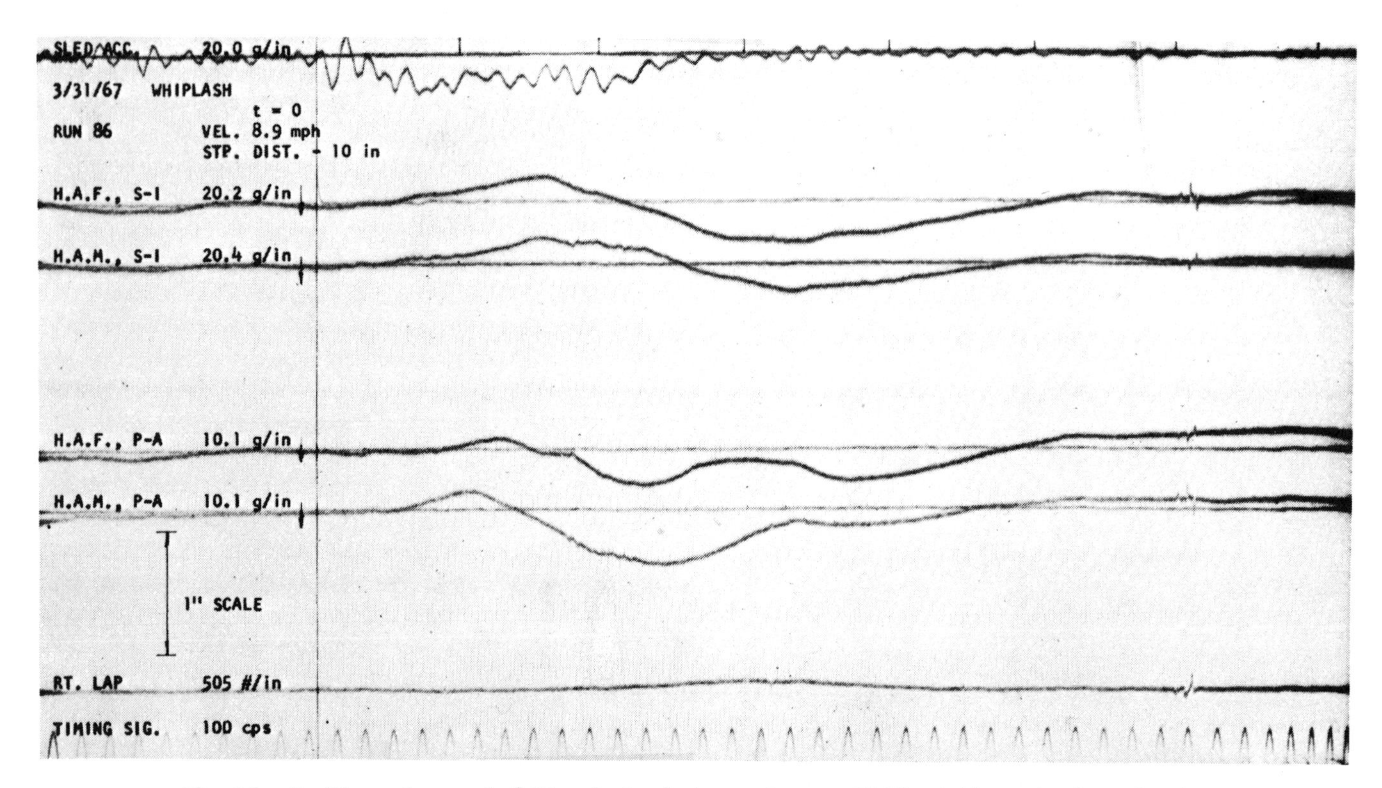

Fig. 10 - Oscillograph record of 10 mph simulation, volunteer (LMP), rigid seat back, no headrest

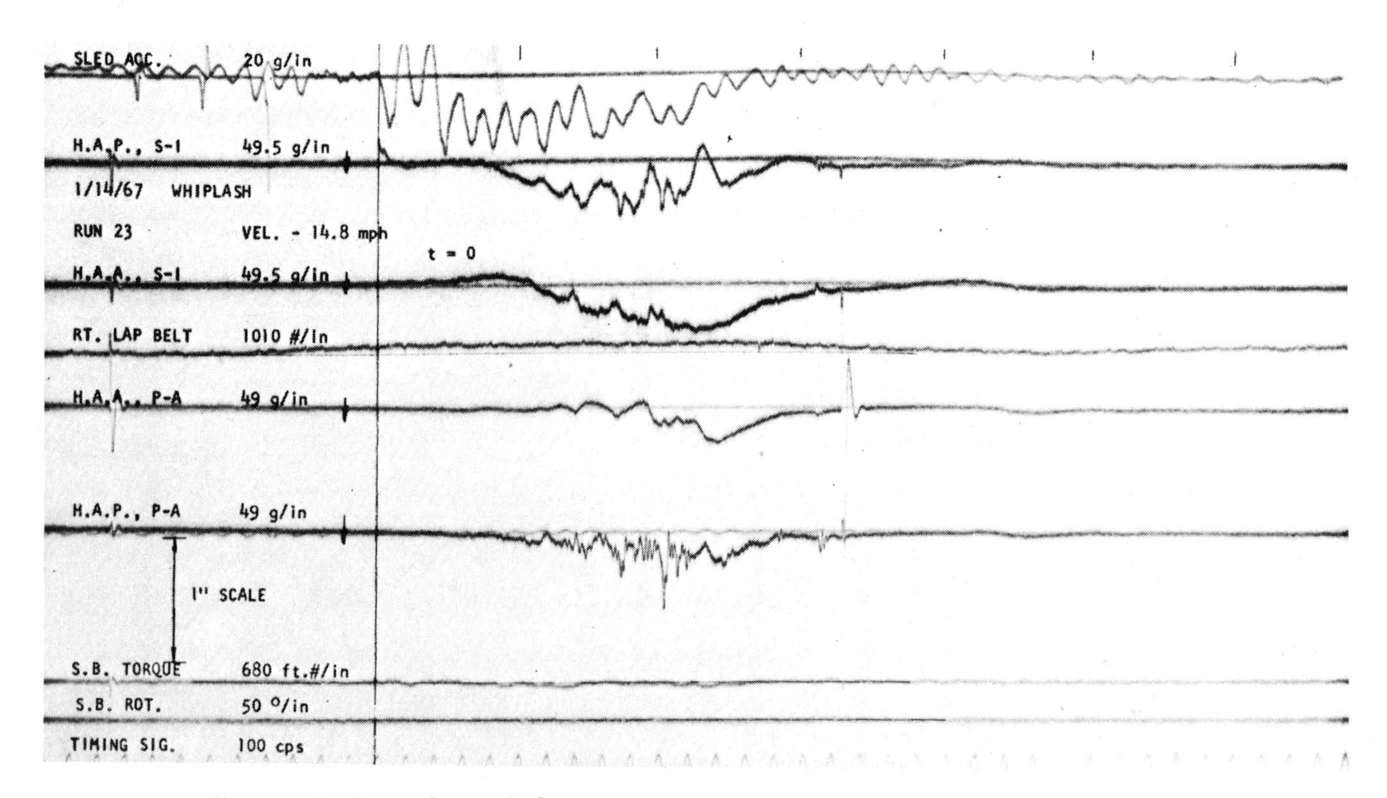

Fig. 11 - Oscillograph record of 23 mph simulation, dummy 1, rigid seat back, no headrest

Another undesirable trait related to dummy construction is the high frequency oscillation due to local vibration at the accelerometer mount which makes the determination of the rigid body acceleration difficult. None of these artifacts occurred on any of the volunteer or cadaver traces.

An artifact common to all traces for all subjects occurs prior to t = 0 and is the result of a switching transient caused by closing the braking valve for the power cylinder.

Also, depicted on the oscillograph records are two different sled acceleration wave shapes. For simulations involving

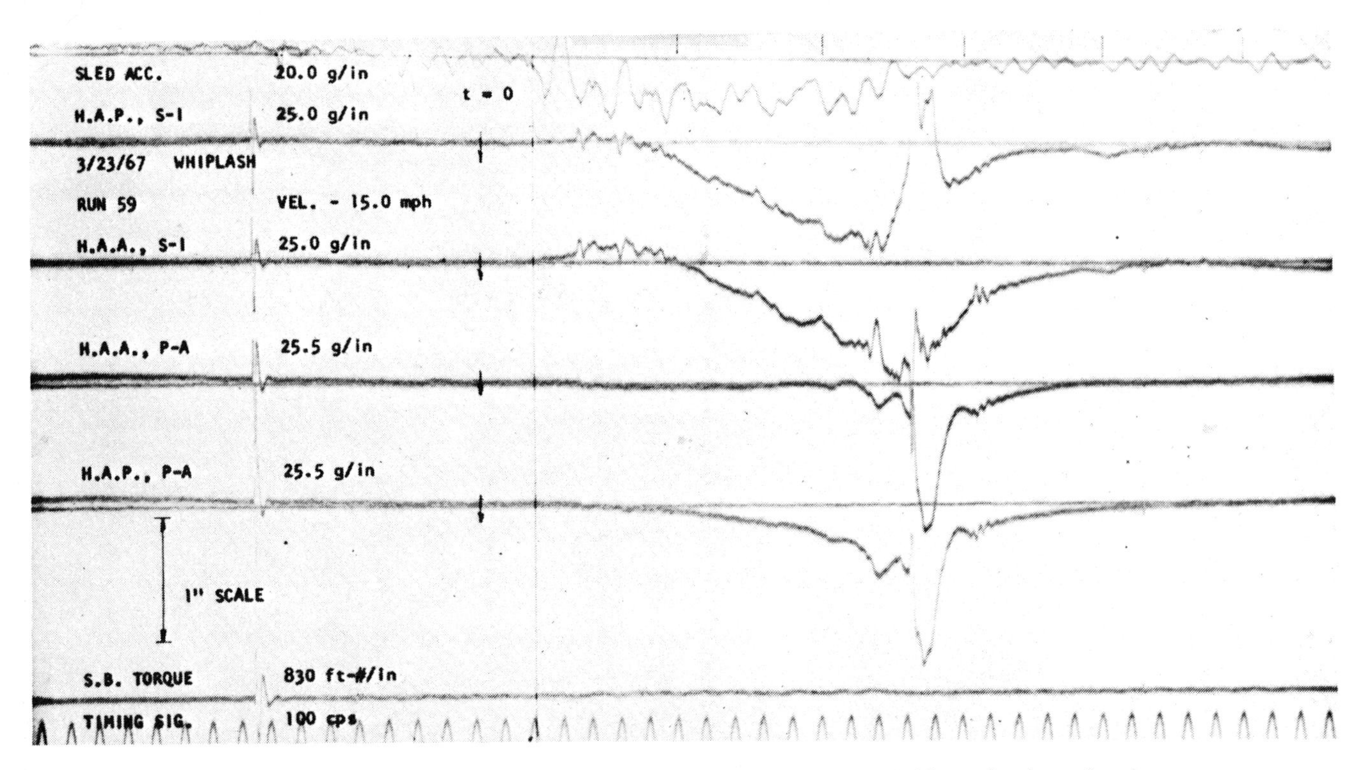

Fig. 12 - Oscillograph record of 23 mph simulation, dummy 2, rigid seat back, no headrest

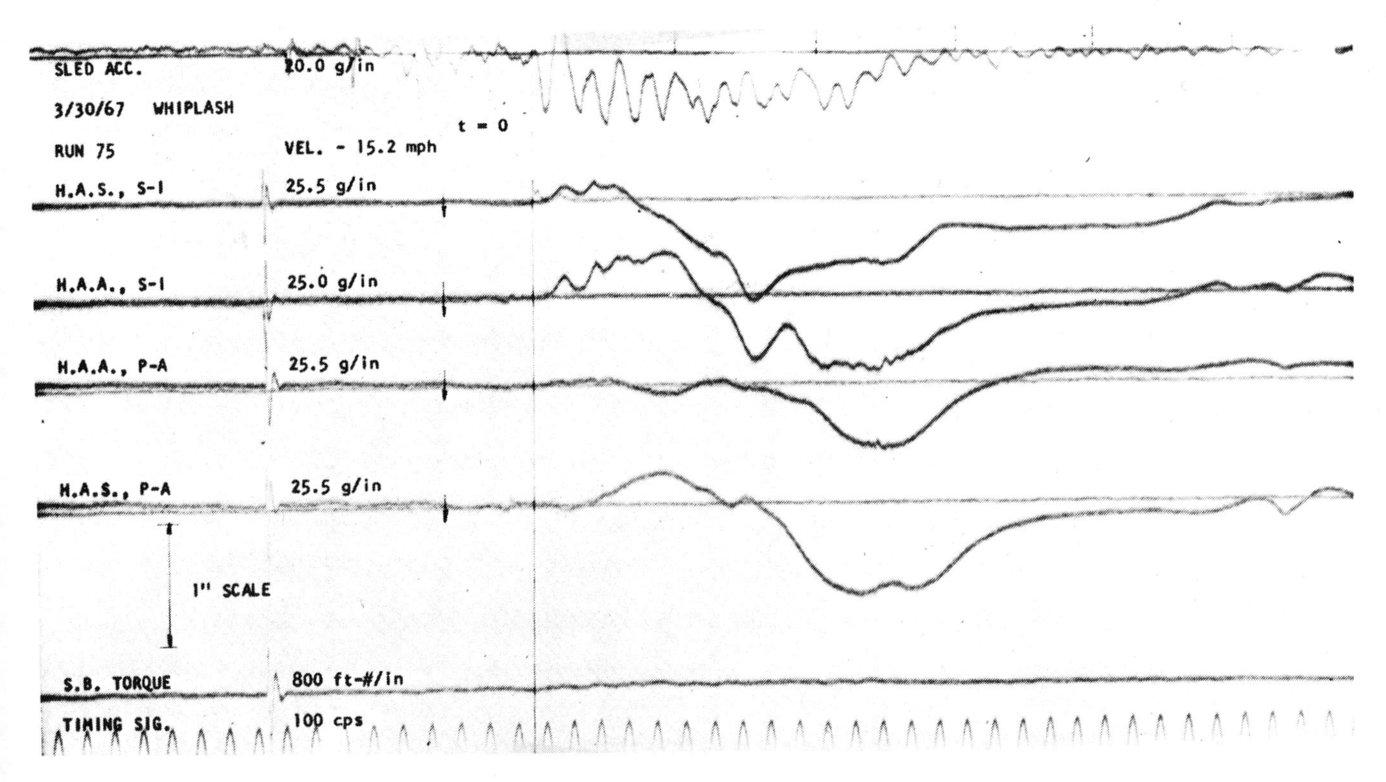

Fig. 13 - Oscillograph record of 23 mph simulation, cadaver 1035, rigid seat back, no headrest

dummies 1 and 2, the volunteer and cadaver 1035 (Figs. 6-8 and 10-13), the sled accelerometer was wrapped in sponge rubber and enclosed in a box which was mounted to the sled. This eliminated the very high g, short time duration structural vibrations of the sled. The sponge rubber caused a lower g, longer time duration vibration to be superimposed on the rigid body acceleration of the sled. The resulting wave shape was satisfactory since a mean sled acceleration could be approximated. For the sequence of simulations using cadaver 1089 (Figs. 9 and 14), the accelerometer was mounted directly to the sled frame. The output of the accelerometer was not amplified but was fed directly to a sensitive, low frequency, critically damped galvanometer which mechanically filtered the high frequency structural vibrations.

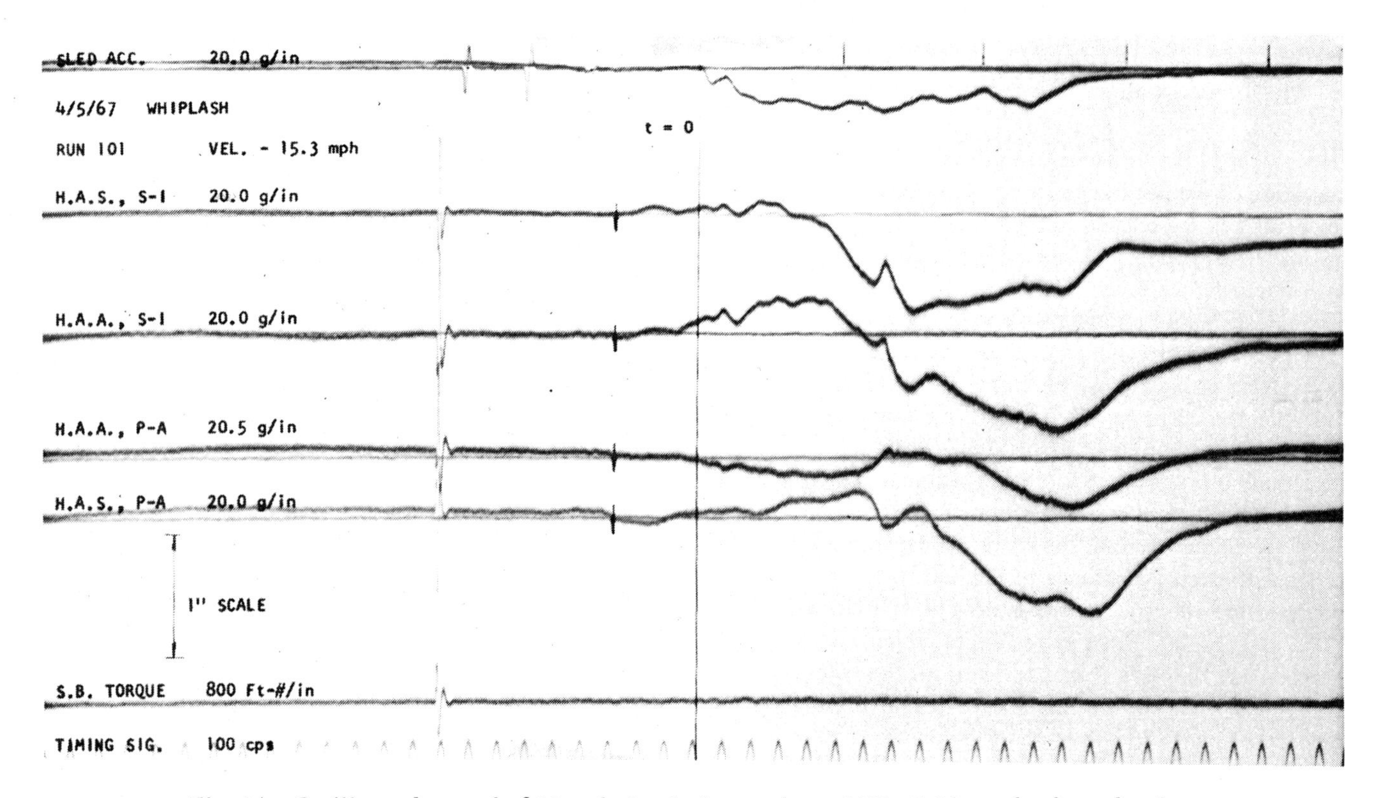

Fig. 14 - Oscillograph record of 23 mph simulation, cadaver 1089, rigid seat back, no headrest

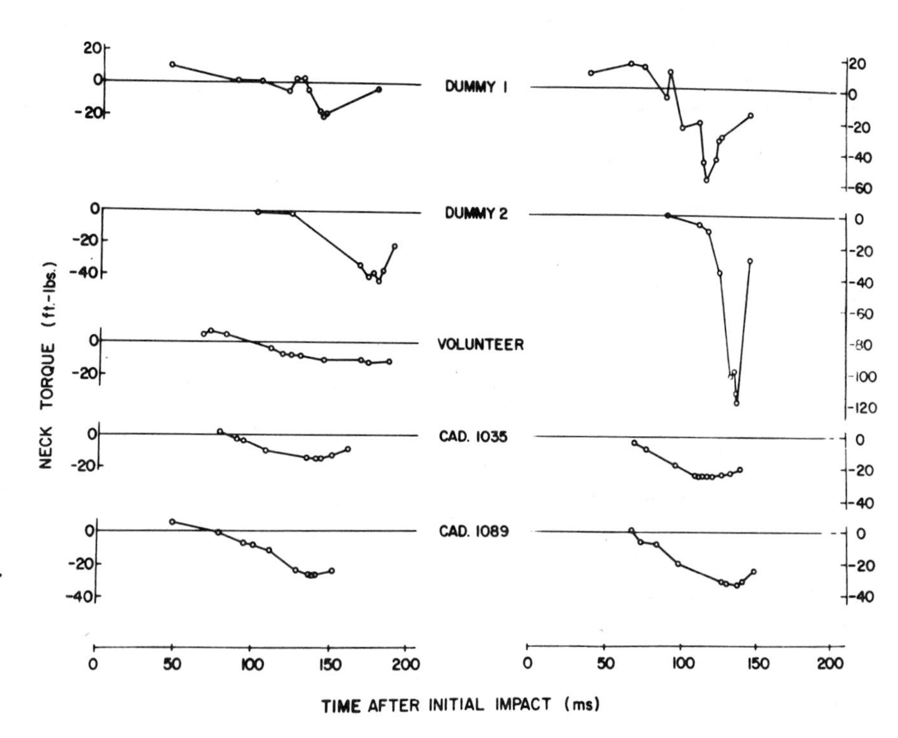

Fig. 15 - Computed neck torques acting on the head at the occipital condyles as function of time for (from left to right) 10 and 23 mph simulations, no headrest, rigid seat back

From these oscillograph records the neck reactions, the acceleration of the c.g. of the head, and the head angular acceleration were computed for various time increments using Eqs. 4-8. The computed neck torques acting on the head at the occipital condyles as a function of time for both the 10 and 23 mph simulations for the various subjects are shown in Fig 15. The dummy's resistive neck torques for both simulation conditions are characterized by high rates of loading, with the response of dummy 2 being more pronounced than that of dummy 1. Considering the neck construction of the dummies, this type of response should be expected. The neck of dummy 1 consists of steel cable. Consequently, the torque needed to produce a change in curvature of the neck depends on the contact surface between segments and the tension in the cable. However, both these quantities vary with neck curvature, resulting in a fluctuation in required torque levels. When relative movement of neck segments is prohibited by metal-to-metal contact (the segments are all bottomed out), further extension of the head takes place primarily by stretching of the steel cable and changing the curvature of the dummy's torso, which requires a large torque increment per degree of rotation. The neck of dummy 2 is constructed of steel segments which are "ball and socketed" together. The torque required to produce a change in curvature of the segments depends upon the rotational frictional resistance of the ball with respect to the socket, which depends on the mating of relative surfaces. Consequently, the torque necessary to produce a change in curvature of the neck fluctuates until all the segments "bottom out." When this occurs, further head extension must produce a change in curvature of the dummy's torso which results in a rapid increase in torque per degree of head extension. Thus, for both dummies if the head still has a velocity relative to the torso when all the neck segments have bottomed out, the neck torque must increase rapidly in order to slow down the head relative to the torso. Comparing the two different types of neck construction with the segments bottomed out, the required torque per degree of extension for dummy 1 will be less than that of dummy 2 because of the increase in neck curvature due to cable extension for dummy 1. This is demonstrated by comparing the maximum resistive neck torque values for dummy 1 and dummy 2, which are 20.3 and 43.3 ft-lb for the 10 mph simulation and 58.5 and 118 ft-lb for the 23 mph simulation, respectively.

Unlike the responses of the dummies, the resistive neck torques for both the cadavers and volunteer build up gradually and form continuous loading curves. In the case of the cadaver, the resistance to head extension is provided by the stretching of the muscles, tissues, and ligaments of the neck. This type of neck loading can be compared to that of a living person who is unexpectedly subjected to a rear-end collision and does not have time for his muscles to react in order to resist the head extension. In this case the head rotates with very little resistive torque until the anterior neck ligaments become stretched and posterior tissue between the spinous process of the neck vertebrae is compressed. At this point resistive forces which are produced are identical in nature, if not in magnitude, to that of the cadavers.

The volunteer represented the case of a person aware of the impending rear-end collision who tensed his neck muscles in anticipation of the impact. In this case the resistive torque is comprised of muscle reaction during the initial portion of the head rotation with the stretching of tissue and ligaments gradually taking up some of the load as the degree of the extension increases. Since the relative rotation of the head is resisted during the entire head extension for the case of tensing prior to the impact, the maximum neck torque should be less than if the person were not tense. This is verified by comparing the maximum neck torques of 14.9, 27.6, and 12.3 ft-lb for cadavers 1035 and

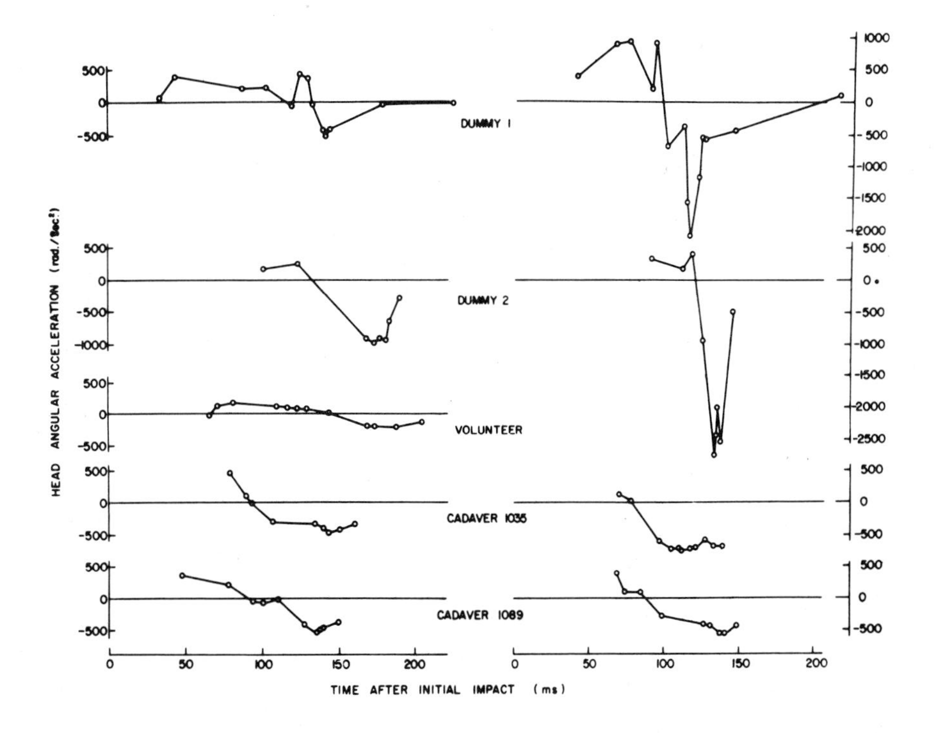

Fig. 16 - Computed head angular accelerations as function of time for (from left to right) 10 and 23 mph simulations, no headrest, rigid seat back

1089 and the volunteer, respectively, for the 10 mph simulation. Hence, tensing prior to a rear-end collision certainly will reduce the relative severity of the impact.

There was a delay in reaching a peak torque for dummy 2 when compared to the other subjects because the angle through which the head could be rotated prior to bottoming out the neck segments was greater and consequently required a longer time period to reach this position.

The angular acceleration of the head as a function of time for the various subjects for the two simulations is shown in Fig. 16. These curves are similar in shape and relative maximum magnitudes to their corresponding torque curves and portray all the characteristics stated for the torque curves.

The P-A accelerations of gravity of the heads of the various subjects, for the 10 and 23 mph simulations are depicted as a function of time in Fig. 17. Again, the smooth, gradually increasing accelerations for the cadavers and volunteer are in contrast to the spiking, variable fluctuating responses of the dummies. The same is true in general for the I-S curves for the acceleration of the c.g. of the heads of the various subjects presented in Fig. 18.

Comparisons of the relative rotations of the heads of the various subjects for the 10 and 23 mph simulations are presented in Figs. 19A and 19B, respectively. The relative position of the head with respect to the torso was obtained from the film analysis and represents the angular position

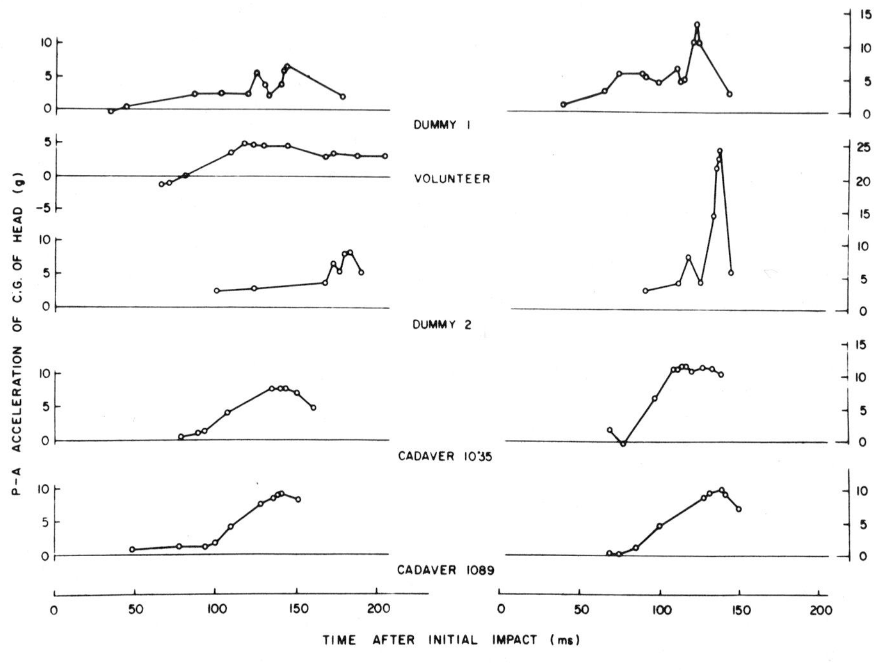

Fig. 17 - Computed P-A acceleration of head c.g. as function of time for (from left to right) 10 and 23 mph simulations, no headrest, rigid seat back

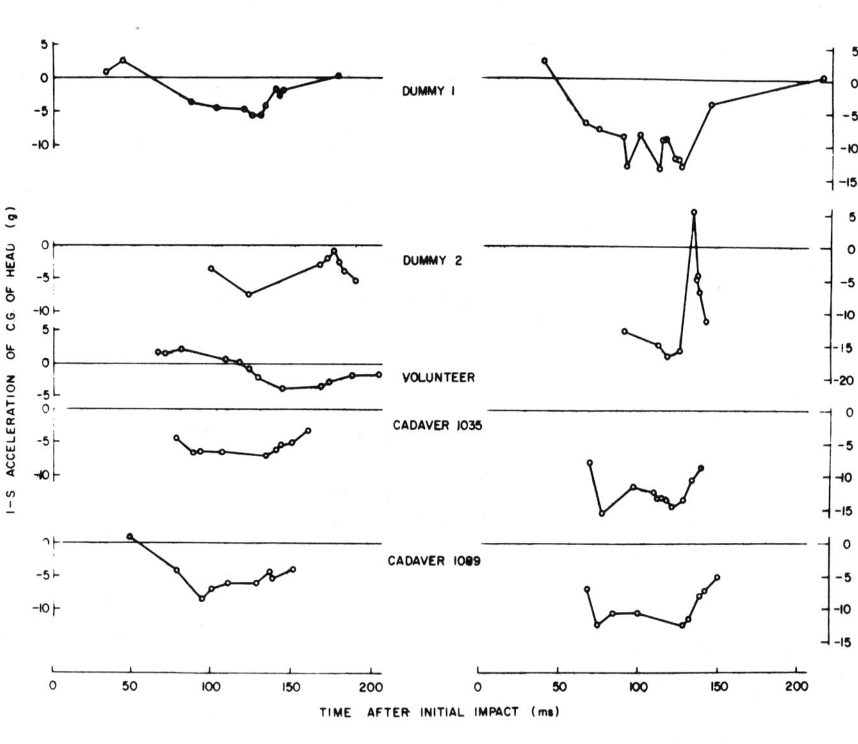

Fig. 18 - Computed I-S accelerations of head c.g. as function of time for (from left to right) 10 and 23 mph simulations, no headrest, rigid seat back

of the X_{112} (I-S) axis of the head with reference to upper torso. The angle for zero extension of the neck was obtained by measuring the angle between the X_{112} axis of the head and the upper torso with the subject seated in a normal position and was taken as -15 deg for all the subjects.

For the 10 mph, nonsevere simulation the maximum head-neck extensions for dummy 1 and cadavers 1035 and 1089 were in good agreement with each other and were 63, 64, and 61 deg, respectively. The maximum head extension for dummy 2 was 93 deg, 30 deg greater than for dummy 1. For the 23 mph simulation, the maximum head extension for dummy 1 and cadavers 1035 and 1089 were approximately the same -- 87, 86, and 84 deg, respectively. The maximum head extension of dummy 2 was 104 deg.

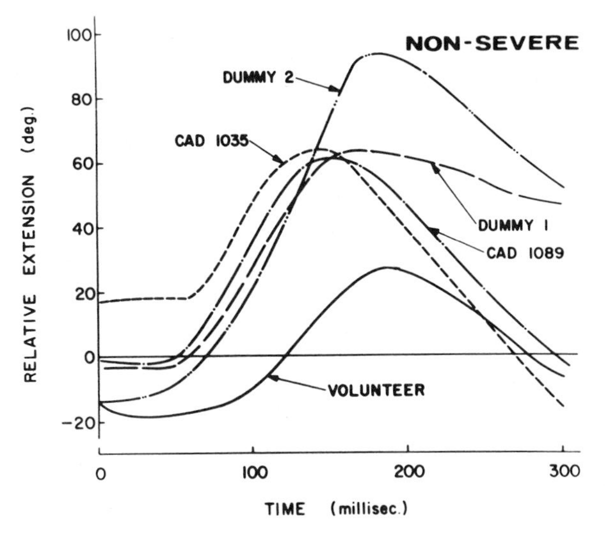

Fig. 19A - Comparison of relative head extension for various subjects for nonsevere 10 mph simulation, no headrest, rigid seat back

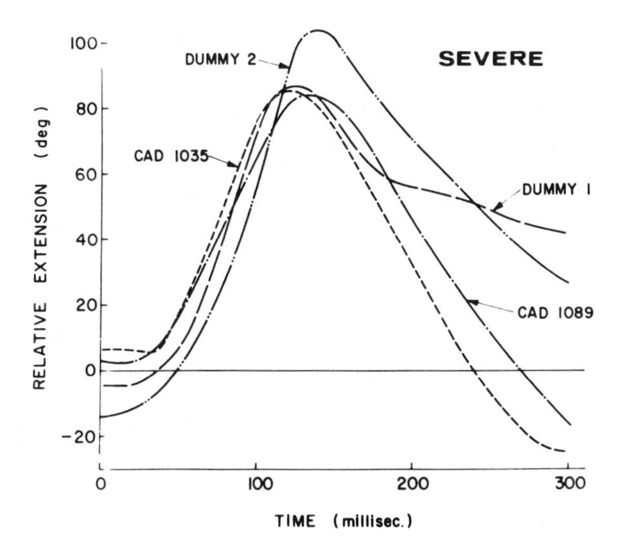

Fig. 19B - Comparison of relative head extension forvarious subjects for severe 23 mph simulation, no headrest, rigid seat back

The greater head extension for dummy 2 which occurred at both simulation levels was due to the fact that the head-neck combination of dummy 2 allowed a greater relative rotation of the head with respect to the torso prior to bottoming of all the individual neck segments. This greater rotation can be reduced by the removal of one of the neck segments.

Comparing the shapes of the extension curves for the time after the maximum head extension has occurred, the response of dummy 1 is more damped than the response of either cadaver. To decrease this damping effect for dummy 1, the neck segments could be encapsulated in a rubber tube capable of storing energy during the extension motion.

The maximum extension of the volunteer's head and neck for the nonsevere simulation was 27 deg which, because of muscle action, is less than the other subjects. In fact, the volunteer rotated his head forward 10 deg in anticipation of the ensuing extending motion. The shape of the head rotation curve for the volunteer after maximum extension has occurred is similar to the rotation curves for the cadavers, indicating that sufficient energy was stored in the neck during extension to cause the head to rebound.

For the 10 mph simulation (Fig. 19A), the head of cadaver 1035 was extended 17 deg at the initiation of the sled deceleration pulse. Consequently, the angular head velocity was less than it would have been if the head were positioned with no initial extension, resulting in a corresponding lower maximum resistive neck torque.

In Table 8 the maximum dynamic head response for the various subjects for the 10 and 23 mph equivalent rear-end collisions are listed. Also, the static voluntary human tolerance levels for the head in an extended position are presented. The indices listed give a method of comparing the responses of the various subjects to identical simulations based on the static voluntary human tolerance levels, taking into account the differences in head weights. Since the dynamic neck reactions are directly proportional to their corresponding head weights, the effect of the differences in head weights was eliminated by dividing the maximum values of the torque, shear, and axial forces by their corresponding head weights. To compare these ratios with the static voluntary human tolerance levels, the volunteer static levels were divided by the volunteer's head weight. The indices which allow for direct comparison between subjects were formed by dividing the dynamic ratios by the corresponding static ratio.

An index number of unity or less indicates that the maximum reaction is either equivalent to or less than its corresponding static voluntary tolerance level and no injury is expected. To evaluate the indices which are greater than unity, the response of the cadavers must be considered. Analysis of X-rays indicated that minor ligamentous damage occurred between the third and fourth cervical vertebrae for cadaver 1035, while no damage was observed for cadaver 1089. Since ligamentous damage occurred for a neck torque

Table 8 - Comparison of Maximum Dynamic Head Responses Based on Voluntary Static Human Tolerance Levels

			Rigid Seat Back						Collapsible Seat Back	
			Torque		Shear Force		Axial Force		Torque	
Subject	Head Weight, lb	Severity[a]	Max. Torque, ft-lb	Index I_1	Max. Shear lb	Index I_2	Max. Axial lb	Index I_3	Min. Torque, ft-lb	Index I_4
Volunteer	10.8	Static	17.5	1.00	190	1.00	255	1.00	--	--
		NS	12.3	0.70	49	0.25	28	0.10	--	--
Dummy 1	12.1	NS	20.3	1.05	89	0.40	61	0.20	--	--
		S	58.5	3.00	172	0.80	159	0.55	34	1.75
Dummy 2	11.4	NS	43.3	2.35	107	0.55	76	0.30	--	--
		S	118.0	6.40	287	1.45	189	0.70	--	--
Cadaver 1035	6.3	NS	14.9	1.45	55	0.50	42	0.30	--	--
		S	25.4	2.50	61	0.55	94	0.65	17	1.65
Cadaver 1089	9.1	NS	27.6	1.85	90	0.55	70	0.35	--	--
		S	33.0	2.25	99	0.60	113	0.55	26	1.75

[a]NS - Nonsevere 10 mph simulation.
S - Severe 23 mph simulation.

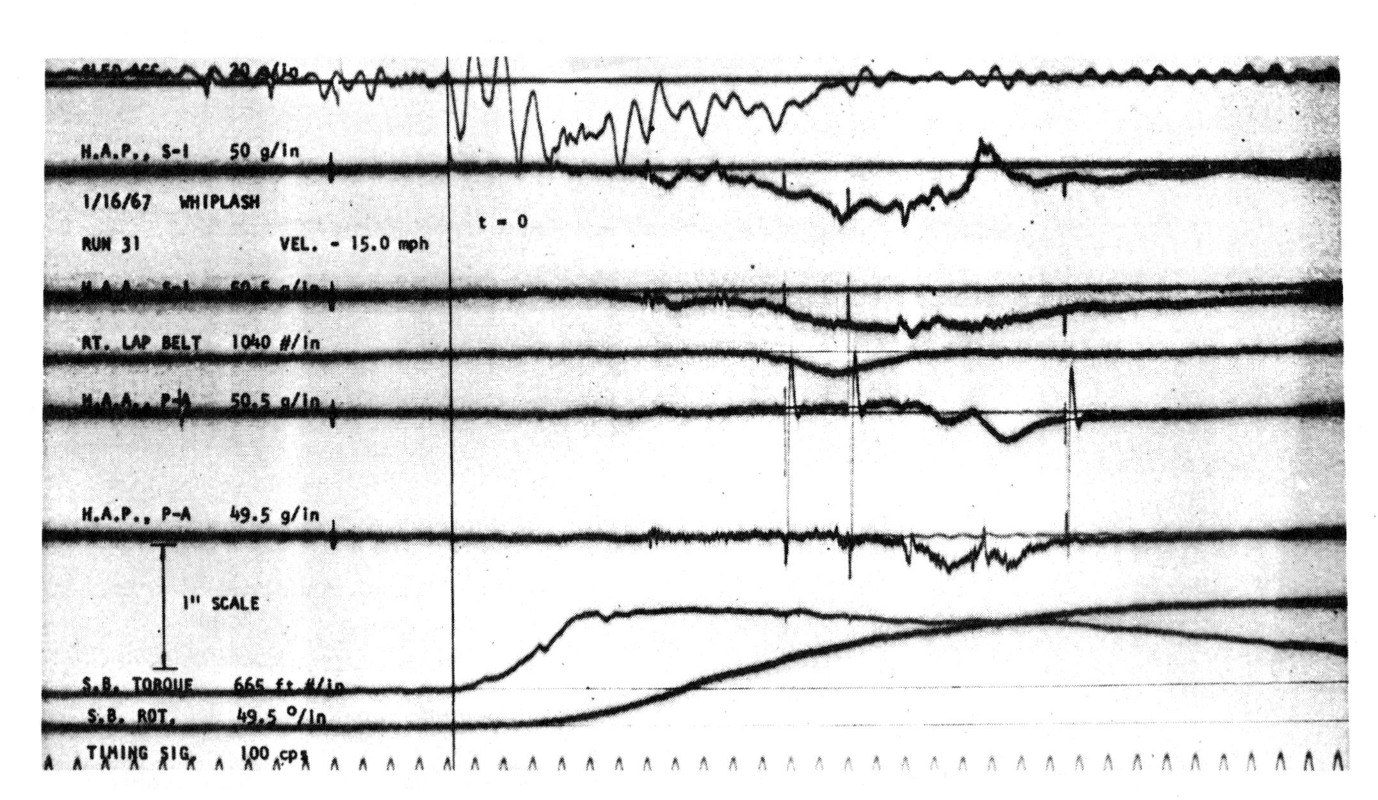

Fig. 20 - Typical oscillograph record (dummy 1, 23 mph simulation, no headrest) for case where seat back was allowed to rotate

index of 2.50 and not for an index of 2.25, a tolerable index is less than 2.50 if the cumulative effects of multiple impacts are neglected. Considering the severity of a 23 mph rear-end collision, where the individual with his head in a normal upright position is not aware of the impending impact, neck discomfort and/or injury would be expected and more so in these simulations since the seat back was rigid and corresponds to rear-end collisions of higher velocity. Consequently, the index value of 2.25 for cadaver 1089 should also be considered higher than a desirable limit. However, the 10 mph rear-end collision should be tolerable even in the case of the unsuspecting individual. Assuming that the neck index of 1.85 (10 mph simulation) for cadaver 1089 is not injurious, a tolerable neck index probably lies between 1.85 and 2.25. Until further data are available, a value of 2.00 is suggested.

The shear and axial force indices for the various subjects and the various simulations are all less than unity except for dummy 2. This implies that these reactions do not play a dominant role in causation of neck injury due to hyperextension and that the neck torque is the predominant factor. Hence, the response of the various subjects to the two simulations will be compared on the basis of their corresponding neck torque indices. For the 23 mph simulations, the indices for cadavers 1035 and 1089 and dummy 1 were 2.50, 2.25, and 3.00, respectively, while the index for dummy 2 was 6.40. Certainly, the response of dummy 2 is not comparable to the response of the other three subjects. For the 10 mph simulation the indices for cadavers 1035 and 1089 are 1.45 and 1.85, respectively, and are greater than the index for dummy 1, which is 1.05. The index of 2.35 for dummy 2 is still greater than those of the other subjects. Based on these indices, the neck structure for dummy 2 gives consistently higher simulated response for hyperextension of the neck, while dummy 1 gives a lower index for the 10 mph and a higher index for the 23 mph simulation when compared to indices of the cadavers. The index of the volunteer, which is 0.70, indicates that this simulation produced a neck torque

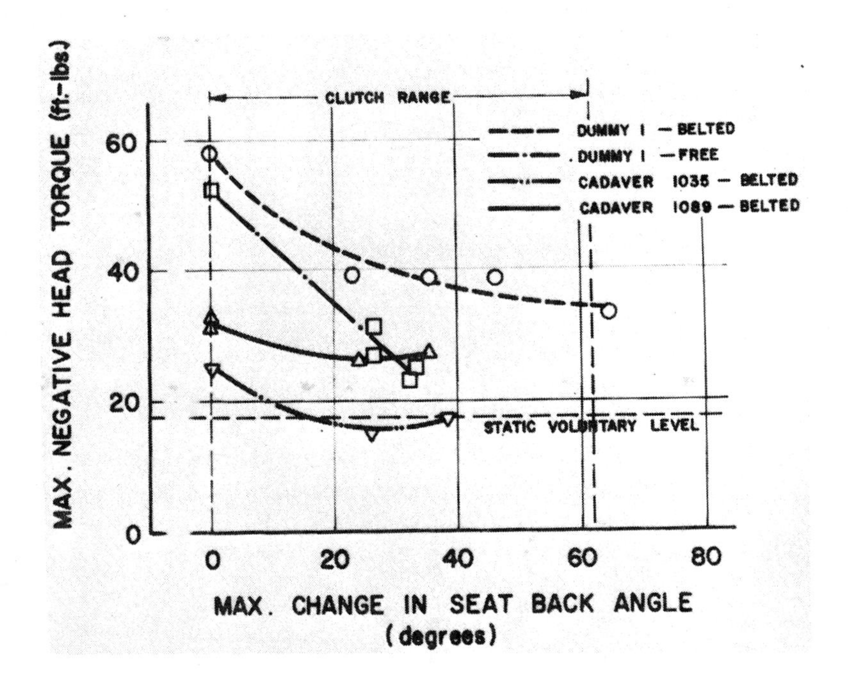

Fig. 21 - Comparison between degree of seat back rotation and resulting maximum negative neck torque for 23 mph simulation, no headrest

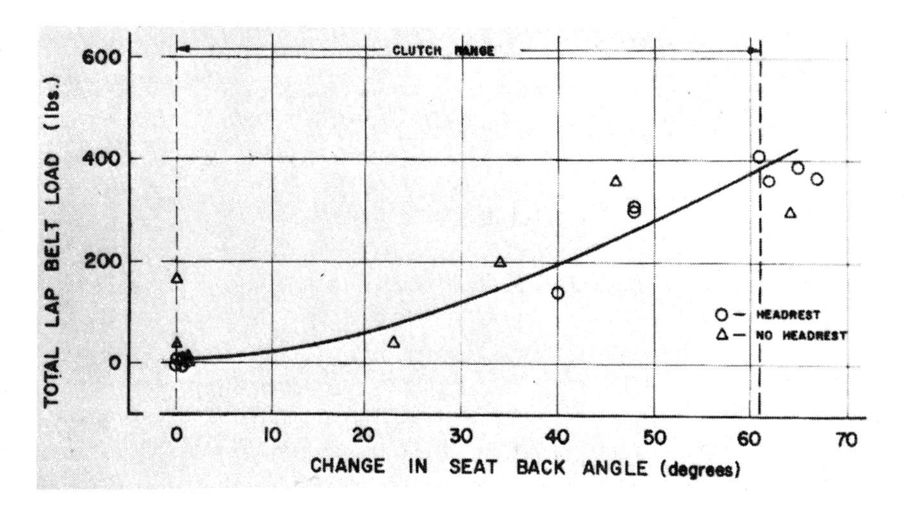

Fig. 22 - Maximum lap belt load on subject as related to maximum change in seat back angle

which was below the static volunteer tolerance level. The cadaver indices were more than double the volunteer's index, which implies that tensing for a rear-end collision reduces the severity of the impact.

EFFECT OF SEAT BACK ROTATION ON REDUCING SEVERITY OF A REAR-END COLLISION WITHOUT EMPLOYING A HEADREST - Fig. 20 depicts a typical oscillograph record (dummy 1, 23 mph simulation, no headrest) for the case where the seat back was allowed to rotate at a prescribed torque controlled by a frictional torque limiter.

The correlation between the degree of seat back rotation and the resulting maximum resistive head torque for various

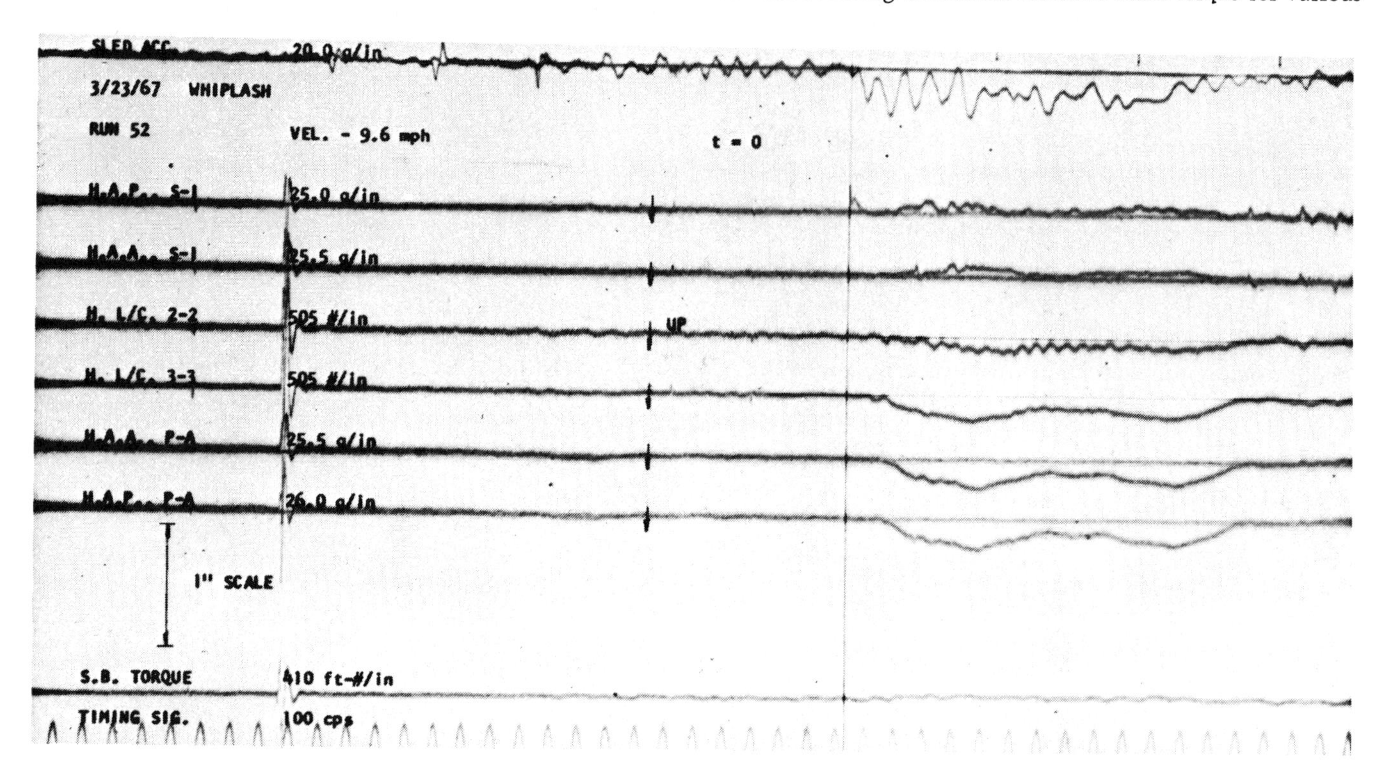

Fig. 23 - Oscillograph record of 10 mph simulation, dummy 2, rigid seat back, headrest

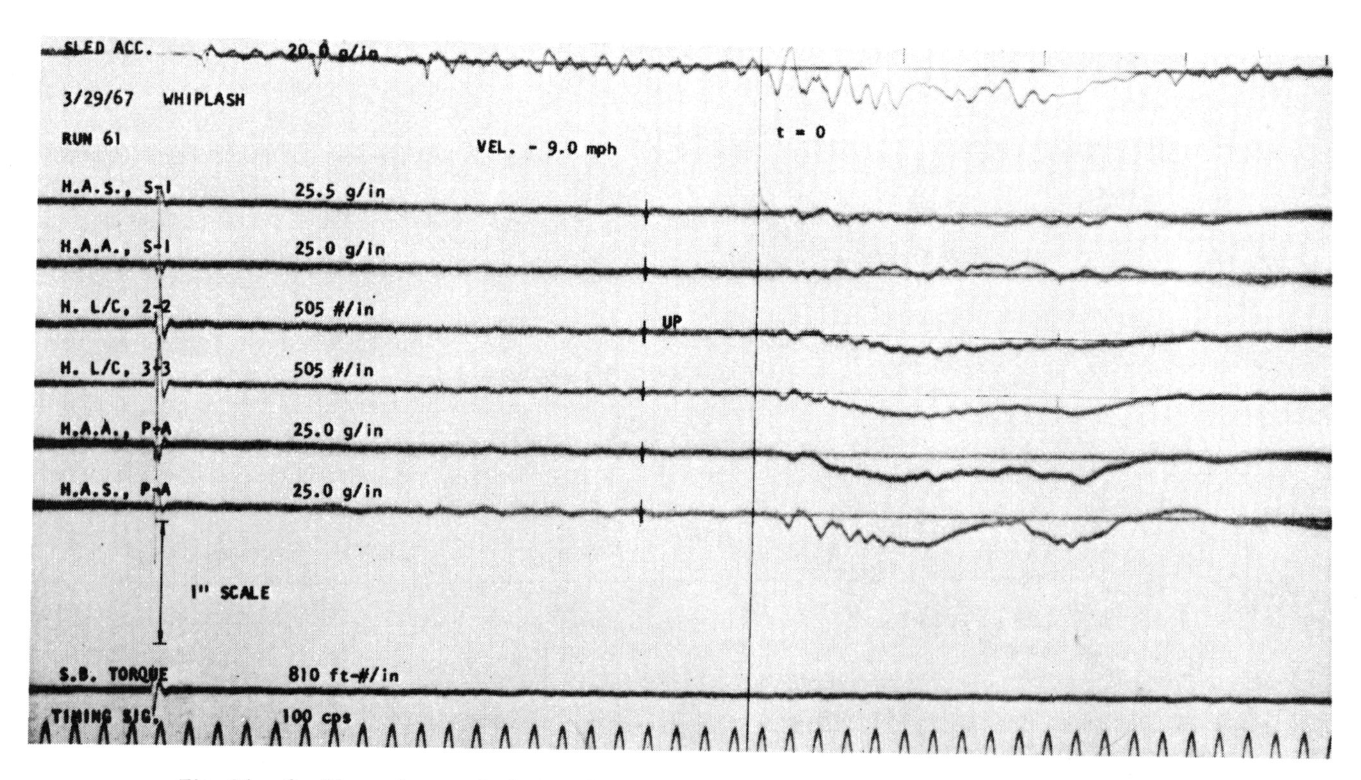

Fig. 24 - Oscillograph record of 10 mph simulation, cadaver 1035, rigid seat back, headrest

subjects for the 23 mph simulation is shown in Fig. 21. All the curves demonstrate a decrease in head torque with a decrease in seat back rigidity. The cadavers' curves indicate that an optimum rotational characteristic may exist.

For the comparative cases where the subjects were lap belted, torque indices based on the minimum neck torque for each subject were computed. These indices were 1.75, 1.65, and 1.75 for dummy 1 and cadavers 1035 and 1089, respectively, which when compared to the tolerable index of 2.00 indicate that with sufficient seat back rotation a 23 mph rear-end collision should not be injurious due to hyperextension of the neck. Simulations were conducted

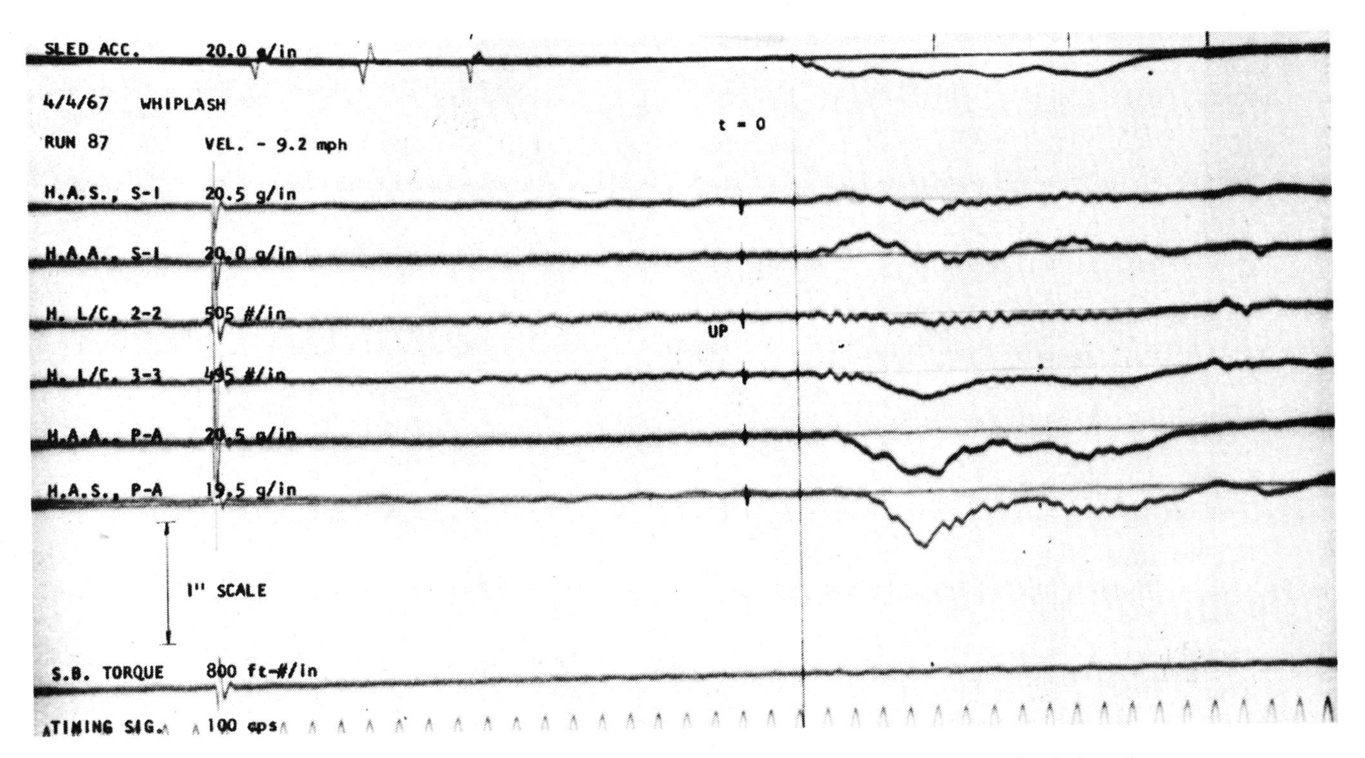

Fig. 25 - Oscillograph record of 10 mph simulation, cadaver 1089, rigid seat back, headrest

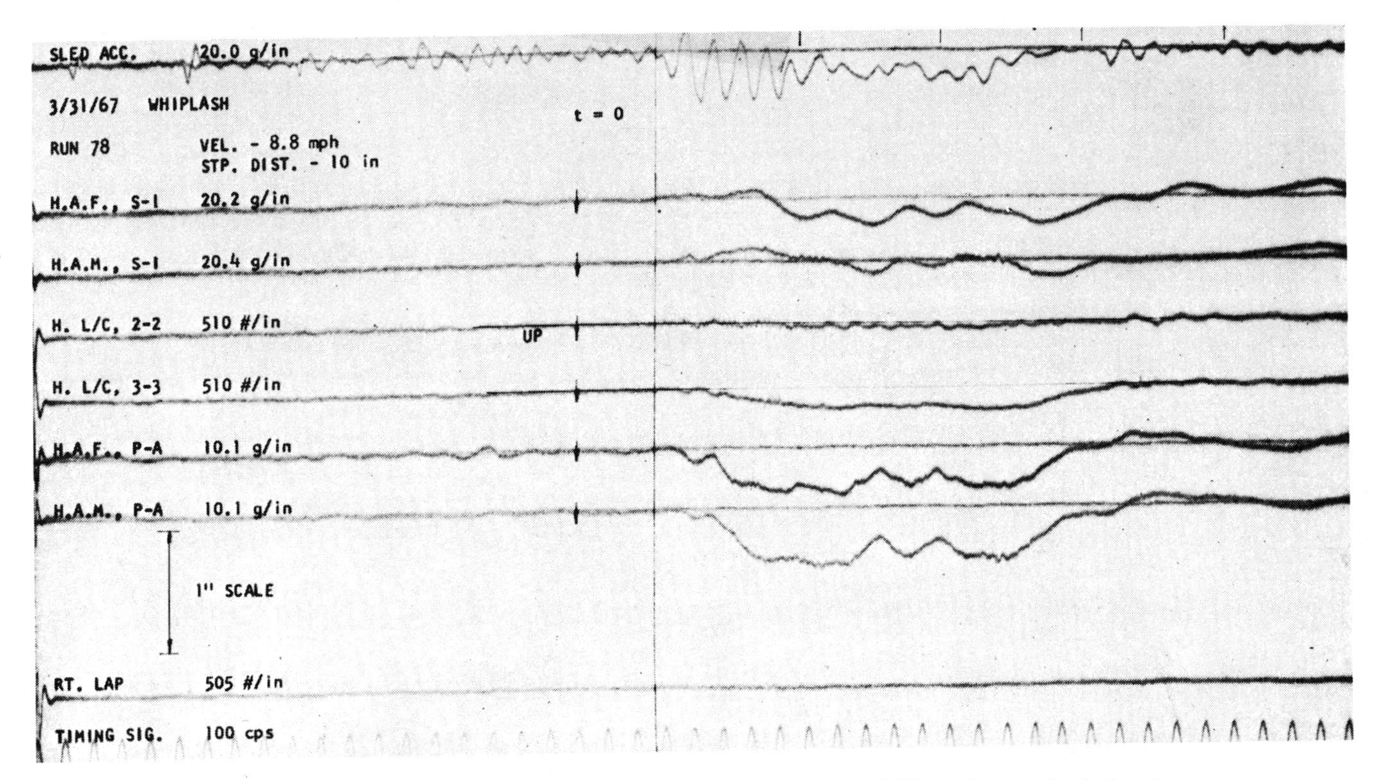

Fig. 26 - Oscillograph record of 10 mph simulation, volunteer (LMP), rigid seat back, headrest

with dummy 1 restrained by a lap belt and unrestrained. Without the lap belt restraint, the head torques were lower since the torso moved up the inclined seat back and rotated, resulting in a larger absolute head angle with no increase in the relative angle between the head and the torso. For small changes in seat back angle, the lap belt has little effect on the maximum head torque since there is very little load (60 lb at 20 deg) in the seat belt, as shown in Fig. 22. However, at larger changes in seat back angle the seat belt load is appreciable, 280 lb at 50 deg, tending to increase the resistive head torque by preventing movement of the lower torso.

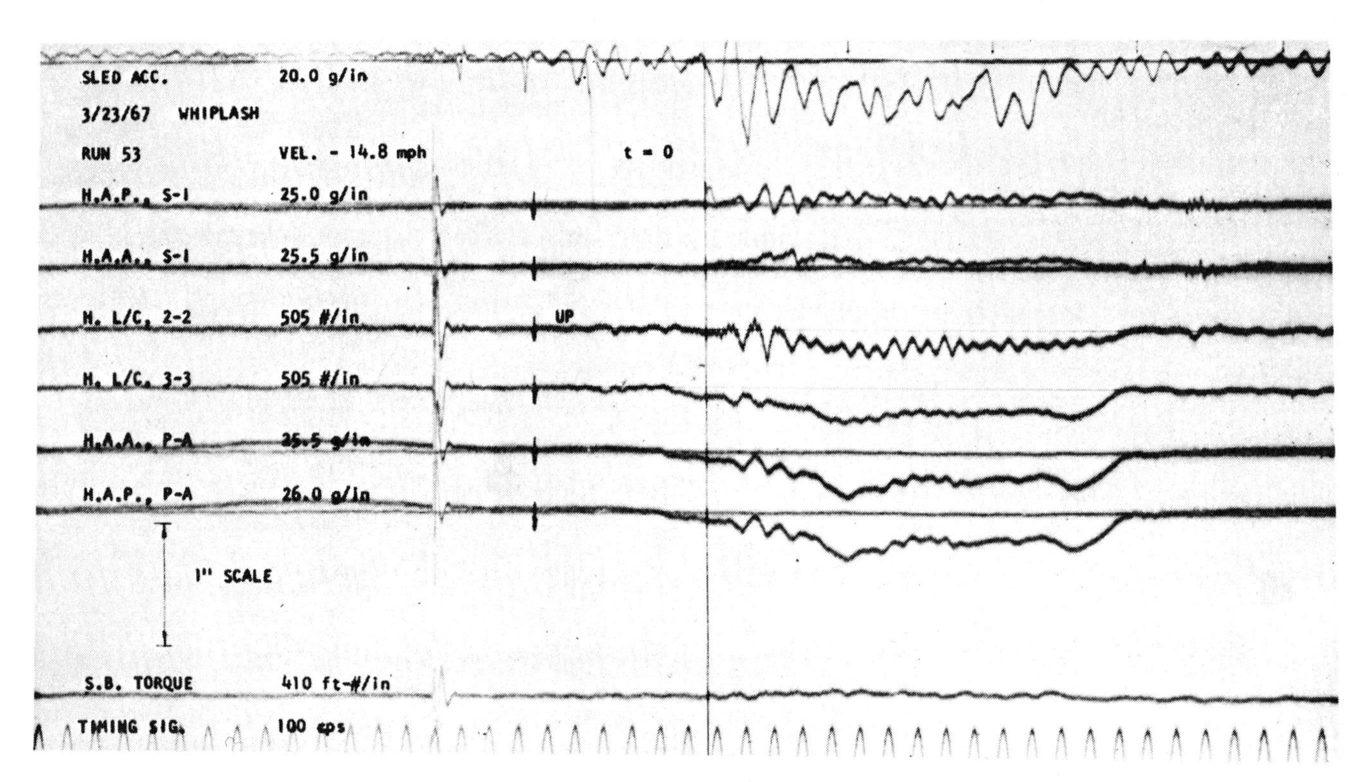

Fig. 27 - Oscillograph record of 23 mph simulation, dummy 2, rigid seat back, headrest

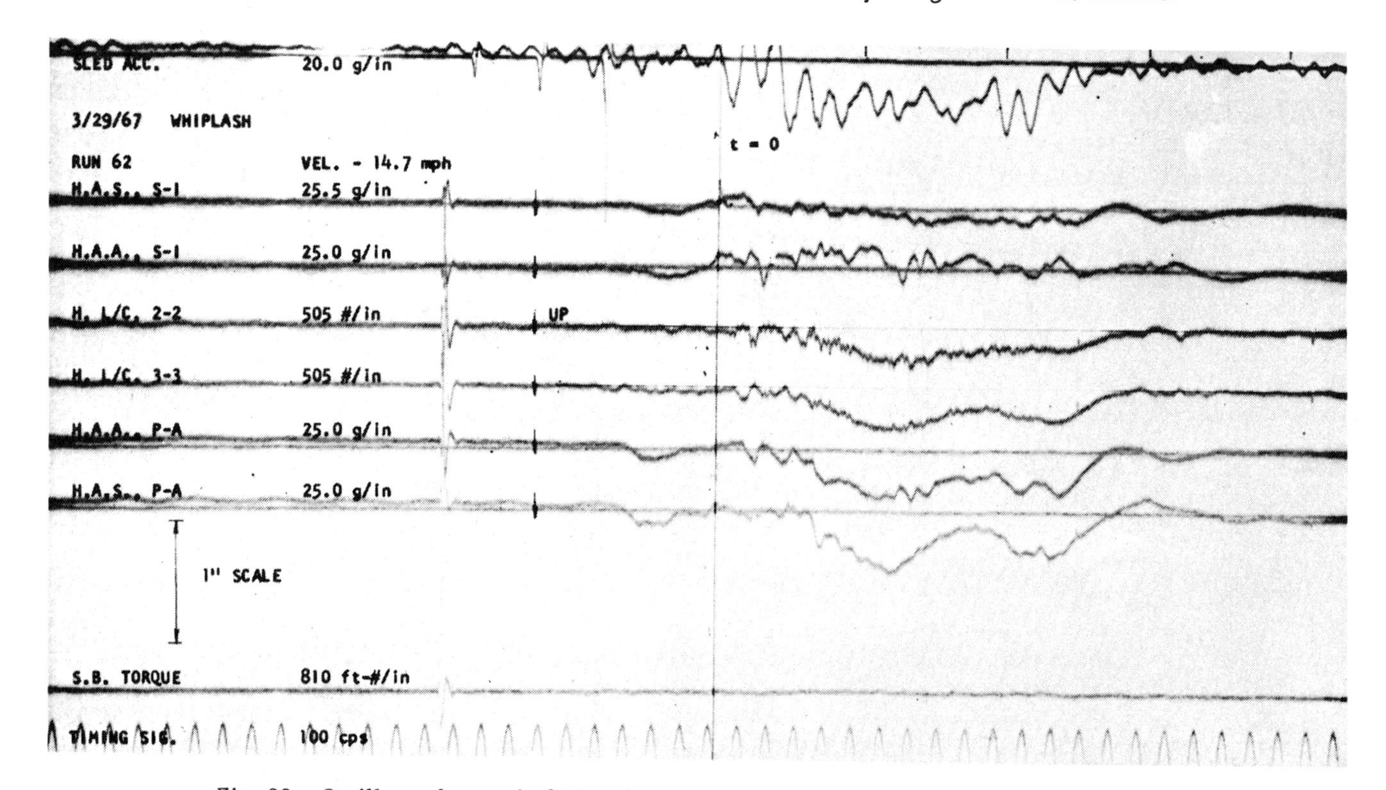

Fig. 28 - Oscillograph record of 23 mph simulation, cadaver 1035, rigid seat back, headrest

COMPARISON OF THE RESPONSES OF THE VARIOUS SUBJECTS WITH HEADREST - The comparative simulations were again the equivalent 10 and 23 mph rear-end collisions. For each simulation the subject was lap belted in the chair with his head initially in contact with the flat headrest and seat back rigid. The oscillograph records for the 10 mph simulation are shown in Figs. 23-26 for dummy 2, cadavers 1035 and 1089, and the volunteer. Comparison with dummy 1 will not be made since a curved headrest was used and is not directly comparable to the flat headrest used with the

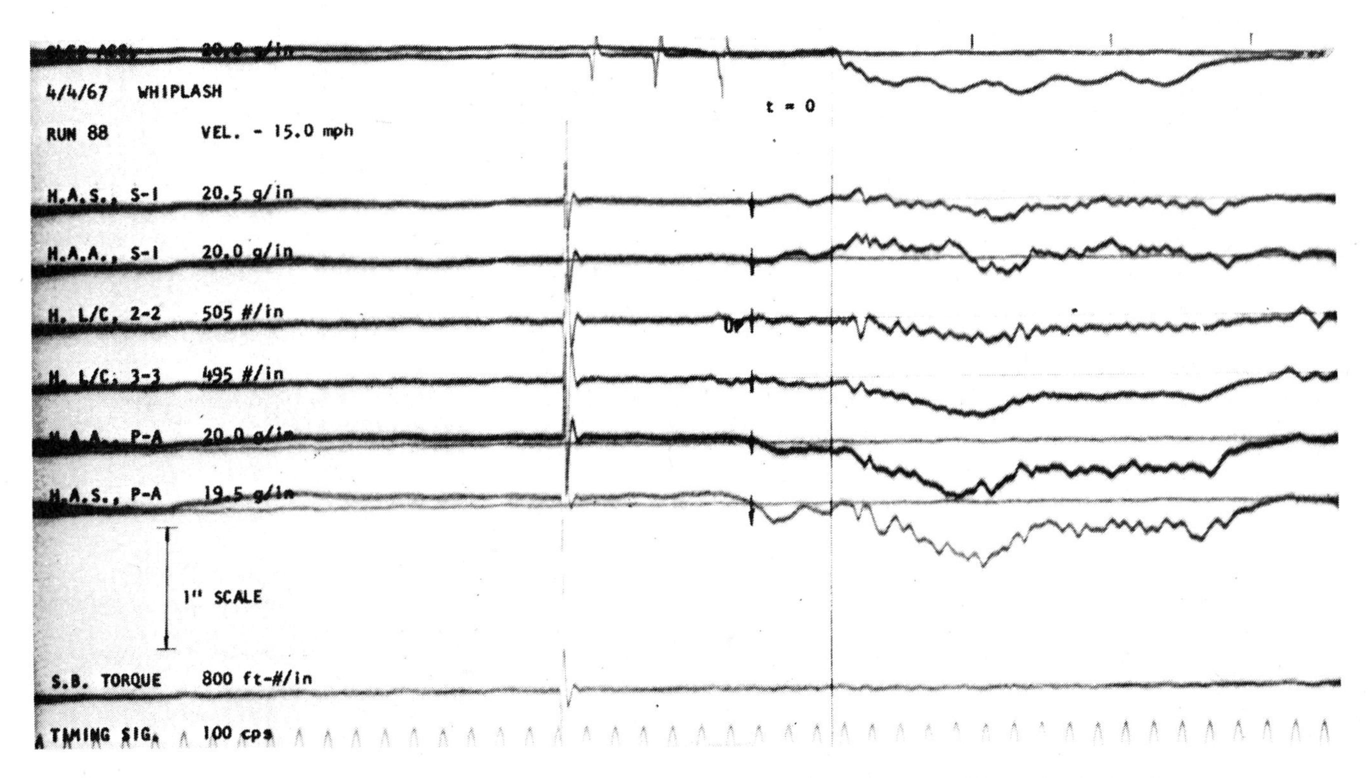

Fig. 29 - Oscillograph record of 23 mph simulation, cadaver 1089, rigid seat back, headrest

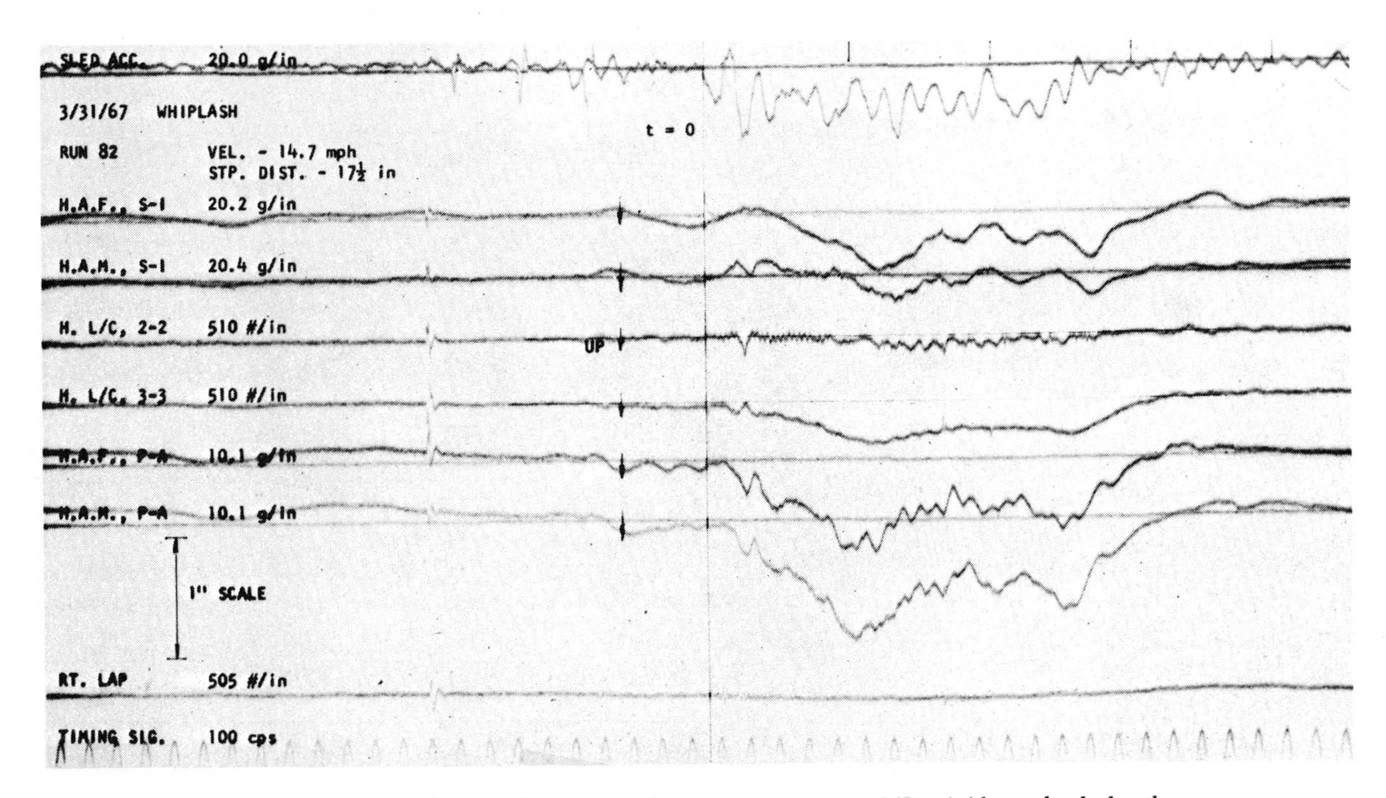

Fig. 30 - Oscillograph record of 23 mph simulation, volunteer (LMP), rigid seat back, headrest

other subjects. H. L/C 3-3 designates the head load cell axis normal to the surface, and H. L/C 2-2 designates an orthogonal axis in a vertical plane. The principal load is in the 3-3 direction and will be used to compare subjects. The same restrictions used for the comparison without a headrest on head acceleration traces and relative scale factors still apply. Noting these restrictions, a good comparison between all subjects on the basis of relative wave shape exists. This is also true for the comparable traces on the oscillograph records for the 23 mph simulation shown in Figs. 27-30.

Based on maximum normal headrest load shown in Fig. 31 for the equivalent 10 and 23 mph impacts, the headrest loads of the cadavers and dummy compared quite favorably with those withstood by the volunteer.

Because of the uncertainty of locating the point of application of the headrest load on the head, the corresponding neck torques were not computed. However, the neck torque and the axial force are not critical parameters when a flat headrest is used, since there is very little head rotation and the principal head load is approximately normal to the axial direction.

A comparison of the maximum shear forces for the various subjects is shown in Fig. 32. For the 10 mph simulation the correlation between the different subjects is good, but for the 23 mph simulation the shear force values of 132 and 107 lb for cadavers 1035 and 1089 are greater than the shear values of 70 and 59 lb for the volunteer and dummy 2. However, the magnitude of the shear load depends on the point of application of the headrest load on the head and, consequently, fluctuation should be expected between subjects. In any case, all the shear loads are below the static voluntary tolerance level of 190 lb.

HUMAN VOLUNTARY SIMULATIONS WITH HEADREST - For these simulations the volunteer was lap belted, his head was initially in contact with the flat surface of the load cell which was padded with one layer of 5/8 in. thick Rubatex, and the seat back was held rigid. The volunteer was subjected to various constant acceleration pulse levels with the maximum being 8.65 g. The curves on Figs. 31 and 32 depict the maximum normal headrest loads and the corresponding maximum shear loads, respectively, which the

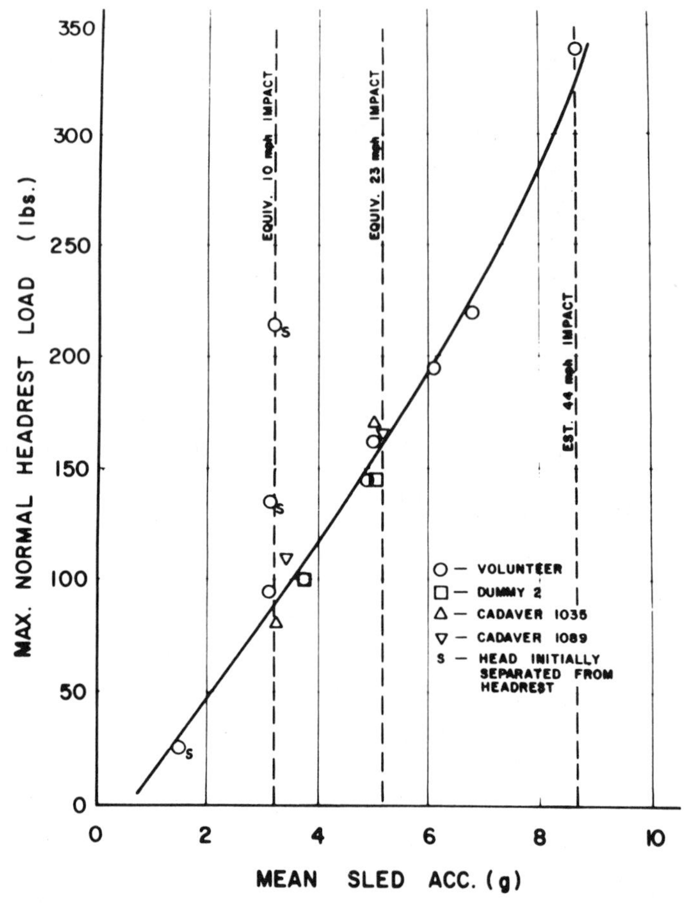

Fig. 31 - Maximum normal headrest loads as function of mean sled acceleration for various subjects, flat headrest, belted, rigid seat back

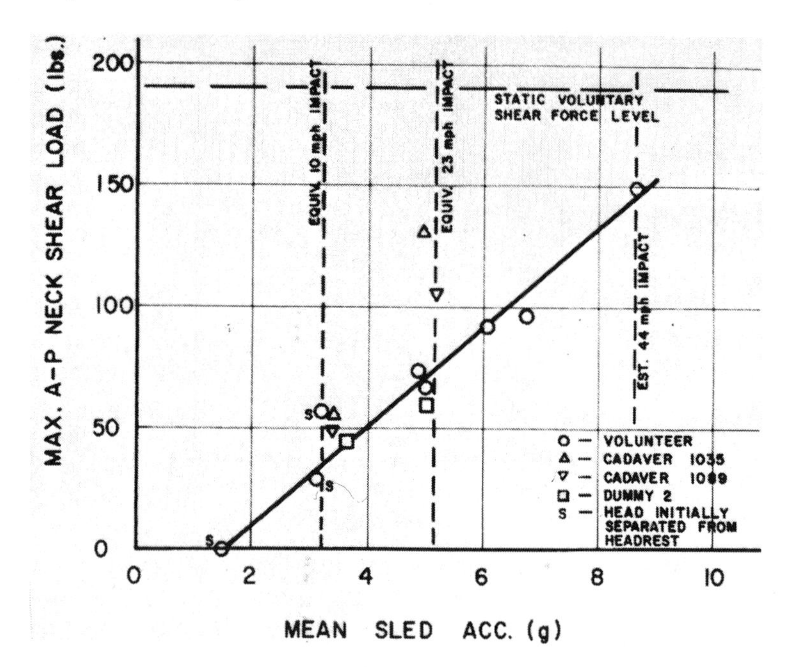

Fig. 32 - Maximum A-P neck shear forces as function of mean sled acceleration for various subjects, flat headrest, belted, rigid seat back

volunteer withstood during these various mean sled acceleration levels.

The curve is linear because the head which is initially in contact with the headrest undergoes a pure translational motion which is identical to the sled's motion. Consequently, for this configuration the headrest load must be directly proportional to the sled acceleration. This dependence on sled acceleration can be demonstrated further by considering the two points on Fig. 31 which represent the responses of the volunteer for an equivalent 23 mph rear-end collision. The headrest loads of 145 and 162 lb were achieved using initial sled velocities of 11.0 and 14.7 mph with corresponding stopping distances of 10 and 17.5 in., which upon applying the relationships for constant accelerations gave approximately the same mean sled accelerations of 4.85 and 4.95 g, respectively.

The maximum mean sled acceleration which the volunteer was subjected to was 8.65 g and only because of fatigue did he stop at this level. The corresponding maximum normal headrest and neck shear loads were 340 and 150 lb, respectively.

To estimate an equivalent rear-end impact at this mean acceleration level, the equivalent car impact velocities of 0, 10 and 23 mph were plotted as a function of mean acceleration as shown in Fig. 33. Since this curve has an increasing slope, a conservative extrapolation can be obtained by extending the chord for the 10 and 23 mph impacts until it intersects the line for 8.65 g which gives a corresponding equivalent car impact velocity of 44 mph. As a further check on this approximation the change in velocity of the struck car (the simulation velocity for the sled) is also plotted as a function of the mean acceleration. A linear relationship exists between the change in velocity of the struck car and its corresponding mean acceleration because the differences

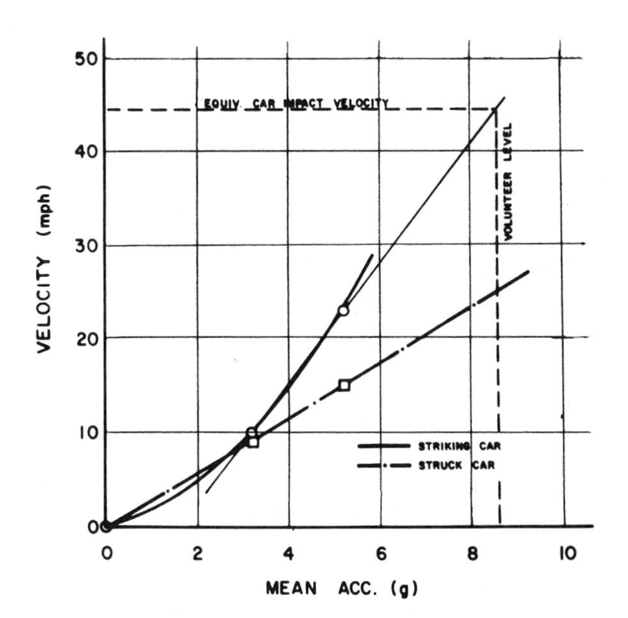

Fig. 33 - Estimation of relative impact velocity of cars of equal weight involved in rear-end collision based on mean car frame acceleration

in pulse time between the two impact conditions (9 and 15 mph changes in velocity) were small when compared to the total pulse duration. Extending this line to 8.65 g gives a change in velocity of the struck car of 25 mph. Since the approximations for the 10 and 23 mph simulations were obtained from impacts between cars of equivalent weights in which no braking action was used, these approximations also applied for this estimation. Consequently, the law of conservation of linear momentum is valid under these conditions and the final velocity of the striking car for any initial velocity can be approximated directly by taking the differences of the ordinates of the two curves for the prescribed impact velocity. Knowing the intial and final velocity of both cars, the coefficient of restitution can be calculated from

$$e = \frac{(u_1 - u_2)}{(v_2 - v_1)}$$

where:

u_1/u_2 = Velocities of the striking and struck cars after impact

v_1/v_2 = Velocities before impact, respectively

For a perfectly elastic impact $e = 1$ and for a perfectly plastic impact $e = 0$.

As the initial relative velocity between impacting cars increases, the percentage of plastic to elastic deformation should increase and this trend will be reflected in a decrease in the coefficient of restitution. The estimated 44 mph rear-end collision fits into this tendency as illustrated by the values of the coefficient of restitution, 0.80, 0.30, and 0.14, for the 10, 23 and 44 mph equivalent rear-end collisions, respectively.

The 44 mph simulation was achieved using a sled velocity of 14.7 mph instead of the 25 mph which is indicated by the change in velocity for the struck car given in Fig. 33. However, as demonstrated previously, the important parameter in the case where the head is initially in contact with the headrest is the mean acceleration. Consequently, the simulation condition which gives a mean sled acceleration of 8.65 g does duplicate the loading during a 44 mph rear-end collision with the only difference being that the pulse duration for the 14.7 mph simulation will be less than for the 25 mph simulation. However, in either case the duration would be classified as "long" (> 100 ms) and have no effect on the volunteer's response to either simulation. Based on these assumptions, it is physically possible for a person to withstand a 44 mph rear-end collision with no injuries, provided his head is initially in contact with a flat headrest which is firmly attached to a rigid seat back.

It should be emphasized that this statement is for a flat headrest. Using a curved headrest, the relative position of the head with respect to the headrest at contact determines the point of application of the applied load. If the load is applied above the c.g. of the head, flexion of the neck will occur, and if the headrest is in the neck area, extension of the neck will occur. Neither of these conditions was evaluated with the volunteer.

EFFECT OF INITIAL SEPARATION OF HEAD WITH RESPECT TO HEADREST - In two instances for the equivalent 10 mph impact, the volunteer's head was not initially in contact with the headrest. The result was that the maximum headrest loads were greater (215 and 135 lb as compared to 90 lb) than when his head was initially in contact. To evaluate this effect, both cadavers were subjected to a sequence of 23 mph simulations identical in setup to the comparison runs, except the distance between the head and the headrest was incrementally increased. The results of these simulations are shown in Fig. 34. For cadavers 1089 and 1035 the normal headrest loads corresponding to a 3-1/2 in. separation were 440 and 310 lb, respectively, compared to approximately 150 and 170 lb for no separation. However, by taping a piece of 1-1/8 in. styrofoam to the back of the head of cadaver 1089, the headrest load for a separation of 2-3/4 in. as indicated by the point with the subscript in Fig. 34 was reduced from 390 to 250 lb.

Consequently, with adequate padding and proper design of the headrest supporting structure, headrest loads for any given head separation can be attenuated with the minimum load for any given mean acceleration being given by the condition of no initial separation.

EFFECT OF SEAT BACK RIGIDITY ON MAXIMUM NORMAL HEADREST LOAD - For these simulations the equivalent 23 mph rear-end collision was used and the subject's head was initially in contact with the headrest. Fig. 35 depicts the effect that seat back rotation has on the resulting maximum normal headrest load.

For dummy 1 only a curved headrest was used and the simulations were conducted with the subject free and lap belted. The response of the dummy with these two types of restraint are quite similar with head loads being approximately 25 lb higher when the dummy is unrestrained.

The relatively large load of 245 lb, which occurred for the lap belted case for a change in angle of 62 deg, resulted because the seat back "bottomed" on mechanical stops which produced an incremental change in the relative velocity between the head and the headrest, resulting in an equivalent impact loading even though the head was in contact with the headrest.

For dummy 2 both a curved and flat headrest were used. The data using the curved headrest are quite scattered compared to response of the dummy using a flat headrest. In general for all subjects, the headrest load decreased as the seat back rigidity decreased which implies that seat back rotation tends to reduce the severity of a given rear-end collision, as was the case for the unsupported head. However, a practical limitation must be placed on the degree of ro-

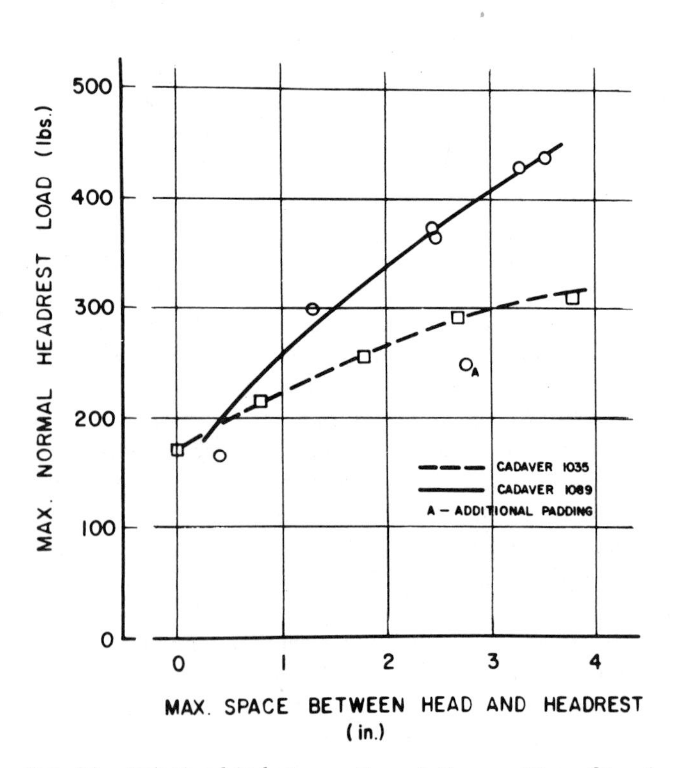

Fig. 34 - Relationship between the relative position of head to headrest and maximum normal headrest load, 23 mph simulation, rigid seat back, flat headrest, lap belted

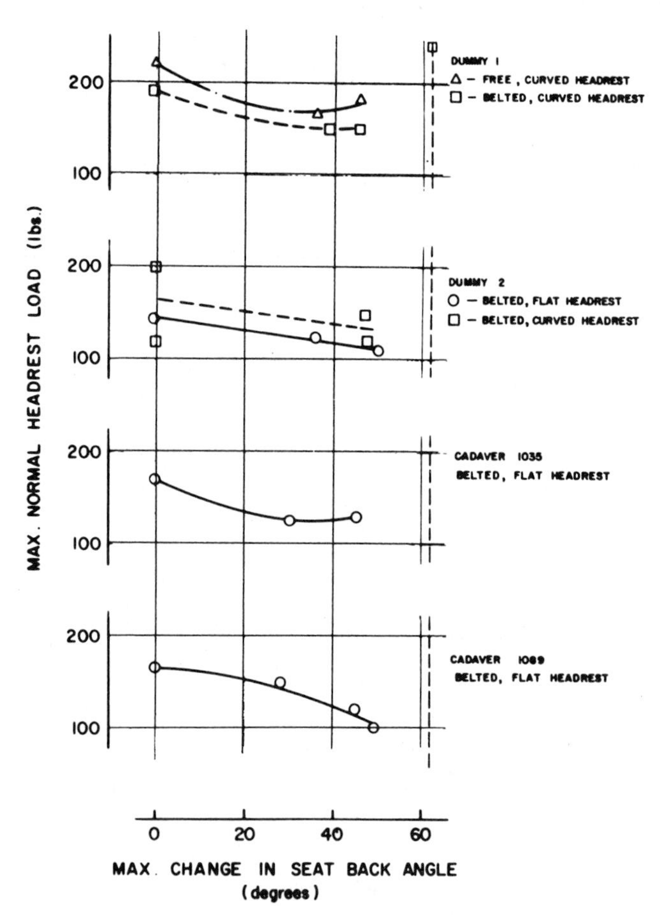

Fig. 35 - Effect of degree of seat back rotation on maximum normal headrest loads for various subjects, 23 mph simulation, head initially in contact with headrest

tation because the driver of the struck car must be in a position to regain control of his car after the collision.

CONCLUSIONS

1. In the case where the car is not equipped with headrests, tensing of neck muscles prior to the impact reduces the possibility of neck injury. The severity of a rear-end collision can be reduced further by flexing the head forward and preventing extension by clasping the hands behind the head.

2. With or without a headrest, controlled seat back collapse reduces the severity of impact. Further study is needed to determine if there is an optimum rotational characteristic.

3. For the unsupported head simulations, cadavers give good representation of the responses of people who are not expecting the rear-end impact. Neither dummy used gave satisfactory responses.

4. With the head initially in contact with the headrest, the responses of all subjects were closely related.

5. With initial separation between the head and the headrest, head loads are higher than with no initial separation. With adequate padding and proper structural design of the headrest, the head loads can be reduced.

6. With the head in contact with a flat headrest and the seat back rigid, a 44 mph rear-end collision can be withstood with little discomfort.

7. Because of the low energy-storing characteristics of the seat used, no appreciable head flexion due to rebound occurred for any of the configurations evaluated.

8. Neck torque at the occipital condyles is the limiting factor in neck injury rather than the shear or axial forces.

9. Results using a curved headrest were scattered. Further evaluation of positioning and optimum curvature of the headrest is needed.

10. The type of restraint (lap belted, lap and diagonal chest belted, or free) had little effect on the response of the subject when the seat back was rigid.

11. With seat back rotation, the lap belt was loaded and increased the severity of the impact in the unsupported head case and had no effect when the head was supported.

12. The diagonal chest strap did not provide any restraint for the configurations in which it was used.

13. Responses of subjects to various degrees of severity of rear-end collisions can be compared on the basis of an index based on the static, voluntary, extended head, neck torque tolerance level with a preliminary tolerable index of 2.00 being given. This value may be changed as further data become available.

ACKNOWLEDGMENTS

The authors wish to express their gratitude to Dr. I. D. Harris for providing the radiological diagnoses of the cadavers, to Clarence Murton for his help in conducting the simulations, and to the members of the Biomechanics Staff of the Engineering Mechanics Department of Wayne State University who contributed to this project.

Also, the services of the Computing and Data Processing Center of Wayne State University for providing the computer time needed for the numerical analysis are gratefully acknowledged.

REFERENCES

1. D. M. Severy and J. H. Mathewson, "Automobile Barrier and Rear-End Collision Performance." Paper 62C presented at SAE Summer Meeting, Atlantic City, June 1958.

2. J. P. Stapp, "Medical Aspects of Safety Seat Belt Development." Proceedings of Sixth Stapp Car Crash and Field Demonstrations Conference, 1963.

3. D. F. Carroll, J. A. Collins, J. L. Haley, Jr., and J. W. Turnbow, "Crashworthiness Study for Passenger Seat Design -- Analysis and Testing of Aircraft Seats." AvSER 67-4, May 1967.

4. Carroll F. Simmons and David N. Herting, "Strength of the Human Neck." Life Sciences Department, Space and Information Systems Division, North American Aviation, Inc., SID 65-1180, Sept. 22, 1965.

5. K. N. Naab, "Measurement of Detailed Inertial Properties and Dimensions of a 50th Percentile Anthropometric Dummy." Proceedings of 10th Stapp Car Crash Conference, 1966.

6. W. T. Dempster and G. R. L. Gaughran, "Properties of Body Segments Based on Size and Weight." The American Jrl. of Anatomy, Vol. 120, No. 1, January 1967.

7. John Martinez, J. Wickstrom, and B. Barcelo, "The Whiplash Injury -- A Study of Head-Neck Action and Injuries in Animals." Paper 65-WA/HUF-6, presented at ASME Meeting, Chicago, November 1965.

710856

A Study of Responses and Tolerances of the Neck

C. W. Gadd and C. C. Culver
General Motors Corp.

A. M. Nahum
Universtiy of California

Abstract

The principal objectives of this study were first to obtain experimental curves of angulation versus moment of resistance of the human neck in hyperextension and lateral flexion, and second to determine angular limits short of significant injury observable in the unembalmed subjects employed in the study. The first of the tests were of the "static" type with load applied over a period of approximately 1s.

To determine the applicability of the data to dynamic conditions, tests were also made of the dissected neck at angulation velocities comparable with those of typical accidental injury. Overall resisting moment and injury threshold were similar under the dynamic loading, but somewhat greater moment of resistance was noted during the (earlier) portion of the loading cycle when angular velocity was greatest.

It is believed the data obtained together with muscular restraint data of other investigators who have used volunteers should be of value in the selection of neck characteristics for anthropometric test devices.

Tolerances of the unembalmed laryngeal cartilages to direct frontal impact in situ were also obtained.

MORE COMPLETE KNOWLEDGE of the flexural properties of the human neck is of practical importance for two reasons. First, a need has existed for more accurate specification of its load-angulation characteristics as a basis for the design of anthropometric test devices which will, as a result, be capable of experiencing more realistic head trajectories and yield more reliable head acceleration Severity Index values. Second, there is a need for a better understanding of the permissible limits of angulation of the neck in the longitudinal and lateral planes.

The following study, using unembalmed subjects, was undertaken to shed more light on these questions. Although it addresses only a portion of the total problem, it is hoped that the results will prove of value in the search for ultimate solutions. It is believed that the angulation characteristics and damage thresholds seen in this study are reasonably representative of a "relaxed" state of the neck. Some consideration is given also to the additional resisting moment which can be exerted by the neck muscles, but it is felt that more study must be given to the question of maximum levels of muscle tension which can typically be exerted in accidents, as well as to such factors as age, stature, and build, before a complete understanding is reached.

Methodology of Angulation Experiments

Prior to conducting the tests reported, exploratory tests were made on another subject to determine the most satisfactory loading and measuring means. Deflection of the torso below the neck level proved to be a problem, and it was determined to bind this securely against a heavy wood block of approximately 20 X 30 X 3 in to which were clamped additional side supports, as shown in Fig. 1. The latter were positioned before clamping so that they pressed firmly against the sides of the body and, in particular, prevented appreciable displacement or rotation of the shoulders. It was found that bending moment could be imposed in a repeatable manner by either of two means: an extension arm

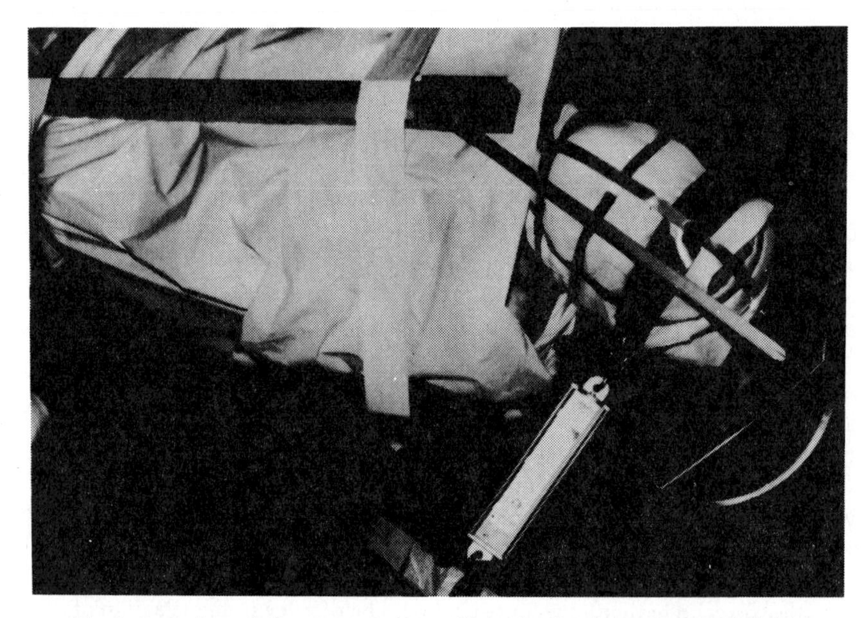

Fig. 1 - Lateral angulation test technique utilizing loading band above orbital ridges

bound to the head, or a flexible band encircling the head at the forehead level. (See Fig. 2.)

The force was applied manually via a spring scale which was held at all times at right angles to the vertical axis of the head as it rotated. Bending moments were calculated about a point within the neck midway vertically between horizontal planes through the inferior surface of the chin and the crest of the shoulders. Head angulation was monitored with respect to the vertical axis of the torso by use of a protractor device. Curvature of the cervical spine was uniform over its length to the degree with which this could be observed in the x-ray (Fig. 2). The subjects, who were 66 years of age and older, were tested between one and several days after death. No evidence of rigor mortis remained, and they were

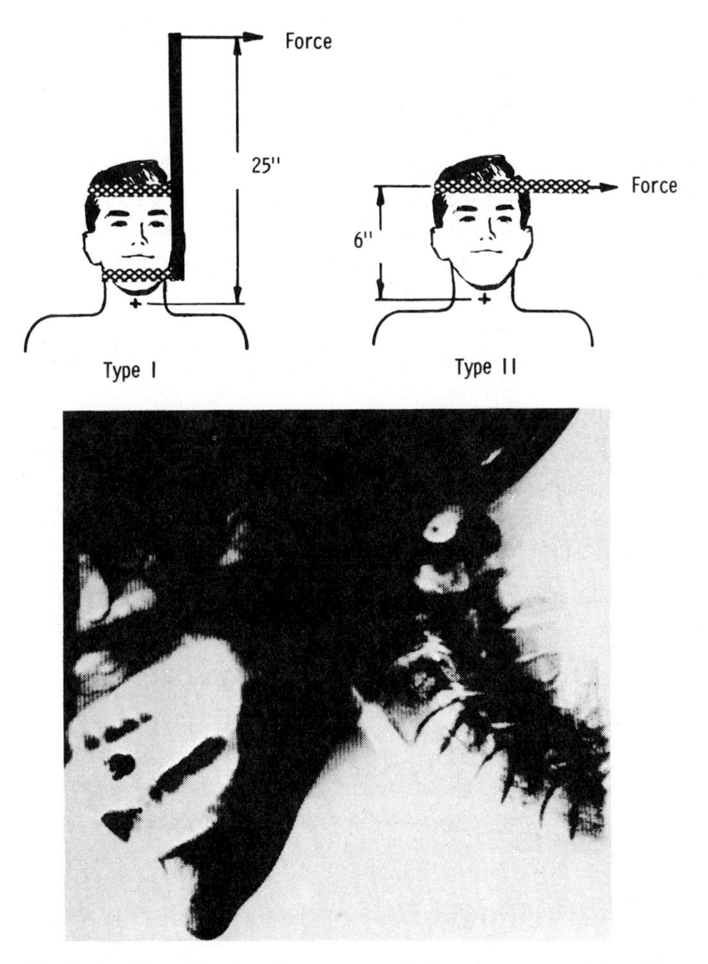

Fig. 2 - Loading of head-neck system, and below, hyperextended neck

considered to represent well a relaxed state, free of appreciable muscular or other soft tissue restraint.

Since the four subjects of the first test series were loaded to the threshold of detectable injury both laterally and in hyperextension, the question arose of possible effect of the injury in one plane upon load-deflection characteristics in the other. This was investigated by reversing the order of testing specimens 15 and 16 as compared with 13 and 14. No bias of results could be discerned from test sequence in keeping with the fact that different restraining ligaments come into play in the two planes.

Investigation of Possible Effect of Loading Rate - In conducting the experiments described above, the load rise time was in the range of 1 s; therefore, a question arose of their applicability to typical accident situations which involve the order of 10-100 ms. Accordingly, a second test series was carried out dynamically, as illustrated in Fig. 3. Here a section of cervical spine was dissected out to include all ligamentous material, but without the bulk of the musculature system of the neck. In this way it was possible to apply hyperextensive loading at high rates of onset and monitor resisting moment unaffected by inertial resisting force or torque of the head. The test was further simplified by employing an active test section of only two joints or vertebral interfaces. To do this, two adjacent vertebrae were selected to be clamped rigidly to a base while a loading beam was inserted sagittally through the body of the second vertebra above for the purpose of applying the bending moment to the two joints which were free to flex. Thus, it was possible to measure angular stiffness of the neck over essentially one-fourth of its length. This was advantageous in two ways:

1. It was necessary to impose only modest beam angles of up to 20 deg to reach the minor injury threshold, thus minimizing inaccuracies attendant to applying loads and monitoring angulation over a large arc.
2. Only modest angular velocities were needed to simulate neck joint angulation rates similar those of field accidents.

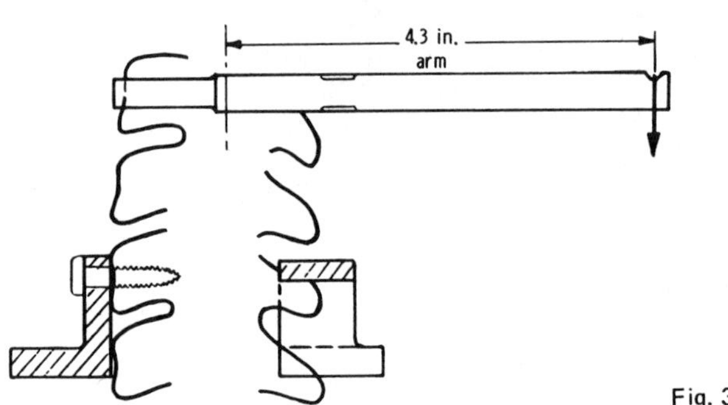

Fig. 3 - Dynamic loading fixture

As a result, a simple falling weight attached by a light chain to the tip of the beam could be employed for applying the dynamic load. Preliminary experiments with this device proved that, as would be expected, the axis of rotation of a vertebra and of the loading beam was on the longitudinal axis of the spinal cord. The geometry of beam movement, further, was such that vertical tip movement measured by a potentiometer type displacement transducer was a linear function of, or proportional to, the angulation imposed upon the upper vertebra of the test section.

The dynamic tests were conducted in hyperextension only, monitoring bending moment by the use of strain gages attached to the upper and lower surfaces of the loading beam immediately posterior to vertebra into which it was inserted. By allowing the falling weight to rebound from its lowerst excursion, it was possible to obtain hysteresis loops of bending moment versus angular deflection of the test section.

It was assumed that the layers of anterior longitudinal ligament of the neck would play a predominant role in resisting the hyperextension; therefore, the portions of the test piece immediately above and below the active section were tightly bound circumferentially with wire. Thus, the first specimen with active interfaces between vertebrae 3, 4, and 5 were bound with wire between vertebrae 2 and 3 and between 5 and 6. To insure against any working of the loading beam relative to the vertebra into which it was inserted, this was oriented so as to rest firmly against the upper surface of the bony portion of the vertebra posterior to the spinal cord. For the dynamic tests of the second dissected neck specimen the active section was chosen to be between vertebrae 4, 5, and 6.

Results

Table 1 lists the bending moment values sustained by the four subjects of the first series tested at low velocity of load application. Load was applied in increasing increments with an interval of load removal of a few minutes between each increment. The resisting moments followed in each case a steadily increasing pattern as a function of increasing angle imposed by the lever arm or forehead strap. When angulations of 80 deg of hyperextension or 60 deg laterally were exceeded, minor strain or injury occurred in some of the specimens detectable audibly or felt by the doctor monitoring the test. The occurrence of such strains was confirmed by reapplying one of the previous load increments and noting failure of the resisting moment to reach its former value for that increment. The load-deflection results are also plotted as curves in Figs. 4 and 5, employing a legend which keys the data points to Table 1.

The moment-of-resistance values found under dynamic loading are summarized in Fig. 6. These are plotted to an abscissa scale which is expanded by a factor of four, so that if one makes the assumption that angulation over two vertebral joints should be approximately one-fourth of that over the entire neck with

Table 1 - Bending Moments and Angulations Sustained by Four Subjects Loaded as Shown in Fig. 2

	Hyperextension Angle, deg	Bending in, lb	Fig. 4 Curve Key	Laterial Angle, deg	Bending in, lb	Fig. 5 Curve Key
Subject No. 13	60	44.0	●	50	61.0	●
	72	77.0		60	130.0	
	80	93.5		70	182.0	
	85	110.0				
	90	121.0				
	95	143.0				
Subject No. 14	60	66.0	+	30	78.0	+
	70	132.0		40	130.0	
	80	154.0		50	169.0	
	90	286.0*		60	188.5*	
	60	55.0	○	70	182.0	○
	100	319.0*				
	70	121.0	□	40	52.0**	□
Subject No. 15	30†	20.0	X	30	100.0	X
	60	35.0		45	125.0	
	70	60.0		60	150.0	
	75	70.0		65	188.0	
	80	120.0		70	200.0	
	70	55.0	⊗			
Subject No. 16	30†	25.0	△	30†	36.0	△
	45	47.0		45	90.0	
	60	99.0		50	135.0	
	70	160.0*		55	162.0	
	75	209.0	▲	60	186.0 shoulder support	
	60	83.0	◬	65	222.0 shoulder support	

*Detectable strain.
**Verification of damage by reapplication of load.
†These series' loaded through band around forehead.

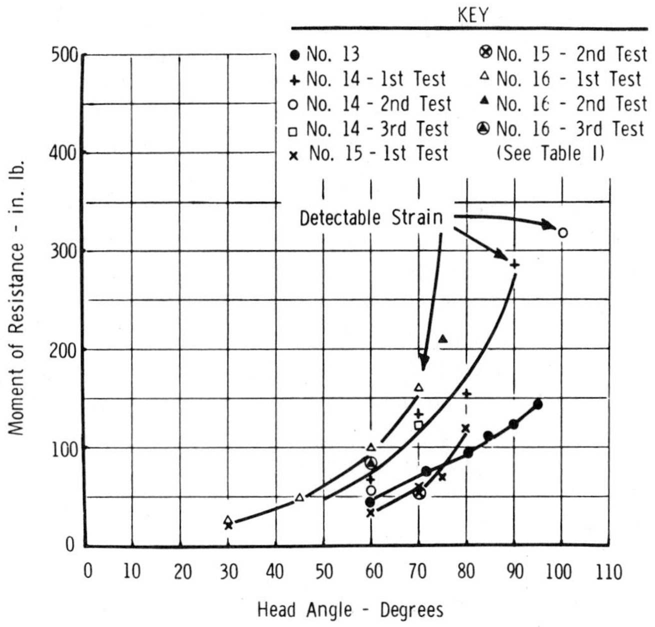

Fig. 4 - Resistance of head-neck system to hyperextension

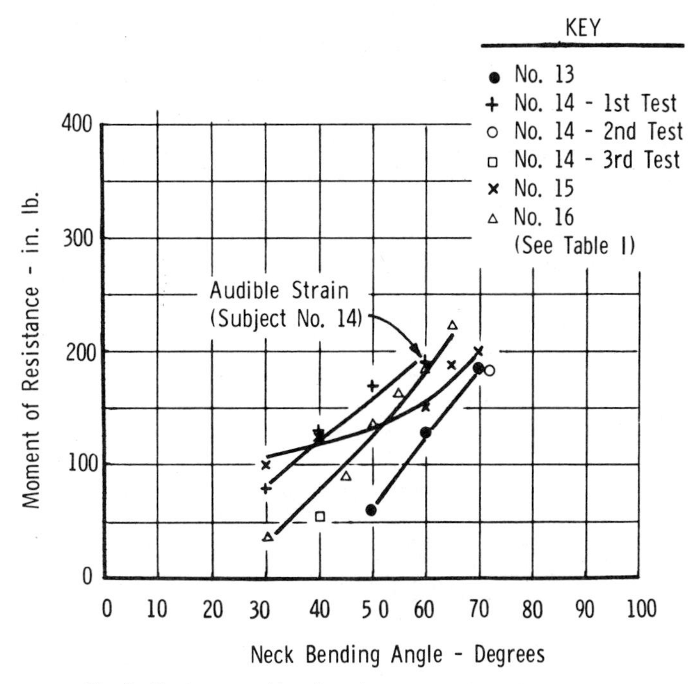

Fig. 5 - Resistance of head-neck system to lateral angulation

eight interfaces, he can make an immediate evaluation of rate sensitivity. Interestingly enough, the low- and high-velocity curves superimpose very well; any real difference appears to be overshadowed by greater differences between test subjects exposed to the same loading.

As a basis for examining rate sensitivity somewhat further, several low-level static loads were applied to the dissected specimens prior to conducting the dynamic experiments. These were applied over a period again of about 1 s, as shown in Fig. 7, and at levels low enough to eliminate any possibility of damage (as evidenced by the fact that when a given static load was repeated an identical value of angulation was recorded). These data points, labeled "S" in Fig. 6, may be compared with instantaneous dynamic deflection at the same load levels on the hysteresis loops labeled "A" taken subsequently. This comparison shows that during the high-velocity portion of the dynamic loading cycle, well short of maximum deflection where the angular velocity goes to zero, the resisting moment is appreciably greater than under static load. The rate sensitivity, however, does not appear to be sustained sufficiently to affect greatly the overall deflection of the neck.

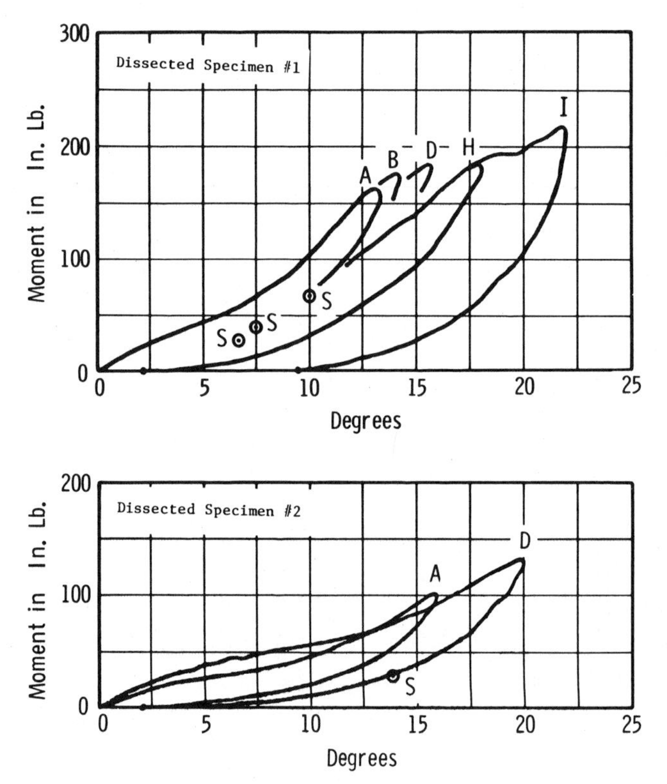

Fig. 6 - Dynamic loading characteristics under repeated increasing load

Following application of the dynamic cycles labeled "A," a succession of repeat cycles (labeled B, C, D, . . .) were applied at slightly higher loading. These confirmed that the threshold for minor damage again lay at approximately the same level of spinal curvature in hyperextension as seen in the first test series conducted on the complete neck; that is at about 20 deg angulation of the short test section. Here the onset of minor damage was evidenced again by a reduction of stiffness and a permanent set developed after several load cycles. Final test "I" conducted at maximum loading and 22-deg angulation exhibited a rather dramatic increase in hysteresis. Finally, Figs. 8 and 9 give several of the dynamic recordings in their original form. Load scales read downward with one major screen division representing a moment equivalent to that produced by 10 lb applied to the tip of the beam. The second trace, also recorded in most instances, is a separate readout of beam deflection as a function of time.

Augmentation of Moment of Resistance by Muscle Tension - There has been some previous investigation (1)* of measured loads which volunteers could sustain against their forehead or occiput by muscle tension. A brief test series of this nature was conducted by the authors to confirm such values and arrive at a rough overall value by which a tensed resisting load might exceed the relaxed. These were conducted with the volunteer lying prone or supine, then arching his body slightly so that his entire weight was divided between the feet and the forehead or occiput; the force exerted by the head being read by an instrumented head band. Males of medium stature were found able to support loadings up to 54 lb against the forehead and 97 lb against the occiput. These values, which equal or somewhat exceed others which have been published, give an indication of how much the resisting moments in extension and flexion might be increased under typical tensed conditions. Studies currently under way by

*Numbers in parentheses designate References at end of paper.

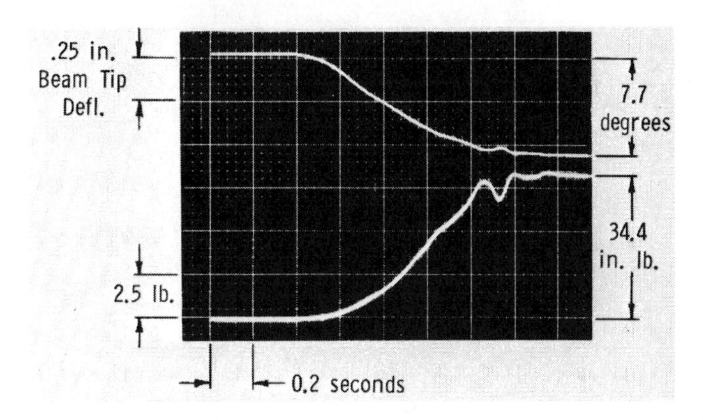

Fig. 7 - Example of static loading of dissected specimen No. 1

others treat the question of muscle tension in more detail on the basis of values sustained by volunteers over a range of head angulations (2, 3).

Impact Tolerance of the Laryngeal Cartilages

Direct impact against the neck is a less frequent problem in accidental injury. Here protection is usually afforded by the chin. The authors, however, have over a period of time collected certain tolerance values for this kind of loading. They have previously (4) reported values in the range of 200-250 lb for marginal fracturing of the thyroid cartilage of embalmed subjects under an impactor of

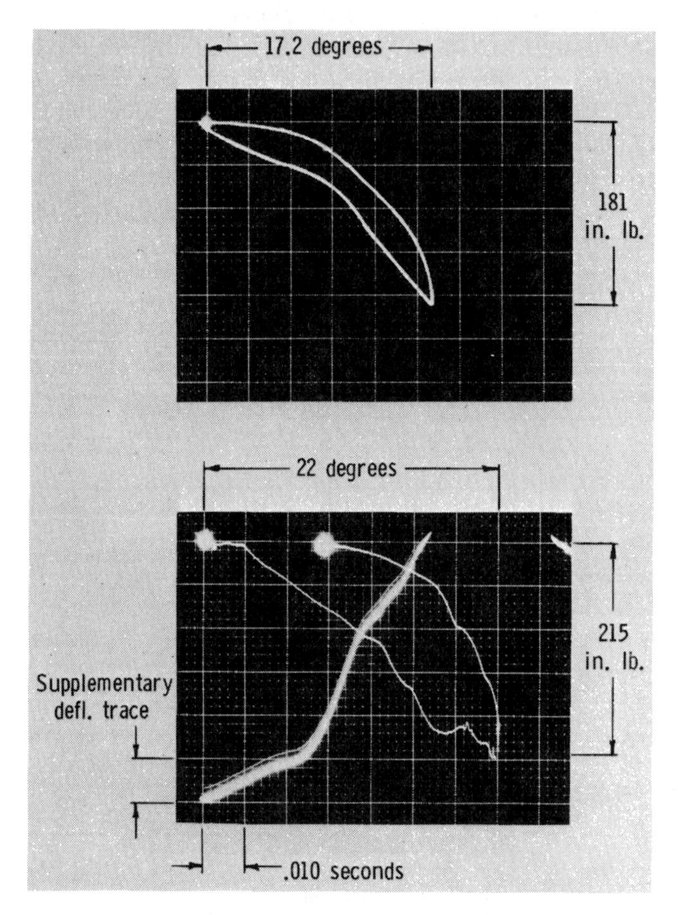

Fig. 8 - Typical oscillograms of dynamic tests of dissected specimen No. 1; final test below shows considerable permanent strain

1-in area against the anterior midline of the neck. A tolerance of 175-225 lb was found at the same time for the cricoid cartilage.

More recent experiments of a similar nature, but using unembalmed subjects, have yielded lower values. Whereas the muscular and superficial tissues surrounding the larynx of the embalmed subjects had firmed up appreciably and shared a great deal of the striker impact, the same tissues of the unembalmed were very soft and life-like, allowing almost the full force of the impact to be absorbed by the larynx itself. As a result, it could absorb only between 90-100 lb for marginal fracture of either the thyroid or cricoid.

As an adjunct to this program, a series of interior impacts were also carried out against dissected laryngeal cartilages of pigs, whose architecture here is very similar to that of the human. Similar fractures occurred in these subjects at load levels in the range of 60-65 lb.

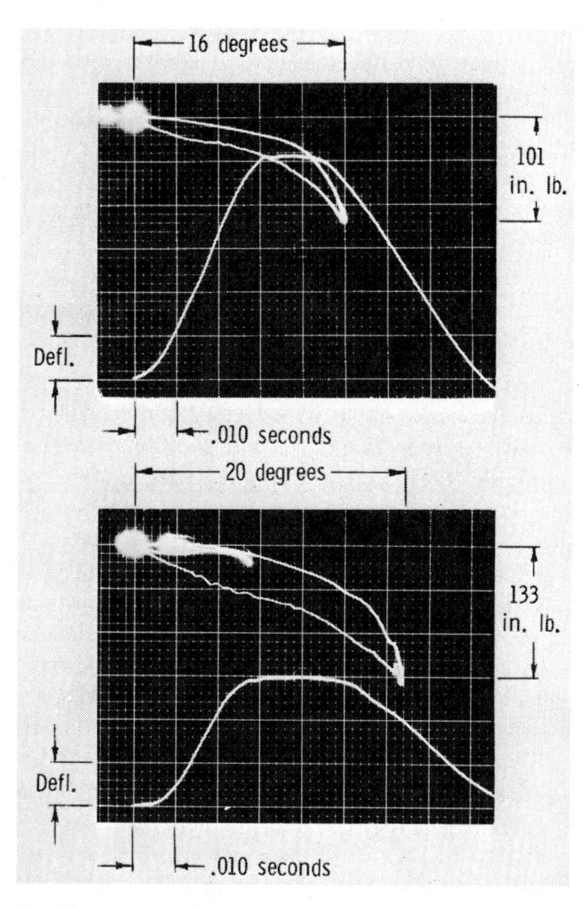

Fig. 9 - Typical oscillograms from dynamic tests of dissected specimens—specimen No. 2

Discussion

Only unembalmed subjects were used in the first test series of intact subjects; it was felt desirable not to complicate the interpretation of these tests by introducing the additional variable of soft tissue stiffening which could have resulted from the embalming. The possibility of such a problem is illustrated by the higher tolerance exhibited by the embalmed subjects of the direct neck impact experiments. From palpation it was evident that in these latter experiments that the embalming stiffened the muscular and superficial tissues on either side of the larynx in such a way as to provide support against lateral deflection of its side walls and resulting concentration of tensile stress inside its apex.

Embalming, however, had no discernable effect upon resisting moment developed by the dissected specimens, wherein only bony and ligamentous materials were involved. An embalmed specimen of this type (for which graphs are not included) responded both statically and dynamically in the same manner as the unembalmed.

Comparing this study with others using volunteers (for example, those of Ewing and Mertz), the latter enable the true moment of resistance contributed by muscle tension to be included. On the other hand, the volunteer tests have the limitation that they cannot go high enough to well delineate the injury threshold. The spectrum of field accidents, of course, includes both a relaxed state as typified by the present study and the tensed state which can be studied by the use of volunteers; both types of study are felt to be valuable. The study of Ewing, et al., (2) gives the overall response of volunteers whose neck muscles are tensed; Mertz and Patrick (1) estimate the separate contribution of muscle tension.

Neck angulation rates are of the same order in the volunteer experiments as in the present study. For example, Ewing, et al., quote angular velocities up to 1784 deg/s; whereas inspection of the deflection-time traces of Fig. 9 shows angulations of the short test sections of up to 20 deg in some 20 ms, corresponding to 1000 deg/s for this short section or roughly four times this rate for a full neck.

The Ewing results are of special interest, among other things, because of the extreme angular accelerations which could apparently be sustained by the head without concussion (evidenced by the slope of the angular velocity curves of their Figs. 1 and 2 (2)). It is not possible to make a significant comparison of neck resisting moments, however, because those of the Ewing study were principally in hyperflexion and have not as yet been separated out from the inertial loadings.

Several aspects of the present study might well be pursued further. The regime of injury was only marginally penetrated, and injury was not sufficiently great for delineation of the true nature of the strains which were observed. While they were thought to be ligamentous for the most part, it would be desirable to extend the loadings to higher levels and look for bony as well as ligamentous damage. Rate sensitivity could also be studied in greater detail and compared with rate sensitivity of companion tests of ligament coupons or samples. It would also be desirable to include younger subjects in the study who would be expected to sustain larger angulations before onset of injury.

Summary

Resistance of the head-neck system to hyperextension and lateral angulation was found to follow concave-upward curves which reached a level of 200 in /lb in the range of 70-90 deg in hyperextension and 60-70 deg laterally. Evidence of minor injury to the neck appeared at approximately 80 deg in hyperextension and 60 deg laterally in the elderly subjects tested.

Supplementary tests conducted dynamically and involving load rise times in the range of 20-30 ms showed little overall differences in angulation.

Attention is called to the substantial additions to angulation resistance which are usually present in accident kinematics (and which are under investigation by others) as a result of muscle tension.

Tolerance of the laryngeal cartilages to localized impact over 1 sq in was found to be in the range of 90-100 lb.

Acknowledgment

The authors wish to express their appreciation to Richard G. Madeira who constructed and made the initial trials of the dynamic loading fixture and to Dennis C. Schneider and George Nakamura who collaborated in the conduct of the experiments.

References

1. H. J. Mertz and L. M. Patrick, "Investigation of the Kinematics and Kinetics of Whiplash." 11th Stapp Car Crash Conference, P-20, paper 670919. New York: Society of Automotive Engineers, Inc., 1967.

2. C. L. Ewing, D. J. Thomas, L. M. Patrick, G. W. Beeler, and M. J. Smith, "Living Human Dynamic Response to -G Impact Acceleration - II, Accelerations; Measured on the Head and Neck." Proceedings of Thirteenth Stapp Car Crash Conference, P-28, paper 690817. New York: Society of Automotive Engineers, Inc., 1969.

3. H. J. Mertz and L. M. Patrick, "Strength and Response of the Human Neck." Proceedings of Fifteenth Stapp Car Crash Conference, P-39, paper 710855. New York: Society of Automotive Engineers, Inc., 1971.

4. Alan M. Nahum, J. D. Gatts, C. W. Gadd, and J. P. Danforth, "Impact Tolerance of the Skull and Face." Proceedings of Twelfth Stapp Car Crash Conference, P-26, paper 680785. New York: Society of Automotive Engineers, Inc., 1968.

710857

Human Head Linear and Angular Accelerations During Impact

Thomas D. Clarke, C. Dee Gragg, James F. Sprouffske and Edwin M. Trout
Holloman Air Force Base

Roger M. Zimmerman
New Mexico State University

William H. Muzzy
Naval Aerospace Medical Research Laboratory

Abstract

Head linear and angular accelerations of humans were investigated during exposure to abrupt linear deceleration ($-G_X$). The 14 subjects were restrained with three different restraints: lap belt only, Air Force shoulder harness-lap belt and air bag plus lap belt. Peak sled decelerations ranged from 7.7-10.3 g.

The results indicated that peak head angular and linear resultant accelerations were elevated with the air bag in contrast to the Air Force shoulder harness or lap belt only restraints. However, the peak angular and linear accelerations may have less traumatic consequences than the degree of head-neck hyperextension.

PREVIOUS RESEARCH HAS REVEALED the favorable impact protection afforded by the air bag (1,2)*. However, at their present stage of development, air bags involve greater uncertainty and risk than restraints previously introduced in the auto safety field (3). Of particular concern is whether the restraining force of the air bag may result in exceeding human tolerance limitations of the head-neck. In this report, the objectives were to determine head linear accelerations plus angular accelerations, velocities and displacements during exposure to abrupt linear deceleration ($-G_X$) while restrained with a lap belt only, standard Air Force shoulder harness-lap belt or air bag plus lap belt**. Correlation of the acceleration, velocity and displacement values with human subject response and trauma will be discussed.

---

*Numbers in parentheses designate References at end of paper.

**The voluntary informed consent of the subjects used in this research was obtained as required by Air Force Regulation 169-8.

Materials and Methods

Thirty-six deceleration tests were performed with 14 adult male volunteers. Peak sled velocities ranged from 23.1-27.2 ft/s resulting in maximum sled decelerations of 7.7-10.3 g. The impact pulse was approximately half sine with a stopping distance of 2 ft. Air bag inflation was mechanically activated at the initiation of the sled deceleration pulse†. A low-pass filter (100 Hz) improved the legibility of the acceleration traces with no appreciable loss of response (1, 2).

Previous investigators (4) computed head angular accelerations via numerical differentiations of photographically derived angular displacement curves. However, envelopment of the head by the air bag precluded this type of analysis (Fig. 1).

Biaxial accelerometer clusters were affixed to anterior and posterior flanges of a lightweight plastic head mount. The mount, restraining straps, and accelerometers weighed 11.3 oz (Fig. 2). The design of the angular acceleration system is dependent upon the principle that the tangential acceleration of point A on a rigid body relative to point B on the body, divided by their separation distance, is the angular acceleration of the body within a spatial reference system

†The air bag systems were gratuitously provided by Eaton, Yale and Towne, Inc.

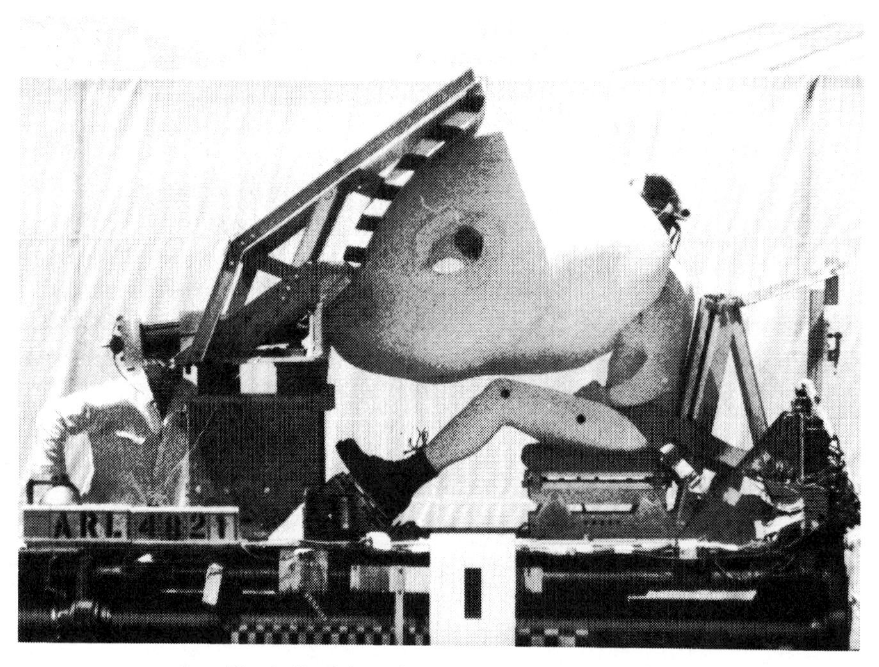

Fig. 1 - Partial envelopment of head by air bag

(5). Only the rotation in the sagittal plane or about the lateral axis of the head (pitch) is described. Integration of head angular accelerations yielded head angular velocity. Angular displacement of the head was computed by an additional integration. Front and rear linear accelerations were computed by vector summation of the X and Z components of the anterior and posterior accelerometer clusters.

Results and Discussion

The head reference axes are depicted using a polar coordinate system where angular displacement of the head was positive with flexion and negative with hyperextension.

Lap Belt Restraint - Head angular acceleration, velocity, and displacement of a human restrained with only a lap belt are graphically displayed for a 9.2 g impact (Fig. 3).

For all tests, the excursions of the accelerometer data correspond closely in magnitude and time phase. The head angular acceleration trace was typically biphasic in shape. The positive peak normally occurred 160 ms after time zero. The acceleration trace reached the maximum negative value 20-50 ms later.

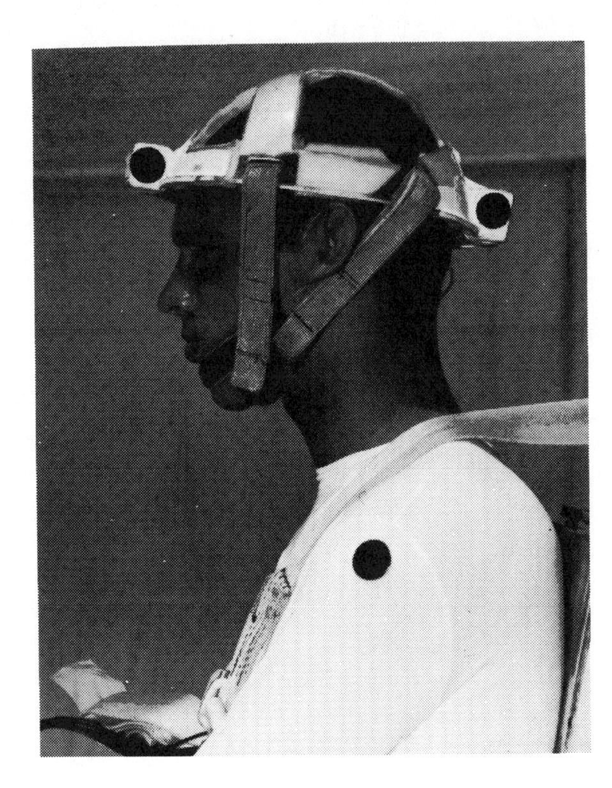

Fig. 2 - Test subject with accelerometer head mount

The positive peak was invariably greater in magnitude than the negative excursion (Table 1). The trace returned to zero by 350 ms.

The head angular velocity trace was primarily positive; therefore the maximum negative value was excluded from Table 1. The head angular displacement curve was invariably positive. This was indicative that the head-neck only underwent flexion. It should be emphasized that the zero position of the head was the actual position of the head at time zero and not necessarily the neutral anatomical position (head erect). However, the head at time zero never exceeded 15 deg positive or negative from the neutral anatomical position.

Although the head angular displacement trace remained at zero for at least the first 100 ms of the impact event, the head was translating forward with the torso in a linear manner. The maximum head angular displacement occurred at approximately 300 ms. While the mean angular displacement of the head was 103 deg (Table 1), the extent of head-neck flexion with respect to the torso never exceeded 50 deg.

The linear resultants (Fig. 4) of the front and rear accelerometer clusters peaked at approximately the same time of the head angular acceleration excursions (Fig. 3). It should be noted that Fig. 3 shows the coplanar vector representation of the front and rear accelerometers and not necessarily the acceleration of the center of gravity of the head. The maximum linear acceleration was invariably recorded with the front accelerometer cluster.

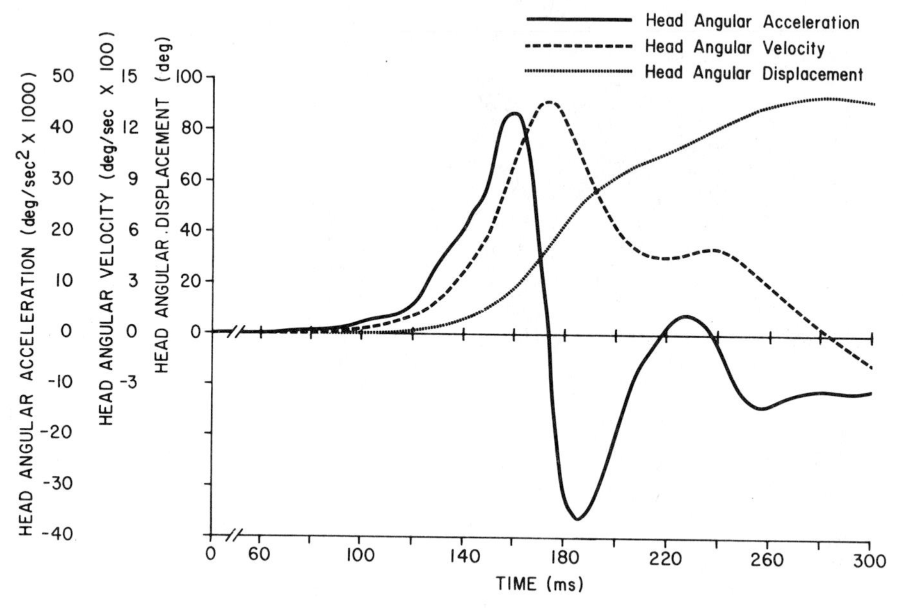

Fig. 3 - Head angular acceleration, velocity and displacement versus time. Lap belt restraint at 9.2 g. Run No. 4931, subject No. 1

Air Force Shoulder Harness Restraint - The head angular acceleration, velocity, and displacement traces when restrained with an Air Force shoulder harness-lap belt (Fig. 5) displayed close similarity to the previously discussed lap belt traces (Fig. 3). The major dissimilarity was the shorter duration from time zero to the initiation of the excursions.

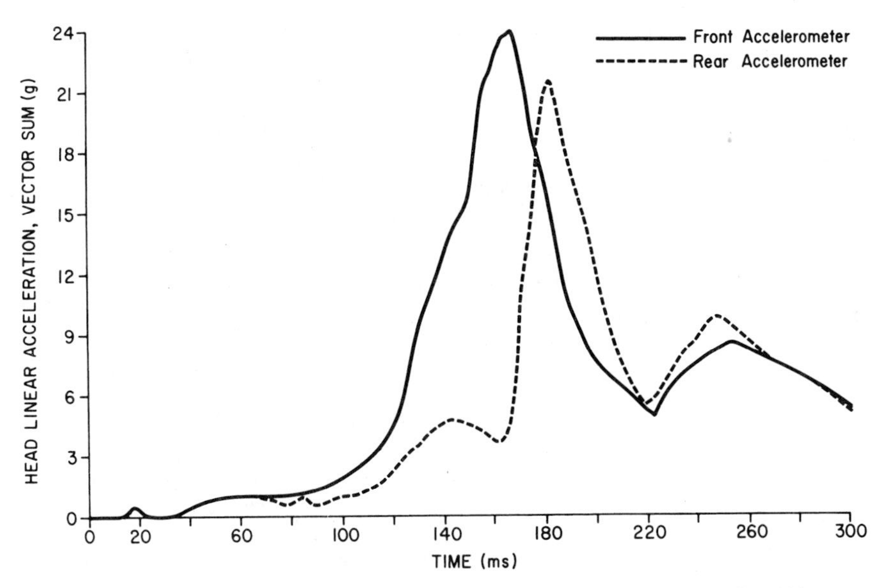

Fig. 4 - Head linear acceleration versus time. Lap belt restraint at 9.2 g. Run No. 4931, subject No. 1

Table 1 - Lap Belt Restraint (10 Subjects)

	A	B	C	D	E	F	G	H
Mean	+45,700	−29,500	+1,012	–	22	14	+103	9.4
S. D.	±20,500	±11,300	±182	–	±7	±5	±10	±0.3

A = maximum head angular acceleration (deg/s^2)
B = maximum head angular deceleration (deg/s^2)
C = maximum positive head angular velocity (deg/s)
D = maximum negative head angular velocity (deg/s)
E = maximum front head linear acceleration (g)
F = maximum rear head linear acceleration (g)
G = maximum flexion of head-neck from position at time zero (deg)
H = maximum sled deceleration (g)
– = highly inconsistent or of minimal value

The mean peak angular accelerations and decelerations while restrained with the Air Force shoulder harness (Table 2) were less than restrained with the lap belt (Table 1), even though the mean sled deceleration with the shoulder harness was slightly higher than comparable tests using the lap belt only.

The mean head angular displacement with the Air Force shoulder harness was 75 deg. However, since the torso was restrained and never rotated foreward more than 20 deg from the erect position, the extent of head-neck flexion with respect to the torso was greatest with this restraint. With this extent of head-neck flexion the subject's mandible often contacted the sternum. Previous researchers have found the maximum angle of flexion to be 53 deg (6).

The peak linear accelerations were slightly less when restrained with the shoulder harness (Table 2). The peak excursions occurred at the same time as the peak head angular accelerations (Fig. 6).

Air Bag Plus Lap Belt Restraint - From 60-70 ms, the rapid excursions of the head acceleration trace were primarily attributable to bag contact with the thorax and head (Fig. 7). The negative spike near 80 ms averaged −178, 200 deg/s^2 (Table 3) and was invariably the maximum head angular deceleration.

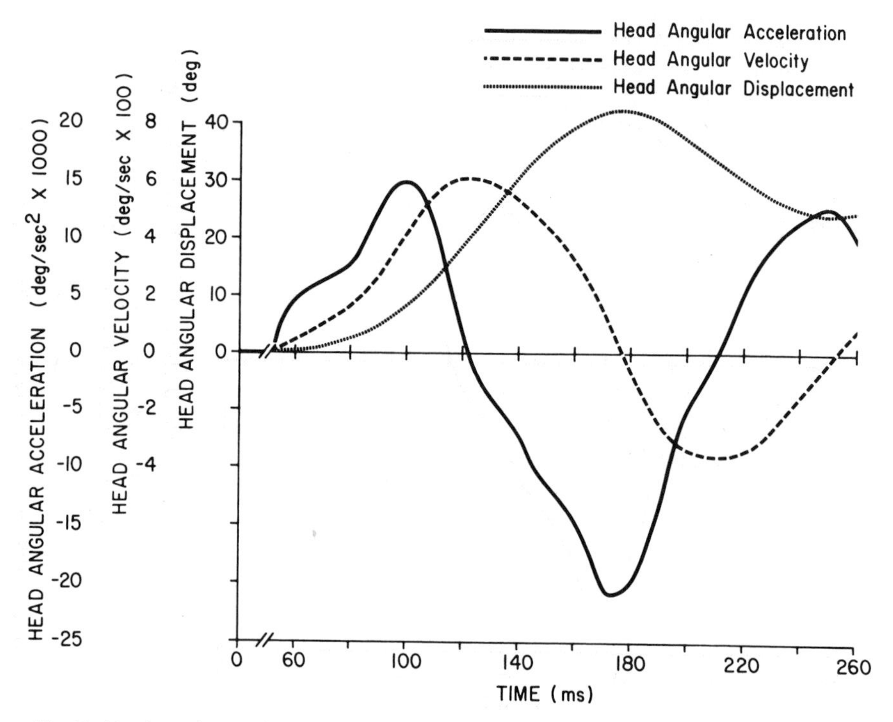

Fig. 5 - Head angular acceleration, velocity and displacement versus time. Shoulder harness restraint at 9.5 g. Run No. 4949, subject No. 1

The duration of this spike was much shorter than the subsequent positive and negative peaks. The positive head angular acceleration peak was extremely inconsistent in magnitude and often of inconsequential value.

The head angular displacement curve was invariably negative, indicating that the head only underwent hyperextension after time zero. Although the mean peak negative rotation of the head was 42 deg, the longitudinal axis of the

Table 2 - Air Force Shoulder Harness Restraint (12 Subjects)

	A	B	C	D	E	F	G	H
Mean	+31,800	−25,500	+1,060	–	20	12	+75	9.6
S. D.	±14,500	±7,900	±290	–	±6	±2	±20	±0.4

A = maximum head angular acceleration (deg/s^2)
B = maximum head angular deceleration (deg/s^2)
C = maximum positive head angular velocity (deg/s)
D = maximum negative head angular velocity (deg/s)
E = maximum front head linear acceleration (g)
F = maximum rear head linear acceleration (g)
G = maximum flexion of head-neck from position at time zero (deg)
H = maximum sled deceleration (g)
– = highly inconsistent or of minimal value

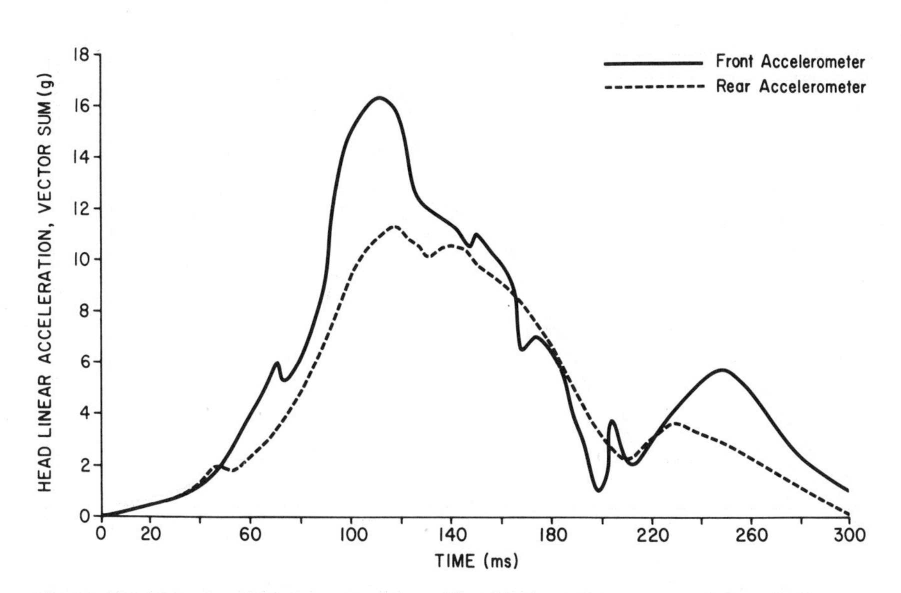

Fig. 6 - Head linear acceleration versus time. Shoulder harness restraint at 9.5 g. Run No. 4949, subject No. 1

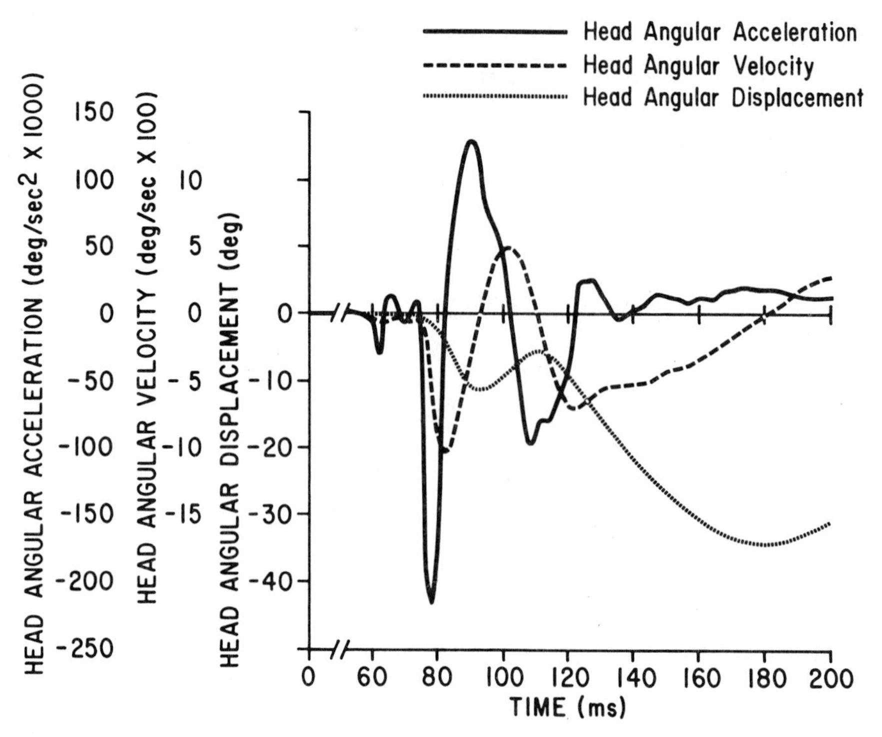

Fig. 7 - Head angular acceleration, velocity and displacement versus time. Air bag and lap belt restraint at 8.4 g. Run No. 4821, subject No. 1

Table 3 - Air Bag Plus Lap Belt Restraint (14 Subjects)

	A	B	C	D	E	F	G	H
Mean	–	−178,200	–	−1,027	71	55	−42	8.9
S. D.	–	±38,900	–	±125	±17	±7	±16	±0.6

A = maximum head angular acceleration (deg/s^2)
B = maximum head angular deceleration (deg/s^2)
C = maximum positive head angular velocity (deg/s)
D = maximum negative head angular velocity (deg/s)
E = maximum front head linear acceleration (g)
F = maximum rear head linear acceleration (g)
G = maximum hyperextension of head-neck from position at time zero (deg)
H = maximum sled deceleration (g)
– = highly inconsistent or of minimal value

torso displaced forward 10-20 deg from vertical. Likewise, the head during many of the tests was hyperextended 15 deg from the neutral anatomical position at time zero. Therefore, at 180 ms the actual head-neck hyperextension relative to the torso for a specific test may average -77 deg or only 2 deg less than the mean voluntary limit of hyperextension (6). This pronounced hyperextension of the head-neck was attributable to the restraining force of the air bag. In essence, the air bag was sufficient to overcome the forward momentum of the head to a much greater extent than the torso of larger mass.

In five tests with the air bag, the peak linear resultants of the front and rear accelerometer clusters exceeded 70 g for a cumulative duration of more than 3 ms. These head accelerations were not appreciably lower than the proposed quantitative occupant injury criteria for a 30 mph barrier impact using anthropomorphic test devices (7). The higher readings of the front accelerometer cluster may be in part due to bag "slap" (Fig. 8).

Although there was no evidence of definitive whiplash injury, six subjects restrained with the air bag plus lap belt developed mild to moderate headaches from 2-24 h post impact (Table 4). This appears to be related to the degree of head-neck hyperextension since headaches resulted with only the subjects who

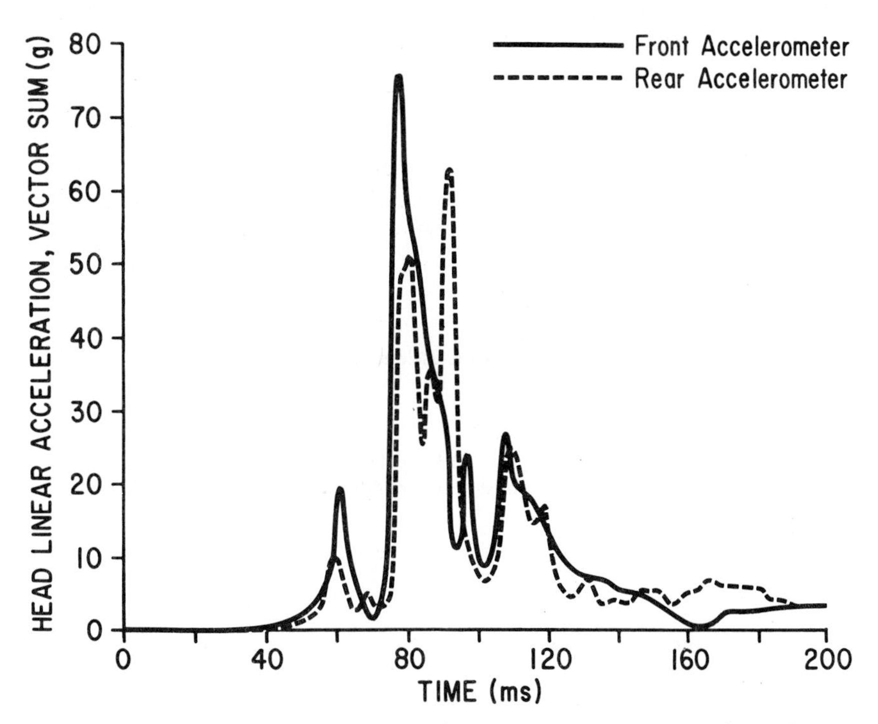

Fig. 8 - Head linear acceleration versus time. Air bag and lap belt restraint at 8.4 g. Run No. 4821, subject No. 1

Table 4 - Correlation
Hyperextension and Headache

Subject	Extent of Head-Neck Hyperextension, deg	Severity of Headache
1	-69	******
2	-61	*****
3	-58	****
4	-56	***
5	-53	*
6	-47	**

*Least severe headache; ****** most severe headache.

experienced the greatest degree of hyperextension. Furthermore, the subject with the most severe headache underwent the greatest head-neck hyperextension.

No correlation was found between headache and angular or linear accelerations.

Conclusions

The results indicated that peak head angular accelerations and linear resultants were elevated with the air bag in contrast to the Air Force shoulder harness-lap belt or lap belt only restraints. However, the angular and linear accelerations may have less traumatic consequences than the degree of head-neck hyperextension.

Acknowledgment

The entire Daisy Deceleration staff of the Land-Air Div., Dynalectron Corp., deserves commendation for its support in the completion of this project.

Portions of this project were funded under contract with the U. S. Department of Transportation.

References

1. T. D. Clarke, J. F. Sprouffske, E. M. Trout, C. D. Gragg, W. H. Muzzy, and H. S. Klopfenstein, "Baboon Tolerance to Linear Deceleration ($-G_x$): Air Bag Restraint." Proceedings of Fourteenth Stapp Car Crash Conference, P-33, paper 700905. New York: Society of Automotive Engineers, Inc., 1970.

2. C. D. Gragg, C. D. Bendixen, T. D. Clarke, H. S. Klopfenstein, and J. F. Sprouffske, "Evaluation of the Lap Belt, Air Bag and Air Force Restraint Systems During Impact with Living Human Sled Subjects." Proceedings of Fourteenth Stapp Car Crash Conference, P-33, paper 700904. New York: Society of Automotive Engineers, Inc., 1970.

3. J. A. Pflug, "Dynamic Problems with an Air Bag Restraint System." Paper 710021 presented at SAE Automotive Engineering Congress, Detroit, January 1971.

4. A. K. Ommaya, A. E. Hirsch, and J. L. Martinez, "The Role of 'Whiplash' in Cerebral Concussion." Proceedings of Tenth Stapp Car Crash Conference, P-12, paper 660804. New York: Society of Automotive Engineers, Inc., 1966.

5. C. D. Bendixen, "Measurement of Head Angular Acceleration During Impact." ARL-TR-70-5, Holloman Air Force Base, New Mexico, 1970.

6. J. J. Defibaugh, "Measurement of Head Motion." Physical Therapy, Vol. 44 (1964), pp. 157-168.

7. "Part 571 - Federal Motor Vehicle Safety Standards." Docket No. 69-7; Notice 9, Federal Register, Vol. 36, No. 47, March 10, 1971.

Appendix

Object - The measurement of linear accelerations at two points on the head will yield after calculations absolute or total angular accelerations of the head with respect to a nonrotating spatial reference system.

Limitations - The angular accelerations of the head with respect to the neck or torso are not being derived. The angular accelerations, velocities, and displacements of the head are only in the sagittal plane (pitch). There was no evidence of lateral head rotation (roll or yaw) from the high-speed film.

Governing Equations and Calculations -

Graphical Computation - Given a subject at time zero referenced to the X-Z coordinate system (Fig. A-1). The head of the subject is defined by two points A and B, which contain accelerometers aligned along the +X and +Z axes. The points are separated by the length L which is of constant value.

After a time "t" the head of the subject has rotated from the original position (Fig. A-2). This is represented by line AB which has rotated through an angle θ (Fig. A-3). Likewise, the accelerometers have rotated through the same angle θ and the rotated axes are designated by X' and Z'.

The accelerometers at A and B measure the instantaneous coplanar accelerations of each point which are typically represented by vectors as shown in Fig. A-4.

The relative acceleration of point A relative to point B is obtained with the vector equation:

$$\mathbf{a}_{A/B} = \mathbf{a}_A - \mathbf{a}_B \tag{A-1}$$

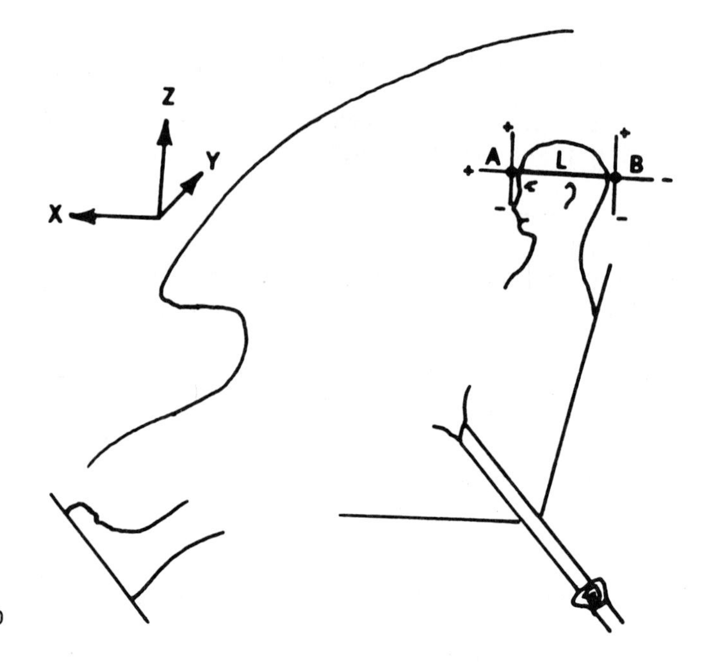

Fig. A-1 - Time: 0

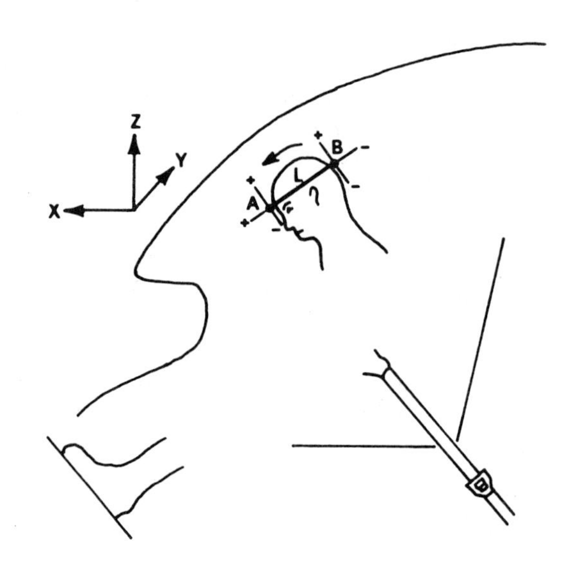

Fig. A-2 - Time: 0 + t

Since the line AB is fixed in length then the relative acceleration of $\mathbf{a}_{A/B}$ can be broken into its normal and tangential components of $\mathbf{a}_{A/B(N)}$ and $\mathbf{a}_{A/B(T)}$. These components are parallel and perpendicular to the line AB, respectively (Fig. A-5).

By definition the magnitudes of coplanar rotation components are:

$$a_{A/B(N)} = -L\dot{\theta}^2 \qquad \text{(A-2)}$$

$$a_{A/B(T)} = -L\ddot{\theta} \qquad \text{(A-3)}$$

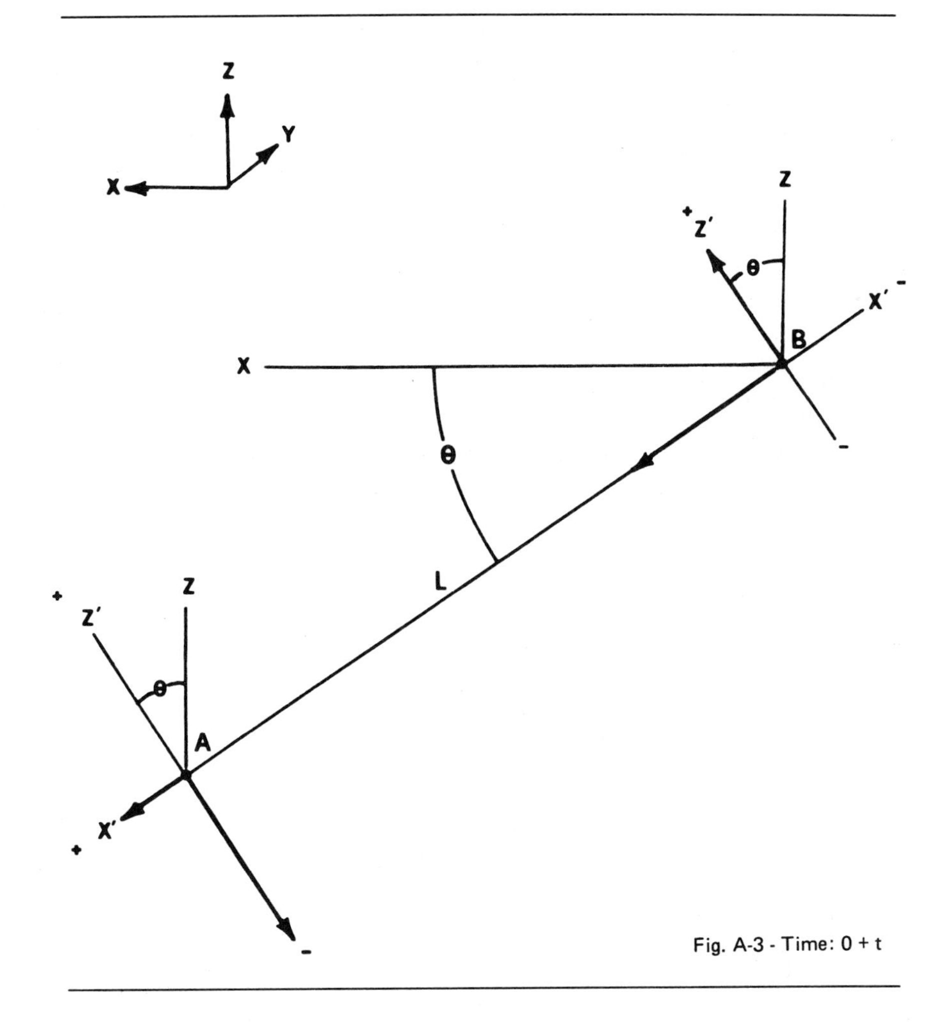

Fig. A-3 - Time: 0 + t

where $\dot{\theta}$ and $\ddot{\theta}$ are the angular velocity and angular acceleration respectively.

It is observed that the magnitude of $\mathbf{a}_{A/B(T)}$ is the difference between the Z′ components of $\mathbf{a}_A$ and $\mathbf{a}_B$ or

$$a_{A/B(T)} = a_{AZ'} - a_{BZ'} = -L\ddot{\theta} \tag{A-4}$$

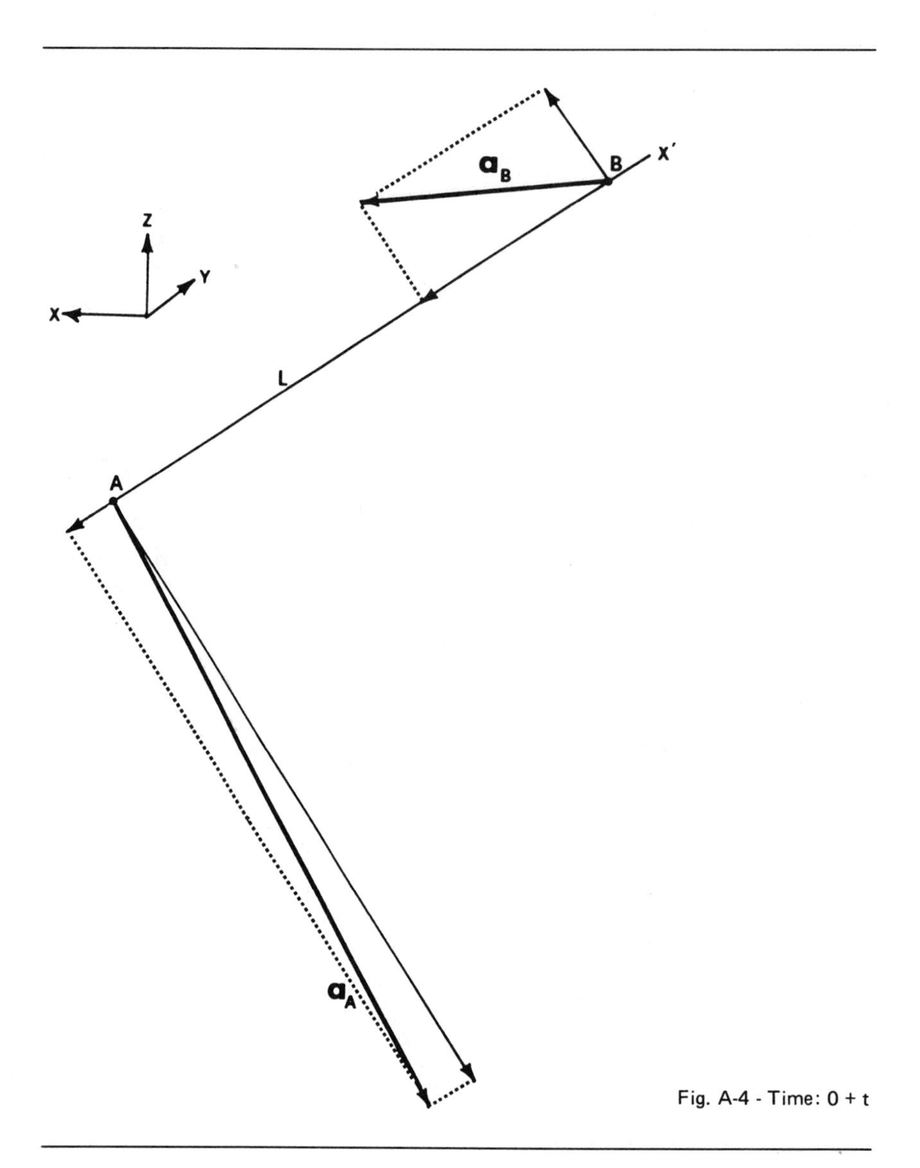

Fig. A-4 - Time: 0 + t

Thus

$$\ddot{\theta} = \frac{a_{BZ'} - a_{AZ'}}{L} \tag{A-5}$$

Where R(1) and R(2) are the readings of the front and rear Z accelerometers respectively:

$$\ddot{\theta} = \frac{R(2) - R(1)}{L} \tag{A-6}$$

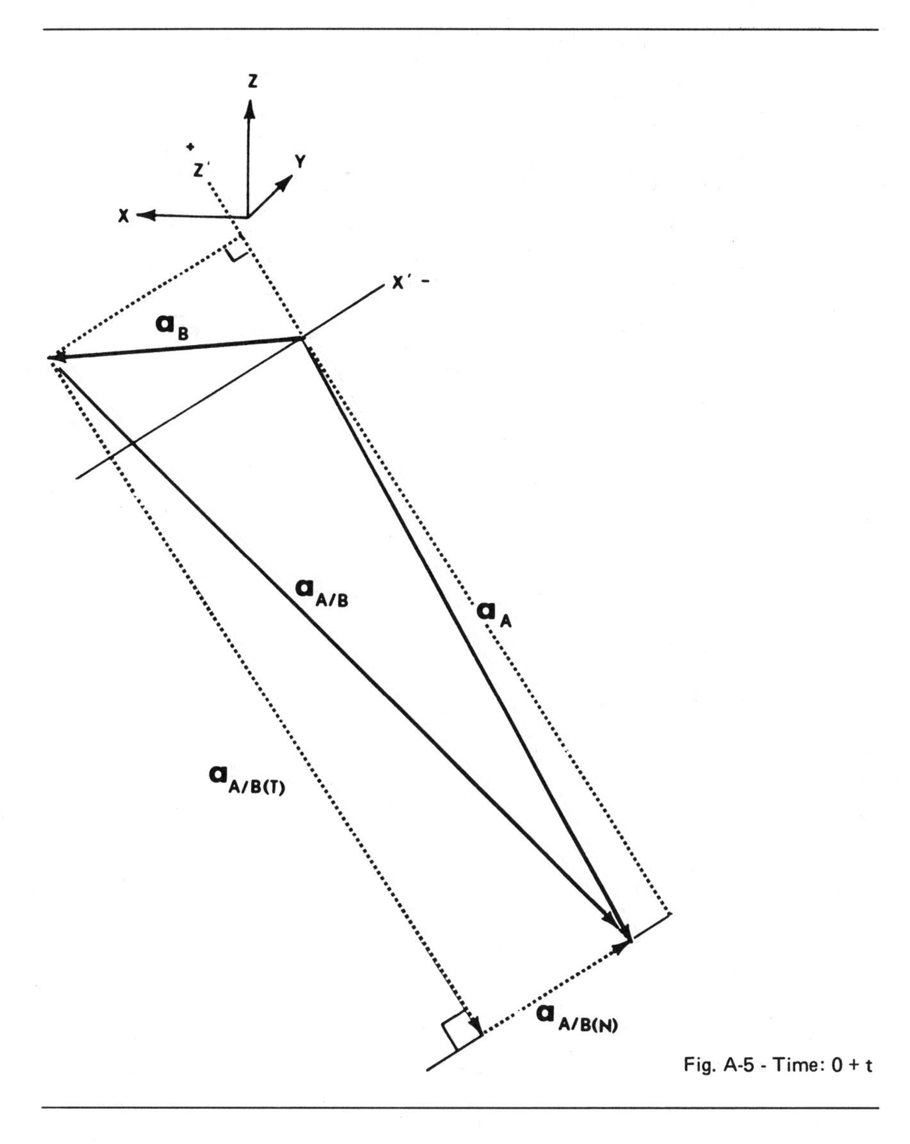

Fig. A-5 - Time: 0 + t

or

$$\ddot{\theta} = \frac{a_{\text{rear } Z} - a_{\text{front } Z}}{L} \tag{A-7}$$

Vector Computation - A compatible vector representation of the coplanar motion can be achieved by using Fig. A-6. The X and Z axes are defined as positive to the left and upward, respectively. Using the left-handed rule, the Y axis is directed into the paper. This reference system, then, is compatible with the series of photographs of the tests.

Mathematically, the motion of line AB may be referenced to a translating reference system with origin at B. Points A and B contain the accelerometers. For this development the unit vector **i** is directed along the X axis, the unit vector **j** along the Y axis, and the unit vector **k** along the Z axis. **r** is the vector representing the distance AB. The general components of the vector **r** are taken as:

$$\mathbf{r} = r_X \mathbf{I} + r_Z \mathbf{K} = \mathbf{r}_A - \mathbf{r}_B \tag{A-8}$$

Since the magnitude of **r** is constant, the relative velocity of point A with respect to point B represents the rotational motion of line AB. Thus, the following results:

$$\mathbf{v}_{A/B} = \frac{d\mathbf{r}}{dt} = r_X \frac{d\mathbf{I}}{dt} + r_Z \frac{d\mathbf{K}}{dt} = \omega \times \mathbf{r} \tag{A-9}$$

where $\omega = \dot{\theta}\mathbf{j}$.

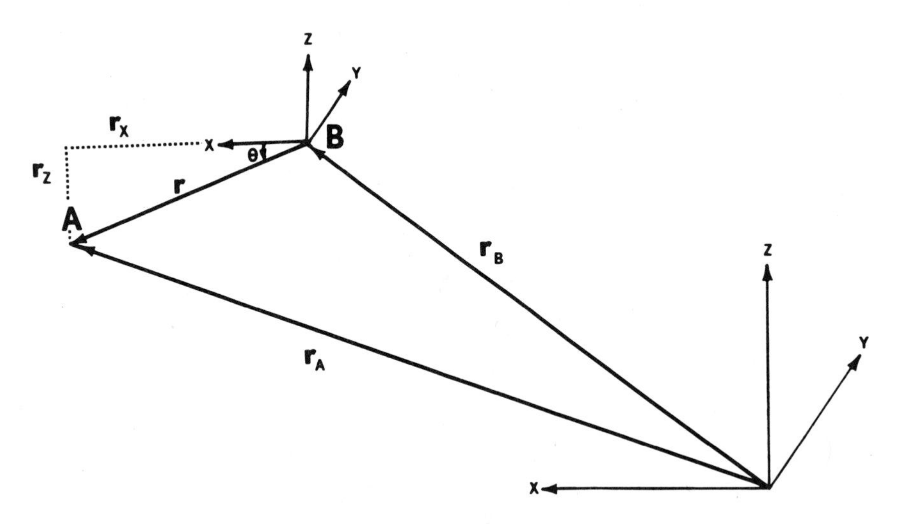

Fig. A-6 - Time: 0 + t

The acceleration of point A relative to point B is defined as:

$$\mathbf{a}_{A/B} = \frac{d\mathbf{v}_{A/B}}{dt} = \omega \times \frac{d\mathbf{r}}{dt} + \frac{d\omega}{dt} \times \mathbf{r} \tag{A-10}$$

where:

$$\frac{d\mathbf{r}}{dt} = r_X \frac{d\mathbf{I}}{dt} + r_Z \frac{d\mathbf{K}}{dt} = \omega \times \mathbf{r} \tag{A-11}$$

$$\frac{d\omega}{dt} = \dot{\theta}\frac{d\mathbf{J}}{dt} + \frac{d\dot{\theta}}{dt}\mathbf{J} \tag{A-12}$$

Since the axis J does not rotate, $\frac{dj}{dt} = 0$

Thus,

$$\frac{d\omega}{dt} = \alpha = \frac{d\dot{\theta}}{dt}\mathbf{J} = \ddot{\theta}\mathbf{J} \tag{A-13}$$

Finally,

$$\mathbf{a}_{A/B} = \omega \times (\omega \times \mathbf{r}) + \alpha \times \mathbf{r} = \mathbf{a}_A - \mathbf{a}_B \tag{A-14}$$

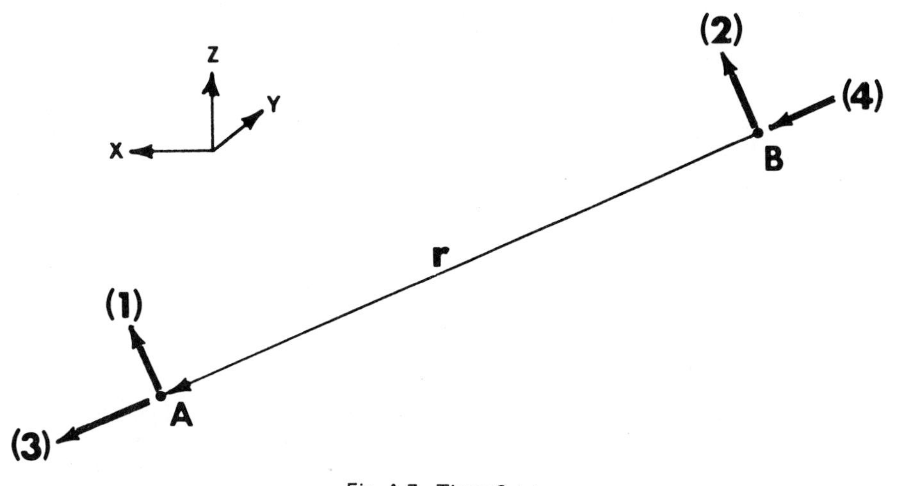

Fig. A-7 - Time: 0 + t

The vector $\omega \times (\omega \times \mathbf{r})$ is directed along the line AB and represents the relative normal acceleration $\mathbf{a}_{A/B(N)}$. The vector $\mathbf{a} \times \mathbf{r}$ is directed perpendicular to line AB and represents the relative tangential acceleration $\mathbf{a}_{A/B(T)}$. The X′ - Z′ coordinate system (Fig. A-2) also defines these normal and tangential components. Accelerometers numbered (1) and (2) (Fig. A-7) measure the Z′ components of the accelerations of A and B which can be represented by:

$$\mathbf{a}_{A/B(T)} = \mathbf{a}_{AZ'} - \mathbf{a}_{BZ'} \qquad \text{(A-15)}$$

In Scalar form

$$a_{A/B(T)} = -r\ddot{\theta} = a_{AZ'} - a_{BZ'} \qquad \text{(A-16)}$$

$$\ddot{\theta} = \frac{a_{BZ'} - a_{AZ'}}{r} \qquad \text{(A-17)}$$

$$\ddot{\vartheta} = \frac{a_{\text{rear Z}} - a_{\text{front Z}}}{r} \qquad \text{(A-18)}$$

741186

Traumatic Distortions of the Primate Head and Chest: Correlation of Biomechanical, Radiological and Pathological Data

Stanley A. Shatsky, William A. Alter III and Delbert E. Evans
Armed Forces Radiobiology Research Institute

Vernon W. Armbrustmacher and Kenneth M. Earle
Armed Forces Institute of Pathology

Gary Clark
Walter Reed Army Medical Center

THE DEVELOPMENT OF PATHOLOGIC LESIONS within the central nervous system and thorax following mechanical trauma has been the subject of extensive clinical and experimental study. Yet there remains a paucity of information regarding the actual movements within the head and thorax during impact injury. The purpose of the present investigation was to examine the role of transient internal mass movements occurring during traumatic events and to relate these to the development of post-injury physiologic and pathologic changes. The studies are as yet preliminary in nature and extent and are limited to small subhuman primates.

Note: Research was conducted according to the principles enunciated in the "Guide for Laboratory Animal Facilities and Care," prepared by the National Academy of Sciences–National Research Council.

The visualization of internal anatomy is a routine procedure in medical radiology, which has only recently been applied to the study of impact trauma (15)*. The prime reasons have been technological, in that conventional radiographic instruments have had neither the frame rate capability nor the short pulse radiation generating capacity necessary for very high-speed x-ray photography. In order to study fast impact mechanisms, it has been necessary to develop new instrumentation and utilize a technique we have termed "flash x-ray cinematography." The procedure combines nanosecond, flash x-ray durations of single exposure ballistic radiographic systems with the cine capability of high-speed medical radiographic instruments. This technique increases the usable

*Numbers in parentheses designate References at end of paper.

ABSTRACT

High speed cinefluorographic studies were performed on anesthetized primates during graded, experimental blunt impacts of the head or chest. Cineframe data were analyzed frame by frame to identify dynamic anatomic movement patterns during each injury. The results indicate that the brain and heart undergo significant displacements within the first few milliseconds (ms) post-impact and these transient interior motions were correlated with physiologic and pathologic changes as well as impact force and deceleration.

frame rate capability by one and one half orders of magnitude. The combination of extremely short exposure times and high frame rates allows study of transient high speed phenomena not previously observable.

METHODS

FLASH X-RAY SYSTEM - To study the movements within the head and chest, a high-speed cinefluorographic system was developed (15). The instrument has seven components (Fig. 1):

1. A high voltage source.
2. A repetitive Marx Surge electronic pulse generator, (Field Emission #815).
3. A remote 150 kVp flash x-ray tube with an effective source size of 2.5 mm.
4. A variable-zoom cesium iodide x-ray image intensifier with a measured brightness gain of 8650 (Varian VXI-155).
5. Coupling optics
6. A high-speed 16 mm pin register camera (Photosonics 16-1W).
7. Trigger, control and synchronization circuitry.

The instrument operates as an x-ray strobe with a 30 nanosecond x-ray pulse exposing each frame. Frame rates are variable up to 1000 frames per second (f/s) with independence of film exposure and frame rate. Radiation exposure at the subject plane is 300 μ R6frame at 150 kVp, with a maximum pulse train number of 100. Images are recorded on Linegraph Shellburst EK-2476 or Cronex SF-2 and developed to a density of 1.3 in Versaflo in a Kodak Versamatt.

EXPERIMENTAL PROCEDURE - In our initial studies, adult Macacca mulatta monkeys weighing 3-6 kg were anesthetized and maintained using sodium pentobarbital. To permit a more complete analysis of the neurologic deficits following trauma, later studies were conducted with the primates maintained under inhalation anesthesia—74% nitrous oxide, 25% oxygen and 1% halothane. To determine the displacement of the cerebral vasculature, a catheter was placed in the left internal carotid artery and served as the route for the injection of radio-dense media—80% Angioconray or thorium dioxide enriched Angioconray. In other animals, the intracranial ventricular system was studied by insertion of a needle stereotaxically into the lateral ventricular atrium and infusion (4 ml) of radio-dense Conray or Pantopaque. A relatively constant intracranial fluid volume was maintained by removal of a similar quantity of CSF via cisternal tap.

Primates prepared for either cerebral angiography or ventricularography were seated on the MB4 acceleration sled (Fig. 1). For left temporoparietal impact studies, the head position was stabilized by attachment of the fin shown above the animal chair. This fin was secured at the bregma—with dental acrylic—and was positioned within the guide rail, thus stabilizing the primate's head during acceleration and restricting impact induced head movement to the impact plane. The animal was then accelerated in the chair by the force of a falling 40 kg weight. Just prior to impact, the chair passed through a velocity trap which detected the terminal velocity of the chair. The chair also passed over a magnetic switch positioned to initiate the cinefluorographic system 10 ms prior to impact. The cinefluorographic system is aligned at the terminal end of the sled to provide radiographs in a projection perpendicular to the chair's direction of travel.

For occipital head impacts, the chair was rotated 90 deg, but the head finguide rail unit was not used. To maintain relatively stable pre-impact head position, a breakaway styrofoam collar was placed around the primate's neck. Post-impact movement of the head was not restricted.

Ten animals were prepared for chest impact studies. A catheter was inserted under direct fluoroscopic control into the left cardiac ventricle in six and into the aortic arch in four. The animal was then seated on the acceleration sled in a modified open chair, and the impactor was positioned to strike the frontal chest in the midline.

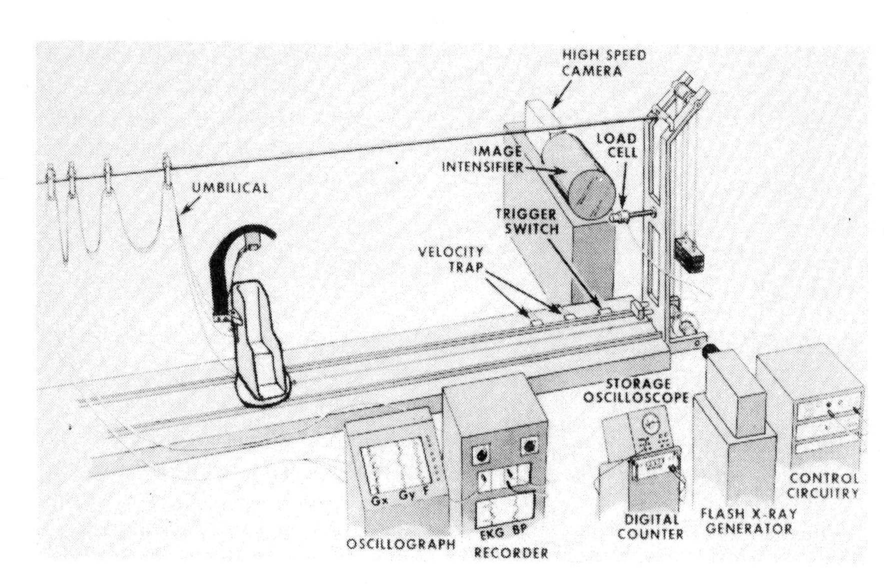

Fig. 1 - MB4 Impact sled and cinefluorographic system

Biomechanical impact parameters recorded during the impact studies included skull deceleration and contact force time history. Skull deceleration was detected by Entran (EGB), miniature, piezoresistive accelerometers secured at the skull bregma, while contact force was detected by a BLH 2CMI load cell mounted in series with the impactor. These data were directly recorded on a Honeywell Visicorder with a maximum frequency response of 3 K Hz.

Primates were accelerated to terminal velocities ranging from 3-8 m/s before striking either the head or chest against a fixed impactor. In the cinearteriographic studies, injection of radio-dense media began ½ s prior to impact. Lateral arteriograms were obtained during frontal thoracic and occipital head impacts, while frontal arteriograms were obtained during temporoparietal head impacts. Physiologic data recorded during these studies included systemic arterial blood pressure and electrocardiogram. Following impact, the animals were allowed to recover to permit physiologic and neurologic examinations. Those animals which appeared to be clinically normal were maintained up to 15 d post-impact, whereas those which demonstrated an obvious clinical deficit received supportive care until death or euthanasia.

CINEFLUOROGRAPHIC DATA ANALYSIS - The data were analyzed using a Vanguard Motion Analyzer, and film from eight of the temporoparietal head impact experiments had sufficient quality to permit quantitative measurements of cerebral vasculature. In addition to the fin-guide rail system for head stability, lead markers were secured to the occipital surface to serve as indicators of any post-impact lateral head rotation. Measurements were made of the locations of the midline anterior cerebral artery, middle cerebral arteries and lateral skull diameter at the level of the pterion. All measurements were obtained relative to external radiodense markers incorporated into the head fin. These markers permitted the data to be scaled to true size. Differences between control and post-impact data were analyzed statistically by the student's paired t-test. Repeated determinations of the same points indicated that measurement errors in our studies were less than 0.5 mm.

Cineventriculograms obtained during temporoparietal and occipital impact and cineangiograms obtained from an occipital impact group were studied only qualitatively for an estimation of the overall movements of the specific structures.

Cineangiograms exposed during thoracic impacts were studied to determine chest compression. Measurements were made of the external radiographic thoracic diameters and comparisons were made between preimpact diameter and diameter during maximal chest compression.

PATHOLOGIC EXAMINATIONS - Primates were euthanatized with intravenous pentobarbital 100 mg/kg. Brains and spinal cords, or heart, lungs and great vessels were removed intact and placed in buffered 10% formalin for a period of at least two weeks. Brains and spinal cords were examined by a neuropathologist and thoracic organs by a general pathologist, each with no prior knowledge of the site, direction or biomechanical level of impact.

CENTRAL NERVOUS SYSTEM EXAMINATION - The brains were examined grossly and coronal sections of the cerebral hemispheres and transverse sections of the cerebellum, brainstem and cervical spinal cord were made. Selected sections were photographed and submitted for microscopic examination after dehydration and embedding in paraffin by standard techniques. Microscopic sections were cut at eight micra for hematoxylin and eosin staining. Twenty-five micra sections were stained with standard Bodian silver technique for axons and standard luxol fast blue technique for myelin.

CARDIOTHORACIC EXAMINATION - Horizontal sections of the heart were made at a minimum of 6 levels, and representative blocks were obtained for embedding in paraffin. The blocks were sectioned at 5-10 micra and stained with hematoxylin and eosin for general histologic examinations. Verhoff's elastic stain was used for selected sections of great vessels.

RESULTS

INTRACRANIAL MASS MOVEMENTS - Five distinct phenomena were observed during these experiments. These were:

1. Transitory, high frequency depression of the skull.
2. Early high frequency movements of cerebral blood vessels.
3. Low frequency movements of cerebral blood vessels.
4. Early high frequency oscillatory movements of the ventricular system.
5. Low frequency movements of the ventricular system.

An initial depression of the skull was characteristic of both fracturing and non-fracturing blunt temporoparietal impacts but was not radiographically detectable following occipital blows. Fig. 2 is a sequence of nine cineframes from a 1000 f/s cineradiographic study of temporoparietal injury.

The primate head was radiographed, moving from right to left at 7.5 m/s. The intracranial arterial system was opacified by thorium dioxide enriched Angioconray and the anterior, middle cerebral and carotid arteries as well as the complete Circle of Willis were visualized in frontal arteriograms. The skull was 60 mm in lateral diameter prior to impact. Two ms after impact, it was 56.6 mm, and five ms post-impact, the skull had returned to its pre-impact diameter. The data in Table 1 summarizes the results obtained in eight experiments. For this group, the mean control lateral skull diameter was 58.7 ± 0.5 mm, and at impact it was reduced 2.0 ± 0.4 mm. This change was statistically significant at the $p < 0.005$ level. The maximum depression of the skull usually occurred within the first 2 ms and returned to near control values by 3-6 ms post-impact. The strain of the skull during temporoparietal impact, calculated as the ratio of the change in the skull diameter to the control diameter, was 3.8 ± 0.7%.

The second transient intracranial movement was a high frequency oscillation of midline blood vessels occurring during the time of maximum skull depression after temporoparietal impact. The movement consisted of small amplitude—2-3 mm—oscillations of the anterior cerebral artery across the midline. These movements were most prominent in the proximal segment of this vessel near its origin from the Circle of

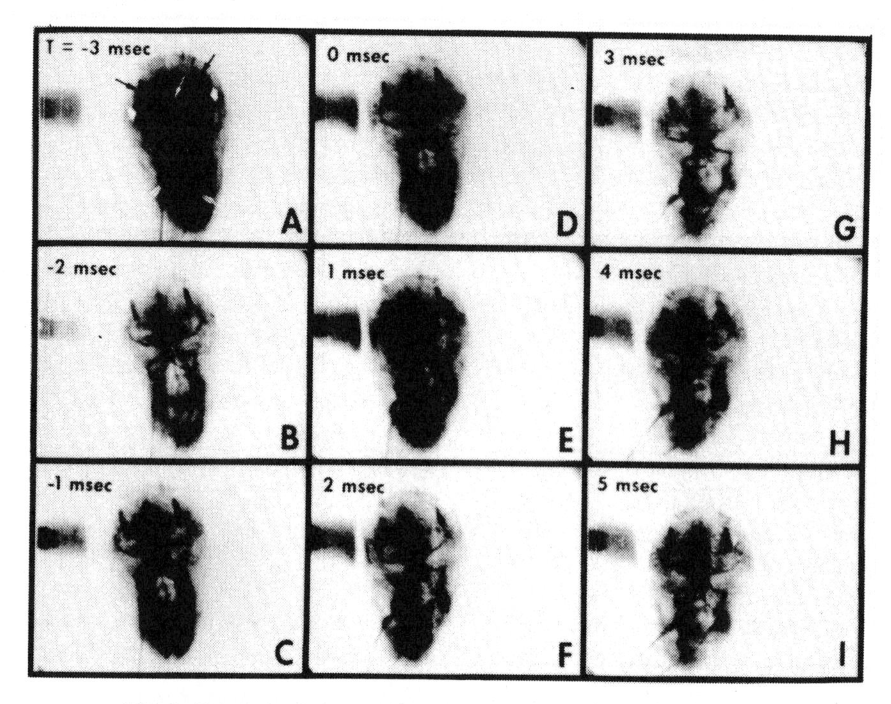

Fig. 2 - Temporoparietal impact sequence, cinearteriogram—An anesthetized primate is traveling from right to left at 7.5 m/s and impacts a fixed load cell in Frame D. A frontally viewed carotid anteriogram was performed at 1000 f/s and demonstrates the midline anterior cerebral artery, (white arrow), bilaterally symmetric middle cerebral arteries, (white arrowheads), bilateral internal carotid arteries, (paired white arrows), and the diamond shaped Circle of Willis. Vertical lead markers (black arrows) cemented to the calvarium demonstrate there was minimal rotation of the skull during the sequence.

Willis but were also recorded on occasion more distally. Occurring simultaneously with this anterior cerebral artery oscillation, there was medial displacement of the middle cerebral artery ipsilateral to the impact site. This displacement data is summarized in Table 1. The control position of 43.5 ± 0.9 mm is relative to the external scale markers. Post-impact it was shifted medially and this change was statistically significant at the $p < 0.025$ level. The time frame for movement of the left middle cerebral artery—2-3 ms—was similar to the time for maximal skull compression—2 ms.

A third transitory event observed during these experiments was a lower frequency oscillatory movement of the midline intracranial arteries during temporoparietal impact. An example of this phenomenon is shown in Fig. 2. Stability of the radio-dense skull markers in Frames D-I indicated minimal lateral head rotation post-impact. During the post-impact period there was a progressive displacement of the anterior cerebral artery towards the impact site. This displacement reached its maximum at 5 ms (Frame I) and then began its return to the pre-impact position. This movement of the anterior cerebral artery was observed in eight animals studied for cerebral vascular displacement after temporoparietal head impact, and this data is summarized in Table 1. The control position of the vessel, 23.6 ± 0.4 mm, was recorded with reference to the external scale marker. Maximum displacement of the anterior cerebral artery 2.2 ± 0.6 mm occurred 3-6 ms post-impact and this change was significant at the $p < 0.005$ level.

To analyze the strain which the brain tissue underwent after temporoparietal impact, changes in the lateral diameter of the left cerebral hemisphere were calculated as the distance between the left lateral skull surface and the midline anterior cerebral artery. These data were summarized in Table 1. Control diameter of the left cerebral hemisphere was 35.5 ± 0.8 mm. Impact transiently reduced this diameter by 3.0 ± 0.7 mm and this change was statistically significant at the p < 0.005 level. This change can be translated into a strain of 8.6 ± 1.9%.

A fourth type of high frequency phenomenon radiographed during blunt head impacts was lateral oscillatory movements of the contrast outlined intracranial ventricular system following temporoparietal impacts. When the ventricles were filled with conray or thorium dioxide enriched conray, they underwent high frequency small amplitude oscillations. The initial displacement was towards the impactor and occured from 1-3 ms post-impact. This was followed by contralateral displacement of the ventricular shadow with a maximum peak to peak displacement of 6 mm and an approximate period of 3-5 ms. At least three oscillations of the ventricular outlines were observed before critical damping took place. However, the magnitude and frequency of ventricular oscillation can be affected by experimental techniques. The extent of ventric-

Table 1—Anatomical Displacements After Left Temporoparietal Impact

	Lateral Skull Diameter	Anterior Cerebral Artery	Compression of Left Cerebral Hemisphere	Left Middle Cerebral Artery
Control in mm.	58.7 ± 0.5*	23.6 ± 0.4	35.5 ± 0.8	43.5 ± 1.0
Post-impact Change in mm.	2.0 ± 0.4	2.2 ± 0.6	3.0 ± 0.7	2.9 ± 0.8
Strain in %	3.8 ± 0.7%	—	8.6 ± 1.9%	—
Time of Post-impact Maximum change in ms	2	3–6	0–5	2–3
n	8	7**	8	4†
P-value††	0.005	0.005	0.005	0.025

*Mean ± S.E.

**Data from one experiment indicated sufficient lateral rotation to affect the anterior cerebral artery movements.

†Only four primates had sufficient contrast media in the left middle cerebral artery to permit analysis.

††Statistical significance determined by a one-tailed paired t-test.

ular filling and density of contrast media modified post-impact results. An example of this effect can be seen in Fig. 13. In this experiment, the ventricular system has been completely filled with contrast media of two densities: air in the bodies of the lateral ventricles, thorium dioxide suspension in the remainder of the ventricular system, subarachnoid cisterns and cervical subarachnoid space. The air appears white in these cinefluorographs and is marked by single white arrows. The thorium suspension appears black against a mottled grey appearing head. The skull was impacted in the left temporo-parietal region (Frame B). The first movement was a contra-lateral shift of low density air to the temporal horn of the right lateral ventricle. This probably resulted from an ipsi-lateral displacement of more dense contrast media. The extent of displacement of contrast media was greater in this experiment than in those in which only contrast media had been injected.

A fifth transitory phenomenon observed during these experiments was a relatively slow oscillatory displacement of the contrast filled ventricular system. This was seen on lateral radiographic projection following occipital impacts and consisted of a slow postero-inferior displacement of the radiographic position of the temporal horn and a postero-superior displacement of the body and atrium.

This phenomenon is illustrated in Fig. 3 which is a sequence of 15 cinefluorographs reproduced at 5 ms spacing from a 1000 f/s cineventriculogram obtained during a 4 m/s non-fracturing occipital head impact. Frame A demonstrates the radiographic contour of the incompletely filled ventricular system 5 ms prior to impact. The temporal horn is marked by arrows, the atrium and body by arrowheads. Frames B through H illustrate the progressive distortion of the ventricular outlines, with inferior and posterior movement of the opacified temporal horn and superior posterior movement of the atrium and body. The distortion appears maximal at 30 ms and is followed by a progressive return to pre-impact position in Frames I through O.

NEUROPATHOLOGY

Temporo-parietal impacts resulted in discrete lesion patterns involving both the brain and spinal cord.

At low, non-fracturing impact levels, either no abnormalities were found or occasionally subarachnoid hemorrhages were seen around the infundibulum and the medial aspects of the temporal lobes and in areas underlying the impact area.

At moderate–diastatic fracture–non-lethal impact levels, petechial hemorrhages were found in the dura around the posterior one-half of the sagittal and medial one-half of each transverse sinus. Subarachnoid hemorrhages were more numerous in this group and were especially concentrated around the lateral surfaces of each parietal and temporal lobe. Hemorrhages were also more numerous around the optic chiasm and infundibulum and were frequently seen around the basis pontis and the medullary pyramids and occassionally on the surface of the corpora quadragemina. Contusions were seen in a contrecoup position along the lateral aspect of the right posterior temporal lobe.

At high–depressed fracture–lethal impact levels, multiple contusions were found. These were most often located in the area of the impact zone–coup location–and were less often located in a position opposite the impact zone on the surface

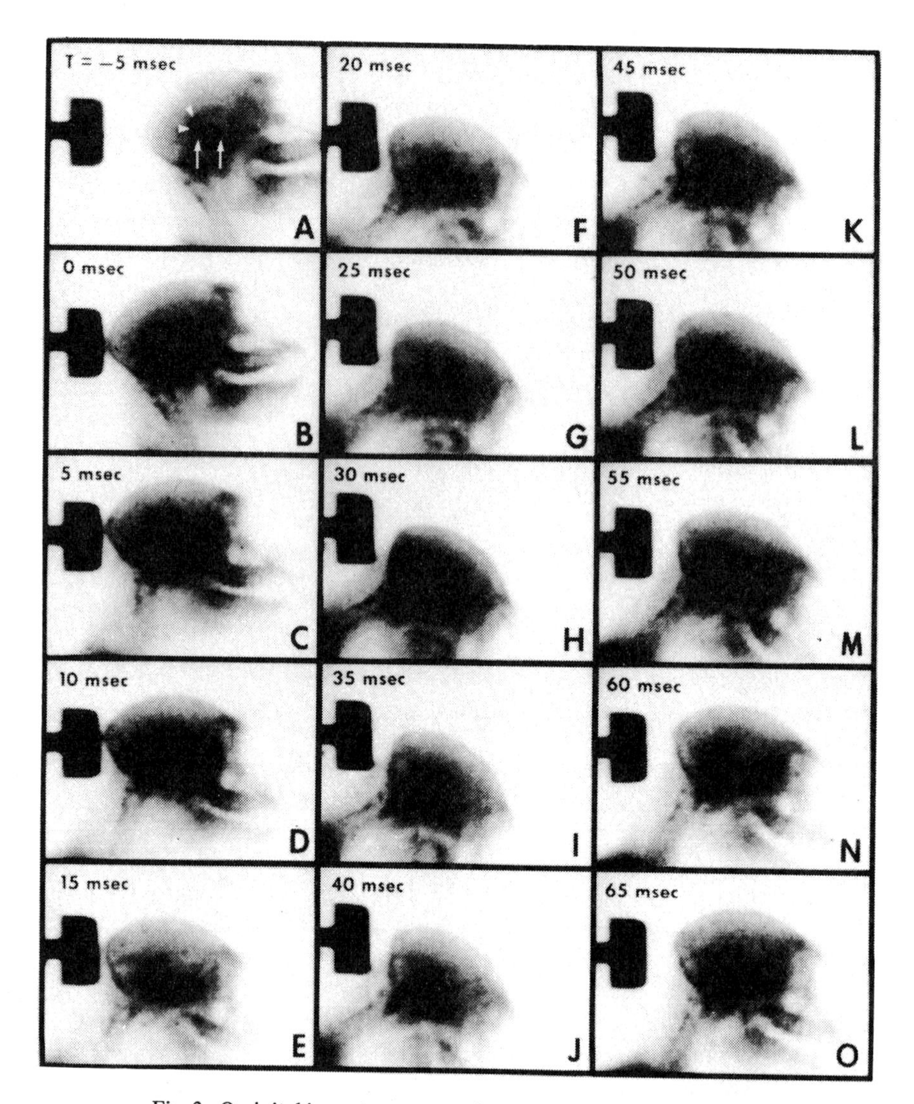

Fig. 3 - Occipital impact sequence, cineventriculogram—An anesthetized primate prepared for ventriculography moved right to left at four m/s prior to impact. Frame A demonstrates the preimpact anatomy with the temporal horn marked by arrows, the incompletely filled atrium and body by arrowheads.

of the right temporal lobe—contrecoup pattern. Again, subarachnoid hemorrhages were seen around the base of the brain. In several cases, hemorrhages and necrosis of the spinal cord were found, especially in segments C3 and C4.

Typical lesions following occipital blows at low impact levels were scattered foci of subarachnoid hemorrhages over the superior surfaces of each frontal lobe and the surfaces of the cingulate gyri.

At moderate–pre-fracture–non-lethal impact levels, these subarachnoid hemorrhages were increased in number and were most prominent over the inferior aspects of each frontal lobe as well as the medial portions of the temporal lobes.

At high–linear fracture–lethal impact levels, numerous contusions sometimes associated with large hematomas were found over both the inferior and lateral surfaces of each temporal lobe. Foci of subarachnoid hemorrhage were extensive, located over the superior and inferior surfaces of the cerebral hemispheres as well as around the optic chiasm and infundibulum. These hemorrhages were also seen around the cerebral peduncles, base of the pons and medullary pyramids, and on the surface of the corpora quadragemina.

Figs. 4-9 are summary diagrams of the neuropathologic lesions tabulated from 43 primates following low, moderate, and high level temporoparietal (Figs. 4-6) and occipital (Figs. 7-9) impacts. Sections from top to bottom are: frontal lobe, level of the optic chiasm, level of the infundibulum, occipital lobe, cerebellum, midbrain, pons and medulla.

CORRELATIONS OF BIOMECHANICAL-PHYSIOLOGICAL-PATHOLOGICAL DATA - Table 2 summarizes biomechanical, physiological and pathological data obtained in the study of temporoparietal impacts. The data includes only primates in which all categories of information were examined and excludes any which experienced toxic reactions to surgical procedures, intra-arterial, or intraventricular contrast media.

The impact level at which skull fracture occurred was well

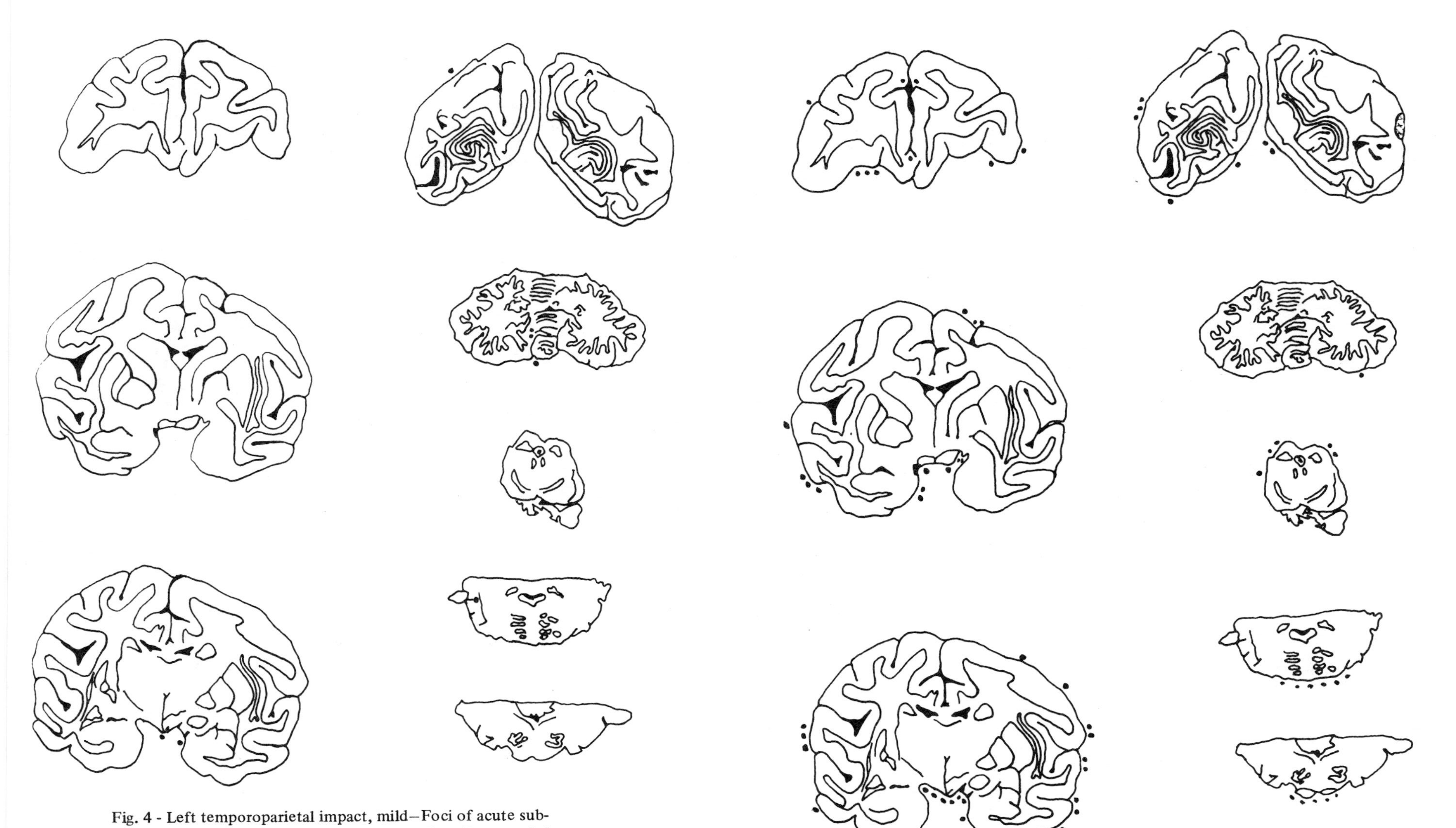

Fig. 4 - Left temporoparietal impact, mild—Foci of acute subarachnoid hemorrhage were noted only occasionally around the infundibulum and over the left posterior temporo-occipital area. These and the following five figures are summary drawings based on results obtained from 43 primates brains. The dots indicate the frequency of subarachnoid hemorrhages. The cross-hatched areas show the locations and extent of contusions.

Fig. 5 - Left temporoparietal impact, moderate—Contusions were seen over the right temporo-occipital area. Scattered subarachnoid hemorrhages were seen as indicated by the dots around the surface of the brain

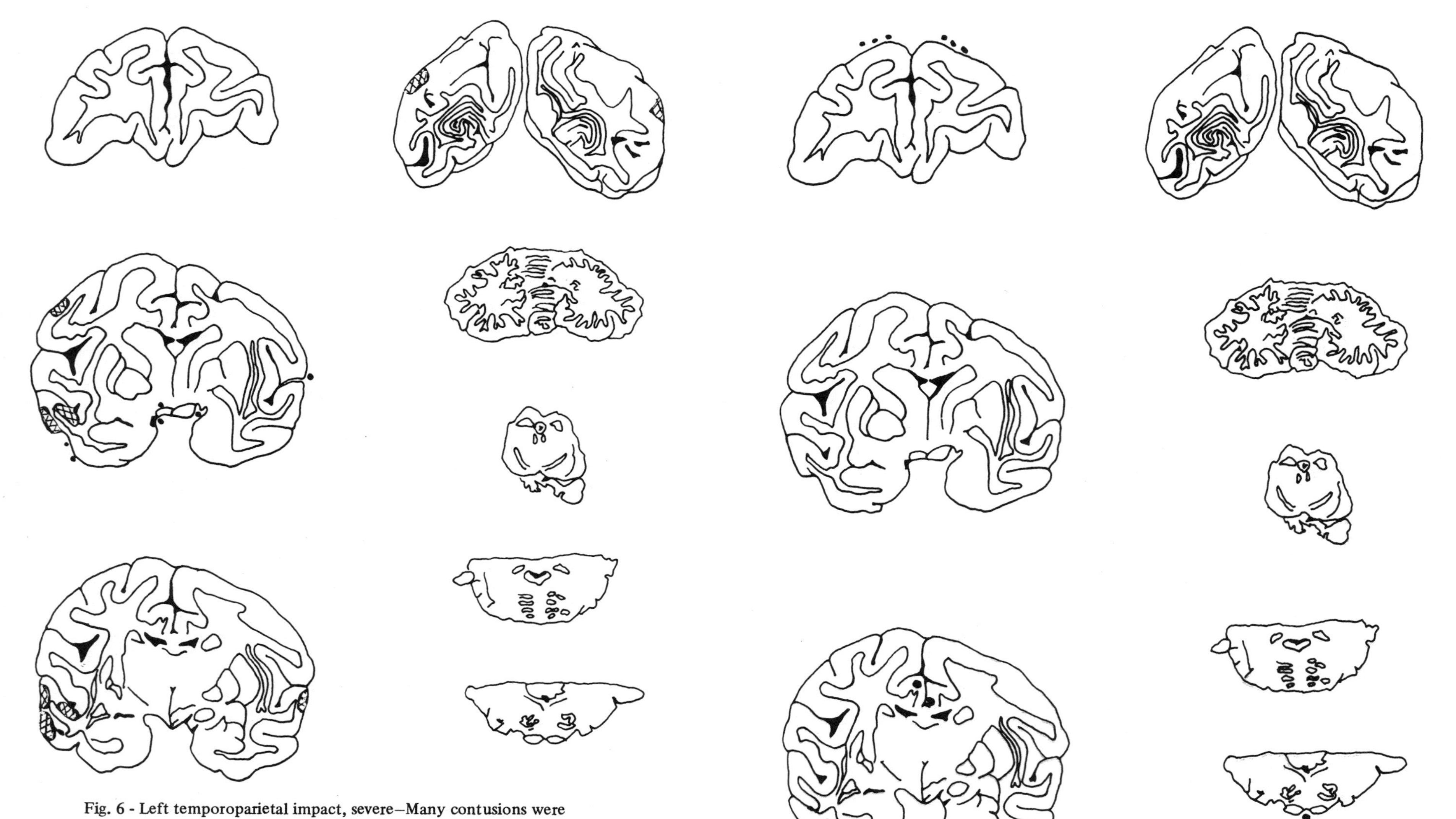

Fig. 6 - Left temporoparietal impact, severe—Many contusions were present along the lateral surface of the left posterior frontal, parietal, temporal, and temporo-occipital areas. Contusions were occasionally noted in a contrecoup location along the right mid and posteriolateral temporal areas. Scattered subarachnoid hemorrhages were noted in the area indicated by the dots.

Fig. 7 - Occipital impact, mild—Foci of subarachnoid hemorrhage were seen occasionally in the area indicated by the markings.

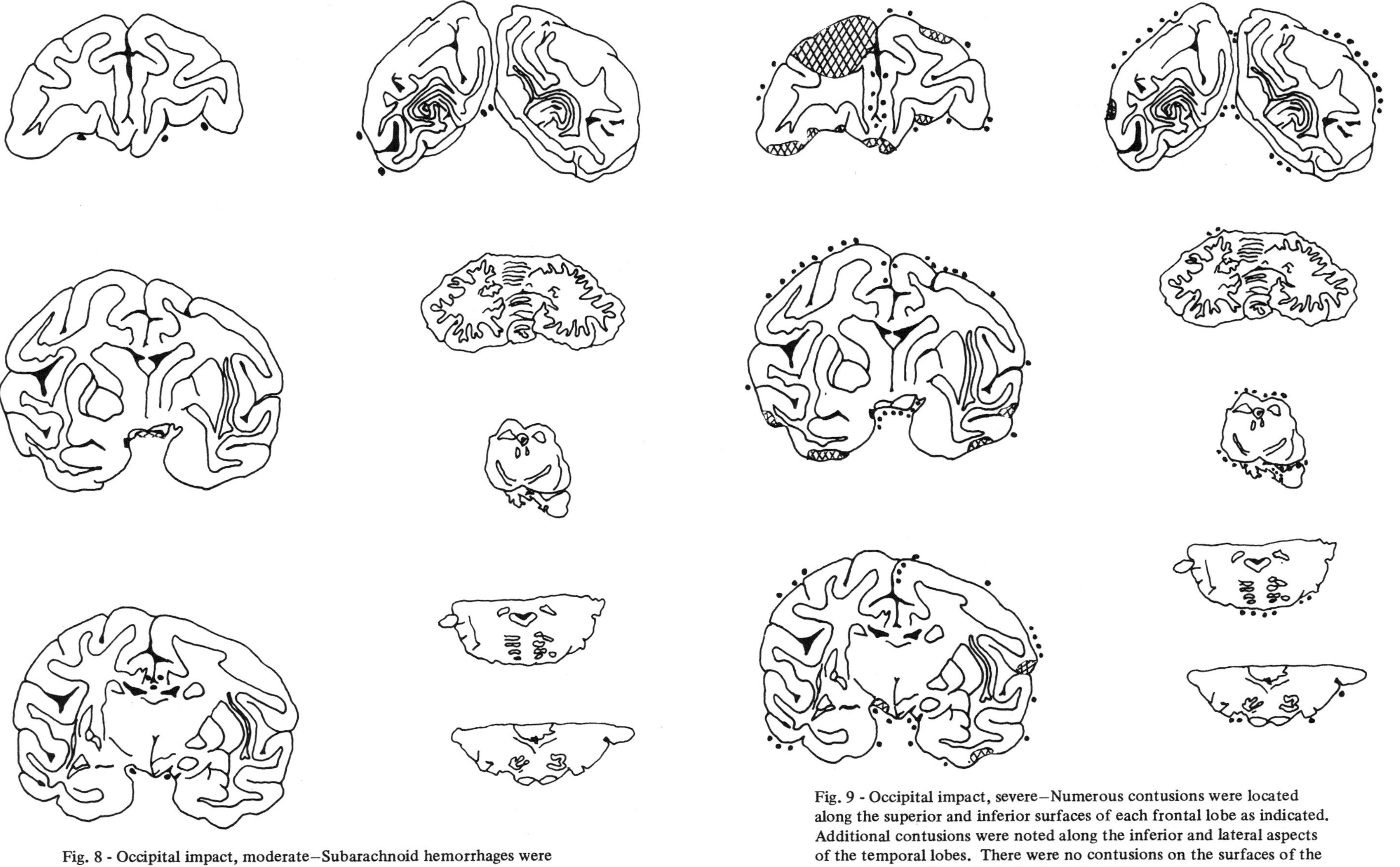

Fig. 8 - Occipital impact, moderate–Subarachnoid hemorrhages were noted as indicated along the inferior frontal lobes and on the surfaces of the cingulate gyrus and around the splenium of the corpus callosum. In addition, occasional contusions were seen along the inferior margin of the optic chiasm.

Fig. 9 - Occipital impact, severe–Numerous contusions were located along the superior and inferior surfaces of each frontal lobe as indicated. Additional contusions were noted along the inferior and lateral aspects of the temporal lobes. There were no contusions on the surfaces of the occipital lobes. Many foci of subarachnoid hemorrhages were seen on all aurfaces of the frontal, parietal, and temporal lobes. In addition, subarachnoid hemorrhages were seen along the brain-stem as indicated by the dots.

Table 2—Temporoparietal Impact Data

Series	Force, peak lbs	Deceleration (Gy)	Physiologic Response	Fracture	Pathology*
H-13	250	435	No changes	No	None
H-14	235	440	Flaccidity, seizures death at 36 hours	No	SAH-brainstem, Left occipital, right hippocampus
H-31	236	350	Transient flaccidity, bradycardia	No	PH-saggital and transverse sinuses; hemorrhage and necrosis-C 3
H-26	388	470	No changes	No	None
H-40	480	430	Transient dilatation of left pupil	Linear	Thrombosis superior sagittal sinus and cortical veins
H-37	515	475	Dilatation of pupils left > right, transient apnea	Suture	SAH-base of brain and brainstem
H-35	576	560	Immediate fixed, dilated pupils, death at 5 minutes	Depressed	SDH-acute left temporal; SAH-bilateral temporal, right occipital, base of brain and brainstem
H-38	585	478	Pupillary dilatation left > right, irregular respirations, death at 18 hours	Suture	SAH-right frontal, parietal, left temporo-occipital, base of brain, and brainstem; PH-grey matter of C3-4
H-46	577	946	Nodal arrhythmia, terminal coma, death at 96 hours	Depressed	EDH-left temporal; contusions-bilateral temporal; hemorrhage and necrosis at C 1-2.
H-45	592	1160	Immediate loss of reaction lower extremities, delayed coma, death at 72 hours	Suture	SAH-right lateral frontal base of brain; PH-grey matter C 3
H-41	608	—	Transient pupillary dilatation, loss of sensation in extremities, bradycardia	Depressed	SAH-small, base of brain; DH-left temporal
H-44	608	—	Transient pupillary dilatation, flaccidity, loss of sensation in extremities, bradycardia	Depressed	SDH-left temporal; SAH-left temporal, base, splenium, right occipital; contusion-left temporal, occipital
H-47	736	903	Immediate apnea, flaccidity, death at 5 minutes	Suture	SAH-splenium of corpus callosum, base of brain
H-48	960	1500	Nodal arrhythmia, immediately euthanized for pathology	Suture	SAH-left frontal, occipital, right parietal, splenium of corpus callosum, base of brain, brainstem; contusion-left temporal; DH-right parietal
H-39	1088	1188	Pupillary dilatation left > right, periodic respirations, delayed opisthotonus, death at 9 hours	Depressed	SAH-bilateral frontal, occipital, temporo-parietal, base of brain, brainstem; contusion-left frontal, right occipital, bilateral temporal

*SAH-subarachnoid hemorrhage SDH-subdural hemorrhage PH-petechial hemorrhage
EDH-extradural hemorrhage DH-dural hemorrhage

defined. Above impacts of 480 lb peak measured force or 430 g peak vertex deceleration, temporoparietal skull fracture occurred. A level for neurologic impairment was less well defined, and increasing neurological deficits were documented with increasing peak force and peak deceleration. In these studies, a peak measured force of 576 lb appeared to be the threshold for fatal injury.

Table 3 summarizes the data obtained on a similar series of occipital head impacts. In this group, the minimum impact force for skull fracture was 550 lb. The degree of physiologic

Table 3—Occipital Impacts

Series	Force, peak lbs	Deceleration (Gx)	Physiologic response	Fracture	Pathology*
H-2	240	640	Transient depressed reaction to stimuli	No	SAH-small, bilateral frontal, left cingulate
H-1	315	—	—	No	SAH-brainstem hemorrhage, softening, C 4-5
H-21	375	—	No changes	No	None
H-6	385	725	Transient pupillary dilatation	No	SAH-corpus callosum, bilateral cingulate, left geniculate; contusion-optic chiasm
H-5	385	985	No changes	No	None
H-56	550	940	Pupillary dilatation, waking period, coma, death at 36 hours	Depressed	SAH-bilateral frontal, occipital; contusion-small, right occipital
H-54	600	1390	Pupils fixed, mid-position, waking period, coma, death at 12 hours	Linear	SAH-bilateral frontal, parietal, temporal, occipital, splenium, brainstem; contusion-bilateral inferior frontal, lateral, and inferior temporal; right oculomotor nerve
H-55	750	1360	Pupillary dilatation, terminal coma, death at 6 hours	Linear	SAH-bilateral frontal, occipital, inferior and lateral temporal, brainstem; contusion-bilateral frontal; left temporal; hematoma-left frontal

*SAH-subarachnoid hemorrhage SDH-subdural hemorrhage PH-petechial hemorrhage
EDH-extradural hemorrhage DH-dural hemorrhage

impairment and the extent of neuropathologic lesions increased with increasing peak force and peak deceleration. Fatal injuries also resulted from occipital impacts with a minimal peak measured force of 550 lb. One notable difference in results obtained from the temporoparietal and occipital impacts was the deceleration and force-time histories. At any given velocity, the deceleration measured at the skull vertex was numerically lower, the rise time slower, and the duration longer for temporoparietal than for lateral impacts. This was believed to be due to two factors: a cushioning effect of the temporalis muscle for lateral impacts, and the elasticity of the thinner temporal bone, compared to the thicker buttressed occipital skull.

THORACIC IMPACT STUDIES - In the thoracic experiments, five monkeys received anterior chest impacts. Of this group, the impact was lethal to two and the others survived two days post-impact, at which time they were euthanatized for pathologic examination.

Measurements of thoracic and cardiac compression were performed in five experiments and these data are summarized in Table 4. The control chest diameter was 82.8 ± 2.5 mm which was reduced 51.9 ± 2.8 mm within the initial 20 ms post-impact and then re-expanded to 64.8 ± 4.4 mm. Three of these survived a 60.2 ± 1.4% strain of the anterior chest diameter, but a 66.2 ± 4.5% strain was lethal to two primates. Significant cardiac lesions were found in four of the five primates and this data also appears in Table 4.

Fig. 10 is a sequence of 15 cineframes selected at 5 ms intervals from a 1000 f/s cineangiographic sequence of a 5 m/s anterior thoracic impact of an anesthetized primate (experiment C 4). Four ml of Angioconray were injected into the left ventricle ½ s prior to injury. Frames A and B demonstrate the pre-impact, lateral radiographic anatomy, showing the anterior thoracic wall, the lateral cardiac silhouette, contrast filled left ventricle, ascending aorta, arch, descending aorta, diaphragm, ribs, and vertebral bodies. Frame C demonstrated initial compression of the thoracic wall with distortion and posterior displacement of the cardiac silhouette. Frame D, at 10 ms post-impact, demonstrates more severe compression of the ventricle and distortion of the aortic arch. Later frames document further compression and transitory alterations of the normal anatomic relationships, with posterior displacement of the thoracic wall against cardiac ventricle and with trapping and occlusion of the descending aorta between the posterior ventricular wall and vertebral bodies. The compression phase lasted approximately 25 ms, which corresponds to the force-time history of the impact. Reexpansion is nearly complete by 65 ms with almost com-

Table 4 — Anterior Thoracic Impacts Data

Experiment	Force, peak lbs	A.P. Initial mm.	Maximum Change mm.	Strain percent %	A.P. final mm.	Pathology
C-3	205	75.5	44.0	58.3	65.1	H*-anterior surface left ventricle, slight
C-4	250	82.4	50.9	61.7	59.9	H-superior anterior mediastinum, severe. Subepicardial H-anterior left ventricular and atrial walls, severe. Septal H-severe. Death at 25 minutes.
C-5	—	81.3	48.3	59.4	74.3	Normal
C-7	300	83.2	58.9	70.8	51.1	H-anterior mediastinum extending around aortic arch and descending aorta. Hemopericardium. Left hemothorax (25cc) Subpicardial H-left anterior atrial wall extending into myocardium. Death at 5 minutes.
C-8	250	91.4	57.6	63.0	73.7	Softening and congestion of anterior left atrial wall; periaortic H-primarily around ascending aorta and arch, without disruption of intima. Marked congestion of both lungs.
Mean	251.2	82.8	51.9	62.6	64.8	
S.E.	±19.4	±2.5	±2.8	±2.2	±4.4	

*H — hemorrhage.

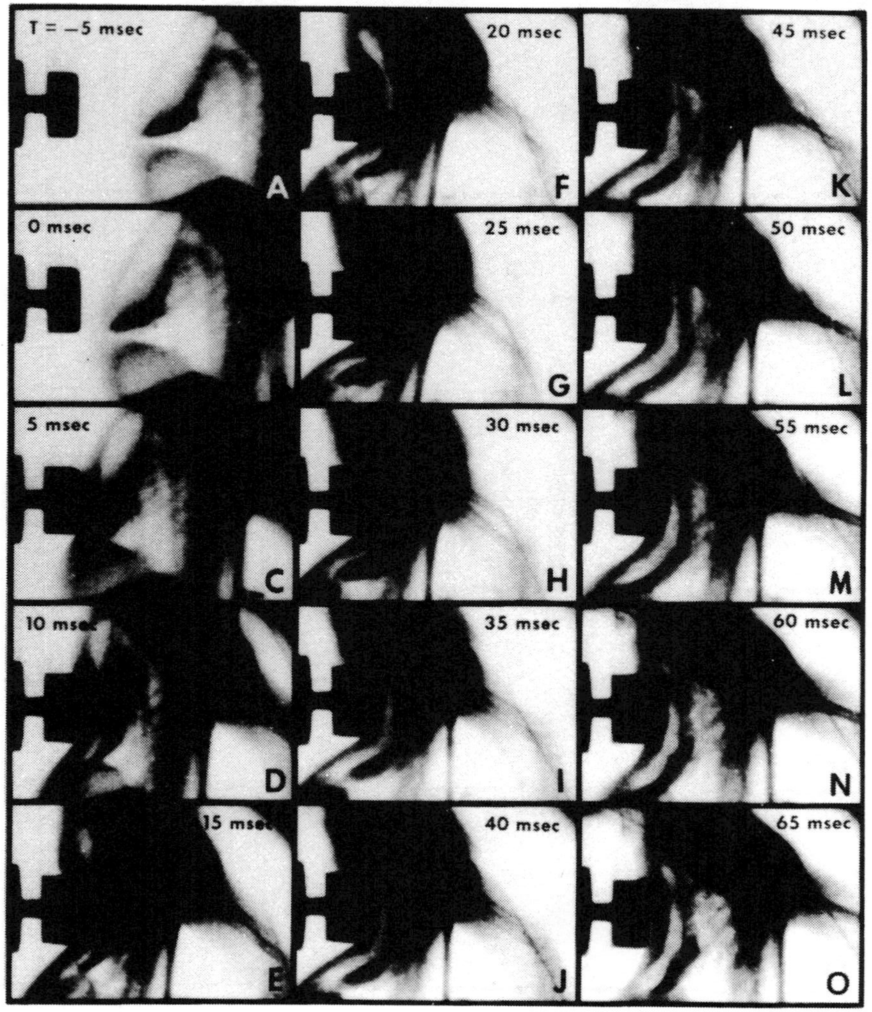

Fig. 10 - Frontal thoracic impact sequence, cinearteriogram

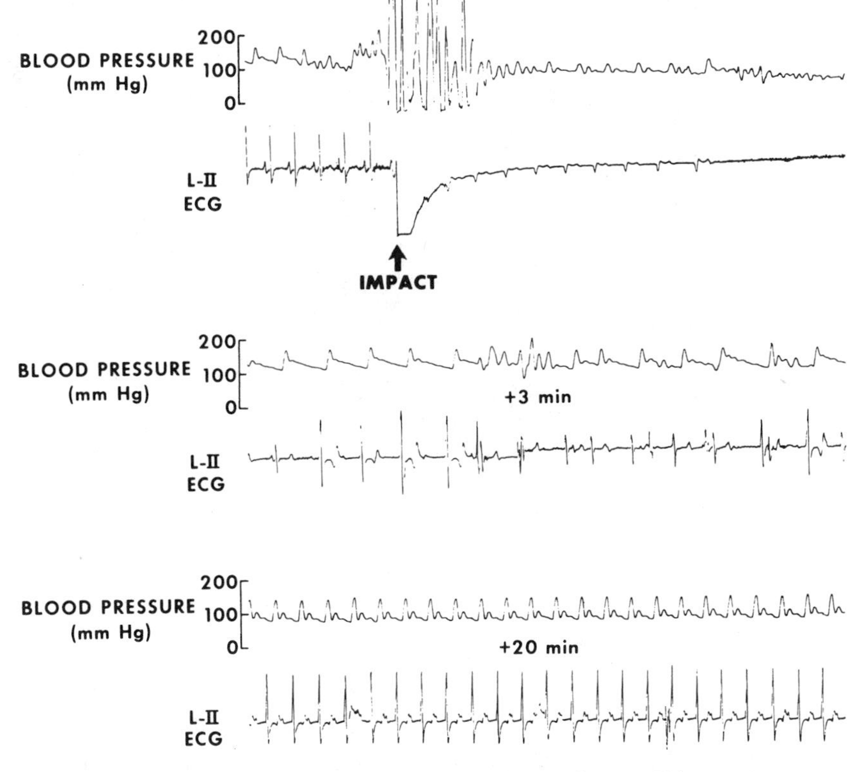

Fig. 11 - Cardiac arrhythmia after frontal thoracic impact–This record is of a primate (C #4) which received a frontal temporoparietal impact. A nodal arrhythmia was observed initially. Three minutes post-impact, P waves returned but were only occasionally transmitted to the ventricles (first complete ECG complex). By 20 minutes post-impact, atrioventricular dissociation had developed.

plete restoration of intrathoracic anatomic relationships, but with sternal fracture. In this animal, a nodal arrhythmia and marked bradycardia were recorded immediately post-impact (Fig. 11). By three minutes post-impact, P waves were recorded but were not synchronous with the QRS complexes. Twenty minutes post-impact, the arrhythmia had progressed to complete atrioventricular dissociation. In this ECG trace, the P wave appears to be superimposed on the preceding T wave.

Fig. 12 illustrates the severe septal hemorrhages found on postmortum examination of this primate.

DISCUSSION

TEMPOROPARIETAL HEAD IMPACT - The present study demonstrates that during closed head injuries, both microscopic and macroscopic pathologic lesions occurred in the subhuman primate brain. These were associated with intracranial movements.

Observation of the intracranial vasculature and ventricles indicated that reproducible movements occurred during and after head impact. Temporoparietal head impact resulted in an initial skull depression and medial displacement of the ipsilateral middle cerebral artery. Both phenomena were transient and these structures recovered towards pre-impact positions. During these first few milliseconds, the midline anterior cerebral artery underwent high frequency, small amplitude oscillation which was followed by displacement of the vessel towards the impact site. This pattern of cerebrovascular displacement after temporoparietal impact led us to conclude that the compressive effects of skull deformation predominate near the impact site resulting in a medial shift of the middle cerebral artery, whereas inertial effects predominate near the head midline resulting in an ipsilateral shift of the anterior cerebral artery. Although the latter lags slightly behind the former, they are sufficiently close temporally to indicate that the left middle cerebral hemisphere's lateral diameter was decreased during this period. The extent of this change can be calculated as strain which was 8.6 ± 1.9%.

In that the brain is a relatively incompressible semiliquid mass with a high bulk modulus, these results led us to conclude that several possible mechanisms may explain this decrease in hemisphere diameter. One possible explanation would be the movement of fluids–blood, CSF–out of the hemisphere; however, the rapidity of this diameter change probably would not allow for sufficient fluid to be displaced. Another more likely explanation is that the hemisphere underwent a concomitant increase in the superior-inferior and/or anterior-posterior hemisphere diameter. The relatively low shear modulus of the brain adds credence to this possible

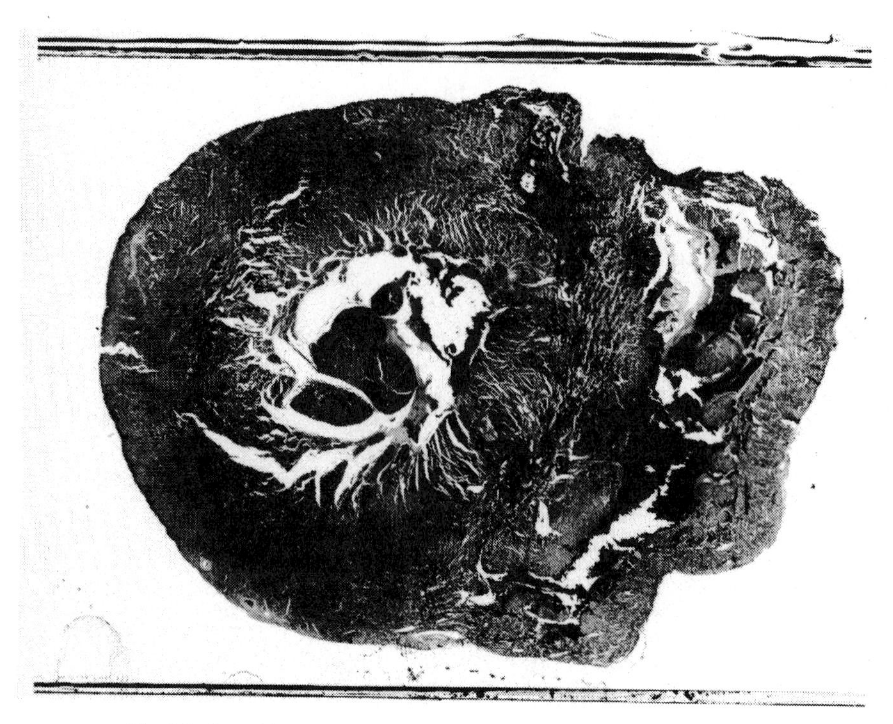

Fig. 12 - Septal hemorrhage in the heart of primate after frontal thoracic impact–The primate in experiment C #4 sustained a severe septal hemorrhage following a frontal impact. The left ventricle is identified by the markedly thicker outer wall. The anterior septal hemorrhage is marked by triple black arrows. There is also more septal hemorrhage posteriorly.

explanation. Our experimental preparation did not permit analysis of changes of the other hemispheric dimensions; therefore, additional studies need to be done to evaluate this hypothesis.

In those primates studied for movements of the intracranial ventricular system after temporoparietal impact, oscillations of the ventricular shadow also occurred during the first few milliseconds post-impact. The extent of these displacements was evaluated qualitatively in these studies, thus preventing any direct comparison between vascular movements. The ventricular shadows appeared to shift initially towards the impact site–similar to the anterior cerebral artery displacement–and then oscillated contralaterally. Conclusions about movement of the ventricles themselves is made difficult because of the possibility that these results may reflect simply a movement of radio-dense media within the ventricles rather than ventricular wall movement. The effect of density differences in ventricular contents on ventricular shadow movements is well demonstrated in Fig. 13 and conclusions based on our present results must be considered with this qualification.

OCCIPITAL HEAD IMPACT - Measurement of intracranial motions after occipital impact were not undertaken because of the lack of suitable head restraint–to minimize lateral head rotation. Qualitative analysis indicated that, at these impact forces, no skull deformation occurred and that the lateral ventricular system was slowly displaced towards the impact site–occiput. Study of vascular displacements during similar occipital impacts is currently underway.

NEUROPATHOLOGY - Very discrete patterns of pathologic changes were recorded following temporoparietal and occipital blunt head impacts. These consisted primarily of localized subarachnoid hemorrhages at low impact levels, with prominent cortical contusions noted following more severe impacts. Following temporoparietal impact, coup lesions were more extensive than contrecoup, although the two were often found to be coexistent. Contrecoup lesions predominated after occipital impacts, and only occasional coup lesions were recorded. Coup lesions were strongly associated with depressed skull fracture. Contrecoup lesions were found to be independent of skull fracture.

The location of contrecoup hemorrhages and contusions was similar but not identical to the clinical data of Courville (2), Freytag (5), Gurdjian (6), Lindenberg (8), Sano (14) and Spatz (18). Ommaya (11), reviewing these authors, found that the anterior frontal and temporal lobes are the most common brain injury sites and that lateral and occipital impacts lead to both coup and contrecoup lesions.

The present experiments reproduced in the monkey brain lesions characteristic of those observed in humans after temporoparietal or occipital impact. In addition, these studies also demonstrated that the lesions were associated with significant deep intracranial movements.

One area of controversy regarding clinical brain injuries has been significance of primary brainstem injuries. Maloney and Whatmore (9) and Crompton (3) have found these to be numerous in fatalities, while Tandon and Kristiansen (20) and others have recorded them rarely. In the present experiments,

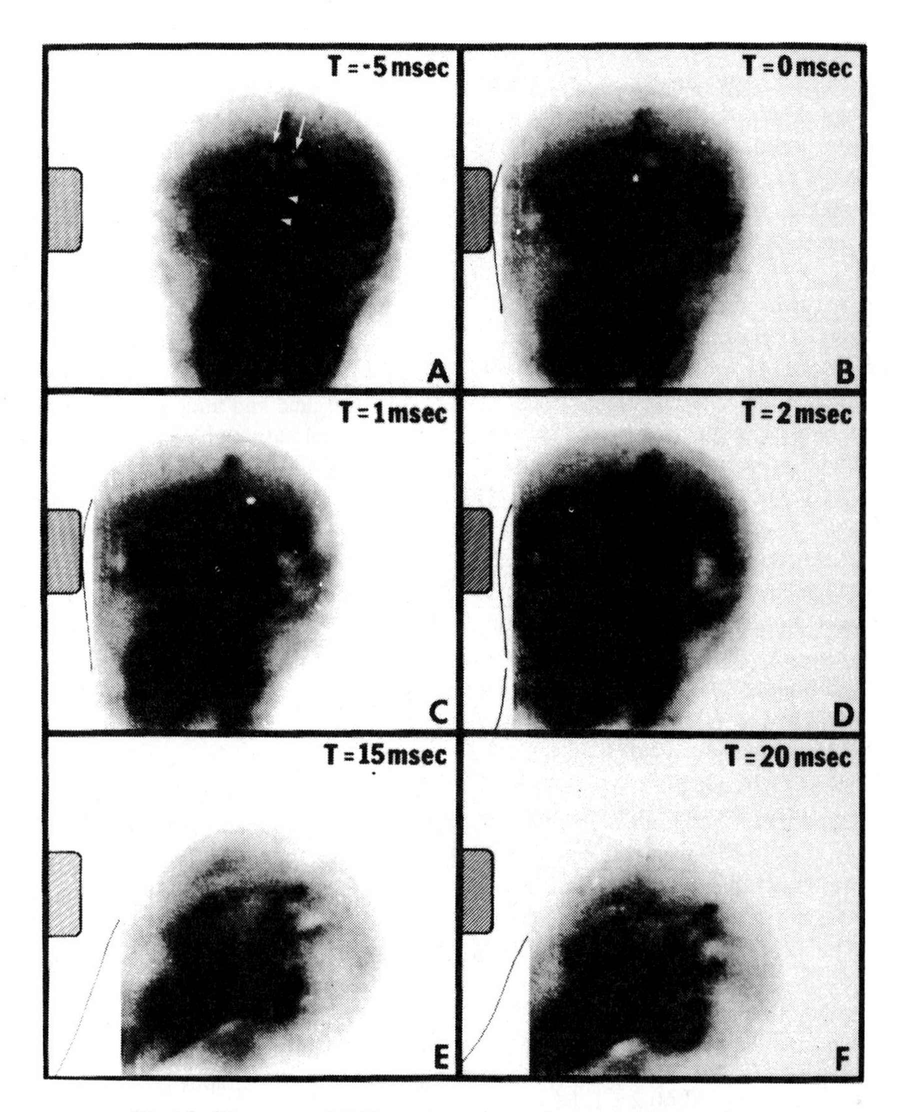

Fig. 13 - Temporoparietal impact sequence, cineventriculogram—An anesthetized primate is traveling from right to left at 4.0 m/s and strikes the temporoparietal region of his head against a fixed impactor in Frame B, and rebounds in succeeding frames. A 1000 f/s frontally viewed air-thorotrast ventriculogram shows air as white bubbles in the bodies of the lateral ventricles (white arrows) and thorotrast as a black shadow in the ventricular frontal and temporal horns, third and dourth ventricles (arrowheads), subarachnoid cisterns, and cervical subarachnoid space.

brainstem injuries occurred occasionally at all impact levels but were most numerous in lethal experiments.

Brain injury mechanisms have been postulated since antiquity and the relevance of diverse short-lived mechanical displacements has long been postulated. However, there has been little physical evidence to verify such movements. The present experimental approach offers a direct way of documenting and quantitating such transient events.

Reversible deformation of the skull during impact is not a new concept. This was proposed by Gurdjian (6), who demonstrated deformation with high-speed cinephotography in human cadaver skulls. The present study, however, is the first demonstration of such deformation in vivo. This illustrates conclusively that indentation of the adult primate skull may occur during impact in the absence of mechanical failure or fracture.

Pudenz and Sheldon (12) demonstrated that mass movements of brain occurred and hypothesized they were etiologic in brain injury. These authors provided the first rigorous documentation of brain movement during trauma by high-speed cinephotography of the brain surface through a lucite calvarium. Their results were later duplicated and extended by Gosch, Gooding and Schneider (4), Ommaya and Hirsch (10), and Gurdjian (6). Pudenz and Sheldon described anterior-posterior movements only and described one artifact of the technique—that draining off cerebrospinal fluid markedly increases the amplitude of such movements. We have observed that different densities of media within the ventricular system affect experimental results. In addition, we have demonstrated that deep movements of cerebral vasculature and ventricular shadows occurred during the first few milliseconds post-impact.

Our observations of transitory movements following impact to the head led to some initial interpretations and point towards areas of further study. First, coup lesions during temporoparietal impacts may result from compression of brain parenchyma due to the indentation of the skull during the first milliseconds. The indentation of the skull physically may compress the brain parenchyma applying the greatest shear forces to the crowns of cortical gyri where coup lesions were characteristically found. In the occipital impact studies, skull deformation was not apparent and coup lesions were not observed. This evidence is also in agreement with the hypothesis of skull deformation being intimately involved in coup lesions. The occipital impact, however, produced significant contrecoup or distal site pathology at the most anterior and inferior surfaces of the frontal and temporal lobes. It is likely that at least some of these lesions resulted from the initial posterior-inferior movement and then rebound of the gelatinous brain over the sharply angulated basal skull surface and the wings of the sphenoid bone. The smoother contour of the internal occipital skull surface would offer much less resistance to such motion, and therefore, produce less shear forces. This theory, that is, that mass movements within the brain result in shearing forces and are etiologic in contrecoup injury, has most recently been maintained by Gurdjian (6).

THORACIC IMPACT - The most striking finding of the thoracic impact studies was the degree of compression that was consistent with relatively unimpaired survival. In the primates from which usable data was obtained, anesthetized animals tolerated thoracic compressions of 60.2 ± 1.4% without lethality. These animals did however sustain significant cardiothoracic pathology and transient bradyarrhythmias. In contrast, an arrhythmia did persist (Fig. 11) in one of the two primates which failed to survive the impact. The finding of a severe septal hemorrhage in this primate suggests that the impact induced injury may have caused the arrhythmia. However, a more definitive conclusion would require further studies to determine the role of specific cardiac lesions in impact induced arrhythmia.

Impact trauma is believed to induce cardiovascular pathology by at least three discrete mechanisms. The first is the direct compression of the heart during impact. This is believed etiologic in myocardial contusions as well as rupture of the heart. The anatomic sequence of the latter is believed to be an initial inward deformation of the sternum and anterior thoracic wall, direct compression, distortion and posterior displacement of the heart with compression of the heart between sternum and the vertebral bodies. Such anatomic sequences were consistently radiographed during anterior chest impacts which produced prominent myocardial contusions, although no radiographic sequence of actual cardic rupture was obtained.

Torsional stresses at the base of the heart have been proposed to cause the aortic ruptures just rostral to the semilunar valve. A second aortic region likely to rupture is the isthmus which lies just distal to the origin of the left subclavian artery (19). As early as 1929, Shennan (17) proposed that this area was particularly prone to rupture and later investigations confirmed the sensitivity of this region (21). Anatomically, this region of the aorta is more constrained by attachments to surrounding tissues–for example, ligamentum arteriosum–compared to the rostral aortic arch and more distal thoracic aorta. In our earlier cineangiograms during anterior chest impact (15), it was demonstrated that the isthmus is selectively stretched and undergoes marked torsion even with diaphragm level blows which caused no direct compression of this region.

The third theory of cardiac injury proposes that increases in pressure within the vascular system during thoracic impact exceed the elastic limit of tissues resulting in rupture. The results of the present studies support the first two mechanisms of cardiothoracic injury, but these studies did not evaluate the last theory.

SUMMARY

High speed cinefluorography was performed on rhesus monkeys subjected to either graded blunt head impacts or blunt thoracic impacts. Marked intracranial and intrathoracic movements were demonstrated at all impact levels.

Temporoparietal impacts demonstrated high frequency reversible indenting of the skull, oscillatory movements of midline and peripheral vascular structures, and of the ventricular shadow.

Occipital impacts demonstrated no inbending of the skull; however, low frequency movements of the ventricular shadow were radiographed.

Frontal thoracic impacts demonstrated anatomic sequences of sternal compression, ventricular distortion and displacement, and trapping of the aorta between the ventricle and vertebral bodies.

ACKNOWLEDGMENTS

The authors sincerely thank Rowena Adamson, Paul Ellis, Donald Godhart and Fritz Miller of the Armed Forces Radiobiology Research Institute (AFRRI); Harold Dixon and his staff of the Walter Reed Medical Audiovisual Department; Francis Chabonnier, Jack Barbour and Lyn Crouser of Hewlitt-Packard; Warren Coon of Varian Enterprises; Vincent Koening and Richard Freeborg of Instrument Marketing Corporation for their excellent technical support; Guy Bateman and Carolyn Armbrustmacher for their meticulous illustrations; and Dr Richard L. Donovan, Charles Gadd and Professor Elisha Gurdjian for their ever helpful suggestions and technical discussions.

REFERENCES

1. W. A. Alter III, S. A. Shatsky, E. D. Evans, R. L. Donovan, and V. W. Armbrustmacher, "Anatomical Displacements in Temporoparietal Impact," AFRRI Scientific Report, in prep.

2. C. E. Courville, "The Mechanism of Coup-Contrecoup Injuries of the Brain," Bull. Los Angeles Neurol. Soc.: 1572-1586, 1950.

3. M. R. Crompton, "Brainstem Lesions Due to Enclosed Head Injury." Lancet: 669-673 April 1971.

4. H. H. Gosch, E. Gooding, and R. Schneider, "Distortion and Displacement of the Brain in Experimental Head Injuries." Surgical Forum 20:425-426, 1969.

5. E. Freytag, "Autopsy Findings in Head Injuries from Blunt Forces." Arch. Path. 75:402-413, 1963.

6. E. S. Gurdjian, "Recent Advances in the Study of the Mechanism of Impact Injury of the Head-A Summary." Clin. Neurosurg. 19:1-42, 1972.

7. J. S. Life, and B. W. Prince, "Response of the Canine Heart to Thoracic Impact During Ventricular Diastole and Systole." J. Biomechanics 1:169-173, 1968.

8. R. Lindenberg, and E. Freytag, "The Mechanisms of Cerebral Contusions." Arch. Path. 69:440-469, 1960.

9. E. J. Maloney, and W. J. Whatmore, "Clinical and Pathological Observations in Fatal Head Injuries." Br. J. Surg. 56, 1:23-31, 1969.

10. A. K. Ommaya, and A. E. Hirsch, "Protection of the Brain from Injury During Impact: Experimental Studies in the Biomechanics of Head Injury. Chapter AGARD Conference, Portugal, 1971.

11. A. K. Ommaya, "Head Injury Mechanisms." Department of Transportation Report No. HS-800:959, 1973.

12. R. H. Pudenz, and C. H. Shelden, "The Lucite Calvarium, A Method for Direct Observation of the Brain:II J. Neurosurg. 3:487-505, 1944.

13. V. L. Roberts, F. R. Jackson, and E. M. Berkas, "Heart Motion Due to Blunt Trauma to the Thorax." Proceedings of Tenth Stapp Car Crash Conference, p-10. New York: Society of Automotive Engineers, Inc., 1967.

14. K. Sano, N. Nakamura, K. Hirakawa, J. Masuzawa, K. Hashizune, T. Hayashi, and S. Fujii, "Mechanisms and Dynamics of Closed Head Injury." Neurologia, Medico-chirrurgica, 9:21-23, 1967.

15. S. A. Shatsky, "Flash x-ray Cinematography During Impact Injury." SAE Paper 730978, Proceedings of the Seventeenth Stapp Car Crash Conference, p-51. New York: Society of Automotive Engineers, Inc., 1973.

16. S. A. Shatsky, and R. Severance, "The AFRRI Flash X-ray Cinematography Systems." AFRRI Technical Note, in prep.

17. T. Shennan, "Traumatic (False) Aneurysms of the Aorta." J. Path. Bact. 32:795-798, 1929.

18. H. Spatz, "Brain Trauma IN: German Aviation Medicine in World War II." Surgeons General's Office, USAF Dept 1950.

19. G. Strassman, "Traumatic Rupture of the Aorta." Amer. Heart J. 33:508-515, 1947.

20. P. N. Tandon, and K. Kristianson, "Clinico-pathological Observations on Brainstem Dysfunction in Craniocerebral Injuries." Proc. IIIrd Intn'l Cong. Neurol. Surg. Copenhagen, 1965.

21. M. A. Zehnder, "Unfallsmechanismus und Unfallmechanik der Aortenruptur in Gesch Lossesen Thoraxtrauma." Thorax-chirurgie 8:47-65, 1960.

791020

Head Impact Response Comparisons of Human Surrogates

Guy S. Nusholtz, John W. Melvin and Nabih M. Alem
University of Michigan

Abstract

The response of the head to impact in the posterior-to-anterior direction was investigated with live anesthetized and post-mortem primates.* The purpose of the project was to relate animal test results to previous head impact tests conducted with cadavers (reported at the 21st Stapp Car Crash Conference (1),** and to study the differences between the living and post-mortem state in terms of mechanical response.

The three-dimensional motion of the head, during and after impact, was derived from experimental measurements and expressed as kinematic quantities in various reference frames. Comparison of kinematic quantities between subjects is normally done by referring the results to a standard anatomical reference frame, or to a predefined laboratory reference frame. This paper uses an additional method for describing the kinematics of head motion through the use of Frenet-Serret frame fields.

The experimental technique used a nine-accelerometer system, mounted rigidly to the head, to measure head motions. Additional measurements included impact force, epidural pressure, and strains in the skull bone. High-speed cineradiography (1000 frames/second) was used during the impact. A total of seven animals were tested in the project, five post-mortem and two live.

The results of the tests are presented to demonstrate the similarities and differences found between animal and human cadaver subjects and between living and post-mortem subjects. The effects of the following factors are discussed:

1. Relative magnitudes of brain mass, skull mass and external soft tissue mass.
2. Head surface geometry at the impact site and mass distribution.
3. Differences in epidural pressures and in head trajectories.

EXPERIMENTAL INVESTIGATIONS IN THE BIOMECHANICS OF HEAD IMPACT RESPONSE have used human cadavers and animals as surrogates of the living human. The parameters commonly used for describing head mechanical response during direct impact have been angular and translational accelerations, velocities, and displacements of the head as a rigid body, skull bone deformations, and internal pressures in the brain. The unembalmed cadaver is often chosen as an experimental model because its geometry and soft tissue distribution is similar to that of the live human. In addition, soft tissue damage can be directly related to injury patterns observed in clinical studies. The disadvantages of the cadaver include the inability to measure pathophysiological response and the susceptibility of some tissues to post-mortem degradation. Also, it has been reported (1) that, during the contact time of direct impact, the motion of the brain of the unembalmed cadaver can be only partially constrained by the skull; the degree of constraint can depend on the time after death and the preparation of the cadaver. This partial decoupling may have marked effects on kinematic time history of the head during and following an impact.

Experimental impact testing of animals, in particular primates, provides basic neurophysiological information related to neuropathology. However, although the primate geometry is the most similar to man's, it is significantly different in anatomic soft tissue distribution and skull morphology. This can present severe prob-

*Animals cared for and handled according to AALAC guidelines.
**Numbers in parentheses indicate reference at the end of the paper.

lems when scaling the test results to human levels. Ultimately these differences lead to complications in the very complex phenomena of head injury (2).

Techniques have been developed in the past few years for accurate determination of three-dimensional motion of the human head (3,4,5,6,7, 8), preparation of the unembalmed human cadaver (1,9,10), and high speed cineradiography of the body (11). This paper discusses techniques used at the Highway Safety Research Institute (HSRI) for conducting posterior-to-anterior head impacts with primates while measuring three-dimensional head motion, skull strain, and epidural pressure. In addition, the results of the series of live and post-mortem posterior-to-anterior impacts with primates are compared to previous posterior-to-anterior direct head impacts with unembalmed human cadavers using the same technique.

METHODOLOGY

THREE-DIMENSIONAL MOTION DETERMINATION - The HSRI method used for measuring the three-dimensional motion of the head is based on a technique used to measure the general motion of a vehicle under a simulated crash (12). In the current application, three triaxial clusters of Endevco Series 2264-2000 accelerometers are affixed to a lightweight rigid magnesium plate (Figure 1) which is then solidly attached to the skull. With this method it is possible to take advantage of the physical and geometrical properties of the test subject as well as the site of impact, in the design of a system for measurement of 3-D motion. In the case of small primates, it is more convenient to design a specific system for each species and site of impact. Two systems were designed and constructed at HSRI, each utilizing a lightweight magnesium plate to mount 9-accelerometers, one for each of the two species tested. The prominent orbital ridges and dental plate found in these species were used to install the rigid plate. Using a multipoint attachment scheme, it is secured at a maximum distance from the point of impact. This minimizes the effect of skull deformation on the 3-D motion analysis during impact. The approximate installed weight of the plate and mounts is 50 grams. Typical distances between accelerometer clusters is about 6.5 cm.

The nine acceleration signals obtained from the three triaxial clusters are used for the computation of head motion using a least squares technique, the details of which are described elsewhere (1). The method takes advantage of the redundancy of nine independent acceleration measurements to minimize the effect of experimental error to produce three angular accelerations and three estimates, in the least square sense, of the true solution.

REFERENCE FRAMES - The impact response of the human body and its surrogates may be described as kinematic quantities derived from experimental measurements and expressed as vectors in reference frames which vary from one instrumentation method to another. In general, comparison of mechanical responses between subjects is achieved by referring results to a "standard" anatomical frame which may be easily identified. On the other hand, it is impractical to require that transducers be aligned with this

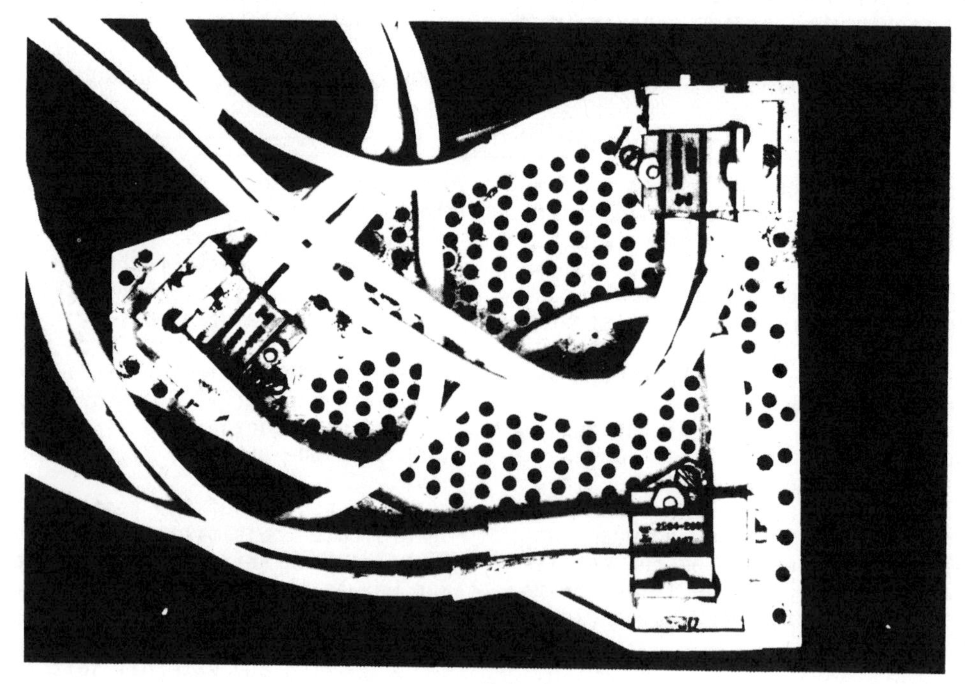

50ths

Fig. 1 - 9-Accelerometer plate

anatomical frame, since this may create unnecessary problems. An alternative is to mount transducers in an arbitrary and convenient reference frame, then describe the transformation necessary to convert the data from this frame to the desired anatomical frame (13).

The three basic reference frames which are used to describe kinematic quantities are: the instrumentation frame, the anatomical frame and the inertial (or laboratory) frame. An additional method to describe motion in space is to utilize the concepts of vector and frame fields. The definitions of the reference frames used at HSRI are given in the following sections.

Anatomical Reference Frame ($\hat{i},\hat{j},\hat{k}$). The $\hat{i}$-axis and the $\hat{j}$-axis of this reference frame lie in the Frankfort plane. The Frankfort plane is defined as passing through the superior edges of the two auditory meati and the two infraorbital notches. The $\hat{i}$-axis is defined along the intersection of the Frankfort and midsagittal planes in the posterior-to-anterior (P-A) direction. The $\hat{j}$-axis is defined along the line joining the two superior edges of the auditory meati, in the right-to-left (R-L) direction. This $\hat{j}$-axis is perpendicular to the midsagittal plane at the "Anatomical Center" (A.C.), which is taken as the origin of the anatomical frame (Figure 2). Finally, the $\hat{k}$-axis is defined as cross-product of the unit vectors of $\hat{i}$- and $\hat{j}$-axes, and therefore, will lie in the midsagittal plane perpendicular to the Frankfort plane, and will be in the inferior-to-superior (I-S) direction.

Thus, the anatomical reference frame ($\hat{i}$, $\hat{j}$, $\hat{k}$) can be completely defined once the four anatomical landmarks are specified.

Instrumentation Reference Frame ($\hat{E}1$, $\hat{E}2$, $\hat{E}3$). This orthogonal frame is embedded in the rigid magnesium plate which carries the 9 accelerometers and is defined by its origin and the plane of ($\hat{E}1$, $\hat{E}2$) which is parallel to the plate (Figure 2). The $\hat{E}3$ axis is defined as the cross-product $\hat{E}1 \times \hat{E}2$. During 3-D radiographic reconstruction, this reference plane is identified by the coordinates of 4 lead pellets permanently installed in the plate to serve as instrumentation landmarks.

The nine accelerometers are arranged in 3 clusters, each forming an orthogonal triad, and are designed to be installed on the plate at precise locations and orientations. Thus, once the instrumentation reference frame ($\hat{E}1$, $\hat{E}2$, $\hat{E}3$) has been determined, the location and direction of all nine acceleration readings may be accurately determined. These readings are immediately transformed to the anatomical reference frame before any 3-D motion computations are carried out.

Laboratory Reference Frame ($\hat{I},\hat{J},\hat{K}$). It is desired to describe the instrumentation reference frame ($\hat{E}1$, $\hat{E}2$, $\hat{E}3$) in terms of the anatomical reference frame ($\hat{i}$, $\hat{j}$, $\hat{k}$) unit vectors:

$$\begin{matrix}\hat{E}1\\ \hat{E}2\\ \hat{E}3\end{matrix} = [E] \begin{matrix}\hat{i}\\ \hat{j}\\ \hat{k}\end{matrix}$$

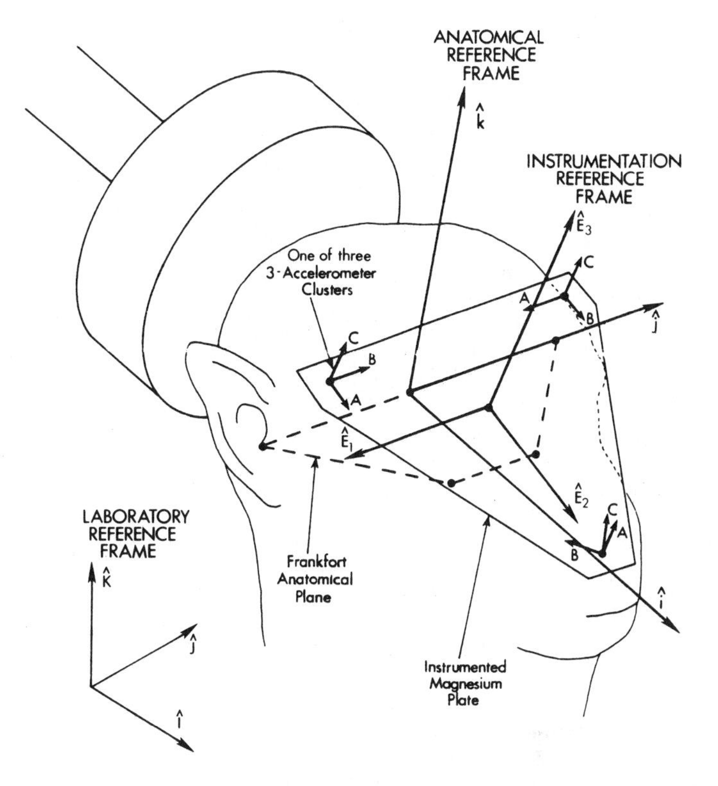

Fig. 2 - Instrumentation reference frame and the three triaxial accelerometer

where [E] is an orthogonal transformation matrix consisting of the nine unknown direction cosines. This matrix may be determined by first expressing each of the ($\hat{E}1$, $\hat{E}2$, $\hat{E}3$) and ($\hat{i}$, $\hat{j}$, $\hat{k}$) in terms of an arbitrary frame ($\hat{I}$, $\hat{J}$, $\hat{K}$):

$$\begin{matrix}\hat{E}1\\ \hat{E}2\\ \hat{E}3\end{matrix} = [U] \begin{matrix}\hat{I}\\ \hat{J}\\ \hat{K}\end{matrix} \quad \text{and} \quad \begin{matrix}\hat{i}\\ \hat{j}\\ \hat{k}\end{matrix} = (V) \begin{matrix}\hat{I}\\ \hat{J}\\ \hat{K}\end{matrix}$$

then eliminating the ($\hat{I}$, $\hat{J}$, $\hat{K}$) between the two expressions to obtain the matrix [E]:

$$[E] = [U] [V]^{-1}$$

Since [U] and [V] are determined from coordinates of several points, the arbitrary frame ($\hat{I}$, $\hat{J}$, $\hat{K}$) will simply be the laboratory frame in which these coordinates are measured. The x-ray method used at HSRI to measure the points automatically defines the laboratory reference frame.

Frame Fields - The description of the motion of the head in response to an impact has always been a central theme in head injury research. With the development of three-dimensional accelerometry techniques for accurate measurement of head motions during impact, it is now possible to produce information on three components of translational acceleration, three components of rotational acceleration and the corresponding components of velocities and displacements. In applying such extensive information to the study of the complex phenomena of head injury it is advantageous to use methods of motion analysis which can simplify and enhance our insight into the events that characterize a particular impact situation. As the head moves through space, any point on the head generates a path in space. In head injury research we are interested in the description of the path of the anatomical center and in events which occur as the head A.C. moves along that path. A very effective method for analyzing the motion of such a point, as it moves along a curved path in space, is to introduce the concept of a frame field. The distance the point travels along the path is a function of time which defines the path and the point of application of any vector on the curve. A vector field is a function which assigns a uniquely defined vector to each point along a path. Similarly, any collection of three mutually orthogonal unit vectors defined on a path is a frame field. Thus, any vector defined on the path (for example, acceleration) may be resolved into three orthogonal components of any well-defined frame field, such as the laboratory or anatomical reference frames. Changes in a frame field with time (for example, angular acceleration of the frame field) are interpreted as vectors defined on the curve and are also resolved into three components.

A frame field which is particularly useful for describing motion along a curved path in space is the Frenet-Serret frame field (discovered by Frenet in 1847, and independently by Serret in 1851). This orthogonal frame is derived from the three-dimensional space curve traveled by the point of interest (the anatomical center, in our case) and the velocity of travel of the point. A complete discussion of this method can be found elsewhere (14) and will only be summarized here.

The curvilinear distance traveled along the path is a function of time and is denoted by s (t) (see Figure 3). For a point moving along the path, the absolute position vector ($\vec{R}$ (t)) can be defined relative to the initial starting position at t = 0. The first and second derivatives of the position vector with respect to time yield the velocity and acceleration vectors of the point, respectively. The time derivative of the path distance, denoted V(t), is called the speed of the curve. At any instant of time, a unit vector ($\hat{T}$) which is tangent to the path at that point on the curve can be defined as the derivative ($d\vec{R}/ds$) of the position vector with respect to the curvilinear distance s. Since the velocity vector of the point ($d\vec{R}/dt$) can be written as $d\vec{R}/ds \cdot ds/dt$, it is possible to substitute the tangent unit vector ($\hat{T}$) for $d\vec{R}/ds$ and the speed of the curve (V(t) for ds/dt resulting in the expression $d\vec{R}/dt = \hat{T} \cdot V(t)$ or $\hat{T} = d\vec{R}/dt/V(t)$. Thus, $\hat{T}$ can be considered a normalized velocity vector. A second unit vector which is perpendicular to $\hat{T}$, called the principal normal unit vector, $\hat{N}$, can be defined to be codirectional with the vector $d\hat{T}/ds$ (since the derivative of a vector is normal to the vector). A third unit vector to complete the orthogonal frame can be defined as the cross product of $\hat{T}$ and $\hat{N}$ and is called the binormal unit vector ($\hat{B} = \hat{T} x \hat{N}$). It can be shown that $\hat{N}$ lies in the plane that contains both the acceleration vector ($\vec{A}$) and the velocity vector

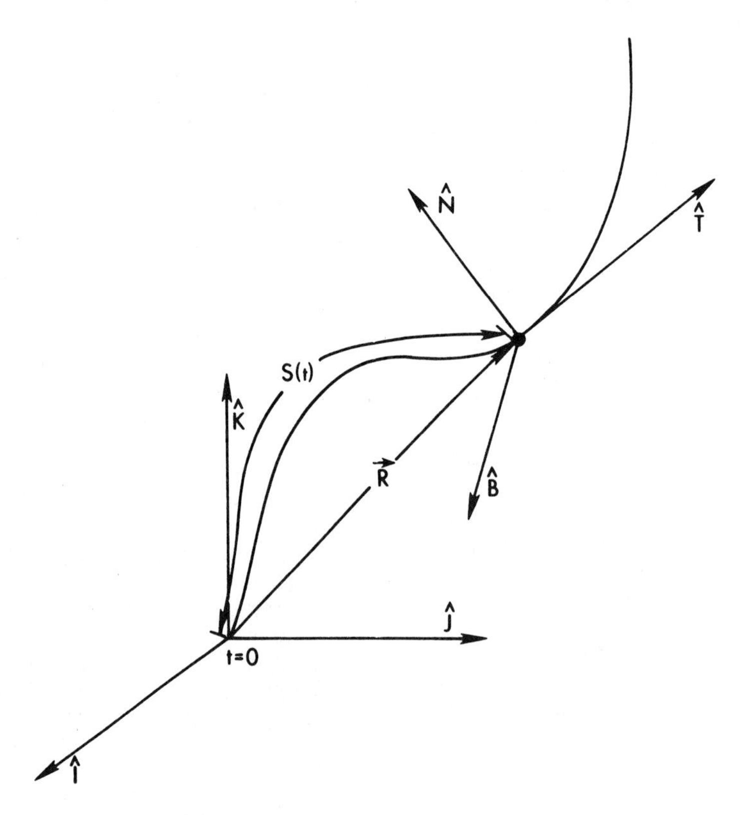

Fig. 3 - Frenet-Serret frame

($\vec{V}$) of the moving point. Since B is perpendicular to both $\hat{T}$ and $\hat{N}$, it is computed as a normalized cross product $\vec{V} \times \vec{A}$ and then $\hat{N}$ is obtained by forming cross product $\hat{B} \times \hat{T}$.

The three orthogonal unit vectors ($\hat{T},\hat{N},\hat{B}$) shown in Figure 3 form a right-handed triad, called the Frenet triad at each point along the space curve. The collection of these triads along a given curve is known as a Frenet-Serret frame field, which is stationary in three-dimensional space. The turning and twisting of a space curve generated by a moving point can be described in terms of curvature, κ, and torsion, τ. Curvature is defined in terms of the Frenet triad as $\kappa\hat{N} = d\hat{T}/ds$, while the torsion is given by $\tau\hat{N} = -d\hat{B}/ds$. The rates of change of ($\hat{T}$, $\hat{N}$, $\hat{B}$) with respect to time may be obtained from the following relations: (T-rate) $d\hat{T}/dt = \kappa V\hat{N}$, $d\hat{N}/dt = -\kappa V\hat{T} + \tau V\hat{B}$, and (B-rate) $d\hat{B}/dt = -\tau V\hat{N}$. Thus, the turning and twisting of a space curve and the rates of turning and twisting are described by the Frenet triad ($\hat{T}$, $\hat{N}$, $\hat{B}$).

In principle, every geometric problem involving motion along a curved path can be solved by means of the Frenet-Serret method. For simple cases, it may be sufficient to record the acceleration data and express it in a convenient form. Examples of such instances are: for zero curvature the motion of a point will be a straight line; for positive curvature ($\kappa>0$) and zero torsion the motion of a point will be in a plane; and for constant positive curvature with zero torsion, the motion of a point will be in a circle. For more complicated motions, which are common even in simple head impact experiments, it is desirable to be able to classify the types of motions in a convenient manner. The Frenet-Serret frame field approach provides such a method.

In many instances, the motion of a point along a path in space can be usefully reparameterized to describe the motion in terms of distance along the path (s) rather than in time (t). Such a procedure can simplify the formulas associated with the description of the curves and can clarify similarities and differences in events that occur as the curve is traversed.

EXPERIMENTAL TECHNIQUES - The installation of the 9-accelerometer plate is accomplished in the following manner. The scalp is removed from the frontal bone over the orbital ridges. Several metal self-tapping screws are attached firmly to the skull through small pilot holes drilled in the orbital ridges and in the dental plate above the eye teeth. Several magnesium feet are attached to the plate. Quick-setting acrylic plastic is molded around each of the screws and the feet are embedded in the plastic. In fifteen minutes when the acrylic has set, the plate is rigidly attached to the skull. The orientation of the plate in this position is shown in Figure 2.

Since this instrumentation frame does not coincide with the standard anatomical reference frame, it is necessary to determine its exact location and orientation in relation to the anatomical frame. A three-dimensional x-ray technique was developed which requires taking two orthogonal radiographs of the instrumented head. The procedure requires the identification of four anatomical landmarks (two superior edges of the auditory meati and two infraorbital notches) with four distinguishable lead pellets, and the identification of four lead pellets, inlaid in the plate to define the instrumentation frame. Lead pellets were located beneath the effective center of gravity of each of the three triaxial accelerometer clusters, and the fourth was placed beneath the effective center of gravity of the three other centroids.

These eight lead pellets are then radiographed twice: once in the x-z plane and once in the y-z plane. On each of the two radiographs, the optical center and the laboratory vertical z-axis are simultaneously x-rayed, the distances between the x-ray film and each target are recorded for each view.

The computations which follow reconstruct the laboratory coordinates (x,y,z) of each of the eight pellets. The anatomical reference frame is reconstructed from the four anatomical points. The instrumentation frame and its origin are determined from the four plate targets. Finally, the transformation matrix between the instrumentation and the anatomical frame is obtained.

The method used for obtaining epidural pressures employs a Kulite model MCP-055-5F catheter tip pressure transducer. A Stryker bone coring tool is used to make a hole with a special 3 mm circular bit. An adjustable set-screwed collet is used which enables the technicians to core into the skull in small increments, preventing damage to the dura mater. The increment of the bone is then removed using a dental scoop, and the resulting hole is tapped for a coarse thread.

A tubular magnesium coupling device is screwed into the tapped hole in the skull. It is anchored into place using quick-setting acrylic plastic molded around the base. A one cm section of rubber tubing is then clamped onto the top of the device and Dow Corning dielectric gel (silicon fluid) is injected into the tubing to act as a coupling medium. The Kulite pressure transducer is then inserted, and secured at proper depth.

Strain gauges are used on the skull bone to record the vibrations of the skull during impact for the purpose of estimating their effect on the accelerometer signals. A Micro-Measurements Type EA-13-015Y-120 strain gauge rosette is bonded with M-bond 200 cyanoacrylate adhesive to the surface of the frontal bone, anterior to the Bregma over the sagittal sinus. After the leads are soldered, the rosette and leads are covered and sealed by a coat of M-coat D paint.

TEST SUBJECT PREPARATION

Seven primate subjects were used in these experiments: five *Macaca mulatta*, one *Macaca assamensis*, and one *Papio cynocephalus*. These were obtained by HSRI from the University of Michigan Unit for Laboratory Animal Medicine

(ULAM). Prior to acquisition, the _Macaca_ subjects had been used in one or more pharmacological research projects and the _Papio cynocephalus_ had been used in a terminal physiology experiment.

The impacts of the _Papio_ and the first four _Macacas_ were conducted on post-mortem subjects. Upon termination, they were stored in a cooler at 4°C for 48 hours before testing. Living _Macacas_ were used in the two final experiments. The protocol for post-mortem primates was less complex than for that of the live primates, which is outlined below. Notes, data , and changes in in procedures were recorded on a chronological checklist.

On the morning of the experiment the primate is given an intramuscular injection of ketamine (dL-2-[O-chlorophenyl]-2-[methylamino] cyclohexanone hydrochloride) before being delivered to the HSRI Biomedical Laboratory by a ULAM technician. A catheter with a three-way valve was inserted into the _saphena parva_ vein in the hind leg, and sodium pentobarbital injected through the valve at a dosage of 15 mg/kg, to effect. An airway is established. The upper body is prepared and the weight and biometrical measurements are taken with a standard anthropometer, a stainless steel tape, a ruler and a Homes Model 51HH beam scale. Body measurements are illustrated in Figure 4; head measurements in Figure 5. Using a cauterizing scalpel, the scalp and muscle mass are removed from the frontal bone and the screws used to moor the nine-accelerometer plate and the epidural pressure transducer skull fitting are screwed into place. The frontal bone is sanded with 200, then 400 weight wet/dry sandpaper and the strain gauge rosette is

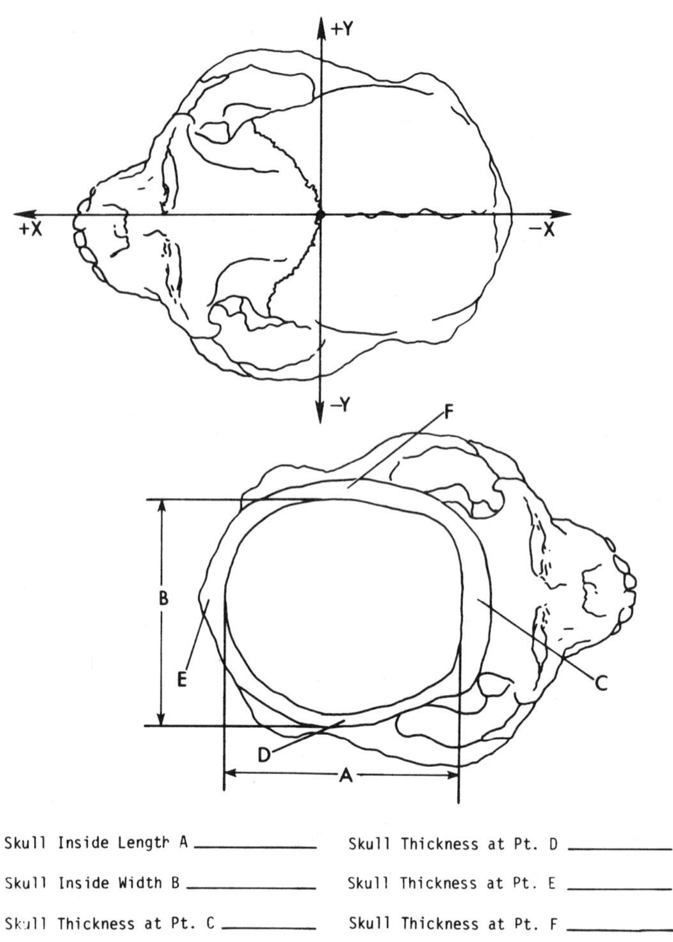

Fig. 5 - Identification of head measurements

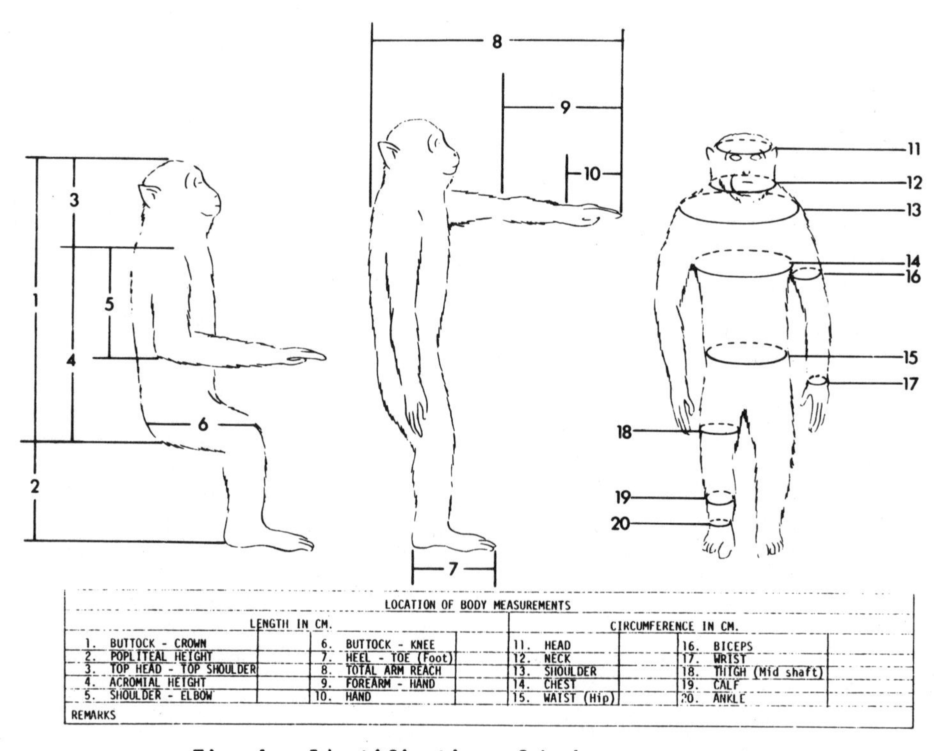

LOCATION OF BODY MEASUREMENTS

LENGTH IN CM.				CIRCUMFERENCE IN CM.			
1. BUTTOCK - CROWN		6. BUTTOCK - KNEE		11. HEAD		16. BICEPS	
2. POPLITEAL HEIGHT		7. HEEL - TOE (Foot)		12. NECK		17. WRIST	
3. TOP HEAD - TOP SHOULDER		8. TOTAL ARM REACH		13. SHOULDER		18. THIGH (Mid shaft)	
4. ACROMIAL HEIGHT		9. FOREARM - HAND		14. CHEST		19. CALF	
5. SHOULDER - ELBOW		10. HAND		15. WAIST (Hip)		20. ANKLE	
REMARKS							

Fig. 4 - Identification of body measurements

then cemented into position. The strain gauge wires are soldered to the leads and the assembly covered with sealant. Quick-setting acrylic is molded around the pressure transducer fitting and nine-accelerometer moorings. When the acrylic is set, the strain gauge wires are strain-relieved by being secured to the nine-accelerometer plate with nylon wire wrap. Figures 2 and 6 show the positioning of the instrumentation on the skull.

Next, eye and ear x-ray targets are positioned. The primate is transported to the x-ray room where two head x-rays (x-z and y-z views) are taken. The distance between each of the eight targets and the surface of the x-ray table is measured and recorded.

The primate is then taken to the impact laboratory and placed in the impact chair. Three triaxial accelerometer clusters are fastened to the nine-accelerometer plate. Silicon fluid is injected into the pressure coupler, thus removing all air, and the pressure transducer is inserted. The primate is positioned in front of the impactor and stabilized with paper tape. All of the transducer wires are then connected and cabled, and the transducers checked for continuity and function.

A Polaroid photograph is then taken through the cineradiograph to check the position of the primate and the x-ray settings. Final adjustments are made on the x-ray settings, amplifier settings, and the position of the primate. The x-ray settings, the distance from the mid-line of the primate to the cineradiograph screen are recorded. At this point setup photographs are taken. Holes are then punched into the paper tape supporting the primate. The Polaroid camera is exchanged for the Photosonics high-speed motion picture camera in the cineradiographic system. One hour after impact, a 5 ml dose of Uthol (concentrated, unpure sodium pentobarbital) is injected via the hind leg I.V. catheter to euthanize the primate.

IMPACT TEST CONDITIONS - The impacts were conducted using the HSRI pneumatic impacting device, which was specifically constructed to give impacts of reproducible velocity. This instrument is comprised of a 0.43 cubic meter air reservoir attached to 167 cm long by 10.2 cm diameter honed steel cylinder with two carefully fitted steel pistons.

Compressed air from the building's air compression system is introduced into the reservoir. The pressure is regulated by a series of hand valves and measured with a gauge having an accuracy of 0.25%. The driver piston is secured at the reservoir end of the cylinder by an electronically controlled locking mechanism. At the opposite end of the cylinder the striker position is rigidly connected to the impact force head. Both steel pistons have interchangeable elastic bumpers. When the air reservoir is pressurized and the locking mechanism released, the driver piston is propelled by the compressed air through the cylinder until its bumper impacts the bumper on the striker piston. Momentum is transferred to the striker piston, which is allowed to travel up to 25.4 cm. The excess kinetic energy is absorbed by a 3003 H14 seamless aluminum inversion tube, 6.35 cm in diameter with a 0.165 cm wall thickness.

Impactor velocity is controlled by reservoir pressure and the ratio of the masses of the driver and striker pistons. The desired impactor stroke can be accurately controlled by the initial positioning of the striker piston with respect to the inversion tube. Both driver and striker pistons have a mass of 10 Kg for these experiments.

The impactor surface is a 10.2 cm rigid metal plate padded with 2.5 cm Ensolite. The impactor force transducer assembly consists of a Kistler 904A piezoelectric load washer with a Kistler 804A piezoelectric accelerometer mounted internally for inertial compensation.

A chair was designed and constructed for positioning of the primates which could be adapted to the varying primates and test conditions. In addition, a moveable overhead arm facilitated suspension of the primate.

The primate is placed in an erect sitting position with its posterior side towards the impactor, so that the line of impact is in the mid-sagittal plane in the posterior-anterior direction. The primate is held in place with paper tape which grips the body under the armpits, suspending the head and torso from the overhead arm. Immediately before testing, holes are punched into the tape to facilitate release upon impact.

The test subjects were impacted at the occiput. In tests 78A232 and 78A234 the eyes were on a horizontal line through the impact site, thus the estimated center of mass was below the line of impact. In tests 78A236 and 78A239 the eyes were raised to a point slightly above a horizontal line through the impact site, thus the line of impact was through the estimated center of mass. In tests 78A238 and 78A241 the eyes were raised higher than the previous two tests, thus the center of mass was above the line of impact (Figure 6).

CINERADIOGRAPHS - Cineradiographs were taken of the impact events at 1000 frames per second. The HSRI high-speed cineradiograph system (11) consists of a Photosonics 1B high-speed 16-mm motion-picture camera which views a 2-inch diameter output phosphor of a high-gain, four-stage, magnetically focused image intensifier tube, gated on and off synchronously with shutter pulses from the motion-picture camera. A lens optically couples the input photocathode of the image intensifier tube to x-ray images produced on a fluorescent screen by a smoothed direct-current x-ray generator. Smoothing of the full-wave rectified x-ray output is accomplished by placing a pair of high-voltage capacitors in parallel with the x-ray tube. A 22 cm diameter circular field was viewed in these experiments.

DATA HANDLING - The impact force and acceleration, epidural pressure, and nine head accelerations are recorded unfiltered on a Honeywell 7600 FM Tape Recorder. The impact force, velocity, and strain are also recorded unfiltered on a Bell and Howell CEC 3300 FM Tape Recorder. A

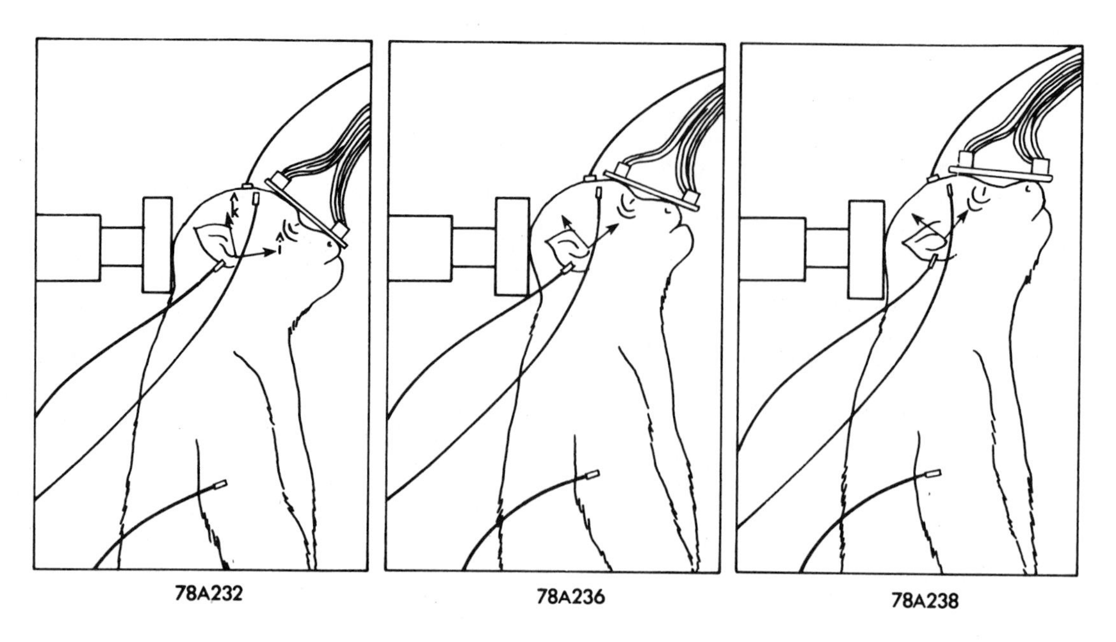

Fig. 6 - Initial conditions of impact test

synchronyzing gate is recorded on both tapes. All data is recorded at 30 ips.

The analog data on the FM tapes is played back for digitizing through proper anti-aliasing analog filters. The A-to-D process for all data, results in a digital signal sampled at a 6400 Hz equivalent sampling rate. Following this step, the power spectrum of the signal is obtained and plotted along with the time history to determine the relevance of the data before being used in the three-dimensional motion analysis computer program.

RESULTS - Full documentation of the experimental results for the primate head impacts discussed in this paper is given in reference 15. The tables and graphs presented here represent the data found to be most pertinent in discussing the test results. A summary of the test conditions are given in Table 1, the primate biometric measurements are given in Table 2, and the time histories are given in the Appendix.

Data obtained from the three-dimensional head motion analysis consisted of translational and angular displacements, velocities, and accelerations resolved into the laboratory, anatomical and Frenet-Serret reference frames. Before impact, the angular velocities of the Frenet-Serret frame are not defined because the path of motion has not been established. However, position, defined as the displacement from the initial condition of the anatomical center, must necessarily be resolved into the laboratory frame.

In order to define the pulse duration, a standard procedure was adopted which determines the beginning and end of the pulse. The procedure is to determine first the peak and the

Table 1 - Primate Head Impact Initial Test Conditions

Test No.	Subject Condition	Subject Positioning C.G. w/r Occipital Impact	Impact Surface	Pressure Transducer Location Relative To Bregma (cm) x	y
78A229	Post-mortem	At	2.5 cm Ensolite		
78A232	Post-mortem	Below	2.5 cm Ensolite	+3.3	+1.3
78A234	Post-mortem	Below	2.5 cm Ensolite	+3.2	-1.1
78A236	Post-mortem	At	2.5 cm Ensolite	+2.8	-0.8
78A238	Post-mortem	Above	2.5 cm Ensolite	+3.0	+0.6
78A239	Live	At	2.5 cm Ensolite	+2.9	-0.8
78A241	Live	Above	2.5 cm Ensolite	+3.6	+0.8

Table 2 - Primate Biometric Measurements

Test No.	Species	Height (Sitting) (cm)	Mass (kg)	Max. Skull Length (cm)	Max. Skull Breadth (cm)	Ave. Skull Thickness (cm)	Head Muscle Mass(g)	Head Mass (g)	Brain Mass (g)	Brain Volume (ml)
78A232	Macaca assamensis	49.1	6.9	6.4	5.1	0.47	125	595	70	
78A234	Macaca mulatta	53.6	9.5	7.4	5.6	0.28	135	625	105	95
78A236	Macaca mulatta	53.4	8.2	6.6	5.6	0.24	140	510	100	90
78A238	Macaca mulatta	53.9	8.2	6.9	5.6	0.23	130	540	115	105
78A239	Macaca mulatta	48.8	8.2	6.4	5.3	0.22	80	510	85	75
78A241	Macaca mulatta	47.0	10.6	8.4	6.4	0.43	190	750	125	115
78A229	Papio cynocephalus		23.0				825	1020	155	

time at which it occurs. Next, the left half of the pulse, defined from the point where the pulse starts to rise to the time of peak, is least-squares fitted with a straight line. This rise line intersects the time-axis at a point which is taken as the formal beginning of the pulse. A similar procedure is followed for the right half of the pulse, i.e., a least squares straight line is fitted to the fall section of the pulse which is defined from the peak to the point where the first pulse minimum occurs. The formal end of the pulse is defined then as the point where the fall line intersects the time axis.

A summary of the test results is given in Table 3 which includes velocity, peak force, force duration, and the greatest positive and negative epidural pressures. The velocity was obtained from a magnetic pickup that produces time pulses at 1.3 cm intervals on the impacting device. The force duration was determined as described above. The peak positive and negative epidural pressures were obtained by measurement from the zero pressure point on the transducer time histories. Two similar cadaver tests (1) are also listed.

DISCUSSION

The work presented in this paper is part of a continuing program on head injury research. A great deal of data has been generated and a complete presentation of the data is beyond the scope of this paper. The results discussed in this section are based on analysis of the data presented in abbreviated form in the Appendix to this paper. Because of the complexity of the experiments, only a limited number of subjects have been tested. Thus, the features of the

Table 3 - Impact Test Summary

Test No.	Description	Velocity (m/s)	Force (N)	Force Duration (ms)	Pressure (mm Hg)
78A229	Post-mortem Occipital Impact	14.0	11500	6 msec	-
78A232	Post-mortem Occipital Impact	11.5	5800	3 msec	-
78A234	Post-mortem Occipital Impact	13.5	5700	4 msec	-
78A236	Post-mortem Occipital Impact	12.5	5600	4 msec	-721/+420
78A238	Post-mortem Occipital Impact	12.5	6600	3 msec	+1041/-410
78A239	Live Occipital Impact	12.5	5000	5 msec	+1977
78A241	Live Occipital Impact	12.5	8700	3 msec	+3682/-710
76A135	Post-mortem (cadaver) Occipital Impact	7.0	5000	13 msec	-
76A137	Post-mortem (cadaver) Occipital Impact	8.0	5500	8 msec	-

Table 4 - Reparameterization of Primate and Cadaver Test Data

PRIMATES

Test No.	Arc Length (mm) Q1	Q2	Q3	Time (ms) Q1	Q2	Q3
78A229	0	6.5	15	0	3.0	4.7
78A232	0	6.7	58	0	2.5	7.5
78A234	0	7.3	50	0	3.0	6.5
78A236	0	7.3	50	0	2.5	7.0
78A238	0	6.8	90*	0	3.0	10.5*
78A239	0	8.0	55	0	3.0	7.0
78A241	0	7.2	56	0	2.5	8.5

*ill-defined due to instrumentation difficulties.

CADAVERS

Test No.	Arc Length (mm) Q1	Q2	Q3	Time (ms) Q1	Q2	Q3
76A135	0	8.3	56	0	5.0	13.1
76A137	0	8.7	60	0	5.0	8.4

data noted in this section represent trends that are felt to be important factors in head impact response. Continued work is necessary to be able to generalize the findings.

HEAD IMPACT RESPONSE DEFINITION - Head impact response may be defined as a continuum of "events" characterized by the path, and all the vectors defined on that path, which are generated by the motion of the anatomical center and by changes of the attached frame fields. Physically this implies that head impact response is interpreted as the response of a material body (the skull) which is in contact with other material bodies (the neck, impactor, soft tissue and brain). The curve and the vectors generated as the anatomical center moves in time are therefore a result of the interactions of the skull with these other material bodies.

Generally, an event that happens in one specific impact will not occur in all other impacts. However, in specific impacts, certain events will always occur. Examples of these events which are important in head impact are: the initiation of head impact response (denoted by Q_1 on the tangential acceleration-time histories in the accompanying data); the positive maximum of the tangential acceleration time history (denoted Q_2 in the accompanying data); and the negative maximum of the tangential acceleration time history (denoted by Q_3 on the accompanying data). These events can then be used to define different types of impacts and to compare the response of one type of surrogate to another.

The motion of a rigid body in space is the result of generalized forces: the total force, and the total torque about a suitable axis. The dynamic problem of the motion of the skull can be interpreted in the same way. However, because of the complex interactions of the skull with the other material bodies, serious problems can arise in determining which of the bodies is producing these generalized forces.

EFFECT OF SKULL DEFORMATIONS ON RIGID BODY MOTION ANALYSIS - The condition of small skull deformations in head impacts is generally not of interest from an injury standpoint. However, in the circumstances of rigid body motion, intractable complications can result if precautions are not taken to minimize the effect of skull deformations. The choice of location of the accelerometers is critical in this respect for primate impacts. The orbital ridges proved to be relatively stable during impact, as the strain gage rosettes placed near the point of attachment of the 9 acceleration plates showed that minimal strains ($<200\mu\varepsilon$) occurred in this area. In addition, a filter, applied to the raw data, can be used to minimize the effects of high frequency vibrations. The choice of filter is based on the power spectrum of the raw data. The power spectrum revealed that the highest frequency content, excluding those components attributed to noise, were in the pressures, where the power spectrum dropped effectively to zero at 600 Hz. The reconstructed time histories, filtered at 600 Hz, show no significant difference from the raw time histories except for the elimination of high-frequency noise.

EFFECT OF SKULL GEOMETRY - The time history of the anatomical center during the time interval

(Q_1-Q_2) is considered to be primarily a result of the interaction of the impactor with the skull. During this period, the angular acceleration found to be principally in the binormal direction (although there were lesser components in the normal and tangential directions) in tests 78A232, 78A234, 78A238 and 78A239. This implies that the skull may be rotating about a point of closest approach of the skull to the impactor (soft tissue prevents actual contact of the skull and impactor surface). Under this condition the rate of change of the tangent unit vector (T-rate) and the translational acceleration in the normal direction will be a result of the torque produced by this rotation. Calculations of the translational acceleration in the normal direction at this point show significant reductions of this acceleration in comparison to that of A.C., confirming that this type of rotational motion is taking place.

The implication of this result relates directly to skull geometry. Figure 7 shows reconstruction of the x-rays from test 76A135, an unembalmed human cadaver, test 78A229 and test 78A239. Examination of this figure shows that the shape of the human skull in the posterior section is considerably different from the other primates. This difference has a direct effect on the rotational stability of the skull during impact.

Figure 6 is a reconstruction of the test set-up for three different types of impacts: Test 78A232 is arranged in a manner similar to the unembalmed human cadaver, the eyes being in frontal direction; Test 78A236 is positioned in an attempt to minimize rotation about the right-left axis during impact; and Test 78A238 is positioned in order to produce such a rotation. These types of set-ups take advantage of the skull geometry of the primate and use the location of the center of gravity relative to the applied impact force axis to adjust the direction of applied torque. This phenomenon is apparent in the evaluation of the angular acceleration for Tests 78A232, 78A236 and 78A238 during the $(Q-Q_2)$ interval. In test 78A232, the angular acceleration is about the right-left axis, while in Test 78A236, during the (Q_1-Q_2) time interval, the angular acceleration is in the opposite direction. Similarly, the angular acceleration in test 78A238 is like that of 78A236, but greater in magnitude.

The top portion of Figure 8 represents the general directions of the tangent and normal vectors at an instant after the Q_2 event. The direction of the binormal vector is determined by the right plane represented in the illustration. Care must be taken in the interpretation of this figure since motion during the (Q_1-Q_2) interval is usually three-dimensional and will not be restricted to a single plane the figure is presented to convey, schematically, the general idea of the differences in head motion produced by the different initial conditions. The three types of tests are distinguished by the Frenet frame and angular acceleration defined on the path during the (Q_1-Q_2) interval. The significant feature is that dynamically similar head motions can be produced in cadaver and lower primate impact tests, but the lower primates require different initial conditions than the cadaver due to skull geometry differences.

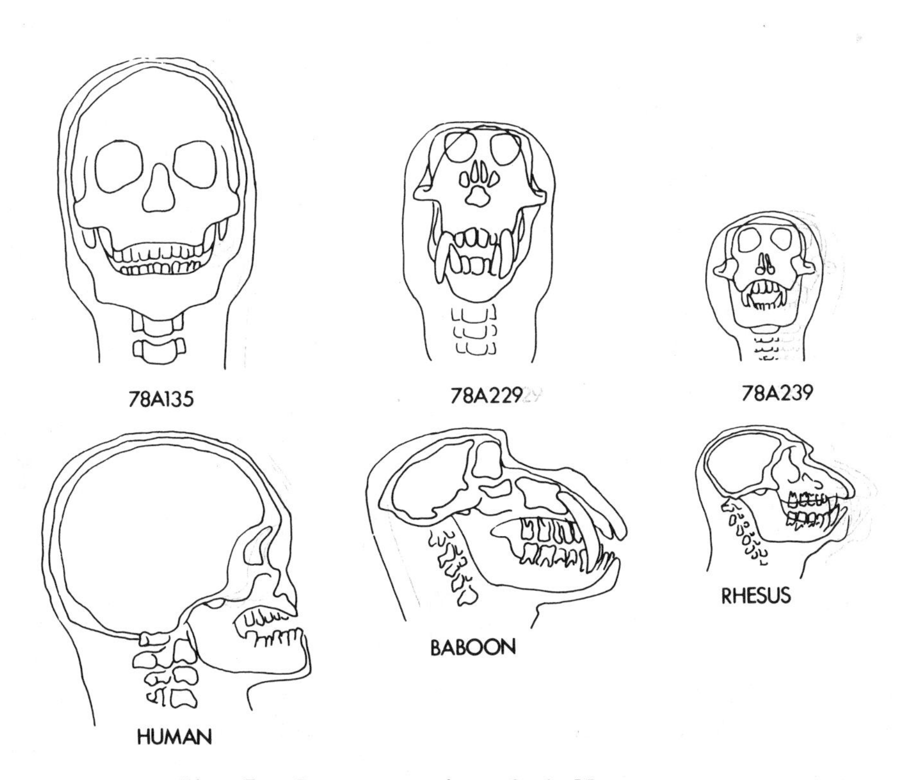

Fig. 7 - Reconstruction of skull x-rays

EFFECT OF SOFT TISSUE - In general, as the impact is increasing, the response, when unopposed by additional generalized forces, will be increasing. Conversely, when the force on the head is decreasing the response should decrease. During the time interval (Q_2-Q_3) this was generally found not to be the case. In this interval, it can be assumed that the impactor no longer dominates the skull's response and that other material bodies contribute input to the path of the A.C. and its vectors. In addition, the events during the interval (Q_1-Q_2) can affect those during (Q_2-Q_3). This conclusion is based on the following findings. For all the tests, the tangential acceleration decreases more rapidly than the force, from the Q_2 event until the time it crosses the zero axis. In tests 76A135 (cadaver), 78A236, 78A238, 78A239 and 78A241 (all Macaca) the normal acceleration and angular acceleration remain significant after the impact force drops to zero and the binormal vector rotates π radians during the (Q_2-Q_3) interval.

Examination of all the tests reported in the Appendix shows that during the (Q_1-Q_2) interval, unless there are rapid changes in the binormal vector direction (large B-rates), the normal acceleration and the binormal vector direction are induced by the angular acceleration of the head. For tests 76A135, 78A238, 78A239 and 78A241 the angular acceleration vector reverses direction around the Q_2 event time with the binormal vector rotating π radians shortly afterwards. This indicates that, in those tests, there are torques being generated that are sufficiently large that they dominate the other factors (such as the impact force) involved in the establishment of the binormal vector direction and the normal acceleration. In addition, the angular acceleration vector is of sufficient duration in its new direction eventually to reestablish it in the binormal vector direction. The rotation of the binormal vector can be very rapid, indicating an inflection point in the path of the A.C. in space. Test 78A239 (see Appendix) shows a large B-rate with the binormal vector rotating π radians almost instantaneously. The fact that in tests 78A232 and 78A234, during the (Q_2-Q_3) interval, there is relatively little rotation of the binormal vector and that the angular acceleration is greatly reduced after the force drops to zero indicates that the torque induced by the impactor is codirectional of much greater than the torque generated by other forces. Therefore the normal and binormal vector directions are basically the same for all tests at the end of the (Q_2-Q_3) interval, as represented for the Macaca subjects by the bottom section of Figure 8.

The fact that torques which result at the beginning and end of the (Q_2-Q_3) interval are opposite in direction is best understood by examining the resultant angular velocity comparison of tests 78A236 and 78A239 to test 78A232 reveals the basic differences. In tests 78A236 and 78A239 the angular velocity reaches a maximum close to the Q_2 time, with the minimum occurring around Q_3. Shortly following Q_3, there is a rapid rise in the angular velocity with rotation in the opposite direction. A similar behavior

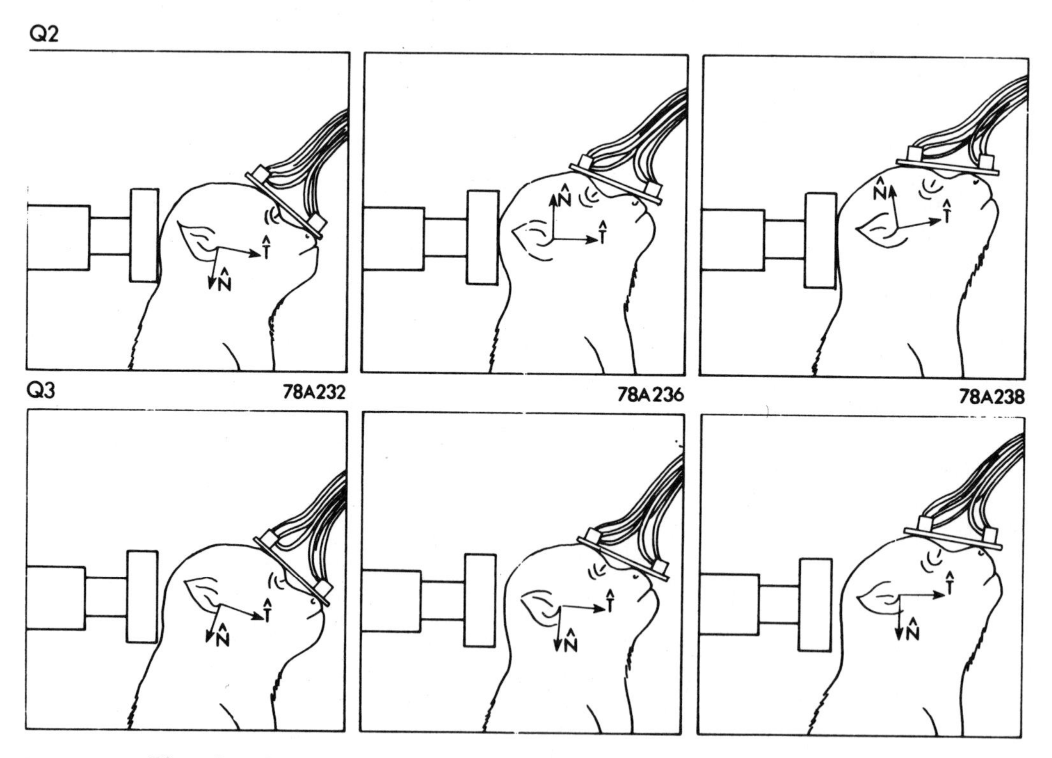

Fig. 8 - Reconstruction of motion from x-ray, cineradiographs, and kinematic time histories for the HSRI 3-D motion measurements

can occur in cadavers (test 76A135) but the first peak is generally significantly smaller than the second. This may be attributed to the differences in skull geometries and soft tissue distributions.

In the unembalmed cadaver, it has been determined (1) that the source of the generalized forces during (Q_2-Q_3) is the soft tissue (i.e., the brain). It is, however, more difficult in primates to separate the effects of the brain from that of the soft tissue in the head and neck. In Figure 7, the x-ray reconstruction compares the soft tissue distribution of the various human surrogates. The brain is the primary source of soft tissue in the head of the cadaver while in the primate, it is the external head and neck masses. The effect of muscle mass can become so great that, during *Papio* Test 78A229, more momentum is transferred to the soft tissue than to the skull. Examination of the data in the Appendix for that test shows that a large positive tangential acceleration occurs after impact, indicating that momentum is being transferred to the skull from an outside source. The amount transferred is, in fact, so great that the maximum velocity is not reached until well after impact. This is consistent with the measured masses of the soft tissue and skull of this subject, given in Table 2.

REPARAMETERIZATION OF EVENT OCCURENCE FROM TIME TO ARC LENGTH - Examination of the primate test data reveals that many of the time histories of the motions are similar in waveform to that of a cadaver (Test 76A135). However, the time intervals between events are necessarily much shorter for primates than for the cadaver, since the inertias of their respective masses are quite different. The events that occur for all tests during the (Q_1-Q_3) interval suggests that, by reparameterization of the data from time (t) to path are length (s(t)), the geometry of the space curves of the A.C. could be made more congruent. This was done for all the primate tests and for two equivalent cadaver tests, reported in Reference (1), by integrating the resultant velocity of the A.C. to obtain arc length. Table 4 lists the results of the reparameterization for both primates and cadavers in posterior-to-anterior impacts. With two exceptions (78A229 and 78A238) the reparameterization shifts the occurence of the Q_2 and Q_3 events into congruence in terms of arc length. Test 78A238 produced a reasonable shift of the Q_2 event but, due to instrumentation difficulties, Q_3 was poorly defined in time and therefore produced a poor definition in terms of arc length. Test 78A229 was conducted with a *Papio* test subject and, as noted earlier, the head motion was dominated by soft tissue effects which produced a different trajectory during the (Q_2-Q_3) interval and thus, is expected to result in a different head impact response.

EFFECTS OF THE POST-MORTEM STATE - Previous work, reported at the 21st Stapp Car Crash Conference (1), has shown that major effects attributable to gross brain motions in head impact with unembalmed human cadavers can be studied using the nine-accelerometer motion analysis technique and with high-speed cineradiography. One of the original purposes of the present study was to investigate such effects in living and post-mortem animal subjects. As discussed in the previous section, there are large soft tissue masses (external muscle masses) in primates whose motions can mask the effects of the motion of the brain on the skull response. Thus, the use of motion analysis of the skull to indicate resulting partially decoupled brain motion is difficult with animal test subjects. Within a particular species comparative effects can be discerned, however. Test 78A236 is a post-mortem test with the same initial conditions as Test 78A239, a live animal test. Examination of the resultant angular velocity traces for both these tests (see Appendix) shows significant differences in the waveform shape and magnitude. For the post-mortem subject the waveform is smoother with less-abrupt changes in angular velocity than for the live subject. The initial peak angular velocity for the post-mortem subject while the final levels were similar. Tests 78A238 and 78A241 (traces not shown in this paper) provide a similar comparison for a different initial test condition.

High-speed cineradiography is a technique which can provide a more direct indication of brain motion. In some of the previously reported cadaver head impacts (1) this method has demonstrated the formation of radiotransparent regions in the brain cavity during impact motion. The appearance of these regions was considered to be related to motion of the brain relative to the skull. In the six tests conducted with *Macaca* subjects, cineradiographs were taken and no such phenomena were noted. This may be due to tangential accelerations which were insufficient to move the skull away from the brain or due to angular accelerations which affected the dynamics of the motion sufficiently to suppress the phenomena. A more severe impact environment than that used in this test series may be necessary to produce conditions in the primates which are analogous to those for cadavers.

Measurement of epidural pressures is another direct measurement of brain response to impact. Figures 9 and 10 present epidural pressure-time histories and their accompanying power spectra for post-mortem and live primate test subjects. The peak pressures for the living state are about two to five times the values for the post-mortem state and, in addition, the post-mortem waveforms contain much lower power than those of the living state. This difference suggests that the living brain is stiffer than the unpressurized post-mortem brain.

CONCLUSIONS

Many features of the data presented in this paper are felt to be indicative of important kinematic factors in head impact. More work is necessary before these findings can be generalized. The following specific conclusions can be drawn:

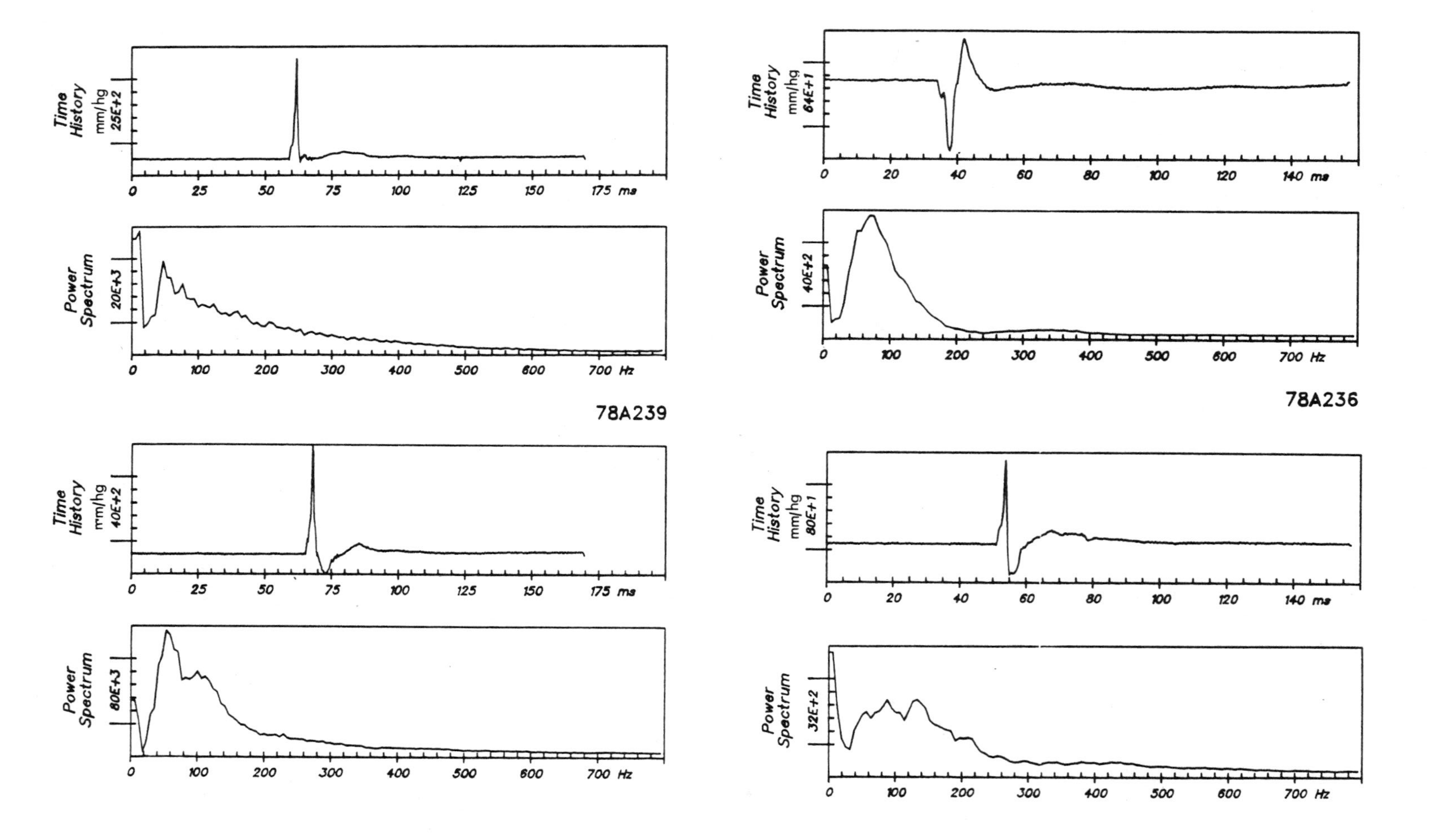

Fig. 9 - Time history and power spectrum of live primate subjects

Fig. 10 - Time history and power spectrum of post-mortem primate subjects

1. Three-dimensional accelerometry techniques can be applied to direct head impacts with lower primate subjects.
2. The Frenet-Serret frame field method for characterizing the motion of a point in space shows promise in describing head impact motion.
3. Reparameterization of head impact motion events may be useful in describing similarities and differences between the head impact responses of human surrogates.
4. The response of the head during the time of contact, in direct impact is affected by the geometry of the skull in the region of contact.
5. The response of the head, during and after the time of contact, is affected by the soft tissues of the head and neck.
6. Comparison of live and post-mortem (unpressurized) Macaca test subjects shows differences in both head impact response and epidural pressure.

ACKNOWLEDGEMENTS

This work was conducted under the sponsorship of the Motor Vehicle Manufacturers Association. The authors would like to acknowledge the contributions of J. Axelrod, J. Benson, J. Brindamour, M. Dunlap, G. Holstein and D. Kashkashian in the performance of this work.

REFERENCES

1. R. L. Stalnaker, J. W. Melvin, G. S. Nusholtz, N. M. Alem and J. B. Benson, "Head Impact Response." Paper No. 770921, Proceedings of the Twenty-first Stapp Car Crash Conference, Society of Automotive Engineers, Inc., 1977.

2. L. M. Patrick, "Head Impact Protection." Head Injury Conference Proceedings, edited by W. F. Caveness and A. E. Walker, pp. 4:41-48, J. P. Lippincott Co., 1966.

3. A. J. Padgaonkar, K. W. Krieger, A. J. King, "Measurement of Angular Acceleration of a Rigid Body Using Linear Accelerations." ASME Preprint -75-APM3, Jule 1975.

4. E. Becker, G. Willems, "An Experimentally Validated 3-D Inertial Tracking Package for Application in Biodynamic Research." Paper No. 751173, Proceedings of the Nineteenth Stapp Car Crash Conference, Society of Automotive Engineers, Inc., 1975.

5. C. L. Ewing, et al., "The Effect of the Initial Position of the Head and Neck on the Dynamic Response of the Human Head and Neck to -Gx Impact Acceleration." Paper No. 751157, Proceedings of the 19th Stapp Car Crash Conference, Society of Automotive Engineers, Inc., 1975.

6. C. L. Ewing, D. J. Thomas, "Human Head and Neck Response to Impact Acceleration." Naval Aerospace Medical Research Laboratory Detachment, New Orleans, Monograph 21, August 1972.

7. C. L. Ewing, D. J. Thomas, "Torque versus Angular Displacement Response of Human Head to -Gx Impact Acceleration." Paper No. 730976, Proceedings of the 17th Stapp Car Crash Conference, Society of Automotive Engineers, Inc., 1973.

8. D. J. Thomas, "Specialized Anthropometry Requirements for Protective Equipment Evaluation." AGARD Conference Proceedings No. 110, Current Status in Aerospace Medicine, Glasgow, Scotland, September 1972.

9. A. Fayon, et al., "Thorax of 3-Point Belt Wearers During a Crash (Experiments for Cadavers)." Paper No. 751148, Proceedings of the Nineteenth Stapp Car Crash Conference, Society of Automotive Engineers, Inc., 1975.

10. A. M. Nahum, R. W. Smith, "An Experimental Model for Closed Head Injury." Paper No. 760825, Proceedings of the Twentieth Stapp Car Conference, Society of Automotive Engineers, Inc., 1976.

11. M. Bender, J. W. Melvin, R. L. Stalnaker, "A High-Speed Cineradiographic Technique for Biomechanical Impact." Paper No. 760824, Proceedings of the Twentieth Stapp Car Crash Conference, Society of Automotive Engineers, Inc., 1976.

12. J. A. Bartz, F. E. Butler, "Passenger Compartment with Six Degrees of Freedom." Auxiliary Programs to "Three Dimensional Computer Simulation of a Motor Vehicle Crash Victim." Final Technical Report for DOT Contract No. FH-11-7592, 1972.

13. D. J. Thomas, D. H. Robbins, R. H. Eppinger, A. I. King, and R. P. Hubbard. "Guidelines for the Comparison of Human and Human Analogue Biomechanical Data." A report of an ad hoc committee, Ann Arbor, Michigan, December 6, 1974.

14. B. O'Neill, "Elementary Differential Geometry." Academic Press, New York, 1967.

15. G. S. Nusholtz, et al., "Direct Head Impacts of Human Surrogates." Final Report for Motor Vehicle Manufacturers Association of the Unites States, Inc., In Preparation.

APPENDIX

The following time-histories are included for reference with this paper. Because of space limitations not all the test variables are shown and some of the tests, which were essentially duplicates of the ones presented, are also deleted.

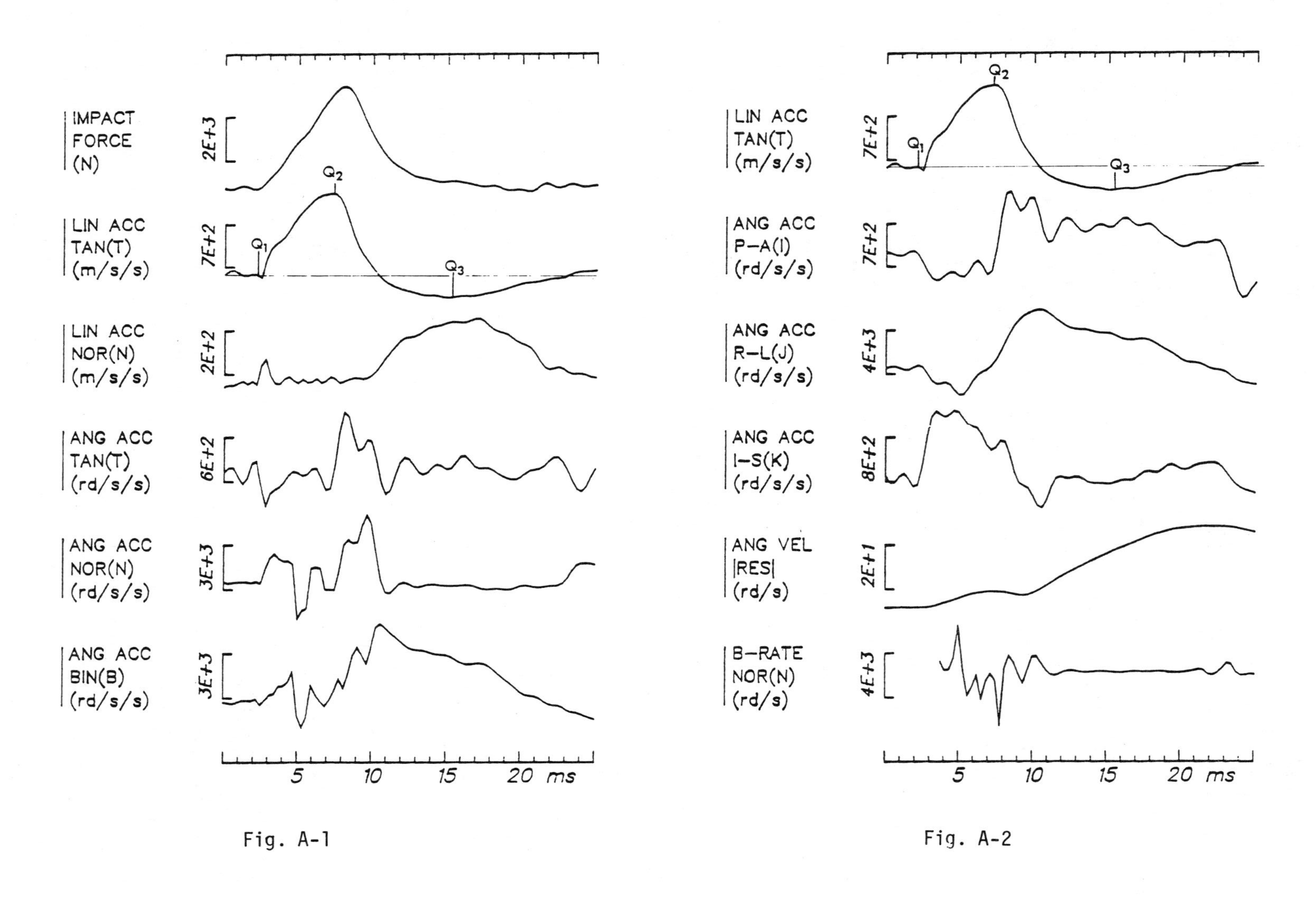
IMPACT FORCE (N)
2E+3
LIN ACC TAN(T) (m/s/s)
7E+2
LIN ACC NOR(N) (m/s/s)
2E+2
ANG ACC TAN(T) (rd/s/s)
6E+2
ANG ACC NOR(N) (rd/s/s)
3E+3
ANG ACC BIN(B) (rd/s/s)
3E+3
Q1
Q2
Q3
5
10
15
20 ms
LIN ACC TAN(T) (m/s/s)
7E+2
ANG ACC P-A(I) (rd/s/s)
7E+2
ANG ACC R-L(J) (rd/s/s)
4E+3
ANG ACC I-S(K) (rd/s/s)
8E+2
ANG VEL |RES| (rd/s)
2E+1
B-RATE NOR(N) (rd/s)
4E+3
Q1
Q2
Q3
5
10
15
20 ms

Fig. A-1

Fig. A-2

791027

Biodynamics of the Living Human Spine During $-G_x$ Impact Acceleration

R. Cheng, N. K. Mital, R. S. Levine and A. I. King
Wayne State University

ABSTRACT

Spinal kinematics of the living human volunteers undergoing $-G_x$ impact acceleration are described along with the experimental procedures followed to acquire such data. There were 4 male and 3 female volunteers who were subjected to impacts in the tensed and relaxed mode from 2 - 8 g, in 1-g increments. Their lower extremities were tightly clamped to the impact seat and the pelvis was restrained by a lapbelt. The biodynamic response of the living spine is quite similar to that of the cadaveric spine, particularly in terms of T1 displacement, acceleration at T1 and flexural resistance. Female volunteers tend to withdraw from the test program at lower g-levels than males due to transient neck pain.

SPINAL RESPONSE TO $-G_x$ IMPACT ACCELERATION was quantified by Mital et al (1)* recently. Cadaveric data were provided to demonstrate the relative motion of the spinal segments and the flexibility of the thoracic and lumbar spine. A surrogate spine in the form of a rigid link pinned at both ends was designed for use in a current anthropomorphic test device (ATD), the Part 572 dummy. The basis for a rigid link was the dual requirement of repeatability and reproducibility in an ATD. These conditions would be difficult to satisfy if a multi-segment spine were used. A preliminary physical model of this surrogate spine was fabricated and tested by Mital et al (2). It was found to perform satisfactorily. Its T1 kinematics relative to the pelvis fell within the corridor of cadaveric data upon which the spine was designed.

In this paper, human volunteer data on spinal kinematics are reported. Both male and female subjects participated in this experimental test series. It describes the experimental instrumentation packages applied to these subjects for consistent data without discomfort. The data are compared with those obtained from cadaveric tests.

*Numbers in parentheses designate References at the end of the paper.

METHODS OF PROCEDURE

INSTITUTIONAL REVIEW BOARD APPROVAL - All procedures involving the use of human subjects were submitted to the Wayne State University Committee on Human and Animal Investigation (the Institutional Review Board for the University) for review and approval. Detailed protocols and assurances were submitted along with the appropriate forms for informed consent. All experiments were carried out within the limitations of the approved procedures and additional tests of procedures required separate approvals from the same board.

EXPERIMENTAL DESIGN - The objective of this study is to acquire data for the design of an ATD spine which has human-like responses. To achieve this, it is necessary to quantify living spinal response to $-G_x$ acceleration in terms of the kinematics of selected spinal segments and the kinetics of resistance to flexion and extension. Moreover, a restraint system needs to be specified to protect the test subject during the impact test.

The experimental plan called for 2 modes of restraint. The first is called the clamped mode in which the lower extremities are tightly clamped to the impact seat and the pelvis is restrained by a lap belt. The upper torso is free to flex forward but is protected from impacting the thighs by a cushion placed across the lap. This mode permitted substantial flexion of the spine without inordinate risk to the test subject. The second mode utilizes the standard three-point belt system which yields a set of norms for spinal kinematics that the ATD spine is expected to match.

The deceleration pulse shape should be representative of an automotive collision. However, the measurement of spinal resistance to bending requires the sled to be brought to rest as quickly as possible. in this way, a static analysis of seat pan loads and moments results in a resisting moment which can be expressed as a function of spinal flexion.

The magnitude of peak deceleration exposure is set at 8 g. It can be lower if the volunteer refuses to go on with the test or if it is deemed by the medical officer that the subject should not go on. The first run is to be made at a level of 2 g to acclimate the subject to the acceleration environment. The incremental level is 1 g and at each g-level, the subject makes 2 runs, first with the muscles tensed and then relaxed.

SUBJECT SELECTION - All subjects were recruited from a pool of over 50,000 students at Wayne State University. Their age range is from 18 to 25. The primary requirement is good health. Individuals with a history of bone fractures or other injuries are disqualified. They can also be disqualified by the examining physician for a variety of pre-existing conditions. The medical examination is scheduled for each volunteer to minimize unnecessary tests in the event that the subject is disqualified during the process. Thus, the x-ray exam is taken last. The medical evaluation consists of the following:

1. Complete history
2. Complete physical
3. ECG
4. Stress ECG
5. Laboratory tests, including
   a) SMA 1260
   b) Electrolytes
   c) Complete hemogram
   d) Urinanalysis
6. Ophthalmologic exam with fundus photography
7. Orthopedic exam
8. X-rays of the skull, spine and chest

The subjects were covered by workmen's compensation as they were hired as student assitants and performed other laboratory tasks in addition to volunteering for the sled runs. A medical insurance policy was purchased for the group from the Continental Casualty Insurance Co. as additional protection.

INSTRUMENTATION - Accelerometer clusters were mounted on the spinous process of T1 and T12 and on the pelvis. Each cluster consisted of 9 uni-axial accelerometers arranged in the configuration described by Padgaonkar et al (3) for the purpose of measuring three-dimensional linear and angular acceleration. The mounts were custom-made to fit the spinal contours at each vertebral level. They were held onto the spinous process by means of a series of elastic straps and belts. Figure 1 shows the T1 mount which is most complex in design. A cubic photographic target is attached rigidly to the mount. Although it is a 7.6 mm cube, it is extremely lightweight, being made out of

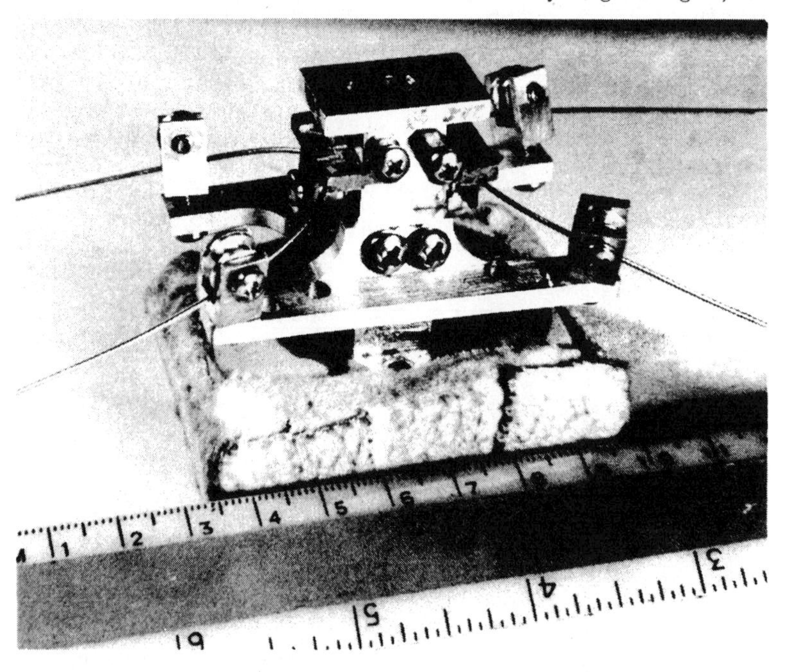

Fig. 1 - T1 accelerometer mount for human volunteers

urethane foam. Figure 2 shows an instrumented volunteer subject with all mounts attached to his spine.

There was photographic coverage from three high-speed cameras placed along 3 orthogonal axes. Other instrumentation included a sled accelerometer and velocity transducer, a six-axis seat pan load cell, an on-board ECG monitor and a time synchronization indicator to match film and transducer data. Belt load cells were used during runs involving the three-point belt restraint system.

SLED FACILITY - All runs were carried out on WHAM III (Wayne Horizontal Accelerator Mechanism), a flat bed sled which was slowly accelerated to a speed of 16 to 17 km/h and was made to impact a programmable hydraulic snubber. The deceleration distance varied from 240 mm to 585 mm. Figure 3 shows a subject seated upright in the impact seat and restrained in the clamped mode. The thighs and legs were held firmly to the seat by clamps which could be released rapidly for emergency egress from the sled. A lapbelt was used to prevent excessive lifting of the pelvis. It was found during the cadaveric experiments that femoral fractures occurred without a lapbelt and were prevented by means of a lapbelt. A piece of 140 mm thick urethane foam was placed over the lap as a cushion to prevent excessive spinal flexion and to protect the rib cage. Head contact with the cushion did not occur. Because of the accelerometer mounts on the back of the subject, two pieces of seat belt webbing were used as a seat back. They were positioned to avoid interaction with the targets and mounts.

Before each run, at least two responsible investigators, including a physician go through a detailed check-list, as required by the test protocol. An ambulance and crew are required to stand-by for transport of the test subject to a predesignated nearby hospital, in the event of an injury. The emergency ward of the hospital was also notified in advance of the scheduled tests. The snubber length, accumulator pressure and all on-board equipment were checked by at least two people. The subject's blood pressure and pulse rate were measured immediately before and after each run and at regular intervals. After each series of runs, the subject was asked to report any pain or discomfort to the physician. Such reports were usually received on the following day but they were all minor and short-lived.

RESULTS

ANTHROPOMETRIC DATA - Table 1 shows the weight, age, sex and sitting height of the seven subjects who participated in this study. There were 3 female subjects and 4 male subjects. Data used for the computation of percentiles were taken from 1962 HEW data (3). Anthropometric data of interest consist of the length of the thoracic and lumbo-sacral spines. They were measured by identifying the coccyx

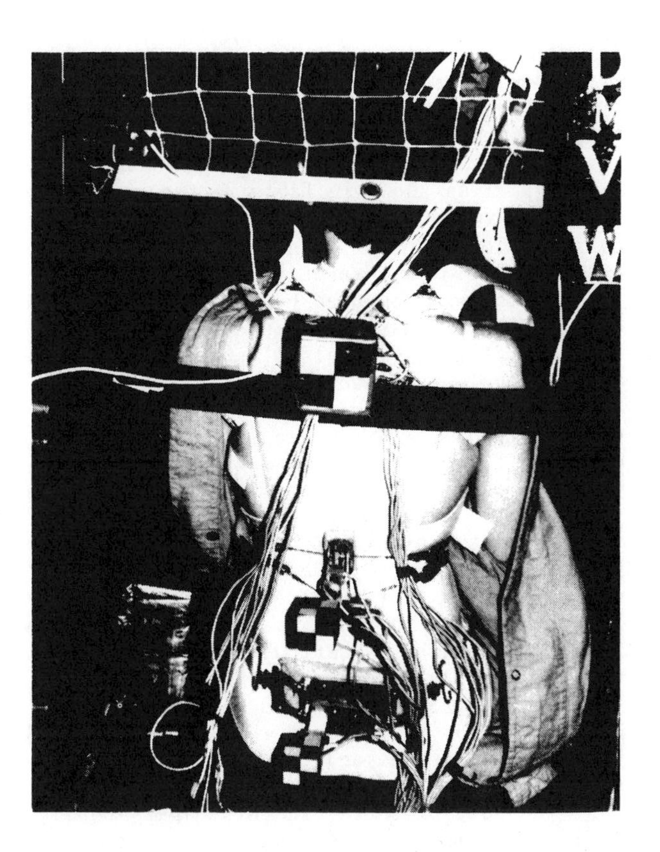

Fig. 2 - Volunteer instrumented with accelerometer mounts

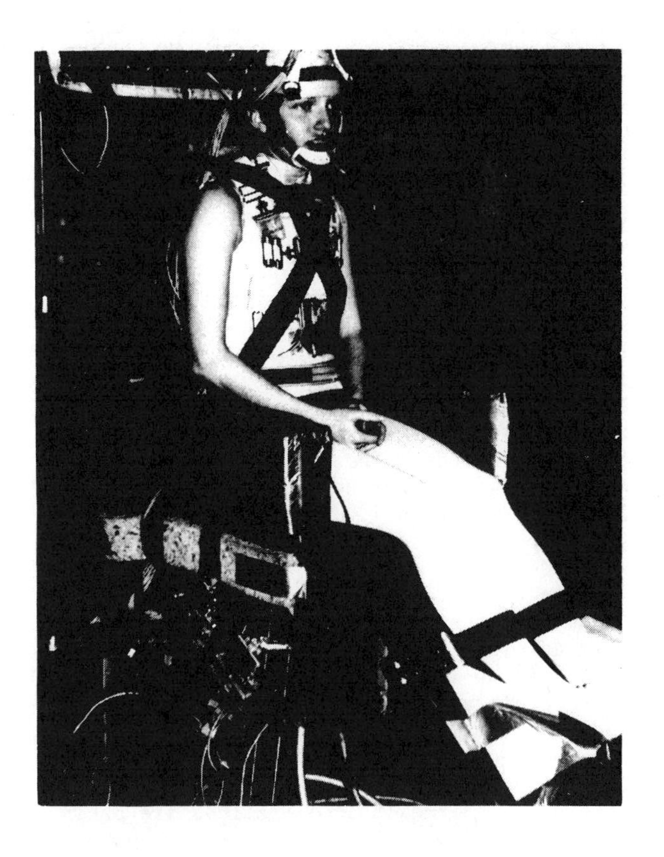

Fig. 3 - Volunteer in impact seal

and spinous process of T12 and C7. The volunteers were measured in a standing position. These dimensions were required to compare the length of the spine of volunteers to the length of the spines of cadavers reported by Mital et al (1). As shown in Table 2, the spine length of volunteers ranged from 440 mm to 592 mm. The mean and standard deviation were 497.7 and 51.3 mm respectively. A normalization factor was computed by dividing the average length of cadaveric spines by the spine length of the individual volunteers as shown in the last column of Table 2.

EXPERIMENTAL RUNS - The volunteer runs were divided in 4 broad groups according to mode and g-level. Runs at 2-4 g (2-3 g for females) were placed in the low-g category and those at 5-8 g (4-8 g for females) were considered as the high-g group. Table 3 summarizes the number of runs conducted on each volunteer in different modes and g-levels. The total number of runs was 88 (25 female runs and 63 male runs). There were a total of 36 belted runs (24 male and 12 female) and 52 clamped runs (39 male and 13 female). Approximately half of the clamped runs were conducted with muscles tensed and the other half with muscles relaxed.

The data presented in this paper are taken from the clamped runs during which there is substantial spinal flexion and extension and with which cadaveric data from (1) can be compared.

ACCELEROMETER DATA - During each run, the acceleration of T1, T12 and the pelvis was measured by a nine-accelerometer module. With these data, the acceleration at the origin of a body-fixed anatomical system can be computed. Using the coordinate system defined by Ewing et al (4) for T1, the resultant acceleration of T1 for runs made at 4-7 g by a male subject is shown in Figure 4. A peak acceleration of approximately 20 g is recorded. The angular acceleration and velocity about the lateral or Y-axis at T1 are shown in Figures 5 and 6 respectively for these runs by the same volunteer. The axis is normal to the mid-sagittal plane and is positive towards the left. The acceleration values are largest about this axis due to the predominately two dimensional motion of the clamped mode. The peak angular acceleration is of the order to 700-900 rad/s^2 for the 7-g case. Figures 7, 8 and 9 show the data from a parallel sequence of tests for a female volunteer.

In terms of data reproducibility, it was found that the corridors for X-axis and Z-axis acceleration (relative to the mount) were quite narrow for each group of runs. The X-axis acceleration at T1 is shown in Figures 10 and 11 for male high-g runs and female low-g runs respectively. There is a similarity in the two negative peaks between these two sets of runs. The Z-axis accelerations are shown in Figures 12 and 13 for the same groups of volunteers. There is again good reproducibility and a characteristic positive peak followed by a negative one. The only difference between the male and female data is the extended duration of the negative peak for the female runs. The Z-axis pelvic acceleration are 2 positive peaks separated by a negative one.

FILM DATA - The rotation and displacement of the cubic film target were measured from high-speed film using a Vanguard film analyzer. A computer program analyzed the data and plotted the position and rotation ot T1, T12 and pelvic target with respect to an inertially-fixed frame or with respect to each other. The data can also be transformed to a specified body fixed reference frame. Figure 16 shows target motion of T1, T12 and the pelvis relative to the inertial reference frame for a high-g male volunteer run in the tensed mode. Relative displacement data for the same run are shown in Figure 17. Figures 18 and 19 show the data for the same volunteer in the relaxed mode. The motion in the relaxed mode is usually larger than that in the tensed mode for the same g-level. Film data for a

Table 1 - Pertinent Data on Volunteers

Volunteer No.	Body Weight (kg)	Age (yrs)	Sex	Sitting Height (mm)	Percentile
1454	40.8	22	F	840	30
1156	50.0	24	F	820	11
1760	45.8	18	F	878	60
2953	59.0	23	M	880	30
1555	60.1	21	M	880	30
0159	71.2	19	M	960	92
0252	61.2	24	M	920	55

Table 2 - Anthropometric Data on Volunteer Spines

Volunteer No.	Spinal Length (mm) Thoracic C7-T12	Lumbo-sacral T12-Coccyx	Total	Normalization Factor
F 1454	245	220	465	1.28
F 1156	280	200	480	1.24
F 1760	300	185	485	1.24
M 2953	265	175	440	1.36
M 1555	321	161	482	1.24
M 0159	370	222	592	1.01
M 0252	280	260	540	1.10

Average Total Length ± S.D. = 497.9 ± 51.3 mm

$$\text{Normalizing Factor} = \frac{\text{Average Cadaveric Spine Length (596.3)}}{\text{Volunteer Spinal Length}}$$

Table 3 - Summary of Experimental Runs

Vol. No. Male	Belted 2-4g	Belted 5-8g	Clamped 2-4g	Clamped 5-8g
2953	6	2	2	7
1555	7			
0159	5	4	6	8
0252			6	12

Vol. No. Female	Belted 2-4g	Belted 5-8g	Clamped 2-3g	Clamped 4-5g
1454	4	8		
1156			4	3
1760			6	

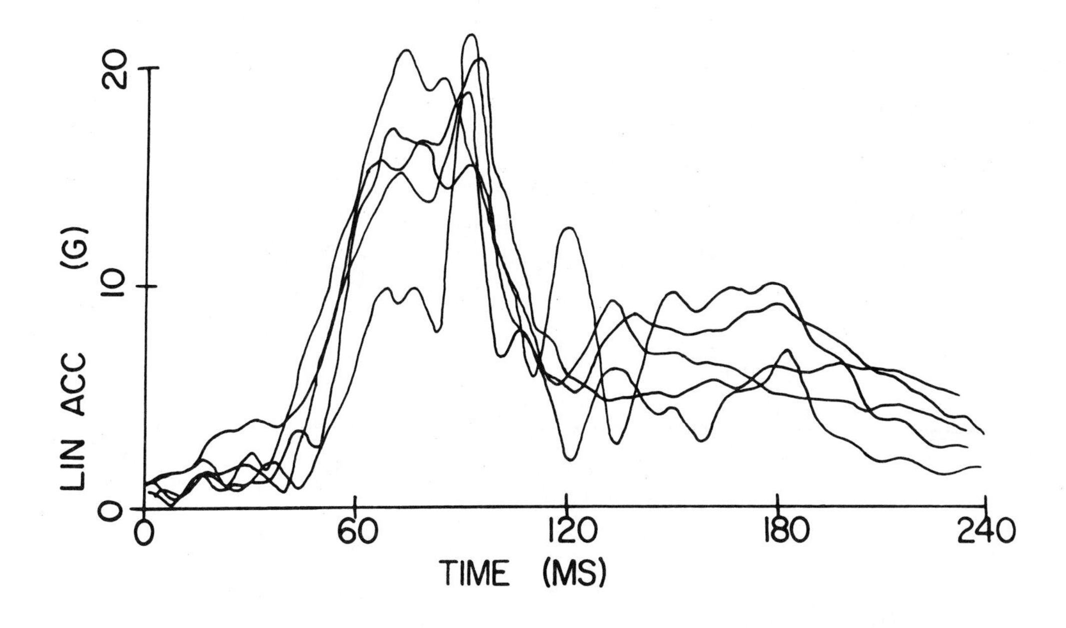

Fig. 4 - Resultant acceleration of T1 at the origin of the T1 coordinate system - male subject (4-7g runs)

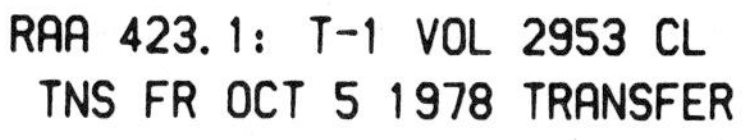

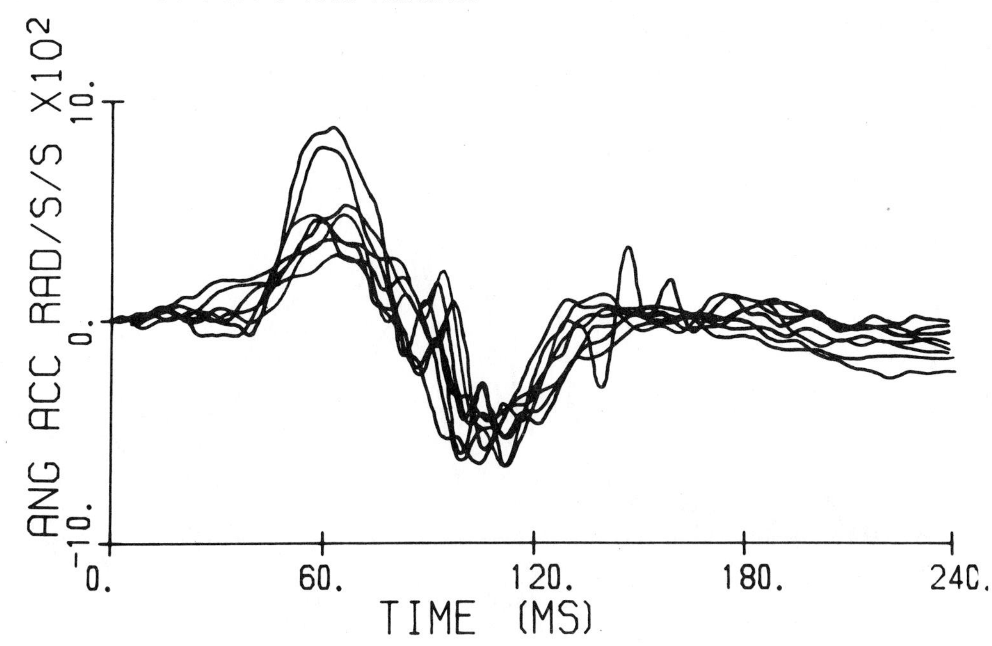

Fig. 5 - Angular acceleration of T1 about the Y-axis - male subject (4-7g runs)

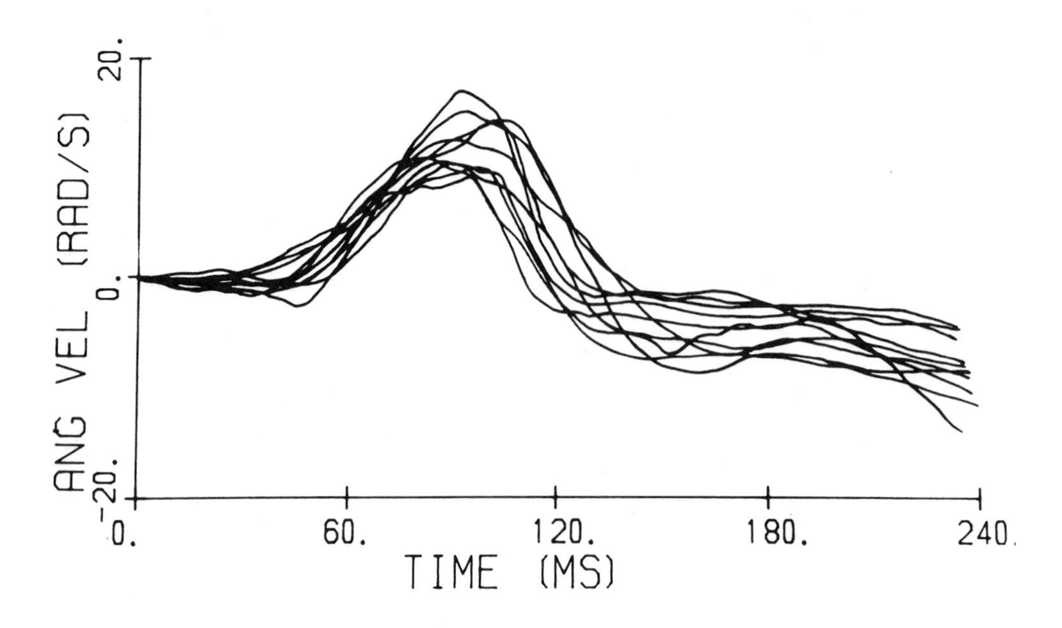

Fig. 6 - Angular velocity of T1 about the Y-axis - male subject (4-7g runs)

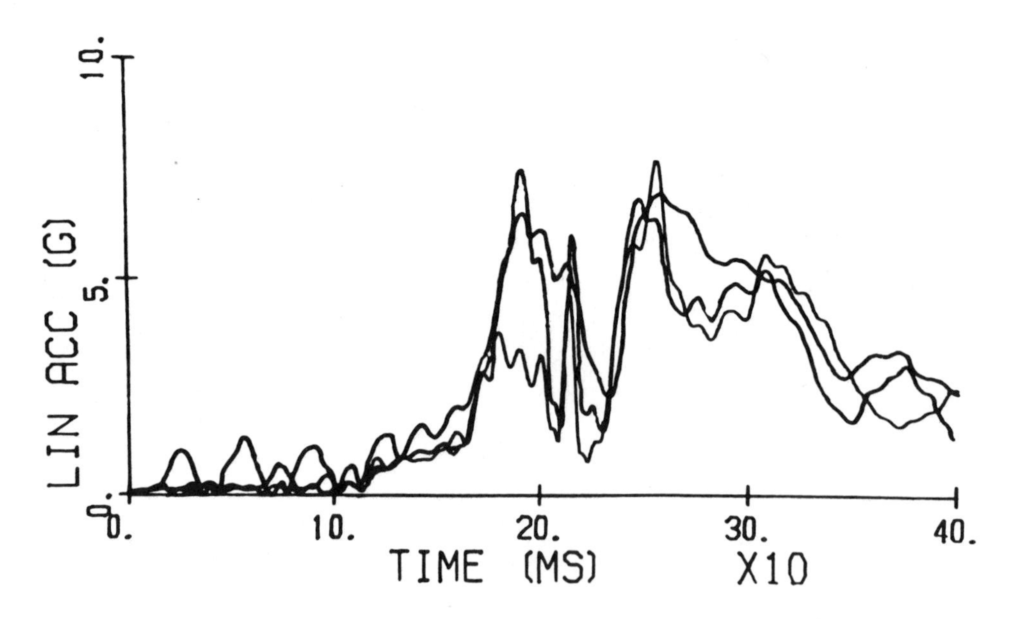

Fig. 7 - Resultant acceleration of T1 at the origin of the T1 coordinate system - low-g runs (female subject)

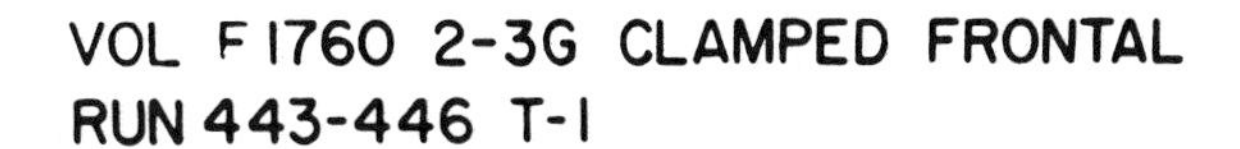

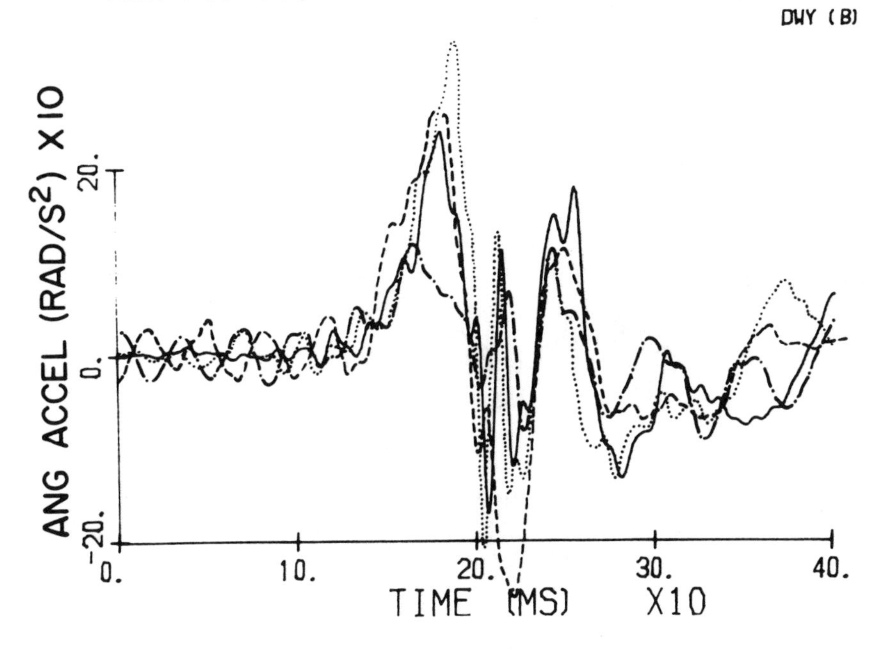

Fig. 8 - Angular acceleration of T1 about the Y-axis - low-g runs (female subjects)

VOL F1760 2-3G CLAMPED FRONTAL
RUN 443-446 T-1

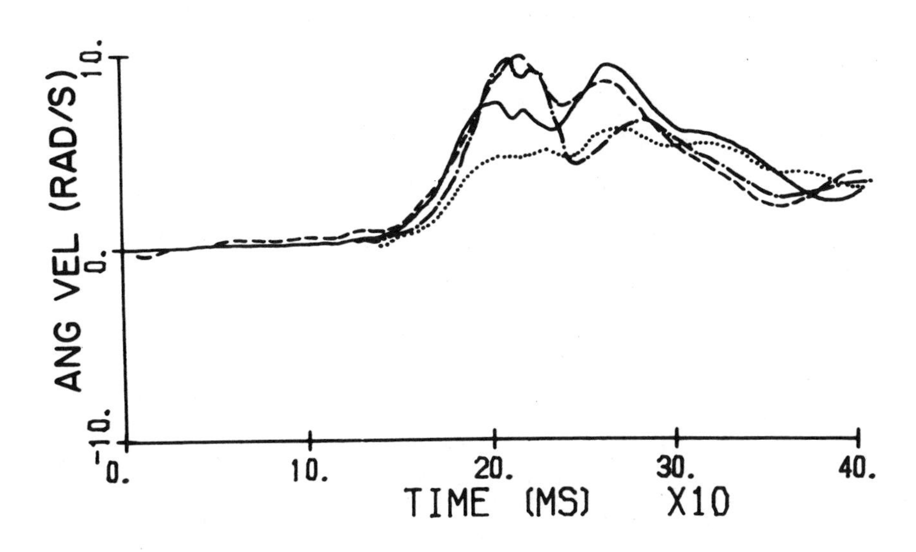

Fig. 9 - Angular velocity of T1 about the Y-axis - low-g runs (female subjects)

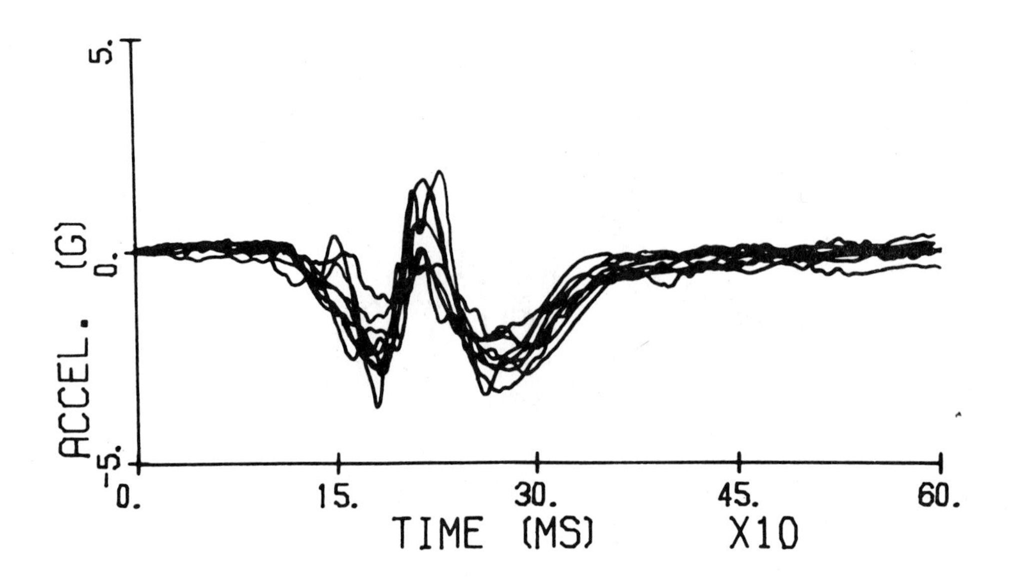

Fig. 10 - T1 X-axis acceleration corridor - male high-g runs

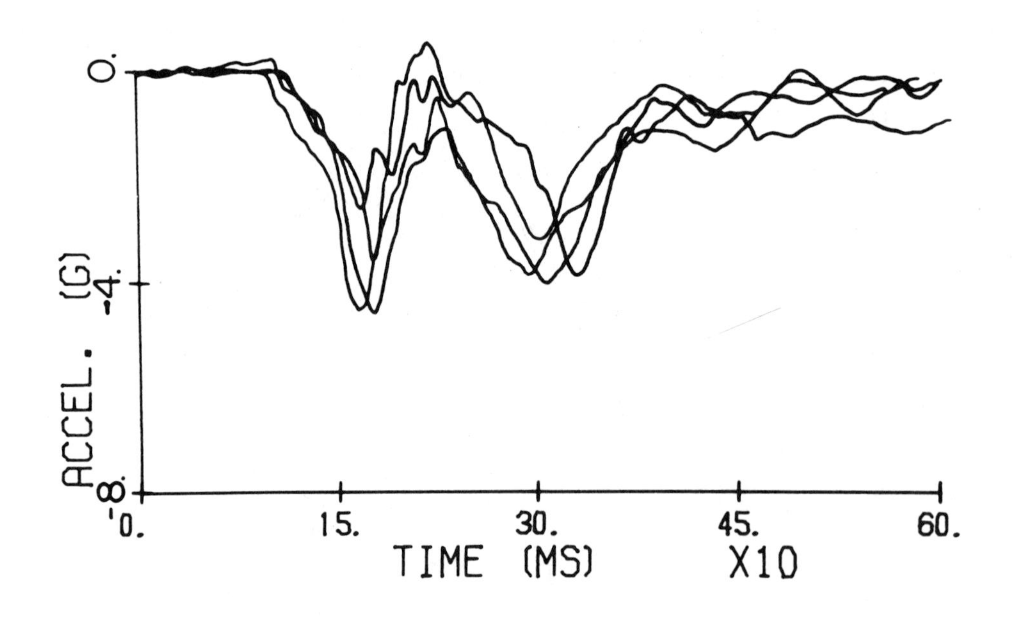

Fig. 11 - T1 X-axis acceleration corridor - female low-g runs

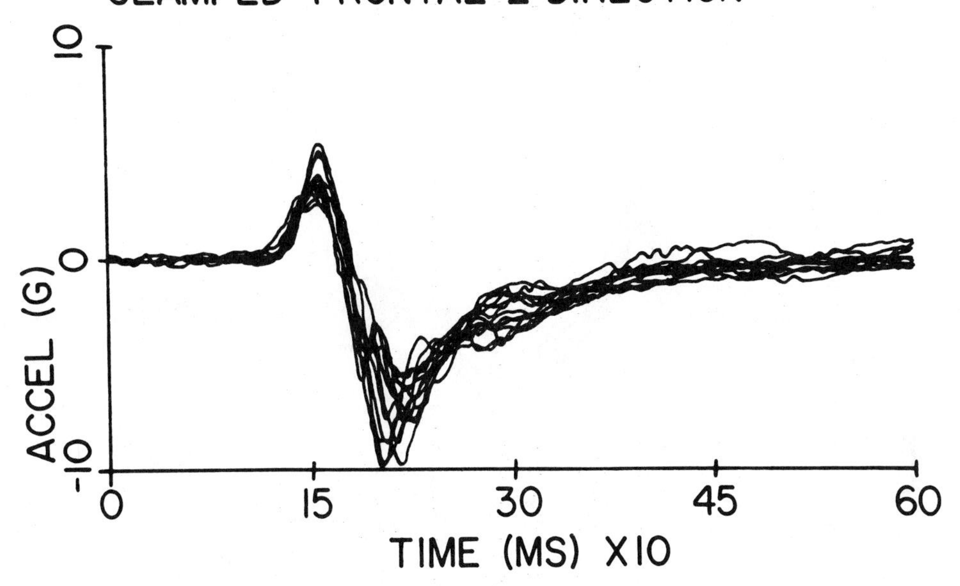

Fig. 12 - T1 Z-axis acceleration corridor - male high-g runs

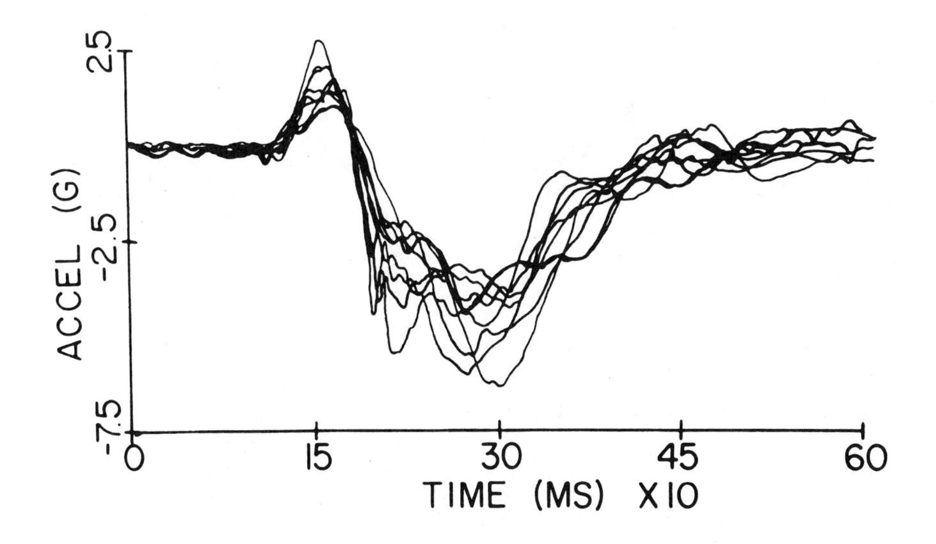

Fig. 13 - T1 Z-axis acceleration corridor - female low-g runs

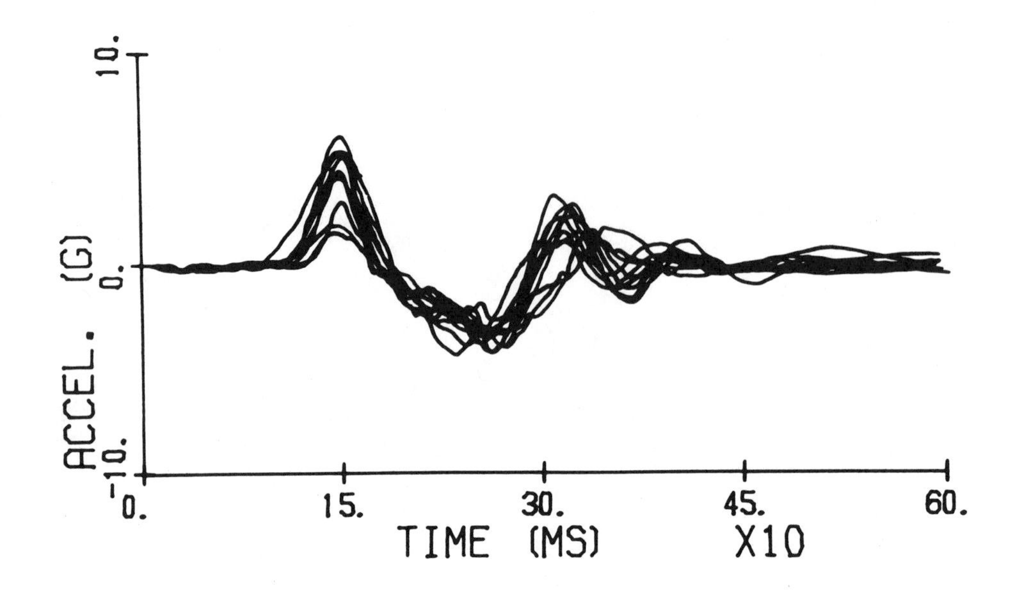

Fig. 14 - Pelvic Z-axis acceleration corridor - male high-g runs

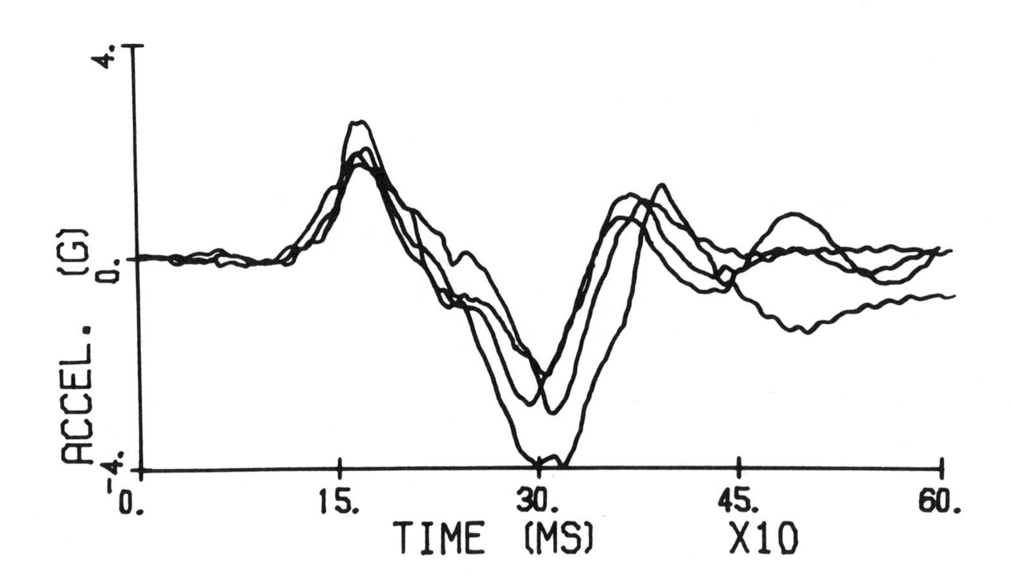

Fig. 15 - Pelvic Z-axis acceleration corridor - female low-g runs

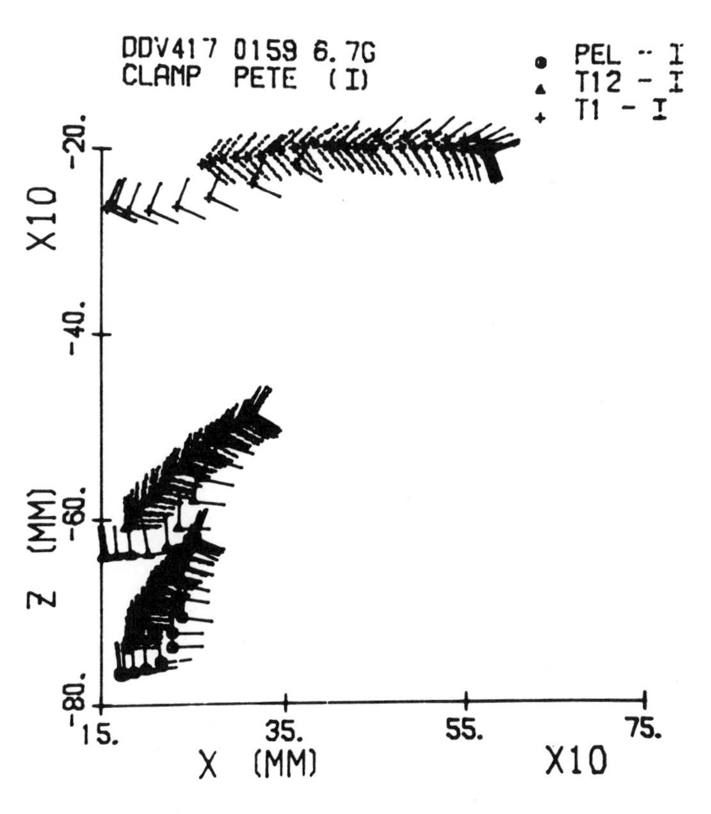

Fig. 16 - Position of T1, T12 and pelvic targets relative to the inertial frame - tensed male volunteer run

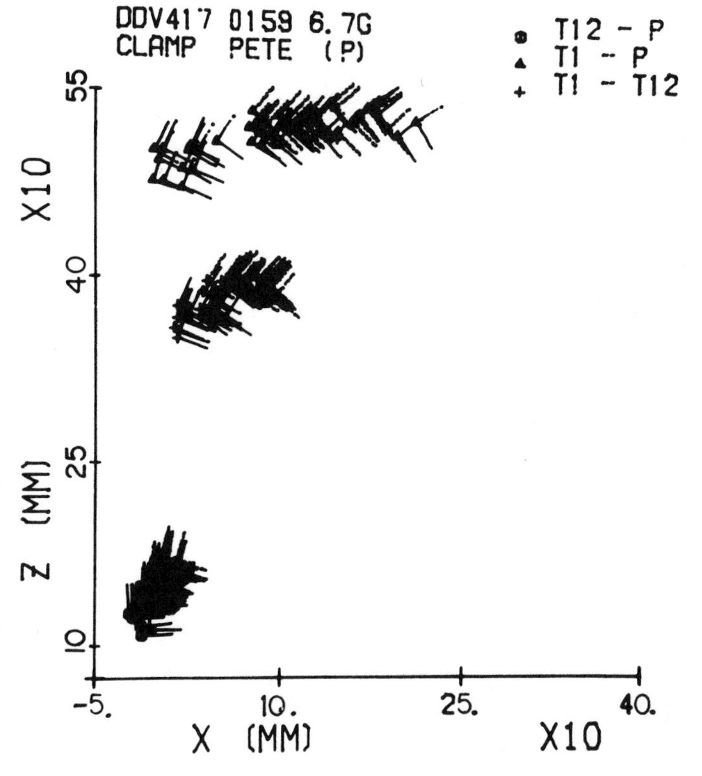

Fig. 17 - Position of T1 and T12 relative to the pelvis and position of T1 relative to T12 - tensed male volunteer run

female volunteer run in the tensed mode are shown in Figures 20 and 21.

LOAD CELL DATA - A six axis load cell was used to measure seat pan forces and moments. A typical output of load cell for a high-g female volunteer run is shown in Figures 22 and 23. The sled-fixed coordinate system was used to define the components of forces and moments measured by this lead cell. Since the impacts were principally 2-dimensional, F_X and F_Z in Figure 22 and M_Y in Figure 23 were used in data analysis. It can be seen from these figures that F_Y, M_X and M_Z are all small.

DATA ANALYSIS

SPINAL FLEXIBILITY - In order to design a surrogate spine which simulates the kinematics of cadaveric spines and volunteer spines, it is necessary to delineate the principal characteristics of vertebral motion during impact. One of the obvious parameters is relative rotation of the targeted vertebra with respect to each other and to the pelvis. The total relative rotation of T1 with respect to T12 was taken to be the algebraic sum of maximum extension and maximum flexion. This represented the total bending of the thoracic spine. A similar rotation was also computed for the lumbar spine; that is the rotation of T12 with respect to the pelvis. A unit rotation is defined as the average rotation of each joint required to attain the total relative rotation and is computed by dividing the total rotation by the number of joints involved, 6 for the lumbar segment and 11 for the thoracic spine. Data for male volunteer run (stensed and relaxed) are listed in Table 4 and for female volunteer runs in Table 5.

RELATIVE ROTATION - Volunteer data of the motion of T1 relative to the pelvis can be compared with cadaveric data reported by Mital et al (1). The corridors in Figure 24 envelope the cadaveric data reported previously. The position plots within the corridors are representative data taken from a large pool of data.

RESISTANCE TO FLEXION - The mathematical formulation to compute the resistance of the spine to flexion was given by Mital et al (1). The free body diagram from computing the resisting moment of the spine at the hip joint is reproduced in Figure 25. The expression for the resultant hip joint moment M_R is given by:

$$M_R = M_Y - F_Z X_L - F_X Z_L$$

where,

M_Y = seat pan moment about the sled fixed Z-axis

F_X and F_Z = seat pan loads along the sled fixed X- and Z-axis

F_L and Z_L = moment arms as shown in Fig. 25

Figure 26 shows the resisting moment and its components for a high-g female volunteer run. The corresponding rotation of T1 with respect to pelvis (Angle Θ) as a function of time is generated from the film data. A cubic spline

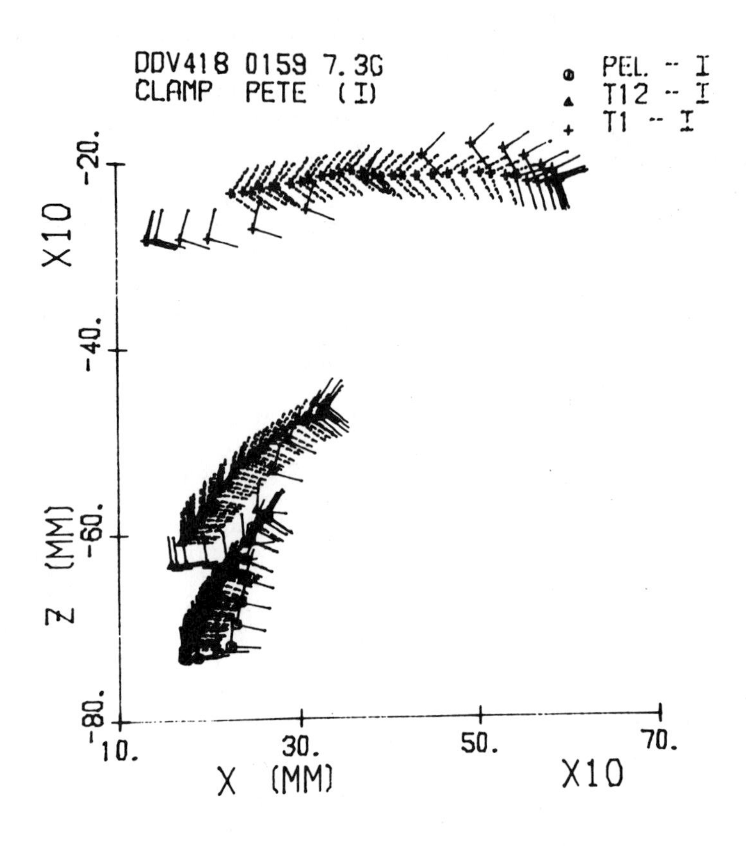

Fig. 18 - Position of T1, T12 and pelvic targets relative to the inertial frame - relaxed male volunteer run

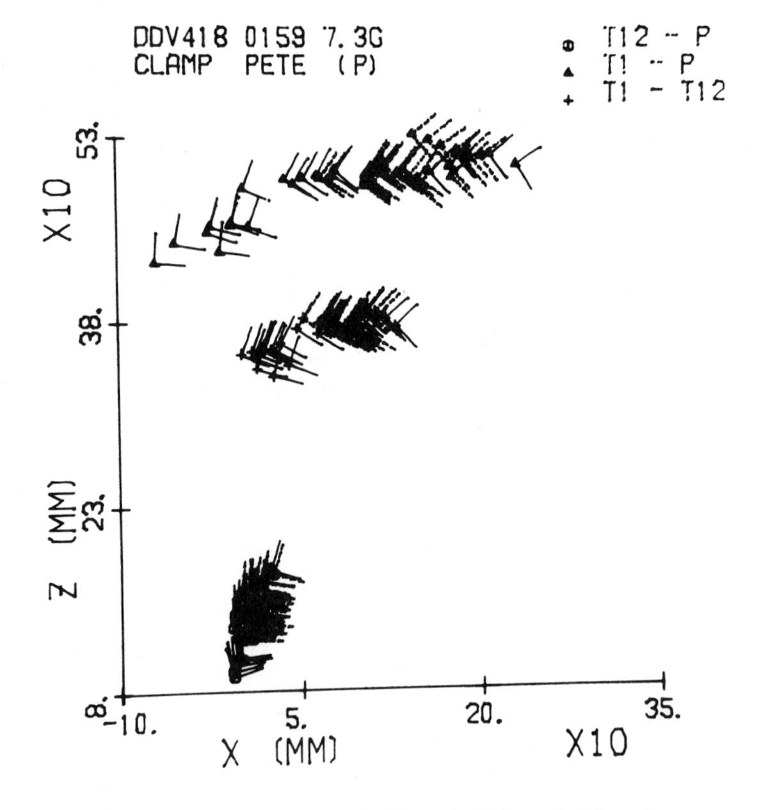

Fig. 19 - Position of T1 and T12 relative to the pelvis and position of T1 relative to T12 - relaxed male volunteer run

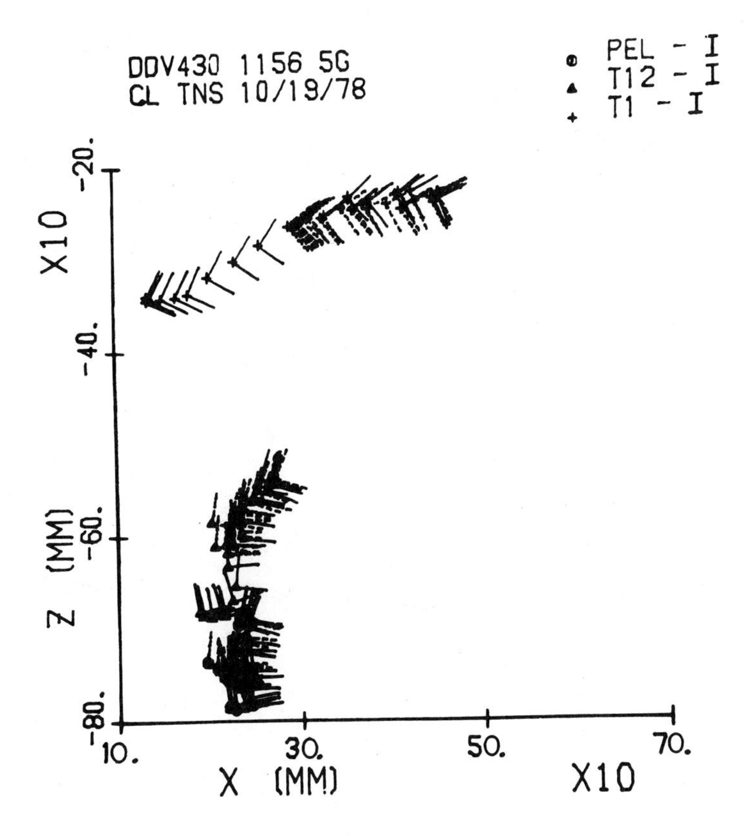

Fig. 20 - Position of T1, T12 and plevic targets relative to the inertial frame - tensed female volunteer run

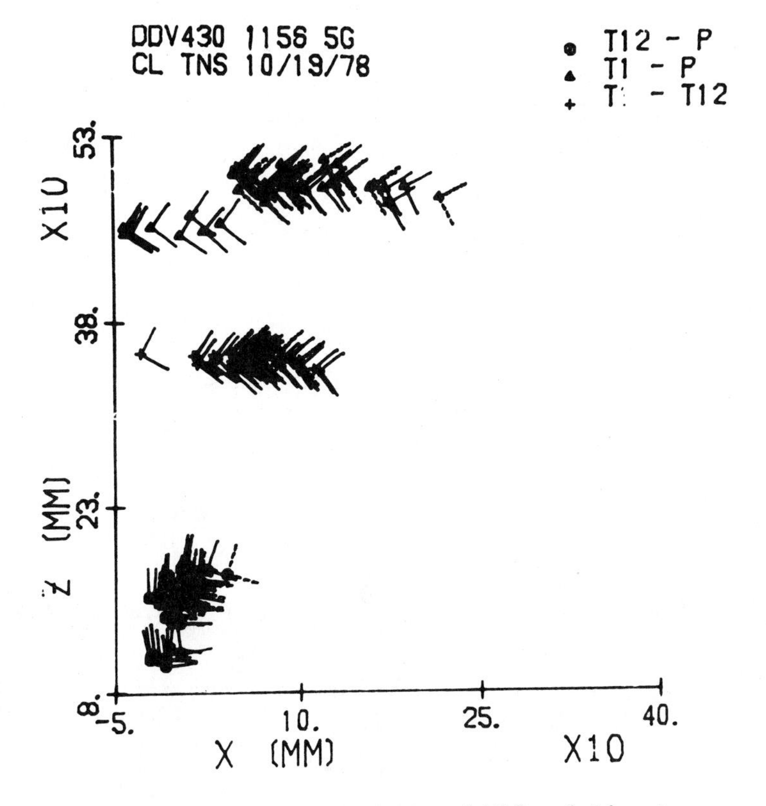

Fig. 21 - Position of T1 and T12 relative to the pelvis and position of T1 relative to T12 - tensed female volunteer run

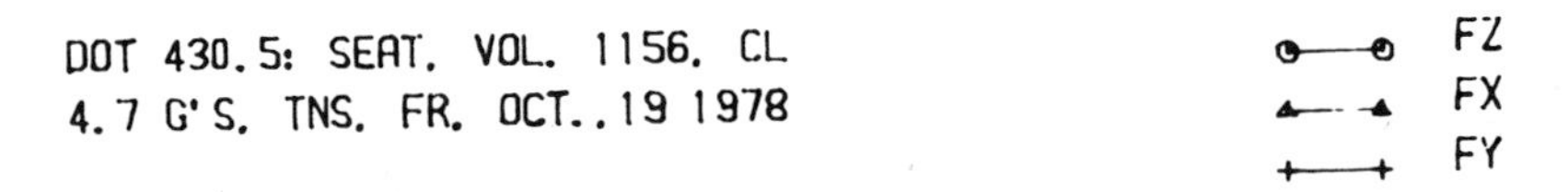

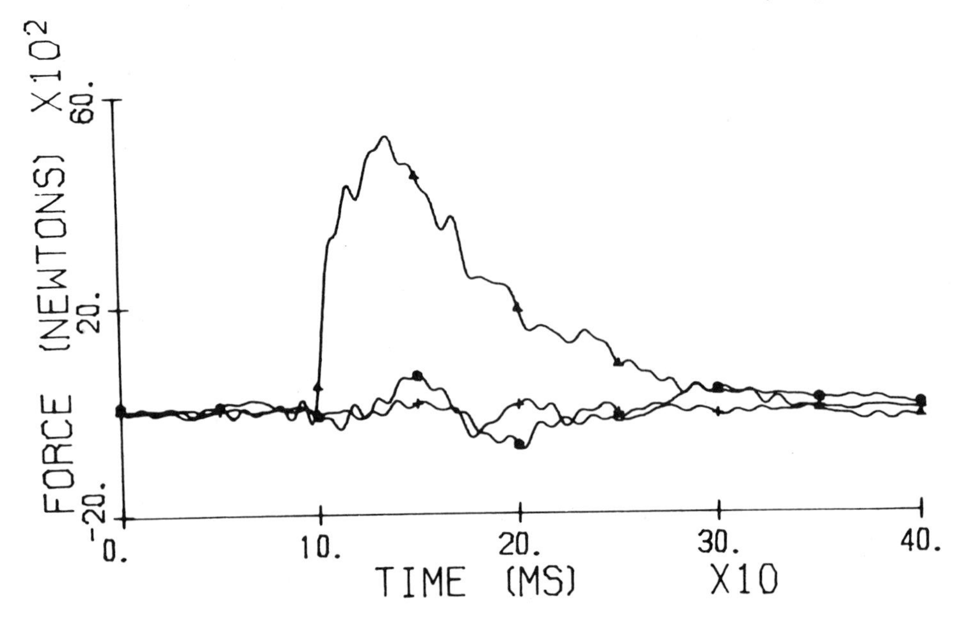

Fig. 22 - Seat pan load cell force output - female 5-g run

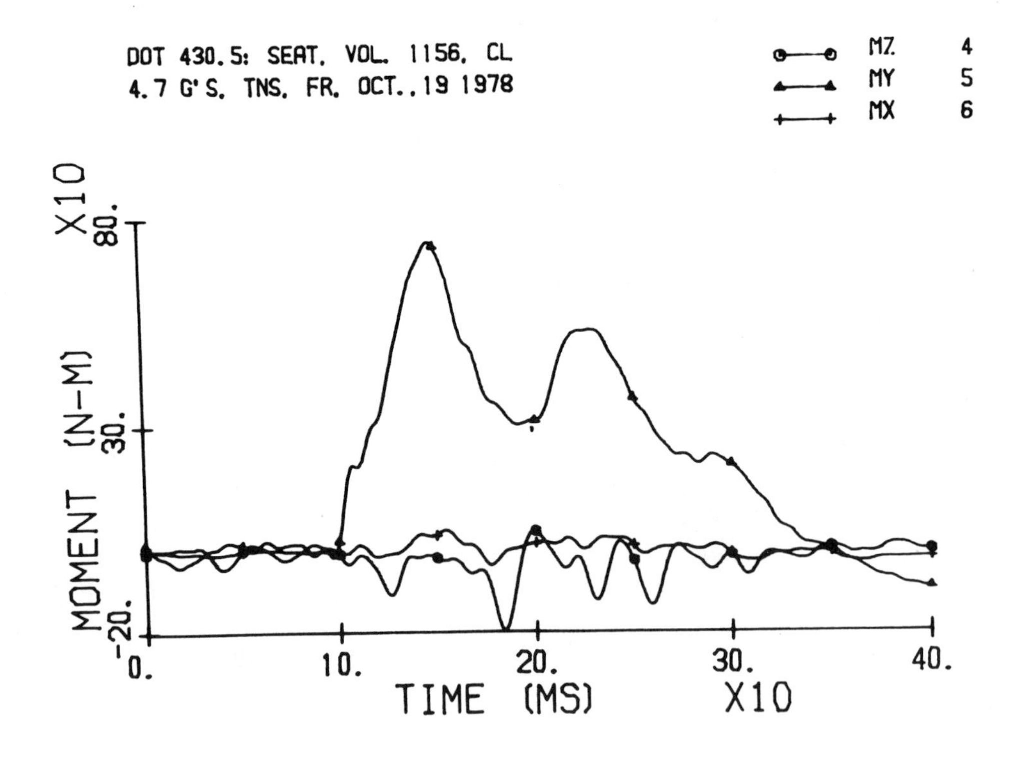

Fig. 23 - Seat pan load cell moment output female 5-g run

Table 4 - Relative Rotation of T1 with respect to T12 and T12 with Respect to the Pelvis

Nominal g-Level	Run No.	Rotation of T1 with Respect to T12 (deg) Max. Ext.	Max. Flex.	Tot. Rot.	Unit Rot.*	Rotation of T12 with Respect to Pelvis (deg) Max. Ext.	Max. Flex.	Tot. Rot.	Unit Rot.†
Male Tense									
1.8	403	-33.2	-59.1	25.9	2.35	1.1	-28.3	29.3	4.88
2.8	433	-34.4	-63.1	28.7	2.60	10.0	-17.6	27.6	4.60
3.2	405	-41.3	-61.4	20.1	1.83	14.8	-10.9	25.7	4.30
3.5	407	-38.4	-55.8	17.4	1.58	25.4	4.7	20.7	3.45
4.0	435	-36.0	-65.6	29.6	2.69	2.8	-25.4	28.1	4.60
4.4	419	-18.9	-39.9	21.0	1.90	11.4	-16.7	28.1	4.68
4.8	409	-23.4	-60.9	37.5	3.40	8.4	-3.6	12.0	2.00
5.0	437	-30.6	-60.3	29.7	2.70	9.4	-21.6	31.0	5.10
5.4	421	-17.4	-47.3	29.9	2.71	7.2	-17.5	24.7	4.12
5.7	453	-32.5	-63.2	30.7	2.79	-55.5	-31.4	24.1	4.01
5.7	455	-34.3	-58.8	24.5	2.22	-1.5	-29.1	27.6	4.60
6.1	411	-17.4	-36.3	18.9	1.72	2.3	-18.6	13.7	2.28
6.5	457	-11.7	-50.3	38.6	3.50	-1.5	-32.1	30.6	5.10
6.6	424	-31.4	-61.6	30.2	2.75	4.8	-26.8	31.6	5.27
6.7	417	-19.4	-45.8	26.4	2.40	4.9	-16.0	10.9	1.81
8.1	426	-24.9	-51.3	26.4	2.40	4.6	-26.9	31.5	5.25
* Average Unit Rotation ± S.D. = 2.47 ± 0.54						† Average Unit Rotation ± S.D. = 4.1 ± 1.15			
Male Relax									
1.5	404	-54.9	-100.0	45.1	4.10	-2.9	-23.3	20.4	3.40
2.0	432	-36.3	-75.7	39.4	3.58	3.9	-22.8	26.7	4.45
2.8	434	-33.0	-83.0	50.3	4.57	-13.0	-31.2	18.2	3.03
3.1	406	-36.1	-64.3	28.2	2.56	10.1	-8.2	18.3	3.05
4.0	436	-33.3	-70.0	36.7	3.33	-2.5	-35.9	33.4	5.56
4.2	408	-35.8	-94.8	59.0	5.36	18.8	-11.7	30.5	5.08
4.4	420	-24.1	-79.5	55.4	5.03	6.8	-16.2	23.0	3.80
4.9	410	-23.4	-82.7	39.3	3.57	2.7	-27.7	30.4	5.06
5.0	438	-41.3	-84.9	43.6	3.96	-2.7	-20.2	17.5	2.92
5.6	422	-21.9	-60.3	38.4	3.50	11.2	-16.6	27.8	4.63
5.7	454	-40.7	-63.2	22.5	2.05	3.2	-30.7	33.9	5.65
5.8	412	-26.9	-50.7	23.6	2.14	6.3	22.2	28.5	4.75
6.2	458	-44.2	-82.0	37.8	3.43	-24.8	-0.4	24.4	4.06
6.6	425	-22.9	-63.4	40.5	3.68	2.7	-28.2	30.9	5.15
7.0	456	-28.9	-60.6	31.7	2.88	-1.7	-33.2	31.5	5.25
7.3	427	-24.3	-54.9	30.6	2.78	9.1	-16.8	25.9	4.32

* Average Unit Rotation ± S.D. = 3.51 ± 0.89

† Average Unit Rotation ± S.D. = 4.4 ± 0.95

Table 5 - Relative Rotation of T1 with respect to T12 and T12 with Respect to the Pelvis

Nominal g-Level	Run No.	Rotation of T1 with Respect to T12 (deg) Max. Ext.	Max. Flex.	Tot. Rot.	Unit Rot.*	Rotation of T12 with Respect to Pelvis (deg) Max. Ext.	Max. Flex.	Tot. Rot.	Unit Rot.†
Female Tense									
2.0	443	-40.9	-68.6	27.7	2.51	8.4	-3.8	12.2	2.03
2.3	413	-24.7	-55.6	30.9	2.80	2.3	-18.6	20.9	3.48
3.2	415	-34.1	-54.4	20.3	1.84	1.2	-21.9	22.9	3.81
4.7	428	-30.1	-48.2	18.1	1.64	2.2	-26.3	28.5	4.75
4.7	430	-28.6	-56.0	27.4	2.49	8.4	-20.4	28.8	4.80
* Average Unit Rotation ± S.D. = 2.25 ± 0.49						† Average Unit Rotation ± S.D. = 3.77 ± 1.13			
Female Relax									
2.0	444	-50.7	-85.1	34.4	3.12	13.7	-16.3	30.0	5.00
2.4	414	-42.7	-67.4	24.7	2.24	-3.9	-30.0	26.1	4.35
3.2	416	-34.5	-59.5	25.0	2.20	-7.6	-29.1	21.5	3.58
4.0	452	-39.0	-65.9	26.9	2.44	8.2	-21.4	29.6	4.90
4.7	429	-28.6	-57.3	28.7	2.60	5.7	-26.9	32.6	5.43

* Average Unit Rotation ± S.D. = 2.52 ± 0.37

† Average Unit Rotation ± S.D. = 4.6 ± 0.37

interpolation program was required to obtain equally spaced film data samples compatible with digital load cell data before a cross plot could be made. This cross plot of M_R against θ shows a curve defining the stiffness of spine and in most cases is a hysteresis loop which is indicative of energy dissipation. Figure 27 shows several M_R - θ curves for male volunteer data at different g-levels in the tensed and relaxed modes. Figure 28 shows M_R - θ curves for female data. The volunteer curves are all within the unembalmed cadaver curve reported by Mital et al (1).

DISCUSSION

The Z-axis acceleration corridors shown in Figures 12, 13 and 14 match well with cadaver data (1). There is similarity in the first positive peak and the second negative peak which occur approximately at the same time. Beyond this time the cadaver torso interacts with the lap cushion and more variation is observed in the cadaver data as compared to volunteer data. In most cases, volunteer data showed only 1 or 2 negative peaks. Load cell data for cadaver and volun-

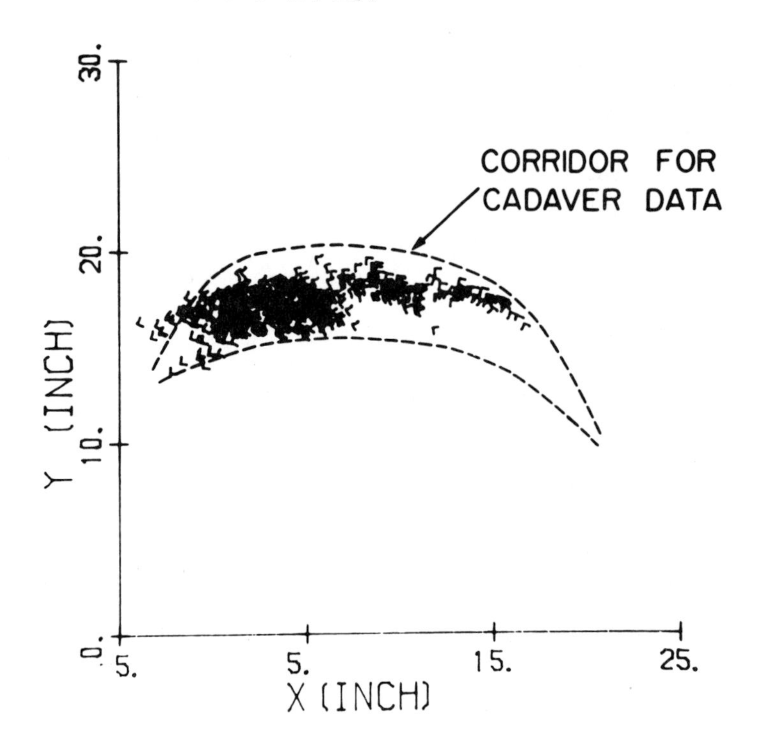

Fig. 24 - T1 motion relative to the pelvis - volunteer data within cadaver data corridors

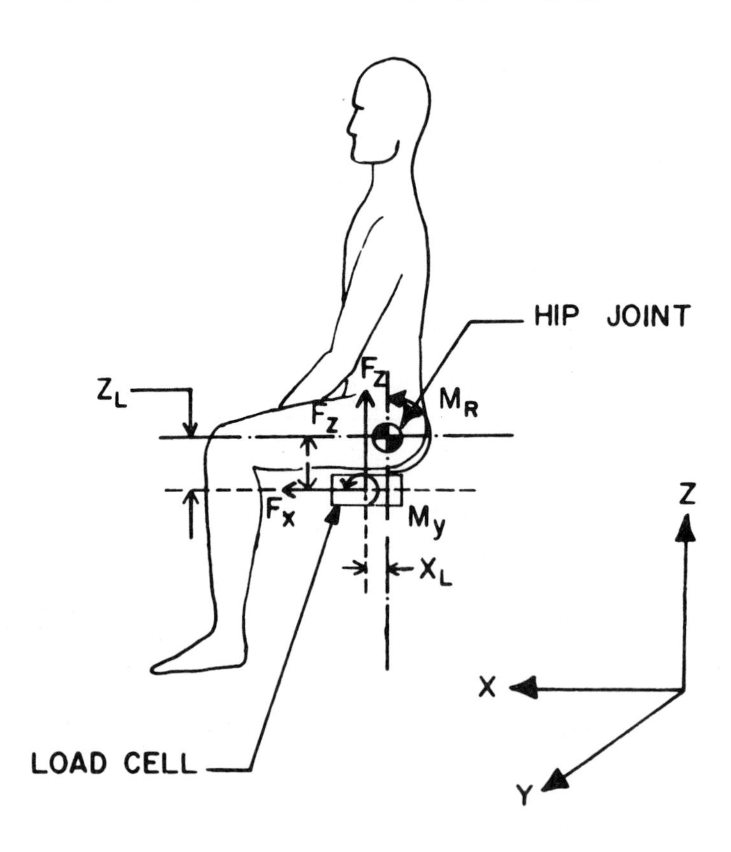

Fig. 25 - Free body diagram for computing resisting moment at the hip

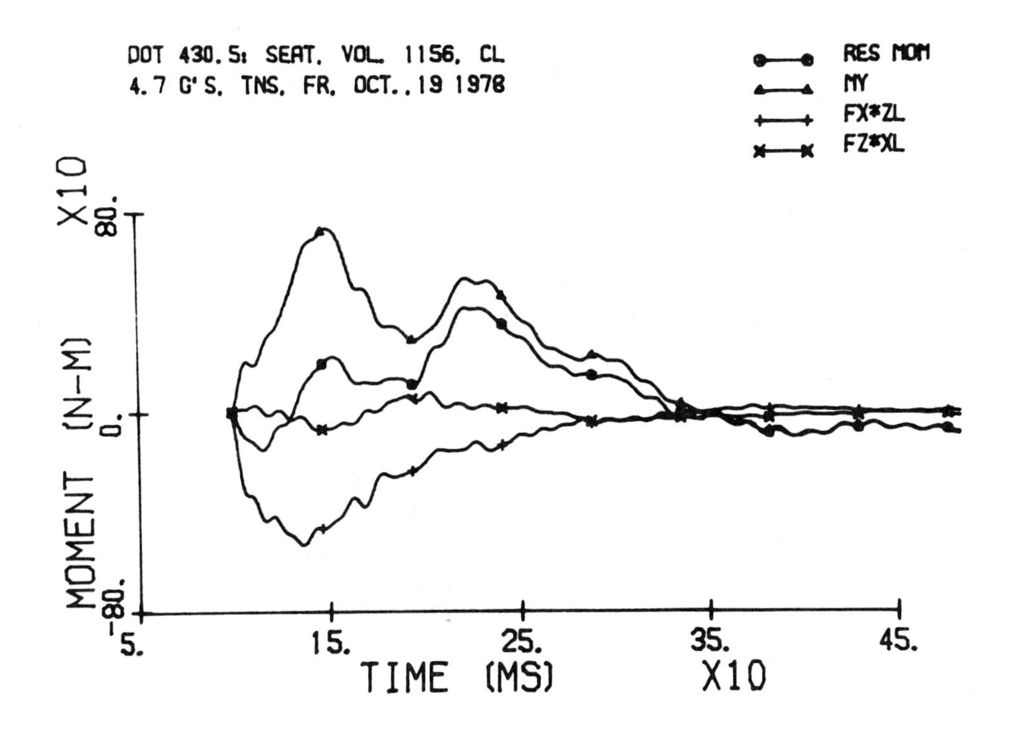

Fig. 26 - Resultant moment computed from seat pan load cell data

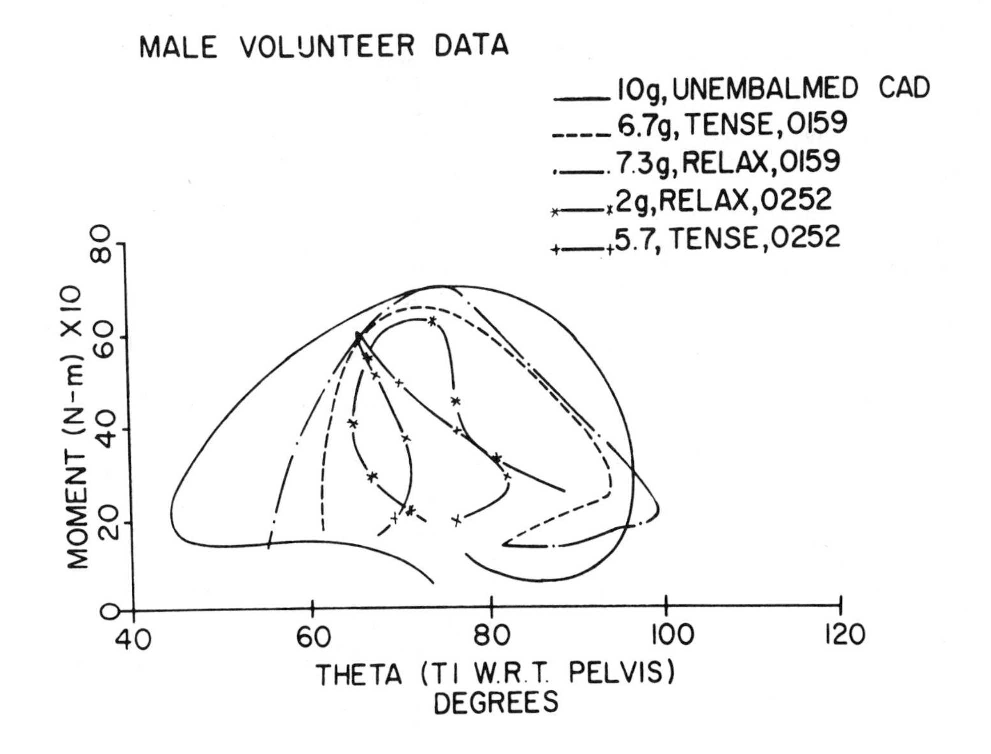

Fig. 27 - Moment - angle curves for spinal flexion - male data

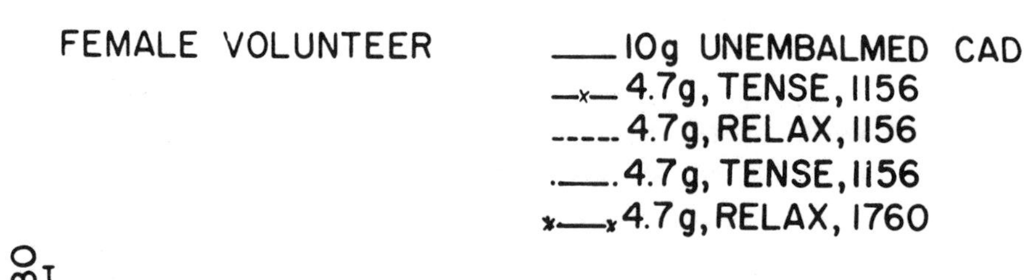

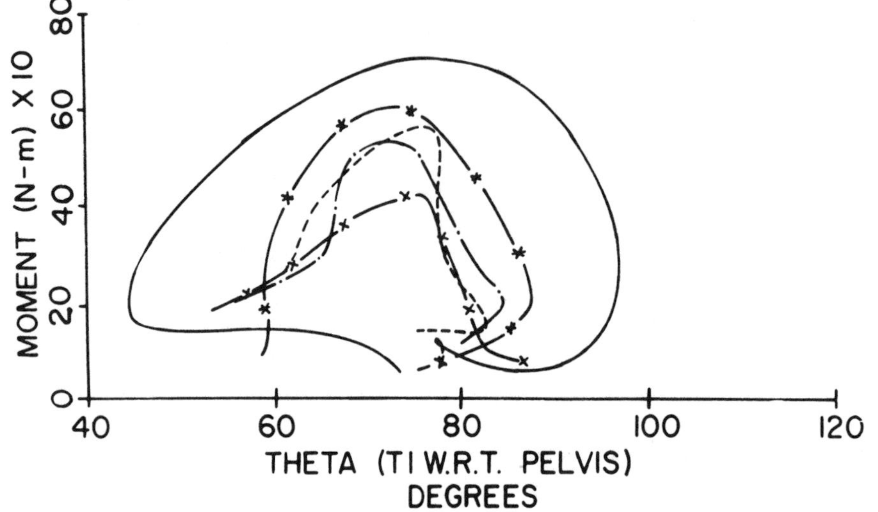

Fig. 28 - Moment - angle curves for spinal flexion - female data

teer runs are also comparable. F_x and M_y are the major components for force and moment and are in the order of 6000 N and 700 N-m, as shown in Figures 27 and 28, are comparable to those plotted from cadaver data. The energy dissipated is less due to the lower g-level of impact. In general, it can be observed from these figures that there is more dissipation with increased g-level and in the relaxed mode in comparison with that for the tensed mode.

The total relative rotation of the thoracic spine appeared to be generally larger than that of the lumbar spine. However, in terms of the flexibility of the two segments, defined as an average or unit rotation at each vertebral joint (Tables 4 and 5) the lumbar spine is apparently more flexible than the thoracic spine for both male and female subjects. As far as cadaver data are concerned, the unit rotation for the thoracic and lumbar segments is comparable for embalmed cadavers. That for the unembalmed cadaver is similar to the data reported for volunteers in this paper. In the relaxed mode, the unit rotation is larger than that in the tensed mode. Male and female unit rotations are comparable. It should be noted, however, that none of the observed differences were found to be statistically significant as determined by the t-test.

The use of a lapbelt to prevent excessive rotation and lifting of the pelvis was the result of cadaveric tests made prior to the human volunteer runs. Bilateral femoral fractures were observed at relatively low g-levels in cadavers due to the excessively high moments applied to the femurs. These moments were reduced by the lapbelt which eliminated femoral fractures up to 20 g. The value of cadaver testing as a prelude to volunteer testing cannot be overemphasized and should be incorporated in all protocols involving human experiments of this type.

Female subjects were more likely to withdraw from the test program than males. Only one of the three females completed an entire sequence of runs up to 8 g. The other two stopped at 4 or 5 g because of transient neck pain. One of the five male subjects was disqualified by the medical officer after 2 runs at 2 g. He had a precipitous rise in blood pressure and pulse rate and was visibly in fear. The data from these 2 runs were not reported and he was not considered as one of the subjects for this experiment.

CONCLUSIONS

1. Spinal kinematics of living human subjects during $-G_x$ acceleration have been quantified up to 8_xg.

2. Displacement data of T1 relative to the pelvis are comparable to those of cadaver runs.

3. There is also a qualitative similarity in T1 acceleration between volunteer and cadaver data.

4. Flexural resistance of the spine in human subjects exhibit similar characteristics to that of unembalmed cadavers. There is appreciable dissipation of energy.

5. Cadavers are a valuable tool in human volunteer experiments of this kind. They point out possible sources of injury to the volunteer and should be used, if available.

6. Female subjects have a weaker neck musculature and are likely to withdraw from the program before reaching the maximum level of 8 g. the principal reason is transient neck pain.

ACKNOWLEDGMENTS

This research was sponsored in part by the National Highway Traffic Safety Administration (NHTSA), under Contract No. DOT-HS-5-01232. Opinions expresed in this paper are those of the authors and are not necessarily of NHTSA or of Wayne State University.

REFERENCES

1. Mital, N.K., Cheng, R., Levine, R.S. and King, A.I. Dynamic Characteristics of the Human Spine During $-G_x$ Acceleration, Proc. 22nd Stapp Car Crash Conf., pp. 141-165, 1978.

2. Mital, N.K., Cheng, R., and King, A.I., A New Design for a Surrogate Spine, Proc. 7th ESV Conf., Paris, France, June, 1979 (In Press).

3. Department of Health, Education and Welfare, Weight, Height and Selected Body Dimensions of Adults, National Center for Health Statistics, June 1965.

4. Ewing, C.L., Thomas, D.J., Lustick, L., Becker, E., Willems, G. and Muzzy, III, W.H., The Effect of the Initial Position of the Human Head and Neck to $-G_x$ Impact Acceleration, Proc. 19th Stapp Car Crash Conf., pp. 487-512, 1975.

841657

Head Impact Response — Skull Deformation and Angular Accelerations

Guy S. Nusholtz, Paula Lux, Patricia Kaiker and Miles A. Janicki
University of Michigan

ABSTRACT

The response of the head to impact was investigated using live anesthetized and post-mortem Rhesus monkeys and repressurized cadavers. The stationary test subject was struck by a guided moving impactor of 10 kg for monkeys; 25 or 65 kg for cadavers. The impactor striking surface was fitted with padding to vary the contact force-time characteristics. The experimental technique used a nine-accelerometer system rigidly mounted on the head to measure head motion, transducers placed at specific points below the skull to record epidural pressure, repressurization of both the vascular and cerebral spinal systems of the cadaver model, and high-speed cineradiography (at 400 or 1000 frames per second) of selected test subjects. The results of the tests demonstrate the potential importance of skull deformation and angular acceleration on the injury produced in the live Rhesus and the damage produced in both the post-mortem Rhesus and the cadaver as a result of impact.

IN THE UNITED STATES roughly 49 percent of head injuries can be attributed to motor vehicle accidents(1-3)*. Investigation of trauma to the head and of mechanisms of injury becomes invaluable for allocating resources and for formulating policy to reduce its incidence, morbidity, and mortality.

Because motor vehicle field accident data do not provide the level of detail necessary to ascertain mechanisms of injury resulting from the interactions of the passenger with the vehicle interior during an accident, bioengineers use trauma experiments with human surrogates (cadavers or animals) to document kinematic parameters so that mechanisms of injury can be better simulated, modeled and verified.

The repressurized cadaver is often chosen as an experimental model because its geometry and soft tissue distribution is similar to that of a live human. Damages to repressurized cadavers which correlate well with clinically observed injuries are those that can be documented by gross autopsy. They include scalp lacerations (linear, flap, stellate), fractures of the cranial vault or base (linear, depressed, comminuted), lesions which are visible to the naked eye and hemorrhage (petechia; contusion; epidural, subdural, subarachnoid and intracerebral hematomas). Mild concussion, cerebral concussion and diffuse brain injury disrupt the functioning of the brain without gross overt structural evidence. Because diagnosis of concussion requires the observation of behavioral responses, animal subjects are necessary in experimental contexts. As a result, subtle distinctions are lost, making scaling of the results to the human level approximate (4).

The bioengineering literature on mechanisms of injury is a rich one (5-43). However, despite conscientious speculation as to the mechanics which produce head trauma and theoretical and experimental scrutiny of various hypotheses about it, considerable controversy still exists. The hypotheses usually cannot wholly fit clinical or pathological observations, assessment of biomaterial properties, or may not agree with predictions of mathematical models.

Several such controversies involve angular and linear acceleration and skull deformation (5-8,12,16-17,21-22,34,38,40). These debates are complicated by the complex mechanical structure of the head and brain. A number of biomechanical analyses (finite element analyses, in particular) have attempted to model the brain with reasonable geometric precision. However, these calculations involve over-simplified modeling of materials and behavior, assuming simple elasticity or homogeneous materials (12-13,23-24,38,41-44). A number of other investigators have attempted to define the

---

*Numbers in parenthesis designate references at end of paper.

pressure-volume relationship of the skull to impact (19,25-26,28-30,32,36,43-44). However, the skull-brain interface as well as the effects of skull deformation on the impact response of the brain is not well understood.

A major difficulty in the investigation of head trauma is designing kinematic experiments which interfere minimally with the biological and and physical systems being observed, yet produce results that correspond well with clinically observed trauma. Some understanding of head injury mechanisms as a result of blunt impact has resulted from relating kinematic parameters to injury modes. However, with the possibility of several injury mechanisms correlations of this type do not always imply causation. The parameters commonly used for describing head mechanical response during direct impact have been angular and translational accelerations, velocities, displacements of the head as a rigid body, skull bone deformations, and internal pressures in the brain. Many investigators have chosen to investigate a single parameter, such as resultant head acceleration for Head Injury Criterion (HIC) calculation, and later use it as an index of severity or tolerance threshold.

This paper discusses techniques developed and used by the Biomechanics Group at the University of Michigan Transportation Research Institute (UMTRI) for measuring three-dimensional head motion, epidural pressure, and internal brain motion of repressurized cadavers and Rhesus monkeys during head impact. The results will be compared to those previously presented (30,40) and a possible injury mechanism will be suggested.

METHODOLOGY

Three live anesthetized and three post-mortem Rhesus monkeys* and nine repressurized cadavers** were tested. Five cadavers were each subjected to a series of up to three head impacts using the UMTRI pneumatic impacting device with a 25 or 65 kg impactor. The remaining four cadavers were subjected to two head impacts each with the UMTRI pendulum impacting device with a 25 kg impactor. The cadavers were instrumented with a nine-accelerometer array located on the head to measure three-dimensional motion. Epidural pressure transducers were used to monitor pressure changes during impact of the brain-skull interface. Both cerebrospinal and vascular systems of the cadaver head-brain complex were repressurized. In addition, high-speed photokinemetrics were obtained using normal photographic or cineradiographic techniques.

Using the pneumatic impacting device with a 10 kg impactor, three live and three post-mortem non-human primates (Macaca mulatta) were each subjected to a single head impact to the occipital region. The six Rhesus subjects used in these experiments were obtained by UMTRI from the University of Michigan Unit for Laboratory Animal Medicine (ULAM). The six Rhesus subjects were instrumented similarly to the cadavers with nine-accelerometer arrays on the head and with epidural pressure transducers to document transient pressures. For post-mortem Rhesus subjects, neither cerebrospinal nor vascular systems of the head-brain complex were repressurized.

IMPACT TESTING - The methods and procedures used in this research are outlined below. Additional information can be found in (30-32,45-46).

Linear Pendulum Impact Device - The UMTRI linear pendulum impact device, using a free-falling pendulum as an energy source, strikes either a 25 kg or 56 kg impact piston. The piston is guided by a set of Thomson linear ball bushings. Axial loads were calculated from data recorded using a Setra Model 111 accelerometer.

Impact conditions between tests were controlled by varying impact velocity and the type and depth of padding on the impactor surface. Piston velocity was measured by timing the pulses from a magnetic probe which senses the motion of the targets in the piston.

A specially designed timer box was used to control and synchronize the impact events during a particular test, such as the release of the pendulum and activation and deactivation of the lights and high-speed cameras.

Ballistic Impact Device - The UMTRI ballistic impact device (Figure 1), consists of an air reservoir, a ground and honed cylinder, and a carefully fitted piston mechanically coupled to a ballistic pendulum. The piston, propelled by compressed air through the cylinder from the air reservoir chamber, accelerates the ballistic pendulum. The mass of the ballistic pendulum can be varied from 10 to 150 kg. The piston is arrested at the end of its travel allowing the ballistic pendulum to become a free-traveling impactor. The ballistic pendulum is fitted with an inertia-compensated load cell.

Nine-Accelerometer Head Plate - For Rhesus subjects, the installation of the nine-accelerometer plate is accomplished as follows.

*Animals were handled according to the American Association for Accreditation of Laboratory Animal Care and National Institutes of Health guidelines.

**The protocol for the use of cadavers in this study was approved by the Committee to Review Grants for Clinical Research of the University of Michigan Medical Center and follows guidelines established by the U.S. Public Health Service and recommended by the National Academy of Sciences/National Research Council.

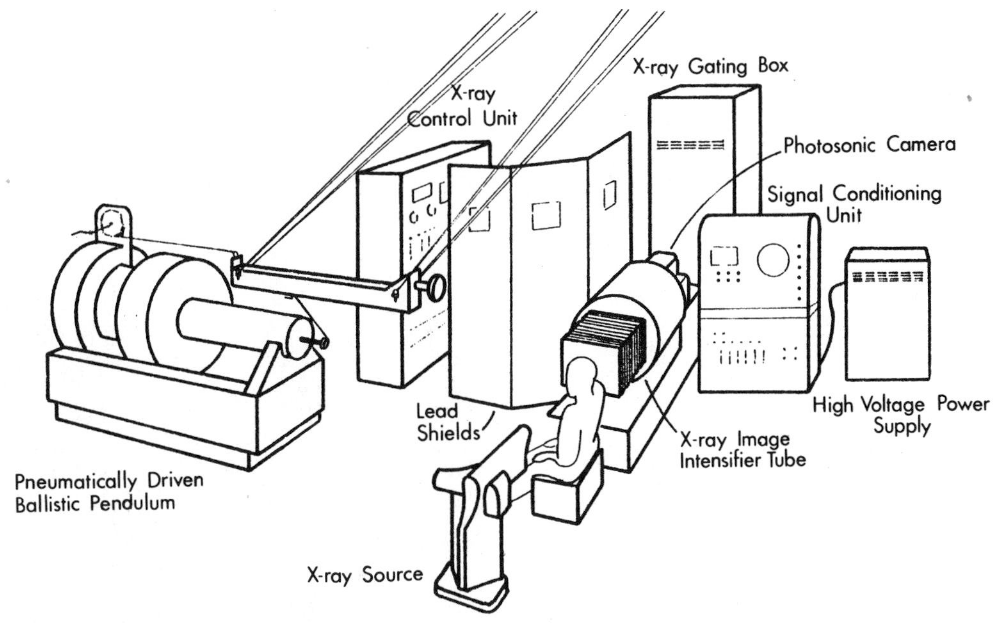

Fig. 1. Pneumatic Ballistic Pendulum and Test Setup for Cadaver Impacts.

The scalp is removed from the frontal bone over the orbital ridges. Several metal self-tapping screws are attached firmly to the skull through small pilot holes drilled into the orbital ridges and into the dental plate above the canine teeth. Quick-setting acrylic plastic is molded around each of the screws and the base of the plate mount embedding it in the plastic. The orientation of the plate in this position is shown in Figure 2.

For cadavers, the nine-accelerometer plate is installed in the following manner. A two-by-two inch patch of scalp is removed from the right occipital-parietal area. Four small screws are then placed in a trapezoidal pattern in the skull within the dimensions of the accelerometer plate mount. Quick setting acrylic plastic is molded around the screws forming a base. The plate mount is then placed in the acrylic base. See Figure 3A for the location of the plate mount.

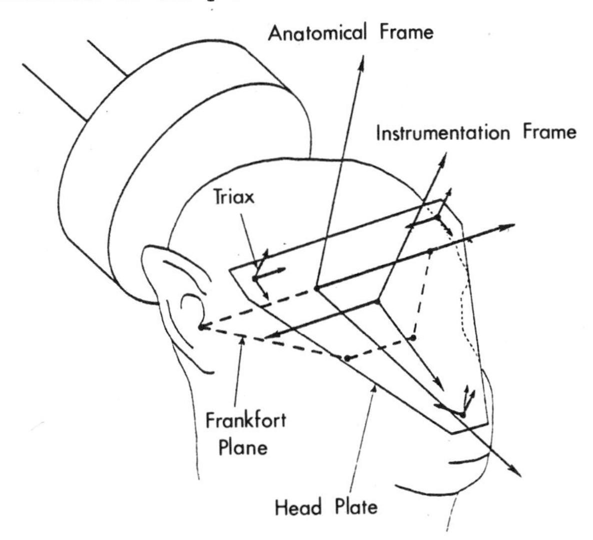

Fig. 2. Nine-Accelerometer Plate Position for Rhesus Monkey Impacts.

It is necessary to determine the instrumentation frame's exact location and orientation in relation to the anatomical frame. A three-dimensional x-ray technique was developed which requires taking two orthogonal radiographs of the instrumented head. The procedure requires the identification of four anatomical landmarks (two superior edges of the auditory meati and two infraorbital notches) with four distinguishable lead pellets, plus the identification of four lead pellets inlaid in the plate to define the instrumentation frame. A similar radiographic technique was used with Rhesus monkeys to determine the orientation and location of the instrumentation frame with respect to the anatomical frame.

Epidural Pressure Measurements - Epidural pressures of Rhesus subjects are obtained with a Kulite model MCP-55-5F catheter tip pressure transducer. For cadavers, Endevco series 8510 piezoresistive pressure transducers are used.

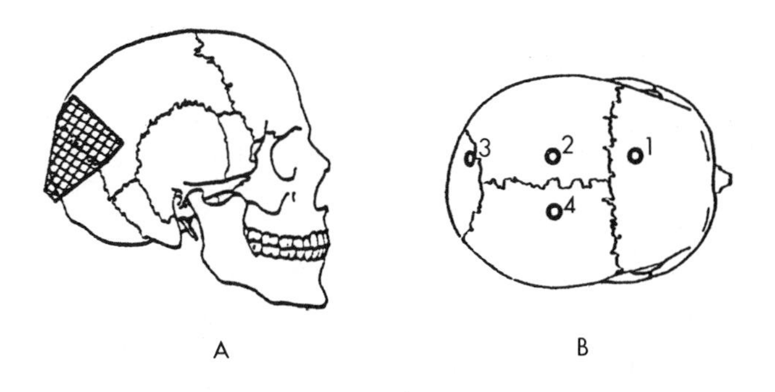

Fig. 3A. Nine-Accelerometer Plate Position for Cadaver Impacts.

3B. Location of Transducers for Epidural Pressures 1, 2, 3, and 4.

For Rhesus subjects a small circle of scalp is removed and a Stryker bone coring tool is used to make a hole in the skull with a circular bit. An adjustable set-screwed collet is used to core into the skull in small increments, preventing damage to the dura mater. The increment of bone is then removed using a dental scoop, and the hole is tapped for a coarse thread. A tabular magnesium coupling device is screwed into the tapped hole in the skull (Figure 4A). It is anchored into place using quick-setting acrylic plastic molded around the base. Dow Corning dielectric gel (silicon fluid) is injected into the tubing to act as a coupling medium. The Kulite pressure transducer is then inserted and secured at proper depth.

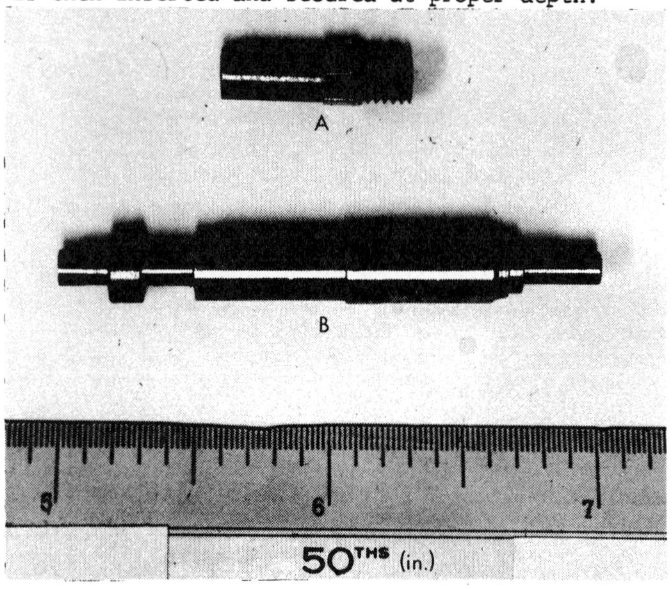

Fig. 4A. Magnesium Coupling Device.
4B. Skull Bone Coring Device.

For cadaver subjects four 1 cm diameter circles of scalp are removed over the frontal, parietal and occipital bones and a Stryker bone coring tool is used to make four holes in the skull with a circular bit (Figure 4B). Holes are placed out of the contact area of the impacting surface and are not drilled into sutures (Figure 3B). The dura mater under these holes is perforated without cutting the brain. The holes are tapped. The brass transducer couplings are inserted into the tapped holes. Two pinhead screws are attached 2 cm from each transducer. Quick-setting acrylic plastic is molded around the pinhead screws and the mouth of each transducer coupling as a mooring device. After checking fluid flow, the Endevco series 8510 piezoresistive pressure transducers are attached to the fittings.

Radiopaque Target Gel - A neutral density radiopaque gel is used to determine motion of the brain during impact. The gel is injected into the brain through the holes used for insertion of the pressure transducers. The injection technique produces lines of radio-contrast in the brain that show up in high-speed cineradiographic movies. See Figure 5.

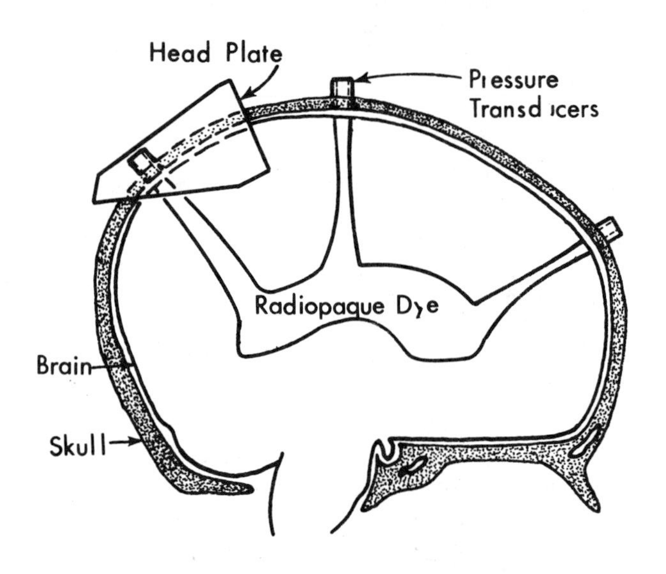

Fig. 5. Radiograph and Reconstruction of Radiograph With Radio-contrast Dye *in situ*.

X-Ray Motion Descriptors - The procedures used for defining x-ray motion descriptors are explained in (45) and briefly outlined below. The cineradiograph system allows non-invasive viewing of internal anatomical structure *in vivo*. In the case of a rigid structure such as bone, the radiopaque targets can be placed on or near anatomical landmarks and motion can be described similarly to that of standard photometric techniques. Problems arise when soft tissue is to be analyzed and rigid body dynamics no longer offer a good approximation.

Several methods could be suggested to produce analytical information describing the soft tissue of the brain. In this paper the motion discriptors chosen are based upon the shadows of objects in a two-dimensional image space produced by a point source of x-rays which are associated with the anatomical structures or the radiopaque dye injected into the brain. The descriptors are at most two-dimensional and do not take into account rotations and translations which move objects in and out of the plane of gross whole body motion. In addition, changes in the x-ray cross-section of objects can lead to changes in the descriptors which do not have a direct relation to rigid body motion. In the impact tests presented in this paper radiopaque gel was injected into the head producing four curved lines in the brain and outlining the ventricles in some tests. Differential motion between the brain and the skull was obtained by comparing the motion of points on the curve closest to the center of the epidural pressure transducer. General characteristics of the motion of the brain were obtained through the changes in shape of the curved lines and ventricles.

Cadaver Vascular Repressurization - To pressurize the cadaver vascular system of the head, the common carotid artery is located at a point in the neck and an incision is made. (See Figure 6.) A balloon catheter is inserted and positioned such that the balloon is in the internal carotid artery just above the point where the external carotid artery branches. A narrow polyethylene tube is inserted at the same point and runs into the internal carotid artery just past the balloon. A Kulite pressure transducer is then fed through this tube so that vascular pressure may be monitored. Finally, the vertebral arteries are tied off above the clavicle such that fluid pressure in the head may be maintained. Just prior to testing, a solution of India ink and saline is released from a tank into the vascular system of the head. A pressure transducer monitors the flow so that the system is brought to normal physiological pressure just prior to impact.

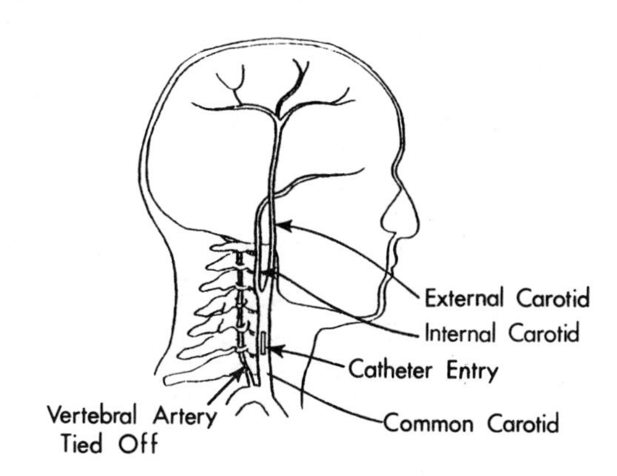

Fig. 6. Vascular Pressurization Procedure and Placement.

Cadaver Cerebrospinal Repressurization - For the cadaver head impacts, the subdural region surrounding the brain and spinal cord is repressurized by coring a small hole into the second lumbar vertebra and inserting a Foley catheter under the dura of the spinal cord such that the balloon of the catheter reaches mid-thorax level. To check fluid flow through the ventricles, saline is injected through the Foley catheter until fluid rises to the top of the piezoresistive pressure transducer couplings. The couplings are capped until the radiopaque sodium iodide gel target has been slowly injected through the couplings into the brain cortex and a setup radiograph has been made of the head. The point at which the catheter passes through the lamina of the second lumbar vertebra is sealed with plastic acrylic. (Figure 7)

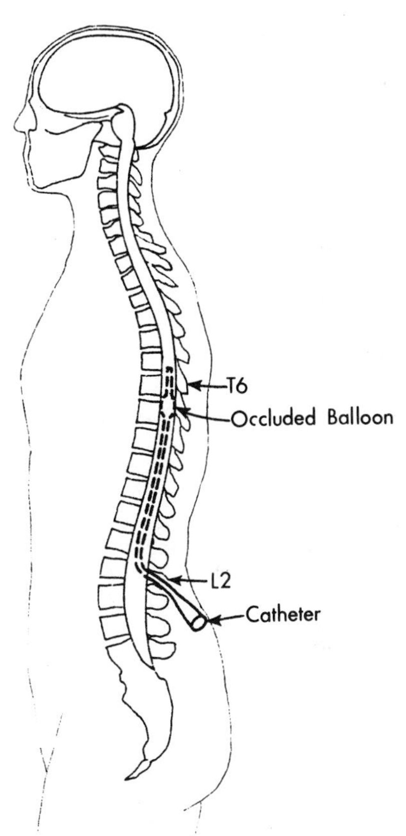

Fig. 7. Cerebrospinal Pressurization Procedure and Placement.

Test Subject Preparation - The unembalmed cadavers were stored at 4 degrees centigrade prior to testing. The cadaver is x-rayed as part of the structural damage evaluation and anthropomorphic measurements are recorded. Next, the cadaver is instrumented, sanitarily, dressed and transported to the testing room where the accelerometers and pressure transducers are attached. The subject is positioned. Next, the radiopaque gel target is inserted, and pretest x-rays and photographs are taken. Pressurization is checked. Then the subject is impacted. Each cadaver received either two duplicate head impacts or one low-

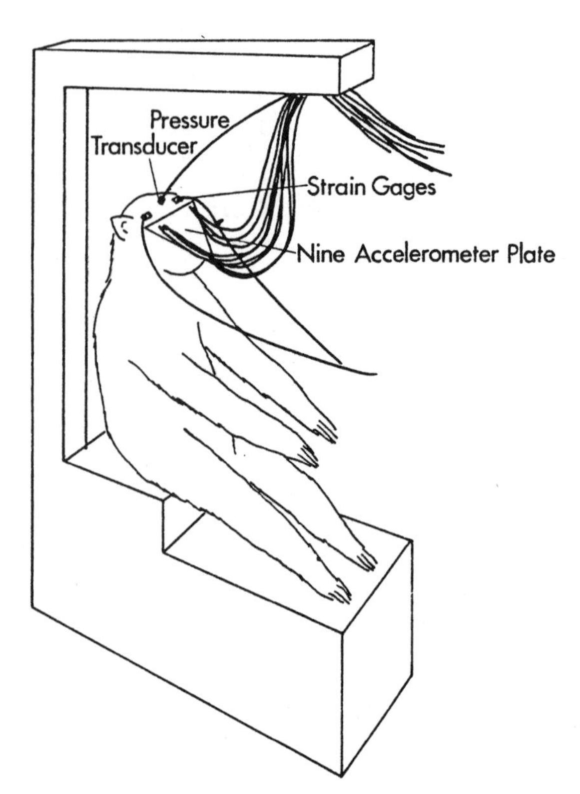

Fig. 8. Test Setup for Posterior-Anterior Rhesus Head Impact.

energy heavy-padded frontal impact and one medium energy slightly-padded frontal impact.

Three Rhesus impacts were conducted on post-mortem subjects. Upon termination, they were stored in a cooler at 4 degrees centigrade for 48 hours before testing. Living Macacas were used in the other three experiments. The protocol for post-mortem animals was less complex than that for the live animals, which was presented earlier (32), outlined below.

The subject is first anesthetized with ketamine and then maintained with ketamine and sodium pentobarbital. An airway is established. The upper body is prepared and the weight and biometric measurements are taken. Using a cauterizing scalpel, the scalp and muscle mass are removed from the frontal bone and the screws used to affix the nine-accelerometer plate and the epidural pressure transducer skull fitting are screwed into place. Figure 8 shows the positioning of the instrumentation on the Rhesus skull.

The Rhesus is placed in an erect sitting position with its posterior side towards the impactor, so that the line of impact is in the mid-sagittal plane in the posterior-anterior direction. The Rhesus is stabilized with paper tape. A polaroid photograph is then taken through the cineradiograph to check the position of the animal and the x-ray settings. Figure 8 also shows the basic test setup. Eight hours after impact, a 5 ml dose of Uthol (concentrated, unpure sodium pentobarbital) is injected via the hind leg I.V. catheter to euthanize the animal.

Rhesus Initial Test Conditions - The tests being reported here are a continuation of Rhesus tests reported earlier (30). Conditions for this set of tests were similar except higher impact energy levels were used and the subjects were positioned so that impact occurred through the estimated center of mass. The impactor was the UMTRI 10 kg pneumatic impacting device and the impacting surface was 10 cm in diameter and padded with 2.5 cm Ensolite.

Cadaver Initial Test Conditions - Tests 82E001 thru 82E062 used the UMTRI 25 kg linear pendulum impacting device with a 15 cm diameter impacting surface padded with 2.5 cm Ensolite. Tests 82E081 thru 84E141D used the UMTRI 25 kg ballistic impacting device fitted with a 15 cm diameter impacting surface padded with 2.5 cm Ensolite, or a sandwich of 2.5 cm styrofoam, 5 cm Dow Ethafoam plus 2.5 cm Ensolite, or one of 0.5 cm Ensolite, 5 cm seating foam plus 0.5 cm Ensolite. Tests 84E151A through 84E161C used the UMTRI 65 kg ballistic impacting device fitted with a 15 cm diameter impacting surface padded with either 0.5 cm Ensolite or a sandwich of 0.5 Ensolite, 5 cm seating foam, and 0.5 cm Ensolite. The target area for all of these impacts was the center of the forehead above the orbits (frontal). Impact occurred in the anterior to posterior direction. All cadavers were seated and positioned with paper tape so that the subject and the impact target were stable (Figure 9).

Cineradiographs - In selected subjects high-speed cineradiographs were taken (some cadavers and all Rhesus). The cineradiographs were taken of the impact events at 1000 or 400 frames per second. The UMTRI high-speed cineradiographic system (47-48) consists of either a Photosonics 1B or Miliken high-speed 16 mm motion-picture camera which views a 5 cm diameter output phosphor of a high-gain, four-stage, magnetically focused image intensifier tube, gated on and off synchronously with shutter pulses from the motion-picture camera. A lens optically couples the input photocathode of the image intensifier tube to x-ray images produced on a fluorescent screen by a smoothed direct-current x-ray generator. Smoothing of the full-wave rectified x-ray output is accomplished by placing a pair of high-voltage capacitors in parallel with the x-ray tube. The viewing field for these experiments was between 20 and 40 cm.

Photokinemetrics - The motion of the subject was determined from the high-speed (1000 frames per second) film by following the motion of single-point phototargets on the head and on the impactor piston. For selected cadaver frontal impacts, a Hycam camera operating at 3000 frames per second provided a close-up

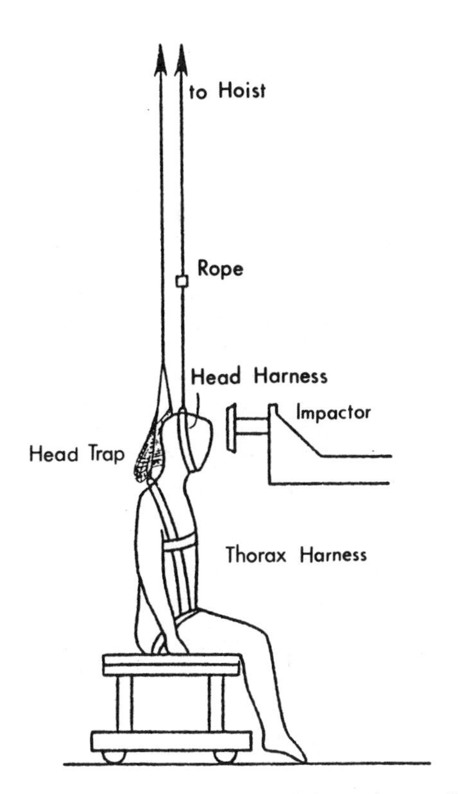

Fig. 9. Initial Conditions for Cadaver Frontal Head Impact.

lateral view of the impact. For selected cadaver frontal impacts the Photosonics provided a lateral 1000 frames per second overall view.

Data Handling - All transducer time histories (impact force, impact acceleration, epidural pressures, nine head-accelerations) were recorded unfiltered on either a Honeywell 7600 FM Tape Recorder or a Bell and Howell CEC 3300/CEC 3400 FM Tape Recorder. A synchronizing gate was recorded on all tapes. All data was recorded at 30 ips. The analog data on the FM tapes was played back for digitizing through proper anti-aliasing analog filters. The analog-to-digital process for all data, results in a digitial signal sampled at 6400 Hz equivalent sampling rate. It has been reported that skull vibrations above 1300 Hz could cause very local motion in the accelerometer mountings (40). To reduce this effect, the raw transducer time histories were filtered, digitally, with a Butterworth filter at 1000 Hz, 6th order.

METHOD OF ANALYSIS - The techniques used to analyze the results are outlined below. Additional information can be found in (30-32,41-42).

Frame Fields - As the head moves through space, any point on the head generates a path in space. In head injury research we are interested in the description of the path of the anatomical center and in events which occur as it moves. A very effective tool for analyzing the motion of such a point (the anatomical center), as it moves along a curved path in space, is the concept of a moving frame (49-52). The path generated as the point travels through space is a function of time and velocity. A vector field is a function which assigns a uniquely defined vector to each point along a path. Thus, any collection of three mutually orthogonal unit vectors defined on a path is a frame field. Therefore, any vector defined on the path (for example, acceleration) may be resolved into three orthogonal components of any well-defined frame field, such as the laboratory or anatomical reference frames. Changes in a frame field with time (for example, angular acceleration of the frame field) are interpreted as vectors defined on the curve and are also resolved into three components.

In biomechanics research, frame fields are defined based on anatomical reference frames. Other frame fields such as the Frenet-Serret frame or the Principal Direction Triad (32,46), which contain information about the motion embedded in the frame field, have also been used to describe motion resulting from impact.

The Frenet-Serret frame (51-52) consists of three mutually orthogonal vectors T, N, B. At any point in time a unit vector can be constructed that is co-directional with the velocity vector. This normalized velocity vector defines the tangent direction T. A second unit vector N is constructed by forming a unit vector co-directional with the time derivative of the tangent vector T (the derivative of a unit vector is normal to the vector). To complete the orthogonal frame, a third unit vector B (the unit binormal) can be defined as the cross product T x N. This procedure defines a frame at each point along the path of the anatomical center. Within the frame field, the linear acceleration is resolved into two distinct types. The tangent acceleration [Tan(T)] is always the rate of change of speed (absolute velocity) and the normal acceleration [Nor(N)] gives information about the change in direction of the velocity vector. The binormal direction contains no acceleration information.

Transfer Function Analysis - With blunt impacts, the relationship between a transducer time history at a given point and the transducer time history of another given point of a system can be expressed in the frequency domain through the use of a transfer function. A Fast Fourier Transformation of simultaneously monitored transducer time histories can be used to obtain the frequency response functions of impact force and accelerations of remote points. Once obtained, a transfer function of the form:

$$(Z)(iw) = (w) \quad \mathcal{F}[F(t)]/\ \mathcal{F}[A(t)]$$

can be calculated from the transformed quantities where w is the given frequency, and $\mathcal{F}[F(t)]$ and $\mathcal{F}[A(t)]$ are the Fourier transforms of the impact forces and acceleration of the point of interest, at the given frequency. This particular transfer function is closely related to a mechanical transfer impedance (53) which is defined as the ratio between simple harmonic driving force and corresponding velocity of the point of interest. Mechanical transfer impedance is a complex valued function which for the purpose of presentation will be described by its magnitude and its phase angle. In addition to a transfer function relating force to velocity, a transfer function can be generated which relates the response of one point in the system to another point in the system, the response being expressed in the frequency domain. Analogous to mechanical impedance, a Fast Fourier Transformation of simultaneously monitored transducer time histories from any two points in the system can be used to obtain the frequency response functions relating those two points. In the case of a force and a pressure, such as impact force and epidural pressure, a transformation of the form:

$$(X)(iw) = F[F(t)]/\ F[p(t)]$$

can be calculated from the transformed quantities, where w is the given frequency, $F[F(t)]$ and $F[p(t)]$ are the Fourier transforms of the impact force time history and the pressure time history.

Correlation Functions - To describe some of the fundamental properties of a time history, such as acceleration or force, two types of statistical measures may be used:

1. Auto-correlation Function. This measure is the correlation between two points on a time history and is a measure of the dependence of the amplitude at time t_1 on the amplitude at t_2.

2. Cross-correlation Function. This is a measure of how predictable, on the average, a signal (transducer time history) at any particular moment in time is from a second signal at any other particular moment in time.

The auto-correlation function is formally defined as the average over the ensemble of the product of two amplitudes:

$$R_x(t_1,T_2) = \int_{-\infty}^{+\infty}\int_{-\infty}^{+\infty} X_1,X_2,P(x_1,x_2,T_1,T_2)dx_1,dx_2$$

where x_1, x_2 are the amplitudes of the time history and $p(x_1,x_2,t_1,t_2)$ is the joint probability density. Normally the above definition cannot be used to generate an auto-correlation function directly. However, it can be shown that for a discrete time history of a finite duration, a close approximation of the auto-correlation function can be obtained through the use of a Fourier transform. (53)

In addition to auto-correlation, cross-correlation can be used to obtain useful information about the relationship between two different time histories. For example, the cross-correlation between acceleration measurements at two different points of a material body may be determined for the purpose of studying the propagation of differential motion through the material body. Cross-correlation functions are not restricted to correlation of parameters with the same physical units; for example, one might determine the cross-correlation between the applied force and the acceleration response to that force. Similar to the auto-correlation function, the calculation of the cross-correlation of two signals begins by taking the Fourier transform of both time histories (Y_1,Y_2). The cross-spectral density is the complex-valued function $(Y_1 \cdot Y_2^*)$. The cross correlation is then the Fourier transform of the cross-spectral density. (53)

Pressure Time Duration Determination - Two different types of pressure-time histories were observed, unimodal and bimodal. The unimodal waveform was characterized by one maximum and the bimodal waveform by two local maxima. In order to define the pressure duration, a standard procedure was adopted which determined the beginning and end of a pulse. This procedure began by determining the peak, or the first peak in the case of a bimodal waveform. Next, the left half of the pulse, defined from the point where the pulse starts to rise to the time of peak, was least-squares fitted with a straight line. This rise line intersected the time axis at a point which was taken as the formal beginning of the pulse. A similar procedure was followed for the right half of this pulse, i.e., a least-squares straight line was fitted to the fall section of the pulse, which was defined from the peak to the point where the pulse minimum occurred. The point where this line intersected the time axis was the formal end of the pulse in the unimodal case, and the formal end of the first peak in the bimodal case. The pressure duration for a unimodal waveform was defined by these points. For a bimodal waveform, these two points were used to determine the first pressure duration. Another least-squares straight line was fitted to the fall section of the second pulse. The point at which this line intersected the time axis was the formal end of the waveform, and the total pressure duration was then defined from this point and the beginning point.

Force Time History Determination - In general the force-time histories were unimodal with a single maximum, smoothly rising, peaking and then falling. Paddings on the striker surface effected different force-time history durations. Force duration was determined using the same techniques for determining pressure duration, that is the same boundary defining and least-squares straight-line fitting techniques were used.

RESULTS

Table 1 lists the initial test conditions for the Rhesus subjects while Table 2 lists those for cadaver subjects. Table 3 summarizes the Rhesus impacts and Table 4 summarizes the cadaver impacts. Table 5 characterizes impact pressures for cadavers. Table 6 reports the injuries/damages for the Rhesus subjects for which kinematic response was reported earlier (30) and Table 7 reports on injuries/damages for Rhesus subjects for which the kinematic response is being reported here. Table 8 reports the damage observed in cadaver subjects. A gross inspection was carried out for each test subject. In addition, representative microscopic sections were examined histologically for each Rhesus brain and included samples from the following areas: frontal cortex, anterior commisure level, mammilary body level, hoppocampus at lateral geniculate level, occipital cortex, cerebellum, midbrain, pons, medulla, and spinal cord. Selected time histories in the appendix are examples of important kinematic factors associated with the research performed in this project. The variables these examples illustrate are tangential and normal acceleration, resultant acceleration, rate of change of the tangential vector (T-rate) and rate of change of the binormal vector (B-rate). In addition, impact force, resultant angular acceleration and velocity, linear velocity, and pressures for both live Rhesus and repressurized cadavers are shown.

Table 1. Rhesus Head Impact
Initial Test Conditions

Test No.	Subject Condition	Subject Positioning Center of Gravity With Respect to Occipital Impact	Impact Surface Padding Thickness+	Pressure Transducer Location Relative to Bregma (cm)		Velocity (m/s)
				X	Y	
78A232	post-mortem	below	2.5 cm Ensolite++	3.2	-1.1	11.5
78A234	post-mortem	below	2.5 cm Ensolite++	2.8	-0.8	13.5
78A236	post-mortem	at	2.5 cm Ensolite++	3.0	0.6	12.5
78A238	post-mortem	above	2.5 cm Ensolite++	2.9	-0.8	12.5
78A239	live	at	2.5 cm Ensolite++	3.6	0.8	12.5
78A241	live	above	2.5 cm Ensolite++	1.4	0.3	12.5
79A249	live	at	2.5 cm Ensolite++	0.0	-2.0	15.4
79A251	live	at	2.5 cm Ensolite++	-2.5	-2.0	13.0
79A253	live	at	2.5 cm Ensolite++	-2.0	2.0	14.8
79A256	post-mortem	at	2.5 cm Ensolite++	-1.0	-2.0	15.4
79A258	post-mortem	at	2.5 cm Ensolite++	-1.5	.5	14.5
79A260	post-mortem	at	2.5 cm Ensolite++	-2.5	2.5	16.0

+10 cm diameter

++10kg striker

Table 2. Cadaver Initial Test Conditions

Test No.	Subject Condition	Impact Surface Padding Thickness+	Velocity m/s
82E001++	repressurized	2.5 cm Ensolite	5
82E021++ 82E022++	repressurized repressurized	2.5 cm Ensolite 2.5 cm Ensolite	5.2 5.7
82E041++ 82E042++	repressurized repressurized	2.5 cm Ensolite 2.5 cm Ensolite	5.5 5.5
82E061++ 82E062++	repressurized repressurized	2.5 cm Ensolite 2.5 cm Ensolite	5.5 5.5
82E081+++ 82E082+++	repressurized repressurized	2.5 cm Ensolite 2.5 cm Ensolite	3.8 3.8
83E102+++	repressurized	2.5 cm styrofoam 5.0 cm Dow Ethafoam 2.5 cm Ensolite	4.5
84E141A+++	repressurized	0.5 cm Ensolite 5.0 cm seating foam 0.5 cm Ensolite	4.5
84E141B+++	repressurized	0.5 cm Ensolite 5.0 cm seating foam 0.5 cm Ensolite	4.5
84E141C+++	repressurized	0.5 cm Ensolite	4.5
84E141D+++	repressurized	0.5 cm Ensolite	3.8
84E151A++++	repressurized	0.5 cm Ensolite 5.0 cm seating foam 0.5 cm Ensolite	3.6
84E151B++++	repressurized	0.5 cm Ensolite	3.8
84E161A++++	repressurized	0.5 cm Ensolite 5.0 cm seating foam 0.5 cm Ensolite	3.8
84E161B++++	repressurized	0.5 cm Ensolite	5.0

+15 cm diameter
++25 kg pendulum
+++25 kg cannon
++++65 kg cannon

Table 3. Rhesus Impact Test Summary

Test No.	Linear Acceleration Tangent m/s/s	Resultant Acceleration m/s/s	Resultant Angular Acceleration r/s/s	Resultant Angular Velocity r/s	Linear Velocity m/s	Kulite Pressure Kpa	Force N	Force Duration ms
78A232	7000	7000	22000	40	12	117	5800	3
78A234	7500	6400	37500	40	14	--	5700	4
78A236	7000	7000	28000	30	13	-9	5600	4
78A238	7100	7500	40000	60	13	100,-55	6600	4
78A239	7500	7500	41200	60	13	260	5000	5
78A241	8400	8000	54000	70	13	490	8700	3
79A249	14000	15200	70000	63	16	-70	7200	4
79A251	13000	13600	63600	75	13	200,-110	7500	4
79A253	9500	9800	40000	75	15	-101	5400	5
79A256	10500	10500	66000	84	16	--	6000	10
79A258	10000	11900	70000	75	15	-35,40	5700	5
79A260	12600	12600	60000	80	16	140,-100	7100	5

Table 4. Cadaver Impact Test Summary

Test No.	Linear Acceleration Tangent m/s/s	Resultant Acceleration m/s/s	Resultant Angular Acceleration r/s/s	Resultant Angular Velocity r/s	Linear Velocity m/s	Force N	Force Duration ms
82E001	3600	4500	42000	52	5	9100	10
82E021	1400	1440	7500	20	5.2	8400	11
82E022	1900	1900	7250	28	7.0	9600	10
82E041	1800	1800	7000	19	6.4	9600	12
82E042	1600	1800	8000	20	7.5	10200	12
82E061	1600	1700	6000	25	6.5	9000	10
82E062	1500	1600	7500	30	6.5	9600	12
83E081	1350	1350	7500	22	3.8	9600	12
83E082	1250	1000	7000	24	3.5	4100	8
83E102	—	—	—	—	—	1800	64
84E141A	500	560	5600	37.5	4.5	3200	30
84E141B	375	420	3900	30	4.5	2400	25
84E141C	1500	1575	20000	45	4.5	7500	10
84E141D	980	1200	16000	45	3.8	7500	8
84E151A	350	350	9000	50	2.6	2600	25
84E151B	2200	2200	25000	44	2	8000	10
84E161A	240	250	840	18	3.8	840	50+
84E161B	780	840	3750	25	5	4800	15

Table 5. Cadaver Test Summary Pressures

Test No.	Location	Type	Maximum Kpa	Time at Maximum ms	Duration ms
82E001	Epidural 1	Unimodal	75	5	10
	Epidural 2	Bimodal	11,3	5/25	10/120+
	Epidural 3	Unimodal	-36	5	15
	Epidural 4	Unimodal	11	5	5
82E021	Epidural 1	Unimodal	161	5	12
	Epidural 2	Bimodal	48,7	5/40	5/80
	Epidural 3	Bimodal	-61,8	5/45	10/80
	Epidural 4	Bimodal	34,6	5/25	5/70
82E022	Epidural 1	Unimodal	180	5	10
	Epidural 2	Bimodal	47,6	5/35	10/80
	Epidural 3	Bimodal	-43,6	5/50	15/100
	Epidural 4	Bimodal	12,51	5/13	5/5
82E041	Epidural 1	Bimodal	22,2	5/40	15/20
	Epidural 2	Bimodal	-20,11	5/45	10/15
	Epidural 3	Bimodal	-55,28	5/50	10/40
	Epidural 4	Bimodal	39,31	5/50	10/70
82E042	Epidural 1	Unimodal	58	5	140+
	Epidural 2	Bimodal	-20,9	5/45	5/20
	Epidural 3	Bimodal	-53,13	5/45	10/40
	Epidural 4	Bimodal	38,42	5/60	5/25
82E061	Epidural 1	Unimodal	97	5	8
	Epidural 2	Unimodal	24	5	5
	Epidural 3	Unimodal	-31	5	8
	Epidural 4	Bimodal	15,7	5/40	10/150
82E062	Epidural 1	Unimodal	55	5	12
	Epidural 2	Bimodal	27,12	5/40	10/35
	Epidural 3	Bimodal	31,14	5/42	10/40
	Epidural 4	Bimodal	37,12	5/45	10/40
83E081	Epidural 1	Unimodal	52	5	150+
	Epidural 2	Bimodal	20,14	5/20	10/135+
	Epidural 3	Bimodal	-18,14	5/20	7/125+
	Epidural 4	Unimodal	25	5	75
83E082	Epidural 1	Unimodal	46	5	15
	Epidural 2	Bimodal	10,5	5/20	10/125+
	Epidural 3	Bimodal	-13,3	5/50	10/100+
	Epidural 4	Bimodal	7,4	5/25	5/50
83E102	Epidural 1	Unimodal	-6	5	—
	Epidural 2	Unimodal	-4	5	—
	Epidural 3	Unimodal	6	5	—
	Epidural 4	Unimodal	3	5	—
84E141B	Epidural 1	Bimodal	30,7	5/65	30/210
	Epidural 2	Bimodal	10,6	5/65	30/200
	Epidural 3	Bimodal	-8,12	5/85	30/250+
	Epidural 4	Bimodal	11,12	5/75	20/250+
84E141D	Epidural 1	Bimodal	46,28	5/50	10/250+
	Epidural 2	Bimodal	21,11	5/55	5/250+
	Epidural 3	Bimodal	-62,28	5/65	5/250+
	Epidural 4	Bimodal	28,13	5/65	5/250+
84E151A	Epidural 1	Bimodal	40,8	5/105	20/200+
	Epidural 2	Bimodal	10,8	5/140	40/200+
	Epidural 3	Unimodal	15	5	30
	Epidural 4	Bimodal	4,12	5/75	50/250
84E151B	Epidural 1	Unimodal	41	5	5
	Epidural 2	Unimodal	30	5	5
	Epidural 3	Unimodal	8	5	10
	Epidural 4	Bimodal	33,17	5/80	20/180
84E161A	Epidural 1	Unimodal	6	80	200+
	Epidural 2	Unimodal	20	70	130
	Epidural 3	Unimodal	2	80	200+
	Epidural 4	Unimodal	15	70	200+
84E161B	Epidural 1	Unimodal	19	140	250+
	Epidural 2	Unimodal	32	15	250+
	Epidural 3	Bimodal	10	135	250+
	Epidural 4	Bimodal	14	140	200+

+Extends beyond the end of sampling.

Table 6. Rhesus Injuries/Damages+

Test No.	Gross Skull	Gross Brain	Micro Brain	Micro Spinal Cord	Gross Other
78A232	No abnormality or injury	No abnormality or injury	No abnormality or injury	No abnormality or injury	Epidural hematoma at C1, dura lacerated at C1
78A234	No abnormality or injury	No abnormality or injury	No abnormality or injury	No abnormality or injury	Epidural hematoma at C1, torn muscle at base of occiput
78A236	No abnormality or injury	No abnormality or injury	No abnormality or injury	No abnormality or injury	No abnormality or injury
78A238	No abnormality or injury	No abnormality or injury	No abnormality or injury	No abnormality or injury	Epidural hematoma at C1 disk
78A239	No abnormality or injury	No abnormality or injury	No abnormality or injury	No abnormality or injury	1/4 cc blood in occiput from epidural hematoma at C1
78A241	No abnormality or injury	No abnormality or injury	No abnormality or injury	No abnormality or injury	Epidural hematoma at C1
78A249	No abnormality. Basilar fracture (ring)	No abnormality. 3/4cc subdural hematoma right frontal lobe (cerebrum) subarachnoid hemorrhage base of cerebellum, pons and medulla	Foci of acute hemorrhage frontal cortex and cerebellum. Subarachnoid hemorrhage base of midbrain, cerebellum, and pons	Subarachnoid hemorrhage around spinal cord	Epidural hematoma at C1, occipital muscles damaged

+The kinetic impact response for these tests is presented in Reference 38.

Table 7. Rhesus Injuries/Damages

Test No.	Gross Skull	Gross Brain	Micro Brain	Micro Spinal Cord	Gross Other
79A251	No abnormality. Linear basilar fracture from foramen magnum to occipital contact point	No abnormality or injury	No abnormality or injury	No abnormality or injury	Epidural hematoma at C1 and C2
79A253	No abnormality. Basliar fracture (quasi-ring), temporal fracture	No abnormality. Petechial lesion left frontal lobe (cerebrum) subarachnoid hemorrhage base, pons, medulla and vermis of cerebellum	Focal hemorrhage frontal cortex, focal hemorrhage cerebellum associated with focal necrosis of purkinje cells,sarachroid hemorrhage cerebellum and base of midbrain plus blood in aqueduct of sylvius, hemorrhage base of midbrain	Subarachnoid hemorrhage over spinal cord and blood in its central canal at C1-C2 level	Damaged neck ligaments
79A256	Abnormality. Very thin skull. Basilar fracture (petrous to petrous). Connected linear fractures to to frontal, temporal, parietal bones	No abnormality. Dura torn along sagittal sinus, emaciated tissue cerebellum, medulla	No abnormality or injury	No abnormality	Lacerated spinal cord
79A258	No abnormality. Basilar fracture (quasi-ring)	No abnormality. Lacerated medulla	No abnormality or injury	No abnormality	Epidural hematoma at C1
79A260	No abnormality. Basilar fracture (right petrous)	No abnormality. Subdural hematoma along sagittal sinus	No abnormality or injury	No abnormality	Epidural hematoma at C1

Table 8. Cadaver Damages

Test No.	Gross Skull	Gross Brain	Gross Other
82E001	No abnormality. Parietal fracture. Basilar fracture.	No abnormality. Subarachnoid hematoma frontal lobes (cerebrum) and on base of occipital lobe (cerebrum)	No abnormality or injury
82E021 82E022	No abnormality or injury	No abnormality. Subarachnoid hematoma right frontal lobe (cerebrum), hemorrhage central area left frontal lobe (cerebrum)	No abnormality or injury
82E041 82E042	No abnormality or injury	No abnormality. Subarachnoid hematoma frontal lobes (cerebrum) and subarachnoid hemorrhage parietal lobe (cerebrum)	No abnormality or injury
82E061 82E062	No abnormality or injury	No abnormality or injury	No abnormality or injury
82E081 82E082	No abnormality or injury	No abnormality or injury	No abnormality or injury
83E101 83E102	No abnormality or injury	No abnormality or or injury Mechanical abnormality of incomplete repressurization	No abnormality or or injury
84E141A 84E141B 84E141C 84E141D	No abnormality or injury	No abnormality. Subarachnoid hemorrhage frontal lobes (cerebrum) and subarachnoid hemorrhage right parietal lobe (cerebrum)	7 cm longitudinal laceration between eyes on forehead
84E151A 84E151B	No abnormality. 2 cm linear fracture of frontal bone continues as left orbital fracture	No abnormality. Subarachnoid hemorrhage left frontal lobe (cerebrum), subarachnoid hematoma right frontal lobe (cerebrum)	Hemorrhage to occipital belly muscle small laceration to side of skull fracture
84E161A 84E161B	Abnormality of very thick skull	Abnormality of massive tumor on the right frontal lobe (cerebrum). Coded "no injury"	No other abnormality

DISCUSSION

The results of a series of head impact research programs conducted during the past five years at UMTRI are presented. The tests entail different initial conditions, human surrogates, impact directions, and locations for the recording instruments. Therefore, in order to compare tests, frame-independent variables and Frenet-Serret vectors are used for examination and analysis. Frame-independent variables include resultant angular and linear velocities and accelerations. Vectors expressed in the Frenet-Serret frame field include tangential acceleration, normal acceleration, T-rate and B-rate. The features of the data discussed in this section in abbreviated form represent trends that may be important factors in head impact response. In particular, the potential effect of skull deformation on head angular acceleration as well as on impact and injury response appears significant.

FORCE TIME HISTORIES - Force time histories of repressurized cadaver tests are divided into two types which correlate well with fracture and non-fracture cases. In non-fracture cases, the force rises smoothly to a maximum and drops smoothly to zero. In fracture cases, although the force rises smoothly to a maximum, the drop to zero has a greater number of inflections or local maxima and is of longer duration. Fracture cases include Tests 82E001 and 84E151B. Test 82E001 is illustrated in the appendix.

Non-fracture impacts of repressurized cadavers can be broken into two groups consisting of long and short-duration impacts. Short-duration impacts are those lasting less than 15 ms; long-duration impacts are defined as 15 ms or longer. In some cases, such as Test 83E102, durations as long as 60 ms were recorded. In the appendix Test 82E041 illustrates a short duration impact and Test 84E161A illustrates a long duration impact.

Force time histories of Rhesus impacts can also be divided into two types which correlate well with non-fracture (or simple linear fracture) and basilar fracture. In the non-fracture/simple linear fracture group (Test 79A251 in the appendix), the force time history smoothly rises to a maximum and drops smoothly to zero. However, as with the repressurized cadaver, in cases with complex basilar fractures (Test 79A253 in the appendix), the force smoothly rose to a maximum, but fell with a greater number of inflections and/or local maxima of longer durations. The energy released from the skull appears to affect the force time history of repressurized cadaver subjects, but not Rhesus subjects, in non-fracture/simple linear fracture cases. Possibly, the Rhesus neck-head soft tissue muscle mass buffers the effect of fracture until "significant fracture" (complex basilar fracture) has occurred.

TANGENTIAL ACCELERATION TIME HISTORIES - The tangential acceleration time histories of repressurized cadavers divide into two groups that correlate well with subarachnoid hemorrhage or the absence of it. For those tests in which no subarachnoid hemorrhage was observed, the tangential acceleration had a single local maximum in the area of maximum acceleration. However, for those tests in which subarachnoid hemorrhage was observed, there were several local maxima in the area of maximum acceleration.

For both live and post-mortem Rhesus, the multimodal tangential acceleration occurred when complex skull fracture (basilar skull fracture) was observed. When there was an absence of complex fracture, the tangential acceleration was unimodal and smooth in the area of the maximum acceleration for both live and post-mortem Rhesus.

COMPARISON OF IMPACTS: CADAVER VARIABILITY - To examine variability within cadaver subjects, some subjects received two similar impacts (Tests 82E001 thru 82E082). Figure 10 is an example of cross-and auto-correlations for Tests 82E021 X 82E022 and 82E061 X 82E062 and 82E021 X 82E061. The figure represents the general trend observed in relating similar tests with different subjects to similar tests with the same subject in terms of force time histories. In general, it seems that force-time history as well as acceleration-time history vary more between subjects than between tests on the same subject. An analogous comparison for epidural pressures (not illustrated) shows equivalent variance between different subjects having similar impacts or between the same subject having similar impacts. This implies that experimental techniques associated with repressurization or with the effects of the post-mortem state may produce as much variance in the pressure time history response as do variations due to the population of test subjects.

IMPACT RESPONSE - The motion of a rigid body in space is the result of generalized forces: the total force and the total torque about a suitable axis. The dynamic problem of the motion of the area of the skull local to the nine-accelerometer array can be interpreted in the same way. However, because of the complex interactions of the area of the skull local to the nine-accelerometer array with the other material bodies, (for example, the muscle soft tissues of the neck, the rest of the skull, the brain, or the impactor), serious problems can arise in determining which of the bodies is producing these generalized forces.

For example, when the head receives an impact, several events occur: 1) stress waves are propagated from the impact site, 2) the skull starts to deform, and 3) the skull begins to move due to the impact, transmitting energy

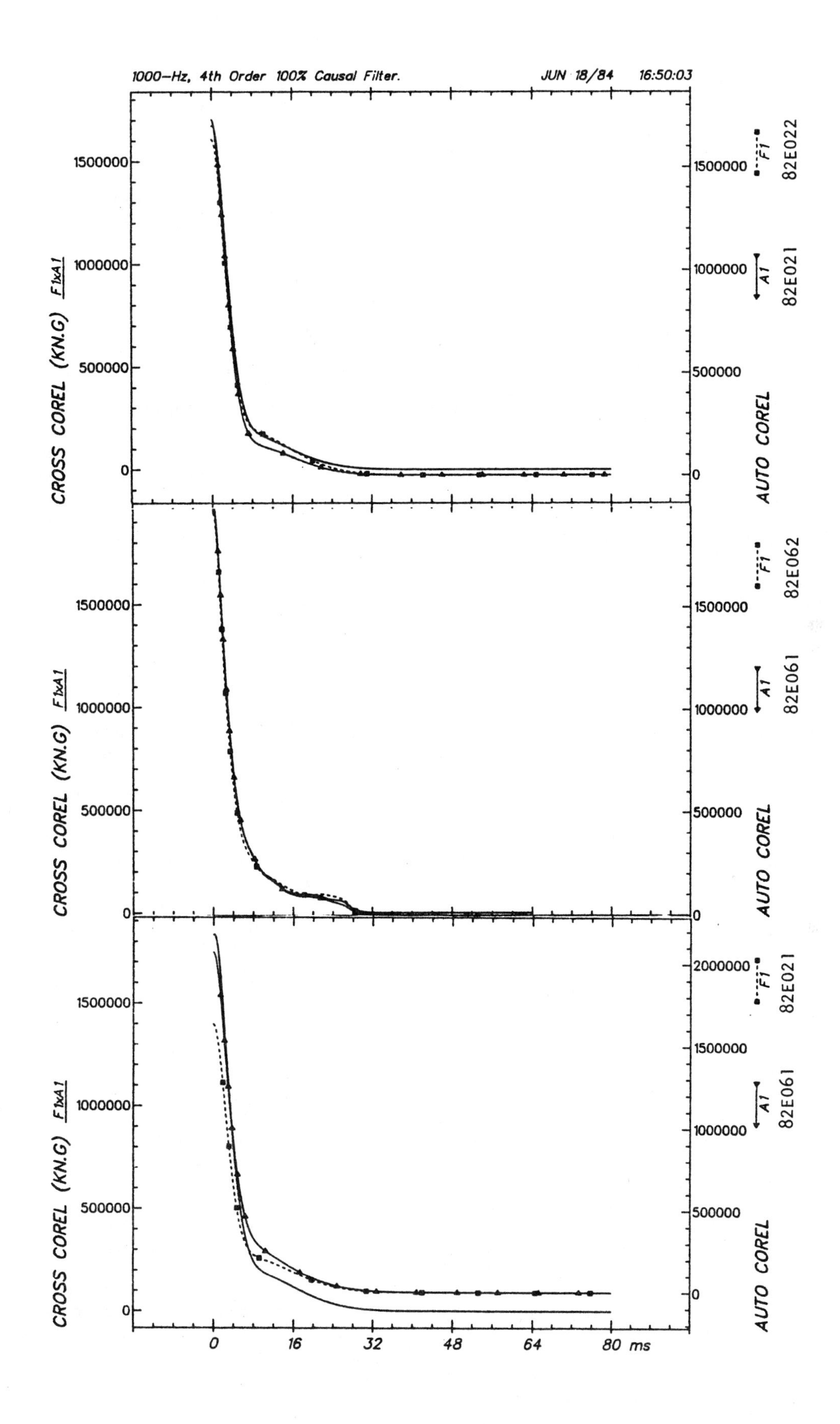

Fig. 10. Auto- and Cross-Correlation for Force Time History.

to the brain via the dura mater. Eventually, the waves are dissipated, the deformation of the skull recovers partially or fully on removal of the impact loads, and the acceleration of the skull comes primarily from forces generated from the brain and neck. If differential skull motion is severe, essentially due to either sufficient energy in the high frequency components of the force time history or sufficient peak force, the stresses at some point in the skull may exceed the failure strength of the bone, thereby producing fracture. The loads producing this type of impact are generally of shorter duration or contain a rise time sufficient to generate the necessary high frequency components to fracture the skull. The motion of the entire skull as a rigid body as estimated by the nine-accelerometer array depends on the degree of skull deformation as well as on the degree of precision being used in the investigation. If the skull deformations are small during and after impact, and the accelerometers are sufficiently far from the impact contact point, then valid rigid body motion can be assumed. However, if skull deformations are significant, then three-dimensional motion of the nine-accelerometer array and of the skull local to its instrumentation mount can only be used to estimate the motion of the rest of the skull through the use of an "estimated anatomical center." Interpretation of the results from the nine-accelerometer array must, therefore, take into account the non-rigid body motion taking place during "significant skull deformation" impacts. Using translations obtained from x-rays, three-dimensional approximate motion of an "estimated anatomical center" can be determined.

Impact Response Definition-With the use of the UMTRI nine-accelerometer array it is possible to record three-dimensional six-degrees-of-freedom motion of an area of the the skull in which the accelerometers are located. Therefore, head impact response can be defined as a continuum of "events" characterized by the path traced by the motion of the "estimated anatomical center," all the vectors defined on that path, and by changes of the associated frame fields. Physically this implies that head impact response is interpreted as the response of a material body (the nine-accelerometer array and area of the skull local to it) in contact with other material bodies. The curve and the vectors generated as the "estimated anatomical center" moves in time are, therefore, a result of the interactions of the skull-mount area with other material bodies.

Examples of events which are used to characterize head impact are: the initiation of head impact response (denoted by Q_1 on the tangential acceleration time histories in the appendix); the positive maximum of the tangential acceleration time history (denoted by Q_2 in the accompanying data); and the negative maximum of the tangential acceleration time history (denoted by Q_3 in the accompanying data). In research reported earlier in which similar Q_1, Q_2 and Q_3 events were defined (38), the tangential acceleration rose smoothly to a single maximum and fell smoothly until crossing zero. In some of the tests being reported here the time interval near Q_2 contains several local maxima, therefore direct comparison is complex. Nevertheless, these defined events can be used to compare different types of impacts for the same human surrogate and to compare the response of one type of human surrogate to another.

THE EFFECTS OF SKULL DEFORMATION ON LINEAR AND ANGULAR ACCELERATION - Inspection of the three-dimensional motion of the skull local to the accelerometers, epidural pressure transducer response, and contact forces showed that skull deformation may have important implications for injury produced in blunt head impact.

For repressurized cadaver tests with time histories having unimodal peaks of the tangential acceleration of the "estimated anatomical center," the time interval between the events Q_1-Q_2 is probably primarily a result of the interaction of the impactor with the skull. During the Q_1-Q_2 interval, the "estimated anatomical center" does not move more than 1 cm and the motion is to some extent three-dimensional. This is indicated by the rate of change of the tangent vector (T-rate) and binormal vector (B-rate). (A positive T-rate implies a curvature of the path or two-dimensional motion; significant T and B rate imply a torsion of the path or three-dimensional motion.) However, the angular acceleration is small or non-existent, and when present lies principally in the binormal direction. The normal acceleration of a point on the skull of closest approach to the impactor was found to be less than that of the "estimated anatomical center." (Reduced normal acceleration implies a "straighter" path of that point.) These measurements of angular and normal acceleration imply that the skull may be rotating about a point of closest approach to the impactor. For those tests with time histories displaying multimodal peaks of the tangential acceleration of the "estimated anatomical center" in the neighborhood of the Q_2 event, the time interval between the events Q_1-Q_2 is also probably primarily a result of the interaction of the impactor and skull. However, in these these tests skull deformations seem to have significant effect on the angular, tangential, and normal acceleration responses. Comparison of this multimodal impact response (Test 82E041 in the appendix) to the unimodal tangential acceleration response (Test 82E061 in the appendix), shows that the following variables are greater during the Q_1-Q_2 interval: angular acceleration, normal acceleration, T-rate and B-rate. This implies that for the multimodal type of impact, the path of the "estimated anatomical center" is, to a greater degree than for the unimodal pattern impacts, moving in a three-dimensional manner and that this increased three-dimensional motion correlates well with the angular acceleration.

Comparison of the ratios of peak angular acceleration and velocity during the Q_1-Q_2 interval to those of peak angular acceleration and velocity during the Q_2-Q_3 interval indicates that for a given test subject there is respectively more angular acceleration during the Q_1-Q_2 interval for the multimodal impacts. In addition, the local maxima of the angular velocities in the multimodal impact as well as the rapid rotation of the binormal and normal vectors of between $\pi/2$ and π radians indicates that the path of the "estimated anatomical center" has passed an inflection point near the Q_2 event. This is most evident when the skull fractures. In a skull fracture test, the head is loaded very rapidly (e.g., Test 82E001, the force drops while the tangential acceleration drops below zero). This is accompanied by a short-lived rotation of the skull which produces a local maximum in the angular velocity. Subsequent to fracture, the skull is in more complete contact with the impactor. The tangential acceleration increases, the angular velocity decreases, and the angular acceleration reverses direction.

In general, the head is modeled as a rigid body when interpreting angular acceleration from nine accelerometers. However, the complex nature of the skull (4,24,54-55) causes asymmetric loading during blunt impact, which leads to an interpretation of an angular acceleration by the nine-accelerometer array that is not directly related to rigid body motion. Therefore, in addition to local skull bending in the area of the nine-accelerometer array, a second mechanism of skull deformation which causes the accelerometers to interpret angular acceleration can be hypothesized.

A schematic display of this type of response is presented in Figure 11 to illustrate the effect of skull deformation on angular acceleration (a rotation is produced). The figure demonstrates the type of motion that might occur and is not necessarily representative of motion actually observed. Also, motion of the skull is not necessarily in the anterior-posterior, inferior-superior plane. Because angular displacement is small, movements are best detected through evaluation of angular acceleration.

Because angular acceleration is an acceleration gradient over displacement at a given instant in time, the results of the linear acceleration are influenced by the angular acceleration. Thus, the differences in the neighborhood of the Q_2 event between the multimodal aspect and the unimodal aspect of the tangential acceleration of the "estimated anatomical center" are a result of the acceleration gradient caused by the angular acceleration.

Figure 12 represents the mechanical impedance corridor of force and tangential acceleration for repressurized cadaver tests in which skull deformation was observed and no skull fracture occurred (81E021, 82E022, 82E041, 82E042, and 84E141). The impedance values for these impacts are similar to driving point impedance tests reported by other researchers (19-20,27,33,39-40,56). This implies that the skull deformation observed could be related to the same type of skull deformation obtained from the driving point impedance tests mentioned above. The results from Tests 84E161A and 84E161B support this conclusion. This test subject had the thickest skull of any of the subjects tested. Both Tests 84E161A and 84E161B were at low severity HIC values of 70 and 400, respectively. If the results of the impedance are primarily a result of skull deformation, then it would be expected that the tests would look more like those of a rigid body. In the frequency range between 10 and 1000 Hz this is,

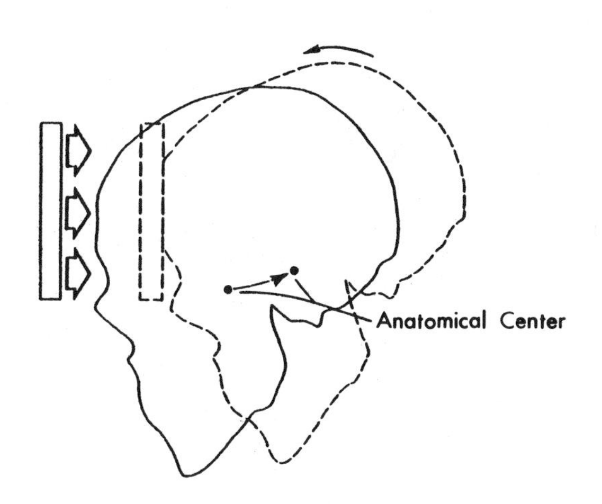

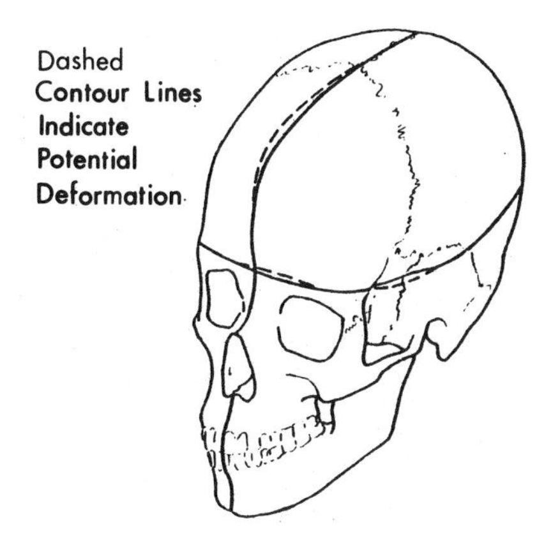

Fig. 11. Schematic Representations of Skull Deformation and Potential Effect on Angular Accelerations.

indeed, the case for Test 84E161A (Figure 13). However, for Test 84E161B there is a slight difference between its response and that of a rigid body at the low frequency. This is, perhaps, the result of the differential movement of soft tissue.

The results reported here extend the studies of the kinematic response of non-human primates (30) and cadavers (40) to blunt impacts. The earlier reports showed some of the differences in kinematic response between non-human primates and cadavers to be attributable to two factors: soft tissue distribution and skull geometry.

In the present study, skull fracture and non-skull fracture tests were analyzed for both Rhesus and cadavers. The difference between fracture and non-fracture impact response was found to be greater for cadavers. This is illustrated by the following tests presented in the appendix: non-fracture Rhesus, 78A241; fracture Rhesus, 79A249 and 79A251; non-fracture cadavers, 82E061; fracture cadavers, 82E001. Differences may be attributed to: 1) the nine-accelerometer array in Rhesus tests are mounted in an area that is to some degree more removed from the calvarium (the accelerometers are mounted on the orbital ridges of the face of the Rhesus), 2) the mounting site in the Rhesus is more massive with respect to the rest of the skull than that of the mounting site for cadavers (the instrumentation is the same size, but the Rhesus head is much smaller than the cadaver head), 3) the skull geometry is significantly different between the Rhesus and cadaver at the impact site, and 4) the differences in the external muscle mass at the impact site affect the interaction of the skull with the load cell.

KINEMATIC RESPONSE AFTER IMPACT: EFFECT OF SOFT TISSUE - Transmission of energy during intervals Q_1-Q_2 and Q_2-Q_3 was analyzed by comparing the acceleration response of the skull to the force-time history of the impactor. The following observations were made. During the Q_1-Q_2 interval, energy was transferred from the impactor to the skull and from the skull to the brain and neck. During the Q_2-Q_3 interval, significant energy was transferred from the brain and neck to the skull. Examination of all the tests shows that during the Q_2-Q_3 interval, unless there were rapid changes in the binormal vector direction (large torsion and large B-rate), the normal acceleration is established by angular acceleration. In addition, the normal and binormal vectors are established first by the angular acceleration during the Q_2-Q_3 interval and then by the angular acceleration and angular velocity after the Q_3 event. In general, for those tests with multimodal/unimodal peaks, the angular acceleration direction changes near the Q_2 event. The extent and amount of rotation changes from test to test. This is probably a result of complex three-dimensional motion of the head during the Q_1-Q_2 interval as well as of the geometry of the head and skull. The rotation tends to be between $\pi/2$ and π radians. The motion past the Q_3 event for multimodal tangential acceleration tests is similar to the unimodal tangential acceleration tests. In other words, the trajectory traced by the "estimated

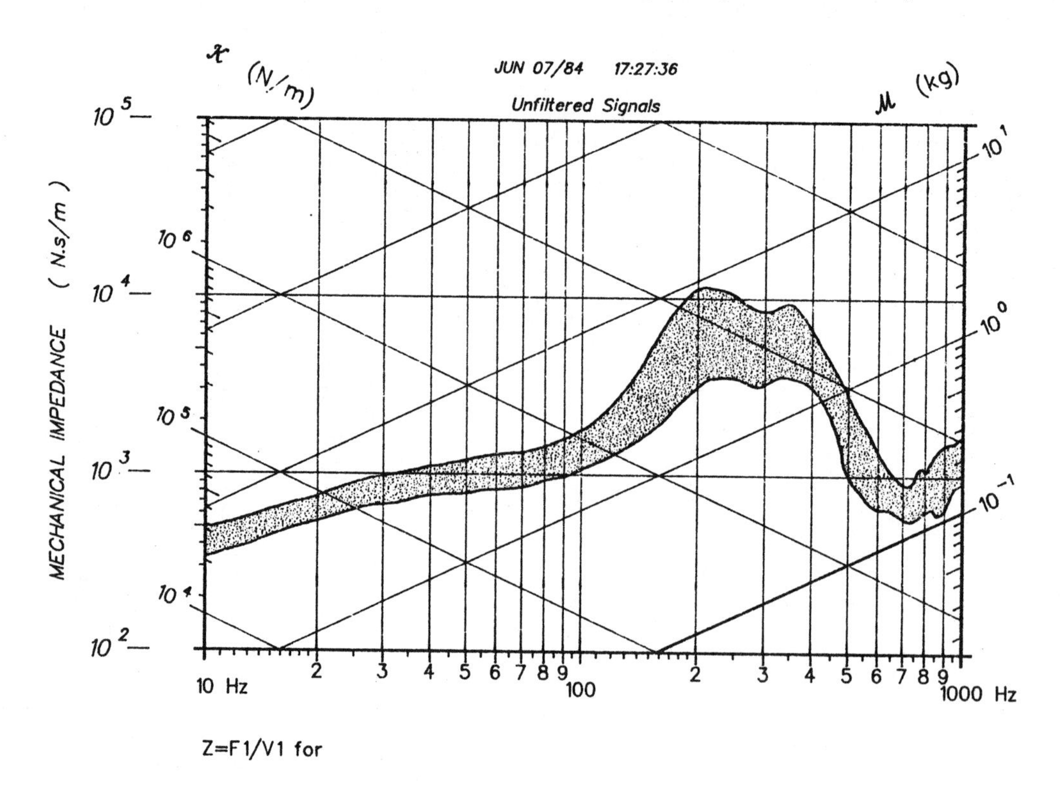

Fig. 12. Mechanical Impedance Corridor of Impact Force and Tangential Accelerations for Tests 82E021, 82E022, 82E041, 82E042, and 84E141.

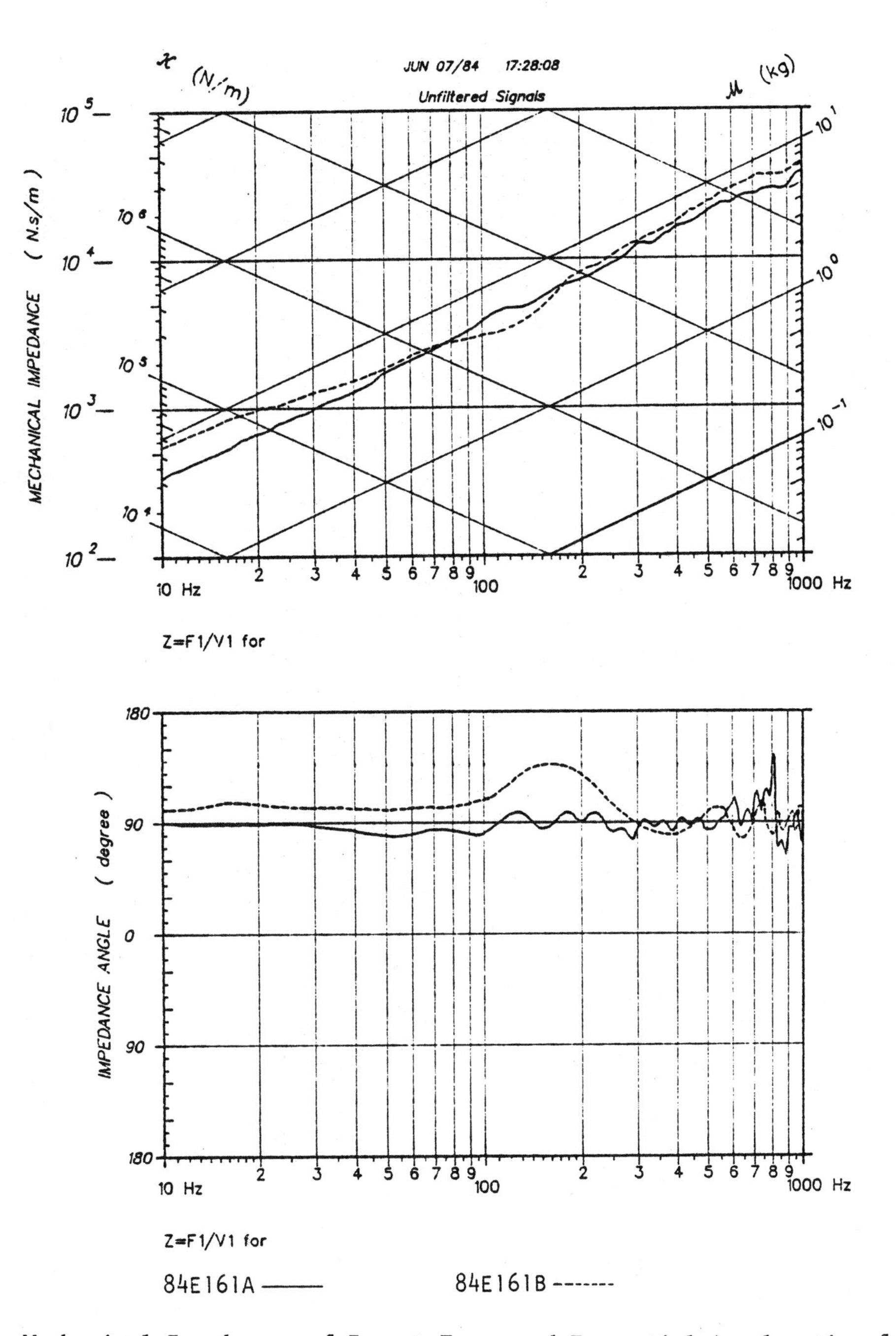

Fig. 13. Mechanical Impedances of Impact Force and Tangential Acceleration for Tests 84E161A and 84E161B.

anatomical center" and its attached frame field during multimodal tangential acceleration impacts is different from that traced during unimodal tangential acceleration impacts. However, the motion after impact is similar and the driving force is obviously other than the impactor. In past research (40) it has been determined that in the unpressurized or partially repressurized cadaver the response of the skull after impact is influenced by differential motion of the brain. In a similar manner, in the data presented here it seems that the brain is driving the skull and that this is manifested in both a linear and rotational manner. Potentially, energy has been transferred from the skull to the brain during impact, was stored as energy and then was released as the impact force dropped below a given level.

PRESSURE TIME HISTORY RESPONSE - The pressure time histories for repressurized cadavers were separated into two significant types, unimodal and bimodal. The unimodal pressure pulses correlate well with short-duration (less than 15 ms) large-valued (1500 m/s/s) tangential accelerations. Bimodal pressure pulses were more commonly observed in longer

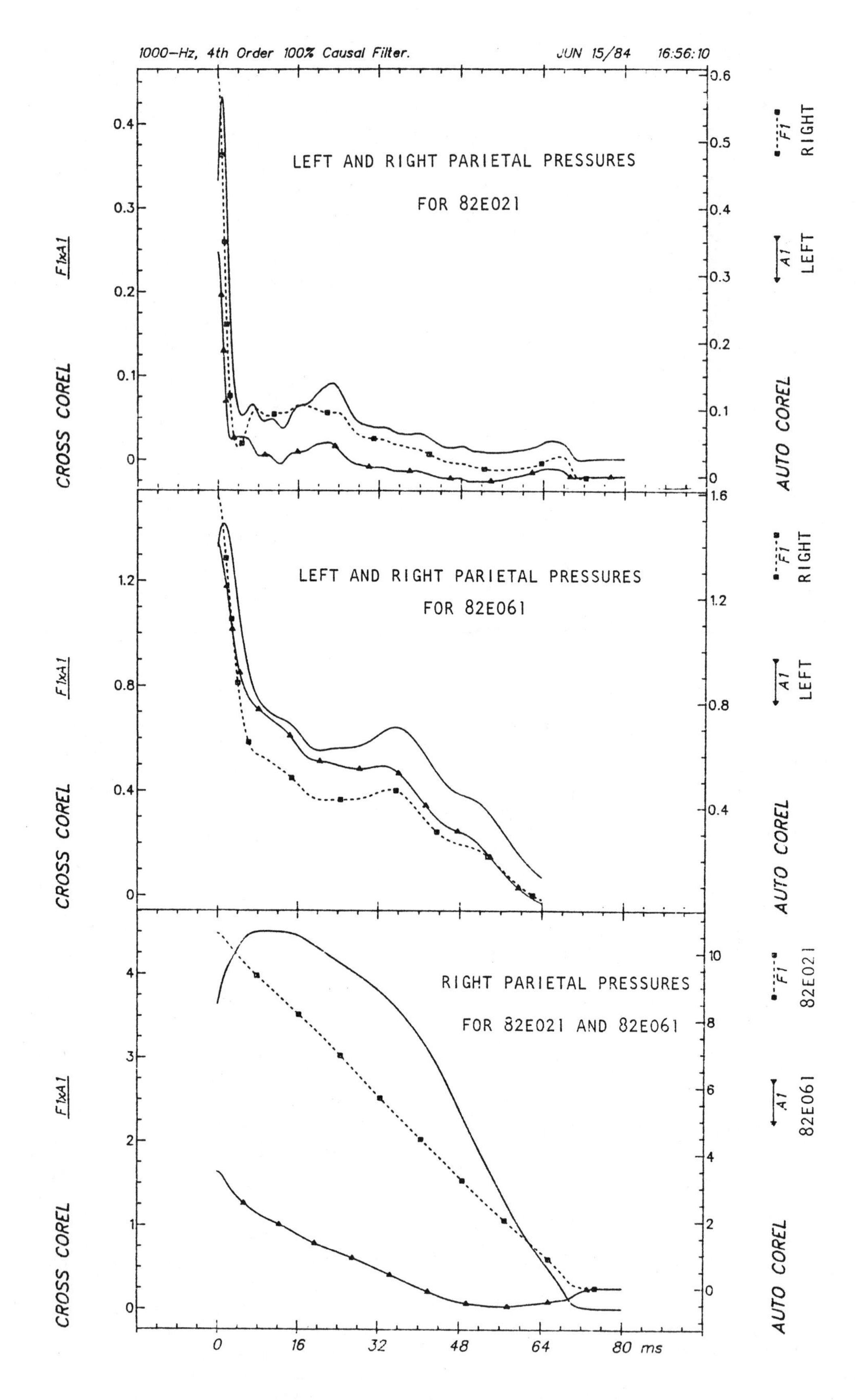

Fig. 14. Comparison of Right and Left Parietal Pressures and Comparison of Right Parietal Pressures for Different Impacts.

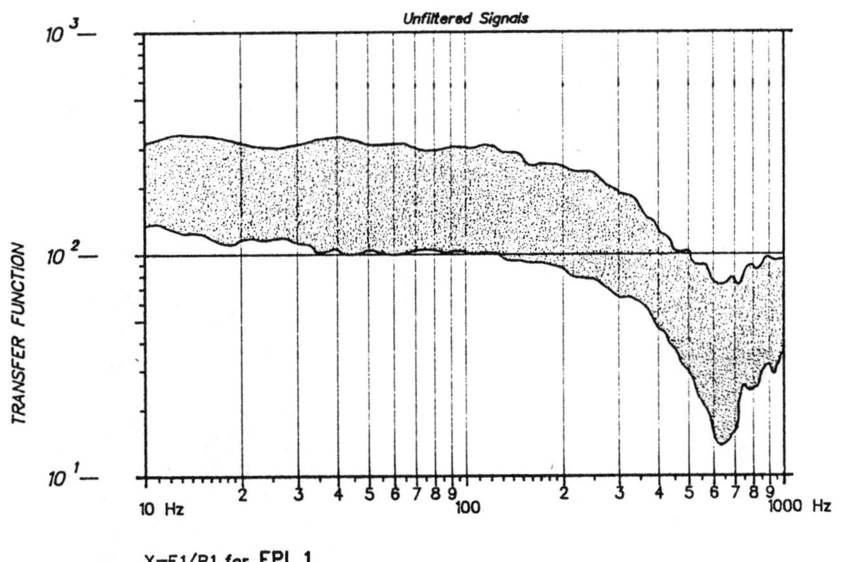

X=F1/P1 for EPI 1

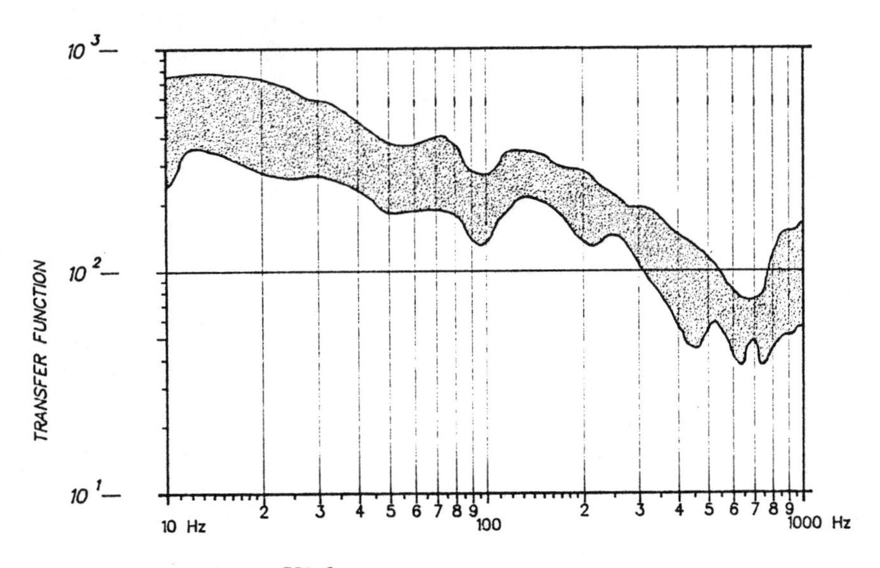

X=F1/P1 for EPI 2

Fig. 15. Transfer Function Corridor of Frontal and Right Parietal Epidural Pressures for Tests 82E021, 82E022, 82E041, 82E042, and 84E141.

duration and lower acceleration impacts. This result seems to be a consequence of the superposition of two different types of mechanisms for producing pressure changes in the head during and after blunt impact.

The first pressure mechanism is associated with impact force time histories which contain sufficient high-frequency components to excite a short-duration loading of the skull on the brain, and probably is primarily a result of inertial loading. When a blunt impact blow is delivered to the head, the skull is initially accelerated. Shortly afterwards the brain compresses on the side closest to impact and is in tension on the side polarly distal to impact. The result is a pressure gradient in the brain going from point of impact to an area opposite from impact. Test 82E021 in the appendix illustrates such pressures for selected impacts and shows that highest magnitudes and positive pressures occur in the frontal lobe (epidural 1) and that negative pressures develop in the occipital lobe (epidural 3). Pressures in the parietal areas (epidural 2, epidural 4) are between the coup and counter coup areas. The pressures in epidural 2 and epidural 4, for most of these repressurized cadaver tests, correlated well indicating that the pressure gradients were, in general, symmetric. However, some differences do exist which may be the result of three-dimensional motion of the head or of some asymmetry associated with the test subject. Figure 14 illustrates a cross- and auto-correlation between epidural 2 and epidural 4 for Tests 82E021 and 82E061 and shows that the auto-correlation for each pressure is similar to the cross-correlation. This is similar to results reported by others (42-43).

Figure 15 represents transfer functions between the force and the epidural 1 and 2 pressures for Tests 82E021, 82E022, 82E041 and 82E042 in which skull deformation occurred in the absence of skull fracture. These transfer functions have a possible resonance in the area for which a resonance was predicted from the impedance transfer function for force and acceleration. This indicates that although the exact amount of the effect of skull deformation on the pressure response is not completely determined, it has some effect which is observable in the pressure time history. Therefore, a reasonable correlation might be found between pressure and acceleration. However, such a correlation would depend on where the accelerometers are placed on the skull.

The second pressure mechanism is associated with impact force time histories which contain low-frequency components or motion of the head after blunt impact. Unlike the first pressure mechanism which rarely produces pressure pulses longer than 15 ms, the second pressure mechanism produces pressure pulses that can last as long as 200 ms. Possibly, the second pressure mechanism is a result of the brain driving the skull as discussed earlier. Since the pressure is positive in all transducers regardless of position, the brain is possibly transferring energy to the skull, thus, accelerating it. This is consistent with the results discussed earlier where the brain stores energy and releases it shortly afterwards in a way that is manifested by skull angular acceleration. The results obtained from the high-speed cineradiograph support this hypothesis. Figure 16 is two frames from 16 mm high-speed cineradiograph movies of Test 84E161 showing the outline of the radio-opaque gel target injected into the brain tissue and ventricles. Inspection of the curves traced out by the gel from the pressure transducers to the ventricles, shows that although no motion could be detected between the skull and the brain, differential motion of parts of the brain was occurring.

All the pressure time histories for Rhesus subjects were unimodal and of short duration. This may be the result of the greater tangential acceleration which was associated with Rhesus impacts as compared to cadaver ones, or it may be the result of the differences between the response of the repressurized cadaver and that of the live unpressurized post-mortem Rhesus. Although there are a limited number of tests in this study, comparisons with pressure time histories from tests reported earlier not having skull fracture (30) indicate that there is a difference in response between those tests and the ones being reported here. Possibly, this is a result of the pressure-volume changes which accompany skull deformation that become acute during basilar skull fracture. The feature of the data that seems to indicate that this is true, is that near the Q_2 event the pressure becomes negative faster and obtains a greater negative maximum. Test 79A249, for example, reaches a negative maximum of about one atmosphere and maintains this for approximately 2 ms. In the appendix, Test 78A241 (non-skull fracture) and Tests 79A249 and 79A251 (skull fracture) illustrate this contrast. In addition, the high-speed cineradiographic film recorded a radio-transparent region forming at the top of the skull for Test 79A253 (Figure 17) indicating that the skull in the area of the pressure transducer had moved completely away from the brain. In previous work (40), there

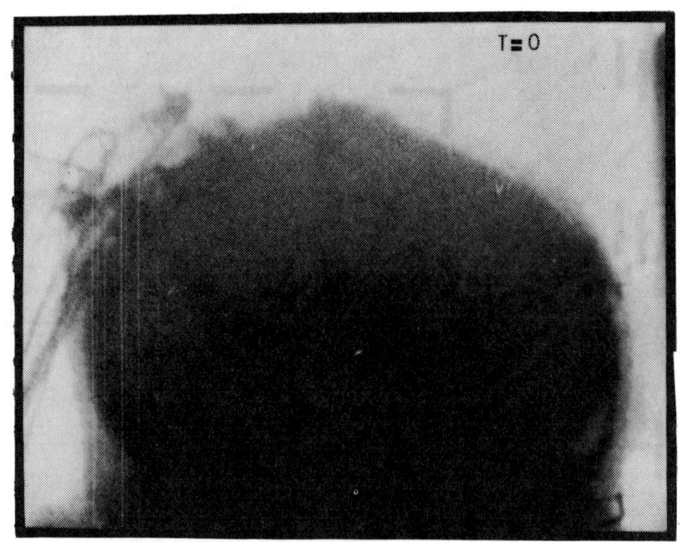

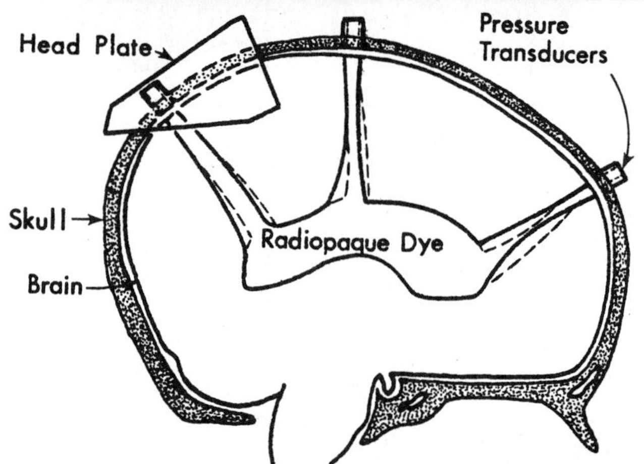

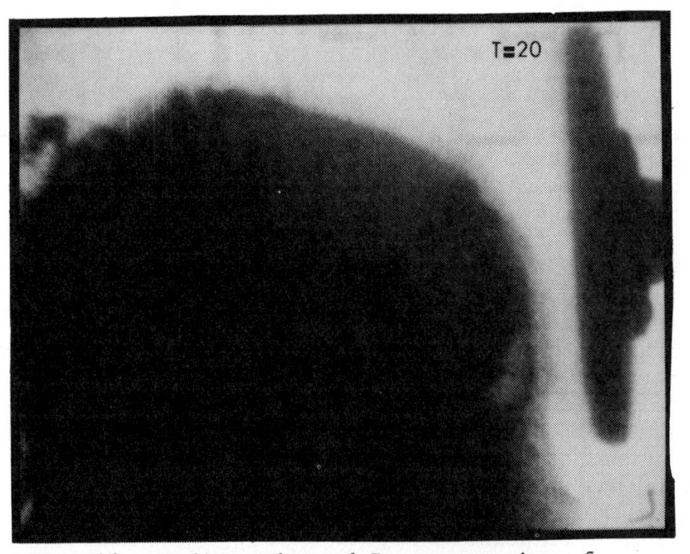

Fig. 16. Radiographs and Reconstruction of Radiographs with Radio-contrast Dye Injection Showing Differential Motion of the Brain at T=0 and T=20 ms.

was a significant difference in response between live and unpressurized post-mortem Rhesus; however, in this study when complex skull fracture occurred, it seems that skull deformation obscurred determination of whether there was such a difference between live, unpressurized post-mortem Rhesus.

INJURY/DAMAGE RESPONSE - The results presented in Tables 6,7 and 8 show that the most common brain injury/damage in the Rhesus and

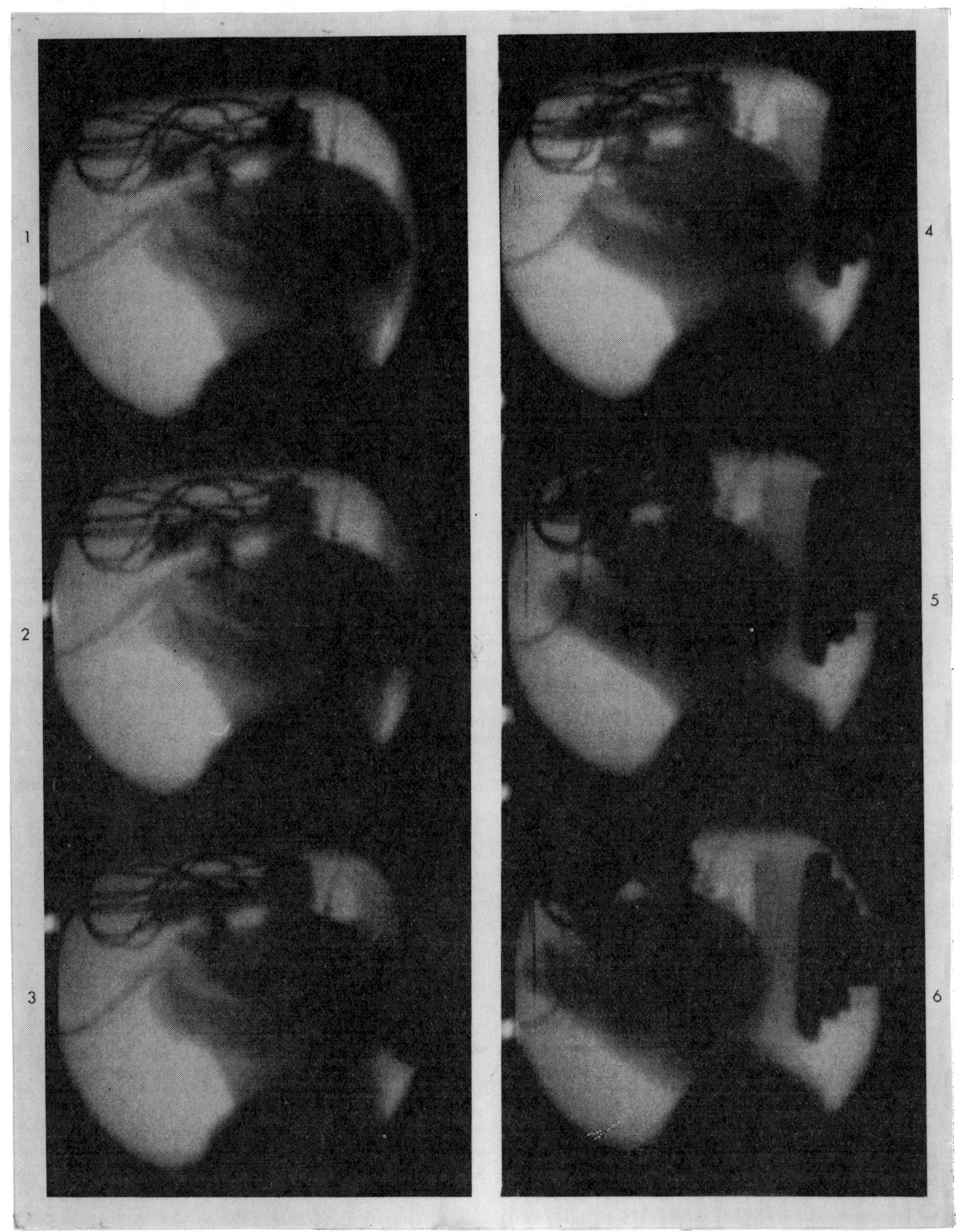

Fig. 17. Radiograph of Test 79A253.

repressurized cadaver is subarachnoid hemorrhage. In the live Rhesus, the injury tends to be in the area at the base of the brain near the cerebellum, pons and midbrain. In the post-mortem Rhesus, no brain damage was observed except damage associated with the skull impinging upon the soft tissue of the brain during skull fracture. In particular, Test 79A256 was the only Rhesus test with this type of damage to the brain: the skull was thin enough so that fracture resulted to a much greater degree than for any other test. Damage occurs for repressurized cadavers in the frontal or parietal lobes of the cerebrum.

Despite the different types of surrogates for humans in these series, the different initial test conditions, and the different injury locations, subarachnoid hemorrhage did not occur unless "significant skull deformation" was present. Except for Test 79A251, no subarachnoid hemorrhage was observed in the absence of skull deformation.

Identifying mechanisms of head injury poses a formidable problem. In head impact response a number of potential injury mechanisms have been proposed (5-43). It is believed that different mechanisms occur for direct head impact than for non-impact (inertial conditions). It is also possible that several mechanisms could be responsible for producing the same injury/damage. The complex nature of the head/skull system under loading implies that under any given impact several mechanisms could be occurring and that they may complement each other to produce injury/damage.

One possible mechanism for production of subarachnoid hemorrhage in both the repressurized cadaver and the live Rhesus human surrogates is induced differential motion between the skull-brain interface. Potentially, there are two types of differential motion of the skull with respect to the brain. One is associated with "local" movement of the skull differentially with respect to the brain. The second requires rotational differential motion of a "significantly large" section of the skull with respect to the brain. In Tests 84E161A and 84E161B the high-speed cineradiograph films show changes in the radiopaque target area that formed curves extending from the pressure transducers to the lateral ventricle, which clearly indicate internal movement of the brain. However, no movement of the skull with respect to the brain was concurrently observed. Yet, in Test 84E151 both skull fracture and differential movement of the brain with the skull of up to 6 mm was observed. This potentially indicates that a "stick-slip" condition occurs, and that a "significant local acceleration" of any part of the skull can initiate differential motion of the brain surface with respect to the skull. However, because only a limited number of tests have been performed using techniques which make such observations possible, more work needs to be done before this hypothesis can be verified.

In repressurized cadaver tests, comparatively large pressure peaks were observed. It is possible that in those tests, high stress in the brain as well as skull deformations and angular accelerations were needed to produce the observed damage.

In several tests, duplicate impacts were made to each subject. It is possible that this enhanced the damage response; and therefore the results presented here should not be used to set tolerance levels. However, it is believed that this did not affect the general trend of damage and/or injury response observed.

CONCLUSIONS

This was a limited study of some important kinematic factors and injury/damage modes associated with direct blunt head impact. Because of the complex nature of the skull-brain interaction during an impact event, more work is necessary before these kinematic factors can be generalized to describe head impact response. However, the following conclusions can be made:

1. "Severe impacts" to the heads of repressurized cadavers can cause local motions in the skull with or without skull fracture. The motions are interpreted as angular acceleration by nine accelerometers mounted in a single array used to determine three-dimensional motion.
2. In repressurized cadavers and live Rhesus subjects skull deformation may cause direct and/or indirect subarachnoid hemorrhage.
3. For live Rhesus subjects, negative pressure peaks during an impact event equal to or greater than one atmosphere do not appear to produce injury.
4. Three-dimensional rigid body motion is not well defined in a "severe head impact" when using accelerometers located on the skull. The acceleration time histories, including the resultant acceleration used to calculate the Head Injury Criterion (HIC), of the anatomical center, depends not only on where the accelerometers have been placed on the skull but also on the biovariability of the test subject's skull.
5. Short duration impacts (less than 15 ms) in the anterior to posterior direction appear to involve two skull-brain interactions. One occurs during impact and is characterized by a transfer of energy from the skull to the brain and a pressure gradient in the brain positive at the frontal bone and negative at the occipital bone. The second interaction occurs during and after impact and is characterized by energy transmission from the brain to the skull and positive pressure in the brain at the frontal, parietal, and occipital bones.

ACKNOWLEDGEMENTS

The results presented in this paper were obtained through series of independently funded research programs conducted during the past five years. The funding agencies were: The Motor Vehicle Manufacturers Association and the United States Department of Transportation, National Highway Traffic Safety Administration, Contract No. DOT-HS-7-01636.

The authors acknowledge the assistance in injury evaluation provided by C.J. D'Amato, P.W. Gikas, Don Huelke and Julian Hoff as well as the encouragement of John C. Scowcroft. The authors also acknowledge the technical assistance of Nabih Alem, John Melvin, Jeff Axelrod, Gary Blair, Gary Holstein, Jeff Lehman, Gail Muscott, Jeff Pinsky, Carol Sobecki, and Bryan Suggitt. A special thanks is given to Jeff Marcus without whom this project could not have been completed.

REFERENCES

1. Anderson, D.W.; Miller, J.D.; and Kalsbeek, W.D. 1983. Findings from a Major Survey of Persons Hospitalized with Head Injuries. Public Health Reports 98(5):475-478.

2. Insurance Institute for Highway Safety, 1982. The Year's Work 1981-1982. Washington, D.C.

3. Kraus, J.F., et al. 1984. The Incidence of Acute Brain Injury and Serious Impairment in a Defined Population. American Journal of Epidemiology. 19(2):186-201.

4. Thomas, D.J.; Robbins, D.H.; Eppinger, R.H.; King, A.I., and Hubbard, R.P. 1974. Guidelines for the Comparison of Human and Human Analogue Biomechanical Data. A report of an ad hoc committee, Ann Arbor, Michigan, December 6.

5. Abel, J.M.; Gennarelli, T.A.; and Segawa, H. 1978. Incidence and Severity of Cerebral Concussion in the Rhesus Monkey Following Sagittal Plane Angular Acceleration. In: 22nd Stapp Car Crash Conference Proceedings, 24-26 October 1978, Ann Arbor, MI, Warrendale, PA: SAE Paper No. 780886, pp. 35-53.

6. Adams, J.H.; Graham, D.I.; and Gennarelli, T.A. 1981. Acceleration Induced Head Injury in the Monkey. II Neuropathology. In: K. Jellinger, F. Gullotta, M. Mossakowski, eds., Experimental Clinical Neuropathology. Acta Neuropathol. (Berl.) Suppl. vii, pp. 26-28.

7. Aldman, B.; Thorngren, L.; and Ljung, C. 1981. Patterns of Deformation in Brain Models Under Rotational Motion. In: DOT, NHTSA, Head and Neck Injury Criteria, Washington, D.C.

8. Alem, N.M. 1974. Simulation of Head Injury Due to Combined Rotation and Translation of the Brain. In: 18th Stapp Car Crash Conference Proceedings, pp. 579.

9. Chan, M. and Ward, C. 1981. Relative Importance of Skull Deformation. Proceedings ASME. Biomechanics Symposium, June 22-24.

10. Engin, A.E. 1969. Axisymmetric Response of a Fluid-filled Spherical Shell to a Local Radial Impulse--A Model for Head Injury. Journal of Biomechanics 2(3):395-341.

11. Engin, A.E. and Akkas, N. 1978. Application of a Fluid-Filled Spherical Sandwich Shell as a Biodynamic Head Injury Model for Primates. Aviation, Space and Environmental Medicine, pp. 120-124.

12. Evans, F.G.; Lissner, H.R.; and Lebow, M. 1958. The Relation of Energy, Velocity and Acceleration to Skull Deformation and Fracture. Surgery, Gynecology and Obstetrics 107:593-601.

13. Ewing, C.L. and Thomas, D.J. 1972. Human Head and Neck Response to Impact Acceleration. Naval Aerospace Medical Research Laboratory Detachment, New Orleans, Monograph 21, August.

14. Ewing, C.L. and Thomas, D.J. 1973. Torque Versus Angular Displacement Response of Human Head to -Gx Impact Acceleration. In: 17th Stapp Car Crash Conference Proceedings, Paper No. 730976.

15. Ewing, C.L., et al. 1975. The Effect of the Initial Position of the Head and Neck on the Dynamic Response of the Human Head and Neck to -Gx Impact Acceleration. In: 19th Stapp Car Crash Conference Proceedings, Paper No. 751157.

16. Gennarelli, T.A., et al. 1979. Differential Tolerance of Frontal and Temporal Lobes to Contusion Induced by Angular Acceleration. In: 23rd Stapp Car Crash Conference Proceedings, 17-19 October, San Diego, pp. 563-586. SAE Paper No. 791022.

17. Gennarelli, T.A.; Adams, J.H.; and Graham, D.I. 1981. Acceleration Induced Head Injury in the Monkey. 1. The Model Its Mechanical and Physiological Correlates. In: K. Jellinger, F. Gullotta, M. Mossakowski, eds, Experimental Clinical Neuropathology. Acta Neuropathol. (Berl.) Suppl. VII, pp. 23-25.

18. Gennarelli, T.A., et al. 1982. Diffuse Axonal Injury and Traumatic Coma in the Primate. Ann. Neurol. 12:564-74.

19. Gurdjian, E.S., et al. 1961. Intracranial Pressure and Acceleration Accompanying Head Impacts in Human Cadavers. Surgery, Gynecology and Obstetrics 113:185-190.

20. Gurdjian, E.S., et al. 1968. Significance of Relative Movements of Scalp, Skull and Intracranial Contents During Impact Injury of the Head. Journal of Neurosurgery 29(1):70-72.

21. Higgins, L.S. and Schmall, R.A. 1967. A Device for the Investigation of Head Injury Effected by Non-Deforming Head Accelerations. In: 11th Stapp Car Crash Conference Proceedings, pp. 35-46.

22. Hodgson, V.R. and Thomas, L.M. 1979. Acceleration Induced Shear Strains on a Monkey Brain Hemisection. In: 23rd Stapp Car Crash Conference Proceedings, 17-19 October, San Diego, Calif, pp. 589-611. SAE Paper No. 791023.

23. Hosey, R.R. and Liu, Y.K. 1982. A Homeomorphic Finite Element Model of the Human Head and Neck. In: R. H. Gallagher, et al., Finite Elements in Biomechanics, NY: John Wiley and Sons, Ltd., pp. 379-401.

24. Khalil, T.B. and Hubbard, R.P. 1977. Parametric Study of Head Response by Finite Element Modeling. Journal of Biomechanics 10(2):119-132.

25. Lissner, H.R.; Lebow, M.; and Evans, F.G. 1960. Experimental Studies on the Relation between Acceleration and Intracranial Pressure Changes in Man. Surgery, Gynecology, and Obstetrics 111:329-338.

26. Lowenhielm, P. 1974. Strain Tolerance of the W. Cerebri Sup. (bridging veins) Calculated from Head-on Collision Tests with Cadavers. Z. Rechtsmedizin 75:131-144.

27. McElhaney, J.H.; Stalnaker, R.L.; and Roberts, V.L. 1972. Biomechanical Aspects of Head Injury. In: W.F. King and H.J. Mertz, eds., Human Impact Response: Measurement and Simulation, N.Y.: Plenum Press.

28. Nahum, A.M. and Smith, R.W. 1976. An Experimental Model for Closed Head Injury. In: 20th Stapp Car Crash Conference Proceedings, Paper No. 760825.

29. Nahum, A.; Smith, R.W.; and Ward, C.C. 1977. Intracranial Pressure Dynamics During Head Impact. In: 21st Stapp Car Crash Conference Proceedings, pp. 337-366.

30. Nusholtz, G.S.; Melvin, J.W.; and Alem, N.M. 1979. Head Impact Response Comparisons of Human Surrogates. In: 23rd Stapp Car Crash Conference Proceedings, 17-19 October, San Diego, Calif., pp. 499-541. SAE Paper No. 791020.

31. Nusholtz, G.S.; Melvin, J.W.; and Lux, P. 1983. The Influence of Impact Energy and Direction on Thoracic Response. In: 27th Stapp Car Crash Conference Proceedings, pp. 69-94.

32. Nusholtz, G.S., et al. 1979. Comparison of Epidural Pressure in Live Anesthetized and Post-Mortem Primates. In: 7th International Workshop on Human Subjects for Biomechanical Research Proceedings, 16 October 1979, Coronado, Calif., pp. 175-200. Washington, D.C.: Distributed by National Highway Traffic Safety Administration.

33. Ommaya, A.K.; Hirsch, A.E.; Flamim, E.S.; and Mahone, R.H. 1966. Cerebral Concussion in the Monkey: An Experimental Model. Science, 153:211-212.

34. Ono, K.; Kikuchi, K.; and Nakamura, M. 1980. Human Head Tolerance to Sagittal Impact Reliable Estimation Deduced from Experimenetal Head Injury Using Subhuman Primates and Human Cadaver Skulls. In: 24th Stapp Car Crash Proceedings, 15-17 October, Troy, MI, pp. 104-160. SAE Paper No. 801300.

35. Padgaonkar, A.J.; Krieger, K.W.; and King, A.J. 1975. Measurement of Angular Acceleration of a Rigid Body Using Linear Accelerations. ASME Preprint-75-APM3, June.

36. Roberts, V.L.; Hodgson, V.R.; and Thomas, L.M. 1967. Fluid Pressure Gradients Caused by Impact to the Human Skull. In: Biomechanics Monograph, pp. 223-235. New York: American Society of Mechanical Engineers.

37. Sances, A., et al. 1984. Biodynamics of Vehicular Injuries. In: G. A. Peters and B. J. Peters, Eds., Automotive Engineering and Litigation, New York: Garland Law Publishing, pp. 449-550.

38. Shugar, T.A. 1975. Transient Structural Response of the Linear Skull-Brain System. In: 19th Stapp Car Crash Conference Proceedings, pp. 581-614.

39. Stalnaker, R.L., et al. Door Crashworthiness Criteria. Final Report. 1971. Contract No. FH-11-7288, U.S. Dept. of Transportation, National Highway Traffic Safety Administration, Washington, D.C.

40. Stalnaker, R.L., et al. 1977. Head Impact Response. In: 21st Stapp Car Crash Conference Proceedings, 19-21 October, New Orleans, pp. 303-335. SAE Paper No. 770921.

41. Thomas, L.M., et al. 1968. Static Deformation and Volume Changes in the Human Skull. In: 12th Stapp Car Crash Conference Proceedings, pp. 260-270.

42. Unterharnscheidt, F. 1983. Neuropathology of Rhesus Monkeys Undergoing -Gx impact acceleration. In: C.L. Ewing, et al., eds., Impact Injury of the Head and Spine, Springfield, IL: Thomas, pp. 94-176.

43. Ward, C.C.; Chan, M.; and Nahum, A. 1980. Intracranial Pressure - A Brain Injury Criterion. In: 24th Stapp Car Crash Conference Proceedings. SAE Paper No. 801304.

44. Ward, C.C.; Nikravesch, P.E.; and Thompson, R.B. 1978. Biodynamic Finite Element Models Used in Brain Injury Research. Journal of Aviation Space and Environmental Medicine, 49(1).

45. Nusholtz, G.S., et al. 1980. Thoraco-Abdominal Response and Injury. In: 24th Stapp Car Crash Conference Proceedings, pp. 187-228.

46. Nusholtz, G.S., et al. 1983. Cervical Spine Injury Mechanisms. In: 27th Stapp Car Crash Conference Proceedings, pp. 179-188.

47. Alem, N.M.; Melvin, J.W.; and Holstein, G.L. 1978. Biomechanics Applications of Direct Linear Transformation in Close-Range Photogrammetry. Proceedings of the Sixth New England Bioengineering Conference, New York: Pergamon Press.

48. Bender, M.; Melvin, J.W.; and Stalnaker, R.L. 1976. A High-Speed Cineradiograph Technique for Biomechanical Impact. In: 20th Stapp Car Crash Conference Proceedings, Paper No. 760824.

49. Bishop, R.L. and Goldberg, S.I. 1968. Tensor Analysis on Manifolds, New York, MacMillan.

50. Cartan, E. 1946. Lecons sur la Geometrie des Espaces de Rieman, Second Edition, Paris, Gautheir Villars.

51. O'Neill, B. 1967. Elementary Differential Geometry, New York Academic Press.

52. Stoker, J.J. 1969. Differential Geometry, New York, Wiley Intersciences.

53. Harris, C.M., and Crede, C.E. 1976. Shock and Vibration Handbook. New York, McGraw-Hill Book Company.

54. Hartman, C.G. and Straus, Jr., W.L., eds. The Anatomy of the Rhesus Monkey (Macaca mulatta), New York: Hafner Publishing Co., 1965 edition.

55. Heimer, L. 1983. The Human Brain and Spinal Cord Functional Neuroanatomy and Dissection Guide, New York: Springer-Verlag.

56. Ommaya, A.K. 1973. Head Injury Mechanisms. Final Report Contract No. DOT-HS-081-1-106IA, U.S. Dept. of Transportation, National Highway Traffic Safety Administration, Washington,D.C.

APPENDICES

Three-Dimensional Motion Time History

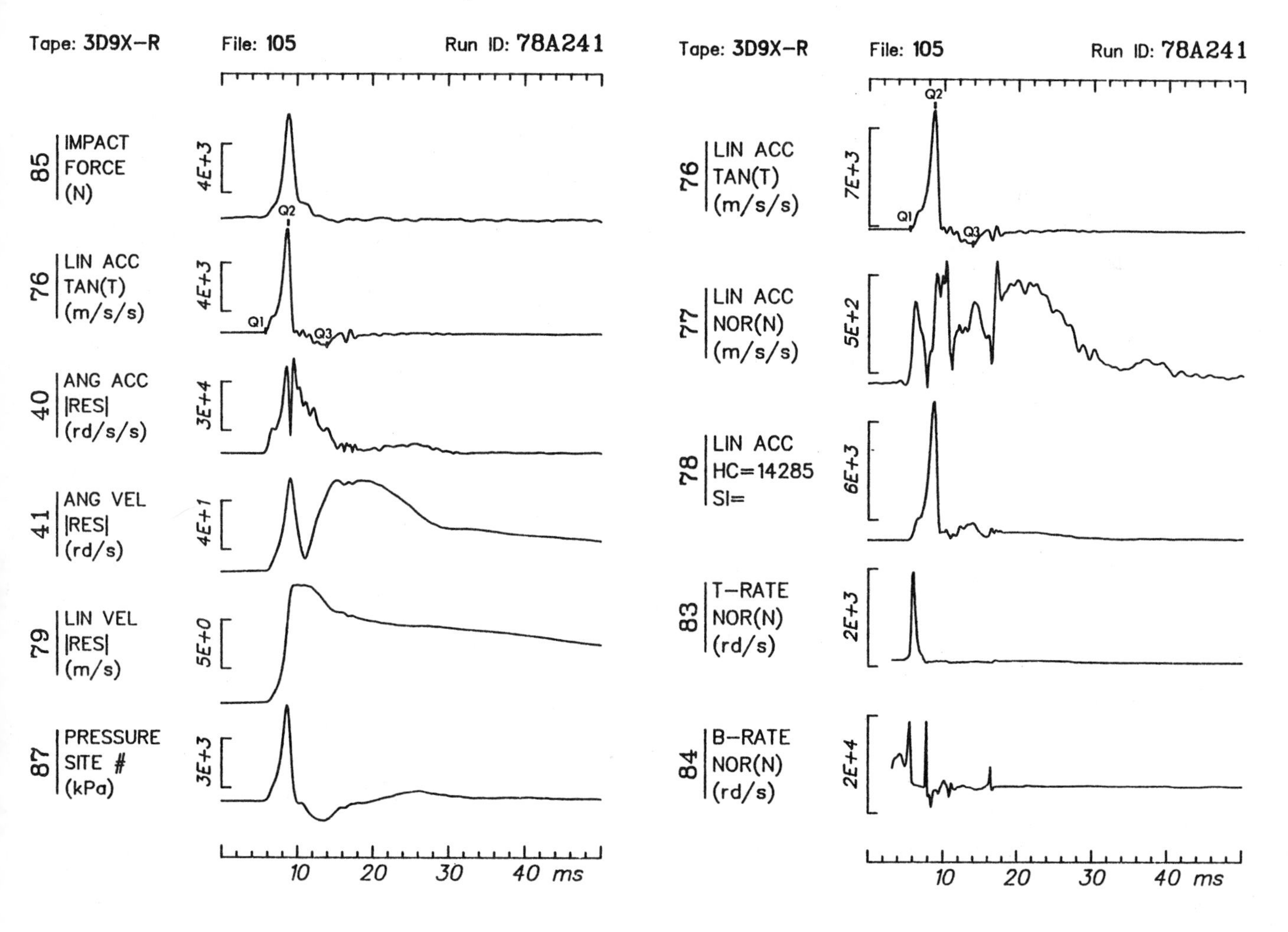

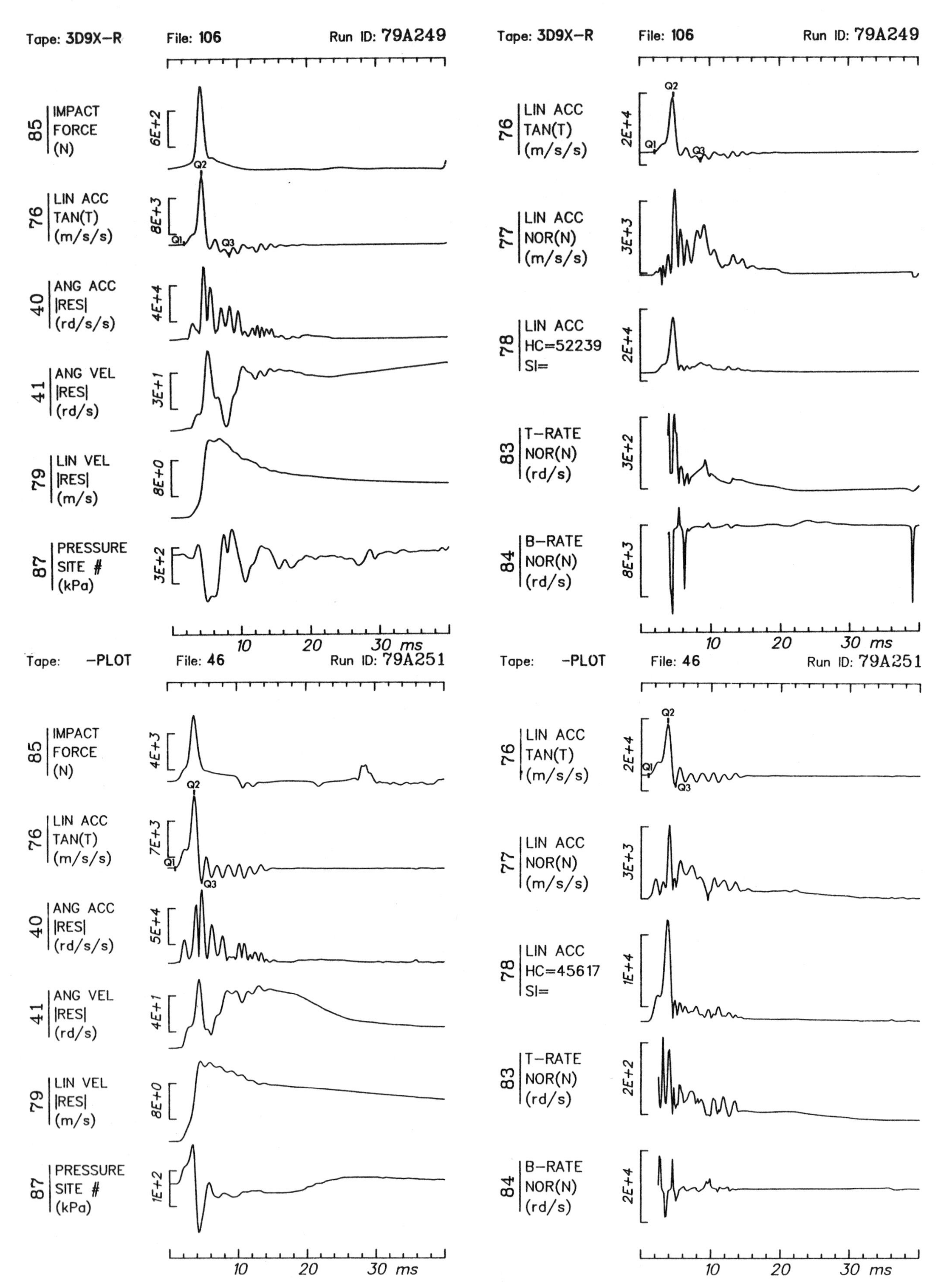

Tape: 3D9X-R
File: 106
Run ID: 79A249
IMPACT
FORCE
(N)
LIN ACC
TAN(T)
(m/s/s)
ANG ACC
|RES|
(rd/s/s)
ANG VEL
|RES|
(rd/s)
LIN VEL
|RES|
(m/s)
PRESSURE
SITE #
(kPa)
LIN ACC
NOR(N)
(m/s/s)
LIN ACC
HC=52239
SI=
T-RATE
NOR(N)
(rd/s)
B-RATE
NOR(N)
(rd/s)
Tape: -PLOT
File: 46
Run ID: 79A251
LIN ACC
HC=45617
SI=
10
20
30 ms

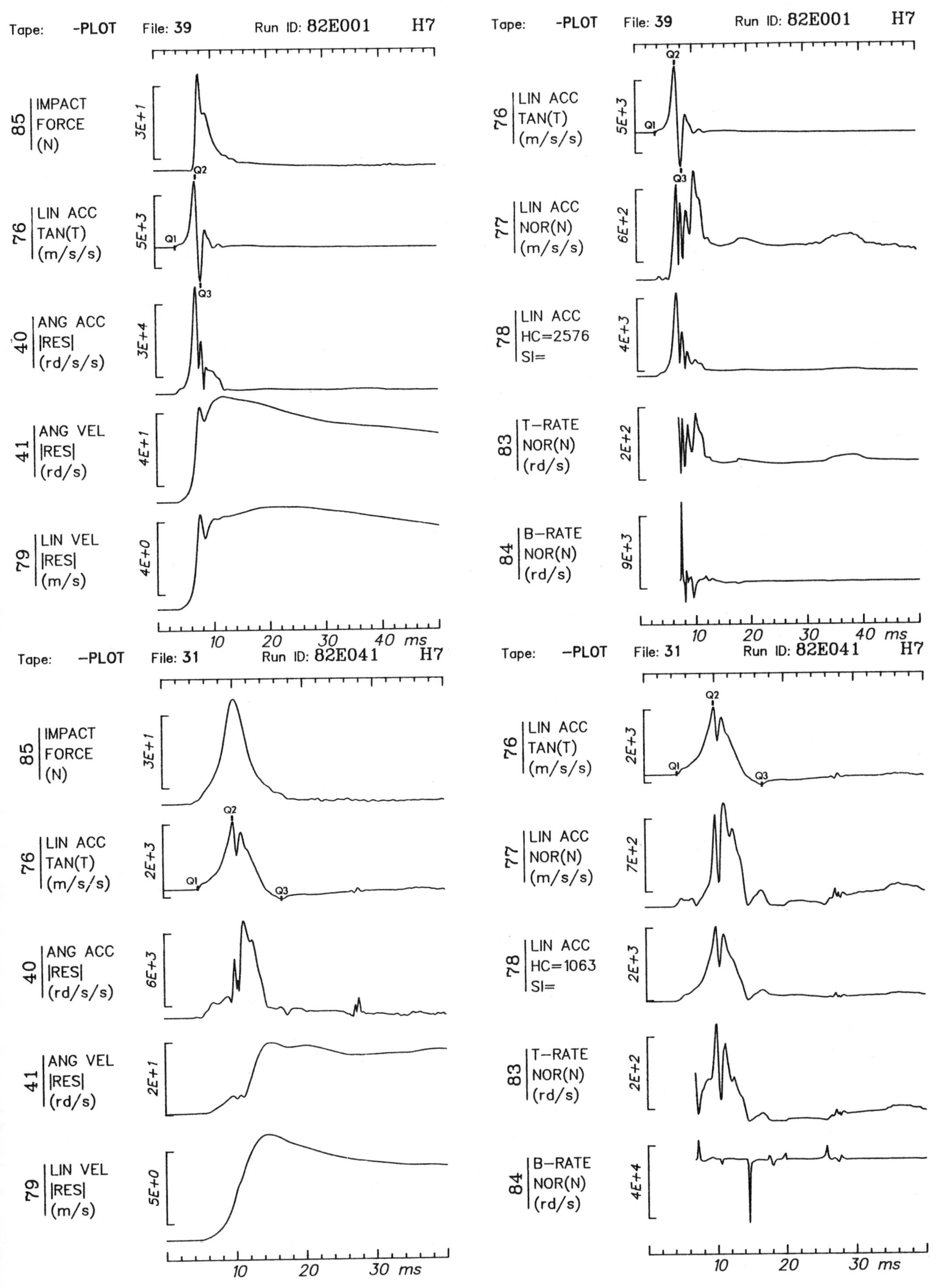

Tape: -PLOT
File: 39
Run ID: 82E001
H7
85 IMPACT FORCE (N)
3E+1
76 LIN ACC TAN(T) (m/s/s)
5E+3
Q1
Q2
Q3
40 ANG ACC |RES| (rd/s/s)
3E+4
41 ANG VEL |RES| (rd/s)
4E+1
79 LIN VEL |RES| (m/s)
4E+0
10
20
30
40 ms
77 LIN ACC NOR(N) (m/s/s)
6E+2
78 LIN ACC HC=2576 SI=
4E+3
83 T-RATE NOR(N) (rd/s)
2E+2
84 B-RATE NOR(N) (rd/s)
9E+3
Tape: -PLOT
File: 31
Run ID: 82E041
H7
2E+3
6E+3
2E+1
5E+0
7E+2
78 LIN ACC HC=1063 SI=
4E+4
10
20
30 ms

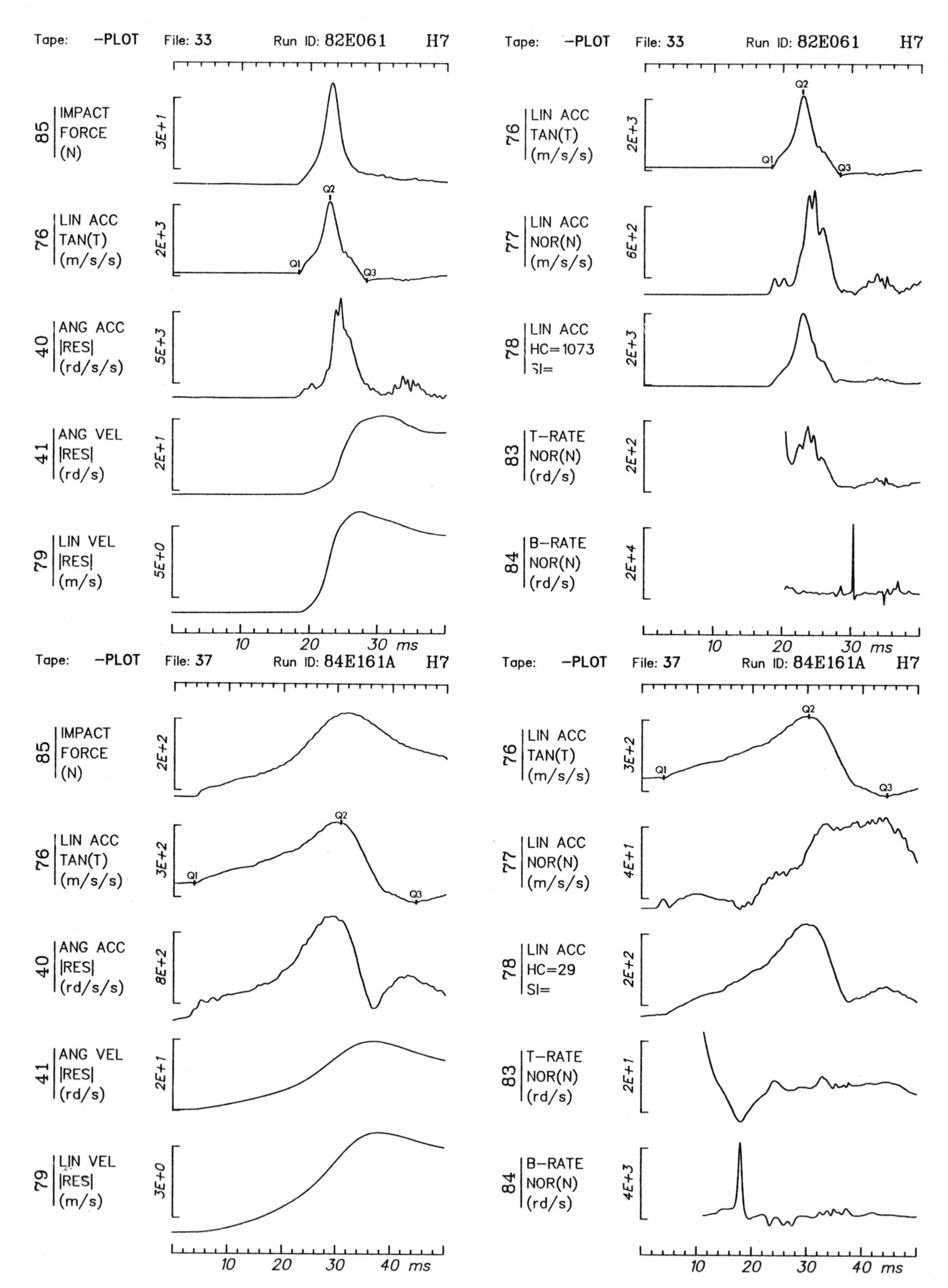

Tape: -PLOT File: 33 Run ID: 82E061 H7
85 IMPACT FORCE (N)
3E+1
76 LIN ACC TAN(T) (m/s/s)
2E+3
Q1
Q2
Q3
40 ANG ACC |RES| (rd/s/s)
5E+3
41 ANG VEL |RES| (rd/s)
2E+1
79 LIN VEL |RES| (m/s)
5E+0
10 20 30 ms
Tape: -PLOT File: 33 Run ID: 82E061 H7
76 LIN ACC TAN(T) (m/s/s)
2E+3
77 LIN ACC NOR(N) (m/s/s)
6E+2
78 LIN ACC HC=1073 SI=
2E+3
83 T-RATE NOR(N) (rd/s)
2E+2
84 B-RATE NOR(N) (rd/s)
2E+4
10 20 30 ms
Tape: -PLOT File: 37 Run ID: 84E161A H7
85 IMPACT FORCE (N)
2E+2
76 LIN ACC TAN(T) (m/s/s)
3E+2
40 ANG ACC |RES| (rd/s/s)
8E+2
41 ANG VEL |RES| (rd/s)
2E+1
79 LIN VEL |RES| (m/s)
3E+0
10 20 30 40 ms
Tape: -PLOT File: 37 Run ID: 84E161A H7
76 LIN ACC TAN(T) (m/s/s)
3E+2
77 LIN ACC NOR(N) (m/s/s)
4E+1
78 LIN ACC HC=29 SI=
2E+2
83 T-RATE NOR(N) (rd/s)
2E+1
84 B-RATE NOR(N) (rd/s)
4E+3
10 20 30 40 ms

Force and Pressure Time History

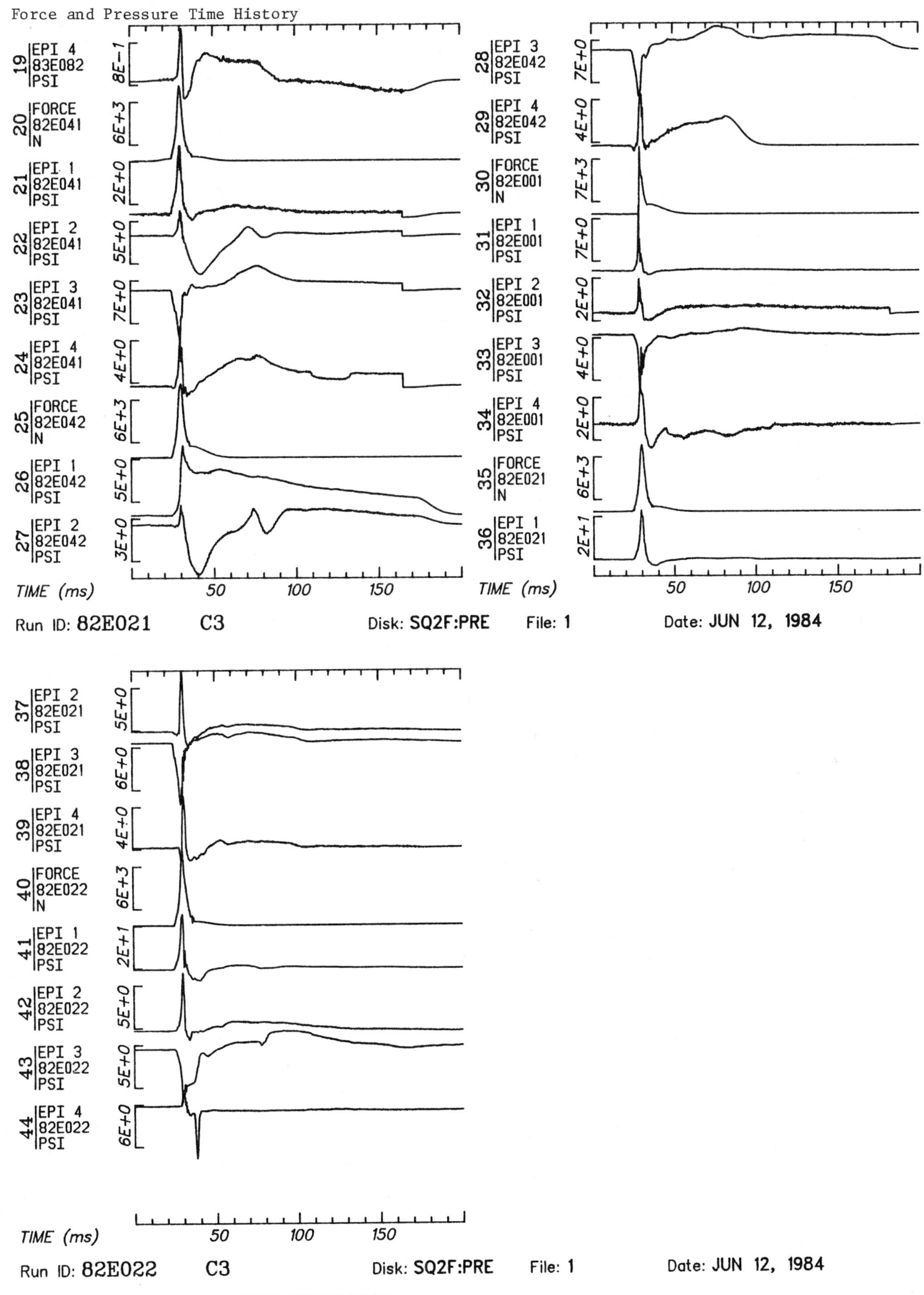

FORCE AND PRESSURE TIME HISTORIES

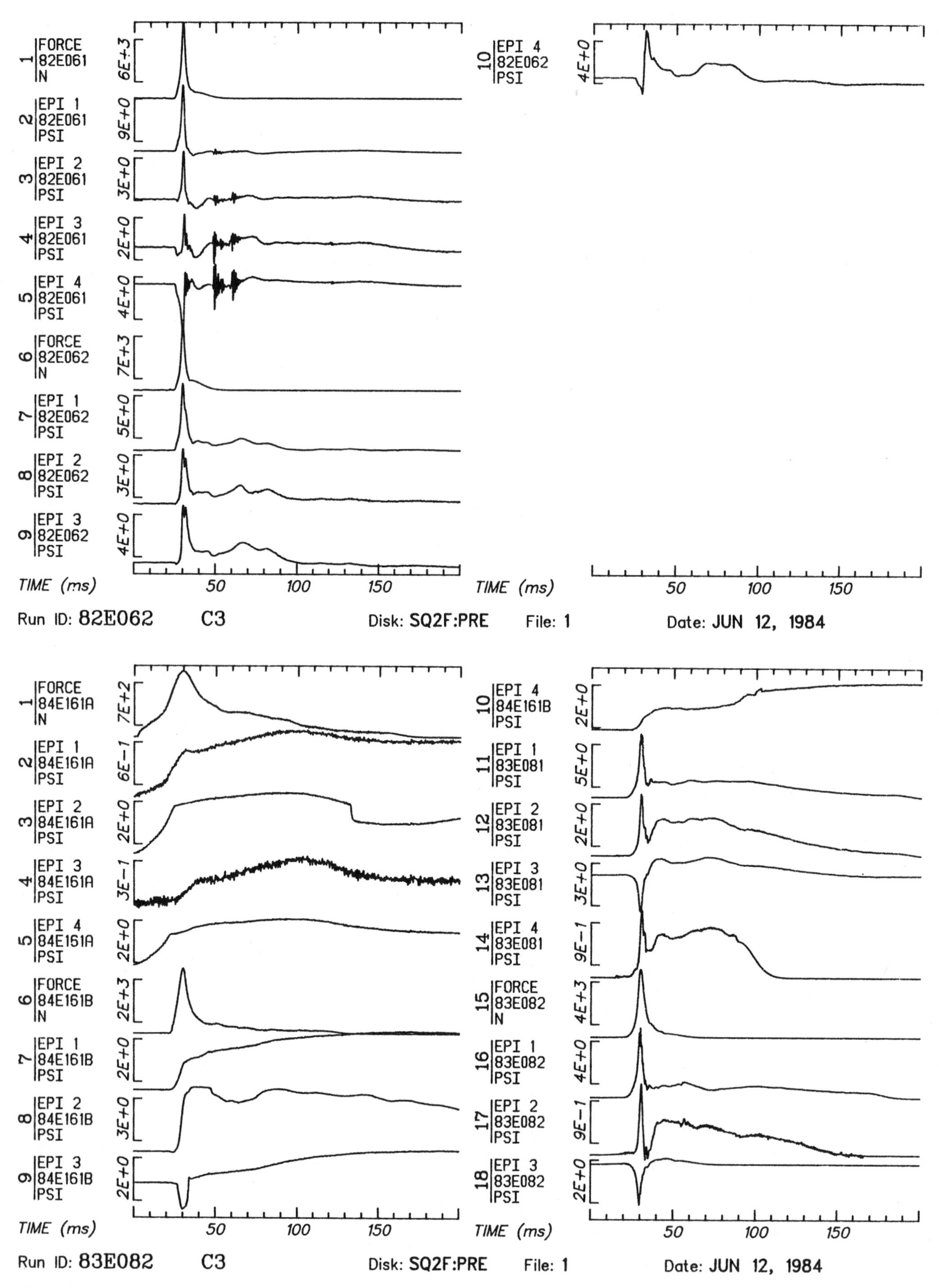

FORCE AND PRESSURE TIME HISTORIES

861893

Omni-Directional Human Head-Neck Response

J. Wismans and H. van Oorschot
TNO Road-Vehicles Research Institute

H. J. Woltring
Biomechanics Consultant

ABSTRACT

The Naval Biodynamics Laboratory (NBDL) in New Orleans has conducted an extensive research program over the past years to determine the head-neck response of volunteer subjects to impact acceleration. These subjects were exposed to impacts in frontal, lateral and oblique directions. An analysis of a limited number of frontal and lateral tests from a test series conducted in the late seventies with two subjects showed that the observed head-neck dynamics can be described by means of a relatively simple 2-pivot analog system (1,2)*.

The present study extends this analysis to a more recent NBDL test program with 16 human subjects. The database consists of 119 frontal, 72 lateral and 62 oblique tests. The research methodology used for this analysis includes a detailed description of three-dimensional kinematics as well as load calculations near T1 and the occipital condyles. A description of this research methodology and a summary of the major test results will be presented. Special attention is given to the influence of impact severity and impact direction on the head-neck dynamics. It will be shown that a similar analog system as proposed earlier for frontal and lateral impacts, is suitable for all impact directions. Geometrical properties of this analog have been determined by means of newly developed numerical techniques rather than through the graphical techniques that were used earlier. Findings of this analysis will be discussed in view of future omni-directional mechanical neck developments.

THE PURPOSE of this paper is to analyse the head-neck response in frontal, oblique and lateral impacts as measured in a large number of human volunteer tests conducted in 1981 and 1982 at the Naval Biodynamics Laboratory (NBDL) in New Orleans. This study is an extension of a previous analysis of a limited set of earlier NBDL tests (1,2,3).

The major aim of this research program is the development of a mechanical neck for crash dummies with omni-directional biofidelity. Realistic simulation of the neck response in a dummy is of particular importance to get a humanlike dynamical behaviour of the dummy head. Trajectories of the head and the nature of the head contact with vehicle interior or exterior are critically dependent on the dummy's neck design.

Neck performance requirements in the literature appear to be mainly related to the neck response in forward flexion and extension. A review of this literature showed that the existing requirements are not sufficient conditions to ensure a humanlike response (3). Performance requirements in this paper will be defined quite differently namely by means of a 2-pivot analog system proposed in the initial phase of this research program (1,2,3). Detailed geometrical and dynamical properties of this analog system will be presented.

DATABASE DESCRIPTION

In the tests the subjects are seated in an upright position on a HYGE Accelerator (0.3048 m) and exposed to short duration accelerations simulating frontal, oblique or lateral impacts. The resulting three-dimensional motions of the head and first thoracic vertebral body (T1) are monitored by anatomically mounted clusters of accelerometers and photographic targets. A detailed description of the instrumentation and test methods is provided in (4,5).

In the frontal impact tests, the subjects are restrained by shoulder straps, a lap belt and an inverted V-pelvic strap tied to the lap belt. Upper arm and wrist restraints were used to prevent flailing (5). In addition a loose safety belt around the chest is employed. The same restraint system is used in lateral and oblique tests along with a 25 cm wide chest

* Numbers in parentheses designate references at end of paper.

strap to minimize the load on the right shoulder. In addition a lightly padded wooden board is placed against the right shoulder of the subject to limit the upper torso motion.

SUBSET SPECIFICATION - Testresults for 253 tests have been obtained from NBDL. Out of these test a subset of 109 tests was selected for further analysis. Selection of this subset was based on the following criteria:

- Tests specified by NBDL with errors in sensor or film data or with possible contact of instrumentation with restraint system or subjects chest have been omitted.
- No large forward or lateral bending of the head in the initial position is allowed.
- Tests with large data gaps or data shifts in the photographically derived variables are omitted.
- Only tests with a more severe impact level will be considered: frontal tests > 8 g, lateral tests > 5 g, oblique tests > 7 g.

Appendix A summarizes the tests incorporated in this subset and the most important test characteristics. The subset contains 46 frontal, 31 lateral and 32 oblique tests with 15 subjects. Three of these subjects were exposed to all three impact directions namely subject H00133, H00135 and H00136. For all tests the impact velocity is larger than or equal to 6 m/s.

SLED ACCELERATION - Mean values of the sled acceleration-time histories for the most severe frontal, lateral and oblique tests in the present database are shown in Fig. 1. The oblique sled pulses appear to be more severe than the lateral ones, while the frontal are more severe than the oblique ones. These test conditions are close to the test conditions in the earlier tests with subjects H00083 and H00093 (1,2,3). Only for the oblique tests a slight difference can be observed: a maximum peak sled acceleration of 11.4 g in the present database compared to 9.7 g in the previous tests.

COORDINATE SYSTEMS - Figure 2 illustrates the location of the head and T1 anatomical coordinate systems as defined by NBDL. Both coordinate systems are orthogonal and right-handed. Three-dimensional X-ray techniques were used to specify in each test these coordinate systems relative to head and T1 anatomical landmarks. The initial nominal orientation (i.e. before moving of the sled) of these systems relative to the laboratory and the sled coordinate systems for the 3 impact directions, is illustrated in Fig. 3. More details on the definition of the NBDL coordinate systems are provided in (1,5).

The distance between T1 and head anatomical origin just before the impact (time = 0) is called initial neck length. In the earlier test series (1,2,3) differences in this parameter up to 0.07 m per subject were observed which was explained partly by errors in specification of the T1 coordinate system. These earlier tests, moreover, showed for different tests with the same subject considerable variations in the initial orientation of the T1 coordinate system. In the present database the variations in neck length are found to be smaller namely less than 0.03 m (6), but variations in initial T1 coordinate system orientation still appear to be large. For instance the angle between T1 and laboratory z-axis projected on the plane of impact (i.e. a plane parallel to the laboratory (x,z)-plane) showed deviations for the same subject and same impact direction up to 24 degrees (6). Such deviations can not be fully explained from variations in the initial torso orientation and are consequently attributed in part to errors in the specification of the T1 coordinate system itself.

On the bases of these findings a new corrected T1 coordinate system will be introduced here similar to the one defined for the previous test series. This coordinate system is obtained in the following way:

- For each subject an average initial neck length is determined on the basis of calculated initial neck length values for each test (see Table 1).
- The origin of the T1 coordinate system is shifted vertically with respect to the laboratory in such a way that the initial neck length becomes identical to the subject's average initial neck length.
- Finally the T1 coordinate system is rotated so that it becomes aligned with the

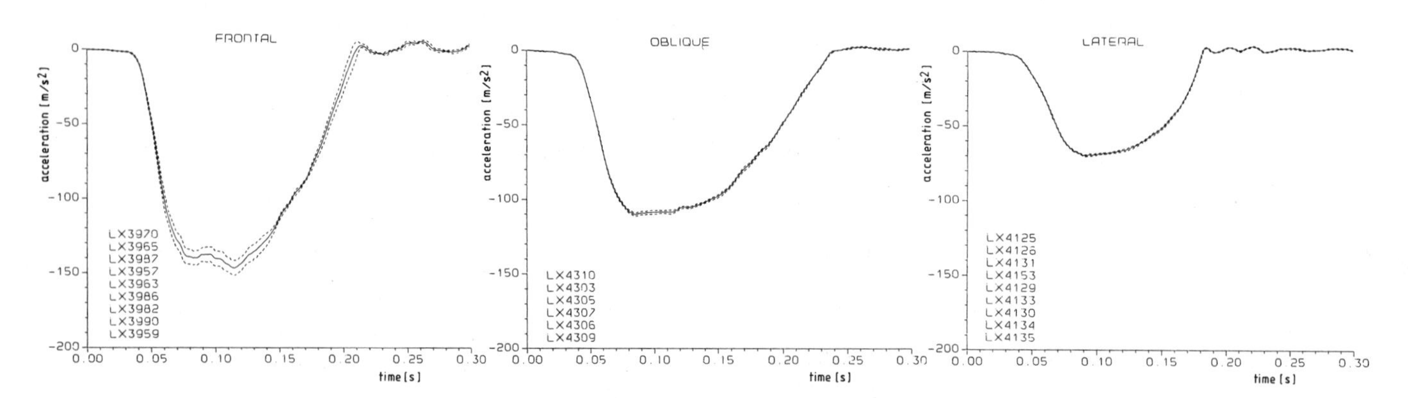

Fig. 1. Mean sled acceleration-time histories for most severe frontal, oblique and lateral impacts (——— mean value, ---- corresponding standard deviation).

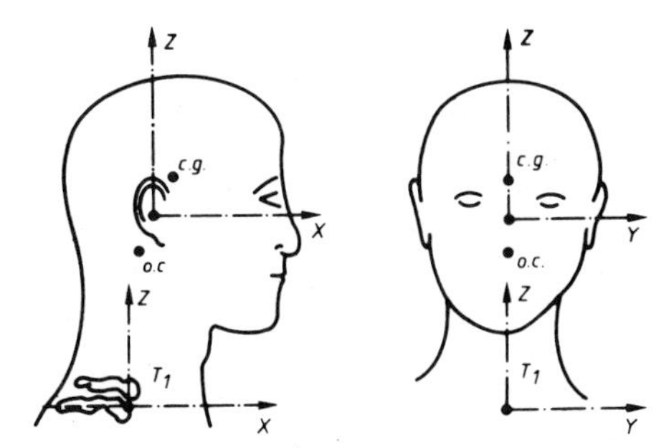

Fig. 2. Location of anatomical coordinate systems according to NBDL (o.c. = occipital condyles, c.g. = center of gravity).

sled (laboratory) coordinate system. As a consequence the orientation of the corrected T1 coordinate system relative to the T1 vertebral body will be dependent on the impact direction. In lateral (oblique) direction it is rotated nominally 90 degrees (45 degrees) with respect to its orientation in frontal direction. Fig. 3 includes the orientation of the corrected T1 coordinate system.

HUMAN SUBJECT ANTHROPOMETRY - At NBDL detailed anthropometric measurements are conducted for each subject as part of the test protocol. Table 1 summarizes the most significant data in this respect. Definitions for these anthropometric variables can be found in (7).

In order to calculate neck load, estimates have to be made for the head mass distribution.

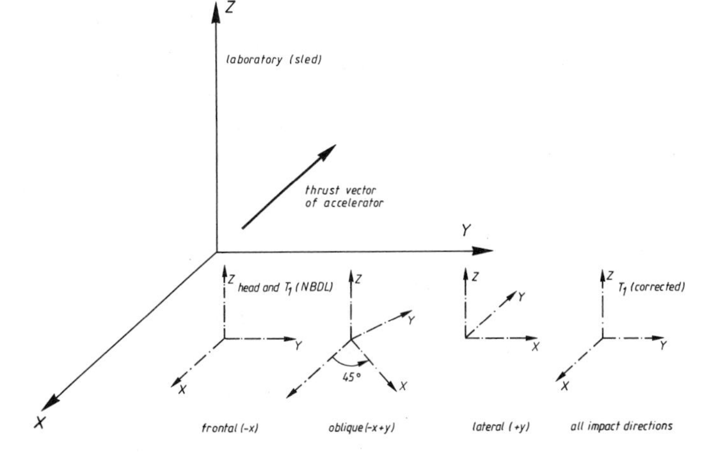

Fig. 3. Nominal initial orientation of head and T1 (NBDL) anatomical coordinate systems relative to laboratory and sled coordinate system in case of frontal, oblique and lateral impacts. Also the orientation of the corrected T1 coordinate systems is included in this figure.

The subjects head mass and the principal moments of inertia will be based on regression equations proposed by McConville et al. (8) using head anthropometry data presented in Table 1. For details on the selected regression equations see (3). The resulting masses and moments of inertia, including a correction for the head instrumentation, are presented in Table 1.

Table 1. Human subject anthropometry.

	Anthropometric measurements at NBDL							Mass distribution Estimates[2]	Principal moments of Inertia		
Subject number	Standing height (cm)	Weight (kg)	Sitting height (cm)	Head circumference (cm)	Head breadth (cm)	Head length (cm)	Initial neck length[1] (m)	Mass (kg)	I_{xx} (kgm^2)	I_{yy} (kgm^2)	I_{zz} (kgm^2)
H00118	185.5	73.8	97.9	57.0	14.4	20.3	0.172	4.79	0.0266	0.0303	0.0149
H00120	172.6	83.0	91.1	58.6	15.6	20.2	0.172	5.14	0.0297	0.0331	0.0169
H00127	172.3	62.1	89.8	54.2	14.9	18.5	0.162	4.40	0.0239	0.0252	0.0129
H00130	180.1	72.6	94.5	56.5	14.9	19.7	0.180	4.75	0.0266	0.0294	0.0148
H00131	167.0	67.6	90.0	57.5	15.4	19.6	0.156	4.98	0.0283	0.0311	0.0160
H00132	172.9	79.8	89.6	57.9	15.7	19.7	0.141	5.05	0.0290	0.0319	0.0164
H00133	161.7	61.2	86.8	56.1	14.7	19.4	0.165	4.70	0.0259	0.0286	0.0145
H00134	178.3	75.3	93.0	56.6	14.4	19.4	0.158	4.81	0.0261	0.0295	0.0151
H00135	171.6	68.9	90.7	53.5	14.6	17.9	0.150	4.32	0.0228	0.0240	0.0125
H00136	185.4	88.9	92.3	56.4	15.0	19.4	0.173	4.77	0.0266	0.0292	0.0149
H00138	186.0	78.9	99.2	57.1	15.4	19.8	0.174	4.87	0.0278	0.0304	0.0154
H00139	174.4	72.6	94.3	57.2	15.7	19.4	0.164	4.94	0.0282	0.0306	0.0158
H00140	177.3	86.2	94.5	56.7	15.4	19.0	0.173	4.88	0.0273	0.0297	0.0155
H00141	183.3	80.7	95.6	55.4	15.2	19.2	0.175	4.57	0.0256	0.0273	0.0138
H00142	182.3	87.5	95.6	56.4	14.9	19.5	0.161	4.75	0.0264	0.0292	0.0148

1) Defined as the average value of the initial distance between T1 and head anatomical origin in several tests.

2) Including correction for instrumentation.

The orientation of the principal inertia axes, location of center of gravity and location of occipital condyles are assumed to be subject independent. The same values are selected as in the earlier analysis on the basis of data identified in the literature (1,2,3). Table 2 summarizes these data.

Table 2. Subject independent anthropometric data estimated on the basis of data in the literature (relative to head anatomical coordinate system).

	x (cm)	y (cm)	z (cm)
center of gravity*)	1.2	0	2.9
occipital condylar point	-1.1	0	-2.6
principal axis system	rotated -36° about anatomical y-axis (backwards)		

*) including correction for instrumentation.

Earlier (see coordinate system section) the anthropometric quantity 'initial neck length' was introduced which was defined as the distance between the T1 and the head anatomical origin. Average values per subject are presented in Table 1 and vary between 0.14 m and 0.18 m. In the previous study (3) one subject (i.e. H00083) showed a significant shorter neck length namely 0.111 m which makes the use of test results from this subject questionable.

T1 DISPLACEMENTS AND ACCELERATIONS - A detailed analysis has been performed of the displacements and accelerations of T1 (6). In agreement with the results of the earlier study (3) the only significant linear displacement of T1 is found in the direction of impact (i.e. along the sled thrust vector). In other words vertical and lateral T1 displacements can be neglected.

Fig. 4 shows for each impact direction separately mean values resulting from the most severe tests for the T1 horizontal acceleration as a function of time. The corresponding standard deviation is incorporated in this figure. It can be seen that the T1 accelerations deviate considerably from the sled accelerations presented in Fig. 1. For all impact directions initially a large spike can be observed due to the interaction between the thorax and the restraint system. As a consequence the peak input acceleration experienced by the head neck system is about twice the peak sled acceleration. Further it follows that the 15 g frontal tests clearly are the most severe tests in terms of input to the head-neck system.

In the previous study a limited rotation of T1 in response to the impact was observed. For the present database such rotations also exist, however, they are considered small enough to be neglected in the remaining part of this paper.

ANALYSIS OF RELATIVE HEAD MOTIONS: THE 2-PIVOT LINKAGE MECHANISM

This section deals with the motions of the head during the loading phase i.e. up to maximum head excursion. The method used for the analysis in general will be similar to the one used for the previous database. Head motions will be expressed relative to a corrected T1 coordinate system as defined in the preceding section. Since T1 rotations will be neglected here, head motions will be presented with respect to a coordinate system which stays aligned with the laboratory coordinate system.

First an analysis will be made of the occipital condyle trajectories. Appendix B shows a projection of the trajectories on the plane of impact for all tests incorporated in the subset. Results are summarized per subject and per impact direction. It follows that:

- trajectories for different tests with the same subject are quite close to each other;
- maximum head excursions in frontal impacts are slightly larger than in oblique impacts and much larger than in the lateral ones;
- the shape of all trajectories is almost circular.

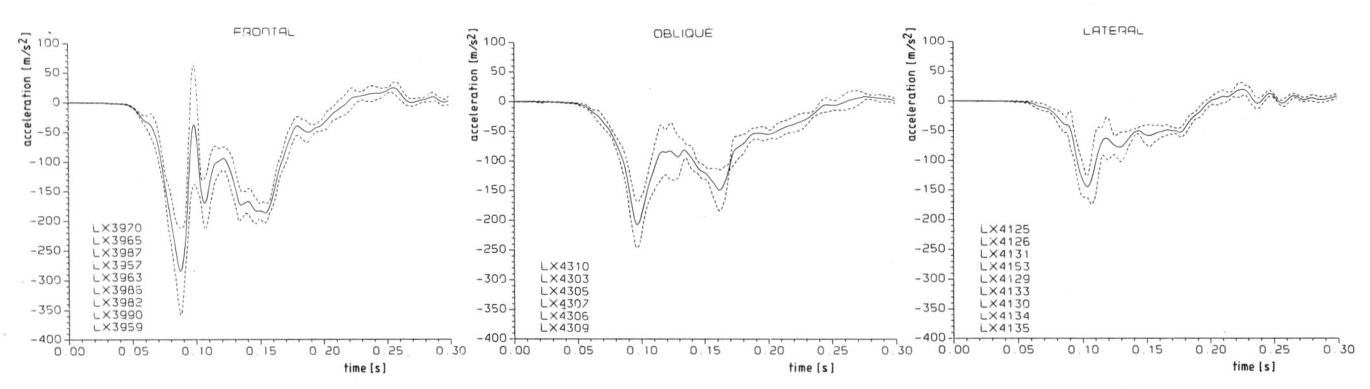

Fig. 4. Mean T1 horizontal acceleration-time histories for the most severe tests in three different impact directions.(—— mean value, --- corresponding standard deviation).

This last finding is the most interesting one. Because displacements of the occipital condyles in a direction perpendicular to the plane of impact are small (less than 0.03 m and without a preferential direction (6)), the 2-pivot linkage mechanism concept proposed in the previous study is also applicable for the present database. Fig. 5 illustrates this linkage mechanism. The upper link represents the head, the middle link the neck and the lower link the torso. The upper pivot is located in the occipital condyles and the lower pivot in the center of the circular arc approximating the occipital condyle trajectories. This lower pivot is a pin joint i.e. a joint with one degree of freedom with the rotation axis perpendicular to the plane of impact. The rotation in this joint is denoted by θ and is defined as the angle between neck link and z-axis of the corrected T1 coordinate system.

The upper pivot is a joint with two degrees of freedom (universal joint). The first degree of freedom of this joint allows the head link to rotate relative to the neck link in the plane of impact. This rotation angle will be denoted by ϕ and is defined here as the angle in the plane of impact between the z-axis of the head anatomical coordinate system and the corrected T1 coordinate system. The second degree of freedom of this upper joint is the rotation ψ of the head about the head anatomical z-axis indicating the head torsion or twist. In frontal impacts this twist motion can be neglected. The rotations of the head link (i.e. head anatomical z-axis) out of the plane of impact will be neglected here because such rotations were found to be very small.

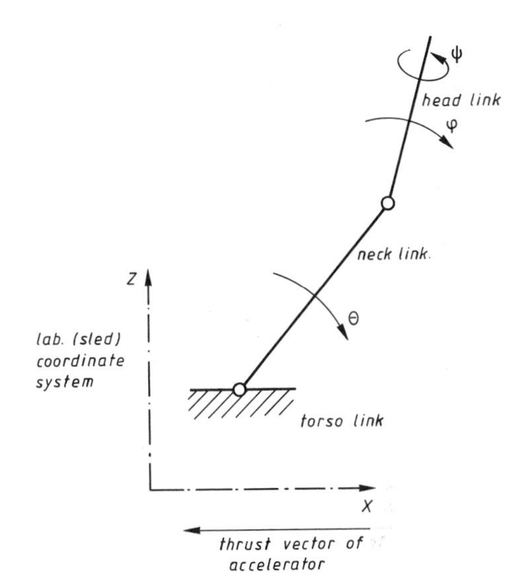

Fig. 5. Analog system for the description of the relative head motions.

NECK LINK LENGTH AND LOWER PIVOT LOCATION - On the basis of the occipital condyle trajectories the neck link length and the lower pivot location will be determined here. Numerical techniques rather than the graphical method applied in the previous study are used in order to improve the accuracy of these estimations. The computer program developed for this purpose allows estimations of an optimal radius and corresponding average center of rotation for a given planar trajectory by means of least-squares estimation techniques (9). The program can also be used to determine the optimal centre of rotation for this given trajectory if the radius has a pre-selected value. Output of the program includes the maximal fitting error in the trajectory (devmax) and the residual standard deviation (sdw).

First the program was applied to estimate the optimal radius for each test separately. Results are presented in Fig. 6 as function of initial neck length. Fig. 7 shows the mean radius as a function of impact severity. Data are given separately for each impact direction. It follows that the radius does not increase for a larger impact severity, whereas the effect of the initial neck length on the optimal radius value is small. The distribution in calculated radii appears to be larger in lateral direction than in oblique or frontal direction which can be explained by the shorter trajectories for lateral impacts. The average optimal radius per impact direction appears to vary slightly: 130.5 mm in frontal, 123.2 mm in lateral and 138.9 mm in oblique direction.

Table 3. Mean values for fitting accuracy in case of free and predescribed radii per impact direction, resulting from numerical estimation techniques.

Impact direction	Free radius: maximal fitting error (devmax) (mm)	Free radius: residual standard deviation (sdw) (mm)	Predescribed radius: 0.129m: maximal fitting error (devmax) (mm)	Predescribed radius: 0.129m: residual standard deviation (sdw) (mm)
Frontal	6.92	1.730	7.05	2.185
Lateral	5.13	1.271	5.41	1.565
Oblique	5.16	0.93	6.00	1.700

In the previous study (3) an average radius of 0.125 m was graphically selected for all tests in all impact directions. The average radius resulting from the present database appears to be surprisingly close to this earlier value, namely 0.129 m.

The effect of this fixed radius on the fitting accuracy has been calculated with the least-squares estimation program and is illustrated in Table 3. This table shows average values for devmax and sdw per impact direction resulting from calculations with and without a predescribed radius. It follows that the influence on the goodness of fit in case of an identical radius for all tests is small: the increase in the average standard deviation is less than 0.8 mm. As analyzed in (9) the radius length and the pivot location are highly correlated; this implies that different values for the radius length are largely

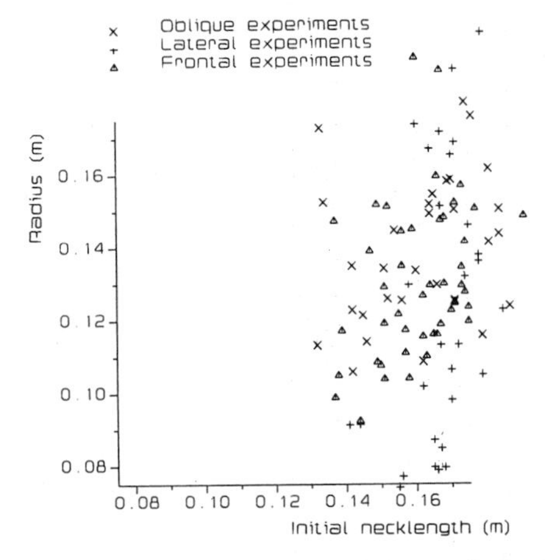

Fig. 6. Optimal radius for occipital condyle trajectories as function of initial neck length and impact direction.

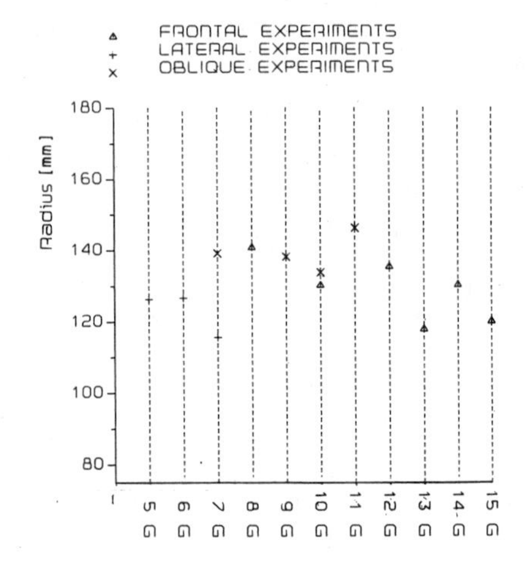

Fig. 7. Mean optimal radius for occipital condyle trajectories as function of impact severity and impact direction.

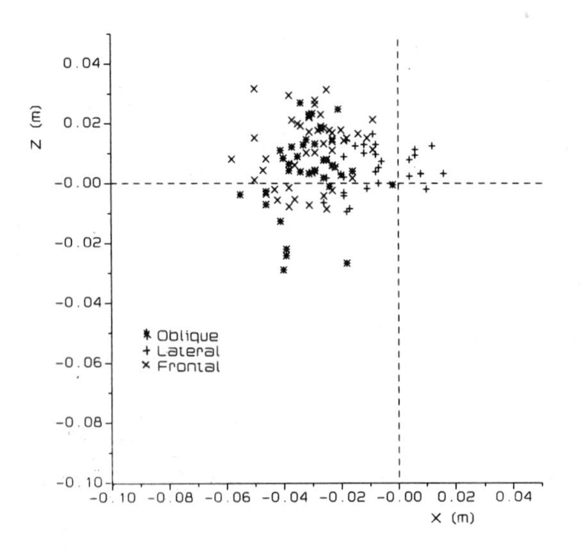

Fig. 8. Lower pivot locations relative to the corrected T1 coordinate system for a fixed neck link length.

compensated for by other optimal pivot values. On the basis of these findings for all tests one fixed value of 0.129 mm will further be used for the neck link length.

The optimal corresponding lower pivot locations for this link length are presented for each test in Appendix B and summarized in Fig. 8. Locations are given per impact direction relative to the (x,z)-plane of the corrected T1 coordinate system. Average values for the x and z coordinates and corresponding standard deviation, per impact direction are summarized in Table 4.

Table 4. Average values and corresponding standard deviations for lower pivot locations per impact direction (relative to the corrected T1 coordinate system).

Impact direction	x (s.d.) (m)	z (s.d.) (m)
frontal	- 0.031 (0.011)	0.011 (0.011)
lateral	- 0.008 (0.012)	0.005 (0.007)
oblique	- 0.032 (0.010)	0.003 (0.014)

The average z-coordinate appears to be close to zero. Variations in this coordinate are mainly due to differences per subject in initial neck length. The x-coordinate has a negative value in most of the tests i.e. opposite to the direction of motion of the head. For lateral impacts the mean value is close to 0.01 m while for frontal and oblique impacts a value close to 0.03 m is found.

ANALYSIS OF RELATIVE HEAD MOTIONS: HEAD AND NECK LINK ROTATIONS

Relative head motions will be analyzed here using the linkage concept presented before. Fig. 5 illustrates this mechanism. Three rotations (i.e. degrees of freedom) can be distinguised: head flexion ϕ, head twist Ψ and neck link rotation θ. The head flexion ϕ and head twist Ψ can be calculated directly from the NBDL test results (i.e. the head rotation matrix) and are independent of the selected geometrical linkage parameters.

The definition used for the head twist motion in the present study is slightly different from the one used in the previous study where the orientation of the head was considered to be the result of three successive Euler rotations about the head anatomical x-axis, the (transformed) head anatomical y-axis and the (transformed) head anatomical z-axis, respectively. This last rotation was defined as the twist angle. In the present study a more adequate definition is used: the relative orientation of the head is considered to be the result of two rotations. The first one transforms the head z-axis directly in its final position and the second rotation i.e. the twist motion is a rotation about the (rotated) z-axis (6).

The neck link rotation θ is calculated from the occipital condyle trajectories and the lower pivot location. Data presented here will be based on the calculated radius of 0.129 m and the corresponding test specific optimal lower pivot location.

The magnitude of the angles ϕ, Ψ and θ in the initial position (time = 0) will be denoted by ϕ_o, Ψ_o and θ_o, respectively. Maximum values for these rotations are denoted by ϕ_{max}, Ψ_{max} and θ_{max}. Appendix A summarizes the initial values as well as $(\phi_{max} - \phi_o)$, $(\theta_{max} - \theta_o)$ and $(\Psi_{max} - \Psi_o)$ for 109 tests.

INITIAL VALUES - Table 5 summarizes the mean values per impact direction for ϕ_o, Ψ_o* and θ_o. It follows that initial head twist (Ψ^o) and head flexion (ϕ_o) are close to zero, indicating that the 'average' initial head orientation is close to the orientation of the laboratory coordinate system. The initial neck link rotation θ_o shows, except for the lateral direction, a significant positive value namely 10.8 degrees in oblique and 17.6 degrees in frontal impacts.

Table 5 Mean values per impact direction for the initial values of the angles ϕ, Ψ and θ.

Impact direction	ϕ_o (deg.)	Ψ_o * (deg.)	θ_o (deg.)
frontal	1.8	-1.0	17.6
lateral	0.8	1.2	0.7
oblique	-2.9	-1.3	10.8

PEAK ROTATIONS AS FUNCTION OF IMPACT SEVERITY - Mean values for $(\phi_{max} - \phi_o)$, $(\theta_{max} - \theta_o)$ and $(\Psi_{max} - \Psi_o)$ as function of impact direction and impact severity are presented in Fig. 9. Both the head flexion and the neck link rotation appear to be strongly dependent on the peak sled acceleration. The influence of the impact direction is most clearly present in the neck link rotation: frontal impacts show larger neck link rotations (for the same impact severity) than oblique ones and oblique impacts show larger neck link rotations than the lateral ones. Finally it can be seen that lateral impacts show a much larger twist than oblique ones.

NECK LINK ROTATIONS AS FUNCTION OF HEAD FLEXION - Fig. 10 presents for the three impact directions the neck link rotation $(\theta - \theta_o)$ as function of head flexion $(\phi - \phi_o)$. Results for the most severe tests are given. In the initial phase of the motion for all impact directions the head flexion is smaller than the neck link rotation illustrating the translational nature of the initial head motion. This response was also observed in the previous study (3). As

* oblique impacts: Ψ_o - 45°; lateral impacts: Ψ_o - 90°

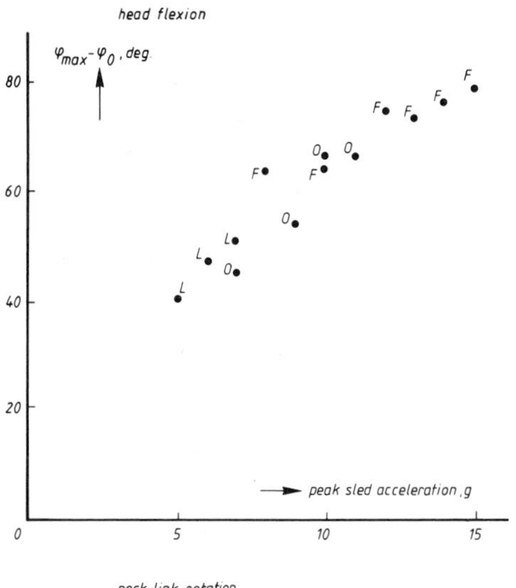

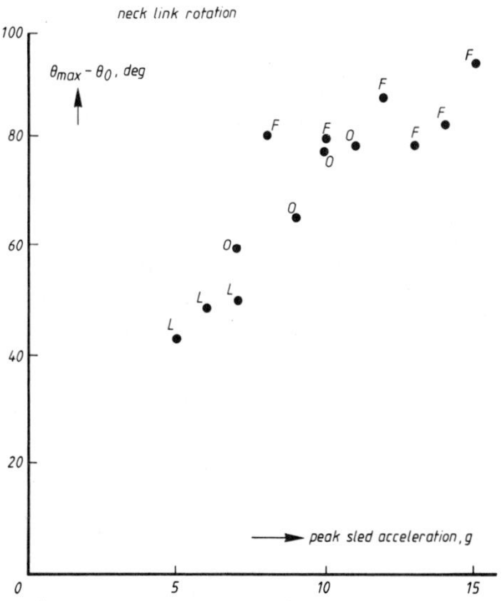

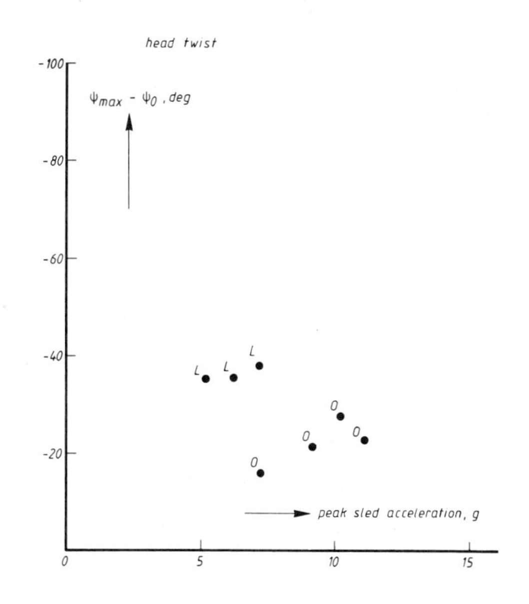

Fig. 9. Mean values for peak rotations as function of impact direction and impact severity.
(F = Frontal, O = Oblique, L = Lateral)

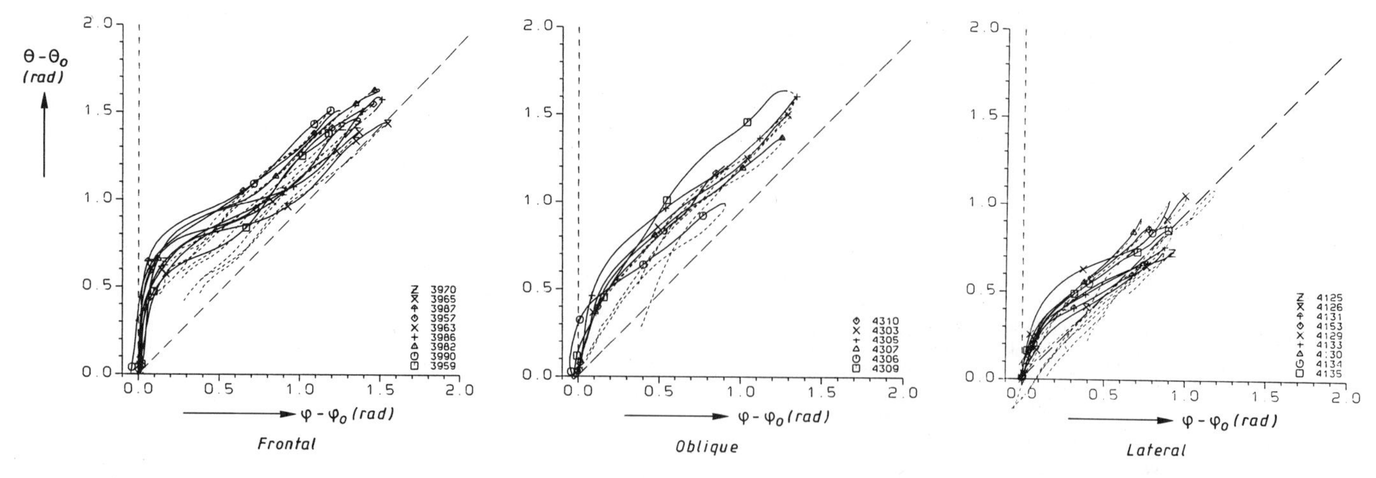

Fig. 10. Neck link rotation $(\theta - \theta_o)$ as function of head flexion $(\phi - \phi_o)$ for the most severe tests as function of impact direction. (——— loading -------- unloading)

soon as the relative angle $(\theta - \theta_o) - (\phi - \phi_o)$ reaches a certain level the head and neck link become more or less locked. For most of the tests this relative angle appears to decrease slightly in the final part of the loading phase.

The maximum relative angle i.e. $((\theta - \theta_o)-(\phi - \phi_o))_{max}$ between head and neck link has been calculated for all tests (see Appendix A). The influence of impact direction and impact severity on this angle is presented in Fig. 11. The impact direction appears to have a significant influence while the effect of impact severity is small. Average values for the upper pivot maximum rotation are 30 degrees in frontal, 20 degrees in oblique and 10 degrees in lateral impacts.

HEAD TWIST AS FUNCTION OF HEAD FLEXION - Fig. 12 shows for the lateral and oblique impact directions the head twist $(\Psi - \Psi_o)$ as function of head flexion $(\phi - \phi_o)$. Results for the most severe tests are presented. It follows that for all tests the head twist is smaller than the head flexion.

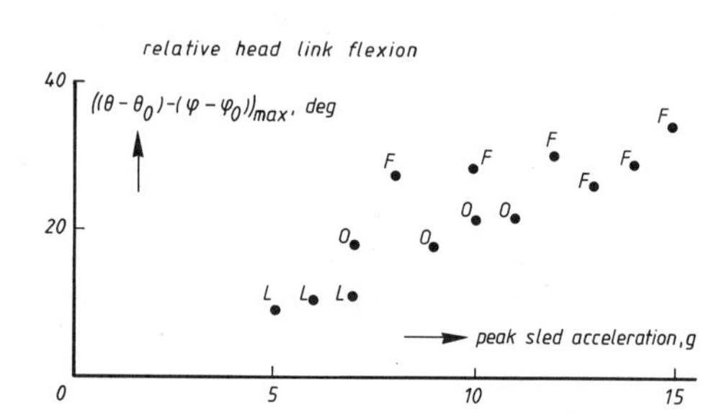

Fig. 11. Influence of impact severity on maximum upper pivot angle (relative head link flexion). (F = Frontal, O = Oblique, L = Lateral).

ANALYSIS OF NECK LOADS

The loads applied by the neck to the head can be calculated using measurements of head acceleration and angular velocity, if the head is regarded as a rigid body and does not come into contact with any other object or body part. Neck load equations used in this study

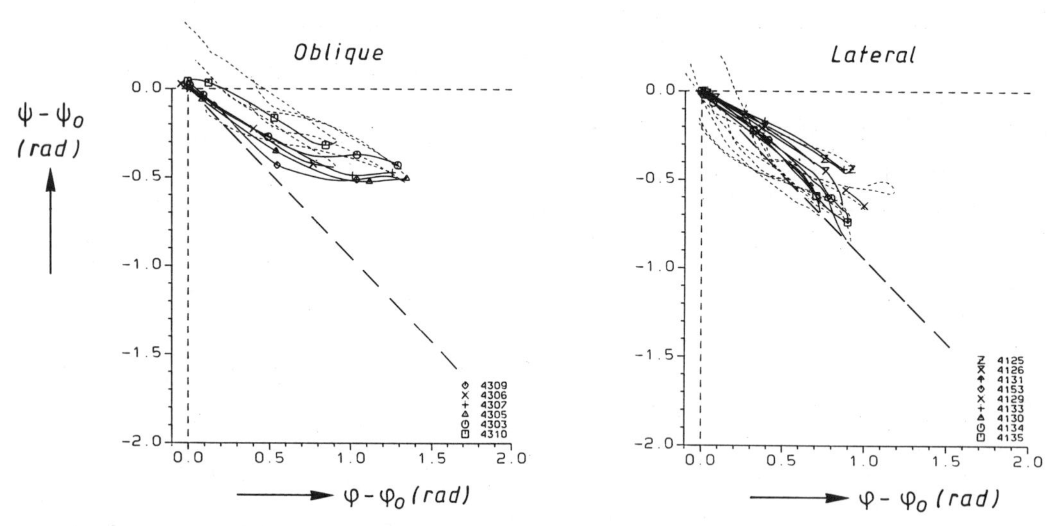

Fig. 12. Head twist $(\Psi - \Psi_o)$ as function of head flexion $(\phi - \phi_o)$ for the most severe tests in lateral and oblique directions. (—— loading --- unloading)

are presented in (1,3,10). Some of the reasons to perform these calculations are:

- they offer an excellent insight in the system's behaviour, for instance with respect to the role of muscle activity;
- load-displacement relations can be used to formulate dummy performance requirements and dummy design specifications;
- it is generally assumed that neck loads correlate quite well with neck injuries.

Another reason to perform such calculations is to estimate the dynamic characteristics of the upper and lower pivot in the analog system proposed in the preceding section. These pivots are located in the occipital condyles and near the T1 origin (i.e. in the center of a circular arc approximating the occipital condyle trajectories). For the calculation of the load in the lower pivot it is assumed that the effect of neck inertia can be neglected.

For all tests incorporated in this study neck loads have been calculated. Detailed results are presented in (6). In this paper some of the most interesting results with respect to the dynamic pivot characteristics will be summarized. Data will be presented for:

- the torque M_ϕ at the upper pivot about an axis perpendicular to the impact plane as function at the relative angle between head and neck link $((\theta - \theta_o) - (\phi - \phi_o))$;
- the torque M_ψ at the upper pivot about an axis parallel to the head anatomical z-axis as function of the head twist $(\Psi - \Psi_o)$;
- the torque M_θ at the lower pivot about an axis perpendicular to the impact plane as function of the neck link angle $(\theta - \theta_o)$.

Torques are defined here as torques applied by the neck to the head (M_ϕ and M_ψ) or by the torso to the head-neck system (M_θ). Fig. 13 presents the sign conventions for the torques and the degrees of freedom of the analog system.

The calculated torque-rotation characteristics for the most severe tests per impact direction are summarized in Fig. 14. The following observations can be made:

- The occipital condyle torques M_ϕ as function of the relative angle between head and neck link confirm the findings in the preceding section for the relative motion in the upper pivot of the linkage: dependent on the impact direction a certain free range of motion exists where the occipital condyle torque is relatively small. As soon as this angle is exceeded the upper pivot gets more or less locked and the occipital condyle torque strongly increases.
- Largest occipital condyle torques M_ϕ (i.e. 60-90 Nm) can be observed for the frontal impact tests. These values appear to be slightly larger than calculated from the previous database (2).
- Peak values for the component M_ψ of the occipital condyle torque appear to be much

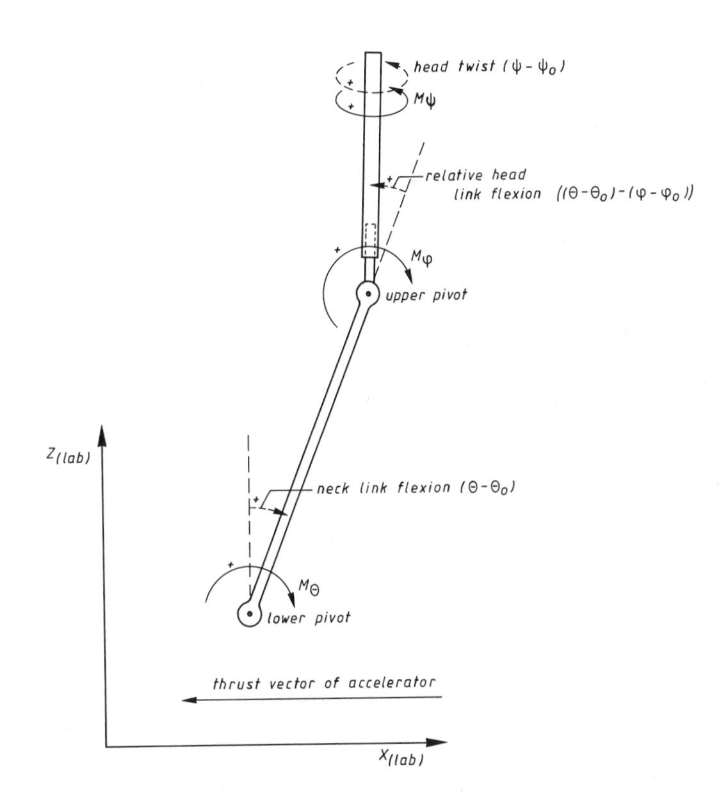

Fig. 13. Analog systems with positive torques and rotations.

smaller (i.e. 4-6 times) than the component M_ϕ.

- For the torques at the lower pivot (i.e. M_θ) peaks close to 200 Nm can be observed. As for the occipital condyle torques, largest torque values are found in the most severe frontal impact tests.
- The lower pivot torque characteristics show for frontal impacts an initial spike. The reason for this behaviour is not yet fully understood.

The test results presented in Fig. 14 relate to the most severe tests in the database. In general, less severe tests showed an identical shape for the torque-rotation characteristics (6). The effect of test severity on joint stiffness was found to be small. Since, moreover, variations in torque-rotation characteristics between subjects are relatively small, selection of one average joint characteristic seems to be appropriate here. In the following sections estimates for these characteristics will be given:

UPPER PIVOT: RELATIVE HEAD LINK FLEXION - A free range of motion can be observed in this joint which varies per impact direction: about 30 degrees in the frontal, 20 degrees in the oblique and 10 degrees in the lateral direction (see section: "analysis of relative head motions: head and neck link rotations"). The joint stiffness after locking of this joint shows no significant differences per impact direction: this joint behaviour can be quite well approximated by a linear function with a slope

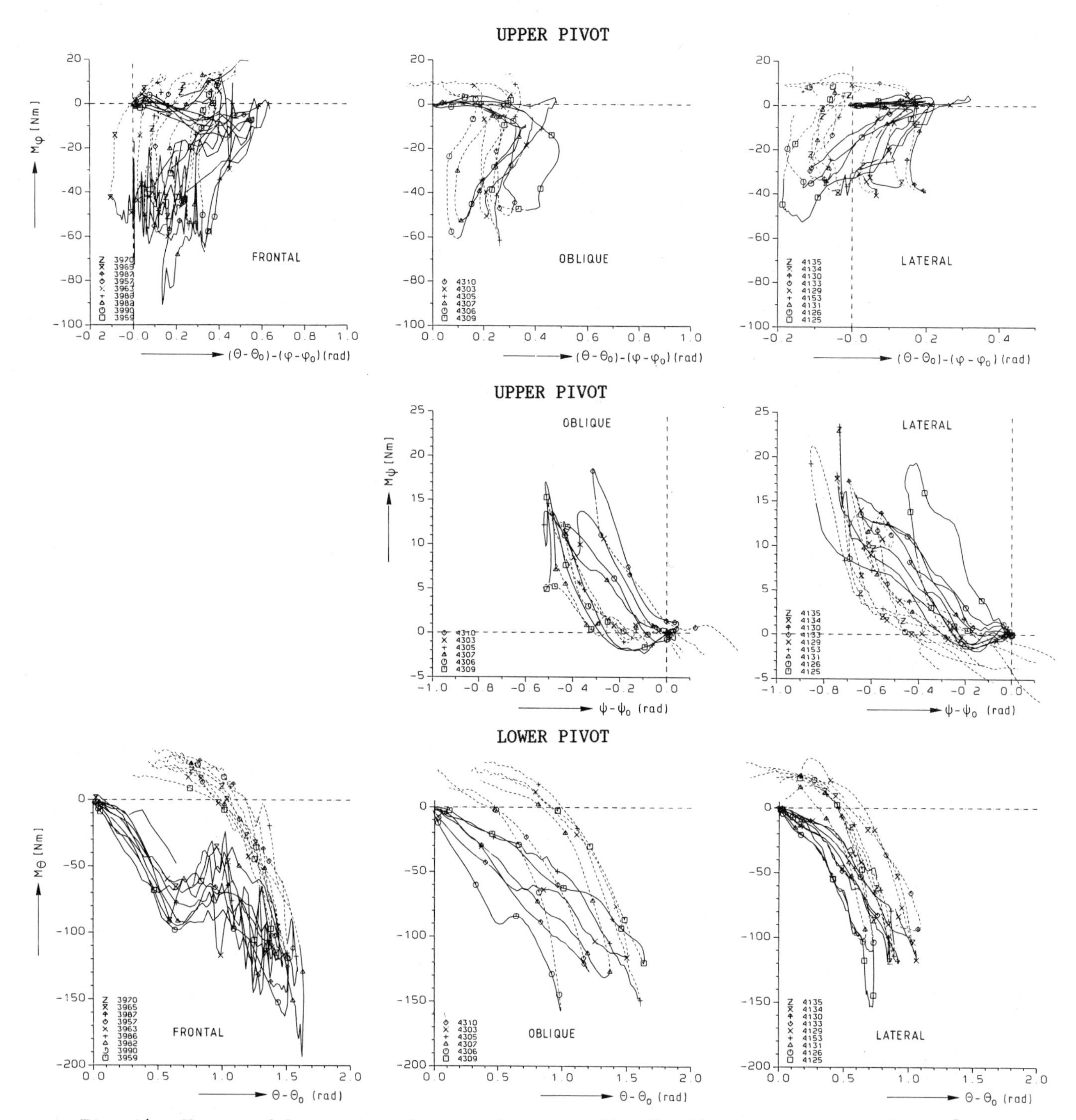

Fig. 14. Upper and lower pivot dynamic characteristics for the most severe tests as function of impact direction (—— = loading --- = unloading).

(i.e. joint stiffness) of 3 Nm/degree.

UPPER PIVOT: HEAD TWIST - A linear function is proposed to characterize the head twist. The stiffness appears to be relatively small and varies per impact direction: about 0.4 Nm/degree in the lateral direction and 0.75 Nm/degree in the oblique direction (note: no torsion motion is present in frontal impacts).

LOWER PIVOT: NECK LINK FLEXION - In order to determine the properties for the lower pivot, first an average torque-rotation characteristic per impact direction has been calculated for the most severe tests. Fig. 15 shows the results of these calculations with corresponding standard deviations. If a linear function is selected to describe this joint behaviour realistic estimates for joint stiffness are as follows:

frontal	:	1.2 Nm/degree
oblique	:	1.5 Nm/degree
lateral	:	2.2 Nm/degree

The initial peak in the frontal impact characteristics has not been taken into account in this linear approximation.

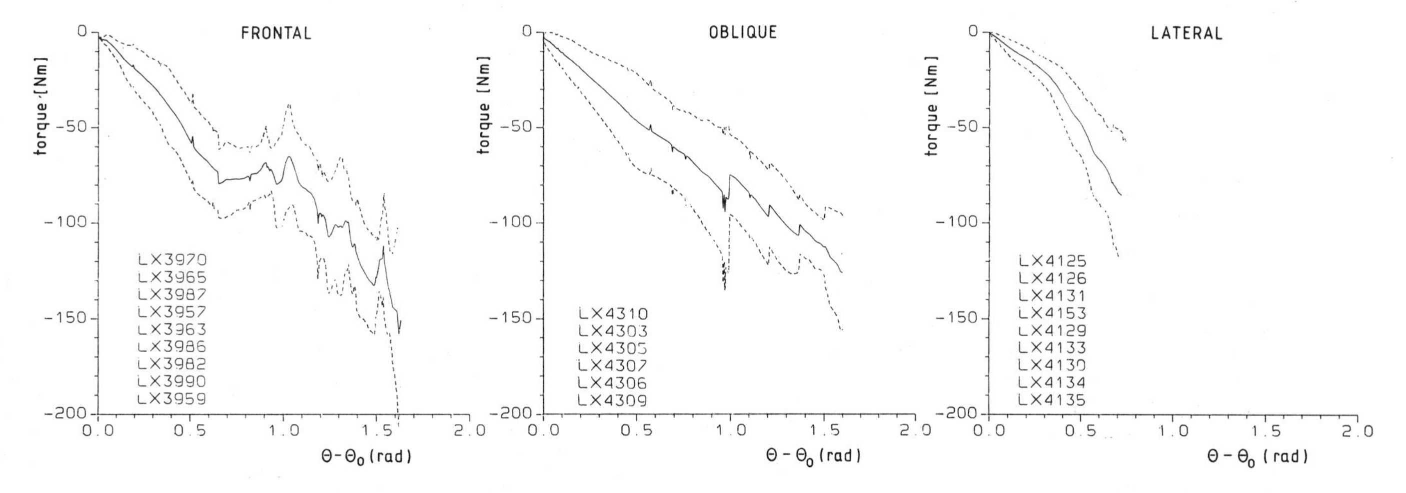

Fig. 15. Average lower pivot torque-rotation characteristics with corresponding standard deviations based on the most severe tests (—— mean value, ---- s.d.).

DISCUSSION

Test results of 253 human volunteer tests have been obtained from the Naval Biodynamics Laboratory in New Orleans. A subset of 109 tests with 15 subjects was selected and partially analyzed in this paper. Special attention was given to the most severe tests in this subset.

This study is an extension of a previous analysis of NBDL tests with two subjects (1,2,3). In this earlier work significant errors were noted in the specification of the T1 coordinate system. In the present database these errors are smaller indicating that the present data are more reliable. On the other hand the present errors are still of such a magnitude that they justified introduction of a corrected T1 coordinate system in a similar way as in the previous study.

The most important finding of the present study is that the observed human head-neck response, like in the previous study, can be represented adequately by a linkage system with 2 pivots. One link represents the head, one link the neck and one link the torso. For each impact direction a separate analog system is proposed. Geometrical parameters of this analog are identical for each impact direction as far as the neck link length and the upper pivot location are concerned. The location of the lower pivot in the torso appears to be slightly different for each impact direction as was illustrated in Table 4 and Fig. 8. Also the initial position of this linkage system varies per impact direction: the initial neck link rotation is almost 18 degrees in frontal impacts, 11 degrees in oblique impacts and close to zero in lateral ones.

For frontal impacts the analog system has 2 and for lateral and oblique impacts 3 degrees of freedom. These degrees of freedom are the neck link rotation in the plane of impact, a head rotation (i.e. head flexion) in the plane of impact and a head twist about the head anatomical z-axis (absent in frontal impacts). Based on load calculations joint stiffness values have been identified. Different characteristics per impact mode can be observed. For the lower pivot the stiffness in lateral direction is larger than in the oblique direction and the stiffness in the oblique direction is larger than in the frontal direction. For the upper pivot rotation in the plane of impact a free range of motion can be observed, which is about 30 degrees in the frontal, 20 degrees in the oblique and only 10 degrees in the lateral direction.

The 2-pivot analog system has been introduced for the purpose of characterizing the three-dimensional head motion in this type of dynamical events. This analog system should not be considered as a representation of the very complex anatomical structure of the cervical articulations and the associated structures (including musculature). See for instance Huelke (11) and Kapandji (12) for a detailed treatment of the neck anatomy and neck motions.

Torques calculated in this study near the occipital condyles and near T1 represent the resultant torque in the head-neck and neck-torso interface, respectively. These torques are developed by tension forces in the musculature, internal neck compression forces, etc. More detailed neck models are needed to study the load contribution of the various neck structures.

As in the previous studies (1,2,3) it is observed that the initial head rotation is smaller than the neck link rotation followed by a lock up between the head and neck link. This behaviour most likely can be explained by the contribution of the various muscle groups. Also the initial peak in the T1 torque (frontal impacts) might be caused mainly by the musculature (i.e. the large posterior muscle groups of the neck).

In this study separate systems are proposed for the three impact directions. As an alternative, a single three-dimensional linkage could have been proposed to simulate all three impact directions. Such a mechanism would have 5 degrees of freedom: 2 degrees of freedom in the lower pivot and three degrees of freedom in the upper pivot. In such a system the neck link rotation is not limited to a planar motion in the plane of impact. The major problem in defining such a system is the out of mid-sagittal plane location of the lower pivot in oblique impacts. If this lower pivot location can be assumed to be located in the mid-sagittal plane, the formulation of a three-dimensional system is not expected to cause significant problems. Attractive aspects of such a system are, among other things, the possibility to predict the head response in arbitrary directions and the possibility to determine out-of-impact plane motions which were neglected in the present analysis.

The proposed analog system is only valid for low severity impacts, i.e. the NBDL human volunteer test conditions. Additional information should be obtained for higher exposure levels from human cadaver tests. Such tests, moreover, are needed for a better understanding of the contribution of muscle activity. If such data become available, adjustment of the proposed analog system might be necessary.

The proposed analog system implicitly constitutes a performance requirement. As an alternative, performance requirements could have been formulated explicitly by means of kinematic requirements like displacements and accelerations for several impact levels. Such a formulation is desirable for instance when establishing a standard to be used in the laboratory for verifying the dummy response. Such criteria can be derived directly from the NBDL test data or indirectly from the analog system.

Although the developed analog system is not intended as a design principle for a future omni-directional mechanical neck it illustrates some important design principles:

- The basic geometrical set-up of the neck design can be identical for the various impact modes. In other words it is expected that one neck design can be used for all impact directions.
- The stiffness of a dummy neck is dependent on the impact direction. In lateral direction for instance the neck should be more stiff (almost twice as stiff) than in forward direction.
- A feature should be incorporated in the design which suppresses large head flexions but allows relatively large neck link rotations. In order to realize this the effect of the locking mechanism which was identified for the upper pivot, should be approximated in a future dummy neck design.

CONCLUSIONS

1. 109 human volunteer test with 15 subjects have been analyzed. Tests were conducted in the frontal, the lateral as well as the oblique impact direction.
2. A 2-pivot analog system appears to be adequate to describe the observed head-neck motions.
3. Numerical techniques were developed and applied to determine the geometrical properties of this analog system.
4. Different analog systems per impact mode are proposed of which the most important geometrical parameters are identical.
5. Load calculations were conducted in order to specify the analog system's dynamic pivot properties.
6. Pivot torque-rotation characteristics appear to be stiffer in the lateral than in the frontal direction.
7. Some preliminary guidelines for future dummy neck designs have been presented.

ACKNOWLEDGEMENT

This study has been supported by the Department of Transportation/National Highway Traffic Safety Administration. All opinions given in this paper are those of the authors and not necessarily those of DOT/NHTSA.

The authors wish to express their special thanks to the staff of the Naval Biodynamics Laboratory in New Orleans for providing the human volunteer test data and the additional information necessary to perform this analysis

Also we would like to acknowledge the University of Nijmegen for preliminary data processing facilities.

REFERENCES

1. Wismans, J. and Spenny, C.H.: 'Performance requirements for mechanical necks in lateral flexion'. In: Proceedings of the 27th Stapp Car Crash Conference. SAE Paper No. 831613, 1983.
2. Wismans, J. and Spenny, C.H.: 'Head-neck response in frontal flexion'. In: Proceedings of the 28th Stapp Car Crash Conference. SAE Paper No. 841666, 1984.
3. Wismans, J.: 'Preliminary development head-neck simulator'. Vol. 1: Analysis of human volunteer tests. Final report Phase I Project SRL-59. Vehicle Research and Test Center, East-Liberty, Ohio, 1985.
4. Ewing, C.L., Thomas, D.J., Lustick, L., Williams, G.G., Muzzy III, W.H., Becker, E.B. and Jessop, M.E.: 'Dynamic response of human and primate head and neck to +Gy impact acceleration'. Report DOT HS-803 058, 1978.
5. Ewing, C.L., Thomas, D.J. and Lustick, L.: 'Multiaxis dynamic response of the human head and neck to impact acceleration'. Aerospace Medical Panel's Specialist's meeting. Paris, AGARD Conference Pro-

ceedings no. 153. North Atlantic Treaty Organisation. Advisory Group for Aerospace Research and Development, 1978.
6. Wismans, J.: 'Human volunteer head-neck motions in frontal, lateral and oblique impacts: analysis of tests with 15 subjects'. Phase II Head-neck Simulator Development (SRL-59). Vehicle Research and Test Center, East-Liberty, Ohio, 1986.
7. Ewing, C.L. and Thomas, D.J.: 'Human head and neck response to impact acceleration'. NAMRL Monograph 21. Naval Aerospace Medical Research Laboratory, Pensacola, Florida, 32512, 1973.
8. McConville, J.T., Churchill, T.D., Kaleps, I., Clauser, C.E. and Cuzzi, J.: 'Antropometric relationships of body and body segment moments of inertia'. Report AFAMRL-TR-80-119. Air Force Aerospace Medical Research Lab. Wright Patterson Airforce Base, Ohio 45433, 1980.
9. Woltring, H.J.: 'An algorithm for weigthed least-squares estimation of rotation pivot and radius in planar point kinematics'. Research Report 85/1 prepared for TNO Road-Vehicles Research Institute Order 13.575.67, 1985.
10. Spenny, C.H. and Wismans, J.: 'Dynamic Analysis of Head Motion'. In: A. Sances, D.J. Thomas, C.L. Ewing, S.J. Larson and F. Unterharnscheidt (eds.). 'Mechanisms of Head and Spine Trauma'. p. 157-185. Aloray, publisher, Goshen New York, 1986.
11. Huelke, D.F.: 'Anatomy of the human cervical spine and associated structures' In: 'The human neck - Anatomy, Injury Mechanisms and Biomechanisms' SAE Warrendale SP-438. Paper 790130, 1979.
12. Kapandji, I.A.: 'The physiology of the joints', Vol. III The trunk and the vertebral column. Churchill Livingstone Edinburgh London and New York, 1974.

APPENDIX A

SUMMARY OF TEST CONDITIONS AND TEST RESULTS

Subject	No.	Peak sled acc. (G)	Rate of onset (G/s)	Sled vel. change (m/s)	ϕ_o (rad)	θ_o (rad)	ψ_o (rad)	$\phi_{max}-\phi_o$ (rad)	$\theta_{max}-\theta_o$ (rad)	$\psi_{max}-\psi_o$ (rad)	$((\theta-\theta_o)-(\phi-\phi_o))_{max}$ (rad)
Frontal experiments.											
H00118	3886	8.2	203.	12.02	-0.186	0.057	-0.030	1.497	1.884	-0.014	0.751
	3903	10.2	284.	13.72	-0.126	0.144	0.005	1.208	1.874	-0.069	0.890
	3985	10.3	292.	13.82	0.002	0.180	-0.024	1.302	1.796	-0.011	0.792
	3920	12.3	392.	15.39	-0.154	0.139	-0.045	1.472	1.964	-0.043	0.867
	3958	14.6	495.	16.85	-0.021	0.281	-0.067	1.490	1.829	-0.079	0.804
	3969	15.4	547.	17.21	-0.013	0.163	0.017	1.433	2.054	-0.104	0.902
H00120	3882	8.2	206.	11.93	-0.024	0.345	0.003	0.997	1.389	-0.045	0.489
	3906	10.2	281.	13.69	0.087	0.539	-0.012	0.921	1.102	-0.033	0.404
	3995	10.2	282.	13.84	-0.001	0.412	0.010	1.067	1.339	-0.055	0.455
	3921	12.1	382.	15.14	0.053	0.423	-0.033	1.167	1.373	-0.028	0.499
	3946	13.6	435.	16.16	0.038	0.394	-0.005	1.174	1.327	-0.048	0.564
	3954	14.1	479.	16.50	0.084	0.431	-0.023	1.243	1.328	-0.060	0.559
H00127	3883	8.2	206.	11.99	-0.035	0.322	-0.008	1.034	1.439	-0.043	0.471
	3904	10.3	277.	13.76	-0.094	0.386	-0.017	1.237	1.380	-0.059	0.431
	3924	12.4	387.	15.43	-0.008	0.352	-0.030	1.200	1.451	-0.057	0.505
	3949	13.6	445.	16.09	0.029	0.447	-0.068	1.186	1.343	-0.047	0.405
	3959	14.8	467.	16.84	0.091	0.393	-0.120	1.179	1.387	-0.013	0.392
H00131	3894	8.4	205.	12.07	-0.136	0.189	0.057	1.044	1.521	-0.111	0.559
	3908	10.2	275.	13.71	-0.040	0.281	-0.022	1.085	1.432	-0.075	0.565
	3999	10.3	287.	13.93	-0.074	0.198	-0.026	1.017	1.453	-0.083	0.503
	3926	12.1	380.	15.03	-0.131	0.136	0.061	1.313	1.755	-0.125	0.552
	3987	14.5	480.	16.76	0.005	0.234	0.043	1.248	1.511	-0.122	0.639
	3990	15.4	527.	17.26	0.011	0.164	-0.001	1.194	1.515	-0.069	0.566
H00132	3997	8.1	203.	12.12	-0.228	0.393	-0.141	0.975	1.204	-0.016	0.332
	3989	10.2	290.	13.78	-0.222	0.313	-0.087	1.107	1.164	-0.035	0.300
	3927	12.2	377.	15.34	-0.257	0.257	-0.107	1.354	1.550	-0.016	0.560
	3950	13.6	441.	16.14	-0.242	0.293	-0.103	1.508	1.597	0.000	0.532
	3957	14.6	509.	16.75	-0.194	0.337	-0.080	1.457	1.552	-0.032	0.552
	3982	15.6	542.	17.47	-0.161	0.322	-0.097	1.493	1.635	-0.020	0.581
H00133	3895	8.2	206.	12.00	-0.135	0.298	-0.012	1.364	1.382	-0.027	0.405
	3913	10.3	274.	13.87	0.060	0.315	-0.035	1.235	1.391	-0.067	0.455
	3998	10.2	285.	13.71	0.104	0.294	-0.018	1.018	1.335	-0.034	0.443
	3939	12.4	382.	15.36	-0.138	0.204	-0.005	1.463	1.488	-0.046	0.451
	3963	14.5	476.	16.69	-0.020	0.259	-0.059	1.368	1.391	-0.008	0.414
	3986	15.6	538.	17.31	-0.054	0.149	-0.004	1.506	1.589	-0.083	0.541
H00135	3898	8.3	207.	12.15	-0.131	0.296	-0.025	1.108	1.362	-0.061	0.553
	3916	10.3	276.	13.83	-0.259	0.302	0.017	1.384	1.422	-0.080	0.487
	3941	12.5	382.	15.52	-0.223	0.423	0.001	1.382	1.382	-0.057	0.479
	3955	13.6	444.	16.18	-0.212	0.414	-0.010	1.416	1.391	-0.054	0.484
	3965	14.6	493.	16.67	-0.233	0.437	-0.004	1.544	1.444	-0.064	0.466
	3970	15.6	534.	17.26	-0.170	0.387	-0.021	1.373	1.470	-0.065	0.481
H00136	3901	7.9	200.	11.88	-0.033	0.366	0.066	1.000	1.128	-0.071	0.432
	3918	10.2	293.	14.03	0.035	0.418	0.035	1.139	1.219	-0.064	0.421
	3942	12.0	382.	15.26	0.029	0.308	0.072	1.236	1.438	-0.113	0.408
	3953	13.3	443.	16.03	0.045	0.415	0.069	1.241	1.251	-0.083	0.362
	3962	14.1	474.	16.59	0.078	0.343	0.067	1.277	1.250	-0.096	0.319

SUMMARY OF TEST CONDITIONS AND TEST RESULTS (cont.)

Subject	No.	Peak sled acc. (G)	Rate of onset (G/s)	Sled vel. change (m/s)	ϕ_o (rad)	Θ_o (rad)	$\Psi_o - \frac{\pi}{4}$ (rad)	$\phi_{max} - \phi_o$ (rad)	$\Theta_{max} - \Theta_o$ (rad)	$\Psi_{max} - \Psi_o$ (rad)	$((\Theta - \Theta_o) - (\phi - \phi_o))_{max}$ (rad)
Oblique experiments.											
H00130	4235	7.0	159.	10.89	-0.012	0.167	-0.002	0.953	1.405	-0.336	0.534
	4301	9.3	252.	13.12	0.022	0.182	-0.044	1.015	1.321	-0.432	0.432
	4286	10.1	281.	13.64	-0.030	0.072	-0.042	1.359	1.766	-0.464	0.663
	4309	11.3	334.	14.64	-0.073	0.061	0.005	1.333	1.638	-0.524	0.492
H00132	4244	7.3	175.	11.40	-0.138	0.215	-0.125	0.689	0.904	-0.313	0.295
	4261	9.0	240.	12.72	-0.083	0.410	-0.089	0.890	0.990	-0.431	0.370
	4287	10.2	288.	13.83	-0.198	0.117	-0.089	1.156	1.175	-0.519	0.275
	4297	10.0	282.	13.78	-0.390	0.096	0.025	1.361	1.266	-0.652	0.156
	4306	11.1	335.	14.54	-0.053	0.308	-0.184	0.905	0.991	-0.440	0.340
H00133	4236	7.3	171.	11.36	-0.051	0.245	0.001	0.778	0.864	-0.169	0.263
	4240	9.1	236.	12.95	-0.028	0.165	0.007	0.959	1.070	-0.241	0.279
H00134	4237	7.2	170.	11.26	-0.181	0.136	0.025	0.840	1.074	-0.348	0.315
	4264	9.3	248.	13.03	-0.107	0.158	0.041	1.036	1.250	-0.349	0.382
	4298	10.1	285.	13.78	-0.155	0.185	0.071	1.219	1.312	-0.455	0.266
	4307	11.4	337.	14.89	-0.089	0.139	0.017	1.264	1.368	-0.488	0.363
H00135	4238	7.3	169.	11.44	-0.062	0.343	-0.047	0.731	0.903	-0.224	0.238
	4314	9.1	244.	12.90	-0.004	0.309	-0.037	0.815	1.093	-0.334	0.316
	4316	10.1	290.	13.62	-0.004	0.347	-0.042	0.915	1.147	-0.393	0.331
H00136	4247	7.1	166.	11.11	0.018	0.242	0.057	0.749	0.832	-0.437	0.160
	4263	9.2	247.	12.98	0.068	0.093	0.068	0.935	1.041	-0.605	0.270
H00138	4241	7.2	167.	11.17	-0.079	0.139	-0.030	0.909	1.236	-0.345	0.393
	4265	9.2	245.	12.98	-0.045	0.127	-0.013	1.163	1.468	-0.403	0.413
	4296	10.1	286.	13.77	-0.075	0.035	-0.064	1.253	1.546	-0.460	0.563
	4305	11.4	342.	14.81	-0.062	0.099	-0.038	1.346	1.607	-0.522	0.482
H00139	4243	7.3	170.	11.31	-0.011	0.151	-0.086	0.989	1.268	-0.268	0.360
	4313	9.1	252.	12.81	-0.005	0.218	-0.079	1.083	1.271	-0.307	0.252
	4291	10.3	290.	13.91	0.013	0.116	-0.021	1.326	1.587	-0.499	0.476
	4303	11.3	339.	14.77	-0.031	0.223	0.000	1.311	1.507	-0.454	0.391
H00140	4259	7.2	172.	11.18	0.066	0.242	-0.040	0.588	0.951	-0.225	0.410
	4302	9.1	248.	12.96	0.104	0.358	0.008	0.677	0.902	-0.356	0.256
	4293	10.1	286.	13.76	0.035	0.198	0.084	0.886	1.145	-0.502	0.335
	4310	11.2	334.	14.69	0.009	0.172	-0.069	0.916	1.210	-0.317	0.324

SUMMARY OF TEST CONDITIONS AND TEST RESULTS (cont.)

Subject	No.	Peak sled acc. (G)	Rate of onset (G/s)	Sled vel. change (m/s)	ϕ_o (rad)	θ_o (rad)	$\psi_o-\frac{\pi}{2}$ (rad)	$\phi_{max}-\phi_o$ (rad)	$\theta_{max}-\theta_o$ (rad)	$\psi_{max}-\psi_o$ (rad)	$((\theta-\theta_o)-(\phi-\phi_o))_{max}$ (rad)
Lateral experiments.											
H00133	4093	5.1	105.	7.11	0.020	0.093	0.016	0.671	0.628	-0.331	0.124
	4111	6.1	132.	7.25	0.033	0.007	-0.003	0.939	0.847	-0.454	0.097
	4151	6.1	130.	7.17	0.057	-0.005	-0.032	0.849	0.743	-0.468	0.168
	4125	7.2	164.	7.02	0.057	0.107	-0.042	0.920	0.734	-0.460	0.168
H00134	4097	5.0	104.	7.06	-0.012	0.134	-0.024	0.648	0.683	-0.691	0.147
	4112	6.1	128.	7.13	-0.064	0.170	0.096	0.772	0.613	-0.758	0.057
	4126	7.1	167.	6.90	-0.030	0.106	0.059	0.864	0.730	-0.695	0.100
H00135	4095	5.2	105.	7.14	0.041	0.144	-0.059	0.630	0.602	-0.572	0.128
	4114	6.1	132.	7.19	0.066	0.098	-0.047	0.651	0.655	-0.564	0.198
	4131	7.3	164.	6.94	0.038	0.082	-0.043	0.728	0.670	-0.622	0.163
H00136	4098	5.1	106.	7.03	-0.030	-0.064	0.099	0.642	0.712	-0.652	0.155
	4142	6.0	131.	7.07	-0.007	-0.069	0.096	0.771	0.820	-0.734	0.259
	4153	7.1	157.	6.86	-0.016	-0.075	0.121	0.912	0.889	-0.861	0.218
H00138	4092	5.1	106.	7.05	-0.015	0.049	-0.081	0.891	0.944	-0.686	0.202
	4115	6.0	127.	7.11	0.012	-0.121	-0.003	0.932	1.011	-0.672	0.251
	4147	6.0	126.	7.06	-0.024	-0.047	-0.084	0.881	0.977	-0.655	0.259
	4129	7.2	162.	6.92	-0.014	-0.074	-0.059	1.000	1.060	-0.666	0.321
H00139	4100	5.1	107.	7.08	0.044	-0.054	0.050	0.917	0.878	-0.590	0.152
	4118	6.1	128.	7.13	0.032	-0.078	0.005	1.075	0.991	-0.570	0.187
	4144	6.1	130.	7.25	0.003	-0.080	-0.016	1.091	1.010	-0.614	0.212
	4133	7.2	165.	6.91	0.004	0.024	0.006	1.179	1.091	-0.579	0.187
H00140	4099	5.1	108.	7.11	0.029	0.069	0.056	0.598	0.784	-0.520	0.186
	4116	6.1	128.	7.20	0.021	0.067	0.109	0.694	0.848	-0.665	0.181
	4145	6.1	127.	7.20	0.005	-0.009	0.105	0.577	0.788	-0.576	0.211
	4130	7.1	161.	6.89	0.018	0.009	0.127	0.727	0.924	-0.701	0.203
H00141	4094	5.1	106.	7.12	0.079	0.042	0.055	0.716	0.832	-0.896	0.126
	4119	5.9	128.	7.06	0.075	-0.046	0.063	0.852	0.975	-0.795	0.127
	4134	7.1	161.	6.85.	0.073	0.021	0.055	0.900	1.033	-0.747	0.198
H00142	4104	5.1	108.	6.99	-0.047	0.008	-0.069	0.759	0.821	-0.705	0.197
	4120	6.1	131.	7.06	0.035	0.095	-0.133	0.843	0.866	-0.683	0.260
	4135	7.2	161.	6.87	-0.046	0.041	-0.017	0.911	0.856	-0.738	0.220

APPENDIX B

OCCIPITAL CONDYLE TRAJECTORIES RELATIVE TO CORRECTED T1 COORDINATE SYSTEM

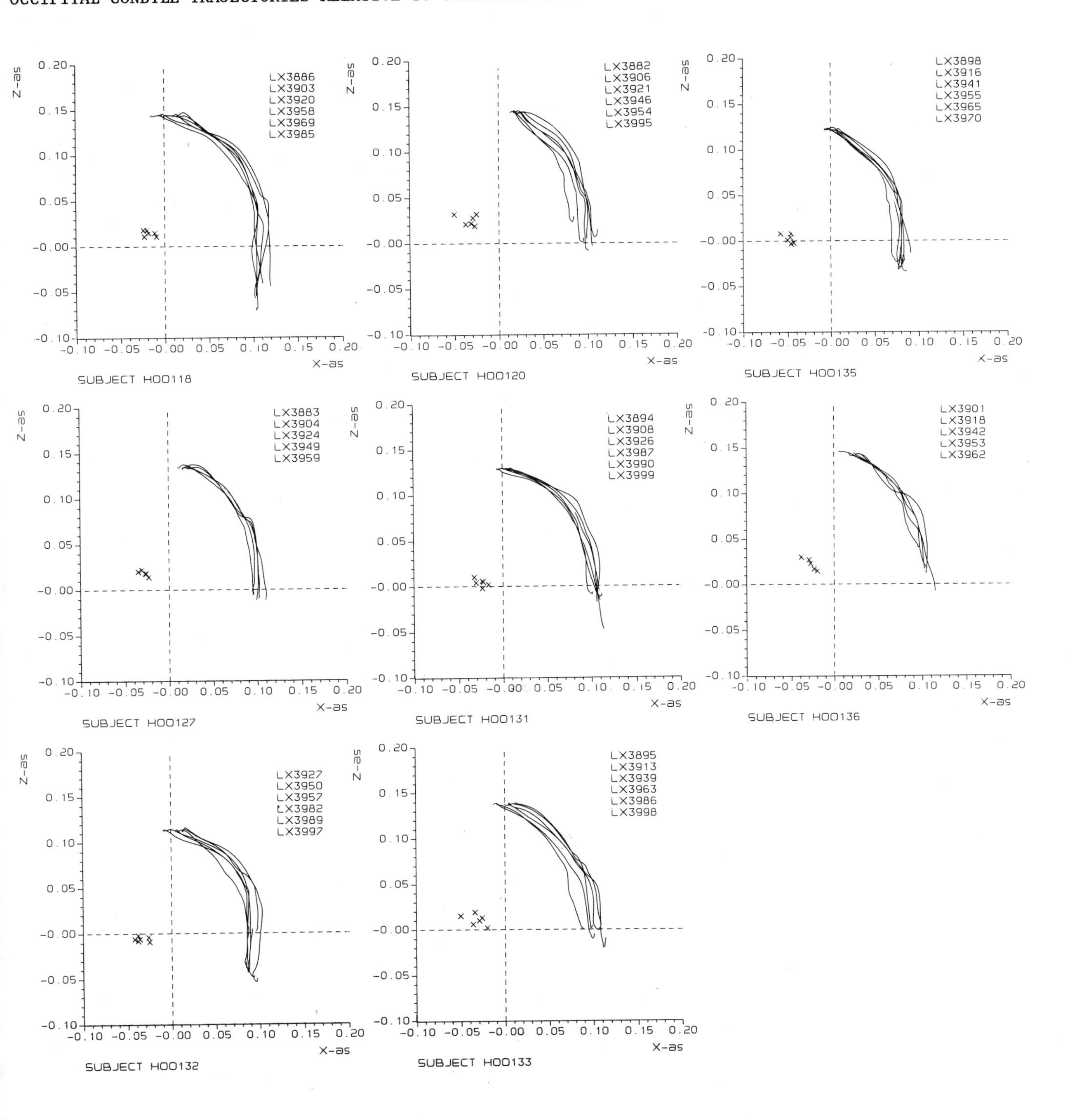

FRONTAL

OCCIPITAL CONDYLE TRAJECTORIES RELATIVE TO CORRECTED T1 COORDINATE SYSTEM (cont.)

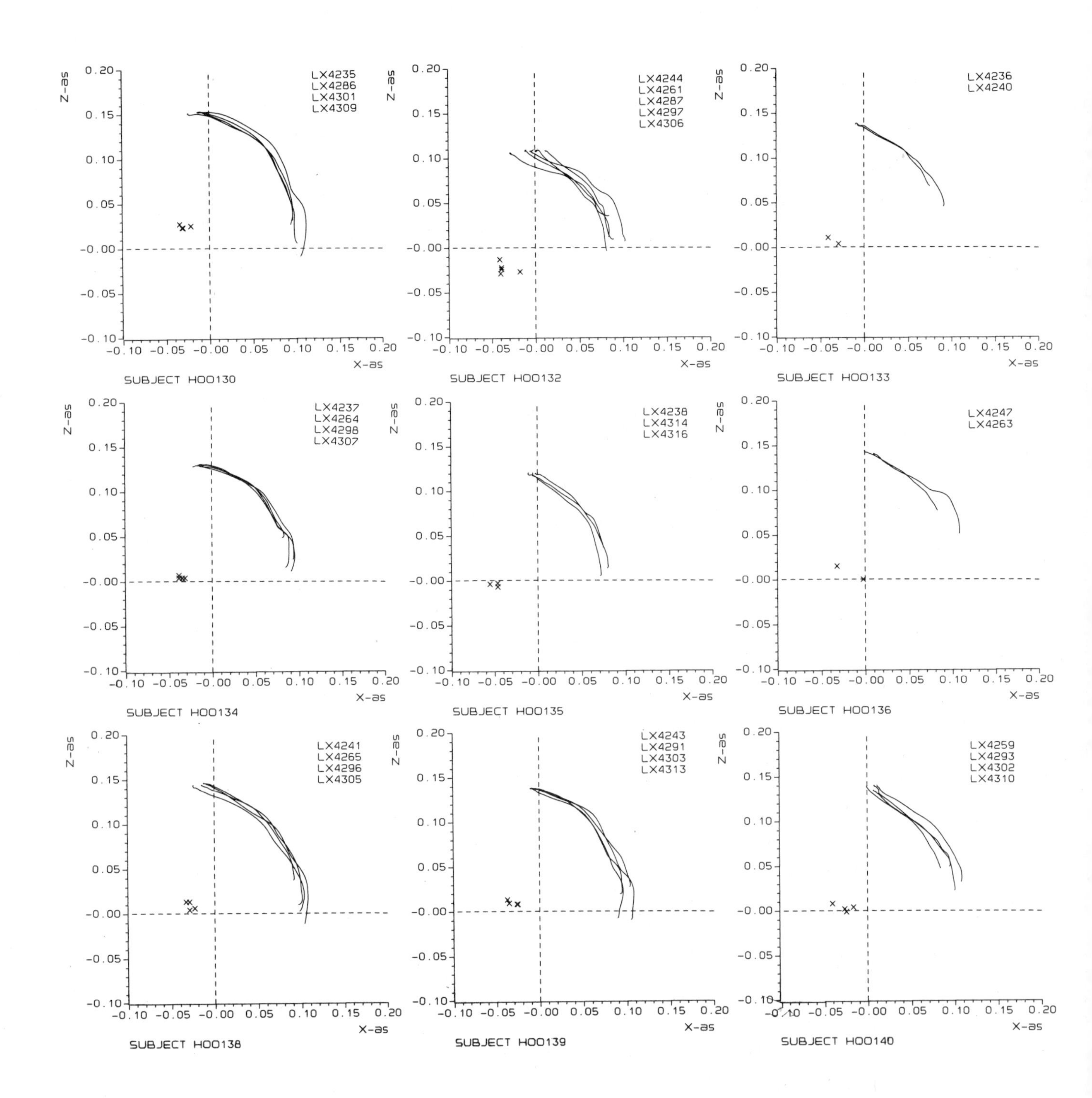

OBLIQUE

OCCIPITAL CONDYLE TRAJECTORIES RELATIVE TO CORRECTED T1 COORDINATE SYSTEM (cont.)

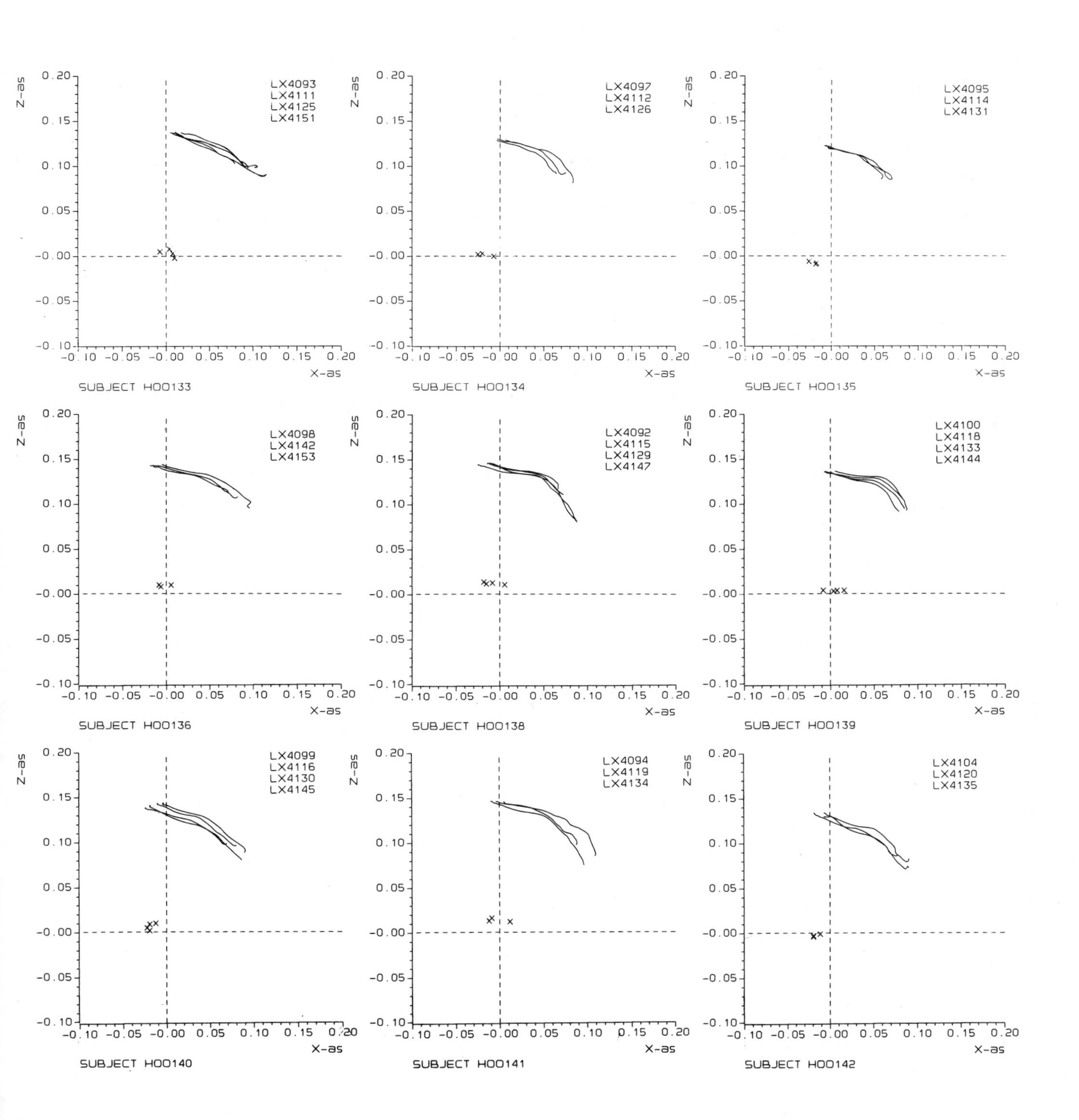

LATERAL

872196

Head and Neck Responses under High G-level Lateral Deceleration

Farid Bendjellal and Claude Tarrière
Laboratory of Physiology and Biomechanics, Peugot/Renault

Didier Gillet and P. Mack
I.R.B.A., Ambroise Pare Hospital

Francois Guillon
Raymond Poincare Hospital

ABSTRACT

Although the neck is one of the least frequently injured body regions, it does play a considerable role in the solicitations of the head in side impact. It can, in fact, be said that the kinematic and dynamic conditions that govern, for instance, a head impact against a vehicle structure depend on the cervical segment.

With a view to characterizing such conditions, i.e. head and neck responses, the LPB-APR conducted a research program including sled tests involving cadavers. These tests were conducted at a low and high G-level sled deceleration, respectively, with the low-violence tests being carried out following collaboration with the Naval Biodynamics Laboratory (New Orleans). Such tests enable direct comparison between volunteer data and cadaver data.

The scope of this paper is to present a synthesis of the data obtained from LPB-APR low and high G-level tests, including, in particular, data obtained from new high severity tests.

The analysis of these data is based on detailed results provided by the processing of both accelerometers and film informations.

Discussion is devoted to the type of kinematics described by the head as well as to the head-neck interactions in terms of the moments acting at the level of the occipital condyles. The results of the autopsies of the subjects, who sustained no injuries, and the linear and angular acceleration levels indicate that these two new tests appear to be satisfactory, as an initial basis, for the defining of specifications for a dummy's neck exposed to violent conditions.

OVERVIEW

Accidentological data for lateral impacts show that the head, thorax, abdomen and pelvis are the body regions that are the most frequently and most severely injured. Compared to the above regions, the neck has a negligible injury frequency. An accidentological study on a French sample (1)* of 1,446 occupants involved in lateral impacts shows that the incidence of neck injury is 1 % for an AIS>1. This level of frequency occurs with occupants located on the impact side with or without intrusion. The neck injuries with this AIS value are principally found in unbelted occupants subjected to impact to the head. This analysis confirms that, considering the level of frequency and severity and the after-effects observed, the neck cannot be viewed as a priority region for lateral impact protection when compared to other body regions. The low level of frequency of neck injury should not, however, obscure the fundamental role that the neck plays in head response. The kinematic and dynamic conditions governing, for example, an impact of the head against a vehicle structure depend mainly on the cervical region. With the aim of characterizing this role, various experiments have been performed that can be classified into two groups :

- global experiments simulating a lateral collision using cadavers in conditions close to real-world road situations (such as the FAT program (2) tests), or identical to real-world road situations (such as the reconstructions of real-world accidents performed in the KOB program (3)). Since the main object of these tests was not solely the study of the behaviour of the head and neck region, the resulting data do not permit a detailed analysis of the responses of this body region ;

---

* Numbers in parentheses designate references at the end of the paper.

- experiments focussed mainly on the dynamic behaviour of the head and the neck when the whole body is subjected to a lateral deceleration. This is the case in the sled tests carried out by the Naval Biodynamics Laboratory using volunteers (4), where the maximum acceleration and velocity of the sled were 7 G and 6 m/s, respectively.

Because of the number of tests performed and the various analyses based on them (4) (5), these tests constitute an important data base for the understanding and quantifying of head and neck responses. The volunteers used were, however, subjected to much lower levels of acceleration than those generally observed in real-world road situations, whence the interest in experiments involving cadavers with higher impact levels.

The aim of our research was first to duplicate the NBDL tests by performing tests using cadavers in an attempt to highlight the difference between the cadaver and volunteer responses, and to follow this with more violent cadaver tests. To ensure accurate duplication of the volunteer tests, a collaboration project was set up between the NBDL and our laboratory using the same HYBRID II dummy instrumented in the same manner as the volunteer, to perform sled tests under HOLD (High Rate of Onset-Long Duration (4)) conditions in both test sites. Once a satisfactory comparison between the dummy and the volunteer responses has been achieved, the LPB-APR test program was started.

The purpose of this paper is to present a summary of the results obtained from the cadaver tests carried out at low and high levels of violence. The cadaver kinematic responses are compared with those of a volunteer subjected to an identical test. A detailed analysis of the new cadaver tests recently carried out is proposed. This analysis is based on the three-dimensional reduction of the kinematographic data and the result of the processing of the head accelerations. Particular attention is paid to new methods, in terms of the preparation and processing of data, set up for high violence tests.

LOW G-LEVEL CADAVER TESTS

TESTS SET UP AND INSTRUMENTATION - A sled similar to that used by Ewing et al. (4) was used. A rigid seat was fixed to the sled in a sideward upright position. In order to limit the translation of the subject as well as the motion of its upper torso, a wooden side board was attached vertically to the seat. The top of this board was on a level with the subject's right shoulder. The subject was positioned on the seat in an upright sitting position and its midsagittal plane was vertical. The subject was restrained by a lap belt and a pelvis stap. In addition, the subject's torso was secured by shoulder straps and a nylon belt around its chest. During the acceleration of the sled, an adhesive tape attached the subject's head to the structure of the sled. This allowed the head's anterior-posterior axis to be maintained approximately horizontal. So as to avoid any influence of the adhesive tape on the kinematics of the head, the tape was partially cut so that the head would be freed as soon as the sled was impacted. It must be stressed that the cadaver tests were performed using a sled that was decelerated by means of polyurethane tubes once the initial velocity of 6 m/s has been reached. The array of instrumentation used for these tests was identical to that developed by the NBDL (6). Two T-shaped plates, each equipped with 6 monoaxial accelerometers, were fitted to the chin and at the level of the first thoracic vertebra (T1) of the subject. In addition, two accelerometers, fixed to the sled and the seat, respectively, were used for the measurement of the sled deceleration.

In order to study head and neck kinematics, 3 cameras, as shown in figure 1, were used. A set of photographic targets were connected to both T-shaped mounts carried by the head and the vertebra T1. The velocity of filming was 500 frames per second. Signals recorded from the various channel measurements were filtered according to channel frequency class of 1,000 for the head and T1, and of 180 for the sled and seat respectively. Furthermore, all the subjects involved were subjected to the injection of the vascular system during the test setting-up procedure in order to re-establish arterial pressure as well as to allow detailed analysis of brain injuries.

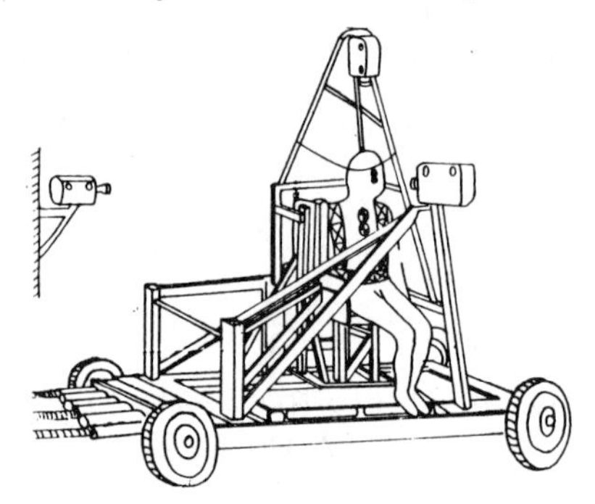

Figure 1. Testing unit for cadaver low G-level experiments according to NBDL methodology

The characteristics of these 4 tests, which were carried out with a nominal level of sled deceleration of 7 G, are summarized in table 1. The initial sled velocity was between 5.9 and 6.3 m/s and the maximum deceleration between 6.6 and 9.2 G. Lateral acceleration measured at T1 reached levels of between 12 G and 32.5 G. In table 1, the term ΔT expresses the period of time during which the sled deceleration is higher than or equal to 75 % of its maximum amplitude.

Table 1. Test characteristics of low G-level cadaver experiments

No. test	Peak sled deceleration (G)	ΔT (ms)	Initial sled velocity (m/s)	Peak lateral acceleration at T1 (G)
MS 234	9.2	72	6.32	32.5
MS 235	7.10	92	6.18	20.5
MS 239	6.60	80	5.87	19
MS 240	6.90	84	5.95	12

COORDINATE SYSTEMS - The orientation of the head and T1 coordinate systems is indicated in figure 2. Both are right-handed orthogonal systems. The subject's head coordinate system is located at the head anatomical origin, while the T1 coordinate system is fixed near the base of the neck at the posterior extremity of the T1 vertebra. The head and T1 y axes are parallel with the direction of the sled acceleration vector, but opposite to the sled displacement direction. The plane of impact is here defined by the YZ plane.

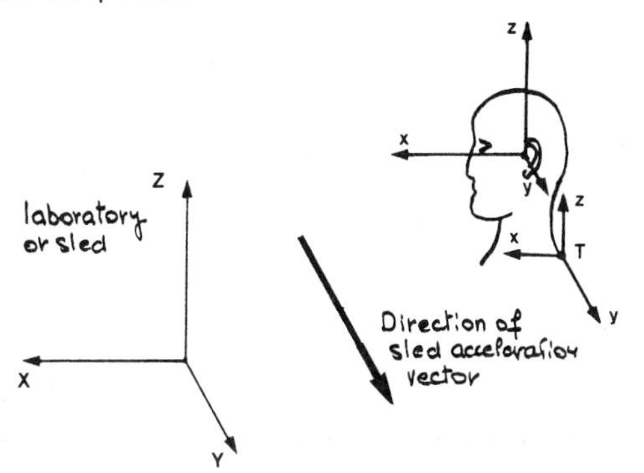

Figure 2. Orientation of head and T1 coordinate systems with respect to the sled

It will be observed that the orientation of the laboratory coordinate system, shown in figure 2, is different from that used in NBDL tests (5), where the laboratory X-axis is parallel with and opposite to the direction of the sled acceleration vector.

Subject's anthropometry : the principal anthropometric data of each subject are summarized in table 2, while the head-neck specific anthropometry is given in table 3. The respective measurements of head and neck masses were performed after the autopsy procedure.

RESULTS - The viewing of the film shows in general that during the initial moments of sled deceleration, the subject's shoulder deflects completely whilst the head describes a movement of pure translation in the impact plane. There follows a three-dimensional movement of the head, in which head rotations around the anterior-posterior and inferior-superior axes predominate.

Figure 3 shows the deceleration curves as a function of time in the 4 cadaver tests. A good level of similarity was observed between the curves, with a maximum mean deceleration level of 6.2 G. The upward phase of the deceleration curve seems to have a significant effect on the response of the first thoracic vertebra T1. The sled deceleration curve in test MS 234 in figure 3, with a peak of amplitude 9.2 G and a pronounced slope, generated large peaks for lateral acceleration at T1, as is shown in figure 4. We can also see that the curves shown in this figure 4 have a high first peak followed by other peaks with a lesser amplitude. It is interesting to note that the volunteer tests also showed this characteristic (4). To compensate for the scatter due to slight differences in subject positioning, the curves in figure 4 were shifted so that the first peaks occur at the same time.

Since we are comparing the cadaver data with volunteer data, we will first compare the input data, i.e. the sled accelerations and the T1 accelerations. We will then analyze the cadaver and volunteer responses in terms of the accelerations and displacements of the head. Figure 5 shows the mean acceleration curves for the sled in the cadaver tests with the corridor defined on the basis of 9 volunteer tests analyzed by Wismans (7). Since the choice of an instant of zero in the NBDL tests (4) differs from that of our tests by approximately 50 ms, the mean curve from the cadaver tests was shifted by the same amount. We see from figure 5 that the sled accelerations in the volunteer and cadaver tests, respectively, are close to each other. This similarity in terms of input data is confirmed by the lateral accelerations at T1 shown in figure 6.

In order to compare the respective responses of the cadaver and the volunteer in terms of kinematics, we shall consider the lateral acceleration calculated at the head c.g. and the linear and angular displacements of the head. In the cadaver tests, these data were obtained by the processing of signals supplied by means of the mounting of 6 sensors fixed to the chin of the subject according to the method called 3-2-1 (6).

Table 2. Principal anthropometric data of subjects involved in low G-level tests

Subject	age years	Sex -	Height (m)	Weight (kg)	Sitting height (m)
MS 324	51	Male	1.70	73	0.95
MS 235	64	Male	1.74	77	0.96
MS 239	50	Male	1.68	71	0.93
MS 240	66	Male	1.68	78	0.90

Table 3. Head and neck anthropometry

Subject	Head				Neck		
	Length (m)	Breadth (m)	Circ.* (m)	Mass (kg)	Breadth (m)	Circ.* (m)	Mass (kg)
MS 234	0.185	0.177	0.53	3.900	0.120	0.410	1.350
MS 235	0.200	0.157	0.59	4.350	0.130	0.435	1.200
MS 239	0.190	0.151	0.56	3.450	0.120	0.412	1.150
MS 240	0.185	0.175	0.57	3.410	0.125	0.410	1.250

*Circumference

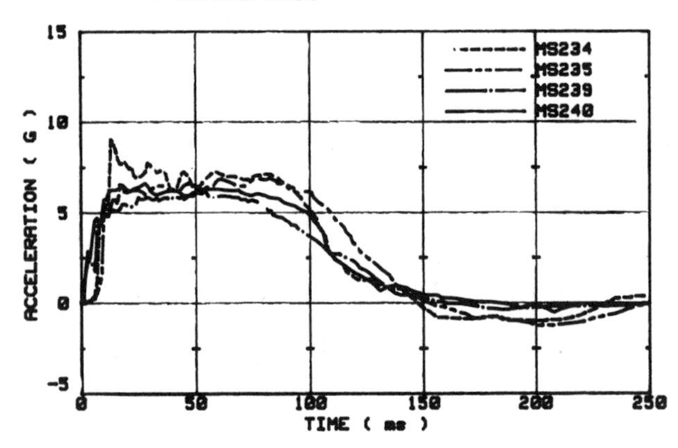

Figure 3. Sled deceleration : time histories obtained from 4 low G-level cadaver tests

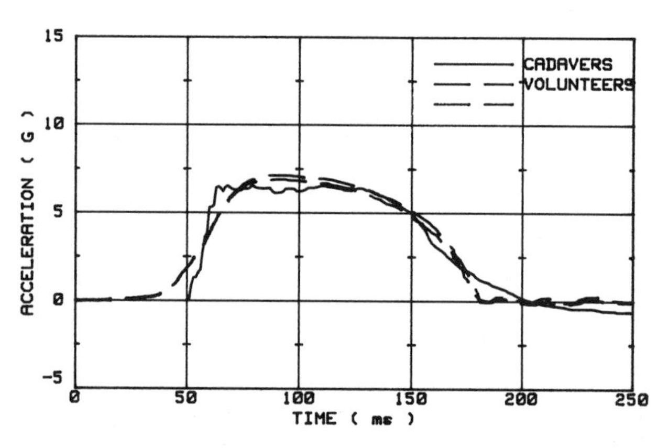

Figure 5. Mean sled acceleration : time history in cadaver tests compared with a corridor obtained from 9 volunteer tests

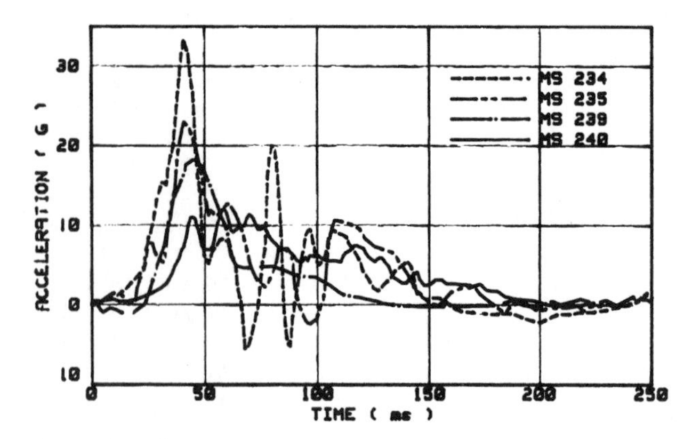

Figure 4. T1 lateral acceleration : time histories obtained from 4 low G-level cadaver tests

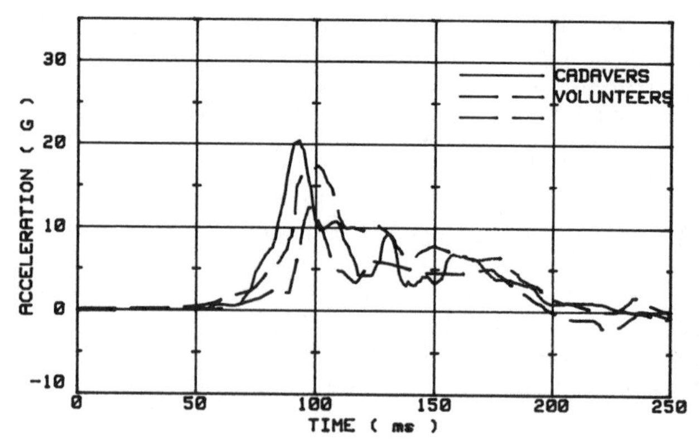

Figure 6. T1 (mean) lateral acceleration : time history in cadaver tests compared with a corridor obtained from 9 volunteer tests

Because of problems that occurred in the recording of certain measurement channels, it was not possible to process all the cadaver tests. Furthermore, vibration problems specific to the instrumentation used made the processing of the data very difficult. This is why, in the comparison that follows, we shall use only the data obtained from one cadaver test. Figure 7 shows respectively the linear acceleration of the head c.g. as a function of time in test MS 239 and the equivalent data from the 9 volunteer tests (7). We can see that the curve for the cadaver has two pronounced peaks, of the order of 19 G and 22 G, respectively, which occur approximately at the same instants as those for the volunteer data.

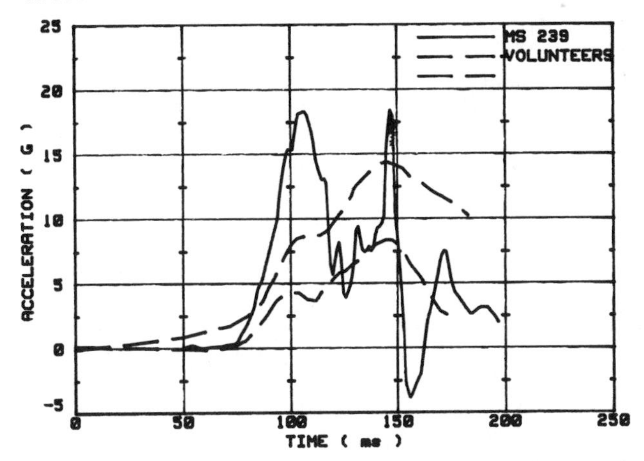

Figure 7. Lateral acceleration of head c.g. in MS 239 test compared with a corridor obtained from 9 volunteer tests

The greater amplitude of the MS 239 curve peaks can be explained, as we shall see in the following sections, by the cadaver's head excursions, which are necessarily greater than those of the volunteer and may be due to the absence of muscular tonicity.

The analysis of head kinematics will be limited to the loading phase, which lasts for 110 ms in the MS 239 test. Film analysis, together with the results derived from accelerometric data processing, show that the movement of the head can be broken down into two phases :

- during the first phase, the head goes through a pure translational movement in the impact plane ;
- in the second phase, the movement is made up of translation and of rotations. Apart from lateral flexion and torsion, anterior flexion of the head (i.e. a rotation of the head about its left-right axis) can also be observed.

Thorax movement is small in this test. For example, the rotation of the T1 z-axis in the impact plane does not exceed 15 degrees. The head c.g. displacements outside the impact plane in the MS 239 test are negligible.

Expressed with respect to the sled, the head c.g. lateral displacement in MS 239 test reached a magnitude of 215 mm, whilst the vertical displacement was approximately 100 mm. Angular displacements of the head were 57°, 18° and 20° for the lateral flexion, the anterior flexion and the torsion respectively. These rotations are defined as follows :

- the lateral flexion is represented by the angle between the vertical and the head z-axis in the impact plane ;
- the anterior flexion is represented by the angle between the vertical and the head z-axis in a plane perpendicular to the impact plane ;
- the torsion is given by the angle between a plane parallel with the impact plane and the head y-axis.

In the case of the volunteers (5), the head is subjected to a pure translation followed by a movement in which lateral flexion and torsion predominate. The magnitudes reached by the head flexion and torsion in the volunteer tests were of the same order, i.e. 50 degrees in the case of the LX 2302 test, for instance. Linear displacements of the head c.g. relative to the sled in this test are 153 mm and 80 mm for the lateral and vertical directions, respectively (8).

The comparison between volunteer and cadaver tests in terms of head movement suggests the following remarks :

- head excursions are greater for the cadaver ; this illustrates the influence of the absence of muscular tonicity in the cadaver ;
- head angular displacements appear to be of the same nature, except for an anterior flexion not seen in the volunteer test ;
- head lateral flexion is slightly greater for the cadaver;
- head torsion for the cadaver is about 50 % less than for the volunteer ;
- in the case of relative head movement with respect to the torso, film analysis of cadaver tests indicates that the head rotates about a centre of rotation probably located in the torso midsagittal plane. No precise estimation of this centre of rotation can be made since the processing of the film data could not be very accurate because of its two-dimensional nature.

Lastly, we note that the autopsy results for the subjects involved in low violence tests show up no cases of injury. In order to perform more violent experiments, tests with a nominal sled acceleration level of 13 G were carried out. The results of these tests are given further on.

HIGH G-LEVEL CADAVER TESTS

Seven tests were performed using unembalmed cadavers, with deceleration and initial sled velocity levels of between 12.2 and 14.7 G and 6.08 and 8.61 m/s, respectively. Table 4 summarizes the characteristics of 4 of these tests which have been analyzed previously (9),

Table 4. Test characteristics of high severity cadaver tests

Test No.	Peak sled deceleration (G)	ΔT (ms)	Initial sled velocity m/s	Peak lateral acceleration at T1 (G)
MS 249	12.2	55	6.08	12.2
MS 297	14.2	48	6.19	44
MS 360	14.6	58	8.61	34.4
MS 361	14.0	46	6.25	46

(10), and which we shall not touch upon here. We can however summarize the main results obtained from these tests as follows :

- the head displacements were greater than those observed in the low violence tests ;
- the levels reached for lateral acceleration at T1 were higher and the curve representing this parameter in time showed the same characteristics as in the low violence tests, with several peaks following the first peak ;
- differences in the subject's positioning or restraint system caused divergences in the accelerometric responses and the amplitude of head movement ;
- no injury was observed in the head-neck region, except in one test where cervical fractures occurred (9), (10).

Two new tests recently performed were chosen for this study. Compared to previous tests, they are subjected to a particularly careful study in terms of methodology and processes of analysis. New methods were developed to ensure that the maximum amount of information was retrieved and to facilitate analysis of the data.

METHODOLOGY - Once the subject has been chosen for this type of test, the next step is to analyze the subject's bone condition by evaluating the mineral condition of the eleventh rib. If the mineral condition is satisfactory (i.e. close to the average level of 0.30 (11)), the subject is selected for the test and one proceeds to the fitting of the sensor mountings and to the preparations for injection of the vascular system.

This injection (12), performed just before the start of the test, allows one to accurately analyze any brain injuries during autopsy of the subject. X-rays are taken of the head and neck before and after the test, so as to, on the one hand, isolate any possible fractures, and on the other hand to locate the occipital condyles and the vertebra T1 in relation to the head anatomical coordinate system.

It is sometimes difficult to highlight the anatomical relationships between the occipital condyles and the first two cervical vertebrae on X-rays of the head taken frontally and in profile.

Frontally (figure 8a), whilst the "open mouth" pictures allow us to see the atlanto-axial articulation (C1-C2), the study of the atlanto-occipital line is obstructed by the superimposition of images of the maxilla and upper teeth. Furthermore, the projection of the mandible onto the 2nd and 3rd cervical vertebrae stops us from being able to locate their position in relation to the antlanto-occipital and atlanto-axial articulations. In profile (figure 8c), the image of the mastoid projection hides the image of the atlanto-occipital line. In the absence of tomography and because of the characteristics of our subjects, we prepared them before the X-ray examination. This preparation consists of the ablation of the whole upper maxillary projection on the palatine plane, the dislocation of the mandible and the excision of the two mastoid processes (figures 8b-8d). The preparation is completed by an opaque radio apparatus which allows us to locate the points of accelerometric measurement and the head anatomical axes, and by a simple system for calculating the scaling coefficients.

To measure head accelerations, an aluminum mounting fitted with four triaxial sensors was fixed to the subject's head, as shown in figure 9. This accelerometer configuration allows one to use either the 3-3-3 method (13) or the method developed by Padgaonkar et al., called 3-2-2-2 (14), for the calculation of angular acceleration of the head, considered as a rigid body.

In order to determine the kinematics of the head and of the base of the neck in relation to the sled from the film, 3 rigid aluminum targets were fitted to the head and at T1 level as shown in figure 10.

Furthermore, the subject instrumentation included acceleration measurements at T1, T4 and at the pelvis respectively.

A method based on the digitization of the photographs of the subject taken from different angles was set up for the three-dimensional calculation of the locations and orientations of the head acceleration sensors and for the locations of the photographic targets in relation to the head anatomical coordinate system. An illustration of this method (called OPT-3D) is given in figure 11, where the Francfort plane of the head is defined by the ends of the threaded rods and the plastic target located between the subject's eyes. At the sensor locations, rigid supports fitted with rods of known dimensions are used for the calibration of the photographs.

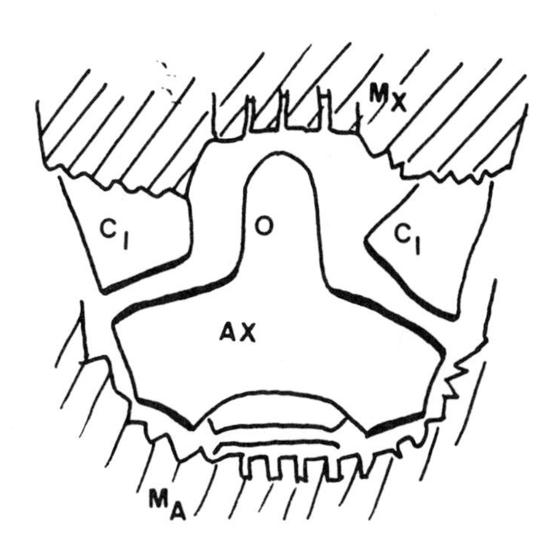

Figure 8a

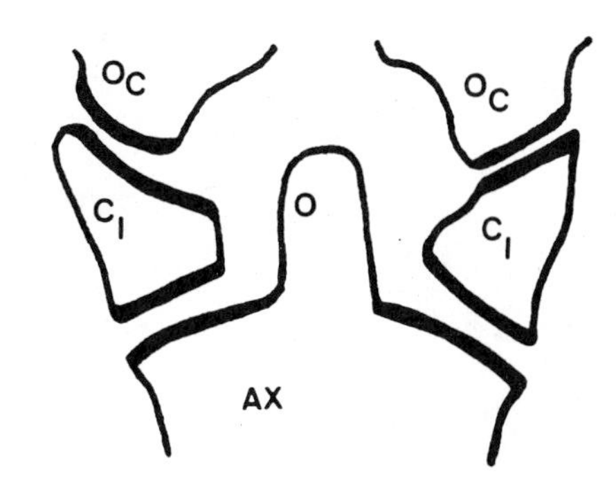

Figure 8b

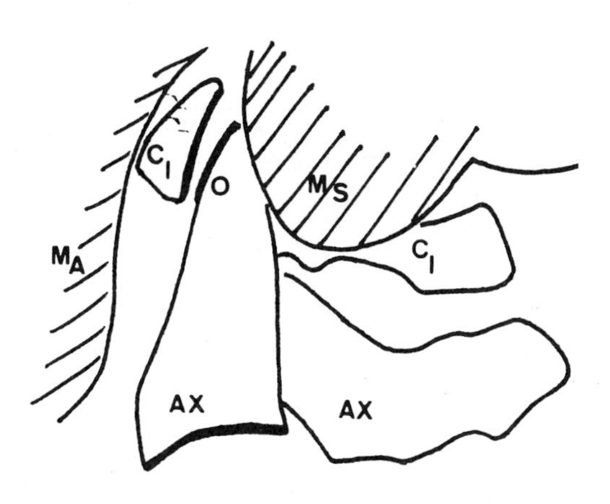

Figure 8c

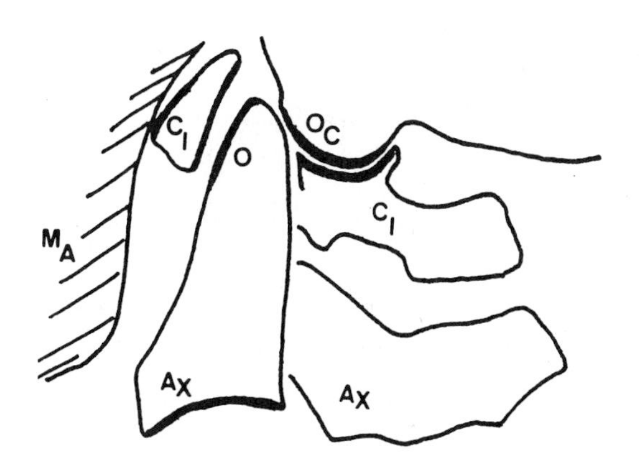

Figure 8d

Oc : Occipital condyle
Cl : First cervical vertebra
O : Odontoïde apophysis
Ax : Axis (second cervical vertebra)

Mx : Maxillary
MA : Mandibule
MS : Mastoïde apophysis

Figure 8. Subject's preparation for the determination of the location of the occipital condyles

8a and 8b : Front view before and after preparation, respectively
8c and 8d : Side view before and after preparation, respectively

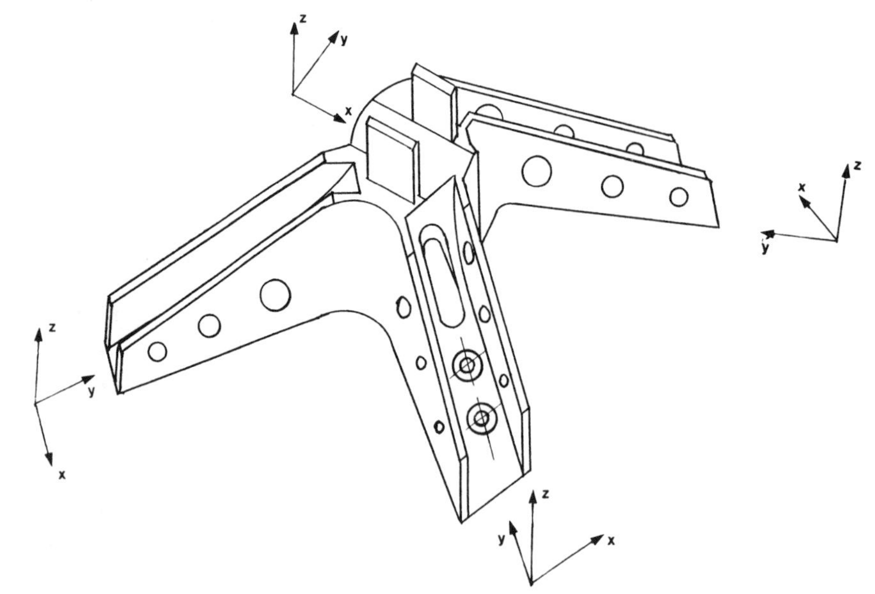

Figure 9. Twelve-accelerometer array used for high G-level cadaver tests

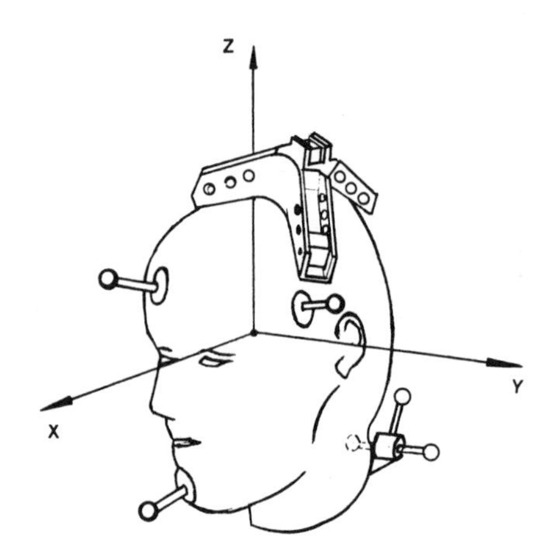

Figure 10. Head and T1 instrumentation

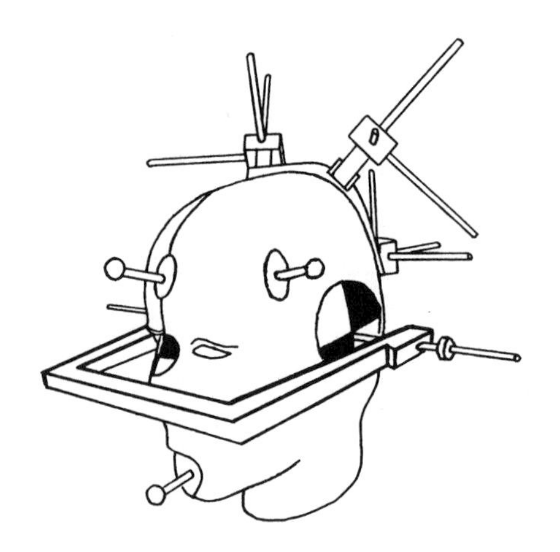

Figure 11. The setting up procedure used for the calculation of the location and orientation of head instrumentation with respect to the head anatomical coordinate system

To enable three-dimensional analysis of head and neck movement in relation to the sled from the film, the cameras are calibrated before testing begins. The camera calibration procedure was performed experimentally, using non-coplanar spherical targets as control points (15) fixed to a cube-shaped mounting. The coordinates of each target relative to the laboratory coordinate system are known. During the test set-up procedure, all targets are filmed by five 16-mm cine cameras. This allows the evaluation of calibration coefficients for each camera and the relationship between digitizer coordinates and object coordinates to be determined. The redundancy of control points and cameras is treated using a linear least-square method.

Once the test preparation is finished, the subject is positioned according to the test set-up procedure described above for the cadaver tests performed at low violence levels.

AVAILABLE DATA - The following parameters as a function of time are available for the MS 375 and MS 376 tests which were carried out recently :

- sled acceleration ;
- frontal, lateral and vertical acceleration for T1, T4 and the pelvis, respectively ;
- three-dimensional displacements of head c.g. and T1, respectively, relative to the sled, obtained from film analysis ;
- head and T1 rotations relative to the sled, expressed in Euler angles or in quaternions, obtained from film analysis ;
- relative head movements with respect to the T1 coordinate system ;
- visualization of head and T1 movements relative to the sled.

From the accelerometric data processing, the parameters shown in table 5 are available.

The subjects' general anthropometry and the head and neck specific anthropometry are also available. These data are summarized in tables 6 and 7, respectively.

RESULTS -

Sled deceleration pulse and T1 acceleration

The type of sled deceleration to which the subjects were subjected is shown in figure 12. The pulse of this deceleration is characterized for the MS 375 test, for example, by a change of 0 to 13 G during the first 17 ms, followed by a plateau of between 12 G and 14.7 G, which is maintained up until 50 ms.

The two subjects sustained a deceleration higher than or equal to 75 % of the maximum sled deceleration for a period of 37 ms in the MS 375 test and 48 ms in the MS 376 test, respectively. The rate of onset of the sled deceleration defined in relation to the sled deceleration change, between 20 % and 50 % of its maximum value and the corresponding time duration (4), was 802 G/s for test MS 375 and 610 G/s for test MS 376. This gives a good illustration of the particularly high level of violence present in these experiments.

The characteristics of these tests are summarized in table 8.

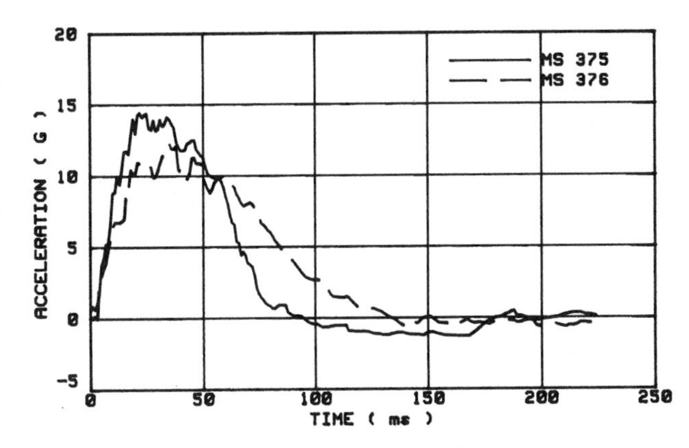

Figure 12. Sled deceleration : time history in MS 375 and MS 376 tests

The lateral acceleration components measured at T1, T4 and on the pelvis, respectively, are shown in figure 13 for the MS 375 test. The order in which the solicitations on the subject occur can be seen. During sled deceleration, it would appear that it is the pelvis which is contacted first, followed by the thorax and the base of the neck. Thus the lateral acceleration in the MS 375 test shows its first peaks at the pelvis, T4 and T1 at 28 ms, 42 ms and 45 ms respectively. We also note the presence of several peaks coming after the first peak in the T1 and T4 curves, as was the case in the preceding tests. It would appear that these peaks are mainly due to dynamic interactions transmitted to the thoracic spine by the head.

The maximum amplitudes for head kinematics in relation to the thorax, which would perhaps have been able to account for the presence of these peaks, occur later on than the secondary peaks for the thoracic accelerations.

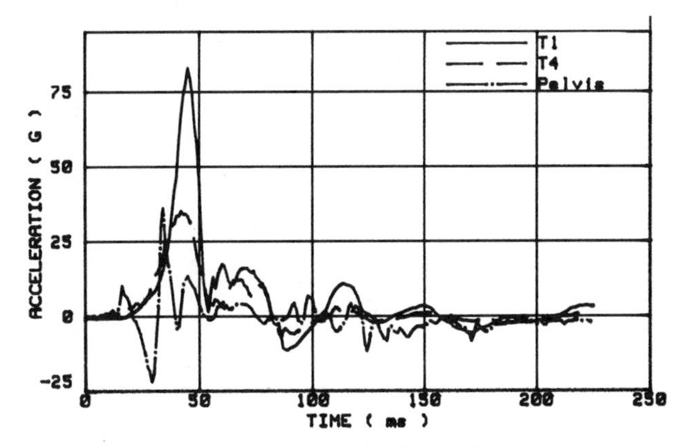

Figure 13. Lateral acceleration : time histories in MS 375 test measured at pelvic, T4 and T1 levels, respectively

Head and T1 movement - The kinematographic data relevant to the movements of the targets attached to the head and to the base of the neck were processed simultaneously with a millisecond time interval via a data-processing program developed at the LPB-APR. On the basis of the photographic input data filmed prior to testing, this program supplied the results of the tri-dimensional kinematics of the head and of T1 in relation to the sled. These results include the translations of the head and T1 as well as their angular orientations expressed in the form of a matrix that can be stated as follows :

$$A = \begin{vmatrix} A_{11} & A_{12} & A_{13} \\ A_{21} & A_{22} & A_{23} \\ A_{31} & A_{32} & A_{33} \end{vmatrix} \quad (1)$$

in which A_{11} to A_{33} are the elements of orientation matrix A, which are actually linear combinations of trigonometric functions of the rotation angles of the head (or of T1).

To describe these rotations, we used three angular coordinates, called Euler angles, which are defined as follows :

- Let xyz be the system of coordinates of the anatomical reference of the head or T1. It is assumed that xyz coincides with the laboratory coordinate system. The xyz axis system can be conveyed onto an x"y"z" system by three successive rotations ;
- the first rotation occurs around the x axis ; the xyz system is conveyed onto the xy'z' system. The second rotation occurs around the new y' axis ; the xy'z' system is thereupon conveyed onto the x'y'z" system. The third rotation occurs around the new z" axis, thereby yielding the new x"y"z" system.

Table 5. Available data in MS 375 and MS 376 tests, respectively, obtained from accelerometric data-processing

- Head linear acceleration in longitudinal, lateral and vertical directions, respectively, with respect to the laboratory coordinate system ;
- the three components of head angular velocity and acceleration, respectively, with respect to the head anatomical coordinate system as well as to the laboratory coordinate system ;
- head c.g. linear displacements and velocities with respect to the laboratory coordinate system ;
- head rotations, expressed in Euler angles with respect to the laboratory coordinate system.

Table 6. Principal anthropometric data of subjects involved in high G-level tests

Subject	Age years	Sex -	Height (m)	Weight (kg)	Sitting height (m)	Mineralization of the 11th rib (g/cm)
MS 375	56	M	1.68	61	0.87	0.31
MS 376	35	M	1.70	77	0.80	0.32

Table 7. Head and neck anthropometry

Subject	Head				Neck		
	Length (m)	Breadth (m)	Circumference (m)	Mass (kg)	Breadth (m)	Circumference (m)	Mass (kg)
MS 375	0.194	0.140	0.560	4.200	0.114	0.375	1.290
MS 376	0.184	0.151	0.550	3.960	0.134	0.395	1.380

Table 8. Test characteristics of new cadaver tests

Subject	Peak sled deceler-ation G	ΔT ms	Rate of onset G/s	Initial sled velocity m/s	Peak T1 lateral acceleration G	Peak T4 lateral acceleration G
MS 375	14.7	37	802	6.3	84	34.6
MS 376	12.2	48	610	6.3	43	20.5

In order to describe the head's lateral flexion (or the rotation of T1), i.e. the angle in the plane of impact between the anatomical axis oz and a vertical axis, we used the first Euler angle.

The head's anterior flexion, i.e. the angle within a vertical plane between the anatomical axis oz and a vertical axis, was defined on the basis of the second Euler angle.

Other combinations of Euler angles can be used. For example, we used another combination that consisted of conveying the xyz axis system onto the x'y'z' by first performing a rotation around the z axis, followed by a rotation around the new x' axis, and, finally, by rotation around the new z' axis.

The torsion of the head, i.e. the head's rotation around its anatomical axis oz, was expressed by means of the third Euler angle. These definitions are illustrated in figure 14.

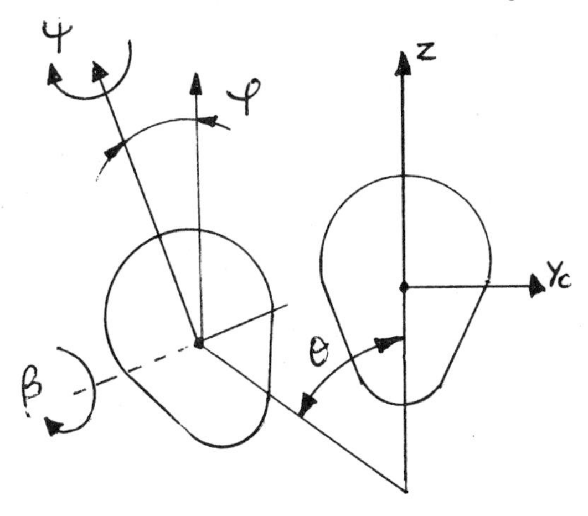

z : head anatomical z-axis

y_c : axis parallel with the head anatomical y-axis

φ : head lateral flexion

β : head anterior flexion

ψ : head torsion

θ : neck flexion

Figure 14. A simple definition of head and neck rotations

The viewing of the film shows that, in general, the kinematics of the head and the base of the neck are here similar to the cadavers' kinematics in the low violence tests. Differences do appear, however, especially in terms of the excursion and rotation of the head in relation to the sled, which are greater in this case. In order to illustrate the head and the neck movements during the sled deceleration, sequences from the film of the MS 375 test are given in figure 15 with the following main phases :

- the translation phase of the head movement, which lasts approximately 42 ms ;
- the rotational phase of the head movement or the intermediary phase of the head, between its translation and the moment of maximum amplitude of its movement ;
- the unloading phase.

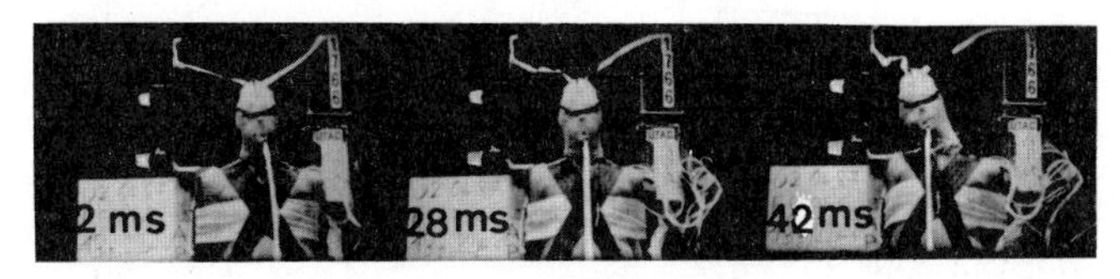

a) Translational phase

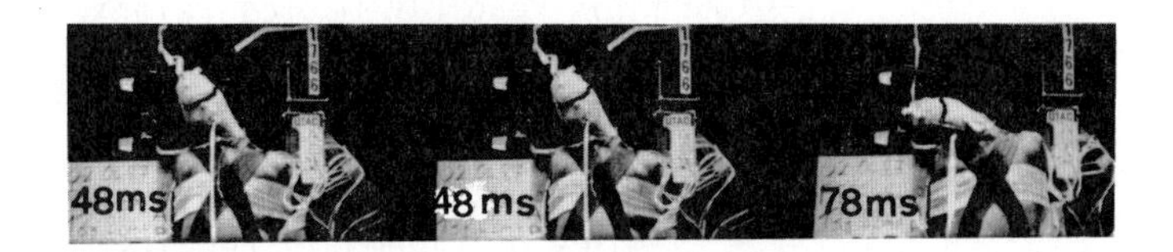

b) Rotational phase

c) Unloading phase

Figure 15. Principal film sequences of head-neck motion in MS 375 test

These main phases are illustrated in another form in figures 16a, 16b and 16c, respectively, where the head kinematics are expressed in relation to the sled ; the head is simulated by a cube in these graphic outputs. This simplified representation of the head generated by a program from the three-dimensional analysis of the film allows the rapid evaluation of the type of kinematics which the head goes through during the test, on the basis of 3 graphic representations in the impacted plane, in a plane vertical and perpendicular to the impact plane and in a plane horizontal and perpendicular to the impact plane.

Thus the orientation of the cube and the anatomical z-axis in figures 16a and 16b, respectively, allow us to follow the passage from the translational phase of the movement to the intermediary phase, where the lateral flexion increases significantly.

Table 9 summarizes the maximum amplitudes of the head and T1 kinematics expressed in relation to the sled, which were obtained for the MS 375 and MS 376 tests.

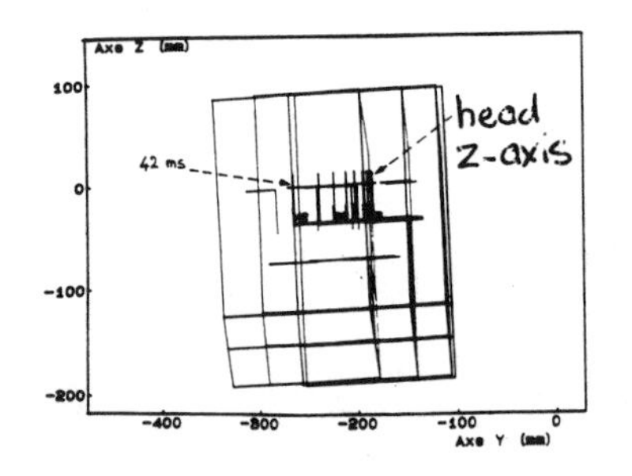

Figure 16a. Head kinematics during the first 42 ms

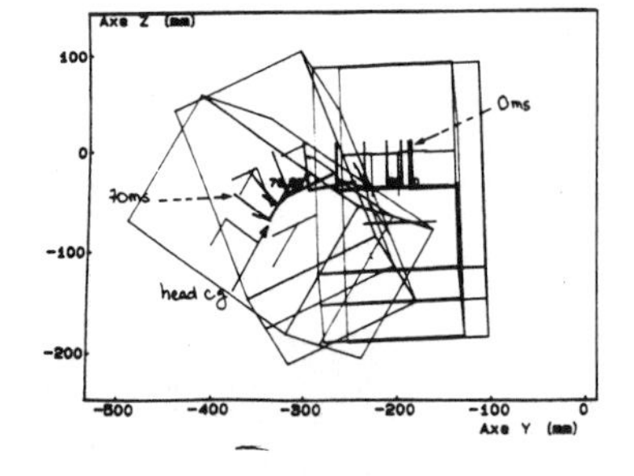

Figure 16b. Head kinematics from 0 ms up to 70 ms

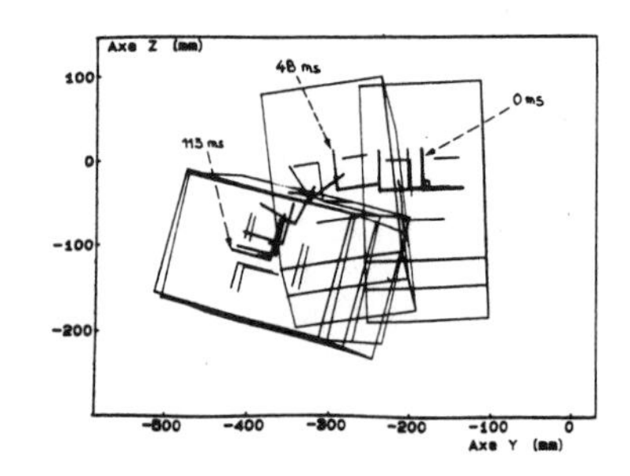

Figure 16c. Head kinematics during the entire loading phase of motion

Figure 16. Head kinematics relative to the sled in the MS 375 test as generated by the Anafilm-3D program

Table 9. Summary of kinematic results with respect to the sled obtained from MS 375 and MS 376 cadaver tests

Test number	T1-kinematics		Head kinematics			
	Lateral Displacement	Rotation of T1	Displacements of c.g.		Head torsion	
			Lateral	Vertical	Lateral flexion	Torsion
	mm	degrees	mm	mm	degrees	degrees
MS 375	63	30	180	82	(2) 75	(0) -19
MS 376	50	18	167	60	(4) 50	(0) -57

Numbers in brackets denote initial value of angle of rotation

From observation of the films and the data in table 9, it would appear that the initial orientation of the head in relation to the sled coordinate system has a significant influence on its kinematic tendencies during impact. Thus, in the MS 375 test, where the head was oriented towards the rear by around 15°, it is the lateral flexion which predominates over the torsion. In the MS 376 test, where the head is leant slightly forward, by about 7° in its initial stage, the tendency is reversed, although less accentuated than in the previous case.

Relative head movements - The relative kinematics of the head in relation to T1 are shown in figures 17 and 18 for tests MS 375 and MS 376, respectively. We can see that the trajectory of the head in the MS 376 test is more circular than in test MS 375. The amplitude of the thorax rotation in this test (as can be seen in figures 19 and 20) has in fact visibly reduced the radius of curvature of the trajectory of the head. The rotation of T1 is expressed in these figures by the rotation of its z-axis in the impact plane.

So as to quantify the tendencies observed in the T1 and head kinematics in the two tests, we give in figures 21, 22 and 23 the angle of rotation of T1 z-axis as a function of time, the angle of torsion of the head as a function of the lateral flexion, and the vertical displacement of the head c.g. as a function of the lateral displacement, respectively.

A study of the curves in figure 21 gives a better understanding of the influence of the T1 rotation, particularly on the displacements of the head c.g. The variation of this rotation as a function of the time in the two tests suggests the following remarks :

- between 0 and 30 ms, the difference in the angle of rotation at T1 is constant ;
- between 30 and 50 ms, the difference increases, with a larger variation in the MS 375 test ;
- lastly, between 50 and 100 ms, this difference is maintained, with a difference of 21 degrees and 14 degrees for the MS 375 and the MS 376 tests

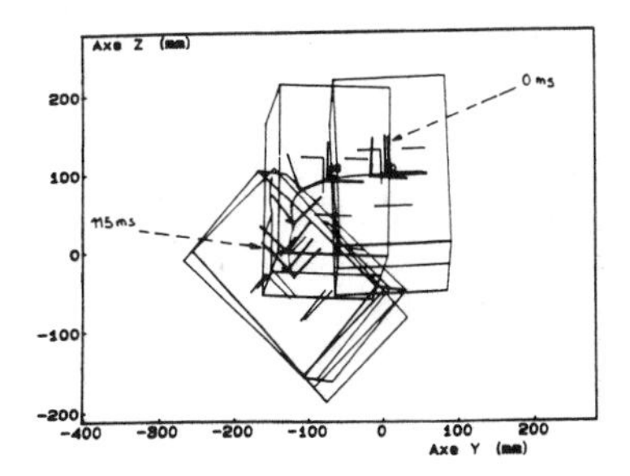

Figure 17. Head-relative kinematics in the impact plane with respect to T1 in MS 375 test

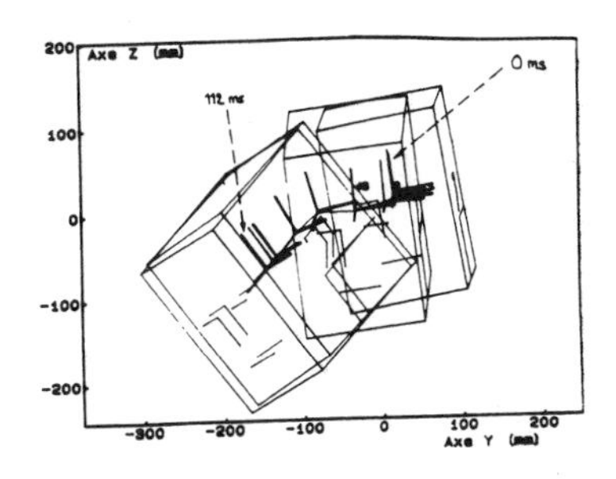

Figure 18. Head-relative kinematics in the impact plane with respect to T1 in MS 376 test

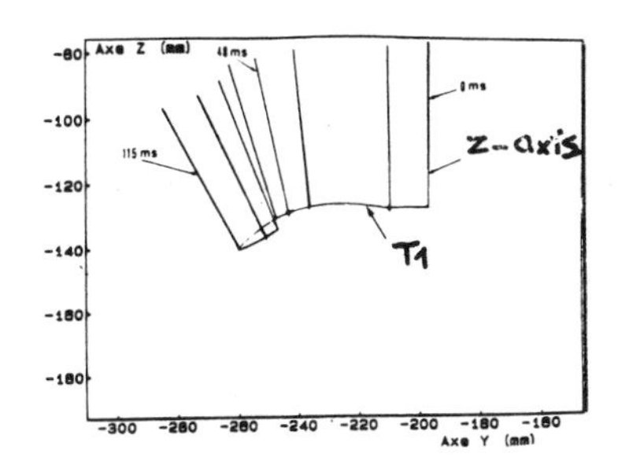

Figure 19. T1-kinematics with respect to the sled in MS 375 test

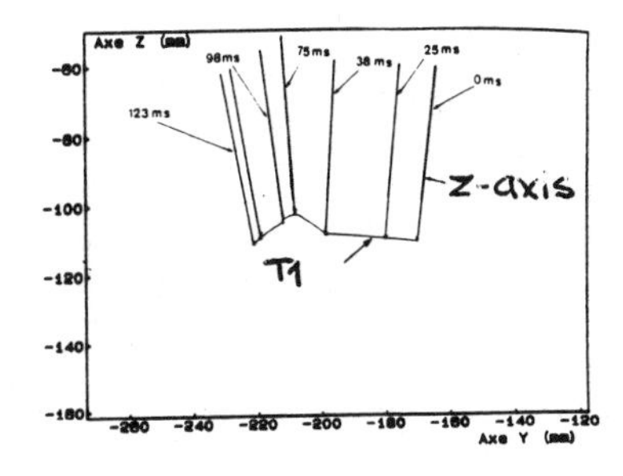

Figure 20. T1-kinematics with respect to the sled in MS 376 test

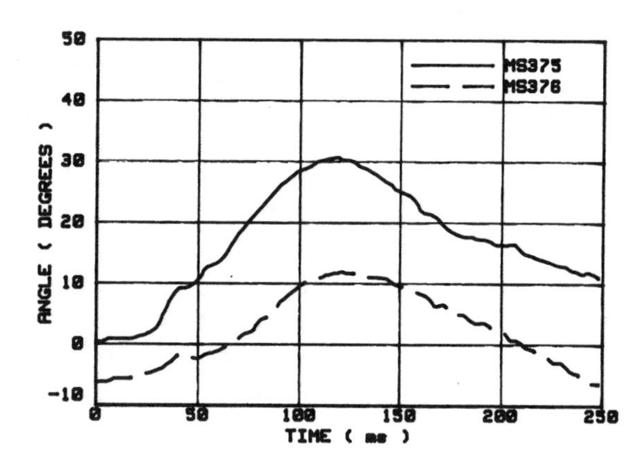

Figure 21. T1 z-axis rotation relative to the sled in MS 375 and MS 376 tests

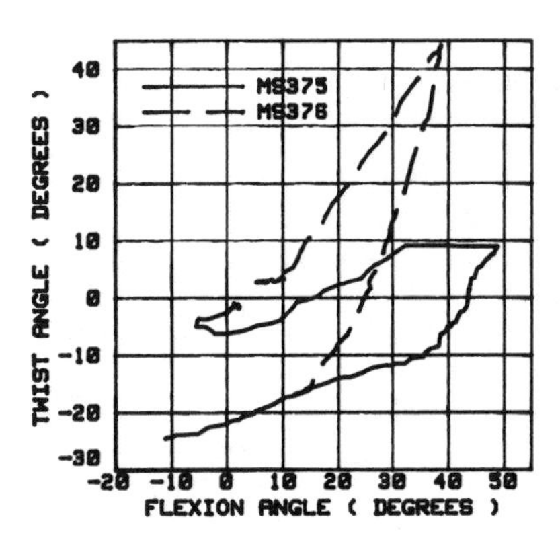

Figure 22. Head torsion angle as a funtion of head flexion angle in MS 375 and MS 376 tests

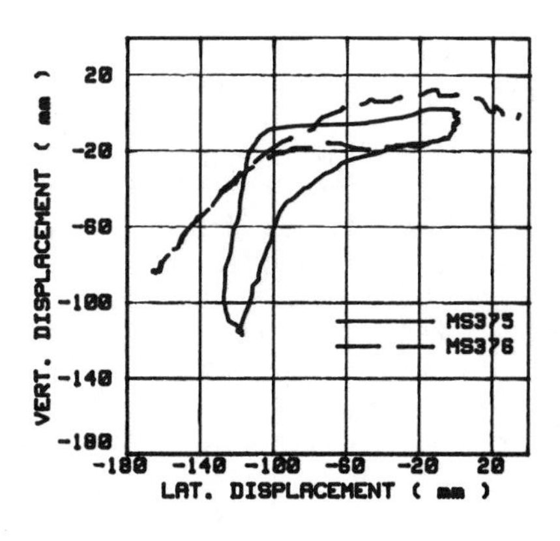

Figure 23. Trajectory of head c.g. relative to T1 in MS 375 and MS 376 tests

This influence of the T1 rotation on the relative kinematics of the head is all the larger in the case of the MS 375 test because the trajectory of the head c.g. takes on a convex form which increases from 51 ms up to 100 ms (that is, during the phase where the rotation of the T1 z-axis increases to a greater extent than that in the MS 376 test).

As far as the rotations of the head in relation to T1 are concerned, the amplitude of the lateral flexion is higher than that of the torsion in the MS 375 test, with maximum values of 46 and 12 degrees, respectively. In the MS 376 test, it is the amplitude of the torsion which is the higher, with maximum levels of 42 and 28 degrees, respectively. These differences in the rotations between the two tests, which we already observed when the kinematics of the head were expressed in relation to the sled, are amplified by the fact that T1 is treated as a rigid body in this analysis.

For the purpose of evaluating head-neck relative behaviour in terms of rotation, we considered the difference in function of the time between the head's lateral flexion and the rotation of T1, i.e. angle $(\psi - \theta)$. The corresponding curve is shown in figure 24. Both tests show the similar behaviour of this angle, with two peaks.

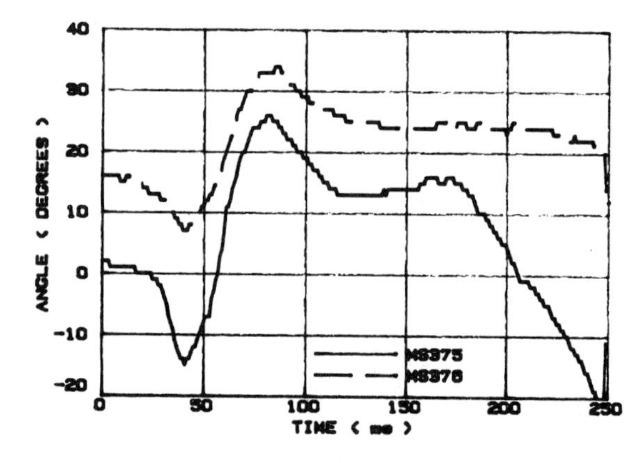

Figure 24. Head-neck relative rotation in MS 375 and MS 376 tests, respectively

The first peak follows an initial phase in which the variation of $(\psi - \theta)$ is negative. The curve then abruptly changes its profile, and the variation of $(\psi - \theta)$ becomes positive. This behaviour can be explained as follows : in an initial phase, the head describes a pure translation, i.e. a rotation of zero amplitude ; during this phase, the neck gradually bends up to a certain limit, at which point a locking appears to occur in the head-neck connection. This phenomenon is thereupon followed by a phase in which head flexion increases faster than neck rotation, as is shown in figure 24. This interesting finding should be confirmed by the accelerometric data, the analysis of which is proposed in the following section.

Lastly, as regards the head anterior flexion, the findings show that its maximum amplitude is, respectively, 3 and 13 degrees for tests MS 375 and MS 376. Since these levels are lower than those observed in low-violence tests with cadavers, it follows that the effect of the head anterior flexion is, in the case of the tests analyzed here, limited by the greater deceleration of the sled.

Head acccelerations - The data supplied by the instrumentation of the subject's head were processed via method 3-3-3 (13). The only elements considered were the measurement channels of the three sensors attached to the peripheral branches of the assembly depicted in figure 9. These data, with those concerning the positions and orientations of the accelerometers in relation to the anatomical coordinate system of the head, were introduced into a processing program called "9-Gamma". This program initially transposes the recorded acceleration channels into the anatomical coordinate system of the head, and then computes the angular velocity and acceleration of the head, considered here as a rigid body. The linear acceleration of the c.g. of the head is established on the basis of the following vectorial equation :

$$\overline{a_G} = \overline{a_R} + \overline{\alpha} \times \overline{GR} + \overline{\omega} \times (\overline{\omega} \times \overline{GR}) \tag{2}$$

in which

- $\overline{a_R}$ is the vector of the linear acceleration of the origin of the head anatomical coordinate system ;
- $\overline{GR}$ is the position vector of the head's c.g. in relation to the head anatomical coordinate system ;
- $\overline{\alpha}$ and $\overline{\omega}$ are the vectors of the head angular acceleration and velocity, respectively.

The transformation matrix that makes it possible to transpose a vector from the anatomical coordinate system to the laboratory coordinate system is expressed on the basis of Euler angles as the transposed matrix of the one defined in section intitled "Head and T1 movement".

In addition to the accelerations and velocities of the head, the graphic outputs generated by the 9-Gamma program include the head orientation as well as the position of its c.g. in relation to the laboratory's coordinate system in function of time.

Generally speaking, the head kinematics obtained from the data-processing yields findings similar to those obtained with the film, with the exception of the head rotations, which display significant differences. This is due, notably, to the errors in the various calculation phases specific to method 3-3-3. Whence the priority assigned to kinematographic data for the analysis of the head displacements.

Figures 25 and 26 present the lateral and vertical components of the linear acceleration of the head c.g., respectively, for tests MS 375 and MS 376.

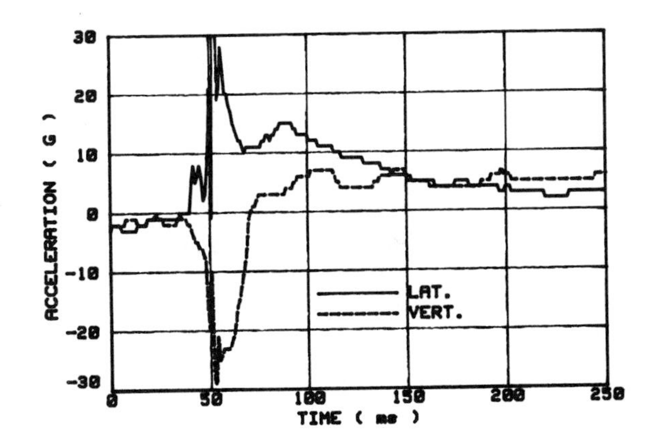

Figure 25. Head linear acceleration in MS 375 test

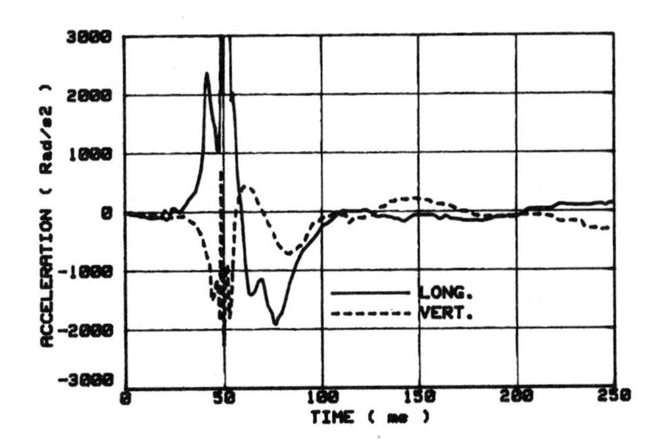

Figure 27. Head angular acceleration in MS 375 test

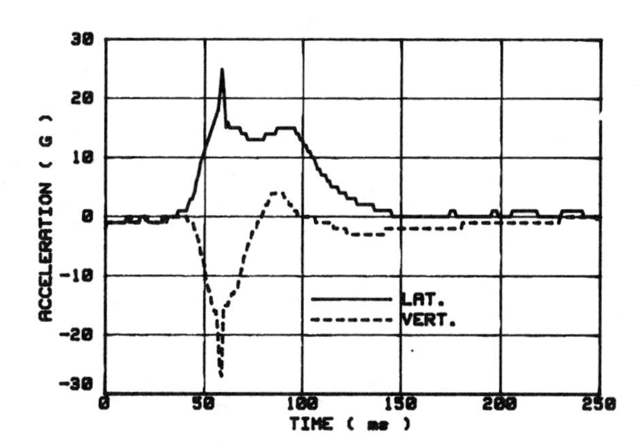

Figure 26. Head linear acceleration in MS 376 test

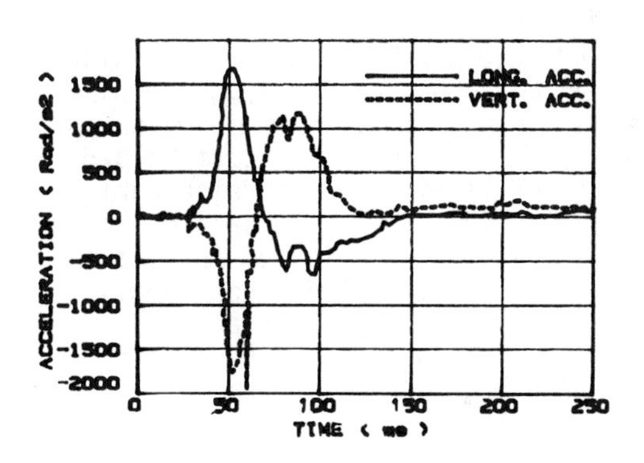

Figure 28. Head angular acceleration in MS 376 test

Figures 27 and 28 illustrate the lateral and vertical components of the angular acceleration of the head. First of all, we see that the curve fairly accurately reflects the type of solicitation to which the cadaver's head is subjected. Thus, the positive sign of the lateral component of the head linear acceleration corresponds to the direction of the impact, i.e. from right to left, whereas the negative sign of the vertical component expresses the retention of the head (by the neck), which is applied in a top-to-bottom direction. The signs of the peaks of the components of the head angular acceleration, expressed in the laboratory coordinate system, are in agreement with the rotations described by the head, i.e. a positive sign for the x component corresponding to the head flexion and a negative sign for the z component corresponding to the head torsion.

In addition, a pronounced, brief peak, which is unusual in this kind of test, in which the head sustains no impact, appears on the components of the head linear and angular accelerations. Whereas the analysis of the film reveals no anomaly, checking of the raw accelerometric data revealed that certain measurement channels showed extremely brief peaks, of a maximum length of one millisecond. The amplitudes of these peaks and their time lengths might be accounted for by a slight vibration in the attachments of support of the head accelerometers.

In consequence, in order to prevent an overestimate of the amplitudes achieved by the various accelerometric parameters, we hence expressed their maximums in terms of 3-ms values, as indicated in table 10. In the light of the data in this table and in figures 25 to 28, the following remarks can be made :

Table 10. Head maximum accelerations expressed in terms of 3-ms values in MS 375 and MS 376 tests, respectively

Test No	Linear acceleration of the head c.g.			Angular acceleration of the head	
	x-component G	y-component G	z-component G	x-component rad/s^2	z-component rad/s^2
MS 375	12.5	25.4	25.2	2526	1451
MS 376	17.2	16.8	15.2	1588	1797

- the ratio between the lateral and vertical components of the linear acceleration of the head c.g. is close to 1 for the two tests. However, the amplitudes of the corresponding peaks are greater for test MS 375 ;
- in relation to the lateral and vertical components, the amplitude of the longitudinal acceleration of the head c.g. is greater in the case of test MS 376 ;
- in the head angular acceleration, the ratio of the longitudinal and vertical components is greater in test MS 375.

The higher levels of linear and angular acceleration can be accounted for by the amplitude of the T1 lateral acceleration, which is approximately twice as great in test MS 375. The predominance of flexion over torsion during the head displacement in the case of test MS 375 is confirmed by the ratios of the peaks, respectively, of the longitudinal acceleration of the head c.g. in relation to the other two components and of the longitudinal component of the angular acceleration in relation to the vertical component.

It will further be noted that the maximum amplitudes of the head linear and angular accelerations occur during the test long before the instant corresponding to the head maximum excursion. If we take, for example, the case of test MS 376, it can be seen that the peaks in the head linear and angular accelerations occur at 52 and 60 ms, respectively. In addition, it was noted earlier that the kinematics of head-neck system in relation to T1 was characterized by a locking mechanism (between the head and neck), which was illustrated by a sudden variation in the head's flexion angle in relation to that of the neck. Since this variation occurs at 42 ms, this locking mechanism may possibly be the cause of the peaks in the accelerations of the head. To illustrate this observation, the head-neck relative angle as a function of time in test MS 376 is shown in figure 29 with the lateral and longitudinal components of the head's linear and angular accelerations, respectively. The data for test MS 375 appear to bear out those of test MS 376, but the occurrence of several peaks in the longitudinal component of the angular acceleration, which however occurred subsequently to the locking mechanism, makes interpretation less easy.

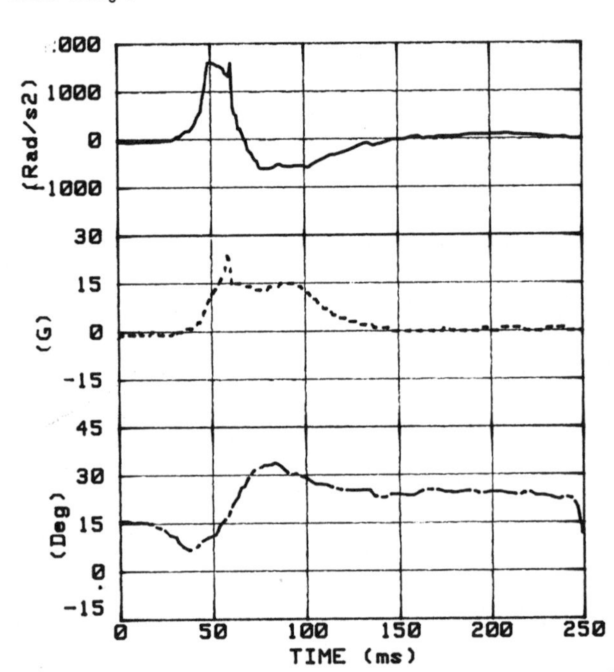

Figure 29. Influence of the head-neck relative angle on the peaks in the head accelerations

Head-neck dynamics - Assuming the head to be a rigid body, and assuming that the type of test presented here does not generate impacts to the head, the moment transmitted by the neck to the head can be calculated by means of the following equation :

$$\overline{M}_C = m_h \cdot \left[\overline{r}_c \times (\overline{a}_g - \overline{g})\right] + I.\overline{\alpha} + \overline{\omega} \times (I.\overline{\omega}) \quad (3)$$

in which :
- $\overline{M}_C$ is the moment transmitted to the head by the neck ;
- m_h is the mass of the head ;

- $\overline{r_c}$ is the position vector of the head's c.g. in relation to the occipital condyles ;
- $\overline{a_g}$, $\overline{g}$ are respectively, the linear acceleration of the c.g. of the head and the acceleration of gravity ;
- I is the moment of the inertia tensor of the head expressed in relation to the c.g. ;
- $\overline{\alpha}$, $\overline{\omega}$ are, respectively, the acceleration and angular velocity of the head ;
- x, . indicates the vectorial and scalar products, respectively.

For the subjects MS 375 and MS 376, the mass of the head m_h was determined by weighing. The position of the occipital condyles in relation to the head anatomical origin was measured graphically in accordance with the procedures described in the section intitled "METHODOLOGY".

The coordinates of the c.g. as well as those of the occipital condyles in relation to the anatomical coordinate system are presented in table 11. The moments of inertia of the head around the principal axes were calculated in accordance with the formulas proposed by Mc Conville et al. (16), on the basis of the anthropometry of the head, i.e. length, width and circumference. The accelerations α, a_g and the velocity ω were calculated earlier.

The results of the calculation of the loads of the head-neck system are presented in figures 30 and 31. The longitudinal and vertical components of the moment that acts at the level of the occipital condyles are called Mx and Mz, respectively. The moment Mx, around an axis parallel to the anatomical axis ox, is shown as a function of the head flexion angle in figure 30. It will be seen that the curves corresponding to tests MS 375 and MS 376 display similar behaviours, with a positive initial phase. The moment Mx during this phase achieves, for tests MS 375 and MS 376, values of 30 and 10 N.m., respectively. The negative phase of moment Mx yields greater amplitudes with maximums of 55 and 38 N.m., respectively. Figure 31 shows the moment Mz around an axis parallel to the anatomical axis oz as a function of the head torsion angle. The curves corresponding to the two tests are similar in their initial phases, where we observe a significant increase in the moment for an almost zero variable in the torsion angle. The maximum amplitudes of the Mz moment are of the same order of magnitude for both tests, i.e. 17 N.m., and are distinctly smaller than those of the Mx moment. The results shown in figures 30 and 31 are presented in their respective loading phases, which, in the case of the tests analyzed here, are different from the unloading phases. This difference appears to be characterized by the absence of muscular tonicity in the cadavers.

In order to illustrate the mechanism of the locking between the head and neck that had emerged earlier, the moment at the level of the occipital condyles around an axis parallel to the anatomical axis ox, i.e. Mx, is shown in figure 32 as a function of the head-neck relative angle for test MS 376. During the initial phase, the head-neck relative angle is observed to increase, whereas the moment at the level of the occipital condyles remains constant and close to zero. This phase ends with a sudden increase in the moment, thereby, confirming the results of the analysis of the kinematics and of the angular accelerations.

Autopsy results - The autopsy results of the subjects revealed no injuries either to brain or cervical region.

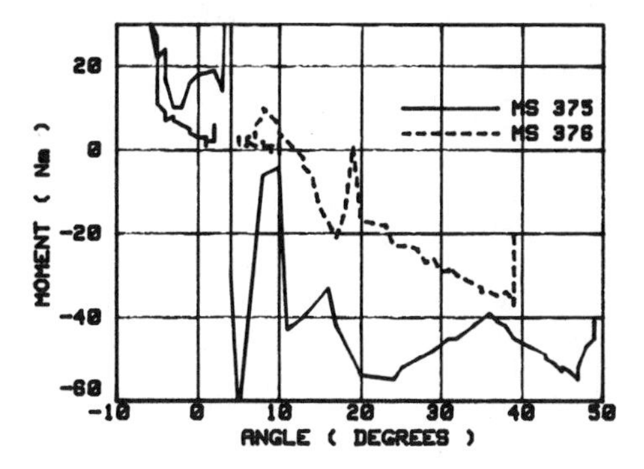

Figure 30. Moment acting at the level of the occipital condyles, around an axis parallel to the anatomical axis ox, as a function of the head flexion angle

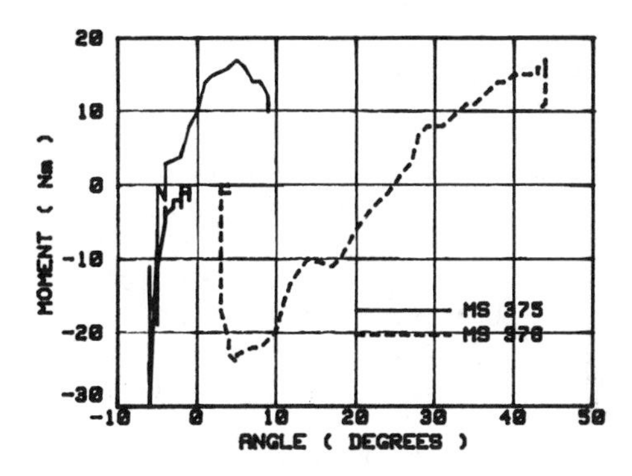

Figure 31. Moment acting at the level of the occipital condyles, around an axis parallel to the anatomical axis oz, as a function of head torsion angle

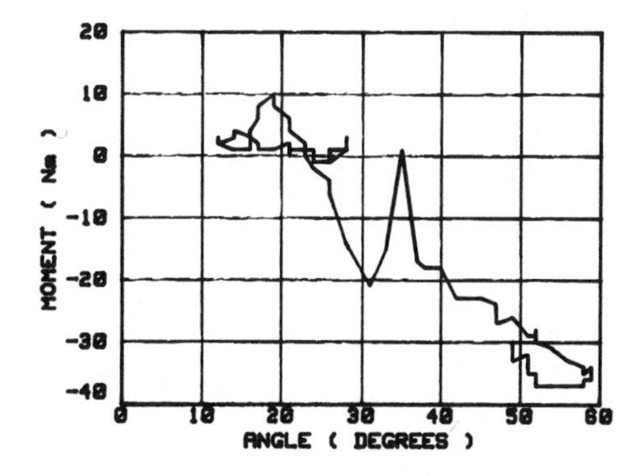

Figure 32. Moment acting at the level of the occipital condyles, around an axis parallel to the anatomical axis ox, as a function of head neck rotation angle in MS 376 test

Table 11. Location of head c.g. and occipital condyles with respect to the anatomical coordinate system of the head

Test no.	Occipital condyles			Head c.g.		
	x m	y m	z m	x m	y m	z m
MS 375	0.0070	0	-0.030	0.0098	0	0.0213
MS 376	0.0082	0	-0.025	0.0098	0	0.0213

CONCLUSIONS

Two experimental phases involving cadavers were carried out in accordance with the sled-type testing environment. The subjects were submitted to decelerations of 7 G and 13 G for an initial sled velocity of 6 m/s. The conclusions of the first phase can be summarized as follows :

- the duplication of tests involving volunteer subjects by experiments with cadavers was satisfactory. The input data in terms of sled acceleration and acceleration of the first thoracic vertebra T1 were comparable ;
- in the initial sled deceleration phase, the kinematics of the cadaver's head and of the volunteer's head are similar. In a second phase, the movement of the cadaver's head is characterized by wider excursions and by a three-dimensional aspect, including, in addition to lateral flexion and torsion, an anterior flexion that had not been observed in the volunteer. The absence of muscular tonicity could account for this rotation.

On the basis of a detailed analysis of two new high-violence tests, the following conclusions can be formulated :

- new means in terms of methodology and analysis have been developed. These appear to be satisfactory since most of the findings of the kinematographic analysis corroborate those supplied by the processing of the measurement data ;
- in general, a high peak of T1 acceleration brings about large amplitudes in the accelerations of the head. For a maximum T1 acceleration of 84 G, amplitudes of 25 G and 2526 rad/s^2 were found for linear acceleration and the longitudinal component of angular acceleration, respectively ;
- the head initial orientation in relation to the sled plays a preponderant role in the relationships of the amplitudes of its rotations. In addition, the trajectory of the head c.g. in relation to T1 depends on a large extent on the rotation of T1. The greater this rotation, the smaller is the curve radius of the head trajectory ;
- a locking mechanism in the head-neck connection appears to occur after a certain level of head-neck relative angle. This finding is in agreement with the data of Wismans et al. (5) concerning tests performed with volunteers. This mechanism seems to determine the instants of the peaks of the head angular acceleration ;
- the moment that acts at the level of the occipital condyles around an axis parallel with the anatomical axis oz is distinctly lesser than the moment around an axis parallel with the anatomical axis ox ;
- the absence of muscular tonicity is characterized in the cadaver by differences between the phases of the loading and unloading of the parameters analyzed, notably with respect to the trajectory of the head c.g., the torsion angle in function of the flexion angle, and the moment-angle characteristics of the head-neck connection ;
- since the autopsies of the subjects reveal no occurrence of injuries in the head-neck area, the data analyzed here appear to be valuable for the defining of a high-violence specification for a dummy's neck. However, it would be desirable for these specifications to be completed by other findings from tests performed under the same conditions.

ACKNOWLEDGEMENTS

We want to express our sincere thanks to Dr Ewing and W.H. Muzzy and the whole team of the NBDL (Naval Biodynamics Laboratory, New Orleans). Without them, without their wish of co-operation, this research could not have been started and brought to a successful issue ; they are, in a way, joint authors of this work, though their scientific liability can be involved only by themselves. Our thanks also go to the UTAC (Union Technique de l'Automobile, du Motocycle et du Cycle) team for their contribution to this study. All opinions given in this paper are those of the authors and not necessarily those of LPB-APR.

This research was sponsored partly by the CEC (Commission of the European Communities) as far as the low-violence tests are concerned and by the CCMC (Committee of Common Market Automobile Constructors) for high-violence tests.

REFERENCES

(1) "Choc latéral : Synthèse Accidentologique", Laboratory of Physiology and Biomechanics Peugeot S.A./Renault, June 1983, (Internal Report).

(2) D. Kallieris, R. Mattern and W. Hardle "Belastbarkeitsgrenzen und Verletzungsmechanick des angegurteten Pkw - Insassen beim Seitenaufprall. Phase II : Ansätze für Verletzungs prädiktionen", FAT Report Number 60.

(3) F. Brun-Cassan, Y. Pincemaille, C. Tarrière, "Reconstructions of Real-Life Side Impact Collisions", International IRCOBI Conference on the Biomechanics of Impacts, September 1987, Birmingham.

(4) C.L. Ewing, D.J. Thomas, L. Lustick, W.H. Muzzy, G.C. Willems and P. Majewski "Dynamic Response of Human Head and Neck to+Gy Impact Acceleration", Proceedings of the 21st Stapp Car Crash Conference, October 1977, New Orleans, Louisiana.

(5) J. Wismans and C.H. Spenny "Performance Requirements for Mechanical Necks in Lateral Flexion", Proceedings of the 27th Stapp Car Crash Conference, October 1983, San Diego, California.

(6) F. Becker and G. Willems "An Experimentally Validated 3-D Inertial Tracking Package for Application in Biodynamic Research", Proceedings of the 19th Stapp Car Crash Conference, November 1975, San Diego, California.

(7) H. Mertz "Lateral Neck-Bending Response Requirements", ISO/TC22/SC12/WG5, Document Number 139, February 1987.

(8) H. Mertz "Lateral Neck-Bending Response Requirements", ISO/TC22/SC12/WG5, Document Number 139, Draft 1, September 1985.

(9) F. Bendjellal et al. "Development of a Dummy Neck for Lateral Collisions", Draft Report to the Tenth International Technical Conference on Experimental Safety Vehicles, Oxford, 1985.

(10) F. Bendjellal et al. "Development of a Dummy Neck for Lateral Collisions, Part 2 : Cadaver tests at a high level of violence, proposed neck requirements", ESV Conference, Oxford, 1985, (Final report not included in the proceedings).

(11) F. Guillon et al. "About bone conditions of subjects involved in biomechanical experiments", Internal Report n° 87/340 in preparation.

(12) C. Tarrière, F. Chamouard "Repressurized Cadaver Head Impacts : Some Findings", Wayne State University Symposium on "Head injury prevention, past and present research", December 1985.

(13) M. Nabih Alem and Holstein, L. Garry "Measurement of 3D-Motion", HSRI, Ann Arbor, Michigan, October 1977.

(14) A.J. Padgaonkar, K.W. Krieger and A.I. Wing "Measurement of Angular Acceleration of a Rigid Body Using Linear Accelerometers", ASME, Journal of Applied Mechanics, September 1975, pp. 552-556.

(15) James S. Walton "Close-Range Cine-Photogrammetry : Another Approach to Motion Analysis", Academic Publishers, Del Mar, Stanford University, Stanford, California.

(16) Mc Conville et al. "Anthropometric Relationship of Body and Body Segment Moments of Inertia", Report AFAMRL-TR-80-119, Air Force Aerospace Medical Research Lab. Wright Patterson Airforce Base, Ohio, 1980.

892435

Some New Data Related to Human Tolerance Obtained from Volunteer Boxers

Y. Pincemaille, X. Trosseille, P. Mack and C. Tarrière
Laboratory of Physiology and Biomechanics

F. Breton and B. Renault
U.R.A. Unité de Recherche en Physiologie Cognitive Pathologie

ABSTRACT

In order to obtain data about human head tolerance, the LPB-APR has conducted some experimentations with volunteer boxers. Five fights, i.e. fifteen rounds were carried out.

Such research was undertaken because they expose themselves, in their normal body activities to direct head impacts.

In an earlier publication, the methodology used for these experimentations was presented. The scope of this paper is to present the results obtained :
- the head accelerations,
- the head kinematics,
- the physiological effects.

The findings showed that the angular accelerations were in all cases higher than 3500 rd/s^2 exceeding the values considered as tolerance limit for volunteers given in the literature already available. The maximum angular velocity was 48 rd/s with a corresponding angular acceleration of 13600 rd/s^2.

In complement to physical measurements, tests of Event Related Potential Assessment of Attention and Orienting Reaction were performed by the CNRS-LENA in Hôpital de la Salpétrière, Paris.

The data recorded were analysed in statistical way for all boxers together. This sophisticated analysis did not indicate any anomaly in the results, except for one case where a small effect was observed, although the behaviour of the boxers remain in the normal range. This particular effect consists on an impairment of the ability to detect and react towards stimuli delivered in the right ear.

OBJECTIVE AND SCOPE OF THE WHOLE PROJECT

This study aims to find out :
- the severity of the blows sustained during a boxing fight, in terms of head accelerations,
- the corresponding head kinematics obtained with the video system,
- the corresponding physiological effects, in medical terms, which may appear on volunteers.

To the degree these results would be achieved it would thus be possible to establish a better understanding of the injury mechanism associated with direct head impact.

ACQUISITION OF BIOMECHANICAL AND PHYSICAL DATA

METHODOLOGY - We present here some recall about the methodology which was previously published (1).

In order to calculate linear and angular accelerations of the head on the basis of peripheral acceleration measurements, a particular head gear fitted to each selected volunteer was developed.

This helmet supports four transducers whose relative positioning must be known with accuracy. It must be snugly fitted to the boxer's head.

Each fight is recorded on video at 500 frames/s.

In addition cameras operating at 40 or 80 frames/second are located around the ring.

COMPUTATION OF HEAD ANGULAR ACCELERATIONS AND HEAD KINEMATICS - Three methods can be used in order to calculate the kinematics of the head (assumed as undeformable solid) from measurements of accelerations at different points. They are :

- the 3-3-3 method ;
- the N * 1 method ;
- the WSU method.

The 3-3-3 method, as developed by N. Alem (2), uses 3 tri-axial transducers. The excessive number of measurement channels (9 instead of the 6 necessary and theoretically sufficient) allows, by means of a least square method, a reduction of the influence of unavoidable measurement errors.

The N * 1 method (3), which is in fact a generalization of the preceding method , was developed in the A.P.R. Laboratory in order to allow the cases with one or several missing channels to be processed. This method requires from six to nine acceleration channels.

The Wayne State University method (4), uses 9 sensitive axes of accelerometers so that the terms which contain components of the angular velocity can be eliminated from the system of equations. Angular accelerations are then directly obtained from the measurements, and angular velocity becomes the integral of angular acceleration.

Advantage of the method : a sophisticated routine of numerical integration of differential equations is not needed, but above all, the sensitivity of the results to measurement errors are considerably reduced, since errors do not accumulate exponentially between angular velocity and acceleration.

ORGANIZATION OF THE FIGHTS - Each fight was organized between one boxer with the head instrumented and one boxer with the hands instrumented. Each fight consisted of three rounds lasting three minutes. The principal characteristics of the boxers are given in table 1.

SELECTION OF BLOWS : REQUIREMENTS - The selection of the blows is made with respect to the accelerometers by reading rough data provided by each triaxial. In order to retain only interesting blows, in terms of violence, we decided to consider the blows where at least one acceleration channel had a magnitude greater or equal to 70 g. When the acceleration level was very high (more than 300 g for instance) we checked on the video tape the exact localization of the impact, to detect possible blows on the helmet.

VIDEO ANALYSIS - We analysed the different views of the films, especially on the video SPIN PHYSICS system, set-up views at 500 f/s in relation with the time base, in order to reject blows which could have occurred on the helmet on one hand, and on the other hand to characterize the blows in terms of the direction of the blow (right or left), the movement described by the head, the localization on the head (temporal...).

Tables 2 to 6 give a description of each blow for each fight.The video analysis allowed 45 validated blows to be classified as follows :

- 31 blows on the LEFT SIDE associated with a right flexion and (or) a torsion movement.
- 6 blows on the RIGHT SIDE associated with a left flexion and (or) a torsion movement.
- 8 other blows associated with antero-posterior flexion.

Table 1. Physical characteristics of the boxers

Boxer with instrumented head			
NB	Age (Years)	Mass (kg)	Size (m)
1	27	83	1.83
2	24	96	1.80
3	27	80	1.90
4	27	78	1.92
5	26	76	1.80
Other boxer			
NB	Age (years)	Mass (kg)	Size (m)
1	19	86	1.95
2	27	80	1.90
3	27	78	1.92
4	26	76	1.80
5	27	78	1.92

Figure 1 shows the different kinds of head rotations.

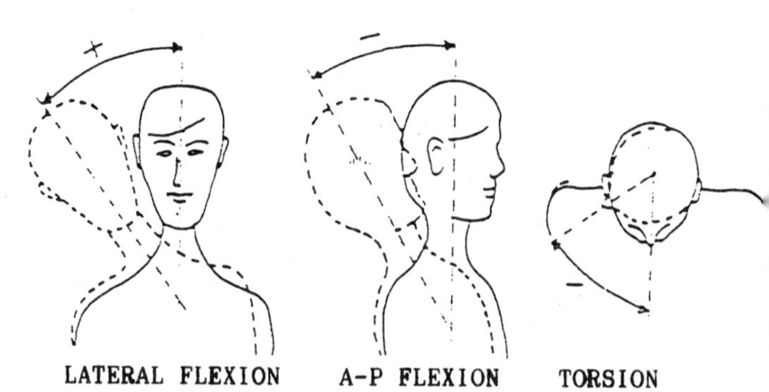

Figure 1. Definition of head rotations with corresponding signs.

Table 2. FIRST FIGHT : VIDEO ANALYSIS

Blow Reference number	Round	Type of blow	Head area involved	Movement of the head
B 51	1	Right Hook	Left Checkbone	Right torsion (-)
B 52	1	Right hook	Left temple	Right torsion (-) and small right lateral flexion (+)
B 53	1	Straight right	Left checkbone	Right torsion (-)
B 54	1	Right Upper	Chin	A-P Flexion
B 55	1	Right Hook	Left temple	Right torsion (-) and small right lateral flexion (+)
B 56	2	Right Hook	Left fronto-temporal	A-P Flexion (-) and right lateral flexion (+)
B 58	2	Right Hook	Left temporal	Left torsion (+) and right lateral flexion (+)
B 59	2	Right Hook	Right fronto-temporal	A-P Flexion (-) and left flexion (+)
B 510	3	Right Hook	Left parietal	Left torsion
B 511	3	Right Hook	Left ear	Right lateral flexion (+)

Table 3. SECOND FIGHT : VIDEO ANALYSIS

Blow Reference number	Round	Type of blow	Head area involved	Movement of the head
601	1	No possibility of analysis	No possibility of analysis	
602	1	Left Hook	Left ear	Torsion (+)
603	1	Right Hook	Left ear	Lateral flexion (+) and small torsion (-)
604	1	Right Hook	Left ear	Lateral flexion
606	2	Right Hook	Left part of the occipital	Small torsion (+)
610	3	Right Uppercut	Chin	A-P Flexion (-) and torsion (+)
611	3	Right uppercut	Left ear	Small lateral flexion (+)

Table 4. THIRD FIGHT : VIDEO ANALYSIS

Blow Reference Number	Round	Type of blow	Head area involved	Observations-Movement of the head
702	1	Straight left	Right temporal	Lateral flexion (-)
703	1	Right Hook	Left temporal	Lateral flexion (+)
704	1	Straight left	Left temporal	Torsion (-) and small A-P flexion (-)
705	1	Straight left	Left temporal	A-P flexion (-)
706	1	Straight left	Temporal	A-P flexion (-)
707	1	Right Hook	Left temporal	Lateral flexion (+)
708	1	Right Hook	Left checkbone	Torsion (-)
709	1	Straight left	Frontal	A-P flexion(-) and small torsion(+)
710	2	Straight left	Left part of the frontal	A-P flexion (-) and lateral flexion(+)
712	2	Left Hook	Right part of the bottom jaw	Torsion (+) and lateral flexion (-)
713	2	Right Hook	Left temporal	Lateral flexion (+)
719	3	Right Hook	Left ear	Lateral flexion (+)
722	3	Right Hook	Left parietal	Small torsion (-), small lateral flexion(+) and small A-P flexion(+)

Table 5. FOURTH FIGHT : VIDEO ANALYSIS

Blow Reference Number	Round	Type of blow	Head area involved	Movement of the head
801	1	Left Hook	Right ear	Lateral flexion (-), torsion (-)
803	1	Right Hook	Left parietal	Torsion (+), lateral flexion (-)
805	2	Right Hook	Left ear	Lateral flexion (+)
807	3	Left Hook	Right ear	Torsion (-), lateral flexion (-), and A-P flexion (+)
808	3	Right Hook	Left ear	Lateral flexion (+)
812	3	Right Hook	Left parietal	Lateral flexion (+)

Table 6. FIFTH FIGHT : VIDEO ANALYSIS

Blow Reference Number	Round	Type of blow	Head area involved	Observations - Movement of the head
903	1	Right Hook	Left ear	Small torsion (-)
904	1	Straight left	Frontal	A-P flexion (-)
907	1	Right Hook	Left temporal	Lateral flexion (+), and torsion (-)
908	1	Right Hook	Left part of the occipital	A-P flexion (+) and lateral flexion (+)
910	2	Right Hook	Left parietal	Lateral flexion (+)
915	3	Left Hook	Occipital (F)	A-P flexion (+)
917	3	Left Hook	Left part of the occipital	A-P flexion (+) and torsion (+)
918	3	Left Hook	Left parietal	Lateral flexion (+)
919	3	Right Hook	Occipital (F)	A-P flexion (+) and torsion (+)

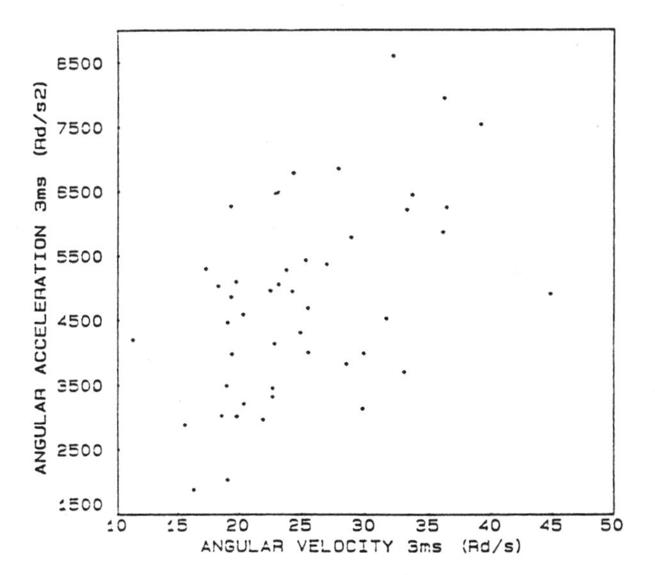

Figure 2. 3ms Angular acceleration versus 3ms velocity for all the blows.

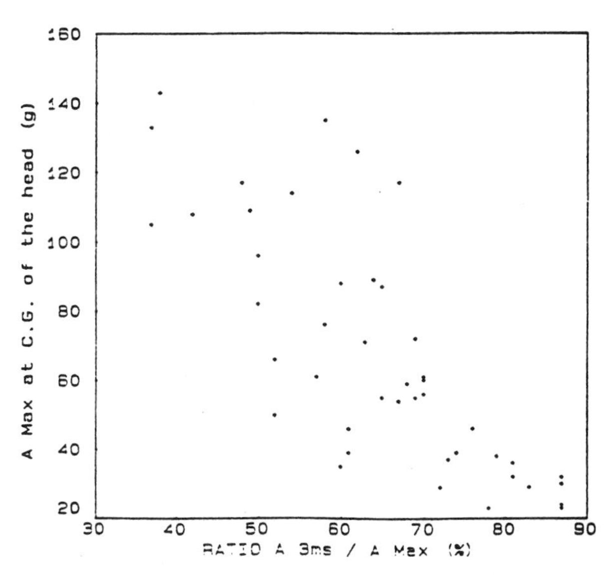

Figure 3. Maximum linear acceleration AR versus A3ms/Amax ratio

RESULTS OF DATA PROCESSING - From the measurement data, the resultant linear accelerations, HIC values, angular accelerations and velocities were determined at the center of gravity of the head. The corresponding values are given in tables 7 through 11. The HIC values range from 5 to 348 and the maximum linear accelerations range from 20 g (B57) to 159 g (705), the 3 ms linear acceleration range from 18 g (602) to 79 g (908).

The maximum peak values of the angular acceleration as well as angular velocity are of a high level 16234 rd/s² (722) and 48 rd/s (702) respectively. These levels can be explained by the type of kinematics observed, which consists of larger angular displacements in comparison with the linear ones. Almost all the blows correspond to angular accelerations higher than 3500 rd/s² as can be seen on figure 2 which gives the 3 ms angular accelerations in relation to the angular velocities. The preceeding tables also display the ratios a3ms/amax in percent, which give an indication of the duration of blow. For instance, ratios of an order of magnitude of 30 % correspond to very brief duration of blows when a ratio of 80 % will correspond to blows of higher duration. This is illustrated on figure 3. The most brief blows produce highest peak linear accelerations whereas less brief blows produce a lesser head acceleration.

From the preceding tables corresponding to the whole sample of blows analysed we have chosen some cases which seemed to be more interesting in terms of angular acceleration especially (see table 12). We have here some examples of important angular accelerations from 7300 rd/s² through 16234 rd/s².

We have chosen one case 722 for instance, for which the following curves are given :

- the components of the linear acceleration at the centre of gravity of the head and the resultant acceleration at the centre of gravity (figure 4),
- the angular acceleration and angular velocity of the head in the laboratory coordinate system (figures 5 and 6),
- the head kinematics in the three planes (figures 7,8 and 9).

This case corresponds to the blow which had the most important head angular acceleration.

EVENT RELATED POTENTIAL ASSESSMENT OF ATTENTION AND THE ORIENTING REACTION IN BOXERS BEFORE AND AFTER A FIGHT

Due to their high resolution in the time domain, Event-Related Potentials of the brain (ERPs) are, at present, one of the best non invasive measurement of functional brain activity during cognitive behaviour (for a recent overview see (5)). Although their physiological basis are not fully explained, ERPs do provide useful assessment of the orienting reaction and the capacity of selective attention which are known to be fundamental, both in animals and in humans, for protection against environmental dangers. This ERP methodology has been therefore used in order to assess boxer's attention mechanisms and their capacity of orienting towards significant stimuli.

Table 7. FIRST FIGHT : COMPUTATION RESULTS

BLOW NB	Resultant Acceleration (G)			Resultant angular acceleration (rd/s^2)		Resultant angular velocity (rd/s)			
	a_{max}	$a_{3\ ms}$	$\frac{a_{3\ ms}}{a_{max}}$ (%)	$\dot{\omega}$ max	$\dot{\omega}$ 3 ms	ω max	ω 3ms	HIC	Δt HIC (ms)
B 51	30	26	87	5784	3709	33.27	33.12	22	8.6
B 52	29	24	83	4557	3020	20.01	19.84	17	8.1
B 53	37	27	73	4484	3145	30.14	29.87	22	5.0
B 54	59	40	68	11108	6850	28.38	27.96	64	7.0
B 55	35	21	60	6304	4592	20.49	20.36	16	15.5
B 56	55	38	69	8655	6219	35.24	33.35	67	11.5
B 58	66	34	52	9334	5875	36.99	36.19	59	6.2
B 59	71	45	63	10572	7545	40.78	39.23	94	6.6
B 510	23	20	87	6641	3995	31.85	29.96	15	19.5
B 511	46	28	61	7154	4689	25.65	25.57	26	7.6

Table 8. SECOND FIGHT : COMPUTATION RESULTS

BLOW NB	Resultant Acceleration (G)			Resultant angular acceleration (rd/s)		Resultant angular velocity (rd/s)			
	a_{max}	$a_{3\ ms}$	$\frac{a_{3\ ms}}{a_{max}}$ (%)	$\dot{\omega}$ max	$\dot{\omega}$ 3 ms	ω max	ω 3 ms	HIC	Δt HIC
601	56	39	70	6545	4863	20.39	19.39	61	11.50
602	23	18	78	4523	2887	16.55	15.63	8.6	6.12
603	133	49	37	9593	6267	24.44	19.36	241	2.51
604	61	35	57	4956	3985	19.56	19.45	58	10.13
606	54	36	67	7690	4533	34.10	31.75	55	12.75
610	61	43	70	3989	3831	29.18	28.59	80	6.62
611	82	41	50	3492	3214	20.57	20.43	98	3.25

Table 9. THIRD FIGHT : COMPUTATION RESULTS

BLOW NB	Resultant acceleration (G)			Resultant angular acceleration (rd/s²)		Resultant angular velocity (rd/s)			
	a_{max}	$a_{3\ ms}$	$\frac{a_{3ms}}{a_{max}}$ (%)	$\dot{\omega}$ max	$\dot{\omega}$ 3 ms	ω max	ω 3 ms	HIC	Δt_{HIC}(ms)
702	117	56	48	13648	4916	48.00	44.91	192	10.7
703	114	61	54	7354	5285	24.30	23.83	252	4.5
704	50	26	52	7581	1885	16.45	16.36	21	9.5
705	159	46	30	11956	4140	23.75	22.88	105	1.4
706	32	28	87	4393	3455	23.12	22.72	30	17.4
707	36	29	81	4529	3030	20.70	18.66	32	16.2
708	143	54	38	11551	5302	21.33	17.33	203	4.9
709	24	21	87	4532	3321	22.99	22.73	15	24.6
710	46	35	76	5169	4311	27.39	24.97	39	9.2
712	39	29	74	3918	2036	19.22	19.13	29	8.25
713	32	26	81	5265	4949	25.01	24.31	20	30
719	60	42	70	6626	5374	28.23	27.05	85	11
.722	89	57	64	16234	4961	25.08	22.55	173	14.5

Table 10. FOURTH FIGHT : COMPUTATION FIGHT

BLOW NB	Resultant acceleration (G)			Resultant angular acceleration (rd/s)		Resultant angular velocity (rd/s)			
	a_{max}	$a_{3\ ms}$	$\frac{a\,3\ ms}{a\ max}$ (%)	$\dot{\omega}$ max	$\dot{\omega}$ 3 ms	ω max	ω 3 ms	HIC	Δ t HIC (ms)
801	88	53	60	6426	5037	18.86	18.33	138	6.51
803	38	30	79	6125	5064	25.27	23.20	41	18.37
805	108	45	42	5565	4005	26.32	25.59	155	2.62
807	109	54	49	9435	8601	35.50	32.28	170	5.25
808	29	21	72	4620	3491	20.37	19.03	22	18.00
812	87	57	65	6184	4200	11.40	11.36	145	6.87

Table 11. FIFTH FIGHT : COMPUTATION RESULTS

BLOW NB	Resultant Acceleration (G)			Resultant angular acceleration (rd/s)		Resultant angular velocity (rd/s)			
	a_{max}	$a_{3\ ms}$	$\frac{a_{3\ ms}}{a_{max}}$ (%)	$\dot{\omega}$ max	$\dot{\omega}$3 ms	ω max	ω3 ms	HIC	Δt (HIC) (ms)
903	135	78	58	10550	6781	29.20	24.39	348	6.75
904	72	50	69	4860	2971	23.88	21.97	135	11.74
907	126	78	62	7008	4467	19.32	19.11	201	8.51
908	117	79	67	11160	7951	38.99	36.28	342	14.63
910	39	24	61	5504	5101	21.97	19.78	27	16.37
915	105	39	37	9817	6257	38.31	36.46	122	16.25
917	96	48	50	9479	6447	35.21	33.75	123	10.75
918	55	36	65	6075	5434	26.69	25.41	44	13.38
919	76	44	58	9933	5791	30.92	28.96	98	13.00

Table 12. Main results for 10 blows which seemed to be the more interesting as regards the violence (* the maximum in each case)

	HIC	W max rd/s²	w max rd/s	A max (g)
B54	64	11108	28	59
B59	94	10572	41	71
603	241	9593	24	133
702	192	13648	48*	117
703	252	7354	24	114
705	105	11956	24	159
708	203	11551	21	143
722	173	16234*	25	89
903	348*	10550	29	135
908	342	11160	39	117

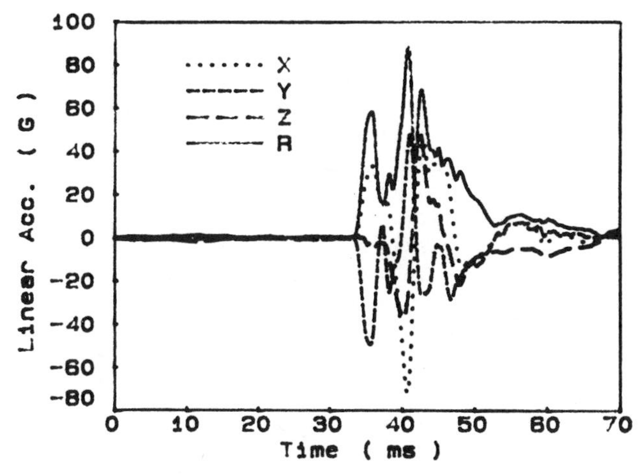

Figure 4. Components and resultant (R) linear acceleration at the centre of gravity of the head (blow 722)

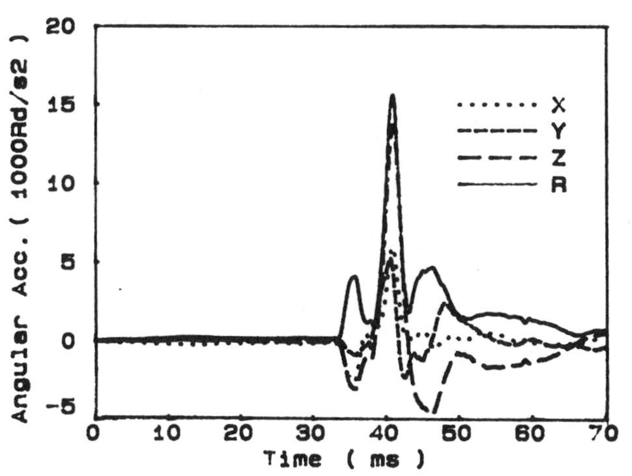

Figure 5. Components and resultant angular acceleration of the head (blow 722)

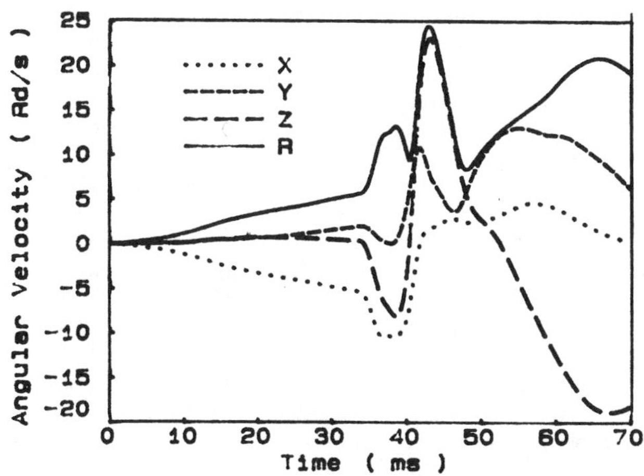

Figure 6. Component and resultant angular velocity of the head (blow 722)

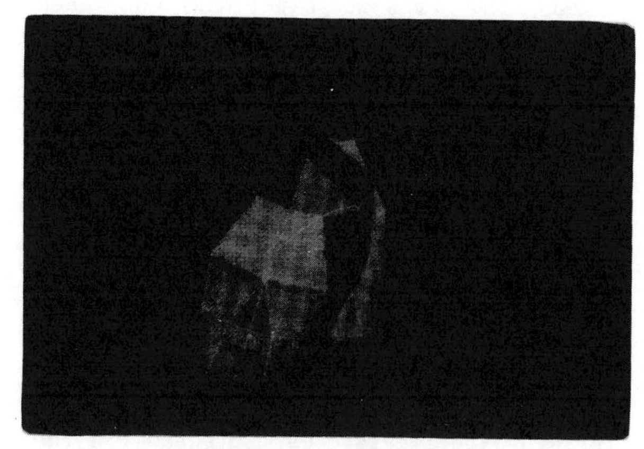

Figure 7. Plane ZOY

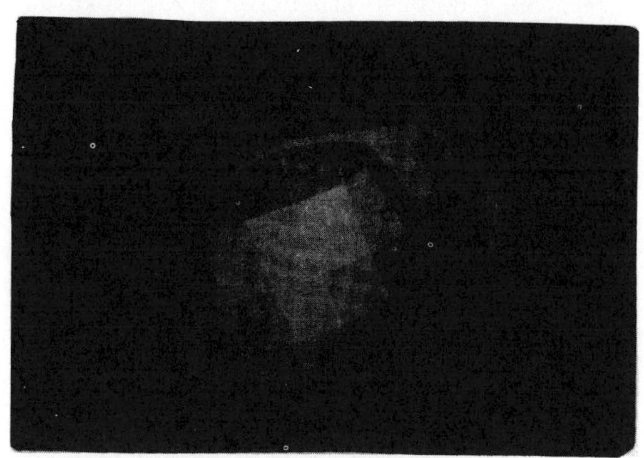

Figure 8. Plane ZOX

Figure 9. Plane YOX

Figures 7-8-9 :

Visualisation of head kinematics during blow as obtained from accelerometric data processing in the three different planes as indicated under figures :

- plane ZOY
- plane ZOX
- plane YOX

In the present experiment we used location-selection conditions in a dichotic listening task. Both attention mechanisms and the orienting reaction were studied before and after the fight by measuring :

1) the processing negativity
2) the mismatch negativity
3) the N2b component

The reaction time and percentage of errors to deviant tones were also measured in order to quantify the level of performance.

METHOD - Five boxers never involved in similar experiments, participated in the study. Their EEG was recorded twice, in the morning and at the end of the afternoon, before and after a fight which took place elsewhere. The time interval between the beginning of each EEG recording was about eight hours. Each recording lasted two hours and the second recording was done about one hour and a half after the fight.

EXPERIMENTAL DESIGN - Standard, or "frequent" (80 %), low pitch tones (1000 Hz) and deviant, or "rare" (20 %), high pitch tones (2000 Hz) were presented randomly, through earphones, to each ear. The interstimulus intervals were at random, between 400 and 800 msec. Each tone lasted 10msec ; its intensity was adjusted for each subject, at the beginning of each experiment, in order to obtain the same subjective intensity for both ears. Two selective attention situations (attend right and attend left) were recorded, during which subjects were required to repond as quickly as possible by a fore-finger displacement of their preferential hand over a photoelectric cell, as soon as they detected a target (high pitch tone) in the attented ear. Subjects were told to whithold their response for all other stimuli : standards (low pitch tone) in both ears and deviants in the non-attended ear. A divided attention condition was also recorded where the subject task was to detect deviants in both ears. Each of these three attention situations comprised 500 stimuli that were presented in blocks of 250 stimuli lasting about 2.5 minutes. The order of these attention situations was distributed across subjects according to a latin-square design. These situations were preceded by a passive situation (same stimuli, no task) during which the subjects were told to pay no attention at all to the sounds. For all situations, subjects were asked to keep their eyes closed and avoid blinking and eye movements.

ELECTROPHYSIOLOGICAL RECORDING AND DATA PROCESSING - The EEG was recorded from 9 equally-spaced electrodes, affixed with collodion along a line going from the right temporal (T4) to the left temporal (T3) region. Electrodes were referenced to linked ears : horizontal and vertical electro-occulograms were recorded simultaneously in order to control eye movements. The data base of this study consisted of 1440 different average ERPs (two types of stimulus x two ears x four situation x before and after the fight x five subjects x 9 electrodes).

For each condition and for each subject, a spatio-temporal map (6) was calculated. On these maps, amplitude variations between two successive electrodes are obtained by a second order interpolation and are represented in the form of isopotential lines as a function of time on the abscissa and of electrode location on the ordinate. The following subtractions of ERP spatio-temporal maps were computed for both standard attention and deviant stimuli, before and after the fight : attention minus passive, attention minus inattention, divided attention minus passive. These 12 subtractions yielded processing negativities. Furthermore, the mismatch negativity (MMN) and the N2b were evidenced, before and after the fight, by subtracting standards from deviants in the passive condition (MMN alone) and in inattention, divided attention and selective attention conditions (MMN plus N2b).

RESULTS -

<u>The data base</u> - Figures 10 and 11 depict spacio-temporal mapping of event-related potentials for standards and for deviants, before and after the fight, in the four experimental conditions. These figures are grand means, across subjects, of individual data. The negativve N1 wave, on going from 40 to 150 msec, was noticeable on all these maps ; the P2 wave positive between 150 and 250 msec, was only seen in response to standard tones. In the case of deviants, this wave disappeared (except in the passive condition) and was replaced by a negative N2 wave that developped between 150 and 500 msec post inattentive and passive conditions and that N2 was larger in selective and divided attention conditions.

Subtraction spatio-temporal maps depict the processing negativity - PN - (figure 12) and the orienting reaction (figure 13). These subtractions were used in order to define several time intervals

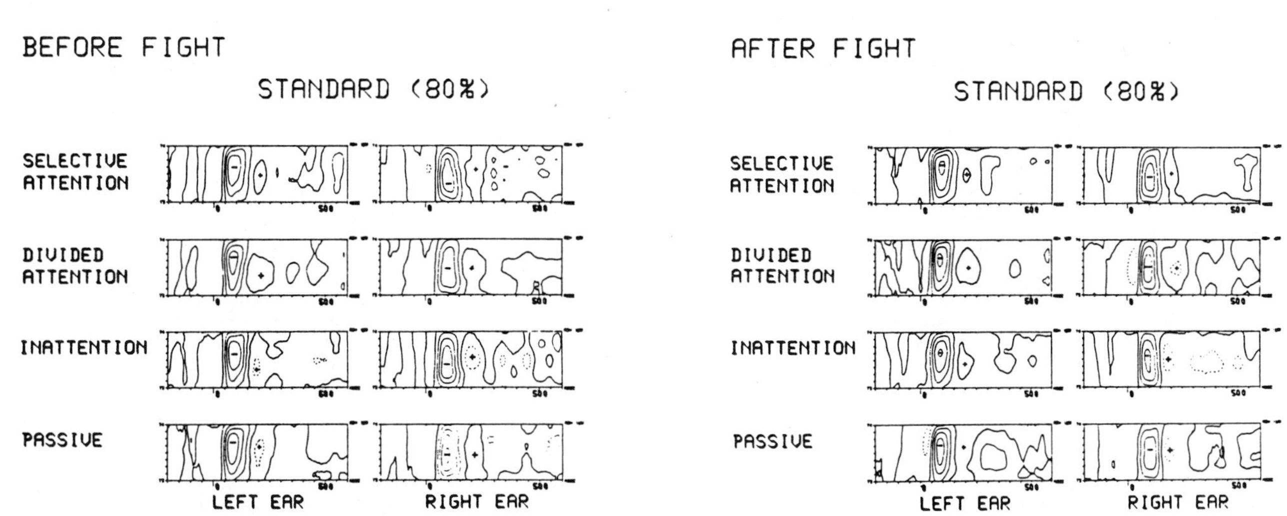

Figure 10. Spatio-temporal maps or ERPS obtained in response to standard tones, before and after the fight. Thin isopotential lines represent the negative potential, thick lines, zero potential, and dotted lines the positive potential. Between two isopotential lines the potential increases by 1uv. Peaks are indicated by sign + or -.

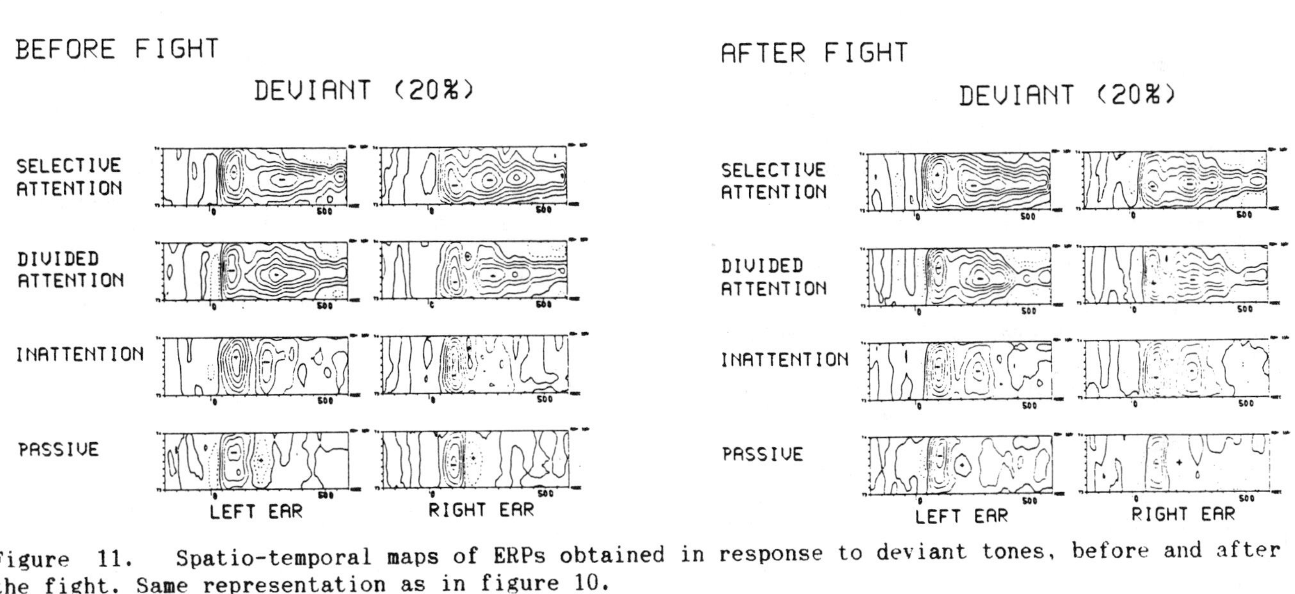

Figure 11. Spatio-temporal maps of ERPs obtained in response to deviant tones, before and after the fight. Same representation as in figure 10.

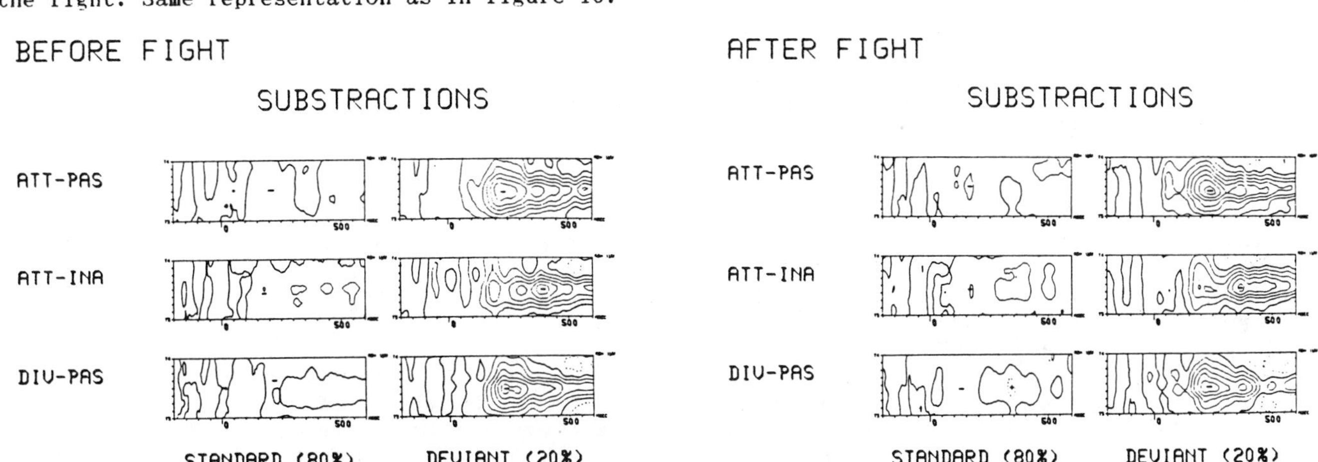

Figure 12. Spatio-temporal maps of the Processing Negativity obtained by substracting passive from attention (ATT-PAS), inattention from attention (ATT-INA) and passive from divided (DIV-PAS). Means of right and left stimuli before and after the fight.

BEFORE FIGHT

AFTER FIGHT

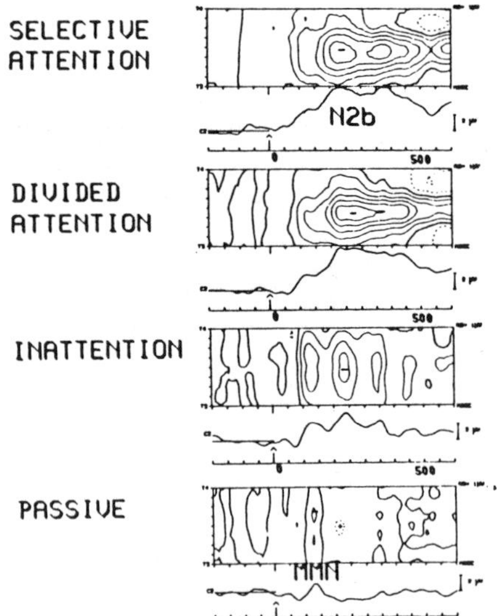

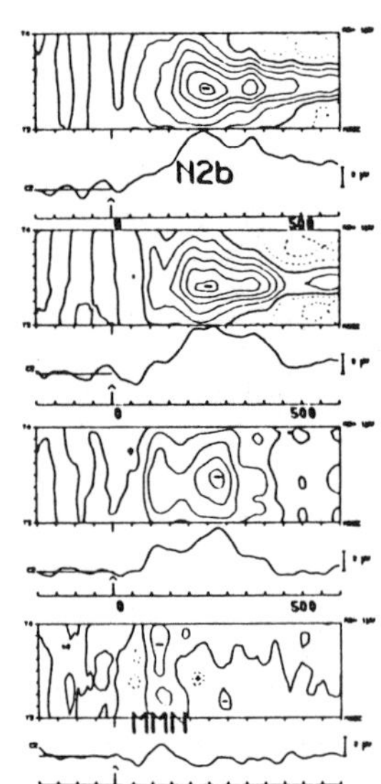

Figure 13. Spatio-temporal maps of MNN and N2b obtained in substracting standards from deviants. Means of right an left stimuli before and after the fight.

in which the amplitude of the resulting waves were measured.

These measurements were made thanks to a computer program taking into account the amplitude values at each electrode site. These measurements obtained for subject and in each condition were submitted to succesive analyses of each variance in order to test the significance level of the differences between conditions and between the before and after -fight situations.

Subsequent analyses for each attention condition confirmed the existence of a N2b. They also revealed an effect of the fight. This effect was a significant interaction between fight, tone pitch and right or left ear of delivery, but only in case of the attention condition ($F1,8 = 5.64$, $p<.05$).

Moreover, it should be pointed out that the differences between ERPs before and after the fight, for the attention condition, were due to a decrease of N2b when the stimuli were delivered to the right ear (Figure 14).

Discussion - Let us first emphasize that the Event-Related-Potential data obtained in this study are in general agreement with other ERP studies of atention and of the orienting reaction (for reviews see Naatanen, 1982 (7) ; Hillyard and Kutas, 1983 (8)). The fact that the processing negativity failed to reach the significance level for standard tones in the attention condition is probably due to the small number of subjects since the waveform is usually of small amplitude (1 or 2 uv) and thus difficult to demonstrate.

The results were not modified by the fight, except in one case : the N2b obtained in response to right deviant tones was of lower amplitude after the fight than before, for the right attention condition only. This result is unlikely to be explained by habituation or learning mechanisms ; being non-selective their effects should then have concerned right and left tones in every experimental condition.

Consequently, this result would rather suggest that the ability to detect right

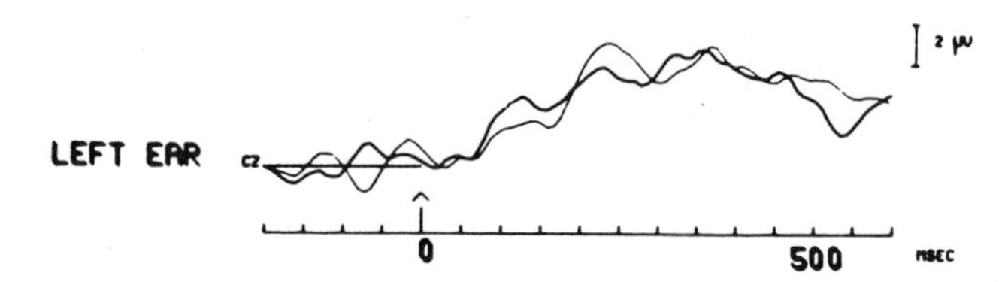

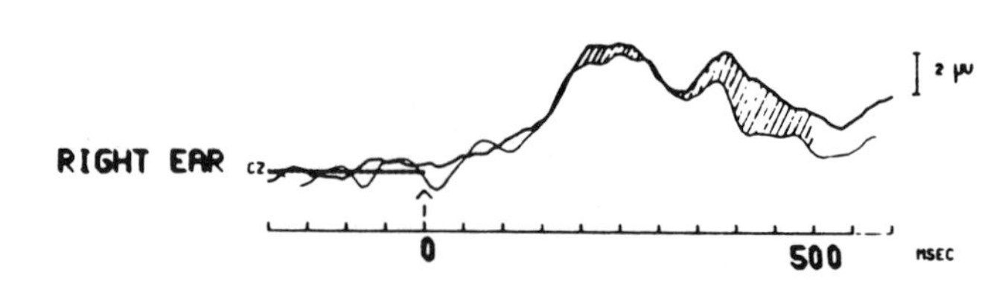

Figure 14. Subtractions of standards from deviants, for right and left stimuli; Cz recording. Thick lines represent MNN and N2b before the fight and thin lines refer to the same waves after the fight.

deviant stimuli and to react to them was partially impaired. Furthermore, right ear stimuli being first processed by the left hemisphere, a relative impairement of this hemisphere in the processing of deviant (rare) stimuli can be suggested.

CONCLUSIONS

In the framework of a research on head tolerance to impact, the Laboratory of Physiology and Biomechanics of Peugeot S.A./Renault Association has engaged a programme with volunteer boxers, in cooperation with I.R.O (France), under contract with the DOT.
Five fights, consisting of 3 rounds each, were performed between volunteer boxers equipped with accelerometric measurement devices. From these fights, 45 blows were selected as "interesting" (that means with a sufficiency preselected violence level) and processed as individual impacts with their own violence and kinematics.
Independently from the severity of blows, it was noted that 69 % were from the left side against 13 % from the right side: their intensities were on an average of the same order of magnitude.
As regards angular accelerations, they were in all cases higher than 3500 rad/s², exceeding the values considered as tolerance limit for volunteers, given in the literature already available. The maximum value obtained was 16000 rd/s², in association with an angular velocity of 25 rad/s.
The maximum angular velocity was 48 rad/s with a corresponding angular acceleration of a rather high level : 13600 rad/s². These values are reported in table 13 together with other data available, related to animals, cadavers, and volunteers.
These results represent sets of values, never seen on volunteers, without any problem : they widely exceed those already given as proposed tolerance limits by other authors.
All these data correspond to angular acceleration peak values, of very short duration. It could be more interesting to take into account the 3ms acceleration value. For instance, in blow 807, the maximum angular acceleration (peak value) was lower than for the cases reported in this table (9435) but the corresponding 3ms value was higher (8601) than all peak values given by the other authors as tolerance limit. There are several such cases, as can be seen in tables 7 through 11 ; this indicates that very high values of 3ms angular accelerations can be tolerated by volunteers without any

Table 13. Summary of proposed tolerance limits for different surrogates in terms of head angular acceleration and velocity.

REFERENCE	TYPE OF TESTS	LIMITS PROPOSED
OMMAYA 1967 (9)	Primate Tests	7500 rd/s²
OMMAYA 1971 (10)	Primate Tests	1800 rd/s² and 60-70 rd/s
LOWEHIELM 1975 (11)	Cadaver Tests Mathematical Model	4500 rd/s² and 50-70 rd/s
EWING 1975 (12)	Volunteers	1700 rd/s² and 32 rd/s
A.P.R. 1988 (13)	Volunteer Boxers	. 16000 rd/s² and 25 rd/s . 13600 rd/s² and 48 rd/s

trouble and so that the tolerance level could be more than this value (8600 rad/s²).
In complement to physical measurements, tests of Event Related Potential Assessement of Attention and Orienting Reaction were performed by the CNRS-LENA, in Hôpital de la Salpétrière, Paris.
The data recorded were analysed in a statistical way for all boxers together. This very sophisticated analysis did not indicate any anomaly in the results, only a small effect in one parameter as a whole, after the fight : the ability to detect and react towards stimuli delivered in the right ear was partially impaired, although the behaviours of the boxers remain in the normal range.
This modification of attention at the right ear could be explained by a perturbation in the left hemispher, which seems to have suffered more than the right one; as a matter of fact there were 31 blows on the left side of the head (i.e. 69 %) against 6 on the right side (i.e. 13 %), all the blows being of about the same intensity.
This particular effect is an illustration of the capability of the Event Related Potential of Attention and Orienting Reaction to reveal a possible very limited disturbance in the neurophysioligical state : false responses, in the right attention

condition, increased from 3,2 % (before) to 5,8 % (after the fight).

This limited effect has to be situated in the whole results, emphasizing that :

- no effect of the fight on the reaction time data,
- no effect of the fight on 32 pairs of mean amplitudes and standard deviations during the time window of the Processing Negativity,
- no effect of the fight on 32 pairs of mean amplitudes and standard deviations or ERPs in the latency window of the MMN (mismatch negativity).

The conclusions of this study are necessarily limited, given the small number of subjects and it would be of great interest in the future to continue along this line of research.

ACKNOWLEDGEMENTS

This study has been supported by the Departement of Transportation/National Highway Traffic Safety Administration. All opinions given in this paper are those of the authors and not necessarily those of DOT/NHTSA. We want to express our sincere thanks to the responsibles of the "Federation Francaise de Boxe", to Mr Guy Cuignard the arbitrator, and particularly to the boxers Mr Fabrice Dorard and Mr Gérard Jongbloët of the "Boxing Club Rodrigues", and their coach Mr Rodrigues; as well as Mr Salah Slimani, Mr Michel Boukorv, Mr Djamel Chebana and Mr Olivier Kemayou of the "Boxing Club de Rouen", and their coach Mr René Fromont. We want also to thank the Technical Staff of the "Laboratoire de Sécurité Automobile", Peugeot S.A., La Garenne, France.

REFERENCE

(1) F. Chamouard, X. Troseille, Y. Pincemaille, C. Tarrière, Laboratory of Physiology and Biomechanics Associated with Peugeot S.A./Renault,"Methodological Aspects of an Experimental Research on Cerebral Tolerance on the Basis of Boxers Training Fights", Proceedings of the 31th Stapp Car Crash Conference, 1987.

(2) N.M. Alem "The Measurement of 3-D Rigid Body Motion", in the proceedings of the 2nd Annual International Meeting of the Ad Hoc Committee on Human Subjects for Biomechanical Research, ed., Hirsch, A. E.

(3) Laboratory of Physiology and Biomechanics APR "Détermination de la cinématique de la tête d'un sujet humain au cours d'un choc, à partir de données accélérométriques" - Programme 9 GAMMA, unpublished.

(4) A.J. Padgaonkar,K.W. Krieger and A.I. King "Measurement of Angular Acceleration of a Rigid Body using Linear Accelerometers", ASME Journal of Applied Mechanics, Vol. 42, September 1975, PP. 552-556.

(5) Renault, B., Kutas, M., Coles M.G.H., and Gaillard, A.W.K. "Event Related Potential Investigation of Cognition" Biological Psychology, 1988, 26, 1-354.

(6) Rémond, A; (1961). Integrated and Topographical Analysis of the EEG. Electroencephalography and Clinical Neurophysiology, 20, 64-67.

(7) Naatamen, R. (1982). Processing Negativity : an Evoked Potential Reflexion of Selective Attention. Psychological Bulletin, 92 605-640.

(8) Hillyard, S.A. and Kutas, M. (1983). Electrophysiology of Cognitive Processing. Annual Review of Psychology, 34, 33-61.

(9) A.K. Ommaya, P. Yarnell, A.E. Hirsh and E.H. Harris, 1967 : "Scaling of Experimental Data on Cerebral Concussions in Sub-Human Primates to Concussion Threshold for Man", Proceedings of the 11th Stapp Car Crash Conference, October 10-11, 1967, SAE, N.Y. USA.

(10) A.K. Ommaya and A.E. Hirsch, 1971, "Tolerances for Cerebral Concussion from Head Impact and Whiplash in Primates", Journal of Biomechanics 4, 13, 1971.

(11) P. Lowenhielm, 1975, "Brain Susceptibility to Velocity Changes, Relative and Absolute Limits for Brain Tissue Tolerance to Trauma and their Relation to Actual Traumatic Situations", Proceedings of an International Interdisciplinary Symposium on traffic Speed and Casualities held at G1 Avernaess, FUNEN, April 22-24, 1975, Denmark.

(12) C.L. Ewing, D.J. Thomas, L. Lustick, E. Becker, G. Becker, G. Willems, and W.H. Muzzy, "The effect of the initial position of the head and neck to - Gx Impact Acceleration", Proceedings of the 19th Stapp Car Crash Conference, November 17-19, 1975, SAE, WARRENDALE, U.S.A.

(13) Contract NHTSA : DTRS-57-86-C-00037, "Investigation of Relationship between Physical Parameters and Neuro-Physiological Response to Head Impact. Final Report, June 1988.

Section 3:
Head-Neck Impact Tolerance and Injury Criteria

660793

Use of a Weighted-Impulse Criterion for Estimating Injury Hazard

Charles W. Gadd
General Motors Corp.

ABSTRACT

This paper describes the usage of an exponential weighting factor for appraising deceleration or force impulses registered on dummies or impacting hammers in safety testing. The proposed impulse-integration procedure, it is shown, takes into account in a more rational way, and in better conformity with published injury tolerance data, the relative importance of time and intensity of the pulse than do the "peak g" or impulse-area criteria. Use of the new Severity Index for assessment of head impact pulses is illustrated. It is shown to be of special value in comparing the relative severity of pulses which differ markedly in shape (because of structural differences in the component being struck) and it is pointed out that without a weighting factor of this nature, laboratory impact tests can yield incorrect ranking of the relative safety merit of alternative designs. Automated methods for quick calculation of the Severity Index are possible.

IN RECENT YEARS there has been an increasing need for more versatile measures or indices with which to judge the degree of injury hazard likely to be associated with impulses which are applied in the laboratory testing of automotive interior structures and components, or with which to draw comparisons between alternate designs proposed for reducing injury. In view of the wide range of locations and angles of impact which can be experienced by the head or other parts of the body in accidents, and recognizing the diverse mechanisms by which injury can occur within the body, it would be unrealistic to assume that we will ever have a single and rigorously quantitative rating system for the hazard inherent in a given pulse applied to a given part of the human body. Simple measures or at least "yardsticks" are nevertheless very much needed and have therefore come into use. It is believed that these can be further improved to yield better approximation of injury hazard than has been obtained in the past and at the same time yield more repeatable and comparable results. The object of this paper is to describe one approach toward this end.

Various terms from the field of mechanics have come into use over the years to characterize the intensity of a blow or typify the manner in which the impulse must be altered to reduce its injury potential. Of these, the term "energy absorption" has been one of the most popular. This concept has, however, been difficult to apply in a specific way because it means different things to different people, and there is not generally any simple relation between energy involved and injury hazard. Variables usually present such as the crush characteristics of the striking and struck objects prevent good correlation of either the kinetic energy of the striking object, or the energy absorbed by the struck object, with injury hazard.

SELECTION OF THE TRANSDUCER

To arrive at a logical index of injury hazard represented by laboratory or field test results, one must first face the question of what type of transducer measurement to make. Under impact, the body may be exposed to acceleration, force, pressure, stress or strain; depending upon the nature of the problem and type of injury, it may be most practicable to select one or another of these parameters for measurement. Here the best choice for the original tolerance measurement on the biological material is of course to select a transducer whose output is believed to be closely related to the mechanism of injury and locate it as close as possible to the actual injury site. It is then preferable to use the same type of measurement in the impact testing of the design which is under study.

In head injury, the problem to which special attention will be given in this paper, it has been impractical except in very limited instances to obtain transducer readings (e.g., pressure, force, or stress) which are directly associated with the injury, and as a result the overall head acceleration, a rather indirect measure, has come into wide use. While acceleration admittedly does not consistently represent the diversity of kinematics and injury mechanisms actually involved, it does provide probably the best currently available basis for judging head impact severity from an internal injury standpoint. The most reliable information which has been obtained is that for impact of the front of the head, and in particular of the forehead.

INTERPRETATION OF PULSE WAVE-SHAPE

Once a pulse depicting a blow has been obtained on the oscillograph, the next question arising is how its severity should be assessed from the standpoint of wave-form or profile. Various investigators have emphasized differing aspects of the wave. The maximum value reached, or "peak g" if it is an acceleration pulse, is the most widely used rule-of-thumb measure of injury hazard inherent in the pulse because it is the simplest, even though it has been pointed out by various people[1] that from the mechanics standpoint a single point on an applied pulse cannot accurately define the response of a physical structure to that pulse. Area under the g-time pulse has also been suggested as a simple way of at least recognizing that injury hazard generally increases with increasing time of exposure to a loading upon the body.

Rate-of-change of acceleration is still another aspect of pulse wave shape which has been suggested as a critical factor in injury. In discussing this subject it is advisable to differentiate first between the input function and the response function; if the transducer monitoring input excitation must (of necessity) be placed on the body at a point apart from the actual site of injury, and there is a mass-elastic system intervening, then both rate-of-change of acceleration and impulse area can under certain conditions become useful as rough indices of how the dynamic characteristics of the intervening system alter the stress intensity at the injury site, and are thus useful in this sense although not as indices of overall injury to be produced by the input function. It is the belief of the author that in such cases the dynamic response should be treated as a prior and separate problem, just as is done in applying shock and vibration theory to determine how the dynamic response of a mechanical structure aggravates the stress at a critical point, before one applies material strength theory to estimate the damage potential of that stress. In head impact studies the transducer cannot usually be placed at the site of injury, nor is the true mass-elastic system usually well simulated in most cases; fortunately, however, the head appears to be designed sufficiently free of resonant effects to enable useful impact evaluations to be made by observing only the input function.

This paper discusses a method of assessing the pulse wave-form in its entirety, in a manner which is in contrast to methods which consider only one point or aspect of the wave. It begins with the premise that injury is some function of both intensity of the loading and its time duration. Assuming then that the investigator has selected and placed his transducer, whether it measures acceleration, force, pressure, so as to be best representative of distress at the injury site, he can then integrate the pulse obtained in such a way as to take into consideration both intensity and time, employing a mathematical weighting factor which best fits the available range of biomechanics data pertinent to injury at the point in question.

EXPONENTIAL WEIGHTING OF THE PULSE

One of the simplest weighting factors which might be selected for trial is one which weights exponentially the intensity scale. This in effect takes into account time dependency of damage as follows. First, one can visualize a hypothetical completely brittle material which fails suddenly if the loading exceeds a certain level. At the other extreme would be a completely viscous material for which percentage increments of load intensity would be just as damaging as corresponding increments of time duration of loading, with failure defined as some excessive degree of shear strain.

Examination of the biomechanics literature indicates that animal tissues fall somewhere between these two extremes in their failure properties and, furthermore, that the use of either of these extreme criteria will lead at times to false ranking of the relative injury hazard between alternative designs.

To the knowledge of the author the first systematic study of the role of load duration in animal impact injury was that at Wayne State University.[2,3] This showed for cranial pressure pulses of similar shape but differing time duration, a trend as shown by the scatter band of Figure 1. As time of exposure to pressure increased, the tolerable intensity decreased. A similar trend is exhibited by the work of other subsequent investigators who have obtained or assembled tolerance data over a range of pulse time durations; for example, in Figure 2 are shown the trends as portrayed by Eiband of NASA for various impact sled tests.[4] It should be pointed out that such curves are very difficult to develop in the face of the many variables involved, and it is doubtful that enough data will ever be obtained over a range of time durations while holding to a particular wave shape (e.g. square, triangular, or trapezoidal) to arrive at a precise mathematical definition of time dependency for blows to particular parts of the body. The data do exhibit the one feature in common, that of downward sloping tolerance curve over the time duration range of vehicle occupant impact. Therefore, it can be concluded that some limiting "g", force or pressure should not serve as the most appropriate criterion for the threshold of injury. A second feature of the threshold curves is that over this time duration range their slope is considerably less than 45 degrees when plotted on a log-log scale, and therefore a pulse-area criterion is also not the best approximation. It follows that some inter-

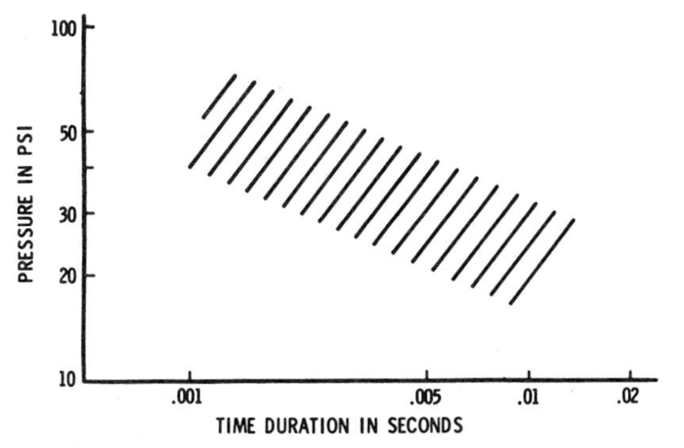

Fig. 1 - Trend of intracranial pressures required for severe concussion in dogs, versus time. From work of Wayne State University

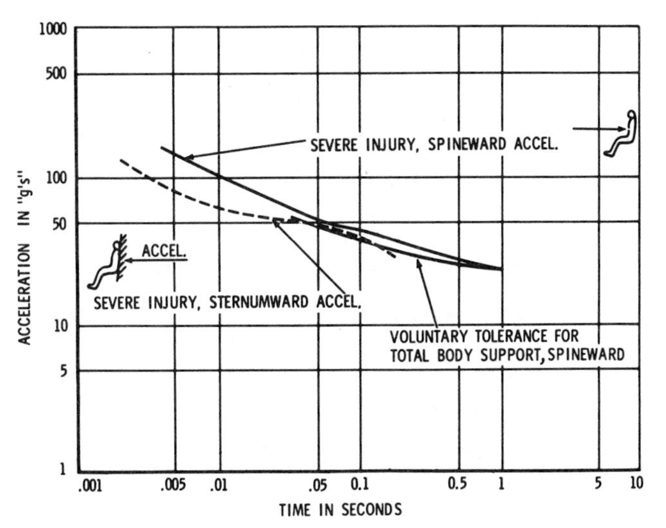

Fig. 2 - Trends of tolerance to square and trapezoidal acceleration from Reference 4

mediate threshold curve is most appropriate. Further, in view of the scatter in the data, it is suggested that a straight line approximation on a log-log plot is sufficient at this time for head injury likely to result from front-to-rear head acceleration over a range of between approximately one and fifty milliseconds. This brackets the pulse time duration range encountered by vehicle interior head impact.

Mathematically the inverse of the slope of such a straight-line threshold corresponds numerically with a simple exponential weighting factor, from which it follows that injury threshold can be defined as a single number,

$$I = \int a^n dt \quad (1)$$

where: a = acceleration, force, or pressure of the response function producing, threshold of injury of a given degree.
n = weighting factor greater than 1.
t = time in seconds.

Integration of this expression yields a severity index which is applicable to a particular class of injury and whose numerical value varies depending upon whether it is developed in terms of acceleration, force, or other indication of loading intensity.

The exponential weighting may be thought of as recognizing that the lower portions of the pulse contribute very little to the injury, but that the more intensive portions contribute to a disproportionately great degree.

The number obtained can be used in two ways: either (1) for comparing different tests for relative severity of impact, or (2) for estimating whether an impact exceeds a safe maximum value.

APPLICATION OF THE SEVERITY INDEX

To use the Index to estimate injury hazard of a given type, two judgments must be made from the available biomechanics data as follows:

A. The appropriate weighting exponent must be selected. If one is interested in only relative ranking of designs, then this one constant is sufficient. For internal injury to the head from frontal blows we have been using an exponent of 2.5, based primarily on the slope of the Wayne animal impact data representing dangerous concussion. The slope selected by Eiband in Reference 4 for spineward (front to rear) acceleration of the seated human also has approximately the same value.

B. The maximum pulse intensity which can be sustained without danger to life must also be selected if absolute rather than relative estimates are to be made. In our work we have been using a numerical value of 1000 for the threshold of serious internal head injury in frontal impact recorded in terms of g's. In other words, if one impacts a structure with a dummy, cadaver, or standard 15-lb headform and finds upon integration of the g-time trace that the above Severity Index exceeds 1000, he assumes that danger to life is indicated for that particular test.

A numerical value of 1000 is in reasonable conformity with the data thus far published. For example, if square waves representing various combinations of g and time are taken from the Eiband tolerance curve (for example 100 g's for .010 sec.), these integrate to approximately 1000. Inasmuch as reading the NASA curve in this manner does not take into consideration the additional damage from the exposure of the test subject to the onset and offset ramps of the acceleration profile, and Index of 1000 may be considered as conservative on the basis of this set of data.

This value is also in reasonable agreement with the Wayne head tolerance curve frequently cited.[5] On first inspection this will not appear to be the case since, for example, this curve passes through a point (Figure 3) whose coordinates are 100 g's for only .005 sec, whereas a square wave of 100 g's for .010 sec would calculate to an injury number of 1000. In discussion with Professor Patrick of Wayne, however, it was verified that the g-time traces used in

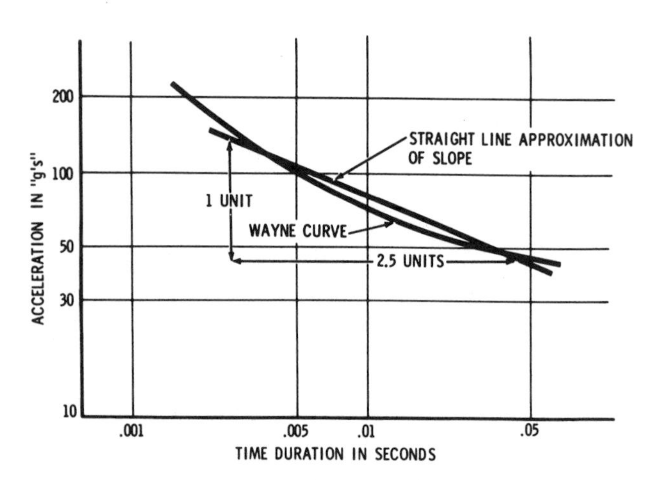

Fig. 3 - Log-log plot of Wayne tolerance curve and straight-line approximation of its slope over time-duration range of automotive interior head impact

developing the intensity scale for the human tolerance curve were not square, and effective values of g were therefore chosen which were less than the peak values. Thus a weighting factor was employed, in effect, on the correct assumption that other investigators using the curve in the future would seldom encounter square waves, and would also need to make a judgment of effective pulse height. Integration of a number of the original Wayne oscillograms shown to the author by Professor Patrick indicated good agreement with the Wayne data.

One of the principal advantages of the Severity Index discussed herein, over a visual weighting which otherwise must be employed, is that it eliminates differences in judgment which are bound to arise even between experienced workers, and thus permits repeatable and comparable test results to be obtained in different laboratories and over an indefinite span of time by different personnel.

A third check against biomechanics data was more recently possible through the cooperation of Mr. Swearingen of FAA, who kindly furnished information regarding the field accident cases, some resulting in fatality, which he simulated in the laboratory. Here an injury number of 1000 fell at approximately the median point between the number of occupants surviving and number who did not survive. It is quite possible that some of the latter received additional injury over and above that from frontal head impact alone, and that on the basis of these tests an Index of 1000 would indicate a survival rate of well over 50%.

It should be pointed out that the value of 1000 for threshold of danger to life for internal head injury in frontal blows is not a fixed quantity; it can be adjusted upward or downward in the future as more survival studies are carried out and if the concensus of the data justifies such an adjustment.

ASSESSMENT OF DIFFERING PULSE WAVE-FORMS

Probably the greatest advantage of an integration procedure is that it can systematically handle or compare widely differing wave-forms. It is known that very high "g" can be tolerated by the head for a few milliseconds, and that only a fraction of this pulse intensity can be withstood in the range of 40 to 50 milliseconds. When, as is usually the case, the pulses are irregular in profile, one cannot accurately compare them visually for relative injury hazard. See, for example, traces a and b in Figure 4. Another dilemma occurs if an extremely sharp "spike" is present as in trace c. If the wave-form is in the range of one to two milliseconds duration, it would be very questionable to select the peak of this spike as one's criterion; yet, it would not be logical to ignore the spike altogether. As integration procedure offers a repeatable and at the same time more rational means to handle a situation of this kind. Again, if a high frequency ringing occurs (as in trace d) which normally represents a spurious vibration in the testing system rather than a real damaging factor, integration will yield a repeatable assessment closer to the true real damage potential represented by the pulse then if one were to select the peak values of the oscillation as the envelope of the curve.

It is of interest to compare a square versus a triangular pulse according to various methods of assessment. The square pulse (trace e) of 54 g for .025 seconds has an injury number of 535 using the 2.5 power weighting factor. Trace f is a triangular pulse of the same time duration, which under a pulse-area criterion would have the same damage potential but under a peak criterion would be twice as damaging. Using 2.5 power integration the triangular pulse of equal area has an injury hazard 1.61 times as great, or 861.

COMPUTATION OF THE SEVERITY INDEX

This is a relatively simple problem. The integration may be done graphically by dividing the pulse into a sufficient number of time increments to define its shape, which usually requires a number in the range of from 20 to 30. At each increment of time the ordinate is raised to the 2.5 power (using a graph or table if one desires) and multiplied by the time duration of the increment. The increments are than added to obtain the injury number.

Modern curve readers and machine computation equipment are advantageous for this purpose and in the author's organization, programs have been set up for this purpose.

PULSE DURATIONS BEYOND .050 SECONDS

It is known that tolerance curves for similar pulse wave-forms tend to asymptote toward a relatively fixed g level at long time duration. It is a matter of some controversy just what value of the latter should be regarded as safe for long time durations, and this is not presently known to any great degree of accuracy even for specific classes of injury. In view of this and because the exponential function

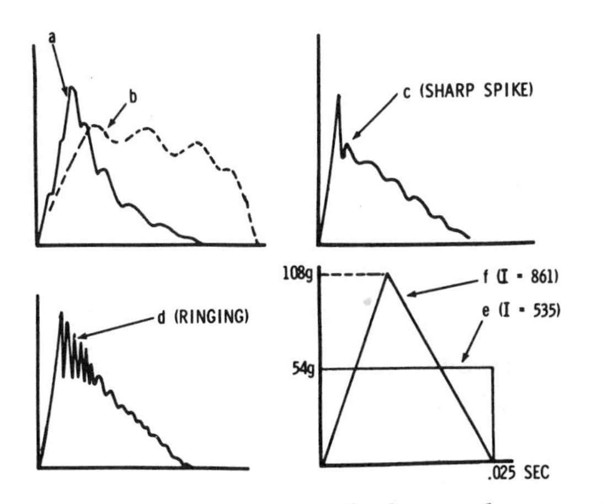

Fig. 4 - Typical wave-forms which must be appraised

fits quite well over the time durations experienced in vehicle head impact, the above Severity Index is suggested for this type of impact. We have recognized, on the other hand, that at some time in the future and for certain purposes it might be well to have available a more complex index designed to extend into the long time-duration range and based upon a larger body of tolerance data than is now available. This would be capable of being fitted to a tolerance curve of arbitrary shape which might be constructed, as were the Eiband and Wayne curves, by noting borderline injury for a range of similar pulses of widely differing time duration.

John P. Danforth* has suggested such an index, in which the exponent n is not a constant but is a prescribed function of acceleration level on the experimentally developed tolerance curve; in other words one does not need to employ a straight-line approximation of the slope of the tolerance curve, but can assume for example that it is steeper in the realm of extremely short time durations and approaches a horizontal asymptote at long duration. A polynomial or a series of two or three simple analytical functions is then employed to express n as a function of g-level over the entire time duration of interest, and this can serve as a basis for machine computation of the Severity Index as before. It is expected that this will be the subject of a future paper.

FURTHER POSSIBLE APPLICATIONS OF SEVERITY INDEX CONCEPT

Thus far our use of this concept has been limited for the most part to the estimation of hazard of internal head injury in frontal impact. It is felt however that there are other classes of impact injury, some to bodily zones other than the head, for which an Index that recognizes time dependency should prove beneficial, as exemplified by the following:

FACIAL INJURY

In contrast with head injury which arises essentially from a disturbance to the head as a whole, there is the important class of frontal injuries which often involves fracture of the facial bone structure or depressed fracture of the forehead. These usually result from severe loading concentrated over a small area. While trauma-indicating headforms serve in a useful way as a relative measure of this hazard, an extention of the use of the Severity Index described herein shows promise as an alternative measure.

As a basis for the development of an Index for facial injury, one can refer to the work of Swearingen[6] and that of Hodgson, Lange, and Talwalker.[7] Both of these independent studies examined the question of time dependency of loading and disclosed that in spite of the fact that relatively brittle bone material was involved, the tolerance to impulses of longer duration was less than that to short duration. Plotting the data of Figure 20 in Reference 6, and that of Figure 11 in Reference 7, on a log-log scale again discloses a trend of time dependency surprisingly close to that for internal head injury. The work of Swearingen included all major areas of the face including the forehead, while the latter plot, Reference 7, summarized a large number of blows of various kinds to the zygoma. In constructing the log-log tolerance plot from the latter reference, only the lower points, that is the lowest striker forces which produced zygoma fracture at a given pulse duration, were employed. Both curves exhibit a slope as great as that represented by a 2.5 power weighting factor.

These data indicate, then, that there is considerable justification for employing at this time a 2.5 power weighting factor for indicating injury hazard inherent in impacts applied to the face. As for a tolerable threshold for damaging, (but usually survivable) injury to the face, one could then employ a Severity Index graded according to the effective contact area. In impact against certain parts of the interior of the body where it is impossible to insure survivability from internal head injury at the higher velocities, such an index could be used as a comparative check of the probability of damage to the facial bones and tissues in the lower speed accidents. The threshold curve for this type of injury would be as given in rudimentary form in **Figure 5. The injury number of 500 for facial bone fracture in impact over an effective area of 3 sq. in. is a preliminary and conservative value based on references 6 and 7, which show a tolerance disparity not as yet resolved.**

CHEST IMPACT

As more experimental tolerance data are obtained in the future it should be possible

---

*Research Laboratories, General Motors Corporation

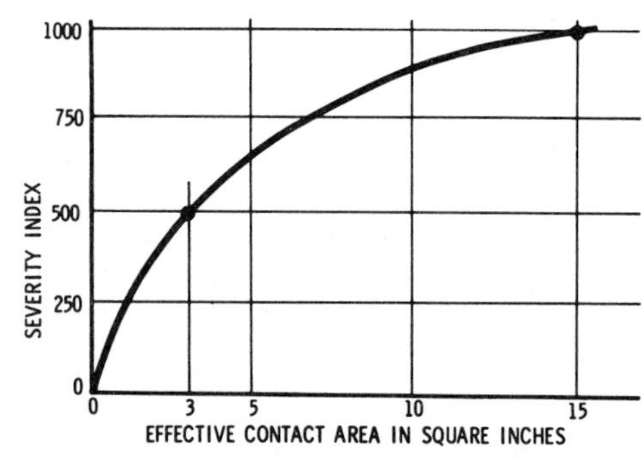

Fig. 5 - General form of a curve expressing lower tolerance of the facial zone of the head to impact over differing areas (damaging, but not fatal, injury)

to recognize time dependency here. Since there is a possibility of transient resonance of the chest, it is necessary to look closely for this in conducting the tests on biological material upon which the Severity Index is to be based. This is done by comparison of the experimentally measured input function with the response function, the first being the force-time trace imposed upon the front of the chest and the second being the chest deflection or other measure of internal injury likely to occur. Chest impact tests with which the author has been associated[8] have not as yet delineated an appreciable dynamic response; if further tests show that this is a small factor, then the input force may be regarded as a reasonable parameter for integration to obtain the Severity Index. (The peak value of input force is the current index for governmental steering system qualification.) If further studies indicate on the other hand that an important dynamic factor is present, it will be necessary to employ a response parameter, e.g. chest deflection, to best reflect the likelihood of internal injury. The procedure for developing a Severity Index is then to apply impacts having similar waveform but differing time durations, and solve for the weighting exponent which best approximates the time dependency.

CONCLUSIONS

A Severity Index has been developed for assessing impact test results which has proven useful in the following respects:

A. It is able to take time dependency of the injury into account in a manner which can be adjusted to that exhibited by the biological material involved, and permits comparisons between pulses of differing time durations experienced by occupants of automotive vehicles in accidents.

B. It permits comparison of the relative hazard between pulses of differing wave-form or profile.

C. It provides a means for different investigators in different laboratories to make numerically consistent interpretations of the hazard represented by a recorded pulse.

REFERENCES

1. H. E. von Gierke, "Biomechanics of Impact Injury." "Impact Acceleration Stress - Proceedings of a Symposium with a Comprehensive Chronological Bibliography." National Academy of Sciences, National Research Council Publication No. 977, 1962.

2. H. R. Lissner, et al, "Experimental Studies on the Relation between Acceleration and Intracranial Pressure Changes in Man." "Surgery, Cynecology and Obstetrics." pp. 329-338, September 1960.

3. H. R. Lissner and E. S. Gurdjian, "Experimental Cerebral Concussion." ASME Annual Meeting Paper No. 60-WA-273, November 27, 1960.

4. A. Eiband, "Human Tolerance to Rapidly Applied Accelerations: A Summary of the Literature." NASA Memorandum 5-19-59E, June 1959.

5. See for example "Human Tolerance to Impact Conditions as Related to Motor Vehicle Design." SAE Handbook Supplement J 885, 1964.

6. J. J. Swearingen, "Tolerances of the Human Face to Crash Impact." Office of Aviation Medicine, F.A.A., Civil Aeromedical Research Institute, Oklahoma City, Okla., July 1965.

7. V. R. Hodgson, W. A. Lange, and Talwalker, "Injury to the Facial Bones." Proceedings of Ninth Stapp Car Crash Conference, University of Minnesota Publication, 1966.

8. L. M. Patrick, C. K. Kroell, and H. J. Mertz, "Forces on the Human Body in Simulated Crashes." Proceedings of Ninth Stapp Car Crash Conference, University of Minnesota Publication, 1966.

9. C. W. Gadd, "Criteria for Injury Potential." Proceedings of Impact Acceleration Stress Symposium, National Academy of Sciences, National Research Council, Publication No. 977, 1962.

670913

Cadaver Knee, Chest and Head Impact Loads

L. M. Patrick and H. Mertz, Jr.
Wayne State University

C. K. Kroell
General Motors Corp.

RECENT RESULTS of a continuing research program to establish human tolerance to impact to the head, chest, and knee are presented. Human tolerance is based upon bone fracture in human cadavers and upon accelerations measured on the head. No attempt is made to determine the effect of the impact on internal organs directly, although some indication of the probable degree of internal injury to the interior of the thorax and the head can be deduced from the injury often accompanying similar skeletal damage in accident cases. For example, a linear skull fracture is generally accompanied by a mild to moderate concussion, according to the neurosurgeons on the interdisciplinary team of engineers, and physicians cooperating in the biomechanics research at Wayne State University. Similarly, simple rib fractures are often observed without internal injuries or other complications, while more extensive rib fractures can produce flail chest, lung, and heart damage. The primary goal of this program is to delineate fracture thresholds of the three areas in terms of quantitative physical measurements of force and/or accelerations, and to present the results in a manner that will permit a rational design of vehicle interior components and other trauma sources to minimize injury.

Embalmed human cadavers are used in this research since they are the closest to living human beings in weight distribution and dynamic reactions, and also embalmed bone has approximately the same strength as living bone. The cadavers are prepared to provide approximately the same degree of flexibility as observed in the living human being. Impacts simulating those encountered in an automobile accident are achieved by seating the cadaver on an automobile seat on the sled and placing load-measuring targets in positions corresponding approximately to the knee-instrument panel, chest-steering wheel, and head-windshield impact sites. When the sled is stopped from a predetermined velocity, the cadaver continues forward, striking the load cells.

ABSTRACT

Human tolerance to knee, chest, and head impacts based upon skeletal fracture of cadavers is reported. The results are based upon unrestrained cadaver impacts in a normal seated position in simulated frontal force accidents at velocities between 10 and 20 mph and stopping distances of 6-8 in.

The head target was covered with 15/16 in. of padding. No skull or facial fractures were observed at loads up to 2640 lb. Extensive facial fractures and a linear skull fracture occurred during the application of the maximum head force of 4350 lb.

The chest target was 6 in. in diameter with 15/16 in. of padding. The padding was rolled over the edge of the target to minimize localized high force areas on the ribs. A 1/8 in. diameter rod was inserted through the chest and fastened through a ball joint and flange to the soft tissue at the sternum. Deflection of the chest was determined by measuring the distance between the back of the cadaver and the end of the probe from the high-speed motion pictures. Four rib fractures were observed at 1340 lb and extensive fractures were observed at 1850 lb. The maximum chest deflection was 2.25 in. at 1545 lb.

The knee targets were padded with 1-7/16 in. of padding. No fractures were observed at forces up to 1970 lb.

A photograph of the experimental setup is shown in Fig. 1. More details of the sled and instrumentation are available in prior publications (1, 2)* In Fig. 1, the cadaver is shown in a seated position on the bench-type automobile seat on a sled. The knee, chest, and head load cells are shown in position, and the accelerometers mounted on the head of the cadaver are visible. A head rest is mounted on the back of the seat to insure proper head position during the acceleration portion of the cycle. The sled is brought up to speed by a pneumatic cylinder, with the speed controlled by the preset pressure in an accumulator tank. At the completion of the acceleration stroke, the sled free-wheels for several feet during which time its velocity is measured. Finally, the sled hits a decelerating cylinder which provides an approximately constant deceleration. The signals from the load cells and accelerometers are taken from the moving sled through a trailing umbilical cord, which in turn feeds the signals into electronic conditioning equipment and thence into a light beam galvanometer recorder.

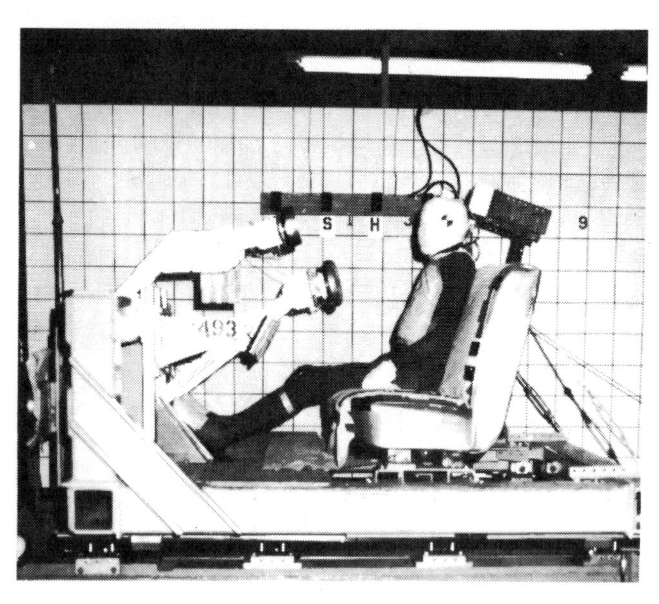

Fig. 1 - Cadaver and load cells on sled prior to run 493

*Numbers in parentheses designate References at end of paper.

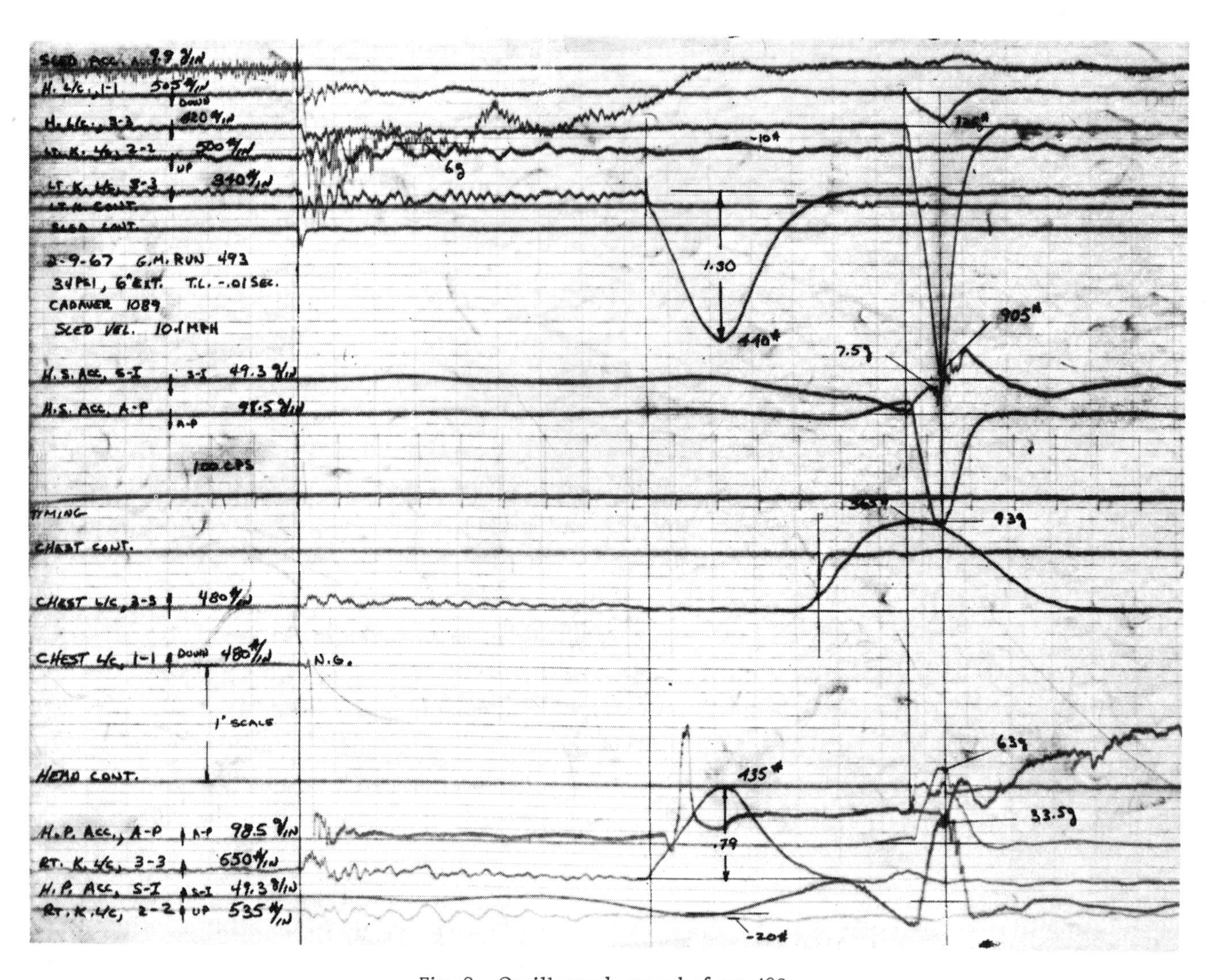

Fig. 2 - Oscillograph record of run 493

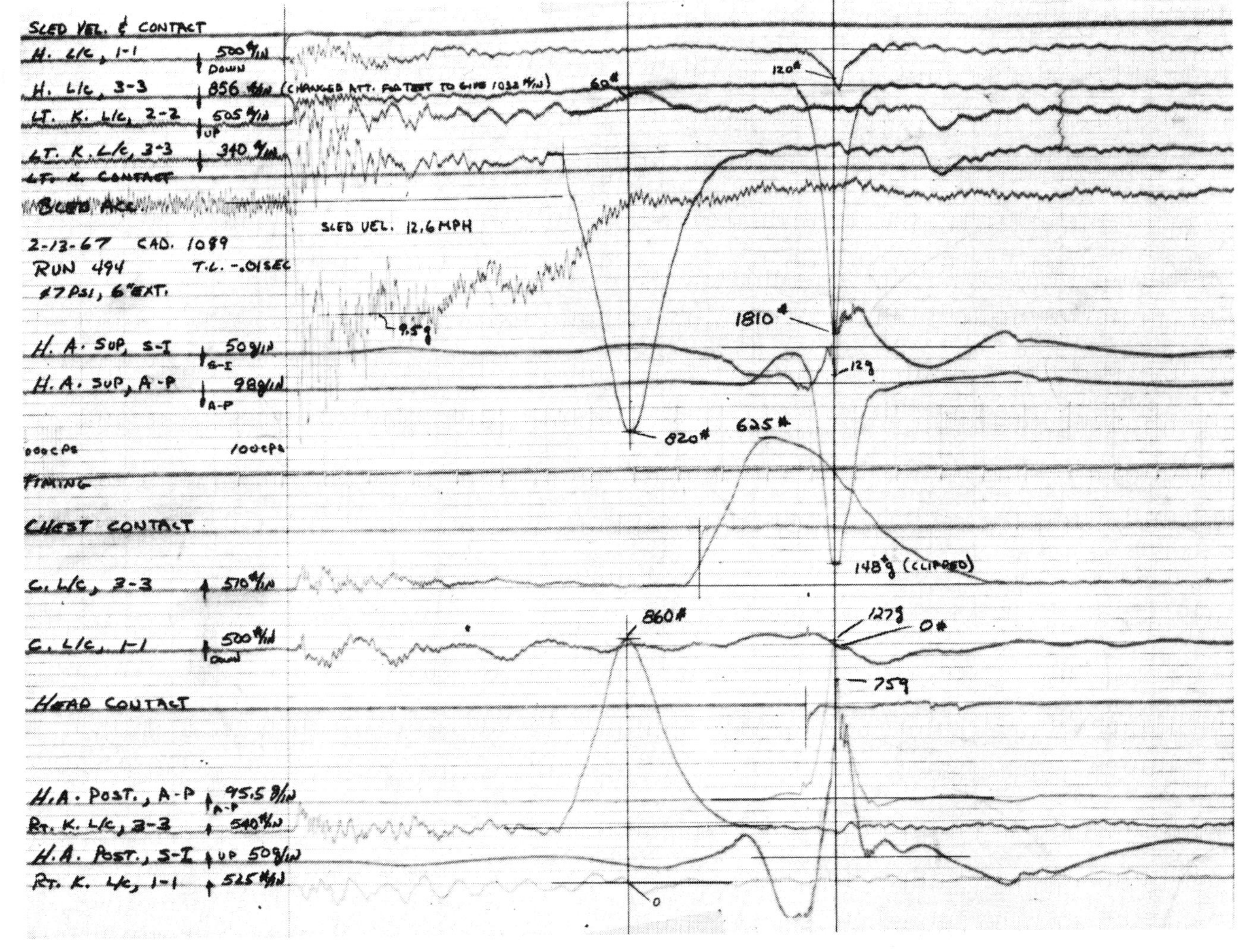

Fig. 3 - Oscillograph record of run 494

Force-time histories of the head, chest, and knee impacts are obtained from triaxial load cells. Biaxial accelerations are recorded from accelerometers on the top at the back of the head.

With the target positions used, the knees struck just before the sled came to a complete stop, while the head and chest impacts occurred after the sled was stopped. The records of the four impacts reported herein are shown in Figs. 2-5. In addition to the forces and accelerations measured, an indication of the deflection of the chest during the impact is obtained by means of a probe extending through the chest and sutured to the soft tissue near the sternum. The deflection is measured from the high-speed motion pictures taken during the impact. A cross-plot of the force-time and deflection-time records provides the chest force-deflection curve under dynamic impact conditions.

Figs. 6 and 7 are a series of photographs taken before and after run 495 and show additional details of the instrumentation. Chalk marks visible in Fig. 7 in the form of crosses on the head and chest load cells are transferred to the head and chest of the cadaver at impact to pinpoint the exact position of the impact. A target mounted on the head is used in analyzing the high-speed motion pictures. The small circle on the chest with a cross in it locates the anterior end of the probe which extends through the chest to indicate deflection. Prior to the impact, small circular indicators are placed on the knee of the cadaver with a small area of adhesive. The other side of the target is completely covered with adhesive. When the knees strike the load cell, the target is transferred from the knee to the load cell, thus providing an accurate indication of the position of the knee on the load cell at impact. In the upper left-hand photograph on the grid of the backdrop there are four flashbulbs labeled with the letters K, S, H, and C. These bulbs indicate the instant of contact of the knee, sled, head, and chest, respectively, and are used to correlate the records with the position of the cadaver in the high-speed motion pictures.

HEAD IMPACT

A rigid load cell for the head impact was covered with 15/16 in. of padding (1/2 in. Rubatex, 7/16 in. Ensolite) which served to distribute the load and prevent localized crushing and fractures from the 3/8 in. thick aluminum plate on the surface of the load cell. The head impact records are included in Figs. 2-5. The trace marked H.

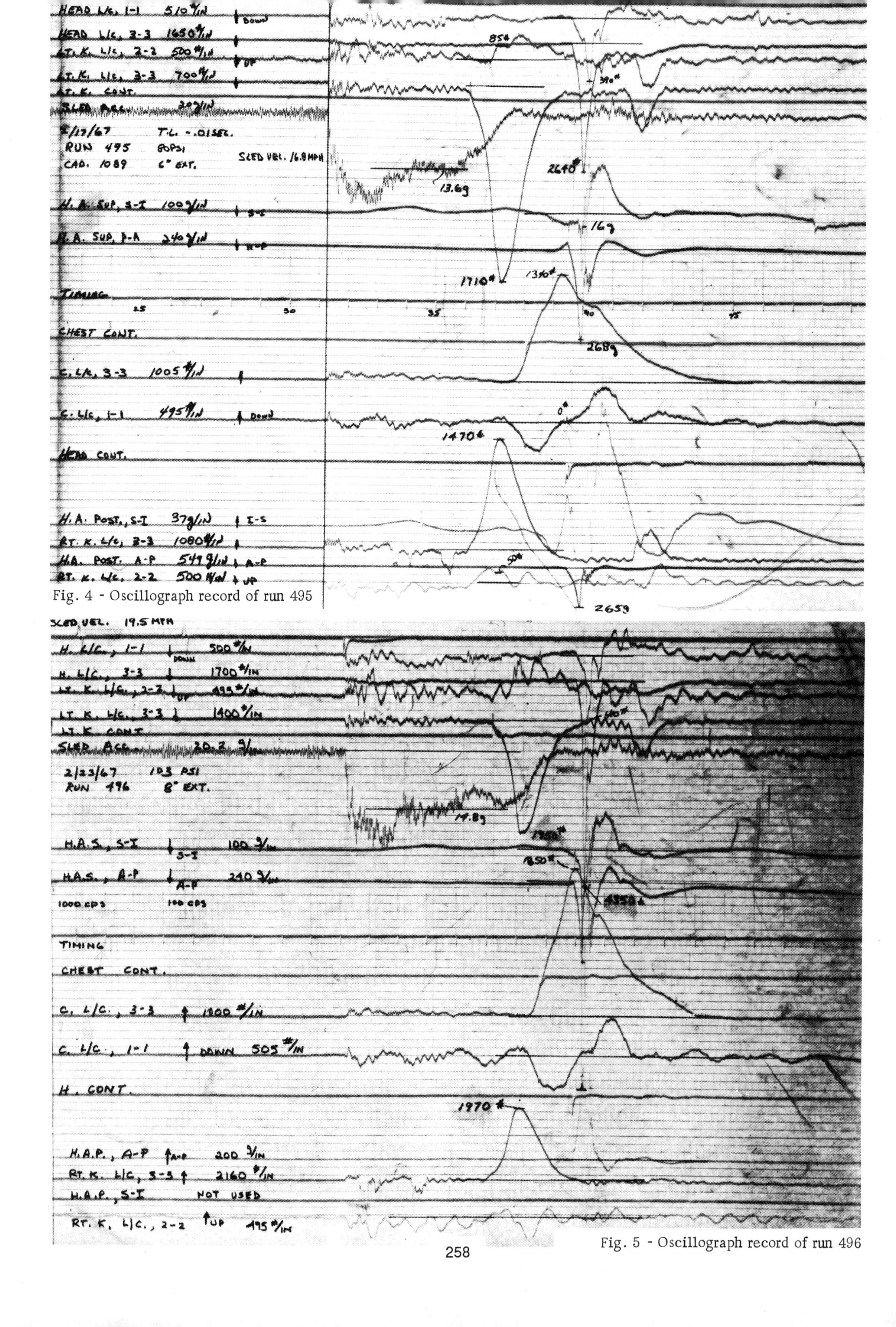

Fig. 4 - Oscillograph record of run 495

Fig. 5 - Oscillograph record of run 496

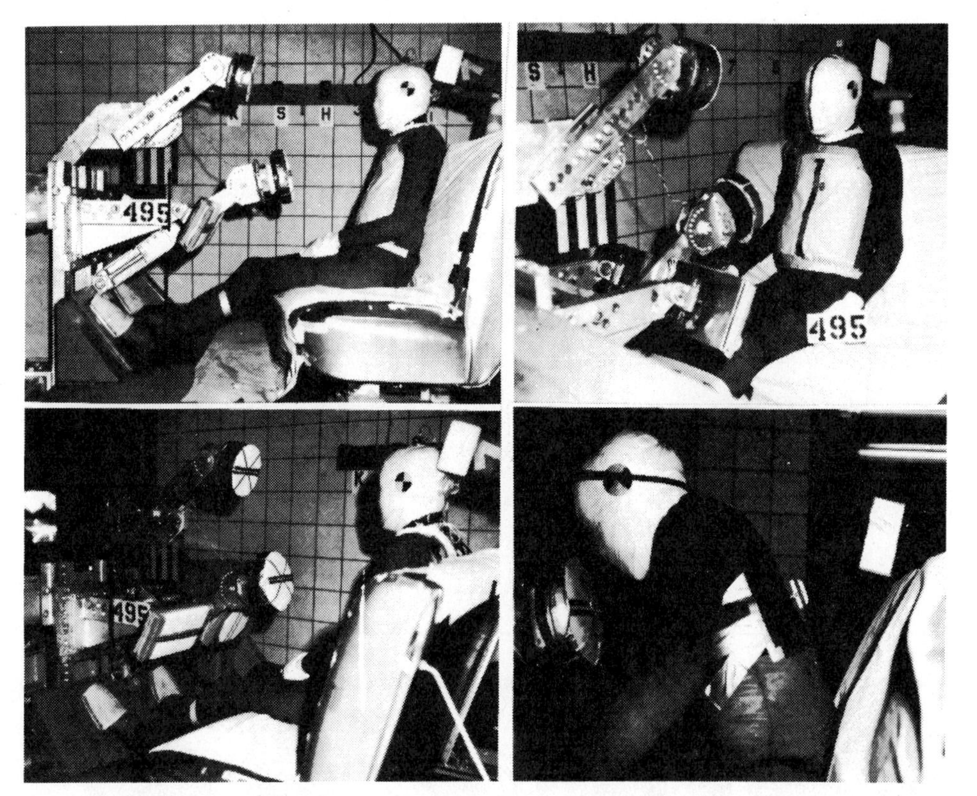

Fig. 6 - Photographs of experimental setup prior to run 495 with cross marks and electrical contacts shown on the head and chest load cells in the lower left photograph and the cadaver pulled forward to show the flag on the end of the chest probe in the lower right photograph

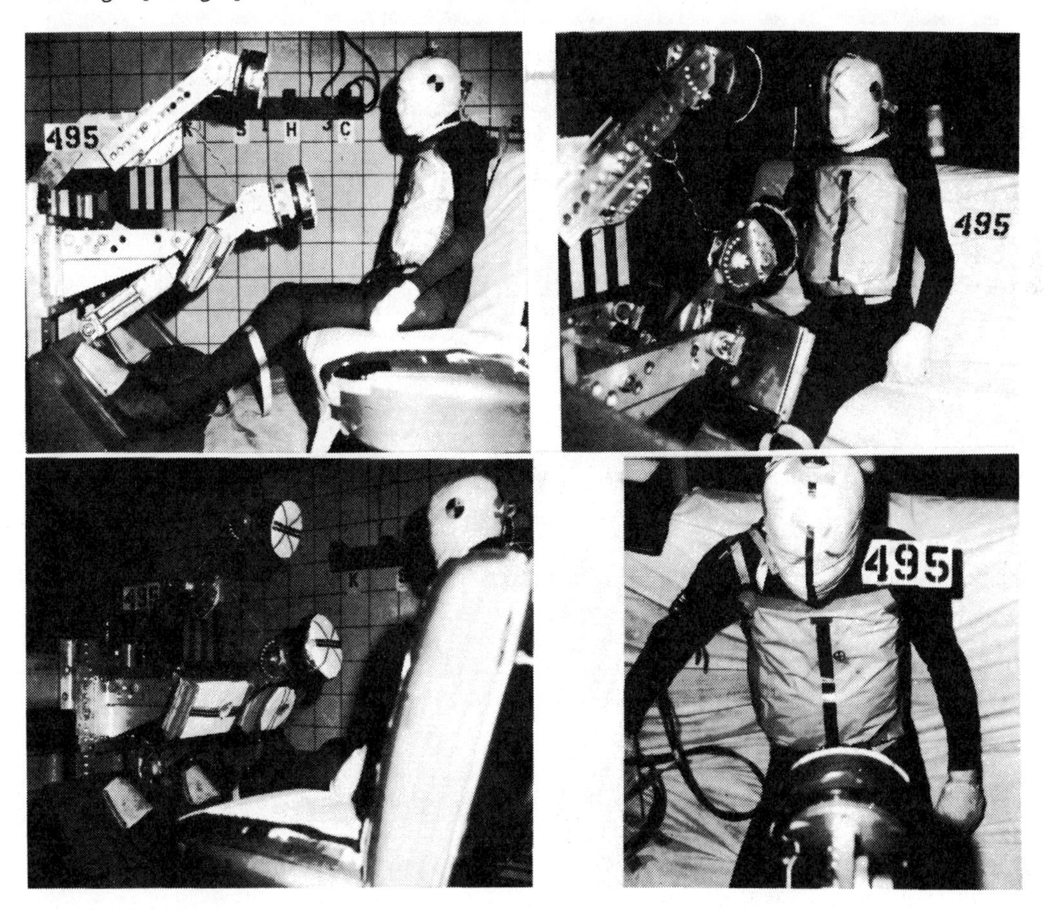

Fig. 7 - Photographs after run 495. The knee targets transferred to the knee load cells are shown in the lower left photograph and the chalk cross-marks are shown in the right photographs

Table 1 - Summary of Results of Head Impacts

Run	Sled Velocity, mph	Normal Head Load, lb	Peak Head Acceleration, g — Post. Loc. A-P	Post. Loc. S-1	Sup. Loc. A-P	Sup. Loc. S-1	Radiological Examination
493	10.1	905	63	34	93	7.5	No abnormalities
494	12.6	1810	127	75	148+	12	No abnormalities
495	16.8	2640	265		268	16	No head abnormalities; 4 rib fracture
496	19.5	4350					Linear temporal fracture Linear fracture of left inferior, lateral orbital margin. Linear fracture of maxilla. Linear fracture of nasal bones. Additional rib fracture

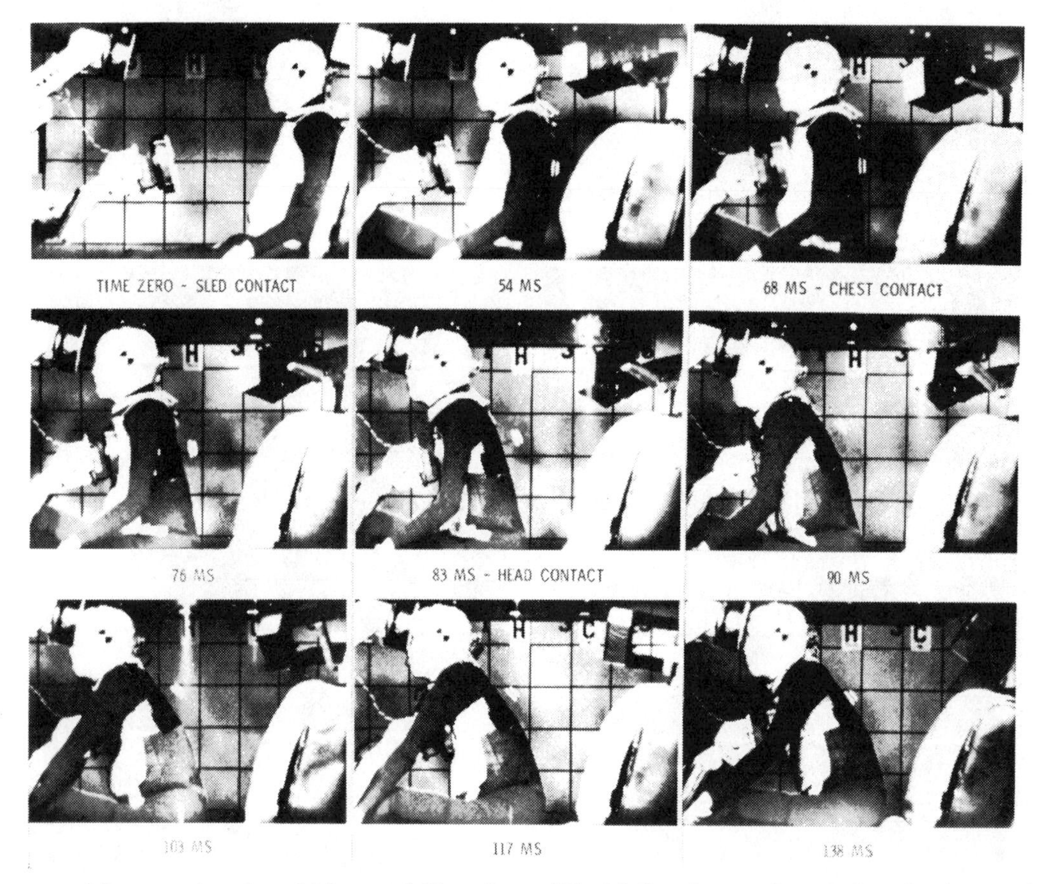

Fig. 8 - Sequence of frames taken from high-speed film of run 495, 16.8 mph, maximum chest load 1340 lb, maximum chest deflection 1.7 in.

L/C 3-3 shows the force normal to the face of the load cell, while the trace marked H. L/C 1-1 shows the force in the plane of the face of the load cell in a vertical plane. Four experimental impacts were conducted at increasing velocities in the range of 10.1-19.5 mph. The results are summarized in Table 1.

No fractures were observed in the head or face until after run 496, where the head load cell recorded a force of 4350 lb. During that run a linear fracture of the left inferior, lateral orbital margin resulted along with linear fractures of the maxilla, nasal bones, and temporal area.

Since no fractures of the head or face were observed in the previous run, the fracture threshold for this cadaver under these conditions was between 2640 and 4350 lb. The corre-

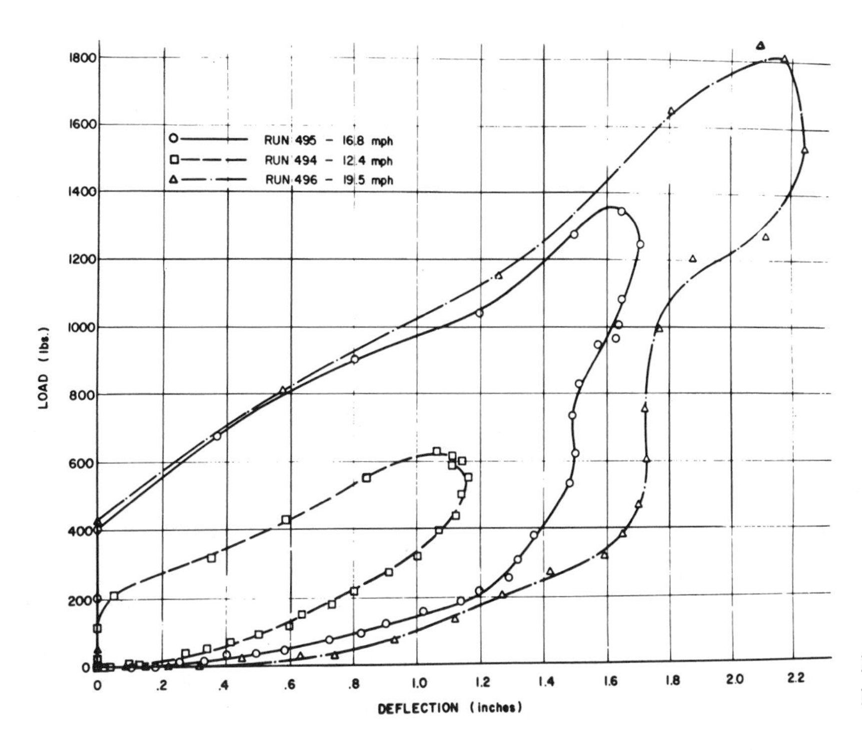

Fig. 9 - Load-deflection curve for the chest. Run 493 could not be read out of the film

sponding head accelerations in the anterior-posterior direction were in the order of 265 g for run 495, where no fractures did occur, the acceleration records were indiscernible. However, since the impact velocity was higher, the accelerations would probably be higher at time of fracture.

A comparison of the head load cell records from the four impacts shows that the head load cell 3-3 axis (perpendicular to the face of the load cell) has a smooth trace for runs 493, 494, and 495, while in run 496 the record shows a distinct break or change in the smooth half-sine or triangular shape observed in the first three impacts. Previous records where fracture occurred show that this is a common indication of fracture on the force-time record.

In previous impact work with dropping cadavers where the head struck a rigid steel plate with no padding, linear fractures occurred at peak accelerations between 225 and 275 g. In the present series of impacts with force-distributing padding, the accelerations in the anterior-posterior direction should be higher prior to skull fracture than for the aforementioned impacts to an unpadded steel plate. Therefore, it is assumed that the facial bone fractures probably occurred near the 2640 lb lower limit while the temporal fracture occurred near the 4350 lb limit. Based upon the Wayne Tolerance Curve (3), some brain damage probably would have occurred during run 494 (148+ g peak, 12 ms), and serious brain damage would be expected during run 495 (268 g peak, 10 ms).

CHEST IMPACT

Establishment of the fracture load of the rib cage is the main objective of the chest impact study. In this respect the dynamic load-deformation characteristics of the thorax, when impacted against the 6 in. diameter target and the maximum deflection of the chest are of importance.

The 6 in. diameter load cell (28 sq in.) is covered with a total of 15/16 in. of padding (1/2 in. Rubatex, 7/16 in. Ensolite). An X is marked on the load cell with chalk prior to the impact. At impact the chalk is transferred to the chest of the cadaver, providing an accurate indication of the point of impact. Fig. 7 contains photographs of the cadaver after impact showing the chalk mark on the chest, while Fig. 6 is a similar series of photographs, taken prior to the impact, showing the position of the cadaver and the chalk mark on the load cell. Horizontal parallel strips of foil on the load cell are connected to an electric circuit. A vertical strip of foil on the chest of the cadaver completes the circuit when the chest contacts the load cell. Completion of the circuit triggers a flashbulb which provides synchronization of the high-speed photographs with the Visicorder records.

A chest probe extending from the front to the rear of the chest is used to measure the deflection during the impact. It is sutured to the front of the chest with a 1 in. diameter flange and consists of a 1/8 in. diameter stainless steel rod attached to the flange through a ball and socket joint. At the back of the chest, a 1 in. diameter bushing is sutured to the cadaver. Thus, the front end of the probe is connected to the ribs with a ball and socket joint, while the probe protrudes through the rear wall of the thorax (between the ribs adjacent to the vertebral column) with a sliding fit through the metal bushing. At the end of the probe where it protrudes through the back, a flag is mounted to serve as a measuring point in the high-speed films.

The flag is shown in Fig. 6 in the bottom right-hand photograph where the cadaver is pulled forward, permitting the flag to be seen. It is also visible in Fig. 8, which is a series of frames taken from the high-speed movies. The displacement from frame to frame is apparent in Fig. 8.

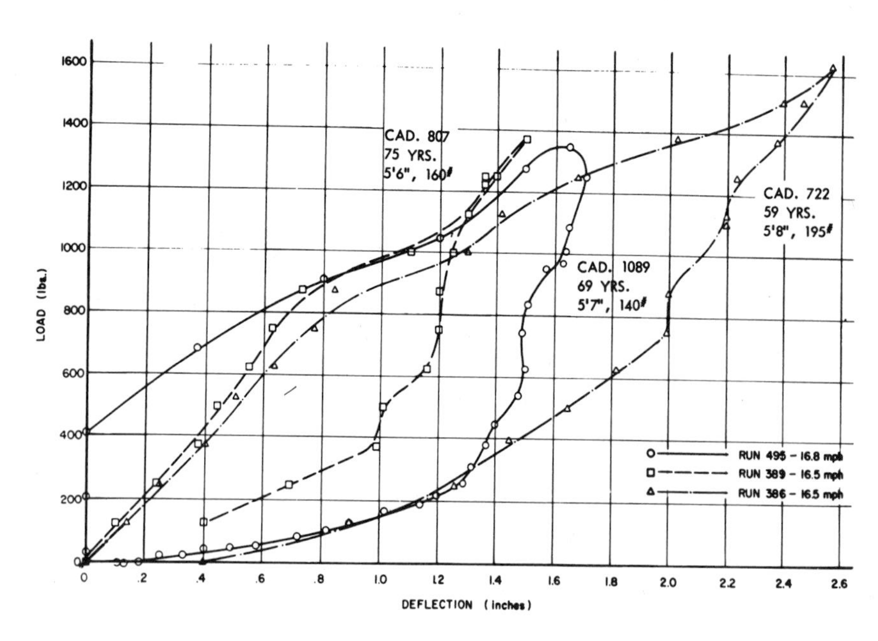

Fig. 10 - Load-deflection curves for three different cadavers at approximately the same impact speed (16.5-16.8 mph)

Table 2 - Summary of Results of Chest Impacts

Run	Sled Velocity, mph	Max Chest Load, lb	Max Chest Deflection, in.	Radiological Examination
493	10.1	365	0.6-0.7[a]	No abnormalities
494	12.6	625	1.16	No abnormalities
495	16.8	1340	1.71	Fracture of left 3rd, 4th, 5th, and 6th ribs in the anterior-axillary line
496	19.5	1850	2.18	Fracture of right 6th and 7th ribs in mid-axillary line. Linear fracture and slight inward depression of sternum. Fracture of facial bones and skull

[a]Best estimate because of poorly defined back reference.

The flash indicating contact of the chest with the load cell foil strips is just barely visible in the second frame of the sequence and gets brighter in succeeding frames. The deflection of the thorax is determined by measuring the distance between the edge of the flag and the back of the body The deflection is first obtained as a function of time from the photographic sequence. From the Visicorder record, the chest load is measured as a function of time. A cross-plot of the two curves eliminates the time factor and results in a force-deflection curve. Fig. 9 shows the force deflection curve for runs 494, 495, and 496 (in run 493 the body covering was loose and prevented accurate deflection measurements). Fig. 10 shows three load-deflection curves at approximately the same velocity (16.5-16.8 mph) for three different cadavers, two of which were reported in the Ninth Stapp Conference (2) and one from run 495.

As the load decreases, there is appreciable hysteresis and a peculiar inflection point. It is conjectured that the inflection point is caused by a comparatively large drag force between the probe rod and the posterior bushing (and possibly the internal body tissues) caused by bending of the rod. The latter apparently was produced by a pronounced upward displacement of the posterior bushing relative to the anterior ball joint, which occurs as the arms and shoulders continued to surge forward after the torso has effectively come to rest. The probe rod exhibited a slight permanent bend after run 494 and somewhat greater deformation after runs 495 and 496 (new rods were used for runs 495

Table 3 - Summary of Results of Knee Impacts

Run	Sled Velocity, mph	Normal Knee Load, lb	
		Left	Right
493	10.1	440	435
494	12.6	820	860
495	16.8	1710	1470
496	19.5	1950	1970

and 496). As the load cell force decays, and the retraction force on the probe thereby increases, a point is reached beyond which the chest relaxes back more rapidly and the load cell force falls off more slowly to zero.

It is thus suggested that the degree of hysteresis exhibited by the resulting load-deflection curves is artificially increased somewhat as the result of the aforesaid combination of deflection measuring technique and natural impact kinematics of the subject. It is not possible to say to what extent the curves are thus influenced. However, it is felt probable that the effect is not a predominating one.

On the other hand, it would appear that the loading portions of the curves are quite well-defined.

Peak chest load appears after the knee load and just prior to the peak head load. It is of longer duration (60-50 ms) than the head impact record (12-8 ms) or the knee impact record (40-20 ms), where the longer time is for the lower velocities.

The results of the chest impact are summarized in Table 2. Linear fracture of the third, fourth, fifth, and sixth ribs on the anterior-axillary line occurred during run 495. Additional fractures of the right sixth and seventh ribs in the mid-axillary line and a linear fracture and slight inward depression of the sternum occurred during run 496. The maximum chest deflection was 2.18 in. at 1815 lb in run 496. The records of the chest load for runs 493 and 494 were smooth, while those of runs 495 and 496 were double humped, which is typical of this type of impact when fracture occurs.

The chest load cell was positioned at an angle which resulted in impact to the lower rim of the load cell at about the lower end of the sternum. This resulted in the load buildup with no appreciable deflection, as shown in Fig. 9. As the impact continued, the body rotated about the load cell until the chest struck in a flat position, causing a deflection. This localized force at the bottom of the sternum probably accounts for the fracture, while the load below the attachment point of the deflection probe does not produce any measurable deflection until a load of about 400 lb is reached in run 496.

KNEE IMPACT

The knee load cells were covered with 1 in. of Ensolite and 7/16 in. of Rubatex. The loads varied from 440 lb at 10.1 mph to 1970 lb at 19.5 mph as shown in Table 3.

No abnormalities of the knee-thigh-hip complex were observed by the radiologists. The record for run 496 (Fig. 5) shows a slight flattening with a slightly ragged top, which when more pronounced has been observed in other knee impacts when fracture occurs. In this run it might show some pending joint, soft tissue, or bone damage.

In a previous publication (2) 1400 lb was recommended as a reasonably conservative value for the injury threshold to the knee-thigh-hip complex from knee impact. With the targets and padding used in these experiments the 1400 lb was reached between 12.6 and 16.8 mph, but no obvious damage occurred at 1950 lb and 19.5 mph. In previous tests much higher loads have not caused fracture in some cadavers, while in one case a mid-shaft femur fracture resulted at 1500 lb. Therefore, the 1950 lb loads without fracture are not unreasonable.

ACKNOWLEDGMENT

The services of Dr. I. D. Harris in providing the radiological diagnoses for this investigation are gratefully acknowledged. The research reported in this paper was sponsored by Research Lab., General Motors Corp.

REFERENCES

1. Charles K. Kroell and Lawrence M. Patrick, "A New Crash Simulator and Biomechanics Research Program." Proc. of Eighth Stapp Car Crash and Field Demonstration Conference, 1964.

2. Lawrence M. Patrick, Charles K Kroell, and Harold J. Mertz, Jr., "Forces on the Human Body in Simulated Crashes." Proc. of Ninth Stapp Car Crash Conference, 1965.

3. Lawrence M. Patrick, Herbert R. Lissner, and E. S. Gurdjian, "Survival by Design -- Head Protection." Proc. of Seventh Stapp Car Crash Conference, 1963.

700401

Comparative Tolerances for Cerebral Concussion by Head Impact and Whiplash Injury in Primates

A. K. Ommaya,
National Institutes of Health

F. J. Fisch and R. M. Mahone
Naval Ship Research and Development Center

P. Corrao and F. Letcher
Naval Medical Research Institute

ABSTRACT

Experimental head impact and whiplash injury experiments have been conducted in 3 sub-human primate species in order to define tolerance thresholds for onset of cerebral concussion. Preliminary analysis of our data support a hypothesis that approximately 50% of the potential for brain injury during impact to the unprotected movable head is directly proportional to head rotation and inversely proportional to head translation by the impact; the remaining brain injury potential of the blow is directly proportional to the contact phenomena of the impact. A scaling method for injurious rotations is presented which predicts that levels of head rotational velocity exceeding 50 Rad/sec and acceleration exceeding 1800 Rad/sec^2 are compatible with a 50% probability for onset of cerebral concussion in man.

RESUME

Des chocs expérimentaux sur la tête et des expériences de lésions en "coup de fouet" ont été menés sur trois espèces de primates non-humains de façon à définir les limites de tolérance avant qu'une commotion cérébrale ne se produise. Une analyse préliminaire des renseignements obtenus soutient l'hypothèse qu'environ 50% du potentiel de lésions cérébrales au moment du choc sur la tête mobile non protégée est directement proportionnel à la rotation de la tête et inversement proportionnel a la translation de la tête causée par le choc; le potentiel restant de lésions cérébrales dues au coup est directement proportionnel au phénomène de contact du choc. Il est presénté une méthode de graduation pour rotations causant des lésions qui prédit que les niveaux de vitesse de rotation de la tête dépassant 50 Rad/sec. et d'une accélération dépassant 1800 Rad/sec^2 sont compatibles avec une probabilite de 50% pour qu'une commotion cérébrale se produise chez l'homme.

AUSZUG

Kopfaufschlag und "whiplash" Verletzungsversuche wurden mit drei Primaten dürchgefuhrt um die Toleranzschwellen des Anfangs der Hirnerschütterung festzustellen. Vorläufige Analyse unserer Daten unterstützen eine Annahme, dass ungefähr 50% der Möglichkeit einer Hirnverletzung während eines Aufpralls des ungeschützten beweglichen Kopfes in direktem Verhältnis zur Kopfdrehung ist, und im umgekehrten Verhältnis zur Kopfbeförderung durch den Aufschlag; die verbleibende Hirnverletzungsmöglichkeit des Schlages ist im direkten Verhaltnis zu den Kontakterscheinungen des Aufpralls. Gezeigt wird eine Stufungsmethode für die verletzenden Drehungen welche vordussagt, dass die Stufen der Kopfdrehungsgeschwindigkeit über 50/Rad/Sek., und Beschleunigung über 1800 Rad/Sek.^{2}, mit einer 50% Wahrscheinlichkeit einer Hirnerschütterung im Menschen angemessen sind

IN EARLIER REPORTS FROM THIS LABORATORY we had observed that angulation of the head on the neck was a necessary common denominator for brain injury in the absence of skull fracture during direct and indirect head impact.[1] Cerebral concussion as well as macroscopic intracranial hemorrhage could be produced by impact to the freely movable head as well as by whiplash injury.[2] These experimental studies of the role of rotational displacement of the head in injuring the brain led us to a direct experimental testing of Holbourn's hypothesis for the mechanism of head injuries. According to this theory, head rotation caused by impact resulted in rotary distortion of the brain with high resultant shear stresses. Cerebral concussion, contre-coup lesions and other effects of head impact were attributed primarily to such rotational effects while the distortion of the skull by the blow (contact phenomena of impact) was thought to produce only local contusions and skull fracture. The translatory component of the impact was considered to be non-injurious. Holbourn also predicted that the response of the head-neck system would be that of a single degree of freedom spring-mass system. Thus short-duration impacts would produce injury proportional to the rotational velocity induced by the blow while long duration impacts would display a dependence on the head's rotational acceleration for injury.[3] We were able to confirm the rotational velocity dependence of short-duration impacts producing cerebral concussion in both direct experimental head injury and whiplash trauma in rhesus monkeys. The main facet of Holbourn's hypothesis was however not proven: thus if head rotation were indeed the crucial brain injury mechanism, cerebral concussion should be produced at an identical threshold for rotational velocity of the head irrespective of how the head rotation were induced i.e. directly or indirectly, and the local effects of impact should have no influence on the threshold for cerebral concussion. Our experimental data did not support this prediction and indeed showed an approximately 50% higher level of rotational velocity for cerebral concussion produced by indirect impact (whiplash).[4,5] This suggested a significant contribution to brain injury by the local effects of impact (i.e. contact phenomena).

Our current hypothesis for brain injury by mechanical trauma to the head is therefore a considerable modification of the rotational theory and is best stated as follows.

Approximately 50% of the potential for brain injury during impact to the unprotected movable head is directly proportional to the amount of head rotation and inversely proportional to the amount of translation of the head; the remaining potential for brain injury will be directly proportional to the contact phenomena of impact (e.g. skull distortion). These phenomena become increasingly significant (i.e. > 50% potential for injury) only if the mobility of the head after impact is progressively reduced. Thus in considering protection of the brain from mechanical trauma to the head two factors have to be controlled; firstly, reduction of the amount of head rotation preferably by its conversion into head translation, and secondly, by diminuition of the contact phenomena of impact (e.g. skull distortion) particularly when the head is relatively immobile. Hitherto, the latter aspect has been considered of prime significance in head protection and insufficient attention has been given to the role of head rotation. In an earlier report, we had also presented our theoretical derivation of a scaling method aimed at extrapolating experimental data on brain injury thresholds from sub-human primates to man in terms of an inverse proportion between the 2/3 power of relative brain weights and levels of rotational acceleration.[6] This method made it possible to compare the injurious effects of head rotations in different primate species but did not include scaling for the contact phenomena of impact.

The present report summarizes all our data on direct and indirect impact induced cerebral concussion in three sub-human primate species and compares the fit of data to theory for our scaling predictions. We will critically examine the value of extrapolating such data to predict the tolerance of man to direct and indirect head impact in the light of our most recent hypothesis as stated above.

METHODS

Cerebral concussion was produced in two ways; by direct impact to the occipital zone of the head and by experimental whiplash injury caused by impact to the base of a mobile chair carrying the seated animal (Fig. 1). Both of these techniques were identical to that described in our earlier publications.[1-7,14] Experiments were performed in three sub-human primate species, the rhesus monkey (brain weight = 70-100 gm), the squirrel monkey (20-27 gm), and the chimpanzee (350-500 gm). The number of each of three sub-human primate species used in these tests are indicated as follows:

Species	Head Impacts	Whiplash Injury
Squirrel monkey	35	35
Rhesus monkey	100	100
Chimpanzee	12	26

A compressed air actuated piston weighing 3.8 pounds was used to impact the head of the rhesus and squirrel monkeys. The force of the impact was controlled by regulating the air pressure behind the piston. Impact force was sensed by a strain gage dynamometer

Note the impacting piston powered by the air compression chamber which accelerates the cart down a track thus subjecting the restrained animal to controlled whiplash trauma. Grid in background is used to measure displacements of monkey's head versus time with high speed cinematography

Fig. 1 - Diagram of whiplash injury apparatus

attached to the striking end of the piston. The velocity of the piston upon impact was measured by recording the voltage generated as a magnet imbedded in the piston passed through a stationary coil. An example of the physical events recorded in rhesus monkeys is shown in Fig. 2. This technique was found to be inadequate to produce significant brain injury in the chimpanzee. A modified humane stunner described previously as well as a specially adapted Hy-G device were utilized for the chimpanzee impacts.[13] Details of this methodology will be published separately.[14] The indirect impact (whiplash injury) experiments in the chimpanzee were also performed differently, and details of this method will be published in a separate report.[15] Animals were usually anesthetized with Nembutal or Sernylan (phencyclidine hydrochloride) prior to preparation for testing. After restraint in the seat had been secured the level of anesthesia was allowed to diminish until the animal exhibited stable vital signs, consistently present corneal and palpebral reflexes, regular respiration and responded to external stimuli with coordinated voluntary movements. The criteria for cerebral concussion onset were similar to that described earlier, and are summarized in Fig. 3.[7] Impacts were given to the occipital zone in all species, attempting to define appropriate 50% concussive levels (e.g. 0.437 lb.sec. for rhesus monkey) in all direct impact tests. Whiplash injury tests were also performed so as to provide inputs at the approximate 50% concussive level (e.g. 200 g. on the rhesus monkey sled with short duration acceleration). Displacement of the head on the neck during direct as well as indirect impact was recorded against a background grid by high speed cinematography at 500-3000 frames/second. Values for rotational velocity and acceleration of the head were calculated from these displacement data. Such values for both

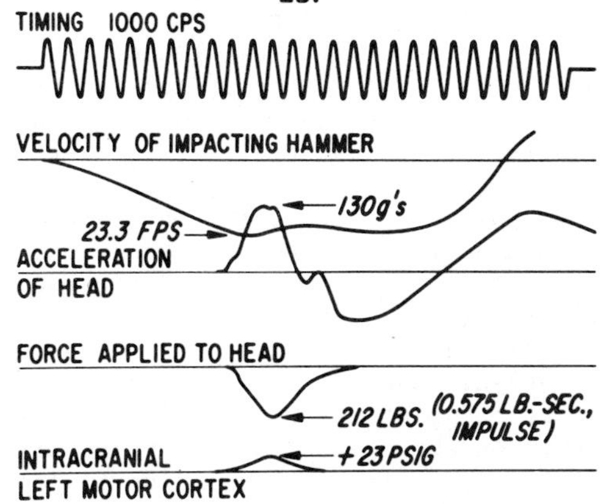

Acceleration was recorded with a miniature piezo-electric accelerometer bonded to skull and intracranial pressure (lowest line) was recorded with a semiconductor strain gage bonded flush with skull surface in a 4 mm diameter water tight hole (diaphragm of gage lying extradurally)

Fig. 2 - Data recorded from impact test in rhesus monkey

CRITERIA FOR EXPERIMENTAL CEREBRAL CONCUSSION IN SUB-HUMAN PRIMATES

(The Basic Phenomenon of Head Injury)

1. Loss of coordinated responses to external stimuli.
2. Apnea >3 seconds, followed by irregular slow respiration.
3. Bradycardia (rate decreased by 20-30 beats/minute).
4. Loss of corneal and palpebral reflexes.
5. Loss of voluntary movements.
6. Pupillary dilatation >15 seconds.

CRITERIA FOR GRADING SEVERITY OF CEREBRAL CONCUSSION

1. Duration of loss of coordinated responses to external stimuli.
2. Abnormalities of EKG pattern and their duration.
3. Duration of pupillary dilatation.

The first criterion (loss of responses to stimuli) is the 'sine qua non' while the others are used as additional factors

Fig. 3 - Criteria for experimental cerebral concussion

impact and whiplash tests were then subjected to two types of statistical evaluation. The first was a probit transformation technique similar to that used earlier.[7] The second was a special parametric technic particularly useful with small sizes wherein the levels of stimulus cannot be precisely controlled.[8] A computer program was developed for this technique employing the method of maximum likelihood to compute estimates of the mean critical level of stimulus and the standard deviation of the critical levels, based on the assumption that the critical levels are normally distributed. The asymptotic approximations for the precision of these estimates was also computed. The computed means and standard deviations for occipital impact and whiplash injury concussive thresholds in the three primate species were used to test the scaling theory referred to above.[6]

In addition to the rotational displacement data, head displacement during head impact in 50 rhesus monkeys in Cartesian coordinates (x, y) derived from the high speed cine films at 1/2 msec intervals was also plotted. A program was written to calculate and plot the x and y velocity and acceleration components and the tangential velocity and acceleration relative to the instant center of rotation for each time increment.

Tolerance curves relating rotational velocity versus acceleration were drawn to provide the best separation of concussed and non-concussed points in the data using the equation

$$\theta_D = \frac{\ddot{\theta}_D}{\omega} \qquad (1)$$

where $\dot{\theta}_D$ is the damaging rotational velocity in radians/second

$\ddot{\theta}_D$ is the damaging rotational acceleration in radians/second^2

and ω is the natural frequency of rotation of the brain in radians/second

Kornhauser has shown that the change over from short duration to long duration impact responses for a single degree of freedom spring-mass system occurs at 1/3 to 1/4 of the period of natural frequency of the system.[9] Since only short duration impact data could be obtained under our experimental conditions, the values for $\ddot{\theta}_D$ were calculated from equation (1) for each species. These values for $\ddot{\theta}_D$ calculated from the <u>experimental</u> data for each species were compared to the theoretically predicted values for $\ddot{\theta}_D$ based on our scaling method from which we derived the following equation:

$$\ddot{\theta}_D = C/M^{2/3} \qquad (2)$$

where M is the mass of brain in grams

and C is a constant derived from the experimental data in gram radians/sec^2

The value of ω, i.e. the natural frequency of rotation of the brain required for equation (1) above was obtained by three separate techniques:[4]

A. High speed cinematography of the brain sulci displacement during impact as seen through a Lexan calvarium;[10]

B. Cineradiographic recording of displacement of opaque iso-density pellets imbedded in the brain of a monkey subjected to head impact;[4]

C. High speed cinematography of brain sulci displacement during vibratory displacement of the animal on a shake table at frequencies equal to 5-15 Hz, with and without total immersion in water to produce immobilization of the head.[11,12]

RESULTS

It is not feasible to tabulate all our data relating onset of cerebral concussion to levels of input impulse and response of the head in either linear or rotational terms. It had been shown in our earlier reports that a correlation between concussion and a measurable index of the input trauma that was statistically significant in rhesus monkeys was the impulse of blows to the back of the head.[7] This correlation failed to hold when the angulation of the head on the neck was reduced by a cervical collar.[1,2] This protective effect of a collar is evident in Fig. 4 which plots the rotational velocity imparted to the heads of rhesus monkeys to the impulse of the blows causing such rotations. This relationship is approximately linear with the probability of concussion onset exceeding 50% as the level of impulse exceeds 0.4 lb. sec. and the level of rotational velocity exceeds 150 Radians/second. However, the non-concussed data points shown as solid triangles indicate that this apparently linear relationship does not hold in the presence of a collar and the rhesus monkeys can sustain much higher impact impulses provided the rotational velocity is kept below threshold levels. The dotted lines drawn on this graph are approximate indicators of the 50% probability level for concussion in rhesus monkeys.

As indicated in the section on methodology, two techniques for statistical analysis of our data were used, probit transformation and the Golub-Grubbs analysis of sensitivity experiments when stimulus levels are not controlled.[8] Both methods gave essentially similar results indicating that the scatter of data was indeed large and that attempts to obtain statistical correlations between input and response measurements of greater precision than that indicated e.g. by the 10%, 50%, and 90% levels of concussion probability ($p < 0.05$) were not practical. Accordingly, the latter derivation of an approximate 50% threshold for onset of cerebral concussion as experimentally defined was chosen as the most realistic comparator for our data.

The measurements of rotational frequency of whole brain movement obtained from the rhesus monkey have not yet provided unequivocal data. The results of two techniques are shown in Figs. 5 & 6 which suggests a frequency of about 5-10 Hz. The third technique has not

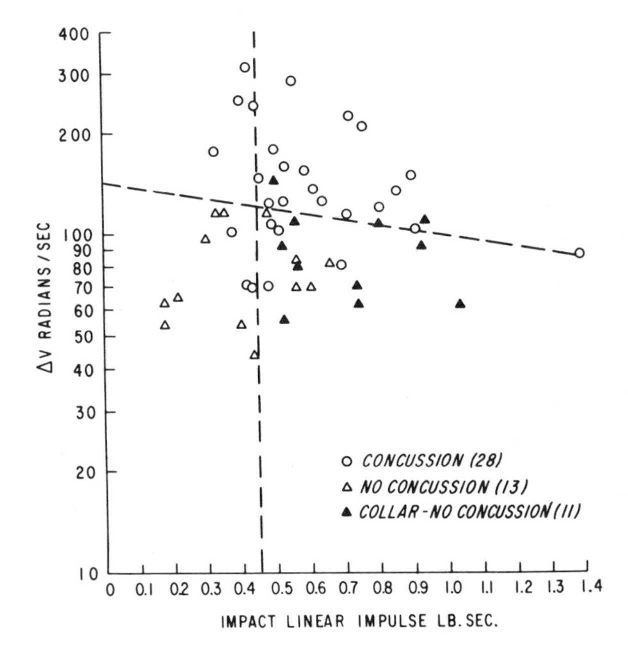

Note how the approximately linear data distribution is deflected to the right when data from monkeys impacted while wearing a cervical collar restraining head angulation are plotted

Fig. 4 - Plot of rotational velocity of head versus linear impulse of impact producing such rotations in the rhesus monkey (52 blows)

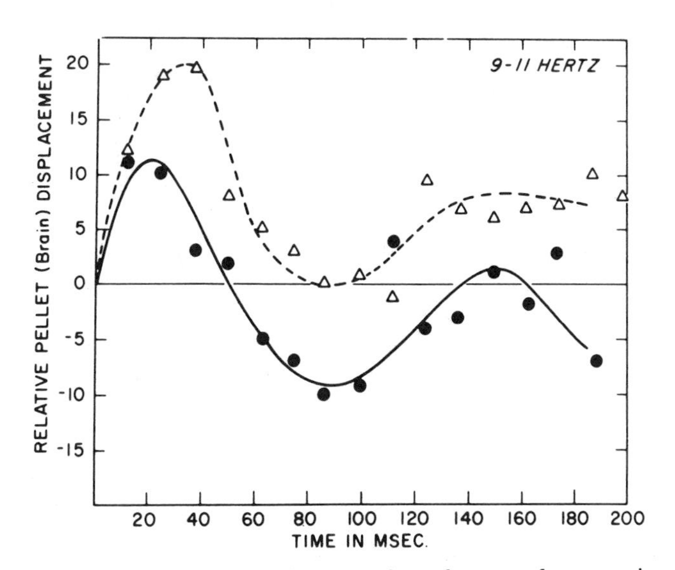

This data allows calculation of an approximate frequency for assumed brain motion during impact induced motion of the head

Fig. 5 - Relative displacement of isodensity pellets in brain of rhesus monkey

yet provided definitive data but preliminary measurements would suggest a value closer to 10 Hz. We have not been able to perform such measurements in the other primate species although we suspect that the frequency should decrease as brain size becomes larger. Thus man's brain should have a frequency around 4-5 Hz, a value suspected from one isolated observation made in a patient with radiopaque silver clips attached

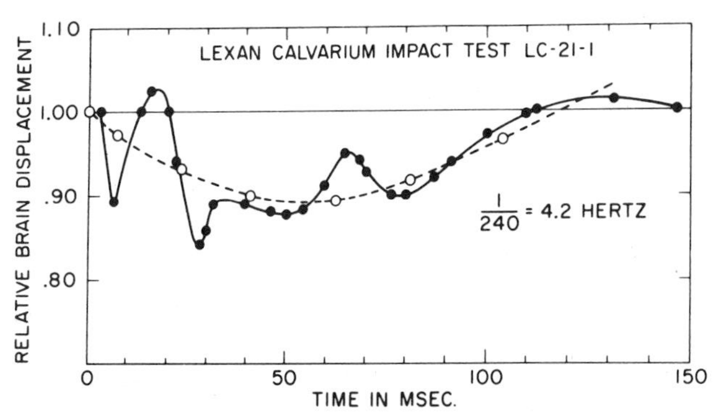

Two frequencies were measured, a relatively slow displacement≃5 Hz, and a superimposed faster frequency≃ 40 Hz. We have assumed that the former is the rotational response, the latter being a type of translational wave over the brain surface

Fig. 6 - Relative displacement of brain surface landmarks (sulci) as viewed through a transparent artificial skull with high speed cinematography

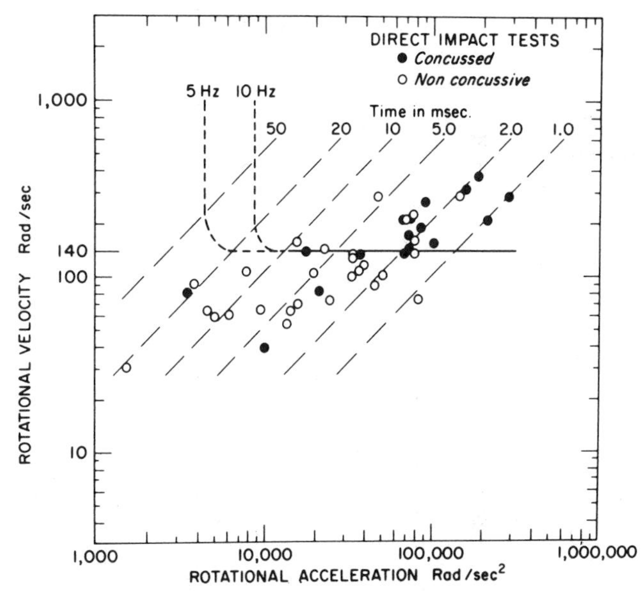

Thresholds for 50% probability of onset of cerebral concussion in terms of rotational velocity (for short duration blows) and rotational acceleration (for long duration blows) after direct head impact are depicted

Fig. 7 - Tolerance curve for rhesus monkey

to the brain surface after removal of a subdural hematoma.

The data on the tolerance to impact and whiplash of the rhesus monkey in terms of the measured rotational velocity and calculated rotational acceleration (from the equation $\dot{\theta} = \frac{\ddot{\theta}}{\omega}$) for the onset of cerebral concussion are shown in Figs. 7 & 8. It will be noted that two thresholds of rotational accelerations for long duration impacts are predicted depending on brain frequencies of 5 Hz and 10 Hz respectively. These acceleration thresholds are derived as shown by the following example for whiplash injury concussion thresholds. Implicit in the rotational velocity threshold of 300 to 350

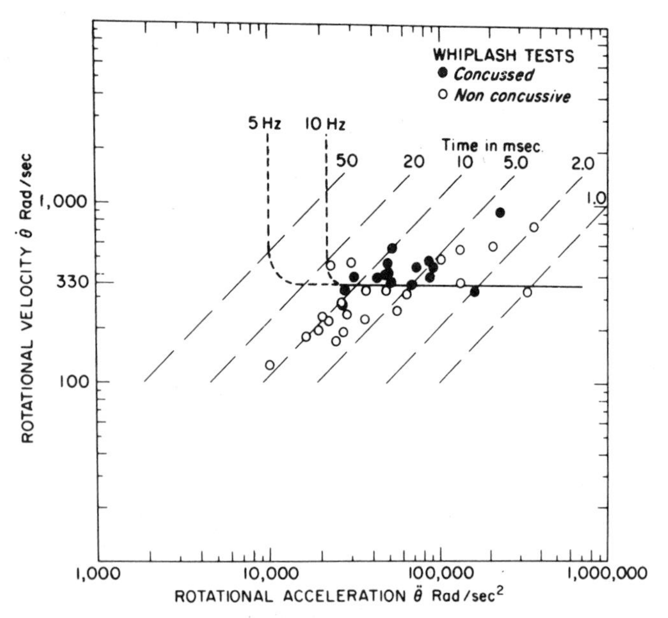

Thresholds for 50% probability of onset of cerebral concussion in terms of rotational velocity (for short duration blows) and rotational acceleration (for long duration blows) after whiplash injury (indirect impact) are depicted

Fig. 8 - Tolerance curve for rhesus monkey

radians/sec is a whole range of angular accelerations from 10,000 to 300,000 radians/ sec^2 and higher. Using equation (1) we have:

$$\dot{\theta} = \frac{\ddot{\theta}}{\omega}$$

$$\ddot{\theta} = \dot{\theta}\omega$$

$$\ddot{\theta} = 2\pi(5 \text{ to } 10) \times (300 \text{ to } 350) \text{ rad/sec}$$

$$\ddot{\theta} = 10{,}000 \text{ to } 20{,}000 \text{ rad/sec}^2 \text{ approximately}$$

Experimentally, it has not been possible to confirm these predictions for long duration impacts for two reasons: the smallness of the monkeys head and the fragility of its skull. The data shown in Figs. 7 & 8 also indicate the almost 50% greater rotation required for cerebral concussion when the head rotation is indirectly induced by whiplash injury.

The data obtained to date on the squirrel monkey during impact and whiplash injury are shown in Figs. 9 & 10. It is evident that our data for this species are not yet adequate to draw the type of tolerance curve obtained from the rhesus monkey data. The scatter of concussed and non-concussed points from both whiplash and the impact data seems to be particularly large. Figure 11 is a probability curve for onset of cerebral concussion in squirrel monkeys obtained by Higgins et al, where the conditions of the experiment were different. Restrained squirrel monkeys were subjected to sudden angulation of the head on the neck through a controlled 45° course while preventing any significant head deformation.[16,17] Thus the conditions of this

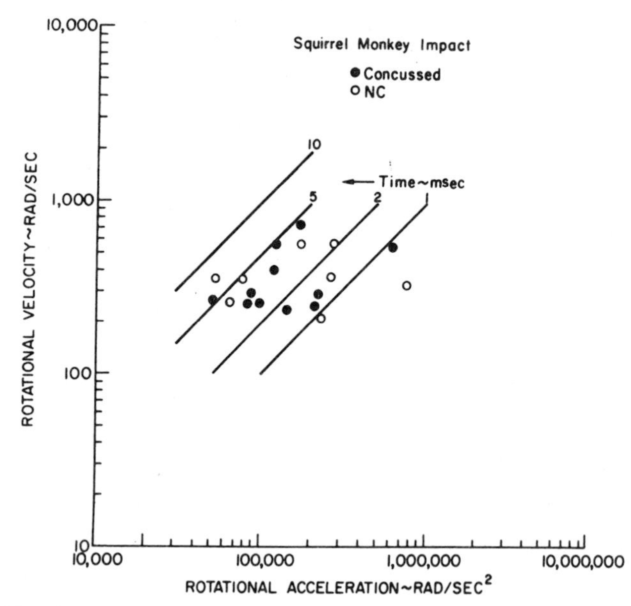

The distribution of concussed and non-concussed data does not yet allow the drawing of tolerance curves as in Figs. 7 and 8

Fig. 9 - Preliminary data plot of rotational velocity versus acceleration during direct head impact in the squirrel monkey

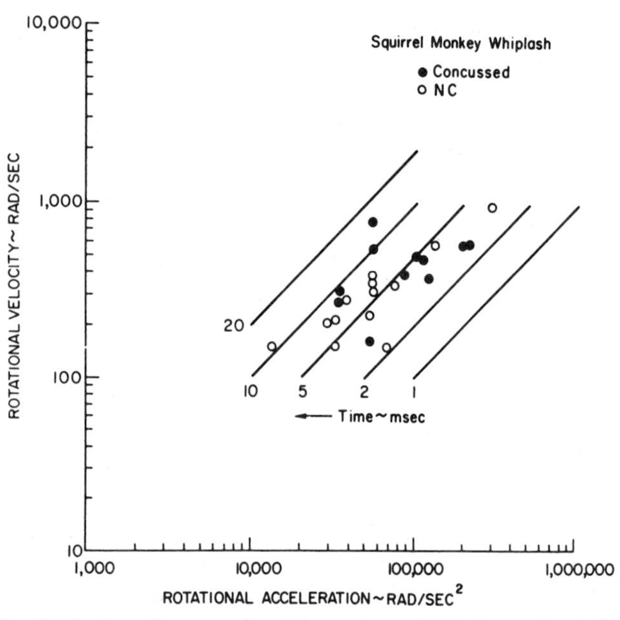

The distribution of concussed and non-concussed data does not yet allow the drawing of tolerance curves as in Figs. 7 and 8

Fig. 10 - Preliminary data plot of rotational velocity versus acceleration during whiplash injury in the squirrel monkey

experiment were analogous to our whiplash experiments where no direct head impact occurred. From Fig. 11 a value of 163,000 Radians/sec^2 for the 50% threshold for onset of concussion in squirrel monkeys is obtainable. When compared to our own whiplash injury data in this species the value of 163,000 Radians/sec^2 would fall on the extreme right of the data points shown in Fig. 10 due to the very short duration of the accelerations depicted in Fig. 11 (i.e. < 2 milliseconds).

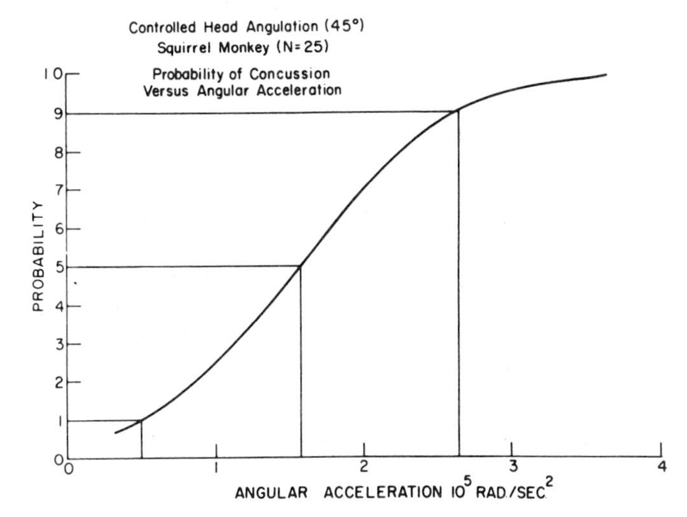

Fig. 11 - Data from Higgins et al. depicting probability of concussion onset in squirrel monkeys subjected to controlled head angulation on neck (45°). Note 50% level is approximately at 163,000 Radians/sec^2

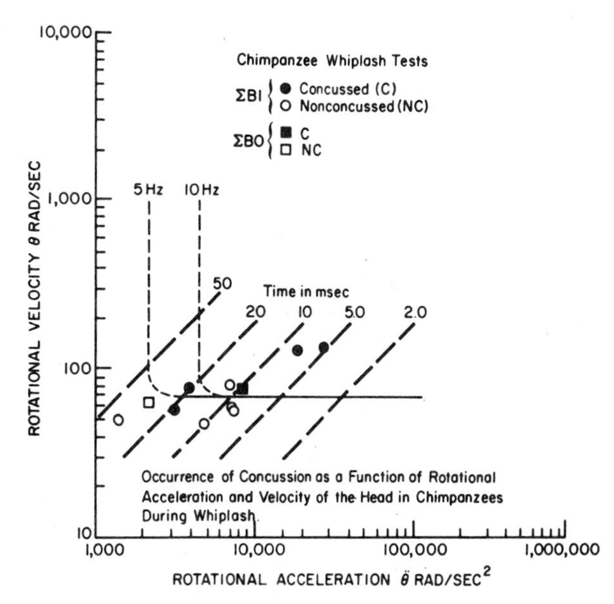

Fig. 12 - Tolerance curve for onset of cerebral concussion in chimpanzee after whiplash injury. ΣB1 and ΣB0 at top of figure refer to "eyeballs in" and "eyeballs out" position of animal relative to direction of impact

The data obtained for whiplash injury in the chimpanzee, the largest brained species in our sub-human primate series, were more susceptible to the type of analysis described above. Thus in Fig. 12 it was feasible to draw a tolerance curve depicting the 50% probability of onset of cerebral concussion in terms of rotational velocity and acceleration during severe whiplash trauma in the lightly anesthetized chimpanzee. Experiments aimed at producing concussion by direct head impact in this species have proven to be extremely difficult to perform successfully. Thus in the eight impacts studied to date, concussion was only produced in one animal in which a relatively high level of anesthesia may have lowered the concussion threshold. Levels of impact impulse as high as 2.0 lb. sec. to the inion have been achieved without producing cerebral concussion and with the durations of force applications not exceeding 4 to 5 milliseconds. The reasons for this failure to concuss chimpanzees by impact are discussed below.

The displacement of the rhesus monkey head in Cartesian coordinates was also analysed in terms of the relation between the linear x, y, and tangential velocity and acceleration of the head and the onset of cerebral concussion. No obvious correlation could be obtained similar to that noted for rotational displacements of the head on the neck. This linear displacement data will be presented in a separate report wherein its significance and possible inverse correlation with rotational displacements for cerebral concussion onset will be reviewed.[18]

In order to compare our data for predictive purposes we have used our scaling method described by equation (2). The constant C in this equation is derived from the rhesus data as follows:

$$\ddot{\theta}_D = C/M^{2/3}$$

$$\ddot{\theta}_D = 10{,}000 \text{ rad/sec}^2, \quad m = 100 \text{ grams}$$

$\therefore$ substituting in equation (2)

$$C = \ddot{\theta}_D \, (m)^{2/3}$$

$$C = 10{,}000 \frac{\text{rad}}{\text{sec}^2} (100)^{2/3} \text{ grams}$$

$$C = 2.16 \times 10^5 \frac{\text{rad(gram)}^{2/3}}{\text{sec}^2}$$

Using this constant it was possible to draw a theoretical plot relating brain mass to rotational acceleration (50% threshold for cerebral concussion) as shown in Fig. 13. The theoretically derived points for squirrel monkey, chimpanzee and man were obtained by simply substituting into equation (2) the brain mass for each species e.g. in the case of the chimpanzee -

$$\ddot{\theta}_D = C/M^{2/3}$$

$$\ddot{\theta}_D = 2.16 \times 10^5 \frac{\text{rad(gram)}^{2/3}}{\text{sec}^2} \Big/ (450 \text{ grams})^{2/3}$$

$$\text{or } \ddot{\theta}_D = 3730 \text{ radians/sec}^2$$

As is evident from Fig. 12 this theoretically derived figure shows reasonable agreement with the experimentally derived value for $\ddot{\theta}_D$ in the chimpanzee = 3000 radians/sec^2. A comparison of the two experimentally derived tolerance curves for rotation induced concussion in rhesus monkey and chimpanzee and the theoretically predicted tolerance

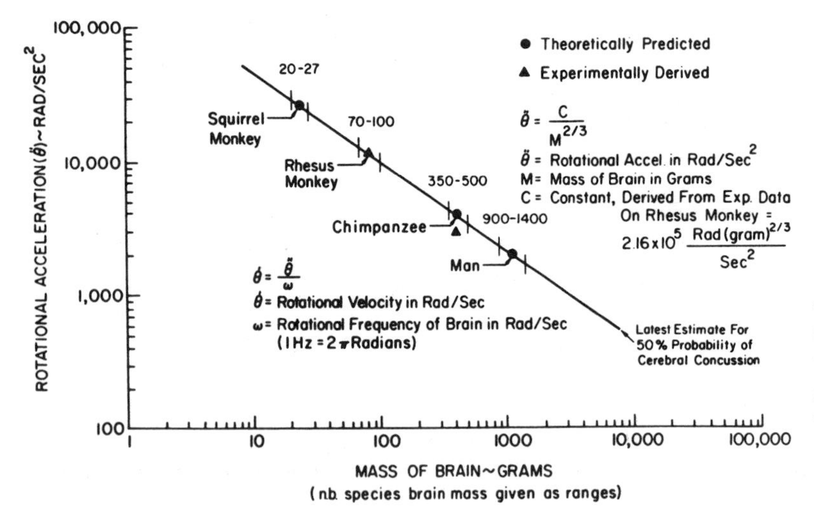

It is assumed that the crucial variable between species is mass of brain and that the crucial injury mechanism is severe shear strain imposed by brain rotation. Note that there are only two experimental data points achieved to date in our continuing attempt to test validity of this curve

Fig. 13 - Theoretical scaling of probability for onset of cerebral concussion in primates

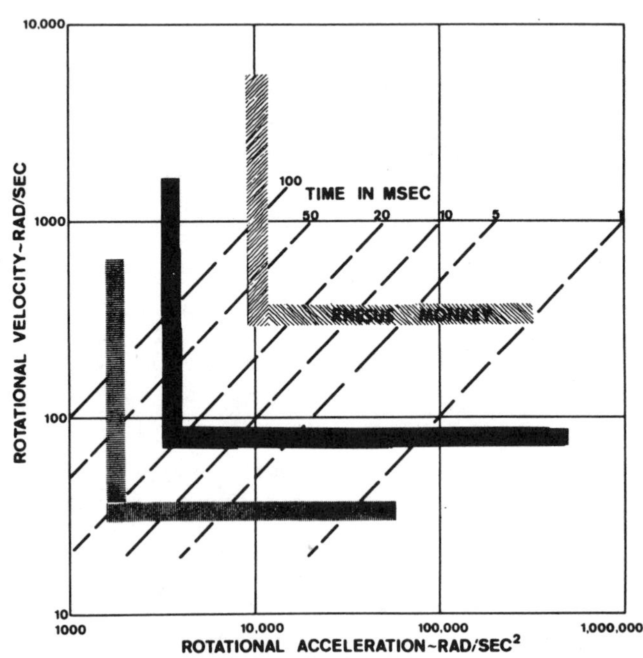

Fig. 14 - Comparative tolerance curves for rhesus monkey, chimpanzee (experimentally derived) and man (theoretically predicted) relating 50% probability for onset of cerebral concussion to rotational velocity and acceleration, assuming that contact phenomena of impact are not contributing significantly to the injury mechanism

curve for man are shown in Fig. 14. It will be noted that this preliminary data suggests approximate values of angular velocity = 50 radians/second and angular acceleration = 1800 radians/sec^2 for a 50% probability of concussion in a man with a 1300 gram brain.

DISCUSSION

It should be emphasized that the predicted relationship between rotational acceleration and brain mass for cerebral concussion shown in Fig. 13 has been validated tentatively for only two species, rhesus monkey and chimpanzee. Our data from the squirrel monkey are not yet sufficient to check the theoretical prediction for this species. If one were to attempt to utilize the data derived by Higgins et al and place the experimental value for 50% concussion in squirrel monkeys from Fig. 11 onto the plot shown in Fig. 13 it is obvious that agreement is not good. However, it must be emphasized that the technique used by Higgins et al was very different from ours and is most probably not comparable in terms of simple rotation of the head in the saggital plane.

It is obvious therefore that our predicted tolerance for the onset of cerebral concussion in man based on our scaling method in terms of rotational velocity and acceleration must be considered only as a useful hypothesis which requires further evaluation. It is possible however to test this theory in two ways, firstly, by further refinement and more precise definition of tolerance levels in the smaller brained primates and secondly by comparing predicted human tolerances for such injurious rotations to actual levels obtained from accident data in selected fortuitious cases. Figure 14 is a summation of our data depicting the scaled relationship between rhesus monkey, chimpanzee, and human tolerance curves for injurious rotational velocity and accelerations when the local effects of impact are not effective (e.g. in whiplash trauma). If the relationship between rotational thresholds for impact and whiplash induced concussion demonstrated in Figs. 7 & 8 holds also for man, then the corresponding values for injurious rotational velocity and acceleration for man after direct head impact would have to be reduced by 50% from that shown in Fig. 14. We must emphasize again that these predicted values are not to be taken as definitive but simply as approximate figures with which we can compare available data. We are currently investigating the restricting set of assumptions for our scaling theory as well as the factual basis for the scaling law stated in equation (2). There is reason to believe that contrary to our simplifying assumptions, brain is not fully elastic, that shape and size of basal foramina and bony protusions in different primate species are not geometrically equal, that the brain is not homogenous and isotropic and that one geometric scale factor between species may not be sufficient. Furthermore, we have not scaled the time factor e.g. for short duration impacts the scaling for rotational velocity would probably be inversely proportional to the 1/3 power of brain mass and not to the 2/3 power as indicated for the rotational acceleration.

There are few situations where one can obtain actual estimates of the rotations sustained by men in accidents. One study undertaken by the U. S. Army and Navy in human volunteers is attempting to measure the linear and rotational accelerations sustained by the head during low level controlled whiplash injury.[19] Preliminary data suggests that as little as 10 g impacts to a mobile carriage can be amplified many fold and rotational velocities of the head as high as 30 radians per second have been observed without obvious ill effects. Such experiments are of necessity limited to very low "tolerable" levels and can only give adequate data on the dynamic response of the head-neck system of man under such loading conditions; data of obvious value in improving scaling laws such as the one outlined above. Only accidents in man can provide the brain injury data required. We have recently described two patients who developed subdural hematomas after indirect impact alone (whiplash injury). One of these cases provided us with valuable information regarding the probable level of rotation required for brain injury in man.[20] Although in this patient, clear-cut cerebral concussion with traumatic amnesia was not described, the production of a large subdural hematoma suggests a level of injury reasonably close to the threshold for cerebral concussion. In our animal experiments, hemorrhages over the brain surface occurred only in the concussed group. It is therefore justifiable to calculate the approximate level of rotational acceleration experienced by the head of the patient as follows:

Weight of patient's car, Mp = 4500 lbs.
Weight of striking truck, Mt = 13000 lbs.
Initial velocity of car, Vp = 0 m.p.h.
Initial striking velocity of truck,
Vt = 30 m.p.h.
Velocity obtained by combined masses = Vf
Acceleration obtained by combined masses = Af

Because this was a nonelastic collision, one object struck a second stationary one and then both obtained some common velocity. This is described as:

$$Mt = 2.89\ Mp \quad \left(\text{i.e. } \frac{13000}{4500}\right)$$

$$MpVp + MtVt = (Mp + Mt)\ Vf$$

$$MpVp + 2.89\ MpVt = (Mp + 2.89\ Mp)\ Vf$$

$$2.89\ MpVt = 3.89\ Vf$$

$$Vf = \frac{2.89}{3.89}\ Vt = \frac{2.89}{3.89}\ (44) = 32.7 \text{ fps}$$

Experimental work on automobile collisions suggests a transfer time, Tt = 0.16 sec (haversine shape). This is the time required to reach peak velocity. Let us assume that the time to reach peak acceleration is one half this value or 0.08 sec.

$$Af = \frac{Vf}{\frac{Tt}{2}} = \frac{2Vf}{Tt} = \frac{2 \times 32.7}{0.08} = 409 \text{ ft/sec}^2$$

According to certain observations made on human volunteers by Ewing, there is an amplification factor of at least 2 which exists between the input acceleration to the base of a seated man and that which is measured at a point on his head.[19] Therefore the head acceleration (Ah) will be:

$$Ah = 2\ Af = 818 \text{ ft/sec}^2$$

$$\text{Rotational acceleration} = \frac{Ah}{R}$$

Where R = pivotal distance = 6 inches

$$\therefore \ddot{\theta} = \frac{818 \text{ ft/sec}^2}{1/2 \text{ ft}} = 1636 \text{ radians/sec}^2$$

From our own experimental work we have predicted that the tolerance levels for onset of cerebral concussion in man in terms of rotational velocity ($\dot{\theta}_m$) are:

$\dot{\theta}_m$ = 50 radians/sec (for short duration) (input where $t \lesssim 20$ msec)

and for rotational acceleration ($\ddot{\theta}_m$) are:

$\ddot{\theta}_m$ = 1800 radians/sec^2 (for long duration) (input where $t > 20$ msec)

Thus it is fairly satisfactory to arrive at a value of $\ddot{\theta}_m$ = 1636 radians/sec^2 from our patient's data when one considers the large number of assumptions made in arriving at this figure. This derived value is just below that of 1800 radians/sec^2 predicted by us to be the onset threshold for cerebral concussion by whiplash injury in man.

Our experiments to date do not allow us to define the role of the contact phenomena of impact and particularly how they would scale to predict tolerance to head impact in man from data in lower primates except in as much as the associated head rotations may be consistently related to the local effects of impact. Our experiments in the chimpanzee strongly indicate that the local factors of skull and scalp thickness, ratio of brain-head weights and strength of the head-neck junction contribute greatly to the effects of impacts. Thus in rhesus monkey and man the fraction of brain mass to head mass is usually greater than 1/5 whereas in the chimpanzee the relatively more massive skull and scalp reduces this fraction considerably. We have been impressed that the thick, movable and springy scalp combined with a thick strong skull in the chimpanzee serve as an excellent "helmet" which raises the threshold for impact and has to date prevented the production of clear-cut cerebral concussion by impacts approaching 4000 lbs of force applied for short durations (< 5 milliseconds, impulse < 2.5 lb.sec.). Contributing to this reduction of the importance of the contact phenomena of impact by the head structure is the reduction of head rotation by impact which is additionally impaired by the massive neck muscles of the chimpanzee. To date our impacts have not succeeded in producing head rotations approaching those achieved in concussive whiplash injury in the chimpanzee.

In conclusion, therefore, it appears that considerable work remains to be done but our current hypothesis remains a valid approach to definition of the comparative tolerances for brain injury by direct and indirect impact to the heads of primates.

We intend to develop more refined theoretical models, both discrete and continuous to further define the mechanism of brain injury as well as the tolerance of man's brain to mechanical trauma.[21]

REFERENCES

1. A. K. Ommaya ,A. Hirsch and J. Martinez, "The role of 'whiplash' in cerebral concussion." Proceedings of the 10th Stapp Car Crash Conference, Society of Automotive Engineers, New York, 197-203, 1966.
2. A. K. Ommaya, F. Faas and P. Yarnell, "Whiplash injury and brain damage: An experimental study." J.A.M.A., 204: 285-289, April 1968.
3. A. H. S. Holbourn, "Mechanics of head injuries." Lancet 2: 245, October 9, 1943.
4. A. E. Hirsch, A. K. Ommaya and R. M. Mahone, "Tolerance of sub-human primate brain to cerebral concussion." Dept. Navy, Naval Ship Research and Development Center, Report 2876, Washington, D.C., August 1968, 19 pp.
5. A. K. Ommaya and P. Corrao, "Pathologic biomechanics of central nervous system injury in head impact and whiplash trauma." Proceedings of the International Conference on Accident Pathology, Washington, D.C.: Government Printing Office, 1969.
6. A. K. Ommaya, A. E. Hirsch, E. Harris and P. Yarnell, "Scaling of experimental data in cerebral concussion in sub-human primates to concussive threshold for man." Proceedings of the 11th Stapp Car Crash Conference, Society of Automotive Engineers, New York, 47-52, 1967.
7. A. K. Ommaya, A. E. Hirsch, E. S. Flamm and R. M. Mahone, "Cerebral concussion in the monkey: An experimental model." Science 153: 211-212 , July 1966.
8. A. Golub and F. E. Grubbs, "Analysis of sensitivity experiment when the levels of stimulus cannot be controlled." J. Amer. Stat. Assoc. 51: 257-265 (NR 274), June 1956.
9. M. Kornhauser, "Prediction and evaluation of sensitivity to transient acceleration." J. Appl. Mechanics 21: 371, 1954.
10. A. K. Ommaya, J. W. Boretos and E. E. Beile, "The Lexan calvarium: An improved method for direct observation of the brain." J. Neurosurg. 30: 25-29, 1969.
11. D. J. Sass, "A discussion of the problems of restraint in experimental investigations of acceleration injury. A lucite water immersion restraint for a vibration injury study with cats." J. Biomechanics 2: 157-162, May 1969.
12. D. Sass, P. Corrao and A. K. Ommaya, "Unpublished observations."
13. F. Faas and A. K. Ommaya, "Brain tissue electrolytes and water content in experimental cerebral concussion in the monkey." J. Neurosurg. 28: 137-144, February 1968.
14. F. Letcher, P. Corrao, R. Mahone and A. K. Ommaya, "Cerebral concussion in the chimpanzee." (To be published).
15. R. Mahone, P. Corrao, E. Hendler, M. Shulman and A. K. Ommaya, "Theory on the mechanics of whiplash produced cerebral concussion in primates." J. Biomechanics (In press).
16. L. S. Higgins and R. A. Schmall, "A device for the investigation of head injuries effected by non-deforming head accelerations." Proceedings of the 11th Stapp Car Crash Conference, Society of Automotive Engineers, New York, 1967.
17. F. Unterharnschiedt and L. S. Higgins, "Traumatic lesions of brain and spinal cord due to non-deforming angular accelerations of the head." Texas Repts. Biol. & Med. 27: 127-166, 1969.
18. A. Hirsch and A. K. Ommaya, "The role of head translation in cerebral concussion." To be presented at the Stapp Car Crash Conference, 1970.
19. C. L. Ewing, D. J. Thomas, L. M. Patrick, G. W. Beeler and M. J. Smith, "Living human dynamic response to $--G_x$ impact acceleration. II. Accelerations measured on the head and neck." Proceedings of the 13th Stapp Car Crash Conference, Society of Automotive Engineers, New York, 400-415, 1969.
20. A. K. Ommaya and P. Yarnell, "Subdural hematoma after whiplash injury." Lancet 2: 237-239, August 1969.
21. A. K. Ommaya, "Trauma to the Nervous System. An Exhibit." 1969.

700899

Protection from Brain Injury: The Relative Significance of Translational and Rotational Motions of the Head After Impact

Arthur E. Hirsch
U.S. Department of Transportation

Ayub Ommaya
National Institutes of Health Dept. H.E.W.

Abstract

The rotational and translational rigid body motions of the head after impact were evaluated by high-speed cinematography in Rhesus monkeys with and without a cervical collar. When a collar was worn, animals displayed increased tolerance to occipital impact for the onset of cerebral concussion. Although head rotations were reduced in this nonconcussed protected group, translational motion of the head *exceeded* that attained by concussed monkeys *not* wearing collars but struck at equivalent impulse levels. These data emphasize the inadequacy of current head impact tolerance criteria which relate the occurrence of brain injury to translational head motions.

AT THE Tenth Stapp Conference we presented our data suggesting that rotational motion of the head on the neck was an extremely important factor in the production of traumatic unconsciousness (cerebral concussion) after head impact (1).[1] Subsequently, we have demonstrated that concussion as well as visible hemorrhages and contusions at the brain surface could be produced during "whiplash" experiments wherein Rhesus monkeys seated in a mobile cart underwent impulsive loading of the head without direct head impact (2). Our observations that such visibly injurious effects on the brain could be produced by head rotation without direct impact were confirmed by Martinez, et al., in Belgian hares subjected to experimental whiplash injury (3). The fact that head rotation due to this type of mechanical input produces traumatic unconsciousness, as well as

[1]Numbers in parentheses designate References at end of paper.

brain hemorrhages and contusions, certainly suggests that rotation would contribute to the injuries caused by impact to the freely movable head. The energy of impact (to the freely movable head) evokes several types of head response: first, local skull bending directly beneath the impact itself; second, excitation of the normal mode oscillations of the skull as a shell, with accompanying pressure oscillations of the head contents; and third, rigid body motions of the head-neck system. The crucial problem in considering the design of protection for the brain during head impact is to determine in quantitative terms the role played by each of these phenomena in the production of brain injury.

In the report referred to above (1), we showed how the tolerance of Rhesus monkeys to concussive head impact was increased when a cervical collar was worn. These data are summarized in Table 1, which indicates that only two cerebral concussions out of 17 impacts occurred at impulse levels above that which had been shown to produce cerebral concussion in at least 50% of a group of Rhesus monkeys not wearing collars (Fig. 1) (4). In the report given at the Tenth Stapp Conference (1), we also discussed the possible ways in which such a neck collar served to reduce the occurrence of cerebral concussion, and concluded that the reduction of head rotational response to the impact was probably the crucial factor. Thus it may be seen from Fig. 2 that if one plots the peak rotational velocity of the head against the impact impulse, angular velocity of heads

Table 1—Impulse Versus Occurrence of Brain Concussion in Rhesus Monkeys Wearing Cervical Collars Exposed to Occipital Impact

Impulse lb • sec[a]	Concussive Effect
0.511	No Concussion
0.535	No Concussion
0.538	No Concussion
0.542	No Concussion
0.561	No Concussion
0.685	No Concussion
0.739	Concussion
0.753	No Concussion
0.753	No Concussion
0.806	No Concussion
0.841	No Concussion
0.894	No Concussion
0.929	No Concussion
0.941	Concussion
0.954	No Concussion
0.958	No Concussion
0.963	No Concussion

[a]Concussion in 50% monkeys occurs at 0.437 lb • sec.

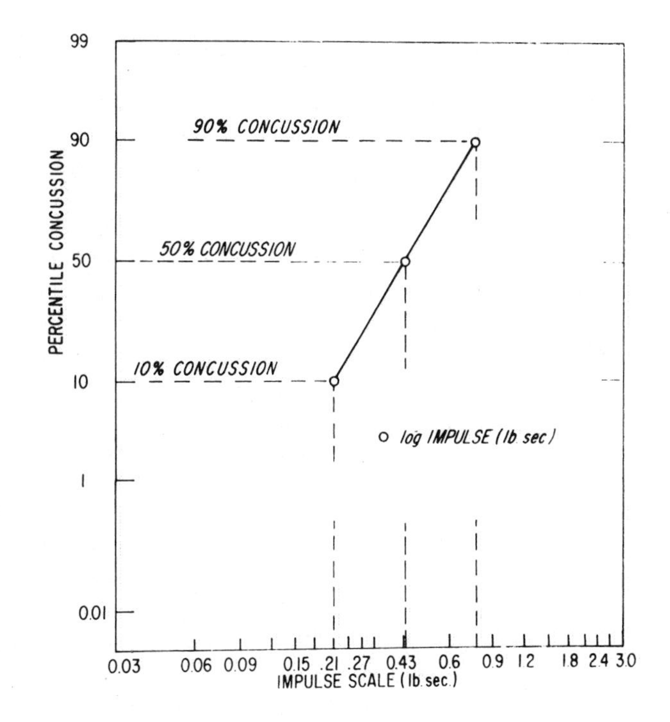

Fig. 1—Logarithmic probability plot of impact impulse versus percentile concussion for Rhesus monkeys subjected to occipital impact of short duration

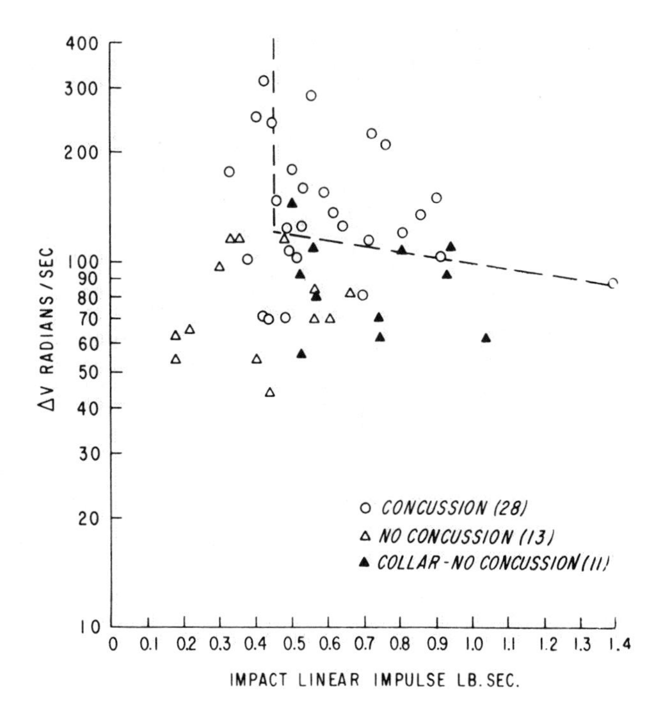

Fig. 2—Sensitivity curve relating impulse, change in angular velocity, and concussive threshold after head impact with and without a cervical collar

of the nonconcussed animals wearing neck collars tends to fall below the level of the majority of the concussed animals not wearing neck collars, in spite of the fact that the impulse delivered to the collar-wearing animals always exceed the 50% concussion threshold of unrestrained animals (1).

In this interpretation of our data, which attributes major importance to rotation of the head on the neck in the production of cerebral concussion, we had not previously considered the possibility that the collar exerted its protective effect by altering the response characteristics as well as the loading pattern of the head so as to reduce the change in angular velocity on impact. The higher impulse levels required for concussion might be explained by the fact that the head was now moving with the trunk and that this considerably larger effective mass would require a larger amount of energy to produce rigid body motions, and that translational as well as rotational responses were therefore reduced in the animals with collars. In evaluating this factor, we have measured the translational accelerations of the heads of such collar-wearing animals and conclude that the data do not support such an explanation of the protection shown. Table 2 summarizes the data, obtained by high-speed cinematography, on the translational head motions of 12 head impacts in Rhesus monkeys. These cinematographic films were obtained at speeds ranging 3000-4000 frames/sec. Digitization of the head displacement data was performed frame by

Table 2—Peak Translational Acceleration, Impulse, and Occurrence of Concussion as a Result of Occipital Blows to Rhesus Monkeys With and Without Collars

Test Numbers	Collars	Impulse lb • sec	Translational Accelerations			C or NC
			A_t (G)	A_x (G)	A_y (G)	
V-681	Yes	0.535	188.13	175.41	167.98	NC
V-175-1	No	0.556	107.39	107.46	9.08	C
V-686	Yes	0.608	700.29	320.42	676.25	NC
V-42-1	No	0.610	153.24	98.28	118.49	C
V-684	Yes	0.646	555.37	260.21	507.76	NC
V-195-1	No	0.672	184.16	184.34	14.17	C
V-685	Yes	0.658	337.62	202.48	336.92	NC
V-183-1	No	0.773	154.85	153.67	20.26	C
V-682	Yes	0.806	322.87	323.70	105.63	NC
V-171-1	No	0.815	136.21	136.20	8.32	C
V-693	Yes	0.958	166.51	168.22	91.20	NC
V-169-1	No	0.952	159.61	134.03	87.55	C

Note: C = Concussed
NC = Not concussed
A_t = Tangential acceleration

frame on an Auto-Trol 3400 linear digitizer. These displacement data were then differentiated once to obtain velocities and twice to obtain accelerations using an extended IBM 1130 computer.

Six of the 12 impacts shown in Table 2 were to animals not wearing collars, all of which were rendered unconscious at impulse levels exceeding 0.5 lb • sec. These six animals were selected at random on the basis of matched impulse value only. The six remaining impact tests at equivalent impulse levels were on animals wearing collars. It should be noted that none of the six collar-protected animals were concussed, while all of the six unprotected animals were concussed. It should also be noted that the translational acceleration response of the collar-wearing animals was considerably higher than the head translations attained by the unprotected animals, rather than lower, as would be predicted from the "larger effective mass" hypothesis for the collar effect. The data from Table 2 are graphically illustrated in Figs. 3–5. As indicated by these data and by the data in Fig. 2, animals wearing a cervical collar are protected from the concussive effects of a blow, not because of the generalized reduction of rigid

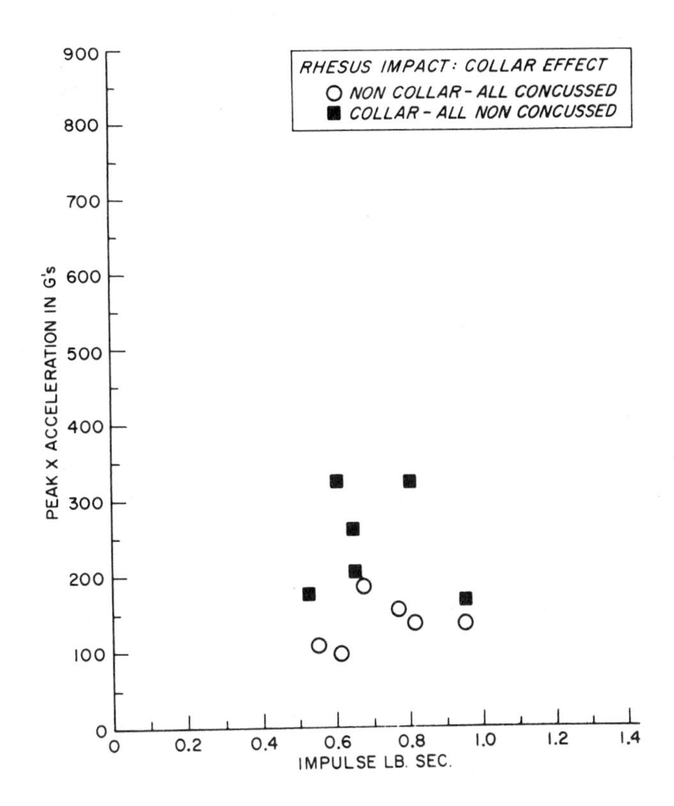

Fig. 3—Peak head A_x versus impact impulse for Rhesus monkeys with and without collars in tests with occipital blows

body motion, but rather by the specific reduction of head rotation. It would thus appear that the Holbourn hypothesis (5), which assigns head rotation a major role in the production of brain injury, is supported (6, 7).

In conclusion, it is our opinion that no convincing evidence has to this date been presented which relates brain injury and concussion to translational motion of the head for short duration force inputs, whether through whiplash or direct impact. Engineers and safety designers should be aware of this when they employ currently popular head tolerance criteria in the design of head impact protection devices (8).

References

1. A. K. Ommaya, A. E. Hirsch, and J. L. Martinez, "The Role of 'Whiplash' in Cerebral Concussion." Proceedings of Tenth Stapp Car Crash Conference, P-12, paper 660804. New York: Society of Automotive Engineers, Inc., 1966.

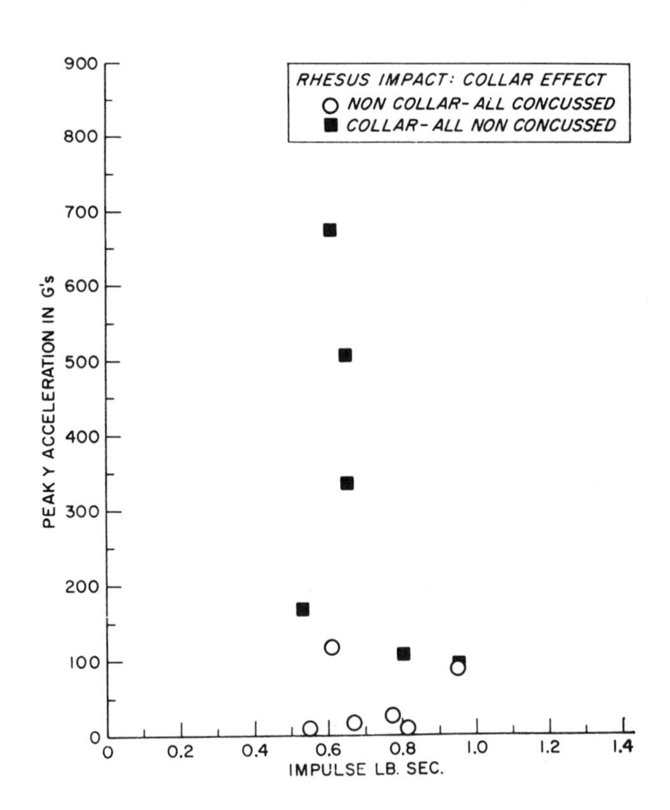

Fig. 4—Peak head A_y versus impact impulse for Rhesus monkeys with and without collars in tests with occipital blows

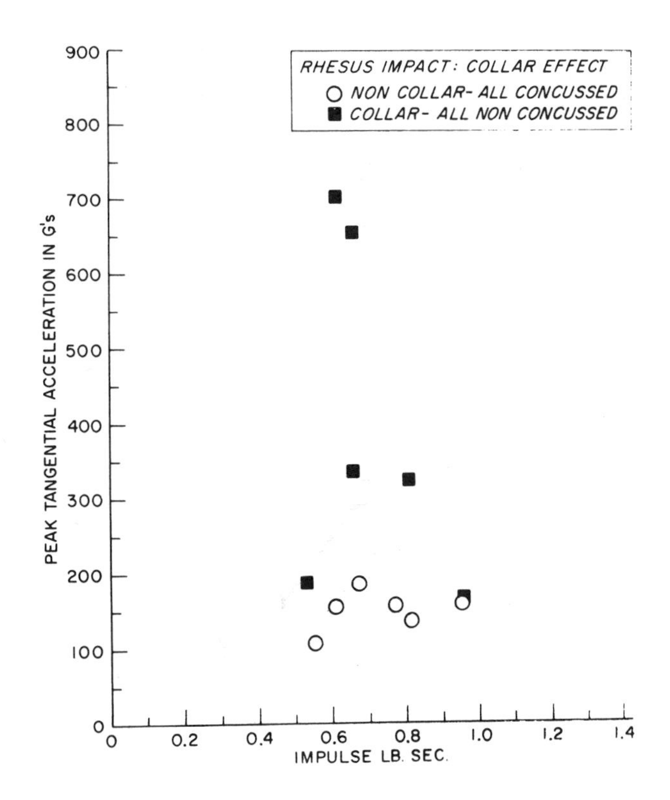

Fig. 5—Peak head A_t versus impact impulse for Rhesus monkeys with and without collars in tests with occipital blows

2. A. K. Ommaya, F. Faas, and P. Yarnell, "Whiplash Injury and Brain Damage: An Experimental Study." JAMA, Vol. 204, April 1968, pp. 285-289.

3. J. L. Martinez, J. K. Wickstrom, and B. T. Barcelo, "The Whiplash Injury — A Study of Head-Neck Action and Injuries in Animals." ASME Paper No. 65-WA/HUF-6, 1965.

4. A. K. Ommaya, A. E. Hirsch, E. S. Flamm, and R. M. Mahone, "Cerebral Concussion in the Monkey: An Experimental Model." Science, Vol. 153, July 1966, pp. 211-212.

5. A. H. S. Holbourn, "Mechanics of Head Injury." Lancet, Vol. 245, 1943, pp. 438-441.

6. A. K. Ommaya, "Trauma to the Nervous System. A Clinical and Experimental Study." Annals of the Royal College of Surgeons of England (Hunterian Lecture), Vol. 39, December 1966, pp. 317-347.

7. A. E. Hirsch, A. K. Ommaya, and R. M. Mahone, "The Tolerance of Sub-Human Primate Brain to Cerebral Concussion." Impact Injury and Crash Protection, Eds: E. S. Gurdjian, et al., Springfield, Ill.: Charles C Thomas, Publisher, 1970, pp. 352-371.

8. C. W. Gadd, "Use of a Weighted-Impulse Criterion for Estimating Injury Hazard." Proceedings of Tenth Stapp Car Crash Conference, P-12, paper 660793. New York: Society of Automotive Engineers, Inc., 1966.

700909

Fracture Behavior of the Skull Frontal Bone Against Cylindrical Surfaces[1]

Voigt R. Hodgson
Wayne State University

Jule Brinn
Chrysler Corp.

L. M. Thomas and S. W. Greenberg
Wayne State University

Abstract

A test program has been conducted to determine the fracture behavior of the human frontal bone against two different rigid cylindrical surfaces; one surface was of 1 in. radius and one was of 5/16 in. radius; both were 6½ in. long. The purpose of this research program was to provide human tolerance data which would:

1. Assist in the design of structures likely to be impacted by the human head.
2. Extend the calibration range of frangible headforms.

Twelve cadavers were tested in this program; seven against the 1 in. radius cylinder and five against the 5/16 in. radius cylinder. The test arrangement employed a guided drop of the test surface against a stationary head which was free to rebound. Drop heights were increased progressively until borderline fractures were obtained.

The large radius shape consistently yielded linear fractures indicating that it is effectively a blunt surface. Fracture loads ranged 950-1650 lb. The small radius shape yielded two linear fractures and three localized elliptical fractures indicating that it is in the transition range between a blunt and concentrated surface. Fracture loads ranged 700-1600 lb.

[1]This study was supported in part by a grant received from the Chrysler Corp.

THERE ARE A NUMBER of relatively narrow structures within automobiles which can be struck by the head. These include:

1. "A" Pillars.
2. Seat structures.
3. Visor brackets.
4. Roof rails.
5. Head restraint supports.
6. Rear view mirror brackets.
7. Side window frames.

Many of these structures would create visibility problems if increased in size to an unrealistic extreme. Therefore, careful analysis will be required in order to obtain an optimum safety design tradeoff. One of the key factors in establishing this balance is knowledge of the strength of the skull against these types of long narrow surfaces. Studies have been reported on the strength of the frontal bone against various flat surfaces; these include a broad flat surface (1)[2]; a 1⅛ in. diameter flat surface (2) and a ½ in. diameter flat surface (3). However, it is not possible to extrapolate this data to determine the strength of the frontal bone against a cylindrical surface. An experimental program was therefore indicated. Two cylindrical shapes were selected for this test work. One was 1 in. in radius and the second was 5/16 in. in radius; both were 6½ in. long to assure a full length contact zone. The 1 in. radius cylinder was considered to approach a blunt surface; the 5/16 in. radius surface was included to obtain some indication of the sensitivity of the frontal bone to different surface radii. These surfaces are shown in Fig. 1. The bone fracture study was undertaken by the Department of Neurosurgery, College of Medicine, Wayne State University. The conventional technique of impacting embalmed cadavers was employed.

Test Procedure

Each of the cadavers was situated with the legs supported by the floor, and the upper torso was suspended by means of shock cord so that the neck was in a slightly extended position as shown in Fig. 2. The frontal bone was positioned to be struck symmetric with respect to the sagittal plane near the minimum radius of the forehead. The cylindrical forms were attached to an assembly which was guided in free fall by two ⅛ in. taut steel cables. Each form-guide assembly weighed 10 lb (approximate head weight) for comparison with previous tests.

[2]Numbers in parentheses designate References at end of paper.

Fig. 1—View of the 1 in. radius and 5/16 in. radius cylindrical surfaces

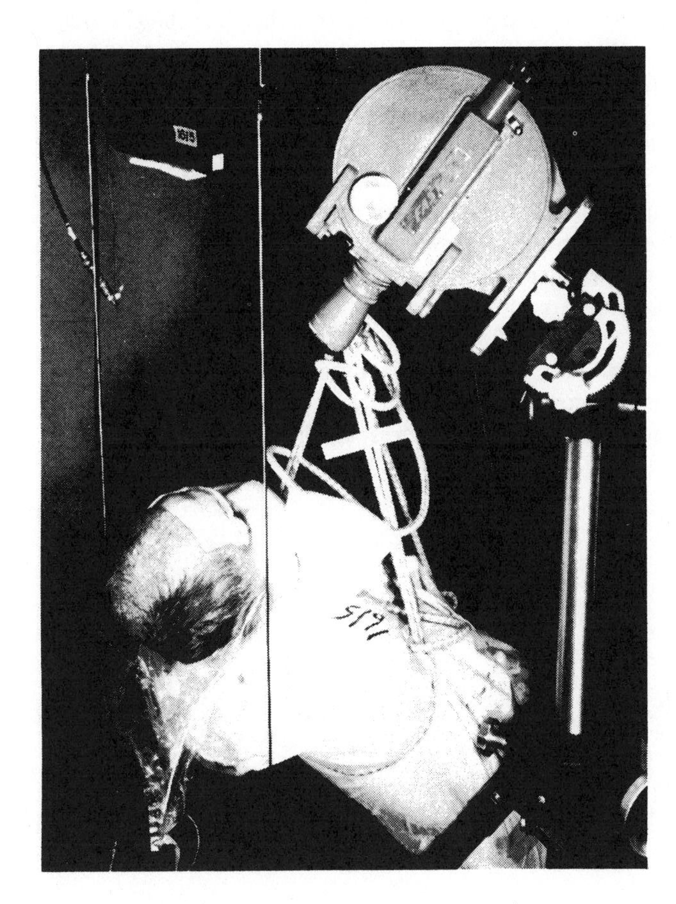

Fig. 2—Cadaver in bungy cord sling preparatory to receiving a blow on the frontal bone with the 1 in. radius cylindrical surface, 0.250 in. thick 1.98 lb/ft^3 dylite pad attached

An Endevco Model 2222 accelerometer was mounted on the form opposite the impact site, from which force of impact was obtained by multiplying form acceleration by form weight of 10 lb. A similar accelerometer was mounted on the occiput opposite the blow to record head accelerations at that point. Both pickups were shock calibrated prior to the tests. A block diagram of the instrumentation is shown in Fig. 3.

The form was first dropped onto the forehead of the intact cadaver from a height of 5 in. Because of the changes which occur with repeated impacts against soft tissue which overlies bone, it was deemed best to replace the soft tissue with a substitute material after the initial blow. Dylite pads of several densities were tested and 1.98 lb/ft^3 density was chosen as being most representative of soft tissue. However, there was a large variation in fat content among cadavers, and the forehead skin thickness varied with location and among cadavers; consequently consistency was the main reason for the choice of dylite. A repeat drop test was performed after replacing the soft tissue at the impact site with ¼ in. thick dylite of 1.98 lb/ft^3 density. Thereafter, the drop height was increased in 5 in. increments up to bone failure which usually occurred at 15 or 20 in. A fresh piece of dylite was used for each increment. Failure was always held to a threshold fracture as evidenced by a departure from a smooth head acceleration trace on the oscillogram. Seven cadavers were struck with the 1 in. radius form and five with the 5/16 in. radius form. Each of the pulses were analysed to determine peak force, peak head acceleration, pulse time, and rise time.

High-speed photographs of selected cases were recorded to determine the crack propagation patterns of the linear skull fractures. The photographs were taken with a Fastax Model WF 17 T camera using a Wallensak Fastax goose control unit WF 301, to time the event near 7000 frames/sec. High-speed photos of elliptical patterns could not be obtained because the test form covered these fractures as they developed.

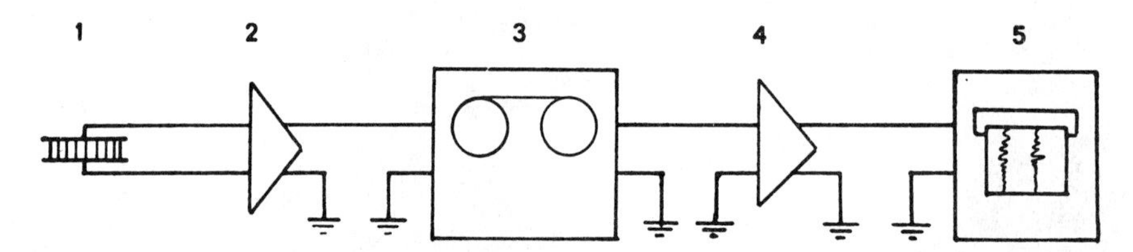

Fig. 3—Instrumentation Diagram: (1) Endevco Model 2222 accelerometer, SN MB-54 Head, SN MB-55 Striker; (2) Endevco 2704 Dynamonitor; (3) Ampex Tape Model SP-300; (4) W.S.U. Galvo Amplifier; (5)Honeywell Model 1108 Visicorder-1650 galvonometers. System Frequency Response Flat to 1000 Hz

After fracture, the heads of the cadavers were removed and weighed. They were then dissected and major diameters, skull thickness, and skull volume were measured.

Cadaver statistics are given in Table 1.

Results

Fig. 4 illustrates the linear fracture of the frontal bone of cadaver 1504 caused by the 1 in. radius striker. Four out of the seven cadavers struck with this surface fractured in this manner. It appears from the high-speed movies that this type of fracture originated at either the right or left supra-orbital notch as shown by the sequence of frames in Fig. 5. Anatomical investigation revealed that the inner table was always involved, with considerable variation among specimens. Of the remaining three fractures, two arose from the frontonasal junction and one ran laterally between the blow site and left temporal bone.

It is understandable that there would be some variation in fracture pattern among specimens in which considerable anatomical difference exists, as shown in Figs. 6A and 6B. Nevertheless, there is enough similarity that the fractures produced by the 1 in. radius striker were in good agreement with predictions made by Gurdjian, Webster, and Lissner

Table 1—Cadaver Statistics

No.	Cadaver	Age	Race	Date of Death	Date of Embalming	Cause of Death
1	1486	82	Caucasian	12/1/68	12/6/68	Cancer of prostate
2	1492	37	Caucasian	12/1/68	12/13/68	Chronic alcohol
3	1582	60	Negro	5/2/69	5/14/69	Pneumonia
4	1581[a]	33	Caucasian	5/5/69	5/5/69	Complications from medications, heart condition
5	1504	83	Caucasian	12/18/68	12/20/68	Heart disease
6	1580	47	Negro	4/30/69	5/12/69	Chronic alcohol; acute hemorrhage of pancreas
7	1536	68	Caucasian	1/20/69	2/3/69	Pulmonary edema
8	1589	59	Caucasian	5/27/69	6/3/69	Gangrene left stump (single amputee)
9	1615	63	Caucasian	7/12/69	7/16/69	Marginal peptic ulcer, septapatic abcess
10	1616	79	Caucasian	7/18/69	7/22/69	Cerebral arteriosclerosis
11	1584	61	Caucasian	5/20/69	5/22/69	(Funeral case)
12	1596	79	Caucasian	6/8/69	6/14/69	Cancer of mouth
13	1471	66	Negro	10/15/68	10/26/68	Pneumonia
14	1614	81	Negro	7/14/69	7/18/69	Arteriosclerosis

[a]Female

(4) on the basis of fracture tests and stress coat analysis of blunt impacts to the frontal bone. In 1953 they wrote, essentially, that there is always an area of inbending immediately beneath and around the point of the blow, with discrete areas of outbending peripheral to the area of inbending. In the outbended bone, tensile stresses of sufficient magnitude may develop to cause a fracture line to start and extend toward the point of impact and in the opposite direction. In their tests of 100 mid-frontal blows to adult skulls, 74% of the fractures were initiated in the neighborhood of the supra-orbital notch, 9% were initiated in the frontonasal junction, and 30% were horizontal fractures in left or right lateral frontal areas — all fractures extending toward the blow site and away from the point of initiation. Since the 1 in. radius striker produced linear fractures similar to those obtained by Gurdjian, et al., it may be considered to behave as a blunt surface for the condition of these tests (velocity = 10 ft/sec).

Of the five cadavers struck with the 5/16 in. radius striker, three received a localized fracture and two received linear fractures originating in the supra-orbital notch. These results indicate that the 5/16 in. radius surface is near the crossover between blunt surfaces, which produce high

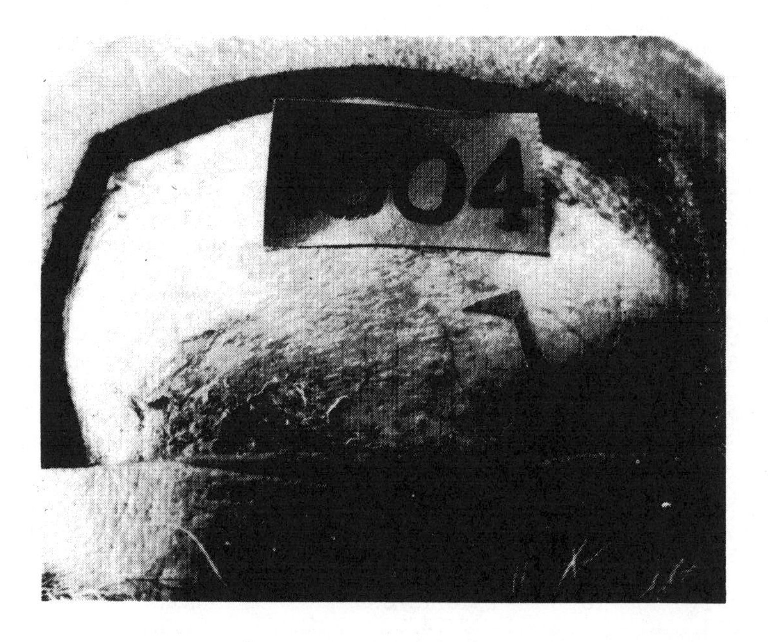

Fig. 4—Linear fracture which originated in or near the left supra-orbital notch and then extended toward the blow site and to the base of the skull

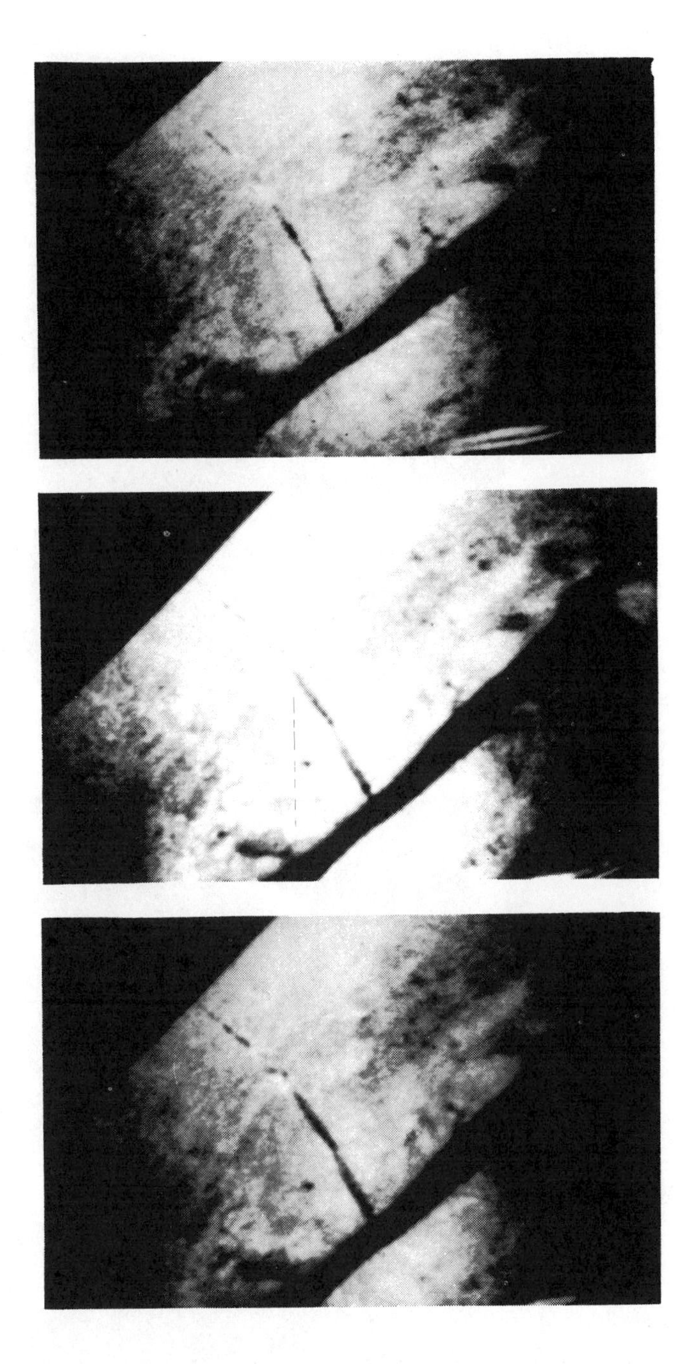

Fig. 5—High speed motion picture frame sequence (7000 fps) showing the development of a fracture beginning in the orbit and extending toward the midfrontal blow site. Read top to bottom

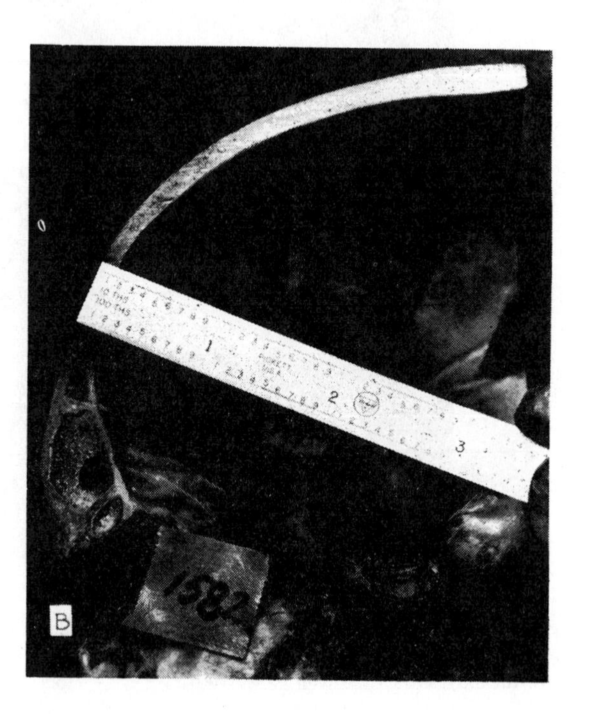

Fig. 6 A & B—Cross section of two skulls to illustrate the anatomical variation among cadavers: A—thick skull 1614; B—thin skull 1582

bending stress, and small sharp area surfaces which cause local failure by crushing or shear. One of the localized elliptically shaped fractures is shown in Fig. 7. The localized fractures also involved the inner table.

Using the 1 in. radius striker, there was a variation of peak force to produce fracture of from 950 lb for female cadaver 1581 to 1650 lb for cadaver 1615, with an average peak force among the seven cadavers of 1260 lb. The 5/16 in. radius striker was more erratic in its performance. It produced three localized fractures at peak forces ranging 700-1730 lb and two linear fractures at peak forces of 1280 and 1600 lb. Average peak force among five cadavers to produce fracture with the 5/16 in. striker was 1230 lb.

Uniaxial head accelerations in the a-p direction were measured opposite the blow. When the peak force obtained from the striker measurements was divided by peak head acceleration, the derived head weight was consistently lower than actual head weight (Table 2). This means that the input force cannot be inferred from such uniaxial acceleration measurements, probably because of skull deformations and/or head rotation. This lack of correlation between force and head acceleration has also been noted by others (2, 5).

Cadaver head measurements are tabulated in Table 3. The dimensions of the frontal bone and its component layers are listed in Table 4 for all cadavers.

Fig. 7—Localized fracture of cadaver 1584 skull due to impact with the 5/16 in. radius cylindrical surface.

Table 2—Frontal Bone Fracture Test Results

No.[a]	Cadaver	Force,[b] lb	Head Acceleration, g	Pulse Time, sec	Rise Time, sec	Drop Height, in.	Velocity, ft/sec	Head Weight, lb	F/A[g] lb	Remarks
1	1486	1320	400	0.0017	0.00040	20	10.3	—	3.3	—
2	1504	1320	300	0.0026	0.00051	15	10.3	10.4	4.5	e
3	1536	1200	260	0.0027	0.00050	15	9.0	8.8	4.6	—
4	1581[f]	950	330	0.0022	0.00050	25	11.6	5.4	2.9	c
5	1582	1350	270	0.0025	0.00062	15	9.0	8.2	5.0	—
6	1589	1650	300	0.0017	0.00060	15	9.0	10.3	5.5	—
7	1615	1070	275	0.0026	0.00045	15	9.0	7.7	3.9	—
	Average	1260	300	0.0023	0.00051	18	9.8	—	—	—
8	1471	1600	360	0.0015	0.00056	20	10.3	9.2	4.4	—
9	1584	700	125	0.0019	0.00032	5	5.2	7.4	5.6	d
10	1596	1730	338	0.002	0.00050	25	11.6	9.6	5.1	d
11	1614	1280	360	0.0022	0.00120	20	10.3	9.6	3.6	e
12	1616	940	150	0.0023	0.00040	15	9.0	7.8	6.3	d
	Average	1230	266	0.0024	0.00059	17	9.6	—	—	—
13	1492[c]	—	260	0.0011	0.00032	15	9.0	—	—	—

[a]Nos. 1-7 struck with 1 in. radius form; 8-13 struck with 5/16 in. radius form.
[b]Peak values.
[c]Cadaver 1492: head dropped onto form. Form dropped on heads of all others.
[d]Localized elliptical fracture.
[e]Head only.
[f]Only female in group.
[g]F = peak force; A = peak head acceleration.

Table 3—Cadaver Head Measurements

No.	Cadaver	Race	Major Outside Skull Dimensions, in.			Major Inside Skull Dimensions, in.			Head Weight, lb	Skull Volume, cc	Forehead Skin Thickness, in.
			a-p	s-i	lat	a-p	s-i	lat			
1	1486	C	7.00	5.65	5.43	6.75	5.40	5.12	—	1400	—
2	1492	C	7.40	5.82	5.34	7.20	5.54	4.98	9.5	1525	0.24
3	1582	N	7.16	5.67	5.68	6.50	5.46	5.34	8.2	1480	0.20
4	1581[a]	C	6.45	5.49	5.51	6.30	5.16	5.03	5.4	1250	0.06
5	1504	C	7.19	6.56	5.85	6.80	5.78	5.53	10.4	1520	0.20
6	1580	N	7.49	5.74	5.80	6.82	5.33	5.49	6.9	1475	0.15
7	1536	C	6.96	6.08	5.53	6.32	5.75	5.20	8.8	1340	0.19
8	1589	C	7.58	6.21	5.76	7.10	5.95	5.33	10.3	1570	0.21
9	1615	C	7.45	5.90	5.65	7.00	5.30	5.35	7.7	1545	0.14
10	1616	C	6.78	5.72	5.64	6.65	5.47	5.38	7.8	1390	0.19
11	1584	C	7.16	5.54	5.86	6.64	5.45	5.80	7.4	1460	0.08
12	1596	C	6.88	5.87	5.98	6.40	5.58	5.65	9.6	1500	—
13	1471	N	7.19	5.35	5.60	6.90	5.00	5.24	9.2	1350	0.22
14	1614	N	6.61	5.97	6.57	6.30	5.47	5.74	9.6	1400	0.26

[a]Female.

Application of the Data to Automotive Design

The above data can be applied to the design of automotive components in two ways:

1. Direct application in guiding design and testing.
2. Indirect application by development and validation of a trauma head form.

Direct Application — This approach is uncomplicated and would apply to simple, relatively rigid surfaces. However, the more typical automotive surface is yielding and deformable to some extent. Thus the contact surface area will generally increase during the deformation process. It will not usually be possible to associate the varying headform forces measured during impact with the varying contact area occurring during the same impact. A realistic evaluation will thus not often be attainable. The cylindrical surfaces chosen in this experimental study were deliberately made hard and unyielding to avoid just this sort of indeterminacy.

Application to a Trauma Headform — A trauma headform is intended to serve as a mechanical analog, which will automatically take into account the varying loads and geometries occurring during an impact, and indicate the gross trauma. The behavior of the headform must be validated over a wide range of impact geometries; it would then be assumed to accurately simulate the human for the unverified "in-between" conditions.

Table 4—Thickness of Frontal Bone Components at the Blow Site

No.	Cadaver	Outer Table	Diploe	Inner Table	Total[a]
1	1471	0.07	0.09	0.06	0.22
2	1486	0.11	0.06	0.08	0.26
3	1492	0.07	0.10	0.06	0.23
4	1504	0.06	0.09	0.09	0.24
5	1536	0.06	0.08	0.06	0.20
6	1580	0.08	0.13	0.06	0.27
7	1581	0.07	0.11	0.06	0.24
8	1582	0.06	0.03	0.08	0.17
9	1584	0.07	0.10	0.04	0.21
10	1589	0.12	0.16	0.09	0.37
11	1596	0.07	0.05	0.10	0.22
12	1614	0.09	0.09	0.09	0.27
13	1615	0.08	0.04	0.06	0.18
14	1616	0.07	0.07	0.08	0.22

[a]All dimensions in inches.

The multi-element headform described at the Thirteenth Stapp Car Crash Conference (6) was tested to determine its correlation to the cadaver behavior described above.[3] Two different headforms were tested against both cylindrical surfaces. These headforms were the type M, which was the model described in the Stapp paper, and the type P, which employed a slightly different plastic formulation. The fracture results were generally encouraging. The type M headform consistently produced elliptical or partial elliptical fractures (Fig. 9); the type P consistently yielded linear remote fractures (Fig. 10). This indicates that the design is near the transition point from linear to elliptical fractures. The fracture loads are

[3]The construction of this headform is shown in Fig. 8. Its fracture behavior against the cylindrical surfaces are given in Table 5 and in Figs. 9 and 10.

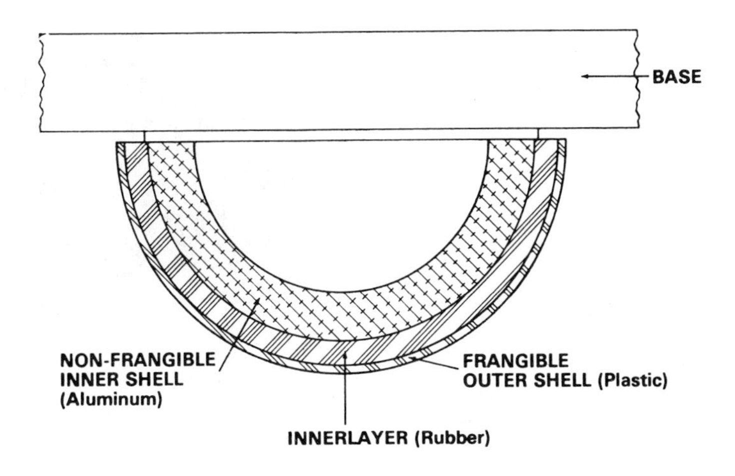

Fig. 8—Cross-section of frangible headform

Table 5—Results of Headform Tests Against Cylindrical Surfaces

Headform	Cylinder Size, in.	Fracture Load, lb.	Fracture Geometry
M	1	1300	Elliptical
M	5/16	1050	Elliptical
P	1	1500	Linear remote
P	5/16	1200	Linear remote

also reasonably close to the cadaver data. The major variation between the headform behavior and the human behavior is the occurrence of the linear crack directly under the site of the blow. This impact site damage was not seen on any cadavers but was seen on every headform fracture. The reason for this qualitative difference is currently under study.

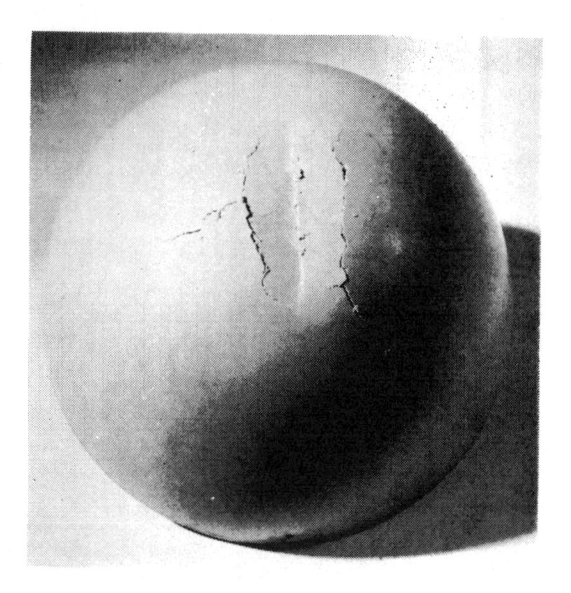

Fig. 9—Localized fracture of type M headform against 5/16 in. radius surface

Fig. 10—Linear fracture of type P headform against 5/16 in. radius cylinder

Conclusions

1. Twelve cadavers were tested to determine the fracture behavior of the frontal bone of the skull against two different rigid cylindrical surfaces. These cylinders were both 6½ in. long: one was of 1 in. radius and one was of 5/16 in. radius. The 1 in. radius consistently yielded remote linear fractures. Fracture loads ranged 950-1650 lb. The 5/16 in. radius cylinder yielded two linear remote fractures and three elliptical fractures. Fracture loads ranged 700-1730 lb.

2. A multi-element frangible headform was tested against these two cylindrical surfaces to determine its match to the cadaver behavior. Its overall simulation was relatively good.

References

1. V. R. Hodgson, "Tolerance of the Facial Bones to Impact." American Journal of Anatomy, Vol. 120 (1967), p. 113.

2. A H. Nahum, J. D. Gatts, C. W. Gadd, and J. Danforth, "Impact Tolerance of the Face and Skull." Proceedings of Twelfth Stapp Car Crash Conference, P-26, paper 680785. New York: Society of Automotive Engineers, Inc., 1968.

3. J. W. Melvin, P. M. Fuller, R. P. Daniel, and G. M. Pavliscak, "Human Head and Knee Tolerance to Localized Pressure." Paper 690477 presented at SAE Mid-Year Meeting, Chicago, May 1969.

4. E. S. Gurdjian, J. E. Webster, and H. R. Lissner, "Observations on Prediction of Fracture Site in Head Injury." Radiology, Vol. 60 (1953), No. 2, pp. 226-235.

5. V. R. Hodgson and L. M. Patrick, "Dynamic Response of the Human Cadaver Head Compared to a Simple Mathematical Model." Proceedings of Twelfth Stapp Car Crash Conference, P-26, paper 680784. New York: Society of Automotive Engineers, Inc., 1968.

6. J. Brinn, "Two Anthropometric Test Forms — The Frontal Bone of the Skull and a Typical Facial Bone." Proceedings of Thirteenth Stapp Car Crash Conference, P-28, paper 690816. New York: Society of Automotive Engineers, Inc., 1969.

710854

Comparison of Head Acceleration Injury Indices in Cadaver Skull Fracture

Voigt R. Hodgson and L. M. Thomas
Wayne State University

THE CEREBRAL CONCUSSION acceleration time tolerance curve due to frontal impact as described in SAE J885a is recognized in vehicular dynamics circles in industry and government, both here and abroad, as the best head injury criterion presently available. Controversy exists about how to apply this information to complex head acceleration pulses experianced by anthropomorphic test devices. Gadd's empirical Severity Index (1)* was the first scheme devised and is the recommended practice in industry standards SAE J885a and government standard MVSS 208.

Mathematically the Severity Index is defined as:

$$I = \int a^{2.5} dt \qquad (1)$$

where:

a = acceleration
2.5 = a weighting factor chosen as the average ratio of time to acceleration over the time range of automotive interior head impacts for the tolerance curve on log-log plot
t = time second

*Numbers in parentheses designate References at end of paper.

Integration of an acceleration time pulse experienced by a model of the human head is assumed critical if a numerical value greater than 1000 is accumulated.

Recently two nearly identical mathematical displacement index models have been proposed. A model by Slattenschek and Tauffkirchen (2) was first proposed followed shortly by that of Brinn and Staffeldt (3). The models are both single degree of freedom, damped spring mass systems represented by second order differential equations, which are uniquely specified by a natural frequency and damping factor. When the model is excited by an acceleration on its base, the response is the relative displacement between the base and the mass which compresses the spring and damper. The model is adjusted to the tolerance curve by assuming a damping factor. Two simultaneous solutions of the differential equation can be obtained in response to triangular acceleration inputs taken from the curve, to yield a natural frequency and critical displacement. A best fit to the curve over the time range of interest is obtained by varying damping and natural frequency. Since the models are similar except for damping factor and natural frequency only the model by Brinn and Staffeldt called the Effective Displacement Index (EDI) will be discussed and compared to the Severity Index (SI).

Because of the dearth of head injury tolerance information

ABSTRACT

Skull fracture was produced in forty cadavers which were dropped with their heads striking rigid, flat, hemispherical, and cylindrically shaped surfaces on the front, side, and rear. The heads were instrumented with biaxial accelerometers and force of impact was measured. Severity Index and Effective Displacement Index are compared at fracture level for all frontal impacts and the frontal flat plate results are compared to the Wayne State Cerebral Concussion Tolerance Curve. Indices calculated for Alderson 50th percentile dummy frontal head impacts onto a rigid flat plate are found to be higher than those for cadaver skull fracture impacts in the same drop height range.

available to be applied in practical design problems, that which is available has often been overextended. The curve is assumed to be valid for a very narrow range of conditions: frontal impact to a rigid flat surface, a-p acceleration pulses of a triangular shape, concussion.

Effects of closely spaced multiple impacts, skull deformation, rise time, reversing acceleration, and long duration acceleration (0.010-0.050 s), some or all of which often accompany head injury accidents are little understood. The SI and EDI treat rise time, multiple pulses, low-level long duration, and fluctuating characteristics of an impact in a very dissimilar manner as has been thoroughly discussed in the 14th Stapp Conference Proceedings. It is because our knowledge of head injury mechanism is so imperfect that two such widely different techniques could be applied to the same tolerance curve.

Comparison of the SI and EDI is further complicated by the fact that human simulators differ in their response to impact because of stiffness and damping differences in comparison with the human. Consequently the SI and EDI will often differ widely in their interpretation of a pulse, in a large part because of the way they differ in the treatment of artifactual resonant frequencies which arise in dummy heads in degree depending upon the stiffness of the impact surface.

The derivation of the Cerebral Concussion Tolerance Curve is discussed in detail in the 14th Stapp Conference Proceedings (4). Linear skull fracture in a cadaver is the condition from which the acceleration time data were obtained in the short duration end of the curve. Clinical statistics reveal that it is generally found to be true that a mild to moderate concussion accompanies a linear skull fracture in the human. Concussion has been noted to have occurred often without fracture. Consequently skull fracture is generally assumed to have occurred under more severe impact conditions than pure concussion alone. Many accidents which cause linear skull fracture result in a serious but not usually fatal injury. Consequently skull fracture represents a useful approximation to concussion and an upper limit which should not be exceeded. In the present state of head injury knowledge, skull fracture data in the human cadaver probably represent the most reliable design guide available. It is for these reasons that the SI and EDI are compared for several conditions of frontal impact which produced linear fracture in a group of 20 cadavers.

TEST PROCEDURE

Embalmed moist human cadavers (Table 1) were strapped to a lightweight honeycomb aluminum pallet which was supported by a pivot point at the feet (see Fig. 1). The cadavers were fitted with two pair of bi-axial accelerometers arranged as shown in Fig. 2, to record essentially transverse and a-p accelerations at two locations on the skull 3 in apart in a plane parallel to the sagittal plane of motion. Mounting, location, and orientation of the pickups were designed to be relatively free of skull vibrations to permit use of rigid body mechanics in calculating translational and angular accelerations as well as the resultant linear acceleration of the head CG from the component measurements. Subminiature Endevco 2222-B accelerometers were glued to either aluminum posts or balsa wood mounts, which were in turn glued to the skull. The balsa wood mounts were the most accurate from the standpoint of positioning the accelerometers relative to each other and of dampening out vibrations, but neither mount was capable of filtering out all skull vibrations, particularly at skull fracture levels. Accelerometers were fed into Endevco 2702 dynamonitor amplifiers and the resultant amplified signal was recorded on Sp-300-C Ampex tape recorder and the signal was played back through a 1650 Hz galvanometer. The entire instrumentation system was flat to 1000 Hz.

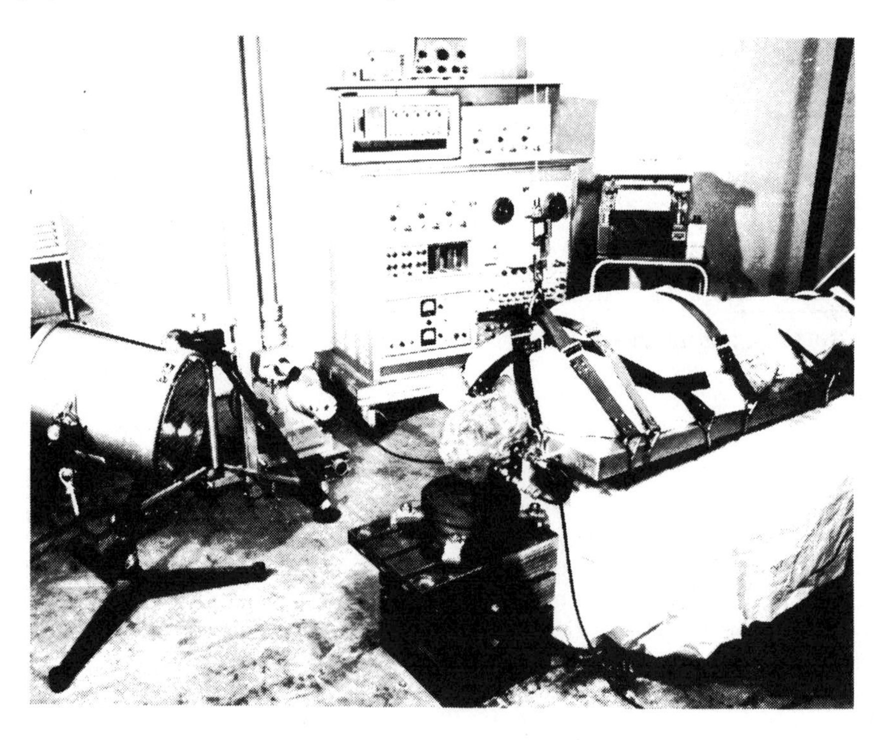

Fig. 1 - Cadaver head drop test setup

The force of impact was recorded by a dynamically calibrated four-arm strain gage load cell mounted under the impact surface, which included a flat plate, 8 in radius hemisphere, 3 in radius hemisphere, and 1 in radius cylinder aligned sagittally. The output of the load cell was fed to a Minneapolis-Honeywell Accudata II Carrier Amplifier and subsequent instrumentation as described above for the accelerometers. Velocity of impact was recorded by means of a probe alongside the head sweeping past two magnet pick-ups, mounted 3 in apart, just prior to impact of the head with the surface.

The aim in most tests was to produce a threshold linear fracture in a minimum number of drops, preferrably one, because of the breakdown in the soft tissue. The cadaver head was raised to the desired height above the surface, the head was restrained by cord from rotating into its preferred position during free fall, and a solenoid quick release was activated. The event was timed to trigger a high-intensity xenon light source and a Hycam camera set at 4000 fps just prior to impact, so that motion of targets on the side of the head could be studied by means of a Vanguard Motion Analyser Model C-11D.

Table 1 - Description of Cadavers

Test No.	Cadaver	Age, Yrs.	Sex	Date of Death	Date of Test	Height, ft-in	Weight, lb	Skin Thickness, in	Head, wt-lb	Circ., in	Head Size, in		$I^{**}_{a\text{-}p}$	$I^{\dagger}_{lat}$
											a-p	lat		
1	1717	67	M	3/28/70	10/9/70	5-6	145	–	10.0	23.0	7.9	6.1	0.20	–
2	1745	49	M	5/10/70	5/11/70	5-11	185	0.19	11.5	22.8	7.8	6.5	0.23	–
3	1701	51	M	2/6/70	10/26/70	5-10	140	–	8.0	20.3	7.1	5.9	0.12	–
4	1699	54	M	2/4/70	10/28/70	5-7	135	0.18	10.0	22.5	8.0	6.3	0.20	–
5*	1638	68	M	9/25/69	11/10/70	5-10	138	0.15	6.1	22.0	7.5	6.0	0.11	–
6	1805	76	M	9/25/70	11/24/70	5-8	163	0.22	10.0	23.5	8.0	6.4	0.21	–
7	1801	60	M	9/7/70	11/25/70	5-7	115	0.18	9.6	21.3	7.4	6.3	–	0.19
8	1820	66	M	10/24/70	12/1/70	5-9	130	0.20	8.3	21.5	7.3	5.8	–	0.17
9	1819	61	M	10/23/70	12/3/70	5-8	125	0.18	10.0	23.0	7.9	6.4	–	0.20
10	1821	69	M	10/30/70	12/4/70	5-11	225	0.30	13.0	24.5	8.4	6.9	–	0.35
11	1829	83	M	11/12/70	12/9/70	5-9	190	0.35	9.3	23.0	7.5	6.4	–	0.22
13	44	48	M	–	12/10/70	5-8	125	–	9.0	22.8	–	–	–	0.21
14	57	45	M	–	12/17/70	5-9	110	–	9.0	22.5	–	–	–	0.20
15	1848	51	M	12/22/70	1/20/71	5-10	182	0.30	9.6	24.0	7.8	6.0	0.21	–
16	1838	50	F	12/7/70	1/22/71	5-6	140	0.18	11.0	21.0	–	–	0.19	–
17	1841	56	M	12/7/70	2/1/71	5-8	230	0.25	14.0	24.0	–	–	0.31	–
18	1862	41	M	–	2/9/71	5-10	160	0.21	10.6	23.5	8.0	6.5	0.23	–
19	1876	61	F	2/6/71	2/16/71	5	110	0.13	7.9	21.0	7.4	5.7	0.14	–
20	1875	77	F	1/29/71	2/19/71	5-2	160	–	10.0	23.5	7.8	6.7	0.21	–
21	1859	84	F	1/7/71	2/23/71	5-3	158	0.22	9.2	22.5	7.8	6.0	0.18	–
22	1871	55	M	1/31/71	2/24/71	5-9	220	0.25	11.7	24.5	7.9	6.8	0.27	–
23	1861	71	M	11/4/71	3/4/71	5-9	170	–	10.5	23.2	7.1	6.3	0.22	–
24	1843	72	M	12/16/70	3/10/71	5-7	128	–	10.7	24.0	8.0	6.6	0.24	–
25	1879	63	M	2/3/71	3/22/71	5-11	210	0.20	12.1	24.0	8.0	6.8	0.27	–
26	188	–	M	–	3/24/71	5-10	145	0.25	9.2	21.8	7.3	6.3	0.17	–
27	1849	67	M	11/16/70	3/25/71	6-2	160	–	13.0	24.4	8.0	6.8	0.30	–
28	154	61	M	–	4/5/71	5-6	110	–	10.6	22.0	7.4	6.0	0.20	–
29	1873	63	M	1/25/71	4/8/71	5-9	140	–	10.0	21.3	7.4	6.0	0.18	–
30	1857	61	M	1/4/71	4/9/71	5-9	183	–	11.8	23.5	7.4	6.4	0.25	–
31	1890	85	M	2/22/71	5/8/71	5-6	150	–	9.7	22.8	7.1	6.4	0.20	–
32	1905	82	F	3/11/71	5/8/71	5-3	140	–	9.2	22.8	7.5	6.1	0.19	–
33	1912	73	F	3/26/71	5/8/71	5-8	238	–	9.7	22.5	8.0	6.1	0.19	–
34	1906	73	M	3/22/71	5/25/71	5-7	238	0.20	13.8	25.0	8.0	7.3	0.33	–
35	1910	91	M	3/17/71	5/26/71	5-9	180	0.20	11.6	23.4	7.6	6.0	0.24	–
36	1938	67	M	5/5/71	6/23/71	5-7	122	0.10	10.0	21.3	7.3	6.3	0.18	–
37	1935	71	M	5/1/71	6/24/71	5-10	180	0.20	14.4	23.8	7.9	6.5	0.32	–
38	1940	62	M	5/5/71	6/25/71	5-4	138	0.10	10.7	22.0	7.4	6.5	0.20	–
39	1936	67	M	5/4/71	6/26/71	5-7	160	0.18	10.3	22.8	7.6	6.3	0.21	–
40	1932	75	F	4/19/71	6/28/71	5-4	130	0.20	10.7	22.5	7.6	6.4	0.21	

*Cadaver dehydrated - test used to check out equipment.

$^{**}I_{a\text{-}p}$ - Mass moment of inertia about transverse axis through CG (units: in-lb-s^2).

$^{\dagger}I_{lat}$ - Mass moment of inertia about a-p axis through CG (units: in-lb-s^2).

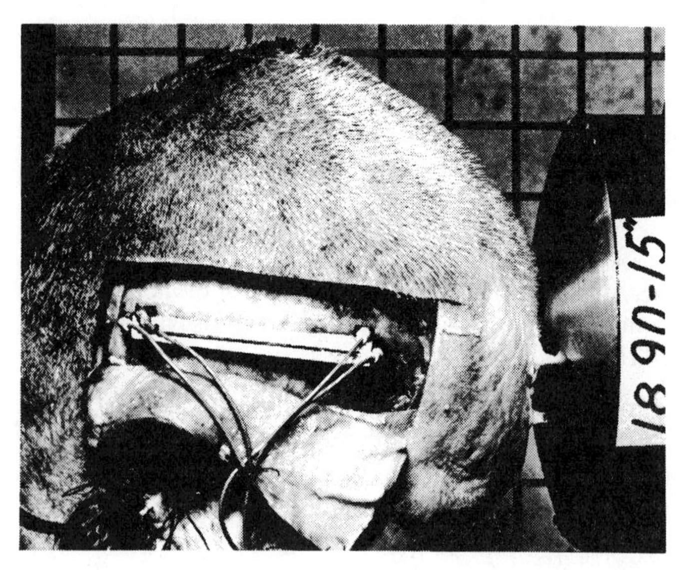

Fig. 2 - Two biaxial accelerometers attached to side of cadaver skull

Generally the oscillograph records gave evidence of a fracture except in cases of severe skin laceration by some of the sharper surfaces, in which case the surface contacted the bone causing vibrations which tended to obscure the jagged traces usually associated with fracture. When a fracture was indicated by the records, a flap at the blow site was lifted back and a visual investigation and a photograph of the fracture, if any, was made. If no fracture was found, the flap was replaced but shifted to apply relatively intact tissue to the blow site, or a piece of 1/4 in thick, 1.98 lb/in^3 dylite (5) was placed over the blow site if the soft tissue was irreparably damaged. The skin condition for the impacts is assumed to be intact unless otherwise noted in Table 2.

For comparison to cadaver data the Alderson 50 percentile dummy was dropped onto the flat plate striking its forehead at the minimum radius point in a manner similar to the cadaver (Fig. 3). The accelerometer mount inside the plate bolted to the back of the head was fitted with an Endevco tri-axial ac-

Table 2 - Results of Head Impacts at Fracture Level

Cad No.	Drop ht., in	Vel., ft/sec	Peak Force, lb	Peak A_{CG} Result, G	Mean A_{CG}, G	Pulse Time, s	Peak A_{a-p}, G	(SI) (A_{CG})	(SI) (A_{a-p})	EDI (A_{CG}) Resp, in	EDI (A_{a-p}) Resp, in
Front-Flat Plate											
1747	10	7.3	1600	230	75	0.0076	195	794	400	0.14	0.11
1701	10	7.3	1450	370	123	0.0045	345	2020	1800	0.18	0.12
1699	10	7.3	1700	270	87	0.0076	230	1280	792	0.17	0.13
1805	10	7.3	1450	190	63	0.0076	150	565	390	0.13	0.11
1873	25	11.5	2100	240	85	0.0061	195	1020	724	0.16	0.13
1857	30	12.7	2000	220	83	0.0076	220	1250	561	0.19	0.11
Average	14	8.9	1720	250	86	0.0068	220	1150	780	0.16	0.12
Front - 8 in Radius Hemisphere											
1890	20	9.4	1150	–	–	0.0078	210	–	370	–	0.11*
1905	20	10.3	1800	–	–	0.0055	390	–	1050	–	0.12*
1912	20	10.3	970	–	–	0.0075	150	–	370	–	0.13
1906	20	10.3	1530	–	–	0.0063	160	–	520	–	0.13
1910	25	11.5	1500	–	–	0.007	200	–	541	–	0.13*
Average	21	10.3	1400	–	–	0.0068	220	–	565	–	0.12
Front - 3 in Radius Hemisphere											
1859	15	9.4	970	–	–	0.006	195	–	765	0.16	–
1871	15	9.4	1140	–	–	0.006	185	–	468	0.13	–
1861	15	9.4	930	–	–	0.0075	170	–	272	0.07	–*
1843	15	10.3	1100	–	–	0.006	225	–	686	0.14	–
Average	15	9.7	1030	–	–	0.0064	190	–	550	0.13	–
Front - 1 in Radius Cylinder - Sagittal Plane											
1848	10	7.3	2000	–	–	0.004	285	–	800	–	0.11**
1938	10	7.3	1150	–	–	0.0075	260	–	822	–	0.11*
1841	10	7.3	2200	–	–	0.0025	230	–	915	–	0.13**
1876	15	9.4	2000	–	–	0.0025	320	–	1285	–	0.11**
1875	20	10.3	2450	–	–	0.0075	310	–	1220	–	0.15*
Average	16	8.5	1960	–	–	0.0048	280	–	910	–	0.12

*Fresh skin patch replaced split skin from previous impact.
**Dylite skin patch replaced split skin from previous impact.

celerometer model 2228B, recording only two channels, anterior-posterior (a-p) and inferior-superior (90 deg to a-p). Force of impact and head acceleration were recorded for 5, 10, 15, and 20 in drop heights, which was the range which produced fracture in most of the cadavers. Other instrumentation details in the dummy were the same as for the cadavers.

RESULTS

Anthropometry and vital statistics of the cadavers are listed in Table 1. Shown in the table is information from an additional 20 cadavers which were not included in the frontal impact tests.

Shown in Table 2 are the results of head impacts at fracture level against four surfaces beginning with the flat plate. The most comprehensive set of data is listed for this surface because it is the condition upon which the concussion tolerance curve is based and because the most reliable measurements were obtained in the present series of tests for this surface. The drop height, velocity, and peak force are measured values. The peak value of the resultant CG accelerations (A_{CG}) were calculated from the four measured acceleration components on the side of the head and distance measurements between pickups and CG using 1 to 1 photographic enlargements. The CG location was obtained by suspending the head from three different points in the sagittal plane in the usual manner of obtaining body segment CG locations. The (effective) head

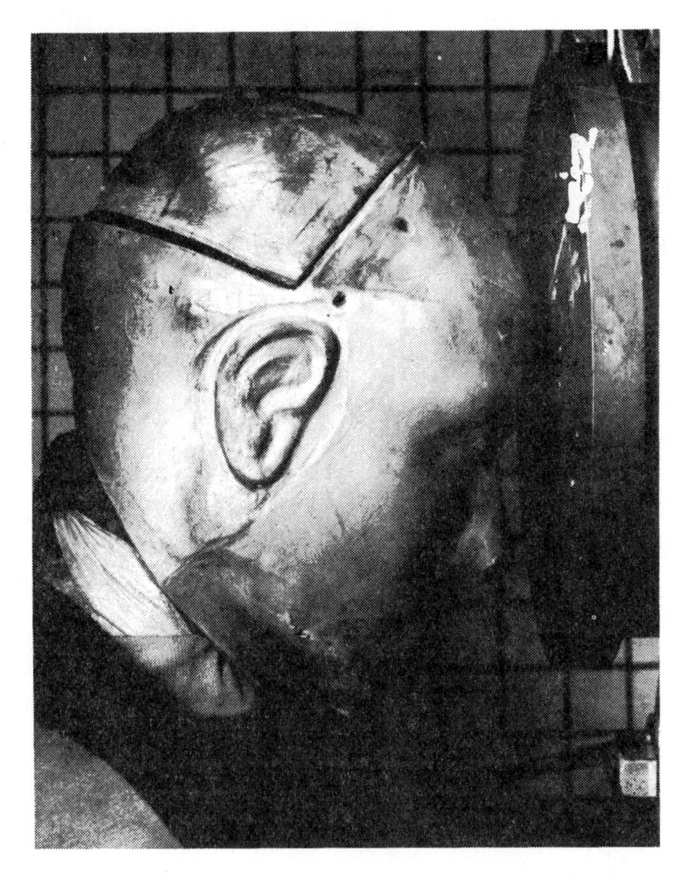

Fig. 3 - Alderson 50th percentile dummy forehead in flat plate impact attitude

acceleration was obtained from integrating the resultant CG head acceleration and dividing this velocity change by the pulse duration, which is the time from initiation of acceleration to return to zero. The peak a-p acceleration ($A_{a\text{-}p}$) is the peak value of the measured acceleration of the a-p accelerometer most remote from the blow site (Fig. 2).

The SI (A_{CG}) was computed from the calculated resultant CG acceleration (A_{CG}). The SI ($A_{a\text{-}p}$) was calculated from the a-p acceleration measured by the pickup most remote from the impact site. The EDI (A_{CG}) was the response calculated in inches using the resultant CG acceleration (A_{CG}) as the excitation to the model. The EDI ($A_{a\text{-}p}$) is the response of the model due to excitation by the a-p acceleration measured by the a-p accelerometer most remote from the impact site.

The average for all the measured and calculated impact parameters for the front-flat plate condition are given in line 7 of Table 2. Fig. 4 shows a typical oscillogram record at fracture from which the measured and calculated values in Table 2 were obtained. Fig. 5 shows a typical linear fracture due to impact with this surface.

The second condition shown in Table 2 is for the front -8 in radius hemisphere surface. As in the case of the front-flat plate condition, drop height, velocity, and peak force are measured values. Because of the skin-splitting effect of this surface and subsequent contact of bone with the surface, the oscillograph records at fracture level were too ragged to evaluate by means of rigid body mechanics and consequently peak and mean CG acceleration were not calculated. The pulse time is the total duration from the initiation of the a-p acceleration measurement until the acceleration subsides to zero. The SI and EDI were calculated only for the a-p acceleration for the pickup most remote from the impact site. Three out of of five cadavers received a linear fracture. Impacts to cadavers 1890 and 1906 produced circular localized fractures.

The measured and calculated values for the front 3 in radius hemisphere impacts were obtained in a similar manner to the 8 in radius hemisphere impacts described above. A linear fracture was produced by impact against this surface for three of the cadavers. Cadaver 1961 suffered a severe compound fracture.

Data for the front 1 in radius cylinder aligned with the sagittal plane was also obtained in a manner as described for the 8 in radius hemisphere impacts. Three cadavers received a linear fracture while cadavers 1876 and 1875 received a combination localized elliptically shaped fracture and linear fracture.

Table 3 shows a comparison of the results for the front-flat plate impact condition in the cadaver as compared to the results obtained for impact of the forehead of the 50th percentile Alderson dummy against the flat plate. The drop height and peak force are measured values. The pulse time (Force) is the time from initiation of impact until the force again reaches zero. The peak acceleration resultant is the same as A_{CG} for the cadaver as given in Table 2. For the dummy it is the vector sum of the a-p and transverse acceleration values. Given next are the a-p accelerations measured at the pickup site most remote from the front of the head for the cadaver and for the a-p oriented pick-up of the bi-axial accelerometer in the dummy. Pulse time (A) is given as the time from initiation of the head accelerations until the subsidence of the acceleration vibrations to zero for both the dummy and the cadaver. The SIs are calculated for the A_{RES}, which for the cadaver is the same as A_{CG} shown in column 9 of Table 2. For the dummy it is the SI calculated for the vector sum of the a-p and transverse acceleration. The SIs for $A_{a\text{-}p}$ are those calculated for the a-p accelerations measured most remote from the impact site in the case of the cadaver and measured for the a-p accelerometer in the dummy. The EDI was calculated for the a-p acceleration ($A_{a\text{-}p}$) for the pickup most remote from the blow site in the cadaver and for the a-p accelerometer in the dummy.

Typical dummy response is shown by the oscillograph records in Fig. 6.

DISCUSSION OF RESULTS

The critical values of SI and EDI are 1000 and approximately 0.15 in (estimated from Fig. 3, Ref. 3) respectively, based upon a-p accelerations recorded for frontal impact to a rigid flat surface in the cadaver. For the SI under these conditions a range of values from 390-1800 was obtained with an average value of 780 (Table 2). The EDI ranged from 0.11-0.13 with an average value of 0.12. Since the MVSS 208 criterion for frontal compact is SI of 1000 for resultant accelerations, these values were also calculated to yield an SI range of 565-2020, average value 1150, and EDI range of 0.13-0.19, average value 0.16. It is to be expected that resultant values should be higher than a-p because the

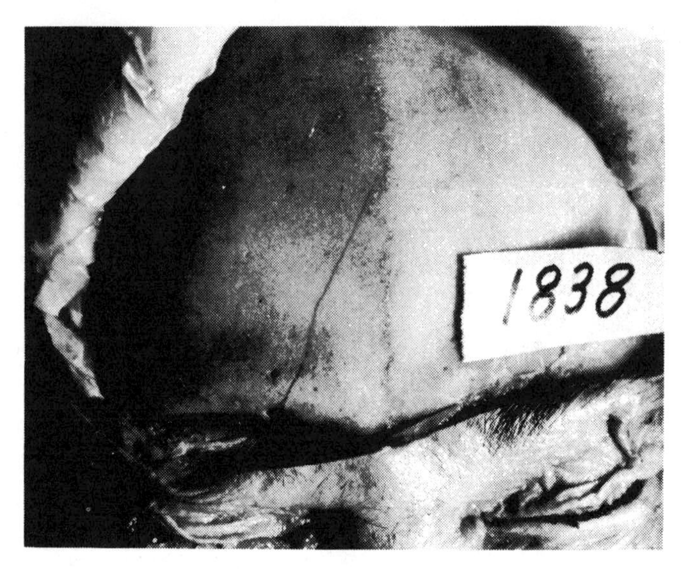

Fig. 5 - Linear fracture originating in right orbit due to frontal impact

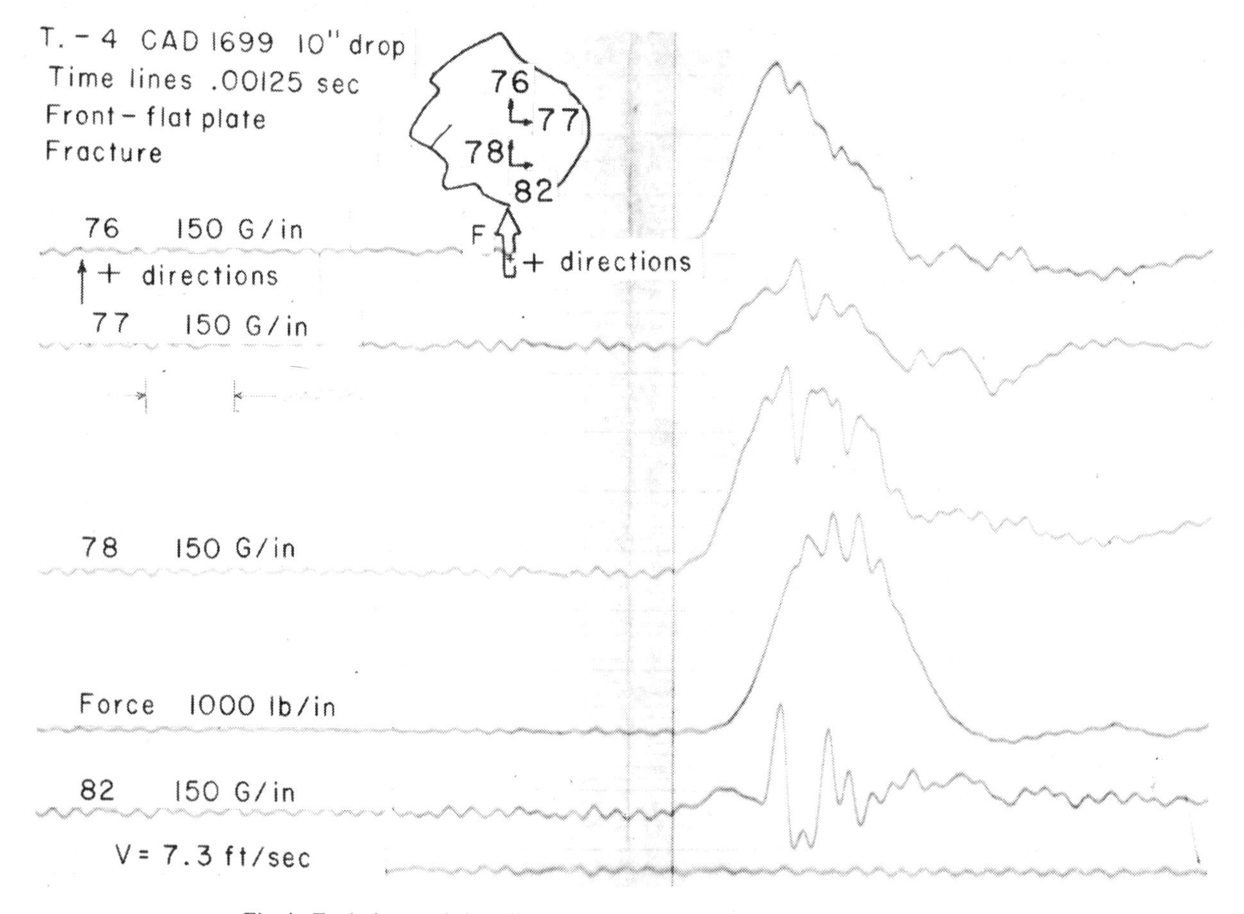

Fig. 4 - Typical expanded oscillographic record of force and acceleration at fracture

resultant values include transverse as well as rotational acceleration components (4).

The six frontal rigid flat plate impacts have been plotted along with the Tolerance Curve in Fig. 7. Cadaver 1701, which produced the highest SI at fracture level, had the highest acceleration for the shortest duration and therefore plots highest on the Tolerance Curve (point 3). Based on the values recorded for the other five it probably represents an atypically stiff cadaver. The SI tends to accent these differences because of the exponential weighting factor, whereas the EDI, especially EDI (A_{a-p}) does not show a significant difference among the six, all of which produced a linear fracture.

For the other impact surfaces the SI and EDI are com-

Table 3 - Comparison of Alderson 50th Percentile Dummy to Cadaver Response at Fracture Level Front-Flat Plate

Subject	Drop Ht., in.	Peak Force, lb	Pulse Time (Force), s	Peak Accel, G res	Peak Accel, G a-p	Pulse Time (accel), s	Severity Index A_{res}*	Severity Index A_{a-p}	EDI (A_{a-p}) Resp, in
Cad 1747	10	1600	0.005	230	195	0.0076	794	400	0.11
Cad 1701	10	1450	0.0025	370	345	0.0045	2020	1800	0.12
Cad 1699	10	1700	0.0034	270	230	0.0076	1280	792	0.13
Cad 1805	10	1450	0.0064	190	150	0.0076	565	390	0.11
Cad 1873	25	2100	0.0056	240	195	0.0061	1020	724	0.13
Cad 1857	30	2000	0.0075	220	220	0.0076	1250	561	0.11
Average	14	1720	0.0051	250	220	0.0068	1150	780	0.12
Dummy	10	2350	0.0034	305	270	0.0072	1680	890	0.10
	15	3070	0.0034	325	290	0.0072	2040	1150	0.12
	20	4000	0.0028	425	384	0.0061	3270	2020	0.14

*A_{res} = Resultant of A_{a-p} and $A_{transverse}$ measurements of dummy.
A_{res} = Resultant CG Acceleration (A_{CG}) for cadavers.

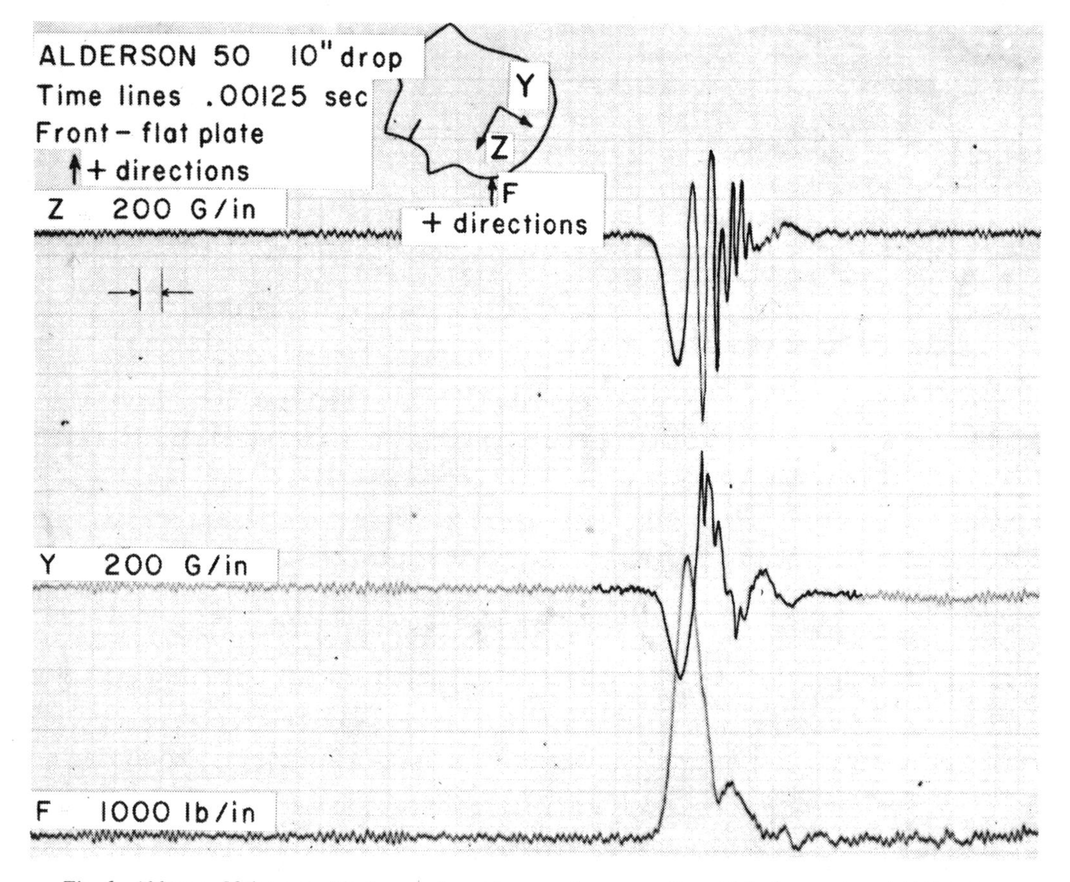

Fig. 6 - Alderson 50th percentile dummy input force and response for a 10 in drop onto a rigid flat plate

pared only on the basis of $A_{a\text{-}p}$ because there is too much high-frequency hash at fracture level against other than flat surfaces. Convex surfaces tend to split the skin and cause surface-to-bone contact which excites bone resonances, making CG acceleration derivations by rigid body mechanics too inaccurate. The SI for the 3 and 8 in radius hemispheres were generally lower than for the flat plate, possibly because these surfaces produce higher localized stresses in the bone thus tending to cause bone failure before high accelerations can develop. The EDI showed little distinction between all the surfaces except for cadaver 1961. The SI was also low for this depressed fracture, which may have had atypically low strength.

The tolerance curve on which the indices are based applies only to flat surface impact. However, the curved surface data are included because the effective area of impact can be more or less than a flat surface impact depending upon other things such as variation in skull shape and rigidity and skin thickness and consistency among cadavers.

The 1 in radius cylinder was aligned with the sagittal plane, which is undergirded by a suture buttress. This may account for the higher forces and acceleration and therefore higher SI for this condition than for the other curved surfaces.

It is interesting to compare the SI and EDI for 3 dummy drops of 10, 15, and 20 in. The peak a-p acceleration increased from 270 to 384, resulting in SI increase from 890 to 2020, whereas the EDI only went from 0.10 to 0.14. As can be seen by Fig. 6 the dummy head response to a rigid flat plate impact oscillates plus and minus. The SI treats the absolute value of the data because negative values of acceleration would lead to imaginary numbers when taken to a fractional exponent. Response of the EDI to alternating acceleration depends upon the algebraic sign of the data. Consequently, it is seen that both methods have mathematical peculiarities which operate on dummy response, which can be artifactual when compared to that of the more highly damped, less rigid human cadaver, to yield two very different pictures of impact severity. It is possible that impact against more resilient surfaces would produce less skull vibrations in the dummy, and better correlation between indices. However, this investigation was limited to rigid surfaces because of the convenience with which fractures can be produced for this condition.

CONCLUSIONS

These test results confirm that frontal bone fracture occurs in the human cadaver at the same level of acceleration as predicted by the Wayne State University Tolerance Curve.

Average values of SI and EDI computed for six impacts which produced linear skull fracture due to frontal rigid flat plate impact are in close agreement with critical values predicted by authors of these two methods.

For the drop height range which produced linear fracture in the cadaver due to frontal impact against a rigid flat plate, the SI for the Alderson 50th percentile dummy head acceleration response was much higher than for the cadaver. The EDI was essentially the same for both cadaver and dummy.

At our present level of head injury knowledge, the results of these tests on cadavers and dummies are difficult to interpret in favor of either index.

ACKNOWLEDGMENT

This work was supported by the U.S. Department of Transportation - National Highway Safety Bureau, Contract Number FH-11-7609.

The contents of this report reflect the views of the Wayne State University, Department of Neurosurgery, School of Medicine which is responsible for the facts and the accuracy of the data presented herein. The contents do not necessarily reflect the official views or policy of the Department of Transportation. This report does not constitute a standard, specification, or regulation.

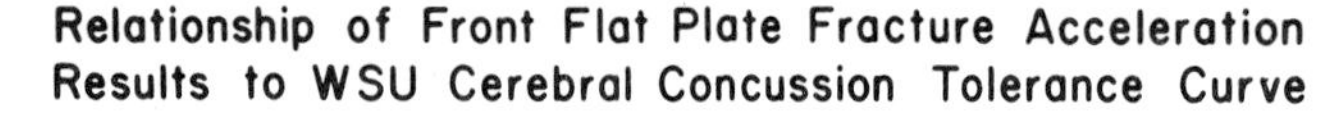

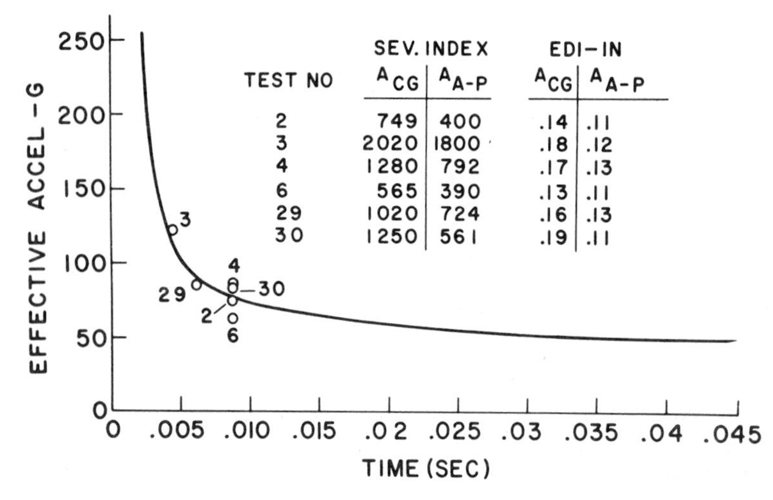

Fig. 7 - Comparison of SI, EDI, and kinematics of six frontal producing linear fracture

REFERENCES

1. C. W. Gadd, "Use of a Weighted-Impulse Criterion for Estimating Injury Hazard." Proceedings of Tenth Stapp Car Crash Conference, P-12, paper 660793. New York: Society of Automotive Engineers, Inc., 1966.

2. A. Slattenschek and W. Tauffkirchen, "Critical Evaluation of Assessment Methods for Head Impact Applied in Appraisal of Brain Injury Hazard, in Particular in Head Impact on Windshields." International Automobile Safety Conference Compendium, P-30, paper 700426. New York: Society of Automotive Engineers, Inc., 1970.

3. J. Brinn and S. E. Staffeld, "Evaluation of Impact Test Accelerations: A Damage Index for the Head and Torso." Proceedings of Fourteenth Stapp Car Crash Conference, P-33, paper 700902. New York: Society of Automotive Engineers, Inc., 1970.

4. V. R. Hodgson, L. M. Thomas, and P. Prasad, "Testing the Validity and Limitations of the Severity Index." Proceedings of Fourteenth Stapp Car Crash Conference, P-33, paper 700901. New York: Society of Automotive Engineers, Inc., 1970.

5. V. R. Hodgson, J. Brinn, S. Greenberg, and L. M. Thomas, "Fracture Behavior of the Frontal Bone Against Cylindrical Surfaces." Proceedings of Fourteenth Stapp Car Crash Conference, P-33, paper 700909. New York: Society of Automotive Engineers, Inc., 1970.

710881

A Review of the Severity Index

John Versace
Ford Motor Co.

Abstract

The SAE Severity Index is supposed to be an approximation to tolerance limit data, but there are incongruities in its derivation which renders the formula unsupportable. The same logic on which the Severity Index appears to be based can be used to support a wide range of possible values for the exponent on the acceleration, including infinity. This inconsistency results because necessary distinctions have not been made between: the formula for a fitted approximation to the tolerance limit data, the scaling of severity as such, and the measure of the acceleration magnitude of a pulse, the "effective acceleration." It is recommended that a formula which more literally follows from the tolerance limit data be adopted. In the long run, however, it is believed that a much more appropriate measure of injury severity would result from processing head impact data in such a way as to reflect the probable degree of brain injury.

Predicting Fatalities and Injuries from Tests

Human Tolerance - The degree of human tolerance to head impact is not well established, although there has been as much or more work in this area of human tolerance as in any other. Laboratory studies to find out how hard an impact can be before it produces a serious injury just cannot be done on people the way other biologists might study, say, insects.

When a toxicologist wants to find out how much chemical stress an insect species can tolerate, he does a very direct experiment. He puts 100 insects in each of 7 jars and then he sprays them with the insecticide, incrementing the dosage from jar to jar. He determines the tolerance level by counting how many insects died for the dosage applied, the test results being known as a "bioassay." Two types of findings make up the bioassay: the tolerance limit or threshold dosage for some designated degree of response to occur, and the scaling of the response relative to the dosage—that is, how many more insects are killed by an x% increase in dosage—once the tolerance limit or threshold has been exceeded. The distinction between these two concepts of response sensitivity should be kept in mind, for one of the main points to be made in this paper will bear upon it.

In contrast to such a direct approach, human tolerance to impact can be studied only by indirect methods. For example, subinjurious tests have been run with human volunteers. Such tests at least reveal that the tolerance is some level greater than what was voluntarily endured. Also, human cadavers (usually of quite old men, long embalmed medical school rejects) and live animals have been assaulted in various ways and the resulting damage inspected. But these methods are of such limited value, and the studies so few and of such small scope, that present knowledge of human tolerance to impact is quite incomplete and tentative.

Calibration of Test Devices - Even if human tolerance information issuing from the biological laboratories were accurate and comprehensive, there would still be significant problems of interpretation because practical testing by a vehicle manufacturer requires simple and repeatable test methods and impacters, not cadavers or live animals; after all, his purpose is to test the vehicle components, not the cadavers or even the mechanical impacters. Ideally, for practical testing purposes there should be a "calibration curve" which relates the test readings obtained with specific impacters to the percent of occupants who will survive when exposed to real accidents.

The kind of information needed to construct such a calibration is almost totally lacking. As a result, there is a strong tendency to act as if the numerical values of force and deceleration measured in vehicle tests have the same significance as those observed in the biological laboratories where the response of animals or cadavers was originally studied. There is little reason to believe that the numbers found in a biological experiment and those registered in a mechanical impacter used for testing instrument panels will correspond. In fact, there is a three-way equivocation: Real world impact experience may be reflected only poorly by the findings from the biological laboratory; the biological laboratory's procedures in turn are likely to be only poorly duplicated in the test engineer's devices. As a result, the test engineer's measurements are not likely to relate to real-world accident experience except by a considerable margin of uncertainty. While test-derived indexes of impact severity have been of practical use in facilitating product design decisions, correlation with the eventual count of traffic injuries and fatalities is largely unknown. Because of the lack of information, methods of measurement put forth in this paper are also subject to much of the same criticism.

Head Injury Tolerance Limit

The basic criterion for most evaluations of head impact trauma is the Wayne State University Tolerance Limit. It is shown in Fig. 1, which is reproduced from SAE J885a. Any point on the curve of the Tolerance Limit is supposed to represent the same threshold of injury as any other point. The curve shows that very intense head acceleration is tolerable if it is very brief, but that much less is tolerable if the pulse duration exceeds 10 or 15 ms.

This tolerance limit is probably not accurate. The impact data on which it is based were sparse and they were not very representative. It resulted from animal tests involving frontal hammer blows and air blasts to the exposed brain, and from drop tests of human cadaver heads. No direct tests using pulses longer than some 25 ms were involved in determining the tolerance limit, but consideration was given to assumed levels of deceleration experienced by humans in events such as free falls.

An alternative tolerance limit criterion might be based on the summary compilation of tolerance data, based mainly on military research, published in 1959 by A. M. Eiband (1)*. His summary for frontal deceleration is reproduced in Fig. 2. Note it is plotted in log-log coordinates; it is otherwise analogous to the Wayne State findings in Fig. 1.

Because Fig. 2 covers a much broader range of pulse durations than do the Wayne State data, it has the appearance of presenting a more valid overview of human tolerance. But the data comprising it are probably less accurate, for head injury severity purposes, than that obtained at Wayne State, as will be discussed below. Proposals have been made, notably by Gadd (2) for a criterion to be based on the Wayne State data and the above findings, with emphasis on the latter. Gadd's suggested criterion is represented by a straight line approximating the data and identified as "$1000 = TA^{2.5}$" shown superimposed on Fig. 2.

The approximation shown in Fig. 2 would seem to be compatible with the

*Numbers in parentheses designate References at end of paper.

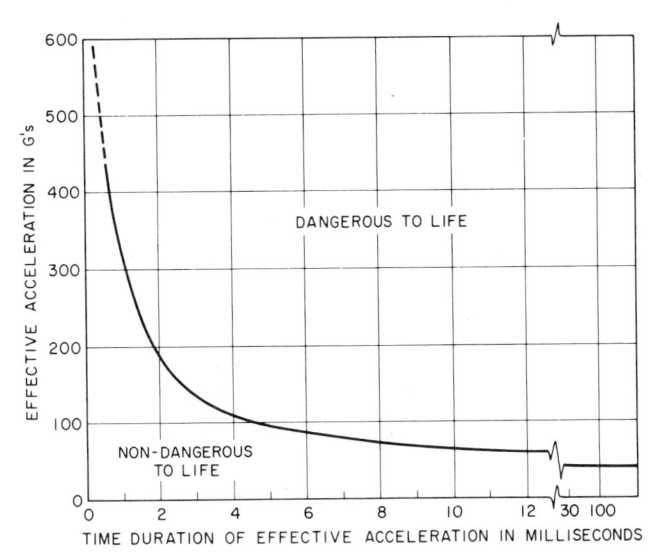

Fig. 1 - Impact Tolerance for the human brain in forehead impacts against plane, unyielding surfaces

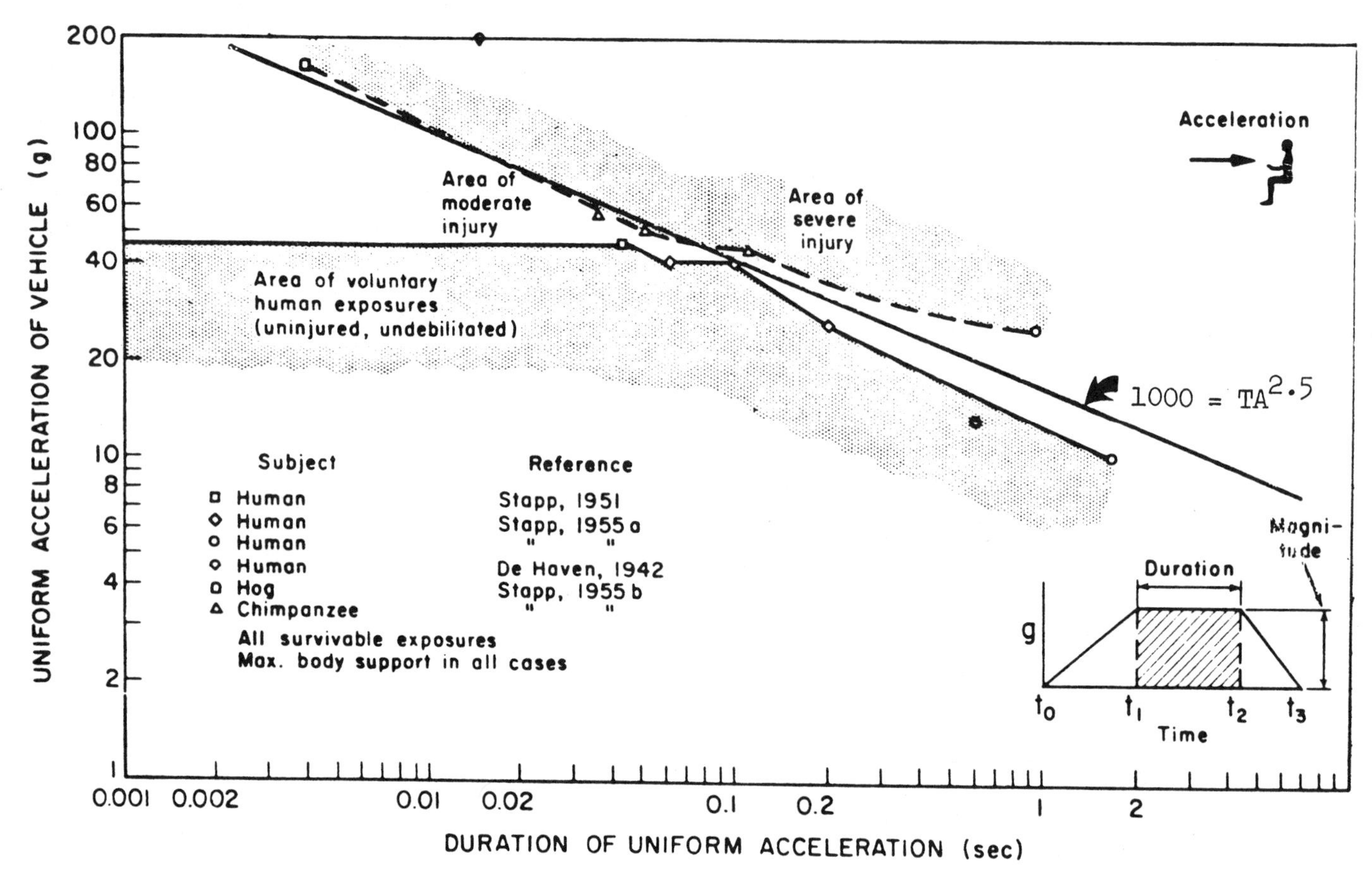

Fig. 2 - Frontal deceleration

Wayne State Tolerance Limit data, but that is not necessarily so because the acceleration measures in the two studies are not commensurate. They are not comparable for at least three reasons:

1. The Wayne State data refer to accelerations measured on the body of the subject experiencing the acceleration. In that respect, these measurements bear a closer relevance to crash testing because the response in the head of a crash dummy or analogous impacter is recorded. By contrast, the military data summarized by Eiband are expressed in terms of the deceleration of the seat on which the test subject rode, not the dynamic response of the test subject. In view of the almost certain amplification of acceleration at the subject's head because of the dynamic action of his body with the restraining devices, the experienced accelerations were undoubtedly greater than as plotted. Consequently, we should expect that the true tolerance is higher and that the fitted approximation should likely have a considerably shallower slope in the 0.1 s region then is shown–if the ordinate of the graph were to represent g experienced by the subject rather than the vehicle, as in the Wayne State data.

2. A second source of difference arises because tolerance data for the very long durations more likely relate to a different mode of injury, fluid displacement, rather than the mechanical type of damage we expect from the shorter duration impacts in car crashes. Therefore, tolerance data recorded for the longer durations should not be given as much weight; to do so would cause the fitted approximation to be skewed away from its appropriate level at the shorter or intermediate durations which are more pertinent to the automotive case. This is a common problem with straight line fits of data over a wide range. Forcing the fit to be best on the average over the whole range can result in a worse fit in a local region of high interest. Wide ranges often span different modes of response.

3. The third source of incomparability to the Wayne State data is the very definition of acceleration level, the value that is plotted on Figs. 1 and 2. The "level" of acceleration must be abstracted from the acceleration pulse. Should it be the peak value? The situation is simple in the case of the military data because all the pulses, being programmed vehicle input rather than occupant response measures, were more or less rectangular waves of constant acceleration level. On the other hand, the accelerations recorded in the Wayne State data tended to be somewhat rounded-off triangular pulses. As a result, the ordinate of the Wayne State graph is stated in units of "effective acceleration." Patrick, et al., state (3):

> "In the general case of impact to a padded surface the acceleration pulse is approximately triangular or sinusoidal, so an effective acceleration value is used. The ordinate of the Tolerance curve of Figure 1 is Effective Acceleration which is based on a modified triangular pulse in which the effective acceleration is somewhat greater than half the peak value. Therefore, triangular or sinusoidal pulses of equal area and higher peak magnitude are

in accord with the experimental evidence from which the Tolerance curve is derived."

In a similar manner, SAE J885 (March 1964 issue) in addition to presenting a weighted-impulse formulation, included the following definition for effective acceleration:

"2.4.2 Impulse Area Criterion - Area under the acceleration-time curve, pressure-time curve, or force-time curve may sometimes provide a useful approximation to its injury potential. This assumes equal importance of the ordinate and its time duration in their contribution to tissue damage.

"Under this criterion, the area under very sharp spikes of the acceleration-time wave is generally negligible and is therefore ignored. The problem is to find the effective acceleration magnitude, which, when multiplied by the

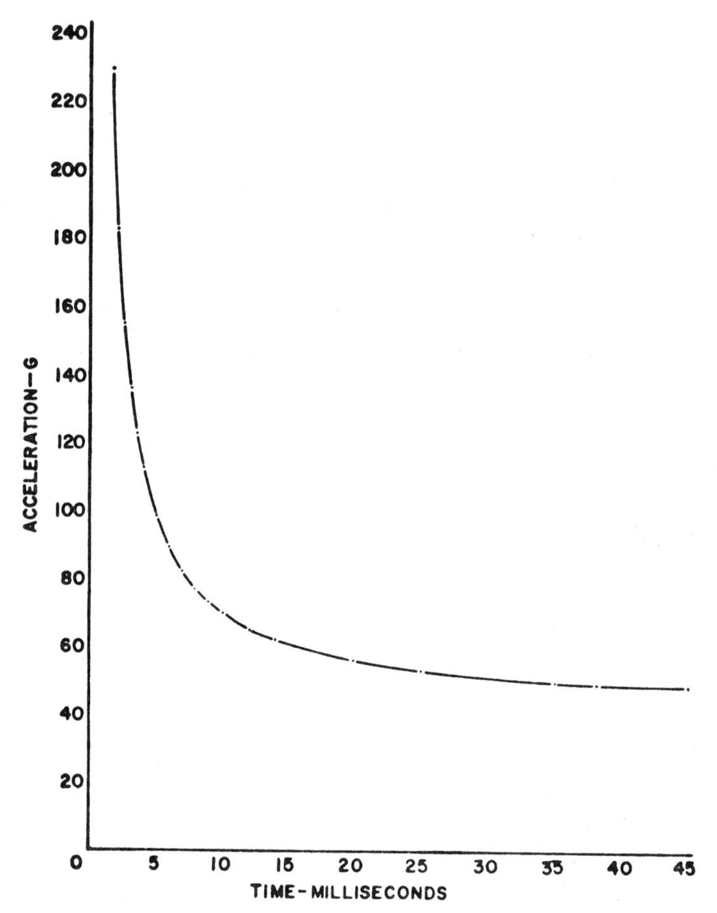

Figure 1 - Acceleration-time tolerance curve for forehead impact to a hard, flat surface

total time duration, will give the area under the curve. Therefore, the following guides should aid in arriving at an effective value for acceleration.

"(a) If the acceleration-time trace has an essentially flat plateau (ignoring any spikes present) the magnitude of the plateau is the effective acceleration.

"(b) For cases in which the acceleration-time trace approximates a half sine or is very irregular in shape, the effective acceleration is the area under the curve divided by its time duration."

Recently, a DOT contract (FH-11-7288) at the University of Michigan (4) revisited the data on which the Wayne State University Tolerance Limit is based. Drawing also upon the results of additional animal studies and on an analytical formulation related to strain in the skull, they concluded that the asymptotic tolerance is greater than is depicted by the Wayne State Limit (Fig. 3). Patrick, et al., had said that too (3): if concentrated loads could be avoided, the effective acceleration asymptote of 42 g was probably unrealistically low—"Consequently, when an impact to padded surface occurs a higher effective acceleration limit is in order. It is estimated that a 60 to 80 g limit is a reasonable value." Presumably, Patrick, et al., refer to the kind of blow which results in an effective acceleration of 60-80 g in real people, not in the heads of crash test dummies.

There are two technicalities associated with the Wayne State Limit which should be noted, as they will be of particular significance in the rest of this paper. The first technical point was covered above, and relates to the meaning of "effective acceleration." The second technical point deals with the distinction

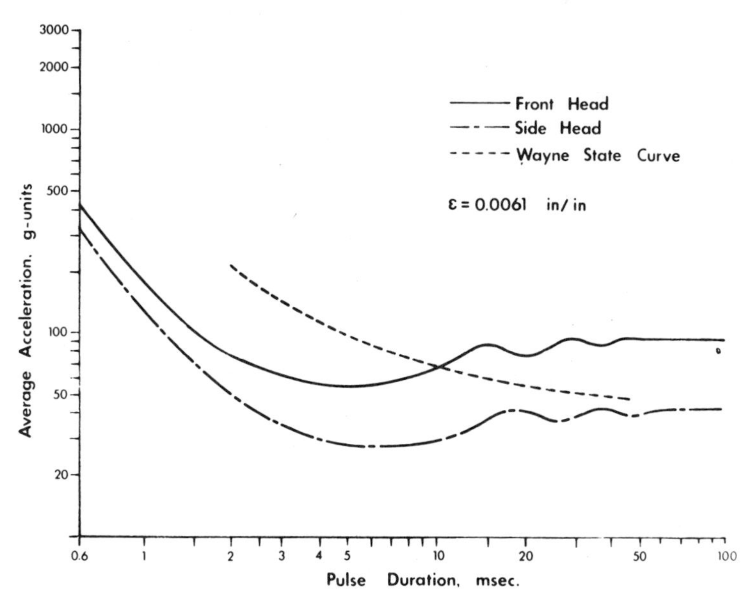

Fig. 3 - Maximum strain criterion for humans

between a tolerance limit or threshold and the scaling of the response once the threshold has been exceeded, a point emphasized in the opening paragraphs of this paper. The Wayne State Limit or any mathematical approximation to it, or any alternative curve brought forth as a replacement for it, represents only the boundary between acceptable and unacceptable levels. It in no way provides a basis for scaling the injury severity or hazard to a victim as the pulse amplitude is varied, for the same pulse duration. A later section of this report will deal with that matter in more detail.

Mathematical Approximation to the Tolerance Limit

The acceleration-time tolerance values, such as seen in the Wayne State Tolerance Limit curve in Fig. 1 and as collected by Eiband, can be plotted on log-log coordinates. In so doing, the data will tend to fall into a somewhat linear configuration. Gadd observed that a straight line on log-log coordinates fits the available data reasonably well.

A straight line in log-log form has the following equation:

$$\log A = m \log T + \log k \qquad (1)$$

the slope of the line is m, and (log k) is a constant intercept. The best-fitting line at which Gadd arrived (as shown in Fig. 2) has a slope m of −0.4. The value for k is determined by setting $T = 1$ (in which case $\log T = 0$); at $T = 1$, the value of A—as seen on the plotted graph—is 15.85 g. Substituting these into Eq. 1 gives the value of k:

$$\log 15.85 = -0.4\,(0) + \log k$$

$$15.85 = k$$

Putting these constant values of m and k back into Eq. 1,

$$\log A = -0.4 \log T + \log 15.85$$

$$\log A = \log T^{-0.4} + \log 15.85$$

$$\log A = \log (15.85\, T^{-0.4})$$

$$A = 15.85\, T^{-0.4}$$

$$TA^{2.5} = 15.85^{2.5} = 1000$$

Thus, the equation for Gadd's line, the *boundary* between acceptable and unacceptable is

$$\boxed{\begin{aligned} 1000 &= TA^{2.5} \\ 15.85 &= AT^{0.4} \end{aligned}} \qquad (2)$$

There can be no cavil with this equation; it is the only equation of that line. The SAE Severity Index, which bears a superficial similarity to this equation, is not the same thing and cannot be derived from these data or from the line which the above equation represents. However, there are two uncertain aspects to the above equation: First, the line it represents is probably not the most appropriate fit and hence the formula would be different if a better fitting line were to be found. And, there is no reason why the fitted line should have to be straight. While all impact tolerance data are rather dubious, this line is skewed by some that are particularly dubious, those for the longer pulse durations. The second uncertain aspect of the equation relates to the definition of the level of "A." Patrick, et al., (3) regard "A" to be some kind of average, the "effective acceleration." It is not certain whether the Wayne State data can even be plotted on the same graph with the military data because of the equivocal identification of the ordinate variable.

The formula can be used in lieu of a tolerance limit graph in establishing whether any observed impact exceeds the tolerance boundary or not. But it can only serve as a "go/no-go" gage. For example, suppose a head deceleration of 46 g and 70 ms were recorded. The point (46, 0.07) could be plotted on the tolerance graph and its position compared to the Tolerance Limit line. Alternatively, either form of the above formula could be used to calculate whether the pulse exceeded the Tolerance Limit:

$$(0.07)(46)^{2.5} = 1005$$

or equivalently,

$$(46)(0.07)^{0.4} = 15.88$$

The first solution exceeds the constant 1000, the second exceeds 15.85; by either version of the formula, the pulse is seen to have a value that places it slightly above the plotted Tolerance Limit line.

Eq. 2 should not be invested with more meaning than it has. It merely specifies the criterion line. That line is supposed to represent a boundary, or threshold, and therefore can in no way represent a scaling of injury severity.

Severity Scaling versus Tolerance Limit

Constants have been taken from the formula of the Tolerance Limit line for severity scaling purposes, a use which cannot be justified conceptually or mathematically. The SAE Severity Index, for example, expropriates the constants "1000" and "2.5" from the boundary formula and inserts them into a different formula:

$$1000 = \int a^{2.5}\, dt \tag{3}$$

It was shown above that there are two different ways of depicting the same line:

$$1000 = TA^{2.5}$$

$$15.85 = AT^{0.4}$$

If the exponent of "A" is somehow indicative of what the severity scaling should be, why not pick the second form of the equation, where the exponent is unity? In fact, the very same Tolerance Limit line can also be depicted by any number of equivalent equations, with the exponent "A" taking on any value desired. For example,

$$10^6 = A^5T^2$$

$$3.98 = \sqrt{A}/T^5$$

are also equivalent, and still refer to the same line.

Now consider another example of this distinction, the distinction between the parameters of a boundary and the scaling of severity, by examining tolerance limit lines that are somewhat different from one another but within the range of normal variation. Assume we have a test result whose acceleration level and duration happen to place it right on the $1000 = TA^{2.5}$ line. We would conclude it just fails being acceptable. Now, suppose that we were to discover some new biological tolerance data, data which cause us to seek a better fitting Tolerance Limit, one whose equation turns out to be, say, $K = TA^{3.6}$, a value which is quite plausible. Let us also suppose the newly fitted line happens to go right through the same test point on the graph, pivoting at that point (not a necessary assumption, but it simplifies the discussion.) Now, in both cases the data point is right on the criterion line, but in the first it involves an acceleration exponent of 2.5 and in the second an exponent of 3.6. The pulse has not changed, so its seriousness is no different just because of a change in shape of the boundary. (To see the effect of exponentiation, note that if $A = 50$, $A^{3.6}$

is about 74 times larger than $A^{2.5}$!) The exponent in the tolerance limit equation has nothing whatsoever to do with the seriousness of the pulse. (See Fig. 4.)

The line corresponding to the equation $1000 = TA^{2.5}$ is not the only plausible line. In fact, it does not fit tightly to the Wayne State Tolerance Limit. The fit is better in some parts of the curve than at others. Table 1 shows constants for several alternative fitted curves.

These are all plausible descriptions of the available tolerance limit data. In fact, in the limited range of 30-100 ms, a near asymptote condition exists in the Wayne State and in the University of Michigan tolerance criteria, as seen in Figs. 1 and 3. Although a flat slope may not be strictly accurate, it is far from being implausible as a simplifying assumption over a narrow range. The exponent gets rapidly larger—approaching infinity—as the curve gets more hori-

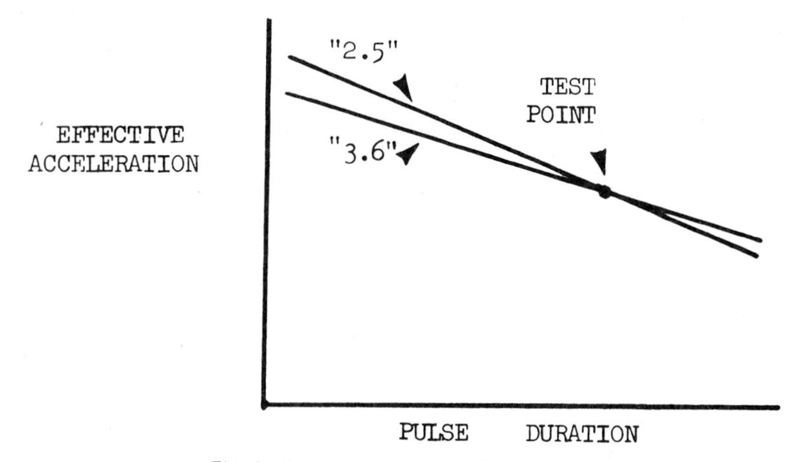

Fig. 4 - Acceleration versus pulse duration

Table 1 - Constants for Several Tolerance Curves

Constant	Power	Comment
1000	2.5	Above the Wayne State Tolerance Curve beyond 2 ms. Goes through a reference point of about 23 g at 400 ms
3780	2.9	Only slightly above Wayne State Tolerance Curve beyond 4 ms. Also goes through same reference point
9580	3.2	Extremely tight fit to Wayne State Tolerance Curve from about 7 ms on. Ditto on reference point
30.8	1.9	Extremely tight fit to Wayne State Tolerance Curve for less than 12 ms. Diverges elsewhere

zontal. The terminal slope of 20 g per decade of time duration implied by extension of the WSU curve corresponds to an exponent of 6. If an exponent is to be expropriated from the formula of the tolerance boundary, and used for severity scaling purposes, which one would be chosen and on what basis would the choice be made? The choice will make quite a difference in practice. But none of the possible exponents are really appropriate because the basis for the scaling of severity must come from some other source than the tolerance limit.

This paper has tried to show that the definition of a Tolerance Limit, or boundary, cannot be used as a definition of a severity scale. In view of the strong intuitive appeal of the SAE Severity Index (5),

$$\int_0^t a^n dt,$$

its apparent simplicity, and the familiarity it has in practice, there may be reluctance to examine critically the logic that produced it. It was stated that the SAE S.I. was not derivable mathematically from the tolerance data. How then, was it produced? The rules which lie behind the SI formula seem to be these:

Rule 1: Fit an approximation to the available tolerance data by drawing a straight line on the graph of log acceleration versus log time duration. (Log A versus log T; the line is drawn so it passes through 100 g at 0.01 s and 300 g at 0.001 s, a plausible fit given that the tolerance data are accepted at face value and that the acceleration levels are established with some consistent method of measurement. This line is shown in Fig. 2. Its equation is $1000 = TA^{2.5}$.)

Rule 2: Find the slope of that line. (Slope is −0.4.)

Rule 3: Take the negative inverse of that slope ($n = 1/-0.4 = 2.5$) and use it as the exponent in the S.I. formula:

$$\int a^{2.5} dt$$

This step appears to follow from postulating the following equivalence

$$\int a^n dt = A^n T$$

which is correct only when "a" is a constant. Here "a" is the instantaneous level of acceleration, "A" is the effective level corresponding to the ordinate of the tolerance data graph of rule 1. Note that if "a" is a constant, it moves outside the integral sign and the formula becomes equivalent to the Tolerance Limit equation. The equations are equivalent only for this case, the rarely encountered square wave.

Rule 4: Use the constant 1000 from the equation of the line (from rule 1), $1000 = TA^{2.5}$, as the criterion level that should not be exceeded.

Rule 4 is a non sequitur. The development of the S.I. from the tolerance data has the appearance of validating the specific value of 2.5 as the exponent. But it is only an appearance. Suppose the equally plausible other Tolerance Limits of Table 1 were to be used. The exponent could be 1.9, or perhaps 3.2. They are equally plausible as Tolerance Limits, but what rule can we invoke in order to make a choice for S.I. purposes?

A choice really cannot be made, which is clearly demonstrated if we now carry this a significant and crucial step further. Consider, as background, that in the limited region of interest to us, for pulses of 50-100 ms, the tolerable acceleration does not change a lot. This is especially apparent if the tolerance data are looked at on a conventional coordinate plot, such as Fig. 1, rather than in log-log form. The available data are so inadequate that they really do not much contradict this idea; a constant or asymptotic value of acceleration level could be taken as a practical approximation in this limited interval. It is roughly about 45 g. Now let us go through the same rules for constructing a S.I., applying them to this not unreasonable case.

Rule 1: Fit a line. The line passes through approximately 45 g at 0.05 s and also at 0.1 s. Its equation is $45 = AT^{\circ}$.

Rule 2: Slope = 0.

Rule 3: Take the negative inverse of the slope ($n = 1/0 = -\infty$) and use it as the exponent in the S.I. formula:

$$\int a^{\pm\infty}\, dt$$

Actually, the exponent in this case is indeterminate, resulting from the attempt to strike the following equivalence

$$\int a^{n}\, dt = AT^{\circ} = A$$

Rule 4: Use the constant from the equation of the line as the S.I. criterion level. The equation $45 = AT^{\circ}$ results in an infinite constant when the attempt is made to obtain the equivalent equation in the form $K = A^{n}\,T$.

The above result may seem too far out to be relevant. However, the accepted line, $1000 = TA^{2.5}$, is not that different from a simplifying approximation of zero slope. The severity index formula becomes nonsensical when the rules are applied to a somewhat different but not unreasonable case. Obviously, the "rules" for developing the S.I. are not trustworthy. The arithmetic may be clearer if we assume a slight slope to the tolerance line, say -0.1, corresponding approximately to a 1 g decline per millisecond decade. The exponent n = 10, and the constant is 44^{10}. The flatter the tolerance curve, the closer to infinity the exponent gets.

The issue which the Severity Index really seeks to resolve is the indefiniteness of the acceleration level, "A." When the pulse is square-topped, there is no problem: A is the amplitude of the top. But what is the acceleration level A when the acceleration pulse is a triangle? The peak? Suppose the triangular wave were to have the same area, the same delta-V as the square-top wave. Are the two waves equivalent in their effect, or is the triangular wave twice as severe? The S.I. formula above can provide no guidance in the matter, which is ultimately one of empirical determination.

Effective Acceleration versus Severity Scaling

The tolerance limit graphs and the formulas based on them suffer from a significant cart-before-the-horse problem because the scale of the ordinate variable "A" is ambiguous in denotation, violating a fundamental prerequisite that measurement must be in consistent terms. It is difficult to know where to plot a point on a graph of acceleration versus duration if the rules for determining the numerical value of the acceleration are vague or ambiguous. ("Consistency" does not mean an impossible freedom from random or experimental error, but uniformity of meaning.) The requirement of consistency precedes the establishment of a reliable tolerance limit. The graphs shown previously—which are the only significant data on this subject—do not satisfy even their avowed purpose of displaying an isoseverity boundary because real-world deceleration pulses do not conform to the narrow definition of acceleration level by which those graphs were plotted.

The scaling of severity and a consistent measure for "effective acceleration" can be easily confused. Severity scaling refers to the idea that a change in severity index level would connote a corresponding increment in real-world injury hazard. Consistency in measuring acceleration level, on the other hand, means that waveforms of quite different shape should be indicated by the same numerical value if they cause the same amount of injury. A consistent measure does not have to produce numbers proportional to the level of injury, only that equal measures refer to equal degree of injury. In only a very limited sense does the establishment of a consistent measure in itself yield a severity scale. Severity scaling goes a big step further, but it is clear that consistency is a prerequisite. Appendix A develops a paradigm for severity scaling based on the proportion of exposed persons who show significant response (say, fatality) to graduated levels of impact.

Summary - Tolerance data exist in the form of graphs showing the tradeoff between acceleration level and pulse duration for a threshold degree of injury. Equations can be derived from fitted approximations to the threshold contour on these graphs, the most familiar criterion being describable by the formula

"$1000 = TA^{2.5}$." The formula for the tolerance limit does not provide any basis whatsoever for the scaling of severity, that being a separate issue requiring presently unavailable data in which incremental degrees of injury would be correlated to incremental levels of input. An even more fundamental defect exists, however, rendering not only the severity scaling an open issue, but frustrating as well attempts to apply the simpler idea of the tolerance limit. That defect is the absence of a measure for acceleration level which is consistent regardless of waveform, the "effective acceleration." The formula for the tolerance limit does not provide any basis whatsoever for establishing a consistent measure of acceleration level either, that too being a separate issue requiring determination of the range of waveform variation resulting in equal degrees of injury.

The remainder of this paper will deal somewhat more closely with the SAE Severity Index, and with an alternative approach based on brain response.

SAE Severity Index

Some of the difficulties with the SAE Severity Index have been pointed out, for example, by Slattenschek and Tauffkirchen (6), by Brinn and Staffeld of Chrysler (7), and by Fan of Ford in two technical reports (8, 9). Different waveforms having the same duration and the same average acceleration yield quite different values on the S.I., as shown in Table 2. The illustrations in Fig. 5 show that the SAE S.I. deviates quite markedly from the Wayne State Tolerance Curve; the discrepancy can exceed 50%.

Table 2 - Severity Index S for Pairs of Values b_{mo} and t_{so} of the Wayne State Curve in Rectangular, Triangular, and Sinusoidal Impact Curves

Wayne State Curve		Severity Index for Different Acceleration Pulses		
t_{so}	b_{mo}	Rectangular $S_{\square}$	Triangular S_{Δ}	Sinusoidal $S_{\frown}$
S	g	$S_{\square} = b_{mo}^{2.5} \cdot t_{so}$	$S_{\Delta} = 1.62 . S_{\square}$	$S_{\frown} = 1.40 . S_{\square}$
0.0025	170	942	1522	1322
0.0050	100	500	808	702
0.0100	71	425	687	596
0.0200	56	469	758	658
0.0250	54	536	866	752
0.0400	52	780	1261	1095
0.0500	51	929	1501	1303
Nominal value		–	1000	–

Source: Ref. 6

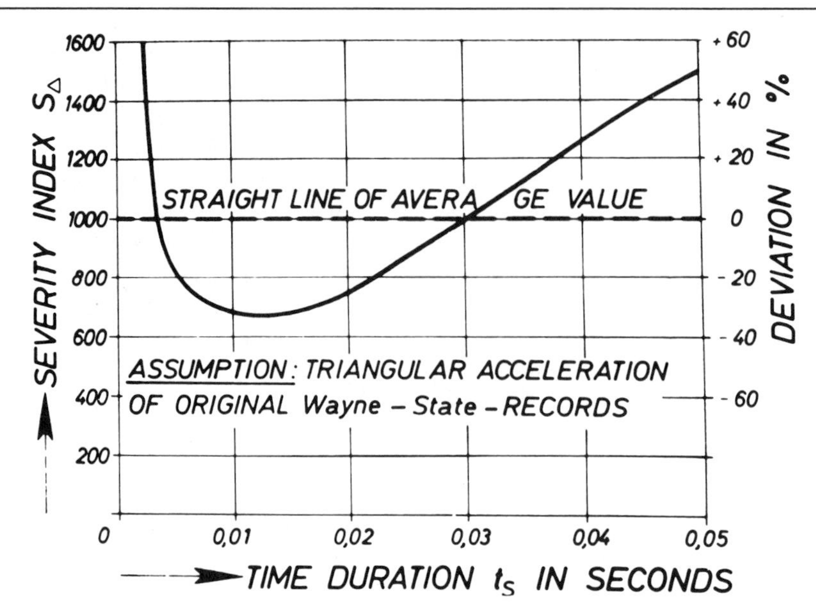

Fig. 4 - Straight line for average value and actual slope of Severity Index S_Δ in the range $2{,}5 \leq t_s \leq 50$ ms. Average value S_Δ = 1,000 equals the theoretical value on the basis of the Wayne State curve. Deviation $A = \frac{S_\Delta - 1{,}000}{1{,}000} \cdot 100\%$

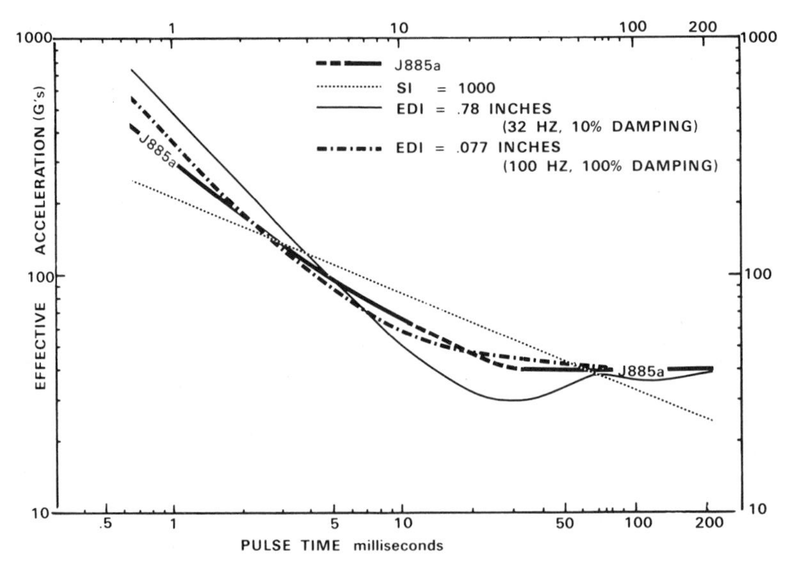

Fig. 2 - Fit of the SI and EDI to the head tolerance curve of SAE Information Report J885a. (Effective acceleration is employed here as defined by Patrick (8). The Severity Index is only recommended for pulse durations ranging 1-50 ms. It has been extended here to cover the range depicted by the curve of J885a)

Fig. 5 - Comparison of S. I.

The SAE S.I. duplicates the Tolerance Limit formula exactly when square waves are involved, so the idea of its general validity tends to be reinforced. But it is only an incidental correspondence, somewhat analogous to the momentarily correct reading displayed at the same time everyday by a stopped clock. For triangular waves, however, the SAE S.I. gives a reading 1.62 times as large as that for the corresponding square wave (that is, square wave of same duration and area). Some would say that is desirable, that peaked waves should read more than flat-topped waves. But, how much more? Earlier in this report it was seen that according to the logic of S.I. definition, any number of equally plausible indexes might be proposed–even one with infinite exponent! Previously, exponents ranging 1.9-3.2 were shown to be consistent with the Wayne State Tolerance Limit, and the extension of that Tolerance Limit in the narrow interval between 50 and 100 ms (the zone of greatest interest to us) has an equation with an exponent of 6. If the Severity Index were to take on the exponent of the associated Tolerance Limit line in each case, the ratio of resultant readouts for triangular versus square waves (of equal area and duration) would be as follows:

n	Ratio
2	1.33
2.5	1.62
3.6	2.64
5	5.35

An example was discussed earlier where a test result had the combination of acceleration and duration which happened to fall at the crossover point of two

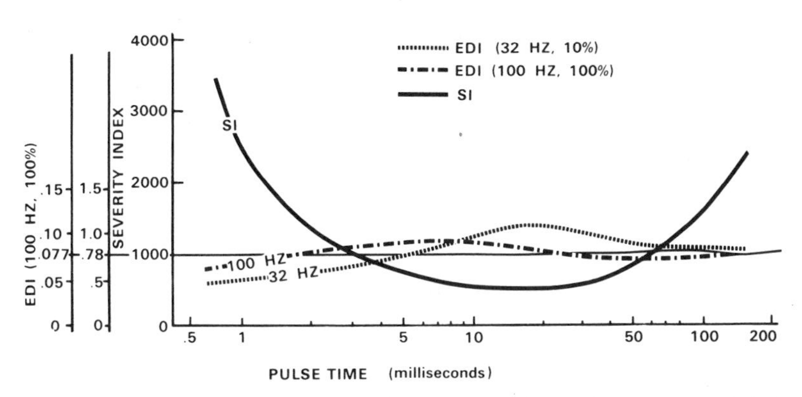

Fig. 4 - Damage indices calculated for each point on the head tolerance curve of J885a and Wayne State Tolerance Curve

alternative and equally plausible Tolerance Limit lines; one line had the 2.5 exponent, the other was a slightly shallower hypothetical line with a 3.6 exponent. That test point would produce S.I. values exactly equal to the criterion value appropriate to each curve (1000 in the first case, some other equally definitive constant in the second) if the wave were rectangular. But if triangular, the S.I. would read 1.62 times as great in the first case, and 2.64 in the second. It is the same pulse and it has not moved on the tolerance limit graph, but if the wave was triangular it would produce a reading which is 63% larger (that is, 2.64/1.62 = 1.63) on the "3.6" S.I. than on the "2.5" S.I.

If the SAE Severity Index cannot be justified on the basis of tolerance limit data, as was amply documented in preceding sections of this paper, then what should replace it? There is no ready answer to that because the human tolerance data required to establish a consistent definition of "effective acceleration" are not available, as was discussed earlier. If a consistent definition were available, the tolerance limit graphs would make more sense as far as practical application goes, and tolerance limit formulas based on them could be used. The tolerance limit formulas developed in this paper were shown to be valid in form, even if there is inadequate data to assure certainty in the precise parameter values, unlike the SAE Severity Index which does not follow from its premises even in form.

What is the SAE Severity Index, really, if it cannot be justified on the basis of the tolerance limit data, and especially since it does happen to be equivalent to the Tolerance Limit formula in the incidental instance of square waves? To clear up the last part of the question first, it is easy to see that for square waves, instantaneous acceleration is constant (at level A), so that

$$\int a^n \, dt = a^n \int dt = A^n T$$

The same kind of relation, of course, holds for the other indexes that might be contrived from the parameters listed earlier. For example,

$$9580 = \int a^{3.2} \, dt = a^{3.2} \int dt = A^{3.2} T$$

This fortuitous relation no longer holds if the waves are not square waves. The integrand is no longer a constant so it cannot be moved ahead of the integral sign. In puzzling over what the S.I. really is, it seemed plausible to consider it to be a variable that reflects some unique definition for the "effective acceleration." To investigate that, consider the general form that the "effective acceleration" might be postulated to take, for example

$$A = \left(\frac{1}{T} \int a^m \, dt \right)^r$$

where m and r are unknown (and strictly speaking, unknowable with the present absence of biomechanical tolerance data that would pertain to a wide range of pulse shapes.) It might be concluded from Patrick's explanation (3) and the representation by the University of Michigan group (see Fig. 2) that they consider the "effective acceleration" to be the average waveform level, implying that the constants m and r are 1.0 or something pretty close to it. If both the S.I. formula and the Tolerance Limit equation are to equal the same constant when at the critical level

$$S.I. = T A^n$$

$$\int a^n \, dt = T\left(\frac{1}{T}\int a^m \, dt\right)^{rn}$$

The equality will hold true only if m = n and if the product of rn = 1. Since n is independently determined from the Tolerance Limit line, and might be, for example, 2.5, then r must be 1/n. From this development, the "effective acceleration" implied by the Severity Index formula would be

$$A = \left(\frac{1}{T}\int a^n \, dt\right)^{1/n}$$

It can be seen that the larger n is, the larger the reading will be, and that non-rectangular pulses will be sensitively affected by the value of n. But it is difficult to see why the slope of the tolerance limit line, from which n is gotten, should have any bearing on this quantity at all. The definition of "effective acceleration" should in no way depend on the slope of the tolerance limit boundary; the definition must come first because the data points from which the tolerance slope is extracted cannot even be plotted until the "effective acceleration" can be measured.

We recommend that the waveform average be used for "effective acceleration." While it is probably not strictly accurate, we do not know that it is significantly inaccurate. On the other hand, if we hold out for an "effective acceleration" based on the SAE Severity Index we would have to find some justification for the choice of n = 2.5; it certainly cannot be sustained on the basis of tolerance limit data, nor can any other alternative value.

Such a choice would ultimately be judgmental. Our recommendation to use n = 1 is reinforced by the fact that there are no data to support any other formulation; in fact, it is the way tolerance limit data have been presented by the researchers at Wayne State University and at the University of Michigan. Furthermore, a conservative approach is more realistic because the existing human tolerance data are not known to bear any significant correspondence to the kind of response recorded with crash dummies and other laboratory impacters. And probably of foremost significance, the kinds of S.I. readings currently obtained on vehicle components and systems known from field experience to be highly effective greatly exceed the levels that are supposedly critical.

Implementation of the Tolerance Limit formula,

$$\text{Constant} = TA^{2.5}$$

using the conventional average waveform level for the "effective acceleration" can be effected as follows: The T.L. formula is equivalent to

$$\frac{(\int a\, dt)^{2.5}}{T^{1.5}}$$

One form of readout is to plot the value of that function as increments of T are accumulated over the pulse duration. The terminal value of this ratio, at the end of the pulse, gives the T.L. value for the whole pulse. However, it is possible that in an extremely irregular pulse, some portion of it could actually have a higher value than the pulse as a whole. In that case, the numerator and denominator would expand at different rates, so the ratio would show a peak. The record is evaluated by noting if the ratio at any point exceeds the criterion value.

There are obvious inadequacies in the above approach, aside from those due to the lack of biomechanical information from which a better estimate of the several exponents appearing in the general form of the Tolerance Limit and "effective acceleration" expression might be gotten. The main inadequacy is that this index represents only an intermediate event in the chain of traumatic events. We need to measure the eventual consequences resulting from such inputs. Of more relevance, then, would be some measure that reflected the response of the brain itself upon head impact. With such a measure, all these concerns for waveforms and severity indices would become unimportant, as we look to brain response for an indication of severity.

Alternative Approaches to an Index

An alternative approach stems from our current understanding, admittedly meager, of the nature of closed head injuries. As described in the paper by Fan (10), other researchers had begun to converge toward the idea that closed head trauma is the result of relative displacement between the brain and the skull. In order to elucidate the mechanism, mathematical models of the brain-skull system have been postulated. Fan has advanced a plausible version of such a model. It is described in both of the referenced reports, and is basically similar to models proposed by Slattenchek and by Brinn (6, 7).

References

1. A. M. Eiband, "Human Tolerance to Rapidly Applied Accelerations: A Summary of the Literature." NASA Memo 5-19-59 E.
2. C. W. Gadd, "Use of Weighted-Impulse Criterion for Estimating Injury Hazard." Proceedings of Tenth Stapp Car Crash Conference, P-12, paper 660793. New York: Society of Automotive Engineers, Inc., 1966.
3. L. M. Patrick, et al., "Survival by Design–Head Protection." Seventh Stapp Car Crash Conference Proceedings. Springfield, Ill.: Charles C Thomas, Publisher, 1963.
4. R. L. Stalnaker, J. H. McElhaney, and V. L. Roberts, "MSC Tolerance Curve for Head Impacts." ASME paper ul-WA/BHF-10, November 1971.
5. Human Tolerance to Impact Conditions as Related to Motor Vehicle Design, SAE J885a. SAE Information Report, October 1966, SAE Handbook.
6. A. Slattenschek and W. Tauffkirchen, "Critical Evaluation of Assessment Methods for Head Impact Applied in Appraisal of Brain Injury Hazard, in Particular Head Impact on Windshields." Paper 700426, 1970 International Automobile Safety Conference Compendium, P-30. New York: Society of Automotive Engineers, Inc., 1970.
7. J. Brinn and S. E. Staffeld, "Evaluation of Impact Test Accelerations: A Damage Index for the Head and Torso." Proceedings of Fourteenth Stapp Car Crash Conference, Paper 700902. New York: Society of Automotive Engineers, Inc., 1970.
8. W.R.S. Fan, "A Technique For Determining Head Impact Severity Index." Ford Motor Co., Automotive Safety Affairs Office Report No. S-71-9, February 16, 1971.
9. W.R.S. Fan, "Comparison of Several Head Injury Severity Indexes." Ford Motor Co., Automotive Safety Affairs Office Report No. S-71-28, June 18, 1971.
10. E. S. Grush, S. E. Henson, and O. R. Ritterling, "Restraint System Effectiveness." Ford Motor Co., Automotive Safety Affairs Office Report No. S-71-40, September 21, 1971.
11. J. Berkson, "A Statistically Precise and Relatively Simple Method of Estimating the Bio-Assay with Quantal Response, Based on the Logistic Function." J. Amer. Stat. Assoc., Vol. 48 (1953), pp. 565-599.
12. System Effectiveness Study.

APPENDIX
THE SCALING OF INJURY SEVERITY

One way to depict the degree of injury is in conventional bioassay terms, the form which biologists customarily use in describing tolerance to stressing agents. The severity is measured by the proportion of exposed individuals who are affected, say by dying or by showing certain signs of injury. Thus, the LD 50, the "lethal dosage for 50% survival," is commonly encountered as a measure of tolerance. A broader view would be achieved by reporting LD 5, LD 10, LD 25, etc. Indeed, the most appropriate scale for depicting severity would be one whose readings are at least proportional to the scale of survival probability. A scale commonly employed for this purpose by toxicologists and other biologists is the logistic function (11).

$$P = \frac{1}{1 + \exp(-(a + b X))}$$

where P is the probability of being affected in the defined manner (say, a fatality) by an applied stress of magnitude X, and a and b are parameters of the function. The logistic function closely resembles the normal distribution. The tolerance curve estimated by Verne Roberts for closed-head injury and shown in Fig. 31 in Ford's Restraint System Effectiveness study (12) fits the logistic function quite well. Fig. A-1 shows the plot on Berkson's (11) logistic paper. An estimated fit, if X-peak g of a triangular pulse, is provided by

$$P = \frac{1}{1 + \exp(6.66 - 0.037 X)}$$

The exponent of e would be (6.66 - 0.074 X) if X = "effective acceleration" = amplitude of the equivalent square wave.

It might be argued that Roberts' survival function, on which this formula is based, is not realistic because it requires the same amount of acceleration to be lethal regardless of pulse duration (but only for pulses exceeding 20 ms). But the critical amount of acceleration does not vary a whole lot, in most reported findings, over the limited range of pulse durations which 30 mph automotive impacts will entail, some 40-100 ms.

However, if a declining tolerance with longer pulses is still deemed more realistic, the logistic function can be modified to take that into account. A plausible adjustment would be to allow the whole logistic function to shift gradually downward in accordance with the acceleration-duration Tolerance Limit discussed in the main text:

$$1000 = T A^{2.5}$$

The critical question which now arises is this: to what LD percentile (or P, in the logistic equation above) does the Tolerance Limit curve correspond? It is likely that it corresponds to a very low LD percentile in view of the data on which that limit was based—the Wayne State experiments, survived free falls, voluntary sled tests, etc. Patrick, et al., say that the WSU Tolerance Limit "is

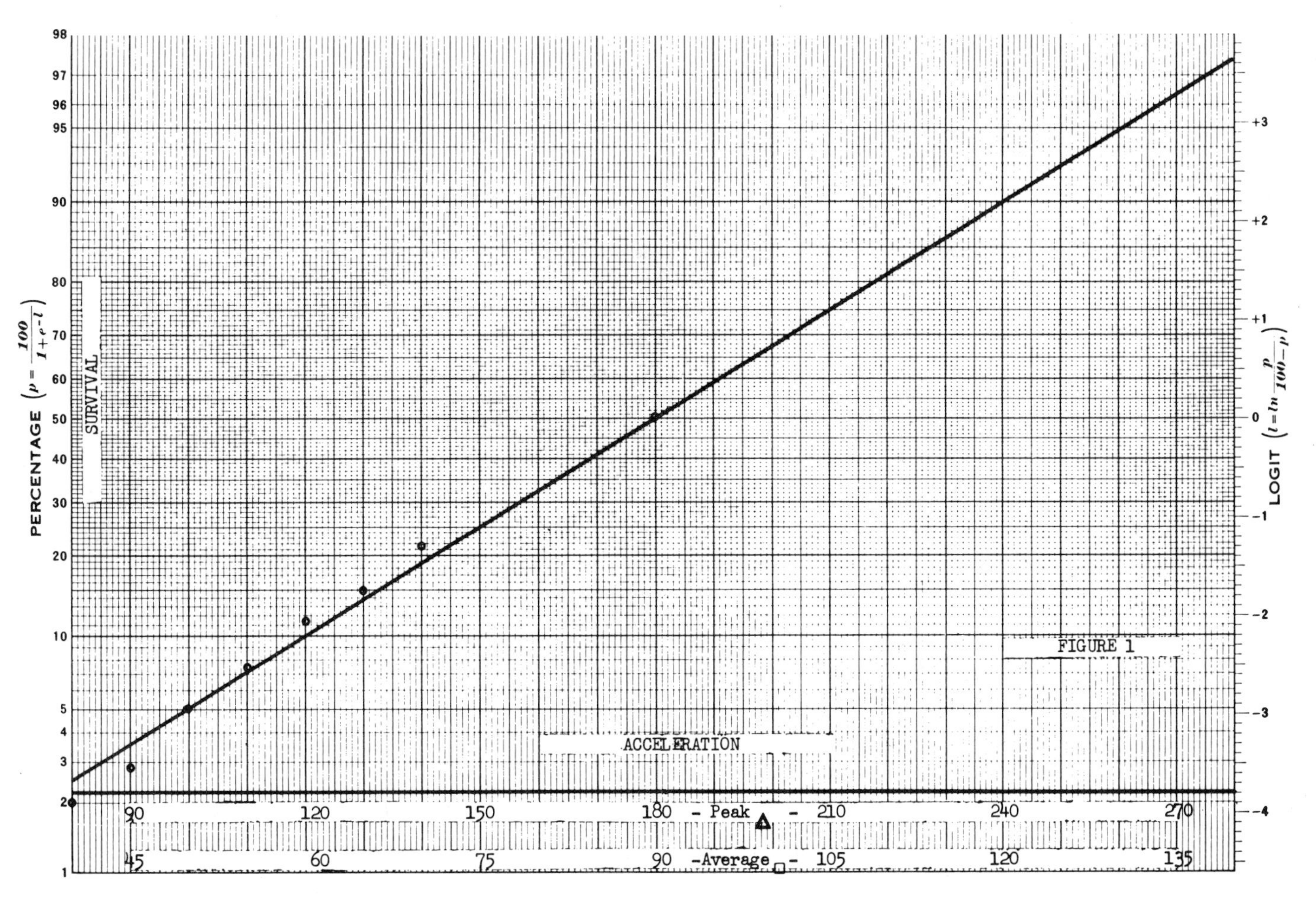
PERCENTAGE $\left(p = \frac{100}{1+e^{-l}}\right)$
SURVIVAL
LOGIT $\left(l = ln\frac{p}{100-p}\right)$
FIGURE 1
ACCELERATION
90 120 150 180 - Peak - 210 240 270
45 60 75 90 -Average - 105 120 135

Fig. A-1

based on a reversible concussion with no after effects" (3). For purposes of further exposition here, it is assumed that the Tolerance Limit curve corresponds to LD 5, the chance of fatality is 5%. While hairline fractures may appear and some degree of concussion may result—assuming no concentrated loads that would cause depressed injury—fatality is likely to be infrequent.

The choice of LD 5 would be consistent with both the Tolerance Limit and with Roberts' survivability curve if we posit the following: Roberts' curve really does not apply uniformly regardless of pulse duration but will be assumed to apply only midway between 20 and 100 ms, at which point the 5th percentile of the survivability curve intersects the (1000; 2.5) Tolerance Limit; the whole survivability curve shifts according to the (1000; 2.5) Tolerance Limit curve; "effective" g is one-half the triangular peak value. On that basis, waves of approximately 45 g (effective) and 50 ms would be 5% lethal, 90 g 50% lethal, and 130 g would be 95% lethal. The transition between 5 and 50% fatality spans 45 g. If we assume the same spread of g for all the pulse durations that might occur in the range being considered, the logistic function becomes:

$$P = \frac{1}{1 + \exp\left(3.33 + 0.074\left(\left(1000/T\right)^{0.4} - \overline{X}\right)\right)}$$

where:

T = duration, s
$\overline{X}$ = "effective acceleration"

Fig. A-2 shows a family of logistic functions corresponding to this expression. An alternative view is seen in Fig. A-3, where an isoseverity family is plotted in the more familiar acceleration-duration graph.

Another question, among others, is this: is the survivability function really symmetric with respect to "effective g" as seen in these figures? In bioassay work, organisms often respond, in logistic curve manner, to the logarithm of the stressing agent. (Because the logarithmic relation is so common, special graph paper is available for plotting the logistic function that way.) In the formula above, X has been taken alternatively as peak g and as average g—should it perhaps be log g, or $g^{2.5}$? No information exists which would allow for a definitive choice. In view of the Wayne State data being expressed as "effective acceleration," and the military human exposure data being very equivocal (square-top time-acceleration *sled input*, not occupant response) it would seem that the most plausible hypothesis at this point is that the survivability function is symmetric and that "effective g," the pulse average measured on the head, is the appropriate pulse descriptor.

Still another question, and perhaps the most crucial of all, bears upon practical application in distinction to scientific findings. The numbers that have been used in this report are supposedly normative, they deal with values that are observed or are at least presumed to be observable in real people—not in the mechanical devices of the vehicle testing workshop. The latter require a *calibration curve* to relate their readings to the normative population figures. (However, do not lose sight of the fact that no true bioassay type experiments have ever been done, or could ever be allowed to be done, with live people. So, a second source of validity erosion occurs in the inference that embalmed cadavers

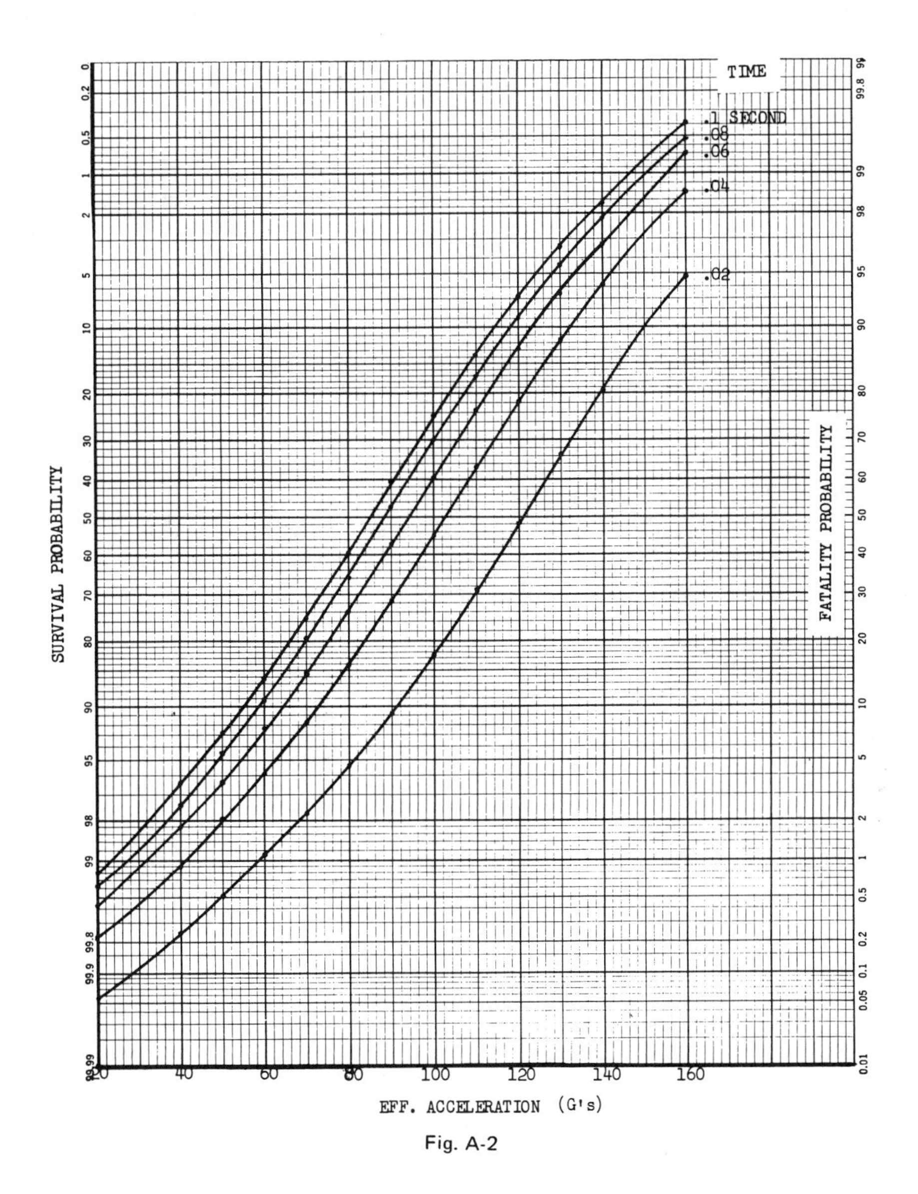

Fig. A-2

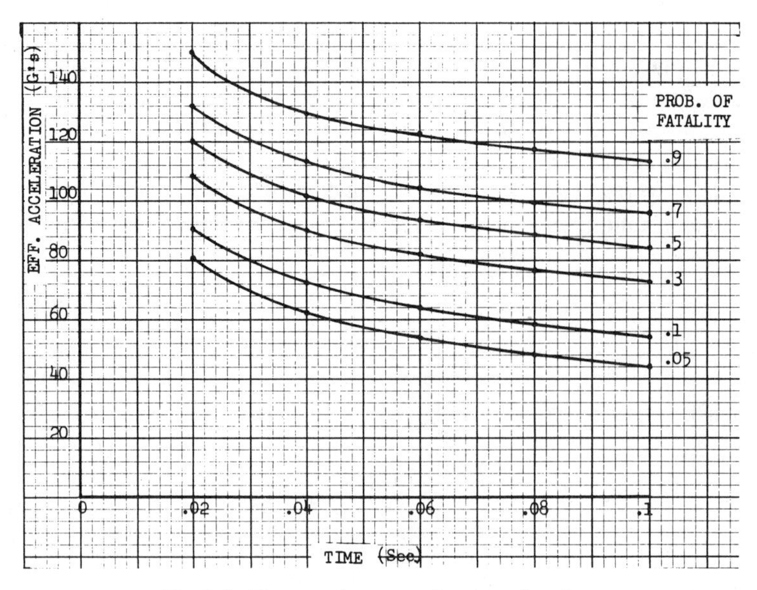

Fig. A-3 - Human tolerance to head acceleration

or live dogs produce responses that are one-to-one normative to people.) Owing to quite different impedances in the testing workshop devices as compared to humans, and to the observed differences in readings recorded when crash test dummies are subjected to the same tests as cadavers, it is likely that the criteria expressed in the functions shown in this paper are quite conservative. Thus, the parameters in the logistic function will no doubt be quite different if that function is supposed to relate the LD of real people to the readings gotten in crash dummies exposed to the same crash. And that relationship is the only valid one.

720956

Effect of Long-Duration Impact on Head

V. R. Hodgson and L. M. Thomas
Wayne State University

Abstract

Impacts have been analyzed in terms of degree of injury, head injury criterion (HIC), and average acceleration as a function of time for frontal impacts against the following surfaces:

1. Rigid flat surface—fractured cadaver skull.
2. Astroturf—head drop of football-helmeted cadaver.
3. Windshield penetrating impact of a dummy.
4. Airbag—dynamic test by human volunteers.

It is concluded that the linear acceleration/time concussion tolerance curve may not exist and that only impacts against relatively stiff surfaces producing impulses with short rise times can be critical. The authors hypothesize that if a head impact does not contain a critical HIC interval of less than 0.015 s, it should be considered safe as far as cerebral concussion is concerned.

ESTABLISHMENT of a head injury acceleration performance standard for a wide range of impact conditions, including direct impact to rigid surfaces, padded surfaces, airbags, and indirect impact through harness assemblies, and using injury indexes based upon the Wayne State University Cerebral Concussion Linear Acceleration-Time Tolerance Curve as published in SAE J885a, has been attempted for several years. The curve was drawn as though the phenomenon of concussion would occur as a result of any combination of average acceleration and pulse duration above the curve from 0.002–0.050 s. As such, the curve has been a convenient design reference based on a conservative and reversible injury, assumed to be a continuous function that could be closely approximated by empirical formulas.

The original data used to construct the short end of the curve were obtained by measuring head accelerations during cadaver frontal skull fracture and have been confirmed as reported in Ref. 1. These data have been the most reliable tie-in available between experimental models and living humans, since clinical experience shows that approximately 80% of those having suffered a linear skull fracture have also experienced a moderate concussion (2)*. However, a skull fracture is evidently only produced during very short-duration impacts in the range of 0.002–0.007 s, or in impacts involving steep rates of onset (20,000–50,000 G/s). Other evidence from the field relative to longer-duration head impacts is beginning to manifest itself and should serve to guide us in the formulation of safety standards and protective system design.

Football Impacts

Fatalities in American high school football have numbered between 7–25 or an incidence of 1.56/100,000 players each year for the past 20 years (3). Of these,

*Numbers in parentheses designate References at end of paper.

65–77% have been due to head injury involving hemorrhage. No national statistics have been recorded for concussion, the most common type of head injury in the game, which has been observed occasionally by most football fans.

Use of telemetry to obtain acceleration data on head impacts during football games has been tried, but so far the attempts have been unsuccessful due to many technical problems associated with safely obtaining accurate measurements. We have therefore resorted to the simulation of severe football head impact situations by the use of instrumented human cadavers. For example, a helmeted cadaver drop of 48 in (122 cm) onto the forehead resulting in an impact velocity of 16 ft/s (4.9 m/s) against a synthetic turf supported below by a rigid load cell produces, in a good helmet, a head injury criterion (HIC) near 1200 (177 G peak, total pulse duration 0.013 s). No one can say for certain that such a fall would be concussive, but it seems reasonable to assume that this exposure may produce close to a critical level of head acceleration in most individuals.

Air Cushion Impacts

Human volunteer testing of air cushions at Holloman Air Force Base has been reported for 35 rigid barrier impacts in the speed range from 15.1–31.5 mph (24.3–50.7 km/h) by Smith, et al., (4). No severe injuries beyond erythema, abrasion, contusion, and blister were received by participants. The National Highway Traffic Safety Admin. (NHTSA) published one record of the Holloman Study in Notice 17, proposed amendment to the head injury criterion in S6.2 of MVSS 208 (5). The oscillogram showed an initial impact against the air cushion (peak acceleration 75 G, duration approximately 0.070 s, Severity Index (SI) approximately 1000) and rebound against the headrest for a total SI of 1500. A maximum HIC of approximately 900 was given as measured during the initial impact against the airbag.

In the report by Smith, et al., (4), the condition of leaning forward toward the instrument panel during static firing of the airbag was briefly investigated, but was terminated due to the development of symptoms associated with mild concussion. No oscillogram records of the data or more complete medical descriptions of the effects on the volunteers have been published for this phase of the testing. Due to the extremely high velocity of the bag when slapping the head in a crouched forward position, the acceleration profile undoubtedly has a much different shape than the impact of the head into the pillowlike fully inflated airbag, particularly with respect to rise time.

Windshield Impacts

NHTSA Notice 17 also published a typical head c.g. acceleration of a test device impacted into a high penetration resistant (HPR) windshield to simulate the type of accident reported in the Cornell Aeronautical Laboratory (CAL) glass damage study (6). The pulse was idealized in the form of a triangular spike due to cracking the glass, followed by a bell-shaped impulse due to bulging the plastic interlayer. The SI and HIC for the whole pulse were given as 806 and 680, respectively. Field evidence of internal head injury caused by impacts that produced only radial cracks in the windshield was given as involving 14%, while internal head injury produced by impacts causing radial cracks with interlayer bulge was given as involving 28% of such accident victims according to the CAL report (6). However, a close study of the report and original data reveals that the evidence is too gross to determine whether it is the effect of the entire pulse or just the initial spike that caused the head injuries. In any event, the percentage of victims that received a concussion involving known unconsciousness in the accident category of radial crack with bulge reduces to at most 6.7%. Duplication of such accidents with instrumented cadavers is now in progress and represents an opportunity to measure head kinematics and kinetics that will help in the understanding of head injury and add to the relatively small store of human tolerance data available.

Analysis of Data Relative to Linear Average Acceleration/Time Tolerance Curve

The computed average acceleration as a function of time was plotted in Fig. 1 for the following frontal impact conditions:

1. Drop of 10 in (25 cm) onto a flat rigid surface—cadaver frontal fracture.
2. Drop of 48 in (122 cm) onto a flat rigid surface—crash helmet on a rigid metal headform.
3. Drop of 48 in (122 cm) onto a synthetic grass surface (rigid load cell beneath) —cadaver.
4. Windshield impact that produced radial crack plus interlayer bulge—dummy.
5. Dynamic airbag impact—human volunteer (pulse not shown beyond peak).

All of the average acceleration values peak out near the tolerance curve.

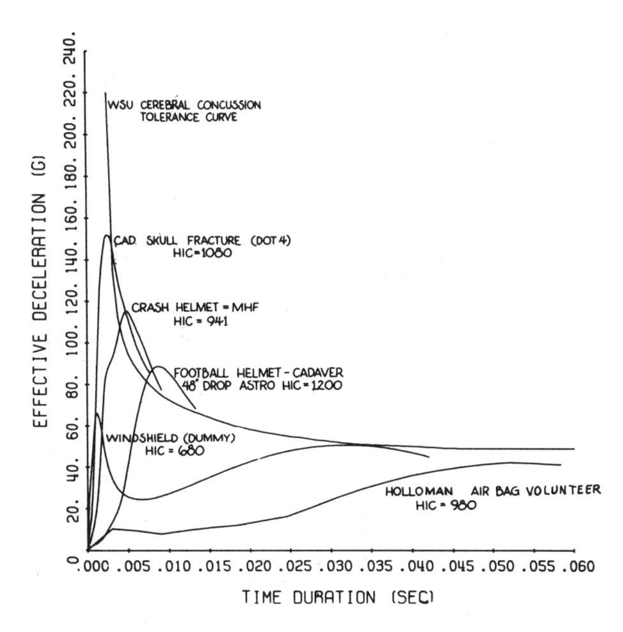

Fig. 1 - Computed average acceleration as function of time for frontal impacts of cadavers, dummies, and volunteers against various surfaces compared to WSU Linear Acceleration-Time Cerebral Concussion Tolerance Curve.

Evidence is mounting that the entire concussion tolerance curve may not exist and that only impacts against relatively stiff surfaces producing impulses with short rise time are critical

1. There is a low incidence of concussion involving unconsciousness in windshield penetration accidents. Injuries occurring may be produced by the initial spike.
2. There were no concussions in the dynamic airbag volunteer studies.
3. Field evidence is that shoulder-belted vehicle occupants who do not strike their heads do not experience serious head injury. The NHTSA has rescinded use of the HIC for the belted dummy that does not strike its head on any portion of the vehicle in compliance tests.
4. In a current series of impact tests on the front of the heads of anesthetized stumptail monkeys, it has been found impossible to render the animals unconscious with anything other than fast-rising, short-duration impact; that is, introduction of any energy-absorbing material to lengthen the pulse time prevents concussion.

Hypothesis

If a head impact does not contain a critical HIC interval of less than 0.015s, it should be considered safe as far as cerebral concussion is concerned.

Acknowledgment

The authors wish to express their appreciation to John E. Kotwick, M.S., Department of Neurosurgery, Wayne State University, for his computer impact analysis.

References

1. V. R. Hodgson and L. M. Thomas, "Comparison of Head Acceleration Injury Indices in Cadaver Skull Fracture." SAE Transactions, Vol. 80 (1971), paper 710854.
2. V. R. Hodgson, L. M. Thomas, and P. Prasad, "Testing the Validity and Limitations of the Severity Index." Paper 700901, Proceedings of Fourteenth Stapp Car Crash Conference, P-33. New York: Society of Automotive Engineers, Inc., 1970.
3. "A Progress Report on the Twenty-Sixth Annual Study of High School Football Fatalities, 1946–1971." Prepared by the National Federation of State High School Associations.
4. G. R. Smith, S. S. Hurite, A. J. Yanik, and C. R. Greer, "Human Volunteer Testing of GM Air Cushions." Paper 720443 presented at 2nd International Conference on Passive Restraints, Detroit, May 1972. (Available from SAE.)
5. Department of Transportation National Highway Traffic Safety Admin. (49 CFR Part 571), Docket No. 69-7; Notice 17, "Occupant Crash Protection Head Injury Criterion."
6. Ronald I. Herman, et al., "A Study of Occupant Injury and Glass Damage Associated with the High Penetration Resistant Windshield." Paper PB 197 569, Cornell Aeronautical Laboratory, Inc., Buffalo, New York, 1970.

720963

Nonlinear Viscoelastic Model for Head Impact Injury Hazard

Roger C. Haut, Charles W. Gadd and Richard G. Madeira
General Motors Corp.

NUMEROUS STUDIES of the tolerance to impulsive loading of either the whole body or the head alone have shown a generally falling trend of tolerable intensity of loading as its time duration is increased. Two NASA summaries of the literature made in 1959 and in 1966 (1,2)* illustrate this trend for whole-body tolerance, and a series of earlier papers from Wayne State University (3) shows the same relationship for head impact tolerance. Similar time dependency of tolerable intensity of load versus exposure duration is common to many nonbiological as well as biological materials, whether exposure is sustained or cyclical. For example, in low-cycle fatigue of metals, the strength plots inversely as approximately the second power of the life. And with respect to central nervous system tissue, Dunn and Fry recently showed (4) that tolerance to ultrasonic dosage varies inversely with the 2.2 power of exposures ranging from less than 1 ms to 100 s (Fig. 1).

These time dependency plots have usually been shown to take the form of skewed hyperbolas in cartesian coordinates or, correspondingly, relatively straight lines if transferred to log-log scales, as illustrated in Fig. 2.

*Numbers in parentheses designate References at end of paper.

Just as in other fields of investigation, the exact form of the right-hand end of the plots may be uncertain. This is not surprising because of the great duration encompassed by a logarithmic timescale, the difficulties of obtaining precise data for a system as complex as the head, the possible effects of load fluctuations or reversals, and the presence of many patterns of stress or strain distribution and mechanisms of injury.

In a paper on the Severity Index (SI) concept in 1961 (5), one of the present authors pointed out that the falling time dependency might be explained on the basis of viscoelastic properties of the materials involved. In addition, it was suggested that the tolerance curve should deviate from this simple trend if more or less discrete masses and elasticities making up the system reacted dynamically, and a family of curves was given indicating a peaking of response (corresponding to a valley in the tolerance curve) if the response of a particular effective mass-spring element predominated.

A number of investigators before and since this time have elected to approximate human tolerance as a function of the single-degree-of-freedom dynamic response, and have suggested a g-time tolerance curve of the form of Fig. 3 containing a shorter-duration realm of slope 45 deg log-log, followed by a horizontal asymptote at long duration. Under this criterion,

ABSTRACT

This study explores the application of viscoelastic modeling for characterization of the response of the brain to impulsive loading with the objective of learning whether such models could exhibit the same time dependency of strain or likelihood of injury, as exhibited by the Severity Index, HIC Index, Wayne Tolerance Curve, and other similar representations of tolerance.

The mathematical relationships between viscoelastic properties and the corresponding time dependency of tolerance are shown for Newtonian, Bingham plastic, and Pseudo-Bingham, as well as more general behavior.

Preliminary static and dynamic tests upon small mammalian material are described with particular attention given to strain in the vicinity of the brainstem as a function of loading profile.

Both the theoretical and experimental results show that the falling time dependency of the above indexes can be interpreted in terms of nonlinear viscoelastic response.

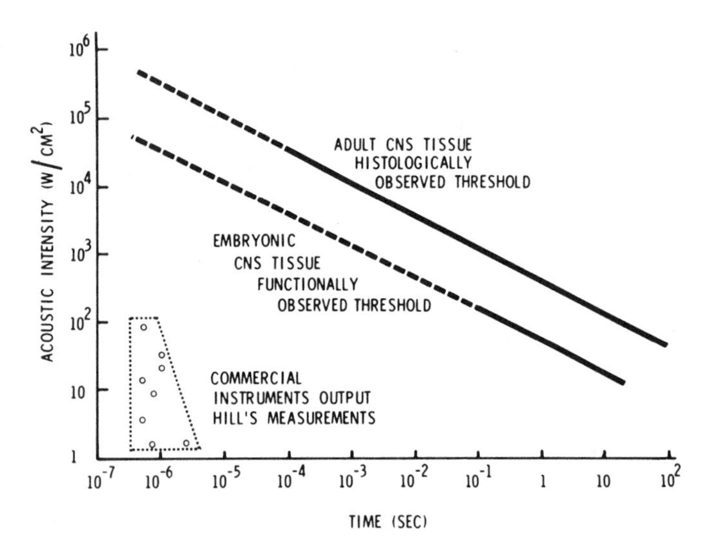

Fig. 1 - Time dependency of central nervous system tissue to ultrasonic exposure (from Dunn and Fry)

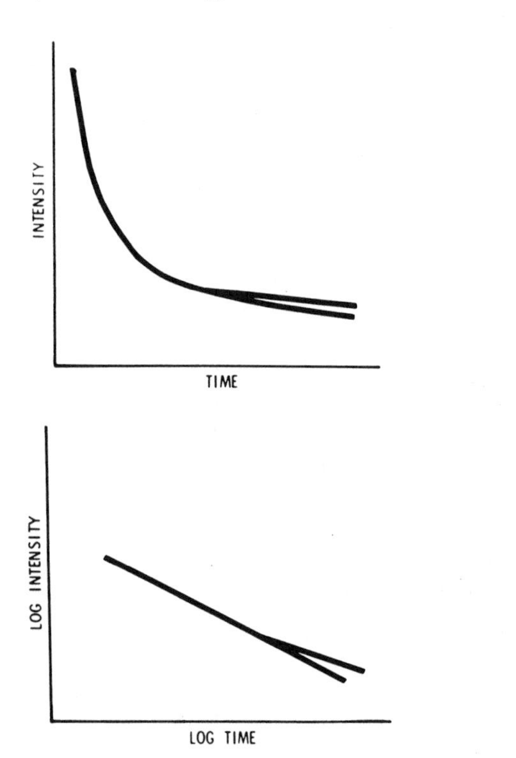

Fig. 2 - Trend of time dependency of head impact tolerance. A - Plotted on cartesian coordinates; B - Plotted on log-log coordinates

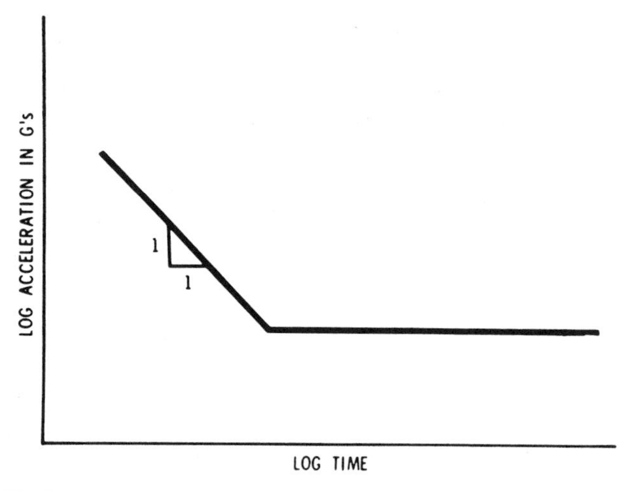

Fig. 3 - Form of tolerance curve employed by Payne, Kornhauser, and others

the tolerance at the shorter duration is limited by the imposed pulse or change in velocity. Kornhauser and Payne both employed this analogy in papers presented in the proceedings of the 1961 conference (6, 7). A limitation of the single-degree analogy having one or two masses with a spring and dashpot between is that it can represent only a 45 deg slope for the tolerance curve at a short duration and a horizontal asymptote at a long duration. The tolerance curve can, however, have a valley or no valley between, depending upon whether one assumes less than critical or critical damping. In another respect, there is an unanswered question, in that to date there has not been a direct observation or measurement substantiating the existence of a single predominating resonant frequency. The difficulty in making such a measurement is not surprising in view of the fact that the head is well balanced in structure and relatively free, through design or natural selection, of any particular kind of damaging resonance.

An alternate and somewhat more versatile analogy for characterizing the falling tolerance expressed by the Wayne Tolerance Curve, the SI, the head injury criterion (HIC) index, or others is to assume a viscoelastic or "passive" structural system. This has an advantage in allowing more flexibility than does the single-degree analogy; it allows the slope to have any value less than 45 deg in the shorter-duration region and permits the tolerance curve to approach to differing degrees a horizontal asymptote.

The objective of this study is to develop further this approach. The purpose will not be to advance viscoelastic response as the sole mechanism by which head injury hazard occurs, but rather to show it to be capable of explaining in a reasonable way the trends that have been observed. It is quite probable that more than one simple mechanism will be required to describe adequately whole-head injury hazard in a general sense. The passive analogy bears similarity to the dynamic analogy in various respects. For example, either may be employed to handle both translational and rotational accelerative hazards, and either may utilize some maximum strain as the point at which significant injury hazard is estimated to be reached.

Following is a step-by-step development of the viscoelastic analogy, together with a citation of related experimental support previously reported by others or obtained by the authors.

NEWTONIAN VISCOSITY

If this kind of response prevailed, rate of strain $d\epsilon/dt$ would be proportional to applied load (F) as given by the expression

$$F = \eta \frac{d\epsilon}{dt} \quad (1)$$

where:

η = coefficient of viscosity.

Integrating both sides, total strain should be proportional to applied impulse or ΔV, and the tolerance curve should assume a 45 deg log-log slope, as in Fig. 4.

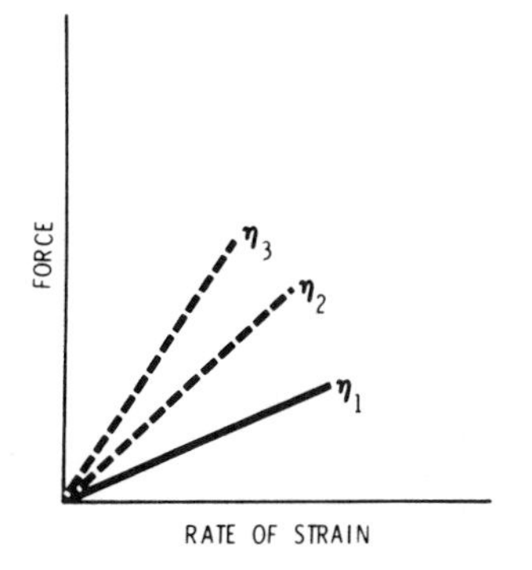

Fig. 4A - Newtonian viscous response

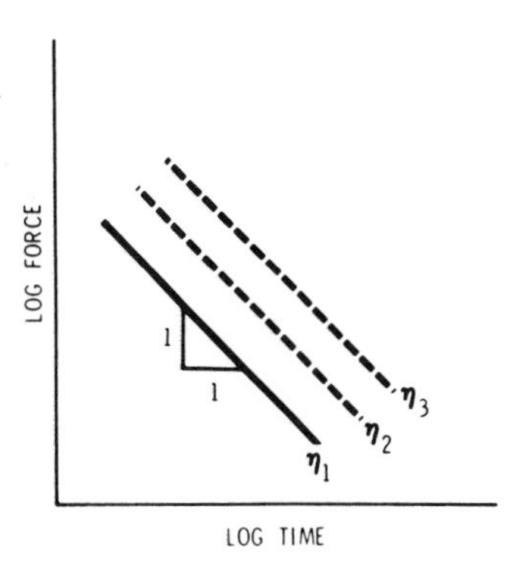

Fig. 4B - Log-log tolerance representation corresponding to Fig. 4A

BINGHAM BEHAVIOR

The next step in complexity of viscoelastic behavior could be taken by assuming behavior of the brain tissue as a Bingham plastic, as illustrated in Fig. 5. Here it is assumed that the brain fully resists flow characteristics up to some finite load level, such that the tolerance curve becomes horizontal, showing that an indefinite load duration can be safely endured below this intensity of load. This relationship may be expressed mathematically by the following:

$$\epsilon \neq \epsilon(t) \text{ for } F \leqslant F_o$$
$$F = \eta \frac{d\epsilon}{dt} \text{ for } F > F_o \qquad (2)$$

PSEUDO-BINGHAM PLASTIC BEHAVIOR

Intuitively, a real material or structure should be expected not to behave in an ideal Bingham fashion, but rather with a less abrupt transition from zero rate of strain to finite values, as shown in Fig. 6. Further, it is reasonable to expect that brain tissue, which is multiphase in character, should contain certain elements of structure (for example, cell walls or fibers), which are more elastic than the intervening matrix materials, and therefore be able to resist viscous or plastic flow effectively at low loads but to a lesser and lesser degree as load is increased.

Actually, some years ago, Ommaya performed an experiment on fresh brain tissue contained in a piston-cylinder arrangement (8). He demonstrated that the sample did, in fact, resist flow (through an escape orifice) to a marked degree up to a finite load level, after which flow became relatively rapid in the manner of a Pseudo-Bingham material.

A Pseudo-Bingham behavior may also be translated mathematically into a corresponding tolerance curve, as follows:

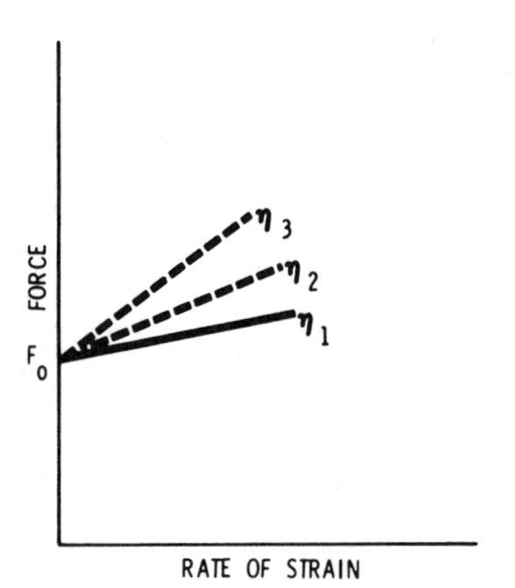

Fig. 5A - Bingham plastic response

$$F^n = C \frac{d\epsilon}{dt} \qquad (3)$$

where:

n = some coefficient greater than 1

Integration of this expression for a constant load and taking the logarithm yields

$$\log F = C' \log \epsilon - \frac{1}{n} \log t \qquad (4)$$

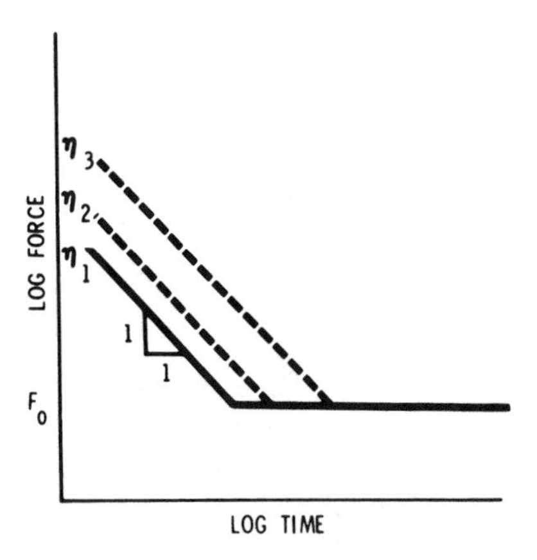

Fig. 5B - Log-log tolerance curve corresponding to Fig. 5A

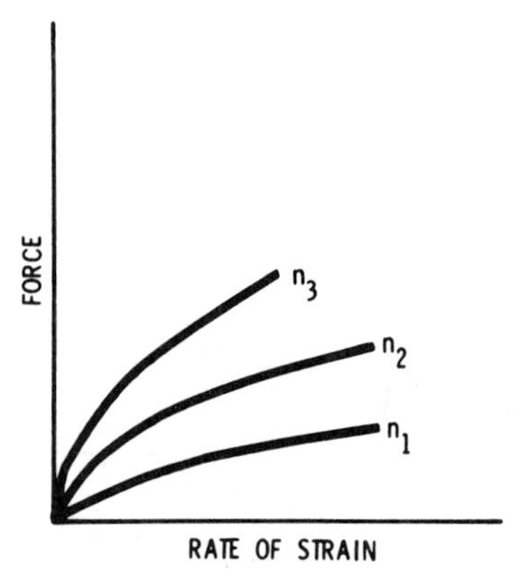

Fig. 6A - Pseudo-Bingham behavior

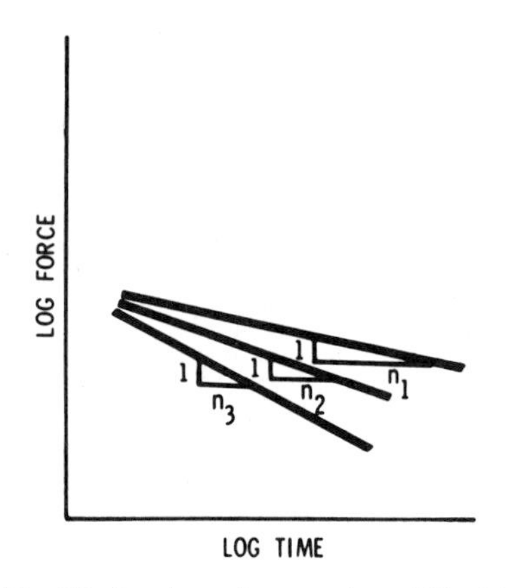

Fig. 6B - Log-log tolerance plot of Fig. 6A

Plotted again on a log-log basis, this corresponds to a tolerance curve having a slope of less than 45 deg. When an elastic region exists before this flow behavior, a horizontal asymptote would follow.

Similar trends in the tolerance curves can be displayed by employing a more general viscoelastic formulation. The time-dependent behavior of viscoelastic materials is usually based on the fact that they exhibit creep, relaxation, and, thus, rate of loading effects. A standard means of determining these material properties is with a creep test at constant load. The response can then be written as

$$\epsilon = Cf(F)\,g(t) \tag{5}$$

where:

ϵ = strain
f = function of load
g = function of time

The creep function g (t), which exhibits the time dependence, has been found to be equal to a linear function of the logarithm of time for many biological tissues (9-11). Although power creep functions have been found for very short-duration approximations, the long-duration tests indicate more of a logarithmic creep response. A number of experiments performed on biological tissues have indicated a nonlinear stress-strain response (9, 11). In this case, instead of the function f (F) being linear, a nonlinear viscoelastic formulation has been proposed (11). The basic response to a creep test can then be written as

$$\epsilon = CF^p\,t^n \tag{6}$$

or

$$\epsilon = CF^p\,(\ln t + C') \tag{7}$$

where:

C and C′ = constants

The response to any load history can then be based on a superposition principle and a hereditary integral. The falling trend of the log-log tolerance curve can be seen by plotting Eqs. 6 and 7 (Fig. 7). As indicated in the introductory remarks above, whether the appropriate form of creep function is a logarithmic or power of time is dependent upon further experimental confirmation.

EXPERIMENTAL PROGRAM

The very interesting result obtained by Ommaya suggested further study of the reaction of the brain to direct loading. The authors' program has, however, differed from Ommaya's and others (12, 13) in that attempts are being made to observe behavior in situ. This approach has been taken in the belief that tests of brain samples removed from the cranial cavity, while valuable, do not fully reflect the mechanics of the entire system. Elastic supporting elements tending to fix the brain in position and resist damaging distortion include not only the cell walls and fibers within the brain, but also such structures as the falx dura and tentorium (partions between sections of the brain), and vascular and arachnoidal structures between the peripheral surface of the brain and the cranial walls. Without the presence of these additional elastic elements, the integrity of the brain under severe loading would in all probability be in more jeopardy.

Since various investigators have shown the brainstem to be a particularly vulnerable zone (as a result of stress concentrations set up by the foramen magnum and other bony ridges, as well as by cord and vascular attachments), the experimental apparatus was designed in such a manner that displacements could be measured indicative of distress in this zone. The device employed in experiments thus far is shown diagrammatically in Fig. 8. The test piece, fresh rabbit material in the tests thus far, included the skull, first cervical vertebra, brain, and sufficient length of spinal cord to extend slightly below the vertebra for monitoring of disturbance at this point. Pressure is applied to the upper region of the brain via a 1/2 in diameter guided piston inserted through the trephined top of the skull, whose diameter was large enough to produce what was thought to be essentially a uniform

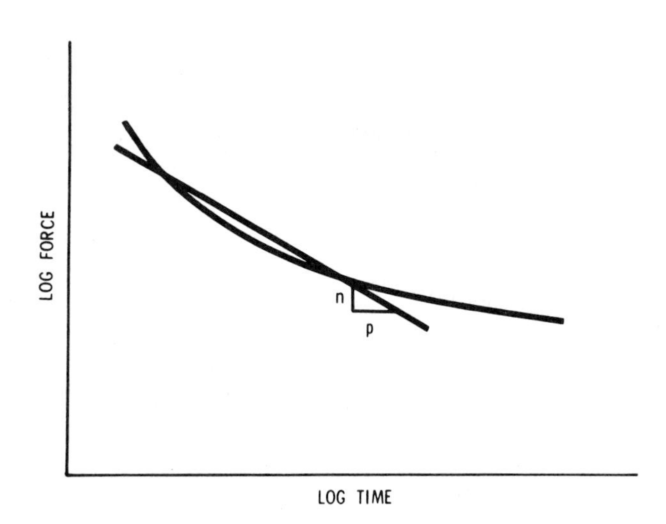

Fig. 7 - Log-log tolerance curves based on response of general viscoelastic materials

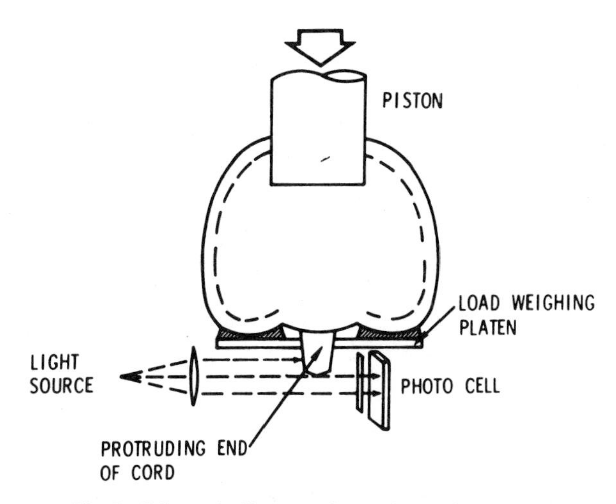

Fig. 8 - Schematic diagram of experimental apparatus

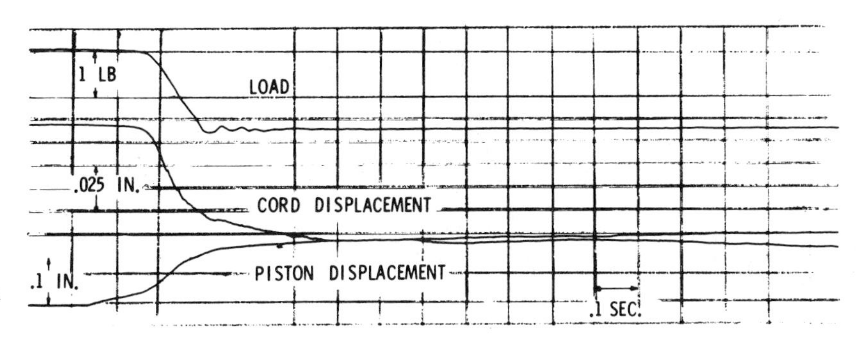

Fig. 9 - Typical oscillograph record from brainstem extrusion tests

pressure in the lower half of the cranial vault. Arbitrary static or dynamic pressure profiles may thus be applied at will to the lower region, that portion under observation in the current study. Applied loads are monitored on a platen immediately beneath the skull and fixed to the latter by means of an intervening layer of quick-setting resin. Movement of the brainstem is monitored by the use of a collimated beam of light passing beneath the severed end of the cord and falling on a photoelectric cell masked to expose only a vertically oriented band 1/32 in in width. It was found necessary to immerse the tip of the cord in a small rectangular transparent container of water to avoid error encountered if a small amount of fluid should otherwise collect on the severed end during the course of a test.

Preliminary static tests verify the existence of elastic components in the system. Elastic action was evidenced to moderate load levels by partial self-retraction of the cord after removal of a given load. As higher loads of longer duration were imposed, the displacements of the cord approached a critical level at which uncontrollable extrusion of the brain through the foramen magnum resulted. Fig. 9 is a typical example of an oscillographic record made of this process. Although only a limited number of tests have been performed, the initial results indicate a definite trend in the mechanics of this extrusion process.

The load-extrusion curves are presented in Fig. 10; they were obtained during the application of various loads. The curves represent a response over a duration of approximately 150 ms required to apply the static weights. The data indicate a force response of the form

$$d = CF^{n} \tag{8}$$

where:

d = cord displacement

From the plot in Fig. 10, the data can be adequately fit with an exponent n of approximately 0.5. Shown in Fig. 11 are the creep curves obtained in these experiments. The tests performed indicate that a creep power function of time for durations of less than 1 s was applicable. The curves are linear when plotted on log-log paper and have an average slope of approximately 0.15. According to the previously presented theory for a nonlinear viscoelastic response, a mathematical model can be written as

$$d = CF^{0.5}\, t^{0.15} \tag{9}$$

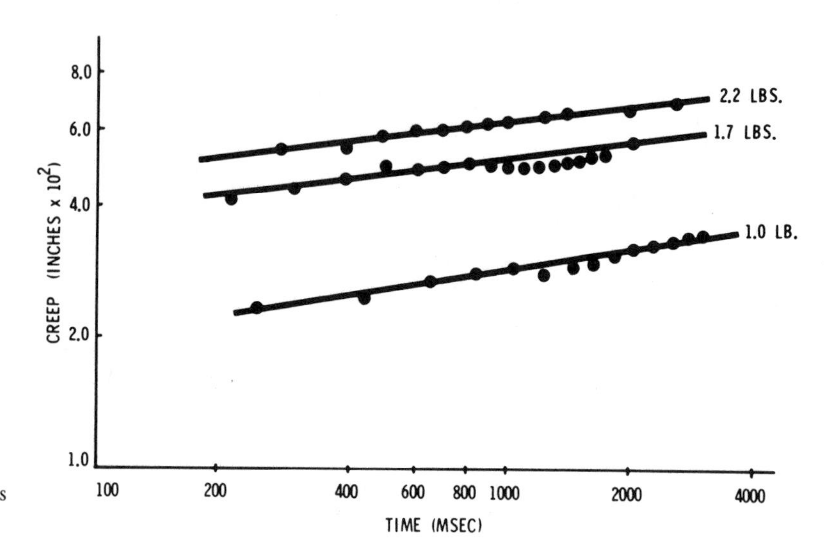

Fig. 11 - Experimental creep curves obtained at various levels of load

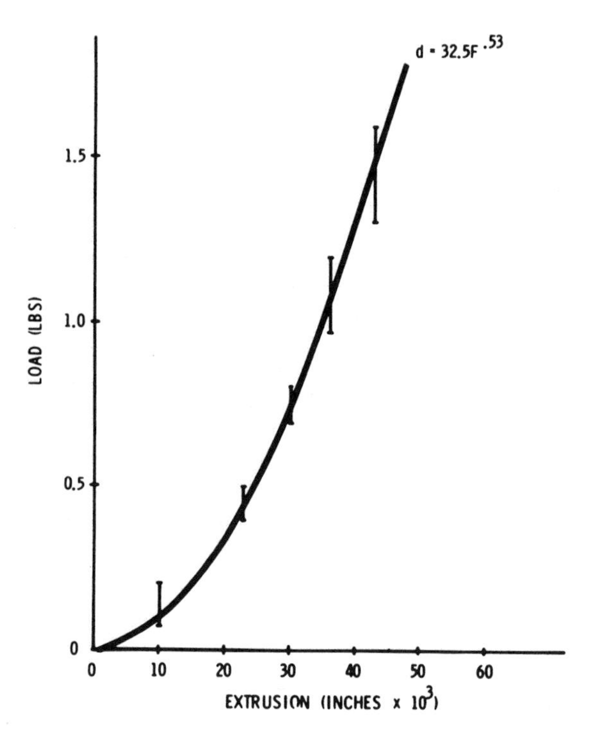

Fig. 10 - Experimental data showing relationship between force and extrusion of brainstem region

to describe this brain extrusion mechanism. A log-log plot of the force versus time to reach a specified level of extrusion indicates a linearly falling tolerance curve whose slope is $-1/3.3$. The data at this point are insufficient to define quantitatively the level of tolerable cord displacement that determines the level of this curve.

SUMMARY

In summary, it is shown that the general shape or time dependence of a frequently plotted head injury tolerance curve is compatible with the expected results of the brain as a viscoelastic structure. Preliminary supporting experiments employing a small mammalian brain in situ are described. It is suggested that no single model can simply describe the variety of stress distributions and resultant sites of injury occurring within the brain, but that a complete model must include both the passive constitutive properties and dynamic response of the system.

ACKNOWLEDGMENT

The authors wish to gratefully acknowledge the assistance given by V. R. Hodgson of Wayne State University.

REFERENCES

1. A. M. Eiband, "Human Tolerance to Rapidly Applied Accelerations: A Summary of the Literature." NASA Memorandum May 19, 1959E, June 1959.

2. T. M. Fraser, "Human Response to Sustained Acceleration." NASA-SP-103, Scientific and Technical Information Div., NASA, Washington, D.C., 1966.

3. E. S. Gurdjian, et al., "Observations on the Mechanism of Brain Concussion, Contusion, and Laceration." Surgery, Gynecology and Obstetrics, December 1955, pp. 680-690.

4. F. Dunn and F. J. Fry, "Ultrasonic Threshold Dosages for the Mammalian Central Nervous System." Transactions on Biomedical Engineering, July 1971, pp. 253-256.

5. C. W. Gadd, "Criteria for Injury Potential." Proc. Symposium on Impact Acceleration Stress, Nat'l Acad. of Science–Natl Res. Council Pub. 977, Washington, D.C., 1962, pp. 141-144.

6. M. Kornhauser and A. Gold, "Application of the Impact Sensitivity Method of Animate Structures." Nat'l Acad. of Science–Nat'l Res. Council Pub. 977, Washington, D.C., 1962, pp. 333-343.

7. P. R. Payne, "An Analogue Computer Which Determines Human Tolerance to Acceleration." Nat'l Acad. of Science–Nat'l Res. Council Pub. 977, Washington, D.C., 1962, pp. 271-296.

8. A. Ommaya, "Mechanical Properties of Tissues of the Nervous System." Jrl. Biomechanics, Pergamon Press, Vol. 1 (1968), pp. 127-138.

9. J. E. Galford and J. H. McElhaney, "Some Viscoelastic Properties of Scalp, Brain, and Dura." ASME Paper 69-BHF-7, 1969.

10. J. B. Koeneman, "Viscoelastic Properties of Brain Tissue." Unpublished M.S. thesis, Cass Inst. of Technology, 1966.

11. R. C. Haut and R. W. Little, "A Constitutive Equation for Collagen Fibers." Jrl. Biomechanics, Vol. 5 (1972), pp. 423-430.

12. M. C. H. Dodgson, "Colloidal Structure of Brain." Biorheology, Vol. 1 (1962), pp. 21-30.

13. J. H. McElhaney, R. L. Stalnaker, and M. S. Estes, "Dynamic Mechanical Properties of Scalp and Brain." Proceedings of Sixth Annual Rocky Mountain Bioengineering Symposium, May 5-6, 1969.

801304

Intracranial Pressure — A Brain Injury Criterion

Carley Ward and Marian Chan
Biodynamics/Engineering Inc.

Alan Nahum
University of California

ABSTRACT

Head impacts in animal and human cadaver tests and in aircraft accidents are simulated using finite element brain models. Brain injury severity is correlated with peak intracranial pressure. The results show that serious and fatal injuries occur when the pressures exceed 34 psi. Considering this value to be the pressure tolerance limit of the brain, a brain injury criterion is proposed. Tolerance curves of head acceleration versus time for frontal and occipital impacts are presented. These brain pressure tolerance curves are compared to existing head injury criteria (the Wayne State tolerance curve, the Vienna Institute model tolerance, the maximum strain criterion from the University of Michigan, the helmet standard, and the head injury criterion used by the Department of Transportation).

MORE SEVERE HEAD INJURIES result from vehicle accidents than from any other cause, according to a recent Mayo Clinic research study (1)*. Their results agree with those from other surveys (2) and (3) which determined head injury to be the major cause of death on the highways. The Mayo Clinic research also showed that head trauma in auto crashes was responsible for more than one-third of all fatalities (from all causes) in the surveyed county. Unlike fatal diseases which are more common in the elderly, head injury affects the young. It is the leading cause of death in the less-than-40 age group. Young men between the ages of 15 and 24 are the most likely victims. This is a tragic loss. Young people have more years to lose, and if severely disabled, the costs to sustain them through their remaining years can be enormous.

The head trauma problem is becoming even more critical with the increasing use of small cars (4) and (5). Fatalities and disabling neurological injuries due to head impact will increase unless better protection is provided. But the design of better, or even adequate, protection requires the following:

(1) An estimate of the human tolerance limits to head impact.

(2) An understanding of the injury mechanisms.

In the past, except for skull fracture, very little factual data were available to safety design engineers. The most disabling injury, brain trauma, was the least understood. Various injury hypotheses were advanced by researchers, such as brain stem shear stresses (6) and (7), pressure differentials in the cerebrum (8) and (9), and negative pressure or cavitation opposite the impact (10). Because each is a partial description of the brain response, proving one to be the primary injury mechanism was impossible.

Brain trauma research is difficult, and progress has been discouragingly slow, primarily because human tolerances cannot be established directly. Experiments must be performed on animal or human cadaver subjects, not on living humans. Animal data cannot be directly extrapolated to the human because of size and shape differences. Animal brain response to a given acceleration is very different from that of the human (11).

Recently, human unembalmed cadaver subjects, which had been repressurized to nominal invivo pressures, were used to establish brain contusion tolerances (12) and (13). But the cadaver subject has limitations. Among these are that they cannot be used to study subtle neurological deficits or cerebral concussion.

Despite this lack of background know-

* Numbers in parentheses designate References at end of paper.

ledge, head protection had to be designed and provided. The first step was to design procedures and numeric measures for evaluating this protection. To obtain these measures, tolerance criteria for head injury were proposed by various research groups. Most have been described in previous Stapp Car Crash Conferences (14) through (19). First, the Wayne State Tolerance (WST) curve was established on the basis of impact tests on animals and human cadavers at Wayne State University (9). Then at General Motors, Charles Gadd proposed a weighted impulse integration procedure, known as the Gadd Severity Index (GSI) (14). The National Highway Traffic Safety Administration, Department of Transportation, selected a Head Injury Criterion (HIC) which was a modified version of the GSI. Their modifications included a maximization procedure. The Vienna Institute based their criterion on the maximum displacement of a single-degree-of-freedom model (17). Another single-degree-of-freedom model was suggested by the Highway Safety Research Institute (HSRI) at the University of Michigan. Their criterion was called the maximum strain criterion (MSC) and was derived from the differences in acceleration between the front and back of the skull (19). All of the above are reviewed in (15) and (18). They were based primarily on measured skull accelerations, and the assumed injury mechanisms were skull fracture or distortion.

In this paper, a brain injury criterion is proposed which is based on the occurrence of brain contusion, hemorrhage, and concussion. It was derived from combined analytical and experimental investigations of intracranial pressures. Animal and pressurized human cadaver head impact tests were simulated using exact geometry brain models. (These models have also been described in previous Stapp Car Crash Conferences (13), (20), and (21).) They predict intracranial pressures throughout the brain. A comparison of the experimental injury results and computed pressures revealed that intracranial pressures above 34 psi can produce severe brain injuries (22). Curves of head acceleration versus time, which produce 34-psi pressure in the brain, are presented. When head accelerations exceed the curve values, the brain is likely to be injured; thus, these traces represent brain tolerance limits. Similar curves for moderate injuries where the maximum pressure is 25 psi are included.

FINITE ELEMENT MODELS

BRAIN MODEL DESIGN - As near as possible, the exact geometries of the human, monkey, and baboon brains were modeled (Figures 1 and 2). The models are composed of three-dimensional finite elements commonly used in structural analyses. In this application, the brick elements represent soft brain tissue, instead of structural continuum. Membrane elements, commonly used in plate and shell analyses, simulate the internal folds of dura: the falx and tentorium.

Only two modifications to accommodate the unique requirements of the brain were made. First, because the brain material is nearly incompressible, a special split energy element was used (23). (Standard structural analysis elements are not accurate for nearly incompressible materials because

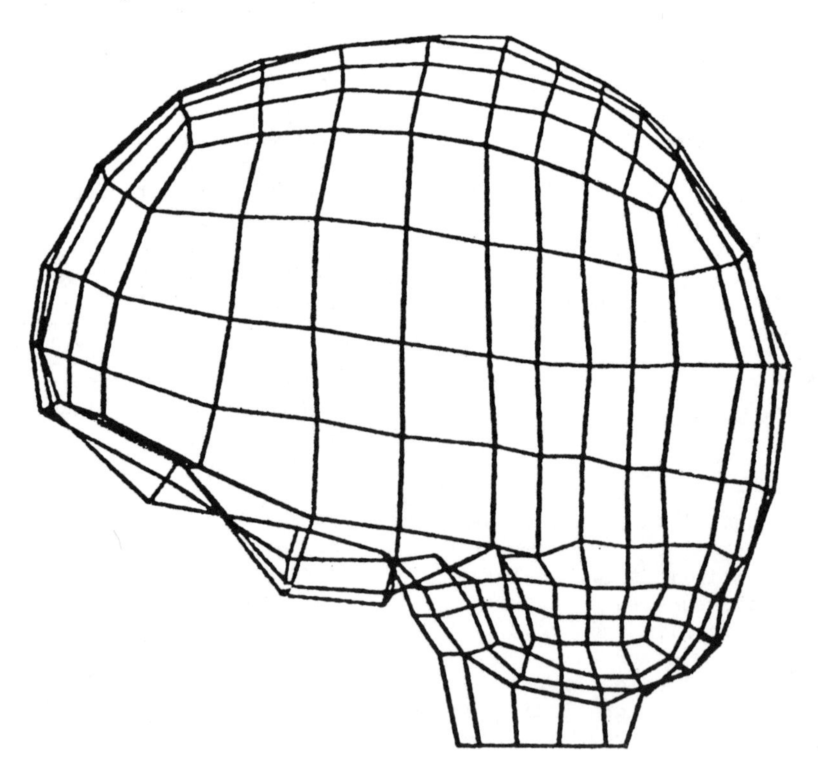

Fig. 1 - Finite element human brain model

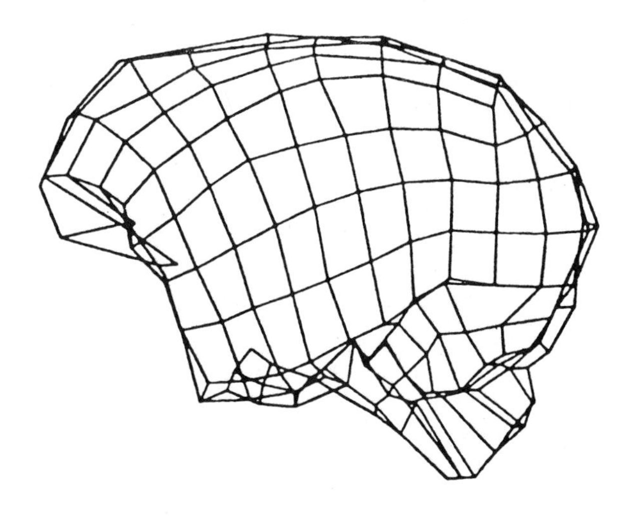

Fig. 2 - Finite element monkey brain model

the stiffness terms approach a numerical singularity.)

The second modification involves a coordinate transformation. Relative coordinates are used to avoid inaccuracies caused by large head displacements and rotations. That is, brain displacements are measured relative to the skull fixed anatomical axis. Known rigid body head accelerations are considered separately and are included in the right side of the equation of motion. The matrix equation of motion becomes

$$M \ddot{X} + K X = - M \ddot{Q} \qquad (1)$$

where
- M = mass matrix of element nodal masses
- K = stiffness matrix derived from the element elasticity equations
- X and $\ddot{X}$ = matrices of nodal displacements and accelerations measured relative to the skull axis
- $\ddot{Q}$ = inertial acceleration matrix

$\ddot{Q}$ contains the skull rotational and translational accelerations and angular velocities (refer to (21) or (23)). In Equation 1, measured head accelerations and velocities are used instead of applied forces. Mathematically, the brain model is forced to move in space and the head moves in the test. The scalp, face, skull, and neck need not be simulated, which greatly simplifies the analysis. Only the shape of the skull's inner surface is simulated to contain the brain.

MATERIAL PROPERTIES - In-situ material properties of the brain and its contained fluids have not been adequately defined for the high strain rate injury-producing event. To determine appropriate values for the models, parametric studies were performed (22) and (24). In these studies, Young's modulus, E, and Poisson's ratio, ν*, were varied. Eighteen tests were simulated and the measured and computed pressures compared. The studies showed that for the composite material (brain, vascular system, and fluid), a Young's modulus of 667,000 Pascal (Pa) gave good results for both animal and human brains. However, a single value for Poisson's ratio could not be used for all tests. Short duration impacts require higher values representing a more nearly incompressible material. When the acceleration pulse durations were less than 2.5 msec, a ν of 0.499 for the human and 0.4999 for the monkey gave the best results. When the duration was 6.5 to 8 msec, a ν of 0.49 for the human and 0.499 for the monkey were best. Between 2.5 and 6.5 msec the value of ν is varied linearly. For long pulses >8 msec, a ν of 0.48 for the human gave good results. (No animal tests had long acceleration pulses.) These pulse range limits are only approximations and need further study.

The different ν values are needed to simulate the pressure release mechanisms or volume elastance. During the short duration, spike-shaped acceleration pulses, these mechanisms have less time to act. The effect is a more incompressible brain. Why the monkey brain needs higher values for ν is unknown. It may be related to the effect of the smaller vasculature and skull orifices, which restrict flow from the animal's brain.

BRAIN STRESSES - A blow to the head produces a pressure (stress) gradient in the brain. The brain tends to lag the skull, compressing near the impact site. Opposite the impact, tension stresses develop as the skull pulls the brain along. In the brain stem and cerebellum, stresses are affected by the opening at the base of the skull, the foramen magnum. Pressure is dissipated by movement of tissue through this opening.

The shapes of the intracranial pressure pulses resemble the skull acceleration traces. When the head impacts a hard surface, the pulse is spike shaped with a short duration. When the impact surface is padded, the pressure pulses are longer with a lower magnitude (22). Pressure pulses are also long in a helmeted head impact. They resemble those produced in an unprotected head impacting a well-padded surface (20).

BRAIN MODEL SIMULATIONS - Head impact tests on human cadavers, baboons, and monkeys, and impacts to helmeted airmen in aircraft accidents were all simulated. The experimental impacts on pressurized unembalmed human cadavers are described in (12), (13), and (20). Descriptions of the baboon and

* In the brain model, ν is an effective Poisson's ratio which includes the volume elastance as well as the tissue compressibility.

monkey tests are in (25) and (26). The aircraft accident data were obtained from reenactments of helmeted impacts (27). Actual helmet damage was duplicated in the laboratory by dropping a helmeted head form.

In these simulations, overall injury severity correlated with intracranial pressure magnitude. At higher pressures, the injuries were more severe, as shown in Figure 3. The maximum stresses are positive (tension) in the animal tests and two crash victims. These were produced by occipital impacts. In the human cadavers and six accident victims, the maximum stresses are negative (compressive) and occur in the frontal lobe. In these tests the frontal bone was impacted. The change in maximum stress from positive to negative is caused by the reversed impact direction.

Figure 3 shows that moderate injuries are likely to occur at pressures above 25 psi. In these cases contusions do not always occur in the high pressure regions. Pressures are below the tissue injury level, and other less severe injuries are produced.

The threshold for serious and fatal injuries is approximately the same, 34 psi. These stresses can be either positive or negative. Most of these injuries are brain contusions or hemorrhages in the high stress regions. However, some animals suffered serious or fatal concussions without contusions. The lowest stress for a severe brain injury, 34 psi, occurred in cadaver test no.

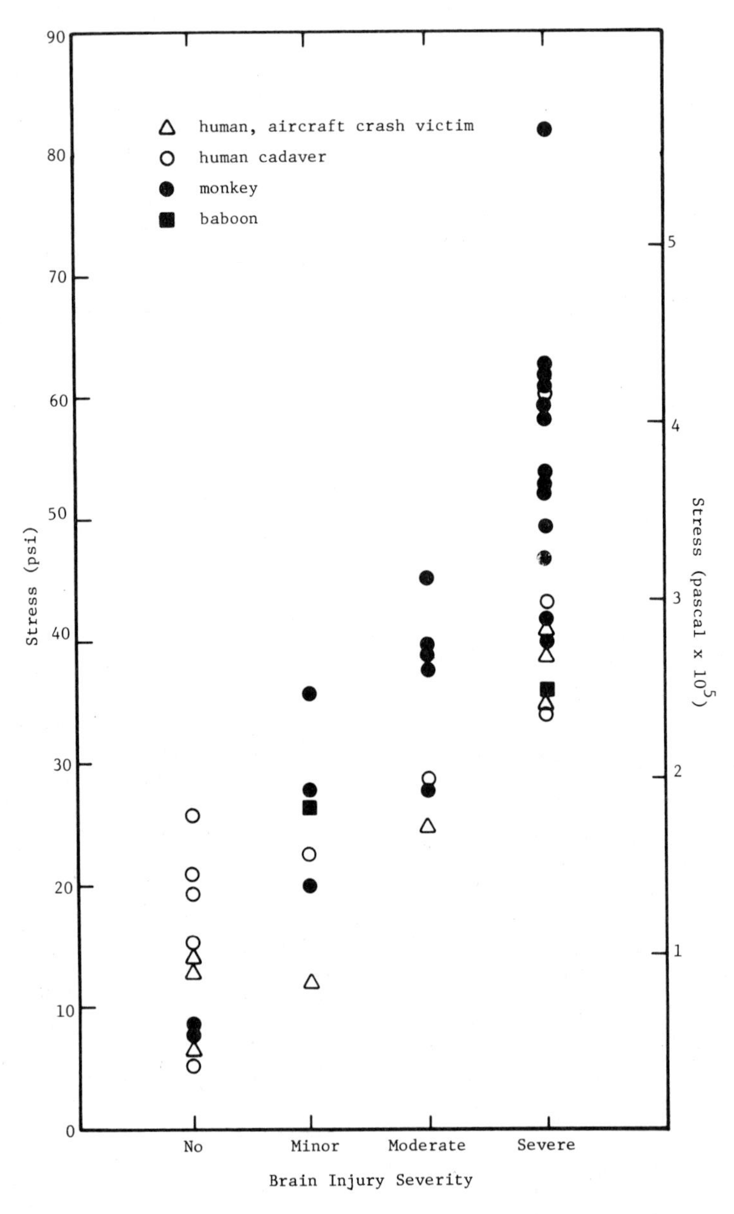

Fig. 3 - Comparison of injury severity and computed maximum pressures (frontal and occipital impacts)

32, where compression produced extensive subpial hemorrhage (12). This stress value was selected as an approximate stress tolerance for brain tissue.

In an analysis of tests reported in (29), it was shown that this stress (34 psi) could contuse the brain. In these tests small 10-gram balance weights were dropped from different heights onto the surface of a living dog brain. When the drop height reached 20 cm, small vessels were ruptured, producing a slight contusion. The calculated impact stress associated with a 20-cm drop height is 203 kPa or 29.4 psi. These values are approximations based on a flat surface impact. The difference between the assumed tolerances (34 psi and 29.4 psi) is not significant considering the assumptions used in the drop test calculations.

INTRACRANIAL PRESSURES AS AN INJURY INDICATOR

BRAIN PRESSURE TOLERANCE CURVES FOR SEVERE INJURY - Assuming a brain tissue tolerance of 34 psi, brain pressure tolerance (BPT) curves for frontal and occipital impacts were determined* (Figures 4 through 7). These curves were calculated as follows. By using the human brain model, intracranial pressures were calculated for a range of head acceleration pulse durations. Four acceleration pulse shapes were considered, beginning with a triangular shape. In the other pulses the peak value was extended to approach a square wave. The acceleration magnitudes which would produce 34 psi were determined and plotted. If the head acceleration exceeds the curve values, the brain is likely to be seriously injured, AIS 5 or 6.

BRAIN PRESSURE TOLERANCE CURVES FOR MODERATE INJURY - Using 25 psi as a threshold for moderate injury, tolerance curves for the same four acceleration pulses were computed. These are the lower curves shown in Figures 4 through 7. Accelerations between the 25 psi and 34 psi curves are likely to produce moderate injuries, with an AIS value

* Side impacts are not included

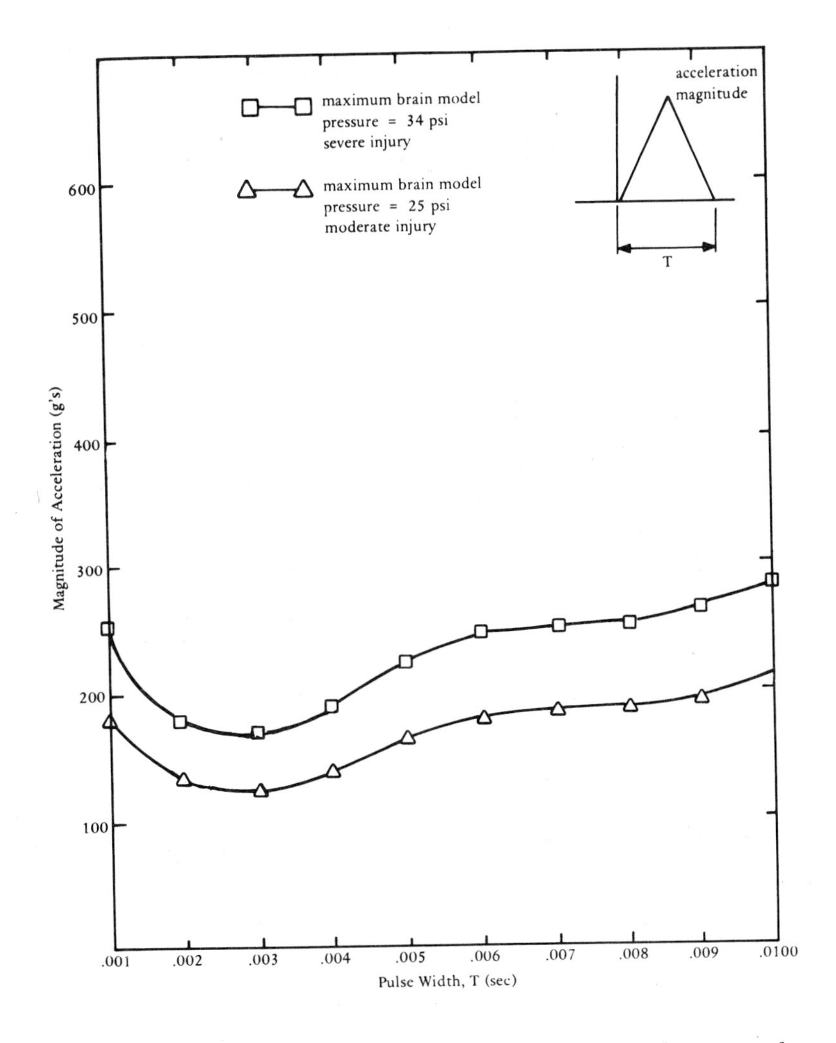

Fig. 4 - Brain pressure tolerance curves for moderate and severe brain injury (frontal and occipital impacts)

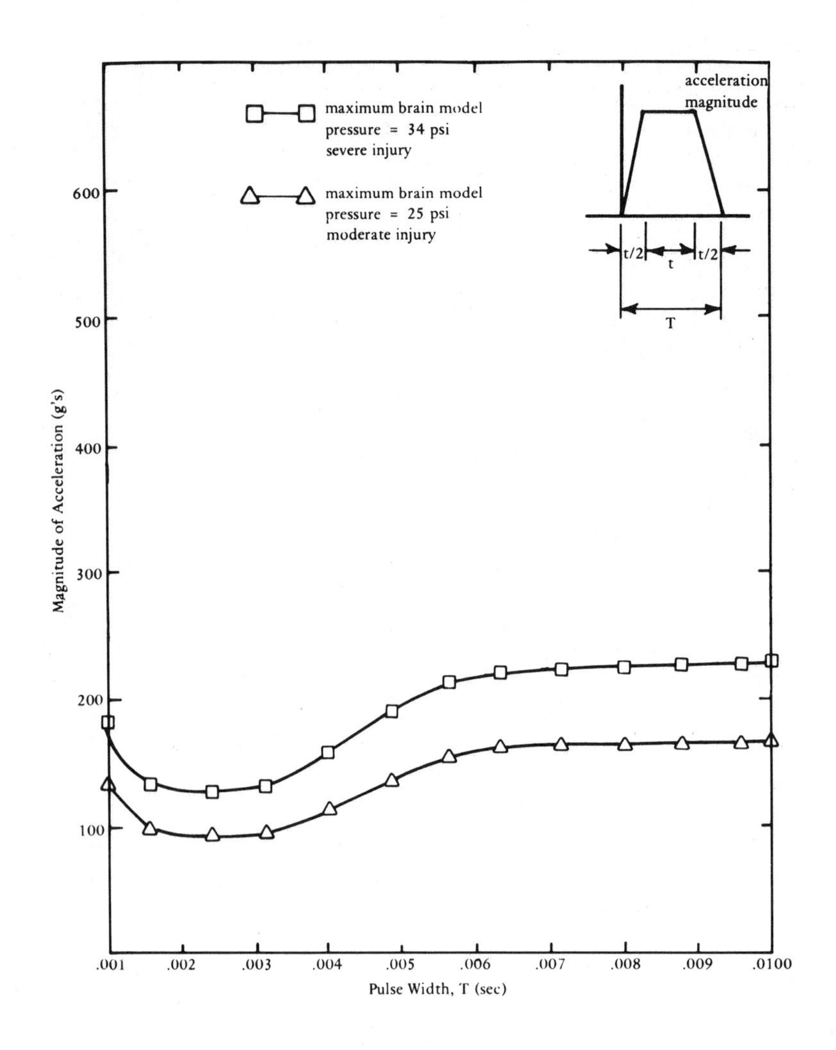

Fig. 5 - Brain pressure tolerance curves for moderate and severe brain injury (frontal and occipital impacts)

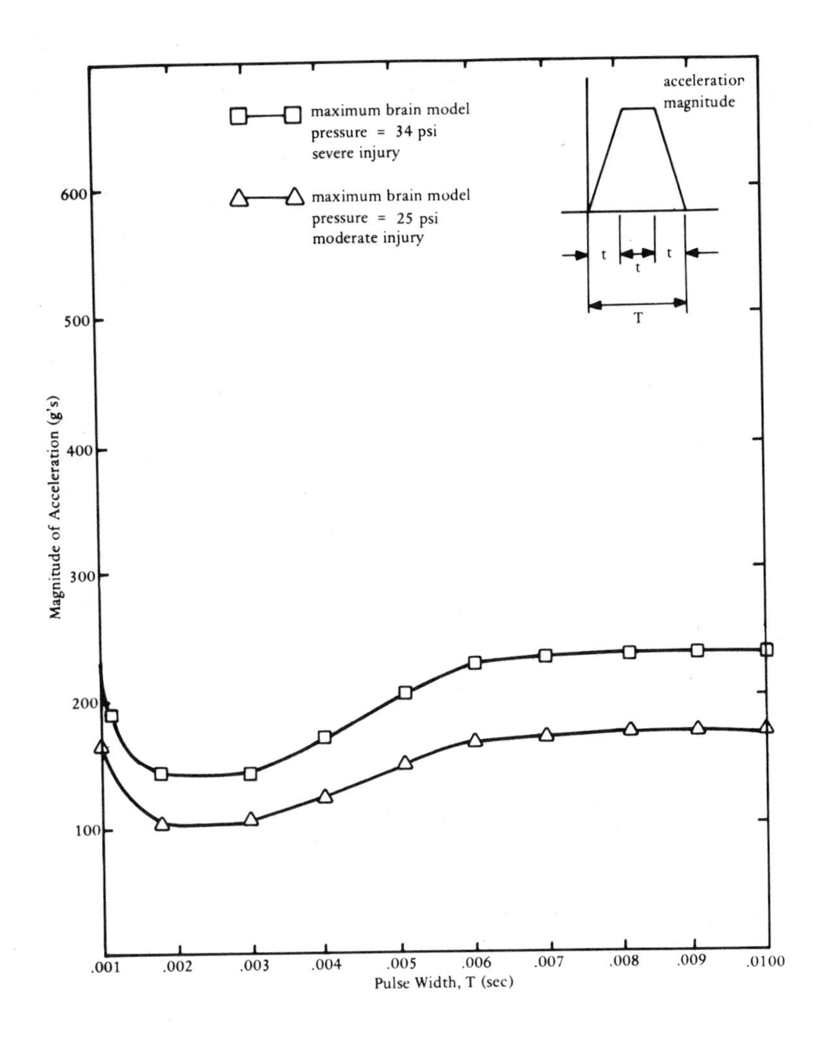

Fig. 6 - Brain pressure tolerance curves for moderate and severe brain injury (frontal and occipital impacts)

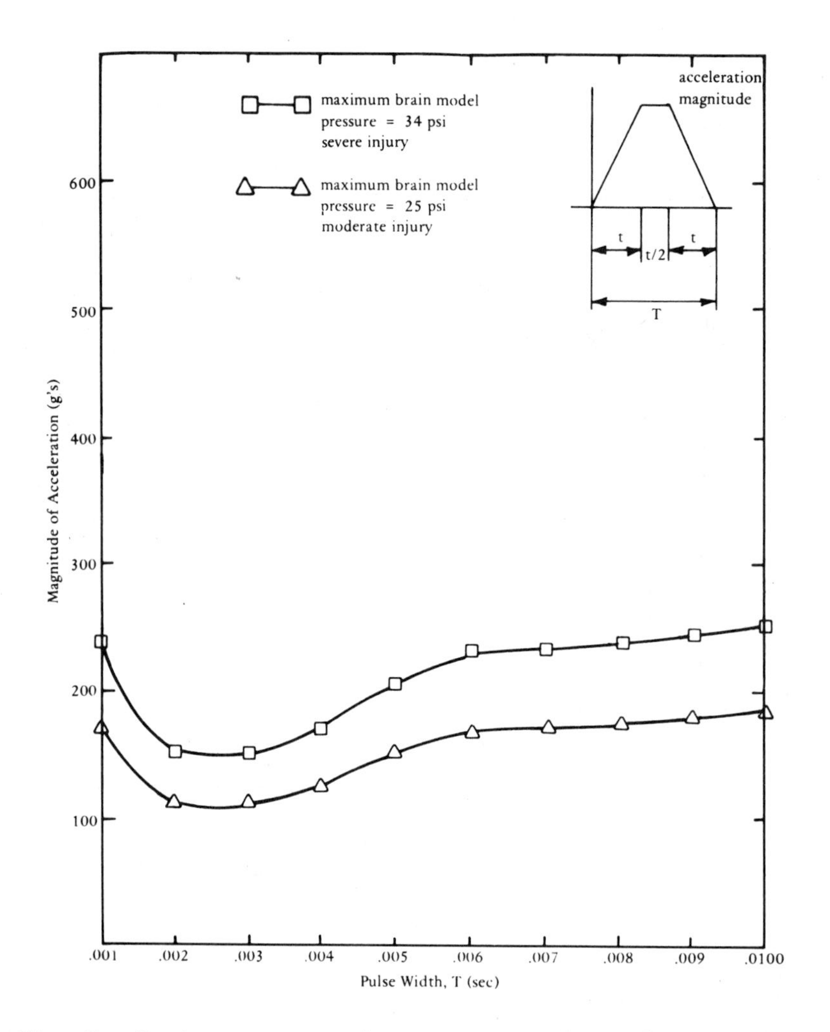

Fig. 7 - Brain pressure tolerance curves for moderate and severe brain injury (frontal and occipital impacts)

of 3 or 4. If the head acceleration is low, below the moderate injury curve, the injury is not likely to be more than minor.

COMPARISON OF THE BRAIN PRESSURE TOLERANCE CURVES WITH OTHER HEAD INJURY CRITERIA

The brain pressure tolerance (BPT) curves are compared to the other head injury indices and criteria in Figures 8 through 10. Curves for the WST, HIC motorcycle standard, and MSC are plotted for the same triangular shaped head accelerations and pulse durations.

THE WAYNE STATE TOLERANCE CURVES - The WST curve uses average acceleration and is plotted for the triangular pulse in Figure 8. This tolerance curve is based on experimental data from human cadaver and animal tests. Tolerances for the short duration impacts were based on cadaver skull fractures. The head was impacted onto an unyielding flat surface, and longitudinal acceleration was measured on the skull opposite the impact. Tolerances for long duration head accelerations were obtained from animal tests.

The brain pressure tolerance (BPT) curves do not agree with the WST curve (refer to Figure 8). The WST would permit much higher accelerations for pulse durations less than 3.5 msec. The deviation is greatest when the impact is of short duration (a head impacting a rigid surface). Agreement is poor for the long duration impacts as well, where the reverse is true. The WST curve is lower than the BPT curve for moderate injury. This BPT curve would permit an acceleration of 206g for a 10-msec triangular-shaped acceleration trace, while the WST curve allows only 140g.

VIENNA INSTITUTE OF TECHNOLOGY MODEL - The Vienna Institute developed a damped spring-mass model to define head injury tolerance. The equation for the system is

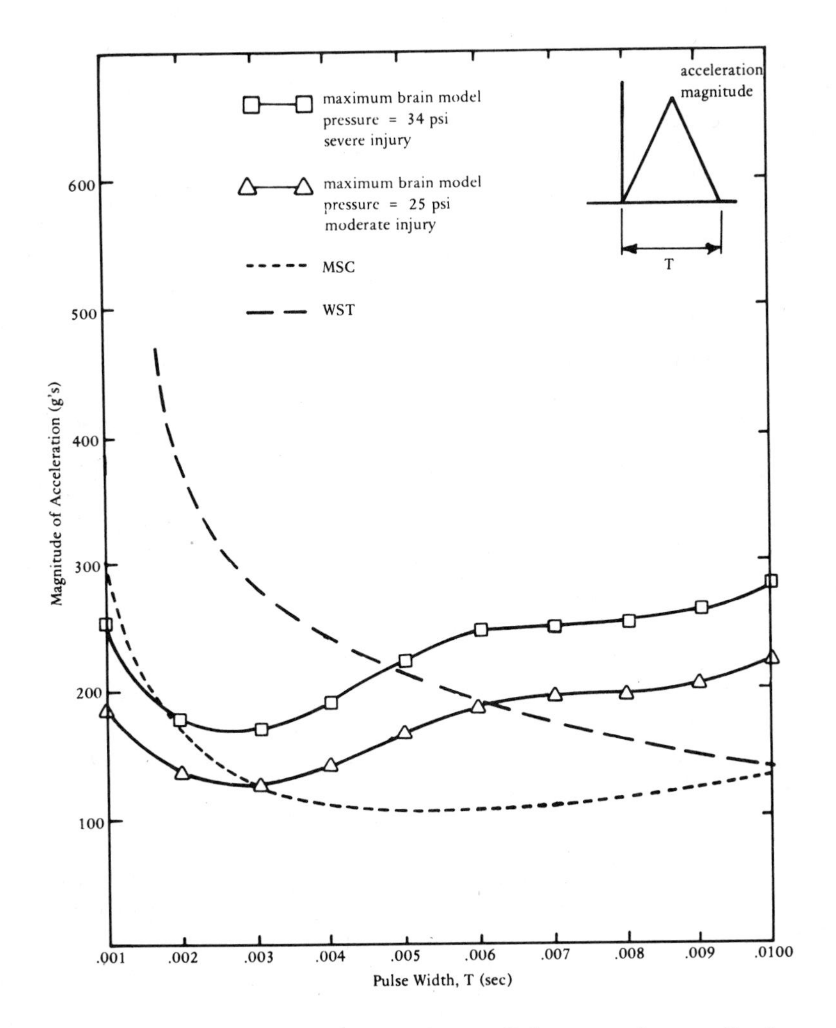

Fig. 8 - Comparison of Wayne State Tolerance Curve, Maximum Strain Criterion, and brain pressure tolerance curves for moderate and severe injury

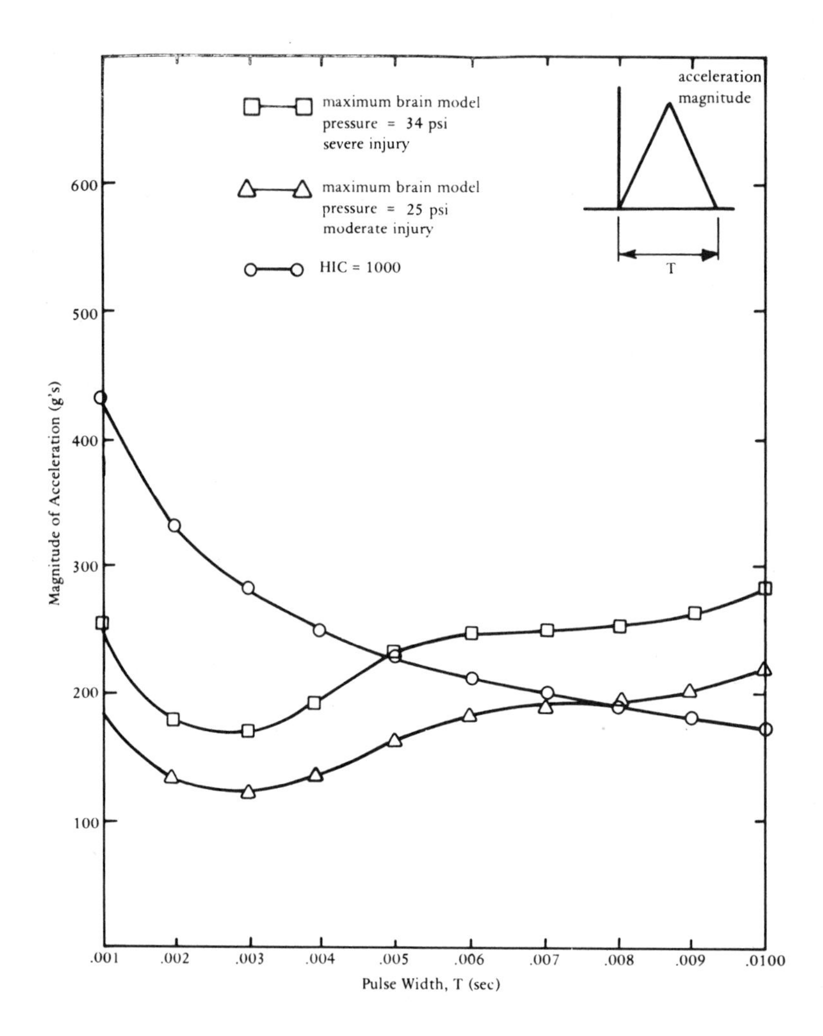

Fig. 9 - Comparison of NHTSA Head Injury Criterion and brain pressure tolerance curves for moderate and severe injury

$$\ddot{X} + 2\,D\,\omega\,\dot{X} + \omega^2\,X = -A(t) \qquad (2)$$

where $D = 1$

ω = natural frequency; $\omega = 635$ rad/sec

$A(t)$ = head acceleration

$X, \dot{X}, \ddot{X}$ = relative displacement, velocity, and acceleration of the mass

A tolerance limit, S, was established for displacement.

$$S = 0.0235 \text{ cm} \qquad (3)$$

In this model, S is assumed to be the maximum displacement that can be tolerated without injury. If for a given head acceleration the mass displacement exceeded 0.0235 cm, injury would be predicted. Using this value, tolerance curves were established. For a triangular pulse the curve approximates the WST curve, the maximum deviation being only 4% (18).

Just as the BPT curves do not agree with the WST curve, they do not agree with the Vienna model injury criteria. A single-degree-of-freedom model approximating the BPT curves would have a natural frequency of 2416 rad/sec and displacement tolerances for severe and moderate injuries of 0.0445 and 0.0327 cm respectively. These values are much higher than the Vienna model frequency (635 rad/sec) and displacement tolerance (0.0235 cm).

MAXIMUM STRAIN CRITERION - HSRI based their maximum strain criterion (MSC) on the mechanical impedance of a human cadaver skull. The input force and output acceleration were measured at the same point on the skull, and a mathematical model was developed for the brain skull system. In this model the maximum tolerable strain was assumed to be 0.00329 in/in. The resulting tolerance curve for a triangular pulse is plotted in Figure 8.

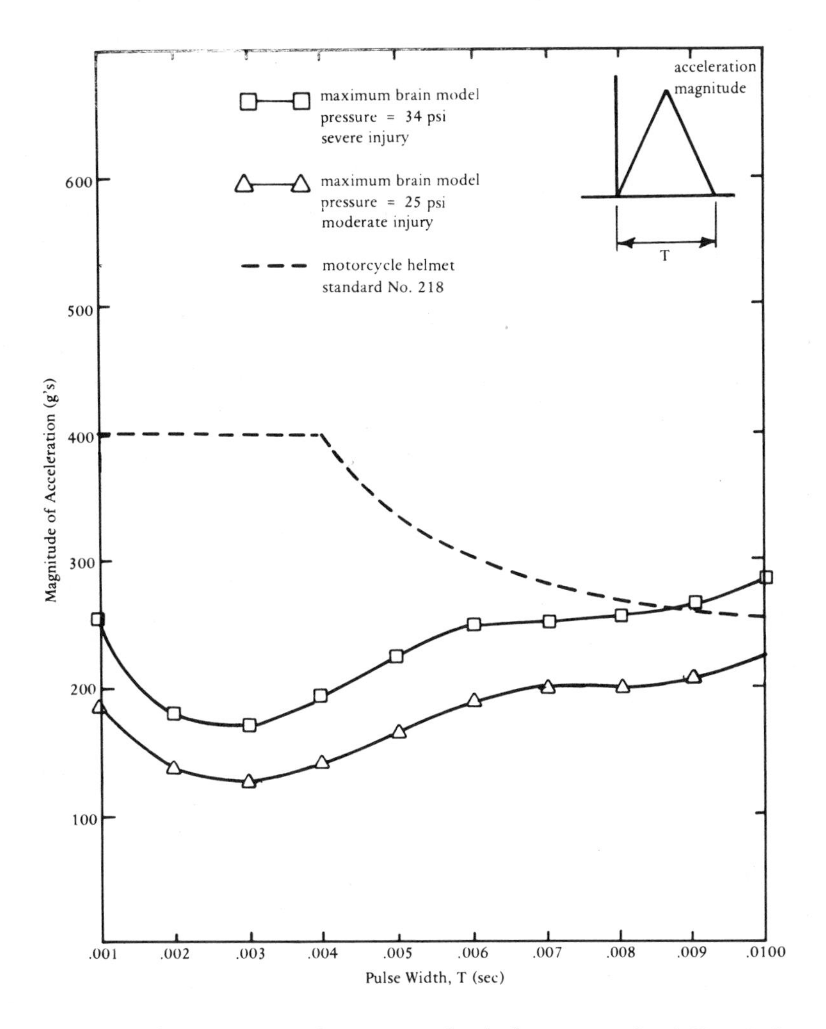

Fig. 10 - Comparison of motorcycle helmet standard No. 218 and brain pressure tolerance curves for moderate and severe injury

The MSC and the BPT curves agree only when the head acceleration duration is between 1.8 and 3 msec (refer to Figure 8). For shorter pulse durations, the MSC allows a higher acceleration. For longer pulses, greater than 3.5 msec, the MSC consistently defines a much lower tolerance than the BPT curve (Figure 8).

THE GADD SEVERITY INDEX AND HEAD INJURY CRITERION - Whole body acceleration data (including Colonel J. Stapp's sled tests) and animal data were reviewed by C. Gadd (14). Acceleration durations were predominantly in the 20-100 msec range. He found that when plotted on log-log coordinates a straight line would fit the data. The equation for the line was

$$A_c T^{0.4} = 15.83 \quad (4)$$

which was changed through multiplication to

$$A_c^{2.5} T = 1000 \quad (5)$$

In these equations, T is the duration of constant acceleration, A_c, and 1000 is the tolerance level. The GSI evolved from Equation 5 to become

$$GSI = \int_{t_1}^{t_2} [A(t)]^{2.5} dt = 1000 \quad (6)$$

where A(t) is the head acceleration in g's expressed as a function of time and the integration is over the pulse duration.

The HIC was developed from the GSI. A maximization procedure was added to help define the pulse duration, thus eliminating contributions from low-level, long-duration accelerations. The following equation is used

$$HIC = \left[\frac{\int_{t_1}^{t_2} A(t)\, dt}{(t_2 - t_1)}\right]^{2.5} (t_2 - t_1) = 1000 \quad (7)$$

where t_1 and t_2 are varied to find the maximum value for HIC. The HIC and GSI are very similar for clearly defined pulses and are equal for a square wave. In general, the HIC tends to be slightly smaller than the GSI because the pulse duration is shorter.

A tolerance curve based on an HIC of 1000 is plotted in Figures 9. Of all the indices, the HIC compares best with the BPT curves. The correlation is best for acceleration pulse durations near 6 msec where it lies in between the two BPT curves. The HIC is more conservative for the long pulses, since it is below the BPT curves. The critical difference between the HIC and BPT curves is in the short impacts, 0 to 3.5 msec. The HIC allows a much higher head acceleration. For a 1.0-msec impact, the HIC tolerance is about one-third higher than the BPT tolerance for severe injury (Figures 9). An HIC of 1000 does not provide sufficient protection for hard surface impacts (0 to 3 msec impacts).

MOTORCYCLE HELMET STANDARD - Motorcycle helmets must attenuate impact accelerations to pass the Motor Vehicle Safety Standard Number 218, (29). This standard requires that during the impact test the headform acceleration does not exceed any of the following:

(a) Peak acceleration of 400g,

(b) Acceleration in excess of 200g for a cumulative duration of 2.0 msec, and

(c) Acceleration in excess of 150g for a cumulative duration of 4.0 msec.

For a triangular head acceleration pulse, the curve defined by the standard is shown in Figure 10. Although the helmet curve is much higher than the BPT curves for 1 to 5 msec impacts, these short duration impacts do not occur. Crush of the liner extends the acceleration pulse to greater than 5 msec. For longer pulses the helmet standard curve approaches the BPT serious injury curve (Figure 10). In the 8 to 10 msec range the agreement is good and the helmet provides protection against serious brain injury.

SUMMARY

Intracranial pressure magnitudes have been shown to correlate with brain injury severity. Simulations of animal and human cadaver tests and aircraft accidents have shown that the peak pressures are, in general, proportional to the trauma severity. These pressures usually produce brain contusions or small vessel hemorrhages on the brain surface. In a few cases, damage was minimal but the animal suffered severe concussions.

Peak pressure (stress) values are measures of the brain's response to the impact. The greater the response, the more serious the trauma. These scalar values can be used to evaluate head protection and establish safe head accelerations.

The injury test simulations show that serious brain injuries occur when the peak intracranial pressure exceeds 34 psi. Using this value as the maximum allowable brain stress, tolerance curves were developed for frontal and occipital impacts. BPT curves for four head acceleration pulse shapes are presented. The impact durations are between 1 and 10 msec.

The BPT curves developed in this study deviate from the WST curve and tolerance curves derived from other indices. This is to be expected, since the others are not based on brain injury. The WST, Vienna model, and MSC all use skull measurements or skull fracture. The HIC and GSI are based primarily on whole body accelerations that are usually greater than 20 msec. The BPT curves are based entirely on brain injury data associated with impact. Acceleration durations in these injury tests are all between 1 and 10 msec. Therefore, they represent only brain tolerance to impact.

The other indices permit higher accelerations for the short impacts and require lower accelerations for the long impacts. Protection based on these indices would protect the brain if the acceleration pulses were long (greater than 4 msec). But a critical difference between the BPT curve and the other curves exists for the short acceleration pulse (less than 3 msec). Protection based on the other criteria would not protect the brain for these short duration impacts. These impacts occur when the head strikes an ineffectively padded or unpadded rigid surface.

The BPT curves developed in this study should not be used to evaluate other head injury criteria. Brain injury is not the only head injury. The other criteria may include tolerances to other injuries like skull fracture. Ideally, tolerance curves for each kind of head injury should be superimposed, and the lowest curves used to define an overall head injury tolerance.

ACKNOWLEDGMENTS

This paper was written at the suggestion of Dr. Rhoads Stephenson, Assoc. Administrator National Highway Traffic Safety Administration (NHTSA). The helpful ideas, encouragement and references provided by Dr. Stephenson are gratefully acknowledged. The study was performed at the Civil Engineering Laboratory, Naval Construction Battalion Center, Port Hueneme, CA for NHTSA under interagency agreement.

REFERENCES

1. J. F. Annegers, Grabow, Kurlona, and Laws. "The incidence, causes, and secular trends of head trauma in Olmstead County, Minnesota 1935-1974." (For copies write to: Dr. J. F. Annegers, Department of Medical Statistics and Epidemiology, Mayo Clinic, 200 First St. S.W., Rochester, MN 55901).

2. J. B. Pedder, S. B. Hagues, and G. M. MacKay. "A study of 93 fatal two wheel

motor vehicle accidents," in Proceedings of Fourth International Conference on Impact Trauma, Götenburg, Sweden, Sep. 1979, IRCOBI.
3. National Safety Council. Accident facts. Chicago, Ill., 1974.
4. "Highway Death Roll Rises Despite Travel Restrictions," The Highway Loss Reduction Status Report, vol. 15, no. 5, Mar 26, 1980.
5. "We .. Will Not Sacrifice Human Safety To Save A Dollar," The Highway Loss Reduction Report, vol. 15, no. 9, Jun 1980.
6. A. K. Ommaya, S. D. Rockoff, and M. Baldwin. "Experimental concussion; A first report," Journal of Neurosurgery, vol. 21, 1964, pp. 249-265.
7. E. S. Gurdjian and H. R. Lissner. "Photoelastic confirmation of the presence of shear strains at the crano-spinal junction in closed head injury," Journal of Neurosurgery, vol. 18, no. 1, 1961, pp. 58-61.
8. A. E. Walker, J. J. Kollros, and T. J. Case. "Physiological basis of concussion," Journal of Neurosurgery, vol. 1, 1944, pp. 102-108.
9. E. S. Gurdjian. Impact head injury. Springfield, Ill., Charles C. Thomas.
10. A. G. Gross. "A new theory on the dynamics of brain concussion and brain injury," Journal of Neurosurgery, vol. 15, 1958, pp. 548-561.
11. C. C. Ward, P. E. Nikravesch, and R. B. Thompson. "Biodynamic finite element models used in brain injury research," Journal of Aviation, Space and Environmental Medicine, vol. 49, no. 1, 1978.
12. A. M. Nahum and R. W. Smith. "An experimental model for closed head impact injury," in Proceedings of Twentieth Stapp Car Crash Conference, Dearborn, Mich., Oct. 1976, pp. 783-813. (SAE paper 760825).
13. A. M. Nahum, R. Smith, and C. C. Ward. "Intracranial pressure dynamics during head impact," in Proceedings of Twenty-first Stapp Car Crash Conference, New Orleans, La., Oct. 1977, pp. 337-365. (SAE paper 770922).
14. C. W. Gadd. "Use of a weighted-impulse criterion for estimating injury hazard," in Proceedings of Tenth Stapp Car Crash Conference, New York, 1966, p. 12. (SAE paper 660793).
15. J. Versace. "A review of the severity index," in Proceedings of Fifteenth Stapp Car Crash Conference, New York, 1971, pp. 771-795. (SAE paper 710881).
16. C. W. Gadd. "Tolerable severity index in whole-head nonmechanical impact," in Proceedings of Fifteenth Stapp Car Crash Conference, Coronado, Calif., Nov. 1971, pp. 809-816.
17. A. Slattenschek, W. Tauffkirchen, and G. Benedikter. "The quantification of internal head injury by means of the phantom head and the impact assessment methods," in Proceedings of Fifteenth Stapp Car Crash Conference, Coronado, Calif., Nov. 1971, pp. 742-765. (SAE paper 710879).
18. W. Fan. "Internal head injury assessment," in Proceedings of Fifteenth Stapp Car Crash Conference, Coronado, Calif., Nov. 1971, pp. 645-665. (SAE paper 710870).
19. R. L. Stalnaker and J. H. McElhaney. "Head injury tolerance for linear impacts by mechanical impedance methods," American Society of Mechanical Engineers. Report 70-WA/BHF-4, 1970.
20. A. Nahum, C. C. Ward, R. Smith, and F. Roasch. "Intracranial pressure relationships on the protected and unprotected head," in Proceedings of Twenty-third Stapp Car Crash Conference, San Diego, Calif., Oct. 1979. (SAE paper 791024).
21. C. C. Ward and R. Thompson. "The development of a detailed finite element brain model," in Proceedings of Nineteenth Stapp Car Crash Conference, San Diego, Calif., Oct. 1975. (SAE paper 751163).
22. C. C. Ward. "A head injury model," in Proceedings of NATO Advisory Group for Aerospace Research and Development Conference, No. 253: Models and Analogues for the Evaluation of Human Biodynamic Response, Performance and Protection, Paris, France, Nov. 1978. (Copies from: National Technical Information (NTIS), 5283 Port Royal Road, Springfield, VA 22151, USA).
23. C. Ward, H. S. Chan, and M. Chan. "Simulation of brain injury tests using a finite element primate brain model," in Proceedings of International Conference on Finite Elements on Biomechanics, Tucson, Ariz. Feb. 18-20, 1980.
24. G. S. Nusholtz, J. Axelrod, J. Melvin, and C. Ward. "Comparison of epidural pressure in live anesthetized and post-mortem primates," Highway Safety Research Institute. Report No. UM-HSRI-79-90, 1979.
25. R. L. Stalnaker, J. B. Benson, and J. W. Melvin. "Dynamic and static load response of the head," in Proceedings of Second International IRCOBI Conference, Biomechanics of Serious Trauma, Birmingham, U.K., Sep. 1975.
26. T. A. Gennarelli, J. M. Able, H. Adams, and D. Graham. "Differential tolerance of frontal and temporal lobes to contusion induced by angular acceleration," in Proceedings of Twenth-third Stapp Car Crash Conference, Coronado, Calif., Oct. 1979. (SAE paper 791022).
27. B. Slobodnik. "Correlation of head injury with mechanical forces based on helmet damage duplication," in Proceedings of NATO Advisory Group for Aerospace Research and Development Conference, No. 253: Models and Analogues for the Evaluation of Human Biodynamic Response, Performance and Protection, Paris, France, Nov. 1978. (Copies from: National Technical Information (NTIS), 5283 Port Royal Road, Springfield, VA 22151, U.S.A.).
28. R. Smith, "The response of unembalmed cadaveric and living cerebral vessels to graded injury - A pilot study," in Proceedings of Twenty-third Stapp Car Crash

Conference, Coronado, Calif., Oct. 1979. (SAE paper 791021).
29. "Motor Vehicle Safety Standard Number 218, Motorcycle Helmets," Part 571 of Title 49 Code of Federal Regulations.

826035

Human Head Tolerance to Lateral Impact Deduced from Experimental Head Injuries Using Primates

Atsumi Kikuchi and Koshiro Ono
Japan Automobile Research Institute, Inc.

Norio Nakamura
Jikeikai University

ABSTRACT

Impacts were applied to temporal regions of 23 subhuman primates, while controlling 1) impact velocity, 2) impactor elasticity and 3) impact stroke. Such impacts were repeatedly applied to test primates until they showed symptoms of brain concussion or other brain injuries. As a result, the range of head accelerations, in which brain concussion and pathological brain injuries such as subdural hemorrhage, brain stem hemorrhage, etc. were likely to occur, was clearly indicated on the coordinates represented by the averaged resultant head acceleration and its duration of impact.

This provided a clear indication of thresholds of severe pathological brain injuries as well as the threshold of brain concussion of subhuman primates against lateral impacts. Based on the experimental results using subhuman primates, the extrapolation was done of human head acceleration and its duration, and thresholds of human brain concussion and severe pathological brain injuries were deduced. From results thus obtained it was estimated that the tolerances to brain concussion and severe pathological brain injuries for humans and subhuman primates were higher for lateral impacts than those of frontal or occipital impacts and that the hemorrhage of callosal, cortical and subcortical was often involved at time of occurrence of brain concussion by lateral impact on the head. According to the above, it was concluded that it should be appropriate to determine the tolerance threshold of the lateral impact, using slight skull fracture, minor subarachnoid hemorrhage or other similar head injury for reference.

INTRODUCTION

Doctors who actually engage in clinical treatment feel that a certain pattern exists among head traumas, and there are clearly some rules concerning the relationship between the direction of force and area of brain injury. Nevertheless, only a few quantitative analyses of the relation between the impact region, injury and impact magnitude have been done, (1,2,3) and the tolerance of each region of the head against impacts has not been clearly developed, whereas relatively more qualitative analyses have been performed.

Gurdjian et al. reported the Wayne State Tolerance Curve (WSTC) regarding tolerance thresholds against frontal impacts in 1965, based on data obtained from a number of animal tests, skulls, cadavers, volunteers and clinical cases. (4,5,6) Since then the WSTC has been used extensively not only as a criterion for motor vehicle safety but also as an index to safety against head impacts. The author et al. recently extrapolated the threshold of human brain concussion against frontal or occipital impacts from the results of head impact tests performed using subhuman primates, (7,8,9) which was reported as the JARI Human Head Tolerance Curve (JHTC) (10,11). As a result, it was confirmed that the WSTC developed by Gurdjian et al. employed injuries similar to brain concussion in terms of severity as the index to tolerance thresholds.

In recent years, safety measures for motor vehicle side collisions have been attracting the attention of people concerned, but safety thresholds of head, thorax, pelvis, etc. against lateral impacts are not recognized yet. The author et al., therefore, carried out lateral impact tests using subhuman primates, in an attempt to clarify tolerance thresholds against lateral impacts through the extrapolation of human brain concussion threshold according to experimental results.

EXPERIMENTAL SUBJECTS

Twenty-three subhuman primates which were morphologically, anatomically and physiologically analogous to humans were used as test subjects. The primates were classified as follows; 7 Macaca fuscatas (Japanese monkeys), 12 Macaca mulattas (Rhesus monkeys) and 4 Macaca fasicularises (crab eating monkeys) consisting of 21 male monkeys and 2 female monkeys. Their weights ranged from 4 to 13 kg, presumed ages from 3 to 9 years old.

EXPERIMENTAL APPARATUS AND IMPACT METHOD

The impactor ejection apparatus shown in Figure 1 was used as the test equipment to apply impacts to temporals of test monkeys. The posture of each test monkey was held by subject restraint device shown in Figure 2. The impactor ejection apparatus was capable of colliding the piston, accelerated by the compressed air, against the impact shaft located at the front end of the cylinder, for ejection of the impact shaft. The impact stroke was con-

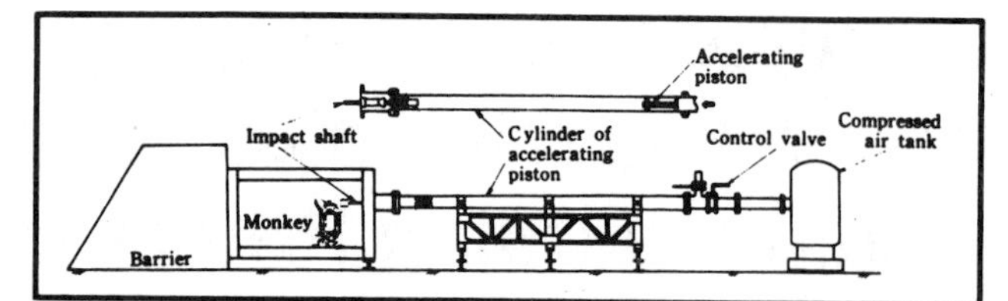

Figure 1. Impactor ejection apparatus.

trolled by the impact shaft stopping position. The maximum capability of the impactor ejection apparatus was 50 m/s where the weight of the impact shaft was 13 kg. The impactor head was made of a solid rubber block of 120 × 80 × 100 mm, and the degrees of hardness of the rubber block were 20, 30 and 70. Elastic characteristics of each impactor head are shown in Figure 3. Each test monkey under light anesthesia was seated on the subject restraint device with its chin held by a urethane foam block, while impacts were applied. The impact direction was vertical to the sagittal plane of the head, with the Frankfurt line of the monkey's head held horizontal (Figure 4). The upper region of the zygomatico arch was selected as the area of impact in order to avoid fracturing of the zygomatico arch. The magnitude of impact was controlled by 1) impactor ejection velocity, 2) impact stroke and 3) impactor head elasticity. After applying impacts, the test monkey's head was held by the supporting net so as to avoid the hyperextension of the neck. At the same time when the body fell against the supporting net, the net started to tilt by 30 to 60° while reducing the velocity, then stopped. Owing to the braking function of the supporting net, the test monkey stopped moving while avoiding secondary collision. At that time, the chair on the subject restraint device did not travel.

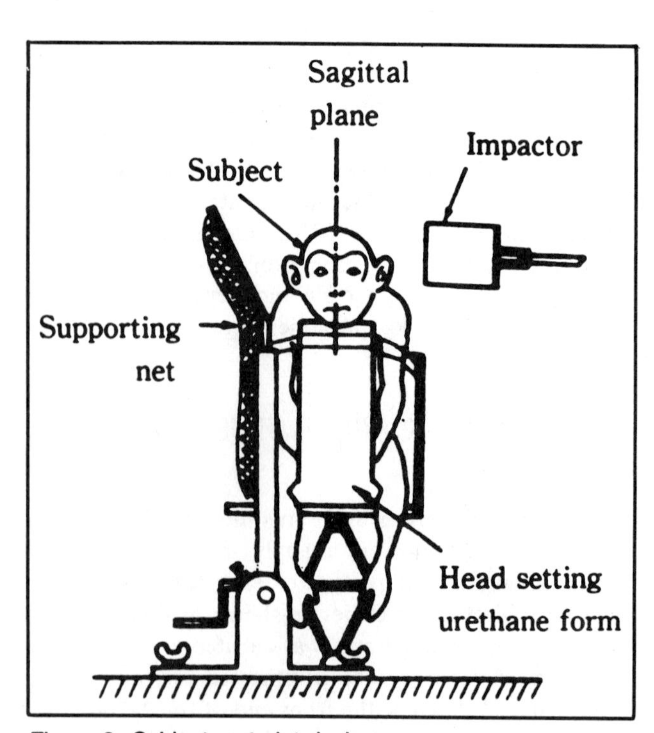

Figure 2. Subject restraint device.

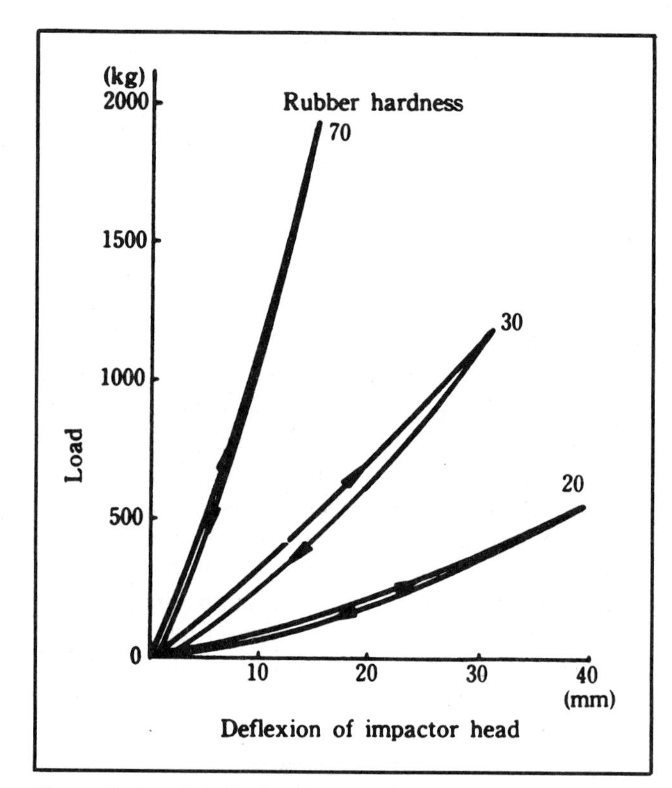

Figure 3. Load-deflexion curve of impactor head.

PHYSICAL MEASUREMENTS AND BIOLOGICAL OBSERVATION

Ketamin hydrochloride for anesthesia was injected into the muscles of each test monkey about 2 hours prior to the test. Under the anesthesia, the test monkey's physical dimensions and weight were measured and various trans-

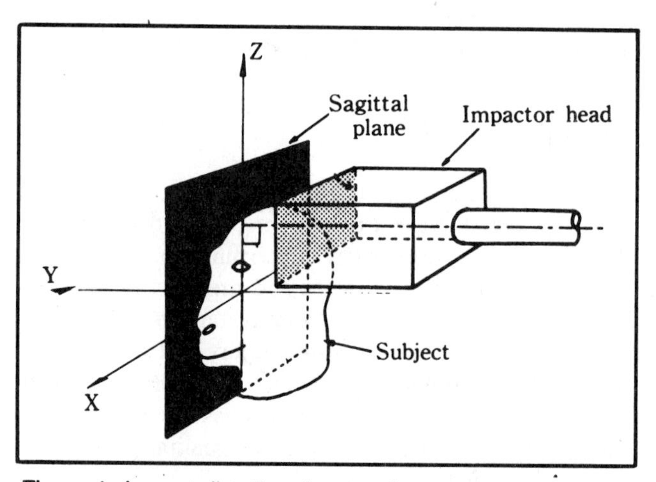

Figure 4. Impact direction; impact direction was set vertical to the sagittal plane, with the Frankfurt line set horizontal.

ducers, electrodes, etc. were installed onto specified locations. The coordinate reference frame of the subject's head was set at the left-hand system. X-axis was set in parallel to the Frankfurt line, while Y and Z axes were set vertical to the X-axis. Head acceleration of each axis was measured by the accelerometer of semiconductor strain gauge type (AH-2000, ST). Values measured by such accelerometers were analyzed and processed through low-path filters by means of a data processing system (NOVA Nihon Minicon). Velocity of impactors was measured optically by laser beams and phototransistors. Behaviors of test monkeys were photographed by high speed cameras at 4000 frames per second (HYCAM, Pedlake) from the direction of X-axis on the head, and the results were studied by a film analyzer (Motion Analyzer, Nack). Electroencephalograms (EEGs) were induced from right and left sides of the frontal-temporal and the parietal scalp, while electrocardiograms (ECG) were induced from the front chest. Respiration measurements were made by means of thermistor-type respiration meters (MTR-2TIS, Nihon Koden). The foregoing biological data were recorded on an electroenceparagraph (EEG-4109, Nihon Koden) and pen-type oscilloscope (RTG-3034, Nihon Koden). Before and after each impact, pupillary light reactions, ciliary reflexes, winks, occular movements, spontaneous movements, etc. were repeatedly observed from time to time. X-ray tests were also done on the head and neck before and after impacts, and skull dimensional measurements were performed and fractures checked, etc. Autopsies were made immediately after death for fatal cases, while surviving monkeys were subjected to euthanasia using nembutal sodium solution 1 hour to 7 days after the impact test, and then given an autopsy. After autopsy, fracture of the skull and cervical vertebra was examined first and extra-intradralle hemorrhage was investigated next. Finally the brain, the cervical spinal cord and upper portion of the thoracic spinal cord were taken out in one, and the skull base duramater was opened to check skull base fractures. After taking photographs and weighing, the brain and spinal cord were held in 10% formalin solution. After 4 weeks or so, they were dissected and studied under an optical microscope.

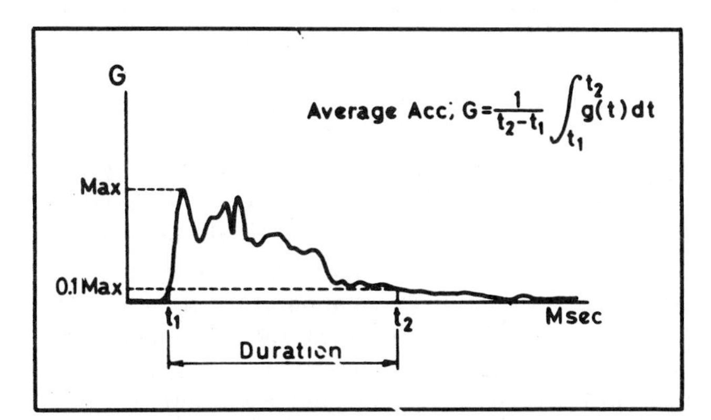

Figure 5. Duration of acceleration; time interval t_1 -t_2 at 10% level of max. acceleration is assumed as acceleration duration.

EVALUATIONS OF IMPACTS AND INJURIES

Magnitudes of impacts on the head were evaluated by head acceleration and its duration, while brain injuries were evaluated by the extent of damage to brain concussion and pathological brain injuries. The methodology and criteria are given in the following.

Magnitude of Impacts

The magnitude of impact from the impactor to the test monkey head is judged by the averaged resultant head acceleration and its duration.

The resultant acceleration is expressed by the absolute value of the sum of vectors of accelerations on X, Y and Z axes. The resultant acceleration is obtained by equation (1), while the duration of the resultant acceleration is expressed by 10% level interval against the maximum of the resultant acceleration (Figure 5).

$$G_R(t) = \sqrt{G_X^2(t) + G_Y^2(t) + G_Z^2(t)} \qquad (1)$$

where,
G_R : resultant acceleration
G_X : X-axis acceleration
G_Y : Y-axis acceleration
G_Z : Z-axis acceleration

The averaged value of the resultant acceleration is the value obtained by the integral value of accelerations within the duration by the duration.

Criteria of Brain Concussion

The extent of damage of brain concussion is judged mainly by 3 indices of 1) disappearance of ciliary reflex, 2) apnea and 3) bradycardia, while making reference to test results of neurological reflex and physiological response.

Criteria:

a. Ciliary reflex disappears for 20 seconds or longer after impact
b. Apnea continues for 20 seconds or longer after impact
c. According to the degree and duration of bradycardia, bradycardia is classified by "none", "slight" and "severe", and judged as "severe".

The degree of brain concussion is classified into 3 grades as listed below, according to the criteria mentioned above.

Brain concussion grade 0: does not fall under either one of a, b or c.
Brain concussion grade I: falls under either one of a, b or c.

Brain concussion grade II: falls under two out of a, b and c.

Brain concussion grade III: falls under all of a, b and c.

Criteria of Pathological Changes

Pathological brain injuries differ greatly according to the area and contents of injury. Although it is difficult to make a general classification and grading of pathological brain injuries in terms of morphology, the location of injury, etc., grading is given in this report as listed on Table 1 for injuries found by this experiment. Consequently, pathological brain injuries were classified into 4 grades, including the grade of "intact".

RESULTS OF EXPERIMENT

Impacts were applied 1 to 5 times per test monkey, which totaled 75 times for 23 monkeys, with impactor velocity of 10 to 27 m/s, impact strokes of 20 to 80 mm and impactor head rubber hardnesses of 20 to 70. As a result, significant brain concussions (grade II and III) were found 12 times (16%) out of total impacts of 75 times, some degrees of pathological brain injuries were found in 18 monkeys (78%) out of 23 monkeys, and skull fractures were found in 6 monkeys (26%).

Head Behaviors

According to results of high speed film analysis, details of head behaviors upon impact differ from impact to impact according to differences in impact conditions. For macroscopic behaviors of test monkeys, however, the following common sequential behaviors were found starting from the moment when the monkey head contacted the impactor to the moment when the monkey head stopped.

(1) Initiation of impact: The impactor head surface contacted the temporal region while keeping itself parallel to the sagittal plane of the test monkey. As the impact progressed, the monkey head started moving toward the direction of impact, while crushing into the impactor head.

(2) Progression of impact: As the impactor further progressed, the monkey head started falling to one side with the neck acting as the fulcrum, while crushing further into the impactor head.

(3) Impactor stop: The impactor stopped with the preset stroke. In some cases, the monkey head was accelerated due to the rebound of the impactor after its stopping, while in other cases the head was not accelerated.

(4) Termination of impact: Even after the monkey head left the impactor head, it fell sidewise further with the neck still acting as the fulcrum, then stopped as the inclination angle became 90° to 100°, and was held by the side net.

(5) Finale: The body fell toward the direction of impact as it was pulled by the head through the neck, then held by the side net. The side net in turn inclined by 60° or so while braking down the speed, then stopped. As mentioned above, nearly reproducible macroscopic head behaviors were obtained throughout the experiment.

Table 1. Classification of pathological brain injury of no-fractural case.

Injury Severity	Injury
0	Intact
I	Subarachnoid hemorrhage
II	Cortical or Subcortical Hemorrhage Callosal hemorrhage Thalamus or Basal ganglia hemorrhage
III	Brain stem hemorrhage Subependymal hemorrhage Subdural hemorrhage

Head Accelerations

Measurements of effective head acceleration for all of X, Y and Z axes were obtained in 56 cases. In the 56 effective measurement cases, accelerations occurred in all directions of X, Y and Z axes. For the processing of acceleration data, a low-pass filter with a cut-off frequency of 1650 Hz was used. Typical examples of time histories of head accelerations are shown in Figure 6. In every impact, the Y-component head acceleration which coincided with the direction of impact showed the maximum acceleration. In case of the time history of Y-component head acceleration, the standard pattern of waves was that a triangular wave with a sharp peak appeared initially, and was followed by a half sine wave or flat wave. The peak value of the sharp triangular wave which appeared first tended to become greater as the impactor velocity increased or the impactor head hardness become harder, ranging from 190 G to 2800 G. The head acceleration wave of Y component that followed the first sharp wave varied its form according to impact conditions. Under conditions where the impactor was soft (rubber hardness 20) and the impact stroke was long (60 to 80 mm), the duration of head acceleration tended to become long, while the duration of head acceleration tended to become shorter where the impactor head was hard (rubber hardness 70) and the impactor strokes were short (20 mm).

Head accelerations of X and Z component were smaller than the Y-component acceleration, but the duration of acceleration was longer for the Z component in most cases. This was due to the rotation of the head around the X axis centering about the neck, and the Z component acceleration continued even after the head had left the impactor head. The acceleration of the X component was smaller than Y and Z components in every case.

The wave form of the resultant acceleration obtained from each component of acceleration in X, Y and Z axes

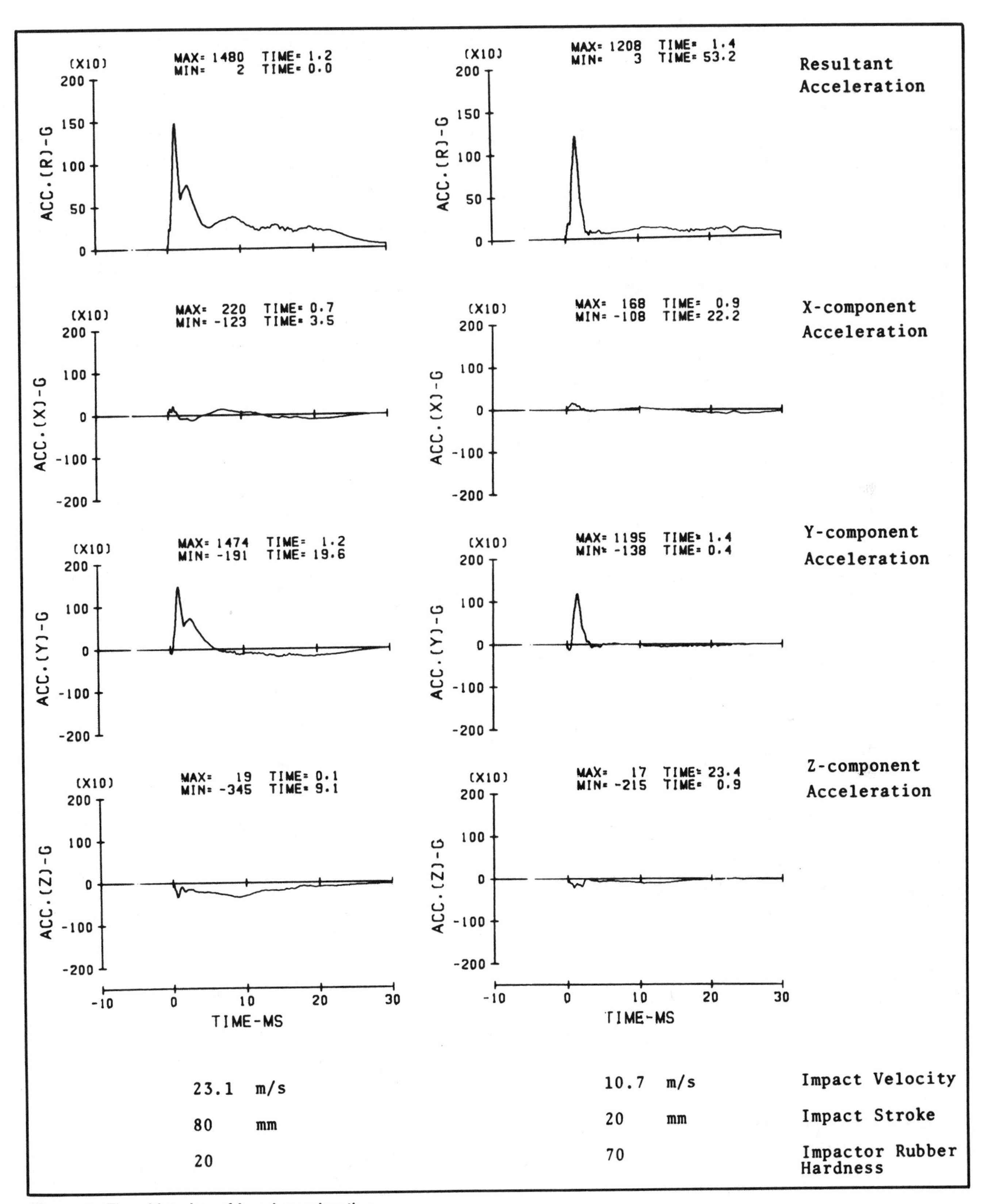

Figure 6. Time histories of head acceleration.

had wave form approximate to the sharp peak wave of the acceleration of the Y component which was the principal component, and it also appeared in the initial stage of impact, and the effect of the Z component on the duration of the resultant acceleration was significant. The averaged resultant head acceleration of 56 effective measurements ranged from 70 to 1310 G, and the duration was in the range of 2.2 to 43.0 ms.

Head Injuries

Physiological and neurological observations: Some physiological and neurological changes were observed in every case. Respiration stopped initially after impact, then resumed as time went by in many cases. In cases where the injury was slight, however, the respiration cycle hardly changed, whereas test monkeys died without recovering from apnea in cases of severe injuries, showing significant variations in respiration. Nevertheless, respiration patterns may be roughly classified into the following 5 types (Fig. 7).

a. No changes in respiration cycle
b. Respiration stopped initially rapidly resumed after a certain time
c. Respiration stopped initially resumed gradually
d. Stopped respiration was initially resumed but deteriorated again
e. Stopped respiration was not resumed

Of 5 types listed above, cases that fell under b and d often showed irregular an weak breathing, but all of them were considered as recovered cases. Therefore, out of 56 effective head acceleration measurements, cases that required more than 20 seconds (a criterion for brain concussion) to recover from apnea were 12, and 10 cases of no-skull fractures were found.

Correlationship was found between blood pressure and heart rate. That is, in many cases, the greater the blood pressure fluctuation, the greater became the variation in heart rate. Blood pressure variation patterns after impact may be classified into the following three types, as shown in Figure 8. Normally, during the fluctuation of blood pressure, bradycardia also appeared, and as the blood pressure resumed its original level, bradycardia also disappeared. Therefore, bradycardia was nearly nil or slight when the blood pressure fluctuation was Type I, but it became more significant for Type II and Type III. The case with the most significant bradycardia showed Type III blood pressure fluctuation and the R-R interval on the ECG was 0.5 sec before impact and 5.6 sec after impact, showing an increase rate of 11 times or so, the original interval then resuming within 5 minutes or so. The recovery time in most cases was between 1 and 5 minutes. Damages appearing in heart rate were judged as "none", "slight" and "severe" using R-R interval and its duration as criteria. Consequently, 17 cases out of 56 impacts of head acceleration effective measurements were judged as "severe", which means that at least one of our criteria for brain concussion, out of 15 cases of non-skull fractures, was found. In most cases ciliary reflex and light reaction were found to be lost, and cases in which ciliary reflex was resumed within 10 sec amounted to 17 cases out of 56, accounting for 30% of the total effective measurements, and those which recovered within 20 sec accounted for 70% or so. The longest recovery time was 154 sec. As for light reaction, on the other hand, 16% were less than 10 sec, and 57% less than 20 sec, with the longest recovery time being 190 sec. The number of cases requiring more than 20 sec (a criterion for brain concussion) for the recovery of ciliary reflex was 17, out of which 12 cases had no-skull fractures.

Using respiration, heart rate and ciliary reflex mentioned in the foregoing as indices, brain concussion judgments for the 56 effective head acceleration measurements were done as follows; brain concussion 0: 30 cases, brain concussion I: 16 cases, brain concussion II: 6 cases, and III: 4 cases. Thus significant brain concussion was observed in 10 cases (18%), out of which 3 monkeys had skull fractures.

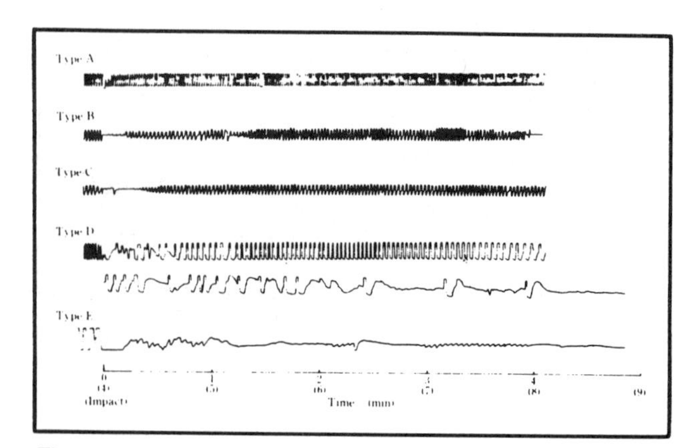

Figure 7. Respiration pattern after impact.

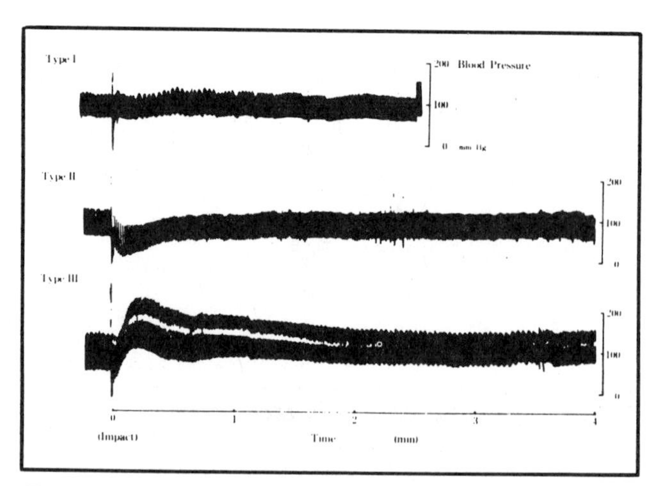

Figure 8. Blood pressure pattern after impact.

Pathological observations: Skull fractures were found in 6 monkeys, out of which one monkey did not have pathological brain injury due to the fact that the fracture was only orbital. The remaining 5 monkeys had skull base fractures, all of which except one were comminuted depressed fractures. This type of fracture caused very serious injuries of brain contusion, brain crush, subdural hemorrhage, etc. as compared with non-fracture cases, and such monkeys died within 15 minutes. Neurological reflex and physiological response of two monkeys of the above resumed within 1 minute, but two other monkeys resumed neither of them.

Of 17 monkeys that had no-skull fracture, on the other hand, one died within 15 minutes and two others died in 6 to 7 days. The 15 minute fatal case had extensive subdural hemorrhage and subarachnoid hemorrhage, and judged as "brain concussion III". Two fatal cases that died in 6 to 7 days did not show any visible injuries, and their brain concussions were below "brain concussion I". Pathological brain injuries of total 18 monkeys consisting of 17 no-fracture cases and 1 case of orbital fracture without any effect to the brain are shown on Table 2. Of the above, subarachnoid hemorrhage was found in 9 monkeys and callosal hemorrhage in 9 monkeys. Cortical hemorrhage and subcortical hemorrhage were also found in 10 monkeys, respectively. Brain stem hemorrhage was observed in 4 monkeys, and subdural hemorrhage was observed in 5 monkeys, out of which 1 fatal case occurred. Moreover injuries were both to the impact side and opposite side of impact, in combinative or separately. No-cortical contusion was found directly beneath the impact.

The classification of pathological brain injury to such no-skull fracture is as follows, according to the injury grades shown on Table 1; pathological injury severity (abbreviated to patho. injury severity) 0: 5 monkeys, severity I: 0, severity II: 7 monkeys and severity III: 6 monkeys. Of these monkeys, effective measurement cases of head acceleration were 3, 0, 4 and 6 monkeys, respectively.

DISCUSSION

Since the mechanism of injuries accompanying skull fractures differs substantially from that of no-fractural cases, the relationship between severity of the no-fractural head trauma and the magnitude of impact will be discussed in this report.

Head Acceleration and Brain Concussion

Assuming that the case of orbital fracture is a no-skull fracture, impacts that did not produce skull fracture were 70 cases, of which 53 cases were effective head acceleration measurements. For the 53 cases, relationship between head acceleration/duration and the severity of brain concussion is shown in Figure 9.

Table 2. Pathological brain injuries no-fractural cases.

Patho. brain injury \ monkey No.	1	2	3	4	5	6	7	8	9	10	11	12	13	14	15	16	17	18
injury severity	0	II	III	II	III	II	III	0	II	0	III	III	0	III	II	0	II	II
subarachnoid hemorrhage			I.L.	C.L.	B.L.		B.L.		I.L.		B.L.	B.L.			B.L.		C.L.	
cortical hemorrhage		I.L.	I.L.		B.L.	I.L.			I.L.						I.L.		C.L.	C.L.
subcortical hemorrhage			B.L.	I.L.	I.L.	C.L.			I.L.		I.L.				I.L.			C.L.
callosal hemorrhage			I.L.	B.L.	B.L.	B.L.			B.L.		B.L.			B.L.	I.L.			B.L.
hemorrhage of thalamus or basal ganglia											I.L.							
brain stem hemorrhage			B.L.		B.L.		I.L.				B.L.							
subependymal hemorrhage			B.L.		I.L.													
subdural hemorrhage			B.L.		B.L.						C.L.	C.L.		C.L.				

I.L.: ipsilateral to the impact side, C.L.: contralateral to the impact side, B.L.: bilateral

The head acceleration is represented by the mean of the resultant acceleration of X, Y and Z components, while the brain concussion is expressed as brain concussion grades of 0 (intact), I (slight) and II or III (significant). In Figure 10, concussion grades 0 and I incidence, and concussion grades I and II or III incidence distributions are partially overlapped, but respective regions of brain concussion grades 0, I and II or III are nearly clarified. In particular, incidence regions of concussion grade 0 and II or III are clearly separated by the boundary which is approximate to the hyperbolic curve, without any overlap. The inclination of the boundary is nearly approximate to inclination of the brain concussion threshold curve on the frontal impact or occipital, and furthermore, the boundary also constitutes approximately the center of the incidence region of brain concussion grade I. From the foregoing findings, it may be deduced that it is necessary to have greater acceleration/duration than the boundary between concussion grade 0 and II or III shown in Figure 9, in the head of a subhuman primate subject to lateral impact in order to induce significant brain concussion in them. It may thus be said that the boundary between brain concussion grades 0 and II or III is the threshold of brain concussion against lateral impact for subhuman primates.

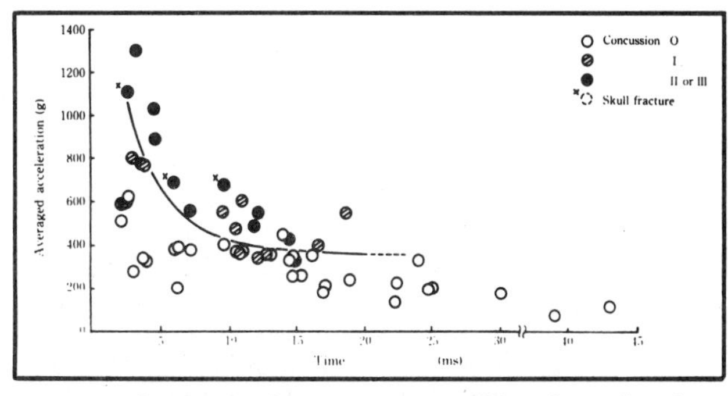

Figure 9. Correlation between averaged head acceleration-duration and grade of concussion in lateral impact.

Head Acceleration and Pathological Brain Injuries

The relationship between the head acceleration/duration and pathological brain injury was investigated, assuming that organic injuries had occurred in impact in which the neurophysiological responses had been the severest in a series of impacts for the same monkey (Fig. 10). In most cases, the severest damage was observed in the final impact. Cases of no-fractural injuries for which head acceleration measurements were effective are as follows; pathological injury severity 0, 4 monkeys; severity I, none (0); severity II, 3 monkeys; and severity III, 6 monkeys, totaling 13 monkeys. As the severity of injury declines from severity III to II and 0, the level of head acceleration/duration also tends to drop. At less than 15 ms of head acceleration's duration in Figure 10, as well as the case of concussion, the pathological injury severity 0 and III incidence distributions are not overlapped, but the underpart of pathological injury severity III incidence and grade II incidence distributions are overlapped. According to the above, the incidence lower limit of pathological injury severity III in Figure 10 which is obtained from these experiment results is the threshold of the subdural hemorrhage, the brain stem hemorrhage and the subependymal hemorrhage, and this threshold is higher than that of the brain concussion by lateral impact on the head as well as the case of frontal and occipital impact. However, it is suggested that the threshold of patho. injury severity II which is the hemorrhage of callosal, cortical or subcortical nearly agrees with or is lower than the threshold of brain concussion.

Extrapolation to Humans

In order to deduce the relationship between the impact and injury of human heads from the relationship between the impacts and injuries of subhuman primates stated above, the dimensional analysis method employed by Stalnaker et al. was applied to extrapolate human brain concussion threshold and pathological brain injury threshold. Assuming, therefore, that shapes of subhuman primates and human heads were analogous, and further that physical characteristics of biological tissues of their brains, skulls, etc. are the same, the head acceleration and duration, which would cause human head injuries equivalent to those of subhuman primates upon lateral impacts, were extrapolated. Parameters necessary for the dimensional analysis were the following five items: 1) head acceleration, a; 2) head acceleration duration, t; 3) impact speed, v; 4) average radius of skull, ℓ and 5) average thickness of skull, h.

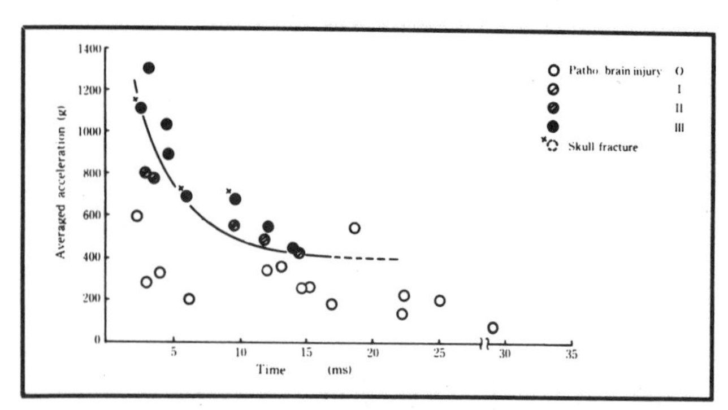

Figure 10. Correlation between averaged head acceleration-duration and grades of pathologica brain injury in lateral; on the patho. brain injury severity O, datas of all impact other than final impact of same subject were contained.

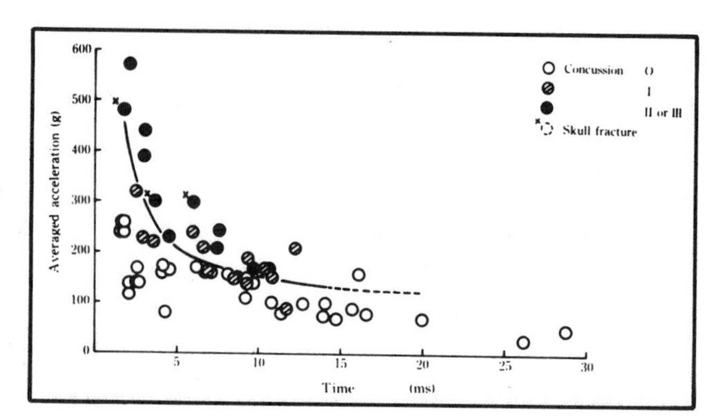

Figure 11. Threshold of concussion in human extrapolated from averaged acceleration-duration by dimensional analysis in lateral impact of the monkey head.

Non-dimensional quantity π is expressed as $\pi_1 = l/h$, $\pi_2 = vt/h$ and $\pi_3 = a \cdot h/v^2$. The relationships among π_1, π_2 and π_3 were determined by using the data (12,13) Stalnaker et al. obtained from subhuman primates of various sizes (sucioroid monkeys, Macaca mulattas, chimpanzees, etc.). The ratio between the average radius and average thickness of human heads in expressed at ℓ /h, and the average weight is assumed to be 1.35 kg.

Human head acceleration and its duration extrapolated from 54 cases of effective measurements of conversion parameters upon lateral impacts of subhuman primates are listed together with the relationship between brain concussion and pathological brain injuries in Figures 11 and 12. Similar to the case of subhuman primates, regions of brain concussions of grades 0 and II or III are also separated for humans, and brain concussion threshold based on grade II or III is also clearly indicated. This brain concussion threshold is represented by the hyperbola passing through 400 G — 2.2 ms and 160 G — 10.0 ms.

Lateral Impact Tolerance Threshold

It was believed from clinical experience, etc. that lateral impact tolerance was lower than the frontal and occipital impact tolerances. However, a comparison between the brain concussion threshold of subhuman primates obtained through this experiment and the results of frontal

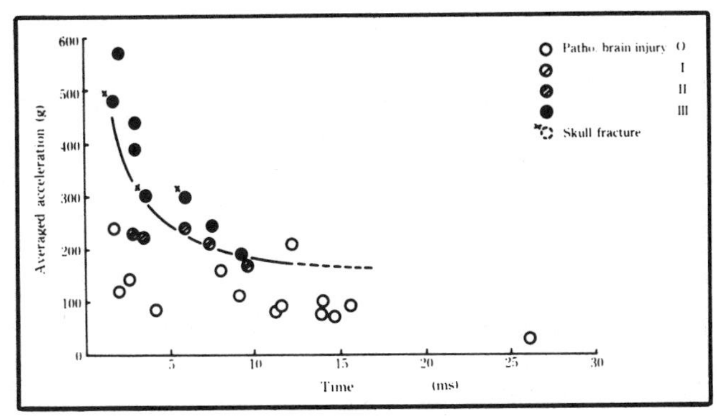

Figure 12. Threshold of pathological brain injury in humans extrapolated from averaged acceleration-duration by dimensional analysis in lateral impact of monkey head.

impact or occipital impact experiment previously reported by the author et al. reveals that the former has markedly higher tolerance than the latter. Gurdjian et al. suggested the WSTC employing brain concussion as the index to the human frontal impact safety threshold, while the author et al. reported the JHTC as the human brain concussion threshold upon frontal or occipital impact, extrapolated from subhuman primates. The comparison between the brain concussion threshold upon frontal or occipital impact indicated by the WSTC and the JHTC and the human brain concussion threshold extrapolated from the results of this experiment shows a clearly high tolerance for lateral impacts.

As regards the threshold of pathological brain injuries, a comparison was also made between the results of frontal or occipital impact already reported by the author et al. and the results of this experiment, which show that the tolerance is higher for the latter as compared with the former similar to the case of brain concussion stated above. According to the results of this experiment described so far, the temporal blow has much higher impact tolerance than the frontal and occipital in terms of brain concussion as well as pathological brain injuries, which agrees with the report published by McElhaney et al., (14) which stated that the impact tolerance to lateral impacts is much higher than that of occipital, according to a head impact experiment performed on primates using the criterion of "whether after effects would remain or not" as the indicator. McElhaney et al., however, reported at the same time that the frontal impact tolerance was higher than the lateral impact tolerance.

As for indices for frontal or lateral impact tolerance threshold, brain concussion is often used as the indicator, as in the JHTC cases and the WSTC. The reasons are that: 1) brain concussion is a transitory and recoverable injury, and 2) brain concussion is likely to occur at the lowest impact level as compared with brain injuries upon frontal or occipital impact. Impacts in the vicinity of the JHTC were such that "at most, subarachnoid hemorrhage might occur (15) (patho. injury severity I)", and it would have been necessary to apply more intense impact to produce skull fracture, subdural hemorrhage, brain stem hemorrhage, etc. (patho. injury severity III). However, the following symptoms were observed upon occurrence of brain concussion of subhuman primates by lateral head impact: 1) subarachnoid hemorrhage (patho. injury severity I) was found in nearly all cases, 2) cortical hemorrhage, subcortical hemorrhage, etc. (patho. injury severity II) were found in more than half the monkeys, and 3) even the concurrence of severe injuries such as subdural hemorrhage, brain stem injury (patho. injury severity III) was found in some cases.

On the other hand, it is reported that the lateral bone fracture load is lower than the frontal bone fracture load according to the studies on skull fracture by head impact, and that lateral bone fracture load is approximately 50 to 80% of the frontal bone fracture load according to a number of experiments. (14,16,17) It is thus deduced, from those studies and skull fracture threshold on frontal or occipital impact as shown in the appendix, that the skull fracture threshold of lateral head impact for humans is the impact level equivalent to or below those of WSTC or JHTC; also it is lower than the threshold for brain concussion on the lateral head impact.

From findings mentioned so far, it was deduced that the thresholds of subdural hemorrhage, brain stem hemorrhage and subependymal hemorrhage (patho. injury severity III) were highest, followed by brain concussion (grade II or III) and cortical hemorrhage and subcortical hemorrhage, etc. (patho. injury severity II) were next, while the threshold of subarachnoid hemorrhage and skull fracture were the lowest for lateral impact, different from frontal or occipital impact. This deduction agrees with a clinical fact that skull fractures are likely to occur in many cases by lateral impacts". Furthermore, this is considered to be one of the reasons for the common belief that the lateral impact tolerance is lower than the frontal impact tolerance due to the difference in the orders of thresholds of injuries as compared with frontal or occipital

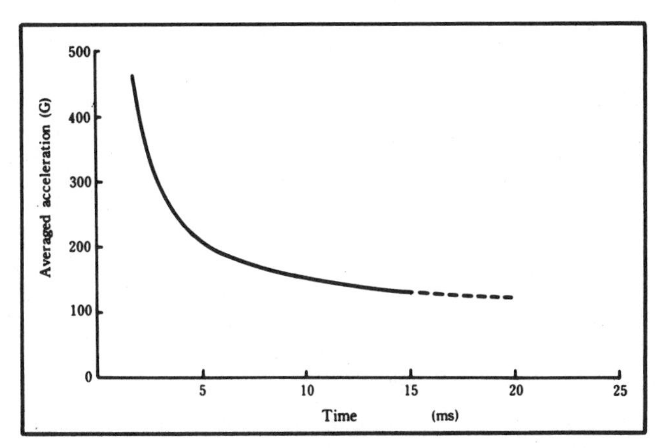

Figure 13. Human brain concussion threshold curve of lateral impact on the head.

impact. The human brain concussion threshold curve of lateral impact on the head extrapolated by the result of this experiment is shown in Figure 13. This is one indicator of judgment on lateral head impact safety. However it was estimated by this experiment that cortical and subcortical hemorrhages and callosal hemorrhage were often involved at time of the occurrence of brain concussion by lateral impact on the head. Consequently it is considered appropriate to determine the safety threshold of lateral impact by referring to skull fractures or subarachnoid hemorrhage of which the tolerance level is deduced to be lower than the tolerance level of brain concussion, brain stem hemorrhage, subdural hemorrhage, etc.

SUMMARY AND CONCLUSIONS

1) In order to clarify thresholds of brain concussion and pathological brain injuries, impacts were applied 75 times to temporals of 23 subhuman primates under various conditions. Of 56 cases of effective measurements, averaged resultant head accelerations of 70 to 1310 G with durations of 2.2 to 43.0 ms were produced, and brain concussion (grade II and III) was found in 10 cases. Of 23 test monkeys, some pathological brain injuries were found in 18 monkeys.

2) For skull fracture cases, the relationship between the head acceleration and brain concussion, and the relationship between head acceleration and pathological brain injury were investigated in order to clarify brain concussion threshold and thresholds of subdural hemorrhage, brain stem hemorrhage, etc. upon application of lateral impact. Extrapolation was also made from data of subhuman primates to humans.

3) Brain concussion threshold of lateral impact, and thresholds of subdural hemorrhage, brain stem hemorrhage, etc. for humans and primates showed higher values than those of frontal or occipital impact.

4) In lateral impact on the head, threshold for hemorrhage of brain stem, subependymal and subdural is higher than that of concussion, and threshold for hemorrhage of cortical, subcortical and callosal nearly agrees with or is lower than the threshold of brain concussion. Comparing these thresholds with JHTC and the skull fracture threshold it is deduced that the threshold of skull fracture is lower than the threshold of brain concussion and cortical hemorrhage, etc.

5) It has been believed from the clinical point of view in craniocerebral injury that the tolerance to lateral impact is lower than the tolerance to frontal impact. It is deduced that one of the reasons for the above concept is that the relationship of skull fracture threshold, brain concussion and various pathological brain injuries threshold differs between the case of frontal or occipital impact and the case of lateral impact.

6) Various pathological brain injuries are often involved at time of occurrence of brain concussion by lateral impact on the head. Therefore slight skull fracture or subarachnoid hemorrhage is an adequate indicator of tolerance by lateral impact on the head.

ACKNOWLEDGEMENTS

The authors received generous assistance from many research and laboratory staff members of the Neurosurgery Departments of the Tokyo Jikeikai University School of Medicine, the University of Tokyo, and Tokyo Women's Medical College and Engineering Department of the University of Tokyo.

REFERENCES

1. K. Sano, N. Nakamura, K. Hirakawa, H. Masuzawa, et al.; Mechanism and dynamics of closed head injuries (preliminary report). Neurol. Medicochir. (Tokyo) 9:(1967)
2. C. B. Courville; Coup-contrecoup mechanism of craniocerebral injuries. Some observations. Arch. Surg. 45:(1942)
3. R. Hooper; Patterns of Acute Head Injury. Edward Arnold Ltd., London (1969)
4. E. S. Gurdjian, H. R. Lissner and L. M. Patrick; Protection of the head and neck in sports. JAMA 182:(1962)
5. E. S. Gurdjian, V. R. Hodgson, W. G. Hardy, L. M. Patrick and H. R. Lissner; Evaluation of the protective characteristics of helmets in sports. J. Trauma 4:(1964)
6. L. M. Patrick, H. R. Lissner and E. S. Gurdjian; Survival by design-----Head protection. Proc. 7th Stapp Conference 7:(1965)
7. H. Sekino, N. Nakamura, A. Kikuchi, K. Ono, et al.; Experimental Head Injury in Monkey Using Rotational Acceleration Impact. Neurologia medico-chirurgica Vol. 20, No. 2: 1979 (Japanese)
8. H. Kohno, N. Nakamura, M. Matuno, et al.; Experimental Head Injury and Concussion—Morphologic Change and Pathophysiologic Following Translational Acceleration in Primates. Neurologia medico-chirurgica Vol. 19, No. 8; 1979 (Japanese)
9. R. Kanda, N. Nakamura, A. Kikuchi, K. Ono, et al.; Experimental Head Injury in Monkeys-Concussion and its Tolerance Level. Neurologia medico-chirurgica, Vol. 21, No. 7:1981
10. K. Ono, A. Kikuchi, M. Nakamura, H. Kobayashi, N. Nakamura; Human Head Tolerance to Sagittal Impact Reliable Estimation Deduced from Experimental Head Injury Using Subhuman Primates and Human Cadaver Skulls. Proc. 24th Stapp Conference 24: 1980

11. A. Kikuchi, K. Ono, H. Kobayashi, N. Nakamura, M. Nakamura; Evaluation of Head Tolerance to Sagittal Impact by Head Acceleration and Duration, and Head Velocity, JSAE Review, No. 3:1980
12. R. L. Stalnaker, V. L. Roberts and J. H. McElhaney; Side Impact Tolerance to Blunt Trauma. Proc. 17th Stapp Conference 17:1973
13. R. L. Stalnaker, J. H. McElhaney, R. G. Snyder and V. L. Roberts; Door Crashworthiness Criteria. U. S. Dept. of Transportation Report No. HS-800-S34; 1971
14. J. H. McElhaney, V. L. Roberts and F. Hilyard; Handbook of Human Tolerance. Japan Automobile Research Institute, Inc.: 1976
15. H. Sekino, N. Nakamura, et al.; Brain Contusion Detected by CT Scan in Minor Head Injury. Neuro Traumatology: 1979
16. J. W. Melvin and P. M. Fuller; Effect of Localized Impact on the Human Skull and Patella. SESA Paper No. 70-491: 1970
17. A. M. Nahum, J. D. Gatts, C. W. Gadd and J. Danforth; Impact tolerance of the skull and force. Proc. 12th Stapp Conference 12: 1968

APPENDIX

It is a figure that shows the threshold of concussion and cadaver skull fracture in human on frontal or occipital impact. (10)

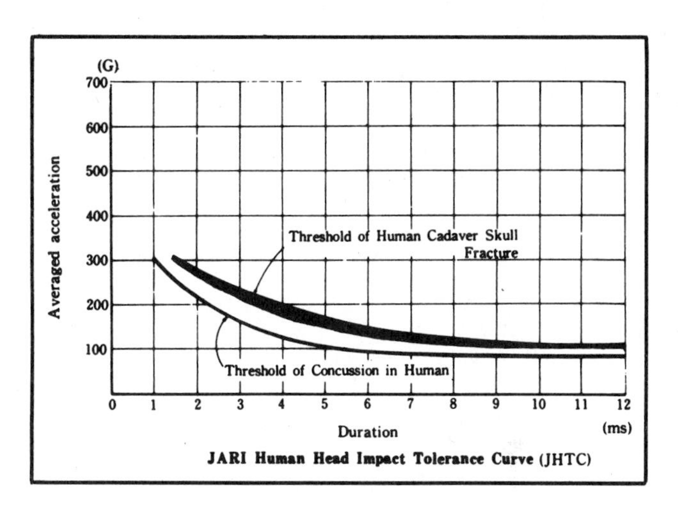

JARI Human Head Impact Tolerance Curve (JHTC)

851246

The Position of the United States Delegation to the ISO Working Group 6 on the Use of HIC in the Automotive Environment

Priya Prasad and Harold J. Mertz
Ford Motor Company and General Motors Corp.

ABSTRACT

A review and analysis of existing cadaver head impact data has been conducted in this paper. The association of the Head Injury Criterion with experimental cadaver skull fracture and brain damage has been investigated, and risk curves of HIC versus skull fracture and brain damage have been developed. Limitation of the search for the maximum HIC duration to 15ms has been recommended for the proper use of HIC in the automotive crash environment.

INTRODUCTION

Working Group 6 (WG-6 -- Performance Criteria Expressed in Biomechanical Terms) of Subcommittee 12 (SC-12 -- Road Vehicle Restraint Systems) of Technical Committee 22 (TC-22 -- Road Vehicles) of the International Standards Organization (ISO) has written a Draft Technical Report (ISO/DTR 7861) entitled "Road Vehicles -- Frontal Collision Protection:Performance Criteria of Biomechanical Origin." They selected the Head Injury Criterion (HIC) as the method for evaluating head protection in frontal automotive crash tests. At a Working Group meeting held in 1982, a HIC value of 1500 was adopted as the level not to be exceeded during the period of head contact for belt-restrained test dummies. Since the level was 50% higher than that specified by the U.S.A. in Federal Motor Vehicle Safety Standard 208, the United States delegate to WG-6 formed a U.S. Advisory Group to evaluate the adequacy of the 1500 HIC level as a head protection limit. The Advisory Group consisted of representatives from the U.S. automotive industry and universities with expertise in biomechanics and/or crash test evaluation, who were actively involved in the activities of the Human Biomechanics and Simulation Subcommittee of the Society of Automotive Engineers. After a thorough review and evaluation of published experimental head injury data, the Advisory Group formulated the following position on the use of HIC for evaluating head impact protection in automotive crash testing.

1. A HIC limit of 1500 should not be supported by the U.S. delegate to WG-6 since this level of HIC may be presumptively associated with too great an apparent risk of brain damage/skull fracture.
2. Any HIC limit chosen for a compliance criterion should be based on the degree of practicable protection sought for the impact environment being evaluated. A tentative risk curve of the percent of population expected to experience a life-threatening brain injury or skull fracture versus HIC values was developed by the U.S. Advisory Group for this purpose.
3. HIC should not be used to define acceptable belt restraint performance in the absence of head impact. In such non-impact cases neck loading should be used.
4. The HIC duration should be limited to 15ms or less for the calculation of the HIC value for a given resultant head acceleration-time history.

This paper discusses the basis for the U.S. Advisory Group Position.

EXPERIMENTAL DATA BASE

The historical development of HIC has been traced by Hess et al.(1)* and Gadd (2), and hence will not be repeated here. It should be noted however that the short duration part of the Wayne State Tolerance Curve (3,4) from which the HIC was derived was based on six data points representing the relationship between head acceleration level and effective impulse duration found to produce linear skull fractures in cadaver skulls. The effective impulse durations ranged from 1 to 6ms. The curve was extended to longer durations with comparative animal, cadaver and human volunteer data. The long duration part of the curve is from human volunteer tests which were obviously not run at injurious levels. The important characteristic of the Wayne State Tolerance curve is that the level of tolerable head acceleration decreases as the effective impulse duration increases. HIC reflects this characteristic.

The validity of the Wayne State Tolerance Curve has been confirmed by Ono et al.(5) for effective impulse durations up to 10ms. They deduced human thresholds for concussion and skull fracture from primate and cadaver skull impacts. Their experimental data ranged between 1-10ms impact durations. Their tolerance curve asymptotes at 90G for 9ms duration and is almost identical to the Wayne State Tolerance Curve in this region. The authors did not extrapolate their data to longer durations. Unfortunately, HIC values for the Wayne State and Ono data were not calculated. Consequently their data were not analyzed by the U.S. Advisory Group.

From a review of the literature four major studies addressing the association of HIC with human skull and brain injuries were identified. The studies were conducted by Got et al.(6), Tarriere et al.(7), Nahum and Smith(8,9) and Hodgson et al.(10,11). All the above studies utilized human cadaver subjects. Many serious doubts are associated with inferring human tolerances from cadaver test data. None the less, these studies were employed because they are considered the best available for making such inferences.

Skull Fracture Data

The skull fracture data base consisted of cadaver head drop tests on flat, rigid and padded surfaces (10), cadaver sled tests against windshields (11) and helmeted cadaver drop tests (6,7). The pooled data, given in Table 1, resulted in 54 cadaver head impacts (25 skull fractures and 29 non-skull fractures) in which the HIC values ranged from 175 to 3400. The HIC durations ranged from 0.9 to 10.1ms. The lowest HIC value associated with a skull fracture was 450 and the highest HIC value associated with non-skull fracture was 2351. Interestingly, the 2351 HIC was associated with the thickest skull in the samples tested by Got et al.(6). A fracture at 516 HIC value was associated with the thinnest skull in the samples tested.

The types of skull fractures ranged from threshold linear to comminuted. Some fractures in the Got et al.(6) series extended to the base of the skull. In the above studies the severities of skull fractures were not reported. Consequently no distinction will be made of the severity of skull fracture in terms of AIS level in this paper. The outcome of a given impact is simply classified as skull fracture or non-skull fracture. It must be stressed that none of these HIC values listed in Table 1 represent the HIC value associated with the level of stimulus required to just produce a skull fracture to the respective cadaver subject. In all tests, the level of stimulus either ex- ceeded or was less than the level of stimulus required to just fracture the cadaver's skull. The implication of this important observation will be discussed further in the " Analysis of Data" section.

Brain Damage Data

The brain damage data base consists of the data reported by Got et al.(6) and Tarriere et al.(7) of the Association Peugeot Renault (APR) of France and Nahum and Smith(8,9) of the United States. The French and the U.S.A. studies were similar in approach but different in the manner in which the cadaver heads were impacted. Both studies utilized fresh cadaveric specimens whose average brain pressures were restored by re-pressurization with fluids containing India ink or dye. The fluid was injected through arteries leading to the brain. After the impact the brain was examined for extravasated ink which was used as an indicator of arterial rupture and hence, brain damage. Both groups attempted to grade the severity of the brain injuries using the extent and location of the extravasated ink. The APR group used the AIS-80 scale for grading the injuries. Nahum and Smith devised their own scale for grading brain injuries.

The APR data, summarized in Table 2, consists of twenty five drop tests with cadavers. Five of these tests were facial impacts and twenty were helmeted or padded impacts. The Nahum and Smith data, summarized in Table 3, consists of eighteen forehead impacts to cadaver heads with a pendulum (rigid and padded). The HIC levels in the APR tests ranged from 516 to 2351, whereas the HIC levels in the Nahum and Smith tests ranged from 31 to 1507 in seventeen tests, with one at 3765. The APR tests in general had longer HIC durations (2.3 to

* denotes references at the end of the paper.

TABLE 1: CADAVER SKULL FRACTURE DATA

Author's Test No.	Source	HIC	Del.t (ms)	Ave.Acc. (G's)	Fracture yes/no
13	2	175	1.5	105	no
4	2	278	3.4	90	no
2	2	291	2.5	106	no
10	2	326	0.9	116	no
17	2	384	3.7	101	no
9	2	411	2.2	128	no
1	2	413	2.3	124	no
DOT6	1	450	2.5	125	yes
3	2	461	4.1	104	no
APR107	3	516	10.1	76	yes
11	2	531	5.7	97	no
8	2	554	0.8	128	no
12	2	554	2.5	138	no
19	2	611	0.8	225	no
DOT2	1	640	2.7	140	yes
APR165	3	692	6.2	104	no
7	2	711	2.5	151	no
APR163	3	750	6.2	108	no
APR110	3	781	8.7	96	yes
14	2	791	1.7	183	no
APR143	3	839	9.2	96	no
16	2	845	2.6	159	no
DOT22	1	850	3.1	150	yes
DOT29	1	870	2.0	175	yes
6	2	892	1.9	185	no
APR144	3	900	2.3	173	no
DOT24	1	965	1.8	200	yes
APR172	3	1042	6.0	125	no
15	2	1050	1.9	197	no
DOT4	1	1070	2.2	185	yes
APR159	3	1078	6.0	126	no
APR251	3	1085	7.5	116	yes
DOT30	1	1130	2.8	175	yes
APR174	3	1156	7.4	120	yes
APR175	3	1200	5.3	139	no
APR166	3	1270	5.8	137	yes
APR162	3	1334	6.2	136	no
DOT26	1	1410	2.9	215	yes
APR160	3	1411	4.8	154	no
APR176	3	1416	4.6	156	yes
APR250	3	1460	9.7	118	yes
APR102	3	1483	3.6	176	yes
18	2	1520	1.9	231	no
DOT3	1	1600	1.7	235	yes
APR108	3	1720	3.5	189	yes
DOT17	1	1930	6.8	150	yes
DOT23	1	2000	4.3	185	yes
APR177	3	2138	4.2	191	no
DOT20	1	2220	5.1	190	yes
APR103	3	2351	3.1	224	no
DOT18	1	2380	3.9	205	yes
DOT19	1	2550	1.9	280	yes
DOT30	1	2780	3.9	295	yes
DOT21	1	3400	2.4	290	yes

SOURCES: 1 Hodgson -- Reference 10
2. Hodgson -- Reference 11
3. APR -- Reference 7: fracture information updated by Reference 24

13.7ms) when compared to those in the Nahum and Smith tests (0.7 to 5.1ms). As was noted for the skull fracture data, none of the HIC values given in Tables 2 and 3 is the HIC value that would be associated with the level of stimulus required to just produce arterial rupture for the cadaver specimen.

The APR tests resulted in a predominance of brain stem injuries. There were fourteen cases of arterial ruptures between HIC levels of 516 and 1720. In two cases no arterial rupture occurred even though the HIC values exceeded 2000. The Nahum and Smith tests resulted in nine cases of arterial ruptures. The associated HIC values ranged from 551 to 3765. In all their tests where the HIC values exceeded 923, arterial ruptures were observed. Hence, the two data sets have comparable injury results at the lower HIC levels, but different results at the higher HIC levels. It is not clear whether these differences are due to the differences in impact conditions or due to differences in the cadaver subjects. Only the APR data were used by WG-6 when they selected a HIC level of 1500 as their head protection limit.

ANALYSIS OF THE DATA

Advisory Group Position on HIC Limit of 1500

Paired points of the average acceleration level and corresponding HIC duration for each HIC value listed in Table 1 are plotted in Figure 1. Also shown are curves of constant HIC levels. Skull fracture data are depicted by solid points; non-skull fracture data by open points. A similar graph for the brain damage and non-brain damage data listed in Tables 2 and 3 is shown in Figure 2. Note that for both data sets there is a trend of increasing frequency of skull fracture and brain damage with increasing HIC level. Note also that nine of the fifteen cadavers that had HIC values be-

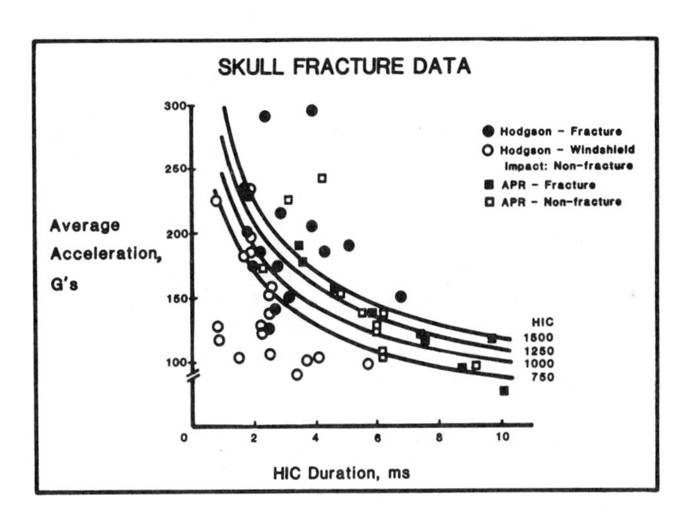

Figure 1. Average Acceleration vs. HIC Duration for Cadaver Skull Fracture.

TABLE 2

APR DATA

Remarks	Test No.	HIC	Del.t (ms)	Ave.Acc. (G's)	Brain AIS	Description of Damage to Cadaver Brain Only.
	107	516	10.1	76	4	extravasation in Pons
facial	89	540	10.	78	2	Corpus Collosum injury
	165	692	6.2	104	0	none
	163	750	6.2	108	2/3	not reported
	110	781	8.7	96	3	frontal contusions and tearings+extravasation in Meninges
	143	839	9.2	96	3	spot in brainstem
	144	900	2.3	173	0	none
	172	1042	6.0	125	0	none
	159	1078	6.0	126	0	none
	251	1085	7.5	116	0	none
facial	95	1150	8.9	111	5	vascular lesions in brain stem and cortical lesions
	174	1156	7.4	120	3/4	not reported
facial	90	1185	8.8	113	0	none
facial	92	1200	6.5	128	3/4	vascular lesions in brain stem
	175	1200	5.3	139	3/4	not reported
	166	1270	5.8	137	3	not reported
	162	1334	6.2	136	0	none
	160	1411	4.8	154	0	none
	176	1416	4.6	157	0	none
	250	1460	9.7	118	5	not reported
	102	1483	3.6	176	5	lesions in brain stem
facial	94	1500	13.7	104	5	lesions in brain stem
	108	1720	3.5	189	5	extravasations in brain stem
	177	2138	4.2	192	0	none
	103	2351	3.1	225	0	none

Notes:

1. Injury descriptions of Test Nos.163,174,175,166 and 250 are not reported by APR.
2. Five facial impact tests reported in the 22nd Stapp have been included in this group but not reported by APR in ISO/TC22/SC 12/GT6, dated June 1982.
3. The Ad-Hoc Group decided to classify any brain arterial bleeding as being a life threatening injury.

tween 1000 and 1500 experienced skull fractures and ten out of seventeen cadavers that had HIC values in this range experienced brain damage. It is because of these relatively high frequencies of skull fracture and brain damage in the HIC range of 1000 to 1500 that the U.S. Advisory Group recommended that the U.S. delegate to WG-6 not support a HIC level of 1500 as a head impact protection limit.

Limit HIC Duration to 15ms or Less

For durations less than 15ms the shapes of the constant HIC curves shown on Figures 1 and 2 are consistent with the Wayne State University(3) and Ono et al.(5) tolerance curves; that is, the tolerable average acceleration increases as the duration decreases. However, the HIC formulation is deficient for average accelerations of relatively long durations. For example, an average acceleration of 1G applied for 1000 seconds gives a HIC of 1000. This deficiency for long durations is the same that was noted for the Gadd Severity Index(12). To overcome this deficiency in the HIC formulation, WG-6 decided to limit the calculation of HIC to that portion of the resultant head acceleration-time history when

TABLE 3

NAHUM and SMITH DATA

Test No.	HIC	Del.t (ms)	Ave.Acc. (G's)	Brain AIS	Description of Damage to Cadaver Brain Only.
27	31	4.2	35	0	none
26	251	2.7	97	0	none
18	366	0.8	184	0	none
44	551	4.2	111	*	not described
31	624	2.7	140	0	none
15	627	2.3	149	*	Subarachnoid hemm. at tip of left occipital lobe. Periventricular hemm. right posterior Corpus Collosum near ventricle.
29	657	0.7	244	*	Diffuse Petechial hemm. of left basal frontal lobe. Petechial hemm. left frontal lobe. Contusion left caudate nucleus, globus pallidus and uncus. Contusion Occi. lobe white matter.
42	703	5.1	114	0	none
37	744	2.4	157	0	none
43	804	2.4	162	0	none
54	820	1.7	187	0	none
19	845	2.2	171	0	none
36	923	2.0	184	*	not described
38	980	1.8	197	*	not described
28	1316	1.9	216	*	Subpial hemm. right and left frontal lobes beneath impact site. Diffuse subarachnoid and some subpial hemm. over brain stem, and medial and base of temporal lobe.
32	1443	1.9	225	*	Subpial hemm. medial tips of frontal lobes, right middle temporal lobe, right occipital pole
17	1507	1.3	266	*	Rupture communicating veins to right side of sup. saggital sinus. Subarachnoid hemm. corpus collosum, etc.
41	3765	2.7	287	*	not described

NOTE: Author's Injury grading not on AIS Scale.
* denotes brain damage

the head was being impacted. This requires a technique for determining the period(s) of head impact.

The U.S. Advisory Group thought that a much simpler way to address this deficiency in the HIC formulation was to place a limit on the HIC duration. The longest HIC duration associated with either a skull fracture or brain damage is 13.7ms. As clearly demonstrated in Figures 1 and 2, the majority of the HIC durations associated with head impacts that produced either brain damage or skull fracture are less than 10ms. No brain damage or skull fracture data exist where the HIC durations are greater than 15ms. Nusholtz et al.(13) have reported two re-pressurized cadaver tests with long impact durations. Both tests resulted in low HIC's (62 and 400) and no apparent brain damage. McElhaney et al.(14) have analyzed and reported the

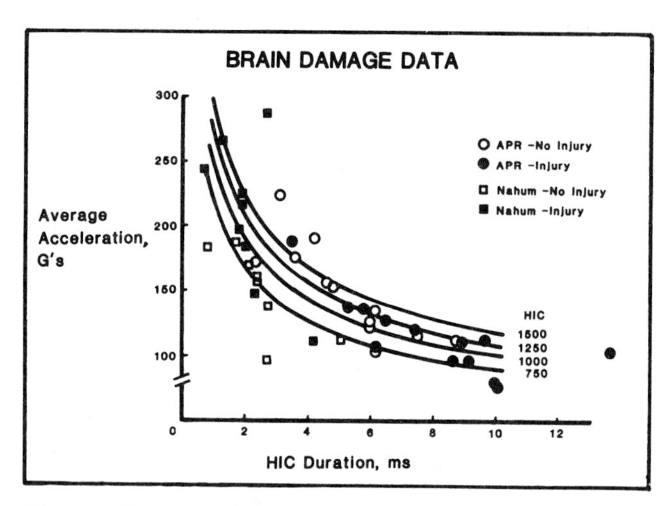

Figure 2. Average Acceleration vs. HIC Duration for Cadaver Brain Damage.

HIC levels experienced by human volunteers restrained by airbags. A HIC as high as 875 with duration of 30ms was observed with no brain injury. Based on these observations, the U.S. Advisory Group proposed that the search for the maximum HIC value for a given acceleration-time history be limited to HIC durations of 15ms or less. This proposal was adopted by WG-6 as an alternative to calculating HIC only during the period of head impact.

HIC in Non-Head Impact Cases

An analysis of field accident data done by WG-6 indicated that there were no cases of brain injuries to the 3-point belt restrained occupants whose head did not impact forward interior components. This observation was their justification for limiting the HIC calculation to the period of head impact. The 15ms or less limit on HIC duration proposed by the U.S. also restricts the HIC value for such non-head impact situations to levels which are consistent with the lack of brain injury noted in the field data. However, cadaver and animal experimental data indicate an increased likelihood of neck injury with increased accident severity level(15,16,17). This trend is also demonstrated by dummy tests and occupant dynamics models that show increased neck loads with increased accident severity when there is no head contact. For these reasons, the U.S. Advisory Group recommended placing a limit on neck loads when the head of the 3-point belt restrained occupant does not contact the forward interior components in crash tests. Specific limits have not yet been proposed.

Skull Fracture and Brain Damage Risk Curves

The data listed in Table 1 were classified into two groups -- skull fracture and non-skull fracture. The HIC values contained in each group are plotted at the top of Figure 1. As indicated previously, none of these HIC values corresponds to the level of stimulus required to just produce skull fracture. The HIC values in the fracture group are all greater than their respective threshold HIC values. The HIC values in the non-fracture group are all less than their respective threshold HIC values. In both cases, the amount that the HIC value exceeds or is below the threshold HIC value for the cadaver specimen is unknown.

Mertz and Weber(18) have addressed the problem of estimating the cumulative distribution curve of injury threshold based on data obtained from specimens which were not tested according to the standard techniques(19,20) for determining threshold information. They note that the distribution of the measured response data provides no information about whether or not the response measure is an indicator of the observed injuries. One must assume apriori that such a correlation exists. If this assumption is made, then an estimate of the range of the threshold values is given by the highest measured response value for the non-injured specimens and the lowest measured response value for the injured specimens. For the data given in Table 1, this estimate of the range of threshold HIC values for skull fracture is 450 to 2351. A first approximation of the cumulative distribution curve of threshold HIC values can be obtained by assuming that it is linear. Such a curve can be constructed by assigning zero percent to the lowest measured response value of the injured specimens and one hundred percent to the highest measured response value of the non-injured specimens. These two points are connected by a straight line as shown in Figure 3 by the curve labeled "Linear Method." Mertz and Weber(18) outline a technique for constructing a cumulative distribution curve if one assumes that the threshold values are normally distributed. Their technique was applied to the data listed in Table 1. The resulting cumulative distribution curve (labeled Mertz/Weber Method) is shown in Figure 3.

The data listed in Tables 2 and 3 were pooled and classified into two groups -- brain damage and no brain damage. The HIC values contained in each group are plotted at the top of Figure 4. Again, none of these HIC values is the HIC value that corresponds to the level of stimulus required to just produce brain damage. The HIC values in the "brain damage group" are all greater than their respective threshold HIC values. The HIC values in the "no brain damage group" are all less than their respective threshold values. In both cases, the amount that a HIC value exceeds or is below its respective threshold HIC value is unknown. Assuming that HIC is an indicator of the threshold of brain damage, then an estimate of the range of threshold HIC values is 516 to 2351. Figure 4 shows the linear method and Mertz/Weber method approxima- tions of cumulative distribution curve for brain damage based on the data of Tables 2 and 3.

The U.S. Advisory Group decided to use the cumulative distribution curves constructed by the Mertz/Weber method as their best estimates of risk curves for skull fracture and brain damage. Comparing these two curves shown on Figures 3 and 4, one notes that they are virtu ally identical. The implication is that for a given level of HIC, skull fracture, brain damage or both are equally likely to occur. The curve labeled "Mertz/Weber Method" on Figure 4 is the risk curve for life-threatening brain injury due to forehead impacts that the U.S. delegate has proposed be used by WG-6 in their deliberations. Note that for a HIC level of 1500, there is a 56 percent risk of life-threatening brain injury, while at a HIC level of 1000 the risk is only 16 percent.

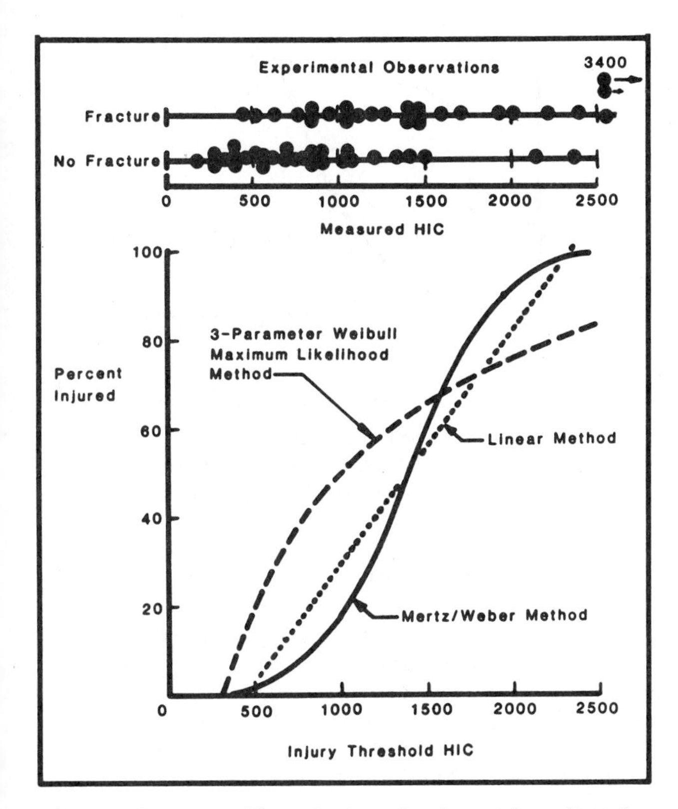

Figure 3. Predicted Cumulative Distribution Curves of Threshold HIC Values for Cadaver Skull Fracture.

DISCUSSION

The preceding analyses of skull fracture and brain arterial damage data show that HIC levels of 1500 may be presumptively associated with apparent high risks of skull fracture and brain injuries. Hence, the U.S. Advisory Group does not recommend that this level be used as a compliance limit for frontal impact environments. The risk curves show that HIC is not a Go/No-Go criterion. The level of HIC that should be used as a compliance limit should reflect the degree of practicable head injury protection sought for the specific environment being considered. Also, the test device used in the test must indicate essentially the same HIC as a human cadaver if these values are to be used for compliance.

Of the two types of data used in the analyses, the skull fracture data were judged to be the more reliable. The test methodology used to obtain the brain damage data has a number of serious limitations. Since only the arterial system is pressurized, damage to the veinous system is not measured. However, veinous ruptures result in subdural hematomas (AIS 4), a major cause of death in severely head-injured patients. Additionally, Diffuse Axonal Injuries (DAI) due to brain cell damage that result in concussion of AIS 2-5, levels can not be detected by this technique. Other brain injuries like cerebral edema and swelling also can not be detected by this method. The implication of these observations is that while the presence of arterial ruptures in the brain in these tests signifies brain damage, the absence of arterial ruptures does not signify an absence of brain injury. Because of these uncertainties the U.S. Advisory Group decided to classify any arterial rupture in these tests as

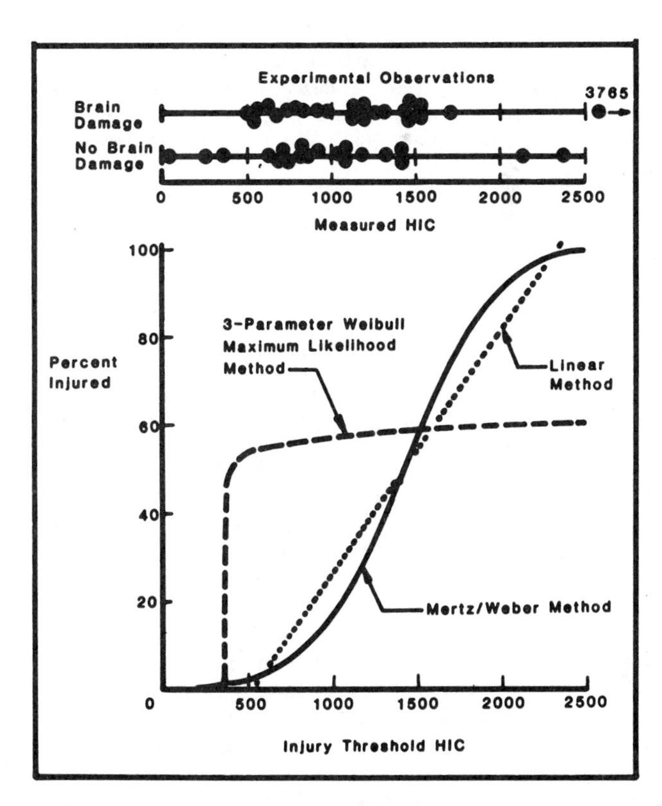

Figure 4. Predicted Cumulative Distribution Curves of Threshold HIC Values for Cadaver Brain Damage.

being a life-threatening injury at the AIS-4 level or greater. In contrast, the arterial rupturing may have occurred at too low a stimulus level due to degeneration of arterial strength properties. Such a degeneration could explain why the risk curves for brain damage and skull fracture were almost identical.

Lowne(21) and Ran et al.(22) have proposed different techniques for constructing risk curves based on the skull fracture data listed in Table 1 and the brain damage data listed in Tables 2 and 3. Lowne suggests that the Probit method(20) be applied to these data bases. This is the standard statistical technique used to analyze threshold data that are obtained under prescribed test conditions. A large number of specimens are exposed to constant stimulus that can be related to their injury threshold. The percent injury is determined for a number of stimulus levels. These data are then analyzed using the Probit method to give a cumulative distribution curve of the threshold values. Unfortunately, the data bases given in Tables 1-3 were not obtained under the conditions required to use the Probit method. Large number of cadavers were not subjected to prescribed levels of constant HIC.

Ran et al.(22) have proposed to use the Maximum Likelihood technique (23) to select appropriate 3-parameter Weibull distributions for the skull fracture and brain damage data bases. The resulting cumulative distribution curves (labeled as "3-parameter Weibull, Maximum Likelihood Method") are shown in Figures 3 and 4. The inability of this technique to provide a good approximation of the actual threshold curve when measured responses are highly scattered is clearly demonstrated by the example given in the Appendix.

Currently, the U.S. Advisory Group is investigating a number of possible techniques for estimating cumulative distribution curves for the skull fracture and brain damage data bases. It may be impossible to predict the cumulative distribution curves for these data bases. Certainly, the data bases cannot be used to determine whether or not HIC is an appropriate indicator of skull fracture or brain damage. This must be assumed apriori. What is required is a test program conducted according to well-established guidelines for determining injury threshold(19,20). The existing data bases were not collected according to such guidelines and therefore are of limited value.

Association Peugeot-Renault has questioned the use of the facial fracture data given in Table 2 in the analysis of brain damage threshold values. All the techniques were applied to the data base given in Tables 2 and 3 with the facial fracture omitted. Figure 5 shows the resulting cumulative distribution curves. Deleting these facial impact data points had no effect on the cumulative distribution curves obtained with either the linear or the Mertz/Weber methods. There was only a very slight change in the cumulative distribution curve predicted by the "3-parameter Weibull, Maximum Likelihood" method. Whether or not these facial fracture points are included in the data base has little influence on the predicted cumulative distribution curve.

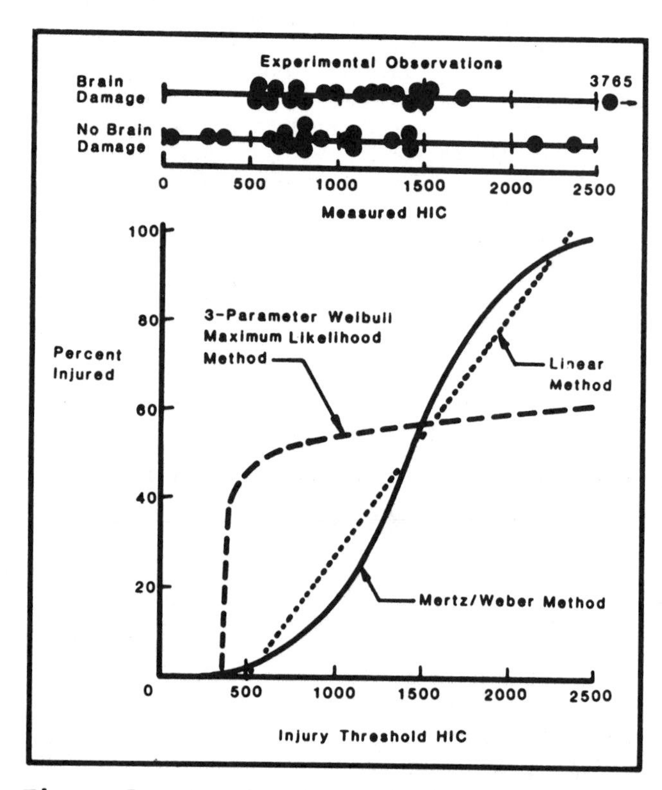

Figure 5. Predicted Cumulative Distribution Curves of Threshold HIC Values for Cadaver Brain Damage. Facial Fracture Data Omitted.

SUMMARY AND CONCLUSIONS

A review of the existing experimental data seeking the association of HIC with cadaver skull fracture and brain damage has been conducted by the U.S. Advisory Group to WG-6 of the ISO. The analysis of the data led to the following conclusions:

1. The HIC limit of 1500 that is specified in ISO/DTR 7861 cannot be supported by the Advisory Group. This value may be presumptively associated with too great an apparent risk of brain damage/skull fracture to be used as a conservative limit for defining acceptable belt restraint performance.

2. A tentative risk curve of the percent of population expected to experience a life-threatening brain injury or skull fracture versus HIC,

developed in this paper, is recommended for use. The HIC limit specified for compliance goals should be based on the degree of practicable protection from life-threatening brain injury or skull fracture sought for the impact environment being evaluated.

3. The use of HIC for defining acceptable belt restraint performance where there is no head impact is not recommended. A measure of neck loading for evaluating belt performance in such cases is recommended. The acceptable neck loads in such situations have not been developed in this paper.

4. The search for the maximum HIC for a given head resultant acceleration-time history should be limited to HIC durations of 15ms or less.

REFERENCES

1. Hess,R.L., Weber,K., and Melvin,J.W., "A Review of Research on Head Impact Tolerance and Injury Criteria Related to Occupant Protection." Head and Neck Injury Criteria, A Consensus Workshop, Edited by A.K.Ommaya, U.S. D.O.T., NHTSA, Wash., D.C. 1981.

2. Gadd,C.W., "Head Injury Discussion Paper." Ibid.

3. Lissner,H.R., Lebow,M. and Evans,F.G., "Experimental Studies on the Relation Between Acceleration and Intracranial Pressure Changes in Man," Surgery, Gynecology and Obstetrics, 111:329-338, 1960.

4. Patrick,L.M., Lissner,H.R. and Gurdjian,E.S., "Survival by Design -- Head Protection," Proc. 7th Stapp Car Crash Confc., SAE, Charles C. Thomas, Springfield, Ill., 1963.

5. Ono,K., Kikuchi,A., Nakamura,M., Kobayashi,H. and Nakamura,N., "Human Head Tolerance to Sagittal Impact: Reliable Estimation Deduced From Experimental Head Injury Using Sub-human Primates and Human Cadaver Skulls," Proc. 24th Stapp Car Crash Confc., SAE, Warrendale, Pa., 1980.

6. Got,C., Patel,A., Fayon,A., Tarriere,C. and Walfisch,G., "Results of Experimental Head Impacts on Cadavers: The Various Data Obtained and Their Relations to Some Measured Physical Parameters," Proc. 22nd Stapp Car Crash Confc., SAE, Warrendale, Pa., 1978.

7. Tarriere,C., Walfisch,G., Fayon,A., Got,C., Guillon,F., Patel,A. and Hureau,J., "Acceleration, Jerk and Neck Flexion Angle: Their Respective Influences on the Occurrence of Brain Injury," ISO/TC 22/SC 12/GT-6 (USA-13) DOC. No. 118, 1982.

8. Nahum,A.M., Smith,R.W., "An Experimental Model for Closed Head Impact Injury," Proc. 20th Stapp Car Crash Confc., SAE, Warrendale, Pa., 1976.

9. Nahum,A.M., Smith,R.W. and Ward,C.C., "Intracranial Pressure Dynamics During Head Impact," Proc. 21st Stapp Car Crash Confc., SAE, Warrendale, Pa., 1977.

10. Hodgson,V.R. and Thomas,L.M., "Breaking Strength of The Human Skull Versus Impact Surface Curvature," Wayne State University, Detroit. DOT- HS-146-2-230, 1977.

11. Hodgson,V.R., Thomas,L.M. and Brinn,J., "Concussion Levels Determined by HPR Windshield Impacts," Proc. 17th Stapp Car Crash Confc., SAE, Warrendale, Pa., 1973.

12. Gadd,C.W., "Tolerable Severity Index in Whole-Head Non-Mechanical Impact," Proc. 15th Stapp Car Crash Confc., Warrendale, Pa., 1971.

13. Nusholtz,G.S., Lux,P., Kaiker,P. and Janicki,M.A., "Head Impact Response- Skull Deformation and Angular Acceleration," Proc. 28th Stapp Car Crash Confc., SAE, Warrendale, Pa., 1984.

14. McElhaney,J.H., Stalnaker,R.H., and Roberts,V.L., "Biomechanical Aspects of Head Injury," Human Impact Response. Edited by W.F.King and H.J.Mertz. Plenum Press, New York-London, 1973

15. Schmidt,G., Kallieris,D., Barz.,J., Mattern,R. and Schulz,F., "Belastbarkeitsgrenze und Verietzungsmechanik des Angurteten Fahrzeuginsassen," FAT Rpt. No. Schriften Reihe Nr. 6, Frankfurt/M, Germany, Jan. 1978.

16. Thomas, D.J. and Jessop, M.E., "Experimental Head and Neck Injury," Impact Injury of the Head and Spine. Edited by C.L.Ewing, D.J.Thomas, A.S.Sances,Jr. and S.J.Larson. Charles C. Thomas, Springfield, Ill., 1983.

17. Unterharnscheidt,F., "Neuropathology of Rhesus Monkeys Undergoing -Gx Impact Acceleration," Ibid.

18. Mertz,H.J. and Weber,D.A., "Interpretations of the Impact Responses of a 3-Year-old Child Dummy Relative to Child Injury Potential," Proc. 9th International Technical

Conference of Experimental Safety Vehicles. U.S. D.O.T., NHTSA, Wash., D.C., 1982.

19. Bliss,C.I., "Calculation of Dosage/Mortality Curve," Annals of Applied Biology, Vol.22, 1935, pp. 134-167.

20. Finney,P.J., Statistical Methods in Biological Assay. 2nd Edition. Griffin Press, London, 1971.

21. Lowne, R., "Estimation of Human Tolerance Distribution," ISO/TC 22/SC 12/WG-6/N173, May 1984.

22. Ran,A., Koch,M. and Mellander,H., "Fitting Injury Versus Exposure Data into Risk Functions," Proc. 1984 International IRCOBI Confc. on the Biomechanics of Impacts.

23. Golub,A. and Grubbs.F.E., "Analysis of Sensitivity Experiments When the Levels of Stimulus Can Not be Controlled," Journal of American Statistical Association, Vol.51, pp. 257-265, 1956.

24. Got,C., Guillon,F., Patel,A., Brun-Cassan,F., Fayon,A., Tarriere,C. and Hureau,J., "Morphological and Biomechanical Study of 146 Human Skulls used in Experimental Impacts in Relation With the Observed Injuries," Proc. 27th. Stapp Car Crash Conference, SAE, Warrendale, Pa., 1983.

APPENDIX

The efficacy of any technique used to estimate the cumulative distribution curve of a given injury threshold can only be judged by determining how well the estimated curve represents the actual curve. In practice, the actual curve is not known, so such comparisons cannot be made. However, hypothetical cases can be constructed where the actual injury threshold curve for a given population is prescribed. A number of specimens with known injury threshold values are selected to be tested by a prescribed test protocol. The result of these tests are analyzed by various techniques giving estimates of the cumulative distribution curve of injury threshold values for the population. These estimated curves can be compared to the actual curve, providing an indication of the ability of each technique to predict the actual curve. This approach is illustrated by the following example.

The injury threshold values of a given population are assumed to be normally distributed. The actual injury threshold of each specimen in the population is known. Ten specimens are selected from the population to be tested. The actual injury threshold value for each specimen is listed in Table A1. The specimens were chosen such that their injury threshold values are normally distributed and their cumulative distribution curve is identical to that of the population. The specimens are given to an experimenter for testing. His test protocol produces the test results listed in Table A1. Note that if the measured response exceeds the

TABLE A1

INJURY THRESHOLD VALUES AND TEST RESULTS FOR TEN SPECIMENS

SPECIMEN NO.	INJURY THRESHOLD	TEST RESULTS RESPONSE	TEST RESULTS INJURED
1	785	800	Yes
2	845	800	No
3	880	900	Yes
4	910	900	No
5	937	950	Yes
6	963	950	No
7	990	1000	Yes
8	1018	1000	No
9	1052	1090	Yes
10	1110	1090	No

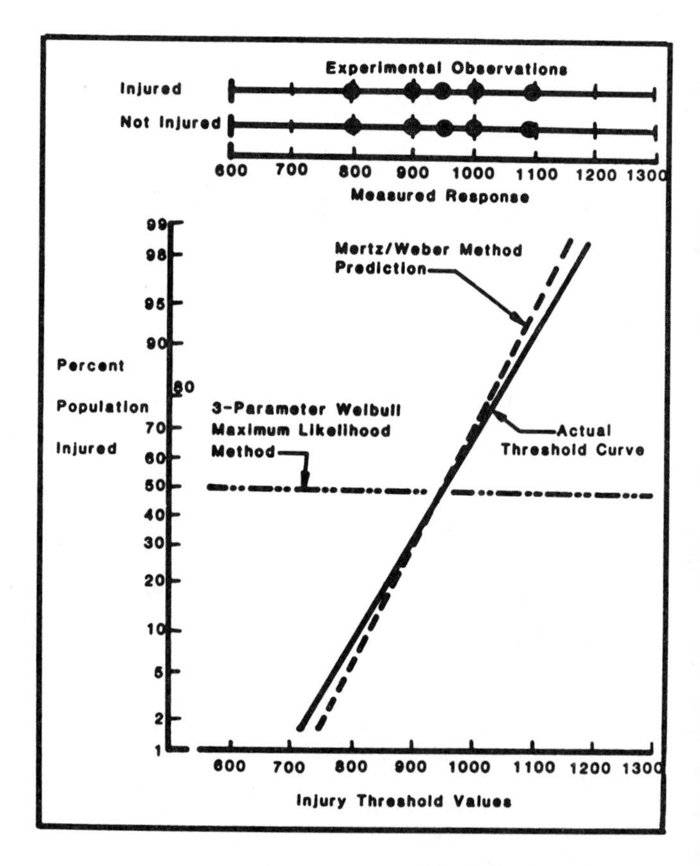

Figure A-1. Predicted and Actual Cumulative Distribution Curves of Injury Threshold Values.

specimen's injury threshold value, injury occurs. If the measured response is below the specimen's injury threshold value, no injury is observed. The test results are graphically depicted at the top of Figure A1. The techniques of Mertz and Weber(18) and Ran et al.(22) were applied to the test results. The predicted and actual injury threshold cumulative distribution curves are shown on the bottom graph of Figure A1. Note the poor prediction given by the "3-parameter Weibull, Maximum Likelihood" technique for these test results.

Acknowledgements

The authors wish to acknowledge the valuable discussions and guidance provided by Mr. Jule Brinn, Dr. John Melvin and Dr. Voight Hodgson who were members of the Advisory Group. The medical advise of Dr. L.M. Thomas, MD, is greatly appreciated. The authors also wish to thank Mr. Stanley Backaitis and Mr. Roger Daniel for their critiques and discussions of the data in this paper. The authors are greatly indebted to Mr. Roger Daniel for the art work and preparation of the manuscript.

Section 4:
Brain Injury

690796

Advances in Understanding of Experimental Concussion Mechanisms[1]

V. R. Hodgson, L. M. Thomas, E. S. Gurdjian, O. U. Fernando, S. W. Greenberg and J. Chason
Wayne State University

Abstract

This paper is an in-depth study of the mechanism of concussion. Prior to 1967, concussion experiments were conducted exclusively on dogs, but after results of trials on dogs, cats, and stumptail monkeys, the authors concentrated on the latter.

Varying numbers of blows, from different angles, were delivered to 12 monkeys. Results, both organic and behavioral, are discussed.

THE PRESENT CONCEPT of the mechanism of concussion held by researchers at Wayne State University had its origins in the work of Gurdjian and Webster in 1943 (8).[2] Following head injury studies in the dog utilizing a hammer, a falling weight, a striker propelled by a spring or pendulum, and by gunshot, they recorded among their conclusions the following:

1. Compression (deformation of the skull), acceleration and deceleration frequently co-exist in the same accident. Pure acceleration or pure deceleration is rare; almost always there is associated compression.
2. The acute physiologic responses and the primary pathologic conditions are caused by mechanical derangement of cells, centers, or supporting structures.
3. The causes of mechanical derangement are:
(a) Deformation of the skull.
(b) Sudden increased intracranial pressure at the time of the blow.
(c) Mass movements of the brain.
(d) Transmitted energy without increased intracranial pressure.
4. The acute physiologic responses to injury may be minimal, moderate, or profound. In moderate and profound injuries, usually there is a sudden increase in blood pressure with respiratory loss, unconsciousness (no reaction to painful stimuli), loss of palpebral or corneal reflexes,

---

[1]This paper has been supported by Army Contracts No. DA 49-193-MD-2924 (Blast Concussion) and No. DA 49-193-MD-2603 (Mechanism of Head and Neck Injury).

[2]Numbers in parentheses designate References at end of paper.

and frequent generalized rigidity followed by limpness. In moderate injuries, the animal survived or died; in the profound group, death almost always occurred.

5. Loss of corneal reflexes represents a more profound response to injury than does unconsciousness. Some animals were observed to be unconscious with active palpebral and corneal reflexes.

6. The acute changes in blood pressure, respiration, and reflexes following trauma to the head are due to stimulation or paralysis of medullary centers.

In 1944, Gurdjian and Lissner (3) delivered hammer blows to the heads of anesthetized dogs through the skulls of which were bonded resistance-wire strain gages to detect deflection, and 2 pressure plugs screwed into the parietal bones to detect intracranial pressure changes during impact. They arrived at the following conclusions:

1. At the time of a hammer blow there is a compression deformation of the outer surface of the skull at the site of the blow, and simultaneously at the side opposite there is a tensile deformation on the outer surface of the skull or a bending outward.

2. Simultaneously with deformation of the skull there is an increase in intracranial pressure on the side of the blow and a decreased pressure at the opposite side.

3. The existence of pressure waves in the cranial cavity may be a possible source of damage to the brain. This damage may be mechanical, with tissue injury due to dynamic stresses set up in the brain by the pressure gradient produced.

4. The traumatic intracranial damage is due to deformation of the skull and changes in intracranial pressure brought about by deformation of the skull and acceleration of the head, after the blow.

Further investigation in the dog was reported by Gurdjian and associates (4) in 1953 in which the intracranial pressure and accelerations and their time durations were measured simultaneously during impact to the head of the anesthetized animal. The time duration of the pressure and the acceleration appear to be the significant factors which explained the clinical effects following impact. It was noted that at the lower values of pressure and acceleration a concussion was produced only if the duration of the acceleration and pressure was of an appreciable length. For the higher accelerations and pressures, it was apparently not necessary that they last for as long a time to produce concussion.

In 1954, Gurdjian, et al (5) performed experiments on the anesthetized dog in which air pulses of controlled amplitude and time duration were delivered through a hole in the skull to the dural membrane. The experiment showed that the shorter the time duration, the higher the pressure necessary to result in a concussive effect. The longer the time duration, the lower the pressure required to effect a concussion. They concluded that concussion resulting from acceleration, deceleration, or compression is caused by an increase in intracranial pressure at the time of impact, based upon these pressure and accleration studies. Subsequently, the brains were studied both macroscopically and microscopically (1). Control animals were used. These studies revealed the following facts:

1. Microscopic abnormalities were present in all but 4 of the 19 experimental animals. The changes consisted of: chromatolysis, swelling and fragmentation of axis cylinders, and petechial and larger hemorrhages.
2. The most significant change following a sudden increase in intracranial pressure in these dogs was chromatolysis involving cells of the reticular substance of the brain stem.
3. The altered cells were in a more medial position with the impacts of lesser magnitude and extended laterally with impacts producing greater pressure changes. These neuropathological changes were observed in animals in which pressure pulses had been delivered to a single port of entry. Subsequent experiments in the dog with portals for air pulse including frontal, parietal, temporal and occipital areas showed results identical qualitatively with those just outlined.

Gurdjian, Webster, and Lissner (6) in discussing the mechanism of brain concussion in 1955 concluded that:

1. Cerebral concussion is an acute post-traumatic state associated with unconsciousness, pallor, and a shock-like state, and is the result of derangement in function of the brain stem. The brain stem involvement may be of varying degree involving both reversible or irreversible damage. A reversible state may result in complete recovery; the irreversible, in unconsciousness with ultimate death. Concussion may be associated with brain contusions and lacerations or it may be unaccompanied by macroscopic evidences of injury to the neural tissue. Contusions and lacerations of the brain may be unaccompanied by concussion.
2. The closed cavity dynamics of the head may be disturbed by injuries resulting in major damage with or without concussion. The closed cavity dynamics in many injuries is the cause of unconsciousness and death. Both are due to physical forces deranging the brain stem function. Concussion occurs as the result of brain stem injury either from increased intracranial pressure at the time of impact, direct injury by distortion, mass movement, shearing, or destruction by a missile.
3. The terms cerebral concussion, contusion, and laceration, should not be used to denote their varying degrees of nervous system damage with concussion identified as representing the mildest form. The evidence indicates that concussion is due to involvement of a specific area in the brain, mainly the brain stem. A contusion or laceration in this area may be fatal. However, contusions and lacerations in other portions

of the nervous system may be present with no associated concussive effect if the brain stem area is not sufficiently involved.

The authors noted that if rotation of the head occurs, portions of the brain may be injured by a butting against bone projections within the skull, thus being compressed. Other portions may be torn because of tension produced as the brain rotates with respect to the skull, and finally shear may occur because of cavitation and pressure gradients, the latter occurring principally in the region of the brain stem. However, it should be noted that Gurdjian, Lissner and Patrick (11) reported in a later paper that concussion has been produced in many experiments where angular acceleration and cavitation or negative pressure have been minimal or absent.

In 1960, Gurdjian and Lissner (9) reported the photoelastic confirmation of the presence of shear strains at the craniospinal junction in closed head injury. They used a 2-dimensional model of the sagittal section of the human head filled with a 1.5% solution of milling yellow which becomes doubly refracting when subjected to shear strain. When

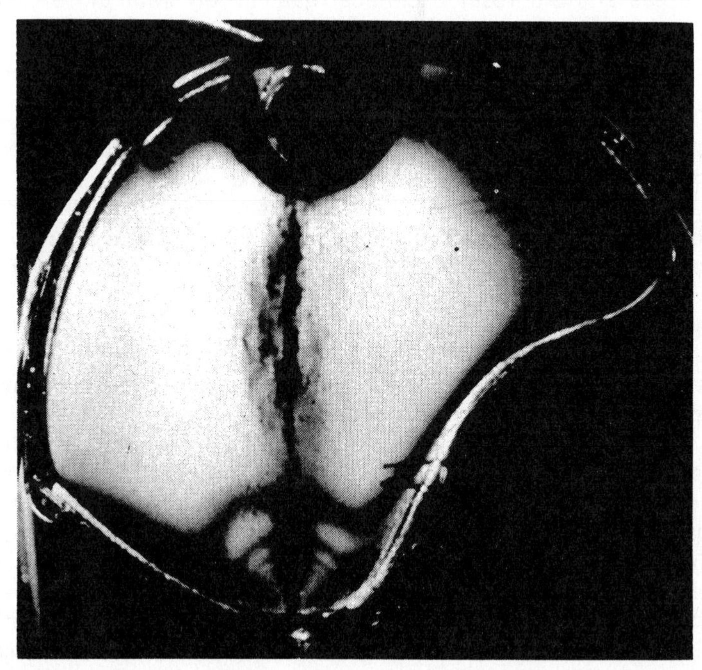

Fig. 1—Plastic model of a sagittal section of the skull including foramen magnum, showing region in which shear strains are developed because of a blow to head. Model is filled with 1.5% solution of milling yellow which becomes doubly refracting when subjected to shear strain, as indicated by alternate dark and light bands. The magnitude of the shear strain is a function of the number of bands produced.

pressure was applied to the walls of the model, shear stresses were obtained in the region of the opening at the base, representing the region of the brain stem. Hammer blows delivered to the circumferential border and to the side walls of the model also resulted in shear stresses, predominantly in the simulated craniospinal junction (Fig. 1).

In 1965, Gurdjian and associates (10) reported on the results of a series of tests involving 34 anesthetized dogs to which blows were delivered to the fixed and free head. The summary and conclusions were as follows:

1. With the head free-to-move or fixed, chromatolysis and cell death occur in greater abundance in the brain stem region.

2. With the less severe impacts, the effects are limited to the lateral medullary reticular formation. With more severe impacts, the

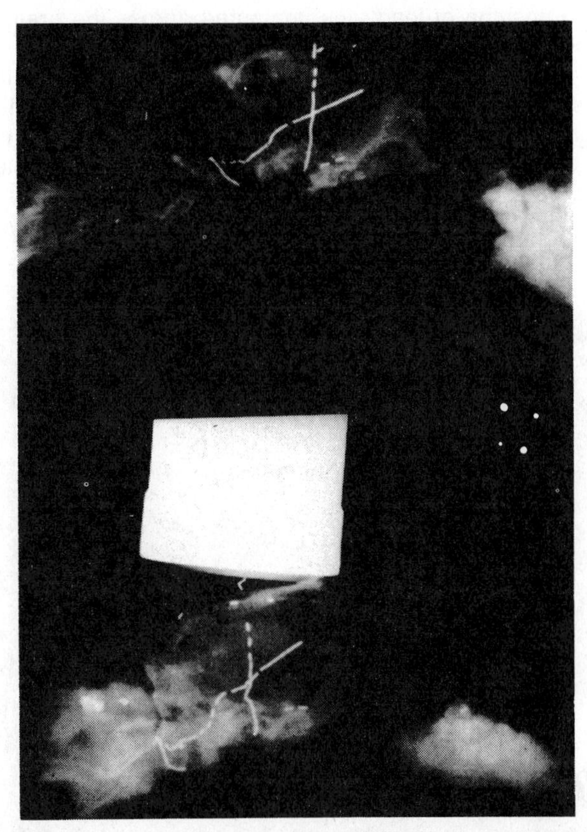

Fig. 2—Flash X-rays of an anesthetized dog's head before (top) and during impact to the fixed head using lead tags to visualize brain movement. Note how the 3 lines of lead tags are distorted from static position. The short line of 7 tags in the lower left part of skull is in brain stem across the foramen magnum

superior portions of the brain stem are also damaged, including the midbrain.

3. A study of the upper cervical spinal cord reveals that cellular damage occurs as frequently with the head fixed as well as with head free-to-move. This raises a reasonable doubt that head-neck movements at the time of impact cause the cellular damage. The cellular damage is caused by shear forces at the craniospinal junction principally resulting from pressure gradients.

In 1966, Thomas, et al. (18) demonstrated the presence of pressure gradients in the water-filled human skull by means of multiple miniature pressure transducers mounted along orthogonal axes during a period of impact acceleration of the head.

Also in 1966, Hodgson, et al. (15) were able to demonstrate evidence of brain movements during impact to the fixed head of the anesthetized dog by means of flash X-ray techniques. The flash X-ray unit was timed to observe the position of lead tags implanted across the brain and brain stem, near the instant of peak force of impact for comparison with static X-rays taken before and after the blow for the same head position. These flash X-ray studies revealed that the line of tags across the foramen magnum were displaced to a much more curved shape during the impact than in the static positions, with maximum displacement at the center, indicative of shear strains in this region (Fig. 2).

In 1967, Gurdjian and associates (13) used the flash X-ray technique

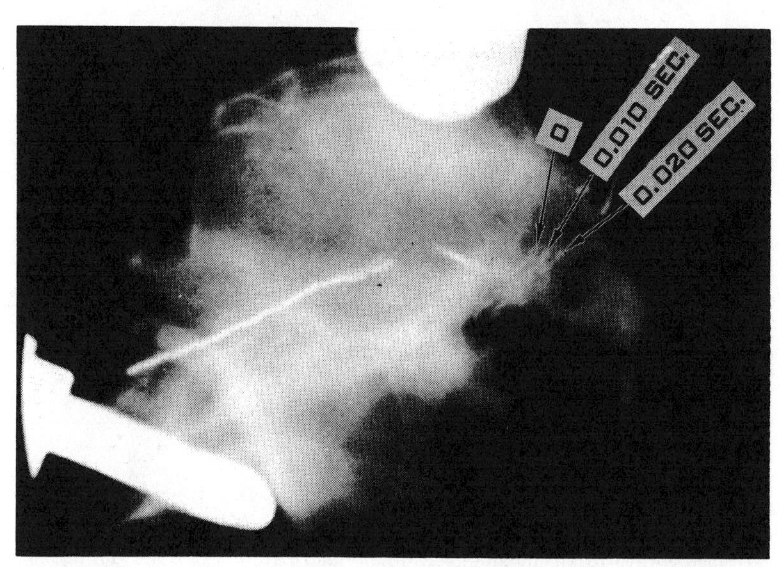

Fig. 3—Overlay of 0.010 and 0.020 sec position flash X-rays upon 0 reference X-ray showing the movement of lead tags in foramen magnum region during a pressure pulse delivered against the dura through a hole in the parietal bone

to demonstrate movements at the craniospinal junction during air pressure impulses through a hole in the skull to the intact dura, using the same techniques previously described to concuss the series of dogs. They also demonstrated downward displacement of lead tags across the foramen magnum of an anesthetized monkey by flash X-ray during occipital impact to the free head of a monkey. (see Figs. 3 and 4).

During the years 1967 and 1968, laboratory tests conducted on the free heads of lightly anesthetized animals including 10 dogs, 10 cats, and 29 stumptail monkeys, demonstrated similar physiological changes in all three species when they were definitely rendered unconscious (16). In the monkey, the best physical index of injury was maximum apparent head acceleration (maximum force/head mass). This apparent head acceleration was found in selected experiments to correspond closely to the center of gravity head acceleration obtained on the basis of accounting for both linear and angular acceleration as calculated from measurements obtained from two triaxial accelerometers mounted on the skull. The most severely injured group of monkeys were associated with pulses which had a combination of high peak amplitude and short rise time.

Following experience gained from these tests a group of 12 stumptail monkeys were struck on the occiput of the free head for the purpose of determining alterations in cell structure of the brain associated with experimental concussion in the primate. These findings, along with physical evidence from high speed photographs of impacts to fresh

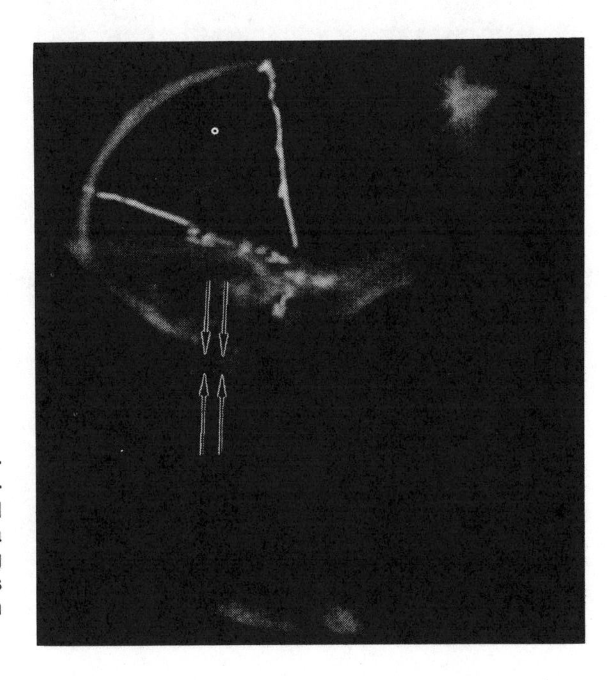

Fig. 4—Flash X-ray during a mid-occipital impact in anesthetized monkey overlayed on a static X-ray showing movement of lead tags toward foramen magnum and spinal canal

sagittal dog and monkey head sections, are herein reported for purposes of comparison with the results of previous works.

Methods and Materials

Twelve stumptail monkeys (Macaca speciosa) weighing an average of 14 lb were used in the neuropathological experiment, one of which was used as control. Each monkey was anesthetized intravenously with sodium pentobarbital, 60 mg per 5 lb of body weight. Single blows to the intact occiput of each monkey were delivered using a 2.2 lb linear striker with a 1 in.^{2} face to which was cemented a ¼ in. thick hard rubber pad. Velocity of the striker ranged from 35 to 52 ft/sec (energy 40-85 ft-lb). The head was free to move as shown in Fig. 5. Each monkey was allowed to recover partially from the anesthesia until all reflexes were present and active prior to delivery of the blow. Reflexes were evaluated immediately after the impact and again for the next 15-20 minutes. The EEG, EKG, direct femoral arterial blood pressure and respiration were monitored with a 6-channel Grass polygraph, Model 7 recording system. The animals were permitted to recover after a single blow. The degree of concussion was graded according to changes in conscious state, reflexes and other physiological parameters into

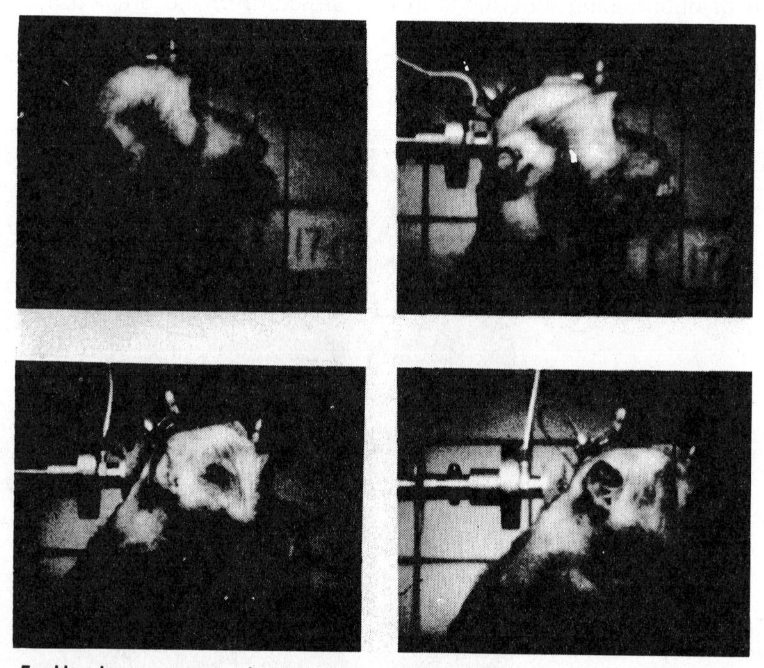

Fig. 5—Head movements in anesthetized stumptail monkey caused by mid-occipital impact. Two triaxial accelerometers are mounted in skull to record translational and angular acceleration

minimal, slight, moderate, moderately severe, and severe. Experimental concussion was produced if the following were observed:

1. Temporary unconsciousness.
2. Temporary loss or depression of the pupillary and corneal reflexes.
3. Transient loss of righting reflex.
4. Depression of the gag and bite reflex.
5. Absent or diminished withdrawal response to pain and irritation.
6. Slowing and flattening of the electroencephalogram.
7. Apnea or bradypnea.
8. Bradycardia.

All monkeys were sacrificed on the seventh day following the blow. This was done since previous experience in agreement with Windle, et al. (19) indicated this to be the best interval to demonstrate cytologic changes. The head was perfused with 10% formalin and the brain and cervical spinal cord were removed and examined by a neuropathologist having no knowledge of the concussion experience of the animals. Sections were taken from the cervical spinal cord, medulla, the pons and overlying cerebellum, the midbrain, corresponding areas in the frontal, parietal, and parieto-occipital areas. In each area, 10 serial sections each 20 microns thick, were studied.

To help visualize gross intracranial strains and mass movements

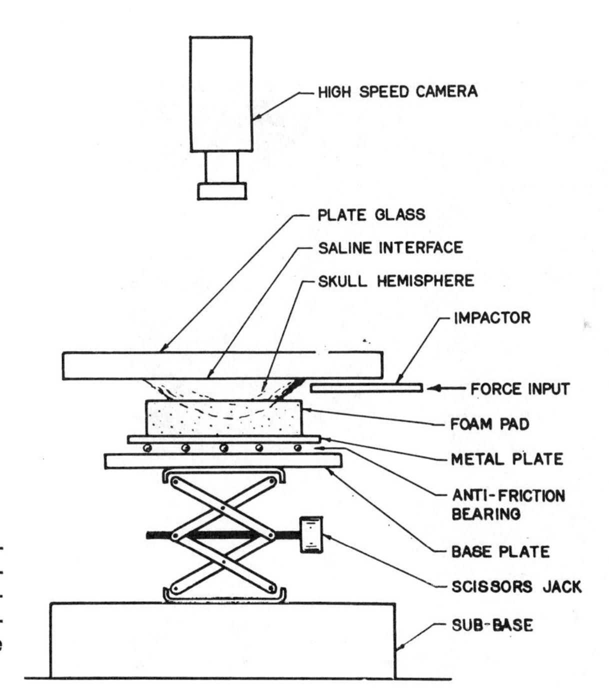

Fig. 6—Diagram of set-up used to record relative motion of components of fresh-like monkey head hemisphere during impact

in the living animal, a model was prepared by the following techniques. The outer soft tissue of the head of an anesthetized animal (dog and monkey) was removed and after 100 cc of heparin was injected to reduce clotting, the animal was sacrificed by a venous injection of 10 cc KCl. Following decapitation at the fourth cervical, the head was quick frozen in liquid nitrogen. A saw cut was made along one side of the sagittal plane and the largest half of the head was applied to the belt sander until a very smooth surface was attained on the sagittal plane. The specimen was then immersed in a 10% saline solution and allowed to thaw overnight in a refrigerator kept at 36 F. After thawing, it was mounted with its sagittal plane pressed against a ½ in. thick panel of plate glass as diagrammatically illustrated in Fig. 6 and shown photographically in plan view in Fig. 7. A film of saline was injected between the fresh-like brain tissue and glass through a filler hole in the glass. The head section was struck by means of a linear striker at several locations to cause general plane motion. Motion of the head was recorded by means of a Fastax Model WF17T high speed camera at a framing rate near 6000 frames/sec.

Results

Pertinent physical impact factors and physiological responses experienced by the twelve anesthetized stumptail monkeys which were struck on the occiput are listed in Table 1. The animals received any-

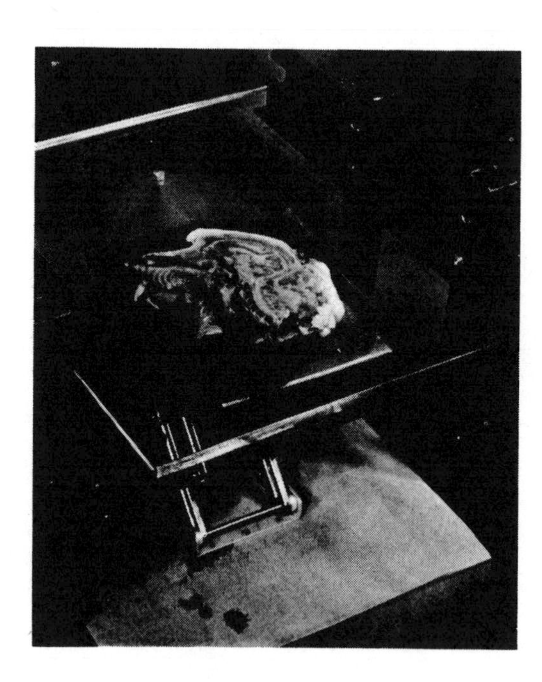

Fig. 7—Plan view of set-up used to record relative motion of components of fresh-like monkey head hemisphere during impact

Table 1—Impact Parameters and Degree of Injury in the Stumptail Monkey

Monkey No.	Sex	Weight, lb	Age	Striker Velocity, fps	Striker Energy, ft-lb	Force, lb	Pulse Time, sec	Clinical Concussion
1	F	13	M	49	74	1070	0.0027	Moderate: Unconsciousness, reflex depression including corneal and pupillary. Slow gradual recovery. Fracture.
2	F	11	Y	35	40	1030	0.0015	Minimal: All reflexes unchanged. No unconsciousness. No fracture.
3	F	16	M	52	85	1430	0.0022	Moderate: Unconscious temporarily. Reflexes depressed. Pupils and corneals unchanged. Quick recovery. No fracture.
4	F	18	Y	50	77	1100	0.0018	Minimal: No change in reflexes. No change in vital signs. Pain response depressed. Fracture.
5	F	12	Y	CONTROL				CONTROL
6	F	14	M	53	86	1800	0.0018	Moderate: No response to pain, corneals absent, disconjugate eye movement, pupillary dilatation and fixation, dazed, Reflexes absent. Fracture.
7	F	11	Y	51	81	1290	0.0029	Severe: Prolonged unconsciousness. Depression of all reflexes. Pupils dilated. Disconjugate eye movement. Apnea and asystole. Gasping respiration. Left corneal remained absent. Right corneal lost momentarily. Fracture.
8	F	9.5	Y	49	78	880	0.0026	Severe: Unconsciousness. Loss of head righting, bite, ear tickle and rectal reflexes. Pupillary and corneal remained. Fracture.
9	M	12	M	49	78	No Record		Severe: Areflexia. Hemorrhage in right ear. Moderate drowsiness. Limited.
10		17	M	42	54	1100	0.0019	Minimal: Reflexes diminished. No pupillary or corneal change. Slightly depressed CSF clear and colorless, microscopic red blood cells. No fracture.
11	F	18	M	47	69	1500	0.0017	Moderate: Dazed. Eyes deviated to right. Pupils dilated. Reflexes diminished. CSF clear and colorless, microscopic red blood cells. No fracture.
12	M	24	O	58	104	1460	0.0018	Slight: Scalp laceration. Reflexes depressed. Pupils dilated (5 mm-8 mm), No focal signs. Fracture.

where from a slight concussion evidenced by depressed reflexes (monkey 12) to severe concussion indicated by unconsciousness and/or areflexia (reflexes absent) (monkeys 7-9). Concussion was accompanied by fracture in at least eight of the twelve animals (monkey 9 extent of fracture uncertain). No relation was observed between degree of concussion and fracture. The force-time history of the pulses is shown in Fig. 8. Failure of the bone is evidenced in the records of monkeys 1, 4, 6-8, and 12 by the rapid fall-off of load after reaching a peak value, thus resulting in a ragged trace.

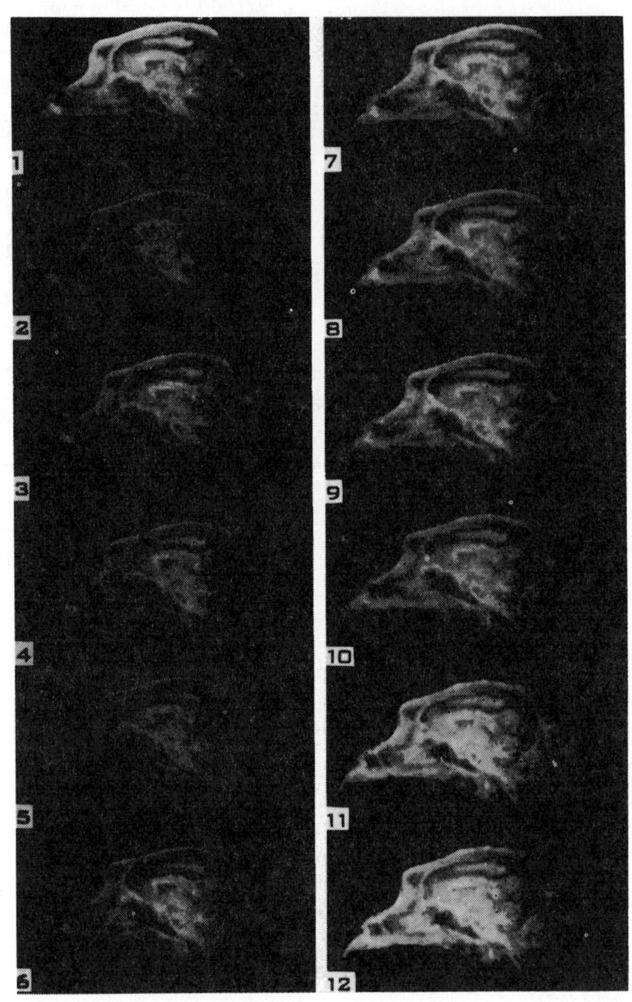

Fig. 8—Force-time records of impact to the occiput of the free intact head of anesthetized stumptail monkeys. 800 lb/in. sensitivity; 1.25 ms time lines except monkey 1, 12.5 ms

Chromatolysis is destruction of the chromatin in the brain cell nucleus, or at least loss of its affinity for the basic dyes. When the cell or axon is injured, the cell can undergo chromatolysis and recover or die. It is a microscopic descriptive term which shows that the cell was not functioning properly when fixation of the tissue occurred. The abnormalities determined by this method are limited to those cells

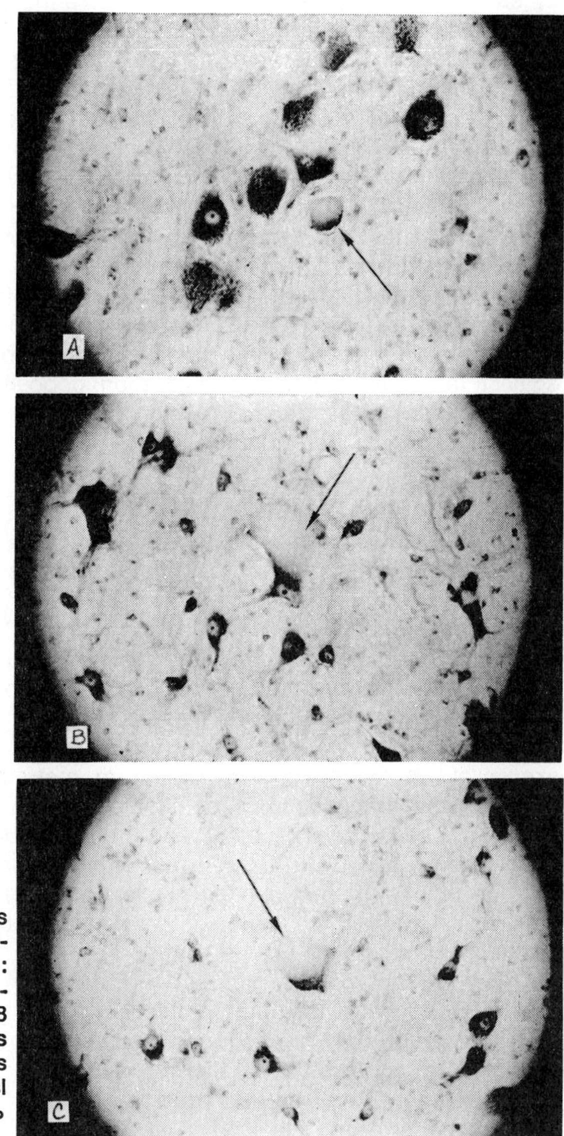

Fig. 9—Cell chromatolysis illustrated by photomicrographs of brain sections in: A-monkey 1 pons; B-monkey 8 pons; C-monkey 8 medullar. The affected cells are located in the centers of the photographs. Nissl stain, 280 x. (Reduced 55% in reproduction)

undergoing recognizable chromatolysis at the time of perfusion, seven days after impact. Photomicrographs showing the cells undergoing chromatolysis are shown in Figs. 9A-9C. The neuropathological examination for chromatolysis comprised a study of seven specific areas representing the cortex and appropriate areas of the brain stem as diagrammatically illustrated in Fig. 10. The diagrams show the total number of affected cells discovered in the 10 sections studied at each level of each animal. Monkeys 9 and 12 were not studied microscopically because of invalid procedures.

Relative motion between parts of the brain tissue and the skull of a hemispherical monkey head section are illustrated in the series of black and white enlargements of frames taken from 16mm color film exposed at approximately 6000 frames/sec, shown in Fig. 11. Because of the short duration of impact (rise time near 0.0005 sec, total pulse time near 0.002 sec), the individual film frames are slighly blurred and do not show the sequence of motion as well as the film projected at 24 frames/sec. In the film there is evidence of pressure gradients which have been observed with other models as is shown by the compression of tissue against the skull near the blow site, the separation of tissue opposite the blow and the forcing of the cerebellum against the brain stem. The separation opposite the blow is the only part of this relative movement which can be readily visualized with the aid of the enlarged frames in Fig. 11.

Discussion

In the brains of previously studied dogs (10) which had received a concussive blow, central chromatolysis was found in the reticular substance of the medulla in 18 of 25 cases. Four of the remaining seven dogs without medullary lesions and 13 other dogs had cells exhibiting central chromatolysis at other levels of the reticular substance, in the cerebral cortex and/or in the cervical cord. Two similarly altered cells were found in the medula of one of the nine control animals.

In the present monkey study, cell chromatolysis was found almost exclusively in the brain stem but more rostrally (toward the nose) in the dog. The preponderance of affected cells were concentrated about the pons. The degree of concussion could not be related to numbers of affected cells except in the severely affected monkeys. In severe cases, such as monkey 7, the damaged cells were more widespread along the cord axis, reaching caudally to the cervical cord and rostrally as far as the lower midbrain.

High speed motion pictures of the heads of the monkeys which were struck on the occiput in this series of experiments shows that both translation and rotation occurred, each part of the head moving in a plane approximately parallel to the sagittal plane in which the blow was delivered. The head was free-to-move on the body and the body was suspended from a point just above the center of gravity of a 3 lb

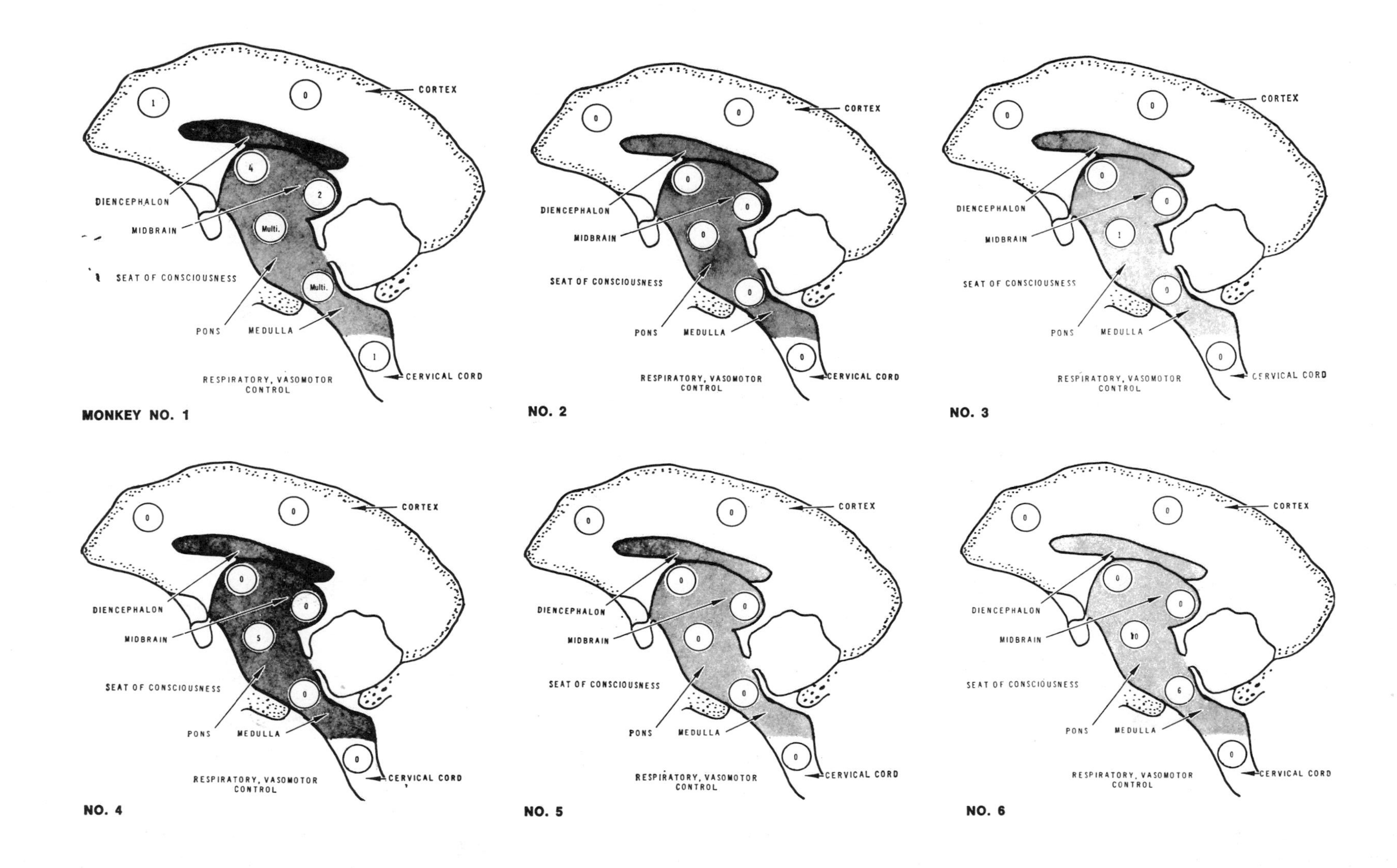
CORTEX
DIENCEPHALON
MIDBRAIN
SEAT OF CONSCIOUSNESS
PONS
MEDULLA
RESPIRATORY, VASOMOTOR CONTROL
CERVICAL CORD
1
0
4
2
Multi.
Multi.
1
MONKEY NO. 1
CORTEX
DIENCEPHALON
MIDBRAIN
SEAT OF CONSCIOUSNESS
PONS
MEDULLA
RESPIRATORY, VASOMOTOR CONTROL
CERVICAL CORD
NO. 2
CORTEX
DIENCEPHALON
MIDBRAIN
SEAT OF CONSCIOUSNESS
PONS
MEDULLA
RESPIRATORY, VASOMOTOR CONTROL
CERVICAL CORD
NO. 3
CORTEX
DIENCEPHALON
MIDBRAIN
SEAT OF CONSCIOUSNESS
PONS
MEDULLA
RESPIRATORY, VASOMOTOR CONTROL
CERVICAL CORD
5
NO. 4
CORTEX
DIENCEPHALON
MIDBRAIN
SEAT OF CONSCIOUSNESS
PONS
MEDULLA
RESPIRATORY, VASOMOTOR CONTROL
CERVICAL CORD
NO. 5
CORTEX
DIENCEPHALON
MIDBRAIN
SEAT OF CONSCIOUSNESS
PONS
MEDULLA
RESPIRATORY, VASOMOTOR CONTROL
CERVICAL CORD
10
6
NO. 6

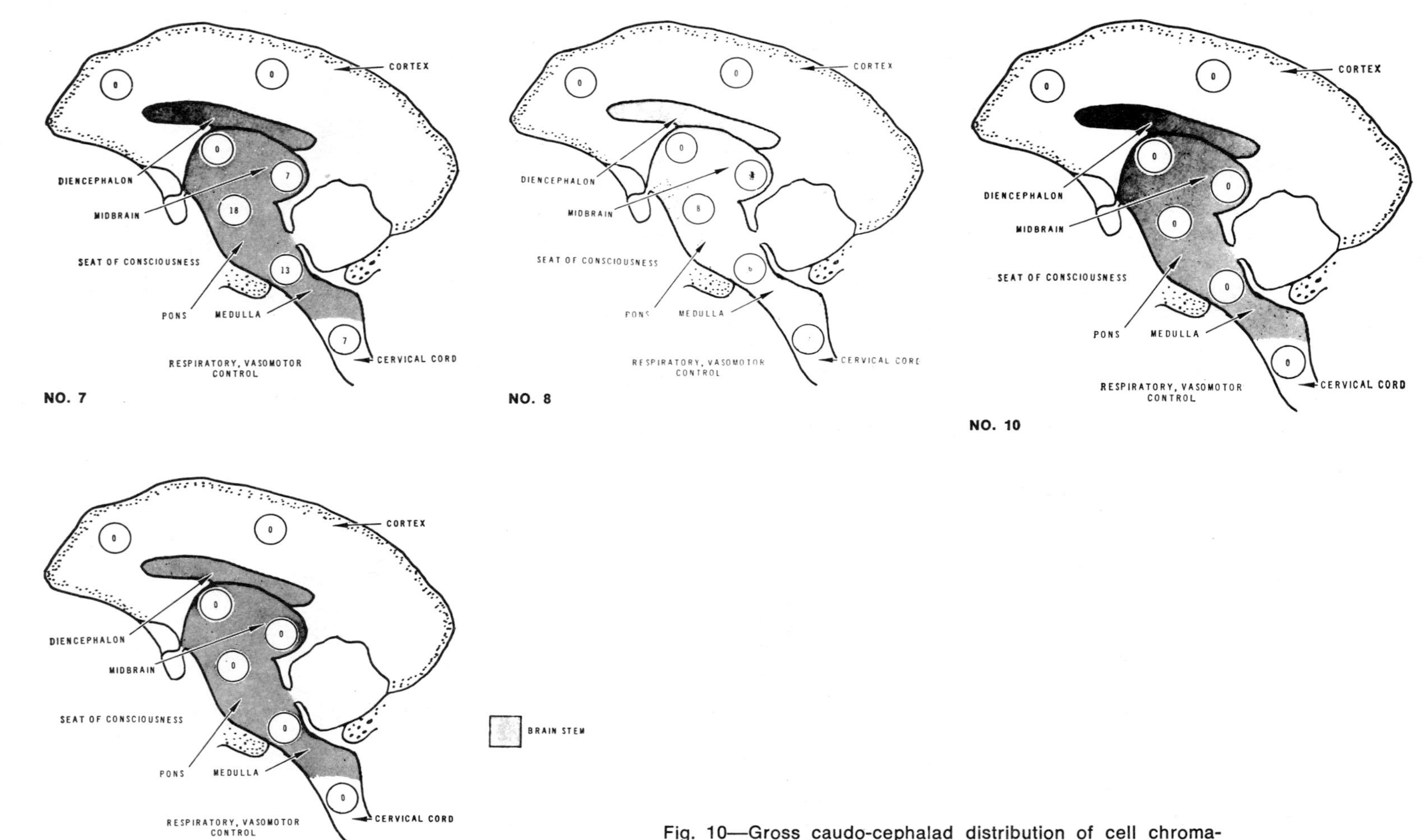

Fig. 10—Gross caudo-cephalad distribution of cell chromatolysis in the monkey.

unconscious by the blow, but only in monkeys 1 and 7 was chromatolysis detected as high as the midbrain. Either some cells were reversibly damaged and recovered before 7 days or affected cells may have been on other levels which were not examined. It seems reasonable to expect that more damaged cells should be found in the cortex (only one in the right frontal lobe of monkey 1) if angular acceleration is an important factor in concussion (14). Similarly more damage should have been seen in the cervical cord if involvement of the brain stem at the craniospinal junction is important as has been claimed (2, 17).

It has been observed by Gurdjian, et al. (12) that the dog is much more sensitive to a blow on the occipital protuberance than to a blow on the sagittal crest as regards changes or cessation in vital functions, that is, respiration, heart rate, and blood pressure. This blow location sensitivity has not been observed in the monkey. As noted above, central chromatolysis was found in the brain stem of both dog and monkey but concentrated more rostrally (toward the nose) in the monkey. To help explain these differences in physiological response and neural tissue stress concentrations, a procedure was devised as outlined earlier for recording the motion of fresh sections of dog and monkey heads under glass. See Figs. 6 and 7. By this technique the skull is more deformable than the intact head, but sufficiently rigid so that low energy blows produce primarily acceleration. The brain tissue that would abut the sagittal section has been removed and whatever forces exist on this plane during a symmetrical sagittal impact have now been replaced by those transmitted through the saline/film as the section slides across the glass. It seems a reasonable assumption that these forces are small compared to the forces generated by impact because the tissue is observed to slide easily in many directions relative to the skull when high speed films are projected.

The preparation has revealed a considerable amount of relative motion between brain components and between the brain components and skull of the dog and the monkey during impact recorded by high speed camera. It has been particularly interesting to observe the interference between cerebellum and the brain stem when an occipital impact causes both to tend to be extruded out of the foramen magnum in the dog. Films of a blow to the sagittal crest of the dog show considerable distortion of the brain stem, but no outward displacement of brain tissue at the foramen magnum, possibly indicating less stress on the brain stem around the vital vegetative functions in the medulla. These visual observations help to account for the blow location sensitivity in the dog discovered by Gurdjian's group as cited previously (12).

The validity of the model as regards its closed system dynamics has not yet been verified. It is planned to compare intracranial pressure and acceleration measurements in the preparation with those obtained in the live animal (dog and monkey). It is also anticipated that a model of the monkey head prepared similarly but with a plastic window screwed to the sagittal edge of the skull is feasible and may prove useful.

Conclusions

1. In concussion, cell chromatolysis was found primarily in the brain stem and in lesser amounts in the cortex of the dog but almost exclusively in the brain stem of the monkey.

2. In the dog brain stem, cell chromatolysis was found concentrated in the medulla and cervical cord.

3. In the monkey, cell chromatolysis was more rostral, being found primarily centered about the pons.

4. The numbers of affected cells could not be correlated with degree of concussion in the monkey except in the more severe cases.

5. In the more severe cases, the number of affected cells were spread out along the brain stem axis, caudally to the cervical cord and rostrally to the lower midbrain.

6. Since the maximum cell chromatolysis seen in these monkeys was found in the brain stem cells, specifically concentrated about the pons with relatively few cells and only one cell found in the cervical cord and cortex, respectively, it increases support for the theory of Gurdjian and Lissner (9) that the mechanism of concussion is shear strain in the brain stem caused by pressure gradients due to closed system dynamics of impact.

7. Although the motion of the head involved both angular and translational acceleration, the preponderance of affected cells found in the brain stem and the almost complete absence of chromatolysis in the cortex, makes it appear likely that translational acceleration is the most important mechanism.

8. The relatively small amount of cell chromatolysis found in the cervical cord makes it unlikely that cerebral concussion is a concommitant of cervical cord involvement.

References

1. J. L. Chason, B. F. Haddad, J. E. Webster, and E. S. Gurdjian, "Alterations in Cell Structure Following Sudden Increases in Intracranial Pressure." Jrl. Neuropath. & Exper. Neurol., Vol. 16 (1), (1957) pp. 102-107.

2. R. L. Friede, "Specific Cord Damage at the Atlas Level as a Pathogenic Mechanism in Cerebral Concussion." Jrl. Neuropath. & Exper. Neurol., Vol. 19 (2), (1960) pp. 266-279.

3. E. S. Gurdjian and H. R. Lissner, "Mechanism of Head Injury as Studied by the Cathode Ray Oscilloscope, Preliminary Report." Jrl. Neurosurg. Vol. 1 (6), (1944) pp. 393-399.

4. E. S. Gurdjian, H. R. Lissner, F. R. Latimer, B. F. Haddad, and J. E. Webster, "Quantitative Determination of Acceleration and Intracranial Pressure in Experimental Head Injury, Preliminary Report." Neurol., Vol. 3 (6), (1953) pp. 417-423.

5. E. S. Gurdjian, H. R. Lissner, J. E. Webster, F. R. Latimer, and B. F. Haddad, "Studies on Experimental Concussion." Neurol., Vol. 4, (1954) pp. 674-681.

6. E. S. Gurdjian, J. E. Webster, and H. R. Lissner, "Observations on the Mechanism of Brain Concussion, Contusion, and Laceration." Surg., Gynec., & Obstet. Vol. 101, (1955) pp. 680-690.

7. E. S. Gurdjian and J. E. Webster: "Head Injuries." Boston: Little, Brown & Co., 1958.

8. E. S. Gurdjian and J. E. Webster, "Experimental Head Injury with Special Reference to the Mechanical Factors in Acute Trauma." Surg., Gynec., & Obstet., Vol. 76 May, 1943, pp. 623-634.

9. E. S. Gurdjian and H. R. Lissner, "Photoelastic Confirmation of the Presence of Shear Strains at the Craniospinal Junction in Closed Head Injury." Jrl. Neurosurg., Vol. 18 (1), (1961) pp. 58-60.

10. E. S. Gurdjian, O. U. Fernando, V. R. Hodgson, L. M. Thomas, and J. L. Chason, "Cellular Changes in the Nervous System Following Compressive and Accelerative Impacts in the Dog." Trabajos del Instituto Cajal de Investigaciones Biologicas, Supplemento al tomo LVII, Madrid, 1965, p. 39.

11. E. S. Gurdjian, H. R. Lissner and L. M. Patrick, "Concussion—Mechanism and Pathology." Proceedings of the Seventh Stapp Car Crash Conference, Springfield, Ill.: Charles C. Thomas, 1965.

12. E. S. Gurdjian, L. M. Thomas, and V. R. Hodgson, "Comparison of Species Response to Concussion." Proceedings of the Ninth Stapp Car Crash Conference, Minneapolis: Univ. of Minn., 1966, pp. 363-382.

13. E. S. Gurdjian, V. R. Hodgson, L. M. Thomas, and L. M. Patrick, "High Speed Techniques in Head Injury Research." Med. Sci., Vol. 18 (11) (1967) pp. 45-56.

14. A. E. Hirsch and A. K. Ommaya, "Tolerance of the Sub-Human Primate Brain to Concussion." Impact Injury & Crash Protection Symp. Proc., Springfield, Ill.: Charles C Thomas (in press).

15. V. R. Hodgson, E. S. Gurdjian, and L. M. Thomas, "Experimental Skull Deformation and Brain Displacement Demonstrated by Flash X-ray Technique." Jrl. Neurosurg., Vol. 15 (5) (1966) pp. 549-552.

16. V. R. Hodgson, "Head Impact Response of Several Mammals Including the Human Cadaver." Ph.D. dissertation, Wayne State Univ., 1968.

17. N. R. Hollister, W. P. Jolley, R. G. Horne, and R. Friede, "Biophysics of Concussion." WADC Tech. Report 58-19f, 1958.

18. L. M. Thomas, V. L. Roberts, and E. S. Gurdjian, "Experimental Intracranial Pressure Gradients in the Human Skull." Jrl. Neurol., Neurosurg., & Psychiat., Vol. 29 (1966) pp. 404-411.

19. W. F. Windle, R. Groat, and C. A. Fox, "Experimental Structural Alterations in the Brain During and After Concussion." Surg., Gynec. Obstet., Vol. 79 (1944) pp. 561-572.

710880

Translational Versus Rotational Acceleration — Animal Experiments with Measured Input

F. J. Unterharnscheidt
The University of Texas

Abstract

Any broadly surfaced impact imparts to the head a force by which it is accelerated. If the impact is directed at the center of mass of a freely movable object, the resulting motion is a translation acceleration. If the impact is directed eccentrically, the result is a combined translational and rotational acceleration. The magnitude of the rotational acceleration is related to the degree of eccentricity of the acting force. The magnitude of the translational acceleration is related to the distance between the point of fixation and the center of gravity of the head.

The distinction between the two types of acceleration is important in view of the different physical processes they initiate in the brain. Pure translational acceleration creates pressure gradients, while pure rotational acceleration produces rotation of the skull relative to the brain. Both processes are the effects of mass inertia of the brain.

It can be expected, according to the physical analysis of translational and rotational trauma, that different mechanisms produce different patterns of lesions.

Experiments with different animal species, which employed translational and rotational accelerations with exactly measured inputs are summarized and the morphological alterations in respect to distribution and quality are discussed.

THIS PAPER is concerned with the effects of translational and rotational acceleration on the brain in closed head injuries.

Any broadly surfaced impact imparts to the head a force by which it is accelerated. If the impact is directed at the center of mass of a freely movable object, the resulting motion is a translational acceleration. If the impact is directed eccentrically, the result is a combined translational and rotational acceleration. The distinction between the two types of acceleration is important in view of the different physical processes they initiate in the brain. Pure translational acceleration creates pressure gradients, while pure rotational acceleration produces rotation of the skull relative to the brain. Both processes are the effects of mass inertia of the brain.

The inertia effect is fundamental for all injuries caused by mechanical forces. The effect can be studied on a rigid, freely movable spherical shell filled with an incompressible, nonviscid and homogeneous fluid.

After the central impact, the shell moves only in translation, carrying with itself the contained liquid. While the blow lasts and the sphere moves, its contents are compressed at the impact side in consequence of the inertia of the liquid particles but are rarefied (expanded) at the side opposite to the impact pole (the "counterpole"). This is the basic pressure distribution picture due to an acceleration motion of the skull. Of major importance is the dependence of the pressure distribution on the time variation and duration of the blow. If the impact is not centrally directed, the skull receives an angular momentum in addition to the acceleration. The farther from the center of rotation the particle of interest is, the greater the acceleration suffered, and thus the greater the potential for injury. Therefore, points near the center of rotation are likely to be damaged only if the acceleration is severe. Rotational traumas involve considerable tensile forces between the accelerated skull and the inert brain.

Translational Acceleration

A single impact of subcommotio strength, at a speed of 7.1 m/s, corresponding to an acceleration of 205 g, imparted to the free head of a cat, caused neither behavioral nor histologic changes in the central nervous system; whereas repeated impacts of the same intensity, without causing primary traumatic lesions, did produce secondary traumatic alterations due to circulatory disturbances. Lesions in the cerebellum included diffuse loss of Purkinje's cells (especially at the summits of the lobuli of the vermis), proliferation of Bergmann's glia, thinning of the granular cell layer with glial reaction, and glial proliferation in the striae medullares and white substance. Alterations in the cerebrum were less severe; they consisted of disseminated ischemic nerve cells and a moderate glial proliferation in the white substance.

Impacts of commotio strength, that is, producing the clinical symptoms of commotio cerebri (unresponsiveness) in cats have a velocity of 8.3-9.4 m/s, corresponding to an acceleration of 280-400 g. After one such impact, the histologic alterations which account for the clinical symptoms of commotio cerebri prove to be traceless with the methods of investigation used today. We found, in particular, no evidence for glial cell proliferation. However, after repeated impacts of equal intensity and at intervals of 1-2 days, the cerebral cortex showed, beside disseminated ischemic nerve cells, extensive focal and pseudolaminary necroses of the parenchyma and loss of nerve cells in various parts of the Ammon's horn formation. Tissue alterations in the cerebellum, although less intense, corresponded in quality to those caused by successive impacts of subcommotio strength.

It follows that blunt impacts of intensities that do not cause noticeable tissue alterations when applied singly may elicit secondary alterations due to circulatory disturbances when applied successively in repeated experiments. A sustained permanent brain injury can therefore result from secondary lesions alone, with no primary traumatic alterations present at all. The time interval of the individual experiments has a distinctive influence on the nature of the morphologic alterations.

Considerable primary traumatic lesions are produced by impacts with a velocity of 10.5 m/s or more, corresponding approximately to more than 400 g. In all instances, there were subarachnoid and subdural hemorrhages, cortical contusions at the counterpole and, at the impact pole, single intracerebral hemorrhages, and traumatic necroses. Speeds of the impacting instrument of 17.2 and 18.3 m/s are fatal to a cat.

Rotational Acceleration

In experiments concerning the effects of rotational acceleration on the central nervous system, controlled nondeforming rotational acceleration was directed through a known path near C7-Th 1 of 24 squirrel monkeys. The equipment used in these studies was designed by Higgins and Schmall. The monkeys were subjected to rotational accelerations of 101.000-386.000 rad/s^2. The result was a continuum of clinical effects from no observable signs through concussion to death.

The lowest rotational accelerations employed (101.000-150.000 rad/s^2) caused apparently no primary or secondary alterations in the cerebrum. However, the next higher accelerations, up to 197.000 rad/s^2, produced in 10 of 13 animals subarachnoid hemorrhages, combined in one instance with primary traumatic hemorrhages in the oculomotor nerve, and tears and avulsions, mainly of veins and capillaries, in superficial cortical layers in 8 animals. Accelerations of more than 200.000 rad/s^2 caused severe primary traumatic hemorrhages in the cortex and white substance.

Rotational accelerations of more than 300.000 rad/s^2 were not survived. These monkeys were the only animals to show additional hemorrhages in more central regions of the brain, that is, very close to the central pivot.

Nearly all the animals tested showed small rhectic hemorrhages in various segments of the spinal cord. Capillary and venous hemorrhages were more frequently found disseminated in the gray substance and were caused by longitudinal and transverse stretching of ascending and descending vessel branches. Primary traumatic alterations were seen in all segments of the cord. These lesions were not fatal and produced no clinical signs in the animals. In two instances subdural hemorrhages were found in the cauda equina.

It must be pointed out that the primary traumatic lesions found in the cortex are venorhectic, and occasionally arterio- or capillary-rhectic hemorrhages of the more superficial cortical layers, as evidenced by torn vessel walls. Also, these hemorrhages are always associated with vessel systems running at right angles to the cortical surface. Pinpoint hemorrhages, so characteristic of cortical contusions in the first stage, were not observed.

Conclusion

In summary, not only does a qualitative difference exist between the primary traumatic cortical hemorrhages produced by rotational acceleration and the so-called cortical contusions found in translational injuries, but there are also different patterns of distribution for the primary traumatic lesions encountered in both types of acceleration, inasmuch as these lesions are arranged in a cylindrically symmetric pattern after translational acceleration, as compared to a radially symmetric pattern located close to the midline after rotational acceleration.

Except for the question of location, these considerations seem to be valid also for the interpretation of findings in the spinal cord, although correlations are not as patently manifest here as they are in the brain. Nevertheless, the relation between severity of primary traumatic lesions and magnitude of acceleration is evident throughout the entire central nervous system.

720970

Pathophysiologic Responses to Rotational and Translational Accelerations of the Head

T. A. Gennarelli, L. E. Thibault and A. K. Ommaya
National Institute of Health and Georgetown University

Abstract

Acceleration-time data in 25 squirrel monkeys subjected to controlled sagittal plane head motions are presented. In 12 of the 25 animals subjected to pure translation of the head at peak positive g levels ranging between 665–1230 g (6–8 ms duration), cerebral concussion was not obtainable. In contrast, 13 of the animals subjected to head rotations at peak positive tangential (at c.g.) g levels ranging between 348–1025 g (5.5–8 ms duration) were all concussed. Visible brain lesions were noted in both translated and rotated groups but with a greater frequency and severity after rotation. An analysis of the lesions produced in both groups is presented, along with our preliminary data on the use of the evoked somatosensory response as an objective, quantifiable index for the onset and severity of brain damage in head injury.

AT THE FIFTEENTH Stapp Car Crash Conference, we presented a report comparing the effects of translational and rotational head motions in producing cerebral concussion (CC). The current study extends and confirms the observations made in that report, describes the patterns of brain injuries produced by the two types of head motion, and presents our use of the somatosensory evoked response (SER) as an objective, quantifiable parameter of the onset and severity of traumatic unconsciousness and brain lesions in vivo.

Method

Twenty-five adult squirrel monkeys were anesthetized with appropriate doses of 1% Brievital. Epidural electrodes were implanted for EEG recording, and subcutaneous electrodes were placed for EKG and for median nerve stimulation from the wrist. After electrode placement, the animal was restrained in a primate chair and allowed to awaken. The animal's head was then encased within a helmet and attached to the HAD-II apparatus. The delivery of rotational and translational motions to the head and measurement of accelerations was carried out as described in our earlier report (1)*. Measured quantities included translational acceleration of the head helmet system, tangential and radial accelerations, and duration of the acceleration pulse.

Because the head is rigidly fastened to the helmet, which is rigidly attached to the HAD-II linkage, the path of the c.g. of the system is precisely controlled. The translational acceleration and the components of the angular acceleration (that is, tangential and radial) are measured equivalently at this point (Fig. 1).

*Number in parentheses designates Reference at end of paper.

Documentation of Head/Neck Motions - The HAD-II apparatus was designed to displace the animal's head and neck within anatomical limits. To verify this and to document where motions were taking place, special lucite helmets and linkages were designed through which x-ray films could be obtained. From these films, line sketches depicting the movements were drawn, and the extent and type of such head/neck movements were found to be within the normal range and anatomic limits of these structures. These findings were supported by the absence of pathologic evidence for visible musculoskeletal injury in the neck structures at autopsy.

Instrumentation - Acceleration, EKG, and high-speed photography were carried out using the apparatus described earlier. For SER recordings, potentials from the epidural EEG electrodes were recorded referentially to linked earlobes with Grass P5 preamplifiers. Stimuli to evoke the SER were delivered to the median nerve at the wrists at 1/s by a Grass 8 stimulator using 0.5 ms pulses at intensities giving a maximal muscle twitch.

Output of the EEG amplifiers was delivered directly to a PDP-12 computer for on-line analysis and simultaneously recorded on FM magnetic tape by a Honeywell Model 7600 recorder. Displayed SER represented the averaged summation of 10 responses. Prolonged EEG recordings were not possible in this experimental preparation because of the need for lesion analysis, and such recordings were therefore continued only until responses stabilized.

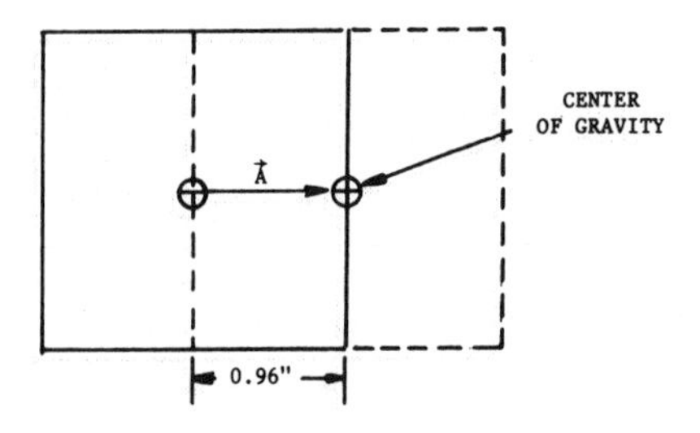

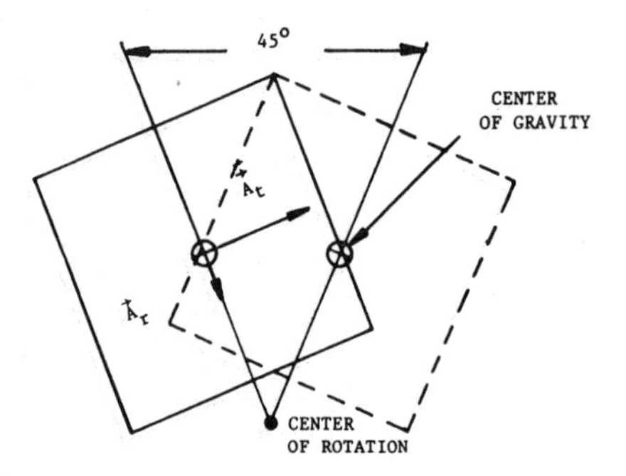

$\vec{A}$ = TRANSLATIONAL ACCELERATION

$\vec{A}_t$ = TANGENTIAL ACCELERATION

$\vec{A}_r$ = RADIAL ACCELERATION

Fig. 1 - Measurement of translational, tangential, and radial acceleration

The 22 animals surviving the impact were observed for varying periods afterwards. Most were sacrificed by perfusion at 24 h, but several animals were observed for periods up to 5 weeks before sacrifice. The gross appearance of the fresh skull and its contents was recorded and photographed and after 7–10 d of fixation, serial sections of the brain were made. Microscopic sections were prepared using hematoxylin-eosin, luxol-cresyl violet, and Bodian reduced silver stains.

Results

Relationship Between Rotational and Translational Motions and Cerebral Conconcussion (CC) -
Translated Group (N = 12) - No animal subjected to pure translational acceleration was concussed; that is, CC could not be produced at these levels of translational acceleration. One animal subjected to a very high G level of translation was awake immediately postimpact but shortly thereafter developed a bradycardia and became unconscious.

Rotated Group (N = 13) - All 13 animals subjected to rotational motions were unequivocally concussed. The duration of concussion ranged from 2–12 min. One animal never regained consciousness and two other animals died within 1 h of impact. The comparative data on accelerations in both groups are summarized in Tables 1 and 2.

Pathological Findings in Translational and Rotational Acceleration - Design demands of the HAD-II required minimization of contact phenomena so as to study more or less pure inertial loading of the head and its contents. This appeared to be the case in all animals because none demonstrated scalp lacerations, scalp contusions, or skull fractures. No fractures, dislocations, or musculoskeletal injuries were noted in the cervical region.

Gross and microscopic lesions were divided into five categories, as shown in

Table 1 - Data on Translated Animals

Monkey No.	rpm	Peak Positive G	Peak Negative G	Acceleration Duration, ms
– SL-2	1500	1230	900	6.0
– SL-1	1500	1140	463	6.5
– SL-12	1300	1058	512	7.0
– SL-3	1200	854	522	7.0
– SL-5	1200	830*	495*	6.5*
– SL-6	1200	—	—	—
– SL-4	1200	812	474	7.0
– SL-8	1000	802	717	7.5
– SL-7	1000	768	665	6.5
– SL-11	1000	734	427	8.0
– SL-9	1000	665	427	7.5
– SL-10	1000	—	—	—

*Estimated values.

Table 2 - Data on Rotated Animals

Monkey No.	rpm	Peak Positive G Tangential at c.g.	Peak Negative G Tangential at c.g.	Peak Radial G (Normal)	Peak Angular Acceleration, 10^5, rad/s^2	Peak Resultant Acceleration, G $A_R = \sqrt{A_T^2 + A_N^2}$	Acceleration Duration, ms
+SR-1	1500	1025	1155	123	3.17	1032	6.5
+SR-2	1500	1025	1155	133	3.17	1033	5.5
+SR-6†	1200	961	731	115	2.97	968	7.0
+SR-12	1200	783	589	94	2.42	786	7.0
+SR-D	1200	728	618	62	2.25	730	7.0
+SR-9†	1200	713	418	85	2.20	715	7.0
+SR-4	1200	710	600	59	2.19	712	7.0
+SR-3	1200	706*	596*	60*	2.18*	708*	7.0*
+SR-5†	1200	700	602	56	2.16	702	7.0
+SR-10	1000	488	545	41	1.51	489	7.5
+SR-11	1000	402	232	35	1.24	403	7.5
+SR-8	1000	387	271	33	1.20	388	8.0
+SR-7	1000	348	248	30	1.08	349	7.5

*Estimated values.
†Animal died.
+Animal concussed.

Table 3, which also summarizes the total number of lesions found in the two groups. Fig. 2 depicts the frequency and distribution of the lesions. Note the marked symmetry of the brain injuries in the rotated group as compared with the asymmetric and much less severe lesions in the translated group (Table E-1).

Lesion Analysis

Subdural Hematomas - Overall, the lesions present in the translated animals (TA) were less frequent and less obvious than those of the rotated animals (RA). Subdural blood in TA was present bifrontally in three animals (six lesions) and unilaterally in two instances for a total of eight occurrences. In all cases where present, only the anterior 5 mm of the frontal pole was affected.

In contrast, subdural accumulations occurred in each animal for a total of 20 times in the RA. In every instance, the blood was thicker and extended further posteriorly than in the TA. Many hematomas covered the entirety of both frontal lobes to the Sylvian fissure, and most were more extensive parasagittally than over the lateral convexity. Additionally, several instances of temporal pole and sub-temporal subdural hematomas occurred, a finding not seen in the TA (Fig. 2).

Subarachnoid Hemorrhage (SAH) - Blood in subarachnoid spaces was categorized into localized, diffuse, and posterior fossa.

1. Localized SAH was present in one TA adjacent to a large surface contusion.
2. Diffuse SAH includes blood either over the cerebral hemispheres or at the base of the brain in the supratentorial compartment. Because the subarachnoid spaces of the hemisphere's convexities are only in indirect intercommunication, SAH on each side is considered a separate lesion.

In the TA, no evidence of diffuse SAH was seen either grossly or microscop-

Table 3 - Gross Pathology—Total Number of Visible Lesions

	Rotation	Translation
Subdural hematoma	20	8
Subarachnoid hemorrhage	27	1
Localized	0	1
Diffuse	18	0
Posterior fossa	9	0
Break in blood-brain barrier (without contusion)	20	8
Cerebral contusions	5	6
Intracerebral blood	22	4
Hematomas	0	2
Petechial hemorrhage	22	2
Total No. lesions	94	27
No. animals	13*	12
No. lesions per animal	7.2	2.2

*Four animals were sacrificed 5 d–10 weeks after impact.

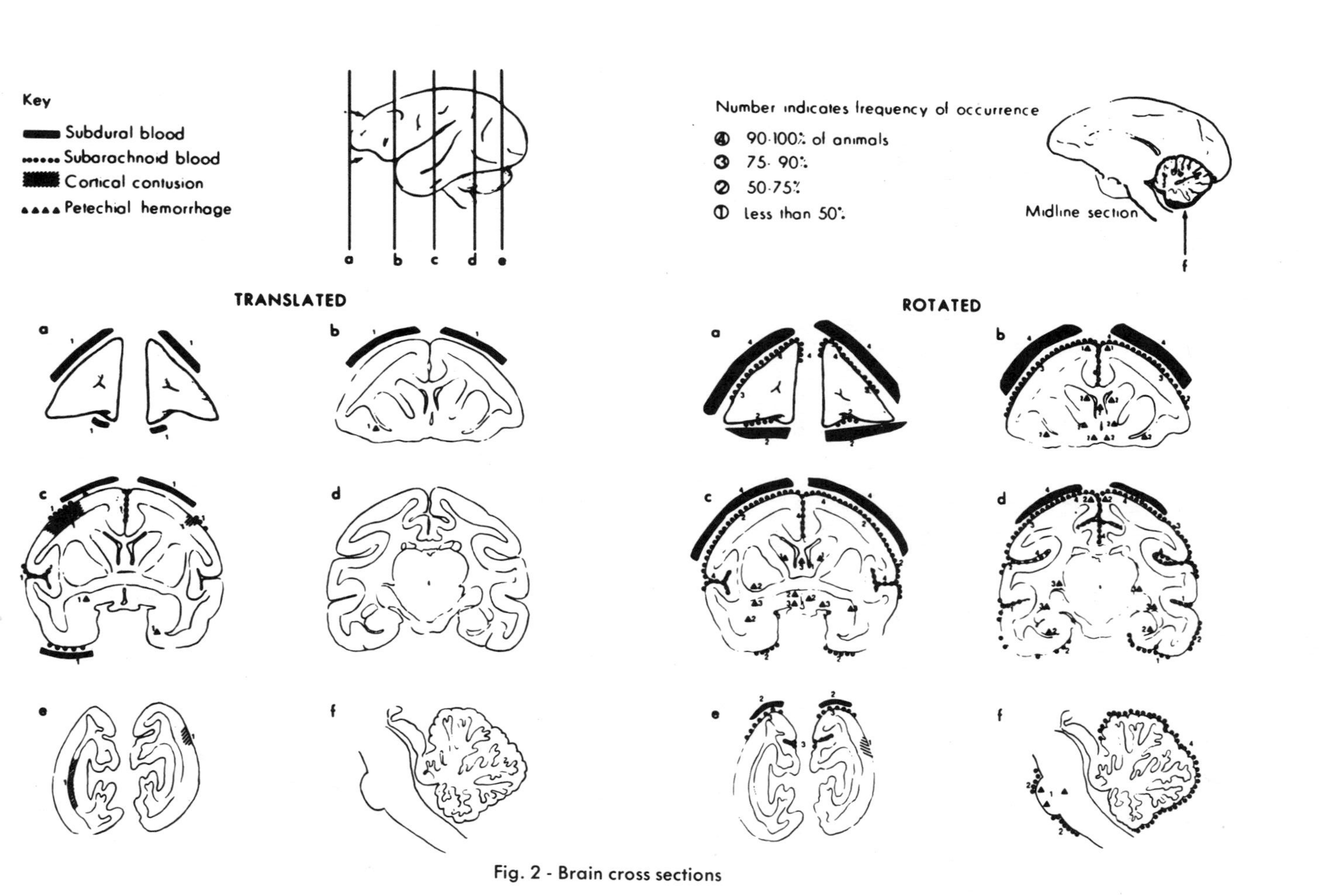

Fig. 2 - Brain cross sections

ically. In contrast, 18 occurrences of diffuse SAH were present in the RA. These occurred with regularity in the callosal and cingulate sulci medially and extended throughout the frontocentral regions bilaterally in close relation to the falx cerebri. Less dense collections of SAH were almost uniformly present over the parasagittal convexities of the hemispheres and decreased in abundance over the lateral convexities. In most cases, SAH was also present in the Sylvian (central) fissure. SAH along the base of the cerebrum and medial temporal regions was less frequent (Fig. 2). In only one instance (SR10) was blood found within the ventricular system.

3. Because the subarachnoid spaces of the posterior fossa intercommunicate, the presence of blood was considered as a single lesion despite bilaterality.

The TA showed no instance of SAH in the posterior fossa even under microscopic examination. In contrast, all the RA that were sacrificed 24 h after impact had posterior fossa SAH. None of the four RA sacrificed 5 d to 5 weeks after impact demonstrated this finding, but one would expect such blood to have dissipated by this time. The SAH in the posterior fossa was seen with greatest frequency over the anterior surface of the cerebellar vermis and hemispheres and less often in the pontomedullary cistern. Even less often was the microscopic occurrence of SAH over the posterior and inferior cerebellum a significant finding (Fig. 2).

Isolated Discontinuities of Blood-Brain Barrier - Most animals were injected with 2% Evan's blue dye several hours before sacrifice. This compound normally remains intravascular and does not stain brain tissue unless anatomical disruption of the capillary-glial blood-brain barrier occurs.

Extravasation of the dye occurred on the cerebral surface in both groups but more than twice as often in the RA. This does not include extravasation accompanying cortical contusion nor accompanying intracerebral blood, which occurred wherever these lesions were seen. The distribution of dye was only slightly different in the two groups. In the TA, extravasated dye was seen mainly in the midfrontal, parasagittal regions, whereas in the RA, dye was seen in the frontotemporal areas more often. The size of the extravasations did not differ in the two groups and tended to be small (1–2 mm) and symmetrical.

Cerebral Contusions - Contusions of the superficial cortex are considered to be hemorrhagic lesions visible on the surface of the uncut brain which on microscopic examination reveal disruption of cortical structure. They are invariably accompanied by extravasation of Evan's blue dye. Cortical contusions were seen in both groups of animals but were more frequent in the TA. The distribution was also similar, being found either at the occipital pole or in the midfrontal region away from the midline. Except in one instance, there was no difference in appearance in the two groups. The contusions were small to moderate size (2–4 mm), circular or ovoid in shape, and were surrounded by a halo of Evan's blue dye. On microscopic examination, these lesions destroyed the upper half of the cortical gray matter and always tapered down and stopped before the grey-white junction. A surrounding area of edema without hemorrhage extended into the superficial white matter. The single exception to these small contusions was in animal SL7. This animal was not concussed and quite normal for the first hour after impact. A marked left hemiparesis then developed which involved hand and face more than leg. At sacrifice, a large, right midfrontal surface contusion measuring 1 × 2 cm was found. Microscopically, despite the large surface size, the hemorrhage extended only through the outer half of the cortex. Several vessels running perpendicular to the surface were ruptured in the deeper gray matter, but no destruction of cortical integrity was seen.

Intracerebral Lesions -

Intracerebral Hematomas - Intracerebral hematomas were seen in two TA and in no RA. Both were large collections of blood deep in the left occipital-parietal regions of animals that had right sided hemiparesis following impact. In neither case was consciousness altered at any time. Both hematomas occurred within the white matter of the centrum semi-ovale distant from the cortex and both separated the fiber tracts with no evidence of myelin disruption on the luxol-cresyl violet stain.

Intracerebral Petechial Hemorrhages (IPH) - IPH are punctate, very small (< 1 mm) hemorrhage lesions found deep within brain substance. With the single exception of animal SL7, these lesions were found only in the RA. Animal SL7 had two IPH in the temporal lobe beneath the very large cortical contusion.

In the RA, IPH tended to be in multiple locations, tended to be in clusters of several petechiae, and tended to be in or near the junction of deep gray matter and large white matter fiber tracts. There was an overall preference for basal portions of the brain (temporal, basal ganglia, diencephalon), although at times more dorsal locations were seen (fornix columns and periventricular caudate). In most instances, no nearby blood vessels could be seen, yet occasionally striking perivascular exudations of blood were present. In one of the chronic animals, perivascular gliosis and spongiosis was seen, perhaps representing the residues of these lesions.

Brainstem Lesions - In only two animals (both were rotated and concussed) were intraaxial lesions of the brainstem seen. One animal had a single petechial hemorrhage in the tectal region of the pons. Its microscopic characteristics were not different than the IPH described above (SR 10).

Animal SR9 was severely concussed, never resumed respiratory activity, and died 10 min after impact. Numerous hemorrhages extending perpendicular to the ventral surface were present in the pontine tegmentum. In general appearance, this lesion was not unlike Duret hemorrhages seen in man.

Cellular Changes - In general, compared with the various hemorrhagic lesions, individual cellular changes were, with one exception, not frequently seen. The exception involves the four chronic RA. All of these animals were concussed but recovered uneventfully and none was hypoxic throughout the postimpact period. Each showed striking loss of cerebellar Purkinje cells. The cell loss was selective for the apical portions of the cerebellar folia and was not seen in the deeper cerebral cortex. No significant cellular changes were found in the brainstem.

Results of SER - In efforts to find an objective measure of the conscious state, we sought a reproducible neurophysiological assay that could withstand the mechanical rigors of an acceleration experiment. The SER is a measure of integration of sensory responses at the level of the cortical neuronal pool. Since this integration involves pathways ascending through the same midbrain reticular system affecting consciousness, it was felt that the SER could be an effective measure of unconsciousness produced by trauma.

The SER was measured before and for varying intervals after impact in both the TA and RA. All preimpact SER were normal and contained the triphasic positive-negative positive waves characteristic of the squirrel monkey. The first positive deflection (P_1) relates to specific and direct sensory projections to the cerebral cortex, while the second positive wave (P_2) represents activity mediated through a nonspecific reticular path. The negative wave (N_1) represents cortical activity.

Immediately upon impact, all waves of the SER were abolished in the concussed (RA) animals. The P_2 wave of the SER remains absent for the duration of concussion and returns as the animal awakens. The return of the P_1 and N_1 waves is not related to the return of consciousness, probably reflecting cortical or subcortical damage (Fig. 3).

The SER of the nonconcussed animals (TA) showed no abolition of the response. Although there is an attenuation of amplitude, all waves are definitely present immediately after impact and remain normal in shape as amplitude is gradually restored (Fig. 4).

Animal SR13 was concussed for a very brief moment but recovered within 1 min. As with the other concussed animals, the P_2 wave disappeared, then reappeared with awakening. Two minutes after impact, the animal again lost consciousness and his P_2 wave disappeared again only to reappear at 9 min as the animal reawakened (Fig. 5).

The SER was also seen to be predictive of cerebral lesions irrespective of the presence or absence of concussion. Animal CL7 previously described had a normal SER until 1 h after impact, but it subsequently became abnormal on the side of the developing cortical contusion. The contusion was 1 cm distant from the electrode site.

Similarly, a depression of parts of the SER in a concussed monkey accurately predicted the side of a large subdural hematoma in one animal.

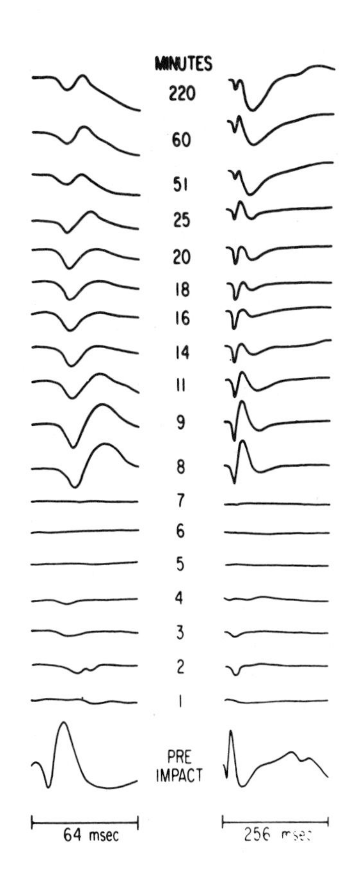

Fig. 3 - SER of concussed SR-10 with no lesion on contralateral medial electrode

Discussion

The data described above provide convincing support for the following statements concerning the mechanism of head injury.

First, measurement of the SER provides the investigator with an objective, reliable, and quantifiable index of the electrophysiologic response of the brain to trauma, which can be correlated with the behavioral abnormalities of traumatic unconsciousness (CC) as well as the postmortem data obtained by lesion analysis. The use of such techniques should obviate the difficulties occasioned in earlier experiments, wherein subjective interpretations of the animal's neurologic and systemic responses to head injury served as the basis for establishing the occurrence of CC and its severity (2).

Second, it has been shown that if head injury is produced in such a way that contact phenomena are eliminated, then the stresses produced by translational head motions alone at the levels examined are not capable of producing CC. In contrast, rotation of the head at equivalent levels of acceleration invariably causes

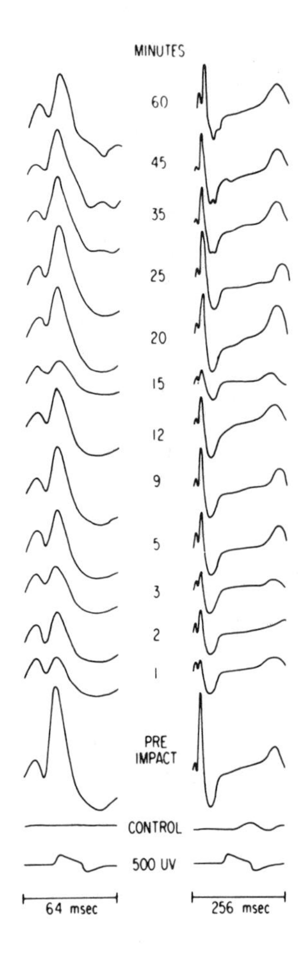

Fig. 4 - SER of SL-11. No concussion or lesion. Right median nerve stimulation on left lateral electrode

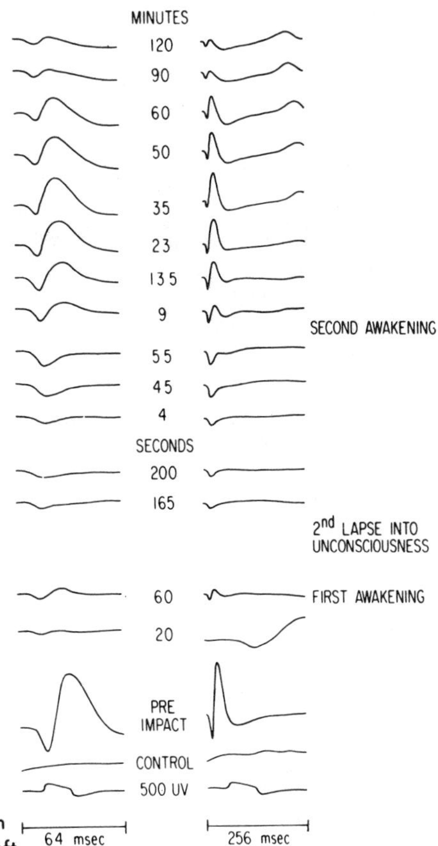

Fig. 5 - SER of SR-13. Concussion and brain lesion. Right median nerve stimulation on left lateral electrode at impact 1

CC as well as more extensive lesions in the brain. We believe that the rotational acceleration experienced by half of the test animals produced shear strains at the brain tissue level. This would be a result of the relative angular acceleration of the skull and brain, which are torsionally coupled. Because of the relatively low dynamic shear modulus of brain, the levels of inertial loading experienced by our test group resulted in gross tissue damage. The lesion distribution very interestingly indicated remarkable symmetry with respect to the saggital plane. Also noted was a predominance of lesions (intracerebral and subdural) at marked discontinuities. Hitherto theoretical models for head injury have been concerned primarily with the analysis of impact phenomena. Our data suggest that attention should also be paid to modeling the significance of inertial loading.

Finally, the pathologic data show the increased severity of damaging effects of inertial loading when the head is allowed to rotate. The shear stresses produced under these conditions appear to be widely distributed and do not indicate a primary concentration at the brainstem, as suggested by other workers. Indeed, the absence of significant vascular or cellular abnormalities in the brainstem in the presence of unequivocal cerebral concussion indicates that the pathology of cerebral concussion is far more complex than its simplistic attribution to "brain-stem damage." (2)

Reference

1. T. A. Gennarelli, A. K. Ommaya, and L. E. Thibault, "Comparison of Translational and Rotation Head Motions in Experimental Cerebral Concussion." Proceedings of Fifteenth Stapp Car Crash Conference, P-39, pp. 797–803. New York: Society of Automotive Engineers, Inc., 1972.
2. A. K. Ommaya, et al., "Head Injury in Chimpanzee, Parts 1 and 2." Jrl. of Neurosurgery, in press.

Appendix A

Representative wave forms of accelerations in rotated (SR) and translated (SL) animals are shown in Fig. A-1. Comparative accelerations at three levels of the two modes of head displacement are shown:

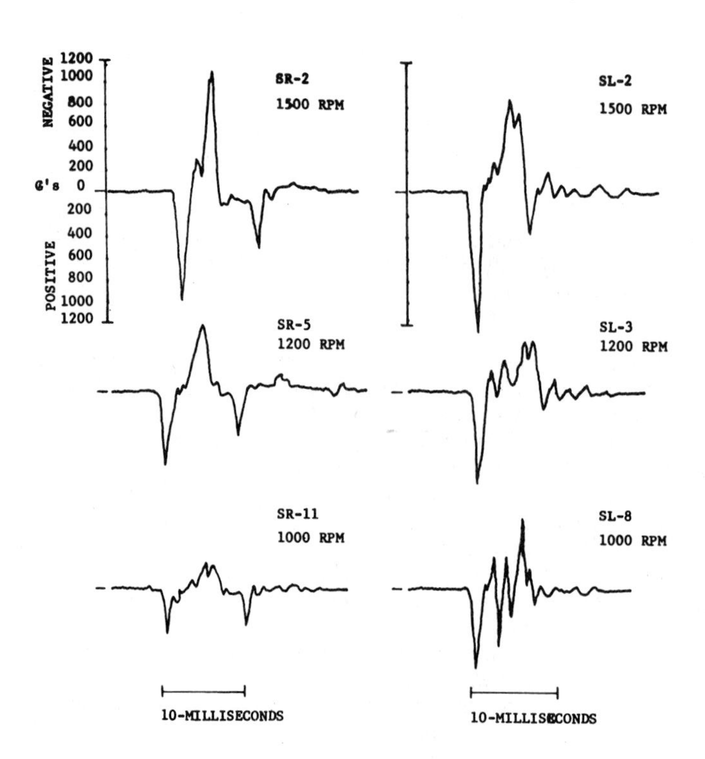

Fig. A-1 - Acceleration wave forms

1. SR-2 and SL-2 at 1500 rpm (1000–1200 G).
2. SR-5 and SL-3 at 1200 rpm (700–850 G).
3. SR-11 and SL-8 at 1000 rpm (450–750 G).

Appendix B

A listing of incidence of primary brain lesions for each animal in both rotated (SR) and translated (SL) groups is shown in Table B-1. Note that all RA became unconscious after the traumatic input, but all TA remained fully conscious in spite of the production of occasional primary brain lesions in some animals.

Table B-1 - Primary Brain Lesions in Rotated (SR) and Translated (SL) Animals—
+ Lesion Present, – Lesion Absent

	Subdural Hematomas	Subarachnoid Hemorrhages	Cortical Contusions	Intracerebral Hematomas	Intracerebral Petechial Hemorrhages	Brainstem Hemorrhages
SR-1	+	+	–	–	+	–
SR-2	+	+	+	+	+	–
SR-3	+	+	–	–	+	–
SR-4	+	+	–	–	–	–
SR-5	+	+	+	–	+	–
SR-6	+	+	–	–	+	–
SR-7, SR-8	Sacrificed 2 to 5 weeks postimpact					
SR-9	+	+	–	–	+	+
SR-10	+	+	+	–	+	+
SR-11	+	+	–	–	–	–
SR-12	+	+	–	–	+	–
SL-1	+	–	–	–	–	–
SL-2	+	–	+	–	–	–
SL-3	+	–	+	+	–	–
SL-4	–	–	–	+	–	–
SL-5	–	–	+	–	–	–
SL-6	+	–	–	–	–	–
SL-7	+	+	+	–	+	–
SL-8	–	–	–	–	–	–
SL-9	–	–	–	–	–	–
SL-10	–	–	–	–	–	–
SL-11	–	–	–	–	–	–
SL-12	–	–	–	–	–	–

770921

Head Impact Response

R. L. Stalnaker, J. W. Melvin, G. S. Nusholtz,
N. M. Alem and J. Benson
University of Michigan

RESEARCH IN THE BIOMECHANICS of head injury has had, for many years, the goal of developing an understanding of brain injury mechanisms. Progress has been slow in relating the mechanical factors associated with head kinematics during an impact to the resulting brain damage or dysfunction in closed head injuries. This is due both to the complexity of the human brain and the complexity of possible mechanical interactions that can take place during a general impact event. Many mechanical parameters have been suggested as the primary injury mechanisms and past research studies have frequently chosen to emphasize one or another of the possible factors while ignoring others.

In terms of mechanical inputs to the head, it has been common to talk of linear and angular accelerations of the head as a rigid body and, in the case of direct head impact, the impact force applied to the head. These variables, when determined in a complete manner, can be used to describe the mechanical response of the head. The response of the brain inside the skull due to the mechanical inputs is quite another matter, however. The creation of pressure waves, pressure gradients, shear strains and normal strains in tissues of the brain can all be the result of various combinations of the mechanical inputs. The effect on the brain tissues of these internal mechanical phenomena is what produces brain injury. Brain injury mechanisms are the internal mechanical phenomena that directly produce the damage to the brain tissues.

It is quite likely that brain tissues can be damaged by more than one injury mechanism and, similarly, a particular injury mechanism may be produced by different combinations of mechanical inputs to the head. In order to understand brain injury completely it will be necessary to have quantified all the complex interactions between brain injury mechanisms and mechanical inputs. This is a formidable task that requires careful step by step development of many methods, both experimental and analytical, necessary to describe these interactions.

Recently, techniques have been developed (1 - 6)* which allow reliable measurement of three-dimensional impact motions of the head using accelerometers. Application of the accelerometry technique of motion measurement eliminates the need of differentiating high-speed photographic displacement data to obtain velocities and accelerations, a method which is subject to inaccuracies. In parallel with the accelerometry developments, techniques for preparing unembalmed human cadavers for head injury research (7, 8) and high speed cineradiographic equipment (9) have been under development. This paper discusses the results of a series of head impacts conducted by the Biomechanics Department of HSRI using these tech-

*Numbers in parentheses designate References at end of paper.

ABSTRACT

A series of head impacts were conducted with 15 unembalmed cadavers. The purpose of the tests was to study the application of three-dimensional motion analysis using accelerometry, brain vascular system pressurization and high speed cineradiography to the understanding of head injury mechanics. The implementation of the techniques is described and their effectiveness is discussed.

The three-dimensional accelerometry technique using nine accelerometers was found to be applicable in direct head impacts. Analysis of the head acceleration data indicates the existence of brain motions which are independent of the motion of the skull. These motions were confirmed by the high speed cineradiographic films. Brain vascular system pressurization and time after death were found to play a role in determining the extent of the brain motions and the resulting brain injuries.

niques. The purpose of the work was to investigate the effectiveness of cadaver brain preparation techniques and to characterize the three-dimensional nature of direct head impacts of the type which have been used in past biomechanical studies of human head injury tolerance.

METHODS

MOTION DETERMINATION WITH ACCELEROMETERS - The HSRI method for measuring the three-dimensional motion of the head is based on a technique used by Calspan (10) to measure the general motion of a vehicle under a simulated crash. By using three triaxial accelerometers (9 readings) properly attached to the skull bones of the head, the complete three-dimensional motion of the head may be computed.

The full analytical details of the method are beyond the scope of this paper. However, for the sake of completeness, the mathematical principles behind this method are briefly outlined in this section.

With 9 acceleration readings, a set of 9 equations may be constructed in the form

$$F(\dot{\vec{\omega}}, \vec{\omega}, \vec{A}_o, \vec{a}_1, \vec{a}_2, \vec{a}_3) = 0$$

where

$\dot{\vec{\omega}}, \vec{\omega}$ are the angular acceleration and velocity vectors (3 unknowns),

$\vec{A}_o$ is acceleration vector of a reference point (3 unknowns), and

$\vec{a}_1, \vec{a}_2, \vec{a}_3$ are the known acceleration vectors of the triaxial accelerometers

Rather than select six independent equations for the six unknowns, it is possible to take advantage of the redundancy of the 9 equations. This is done by realizing that, because of experimental errors inevitably present in the accelerations measurements, the right hand side of the above equation is not exactly zero, but rather is a 9-element small error vector,

$$F(\dot{\vec{\omega}}, \vec{\omega}, \vec{A}_o, \vec{a}_1, \vec{a}_2, \vec{a}_3) = \vec{\varepsilon}.$$

Once this fact is established, the procedure is to search for the 6 unknowns which minimize the sum of squares of these errors. This is accomplished by forming the squares, adding them to form an expression

$$G(\dot{\vec{\omega}}, \vec{\omega}, \vec{A}_o, \vec{a}_1, \vec{a}_2, \vec{a}_3) = \sigma^2,$$

then differentiating this expression with respect to each of the 6 unknowns to produce the 6 equations to be solved. Thus, the partial derivatives of G with respect to the three components of $\dot{\vec{\omega}}$ form a set of 3 differential equations:

$$\frac{\partial G}{\partial \dot{\vec{\omega}}} = H(\dot{\vec{\omega}}, \vec{\omega}, \vec{a}_1, \vec{a}_2, \vec{a}_3) = 0.$$

The solution to this equation produces 3 angular accelerations and 3 angular velocities which are the "best" estimates, in the least square sense, of the true solution.

INSTRUMENTATION FOR 3-D MOTION MEASUREMENT - The cadaver was sanitarily prepared for testing by sealing off all body openings. The installation of the nine accelerometers on the head was accomplished in a systematic manner. An aluminum installation fixture was used to locate accelerometer mounting areas on the head which would be away from the intended impact site and provide the maximum separation between the triaxial assemblies. The use of this fixture is shown in Fig. 1. Where the three triaxial accelerometer units were attached, the scalp was removed from the head exposing the skull. The surface of the skull was cleaned and dried at these locations. Three metal screws were attached to the skull through small pilot drill holes (making sure the drill did not go into the brain). Stainless steel wire was wrapped around the screws and dental acrylic encased the screws and wires at each location. The accelerometer mounts were then placed in the dental acrylic and positioned by the aluminum jig, which insures the mutual orthogonality of the 3 accelerometer axes. The locations of the accelerometer mounts were measured on the aluminum jig to determine accelerometer axis dimensions and then the jig was removed when the acrylic had set, leaving the accelerometer mounts bonded to the skull in an orthogonal relationship at precisely known coordinates as shown in Fig. 2. Since this instrumentation frame does not, in general, coincide with the standard anatomical reference frame, the next step was to determine its exact location and orientation with respect to the anatomical frame.

A three-dimensional x-ray technique was developed which requires taking two orthogonal radiographs of the instrumented head. The procedure requires the identification of four anatomical landmarks (two superior edges of the auditori meati, and two infraorbital notches) with four distinguishable lead pellets; and the identification of each triaxial accelerometer center of mass, also with lead pellets carried by special aluminum blocks which fit the three accelerometer mounts. These seven lead pellets were then radiographed twice - once in the x-z plane and once in the y-z plane. On each of the two radiographs, the optical center and the laboratory vertical z-axis were simultaneously x-rayed, and the distances between the

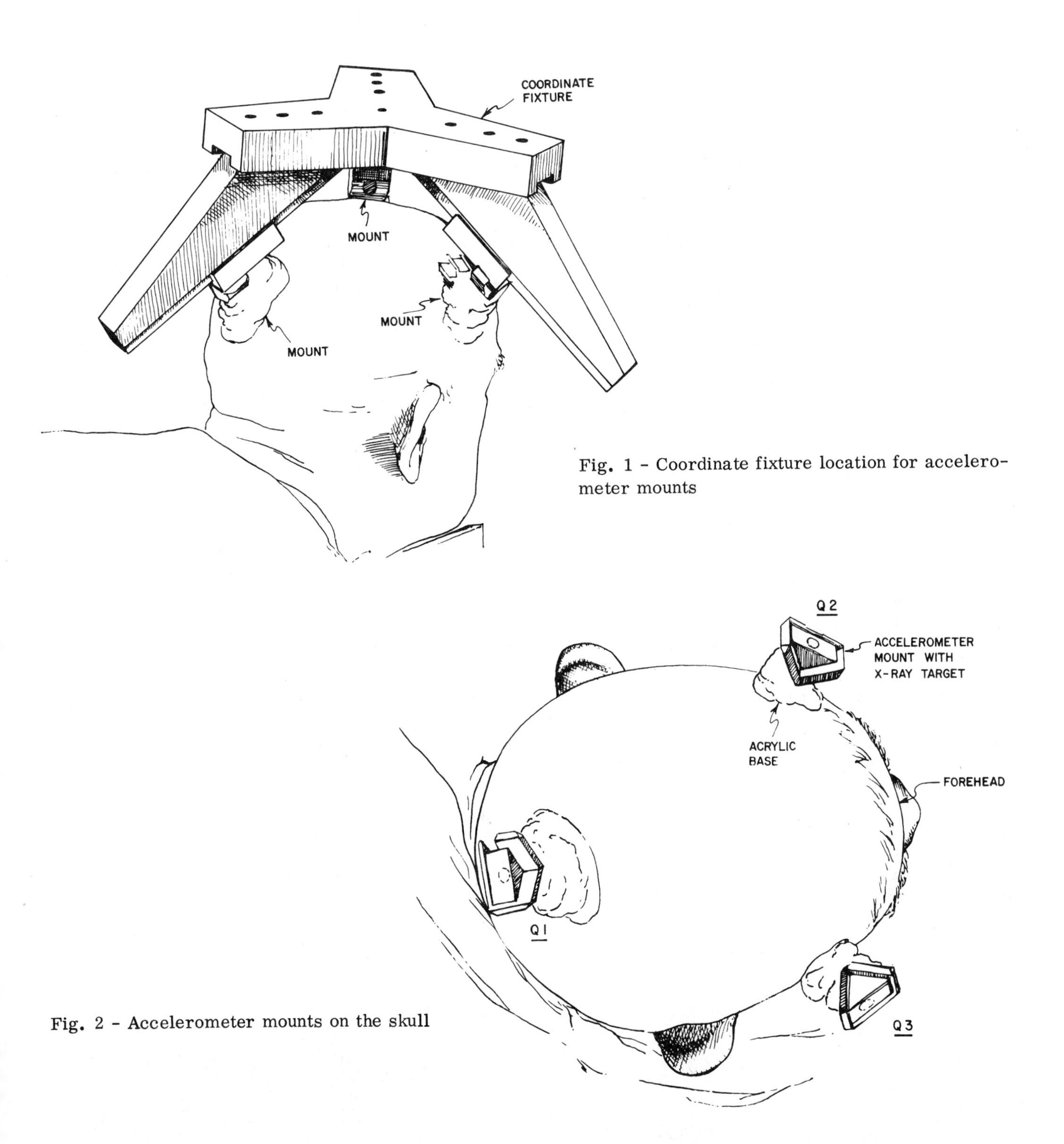

Fig. 1 - Coordinate fixture location for accelerometer mounts

Fig. 2 - Accelerometer mounts on the skull

x-ray film and each target were recorded for each view.

The computations which follow reconstruct the laboratory coordinates (x, y, z) of each of the seven pellets. The Frankfort plane was determined and the anatomical reference frame reconstructed from the four anatomical points. The instrumentation frame and its origin were determined from the three triaxial accelerometer centers. Finally, the transformation matrix between the instrumentation and the anatomical frame was obtained.

PRESSURIZATION - Brain vascular pressurization was not carried out in all tests. In the earlier tests pressurization was tried with varying degrees of success. The following method of pressurization was developed in the latter portions of this study and good vascular pressurization was found to have been produced. The technique is based on modifications of the method developed by Fayon, et al., of the Peugeot-Renault Association, Paris, France. The technique involves catheter construction, and surgical installation.

Catheter construction starts by taking a Foley balloon catheter, and replacing the center section with suitable lengths of polyethylene tubing. Two tubes were necessary, one for pressurizing fluid and one for balloon inflation. The two polyethylene tubes were connected to the inlet and tip ends of the catheter. A small lead pellet was inserted in the end of the exit port of the tip and the port was then sealed with epoxy. This was done so that the precise location of the catheter would be known by x-ray during insertion. Holes were then made in the end of the pressurization tube but before the occluding balloon. This would allow the fluid to flow into the occluded blood vessels. The catheter system is shown in Fig. 3.

In order to measure pressure in the internal carotid artery as close to the brain as possible, the balloon catheter was inserted through the carotid artery and down into the descending aorta just past the branch of the left subclavian artery to provide for pressurization through both of the left and right carotid arteries. One end of a second section of polyethylene tubing (fill tube) was inserted along the outside edge of the balloon catheter and into the ascending aorta while the other end was inserted into the internal carotid artery above the incision. A third tube was inserted into the internal carotid artery above the incision to allow the miniature pressure transducer to be placed up the internal carotid artery and located near the base of the brain.

Once all the tubes have been placed in the appropriate arteries, the carotid incision was sealed with quick setting plastic to prevent fluid leakage. The pressurization fluid consisted of water and black india ink. The purpose of the ink was to mark ruptures due to impact.

The cadaver was then fitted with a vinyl suit, gloves, socks, and face hood as final preparation before testing.

IMPACT TESTING - The cadaver was transported to the impact laboratory and was seated in an upright position in an upright position in an adjustable chair. The head and torso were stabilized by fastening waxed cord to each ear and attaching the cords (which broke easily at impact) to an overhead support structure. The cadaver was then positioned in front of the HSRI pneumatic cannon with the head impact surface oriented approximately parallel to the cannon impacting surface, and the centerline of the impact approximately through the head center of gravity. The nine accelerometers used in this study were Endevco Model 2264-2000, with a Kulite Model MCP 055-5F (006) pressure transducer for measuring intracranial vascular pressure. The accelerometers were attached to the accelerometer mounts on the skull. The pressurization lines and pressure transducers were connected.

The final step before impact was to pressurize the cadaver. A 5 L fluid-filled container was connected to the cadaver's pressurization tube and to an air supply. When air pressure was applied to the fluid container, the fluid was forced into the cadaver. Prior to pressurization, the catheter balloon was inflated to block off the descending aorta. The air pressure was increased until the vascular pressure of 120 mm Hg was indicated and was steady. The fluid could be seen flowing through the fill tube in the neck when good pressurization was achieved.

The impacts were conducted using a pneumatically operated testing machine specifically constructed for impact studies. The machine consists of an air reservoir and a ground and honed cylinder with two carefully fitted pistons. The transfer piston which was propelled by compressed air through the cylinder from the air reservoir chamber, transfers its momentum to the impact piston. A striker

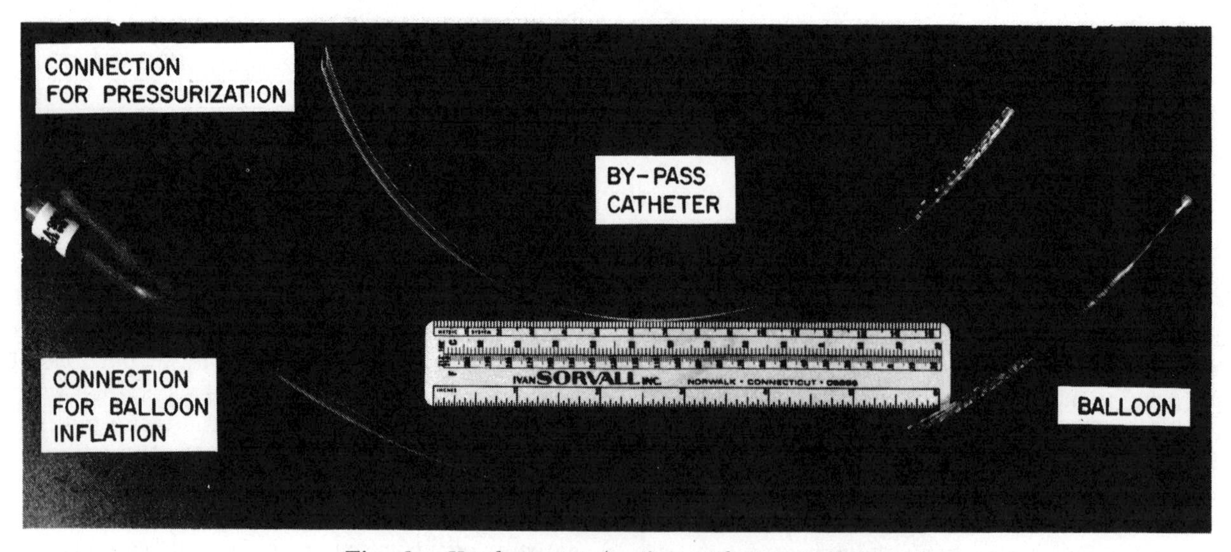

Fig. 3 - Head pressurization catheter system

surface with an inertia-compensated load cell is attached to the impact piston. This piston is allowed to travel up to 25.4 cm and then its motion is arrested by an inversion tube which absorbs the remaining kinetic energy of the piston. The desired impactor stroke can be controlled by the initial positioning of the impact piston with respect to the inversion tube. The impactor velocity is controlled by reservoir pressure and the ratio of the masses of the transfer and impact pistons. The load cell is a Kistler 904A piezoelectric load washer with a Kistler 805A piezoelectric accelerometer mounted internally for inertially compensating the load cell for the striker mass between the load cell and the impact surface. The impact-piston and load cell striker assembly for all these tests had a mass of 10 kg. The striker used was 15.2 cm in diameter.

GROSS AUTOPSY AND INJURY SCALING - The gross autopsies were conducted in the autopsy laboratory, specially equipped for dissection. Careful anatomical dissection of the head and brain was performed to allow discrete identification of all injuries. When gross trauma was found, it was photographically recorded, using a specially modified Pentax camera with a closeup lens, to provide a permanent record of the injury.

The injuries were evaluated using the Abbreviated Injury Scale (AIS) (11). Each individual injury was assigned an AIS value and the overall AIS injury was assigned by summing the cubes of the individual AIS numbers and then taking the cube root (12). Only injuries directly related to the impact are reported on here. Any injury related to instrumentation impact are reported on here. Any injury related to instrumentation installation was ignored, where possible, in injury assessments.

DATA HANDLING - The impact force and acceleration, velocity, pressure, and the nine head accelerations were all recorded unfiltered on a Honeywell 7600 FM Tape Recorder. The analog data on the FM tape was played back for digitizing through proper pre-digitizing analog filters. The A-to-D process resulted in a digital signal sampled at a 6400 Hz equivalent sampling rate. Anti-alias digital filters were then used before reducing the sampling rate to 3200 Hz. Following this step, the signals were digitally filtered and plotted in a strip-chart type plot to determine the usefulness of the data before being used in the three-dimensional motion analysis computer program.

RESULTS

A series of 15 head impact tests were conducted - five side, five front, and five rear. Most of the impacts were conducted with a 2.5 cm Ensolite padded striker, but four were conducted with a rigid impacting surface. Though pressurization was attempted in 11 tests, only four were considered to be good enough for analysis. Leakage and direct impact of tissue on the pressure transducer were the main causes of failures. A summary of the test conditions is shown in Table 1.

The anthropometric measurements and medical history for each cadaver are given in Table 2. The lengths and breadths were maximum inside dimensions of the skull. The average thickness of the skull was determined from measurements of the front, sides, and back. Skull circumference was

TABLE 1. TEST CONDITIONS

TEST NO.	IMPACTOR TYPE	DIRECTION OF IMPACT	PRESSURIZATION	NO. OF DAYS AFTER DEATH BEFORE TESTING
75A113	2.54 cm Ensolite	Left Side	Poor	3
75A116	2.54 cm Ensolite	Left Side	Poor	4
76A126	2.54 cm Ensolite	Frontal	Poor	3
76A133	2.54 cm Ensolite	Frontal	Not done	5
76A134	2.54 cm Ensolite	Left Side	Not done	1
76A135	2.54 cm Ensolite	Rearward	Not done	1
76A136	2.54 cm Ensolite	Frontal	Poor	5
76A137	2.54 cm Ensolite	Rearward	Poor	8
76A144	Rigid	Frontal	Poor	11
76A145	Rigid	Left Side	Good	6
76A152	Rigid	Left Side	Good	9
76A159	2.54 cm Ensolite	Rearward	Not done	6
76A167	2.54 cm Ensolite	Rearward	Good	7
76A169	Rigid	Rearward	Poor	8
76A171	2.54 cm Ensolite	Frontal	Good	12

TABLE 2 CADAVER SPECIFICATION

TEST NO.	SEX	AGE (yr)	HEIGHT (cm)	WEIGHT (kg)	SKULL DIMENSIONS MAX. LENGTH (cm)	MAX. BREADTH (cm)	AVE. THICKNESS (cm)	MAX. CIRCUMFERENCE (cm)	CAUSE OF DEATH
75A113	M	54	179	83.0	18.0	15.5	0.60	53.3	Pneumonia
75A116	F	66	152	44.0	15.7	14.0	0.50	46.4	Myocardial Infarction
76A126	M	65	180	81.2	18.8	15.2	0.57	57.2	Respiratory Failure
76A133	M	54	175	81.6	18.5	14.7	0.57	55.2	Strangulation
76A134	F	72	170	47.2	18.0	14.7	0.58	53.3	Pneumonia
76A135	F	58	160	63.5	18.0	13.5	0.77	55.2	Pneumonia
76A136	M	88	170	76.2	19.1	15.0	0.56	57.8	Septicemia
76A137	F	89	152	21.8	17.3	14.5	0.43	52.1	Myocardial Degeneration
76A144	M	45	168	75.3	18.5	14.0	0.60	56.3	Embolism
76A145	F	78	160	80.3	17.5	14.5	0.53	54.6	Lung Cancer
76A152	F	66	160	-	17.0	13.7	0.64	31.4	Acute Cardiac Arrest
76A159	M	41	177	49.0	18.8	14.7	0.64	61.5	Acute Pulmonary Edema
76A167	F	55	161	43.1	18.3	14.5	0.39	52.6	Renal Failure
76A169	F	75	169	76.2	18.5	14.5	0.65	54.6	Ventricular Fibrillation
76A171	M	57	168	75.7	17.8	14.2	0.52	-	Lung Cancer

taken on the outside of the skull at the level that the length and breadth measurements were taken.

The data obtained from the 3-D head motion analysis consisted of translational and angular displacements, velocities and accelerations. Because of the complexity of this data, it was summarized by computing the resultants of head accelerations and velocities. With better understanding of the resultants, future studies will be made on their components. Four examples of the summaries of the nine accelerometer 3-D analysis are given in Figs. 4 - 7. The complete data set is given in Reference 13.

The significant portion of the impact force, as well as the resultant translational and angular acceleration pulses, occur during the first few milliseconds of impact. The waveform of this portion is essentially triangular with a rise, a peak and a fall. In order to define the pulse duration, a standard procedure was adopted which determines the beginning and end of the pulse, along with the onset rate of its rise portion.

The procedure is to determine first the peak and the time at which it occurs. Next, the left half of the pulse, defined from the point where the pulse starts to rise to the time of peak, is least-squares fitted with a straight line. This rise line intersects the time-axis at a point which is taken as the formal beginning of the pulse, while the slope of this line is taken as the onset rate of the pulse. A similar procedure is followed for the right half of the pulse, i.e., a least squares straight line is fitted to the fall section of the pulse which is defined from the peak to the point where the first pulse minimum occurs. The formal end of the pulse is defined then as the point where the fall line intersects the time axis.

Summary of the results is given in Table 3, which includes the peak values for the pressure, force and accelerations. The velocity, both angular and translational were determined by reading their magnitudes off the summary sheets (e.g., Figs. 4 - 7) at the end of the force pulse duration. The force pulse duration and the rate of onset of the acceleration were obtained as described above.

The injury scaling is presented in Table 4. Any injuries due to instrumentation were not considered in the injury rating. One test subject (76A171) was found to have a benign brain tumor in the left occipital lobe. Injuries in the vicinity of this tumor were ignored. In general, black ink was found throughout the brain where good pressurization had been accomplished, and injuries were easy to observe.

DISCUSSION

A complete detailed analysis of the data generated from the 3-D accelerometer system is beyond the scope of this paper. Nevertheless, the usefulness of the method can be illustrated, and insight into the design of future experiments can be gained by conducting a preliminary analysis of the data.

FREQUENCY ANALYSIS - One of the first points of interest in analyzing this data was to determine the various frequency components of the signals and their origins. The highest frequencies, observed in the raw analog signals were around 30 ±2 kHz. These frequencies were attributed to resonances in the accelerometers themselves.

The next significant frequencies in the raw signals were noted at approximately 10 ±3 kHz, and occurred only in rigid impacts. The magnitudes of these frequencies were highest on the accelerometers mounted closest to the impact site,

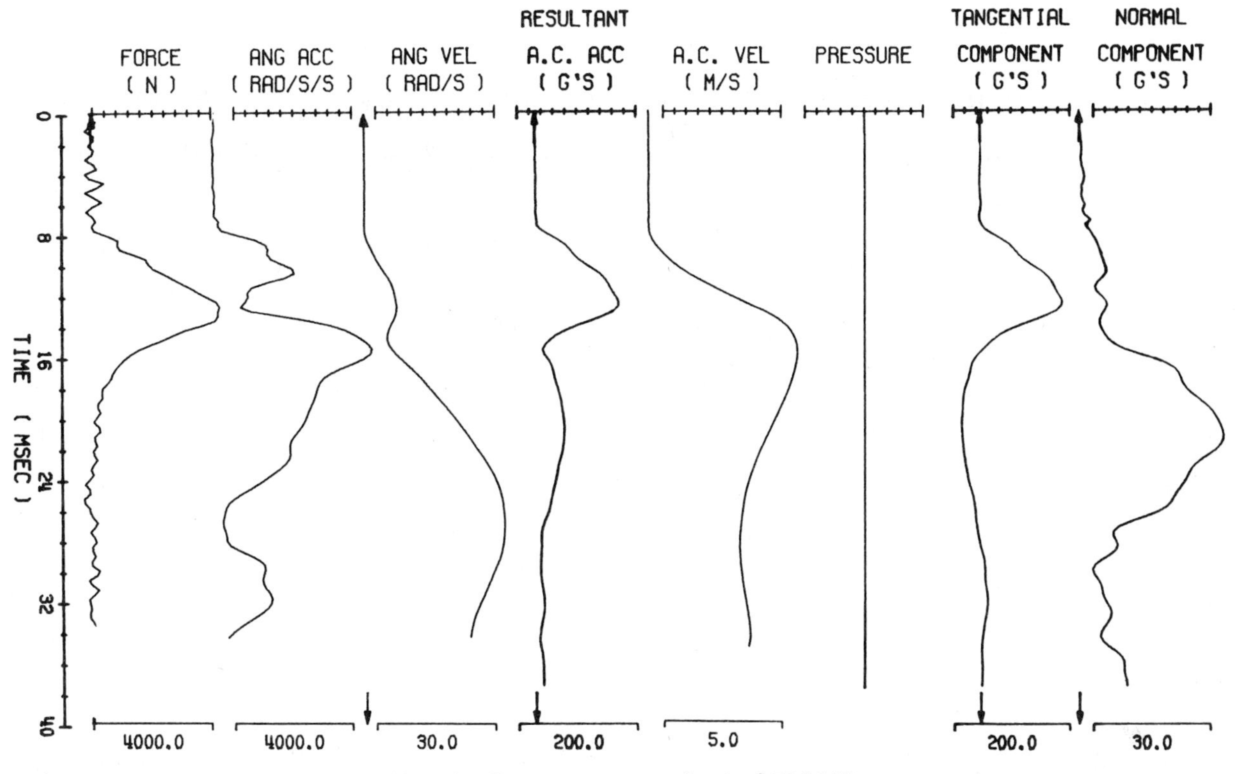

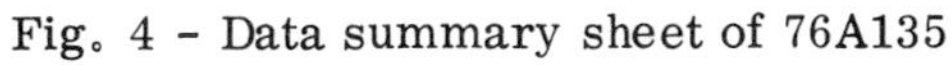

Fig. 4 - Data summary sheet of 76A135

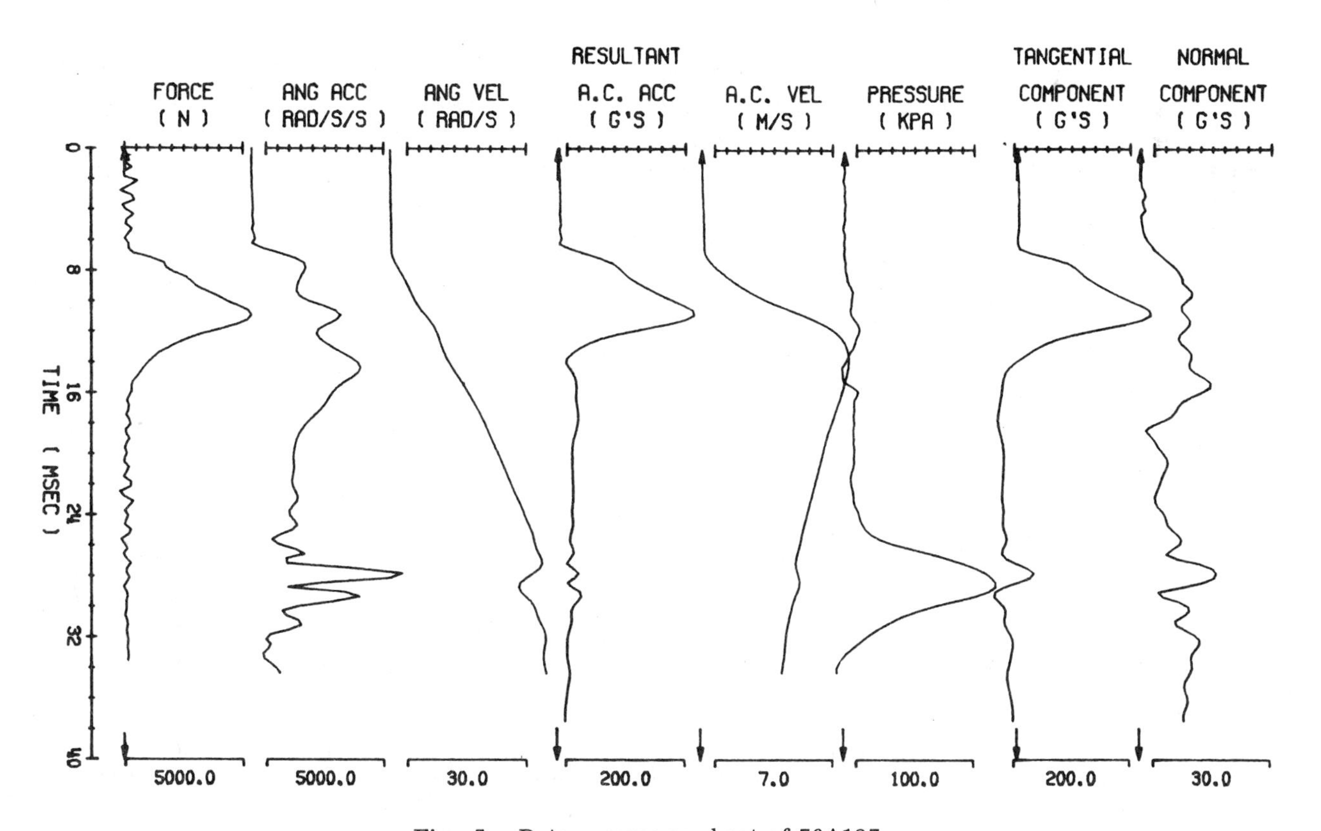

Fig. 5 - Data summary sheet of 76A137

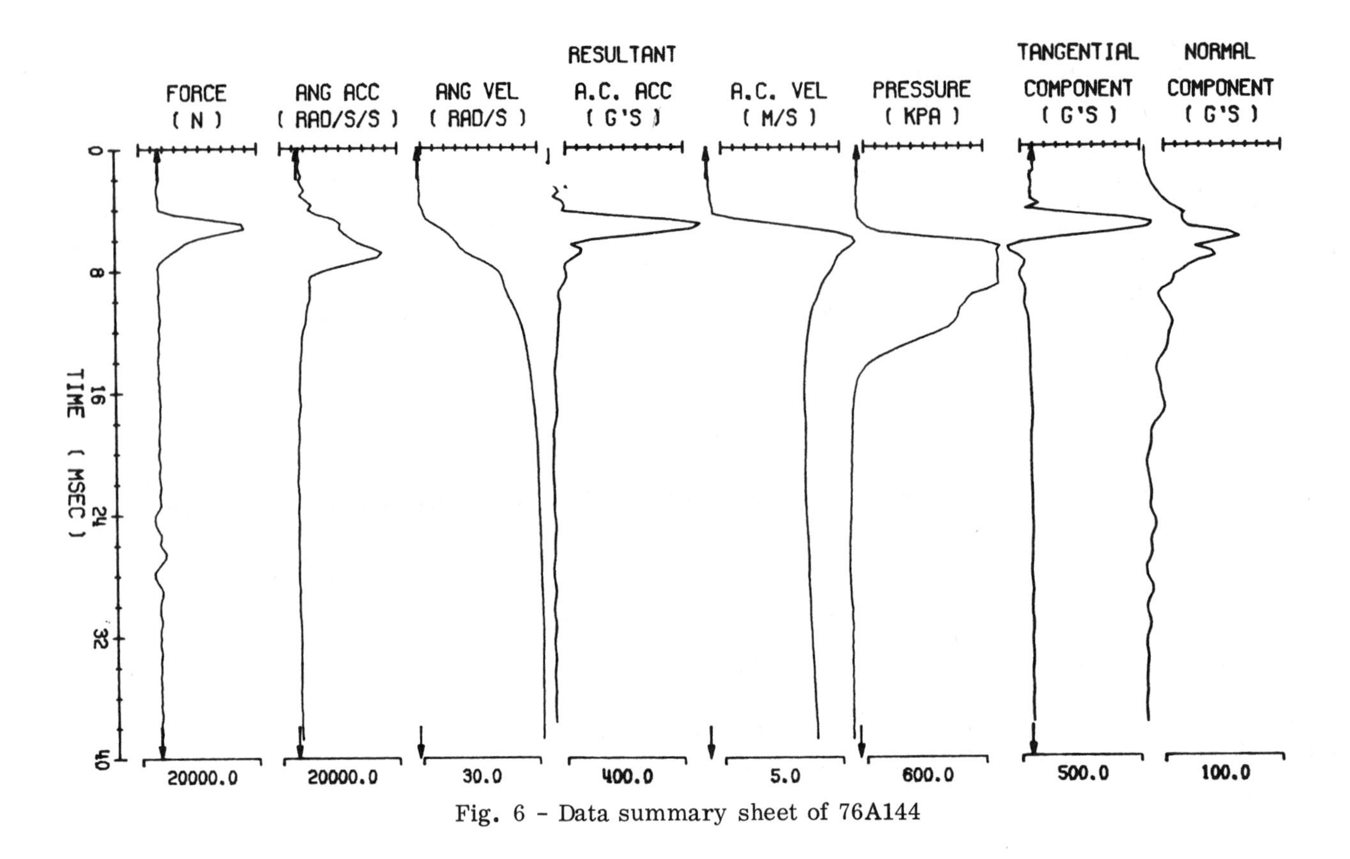

Fig. 6 - Data summary sheet of 76A144

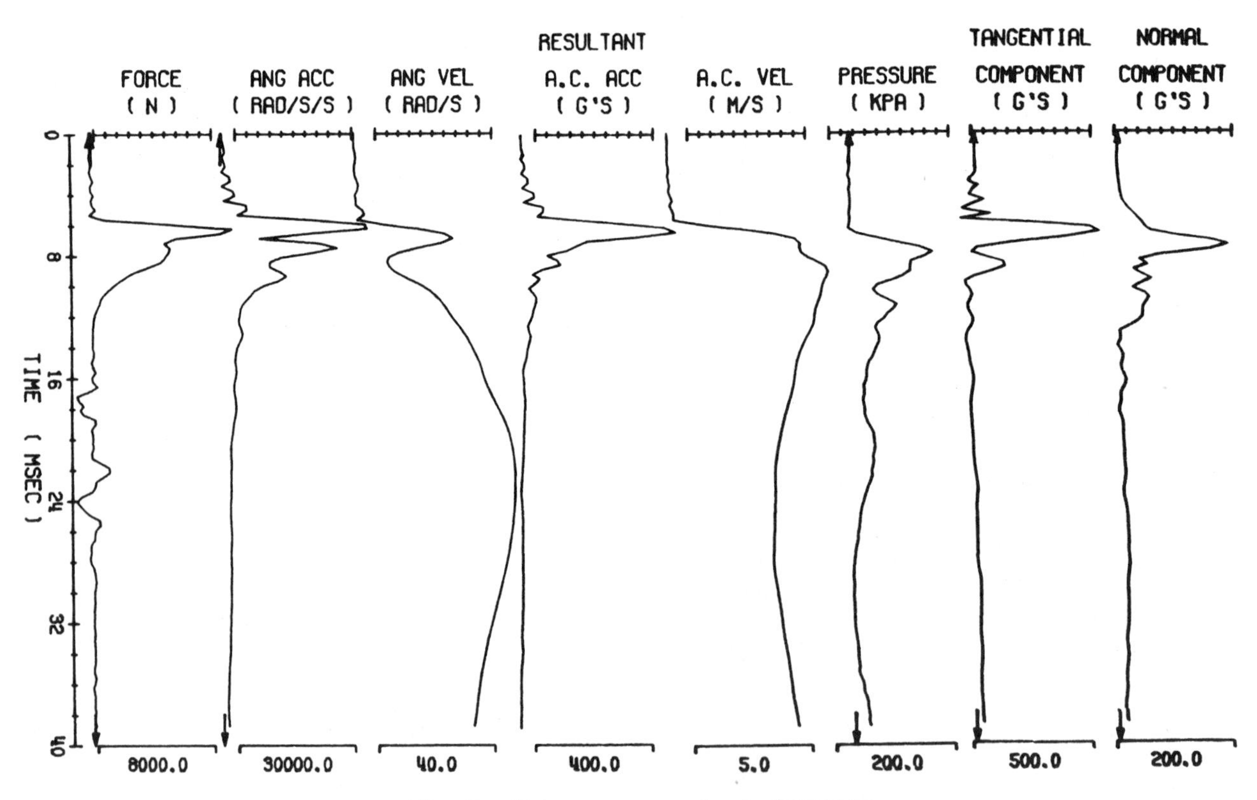

Fig. 7 - Data summary sheet of 76A145

TABLE 3. TEST SUMMARY

TEST NO.	PEAK FORCE (kN)	FORCE DURATION (msec)	IMPULSE (N·sec)	PEAK RESUL. LINEAR VEL. (m/sec)	PEAK RESUL. LINEAR ACC. (g's/m/sec²)	RESUL. LIN. ACC. RATE OF ONSET (k g/sec)	PEAK RESUL. ANG. VEL. (rad/sec)	PEAK RESUL. ANG. ACC. (k rad/sec²)	PEAK PRESSURE kPa x 10^{-2}	AIS NO
75A113	4.82	9.7	24.7	5.97	124.8/1251	31	21.2	5.00	--	4
75A116	4.65	9.1	21.4	7.08	178.8/1784	28	36.5	8.42	--	0
76A126	6.64	9.4	27.7	6.07	190.9/1900	34	15.0	9.57	--	4
76A133	7.70	8.4	24.1	5.92	209.1/2060	30	22.0	6.65	--	3
76A134	4.21	10.6	20.7	6.20	137.6/1371	30	32.6	6.65	--	2
76A135	4.38	10.9	23.2	6.56	141.4/1391	27	19.1	5.45	--	0
76A136	6.83	9.4	28.7	6.10	162.0/1588	30	23.3	7.89	--	3
76A137	5.36	9.1	22.5	8.83	229.9/2276	48	18.5	6.50	--	4
76A144	14.6	3.75	21.5	6.26	515.3/4816	516	19.8	14.62	--	2
76A145	9.59	6.87	22.3	6.82	532.0/5188	516	33.7	37.55	140	5
76A152	7.15	6.87	19.5	7.17	262.3/2538	266	38.1	22.81	70	5
76A159	13.0	9.4	38.6	6.30	237.3/2348	38	43.6	10.24	--	4
76A167	5.39	10.3	22.6	7.96	207.5/2076	36	38.4	8.93	29	0
76A169	9.61	2.19	11.7	6.35	359.8/3489	290	12.3	13.62	--	4
76A171	7.56	8.4	26.9	5.50	161.26/1620	31	26.8	8.62	52	4

TABLE 4. INJURY SCALING

Test No.		AIS
75A113	1. No external damage	0
	2. Hemorrhaging in the arachnoid-left frontal lobe	4
	3. Highly localized hemorrhages on left and right mid-parietal lobes	3
	OVERALL	4
76A116	1. No external damage	0
	2. No visible brain injury	0
	OVERALL	0
76A126	1. Contusion on forehead 5 cm long	1
	2. Highly localized subarachnoid hemorrhage - right front temporal lobe	3
	3. Highly localized hemorrhage - right frontal lobe	3
	4. Highly localized subarachnoid hemorrhage - mid-right parietal lobe	3
	OVERALL	4
76A133	1. No external damage	0
	2. Focussed hemorrhage - right occipital lobe (contre-coup)	3
	OVERALL	3
76A134	1. No external damage	0
	2. Diffuse subarachnoid hemorrhage - left, mid-right parietal lobe	2
	OVERALL	2
76A135	1. No external damage	0
	2. No visible brain injuries	0
	OVERALL	0
76A136	1. No external damage	0
	2. Epidural hemorrhage - midsagittal plane	2
	3. Highly localized hemorrhage - mid-left parietal lobe	3
	OVERALL	3
76A137	1. No external damage	0
	2. Subarachnoid hemorrhage - right and left posterior parietal lobes	3
	3. Highly localized hemorrhage - right temporal, parietal and cerebellum	4
	OVERALL	4

TABLE 4. INJURY SCALING (Continued)

Test No.		AIS
76A144	1. Deep forehead laceration	2
	2. No visible brain injury	0
	OVERALL	2
76A145	1. Blood from external acoustic meatus, squamosa fracture	2
	2. Comminuted fracture of temporal bone into ear canal	3
	3. Highly localized hemorrhage - right frontal lobe	3
	4. Diffuse subdural hemorrhage - left anterior tip of temporal lobe	4
	5. Highly localized hemorrhage - cerebellum	4
	OVERALL	5
76A152	1. Blood from external acoustic meatus, squamosa fracture	2
	2. Compound fracture of temporal bone	3
	3. Simple fracture of left side of occipital bone	3
	4. Diffuse subarachnoid hemorrhage over entire brain	3
	OVERALL	5
76A159	1. No external damage	0
	2. Highly localized hemorrhage - right and left frontal lobes	4
	3. Diffuse subarachnoid hemorrhage over entire brain	3
	OVERALL	4
76A167	1. No external damage	0
	2. No visible brain injury	0
	OVERALL	0
76A169	1. No external damage	0
	2. Simple fracture of occipital bone	3
	3. Highly localized hemorrhage - right frontal lobe	3
	4. Diffuse subarachnoid hemorrhage over right side of brain	3
	OVERALL	4
76A171	1. Bruise on forehead	1
	2. Subdural hemorrhage - left frontal lobe	4
	3. Subdural hemorrhage - left or occipital lobe brain tumor just under the lesion	-
	OVERALL	4

and decreased in magnitudes the farther away they were mounted. Thus, these frequencies may be attributed to vibrations of the skull structure.

The next major frequency range appearing on the raw data was 1300 ±300 Hz and appeared in some accelerometer clusters but not in others. This vibration was attributed to resonances in the accelerometer mounting systems.

The last frequency of interest occurred around 300-600 Hz and had a very small amplitude, and was attributed to instrumentation noise. This frequency noise was found on some channels and not on others, and in all cases occurred before and after the test.

Because the above frequencies were not attributable to the rigid body motion of the head, a filtering rate of 450 Hz was selected to obtain the data to be analyzed in this paper. A careful comparison of the data after filtering with the unfiltered data showed no difference in the basic wave form except for the removal of most of the higher frequency noise mentioned above.

BRAIN MOTION - The acceleration of a rigid body should follow the force in time varying only by a constant, i.e., varying by only the mass of the system. When this does not occur, pure rigid body motion has not been observed. Upon close examination of the translational acceleration and the impact force in the tests conducted for this study, certain variations in this relationship were seen. The rate of onset and decay as well as major peaks did not coincide precisely with each other. The overall basic shapes were similar but not identical as one would expect. In order to gain an understanding of this phenomenon, the head mass for each test subject was estimated by dividing the peak translational acceleration into the peak impact force. This gave effective head masses ranging from 1.9 kg - 5.6 kg with an average of 3.3 kg. The correlation between the calculated skull mass and the measured skull volume obtained from anthropometric measurements of the head was found to be 0.8, much lower than one would expect from these types of measurements.

A possible explanation of this phenomenon might be a fluctuation in the acceleration due to curvature in the motion path of the head. This component was calculated and used to adjust the acceleration around the peak for padded impacts only. This changed the head mass calculation between 1 and 5%, which was not considered significant.

Assuming that the brain was not completely coupled to the skull during impact, variations in the force acceleration traces and in the calculated mass would occur. Should this occur, then one might suspect that the brain would be fairly well coupled to the skull toward the end of impact. The brain mass could then be computed by dividing the impulse of the force by the terminal velocity. With this method, the head mass ranged from 3.4 - 5.7 kg averaging 4.16 kg and correlation with head volume was 0.92. This would indicate that the brain motion lags the skull's during impact by varying amounts. This decoupling of the brain from the skull was observed during a series of impact tests in which the HSRI High Speed Cineradiography system (7) was employed. In the first test, holes were drilled in the skull to allow for lead targets to be placed in the brain. In this experiment, brain-skull decoupling was first observed. The next test was done the same way, but care was taken to seal the holes used for insertion of the lead targets so as not to allow air to enter during impact. In this experiment, brain-skull decoupling was still observed but to a lesser extent. Finally, another test was conducted in which no preparation of the skull was done but pressurization was used. In this test, brain-skull decoupling was again observed as indicated in Fig. 8.

For a more detailed look at this phenomenon it was necessary to break up the translational acceleration into its normal and tangential components. This procedure was applied for example to test 76A135 to study the brain-skull decoupling. At the onset of impact, the effective mass, calculated by dividing the impact force by the translational acceleration, was approximately 2 kg, indicating that the mass being accelerated is primarily that of the skull. After about three-quarters of a millisecond, the brain starts to load the skull and impactor; at this point, the head effective mass becomes a function of time. At approximately peak acceleration, i.e., about 1 ms before peak force, the apparent mass of the head is about 3 kg, and by the time peak force occurs, this mass has increased to approximately 5 kg. Because the force lags the acceleration, the head mass-system consists of at least two masses, i.e., the mass of the skull coupled with the mass of the brain as a function of time. It is believed that the brain is rebounding off the back of the skull, producing an increased force and a decrease in acceleration, as indicated in Fig. 4. This response, seen in the tangential and normal accelerations is therefore an indicator of the brain elastic modulus. Furthermore, it is clear now that the assumption made earlier, that the brain is fairly well coupled to the skull toward the end of impact, is incorrect.

If time-after-death has any effect on the apparent stiffness of the brain, it would show up as a variation in the post-impact tangential acceleration. This portion of the curve was examined for one PMD (Post Mortem Days) test 76A135, Fig. 4, and eight PMD test 76A137, Fig. 5. Several differences were observed. The first difference is in the magnitudes of the impact forces when the tangential accelerations were zero - 2000 N in test 76A135, and considerably lower 1150 N in test 76A137. The second difference is in the oscillation of the curves: while the acceleration curve in test

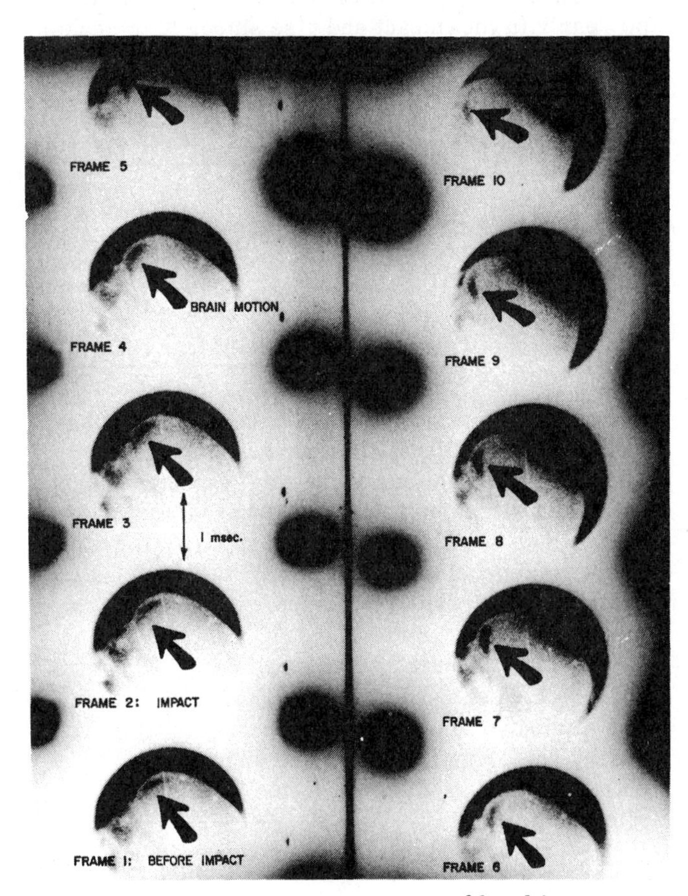

Fig. 8 - X-ray movie sequence of head impact

76A135 oscillates, that of test 76A137 does not. The third observed difference is the settling time of the two acceleration curves - test 76A135 acceleration settles (returns to zero) much faster than that of test 76A137. From these observations, one may conclude that the brain in test 76A135 (1 PMD) is much stiffer than the brain in test 76A137 (8 PMD). In the padded side impacts reported here, there was only a slight difference in brain response in tests conducted up to four days after death.

BRAIN PRESSURIZATION - The effects of pressurization can also be determined by studying the brain motion during impact. Because of the many variables, e.g., time-after-death, padded, unpadded, front-to-back impacts, etc., a valid comparison cannot be made, but trends were observed in several cases where variations could be minimized.

In a direct head impact, one would expect that a brain which had been pressurized to be coupled to the skull more effectively than a non-pressurized brain. If this is true, then a plot of effective head mass (impact force divided by resultant acceleration) versus time would indicate the amount and rate at which the brain loads the impactor and the skull. This parameter is shown in Figs. 9-12, along with impact force. As suspected, tests 75A167 (rear impact) and 76A171 (front), both pressurized, have an effective head mass versus time curves which start early in the impact and rise smoothly to peaks of 5 kg and 8 kg, respectively. In tests 76A137 (rear impact) and 76A126 (front), both unpressurized, the effective head mass versus time curves start to rise late in the impact and reach peaks of 8.5 kg and 16 kg, respectively. From this limited number of samples, brain pressurization does improve the coupling between the brain and the skull, allowing thus for a more realistic test and better injury evaluation.

SKULL FRACTURE-Most of the padded impacts resulted in essentially translational motions; that is, angular accelerations and velocities did not come into play until the head loaded the neck causing it (the head) to rotate. This was observed from both regular and x-ray high-speed photography. This was not, however, the case in rigid head impacts, illustrated in Figs. 6 and 7.

In these rigid impacts, the skull is loaded very rapidly, and in the case of skull fracture (depression under the impactor) as in Fig. 7, the force drops, during which the tangential acceleration drops to zero, indicating that the large unfractured portion of the skull is not in complete contact with the impactor. This is accompanied by a short-lived rotation of the skull until it comes completely in contact with the impactor, evidenced by a rapid increase of the angular velocity. Once the unfractured portion of the skull is back in complete contact with the impacting surface, the tangential acceleration increases and the rotation of the skull reverses its direction.

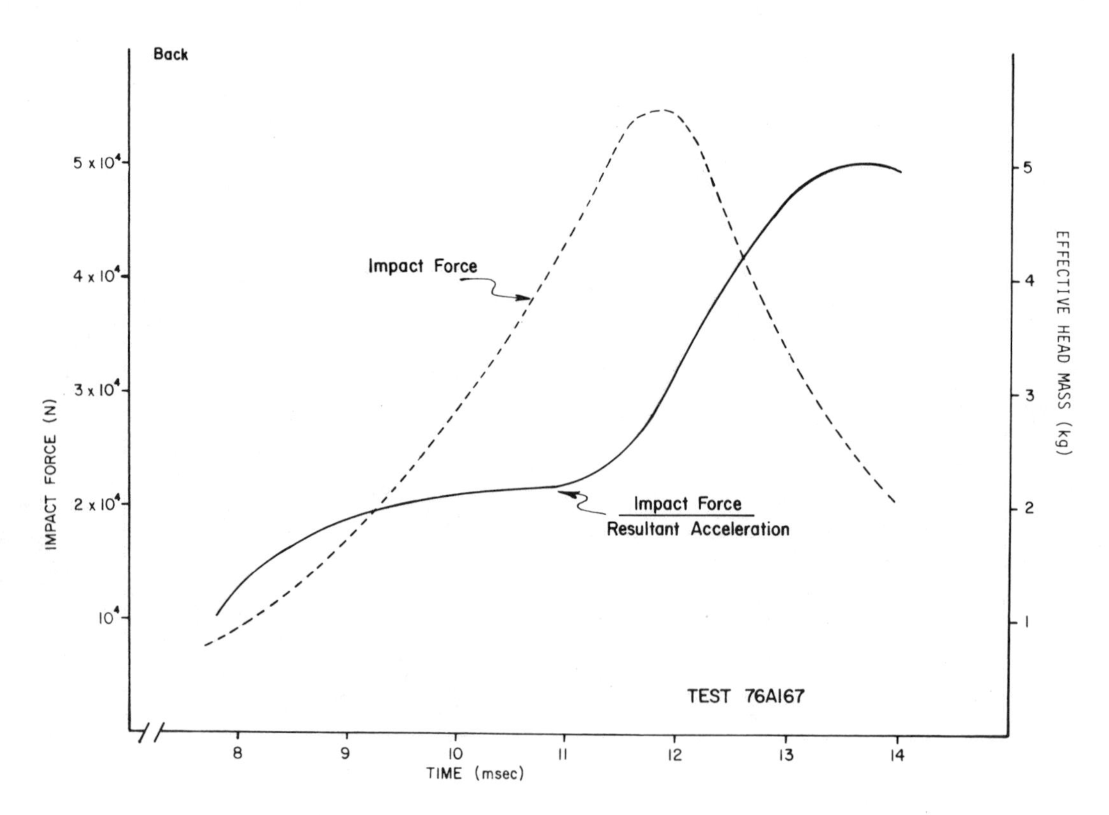

Fig. 9 - Force and effective head mass versus time

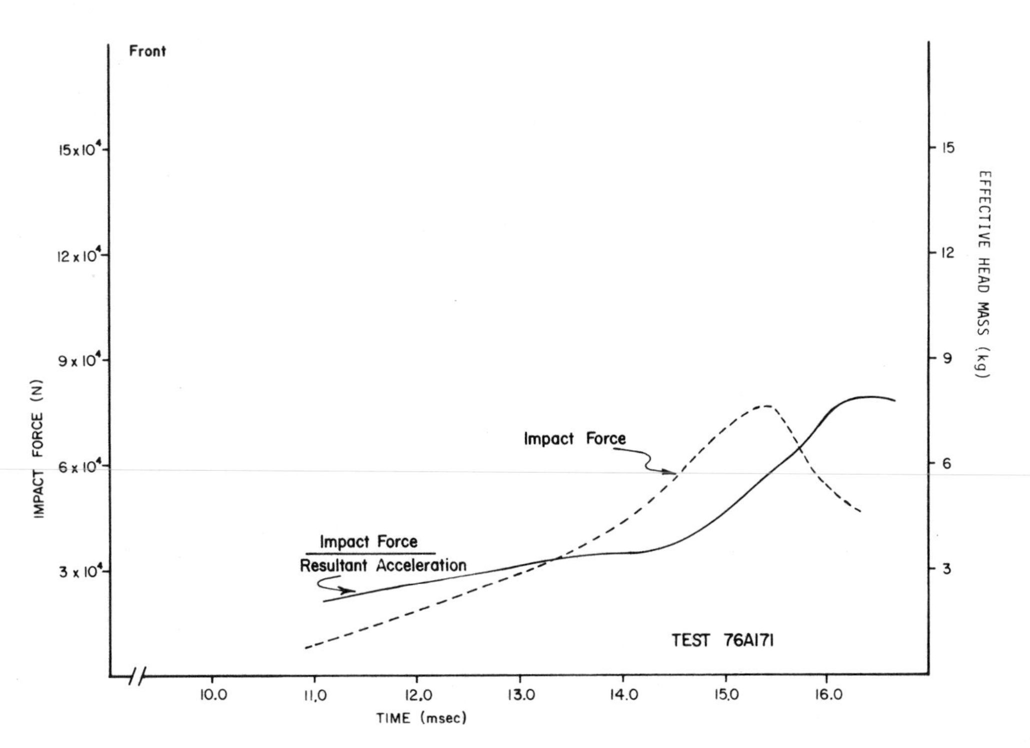

Fig. 10 - Force and effective head mass versus time

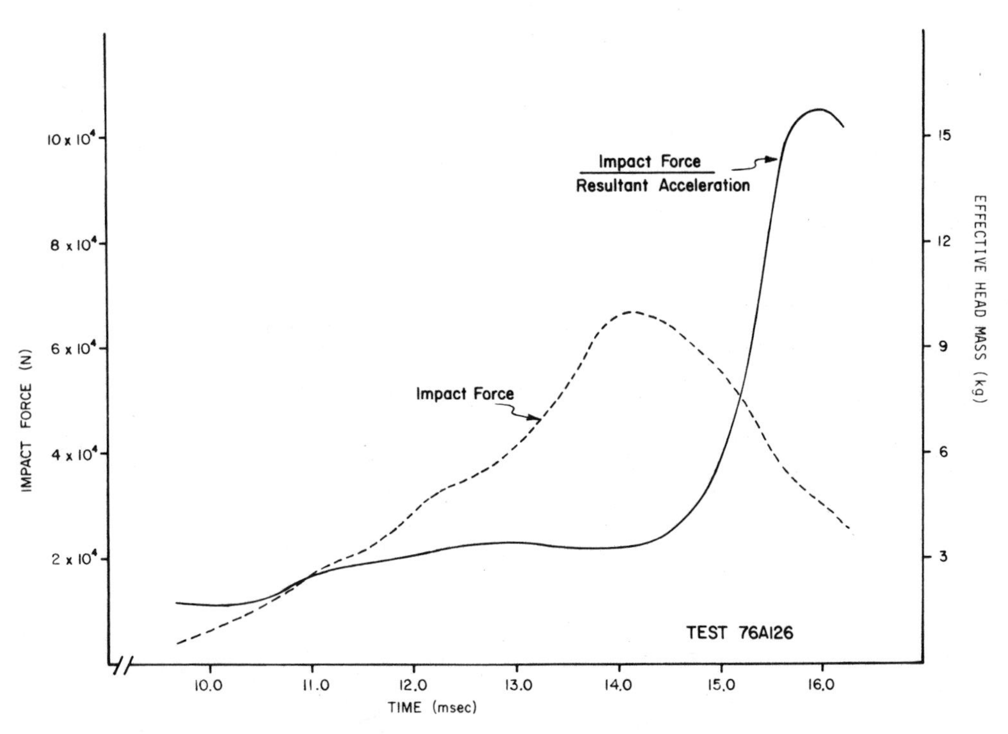

Fig. 11 - Force and effective head mass versus time

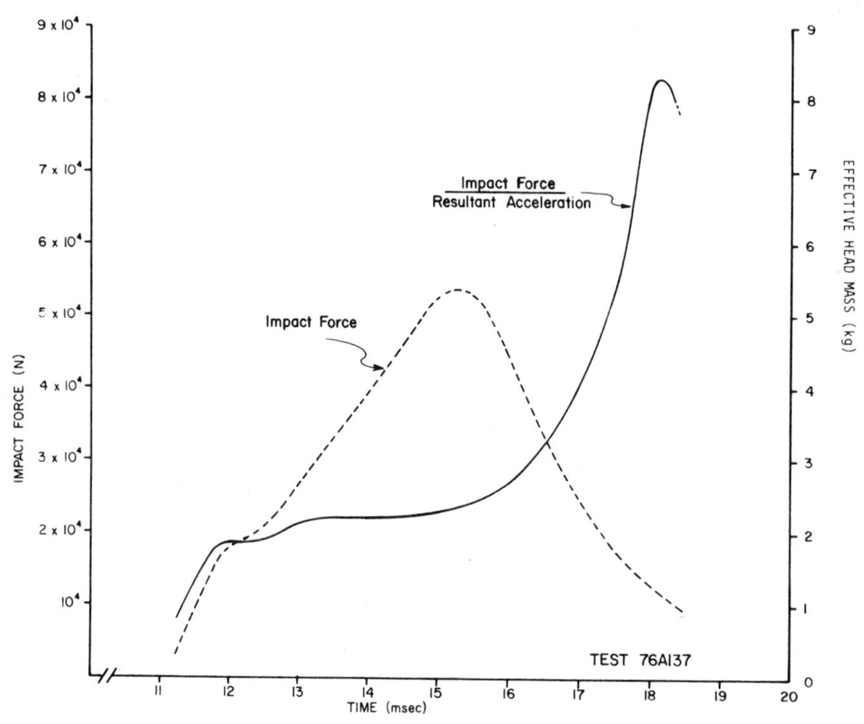

Fig. 12 - Force and effective head mass versus time

It should be noted that as long as the accelerometer mounts remain rigidly attached to the unfractured large portion of the skull, 3-D rigid body motion of the skull continues to be measured. It should also be noted that effective head mass calculations are no longer valid when angular velocities become substantial. Other possible sources of errors in obtaining a realistic head mass include the frictional forces between the interfacing impactor surface and impacted tissue, and the tissue deformation at the impact site.

Friction between the impactor and the skull would result in a higher force and a lower acceleration, thus tending to make the apparent head mass higher than it really is. This was not the case in our tests. The effects of scalp between the impactor and the skull has been shown from past studies (14) to be primarily load distributing with very little energy-absorbing properties and would have little or no effect on the apparent mass.

Taking into consideration all of the above factors, and examining all of the 15 tests reported here, five head impacts were considered to be suitable for calculating head mass: 76A135, 76A144, 76A152, 76A167 and finally 76A169. The head masses calculated from these five tests ranged from 4.8 - 5.8 kg, and averaged 5.5 kg.

INJURY CORRELATION - A correlation of all the parameters given in Table 3 with the resulting AIS values was made, and found to be low. Then an analysis was made after separating the pressurized tests from the non-pressurized tests. The pressurized data showed a much higher correlation than the non-pressurized data. Because of the small number of tests, the correlations were not statistically significant, but a definite trend was observed.

CONCLUSIONS

Although this has not been an exhaustive study of the data obtained from these tests, it may be concluded that:

1. Three-dimensional accelerometry techniques using nine accelerometers in three triaxial clusters was found to be applicable in direct head impacts. The complete description of 3-D motion proved to be invaluable to the understanding of skull-brain kinematics.
2. During the contact time of direct impact, the brain of the cadaver was found to be only partially constrained by the skull.
3. Pressurization of the brain vascular system assists in the evaluation of injury by straining the injury sites and by achieving a more realistic physical state of the brain.
4. At periods of longer than four days after death for this series of cadavers, noticeable brain property degradation appears to occur, as evidenced by excessive brain motions during impact.
5. When skull fracture occurs, a rapid increase in both angular velocity and the normal component of the translational acceleration results.

ACKNOWLEDGMENTS

This work was conducted under the sponsorship of the Motor Vehicle Manufacturers Association.

REFERENCES

1. Padgaonkar, A. J., K. W. Krieger, A. J. King, "Measurement of Angular Acceleration of a Rigid Body Using Linear Accelerations." ASME Preprint-75-APMB-3, June 1975.

2. Becker, E., G. Willems, "An Experimentally Validated 3-D Inertial Tracking Package for Application in Biodynamic Research." The 19th Stapp Car Crash Conference, 1975.

3. Ewing, C. L. et al., "The Effect of the Initial Position of the Head and Neck on the Dynamic Response of the Human Head and Neck to -Gx Impact Acceleration." Paper 751157, Proceedings of the 19th Stapp Car Crash Conference, 1975.

4. Ewing, C. L., D. J. Thomas, "Human Head and Neck Response to Impact Acceleration." Naval Aerospace Medical Research Laboratory Detachment, New Orleans, Monograph 21, August 1972.

5. Ewing, C. L., D. J. Thomas, "Torque versus Angular Displacement Response of Human Head to -Gx Impact Acceleration." Paper 730976, Proceedings of the 17th Stapp Car Crash Conference, Society of Automotive Engineers, Inc. 1973.

6. Thomas, D. J., "Specialized Anthropometry Requirements for Protective Equipment Evaluation." AGARD Conference Proceedings No. 110, Current Status in Aerospace Medicine, Glasgow, Scotland, September 1972.

7. Fayon, A. et al., "Thorax of 3-Point Belt Wearers During a Crash (Experiments for Cadavers)." The 19th Stapp Car Crash Conference, 1975.

8. Nahum, A. M., R. W. Smith, "An Experimental Model for Closed Head Injury." The 20th Stapp Car Crash Conference, 1976.

9. Bender, M., J. W. Melvin, R. L. Stalnaker, "A High-Speed Cineradiographic Technique for Biomechanical Impact." The 20th Stapp Car Crash Conference, 1976.

10. Bartz, J. A., F. E. Butler, "Passenger Compartment with Six Degrees of Freedom." Auxiliary Programs to "Three Dimensional Computer Simulation of a Motor Vehicle Crash Victim." Final Technical Report for DOT Contract No. FH-11-7592, 1972.

11. Joint Committee on Injury Scaling: The Abbreviated Injury Scale (AIS), 1976 Revision, Morton Grove, Illinois, American Association for Automotive Medicine, 1976.

12. Stalnaker, R. L., D. Mohan, J. W. Melvin, "Head Injury Evaluation: Criteria for Assessment of Field, Clinical and Laboratory Data." Proceedings of the 19th Conference of the American Association for Automotive Medicine, San Diego, California, November 17-21, 1975.

13. Stalnaker, R. L. et al., "Human Cadaver Head Impacts." Final Report for Motor Vehicle Manufacturers Association of the United States, Inc., In Preparation.

14. McElhaney, J. H., J. W. Melvin, V. L. Roberts, "Dynamic Characteristics of the Tissues of the Head. Presented at the Symposium on Perspectives in Biomedical Engineering, The University of Strathclyde, Glasgow, Scotland, June 19-20, 1972.

780887

Results of Experimental Head Impacts on Cadavers: The Various Data Obtained and Their Relations to Some Measured Physical Parameters

C. Got and A. Patel
Raymond Poincaré Hospital

A. Fayon, C. Tarrière, G. Walfisch
Laboratory of Physiology and Biomechanics, Peugeot/Renault

ABSTRACT

This report describes the results of 42 tests involving direct impacts on the head, performed on fresh, unembalmed, perfused cadavers, helmeted or not helmeted, by means of a free-fall procedure.

Three main kinds of impact were investigated: frontal, temporal-parietal, and frontal-facial.

The results yield a typology of lesions (associated with various test conditions) that differs from the one described in earlier, similar reports published by A.M. Nahum and R.L. Stalnaker.

The measurements confirm a tolerance level of HIC $>$1500 in the case involving skull impacts under the conditions specified in the text.

THE STATE OF KNOWLEDGE concerning human tolerance of cranial-cerebral traumatisms governs the design of suitably adapted protective devices, whether these are helmets for motorcyclists or the parts of a car that may strike the occupants, or other road users.

Investigations of real-life accidents supply highly accurate information regarding the clinical evolution and nature of injuries in particular when an autopsy is performed; however, these investigations do not permit quantitative knowledge of the physical characteristics of the impacts that caused the injuries. Only in highly special cases of skull impact on deformable surfaces, in particular on automobile hoods, it is possible to reproduce the deformation with anthropomorphic heads or cadaver heads, and to acquire these quantitative data indirectly.

Investigations carried out on volunteers are of relatively limited value, since the impacts must be incapable of resulting in injuries. However, under certain circumstances (American football, boxing), the impacts received render a quantitative study of loss of consciousness possible.

The use of animals is limited by the differences in sizes and shapes between the human skull and those of the animals used, resulting in unreliable extrapolations.

These difficulties led us to develop, in 1972, a method for investigating the resistance of the human brain to impacts by using unembalmed, perfused cadavers. The present paper includes all the experimental cranial impacts carried out via a free-fall procedure using this research during the last three years. The results of the initial impacts have been described in previous publications (7)(8); they are noted here for purposes of synthesis.

METHODOLOGY

Description of standard test - The subject lies prone in a metal cradle. His head, with or without a helmet, protrudes beyond the cradle, as does also part of the upper thorax (fig. 1). The unit is released and allowed to fall freely. The head is maintained by means of a suitable device in alignment with the trunk until impact. The surface against which the head strikes is, in most cases, flat, metallic and rigid.

The cradle containing the body comes to rest against a thick mattress of densely-packed shock-absorbent material sinking slightly into it. The head undergoes a relative movement in relation to the thorax, which will be described later.

Each test is accompanied by anthropometric measurements, acceleration measurements and measurements of the

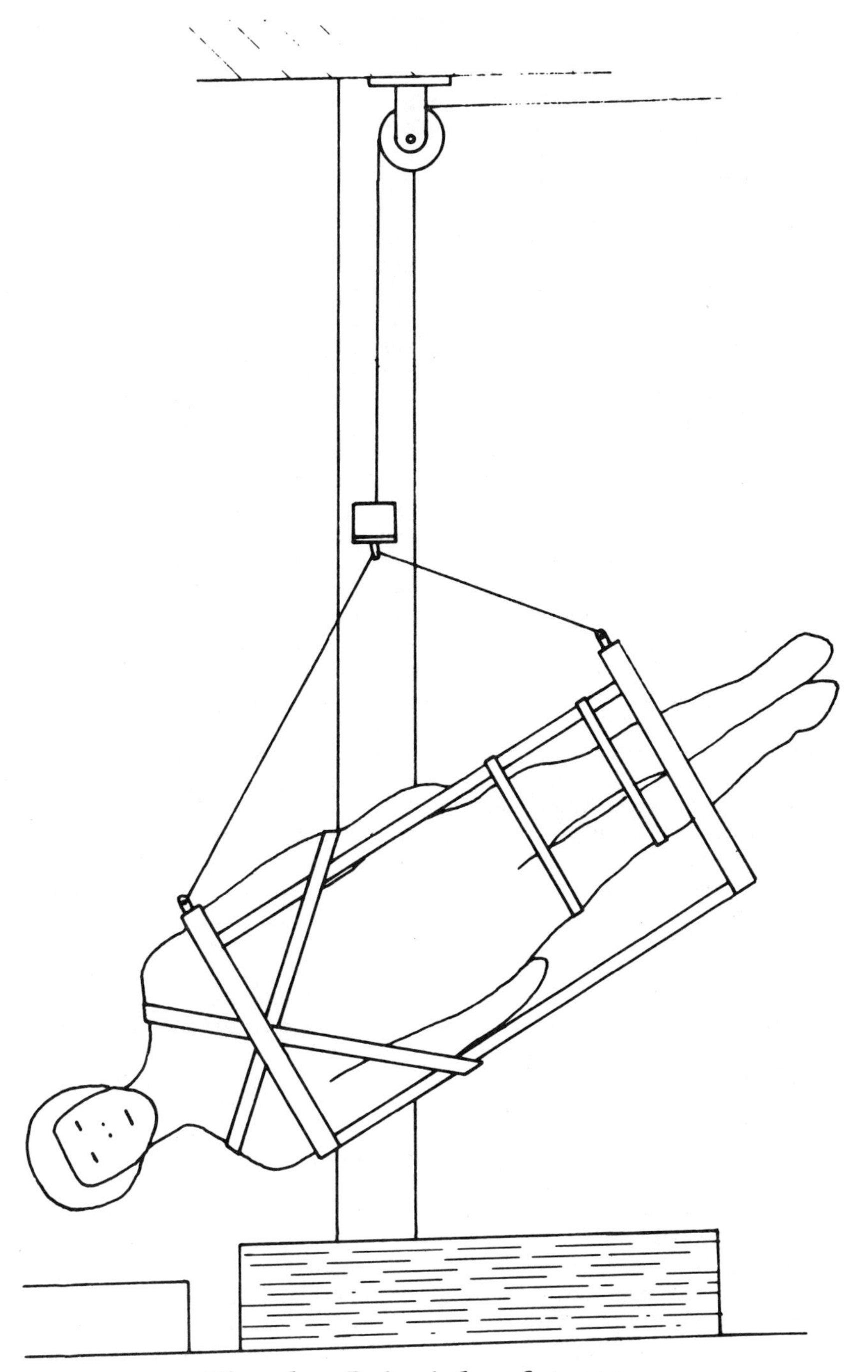

Fig. 1 - Principle of tests

percussion force, and of the pressurization of the encephalon described below.

The impact is recorded on film at a speed of one thousand frames per second by two Stalex cameras.

ANTHROPOMETRIC DATA - Before each test, the age, sex and principal head measurements of each subject are noted. For instance, the circumference and length of the head, measured over the lowest part of the frontal bone and the outer occipital protuberance. The width of the head is considered as the maximum distance between the right and left parietal temporal parts. After the test, the head, neck and skull, and, frequently the brain, are weighed. The cross sections are those of L.B. Walker (9). The results are listed in the tables pertaining to the individual subjects.

MEASUREMENTS RECORDED - Three accelerometers are attached to light alloy plates screwed into the subject's skull. Their position and orientation are defined with relation to the Ewing anatomical reference (10), and hence also to the Frankfort plane.

The positions of the accelerometers vary with each type of test. In the event of left temporal parietal impact, there is an accelerometer on the frontal bone, one on the occiput and another on the right temple.

In cases of frontal impact, there is an accelerometer on each temple and on the occiput.

During the initial tests, two bidirectional accelerometers and one triaxial accelerometer were used; beginning with subject No 83, three tridirectional accelerometers were mounted. The temporal accelerometers lie fairly close together as seen from the side to the center of gravity of the head; similarly, the frontal accelerometer lies often quite close to the center of gravity as seen from the front. This is not always the case, since the frontal sinuses may make it necessary to shift the position of this accelerometer.

The orientation of the sensitive axes of the transducers is generally close to those of the anatomical coordinates of the head. The exception is the occipital accelerometer the attachment plate of which follows the curve of the skull. The measurements of acceleration meet the requirements of Federal Standard 208 and of the SAE J 211 b procedure.

The tables of results list the data for each measurement channel. The helmets of the subjects were hollowed in order to house the accelerometers.

ACCELERATION AT THE CENTER OF GRAVITY - On the basis of the acceleration measurements performed as above, on the periphery of the skull, it is possible to compute the acceleration at the center of gravity of the head. A method of solution was used similar to that of Stalnaker, Alem (4).

The head is assumed to undergo only slight deformation; the inclination of the occipital transducer was rectified by means of a computer programming. The method used for attaching the accelerometer does not permit as accurate an orientation of the captors as with the HSRI, and it will be modified in later experiments. However, fairly satisfactory results were achieved when nine correct $\gamma(t)$ pulses were available. In fact, we obtained $\gamma(t)$ pulses on the center of gravity that were fairly similar for different methods of calculation, and good correspondence with the head movement observed on the films of the test.

Application of the computing program was performed with the use of a sampling range at 5,000 Hz.

When it was not possible to use the nine γ method, an estimate of HIC at C.G. was made on the basis of the results of the accelerometers closest to the C.G., while also checking our estimate by means of various comparisons.

The use of HIC here does not mean that we are assuming a priori a close connection between the values of the HIC and the gravity of the injuries. The purpose is only to endeavour to make optimum use of the routine tool represented by the HIC, the better to define its possibilities and limitations.

MEASUREMENT OF IMPACT FORCE - The vertical load exerted on the plate against which the head strikes was recorded against time; the plate was resting on Kistler cells. The results should however be considered with reservation, since the presence of the helmet, the tangential effect, and the interference of the shoulder can affect the maximum, which is the only item reported on. In addition, the F_H frequency is 600 Hz.

PRESSURIZATION - A perfusion of the encephalon was performed on each subject.

The reproduction of a blood pressure close to the average blood pressure of a live person has multiple effects: it restores to the brain tissue the relative rigidity that characterizes tissues perfused with sufficient pressure; in addition, perfusion eliminates the separation of the brain from the skull caused by the collapsing of unpressurized brain tissue (4). Lastly, it facilitates the display of arteriolar or capillary ruptures through the use of an injection mass containing carbon particles in suspension. The size of these particles does not enable them to move out of vessels when no rupture has occured. Vascular damage is hence marked by black extravasations around the ruptures similar in appearance to hemorrhages observed in live persons. The composition of the liquid injected is as follows:

- 1 volume of India ink
- 3 volumes of formaldehyde at 30 %
- 6 volumes of water.

The presence of formol permits fast fixing of the brain, thereby preventing the progress of autolysis; it facilitates the collection of samples and reduces artifacts. The action of the formol is too short prior to impact to modify the physical characteristics of the perfused tissue.

In order to obtain proper stability of the pressure in the arteries that vascularize the encephalon, the entire arterial system is placed under pressure, via a tubulure inserted through the left ventricle and extending to the base of the aorta. Airtight sealing is achieved through a ligature of the aorta onto the rigid extremity of this tubulure. In certain cases, perfusion was performed via catheterization of a femoral artery or of the abdominal aorta.

The perfusion starts approximately twenty seconds prior to impact; this represents a flow of about 1.5 liters of liquid; it is continued after impact. The total volume injected is three liters.

The head and neck are separated from the trunk after removal of the encephalon; the occipital is sawn off around the occipital aperture to enable the neck to be separated from the head without causing artifactual lesions at the level of the occipital-atloidean articulation. The neck is then frozen, and is cut along a sagittal median plane. The bony medullary lesions can then be observed. After thawing, the condition of the articular apophyses can be readily verified.

The encephalon undergoes additional fixation through immersion in formol. It is then cut into a section one centimeter thick, parallel to a frontal plane or to the Frankfort plane (a plane passing along the lower edge of the rear commissure and the highest point of the floor of the third ventricle is considered as the best encephalic equivalent of the Frankfort plane).

The sections are examined under a binocular magnifying glass, which makes it possible to distinguish the capillary injection zones from the extravasations caused by the rupturing of the vascular walls. Samples for microscopic examination are collected in the dubious zones, and are systematically collected at the level of the cerebral stem (two sections at the level of the cerebral peduncles, one isthmic section, three pons sections and three bulb sections).

The inclusion of the various parts is done with paraffin; dyeing is done with Hematein-Eosine and fast-blue Luxol. In certain dubious cases, non-deparaffined sections can be examined directly, so as to prevent formation of artifacts connected with the passage of the carbon particles outside the vessels during the processes of paraffin removal and dyeing.

RESULTS

Classification of tests performed - The tests performed fall into several categories, as follows:

- depending on the main direction of the impact (fig. 2),
- depending on the nature of the head-protection gear,
- depending on the height from which the fall takes place.

Nineteen of the tests involved falls in which the point of impact was temporal-parietal. These were hence not pure lateral impacts along a G_y anatomical direction: the vertical component is by no means negligible. This point of impact was used because of the frequency of its occurence in accidents involving two-wheeled vehicles or collisions between automobiles. Fourteen of the tests involved a frontal point of impact. These tests are similar to antero-posterior impacts. The face may or may not be involved.

Several types of helmets were used. The helmets are identified by the letters "A", "B", and "C" in the tables of results.

"A" is a "jet" type helmet, lined with expanded polystyrene about twenty millimeters thick along the entire surface covering the skull.

"B" is of the "fully-envelopping" type. The shock absorbent material is thicker and denser than in "A" helmet, but it is arranged in the form of a crown around the head and does not cover the top of the latter.

"C" is a "jet" type helmet characterized by the extremely wide thickness of polystyrene covering the frontal bone.

The heights from which the subjects were dropped were as follows :

- 1.83 meters as in Safety Standard 218,
- 2.5 meters as in other standards, including the French standard,
- 3 meters.

As the height was increased, relative motion between the head and upper thorax rose to levels resulting in injuries owing to the greater displacement of the body compared with that of the head.

The most recent tests, beginning with N°134, were performed under conditions designed to reduce the relative motion between head and trunk.

The cradle was removed and the subject was suspended from several points with no intermediate support. The height of the mattress of shock-absorbent material onto which the body dropped was modified.

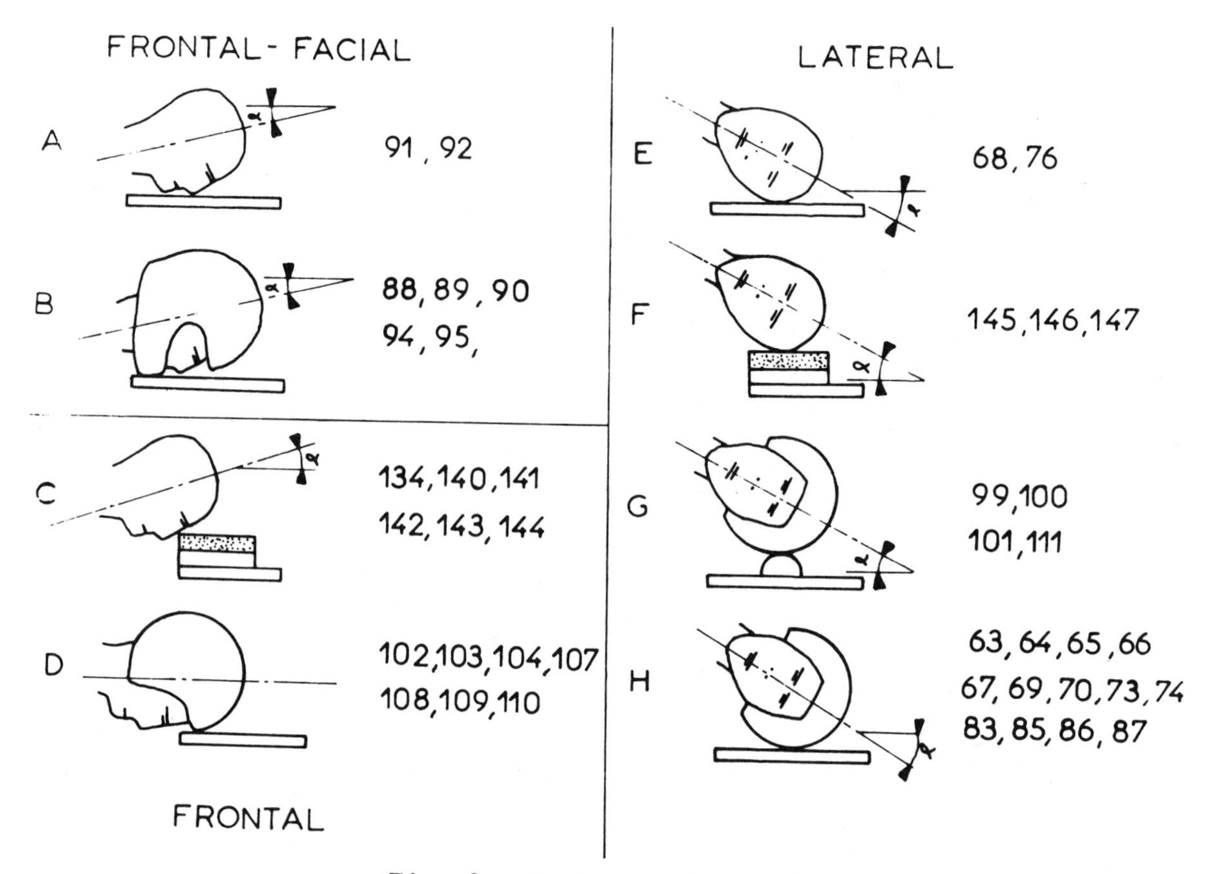

Fig. 2 - Impact configurations

Improvements were also made to the conditions governing observation of the impact of the head and of the stress undergone by the neck. For this purpose, the helmet was removed, and its shock-absorbent function was replaced by using a shock-absorbent material to cover the plate against which the head struck.

KINEMATICS

Generally speaking, the kinematics of the head and neck can be described in a simplified manner, in two phases, as follows : (Fig.3)

A. Vertical descent of the head into the helmet or the shock absorbent material, with no notable rotations in relation to the head at the time of the impact.

B. The start of rebound of the head triggers considerable rotation of the head with relation to the trunk, since the head tends to move upwards while the rest of the body continues to move downwards into the shock-absorbent mattress. This pattern is followed by extremely wide movements at the level of the neck.

In the case of temporal-parietal impact, for example, the translation phase lasts for six to seven ms, while the rotation phase lasts for 15 to 20 ms, and corresponds to a rotation of the vertical of the head in relation to the median axis of the trunk to a variable extent, though frequently greater than 70°. The corresponding angle is recorded as β in the tables of results.

In cases of frontal-parietal impact, with a fully-envelopping helmet, the extent of rotation is considerably decreased.

It will be noted that angular accelerations achieve their maxpoint during the translation phase. In the present state of measurements, the associated acceleration computing program, the $\ddot{\theta}$ curves that were calculated are extremely irregular, and we shall proceed on the assumption that the true maximum angular acceleration does not achieve 7500 rd/s^2 for any significant length of time as concerns the risk of occurence of injuries. Since, on the other hand angular speeds would not appear to exceed 50 rd/s , it will be assumed, on the basis of published data (12)(13), that angular speeds and accelerations do not lead to injury. There is still doubt concerning the consequences of the amplitude of neck movements on brain injuries, though their effect as regards the cervical injuries described below is obvious (5) (6).

In conclusion, a series of head impact tests was performed in which cerebral injuries resulted from the linear accelerations sustained.

As concerns frontal-facial impacts, it should be noted that the kinematics of the subjects were such that in every

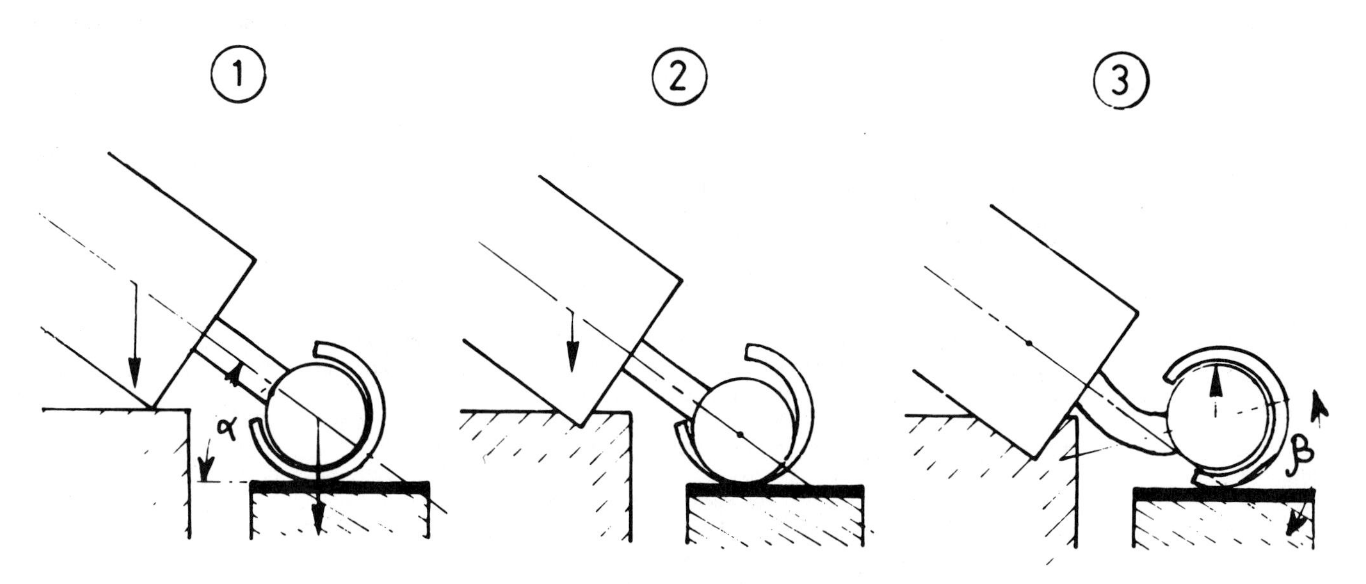

Fig. 3 - Simplified kinematics

case, it was the chin that was struck first (cf. angle α in the tables of results).

A summary of the tests results for each subject is given at the end of the paper. (Fig. 10 to 23).

NECK INJURIES

In the cases of frontal impacts (configurations C and D of Figure 2), seven subjects out of thirteen sustained osteoligamentary injuries of the cervical column, for hyperextensions of the head greater than 65°.

In cases of lateral impact, out of twenty two subjects we found two cases of lesions of the same type occurring for inclinations of the head of 55° and 89° in the film projection plane. In several cases, the lateral inclinations of the head in relation to the trunk went as high as 70° under the same conditions.

In the tests from subject N° 134, injury of the medulla occurred. This cannot be claimed for the earlier tests, in view of the sample collection technique.

FRACTURES OF THE SKULL AND FACE

In cases of impacts sustained by the skull, the absence of fractures was normal for subjects wearing helmets for falls from heights of three meters or less.

However, one exception was noted (case N° 83), in which a fracture of the dome of the skull occurred. For this subject, aged 74, as well as for the others, the ratios

$$\frac{\text{Skull mass}}{(\text{circumference of head})^2} \quad \text{and} \quad \frac{\text{Skull mass}}{(\text{circumference of head}^3}\ ,$$

were calculated.

They are respectively clearly related to the average thickness of the skull and also to its relative average thickness.

Of the entire set of subjects, N° 83 is the one displaying the lowest values for these ratios : in N° 83, the skull was virtually translucent in the vicinity of the point of impact.

In cases of frontal-facial impact of "fully-helmeted" subjects, fractures occurred as a result of falls from a height of 2.5 meters : this finding is not surprising in view of the low shock-absorbent capacity of the maxillary part of the "fully-envelopping" helmets. These fractures lower the acceleration levels.

The subject who sustained the severest injuries under the lowest acceleration conditions (N°89, $\gamma < 150$ g) was, as before, the one with the most fragile skull in the subgroup of subjects who underwent frontal-facial impacts, on the basis of the ratios defined above.

The non-helmeted subjects sustained a miscellaneous array of fractures.

INTERNAL INJURIES

Table 1 shows the injuries that were found.

Regardless of the violence of the impacts, whenever injuries occurred, they were found most frequently in the brainstem : fifteen cases of injuries of the brainstem were found in twenty subjects that had sustained various cerebral injuries. The injuries observed in the cerebral hemispheres - eight subjects out of the twenty injured - were located on the surface and, in most cases, in the impact zone. In two of the injured subjects, the corpus callosum was the seat of injuries. In view of the small number of cases available and the absence of identifications of the violence of the shock in the various categories of impacts investigated (temporal-parietal, frontal-facial, and frontal), it did not prove possible to perform an analysis for each type of impact. Figure 4 shows an example of a microscopic view of superficial injuries (case N° 110).

The topography of the vascular ruptures produced may appear surprising. The brainstem injuries are more frequent than cortical contusions in the zone subjacent to the point of impact. It should be noted that this brainstem damage is clearly predominant at the level of the protuberance and the isthmus, and that it is more deep than it is superficial at the level of these structures. Such injuries are observed in human pathology, and it is primarily their frequence in our experiments that raises a problem. There is no occurence of injection artifacts, since the adjacent structures, in particular the bulb and the cerebellum, are clearly injected with no observable vascular rupture.

The situation of the protuberance, which projects beyond the other elements of the brainstem and is in contact with bony structures (body of the occipital, in front and rear part of the petrosal bone laterally), makes if susceptible to damage, whether the impact moves from front or crosswise, (Fig. 5).

RESULTS OF MEASUREMENTS, HELMETED SUBJECTS OR DROPPED ONTO PADDING

The levels of acceleration achieved depend on the helmet used, on the point of impact, on the mass of the subject's head, and on the position of the body and head at the moment of impact. If the point of impact is near the edge of the helmet, the shock-absorbent action is impaired.

In the frontal impact investigations, all the tests were performed by dropping subjects from a height of three meters. Maximum accelerations ranged from 125 g (N°144 : head falling onto a polystyrene surface 60 mm thick) to 500 g (short peak, approximate measurements : impact on edge of helmet).

The H.I.C. ranged from 900 to over 2500, limited to subjects in which the brain injuries were interpretable.

TABLE 1 - Summary of Cerebral Lesions

Type of impact	N. of Subjects	N. of Serviceable Subjects	Injured	Brain stem lesions	Corpus Callosum	Peripherical Impact area	Peripherical Opposite area	Others
side/temporal	22/20	17/16	9/9	7/6	1/0	1	1	1
face and forehead	7/5	7/5	5/5	4/3	1/1			
forehead (3 m. of free fall)	13/13	7/7	6/6	4/4	2/2			

Note : first figure : all subjects
second figure : helmeted subjects, or equivalent cases

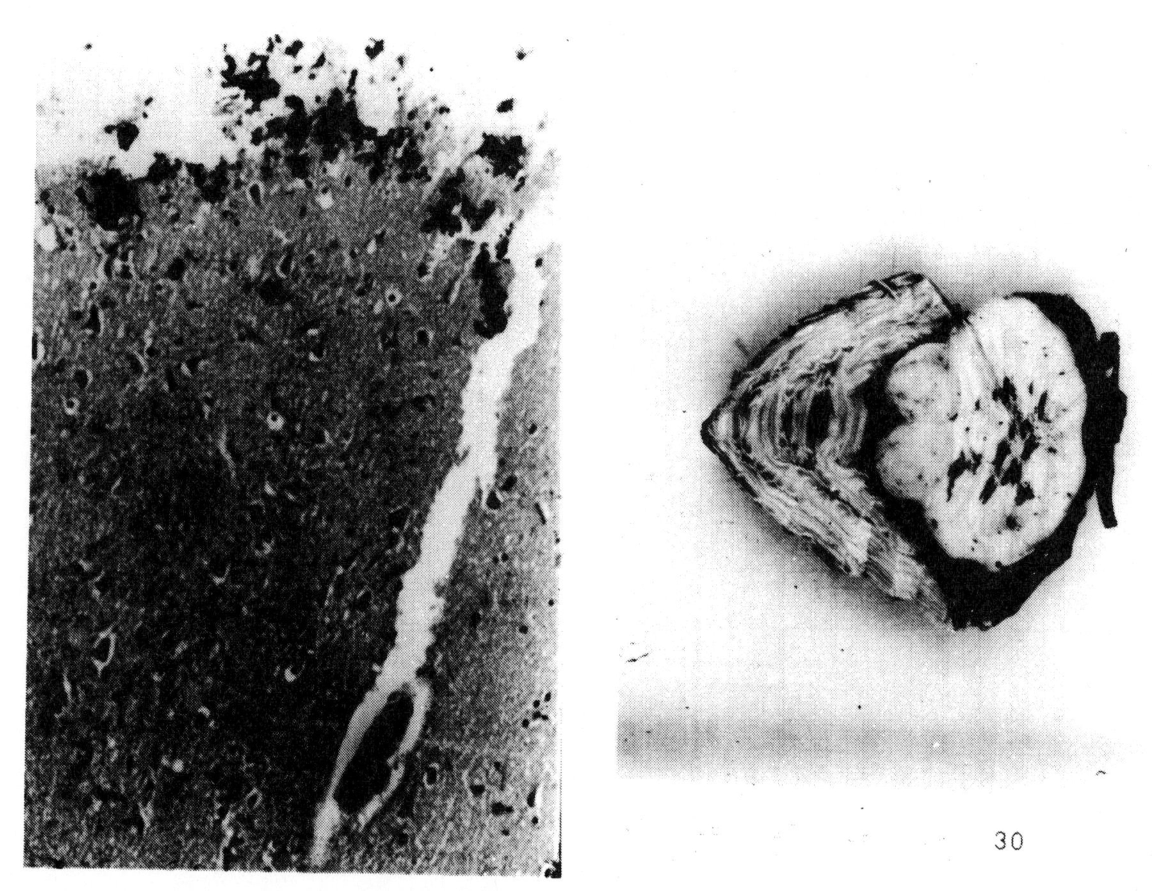

Fig. 4 - Microscopic view of superficial lesions (above) and typical view of brain stem

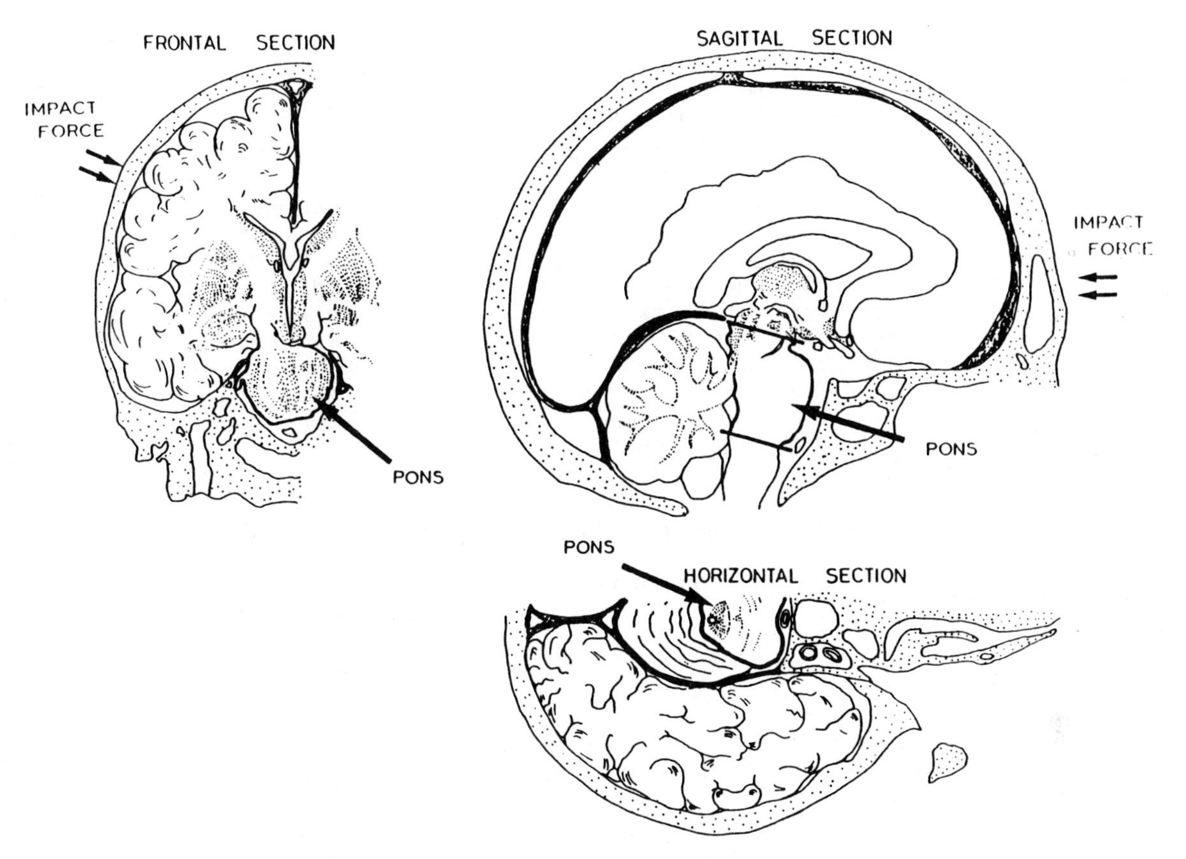

Fig. 5 - Location of pons versus long structures

In the cases of lateral impact, the difference in maximum accelerations is smaller , i.e. by 150 to 260 g, with the H.I.C. ranging from 900 to 2300.

Likewise, in cases of frontal-facial impact, accelerations are lower by about 120 to 180 g whilst the H.I.C. ranges from 540 to 1480.

Generally speaking, the H.I.C. increases with the potential energy of the head, in accordance with a law peculiar to each type of helmet (figure 6 and figure 7). This phenomenon does not occur with the fully-envelopping helmet in temporal-parietal impacts, for various reasons, including the number of points available.

For this shock-absorbent process utilizing polystyrene, the H.I.C. is clearly correlated with acceleration. However, this result cannot be generalized to all the shock-absorbent materials or systems.

RELATIONSHIPS BETWEEN INJURIES AND MEASUREMENTS (Fig.8 & 9)

In the cases of frontal bone impact, two subjects displayed presumably severe injuries and a H.I.C. of $<$ 1500 : N° 107 and 143.

Subject N° 107 had a ratio of 1.41 for $\frac{\text{skull mass}}{\text{circumference of head}^2}$ as compared with subject N°83 (ratio 1.39), noted above. The mean figure for our subjects was 2.28.

In subject N°143, isolated extravasations of the product injected were visible in a single zone - the median part of he protuberance. Medical prognosis is difficult in this case.

As concerns frontal-facial impacts, let us again consider subjects afflicted with injuries with an A.I.S. $\geq$ 3, in which the H.I.C. is less than 1500. Subject N° 92 who was without a helmet and was dropped from a height of 2.5 meters, sustained fractures of the Lefort I, II, III type as well as of other types; his A.I.S. is hard to evaluate (3 to 4), but *a priori* it was not fatal, and he had a H.I.C. of approximately 1200. We do not know to what extent the multiplicity of fractures was the cause of more severe injuries. Subject N° 95 (H.I.C. = 1150), in contrast, can be considered without reserve.

An investigation of the gravity of the cerebral injuries described in terms of A.I.S., situating them in relation to the H.I.C. or to the accelerations measures, does not *a priori* reveal any relationship between these two parameters, (Fig.8 & 9). This is not surprising, since the phenomena causing injuries are not necessarily of identical nature depending on the type of impact - frontal or lateral impact, for example : they hence correspond to different appearance thresholds. In addition, in an identical injury-

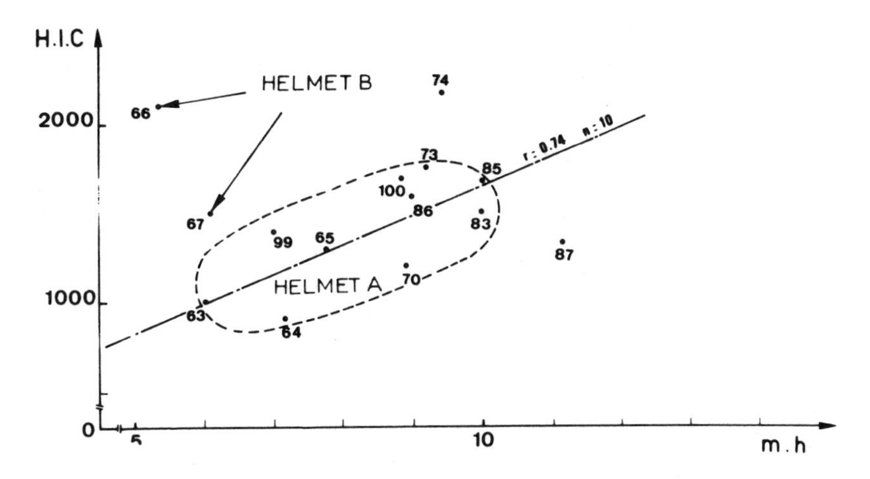

Fig. 6 - H.I.C. versus (head mass x height of fall)-side impacts

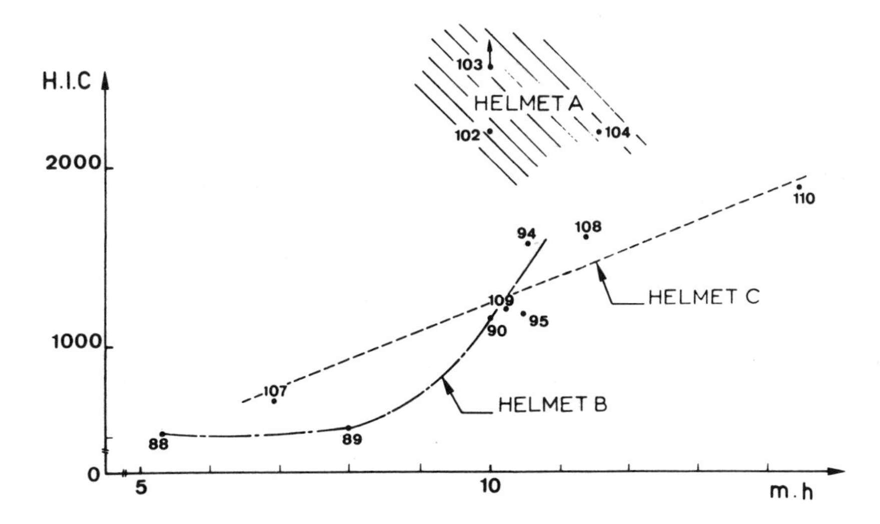

Fig. 7 - H.I.C. versus (head mass x height of fall)-frontal and frontal/facial impacts

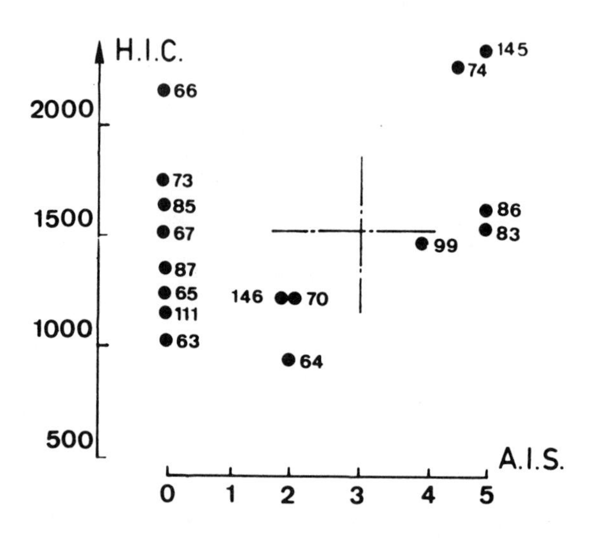

Fig. 8 - H.I.C. versus A.I.S.

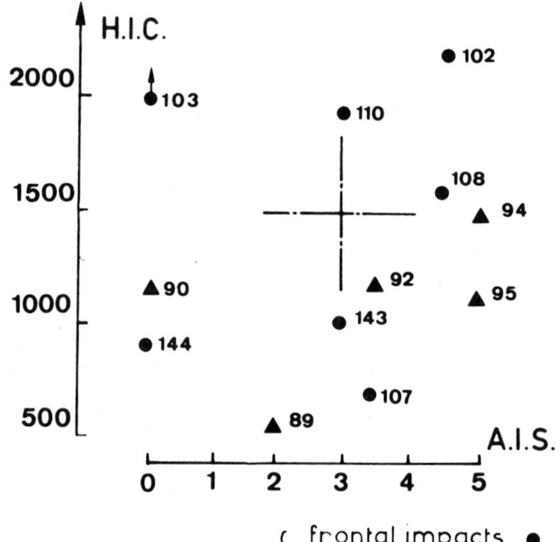

Fig. 9 - H.I.C. versus A.I.S.

causing process, tolerance is a function of numerous parameters related to the subject; last but not least, the A.I.S. is not a proportionate scale.

In all likelihood, therefore, as indicated by Stalnaker (4), we can only progress step-by-step, in pace with the accumulating of available results. Moreover, a statistical analysis allowing the different acting parameters to be separated is necessary.

We can however distribute the injuries found according to types of impact and zones of the encephalon, and endeavor to eliminate particular cases that would lead to misleading conclusions.

In a temporal-parietal impact, serious (A.I.S. $\geqslant$ 3) injuries appear in extremely low H.I.C. values when the subject's brain is in a poor state of conservation (case N° 147, H.I.C. $\sim$ 800), but this result is not meaningful. The lowest H.I.C.s associated with serious injuries are those of N° 88 and N° 99, which are respectively, 1500 and 1400. It will be noted incidentally that these two subjects have skulls that are relatively light for their volumes. The insufficiency of the data prevented us from exhaustively investigating the effect of skull thickness on tolerance.

Conversely, it is possible to obtain high H.I.C. values without injuries. It should be noted that we found no serious injury except when the accelerations maintained during three milliseconds at the most were more than 130 g.

SUMMARY OF THE FOREGOING

To summarize the foregoing, it was found that during skull impacts, there is but slight likelihood of the occurrence of a serious injury for a value of H.I.C. $<$ 1500. It did not occur in practice in temporo-lateral impacts.

Not even one single medium-skulled subject that had simultaneously a medium skull associated both with a serious injury and a H.I.C. $<$ 1500 was found among frontal impact cases.

As regards impacts sustained on both face and skull, the phenomena appear more complex. All the subjects that underwent the frontal-facial impacts reported on here sustained fractures, except N° 88 (dropped from a height of I.83 meters and not considered in the analysis because of the condition of his brain). Moreover, we have here a case in which the deviation between the accelerations obtained with cadavers and those obtained by performing tests with dummies is quite high, since the faces of dummies do not shatter.

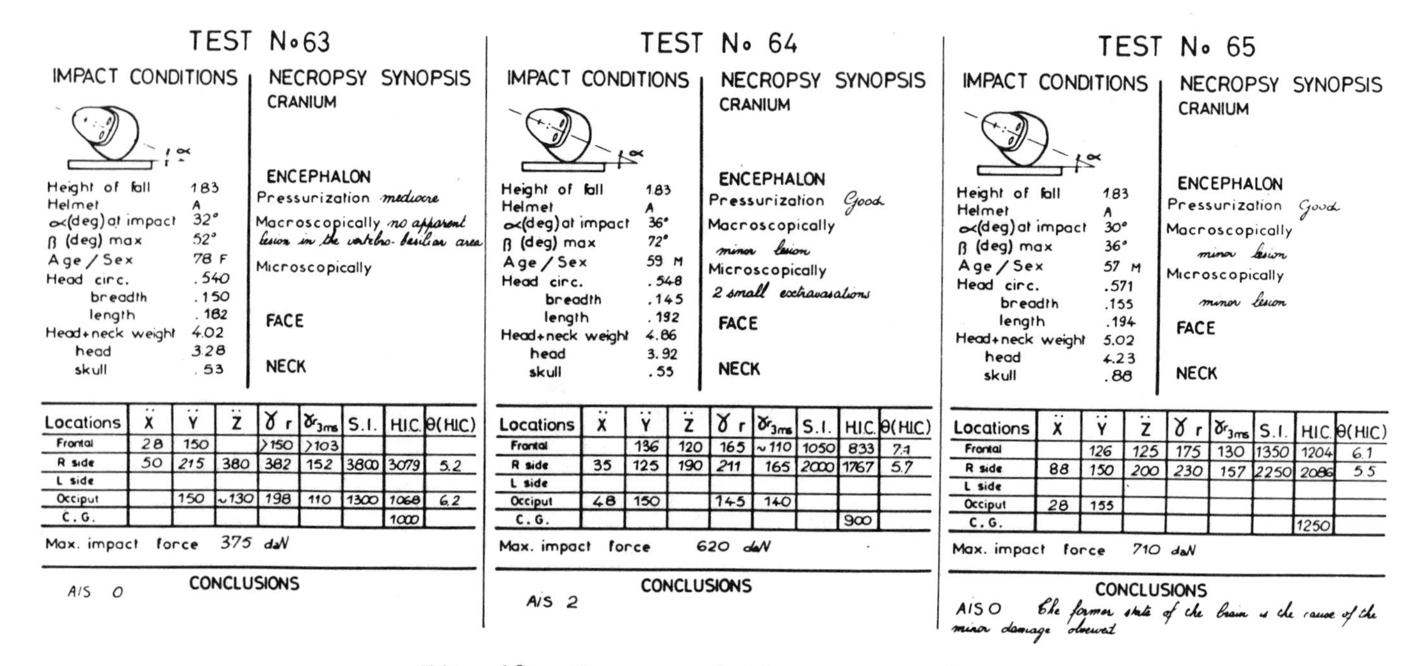

TEST No 63

IMPACT CONDITIONS

- Height of fall: 1.83
- Helmet: A
- α (deg) at impact: 32°
- β (deg) max: 52°
- Age / Sex: 78 F
- Head circ.: .540
  - breadth: .150
  - length: .182
- Head+neck weight: 4.02
  - head: 3.28
  - skull: .53

NECROPSY SYNOPSIS

CRANIUM

ENCEPHALON

Pressurization *mediocre*

Macroscopically *no apparent lesion in the vertebro-basilar area*

Microscopically

FACE

NECK

Locations	$\ddot{X}$	$\ddot{Y}$	$\ddot{Z}$	γ_r	γ_{r3ms}	S.I.	H.I.C.	θ(H.I.C.)
Frontal	28	150		>150	>103			
R side	50	215	380	382	152	3800	3079	5.2
L side								
Occiput		150	~130	198	110	1300	1068	6.2
C. G.							1000	

Max. impact force 375 daN

CONCLUSIONS

AIS 0

TEST No 64

IMPACT CONDITIONS

- Height of fall: 1.83
- Helmet: A
- α (deg) at impact: 36°
- β (deg) max: 72°
- Age / Sex: 59 M
- Head circ.: .548
  - breadth: .145
  - length: .192
- Head+neck weight: 4.86
  - head: 3.92
  - skull: .55

NECROPSY SYNOPSIS

CRANIUM

ENCEPHALON

Pressurization *Good*

Macroscopically *minor lesion*

Microscopically *2 small extravasations*

FACE

NECK

Locations	$\ddot{X}$	$\ddot{Y}$	$\ddot{Z}$	γ_r	γ_{r3ms}	S.I.	H.I.C.	θ(H.I.C.)
Frontal		136	120	165	~110	1050	833	7.1
R side	35	125	190	211	165	2000	1767	5.7
L side								
Occiput	48	150		145	140			
C. G.							900	

Max. impact force 620 daN

CONCLUSIONS

AIS 2

TEST No 65

IMPACT CONDITIONS

- Height of fall: 1.83
- Helmet: A
- α (deg) at impact: 30°
- β (deg) max: 36°
- Age / Sex: 57 M
- Head circ.: .571
  - breadth: .155
  - length: .194
- Head+neck weight: 5.02
  - head: 4.23
  - skull: .88

NECROPSY SYNOPSIS

CRANIUM

ENCEPHALON

Pressurization *Good*

Macroscopically *minor lesion*

Microscopically *minor lesion*

FACE

NECK

Locations	$\ddot{X}$	$\ddot{Y}$	$\ddot{Z}$	γ_r	γ_{r3ms}	S.I.	H.I.C.	θ(H.I.C.)
Frontal		126	125	175	130	1350	1204	6.1
R side	88	150	200	230	157	2250	2086	5.5
L side								
Occiput	28	155						
C. G.							1250	

Max. impact force 710 daN

CONCLUSIONS

AIS 0 *The former state of the brain is the cause of the minor damage observed*

Fig. 10 - Summary of the test results

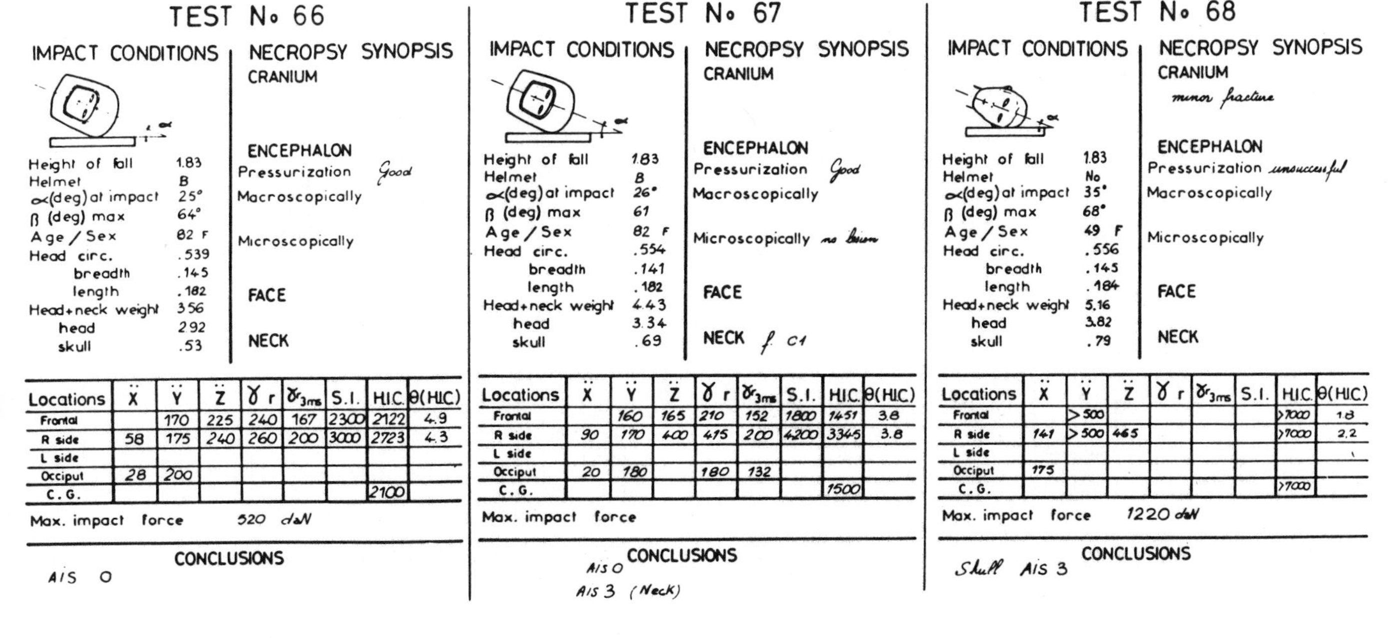

TEST No 66

IMPACT CONDITIONS

Height of fall 1.83
Helmet B
α(deg) at impact 25°
β (deg) max 64°
Age / Sex 82 F
Head circ. .539
breadth .145
length .182
Head+neck weight 3.56
head 2.92
skull .53

NECROPSY SYNOPSIS

CRANIUM

ENCEPHALON
Pressurization Good
Macroscopically
Microscopically

FACE

NECK

Locations	$\ddot{X}$	$\ddot{Y}$	$\ddot{Z}$	γ_r	$\gamma_{r\,3ms}$	S.I.	H.I.C.	θ(H.I.C.)
Frontal		170	225	240	167	2300	2122	4.9
R side	58	175	240	260	200	3000	2723	4.3
L side								
Occiput	28	200						
C. G.							2100	

Max. impact force 520 daN

CONCLUSIONS

AIS 0

TEST No 67

IMPACT CONDITIONS

Height of fall 1.83
Helmet B
α(deg) at impact 26°
β (deg) max 61
Age / Sex 82 F
Head circ. .554
breadth .141
length .182
Head+neck weight 4.43
head 3.34
skull .69

NECROPSY SYNOPSIS

CRANIUM

ENCEPHALON
Pressurization Good
Macroscopically
Microscopically no lesion

FACE

NECK f. C1

Locations	$\ddot{X}$	$\ddot{Y}$	$\ddot{Z}$	γ_r	$\gamma_{r\,3ms}$	S.I.	H.I.C.	θ(H.I.C.)
Frontal		160	165	210	152	1800	1451	3.8
R side	90	170	400	415	200	4200	3345	3.8
L side								
Occiput	20	180		180	132			
C. G.							1500	

Max. impact force

CONCLUSIONS

AIS 0
AIS 3 (Neck)

TEST No 68

IMPACT CONDITIONS

Height of fall 1.83
Helmet No
α(deg) at impact 35°
β (deg) max 68°
Age / Sex 49 F
Head circ. .556
breadth .145
length .184
Head+neck weight 5.16
head 3.82
skull .79

NECROPSY SYNOPSIS

CRANIUM
minor fracture

ENCEPHALON
Pressurization unsuccessful
Macroscopically
Microscopically

FACE

NECK

Locations	$\ddot{X}$	$\ddot{Y}$	$\ddot{Z}$	γ_r	$\gamma_{r\,3ms}$	S.I.	H.I.C.	θ(H.I.C.)
Frontal		>500					>7000	1.8
R side	141	>500	465				>7000	2.2
L side								
Occiput	175							
C. G.							>7000	

Max. impact force 1220 daN

CONCLUSIONS

Skull AIS 3

Fig. 11 - Summary of the test results

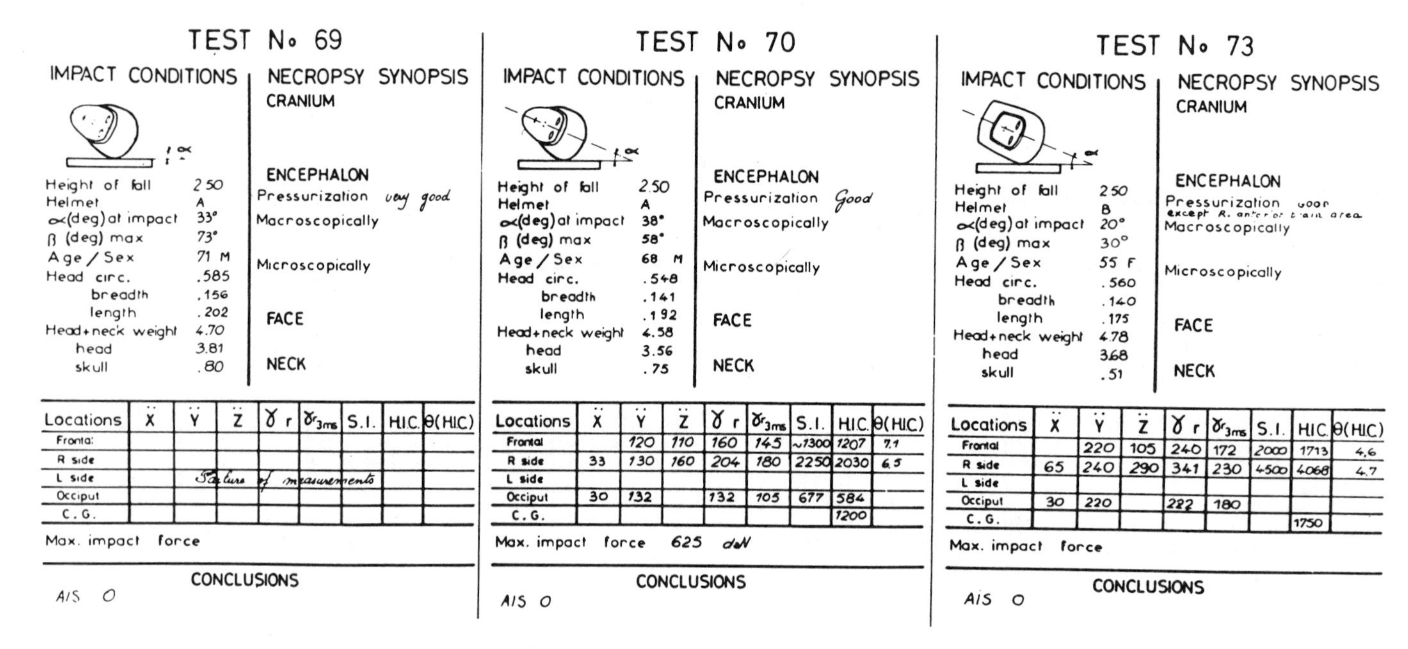

TEST No 69

IMPACT CONDITIONS

Height of fall	2.50
Helmet	A
α (deg) at impact	33°
β (deg) max	73°
Age / Sex	71 M
Head circ.	.585
breadth	.156
length	.202
Head+neck weight	4.70
head	3.81
skull	.80

NECROPSY SYNOPSIS

CRANIUM

ENCEPHALON

Pressurization very good

Macroscopically

Microscopically

FACE

NECK

Locations	$\ddot{X}$	$\ddot{Y}$	$\ddot{Z}$	γ_r	$\gamma_{r_{3ms}}$	S.I.	H.I.C.	θ(H.I.C.)
Frontal								
R side								
L side		Failure of measurements						
Occiput								
C.G.								

Max. impact force

CONCLUSIONS

AIS 0

TEST No 70

IMPACT CONDITIONS

Height of fall	2.50
Helmet	A
α (deg) at impact	38°
β (deg) max	58°
Age / Sex	68 M
Head circ.	.548
breadth	.141
length	.192
Head+neck weight	4.58
head	3.56
skull	.75

NECROPSY SYNOPSIS

CRANIUM

ENCEPHALON

Pressurization Good

Macroscopically

Microscopically

FACE

NECK

Locations	$\ddot{X}$	$\ddot{Y}$	$\ddot{Z}$	γ_r	$\gamma_{r_{3ms}}$	S.I.	H.I.C.	θ(H.I.C.)
Frontal		120	110	160	145	~1300	1207	7.1
R side	33	130	160	204	180	2250	2030	6.5
L side								
Occiput	30	132		132	105	677	584	
C.G.							1200	

Max. impact force 625 daN

CONCLUSIONS

AIS 0

TEST No 73

IMPACT CONDITIONS

Height of fall	2.50
Helmet	B
α (deg) at impact	20°
β (deg) max	30°
Age / Sex	55 F
Head circ.	.560
breadth	.140
length	.175
Head+neck weight	4.78
head	3.68
skull	.51

NECROPSY SYNOPSIS

CRANIUM

ENCEPHALON

Pressurization Good except R. anterior brain area

Macroscopically

Microscopically

FACE

NECK

Locations	$\ddot{X}$	$\ddot{Y}$	$\ddot{Z}$	γ_r	$\gamma_{r_{3ms}}$	S.I.	H.I.C.	θ(H.I.C.)
Frontal		220	105	240	172	2000	1713	4.6
R side	65	240	290	341	230	4500	4068	4.7
L side								
Occiput	30	220		222	180			
C.G.							1750	

Max. impact force

CONCLUSIONS

AIS 0

Fig. 12 - Summary of the test results

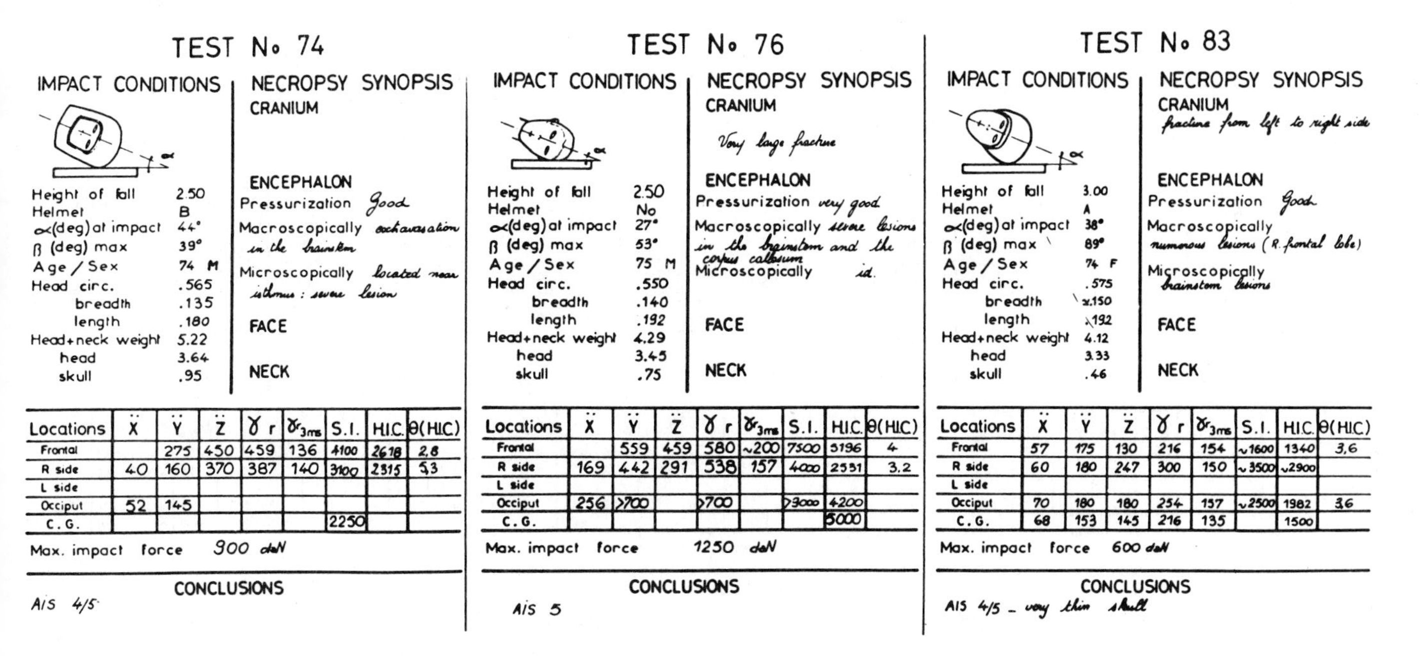

TEST No 74

IMPACT CONDITIONS

Height of fall	2.50
Helmet	B
α(deg) at impact	44°
β (deg) max	39°
Age / Sex	74 M
Head circ.	.565
breadth	.135
length	.180
Head+neck weight	5.22
head	3.64
skull	.95

NECROPSY SYNOPSIS

CRANIUM

ENCEPHALON

Pressurization *Good*

Macroscopically *extravasation in the brainstem*

Microscopically *located near isthmus : severe lesion*

FACE

NECK

Locations	$\ddot{X}$	$\ddot{Y}$	$\ddot{Z}$	γ_r	$\gamma_{r\,3ms}$	S.I.	H.I.C.	θ(H.I.C.)
Frontal		275	450	459	136	4100	2618	2.8
R side	40	160	370	387	140	3100	2315	5.3
L side								
Occiput	52	145						
C. G.						2250		

Max. impact force *900 daN*

CONCLUSIONS

AIS 4/5

TEST No 76

IMPACT CONDITIONS

Height of fall	2.50
Helmet	No
α(deg) at impact	27°
β (deg) max	53°
Age / Sex	75 M
Head circ.	.550
breadth	.140
length	.192
Head+neck weight	4.29
head	3.45
skull	.75

NECROPSY SYNOPSIS

CRANIUM

Very large fracture

ENCEPHALON

Pressurization *very good*

Macroscopically *severe lesions in the brainstem and the corpus callosum*

Microscopically *id.*

FACE

NECK

Locations	$\ddot{X}$	$\ddot{Y}$	$\ddot{Z}$	γ_r	$\gamma_{r\,3ms}$	S.I.	H.I.C.	θ(H.I.C.)
Frontal		559	459	580	~200	7500	5196	4
R side	169	442	291	538	157	4000	2551	3.2
L side								
Occiput	256	>700		>700		>3000	4200	
C. G.							5000	

Max. impact force *1250 daN*

CONCLUSIONS

AIS 5

TEST No 83

IMPACT CONDITIONS

Height of fall	3.00
Helmet	A
α(deg) at impact	38°
β (deg) max	89°
Age / Sex	74 F
Head circ.	.575
breadth	≈.150
length	≈.192
Head+neck weight	4.12
head	3.33
skull	.46

NECROPSY SYNOPSIS

CRANIUM

fracture from left to right side

ENCEPHALON

Pressurization *Good*

Macroscopically *numerous lesions (R. frontal lobe)*

Microscopically *brainstem lesions*

FACE

NECK

Locations	$\ddot{X}$	$\ddot{Y}$	$\ddot{Z}$	γ_r	$\gamma_{r\,3ms}$	S.I.	H.I.C.	θ(H.I.C.)
Frontal	57	175	130	216	154	~1600	1340	3,6
R side	60	180	247	300	150	~3500	~2900	
L side								
Occiput	70	180	180	254	157	~2500	1982	3,6
C. G.	68	153	145	216	135		1500	

Max. impact force *600 daN*

CONCLUSIONS

AIS 4/5 – *very thin skull*

Fig. 13 - Summary of the test results

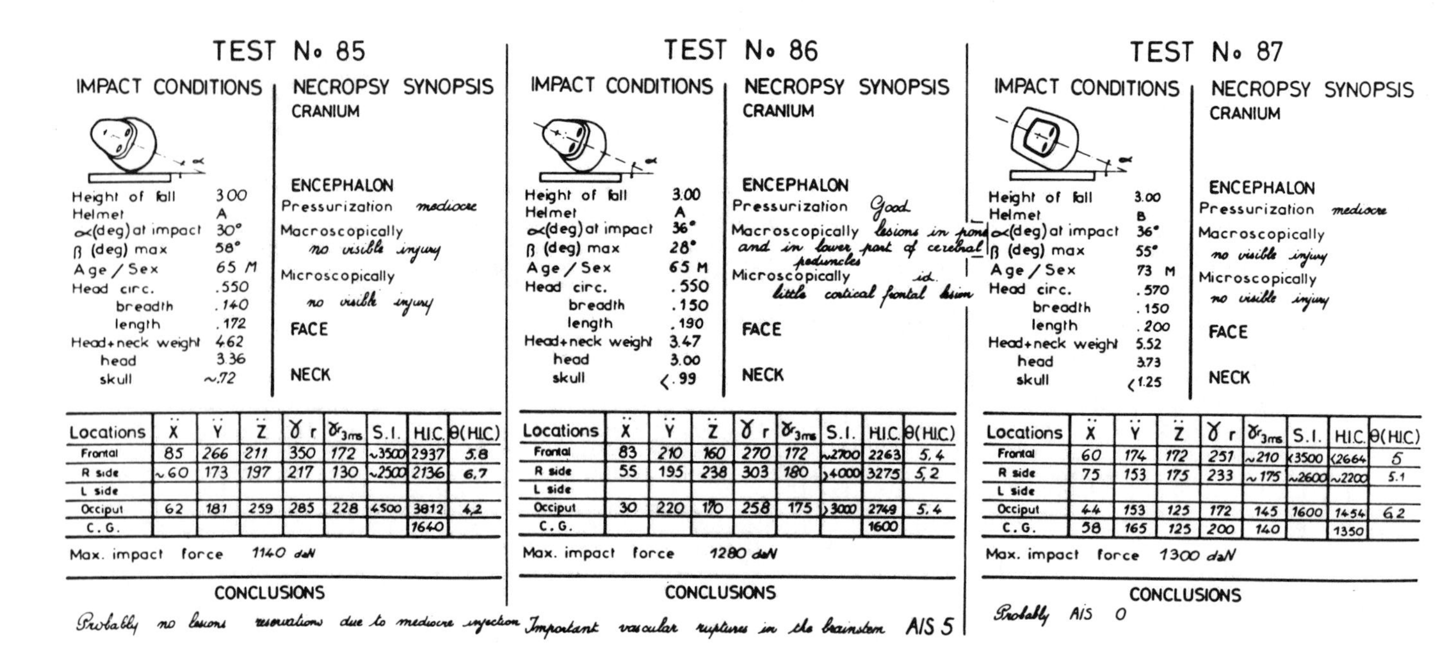

TEST No 85

IMPACT CONDITIONS

Height of fall	3.00
Helmet	A
α(deg) at impact	30°
β (deg) max	58°
Age / Sex	65 M
Head circ.	.550
breadth	.140
length	.172
Head+neck weight	4.62
head	3.36
skull	~.72

NECROPSY SYNOPSIS

CRANIUM

ENCEPHALON

Pressurization mediocre

Macroscopically no visible injury

Microscopically no visible injury

FACE

NECK

Locations	$\ddot{X}$	$\ddot{Y}$	$\ddot{Z}$	γr	γr_{3ms}	S.I.	H.I.C.	θ(H.I.C.)
Frontal	85	266	211	350	172	~3500	2937	5.8
R side	~60	173	197	217	130	~2500	2136	6.7
L side								
Occiput	62	181	259	285	228	4500	3812	4.2
C.G.							1640	

Max. impact force 1140 daN

CONCLUSIONS

Probably no lesions reservations due to mediocre injection

TEST No 86

IMPACT CONDITIONS

Height of fall	3.00
Helmet	A
α(deg) at impact	36°
β (deg) max	28°
Age / Sex	65 M
Head circ.	.550
breadth	.150
length	.190
Head+neck weight	3.47
head	3.00
skull	<.99

NECROPSY SYNOPSIS

CRANIUM

ENCEPHALON

Pressurization Good

Macroscopically lesions in pons and in lower part of cerebral peduncles

Microscopically id. little cortical frontal lesion

FACE

NECK

Locations	$\ddot{X}$	$\ddot{Y}$	$\ddot{Z}$	γr	γr_{3ms}	S.I.	H.I.C.	θ(H.I.C.)
Frontal	83	210	160	270	172	~2700	2263	5.4
R side	55	195	238	303	180	>4000	3275	5.2
L side								
Occiput	30	220	170	258	175	>3000	2749	5.4
C.G.							1600	

Max. impact force 1280 daN

CONCLUSIONS

Important vascular ruptures in the brainstem AIS 5

TEST No 87

IMPACT CONDITIONS

Height of fall	3.00
Helmet	B
α(deg) at impact	36°
β (deg) max	55°
Age / Sex	73 M
Head circ.	.570
breadth	.150
length	.200
Head+neck weight	5.52
head	3.73
skull	<1.25

NECROPSY SYNOPSIS

CRANIUM

ENCEPHALON

Pressurization mediocre

Macroscopically no visible injury

Microscopically no visible injury

FACE

NECK

Locations	$\ddot{X}$	$\ddot{Y}$	$\ddot{Z}$	γr	γr_{3ms}	S.I.	H.I.C.	θ(H.I.C.)
Frontal	60	174	172	251	~210	<3500	<2664	5
R side	75	153	175	233	~175	~2600	~2200	5.1
L side								
Occiput	44	153	125	172	145	1600	1454	6.2
C.G.	58	165	125	200	140		1350	

Max. impact force 1300 daN

CONCLUSIONS

Probably AIS 0

Fig. 14 - Summary of the test results

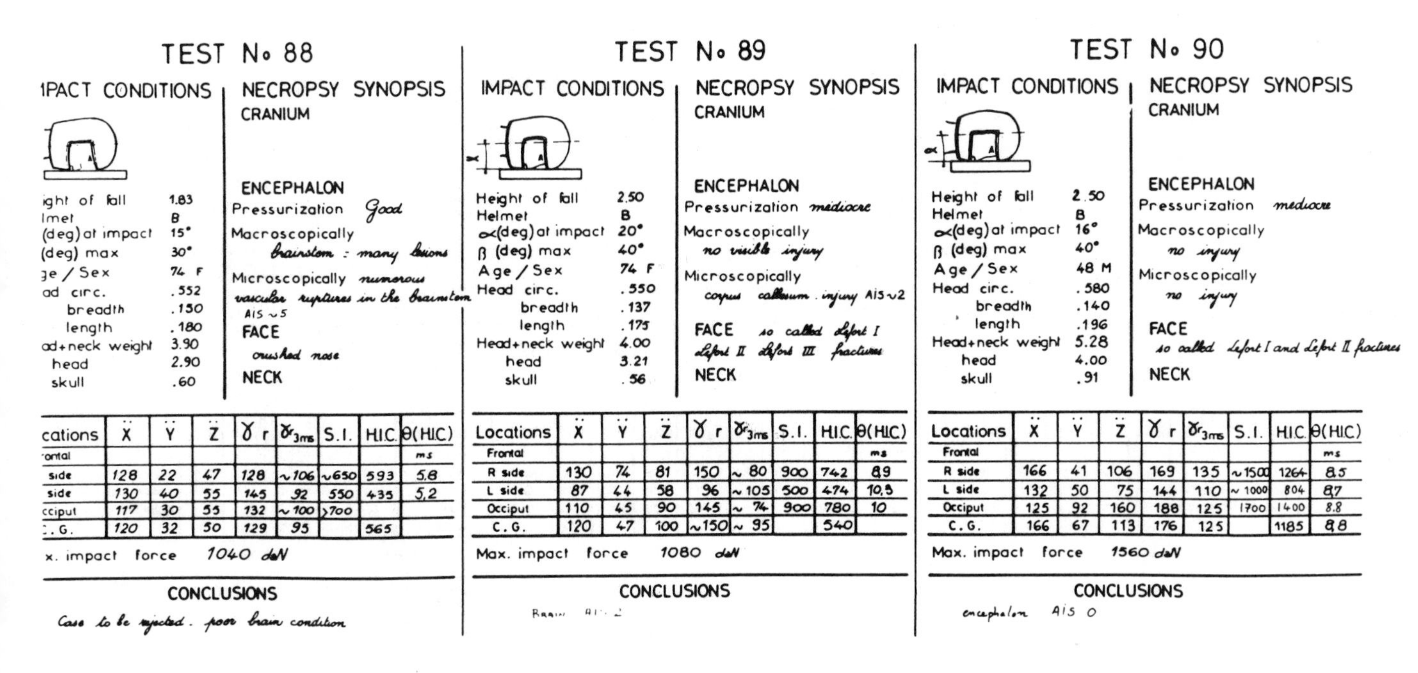

TEST No 88

IMPACT CONDITIONS

ight of fall	1.83
lmet	B
(deg) at impact	15°
(deg) max	30°
ge / Sex	74 F
ad circ.	.552
breadth	.150
length	.180
ad+neck weight	3.90
head	2.90
skull	.60

NECROPSY SYNOPSIS

CRANIUM

ENCEPHALON

Pressurization *Good*

Macroscopically *brainstem : many lesions*

Microscopically *numerous vascular ruptures in the brainstem* AIS ~5

FACE *crushed nose*

NECK

cations	Ẍ	Ÿ	Z̈	γ r	$\gamma_{r\,3ms}$	S.I.	H.I.C.	θ(H.I.C.)
ontal								ms
side	128	22	47	128	~106	~650	593	5.8
side	130	40	55	145	92	550	435	5,2
cciput	117	30	55	132	~100	>700		
. G.	120	32	50	129	95		565	

x. impact force 1040 daN

CONCLUSIONS

Case to be rejected. poor brain condition

TEST No 89

IMPACT CONDITIONS

Height of fall	2.50
Helmet	B
α (deg) at impact	20°
β (deg) max	40°
Age / Sex	74 F
Head circ.	.550
breadth	.137
length	.175
Head+neck weight	4.00
head	3.21
skull	.56

NECROPSY SYNOPSIS

CRANIUM

ENCEPHALON

Pressurization *mediocre*

Macroscopically *no visible injury*

Microscopically *corpus callosum . injury* AIS ~2

FACE *so called Lefort I Lefort II Lefort III fractures*

NECK

Locations	Ẍ	Ÿ	Z̈	γ r	$\gamma_{r\,3ms}$	S.I.	H.I.C.	θ(H.I.C.)
Frontal								ms
R side	130	74	81	150	~ 80	900	742	8.9
L side	87	44	58	96	~ 105	500	474	10,5
Occiput	110	45	90	145	~ 74	900	780	10
C. G.	120	47	100	~150	~ 95		540	

Max. impact force 1080 daN

CONCLUSIONS

Brain AIS 2

TEST No 90

IMPACT CONDITIONS

Height of fall	2.50
Helmet	B
α (deg) at impact	16°
β (deg) max	40°
Age / Sex	48 M
Head circ.	.580
breadth	.140
length	.196
Head+neck weight	5.28
head	4.00
skull	.91

NECROPSY SYNOPSIS

CRANIUM

ENCEPHALON

Pressurization *mediocre*

Macroscopically *no injury*

Microscopically *no injury*

FACE *so called Lefort I and Lefort II fractures*

NECK

Locations	Ẍ	Ÿ	Z̈	γ r	$\gamma_{r\,3ms}$	S.I.	H.I.C.	θ(H.I.C.)
Frontal								ms
R side	166	41	106	169	135	~1500	1264	8.5
L side	132	50	75	144	110	~ 1000	804	8.7
Occiput	125	92	160	188	125	1700	1400	8.8
C. G.	166	67	113	176	125		1185	8.8

Max. impact force 1560 daN

CONCLUSIONS

encephalon AIS 0

Fig. 15 - Summary of the test results

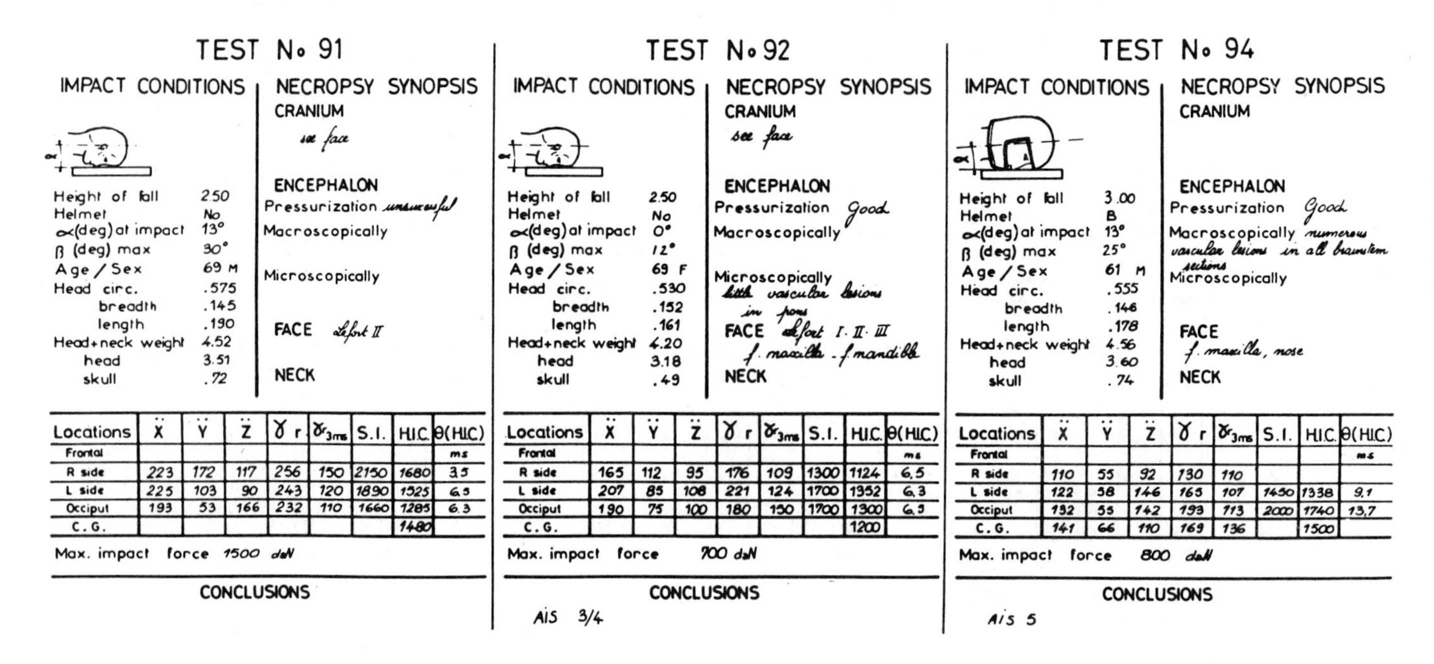

TEST No 91

IMPACT CONDITIONS

Height of fall	2.50
Helmet	No
α (deg) at impact	13°
β (deg) max	30°
Age / Sex	69 M
Head circ.	.575
breadth	.145
length	.190
Head+neck weight	4.52
head	3.51
skull	.72

NECROPSY SYNOPSIS

CRANIUM
see face

ENCEPHALON
Pressurization unsuccessful
Macroscopically
Microscopically

FACE Lefort II

NECK

Locations	$\ddot{X}$	$\ddot{Y}$	$\ddot{Z}$	γ r	γ_{3ms}	S.I.	H.I.C.	θ(H.I.C.)
Frontal								ms
R side	223	172	117	256	150	2150	1680	3.5
L side	225	103	90	243	120	1890	1525	6.5
Occiput	193	53	166	232	110	1660	1285	6.3
C.G.							1480	

Max. impact force 1500 daN

CONCLUSIONS

TEST No 92

IMPACT CONDITIONS

Height of fall	2.50
Helmet	No
α (deg) at impact	0°
β (deg) max	12°
Age / Sex	69 F
Head circ.	.530
breadth	.152
length	.161
Head+neck weight	4.20
head	3.18
skull	.49

NECROPSY SYNOPSIS

CRANIUM
see face

ENCEPHALON
Pressurization Good
Macroscopically
Microscopically
little vascular lesions in pons

FACE Lefort I · II · III
f. maxilla - f. mandible

NECK

Locations	$\ddot{X}$	$\ddot{Y}$	$\ddot{Z}$	γ r	γ_{3ms}	S.I.	H.I.C.	θ(H.I.C.)
Frontal								ms
R side	165	112	95	176	109	1300	1124	6,5
L side	207	85	108	221	124	1700	1352	6,3
Occiput	190	75	100	180	150	1700	1300	6,3
C.G.							1200	

Max. impact force 700 daN

CONCLUSIONS

AIS 3/4

TEST No 94

IMPACT CONDITIONS

Height of fall	3.00
Helmet	B
α (deg) at impact	13°
β (deg) max	25°
Age / Sex	61 M
Head circ.	.555
breadth	.146
length	.178
Head+neck weight	4.56
head	3.60
skull	.74

NECROPSY SYNOPSIS

CRANIUM

ENCEPHALON
Pressurization Good
Macroscopically numerous vascular lesions in all brainstem sections
Microscopically

FACE
f. maxilla, nose

NECK

Locations	$\ddot{X}$	$\ddot{Y}$	$\ddot{Z}$	γ r	γ_{3ms}	S.I.	H.I.C.	θ(H.I.C.)
Frontal								ms
R side	110	55	92	130	110			
L side	122	58	146	165	107	1450	1338	9,1
Occiput	132	55	142	193	113	2000	1740	13,7
C.G.	141	66	110	169	136		1500	

Max. impact force 800 daN

CONCLUSIONS

AIS 5

Fig. 16 - Summary of the test results

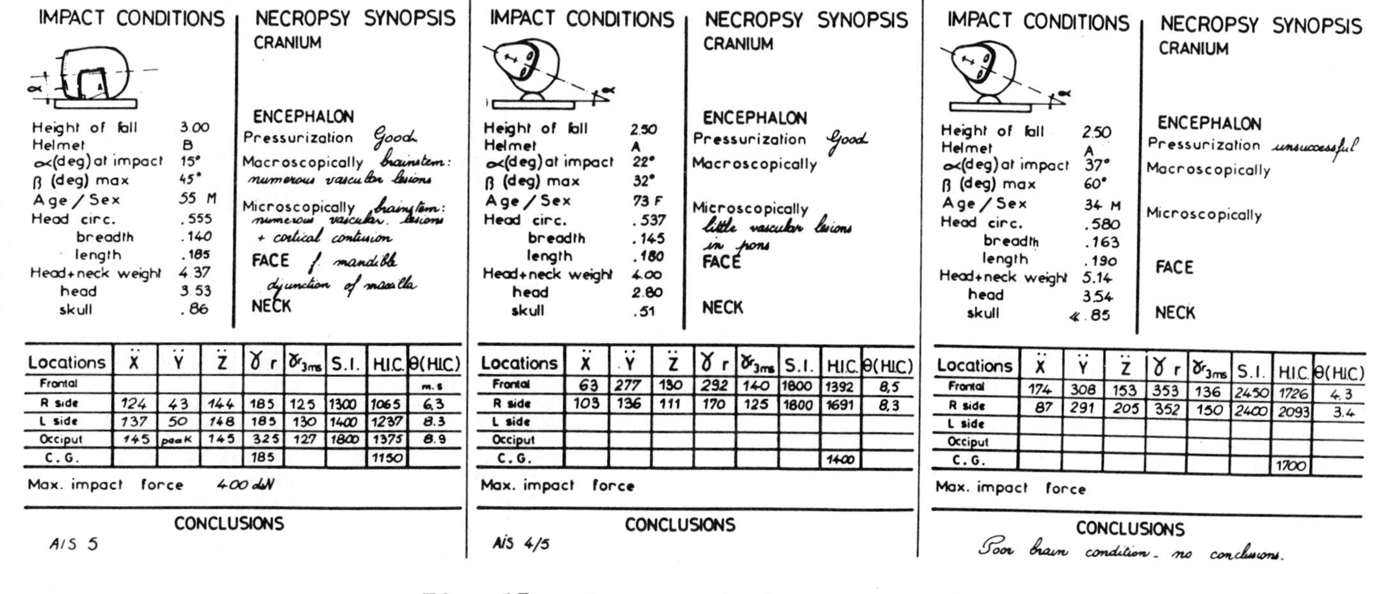

TEST No 95

IMPACT CONDITIONS

Height of fall	3.00
Helmet	B
α(deg) at impact	15°
β (deg) max	45°
Age / Sex	55 M
Head circ.	.555
breadth	.140
length	.185
Head+neck weight	4.37
head	3.53
skull	.86

NECROPSY SYNOPSIS

CRANIUM

ENCEPHALON

Pressurization Good

Macroscopically brainstem: numerous vascular lesions

Microscopically brainstem: numerous vascular lesions + cortical contusion

FACE f. mandible dyunction of maxilla

NECK

Locations	$\ddot{X}$	$\ddot{Y}$	$\ddot{Z}$	γr	γr_{3ms}	S.I.	H.I.C.	θ(H.I.C)
Frontal								m.s
R side	124	43	144	185	125	1300	1065	6,3
L side	137	50	148	185	130	1400	1237	8.3
Occiput	145	peak	145	325	127	1800	1375	8.9
C. G.				185			1150	

Max. impact force 400 daN

CONCLUSIONS

AIS 5

TEST No 99

IMPACT CONDITIONS

Height of fall	2.50
Helmet	A
α(deg) at impact	22°
β (deg) max	32°
Age / Sex	73 F
Head circ.	.537
breadth	.145
length	.180
Head+neck weight	4.00
head	2.80
skull	.51

NECROPSY SYNOPSIS

CRANIUM

ENCEPHALON

Pressurization Good

Macroscopically

Microscopically little vascular lesions in pons

FACE

NECK

Locations	$\ddot{X}$	$\ddot{Y}$	$\ddot{Z}$	γr	γr_{3ms}	S.I.	H.I.C.	θ(H.I.C)
Frontal	63	277	130	292	140	1800	1392	8,5
R side	103	136	111	170	125	1800	1691	8,3
L side								
Occiput								
C. G.							1400	

Max. impact force

CONCLUSIONS

AIS 4/5

TEST No 100

IMPACT CONDITIONS

Height of fall	2.50
Helmet	A
α(deg) at impact	37°
β (deg) max	60°
Age / Sex	34 M
Head circ.	.580
breadth	.163
length	.190
Head+neck weight	5.14
head	3.54
skull	« .85

NECROPSY SYNOPSIS

CRANIUM

ENCEPHALON

Pressurization unsuccessful

Macroscopically

Microscopically

FACE

NECK

Locations	$\ddot{X}$	$\ddot{Y}$	$\ddot{Z}$	γr	γr_{3ms}	S.I.	H.I.C.	θ(H.I.C)
Frontal	174	308	153	353	136	2450	1726	4.3
R side	87	291	205	352	150	2400	2093	3.4
L side								
Occiput								
C. G.							1700	

Max. impact force

CONCLUSIONS

Poor brain condition - no conclusions.

Fig. 17 - Summary of the test results

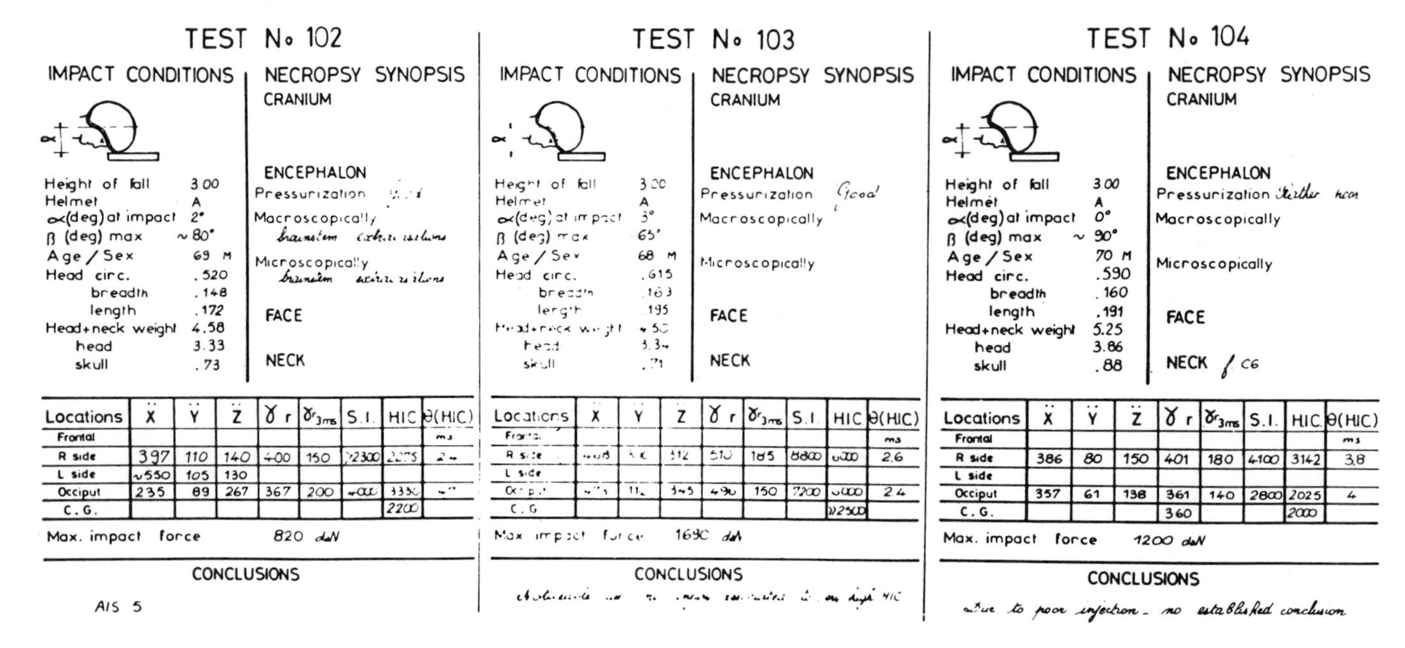

TEST No 102

IMPACT CONDITIONS

Height of fall	3.00
Helmet	A
α(deg) at impact	2°
β (deg) max	~ 80°
Age / Sex	69 M
Head circ.	.520
breadth	.148
length	.172
Head+neck weight	4.58
head	3.33
skull	.73

NECROPSY SYNOPSIS

CRANIUM

ENCEPHALON

Pressurization [illegible]

Macroscopically [illegible]

Microscopically [illegible]

FACE

NECK

Locations	$\ddot{X}$	$\ddot{Y}$	$\ddot{Z}$	γ_r	γ_{r3ms}	S.I.	HIC	θ(HIC)
Frontal								ms
R side	397	110	140	400	150	>2300	2275	[illegible]
L side	~550	105	130					
Occiput	235	89	267	367	200	[illegible]	3330	[illegible]
C.G.							2200	

Max. impact force 820 daN

CONCLUSIONS

AIS 5

TEST No 103

IMPACT CONDITIONS

Height of fall	3.00
Helmet	A
α(deg) at impact	3°
β (deg) max	65°
Age / Sex	68 M
Head circ.	.615
breadth	.163
length	.195
Head+neck weight	[illegible]
head	[illegible]
skull	[illegible]

NECROPSY SYNOPSIS

CRANIUM

ENCEPHALON

Pressurization Good

Macroscopically

Microscopically

FACE

NECK

Locations	$\ddot{X}$	$\ddot{Y}$	$\ddot{Z}$	γ_r	γ_{r3ms}	S.I.	HIC	θ(HIC)
Frontal								ms
R side	[illegible]	[illegible]	312	510	185	8800	[illegible]	2.6
L side								
Occiput	[illegible]	[illegible]	[illegible]	490	150	7200	[illegible]	2.4
C.G.							>>2500	

Max. impact force 1690 daN

CONCLUSIONS

[illegible]

TEST No 104

IMPACT CONDITIONS

Height of fall	3.00
Helmet	A
α(deg) at impact	0°
β (deg) max	~ 90°
Age / Sex	70 M
Head circ.	.590
breadth	.160
length	.191
Head+neck weight	5.25
head	3.86
skull	.88

NECROPSY SYNOPSIS

CRANIUM

ENCEPHALON

Pressurization [illegible]

Macroscopically

Microscopically

FACE

NECK f C6

Locations	$\ddot{X}$	$\ddot{Y}$	$\ddot{Z}$	γ_r	γ_{r3ms}	S.I.	HIC	θ(HIC)
Frontal								ms
R side	386	80	150	401	180	4100	3142	3.8
L side								
Occiput	357	61	138	361	140	2800	2025	4
C.G.				360			2000	

Max. impact force 1200 daN

CONCLUSIONS

Due to poor injection - no established conclusion

Fig. 18 - Summary of the test results

TEST No 107

IMPACT CONDITIONS

Height of fall	3.00
Helmet	C
α(deg) at impact	0°
β (deg) max	
Age / Sex	55 F
Head circ.	.520
breadth	.132
length	.175
Head+neck weight	3.57
head	2.32
skull	.38

NECROPSY SYNOPSIS

CRANIUM

ENCEPHALON

Pressurization *Good*

Macroscopically *nothing*

Microscopically *little extravasations in pons*

FACE

NECK

Locations	$\ddot{X}$	$\ddot{Y}$	$\ddot{Z}$	γ_r	$\gamma_{r_{3ms}}$	S.I.	H.I.C.	θ(H.I.C.) ms
Frontal								
R side	167	60	52	170	116	~800	712	7.1
L side		55	53					
Occiput	125	39	150	195	147	1200	1123	4.5
C.G.							700	

Max. impact force *520 daN*

CONCLUSIONS

AIS *unknown* . 3/4 ?

TEST No 108

IMPACT CONDITIONS

Height of fall	3.00
Helmet	C
α(deg) at impact	
β (deg) max	
Age / Sex	64 M
Head circ.	.558
breadth	.147
length	.188
Head+neck weight	5.30
head	3.78
skull	≪.97

NECROPSY SYNOPSIS

CRANIUM

ENCEPHALON

Pressurization *very good*

Macroscopically *nothing*

Microscopically *some extravasations in brainstem*

FACE

NECK

Locations	$\ddot{X}$	$\ddot{Y}$	$\ddot{Z}$	γ_r	$\gamma_{r_{3ms}}$	S.I.	H.I.C.	θ(H.I.C.) ms
Frontal								
R side	239	48	76	256	150	1800	1587	3,9
L side								
Occiput	206	28	276	346	170	4000	3351	3,7
C.G.							1600	

Max. impact force *1280 daN*

CONCLUSIONS

AIS 5: *possible, not sure*

TEST No 109

IMPACT CONDITIONS

Height of fall	3.00
Helmet	C
α(deg) at impact	5°
β (deg) max	80°
Age / Sex	68 F
Head circ.	.545
breadth	.143
length	.190
Head+neck weight	4.20
head	3.41
skull	.50

NECROPSY SYNOPSIS

CRANIUM

ENCEPHALON

Pressurization *unsuccessful*

Macroscopically

Microscopically

FACE

NECK

Locations	$\ddot{X}$	$\ddot{Y}$	$\ddot{Z}$	γ_r	$\gamma_{r_{3ms}}$	S.I.	H.I.C.	θ(H.I.C.) ms
Frontal								
R side	246	100	64	265	205	1300	1180	3,6
L side								
Occiput	200	60	105	238	160	1690	1337	4,8
C.G.							1200	

Max. impact force *1010 daN*

CONCLUSIONS

unserviceable.

Fig. 19 - Summary of the test results

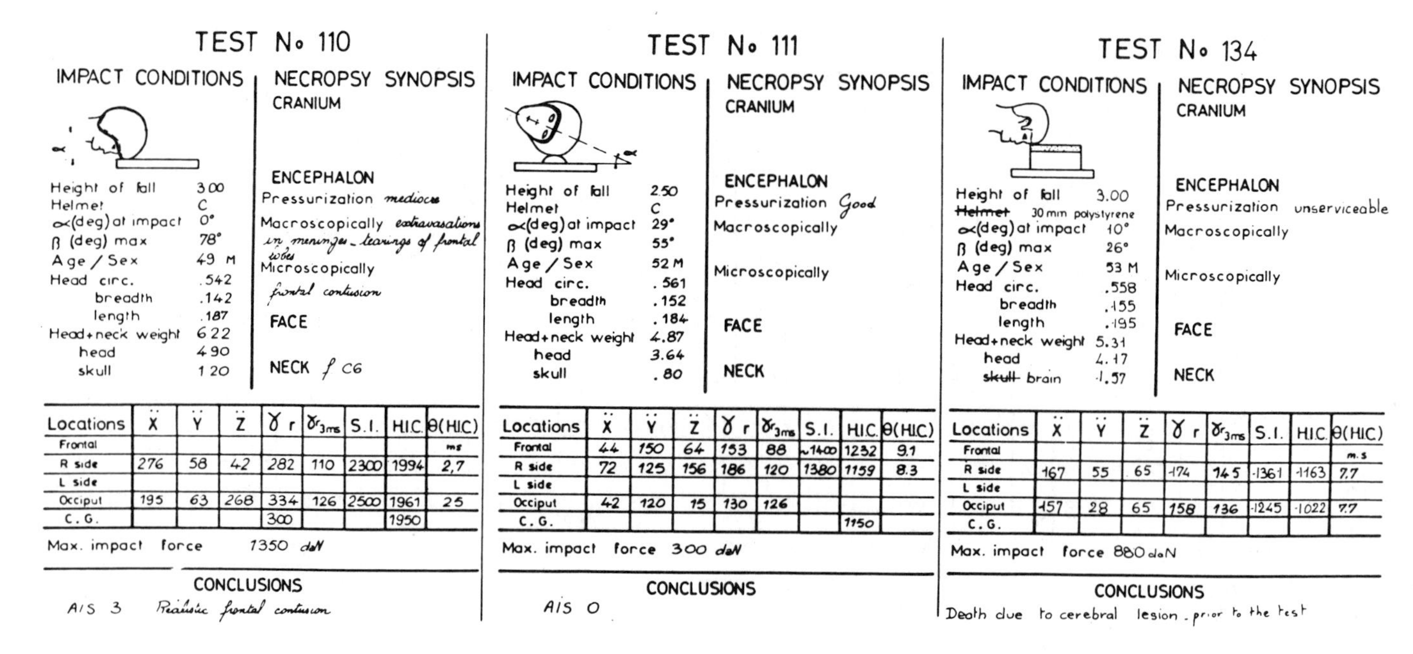

TEST No 110

IMPACT CONDITIONS

Height of fall	3.00
Helmet	C
α(deg) at impact	0°
β (deg) max	78°
Age / Sex	49 M
Head circ.	.542
breadth	.142
length	.187
Head+neck weight	6.22
head	4.90
skull	1.20

NECROPSY SYNOPSIS

CRANIUM

ENCEPHALON

Pressurization mediocre

Macroscopically extravasations in meninges – tearings of frontal lobes

Microscopically frontal contusion

FACE

NECK f C6

Locations	$\ddot{X}$	$\ddot{Y}$	$\ddot{Z}$	γ_r	$\gamma_{r_{3ms}}$	S.I.	H.I.C.	θ(H.I.C.)
Frontal								ms
R side	276	58	42	282	110	2300	1994	2,7
L side								
Occiput	195	63	268	334	126	2500	1961	25
C. G.				300			1950	

Max. impact force 1350 daN

CONCLUSIONS

AIS 3 Realistic frontal contusion

TEST No 111

IMPACT CONDITIONS

Height of fall	2.50
Helmet	C
α(deg) at impact	29°
β (deg) max	55°
Age / Sex	52 M
Head circ.	.561
breadth	.152
length	.184
Head+neck weight	4.87
head	3.64
skull	.80

NECROPSY SYNOPSIS

CRANIUM

ENCEPHALON

Pressurization Good

Macroscopically

Microscopically

FACE

NECK

Locations	$\ddot{X}$	$\ddot{Y}$	$\ddot{Z}$	γ_r	$\gamma_{r_{3ms}}$	S.I.	H.I.C.	θ(H.I.C.)
Frontal	44	150	64	153	88	~1400	1232	9.1
R side	72	125	156	186	120	1380	1159	8.3
L side								
Occiput	42	120	15	130	126			
C. G.							1150	

Max. impact force 300 daN

CONCLUSIONS

AIS 0

TEST No 134

IMPACT CONDITIONS

Height of fall	3.00
~~Helmet~~ 30 mm polystyrene	
α(deg) at impact	10°
β (deg) max	26°
Age / Sex	53 M
Head circ.	.558
breadth	.155
length	.195
Head+neck weight	5.31
head	4.17
~~skull~~ brain	1.57

NECROPSY SYNOPSIS

CRANIUM

ENCEPHALON

Pressurization unserviceable

Macroscopically

Microscopically

FACE

NECK

Locations	$\ddot{X}$	$\ddot{Y}$	$\ddot{Z}$	γ_r	$\gamma_{r_{3ms}}$	S.I.	H.I.C.	θ(H.I.C.)
Frontal								m.s
R side	167	55	65	174	145	1361	1163	7.7
L side								
Occiput	157	28	65	158	136	1245	1022	7.7
C. G.								

Max. impact force 880 daN

CONCLUSIONS

Death due to cerebral lesion - prior to the test

Fig. 20 - Summary of the test results

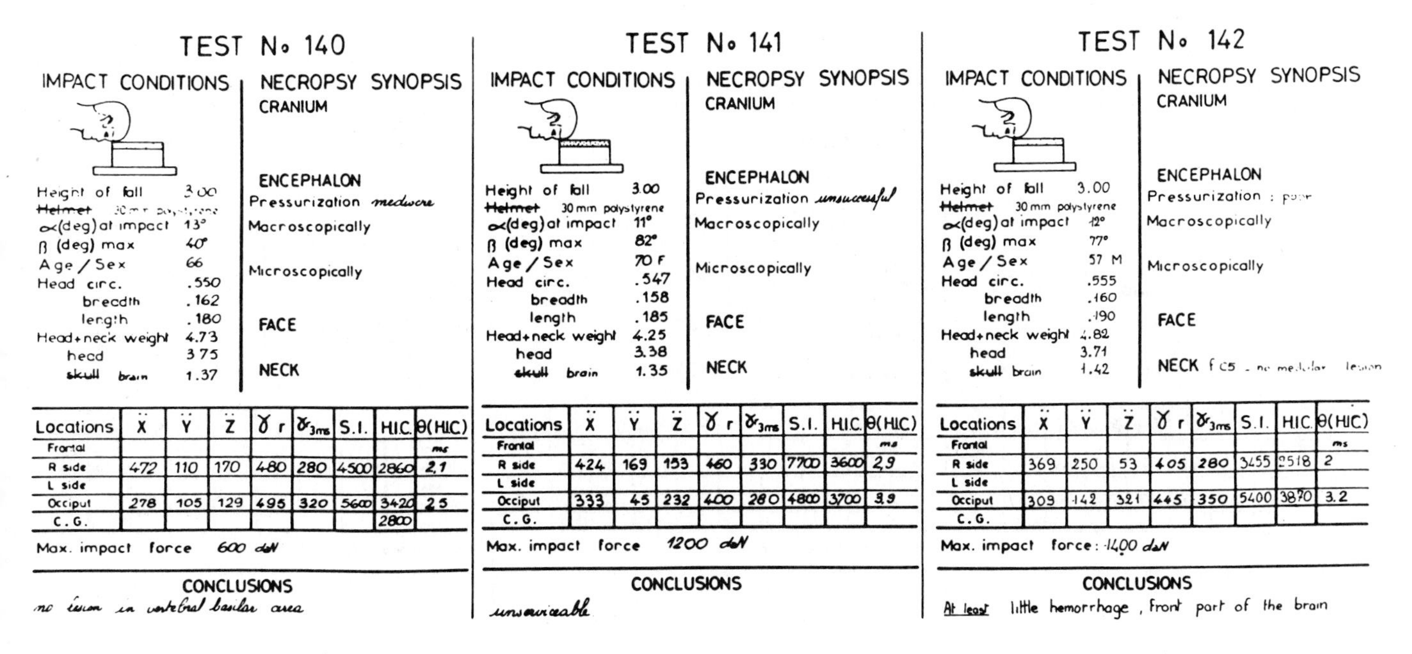

TEST No 140

IMPACT CONDITIONS

- Height of fall: 3.00
- ~~Helmet~~ 30 mm polystyrene
- α (deg) at impact: 13°
- β (deg) max: 40°
- Age / Sex: 66
- Head circ.: .550
- breadth: .162
- length: .180
- Head + neck weight: 4.73
- head: 3.75
- ~~skull~~ brain: 1.37

NECROPSY SYNOPSIS

CRANIUM

ENCEPHALON

Pressurization mediocre

Macroscopically

Microscopically

FACE

NECK

Locations	$\ddot{X}$	$\ddot{Y}$	$\ddot{Z}$	γ r	γ_{3ms}	S.I.	H.I.C.	θ(H.I.C.)
Frontal								ms
R side	472	110	170	480	280	4500	2860	2,1
L side								
Occiput	278	105	129	495	320	5600	3420	2,5
C.G.							2800	

Max. impact force 600 daN

CONCLUSIONS

no lesion in vertebral basilar area

TEST No 141

IMPACT CONDITIONS

- Height of fall: 3.00
- ~~Helmet~~ 30 mm polystyrene
- α (deg) at impact: 11°
- β (deg) max: 82°
- Age / Sex: 70 F
- Head circ.: .547
- breadth: .158
- length: .185
- Head + neck weight: 4.25
- head: 3.38
- ~~skull~~ brain: 1.35

NECROPSY SYNOPSIS

CRANIUM

ENCEPHALON

Pressurization unsuccessful

Macroscopically

Microscopically

FACE

NECK

Locations	$\ddot{X}$	$\ddot{Y}$	$\ddot{Z}$	γ r	γ_{3ms}	S.I.	H.I.C.	θ(H.I.C.)
Frontal								ms
R side	424	169	153	460	330	7700	3600	2,9
L side								
Occiput	333	45	232	400	280	4800	3700	3,9
C.G.								

Max. impact force 1200 daN

CONCLUSIONS

unserviceable

TEST No 142

IMPACT CONDITIONS

- Height of fall: 3.00
- ~~Helmet~~ 30 mm polystyrene
- α (deg) at impact: 12°
- β (deg) max: 77°
- Age / Sex: 57 M
- Head circ.: .555
- breadth: .160
- length: .190
- Head + neck weight: 4.82
- head: 3.71
- ~~skull~~ brain: 1.42

NECROPSY SYNOPSIS

CRANIUM

ENCEPHALON

Pressurization : poor

Macroscopically

Microscopically

FACE

NECK f C5 - no medullar lesion

Locations	$\ddot{X}$	$\ddot{Y}$	$\ddot{Z}$	γ r	γ_{3ms}	S.I.	H.I.C.	θ(H.I.C.)
Frontal								ms
R side	369	250	53	405	280	3455	2518	2
L side								
Occiput	309	142	321	445	350	5400	3870	3.2
C.G.								

Max. impact force: 1400 daN

CONCLUSIONS

At least little hemorrhage, front part of the brain

Fig. 21 - Summary of the test results

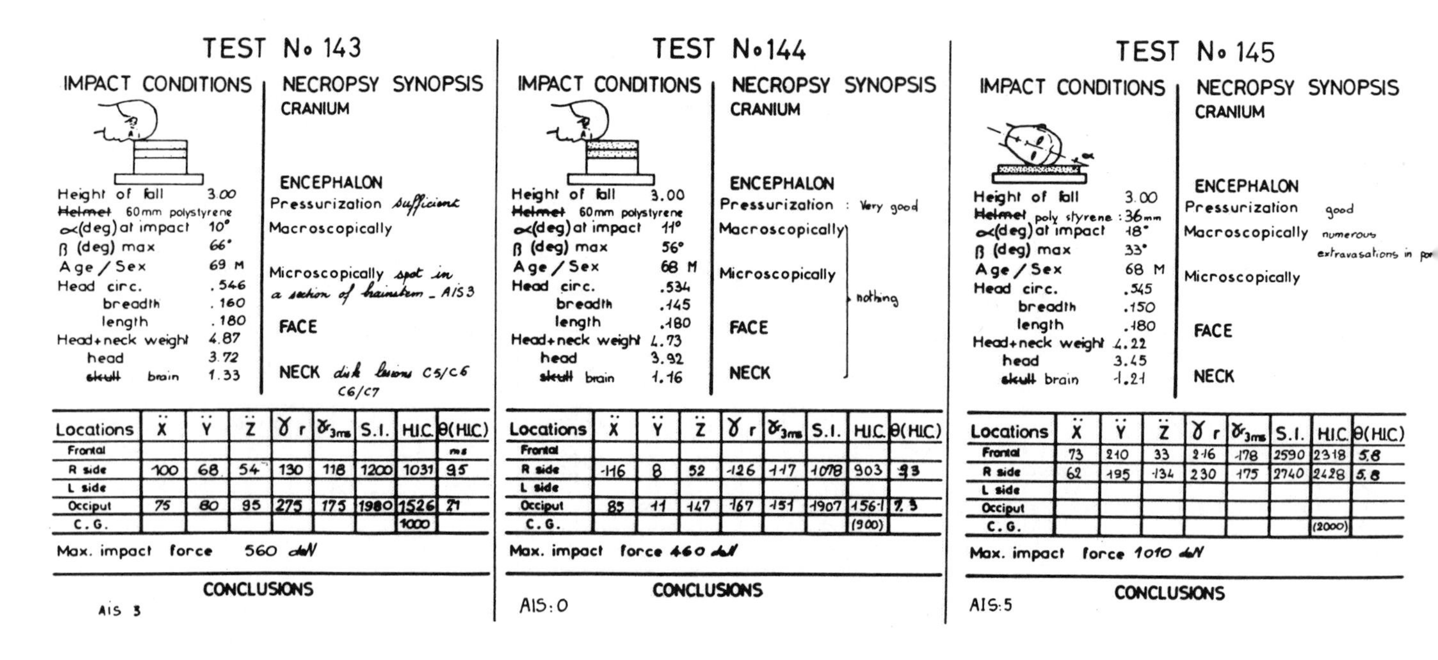

TEST No 143

IMPACT CONDITIONS

Height of fall 3.00
~~Helmet~~ 60mm polystyrene
α(deg) at impact 10°
β (deg) max 66°
Age / Sex 69 M
Head circ. .546
breadth .160
length .180
Head+neck weight 4.87
head 3.72
~~skull~~ brain 1.33

NECROPSY SYNOPSIS

CRANIUM

ENCEPHALON
Pressurization sufficient
Macroscopically
Microscopically spot in a section of brainstem _ AIS 3

FACE

NECK disk lesions C5/C6 C6/C7

Locations	$\ddot{X}$	$\ddot{Y}$	$\ddot{Z}$	γ_r	γ_{3ms}	S.I.	H.I.C.	θ(H.I.C.)
Frontal								ms
R side	100	68	54	130	118	1200	1031	9.5
L side								
Occiput	75	80	95	275	175	1980	1526	21
C. G.							1000	

Max. impact force 560 daN

CONCLUSIONS

AIS 3

TEST No 144

IMPACT CONDITIONS

Height of fall 3.00
~~Helmet~~ 60mm polystyrene
α(deg) at impact 11°
β (deg) max 56°
Age / Sex 68 M
Head circ. .534
breadth .145
length .180
Head+neck weight 4.73
head 3.92
~~skull~~ brain 1.16

NECROPSY SYNOPSIS

CRANIUM

ENCEPHALON
Pressurization : Very good
Macroscopically
Microscopically
FACE
NECK
(Macroscopically to NECK:) nothing

Locations	$\ddot{X}$	$\ddot{Y}$	$\ddot{Z}$	γ_r	γ_{3ms}	S.I.	H.I.C.	θ(H.I.C.)
Frontal								
R side	116	8	52	126	117	1078	903	9.3
L side								
Occiput	85	11	147	167	151	1907	1561	7.5
C. G.							(900)	

Max. impact force 460 daN

CONCLUSIONS

AIS: 0

TEST No 145

IMPACT CONDITIONS

Height of fall 3.00
~~Helmet~~ poly styrene : 36mm
α(deg) at impact 18°
β (deg) max 33°
Age / Sex 68 M
Head circ. .545
breadth .150
length .180
Head+neck weight 4.22
head 3.45
~~skull~~ brain 1.21

NECROPSY SYNOPSIS

CRANIUM

ENCEPHALON
Pressurization good
Macroscopically numerous extravasations in pon
Microscopically

FACE

NECK

Locations	$\ddot{X}$	$\ddot{Y}$	$\ddot{Z}$	γ_r	γ_{3ms}	S.I.	H.I.C.	θ(H.I.C.)
Frontal	73	210	33	216	178	2590	2318	5.8
R side	62	195	134	230	175	2740	2428	5.8
L side								
Occiput								
C. G.							(2000)	

Max. impact force 1010 daN

CONCLUSIONS

AIS: 5

Fig. 22 - Summary of the test results

TEST No 146

IMPACT CONDITIONS

Height of fall	3.00
Helmet	36mm polystyrene
α(deg) at impact	28°
β (deg) max	
Age / Sex	68 M
Head circ.	.548
breadth	.150
length	.180
Head+neck weight	4.96
head	3.91
skull	

NECROPSY SYNOPSIS

CRANIUM

ENCEPHALON

Pressurization *Good*

Macroscopically

Microscopically *cortical lesion, left side*

FACE

NECK

Locations	$\ddot{X}$	$\ddot{Y}$	$\ddot{Z}$	γ_r	γ_{r3ms}	S.I.	H.I.C.	θ(H.I.C.)
Frontal	42	150	68	166	140	1579	1315	6
R side	40	139	138	200	165	2258	1911	5.8
L side								
Occiput	12	122	95	149	130	1250	1114	6.8
C.G.							1200	

Max. impact force 690 daN

CONCLUSIONS

AIS 2/3 - flabby brain

TEST No 147

IMPACT CONDITIONS

Height of fall	3.00
Helmet	36mm polystyrene
α(deg) at impact	35°
β (deg) max	26°
Age / Sex	57 F
Head circ.	.538
breadth	.140
length	.190
Head+neck weight	4.42
head	3.59
skull	

NECROPSY SYNOPSIS

CRANIUM

ENCEPHALON

Pressurization *mediocre*

Macroscopically *brainstem lesions, mostly in pons*

Microscopically *brainstem lesions, mostly in pons*. AIS 5

FACE

NECK

Locations	$\ddot{X}$	$\ddot{Y}$	$\ddot{Z}$	γ_r	γ_{r3ms}	S.I.	H.I.C.	θ(H.I.C.)
Frontal	10	115	95	149	129	1200	1045	6.4
R side	23	80	166	183	157	1900	1695	6.2
L side	?							
Occiput	38	80	79	116	101	700	592	5.6
C.G.							800	

Max. impact force 500 daN

CONCLUSIONS

very flabby brain. case to be rejected

Fig. 23 - Summary of the test results

DISCUSSION

Discussion of validity of method of brain-injury detection

The criticisms that can be made of such a method are numerous, the most important being the following :

1. Perfusion of the encephalon is extremely irregular, and it is indispensable to eliminate those subjects who have undergone assisted ventilation and in whom brain injuries were present prior to stopping the reanimation procedures. In cases in which brain death has occurred at the same time as the "legal death", it is also possible to observe wide variations in perfusion; in privileged cases, virtually the entire capillary system will be injected, and, very often, the arteriolar injection is excellent but only small capillary areas have been perfused. In this case, the validity of the experiment is limited to exploration of the arteriolar injuries.

2. It is possible for the resistance of the vessels in a cadaver not to be the same as in a living individual. In fact, alterations by autolysis occur late in the vascular walls. It should be noted that experimental surgery have demonstrated the proper conservation of vessels by freezing. (Grafts can be performed after several days of conservation in a cold atmosphere). Clearly, is is imperative to have fast refrigeration of the cadavers used for these experiments.

3. The cadavers used do not belong to the same age groups as those of individuals who have died from accidents in real pathology. Older individuals can display lower vascular resistance than that of younger subjects, and the thresholds of tolerance will be lower than those of persons exposed to accidents. However, it is useful to know the thresholds for individuals who are a priori the least resistant.

4. The absence of muscular contraction in the neck facilitates large-scale relative movements between the head and neck. This relaxing of the cervical musculature can clearly foster the occurrence of fractures of the cervical column. In the absence of these fractures, it is difficult to assert that this relative movement can generate injuries of the bulbo-medullary junction of above it.

5. Recent publications have emphasized the frequency of ruptures of the nerve fibres in cervical traumatisms; the morphological malfunctions that occur, in particular the "ball injuries" are not found in cadavers. Tolerance thresholds that have vascular ruptures as their sole criterion are hence incomplete.

6. The injuries that occur at secondary level are not observable in cadavers, and we know the significance of the cerebral oedema in the development pattern of cerebral traumatisms.

7. Nothing is known concerning the clinical correspondence of minimum experimental injuries. The pathologists are familiar with lethal injuries that can be investigated during autopsies; however, less is known about encephalic injuries that justify keeping an individual alive by means of intensive care. They can be observed morphologically, in in particular as concerns the brain stem, only in cases in which death occurs secondarily as the result of an injury outside the skull.It is therefore difficult to indicate an A.I.S. value for minimal experimental injuries. There is reason to hope that the continual development of techniques for exploring the central nervous system, (scanning, e.g.) will enable comparisons between non fatal real injuries and small-scale experimental injuries. Taking into account the criticisms, one may conclude that the utilized method is particularly interesting when perfusion succeeds and when no evidence of injuries occurs in the same test.

Comparison of the present results against those of A.M. NAHUM

The reference is to publications (3) and (14), describing frontal impacts. The direction of these impacts differs a little from those presented in this paper. The pressurization method is very similar, but in addition, there was injection of a normal saline solution in order to restore the pressure and level in the space allotted to the cerebrospinal fluid, and specifically in that of the lateral ventricles (3).

Nahum performed tests with the heads raised, whereas the tests described in this report were performed with heads lowered, a fact that perhaps minimizes the problems related to the cephalo-rachidian liquid.

At the moment of impact, the heads of the subjects tested by Nahum were inclined approximately 45° forward. This procedure reduces neck extension and changes the direction of the principal accelerations as compared with those in the present tests. In (3), injuries were found frequently, which occurred on the surface of the encephalon of varying degrees of gravity. No injuries of the brain stem were found.

The highest H.I.C. without injuries was 845.

The highest H.I.C. with moderate injuries was 1316.

The lowest H.I.C. with severe injuries was 657. This value is associated with a high acceleration peak (290 g) and a low period of calculation of the H.I.C.(0.7ms), due to a striker that was only thinly padded.

In (14), there is no description of injuries.

The highest H.I.C. without injuries is 820.

The highest H.I.C. with moderate injuries is 3765.

There are no cases of severe injuries.

The H.I.C. were calculated on the basis of the two largest acceleration components. The third component, which is transversal in relation to the head, was not measured. The H.I.C. can hence be slightly higher than the findings indicated.

The small number of test results available, as in the present report, means that caution must be exercised. Nahum's tests were performed under various conditions that apparently led to localizations of injuries different from those in that paper. It is possible that tolerance may be lower in such a case, notably for impacts of shorter duration. The impact time-period to be considered first is the one that intervenes in accidents against which protection has been provided. The time periods for calculating the H.I.C. of (3) are all under 4.2 ms; in the present paper, for helmeted subjects, they are approximately the same in cases of frontal impact, and longer in cases of temporal-parietal impact.

Comparison of the present results against those of R.L. STALNAKER

The reference is (4).

The impacts sustained by fifteen fresh cadavers were distributed as follows :

- five frontal,
- five lateral,
- five occipital.

All impact directions were perpendicular to a vertical of the head and so differ a little from ours, as regards lateral ones.

As in Nahum's cases, the subjects were in a sitting position. There was pressurization, but there was no intervention at the level of the cephalo-rachidian liquid.

The H.I.C. was not calculated. However, the γ (t) acceleration curves calculated for the center of gravity of the heads have been published for four tests, thereby, enabling us to make the evaluations listed below.

Stalnaker reference	Type of impact	G.S.I.	H.I.C.	A.I.S. head
75 A 135	occipital	650	530	0
76 A 137	occipital	1800	1600	4
76 A 144	frontal	3400	3200	2
76 A 145	lateral	4000	2700	5

The figures listed above correspond to filtering at 450 Hz.

We did not perform occipital impacts, which seldom occur in real-life accidents. Test 76 A 144 is another case of high H.I.C. with moderate injuries. The A.I.S. 5 obtained with the parietal-temporal impact 76 A 145 shows a value considerably higher than the level of probable tolerance.

There is therefore no contradiction between our results and those published by Stalnaker.

GENERAL REMARK

In the view of the writers, the scatter of the H.I.C. likely to produce experimental vascular ruptures is not of a nature such as to cast doubts on the value of these investigations. The variations in the skull skeleton (shape, masses, mineralization), in the mass of the encephalon, and in the stress exerted on the structures (direction, time length) afford no hope of finding a simple solution to the problem in the form of some "magic" formula that would determine the threshold of the onset of injuries.

Investigations using injections of the cerebral vascular system before, during, and after an experimental traumatism can be interpreted restrictively. A good-quality injection, without occurrence of vascular ruptures, situates the impact in the tolerance zone for this criterion of integrity of vascularization. However, we find that acceptance of an experimental injury of the vessels as the criterion for exceeding the injury threshold seems insufficiently proved in the present state of knowledge. This allows us to be optimistic in estimating the tolerance levels.

Relations with experiments carried out with dummies

Some of the tests effected with cadavers and which are reported here have been reproduced with "Part 572" dummies under conditions as near as possible to those of the tests with cadavers. The small amount of tests that have been made and the variety of the results prevent us from being too systêmatic, nevertheless, it has been possible to make a few observations.

The degree of similarity of kinematics and acceleration between the tests with dummies and the tests with cadavers differs in accordance with the three categories of impacts considered in this study (temporo-parietal, frontal, frontal-facial). Accelerations measured are shown in the table 2
In the case of temporo-parietal impact, when the severity of the impact is near to the limits of the capacity of attenuation of the impact by the helmet, results are very similar in terms of acceleration, whether with a "Part 572" dummy or with cadavers; (e.g. impacts on steel hemisphere of safety standard 218). The same does not apply to less severe impacts (impacts on flat surfaces) with regard to the helmet. The H.I.C. measured on "Part 572" dummies are then less than the H.I.C. of the subjects; nevertheless, the γ 3ms remain close.

TABLE 2 - Compared measureaments results of dummies and cadavers

Type of test	Test N°	Type of helmet	Height of fall (m)	Results of measurements			
				Cadavers		Part 572	
				γ 3ms	HIC	γ 3ms	HIC
	69	A	2.5	145	1200	138	1125
	70						
	73	B	2.5	170	1750	138	1125
	74			140	2250		
	83	A	3	135	1500		
	85			180	1640	140	1169
Temporo -	86			175	1600		
	87	B	3	140	1350	158	1514
parietal	99	A	2.5 against steel hemisphere	140	1400		
	100			140	1700		
	111			85	1150	83	1594
	145	polystyrène 36 mm	3	135	1200	124	995
	147			130	800	120	950
	102	A	3	150	2200		
Frontal	103			185	2500	30	3454
	104			180	2000		

Frontal	107	C	3	120	700	124	939
	108			150	1600		
	109			130	1200		
	110			115	1250		
Fronto-facial	88	B	1.83	95	565	120	1406
	89		2.50	95	540	130	3400
	90		2.50	125	1185	150	4150
	94		3	136	1480	130	4300
	95		3	125	1150		

With regard to frontal impacts, results of the comparison vary widely depending upon the helmet and the test, for the phenomena which occur at a point of impact near to the edge of the helmet have a poor repeatability.

When the helmet is subject to a severe test, acceleration of the head of the dummy looks higher; otherwise, results of the same order are found for "Part 572" dummies and for cadavers.

The above results become coherent when considering the severety of the tests applied to the helmet, frontal impacts are placed in the same class as temporo-lateral impacts for which a smaller quantity of shock-absorbent material is used.

They are explained if allowance is made for the greater mass of the "Part 572" dummy head and its lack of deformability?

With regard to frontal-facial impacts, no comparison is possible because of the difference of rigidity between the human face (which breaks) and the face of the dummy. Accelerations and indices are much higher with a "Part 572" dummy (= 4150, for the "Part 572" dummy instead of 1480, for a subject, for example).

The "Part 572" dummy is not convenient for simulation of impacts where the face is concerned.

These few results confirm that there is no constant ratio between acceleration measured on a dummy and those measured on a cadaver in an identical test but that the relation is variable depending upon the conditions of the test.

CONCLUSIONS

1. Injection into the circulatory system provides a means of detecting injuries, resulting from vascular rupture on fresh cadavers submitted to experiment. This kind of technique has been used successfully by the writers since 1973 and is a privileged means of evaluating risk of cerebral injury.
2. A high frequency of injury to the brain stem has been revealed in the case of cadavers which were injured during experiments, mainly for the subjects falling from a height of 3 meters.
3. Research for the correlation between A.I.S. and H.I.C. has met with a variety of difficulties.
   - Firstly, individual tolerances are very scattered.
   - Secondly, the type of variables raises special problems due to definition of A.I.S. by a figure, approximate determination of gravity of injuries, mathematical definition of H.I.C., etc...
   - Finally, the analysis that would provide the means of separating the influence of various parameters available after experiments has not yet started.

4. It has nevertheless been possible to set a threshold for the appearance of severe injuries during temporo-parietal impacts without fractures because no grave injuries have been found in this type of impact - beneath an H.I.C. value of about 1500.

It should be noted that this threshold at which injuries appear is different from the "tolerance level", this last magnitude is necessarily higher.

5. With regard to pure frontal impact without fractures, it would seem that an acceleration pulse linked to an H.I.C. exceeding 1500, could be supported by the head of a living person (a priori, more tolerant than that of a cadaver) without severe injury in the majority of cases. A greater number of experiments are needed to confirm this result.

6. Impacts that give rise to injury of the facial bones, injuries that occur at lower overall acceleration levels and under conditions of impact for which simulation by "Part 572" dummies is unsatisfactory, must be left aside.

7. From the experimental point of view, the state of the brain of the subject before the impact must be checked as being suitable, by appropriate tests which will only be possible to carry out after the impact. The characteristics of condition of the cranium must be determined because it is possible that the mechanical characteristics of the cranium are linked with cerebral tolerance. Finally because this is due to the variety of typologies of injuries in connection to the conditions under which they appear, a large number of directions and conditions of impact shall have to be considered.

This research was performed with the collaboration of Anatomy Laboratory of the Biomedical Research and Teaching Dept. (UER) of René Descartes University, Paris, Head: Pr. Delmas.

This work was supported as part of the French Thematic Action Program; Opinions of the Authors

REFERENCES

1. H. Metz, J. McElhaney, A. Ommaya : "A Comparison of the Elasticity of Live, Dead and Fixed Brain Tissue". J. Biomechanics, 3, No 4, July I970.

2. D.R. Foust, B.M. Bowan, R.G. Snyder : "Study of Human Impact Tolerance Using Investigations and Simulations of Free Falls". Proceedings of 21st Stapp Car Crash Conference, SAE paper 770 915.

3. A.M. Nahum, R.W. Smith : "An Experimental Model for Closed Head Impact Injury". Proceedings of 20th Stapp Car Crash Conference, SAE paper 760 825.

4. R.L. Stalnaker, J.W. Melvin, G.S. Nusholtz, N.M. Alem, J.B. Benson : "Head Impact Response"". Proceedings of 21st Stapp Car Crash Conference, SAE paper 770 921.

5. F.J. Unterharnscheidt : "Die Traumatischen Hirnschäden. Mechanogenese, Pathomorphologie und Klinik".Z. für Rechtsmedizin, 71, 153-221 (1972) - J.F. Bergmann, München.

6. R.L. Friede : "The Pathology and Mechanics of Experimental Cerebral Concussion". W.A.D.D. Technical Report, 61-256.

7. A. Fayon, C. Tarrière, G. Walfisch, C. Got, A. Patel : "Performance of Helmets and Contribution to the Definition of the Tolerances of the Human Head to Impact". Proceedings of the 3rd International Conference IRCOBI, IRCOBI Secretariate, 109 Avenue Salvador Allende, (69) Bron, France.

8. A. Fayon, C. Tarrière, G. Walfisch, C. Got, A. Patel : "Contribution to the Definition of the Tolerances of the Human Head to Impact". Proceedings of the 6th ESV Conference, Washington, Oct. I976.

9. L.B. Walker, E.H. Harris, V.R. Pontius :"Mass, Volume, Center of Mass and Mass Moment of Inertia of Head and Head and Neck of Human Body". Proceedings of 17th Stapp Car Crash Conference, SAE, 1973.

10. C.E. Ewing, D.J. Thomas :"Torque Versus Angular Displacement Response of Human Head". Proceedings of 17th Stapp Car Crash Conference, SAE, I973

11. J.M. Mc Elhaney, R.L. Syalnaker, V.L. Roberts : "Bio-mechanical Aspects of Head Injury", in Human Impact Response Measurement and Simulation, W.F. King and H.J. Mertz, editors, Plenum Press, 1973.

12. A.K. Ommaya, P. Yarnell, A.E. Hirsch, E.H. Harris : "Scaling of Experimental Data on Cerebral Concussions in Sub-Human Primates to Concussion Threshold for Man". Proceedings of 11th Stapp Car Crash Conference, SAE, 1967

13. A.K. Ommaya, A.E. Hirsch : "Tolerances for Cerebral Concussion from Head Impact and Whiplash in Primates". J. of Biomechanics 4, 13, 1971.

14. A.M. Nahum, R. Smith, C.C. Ward : "Intracranial Pressure Dynamics During Head Impact". Proceedings of 21st Stapp Car Crash Conference, SAE, 1977.

791023

Acceleration Induced Shear Strains in a Monkey Brain Hemisection

V. R. Hodgson and L. M. Thomas
Wayne State University

ABSTRACT

A freshly dead Stumptail (Macaca speciosa) monkey brain hemisection model has been subjected to translation, pure rotation and a combination motion. Linear and angular head accelerations were measured as well as brain displacement relative to the skull and shear strain at several locations. Much higher than previously predicted shear strain was measured at acceleration levels which have been recorded during impacts which produced concussion in live monkeys. Pure rotation produced the highest, most diffuse and long lasting shear strain and brain displacement, while translation produced very low shear strain. Highest shear strain during rotation was recorded in the brainstem rather than on the periphery as many have predicted. Results suggest that the mechanism of brainstem injury, regardless of head motion, is due to shear caused by stretching of the cervical cord.

LINEAR ACCELERATION HEAD injury criteria are in use partially because of convincing arguments in their favor, convenience of linear acceleration measurements, nothing else is available for practical application and because of the difficulties in refuting the experimental evidence on which they are largely based.

Holbourn (1)*, and later principally Ommaya and associates, (2, 3, 4) have presented evidence and argued for rotational criteria, convinced that shear strains arise to disrupt brain tissue only during rotation. Much of Ommaya's evidence was developed on experiments designed to separate translational and rotational motion of monkeys' heads. Cerebral concussion (cc) tests were conducted on monkeys with and without collars designed to immobilize the cervical spine (3), or with monkeys clamped into the HAD II device (4).

Unterharnscheidt and Higgins, (5) have also presented strong evidence for rotational criteria using the HAD II machine. Abel, et al, (6) have produced serious injuries in the monkey using a HYGE® actuated head rotation machine. The HYGE® machine rotation device produced pure rotation of the head about a fixed axis through the neck in a manner similar to the HAD II machine.

The Japanese team of Tsubokawa, et al, on the other hand have encased monkeys in a HYGE translation sled and produced concussion in degrees up to lethal (7).

*Numbers in parentheses designated References at end of paper.

Hodgson, et al, have found in two recent series of cc experiments with Stumptail (Macaca Speciosa) monkeys, wearing protective caps with head free to move that there was no correlation between angular acceleration and duration of unconsciousness (8, 9). They also found it possible to obtain correlation between linear acceleration parameters and cc only on side impacts (when compared to front, rear and top blows).

It is felt that head motion constraint has been an important difference in all of these experiments and has been largely responsible for the conflicting results. Separating translation and rotation in a living animal experiment is virtually impossible. The purpose of this research is to measure shear strain in a freshly dead hemisection monkey brain model, not subject to skull deformation and constrained to move in distinctly separated translational and rotational modes of accelerated motion.

METHODS

A. CONTAINER - An aluminum piece measuring 152 x 167 x 76 mm was machined out to form a hollow rigid block with wall thickness of 16 mm. The hole in the block was large enough to accept half a monkey head cut along the medial sagittal plane, with the soft outer tissue removed from the skull and the neck cut off at C7.

B. MOTION CONSTRAINTS

1. Translation - The container (block) can be constrained by a guide to

slide in translation (Fig. 1) when struck by a linear air propelled striker.

2. <u>Rotation Free</u> - When attached to the end of an arm, the block can be free to rotate not only about a fixed axis, but also about an axis on the end of the arm as shown in Fig 2. When the model is struck a blow tangential to the radius, it rotates about a fixed pillow block axis but not about an axis through its own center of gravity (CG).
3. <u>Rotation Fixed</u> - A pin can be inserted between the block and arm to prevent the block from turning with respect to the arm. When struck a blow tangential to the radius arm, the model rotates about the shaft mounted in the pillow block and about its own CG (see Fig. 3).

C. IMPACTOR - A 37 mm diameter steel cylinder containing a rubber tipped load cell on its striking end (Fig 1), was used to deliver blows to the aluminum block for purposes of producing accelerated motion. Weight of the striker was 22 N, approximately that of the block and contents. An air cylinder was used to propel the striker up to the desired speed and allow it to free-wheel before striking the test specimen.

D. HEMISECTION MODEL PREPARATION - The head of a female Stumptail monkey, which had to be sacrificed for other reasons, was decapitated at C7 and soft tissue was removed down to the

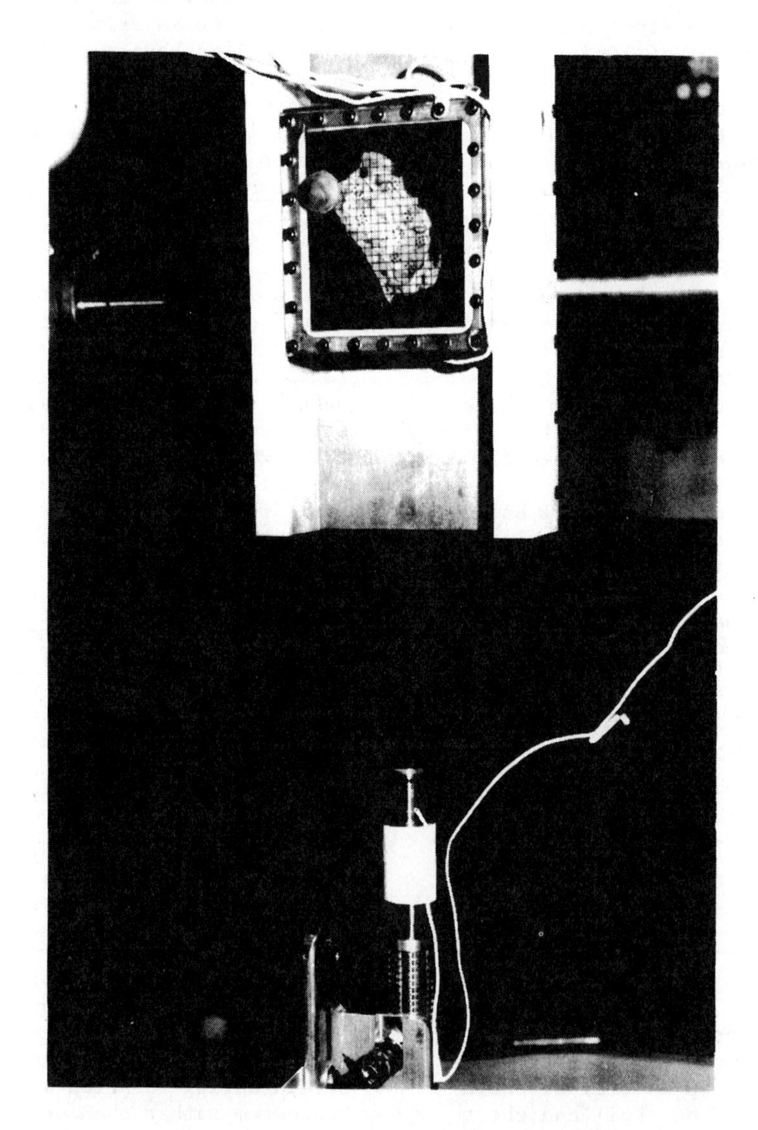

Fig. 1 - Hemisection model of a dog head showing strain markers (similar to the monkey model) in position for a front translational impact. Fluid filled cistern with elastic cover simulates the spinal cord-CSF-dura system

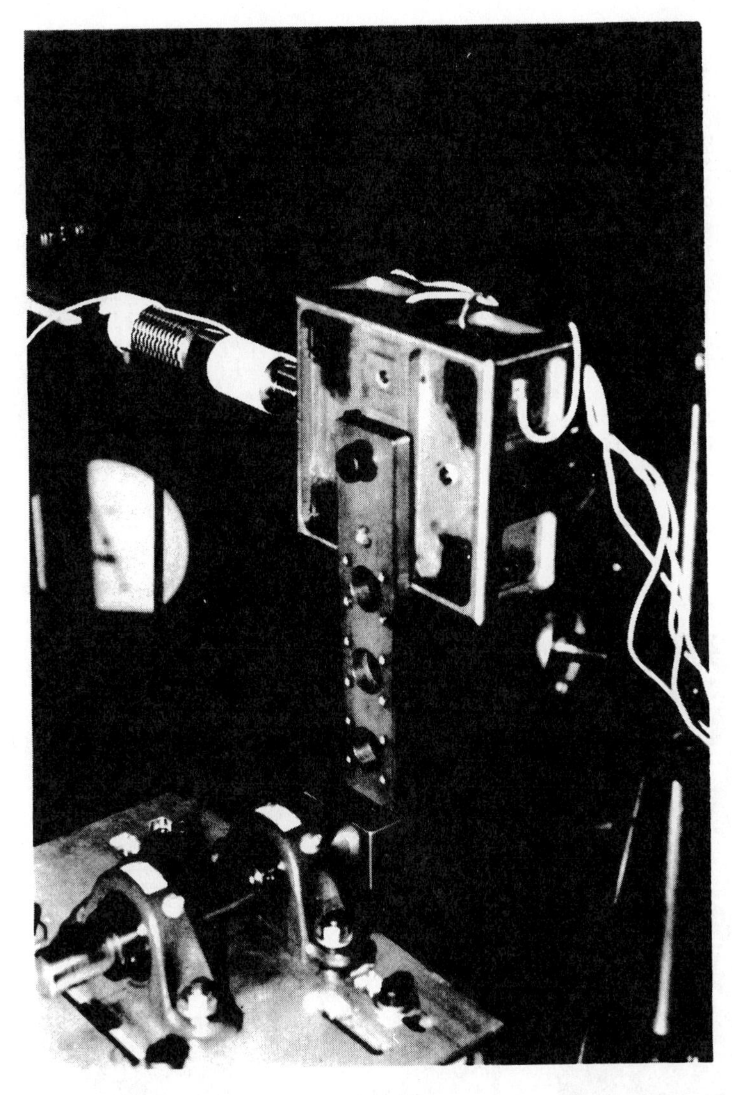

Fig. 2 - Rear view of hemisection model showing the mechanism which allows motion at several radii above the pillow block shaft but free movement about a bearing in the end of the arm and consequently no rotation about its own CG

skull and cervical spine.* The specimen was quick frozen and sawed to one side of the mid-sagittal plane. In the brain the largest of the two parts was sanded down to the falx cerebri which was then cut along the periphery and removed, exposing the left half of the brain and spinal cord.

Nylon brush bristles were exposed to heat, melting the ends to form a ball. The bristles were cut, leaving a stem of 3 mm on the balls. The stems were inserted into the brain surface to the level of the balls in tic-tack-toe patterns, with 5.1 mm spacing at several locations as shown in Fig. 4.

The hemisection was placed into the block using a resilient foam spacer to bring the sagittal surface into the viewing plane of the block. RTV rubber was poured around the specimen to keep it in position. A thin film of melted beeswax was layed down on the skull and spine edges. A 6.4 mm plexiglass window was clamped to the block using a gasket to seal its edges, and a gentle pressure was applied by the window to the soft wax on the bony edges of the skull and cervical spines to seal the brain cavity and spinal canal. The cervical cord

*Wayne State University is approved by the Animal Association of Accreditation for Laboratory Animal Care and the protocol for this experiment was reviewed and approved by the Animal Protection Subcommittee.

terminated in a location of the window in which a hole had been cut and out of which a 25 mm diameter plastic tube protruded 5 mm to form a cistern. A 10% saline solution was injected into a filler hole near the top of the window until all voids in the specimen were filled including ventricles and subdural spaces from which spinal fluid had leaked during the preparation. When the cistern was filled, it was capped with a surgical glove diaphram to simulate a near atmospheric fluid pressure which exists within the distensible spinal dura mater.

E. INSTRUMENTATION - Miniature linear accelerometers were mounted on the block to record linear acceleration in the line of force for translational impacts and to record tangential and radial acceleration for rotational impacts.

Velocity of the striker was recorded by measuring the time for the free-wheeling missile to pass two magnetic probes. A HYCAM rotating prism camera was focused on the surface of the specimen to record motion of the strain markers while operating at 4000 fps.

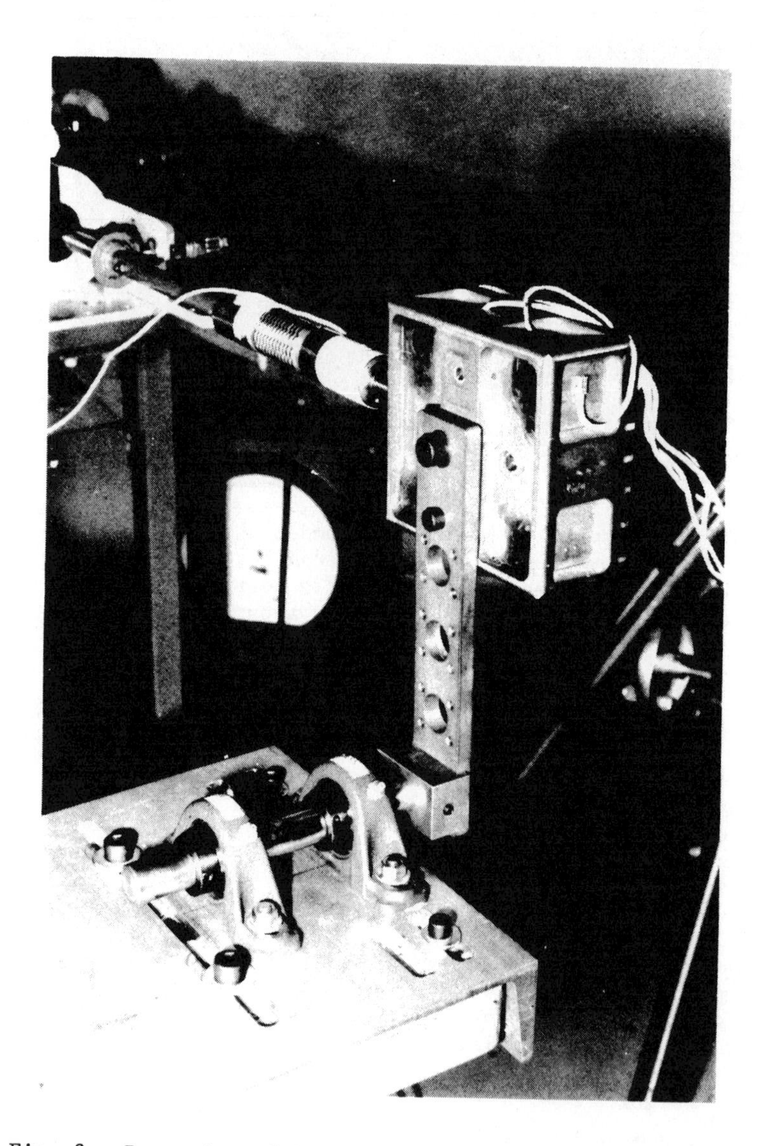

Fig. 3 - Rear view of hemisection model showing the mechanism which allows motion at several radii about the shaft with a bolt locking the arm and model thereby causing pure rotation

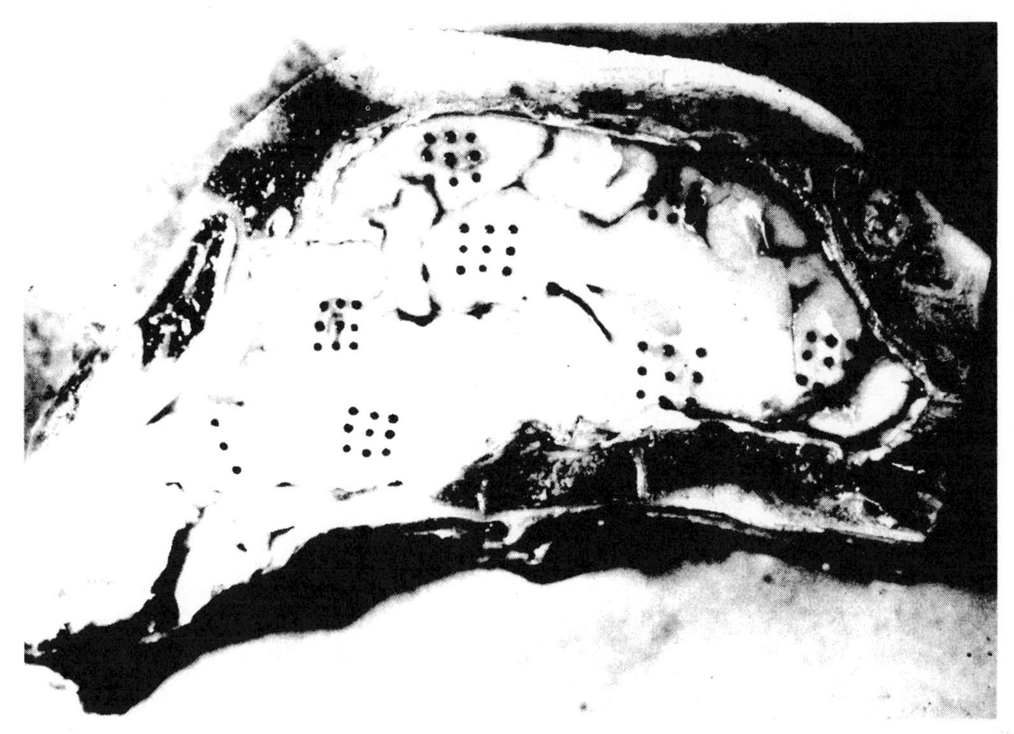

Fig. 4 - Closeup of shear strain markers implanted in a dog hemisection model typical of monkey brain implantation

RESULTS

The input to the hemisection monkey brain in terms of energy, force, linear acceleration, angular velocity and angular acceleration are given in Table 1 along with brain displacement and shear strain response for all tests.

ENERGY - Each of the three modes of motion were produced by an impact of the striker through the CG of the block. Three levels of energy were delivered by the 22 N rubber tipped striker, each level being approximately equal in the three modes of motion as indicated in column 3. The original intent had been to match the hemisection block mass, which turned out to be 24 N for the translation case and slightly greater for rotation, when the constraint arm is attached.

IMPULSE - Approximately equal force magnitude and pulse duration impulses were delivered at each energy level as listed in column 4 and 5, and illustrated in Fig. 5 for the maximum energy impact in each motion mode.

LINEAR ACCELERATION - The three levels of striker energy were designed to produce linear acceleration of the block, beginning at what is considered to be below injury level (380G) and proceding through an intermediate step (750 G), up to the 1200 G range which has been recorded during injury producing experiments with this species (9). These values are given in column 6.

ANGULAR VELOCITY AND ACCELERATION - Angular velocity values given in column 7 were obtained by measuring changes in angular displacement of the block for each frame of film which had been recorded at 4000 fps. The angular displacement data was then numercially differentiated in time increments of 0.00025 sec. Characteristically the angular velocity peaked within 1-2 ms and dropped to the average values listed within 4-5 ms for the rotation free cases. For rotation-fixed tests the angular velocity did not level off to the average values until 6-10 ms.

Angular acceleration values to the left of the slash mark in column 8 were obtained by numerical differentiation of the angular velocity, while those to the right were obtained as the quotient of tangential acceleration of the block CG divided by the 229 mm radius of rotation. Most confidence is placed in the angular acceleration values obtained from peak tangential linear acceleration measurements. These were obtained as the average of two linear accelerometer measurements made on the front and rear of the block by accelerometers which had been shock calibrated against the laboratory standard into the 1200 G range, with an accuracy of $\pm$ 5 percent.

Angular acceleration levels obtained by film analysis were limited in accuracy by the resolution of angular displacement changes made between frames on the film analyzer. This resolution was $\pm$ 0.003 rad for peak angular changes of 0.012 rad, giving an estimated error of $\pm$ 25 percent for angular acceleration measurements calculated by double numerical differentiation of the angular changes per 0.00025 sec time increment.

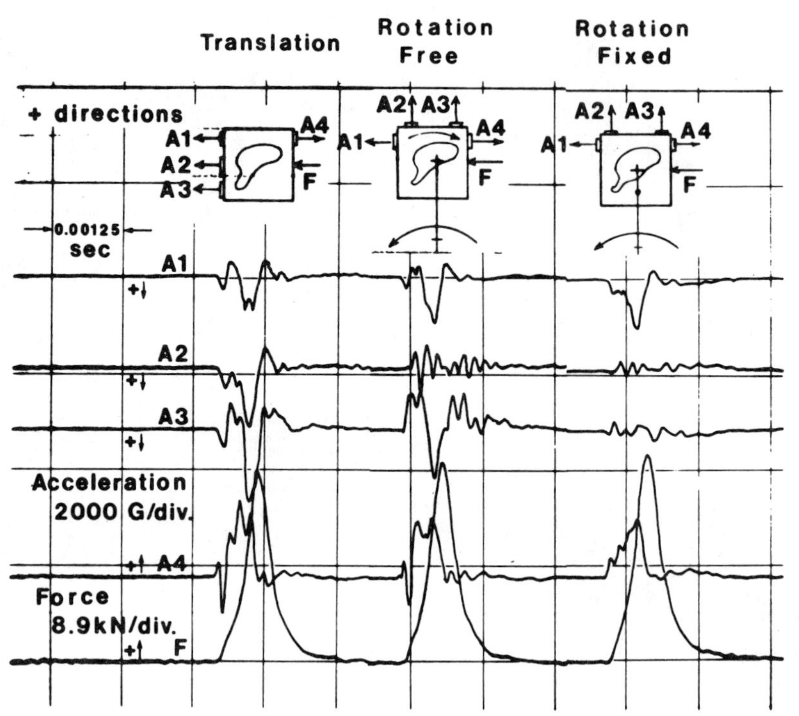

Fig. 5 - Comparison of impulse and linear acceleration response for the modes of motion at highest energy level

Table I

INPUT AND RESPONSE OF MONKEY HEAD HEMISECTION FOR THREE LEVELS OF ENERGY IN EACH OF THREE MODES OF MOTION

1	2	3	4	5	6	7	8	9	10	11	12	13	14
									SHEAR STRAIN - RADIANS (5)				
TEST	MODE	ENERGY - J -	FORCE - kN -	IMPACT DURATION - ms -	LINEAR ACCELERATION - G -	ANGULAR (1) VELOCITY - r/sec -	ANGULAR (2) ACCELERATION - r/sec² x 10³ -	BRAIN DISPLACE-MENT (3,5) - ms -	□ BRAIN STEM	△ CONTRE COUP	○ CORTEX	☆ FRONT LOBE	⬡ CEREBELLUM
1	Trans.	19.7	6.5	1.8	350	---	---	neg.	............MISSED DATA............				
2	Trans.	38.9	12.8	1.6	750	---	---	neg.	0.10	0.20	0.07	0.05	0.09
3	Trans.	55.8	18.0	1.4	1250	---	---	0.51	0.05	0.23	0.12	0.05	0.07
4	Rot. Free	22.2	7.4	1.7	380	$^{10}/_{9}$	$^{16}/_{16}$	0.6	$^{0.13}/_{0.10}$ (4)	$^{0.06}/_{0.11}$	0.07	0.09	0.05
5	Rot. Free	40.7	12.9	1.6	750	$^{18}/_{13}$	$^{35}/_{32}$	0.9	$^{0.13}/_{0.14}$	$^{0.15}/_{0.17}$	0.23	0.09	0.09
6	Rot. Free	55.6	18.7	1.4	1200	$^{22}/_{15}$	$^{62}/_{51}$	1.9	$^{0.13}/_{0.19}$	$^{0.16}/_{0.18}$	0.12	$^{0.09}/_{0.11}$	$^{0.06}/_{0.09}$
7	Rot. Fixed	21.1	6.9	1.7	380	$^{20}/_{8.5}$	$^{16}/_{16}$	1.8	0.19	$^{0.17}/_{0.20}$	0.08	0.06	0.07
8	Rot. Fixed	40.9	13.8	1.6	800	$^{30}/_{14}$	$^{48}/_{34}$	2.2	0.19	0.25	0.10	0.07	0.11
9	Rot. Fixed	57.4	19.4	1.4	1200	$^{36}/_{17}$	$^{96}/_{51}$	2.5	0.22	0.21	0.15	0.08	0.07

(1) Maximum amplitude attained in 1-2 ms/average amplitude after impulse.

(2) Values obtained by double differentiation of film/values obtained from linear tangential acceleration ÷ radius.

(3) Maximum amplitude of brain with respect to the skull in the superior cortex.

(4) Late surge at 15-25 ms indicated by shear strain level under diagonal.

(5) Brain displacement and shear strain measurement locations shown in Figure:

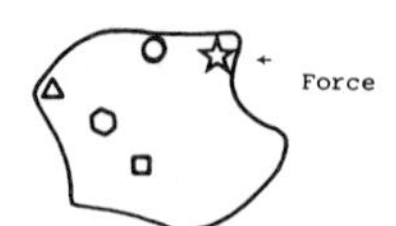

The angular acceleration obtained from the film analysis is higher in all except lower energy level impacts than the values obtained from linear acceleration measurements, because the latter are for the block CG while the film analysis included rotation of the block about its CG also. These differences are greater for fixed rotation than for the rotation free case. The range of peak angular acceleration values of

$$1.6 \times 10^4 - 9.6 \times 10^4 \text{ rad/sec}^2$$

compared to that reported by Abel, et al, (6) of

$$1.8 \times 10^4 - 1.2 \times 10^5 \text{ rad/sec}^2$$

BRAIN DISPLACEMENT - Brain displacement with respect to the skull is most critical around the superior sagittal sinus area where bridging veins between the arachnoid and dura may be ruptured due to stretching. Consequently, displacement of one of the brain surface markers located in the top of the model was monitored on film by means of the film film analyzer. Maxiumum excursion relative to the skull are given in column 9, the greatest being 2.5 mm for the high energy rotation-fixed case. Resolution of these movements was considered to be $\pm$ 0.5 mm.

SHEAR STRAIN - Listed in the last five columns are maximum values of shear strain measured at five locations in the model for all but the lowest level translation test. These represent maximum changes in an approximate right angle formed by three markers implanted in the brain tissue. Principally in the rotation-free tests, the motion of the block was such as to cause an initial shear distortion which is listed to the left of the diagonals in columns 10-14, followed by a later surge at some locations, as the model began rotating about its CG after becoming free from the striker. Maximum value of shear for these late surges is shown to the right of the diagonal in columns 10-14.

IMPLICATIONS OF THE RESULTS

BRAIN DISPLACEMENT MEASURMENTS - Viewing the films of the brain motion during and following impact leaves the viewer with the impression that there is little displacement or distortion in the translation tests, more activity for the rotation-free tests and maximum movement relative to the skull for the fixed rotation case. A comparison of the relatively similar impulse and linear acceleration oscillograms, shown in Figure 5 for the highest energy blow in each of the three modes of motion, does not give an indication of the graphic differences in internal motion viewed on film. The maximum displacements relative to the skull listed in the Table, column 9, show in a quantitative way that at least for the superior cortex location, the displacement becomes gradually more severe in the same order as the impressions left by viewing the film.

It is not certain what limits of stretch are critical for veins in the vicinity of the superior sagittal sinus of the female Stumptail monkey, but Abel, et al (6), produced subdural hematomas resulting from ruptured parasagittal bridging veins in 16 out of 50 impact tests using Rhesus (Macaca Mulatta) monkeys. Their setup produced similar initial head motion as the rotation-fixed case of the present series. The injuries were reported as correlating with peak values of tangential acceleration with onset occurring at values of 700 G. Angular acceleration ranges of the two experiments overlapped as shown above, however other differences in motion, as discussed below, may preclude correlating results.

Liu, et al (10), devised a mathematical model of viscoelastic material in a rigid spherical shell subject to rotation, similar to that of Bycroft (11), with the same no-slip condition at the brain-skull interface. From this they deduced that higher shear stresses are at the surface of the sphere and are attenuated toward the center. Their model predicts the possibility of subdural and subarachnoid injuries without subcortical involvement but the reverse is impossible. They also reported that their findings confirmed the neuropathological study of Unterharnscheidt and Higgins (5), using Squirrel (Saimiri Sciureus) monkeys encased in a fixed rotation setup designated as the HAD II apparatus. Unterharnscheidt and Higgins did not publish the shape of the pulse which they described as a triangular waveform. However, Gennarelli, et al (4), using the same equipment, did publish the waveforms and the positive acceleration waveform is followed by a negative acceleration of the head which is sometimes higher than the positive phase. The waveforms published by Abel, et al (6), on the HYGE machine are also rapid reversals of nearly equal tangential acceleration.

Both Unterharnsheidt and Higgins (5) as well as Abel, et al (6), produced subdural hemorrhages which they each attributed to tearing of bridging veins. Unterharnsheidt and Higgins stated that the range of available rotational acceleration was found to produce a continuum of clinic effect from no observable signs through concussion to death. Abel, et al, concluded that mid-sagittal plane angulations, resulted in injuries, the severity of which is strongly related, not only to the magnitude of peak positive angular acceleration, but also to peak positive tangential acceleration and to peak positive tangential force. It is difficult to understand how the findings on a mathematical model with a no-slip condition at the skull-brain interface can correlate with those obtained on experimental animals which suffered subdural hematomas due

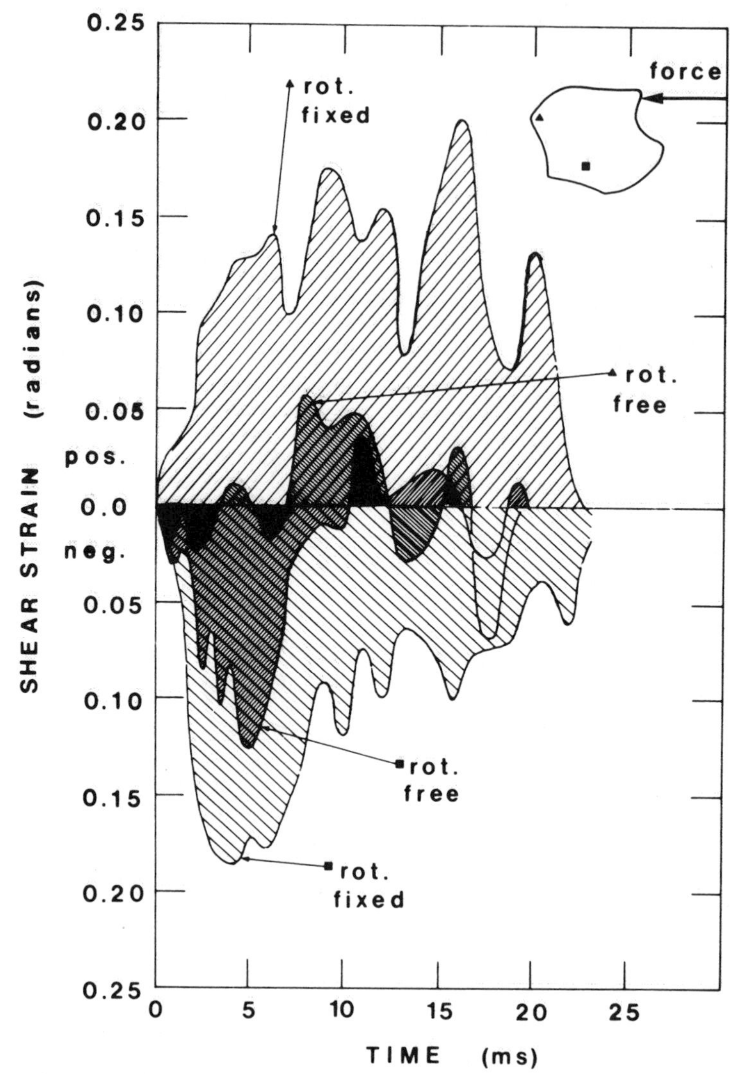

Fig. 6 - Comparison of shear strain in the contre coup location (top) and brainstem (lower) for rotation fixed and rotation free motions

to stretching of bridging veins between the brain and dura which is firmly adherent to the skull. Furthermore, until the effects of rapid reversals of nearly equal amplitude impulses to the head are investigated, conclusions about the relationships between kinematic and/or force parameters and pathology accomplished on these machines are premature.

SHEAR STRAIN - The present study may indicate why the HAD II and HYGE® head rotation machines are so productive of not only subdural hematomas but diffuse brain damage. Heads of animals in both cases are constrained to move in pure rotational motion. A study of Fig. 6 shows that when the hemisection block is allowed to rotate free on the end of the radius arm, at the lowest energy level, the shear strain both in the brain stem and contrecoup locations, are greatly reduced. (Also as shown in Table 1, column 9, the relative displacement between brain and skull in the region of the superior cortex is reduced). Figure 7 shows the shear strain as a function of time for all three modes of motion. Translation further relieves shear strain everywhere measured and as observed on film, except at the contrecoup location. The shear strain produced at the contrecoup measuring site in translation was due to rapid rarefaction and compression of entrained gases or air in the system and diminished to negligible amounts after 5 ms. Shear at other measuring sites in translation was relatively low. Rotation on the other hand, excited a shear response in the brain tissue more diffuse and much longer lasting than in the case of translation.

As has been pointed out by many workers beginning at least as early as Holbourn (1), the shear modulus of brain tissue (ratio of tangential force to change in angular deforma-

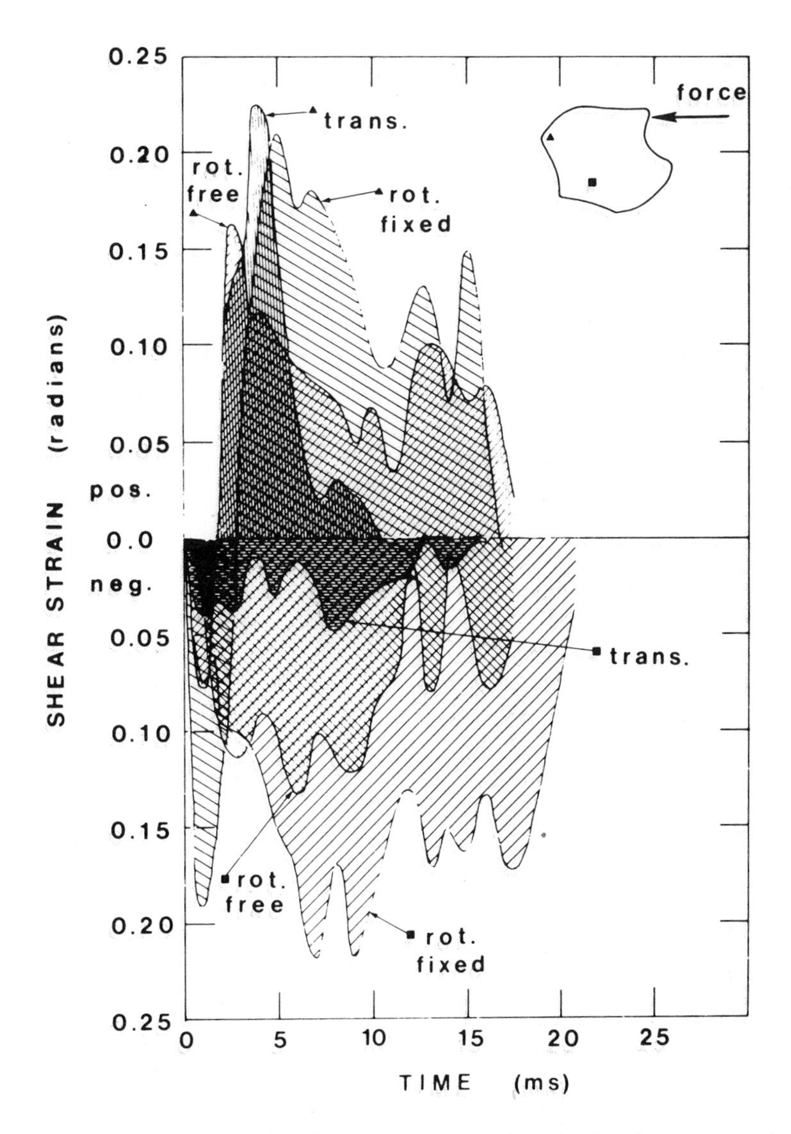

Fig. 7 - Comparison of shear strain in the brain stem and contre coup for the three modes of motion

ation) is very much lower than the bulk modulus (ratio of a compressing or distending force to volume change) and is therefore the most likely mode of failure. This great difference in moduli is also the reason for the difference in time response of brain tissue in translation compared to rotation motion. Shear strain produced at the contrecoup site in translation rises and diminishes rapidly because the brain-entrapped gas and air is a relatively stiff system in tension or compression. Shear strain produced in rotation, at other than the contrecoup site (where the degree of shear strain depends on both tangential linear acceleration and rotation) was relatively slow to rise and diminish because the brain is a soft system in this mode of distortion. Although the impulses were only 1.4 - 1.8 ms duration, the shear strain in rotation was active even as the measuring site went out of the camera field between 0.025 - 0.035 sec.

Some of this longer response may also be due to the radial acceleration, which exists in the system at a much lower level as long as the model rotates free after the impulse.

Fig. 8 demonstrates the distribution of shear strain as a function of time at the five measuring sites during the highest energy level of the fixed rotation case. The brain stem location, as at the other energy levels, generally displayed the highest and most sustained levels of shear strain. The contrecoup site was also high but part of this may have been artifactual distortion caused by more entrained gas being sucked to this area, due to autolysis, than would be present in live brain tissue. Shear strain at the peripheral (superior cortex) site which would be the predictable location of maximum strain by no-slip mathematical models, was generally lower than the brain stem.

Holbourn (1) reasoned and used a two

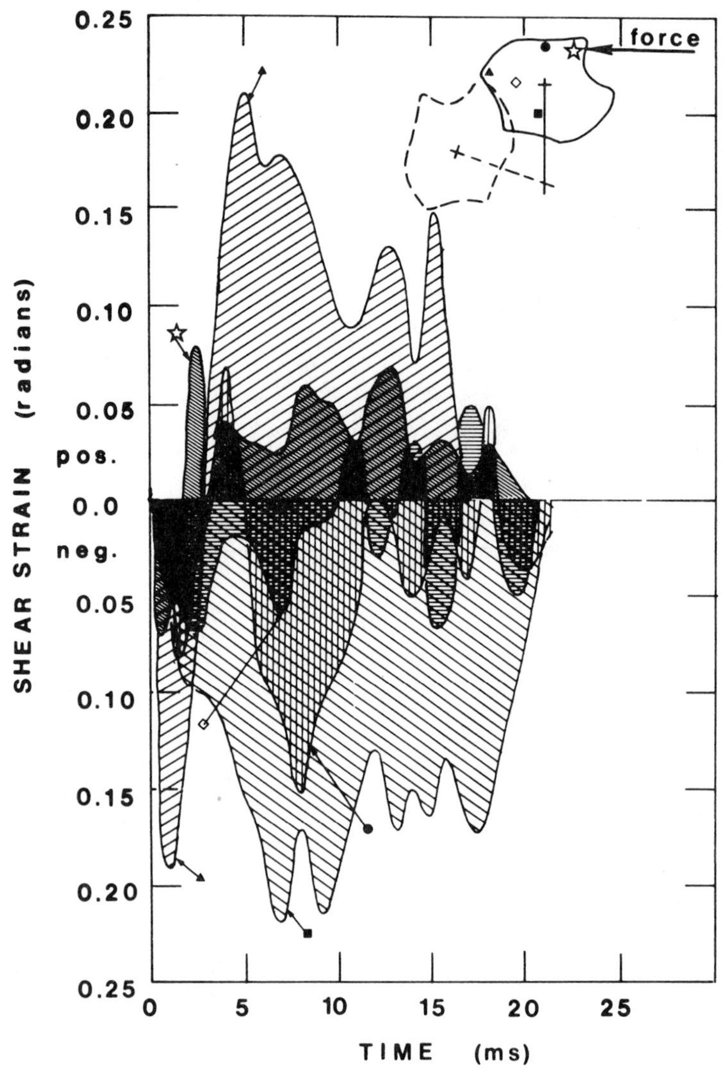

Fig. 8 - Distribution of shear strain for the rotation fixed case at highest energy level

dimensional gel model, to predict highest shear strain in locations where the skull gets a good grip on the brain and can rotate it. As such the frontal lobe location in the present model should have shown high shear strain, but it was generally, along with the cerebellum location, the lowest. The results do support his deduction that maximum sliding between pia and arachnoid would occur on the vertex since it is furthest removed from gripping areas, thereby making this the most likely site for hemorrhages due to stretched bridging veins. He recognized the deficiencies of his model in terms of its uniform elastic properties and lack of foramen magnum, hence acknowledging that there may be considerable deviation from his predictions.

The results also support those beginning at least as early as Holbourn, who cited rotation as being the most injurious mode of motion of the head. Fortunately, the fixed rotation case demonstrated in this study as well as on the HAD II and HYGE® rotation machines, is probably an extremely rare occurrence in human experience. The gripping action of impacting surfaces, as well as the neck linkage both tend to relieve the head from experiencing the pure rotation (fixed) condition, by constraining or allowing the head to move more like the head free or even the translation condition.

In two recent series of head injury research in the Stumptail monkey, the heads were protected against skull fracture and allowed to move free other than from neck constraint. Multiple blows spaced in days or weeks were delivered to several locations in each animal. Higher forces and linear and angular accelerations were experienced than in the present or other cited tests. Only short duration concussion with 15-100 sec periods of unconsciousness were produced as far as could be observed (9).

The experience in American football has been that concussions are commonplace, while subdural hematomas are extremely rare. In 1977 only three deaths resulted from subdural hematomas among 1,500,000 high school and junior high players (12). It is not known how many subdural hematomas were produced in total but it cannot be very many because as Gurdjian and Gurdjian have pointed out, acute subdural hematomas have a high mortality (13).

MAXIMUM VALUES OF SHEAR STRAIN - The values listed in Table 1 columns 10-14 are useful as a comparison with predicted brain injury levels of other workers. Bycroft (11) has predicted a value of 0.05 rad in the region of the upper reticular formation (Thomas 1970 (14)) as that necessary to cause concussion in man. He deduced this level from a mathematical model of a spherical viscoelastic medium attached to the inside of a rigid spherical shell (no-slip) subject to angular acceleration. He correlated his model analysis with the experimental head impacts of Ommaya, et al (2, 3, 4), and Unterharnscheidt and Higgins (5) who used Squirrel and Rhesus monkeys.

The brain stem shear strain location in the present series is approximately at the upper reticular formation level and the peak values recorded are around four times higher than those predicted by Bycroft. There were possibly higher shear strains than those locations at which markers were located in the brain hemisection. Bycroft's no-slip model is analogous to assuming that the outer surface of the brain moves with the skull.

MECHANISM OF CONCUSSION - Results of these tests are supportive but also refute part of the theories of both sides of the debate over the mechanics of injury to brain tissue which results in concussion. Pure rotation has been found to be most certainly the cause of higher, longer lasting and more diffuse shear strain and displacements than the other motions. However, both the experimental and analytical proponents of rotation have predicted that the shear strains are initiated at threshold energy levels on the periphery and procede inward as the impact intensity increases (10, 11, 17). The hemisection model results also support those such as Gurdjian, et al (15, 16), who deduced that the mechanism was due to the production of shear stress along the brain stem axis. However they assumed the shear stress to arise as the result of flow generated from intracranial pressure gradients. If this theory is correct it should hold for translation as well as rotation. In this model there was negligible movement or distortion of the brain stem in translation.

It is deduced from watching films of the present series that the reason there are higher shear strains in the brain stem during rotation is because when the brain is rotated with respect to the skull, it must pull on the spinal cord through the brain stem. It is well known that tension produces shearing stress which in many materials is the determining factor in failure rather than the tensile stress. The present model may be inadequate to show that such a shearing action may also be present in a live monkey subject to translation of the head, such as in the seris of tests by Tsubokawa, et al (7), and Komaki, et al (18). They accelerated the whole body of their primates and stopped the head first, followed by a more gradual deceleration of the body. No relation was found between brain injury and intracranial pressure, but they produced many neck injuries including one transection at C1 due to "severe jackknife effect at the neck joint (18)." Such a motion will move the head with respect to the neck, stretching the brain stem and thereby setting up shear strains to produce tissue damage. Tsubokawa, et al (7), reported that in dead animals, primary injury consisted of narrowing of lumen and stoppage of blood flow in perforating arteries of the brain stem, and small hemorrhages in the same area were noted.

In the present series, the spinal cord to C7 moved along with the block, preventing the possibility of such stretch in the brain stem suring translation, but allowing it to occur in rotation because there is some resistance of even the partial cord of this model to stretch when the brain rotates within the cranial cavity.* Friede (19) and Hollister, et al (20), have stressed the importance of cerebral concussion in the cat.

LINEAR HEAD INJURY INDICES - As indicated by the oscillograph record comparisons for the three modes of motion used in these tests, Fig. 5, a similar Head Injury Criterion (HIC) could be recorded for each case but with very different intracranial results. Peak shear strain values listed in the table do not show a definite pattern of increasing with either linear or angular acceleration. Some of this lack of pattern could be due to the fact that nine tests at such high shear strain levels were too destructive and the tissue could not transmit higher shear stress with increasing input.

Linear indices are based upon accelerometer measurements which reflect at least in some degree the intensity of both translation and rotation by picking up components of both. Also, it should be noted in the table that brain displacement with respect to the skull increased with increasing linear acceleration for both rotational motions. Therefore, these

*The denticulate ligament is a fibrous band of pia mater extending the entire length of the spinal cord on each side and is attached to the dura at 21 points beginning at the foramen magnum (21).

results serve to both justify continued use of linear indices in estimating the degree of trauma indicated by an instrumented human surrogate, at the same time pointing out possible shortcomings.

CONCLUSIONS

1. Fixed (pure) rotation produced the highest, most diffuse and longest lasting shear strain among the three modes of motion.

2. Maximum levels of shear strain

$$(0.20 - 0.25 \text{ rad}) \qquad (1)$$

measured at linear and rotational acceleration levels recorded during concussion producing live monkey experiments were about four times higher than predicted by mathematical models of a rigid hollow sphere containing a viscoelastic medium with no-slip at the interface.

3. Shear strain measurements for rotational motion were maximum in the brain stem rather than on the periphery as has been predicted by rotational proponents.

4. Shear strain measurements during translation were negligible in the brain stem.

5. It is deduced that the shear strain in the brain stem during rotation is due to stretching of the spinal cord.

6. The model is inadequate to show that stretching can occur in translation if relative motion between neck and head occurs.

ACKNOWLEDGEMENTS

This research was supported by the Office of Naval Research Contract Number N00014-75-C-0975. Matthew W. Mason was responsible for building the head holding fixtures.

REFERENCES

1. A.H.S. Holbourn, "Mechanics of Head Injuries." The Lancet, p. 438-440, October 9, 1943.

2. A.E. Hirsch, A.K. Ommaya and R.M. Mahone, "Tolerance of Subhuman Primate Brain To Cerebral Concusion." Report 2876, Department of the Navy, Navel Ship Research and Development Center, Washington, D.C., August 1968.

3. S.D. Rockoff and A.K. Ommaya, "Experimental Head Trauma" Cerebral Angiographic Observations in the Early Post-Traumatic Period." Am. J. Roentgenology, May, 1964.

4. T.A. Gennarelli, L.E. Thibault, and A. K. Ommaya, "Pathophysiological Responses to Rotational and Translational Accelerations of the Head." Sixteenth Stapp Car Crash Conference, Detroit, Michigan: Society of Automotive Engineers, Inc., p. 296-308, 1972.

5. F. Unterharnscheidt and L.S. Higgins, "Traumatic Lesions of Brain and Spinal Cord Due to Non-Deforming Angular Acceleration of the Head." Texas Report on Biology and Medicine, Volume 27, Number 1, Spring, 1969.

6. J.M. Abel, T.A. Gennarelli, H. Segawa, "Incidence and Severity of Cerebral Concussion in the Rhesus Monkey Following Sagittal Plane Angular Acceleration." Twenty-Second Stapp Car Crash Conference. Warrendale, PA: Society of Automotive Engineers, Inc., 1978.

7. T. Tsubokaqa, S. Nakamura, N. Hayashi, M. Miyagami, N. Taguma, J. Yamada, M. Kurishaka, T. Sugawara, H. Shinozaki, T. Goto, T. Takeuchi and N. Moriyasu, "Experimental Primary Fatal Head Injury Caused by Linear Acceleration-Biomechanics and Pathogenesis." Tokyo, Japan.

8. V.R. Hodgson, T.B. Khalil and L.M. Thomas, "Reversible Concussion in the Monkey Compared at Front and Side Blow Locations." Progress Report, Department of Neurosurgery Wayne State University, Detroit, Michigan, November 1978.

9. V.R. Hodgson, T.B. Khalil and L.M. Thomas, "Reversible Concussion in Monkeys Compared at Four Impact Locations." Progess Report, Department of Neurosurgery, Wayne State University, Detroit, Michigan, November 1978.

10. Y.K. Kiu, K.B. Chadran and D.U. von Rosenberg, "Angular Acceleration of Viscoelastic (Kelvin) Material in a Rigid Spherical Shell - A Rotational Head Injury Model." Journal of Biomechanics, 1974.

11. G.N. Bycroft, "Mathematical Model of a Head Subjected an Angular Acceleration." Journal of Biomechanics, 1972.

12. F.O. Mueller and D.C. Arnold, "1977 Survey of Catastrophic Head and Neck Football Injuries."

13. E.S. Gurdjian and J.E. Webster, "Head Injuries, Mechanism, Diagnosis and Management." Boston - Toronto; Little, Brown and Company, 1958.

14. L.M. Thomas, "Mechanisms of Head Injury." Impact Injury and Crash Protection, Springfield, Illinois: Charles C. Thomas, p. 27-62, 1970.

15. E.S. Gurdjian and H.R. Lissner, "Mechanism of Head Injury as Studied by Cathode Ray Oscilloscope." Journal of Neurosurgery, Volume 1, p. 393-399, 1944.

16. E.S. Gurdjian and H.R. Lissner, "Photoelastic Confirmation on the Presence of Shear Strains at the Craniospinal Junction in Close Head Injuries." Journal of Neurosurgery, Volume 18, p. 58-60, 1961.

17. A.K. Ommaya and T.A. Gennarelli, "Cerebral Concussion and Traumatic Unconsciousness." Correlation of Experimental and Clinical Observations on Blunt Head Injuries. Brain, Volume 97, p. 633-645, 1974.

18. U. Komaki, A. Kikuchi, M. Horii, K. Ono, M. Kitagawa and M. Matsuno, "The Research on the Fatal Level of the Primates Head Impact Tolerance." The Japan Automotive Research Institute, Inc. (JARI) October 11, 1978.

19. R. Friede, "Specitic Cord Damage at the Atlas Level as a Pathogenic Mechanism in Cerebral Concussion." Journal of Neuropathology Exp. Neurology., Volume 19, p. 266-279, 1960.

20. N.R. Hollister, W.P. Jolley and R.G. Horne, "Biophysics of Concussion." WADC Tech-

nical Report 58-198, Wright-Patterson Air Force Base, Ohio, September 1978, pt. I.

21. H. Gray, "Anatomy of the Human Body." Philadelphia, Pennsylvania: C.M. Goss, Editor, Lea and Febiger, 1975.

791024

Intracranial Pressure Relationships in the Protected and Unprotected Head

Alan M. Nahum, Randall Smith and Frank Raasch
University of California

C. Ward
Naval Const. Batt. Center

THIS PAPER REPRESENTS a continuation of previous research (1)* (2) on closed impact in the human cadaver and an associated mathematical model. The long term goal of this study is to describe the relationships between head impact events which might be useful in understanding what takes place in the living human. An ultimate goal is the reduction and prevention of head impact injury. The most recent paper (2) presented data from a series of experiments upon unprotected heads with particular attention to the relationships between impact parameters and intracranial pressures. The present paper represents an extension of this work to protected (helmeted) heads and a comparison with the unprotected (unhelmeted) situation.

*Numbers in parentheses designate References at end of paper.

METHODOLOGY

In the current study two different sets of experiments were conducted which will be subsequently referred to Series I and Series II. Seated cadavers were impacted in the frontal region by a rigid mass traveling at a constant velocity. The impact was delivered in the mid-saggital plane in an anterior-posterior direction. The head was rotated forward so that the Frankfort anatomical plane was inclined 45° to the horizontal. Various padding materials were used between the skull and impactor to vary the duration of the applied load and to prevent skull fracture. The input force and the biaxial acceleration-time history was recorded during impact. Pressurization of both the intravascular and cerebrospinal fluid compartments were produced to normal levels just prior to impact.

Abstract

This paper represents a continuation of previous research on closed head impact in the human cadaver and an associated mathematical model. The long term goal of the study is to describe the relationships between head impact events which might be useful in understanding what takes place in the living human. In the current study, two different sets of experiments were conducted 1) sequential impacts on a single embalmed helmeted specimen and 2) impact experiments on individual helmeted unembalmed specimens. Impact parameters and intracranial pressures were measured and discussed. A finite element model is presented which can predict the intracranial pressures throughout the brain for the first 8 msec. It is apparent that the helmet prevents high magnitude, short duration intracranial pressures and that posterior pressures develop after the acceleration phase and helmeted impacts.

Endevco model 8510 piezo resistive pressure transducers (resonant frequency 180 KHz) were used to monitor the intracranial pressure events. Recording sites were obtained by removing a bone plug and sealing the transducer in the subdural space. The intravascular pressures were recorded by the insertion of a kulite (model MCP-808-9R; resonant frequency 150 KHz) catheter in the internal carotid artery at the level of the siphon. Head protection was obtained by the addition of a Bell RT model DOT approved helmet with an expanded polystyrene liner (normal density of 40 gms/liter). Circular plugs were removed from the helmet at appropriate locations to accomodate the pressure transducers and to prevent potentially damaging contact during impact. Because of concerns about possible changes in helmet impact responses when plugs were removed, several test impacts were conducted prior to the full-scale experimental series. This disclosed that the drilled out helmets did not have significantly different force-time responses to impact in the frontal mode than did intact helmets.

New helmets and liners were used for each of the individual impact experiments.

EXPERIMENTAL RESULTS

Series I consisted of 7 sequential impacts on a single embalmed helmeted specimen. The experiments were designed to explore the range of possible impact exposures and resulting severity indices. Table I displays the range of impact conditions. Table II displays the force-time parameters and severity indices for the helmeted single specimen. The bottom of the table shows values for comparison from prior multiple impacts upon unprotected heads. It can be readily seen that the helmet reduced the severity indices for comparable impacts and also allowed the exploration of greater kinetic energies without the fear of producing skull fracture.

Series II consisted of 5 impact experiments on individual helmeted unembalmed specimens. Table III displays the impact conditions for this series. Table IV displays the force-time and severity indices at the top of the page for the current protected series and at the bottom of the page for the prior unprotected series. Again the data indicates reduction in severity indices of helmeted specimens for comparable impact exposures.

Table V presents the force-time and pressure data at the top for the current helmeted series and at the bottom for the prior unprotected series. Because of concern for possible impact damage to the pressure transducers, frontal locations were not used for the protected series. When two valid pressure measurements were obtained from comparable opposite sides of the skull, these values were averaged to present a single figure in the tables. Ordinarily the two values were relatively identical. A single foramen magnum pressure was recorded in the final experiment.

A comparison of experiments having equivalent peak head accelerations e.g. experiments 65W (protected) and 37 (unprotected) shows that pressures are somewhat lower in the protected head but that the HIC values (1542 in the former and 744 in the latter) are higher be-

Table III. Helmeted Single.

Experiment	64W	65W	66W	67W	68W
Impactor Velocity (M/sec)	9.85	8.47	5.81	7.24	7.86
Impact Mass	14.83 (KG)				

Table I. Helmeted Multiple.

Experiment	055W	056W	057W	058W	059W	060W	063W
Velocity (M/sec)	7.51	10.30	12.70	6.44	7.38	9.97	9.95
Impactor Mass (KG)	5.30	5.30	5.30	14.92	14.92	14.92	14.82

Table II. Helmeted Multiple.

Experiment	KE ($kg.M^2/sec.^2$)	Peak Head Accelerations ($m/s^2 \times 10^3$)	GSI	HIC	t_1	t_2
055W	149	1.65	668.6	563	3.3	6.3
056W	281	1.13	251.5	221	3.8	6.8
057W	427	1.43	420.8	367	3.9	6.6
058W	303	1.19	350.7	298	5.1	10.0
059W	406	1.51	640.8	546	4.8	8.7
060W	741	1.67	1091.5	1010	2.9	7.6
063W	733	1.98	1087.1	1000	5.5	9.9
		Unhelmeted Multiple.				
46	51	0.31	36	32	2.7	13.0
47	51	0.29	24	21	4.5	12.7
48	46	1.28	342	297	2.2	4.8
49	182	3.42	1153	1008	1.3	2.2
50	182	1.42	675	539	4.4	9.5
51	179	5.39	4394	3895	3.6	4.4
52	179	4.29	3687	3182	1.9	3.2

Table IV. Helmeted Single.

Experiment	KE ($kg.M^2/sec.^2$)	Peak Head Accelerations ($m/s^2 \times 10^3$)	GSI	HIC	t_1	t_2
64W	719	2.69	2819.6	2684.9	2.5	6.7
65W	532	1.97	1626.8	1542.1	2.6	7.4
66W	250	1.26	543.9	465.3	4.1	9.3
67W	388	1.96	1584.8	1581.1	3.4	8.0
68W	458	1.43	857.2	806.6	3.1	8.9
		Unhelmeted Single.				
36	205	2.30	1068	923	2.4	4.4
37	276	2.00	861	744	3.1	5.5
38	245	2.42	1153	980	1.1	2.9
41	1900	3.90	4756	3765	5.9	8.6
42	438	1.59	842	703	2.0	7.1
43	438	2.23	1008	804	3.7	6.1
44	50	1.52	675	551	2.8	7.0
54	184	2.34	1061	820	2.8	4.5

Table V. Helmeted Single.

Experiment	KE ($kg.^2/sec.^2$)	Peak Head Acceleration ($M/sec^2 \times 10^3$)	Parietal	Occipital	Posterior Fossa	Carotid Siphon	Foramen Magnum
64W	719	2.69	887	-487	-201	5815	-
65W	532	1.97	436	-157	-536	2676	-
66W	250	1.26	398	- 79	-139	1365	-
67W	388	1.96	479	- 45	-184	6749	-
68W	458	1.43	365	-268	-332	1518	1193
			Unhelmeted Single.				
36	205	2.30	594	-205	-480	-	-
37	276	2.00	552	-341	-452	-	-
38	245	2.42	494	-205	-485	-	-
41	1900	3.90	1414	857	-426	352	-
42	438	1.59	70	-	-329	550	-
43	438	2.23	1664	482	-136	811	-
44	50	1.52	150	109	- 19	860	-
54	184	2.34	1354	248	-483	356	-

cause the acceleration pulses are longer in duration for the protected heads. However, pulse duration has little or no effect upon intracranial pressure magnitudes and influences only the pressure pulse durations. Intracranial pressures are system response parameters and are related to the rise time and frequency content of the acceleration forcing function. Pressures in the protected and unprotected head are different because the acceleration traces have a different shape. Helmeted head accelerations do not have the high frequency components and fast rise times which typify unprotected head responses (3).

A blow on a helmeted head initiates a series of dynamic reactions. First the helmet, then the skull, and finally the brain is accelerated. The helmet liner compresses, dissipating energy and filtering the high frequency components from the transmitted force. The brain, last to be accelerated, compresses on the side close to the impact and elongates on the side opposite the impact. Figures 1-4 are data samples from expt. 68W. During the acceleration phase, positive pressures develop in the frontal and parietal lobes, Figure 1, and negative pressures (pressures less than atmospheric), develop in the cerebellum,

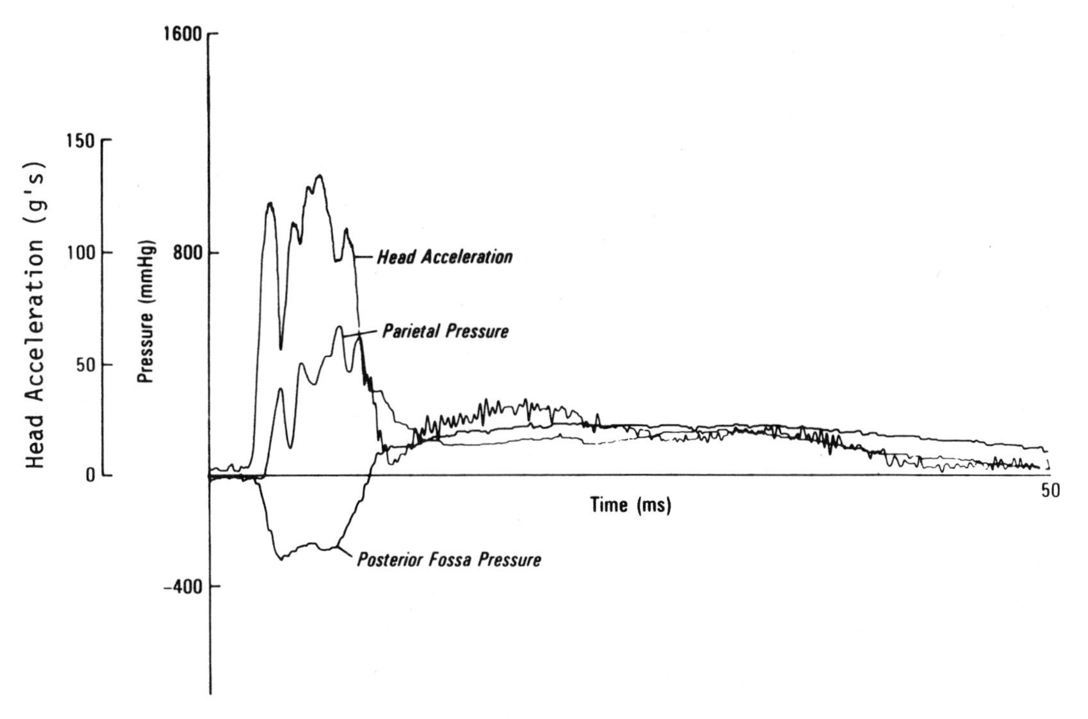

Fig. 1 - Data sample from impact on protected head in Experiment 68W

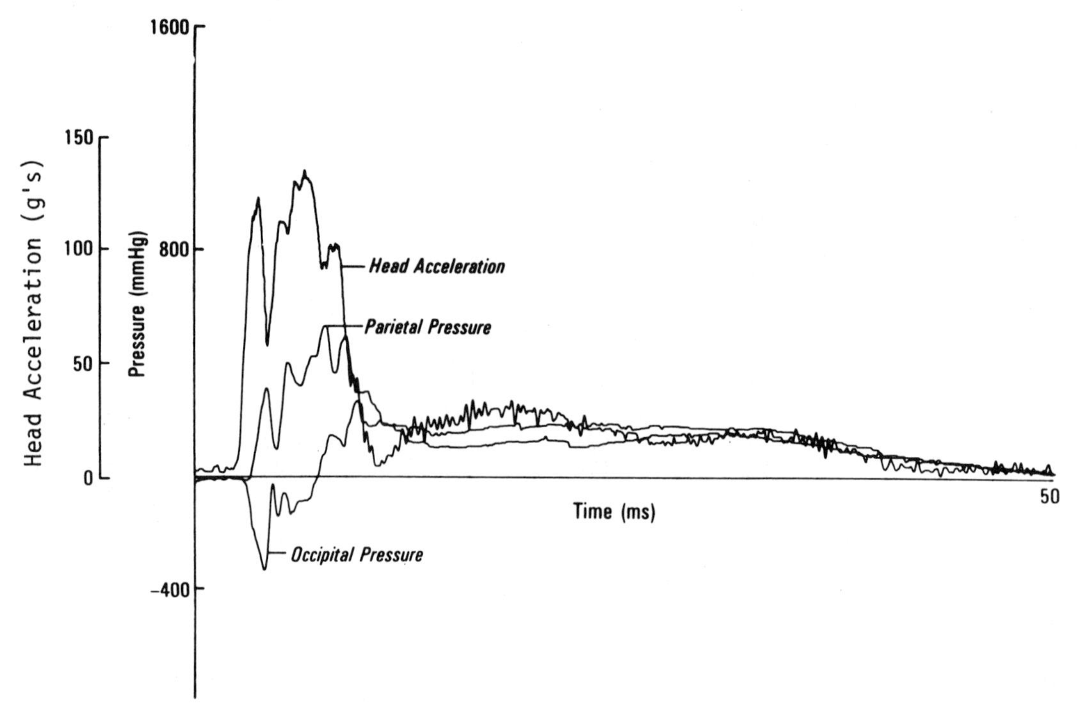

Fig. 2 - Data sample from impact on protected head in Experiment 68W

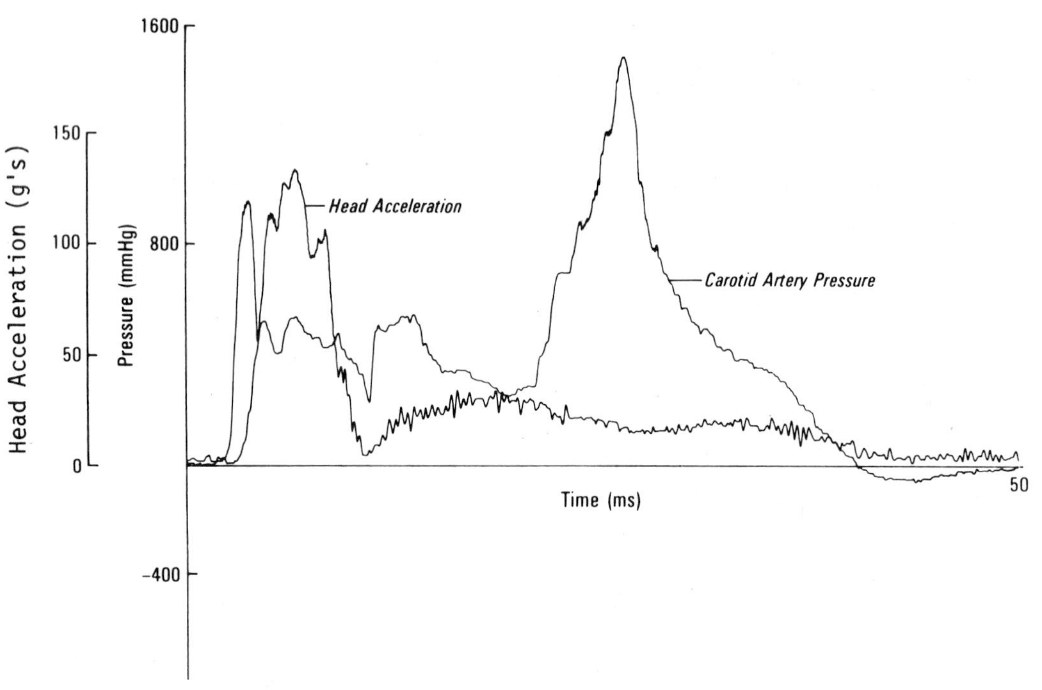

Fig. 3 - Data sample from impact on protected head in Experiment 68W

Figure 2. The highest magnitude positive and negative pressures occur in the frontal lobe and posterior cerebellum, respectively, the coup and contra coup sites in this frontal impact series. Lower magnitude pressures develop in the parietal and occipital lobes, Figures 1 and 2. Such pressure differences or gradients exist whenever the head is accelerated even though the displacements of the brain relative to the skull are small.

Pressures developing after the acceleration phase are not as well understood. Unusually high pressures were measured at the foramen magnum (the opening at the base of the skull), and in the carotid artery, Figures 3 and 4. Carotid artery pressure

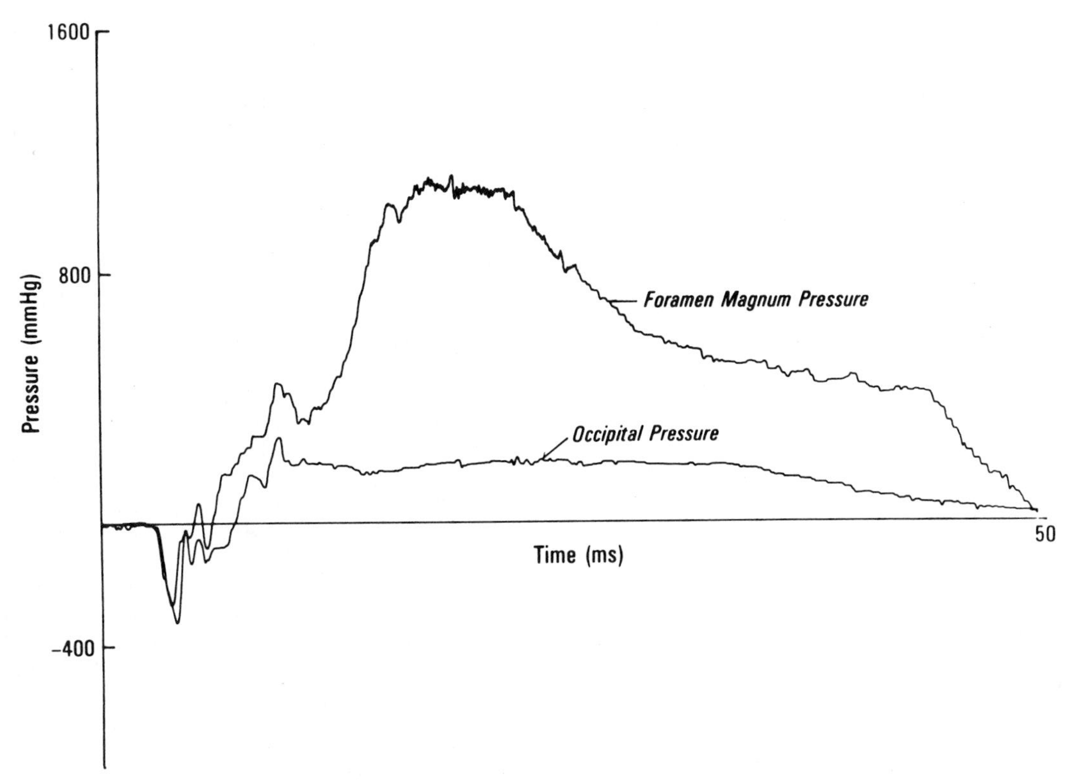

Fig. 4 - Data sample from impact on protected head in Experiment 68W

magnitudes are listed in Table V for both protected and unprotected head experiments. In the protected head, these pressures are significantly higher and far exceed the pressures measured in the brain. This suggests that the pressure may be due to a local compression of the artery, or related to a neck response which is more severe in the protected head. Foramen magnum pressure was measured in only one experiment. Initially it resembles the occipital pressure, Figure 4, indicating a similarity between the responses at the two locations. But the timing and magnitude of the peak value do not correlate with any measured or predicted brain response. The unknown forces or local responses producing these high pressures late in the event require additional investigation.

FINITE ELEMENT MODEL

The response of the brain is simulated using a structural analysis model to determine the pressure distribution, displacements, and stresses. Just as the brain is physically separated from the helmet by the skull, the response of the brain can be studied separately from the helmet by using the skull acceleration as the forcing function. With this procedure a mathematical idealization of the helmet is not required, provided skull acceleration values are known.

The same brain model, Figure 5, validated in unhelmeted impacts (2 and 3), is used in the helmeted tests. (Thirty-two human cadaver impact tests have been simulated with versions of this model.) All calculations are performed in relative coordinates to eliminate inaccuracies due to large displacements and

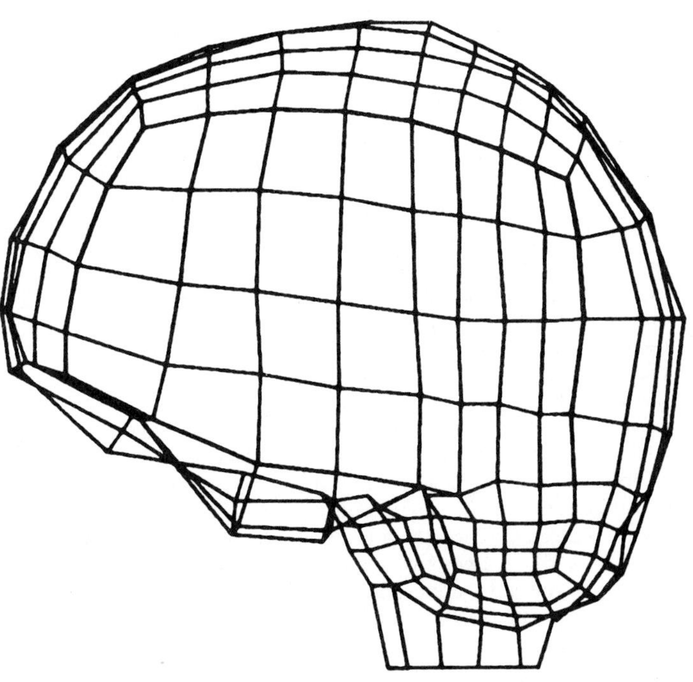

Fig. 5 - Finite element brain model

rotations of the head. Thus the coordinates and node displacements are measured relative to the skull fixed reference frame. External geometry of the brain model is maintained throughout the event just as the external shape of the brain is maintained by the skull. Mathematically the brain model is compelled to move as the impacted skull moves in the test. The foramen magnum and a short section of the cervical cord are modeled. This opening permits an in and out motion of the adjacent soft tissue elements, simulating an intracranial pressure release mechanism (4).

In the finite element model, the brain is separated into eight node brick and four node membrane elements. One hundred eighty-nine brick elements simulate the soft tissue and eighteen membrane elements simulate the partitioning folds of dura, the falx cerebri and tentorium. Brain material is assumed to be isotropic, homogeneous, and linearly elastic in the finite element idealization. In recent years the model has been modified and improved. The elastic brick elements (5) were converted to split energy brick elements greatly improving the accuracy of the shear stress calculation. In this element the dilatation and distortion energies are separated, and the dilatation terms computed with a reduced integration order. Using this technique, the ill conditioning encountered in the stiffness terms when Poisson's ratio (ν) approaches .5 is avoided.

The matrix equation of motion for the finite element model is written,

$$M \ddot{X} = K X = M \ddot{Q} \qquad (1)$$

where M is the diagonal lumped mass matrix, and K is the structural stiffness matrix. The terms $\ddot{X}$ and X are the node translational accelerations and displacements, respectively, measured relative to the skull fixed anatomical axis. The translational acceleration, rotational velocity and acceleration of the skull are contained in $\ddot{Q}$ (the acceleration function in the inertial load matrix, $-M\ddot{Q}$). Eq. 1 contains four hundred and six separate equations (degrees of freedom) and can be solved using matrix analysis procedures. But it is difficult to formulate $\ddot{Q}$. The $\ddot{Q}$ term requires an unusual vector differentiation especially developed for the brain model. In the past, the $\ddot{Q}$ terms were computed separately and input to the structural analysis program (SAP). Two computer programs were required. In order to simplify the solution, the required modifications have been combined into one program. Calculation of $\ddot{Q}$, solution of Eq. 1, and the split energy element equations have been added to the EASE 2 program. Computational accuracy is improved and computational efficiency greatly increased. Head translational and rotational accelerations, and rotational velocity are input as time functions to the new program. Access to the model is also improved, since the EASE 2 program is maintained by the Control Data Corporation (CDC). It is available to all U.S. and foreign users of the CDC cynernet system.

Selection of material properties is often the most troublesome task in biodynamic modeling. Soft tissue is expecially difficult to define. The material is non-isotropic, nonhomogeneous, highly damped and strain rate dependent. In the absence of reliable brain material properties for the high strain rate injury event, an inverse approach was taken. A parametric study was conducted to determine which material properties provide the correct response (3). It was shown that the appropriate value for ν depends on the acceleration pulse shape; sharper head acceleration traces require higher values of ν (values approach .5). The selected value is an effective value. It must include the compressibility provided by the blood and cerebrospinal fluid flowing out of the cranial cavity. During the longer pulse, there is more flow and greater pressure dissipation, requiring a lower Poisson's ratio. For the helmeted head acceleration pulse, the parametric study (3) suggests the following values for ν and Youngs modulus (E).

$$\nu \text{ (brick el.)} = .48$$
$$E \text{ (brick el.)} = 6.67 \times 10^{6} \text{ dynes/sq cm}$$

The dura has an elastic stress-strain relationship (6) and is independent of the acceleration pulse shape. The Youngs modulus obtained from (6) is

$$E \text{ (brick el.)} - 3.152 \times 10^{8}$$

The Poisson's ratio of the membrane elements has little or no effect on the response of the brain.

The model can accurately predict pressures throughout the brain for the first eight milliseconds of the event. A comparison of measured and computed pressures is shown in Figures 6 and 7. Measured (right and left) and computed posterior fossa pressures are compared for four tests. In Figure 8 the computed frontal lobe pressure is presented (frontal pressures were not measured). The results show that a positive pressure pulse develops near the frontal bone and propagates in the posterior direction (shown by Figure 8). At the same time a low magnitude, negative pressure pulse propagates from the posterior fossa in the anterior direction, Figure 9. Both the negative and positive pressures dissipate as they propagate. Pressures are lowest near the brain center of mass, the superior mid-brain region.

Correlation between the measured and computed pressures begins to degrade after eight milliseconds. The predominate pressure response in the parietal and occipital lobe occurs within the eight milliseconds. But the posterior fossa, Figure 1, a positive pressure occurs after eight milliseconds which is not computed. This post impact positive pressure appears to be characteristic of helmeted impacts. Positive pressures of this magnitude rarely develop

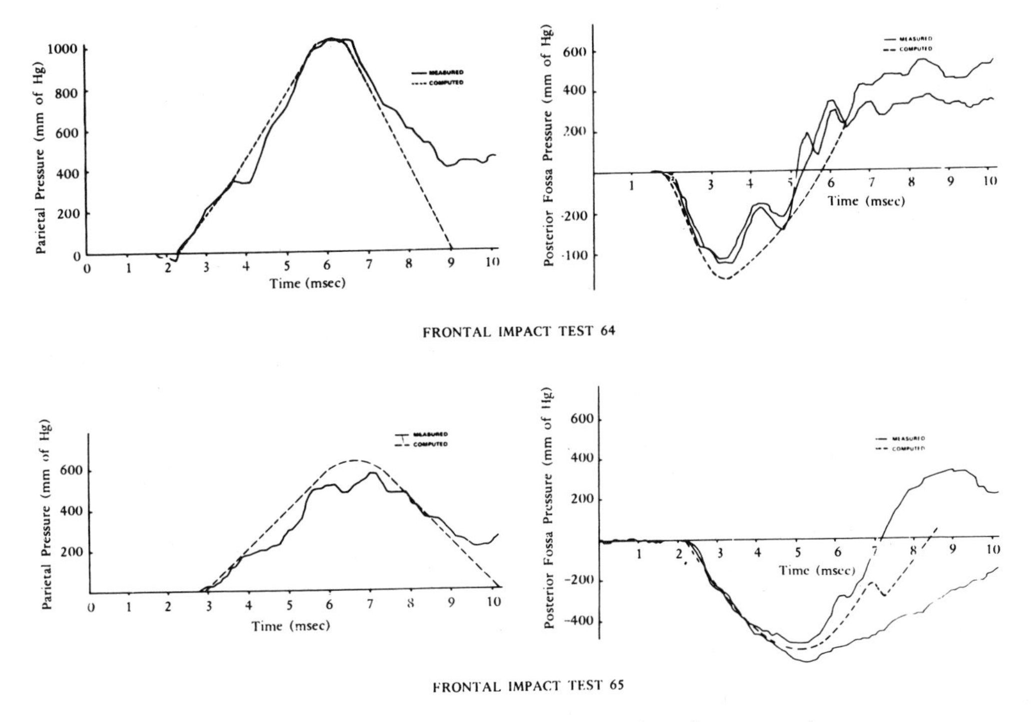

Fig. 6 - Comparison of computed and measured pressures

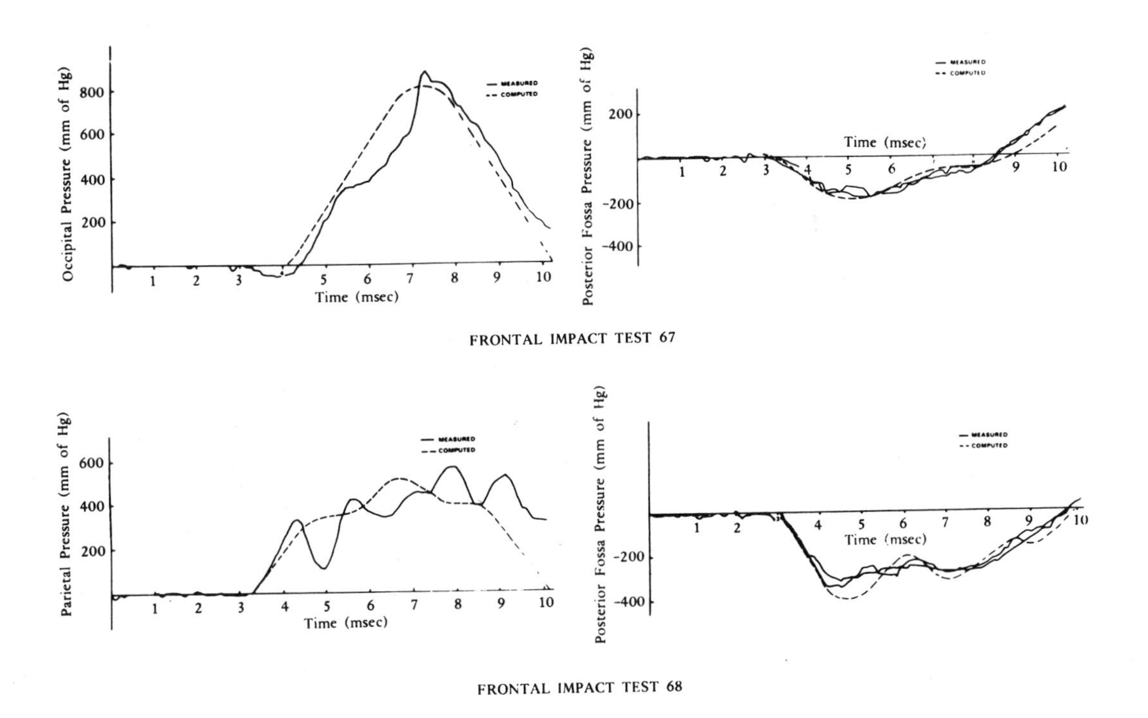

Fig. 7 - Comparison of computed and measured pressures

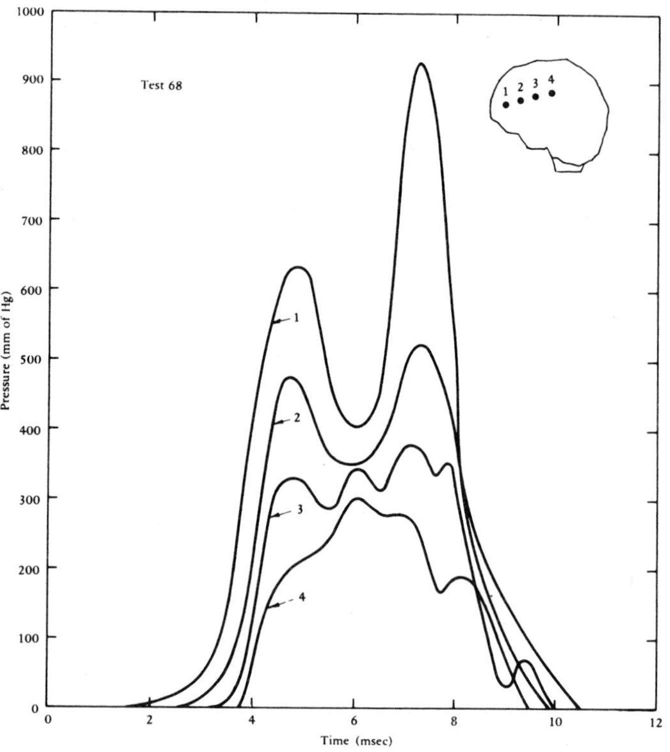

PRESSURES IN FOUR ELEMENTS EXTENDING FROM MID-BRAIN TO POSTERIOR OCCIPITAL LOBE

Fig. 8 - Computed frontal pressures

in unhelmeted head impacts. It is conceivable that the pressures are produced by a nonlinear phenomenon not simulated by the model, such as the compression of the cervical cord. (The cervical cord compresses as the head rotates back in neck extension.)

CONCLUSIONS

1. The model can predict the intracranial pressures throughout the brain for the first eight milliseconds.
2. The helmet prevents high magnitude short duration intracranial pressures.
3. The highest positive pressures develop in the frontal lobe and propagate in the posterior direction.
4. Low magnitude negative pressures develop in the posterior fossa during head acceleration and propagate in the anterior direction.
5. Positive posterior pressures develop after the acceleration phase in helmeted impacts.

ACKNOWLEDGEMENTS

The authors wish to express their appreciation to Bell Helmets for providing helmets used in these experiments. This work was supported by Grant No. OH-00404-04 from the National Institute of Occupational Safety and Health, Department of Health, Education and Welfare.

REFERENCES

1. A.M. Nahum and R.W. Smith. "An Experimental Model for Closed Head Impact Injury." Proc. 20th Stapp Car Crash Conf. SAE 760825, 1976.
2. A.M. Nahum, R.W. Smith and C.C. Ward. "Intracranial Pressure Dynamics During Head Impact." Proc. 21st Stapp Car Crash Conf. SAE 770922, 1977.
3. C.C. Ward. "A Head Injury Model." NATO Advisory Group for Aerospace Research and Development Conference Preprint No. 263, Models and Analogues for the Evaluation of Human Biodynamic Response, Performance and Protection, 1978.
4. Y.K. Liu. "Biomechanics of Closed Head Impact." ASCE Journal, Engineering Mechanics Division, February 1978.
5. D.S. Malkus and T.J.R. Hughes. "Mixed Finite Element Methods Reduced and Selective Integration Techniques: A Unification of Concepts." Submitted to Computer Methods in Applied Mechanics and

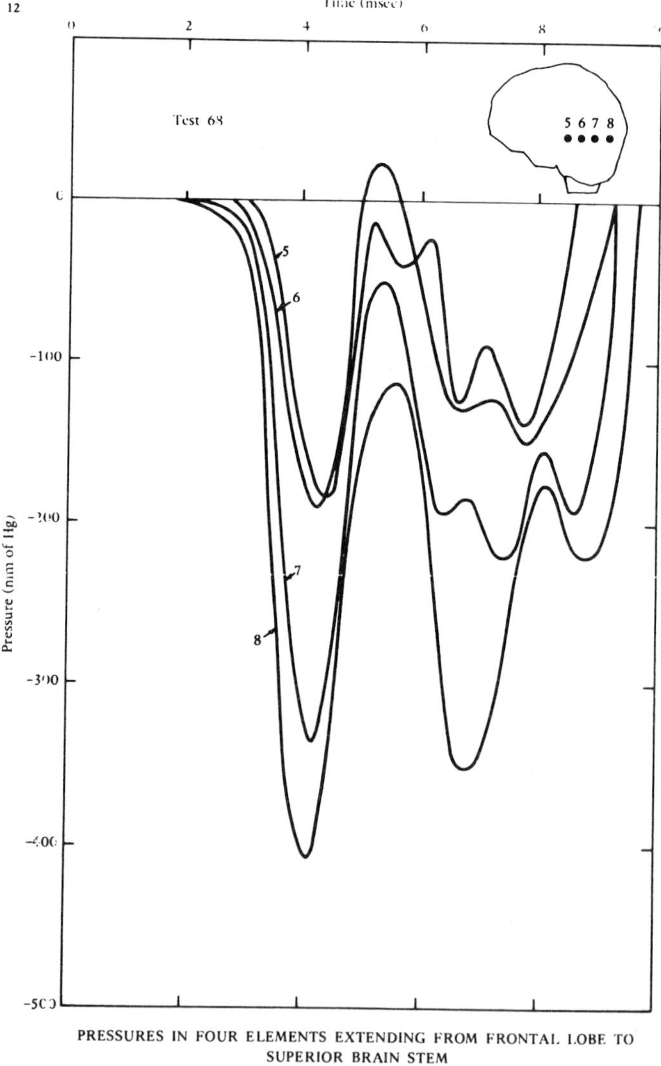

PRESSURES IN FOUR ELEMENTS EXTENDING FROM FRONTAL LOBE TO SUPERIOR BRAIN STEM

Fig. 9 - Computed negative pressure pulses

Engineering.
6. "Determination of the Physical Properties of Tissues of the Human Head-Final Report." Biomechanics Laboratories of Department of Theoretical and Applied Mechanics College of Engineering, West Virginia University. Contract No. PH-43-6-1137, May 1970.

801303

Human Head Tolerance to Sagittal Impact — Reliable Estimation Deduced from Experimental Head Injury Using Subhuman Primates and Human Cadaver Skulls

Koshiro Ono, Astumi Kikuchi and Marumi Nakamura
Japan Automobile Research Institute

Hajime Kobayashi
University of Tokyo

Norio Nakamura
Jikei-kai University

ALTHOUGH THERE have been many studies on the mechanism by which an external force impinging on the head causes injury, and on the tolerance limit of the head to such a force, the problem has still not been fully clarified [1]∿[3]*. In particular, there is a lack of systematic studies on the patterns of head injury likely to result from difference patterns of impact to the head. This question is of special concern in devising safety measures for protection in automobile accidents.

In order to clarify these points and to find an indication of the depression of vital functions, which is necessary to determine the human head impact tolerance threshold, the authors conducted a series of impact experiments using monkeys as subject. [4]∿[8] They also experimented to find the fracture threshold of human cadaver skulls as an aid to extrapolating the human head impact tolerance threshold from the results of the preceding experiments.

In the head impact experiments on monkeys, four items were selected for study from the pos-

* Numbers in brackets designate References at end of paper.

Abstract

To investigate the human head impact tolerance in terms of changes in vital functions, a series of head impact experiments was performed using live monkeys, which are morphologically analogous to humans. To find a causal relationship between the impact and changes in vital functions, three kinds of experimental conditions were used: translational acceleration impact and rotational acceleration impact (both using a head restraint mask with broad contact area), and impact of the unrestrained head against a padded flat surface.

The results indicated that the concussion, cerebral contusion and skull fracture in the monkeys depended on: i) the translational and rotational acceleration impact; ii) the contact area of the impact; iii) the amplitude and duration of the imposed head acceleration; iv) the direction of the impact region (whether frontal or occipital). It was also determined that, of these three patterns of injury, the threshold for the occurrence of concussion is a tolerance threshold (i.e., it indicates transitory and reversible effects). A curve was drawn up for the threshold of concussion occurrence (TCO) in monkeys. Further, the fracture threshold curve for human cadaver skulls was obtained experimentally.

Dimensional analysis and the similarity principle were then used to extrapolate the human threshold for the occurrence of concussion from that obtained experimentally for monkeys, and thus to derive the human head impact tolerance curve. At the same time a comparison was made between the human cadaver skull fracture threshold curve and the monkey skull fractures to confirm the reliability of the estimation of the human head impact tolerance.

sible impact conditions. These were: i) the translational and rotational acceleration impact; ii) the size of the contact area on impact between the impactor and the head; iii) the amplitude and duration of the imposed head acceleration; and iv) the impact region (frontal or occipital). These conditions were applied in different combinations and their influence on the occurrence and severity of head injury was studied.

The results showed that in the head impact experiments on monkeys the TCO is a tolerance threshold (i.e., it indicates transitory and reversible effects). It was also shown that the occurrence of skull fracture is closely related to the severity of head injury and serves as an indication of the tolerance threshold at which severe brain injury is likely to result.

The authors applied the dimensional analysis method of Stalnaker et al. [9]∿[12] to the TCO in monkeys to estimate the TCO in humans. By combining this with the human cadaver skull fracture threshold they derived the human head impact tolerance threshold.

HEAD IMPACT EXPERIMENTS ON MONKEYS

The impact conditions which affect the head injury outcome were classified into the following three groups:

Group A - experiments on head injury at the fatal level, under the conditions of pure translational acceleration impact over a broad contact area using a head restraint mask (TAIRH);

Group B - experiments on brain contusion at the fatal level, under the conditions of rotational acceleration impact over a broad contact area using a head restraint mask (RAIRH);

Group C - experiments on concussion, under the conditions of milder impact without head or neck restraint, over a narrow contact area, using of a rotational component (DIUH).

The experimental subjects were a total of 63 monkeys of the species Macaca fuscata (Japanese monkey), Macaca mulatta (Rhesus monkey), Macaca fasicularis (Crab eating monkey), and Papio cynocephalus (Baboon). Table 1 shows the combinations of impact conditions used in the three groups of experiments, and the age, weight, and sex of the subjects used in each group.

EXPERIMENTAL APPARATUS AND IMPACT SYSTEM

GROUP A: TAIRH - A newly developed impact apparatus consisting of a slider mounted on a 12" HYGE sled (Fig. 1) was used. The slider, which is part of a unit also including the mask, pin and body protector, can move freely and smoothly along the slider guide. When the sled is launched, the slider travels along the slider guide and collides with the barrier above the sled. On collision, the pin impinges on the lead block and the impact is passed on to the head of the subject through the slider and mask. The impact

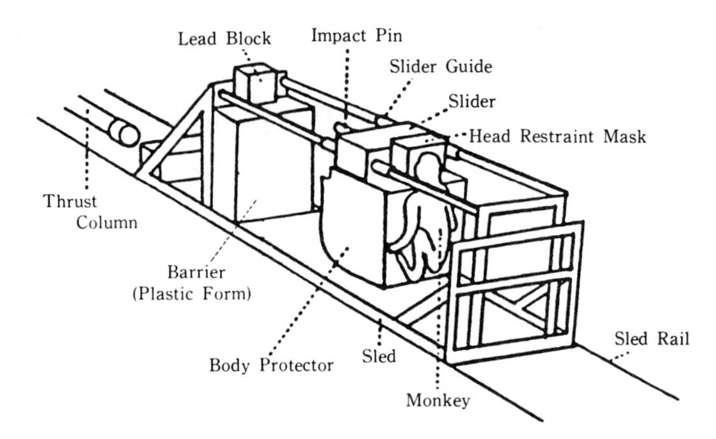

Fig. 1 Overall view of experimental set-up of newly developed impact system on HYGE sled

to the body is cushioned as far as possible by means of a body protector made of a bulky foamed urethane mattress. The impact on the monkey head is aligned parallel to its line of action, i.e., parallel to the Frankfort line. [13],[14] Masking is done by means of a plaster-of-Paris mask (cast individually for each monkey) in an iron box (Fig. 2(a), (b)). The head impact is controlled by varying the sled velocity and the shape of the pin.

GROUP B: RAIRH - The apparatus for rotational acceleration impact consists of a monkey chair device (restraint system) and an impactor ejection apparatus (impactor system) driven by compressed air which delivers the impact to the monkey head. The former is shown in Fig. 3 and the latter in Fig. 4. The accelerating piston of the impactor system is accelerated by compressed air and thrust against the impact shaft. Upon collision the accelerating piston and the impact shaft, acting as a single unit, deliver an impact to the subject. The subject is seated in the chair device of restraint system with its head immobilized by a head impact mask consisting of an iron box filled with plaster of Paris, which is individually cast for each monkey. The impact is aligned parallel to its line of action, i.e., parallel to the Frankfort line. The body is fixed loosely to the chair device by a back-plate and restraining net around the thorax. Upon impact, the head mask rotates around the neck region ($C_{6\sim7}$) through about 45° - 60°; this rotation is then transmitted to the body, pivoting around the hips. The body undergoes a gradual braking action and comes to rest at an angle of approximately 60°.

GROUP C: DIUH - The apparatus in this case consists of an impactor ejection apparatus driven by compressed air and a restraining device which holds the head of the subject in position for the impact and then gradually absorbs the kinetic energy imparted to the subject, and brings it to rest. The impactor ejection apparatus is the same as that used in the rotational acceleration impact experiments (Fig. 4). The impactor head, which delivers a direct head im-

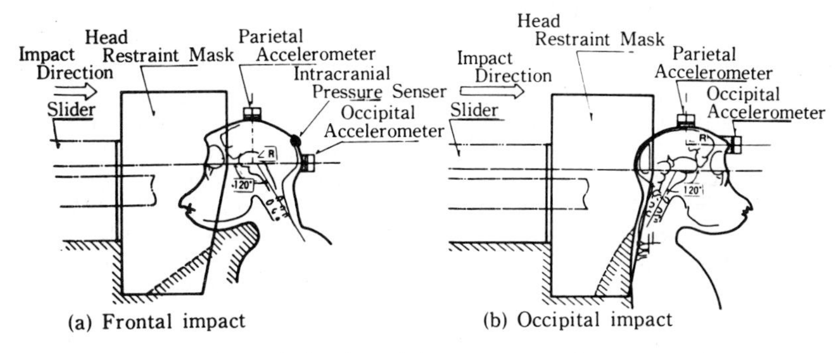

Fig. 2 Translational head impact test configuration and set-up for masking of subject's head

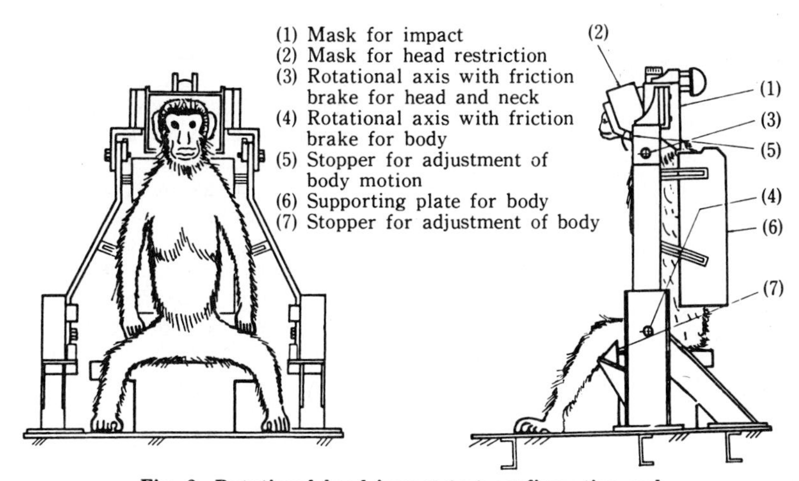

Fig. 3 Rotational head impact test configuration and set-up for masking of subject's head

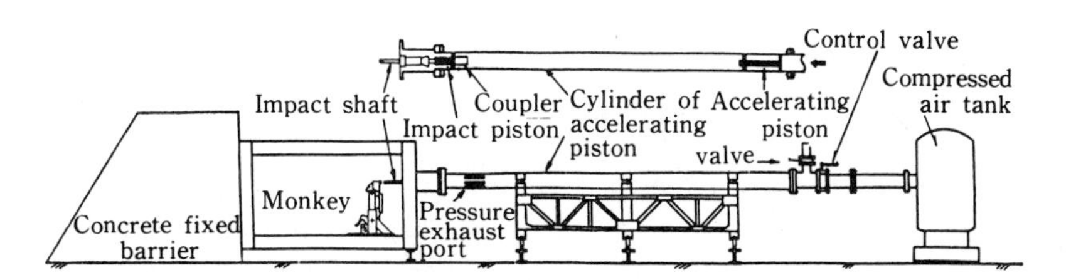

Fig. 4 Illustration of the impactor system for rotational and direct head impact

pact, consists of a rubber block measuring 120 x 80 x 100 mm, and having four degrees of hardness, namely, 20, 40, 60 and 80. The rubber impactor was so designed that the area of contact with the monkey head would be larger than the projected area of the latter, i.e., so that its surface area could adequately accommodate the whole head of the monkey. The elasticity characteristics of the rubber blocks are shown in Fig. 5. In the case of occipital impact, the rubber block strikes the head directly, but in the case of frontal impact, the impact is received through a mask so as to protect the eyes. The monkey head is fastened at the impact position by surgical threads stitched to the skin. During the impact, the surgical threads are released when the force exerted on them exceeds a certain limit, freeing the head to move. The direction of the impact is almost parallel to the Frankfort line. To prevent cervical hyperextension after the impact, a supporting net is employed to cushion and stop the head movement at an angle of approximately 60° - 90°. For further absorption of the head impact energy, the frame which supports the monkey head and body is designed to rotate around the hip point (so that the rotational force could be absorbed), and to stop at an inclination of 30° - 60°. Fig. 6 illustrates the position of the subject in the restraint system.

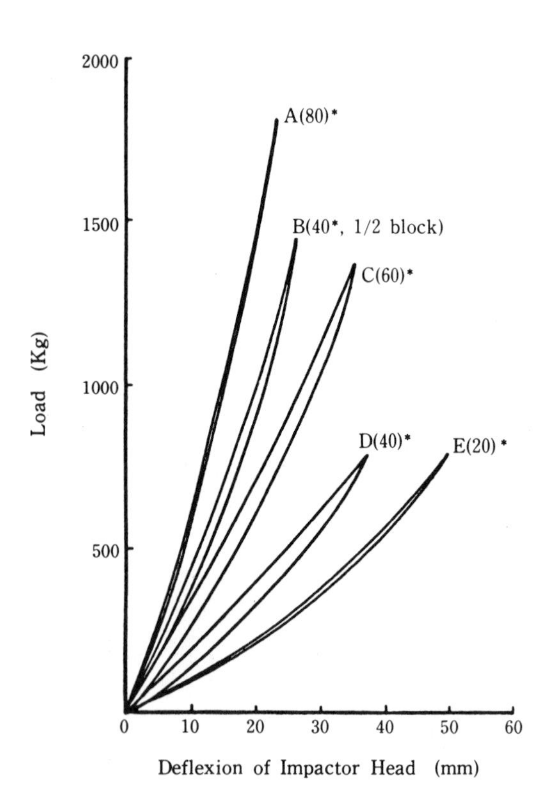

Fig. 5 Load-deflexion curve of impactor head
***Numbers in parentheses show the type of impactor head**

PHYSICAL MEASUREMENTS AND BIOLOGICAL OBSERVATIONS

The subjects were kept under anesthesia for about 2 to 6 hours (during the preparations and up to the time of impact) by intramuscular injection of 5 - 10 mg/kg ketamine hydrochloride every 30 to 60 minutes. During this time measurements of the body were taken and sensors were placed in position. The impact experiment was performed after waiting for a condition of light anesthesia in which the EEG waves and neurophysiological responses had been recovered.

In Group A (TAIRH) experiments, four accelerometers (two X-axis and two Y-axis) were mounted on the skull parietally, occipitally or frontally. The intracranial pressure sensor was placed in the epidural space of the counterpart of the impact. In Group B (RAIRH) and Group C (DIUH) experiments, nine sets of accelerometers were mounted frontally or occipitally in order to measure the rotational acceleration of the subject's head. (Details of the measurement method are given in Appendix A.)

The velocities of the impactor and the monkey head were measured optically by laser beam and photo conductive cell. The measurements were analyzed by a data processing system [15] (NOVA, Nihon Minicomputer Ltd.). The behavior of head and body were also filmed for analysis with 16-mm high-speed cameras (HYCAM, Redlake; STALEX, Weinberger) at 4,000 and 2,000 frames per second, respectively.

Electroencephelograms (EEGs) were taken from leads attached to both frontal-temporal and parietal regions. Electrocardiograms (ECGs) were taken from chest leads, and respiration from a strain-gauge or thermistertype pneumotachometer (MTR-ITA or MTR-2TIS, Nihonkohden Inc.). Systemic blood pressure was recorded from an indwelling femoral arterial catheter through an external pressure transducer.

Consecutive measurements of the neurological status, particularly the pupillary light reflex, ciliary reflex, spontaneous blinking and spontaneous behavioral changes, were made prior to and after the experiment (generally commencing 10 seconds after impact). X-rays of the head, neck and chest were taken in two planes prior to and after the experiment to check for fractures. Monkeys which survived the experiment were sacrificed from 18 hours to 1 week later by perfusion of nembutal sodium solution, and autopcies were performed immediately both in these cases and in those which died in the experiment. After inspection for surface contusions, bleeding, injury to the visceral organs and fractures of the cranium and cervical vertebrae, the skull and dura mater were opened and the brain and cervical spinal cord were removed as a unit. An inspection was then made for basal fractures. After weighing the brain and cervical spinal cord, they were preserved along with the thoracic and abdominal organs in 10% formalin solution. Then, about 4 weeks later, they were sectioned and studied under the optical and/or electron microscope. The items of physical

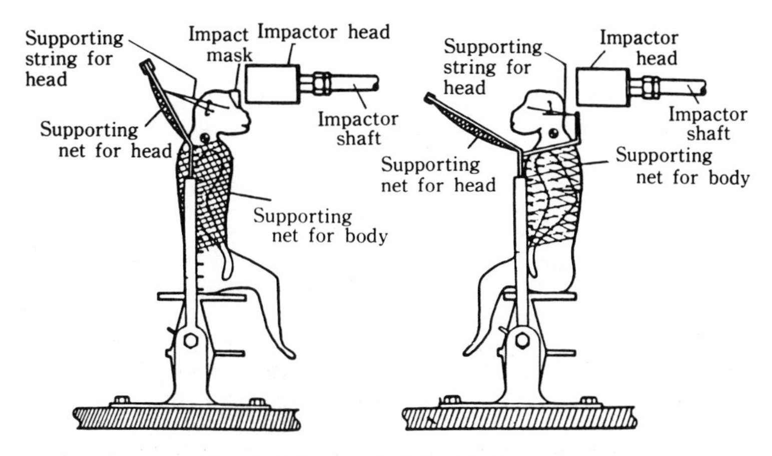

Fig. 6 Set-up restraint system

Table 1 Impact Conditions and Subjects in the Classified Experiments

Grouping \ Items	Impact Condition			Location of Impact	Subject				
	Impact Velocity (m/s)	Component of Head Motion	Contact Area		Species	Age (average)	Body Weight (Kg) (average)	Sex M	Sex F
Translational Impact (Group A)	27 <	Translation	Mask (Board)	Frontal	Macaca fuscata	5.7	8.2	17	3
					Papio cynocephalus	20	25	1	0
				Occipital	Macaca fuscata	7.8	7.7	2	2
					Papio cynocephalus	9	35.8	2	0
Rotational Impact (Group B)	12 ~ 28	Translation + Rotation	Mask (Board)	Occipital	Macaca fuscata	10.1	7.9	5	4
					Macaca fasicularis	6	8.1	8	0
					Macaca mulatta	6	9	1	0
Direct Impact (Group C)	5 ~ 15	Translation + Rotation	Padded flat surface (narrow)	Frontal	Macaca fuscata	12	8.8	1	1
					Macaca mulatta	6.7	7.1	3	4
				Occipital	Macaca fuscata	6.8	6.9	0	5
					Macaca mulatta	5.8	6.1	1	3

M : Male F : Female

measurement are listed in Table 2, and those of biological observation in Table 3. It should be noted that in the Group C (DIUH) experiments, individual monkeys were subjected to one or more experiments until concussion occurred.

CRITERIA OF CONCUSSION

The severity of concussion was judged according to the following criteria immediately after the impact: (i) apnea and (ii) loss of the corneal reflex are observed for at least 20 seconds; and (iii) blood pressure decreases immediately after the impact but rises a few seconds later to above its pre-experimental level, and brady-cardia is observed.

SEVERITY CRITERIA:

a. Loss of the corneal reflex persists for at least 20 seconds after the impact.
b. Respiration ceases for at least 20 seconds after the impact.
c. Two levels of severity of blood pressure disturbance occurs: "absent-mild" on one level, and "severe" on another level. These level are distinguished according to three indices, namely, the bradycardia, the duration and the overall patterns of the blood pressure.

These three criteria were used to grade the level of concussion as follows:

0: none of the three criteria applies.
I: one of the three criteria applies.
II: two of the three criteria apply.
III: all three criteria apply.

EXPERIMENTAL RESULTS

The experiments, using 63 subjects, comprised 60 frontal and 59 occipital impacts. Since all subjects in the TAIRH and RAIRH experiments showed concussion on the first impact, none were subjected to two or more impacts.

GROUP A: TAIRH - The resultant average head acceleration ranged from approximately 240 to 1,100 G, and its duration from approximately 3 to 18 ms. The slider acceleration and the X-axis acceleration of the head had very similar values. Analysis of the high-speed movie films showed almost no rotational head movement during

Table 2 Physical Measurements

	Velocity	Acceleration	Intracranial pressure	Motion	Dimensions recorded by X-ray
Group A	Sled	Head (X, Z axes) Slider (X axis) Sled (X axis)	Epidural space	Head (4000 f/s) Head and body (2000 f/s)	Location and direction of the attached accelerometers Skull diameter Thickness of skull bone
Group B	Impactor Head Mask	Head (X, Y, Z axes; 9 ch) Head Mask	–	Head (4000 f/s) Head and body (2000 f/s)	Location and direction of the attached accelerometers Skull diameter Thickness of skull bone
Group C	Impactor	Head (X, Y, Z axes; 9 ch)	–	Head (4000 f/s) Head and body (2000 f/s)	Location and direction of the attached accelerometers Skull diameter Thickness of skull bone

Table 3 Biological Observations

1) Physiological survey
   Vital sings (respiration, pulse rate, blood pressure, ECG)
   EEG
2) Biochemical survey
   Blood gas analysis (PO_2, PCO_2, PH)
3) Neurological survey
   Eye movement, blinking, pupillary response
   Light reflex, corneal reflex, behavioral response to pain
   Oculocephalic response
4) Pathological survey
   Autophy examination
   Optical microscopic or electron microscopic examination

the time of impact. Fig. 7 is an example of the time history of head acceleration and intracranial pressure. Table 4 shows survival and mortality for each direction of impact. The conditions of the subjects, head accelerations and durations, and the pathological and neurophysiological findings are summarized in Appendix B.

Concussion occurred in all subjects; among the survivors it lasted from 20 seconds to 7 minutes. Of the 26 monkeys tested (i.e., excluding one case where transection of the spinal cord was found traumatically severed), 15 survived and 11 died. A linear fracture of the skull was seen in one case. Although negative intracranial pressure of -1 (close to atmospheric pressure) developed, only slight subarachnoid hemorrhage was seen in seven cases including the fracture case; no other intracerebral hemorrhages were seen on visual examination. In particular, no contusions were visible in the impact area or the counterpart of the impact. Slight brain stem hemorrhage was seen under the microscope, however. Substantia gelatinosa and rarefaction and sponginess of the white matter were observed in the cervical spinal cord; these findings were particularly marked in the fatal cases. Furthermore, all 12 of the fatal cases showed chest injuries, including rib fractures, mediastinal hemorrhage, and pulmonary contusion. However, in no case were there any chest injuries serious enough to be fatal lesions of primary significance, such as multiple rib fractures, pneumothorax or injury of large vessels.

GROUP B: RAIRH - The impact velocity was approximately 20 - 36 m/s in terms of impactor velocity and approximately 12 - 28 m/s in terms of the tangential velocity in the parietal skull region. The resultant average head acceleration was approximately 500 - 1,100 G, and duration ranged from approximately 1 to 5 ms. The resultant average rotational head acceleration was in the range of approximately 5×10^4 - 29×10^4

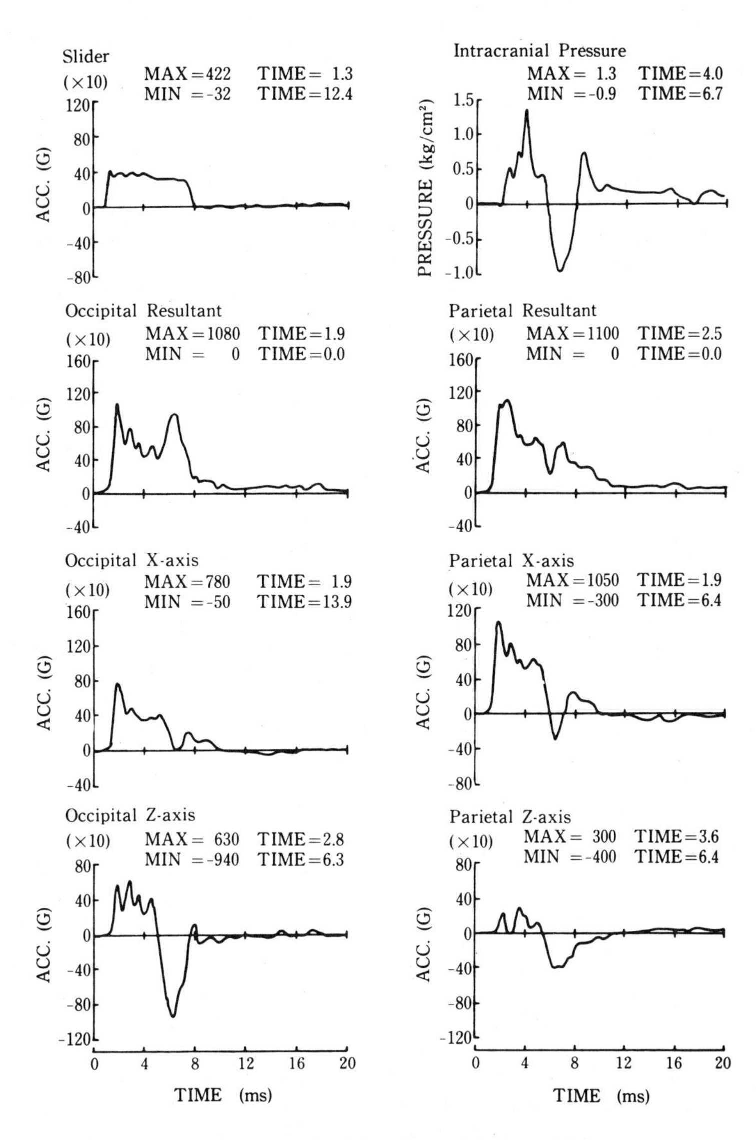

Fig. 7 Head acceleration and intracranial pressure-time history in frontal impact. (Exp. No. 102, Group A)

Table 4 Number of Outcomes in Translational Impact (Group A)

	Species	Survied	Died	Total
Frontal Impact	Macaca fuscata	10	10	20
	Papio cynocephalus	0	1	1
Occipital Impact	Macaca fuscata	4	0	4
	Papio cynocephalus	1	1	2
	Total	15	12	27

rad/s^2, and its duration was in the range of approximately 1 - 4 ms. Fig. 8 shows a sample time history of head acceleration and rotational acceleration. The results are summarized in Appendix C.

Of the 18 subjects, 12 survived and six died. Table 5 shows the number of outcomes, skull fractures and brain contusions. Five of the deaths were instantaneous; the other occurred 14 hours after the experiment. In four of the fatal cases, comminuted depressed fractures had occurred in the occipital, parietal or temporal skull or the skull base. The depressed fragmented bone had caused laceration of the medulla oblongata and/or the main trunk of the intracranial vessels, resulting in instant death. Examination of the neurophysiological responses indicated that in all but one of the monkeys that died instantly, the neurological responses were not regained before death. In the survivors, the average recovery time of the pupillary light reflex was 2 minutes 20 seconds, that of the ciliary reflex was 2 minutes 35 seconds, and that of spontaneous blinking was 3 minutes 30 seconds.

Pathological findings were significant. Brain laceration and contusion were seen in all six cases where fractures occurred; of the 12 non-fracture cases, seven showed brain contusion and/or subarachnoid hemorrhage in i) the parasagittal region, ii) the basal surface and tip of the frontal and temporal lobes, and iii) the brain stem, etc. In the case of ii), these were contre-coup injuries due to occipital impact. No coup injuries in the impact region were seen in the non-fracture cases, but brain contusions were seen in the case of i), where the impact is believed to have been received against the edge of the mask.

In the RAIRH, concussion of grade III occurred in 94% of the cases except one case of concussion of grade II, while brain contusions occurred in 67% of the cases (Fig. 9). Both concussion and contusion cases were below the concussion threshold usually proposed for rotational acceleration. However, no correlation was seen between the resultant average rotational head acceleration and its duration, and between the incidence of concussion and skull fractures. As for the relationship between the resultant average head acceleration and its duration, and the incidence of head contusions, a threshold for the occurrence of skull fractures and brain contusions was recognized, although the number of cases was small (Fig. 10).

GROUP C: DIUH - In these experiments, eighteen subjects were used; of which nine were subjected to a total of 39 frontal impacts, and the other nine to a total of 35 occipital impacts.

The velocity of the impactor ranged from approximately 5 to 15 m/s, the resultant average head acceleration from 140 to 1,100 G, and the resultant rotational head acceleration from 6,000 to 120,000 rad/s^2, with duration from 1 to 14 ms. Fig. 11 shows a time history of head acceleration and rotational acceleration. The results are tabulated in Appendix D.

Of the 18 subjects, four showed skull fractures, and of these, three died. Five of the remaining 14 cases without skull fractures also died. From the neurophysiological findings (excluding the fracture cases), concussion was found to occur at the following grades and frequencies: on frontal impact--0 (no concussion), 17 times; I, eight times; II, five times; III, two times; on occipital impact--0, 12 times; I, seven times; II, seven times; III, five times. The pathological findings showed no brain contusion in the parasagittal region (such as appeared in rotational acceleration impact), but instead, contusions were present in the basal surface and elipse of the frontal lobe, the occipital lobe, and the tip of the temporal lobe. In the non-fracture cases, contre-coup injuries in the form of contusions were seen on frontal impact in one case, and on occipital impact in two cases. No coup injuries in the form of contusions were seen on either frontal or occipital impact. Fig. 12 shows the distribution of brain contusions. Subarachnoid hemorrhage was seen in six of the 14 cases without skull fractures, and in all four of the cases with skull fractures. In the non-fracture cases, subarachnoid hemorrhage occurred in the contre-coup region on frontal impact in one case, and on occipital impact in two cases, but did not occur in the coup region on either frontal or occipital impact (Fig. 13). Brain stem injury accompanied by deep-seated hemorrhage was seen in one fracture case and four non-fracture cases. Three of the latter were cases occurring on occipital impact.

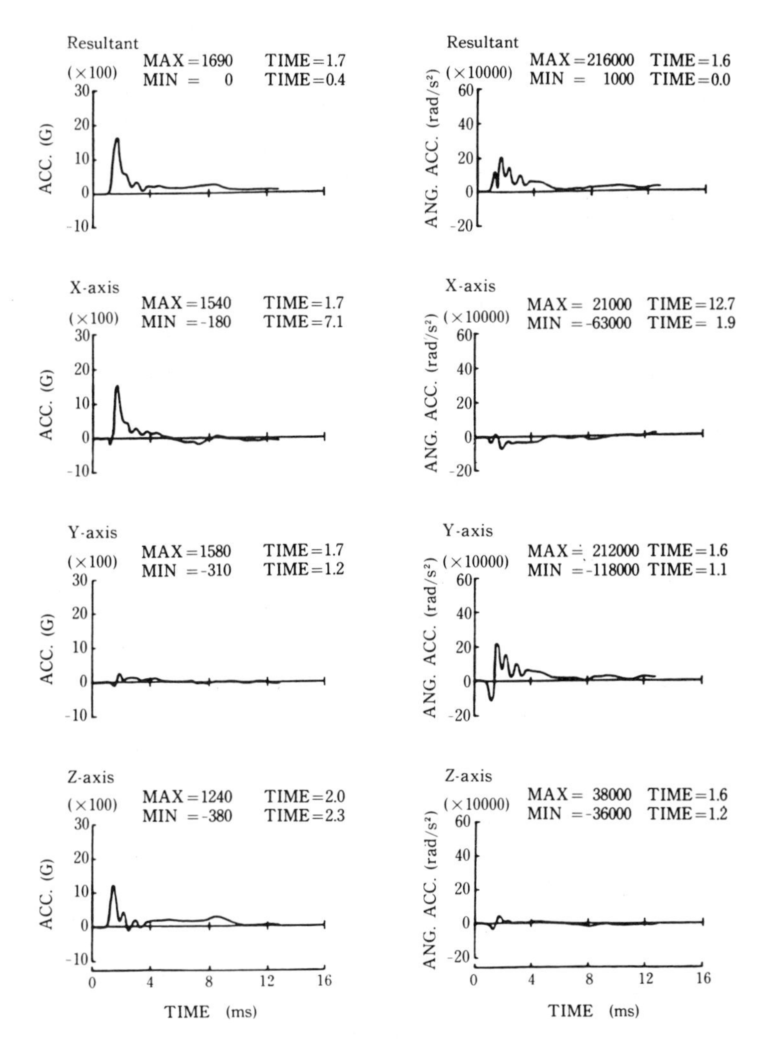

Fig. 8 Head acceleration and angular acceleration-time history in occipital impact. (Exp. No. 211, Group B)

Further, four of the five subjects with brain stem injury died within one hour. Of the total of eight fatal cases, four (50%) showed cervical injury without skull fracture. The mortality rate among cases of intramedullary hemorrhage alone was high, namely four out of six cases or 67%. Photograph 1 shows a specimen from a suvvivor with concussion grade III, injected with Evans Blue. The dye can be seen not only in the area of brain contusions on the undersurface of the frontal lobe, but also in the white matter of the frontal lobe, the cortex of the occipital lobe, and the cerebellum. Photograph 2 shows the cervical spinal cord of a subject which died of respiratory disorder. Hemorrhage can be seen distinctly on the dorsolateral side.

FREE-DROP IMPACT EXPERIMENTS ON HUMAN CADAVER SKULLS

Fifteen dried human cadaver skulls were used. They averaged 173 mm in length, 130 mm in breath, and 0.59 kg in weight. To obtain a closer approximation to the actual condition of the human head, the skulls were prepared as follows.

The skull weight was increased by filling the intracranium with a 10% aqueous solution of gelatin, plastering the outer surface with clay, and covering the entire skull with a rubber skin. Since the weight of the human head is generally taken to be 5 kg [16],[17], the total weight of each skull was adjusted to this figure. The rubber skin used was Hybrid II dummy head skin.

Points of impact were selected in the frontal and occipital skull regions. In the frontal region, the point was on the frontal sagittal line at a distance of 55 mm from the nasion. In the occipital region, it was on the sagittal line near the parietal at a distance of 30 mm from the external occipital protuberance. The skulls were suspended along a line passing

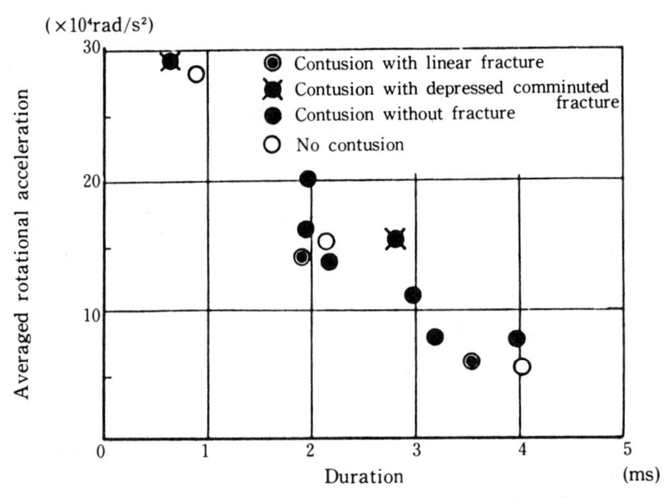

Fig. 9 Averaged rotational acceleration-duration of the monkey head v.s. head injuries

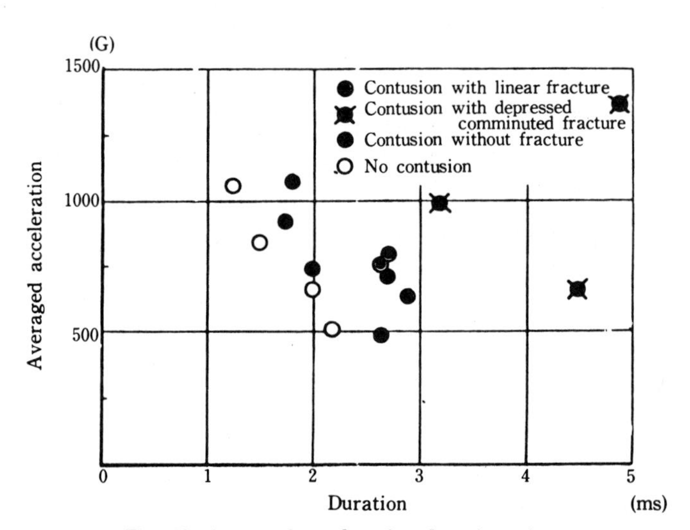

Fig. 10 Averaged acceleration-duration of the monkey head v.s. head injuries

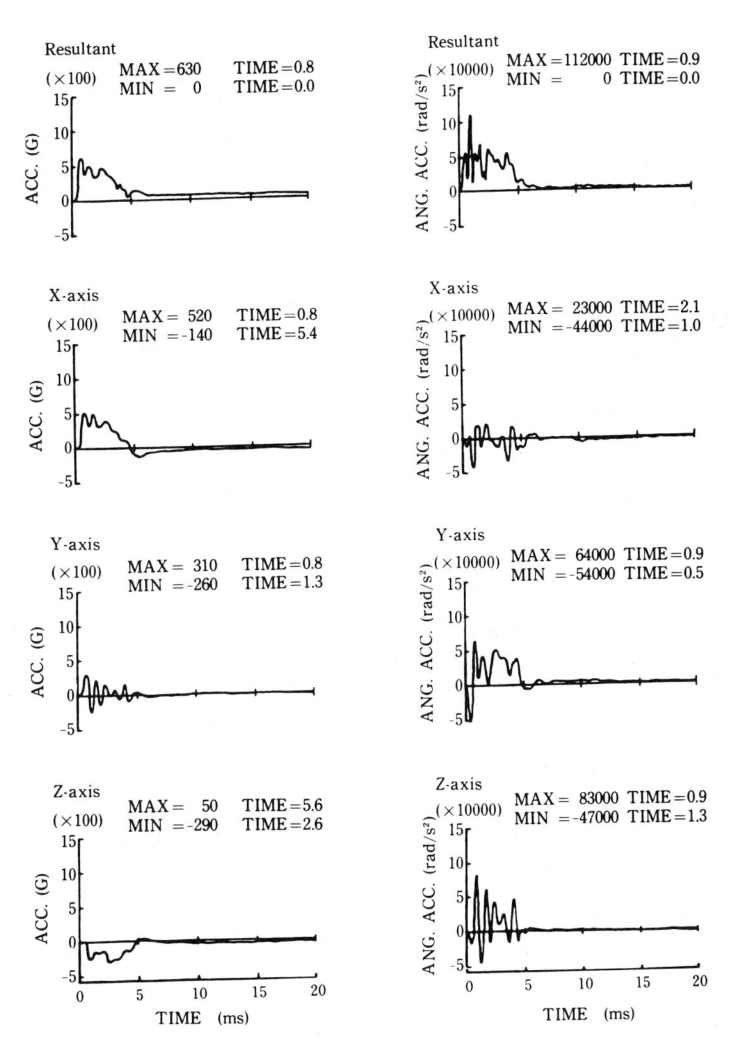

Fig. 11 Head acceleration and angular acceleration-time history in frontal impact. (Exp. No. 336, Group C)

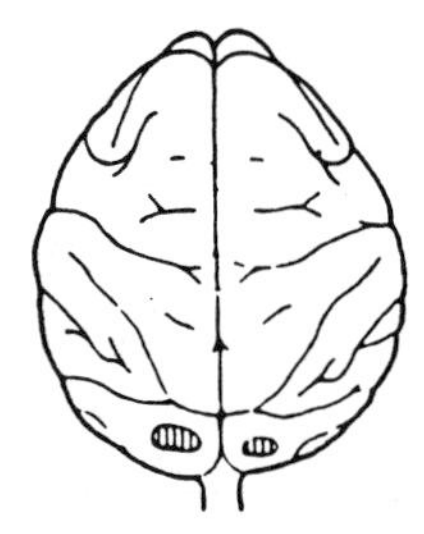

Frontal impact

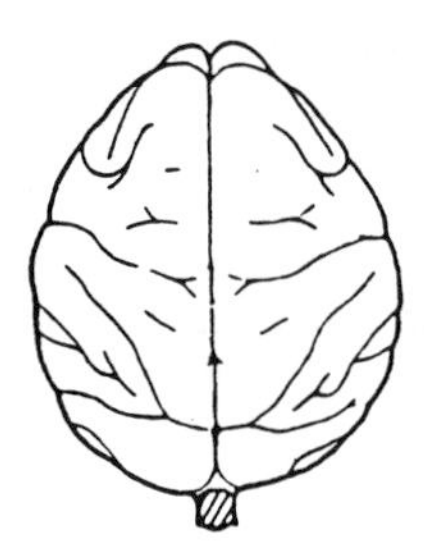

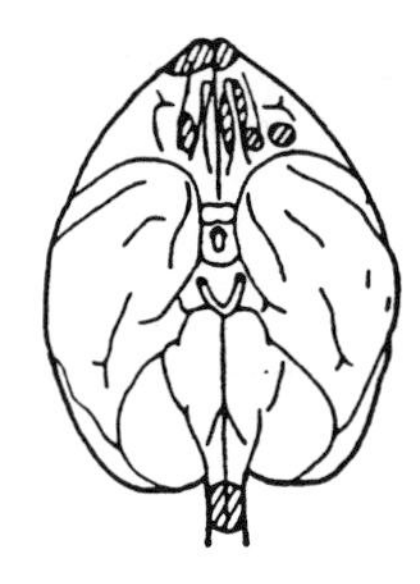

Occipital impact

Fig. 12 Location of contusion of the monkey brain without skull fracture in direct impact (Group C)

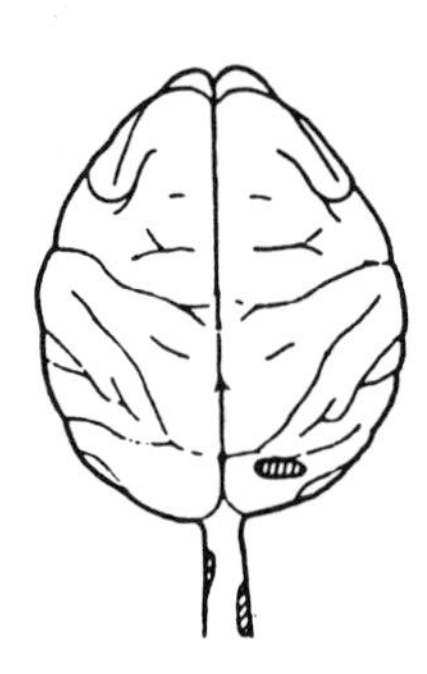

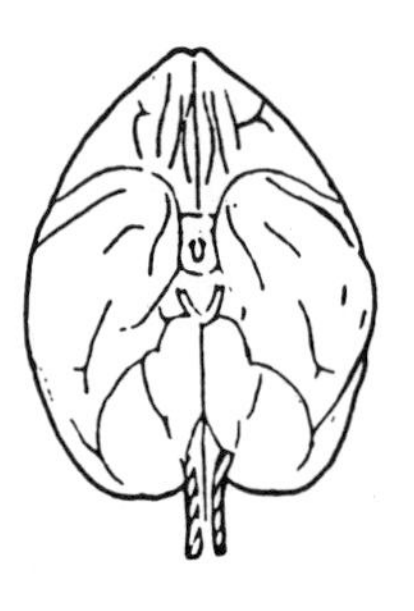

Frontal impact

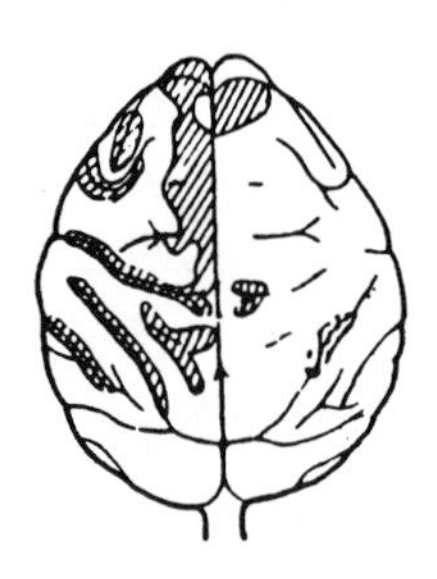

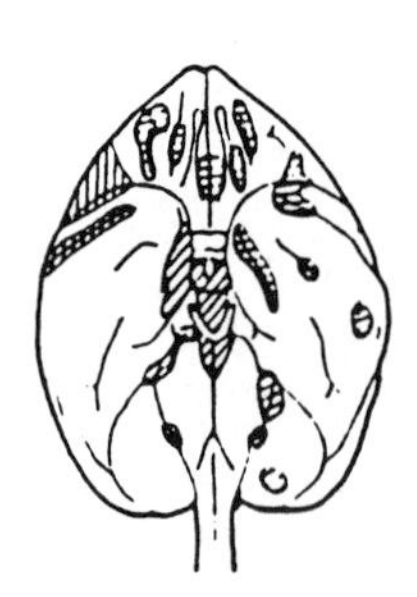

Occipital impact

Fig. 13 Location of subarachnoid hemorrhage of the monkey brain without skull fracture in direct impact (Group C)

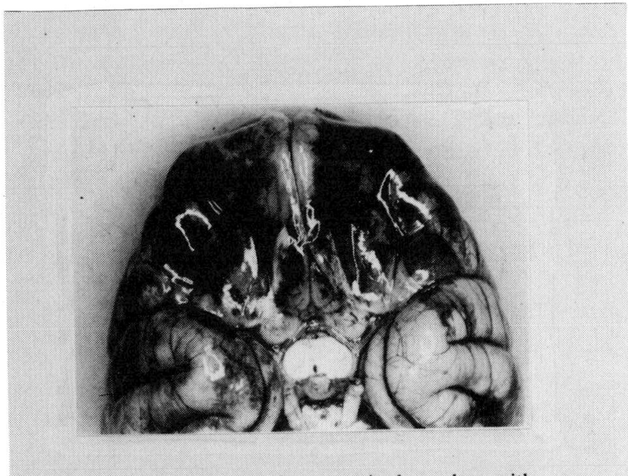

Photo. 1 A specimen from survival monkey with concussion grade III, injected with Evans Blue. Visual contusions on the base of frontal lobe of the monkey brain in occipital impact. (Exp. No. 353, Group C)

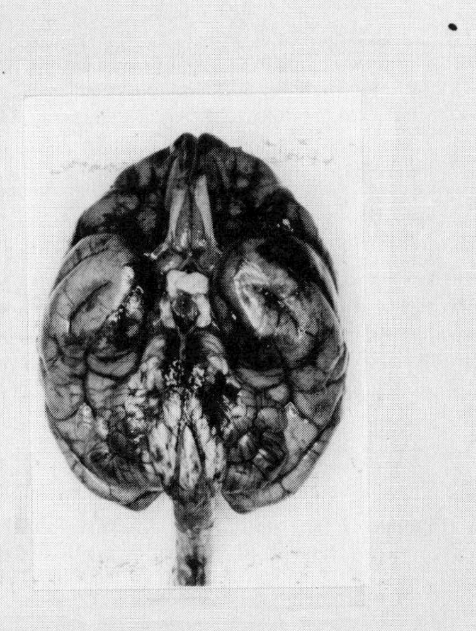

Photo. 2 A specimen from fatal monkey with concussion grade II, injected Evans Blue. Subdural hemorrhage of the upper cervical spinal cord and gross contusion on the base of frontal lobe of the monkey brain in occipital impact. (Exp. No. 348, Group C)

through the impact point and the center of gravity. They were dropped from a gradually increasing height until fracture occurred, with the impactor base being replaced each time (see Fig. 15). The impactor base was the same as that used for the impactor head in the DIUH experiments on monkeys.

The skull acceleration was measured by mounting three piezo-type accelerometers (X-, Y- and Z-axis) between the basal vomer and the foramen magnum (Fig. 14). The impact load was measured by a load transducer mounted on the free-drop platform. The experimental apparatus and method are outlined in Fig. 15.

The fractures were classified by degree into: 1) rudimentary fractures; 2) linear or radiating fractures; and 3) multilinear comminuted fractures with deflection or detachment of bone fragments. Where skull base fractures occurred, these were also recorded in addition to the above classification.

EXPERIMENTAL RESULTS

A total of 84 impacts were delivered to the frontal and occipital regions. Six results had to be discarded, leaving a total of 78. The results are summarized in Appendix E(a), (b).

In cases where skull fractures occurred, the basic impact wave of the skull impace acceleration shows spike-shaped, higher-harmonic, damped oscillations (Fig. 16). This phenomenon, which was seen in all fracture cases, including basal fractures, results from changes in the

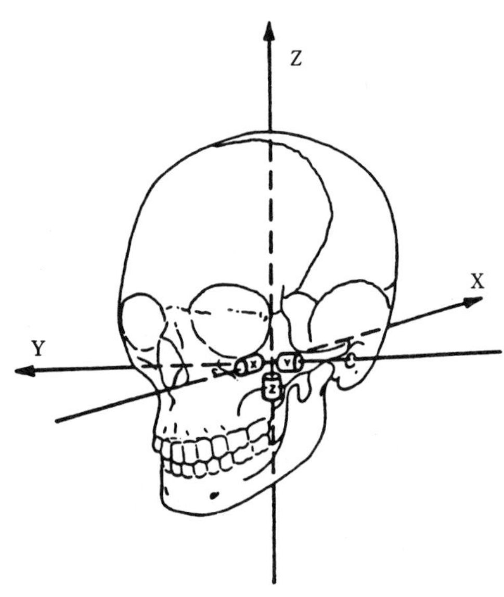

Fig. 14 Instrumentation reference frame and location of the tri-axial accelerometer (Combination of three uniaxial accelerometers)

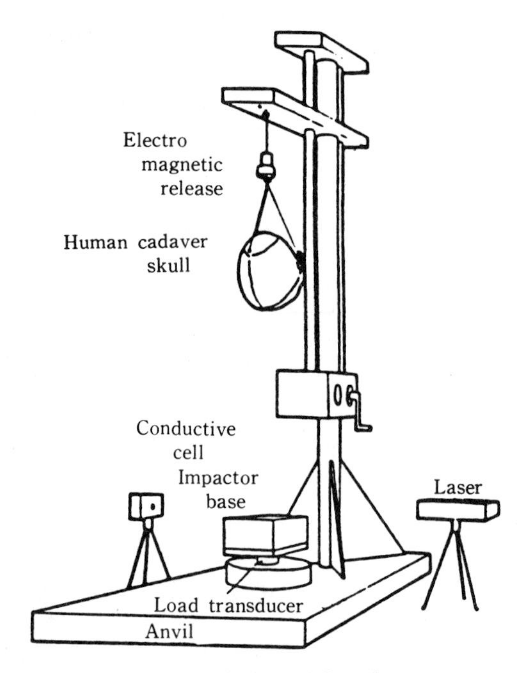

Fig. 15 Overall view of free drop test on human cadaver skull

Table 5 Number of Outcomes, Skull Fractures and Brain Contusions in Rotational Impact (Group B)

Head Injury			Survived	Died	Total
Fracture	Linear	Contusion	2	0	2
		Non-contusion	0	0	0
	Depressed and/or Comminuted	Contusion	0	3	3
		Non-contusion	0	1	1
Non-fracture		Contusion	5	2	7
		Non-contusion	5	0	5
Total			12	6	18

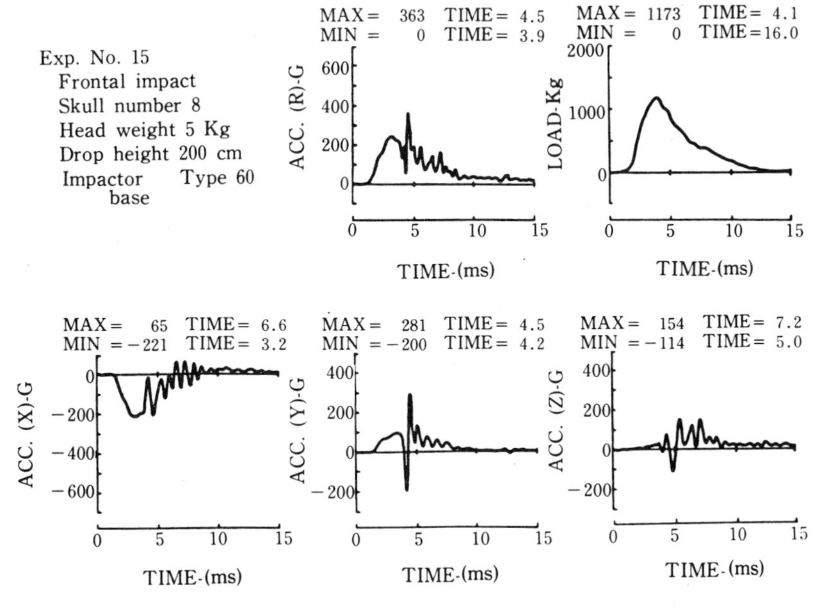

Fig. 16 Skull acceleration and load from frontal impact on human cadaver skull in experiment No. 15

vibration pattern of the skull itself due to the fracture. The phenomenon of skull fracture can therefore be seen in terms of a change in pattern in the time history of the skull acceleration. Skull fracture is more pronounced in the case of impact on an anvil or rigid impactor base from a relatively great height. In the special case of "over fractures" (where large bone fragments are broken off), spike-shaped higher-harmonic damped oscillations appear, then fall off rapidly before the time history of skull acceleration reaches its maximum amplitude (Fig. 16). Fig. 17 shows the relationship between skull fractures and resultant average skull acceleration, and its duration in the case of frontal impact. In this figure the symbol "◎" indicates over-fractures, and "●" indicates fractures on visual examination. Fractures in the intracranium (in the region of the sella turcica or canalis opticus, for example) are difficult to detect visually, however. In one case no fracture was detected until the skull was opened. It was therefore decided to review the time history of skull acceleration patterns even in cases where fractures had been detected by visual examination. This review resulted in the detection of "subfracures", which the figure by the symbol "◍". This review ascertained a distinct boundary between the fracture and nonfracture groups, and clarified the frontal fracture threshold. The fracture threshold in the case of occipital impact is almost the same as that of frontal impact (Fig. 18).

DISCUSSION

IMPACT CONDITIONS AND INJURIES; SURVIVAL AND MORTALITY OF SUBJECTS - Mortality rates in the three groups of head impact experiments on monkeys was as follows: TAIRH, 11 out of 26 (excluding one case of transection of the spinal cord junction); RAIRH, six out of 18; DIUH, eight out of 18. Under all three impact conditions, autopsy revealed that the macroscopic lesions most highly correlated with mortality were the depressed fractures accompanying skull base fractures, brain stem injuries, cervical injuries and thoracic injuries. Each type of injury showed a high mortality rate; namely, 100, 83, 61 and 61%, respectively. (Table 6 shows the number, percentage and mortality rate of each type of injury under the different impact conditions). In the TAIRH head restraint mask with broad contact area was employed to impart a purely translational motion to the head, but unfortunately, thoracic injuries also resulted. In the RAIRH experiments, the same restraint system was employed to impart a purely rotational motion to the head, but again unintended injuries resulted. This time, due to slippage of the skin at the time of impact, the tip of the mask and other points delivered localized impacts which depressed the skull or injured the cervical region.

Nevertheless, the experiments throw a clear light on the possible cause of death. In the fatal cases there was a high incidence of depressed fractures accompanying skull base fractures (20%), cervical injuries (78%), thoracic injuries (61%), and brain stem injuries (28%). In the case of survivors, these cases appeared at the low incidence rates of 0, 26, 20 and 3%, respectively. Taking these results together with the neurophysiological findings after the impact, it can be seen that in both fatal and surviving cases, the effect of the impact on the

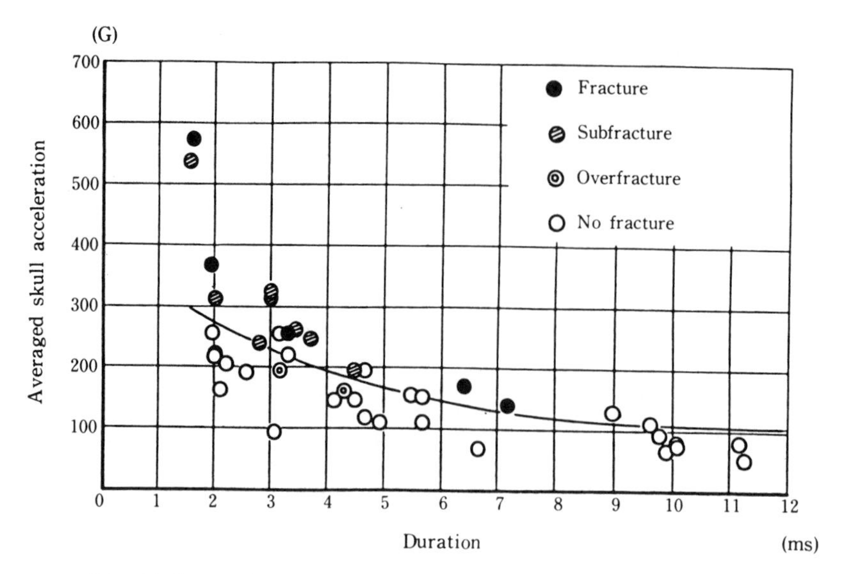

Fig. 17 Correlation between averaged acceleration-duration and skull fracture in frontal impact of human cadaver skulls

brain stem inhibits the heart and respiratory system, which this brain stem control. This results in a vicious circle which further suppresses the brain functions. It is thought that in the fatal cases seen in the study, recovery from this vicious circle was blocked by the effects of the cervical and thoracic injuries, so that ultimately a fatal condition ensued.

HEAD ACCELERATION AND BRAIN CONTUSIONS - In the TAIRH experiments, there were no visible brain contusions, nor any brain injuries even in cases where a negative intracranial pressure of -1 (lower than atmospheric pressure) was thought to have developed [18],[19]. In the RAIRH and DIUH experiments, on the other hand, brain contusions occurred which, although relatively slight (apart from fracture cases), were clearly visible in the contrecoup or coup region. In the non-fracture groups, brain contusions occurred at a rate of 35%, and under rotational acceleration impact in particular, they occurred at the high rate of 50%. Fig. 19 shows the relationship among the resultant rotational head acceleration, duration, and brain contusions in the non-fracture groups under the RAIRH and DIUH impact conditions. This relationship indicates a tendency for brain contusions to appear at a rotational velocity of at least 300 rad/s, and suggests that a rotational component is necessary for the occurrence of brain contusions. This is in agreement with the view of Ommaya et al. [20],[21].

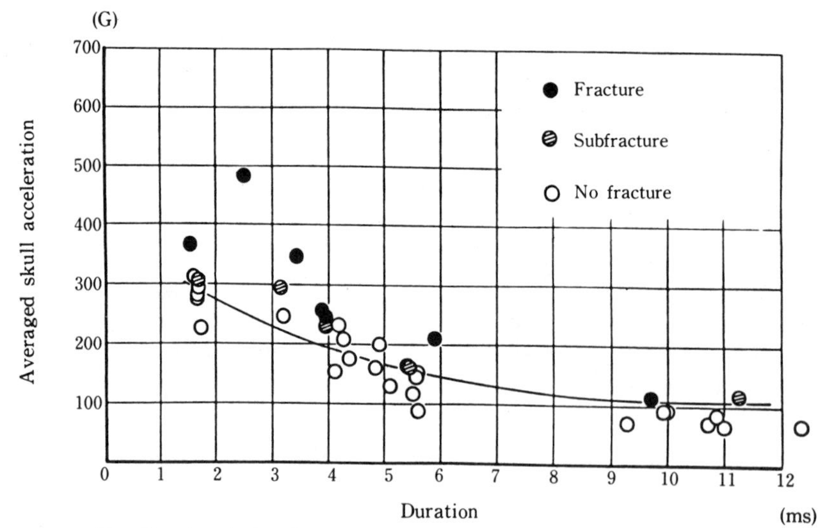

Fig. 18 Correlation between averaged acceleration-duration and skull fracture in occipital impact of human cadaver skulls

Table 6 Comparison of the Rate and the Number of Injuries in Groups A, B and C

Grouping	OC	TS	Head Injury									Cervical Injury	Thoracic Injury
			STF	Skull Fracture				SOF	No Skull Fracture				
				TF	Linear L(L/TS%)	Depressed and/or Comminuted DC(DC/TS%)	Basal B(B/TS%)		TI	Contusion C(C/SOF%)	Brain Stem Injury R(R/SOF%)	CI (CI/SOF%)	TI (TI/SOF%)
Translational Impact (Group A)	S	15	1	1	1(7)	0(0)	0(0)	14	0	0(0)	0(0)	3(21)	6(32)
	D	11	0	0	0(0)	0(0)	0(0)	11	1	0(0)	1(9)	8(73)	11(100)
Rotational Impact (Group B)	S	12	2	4	2(17)	0(0)	2(17)	10	5	5(50)	0(0)	2(20)	1(10)
	D	6	4	4	0(0)	0(0)	4(67)	2	3	2(100)	1(100)	2(100)	0(0)
Direct Impact (Group C)	S	10	1	1	1(10)	0(0)	0(0)	9	3	2(22)	1(11)	4(44)	0(0)
	D	8	3	4	2(25)	1(13)	1(13)	5	4	1(20)	3(60)	4(80)	0(0)
Total	S	37	4	6	4(11)	0(0)	2(5)	33	8	7(20)	1(3)	9(26)	7(20)
	D	25	7	12	2(8)	5(20)	5(20)	18	8	3(17)	5(28)	14(78)	11(61)
Mortality Rate	D/(D+S)				33%	100%	71%			30%	83%	61%	61%

OC: Outcome S: Survived D: Died
TS: Total number of subjects
STF: Subtotal number of subjects with skull fracture
TF: Total number of occurrences of linear, depressed comminuted and basal fractures
L: Total number of subjects with linear fracture
DC: Total number of subjects with depressed comminuted fracture
B: Total number of subjects with basal fracture
C: Total number of subjects with brain contusion
R: Total number of subjects with brain stem injury
CI: Total number of subjects with cervical injury
TI: Total number of subjects with thoracic injury
SOF: Total number of subjects with no skull fracture

However, the present study yielded several findings on the causes of brain contusion under different conditions of impact, namely:
(1) Although the rotational velocity of the head was almost the same under rotational acceleration impact and under direct impact on the unrestrained head, only the former showed a high incidence of brain contusion.
(2) Brain contusion did not occur in the TAIRH experiments in spite of the large impact delivered.
(3) In the RAIRH experiments, the imperfect fit between the head mask and the head resulted in a 33% incidence of skull fractures. Hence it can be deduced that, compared to the other groups, the non-fracture cases in this group included more cases of deformation of the skull that were nearly, but not really, fractures.

The above findings suggest that, brain ccntusion is caused not only by rotational acceleration as generally proposed, but also by the contact area between the impactor and the head, which determines the deformation of the skull on impact.

HEAD ACCELERATION AND CONCUSSION; THE CONCUSSION THRESHOLD IN MONKEYS - While it is difficult to determine the presence, degree and duration of concussion in animals, as a rule the loss and subsequent recovery of the pupillary light reflex, ciliary reflex, blinking, spontaneous movement and respiration, etc., should be sufficient criteria to indicate concussion in the clinical aspect.

Of the three groups of impact conditions, those showing the most cases of severe concussion (i.e., almost all at grade III or above) were translational and rotational acceleration impact. The severest cases occurred in the former. DIUH resulted in relatively mild concussion of grades II-III or below. In Fig. 20 (a), (b), which illustrates these results in relation to resultant average head acceleration and its duration, the resultant average head acceleration is seen to fall on almost the same velocity line in the case of both RAIRH and DIUH, but the duration of the former is relatively short, ranging from 1.5 to 2.5 ms. Further, a comparison of concussion of grades II-III in non-fracture cases showed that after the RAIRH experiment, it required an average of three times longer time for recovery of the corneal reflex and 1.5 times longer for recovery of respiration, than in the case of the DIUH experiment. Also, brain contusions were found in 17% of all fatal cases and 20% of all survivors (Table 6).

The above findings on the course of concussion and the occurrence of brain contusions suggest that not only pathological findings in the brain but also neurophysiological findings must be taken into consideration as indications of the head impact tolerance threshold under different impact conditions.

Next, attention was focussed on relatively mild impacts of the unrestrained head against padded flat surfaces, since of the three types of impact this comes closest to the conditions found in actual collisions. The relationship between concussion and head acceleration was studied. Fig. 21(a), (b) shows the relationship between the resultant average head acceleration and the grade of concussion. The incidence of grades 0, I, II and III on the occipital impact can be divided into zones along the iso-velocity lines. The particular zone in which concussion occurs is clearly delineated. The zone of occurrence under frontal impact is not as well

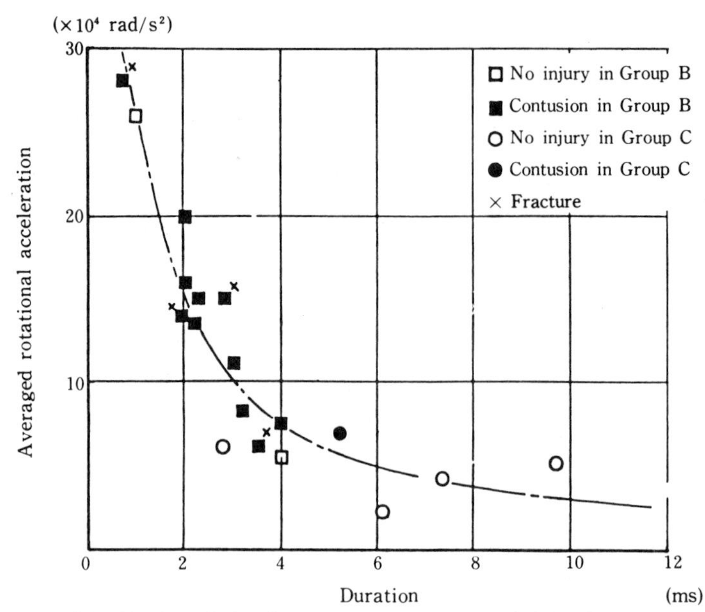

Fig. 19 Correlation between averaged rotational head acceleration-duration and brain concussion of the monkey in occipital impact (Groups B and C)

defined, but can be estimated by taking the occurrence of concussion grades II-III as the criterion. These results suggest that the conditions under which concussion occurs in frontal or occipital impact against a padded flat surface may be determined primarily by the resultant average head acceleration and its duration. Further, comparison of the concussion thresholds in frontal and occipital impacts shows that the tolerance threshold is higher in frontal impact. As Fig. 21 shows, frontal impact results in fewer concussions of grades II and III above the threshold line drawn.

From the viewpoint of the safety threshold, however, the authors regarded the concussion threshold as virtually the same for both frontal and occipital impacts.

ESTIMATION OF THE CONCUSSION THRESHOLD IN HUMANS - The head impact experiments on monkeys provided various kinds of information on how the force and conditions of a head impact are related to the ensuing head and brain injuries. The results showed that the occurrence of concussion in the subjects is an indication of the tolerance threshold, representing a disturbance of vital functions which is transitory and reversible. Moreover, the close correlation between skull fractures and survival or mortality, serve as an indication of the tolerance threshold at which hemorrhage due to brain contusion is likely to result. Thus, by applying the dimensional analysis method of Stalnaker et al. [9],[10] to the experimentally determined concussion threshold in monkeys, the authors are able to estimate the concussion threshold in humans.

To further clarify the reliability of the estimate obtained by this dimensional analysis, estimation by this method will be performed on one case of frontal impact and four cases of occipital impact on the monkey head impact experiments. The results will then be compared with the fracture threshold in these fracture cases and in human cadaver skulls. In the previous dimensional analysis, the five parameters used for conversion from the monkey to the human case were: 1) head acceleration; 2) duration of head acceleration; 3) impact velocity; 4) average skull radius; 5) average skull bone thickness. It was assumed that the geometrical forms of the monkey and human head are analogous, and that the vital tissues have the same characteristics. As π numbers, the relationship of π_1, π_2 and π_3 as determined by Stalnaker et al. from several species of sub-human primates of all sizes (squirrel monkeys, rhesus monkeys, chimpanzees, etc.) was employed. The ratio of the average human skull bone thickness (h) to the average human skull radius (ℓ) was taken to be $h/\ell=13.3$, and the brain weight (w_b) was taken to be 1.74 kg.

For frontal and occipital impacts, respectively, Figs. 22(a), (b) show the relationship between head acceleration, duration, and concussion as determined by dimensional analysis. The concussion thresholds for the frontal and occipital regions of the human head can be indicated here, as in the case of the monkey head, by means of the hyperbolic curve which links the resultant average head accelerations and durations of 220 G, 2 ms and 90 G, 9 ms.

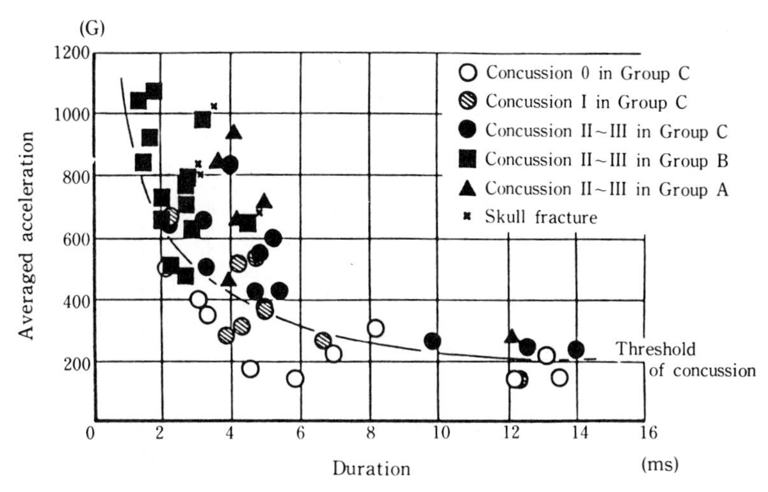

Fig. 20-a Comparison between averaged acceleration-duration of the monkey head and grades of concussion in occipital impact among three groups (A, B, C)

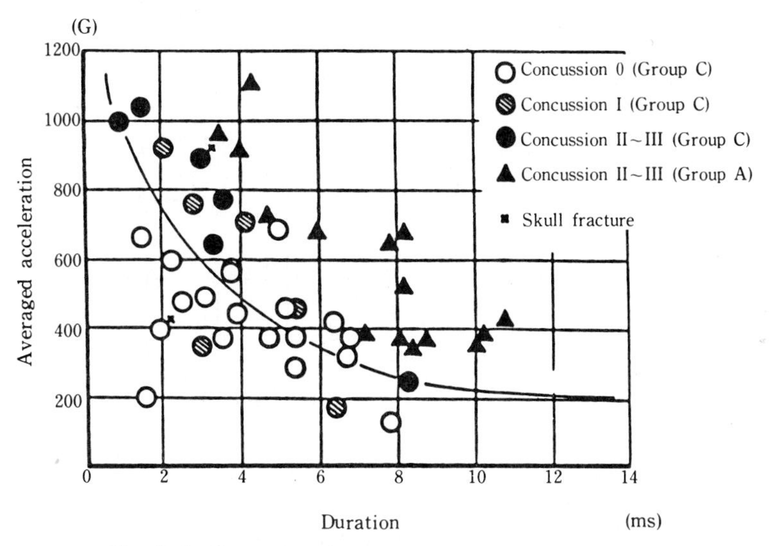

Fig. 20-b Comparison between averaged head acceleration-duration and grades of concussion in frontal impact (Groups A and C)

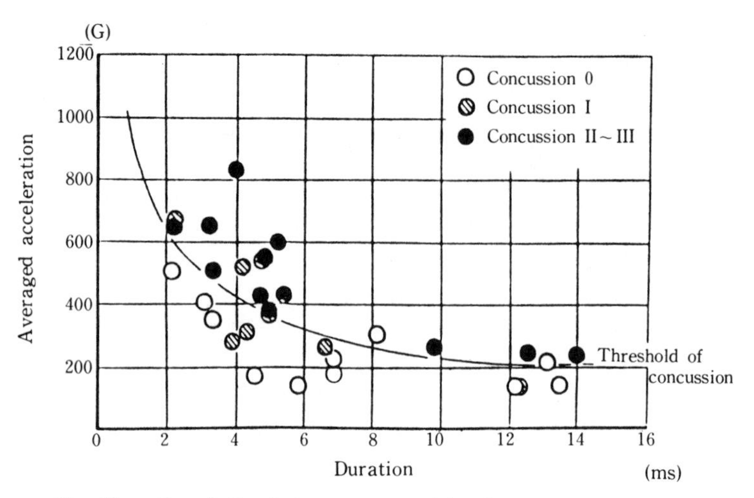

Fig. 21-a Correlation between averaged head acceleration-duration and grades of concussion in occipital impact (Group C)

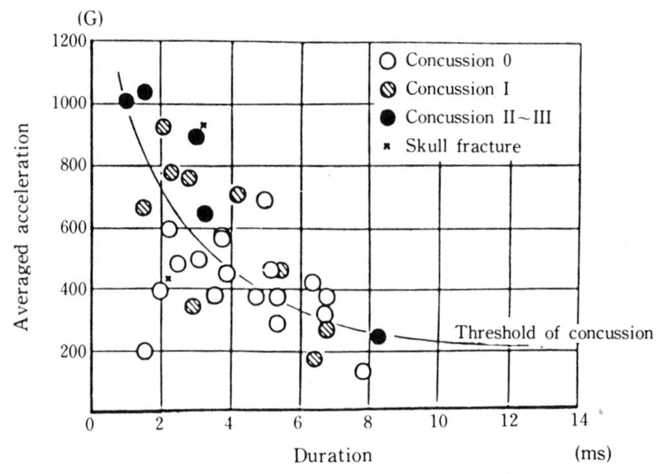

Fig. 21-b Correlation between averaged head acceleration-duration and grades of concussion in frontal impact (Group C)

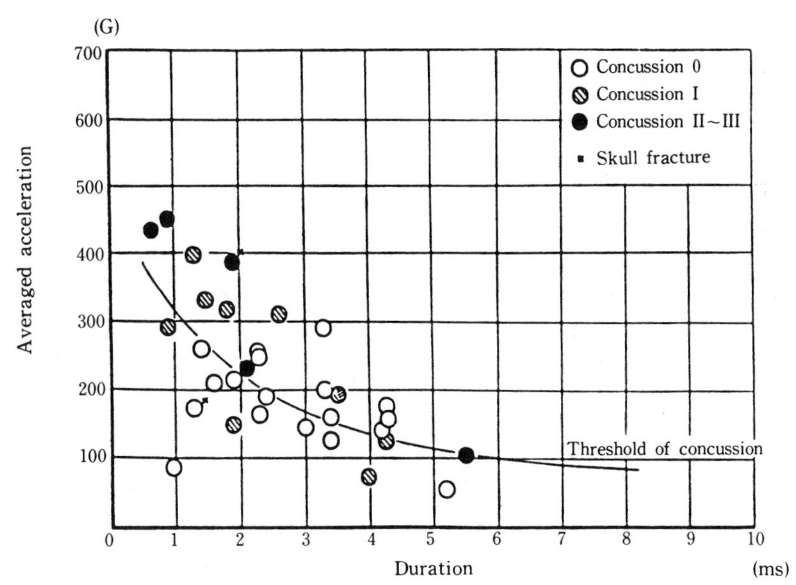

Fig. 22-a Threshold of concussion in humans extrapolated from averaged acceleration-duration by dimensional analysis in frontal impact of the monkey head

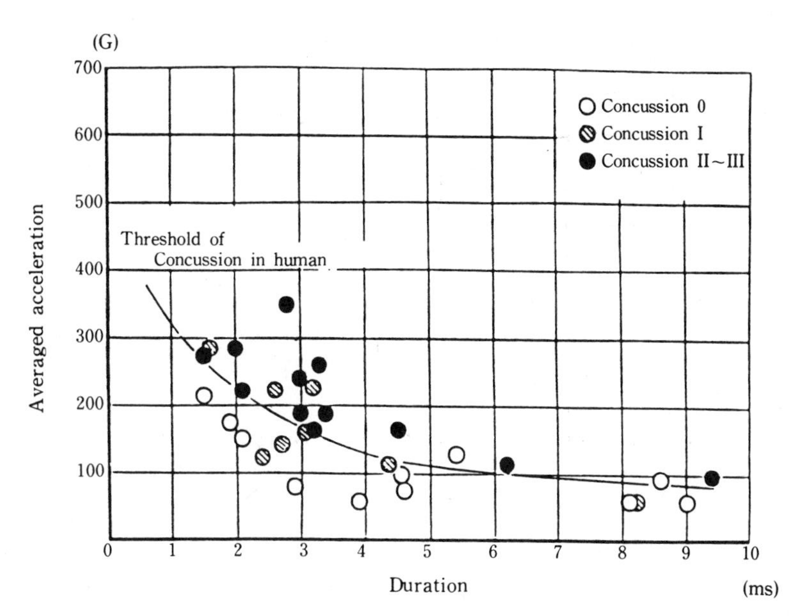

Fig. 22-b Threshold of concussion in humans extrapolated from averaged acceleration-duration by dimensional analysis in occipital impact of the monkey head

Fig. 23(a), (b) shows a comparison between the fracture cases in the experiment on human cadaver skulls and fracture cases estimated by dimensional analysis from experiments on monkey skulls. (The symbol "▲" indicates the fracture cases in monkeys.) No significant difference was found between the human skull fracture level estimated from monkey skull fractures and that determined from human cadaver skulls.

It was therefore concluded that the estimates of human levels obtained by dimensional analysis are sufficiently reliable.

TOLERANCE THRESHOLD FOR HUMAN HEAD IMPACT - The authors obtained the concussion threshold curve for the monkey head and, using this as a basis, estimated the concussion threshold curve of the human head. Further, using human cadaver skulls they were able to experimentally determine the human skull fracture threshold curve.

In autopsies on monkeys which had shown concussion, subarachnoid and subpial hemorrhage were visible in 56% of all cases. Also, of the 45 monkeys which recovered from concussion, 12 showed brain contusions. This fact suggest that, from the clinical aspect, cases of slight lesions to the head which have previously been regarded as "concussion only" may include some cases of brain contusion, even if these do not present obvious symptoms. In fact, as one of the authors (N. N. et al.) previously reported [22], CT scanning has revealed brain contusions in 4-5% of cases clinically diagnosed as "concussion only."

The human concussion threshold curve estimated in this study from that of monkeys, when considered in the light of these clinical cases, can be seen to represent a strict tolerance limit which tends towards the conservative side. The fracture threshold, as can be seen from the monkey experiments also, is a threshold indication at which there is a high probability of fatality. The human concussion threshold curve derived both from that of monkeys and from the human cadaver skull fracture threshold curve is shown in Fig. 24. This is taken to be the head impact tolerance curve for humans, and has been named the JARI Human Head Impact Tolerance Curve (abbreviated to JHTC).

A head impact tolerance threshold known as the Wayne State Tolerance Curve (WSTC) [23] has been proposed by researchers a Wayne State University. This forms the basis for the head injury criteria of FMVSS [24]. In the tolerance zone of the WSTC, injuries are said to be limited to concussion with no sequelae. According to the literature [25], however, the WSTC has been compiled from various experimental results using cadavers, animals, dummies, volunteers, etc., and the indices of injury in the separate experiments have not been made entirely clear. The values obtained in the present study differ from those of WSTC due to a difference in head acceleration amplitude within long duration. In spite of this difference, it can be inferred that WSTC also uses concussion as a tolerance indication.

SUMMARY AND CONCLUSIONS

1. In order to investigate the head impact tolerance threshold, a series of experiments was conducted on 63 monkeys under conditions of translational and rotational acceleration impact using a head restraint mask with broad contact area, and direct impact to the unrestrained head using a padded flat surface. The results have been presented and discussed. Free-drop impact experiments were also conducted on human cadaver skulls to determine the human skull fracture threshold, and the results were compared with the monkey skull fracture threshold.

2. Fatal visible brain injury could not be produced simply by means of a translational acceleration impact to the head at around 1,000 G. What actually produces the injury is brain-sustained functional damage due to primary direct effects, and vicious circle complication caused by injury to the heart and respiratory system.

3. In rotational acceleration, impact and direct impact to the unrestrained head (where a rotational compornent is also present), produce a higher rate of both brain contusions and skull fractures than translational acceleration. This suggests that the rotational acceleration of the head is not the only indicative factor in brain contusions. Deformation of the skull at impact, and therefore the contact area of the head at impact, also play a part.

4. The occurrence of concussion in monkeys showed no correlation with the rotational acceleration of the head, but was highly correlated with the head acceleration. Accordingly, the threshold of concussion in the monkey was determined primarily from the resultant average head acceleration and duration.

5. It was found that the occurrence of concussion in monkeys is an indication of the tolerance threshold (i.e., effects are transitory and reversible), and that the occurrence of skull fractures is an indication of the danger threshold in which severe, and even fatal, brain injury is likely to result.

6. Comparison of the concussion thresholds on frontal and occipital impact in monkeys showed a tendency to slightly higher levels on frontal impact, but in terms of the tolerance limit, the two could be regarded as having almost the same level.

7. The human head impact tolerance threshold was derived by applying the dimensional analysis method of Stalnaker et al. to the concussion threshold in the monkey. This threshold curve was named the JARI Human Head Tolerance Curve (JHTC). Further, the reliability of the JHTC estimated by dimensional analysis was confirmed by comparing observed fractures in monkey skulls with those in human cadaver skulls.

8. From the estimated human head impact tolerance curve it was inferred that the criterion for the tolerance threshold used in the human head impact tolerance curve of SAE Recommended Practice J885a [26] is the level of concussion.

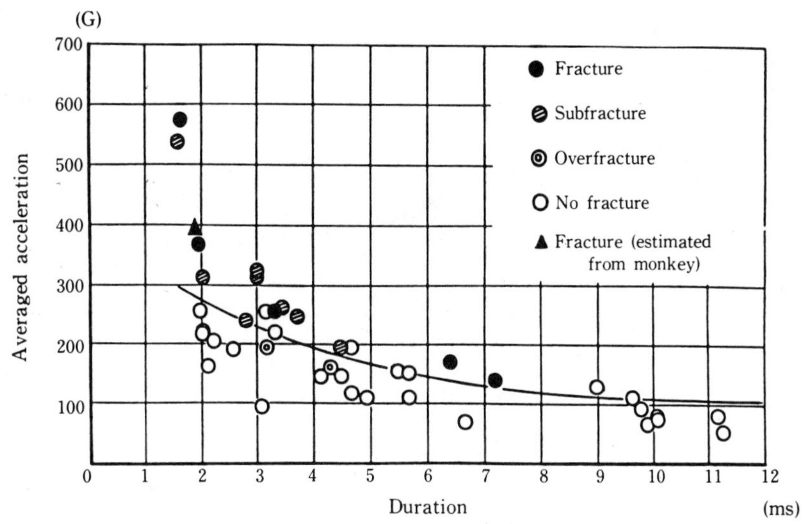

Fig. 23-a Comparison between averaged acceleration-duration of human cadaver skulls and skull fracture extrapolated from monkey in frontal impact

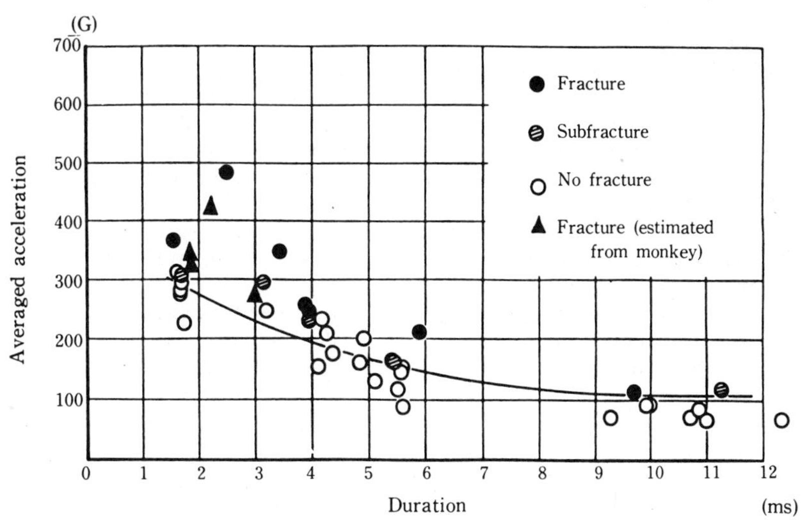

Fig. 23-b Comparison between averaged acceleration-duration of human cadaver skulls and skull fracture extrapolated from monkey in occipital impact

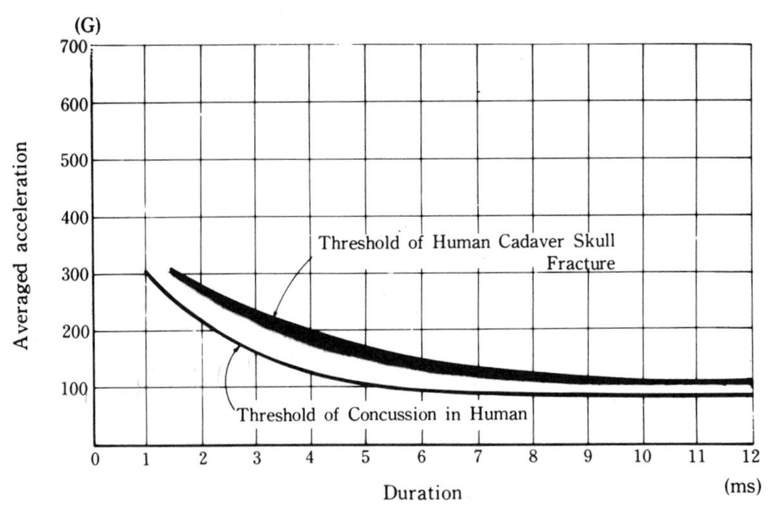

Fig. 24 JARI Human Head Impact Tolerance Curve (JHTC)

ACKNOWLEDGMENTS

This study was supported over a three-year period (1976-1978) by grants-in-aid Nos. 163-2, 173-9, and 193-10 of the Automobile Bureau Safety Section, under the Automobile Safety Standards Research Program, Automobile Accident Countermeasures Fund, and Project Subsidies of the Ministry of Transportation.

The authors received generous assistance from many members of the research and laboratory staffs of the Neurosurgery Departments of the Tokyo Jikei-kai University School of Medicine, the University of Tokyo, and Tokyo Women's Medical College. They are grateful also for the advice provided by Dr. Masanori Matsuno, former director and head of the 1st Research Division, Japan Automobile Research Institute; Dr. Sumiji Fujii, Professor of the Faculty of Engineering, University of Tokyo; Dr. Umio Soejima, Professor of the Department of Engineering, Nihon University; and Dr. Tsuyoshi Inoue, Professor Emeritus of the School of Medicine, Kanazawa University.

REFERENCES

1) R. H. Pudenz and C. H. Shelden, "The Lucite Calvarium - A Method for Direct Observation of the Brain. II. Cranial Trauma and Brain Movement", Journal of Neurosurg. 3, 487-505 1946

2) A. G. Gross, "A New Theory on the Dynamics of Brain Concussion and Brain Injury", Journal of Neurosurg. 1, 190-200 1944

3) R. L. Stalnaker, J. W. Melvin, G. S. Nusholtz, N. M. Alem and J. B. Benson, "Head Impact Response", Paper 770921 Proceedings of the 21st Stapp Car Crash Conference, 1977

4) M. Masuzawa, N. Nakamura, et al., "Experimental Head Injury and Concussion in Monkey Using - Pure Linear Acceleration Impact" Neurologic medico-chirurgica Vol. 16, Part I, 1976

5) Y. Komaki, A. Kikuchi, M. Horii, K. Ono, M. Kitagawa, and M. Matsuno, "The Research on the Fatal Level of the Primates Head Impact Tolerance", HOPE International JSME Symposium, Hazard-free Operation against Potential Emergencies, October 1977

6) T. Tsubokawa, N. Moriyasu, M. Matuno, et al., "Fatal Brainstem Damage Caused by Linear Acceleration Impact - Relationship between Pathophysiological Findings and Far Field Acoustic Response -", Neurologica medico-chirurgica Vol. 19, No. 3, 1979 (Japanese)

7) H. Kohno, N. Nakamura, M. Matsuno, et al., "Experimental Head Injury and Concussion : Morphologic Change and Pathophysiologic Following Translational Acceleration in Primates", Neurologia medico-chirurgica Vol. 19, No. 8, 1979 (Japanese)

8) H. Sekine, N. Nakamura, A. Kikuchi, K. Ono, et al., "Experimental Head Injury in Monkey Using Rotational Acceleration Impact", Neurologia medico-chirurgica Vol. 20, No. 2, 1980 (Japanese)

9) R. L. Stalnaker, V. L. Roberts, and J. H. McElhaney, "Side Impact Tolerance to Blunt Trauma", Paper 730979 Proceedings of the 17th Stapp Car Crash Conference

10) R. L. Stalnaker, J. H. McElhaney, R. G. Snyder, and V. L. Roberts, "Door Crashworthiness Criteria" U. S. Dept. of Transportation Report No. HS-800-S34, June 1971

11) J. H. McElhaney, R. L. Stalnaker, V. L. Roberts, and R. G. Snyder, "Door Crashworthiness Criteria", Paper 710864 Proceedings of the 15th Stapp Car Crash Conference

12) R. L. Stalnaker and J. H. McElhaney, "Head Injury Tolerance for Linear Impacts by Mechanical Impedance Methods", ASME Paper No. 70-WA/BHF -4, 1970

13) L. B. Walker, E. H. Harris, and U. R. Pontinus, "Mass, Volume, Center of Mass, and Mass Moment of Inertia of Head and Head and Neck of Human Body", Paper 730985 Proceedings of the 17th Stapp Car Crash Conference

14) A. M. Nahum and R. W. Smith, "An Experimental Model for Closed Head Impact Injury", Paper

760825 Proceedings of the 20th Stapp Car Crash Conference
15) M. Sakimura, "Reviews on Accuracies and Results of ESV Collision Tests", Fifth International Technical Conference on Experimental Safety Vehicles, London, England, 1974
16) E. B. Becker, "Measurement of Mass Distribution Parameters of Anatomical Segments", Paper 720964 Proceedings of the 16th Stapp Car Crash Conference
17) R. P. Hubbard and D. G. Mcleod, "Geometric, Inertial, and Joint Characteristics of Two Part 572 Dummies for Occupant Modeling", Paper 770937 Proceedings of the 21st Stapp Car Crash Conference
18) V. L. Roberts V. R. Hodgson, and L. M. Thomas, "Fluid Pressure Gradients Caused by Impact to the Human Skull", ASME 1966
19) T. Hayashi, "Theory of Human Tolerance Limit for Head Impact and its Cushioning Method", Journal, Japan Soc. Auto. Engrs., Vol. 24, No. 7, January 1970 (Japanese)
20) A. K. Ommaya, P. Yarnell, A. E. Hirsch, and E. H. Harris, "Scaling of Experimental Data on Cerebral Concussion in Sub-Human Primates to Concussion Threshold for Man", Paper 670906 Proceedings of the 11th Stapp Car Crash Conference
21) T. A. Gennarelli, A. K. Ommaya, and L. E. Thibault, "Comparison of Translational and Rotational Head Motions in Experimental Cerebral Concussion", Proceedings of the 15th Stapp Car Crash Conference, November 1971
22) H. Sekino, N. Nakamura, et al., "Brain Contusion Detected by CT Scann in Minor Head Injury", Neuro Traumatology 1979, 31p
23) L. M. Patrick, H. R. Lissner and E. S. Gurdjian, "Survival by Design - Head Protection", Proceedings the 7th Stapp Car Crash Conference 1963
24) Federal Motor Vehicle Safety Standard, Standard No. 208, Occupant Crash Protection, January 31, 1967
25) R. G. Snyder, "State-of-the-Art Human Impact Tolerance" SAE Paper No. 770398
26) SAE Recommended Practice J885a, "Human Tolerance to Impact Conditions as Related Motor Vehicle Design"
27) K. Ono, K. Miyazaki, M. Sakimura, S. Suzuki, and H. Ohashi, "Total Motion Measurement and Analysis on Dummy Head-Neck System" Transactions of The Society of Automotive Engineers of Japan, Inc. No. 14, 1978 (Japanese)
28) E. Becker and G. Willems, "An Experimentally Validated 3-D Internal Tracking Package for Application in Biodynamic Research", Paper 751173, Proceedings of the 19th Stapp Car Crash Conference

APPENDIX A

ANATOMICAL COORDINATE SYSTEM OF SUBJECT HEAD - The coordinate reference frame of the subject head is a right-handed system; its reference axes are shown in Fig. A-1. The X-axis of the head runs parallel to the Frankfort line which connects the orbital notch with the external autitory meatus (an anatomical reference line of the head) and passes through the center of gravity of the head. The Y- and Z-axes are the lines which pass through the center of gravity of the head and intersect the X-axis at right angles to it and to each other.

MEASUREMENT OF HEAD ACCELERATION - To measure the translational and rotational accelerations of the head using a set of uniaxial accelerometers, the component measurement method with six degrees of freedom using nine-channel accelerometers was applied [27],[28].

The acceleration values measured by the accelerometers can be expressed by the following equation:

$$\alpha_K = \vec{\alpha}_G \cdot \vec{e}_K + (\ddot{\vec{\theta}} \times \vec{r}_K \cdot \vec{e}_K) + (\dot{\vec{\theta}} \times (\vec{r}_K \times \dot{\vec{\theta}}) \cdot \vec{e}_K) \quad (1)$$

where:

α_K = acceleration sensed at the attachment point of the k'th accelerometer

$\vec{\alpha}_G$ = translational acceleration

$\ddot{\vec{\theta}}$ = rotational acceleration

$\dot{\vec{\theta}}$ = angular velocity

$\vec{r}_K$ = attachment point of the k'th accelerometer

$\vec{e}_K$ = direction of attachment of the k'th accelerometer.

To calculate the translational and rotational acceleration components from the measurements of the nine-channel accelerometers, the three accelerometers aligned on the same axis must be parallel to the same axis and must lie on the same plane. Also, the sensitivity axes of the accelerometers aligned in each of the three directions must lie at right angles to that direction.

Fig. A-1 shows the nine accelerometers attached to the frontal region of the subject head, showing also the alignment of the sensitivity axis of each. This is the standard arrangement of nine-channel accelerometers on the subject head. The distances between the accelerometers in this arrangement can be represented by equation (2).

$L_1 \sim L_6$: distances between the nine accelerometers

$$\left.\begin{array}{ll} L_1 = \ell_{2y} - \ell_{1y} & L_2 = \ell_{3z} - \ell_{1z} \\ L_3 = m_{2z} - m_{1z} & L_4 = m_{3x} - m_{1x} \\ L_5 = n_{2x} - n_{1x} & L_6 = n_{3y} - n_{1y} \end{array}\right\} \quad (2)$$

where ℓ, m and n are the attachment pointed of the nine accelerometers, as follows

$$\vec{\ell}_1(X) = (\ell_{1x}, \ell_{1y}, \ell_{1z}), \quad \vec{m}_1(Y) = (m_{1x}, m_{1y}, m_{1z}), \quad \vec{n}_1(Z) = (n_{1x}, n_{1y}, n_{1z})$$

$$
\begin{aligned}
\vec{\ell}_2(TX)&=(\ell_{2x},\ell_{2y},\ell_{2z}), \quad \vec{m}_2(FY)=(m_{2x},m_{2y},m_{2z}),\\
&\qquad\qquad \vec{n}_2(PZ)=(n_{2x},n_{2y},n_{2z})\\
\vec{\ell}_3(FX)&=(\ell_{3x},\ell_{3y},\ell_{3z}), \quad \vec{m}_3(PY)=(m_{3x},m_{3y},m_{3z}),\\
&\qquad\qquad \vec{n}_3(TZ)=(n_{3x},n_{3y},n_{3z})
\end{aligned}
\tag{3}
$$

Using equations (1), (2) and (3) above, the rotational acceleration can be calculated by the component measurement method with six degrees of freedom according to equation (4).

$$
\vec{\ddot{\theta}}=\begin{bmatrix}\ddot{\theta}_x\\ \ddot{\theta}_y\\ \ddot{\theta}_z\end{bmatrix}=\frac{1}{2}\left[\begin{bmatrix}0 & \frac{1}{L_3} & \frac{-1}{L_6}\\ \frac{-1}{L_2} & 0 & \frac{1}{L_5}\\ \frac{1}{L_1} & \frac{-1}{L_4} & 0\end{bmatrix}\begin{bmatrix}X\\ Y\\ Z\end{bmatrix}-\begin{bmatrix}0 & \frac{1}{L_3} & 0\\ 0 & 0 & \frac{1}{L_5}\\ \frac{1}{L_1} & 0 & 0\end{bmatrix}\begin{bmatrix}TX\\ FY\\ PZ\end{bmatrix}+\begin{bmatrix}0 & 0 & \frac{1}{L_6}\\ \frac{1}{L_2} & 0 & 0\\ 0 & \frac{1}{L_4} & 0\end{bmatrix}\begin{bmatrix}FX\\ PY\\ TZ\end{bmatrix}\right]
\tag{4}
$$

where X, Y, Z, TX, FY, PZ, FX, PY and TZ are the accelerations obtained from the respective accelerometers attached to the head.

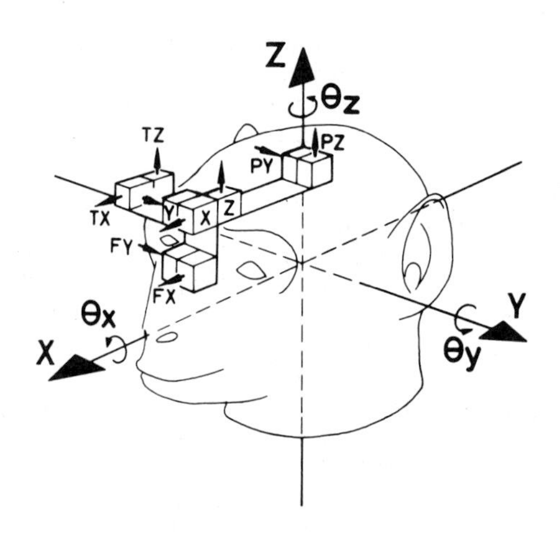

Fig. A-1 Coordinate reference frame and location of nine-channel accelerometers for subject head.

Appendix B Summary of Translational Head Impact (group A)

Experimental No.	Species	Age	Sex	Body Weight (Kg)	Brain Weight (g)	Skull Diameter A - P (mm)	Skull Diameter Lateral (mm)	Skull Diameter Thickness (mm)	Impact Location	Impact Velocity (m/s)	Frontal Acceleration Peak G_x (G)	Frontal Peak G_z (G)	Frontal Peak Average G_r (G)	Frontal Duration (ms)	Occipital Acceleration Peak G_x (G)	Occipital Peak G_z (G)	Occipital Peak Average G_r (G)	Occipital Duration (ms)	Parietal Acceleration Peak G_x (G)	Parietal Peak G_z (G)	Parietal Peak Average G_r (G)	Parietal Duration (ms)	Intracranial Pressure (Kg/cm^2)	Outcome	Absence Duration Light React. (m)	Absence Duration Blink (m)	Skull Fracture	SAH	Autopsy Findings
101	J	5	F	7.8	114	72	60	2.7	F	27.1	/	/	/	/	600	-640	760 350	10.0	890	220	900 600	10.4	1.29/-1.07	S	5	5	—	—	C_6-Th_5 EDH, adrenal gland slight contusion
102	J	5	M	7.1	107	75	60	2.6	F	27.2	/	/	/	/	780	-940	1080 520	8.2	1060	-400	1110 500	9.6	1.33/-0.98	S	8	8	—	+	Mediastinal slight hemo., liver hemo.
103	J	4	M	6.6	103	—	—	—	F	27.1	/	/	/	/	470	-740	750 430	10.7	—	—	—	—	0.40/0.0	D	2	2	—	+	C_6-Th_5 EDH, mediastinal and pulmonary hemo.
104	J	8	M	8.3	109	79	62	2.8	F	27.1	/	/	/	/	600	-690	750 370	10.2	750	400	770 370	10.6	0.54/-0.75	S	1	1	—	—	Pulmonary hemo.
105	J	13	F	8.5	98	74	60	2.8	F	27.2	/	/	/	/	510	-360	610 390	7.2	670	-230	710 470	6.7	0.18/-0.76	D	∞	∞	—	—	C_6-Th_2 EDH, mediastinal hemo., rt. and lt. clavicls fracture
106	J	5	M	8.5	121	75	63	2.8	F	27.3	/	/	/	/	410	-660	720 370	8.8	630	190	690 360	10.7	1.01/-0.58	S	3	3	—	—	Brain congestion, pulmonary and liver congestion
107	J	4	M	5.5	98	73	61	2.3	F	27.4	/	/	/	/	540	400	540 340	8.4	660	220	660 350	9.2	0.24/-0.26	S	3	3	—	+	Mediastinal congestion, basal cistern SAH, C-Th EDH
108	J	5	M	7.8	114	76	66	2.7	F	27.1	/	/	/	/	510	430	680 370	8.1	830	220	830 400	9.4	0.18/-0.45	D	6	8	—	+	Occipital and temporal PTH, brain stem hemo., C_7-Th_2 EDH, pulmonary congestion
109	J	5	M	8.3	120	78	64	—	F	28.1	/	/	/	/	—	—	—	—	—	—	—	—	0.34/-0.36	S	3	3	—	+	Frontal lobe extensive SAH
110	J	6	M	8.4	114	85	72	2.8	F	28.1	/	/	/	/	—	—	—	—	850	-280	860 490	7.4	1.01/-0.79	D	5	6	—	+	Occipital lobe PTH, C_7-Th_5 EDH, media-stinal hemo., rib fractures
111	J	5	M	7.4	130	80	60	—	F	28.1	/	/	/	/	—	—	—	—	—	—	—	—	0.29/-0.06	D	5	—	—	—	Th_{2-8} and L-S EDH, mediastinal hemo., clavicle fracture, kidney congestion
112	J	5	M	8.6	122	—	—	—	F	28.2	/	/	/	/	—	—	—	—	—	—	—	—	0.67/-1.26	S	2	2	—	—	Ribs fractures
113	B	20	M	25.0	175	94	83	—	F	27.2	/	/	/	/	—	—	—	—	—	—	—	—	/	D	∞	∞	—	—	C_6-Th EDH, mediastinal hemo., clavicles fractures, pulmonary slight congestion
114	J	5	M	8.0	99				F	26.7	/	/	/	/	1200	690	1200 680	5.9	/	/	/	/	/	D	∞	∞	—	—	clavicles fractures, ribs fractures pulmonary hemo.
115	J	5	M	7.2	97				F	26.9	/	/	/	/	990	600	990 730	4.7	/	/	/	/	/	D	∞	∞	—	—	C_6-Th_2 SDH, Th_{7-8} EDH, ribs fractures, pulmonary contusion
116	J	5	M	9.6	1o2				F	26.9	/	/	/	/	900	1910	2110 680	8.2	/	/	/	/	/	D	∞	∞	—	—	Clavicles, rib and shoulder blade frac-tuers, plumonary hemo., mediastnal hemo.
117	J	5	M	13.5	113				F	26.9	/	/	/	/	1200	1750	1360 650	7.8	/	/	/	/	/	D	∞	∞	—	—	C_5-Th_2 SAH, rib and clavicles fractuers, pulmonary hemo.
118	J	4	M	7.6	112				F	23.6	/	/	/	/	1130	-1340	1600 960	3.5	/	/	/	/	/	S	3	7	—	—	Eyelid and eyeball PTH
119	J	4	M	7.5	96				F	26.7	/	/	/	/	1340	1150	1590 920	4.0	/	/	/	/	/	S	—	5	—	+	C_6-Th_2 pia mater hemo. with SAH
120	J	4	M	7.6	84				F	27.0	/	/	/	/	—	—	—	—	—	—	—	—	/	S	4	3	—	—	Pericardium hemo.
121	J	4	M	6.4	92				O	26.9	1110	-400	1180 720	4.9	/	/	/	/		—	—	-	/	S	3	3	+	—	Lineal fractuer from occipital to cranial base
122	J	5	F	7.4	98				O	27.0	1520	-300	1520 940	4.1	/	/	/	/	1430	-340	1490 900	4.3	/	S	0.3	0.3	—	—	Occipital base slight EDH
123	J	12	F	9.6	107				F	26.9	/	/	/	/	1290	1010	1610 1110	4.3	1430	-820	1520 920	4.4	/	D	∞	∞	—	—	Cervical spinal cord SDH, mediastinal hemo., pulmonary contusion
124	B	8	M	31.5	208				O	22.4	760	-430	1160 640	4.4	/	/	/	/	1510	530	1570 780	3.4	/	S	3	2	—	—	
125	J	10	F	6.8	110				O	27.0	1190	-490	1190 870	3.7	/	/	/	/	1360	570	1360 850		/	S	—	1	—	—	
126	J	12	M	10.0	126				O	27.2	610	-200	620 280	12.1	/	/	/	/	1070	570	1130 320	9.3	/	S	—	—	—	—	
127	B	10	M	40.0	193				O	26.9	510	-160	520 240	17.9	/	/	/	/	—	—	—	—	/	D	∞	∞	—		Transection of spinal cord at C_1

J: Macaca fuscata
B: Papio cynocephalus
F: Female
M: Male

F: Frontal
O: Occipital

D: Died
S: Survived

EDH: Epidural hemorrhage
SDH: Subdural hemorrhage
SAH: Subarachnoid hemorrhage
PTH: Petechial hemorrhage

Appendix C Summary of Rotational Head Impact Data (Group B)

Experimental No.	Species	Age	Sex	Body Weight (Kg)	Brain Weight (g)	Skull Diameter A - P (mm)	Skull Diameter Lateral (mm)	Skull Diameter Thickness (mm)	Impact Location	Impact Velocity ($^m/_s$)	Frontal Acceleration Peak Gx (G)	Frontal Acceleration Peak Gz (G)	Frontal Acceleration Peak Gr (G)	Frontal Acceleration Average Gr (G)	Frontal Acceleration Duration (ms)	Rotational Acc. Average Gr ($^{rad}/_{s^2}$ x10^4)	Rotational Acc. Duration (ms)	Outcome	Duration of Absence Light React. (s)	Duration of Absence Wink (s)	Duration of Absence Ciliary Reflex (s)	Duration of Absence Spont. Movement (s)	Duration of Absence Apnea (s)	Subarachnoid Hemorrhage	Skull Fracture	Autopsy Findings
201	J	13	F	7.5	97	—	—	—	O	31.6	2420	-1110	2640	990	3.2	29.0	0.6	D	∞	∞	∞	∞	∞	+	^+c	Pulmonary hemorrhage, transection of medulla
202	J	18	F	9.5	87	—	—	—	O	20.2	4040	-2440	4640	1380	4.9	—	—	D	∞	∞	∞	∞	∞	—	^+c	Hematoma in the pituitary gland, lt. ext. jugulan vein laceration
203	R	6	M	9.0	94	97	75	3.4	O	32.0	1080	-1840	2150	1060	1.3	—	—	S	90	60	—	10	35	—	—	Pulmonary congestion
204	J	13	F	6.5	85	—	—	—	O	32.6	—	—	—	—	—	—	—	D	∞	∞	∞	∞	∞	+	^+c	Cerebral hemorrhage, subdural hemorrhage medulla, cerebral herniation
205	J	5	M	10.5	97	—	—	—	O	29.1	—	—	—	—	—	—	—	S	300	360	210	390	20	—	—	
206	C	8	M	9.0	100	99	80	2.5	O	35.8	1590	-1500	2190	920	1.7	—	—	D	110	50	220	270	35	+	—	
207	J	10	M	8.2	112	105	80	3.5	O	28.4	2090	-1240	2420	1080	1.8	7.7	3.2	S	120	310	180	—	45	+	—	Intramedullary hemorrhage of cervical spinal cord, contusion of rt. frontal tip
208	J	9	M	7.8	134	97	78	3.7	O	28.3	1280	-620	1400	720	2.7	20.1	2.0	S	90	40	105	—	45	+	—	
209	J	8	M	8.3	104	—	—	—	O	32.7	1700	-430	1970	790	2.7	6.5	3.5	S	125	180	180	—	55	+	+	Intramedullary hemorrhage of upper cervical spinal cord
210	J	8	F	5.9	98	90	73	2.4	O	34.0	1240	-740	1310	650	4.5	15.3	2.8	D	∞	75	∞	∞	∞	+	^+c	Intramedullaryhemorrhage of C_{1-5}, subdural hemorrhage of C_4-Th_1
211	C	10	M	10.4	85	99	77	2.8	O	33.9	1540	-1240	1690	660	2.1	15.4	2.1	S	170	195	125	10	75	—	—	
212	J	7	M	6.7	109	—	—	—	O	28.8	1880	-880	2040	770	2.7	13.9	1.9	S	—	330	330	10	10	+	+	Slight contusion of rt. pulmonary, mandibular fractre
213	C	5	M	6.4	80	—	—	—	O	30.8	—	—	—	—	—	15.9	1.9	S	130	225	90	315	60	+	—	
214	C	5	M	8.5	71	90	73	2.7	O	32.4	1720	-1220	1330	510	2.2	28.1	0.9	S	105	270	175	200	50	—	—	
215	C	5	M	7.3	72	89	72	2.5	O	36.2	1250	-1500	1550	740	2.0	10.9	3.0	S	90	190	110	11	40	+	—	
216	C	5	M	7.4	75	93	73	2.8	O	36.4	820	-1520	1730	630	2.9	7.9	4.0	D	85	∞	∞	∞	20	+	—	Petechial hemorrhage of medulla
217	C	5	M	7.8	82	88	71	2.6	O	34.0	800	-1340	1420	840	1.5	5.2	4.0	S	90	—	130	720	50	—	—	Zygomatic fracture
218	C	5	M	8.0	75	91	72	2.2	O	33.3	760	-1010	1100	480	2.7	13.6	2.3	S	60	120	65	270	15	+	—	

J: Macaca fuscata
R: Macacamulatta
C: Macaca fascicularis

O: Occipital
M: Male
F: Female

D: Died
S: Survived

^+c: comminuted + linear

Appendix D Summary of Direct Head Impact Data (Group C)

Experimental No.	Subject No.	Species	Sex	Age	Body Weight (Kg)	Brain Weight (g)	Skull Diameter A - P (mm)	Skull Diameter Lateral (mm)	Skull Diameter Thickness (mm)	Impact Location	Impact Velocity ($^m/_s$)	Impact Mask	Impacter Head	Impact Stroke (mm)	Head Acceleration Peak (G)	Head Acceleration Average (G)	Head Acceleration Duration (ms)	Head Rotational Acceleration Peak (x10^4 $^{rad}/_{s^2}$)	Head Rotational Acceleration Average	Head Rotational Acceleration Duration (ms)	Outcome	Duration Absence Light React. (s)	Duration Absence Apnea (s)	Duration Absence Blink (s)	Bradycardia Grade	Concussion Grade	SAH	SDH	Contusion	Skull Fracture	Autopsy Findings
301	5	R	M	5	7.8	113	101	82	5.6	F	—	No	D	20	—	—	—	—	—	—	S	—	0	10	I	0					
302											—				—	—	—	—	—	—		—	0	12	I	0	-	-	-	-	
303											—				340	200	1.5	—	—	—		—	5	10	I	0					
304											—				830	440	3.9	—	—	—		—	5	19	I	0					
305	1	R	F	5	5.7	81	97	79	5.8	F	10.5	No	C	20	580	350	3.0	—	—	—	S	75	5	22	I	I					
306											12.9				520	380	3.6	10.49	2.82	4.5		36	8	8	II	0					No autopsy
307											15.5				710	480	2.5	4.51	2.33	9.8		600	5	16	II	0					(The subject was used to No 331-335
308											17.5				1010	670	1.5	14.20	5.28	7.6		600	15	13	II	I					impact test)
309											18.8				1290	490	3.1	25.69	5.49	3.7		120	3	10	II	0					
310											21.6				1980	600	2.3	83.32	3.97	8.9		—	15	16	I	—					
311	2	R	F	5	4.6	83	102	83	6.5	O	10.9	No	B	20	300	180	4.6	5.07	1.73	5.1	S	13	0	16	I	0					
312											14.4				580	350	3.3	6.40	2.30	4.0		14	0	18	I	0					Occipital base congestion
313											16.3				500	320	4.3	5.08	2.04	4.9		9	0	30	I	I					Parietal lobe and upper cervical spinal
314											18.0				650	370	4.9	11.53	—	—		10	0	34	I	I					cord SPH
315											22.6				—	—	—	—	—	—		8	0	7	I	0	-	-	-	-	
316											19.8				690	400	3.1	8.26	2.80	4.2		12	0	11	I	0					
317											21.0				1020	650	3.2	7.25	3.00	4.0		15	8	22	III	II					
318											21.7			30	430	260	9.9	17.82	5.07	9.6		44	50	15	III	II					
319	9	R	F	5	5.6	112	103	84	7.1	O	14.2	No	A	20	760	510	3.3	6.04	2.30	4.5	S	15	26	29	I	II					
320											16.3			30	780	430	4.7	7.01	3.09	5.0		11	30	13	III	II	-	-	-	-	
321											21.7				880	370	4.9	10.52	4.28	7.4		9	36	12	III	II					
322	10	R	F	5	5.8	80	—	—	—	O	15.5	No	D	30	690	290	3.9	17.10	5.15	9.9		13	32	13	I	I					
323											20.3				950	550	4.8	11.30	5.26	9.2		15	70	9	III	II	+	-	-	-	Parietal lobe and upper cervical spinal cord SDH
324											20.0				—	—	—	—	—	—	D	—	—	32	III	III					
325	7	R	F	5	4.3	80	98	76	6.6	F	15.7	Use	C	30	930	460	5.2	18.94	4.23	6.5	S	5	0	12	I	0					
326											20.0				1660	1040	1.5	21.99	9.95	1.9		55	30	30	III	III	+	-	+	-	Occipital lobe contusion
327											—				1630	1000	1.0	—	—	—		140	47	110	II	II					
328	4	R	F	10	8.4	85	97	77	6.8	F	15.5	Use	E	30	600	290	5.4	—	—	—	S	14	0	15	I	0					
329											19.6				340	175	6.4	3.50	1.10	10.0		13	2	20	II	I	-	-	-	-	
330											—			45	—	—	—	—	—	—		24	8	36	II	I					
331	1	R	F	5	4.7	81	90	79	5.4	F	15.5	Use	C	30	1470	570	3.7	25.68	7.48	3.3	S	75	0	10	I	0					
332											19.0				1140	580	3.7	12.54	4.33	6.6		52	0	16	II	0					
333											20.1			45	1340	710	4.1	19.60	6.75	5.0		18	5	20	II	I	+	-	-	-	
334											—			60	—	—	—	—	—	—		24	30	12	III	II					
335											22.5			45	—	—	—	—	—	—		190	10	220	III	II					
336	6	R	M	7	8.7	105	85	73	4.0	F	19.4	Use	E	30	630	330	4.7	11.21	3.67	5.8	S	8	8	15	I	0					
337											18.9			60	1020	370	5.4	10.55	4.03	6.3		15	8	12	II	0					No autopsy
338											18.4			90	1190	360	6.8	10.29	4.45	7.8		22	10	10	II	0					(The subject was used to No 341-343
339											21.7			60	990	320	6.7	13.39	3.63	6.9		15	4	9	II	0					impact test)
340	8	R	M	7	7.4	99				F	10.1	Use	A	30	2310	890	3.0	32.44	4.05	3.7	S	76	—	750	III	III	+	+	-	+	Occipital lobe SDH

R:Macaca mulatta
J;Macaca fuscata
F:Female
M:Male

F:Frontal
O:Occipital

S: Survived
D: Died

SDH: Subdural hemorrhage
SPH: Subpial hemorrhage
SAH:

Appendix D Summary of Direct Head Impact Data (Group C, Continued)

							Skull Diameter								Head Acceleration			Head Rotational Acceleration				Duration Absence									
Experimental No.	Subject No.	Species	Sex	Age	Body Weight (Kg)	Brain Weight (g)	A - P (mm)	Lateral (mm)	Thickness (mm)	Impact Location	Impact Velocity (m/s)	Impact Mask	Impacter Head	Impact Stroke (mm)	Peak (G)	Average (G)	Duration (ms)	Peak (10^4rad s^2)	Average	Duration (ms)	Outcome	Light React. (s)	Apnea (s)	Blink (s)	Bradycardia Grade	Concussion Grade	SAH	SDH	Contusion	Skull Fracture	Autopsy Findings
341	6	R	M	7	8.7	105	101	82	5.6	F	24.7	Use	E	60	1470	450	5.5	17.7	4.9	6.8	S	30	0	20	II	I					No autopsy
342											—			90	—	—	—	—	—	—		10	24	9	III	II					(the subject was used to No 336-339
343											24.6				2060	650	3.3	21.2	6.1	7.1		14	20	23	III	III					and No 349 impact test)
344	18	J	F	7	7.8	112	97	79	5.8	O	—	No	E	40	460	230	6.9	—	—	—	S	19	0	5	I	0					Frontal tip contusion
345											17.0				710	270	6.6	8.9	3.2	7.6		35	0	30	II	I					
346											19.7			60	510	310	8.1	7.9	2.6	11.7		20	0	8	II	0	+	+	+	-	
347											24.5				1010	540	4.8	10.5	3.7	7.1		20	5	15	III	I					
348											24.7			90	1670	830	4.0	19.8	5.9	5.2	D	105	∞	110	I	II					
349	6	R	M	7	8.7	—	102	83	6.5	F	—	Use	E	90	—	—	—	—	—	—	D	24	0	10	III	I	-	-	-	-	
350	3	R	M	8	8.4	100	103	84	7.1	O	25.3	No	E	60	1020	430	5.4	14.7	3.7	5.9	S	20	6	28	III	II					
351											26.2				980	520	4.2	13.0	5.1	3.0		20	10	15	III	I					Frontal tip contusion
352											24.3			90	1100	600	5.2	14.6	5.2	4.6		20	24	33	III	III	+	+	+	-	Frontal and temporal base SDH
353											—				—	—	—	—	—	—		13	40	20	III	III					Cervical spinal cord EDH
354	55	R	F	8	7.5	105	—	—	—	F	11.1	Use	A	20	1800	920	2.1	36.0	11.6	2.2	S	140	18	∞	III	I	+	+	+	+	Blow out fracture of frontal base
355											5.0				850	400	2.0	54.4	2.0	3.0	D	13	0	∞	II	0					
356	12	J	F	10	9.2	102	98	76	6.6	O	6.8	No	E	60	220	140	12.2	2.0	0.8	8.9	S	29	0	19	I	0					
357											7.5			90	210	140	12.3	1.3	0.6	9.1		44	10	33	I	I					
358											11.5				380	220	13.1	3.8	1.2	8.1		56	0	18	I	0	-	-	-	-	
359											13.7			60	470	235	14.0	4.7	1.6	10.1		27	0	30	III	II					
360											16.4				580	255	12.6	9.5	2.2	6.1		22	20	14	III	II					
361	111	J	F	7	7.5	104	97	77	6.8	O	11.2	No	E	10	1100	650	2.2	17.5	6.0	2.8	D	140	∞	∞	II	II	+	-	-	-	
362	777	J	F	6	5.6	108	90	79	5.4	O	10.8	No	E	10	830	510	2.2	5.5	2.0	6.3	S	10	0	10	I	0					
363											13.4				1340	680	2.4	17.9	7.6	4.5		37	19	10	III	I	+	+	+	+	Frontal and temporal tips contusion
364											—				—	—	—	—	—	—	D	∞		22	II	II					
365	555	J	F	6	4.5	92	85	73	4.0	F	19.9	Use	D	60	1370	690	5.0	29.2	7.1	6.9	S	12	10	10	—	—					
366											19.4			80	1300	760	2.7	18.9	9.0	2.6		20	2	10	III	I					Frontal base and occipital lobe
367											25.3			40	940	420	6.4	16.9	5.2	7.9		14	5	11	II	0	+	+	+	+	contusion
368											28.7				1100	790	2.2	29.2	9.0	2.5		25	2	10	III	I					SDH
369											—			80	—	—	—	—	—	—	D	∞	4	400	III	II					
370	44	J	F	4	4.5	114	93	84	5.0	O	8.0	No	E	30	290	140	13.5	4.6	1.2	8.7	S	10	0	11	I	0					
371											7.3			20	220	140	5.8	1.4	0.8	6.8		10	0	6	I	0					No autopsy
372											8.5			25	300	180	6.9	4.1	1.0	8.2		5	0	10	I	0					
373	88	J	M	18	13.0	124	104	85	6.1	F	8.4	Use	E	50	210	130	7.8	4.1	2.3	7.6	S	8	6	12	I	0	-	-	-	-	C_7-Th_1 subluxation
374											15.5			80	450	250	8.3	9.6	3.4	—	D	5	40	9	III	II					

R:Macaca mulatta
J:Macaca fuscata
F:Female
M:Male

F:Frontal
O:Occipital

S:Survived
D:Died

SDH: Subdural hemorrhage
EDH: Epidural hemorrhage
SAH: Subarachnoid hemorrhage

Appendix E-(a) Results of Free-drop Tests on Human Cadaver Skulls

No.	SN No.	HW (Kg)	LI	IM	DH (cm)	IV (m/s)	Load (Kg) Max.	Load (Kg) Ave.	SACC (G) Res.	SACC (G) Ave.	DR (ms)	Fracture
1	7	5.00	F	60	100	4.43	741	437	121	70	6.67	
2					250	7.00	1393	754	282	151	5.67	
3				80	150	5.42	1100	683	185	119	4.65	
4					250	7.00	1596	952	311	198	4.47▲	
5				Anvil	40	2.80	640	311	171	94	3.07	
6					100	4.43	1184	651	310	190	2.57	
7					150	5.42	1572	660	480	240	2.77▲	
8					200	6.26	1187	363	422	198	3.17◎	3 +, B 3 +
9	3	5.01	F	60	250	7.00	1500	578	349	156	4.27◎	3 +, B 3 +
10	6	5.00	F	60	150	5.42	1104	645	234	148	4.47	
11					200	6.26	–	–	–	–	–	
12					200	6.26	1340	741	275	172	6.40	2 +, B 2 +
13	8	5.00	F	20	250	7.00	985	601	143	82	11.15	
14				60	150	5.42	1043	611	208	113	5.65	
15					200	6.26	1173	568	363	140	7.17	2 +, B 2 +
16	1	5.05	O	80	250	7.00	1635	646	866	348	3.45	3 +, B 2 +
17	5	5.01	O	60	100	4.43	1035	643	136	88	5.62	
18					200	6.26	1331	801	213	92	9.92	
19					250	7.00	1603	925	238	145	5.57	
20				80	150	5.42	2000	1161	247	154	4.12	
21					200	6.26	1946	731	439	210	5.90	3 +, B 2 +
22	4	5.00	O	60	150	5.42	1043	482	524	247	3.95	2 +
23	2	5.05	O	20	250	7.00	–	–	–	–	–	
24					250	7.00	831	453	572	113	9.70	2 +, B 2 +
25	12	5.07	O	Anvil	40	2.80	1114	589	398	225	1.75	
26					60	3.43	1292	713	511	281	1.67	
27					60	3.43	1182	632	497	273	1.67	
28					80	3.96	–	–	–	–	–	
29					80	3.96	1316	710	609	313	1.60	
30					80	3.96	1423	821	534	306	1.72▲	
31					100	4.43	1430	824	693	369	1.55	1 +
32	11	5.00	F	Anvil	40	2.80	797	338	287	163	2.10	
33					60	3.43	880	408	391	219	2.00	
34					80	3.96	1058	292	467	257	1.97	
35					100	4.43	1220	495	625	312	2.02▲	
36					120	4.85	1220	474	732	367	1.95	1 +
37	14	5.05	O	Anvil	40	2.80	–	–	–	–	–	
38					60	3.43	1293	796	510	293	1.70	
39					80	3.96	1392	413	1089	483	2.50	2 +, B 2 +
40	9	5.05	F	20+ C	150	5.42	513	436	120	55	11.25	
41				20	150	5.42	593	371	117	68	9.90	
42				20+ C	200	6.26	674	383	139	75	10.62	
43				20+ C	250	7.00	766	486	147	84	10.87	
44				20	250	7.00	788	471	155	94	9.75	
45				60+ C	100	4.43	610	429	169	110	4.90	
46				60+ C	150	5.42	821	498	199	147	4.12	
47				60+ C	200	6.26	1008	593	254	112	9.62	
48				Anvil	60	3.43	814	355	374	205	2.20	
49					80	3.96	846	339	553	256	3.32	2 +, B 2 +
50	13	5.00	O	20+ C	150	5.42	495	292	137	68	11.00	
51				20	150	5.42	512	306	134	73	9.27	
52				20+ C	200	6.26	577	354	129	72	10.70	
53				20	200	6.26	627	372	168	74	12.37	
54				60+ C	100	4.43	738	430	190	118	5.50	
55				60	100	4.43	754	470	211	130	5.12	
56				60+ C	150	5.42	911	555	253	160	4.85	
57				60+ C	250	7.00	683	407	162	85	10.87	
58				20	250	7.00	700	391	178	93	10.05	
59				60	200	6.26	989	640	315	119	11.25▲	
60				60+ C	200	6.26	998	599	276	162	5.45▲	
61				60	200	6.26	948	575	298	166	5.47	1 +
62	10	5.04	O	60+ C	150	5.42	1006	544	257	153	5.62	
63				60	150	5.42	1014	582	282	177	4.37	
64				60+ C	200	6.26	–	–	–	–	–	
65				60	200	6.26	1232	705	337	210	4.27	
66				60+ C	250	7.00	1208	711	320	200	4.92	
67				60	250	7.00	1304	775	375	234	4.20	
68				80	150	5.42	1183	673	420	249	3.22	
69				80+ C	200	6.26	1257	702	372	230	3.97▲	
70				80	200	6.26	1442	780	485	296	3.15▲	
71				80+ C	250	7.00	1378	764	400	269	3.90	1 +
72	15	5.00	F	80	150	5.42	1031	585	365	154	5.47	
73				80+ C	200	6.26	1077	610	343	230	8.97	
74				80	200	6.26	1192	606	472	249	3.70▲	
75				80+ C	250	7.00	1257	759	321	197	4.65	
76				80+ C	250	7.00	1225	676	426	264	3.45▲	
77				80	250	7.00	1305	681	561	329	3.00▲	
78				Anvil	80	3.96	891	383	399	237	2.07	
79					100	4.43	–	–	–	–	–	
80					100	4.43	900	382	504	220	3.32	
81					120	4.85	916	418	658	256	3.15	
82					150	5.42	1062	446	825	317	3.00▲	
83					180	5.94	1268	483	887	540	1.55▲	
84					200	6.26	1346	584	920	585	1.57	3 +, B 3 +

SN: Skull No.
HW: Head weigt
LI: Location of impact
IM: Impactor base
DH: Drop height
1+: Rudimentary fracture
3+: Multilinear comminuted fracture
C: Buffer material (Viscol A-30. Thiokol)

IV: Impact velocity
SACC: Skull acceleration
DR: Duration
F: Frontal impact
O: Occipital impact
2+: Linear and/or radiating fracture
B: Basal fracture

▲: Subfracture
◎: Overfracture

Appendix E-(b) Human Cadaver Skull Measurements

Items \ Skull No.	1	2	3	4	5	6	7	8	9	10	11	12	13	14	15
Impact Direction	O	O	F	O	O	F	F	F	F	O	F	O	O	O	F
Skull Weight (gr.)	467	470	494	557	570	574	648	664	595	650	690	585	665	695	653
Skull Breadth (mm)	126	129	127	127	133	128	128	135	125	125	132	129	135	136	132
Skull Length (mm)	173	168	167	168	157	180	174	179	172	169	184	165	186	177	179
Thickness of Frontal Bone (mm)	5.5	5.3	7.8	8.5	8.0	7.0	7.0	7.5	5.5	5.0	8.2	5.7	8.1	7.2	7.8
Thickness of Frontal SKin (mm)	7.0	7.0	7.0	7.0	7.0	7.0	7.0	7.0	7.0	7.0	7.0	7.0	7.0	7.0	7.0
Thickness of Occipital Bone (mm)	6.8	6.7	6.0	9.9	5.5	5.9	7.3	5.5	7.2	6.1	6.5	5.8	7.2	.7.4	7.3
Thickness of Occipital Skin (mm)	7.0	7.0	7.0	7.0	7.0	7.0	7.0	7.0	7.0	7.0	7.0	7.0	7.0	7.0	7.0
Total Weight of Head (Kg)	5.05	5.00	5.01	5.00	5.01	5.00	5.01	5.00	5.05	5.04	5.00	5.07	5.00	5.05	5.00
Moment of Inertia of Head $(Kgm^2) \times 10^{-2}$	4.63	2.86	3.42	2.46	4.70	2.54	2.77	3.20	4.71	3.77	3.90	4.47	4.96	3.70	4.86

F: Frontal Impact
O: Occipital Impact

831618

The Role of Impact Location in Reversible Cerebral Concussion

Voigt R. Hodgson and L. Murray Thomas
Wayne State Univ. School of Medicine

Tawfik B. Khalil
General Motors Research Laboratories

ABSTRACT

Mechanical impacts were delivered by an air propelled striker to the front, side, rear and top of rigid protective caps worn by six anesthetized monkeys. These tests were to produce reversible concussion and to determine differences in tolerance to concussion among the four impact sites. Striker force and cap accelerations were measures of the impact severity and animal blood pressure, respiration and ECG changes were measures of the physiological effects.

By distributing the blow with a protective cap, allowing free head movement after impact, skull fracture was eliminated and simple reversible concussion could be produced without symptoms of residual neurological deficit. Higher linear and angular accelerations produced longer periods of unconsciousness (more than 3 times) on the side than at any of the other locations. It is hypothesized that the decrease in concussion tolerance accompanied by higher accelerations for side impacts may be the result of lower mechanical impedance due to the oval shape of the animal head.

MANY INVESTIGATORS HAVE ATTEMPTED to explain the events that follow the application of a blow to the head. Of particular interest has been the phenomenon of cerebral concussion because of its usually reversible nature and the fact that it appears to be a conservative end point for design of head protection. The committee of the Congress of Neurological Surgeons has defined cerebral concussion as "a clinical syndrome characterized by immediate transient impairment of neural function such as alteration of consciousness, disturbances of vision, equilibrium, etc., due to mechanical forces" (1)*.

Gurdjian and associates (2) from Wayne State University have conducted numerous cerebral concussion experiments utilizing dogs, cats and primates. Their conclusions concerning the pathogenesis of this injury can be summarized as follows:

1. Cerebral concussion is attributable to brain stem damage.
2. The brain stem damage is primarily caused by a combination of total head motion, skull deformation and relative motion of the central nervous system with respect to the skull.
3. The observed changes in blood pressure, heart rate and respiration rate are attributable to medullary receptors.

Ommaya and his co-workers (3) have extensively investigated cerebral concussion in three subhuman primates (Rhesus monkeys [Macaca mulatta], squirrel monkeys [Saimiri sciureus], and Chimpanzee) due to direct occipital head impacts and whiplash. Many of the studies were conducted to test Holbourn's hypothesis which primarily attributes brain injuries to head rotations (4). Ommaya's data indicates that about fifty percent of the potential brain injury is attributable to head rotations. The remaining potential injury is conceivably due to direct impact effects. Comparative tolerance curves of the tested species and a theoretically derived curve for man, relating fifty percent probability for cerebral concussion as a function of rotational velocity and acceleration have been constructed (5).

In 1974, Ommaya and Gennarelli published a contemporary definition of concussion which suggests that a syndrome is attributable to focal and diffuse injury of nerve cells not only involving the brain stem but also occurring at other brain loci (6). The diffuse injuries are mainly caused by rotational acceleration while translational accellerations cause mainly focal injuries. The authors hypothesized that cerebral concussion is a graded set of clinical syndromes caused by mechanically induced strains. These strains always begin at the surface of the brain in the mild cases and extend inward to affect the diencephalic-mesencephalic core in severe cases.

Gennarelli, et al., (7,8) have used squirrel primates (Saimiri sciureus) to perform experiments

* Numbers in parentheses designate references at end of paper.

on the head accelerating device (HAD II) in which they compared translational and rotational head motions. They found that in contrast to translational head motions, which did not produce cerebral concussion at the levels examined, rotation of the head at equivalent levels of acceleration invariably causes cerebral concussion as well as more extensive lesions in the brain.

Abel, et al. (9), have used a rotational acceleration device to produce head injury in the Rhesus species (Macaca mulatta). Peak values of angular acceleration range from 1.8×10^4 to 1.2×10^5, rad/sec^2 with total pulse durations between 6 and 11 ms. Peak values of tangential acceleration at the center of mass of the brain reached 1300 g's. The pulse was described as "a controlled single, approximately sinusoidal, pulse of angular acceleration about a fixed axis perpendicular to the sagittal plane." However, the device used for propulsion accelerated and brought the head to rest in 60 degrees, producing a deceleration with shorter rise and higher amplitude than the acceleration which was used in the analysis. Abel, et al., used an injury classification scale from 0-6, in which grade zero exhibited no physiological changes; grade 3 animals were traumatically unconscious (absent corneal reflex) for 30 sec or less; more severe concussion and small amounts of subdural blood was frequently present in grade 4; large size subdural hematomas were present in seven of nine cases in grade 5 resulting in neurological death within six hours. Grade 6 animals suffered instantaneous death from pontomedullary lacerations and gross brain disruption. By a statistical analysis, the injury severity was related to peak positive angular acceleration, peak tangential force and peak positive tangential acceleration. The authors concluded that angular accelerations alone does not account for the incidence of concussion as predicted by Holbourn's theory, but rather, a combination of translational and rotational phenomena involved.

Unterharnscheidt and Higgins (10) also investigated the effects of rotational acceleration on the heads of squirrel monkeys using HAD II device and found that rotational acceleration in the range of 101,000-150,000 rad/sec^2 produced subarachnoid hemorrhage and tear of veins and capillaries in the cortex in ten of thirteen animals. Rotational acceleration higher than 300,000 rad/sec^2 was fatal to all subjects, causing additional hemorrhages in the central region of the brain.

Historically, most animal experiments have involved direct impact on the occipital region of the head because it is the most accessible place to contact. In contrast, human tolerance criteria have been published for frontal impact. However, recently McElhaney, et al. (11), published a curve showing significantly higher tolerance for the frontal compared to temporoparietal head impacts after experiments on primates (squirrel, cynomolgus, Rhesus, Chimpanzee), and human cadavers.

This paper describes concussion-producing experiments conducted on six female stumptail monkeys (Macaca speciosa). It is an attempt to determine the relative sensitivity of this subhuman species to concussion-producing impact against the frontal, temporoparietal, occipital and cranial regions of the cap-protected head.

METHODS

Six mature female subhuman primates* (Macaca speciosa), weighing 6-10 kg, were used in the experiments conducted at the Department of Neurosurgery, Wayne State University, School of Medicine. The primates were sedated with Sernalyn R for safe handling and anesthetized by Nembutal R administered intravenously as a general anesthetic for installing monitoring pickups. ECG was recorded with needle electrodes in the Lead II (right arm-left leg) position. Blood pressure was monitored by a Statham model 23 Pc transducer connected via an indwelling catheter inserted in the right femoral artery. Respiration rate was recorded through a strain gage transducer in an elastic cord around the chest cavity.

An individual close fitting fiberglass cap approximately 5 mm thick and weighing between 200-300 g was used to protect the scalp and prevent skull fracture. The cap was fastened to the head by screws connected to a vertical arm of the bite plate extended under the zygomatic arches. The bite plate was equipped with register pins for impact location purposes. Four Endevco 2264-4500 G accelerometers were arranged in biaxial configuration on an aluminum block attached to the protective skull cap. The animal was loosely taped in the seated position to a magnesium seat framework which was free to slide on linear bearings. A 1.1 kg rubber-tipped air propelled metal striker with an impact area of 6.54 cm^2 was used to deliver the impacts.

Preliminary tests indicated that the velocity of 19.5 m/s (64 ft/s) and the kinetic energy of 217 J (160 ft-lb) would result in extremely brief or no unconsciousness for all impact locations except the lateral where short periods of unconsciousness up to two minutes were produced. All experimental impacts were therefore administered using this level of energy. The 1.1 kg mass of striker was chosen because it was equal to the combined approximate weight of the head and cap and essentially exhausted its kinetic energy in the experimental impact. Impact force was measured by a Kistler quartz crystal load cell mounted on the impacted side of the striker which had a 0.32 cm thick, 60-durometer rubber pad glued to its surface for force distribution.

TEST PROCEDURE

After installation of the monitoring pickups

* The rational and experimental protocol for the use of an animal model in this program have been reviewed by Wayne State University's Animal Research Committee. The research follows procedures outlined in publications by the U.S. Department of Health, Education, and Welfare, "Guide for the Care and Use of Laboratory Animals," or the U.S. DHEW National Institutes of Health (NIH), "Guidelines for the Use of Experimental Animals," and complies with U.S. Department of Agriculture (USDA) regulations as specified in the Laboratory Animal Welfare Act (PL 89-544), as amended in 1970 and 1976 (PL 91-579 and PL 94-279).

and closely fitted skull cap, the animal was allowed to come up to a level of anesthesia at which gag, pinna, and lid reflexes were present. The head of the animal was properly positioned with the help of registering pins on the impact side of the skull cap and supported by a fall-away styrofoam block on the opposite side Pressurized air propelled the striker up to the point immediately before the impact and the impact velocity was measured by two magnetic probes, spaced 7.62 cm apart during the period of free-wheeling striker before the contact. The head of the animal was free to move after the impact with a maximum of 5-8 cm total travel limited by a styrofoam cushion. Immediately after the impact, the animal was continuously checked for signs of reflex response to mechanical stimulus, ammonia odor perception and the presence of neck muscle tone. The cap was removed from the head within thirty seconds of the impact exposure so that lid or corneal reflexes could be checked for signs of consciousness as well. None of the periods of unconsciousness was longer than 120-180 sec. and often the return of consciousness occurred within a period of 5-15 sec from the time of first response. The ECG, blood pressure and respiratory frequency were monitored continuously prior to the test and until at least five minutes following arousal from unconsciousness.

All impacts were filmed at 4000 frames per sec (FPS) with a high-speed camera. Analysis of films indicated essentially planar motion of the head and of the protective cap.

The protective cap was fastened by four-bolt attachments to a bite plate having arms extending under the zygomatic arches. Films taken at a constant speed of 4000 fps revealed that the bite plate moved synchronically with the head after the third frame following initial contact (0.25 ms/frame) in the majority of tests. Exceptions were noted in one case of frontal impact in which the cap was torn off its bolts, and by the cranial and occipital impacts, which tended to separate cap and bite plate.

Following the test, the animals were observed and appropriate doses of analgesics (Meperidine HCl, 11 mg/Kg IM of body weight) were administered as necessary for a minimum period of six hours. Extreme level of care was provided to assure that no unnecessary pain was perceived by the animal subjects even in the remaining postexperimental period. The animals were then daily monitored in their cages. All of the animals were impacted at least once on the front, side, rear and top of the head cap, allowing a minimum of one week period between impacts. On several occasions, the order of blow location was reversed to eliminate the possibility that cumulative effects might bias the differences in the direction of blow location.

RESULTS

HEAD MOTION FOLLOWING IMPACT --- High-speed film analysis show that primarily translational and rotational movements are involved in the head kinematics following impacts to all four locations since the impact loads did not coincide with the center of mass of the head and because of possible neck force constraints.

ACCELERATION AND FORCE ANALYSIS --- Figure 1 presents the oscillograms for impact to each of the four blow locations. The acceleration records are in the top four traces and the force (F) is in the bottom trace. The impulse in all cases is of approximately 1 millisecond (ms) duration except for the cranial impact which is about 1.5 ms long. Evidently the cranial impulse is longer because there is more effective mass when the action force is aligned with the neck, and the cap may be more deformable on the crown than in the other locations. Primary acceleration response is of 1 ms or less duration in all cases with some fluctuations lasting as long as 3-4 ms.

Impact records for all six animals for each of the four blow locations were superimposed to produce the envelopes of impulse and responses shown in Figure 2. There is a similarity in direction of excursions for planar inline acceleration (X) and distal transverse acceleration (Y). The proximal transverse accelerations (Y) for frontal and cranial locations are similar and opposite in direction to those due to temporoparietal and occipital impacts, indicating that the impacts did not produce identical vectors of rotation for all impacts.

The peak values of force and acceleration for each impact, as typified in Figure 1 are shown in Table 1. The body and head weights of the animals as well as striker velocity, impact duration and duration of unconsciousness are also included to permit the correlations. Table 2 summarizes the average data for all impact locations and their statistical differences. On the average, the peak force for temporoparietal impact was highest followed by frontal, occipital and cranial, respectively. Temporoparietal impacts produced highest peak acceleration on the average and the cranial impacts the lowest.

Table 3 lists the recorded and computed peak values, SI, HIC, angular velocity and angular acceleration for all impacts for which values are available. Column three shows the rigid body head acceleration obtained by dividing peak force by head weight. In column four, the distal inline to rigid body acceleration are presented. Each is a measure of departure from a central rigid body impact (r = 1) which evidently is the most closely approximated in top impacts and the least approximated in side impacts (shown more clearly on Table 3). The Head Injury Criteria (HIC) and Severity Indexes (SI) in columns six and seven were computed from the recorded distal inline accelerations.

ANGULAR KINEMATICS --- The original intent in the tests was to derive values of angular acceleration using rigid body mechnics application and the readings from the linear accelerometers arranged in biaxial configuration. The effect of time duration of the impulse by interposing various kinds of energy absorbing materials between the striker and head cap was also expected to be investigated in a similar way. However, only impacts which produced pulses on the order of 1-2 ms

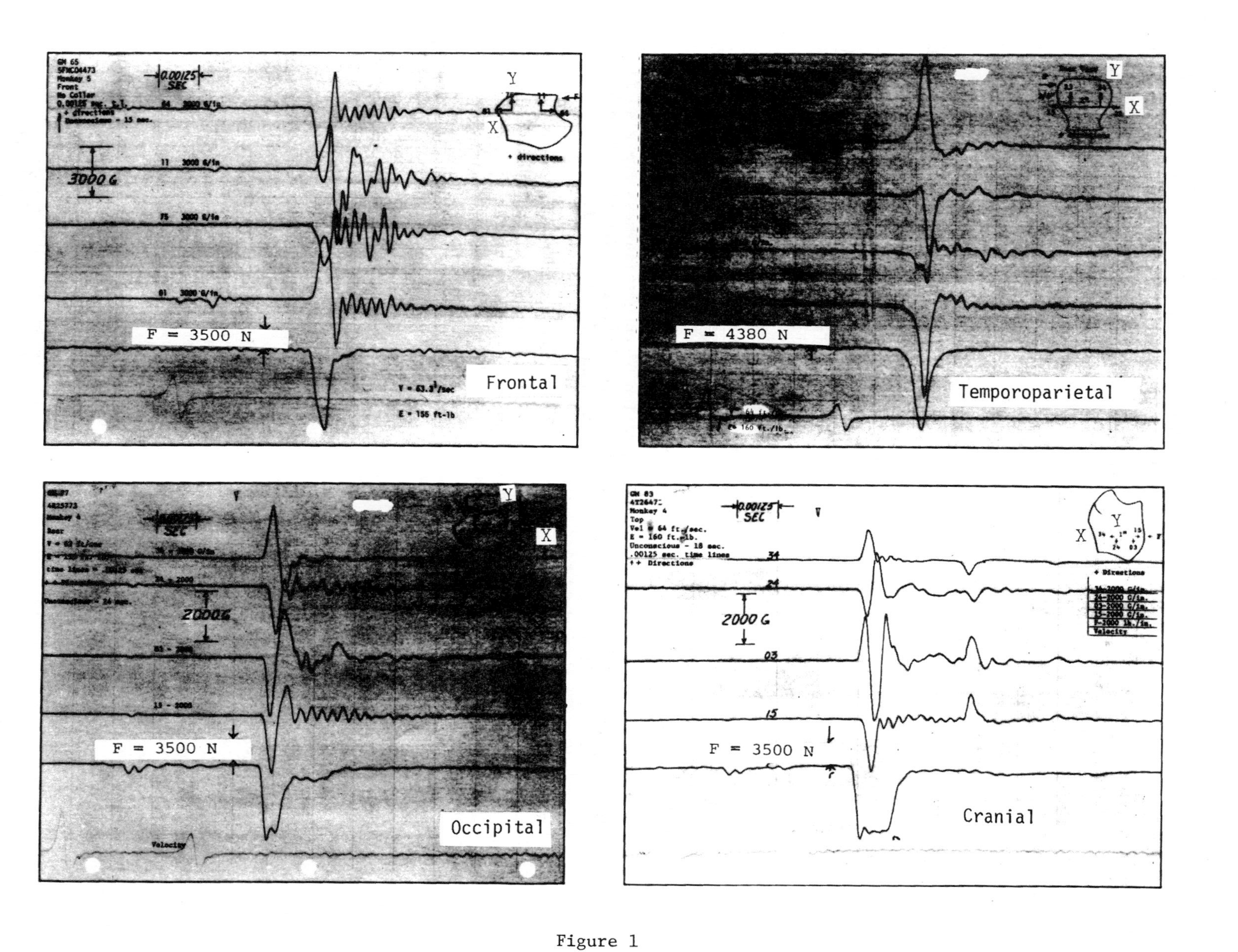

Figure 1

Typical oscillograms of force and acceleration response for the four blow locations

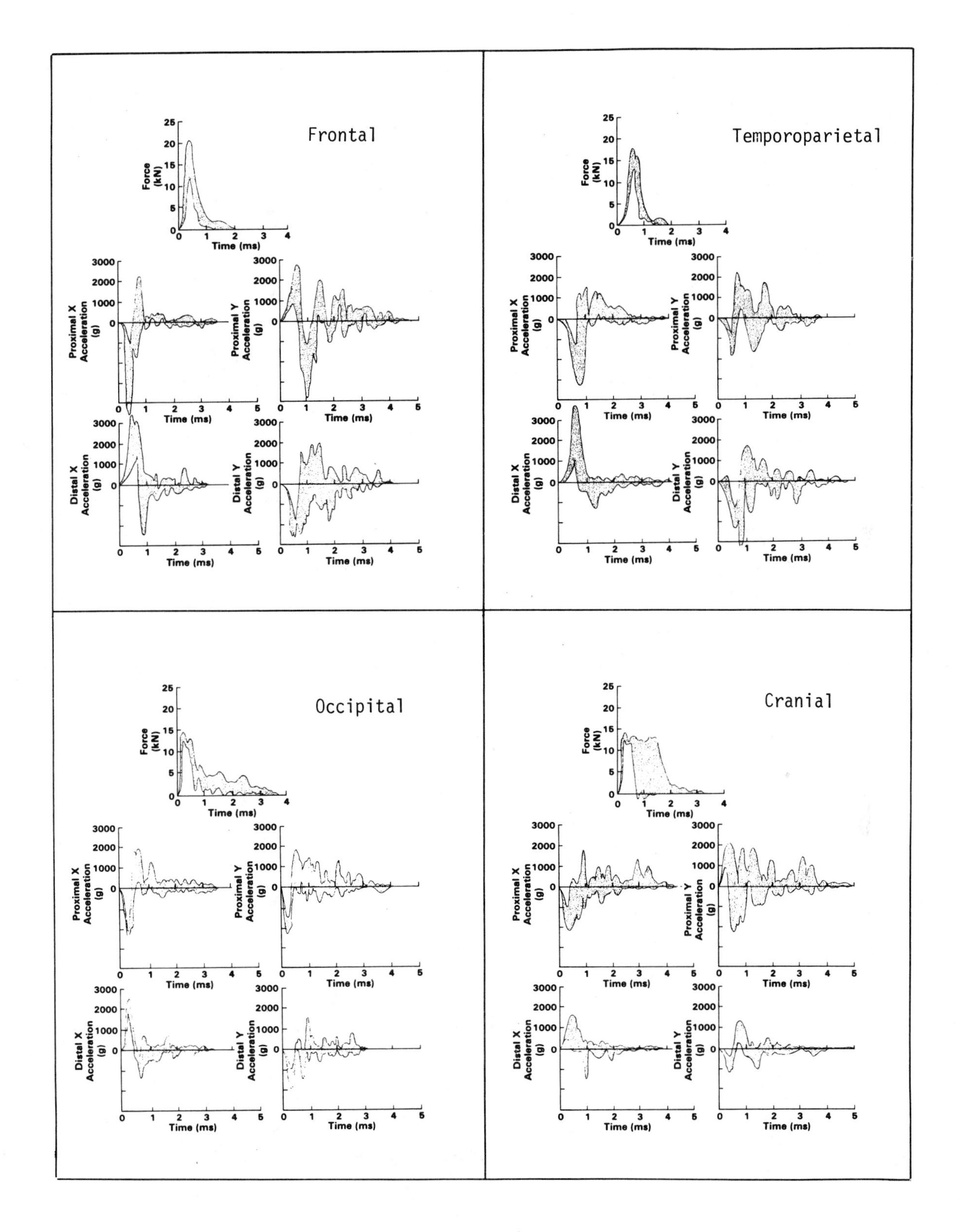

Figure 2

Superposition of impulse and response records for all six monkeys struck at four locations on the head.

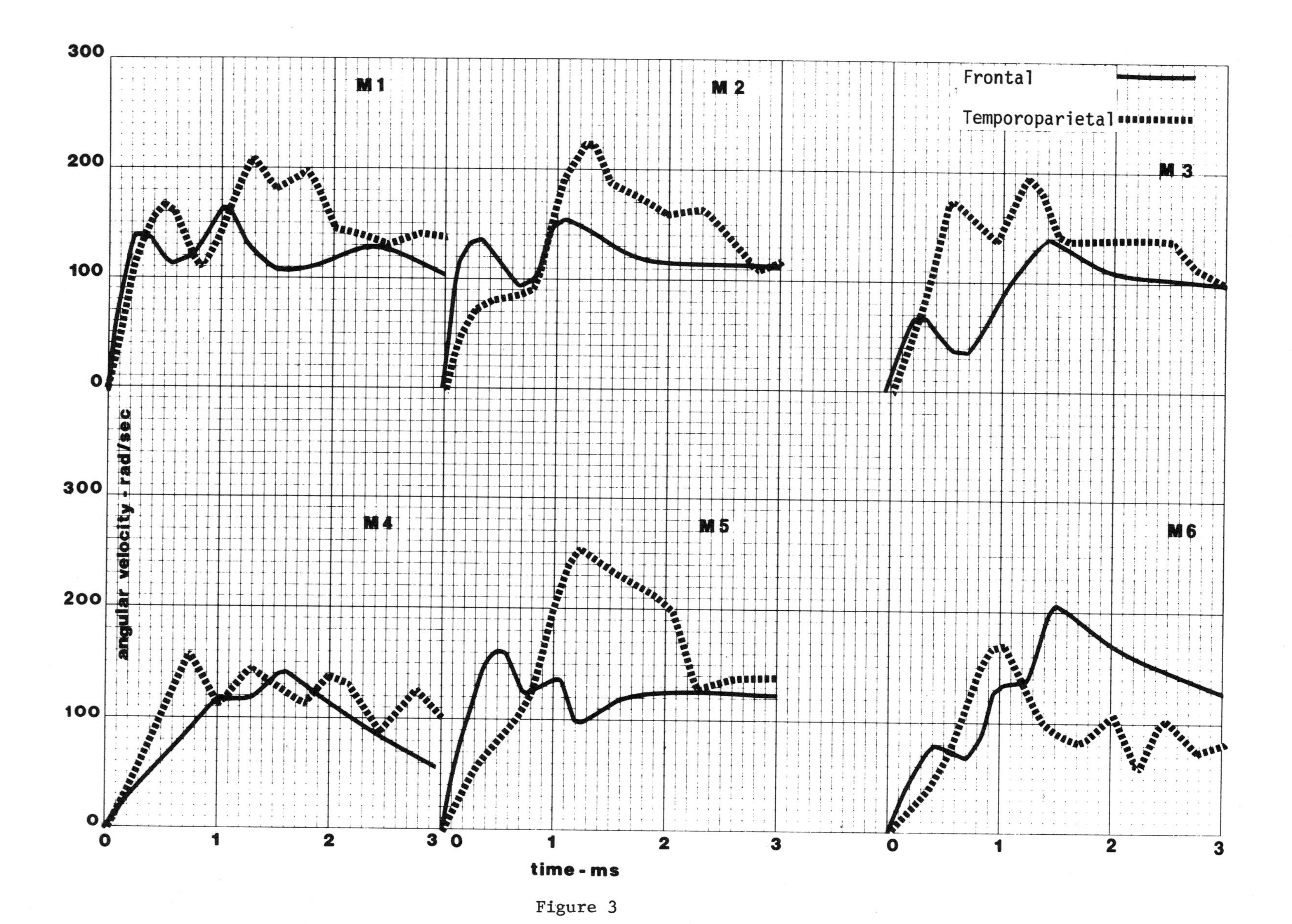

Figure 3

Angular velocity amplitude versus time plots for the first 3 ms following head impact to front and side monkey heads (M1-M6). Data obtained from frame-by-frame analysis of bite plate movement recorded on film at 4000 frames per second.

Table 1

Summary of Individual Experimental Data

No	Body Mass kg	Head* Mass kg	Impact Location	Peak Force N X10³	Striker vel. ** m/s	Duration ms	Peak Proximal Accel., (g) X	Peak Proximal Accel., (g) Y	Peak Distal Accel., (g) X	Peak Distal Accel., (g) Y	Time of Unconsciousness - sec -
1	5.9	0.64	F	15.6	19.1	1.1	2100	3000	3000	1800	5
			S	18.4	19.7	1.1	3000	1400	3060	2050	84
			R	13.4	18.8	0.80	2250	2250	2100	950	5
			T	12.9	18.8	0.70	2100	1940	1620	1250	40
2	7.7	0.86	F	17.8	20.3	1.1	2250	****	1970	2060	15
			S	20.9	19.3	0.80	****	****	3750	****	100
			R	13.4	19.5	1.1	2200	2200	1750	1440	12
			T	12.9	18.8	2.3	2000	2120	1000	1130	5
3	6.8	0.77	F	10.2	19.5	****	****	****	****	****	0
			S	21.1	20.3	0.80	****	****	****	****	48
			R	13.4	19.4	1.8	2350	1800	2100	1880	5
			T	13.1	19.1	1.3	1630	1400	1000	750	40
4	10.4	1.13	F	13.9	20.3	0.86	1150	2600	1400	2180	5
			S	17.8	19.4	1.1	3500	1200	3400	2200	86
			R	12.9	18.7	1.1	2250	2300	2000	1880	30
			T	12.4	19.5	1.9	2100	2380	1140	1400	20
5	5.0	0.54	F	17.8	20.8	1.3	3150	1600	3360	1500	30
			S	15.3	19.3	0.94	2050	1050	2160	2100	108
			R	13.3	19.3	0.70	2250	2060	2120	1860	60
			T	13.1	18.9	0.78	1560	900	1000	450	30
6	7.7	0.86	F	11.3	19.7	1.6	2400	900	2300	2000	15
			S	16.1	19.1	0.94	3050	1000	3100	3100	90
			R	13.3	19.3	1.3	2320	2100	2500	1800	42
			T	12.9	19.5	2.3	1760	2000	900	1400	5

* Protective cap and accessories mass = 0.46 kg.
** Average kinetic energy = 207 ±0.16 J (155 ±0.12 ft. lb.)
F = Frontal
S = Temporoparietal
R = Occipital
T = Cranial

Table 2

Average Data of Impact Parameters From All Experiments

Impact Location	Peak Force N $X10^3$	Striker vel.* m/s	Duration ms	Peak Proximal Accel., (g) X	Peak Proximal Accel., (g) Y	Peak Distal Accel., (g) X	Peak Distal Accel., (g) Y	Time of Unconsciousness - sec -
Frontal	14.4 ±3.2	20.0 ±0.6	1.2 ±0.3	2210 ±720	2030 ±950	2410 ±790	1910 ±270	13 ±10
Temporoparietal	18.3 ±2.4	19.5 ±0.4	0.95 ±0.13	2900 ±610	1160 ±180	3090 ±590	2360 ±496	86 ±21
Occipital	13.3 ±0.20	19.2 ±0.3	0.97 ±0.23	2270 ± 55	2120 ±180	2100 ±240	1640 ±380	26 ±22
Cranial	12.9 ±0.3	19.2 ±0.9	1.2 ±0.9	1860 ±239	1790 ±540	1110 ±260	1060 ±380	23 ±16

* Average kinetic energy = 207 ±0.6 J (155 ±0.12 ft. lb.)

Table 3

Measured and Computed Linear Acceleration Indices and Rotational Velocity and Acceleration from Film Analysis

1	2	3	4	5	6	7	8	9	10	11	
Monkey Number	Impact Location	Rigid Body Accel (g)	Distal in-line Accel (g)	Normal-ized Accel***	HIC* $X10^3$	SI* 10^3	$\dot{\theta}$ r/sec	$\ddot{\theta}$ r/sec^2 $X10^3$**	time to peak $\dot{\theta}$ -m-	Unconscious Duration	Remarks
1	F****	1450	3000	2.07	81.6	107	163	164	1.0	5	** Estimated accuracy of angular measurements per frame:
2	F	1380	1970	1.43	54.1	61.4	157	157	1.0	15	θ = 20 min (0.006 rad)
3	F	850	****	****	****	****	140	101	1.5	0	$\dot{\theta}$ = 23 r/sec
4	F	893	1400	1.57	15.6	21.0	140	93	1.5	5	$\ddot{\theta}$ = 93 r/sec^2
5	F	1820	3360	1.85	108	148	140	140	1.0	30	
6	F	876	2300	2.63	51.9	70.2	209	139	1.5	15	
1	S	1710	3060	1.79	97.6	116	210	167	1.3	84	
2	S	1620	3750	2.32	187	252	220	177	1.3	100	
3	S	1760	****	****	****	***	200	160	1.3	48	
4	S	1140	3400	2.98	127	148	140	112	1.3	86	
5	S	1560	2160	1.39	58.5	65.4	256	205	1.3	108	
6	S	1250	3100	2.48	97.3	134	174	174	1.0	90	
1	R	1250	2100	1.68	28.4	35.1				5	
2	R	1040	1750	1.68	17.4	25.6				12	
3	R	1120	2100	1.88	28.5	31.6				5	
4	R	828	2000	2.42	28.3	42.6				30	
5	R	1360	2120	1.56	36.6	40.7				60	
6	R	1030	2500	2.43	56.0	80.5				42	
1	T	1200	1620	1.35	30.3	41.6				40	
2	T	1000	1000	1.00	7.22	11.4				5	
3	T	1070	1000	0.94	3.76	6.91				40	
4	T	797	1140	1.43	7.84	10.1				20	
5	T	1340	1000	0.75	9.22	13.7				30	
6	T	997	900	0.90	5.68	7.21				5	

* Computed from distal in-line acceleration
*** Ratio Col. 4 ÷ Col. 3
**** F = Frontal
S = Temporoparietal
R = Occipital
T = Cranial

and linear accelerations of 2000-5000 G produced concussion in the monkeys wearing a protective cap. The high intensity impacts occasionally resulted in transducer failures, or vibrations of the cap-transducer assembly which particularly affects the transverse acceleration values used in angular acceleration computation. Consequently, the in-line distal accelerometer responses were used to evaluate linear acceleration and to calculate linear acceleration indices, while films were used to obtain angular velocity and acceleration.

Films were analyzed by measuring the angular changes at the bite plate using all frames obtained during a period of 3-5 ms. Numerical differentiation averaged over each frame increment of 0.25 ms was then used to evaluate angular velocity. Obtained values for front and side rotational response are compared in Figure 5. As a general rule, the protective cap itself was accelerated first for 0.5 to 0.75 ms and the cap, bite plate and head acceleration followed up to peak level of angular velocity. Afterwards, the velocity decreased as the striker separated from head contact and neck restraining forces decelerated the head. On the average, peak angular velocity was reached in 1.3 ms for front impacts, and 1.2 ms for side impacts.

Angular acceleration was characterized by the slope of the line from time zero to peak angular velocity are listed in columns eight, nine and ten of Table 3. For comparison, the duration of unconsciousness is given in column eleven. Calculated average data and standard deviations are shown in Table 4.

PHYSIOLOGICAL CHANGES DURING CONCUSSION --- Figure 4 shows six oscillogram excerpts of typical changes in respiration, blood pressure, and ECG following concussion-producing impact. An initial drop in blood pressure (from 150/85 to 130/60 mm Hg) is followed by a sustained increase in blood pressure (210/120 mm Hg), particularly in the systolic pressure, until the time of regaining consciousness. Bradycardia and transient arrhythmia in the ECG also typify this mode of experimental concussion. Bradycardia is, however, sustained even after the animal is fully aroused. Respiratory apnea is followed by a period of rapid shallow breathing, and accompanies the cardiovascular changes late into the period of full consciousness.

Figure 5 summarizes the time pattern of recorded physiological changes during the entire period of typical concussive impact experiment. Immediately following impact, apnea or depression of respiration is noted, lasting for approximately 60 seconds. This is followed by shallow breathing at higher respiration rate, reaching a level of forty per minute. A slight drop in blood pressure is also observed during the first 10 seconds after impact. Then the blood pressure rises for a period of 50 to 60 seconds, followed by a prolonged period of hypotension lasting about 20 minutes. The electrocardiogram records shows an immediate drop in heart rate from 100/min to about 50/min. The bradycardia persisted for an approximate period of 60 seconds after which the heart rate increased to its normal rate. In the tests, bradycardia usually accompanied cerebral concussion. Changes in the ECG record included increase in the amplitude of the T wave, inversion of QRS complex and absence of P wave. In general, most of the myocardial dysfunctions associated with head impact may be attributed to vagal stimulation due to central nervous system irritation.

DURATION OF UNCONSCIOUSNESS VERSUS BLOW LOCATION --- Bar graphs in Figure 6 indicate the duration of unconsciousness for the four blow locations in each animal. In all animals the duration of unconsciousness for the lateral impact was longer than for any other location. The average duration of unconsciousness for frontal, cranial, occipital and temporoparietal impacts was 13 ±10, 23 ±16, 26 ±22 and 86 ±21 sec, respectively. According to the Abbreviated Injury Scale (12), the duration of unconsciousness reflects directly to the severity of head injury. When the clinical experience is extrapolated to these experiments, the level of induced injury would indicate the level of AIS = 2 for all impact sites.

DISCUSSION OF RESULTS

The temporoparietal impacts produced the longest duration of unconsciousness among the animals, when similar levels of energy input are compared. The high-speed film analysis indicates that there was essentially equal energy transfer for all locations. Head accelerations, SI and HIC were, in general, higher on the temporoparietal location than at the other locations and the pulse duration was on average lower for that location (Table 2). Angular velocity and acceleration levels were about twenty-six percent higher on the temporoparietal site than for frontal impacts. It is assumed that the protective cap virtually eliminated skull deformation as a significant factor in the tests and that the resulting acceleration of the head was the main injury factor. A linear regression analysis was performed and low and insignificant correlation coefficients were found for all parameters listed in Table 3. It could be assumed that the reading obtained from the caps were unsatisfactory indicators of head motion or that the sample size is too small to show significant differences because of the variable resistance against cerebral concussion among the used individual animal subjects.

There is no apparent intracranial anatomic reason why the temporoparietal region of the head in this primate species should be more sensitive to concussion than the frontal or occipital area. However, the effect may be biomechanically related to the oval shape of the head. All three locations were struck at the same level above the center of gravity. On this basis, the side presents a lower mechanical impedance to a striking body because of the lower mass moment of inertia about the anterior-posterior axis through the CG than the lateral axis about which the front or rear blows act. Consequently, temporoparietal impacts produce higher acceleration levels of

Table 4

Calculated Average Kinematic Data at Each Location*

1 IMPACT LOCA-TION**	2 RIGID BODY ACCELERATION (g)	3 DISTAL IN-LINE ACCELERATION (g)	4 NORMALIZED ACCELERATION	5 HIC X10^3	6 SI 10^3	7 $\dot{\theta}$ RAD/SEC	8 $\ddot{\theta}$ RAD/SEC^2 X10^3	9 TIME TO PEAK $\dot{\theta}$ ms	10 UNCONSCIOUS DURATION SEC
F	1210 ±400	2410 ±790	1.91 ±0.47	62.2 ±34.7	81.5 ±48.1	158 ±26.8	132 ±28.9	1.3 ±0.27	13 ±11
S	1510 ±250	3090 ±590	2.19 ±0.62	113 ±47.8	143 ±68.4	200 ±39.8	166 ±30.5	1.3 ±0.12	86 ±21
R	1100 ±190	2100 ±240	1.94 ±0.39	32.5 ±13.0	42.7 ±19.5				26 ±22
T	1070 ±190	1110 ±260	1.06 ±0.27	10.7 ±9.8	15.2 ±13.2				23 ±16

* Calculated from Table 2 data

** F = Frontal
S = Temporoparietal
R = Occipital
T = Cranial

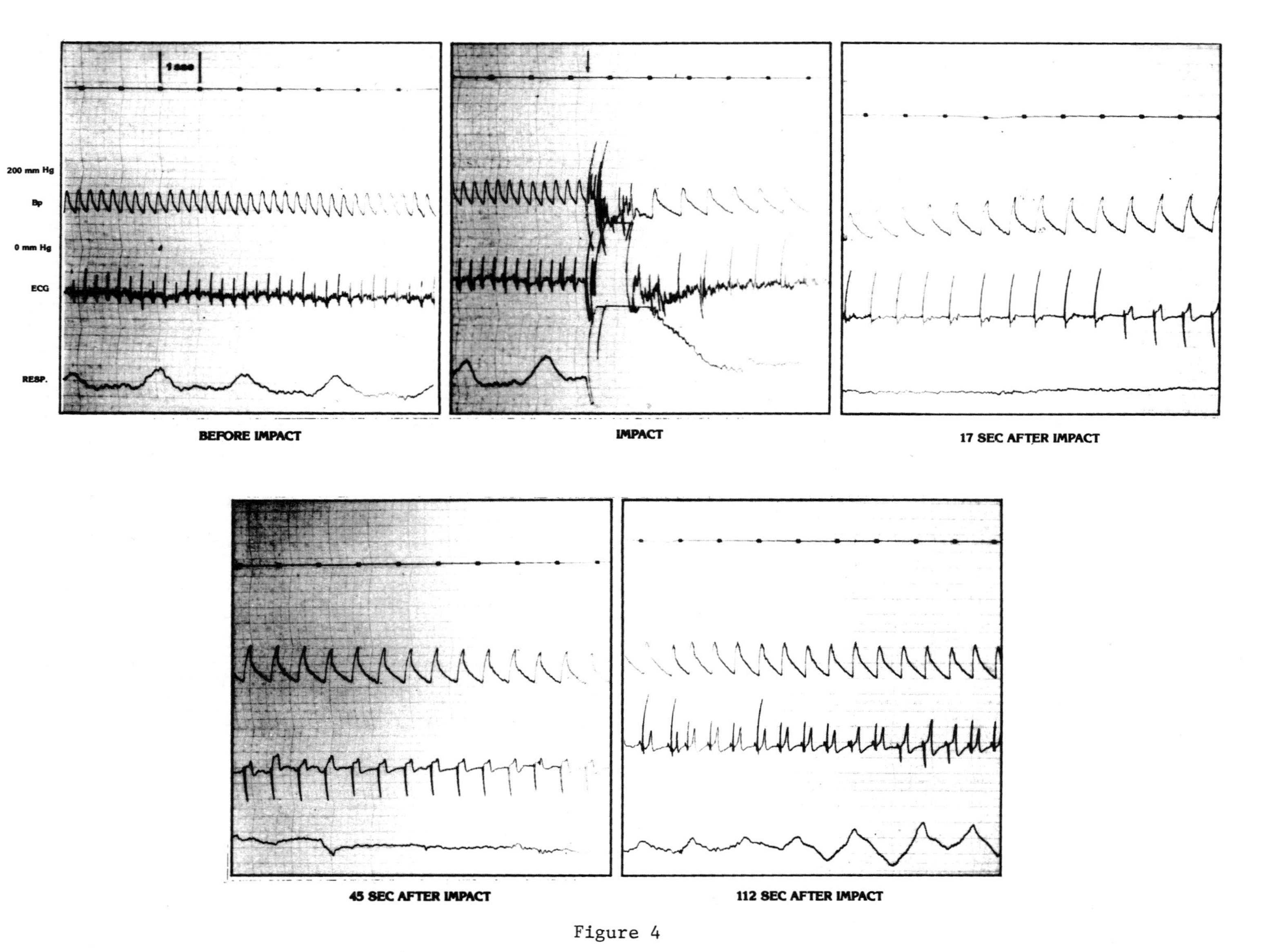

Figure 4

Typical oscillograms excerpts showing changes in blood pressure (BP), heart rate (ECG), and respiration (Resp.), following a concussion-producing impact to a monkey's head.

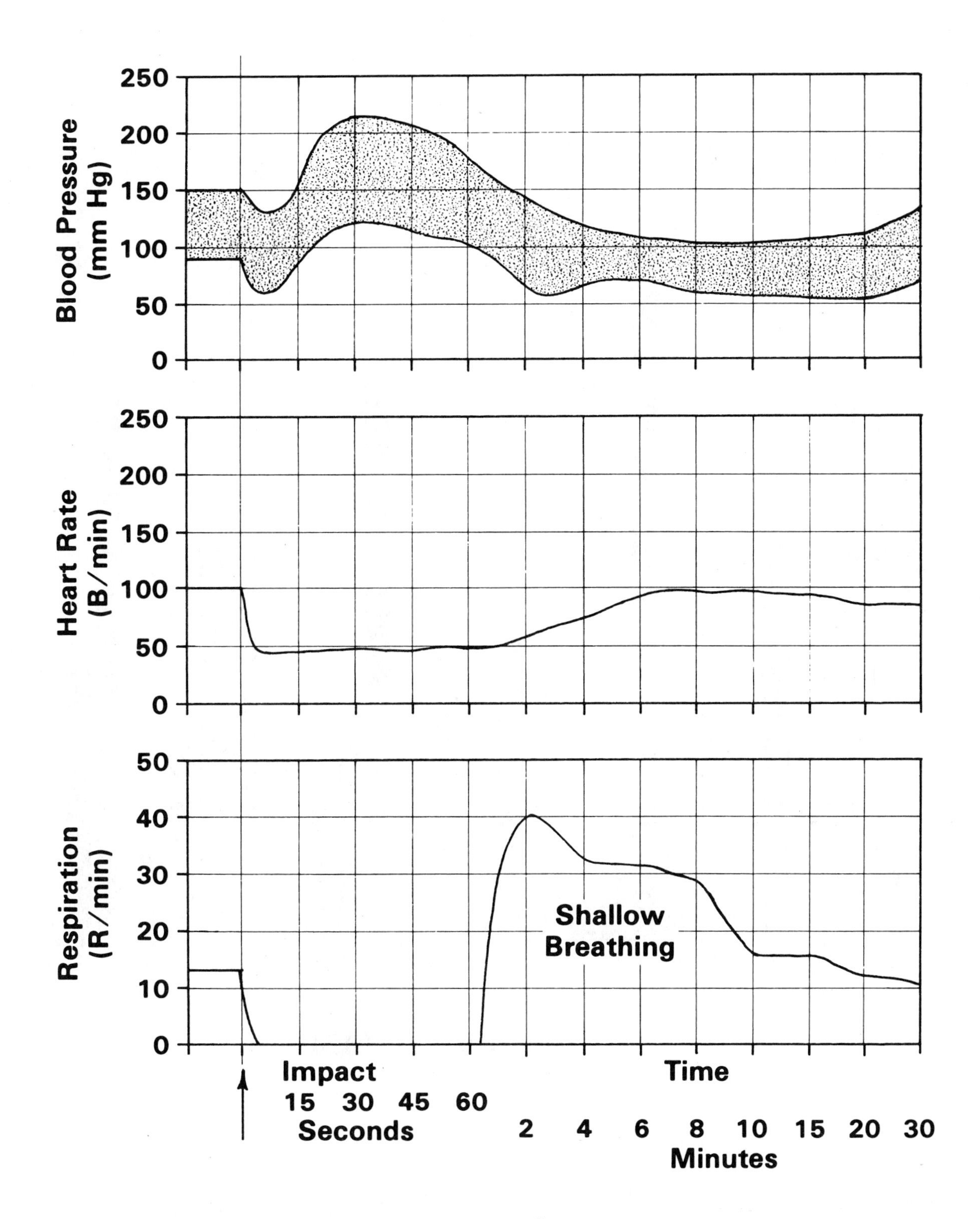

Figure 5

A continuous plot of rate and amplitude of physiological changes accompanying a concussion-producing impact to the head of a monkey.

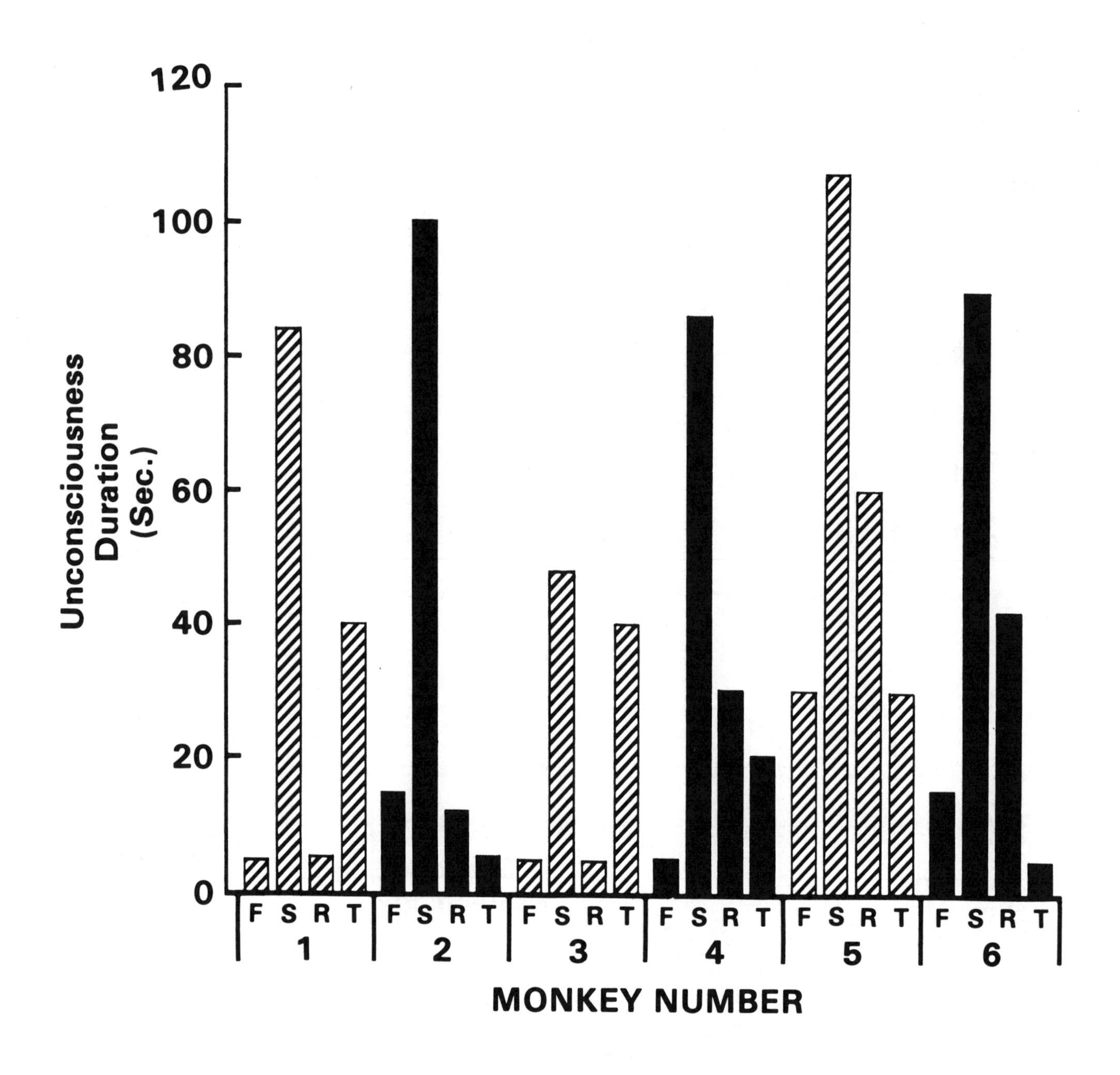

Figure 6

Duration of unconsciousness versus impact location on frontal (F), temporoparietal (S), occipital (R) and cranial (T), for concussive impacts to six monkeys.

shorter duration, which presumably could have caused the longer periods of unconsciousness.

The concussions produced in this experiment are consistent with the injury definition as outlined by the Congress of Neurological Surgeons and are in marked contrast with the experimental data produced by other authors using head holders designed to separate translatory and rotatory head motions. The HAD II apparatus has been used to produce head injury in squirrel monkeys both by Unterharnscheidt and Higgins (10) as well as Gennarelli, et al. (7,8). The apparatus produces either a purely translational motion or a purely rotational motion, with rotation about an axis at the approximate level of the seventh cervical vertebra. The head of the animal is closely connected with a metal cap to eliminate skull deformation.

Gennarelli's group found that no cerebral concussion was produced by the translational mode in twelve tested animals while thirteen animals exposed to the rotational mode were all unconscious for two to twelve minutes. One animal in the rotational mode group never regained consciousness and two others died within one hour of impact. The visible lesions included twenty subdural hematoma among rotated and eight among translated primates. Twenty-seven subarachnoid hemorrhages, numerous contusions and intra-cerebral hemorrhages were recorded. A total number of ninety-four lesions were reported in the rotated group with an average of 7.2 lesions per animal and twenty-seven lesions occurred in the translated group with an average of 2.2 lesions per animal. Peak angular acceleration in the Gennarelli experiments was calculated from a tangential acceleration on the top of the head holder and ranged from 108,000 to 317,000 rad/sec^2. The form of the acceleration wave from tangential or inline acceleration measurements in both rotated and translated animals was initially positive followed by a negative acceleration of almost equal amplitude for a period ranging from 6-8 ms.

Similar results have been reported by Unterharnscheidt and Higgins on the HAD II apparatus with angular accelerations ranging up to 386,000 rad/sec^2. As noted earlier, acceleration of about 200,000 rad/sec^2 produced cerebrovascular injury to most of the subjects. Animals exposed to accelerations above 300,000 rad/sec^2 did not survive. The head injury study conducted on the Hyge crash simulator by Abel,et al. (9), also produced more severe cerebral pathology than the present study.

As indicated in Table 3, only one from twelve frontal and temporoparietal impacts produced angular accelerations above 200,000 rad/sec , which may account for the absence of serious cerebral injuries observed in the present study. (Both HAD II and the Hyge experiments mentioned above used impacts in the sagittal plane.) None of our animals showed symptoms of residual neurological pathology. The animals were alive, alert and moving about well in their cages upon recovery from the analgesia. It should be noted, however, that the accelerations in the present experiment were administered to primate species with brains relatively larger than those of squirrel monkeys and, therefore, lower resistance to head injury among our larger species should be expected.

It is plausible to assume that the excessive brain injuries produced in HAD II and Hyge experiments might have been produced by different acceleration wave forms which have not been experienced by animals in the present study. The HAD II produced a rapid acceleration in the 600-1230 G range followed by deceleration of even greater magnitude; the entire motion occurs within a period of 6-8 ms. This contrasts with the short but higher acceleration (2000-5000 G) lasting only about 1 ms and followed by gradual deceleration when our head impacts were produced. Masuzawa, et al. (13), produced fatal concussions in primates exposed to almost square wave decelerations in the 300-600 G range, lasting 6-8 ms. This seems to indicate that the impacts of different duration might produce different forms of cerebral cell damage or that the rates of the acceleration onset might play a significant role in the mechanism of cerebral concussion and permanent brain injury. Obviously, more research is needed to clarify the role of cerebral injury mechanisms. Until the effects of the rapid reversal of accelerations produced by the HAD II and Hyge machines are explained, it would be premature to attribute the different injuries produced by them solely to the types and levels of onset of acceleration. Their principal value appears to be as producers of severe brain injury, somewhat repeatedly, for the study and treatment of their sequelae.

CONCLUSIONS

1. It was found that by distributing a blunt impact blow with a protective head-cap, and allowing free head movement after impact, skull fracture was eliminated and simple reversible cerebral concussion could be produced without symptoms of residual neurological deficit.

2. On the average, similar energy impacts produced higher force and shorter duration impulses for temporoparietal blows than at another location, resulting in higher linear and angular head accelerations.

3. Higher acceleration and longer concussions due to temporoparietal compared to frontal and occipital impacts could be related to lower mechanical impedance at the temporoparietal blow location because of the oval shape of the animal's head (see discussion of results).

5. Although four reversible concussive impacts were administered to each animal within a two month period, no evidence of neurological deficit or neurogenic dysfunction was observed in any animal.

REFERENCES

1. Clinical Neurosurgery Proceedings of the Congress of Neurological Surgeons, 61:14. The Williams and Williams Company, Baltimore, 1966.
2. E. S. Gurdjian, Impact Head Injury, Charles C. Thomas, Publisher, Springfield, Illinois, 1975.
3. A. K. Ommaya, F. J. Fisch, R. M. Mahone,

P. Corrao and F. Letcher, Comparative Tolerances for Cerebral Concussion by Head Impact and Whiplash Injury in Primates, 1970. International Automovile Safety Conference Compendium, Detroit, Michigan. May 13-15, 1970 and June 8-11, 1970, pp. 808-818.
4. A. H. S. Holbourn, Mechanics of Head Injuries, Lancet 2:438-41, 1943.
5. A. K. Ommaya, A. E. Hirsch, Tolerance for cerebral concussion from head impact and whiplash in primates, J. of Biomechanics, 4:12-31, Pergamon Press, 1971.
6. A. K. Ommaya and T. A. Gennarelli, Cerebral concussion and traumatic unconsciousness - correlations of experimental and clinical observations on blunt head injuries, Brain, 97:633-654, 1974.
7. T. A. Gennarelli, A. K. Ommaya and L. E. Thibault, Comparison of translational and rotational head motions in experimental cerebral concussion, Proceedings of the Fifteenth Stapp Conference, Coronado, California, November, 1971.
8. T. A. Gennarelli, L. E. Thibault and A. K. Ommaya, Pathophysiologic responses to rotational and translational accelerations of the head, Proceedings of the Sixteenth Stapp Conference, Detroit, Michigan, November, 1972, SAE 720970.
9. J. M. Abel, T. A. Gennarelli and H. Segawa, Incidence and severity of cerebral concussion in the Rhesus monkey following sagittal plane angular acceleration, Proceedings of the Twenty-Second Stapp Conference, Ann Arbor, Michigan, October, 1978.
10. F. Unterharnscheidt and L. S. Higgins, Pathomorphology of experimental head injury due to rotational acceleration, Acta-Neuropathology, 12: 200-204, 1969.
11. J. H. McElhaney, R. L. Stalnaker, V. L. Roberts and R. G. Snyder, Door Crashworthiness Criteria, Proceedings of the Fifteen Stapp Conference, Coronado, California, November, 1971, SAE 710864.
12. Proceedings of Nineteenth Conference of American Association for Automotive Medicine, AIS, p. 438, November 20, 1975. (D. Huelke, Ed.)
13. H. Masuzawa, N. Nakamura, K. Hirakawa, K. Sand, M. Matsuno, H. Sekino, K. Mii, and Y. Abe. Experimental head injury and concussion in monkeys using pure linear acceleration impact. Symposium of 34th Annual Meeting of Japan Neurosurgical Society, Nagoya, October 24, 1975.

856022

Biomechanics of Diffuse Brain Injuries

Lawrence E. Thibault and Thomas A. Gennarelli
University of Pennsylvania

Abstract

This report discusses the development of brain injury tolerance criteria based on the study of three model systems: the primate, inanimate physical surrogates, and isolated tissue elements. Although we are equally concerned with the neural and neurovascular tissue components of the brain, the report will focus on the former and, in particular, the axonal elements. Under conditions of distributed, impulsive, angular acceleration loading, the primate model exhibits a pathophysiological response ranging from mild cerebral concussion to massive, diffuse white matter damage with prolonged coma. When physical models are subjected to identical loading conditions it becomes possible to map the displacements and calculate the associated strains and stresses within the field simulating the brain. Correlating these experimental models leads to predictive levels of tissue element deformation that may be considered as a threshold for specific mechanisms of injury. Isolated tissue studies in the axon then serve to confirm the relationship between ultimate strain, for example, and the pathophysiological consequences. At this time, we have found that elongating strains of between 5 and 10 percent at strain rates of greater than $50 sec^{-1}$ produce membrance depolarization and a concomitant decrease in excitability that recovers in minutes. The degree of depolarization is directly proportional to the magnitude of the stretch. Further, total structural failure of the axonal membrane (diameters 400-750.jmm) occurs between 25 and 50 percent elongation at comparable strain rates. These observations are in reasonable agreement with the estimations of critical strains obtained from the primate and physical model studies.

Introduction

The development of head injury tolerance criteria for specific forms of brain injury remains one of the many challenging problems in the biomechanics of central nervous system trauma. Part of this problem is to describe the variety of head injuries that occur and to associate with each injury a severity index. Additionally, it becomes necessary to elucidate the mechanisms that produce each injury and to develop tolerance (or failure) criteria for the components that constitute the brain. Finally, a method to relate the kinematics of dynamic mechanical loading of the head to the deformations of the brain tissue is required.

This report will discuss the use of three model systems selected to address these issues and, further, it will focus on diffuse brain injury. Although the neural and neurovascular tissue injuries may be of equal importance in terms of injury severity, the scope of this study is restricted to the axonal structures of the brain. The three models therefore include the primate, simplified, inanimate, structural models of the skull brain system, and isolated axonal tissue preparations. Previous studies have found(1,2,3) that inertial loading, and in particular rotational acceleration, is capable of producing diffuse injury to the brain of the primate. We have more recently shown that the severity of this injury can be varied to include mild cerebral concussion as well as prolonged coma with axonal damage bilaterally in the hemispheres and in the corpus colossum and brain stem. Results to date suggest that as the brain experiences inertial loading with significant rotational components it undergoes a deformation that depends not only on the details of the loading but obviously on the local geometry and the tissue material and structural properties.

Simple physical models of the skull-brain structure under carefully controlled inertial loading experiments allow one to study the temporal and spatial variation of these deformations as a function of the magnitude, direction, and time history of the rotational accelerations. These deformations result in strains and stresses that, when excessive, are manifested in functional and structural changes in the tissue elements. To demonstrate the latter, it is most practical to use isolated tissue elements. We will discuss the giant axon of the squid as a model of axonal structures of the brain. Electron miscroscopic examination of the brain's deep white matter reveals intracellular injury to the axons in the regions of the nodes of Ranvier, and this particular injury has been associated with prolonged coma(4). Although the squid axon is non-myelinated, we feel it is a reasonable model for this nodal region that is probably the most vulnerable from a structural point of view. Changes in the electrophysiology of the axon under conditions of variable strain-rate uniaxial extension are then related to the animal and physical model results.

Methods

Primate Model

A standard 6in diameter HYGE® (Bendix) shock tester has been modified in the following manner:

- The stroke length was reduced to 3.125in.
- The internal acceleration and deceleration metering pins were redesigned to produce the desired acceleration-time characteristics.
- A kinematic linkage was designed that converts the translational motion of the HYGE thrust column into an angular motion.
- A helmet system was designed to rigidly couple the primate head to the kinematic linkage.

This system is used to provide controlled inertial loading to the primate as well as the physical models. Figure 1 depicts the kinematic linkage whose angular displacement is set at approximately 65°. Also shown in the figure is a typical acceleration-time history expressed as tangential acceleration.

The primates used in these studies were baboons ranging in weight from approximately 5 to 10kg. Standard physiological monitoring was performed including intra-cranial pressure, arterial pressure, respiration, respiratory carbon dioxide, respiratory oxygen, EKG, EEG, and multimodality evoked potentials. Upon sacrifice, both light and electron microscopic studies were performed on the brain tissue. With particular emphasis on the axonal structures, the lesions are mapped according to anatomic location and degree of severity.

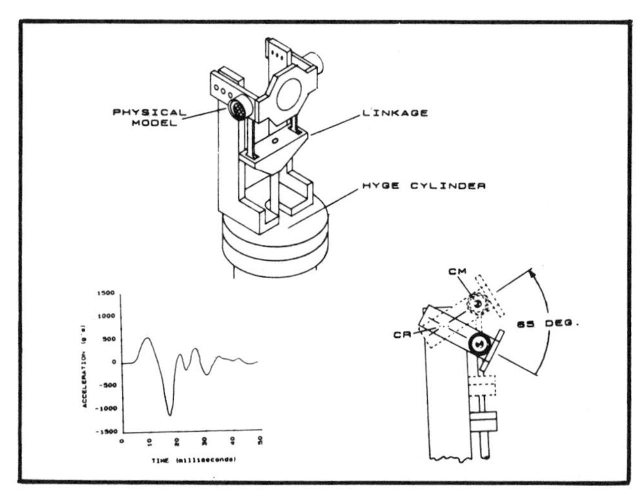

Figure 1. Experimental linkage and typical acceleration profile

Fresh brain weights are recorded in all cases immediately post-mortem and in this study range from approximately 110 to 170g.

The rotational axis was varied to include the three mutually perpendicular planes of sagittal, coronal, and horizontal.

Physical Models

A surrogate brain material in the form of a silicone-based gel, known as Silicone Gel System (Dow Corning), was used for these studies. The material is supplied as a two-part mix of polymer and catalyst. By varying the relative concentrations of these two components, it is possible to alter the modulus of elasticity of the gel by several orders of magnitude. For these experiments, the material properties were closely matched to brain tissue. The gel is an optically transparent substance cast at room temperature and can be layered with no mechanical discontinuity (i.e., the gel in a liquid state will adhere to a polymerized surface upon curing). It is therefore possible to cast the surrogate brain in layers and print or paint an orthogonal grid on selected surfaces that can be entrained in depth within the model. As depicted in Figure 2, both idealized right circular cylindrical geometries, as well as realistic skull boundaries, can be studied.

The physical models are mechanically fastened to the kinematic linkage of the HYGE system in such an orientation that deformations of the grid within the gel can be filmed in the plane of interest. Photography is performed at 4,000 frames per second on 16mm film. Individual frames of interest are printed and then digitized using a Hewlett Packard 9836 computer and bit pad. From the digitized data, one can then compute the field parameters of displacement, strain, and stress at node points within the grid. Therefore, both the strains and strain-rates at a particular location can be determined for any given loading condition.

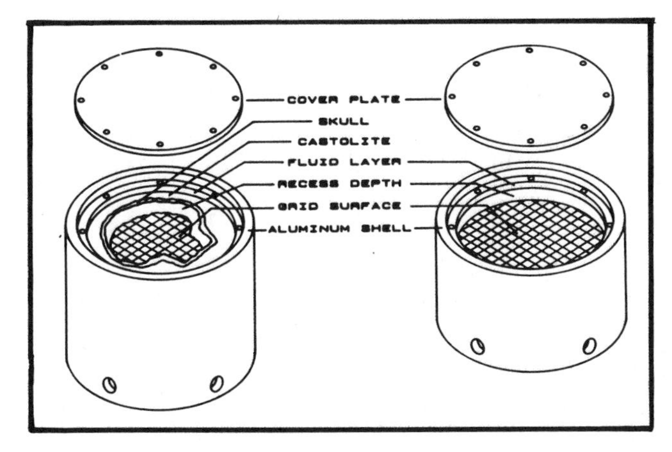

Figure 2. Construction of the physical models

Isolated Tissue

A system has been developed that permits one to study the electrophysiology, morphology, biochemistry, and transport properties of the giant axon of the squid as a function of variable strain-rate and variable magnitude of stretch in uniaxial extension.

Approximately 5cm of dissected axon is mounted in a chamber where glass cannulae are inserted through small cuts in the membrane. As portions of the membrane are allowed to dry to the glass cannula, a tight grip is formed between the glass and the internal circumference of the axon. This procedure produces a viable length of axon of approximately 1.5cm, which is immersed in the physiologically suitable fluid. The glass cannulae are part of platens that on one end attach to an isometric force transducer and on the other to a movable linkage. The linkage is driven by an audio speaker that can be programmed to move a predetermined displacement at a given velocity. This constitutes a small materials testing machine with a specimen environmental chamber. For purposes of this study, a wire electrode is inserted axially into the preparation to obtain membrane potential measurements. The system is depicted schematically in Figure 3. The entire device is integrated onto the movable

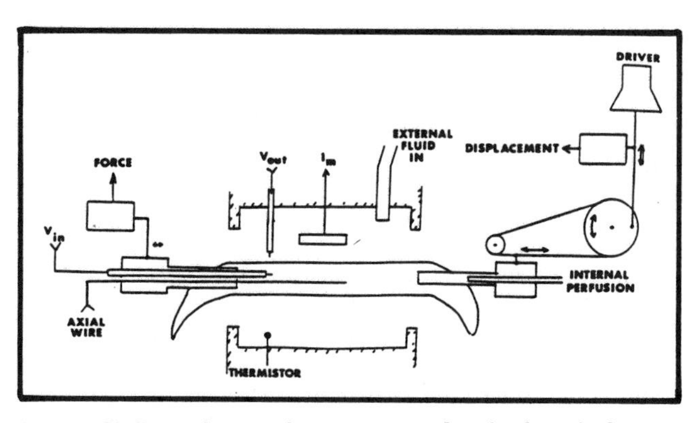

Figure 3. Experimental apparatus for isolated tissue studies

stage of an inverted phase contrast microscope. With this arrangement, it is then possible to obtain axonal mechanical properties as well as the neural tissue element electrophysiological response to mechanical stimulation.

Results

Primate Model

Thus far we have been able to programmably reproduce the following brain injuries in the primate model:

1. The acute subdural hematoma with varying degrees of severity.
2. Cerebral concussion from mild to periods of unconsciousness of up to 2 hours.
3. Diffuse axonal injury with prolonged coma.

The subdural hematoma as part of a class of neurovascular injuries will not be presented here.

The cerebral concussion and diffuse axonal injuries are shown in Figure 4. These same data are presented in Figure 5 where the peak rotational acceleration is shown plotted against the change in angular velocity. The data are scaled based on Holbourn's(5) suggested relationship as a function of brain mass to an average human brain of approximately 1,400g.

In the case of cerebral concussion, the period of unconsciousness varied from less than 1 minute to approximately 2 hours. For those animals listed as having diffuse axonal injury with prolonged coma, the longest survivor was sacrificed at 6 weeks. At this point in our experiments, the threshold levels for cerebral concussion and diffuse axonal injury when scaling the primate data to man appears to be 5,000 rad/s^2 (75 rad/s) and approximately 15,000 rad/s^2 (150 rad/s), respectively. The symbols □ and x in both figures refer to diffuse axonal injury and cerebral concussion.

Physical Model

Depicted in Figure 6 are the digitized and computer-reconstructed still frames of the grid deformations within an idealized right circular cylindrical geometry. Time is shown beneath each frame and is expressed in milliseconds. As the frames are presented, the cylinder is rotating in a counterclockwise direction. Recalling the asymmetric acceleration-time history character of the loading, the peak deformation is seen to occur at a point that temporally corresponds to the peak of the tangential deceleration waveform. Once the deformation field is digitized, the computation of the field parameters can be performed. Shown in Figure 7 is a plot of one component of the nonlinear strain tensor, E_{12}, as a function of the nondimensional cylinder radius. This type of analysis is

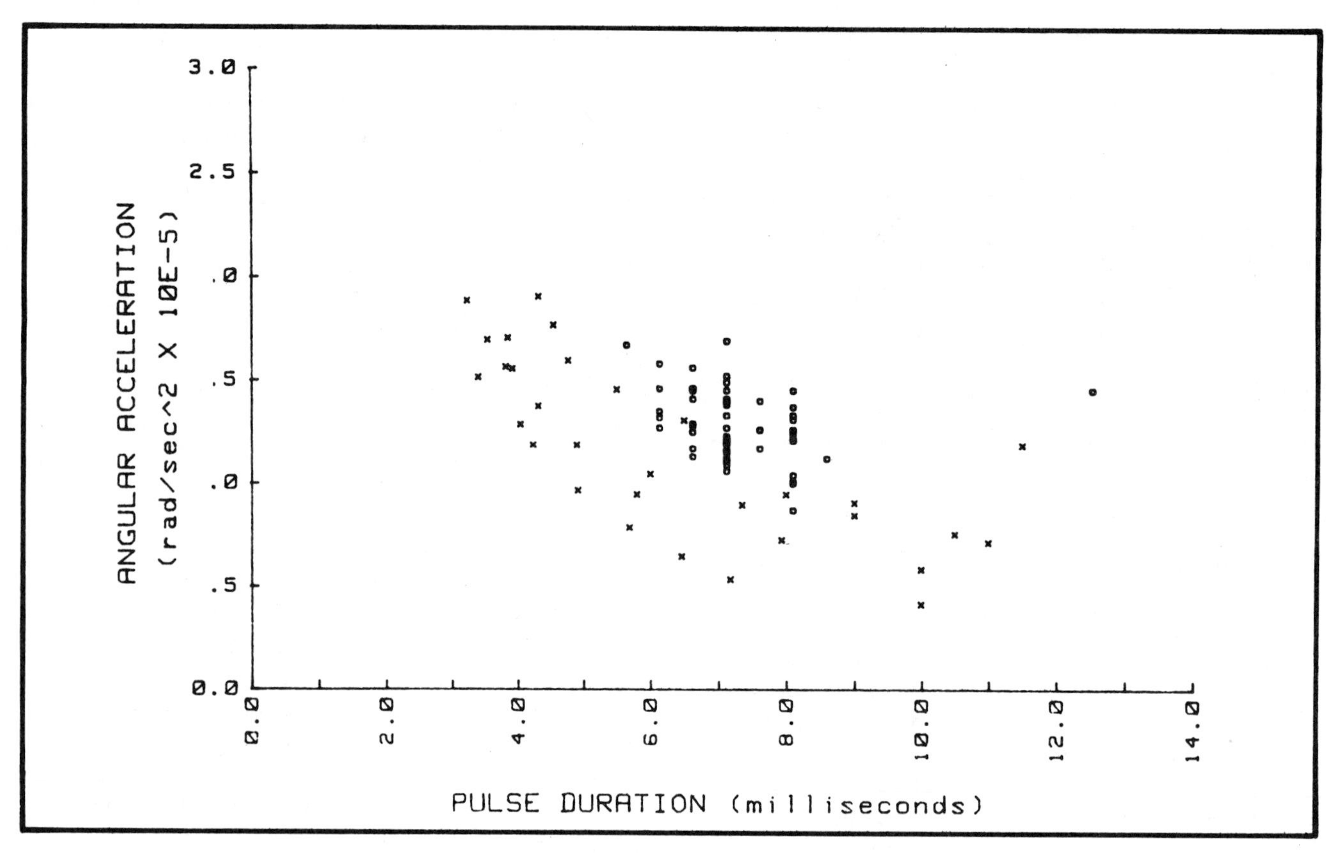

Figure 4. Primate experiments—angular acceleration versus pulse duration

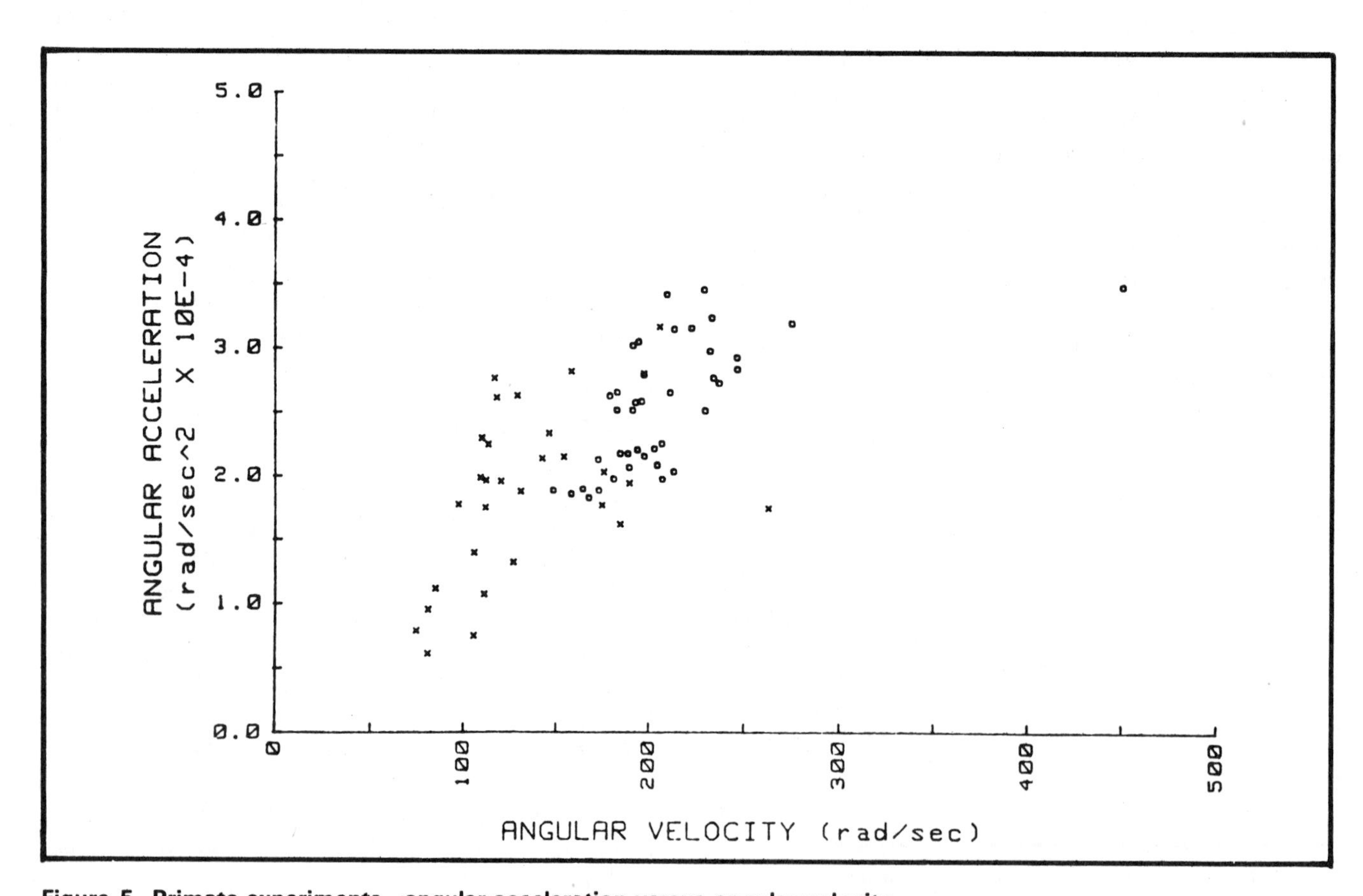

Figure 5. Primate experiments—angular acceleration versus angular velocity

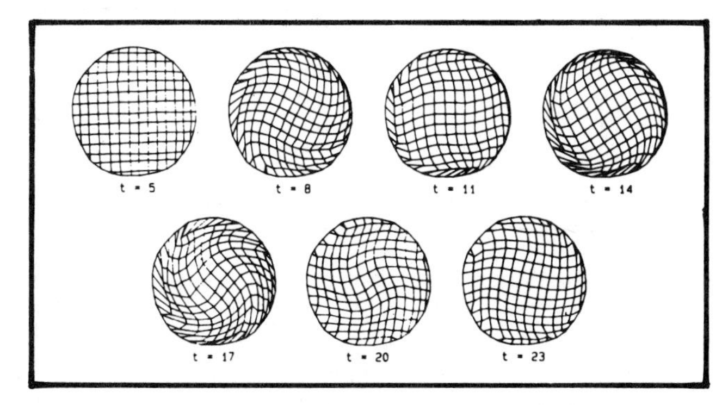

Figure 6. Digitized reconstruction of still frames for the right circular cylinder model

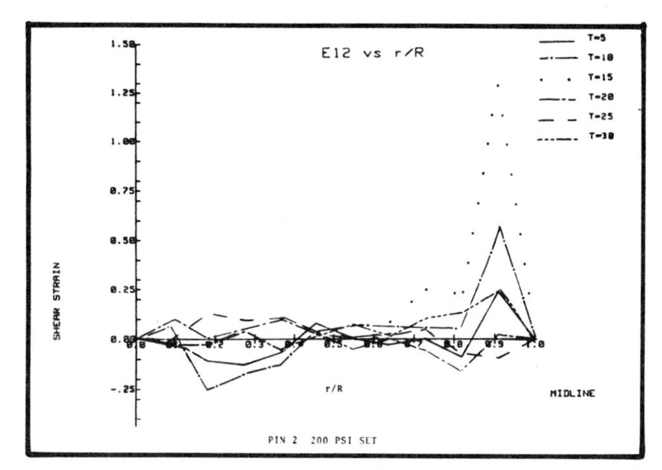

Figure 7. Strain versus nondimensional radius

presented to demonstrate the somewhat asymmetric strain field expected in the case of a noncentroidal rotation and to show an example of the many kinds of analysis that can be performed with the physical model data.

Figure 8 represents a comparison of the maximum elongating strains in three cylindrical models of different radii as a function of the peak rotational acceleration. The curves are labeled A, B, and C, representing cylinder diameters of 11.12cm, 7.94cm, and 6.35cm. Experiments such as these enable one to test scaling relationships such as the one proposed by Holbourn given as

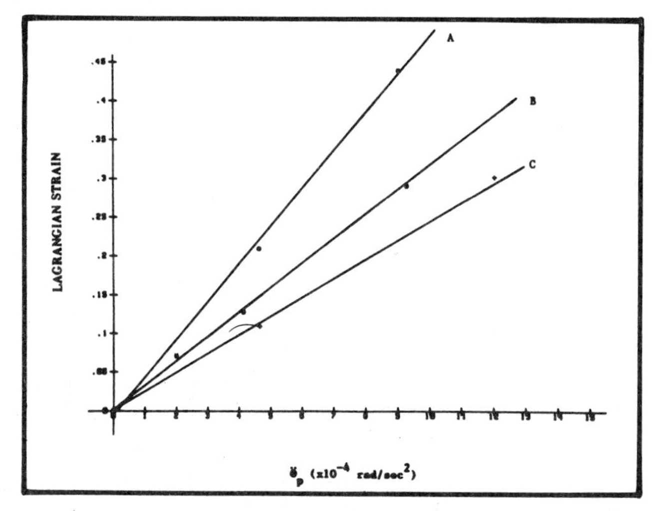

Figure 8. Strain versus angular acceleration for three cylindrical models A, B, and C of mass m_1, m_2, and m_3

$$\frac{\ddot{\Theta}_m}{\ddot{\Theta}_p} = \left(\frac{M_p}{M_m}\right)^{2/3}$$

Figure 9 shows the relationship between the maximum strain and the peak rotational acceleration when this scaling relationship is applied. The data in this case are scaled to the average brain mass of the baboon from the mass of each of the three cylinder models.

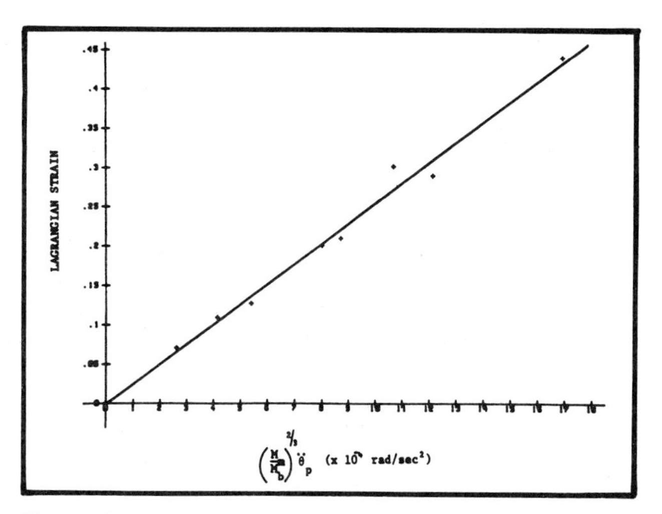

Figure 9. Strain versus angular acceleration from the three cylinder experiments scaled to the primate brain mass

Isolated Tissue

Figure 10 shows examples of the mechanical testing performed on the giant axon of the squid. The load-elongation curve exhibits the nonlinear behavior characteristic of soft tissue. Hysteresis and stress-relaxation studies also depicted are representative of the viscoelastic behavior of this class of biological materials. Under conditions of high strain-rate uniaxial extension, there is an associated depolarization of the axon. Figure 11 shows two such experiments where the magnitude of the axial elongation is increased from approximately 15 percent in the first case to 30 percent in the second study. A proportionate increase in the depolarization is observed. We believe that this phenomena and the accompanying decrease in membrane excitability are analogous to the cerebral concussion observed in the primate model. Figure 12 shows the time course of recovery of the resting membrane potential following a high strain-rate ($50s^{-1}$) uniaxial extension (20 percent). The exponential

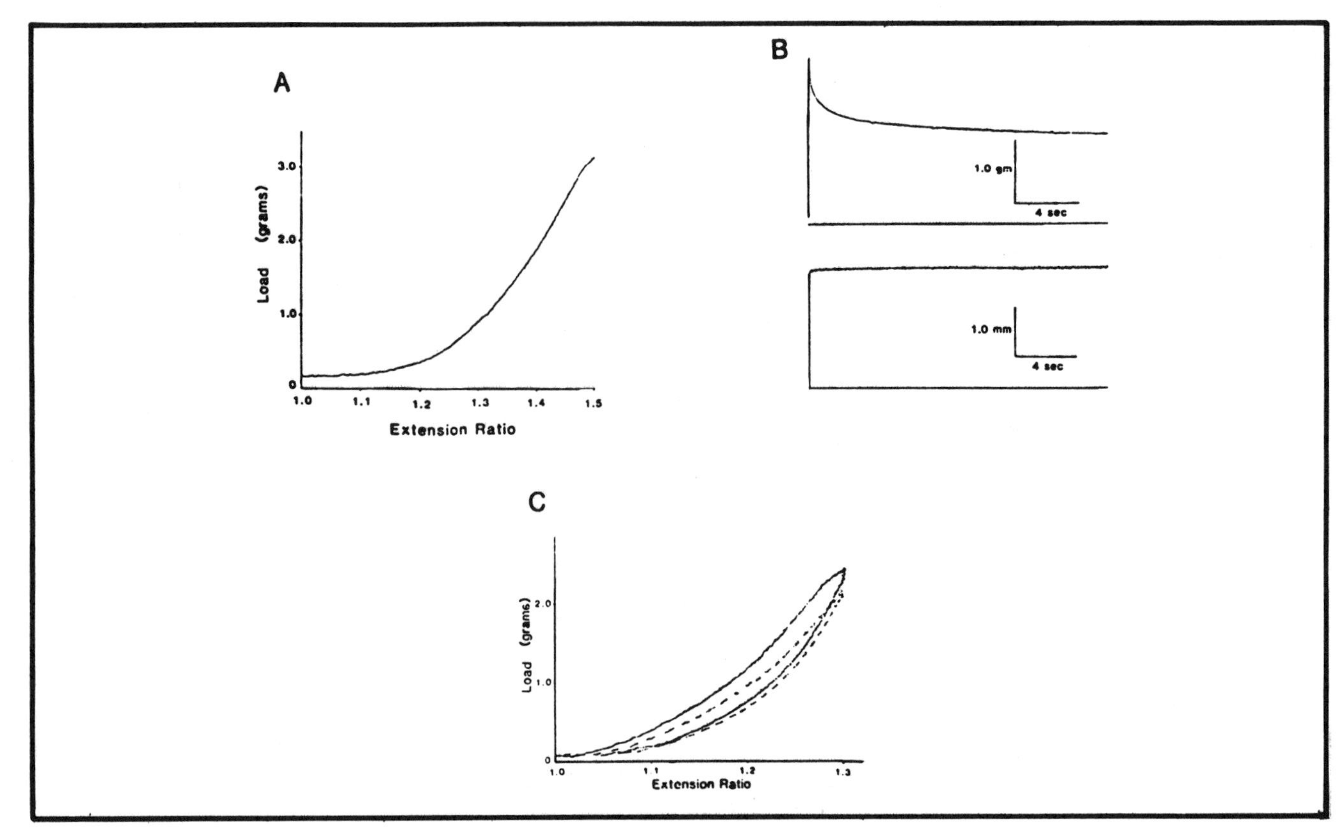

Figure 10. Mechanical properties of the isolated squid giant axon

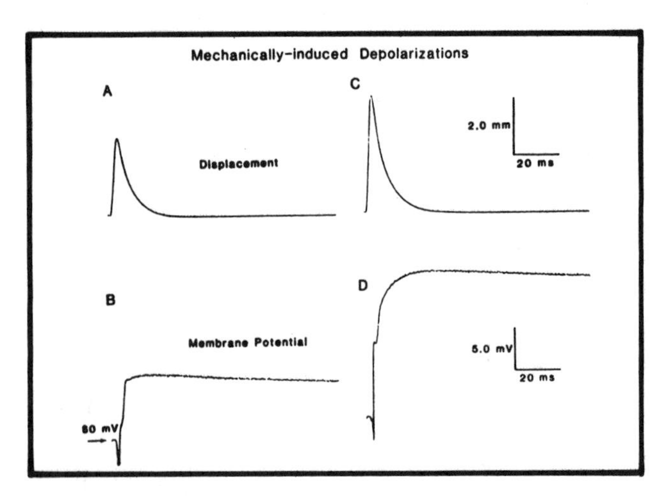

Figure 11. Changes in membrane potential associated with high strain-rate uniaxial elongation

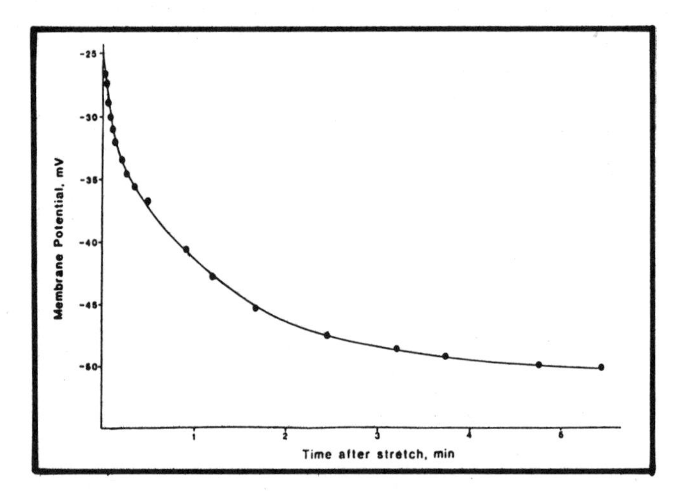

Figure 12. Membrane potential time course of recovery following mechanical stimulation

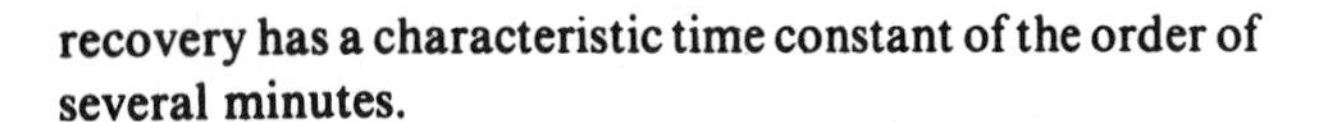

recovery has a characteristic time constant of the order of several minutes.

Discussion

The experiments described herein are designed to provide methods of improving on existing head injury tolerance criteria and to develop model systems whereby it is possible to study the mechanisms of injury with the ultimate goal of therapeutic intervention. These two distinctly different objectives share a common approach that we describe as the cellular basis for injury from both the biomechanics and pathophysiological points of view.

Figure 13 represents our first approach to developing injury criteria based on tissue response to mechanical strain. The physical models, which are highly idealized at this point, enable one to transform the inertial loading data into tissue deformation. The predicted thresholds for cerebral concussion and diffuse axonal injury ap roximate the isolated tissue response.

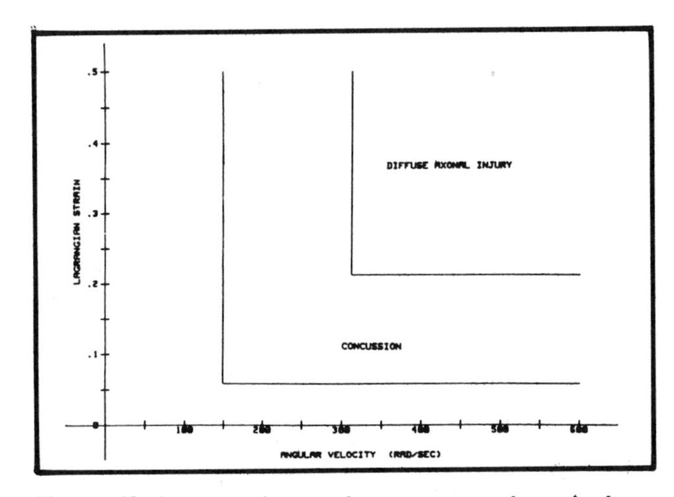

Figure 13. Langrangian strain versus angular velocity

Acknowledgement

This work was supported in part by NHTSA of the U.S. Department of Transportation, Contract DTNH 22-82-C07187.

References

1. Higgins, L.S., and R.A. Schmall, "A device for the investigation of head injury effected by non-deforming head accelerations," Proceedings 11th Stapp Car Crash Conference, Soc. Auto. Eng., New York, pp. 57-72, October 1967.
2. Ommaya, A.K., F. Taas, and P. Yarnell, "Head injury in the chimpanzee. 1. Biodynamics of traumatic unconsciousness," *J. Neurosurg*, 39:2:152-166, August 1973.
3. Gennarelli, T.A., L.E. Thibault, and A.K. Ommaya, "Pathophysiological response to rotational and translational accelerations of the head," Proceedings 16th Stapp Car Crash Conference, Soc. Auto. Eng., New York, pp. 296-308, 1972.
4. Gennarelli, T.A., L.E. Thibault, J.H. Adams, D.I. Graham, C.J. Thompson, and R.P. Marcincin, "Diffuse axonal injury and traumatic coma in the primate," *Annals of Neurology*, Vol. 12, pp. 564-574, 1982.
5. Holbourn, A.H.S., "Mechanics of head injuries," Lancet (2), pp. 438-441, 1943.

861890

Critical Limitations on Significant Factors in Head Injury Research

Guy S. Nusholtz, Patricia S. Kaiker and Richard J. Lehman
UMTRI

ABSTRACT

The response of the head to blunt impact was investigated using anesthetized live and repressurized- and unrepressurized-postmortem Rhesus. The stationary test subject was struck on the occipital by a 10 kg guided moving impactor. The impactor striking surface was fitted with padding to vary the contact force-time characteristics. A nine-accelerometer system, rigidly affixed to the skull, measured head motion. Transducers placed at specific points below the skull recorded epidural pressure. The repressurization of postmortem subjects included repressurization of both the vascular and cerebrospinal systems. The results of the tests demonstrate that: 1) Head impact and injury response are complex phenomena and require careful attention to the experimental techniques used to obtain the impact data as well as the interpretation of the results; 2) A set of initial conditions has been found such that the repressurized-postmortem Rhesus was more similar in head impact response to the anesthetized Rhesus than was the unrepressurized-postmortem Rhesus (in terms of the transfer function relationships between tangential acceleration and epidural pressure; 3) In terms of subarachnoid hemorrhage in the *medulla-pons* area associated with skull fracture, the repressurized-postmortem and anesthetized Rhesus were similar in injury response; 4) The initial position of the head-neck system was a critical factor associated with the brain-cerebrospinal fluid system's response to head impact; 5) The thermodynamic response (fluid vaporization) of the head-brain system was an important consideration when the impact produced significant tension; and 6) No relationships could be found between injury and negative pressures equal to one atmosphere.

[1]Numbers in parentheses refer to references found at the end of the text.

INTRODUCTION

EXPERIMENTAL IMPACT TESTING of human surrogates provides useful information about biomechanics data-gathering techniques, the general nature of human injury mechanisms, kinematic and injury response parameters, and injury tolerance values. One body region that has been investigated through the use of the human surrogate is the head (1,5,6,9,11-21,23,25-28)[1]. In an attempt to reduce the injuries that can occur as a result of blunt impact to the head, researchers have used a number of procedures to deliver a calibrated amount of energy to the head of a human surrogate. The resulting kinematic motion and injury, then, potentially, can be used to obtain information to address head injury mechanisms, providing that the experimental techniques (including the choice and preparation of a surrogate) are appropriate.

A major difficulty in the investigation of head trauma is designing kinematic experiments which minimally interfere with the biological and physical systems being studied, yet produce results that are accurate and can be used effectively to describe the physical phenomena which occur as a result of impact. The goal of such research is a paradigm of injury mechanisms which corresponds well with clinically observed trauma. Some understanding of head injury mechanisms as a result of dynamic biomechanics blunt impact experiments has resulted from relating kinematic parameters to injury modes. Correlations of this type do not always imply causation, however, because of 1) the possibility of several injury mechanisms; 2) experimental inaccuracy associated with the choice of human surrogate; and 3) misinterpretation of the results as a consequence of faulty assumptions often compounded by errors which occur because of experimental techniques that do not (or do not or accurately) measure the needed parameters. In this regard it is critical to understand the problems and limitations associated with the experimental techniques used in head injury research.

For example, the Head Injury Criterion (HIC) is derived from an acceleration response. One method of obtaining the acceleration information required to generate the HIC for human surrogates, such as the cadaver, is to mount accelerometers on the skull and assume that rigid body motion will occur during impact. In a previous study in which a nine-accelerometer array was rigidly affixed to the human cadaver skull, it was concluded that three-dimensional rigid body motion of the head is not well defined during a "severe" head impact when motion is measured by accelerometers affixed to the skull (14). In such impacts, skull deformation was interpreted by the nine-accelerometer array as an angular and linear acceleration not directly related to rigid body motion. Therefore, when the skull deformations (with or without skull fracture) were significant in severe impacts, the three-dimensional motion derived from the nine-accelerometer array could only be used to estimate the motion of the rest of the skull. This result implies that even when three-dimensional motion (consisting of both linear and angular motion) is obtained using a nine-accelerometer array in one lab, the results may not be closely comparable to those obtained in other labs, and may not truly represent the desired three-dimensional motion. Different experimental techniques (e.g. transducer, mount type, and location) employed for obtaining the three-dimensional motion may produce uncomparable data. Therefore, the HIC may not be well-defined for "severe" head impacts.

A second example of the problems that can be associated with experimental techniques comes from human cadaver blunt crown head impact observations (13). The experimental testing showed that peak forces are not a good indicator of cervical spine injury. Forces as high as 3000 lbs did not produce cervical spine injury in some tests, while those as low as 500 lbs produced spinal cord transection in others. The faulty assumption that the kinematic parameter of peak force correlates with injury to the cervical spine allowed an experimental design in which a negative correlation between peak force and trauma to the cervical spine was obtained.

A third example comes from the earlier observations made on some of the Rhesus being presented in this report (12,16). A particular set of initial conditions produced a very significant difference in blunt head impact response between the unrepressurized-postmortem and anesthetized Rhesus. However, we realize that other sets of initial conditions may produce a very small difference between these two models. It should be possible to conduct two series of impact experiments having two slightly different sets of initial conditions in which one set would produce no difference in response between the unrepressurized-postmortem and anesthetized Rhesus, while the other would produce a significant difference. Therefore, it would be possible for two different labs using seemingly similar procedures to produce contradictory conclusions. Such variations in conclusions seem to stem from a lack of knowledge of the subtle aspects of head impact.

The purpose of the dynamic biomechanics blunt head impact study being reported here was to qualify, in a limited sense, a few of the critical factors associated with head trauma research. In particular, the effects of fluid vaporization, the effects of material flow through the foramen magnum, the potential effects of the postmortem state, the effects of vascular and cerebrospinal repressurization on the postmortem state, the effect of the initial positioning of the head and neck on kinematic/injury response, and the effect of constraints on results produced by experimental techniques (i.e. measurement and analytical procedures) will be discussed. The techniques developed and used by the University of Michigan Transportation Research Institute (UMTRI) for measuring three-dimensional head motion and epidural pressure will be presented. Anesthetized live Rheus and repressurized- and unrepressurized-postmortem Rhesus were subjected to a posterior-to-anterior direction occipital head impact. The results presented here are in addition to, build upon, and are compared with those previously presented (See 12,14,16) so that they cover the data from related research programs conducted over the last nine years.

METHODOLOGY

The methods and procedures that were used in this dynamic biomechanics trauma research are outlined below. Additional information about the methods and procedures can be found elsewhere (2,3,8,12,14,16).

Impact Testing - The twenty-two laboratory animal subjects used in these experiments were obtained by UMTRI from the University of Michigan Unit for Laboratory Animal Medicine. The Rhesus were instrumented using procedures similar to those used to instrument human cadaver subjects. A nine-accelerometer array was rigidly affixed to the skull and epidural pressure transducers were used to document pressure changes. The cerebrospinal and vascular systems of five of the postmortem Rhesus were repressurized.

Using the pneumatic impacting device with a 10 kg impactor (12), ten anesthetized and twelve postmortem Rhesus were each subjected to a single head impact to the occipital region. Preliminary findings on five of the anesthetized Rhesus and seven of the postmortem Rhesus have been presented earlier elsewhere (12,14,16).

The impactor surface was a 10.2 cm rigid metal plate padded with 2.5 cm Ensolite. The impactor force transducer assembly consisted of a Kistler 904A piezoelectric load washer with a Kistler 804A piezoelectric accelerometer mounted internally for inertial compensation.

A specially designed timing unit was used to synchronize the impact events during each test, such as the release of the striker and activation and de-activation of the high-speed photographic equipment. The impact conditions between tests were controlled by varying the impact velocity and the initial positioning of the test subject. The piston velocity was measured by timing the pulses from a magnetic probe which sensed the motion of targets in the piston.

Triaxial Acceleration Measurement - An interrelated set of three triaxial accelerometers (the nine-accelerometer array) recorded head accelerations. The accelerometers were either Endevco 2264-2000 piezoresistive ones or Kistler Model 8694 piezoelectric ones.

Epidural Pressure Measurements - Epidural pressures were measured with piezoelectric Kulite model MCP-055-5F catheter tip pressure transducers or with Endevco 8507-50 pressure transducers.

Cineradiographs - High-speed cineradiographs were taken of the impact events at 400 or 1000 frames per second. The UMTRI high-speed cineradiographic system (24) consists of either a Photosonics 1B or Milliken high-speed 16 mm motion-picture camera which views a 5 cm diameter output phosphor of a high-gain, four-stage, magnetically focused image intensifier tube, gated on and off synchronously with shutter pulses from the motion-picture camera. A lens optically couples the input photocathode of the image intensifier tube to x-ray images produced on a fluorescent screen by a smoothed direct-current x-ray generator. Smoothing of the full-wave rectified x-ray output is accomplished by placing a pair of high-voltage capacitors in parallel with the x-ray tube. The viewing field for these experiments was between 20-by-40 cm (See Figure 1).

Data Handling - All transducer time-histories (i.e. impact force, impact acceleration, epidural pressures, nine head accelerations) were recorded unfiltered on a Honeywell 9600 FM Tape Recorder with an EMI multiplex unit or a Honeywell 7600 FM Tape Recorder. A synchronizing gate was recorded on all tapes. The analog data on the FM tapes were played back for digitizing through proper anti-aliasing analog filters. The analog-to-digital process for all data resulted in a digitial signal sampled at 6400 Hz equivalent sampling rate.

Test Synopsis

Twelve head impacts were conducted on post-mortem Rhesus. Upon termination, they were stored in a cooler at 4°C for 72 hours (78A232-79A260)/5 days (85R005-86R015)/3 weeks (85R002) prior to testing. Anesthetized Rhesus were used in the other ten experiments. The protocol for postmortem animals, presented in detail elsewhere (12), was less complex than that for the anesthetized laboratory animals, which is outlined below.

Test Subject Preparation - The animal was first anesthetized with an injection of ketamine and then maintained with ketamine and sodium pentobarbital by means of a catheter with a three-way valve which had been inserted into the *long saphenous* vein of the leg. An airway was established to aid breathing when necessary. Then the subject was weighed and measured. Next, using a cauterizing scalpel, the scalp and muscle mass were removed from the frontal bone and the screws used to secure the epidural pressure transducer coupling devices were screwed into place. Then the nine-accelerometer plate was installed. Next, a strain gauge was attached to the skull bone to record the vibrations of the skull during impact for the purpose of estimating their effect on the accelerometer signals.

Eye and ear x-ray targets were positioned and the Rhesus was transported to the x-ray room where two head x-rays (x-z and y-z views) were taken, or were taken later as part of the necropsy examination (85R002-86R015). The subject was then taken to the impact laboratory and placed in an erect sitting position with its posterior side towards the impactor, so that the line of impact was in the mid-sagittal plane in the posterior-anterior direction.

Subject Positioning - The subject was held in place with paper tape which secured the body under the armpits, suspending the head and torso from an overhead hoist. The test subjects were impacted at the occiput. In tests 78A232 and 78A234 the eyes were on a horizontal line through the impact site, thus the estimated center of mass was below the line of impact. In tests 78A236 and 78A239 the eyes were raised to a point slightly above a horizontal line through the impact site, thus the line of impact was through the estimated center of mass. In tests 78A238 and 78A241 the eyes were raised higher than the previous two tests, thus the center of mass was above the line of impact (See Figure 2). Tests 79A249-79A260 were similar to those just de-

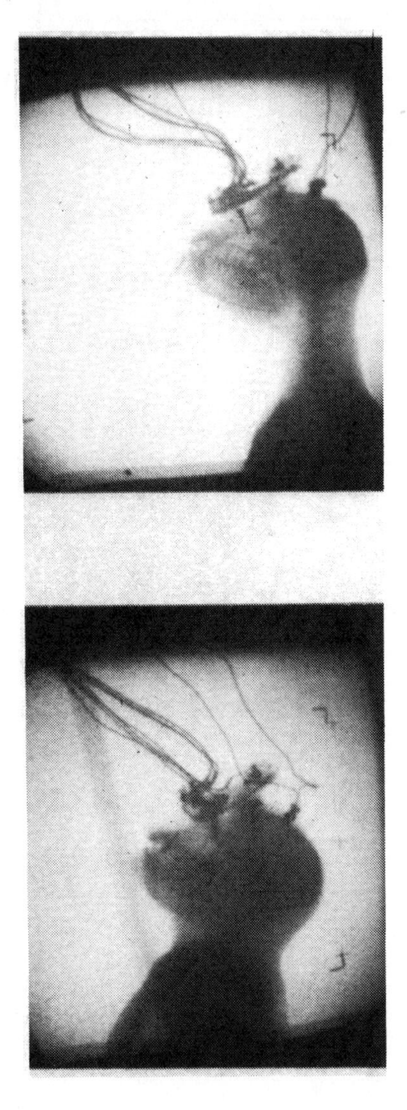

FIGURE 1:

CINERADIOGRAPH PHOTOS OF INITIAL POSITIONING OF THE HEAD-NECK: NECK-STRETCHED (TOP) AND NECK-RELAXED (BOTTOM)

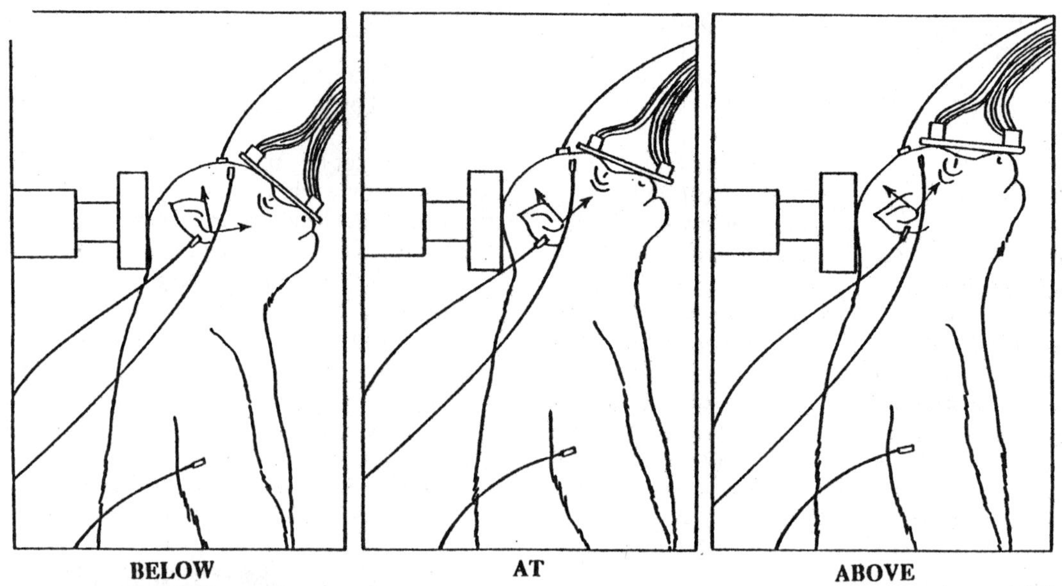

FIGURE 2: SUBJECT POSITIONING - CENTER OF HEAD MASS RELATED TO LINE OF IMPACT

scribed except that higher impact energy levels were used and the subjects were positioned so that impact occurred through the estimated center of mass. For Tests 85R002-86R015, the eyes were raised, positioning the subjects similar to those in Tests 79A238 and 78A241.

The initial positioning of the subject also included placing the subject in either the neck-stretched (86R011-86R013) or neck-unstretched position (78A232-79A260, 85R002-8, and 86R014-15). See Figure 1 for illustration of the initial positioning of the head-neck. To achieve the neck-stretched configuration, strings were placed through the ears and the subject was suspended by an overhead hoist so that approximately one-half of its entire body weight was supported by the neck. To achieve the neck-unstretched configuration, paper tape was used to support the head via the overhead hoist so that only one-half the weight of the head and none of the remaining body weight was supported by the neck.

After the positioning, three triaxial accelerometer clusters were fastened to the nine-accelerometer plate. Silicon fluid was injected into the pressure couplings, thus removing all air, and the pressure transducers were inserted. The subject was positioned in front of the impactor and stabilized with paper tape. All of the transducers were then connected, cabled, and checked for continuity and function.

Figure 3 shows the basic dynamic biomechanics impact test setup. One and a half-hours after impact for Tests

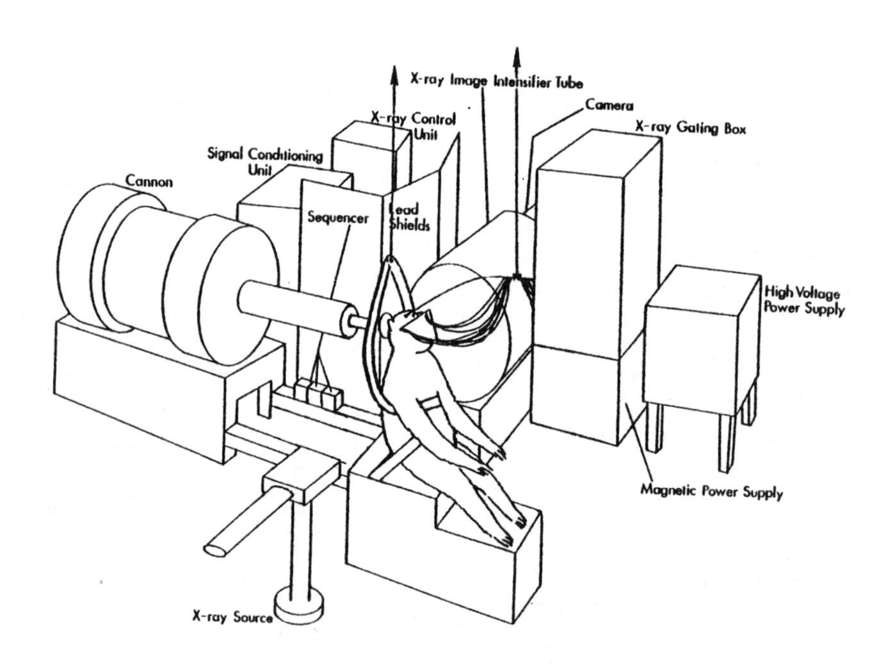

FIGURE 3: TEST SET-UP FOR DYNAMIC BIOMECHANIC IMPACT TESTING

85R002-86R015, one hour after impact for tests 78A232-78A249, and eight hours after impact for tests 79A251-79A260, a 5 ml dose of Uthol (concentrated, unpure sodium pentobarbital) was injected via the hind leg intravenous catheter to euthanize the anesthetized laboratory animal subjects. A bilateral pneumothorax was performed to insure termination.

Surgical Instrumentation

Nine-Accelerometer Head Plate - To install the nine-accelerometer plate the scalp was removed from the frontal bone over the orbital ridges. Five metal self-tapping screws were attached firmly to the skull through small pilot holes drilled into the orbital ridges and into the dental plate above the canine teeth. Quick-setting acrylic plastic was molded around each of the screws and the nine-accelerometer plate mount, embedding the mount in the plastic. After the acrylic set, the plate was rigidly attached to the skull. The orientation of the plate in this position is shown in Figure 4.

Since it was necessary to determine the instrumentation frame's exact location and orientation in relation to the anatomical frame, a three-dimensional x-ray technique was developed. Two orthogonal radiographs of the instrumented head were taken. The procedure required the identification of four anatomical landmarks (i.e. two superior edges of the auditory meati and two infraorbital notches) with four distinguishable lead pellets, plus the identification of four lead pellets inlaid in the plate which defined the instrumentation frame. The targets were digitized and related using a mathematical algorithm which resolved the instrumentation center of gravity into the translations and rotations of the anatomical center of gravity.

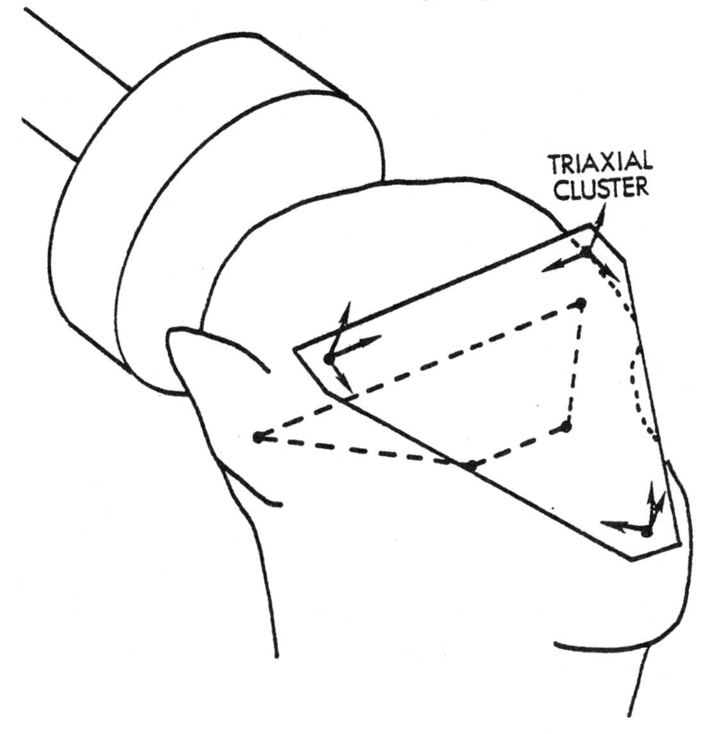

FIGURE 4: ORIENTATION OF NINE-ACCELEROMETER PLATE

Epidural Pressure Couplings - After the scalp was removed from a small area over the right and left sides of the frontal bone, the surface was sanded. Next, two 0.3 cm diameter holes were drilled using a Stryker bone coring tool, the bone was tapped, and the *dura mater* was punctured under each hole. Aluminum couplings for the pressure transducers were screwed into the tapped holes (See Figure 5). Dental acrylic was applied around the base of the couplings to secure them.

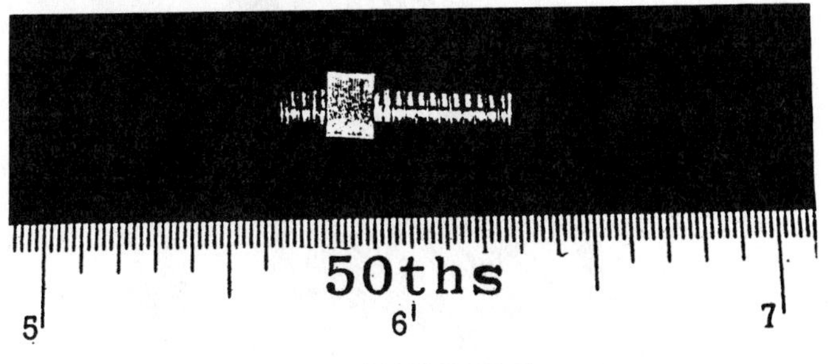

FIGURE 5:

EPIDURAL PRESSURE COUPLER

Strain Gauge - A Micro-Measurement Type EA-13-015Y-120 strain gauge rosette was bonded with M-bond 200 cyanoacrylate adhesive to the surface of the frontal bone, anterior to the bregma over the sagittal sinus. After the leads were soldered, the rosette and leads were covered and sealed with M-coat D paint.

Cerebrospinal Repressurization - A small hole was cored in the L6 lumbar vertebra and a Foley catheter was inserted under the dura of the spinal cord such that the balloon of the catheter reached mid-thorax level (See Figure 6). The point at which the catheter passed through the lamina of the sixth lumbar vertebra was sealed with plastic acrylic. To check fluid flow through the ventricles, saline was injected through the Foley catheter until it rose to the top of the couplings. The couplings were capped until the pressure transducers were attached in the impact laboratory. Dow Corning dielectric gel (silicon fluid) was injected into each coupling device to act as a securing medium. The pressure transducer was then inserted and secured at the proper depth. Then a setup radiograph was made of the head.

Vascular Repressurization - The common carotid artery was located at a point in the neck and an incision was made. String was looped around the common carotid and it was cut lengthwise. A polyethylene tube was inserted into the ascending common carotid, and the descending common carotid was ligated. The opposite common carotid was located and similarly cut. Vascular flow was checked. Then a Millar catheter tip pressure transducer was inserted into the ascending common carotid to the point of its branching with the external carotid, secured with tape and sewn in place (See Figure 7).

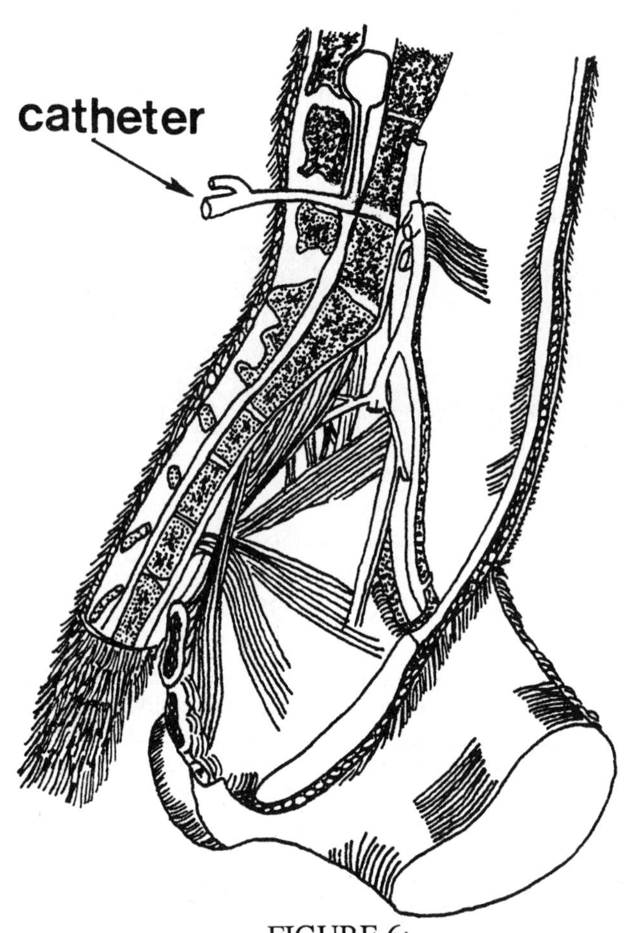

FIGURE 6:
RHESUS CEREBROSPINAL
REPRESSURIZATION CATHETER

METHOD OF ANALYSIS - The techniques used to analyze the results are outlined below. Additional information can be found elsewhere (2,3,8,12,14,16,24).

Tangential Acceleration - As the head moves through space, any point on the head generates a path in space. In head injury research interest is in the description of the path of the anatomical center and in events which occur as it moves. A very effective tool for analyzing the motion of the anatomical center as it moves along a curved path (a one-dimensional manifold) in space is the concept of a moving frame (2,3,19,24). The Frenet-Serret frame (3), which contains information about the motion embedded in the frame field, is one such moving frame. The tangential acceleration is a quantity which can be derived from the Frenet-Serret frame. The tangential acceleration is the rate of change of the absolute value of the velocity of a given moving point. It is co-directional with the velocity. The tangential acceleration is a very useful tool when working with one-dimensional transfer functions. A short summary of this method is included elsewhere (14).

Transfer Function Analysis - With blunt impacts, the relationship between a transducer time-history at a given point and the transducer time-history of another given point of a system can be expressed in the frequency domain through the use of a transfer function. A Fast Fourier Transformation of simultaneously monitored transducer time-histories can be used to obtain the frequency response functions of the impact force and accelerations of remote points. Once obtained, a transfer function of the form:

$$(Z)(i\omega) = (\omega)\ \mathbf{F}[F(t)]/\mathbf{F}[A_t(t)]$$

can be calculated from the transformed quantities where ω is the given frequency, and $\mathbf{F}[F(t)]$ and $\mathbf{F}[A_t(t)]$ are the Fourier transforms of the impact force and tangential acceleration of the point of interest, at the given frequency. This particular transfer function is closely related to a mechanical transfer impedance (8). Mechanical transfer impedance is defined as the ratio between the harmonic driving force and the corresponding velocity of the point of interest, and is a complex-valued function which can be described simply by its magnitude and phase angle. In addition to a transfer function relating force to velocity, a transfer function can be generated which relates the response of one point in the system to another point in the system, the relationship being expressed in the frequency domain. Analogous to mechanical impedance, a Fast Fourier Transformation of simultaneously monitored transducer time-histories from any two points in the system can be used to obtain the frequency response functions relating those two points. In the case of an acceleration and a pressure, such as tangential acceleration and epidural pressure, a transformation of the form:

$$(X)(i\omega) = \mathbf{F}[A_t(t)]/\mathbf{F}[p(t)]$$

can be calculated from the transformed quantities, where ω is the given frequency, and $F[A_t(t)]$ and $F[p(t)]$ are the Fourier transforms of the tangential acceleration time-history and the epidural pressure time-history, respectively. The data will be presented in terms of a transfer function generated for tangential acceleration divided by epidural pressure.

Transfer Function Spectral Coherence - The coherence function $Cxy^2(\omega)$ is not strictly speaking a transfer function, but instead is a measure of the quality of a given transfer function at a given frequency.

$$Cxy^2(\omega) = |Gxy(\omega)|^2 / (Gxx(\omega)Gyy(\omega))$$

where $Gxx(\omega)$ and $Gyy(\omega)$ are the power spectral densities of the two signals, respectively. Power Spectral Density is a Fourier Transform of each signal's auto-correlation. $|Gxy(\omega)|^2$ is the Cross-Spectral Density function squared. Cross-Spectral Density is the Fourier Transform of the cross-correlation of the two signals and ω at the given frequency. By definition, $0 \leq Cxy^2(\omega) \leq 1$. Values of $Cxy^2(\omega)$ near 1 indicate that the two signals may be considered causally connected at that frequency. Values sig-

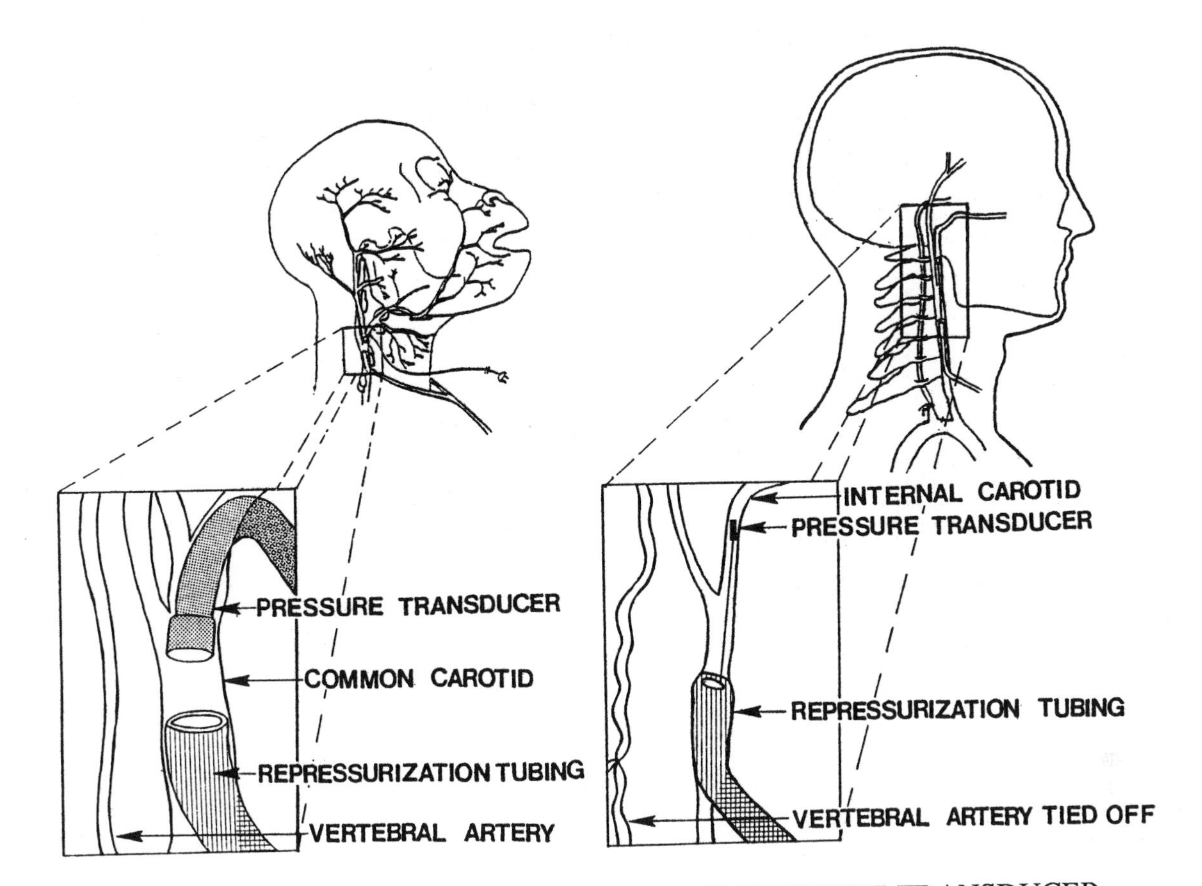

FIGURE 7: LOCATION OF VASCULAR PRESSURE TRANSDUCER IN RHESUS AND CADAVER MODELS

nificantly below 1 at a given frequency indicate that the transfer function at that frequency cannot accurately be determined. In the case of an input-output relationship, values of $Cxy^2(\omega)$ less than 1 indicate that the output is not attributable to the input and is perhaps due to extraneous noise. The coherence function in the frequency domain is analogous to the correlation coefficient in the time domain. The coherence function can be used to determine the useful range of the data in the frequency domain.

Correlation Functions - Correlation function statistical measures (8) were also used in the analysis of the data to describe some of the fundamental properties of a time-history, such as linear acceleration or epidural pressure as shown in the appendix.

Auto-correlation Function. This measure is the correlation between two points on a time-history and is a measure of the dependence of the amplitude at time t_1 on the amplitude at t_2 and is defined as:

$$R_{xx}(t_1,t_2)=\int_{-\infty}^{\infty}\int_{-\infty}^{\infty} x_1 \cdot x_2 \cdot p(x_1,x_2,t_1,t_2)dx_1 \cdot dx_2$$

where x_1,x_2 are the amplitudes of the time-history and $p(x_1,x_2,t_1,t_2)$ is the joint probability density.

Cross-correlation Function. This is a measure of how predictable, on the average, a signal (transducer time-history) at any particular moment in time is from a second signal at any other particular moment in time and is defined as:

$$R_{xy}(t_x,t_y)=\int_{-\infty}^{\infty}\int_{-\infty}^{\infty} x \cdot y \cdot p(x,y,t_x,t_y)dx \cdot dy$$

where x,y are the amplitudes of the two time-histories, respectively, and $p(x,y,t_x,t_y)$ is the cross probability density. Cross-correlation can be used to obtain useful information about the relationship between two different time-histories. For example, the cross-correlation between the acceleration measurements at two different points of a material body may be determined for the purpose of studying the propagation of differential motion through the material body. Cross-correlation functions are not restricted to correlation of parameters with the same physical units; for example, we determined the cross-correlation between the linear acceleration and the epidural pressure for the three Rhesus models. Cross-correlations, in general, do not commute, i.e. tangential acceleration cross epidural pressure does not equal epidural pressure cross tangential acceleration.

RESULTS

Table 1 lists the initial test conditions. Table 2 lists the Rhesus biometric measurements. Table 3 summarizes the head impacts. Table 4 reports the injuries/damages for which the preliminary analysis of the kinematic response was reported earlier (12) and Table 5 reports on the injuries/damages for which the kinematic response is being reported here. A gross pathological inspection was carried

Table 1. Rhesus Head Impact
Initial Test Conditions

Test No.	Subject Condition	Repressurization Condition	Subject Positioning Center of Gravity With Respect to Occipital Impact	Neck Condition	Velocity (m/s)
78A232	postmortem	unrepressurized	below	unstretched	11.5
78A234	postmortem	unrepressurized	below	unstretched	13.5
78A236	postmortem	unrepressurized	at	unstretched	12.5
78A238	postmortem	unrepressurized	above	unstretched	12.5
78A239	anesthetized	--	at	unstretched	12.5
78A241	anesthetized	--	above	unstretched	12.5
79A249	anesthetized	--	at	unstretched	15.4
79A251	anesthetized	--	at	unstretched	13.0
79A253	anesthetized	--	at	unstretched	14.8
79A256	postmortem	unrepressurized	at	unstretched	15.4
79A258	postmortem	unrepressurized	at	unstretched	14.5
79A260	postmortem	unrepressurized	at	unstretched	16.0
85R002	postmortem	repressurized	above	unstretched	11.0
85R005	postmortem	repressurized	above	unstretched	12.2
85R006	postmortem	repressurized	above	unstretched	11.9
85R008	postmortem	repressurized	above	unstretched	11.0
86R010	postmortem	unrepressurized	above	unstretched	11.0
86R011	anesthetized	--	above	stretched	10.0
86R012	anesthetized	--	above	stretched	12.2
86R013	anesthetized	--	above	stretched	11.9
86R014	anesthetized	--	above	unstretched	11.9
86R015	anesthetized	--	above	unstretched	12.1

Table 2. Rhesus Biometric Measurements*

Test No.	Height (Sitting) (cm)	Mass (kg)	Max. Skull Length (cm)	Max. Skull Breadth (cm)	Ave. Skull Thickness (cm)	Head Muscle Mass (g)	Head Mass (g)	Brain Mass (g)	Brain Volume (ml)
78A232	49.1	6.9	7.4	6.1	0.47	94	276	70	N/A
78A234	53.6	9.5	7.4	5.6	0.28	135	907	105	95
78A236	53.4	8.2	6.6	5.6	0.24	149	510	100	100
78A238	53.9	8.2	6.9	5.6	0.23	130	540	115	90
78A239	48.8	8.2	6.4	5.3	0.22	80	510	85	75
78A241	47.0	10.6	8.4	6.4	0.43	190	750	125	115
79A249	55.0	5.8	6.5	5.5	0.40	128	454	85	80
79A251	66.1	7.7	8.0	6.0	0.45	113	726	113	105
79A253	58.5	5.0	6.8	5.5	0.25	99	610	92	85
79A256	61.3	6.9	7.0	6.0	0.10	106	823	99	80
79A258	62.0	7.6	7.0	5.5	0.26	85	595	110	90
79A260	61.8	7.1	7.3	6.0	0.41	113	957	85	90
85R002	71.0	5.0	6.2	5.8	0.38	116	475	71	75
85R005	72.3	9.1	7.5	6.0	0.38	287	N/A	99	100
85R006	67.3	8.4	7.2	6.0	0.31	191	481	99	90
85R008	65.0	6.8	N/A	N/A	0.25	156	511	N/A	70
86R010	70.0	7.3	7.2	5.5	.0.25	198	595	99	80
86R011	66.0	4.9	6.5	5.0	0.41	142	425	99	80
86R012	68.5	6.9	6.5	5.5	0.35	170	479	99	70
86R013	86.5	8.7	6.8	5.6	0.30	225	709	113	100
86R014	71.0	8.6	6.5	5.5	0.35	298	759	113	95
86R015	69.0	8.0	7.0	6.0	0.25	255	730	113	90

*All test subjects were Macaca mulatta except 78A232 which was Macaca assamenis.

Table 3. Rhesus Impact Test Summary

Test No.	Linear Acceleration Tangent m/s/s/	Resultant Acceleration m/s/s/	Resultant Angular Acceleration r/s/s	Resultant Angular Velocity r/s	Linear Velocity m/s	Epidural Pressure Kpa	Force N	Force Duration ms
78A232	7000	7000	22000	40	12	117	5800	3
78A234	8000	6400	38000	40	14	N/A	5700	4
78A236	7000	7000	28000	30	13	-9	5600	4
78A238	7100	8000	40000	60	13	-100,55	6600	4
78A239	8000	8000	41000	60	13	260	5000	5
78A241	8400	8000	54000	70	13	490	8700	3
79A249	14000	15000	70000	63	16	-70	7200	4
79A251	13000	14000	64000	75	13	200,-110	7500	4
79A253	10000	10000	40000	75	15	-101	5400	5
79A256	11000	11000	70000	84	16	N/A	6000	10
79A258	10000	12000	70000	75	15	-35,40	5700	5
79A260	13000	13000	60000	80	16	140,-100	7100	4
85R002	5300	5300	7900	30	12	270	4000	4
85R005	20000	18000	28000	45	15	-100,-105	N/A	N/A
85R006	11000	10000	65000	75	13	375	7500	5
85R008	11000	11000	60000	70	16	-98,-100	5800	5
86R010	7900	7900	16000	33	11	-375	7000	4
86R011	6000	5300	33000	24	13	11	N/A	N/A
86R012	18000	18000	110000	130	16	45	7000	4
86R013	9000	8800	80000	105	13	75	4900	5
86R014	12000	11000	120000	105	15	250	7000	4
86R015	13000	11000	120000	90	16	140,190	6500	4

Table 4. Rhesus Injuries/Damage Summary

Test No.	Gross Skull	Gross Brain	Gross Other
78A232	No injury	No injury	Epidural hematoma at C1, dura lacerated at C1
78A234	No injury	No injury	Epidural hematoma at C1, torn muscle at base of occiput
78A236	No injury	No injury	No abnormality or injury
78A238	No injury	No injury	Epidural hematoma at C1 disk
78A239	No injury	No injury	1/4 cc blood in occiput from epidural hematoma at C1
78A241	No injury	No injury	Epidural hematoma at C1
79A249	Basilar fracture (ring)	3/4 cc subdural hematoma right frontal lobe (cerebrum) subarachnoid hemorrhage base of cerebellum, pons, and medulla	Epidural hematoma at C1, occipital muscles damaged
79251	Linear basilar fracture from foramen magnum to occipital contact point	No injury	Epidural hematoma at C1 and C2
79A253	Basilar fracture (quasi-ring), temporal fracture	Petechial lesion left frontal lobe (cerebrum) subarachnoid hemorrhage base, pons, medulla and vermis of cerebellum	Damaged neck ligaments
79A256	Basilar fracture (petrous to petrous). Connected linear fracture to parietal bone*	Dura torn along sagittal sinus, emaciated tissue cerebellum, medulla	Lacerated spinal cord
79A258	Basilar fracture (quasi-ring)	Lacerated medulla	Epidural hematoma at C1
79A260	Basilar fracture (right petrous)	Subdural hematoma along sagittal sinus	Epidural hematoma at C1

*Abnormality - Test subject had very thin skull.

Table 5. Rhesus Injuries/Damage Summary

Test No.	Gross Skull	Gross Brain	Gross Other
85R002	No injury	No injury	No injury
85R005	Basilar skull fracture	Subarachnoid hemorrhage at medulla and brain stem	No injury
85R006	No injury	No injury	No injury
85R008	Basilar ring fracture*	Subarachnoid hemorrhage at medulla and brain stem	No injury
86R010	Basilar ring fracture	No injury	No injury
86R011	No injury	No injury	Subdural hemorrhage of spinal cord at base of brain
86R012	No injury	Subdural hemorrhage left frontal lobe	No injury
86R013	No injury	Subdural hemorrhage left frontal lobe	No injury
86R014	No injury	No injury	Subdural hemorrhage of spinal cord at base of brain
86R015	Linear fracture of superior right petrous Linear fracture of left occipital, Linear fracture of right temporal	No injury	Bilateral hemorrhage to side of spinal cord at base of medulla

*Abnormality - Test subject had a thin skull

out for each test subject. Selected time-histories in the appendix are examples of important kinematic factors associated with the observed head impact response.

DISCUSSION

The results of a series of head impact experiments conducted over the past nine years at UMTRI using anesthetized and postmortem Rhesus are presented. The tests entail different initial conditions and are compared to tests of another human surrogate (i.e. the repressurized human cadaver) reported in previous Stapp Conferences (12,14 and 16). Frame-independent variables and vectors were used to compare these different tests. The significant frame-independent vector tangential acceleration will be presented in detail throughout this discussion.

It would be preferable to run a limited set of experiments and generalize the results. However, because of the geometry of the skull-brain-spinal cord, the possible different initial conditions, the complex interaction of the skull with the brain, and of the different injury modes that can result from blunt head impact, *the results presented here apply only to the test conditions in which the experiments were run and may not apply to all situations*. The features of the data, discussed in this section in abbreviated form, represent trends that are felt to be important factors in head injury research.

The experiments can be broken into three significant classes: repressurized-postmortem subjects, unrepressurized-postmortem subjects, and anesthetized live subjects. Each class can be divided into two groups, skull fracture and no skull fracture. In addition, the anesthetized test subjects can be divided into two groups, neck-stretched and neck-unstretched. Although most of the subjects were positioned so that the estimated center of mass of the head was above the contact point of the head with a line running horizontally through the center of the impactor, a few were positioned "at" or "below." This means that, in general, there were two-four subjects in each grouping. Therefore, when comparing between groupings it is necessary to use only the most robust aspects of the data.

In head impact experiments it is common to find variations in the input force time-histories as well as variations in the acceleration (both linear and angular) responses. The differences and similarities in the force time-histories as well as in the linear and angular accelerations for the impact conditions used in these tests have been described elsewhere (12,16). The two initial test conditions (neck-stretched positioning and repressurization), absent in the tests reported earlier (12,16) and part of the entire sample being reported here did not affect significantly the response in terms of force and angular/linear acceleration. for other similar initial conditions, the repressurized-postmortem and the neck-stretched anesthetized Rhesus subjects were not distinguishable in response from the class, anesthetized Rhesus. For a given set of initial conditions, these variations can be addressed when the sample size is large for each class and group and the statistics are appropriate. However, for the small sample size that each of the different classes of this study entail, it is helpful to normalize the data through the use of transfer functions and to address only the most robust aspects of the differences. In general, the force time-histories were similar in terms of wave shape and magnitude for the no-skull-fracture case. A general idea of the range of peak forces and durations can be obtained by referring to Table 3 which shows an average peak force of 6300 N with a standard deviation of 1100 N and a duration of 4.5 ms with a standard deviation of 1.5 ms. Because of the small sample size and the variations among the tests, there was no significant difference in response in terms of peak values for linear/angular velocity or linear/angular acceleration between the different group of tests for the no-skull-fracture case.

Impact Response Definition - Using the UMTRI nine-accelerometer array, it is possible to record three-dimensional six-degrees-of-freedom motion of an area of the skull in which the accelerometers are located. Since the nine-accelerometer motion analysis assumes rigid body motion and the skull and brain deform during impact, the output from this three-dimensional motion experimental technique can only be used to estimate the motion of the head. Thus, head impact response can be defined as a continuum of "events" characterized by the path traced by the motion of the "estimated anatomical center," by all the vectors defined on that path, and by the changes of the accelerated frame fields. Physically, head impact response is interpreted as the response of a material body (i.e. the nine-accelerometer array) in contact with other material bodies. The curve and the vectors generated as the "estimated anatomical center" moves in time are, therefore, a result of the interactions of the skull-mount area with other material bodies.

Three-Dimensional Motion Technique - When the head receives an impact, several events occur: 1) stress waves are propagated from the impact site, 2) the skull starts to deform, and 3) the skull begins to move due to the impact, transmitting energy to the brain via the *dura mater*. Eventually, the stress waves are dissipated, the deformation of the skull recovers partially or fully on removal of the impact loads, and the acceleration of the skull continues primarily from forces generated by the brain and neck. If the differential skull motion is "severe," due to either sufficient energy in the high frequency components of the force time-history or to sufficient peak force, the stresses at some point in the skull may exceed the failure strength of the bone, thereby producing fracture. Dynamic human cadaver impact experiments have shown that the motion of the entire skull as a rigid body as estimated by the nine-accelerometer array depends on the degree of skull deformation (14). The previous study showed that cadaver skull deformation, with or without skull fracture, correlated well with injury. In addition, whenever skull deformation was present, it significantly affected the linear and angular acceleration response. For the Rhesus model, however, the accelerometers are located on the orbital ridges and dental palate, sufficiently far from the impact site so that the local perturbations are small even when the skull fractures (12,14). If the skull near the nine-accelerometer array had been significantly perturbed by skull fracture, then the mechanical impedance of the skull-fracture tests should have been different from that of the no-skull-fracture tests. This, in general, was not the case. The mechanical impedance transfer functions in the appendix show that the mechanical impedance for both the skull-fracture and the no-skull-fracture tests were the same. Based upon human cadaver data, that result is different from what would be expected for the live human. It is believed that the mechanical impedance observed for the Rhesus tests was a result of the different skull geometry and soft-tissue distribution between the Rhesus and human cadaver models. The observation implies that the effect of skull deformation on the nine-accelerometer recording is different for the Rhesus as compared to the human cadaver models. Because skull deformation can be an important aspect of mechanisms of focal injury and kinematic response (14), and because the experimental techniques used to simulate the three-dimensional motion of the head of the human cadaver and lower primate models produce significantly different results with regard to skull deformation, this difference implies that the scaling of impact and injury response between human cadaver and Rhesus models for some types of impacts may not be possible. As stated for the no-skull-fracture case, there was no significant difference in terms of peak values for linear/angular velocity and linear/angular acceleration; the no-skull-fracture and skull-fracture cases had the same mechanical impedance which further supports that finding significant difference in terms of force and linear acceleration was difficult for this sample.

Vascular Repressurization Technique - The results presented here regarding the repressurized-postmortem Rhesus are potentially a function of the repressurization technique used. Different repressurization techniques may give different results. The repressurization techniques developed for Rhesus subjects are similar to those used for repressurized human cadavers (12,14,16). However, because of the smaller size of the vessels in the vascular system of the Rhesus as compared to that of the human cadaver, somewhat different procedures had to be used (See Figure 7). The Rhesus vertebral artery was not ligated. In the Rhesus model the pressure transducer in the external carotid artery restricts the flow of the internal carotid artery at the measurement site. The pressure transducer used to monitor the carotid pressures, and, therefore, used to determine the vascular pressure in the brain, was placed in different locations in the two species. In the cadaver model the transducer was in the internal carotid close to the entrance of the carotid into the brain. In the Rhesus model the transducer was in the external carotid distal to the brain. The precise difference this makes between the results is unknown. The initial repressurization levels may not be comparable. In addition, it was more difficult to remove the air in the Rhesus skull-brain area than in the cadaver model; thus, a greater proportion of the Rhesus brain cavity may have contained air than that of the cadaver brain cavity.

The No Skull Fracture Impacts - Comparisons of the three classes of no-fracture head impacts for experiments, in which the initial conditions were impact velocity of 11.5-12.5 m/s, neck-unstretched positioning, impact below the head center of mass, 2.5 cm Ensolite padding, and the pressure transducer located at the front of the skull, imply that:

1. The unrepressurized-postmortem Rhesus is significantly different than the anesthetized Rhesus as shown by the transfer function relationships of tangential acceleration divided by epidural pressure.

2. In terms of peak pressure (largest absolute value), the repressurized-postmortem Rhesus' response is more similar to the anesthetized Rhesus' than the unrepressurized-postmortem Rhesus'.

3. The impact responses of the anesthetized, unrepressurized-postmortem and repressurized-postmortem Rhesus are all different as shown by the transfer function relationships of tangential acceleration divided by epidural pressure.

Inspection of the peak pressures presented in Table 3 shows that the peak pressures of the unrepressurized-postmortem Rhesus were between 95 and 150 kPa, and those for the anesthetized and repressurized-postmortem Rhesus were between 250 and 500 kPa. Although there were a limited number of tests in each category, the results were consistent enough to show that, in terms of peak pressures, the anesthetized and the repressurized-postmortem subjects were similar. In addition, for some of the tests, the peak pressures for the unrepressurized-postmortem Rhesus were opposite in sign to those of the anesthetized Rhesus and the repressurized-postmortem Rhesus (see Test 78A236 in the appendix). Figure 8 represents the transfer function corridors for the tangential acceleration divided by epidural pressure for the no-fracture class of Rhesus subjects. The key defines the corridors for the anesthetized, the unrepressurized-postmortem, and the repressurized-postmortem Rhesus subjects.

Successful understanding of the performance of a given system depends on being able to predict accurately the output response for a given input. In the case of a time invariant linear system, a transfer function can be generated that characterizes the system and is independent of the input to the system, i.e. it is an invariant of the system. Comparisons of outputs for different inputs will show some variations, but transfer functions that characterize the system will not. In general, systems in the real world are inherently non-linear and that is probably also true for the head when struck by a blunt impactor. We can, however, assert that a linear system is a valid approximation of a non-linear one over a limited range such as the range seen in these experiments. For example, if a transfer function generated for tangential acceleration divided by epidural pressure for an impact to an unrepressurized Rhesus subject is compared to another transfer function of the same parameters for an anesthetized Rhesus subject and is found to be similar, then the transfer functions correlated well and represent the same biological system; when they are found to be different, then the transfer functions do not correlate well and are said to be representing different biological systems. For the tests being reported here the force time-histories are similar enough so that a valid linear range can be assumed.

For the unrepressurized-postmortem versus the anesthetized Rhesus, the magnitude differed by at least a

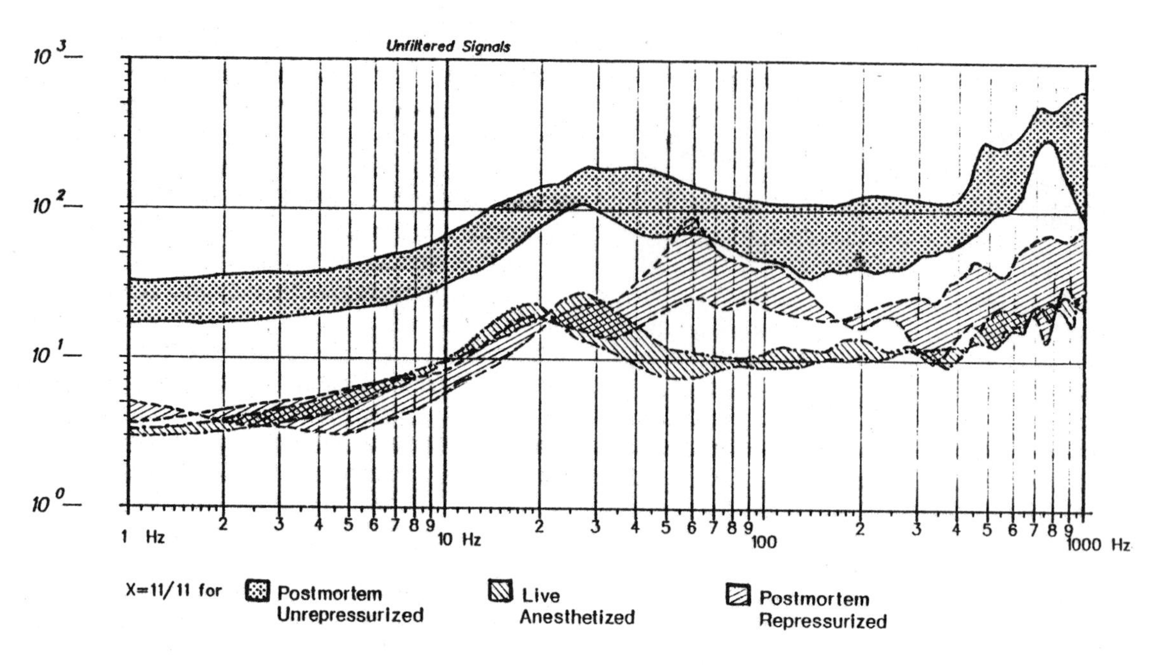

FIGURE 8: TRANSFER FUNCTION CORRIDORS TANGENTIAL ACCELERATION/EPIDURAL PRESSURE NO-FRACTURE RHESUS

factor of 3 at all frequencies, indicating that there is more energy at any given frequency in the output (pressure) for a given input (tangential acceleration) for the anesthetized subject than for the unrepressurized-postmortem subject. However, it is also evident that when comparing the transfer function corridors between the repressurized-postmortem Rhesus and the anesthetized Rhesus, the differences suggested by the comparison of peak pressure are not entirely substantiated. For example, although the magnitude of the transfer function of the repressurized-postmortem Rhesus is similar to that of the anesthetized Rhesus up to about 35 Hz, above that frequency the transfer function magnitude indicates that the repressurized-postmortem, the unrepressurized-postmortem, and the anesthetized Rhesus all differ to some degree. In Figure 8 the greatest energy in the pressure for a given acceleration is in the anesthetized Rhesus, while that for the repressurized-postmortem Rhesus has its magnitude between those of the anesthetized and the unrepressurized-postmortem subjects.

It can be concluded that even though the repressurized-postmortem Rhesus was more similar than the unrepressurized-postmortem Rhesus to the anesthetized Rhesus in response as shown by the transfer function relationships, there were still detectable differences between them. **Note:** It was also realized that this difference only applied to experiments of those initial conditions and would not apply in a general sense. Potentially it would be possible to choose a different set of initial conditions and run a series of blunt head impact experiments to the unrepressurized-postmortem and anesthetized Rhesus and show that there is no difference in the pressure response for similar force and acceleration time-histories between the two models. In general it cannot be stated that there is a difference in pressure response between the unrepressurized-postmortem and anesthetized Rhesus for similar force and acceleration responses, but there exists an initial condition in which such a difference will manifest itself. By comparing blunt head impact experiments conducted (12) on both unrepressurized-postmortem and anesthetized Rhesus plus the additional subjects being reported here, and by carefully choosing the initial position of the head and neck as well as the impact velocity and padding on the impactor surface, it can be shown that in terms of epidural pressure there is a significant difference in the pressure response of the two types of surrogates for similar force and acceleration time-histories.

Transfer functions are complex-valued, including both magnitude and phase. The phase relationship was evaluated with the cross-correlation function. The analysis obtained from the cross-correlation function between the tangential acceleration and the epidural pressure shows that: 1) Effectively there was no phase lag between the tangential acceleration and the epidural pressure for the anesthetized Rhesus (See Figure 9 which illustrates the maximum positive cross-correlation at zero lag and the appendix for other tests); 2) An effective phase lag between the tangential acceleration and the epidural pressure of one or two milliseconds for repressurized-postmortem Rhesus was observed (See Figure 10 which illustrates the maximum positive cross-correlation at one to two milliseconds and the appendix for other tests); and 3) The greatest phase lags between tangential acceleration and the epidural pressure were observed for unrepressurized Rhesus (See Figure 11 which illustrates the maximum negative cross-correlation for these tests and the appendix for other tests). The anesthetized and the repressurized-postmortem Rhesus had more similar phase lags between tangential acceleration and epidural pressure than the unrepressurized- and repressurized-postmortem Rhesus. Therefore, repressurization seemed to make the postmortem subject a better model of the anesthetized subject. However, it seems that the

FIGURE 9: ANESTHETIZED RHESUS, NO SKULL FRACTURE, NECK UNSTRETCHED

Cross– and Auto– Correlations

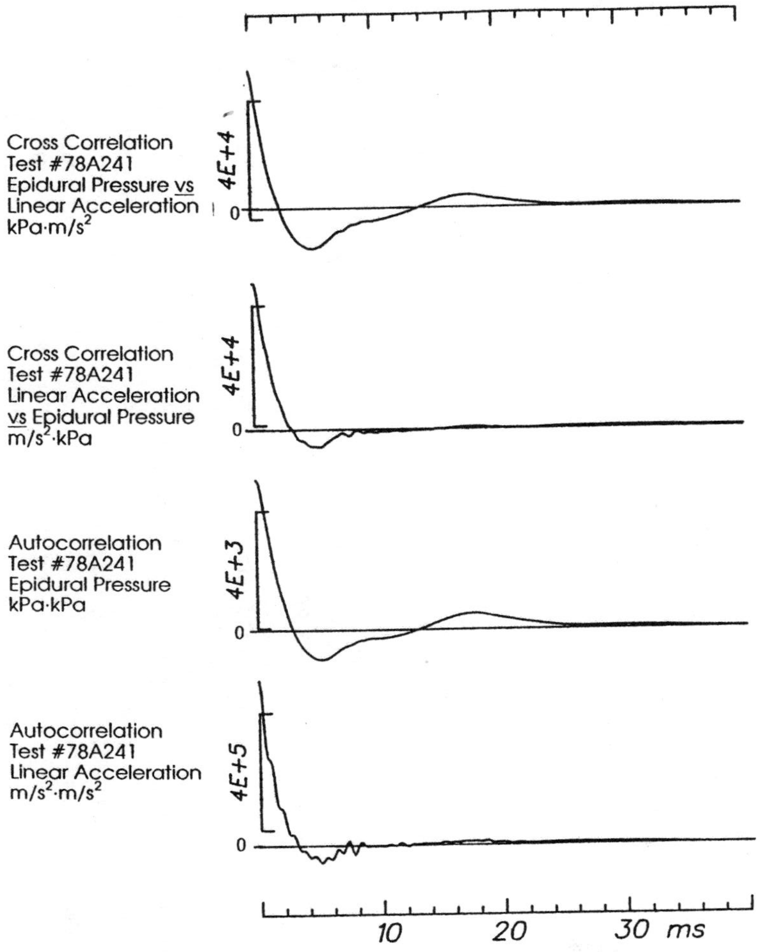

HEAD CENTER OF GRAVITY WITH RESPECT TO LINE OF IMPACT = ABOVE

FIGURE 10: REPRESSURIZED POSTMORTEM RHESUS, NO SKULL FRACTURE, NECK UNSTRETCHED

Auto– and Cross– Correlations

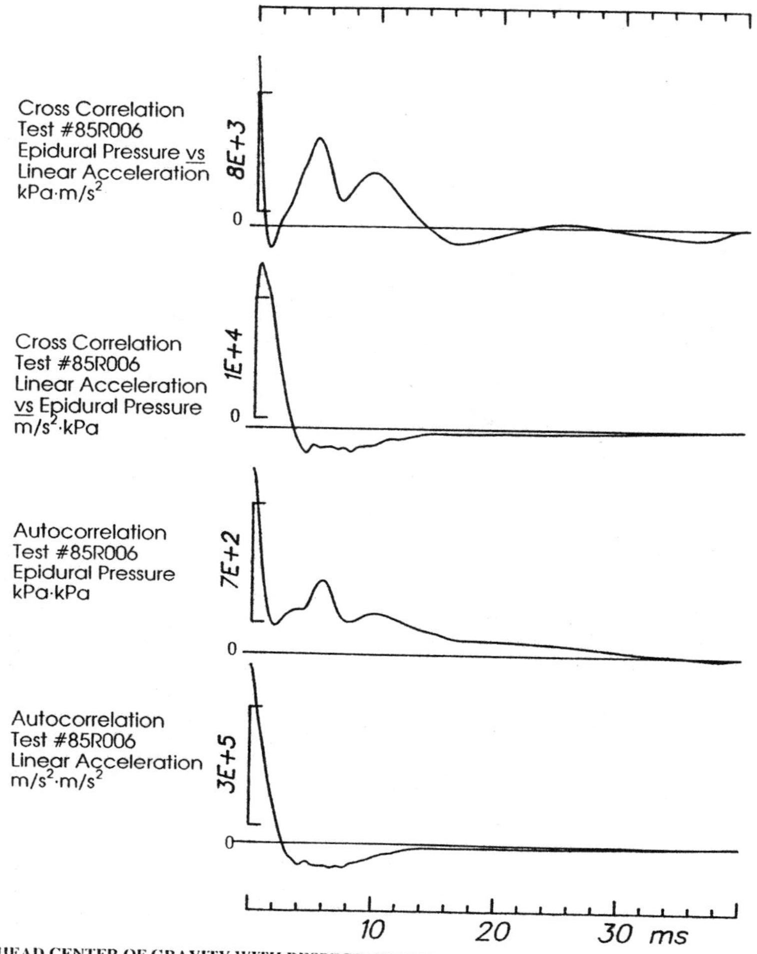

HEAD CENTER OF GRAVITY WITH RESPECT TO LINE OF IMPACT = ABOVE

FIGURE 11: UNREPRESSURIZED POSTMORTEM RHESUS, NO SKULL FRACTURE, NECK UNSTRETCHED

Cross– and Auto– Correlations

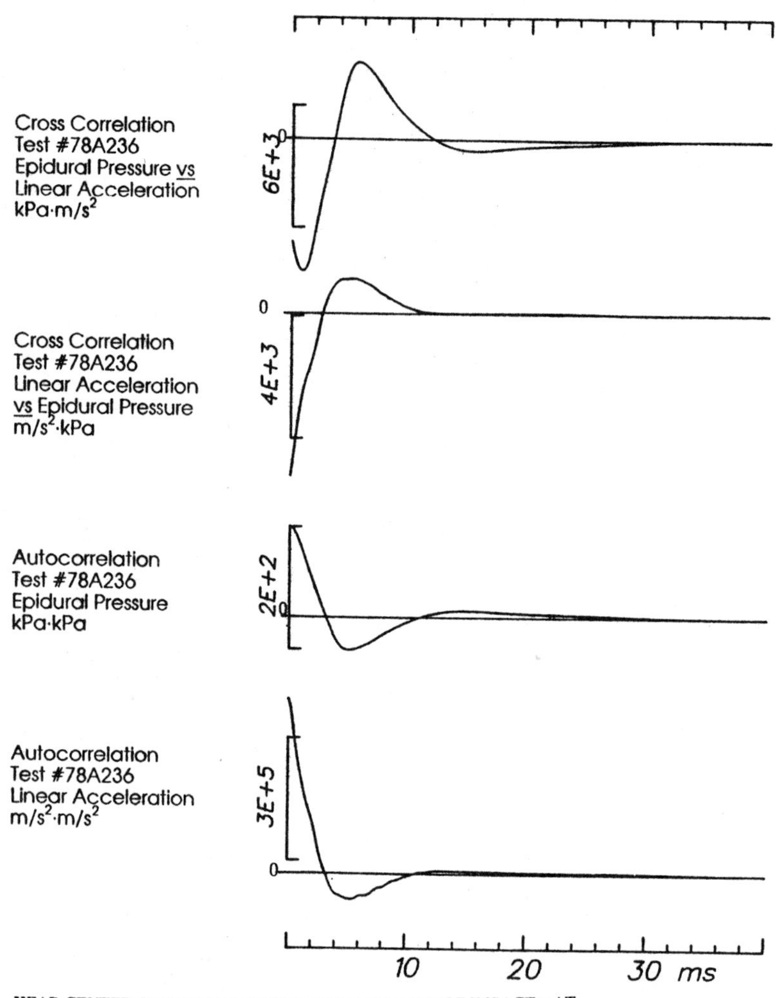

HEAD CENTER OF GRAVITY WITH RESPECT TO LINE OF IMPACT = AT

postmortem Rhesus brain, whether repressurized or unrepressurized, was not as "stiff" as that of an anesthetized subject. This may have been due to the postmortem degradation of the brain tissues, incomplete repressurization, or the repressurization process not removing all of the air from the skull-brain area.

The results reported elsewhere (14) indicated that when basal skull fracture occurred in Rhesus subjects, subarachnoid hemorrhaging was common in the *medulla-pons* area. In addition, this injury did not occur in the unrepressurized-postmortem Rhesus. For the repressurized-postmortem Rhesus, focal injuries similar to those of the anesthetized subject were observed. The implications were that in terms of injury response *for this type of injury mode,* the repressurized-postmortem subject was similar to the anesthetized subject. Caution must be exercised in interpreting these results because one injury mode for a very particular type of impact was involved and the sample size was small. *The results may not occur for other injury modes or impact conditions.*

Head-Neck Positioning Parameter - The results reported elsewhere (16) indicated that the interaction of the spinal cord and the *foramen magnum* with the rest of the brain - cerebrospinal fluid system was a critical factor in certain types of head impact response. In an attempt to determine the effect of the interaction of the head with the cervical spine, a series of impacts were run with the neck stretched. Figure 3 is an x-ray of the two types of initial neck conditions. One shows the neck-stretched configuration, and the other the neck-unstretched configuration. Although the impedance values between force and tangential acceleration as well as the time-histories of the linear and angular accelerations for the two neck conditions were similar (See appendix), the transfer functions for the tangential acceleration divided by epidural pressures were not (Figure 12). This implies that the flow of material through the *foramen magnum* as well as the inertial

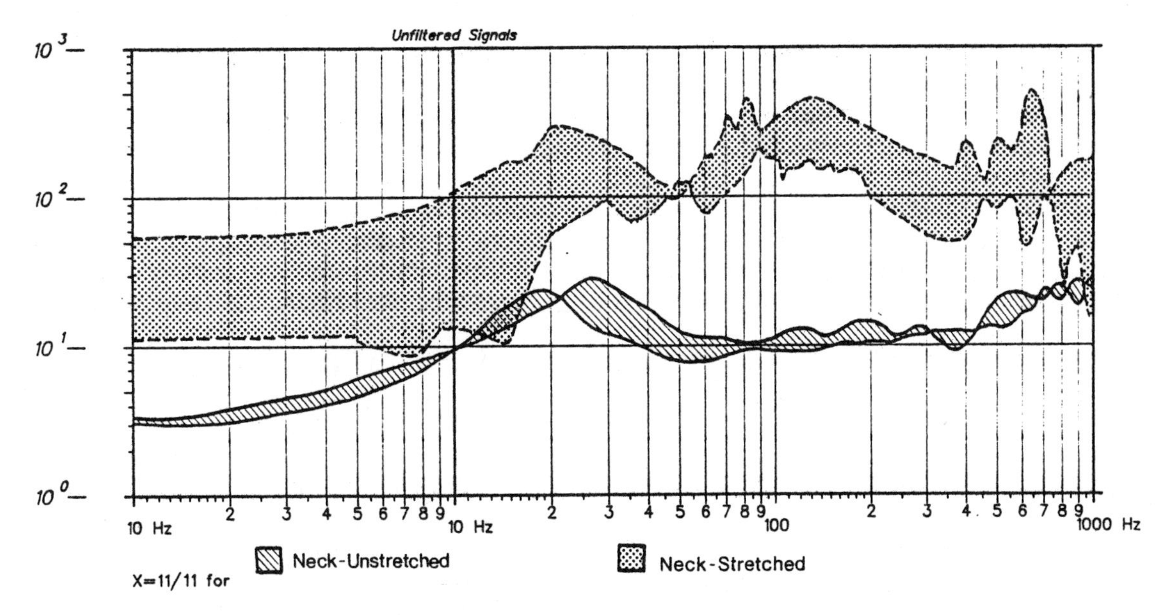

FIGURE 12: TRANSFER FUNCTION CORRIDORS TANGENTIAL ACCELERATION/EPIDURAL PRESSURE NECK-STRETCHED VS. NECK UNSTRETCHED RHESUS

properties of the cord may have been critical factors in determining the type of pressure and stress that developed in the brain during impact. In addition, the neck-stretched tests were the only ones in which subdural hemorrhaging was observed in the frontal lobes of the brain. Therefore, not only was the impact response different, in terms of transfer functions for tangential acceleration and epidural pressure, but the injury response also changed. The difference in initial positioning may have changed the hydrodynamic lubrication and the mechanical restriction associated with the skull-brain interface. In addition, the large pressures that developed during impact in the neck-unstretched anesthetized Rhesus subject and the inertial properties of the restricted flow of the spinal cord-cerebrospinal fluid through the *foramen magnum* may have helped to reduce the differential motion between the brain and the skull during impact. The observed injuries (subdural hematomas in the frontal lobes) in the neck-stretched case with angular acceleration similar to the neck-unstretched case imply that the interaction of the cervical cord and the *foramen magnum* may be critical for these types of injury, and perhaps is more important than the angular or linear velocity/acceleration parameters for determining head-impact tolerance levels.

The Skull-Deformation Impacts - To understand the effects of skull deformation on pressure it is first instructive to look at small deformation experiments, i.e. no skull fracture ones. In a manner similar to that of the repressurized human cadaver, skull deformation played a small part in the pressure response of the anesthetized Rhesus brain up to the level where the skull fractured (see also 14,16). The features of the data that indicated this for anesthetized subjects are:

1) The auto-correlations of the tangential acceleration and epidural pressures were similar to those of the cross-correlation between the two signals for the anesthetized no-skull fracture Rhesus. (See Figure 9 and the appendix for other tests. When the auto-correlation of two given signals have the same magnitude, lag, and wave shape as the cross-correlation of those two signals, then statistically the two signals correlate well. Therefore, it is reasonable to assume that when the correlations follow this pattern as the tangential acceleration ones shown in Figure 8, the pressure response is most likely due to the inertial loading of the brain.); and
2) The greatest ***absolute*** maximum value for severe skull-fracture subjects was generally a negative quantity (i.e. a negative cross-correlation near zero lag as shown in Figure 13. Other examples are given in the appendix), whereas for simple skull fracture subjects the tangential acceleration correlated better with epidural pressure than for severe skull fracture subjects. The negative correlation implies the pressure did not follow the inertial loading of the skull. It is reasonable to assume that because of the geometry of the skull, deformation of the skull as a result of impact to the occiput would increase the volume and decrease the pressure. The negative pressure and the negative cross-correlation seen in the severe skull fracture case indicates the pressure followed the increase in volume associated with skull deformation.

The result that skull deformation played a small part in the epidural pressure response of the Rhesus brain up to the level where the skull fractured has important implications in terms of head trauma modeling. If the head is viewed as a closed shell and the brain and its surrounding tissues and fluids as an incompressible, or nearly incompressible, vis-

FIGURE 13: ANESTHETIZED RHESUS, SKULL FRACTURE, NECK UNSTRETCHED

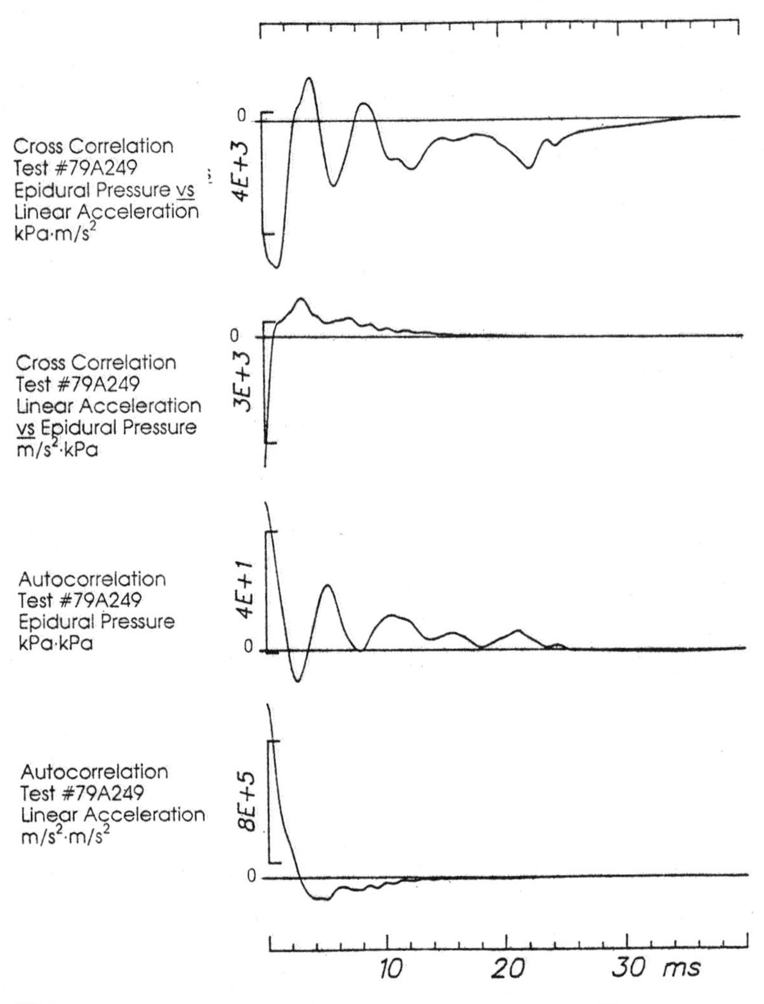

HEAD CENTER OF GRAVITY WITH RESPECT TO LINE OF IMPACT = ABOVE

coelastic material (i.e. as having a bulk modulus near that of water), then small changes in volume would be expected to produce larger pressure changes than those observed in the response of the anesthetized test subjects. In addition, one would expect negative pressures and a negative cross-correlation between tangential acceleration and epidural pressure for the no-skull fracture subjects, because the pressure volume change mechanism would dominate any of the inertial phenomena. The results imply three possibilities for head trauma modelers: 1) The *foramen magnum* was significantly large enough so that it needs to be included in modeling of head blunt impact response; 2) The *effective bulk modulus* may need to be viewed as significantly less than that of water or an equivalent system should be used (Rather than attempting to model all of the foramina and canals of the skull, an equivalent system model might be useful which can reproduce the same kinematics occurring in blunt head impacts by incorporating a "skull" having an evenly distributed set of holes.); and 3) or both.

Although the bulk modulus of the brain may be very close to water, the concept of an effective bulk modulus may be useful in describing the blunt impact response of the brain. The need for an effective bulk modulus concept results from the fact that although the foramen magnum is the largest entrance and exit port of the skull, other holes may have some small effect on the pressure response of the brain. Therefore, to estimate this effect as well as the effect of small amounts of dissolved gas in the brain-cerebrospinal fluid system, an effective bulk modulus may be needed for successful modeling. An alternative approach might be to leave the bulk modulus similar to that of water and use an equivalent system in which the skull is idealized as a structure with small openings.

Fluid Vaporization - Cavitation - In any situation where fluid is involved (e.g. water) and the tensile stress is reduced below the vapor pressure, vaporization occurs almost instantaneously (7,29,30). Therefore, in any blunt head impact test in which the epidural pressures are less than 0.13 kPa *absolute* fluid vaporization occurs.

During skull-fracture Rhesus tests, pressures slightly below one negative atmosphere gauge were observed

(See 14 and appendix). These were the only tests in which negative pressures of that magnitude were produced. Therefore, it seems that vaporization of the fluid near the pressure transducer was rapid enough to effectively act as a tension[2] (negative pressure) limiting function. This thermodynamic phenomenon would have affected significantly the stresses and strains (8) that were being produced in other parts of the brain during the severe impacts. The earlier observations indicated that when negative pressures were equal to or slightly less than one negative atmosphere for anesthetized Rhesus, no injuries were observed in the brain near the pressure transducer measuring sites (14). The same observation was made for all the anesthetized Rhesus in this study in which skull fractures produced epidural pressures of negative one atmosphere. Therefore, it seems that cavitation[3] was not a mechanism of injury for these experiments. Cavitation as a potential mechanism of injury for these tests would require the non-repetitive formation of bubbles or cavities within the fluid surrounding and within the brain tissue, displacing the tissue. We believe that cavities did not form within the fluid within the brain tissue, but rather occurred within the cerebrospinal fluid located at the dura-cerebrospinal fluid interface because no focal injuries or evidence of cavities were observed within the brain tissue.

Injury prediction based on stress or the differential motion between the skull and the brain should address the phenomenon of fluid vaporization at the dura-cerebrospinal fluid interface. It, therefore, may be essential to include this phenomenon in any modeling effort of head impact response.

CONCLUSIONS

This has been a limited study, using anesthetized, unrepressurized-postmortem, and repressurized-postmortem Rhesus, of a few important kinematic factors and injury modes associated with blunt impact to the head. Because of the complex nature of the head during impact, more work is necessary before these kinematic factors can be generalized to describe the responses of the skull, brain, cerebrospinal fluid, and the neck to blunt head impact. However, the following conclusions are drawn.

1. A set of initial conditions can be found such that in terms of peak pressure, the repressurized-postmortem and the anesthetized Rhesus responses are significantly different from that of the unrepressurized-postmortem Rhesus.

2. A set of initial conditions exists in which the repressurized-postmortem Rhesus, when compared to the unrepressurized-postmortem Rhesus, will be more similar in response to that of the anesthetized Rhesus (as shown by transfer function relationships for tangential acceleration divided by epidural pressure). The differences detected between the repressurized-postmortem and the anesthetized Rhesus may be important in terms of kinematic response.

3. The repressurized-postmortem Rhesus and the anesthetized Rhesus produce similar subarachnoid hemorrhaging in the *medulla-pons* area associated with skull fracture.

4. The initial position of the head-neck system is a critical factor associated with the brain-cerebrospinal system's response to head impact. For given linear and angular acceleration-time histories, the initial position of the head-neck system was, in general, successful in predicting the pressure and injury response. Unless the initial position of the head-neck system is included in the head tolerance criteria function (unlike those functions solely based on linear and angular accelerations, intracranial pressure, and skull strain), the function may not prove useful.

5. The thermodynamic response (fluid vaporization) of the head-brain system is an important consideration when the impact produced significant tension.

6. For the anesthetized Rhesus presented here and in the results presented elsewhere (12,16), no relationships could be found between injury and negative pressures equal to one atmosphere. This implies that cavitation does not seem to be a mechanism of injury in head impact response.

7. The above conclusions are important for both the mathematical modeling and the determination of the injury tolerance of the head. The following factors significantly affect the stress and possibly the strain in the brain as well as the interaction of the brain with the skull.

- The tension in the brain, limited by vaporiztion in "severe" impacts;
- The inertial properties of the cervical cord as well as the flow of material through the foramen magnum;
- The initial position of the head-neck system; and
- The repressurization of postmortem surrogates.

ACKNOWLEDGEMENTS

The results presented in this paper were obtained through a series of independently funded research programs conducted during the past nine years. The funding agencies were: The Motor Vehicle Manufacturers Association and the United States Department of Transportation, National Highway Traffic Safety Administration. The assistance in injury evaluation provided by C. J. D'Amato, P. W. Gikas, D. F. Huelke, and J. Hoff is gratefully acknowledged. The authors are indebted to Nabih Alem,

[2]Tensile Stress or Tension is the internal force that resists the action of external forces tending to increase the length of the body. Compression is the internal force that resists the actions of external forces tending to decrease the length of a body (4).

[3]When fluid pressure is reduced to the vapor pressure at the particular temperature, then the liquid will vaporize and a cavity forms. As the fluid flows into a region of higher pressure, the bubbles of vapor suddenly condense. This may produce very high dynamic pressure upon adjacent structures. When this action is continuous and has a high frequency, the material of the adjacent structure may be damaged (7. See also 22,30).

Joseph Benson, Miles Janicki, John Melvin, Richard Stalnaker, Paula Lux, Wendy Gould, Valerie Karime, Lilly Kim, and Shawn Cowper for their technical assistance. A special thanks is given to Jeff Marcus for his support.

In conducting the research on the laboratory Rhesus subjects described in this study, the investigators adhered to the "Guide for the Care and Use of Laboratory Animals," as prepared by the Committee on the Care and Use of Laboratory Animals of the Institute of Laboratory Animal Resources, National Research Council.

REFERENCES

1. Atluri, S., Kobayashi, A.S., and Cheng, J.S. 1975. Brain Tissue Fragility - A Finite Strain Analysis by Hybrid Finite-Element Method. Journal of Applied Mechanics 42:269-273. June.

2. Bishop, R.L. and Goldberg, S.I. 1968. Tensor Analysis on Manifolds, New York: MacMillan.

3. Cartan, E. 1946. Lecons sur la Geometrie des Espaces de Rieman, Second Edition, Paris: Gautheir Villars.

4. Esbach, O.W. and Souders, M., editors, 1975 ed. Handbook of Engineering Fundamentals. NY: John Wiley and Sons, especially p. 488.

5. Ewing, C.L., et al. 1975. The Effect of the Initial Position of the Head and Neck on the Dynamic Response of the Human Head and Neck to -Gx Impact Acceleration. 19th Stapp Car Crash Conf. Proc., Paper No. 751157.

6. Flexner, L.B. and Weed, L.H. 1983. Note on Cerebrospinal Elasticity in a Chimpanzee. Am. J. Physiol. 105:571-573.

7. Granger, R.A. 1985. Fluid Mechanics. NY: Holt, Rinehart and Winston, p. 281.

8. Harris, C.M. and Crede, C.E. 1976. Shock and Vibration Handbook. New York: McGraw- Hill Book Company.

9. Hashizume, K. 1972. A Study of the Experimental Brain Injury. Brain and Nerve, 14:991-1002.

10. Nusholtz, G.S. 1985. Critical Factors in Head Injury Research. Head Injury Prevention - Past and Present. Publisher: Harbo. In press.

11. Nusholtz, G.S., Bender, M. and Kaiker, P.S. 1986. Photogrammetric Techniques Using High-Speed Cineradiography. Optical Engineering, 25(6):791-798.

12. Nusholtz, G.S., Melvin, J.W., and Alem, N.M. 1979. Head Impact Response Comparisons of Human Surrogates. 23rd Stapp Car Crash Conf. Proc., 17-19 October, San Diego, Calif., pp. 499-541. SAE Paper No. 791020.

13. Nusholtz, G.S., et al. 1981. Response of the Cervical Spine to Superior-Inferior Head Impact. 25th Stapp Car Crash Conference Proc., pp. 197-237.

14. Nusholtz, G.S., et al. 1984. Head Impact Response-Skull Deformation and Angular Accelerations. 28th Stapp Car Crash Conf. Proc., pp. 41-74.

15. Nusholtz, G.S., et al. 1986. Evaluation of Experimental Techniques in Head Injury Research. Final Report, UMTRI-86-30. DOT-NHTSA Contract No. DTNH22-83-C-07095.

16. Nusholtz, G.S. and Ward, C.C. 1986. Comparison of Epidural Pressure in Live Anesthetized and Postmortem Primates. Journal of Aviation, Space and Environmental Medicine. In press.

17. Ommaya, A.K., et al. 1967. Scaling of Experimental Data on Cerebral Concussion in Sub-Human Primates to Concussion Threshold for Man. 11th Stapp Car Crash Conf. Proc., pp. 45-52.

18. Ommaya, A.K. 1973. Head Injury Mechanisms. Final Report Contract No. DOT-HS-081-1-1061A, U. S. Dept. of Transportation, National Highway Traffic Safety Administration, Washington, D.C.

19. O'Neill, B. 1967. Elementary Differential Geometry, New York: Academic Press.

20. Portnoy, H.D., et al. 1970. Intracranial Pressure and Head Acceleration During Whiplash. 14th Stapp Car Crash Conf. Proc., pp. 152-168. SAE Paper No. 700900.

21. Saczalski, K.J. 1976. A Critical Assessment of the Use of Non-Human Surrogates for Safety System Evaluation. 20th Stapp Car Crash Conf. Proc., pp. 159-187.

22. Schweitzer, D.H. and Szebehely, V.G. 1950. Gas Evolution in Liquids and Cavitation. J. of Applied Physics 21:1218-1224 (Dec.).

23. Smith, R.W. 1979. The Response of Unembalmed Cadaveric and Living Cerebral Vessels to Graded Injury - A Pilot Study. 23rd Stapp Car Crash Conf. Proc., pp. 543-560. SAE Paper No. 791021.

24. Stoker, J.J. 1969. Differential Geometry, New York: Wiley Intersciences.

25. Symon, L. 1967. A Comparative Study of Middle Cerebral Pressure in Dogs and Macaques. Journal of Physiol. 191:449-465.

26. Thomas, D.J., Robbins, D.H., Eppinger, R.H., King, A.I., and Hubbard, R.P. 1974. Guidelines for the Comparison of Human and Human Analogue Biomechanical Data. A report of an ad hoc committee, Ann Arbor, Michigan, December 6.

27. Unterharnscheidt, F. and Higgins, L.S. 1969. Traumatic Lesions of Brain and Spinal Cord Due to Nondeforming Angular Accelerations of the Head. Texas Report on Biol. and Med. 27:127-166.

28. Weed, L.H. and Flexner, L.B. 1932. Cerebrospinal Elasticity in the Cat and Macaques. American Journal of Physiol. 101:668-667.

29. Wiggert, D.C. and Sundquist, M.J. 1979. The Effect of Gaseous Cavitation on Fluid Transients. J. of Fluid Engineering 101:79-88 (March).

30. Wylie, E.B. and Streeter, V.L. 1985 edition, Fluid Transients. Ann Arbor, MI: FEB Press, especially pp. 12-13 and 136-155.

MECHANICAL IMPEDANCE

TYPE: UNREPRESSURIZED POSTMORTEM RHESUS, NO SKULL FRACTURE, NECK UNSTRETCHED

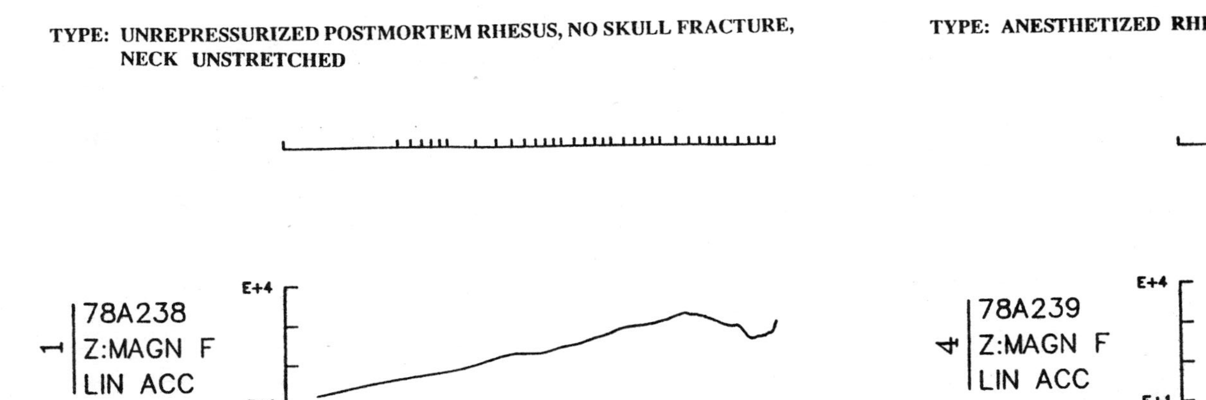

HEAD CENTER OF GRAVITY WITH RESPECT TO LINE OF IMPACT = ABOVE

MECHANICAL IMPEDANCE

TYPE: ANESTHETIZED RHESUS, NO SKULL FRACTURE, NECK UNSTRETCHED

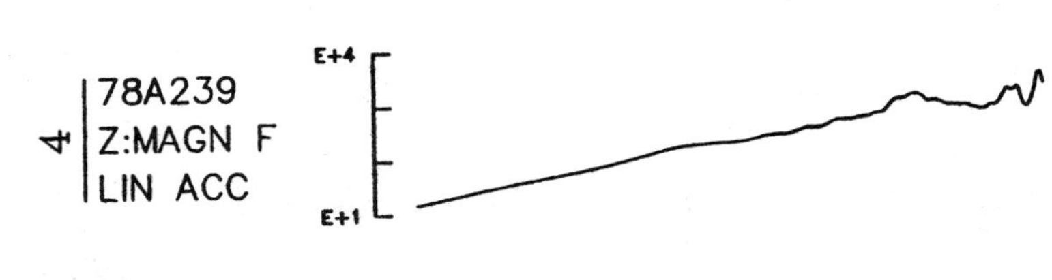

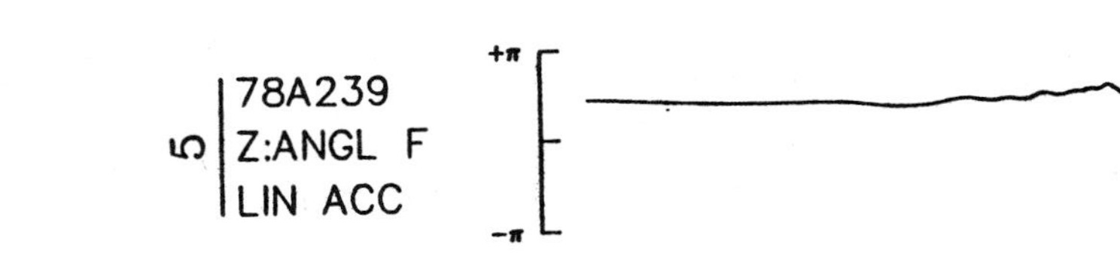

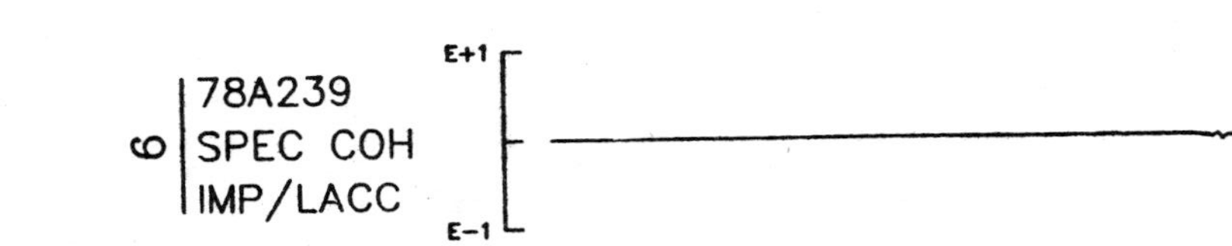

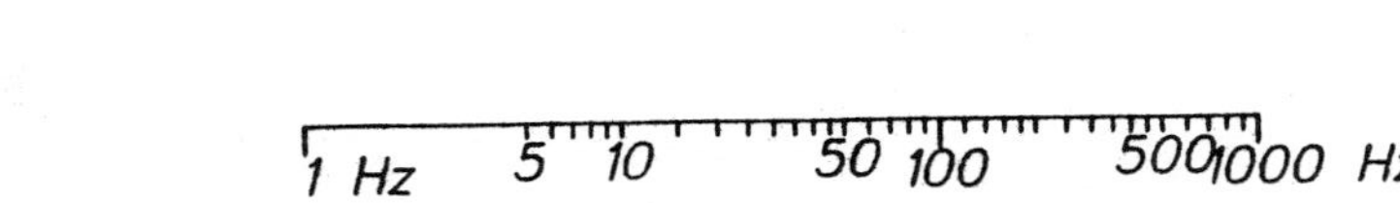

HEAD CENTER OF GRAVITY WITH RESPECT TO LINE OF IMPACT = AT

MECHANICAL IMPEDANCE

TYPE: ANESTHETIZED RHESUS, SKULL FRACTURE, NECK UNSTRETCHED

10 | 79A249 Z:MAGN F LIN ACC

E+4

E+1

11 | 79A249 Z:ANGL F LIN ACC

+π

−π

12 | 79A249 SPEC COH IMP/LACC

E+1

E−1

1 Hz 5 10 50 100 500 1000 Hz

HEAD CENTER OF GRAVITY WITH RESPECT TO LINE OF IMPACT = ABOVE

MECHANCIAL IMPEDANCE

TYPE: ANESTHETIZED RHESUS, SKULL FRACTURE, NECK UNSTRETCHED

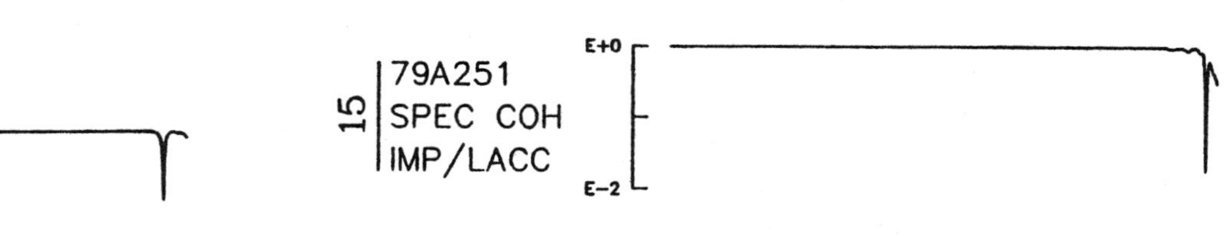

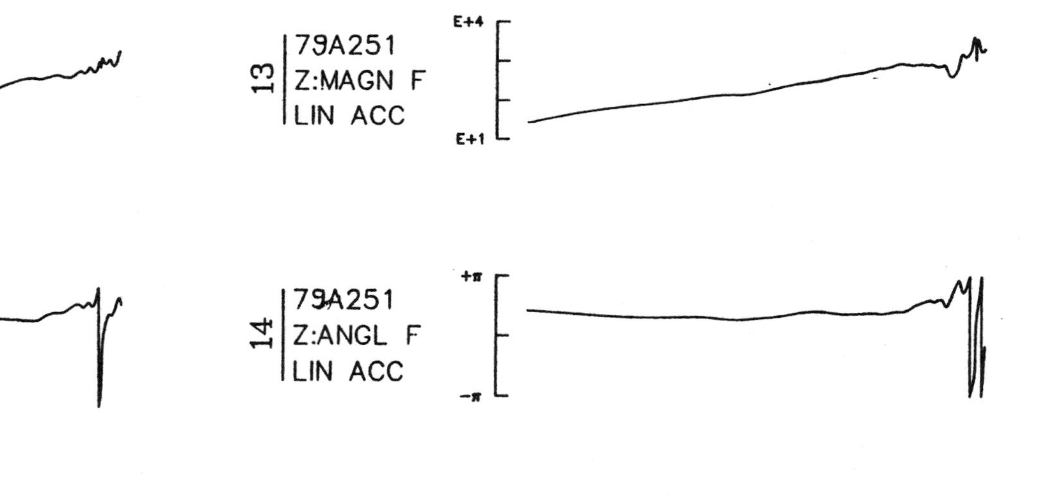

HEAD CENTER OF GRAVITY WITH RESPECT TO LINE OF IMPACT = AT

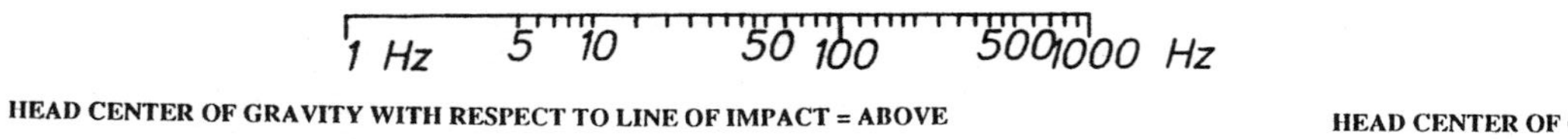

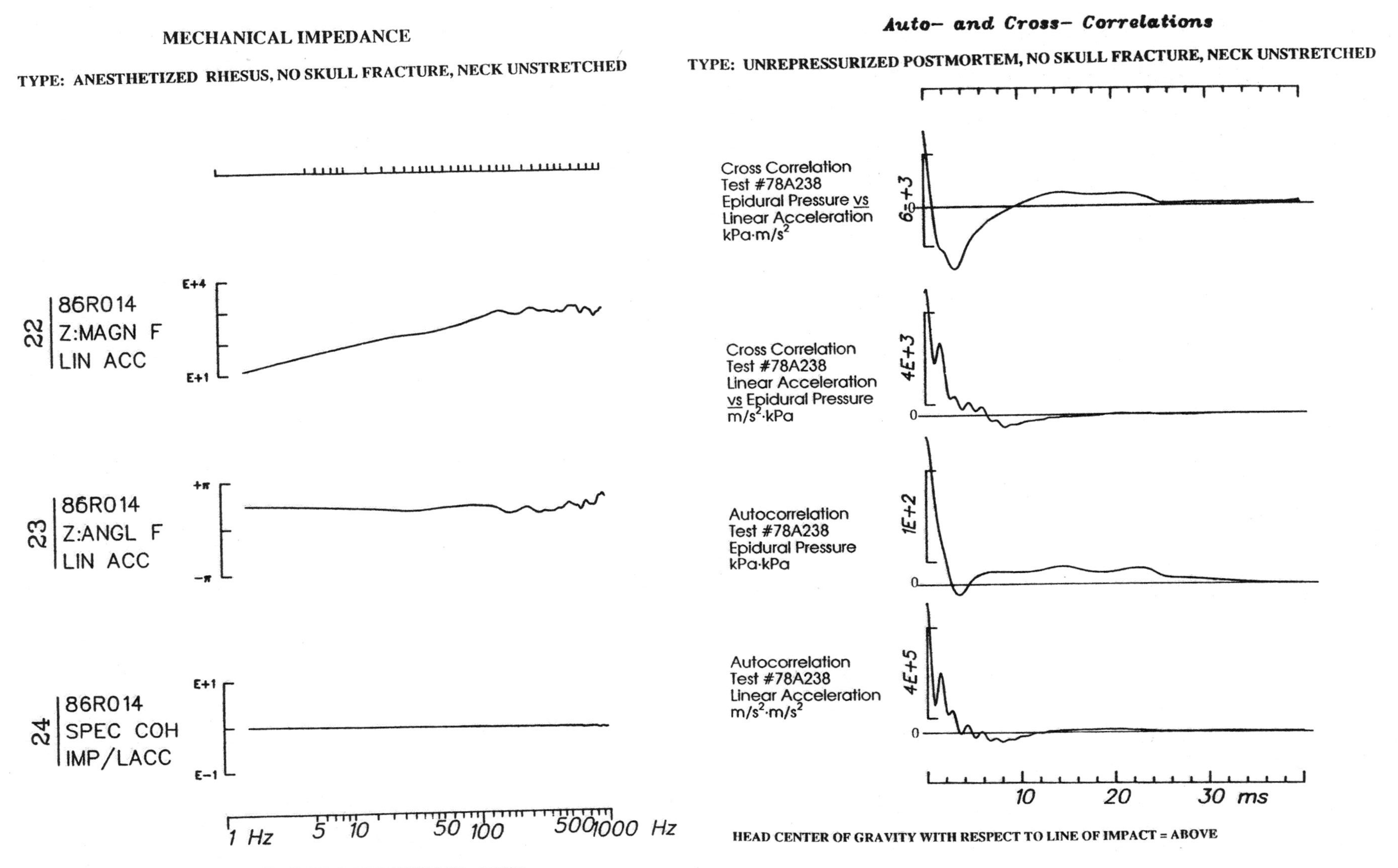
MECHANICAL IMPEDANCE
TYPE: ANESTHETIZED RHESUS, NO SKULL FRACTURE, NECK UNSTRETCHED
22
86R014
Z:MAGN F
LIN ACC
E+4
E+1
23
86R014
Z:ANGL F
LIN ACC
+π
-π
24
86R014
SPEC COH
IMP/LACC
E+1
E-1
1 Hz
5
10
50
100
500
1000
Hz
HEAD CENTER OF GRAVITY WITH RESPECT TO LINE OF IMPACT = ABOVE
Auto- and Cross- Correlations
TYPE: UNREPRESSURIZED POSTMORTEM, NO SKULL FRACTURE, NECK UNSTRETCHED
Cross Correlation
Test #78A238
Epidural Pressure vs
Linear Acceleration
kPa·m/s²
6=+3
0
Cross Correlation
Test #78A238
Linear Acceleration
vs Epidural Pressure
m/s²·kPa
4E+3
0
Autocorrelation
Test #78A238
Epidural Pressure
kPa·kPa
1E+2
0
Autocorrelation
Test #78A238
Linear Acceleration
m/s²·m/s²
4E+5
0
10
20
30 ms
HEAD CENTER OF GRAVITY WITH RESPECT TO LINE OF IMPACT = ABOVE

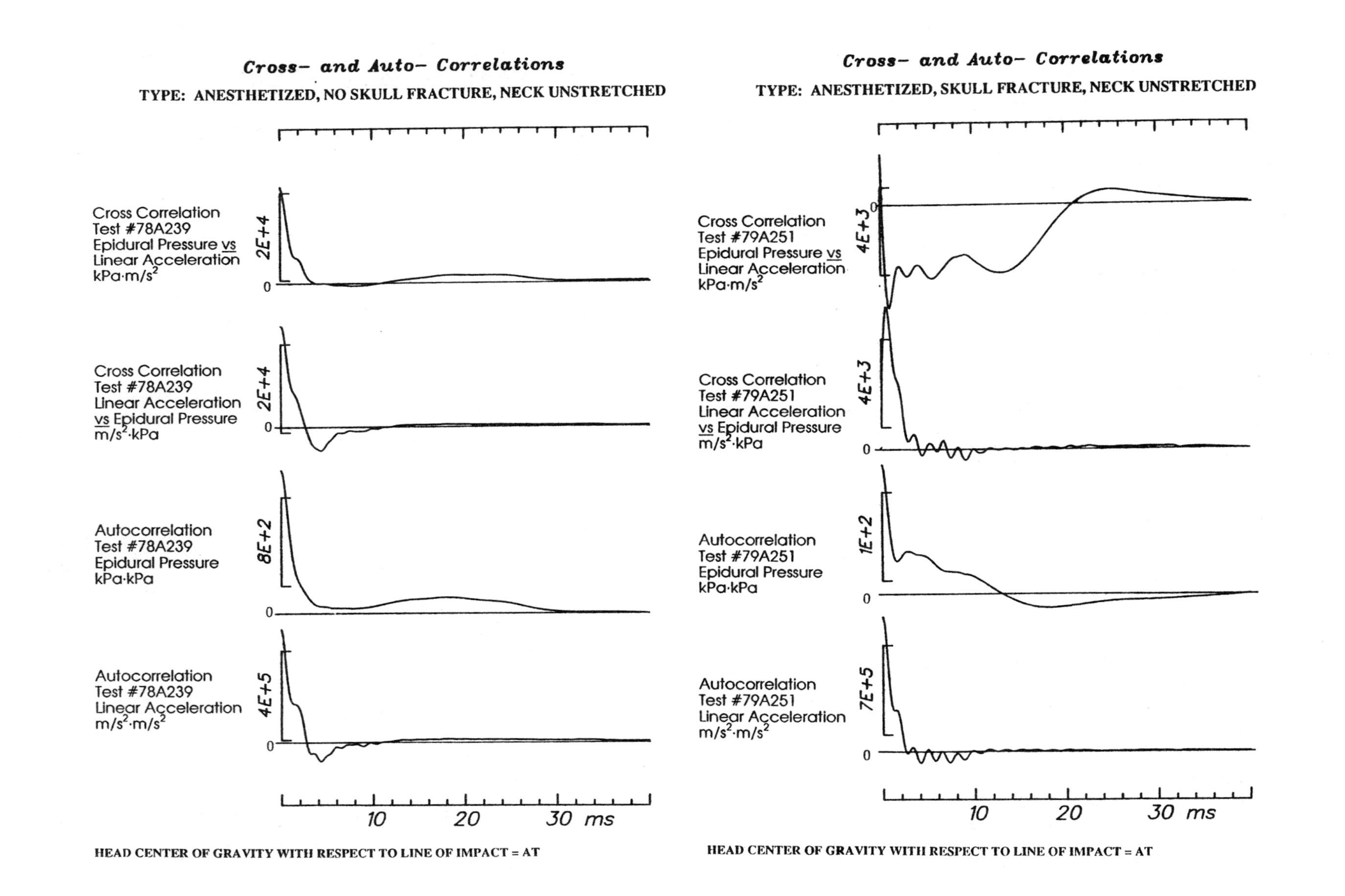
Cross- and Auto- Correlations
TYPE: ANESTHETIZED, NO SKULL FRACTURE, NECK UNSTRETCHED
Cross Correlation
Test #78A239
Epidural Pressure vs
Linear Acceleration
kPa·m/s²
2E+4
0
Cross Correlation
Test #78A239
Linear Acceleration
vs Epidural Pressure
m/s²·kPa
2E+4
0
Autocorrelation
Test #78A239
Epidural Pressure
kPa·kPa
8E+2
0
Autocorrelation
Test #78A239
Linear Acceleration
m/s²·m/s²
4E+5
0
10
20
30 ms
HEAD CENTER OF GRAVITY WITH RESPECT TO LINE OF IMPACT = AT
Cross- and Auto- Correlations
TYPE: ANESTHETIZED, SKULL FRACTURE, NECK UNSTRETCHED
Cross Correlation
Test #79A251
Epidural Pressure vs
Linear Acceleration
kPa·m/s²
4E+3
0
Cross Correlation
Test #79A251
Linear Acceleration
vs Epidural Pressure
m/s²·kPa
4E+3
0
Autocorrelation
Test #79A251
Epidural Pressure
kPa·kPa
1E+2
0
Autocorrelation
Test #79A251
Linear Acceleration
m/s²·m/s²
7E+5
0
10
20
30 ms
HEAD CENTER OF GRAVITY WITH RESPECT TO LINE OF IMPACT = AT

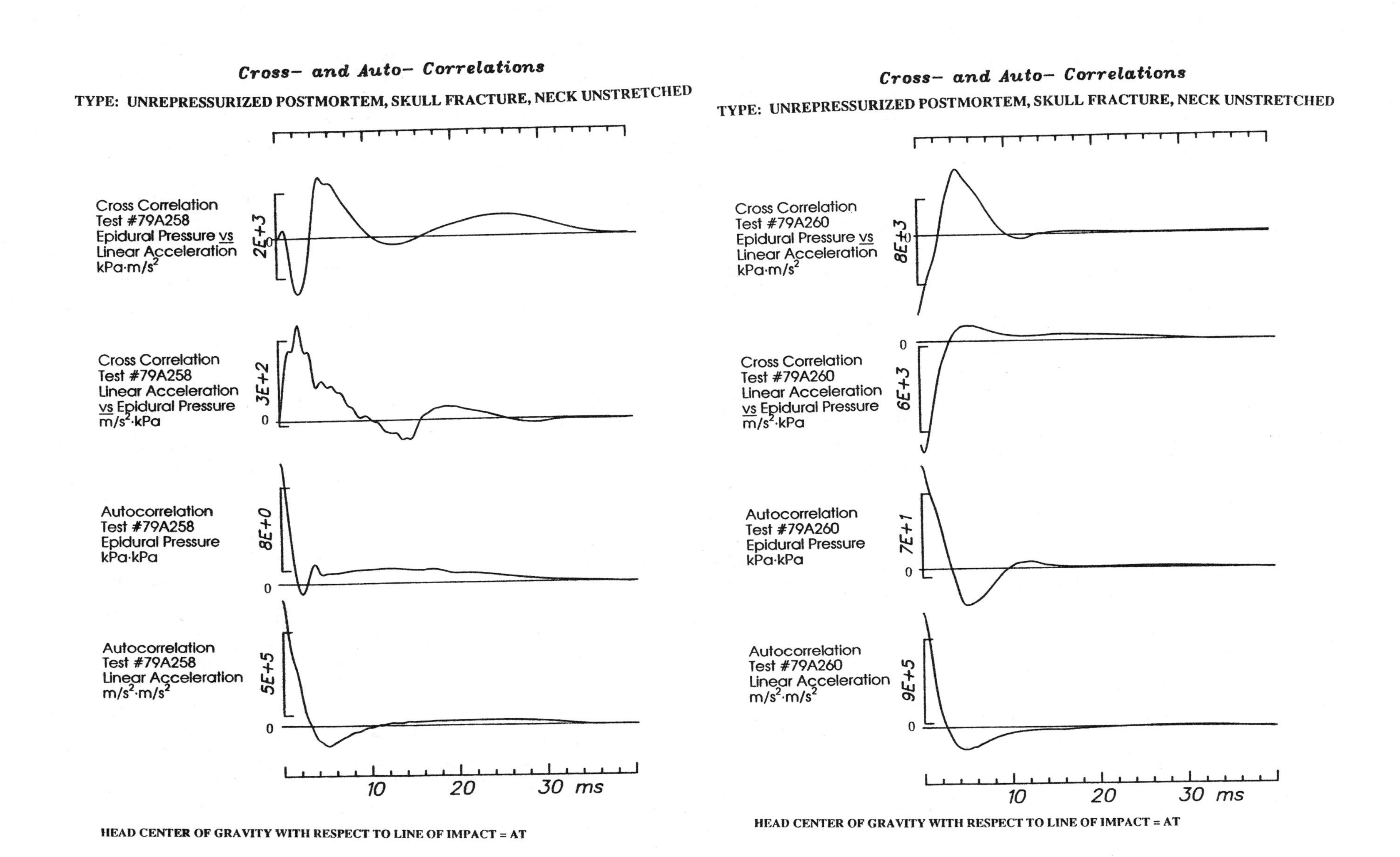
Cross- and Auto- Correlations
TYPE: UNREPRESSURIZED POSTMORTEM, SKULL FRACTURE, NECK UNSTRETCHED
Cross Correlation
Test #79A258
Epidural Pressure vs
Linear Acceleration
kPa·m/s²
2E+3
0
Cross Correlation
Test #79A258
Linear Acceleration
vs Epidural Pressure
m/s²·kPa
3E+2
0
Autocorrelation
Test #79A258
Epidural Pressure
kPa·kPa
8E+0
0
Autocorrelation
Test #79A258
Linear Acceleration
m/s²·m/s²
5E+5
0
10
20
30 ms
HEAD CENTER OF GRAVITY WITH RESPECT TO LINE OF IMPACT = AT
Cross- and Auto- Correlations
TYPE: UNREPRESSURIZED POSTMORTEM, SKULL FRACTURE, NECK UNSTRETCHED
Cross Correlation
Test #79A260
Epidural Pressure vs
Linear Acceleration
kPa·m/s²
8E+3
0
Cross Correlation
Test #79A260
Linear Acceleration
vs Epidural Pressure
m/s²·kPa
6E+3
0
Autocorrelation
Test #79A260
Epidural Pressure
kPa·kPa
7E+1
0
Autocorrelation
Test #79A260
Linear Acceleration
m/s²·m/s²
9E+5
0
10
20
30 ms
HEAD CENTER OF GRAVITY WITH RESPECT TO LINE OF IMPACT = AT

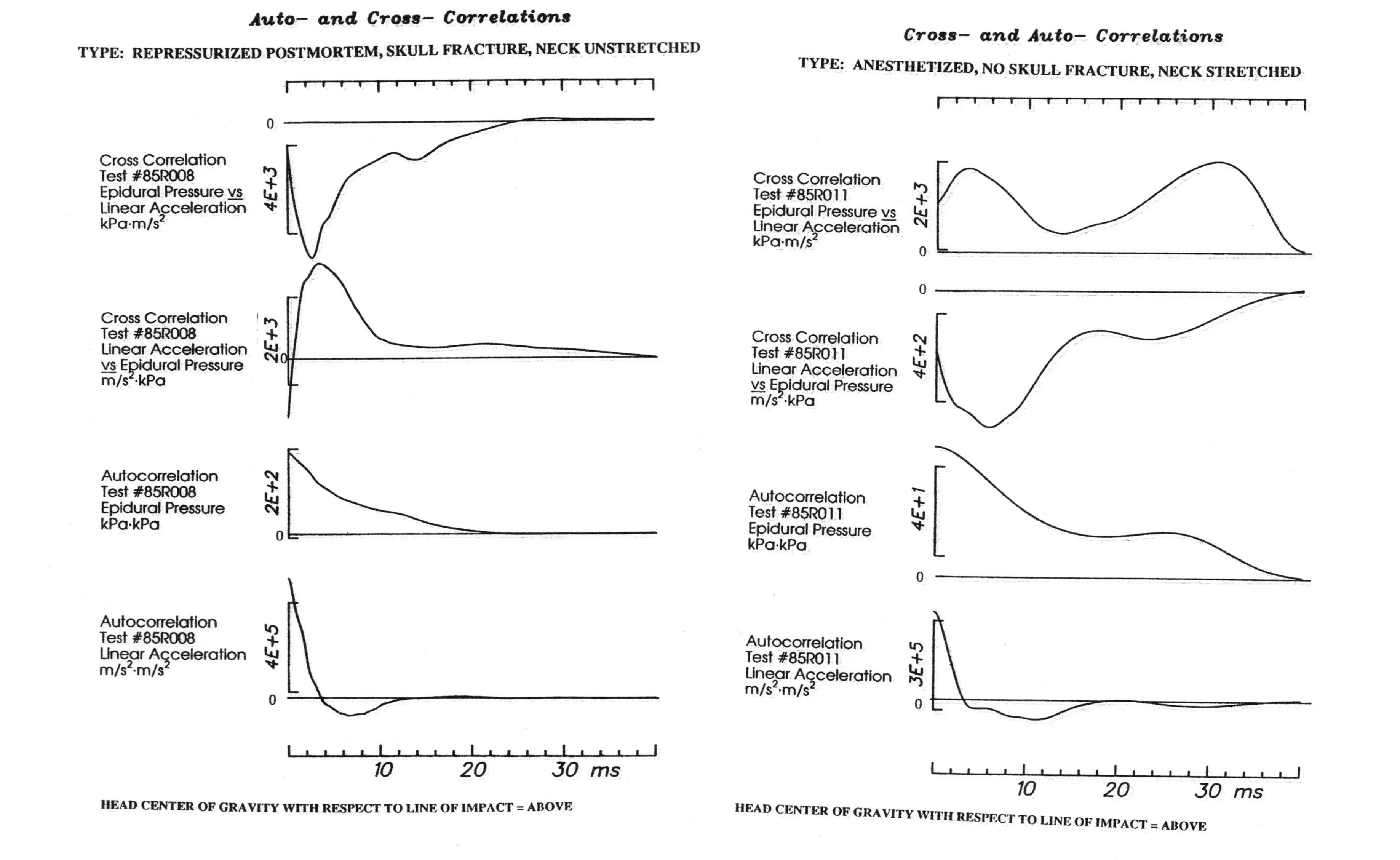
Auto- and Cross- Correlations
TYPE: REPRESSURIZED POSTMORTEM, SKULL FRACTURE, NECK UNSTRETCHED
Cross Correlation
Test #85R008
Epidural Pressure vs
Linear Acceleration
kPa·m/s²
4E+3
0
Cross Correlation
Test #85R008
Linear Acceleration
vs Epidural Pressure
m/s²·kPa
2E+3
0
Autocorrelation
Test #85R008
Epidural Pressure
kPa·kPa
2E+2
0
Autocorrelation
Test #85R008
Linear Acceleration
m/s²·m/s²
4E+5
0
10
20
30 ms
HEAD CENTER OF GRAVITY WITH RESPECT TO LINE OF IMPACT = ABOVE
Cross- and Auto- Correlations
TYPE: ANESTHETIZED, NO SKULL FRACTURE, NECK STRETCHED
Cross Correlation
Test #85R011
Epidural Pressure vs
Linear Acceleration
kPa·m/s²
2E+3
0
Cross Correlation
Test #85R011
Linear Acceleration
vs Epidural Pressure
m/s²·kPa
4E+2
0
Autocorrelation
Test #85R011
Epidural Pressure
kPa·kPa
4E+1
0
Autocorrelation
Test #85R011
Linear Acceleration
m/s²·m/s²
3E+5
0
10
20
30 ms
HEAD CENTER OF GRAVITY WITH RESPECT TO LINE OF IMPACT = ABOVE

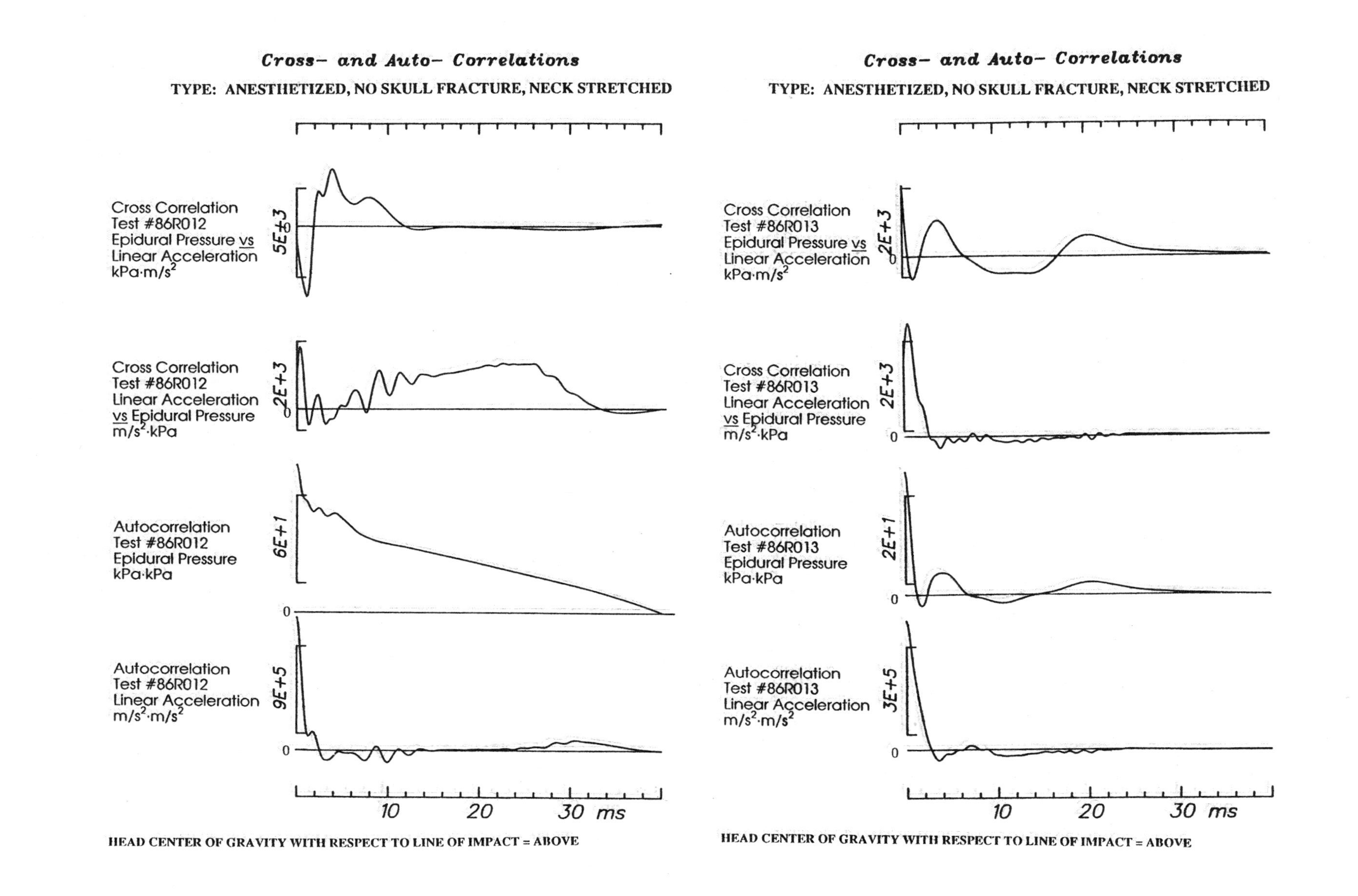
Cross- and Auto- Correlations
TYPE: ANESTHETIZED, NO SKULL FRACTURE, NECK STRETCHED
Cross Correlation
Test #86R012
Epidural Pressure vs
Linear Acceleration
kPa·m/s²
5E+3
0
Cross Correlation
Test #86R012
Linear Acceleration
vs Epidural Pressure
m/s²·kPa
2E+3
0
Autocorrelation
Test #86R012
Epidural Pressure
kPa·kPa
6E+1
0
Autocorrelation
Test #86R012
Linear Acceleration
m/s²·m/s²
9E+5
0
10
20
30 ms
HEAD CENTER OF GRAVITY WITH RESPECT TO LINE OF IMPACT = ABOVE
Cross- and Auto- Correlations
TYPE: ANESTHETIZED, NO SKULL FRACTURE, NECK STRETCHED
Cross Correlation
Test #86R013
Epidural Pressure vs
Linear Acceleration
kPa·m/s²
2E+3
0
Cross Correlation
Test #86R013
Linear Acceleration
vs Epidural Pressure
m/s²·kPa
2E+3
0
Autocorrelation
Test #86R013
Epidural Pressure
kPa·kPa
2E+1
0
Autocorrelation
Test #86R013
Linear Acceleration
m/s²·m/s²
3E+5
0
10
20
30 ms
HEAD CENTER OF GRAVITY WITH RESPECT TO LINE OF IMPACT = ABOVE

Cross- and Auto- Correlations

TYPE: ANESTHETIZED, NO SKULL FRACTURE, NECK UNSTRETCHED

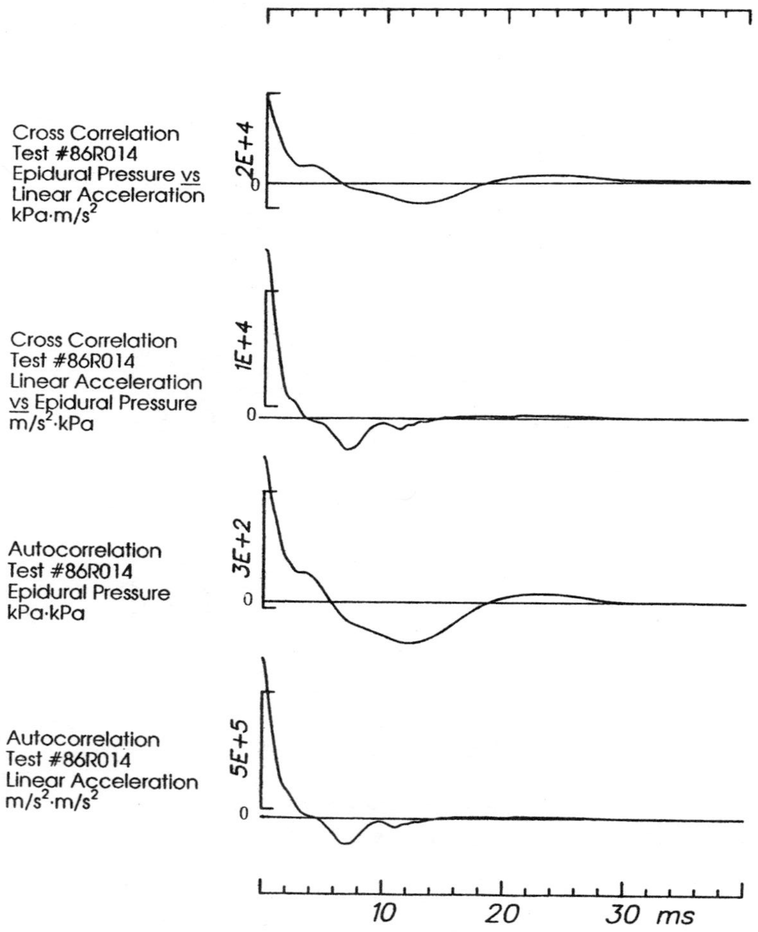

HEAD CENTER OF GRAVITY WITH RESPECT TO LINE OF IMPACT = ABOVE

KINEMATIC TIME-HISTORIES

TYPE: UNREPRESSURIZED POSTMORTEM, NO SKULL FRACTURE, NECK UNSTRETCHED

Run ID: 78A236 *Filter: 1600*4C* *Smooth: 3SD*

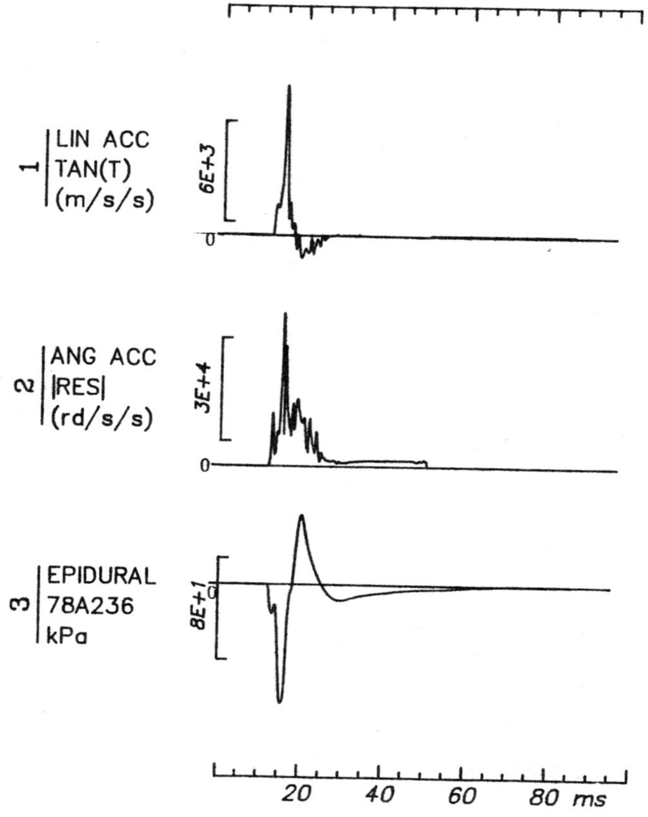

1=Tangential Acceleration
2=Angular Acceleration (Resultant)
3=Epidural Pressure

HEAD CENTER OF GRAVITY WITH RESPECT TO LINE OF IMPACT = AT

KINEMATIC TIME-HISTORIES

TYPE: ANESTHETIZED, SKULL FRACTURE, NECK UNSTRETCHED

Run ID: 79A249 *Filter: 1600*4C* *Smooth: 3SD*

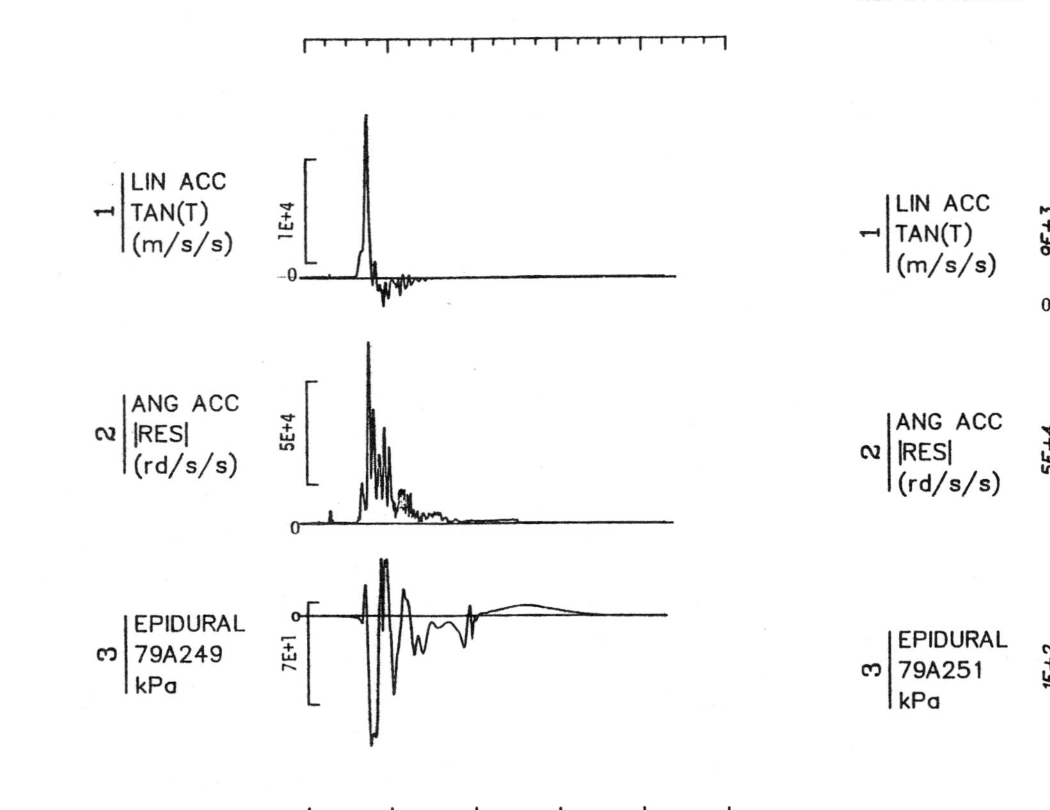

1=Tangential Acceleration
2=Angular Acceleration (Resultant)
3=Epidural Pressure

HEAD CENTER OF GRAVITY WITH RESPECT TO LINE OF IMPACT = ABOVE

KINEMATIC TIME-HISTORIES

TYPE: ANESTHETIZED, SKULL FRACTURE, NECK UNSTRETCHED

Run ID: 79A251 *Filter: 1600*4C* *Smooth: 3SD*

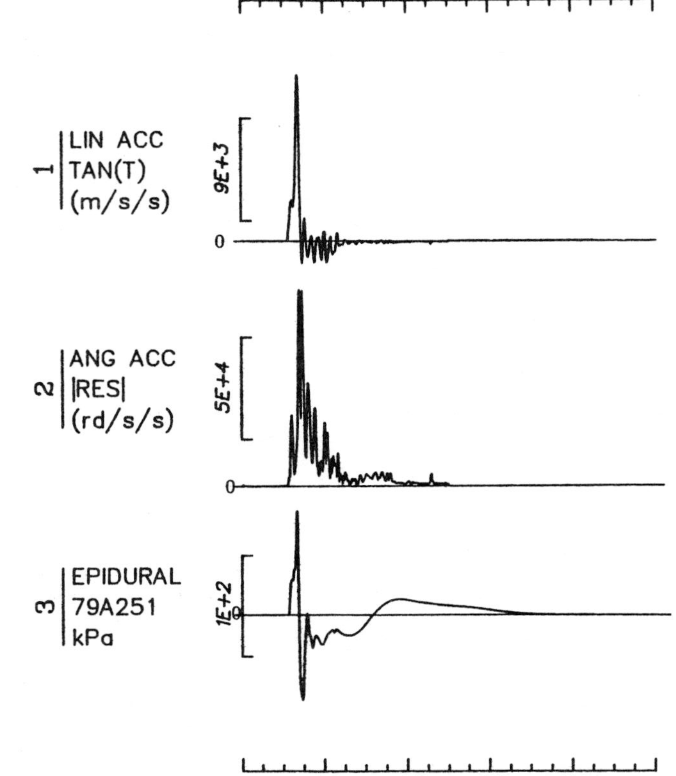

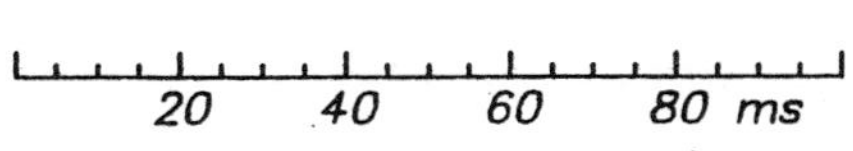

1=Tangential Acceleration
2=Angular Acceleration (Resultant)
3=Epidural Pressure

HEAD CENTER OF GRAVITY WITH RESPECT TO LINE OF IMPACT = AT

KINEMATIC TIME-HISTORIES

TYPE: REPRESSURIZED POSTMORTEM, NO SKULL FRACTURE, NECK UNSTRETCHED

Run ID: 85R006 H9 *No Filtering*

KINEMATIC TIME-HISTORIES

TYPE: ANESTHETIZED, NO SKULL FRACTURE, NECK STRETCHED

Run ID: 86R013 H7 *No Filtering*

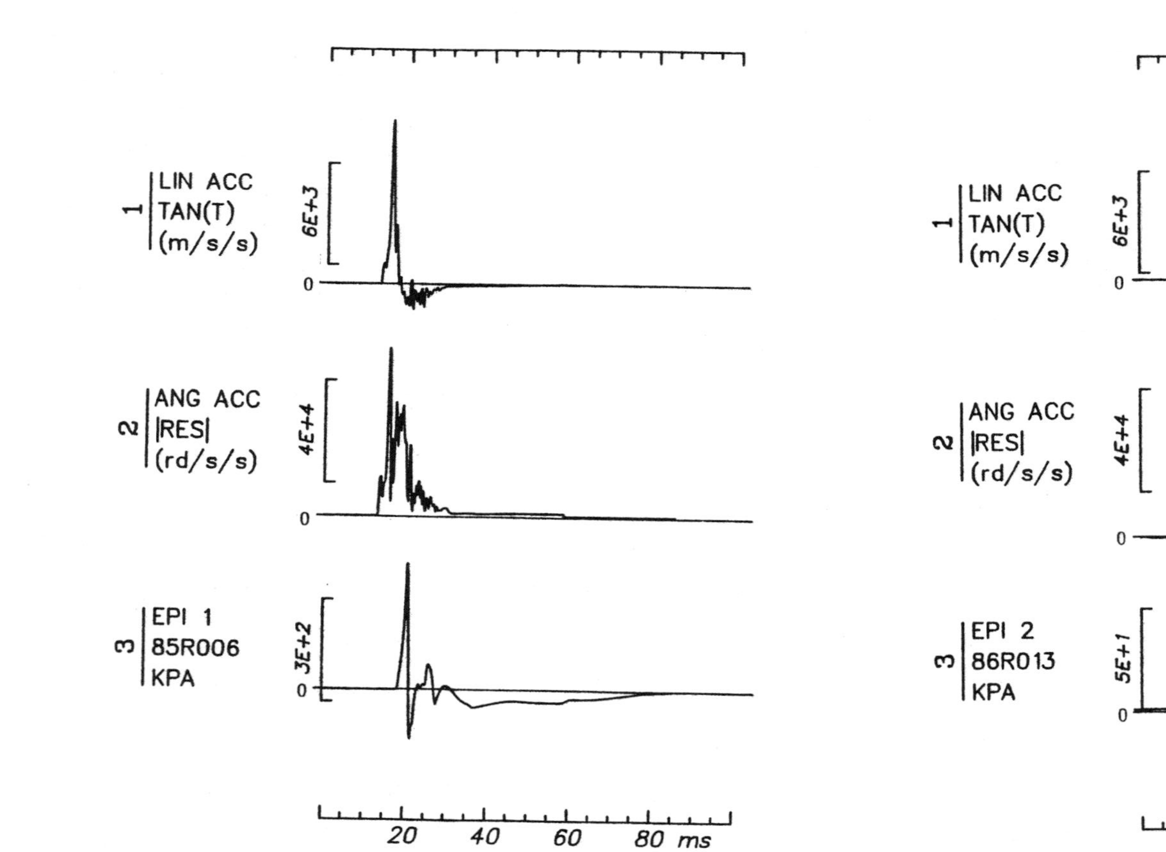

1=Tangential Acceleration
2=Angular Acceleration (Resultant)
3=Epidural Pressure

HEAD CENTER OF GRAVITY WITH RESPECT TO LINE OF IMPACT = ABOVE

1=Tangential Acceleration
2=Angular Acceleration (Resultant)
3=Epidural Pressure

HEAD CENTER OF GRAVITY WITH RESPECT TO LINE OF IMPACT = ABOVE

KINEMATIC TIME-HISTORIES

TYPE: ANESTHETIZED, NO SKULL FRACTURE, NECK UNSTRETCHED

Run ID: 86R014 H7 *No Filtering*

KINEMATIC TIME-HISTORIES

TYPE: ANESTHETIZED, SKULL FRACTURE, NECK UNSTRETCHED

Run ID: 86R015 H7 *No Filtering*

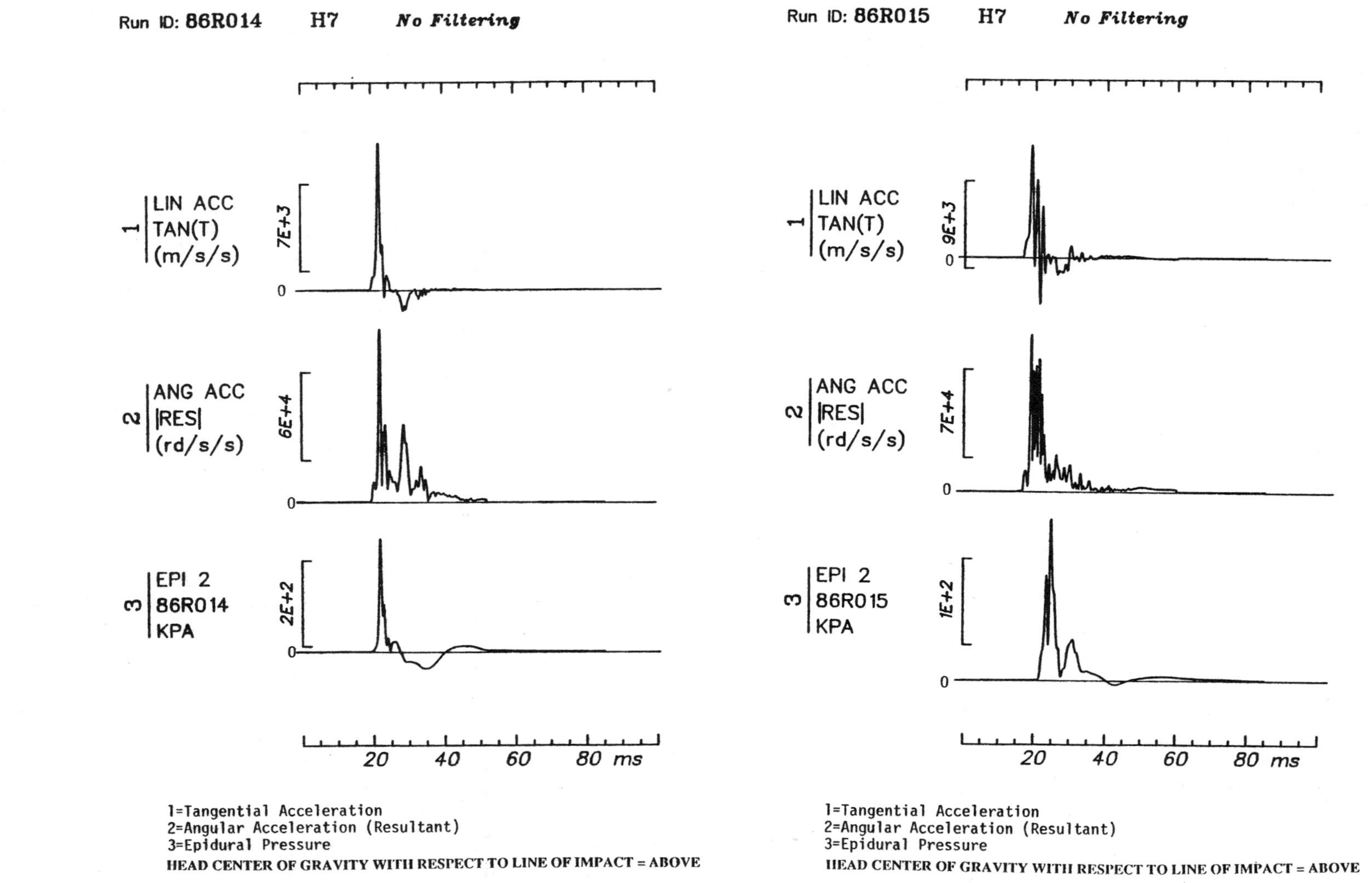

872197

Directional Dependence of Axonal Brain Injury due to Centroidal and Non-Centroidal Acceleration

Thomas A. Gennarelli and Lawrence E. Thibault
University of Pennsylvania

G. Tomei, R. Wiser, D. Graham and J. Adams
University of Glasgow

ABSTRACT

DIFFUSE AXONAL INJURY (DAI) is a brain injury characterized by prolonged traumatic coma not due to mass lesions that has dysfunction or structural damage to brain axons. DAI can be produced by inertial loading of the head in a centroidal or non-centroidal manner. This paper compares the effect of varying the direction of head movement on the severity of DAI. Three groups of 13 monkeys are presented, each subjected to a single non-impact distributed inertial acceleration pulse with head motion constrained to a single plane. In groups 1 and 3, non-centroidal acceleration was produced in the sagittal (rotation about the y axis) and coronal (about the x axis) planes respectively, with the center of rotation in the lower cervical spine. Group 2 was subjected to centroidal acceleration in the horizontal plane (z axis). Deceleration pulse duration (6-8 msec), peak angular deceleration (1-2 x 10^5 rad/sec^2) and angular velocity (475-510 rad/sec) were comparable in each group. Results show that traumatic coma is longest in group 3, shortest in group 1, with group 2 intermediate. The percent of animals still in coma at 1 hour was 0, 30 and 100 in groups 1, 2 and 3 respectively. Pathologic analysis of axonal damage showed a similar difference and was more prevalent and more widespread in group 3 than in group 2, with group 1 having the least damage. Axonal damage of mild, moderate and severe degree was present respectively in 56%, 0, 0 of group 1, 10%, 90%, 0 in group 2 and 0, 12%, 88% of group 3. Brain stem damage was seen almost exclusively in group 3. All differences were significant at the $p<0.05$ level. These findings demonstrate 1) that the direction of brain motion is important in the amount of axonal brain damage produced by inertial loading, 2) that lateral head motions are more injurious than horizontal or sagittal, 3) that pure centroidal acceleration can be produced experimentally.

DATA ON TOLERANCE of the brain to various directions of inertial loading is virtually unavailable. For this reason we present data from primate experiments involving controlled non-impact distributed inertial acceleration loading of the head in each of three single planes. We assessed the influence of the direction of head motion on the amount and distribution of demonstrable axonal damage within the brain. Severity of injury was assessed by clinical examination and the amount of axonal damage as determined by neuropathological examination. These were compared to input parameters of tangential and angular acceleration, angular velocity and acceleration duration. We found that for comparable acceleration levels, the brain in most susceptible to axonal damage in coronal plane acceleration, while horizontal and sagittal plane acceleration produce less damage.

METHODS:

<u>BIOMECHANICS</u> - The methodology of these experiments has been previously reported (1)*. Briefly, adult monkeys (M. mulata or P. nubis) are subjected to a single acceleration deceleration pulse while under phencyclidine or ketamine anesthesia. The head is

constrained within an aluminum helmet to which it is attached by dental cement. A 6" Hyge pneumatic actuator delivers a programmable pulse to a linkage that is attached to the helmet that converts linear thrust into a 60 degree angular displacement around a center of rotation that can be varied.

Three groups of animals were studied:

Group 1 Sagittal plane acceleration: Non-centroidal sagittal plane motions were produced while placing the animal supine in the injury apparatus and moving the head in an angular manner from posterior to anterior with the center of rotation in the mid to lower cervical spine. Head motions were from 30 degrees of extension to 30 degrees of flexion. Rotation occurred through the y axis.

Group 2 Horizontal plane acceleration: Centroidal horizontal plane motions were produced with a different linkage from groups 1 and 3. Here, the supine animal's head was rotated about the Z axis, through the center of gravity of the brain, such that the head began with the nose pointed 30 degrees to the right and ended 30 degrees to the left.

Group 3 Coronal plane acceleration: Coronal plane non-centroidal motions were produced in a manner similar to group 1. The animal was placed on its left side with the center of rotation occurring through the X axis in the mid to lower cervical spine. The head was moved laterally from left to right equidistant about the midline. The head began 30 degrees to the left of the midline and was laterally bent to end 30 degrees to the right of the midline.

In the remainder of this paper these three groups are referred to as sagittal, horizontal and coronal respectively. For obvious anatomic reasons the head accelerations were non-centroidal in the sagittal and coronal groups. However, we have shown that there is no difference between centroidal and non-centroidal acceleration as measured by displacement and strain in physical models using identical loading conditions (2).

In all cases, acceleration-time histories were recorded by an Endevco uniaxal accelerometer mounted on the helmet tangential to the motion. In all cases the input was an asymmetric pulse where the acceleration and deceleration each described a half-sine shape and the deceleration phase was approximately three times the peak magnitude of the acceleration phase.

NEUROLOGICAL OBSERVATIONS - Neurological observations were made by clinical examination. Coma was defined as the lack of eye opening spontaneously or to noxious stimuli and the lack of behavioral contact with the environment. The previously described primate coma scale was used to document behavior, motor activity, eye opening ability, pupillary function and corneal reflexes (3).

NEUROPATHOLOGY - Animals were sacrificed under deep anesthesia and had transcardiac intravascular perfusion of fixatives appropriate for either light or electron microscopy according to previously described protocols (1). The distribution and amount of axonal damage was graded as follows:

Grade 0: No axonal damage was seen

Grade 1: Axonal damage was confined to the white matter of the cerebral hemispheres.

Grade 2: White matter axonal damage plus tissue tears in the corpus callosum or central brain area were present. Tissue tear hemorrhages were characterized by axonal and small vascular damage.

Grade 3: There was axonal damage with or without tissue tears in the brainstem in additional to axonal damage in the white matter, corpus callosum and central brain regions.

RESULTS

BIOMECHANICAL INPUT PARAMETERS - 13 animals were studied in each group. Individual values for each animal are presented in Figure 1 and Figure 2 and are summarized in Table 1. Since, because of the asymetric wave-shape, the deceleration phase was approximately three times larger than the acceleration phase, we have chosen to use the peak deceleration values for correlative purposes. We realize other descriptors could be used and in fact, physical model experiments are being performed in order to determine the best descriptors of this waveform. Angular deceleration values ranging from 7 to 18 x 10^4 rad/sec^2 were achieved in these experiments. Although the range was slightly more wide in the sagittal than in the horizontal or coronal groups, there was no statistical difference between the mean values presented in Table 1 for angular acceleration. Deceleration duration ranged from 5.8 to 8.6 msec and was not statistically different among the three groups. Angular velocity (Figure 2) was slightly lower in four sagitally injured animals than in the majority of cases but was not statistically different as a group. Angular velocities ranged from 250 to 750 rad/sec. Tangential deceleration of the sagittal and coronal groups was not statistically different but was slightly higher in the horizontal injuries.

FIGURE 1

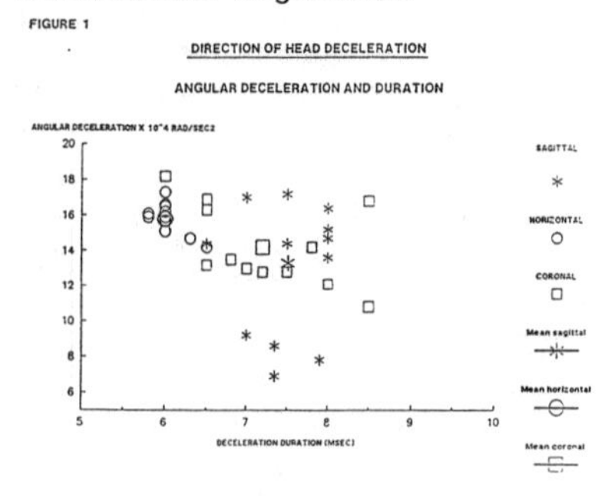

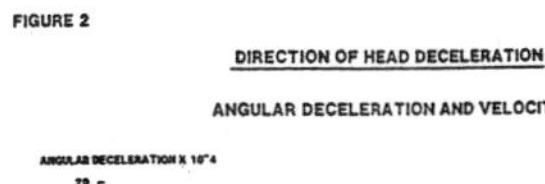

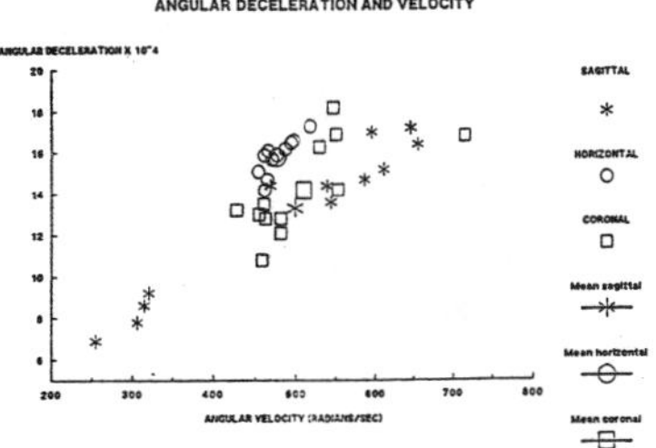

TABLE 1

DIRECTION OF HEAD MOTION	SAGITTAL	HORIZONTAL	CORONAL
ANGULAR DECELERATION (RAD/SEC/SEC X 10^-5)	1.33	1.58	1.42
ANGULAR VELOCITY (RADIANS/SEC)	499	477	510
DECELERATION DURATION (MSEC)	7.51	6.0	7.2
RADIUS (INCHES)	3.65	0	3.62
TANGENTIAL DECELERATION (G)	1185	1486	1165

NEUROLOGICAL OBSERVATIONS - All 39 animals exhibited immediate tramatic coma upon acceleration. The duration of coma is shown in Figure 3 and was less than 5 minutes in all of the sagittally injured animals and in 7 of 13 horizontally injured animals. In contrast, all 13 coronally injured animals were still comatose at 6 hours after injury. At 1 hour the percent of animals in coma in the sagittal, horizontal and coronal groups was 0, 30 and 100 respectively. Eight of the 13 remained permanently comatose until sacrificed. Those coronally injured animals who awakened remained severely disabled. Moderate degrees of neurological impairment were seen in 6 of the 13 horizontally injured animals but all sagittally injured animals made a good recovery. Comparable changes in intracranial pressure and blood pressure have previously been reported (4).

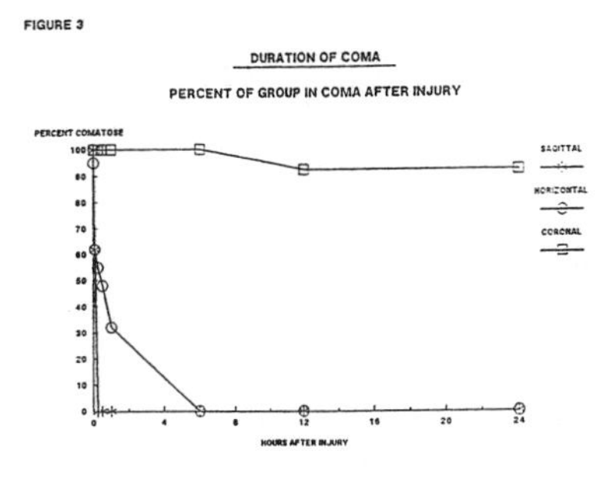

NEUROPATHOLOGICAL OBSERVATIONS - The amount and distribution of axonal damage is shown in Table 2. It should be noted that not all of the animals were available for comprehensive neuropathology but the number undergoing comprehensive examination was comparable in each group. Four of the nine sagittally injured animals had no evidence of axonal damage morphologically. Those five that did have axonal damage had damage confined only to the cerebral hemispheres. In contrast only one horizontally injured and none of the coronally injured animals had grade 1 axonal damage confined to the hemispheres. Nine of the ten horizontally injured had evidence of hemispheric and corpus callosum/central brain axonal damage. This was also true of one of nine coronally injured animals. Most striking was the distribution of severe axonal injury. No sagittal or horizontal animal had severe axonal damage whereas eight of nine coronally injured animals had axonal damage in the brainstem as well as in the central brain corpus callosum and cerebral hemispheres. These differences were highly significant at the $p<0.01$ level.

TABLE 2

GRADE OF DAI	SAGITTAL	HORIZONTAL	CORONAL
0	4	0	0
1	5	1	0
2	0	9	1
3	0	0	8

DISCUSSION

Previous work has demonstrated that inertial loading of the brain in a non-centroidal (previously called angular acceleration) fashion is a mechanism that produces the diffuse brain injuries of cerebral concussion and diffuse axonal injury (DAI)[1,5,6]. Accumulating evidence is compelling that axonal damage is the pathologic subrate of this group of brain injuries[1,7-10]. Cerebral concussion and DAI appear to the less and more severe ends of a continuous spectrum of brain dysfunction characterized by increasing amounts and distribution of axonal damage throughout the brain and brainstem. This paper demonstrates that the brain has directional sensitivity to the amount and distribution of axonal damage that occurs under loading conditions that differ principally in the direction that the head is moved. The data demonstrate that coronal head motions produce the most serious neurological disturbances in terms of depth and duration of coma and in terms of outcome from injury. These neurological observations are highly correlated to the amount and distribution of axonal damage within the brain which was worse than any of the other two groups. In particular, coronal plane head motions were the only type of head motion that commonly produced axonal damage in the brainstem.

In contrast, horizontal plane head motions produced slightly more neurological abnormality, and a slightly worse outcome than those subjected to sagittal plane motions. The amount of axonal damage in the brain was greater in the horizontal than in the sagittal group but was far below that seen in the coronal group.

There then appears to be a gradation of neurological and corresponding axonal damage that is progressively more severe as head motions are changed from sagittal to horizontal to coronal. Since the input parameters of angular acceleration, angular velocity and acceleration duration were comparable in these three groups we are confident that input variations are not responsible for this directional sensitivity. Rather, we feel that geometric constrains of the internal structure of the skull and brain are responsible for these differences. For example, coronal plane acceleration is markedly influenced by the presence of the falx, a firm partition between the two hemispheres. Since the falx runs exactly in the anterior posterior plane, it may also play a role in horizontal motions but plays little or no role in modifying strain within the brain when head motions are confined to the sagittal plane. Further physical modeling experiments may serve to demonstrate and quantitate the differences in these neurological and pathological findings. We anticipate that these models will enable us to determine the influence of these geometric non-linearities.

SUMMARY AND CONCLUSIONS

1. Single plane inertial loading of the head was produced non-centroidally about centers of rotation in the X and Y axes and centroidally about the Z axis.
2. Acceleration and deceleration magnitudes, time durations and angular velocities were comparable in the three groups.
3. The duration of coma, the recovery from coma and the amount of axonal damage in the brain varied according to the direction of head acceleration.
4. The brain appears to have a differential sensitivity to the direction of head motion with regard to the amount of axonal damage produced. More damage occurs as the head motions are changed from sagittal to horizontal to coronal.
5. Pure centroidal motions do not necessarily produce more axonal damage than non-centroidal.
6. The reason for the directional sensitivity of the brain is postulated to be related to geometric properties of the skull and brain.

REFERENCES

1. Gennarelli, TA, Thibault, LE, Adams, JH, Graham, DI, Thompson, CJ and Marcincin, RP: Diffuse Axonal Injury and Traumatic Coma in the Primate, Ann. Neurol., 12:564-574, 1982.
2. Thibault, LE, Margulies, S, and Gennarelli, TA: Temporal and Spatial Deformation Response of a Brain Model in Inertial Loading, 31st Stapp Car Crash Conference, this volume, 1987.
3. Gennarelli, TA, Thibault, LE, Adams, JH, Graham, DI, Thompson, CJ and Marcincin, RP: Diffuse Axonal Injury and Traumatic Coma in the Primate, in Trauma of the Central Nervous System, Dacey, RG (ed), Raven Press, New York, 1985, pp. 447-463.
4. Gennarelli, TA, Pastusko, M, Sakamoto, T, Tomei, G, Duhaime, A, Wiser, R, and Thibault, LE: ICP after Experimental Diffuse Head Injuries, in Intracranial Pressure VI, Teasdale, G. (ed), Springer, Berlin, 1985, pp. 15-20.
5. Gennarelli, TA, Head Injury in Man and Experimental Animals: Clinical Aspects, Acta Neurochirurg., Suppl. 32:1-13, 1983.
6. Gennarelli, TA, Cerebral Concussion and Diffuse Brain Injuries, in Head Injury, Cooper, P. (ed), Williams and Wilkins, Baltimore, Chapter 7, 1987, pp. 108-124.
7. Jane, JA, Stewart, OD and Gennarelli, TA: Axonal Degeneration Induced by Experimental Non-invasive Minor Head Injury, J. Neurosugr., 62:96-100, 1985.
8. Adams, JH, Graham, DI, and Gennarelli, TA: Head Injury in Man and Experimental Animals: Neuropathology. Acta Neurochirurg. Suppl 32, 15-30, 1983.
9. Povlishock, JT and Kontos, HA: Continuing Axonal and Vascular Change Following Experimental Brain Trauma. Central Nervous System Trauma 2:285-298, 1985.

10. Povilshock, JT, Becker, DP, Cheng, CL and Vaughan, GW: Axonal Change in Minor Head Injury. J. Neuropathol. Exp. Neurol., 42:225-242, 1983.

*Numbers in parentheses designate references at end of paper.

881708

Biomechanics of Head Injury — Toward a Theory Linking Head Dynamic Motion, Brain Tissue Deformation and Neural Trauma

David C. Viano
General Motors Research Labs.

ABSTRACT

A "central" theory for the biomechanics of brain injury is proposed that includes the construct that acceleration of the head, per se, is not the proximate cause of injury. Rather, rapid motion of the skull causes displacement of the hard bony structures of the head against the soft tissues of the brain, which lag in their motion due to inertia and loose coupling to the skull. Relative displacement between brain and skull produces deformation of brain tissue and stretching of bridging veins, which contribute to the tissue-level causes of brain injury.

The first step in an accurate interpretation of brain injury risk in dummies involves the measurement of the three-dimensional components of translational and rotational acceleration of the head. The In-Line accelerometry method provides an accurate three-dimensional dynamic measurement in the Hybrid III dummy and uses a row of uniaxial accelerometers to interpret the gradient of rotational acceleration by a linear least-square fit through acceleration responses. Utilizing three orthogonal In-Line rows, and a computer program to post-process the accelerometer data, the six independent components of rigid body dynamics are accurately measured during impact.

In this preliminary study, 2D head dynamics data are used as input to a brain compliance model, which interprets the effect of rapid skull motion as tissue-level deformations of the brain. The brain compliance approach interprets brain deformation by the Viscous response (VC or the product of strain and strain rate at the tissue-level) which recent experiments show may be the underlying cause of neural trauma. It excludes the effects of skull deformation. The approach identifies the time of injury risks within the brain, and offers a different and potentially better assessment of head injury than currently available with the HIC criterion.

REGULATORS AND SAFETY ENGINEERS continue discussing the most appropriate tolerance level for the head injury criterion (HIC). Such discussions tacitly accept the validity of the HIC as a measure of injury risk and overshadow a fundamental issue: acceleration, per se, is not the cause of brain injury. This paper addresses the tissue-level causes of neural trauma, presents a "central" theory for the biomechanics of brain injury, and gives a framework for future research on the biomechanics and assessment of head injury.

CURRENT ASSESSMENT OF HEAD INJURY

The pioneering research of Holbourn (1,2) resulted in an engineering interpretation of the mechanism of head injury for blows of short duration or impact. He proposed that injury is proportional to the force multiplied by the time over which it acts. For closed head impact, without skull fracture, a transient force produces a proportional acceleration of the head. Since acceleration is a measurable quantity on a moving body during impact, the earliest experimental determinations of head injury were related to translational acceleration. Although the important role of rotational acceleration of the head was identified in Holbourn's early work, the lack of procedures to measure it during impact has delayed its emergence as a principal factor in brain injury assessment.

Holbourn's focus on the importance of the duration of impact predated its inclusion in the analysis by Patrick (3) of head acceleration data from Wayne State University and later by Gadd (4) in the acceleration-weighted tolerance criterion called the Gadd Severity Index (GSI). The GSI related injury risk to the integral over the impact duration of head acceleration raised to the 2.5 power (Figure 1). A tolerance level of GSI = 1,000 was established for injury assessment. This index was widely used in crash injury research and was the forerunner of the current Head Injury Criterion (HIC), which is a mathematical refinement of the original acceleration-weighted analysis (5).

The single-axis translational acceleration basis for HIC was later replaced by the resultant acceleration of a triaxial accelerometer array near the center of gravity of the dummy's head. The HIC has been widely used since its introduction in the early 1970's. Recently, the HIC procedure has been modified by either limiting the maximum duration of the calculation or basing the calculation only on the interval of head contact (6). This avoids a misleading interpretation of injury risk for some non-contact situations.

TRADITIONAL VIEW OF HEAD INJURY

Concurrent with the engineering development of a head injury criterion was pioneering medical research. Gurdjian (7,8), Ommaya (9-11), Gennarelli (12) and their co-workers initiated experimentally-based studies on anesthetized animals. This work provided a foundation to interpret the underlying physiology of brain injury after impact. What again emerged was the importance of acceleration as a measurable correlate with injury from head impact. Ommaya (13) later refined the concept that translational and rotational acceleration are coupled during impact and need to be assessed in tandem to interpret the wide range of brain vascular and axonal injury risks from head impact.

The medical and engineering views have jelled into the current "traditional" view of head injury (Table 1). It conceptually frames the biomechanics of head injury as resulting from impact in which translational and rotational acceleration of the head produce either skull fracture or concussive/contusive brain injury. This approach relies on the long-standing view that impact causes acceleration, which causes injury, in sequence. The logic implies that minimizing injury risk is best achieved by limiting both components of acceleration.

But, to further complicate predictive analysis, more recent evaluations of the original data of the GSI and HIC (14) indicate a high degree of overlap between injury and non-injury outcomes (Figure 2). These data really show that HIC is not strongly correlated with either skull fracture or brain vascular injury when the head is subjected to impact, and that HIC is not an adequate biomechanical predictor of injury. However, lack of correlation is somewhat understandable when the wide variety of head impact and acceleration types, and differing injuries are viewed with the expectation that a single measure of head dynamics should relate to all outcomes. In the broad view reliable predictions should not be expected from a measurement of resultant translational acceleration of the head and analysis by a mathematical routine that gives results in a single HIC number.

The current HIC approach to injury prediction also ignores the importance of an injury criterion pin-pointing the time of greatest risk during impact. Knowledge of the instant of greatest injury risk may enable more focused engineering evaluations of potential safety improvements. For example, head impact on the high-penetration resistant (HPR) windshield involves a series of separate but linked events: force build-up, fracture of the glass layers, stretching of the plastic interlayer, and plowing of the head along the windshield (see subsequent discussion and Figure 13).

Each phase and the specific sequence of a head impact may be associated with different types and risks of injury. Dynamic loading to fracture involves a sharp rise in force from the elastic and inertial resistances of the glass with associated head acceleration. Stretching of the plastic interlayer and plowing causes lower-level accelerations of longer duration and rotations of the head. If concussive or contusive injury occurs after impact, an understanding of what specifically caused it would direct attention to potential engineering interventions. In addition, details of the individual events in the head dynamics sequence may be pivotal to the build-up of responses leading to a specific injury.

Because of the diversity of head injuries associated with impact (Table 2), more specific biomechanical interpretations need to be given for the mechanisms of skull fracture, vascular laceration, cortical contusion, and neural and axonal injury. The biomechanics of specific injuries should be based on the fundamental mechanical processes involved with the injury. They should be related to the severity of impact by a method that encompasses the concept that mechanical input is the cause and disruption of physiologic function or tissue injury the consequence. Such an interpretation of the injury process should have as its main focus the tissue-level biomechanical cause of neural or vascular disruption. From that understanding, linkages can be derived to understand the mechanical parameters of head impact response and the relation to brain injury.

CENTRAL THEORY OF SKULL/FACIAL INJURY BIOMECHANICS

Sharp or blunt impact of the head can cause skull fracture with the potential for fragments of bone penetrating brain tissue. Injury to the brain can result in hemorrhage and swelling. Equally important, skull fracture can breach the dura and expose the normally isolated central nervous tissues.

A "central" theory for the biomechanics of skull/facial injury (Figure 3) would recognize that impact produces force as the head is accelerated to a common velocity with the impactor. The force of contact on the skull produces deformation of the skull and strain in the bony layers. Tensile strain is the most common mechanism of compact bone failure and can initiate an area of fracture when it

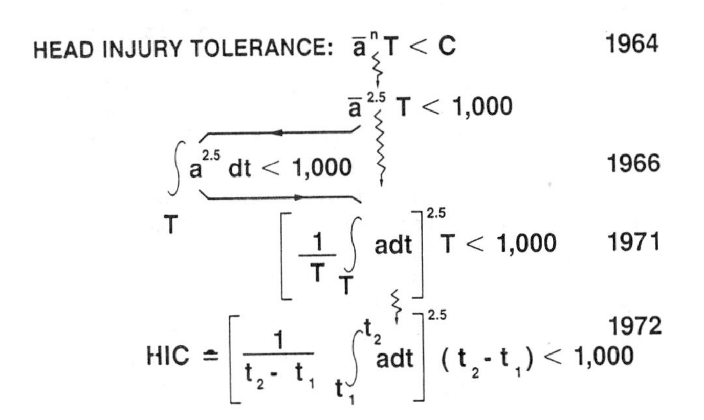

Figure 1: Evolution of criteria for head impact assessment (courtesy of J. Newman Bio-kinetics Inc.).

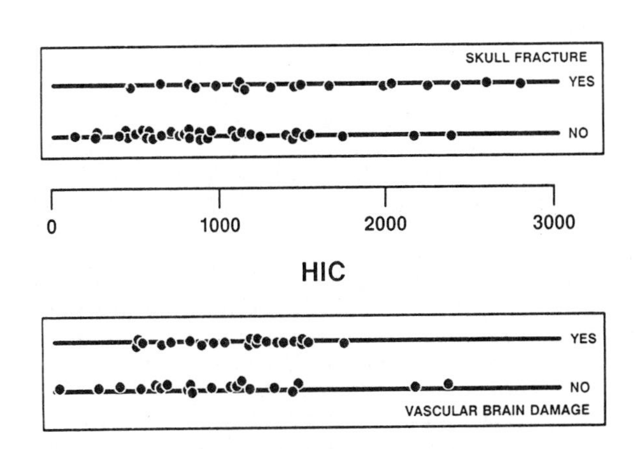

Figure 2: Relationship between measured HIC and the occurrence of skull fracture or the extravasation of fluid from blood vessels (redrawn from 6).

Table 1

BIOMECHANICS OF HEAD INJURY
(Traditional View)

IMPACT

ACCELERATION
- Translation
- Rotation

INJURY
- Fracture
- Concussion
- Contusion

Table 2

HEAD INJURIES

Skull Injuries	Focal Injuries	Diffuse Brain Injuries
Vault Fracture	Epidural Hematoma	Mild Concussion
Linear	Subdural Hematoma	Classical Cerebral Concussion
Depressed	Contusion	Diffuse Injury
Basilar Fracture	Intracranial Hemorrhage	Shearing Injury

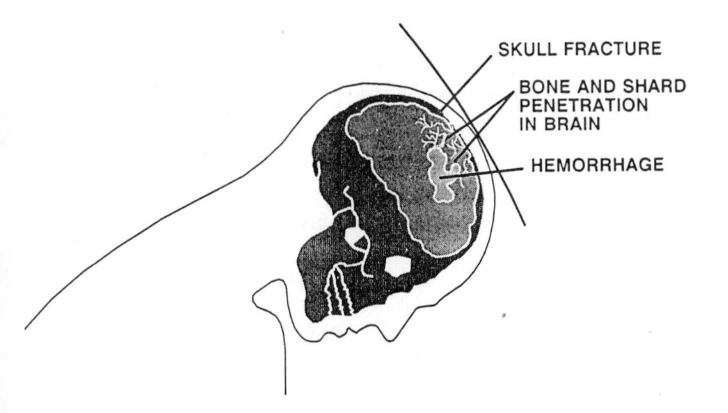

Figure 3: Central theory for the biomechanics of skull/facial injury.

BIOMECHANICS OF SKULL/FACIAL INJURY
(Central Theory)

IMPACT

FORCE
- Strain

INJURY
- Fracture

exceeds 1-3 percent in the tissue (15). The extent and depth of bone fracture is associated with the magnitude and duration of force as well as the area and shape of the skull contact (16). Sufficient load can result in linear, depressed or basilar skull fracture. In this case, the biomechanical mechanism of injury is strain or deformation of the compact bone in the skull.

In the case of impacts with force applied to the face, strain is also the biomechanical cause of fracture. In addition, brain injury is highly dependent on both the normal compliance of facial structures and any resulting fracture/dislocation of bony elements occurring with impact. Recent experiments (17) conducted under contract to our laboratory demonstrate that the human cadaver face is more compliant than the current Hybrid III face. This results in a much lower force plateau and more deflection of facial structures in the human with associated lower induced accelerations of the skull. This markedly alters the interpretation of brain injury risk based on HIC.

For example, full-face impact of the Hybrid III produced 17 kN force and an associated HIC = 1100 (HIC = 1600 computed from acceleration at the top of head); similar tests with human cadavers resulted in a peak force of 5 kN and an HIC = 200 at the top of the head. These data underscore the need for a more sophisticated criterion to reflect actual biomechanical compliance of the dummy face to accurately interpret facial injury risks, head dynamics, and brain injury risk.

CENTRAL THEORY OF BRAIN INJURY BIOMECHANICS

Closed head impact can result in a wide range of injury types and locations within the cranium. Brain injuries can be subdivided into two main categories: focal injuries, which are primarily the observable vascular hemorrhage and contusion of brain tissue; and diffuse brain injury, which is commonly associated with concussion and axonal injury, and can be identified by microscopic evaluation of neural tissue. Focal brain contusion injuries are primarily related to adjacent bony tissues and stiff membranes, particularly the grooves of the anterior and middle fossa supporting the frontal and temporal lobes.

Gurdjian (8) showed the inferior surfaces of the frontal and temporal lobes as common sites of contusive brain injury (Figure 4). In order to capture the anatomical features of the interior skull geometry, our research program includes a contract (18) to compare the anterior and middle fossa anatomy and associated brain injury patterns. This type of data is pivotal to a more complete interpretation of the biomechanics of brain injury.

A "central" theory (Figure 5) for the biomechanics of brain injury was outlined in a recent paper on head injury perspectives (19). In this view, acceleration, per se, is not the cause of injury. Rather rapid motion of the skull (Figure 6) causes displacement of the hard bony structures of the head against the soft tissues of the brain, which lag in their motion due to inertia and loose coupling to the skull (see section on brain compliance modeling for proof of this assertion). This approach more completely identifies the mechanical responses causing injury.

As in the traditional view, closed head impact causes acceleration of the skull and brain, and a complete interpretation of head dynamics, including the three dimensional components of the translational and rotational acceleration is needed. However, the central theory interprets head accelerations as the cause of rapid skull motion producing a transient displacement and velocity of the head. The motions produce relative displacement between brain and skull structures which deform brain tissues and stretch bridging veins (20). These deformations are proposed as the tissue-level cause of brain injury.

Brain and spinal cord tissues are viscoelastic, as are other soft tissues and organs in the body (21,22), and the rate-dependent properties of the soft tissues are suggested here to be the critical parameters of injury. Dynamic brain-skull interactions produce strain and strain-rate effects within brain tissue and the associated vascular system. Here, it is of critical importance that the fundamental biomechanical criterion for brain tissue failure must be identified and used to correctly interpret injury risks associated with brain tissue deformation.

BIOMECHANICS OF BRAIN INJURY

Our research has identified a Viscous mechanism of injury in many of the soft tissues and organs of the body. The early work focused on laceration of liver, rupture of the heart, disruption of cardiac function, and contusion of the lung, and we have developed a large data base on the mechanisms of soft tissue injury (21,22). The Viscous response is a measure of the viscoelastic reaction of tissue to dynamic deformation and combines two important parameters of soft tissue injury: strain (or C-compression) and strain rate (V- velocity of deformation). The Viscous response is merely the instantaneous product of strain and strain rate at the tissue level. VC, or the Viscous response, is an important measure of injury probability for soft tissues undergoing dynamic deformations.

We are continuing our research to determine the applicability of the Viscous response as an injury predictor for brain and spinal cord tissues. Recent experiments in our department (23-26) have involved dynamic compression of brain and spinal cord and have demonstrated a significant correlation of functional (spinal cord) and anatomical (brain) injury to the Viscous response (Figure 7). This data is extremely encouraging as it provides, for the first time, evidence that the Viscous response is a biomechanical correlate with neural trauma. In addition to a valid

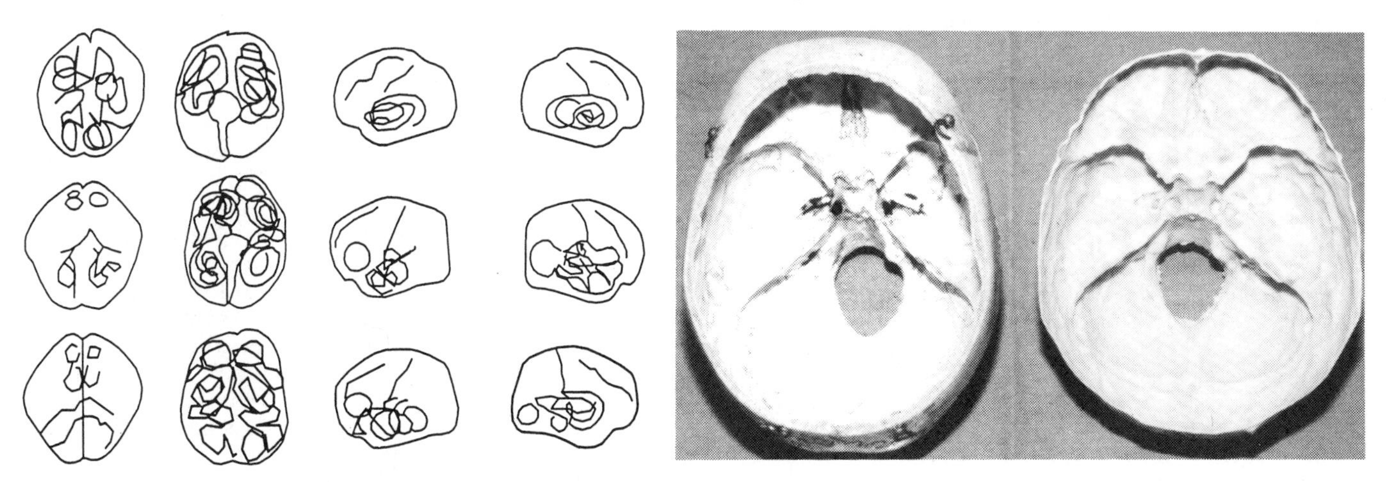

Figure 4: (a) Distribution of contusion injury at the base of the frontal and temporal lobes of the brain (redrawn from 8), (b) photograph and cast of the interior anatomy of the anterior, middle, and posterior fossa at the base of the skull (from (18)).

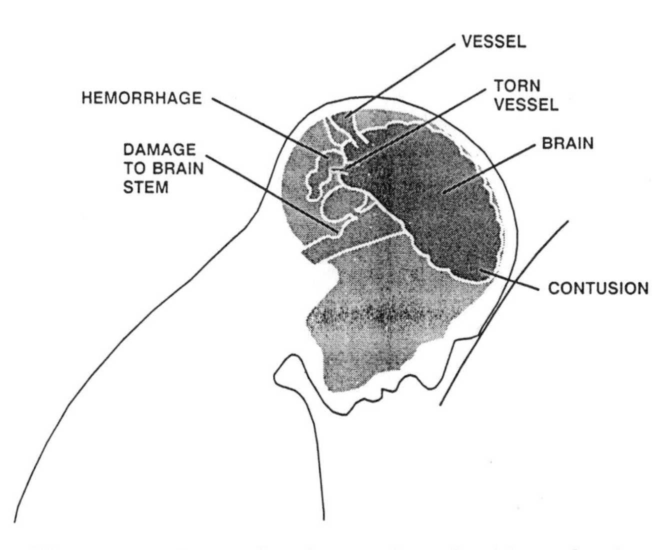

Figure 5: Central theory for the biomechanics of brain injury emphasizing the importance of rapid skull motion causing brain deformations and injuries.

BIOMECHANICS OF BRAIN INJURY
(Central Theory)

IMPACT

ACCELERATION
- Translation
- Rotation

RAPID SKULL MOTION
- Displacement
- Velocity

BRAIN-SKULL INTERACTIONS
- Strain
- Strain-Rate

INJURY
- Concussion
- Contusion

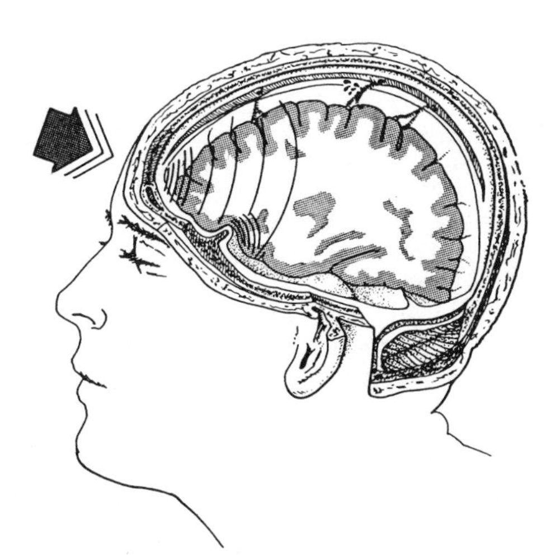

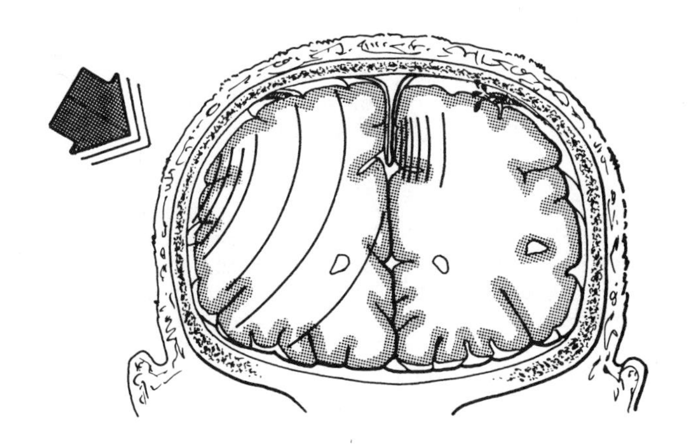

Figure 6: Representation of the dynamics of brain injury associated with rapid skull motion (from 20).

criterion of neural trauma, it is necessary to have a complete interpretation of skull dynamics upon which to compute the rapid displacements and velocities of the skull caused by impact so that the subsequent Viscous response of the brain can be determined.

COMPLETE DETERMINATION OF HEAD DYNAMICS

THREE DIMENSIONAL APPROACH - A new procedure has been developed that provides accurate three-dimensional dynamics of the Hybrid III dummy head during impact. The so-called In-Line accelerometry method (27) uses a row of uniaxial accelerometers to interpret the gradient of rotational acceleration by a linear least-square fit through the accelerometer responses as a function of time. Utilizing three orthogonal In-Line rows (Figure 8) and a computer program to post-process the acceleration data, the six independent components of rigid body dynamics of the Hybrid III head can be accurately measured during a wide range of severe impacts.

An example of the two-dimensional aspects of head dynamics is shown in Figure 9 for a blunt impact on the forehead. Even for this relatively simple loading condition, there are a series of dynamic responses in the translational and rotational accelerations of the head. These data also show the complexity of the dynamic events in which injury may occur and for which injury interpretations need to be made. By attaching the Hybrid III head through a six-axis neck load cell, the impact force on the head can be accurately determined by subtracting the product of the translational acceleration at the center of gravity of the head and head mass from the neck reaction force (Figure 10). In addition, the three moment measurements at the neck can be used to locate the line of action of the external force on the head.

When an oblique impact of the head is measured by the full three-dimensional In-Line accelerometry method, the six independent accelerations of the head show even more complicated acceleration dynamics (Figure 11). In this example, an unrestrained right-front automotive passenger was rotated in-board twenty degrees on the seat and subjected to a 30 mph barrier crash simulation on a Hyge sled. The resulting head dynamics demonstrate oblique contact of the head with the windshield and accelerations transmitted through dummy structures from other body contacts. Although our familiarity with the complex dynamics of the head is developing, a key first-step is to interpret these data in terms of rapid skull displacements and velocities through which the brain is loaded.

TWO DIMENSIONAL APPROACH - Recent research has focused on a two-dimensional In-Line measurement and analysis of data to develop a familiarity with translational and rotational accelerations experienced by the Hybrid III dummy. In this effort, a single row of accelerometers is used from the 3D In-Line package or a row of accelerometers is attached to the back-cap of the Hybrid III head (Figure 12) and the resulting acceleration data is post-processed to determine the best linear fit to the transient acceleration gradient. With an accurate calculation of the rotational acceleration of the head, the full 2D head kinematics, including velocity and displacements can be determined with respect to a fixed coordinate system or a moving coordinate system on the sled fixture (28). The latter allows a determination of head motion with respect to the interior components on the sled by accounting for Hyge sled acceleration. In addition, external forces on the head can be calculated by subtracting the inertial forces due to head acceleration from the neck loads. All of these calculations define the important parameters of head impact, and allow an analysis of brain responses and potential injury risks.

We have used the 2D In-Line approach in a series of sled tests using an automotive interior buck in a simulated 30-35 mph barrier crash. Response data from these tests are summarized in Table 3 and head dynamics responses shown in Figure 13. In Test 1338 the lap-shoulder belted dummy's head interacted with the steering wheel hub and in an otherwise identical Test 1350, the head did not. Other tests simulated an unrestrained passenger interacting with the windshield and instrument panel at two different crash speeds, and either a lap-shoulder belted or unrestrained driver interacting with an air bag. Typical full-waveform HIC calculations are given for the resultant head acceleration, and values for a 36 ms and 15 ms time interval cut-off. Similar calculations were made for the single-channel AP (anterior-posterior head direction) acceleration.

Using the 2D In-Line and kinematic analysis computations, peak head velocities with respect to the moving sled and interior are summarized for the x-component, z-component, and resultant (note that these are not the AP and SI directions of the head which change orientation with respect to the sled during head dynamics). Rotational responses are also summarized by the maximum positive and negative angular acceleration, angular velocity, and displacement. Finally, the components of external contact force acting on the head are given. In the case of the unrestrained passenger, the first value indicates the peak force occurring at contact and fracture of the windshield, and the second value is the maximum force during restraint by stretch of the interlayer of the windshield or subsequent head contact on the instrument panel.

The response data on Test 1350 and 1338 indicate that the full wave-form HIC is actually higher in the non-contact belt restraint test (HIC = 1638 versus 889). This is primarily a result of the sharper acceleration spike and substantially shorter HIC interval with head contact than for the longer-duration, non-contact HIC calculation. In contrast, the peak translational and rotational accelerations were

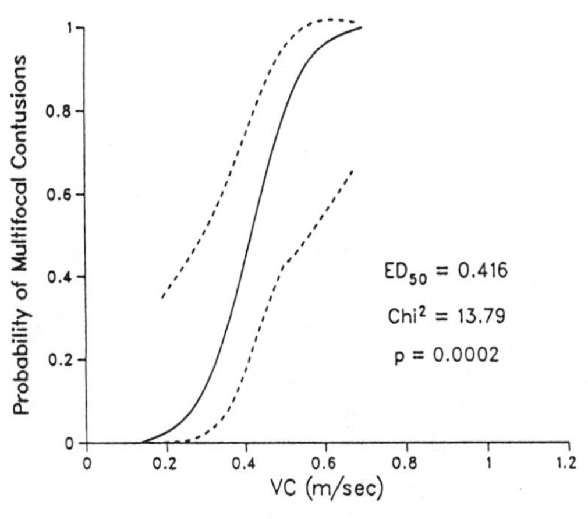

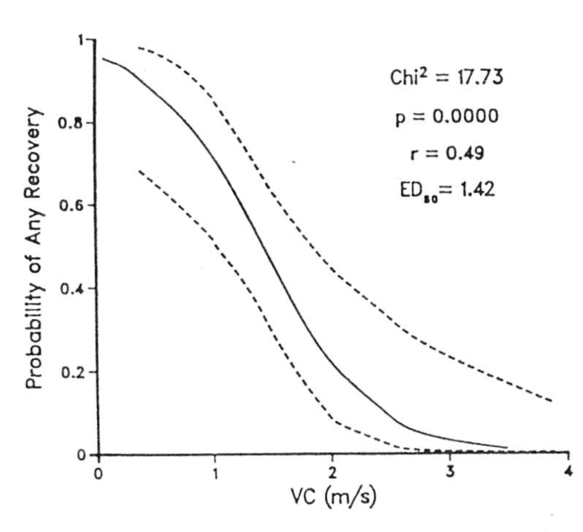

Figure 7: (a) Logist analysis of the probability of neural trauma from direct brain loading as a function of the Viscous response and (b) logist analysis for functional and anatomic injury from direct compressive loading of the cervical spinal cord as a function of the viscous response (courtesy J. Lighthall (24) and P. Kearney (26)).

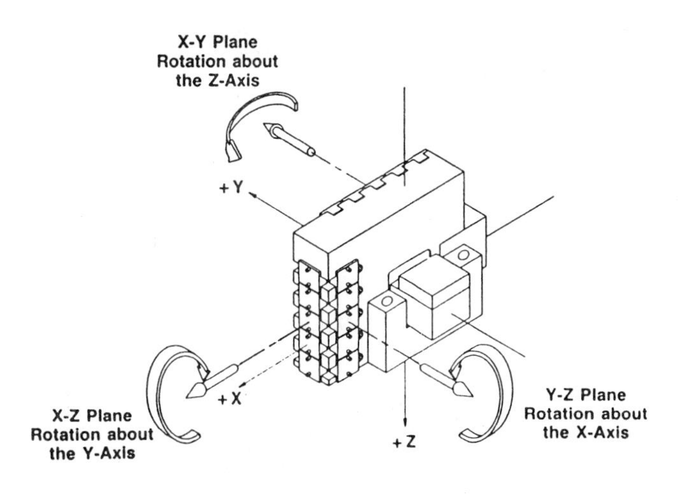

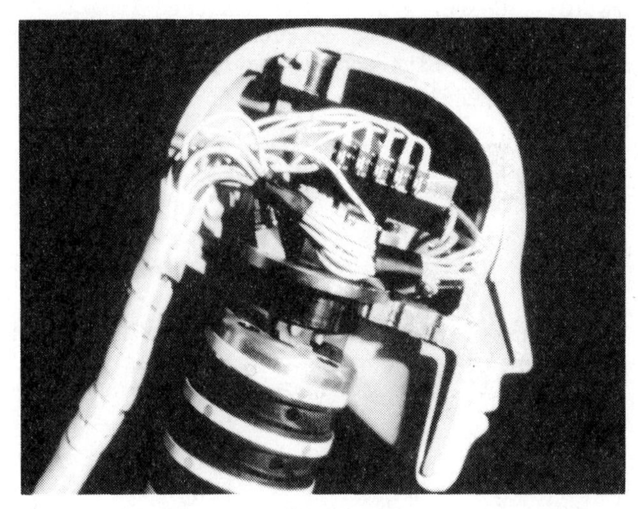

Figure 8: In-Line accelerometry method measures the three-dimensional components of translational and rotational acceleration of the Hybrid III dummy (from 27).

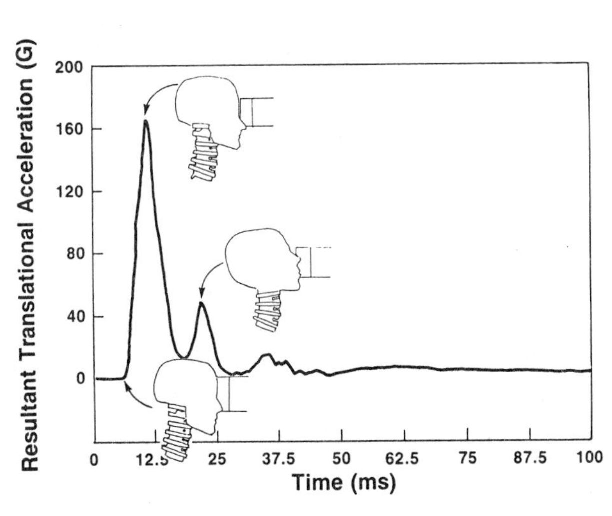

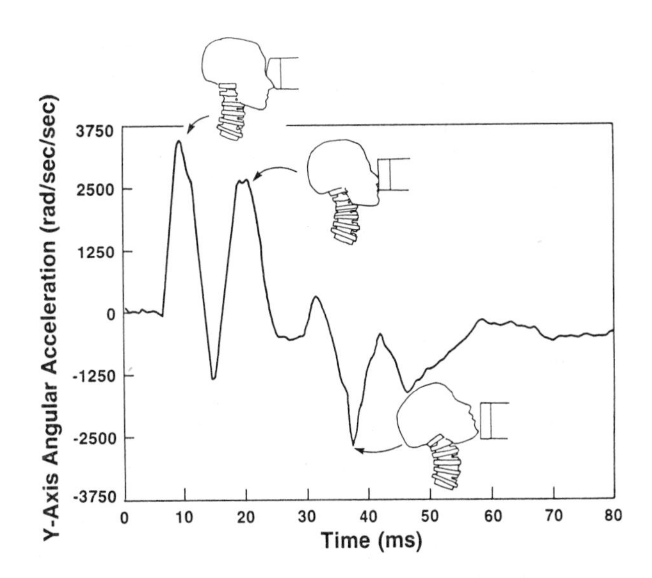

Figure 9: Translational and rotational acceleration of the Hybrid III head subjected to a forehead impact (from 27).

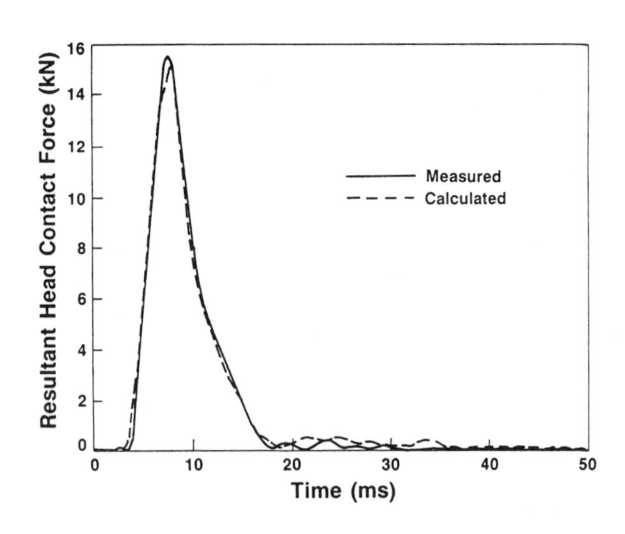

Figure 10: Comparison of the measured impact force on the forehead and computation of impact force from the In-Line accelerometry method and inertial compensation of neck reaction loads (from 27).

Figure 12: Two-dimensional version of the In-Line accelerometry method using a skull-cap mounting of a single In-Line row in the Hybrid III dummy.

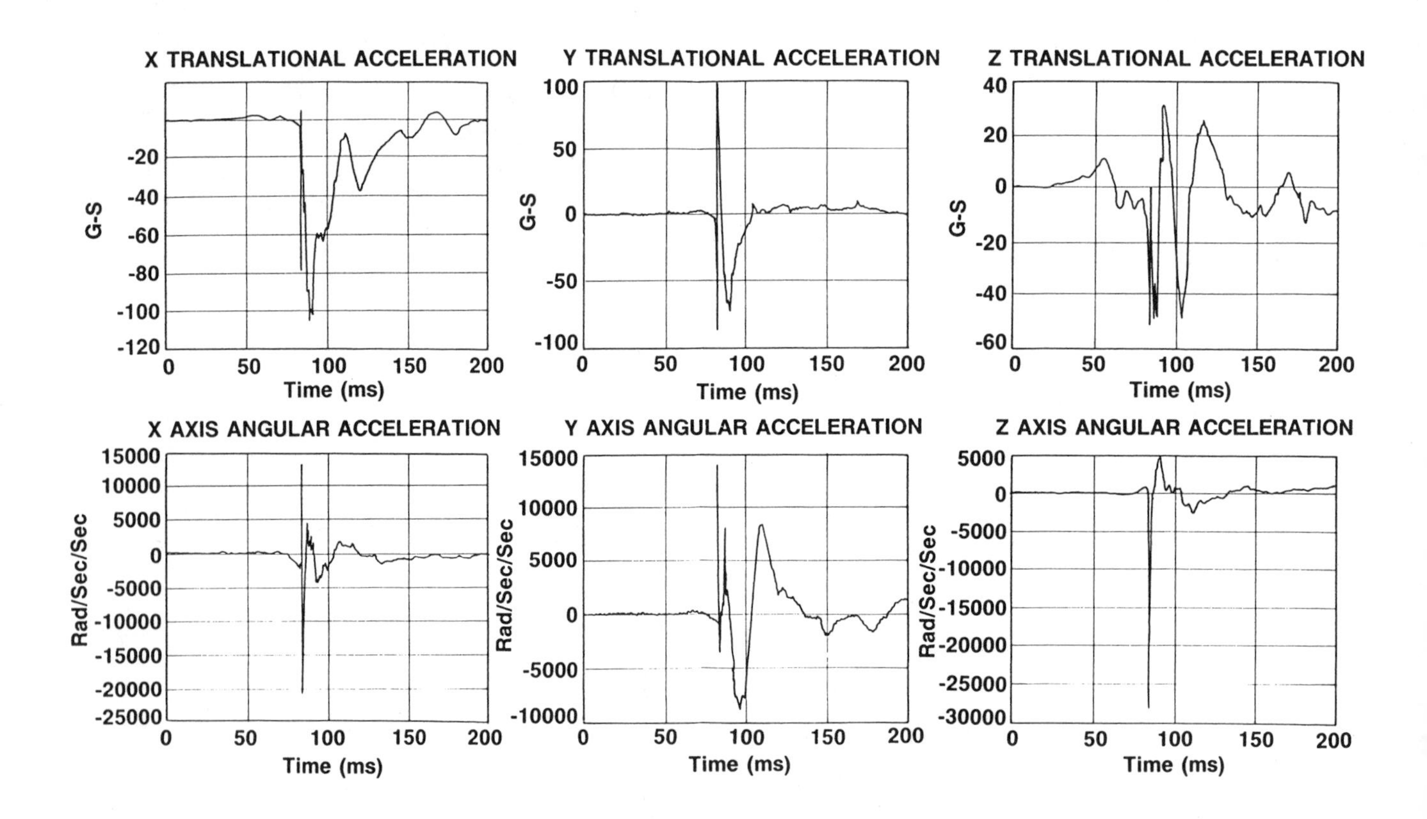

Figure 11: Three-dimensional components of head acceleration for oblique impact of the head (from 27).

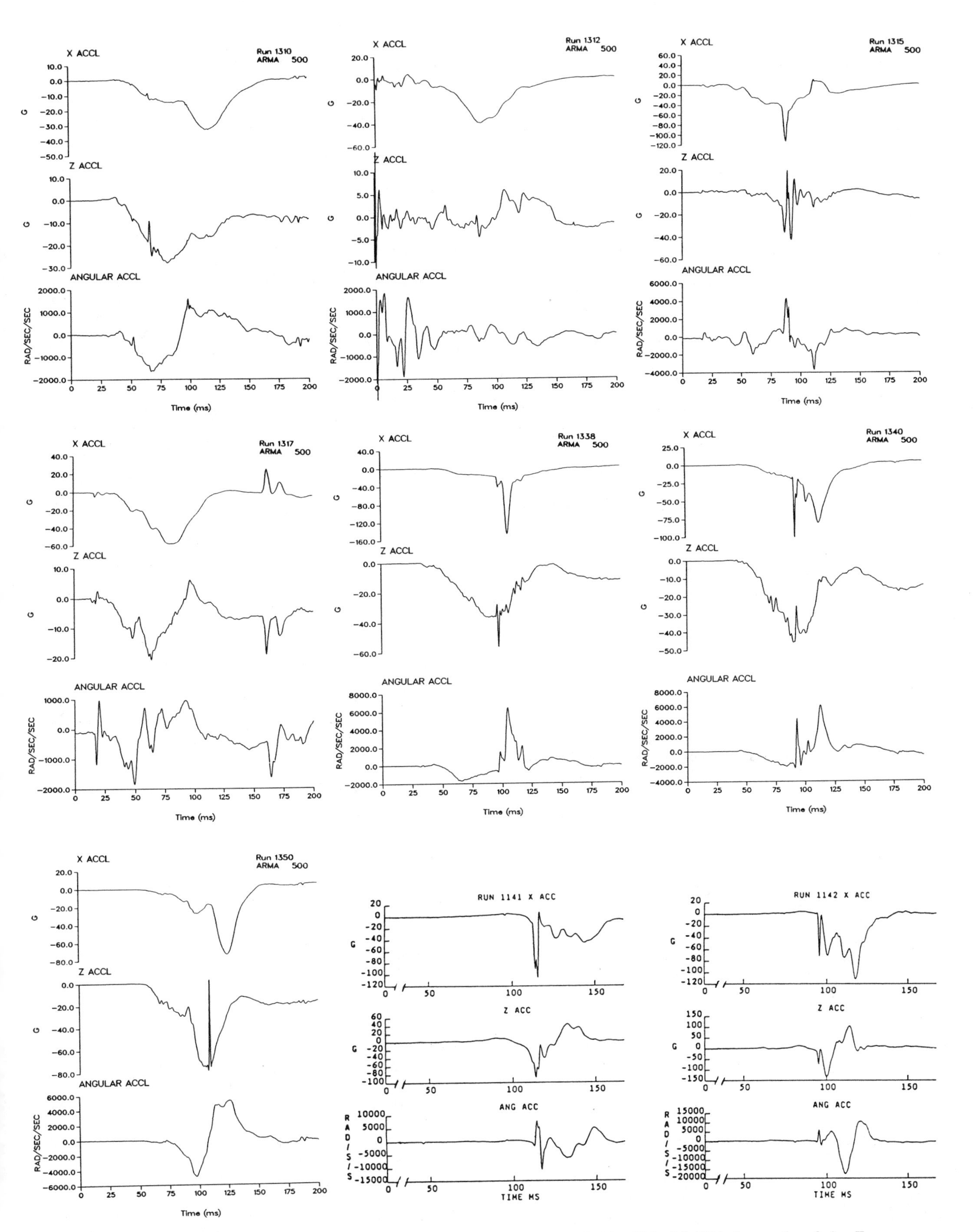

Figure 13: 2D translational and rotational accelerations of the Hybrid III dummy head in Hyge sled tests.

higher with head contact (see Figure 13). So what is the proper interpretation of brain injury risk and when would injury occur?

A 36 ms or 15 ms cut-off on the HIC interval reduces the peak HIC value for head acceleration without steering wheel contact. In fact, a 15 ms cut-off indicates that the exposure with head contact is more severe. However, the full wave-form HIC is higher than that for the unrestrained dummy with direct contact on the windshield for the 11.7 m/s (25 mph) test and of similar magnitude as the 15.0 m/s (30 mph) barrier crash simulation. In contrast, the peak head acceleration is significantly lower for Test 1350 than in similar belt restraint tests with head contact or in the tests with an unrestrained passenger and windshield contact.

Figure 14 shows the resultant head velocity with respect to the sled with peak values of 9-12 m/s (20-27 mph) for belted occupants (this is reduced to 5 m/s (11 mph) for belt plus air bag restraint). Peak velocity of the head is higher for the belted occupant without head contact because contact with the steering wheel rim reduces the maximum velocity and forward excursion of the dummy's head. However, in either case, there is a rapid drop-off in velocity at approximately 100 ms by the restraining effect of the lap-shoulder belt or by steering wheel contact.

By looking at the various responses in Table 3 that may be indicative of brain injury risk, there is wide variability in the interpretation of the most severe exposure for the tests conducted in this demonstration series. What we need to do is interpret the dynamic motion of the skull, the skull-brain interactions, and finally, the strain or deformation response of brain and supporting tissues to more clearly assess brain injury risks.

BRAIN RESPONSES TO HEAD DYNAMICS

Two approaches are being investigated to determine brain tissue responses to the measured head dynamics from Hybrid III dummy tests. One involves brain compliance modeling and the other finite element modeling. Before discussing the current approaches to modeling the elastic and viscous compliance of the brain-skull interaction, a review of the previously developed Mean Strain Criterion (MSC) will be given to put perspective on the current work.

SKULL COMPLIANCE MODELS (MSC AND NMSC) - In the late 1960's and early 70's, experiments were conducted on the forced-vibration response of human and animal heads. These tests involved attachment of a driver piston to the skull and vibrating the head over a range in frequency. The resonance or impedance response defines the vibratory characteristics of a portion of the skull and brain interacting with the driving force (see Figure 15a). The resonance data were used to develop a lumped-mass model called the Mean Strain Criterion (MSC) (29). The maximum strain due to input velocity and displacement from measured acceleration was interpreted as a risk of head injury. The MSC model primarily characterizes the stiff compliance of the skull due to a concentrated loading.

More recently, Stalnaker (30,31) has improved the biofidelity of the original MSC model. In this approach, the skull and brain masses are assumed coupled to the driving point on the skull through parallel dashpot and spring-dashpot elements representing the compliance of the system. Relative displacement between the driving point and skull-brain masses is interpreted as a strain by normalizing with the internal AP dimension of the skull (16.1 cm is used in this paper (32)). As shown by Khalil and Hubbard (32), deflection of the skull can be interpreted as circumferential surface strain and hoop surface strain in the layered skull. In their work, 1 mm deflection of the skull at the site of impact produces .39% tensile strain on the surface of the skull.

In the current use of the NMSC model, dummy test responses are used to drive the heavy skull-brain mass (m2) to determine an effective input force on the frontal skull mass m1 for an AP direction simulation. Stalnaker has reported that the energy dissipated by the viscous damping elements in the NMSC model correlate well with brain injury risk. In a way, this new effort indirectly applies the Viscous response to assess injury, since VC is a measure of energy dissipated during the rapid loading phase of impact. However, the NMSC damping elements are primarily related to the compliance of the skull system with some influence from the mass and viscoelasticity of the brain.

Stalnaker (31) has also used the model to calculate impact force on the head; but, this approach may not correctly account for the force between the head and neck which can be large in comparison to the external forces of contact for some impact situations (this is also clearly the case when the head of the lap-shoulder belted occupant does not experience external contact). A better application of the NMSC model may be possible using the external contact force calculated from the In-Line analysis procedure. The calculated contact force components can be applied directly to the light-weight skull mass (m1) and deflection and strain computed from the NMSC lumped-mass model.

BRAIN COMPLIANCE MODEL (BCM) - Another approach is under investigation at GM to assess brain injury risks from Hybrid III head dynamics. Brain compliance modeling assumes that the Hybrid III dummy has a rigid skull and dynamics accurately characterized by the In-Line accelerometry methods. The integrated velocity and displacement responses of the skull are used as input to a lumped-mass model that simulates the compliance of brain tissue in the skull (see Figure 15b).

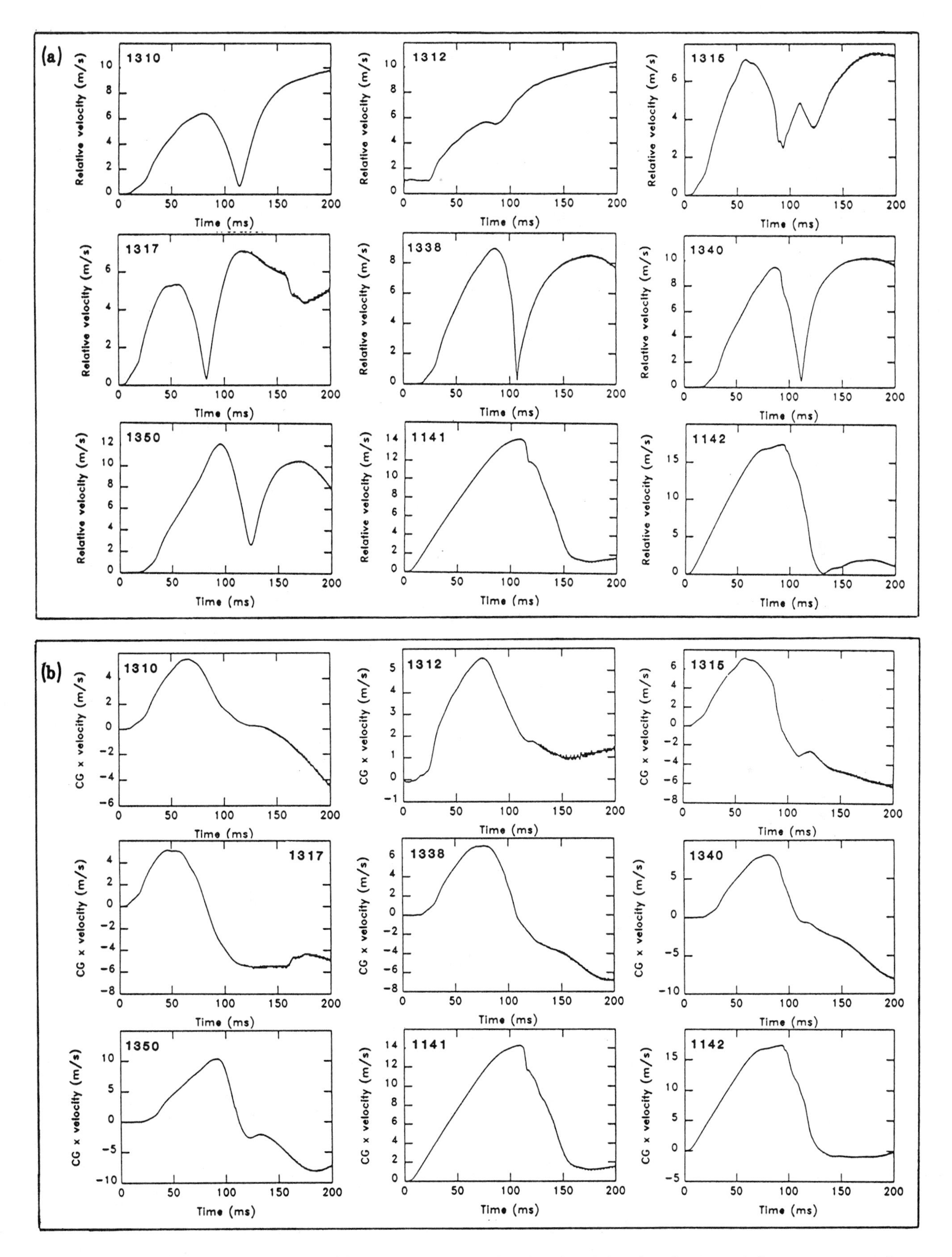

Figure 14: (a) Resultant and (b) x-component of Hybrid III head velocity with respect to the vehicle interior in Hyge sled tests.

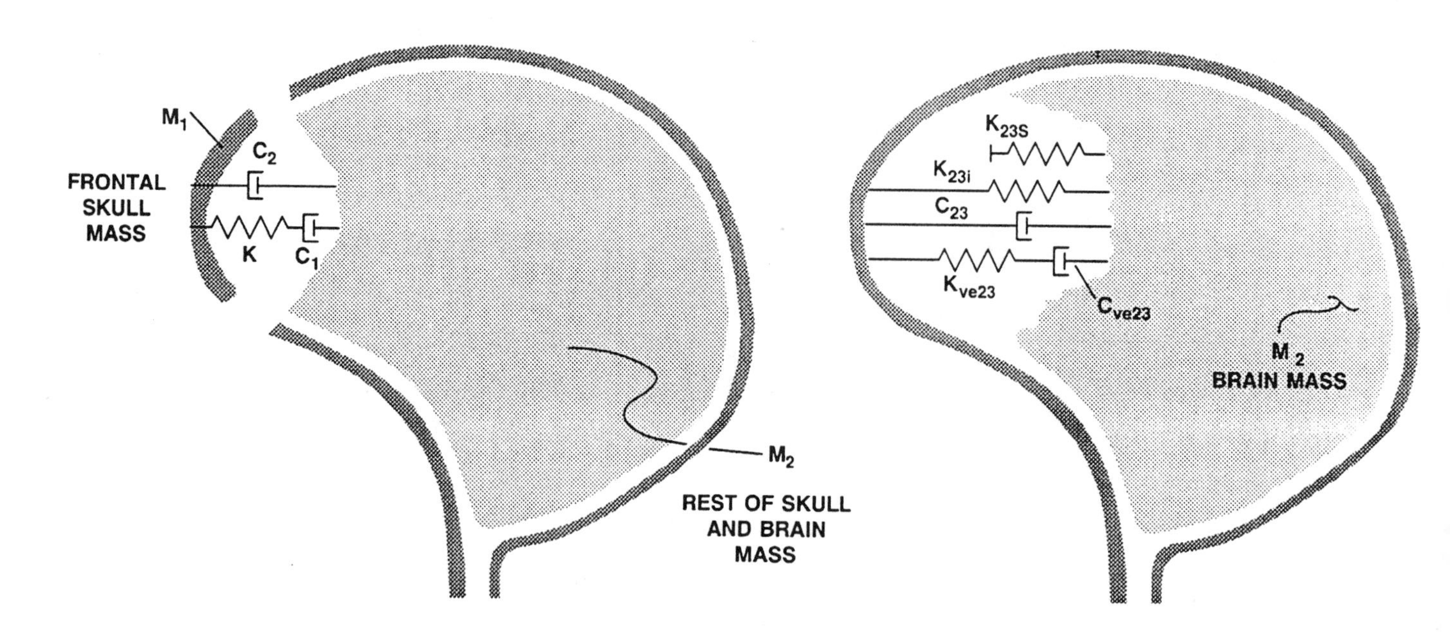

Figure 15: (a) New Mean Strain Criterion Model (NMSC) and (b) Brain Compliance Model (BCM) representations of the spring, dashpot, and mass of the skull-brain system for head dynamics analysis and brain injury assessment.

Table 3

Hyge Sled Test Data with Hybrid III Dummy and 2D Inline Accelerometers

	1310	1312	1315	1317	1338	1340	1350	1141	1142
Sled Speed (m/s)	14.7	17.7	17.0	17.1	17.2	17.2	17.2	11.7	15.0
Restraints	L/S	UNB+ARCS	UNB+SIR	L/S+SIR	PRT:L/S	L/S	L/S	UNB	UNB
Head Contact	NO	NO	WS	NO	HUB	HUB	NO	WS	WS+IP
HIC Result.									
Full	354	215	476	548	889	993	1638	556	1771
36 ms	199	211	476	533	889	983	1577	445	1584
15 ms	122	123	455	328	889	560	712	400	768
HIC AP									
Full	176	212	411	504	816	643	604	209	609
36 ms	150	209	411	498	816	643	599	209	609
15 ms	91	122	355	319	816	430	454	209	182
Head Accel (g)									
Result.	38	40	114	58	150	130	80	125	148
AP	34	38	110	58	145	120	75	105	112
SI	28	6	40	20	-60	-45	-70	86	130
Head Velocity (m/s)									
X	5.8	5.5	7.2	5.0	7.2	8.0	10.6	14.2	17.0
Z	5.2	3.0	0.2	2.2	6.8	7.2	9.0	2.0	0.4
Result.	6.5	5.8	7.2	5.3	9.2	9.5	12.2	14.3	17.2
Rotational									
Accel (r/s^2)	1600/-1300	300/-300	5000/-4000	1000/-1800	1800/-6200	2000/-6000	4600/-5200	12000/-8000	10000/-17000
Vel (r/s)	42	12	-22	10/-18	45	45	76	-52	-80
Displ (deg)	-110	57	-60	-32	-82	-95	-115	14/-30	12/-30
Contact Force (kN)									
Result.	-	0.71	4.45	3.92	6.10	4.00	1.07	5.79/1.60	5.79/5.79
AP	0.28	0.70	4.23	3.65	5.96	3.92	0.89	4.90/1.60	3.47/2.10
SI	-	0.26	1.87	0.94	1.67	0.89	0.22	2.94/1.16	5.56/5.56

L/S Lap-Shoulder Belt; PRT Pretensioned Belts; UNB unbelted, ACRS 1970's Driver Airbag, SIR 1988 Supplemental Driver Airbag
WS Head Impact of Windshield, HUB Head Impact of Steering Wheel Hub, IP Head Impact of Instrument Panel

Head velocity and displacement are the inputs that drive the model, since force is developed in the elastic spring and viscous damper by relative displacement and velocity between the brain and skull, not skull acceleration. As the skull dynamically moves, the inertial, elastic, and viscous compliance of brain tissue respond by deforming and displacing the brain with respect to the rigid skull. For the 2D approach, the three components of motion (AP and SI translational acceleration, and rotational acceleration) are evaluated independently as in the NMSC approach and the total effects determined by the combination in time of the individual responses. Further development of the model is planned to allow use of the resultant acceleration as input for trend comparisons with the HIC.

A preliminary brain compliance model is presented here as an example of the approach and type of human responses that are associated with the range of impacts of the Hybrid III dummy in this study. It is primarily shown for trend comparisons. The static, elastic compliance of human brain was approximated from published data on the intracranial pressure-volume characteristics of the dog (33,34). As a volume of fluid is injected into the CSF space intracranial pressure increases. The relationship was used for the human head without scaling in the preliminary study as an increase in force on the anterior portion of the skull (assuming a half cylinder profile area of 42.5 cm^2 (32)) for an axial displacement of brain tissue equivalent to the associated volume injected assuming comparable compliance for the human and canine cranial cavity (see Figure 16). The response is bi-linear and can be characterized by an initial stiffness of 10 kN/m for deflections up to 1.35 mm and a secondary stiffness of 50 kN/m for deflections beyond 1.35 mm.

The bilinear spring approximates the static compliance of the brain. It is stiffer than the compression response of isolated brain tissue, which have been shown to undergo very large strains (35-40), but is considerably softer than what would be predicted from the bulk modulus of fluid. Using water, the elastic stiffness would be about 58,000 kN/m. Although the static stiffness may become stiffer than predicted by the pressure-volume data for deflections above 3 mm and a tertiary spring may be warranted, the skull is not a closed system and fluids and tissue can be forced out openings. Thus a tertiary spring would have a value much lower than the bulk modulus of brain.

The viscous damping characteristic of brain is approximated by a dashpot value of 0.20 kN/m/s which was selected to give relatively large deflections between brain and skull in this preliminary study. However, trends in relative severity of predicted responses among sled tests were not effected by increasing the dashpot values an order in magnitude. This reduced the maximum deflections to below 7 mm but increased the force developed by the damper. The more-compliant viscous model parameters were selected in this example of the analysis approach, and a more refined data-set is needed for a validated model.

The important biomechanical responses of the brain compliance model are the Viscous response (VC) and the displacement of the brain with respect to the skull. The former can be generally interpreted as the risk of soft neural tissue injury and the latter as the risk of rupture of tethering vessels by relative displacement between the brain and skull. The fact that the brain is loosely coupled to the skull and may undergo relative deflections during impact has been experimentally proven (41-44). The soft compliance of isolated brain tissue is also well known (35-40). Lumped-mass modeling similar to that presented here has been successfully used in a recent analysis of spinal cord injury (26) and has been proposed for the head (45). Brain mass is assumed to be 1.36 kg.

Responses from the NMSC skull compliance model are shown in Figure 17a for several of the sled tests conducted in this series. The deflection response and energy dissipated by the viscous damping elements of the NMSC show a gradual rise to peak value after the primary acceleration phase of impact. In contrast, the brain compliance model shows (Figure 17b) much more variation in its responses to skull motion. The BCM allows restoration of the brain by the elastic elements in the model; the NMSC does not. The BCM also has much lower parameter values than those of the NMSC and thus is more related to the brain response.

Brain compliance modeling may help pinpoint more specifically the time of high injury risks associated with the Viscous loading of brain tissue and the deflection of the brain mass. In all cases, maximum deflection occurs after the significant peaks in Viscous response. This is identical to what we have seen in other soft tissue impacts in the body (20-22). Peak values from the modeling responses are shown in Table 4 and, as in the previous data, show differing levels of brain injury risk depending on the criteria and analysis approach. Because the BCM responses depend on the velocity and displacement of the skull, high frequency content in the input is not needed for the calculation. We used 180 Hz for the analysis, and cutoffs of 500 Hz or 1000 Hz did not change the responses.

COMPARISON OF HEAD RESPONSE TRENDS

Table 5 presents the relative ranking of the Head impact responses for the tests conducted in this study. Tests are ranked by the severity of predicted responses. The full wave-form HIC and peak resultant acceleration measured from the Hybrid III dummy are shown for contrast to the peak Viscous and deflection response predicted by the BCM using the resultant acceleration as input. These analysis approaches exclude the potential effects of

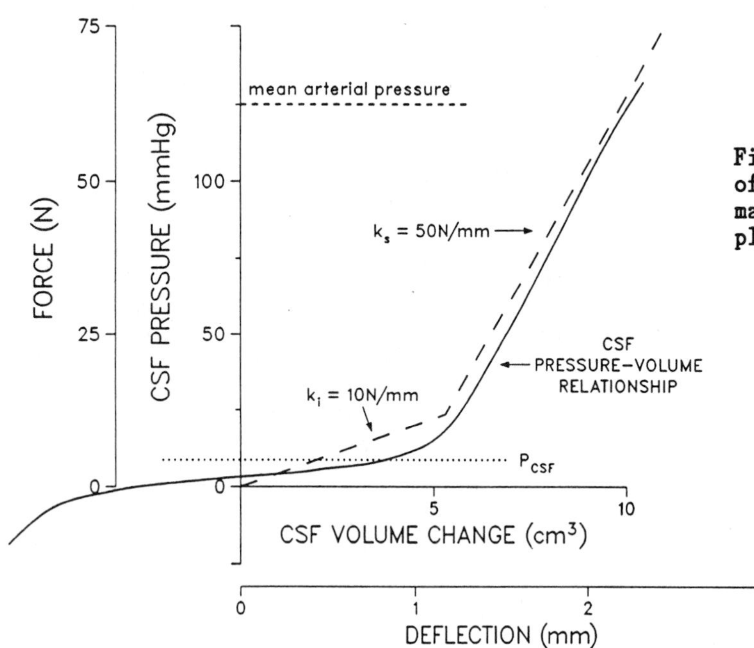

Figure 16: Static pressure-volume relationship of the canine brain (from 33,34) and approximation by elastic components in the Brain Compliance Model.

Table 4

New Mean Strain Criterion, Brain Compliance and Finite Element Modeling 2D Head Dynamics (AP Component Responses)

	1310	1312	1315	1317	1338	1340	1350	1141	1142
Sled Speed (m/s)	14.7	17.7	17.0	17.1	17.2	17.2	17.2	11.7	15.0
Restraints	L/S	UNB+ACRS	UNB+SIR	L/S+SIR	PRT:L/S	L/S	L/S	UNB	UNB
Head Contact	NO	NO	WS	NO	HUB	HUB	NO	WS	WS+IP
NMSC (Stalnaker)									
Defl (mm)	-0.36	-0.45	-0.52	-0.55	-0.57	-0.66	-0.70	-0.27	-0.42
Strain (%)	0.22	0.28	0.33	0.34	0.35	0.41	0.44	0.17	0.26
Force (kN)	1.32	1.58	4.13	2.27	5.77	3.38	2.91	3.36	3.74
Diss Energy (J)	0.04	0.05	0.10	0.11	0.14	0.14	0.15	0.05	0.11
NMSC (Contact)									
Defl (mm)	0.42	0.45	0.52	0.55	2.07	2.10	0.69	2.60	1.71
Strain	0.26	0.28	0.33	0.34	1.30	1.30	0.42	1.62	1.02
Brain Compliance Model									
VC (m/s)	0.011	0.016	0.108	0.026	0.227	0.072	0.049	0.039	0.076
Defl (cm)	-0.82	-0.95	-1.58	-1.34	-2.11	-1.70	-1.65	-0.75	-1.59
Comp. (%)	5.09	5.94	9.88	8.38	13.11	10.56	10.25	4.69	9.94
Accel. (g)	-35.9	-43.0	-87.2	-60.8	-127.9	-84.8	-78.6	-52.1	-81.8
Force (kN)	-0.48	-0.57	-1.16	-0.81	-1.70	-1.13	-1.05	-0.69	-1.09
(Resultant Translational Acceleration)									
Brain Compliance Model									
VC (m/s)	0.012	0.016	0.119	0.025	0.226	0.069	0.109	0.084	0.246
Defl (cm)	-0.92	-0.95	-1.69	-1.35	-2.23	-1.75	-2.02	-1.07	-2.09
Comp. (%)	5.75	5.94	10.56	8.44	13.94	10.94	12.63	6.69	13.06
Accel. (g)	-39.6	-43.1	-92.9	-60.6	-129.3	-83.0	-103.3	-72.6	-133.5
Force (kN)	-0.53	-0.57	-1.24	-0.81	-1.72	-1.11	-1.38	-0.97	-1.78

L/S Lap-Shoulder Belt; PRT Pretensioned Belts; UNB unbelted, ACRS 1970's Driver Airbag, SIR 1988 Supplemental Driver Airbag
WS Head Impact of Windshield, HUB Head Impact of Steering Wheel Hub, IP Head Impact of Instrument Panel

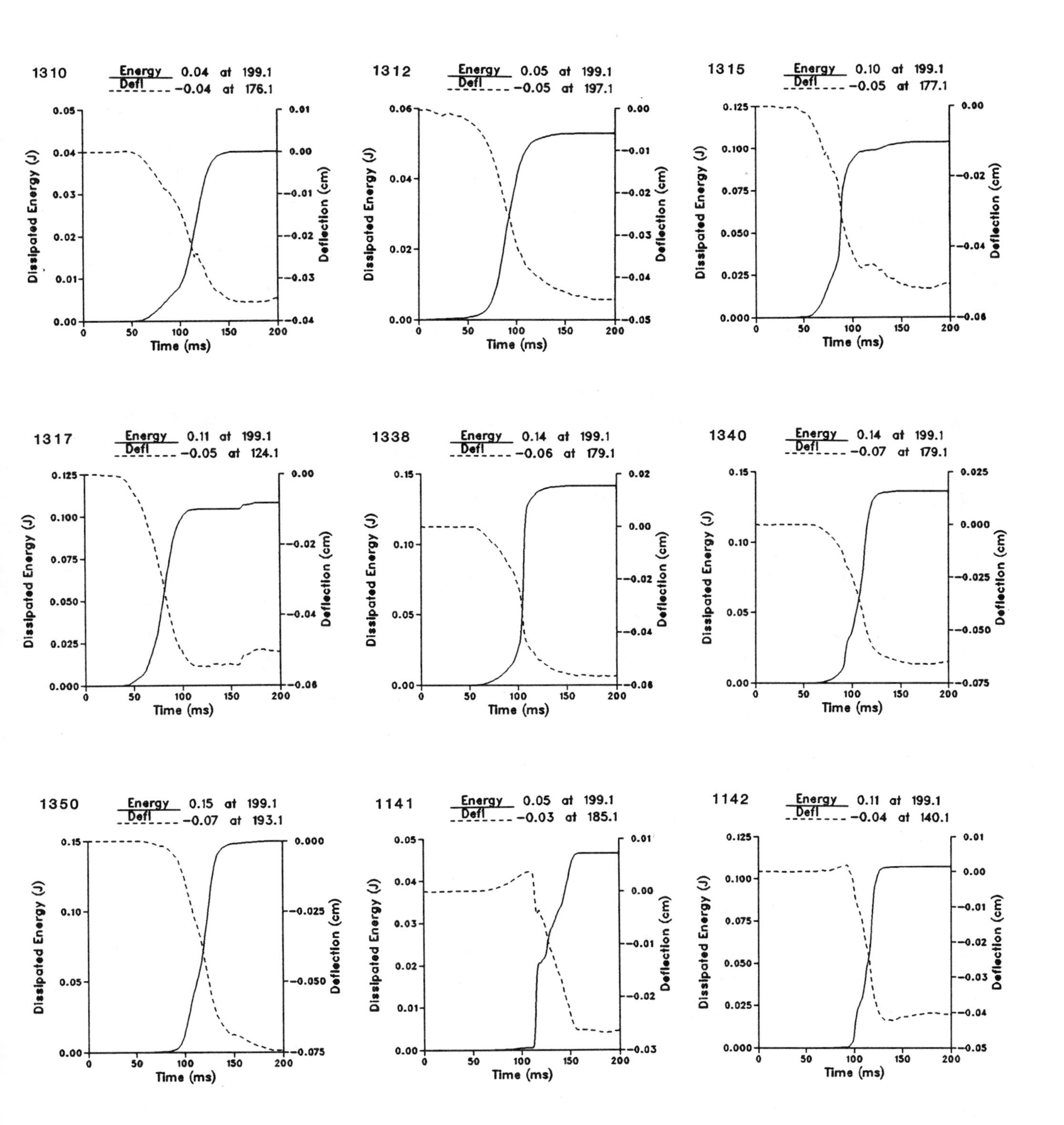

Figure 17: (a) Mean Strain Criterion Model deflection and energy dissipation responses for the Hyge sled tests.

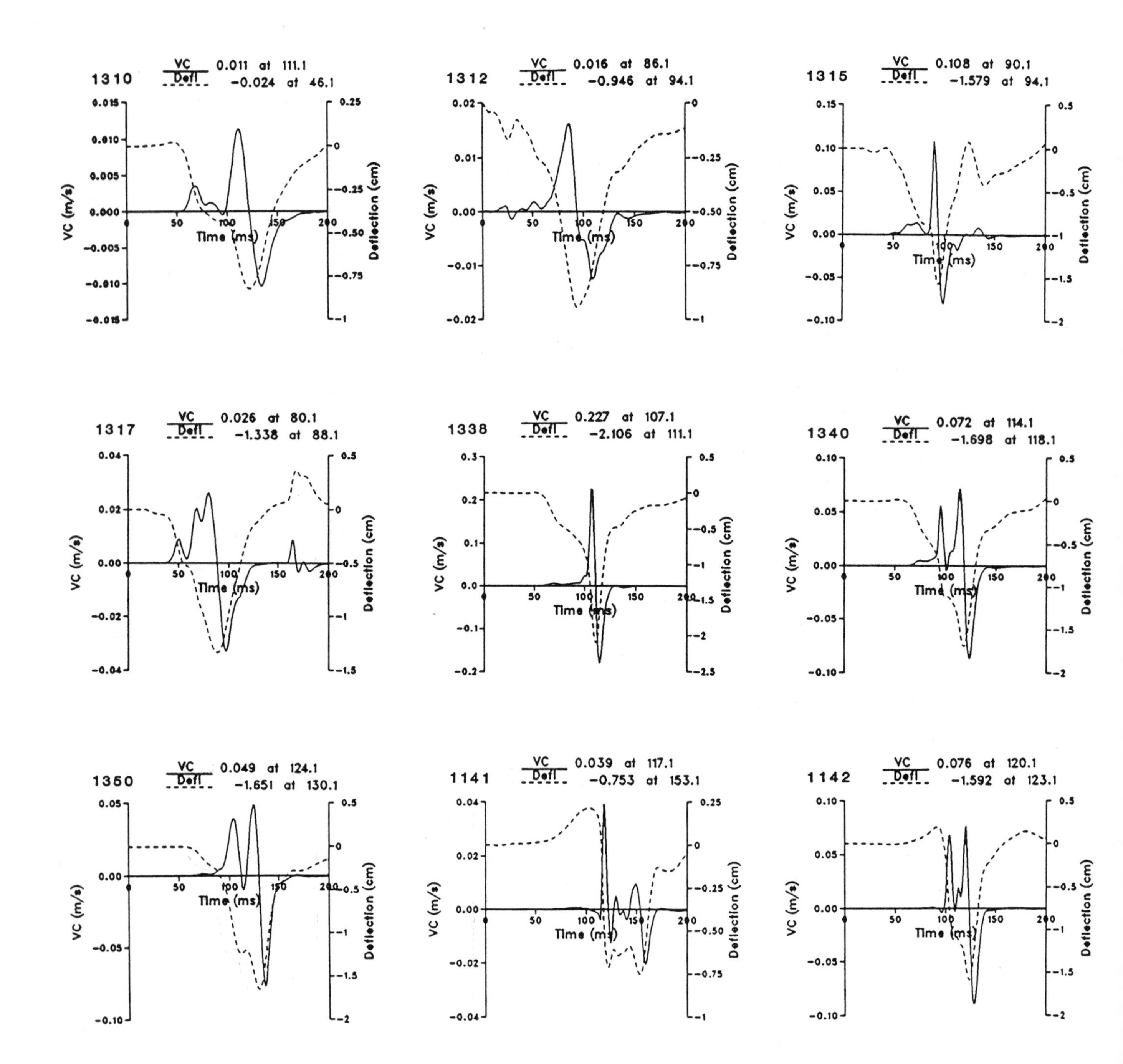

Figure 17: (b) Brain Compliance Model deflection and viscous responses for the sled tests.

rotational acceleration on ranking the severity of impacts for brain injury risk, but they do provide a point of discussion for the contrast in ranking they provide.

For example, the HIC and BCM do a comparable job of ranking the impacts except for the two cases of short duration head impact superimposed on a longer duration acceleration, and the one case of non-contact acceleration. In Test 1338, the belted driver experiences head acceleration and a sharp impact on the steering wheel hub. In Test 1315, the unrestrained driver experiences head acceleration by the steering wheel air bag but rides up and over the bag. There is eventually a head impact with fracture of the windshield close to the header. In both cases, the full wave-form and 36 ms HIC identifies the long-duration, lower level acceleration for calculating the maximum HIC value; whereas, the brain compliance model predicts that the early phase of head acceleration deforms the brain against the skull so that the sharp subsequent impact causes high risk of injury by a viscous mechanism. In contrast, the belted passenger in Test 1350 does not have a head impact, yet the HIC calculation with long duration indicates more risk than predicted by the brain compliance model or by a 15 ms HIC calculation cutoff.

The brain compliance modeling indicates that Test 1141 has a higher viscous injury risk than from brain deflection whereas the longer duration head accelerations with belt restraints (Tests 1350 and 1340) show a higher relative deflection response. This type of response information may provide focus on the underlying type or mechanism of brain injury expected from a head impact, especially when the full 2D responses associated with head translational and rotational acceleration are included in the final brain compliance model.

In contrast, the NMSC seems to give a very different perception of risk. Tests 1141 and 1142 cause a relatively low deflection response probably because of the high compliance of the system representing primarily the skull stiffness. Tests 1350 and 1340 give a high deflection response and may accentuate the relative severity of the impact because of the lack of restoring force without a spring coupling between the head masses.

As refinements in the brain compliance model are made and appropriate responses are used to interpret brain injury risks, the utility of the analyses method as a post-processing program will improve. It does, however, exhibit the important characteristic of a compliant coupling between the brain and skull and uses a response that shows the cumulative effects of head dynamics, such as the deflection or the Viscous response between the brain and skull.

The BCM approach should simplify analysis of the complex head dynamics measured by the In-Line accelerometry method and should provide complementary and more specific injury probability assessment of alternative safety designs. The Viscous mechanism of brain injury may lead to a better interpretation of injury risk from lumped-mass modeling of Hybrid III dummy head dynamics. A correct biomechanical compliance of the Hybrid III face will also help give a more human-like dummy response. Thus, our research is focusing on a better assessment of head injury risks by procedures based on an accurate understanding of the biomechanics of central nervous system trauma, an accurate interpretation of head dynamics, and biofidelity in the dummy's response to impact loads.

This effort is also complementary to a wide range of international research on head injuries. Two recent symposium publications (46,47) attest to the interest in advancing our understandings of brain injuries.

ACKNOWLEDGEMENTS

Sincere appreciation is extended to the many people in the Biomedical Science Department who helped in this research effort, particularly John Horsch and Joe McCleary for conducting sled tests and providing dummy response data from the In-Line accelerometer package, Ken Baron and Sudhaker Arepally for their help in developing the computer program for the brain compliance model, T. Rex Shee for calculating head velocities with respect to the moving coordinate system on the Hyge sled, Albert King and John Melvin for providing preliminary data on the facial impact response of the human cadaver and Hybrid III dummy, and Janis Georgen and Mary Foster for their help in preparing the paper.

With the approval of AAAM (Association for the Advancement of Automotive Medicine) and SAE (Society of Automotive Engineers), portions of this paper were presented and distributed in a report from the September 30, 1987 Symposium on Head Injury Mechanisms hosted by AAAM and the Volvo Car Corporation.

Table 5

Comparison of Head Injury Responses
(Based on Resultant Translation Acceleration)

Hybrid III Response		Brain Compliance Model	
HIC	Acceleration	Viscous	Deflection
1142 (1.00)*	1338 (1.00)	1142 (1.00)	1338 (1.00)
1350 (0.92)	1142 (0.99)	1338 (0.92)	1142 (0.94)
1340 (0.56)	1340 (0.87)	1315 (0.48)	1350 (0.91)
1338 (0.50)	1141 (0.83)	1350 (0.44)	1340 (0.78)
1141 (0.31)	1315 (0.76)	1141 (0.34)	1315 (0.76)
1317 (0.31)	1350 (0.53)	1340 (0.28)	1317 (0.61)
1315 (0.27)	1317 (0.39)	1317 (0.10)	1141 (0.48)
1310 (0.20)	1312 (0.27)	1312 (0.07)	1312 (0.43)
1312 (0.12)	1310 (0.25)	1310 (0.05)	1310 (0.41)

* Test number is followed by the ratio of the response amplitude relative to the maximum observed in the tests.

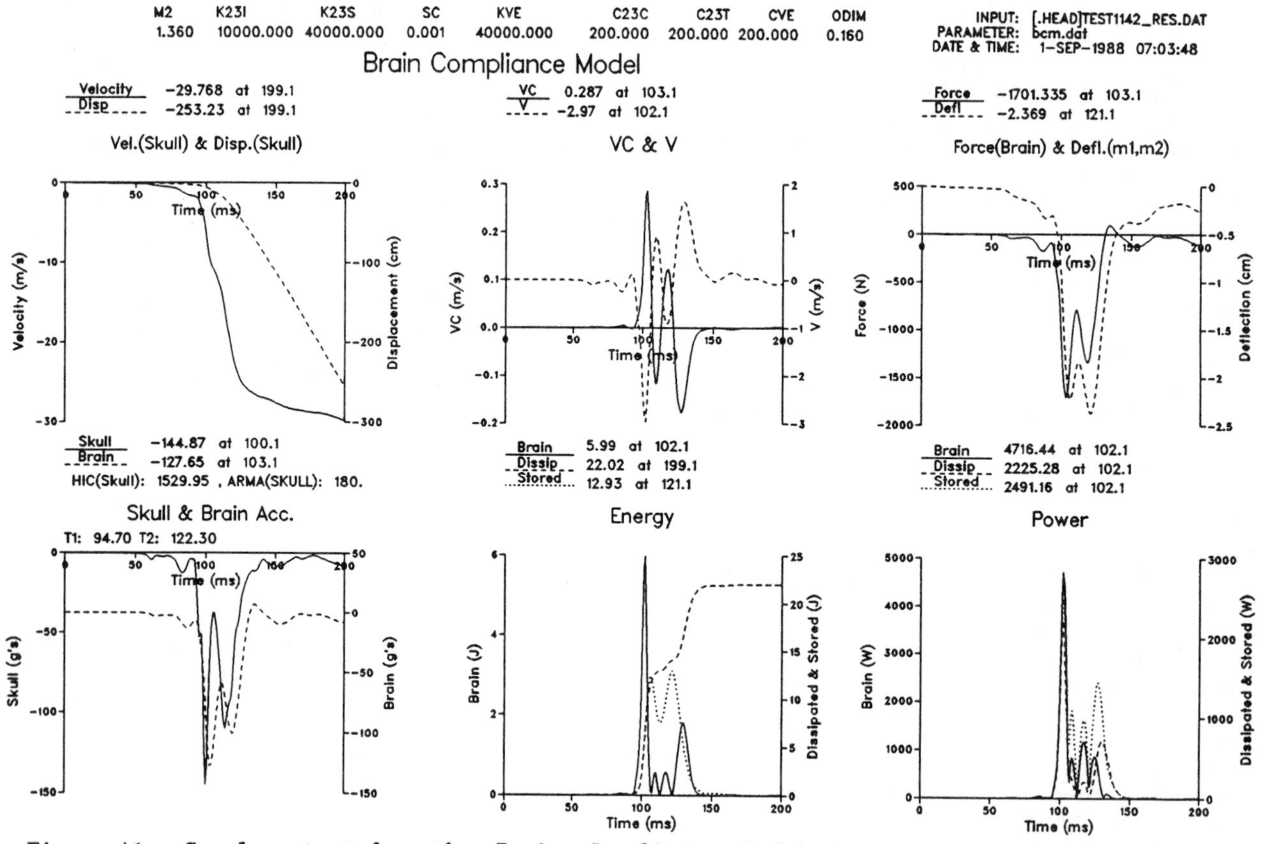

Figure A1: Sample output from the Brain Compliance Model showing the dynamics of the skull velocity and displacement input, the Viscous and displacement response of the brain, force on the brain, accelerations, and energy and power associated with brain responses. The input parameters at the top relate to parameters described in (48,49), with the exceptions that m2 in the BCM is m3 in the Lobdell model and m1 is m2, and energy and power of the brain mass are based on relative velocity and displacment with respect to the skull.

REFERENCES

1. Holbourn, A.H.S., "Mechanics of Head Injury." *Lancet* 2:438-441, 1943.
2. Holbourn, A.H., "The Mechanics of Brain Injuries." *British Medical Bulletin* 3:144-149, 1945.
3. Patrick, L.M., Lissner, H.R., and Gurdjian, E.S., "Survival by Design -- Head Protection." *Proceedings of the 7th Stapp Car Crash Conference* , SAE, Charles C. Thomas, Springfield, Illinois, 1963.
4. Gadd, C.W., "Use of a Weighted Impulse Criterion for Estimating Injury Hazard." *Proceedings of the Tenth Stapp Car Crash Conference,* New York, Society of Automotive Engineers, 1966.
5. Versace, J., "A Review of the Severity Index." *Proceedings of the Fifteenth Stapp Car Crash Conference* , New York, Society of Automotive Engineers, 1971.
6. Prasad, P. and Mertz, H.J. "The Position of the United States Delegation to the ISO Working Group 6 on the Use of HIC in the Automotive Environment." SAE Technical Paper No. 851246, Society of Automotive Engineers, Warrendale, MI, 1985.
7. Gurdjian, E.S. and Gurdjian, E.S., "Acute Head Injuries." *Surgery, Gynecology and Obstetrics* 146:805-820, 1978.
8. Gurdjian, E.S. and Gurdjian, E.S., "Cerebral Contusions: Re-Evaluation of the Mechanism of Their Development." *Journal of Trauma* 16(1):35-51, 1976.
9. Ommaya, A.K., Grubb, R.L., and Naumann, R., "Coup and Contre-coup Injury: Observations on the Mechanics of Visible Brain Injuries in the Rhesus Monkey." *Journal of Neurosurgery* 35(8):503-516, 1971.
10. Ommaya, A.K. (ed), *Proceedings of the Consensus Workshop on Head and Neck Injury Criteria.* Washington, DC, US Government Printing Office, 1983.
11. Ommaya, A.K., and Gennarelli, T.A., "Cerebral Concussion and Traumatic Unconsciousness." *Brain* 97:633-654, 1974.
12. Gennarelli, T.A., Thibault, L.E., Adams, J.H., et al, "Diffuse Axonal Injury and Traumatic Coma in the Primate." *Annals of Neurology* 12:564-574, 1982.
13. Ommaya, A.K., "Biomechanics of Head Injury." *The Biomechanics of Trauma.* A.M. Nahum and J. Melvin (eds), Appleton-Century-Crofts, Norwalk, 1984.
14. Ran, A., Koch, M., and Mellander, H., "Fitting Injury Versus Exposure Data Into a Risk Function." 1984 International IRCOBI Conference on the Biomechanics of Impacts, September, 1984.
15. Viano, D.C., "Biomechanics of Bone and Tissue: A Review of Material Properties and Failure Characteristics," SAE Technical Paper 861923, in SAE Special Publication, *Biomechanics and Medical Aspects of Lower Limb Injuries,* P-186, pp. 33-64, 1986.
16. Khalil, T.B. and Viano, D.C., "Comparison of Human Skull and Spherical Shell Vibrations - Implications for Head Injury Modeling." *J Sound and Vibration,* 82(1):95-110, 1982.
17. King, A.I. and Melvin, J.W., Personal Communication.
18. MacLean, A.J., Personal Communication.
19. Viano, D.C. "Perspectives on Head Injury Research," 1985 International Research Conference on the Biokinetics of Injury, IRCOBI Secretariat, Lyon, France, 1985.
20. Viano, D.C., A.I. King, J.W. Melvin and K. Weber, "Injury Biomechanics Research: An Essential Element in the Prevention of Trauma." Accepted for publication in the *Journal of Biomechanics,* October, 1987.
21. Lau, I.V., and Viano, D.C., "The Viscous Criterion: Bases and Applications of an Injury Severity Index for Soft Tissues." *Proceedings of the 30th Stapp Car Crash Conference,* SAE Technical Paper 861882, pp. 123-142, October, 1986.
22. Viano, D.C., and Lau, I.V., "A Viscous Tolerance Criterion for Soft Tissue Injury Assessment." Accepted for Publication in the *Journal of Biomechanics,* 21(5):387-399, 1988.
23. Lighthall, J.W., "Controlled Cortical Impact: A New Model of Mechanical Brain Injury." *J Neurotrauma,* 5(1):1-12, 1988.
24. Lighthall, J.W., Personal Communication.
25. Anderson, T.E., "Physical Parameters of Cord Contusion Differentiate Between Neural and Vascular Injury." *J Neurosurgery,* 62:115-119, 1985.
26. Kearney, P., Personal Communication.
27. Viano, D.C., Melvin, J.W., McCleary, J.D., Madeira, R.G., Shee, T.R., and Horsch, J.D., "Measurement of Head Dynamics and Facial Contact Forces in the Hybrid III Dummy." *Proceedings of the 30th Stapp Car Crash Conference,* SAE Technical Paper 861891, pp. 269-290, October, 1986.
28. Shee, T.R., Personal Communication.
29. Stalnaker, R.L., J.H. McElhaney, and V.L. Roberts, "MSC Tolerance Curve for Human Head Impacts." The ASME Winter Annual Conference, 71WA/BHF-10, 1971.
30. Stalnaker, R.L., Low, T.C., and Lin, A.C., "Translational Energy Criteria and Its Correlation with Head Injury in the Sub-Human Primate." *Proceedings of the 1987 International IRCOBI Conference on the Biomechanics of Impacts,* pp. 223-238, September, 1987.
31. Stalnaker, R.L., Lin, C.A., and Guenther, D.A., "The Application of the New Mean Strain Criterion (NMSC)," *Proceedings of the 1985 International IRCOBI/AAAM Conference on the Biomechanics of Impacts,* Goteborg, Sweden, June, 1985.
32. Khalil, T.B. and Hubbard, R.P., "Parametric Study of Head Response by Finite Element Modeling." *J Biomechanics* 10:119-132, 1977.

33. Lofgren, J., von Essen, C., and Zwetnow, N.N., "The Pressure-Volume Curve of the Cerebrospinal Fluid Space in Dogs." *Acta Neurol. Scandinav* 49:557-574, 1973.
34. Lofgren, J. and Zwetnow, N.N., "Cranial and Spinal Components of the Cerebrospinal Fluid Pressure-Volume Curve." *Acta Neurol. Scandinav,* 49:575-585, 1973.
35. McElhaney, J.H., Melvin, J.W., Roberts, V.L., and Portnoy, H.D., "Dynamic Characteristics of the Tissues of the Head." *In Perspectives in Biomedical Engineering,* Proceedings of a Symposium Organized in Association with the Biological Engineering Society and Held in the University of Strathclyde, Glasgow, (ed.) R.M. Kenedi, June, 1972.
36. Estes, M.S. and McElhaney, J.H., "Response of Brain Tissue of Compressive Loading." American Society for Mechanical Engineers Paper, 1971.
37. Galford, J.E. and McElhaney, J.H., "A Viscoelastic Study of Scalp, Brain and Dura," *J Biomechanics,* 3:211-221, 1970.
38. Demiray, H., "A Note on the Elasticity of Soft Biological Tissues." *J Biomechanics,* 5:309-311, 1972.
39. Pamidi, M.R. and Advani, S.H., "Nonlinear Constitutive Relations for Human Brain Tissue." *Transactions of the ASME,* 100:44-48, 1978.
40. Shuck, L.Z. and Advani, S.H., "Rheological Response of Human Brain Tissue in Shear." *J Basic Engineering,* 905:911, 1972.
41. Shatsky, S.A., Alter, W.A., Evans, D.E., Armbrustmacher, V., and Clark, G., "Traumatic Distortions of the Primate Head and Chest: Correlation of Biomechanical, Radiological and Pathological Data." *SAE Technical Paper 741186,* 1974.
42. Pudenz, R.H. and Shelden, C.H., "The Lucite Calvarium, A Method for Direct Observation of the Brain:II." *J Neurosurg,* 3:487-505, 1944.
43. Gosch, H.H., Gooding, E., and Schneider, R., "Distortion and Displacement of the Brain in Experimental Head Injuries." *Surgical Forum,* 20:425-426, 1969.
44. Stalnaker, R.L., Melvin, J.W., Nusholtz, G.S., Alem, A.M., and Benson, J.B., "Head Impact Response." *SAE Technical Paper 770921,* Society of Automotive Engineers, Warrendale, PA, 1977.
45. Alem, N.M., "Simulation of Head Injury Due to Combined Rotation and Translation of the Brain." *SAE Technical Paper 741192,* Society of Automotive Engineers, Warrendale, PA, 1974.
46. *Injury Biomechanics,* Special Publication SP-731, Government/Industry Meeting and Exposition, Washington, DC, May 18-21, 1987, Society for Automotive Engineers, Warrendale, PA, 1987.
47. *Head Injury Mechanisms,* Symposium Report, September 30, 1987, New Orleans, LA, Association for the Advancement of Automotive Medicine, Des Plains, IL, 1987.
48. Viano, D.C., "Evaluation of Biomechanical Response and Potential Injury from Thoracic Impact." *Journal of Aviation Space and Environmental Medicine,* 49(1):125-135, January, 1978.
49. Viano, D.C., "Evaluation of the Benefit of Energy-Absorbing Material for Side Impact Protection: Part I." *Proceedings of the 31st Stapp Car Crash Conference,* pp. 185-204, SAE Technical Paper #872212, Society for Automotive Engineers, Warrendale, PA, 1987.

896074

Toward a Biomechanical Criterion for Functional Brain Injury

J. W. Lighthall, J. W. Melvin and K. Ueno
General Motors Research Laboratories

Abstract

This paper describes a clinically relevant experimental brain injury model developed at GMRL for biomechanical and neurophysiological studies to determine tissue-level criteria for functional neural injury. The controlled cortical contusion technique has several advantages over previous techniques for development of brain injury criteria: (1) the mechanical input to the brain tissue is controlled and can be well-characterized, a necessary step toward validation of finite element analysis procedures; (2) the neuropathologic features simulate key aspects of clinical brain injury; and (3) the functional outcome can be varied across the clinical spectrum, from minimal effect through prolonged coma to fatal, a feature critical to successful development of functional brain injury criteria.

Finite element model (FEM) development and analysis is proceeding in parallel for the experimental injury model and for human brain, since geometric differences in brain structures prevent a straightforward scaling of injury thresholds based on dimension or mass. Analysis of the experimental injury event using the FEM presently allows qualitative (better/worse) assessments, and validation of the FEM model and its interpretation against observations in experimental models of brain injury will be used to define injury probability thresholds.

Introduction

The evolution and development of any experimental brain injury model is dependent on the scientific objectives and goals, and may result in different models, each appropriate for specific questions. Despite differing objectives, certain experimental requirements are shared by any acceptable brain injury model. The injury response, whether specified in physiological, behavioral or anatomical terms, must be reproducible and quantifiable, clinically relevant, and produce a continuum of injury severities. The objective for any experimental model of brain injury is to create specific injury response in a reproducible manner (1,2)*.

Experimental injury models that can be characterized biomechanically are required for physical and analytical modeling of tissue deformation. These modeling efforts can be used for correlating experimental results and injury response with human injury. The validation of analytical head impact models using experimental data requires correlation of the mechanical response of the system in terms of pressures, displacements and local accelerations of the important regions of the brain with pathophysiological and functional outcome of the impact. This approach leads to tissue-level injury criteria, independent of overall brain geometry. Once validation has been accomplished the analytical method can be applied to the analysis of crash test data for injury risk assessment. Routine application of this method will require the development of data presentation techniques that enhance the understanding of the results by the test engineer.

Analytical modeling of the tissue response that produces injury requires that mechanical input to the brain be quantifiable and reproducible. Simplification of the input is required in designing a model that can be used to address the mechanics of brain trauma at the tissue level. A simplified model, once characterized biomechanically and physiologically, can be combined with analytic and physical models of the brain and skull to define an injury tolerance criterion at the tissue level.

*Numbers in parentheses designate references at end of paper.

If the experimental brain injury model technique produces structural changes, such as contusion, hemorrhage or axonal injury, in regions similar to those observed clinically, then the mechanics of the injury, at a tissue level, may be assumed to be directly comparable and the technique used to load or deform the brain is unimportant. Direct brain deformation reproduces only some of the dynamics of closed head impact trauma, but it can be readily characterized biomechanically.

We hypothesized that if the injury response were demonstrated to be relevant to hypothesized tissue-level biomechanics and pathophysiology of clinical injury, then a direct cortical deformation technique would provide an injury model with well-controlled mechanics the brain deformation parameters could then be scaled to various species, eliminating the need to restrict studies to primates on the basis of brain mass and geometry as has sometimes been required for clinical relevance in head-impact models or acceleration techniques (1,2).

GMRL cortical impact injury model

Acute brain injury experiments from this laboratory have previously demonstrated that dynamic mechanical compression of the brain produces pathologic changes that are seen clinically in human head injury (3). The technique elicits a range of pathological responses similar to those described for other brain injury models, and allows independent control over contact velocity and amount of brain compression to enable biomechanical studies. We wanted to determine if our technique would also produce axonal damage and prolonged functional coma, as seen in human head injury.

The controlled cortical impact model uses a pneumatic impactor consisting of a small bore, stroke-constrained, pneumatic cylinder with a 5.0 cm stroke (figure 1A). The cylinder is rigidly mounted in a vertical position on a cross bar which can be precisely adjusted in the vertical axis. The velocity of the impactor shaft can be adjusted between 1.0 and 10.0 m/s by controlled air pressure. The experiments were performed using aseptic technique in fully anesthetized laboratory ferrets.[1] Impacts were performed on midline between bregma and lambda through a central craniotomy to the intact dura mater at 4.3m/s or 8.0m/s, with $\simeq$ 10% compression (2.5mm).

The level of outcome severity was evaluated at the gross anatomical and microscopic level. Contusion of the cerebral cortex was seen in every case (figure 1B and 1C). Extensive axonal injury was observed in both the 3 and 7 day post-injury groups using both velocity/compression combinations. Regions displaying axonal injury were: the subcorti-

[1] The Research Biomedical Laboratory of GM Research Laboratories is accredited by the American Association for the Accreditation of Laboratory Animal Care (AAALAC). The rationale and experimental protocol for use of an animal model in this study have been reviewed and approved by both the Research Laboratories' and Wayne State University School of Medicine Animal Research Committees and are in compliance with federal, state and local laws and regulations, and in accordance with the NIH Guide (DHEW Publication # NIH 85–23).

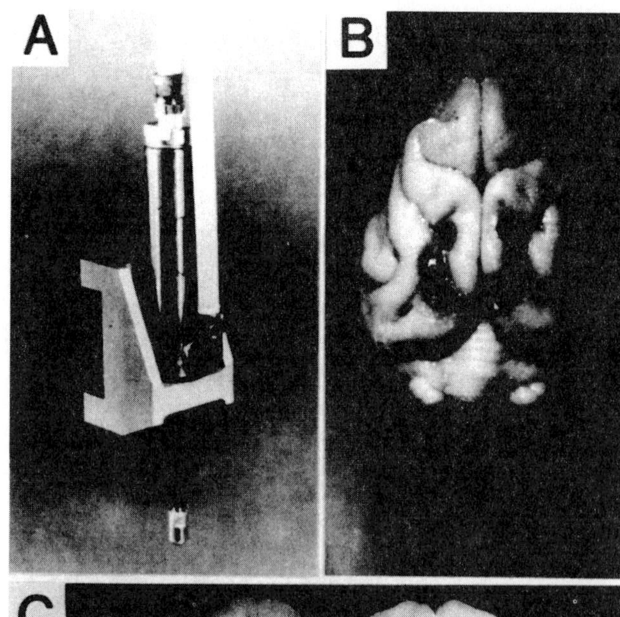

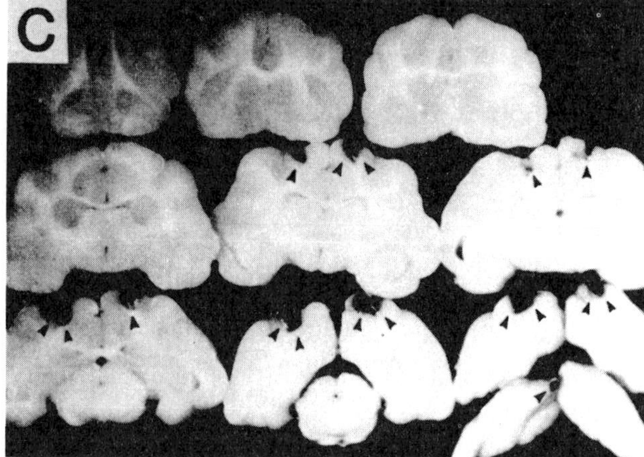

Figure 1. (A) Pneumatic cylinder used in cortical impact experiments, tip diameter = 1.25cm. (B) Cortical contusion produced by a 4.3 m/s—2.5 mm impact (3 day observation period). (C) Serial 2mm frontal sections of the brain shown in figure 2B, arrow heads indicate the margins of the contusion.

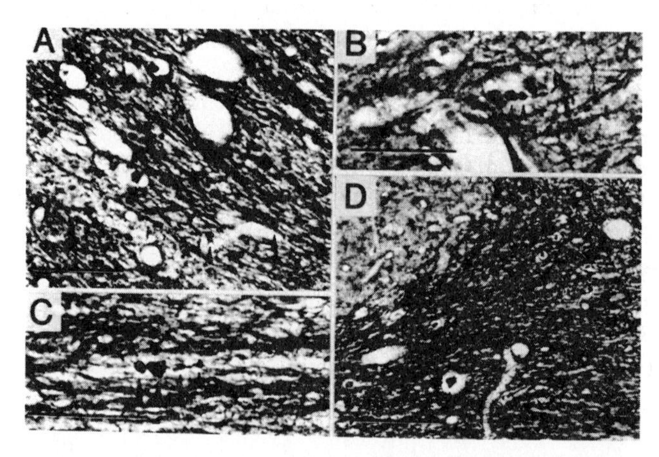

Figure 2. Photomicrographs from brain regions displaying axonal injury. (A) Corpus collosum: arrow heads indicate axonal retraction balls, cal bar = 100 μm; (B) Dorsal thalamus: arrow heads indicate beaded axon, cal bar = 50 μm; (C) Pontine region: arrow heads indicated beaded axon, cal bar = 100 μm; (D) Anterior lobe of the cerebellum: arrow heads and circles indicate the location of axonal injury in this field of cerebellar white matter, cal bar = 100 μm.

cal white matter including the corpus collosum (figure 2A); dorsal thalamus (figure 2B), and brainstem regions including the pons (figure 2C) and medulla. Axonal injury was also evident in the white matter of the cerebellar folia (figure 2D) and the region of the deep cerebellar nuclei. Behavioral assessment showed initial functional coma consisting of decorticate and decerebrate rigidity, followed by muscle flaccidity and inability to maintain sternal recumbancy lasting up to 36 hrs following 8.0m/s impacts, concluding with impaired movement and control of the extremities. It is clear that a wide spectrum of responses can be produced by controlling the velocity and the depth of the cortical impact.

Finite element brain injury models

The goal of our analytical modeling of head and brain impact response is to provide a rational method for assessing injury risk in crash tests based on the head impact forces and accelerations measured in test dummies. At present, our modeling addresses brain response due to accelerations of the head as a rigid body. From a theoretical point of view, since brain tissue injury is initiated by induced stresses and strains within the brain, direct cortical deformation may be used to study acceleration or impact induced brain injury if it results in comparable levels of stress and strain in the cortical and subcortical structures and comparable anatomic and functional injury outcome.

As an overview of the finite element modeling (FEM) work now in progress, and as an indication of how we are addressing the issues of model development, use, and validation, two examples will be given. The first will apply simple linear elastic FEM analysis to laboratory dummy head impact data to demonstrate procedures necessary for acquiring and post-processing data to provide information useful to the test engineer. The second example will show how the method can be used to evaluate the appropriateness of using experimental brain impact data to estimate human head injury tolerance through direct scaling.

Analysis of Hybrid III head acceleration data

The time history of accelerations measured at the center of gravity (CG) of a dummy head is used in crash testing to assess the injury risk caused by head impact. In two-dimensional analysis of head impact, anterior-posterior (AP), superior-inferior (SI), and rotational acceleration measured at the CG establish the complete motion of the head as a rigid body in the sagittal plane. The AP and SI accelerations combine to form a resultant acceleration that represents the magnitude of translational acceleration in the sagittal plane. As the head rotates, the direction of translational acceleration changes.

One of the methods for assessing head injury with acceleration data is the Head Injury Criterion (HIC) (4). Since the HIC is based on resultant translational acceleration and does not include rotational acceleration and the direction of resultant acceleration, it is not a complete analysis of the deformable body in two-dimensional space, but rather, an analysis based on a particular point of interest, i.e., the CG of the head.

The approach of using a finite element model of the head, on the other hand, is an analysis of a deformable body in space, where a motion prescribed by complete acceleration and boundary conditions causes stresses and strains in the brain model and the critical sites of potential injury may be indicated by maximum pressure, stress, strain, or displacement relative to the boundary.

It has been proposed that rotational acceleration rather than translational acceleration may be the likely cause of structural failure in brain tissue (5,6,7). However, the theory that only shear deformation causes injury in the brain has not been established, and rotation is not the sole cause of shear deformation since translational motion can cause shear strain at 45 degrees to normal strain directions. Therefore, the importance of translational acceleration in brain injury should be considered in any analysis.

Two different cases of Hybrid III sled tests were used to analyze the stress and strain field of a 2D finite element model of the head using linear direct transient dynamics in NASTRAN. The values of peak stresses and the HIC for the two different cases are compared with each other.

The accelerations in the midsagittal plane (AP translational, SI translational, and rotational) were measured with uniaxial accelerometers (arranged as a triaxial set and an inline set of five) located at the center of the Hybrid III dummy head in tests simulating frontal impacts. The dummy was restrained by a three point lap/shoulder belt that was adjusted so that in one case (Test 1338) the head hit the steering wheel hub and in the other case (1350) no head contact occurred. The data were filtered by a second order Butterworth low-pass filter with a cut-off frequency of 1000 Hz. Figures 3 and 4 show the time histories of the head accelerations for the two tests.

A two dimensional human brain finite element model was built with SMUG, a pre-processing program at GM, from a sagittal section photograph of a human brain (figure 5). This plane strain model has 510 nodes, 437 isoparametric quadrilateral elements and 74 triangular elements. The model is homogeneous, isotropic, and linear elastic with structural damping characterized by a constant loss tangent. The material properties were chosen from the literature as discussed by Lee (7) and are:

Modulus of Elasticity—0.24MPa
Poisson's ratio—0.49
Mass density—1×10^{-6} kg/mm^3
Total sectional area—1.83×10^4mm^2
Damping loss tangent—0.2

All the surface boundary nodes of the model except the spinal canal opening were rigidly connected, representing an ideal rigid skull boundary. The spinal canal opening was left free to allow a force-free opening along the spinal canal.

The model was executed using acceleration input for the period of 200 ms in NASTRAN. Linear direct transient

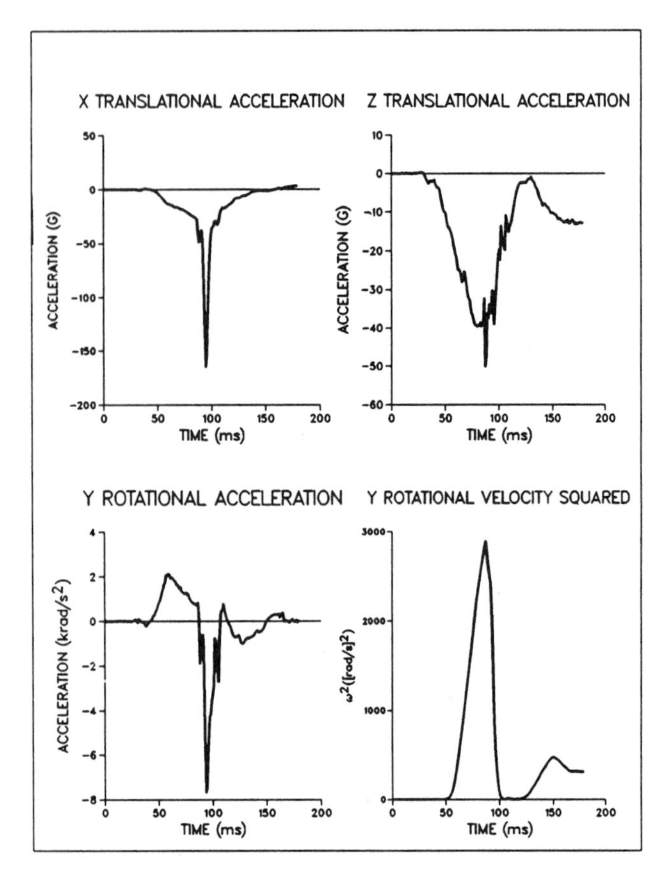

Figure 3. The kinematic input for the contact case (Test 1338).

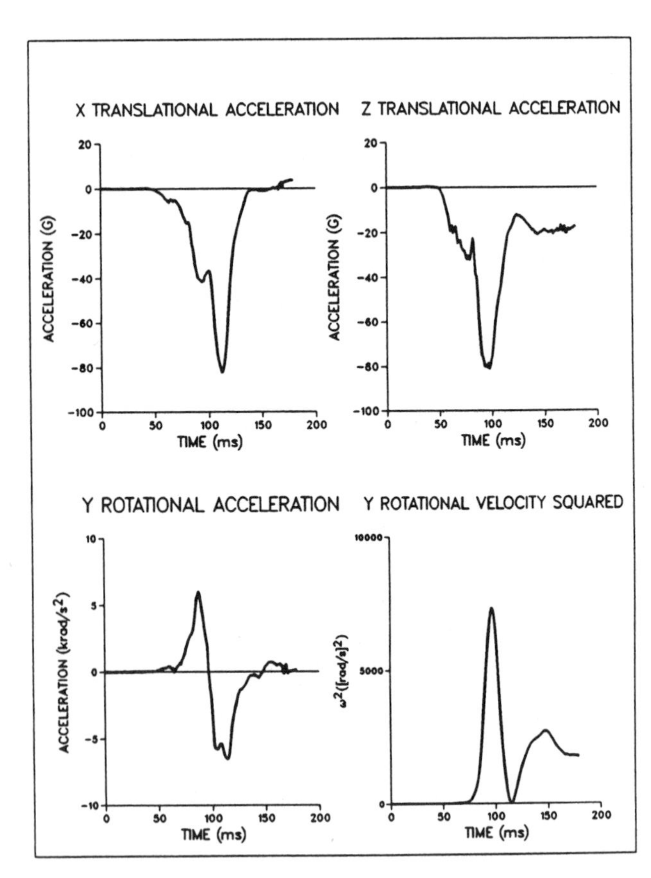

Figure 4. The kinematic input for the non-contact case (Test 1350).

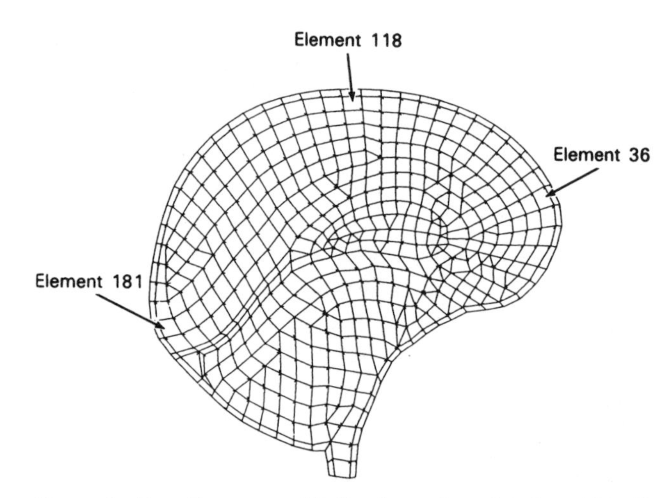

Figure 5. Two dimensional finite element model of a midsagittal section of the human brain. The nodes at the outer boundary are rigidly connected to the skull. The two nodes at the spinal opening are left free. Although the mesh geometry retains the shapes of the brainstem, cerebellum, fluid spaces and tentorium, the materials properties are assigned as homogeneous at this time.

dynamics solution procedures with a 1ms time step were used with equivalent viscous damping in NASTRAN (8). The rotational acceleration at the center of rotation was applied through the method of enforced motion with a large mass. Centrifugal forces were also applied by radial forces with the magnitude of $m_i r_i \omega^2$, where m_i, r_i, and ω represent lumped mass at the node i, distance from the center of rotation to node i, and the angular velocity at the center of rotation, respectively.

Two principal stresses (σ_{max}, σ_{min}) in plane strain elasticity were obtained as results from NASTRAN. From these principal stresses, the maximum shear stress (τ_{max}), and pressure (p) are calculated as follows (9):

$$\tau_{max} = \frac{\sigma_{max} - \sigma_{min}}{2} \tag{1}$$

$$p = \frac{\sigma_{max} + \sigma_{min}}{2} \tag{2}$$

For the purpose of comparison of the stresses produced in the brain model for the two different head acceleration cases, specific points in the brain have been chosen. As indicated in figure 5, these points are Element 36 in the front of the head, Element 181 in the back of the head, and Element 118 at the top of the head. These locations were selected because the pressures reached high values in the front and back of the head and shear stresses were high near the top of the head. Figure 6 shows the time history of the pressure in Element 36 and figure 7 shows the same information at Element 181 for the two tests. At both elements the contact case produces over 70% greater peak pressures than the non-contact case. Comparison of the waveform of the pressures with the acceleration curves in figures 3 and 4 indicates that the pressures in both cases follow the anterior-posterior acceleration.

Figure 8 shows similar data for the maximum shear stress at Element 118. The waveforms of the shear stresses for both cases follow the angular acceleration. The peak shear stress for the contact case is 27% greater than for the non-contact case. Although the magnitudes of the shear stresses are lower than the pressures, they are indicative of large strains because of the low elastic shear modulus of brain.

Figure 9 presents a comparison between the maximum pressures at the front and back of the head and the HIC (at 36 ms) value for both cases. Figure 10 presents the same comparison between maximum shear stress and HIC. The relative severity of the two tests as judged by HIC is opposite that judged by either the resultant pressures or shear stress. The pressures indicate a much greater difference between the tests than the shear stress does.

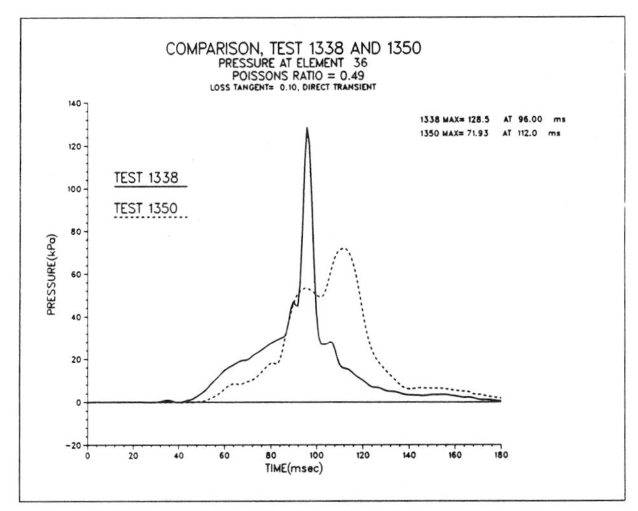

Figure 6. Pressure-time histories in Element 36, located near the front of the human brain model, for the contact case (Test 1338) and the non-contact case (Test 1350).

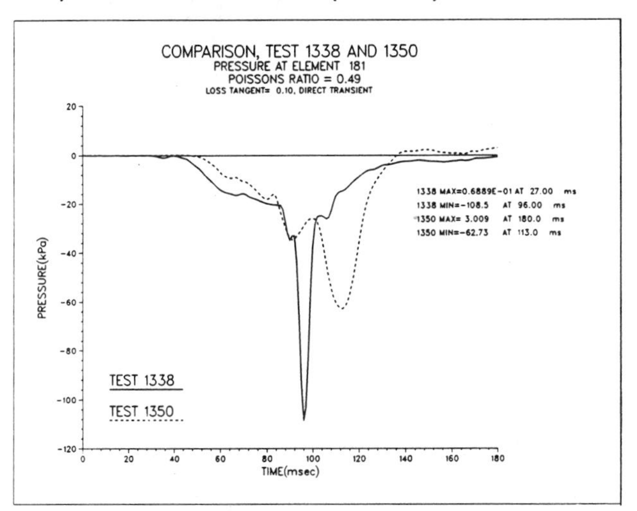

Figure 7. Pressure-time histories in Element 181, located near the back of the human brain model, for the contact case (Test 1338) and the non-contact case (1350).

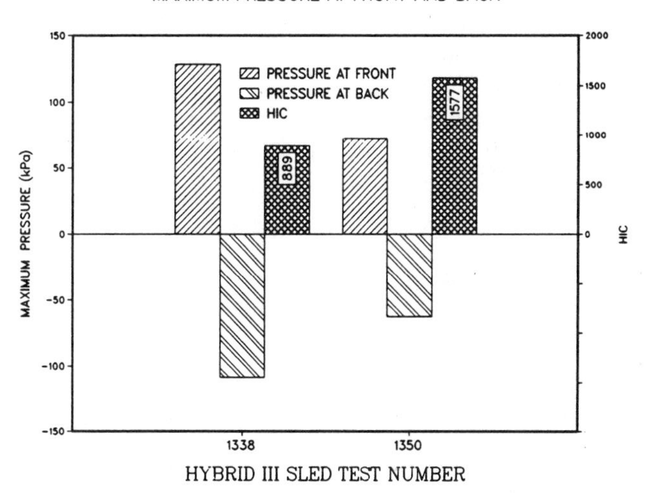

Figure 9. Comparison of the maximum pressures at the front and back of the human brain model and the 36 mc HIC for the contact case (Test 1338) and the non-contact case (Test 1350).

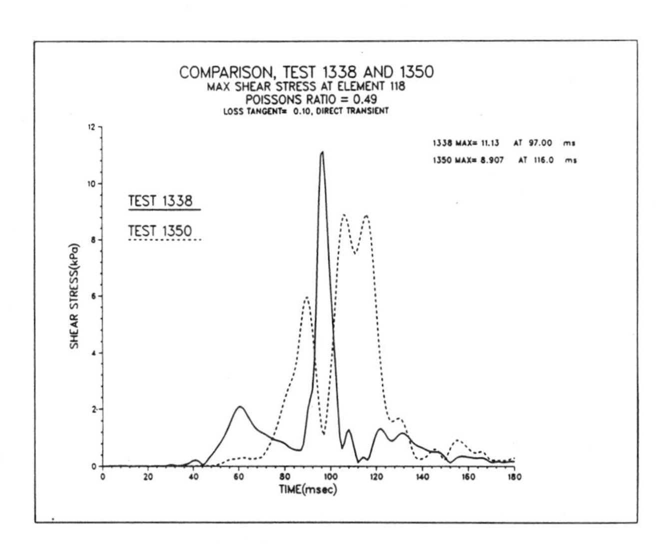

Figure 8. Maximum shear stress-time histories in Element 118, located near the top of the human brain model for the contact case (Test 1338) and the non-contact case (Test 1350).

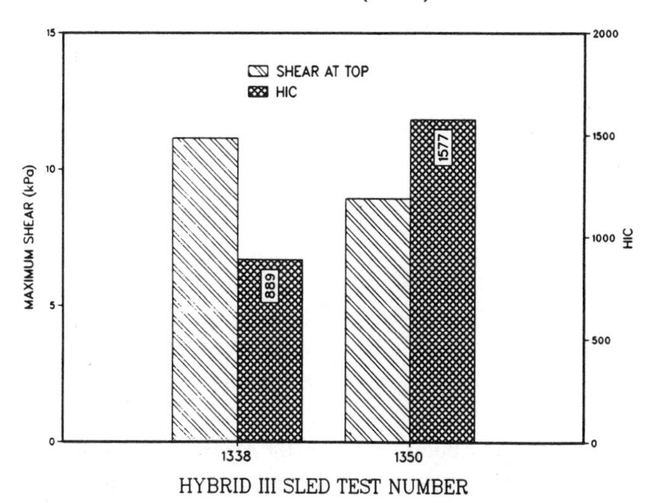

Figure 10. Comparison of the maximum shear stress at the top of the human brain model and the 36 ms HIC for the contact case (Test 1338) and the non-contact case (Test 1350).

Scaling of animal head impact data

Head injuries due to impact have been studied experimentally using animals such as the rhesus monkey with the objective of establishing criteria to determine what types and magnitudes of loading cause injuries in brain (5,10). In order to apply the resulting head injury criteria to human subjects, where the brain size and shape are different from that of the experimental animal, a scaling law must be applied.

Holbourn (11) proposed a scaling law which was based on the mass ratio of brains in the human and the experimental animal. This approach has been used by Ommaya (12) and Margulies (13). According to this law, the angular acceleration α_e that is applied to the experimental animal brain of mass M_e is scaled for the human brain with a mass M_h by the following rule:

$$\alpha_h = \alpha_e * (M_e/M_h)^{2/3} \quad (1)$$

where α_h is a rotational acceleration that produces the same stress level in human brain.

Since mass is proportional to the cube of length, if the scale factor λ represents the ratio of a characteristic length in the human brain to that in the animal brain, this law suggests alternatively

$$\alpha_h = \alpha_e * (1/\lambda)^2 \quad (2)$$

A more general application of dimensional analysis (14), in which the modulus of elasticity and density of the tissues of the brain are assumed to be the same for both the animal and human, yields relationships for scaling all mechanical parameters associated with impact. This scaling procedure, referred to as equal stress-equal velocity scaling, produces scaling of translational acceleration by $1/\lambda$ and time by λ as well as angular acceleration by $1/\lambda^2$. Such scaling is valid only for similarly-shaped structures, however.

To clarify this potential ambiguity in scaling law application, the two dimensional human brain finite element model described above was scaled down and the results compared with those of a rhesus monkey model (shown in figure 11) reported by Lee (7). Since both the monkey model and the human model are two-dimensional, a characteristic length scale factor, λ_A, was taken as the square root of the area ratio as follows:

$$\lambda = (A_H/A_M)^{1/2} = 2.69$$

where

A_H = total area of the human brain model = 183.4 cm^2

A_M = total area of the monkey brain model = 25.27 cm^2

This approach produces a length ratio which represents an average characteristic of the sagittal cross-sections. In accord with the scaling law, the magnitude of the acceleration and the time duration used in the monkey model (shown in figure 12) were scaled in the human model using $1/\lambda_A^2$ and λ_A respectively. The center of rotation was also scaled with λ_A, resulting the distance between the center of mass and center of rotation of 21.5 cm.

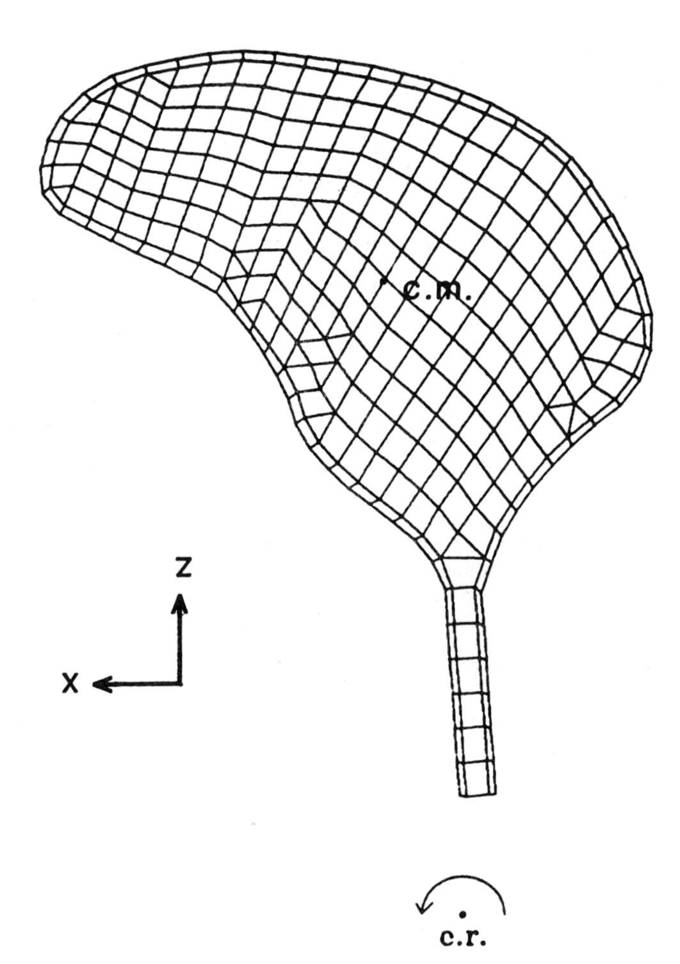

Figure 11. Mesh of the two dimensional finite element model of a rhesus monkey brain. c.m.: center of mass of the model, c.r.: center of rotation.

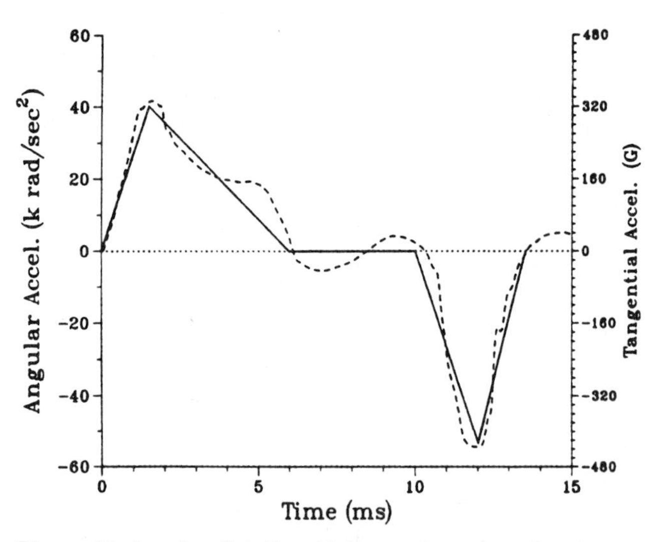

Figure 12. Acceleration-time history enforced on the rhesus monkey brain model (solid curve). The original waveform from (10) is also shown (dashed curve).

Modal transient analysis was used in NASTRAN. A truncation frequency of 400 Hz was used because the Fourier power spectrum of the input data showed the significant frequency range to be well under this frequency.

Overall patterns of stress distribution were very similar between the human and monkey models. However, the maximum shear stress which occurred at the time of deceleration peak was 12 kPa for the monkey and 20 kPa for the human counterpart. Thus, the application of Holbourn's law, which lacks consideration of time scaling and geometrical differences, could lead to a misinterpretation of tolerable translational and rotational accelerations, angular velocity and time duration when extrapolated to man.

Conclusions

The experiments using the GMRL cortical impact model examined the pathogenesis of traumatically-induced changes in the microscopic anatomy of the central nervous system. Confirmation of long term functional outcome including coma and of progressive development of axonal pathology argue strongly in favor of the clinical relevance of this experimental injury model for studies of human brain injury mechanisms. The experimental results confirmed that the cortical impact model of brain injury can be used to investigate the mechanisms underlying axonal damage and prolonged coma, and to more accurately define the threshold and probability of injury in physical and analytical models.

There are several advantages to using a single, controlled mechanical input to produce a brain injury. If the mechanical input is designed to be quantifiable and graded, then correlations can be made between the brain deformation parameters including applied force, the amount of deformation and its time-history and the resultant pathology and functional changes. Such analysis will ultimately lead to enhanced understanding of the interaction between the physical input, the severity of the physiologic injury response and the functional outcome. The restriction to one quantifiable mechanical input variable facilitates biomechanical analysis of the experimental brain injury. Parallel analysis using analytical modeling of tissue deformation is being used to correlate tissue mechanics to the more complex dynamics of human brain injury through derivation of brain tissue injury parameters which produce transient neurological changes, coma or fatality.

The analytical models described here are homogeneous, linearly elastic and damped, but are two dimensional and lack representation of associated structures including membranes, which may be vital to the accurate modeling of detailed brain response. As a result, comparison of the models gives a valid relative indication of the effects of the skull/brain geometry on scaling from animal to man, but absolute comparison will require the development of more detailed brain models. The relative comparison demonstrates an encouraging possibility to scale test results on brain injury between species with different brain geometry.

References

(1) Gennarelli, T.A. and L.E. Thibault, "Biological models of head injury," In Becker, D.P. and J.T. Povlishock (eds). Central Nervous System Status Report. NINCDS publication, pp. 391–405, 1985.

(2) Anderson, T.E. and J.W. Lighthall, "An Evaluation of Experimental Brain Injury Models: Need for Continuing Development," In J. Hoff et al, (eds). Blackwell Scientific Publications, Boston, MA, pp. 77–91, 1989.

(3) Lighthall, J.W., "Controlled cortical impact: A New Experimental Brain Injury Model," J. Neurotrauma 5:1–15, 1988.

(4) Chou, C.C. and G.W. Nyquist, "Analytical Studies of the Head Injury Criterion (HIC)," Automotive Engineering Congress, SAE Technical Paper No. 740082, 1974.

(5) Gennarelli, T.A., J.M. Abel, G. Adams, and D. Graham, "Differential Tolerance of Frontal Temporal Lobes to Contusion Induced by Angular Acceleration," Proc. 23rd Stapp Car Crash Conference, SAE, Warrendale, PA., Paper No. 791022, pp. 561–586, 1979.

(6) Gennarelli, T.A., L. Thibault, G. Tomei, R. Wiser, D. Graham and J. Adams, "Directional Dependence of Axonal Brain Injury due to Central and Non-centroidal Acceleration," Proc. 31st Stapp Car Crash Conference, SAE, Warrendale, PA., Paper No. 872197, pp. 49–53, 1987.

(7) Lee, M.C., J.W. Melvin, and K. Ueno, "Finite Element Analysis of Traumatic Subdural Hematoma," Prod. 31st Stapp Car Crash Conference, SAE, Warrendale, PA., Paper No. 872201, pp. 67–77, 1987.

(8) MSC/NASTRAN, User's Manual 1 and 2, the MacNeal-Schwendler co., 1983.

(9) Timoshenko, S.P. and J.N. Goodier, "Theory of Elasticity," McGraw-Hill, Inc., 1970.

(10) Abel, J.M., T.A. Gennarelli and H. Segawa, "Incidence and Severity of Cerebral Concussion in the Rhesus Monkey following Sagittal Plane Angular Acceleration," Pro. 22nd Stapp Car Crash Conference, SAE, Warrendale, PA., Paper No. 780886, pp. 35–53, 1978.

(11) Holbourn, A.H.S., "Mechanics of Head Injuries," Lancet 2, pp. 438–441, 1943.

(12) Ommaya, A.K., P. Yarnell, A.E. Hirsch and E.H. Harris, "Scaling of Experimental data on Cerebral Concussion in Sub-Human Primates to Concussion Threshold for Man," Proc. 11th Stapp Car Crash Conference, SAE, Warrendale, PA., Paper No. 670906, pp. 47–52, 1967.

(13) Margulies, S.S., P. Yarnell, A.E. Hirsch and E.H. Harris, "A Study of Scaling and Head Injury Criteria using Physical Model Experiment," Proc. 1985 International IRCOBI/AAAM Conference on the Biomechanics of Impacts, pp. 121–234, 1985.

(14) Langhaar, H.L., "Dimensional Analysis and Theory of Models," John Wiley and Sons, Inc. 1951.

Section 5: Face-Skull Injury

680785

Impact Tolerance of the Skull and Face

Alan M. Nahum and James D. Gatts
University of California

Charles W. Gadd and John Danforth
General Motors Corp.

Abstract

Forces necessary for fracture under localized loading have been obtained experimentally for a number of regions of the head. Three of these, the frontal, temporoparietal, and zygomatic, have been studied in sufficient detail to establish that the tolerances are relatively independent of impulse duration, in contrast with the tolerance of the brain to closed-skull injury. Significantly lower average strength has been found for the female bone structure.

Other regions reported upon more briefly are mandible, maxilla, and the laryngotracheal cartilages of the neck.

Pressure distribution has been measured over the impact area, which has been 1 sq in. in these tests, and the relationship between applied force as measured and as predicted from a head accelerometer is examined.

RESEARCH ON the problems of head injury has been concentrated upon the effects of closed head injury to the brain as manifested by concussion and other phenomena. Tolerance to impact has usually been defined as a function of the deceleration of the head occurring either from inertial loading as the subject is suddenly decelerated, or when the head strikes a surface in such a manner that contact area is great enough to preclude a fracture.

An equally important injury occurs when the impact is localized over a sufficiently small area to produce damage to bone and overlying soft tissues. In this event the primary problem is the injury at the impact site. Although brain injury can occur without skull fracture, the hazard increases with skull fracture, particularly if it is depressed or comminuted. Pioneer studies, in particular those of Hodgson (1-3),[1] Swearingen (4), and Gurdjian (5), identified forces that were likely to produce significant injuries to the bones of the skull and face. The research reported here is a further attempt to delineate the tolerances of these

[1]Numbers in parentheses designate References at end of paper.

bones in human cadaver material with special attention to the following variables:

1. Variation among human cadaver test subjects
   (a) Age at time of death.
   (b) Pre-existing diseases at time of death.
   (c) Sex.
   (d) Effects of embalming.
   (e) Effects of overlying soft tissues.
   (f) Characteristics of bone structure.

2. Effects of impulse time history—Fracture properties of biologic materials as a function of such factors as impulse duration, rate of onset, peak force, and deceleration. (Acceleration rather than force was used in many earlier studies, raising the question of any correlation between force and acceleration and which was a more accurate measurement for tolerance descriptions.)

3. Degree of injury—In real life there are many degrees of trauma which range from the barely detectable and not clinically significant to the fatal or near fatal. A range of tolerance values must be established, particularly when it may be possible or desirable to protect for only certain degrees of injury.

4. Injury tolerance sites—Tests have been carried out by others for both skull and facial bones, but it has seemed that more data points with less scatter were needed. Sites were selected which represented areas most commonly impacted in real life situations. In addition, an attempt was made to provide the kind of data that engineers could utilize to select better structural or cushioning thresholds and to make the most efficient use of the crush depths available in vehicle engineering. Another objective is to specify the load-deflection and damage properties of the human head and face so that improved trauma-indicating headforms can be developed.

Methodology

Experimental Apparatus—The device illustrated in Fig. 1A is a scaffold which permits the controlled release of an impacting mass. Both the height and weight can be varied. The tip of the striker is instrumented with a load cell which records the impulse directly on the oscilloscope. A tip contact area of 1 sq in. was employed for all of the tests reported. To distribute the loading over the tip as effectively as possible, MetNet crushable nickel pad (6), 0.2 in. thick, was placed over the tip. Densities of 9, 7, and 3% of this material, by weight, were used for the frontal, temporoparietal, and zygoma areas, respectively. This crushable metal was chosen because it virtually eliminates the time dependency of the yielding of the load-distributing pad which might confuse an interpretation of the crush properties of biologic materials. A second benefit is that, since the pad is calibrated for pressure measurement, it provides a check on the accuracy of the load cell recording

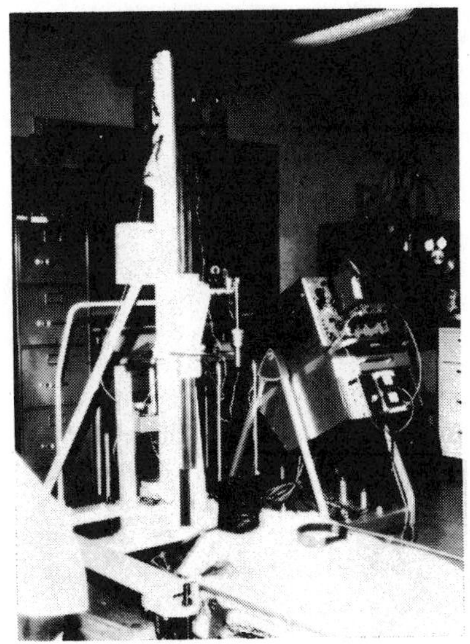

Fig. 1A—Experimental apparatus

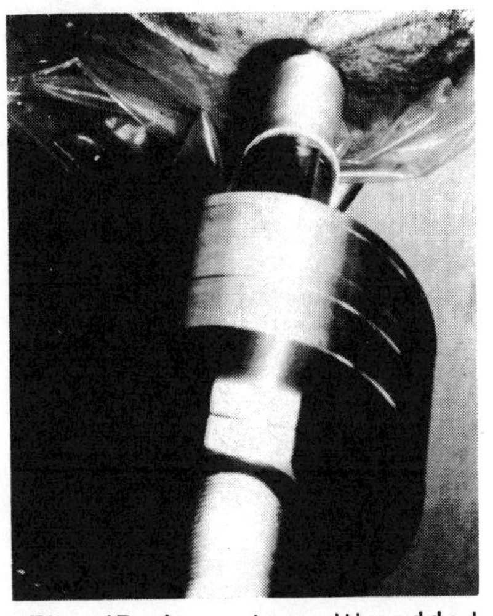

Fig. 1B—Impactor with added weights and dilator tube at lower end for extending pulse duration

and also provides a body of pressure distribution data for investigation of load pressure variation as a factor in tolerance.

Experimental Techniques

Impact Locations—The following locations (Figs. 2A and 2B) were studied on the skull and face:

1. Frontal bone (right and left).
2. Temporoparietal junction (right and left).
3. Zygoma (right and left).
4. Mandible (either midsymphysis or body).

In addition, measurements were made of nasal bones and maxilla when it was felt that the results would not be influenced by the presence of zygomatic fractures. Measurements were also made on the anterior neck of tolerances of the thyroid and cricoid cartilages.

Bone Fracture Identification—Fractures of bones were identified by the following techniques:

1. Palpation visual inspection of the specimen.
2. X-rays of the bones.
3. Acoustic impression at the time of impact.
4. Characteristic changes in the force-time oscillogram which occur in association with various degrees of fracture.
5. Careful anatomic dissection under magnification.
6. Use of dye penetrants used in engineering stress analysis.

Both fractures and soft tissue injuries were photographed and recorded with an anatomic description of the abnormalities noted.

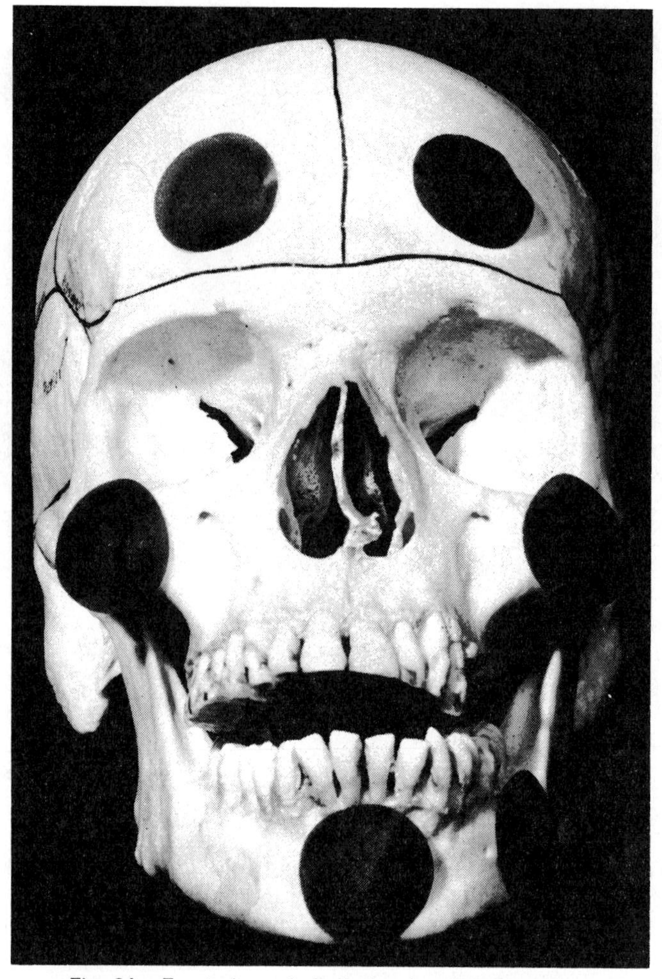

Fig. 2A—Front view of skull showing impact points

Fractures were divided in the following categories:

0 = none

1+ = minimal detectable change, usually a hairline crack, not clinically significant

2+ = readily detectable fracture which is clinically significant

3+ = comminuted and/or depressed fractures which represents the extreme of clinical significance

Even the 1+ fractures, which may not be of clinical significance, are important in tolerance experiments. They serve as stress concentrators in bone and can greatly reduce tolerance as judged by a subsequent blow. Therefore, successive blows should not be delivered for

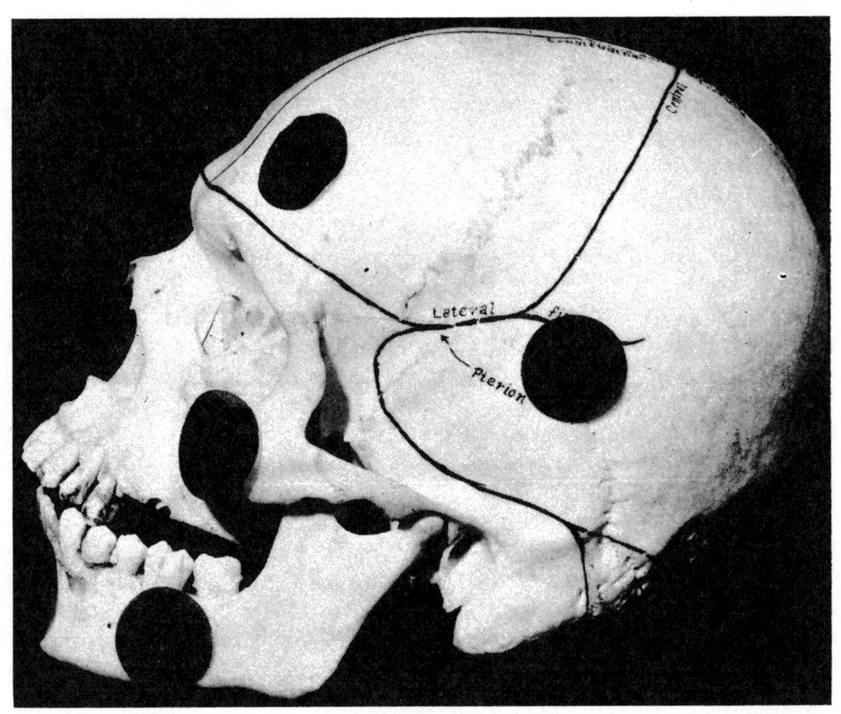

Fig. 2B—Lateral view of skull

that experiment. For this reason, the authors decided that tolerance would be judged for the most part by only single impacts to a particular anatomic location.

Various levels of fracture severity could be distinguished on the oscillogram. This varied from the 3+ fracture characterized by the sudden drop in load (see Fig. 11A) to the 1+ minimal variety best seen on an otherwise smooth tracing as a small notching (see Figs. 10A and 10B).

Paired Comparison Techniques—Data early in the experimental series confirmed that the skull and face were quite symmetrical and that right and left frontal, temporoparietal, and zygomatic areas could be compared in the same specimen. This symmetry was exploited by introducing variables, the effects of which could be measured by the opposite control side for the three above-named locations.

Head Support—The majority of the experiments reported here simulated a relative lack of head support because the head is not supported in real life impacts except by its connection with the neck. Preliminary tests conducted with a headform demonstrated that a rigid support at the back of the head increased the impact force by a range of 20-30% over the freely supported condition. But no measurable error was observed when the headform was supported relatively freely. In most of the experiments, therefore, the specimens were supported by

4-5 in. of soft foam rubber. This compensated for the downward force of gravity on the head, but allowed essentially free inertial reaction during the relatively brief impact. In a few of the most recent paired tests, a stiff moldable material (Duxseal) was used as a head support to facilitate the production of longer pulse durations.

Variation of Impulse Time History—Previous reports of tests of engineering materials and bone coupons have usually shown that strength is somewhat higher when the rate of loading is increased. McElhaney (7) showed human and bovine bone strength in compression to be about 12% higher when the rate of straining was increased by a factor of 10:1 in the pulse range reported. Bird (8) showed even less rate of increase in going from static up to 5 times 10^5 psi/sec. This increase may be a manifestation of the shorter time available for flow or other microeffects leading to fracture to take place under shorter exposures to such stresses. Unpublished flexural tests of bone coupons by one of the authors (Gadd), using a plastic insert to extend pulse duration at the same rate of onset, have shown similarly low sensitivities to rate of onset.

When a more elaborate structure with possible modes of vibration is tested, in which a possible dynamic response is involved, it is conceivable that a high rate of loading could have the opposite effect of decreasing rather than increasing fracture tolerance. For example, in a single-degree-of-freedom system the stress might be doubled if a long pulse is applied which has a rise time that is short in comparison with the natural period of the system.

When impulses of similar shape but various durations are applied, it is important to look for somewhat higher tolerance to short pulses, assuming that the impulse is resisted by the strength of the soft tissue and bone without dynamic effect. But if a predominantly dynamic response occurs, tolerance can vary and be lowest for a pulse which might tend to "tune in" with the natural period of the structure under study. Overlying soft tissues may also have an attenuating effect which should increase duration but decrease peak intensity of the impulse.

The rate of onset was varied in these experiments by selecting a large weight and dropping it from a lower height, or by using a smaller weight dropped from a greater elevation. In this manner, it was possible to obtain pulses which were approximately half sine in shape but which had a variable rate of onset which could be predetermined and compared for corresponding halves of the head and face.

Pulse duration was controlled by adding a 1½ in. section of aluminum tubing (Fig. 1B.) to the impactor tip which was slightly smaller in diameter than the tip. During impact, it dilates and rides up over the tip, thus increasing the duration of the pulse and maintaining the impact force at the desired level.

A uniaxial accelerometer was attached to the back of the head and positioned directly opposite the planned area of impact. Values were recorded on the oscillogram adjacent to the impulse recording.

Biologic Specimens—Both embalmed and unembalmed postmortem specimens were studied (Table 1). Variations in age, pre-existing disease, sex, and preservation were studied. In one instance (not detailed here), a test series was performed on half of an unembalmed specimen which was then embalmed so that identical tests could be performed on the opposite side. Measurements were made of soft tissue thickness overlying the test site. In all the tests reported, the soft tissues were intact over the area of bone to be tested. Some comparative tests were made, however, with and without the presence of overlying soft tissues.

Table 1—Specimens Studied

Specimen Number	Sex	Age	Embalmed	Elapsed Time[a]	Cause of Death
1	M	81	Yes	30	Arteriosclerotic heart disease
2	M	81	Yes	150	Arteriosclerotic heart disease
3	M	75	No	2	Arteriosclerotic heart disease
4	F	68	Yes	34	Multiple myeloma
5	M	70	No	4	Disseminated neoplasm
6	F	71	Yes	360	Cerebral arteriosclerosis
7	F	70	No	2	Arteriosclerotic heart disease
8	F	60	Yes	350	Hypertensive heart disease
9	F	55	No	3	Disseminated neoplasm
10	F	61	No	2	Arteriosclerotic heart disease

[a]Number of days from time of death until experiment.

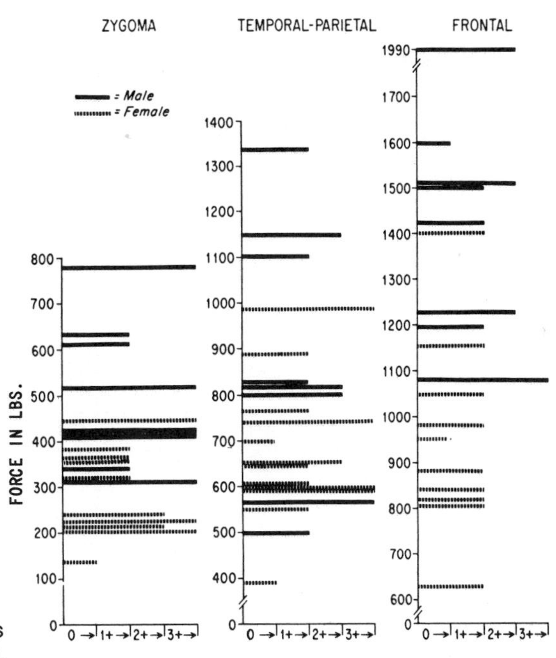

Fig. 3A—Force tolerances for frontal, temporoparietal, and zygomatic sites

Tolerance Thresholds—The fracture data on frontal, temporoparietal, and zygomatic areas are summarized in Table 2. These have been tabulated separately for males and females (Fig. 3A). In each of these test sites, the majority of fractures cluster in groups near the lower end of the observed bands of variance. It is necessary to select tolerance values for engineering purposes which are on the low side of the variance seen in the majority of test subjects, but not so low that a cushioning medium is selected which is so soft that it "bottoms out" under average impact conditions. The following values are suggested for clinically significant fractures when the effective contact area is approximately 1 sq. in.:

Frontal area	1100 lb
Temporoparietal area	550 lb
Zygomatic area	225 lb

Minimal tolerance values for these areas could be specified at:

Frontal area	900 lb
Temporoparietal area	450 lb
Zygomatic area	200 lb

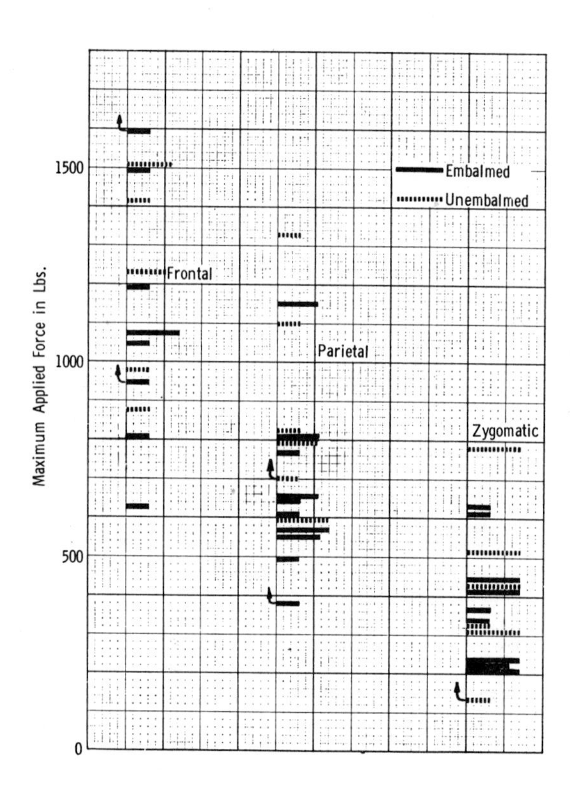

Fig. 3B—Force tolerances showing effects of embalming (first eight test subjects)

Fracture patterns varied from the smallest indication seen on the oscillogram through various degrees of severity. Patterns were also a function of the geometry and composition of the bone in a particular location. For example, a 1+ fracture in the frontal area (Fig. 4) was detectable only with the use of dye penetrant. A 3+ fracture (Fig. 5) of the zygomatic area was comminuted and displaced and quite readily recognized.

Effects of Biologic Variations—Sex of test subject was the most important variable observed. Six of the ten subjects studied were females. The test values for these subjects consistently grouped together at the lower end of the tolerance scale. This suggests that values for all tolerances in females will require important consideration in the

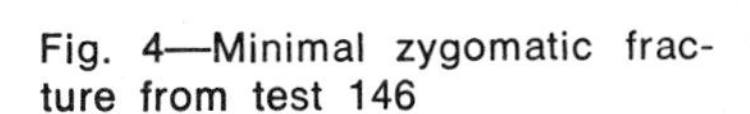
Fig. 4—Minimal zygomatic fracture from test 146

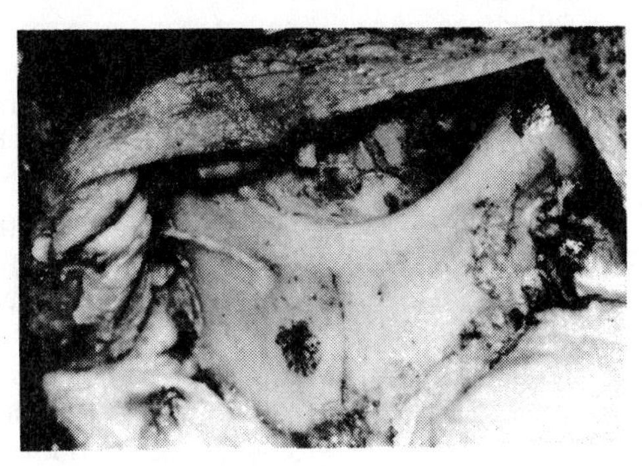

Table 2—Fracture Data

Specimen	Sex	Embalmed	Frontal Test No.	Frontal Force	Frontal Frac.	Temporoparietal Test No.	Temporoparietal Force	Temporoparietal Frac.	Zygoma Test No.	Zygoma Force[a]	Zygoma Frac.
1	M	Yes	3	1194	1+	7	566	3+	1	411	3+
			4	1078	3+	8	498	1+	2	332	1+
2	M	Yes	16	1600+	0	18	1150	2+	13	616	1+
			17	1500	1+	34	817	2+	14	633	1+
3	M	No	25	1420	1+	21	828	1+	20	316	3+
			94	1230	2+	24	798	2+	22	425	3+
4	F	Yes	47	848	3+	56	767	1+	49	443	3+
			50	600	3+	59	551	1+	60	363	1+
5	M	No	74	1990	2+	64	1334	1+	67	780	3+
			76	1510	2+	66	1102	1+	68	518	3+
6	F	Yes	89	1050	1+	84	655	2+	80	228	3+
			91	630	1+	86	608	1+	83	205	3+
7	F	No	102	980	1+	106	594	3+	104	138	0
			103	880	1+	109	594	3+	105	314	1+
8	F	Yes	117	948	0	121	642	1+	120	240	2+
			124	806	1+	129	382	0	123	212	2+
9	F	No	141	1400	1+	142	742	3+	136	348	1+
			147	816	1+	152	700	0	146	382	1+
10	F	No	160	1160	1+	164	882	1+	159	418	3+
			173	848	1+	175	986	3+	172	NG	—

[a] Force values from the local cell have been corrected for the mass of the striker tip and therefore are not numerically equal to the values calculated from the oscilloscope tracings.

final establishment of tolerances for the entire population at risk (excluding children). The effects of individual structural properties of bone were not determined, except that in specimen 4 a pre-existing disease had decreased the strength of the bone in the frontal area. For this reason, the two low values for the frontal area of specimen 4 were not included in Figs. 3A and 3B.

Embalming did not appear to affect results for the areas studied. This is illustrated by the distribution plots of Fig. 3B. It should be noted, however, that no embalmed specimen was used more than 1 year after embalming. In addition, the soft tissue responses appear to be quite different so that when soft tissue properties play a greater role, as in tests of mandible, maxilla, and neck (cricoid and thyroid cartilages), embalming adversely affects the test results.

Thickness of soft tissues and, in particular, the presence or absence of soft tissue was shown to play an important role. For example, test 117 failed to produce any evidence of fracture to the frontal area of specimen 8 at a force of 948 lb. When the soft tissue was removed, a 3+ fracture resulted with a peak force of 708 lb.

Soft tissues over the frontal area (Fig. 6) usually were crushed or lacerated and displaced in a characteristic fashion. Soft tissues over the other impact locations were depressed and contused but rarely lacerated.

Selected tests were performed on both sides of the lower maxilla. This was done only when no zygomatic fractures were present. Values in the range of 175-210 were obtained for clinical fracture tolerances. This is a difficult area to test and requires more investigation.

Tests of the mandible were performed in two locations, the sym-

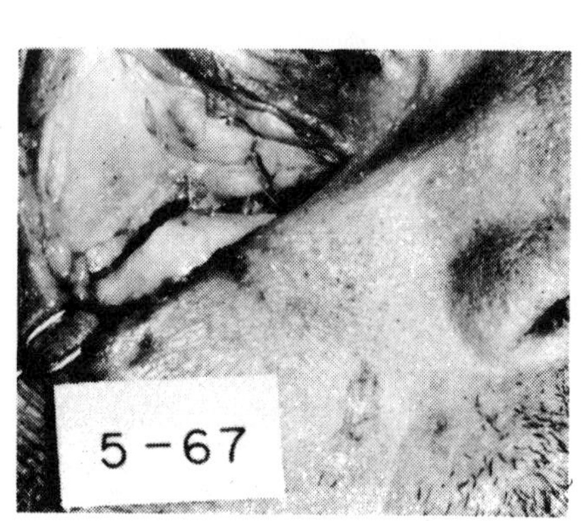

Fig. 5—Severe zygomatic fracture from test 67

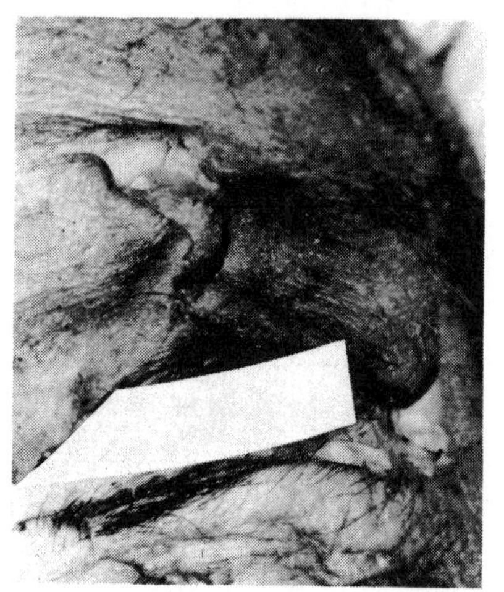
Fig. 6—Example of soft tissue damage noted for frontal impacts

physis and the midbody, on either the left or right side. Variations were observed between embalmed and unembalmed specimens, as a consequence not only of the effects on the overlying soft tissues but also the effects of the ease with which dislocation or movement at the temporomandibular joints could attenuate the forces. At the symphysis a clinical fracture range of 350-400 was obtained. For the mandibular body lower fracture values of 290-325 were obtained. Females displayed significantly lower tolerances than did males.

Tests were also performed on cricoid and thyroid cartilages at the anterior midline of the neck. There appeared to be great individual variations as a function of overlying soft tissue and amount of calcification of the cartilages. Embalmlng had an adverse effect upon the tests, both in fixing the larynx and in increasing the attenuating effects of overlying soft tissues. Values in the range of 200-250 were obtained for the thyroid cartilage and 175-225 for the cricoid cartilage of embalmed subjects. While the thyroid cartilage tends to protrude further, it appeared to have more backward mobility, which would be expected from its shape. On the other hand, the completely circular cricoid cartilage is soon backed up against the vertebral bodies.

In general, tolerance ranges for a particular subject appeared to be quite consistent. For example, if a particular subject showed considerable strength in one area, the tolerance values for the other areas tested tended to be higher as well.

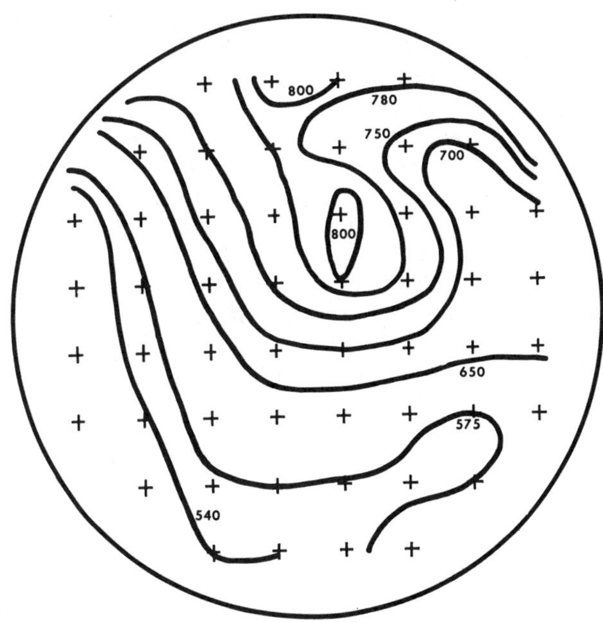

Fig. 7—Pressure distribution calculation from MetNet

Figure 7 - Pressure distribution

Effects of Impulse Variations

Load Distribution—Even with the relatively small impactor tip used and with the load distributing ability of the soft tissues and MetNet pad, considerable variations in local pressures were observed in individual tests. For example, in test 84 on the temporoparietal zone of specimen 6 the load cell recorded 655 lb. A pressure distribution calculation (Fig. 7) revealed local pressure variations between 540 and 810 with an average of 653 psi. In many tests the variation was greater than this. Preliminary study of these has indicated two principal effects tending to cause this:

1. The contour and rigidity of the head material at the zone of contact, which in these tests sometimes increased local pressure 200% over the theoretical average.

2. A time sequencing of the load application over different portions of the contact area during a single impact, which increased average pressure over the area by as much as 30% above the load cell reading.

Further studies must be made before possible relationships between tolerance, on the one hand, and pressure, area, and deflection, on the other, can be developed.

Pulse Time History—The tests conducted to date indicate that pulse duration and rate of onset are not critical factors and that tolerance can be described reasonably well in terms of peak force. For example, tests 21 and 24 on specimens 3 (Fig. 8) utilized 15.5 lb dropped from 20 in. and 2.64 lb dropped from 114 in. to produce measured forces of 828 and 798 psi. Both produced fractures with rise times of about 1 ms for the high velocity and several milliseconds for the lower.

There has been no evidence that transient resonance of the skull structure with soft tissues present is great enough to reduce fracture tolerance materially. Oscillograms of acceleration recorded from the back of the head in some tests show oscillation, but this does not necessarily indicate high stress in the bone structure of the head.

The influence of pulse duration appears to be small and of a degree that would be expected from the fracture properties of the bone as discussed above. That is, a several-fold increase in the duration of the intensive portion of the pulse produced on the average only a small increase in the damage. Citing two examples from the paired cases of Figs. 9-12, the impact of Fig. 9B produced less damage for a little

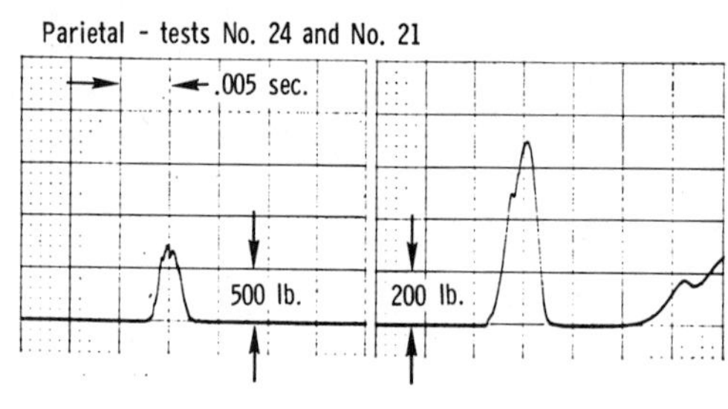

Fig. 8—High and low rates of onset obtained by high and low velocities. Low rate test at right produced similar fracture at approximately same load

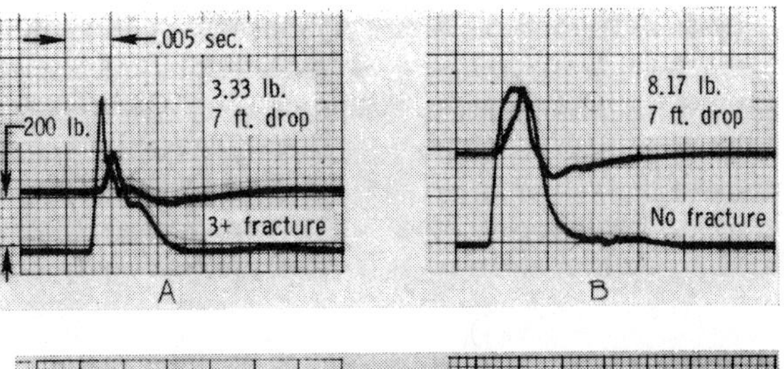

Fig. 9—Paired tests with short and long exposure to high loading, parietal—tests 142 and 152

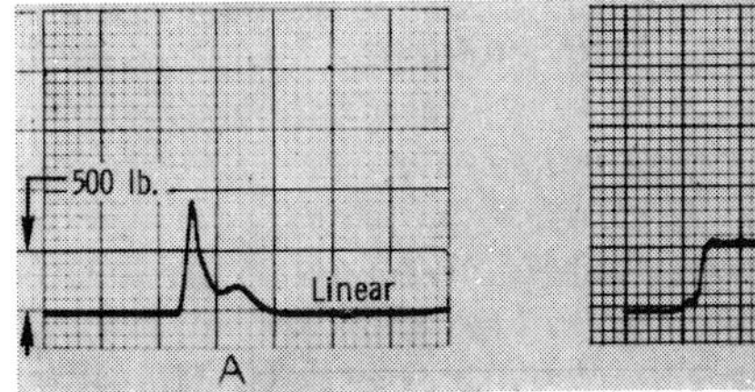

Fig. 10—Short and long exposure, frontal—tests 89 and 91

lower loading of much greater duration than did the impact of Fig. 9A. Those of Figs. 12A and 12B produced fractures of the same degree over a similarly great duration range. In making these tests, a short duration failure was obtained first. Then a companion test was made at longer duration by adding a dilator tube to the impact tip. A tube is selected with a calibrated dilation load which is the same or somewhat less than the previous value of load obtained without the tube. Extra mass is also added when the dilator tube is used to increase the available kinetic energy, but the dilator tube nevertheless succeeds in limiting the applied force to a fixed level. Neither the applied kinetic energy nor the momentum have been found to be accurate criteria of fracture hazard.

Similar differences are noted in the oscillograms (Fig. 10A and 10B) of tests 89 and 91 in the frontal area and tests 49 and 60 in the zygomatic area (Figs. 11A and 11B).

Correlation of Measured Forces with Head Acceleration—Because accelerometers are in wide use for assessing brain injury hazard in studies which use anthropomorphic dummies or impacting head forms, it is of interest to examine whether head acceleration in these tests is compatible on the basis of simple theory with known forces applied to the front of the head. Whereas the head accelerometer has usually been employed to assess the inertial effect upon the brain (closed-skull injury hazard), it was felt of interest also to learn whether an accurate structural simulation of the head should be expected to yield acceleration readings which might be interpreted approximately in terms of localized injury hazard to either the front or side of the head. Thus, if a marking coating on the dummy's head indicated heavy contact over roughly 1 sq in., the question would be whether one might estimate the force

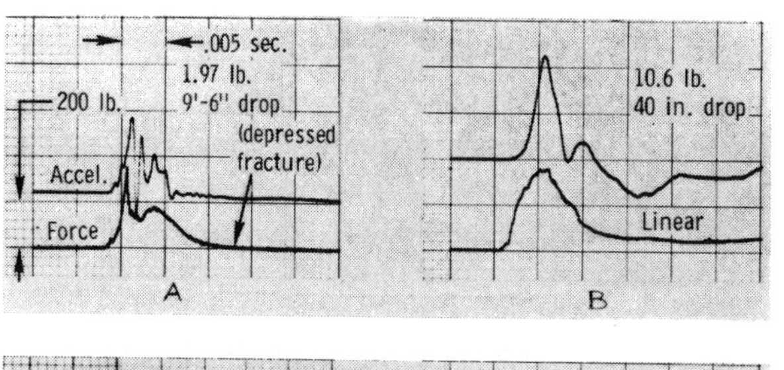

Fig. 11—Short and long exposure, zygoma—tests 49 and 60

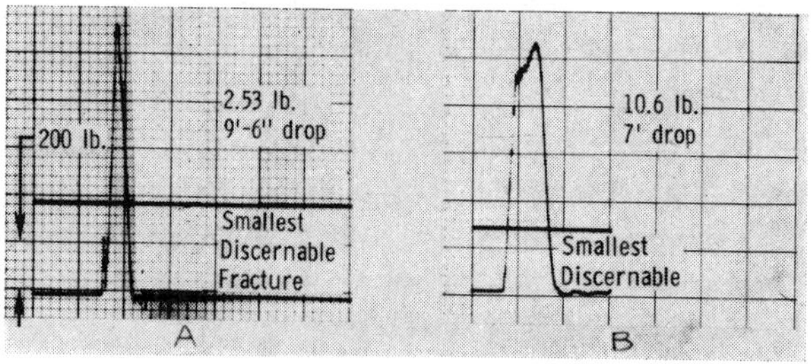

Fig. 12—Short and long exposure, parietal—tests 64 and 66

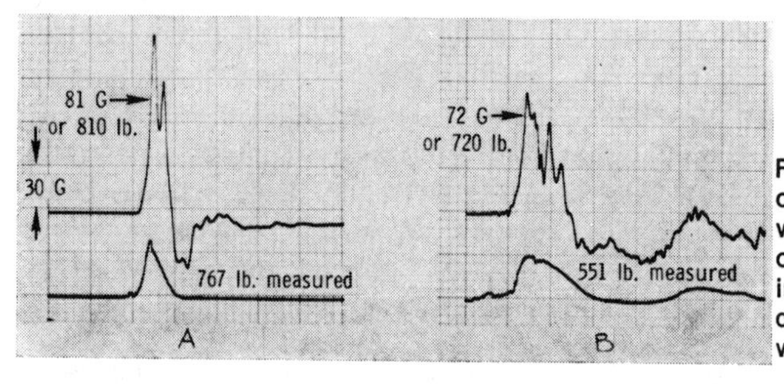

Fig. 13—Comparison of measured force with apparent force obtained by multiplying gross head acceleration by head weight of 10 lb

applied here by multiplying the weight of the head by the number of g's. This calculation, if valid, would give an indication of whether facial fracture tolerance was exceeded in the test and serve as an alternative or supplement to a trauma-indicating head structure.

A uniaxial accelerometer was clamped to the occiput with scalp removed for these tests, such that reasonably accurate readings could be made before the head rotated appreciably. A number of representative oscillograms are given in Figs. 9, 11, and 13. Transient vibration was often seen of the order of 1000 Hz on these traces, even with considerable filtering introduced so that gross movement of the head could more easily be distinguished. Such vibration was often seen both below and above the fracture level and was undoubtedly due to structurla resonances of the head, which in most instances did not represent high amplitude of displacement or stress but, nevertheless, made it difficult to estimate input force by visual inspection of the traces. (A sinusoidal movement of 0.001 in. either side of the mean, at 1000 Hz, would produce about 100 g.)

Only moderate success has been achieved thus far with the attempted correlation.

In a number of cases, very good agreement has been seen while in others the calculated (assuming a head weight of 10 lb) forces have been up to 50% greater than the input forces, evidently as a result of an indeterminate amount of transient overshoot in the response at the occiput. Acceleration traces such as those of Fig. 11 were difficult to interpret accurately. On the other hand, when a more clear-cut ringing occurred, as in Fig. 13, a better visual estimate could be made of probable gross or bodily movement of the head as a whole and exclusive of the ringing. Estimates of this kind have been entered alongside the acceleration traces of this figure. Again assuming a head weight of 10 lb, the estimated forces are 810 and 720 lb in comparison with 767 and 551 lb input force, respectively, for the two tests.

On the average, the g readings at the rear of the head were higher than the $F = ma$ relation would indicate, but when a rigid metal block was substituted for the head and the force applied as before through a 0.2 in. thickness of MetNet, the agreement was within a few per cent. More study will be needed before the optimum dummy head design (or designs) can be determined for estimating both facial and closed skull injury hazard.

References

1. V. R. Hodgson, G. S. Nakamura, and R. F. Talwalker, "Response of the Facial Structure to Impact." Eighth Stapp Car Crash Conference Proceedings. Detroit: Wayne State University Press, 1964.

2. V. R. Hodgson, W. A. Lange, and R. F. Talwalker, "Injury to the Facial Bones." Ninth Stapp Car Crash Conference Proceedings. Minneapolis: University of Minnesota Press, 1965.

3. V. R. Hodgson, "Tolerance of the Facial Bones to Impact." American Jrl. Anat., Vol. 120, 1967, p. 113.

4. J. J. Swearingen, "Tolerance of the Human Face to Crash Impact." Office of Aviation Medicine, FAA, Civil Aeromedical Research Institute, Oklahoma City, July 1965.

5. H. R. Lissner, et al, "Mechanics of Skull Fracture." Proceedings of Experimental Stress Analysis, Vol. 7, No. 1, 1949. See also other papers by Lissner, Gurdjian, and other colleagues.

6. W. K. Miller and S. Katz, "A Technique for Measuring Maximum Local Pressures which Occur During Impact." Proceedings of General Motors Automotive Safety Seminar, Milford, Mich., July 1968.

7. J. H. McElhaney and E. F. Byars, "Dynamic Response of Biological Materials." Biomechanics Monograph. New York: American Society of Mechanical Engineers, New York, 1967.

8. F. Bird, et al, "Experimental Determination of the Mechanical Properties of Bone." Aero. Med., January 1968.

9. C. W. Gadd, A. M. Nahum, James Gatts, and John Danforth, "A Study of Head and Facial Bone Impact Tolerances." Proceedings of General Motors Automotive Safety Seminar, Milford, Mich., July 1968.

700910

Tolerance and Properties of Superficial Soft Tissues in Situ

Charles W. Gadd, Dennis C. Schneider and Richard G. Madeira
General Motors Corp.

Alan M. Nahum
University of California

Abstract

Utilizing unembalmed cadaver test subjects, a series of tests was carried out to characterize quantitatively the resistance of the skin, the soft underlying tissue of the scalp, and certain other typical areas of the body to impact loading. The impacts were delivered by the use of an instrumented free-fall device similar to that previously employed for facial bone fracture experiments. In one group of tests, metal and glass edges were affixed to the impacting device to produce localized trauma under conditions which were standardized with respect to variables affecting the degree of the injury. In the second group of experiments, specimens of skin, together with underlying tissue of uniform thickness, were subjected to compressive impact between the parallel surfaces of the impacting weight and a heavy metal platen. From these latter experiments the force-time histories, coefficient of restitution, and hysteresis loops of load versus deflection were obtained for the specimens. It is suggested that the data obtained should be applicable to the evaluation of soft tissue injury hazard in accidents and to the appraisal of artificial soft tissue simulations for impact headforms and anthropometric dummies.

WHILE MUCH PROGRESS has been made over the last decade in the prescribing of forces or accelerations which major zones of the human body are able to withstand short of sustaining dangerous internal injury, relatively little has been done to quantify thresholds for significant injury hazard to the superficial tissues. Thus, whereas useful measures have been developed for whole-head acceleration tolerance, gross chest force or deflection tolerance, and fracture resistance of specific skeletal areas, the specification of injury hazard to the soft tissues overlying the body has been a more elusive task. No doubt this has largely been because superficial tissue injuries have an almost infinite variety; severity of their injury is a function of numerous variables, including geometry of the soft tissue surface and of the underlying bone, mass, and consequent inertia force of the bodily zone involved, shape of the target area, and others.

In view of this great variety of possible superficial wounds, it was felt logical in this study to consider only a limited number of "standardized" impacts which might nevertheless fulfill two objectives as follows:

1. Provide a limited number of specific tolerance values representative of practical areas of interest, including impact against flat surfaces, glass edges, and other straight edges of differing radii of curvature.
2. Give a basis for the evaluation of artificial skin and underlying tissues used for the construction of anthropomorphic dummies and impact headforms.

The first objective here was to set forth specific test conditions to produce a particular depth and extent of trauma to the biological material which if duplicated in the artificial material should validate it as a trauma-indicating material. Apart from trauma indication, however, an equally important and long-standing question with regard to the superficial tissues of the dummy has been the force or acceleration-attenuating characteristics of the tissues. If these are not correct, considerable doubt is cast particularly upon the validity of head impact Severity Index values or G values obtained in crash experiments. If the crushability of the scalp material — and to a lesser degree the deformation characteristics of the skull — is not true, the G-time histories obtained within the head will be influenced. While this is not apt to be a significant problem in impact of the head into a target area whose deformation is large compared with those of the scalp and skull, it *can* be significant when more rigid objects are struck. The program described herein was accordingly expanded beyond the investigation of common separational wounds to include considerable emphasis upon simple crush between parallel surfaces.

The general approach of this study was empirical, in the sense that overall injuries resulting from particular types of blows representative of accident situations were measured, instead of studying the resistance of tissue samples to simple and uniform states of stress. Both approaches have important application.

Methodology

All tests centered around the use of impactors of the type described in Ref. 1 and illustrated in Fig. 1. Total mass was prescribed and released in free fall from prescribed heights. Since the soft tissue involved was in all cases backed up by rigid and massive material — the skull in most of the cadaver impacts and a steel bed plate in the flat specimen tests — it was easily possible to obtain very high rates of loading and very short exposure durations using the simple rigid test device dropped from modest elevations.

All tissues evaluated were unembalmed, ranging from one and a half to several days old. There was no evidence of initial stiffening of the tissues when the earliest tests were performed, and no change was noted over the several days of testing, during which the material was refrigerated except when under actual test. To the touch, the tissues at all times felt very lifelike.

The complete drop weight assembly — including a built-in load cell for monitoring the force-time history of the impact — acted essentially as a rigid body, as confirmed by accelerometer measurements. It should therefore be possible to duplicate the normal-incidence tests in another laboratory and observe whether the trauma pattern in a soft issue stimulation is consistent with that of the biological tissue, provided only that total mass of impactor, drop height, and impacting tip configuration are

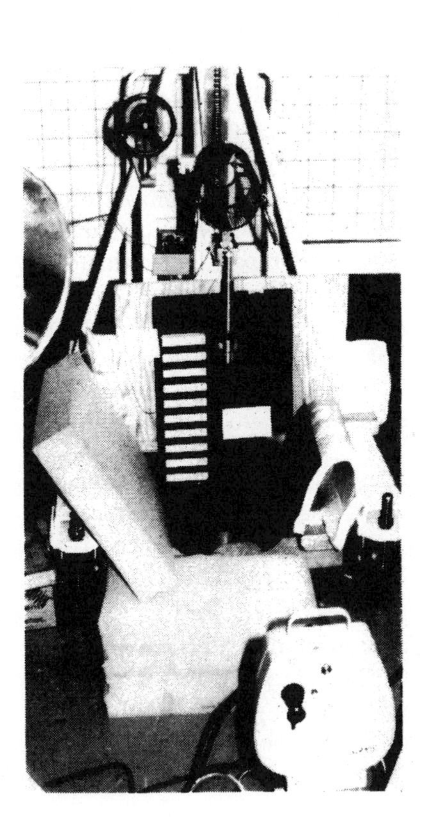

Fig. 1—Free-fall test device, suspended over scalp specimen resting on 100 lb steel block

duplicated. An additional requirement in duplicating *in situ* tests will be essentially to support the dummy head freely by resting it upon at least 3 in. of soft foam padding as was done for the cadaver tests. Discussion of this factor is in Ref. 1.

Soft Tissue Trauma in Situ — These tests — which had earlier been confined to the scalp areas, producing what may be characterized as lacerative injury — were obtained by affixing edge blocks to the lower tip of the drop weights (as sketched in Table 1) ranging in radius of curvature of cross section from that of fractured glass to that of hardened aluminum edges of 1/64 and 1/16 in. Both normal and 45 deg angles of incidence were employed, the latter in an attempt to reproduce "half-moon" shaped lacerations seen at times, and which brought into play the resistance of the scalp materials to tensile and tearing forces as well as their resistance to compressional loading. The glass edges were produced by fracturing common double-strength window glass at 90 deg. In preparing the samples, care was taken to achieve a true square break, free of any lip along the edge.

The initial angular impact tests using the metal edges disclosed a sensitivity of the trauma to the magnitude of inertial resistance of the drop weight assembly to rotational acceleration. The skin involved proved to be sufficiently tough to deflect the metal edge laterally at the moment of impact and at times caused it to skid off the scalp with no penetration. Accordingly, weighting discs were added near the top and bottom of the drop weight assembly to increase its resistance to rotation at impact so that the half-moon lacerations could be obtained. A complete description of angular impact, sufficient to produce the same tendency to lacerate in both the biological and the artificial tissues, thus requires an additional specification of the center of gravity of the drop weight, as well as its moment of inertia about a horizontal axis through this point.

Force-time histories along the axis of the drop weight were recorded as an adjunct to most of the *in situ* experiments. While not essential for specification of the test conditions, they served to establish that the impacts were in a time duration realm of practical interest for accident injury research. They were recorded from a strain-gaged diaphragm mounted within the enlarged portion of the drop weight assembly just above the contacting tip. The records required a correction for the inertia force contributed by the mass of the contacting tip, as is usual in conducting tests with equipment incorporating load cells.

Biological tissue parameters affecting magnitude of trauma to be expected in these tests included scalp thickness, radius of curvature along the line of contact of the metal or glass edge, and strength of the skin and underlying scalp material as a function of the particular area

Table 1—Impact Intensities Producing Marginal Laceration Through Scalp; Normal (right angle) Incidence of Metal and Glass Edges

Location	Drop Weight, lb	Height, in.	Force, lb	Scalp Thickness, mm	Radius of Curvature, in.	Edge Cross Section	Length and Depth of Laceration
Frontal — average results from three subjects, 1/16 in. radius	0.897	36	405	4-5	3.1	90° 1/16 in. rad. metal edge	1.4 cm av *
Subject with unusually thin frontal scalp	0.897	24	260	2.5-3	3		1.5 cm av 1 cm.
Frontal — average of three subjects; 1/64 in. radius edge.	0.897	24	286	4-5	3	90° 1/64 in. rad. metal edge	2.5 cm av. *
Frontal, glass edge	0.897	12	90	4.0	2.6	60 sq. glass edge	2.1 cm av. * * marginally through scalp
Supplementary transverse impacts against mid anterior ridge of tibia	0.897	36	240	2.5	—	Same 1/16 in. metal edge	2 mm
	0.897	15	—	2.5	—	Same square glass edge	1 cm 2 mm

of the scalp under test. Tests thus far have indicated the first two of these to be most significant, and they have accordingly been recorded for reference. Penetrometer readings of the type described in Ref. 2 were taken after performing the impacts, at points near the impact sites, as an independent check of puncture resistance of the skin and underlying soft tissue in the various areas tested.

Impact Between Parallel Surfaces — These tests were conducted primarily as an additional way of validating the performance of artificial tissues for application to dummy or headform construction. They served secondarily for observation of approximate tolerance to flat surface impact. The tests were made by allowing the bare 1 in. area metal impacting tip to fall vertically against 1½ sq in. square specimens of uniform thickness dissected from the scalp, facial, and arm areas and laid on the surface of a 100 lb metal block. The specimens from the frontal and parietal regions of the scalp were left in their full thickness, and those from the zygomatic area were trimmed on their underside to a uniform thickness equal to that originally existing over the zygomatic bone prominence. Those from the arm were arbitrarily trimmed on the underside to a thickness of ¼ in. (6.35 mm), for direct comparison with certain artificial materials which have been used in this thickness. In general the skin was left intact to form the upper surface of the specimens as tested. The specimens were considered large enough in area to eliminate edge effects proximal to the impactor as significant variables from test to test.

The first experiments disclosed that the specimens fell in two categories: first, the relatively tough and somewhat elastic scalp material, and second, the more easily crushed and less resilient tissue from the facial area and from other areas of the body where the material below the skin is essentially fatty in its makeup.

The impactor was dropped from heights sufficient to produce varying degrees of visible damage. While this visually observed damage served as a qualitative measure of injury hazard, it was considered desirable in this portion of the study to describe the tissue reaction more precisely; therefore, in all of the flat specimen testing two additional measurements were made: force-time history and rebound height.

With the force-time history it was possible, assuming the drop weight to be a rigid body and employing a computer program to integrate for deflection at each increment of time[1], to obtain a plot of the hysteresis loop of compressive load versus deflection of the specimen. This plot included a value for the maximum instantaneous crush of the specimen.

[1]A technique developed by J. P. Danforth.

The rebound height of the drop weight, obtained by photographing its vertical motion as a trace on camera film picked up from an attached small convex mirror surface, served as a measure of the coefficient of restitution of the material under the particular impact condition, as well as an independent check of the accuracy of the computed hysteresis loop. Agreement within a few percent was consistently obtained between the two experimental techniques, provided both the load cell and oscilloscope time axis were maintained in proper calibration.

Results

Soft tissue Trauma *in Situ* — Representative results for normal and angular blows are summarized in Tables 1 and 2 respectively, and these are supplemented by the trauma photos of Figs. 2–4. In selecting typical examples for inclusion herein, impacts were chosen which were marginally able to penetrate through the frontal region of the scalp forward of the frontal plane of the head. These examples, with the exception of certain supplementary tests which have been included, were from subjects having scalp thickness essentially equal to the average of some 20 subjects used in this and the previous study of Ref. 3. The average soft tissue thicknesses in various zones of these subjects were as follows:

Frontal	4.0 mm
Parietal	4.7
Top of head	5.4
Zygomatic	8.0
Maxillary	8.75

Table 2 and the corresponding photos depict the magnitude of trauma obtained under 45 deg angular impact in the a-p direction against the frontal region. Accompanying the 45 deg metal edge test results are those of two additional tests which were similar except that the angle of incidence was varied. In one instance the angle was increased to 50 deg to produce a more direct blow which penetrated completely through the scalp over the entire 3 cm length of the cut, and in the other case decreased to 30 deg for a more acute or grazing impact which failed to penetrate completely through the skin proper. Angle of incidence is thus seen to be a critical variable in soft tissue trauma.

Location and orientation of the impacting edge, on the other hand, were not found to be critical factors over the forward half of the scalp. Illustrated in Fig. 2, for example, are two of the lacerations of Table 1, taken longitudinally and transversely in the left frontal region, which were of equal length and extended to equal depth, thus indicating no significant dependence upon edge orientation. This result was in keeping with the fact that spade-tip penetrometer readings over various portions

Table 2—Angular Impacts of Metal and Glass Edges against Frontal Region

Edge	Drop Weight, lb	Height, in.	Scalp Thickness, mm	Scalp Radius of Curvature, in.	Edge Cross Section	Center of Gravity and Moment of Inertia	Angle of Incidence, deg.	Length and Depth of Laceration
1/64 in. metal edge against mid and side frontal (3 tests)	5.1	36	4.5	2⅜	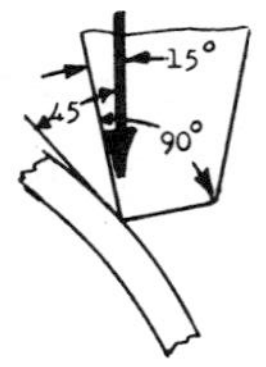	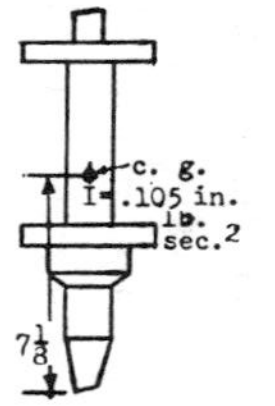	45	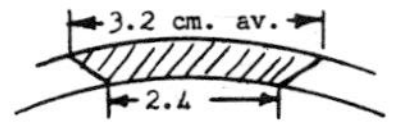
Less acute angle	5.1	36	—	—			50	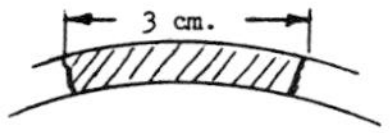
More acute angle	5.1	36	—	—			30	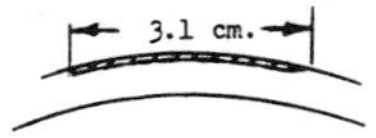
Glass edge against mid and side frontal (3 tests)	3.33	24	4.5	2⅞	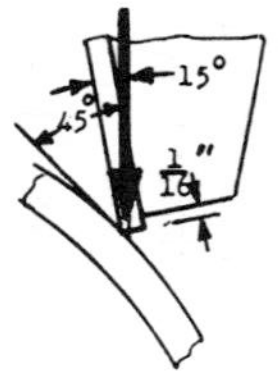	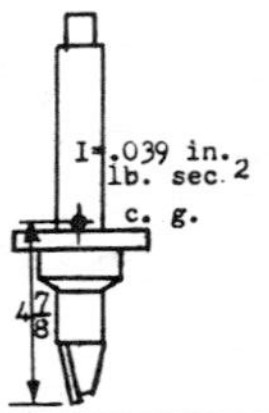	45	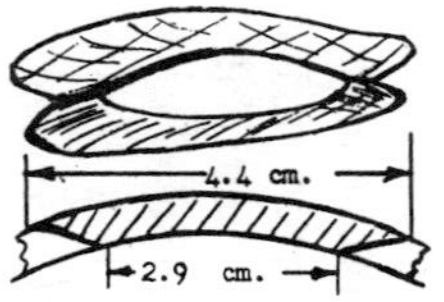

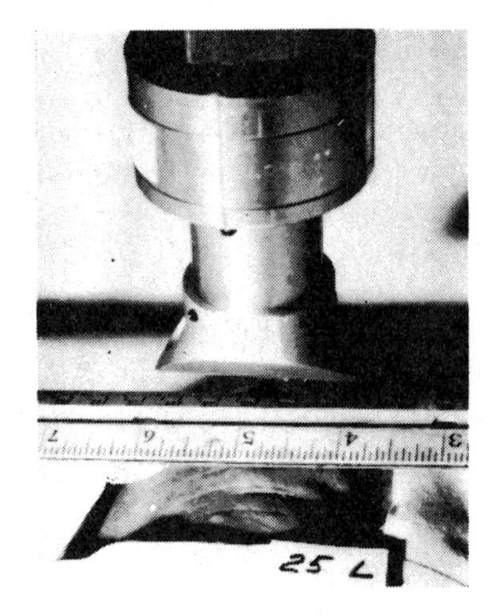

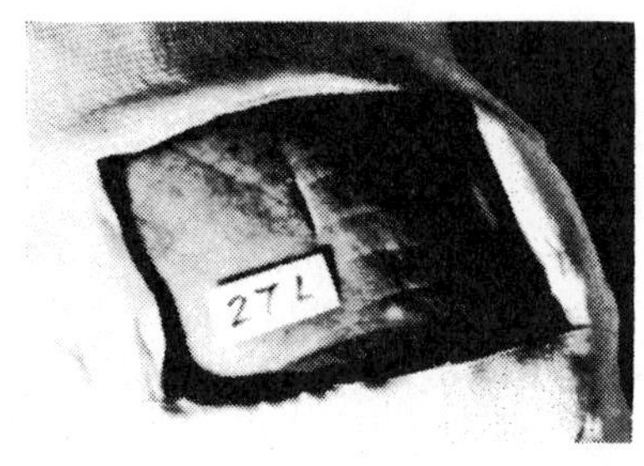

Fig. 2—Above, transverse, and longitudinal laceration in frontal area from 1/16 in. radius metal edge. Below, two "half-moon" lacerations from 1/64 in. metal edge at 45 deg. angle of incidence. Second scar in lower right-hand photo is from impact at 30 deg. from surface, which failed to penetrate completely through skin

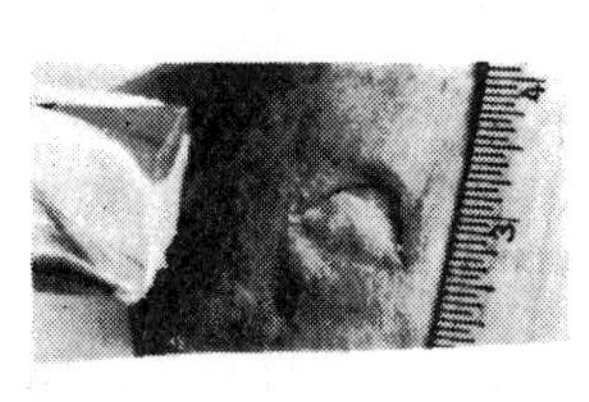

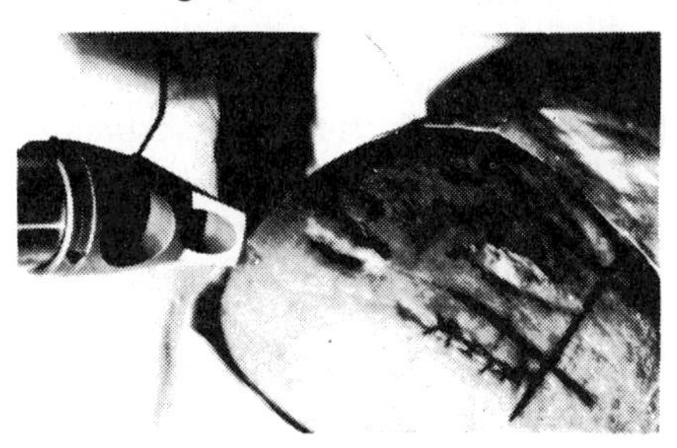

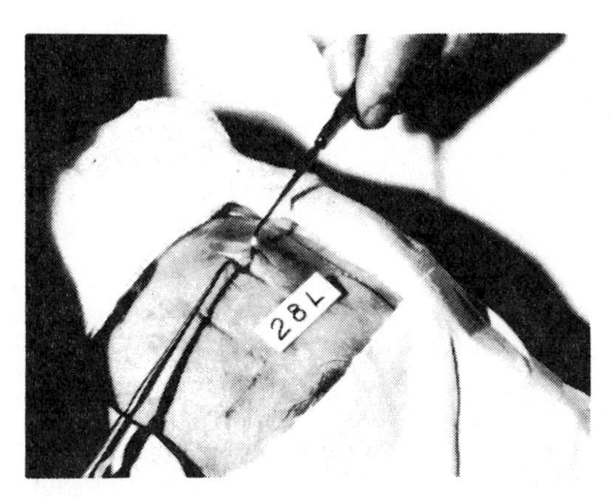

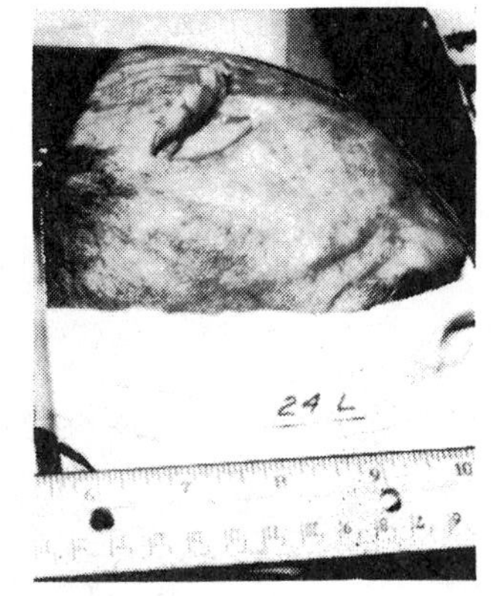

Fig. 3—Glass edge impact in frontal area. Left: at normal angle of incidence. Right: at 45 deg

of the scalp failed to disclose significant dependence upon orientation, in spite of the preferred orientation of certain of the tissues of the scalp.

The penetrometer readings, taken with the standard 0.010 x 0.0625 in. spade tip, usually fell between 14-17 lb for puncture, in keeping with the values previously reported in Ref. 3 for living scalp. This range was also consistent with that reported in Ref. 4, where little variation was seen in unembalmed subjects over an age span of 25-80 plus. The test subjects of this study were 66 years of age and older. Ref. 5 showed a diminution of certain of the physical properties of the skin and soft tissues of some areas of the body between the prime middle period and the elderly; therefore, to the extent that the tensile and elongation properties reported bear upon resistance to trauma, the results given herein may be conservative.

It is of interest to note the remarkably high compressive strength of the skin, as revealed by both the static penetrometer readings and the dynamic metal edge impacts. Thus, on the basic load-area calculation, the nominal stress under the 0.010 x 0.0625 in. penetrometer tip exceeds 20,000 psi when the puncture occurs at 14 lb loading. Similarly high compressive strength can also be estimated from the metal edge test results.

Specimens loaded between parallel surfaces — Representative results of these tests are summarized in Table 3 and are shown in the hysteresis loops of Figs. 5 and 6. Examples are given of scalp material in full thickness, found to be the most resistant to crushing action by virtue of tougher galea tissue, and specimens from other areas where the material under the skin was essentially fatty and weak. The latter material tended

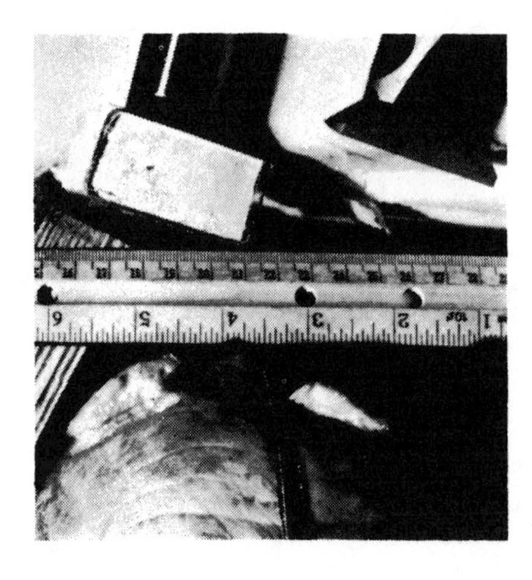

Fig. 4—a-p views of trauma from 45 deg incident angle. Left: the right-hand test of Fig. 3. Right: the lower right-hand test of Fig. 2.

to break down much more easily under impact and produced markedly lower rebound. Intermediate between these two extremes, regarding strength and coefficient of restitution, was the tissue of the zygomatic area. However, this material varied widely between different individuals in both thickness and properties.

Force developed under the impactor, rebound height, and visible residual crush are recorded for typical tests in each of these areas, and corresponding hysteresis graphs are plotted for both 12 and 24 in. drops.

Table 3—Compressive Impact Between Parallel Visible Surfaces (circular, sq in. area)

Location	Drop Weight, lb	Height, in.	Peak Force, lb	Total Specimen Thickness, mm	Rebound Height, in.	Visible Crush, mm	Test No.
Frontal scalp	0.65	12	445	5	3.5	None	76S
	0.65	12	485	3.5	2.25	None	77S
Parietal scalp	0.65	12	390	5.0	2.5	None	42S
	0.65	24	610	5.3	5.9	0.5-1	50S
Zygoma	0.65	12	208	5.5	1.4	None	44S
		24	338	5.5		1-1.5	52S
Fatty tissue,	0.826	12	136	6.35	0.01	4	7S
anterior upper	0.826	12	128	6.35	0.02	4	22S
arm	0.65	24	140	9.5	0.60		83S

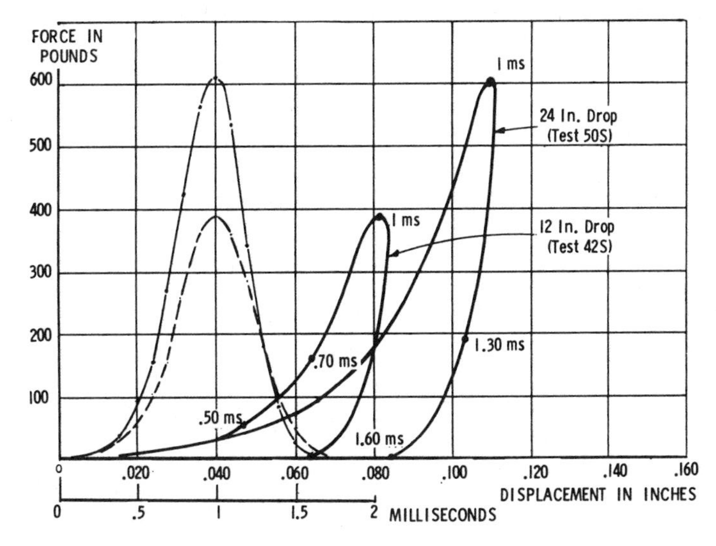

Fig. 5—Hysteresis characteristics of scalp tissue, with corresponding plots of load versus time at left. (Time from initial contact given in milliseconds)

Typifying the strength of scalp material, one specimen sustained a maximum force of 445 lb distributed uniformly over the 1 sq in. impactor area without visible damage. When drop height was increased from 12 to 24 in., peak force increased to 610 lb and marginal damage resulted. A loading between these two values may be looked upon as an approximate tolerance of the scalp material below the skin proper, since the tolerance of the skin itself to compressive loading is much greater — as pointed out above.

Tissue over the zygoma was weaker, depending to a varying degree upon the proportion of fatty material. Two typical specimens (Fig. 6) of approximately the same thickness sustained 208 and 338 lb during impacts from 12 and 24 in. height, respectively. The specimens which were completely fatty under the skin experienced major visible crush even under the least severe impacts. Energy absorption was excellent, as indicated by the minute rebound height, but loads of 100-150 lb over the sq in. area produced over 90% crush.

The hysteresis loops, selected to be typical for the three tissue categories, gave the dynamic response in more quantitative terms. For a given intensity of impact, the load sustained and the recovery during the unloading cycle were seen to decrease progressively in the weaker tissues.

Summary and Conclusions

Effects of selected impacts to localized areas of the superficial tissues are presented. Test conditions, including time of exposure to the loading, have been chosen in such a way that it is believed the results may be of practical value in indicating approximate tolerances of the superficial tissues to common types of accidental loading.

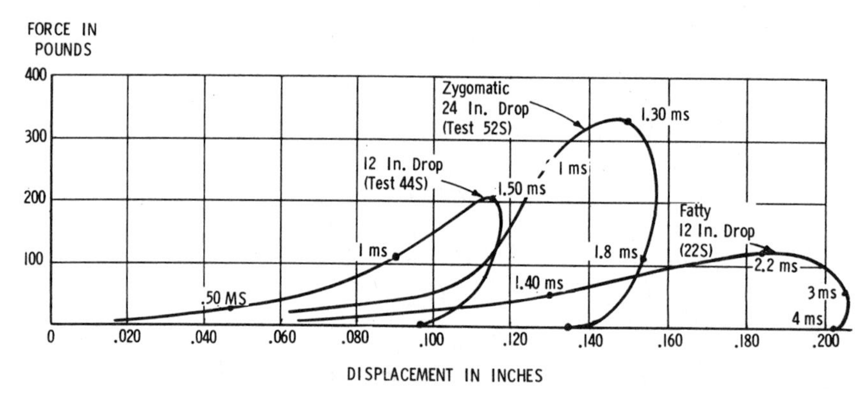

Fig. 6—Hysteresis of zygomatic and fatty tissue. (Time from initial contact given in milliseconds)

The geometrical shapes of the impacting tips, the intensities of impact, and the magnitude of the resulting tissue damage are given in sufficient detail to enable the tests to be duplicated on artificial simulations of the biological materials, and thus to determine their suitability for the construction of anthropometric dummies.

References

1. C. W. Gadd, A. M. Nahum, James D. Gatts, and J. P. Danforth, "A Study of Head and Facial Bone Impact Tolerances." Proceedings of General Motors Automotive Safety Seminar, Milford, Mich., July 1968.
2. C. W. Gadd, W. A. Lange, and F. J. Peterson, "Strength of Skin and its Measurement." Biomechanics Monograph, ASME, New York, 1967.
3. A. M. Nahum, James D. Gatts, C. W. Gadd, and J. P. Danforth, "Impact Tolerance of the Skull and Face", Twelfth Stapp Car Crash Conference Proceedings, P-26, paper 680785. New York: Society of Automotive Engineers, Inc., 1968.
4. Peter Fuller and Donald Huelke, "Skin Puncture Tests." Vol. IV, App. F, Interim Report on Contract PH-43-67-1136, Highway Safety Research Institute, University of Michigan, 1968.
5. H. Yamada, "Human Biomechanics." Kyoto Prefectural University of Medicine, Kyoto, Japan, 1963.

710871

A Strain Energy Approach to the Mechanics of Skull Fracture

J. W. Melvin and F. G. Evans
The University of Michigan

Abstract

The mechanics of skull fracture in humans has been investigated by many people for over 90 years. A variety of techniques has been used in past studies. Test specimens have been whole cadavers, cadaver heads, skulls and sections of skulls with material conditions including both fresh and embalmed tissue, both dried and moist. Test techniques have incorporated cadaver drop tests, drop towers, and universal testing machines with the impacting surfaces including large surfaces, both flat and curved, and localized flat and curved surfaces. Some of the studies used impact energy as the measured test parameter, others used impact load and some studies used both quantities to describe the impact.

The results of recent studies on the mechanical properties of cranial bone suggest that local values of strain energy density present in the bone of an impacted skull may be the critical parameter in the initiation of skull fracture. It is the purpose of this paper to summarize the results of the large body of information on the phenomenology of skull fracture and then to analyze the results in terms of local strain energy density considerations. Based on this reasoning, the differences and similarities between the results of the many different studies are discussed in a qualitative manner. Conclusions are drawn on the influences of such factors as impactor size and shape, skull geometry, and soft tissue effects.

ALTHOUGH THE MECHANICS OF SKULL FRACTURE has been of interest to members of the medical profession for a long time, engineers have also recently become involved with the problem. This is particularly true of the safety engineers in the automobile industry, since the head is the most frequently injured part of the body in automobile accidents. Engineers in the aircraft and aerospace industries, in addition to the designers of safety gear and equipment, also need factual data on the limits of tolerance of the human skull to impacts.

During the last century and the early part of this one, several theories of the mechanics of skull fracture were advanced. Unfortunately, little data on the limits of tolerance of the skull were offered in their support.

According to Aran (1)*, a fracture of the skull vault, arising from an impact to the vertex, reached the base of the skull by the shortest route. This theory implied that the fracture was initiated at the site of impact.

Elasticity of the skull was studied by Bruns (2) who placed undamaged heads between two small boards in a vise and slowly compressed them in either the longitudinal (front to back) or the transverse (side to side) direction. Changes in the longitudinal (sagittal) and transverse diameters of the skull at various stages of compression were measured from markers on the occiput, the forehead, and the two parietal bones. Bruns demonstrated that the skull can undergo a considerable compression in any direction without fracture and that a decrease in diameter in one direction is accomplished by an increase in diameter in other directions.

As a result of his experiments, Bruns theorized that force along one diameter of a skull created tensile forces in parts of the skull perpendicular to the direction of the pressure. Fracture eventually resulted in an area of the skull where the radius of curvature decreased or where outbending occurred. If the compression was too great, fracture and depression occurred at the point of application of the force. Bruns believed the elasticity of the skull was quite high.

In 1873 Félicet (cited by Messerer, 1880) (3) described the buttresses of the human skull and noted that when they were separated by an impact the fracture might pass between them. According to him, a fracture was initiated when an impact flattened out the curved surface of the skull.

Baum (4) repeated the experiments of Bruns, but instead of applying forces to the skull with a vise he used an iron ring with two pads on opposite sides—one fixed and the other movable. The soft tissues were removed from the skull before testing so they would not obscure cracks and small fissures produced in the skull during a test. Three skulls were tested but the results were quite different from those reported by Bruns in that the diameter in a direction perpendicular to the load did not change.

The behavior of the human skull found in Baum's experiments differed from the experiences of nearly all other investigators of the mechanics of skull fracture. Hyrtle (5) concluded that the skull was elastic because a fresh head thrown on the floor bounces several times. A similar conclusion was reached by Félicet (3) who dropped charcoal-blackened skulls from various heights onto a white surface and found oval black spots, as in similar experiments with an ebony ball.

---

*Numbers in parentheses designate References at end of paper.

Cohstein (5), studying the effects of forceps compression during delivery, measured the changes produced in the longitudinal and transverse diameters of the skull of newborn infants. He found that in different cases the diameter perpendicular to the direction of compression either decreased, increased, or remained unchanged. However, as pointed out by Messerer (7), the conditions in the membraneous skull of a newborn infant are quite different than in those of an ossified skull of an adult.

Weber (8) appears to be the first investigator of the mechanics of skull fracture to publish quantitative data on the magnitude of the force required for fracture. He found that a force of 1113 lb was required to fracture the skull of a small-boned tuberculous girl 27 years of age, while the skull of a robust woman 37 years of age did not fracture under a force of 1375 lb.

The classic study on the force required to fracture bones is that of Messerer (7) who determined the breaking load of 500 fresh bones or bone combinations from 90 individuals by loading them to failure in a Werder materials testing machine. Before testing a skull, the soft parts were removed.

The skulls of seven men, with an average age of 42.7 years (18-69), and of six women, with an average age of 48.2 years (22-82) were loaded in the transverse direction. Tests by loading in the longitudinal direction were made on skulls of seven men with an average age of 41.4 years (19-61), and five women with an average age of 41.8 years (20-74).

Messerer found that the load required for fracture, when applied in the transverse direction, was greater for skulls of women than for those of men, while the reverse was true for loads applied in the longitudinal direction. In both sexes, a higher load was supported before failure if applied in the longitudinal direction. Disregarding sex and age, the average breaking load and range of variation was 1142 lb (770-1760 lb) for transverse loading and 1430 lb (880-2650 lb) for longitudinal loading.

In addition to his tests on the effects of longitudinal and transverse compression on skull fracture, Messerer investigated the effects of compression in a direction perpendicular to the skull base. This was done in eight skulls, with three or four attached vertebrae, by loading the skull in the Werder testing machine through the crown on one hand and the inferior surface of the last attached vertebra on the other.

Fracture occurred at relatively low loads and without any perceptible change in the various skull diameters. Because of its weak and brittle character the base of the skull was destroyed before the compression had much effect on the entire skull. In many cases the first or second vertebra fractured before the skull did.

The individuals whose skulls were tested ranged from 24-60 years of age. Six were men and two were women (44 and 60 years of age). The breaking load, regardless of sex, varied from 495-660 lb, with an average of 594 lb.

In a third series of experiments, Messerer studied the effects of concentrated pressure at various sites in the skull. The localized pressure was applied via a cylindrical bolt 0.67 in in diameter, somewhat rounded in front. As before, the tests were made in the Werder testing machine.

Except for three cases in which longitudinal cracks were found, the concentrated pressure produced perforations on the same diameter as the bolt without involving the adjacent area of the skull. Thus, the results of a test were not influenced by pressure at other sites on the same skull.

The penetration load varied from 616-1815 lb for the center of the frontal bone, from 396-1100 lb for the center of the parietal bone, from 1155-2145 lb for the external occipital protuberance, from 374-418 lb for the squamous portion of the temperal bone, and from 55-77 lb for the zygomatic arch.

The total thickness of the skull had no significant effect on the resistance to perforation, thick-walled skulls frequently having small rupture loads and vice versa.

Messerer believed that the massiveness of the diploë was particularly important in skull perforation, although, as he pointed out, it had little influence on strength.

In three cases, pressure on the external occipital protuberance, a thick area, produced a longitudinal fracture, as in the compression tests between plane surface, instead of a perforation. In a 27-year old and a 50-year old man a longitudinal fracture of the skull bases was produced while in a 28-year old woman a diastasis of the cranial suture occurred.

According to Rawling (9), fractures of the base of the skull result from impact that actually splits the skull in much the same manner as a hatchet can split a piece of wood along its grain. It is implied that the fracture was initiated at the point of impact.

The work of LeCont and Apfelbach (10) more or less substantiates the concepts enunciated by Aran (1), Félicet (3), Bruns (2), and Rawling (9). In Rowbotham's textbook on head injury (11), the idea is advanced that depressed fractures result from local deformations while linear fractures result from generalized deformations.

In 1945 Gurdjian and Lissner published the first of their famous series of studies on skull deformation and fracture by means of the "Stresscoat" technique (12). "Stresscoat" is a trade name for a brittle strain sensitive lacquer which cracks in response to tensile strain occurring in the material upon which it is sprayed. The cracks in stresscoat lacquer only arise in tensile strain in the underlying material, the site of their first appearance indicating the area of highest tensile strain where failure may be expected to occur with sufficient load. This study was made with animal skulls.

Gurdjian and Lissner then extended their studies to the mechanism of skull fracture in human skulls (13). The close correspondence between the site where the stresscoat cracks first appeared under known conditions of loading and the direction of the crack, indicated that linear fractures of the skull arose from failure of the bone from tensile stresses created within it by bending. A fracture may be initiated on either the external or the internal surface of the skull. Variations in the shape and thickness of the skull influence the strain propagation characteristics and the stresscoat patterns.

In their second study with the stresscoat technique, Gurdjian and Lissner (14) determined the energy required for threshold deformations of six defleshed skulls obtained from embalmed dissecting room cadavers. Each skull was weighted and then suspended by means of a string a measured distance over a polished steel block weighing several hundred pounds. The skull was dropped upon the steel block by burning through the supporting string.

In this study, Gurdjian and Lissner found that the amount of absorbed energy necessary for threshold deformation of the skull differed in various regions of the skull. Thus, in the midfrontal region, blows of 14-18 in lb of energy caused deformations similar to those obtained with 8 in lb of energy in the mid-occipital region. The amount of energy dissipated in deformation and fracture of the skull was considered to be small.

The distribution and the direction of the deformation patterns produced in the experimental skulls were generally similar to the fracture lines seen in clinical cases. It was pointed out that the shape and velocity of the object causing the injury may influence the strain patterns and give them a directional quality. In these experiments, the velocities of the skull at the time of impact varied from 4.5-9.9 ft/s.

These studies showed that the path of the strain produced in a skull depended on the shape, contour, and thickness of the bone in the area of impact. Deformation patterns were more common in the weaker buttress regions of the skull, and relatively few patterns were found in the region of zygomatic bone, frontal, and petro parietal buttresses. According to the authors, the ease with which deformations were produced in the region of the frontal, foramen magnum, and the parietal-temporal area may account for many fractures in this region without loss of consciousness.

The amount of energy and the time for its absorption required to fracture 55 intact human cadaver heads was investigated by Gurdjian, et al. (15). Tests were made by dropping the head on the steel block so that energy was applied to the midfrontal, the anterior interparietal, the midoccipital, and the right or left posterior parietal regions.

Data obtained in the tests showed that energy varying from 400-900 in lb was required to produce a single linear fracture. An average of 571 in lb was necessary to produce a fracture in the midfrontal regions, 517 in lb in the midoccipital regions, 710 in lb in the anterior interparietal region, and 615 in lb in the regions above each ear. Because of the wide range of variation in each region, the authors considered average differences to be of little significance.

The great range of variation in the amount of energy required to fracture the skulls was a result of individual variations in the shape and thickness of the skull and scalp. They found that a relatively thin layer of soft tissue, for example, the scalp, made a considerable difference in the amount of energy required for fracture. For example, in an intact cadaver head, 400 in lb of energy were required to produce a single linear fracture while as little as 40 in lb would produce a similar fracture in a dry skull. It was found in their tests that after a single linear fracture was initiated very little energy was required to produce multiple fractures and complete destruction of the skull. Furthermore, the average energy required for complete skull destruction often was close to those necessary to produce a single linear fracture. In these tests, the average impact velocity for the midfrontal region test was 15.0 ft/s; for the anterior interparietal test, 19.3 ft/s; for the occipital deceleration impact, 15.8 ft/s; and for the posterior parietal deceleration impact, 18.8 ft/s.

Gurdjian, et al. (16) summarized their studies on the prediction of fracture site in head injury on the basis of test results from 100 randomly selected adult skulls. In these tests deceleration blows were applied to the midfrontal, the anterior interparietal, the posterior interparietal, the midoccipital, the left frontal, the left anterior interparietal, the left posterior parietal, and the left parietal-occipital area. The frontal, anterior parietal, posterior parietal, and parietal-occipital areas were bilateral in position. Because of the symmetry of the skull, the left side and midline areas were used, special care being taken that every part of each area was tested many times. The fractures produced in the test in each area were then analyzed with respect to site and direction. It was found, in general, that frontal blows produced more vertically oriented fractures slanting slightly forward while blows in the anterior and the posterior parietal and in the parietal-occipital areas produced more horizontal fractures. Many fractures on one side or the other of the skull were initiated in the temporo-parietal areas. Ipsilateral fractures arose from blows on one or the other side of the midline. When a blow is applied near the border of an area, the fracture position in the adjacent area aided in predicting a horizontal fracture in the temporal and lower parietal regions created by blows in the anterior and inferior regions of the left parietal-occipital area.

Evans, et al. (18) studied the relation of energy, velocity, and acceleration to skull deformation and fracture in intact human heads taken from adult embalmed cadavers. The test consisted of dropping the heads from various measured distances upon a 1954 model automobile instrument panel mounted in a way similar to that in an intact car. The blow was applied to the forehead such as frequently occurs in automobile accidents.

Fractures were produced with peak impact accelerations of 337, 344, 555, and 724 g, having a total time duration of 0.01125, 0.00488, 0.00903, and 0.00338 s, respectively. In some cases, the head tolerated, without fracture, peak impact acceleration as high as 686 g and available kinetic energy as great as 577 ft lb. The magnitude of the energy producing the fracture was estimated to be approximately 33-75 ft lb and the major portion of the kinetic energy available in each test was expended in bending and deforming the instrument panel upon which the head was dropped. However, in each test, the heads also rebounded to a considerable height but the energy responsible for the rebounding was not considered.

Hodgson, et al. (19) conducted tests on facial impact with human cadavers using a 1 sq in impactor surface. The resulting fracture forces for impact to the supra-orbital ridge of the frontal bone were, 859 lb with skin intact and padding on the striker resulting in a depressed fracture, 940 lb with skin removed and no striker padding resulting in a linear fracture, and 1514 lb with skin removed and striker padding resulting in a depressed fracture.

In a similar study in 1967 Hodgson (20) reported data on blunt impact to the frontal bone using a 5.2 sq in impactor with a 1 in polyurethane padding thickness and skin intact. The fracture forces were 943, 1660, 1980, and 2050 lb with 4-5 ms durations. Linear fractures were obtained in all cases.

The impact tolerance of the human skull and face was investigated by Nahum, et al. (21) in ten human skulls varying from 55-81 years of age. Four of the skulls were from men and the rest from women. Five of them also were from embalmed bodies and five were not.

Drop tower impacts were applied to the frontal bone and the tempero-parietal junction, using a 1 sq in impactor area.

Fractures were verified by visual palpation, x-rays, acoustic impression at time of impact, changes in the force-time oscillogram, anatomical dissection under magnification, and the use of dye penetrants. Fractures were classified into three categories: minimal detectable change (1+); readily detectable fracture (2+); and comminuted and/or depressed fractures (3+), representing the extreme of clinical significance. Only single impacts were applied to a particular anatomical location.

The data obtained from their experiments led the authors to suggest the following values for clinically significant fractures with an effective contact area of approximately 1 sq in: frontal area, 1100 lb; temporo-parietal area, 550 lb; zygomatic area, 225 lb. The minimum tolerance values for these areas were specified as 900 lb for frontal area, 450 lb for temporoparietal area, and 200 lb for the zygomatic area. Comparison of the data according to sex revealed a significantly lower average strength in the female subjects.

The range of variation in the fracturing force for the frontal area regardless of sex was from a minimum of 806 to a maximum of 1990 lb; for the temporoparietal area from a minimum of 551 lb to maximum of 1334 lb; and for the zygoma from a minimum of 314 lb to a maximum of 633 lb. The thickness of soft tissues and particularly its presence or absence was found to play an important role. Thus, no evidence of fracture of the frontal area of a particular skull was produced by a force of 948 lb when soft tissue was present. But, a fracture was produced with a peak force of 708 lb when the soft tissue was removed.

In 1969 Melvin, et al. (22) investigated the tolerance of frontal and parietal bones to local penetration using impactor diameters of 0.432, 0.500, and 0.612 in. Skull caps from embalmed cadavers were mounted in a holding fixture that allowed the tests to be performed in a high-speed universal materials testing machine. Due to the local nature of the loading it was determined that this technique was basically equivalent to whole skull tests for the regions impacted. Tests were performed both statically and at 14 and 28 ft/s. The average penetration forces, ordered by increasing impactor diameter were, for frontal bone; 1030 lb (470-2000), 1480 lb (1000-2100), and 1710 lb (920-2200), and for parietal bone; 780 lb (140-1500), 880 lb (400-1300), and 1290 lb (500-2200). No significant rate effects were found.

Hodgson, et al. (23) investigated the behavior of the frontal bone of intact human heads when subjected to impacts delivered by two rigid cylindrical surfaces—one with a 1 in radius and the other with a 5/16 in radius. Both objects were 6 1/2 in long. A total of 12 cadavers were tested, seven against the 1 in radius cylinder and five against the 5/16 in radius cylinder. A guided drop of the test object against a stationary head which was free to rebound was part of the test. Drop heights were increased progressively until borderline fractures were obtained.

Remote linear fractures were consistently produced with a large radius shape suggesting that it is effectively a blunt surface. Fracture loads varied from 950-1650 lb. With the small radius shape, two remote linear fractures and three localized elliptical fractures were produced indicating that the cylinder was in the transition phase between the blunt and concentrated surfaces. Fracture loads of the smaller radius cylinder ranged from 700-1730 lb.

Properties of The Human Skull

It is evident from the introduction section of this paper that experimental skull fractures have been achieved by a variety of techniques which have produced a wide range of results. The basis of continuity throughout all the test techniques has been the use of the human skull. The three important features of the human skull with respect to its fracture behavior are the overall structure of the skull, the local structure of the bones of the skull, and the mechanical properties of the basic bone material in the skull.

The overall structure of the skull that is of primary interest in head injury is the cranium which encloses the brain. The cranium is composed of eight bones (Fig. 1), four of which (frontal, occipital, and left and right parietal) form the upper surface of the skull (Figs. 1A and 1B). The left and right temporal bones form the side walls of the cranium and along the occipital and frontal bones they also form part of the floor of the cranium (Figs. 1B and 1C). The sphenoid and ethmoid bones (Figs. 1C and 1D) complete the floor of the cranial vault with the sphenoid bone extending up both sides of the cranium between the frontal bone and the temporal bones. The various bones of the cranium are joined together along their boundaries by junctions known as sutures. The sutures are narrow regions of collagenous material and they possess a complex geometry similar to an irregular dovetail joint as shown in Fig. 1.

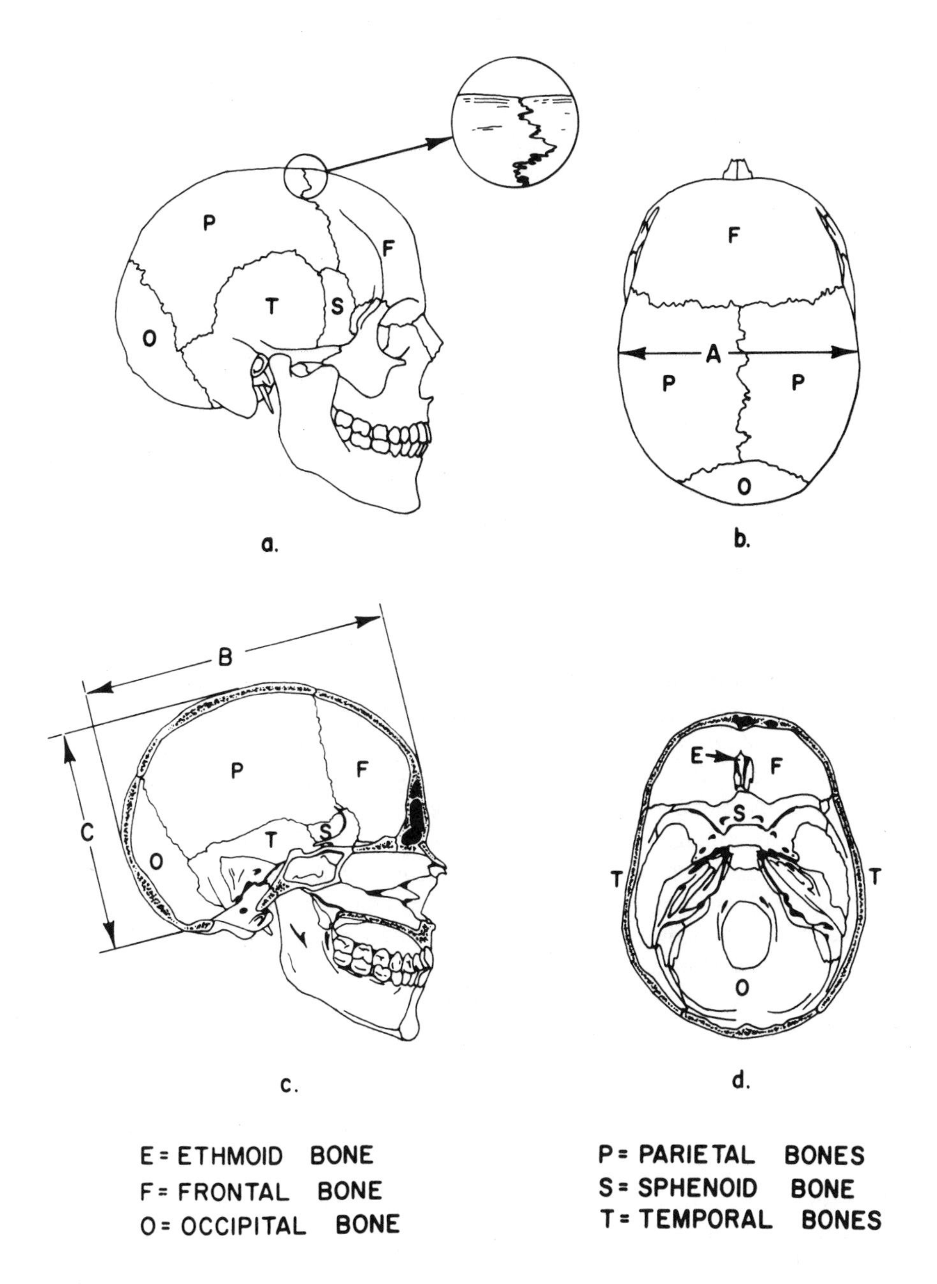

Fig. 1 - Skull geometry and structure

In overall geometry, the cranium is somewhat egg-shaped, with the posterior portion being larger than the anterior portion. Average values for the dimensions shown in Fig. 1 for an adult caucasian male cranium (24) are 5.51 in wide (dimension A), 7.12 in long (dimension B), and 5.34 in high (dimension C). The portion of the cranium above a plane through the temporo-parietal sutures tends to be a fairly uniform domelike structure, while that portion of the cranium below the plane is comparatively flatter and highly irregular in structure, with many openings such as the foramen magnum and other smaller foramina.

The bones of the cranium each have their own characteristic structural variations; their basic structure, however, as shown in Figs. 1E and 1D, is that of an inner and outer layer of compact bone separated by a layer of porous bone known as the diploë layer. In some cranial bones, the diploë layer is not present and the structure is then just a layer of compact bone. The parietal bones and the upper portions of the frontal and occipital bones in many cases possess a diploë layer thickness approximately equal to the sum of the thicknesses of compact bone on each side of it resulting in total thicknesses in the 0.2-0.3 in range. The temporal bones and the lower portions of the frontal and occipital bones commonly have little or no diploë layer and are generally quite thin in such regions as the orbital plates of the frontal bone, the squamous portion of the temporal bones, the posterior cranial fossae of the occipital bone, regions of the ethmoid bone and the greater wings of the sphenoid bone.

Mechanical Properties of Skull Bone - Although skull fracture has been studied for over 100 years, it has been only recently that experiments to determine the mechanical properties of skull bone as a material have been performed. The first such study was that of Evans and Lissner (25) in 1957, in which the static tensile and compressive strengths of embalmed human parietal bone were determined. The average ultimate tensile strength of 15 specimens of parietal compact bone was found to be 10,230 psi (6,030-15,800) while the average compressive strength of 69 specimens of parietal compact bone was 22,080 psi (12,400-47,800) for the same load direction and the average compressive strength of 56 specimens loaded perpendicular to tensile load direction was 24,280 psi (4,500-46,900). The average compressive strength of 23 specimens of the cancellous diploë layer bone was found to be 3640 psi (1700-5770).

Most recently, Wood (26) has reported on a study of the mechanical properties of unembalmed human compact cranial bone in tension based on tests of over 120 specimens from 30 subjects. The specimens were taken from the compact layers of parietal, temporal, and frontal bone at autopsy and the tests were performed at strain rates ranging from 0.005-150 s^{-1} The modulus of elasticity, the breaking stress, and the breaking strain were found to be strain rate sensitive while the energy absorbed to failure did not change with strain rate. The properties showed no important variation with respect to type of bone, side of body, or age of the individual, and there was no apparent variation of properties with respect to direction tangent to the surface of the skull. The average tensile strength of 10,000 psi found at a strain rate of 0.01 s^{-1} agrees closely with the value obtained by Evans and Lissner. For the strain rate range studied by Wood the average modulus of elasticity ranged from approximately 1.8-2.9 $\times 10^6$ psi, the average tensile strength ranged from 10,000-14,000 psi, and the average failure strain ranged from 0.7-0.55%. The average value of energy absorbed to failure was 42.6 in lb/in^3 with the bulk of the data ranging between 20-80 in lb/in^3.

In 1970, Melvin, et al. (27) presented the results of compression tests of 215 unembalmed cancellous bone specimens from the diploë layer of 46 bone plugs from the parietal and frontal regions. The tests were performed over a strain rate range from 0.002-500 s^{-1}. The highly variable trabecular structure of the diploë layer was found to strongly influence the mechanical properties to such an extent that strain rate effects were not significant. The most frequent form of failure of the specimens was an abrupt collapse of the trabecular structure. This occurred at compressive stresses as low as 1300 psi. For specimens whose trabecular structure was massive the compressive behavior became more like compact bone with compressive strengths as high as 31,900 psi reported. The median strength was approximately 5000 psi.

The preceding studies determined the mechanical properties of skull bone from a classic strength of material viewpoint. In order to further understand skull fracture phenomena, it is necessary to consider the fracture mechanics aspect of bone failure. In its simplest form, fracture mechanics as proposed by Griffith (28) is based on the concept of a crack-like flaw of a critical length occurring in a brittle elastic material. The flaw can either be inherent in the material or it can be created by prior loading. If this crack is to propagate in a fast uncontrolled manner under load then the rate of release of strain energy in the loaded material as the crack propagates through it must be greater than the rate of creation of surface energy associated with the fracture surface as it grows. Knowledge of the tensile strength and elastic modulus of the material and the surface energy or work of fracture allows the calculation of the critical flaw size for a particular material. The work of fracture must be determined using special specimens and test techniques.

Piekarski (29) has performed such experiments on beef femur and, for rapid crack propagation, the work of fracture was found to have an average value of 5.5 in lb/in^2 (2.34-20.6). While these values cannot be expected to be exactly representative of human skull bone, they can be useful in providing an order of magnitude estimate of the corresponding work of fracture values for skull bone. Based on typical compact skull bone mechanical properties the above values of work of fracture predict an average critical crack length on the order 0.05 in assuming skull bone to be a perfectly brittle elastic material. In elastic deformations at the tip of a crack will modify the fracture mechanics analysis of bone fracture so that this analysis must be considered as approximate.

Mechanical Properties of Scalp - The scalp consists of two basic layers of tissue (the dermal layer and the subcutaneous layer) which are of importance to its mechanical behavior in head impact. The dermal layer is the outer layer of the skin and consists of tough fibrous connective tissue. Underlying the dermal layer is the subcutaneous layer, which consists of loose fatty connective tissue. When scalp is loaded in compression perpendicular to its surface the two layers deform in series with the greatest part of the initial deformation occurring in the soft subcutaneous layer. The result is the typical compressive stress-strain curve for scalp shown in Fig. 2A. As the strain in the subcutaneous layer apprcaches high values the resistance to further deformation begins to rise rapidly resulting in an exponential increase in stress following the initial low stress region of the curve.

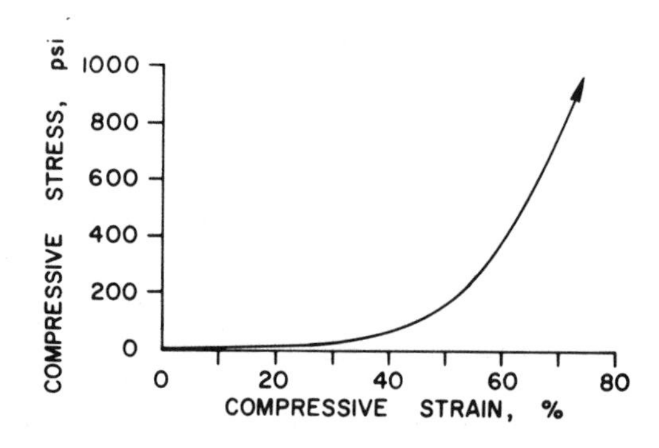

Fig. 2A - Scalp compressive stress-strain behavior

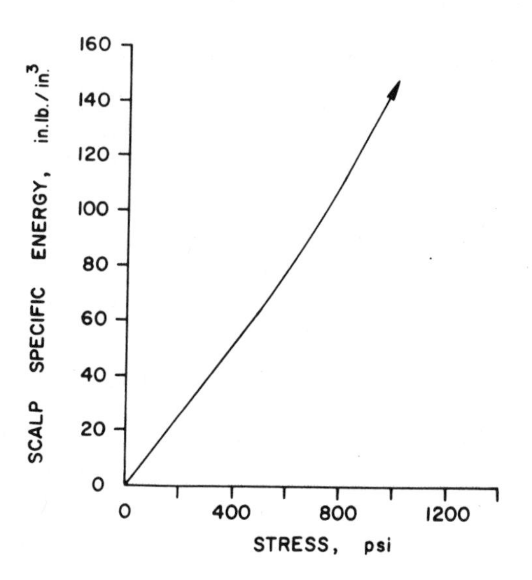

Fig. 2B - Scalp specific energy absorption as a function of stress level

In head impact the characteristic concave-upward shape of the scalp compressive stress-strain curve with the low stress level during the initial portion of the deformation serves to distribute the load over the surface of the skull and to prevent the occurrence of sharp load spikes. As the subcutaneous layer is compressed it tends to preserve its volume by expansion perpendicular to the load direction. Restriction of this lateral growth by the tissue surrounding the impact site tends to make the pressure distribution under the impactor even more uniform. The deformation of the scalp also absorbs some of the kinetic energy of the impactor. The amount of energy absorbed depends on the peak stress level developed in the scalp and the volume of scalp under load. Fig. 2B shows the specific energy absorbed (that is, the area under the stress-strain curve) as a function of stress level for the scalp stress-strain curve of Fig. 2A).

Head Impact

The collision of a head with an impactor or impact surface is governed by the principles of impulsive motion whereby sudden changes in momentum are related to impulsive forces. The magnitude and duration of such forces depend primarily on the relative impact speed and the mechanical and geometric properties of both the head and the impactor. In general, the law of conservation of mechanical energy does not hold when impact phenomena occur; however, it is of interest to consider the storage and dissipation of energy during head impact. The forces generated during impact produce both local deformations in the skull at the impact site and stress waves which travel throughout the structure of the skull. The combined effect of the forces transmitted to the skull and the resulting deformations of the skull structure create a strain energy distribution throughout the material of the skull which, under some conditions, will lead to skull fracture.

Thus, for skull fracture to occur in an impact, the momentum change of the impactor must be sufficient to cause an impact force high enough to produce critical strain energy levels in the skull. It is also necessary for the impactor to possess kinetic energy adequate to fulfill the energy storage and dissipation requirements due to deformation of the impactor, deformation of the scalp, and the deformations of the skull. Impactor kinetic energy alone cannot cause skull fracture if the impact forces developed are not high enough for the particular impactor and head mechanical and geometric properties present. In discussing human skull fracture tolerance, primary consideration is usually focused on the lower portion of the statistical distribution of experimentally obtained values. The following discussion of skull fracture phenomena continues this principle.

For head impacts occurring at velocities typical of automotive accidents the three basic types of skull fracture (shown in Fig. 3) are penetration-type depressed fractures, comminuted depressed fractures, and remote linear fractures (15). These three types of fractures are primarily dependent on the size and geometry of the impactor surface with combinations of more than one type fracture possible in transition regions between the characteristic impactor sizes associated with each type of fracture. As a framework for discussion of the various types of fracture, the fracture force data reviewed in the introduction for the most commonly studied cranial bone, the frontal bone, have been plotted versus impactor area in Fig. 4. The following is an analysis of the general

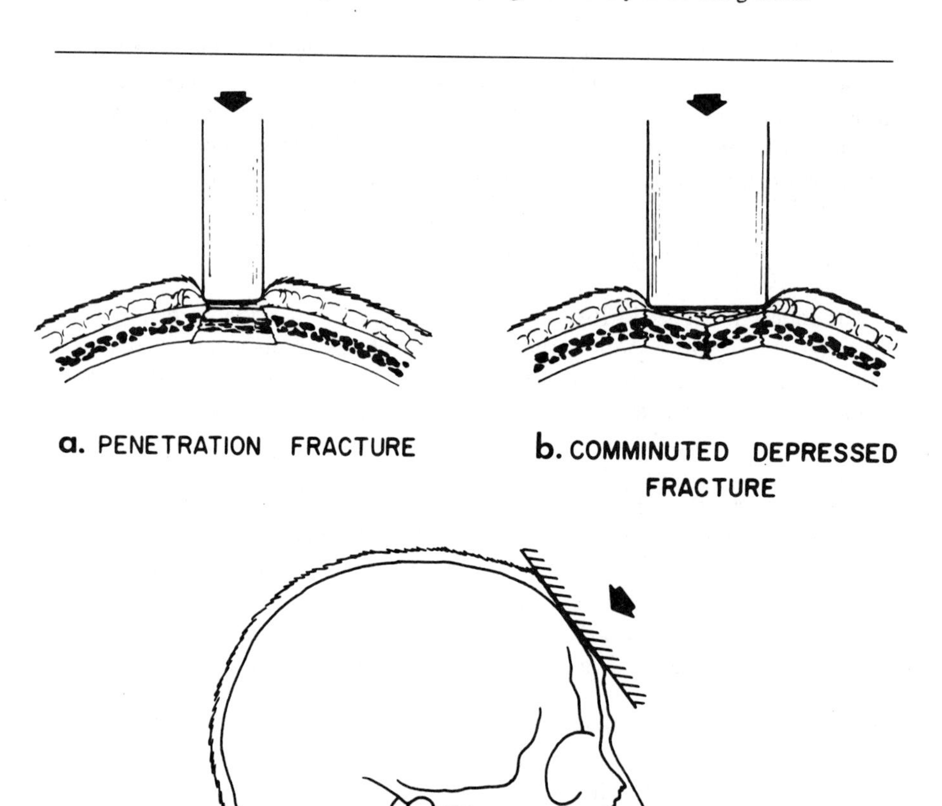

Fig. 3 - Characteristic types of skull fractures

features of the types of skull fracture and it must be kept in mind that the inherent structural variations in skulls can cause some modification and overlapping of fracture phenomena.

Penetration Fractures - Penetration-type depressed fractures are the result of highly localized loading of the skull by impactors with less than 1 sq in of area. The basic features of such a skull fracture are depicted in Fig. 3A, and usually consist of shearing out of a plug the diameter of the impactor with little or no disturbance of the surrounding skull structure. Most of the data obtained on this type of loading (7, 22) has been with frontal and parietal bones which usually possess pronounced diploë layers. The failure initiation appears to occur directly under the impactor on the outer surface of the skull. During the initial loading, the skull deforms as a sandwich shell with the inner and outer tables acting as a unit with the diploë layer core. However, the deformations become more and more localized as the load builds up and the skull no longer behaves as a unit as the compressive stresses increase on the diploë layer directly under the impactor. In view of the abrupt collapse characteristics exhibited by the diploë layer in compression it is probable that, in skulls with moderate to low density trabecular structures directly under the impact site, the peak penetration load is determined primarily by the compressive strength of the diploë layer. Collapse of the diploë layer causes a loss of structural integrity in the layered bone structure with concomitant failure of the outer table bone

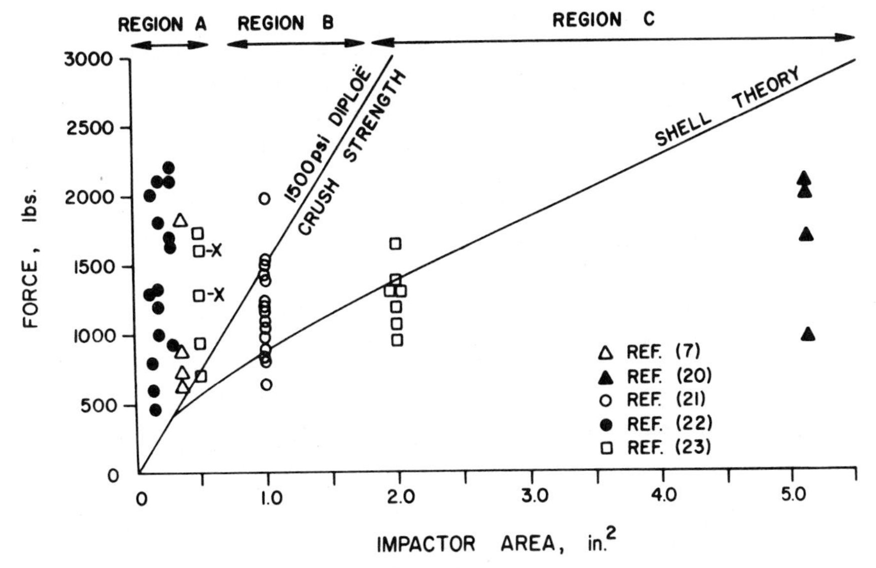

Fig. 4 - Frontal bone tolerance to force
Region A - penetration fractures
Region B - comminuted depressed and local linear fractures
Region C - remote linear fractures

around the circumference of the penetration. Based on this reasoning, a minimal force tolerance for penetration fractures can be calculated for any area impactor by consideration of diploë layer compressive strength data. Such a line has been drawn in Fig. 4 for a diploë strength value of 1500 psi, which lies at the lower end of the strength values reported in the previous section on diploë mechanical properties. Note that all of the experimental penetration data fall above this line whereas for larger area impactors the minimum force tolerance for penetration rapidly reaches values that cannot be obtained due to other modes of failure causing the other types of fractures. The data shown in Fig. 4 for the cylindrical impactor (23) is plotted at an estimated effective area of 0.5 sq in. The two data points marked with an X were remote linear fractures which were most likely due to an interaction of the frontal bone geometry with the flat direction of the impactor resulting in an increased effective impact area.

Comminuted Depressed Fractures (Including Local Linear Fractures) - Comminuted depressed fractures (Fig. 3B) are the result of localized loading of the skull by impactors with areas in the range of 1 sq in (19, 21). The larger impact area generally precludes penetration as shown in Fig. 4; instead, the skull structure in the vicinity of the impactor deforms as a structural unit as the load builds up. This type of loading produces a concentrated bending moment in the skull structure directly under the load region and produces a tensile strain field in the inner table compact bone. As the strain energy density in the inner table material reaches a critical level (20-80 in lb/in^3) fracture occurs in the inner table. If the kinetic energy of the impactor was sufficient to produce only this initial linear fracture it is the threshold energy. Usually the impactor possesses more than a threshold amount of energy with the result that comminuted fractures are produced subsequent to the initial fracture. The loss of structural integrity due to the initial fracture allows the impactor to essentially cave-in the skull bone structure directly under it.

Since the loaded region of the skull and the subsequent failure zone are in the relatively uniform shell-like region of the frontal bone it is possible that this type of skull fracture could be analyzed theoretically on the basis of shell theory. The results of a linear small deflection analysis of stresses in spherical homogeneous shells due to local loadings (30, 31) were specialized for the case of a typical human skull in the following manner; a shell thickness was chosen at 0.25 in , with a radius of curvature of 3.125 in , dynamic loading was assumed and a compact table bone modulus of elasticity of 3.0×10^6 psi was chosen (26); in order to be able to convert from uniform shell analysis to layered shell analysis an equivalent uniform shell modulus of elasticity of 2.25×10^6 psi was chosen (32). The resulting state of stress in the uniform shell material as a function of applied load was found for a series of loading areas. These states of stress were then converted to equivalent strain energy density levels in the layered shell structure and using a minimum critical value of strain energy density of 20 in lb/in^3 a relationship between minimum failure load and loading area was obtained. This relationship has been plotted in Fig. 4. The predicted minimum failure load of 886 lb for a 1 sq in impactor agrees quite well with the suggested level of 900 lb for frontal bone (21). Note that, as in the case of the predicted local penetration failure loads, the predicted failure load for the comminuted depressed fracture mechanism rises above failure loads experimentally obtained for larger impactor areas as the mechanism of failure associated with remote linear fractures comes into effect.

Remote Linear Fractures - Remote linear fractures are the result of general deformations of the skull structure due to distributed loading by impactors with areas generally greater than 2 sq in (12-17, 20, 23). For such distributed loading, the mechanism of penetration fracture does not come into play; however in some cases the bending moment in the skull structure in the region of load application may, in some skulls, lead to initiation of threshold linear fractures and depressed fractures at the impact site. For most cases of blunt impact to the cranium remote linear fractures, originating from the lower bones of the cranial vault, are the primary mode of failure. Blunt impacts to the frontal bone, for instance, result in linear fractures involving the supra-orbital notch areas, the temporal areas or the orbital areas with a frequency of over 90% (16).

In most skulls the generally uniform dome-like structure of the upper half of the cranium (skull cap) is ideal for carrying distributed loads as a sandwich shell. Thus, as the load on the skull builds up during an impact the resulting deformations are distributed throughout the skull structure leading to a much more uniform distribution of strain energy in the skull bone material as compared to the situation where more concentrated loading is applied. The loads and resulting bending moments developed in the skull cap are transmitted into the lower bones of the cranium where the highly nonuniform structure results in concentration of stresses and deformations, particularly in the thin bone regions between the massive buttress regions, with resulting concentrations of strain energy in such areas as the orbital plates of the frontal bone and at sharply irregular areas like the supra-orbital notches. When the total load on the skull reaches a level sufficient to create a critical strain energy density level (20-80 in lb/in^3) in such regions a linear fracture may initiate and providing there is sufficient strain energy density in the surrounding bone material the fracture will propagate. The loads involved in remote linear fractures are large (Fig. 4) and, thus, the general state of strain energy in the skull cap is usually high enough that once a crack is initiated it will propagate up into the upper half of the skull as well as propagating in the lower regions. The use of shell theory to predict this type of skull fracture does not seem likely unless the influence of the highly irregular lower bones of cranium are included as some type of effective stress concentration.

Scalp Energy Absorption in Head Impact - The main function of the scalp in direct head impact is that of load distribution and the prevention of load spikes that would result if a rigid impactor struck the hard bone of the skull. The scalp does absorb some energy as it deforms (Fig. 2B) and the amount depends on the stress level and the volume of scalp being loaded. Typical experimental head impact situations involve impactor kinetic energies in the range of 200-300 in lb for 1 sq in impactor areas and 400-900 in lb for flat surface impacts. Calculation of the average pressure produced by such impactors yields a compressive stress level in the scalp of 800 psi for a typical 1 sq in impactor fracture level of 800 lb (Fig. 4) and a compressive stress of 400 psi for a typical 5 sq in impactor fracture level of 2000 lb. (See Fig. 4.) Referring to Fig. 2B and assuming an average scalp thickness of 0.2 in results in 21.8 in lb of energy being absorbed for the smaller impactor and 51 in lb of energy being absorbed by the large impactor. These values represent fractions on the order of 6-13% of the available impactor energy at impact. For the case of head impacts where there is a glancing or tangential component to the blow, the loose coupling of the scalp to the skull may produce additional energy absorption.

Conclusions

The overall structural geometry of the cranium and the structural characteristics of the individual bones of the cranium interact with the geometry of impactors to produce three general types of skull fracture in moderate velocity impacts. The force level associated with these fracture types depends on the mechanical properties of skull bone as a material. In penetration depressed fractures associated with highly local loading of the skull cap the fracture force depends on the compressive strength of the diploë layer. For comminuted depressed fractures which are associated with impactor areas in the range of 1 sq in the fracture force depends on the strain energy density in the skull bone in the region of the impact. Blunt impact to the head produces remote linear fractures which are associated with high strain energy distributions in irregular and thin regions in the lower portions of the cranium.

The scalp serves primarily to distribute impact loads and to attenuate load spikes. In typical direct head impacts, the scalp absorbs less than 13% of the available impactor energy.

Acknowledgments

This paper was supported in part by National Institutes of Health Grants No. AM-03865 and No. AM-15044.

References

1. F. A. Aran, "Recherches sur les fractures de la base du crane." Archiv. Générales de Medecine, 4th Series, T. VI, 1844, pp. 180-209, 309-347.

2. V. Bruns, "Handbuch der praktischen Chirurgie für Aerzte und Wundärzte." H. Laupp, Tübingen, Vol. 1, 1854-1859.

3. G. M. Félicet, "Researchers anatomiques et experimentals sur les fractures du crane." A Delahaye, Paris, 1873.

4. W. Baum, "Beitrag zur Lehre von den indirecten Schädel fracturen." Archiv f. Klin. Chirurgie, Vol. 19 (1876), pp. 381-399.

5. Hyrtl, "Topographische Anatomie, Vol. 1, 1857, p. 68.

6. Cohnstein, "Ueber Zangenapplication bei Beckenenge." Archiv für path. Anatomie und Physiologie, Vol. 64 (1875), pp. 82-101.

7. O. Messerer, "Über Elastiticät und Festigkeit der Menschlichen Knochen." J. G. Cotta, Stuttgart, 1880.

8. C. O. Weber, "Chirurgische Erfahrunger und Untersuchungen nebstzahlreichen Beobachtungen aus der chirurgischen Klinikund dem evangelischen." Krankenhause zu Bonn, G. Reimer, Berlin, 1859.

9. L. B. Rawling, "Fractures of the Skull." Lancet, Vol. 1 (1904), pp. 973-979, 1034-1039, 1097-1102.

10. E. R. LeCount and C. W. Appelnach, "Pathologic Anatomy of Traumatic Fractures of Cranial Bones and Concomitant Brain Injuries." J. of the American Medical Association, Vol. 74 (1920), pp. 501-511.

11. G. F. Rowbotham, "Acute Injuries of the Head." Baltimore: Williams and Wilkins Co., 1942.

12. E. S. Gurdjian and H. R. Lissner, "Deformation of the Skull in Head Injury. A Study with the "Stresscoat" Technique." Surg., Gyn. Obst., Vol. 81, (1945), pp. 679-687.

13. E. S. Gurdjian, H. R. Lissner, and J. E. Webster, "The Mechanism of Production of Linear Skull Fractures." Surg., Gyn. Obst., Vol. 85 (1947), pp. 195-210.

14. E. S. Gurdjian and H. R. Lissner, "Deformation of the Skull in Head Injury Studied by the "Stresscoat" Technique, Quantitative Determinations." Surg., Gyn. Obst. Vol. 83, (1946), pp. 219-233.

15. E. S. Gurdjian, J. E. Webster, and H. R. Lissner, "Studies on Skull Fracture with Particular References to Engineering Factors." Amer. J. of Surg., Vol. 78 (1949), pp. 736-742.

16. E. S. Gurdjian, J. E. Webster, and H. R. Lissner, "Observations on Prediction of Fracture Site in Head Injury." Radiology, Vol. 60 (1953), pp. 226-235.

17. E. S. Gurdjian and J. E. Webster, "Recent Advances in the Knowledge of the Mechanism, Diagnosis, and Treatment of Head Injury." Am. J. Med. Sci., Vol. 226 (1953), pp. 214-220.

18. F. G. Evans, H. R. Lissner, and M. Lebow, "The Relation of Energy Velocity, and Acceleration to Skull Deformation and Fracture." Surg., Gyn. Obst., Vol. 107 (1958), pp. 593-601.

19. V. R. Hodgson, W. S. Lange, and R. K. Talwalker, "Injury to the Facial Bones." Proceedings of Ninth Stapp Car Crash Conference. Minneapolis: University of Minnesota, 1966.

20. V. R. Hodgson, "Tolerance of the Facial Bones to Impact." Am. J. Anat., Vol. 120 (1967), pp. 113-122.
21. A. M. Nahum, J. D. Gatts, C. W. Gadd, and J. Danforth, "Impact Tolerance of the Skull and Face." Proceedings of Twelfth Stapp Car Crash Conference, P-26, paper 680785. New York: Society of Automotive Engineers, Inc., 1968.
22. J. W. Melvin, P. M. Fuller, R. P. Daniel, and G. M. Pavliscak, "Human Head and Knee Tolerance to Localized Impacts." SAE transactions, Vol. 78 (1969), paper 690477.
23. W. R. Hodgson, J. Brinn, L. M. Thomas, and S. W. Greenberg. "Fracture Behavior of the Skull Frontal Bone Against Cylindrical Surfaces." Proceedings of Fourteenth Stapp Car Crash Conference, P-33, paper 700909. New York: Society of Automotive Engineers, Inc., 1970.
24. E. F. Byars, D. Haynes, T. Durham, and H. Lilly, "Craniometric Measurements of Human Skulls." ASME Winter Annual Meeting, Paper 70-WA/BHF-8, 1970.
25. F. G. Evans and H. R. Lissner, "Tensile and Compressive Strength of Human Parietal Bone." Jrl. of Applied Physiology, Vol. 10 (1957), pp. 493-497.
26. J. L. Wood, "Dynamic Response of Human Cranial Bone." Jrl. of Biomechanics, Vol. 4 (1971), pp. 1-12.
27. J. W. Melvin, P. M. Fuller, and I. T. Barodawala, "The Mechanical Properties of the Diploë Layer in the Human Skull." Presented at 1970 Society for Experimental Stress Analysis Spring Meeting, May 1970.
28. A. A. Griffith, Phil. Trans. Royal Soc., 221, p. 163, 1920 [A].
29. K. Piekarski, "Fracture of Bone." Jrl. of Applied Physics, Vol. 41 (1970), pp. 215-223.
30. R. Kao and N. Perrone, "Stresses in Spherical Shells Due to Local Loadings." Report 71-4, The Catholic University of America, Washington, D.C., September 1970.
31. K. R. Wichman, A. G. Hopper, and J. L. Mershon, "Local Stresses in Spherical and Cylindrical Shells Due to External Loadings." Welding Research Council Bulletin No. 107, August 1965.
32. R. P. Hubbard, "Flexure of Cranial Bone." Ph.D. Dissertation, University of Illinois, 1970.

720965

Impact Studies of Facial Bones and Skull

D. C. Schneider
General Motors Corp.

A. M. Nahum
University of California

Abstract

The dynamic responses of the human skull and facial bones have been determined by a series of impact experiments. A preliminary report was issued on this subject with particular reference to three impact sites of the skull—the frontal, temporo-parietal, and zygomatic. This work has been extended to include more experiments in these areas to delineate further the nature and reliability of the earlier data. In addition, new data have been obtained for the maxilla, mandible, and zygomatic arch in order to include additional anatomic sites that are frequently involved in accidental impacts. Relationships to impulse duration, peak force, and various anatomic characteristics have been studied and will also be presented. Additional data have been obtained for unembalmed anatomic specimens to provide new information to supplement and expand the tolerance base reported in the preliminary study.

IN A PREVIOUS STUDY (1)*, one of the authors investigated the tolerance characteristics of the frontal, temporo-parietal, and zygomatic areas of the human skull. The purpose of the present study was to collect additional data concerning the above-mentioned facial bones, and also to generate new force-tolerance information on the mandible, maxilla, and zygomatic arch.

A total of 106 experiments were performed on 17 human cadavers (see the Appendix). The specimens were both embalmed and unembalmed, male and female. Methods of loading included A-P and lateral exposures of the mandible, oblique impacts to the maxillary air sinus, and lateral loading of the zygomatic arch.

Shown in Fig. 1 are the impact locations discussed in the report. A description of the contact interface and its orientation with respect to the bone surface is contained in the "Results" section.

Questions pertinent to the experimental methodology were also addressed. These included the test variable effects of specimen sex, preservation status (embalmed or unembalmed), and duration of applied force pulses.

Data scatter is to be expected in any experimental study involving biological material. Substantial variations are found when investigating material properties of uniform cranial bone coupons. Evans and Lissner (2) found the compressive strength of 69 specimens of embalmed parietal compact bone ranged from 12,400–47,800 psi. In a later study, Wood (3) examined 120 specimens of unembalmed human compact cranial bone. Within this sample, the bulk of the data showed the energy absorbed to failure extended from 20–80 in lb/in^3. These differences, caused by normal anatomical variations, are multiplied when a system study of the skull is undertaken.

---

*Numbers in parentheses designate References at end of paper.

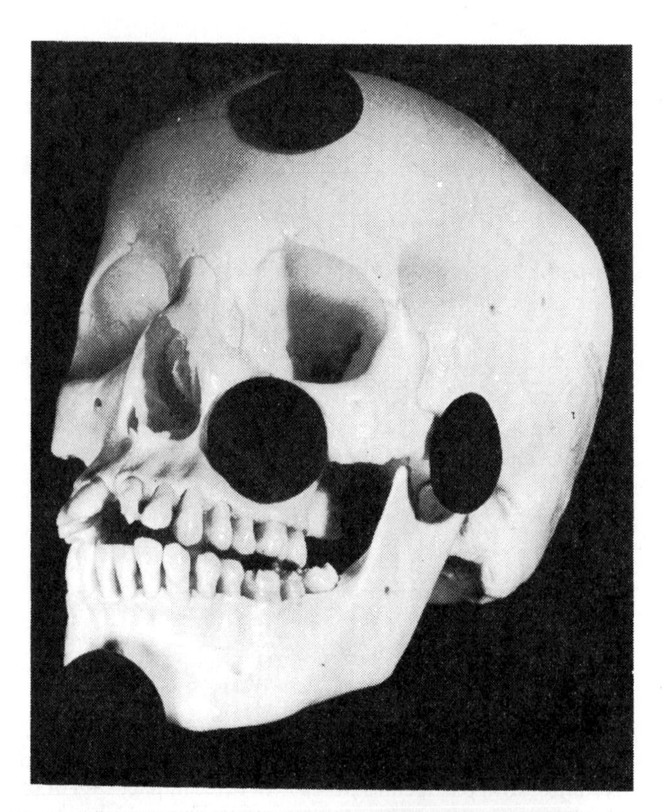

Fig. 1A - Oblique view of skull showing impact sites

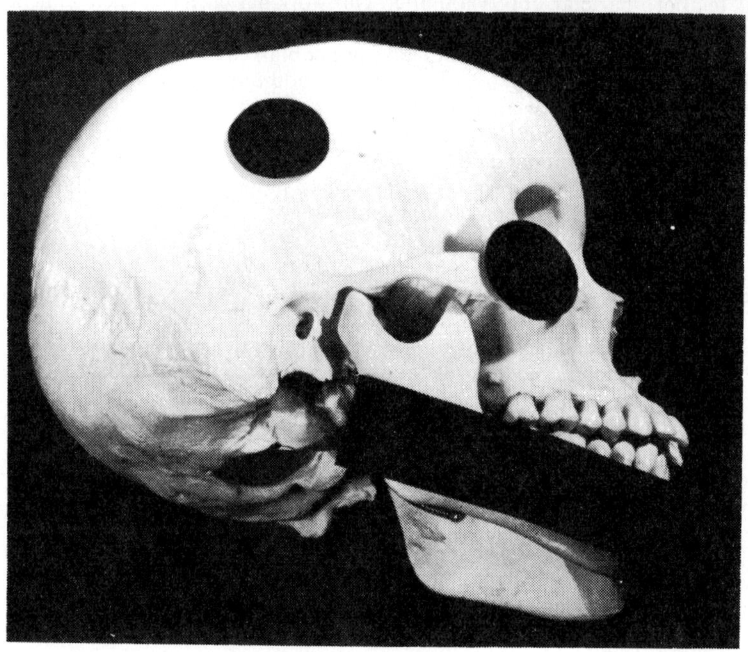

Fig. 1B - Lateral view of skull

Therefore, data offered in this report should be considered preliminary due to the relatively small statistical sample and subject to further refinement as additional research results become available.

Methodology

Impacts were administered to the various anatomic sites by means of a known weight dropped from a preselected height. Incorporated within the drop weight assembly was a diaphragmatic force transducer that provided the load-time history of the impact event. The resonant frequency of the force transducer was a function of its sprung to unsprung mass ratio. Therefore, its natural frequency changed as its total assembly weight varied. The minimum resonant frequency recorded was 3400 Hz.

Data were conditioned via a Honeywell d-c amplifier, permanently recorded on an Ampex FM magnetic tape deck and displayed on a Honeywell light-beam oscillograph. Fig. 2 shows a block schematic of the instrumentation employed. Below each component is listed the upper bound of its linear frequency response.

Both embalmed and unembalmed, male and female human cadavers were used in the study. With the exception of specimens 30 FF through 42 FF, all experiments were performed on intact cadavers, lying supine, utilizing an unconstrained free-falling mass. In the latter specimens (30 FF through 42 FF), the drop weight assembly was guided by a pair of braided steel wires. Two plexiglass cross members were affixed to the drop assembly and contained nylon bushings to minimize any frictional reaction with the steel guidewires (Fig. 3). In addition, the heads were severed at the seventh cervical vertebra in the latter experiments to facilitate the placement of the skull necessary for A-P mandibular impacts. The head, in all cases, was supported by wedges of soft polyurethane padding.

The above-mentioned guidewire system was employed to ensure a high degree of impact location accuracy and impacting mass stability required for the maxilla and mandibular impact sites. This topic will be discussed further in the "Results" section.

To obtain the applied pulse shapes and duration needed, a range of impacting masses, 2.38–8.40 lb (1.08–3.82 kg), and velocities, 9.8–19.7 ft/s (2.99–5.97 m/s), were used.

The contact interface was a 1 in^2 (6.45 cm^2) circular disc covered with a 0.100 in (2.45 mm) layer of crushable nickel foam (5) for all tests, with the exception of the lateral mandible impacts in which a 1 × 4 in (2.45 × 10.16 cm) flat, rectangular surface covered with a 0.200 in (3.08 mm) layer of nickel foam was used. All experiments were performed with the existing soft tissue in situ.

For paired tests of long- and short-duration pulses, an aluminum tube was attached, which rolled over the impactor tip, thereby limiting the force and lengthening the pulse duration.

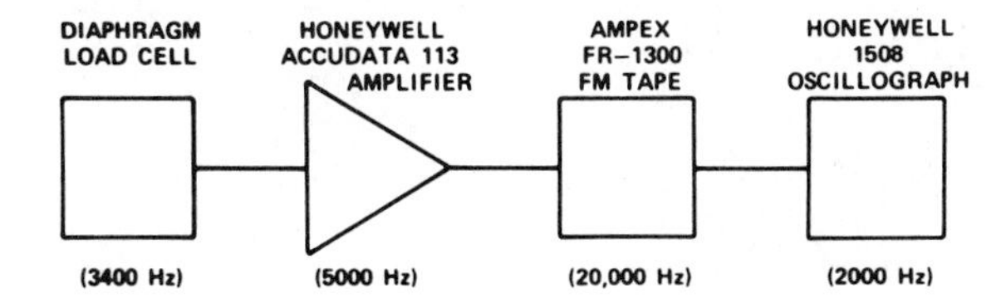

Fig. 2 - Instrumentation block diagram

All fracture data were based on a single impact to the area in question. This was done to preclude any possibility of an undetectable hairline fracture acting as a stress concentration point for a subsequent impact and thereby lowering the required force to fracture.

Following the experiments, the individual impact sites were dissected and any fractures, if present, were photographed. To assist in the detection of hairline fractures, india ink was applied to the area in question. After wiping the bone surface clear, the fracture crevasses retained the ink, providing visual contrast.

The scale used for rating the severity of a fracture was as follows:

0 none
1 minimal detectable change, not clinically significant
2 readily detectable fracture, clinically significant
3 comminuted, and/or depressed fractures

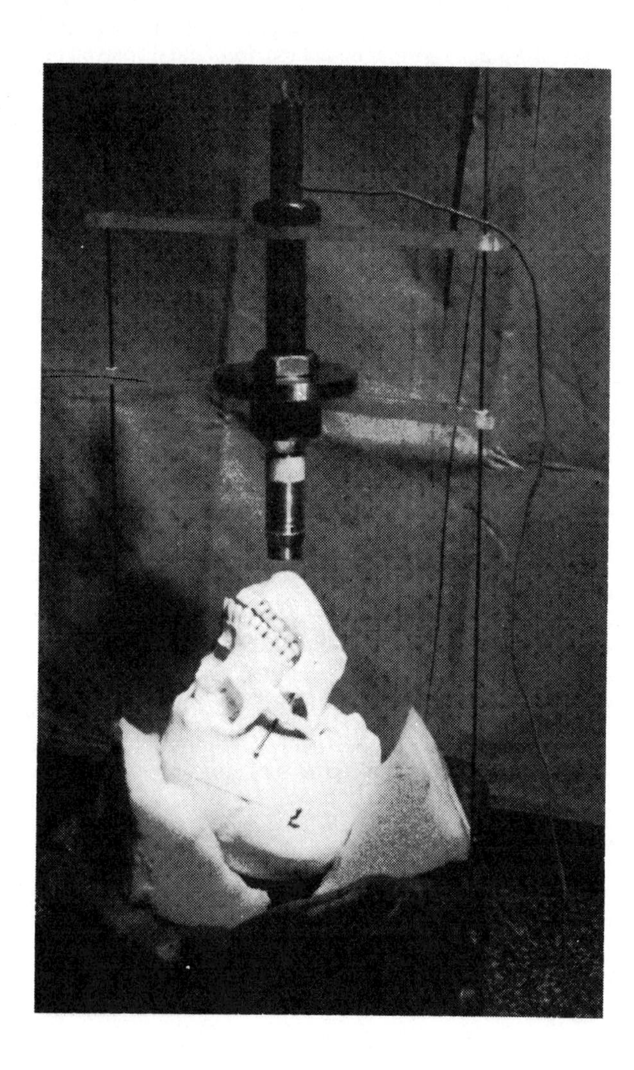

Fig. 3 - Drop weight assembly with guidewire system

Experimental Results

As previously outlined, in addition to expanding the data base of the temporo-parietal, frontal, and zygomatic areas, investigation included the mandible, maxilla, and zygomatic arch.

Of note are certain anatomical features involving the mandible and maxilla. The mandible approximates a rigid semicircular link with pinned joints at its free ends. When applying an A-P load in the sagittal plane to the symphysis, instability is encountered unless the line of action of the force passes through the condylar processes. For this reason, the skulls were removed from the intact cadavers to obtain the exact orientaton required for proper load development. Montgomery (4) has found that 70% of the mandibular fractures occur in either the symphysis, body, or condylar process of the mandible. Each of these three types of trauma was produced by A-P loading of the mandible. Fig. 4 includes a sample oscillograph record with the associated dissection photographs of the symphysis and condylar process fractures. (Note: all oscillograph records have the sprung mass inertial correction factor incorporated into the listed force sensitivity.)

As can be seen in Fig. 7, there is considerable variance in the force required to produce mandibular fracture when the load is applied in the sagittal plane. Unlike the frontal or temporo-parietal areas, whose geometry is described primarily by bone thickness and local radius of curvature, the mandible is a composite of regional geometries. Each of these regions has its own mode of failure and associated tolerance level. Beginning with specimen 41 EM where the tolerance level was determined by the fracture of a single condylar process, the tolerance to peak applied force increased in proportion to the relative size of the mandibular area involved in failure. Specimens 30 FF and 39 EM exhibited fractures of both condyles at 535 lb (2380 N) and 550 lb (2450 N), respectively. At the upper end of the scale, specimens 35 EM and 38 EM sustained fractures of the symphysis.

It is the authors' hypothesis that the method by which force is transmitted through the mandible to the skull determines which areas of the mandible will fail. If the dentures are in contact and the direction of the impact has a sufficient I - S component, then the greater component of the force will be developed across the symphysis and denture contact line causing eventual failure at high force levels of the mandibular body and/or symphysis. In the case of poor denture alignment and an adequate A-P load application, slippage occurs along the denture contact line and the bulk of the load transmission is through the condylar processes that fail at significantly lower magnitudes of peak force. Although care was taken to duplicate the preimpact skull positioning for each specimen, it is believed that normal variations in individual anatomical structure could account for the differences in loading response mentioned above.

Fig. 7, therefore, represents the response of the mandible in a number of different failure modes.

Lateral impacts to the body of the mandible using the 1 × 4 in (2.45 × 10.16 cm) interface produced transverse fractures of the body (Fig. 5). In this series, the long axis of the body was aligned with the 4 in dimension of the impact interface, and the load cell axis was normal to the surface of the mandibular body.

Fracture tolerance data gathered on the maxilla indicate it to be one of the more frangible of the facial bones. The bone covering the maxillary air sinus is thin and shell-like in nature and susceptible to depressed, comminuted fracture patterns. Although covered by relatively thick soft tissue material, 1/4-1 in (6.35–25.4 mm)

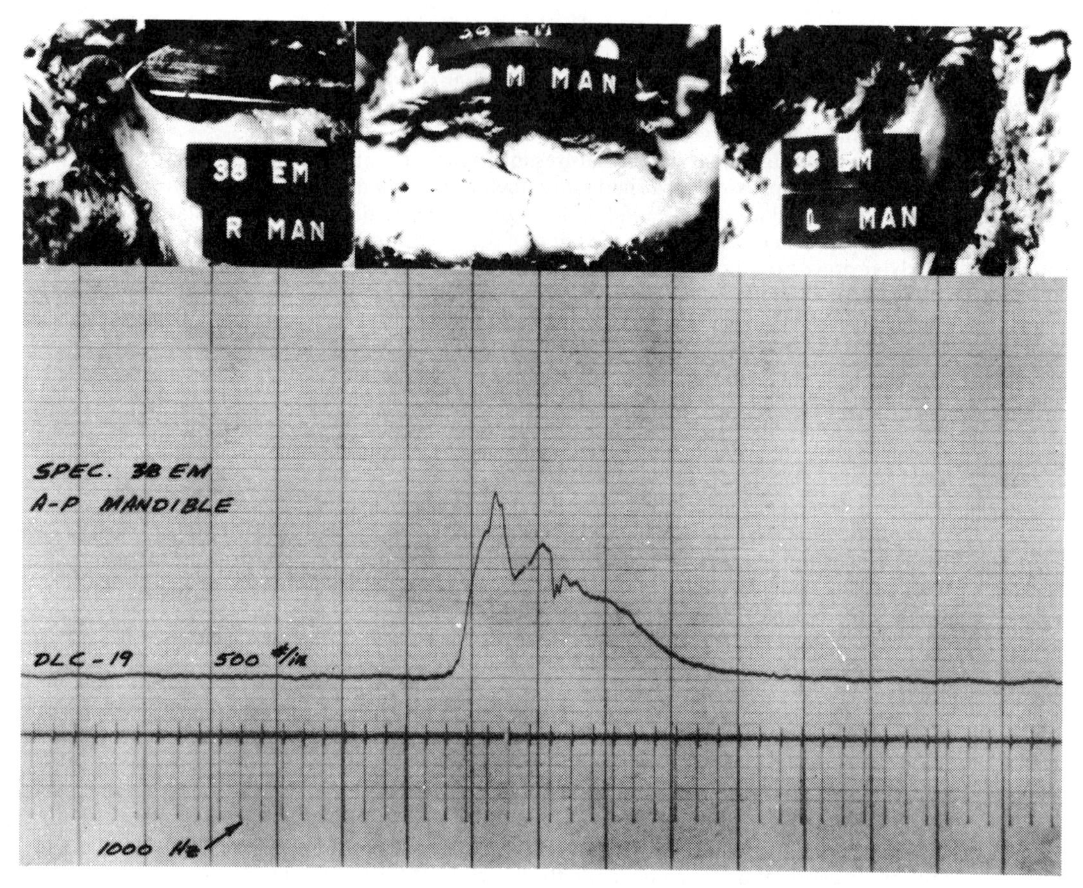

Fig. 4 - A-P impact of mandible (fracture severity: 3)

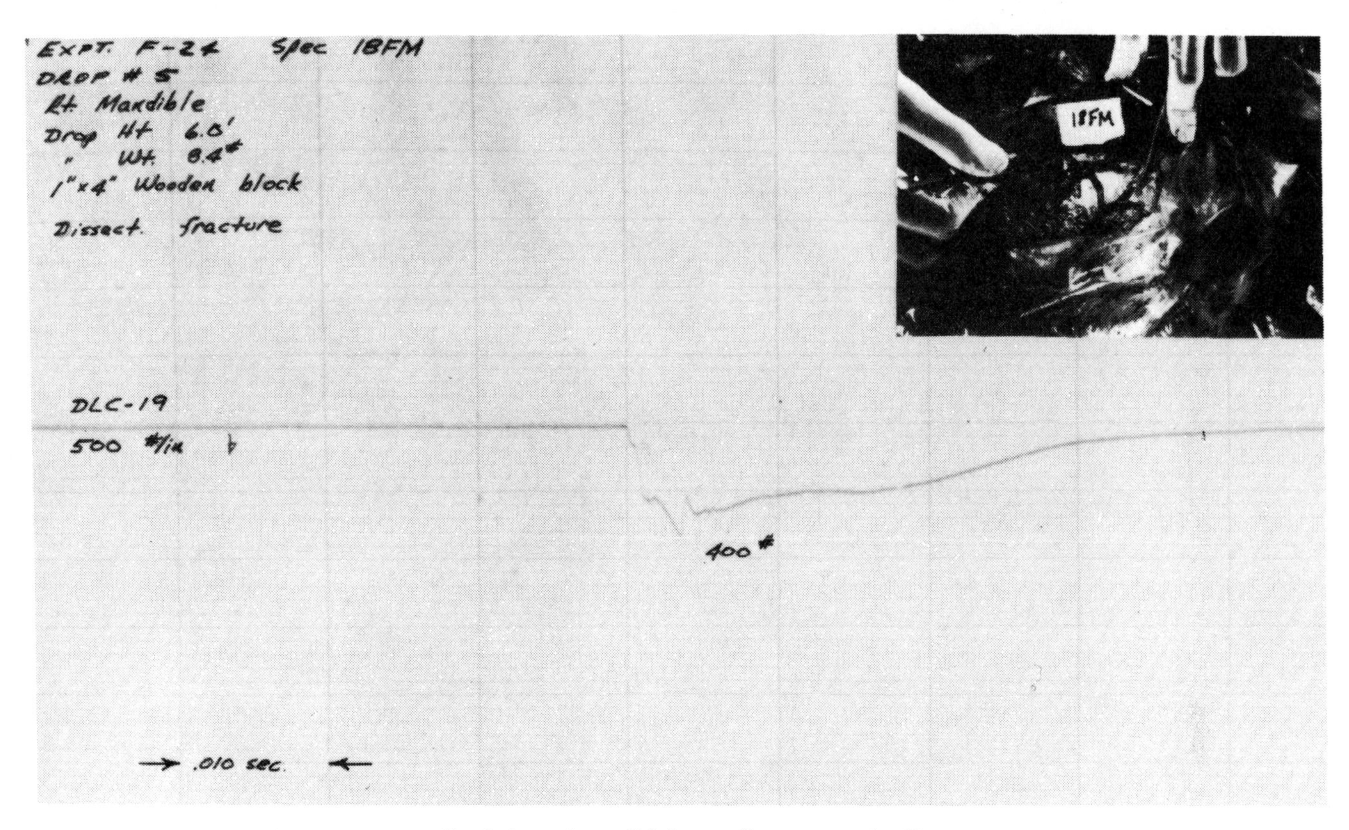

Fig. 5 - Lateral mandible impact (fracture severity: 3)

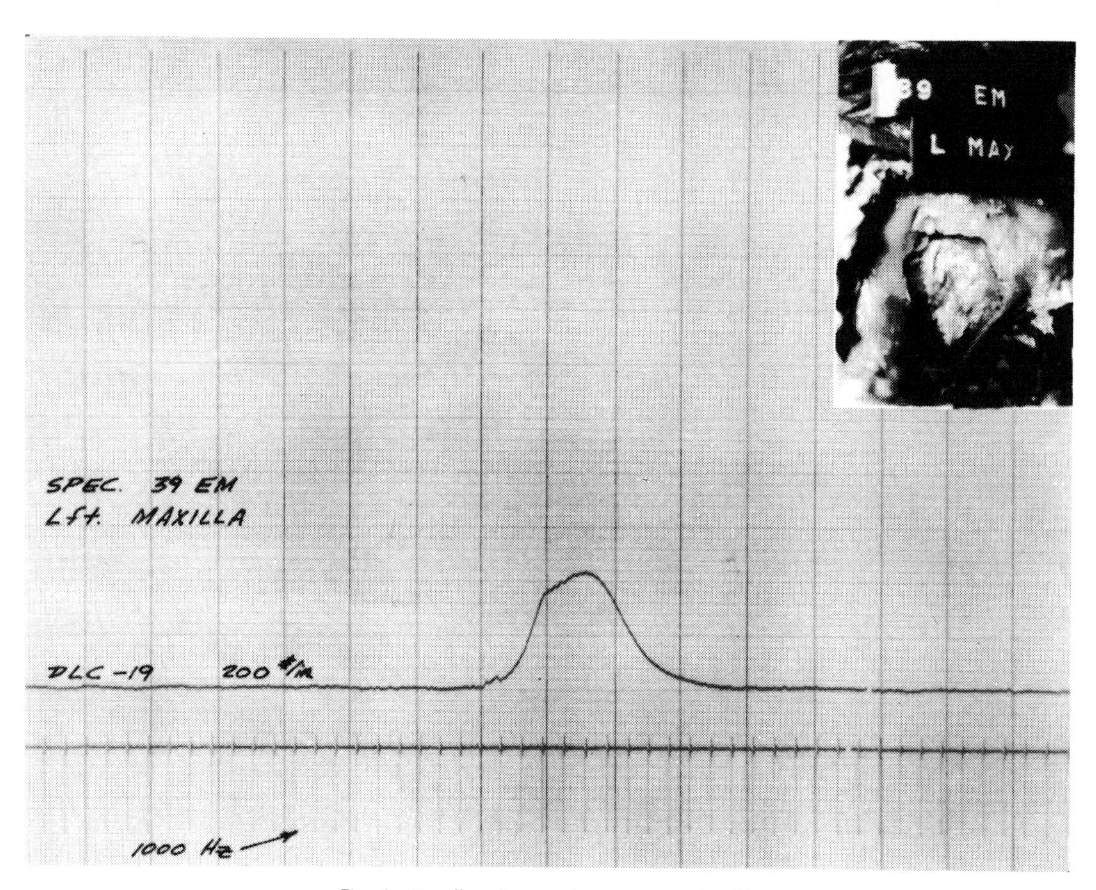

Fig. 6 - Maxillary impact (fracture severity: 3)

on embalmed specimens, the fractures produced were at all times depressed and comminuted (Fig. 6). Figs. 7–9 show force tolerance versus degree of injury for the A-P mandible, lateral mandible, and maxillary exposures, respectively.

Results of tests carried out on the zygomatic arch appear in Fig. 10. Failure occurred at midarch, where the cross-sectional area is at minimum, and also at the zygomatico-temporal suture. Zygomatic tolerance to peak force data is in close agreement with Hodgson (6) for the range of pulse durations studied.

Further investigation of the frontal, temporo-parietal, and zygomatic bones reflects a force-tolerance spectrum paralleling the authors' original work. Figs. 11A, 12A, and 13A present these new data. Results of the authors' original work (1) are reproduced in Figs. 11B, 12B, and 13B for purposes of comparison.

Examination of pulse duration effects utilized the temporo-parietal area. After establishing the load to fracture characteristic on one side of a skull specimen, the opposing side was impacted using an aluminum tube whose square-wave crush load was slightly lower than the force necessary to fracture in the first test. The results of these experiments indicated that no more, and often less, damage resulted from an impact of considerably longer duration than the initial test. An

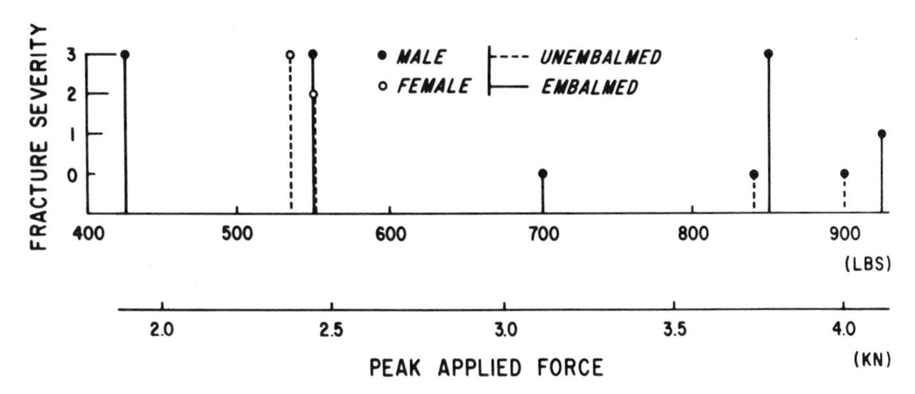

Fig. 7 - Force tolerance of mandible for A-P loading

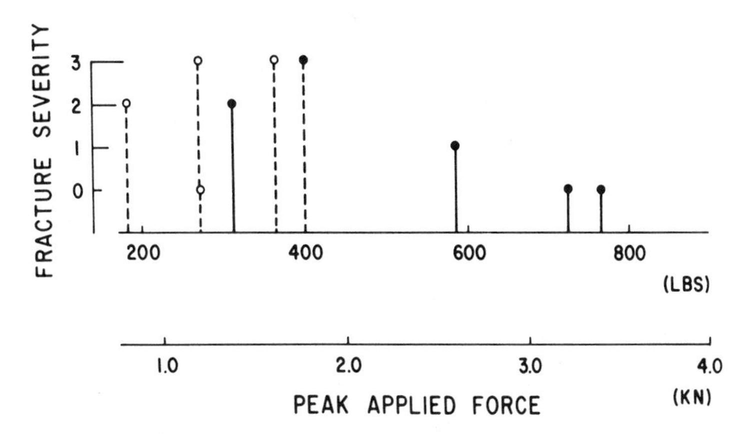

Fig. 8 - Force tolerance of mandible for lateral loading

example of this finding is illustrated in Fig. 14. Specimen 41 EM sustained a depressed, comminuted fracture of the right temporo-parietal bone less than 3 ms after initiation of the loading pulse at a force level of 475 lb (2160 N). The left side of the skull exhibited no evidence of fracture after applying a pulse of over 12 ms in length and 440 lb (1960 N) in average intensity. Those experiments where the pulse-extending technique was employed are noted in the data compilation in the Appendix. Experimental difficulty prevented use of specimen 31 FM in this group.

No trend developed indicating that embalmed bone material is significantly stronger than fresh bone. The distribution of tolerance points is random in this respect, and in some cases the fresh material proved to have the highest tolerance in a given series (that is, zygomatic arch and temporo-parietal).

Contained in the Appendix is the impactor kinetic energy for each experiment. Because the skull was supported by padding and not rigidly fixed, it is not possible to state that all available impactor energy was converted to strain energy of soft tissue and bone deformation. The amount of kinetic energy that contributed to whole head movement varied for each impact and is unknown. For this reason, a correlation between impactor kinetic energy and bone fracture tolerance becomes questionable.

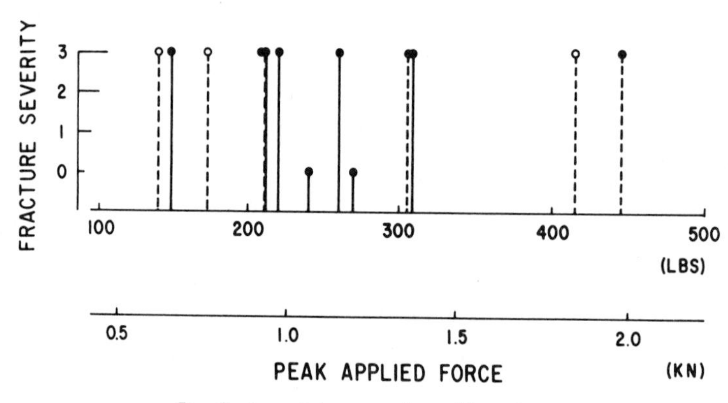

Fig. 9 - Force tolerance of maxillary sinus

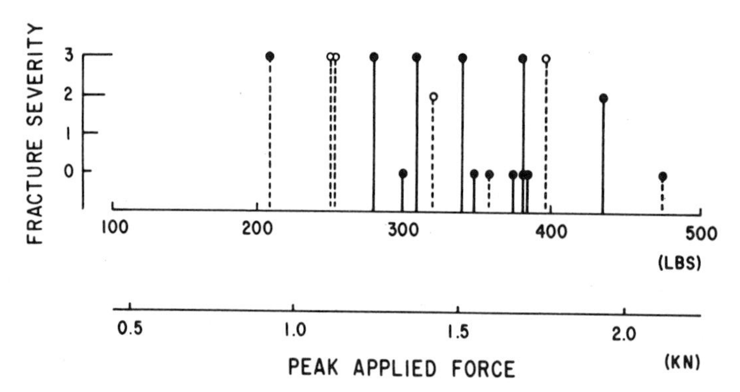

Fig. 10 - Force tolerance of zygomatic arch

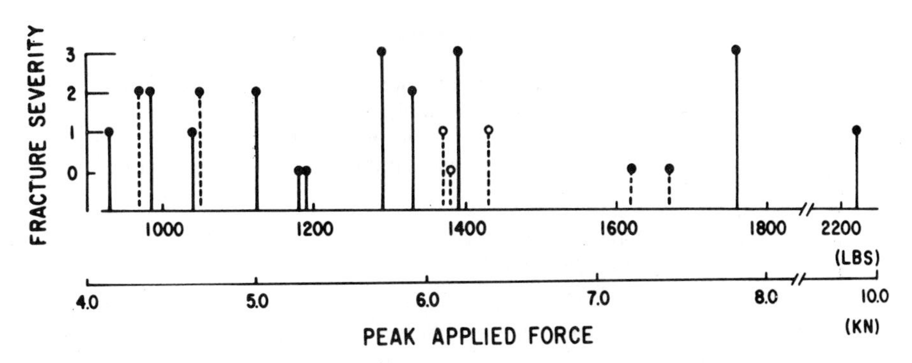

Fig. 11A - Force tolerance of frontal bone

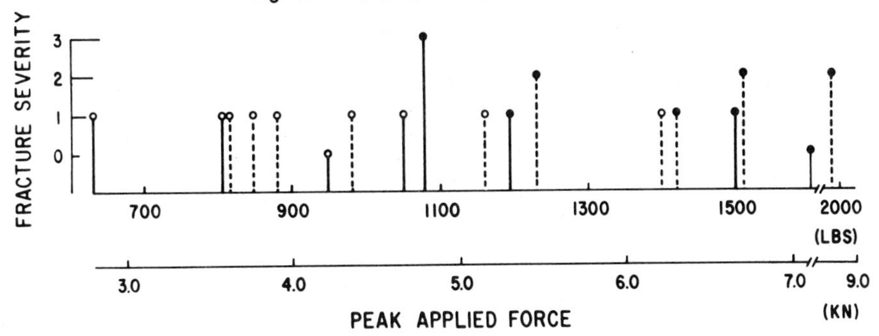

Fig. 11B - Results of authors' original work (1)

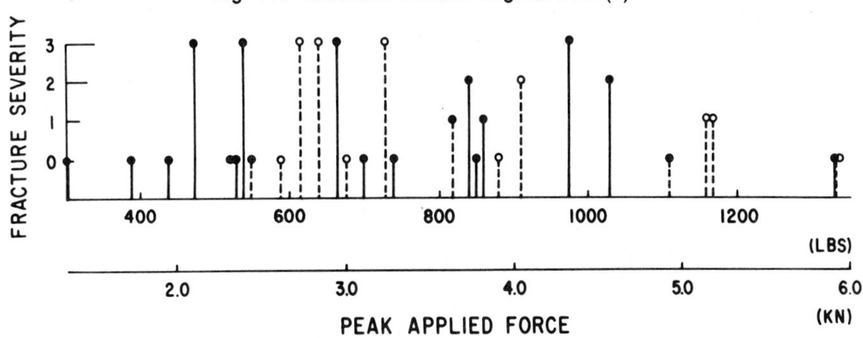

Fig. 12A - Force tolerance of temporo-parietal area

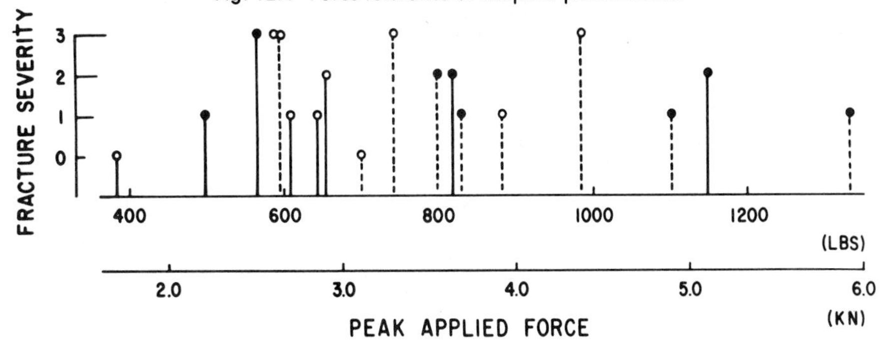

Fig. 12B - Results of authors' original work (1)

Conclusions

1. Previously reported minimal force tolerances for the frontal, temporoparietal, and zygomatic areas of 900 lb (4000 N), 450 lb (2000 N), and 200 lb (890 N), respectively, are corroborated by additional data.

2. Minimal force tolerances for most recent impact locations investigated could be described as follows:

A-P mandible— 400 lb (1780 N).
Lateral mandible—200 lb (890 N).
Maxilla— 150 lb (668 N).
Zygomatic arch— 200 lb (890 N).

3. Females, as a group, tended to have a lower tolerance to applied force than did the male specimens.

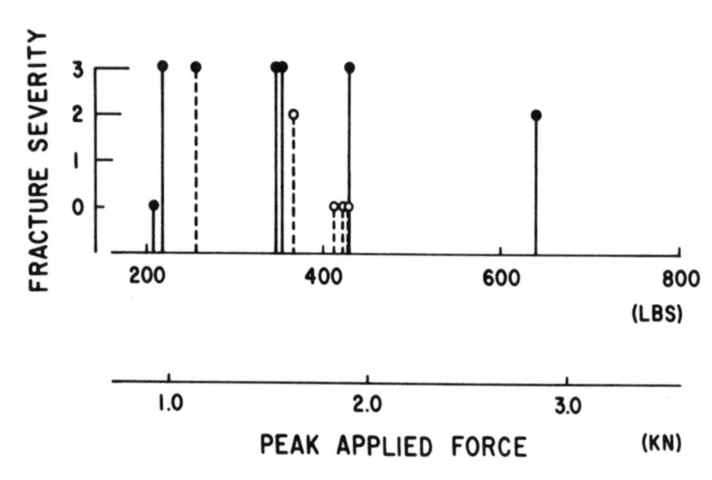

Fig. 13A - Force tolerance of zygoma

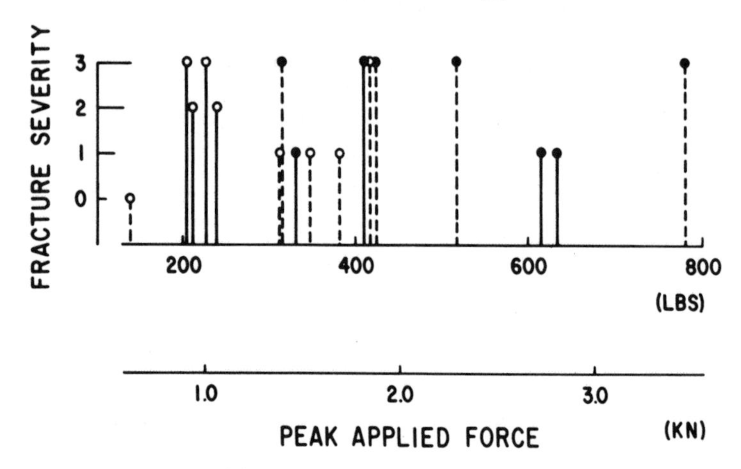

Fig. 13B - Results of authors' original work (1)

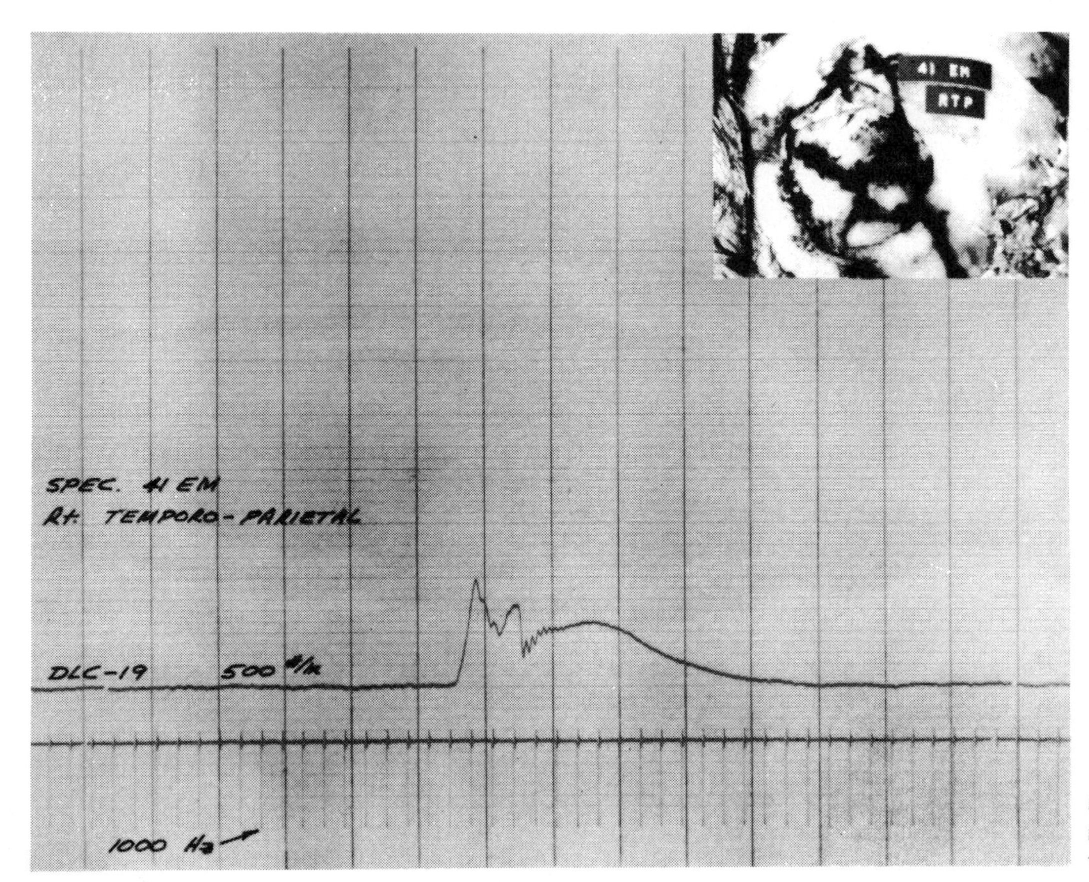

Fig. 14A - Right temporo-parietal impact—specimen 41 EM—(fracture severity: 3)

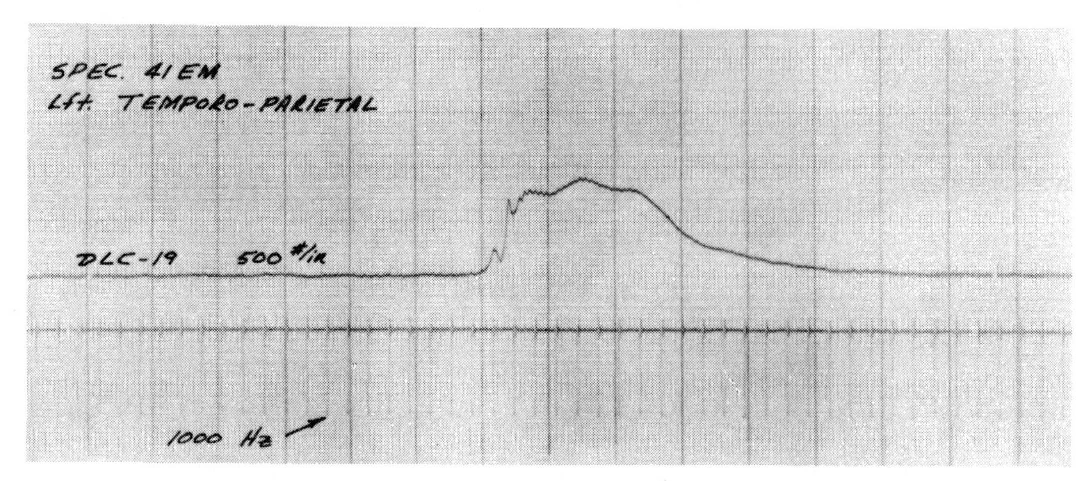

Fig. 14B - Left temporo-parietal pulse extended impact—specimen 41 EM—(fracture severity: 0)

4. Unembalmed material showed no tendency to be weaker in general than embalmed material.

5. In the range of pulse durations studied, there was no evidence to indicate that fracture tolerance to force decreases as pulse duration is increased.

Acknowledgment

The authors wish to express their appreciation to Robert Barton and George Nakamura who assisted in the conduct of the experiments.

References

1. A. M. Nahum, J. D. Gatts, G. W. Gadd, and J. Danforth, "Impact Tolerance of the Skull and Face." Paper 680785, Proceedings of Twelfth Stapp Car Crash Conference, P-26. New York: Society of Automotive Engineers, Inc., 1968.

2. F. G. Evans and J. R. Lissner, "Tensile and Compressive Strength of Human Parietal Bone." Jrl. Applied Physiology, Vol. 10 (1957).

3. J. L. Wood, "Dynamic Response of Human Cranial Bone." Jrl. Biomechanics, Vol. 4 (1971).

4. W. Montgomery, "Surgery of the Upper Respiratory System." Lea & Febiger, 1972.

5. W. K. Miller and S. Katz, "A Technique for Measuring Maximum Local Pressures which Occur During Impact." Proceedings of General Motors Automotive Safety Seminar, Milford, Mich., July 1968.

6. V. R. Hodgson, "Tolerance of the Facial Bones to Impact." American Jrl. Anat., Vol. 120 (1967).

Appendix

Specimen	Age	Weight, lb (kg)	Height, ft, in (m)	Cause of Death
14FF*	75	127 (57.6)	5, 2 (1.57)	Myocardial congestive failure
18EM	73	150 (68.1)	—	—
18FM	78	145 (65.9)	5, 10 (1.78)	Strangulation
19EM	64	120 (54.5)	—	—
20EM	80	175 (79.5)	—	—
21FF	45	151 (68.6)	5, 9 (1.75)	Cerebrovascular hemorrhage
27FF	63	110 (50.0)	5, 1 (1.55)	Toxemia
29FF	57	120 (54.5)	5, 4 (1.63)	Metastatic carcinoma
30FF	52	90 (40.8)	5, 1 (1.55)	Cardiac-respiratory failure
31FM	51	165 (74.9)	6, 0 (1.83)	Hepatic cirrhosis
34FM	64	130 (59.0)	5, 10 (1.78)	Cerebral thrombosis
35EM	64	177 (80.4)	5, 6 (1.68)	Metastatic carcinoma
38EM	72	183 (83.1)	5, 9 (1.75)	Pulmonary hemorrhage
39EM	76	187 (84.9)	—	Myocardial infarct
40EM	58	216 (98.0)	5, 10 (1.78)	Cerebrovascular accident
41EM	71	267 (121.1)	6, 2 (1.88)	Pneumonia
42FF	49	96 (43.6)	5, 3 (1.60)	Myocardial infarct

*Specimen designation: first letter refers to preservation status: E = embalmed, F = unembalmed. Second letter designates sex.

Specimen	Drop Weight, lb (kg)	Velocity, ft/s (m/s)	Kinetic Energy, in-lb (J)	Peak Force, lb (N)	Fracture Severity
			A-P Mandible		
30 FF	6.88 (3.12)	17.0 (5.17)	371 (41.9)	535 (2380)	3
31 FM	6.88 (3.12)	17.0 (5.17)	371 (41.9)	900 (4000)	0
34 FM	6.88 (3.12)	16.0 (4.87)	330 (37.3)	840 (3740)	0
35 EM	6.88 (3.12)	17.9 (5.44)	413 (46.7)	925 (4120)	1
38 EM	6.88 (3.12)	17.9 (5.44)	413 (46.7)	850 (3780)	3
39 EM	6.88 (3.12)	17.9 (5.44)	413 (46.7)	550 (2450)	3
40 EM	6.88 (3.12)	17.9 (5.44)	413 (46.7)	700 (3120)	0
41 EM	6.88 (3.12)	17.9 (5.44)	413 (46.7)	425 (1890)	3
42 FF	6.88 (3.12)	16.0 (4.87)	330 (37.3)	550 (2450)	2
			Lateral Mandible		
14 FF	8.40 (3.81)	17.9 (5.44)	504 (56.9)	273 (1210)	0
18 EM	8.40 (3.81)	17.9 (5.44)	504 (56.9)	584 (2600)	1
	8.40 (3.81)	18.8 (5.72)	564 (63.7)	723 (3220)	0
18 FM	8.40 (3.81)	19.7 (5.97)	605 (68.4)	400 (1780)	3
19 EM	8.40 (3.81)	19.7 (5.97)	605 (68.4)	765 (3400)	0
20 EM	8.40 (3.81)	17.9 (5.44)	504 (56.9)	312 (1390)	2
21 FF	8.40 (3.81)	19.7 (5.97)	605 (68.4)	272 (1210)	3
27 FF	8.40 (3.81)	18.8 (5.72)	564 (63.7)	184 (820)	2
29 FF	8.40 (3.81)	18.8 (5.72)	564 (63.7)	364 (1620)	3
			Maxilla		
30 FF	3.88 (1.76)	15.0 (4.56)	163 (18.4)	415 (1850)	3
31 FM	3.88 (1.76)	17.9 (5.44)	233 (26.3)	445 (1980)	3
34 FM	2.38 (1.08)	12.7 (3.86)	71 (8.0)	307 (1370)	3
	2.38 (1.08)	9.8 (2.99)	43 (4.9)	212 (940)	3
35 EM	2.38 (1.08)	9.8 (2.99)	43 (4.9)	220 (980)	3
38 EM	2.38 (1.08)	9.8 (2.99)	43 (4.9)	270 (1200)	0
	2.38 (1.08)	11.3 (3.45)	57 (6.4)	260 (1160)	3
39 EM	2.38 (1.08)	11.3 (3.45)	57 (6.4)	212 (940)	3
40 EM	2.38 (1.08)	11.3 (3.45)	57 (6.4)	240 (1070)	0
	2.38 (1.08)	13.9 (4.22)	86 (9.7)	308 (1370)	3
41 EM	2.38 (1.08)	9.8 (2.99)	43 (4.9)	148 (660)	3
42 FF	2.38 (1.08)	9.8 (2.99)	43 (4.9)	140 (625)	3
	2.38 (1.08)	9.8 (2.99)	43 (4.9)	172 (765)	3

Specimen	Drop Weight, lb (kg)	Velocity, ft/s (m/s)	Kinetic Energy, in-lb (J)	Peak Force, lb (N)	Fracture Severity
Zygomatic Arch					
30 FF	3.88 (1.76)	12.7 (3.86)	117 (13.2)	396 (1760)	3
	3.88 (1.76)	11.3 (3.44)	93 (10.5)	250 (1110)	3
31 FM	3.88 (1.76)	12.7 (3.86)	117 (13.2)	208 (930)	3
34 FM	2.38 (1.08)	11.3 (3.44)	57 (6.4)	358 (1590)	0
	2.38 (1.08)	13.9 (4.22)	86 (9.7)	475 (2120)	0
35 EM	2.38 (1.08)	11.3 (3.44)	57 (6.4)	376 (1670)	0
	2.38 (1.08)	12.7 (3.86)	71 (8.0)	435 (1940)	2
38 EM	2.38 (1.08)	11.3 (3.44)	57 (6.4)	340 (1510)	3
	2.38 (1.08)	9.8 (2.99)	43 (4.9)	300 (1390)	0
39 EM	2.38 (1.08)	15.0 (4.56)	100 (11.3)	380 (1690)	3
	2.38 (1.08)	12.7 (3.86)	71 (8.0)	280 (1250)	3
40 EM	2.38 (1.08)	11.3 (3.44)	57 (6.4)	374 (1660)	0
	2.38 (1.08)	13.9 (4.22)	86 (9.7)	384 (1710)	0
41 EM	2.38 (1.08)	17.9 (5.44)	413 (46.7)	425 (1890)	3
	2.38 (1.08)	11.3 (3.44)	57 (6.4)	308 (1370)	3
42 FF	2.38 (1.08)	11.3 (3.44)	57 (6.4)	320 (1420)	2
	2.38 (1.08)	11.3 (3.44)	57 (6.4)	252 (1140)	3
Temporo-Parietal					
18 EM	5.30 (2.41)	17.9 (5.44)	318 (36.0)	740 (3290)	0
	6.40 (2.90)	19.7 (5.97)	384 (43.4)	1330 (5920)	0
19 EM	3.30 (1.50)	16.0 (4.87)	159 (17.9)	530 (2360)	0
	5.30 (2.41)	19.7 (5.97)	382 (43.2)	1030 (4580)	2
20 EM	5.30 (2.41)	19.7 (5.97)	382 (43.2)	973 (4330)	3
	3.30 (1.50)	12.7 (3.86)	99 (11.2)	302 (1340)	0
21 FF	6.40 (2.90)	19.7 (5.97)	461 (52.1)	1330 (5920)	0
	3.30 (1.50)	17.9 (5.44)	198 (22.4)	880 (3920)	0
27 FF	5.30 (2.41)	19.7 (5.97)	382 (43.2)	615 (2740)	3
	3.30 (1.50)	19.7 (5.97)	238 (26.9)	728 (3240)	3
29 FF	3.30 (1.50)	19.7 (5.97)	238 (26.9)	1170 (5200)	1
	3.30 (1.50)	17.9 (5.44)	198 (22.4)	1160 (5160)	1
30 FF	3.88 (1.76)	17.9 (5.44)	233 (26.3)	640 (2850)	3
	3.88 (1.76)	17.0 (5.17)	210 (23.7)	590 (2620)	0
31 FM*	3.88 (1.76)	17.9 (5.44)	233 (26.3)	1110 (4940)	0
	6.94 (3.15)	19.7 (5.97)	500 (56.5)	665 (2960)	3
34 FM*	2.38 (1.08)	17.9 (5.44)	143 (16.2)	818 (3640)	1
	6.94 (3.15)	16.0 (4.87)	416 (47.0)	550 (2450)	0
35 EM	2.38 (1.08)	17.9 (5.44)	143 (16.2)	700 (3120)	0
	4.68 (2.12)	17.9 (5.44)	281 (31.8)	860 (3820)	1
38 EM	2.38 (1.08)	19.7 (5.97)	171 (19.3)	390 (1740)	0
	4.68 (2.12)	17.9 (5.44)	281 (31.8)	540 (2400)	3
39 EM*	4.68 (2.12)	17.0 (5.17)	253 (28.6)	840 (3740)	2
	6.94 (3.15)	17.0 (5.17)	374 (42.5)	530 (2360)	0
40 EM	4.68 (2.12)	17.9 (5.44)	281 (31.8)	850 (3780)	0
41 EM*	4.68 (2.12)	17.0 (5.17)	253 (28.6)	475 (2120)	3
	6.94 (3.15)	18.8 (5.72)	458 (51.8)	440 (1960)	0
42 FF*	3.88 (1.76)	16.0 (4.87)	186 (21.0)	910 (4050)	2
	6.94 (3.15)	18.8 (5.72)	458 (51.8)	675 (3000)	0

*Extended pulse duration impact.

Specimen	Drop Weight, lb (kg)	Velocity, ft/s (m/s)	Kinetic Energy, in-lb (J)	Peak Force, lb (N)	Fracture Severity
Frontal					
18 EM	6.88 (3.12)	17.9 (5.44)	413 (46.7)	1290 (5750)	3
	6.88 (3.12)	19.7 (5.97)	495 (56.0)	1390 (6190)	3
19 EM	6.88 (3.12)	16.0 (4.87)	330 (37.3)	1040 (4630)	1
	6.88 (3.12)	19.7 (5.97)	495 (56.0)	2220 (9880)	1
20 EM	6.88 (3.12)	19.7 (5.97)	495 (56.0)	1330 (5920)	2
30 FF	6.88 (3.12)	16.0 (4.87)	330 (37.3)	1380 (6150)	0
	6.88 (3.12)	17.9 (5.44)	413 (46.7)	1430 (6360)	1
31 FM	6.88 (3.12)	19.2 (5.85)	475 (53.6)	1050 (4680)	2
	6.88 (3.12)	19.2 (5.85)	475 (53.6)	970 (4320)	2
34 FM	6.88 (3.12)	16.0 (4.87)	330 (37.3)	1670 (7440)	0
	6.88 (3.12)	17.9 (5.44)	413 (46.7)	1620 (7210)	0
35 EM	6.88 (3.12)	17.9 (5.44)	413 (46.7)	1760 (7840)	3
38 EM	6.88 (3.12)	17.9 (5.44)	413 (46.7)	985 (4380)	2
39 EM*	6.88 (3.12)	17.9 (5.44)	413 (46.7)	1125 (5010)	2
	6.94 (3.15)	19.4 (5.89)	485 (54.8)	930 (4140)	1
40 EM	6.88 (3.12)	18.8 (5.72)	455 (51.4)	1190 (5300)	0
	6.88 (3.12)	19.7 (5.97)	495 (56.0)	1180 (5250)	0
42 FF	6.88 (3.12)	16.0 (4.87)	330 (37.3)	1370 (6100)	1
Zygoma					
14 FF	3.10 (1.41)	15.0 (4.56)	130 (14.7)	429 (1910)	0
	3.10 (1.41)	15.0 (4.56)	130 (14.7)	412 (1830)	0
18 EM	3.10 (1.41)	13.9 (4.22)	112 (12.6)	354 (1580)	3
18 FM	3.30 (1.50)	15.0 (4.56)	139 (15.7)	256 (1140)	3
19 EM	3.30 (1.50)	13.9 (4.22)	119 (13.5)	218 (970)	3
	3.30 (1.50)	17.0 (5.17)	178 (20.1)	640 (2850)	2
20 EM	3.30 (1.50)	13.9 (4.22)	119 (13.5)	208 (930)	0
	3.30 (1.50)	15.0 (4.56)	139 (15.7)	430 (1910)	3
21 FF	3.30 (1.50)	15.0 (4.56)	139 (15.7)	424 (1890)	0
	3.30 (1.50)	17.0 (5.17)	178 (20.1)	366 (1630)	2

*Extended pulse duration impact.

811013

Field Facial Injuries and Study of Their Simulation with Dummy

C. Tarrière, Y.C. Leung and A. Fayon
Laboratoire de Physiologie et de Biomécanique, Peugot/Renault

C. Got and A. Patel
Hôpital Raymond Poincaré

P. Banzet
Hôpital Saint-Louis

Abstract

With drivers wearing 3-point seat belts, the head-steering-wheel impact occurs in most serious accidents, so inducing mainly face injuries.

In a first part, the authors analyze the injuries observed in a sample of 1180 belted drivers involved in frontal collisions, making a distinction, mainly for facial impacts, between injuries related to the properly so-called face and those to the skull and brain and the different possible lesional correlation.

In a second part are presented the results of work carried out in order to define a human face model adaptable to any type of Hybrid II or Hybrid III dummies' heads. The use of this model allows one to elaborate a new protection criterion the head and skull protection criterion, such as the HIC (or another equivalent criterion which could possibly replace it).

BY WAY OF INTRODUCTION, it is important to point out the essentially protective role of seat-belts in the occurence of accidents.
In recent publications (1)(2)*, in which a sample of over 1600 3-point belt wearers is compared with some 5000 non-belt wearers, it was shown that the risk of fatalities is, for belt-wearers, five times less when cars roll-over, 2.3 times less in head-on collisions, 2.1 times less in rear impacts and 1.7 times less in side impacts.

Before investigating the head injuries sustained by belt-bearers and non-belt wearers, we must especially not overlook the fact that the head is the part of the body that is best protected by seat-belts. The risk of serious or fatal head injuries is nine times less for front-seat passengers who are belted (1). The situation is less favorable for the driver because of the presence of the steering-wheel. It is specifically the purpose of this report to analyse the driver's situation, to examine the limits of the protection afforded by seat-belts and to recommend improvements .

FREQUENCY AND GRAVITY OF FACIAL, SKULL AND BRAIN INJURIES

The data analysed here come from the general files of the Laboratoire de Physiologie et de Biomécanique Peugeot S.A./Renault. These records contain data on 3656 automobiles involved in accidents and 7373 passengers involved in accidents, including 1956 3-point belt wearers.

Out of this total sample, we extracted a subsample of 1180 drivers involved in frontal impacts, the violence of which is known,and in which 405 drivers were belted. Among this latter, 74 belted wearers, i.e. 18%, sustained head-against-steering-wheel impacts.

DESCRIPTION OF THE SAMPLE OF BELTED DRIVERS INCLUDING THOSE WHO SUSTAINED HEAD IMPACTS AND THOSE WHO DID NOT -

Variations in Impact Velocity - It was not possible to evaluate impact velocities. However, as compared with test impacts performed at known speeds, the variations in velocity can be computed on the basis of officially, internationally recognized methods described later-on in this report.

Table 1 lists the distribution of these speed variations in km/h for belt-wearers who sustained head impacts (analyzed sample) and for those who did not sustain head impacts (overall sample). The sample of those drivers who sustained head impact represents accidents that were moresevere (by an average of 10 to 16 km/h of Δv) than in the overall sample.

Overall Severity Injuries - The severity of injuries was evaluated on the basis of the international Abbreviated Injury Scale (AIS), most recently updated in 1980. Examples of this scale are given in Table 2.

The term AIS is applied separately to each individual bady segment. It becomes OAIS (Overall

* Numbers in parentheses designate References at end of paper

Table 1 - Variations in Impact Velocity (ΔV).

ΔV in km/h		< 25	26-35	36-45	46-55	56-65	66-75	Total
Analyzed Sample	Number	5	18	22	21	6	2	74
	%	6.7	24.3	29.7	28.4	8.1	2.7	100
Overall Sample	Number	194	96	50	42	17	6	405
	%	47.9	23.7	12.3	10.4	4.2	1.5	100

Table 2 - Examples of Injuries Sustained in Association with Various Degrees of AIS Severity.

Degree 0	No injury.
Degree 1 (Minor)	Skull trauma without loss of consciousness. Fracture of nose, of tooth, (or teeth). Superficial facial injuries.
Degree 2 (Moderate)	Skull trauma with or without dislocated skull fracture and brief loss of consciousness. Fracture of facial bones without dislocation. Deep wound(s).
Degree 5 (Critical, survival uncertain)	Cerebral contusion, loss of consciousness for more than 12 hours with intracranial hemorrhaging and other neurological signs.

Table 3 - Overall Severity of Injuries (OAIS)

	OAIS	C	1	2	3	4	5	6	Total
Analyzed Sample	Number	1	26	21	17	5	3	1	74
	%	1.3	35.1	28.4	23.0	6.8	4.1	1.3	100
Overall Sample	Number	155	157	34	39	5	5	10	405
	%	38.3	38.8	8.4	9.6	1.2	1.2	2.5	100

(OAIS = Overall Abbreviated Injury Scale).

AIS) when there is a synthesis of the total body injuries. Table 3 lists descriptions of our two subsamples (analyzed and overall ones) in accordance with the application of the OAIS.

The degree of severity of the injuries sustained by drivers whose head strucks the steering -wheel (D.S.=35) was three to four times higher than that of the injuries of drivers who did not sustain the impact (D.S.=10).

$$\text{Degree of severity (DS)} = \frac{\text{Number of (SI+FI)}}{\text{Number of involved}} \times 100$$

in which S.I.=severely injured individuals, i.e. OAIS $\geqslant$3, and F.I.= fatally injured individuals.

Age of the drivers - The drivers'age distribution is listed in Table 4. It points up no meaningful difference between the two samples compared.

DISTRIBUTION OF INJURY SEVERITY BY BODY SEGMENT - The data are listed in Tables 5 and 6. The far right-hand column contains the indication Σ AIS^3, the significance of which is that it associates frequency and severity of injuries by assigning greater weight to severe injuries through raising them to third power. This indicator used by a large number of research teams.

We find that head injuries emerge foremost when we consider the aspects of frequency and severity in both the two samples compared.

SEVERITY OF HEAD INJURIES CAUSED BY IMPACT AGAINST STEERING-WHEELS - Head injuries occurring in connection with steering-wheel impacts are responsible for the severity of the subjects' condition in :

(a) 67 % of the cases (with the head AIS possibly associated with injuries of similar severity in other body segments),

(b) 37 % of the cases (the AIS for the head was the highest AIS for isolated injuries).

FREQUENCY OF OCCURENCE OF HEAD INJURIES AND THEIR SEVERITY - Three areas were delimited, as follows :

(A) Face
- A.1 Soft tissue (face + forehead)
- A.2 Fracture

(B) Skull
- B.1 Soft tissue (scalp)
- B.2 Fracture

(C) Brain lesions

Table 7 shows the distribution of injuries by various areas (for five of the drivers, it was not possible to specify the site of impact on the skull soft tissue).

The overall findings were as follows :

(A) Face injury (soft tissue+fracture) in at least 92% of cases,

Table 4 - Ages of the Drivers

	Age	>20	21-30	31-40	41-50	51-60	61-70	Total
Analyzed Sample	Number	2	31	15	14	11	1	74
	%	2.7	41.9	20.3	18.9	14.9	1.3	100
Overall Sample	Number	13	102	85	65	31	14	310*
	%	4.2	32.9	27.4	20.9	10.0	4.5	100

* : 95 Uninjured Drivers Whose Ages are Unknown.

Table 5 - Distribution of Degrees of Injury Severity by Body Segments : 74 Drivers who Sustained Head-Against-Steering Wheel Impacts.

Body Segment	A.I.S. 0	1	2	3	4	5	Total	ΣAIS^3
Head	1	39	25	5	-	4*	74	874
Neck	69	2	1	1	-	1	74	162
Thorax	39	19	9	5	2	-	74	354
Upper Limbs	51	19	2	2	-	-	74	89
Dorso-Lumbar Column	69	4	-	1	-	-	74	31
Pelvis	60	9	3	2	-	-	74	87
Abdomen	61	6	-	1	5	1	74	478
Lower Limbs	28	26	5	15	-	-	74	471

* : Including One Fatal Case (Intrusion++).

Table 6 - Distribution of Degrees of Injury Severity by Body Segments : 398 Drivers in the Overall Sample (7 of the 10 Fatally Injured not Having Been Autopsied).

Body Segment	A.I.S. 0	1	2	3	4	5	Total	ΣAIS^3
Head	267	85	32	7	-	7	398	1405
Neck	372	21	2	2	-	1	398	216
Thorax	286	88	13	7	2	2	398	759
Upper Limbs	339	46	8	5	-	-	398	245
Dorso-Lumbar Column	385	10	1	2	-	-	398	72
Pelvis	368	18	6	6	-	-	398	228
Abdomen	375	13	1	1	7	1	398	621
Lower Limbs	251	104	10	33	-	-	398	1075

Table 7 - Distribution of Head Injuries Among Belt-Wearing Drivers

Site of Injury	1	2	3	4	5	Total	ΣAIS^3
FACE-Soft tissue	38	16	1	-	-	55	
-Soft tissue + fracture	-	6	4	-	3	13	
TOTAL FACE	38	22	5	-	3	68	724
SKULL-Soft tissue	3	2	-	-	1	6	
-Soft tissue + fracture	-	1	-	-	1	2	
TOTAL SKULL	3	3	-	-	2	8	277
BRAIN	-	18	3	-	3	24	600

(B) Skull injury (scalp+fracture) in 11% of cases,

(C) Brain lesion in 32% of cases.

The sixteen cases of very severe injuries found (AIS 3) are distributed as follows:

(A) face, 8 cases

(B) skull, 2 cases

(C) brain, 6 cases

DISTRIBUTION OF FACIAL, SKULL AND BRAIN INJURIES (ASSOCIATED AND NON-ASSOCIATED) - This distribution is shown in Tables 8 and 9. The values between parentheses concern the cases of skull injury (soft tissue) for which the site of impact (scalp or forehead) is not specified.

Overall, out of 74 cases of head-against-steering wheel impact, we found the following:

(A) a single facial injury in 56 % to 58 % of cases,

(B) a single skull injury in 1 case at the most,

(C) association of skull and face injuries (without brain lesion) in 4 % of cases.

(D) face injury associated with brain lesion in 26 to 32 % of cases.

Table 8 - Distribution of Facial-, Skull- and Brain Injuries (Associated and Non-Associated) for Drivers Wearing a Three-Points Seat Belt.

Body Segment/Type of Injury	AIS 1	AIS 2	AIS 3	AIS 4	AIS 5	TOTAL
Face alone, soft tissue	35 (+1)	1	-	-	-	36 (+1)
Soft tissue + fracture	-	3	2	-	-	5
Skull alone, soft tissue	(1)	-	-	-	-	(1)
Soft tissue + fracture	-	-	-	-	-	(1)
Skull + face, soft tissue	3	-	-	-	-	3
Soft tissue + fracture	-	-	-	-	-	-
Face (soft tissue + fracture) + skull (soft tissue)	-	-	-	-	-	-
Skull (soft tissue + fracture) + face (soft tissue)	-	-	-	-	-	-
Face, soft tissue + internal lesions	-	12 (+2)	1	-	(1)	13 (+3)
Skull, soft tissue + internal lesions	-	(1)	-	-	-	(1)
Face, soft tissue + fracture + internal lesions	-	3 (+1)	2	-	1	6 (+1)

Table 9 - Distribution of Associated Facial-, Skull- and Brain Injuries. 3 pts belted drivers.

Site of Injury	AIS 1	AIS 2	AIS 3	AIS 4	AIS 5	TOTAL
Skull : soft tissue + fracture + internal lesions	-	-	-	-	-	-
Face + skull : soft tissue + internal lesions	-	2 (+1)	-	-	(1)	2 (+2)
Face : soft tissue + fracture + Skull : Soft tissue + internal lesions	-	(1)	-	-	1	1 (+1)
Skull + face : soft tissue + fracture + internal lesions	-	-	-	-	1	1
Skull : soft tissue + fracture + Face : soft tissue + internal lesions	-	1	-	-	-	1

(E) Skull injury associated with brain lesion in 1 case at the most.

(F) Skull and face injuries associated with brain lesions in 7 to 11 % of cases.

DISTRIBUTION OF FACIAL, SKULL AND BRAIN INJURIES (ASSOCIATED AND NON-ASSOCIATED) FOR UNBELTED OCCUPANTS - This distribution, given in Table 10 is the part of our analysis which completes, for unbelted people, the data given in Tables 8 and 9 for belted people.

Among 947 unbelted drivers, one found 384 cases with an AIS $\geqslant 1$, i.e. 40 % (against 18 % for belted drivers).

Concerning this sample of 384 cases, the head impact occurs against diverse parts of the passenger compartment. In the previous study here-above mentioned (1), 24,1 % of the observed impacts are impacts against the windshield,9,5 % against the steering-wheel, 3,9 % against windshield upper cross-member, 3,2 % against A-pillar 0,8 % against dashboard.

The following indications arise from data given in Table 10:

(A) a single facial injury in 48 % of cases

(B) a single skull injury in two cases only

(C) associated skull and face injuries (without brain lesion) in 8 % of cases

(D) face injury associated with brain lesion in 28 % of cases,

(E) not only a single case of skull injury+ brain injury

(F) skull and face injuries associated with brain lesions in 5 % of cases.

It is an interesting finding that the dis-

Table 10 - Distribution of Facial, Skull and Brain Injuries (Associated and Non-Associated) for Unbelted Drivers.

Body Segment	AIS 1	AIS 2	AIS 3	AIS 4	AIS 5	AIS 6	TOTAL
Face alone, soft tissue	141	7	1	-	-	-	158
Soft tissue + fracture	14	20	2	-	-	-	27
Skull alone, soft tissue	38	1	-	-	-	-	39
Soft tissue + fracture	1	-	-	-	1	-	2
Skull +face, soft tissue	21	4	-	-	-	-	25
Soft tissue + fracture	-	-	-	-	-	-	0
Face (soft tissue+fracture) + skull(soft tissue)	4	2	-	-	-	-	6
Skull (soft tissue+fracture) + face (soft tissue)	-	-	1	-	-	-	1
Face : soft tissue + internal lesions	1	69	3	2	1	-	76
Skull : soft tissue + internal lesions	-	23	-	1	3	-	27
Face : soft tissue + fracture+internal lesions	-	20	5	3	1	-	32
Skull : soft tissue + fracture+internal lesions	-	-	-	-	-	-	0
Face + Skull :soft tissue + internal lesions	-	9	2	-	-	-	11
Face: soft tissue+fracture + skull : soft tissue + internal lesions	-	1	-	1	2	-	4
Skull + Face :soft tissue + internal lesions	-	-	-	1	1	-	2
Skull: soft tissue+fracture + face : soft tissue + internal lesions	-	1	-	-	-	-	1

tribution of head injuries is very similar in either belted or unbelted drivers.

CONCLUSIONS - Thus, for the belted drivers, the following were found:

- Head-against-steering wheel impact occur--red in 18 % of cases (those of which the violence was, in most cases, the severest: nearly 70 % had a ΔV of $>$ 40 km/h).
- The degree of injury severity was three to four times hugher when head impact occurred.
- Head injuries were the severest injuries in 67 % of cases.

For the belted drivers who sustained head-against-steering wheel impact:

- The face was involved in 92 % of cases.
- Injuries involved the face alone in 56 % of cases.
- Combined face injury and brain lesion occurred in 26 to 32 % of the cases and constituted virtually all the extremely serious injuries.
- The head of the unbelted driver strikes a part of the vehicle in 40 % of cases.
- The face was involved in 82 % of cases.
- The injuries involved the face alone in 48 % of cases.
- Associated injuries to the face and brain occurred in 28 % of cases. They constitute the three-quarters of all severe injuries.

DESIGNING AND BUILDING OF A BIOMECHANICAL MODEL OF THE HUMAN FACE FOR TESTING PURPOSES

The conclusions of the preceding chapter

clearly show that the face is the site of the most frequently occurring injuries and of the injuries that are the most serious through their association with brain lesions.

It is hence important that the dummies used for simulating human behaviour patterns in accidents, for anticipating the risks incurred and for designing protection systems should be equipped with sufficiently realistic faces. On the currently available dummies, the face is a light, moulded rigid alloy covered with a run-of-the-mill "skin" that is unsuited for providing any idea of the risk of laceration (Fig. 1).

In view of the foregoing, research was devoted to human facial features and to designing a crushable, removable model of a face which is adaptable to the heads of existing dummies and is replaceable after every test.

SIMULATION OF THE HUMAN FACE - Stated simply, the face is composed of soft tissues covering the facial skeleton.

Simulation of the Soft Tissue - We investigated the human skin's resistance to laceration by means of dropping sharp-edged impactors of various shapes.

This investigation of the mechanical characteristics of the human skin showed that laceration severity depends on force of impact, on the impactor's geometrical shape and on the different body regions. It further showed that the human skin's laceration resistance depends on the orientation of the laceration with respect to the Langer lines. *

This finding bears out those of automobile accident statistical data: the facial lacerations occurring parallel to the Langer lines are most frequent than those occurring perpendicular to these lines (4). For this reason, the laceration resistance value of the artificial skin used for the model is that of the human skin in the case of lacerations occurring parallel to the Langer lines.

Under the same testing conditions as for human skin (4) - free fall of guided impactor, various dropping heights and different impactors used - 250 tests were performed with five different types of artificial skins two millimeters thick. The general shape of these impactors is illustrated in Figure 2.

Our impactors (DiØj) were designed so as to have different cutting diameters (Di) and different cutting angles.

(Øj): i=1, 2, 3 = 15, 20, 25 mm;
j=1, 2, 3 = 30°, 60°, 90°.

The mean findings are listed in Table 10. When they are compared with the findings for human skin, the efforts appliqued in terms of laceration severities are represented in Figures 3 and 4. The definition of these indicators is identical for both human and artificial skins : " 0 " corresponds to non-occurrence of laceration, " 1 " corresponds to a laceration of one-third of the skin's thickness and " 3 " corresponds to a laceration of the full thickness of the skin.

Figure 3 displays the finding yielded with various different materials. Finding close to those for human skin were obtained with LAB 265 (polyurethane) and RTV 1502 (silicon) materials.

When these findings were compared with those for the simulations, the stability of the materials and the possibility of adherence to the plastic foam, the RTV 1502 silicon emerged as the most suitable component for manufacturing artificial skin, and was therefore used.

In order to define a plastic foam would accurately simulate adipose tissue, we did tests on artificial skins with various foams and compared the findings with those for facial skin with adipose tissue.

Figure 4 shows the force curves in terms of skin laceration with various foams having densities of 0.16 and 0.32.

Simulation of the facial Skeleton - Biomechanical data are available for the fracture resistance of the various facial bones, but such data were lacking concerning fracture tolerance when forces are applied to the face overall (exclusive of the forehead).

Thanks to the work of specialized medical institutes, such as the I.R.O. in Garches,France (5) (6), it was possible to perform facial impact tests with and without rigid masks on the unembalmed cadavers of recently deceased individuals. This made it possible to determine energy absorption by the face, as well as the facial resistance in tests performed without masks; the proper distribution of impact force over the face alone, without touching the forehead, during the tests with masks, enable determination of a mean resistance. Table 11 summarizes the main findings.

This study showed that the facial skeleton undergoes elastic and plastic deformations during occurrence of impact.

Facial resistance to fracture is 770 daN; the average facial surface area is 75 cm2; the average resistance is hence close to 10 daN/cm2. Such a definition suffices for the artificial facial skeleton.

In addition, the average crushing of the facial skeleton is 0.73 cm for a dropping height of 2.5 meters.

We performed a series of impact tests on several kinds of light alloy honeycomb for the purpose of ascertaining the materials best suited to simulations of the facial skeleton. The honeycomb material chosen was AMI 4-40. This honeycomb was designed along the general lines of the facial skeleton (Cf. Fig. 6). It is characterized by an impact resistance of 856 daN and a crushing of 0.8 cm for a dropping height of 2.5 meters. As compared with the biomechanical data, the resistance is 11 % hugher than that of the human face, but is consistently included in the

* Corresponding to the crease lines on the surface of the majority of the body segments.

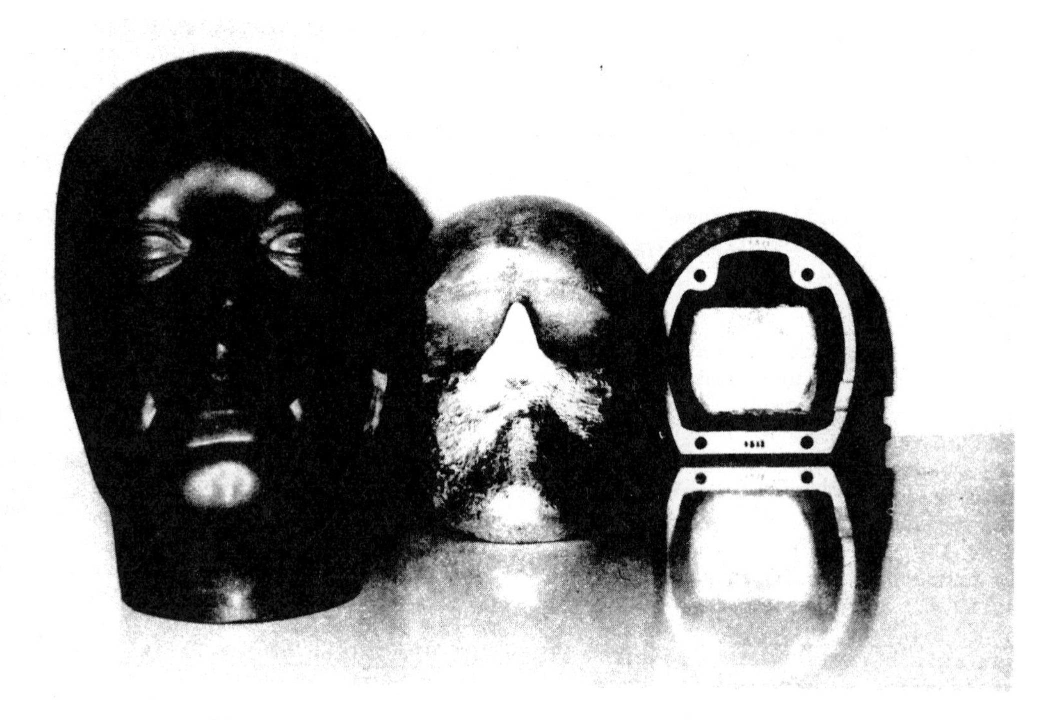

Fig. 1 - Facial elements of the Part 572 dummy

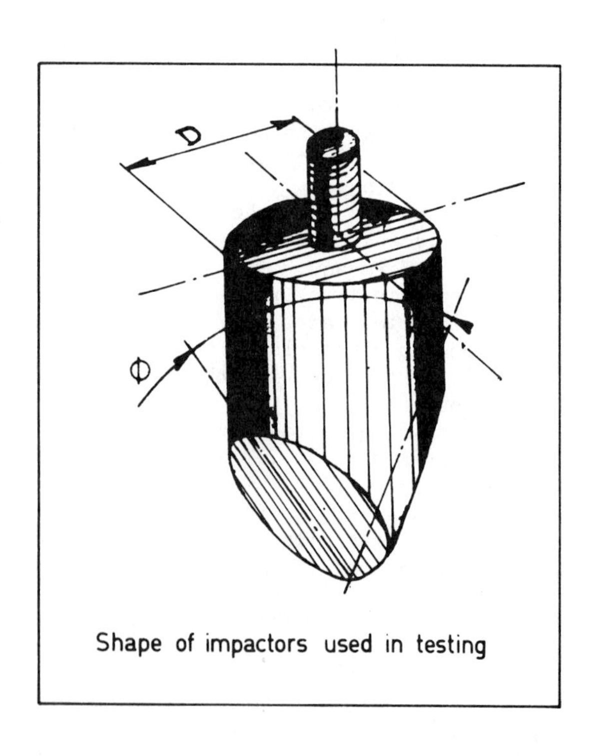

Fig. 2 - General shape of the impactors used in laceration tests

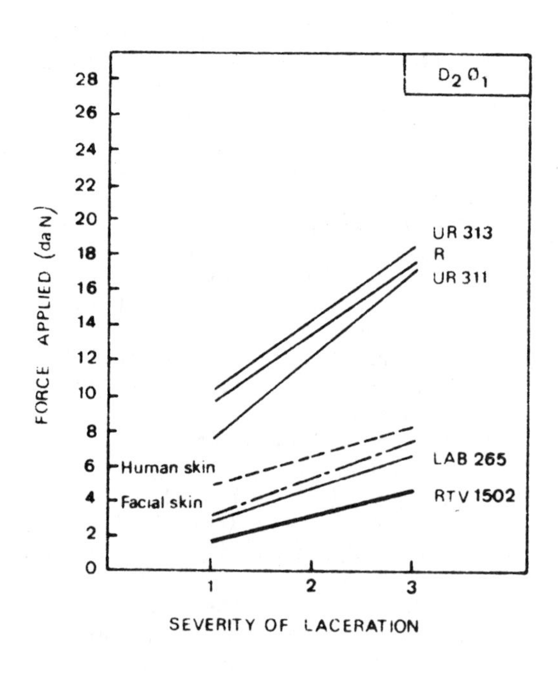

Fig. 3 - Force applied versus severity of laceration. Comparison between human skin and artificial skin

data scatter. In addition, the impact sustained by the skeleton model is appreciably close to that of the human face .

BUILDING OF THE MODEL FACE - Using the simulation of the human face as the basis for design, the model face was produced by making modifications to the skull and skin of the Part 572 dummy. The dimensions, weight and center of gravity of the head and its moments of inertia are identical to those of the Part 572 dummy, in order to obviate any change in its behaviour during occurrence of impact simulation.

The metal support of the Hybrid II dummy's face was replaced by an AS-13 aluminum-alloy plate welded to the skull in an area passing through the temple and running perpendicular

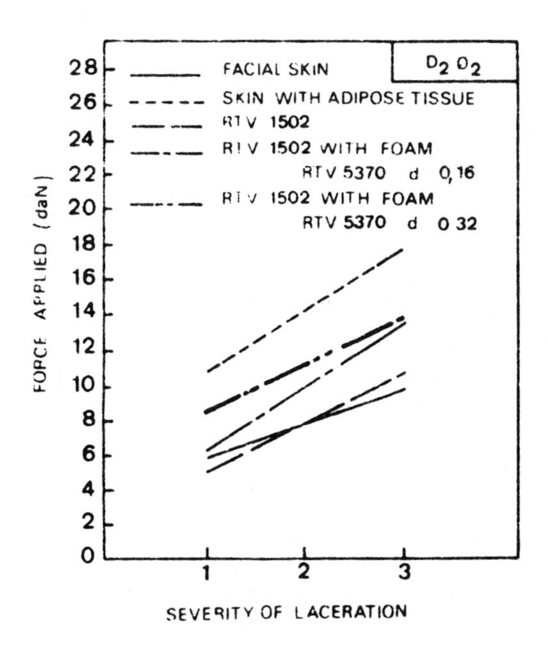

Fig. 4 - Force applied versus severity of laceration. Comparison between human skin and artificial skin

Table 11 - Laceration Resistance of Artificial Skins Compared to That of Human Skin

Material	Reference or description	Thickness of samples (mm)	Impactor used	Force applied (daN) Indicator of severity		
				1	2	3
Polyurethane	R	2	$D_2\Phi_1$	9.9	13.7	17.6
			$D_2\Phi_2$	27.5		
Polyurethane	UR 313	2	$D_2\Phi_1$	10.0	14.0	18.0
			$D_2\Phi_2$	21.6	27.6	
Polyurethane	UR 311	2	$D_2\Phi_1$	7.7	12.4	17.7
			$D_2\Phi_2$	16.6.	20.5	24.4
Polyurethane	L AB 265	2	$D_2\Phi_1$	3.0	5.0	6.9
			$D_2\Phi_2$	9.9	12.0	14.3
			$D_1\Phi_1$	4.5	5.7	11.0
			$D_3\Phi_1$	5.4	7.3	9.2
Silicon	RTV 1502	2	$D_2\Phi_1$	1.9	3.7	5.5
			$D_2\Phi_2$	4.5	7.7	11.0
			$D_2\Phi_3$	10.2	13.0	16.0
			$D_1\Phi_1$	1.3	3.0	4.7
			$D_3\Phi_1$	2.0	4.0	6.0
Silicon	RTV 1502 + RTV Foam 5370 d= 0.16	10	$D_2\Phi_1$	3.0	4.9	6.9
			$D_2\Phi_2$	6.0	9.6	13.0
Silicon	RTV 502 + RTV Foam 5370 d= 0.32	10	$D_2\Phi_1$	3.8	5.2	6.6
			$D_2\Phi_2$	6.6	7.8	9.0
Facial skin	P 82	2.3	$D_2\Phi_1$	3.0	5.4	7.6
Facial skin + adipose tissue	P 82	8.9	$D_2\Phi_1$	6.7	8.4	10.0
			$D_2\Phi_2$	10.8	14.2	17.8
Human skin	F 22-39	2.3	$D_2\Phi_1$	4.9	6.2	7.8
			$D_2\Phi_2$	6.8	10.6	14.2

Table 12 - Findings for Facial Impacts

Test No.	Subject Sex/Age	Dropping Height (m)	Deceleration of the head (g)	Force applied (daN)*	Time (ms)	Mask
91	M 69	2.5	165	880	3.5	no
92	F 69	2.5	115	500	2.5	no
96	M 53	3.0	88	680	2.0	yes
105	M 47	3.5	135	870	2.0	yes

(*) facial fracture strength

to the Frankfurt plane.

The facial skeleton model is attached as follows: the honeycomb is glued onto a fine metal plate, which is then screwed onto the metal support, a space being hollowed out in this plate to accomodate the facial skeleton model. The soft facial tissue part (the skin and adipose tissue) is screwed onto the skull's sides by two fastening clamps which are covered by the scalp of the dummy.

The elements of the facial model are depicted in Figures 5 and 6.

VALIDATION OF THE MODEL FACE - Characteristics of Acceleration Versus Impact Time Length - This check was performed by means of tests involving free falls of the head alone, both modified and unmodified, from a height of 2.5 meters (25km/h velocity of impact against a rigid flat surface. The time accelerations were recorded for these heads, and the curves pertaining thereto are plotted in figure 7.

This latter figure shows that the acceleration versus impact time-length for the Part 572 dummy's head is far higher than those of the model head and the human head, whereas the acceleration versus impact time-length of the model head and the human head are appreciably of the same magnitude.

This comparison enabled us to conclude that in the case of facial impact, the protection criteria (S.I., H.I.C. etc.) calculated on the basis of the acceleration of the Part 572 dummy's

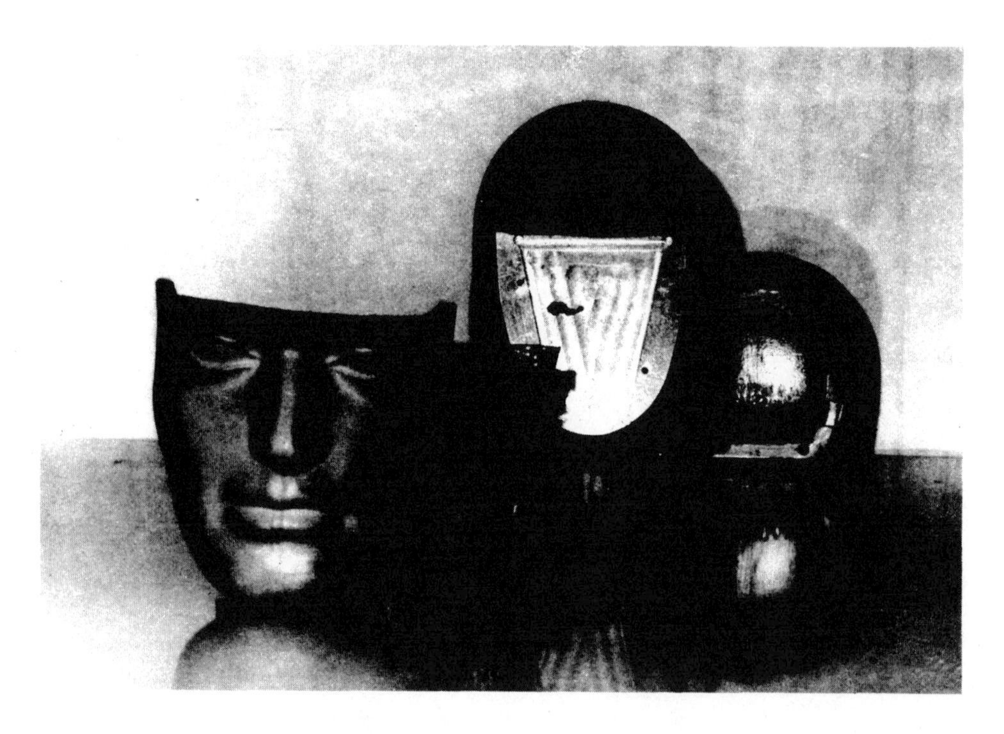

Fig. 5 - Whole elements of the face model

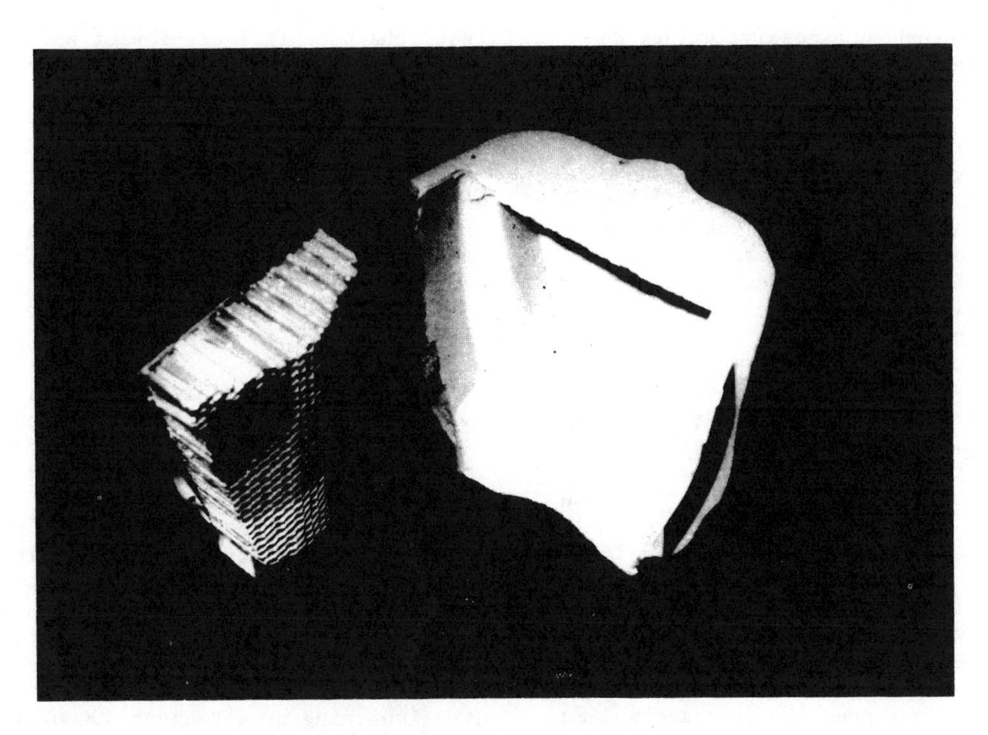

Fig. 6 - Partial elements of the face model

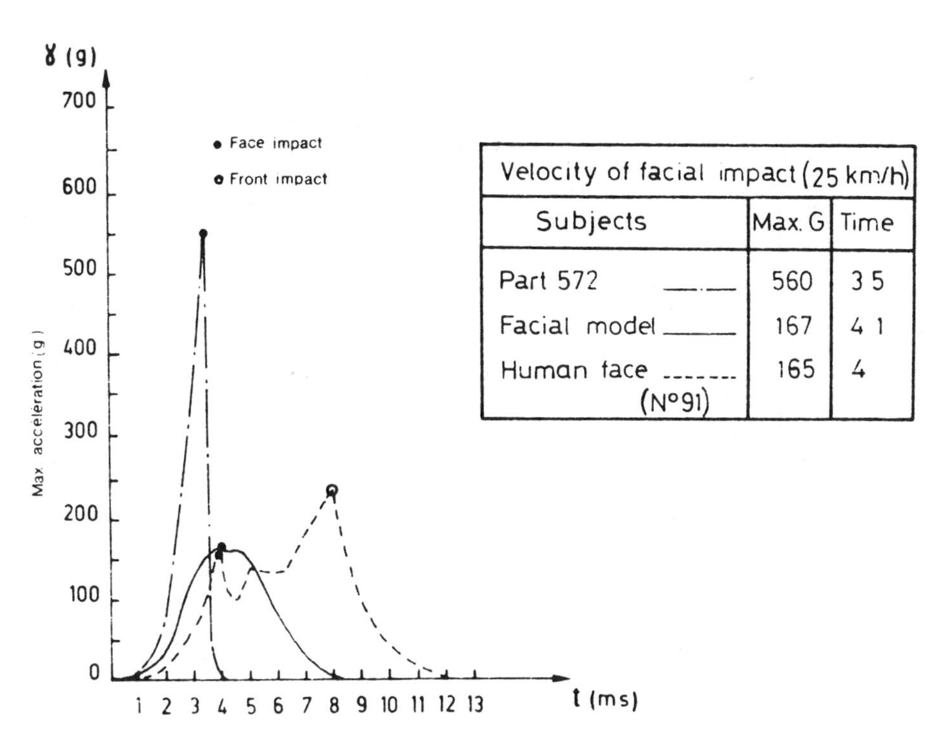

Velocity of facial impact (25 km/h)		
Subjects	Max. G	Time
Part 572 ——.——	560	3 5
Facial model ————	167	4 1
Human face ------- (N°91)	165	4

Fig. 7 - Comparison of facial impact between Part 572 dummy, facial model and human subject

head would correspond to excessive values devoid of biomechanical significance, and that the model yields a proper simulation of facial impact.

Energy Absorption Capacity - A crosscheck was also made for energy absorption by the artificial face. The mean findings were 80 % for the artificial face and 77 % for the human face for a 2.5-meter drop height. These comparative values clearly show that the model's energy absorption capacity is quite similar to that of the human face.

On the basis of this study, we plotted an " absorbed energy versus potential energy curve" (fig.8), which characterizes the model's energy absorption capacity for various dropping heights and for similar platform impacts. This curve emerges as representative of that for the human face.

PROCEDURE FOR USING THE MODEL FACE - To conclude, during the occurrence of facial impact on the basis of the deformation sustained by the model face, we performed a series of impact tests of the model face in free falls against a rigid surface. We recorded the acceleration for each test and also measured the deformed volume of the honeycomb subsequent to impact.

These test findings enabled us to establish a relation-ship between, on the one hand, the deformed volume of the honeycomb,on the other hand,the potential pre-impact energy, the energy absorbed by the face, impact velocity and facial resistance (fig. 9).

If one knows the post-impact deformed volume of the honeycomb, one can judge the other parameters. On this figure, for example, if the honeycomb's deformed volume is 40 cm3, one can find a potential energy of 67 J, a facial resistance of 770 daN and an impact velocity of 15 km/h. The critical velocity of facial impact is 9 km/h, corresponding to a deformed volume of 1.5 cm3, which in turn corresponds to a nose fracture.

In this same manner, one can, specifically, plot curves for impacts against the steering wheel and impacts against the windshield (research devoted to this matter in currently under way).

The model face was assembled onto Part 572 dummies for the purpose of tests involving impacts against steering wheel and windshield in experiments simulating head-on-vehicle impacts. The accompanying photograph (Fig. 10) shows the results of impacts against a hardened-glass wind-shield. The accident simulation involved the launching of a Peugeot 504 against a rigid wall at a speed of 36.4 kph ; we can see directly on the photograph, for example, three serious and several slight lacerations, and glass splinters that have remained stuck in the skin;

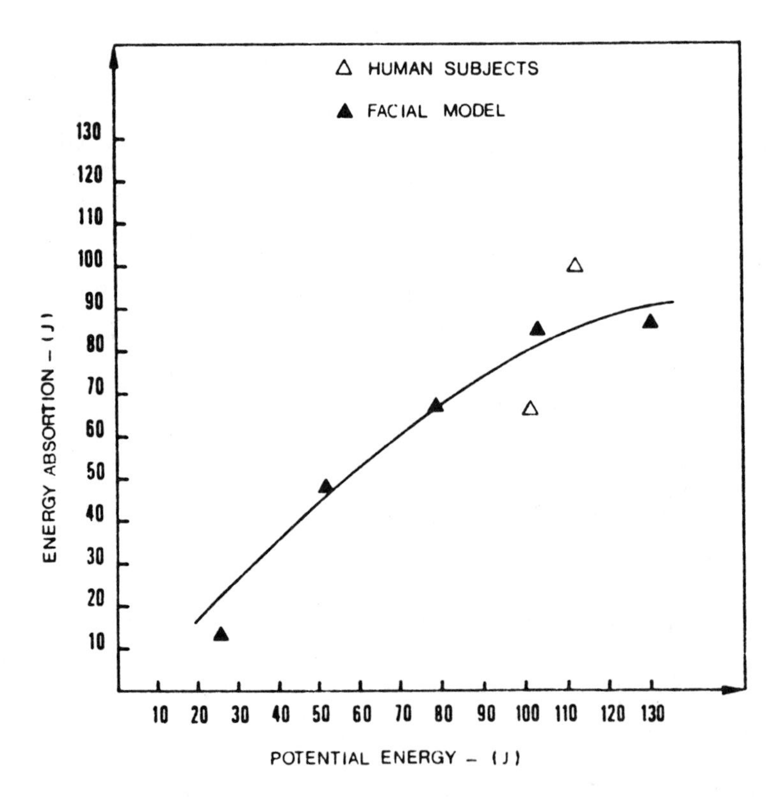

Fig. 8 - Relationship between potential energy and energy absorbed by human face and face model

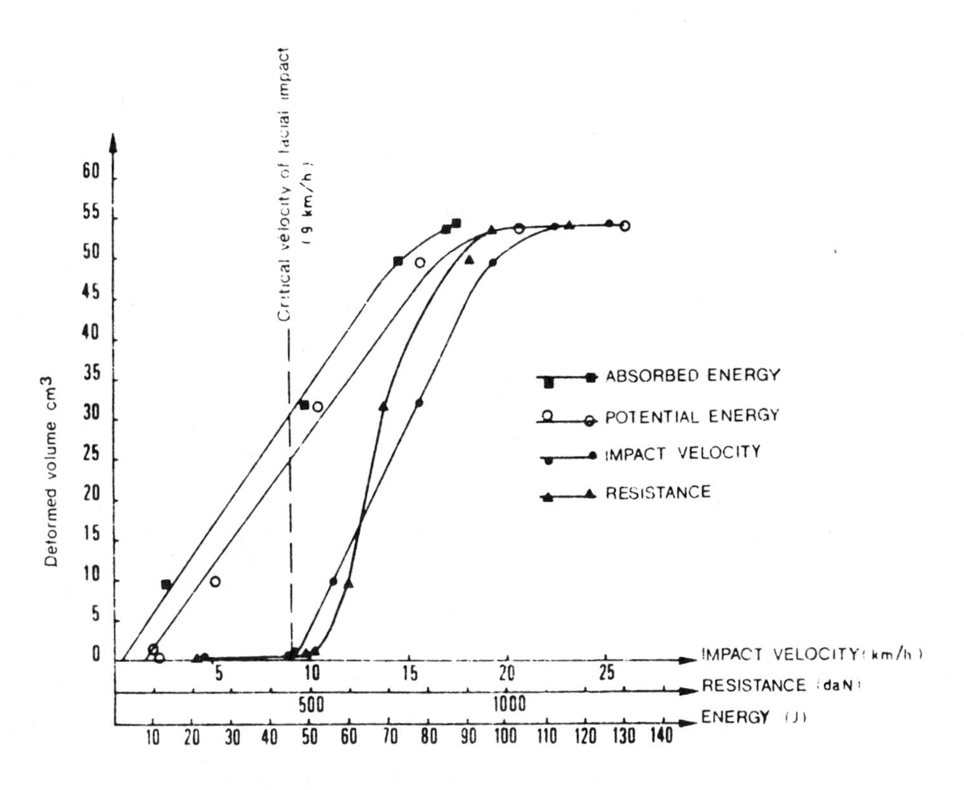

Fig. 9 - Deformed volume of the honeycomb of the face model versus absorbed energy, potential energy, impact velocity and resistance

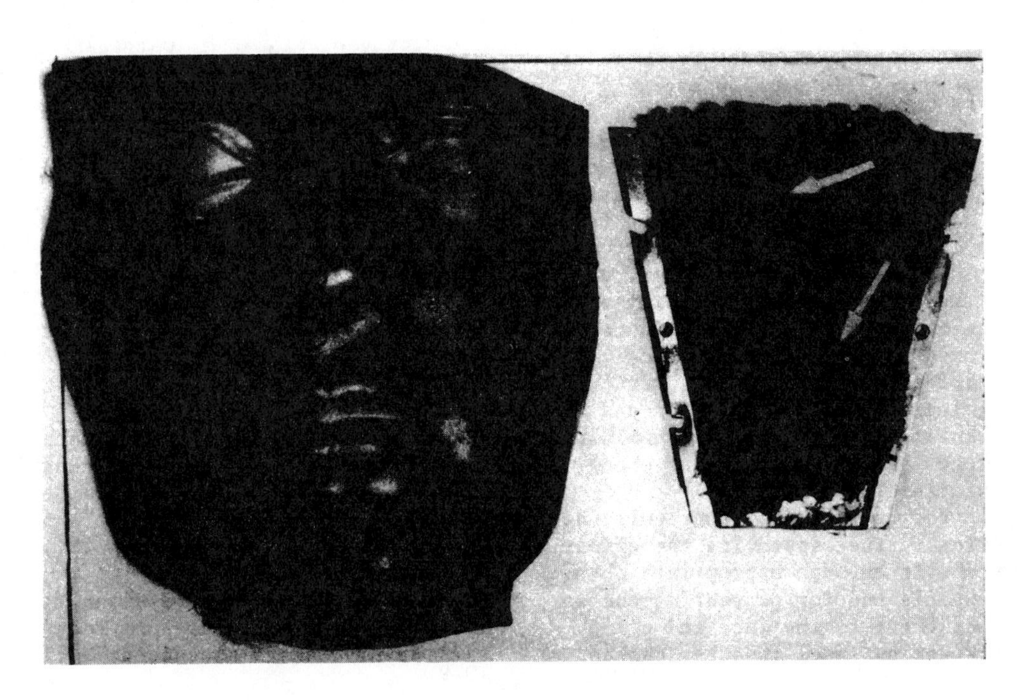

Fig. 10 - Results of a test with the face model: indications of the lacerations and deformations

two impactions on the honeycomb are visible at the level of the right eye socket and of the teeth.

CONCLUSIONS

A model face has been designed on the basis of studies of the laceration of human skin and of responses to facial impact. This model face has been used in tests involving impacts against steering-wheels, against windhields, and in vehicle-impact tests. The advantages of using this model have been confirmed, at least by the following findings :

1. Similitary with the characteristics of laceration, facial resistance and energy-absorption capacity of the human face.

2. facitily of repeated tests thanks to the interchangeability of the deformable elements.

3. Possibility of interpreting the facial impact even in the absence of measurements, through analysis of the degree of crushing of the facial skeleton model.

GENERAL CONCLUSIONS AND PROSPECTS

1. The wearing of seat-belt is the most effective way to reduce the risk of facial injury. The risk of serious, fatal head injuries is reduced nine times for front-seat belt-wearing passengers. However, this protection is less effective for the driver because of the presence of the steering wheel, and 18 % of the seat-belted drivers still sustained a head-against-steering wheel impact.

2. The facial injuries caused by steering-wheels are serious because of the brain lesions with which they are associated in 30 % of cases.

3. In order to anticipate this facial injury risk and to protect car occupants by means of suitable countermeasures (deformable steering wheels fitted with shock-absorbent materials,for example), the dummy used in the acceptance tests should be equipped with a realistic model face. The mecanical characteristics of the bony mass of the face and of the human skin have been investigated and a model simulating the behaviour of the face have been designed. This model, which is adaptable to the heads of existing dummies, enables detection of the risks of facial bone fractures or of skin lacerations. This model, which has been produced in small series, is currently undergoing tests for evaluation and validation in simulations of accidents involving impacts of the face against the steering-wheel, dashboard, and windshield.

Thus, within the near future, there will be a tool suitable for further improving the effectiveness of seat-belts through appropriate changes in steering-wheels and for preventing the severe injuries that occur to the face and brain in head-against-steering wheel impacts. Another solution may be to improve the seat-belt itself, by equipping it with the so-called pre-tension system, which reduces the extent of movement sustained by car occupants. The model face would further be useful for evaluating the risk of contact and its severity, because the approach recommended by us does not require a specific solution; it is confined to evaluating the findings in terms of the non-occurrence of serious injuries.

REFERENCES

(1) A. Fayon, F. Hartemann, C. Tarrière, C. Rhomas, C. Got, A. Patel, " La ceinture de Sécurité. Théorie, Expérimentation, Mesure de son Efficacité sur 3000 Accidents Réels ",published by the C.I.S.R., 34 avenue Marceau, 75008 Paris, Imprimerie Landais, 75010,Paris, July 1979.

(2) C. Thomas, G. Fayon, C. Henry, C. Tarrière, C. Got, A. Patel, " Comparative Study of 1624 Belted and 3242 Non-Belted Occupants : Results on the Effectiveness of Seat-Belts ", in Proceedings of the 24th Annual Conference of the American Association for Automotive Medecine, published by the A.A.A.M., October 1980.

(3) C. Tarrière, A. Fayon, F. Hartemann, P. Ventre, " The Contribution of Physical Analysis of Accidents Towards Interpretation of Severe Traffic Trauma ", in Proceedings of 19th Stapp Car Crash Conference, San Diego, Calif., November 17/19, 1975, published in SAE papers.

(4) Y.C.Leung, E. Lopat, A. Fayon, P.Banzet and C. Tarrière, " Lacerative Properties of the Human Skin During Impact ", 3rd IRCOBI Conference Berlin, 1977, published by the IRCOBI Secretariat ONSER, 109 ave. Salvador Allende, 69500 Bron, France.

(5) Y.C. Leung, G. Walfisch, C. Got, A.Patel P. Banzet and A. Delmas, " Etude d'une Face pour Tête de mannequin ", Report submitted at S.I.A. Symposium (journées de la S.I.A.), published in " Ingenieurs de l'Automobile ", N°.3, 1979, pp. 210-214.

(6) Y.C. Leung, C. Tarrière, A. Fayon and P. Banzet, " Simulation de la Face Humaine sur un Modèle de Mannequin ", in " Annales de Chirurgie Plastique ", published by the Editions de l'Association d'Enseignement Médical des Hôpitaux de Paris, 46, Bd de Latour-Maubourg, 75340 Paris Cédex 07.

(7) C. Tarrière, F. Hartemann, A. Fayon, P. Banzet et Y.C. Leung, " Prévention des Lésions de la Face. Le Rôle de la Ceinture de Sécurité et ses limites ", in " Journal de Médecine Legale, Droit Médical", 1981, 24, N°2, pp. 193-212.

(8) C.Y.Warner and J.Niven,"A Prototype Load Sensing Dummy Face Form Test Device for Facial Injury Hazard Assessment", 23rd A.A.A.M. Conference, Louisville, Kentucky, October 4-6, 1979.

831619

Morphological and Biomechanical Study of 146 Human Skulls Used in Experimental Impacts, in Relation with the Observed Injuries

C. Got, F. Guillon, A. Patel and P. Mack
I.R.B.A., Hôpital R. Poincaré

F. Brun-Cassan, A. Fayon and C. Tarrière
Laboratory of Physiology and Biomechanics, Peugot/Renault

J. Hureau
Anatomical Laboratory, Saint Pères UER

ABSTRACT

Biomechanical studies related to the head have been mainly directed towards the determination of cerebral tolerance to impact in the absence of fracture. However, the frequency of skull trauma producing complex fractures and cerebral lesions linked to these fractures should be taken into consideration. On a human being, impacts under similar mechanical conditons can produce either fatal encephalic lesions without fractures or skull fractures with encephalic lesions if the subject has a different skull morphology. A sample of 146 subjects has been studied to determine the relation between the morphological characteristics of the skulls (weight of the skull cap, thickness, weight of the cranial skeleton...), their mineralization. The mechanical tests were performed on bone fragments (bending and shearing tests). Nine accelerometers were used during the experiments of various types of impacts. The results were computerized. The skull fractures observed (a total of 45) are described. An analysis of the results has been made and enables us to make a correlation between the characteristics of the skull and those of the impacts with the type of fractures.

Studies of real accidents, experiments with volunteers and experiments with human cadavers are the three complementary methods which can help us to improve our knowledge of human tolerance levels to impacts.

The development of experiments with cadavers is justified as it is impossible to measure real accidents and there is a lack of injuries with volunteers.

In order to obtain the best "output" from such experiments, two major problems must be solved :

- The integration of variations among human cadavers, not only the criteria of sex and age, but also the individual variations in morphology and structural properties.
- The selection of pertinent criteria for the injury scale of reference and the definition of adapted parameters to quantify the physical violence of the impact.

Previous studies concerning tolerance levels of thoraxes emphasised the individual variations of the mechanical properties of bone (1). A determination of these variations could be achieved with biological or mechanical tests, and a similar attempt is made here to determine those of the skull, thus trying to solve the two major problems underlined above.

METHODS

Experiments with dead bodies constitute the basis of the data regarding skulls analysed in this study. After examination, skulls of the subjects are submitted to characterization tests (detailed later on). At this stage, fractures resulting from experiments, if any, and measurement results are supposed to be known. Just for memory, the procedure is as follows.

Subjects are fresh unembalmed dead bodies. If brain injuries have to be studied, death should have occurred less than 4 days before tests. A preliminary selection, based on medical files, aims at discarding bone diseases (if any) and cerebral causes of death. Subjects are stored as soon as possible in a cold room, at about 2°C. Preparation before the test includes restoring blood pressure and air volume inside the lungs. This preparation is not adapted to the skull and will not be reported here, since more details are given in previous publications (cf 2 & 3). However, the preparation which concerns the present study is described below.

METHOD FOR THE SKULL - Head and neck are separated from the torso ; the spine is cut between the C7 and T1 vertebrae. To obtain skull caps, the calvaria are sawn along a plane perpendicular to the sagittal plane containing a point located 30 mm above the nasion and another one 20 mm above the external occipital protuberance. After the brain has been removed, the head is separated from the neck by cutting around the foramen magnum. Doing so, occipital condyles remain attached to C1 and the injuries which might occur while separating the top of the spine are eliminated.

The whole skull (in 105 cases), and the calvaria (in all the cases), were cleaned of soft tissues after they had been dissected and boiled. They were weighed before drying. The volume of the skull caps was measured by comparing its apparent weight in air and water. Overall A.P. and R.L. dimensions of the skull caps were taken. The heights were defined by the distance between the apex and the cut plane.

A surface was also defined for the skull caps ; it was obtained in 28 cases by weighing little metal balls (diameter : 4,5 mm) making a single layer against the internal face of the skull cap ; for all the subjects, these surfaces were estimated considering that the skull cap is similar to a spherical area, the radius of which is the average between the 2 available half-diameters and the skull cap height. Moreover, a correction was made for both methods taking into account the average half-thickness of the skull, defined hereafter. The thickness is obtained by taking an average of the measurement results at the level of the edges and for the 4 skull plugs, detailed later and used for the mineralization studies. The thickness of the edges was measured on both sides of the frontal sinuses and on both sides of the superior longitudinal sinus and laterally at the level of the maximal transversal diameter. The skull mineralization was studied by a method similar to that already used for the ribs (1). Circular skull pieces were taken from the frontal and parietal bones and then calcinated in the open air at 700° C. The weight of the ashes for 1 cm^2 of the skull surface is noted. Samples of circular fragments were made with a hole saw, (the interior diameter being : 32,5 mm (Greenlee N° 825) ; exterior diameter : 37 mm). For a more precise localization of the skull plugs, a 6,3 mm hole was previously drilled. The 2 frontal skull plugs were located 20 mm before the frontal-parietal suture, and they were separated by a similar distance, as shown on Fig. 1. Parietal plugs coincide with the tops of the parietal humps. The skull pieces used for mechanical testing were obtained from transverse sections of the skull caps. 10 mm wide bone stripes were cut, as indicated on Fig. 1. Tests were performed on 28 available skull caps. The skull bone stripes were submitted to bending and shearing tests, with the same procedure as for rib tests (1) Fig. 2. The cylindrical plunger goes down at a 1 mm/mn rate. Force and displacement were recorded until the rupture of the bone piece took place. As regards bending, the maximum of the slope force versus displacement is noted. We have the ultimate bending stress, the Young modulus, and an energy for the bending tests. The sections of the bone strips are supposed to be rectangular. In shearing tests, two bone sections separated by a 10 mm distance are cut. In a similar manner, force and displacement are recorded and the energy used for the shearing process is noted. At the beginning of the research, an attempt was made to use the results of punching tests, but it proved to be unserviceable.

TYPE OF TESTS - The skulls we have studied come from dead bodies submitted to various experiments related to several problems of traffic safety. These experiments may be classified into two categories :

1) Drop tests in various configurations where the head strikes more or less padded surfaces (2). Depending on the test, the frontal bone, the temporo-parietal bone or the face, first contact the surface. Forces transmitted to the skull from the impact point are sufficiently well described by accelerations.

2) Other experiments, where a victim is simulated either inside or outside a vehicle ; frontal collisions with restraint, lateral collisions or pedestrians accidents. The analysis of forces transmitted to the skull is quite difficult because the neck is an important path of loading, mainly in frontal collisions when belts are worn. This second category is not studied in the present paper. However, the corresponding skulls were examined in the morphological studies. An attempt to separate the forces transmitted to the skull by the neck and the impact forces, by means of a mathematical model, exists in (8), as regards restrained occupants involved in frontal collisions. In each case, measurements of the head aim at a complete knowledge of the accelerations of the brain envelope (i.e. the skull). 9 accelerometers set on

3 places of the skull are used. Assuming that, during the collision, their distances vary very slightly, this accelerometer array (3,3,3,) can compute the linear accelerations at any point of the head. The accelerometers were not as numerous as in previous tests, but there were at least 6, in 3 different places. Measurement channels meet the requirement of MVSS 208/SAE J 211 b, class 1000. Since the first paper about these tests (2), data were processed in order to give the best C.G. acceleration estimates (8). Moreover, some fractures do not figure in (2). The impact force was also recorded in the drop tests. Table 1 gives a summary of the corresponding results. High speed are available for each case.

RESULTS

MORPHOLOGY AND MINERALIZATION - The morphological and mineral characteristics of 146 skulls are gathered in Table 2. They belong to a series of 301 experimental impacts or collisions performed between 1972 and 1983. The surface of the vault measured with iron balls is well correlated with the calculated value (R = .93) for the part of sample (27 cases) used for this comparison and we retained only the calculated value in the whole sample to obtain the mean thickness.

Age is a poor indicator of bone mineralization (R = -.19) and must be discarded in any comparative studies using cadavers (Fig. 3) : we also had to give up using the bone condition of ribs as a reference value (R = .18) (Fig. 4). There is no relationship between skull and rib mineralization. Variations of vault dimensions (antero-posterior, transversal diameters and vertical radius) are not important, and consequently the surface variations are limited, and the correlation between vault weight and mineralization expressed by surface unit is high (R = .74-). The mean thickness of the vault and the mineralization are fairly well correlated (R = .68) (Fig. 5) but the thickness of the cortical bone is not proportional to the thickness of diploe which explains the poor correlation between density and mineralization (R = .44) (Fig. 6). The relatively bad correlation between mean thickness calculated with volume and surface and edge thickness (R = .61) (Fig.7) seems very important ; it is a consequence of the big variations in thickness at the inferior part of the parietal bones even if the thickness of frontal bone and/or the convexity of the vault are almost identical. Any experiment with lateral impacts using cadavers needs a specific characterization of the lateral part of the skull (thickness and mineralization). The correlation between the frontal bone and the most convex part of the parietal bone is high (R = .73) (Fig. 8) ; two cases with a frontal mineralization, very different from the parietal one, are typical frontal hyperostosis.

We have tried to determine the differences between men and women ; this work was done retrospectively after having dried the skulls, and data are different from those figuring Table 1. The ratio : weight of the vault/weight of skull basis + face (without inferior maxillary) is 1.1 for women and 1.12 for men (a difference which is not significative) but the mean weight of the skull is 584 g. for men and 479 g. for women (dried skull without inferior maxillary).

DEFINITION OF PARAMETERS IN ORDER TO CHARACTERIZE SUBJECTS - The available parameters that could be used for characterizing skulls - and more generally, heads - are rather numerous. They can be classified into three groups :

- Dimensional parameters : such as the AP an RL diameters and the heights of the skull caps which are examined here.
- Those parameters which define the quality and quantity of bone : the degree of skull mass mineralizations, skull thicknesses for example.
- For a smaller sample of subjects, parameters related to the previous description of mechanical tests, such as the energies and the forces recorded in bending or shearing tests.

Although mechanical testing was performed on a small sample, the result was not changed. In fact, the analysis of the results concerning the 28 skulls submitted to mechanical tests indicated a strong correlation between some mechanical results and some other results. The last results are available for all subjects. More precisely, the correlation of the ultimate bending force and the mineralization is used in the following. However, the number of parameters necessary to characterize a subject remains too high ; it makes the analysis more confused. In order to keep the minimum number of independent parameters, the matrix that summarizes the correlations between all the parameters was studied (Tables 3 a and b). It shows that the

3 mineralizations are well correlated ; only one out of these 3 was retained : the averaged mineralization. The thicknesses were considered in the same way and the thickness of the skull cap edge was selected. The density was ne-

glected as it did not give any further detail. The 3 skull dimensions appeared to be poorly correlated between themselves and to many other parameters ; they were retained. As regards the masses, the whole head mass was preferred to the skull cap mass, because the latter is actually correlated to the other parameters which are related to bone resistance.

Finally, 6 parameters were retained at this stage, in order to characterize a subject's head. They are :
- the head mass (HDW).
- the averaged skull cap mineralization (MIM).
- the thickness of the edge (LTH)
- the 2 diameters and the height of the skull cap (APD, RLD, SDP).

This number of parameters is still high ; it makes the observed phenomenon difficult to explain, since they are due to several causes at the same time. In order to reduce the number of parameters, an analysis of principal components was performed to find out combinations of varia bles. Such a method is already known and described (4 & 5) and was previously used successfully for problems concerning the characterization of the human rib cage (1).

- Figure 9 displays the results of the principal components analysis made on 137 subjects for whom the 6 selected parameters were known. These 6 parameters were all used as active variables in the analysis.

- Figure 9 concerns a sample of individuals. According to the validity of such a method, the points corresponding to the variables were also projected onto the plane constituted by the first two factors which emerge from the analysis. These 2 factors, when compared to other couples of factors, clearly describe the sample constituted by the individuals.

3 well separated groups of variables appear :
- APD ans HDW (AP diameter and head mass), which are highly correlated to the artificial variable that the axis 1 shows on figure 9.
- LTH ans MIM (skull cap edge thickness and averaged mineralization), which are highly correlated to the variable defined by the axis 2.
- RLD ans SDP (RL diameter, and height of the skull cap), which have a very small correlation with axis 1, and a poor one with axis 2.

Looking at this table which describes the meaning of these 6 parameters through only 2 of them, (namely the 2 principal components), seemed very difficult. This was due to the last group of parameters, as they are insufficiently correlated to either axis. But it could be simplified : EPB ans MIM could be replaced by the variable defined by axis 2. Actually, the analysis of the results obtained with the 28 skulls, (the mechanical characteristics of which were known), shows that this variable is related to skull resistance (Fig. 10). Considering the total sample of 137 subjects, one can find that the correlation coefficients between MIM, LTH and the artificial variable 2 (VAR 2) are .713 and .791 respectively. The complete expression of VAR 2 is on annexe 1.

Multiplicative factors illustrate the respective weights of the experimental parameters which constitute artificial variable VAR 2. The preponderant effects of thickness and mineralization appear. Such a preponderance differs from the expression of the Bone Condition Factor (BCF) (1), related to the rib cage, and where the various weights were of the same order of magnitude. As regards the BCF, all the parameters used obviously had some influence on rib resistance. On the contrary, when VAR 2 is considered, diameters, for instance, have a secondary effect. Incidentally, the gathered data indicated that rib mineralization and skull mineralization have no correlation. VAR 2 is labelled Skull Bone Condition Factor (SBCF) in the analysis presented hereafter. Consequently, this analysis uses the following parameters :
- the 3 dimensions and the head mass.
- the skull resistance factor SBCF.

To interpret the results, the age of the subjects and their sex will occasionally be taken into account. To explain the occurrence of fractures, and possibly their type, or the absence of fractures when cadavers are submitted to head impacts under similar conditions, some parameters describing impact severities are needed. These parameters are given by the measurements made (Mainly accelerations). Of course, the paths of loading through the skeleton must also be considered.

CLASSIFICATION OF SKULL FRACTURES - The same impact on a subject's head may produce a more or less severe fracture, according to the characteristics of the skull. Therefore, fractures will be classified in the analysis, taking into account the extent and the localization of the damages to the bones.

A first approximate classification may be done to summarize the set of available fracture cases. (cf : Table 4). Furthermore, for a complete information, the 45 fracture cases recorded are schematized on adequate graphs, subject after subject. (Fig. 11 a, b, c, d,).

UTILIZATION OF THE CHARACTERIZATION OF THE SUBJECTS AND THE MEASUREMENT RESULTS IN ORDER TO DEFINE SKULL TOLERANCE TO FRACTURE - Results were analysed while simultaneously taking into account the previously defined SBCF as a resistance index, the type of fracture, and one parameter likely to be related to the fracture tolerance.

Using the maximum of acceleration at first is an attractive approach, since it is measureable, and linked to the loads transmitted to the bone structure. In the small sample of direct blows on the head (drop tests), the accelerations give an accurate account of these loadings, because small forces are transmitted by the neck, due to the test configurations (it was not the case for a vertical fall, head first). The analysis will therefore be performed, first of all, on this small sample. All the tests analyzed here correspond to a good distribution of the impact loads, by means of a liner of shock-absorbing material about one inch thick. This liner is either on the impacted surface, or inside a helmet worn by the subject. The impacted surface is a flat rigid one. More details in (2). The designs on Fig. 12 to 14 illustrate the impact conditions.

Figure 12 concerns falls with a deceleration of the head, due to a set of forces mainly applied to the lower part of the frontal bone ; however, the loading may extend to the nasal area. One may distinguish on this figure a first area where no fracture had occurred, (which corresponds to the most resistant skulls) and a second area where fractures appear with similar acceleration levels . The SBCF allows this ranking. A deeper examination from the right side to the left side of the figure indicates, a deterioration in the resistances of the skulls, which successively corresponds to uninjured people's face fractures, face fractures extended to the basis of the skull, and, finally, to a complete dislocation. According to Fig. 12 and under the illustrated impact conditions, when this particular type of loading is considered, 200 g. is an order of degree for the tolerance to fracture. One may say that the supporting structures of the frontal bone yield, instead of saying that the frontal bone yields. The g-level mentioned is an intermediate force between the skull tolerance to fracture in frontal bone area, and the facial tolerance to fracture, with some load distribution.

Figure 13 concerns the blows against the face, with a distributed loading and a small amount of padding as provided by full-face covering helmets. It indicates a lower level close to 150 g. The limit between the impact conditions of the tests on Fig. 12 and Fig. 13 may be unprecise in particular cases.

Figure 14 refers to impacts against the parieto-temporal area, with a distributed loading and about one inch of polystyrene liner. It shows that no fracture occurs under these conditions, except for very high g level (300 g), or with exceptionally weak subjects. 250 g appear to be tolerable to the greatest part of the experimental population (which we supposed to be more vulnerable than the accident involved population).

In figure 15, all the fracture cases in drop tests are shown ; coordinates are the "resistance" SBCF and the acceleration level (less than 3 mn). Figure 15 shows that this latter parameter varies only slightly when a fracture occurs, whatever its type and whatever the subject. Furthermore, one may also conclude that the previously common limit of 80 g (3 ms) is a very conservative one when distributed padded impacts are considered, and even when the face is involved.

DISCUSSION

RELATIONSHIP BETWEEN SKULL FRACTURES AND CEREBRAL LESIONS - A fracture is not invariably found in fatal head injuries, and, moreover a patient with a fracture may have minimal clinical evidence of brain lesion, if any ; but there is a relationship among car occupants involved in accidents between the A.I.S. level and the frequency of fracture.

In a sample considering 9899 car occupants (+) the frequency of skull fractures was 3 % for head A.I.S. level 2, 14 %for A.I.S. level 3, 26 % for A.I.S. 4, 28 % for A.I.S. 5 (survival), 54 % for A.I.S. 5 (death) and 6.

The fracture may be directly related to the mechanism of brain lesion and death, especially when a fracture of the vault tears a meningeal artery and produces an extradural haematoma, also when a ring fracture of the base of the skull produces a laceration of the brain stem, and when the dislocation of the skull brings major lacerations or crushing of the brain. (24 among 47), as shown in Fig. 16.

In a second group of these cases, there is primary brain damage directly related to the fracture (cortical contusion, cortical laceration) or to the

(+) : IRO-Peugeot-Renault multidisciplinary investigation of road accidents.

deformation of the skull which is the consequence of extensive fractures. But "acceleration" lesions or secondary lesions (brain swelling, raised intracranial pressure, hypoxic lesions), are associated ; so we are unable to specify the exact role of the fracture in the fatal issue (14/46).

In the third group, consequences of skull fractures and brain lesions are independent of head impacts and the fracture can be an indicator of the characteristics of those impacts ((8/46). In a series of autopsies made by one of us, a skull fracture was observed in 63 % of lethal head lesions (47/74 - Fig. 16). The effect of these lesions differs according to the type of medical institution. When J. Hume Adams, et al. indicate (7) that "there is therefore ever increasing evidence that acute subdural haematomas and diffuse axonal injury, the principal causes of death and disability brought about by head injury on man, are caused by angular acceleration of the head", they refer to neuro- surgical centers without immediate deaths.

Studies including immediate deaths are necessary ; in the opinion of the authors, if there is a place for angular acceleration in the production of head injury, the major cause of death for car occupants, two-wheel users and pedestrians is "an impact or blow involving a collision of the head with another solid object at an appreciable velocity. This situation is generally characterized by large linear acceleration and small angular acceleration during the impact phase" (6).

One has to keep in mind that the individual variations are so important that, considering almost identical impacts, we can observe in one case a death secondary to inertial lesions of the brain without any fracture, and, in other cases, a ring fracture of the basis of the skull, a fracture of the odontoïd or a depressed fracture of the frontal bone. We think that the distribution of tolerance levels for those different lesions are not very different and that one of the object of experimental research using cadavers is to specify the similarities and the differences between skull and brain tolerance levels ; different parameters for these two tolerances may be considered.

UTILISATION OF TOLERANCE RESULTS IN FULL SCALE TESTS WITH ANTHROMORPHIC TEST DUMMIES - The transpositions between the mentioned fracture tolerance estimates and some injury criteria to be found on dummies are difficult. The perfect simulation of the skull's response is not the major problem, especially when a padding increases the impact duration. When the face is concerned, the discrepancies in the responses of a fragile human face and a unbreakable dummy face appear. There is an attempt to develop a deformable dummy face (12) and further study is highly desirable. Fracture and brain injury criteria have not been compared here. They are different, since they correspond to two different injury mechanisms. Probably, the use of two distinct head criteria could be examined. Another unsolved problem concerns the restrained occupant, whose neck is an important point of overloading ; therefore, head accelerations are not directly related to impact forces (8). These points have to be considered, taking into account measurement possibilities.

A procedure with two different criteria could be more adequate.

CONCLUSIONS

To be used, the results of experiments performed with cadavers have to be characterized as regards to their morphology and their bone condition. In this study, the necessary morphological data appear to be :

- the weight of the head,
- its dimensions,
- the thickness of the vault - particularly impacted area,
- the weight of the skull part taken apart under normalized conditions.

With previous results concerning ribs, an index describing the skull bone condition was defined and used. This index (SBCF) proved to be well correlated with the mechanical tests performed on skull fragments. The measurements and fracture results obtained in experiments could thus be interpreted, and tolerance levels be specified under particular loading conditions. Generalization of such morphological and bone characterizations are likely to help reducing the number of tests required to solve a given problem, and achieve a better knowledge of tolerances when a population, which runs well defined risks, is involved.

FIG 1 LOCALIZATION OF BONE FRAGMENTS USED FOR SKULL BONE CHARACTERIZATION

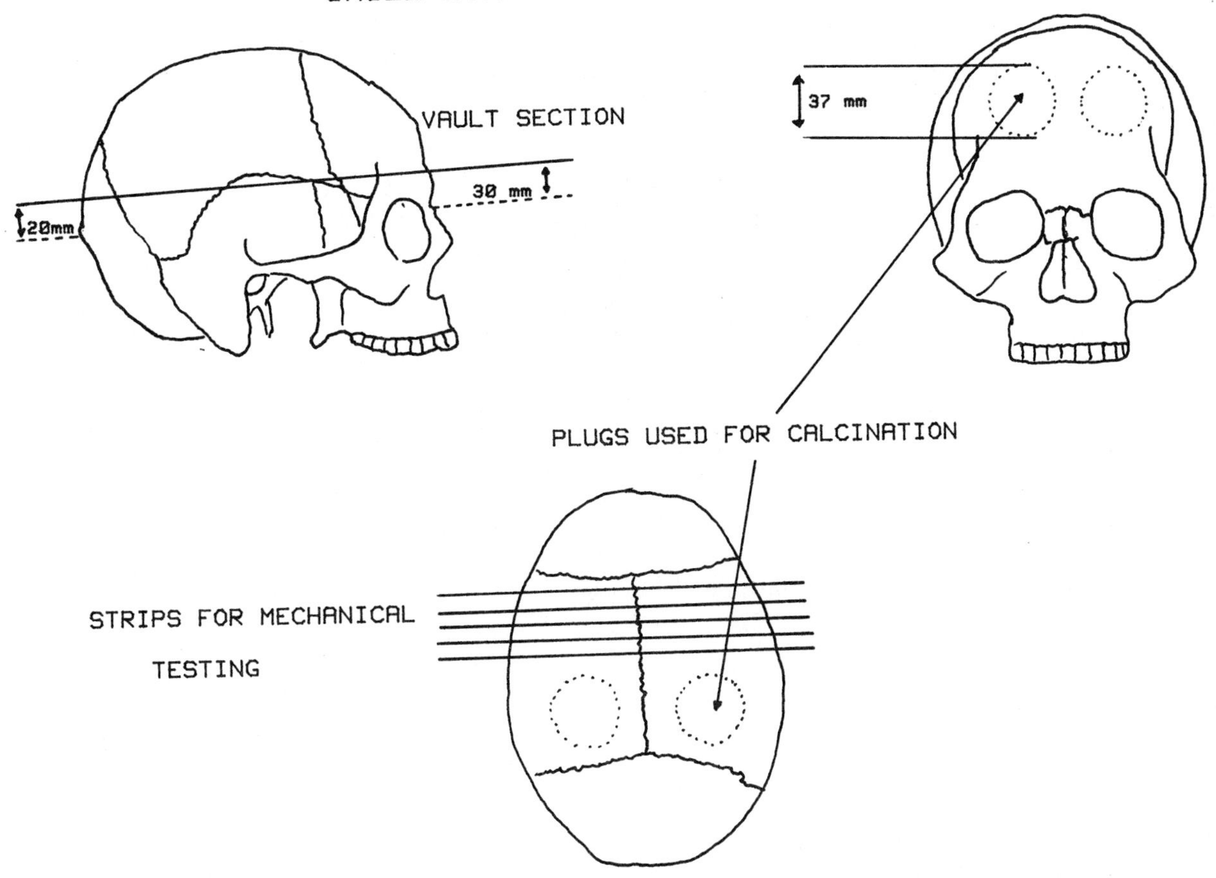

FIG 2 MECHANICAL TESTS ON SKULL STRIPS

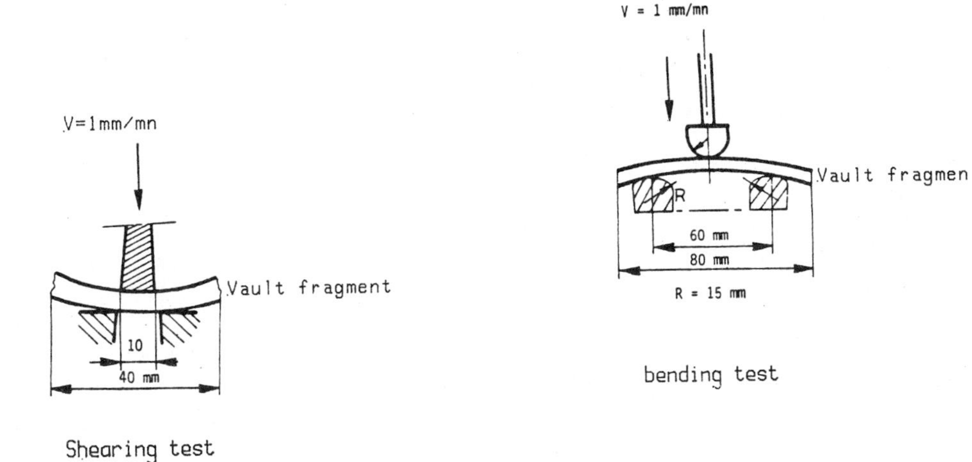

TABLE 1 CHARACTERISTICS OF IMPACTS

Padded γ_x impacts on frontal bone (see fig 12)

Number	γ_{max} (g)	γ_{3ms} (g)	Impact force (daN)	HIC	SBCF
102	232	147	780	1483	.190
103	349	149	1580	2351	.999
107	154	90	516	500	-.005
108	313	144	1720	1200	.402
110	194	109	781	1260	.003
159	214	117	1000	1078	.342
160	187	150	880	1411	.670
162	192	132	850	1334	-.085
163	150	102	580	750	-.110
165	163	109	1000	692	-.050
166	213	133	980	1270	-.510
172	209	102	950	1042	.420
174	182	122	540	1156	-.090
175	190	130	600	1200	-.090
176	257	127	700	1416	-.206
177	255	169	680	2138	.202
250	202	129	/	1460	-1.000
251	252	129	720	1085	.260

Padded impacts on the temporo parietal area (see fig 14)

Number	γ_{max} (g)	γ_{3ms} (g)	Impact force (daN)	HIC	SBCF
63	171	113	/	937	.202
64	151	118	/	832	-.370
65	175	130	700	1250	.470
66	174	137	560	1166	.510
67	251	151	/	1584	1.093
70	121	107	600	641	.350
74	260	109	840	1082	1.000
83	208	142	640	1240	-.720
85	254	150	1160	1665	-.490
86	206	146	1300	1571	.039
87	249	141	1300	1374	.508
273	230	124	800	1223	-.550
274	253	118	940	1508	.098
275	326	104	1150	1307	-.111

γ_x padded impacts on the whole face (see fig 13)

Number	γ_{max} (g)	γ_{3ms} (g)	Impact force (daN)	HIC	SBCF
88	129	95	1040	565	.099
89	150	95	1080	540	-.016
90	176	125	1560	1185	.423
94	169	136	800	1500	-.228
95	185	125	400	1150	.610

TABLE 2

CHARACTERISTICS OF THE SAMPLE – MORPHOLOGY/MINERALIZATION/AGE

VARIABLE	OBSERV.	MEAN	VARIANCE	ST.DEV.	MAX.	MIN.	SKEWNESS	CURTOSIS
HEAD WEIGHT	140	3846	270616	520	5210	2320	.352	.149
SKULL WEIGHT (VAULT)	120	387.3	6271.8	79.19	301	158	.229	.435
VOLUME (VAULT)	146	226.1	2223.8	47.15	393	115	.29	.434
DIAMETER (ANT–POST)	146	180.2	51.29	7.16	203	164	.191	–.003
DIAMETER (TRANSV.)	146	142.7	40.24	6.34	158	122	–.148	.127
RADIUS (VERTICAL)	146	86	80	8.94	105	61	–.486	–.231
EDGES THICKNESS	146	4.3	.632	.795	6.6	2.6	.581	.0151
FRAGMENTS THICKNESS	146	6.47	1.7	1.304	11.12	3.5	.737	1.542
SURFACE (cm2)	146	400.95	1417.3	37.64	472.7	301.6	–.46	–.275
THICKNESS (VOL/SURF)	146	5.64	1.107	1.052	9.64	2.94	.31	.64
DENSITY	120	1.676	.0186	.136	1.99	1.17	–.60	1.052
MINERALIZATION:FR.	146	.605	.0232	.1523	1.171	.222	.865	1.795
MINERALIZATION:PAR.	146	.515	.018	.134	.917	.167	.381	.41
MINERAL.MEAN (g/cm2)	146	.56	.0178	.1334	1.01	.197	.557	.881
AGE	138	57.9	128.15	11.32	82	24	–.558	.376

DIAMETERS, RADIUS AND THICKNESS ARE EXPRESSED IN MILLIMETERS , WEIGHT IN GRAMS, VOLUME IN CUBIC CENTIMETERS

FIG 3a CORRELATIONS BETWEEN MORPHOLOGICAL AND MECHANICAL PARAMETERS (27 SKULLS)

	BENDING FORCE	BENDING ENERGY	ULTIMATE BENDING STRESS	YOUNG MODULUS	SLOPE F/D (BENDING)	MAX. SHEARING FORCE	SHEARING ENERGY
HEAD WEIGHT	–.11	–.13	–.23	–.03	.29	.09	.02
SKULL WEIGHT	.34	.34	.10	.21	.49	.55	.53
VAULT WEIGHT	.09	.39	.51	.13	.36	.53	.60
DIAMETER (A.P.)	–.30	–.15	–.43	–.22	.08	–.01	–.08
DIAMETER (TR.)	–.15	–.08	–.24	.13	.32	–.24	–.21
RADIUS (VERT.)	–.13	.15	–.21	–.06	.10	.07	.08
VAULT VOLUME	.19	.25	–.09	.52	.50	.32	.38
VAULT SURFACE	–.32	.01	–.44	–.07	.22	–.02	–.04
THICKNESS:FRAGMENTS	.30	.21	.02	–.25	.31	.27	.31
THICKNESS:EDGES	.38	.23	.29	.15	.34	.22	.34
THICKNESS:SURF/VOL.	.43	.31	.16	.11	.46	.27	.31
MINERALIZATION	.71	.71	.57	.66	.62	.71	.78
DENSITY	.40	.51	.48	.67	.30	.45	.45

TABLE 3b CORRELATIONS BETWEEN MECHANICAL TESTS AND MORPHOLOGICAL PARAMETERS (27 SKULLS)

	HDW	SKW	SUR	DEN	LTH	TVS	FT	MIF	MIP	MIM	RLD	APD	SDP
HEAD WEIGHT	1												
SKULL WEIGHT	.53	1											
SURFACE (VAULT)	.44	.36	1										
DENSITY (VAULT)	.04	.15	.13	1									
THICKNESS (EDGES)	.15	.40	−.09	−.16	1								
THICKNESS (VOL/SURF)	.31	.69	−.30	−.22	.43	1							
THICKNESS (FRAGM.)	.28	.72	−.04	−.15	.44	.73	1						
MINERALS (FRONTAL)	.14	.70	−.04	.25	.34	.57	.71	1					
MINERALS (PARIETAL)	.23	.74	.02	.28	.34	54	.66	.76	1				
MINERALS (MEAN)	.19	.76	−.02	.28	.36	.59	.73	.94	.93	1			
R.L. DIAMETER	.38	.17	.31	.01	.08	.07	−.01	−.08	−.06	−.08	1		
A.P. DIAMETER	.62	.37	.49	−.06	.12	.13	.16	.02	.03	.01	.21	1	
SKULL CAP DEEPNESS	.40	.48	.70	.20	−.12	.07	.11	.12	.17	.15	.05	.19	1

TABLE 4

CLASSIFICATION OF 45 EXPERIMENTAL SKULL FRACTURES

	SKULL	FACE	SKULL + FACE
LINEAR LOCALIZED	7	0	1
LINEAR EXTENSIVE (INCLUDING MULTIPLE LINES)	7	1	5
DEPRESSED LOCALIZED	1	0	0
DEPRESSED EXTENSIVE	4	0	* 11
DISLOCATION	0	4	4

* FACIAL IMPACTS WITH EXTENSIVE LINEAR OR DEPRESSED FRACTURE OF THE SKULL

Fig 3 AGE/MINERALIZATION (g/cm2)

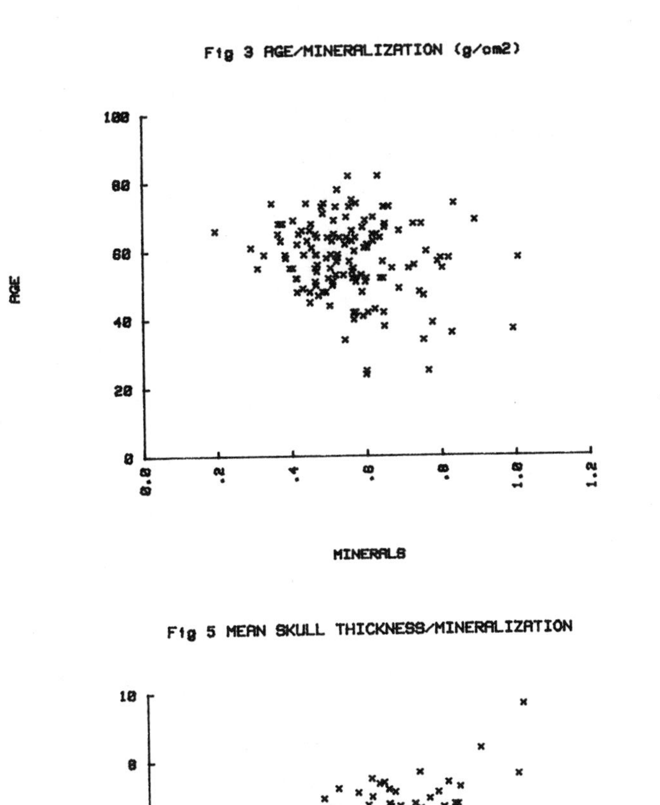

Fig 4 MINERALS : SKULL (g/cm2)/RIBS (g/cm)

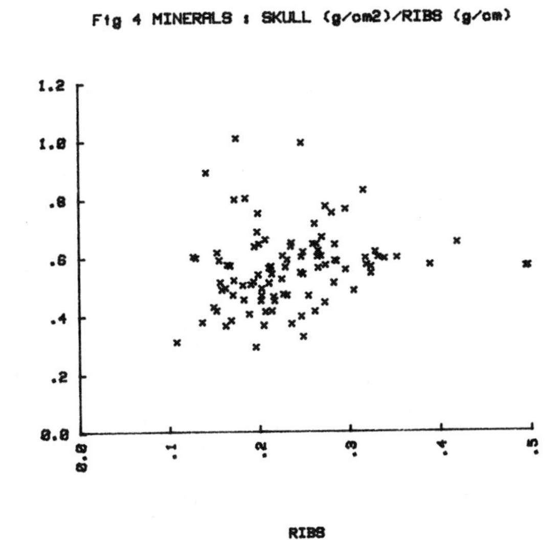

Fig 5 MEAN SKULL THICKNESS/MINERALIZATION

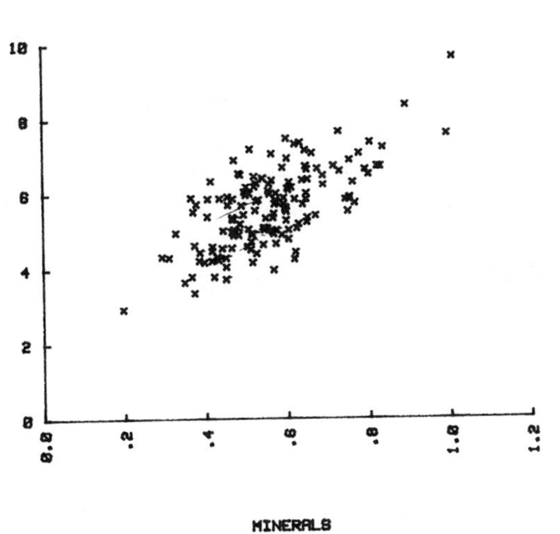

Fig 6 SKULL DENSITY/MINERALIZATION

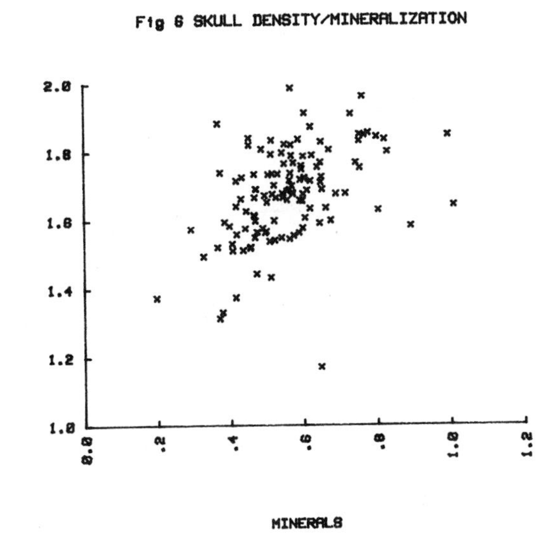

Fig 7 EDGES THICKNESS/MEAN THICKNESS

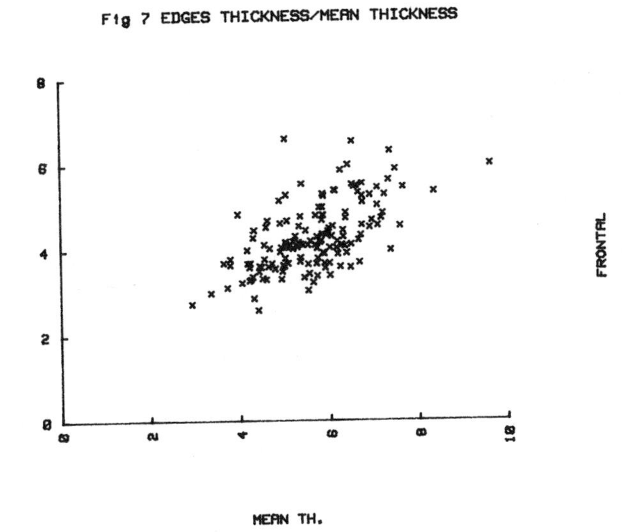

Fig 8 MINERALS(FRONTAL)/MINERALS(PARIETAL)

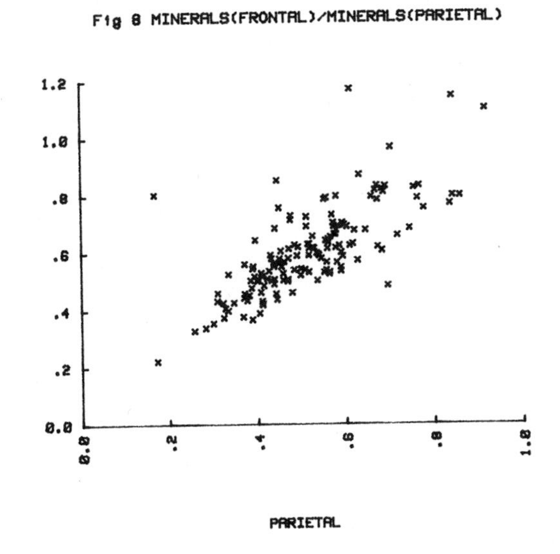

FIG 9 RELATIVE POSITIONS OF MORPHOLOGICAL PARAMETERS IN THE TWO FIRST AXES PARAMETERS (137 SUBJECTS)

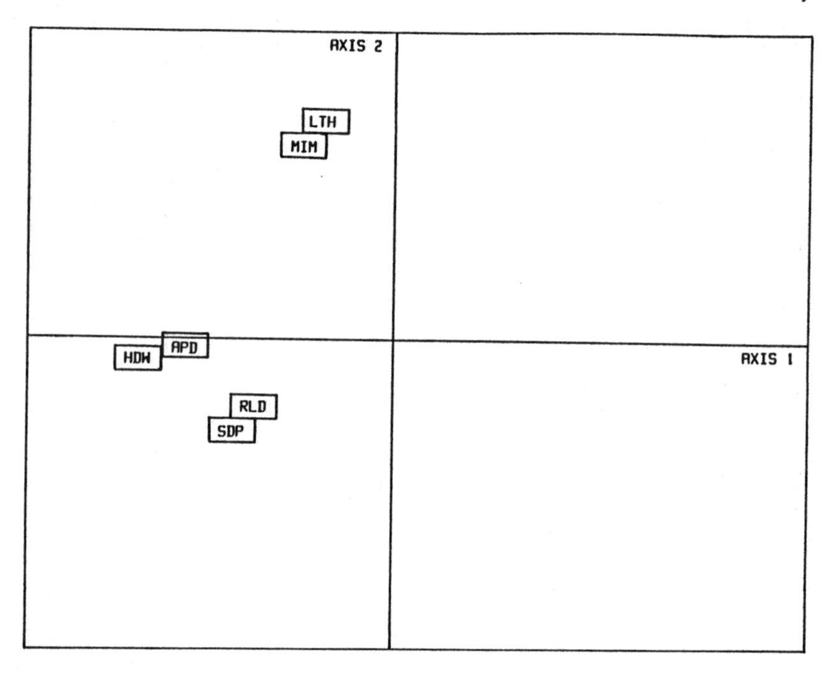

FIG 10 ANALYSIS OF 28 SKULLS – INTERPRETATION OF VAR 2

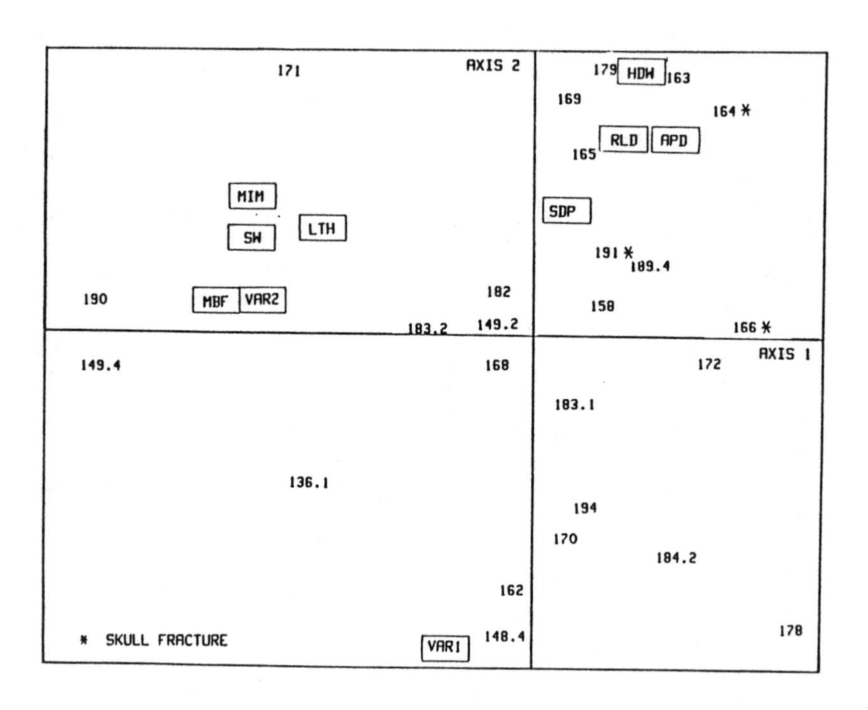

FIG 11a EXPERIMENTAL FRACTURES

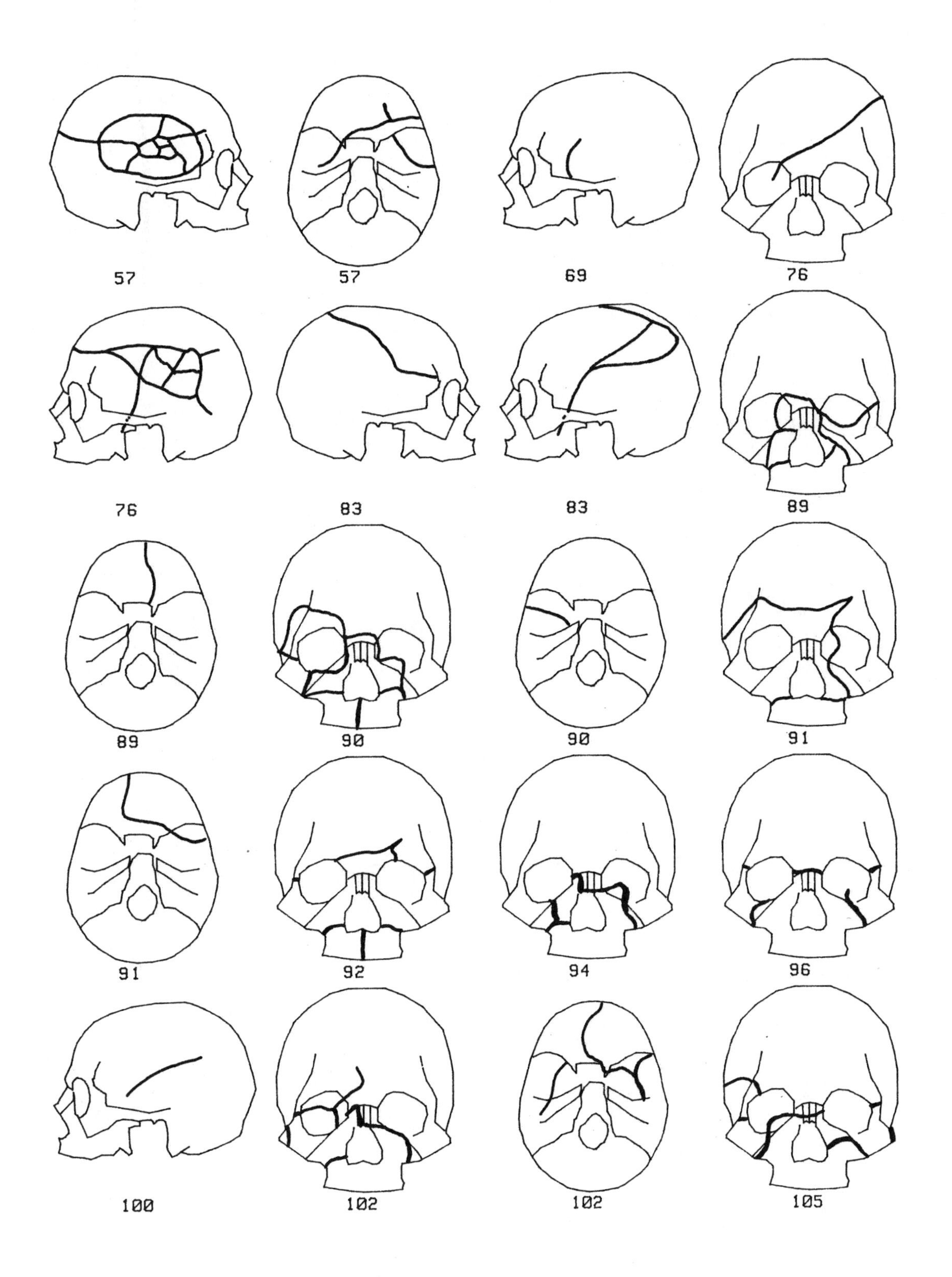

FIG 11b EXPERIMENTAL FRACTURES

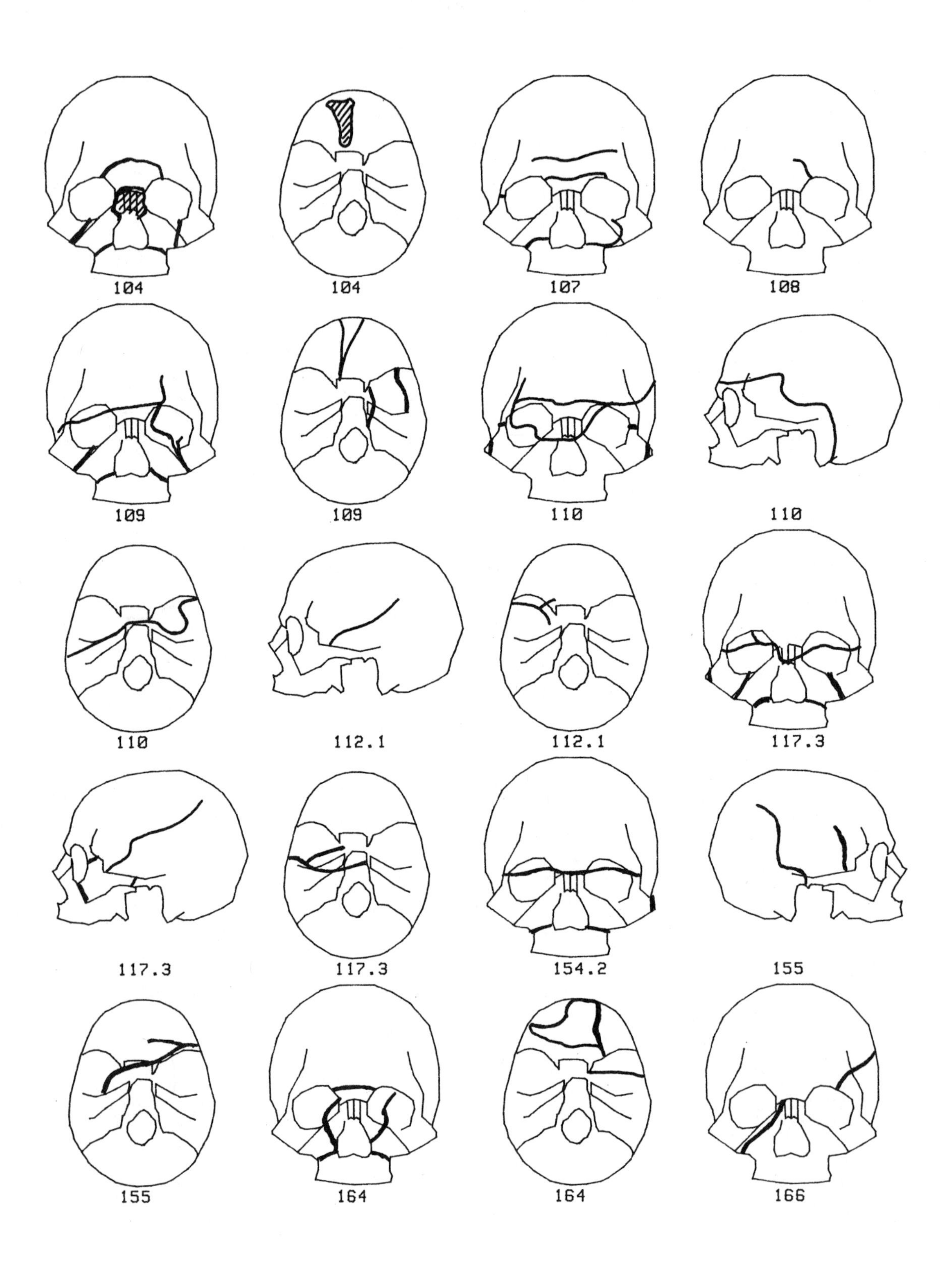

FIG 11c EXPERIMENTAL FRACTURES

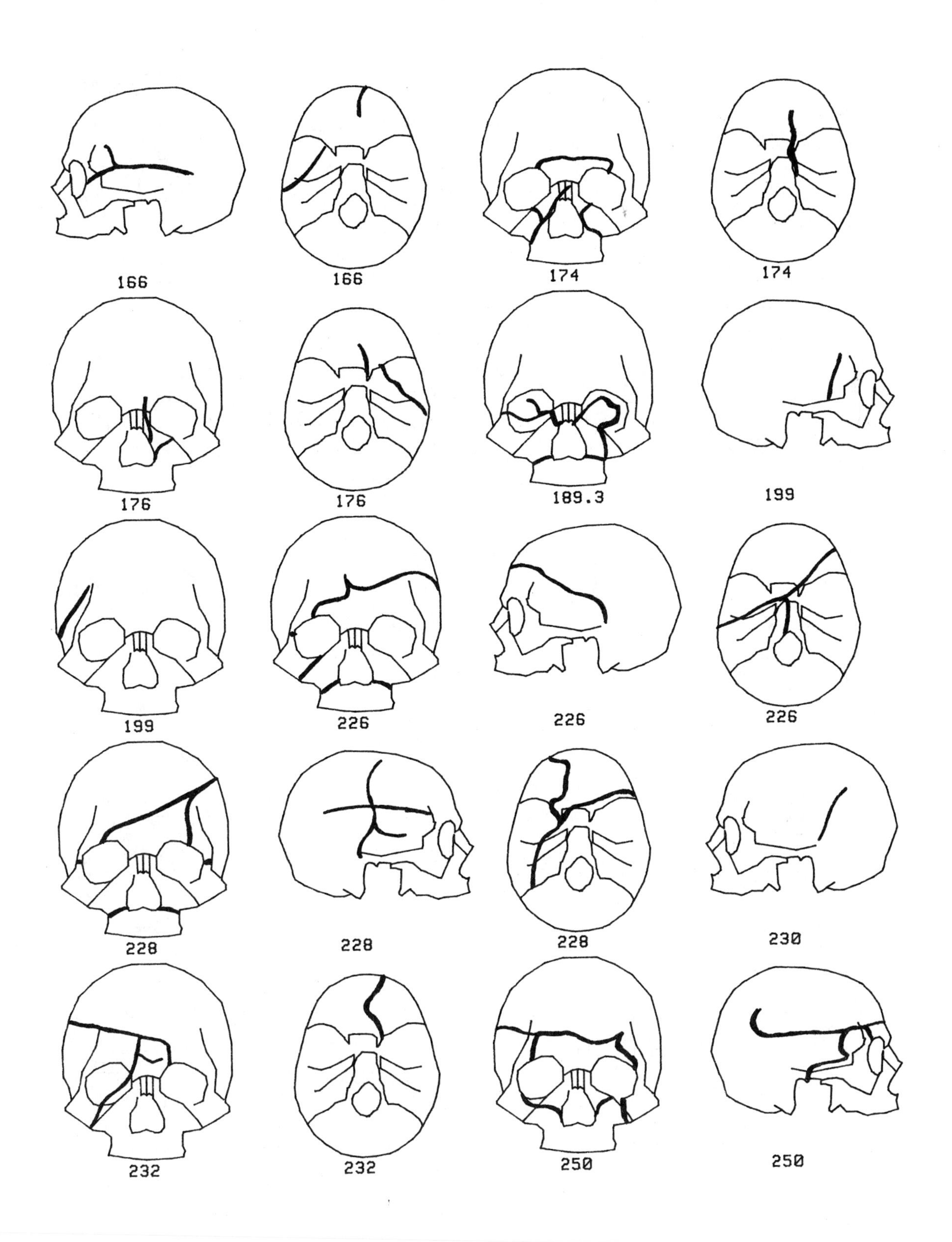

FIG 11d EXPERIMENTAL FRACTURES

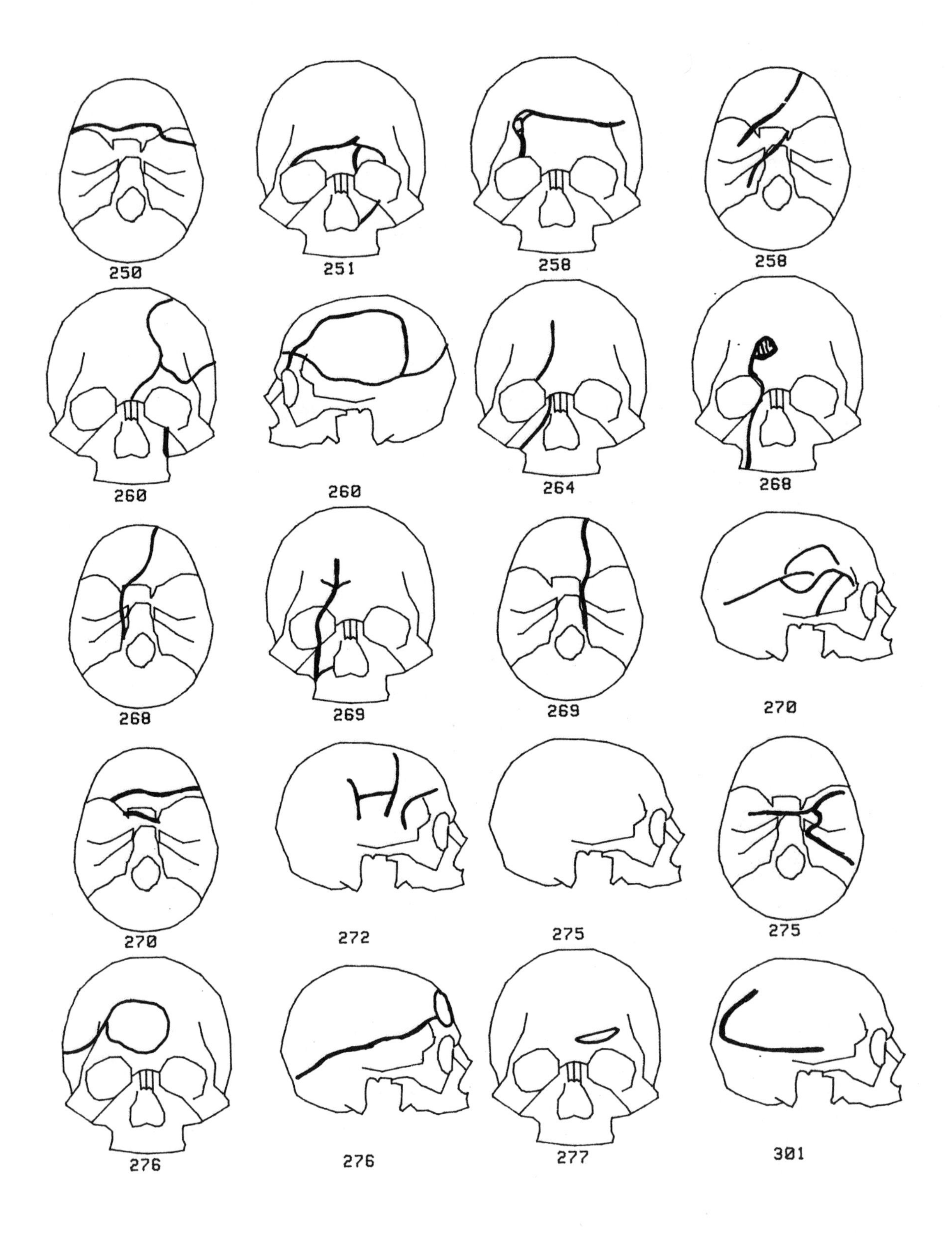

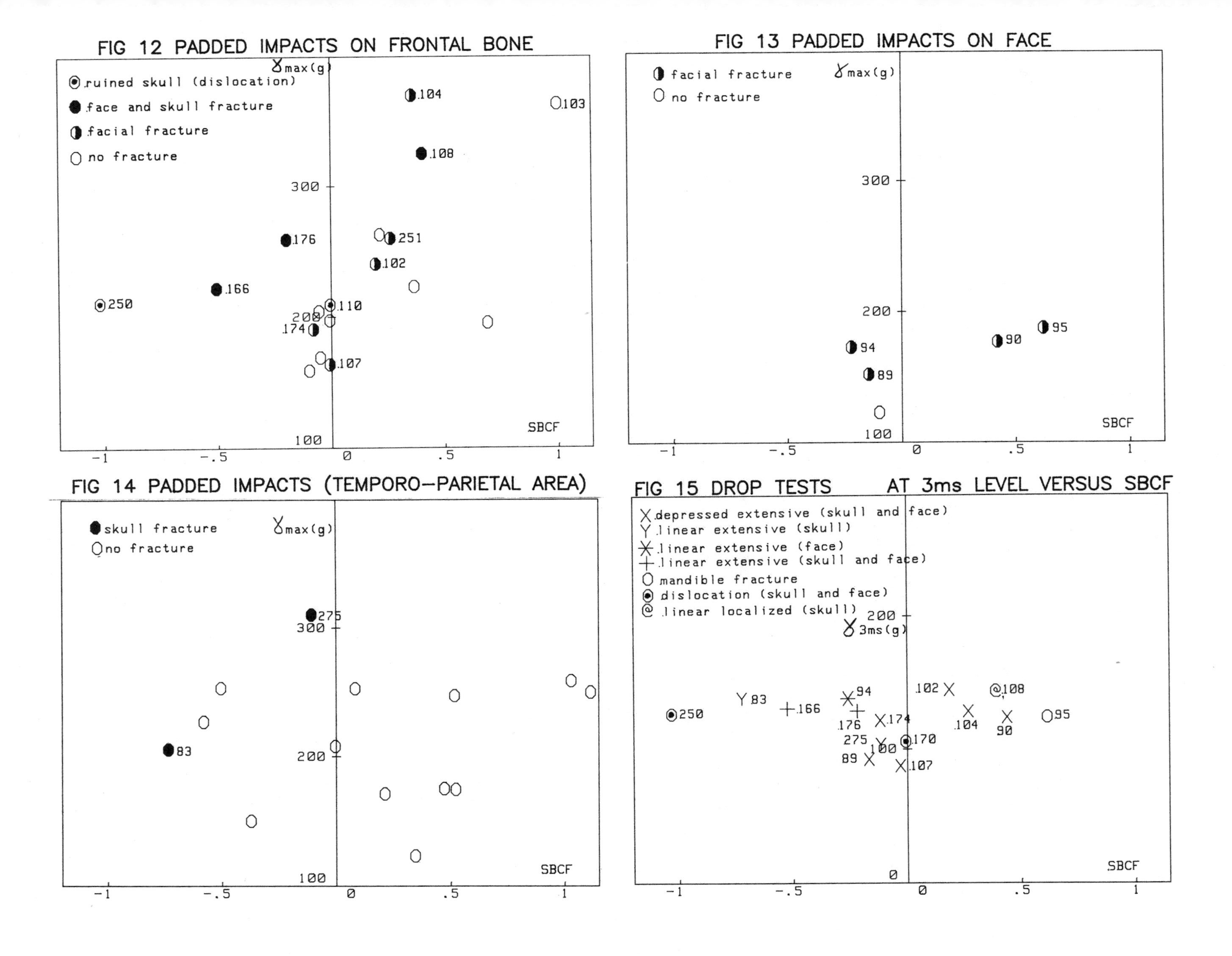

FIG 12 PADDED IMPACTS ON FRONTAL BONE

FIG 13 PADDED IMPACTS ON FACE

FIG 14 PADDED IMPACTS (TEMPORO–PARIETAL AREA)

FIG 15 DROP TESTS AT 3ms LEVEL VERSUS SBCF

FIG 16 SKULL FRACTURES IN TRAFFIC ACCIDENTS

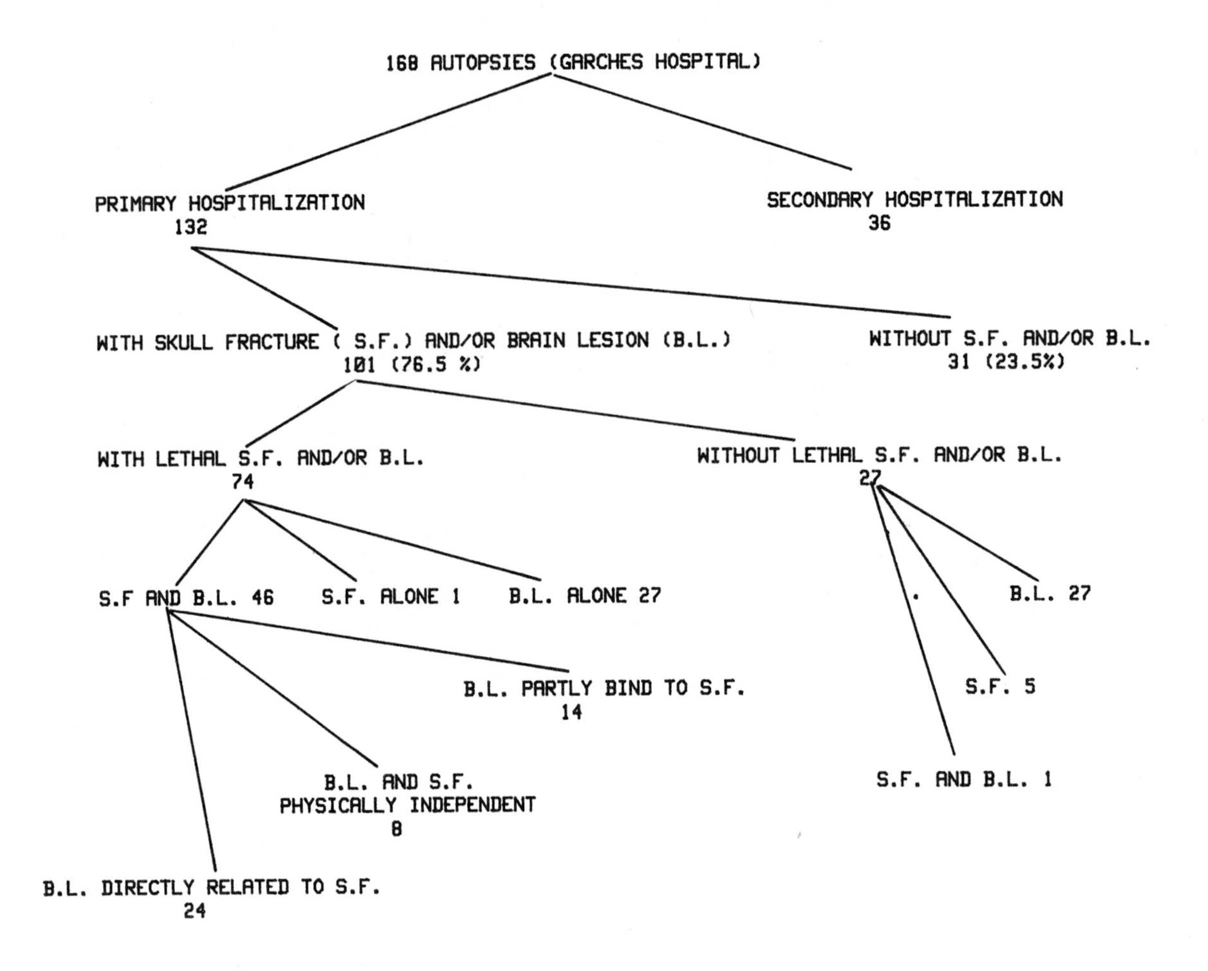

BRAIN LESIONS INCLUDE SUBDURAL HAEMATOMAS
TRAFFIC ACCIDENTS ONLY (CAR OCCUPANTS, PEDESTRIANS AND TWO WHEELS USERS)

ANNEX 1 : EXPRESSION OF VARIABLE 2 (VAR2)

$$VAR2 = \sum_{j=1}^{6} \frac{P_j}{\sqrt{\lambda 2}} \frac{r_{ij} - r_j}{s_j \sqrt{6}}$$

$$VAR2 = \frac{0.791}{1.307}\frac{LTH-4.32}{0.80\sqrt{6}} + \frac{-0.236}{1.307}\frac{RLD-143}{6.48\sqrt{6}} + \frac{0.713}{1.307}\frac{MIM-0.56}{0.13\sqrt{6}}$$

$$+ \frac{-0.078}{1.307}\frac{HDW-3.79}{0.49\sqrt{6}} + \frac{-0.034}{1.307}\frac{APD-180}{7.18\sqrt{6}} + \frac{-0.331}{1.307}\frac{SDP-86}{8.1\sqrt{6}}$$

REFERENCES

1. J. Sacreste, F. Brun Cassan, A. Fayon, C. Tarrière, C. Got, A.Patel, "Proposal for a Thorax Tolerance Level in side impact based on 62 tests performed with cadavers having known bone conditions". S.A.E. Ref. 821 157 - Proceedings of the 26th Stapp Car Crash Conference.

2. C. Got, A.Patel, A. Fayon, C. Tarrière, G. Walfisch, "Results of experimental Head Impacts on Cadavers". S.A.E. Ref. 780 887 - Proceedings of the 22nd Stapp Car Crash Conference.

3. A. Fayon, C.Tarrière, G. Walfisch, C. Got, A. Patel, "Thorax of 3-Point belt wearers during a Crash". S.A.E. Ref. 751 148 - Proceedings of the 19th Stapp Car Crash Conference.

4. L. Lebart, J.P. Fenelon, "Statistique et Informatique appliquée" Dunod 1975 - Paris.

5. J.P. Benzecri and al., "L'analyse des Données" Dunod - Paris.

6. J.M. Elhaney, R.L. Stalnaker, V.L. Roberts, "Biomecanical aspects of Head Injury IN : Human Impact Response measurement and Simulation". Proceedings of the Symposium on Human Impact Response - GMRL, Warren, 1972 - Edited by WF. King and H.J. Mertz - Plenum Press, 1973.

7. J. Hume Adams, T.A. Gennarelly, D.I. Graham, "Brain damage in non missile Head Injury in : Recent advances in Neuropathology, Vol. II" - Edited by W. Thomas Smith, J.B. Cavanagh - Churchill Livingstone - 1982.

8. D. Lestrelin, C. Tarrière, G. Walfisch, A. Fayon, C. Got, A. Patel, J. Hureau, "Proper use of HIC under different typical collision environments" given at the Ninth Inal Technical Conference on E.S.V. - November 82, Kyoto - Japan.

9. S.W. Greenberg, D. Gonzalez, E.J. Gurdjian, L.M. Thomas, "Changes on physical properties of bone between the in vivo, freshly dead and embalmed conditions". S.A.E. Ref. 680 783 - Proceedings of 12th Stapp Car Crash Conference.

10. V.R. Hodgson and L.M. Patrick, "Dynamic response of the human cadaver head compared to a simple mathematical model". S.A.E. Ref. 680 784 - Proceedings of 12th Stapp Car Crash Conference.

11. V.R. Hodgson, J. Brinn, L.M. Thomas, S.W. Greenberg, "Fracture behavior of the skull fontal bone against cylindrical surfaces". S.A.E. Ref. 700 909 - Proceedings of 14th Stapp Car Crash Conference.

12. C. Tarrière, Y.C. Leung, A. Fayon, C. Got, A. Patel and P. Banzet. "Field facial Injuries and study of their simulation with dummy". S.A.E. Ref. 811 013 - Proceedings of the 25th Stapp Car Crash Conference.

861896

Facial Impact Tolerance and Response

Gerald W. Nyquist, John M. Cavanaugh, Sarah J. Goldberg and Albert I. King
Wayne State University

ABSTRACT

Facial impact experiments were conducted on eleven unembalmed human cadavers. A 32 kg or 64 kg impactor with a 25 mm diameter, rigid, cylindrical contact surface was oriented in the left-right direction relative to the face and contacted the nose at the elevation of the infraorbital margins. The impactor was propelled toward the face along an anterior-to-posterior path, with contact velocities ranging from 10 to 26 km/h. Accelerometers mounted on the impactor and the occiput provided data for analyzing the dynamics of the impacts. While the threshold for nasal bone fractures was not determined, it appears that a peak force of about 3 kN (filtered 180 Hz) is a representative threshold for more severe fracture patterns. A preliminary dynamic force vs penetration response specification for the above mode of loading is offered. Analyses of the acceleration data suggest that current typical dummy heads (vinyl flesh over aluminum skull) will not provide proper acceleration data in facial impact environments; the face is too stiff.

WAYNE STATE UNIVERSITY, under contract with the National Highway Traffic Safety Administration, is engaged in a study of the mechanical response and injury tolerance characteristics of three body regions of human cadaver subjects under various dynamic loading conditions.* The regions of study are the face, abdomen and lower leg. This Paper provides an interim summary of the research associated with the facial impacts; a companion Paper by Cavanaugh et al. (1)** covers the status of the abdominal impact study. Nyquist et al. (2) have previously reported on the lower leg study. Since this Paper deals with a topic where little information is available in the literature, discussions judged to be unnecessary for an understanding of the data are brief, thereby enabling detailed test data to be presented within the available space.

Facial impacts to vehicle interior components has long been (and continues to be) a relatively common occurrence. A familiar scenario is facial bone fractures as a consequence of impact to the steering wheel by an unrestrained or lap-shoulder belted driver. While significant discomfort and temporary disability is experienced, the norm is that full recovery is enjoyed when the impact severity is below levels resulting in brain injury. While admittedly not a pleasant thought, the facial structure can serve as a shock absorber for blows to the front of the head. Crushing of the facial bones attenuates the accelerations experienced by the brain. Consequently, the fracture tolerance and dynamic mechanical response characteristics of the face are topics of interest.

Surrogates of the human face, whether mathematical or mechanical, are of limited usefulness without associated tolerance and response data. A realistic contact force during a facial impact will occur only if the dynamic force-penetration behavior of the face at the loading site is humanlike. Furthermore, given that realistic forces are generated, the injury consequences of a blow can be assessed only if facial fracture tolerance data are available. All human surrogates commonly used in impact testing, with the exception of the cadaver, have had profound shortcomings in this regard. The norm is that the face will not fracture (and is therefore too stiff) and that recognized fracture tolerance loads are unavailable for a particular type of impact. An obvious and important example is facial impact to any modern anthropomorphic test dummy. The face is an aluminum casting covered by a vinyl flesh and skin simulation.

----

* This research complies with the provisions of the Uniform Anatomical Gift Act and guidelines established by the National Academy of Sciences and others.

** Numbers in parentheses designate references at the end of the Paper.

Severe blows to the face will result in unrealistically high head accelerations. Yet, facial impacts are common in vehicle crash tests using dummy occupants.

There have been limited studies of facial bone fracture tolerance and response; more of the former than the latter. For example, Hodgson et al. (3) delivered hammer-like blows to the supraorbital ridge, the zygoma and the zygomatic arch of embalmed cadavers of the elderly. Similarly, Nahum et al. (4) used a drop weight striker device to study the fracture tolerance of the frontal bone, the zygoma and mandible of embalmed and unembalmed subjects. There have also been some limited efforts toward the development of better facial simulations for anthropomorphic dummies. McLeod and Gadd (5) developed a frangible skull and face with a simulated flesh covering. Warner and Niven (6) produced a prototype load-sensing faceform. Finally, Tarriere et al. (7) have reported on a modification of a dummy head that incorporates a crushable, light alloy honeycomb insert to represent the facial skeleton. Each of these facial concepts is based on minimal biomechanical data, and to date none are commercially available or in widespread use. More basic biomechanical data are needed; hence this study.

TESTING PROTOCOL

Thus far in the study, facial impacts have been performed in a manner analogous to (but not identical with) loading by a steering wheel rim. The testing protocol resembled that previously described for tibia testing (2) and utilized in the companion paper (1) covering abdominal impacts, in that the so-called Wayne State Translational Impactor was utilized to deliver the impacts. This facility featured a 32* or 64 kg (including instrumentation) rigid impactor that slid horizontally on nylon bearings and was propelled by a 152 mm stroke pneumatic cylinder. Speed was measured during a brief coasting phase immediately prior to impact. A 25 mm diameter aluminum bar on the end of the impactor, orientated with its longitudinal axis parallel to a left-right axis on the head, contacted the face. The bar was longer than the width of the face. The test subject was positioned in an upright seated posture in front of the impactor with the head held by weak tape tethers such that the Frankfort plane was horizontal. Furthermore, the seat height was adjusted to align the palpated infraorbital margins (i.e. bottom of eye sockets) to the same elevation as the axis of the aluminum bar (Fig.1). Under these conditions the head essentially translated rearward (with minimal rotation) during the portion of the event of interest.

The instrumentation utilized in these tests included the following:

- Uniaxial accelerometer on impactor, aligned parallel to path of travel.
- Uniaxial accelerometer on occiput (i.e. back of head) of cadaver, aligned parallel to the anteroposterior direction.
- Magnetic pick-up for speed determination.
- High-speed 16mm camera. (100 Hz timing signal exposed along edge of film.)
- Electronic flash and photovoltaic cell for synchronizing cinematographically and electronically recorded data.

The accelerometers were Endevco ® Model 7264-200. The analog data were prefiltered at 1000 Hz, digitized at a sampling rate of 2000 Hz and further analyzed using the University's computer and plotting facilities. Filters fit the centers of SAE channel class corridors. Phototargets were attached midsagitally at the occipital accelerometer and on the impactor.

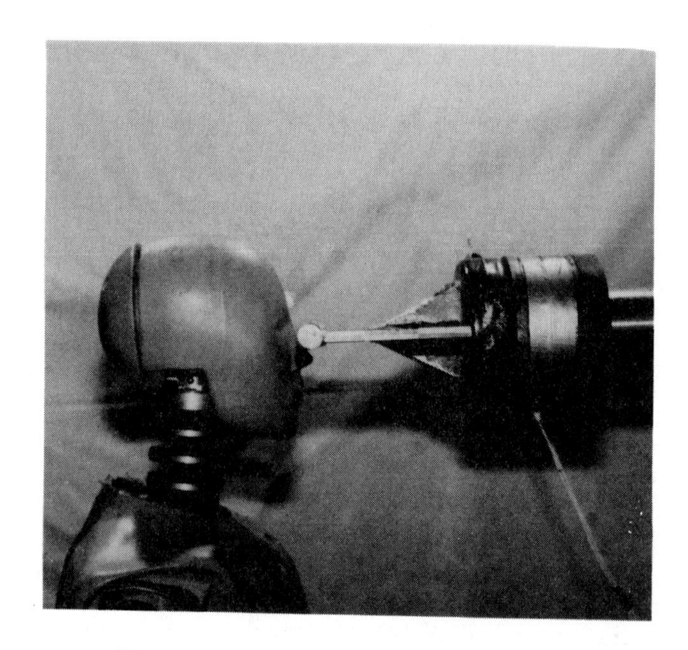

Figure 1. Impactor Aligned with Face

TEST SUBJECTS

Data are presented for one test on each of eleven cadavers. The subjects were obtained through the University medical college shortly after death and were unembalmed. Table 1 provides a summary of their characteristics. There were four females, with ages ranging from 43 to 57 years and body mass ranging from 53 to 75 kg. The seven males ranged from 43 to 66 years and 45 to 92 kg. The subjects were clothed in disposable coveralls and had a thin knit cotton cloth covering the face during the test. The injuries sustained as a result of the impact were diagnosed during dissection at autopsy by a suitably experienced pathologist.

RESULTS

The results of the tests are summarized

----
* Except for the first test, that was 31 kg.

Table 1

Cadaver Characteristics

Test No.	Cadaver No.	Sex	Age (Years)	Stature (m)	Body Mass (kg)	Cause of Death
13	404	M	56	1.76	79	Cardiorespiratory arrest, cerebellar and cerebral infarct.
15	458	M	56	1.82	68	Small cell carcinoma of the lung.
20	473	F	43	1.59	53	Asphyxia due to carbon monoxide poisoning.
25	525	M	57	1.87	45	Ischemic anoxic brain injury, caustic material ingestion, diabetes mellitus, pneumonia.
29	578	F	57	1.63	75	Cardiopulmonary arrest.
34	590	F	51	1.63	68	Congestive heart failure, arterioscl-erotic heart disease, renal insufficiency.
42	712	F	51	1.59	55	Carbon monoxide poisoning.
44	721	M	66	1.70	70	Cardiopulmonary arrest, arterioscl-erotic heart disease, diabetes mellitus.
46	731	M	58	1.76	92	Cardiac arrest.
48	739	M	43	1.72	61	Cardiac arrest, end stage heart failure, cardiomyopathy.
58	751	M	64	1.84	92	Cardiac arrest.

in Table 2. Figures A1 through A11 of the Appendix provide filtered and unfiltered time histories of the impactor and occipital accelerations. Furthermore, Figures A12 and A13 provide plots of impactor penetration into the face as a function of time, based on analyses of the high-speed films and/or double integration of the accelerometer signals. Penetration is defined as the relative displacement of the aluminum cylinder into the face following initial contact (with the nose).

DISCUSSION OF RESULTS

An initial observation from Table 2 is that nasal bone fractures were sustained in every test. This is not surprising, since the impactor alignment utilized in this study results in initial contact at the nose. Referring to Table 2, for fracture limited to the nasal bones, the minimum speed at contact was 10.0 km/h (Test Nos. 44, 46 and 48); the minimum kinetic energy of the impactor was 241 J (Test No. 13). These results

Table 2

Summary of Impact Severity and Injury

Test No.	Nature of Impact			Impactor Peak Penetration into Face (mm)	Facial Injury
	Impactor Mass (kg)	Speed @ Contact (km/h)	Kinetic Energy (J)*		
13	31	14.2	241	15.0	Comminuted fractures of the nasal bones.
15 **	32	24.6	747	30.7	Bilateral vertical fractures of the nasal bones. Two vertical fractures of the frontal process of the maxilla at the medial wall of the right orbital cavity, and three such fractures on the left Vertical fractures of the orbital process of the zygomatic bone at the lateral wall of the left orbital cavity.
20	32	17.1	361	23.6	Two oblique fractures of the nasal bones in the shape of an inverted V. Small horizontal fracture across nasal bones below V.
25	32	16.1	320	6.4	Comminuted fractures of the nasal bones bilaterally.
29 **	32	22.8	642	27.9	Comminuted fractures of the nasal bones and frontal processes of the maxillary bones bilaterally.
34 **	32	25.7	815	29.2	Fractures of nasal bones, frontal process and orbital surface of maxilla, zygomatic bone, greater wing of sphenoid and orbital plate of frontal bone.
42	64	12.4	380	28.2	3/4 inch laceration of bridge of nose, fracture of the nasal bones bilaterally.
44	64	10.0	247	8.1	Abrasions of the nose, comminuted fracture of the nasal bones bilaterally.
46	64	10.0	247	15.2	Abrasion of the nose with bilateral comminuted fractures of the nasal bone.
48	64	10.0	247	29.5	Abrasion at bridge of nose with underlying horizontal linear fracture of nasal bone.
58 **	32	14.0	242	18.8	Laceration and abrasion at bridge of nose with transverse fractures of the frontal bone about the nasal notch and multiple fractures of the nasal bones.

* Kinetic energy of the impactor at instant of contact to face.
** Fracture pattern more extensive than nasal bones-only.

are consistent with the common knowledge that a "broken nose" is relatively easily sustained. On the other hand, with the exception of Test No. 58, the impactor energy associated with tests resulting in more extensive fractures was considerably larger than the above value, ranging from 646 to 815 J. Interestingly, the energy associated with Test No. 58 was 242 J, which is essentially equal to that of Test No. 13 (which had the lowest kinetic energy experienced in the series of tests). No explanation is available for the unexpected results of Test No. 58.

Figure 2 provides a plot of impactor penetration versus impactor speed. The darkened data points represent tests where fractures were not limited to the nasal bones. There is a clustering of darkened points at the upper right (representing high speed, high penetration impacts). Figure 3 provides a similar plot, except that the abscissa is impactor kinetic energy instead of speed. Since impactor mass is accounted for in the kinetic energy computation, there are no longer separate sets of points for the two impactor masses. The separation of the points for the two levels of fracture severity is even more obvious here. The large variability of peak penetrations in Figures 2 and 3 and the fact that nasal bone-only fractures and the more extensive fractures occurred at similar penetrations is suggestive that peak penetration is not a meaningful measure of impact severity. This is postulated to be a result of variability in facial anthropometry among test subjects. While peak penetration for an individual subject undoubtedly would increase with increasing impactor speed and kinetic energy, this trend probably becomes obscure in a data set comprised of single impacts to multiple subjects. A subject with relatively less prominent nasal bones and/or that is well endowed with nasal cartilage and soft tissue should sustain larger peak penetrations at a given level of applied force (or given severity of fracture) than could be tolerated by a subject with more prominent nasal bones and/or relatively less cartilage and soft tissue. Unfortunately, detailed facial anthropometry was not documented for the test subjects in this study. (It is cautioned that the above remarks should not be interpreted to imply that peak penetration may not be an appropriate measure of impact severity for a standardized test tool such as an anthropomorphic dummy head.) One is led to conclude that impactor kinetic energy should correlate with fracture severity better than does peak penetration. The energy increases with increasing impactor mass and impactor speed. Clearly, increases in the magnitudes of either of these individual parameters logically equates to the application of higher loads on the facial bones. Energy may not correlate with injury for limiting cases of very high speed and small mass or very large mass and small speed; however, such impacts fall outside of the realm of interest in this study.

A summary of the impactor peak deceleration magnitudes and related information is provided in Table 3. Peak decelerations were scaled from the

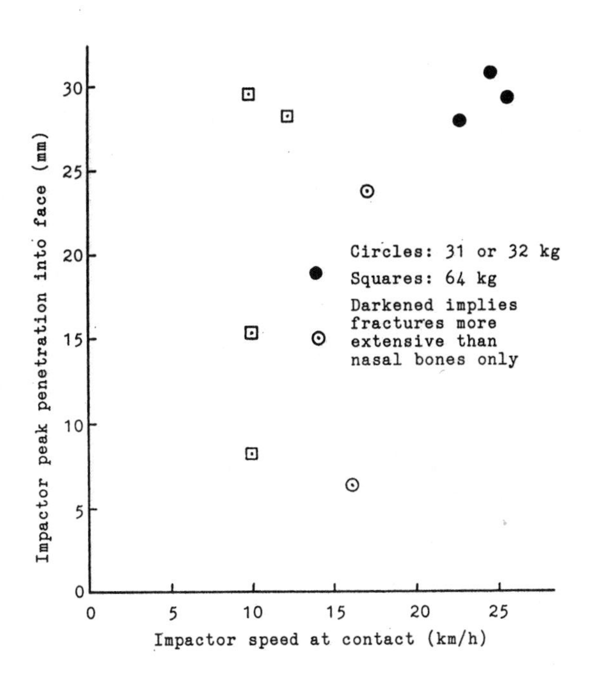

Figure 2. Impactor Penetration Versus Speed

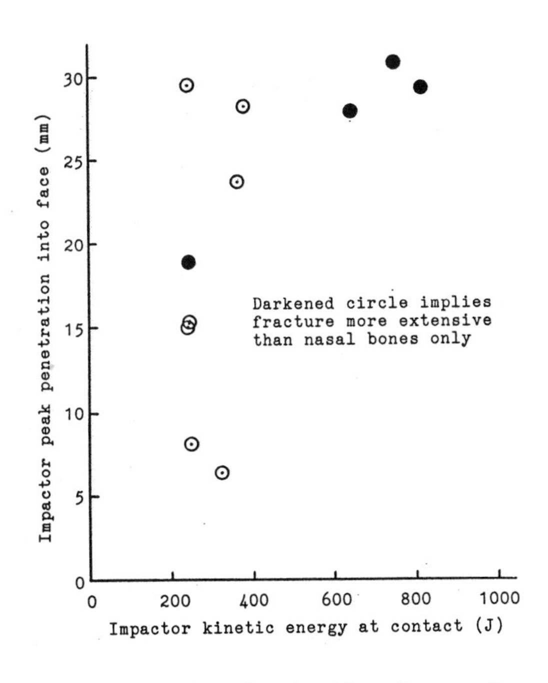

Figure 3. Impactor Penetration Versus Energy

profiles included in the Appendix and the deceleration rise times (zero to peak value) were scaled from the haversine-like 180 Hz filtered traces. Noise problems associated with the 1000 Hz data limit its usefulness. (This is further discussed later in the Paper.) Consequently, the 180 Hz filtered data were relied upon for characterizing the overall nature of the impacts. The peak applied forces provided in Table 3 were computed using the known impactor mass and peak deceleration together with Newton's law. Larger peak forces clearly would result if the 1000 Hz data were used; however the physical meaning of the numbers would be questionable. The peak value based on filtered data are judged to be representative of results that would be achieved had noise-free traces been filtered. While the ensuing discussion of the biomechanics of the fracture process relies on details that are evident only as a result of analyzing the 1000 Hz occipital acceleration profiles, with the exception of Test No. 13 the more heavily filtered signals are nevertheless believed to provide meaningful response data for use in establishing dynamic force-deflection specifications for a surrogate face. (Duplication of typical high-frequency responses in a surrogate face is probably impractical.)

The rise time and peak applied force data of Table 3 may be thought of as the half-period and amplitude of haversine-like force versus time profiles, respectively, for 180 Hz filtered signals. The peak force for Test No. 13 seems pathologically large compared to that of the other tests, and should not be relied upon. This test was performed early in the study, during a period when impactor erratic vibrational problems were experienced. The dramatic effect of filtering for Test No. 13, compared to all other tests, is evident in Table 3. Exclusive of Test No. 13, the mean value of peak force (Table 3) for tests with fractures limited to the nasal bones was 2.77 kN and the mean for tests with more extensive fractures was 3.07 kN. While it is logical that the latter mean force be higher than the former, it is important to keep dose-response principles in mind. Since only four of the ten subjects (Test 13 omitted) sustained more extensive fractures than the remaining six subjects with only nasal bone fractures, and since the peak force for all ten tests ranged from 2.01 to 3.89 kN, the above mean value of 3.07 kN should be a fairly representative measure of the threshold for the more extensive fractures. To the contrary, the above mean value of 2.77 kN for nasal bone-only fractures is an upper bound, not a threshold, since none of the tests were of sufficiently low severity to avoid this injury. (There is no reason to believe that this upper bound is the so-called least upper bound.)

The cadaver head occipital acceleration data have proved to be useful in understanding some of the details of the facial responses during the impacts. It appears that the 1000 Hz data are necessary for studying this detail. The signal-to-noise concerns associated with the 1000 Hz filtered impactor deceleration data are, relatively speaking, absent from the occipital acceleration signals. This is because the head is initially at rest, totally uncoupled from the impactor. Once the impactor contacts the face, it is conceivable that some of the noise is introduced into the occipital accelerometer signal through the head during the course of the impact.

Table 3

Summary of Impactor Deceleration and Applied Force Characteristics

Test No.	Impactor Peak Decel. (G) Filt. 1000 Hz	Filt. 180 Hz	Rise Time (ms)*	Peak Applied Force (kN)**
13	33.8	13.7	8	4.15
15***	12.8	10.7	12	3.35
20	16.5	10.0	11	3.14
25	14.0	11.2	16	3.51
29***	10.4	9.6	6	3.01
34***	13.9	12.4	9	3.89
42	7.2	4.9	5	3.07
44	4.5	3.7	9	2.32
46	5.0	3.5	5	2.20
48	6.5	3.8	6	2.39
58***	10.6	6.4	7	2.01

* From zero to peak value, based on 180 Hz filtered data.

** Force =(Impactor mass) x (180 Hz impactor peak decel.)

*** Fracture pattern more extensive than nasal bones-only.

The noise, however, tends to be of high frequency; consequently one might expect it to be attenuated measurably by the time it reaches the occiput, considering the nature of the impactor engagement with the face and the inherent damping characteristics of the head.

A careful look at the 1000 Hz filtered occipital acceleration traces in the Appendix reveals a strong trend during approximately the first 4.5 ms. There is an initial acceleration peak. (Each of these initial peaks is identified by an arrow on the trace.) The data of Table 4 were scaled from the trace and provide a summary of the nature of the initial peaks as well as the subsequent maximum amplitude later in time. It is often possible to distinguish some form of perturbation in the impactor acceleration trace that appears to correlate with the initial occipital peak; however, the noise associated with the 1000 Hz impactor accelerations obscures any strong trends. The early occipital acceleration peaks are postulated to be the result of nasal bone fracture. It is reasonable to expect a momentary drop in the force level applied to the head as the nasal bones fail. This momentary load loss would cause a concomitant occipital acceleration peak, immediately followed by an again increasing trace as the impactor continued to bear down the on face.

Table 4

Summary of Occipital Acceleration Data

Test No.	Acceleration (G) Initial Peak (1000 Hz)	Subsequent Maximum (Later in Time) (1000 Hz)	(180 Hz)	Time (ms) from Accel. Onset to Initial Peak (1000 Hz)*
13	38	81	51	3.0
15	77	62	55	1.5
20	9	>29 **	20	1.0
25	68	>72 **	66	1.0
29	25	88	63	2.0
34	83	76	47	3.0
42	53	73	56	1.0
44	14	13	12	4.5
46	22	68	47	2.5
48	21	62	45	3.1
58	39	49	19	2.0

* Rounded to nearest 0.5 ms.
** Peak was clipped during recording.

The above postulated explanation for the early peaks is corroborated by the impactor penetration versus time plots of Figures A12 and A13. There are break-points early in these plots where the slope decreases. Since the slope represents rate of penetration, it stands to reason that the slope will decrease rather suddenly when the impactor has penetrated through (or compressed) the soft tissue and cartilage and begins bearing down on the nasal bones. The break-points in the penetration plots tend to occur slightly later in time than the acceleration peaks; however, since displacement is the double integral of acceleration, it takes time for observable penetration phenomena to occur. Furthermore, the resolution associated with determining penetrations from high-speed films was marginal for such details.

Table 4 indicates that the norm is for the initial occipital acceleration peak to be of lesser magnitude than the subsequent maximum; however, this was not the case for Test Nos. 15, 34 and 44. One can question what is different about these three tests compared to the others. First, it should be noted that Test Nos. 15 and 34 had the highest two impactor speeds and the largest two facial penetrations. Furthermore, the fracture patterns associated with these two 32 kg impactor tests were more extensive than in the remaining 32 kg tests. High loads appear to have been sustained during nasal bone fracture, followed by large additional penetration of the impactor into the face and a relatively extensive fracture pattern with concomitant lesser acceleration peak as a result of the larger distance involved. The quantity of data is insufficient at this point in the study to conclude that this behavior is, in fact, caused by the higher impactor speeds (i.e. that the nasal bone fracture tolerance increases dramatically with rate of loading). In fact, Test No. 44 (with 64 kg impactor) had a relatively low speed and one of the smallest peak penetrations, yet exhibited the same phenomenon. (It should be noted that the tabulated occipital accelerations, for unexplained reasons, seem small for Test No. 44 compared to those of similar tests.) Referring to the kinetic energies in Table 2 leads one to conclude that Test No. 44 also was not similar to Test Nos. 15 and 34 with regard to this parameter. Furthermore, based on the cadaver characteristics of Table 1, the subject for Test No. 44 is not an outlier compared to the other two subjects. Details of facial anthropometry, however, could be remarkably different.

An analysis of the pairs of peak impactor deceleration and head acceleration was performed that led to a conclusion regarding the need for a more realistic facial structure on anthropomorphic dummies. Table 5 provides a summary of the ratio of these peaks (occipital divided by impactor). The ratios are tabulated for data filtered at both 1000 Hz and 180 Hz. Under rigid body dynamics (ignoring noise problems and assuming the cadaver neck does not apply significant loads in this time frame) the above ratio of accelerations should be numerically equal to the ratio of the impactor mass divided by the head mass. While the actual head masses are unknown, a well known statistical approximation is that the head mass is 6.1 percent of the total body mass for adults (8). This criterion was used along with information from Tables 1 and 2 to compute the entries in the fourth column of Table 5. The remaining two columns of the Table express percent differences in this ratio, computed as stated in the footnote below the table, for the two levels of filtering of accelerations. Note that these differences are predominantly of negative sign, ranging from -79 to +21 percent for the 1000 Hz data and from -80 to +18 for the 180 Hz data, with averages of -29 and -40 respectively. The nature of the calculations is such

Table 5

Acceleration Ratios Compared to
Predictions Based on Rigid Body Mechanics

Test No.	Ratio of Peak Accelerations (Occipital/Impactor)		Ratio of Masses * Impactor/Head	Difference of Ratios (Percent) **	
	Filtered 1000 Hz	Filtered 180 Hz		Filtered 1000 Hz	Filtered 180 Hz
13	2.4	3.7	6.4	-63	-42
15	6.0	5.1	7.7	-22	-34
20	-	2.0	9.9	-	-80
25	-	5.9	11.7	-	-50
29	8.5	6.6	7.0	+21	-6
34	6.0	3.8	7.7	-22	-51
42	10.1	11.4	19.1	-47	-40
44	3.1	3.2	15.0	-79	-79
46	13.6	13.4	11.4	+19	+18
48	9.5	11.8	17.2	-45	-31
58	4.6	3.0	5.7	-19	-47
Average:				-29	-40

* Head mass estimated as 6.1 percent of body mass (8).
** 100 (Accel. ratio minus mass ratio)/(Mass ratio).

that a negative sign implies an occipital acceleration magnitude that is smaller than would be present under rigid body conditions. Fracturing of the face and damped elastic deformation of the skull are logical explanations to account for this phenomenon. To the contrary, displacements of soft tissues relative to the skull (both inside and outside of the skull) during impact would logically lead to a lower effective head mass. This would cause a larger occipital acceleration than would be present for a rigid body, and hence a positive error. The preponderance of negative signs in this analysis suggests that the combined effect of the two phenomena is dominated by the face fracturing and skull damped elastic deformation. Current typical anthropomorphic dummy heads, composed of a vinyl flesh over a cast aluminum skull, in this impact environment would undoubtedly mimic rigid body behavior much more closely than has been observed for the cadavers in this study. That is, the occipital acceleration magnitudes would be too large. Assuming that facial crush is a more important factor than skull deformation, it follows that acceleration magnitudes measured at the center of gravity of the dummy head would also be too large. The need for some type of deformable facial structure on dummies is thus demonstrated.

While the data already discussed provide a basis for defining performance requirements for a surrogate face, dynamic force vs penetration plots would be helpful. An effort has been made to crossplot force vs time and penetration vs time data to provide the desired results. (The force data used in this analysis were filtered at 100 Hz.) The analysis has been complicated by the impactor deceleration noise problem, time synchronization problems between data sets and the limited precision of the penetration data from the high-speed films. (Double integration of accelerations proved unreliable). The results of these analyses are depicted in Figure 4. While there is considerable variability among the plots, there is a clear trend toward the typical concave-up force-deflection behavior commonly observed for biological materials. Figure 4 includes three candidate curves that tend to be representative of the average response. The curves are labeled with their equations:

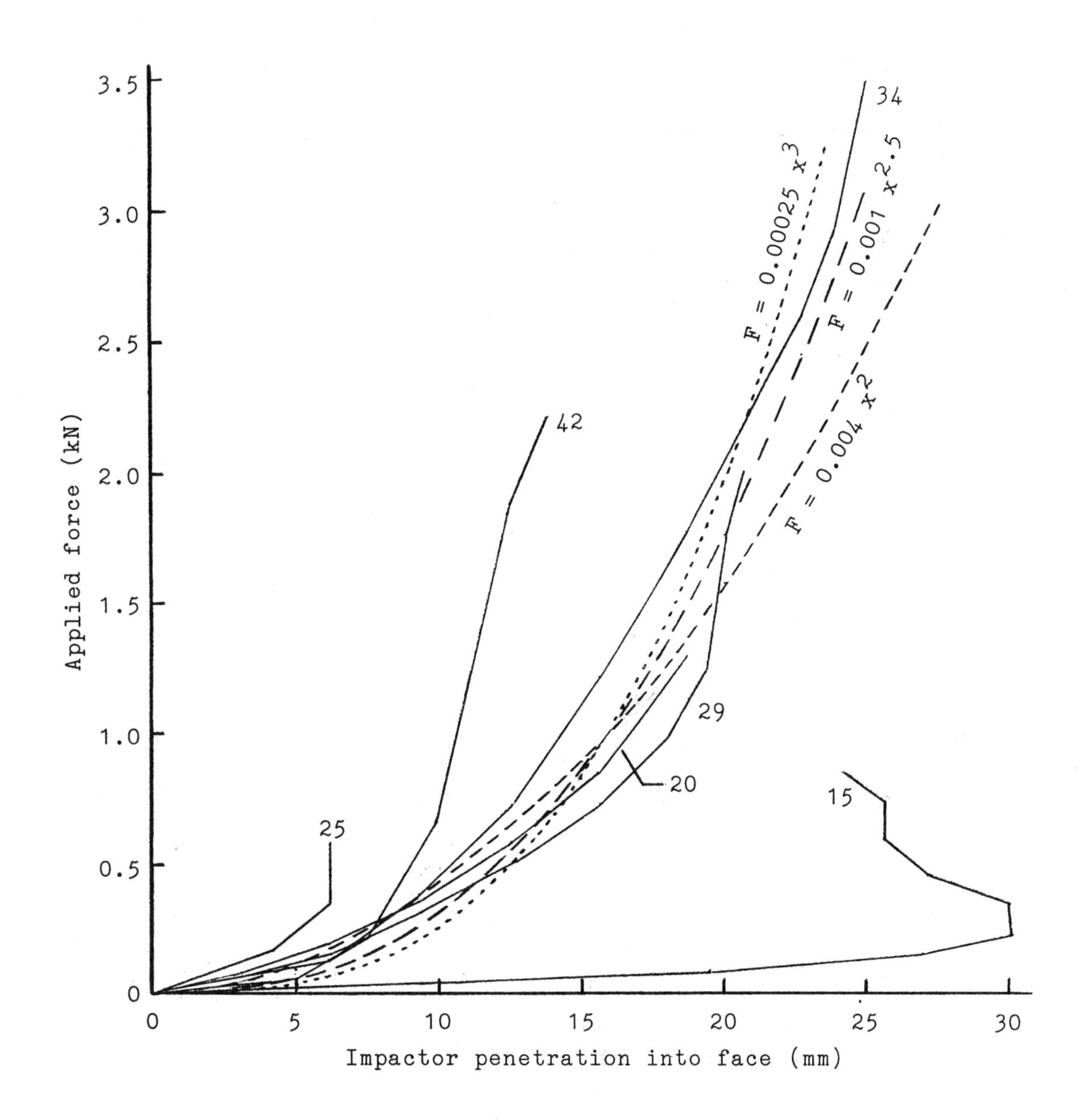

Figure 4. Force vs Penetration for Tests 15, 20, 25, 29, 34 and 42

$$F = 0.00025\ x^3 \qquad (1)$$
$$F = 0.001\ x^{2.5} \qquad (2)$$
$$F = 0.004\ x^2 \qquad (3)$$

where F is the applied force expressed in kN and x is the facial penetration expressed in mm. The cubic fit of Eq.(1) tends to provide too small a force at small penetrations; the quadratic of Eq.(3) tends to provide too large a penetration at large forces. Interestingly, the ubiquitous exponent of 2.5 in Eq.(2) seems to provide a compromise. Considering the paucity of data and concerns with the precision of the crossplots, Equation (2) should be considered the best descriptor of facial response only in a preliminary sense. Further information is needed.

CONCLUSIONS

Based on the data and analyses provided above, the following conclusions are offered:

1. The nasal bones appear to fracture relatively easily, but dramatically larger forces can be sustained with only limited fracture patterns involving one or more of the maxilla, zygoma, sphenoid and frontal bones in proximity of the orbits.
2. Impactor kinetic energy tends to be predictive of fracture pattern severity among subjects; impactor peak penetration into the face is not. A force of about 3 kN appears to be a representative threshold for fracture patterns more extensive than nasal bones-only.

3. The study provides evidence that there is a need for anthropomorphic dummies to have some form of deformable face if head accelerations are to be realistic during facial impacts.
4. The dynamic force-penetration response of the face in this impact environment is characterized by a concave-up shape. A preliminary best estimate of a single dynamic response characterizing the data is provided by Eq. (2).
5. Data collection needs for facial impact experiments of this nature include relatively noise-free acceleration signals at SAE Channel Class 1000 and high-speed films exposed at at least 1000 frames per second.
6. Additional data collected using essentially these same impact conditions are needed to fully understand the facial response and fracture tolerance for this mode of loading.
7. Details of the facial anthropometry should be documented in all future work, in an effort to understand variability in the data. Cadaver head mass should be measured at autopsy.

ACKNOWLEDGMENT

The authors wish to express their gratitude to F. DuPont, G. Locke and J. Ryan for their efforts in conducting the test program and to N. Blevins for the preparation of the manuscript. This research was supported by the National Highway Traffic Safety Administration (Contract Number DTNH-22-83-C-27019).

REFERENCES

1. J.M. Cavanaugh, G.W. Nyquist, S.J. Goldberg and A.I. King, "Lower Abdominal Impact Tolerance and Response," 30th Stapp Car Crash Conference Proceedings, Society of Automotive Engineers, 1986.

2. G.W. Nyquist, R. Cheng, A.A.R. El-Bohy and A.I. King, "Tibia Bending: Strength and Response," SAE Paper No. 851728, 29th Stapp Car Crash Conference Proceedings, Society of Automotive Engineers, 1985.

3. V.R. Hodgson, G.S. Nakamura and R.K. Talwalker, "Response of the Facial Structure to Impact," 8th Stapp Car Crash and Field Demonstration Conference Proceedings, Wayne State University Press, Detroit, 1966.

4. A.M. Nahum, J.D. Gatts, C.W. Gadd and J. Danforth, "Impact Tolerance of the Skull and Face," SAE Paper No. 680785, 12th Stapp Car Crash Conference Proceedings, Society of Automotive Engineers, 1968.

5. D.G. McLeod and C.W. Gadd, "An Anatomical Skull for Impact Testing," contained in Human Impact Response, ed. by W.F. King and H.J. Mertz, Plenum Press, New York, 1973.

6. C.Y. Warner and J. Niven, "A Prototype Load-Sensing Dummy Faceform Test Device for Facial Injury Hazard Assessment," Proceedings of 23rd Conference of American Association for Automotive Medicine, Morton Grove, Illinois, 1979.

7. C. Tarriere, Y.C. Leung, A. Fayon, C. Got, A. Patel and P. Banzet, "Field Facial Injuries and Study of Their Simulation with Dummy," SAE Paper No. 811013, 25th Stapp Car Crash Conference Proceedings, Society of Automotive Engineers, 1981.

8. H.M. Reynolds, C.E. Clauser, J. McConville, R. Chandler and J.W. Young, "Mass Distribution Properties of the Male Cadaver," SAE Paper No. 750424, Automotive Engineering Congress and Exposition, Society of Automotive Engineers, 1975.

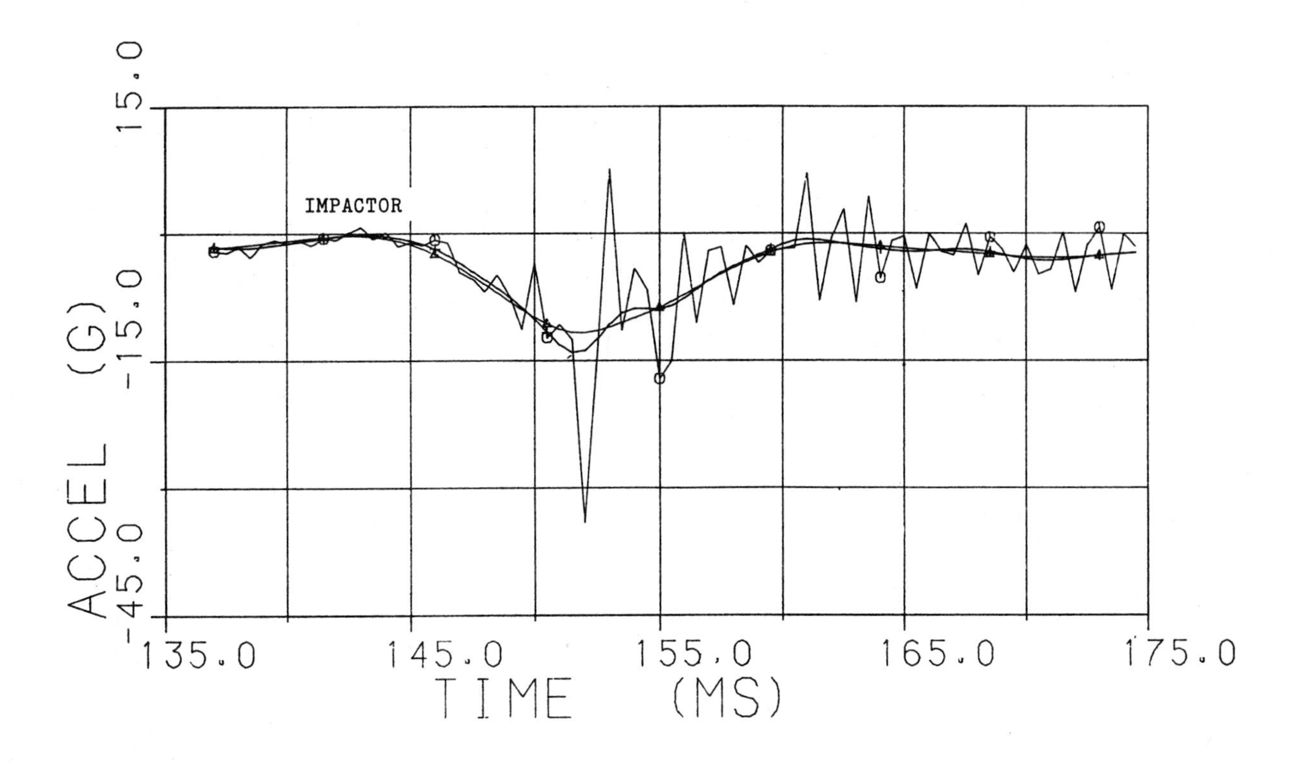

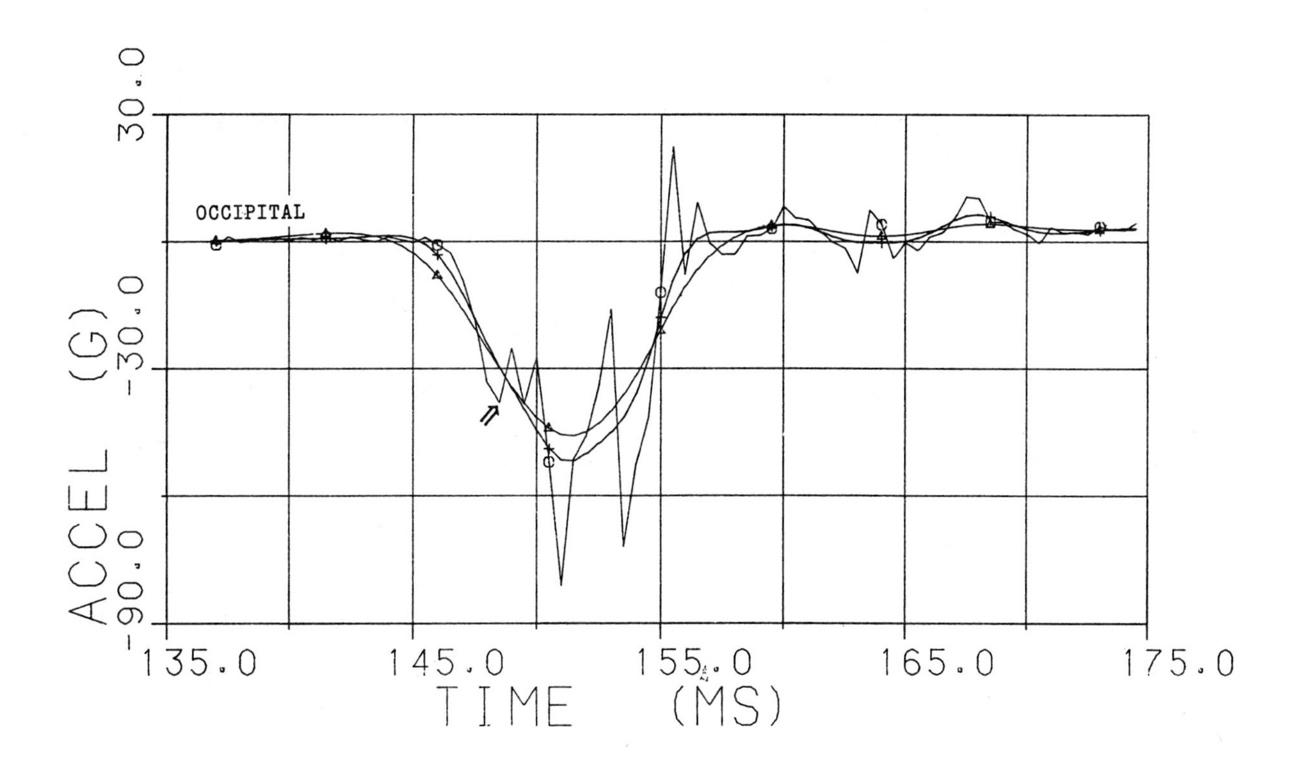

Figure A1. Acceleration Traces For Test No. 13, Filtered 1000, 180 and 100 Hz.

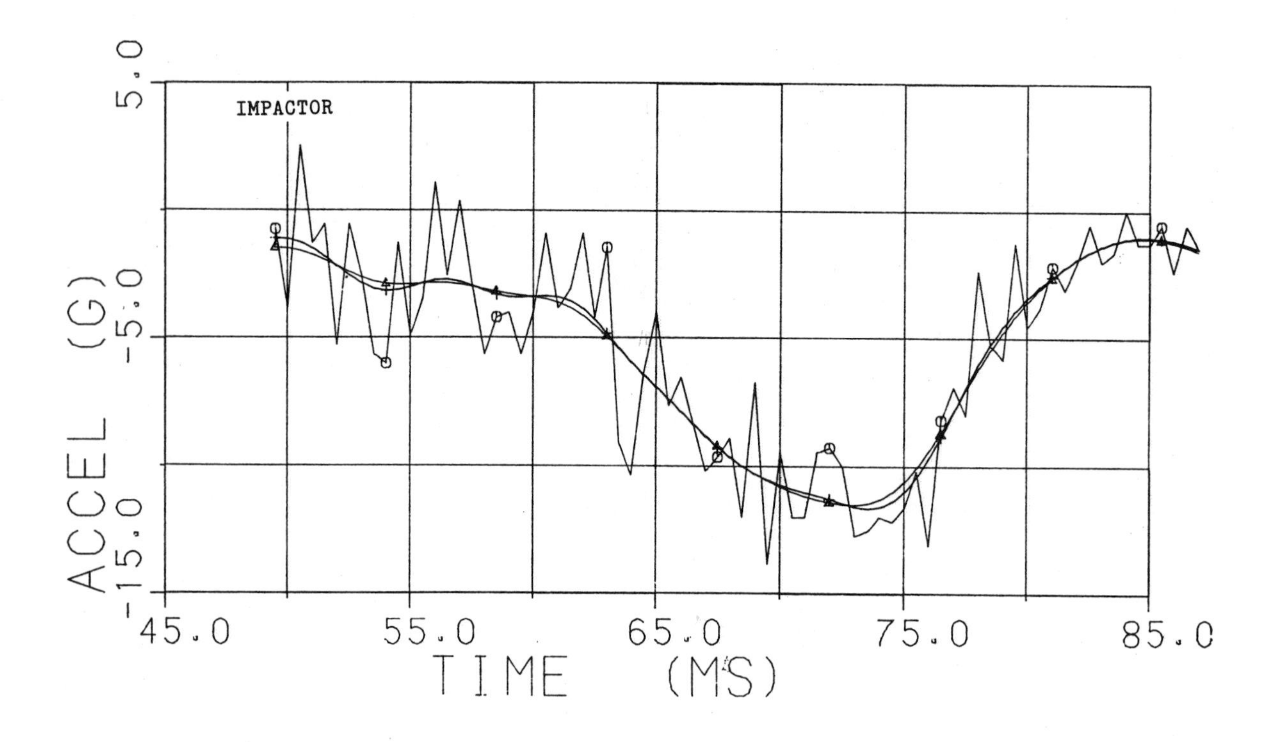

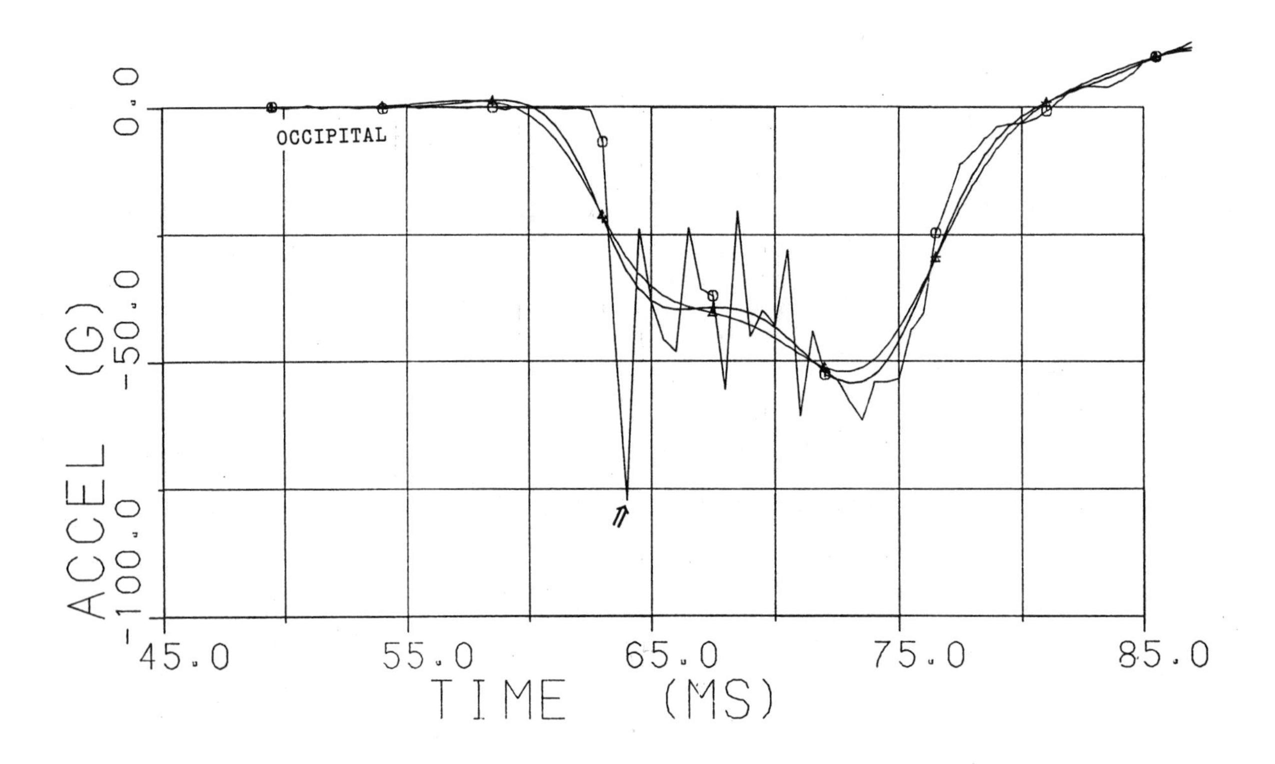

Figure A2. Acceleration Traces For Test No. 15, Filtered 1000, 180 and 100 Hz.

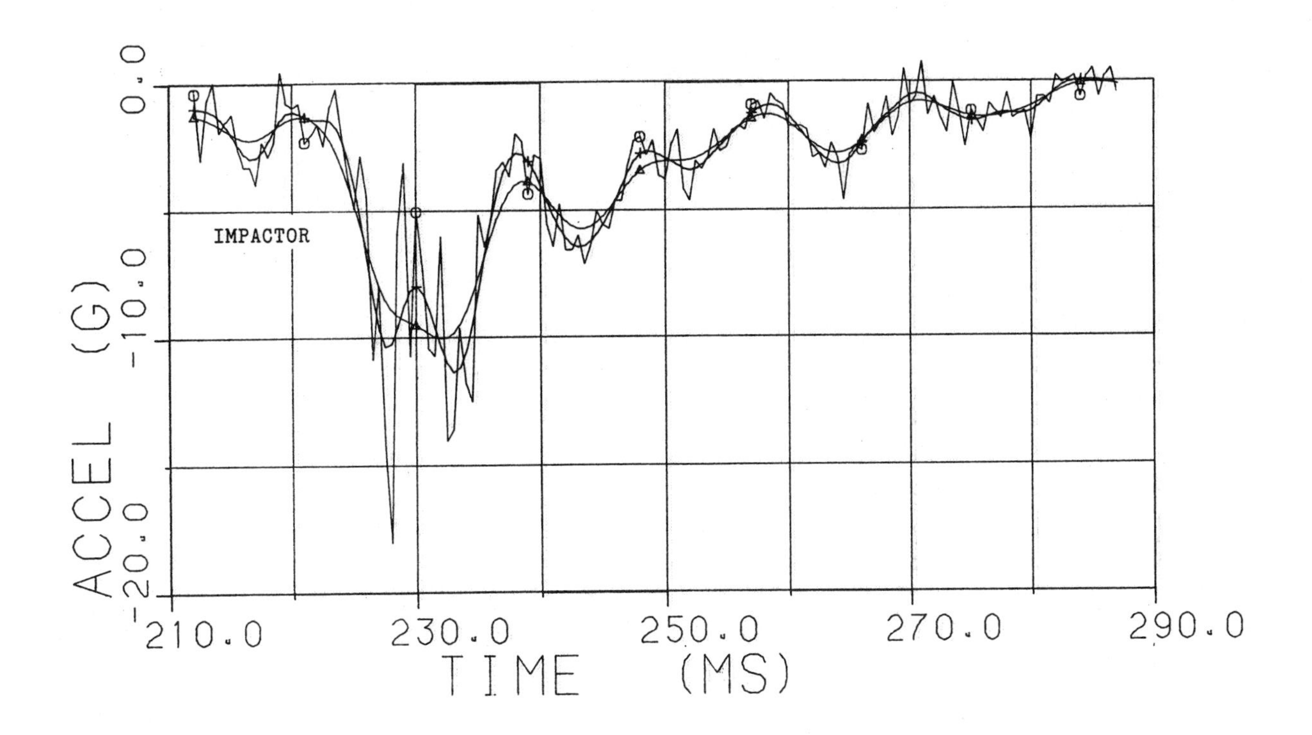

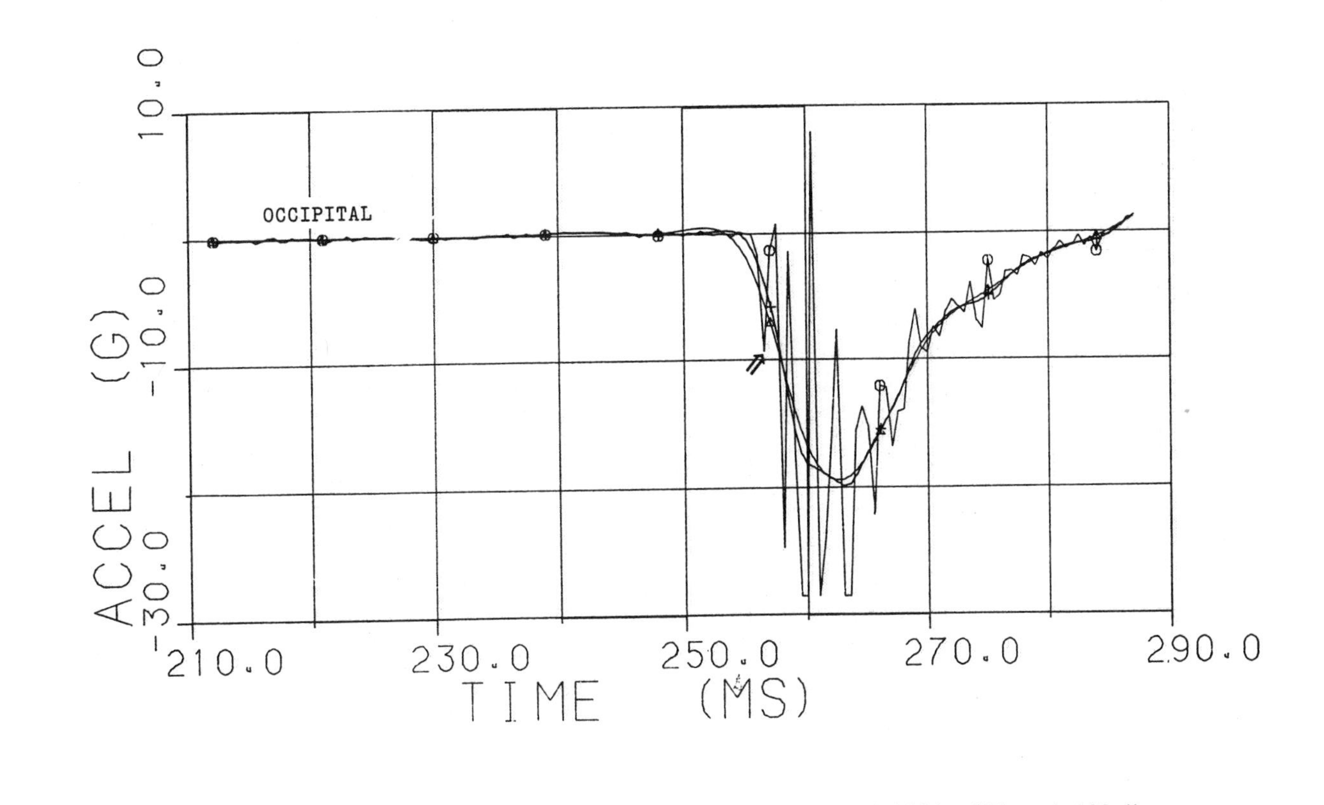

Figure A3. Acceleration Traces For Test No. 20, Filtered 1000, 180 and 100 Hz.

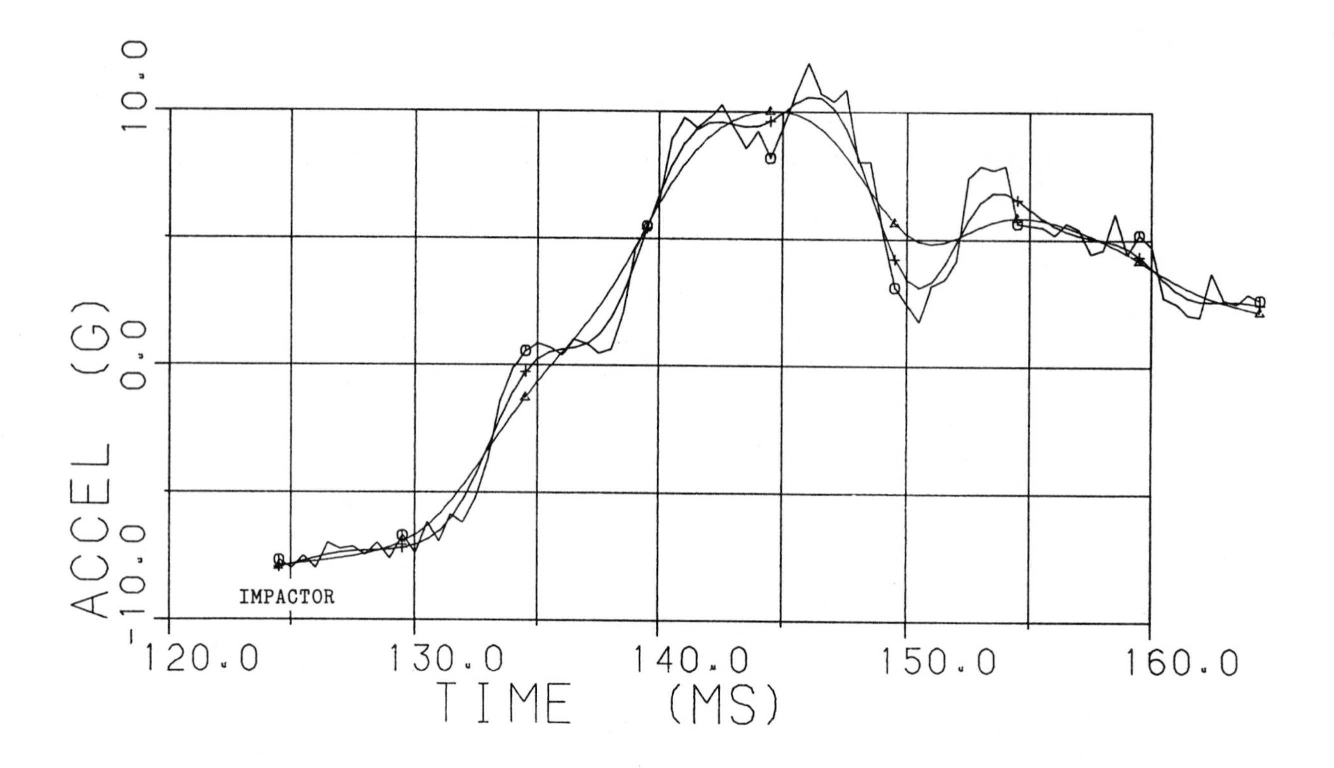

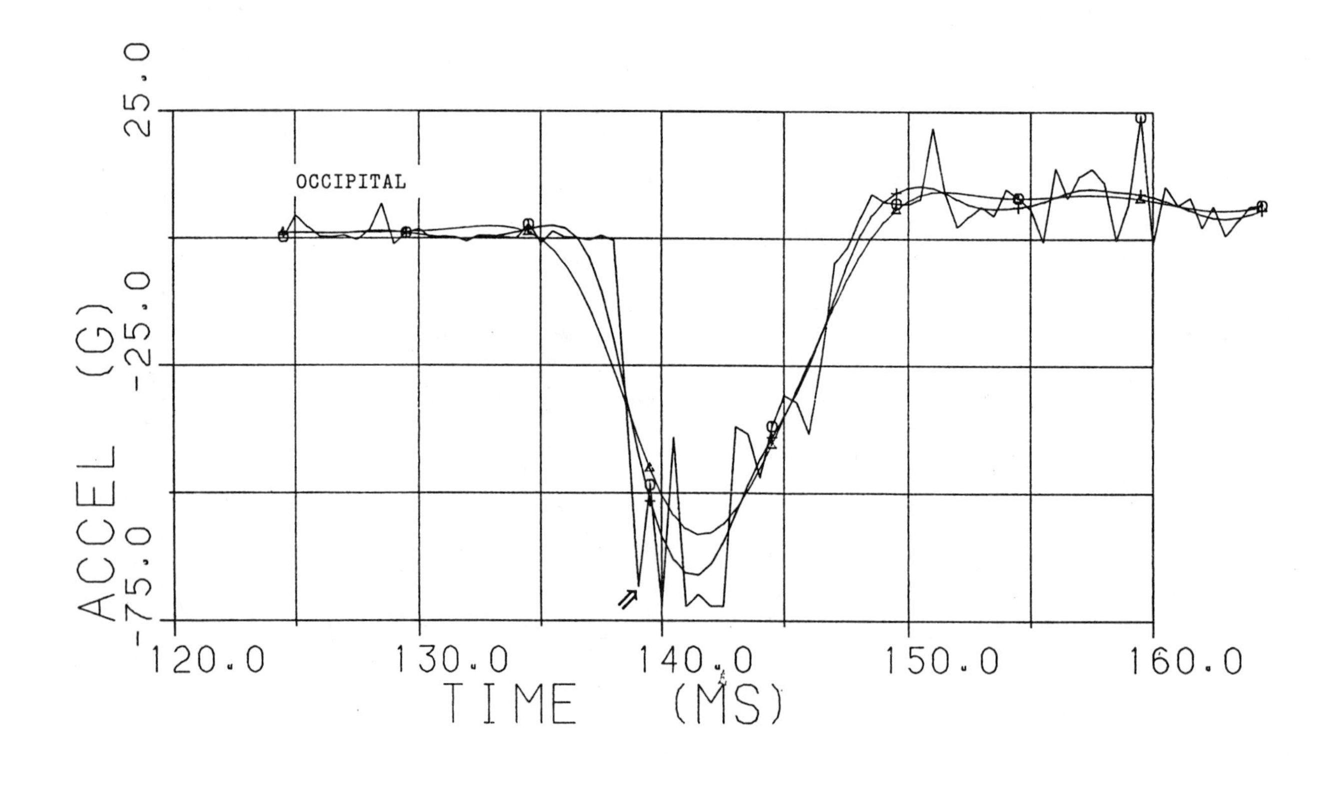

Figure A4. Acceleration Traces For Test No. 25, Filtered 1000, 180 and 100 Hz.

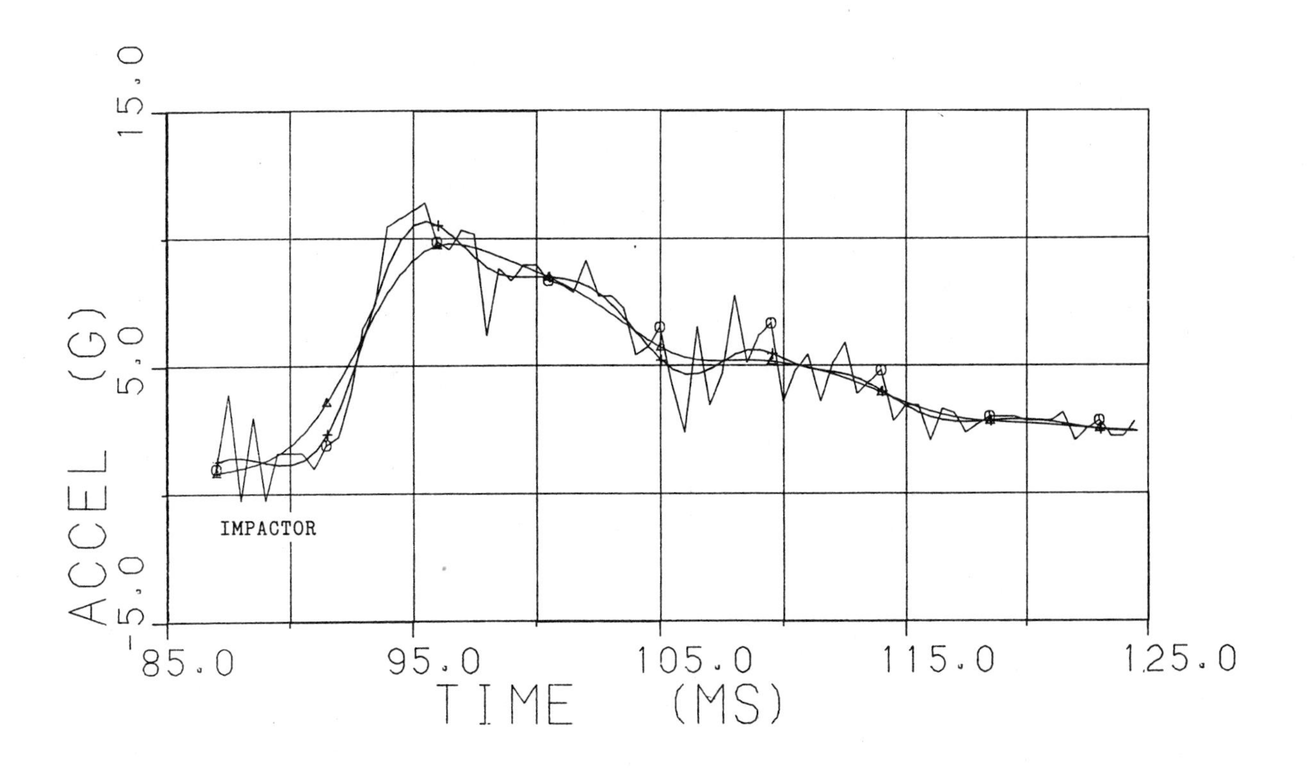

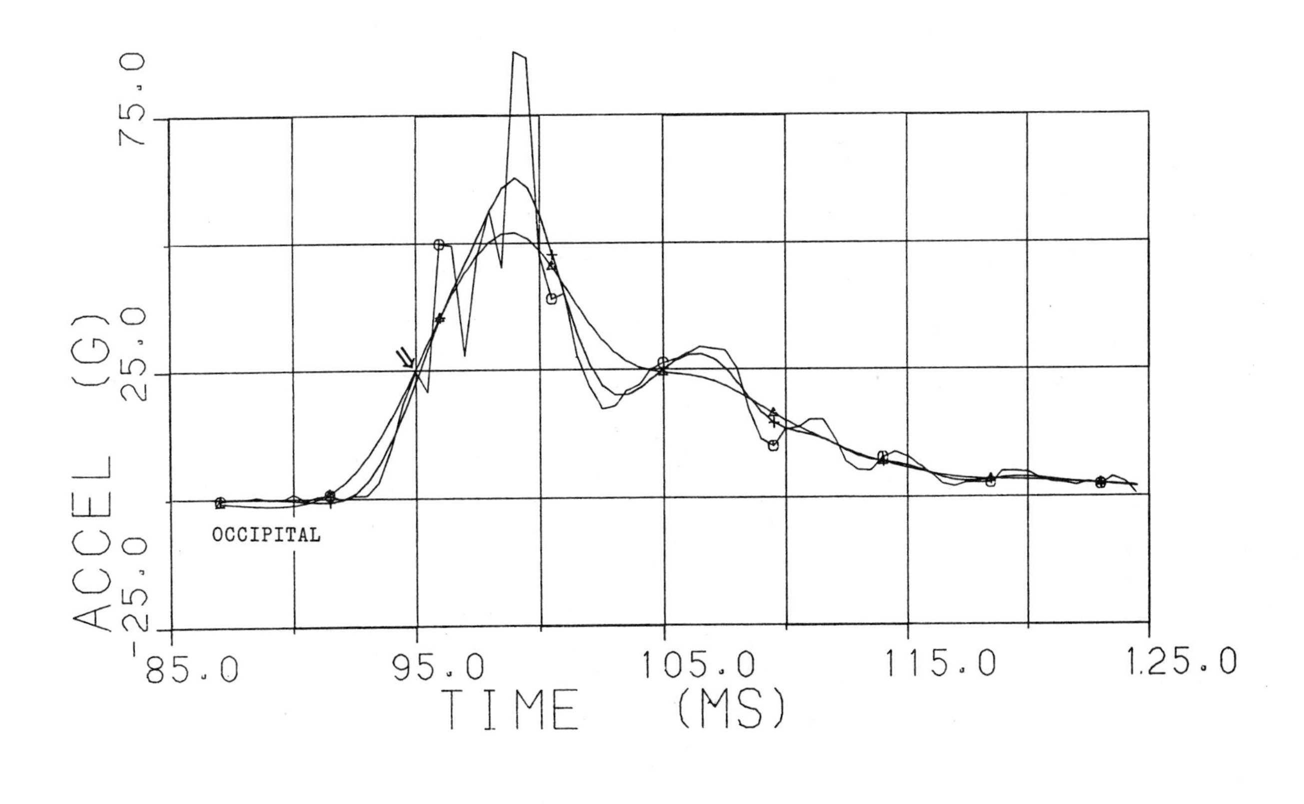

Figure A5. Acceleration Traces For Test No. 29, Filtered 1000, 180 and 100 Hz.

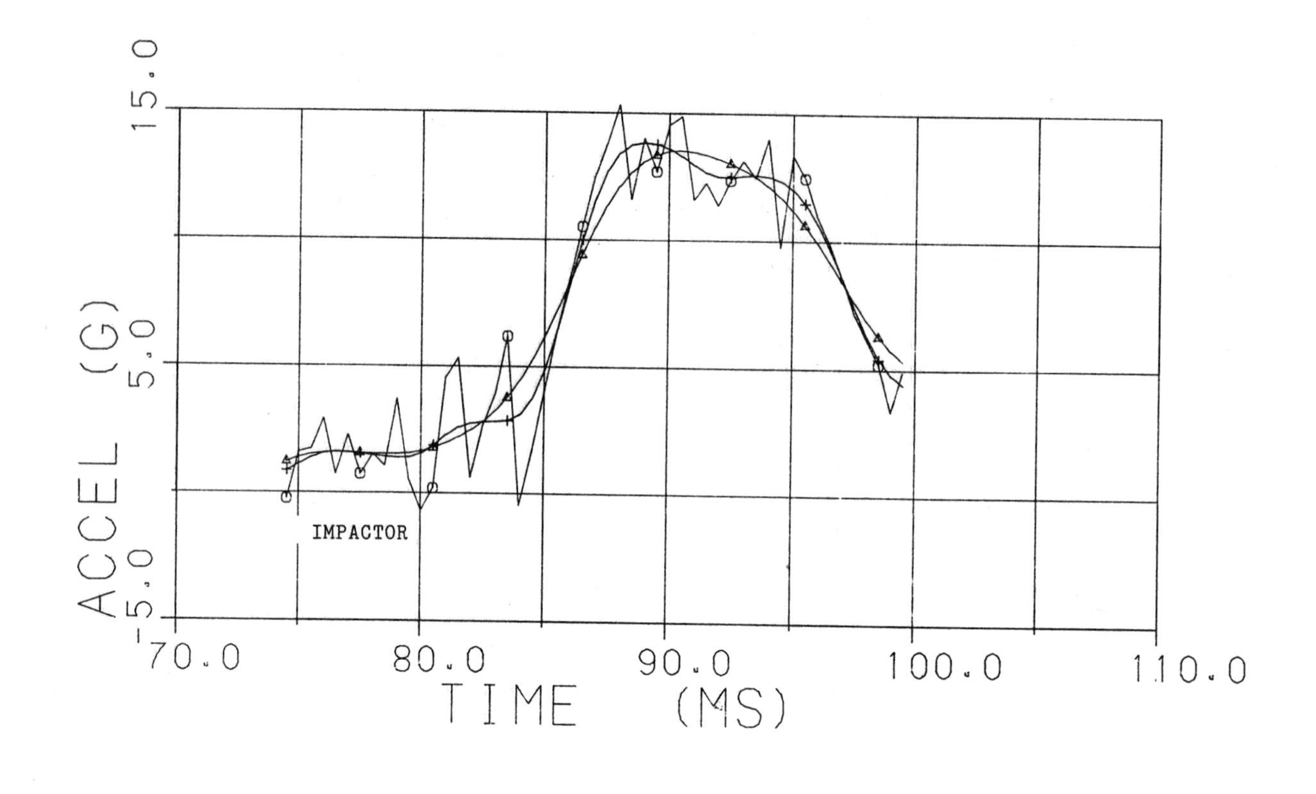

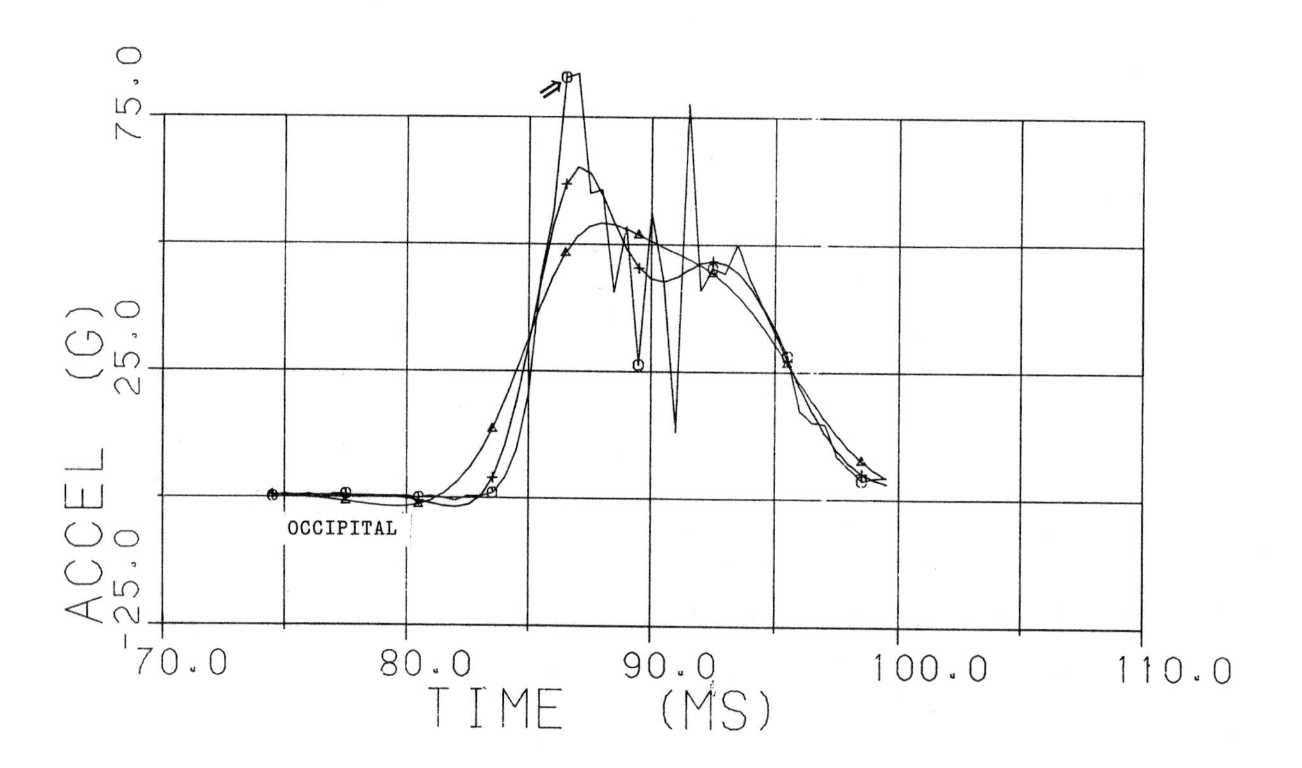

Figure A6. Acceleration Traces For Test No. 34, Filtered 1000, 180 and 100 Hz.

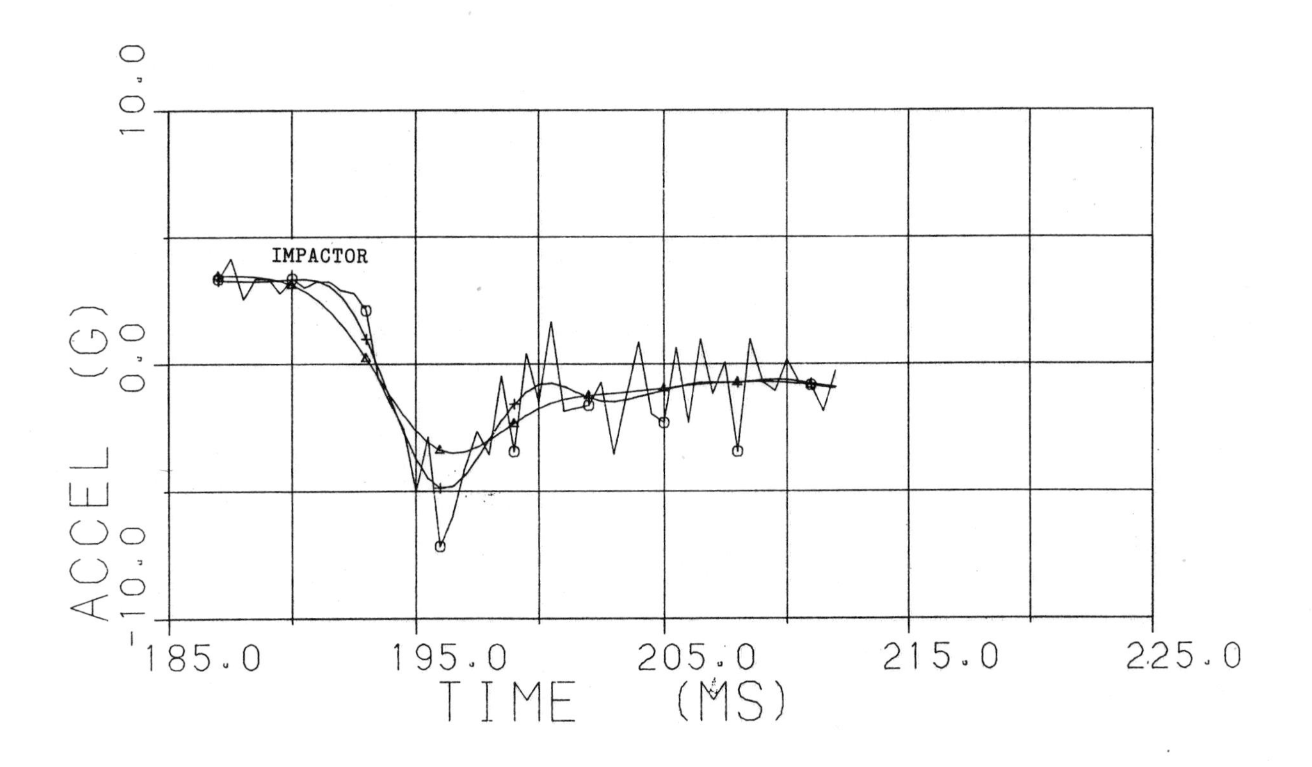

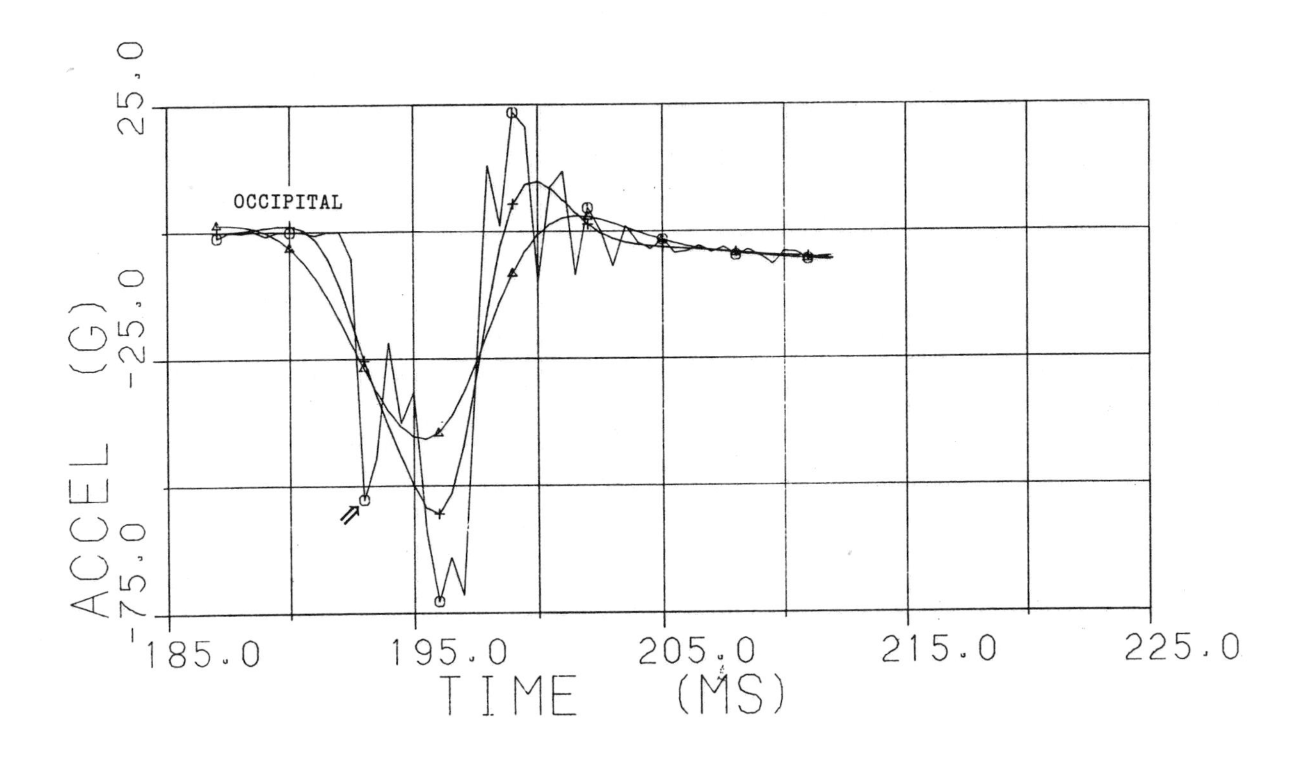

Figure A7. Acceleration Traces For Test No. 42, Filtered 1000, 180 and 100 Hz.

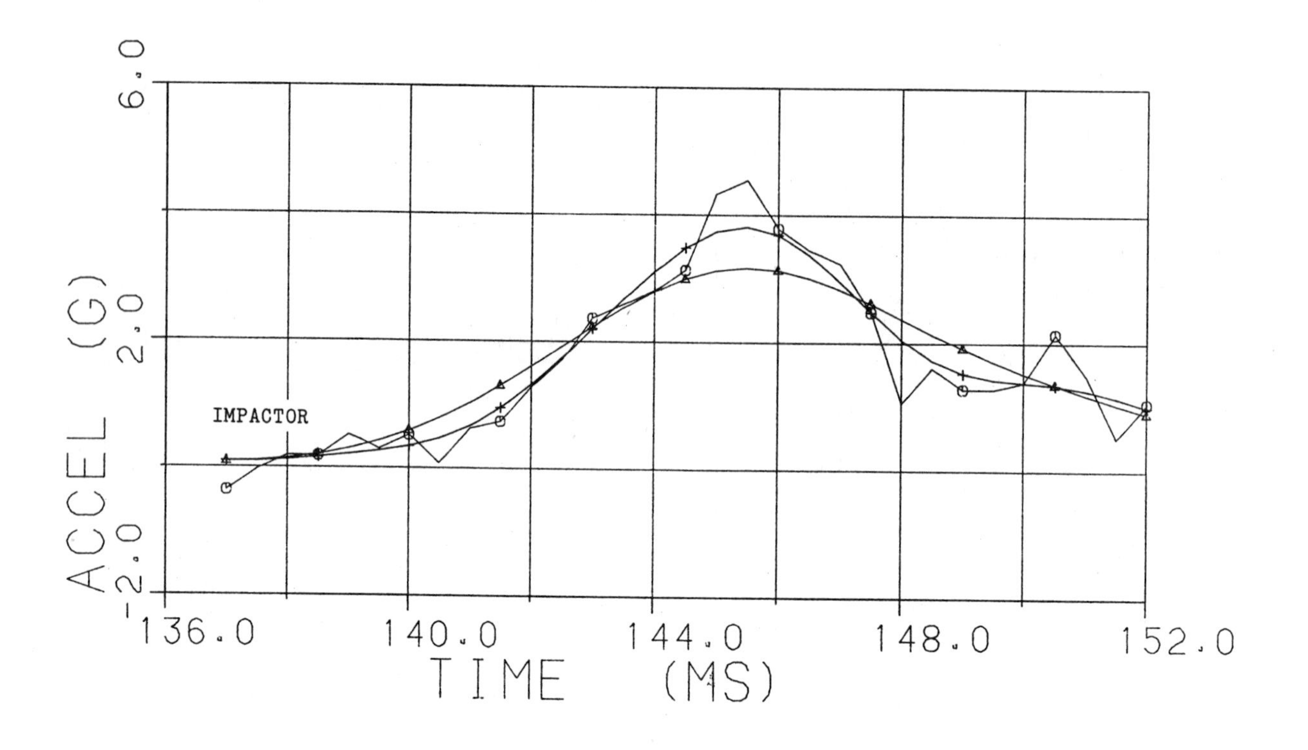

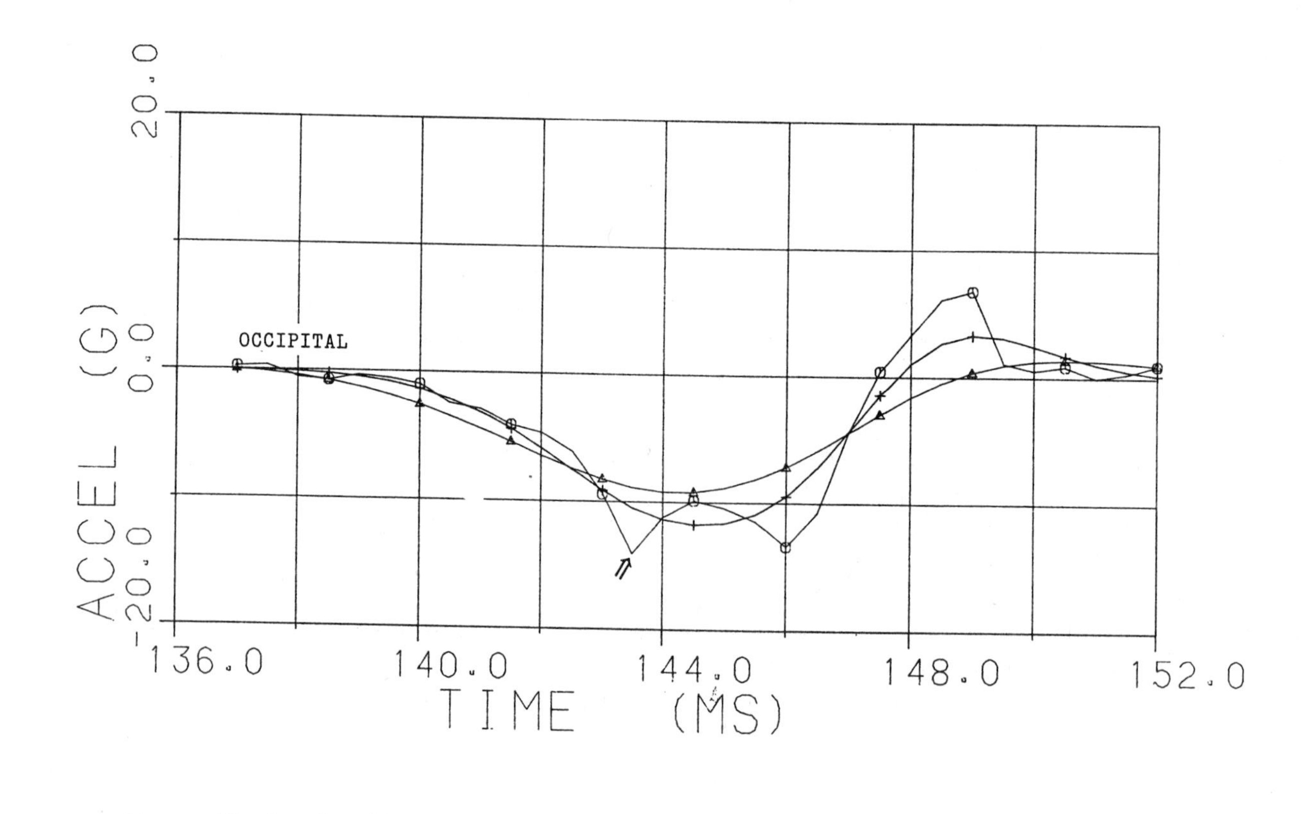

Figure A8. Acceleration Traces For Test No. 44, Filtered 1000, 180 and 100 Hz.

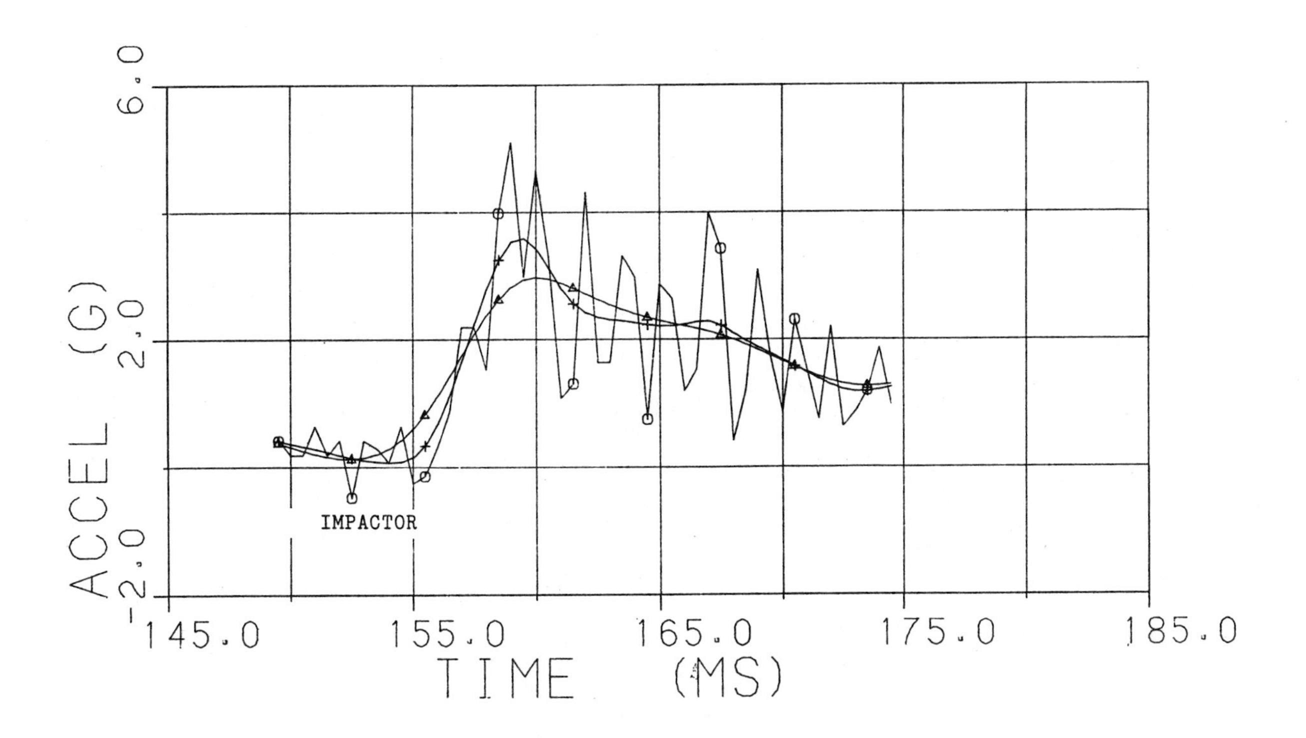

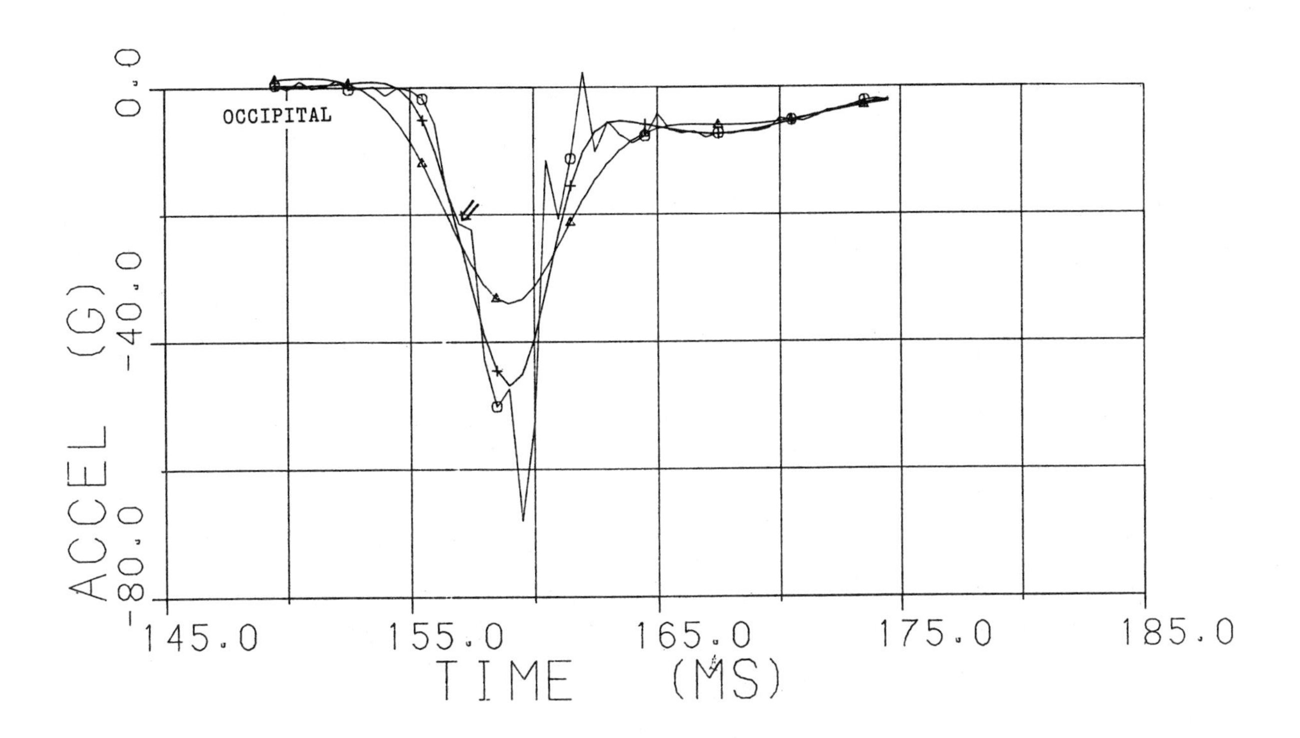

Figure A9. Acceleration Traces For Test No. 46, Filtered 1000, 180 and 100 Hz.

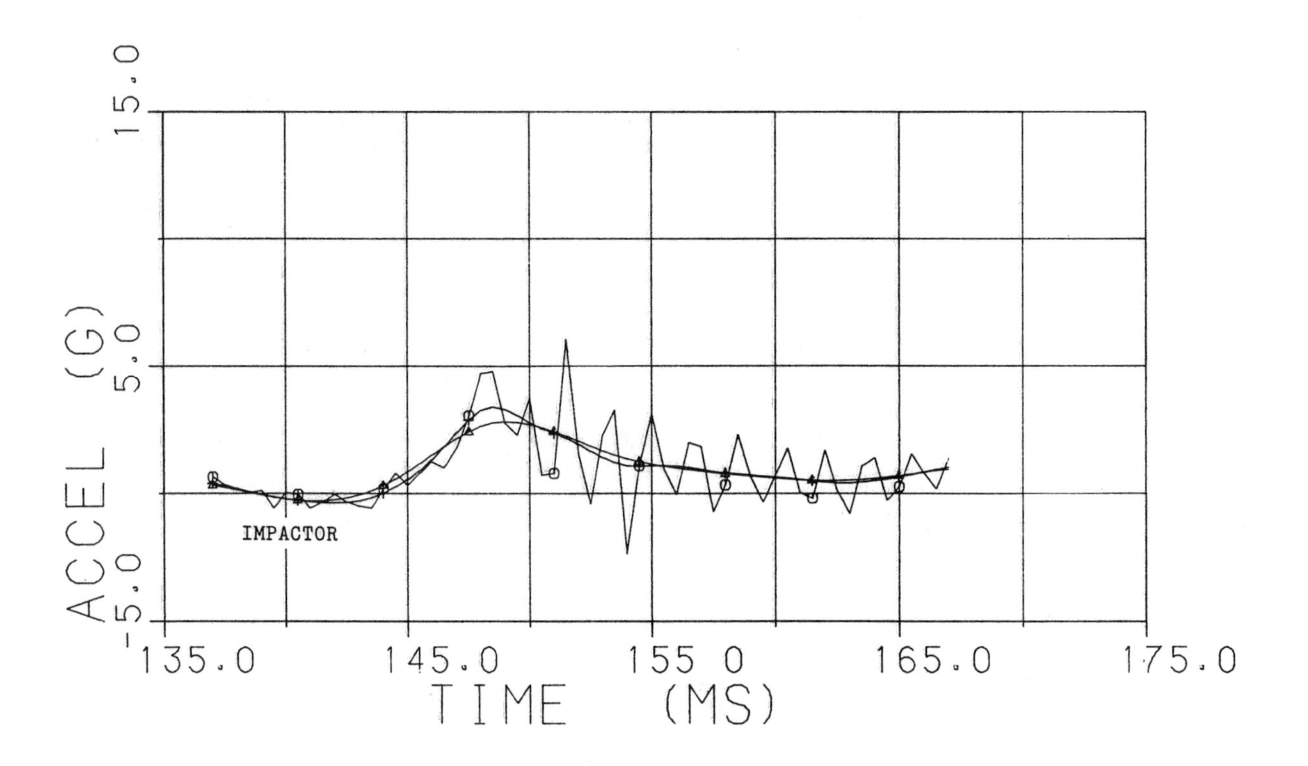

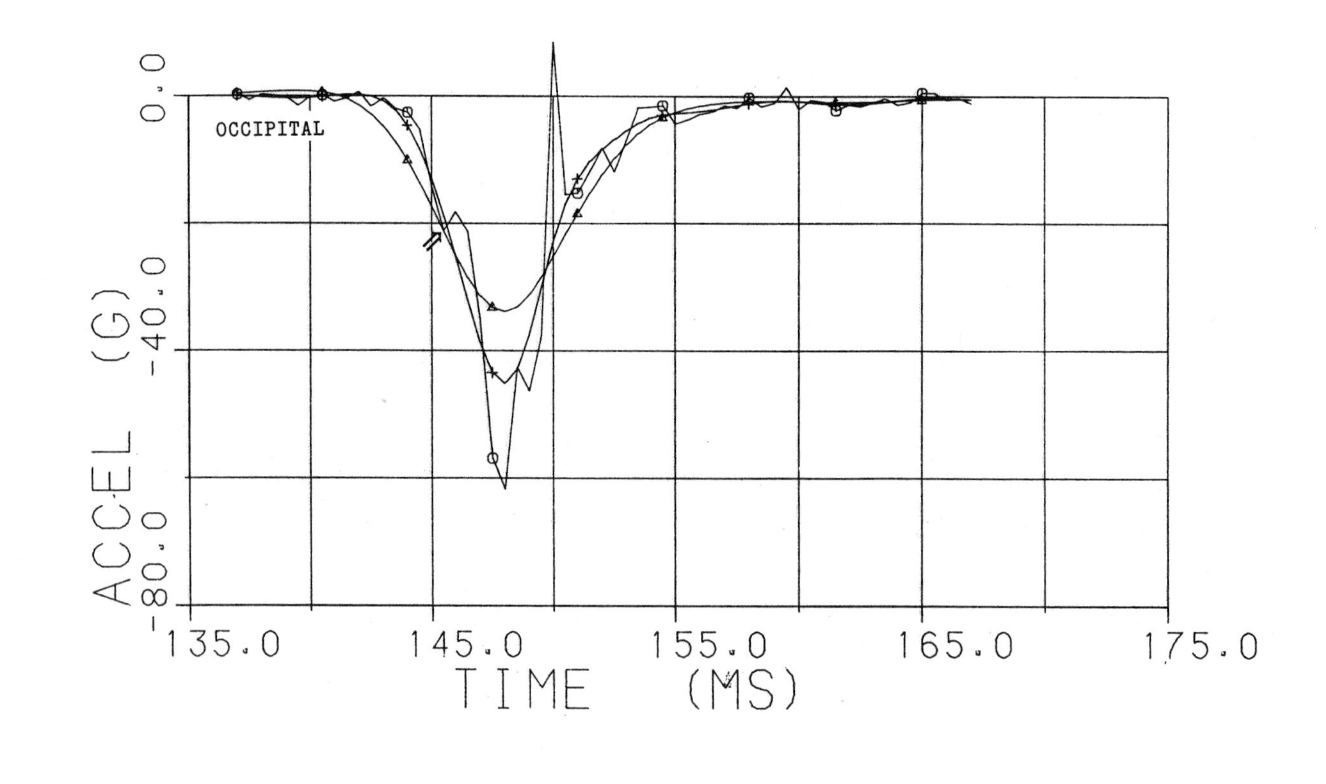

Figure A10. Acceleration Traces For Test No. 48, Filtered 1000, 180 and 100 Hz.

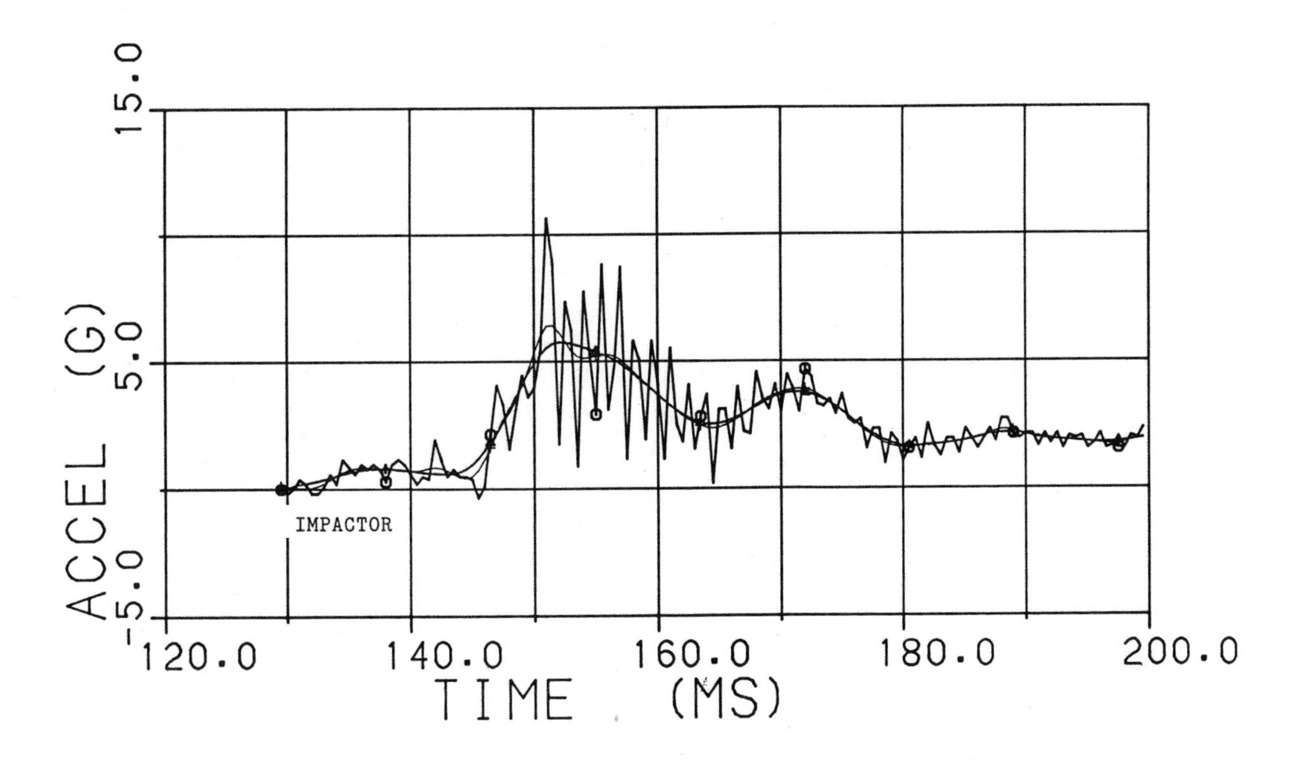

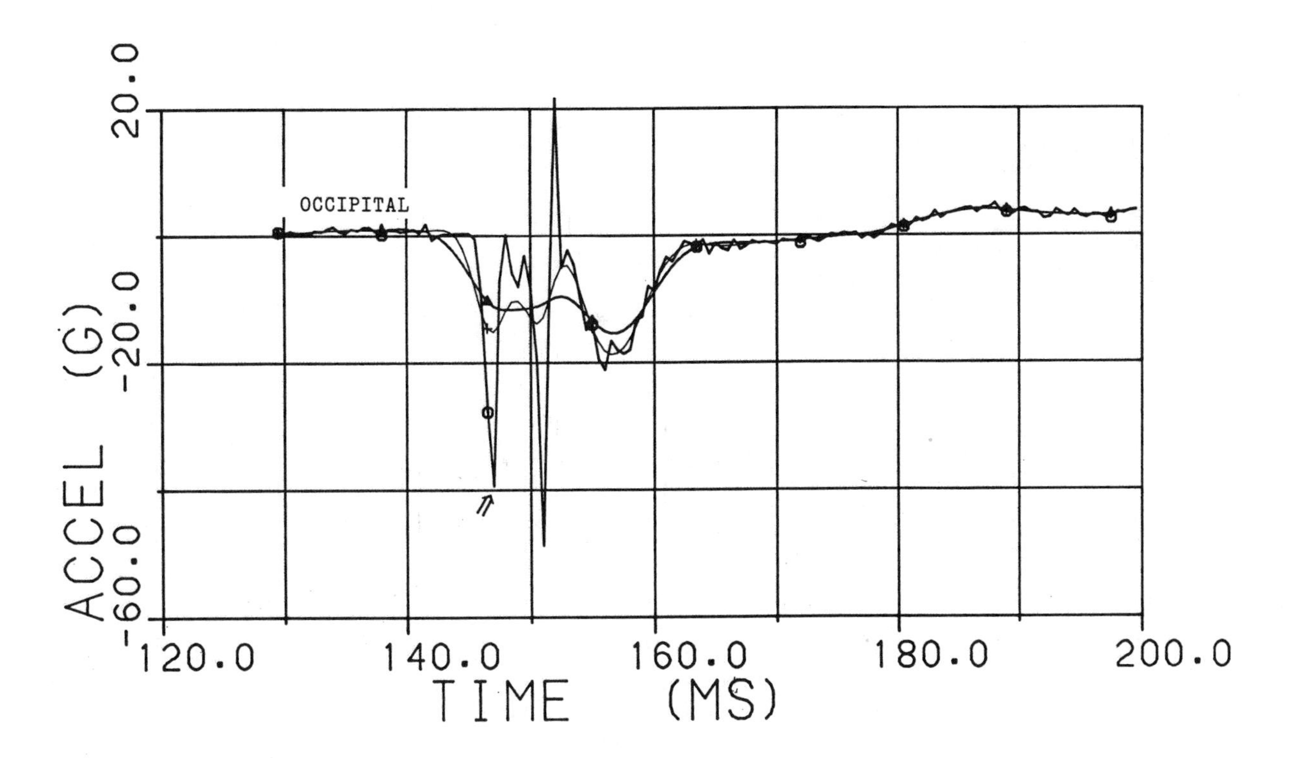

Figure A11. Acceleration Traces For Test No. 58, Filtered 1000, 180 and 100 Hz.

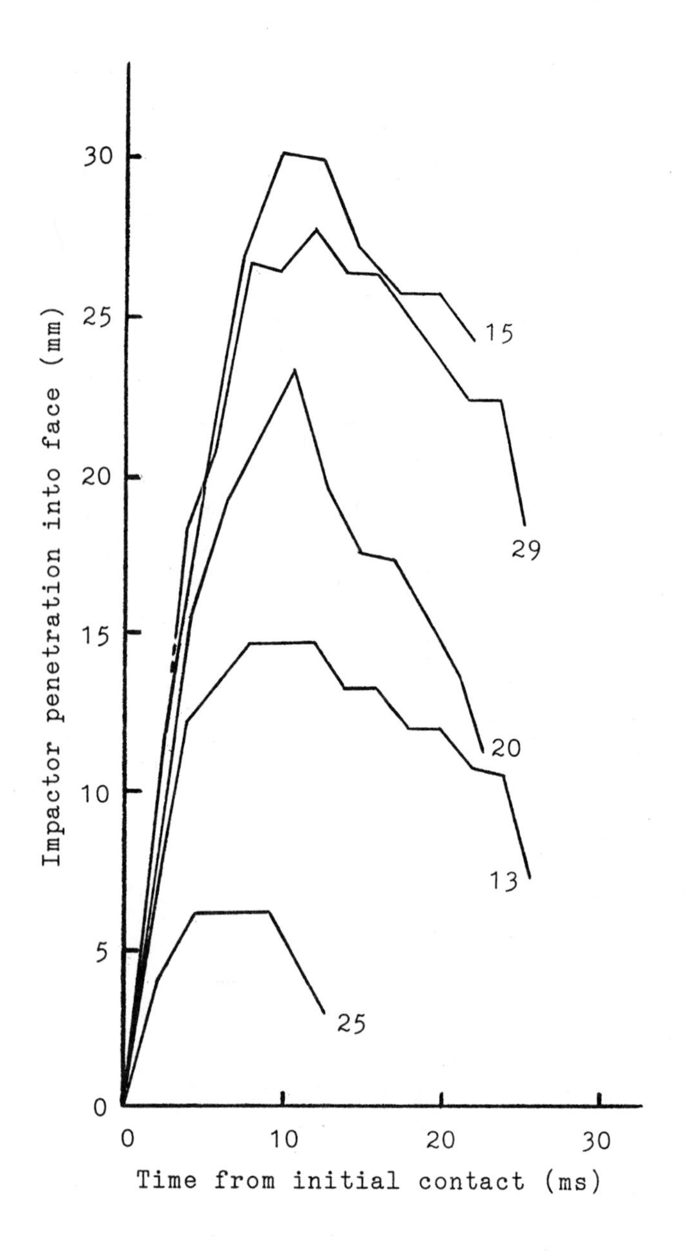

Figure A12. Penetration vs Time, Tests 13, 15, 20, 25 and 29.

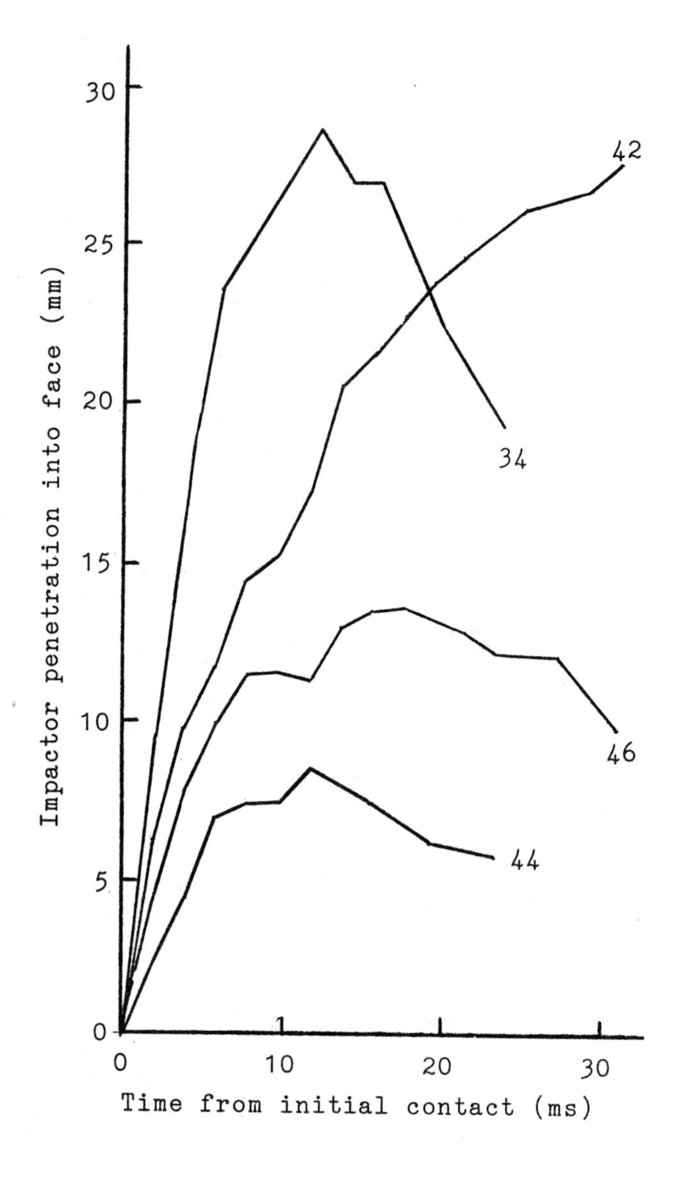

Figure A13. Penetration vs Time, Tests 34, 42, 44 and 46.

881712

Steering Wheel Induced Facial Trauma

Narayan Yoganandan, Frank Pintar, Anthony Sances, Jr., Joel Myklebust, Dale Schmaltz, John Reinartz, John Kalbfleisch and Sanford Larson
Veterans Admin. Medical Center

Gerald Harris
Marquette University

Kedar Chintapalli
Medical College of Wisconsin

ABSTRACT

Studies were conducted on twenty-two fresh human cadavers to determine the probability of facial bone fracture following dynamic contact with steering wheel assemblies of both standard (a commercially available) and energy absorbing (EA) types. Using a specially designed and validated vertical-drop impact test system, either zygoma was impacted once onto the junction of the lower left spoke and rim with velocities ranging from 2.0 to 6.9 m/s. Generalized force histories were recorded with a six-axis load cell placed below the hub. The wheel was inclined 30 degrees to the horizontal. Steering wheel deformations were recorded with a system of potentiometers placed below the impact site on the wheel. Dynamic forces at the zygoma (impact site) were computed using transformation principles. A triaxial accelerometer was placed at the posterior parietal region of the specimen opposite to the impact site to record acceleration histories. High speed photography documented the kinematics. Tissues were evaluated using palpation, gross dissection, plain radiography, 2-D and 3-D CTs, and defleshing techniques. Fracture severity was graded on a scale of 1-4. Quasistatic responses of the cadaver zygoma and the steering wheels were also studied.

At impact velocities of 6.93 to 3.58 m/s for the EA wheel and 3.13 to 2.24 m/s for the standard wheel, severe fractures of the zygoma, zygomatic arch, maxilla and orbit were observed. These clinically significant fractures often require surgical intervention. A velocity of approximately 2.68 m/s for the EA wheel and 2.01 m/s for the standard wheel produced no clinically significant injury to the facial structure.

FACIAL INJURIES IN VEHICULAR CRASHES continue to be a significant problem despite the introduction of seat belt legislation (4,29,39)*. The steering assembly has been implicated as a causal agent inducing facial injury in countries where seat belt requirements have been in effect for a number of years such as in Europe and Canada. In a German belt-wearing community (27), impact with steering wheels produced a significant source of facial trauma. A similar study by Marcus and Blodgett (21), based on 1982-1985 National Accident Sampling System (NASS) files reporting injuries to the head and face, implicated the steering assembly as a leading cause of short-term sensory impairment and medium-term pain impairment concomitant to short-term cosmetic alteration. A review of 397 individual facial injuries in 222 NASS accident cases involving the steering assembly from 1981-1985 indicated that facial injuries included contusions 32%, abrasions 9%, lacerations 36%, avulsions 2%, fractures 17% and other injuries 4%. Seventy percent of all serious (AIS $\geq$ 2) steering assembly induced facial injuries are skeletal, while lacerations and contusions represent less than 25% of the more serious trauma (39).

Complex mid-facial fractures are routinely observed in vehicular crashes (1,16,17). One of the most widely used systems of classifying these fractures was described in 1901 by LeFort who produced lesions in adult cadavers with impact to the lateral and anterior region of the face (15). LeFort described various regions of the face in terms of great lines of weakness and considered many parts of the face as merely a link to the cranium. Most maxillary fractures are severely comminuted and consist of combinations of the LeFort types. A LeFort I fracture is horizontal and segmented across the maxilla. The teeth are often contained in the detailed bony portion. This fracture responds well to intramaxillary fixation. LeFort II is also a fracture of the maxilla. The body of the maxilla is separated from the facial skeleton, across and through the infra-orbital foramen, and across the dorsum of the nose. Surgical fixation is required. LeFort III is a fracture causing craniofacial dissociation. The entire maxilla and one or more bones are completely separated from the cranium through the front zygomatic region, the orbital floors, and across

* Numbers in parentheses designate references at end of paper.

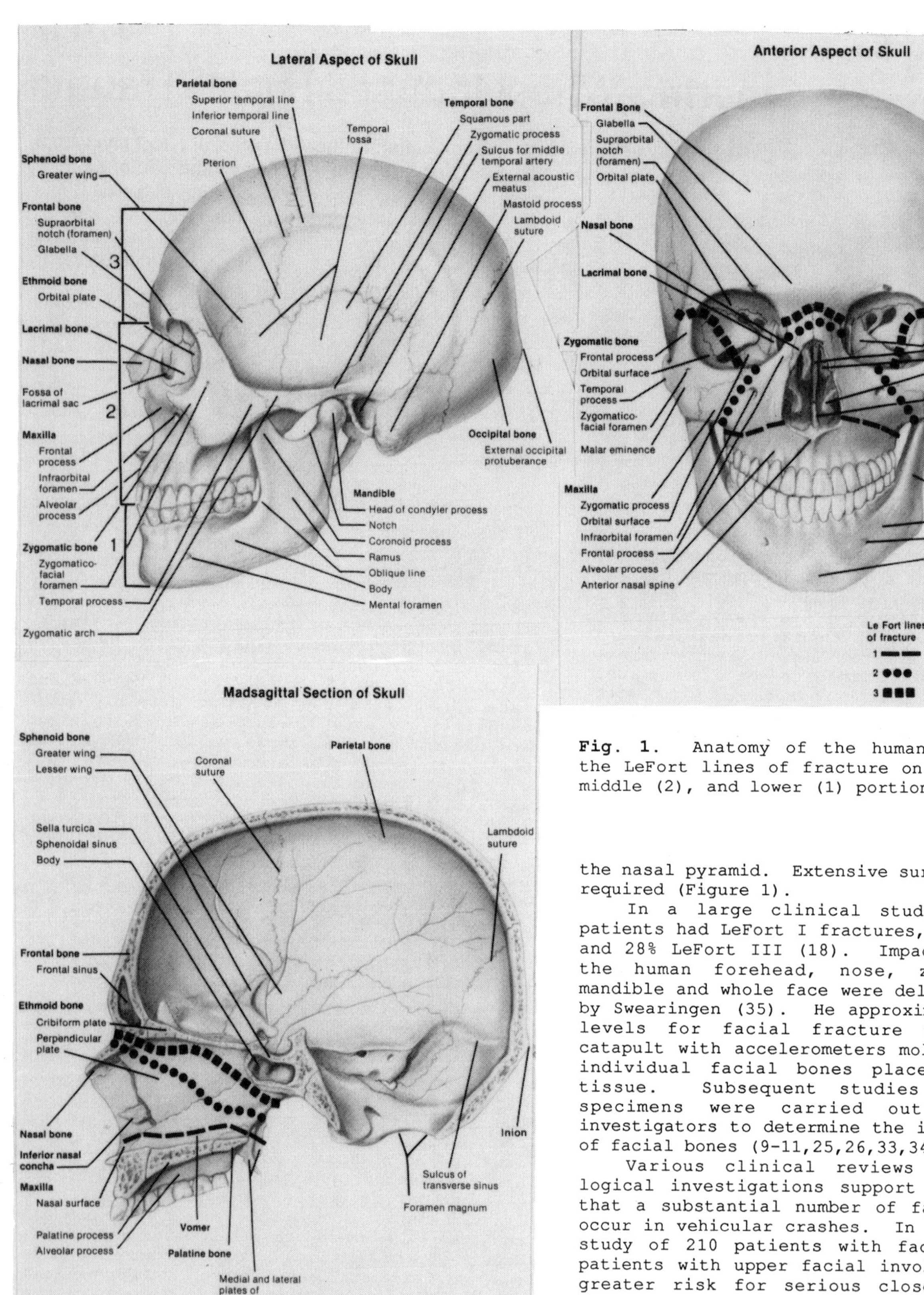

Fig. 1. Anatomy of the human skull showing the LeFort lines of fracture on the upper (3), middle (2), and lower (1) portions of the face.

the nasal pyramid. Extensive surgical repair is required (Figure 1).

In a large clinical study, 30% of the patients had LeFort I fractures, 42% LeFort II, and 28% LeFort III (18). Impact tolerance of the human forehead, nose, zygoma, teeth, mandible and whole face were delineated in 1965 by Swearingen (35). He approximated various G levels for facial fracture using a small catapult with accelerometers molded to fit the individual facial bones placed against the tissue. Subsequent studies in cadaveric specimens were carried out by numerous investigators to determine the impact tolerance of facial bones (9-11,25,26,33,34).

Various clinical reviews and epidemiological investigations support the contention that a substantial number of facial fractures occur in vehicular crashes. In a neurosurgical study of 210 patients with facial fractures, patients with upper facial involvement were at greater risk for serious closed head injury while those with mandibular and mid-face trauma usually had mild closed head injury (3,7,12-14,17-20,22,23,28,37).

The three most common isolated facial fractures are the nasal, the tripod, and blowout fractures of the orbit (1). Zygomatic fractures routinely occur because of the prominence of this mid-facial structure. The tripod fracture associated with the malar, zygomatic maxillary and zygomatic complex is often quoted as the second most common isolated facial fracture, and the most common fracture involving the maxillary sinus (1). Prominent areas of evaluation in facial fractures are the zygoma, nose, orbit, maxilla, mandible and maxillary sinus. Blowout fractures of the orbit which are often associated with entrapment of the inferior rectus muscle can cause blindness (1,37). Detailed computerized tomography (CT) studies are routinely conducted to determine both hard and soft tissue alterations (2,5,6,17,37). Three-dimensional (3-D) CT reconstructions of the craniofacial area in concert with two-dimensional (2-D) CT imaging provides a unique opportunity to noninvasively identify hard tissue alterations (8,36).

Because of these findings and the growing concern for automotive safety, studies were conducted with fresh human cadaveric tissues to determine the probability of facial bone fracture following dynamic contact with steering wheel assemblies of both standard (a commercially available) and energy absorbing (EA) types. The EA wheel is the one used in the earlier study (29).

METHODS

Twenty-two fresh adult human cadaver heads (six female, sixteen male) were used in the study. The subjects ranged in age from 57 to 76 years, height 155 to 196 cm, and weight 45 to 109 kg (Table 1). Medical records were reviewed and plain radiographs taken to select subjects without metastatic bone disease or prevalent head and facial trauma. The specimens were isolated at OC-C1 level within 24-48 hours after death, sealed in plastic bags and deep frozen at -70 degrees Centigrade. Two-dimensional computerized tomography was done using a General Electric Scanner (Model 9800) with 1.5 mm sections. The specimens were thawed at room temperature 12-24 hours prior to experimentation. Radiographs were also taken in the anteroposterior, superior-inferior, lateral, and occipitomental (Waters) views prior to test. Three-dimensional CT's were also done on the tissues with the Dimensional Medicine Model 3200 System.

QUASISTATIC STUDIES

Quasistatic Studies at the Zygoma

Quasistatic studies were conducted on seven isolated cadaver specimens (four with the standard steering wheel and three with the EA). Physical measurements were taken (Table 2). The specimens were rigidly fixed in a halo ring with eight to twelve stainless steel screws 6.35 mm in diameter.

Initially, three specimens (two on the standard and one on the EA) were tested as follows. The specimens were firmly clamped onto the frame of the specially designed MTS testing device. The standard steering wheel was cut to approximately 150 mm in length so that the junction of the lower left spoke and rim could be used as the loading surface. An approximately 300 mm length of a similar region of the EA steering wheel was used. A fixture was designed so that this part of the wheel could be attached to the piston and angulated at approximately 30 degrees with respect to the horizontal so as to approximate dynamic contact studies. The specimens were contacted in the region of the malar eminence of the zygomatic area. Two tests were done on each specimen with the standard wheel; one test with the skin remaining intact on one side and the next test with the skin removed so as to expose the zygomatic area on the other side. On the EA wheel, the right zygoma with the skin remaining intact was tested.

The remaining four specimens were tested with the entire steering wheel. Two specimens were loaded with the EA wheel and the remaining

Table 1
Specimen Data

SPECIMEN #	SEX	AGE (YEARS)	HEIGHT (cm)	WEIGHT (kg)	CAUSE OF DEATH
5149	M	64	175	79	Cardiac decompression
7690	M	65	170	91	Heart failure
5421	M	60	178	82	Myocardial infraction
5812	M	68	178	75	Cardiopulmonary failure
4645	F	71	178	76	Heart Attack
9234	F	65	157	45	Respiratory arrest
3905	M	57	180	57	Congestive heart failure
6216	M	62	180	86	Myocardial infarction
8947	F	68	162	91	Cardiac failure
7628	M	59	178	80	Heart Attack
7589	F	67	163	82	Cardiac failure
8250	M	65	178	75	Internal gastric hemorrhage
8151	M	69	180	79	Stroke
8888	M	63	178	79	Cardiac arrest
5912	M	69	196	77	Heart Attack
7643	F	76	155	64	Cardiac failure
9274	M	71	175	68	Pneumonia
6025	M	66	175	75	Heart Attack
8228	M	66	178	84	Pulmonary embolism
8891	F	71	160	45	Heart failure
8902	M	60	183	109	Heart failure
8973	M	74	183	104	Heart failure

Table 2
Specimen Physical Parameters

SPECIMEN #	CHIN-VERTEX (mm)	LATERAL-LATERAL (mm)	NASION-OCCIPUT (mm)	MAXIMUM CIRCUMFERENCE (mm)	ZYGOMA SKIN THICKNESS (mm)	STUDY DYNAMIC/STATIC	WHEEL TYPE
5149	230	153	191	578	15	Dynamic	EA
7690	228	155	194	560	8	Dynamic	EA
5421	191	163	193	590	10	Dynamic	EA
5812	165	155	189	550	15	Dynamic	EA
4645	190	151	200	575	14	Dynamic	EA
9234	206	142	171	525	7	Dynamic	EA
3905	146	153	172	560	11	Dynamic	EA
6216	290	140	191	---	12	Static	Standard
8947	245	165	197	---	20	Static	Standard
7628	216	162	194	575	16	Static	EA
7589	226	153	187	561	11	Dynamic	EA
8250	235	145	184	545	15	Dynamic	EA
8151	208	157	195	563	6	Dynamic	Standard
8888	242	153	202	577	12	Dynamic	Standard
5912	196	137	203	565	12	Dynamic	Standard
7643	218	147	165	539	11	Dynamic	Standard
9274	232	141	169	532	7	Dynamic	Standard
6025	186	132	180	566	8	Dynamic	Standard
8228	228	157	213	588	12	Static	EA
8891	227	145	182	518	5	Static	EA
8902	250	140	199	561	14	Static	Standard
8973	227	157	209	602	8	Static	Standard

two specimens were loaded with the standard wheel. In all these four specimens, the skin on the zygoma was not removed. The junction of the left lower spoke and rim contacted the zygoma during the test. For this purpose, a fixture was designed one end of which was attached rigidly to a specially designed electrohydraulic Materials Testing System (MTS) piston, and the other end rigidly held the specimen fixated with the halo ring so that the malar eminence region of the zygoma contacted the steering wheel (EA or standard type). The steering wheel mounted onto the MTS platform was inclined at an angle of 30 degrees from the horizontal. The same orientation was used in dynamic drop tests (discussed later). A Denton six-axis load cell (Model # 1994) was placed directly under the hub of the steering wheel to record the generalized force histories.

All specimens were loaded at a constant displacement rate of 2.5 mm/s. Gross dissection, radiography and CT's were done after the test to determine the pathology.

Quasistatic Studies of the Steering Wheels

The energy and force-deflection characteristics of the two types of steering wheels were determined by loading at a constant velocity of 2.5 mm/s with the MTS device. The steering wheels were inclined at 30 degrees from the horizontal. A specially designed, steel cylindrical fixture with curvature approximating the zygomatic region was attached to the piston of the MTS which contacted the junction of the lower left spoke and rim of the steering wheel during loading. This fixture did not bind the steering wheel during the loading process. Studies were also conducted at the center of the largest unsupported part of the rim to compare the responses obtained from tests at the junction of the lower left spoke and rim. Tests were videotaped at 60 frames/s.

DYNAMIC STUDIES

Design and Validation of the Vertical-Drop Test System - The impact test system consisted of a 7.6 m vertically supported stainless steel monorail. A cart with precision bearings was designed to traverse the 19 mm diameter monorail. Linear bearings with a stabilizing outrigger bearing mechanism prevented the rotation of the cart about the vertical axis. Rotation about the other two coordinate axes was constrained by the linear bearing design. A special positioning fixture was designed to accept cadaveric specimens as well as the magnesium Z90.1 headforms. This fixture was designed to allow for 360 degree rotation of the specimens in two mutually orthogonal directions (lateral-lateral and superior-inferior). This positioning insured that the impact location was in the region of the malar eminence of the zygomatic bone.

A floor-mounted fixture was designed for impaction of the specimens with the junction of the lower left spoke and rim of both steering wheels (Figure 2a). The steering wheel was mounted on this fixture so that the plane of the wheel was inclined at 30 degrees with respect to the horizontal. A six-axis Denton load cell was placed directly under the hub of the steering wheel to record the impact force and moment histories. A specially designed system incorporating a combination of linear and rotary potentiometers was placed below the impact point on the steering wheel (junction of the lower left spoke and rim) to record the dynamic steering wheel deformations. The positioning of this instrument did not interfere with the impact characteristics of the specimen, nor did it alter the mechanical properties of the steering wheel.

The vertical-drop impact test system (slider) was validated by comparing results

Fig. 2a. Steering wheel fixture at an angle of 30 degrees with respect to the horizontal. The six-axis Denton load cell is mounted directly beneath the hub of the steering wheel. The steering wheel deformations are recorded by the potentiometers fixed directly below the impact site on the wheel.

obtained with free fall conditions. For this purpose, a Z90.1 magnesium rigid headform (size C) which was mounted in a halo ring was impacted with the junction of the lower left spoke and rim of both steering wheels under free fall conditions as well as under the slider system. A uni-axial accelerometer (Endevco Model 2264A) mounted at the Center of Gravity (CG) of the headform was used for acceleration measurements. The metal headform weighed 4.95 kg resulting in a total weight of 7.14 kg for the headform and the fixtures. The weight for the free fall tests was ballasted to the same amount. The six-axis load cell and the system of potentiometers were used to record the generalized force and deformation data at 8 kHz sampling rate. The tests were conducted at an impact velocity of 3.13 m/s. A new steering wheel was used for each test.

Vertical Drop Tests on the Zygoma - Nine specimens were impacted on the EA wheel and six specimens on the standard wheel. Physical data such as the nasion-inion dimension, maximum lateral-lateral measurement at the level of the auditory meatus, chin-vertex distance, and the maximum circumference of the head were recorded. The thickness of the skin at the impact site in the region of the malar eminence of the zygoma was also measured. The physical parameters of all the specimens are given (Table 2).

After obtaining these measurements, the head was rigidly mounted in a neurosurgical stainless steel halo ring with fixation into the skull using eight to twelve screws 6.35 mm diameter. The mounted specimen was then attached to the positioning fixture of the specially designed vertical-drop impact test system. A triaxial accelerometer (Endevco Model # 7237A) was firmly secured to the skull with four orthopedic stainless steel screws in the region of the posterior parietal skull opposite to the region of impact. The Z-axis of the accelerometer coincided with the vertical axis of the vertical-drop impact test system. The combined weight of the fixture and mounted specimen was ballasted to 6.804 kg.

Mounting and proper positioning of the specimen was made for perpendicular impact to the malar region of the zygomatic bone. The specimen was dropped from a predetermined height onto the junction of the lower left spoke and rim of the standard or EA steering wheels.

The impact event was photographed with a Hycam 16 mm camera (Model # K20S4AE-115, Red Lake Laboratories, CA) at 1000 frames/s and a video camera at 180-240 frames/s. The preparations were impacted on the right or left zygoma at velocities ranging from 2.0-6.9 m/s. Immediately following the drop, colored photographs were taken and the tissues were palpated for fracture. The contact area at the impact site was determined by the residual deformations of the tissue in the zygomatic region. In all specimens, the contact area was approximated to be an ellipse for which the major and minor axes dimensions were measured. Gross dissection at the impact site was then performed to observe the pathology. Radiographs were taken in the anteroposterior, lateral and superior-inferior directions, and Waters view. computed tomography was performed and the specimens defleshed. For defleshing, the specimen was immersed in a caustic liquid and then washed with a strong solution of hydrogen peroxide. Three-dimensional CT images were reconstructed from the pre and post experiment 2-D CT's. The pathology was determined with palpation, gross dissection, plain radiography, 2-D and 3-D CT, and from the defleshed skulls. Fracture severity was graded on a scale of 1-4 using the method reported in literature (14).

BIOMECHANICAL DATA

During dynamic drop studies, biomechanical data were collected according to NHTSA requirements using a modular data acquisition system (TransEra MDAS Model #7000) and specially designed preamplifiers and filters. Class 1000 filtering (SAE J2116) at a sampling rate of 8 kHz was used to collect all the biomechanical data. The sampled data included the force and moment histories from the six-axis load-cell, three linear accelerations from the tri-axial accelerometer, and deformation data from the system of linear potentiometers. All channels of information were then transferred to an IBM PC-AT computer from the MDAS system for biomechanical analysis.

For the vertical drop tests, the dynamic forces at the impact site of the specimen with the junction of the lower left spoke and rim contact were derived using the force and moment records from the six-axis load cell, and deformations of the steering wheel. The resultant transformed peak forces are given in the tables. Because the accelerations recorded on the preparations were oscillatory, a range of average values coincident with the peak forces over a duration of 4 to 5 ms were estimated and are given in the tables. A series of transformations including conversion, translation and rotation were carried out to compute the dynamic force history (Appendix 1). Principles of rigid body mechanics were used to transform the generalized forces from the six-axis load cell to impact site. The methodology included a transformation of the translational and rotational components of steering wheel deformations during zygomatic loading of the impacting specimen (Figure 2b).

Applied force and deformation data for the quasistatic steering wheel characteristic studies were obtained by a force gauge (Kistler Model # 9352-A) and a linear variable differential transducer (LVDT) attached to the piston of the MTS testing device. A digital oscilloscope (Norland Model 3001) capable of acquiring four channels of data was also used to record the signals from the force transducer and LVDT. Both the loading and unloading portion of the responses were obtained, and from the force-time and deformation-time traces, force-deflection characteristics were computed. Stiffness and energy absorbed by the structure were determined.

RESULTS

QUASISTATIC STUDIES

Quasistatic Studies at the Zygoma - Results of the quasistatic loading experiments on the cadaver zygoma using a cut portion of the standard and EA steering wheels as the surface of load application are summarized in Table 3. A comparison of the mechanical response on specimen #6216 with the skin intact on the left side and removed on the opposite side is given (Figure 3a). This specimen was tested on the standard wheel. Failure was defined as the point on the load-deflection curve at which an increasing deflection produced a decreasing force. This coincided with a depressed fracture of the zygomatic bone in the malar eminence area. The mean force and deformation for tests with the intact skin was 1455 N and 14.43 mm, respectively. The corresponding values when the skin was removed were 1211 N and 5.80 mm, respectively.

The deformations and forces indicated in Table 3 represent the variables recorded by the force gauge and LVDT of the MTS device. A displacement of 48.25 mm resulted in a force of 3226 N on the right zygoma of specimen #7628 tested with the EA wheel. The skin thickness at the zygomatic region was 16 mm, and the foam thickness of the EA wheel at the junction of the

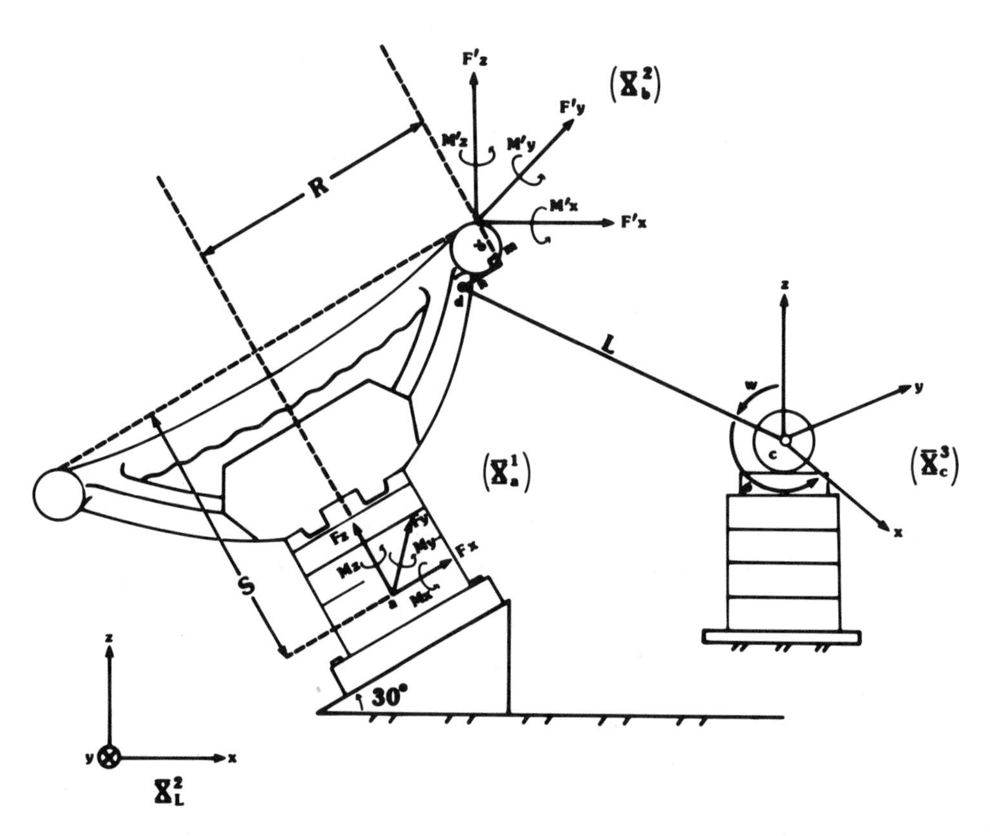

Fig. 2b. Cross-section of the steering wheel fixture which is impacted at the junction of the lower left spoke and rim. L denotes the variable arm length of the potentiometer system. The six-axis load cell is placed under the center of the hub. R denotes the radius of the wheel and S denotes the perpendicular distance to the measuring point on the load cell from the top plane of the steering wheel. The arrows indicated for the forces and moments at the Point a in the hub of the steering wheel corresponds to the forces and moments transferred from the steering wheel and acting on the six-axis load cell. Coordinate transformation procedures were adopted to determine the forces at the impact site on the zygoma (Appendix I).

Table 3
Quasistatic Studies at the Zygoma

SPECIMEN #	WHEEL TYPE	LOADING SURFACE	CONTACT SITE	SKIN PRESENT	SKIN THICKNESS (mm)	FORCE (N)	DEFLECTION (mm)	ZYGOMA FRACTURE
6216	Standard	C	Right Zygoma	No	--	1134	5.06	Yes
6216	Standard	C	Left Zygoma	Yes	12	1512	13.74	Yes
8947	Standard	C	Right Zygoma	No	--	1289	6.55	Yes
8947	Standard	C	Left Zygoma	Yes	20	1398	15.12	Yes
7628	EA	C	Right Zygoma	Yes	16	3226	48.25	No
8228	EA	F	Right Zygoma	Yes	12	2108	114.42	No
8891	EA	F	Right Zygoma	Yes	5	1172	46.14	Yes
8902	Standard	F	Right Zygoma	Yes	14	1539	32.37	Yes
8973	Standard	F	Left Zygoma	Yes	8	2352	38.08	Yes

Note: Piston deflection values for specimen #7628 and #8228 represent maximum values whereas all other values correspond to the fracture point.
C = Cut portions of the junction of the left lower spoke and rim.
F = Junction of the left lower spoke and rim of the entire steering wheel contacted the zygoma.
The load and deflections were recorded by the force gauge and LVDT on the MTS piston, respectively.

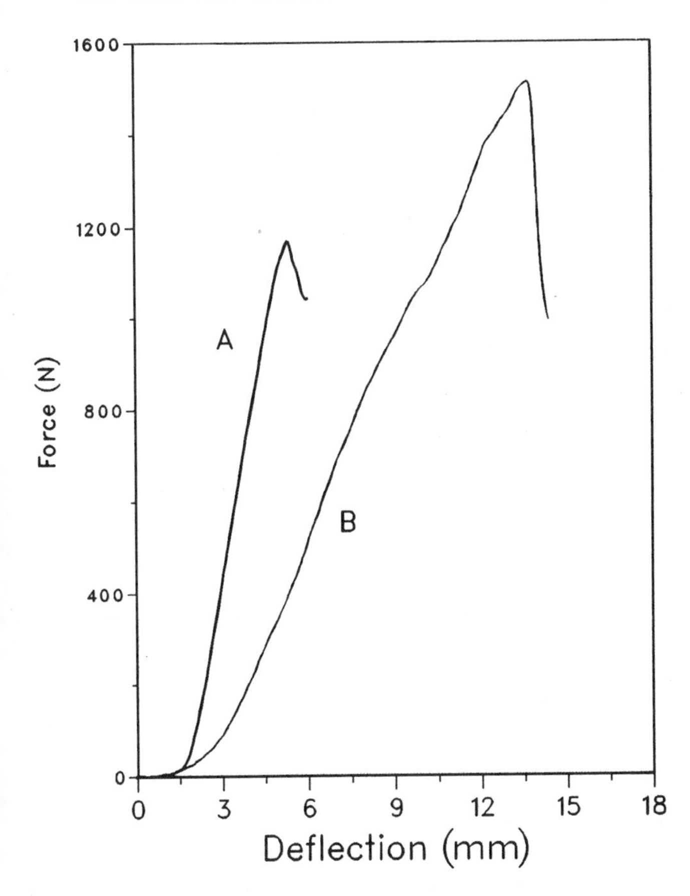

Fig. 3a. Comparison of the load-deflection response of specimen #6216 with the standard steering wheel as the loading surface. Load and deflections were recorded by the force gauge and the LVDT on the MTS piston, respectively. The specimen was tested on the left side without the skin (A). Right zygoma with skin intact (B).

spoke and the rim (at the loading site) was 18 mm. During the experiment, the loading surface initially contacted the zygoma, deformed the skin on the specimen with a simultaneous compression of the foam, and with further displacement the steering wheel wrapped around the zygomatic bone and nasal region. The deformability characteristics of the EA wheel, the compression of the skin, and the mounting configuration of the cut wheel all contributed to the wrapping around phenomenon of the test. This resulted in a minimal fracture of the right nasal bone with no accompanying fractures to the zygomatic bone.

Table 3 also includes the results from quasistatic experiments on the cadaver zygoma tested with the entire standard and EA steering wheels as contacting surfaces. A comparison of the mechanical response between the EA and standard wheels on specimens #8891 and #8902, respectively, is given in Figure 3b. An MTS piston force of 1539 N at a deformation of 32.37 mm for specimen #8902, and a force of 2352 N at a deformation of 38.08 mm for specimen #8973 produced tripod fractures of the zygoma. These two specimens were tested with the entire standard wheel. Out of the two specimens tested with the EA wheel, only one (specimen #8891) resulted in a tripod fracture of the zygoma which occurred at a force and deformation of 1172 N and 46.14 mm, respectively (Figure 3b). However, specimen #8228, also tested with the EA wheel, did not produce fracture of the zygomatic area. This specimen resisted a force of 2108 N at a deformation of 114.42 mm. The forces reported in Table 3 and the data for Figures 3a and 3b were recorded by the MTS piston. The peak transformed force at the loading site (junction of the left lower spoke and rim of the steering wheel with the cadaver zygoma) obtained using the six-axis Denton load cell data were within 5-8 percent of the peak external applied force as measured by the MTS piston. In general, the contact areas of the specimens measured at the loaded site were approximately twice for the EA wheel compared to the standard wheel. The structural properties of the entire wheel and a cut portion of the wheel were found to be dissimilar.

Quasistatic Studies of the Steering Wheels-
Force-deformation characteristics of the standard and EA steering wheels at the center of the largest unsupported part of the rim, and at the junction of the lower left spoke and rim are illustrated (Figure 4). The mechanical response depicted non-linear characteristics both at the

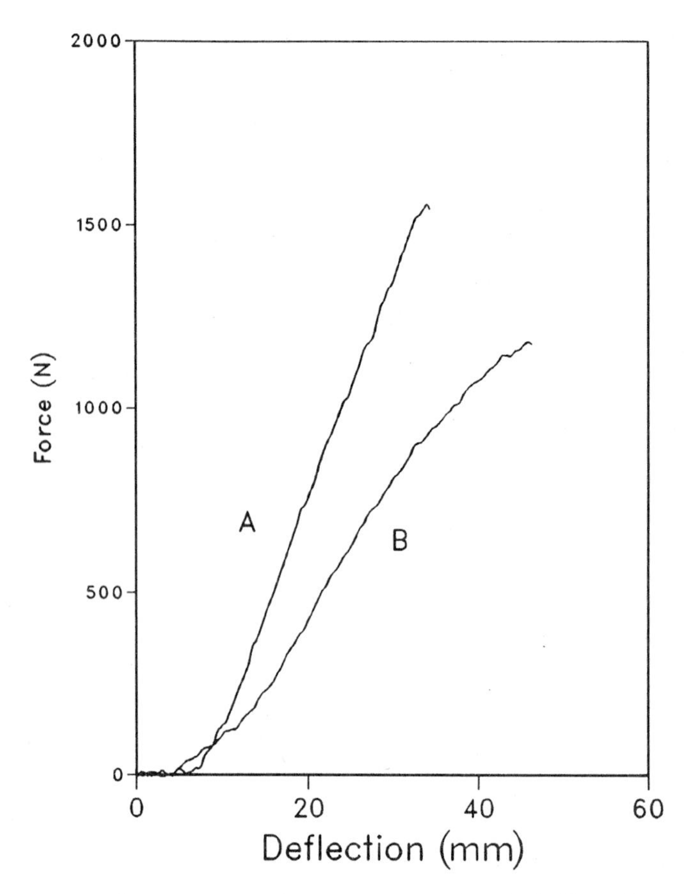

Fig. 3b. Comparison of the load-deflection response of specimen #8902 (Curve A) tested with the entire standard wheel and specimen #8891 (Curve B) tested with the entire EA wheel. Load and deflections were recorded by the force gauge and the LVDT on the MTS piston, respectively.

junction of the lower left spoke and rim (hardest part), and at the central region of the most unsupported part on the rim. The EA steering wheel absorbed 40.68 J (Joules) of energy compared to 67.39 J of energy absorbed by the standard steering wheel when subjected to 101 mm of deformation at the center of the largest unsupported part of the rim. The EA wheel absorbed 56.33 J of energy at a deformation of 76.6 mm at the hardest part. The corresponding force was 1609 N. The standard steering wheel in this region produced a force of 2989 N for 38.04 mm of deformation with an energy absorption of 51.50 J (Table 4). The energy absorbing capacity of the EA steering wheel was approximately 60% of the standard steering wheel at the softest part; at the hardest part, for a 25.58 mm deformation, it absorbed about 50% the energy of the standard steering wheel. Comparison of the energy absorbing characteristics at higher deformation levels was not possible because the standard steering wheel could not be subjected to larger magnitudes of displacement due to the limitations of the six-axis load cell in recording the forces. The EA steering wheel had approximately 50% of the stiffness compared to the standard wheel at the soft part and about 75% at the hard part. The stiffness was taken as the slope in the most linear part of the load-deflection curve. In all tests conducted with the EA steering wheel, permanent deformation occured at the hub where the spokes joined the central mounting ring.

DYNAMIC STUDIES

System Validation Tests - Validation results for the vertical-drop impact test system are included in Table 5. Figure 5a includes a plot of the force history in the Z direction at the impact site of the Z90.1 headform for the free fall and slider tests on the EA and standard steering wheels. The corresponding acceleration histories are illustrated in Figure 5b. From the recorded acceleration history at the CG of the headform, the theoretical peak dynamic forces in the Z-direction (Table 5, column 3) were calculated by multiplying the recorded acceleration by the computed mass of the specimen. This force was compared with the

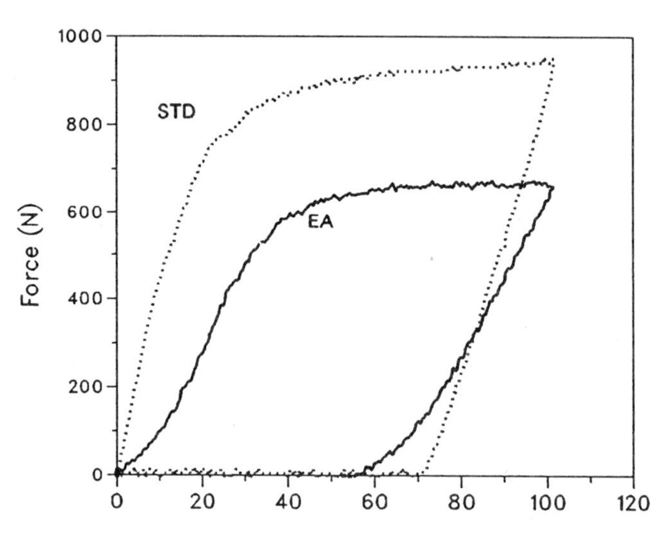

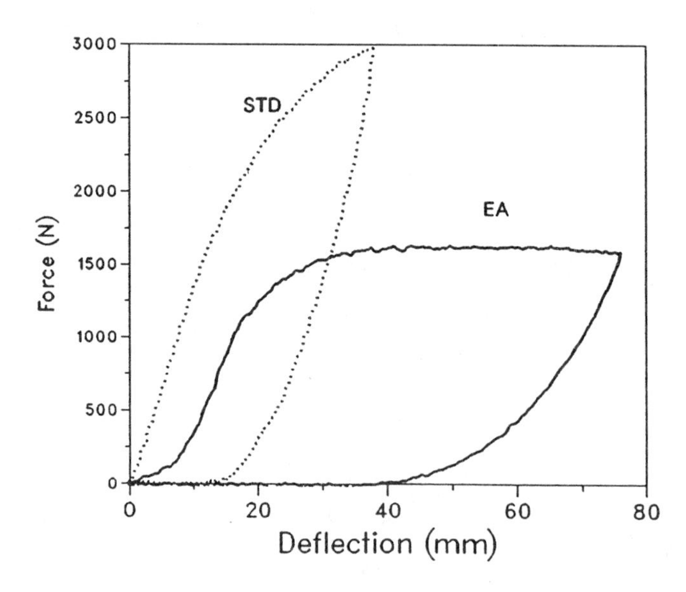

Fig. 4. Force-deformation characteristics of the standard (STD) and EA steering wheels tested at a quasistatic displacement rate of 2.5 mm/s using the MTS device. Top figure: Test at the mid-point on the unsupported rim. Bottom figure: Test at the junction of the lower left spoke and rim.

Table 4
Quasistatic Studies of the Steering Wheels

STEERING WHEEL TYPE	FORCE APPLICATION AT	FORCE (N)	DEFLECTION (mm)	STIFFNESS (N/mm)	ENERGY (J)
EA	Spoke/rim	1609	76.20	110.25	56.33
EA	Rim	664	101.20	21.36	40.68
Standard	Spoke/rim	2989	38.04	139.06	51.50
Standard	Rim	951	101.67	41.27	67.39

NOTE: Stiffness is taken as the slope of the load-deflection diagram in the most linear part of the curve. Energy is calculated by taking the area of the hysteresis curve of each test.

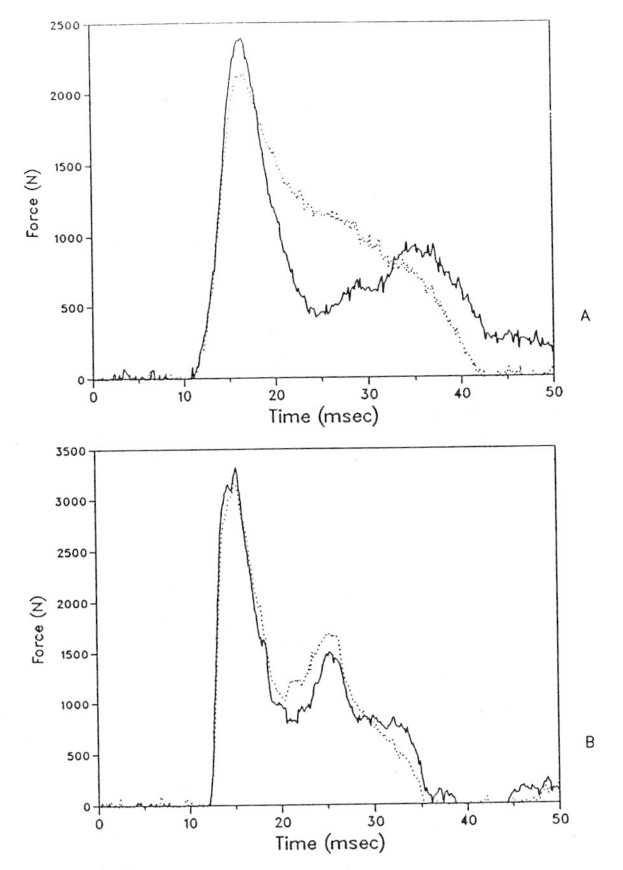

Fig. 5a. Transformed forces (Z-direction) at the impact site on the Z90.1 head form. The vertex of the Z90.1 headform impacted the junction of the left lower spoke and rim of the steering wheel. Solid lines, free-fall tests and dotted lines, slider. Upper: EA wheel. Lower: standard wheel, at impact velocity of 3.13 m/s.

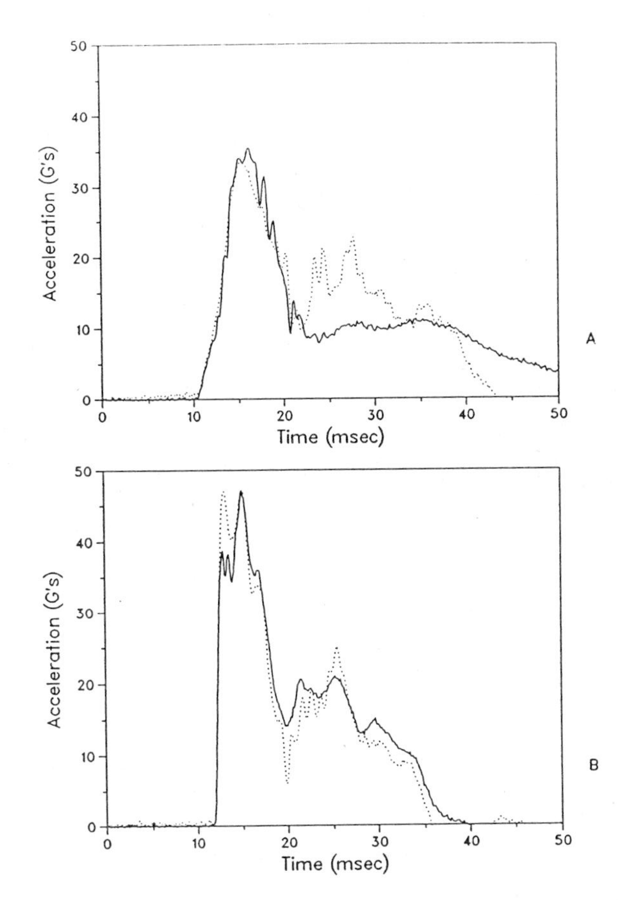

Fig. 5b. Acceleration response (Z-direction) for free-fall (solid lines) and slider (dotted lines) onto the EA wheel (top) and standard wheels (bottom) with Z90.1 head form, impact velocity 3.13 m/s. The vertex of the Z90.1 headform impacted the junction of the left lower spoke and rim of the steering wheel.

transformed force in the Z-direction at the impact site (Table 5, column 4). Impact on the standard steering wheel gave an average difference of 2.78% in the forces while impact with the EA steering wheel resulted in an average difference of 6.66%. Impact with the standard steering wheel gave a 0.23% difference in the peak accelerations between free fall (47.10 G) and slider (47.21 G) conditions, whereas, the EA steering wheel demonstrated a 4.36% difference between the free fall (35.56 G) and slider (34.01 G) experiments.

Although the transformed resultant force at the impact site was computed, it was not used for comparison purposes because the accelerations were recorded at the center of the Z90.1 headform with a uniaxial accelerometer. The validation studies indicated that the slider

Table 5
Validation Studies with Z90.1 Headform
(impact velocity 3.13 m/s)

WHEEL TYPE	DROP CONDITION	THEORETICAL FORCE (N)	TRANSFORMED FORCE (N)	DIFFERENCE BETWEEN FORCES (%)	PEAK ACCELERATION (G)	DIFFERENCE BETWEEN ACCELERATIONS (%)
Standard	Free fall	3307.3	3324.8	0.53	47.10	----
	Vertical drop test device	3334.9	3166.8	5.04	47.21	0.23
EA	Free fall	2492.2	2400.4	3.68	35.56	----
	Vertical drop test device	2384.3	2154.3	9.65	34.01	4.36

Note: Theoretical force is computed using the relation $F = m \cdot a$, where a denotes the acceleration and m denotes the mass.
The acceleration (in G's) is given in Column 6.

system was similar in characteristics to that of the free fall system.

Vertical Drop Tests on the Zygoma - Figures 6a and 6b illustrate respectively the generalized force histories recorded by the six-axis load cell and the triaxial accelerometer data for specimen #7589 which was impacted at the right zygoma onto the EA steering wheel at a velocity of 2.68 m/s. The Y component of the force is of a significantly less magnitude. Because the steering wheel was placed at an angle of 30 degress with respect to the horizontal and the load cell was mounted under the center of the hub orthogonal to the plane of the wheel, Fx and Fz forces are of comparable magnitude. The Mx and Mz components are also small compared to the My component. The transformed generalized force histories are depicted in Figure 6c. Forces in the X' and Y' directions are smaller compared to the Z' (vertical) direction. Moments along the X' and Z' axes are also smaller in magnitude. Only after the steering wheel starts to deflect (after reaching the peak impact force in the Z' direction), bending of the wheel results in a buildup of the moment about the Y' axis. The resultant acceleration at the left posterior parietal region, the transformed resultant force at the impact site, and the dynamic steering wheel deformations exclusive of foam compression are given (Figure 6d).

A summary of the biomechanical data and fracture severity data for specimens impacted on EA and standard steering wheels is given (Table 6). Nine drops were done at impact speeds ranging from 2.68 to 6.93 m/s onto the EA steering wheel. The peak transformed resultant impact forces at the zygoma ranged from 1523 to

Table 6
Summary of Biomechanical Data from Drop Tests

SPECIMEN #	IMPACT VELOCITY (m/s)	IMPACT SITE LEFT/RIGHT	CONTACT AREA (mm^2)	RESULTANT IMPACT FORCE (N)	DYNAMIC STEERING WHEEL* DEFORMATION (mm)	RESULTANT IMPACT ACCELERATION (G)	HEAD INJURY CRITERIA (HIC)	FRACTURE[b] SEVERITY
				EA WHEEL				
5149	6.93	Left	1177	4604	60.5	70-75	383	4
7690	6.93	Right	1390	4565	60.3	70-75	341	4
5421	5.59	Right	1296	2289	47.6	45-50	143	4
5812	4.47	Right	1104	2447	35.5	25-30	123	4
4645[a]	3.58	Right	----	2022	----	-----	---	4
9234	3.58	Right	888	1770	18.9	20-25	58	4
3905	3.13	Right	891	2029	20.6	30-35	63	3
8250	3.13	Left	1062	2113	18.6	25-30	65	2
7589	2.68	Right	629	1523	18.8	15-20	33	0
				STANDARD WHEEL				
8888	3.13	Left	594	2571	22.9	40-45	109	4
8151	2.68	Right	507	1564	16.1	25-30	42	3
5912	2.68	Right	407	1733	10.0	25-30	45	2
7643	2.24	Left	302	1909	----	25-30	45	4
9274	2.24	Left	352	1499	8.2	15-20	23	4
6025	2.01	Left	363	1359	10.6	15-20	30	0

*: The deformations are exclusive of foam compressions of the EA wheel.

a: Resultant from the X and Z channels of force-time data. The Y channel data was lost during transferring.

b: See text for fracture classification and description.

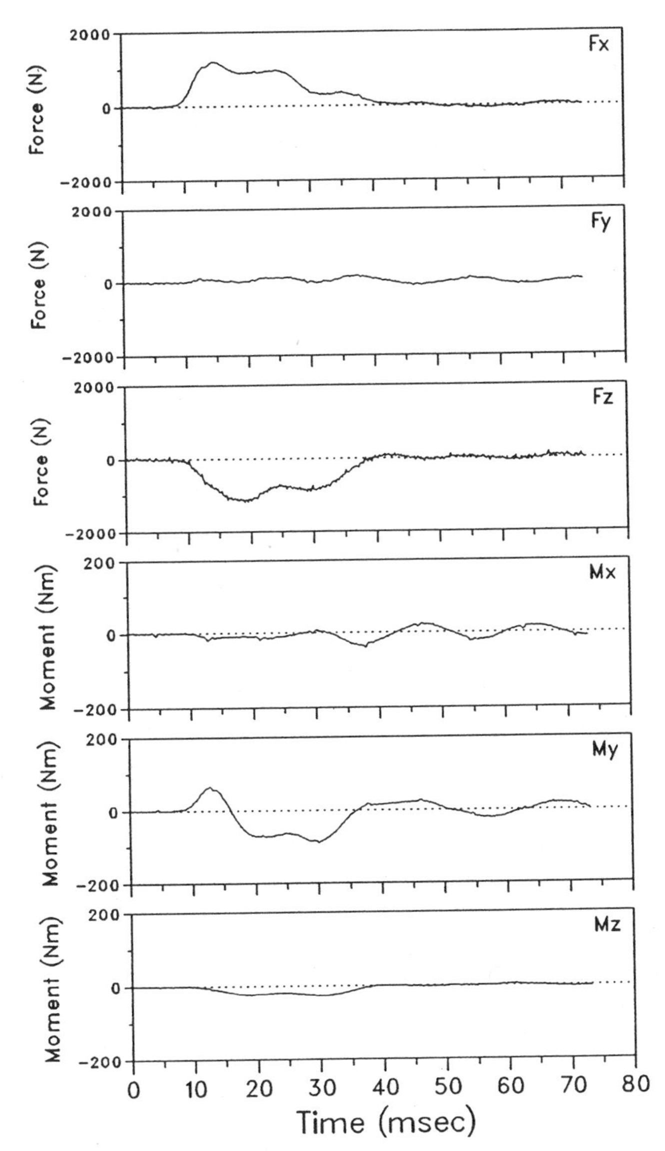

Fig. 6a. Generalized force histories recorded by the six-axis load cell. The top three traces correspond to the forces and the bottom three correspond to the moments along the X, Y, and Z axes respectively. This data corresponds to specimen #7589 which was impacted on the EA wheel at a velocity of 2.68 m/s. Note that the Y component of the force is of a significantly less magnitude. Because the steering wheel was placed at an angle of 30 degrees with respect to the horizontal and the load cell was mounted under the center of the hub orthogonal to the plane of the wheel Fx and Fz forces are of comparable magnitude. The angle between these two principal components was approximately 30 degrees. Note that the Mx and Mz components are also small compared to the My component. Coordinate system response is shown in Figure 2.

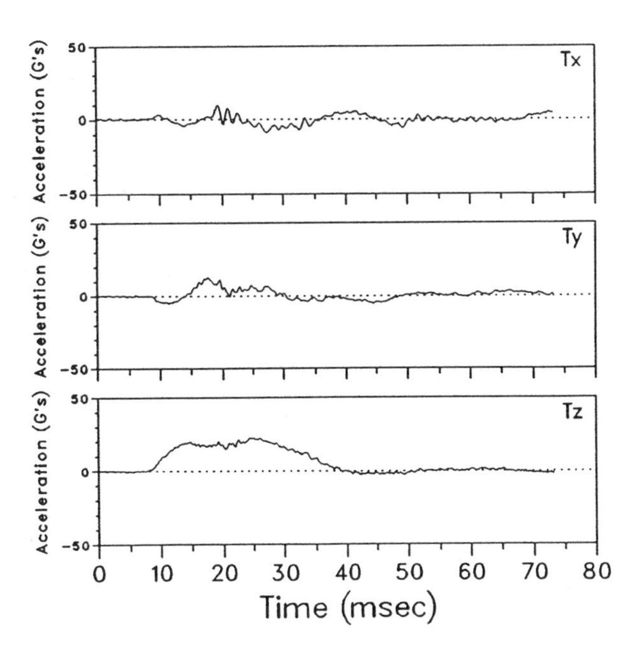

Fig. 6b. Acceleration in X (top), Y (middle), and Z (bottom) directions recorded by the triaxial accelerometer placed at the posterior parietal region opposite to the impact of the specimen #7589. The specimen was impacted onto the EA wheel at a velocity of 2.68 m/s.

4604 N with resultant peak accelerations varying from 15 to 75 G's. Many of the studies showed multiple force and acceleration excursions (Figure 7). Dynamic steering wheel deformations ranged from 19 to 60 mm. Five specimens were tested on the standard steering wheel at impact speeds ranging from 2.01 to 3.13 m/s. Transformed resultant peak forces ranged from 1359 N at 2.01 m/s to 2571 N at 3.13 m/s velocity; resultant peak accelerations at these velocities varied from 15 to 45 G's. Dynamic peak deformations of the steering wheel ranged from 8 to 23 mm. Analysis of the 16 mm films (1000 frame/sec) confirmed the impact velocities. The films suggest that the specimen initially compresses at the impacting zygomatic bone region. The wheel is then deformed and the tissue is loaded to a greater extent when fracture probably occurs. For example, in a 3.1 m/s drop, the tissue and steering wheel compressed in approximately 5 to 6 ms without steering wheel displacement. This was followed by the deformation of the wheel and the specimen with the maximum force occurring at about 10 to 12 ms. The tissue was in contact with the steering wheel throughout this time and rebounded at approximately 20 ms as the wheel unloaded. The specimens rebounded approximately 25 to 30% of the drop height for impacts of 2.68 to 3.1 m/s. The second contact occurred approximately 300 ms following the first. At these speeds, displacements between the steering wheel surface and the tissue were in the range of 20 to 30 mm. The greatest displacements were observed with the EA wheels. No lacerations of the tissue occurred in any of the tests. Maximum permanent deformations of the EA wheel occurred in the hub region at the junction of the spoke and hub ring.

The contact areas for the EA wheel studies were approximately two to three times those measured with the standard wheel (Table 6). The HIC ranged from 23 to 383 for all specimens.

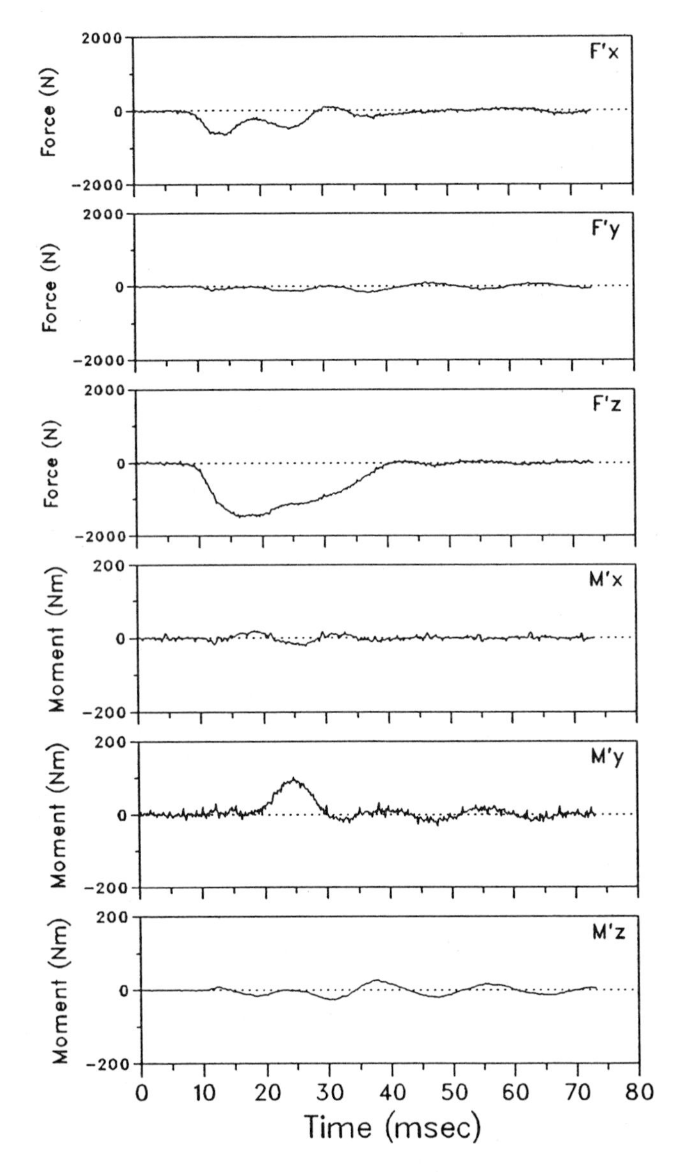

Fig. 6c. Transformed generalized force history at the impact site. Top three traces denote the X, Y, and Z components of the force and the bottom three traces denote the moments along X, Y, and Z axes respectively. Forces in the X and Y directions are smaller compared to the Z (vertical) direction. Moments along the X and Z axes are also smaller in magnitude. Only after the steering wheel starts to deflect (after reaching the peak impact fore in the Z direction), bending of the wheel results in a build up of the moment in the Y direction. (See text for details).

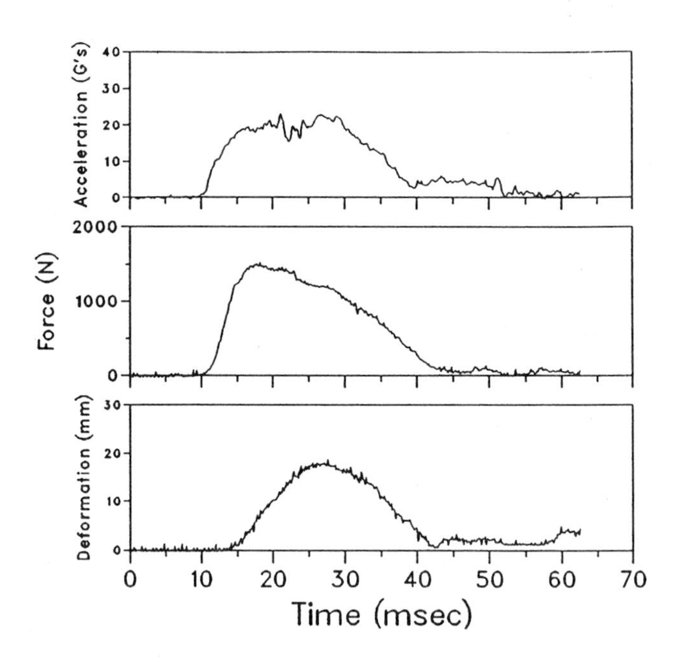

Fig. 6d. Resultant acceleration (top), transformed resultant force (middle) and the dynamic steering wheel deformation exclusive of foam compressions (bottom) for specimen #7589 which was impacted at a velocity of 2.68 m/s.

The forces and accelerations at the higher velocities tended to be more complex than those at the lower levels. The force for fracture could not be determined at the higher velocities since the preparations were probably loaded beyond their fracture threshold levels. Furthermore, many of the preparations sustained bilateral fractures. Specimen #3905 impacted at 3.13 m/s sustained a sphenoid fracture of the skull without facial fracture. The sphenoid fracture is a serious injury. Specimen #7588 sustained an impact velocity of 2.68 m/s without fracture. These two specimens had less complex force time curves. It was difficult to assign an exact value to the head accelerations because of their oscillating nature. However, the higher impact velocities produced larger accelerations. The calculated forces based on the system mass and accelerations did not closely predict the measured forces. Steering wheel displacement routinely commenced several milliseconds after the initial force peak. The standard steering wheel test results are shown in Figure 8. The force and acceleration traces also tend to be more complex as the fractures are more severe. Specimen #6025 with a drop velocity at 2.01 m/s did not sustain a facial fracture. The steering wheel deformations, accelerations and HIC values tended to increase with impact velocity.

PATHOLOGICAL OBSERVATIONS - Patho-anatomic data obtained from gross dissection (at the impact site), 2-D CT axial images (of the specimen), and defleshed skulls are included (Table 7a-c). The fracture severity based on the findings of Tables 7a-c is graded on a scale of 1-4: 1-minor, 2-moderate, 3-severe and 4-most severe. Table 8, adapted from a study on the incidence of hospital-treated facial injuries from vehicles by Karlson, includes a description of the severity groups for facial fractures (14). Figure 9 shows a 3-D CT facial reconstruction of an automobile crash victim. Bilateral fractures of the maxilla, zygoma, nasal and orbital regions are observable. Three-dimensional CTs are routinely used in our institution for reconstruction of facial malformations and for the evaluation of fractures secondary to trauma. Figure 10 shows

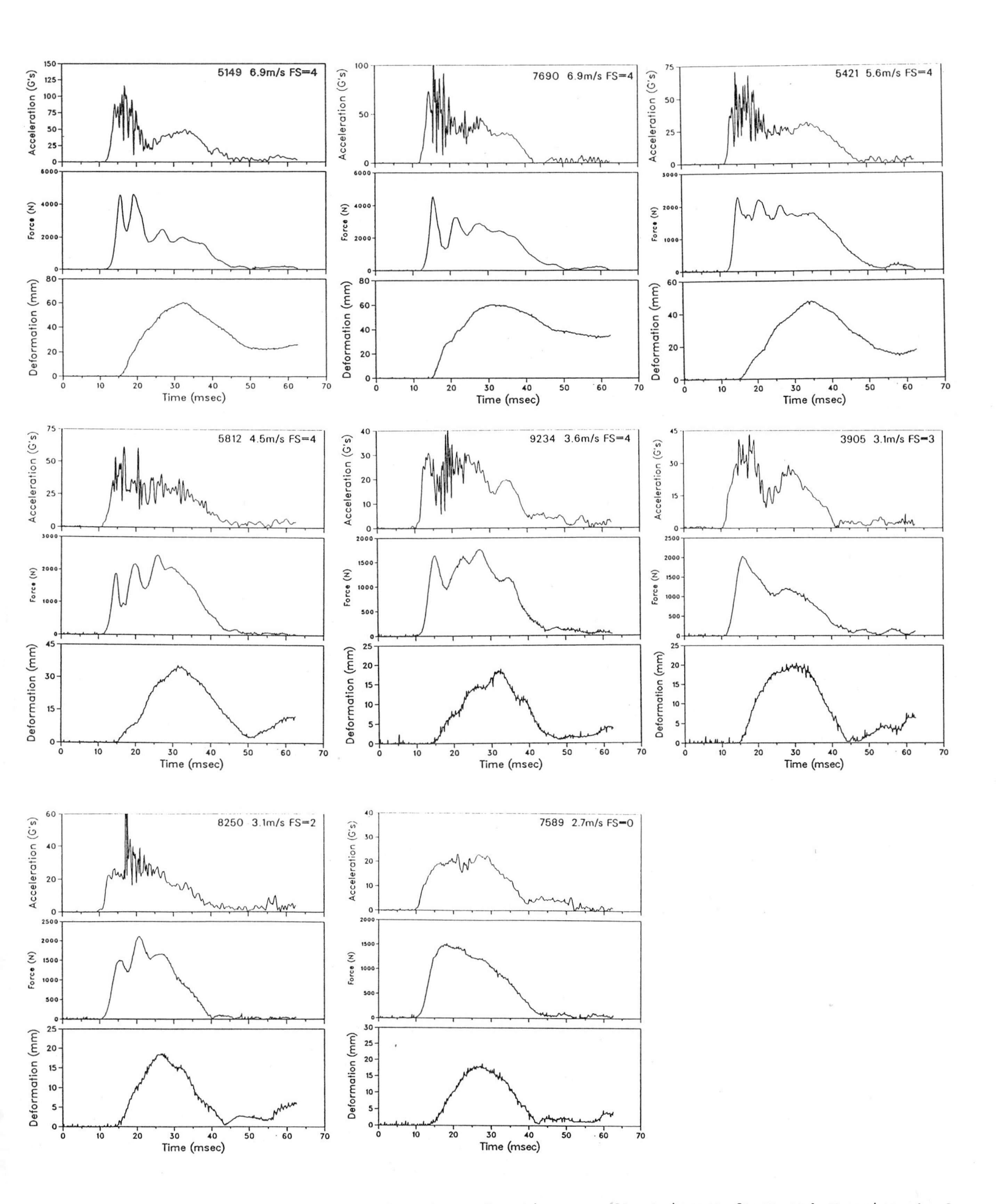

Fig. 7. EA wheel studies. Resultant impact acceleration, resultant impact force and steering wheel deformation for specimen #'s 5149, 7690, 5421 (top left to right), #'s 5812, 9234, 3905 (middle left to right), and #'s 8250, 7589 (lower left to right). Impact velocity and fracture severity (FS) are indicated for each test.

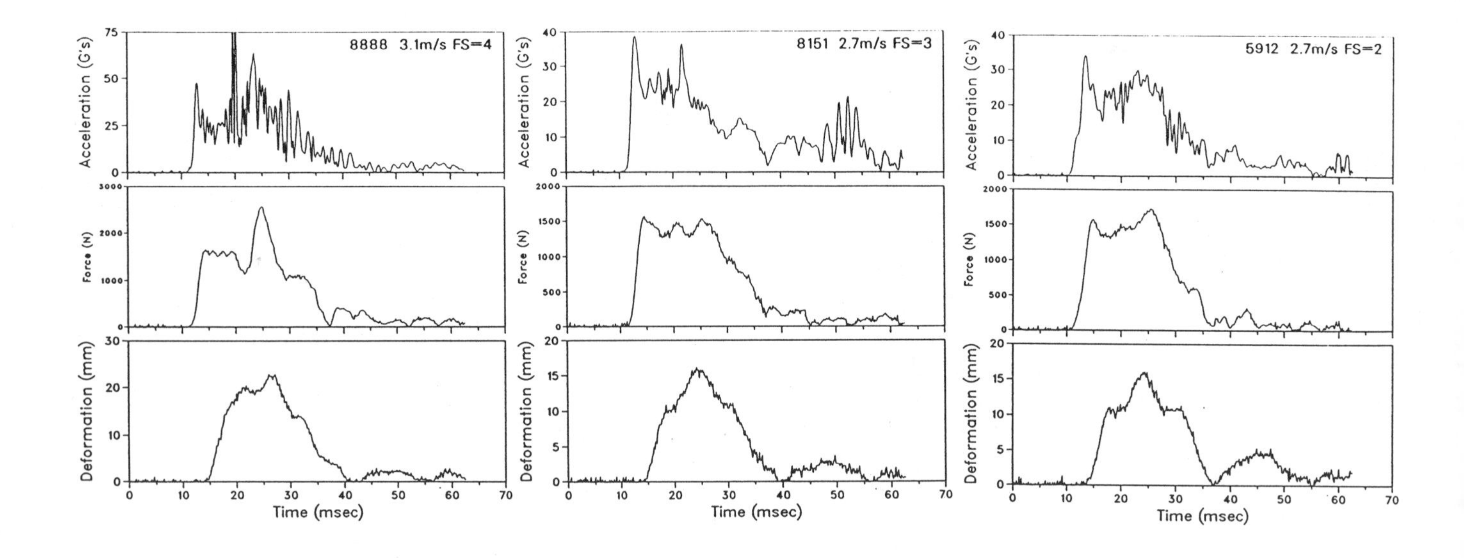

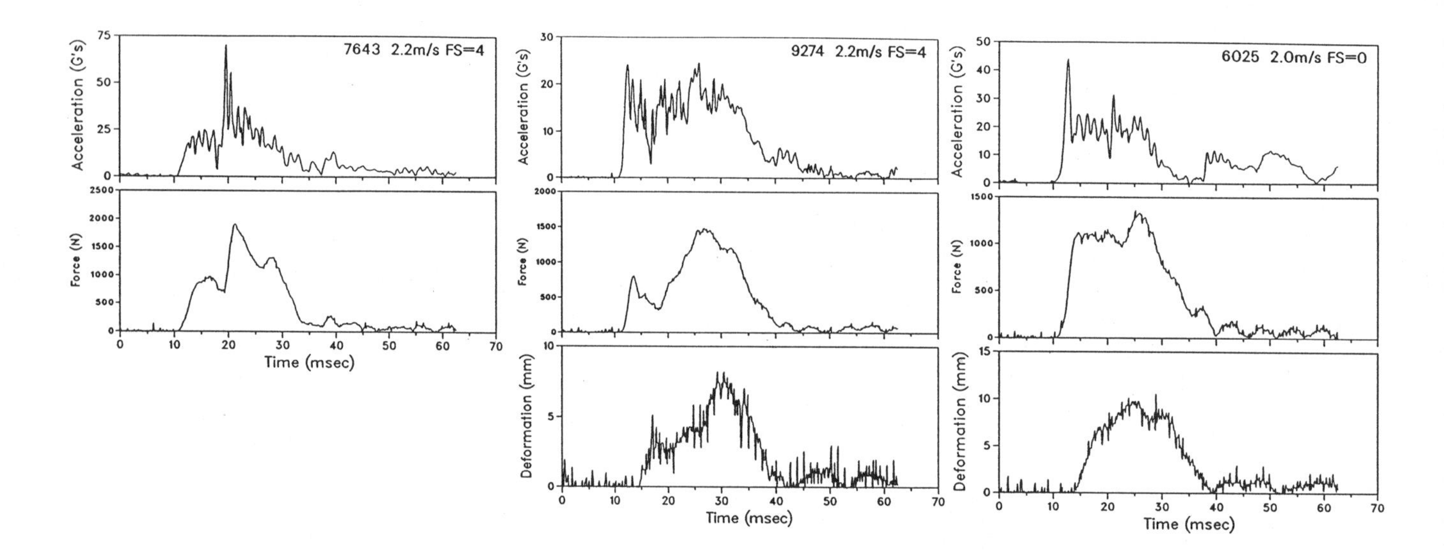

Fig. 8. Standard wheel studies. Resultant impact acceleration, resultant impact force and steering wheel deformation for specimen #'s 8888, 8151, 5912 (top left to right) and #'s 7643, 9274, 6025 (bottom left to right). Impact velocity and fracture severity (FS) are indicated for each test.

the right lateral and left lateral oblique and frontal views of the defleshed skull of specimen #8891. The fracture lines are highlighted with black lines. The three views of the defleshed specimen #8902 tested on the standard steering wheel is shown in Figure 11. For both of these specimens, tripod fractures can be seen on the impacted side with fractures on the opposite side as well.

In the EA steering wheel experiments, fractures occurred in the zygomatic region and orbits at impact velocities of 6.93, 5.59, 4.47 and 3.58 m/s (Figures 12 a,b). Specimen #3905 had a sphenoid fracture without facial fractures (Figure 12b). All these specimens had bilateral fractures. Specimen #8250 impacted at a velocity of 3.13 m/s showed a non-displaced fracture of the left zygomatic arch and maxilla. An impact velocity of 2.68 m/s did not induce fractures in specimen #7589 (Figure 12c).

Studies with the standard steering wheel resulted in depressed fractures of the zygomatic region at 3.13 m/s velocity but no fractures with an impact velocity of 2.01 m/s. In specimens #8151 and 5912 less severe fractures were found at 2.68 m/s velocity, but more severe maxilla and tripod fractures were observed at a velocity of 2.24 m/s (Figures 13a,b). Specimen #'s 9274 and 7643 had severe unilateral fractures to the impacted side of the face (Figure 13b). Specimen #6025 did not sustain fractures. A widened suture line is seen in Figure 13b which is not pathological. Some differences were observed in the 2-D CTs since the images may have been parallel to the fracture plane. The gross dissection findings were similar to the defleshed skull findings.

TABLE 7a
Pathology from Dynamic Drop Tests:
Gross Dissection Findings

	IMPACT SITE		
SPECIMEN #	Zygomatic Arch	Tripod	Maxilla
5149		X	X
7690	X	X	X
5421	X	X	X
5812	X	X	X
4645	X	X	X
9234	X	X	X
3905			Hairline
8250	X		X
7589	NONE	NONE	NONE
8888	X	X	X
8151			Hairline
5912			X
7643		X	X
9274	X	X	X
6025			Hairline

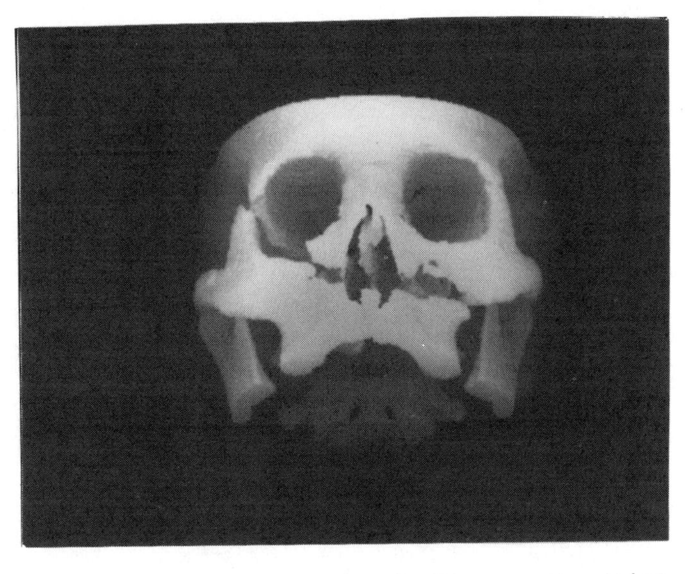

Fig. 9. Three-dimensional CT reconstruction of a patient injured in an automobile collision. Bilateral fractures of the maxilla, zygoma, nasal and orbit regions are observed.

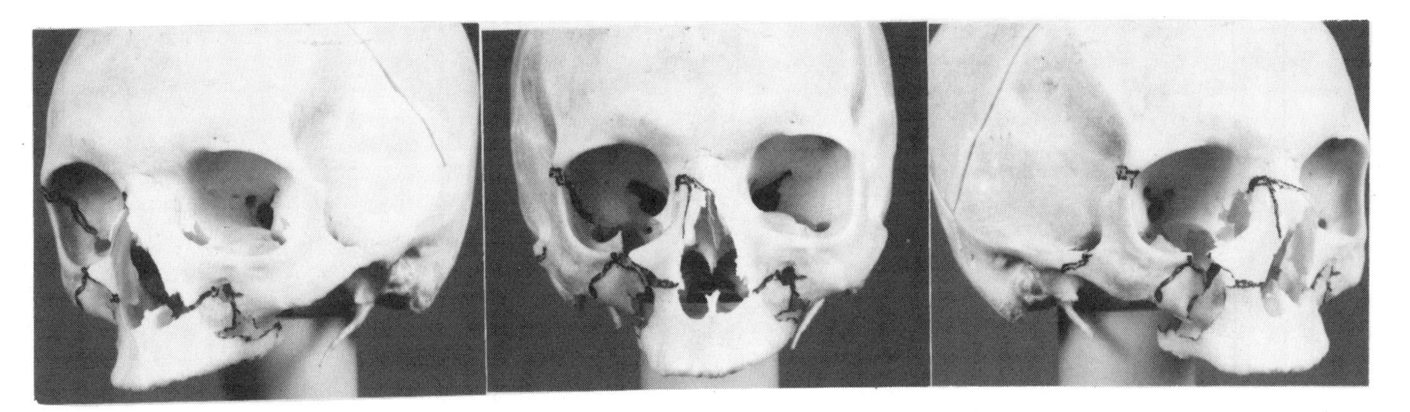

Fig. 10. Defleshed skull photographs of specimen #8891 which was tested quasistatically using the entire EA steering wheel.

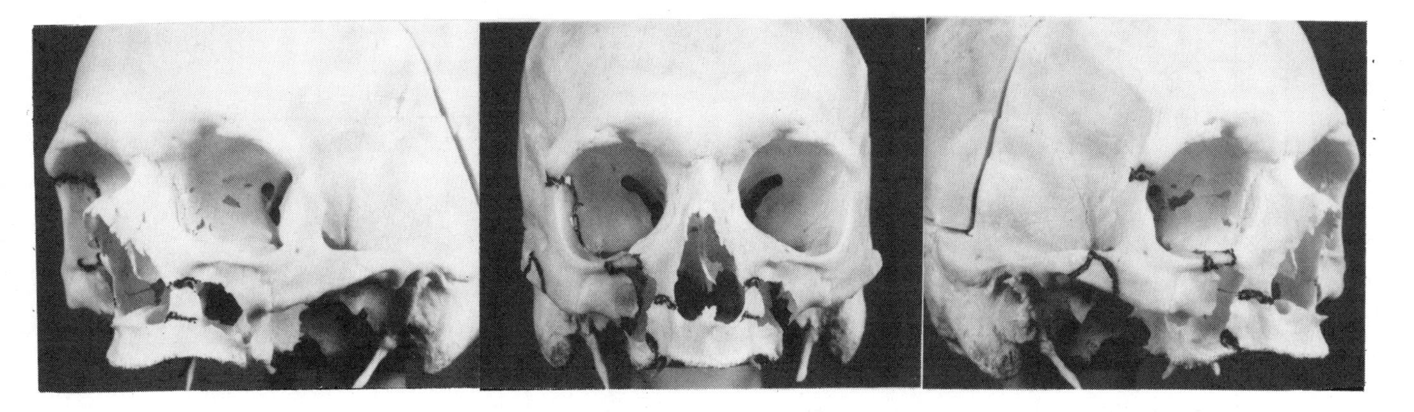

Fig. 11. Defleshed skull photographs of specimen # 8902 which was tested quasistatically using the entire standard steering wheel.

Table 7b
Pathology from Dynamic Drop Tests: 2-D Computed Tomography Findings

SPECIMEN #	Impact Site Left/Right	IMPACT SITE Zygomatic Arch	Tripod	Orbits	Maxilla	Other	OPPOSITE TO THE IMPACT SITE Zygomatic Arch	Tripod Zone	Orbits	Maxilla	Other
5149	L		D			Nasal (LW) N/D				N/D (LW)	
7690	R	D	D		N/D (LW)		D				
5421	R										
5812	R				D (LW) (AW) (PW)					N/D (LW) (ALW)	
4645	R	D	D		D (LW)		None				
9234	R		N/D		N/D (LW)		None				
3905	R	None				Sphenoid Bone	None				
8250	L	N/D					None				
7589	R	None					None				
8888	L	N/D		N/D (LW)	D (AMW)	Vertical fx at base of skull	None				
8151	R		N/D		N/D (LW)			N/D	N/D (LW)		
5912	R				N/D (AW)						
7643	L	N/D	D		N/D (LW) (MS)	Temporal bone N/D					
9274	L	N/D	N/D		N/D (LW) (MS)	Temporal bone N/D			N/D (LW)		
6025	L	None					None				

D: Displaced PW: Posterior wall N/D: Non-displaced ALW: Anterior lateral wall LW: Lateral wall MS: Maxillary sinus

Table 7c
Pathology from Dynamic Drop Tests: Defleshed Skull Findings

SPECIMEN #	Impact Site Right/Left	IMPACT SITE Zygomatic Arch	Tripod	Orbit	Maxilla	Other	OPPOSITE TO IMPACT SITE Zygomatic Arch	Tripod	Orbit	Maxilla	Other
5149	L		X	X	X				X	X	
7690	R	X	X	X	X		X		X		Nasal
5421	R	X	X	X	X				X	X	
5812	R	X	X	X	X	Palate				X	Palate
4645	R	X	X	X	X	Nasal				X	Palate
9234	R	X	X	X	X	Nasal				X	Palate
3905	R					Body of Sphenoid	None				
8250	L	X			X		None				
7589	R	None					None				
8888	L	X	X	X	X		None				
8151	R				Hairline		None				
5912	R				X		None				
7643	L	X	X	X	X		None				
9274	L	X	X	X	X		None				
6025	L	None					None				

Table 8
Severity Groups for Facial Fractures
(Adapted From Ref 14)

GROUP 1 MINOR	GROUP 2 MODERATE	GROUP 3 SEVERE	GROUP 4 MOST SEVERE
Nasal bones Mandible - ramus Mandible - unspecified Maxilla - nasal spine	This group if open, comminuted and/or displaced		
	Maxilla - LeFort I Maxilla - unspecified Mandible - body Mandible - condyle Mandible - alveolus Mandible - subcondylar Orbit Zygoma	This group if open, comminuted and/or displaced	
		Base of skull (basilar, ethmoid orbital roofs sphenoid)	Maxilla, LeFort III More than one group 2 fracture One Group 3 fracture and one Group 1 or Group 2 fracture
	More than one Group 1 fracture Bilateral Group 1 fractures	Maxilla, LeFort II One Group 2 fracture and one Group 1 fracture	This fracture if open, comminuted, and/or displaced

Table 9
Comparison of Cadaver Zygomatic Fracture Loads (Ref #9)
with 645 mm^2 and 3355 mm^2 Impaction Areas
(Ratio of Areas 1:5.2, Ratio of Equivalent Diameters 1:2.27)

SPECIMEN NUMBER*	FRACTURE FORCE (N) WITH IMPACTOR AREA OF 645 mm^2	3355 mm^2	RATIO OF FRACTURE FORCES
772-R	1120.9	----	----
772-L	----	2593.3	2.31
812-L	1663.6	----	----
812-R	----	3362.8	2.02
905-L	1120.9	----	----
905-R	----	1761.5	1.57
907-L	1548.0	----	----
907-L	----	2882.4	1.86

*Specimen numbers used in Ref. 9 are retained in this table.

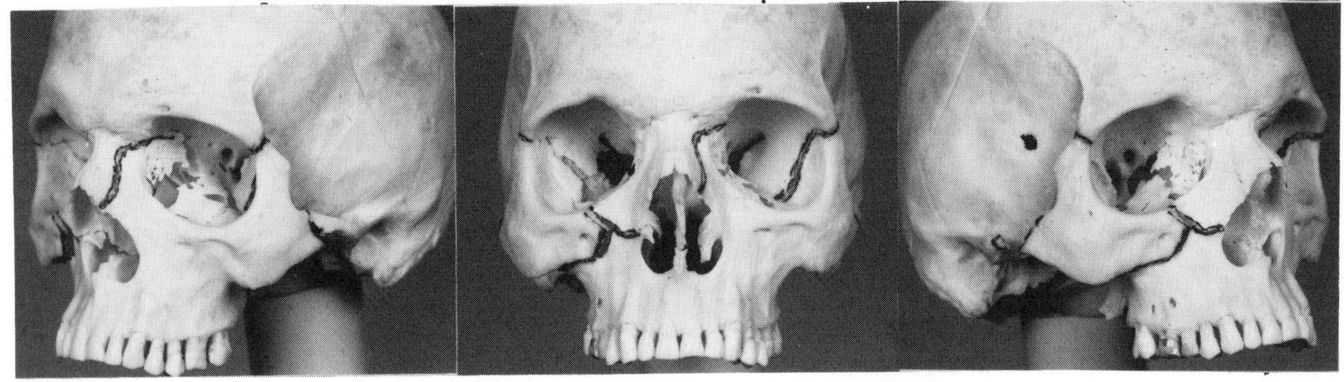

SPECIMEN #7690

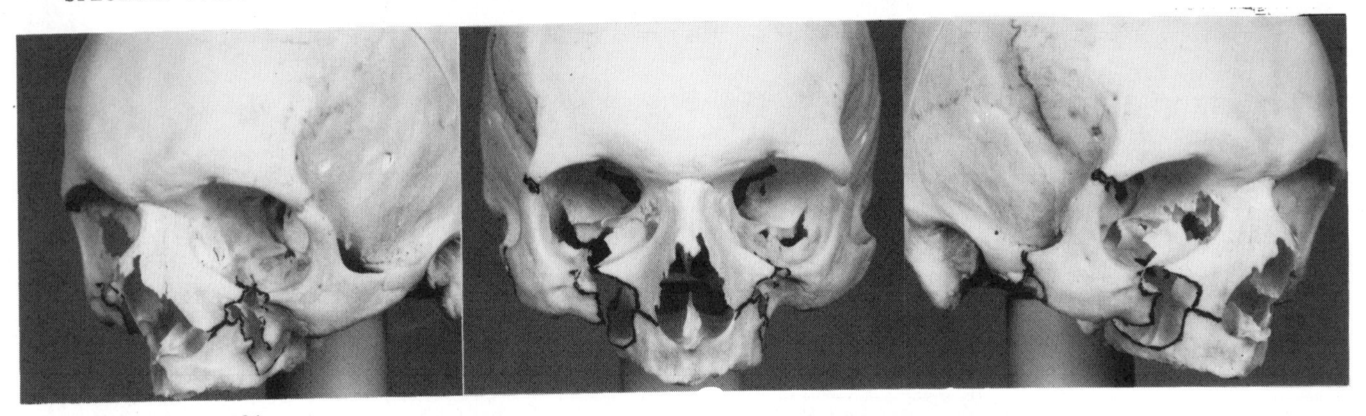

SPECIMEN #5421

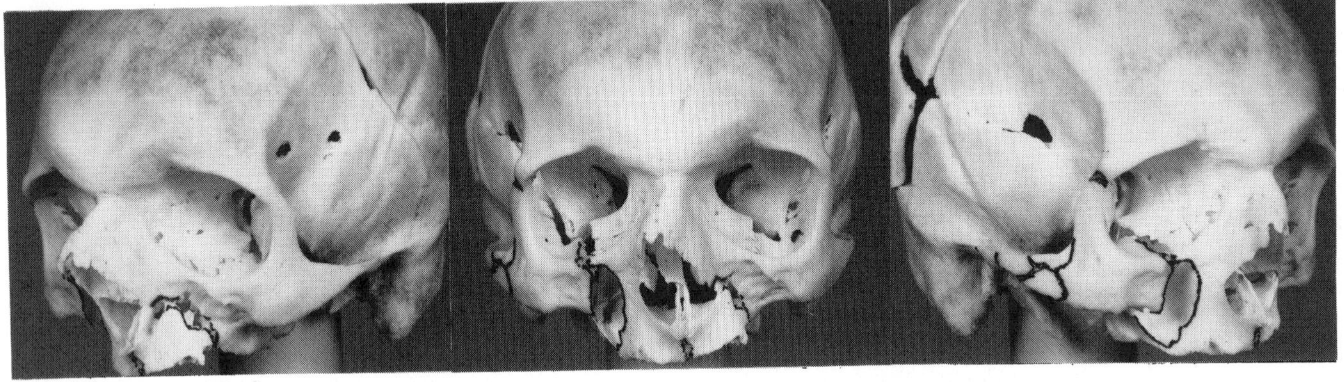

SPECIMEN #5812

Fig. 12a. Defleshed skull photographs of specimens #7690 , 5421, and 5812 which were tested at impact velocities of 6.93, 5.59, and 4.47 m/s respectively with the EA wheel.

DISCUSSION

Previous studies to determine mitigation of injury with steering systems indicate that a large fraction of facial harm occurs at speeds below 8.94 m/s velocity change. For comparison, the fraction of abdominal and chest harm at 13.40 m/s is about the same as that of facial harm at 8.94 m/s (4). The original studies of LeFort, which produced the classic fractures associated with his name are routinely cited in the literature (15). Unfortunately, no biomechanical measurements were made; however, the great lines of weakness in the face were documented. Various attempts have been made in the past to determine the tolerance levels for brain injury. Briefly, the early studies of Gurdjian, Lissner, Eiband and Stapp provide the basis for much of our understanding of human head injury tolerance. These studies have been recently reviewed (30,31). However, many of these studies were conducted with impact to the frontal region of the head because of the interest in the automotive injury consistent with injuries to the head in frontal collisions. Swearingen advanced an interesting model to determine various fracture levels of facial components with molded accelerometer blocks to fit various portions of the face and finally the entire surface of the face (35). He indicated that the weakest part of the face is the nose followed by the zygomatic prominence, condyle of the mandible, and finally the forehead has the greatest tolerance. In 1967 Hodgson showed a

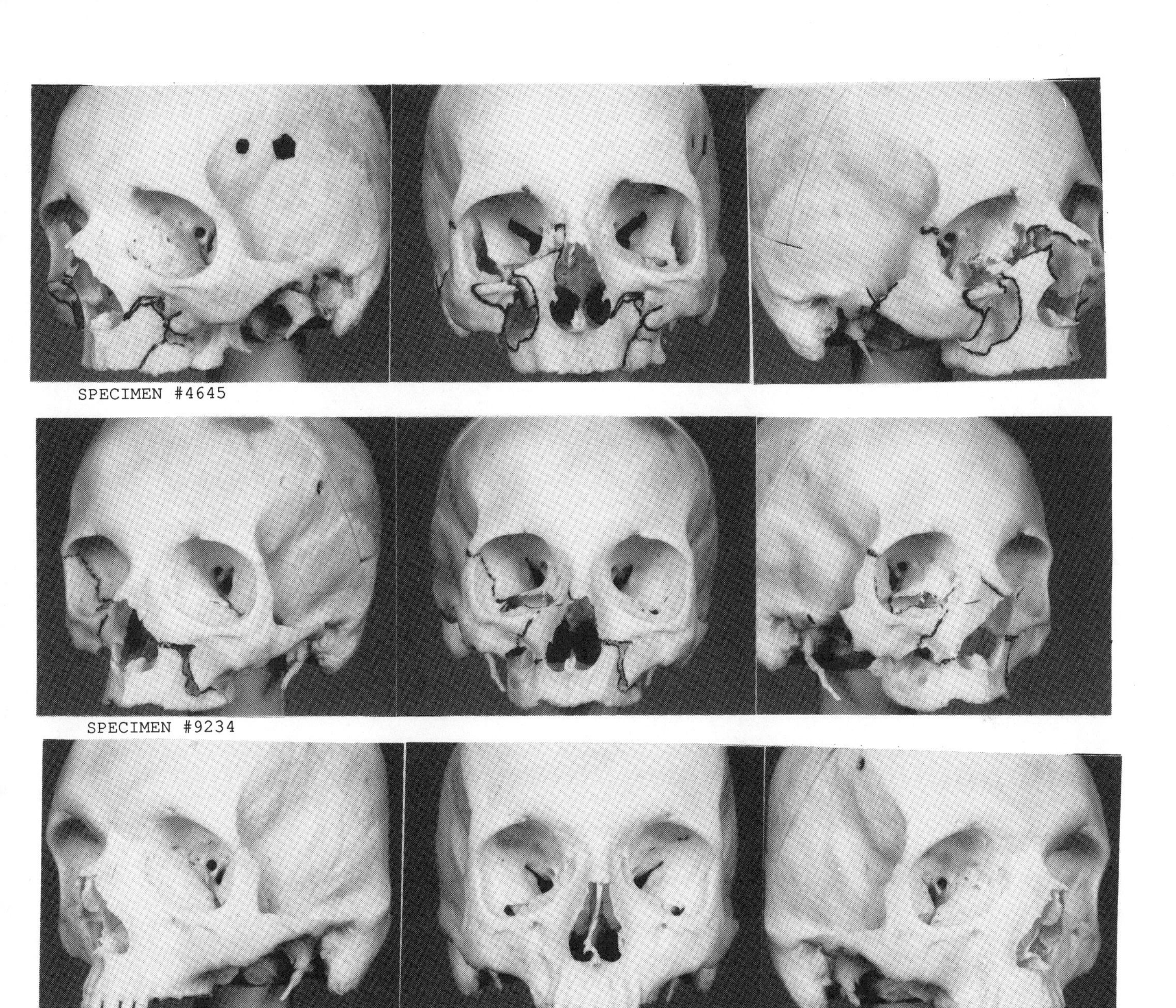

Fig. 12b. Defleshed skull photographs of specimens # 4645, 9234 and 3905 which impacted onto the wheel at velocities of 3.58, 3.58 and 3.13 m/s respectively.

temporal dependence for facial fracture based on the time duration of the impact(9). Impacts lasting beyond 4 ms produced fracture of the zygomatic bone near 880 N with a 645 mm^2, 28.7 mm diameter impactor. However, forces up to approximately 4450 N were tolerated for durations of 3 ms or less. It was found that for impulses lasting longer than 4 ms, the tolerance of the zygomatic bone was raised by a factor of 1.5 to 2.5% if the load was distributed over an area of 3355 mm^2 with a 65 mm diameter striker. An accelerometer was mounted on the rear of the striker to determine the force as a function of time. An accelerometer was mounted opposite the impact region to record the head response. The accelerations recorded from the head were oscillatory, and consistent with the study of Hodgson and Patrick (11) who indicated that the response of the cadaver occiput to a sinusoidal vibration input to the frontal bone corresponds closely to that of a damped spring-mass system. The first and third bending mode of the skull occurred near 300 and 900 Hz for both the cadaver preparation with silicone gel-filled cranial cavity and the living human head. The second mode was found near 600 Hz in the living human. They found accelerations on the occipital bone up to twice those measured in response to forehead impact. Studies by Nahum (25) also indicated that oscillatory responses recorded from an accelerometer opposite the

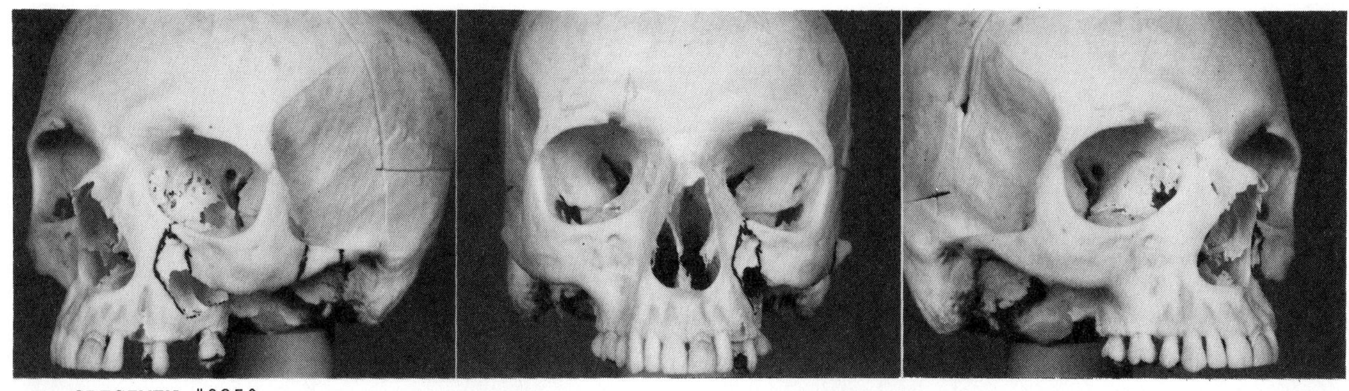

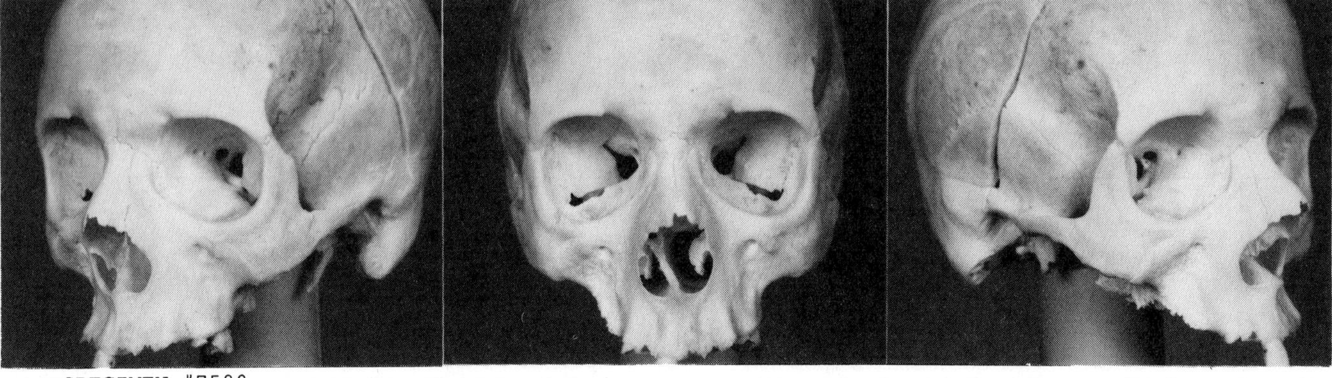

Fig. 12c. Defleshed skull photographs of the specimen #8250 (top) which was tested at a velocity of 3.13 m/s and #7589 (bottom) which was impacted at 2.68 m/s velocity. Both the specimens were tested with the EA wheel.

impacting area of the face confounded the correlation between the impacting force, and the calculated force based on the head acceleration and the weight of the head. Zygoma fractures were produced with a striker instrumented with a load cell which recorded the impulse directlv on the oscilloscope. Tip contact area of 645 mm^2 was employed for all tests. Severe comminuted or depressed fractures with clinical significance were given. Females as a group tended to have a lower tolerence than males. Naham reported severe comminuted or depressed fractures of the zygoma in cadaveric specimens with a force of 912 N to 3470N. The tolerance threshold at the zygomatic area for clinically significant fractures was given as 1000 N.

In Hodgson's studies the impact velocities ranged from 7.87 to 2.62 m/s. Table 9 shows the correspondence between fracture forces with different impactors and the corresponding diameter correlation. These studies indicate that the forces for facial fracture are more closely proportional to the diameter of the impactor than to the area. The penetration forces of the parietal skull with impactors of 95 mm^2 to 200 mm^2 tended to vary more closely with diameter than with area (24). A review of skull fracture is recently reported (31,32). The studies of Nahum and Hodgson were done on both sides of the face. Nyquist conducted facial impact experiments with a 25 mm diameter rigid cylindrical contact surface oriented across the front of the face to contact the nose at the elevation of the infra-orbital margin. The impactor was moved into the face in the anterior posterior direction with velocities of 2.8 to 7.3 m/s. Accelerometers were mounted on the impactor and the occiput. A peak force of approximately 3000 N filtered at 180 Hz was found to be a threshold for more severe facial fractures.

The fracture forces in our study are within the reported range in literature. For the EA wheel severe clinically significant facial fractures were observed at and above 3.58 m/s impact velocities. The demarcation between fracture and no fracture occurred approximately at 3.13 m/s. The corresponding areas at 3.13 m/s were 891 mm^2 and 1062 mm^2 with forces of 2029 and 2113 N. In contrast, the standard steering wheel produced severe clinically significant fractures at 3.13 m/s and 2.24 m/s velocities. The average area for the 2.24 m/s impacts with the standard wheel is 327 mm^2 with an average force of 1704 N. For comparison, the average area for the 3.13 m/s drop on the EA wheel is 976 mm^2 with an average force of 2071 N. These findings suggest that the observed fracture levels are weakly coupled to area. The corresponding dynamic steering wheel deformation (Table 6) is greater for the EA wheel than for the standard wheel at these impact velocities. The foam deformations of the EA wheel are not accounted for in these dynamic steering

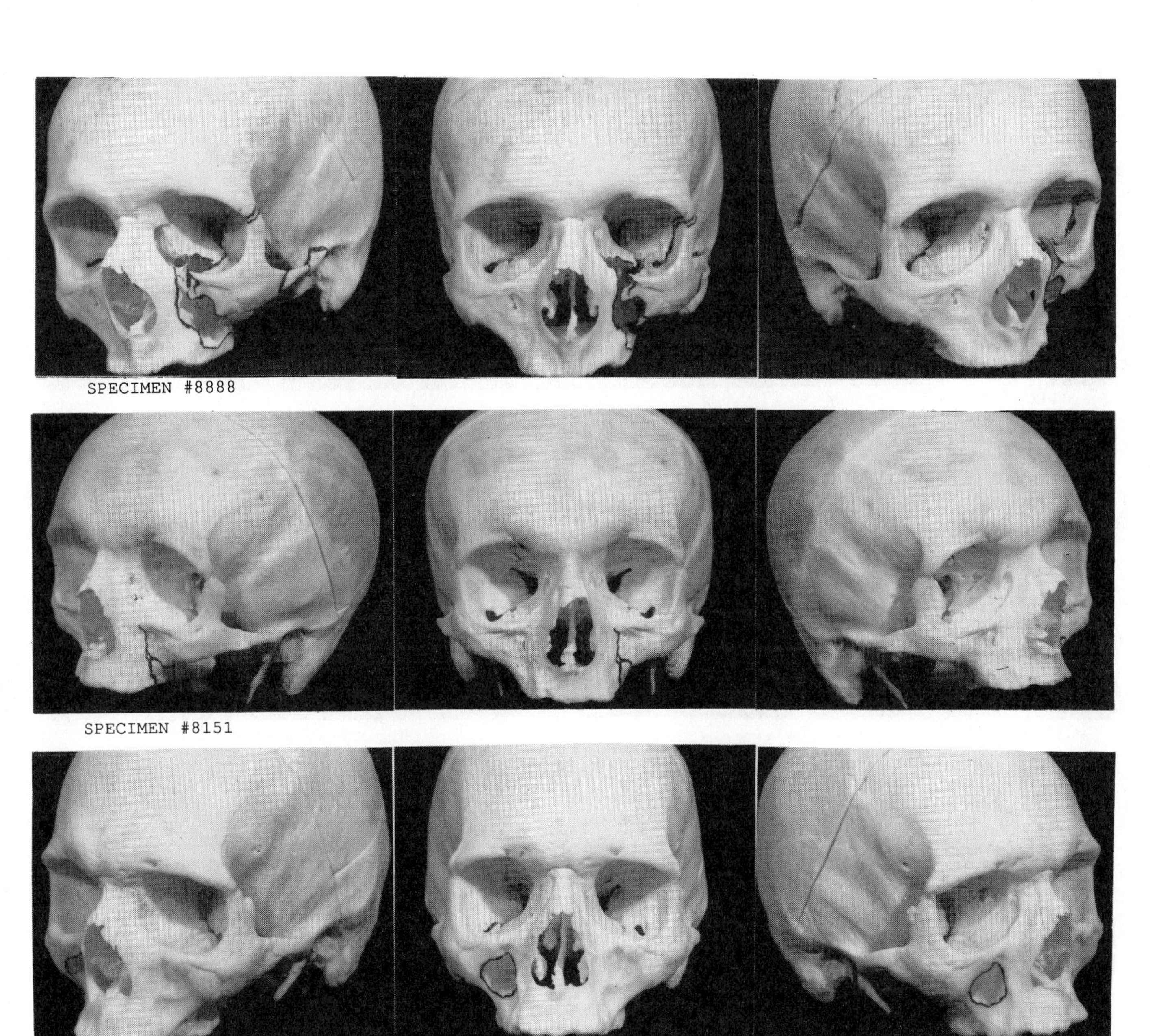

Fig. 13a. Defleshed skull photographs of specimen #'s 8888, 8151, and 5912, which were impacted on to the standard steering wheel at velocities of 2.68, 2.68, and 3.13 m/s respectively.

deformations since the potentiometer is fixed to the underside of the wheel.

The observed differences in the mechanical response between dynamic drops and quasistatic MTS tests on the steering wheels may be due to inertial effects, viscous effects, vibrations and stiffening of the wheel at the junction of the lower left spoke and rim. Validation studies using the rigid Z90.1 headform, at a velocity of 3.13 m/s, which indicated a relatively small percentage difference (5.04 for the standard wheel) between the theoretical force (computed using the F=m·a relation) and the transformed dynamic force at the impact site, and the similarity in these two histories supports the assumption regarding the neglect of the inertial effects under dynamic loading. However, the evaluation of the velocity dependence needs further study. In dynamic tests, the cadaver zygoma essentially bottomed out the skin on the preparations which could induce viscous effects. Secondary oscillations due to vibration in the steering wheel following impact from the cadaver zygoma may have exacerbated the steering wheel deformations in dynamaic drop tests.

Because of the lack of evidence in the literature regarding the promulgation of

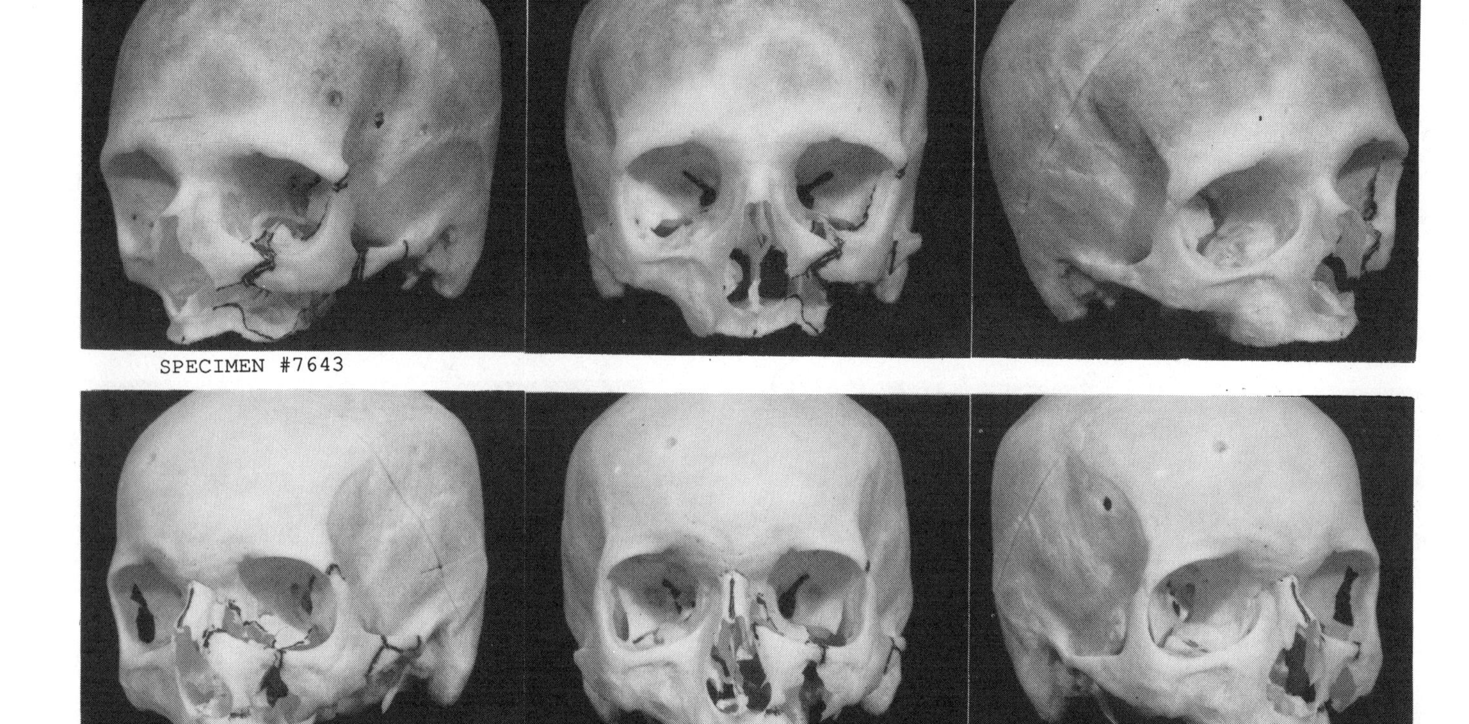

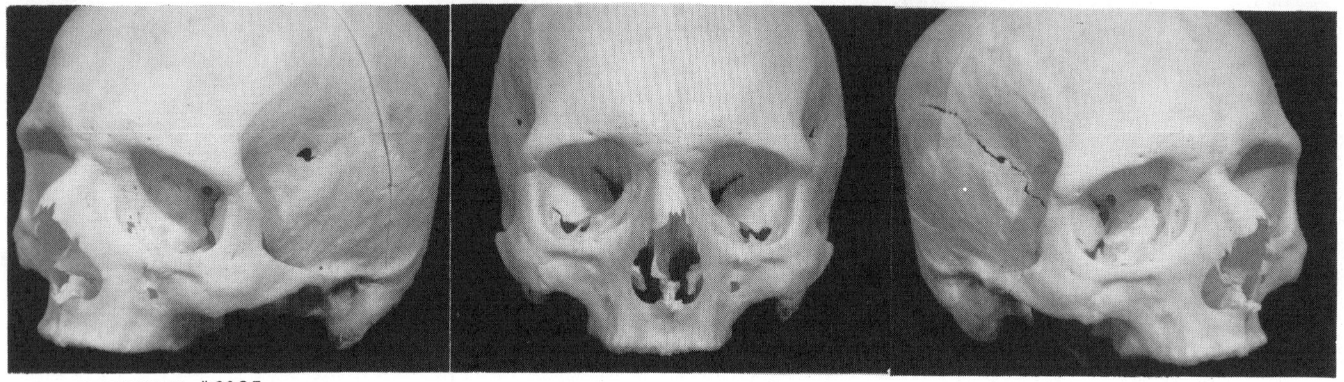

Fig. 13b. Defleshed skull photographs of specimen #'s 7643, 9274, and 6025, which were impacted on to the standard steering wheel at velocities of 2.24, 2.24 and 2.01 m/s respectively.

fractures with unilateral impact to the face, preliminary studies were conducted on four specimens. These specimens were impacted on one side first followed by impact on the other side at impact speeds from 2.01 to 3.8 m/s on the standard wheel. Upon palpation and gross dissection after the first impact on one side, no fracture was observed. However, after a subsequent impact to the opposite side, bilateral fractures were observed with 2-D and 3-D CTs and confirmed with defleshing. This finding emphasized the requirement for unilateral experimentation so as to preclude structural alteration with bilateral impact to the same preparation.

A review of the high-speed photography suggests that the initial contact phase was associated with the bottoming out of the soft tissue, followed by the deformation of the foam for the EA wheel. This occurs at approximately the first 5 to 6 ms with drops in the region of 2.68 to 3.58 m/s velocities. A buildup of force then occurs concomitant with deflection of the steering wheel which is measured by the potentiometer system. It appears that fractures occur during this phase following bottoming out of the foam and skin, and loading of the steering wheel. Unloading of the steering wheel appears to occur at approximately 20 ms with a rebound of approximately 25 to 30% of the original drop height. However, the substantial fractures most likely occurred prior to the rebound.

Because the preparations were ballasted to 6.8 kg and dropped on the junction between the lower left spoke and rim, these studies probably represent a severe test of the biological structure. However, this type of impact can be anticipated in predictable crash sequences. In studies conducted in vertically dropped intact cadavers, the head was often observed to move in a disarticulated fashion with respect to the remainder of the body unless little motion of the head and contact surface occurred (38). Therefore, if one assumes the average weight of the head is 4.5 kg, the available impact energy could be reduced by one third.

In static studies using the MTS device and dynamic drop tests with the slider system, contact areas at the loading site were 2 to 3 times greater in the EA wheel compared to the standard wheel. However, within the limits of biological variability in these experiments, force for fracture did not demonstrate this variation. Furthermore, the most consistent values predicting fracture of the facial elements appear to be the forces measured at the interface between the steering wheel and the face in contrast to those calculated based on the acceleration determinations at the point opposite to impact. However, the contribution of the specimen variation and steering wheel manufacturing differences are difficult to determine at this time. The exact coupling between the impact surface and the surface-mounted accelerometer in the parietal area has not been fully determined. Furthermore, the predicted HIC values are substantially below a value of 1000 for clinically significant fractures to the face.

ACKNOWLEDGEMENT

We respectfully acknowledge the consultation and advice of Drs. Rolfe Eppinger, Mark Haffner, Kennerly Digges, Lee Stucki, Roger Saul of NHTSA and David Zuby of TRC in Ohio. This research was supported in part by PHS, CDC grant #R49 CCR502508-01 and the VA Medical Research Funds.

REFERENCES

1. Angelo Del Balso, ed: Maxillo Facial Injury. In Press, W.B. Saunders, 1989.
2. Brant-Zawadzki MN, Minagi H, Federle MP, Rowe LD: High resolution CT with image reformation in maxillofacial pathology. AJR 138:477-483, 1982.
3. Campbell BJ: Safety belt injury reduction related to crash severity and front seated position. J Trauma 27(7):733-739, 1987.
4. Digges K, Cohen D, Eppinger R, Hackney J, Morgan R, Stucki L, Saul R: Evaluation of devices to measure the injury mitigation properties of steering systems. Intl Research Council on Biokinetics of Impacts, 1987 Intl IRCOBI Conf on the Biomechanics of Impacts, Birmingham (United Kingdom), Sept 8-10, 1987, pp 91-102.
5. Dolan K, Jacoby C, Smoker W: The Radiology of Facial Fractures. RadioGraphics 4(4): 577-668, July 1984.
6. Fujii N, Yamashiro M: Computed tomography for the diagnosis of facial fractures. J Oral Surg 39:735-741, Oct 1981.
7. Gallup BM, Newman JA: The assessment of facial injury to fully restrained drivers through full-scale car crash testing. J Trauma 27(7):711-718, 1987.
8. Hemmy DC, David DJ, Herman GT: Three-dimensional reconstruction of craniofacial deformity using computed tomography. Neurosurgery 13(5):534-541, 1983.
9. Hodgson VR: Tolerance of the facial bones to impact. Am J Anat 120:113-122, 1967.
10. Hodgson VR, Nakamura GS: Mechanical impedance and impact response of the human cadaver zygoma. J Biomechanics 1:73-78, 1968.
11. Hodgson VR, Patrick LM: Dynamic response of the human cadaver head compared to a simple mathematical model. Proc 12th Stapp Car Crash Conf, Society of Automotive Engineers, New York, 1968, pp 280-301.
12. Huelke DF, Compton CP: Facial injuries in automobile crashes. J Oral Maxillofac Surg 41:241-244, 1983.
13. Kahnberg KE, Göthberg AT: LeFort fractures. (I) A study of frequency, etiology and treatments. Int J Oral Maxillofac Surg 16:154-159, 1987.
14. Karlson TA: The incidence of hospital-treated facial injuries from vehicles. J Trauma 22(4):303-310, 1982.
15. LeFort R: Experimental study of fractures of the upper jaw, Part III. Plast Reconstr Surg 50(4):600-607, 1972.
16. Lee KF, Wagner LK, Lee YE, Suh JH, Lee SR: The impact-absorbing effects of facial fractures in closed-head injuries. An analysis of 210 patients. J Neurosurg 66:542-547, 1987.
17. Luce EA, Tubb TD, Moore AM: Review of 1,000 major facial fractures and associated injuries. Plastic Reconstr Surg 63(1):26-30, 1979.
18. Manson PN, Hoopes JE, Su CT: Structural pillars of the facial skeleton: an approach to the management of Le Fort fractures. Plastic Reconstr Surg 66(1):54-61, 1980.
19. Manson PN, Crawley WA, Yaremchuk MJ, Rochman GM, Hoopes JE, French JH Jr: Midface fractures: advantages of immediate extended open reduction and bone grafting. Plastic Reconstr Surg 76:1-10, 1985.
20. Manson PN: Some thoughts on the classification and treatment of Le Fort fractures. Ann Plastic Surg 17(5):356-362, 1986.
21. Marcus JH, Blodgett R: Priorities of automobile crash safety based on impairment. Eleventh International Technical Conf on Experimental Safety Vehicles, Washington, D.C., May 1987.
22. McCoy FJ, Chandler RA, Magnan CG Jr, Moore JR, Siemsen G: An analysis of facial fractures and their complications. Plast Reconstr Surg 29(4):381-391, Apr 1962.

23. McDermott FT, Klug GL: Injury profile of pedal and motor cyclist casual ties in Victoria. Aust NZ J Surg 55:477-483, 1985.
24. Melvin JW, Fuller PM, Roberts VL: Frangible head form development Phase I: A six month study of the effects of localized impact on tissue, Report PO No: NP-47-356319, Highway Safety Research Institute, Ann Arbor, MI, February, 1969.
25. Nahum AM, Gatts JD, Gadd CW, Danforth J: Impact tolerance of the skull and face. Proc 12th Stapp Car Crash Conf, Society of Automotive Engineers, New York, 1968, pp 302-316.
26. Nyquist GW, Cavanaugh JM, Goldberg SJ, King AI: Facial impact tolerance and response. Proc 30th Stapp Car Crash Conf, Society of Automotive Engineers, Warrendale, PA, 1986, pp 379-400.
27. Otte, Dietmar: Residual injuries to restrained car-occupants in front and rear-seat positions. Eleventh International Technical Conf on Experimental Safety Vehicles, Washington, D.C., May 1987.
28. Peck RHL: The pattern of maxillo-facial injuries in Singapore. Ann Acad Med 9(3):374-379, 1980.
29. Petty SPF, Fenn MA: A modified steering wheel to reduce facial injuries and an associated test procedure. Tenth International Transportation, National Highway Traffic Safety Administration, Oxford, England, July 1-4, 1985, pp 342-347.
30. Sances A Jr, Yoganandan N: The societal impact of biomechanics. Proc IEEE/EMBS 10th Ann Intl Conf, New Orleans, LA, Nov 4-7, 1988, (In Press).
31. Sances A Jr, Yoganandan N: Human head injury tolerance. In Mechanisms of Head and Spine Trauma, A Sances, Jr, DJ Thomas, CL Ewing, SJ Larson, F Unterharnscheidt, editors, Aloray Publisher, Goshen, NY, 1986, pp 189-218.
32. Sances A Jr, Weber RC, Larson SJ, Cusick JF, Myklebust JB, Walsh PR: Bioengineering analysis of head and spine injuries. CRC Critical Reviews in Bioengineering 5(2):79-122, 1981.
33. Schneider DC: Biomechanics of Facial Bone Injury: Experimental Aspects. In The Biomechanics of Trauma, Appleton-Century-Crofts, Publisher, 1985, pp 281-299.
34. Schneider DC, Nahum AM: Impact studies of facial bones and skull. Proc 16th Stapp Car Crash Conf, Society of Automotive Engineers, New York, 1972, pp 186-203.
35. Swearingen JJ: Tolerances of the human face to crash impact. Office of Aviation Medicine, Federal Aviation Agency, Oklahoma City, OK, July 1965, 24 pp.
36. Tessier P, Hemmy D: Three dimensional imaging in medicine. A critique by surgeons. Scand J Plast Reconstr Surg 20:3-11, 1986.
37. Weymuller EA Jr: Blindness and LeFort III fractures. Ann Otol Rhinol Laryngol 93:2-5, 1984.
38. Yoganandan N, Sances A Jr, Maiman DJ, Myklebust JB, Pech P, Larson SJ: Experimental spinal injuries with vertical impact. Spine 11(9), 1986, pp 855-860.
39. Zuby D: Steering Assembly Induced Facial Injury. ASME Winter Ann Mtg, American Society of Mechanical Engineers, Boston, MA, Dec 13-18, 1987, 9 pp.

APPENDIX I

WHEEL RIM FORCE COMPUTATION

Reference System

Cartesian references are depicted in Fig. 2b as X_n^m, where m denotes rotational orienta-tion, and n denotes the location of the coordinate origin (n = L at the laboratory reference). The principal assumptions used in the transformation process is that the mass of the steering wheel is considered negligible. This results in the assumption of the neglect of the mass moment of inertia in the computation. The validity of this assumption is well supported by the studies conducted with the rigid Z90.1 headform tests which are explained in the text. Briefly, Figure 5 illustrates a similarity between the acceleration and the transformed force histories for both free fall and slider drop tests. The quantitated data from these four tests is included in Table 5 which gives the theoretical peak forces (force computed by the relation F=m·a) and transformed forces (forces computed using the load cell data). For the standard wheel and the EA wheel the difference between these force magnitudes was 5.04% and 9.65%, respectively. These results are discussed in the text.

Transformation

The transformation procedure includes conversion, translation, and coordinate rotation. Position vectors relative to the respective references are describes as: $M_p^{m,n}$, where M denotes the reference axes, m and n are as defined above, and p denotes the point of position vector termination.

1. M_b^{2a} and M_d^{2a} are determined prior to impact.

2. M_d^{2c} is determined from angular and linear potentiometer position measurements.

3. $M_c^{2a} = M_d^{2a} - M_d^{2c}$

4. $M_{dt}^{2a} = M_c^{2a} + M_{dt}^{2c}$ where dt denotes the position of point d immediately post-impact.

5. $M_{bt}^{2a} = M_{dt}^{2a} + M_{bt}^{2dt}$

Force and Moment Computation

Experimentally determined forces and moments are the 6-axis transducer, $\{N_a'\}$, together with a rotation matrix, [R], and translation matrix, [T], allow determination of wheel rim forces, $\{N_b^2\}$:

$$\underset{6x1}{\{N_a^2\}} = \underset{6x6}{[R]} \; \underset{6x1}{\{N_a'\}} \qquad (6)$$

$$\underset{6x1}{\{N_b^2\}} = \underset{6x6}{[T]} \; \underset{6x1}{\{N_a^2\}} \qquad (7)$$

(Values for [T] are determined from M_{bt}^{2a} defined above).

881719

Facial Impact Response — A Comparison of the Hybrid III Dummy and Human Cadaver

Douglas "L" Allsop and Charles Y. Warner
Collision Safety Engineering

Milton G. Wille
Brigham Young University

Dennis C. Schneider and Alan M. Nahum
University of California

ABSTRACT

Results indicate the need for a redesigned Hybrid III face capable of accurate force and acceleration measurements. New instrumentation and methods for facial fracture detection were developed, including the application of acoustic emissions. Force/deflection information for the human cadaver head and the Hybrid III ATD were generated for the frontal, zygomatic, and maxillary regions.

THE BIOMECHANICS OF FACIAL IMPACT INJURY has been the subject of research for more than two decades. Pioneering efforts by Swearingen (1)*, Hodgson (2), Nahum, et al (3) and others provided reference data regarding impact fracture tolerance of major facial bones. However, little has been published in biomechanical literature regarding overall facial response to impact. Lap and shoulder belt use in severe frontal crashes has increased interest in facial fracture injuries. Thus, the dynamic response of the face has become important as it relates not only to facial trauma per se, but as it affects the overall head dynamics and associated potential brain injury. The work of Nyquist, et al (4) was designed to derive biomechanical data from impact loadings with a round circular rod, similar to a steering wheel rim.

The work reported here had two biomechanical objectives. Recent work on load-and-pressure-sensing anthropomorphic dummy faces indicates that although forces were measured with the dummy face, (5, 6) it is unlikely that those measurements represent forces on the human head during facial impact. The dynamic deflection responses of the stiff dummy skull, through which these forces are transmitted, are significantly different from those of the human skull. Hence, the first goal was to identify the large-deformation response of the human skull.

The deformation response was determined experimentally by application of a perpendicular transverse impact along the mid saggital planes of the skull of severed cadaver heads, with impact sites at the maxillary, zygomatic and supra-orbital frontal bone locations. The specimens were subjected to complete autopsy examinations after the impacts had been performed.

The second objective of this work was to augment the data for small diameter cylindrical facial impact, but with more detailed instrumentation for force distribution over the skull.

The design of an experiment to accommodate these objectives with a finite number of biomechanical specimens proved challenging. The identification of the onset of fracture by any kind of bracketing technique appeared to be prohibitive from a cost and biomaterials standpoint in view of the rather broad variation in facial anatomy in humans. Establishing fracture thresholds by utilizing multiple, gradually-increasing impacts at a given impact site was discarded, since the influence of prefracture impacts on a subsequent fracture threshold would be difficult to determine. Furthermore, the identification of a hairline fracture in a test specimen requires complete dissection of the overlying tissue, thereby rendering the specimen unusable for further testing.

This dilemma was resolved by a major-force impact technique. A special array of instrumentation was used to monitor force level as a heavy weight impactor drove substantially into the skull structure at successive impact sites on each specimen. This instrumentation incorporated a specially built set of force tranducers to measure the time history of the interface force as the

---

* Numbers in parentheses designate references at the end of the paper.

impactor penetrated the specimen. This tranducer array consisted of low mass semi-cylindrical discs backed up by piezoelectric load cells arranged such that for each one half inch of length along the face of the impactor a separate electronic force/time history of the impact was generated. The data thus produced from each impact gave detailed force/time and force/displacement histories along the length of the impactor. This allowed detailed characterization of fracture by identifying key points on the force/time and force/deflection curves.

Additional instrumentation was applied in these experiments to reinforce the force/time observations regarding fracture. An acoustic emissions system was employed to identify bursts of stress wave energy associated with fracture in specimens during impactor penetration.

PROCEDURE

CADAVER HEAD TESTS

INSTRUMENTATION. An impactor capable of detecting fracture during the event was required, since the best use of the cadaver material would require impacting each head with more than enough energy to cause significant fracture. The relatively large mass required for the impactor and the associated large distributed forces associated with fracture precluded the determination of fracture force from accelerometer data. Therefore, an impactor was designed and fabricated for this project (see Figure 1). It consisted of eighteen miniature force transducers with piezoelectric sensors mounted to solid aluminum blocks which were in turn mounted to a falling mass on a drop tower. The force transducers were made by sandwiching piezoelectric sensors between aluminum and steel plates. The forces were transmitted to the sensors by steel rods that connected to semi-cylindrical disks. As previously noted, eighteen semi-cylindrical disks lined up side by side created a semi-circular rod-shapped impactor 20 mm in diameter and 230 mm in length. The masses between the impacted specimen and the loadcells were kept low (17.6 gm) in order to allow detection of fracture from the force/time curves.

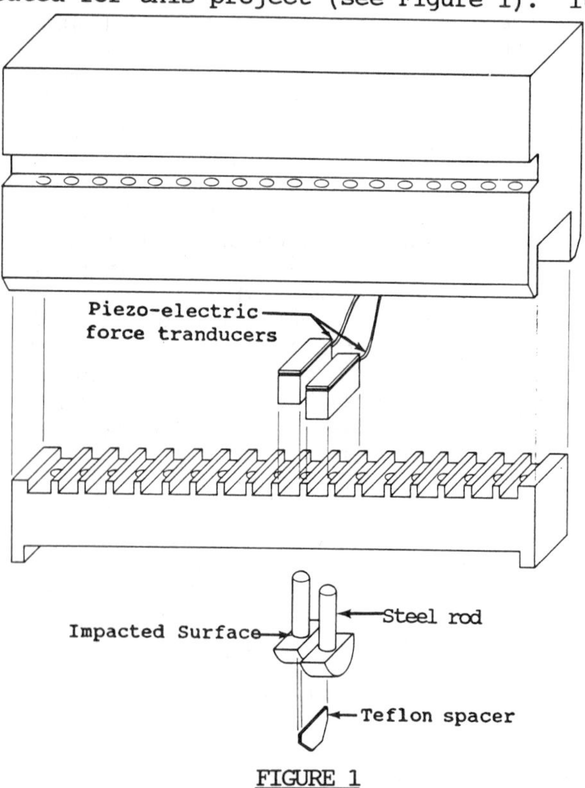

FIGURE 1
Impactor Design

Before the cadaver tests, the impactor and other instrumentation were tested on other materials. First wood slats, then sheep heads were tested with very encouraging results. An additional medium of fracture detection was desired in order to substantiate what was being observed on the force/time histories. Acoustic emissions monitoring was selected as this additional method of detecting fracture.

The basic theory behind the acoustic emissions technique is that fractures create sound or deflection waves, (such as when snapping a pencil or a match stick). While not always in the human audible range, most brittle or semi-brittle materials emit acoustic waves upon fracture. This technique has been used for several years in the composite-materials industry to detect microscopic fractures before macroscopic fractures occur (7). It has also been used in some static bone testing (8, 9). This technique was applied to obtain a secondary fracture indication. (See Appendix #1 for further discussion of acoustic emissions.)

The most troublesome problem encountered in applying acoustic emissions to dynamic bone

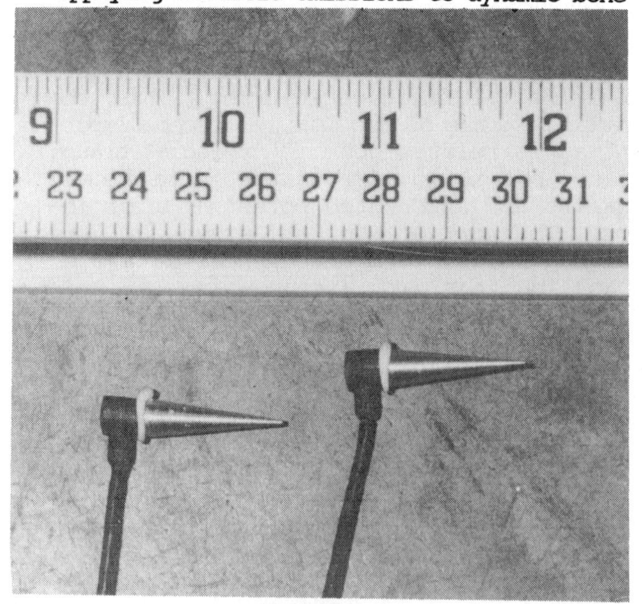

FIGURE 2
Acoustic Emission Sensors and
Wave Guides

testing was the application of the sensors to the impact specimen. Several methods were attempted in sheep head tests. It was discovered that a conical-shaped wave guide (Shown in Figure 2) glued to an acoustic emissions sensor gave good results. This sensor was attached to the specimen by making a very small incision in the skin and then wedging the wave guide into the flesh adjacent to the bone to be impacted.

The instrumentation setup is shown in Figure 3. The drop tower was instrumented with the impactor described above, a potentiometer connected to a continuous loop cable, and a 100G string pot to measure displacement. All instrumentation on the drop tower was sampled digitally at a rate of 5K Hz through a TransEra MDAS 7000 shown on the left side of Figure 3. Acoustic emissions were sampled digitally at 5 MHz by a Soltec SDA 2000 shown

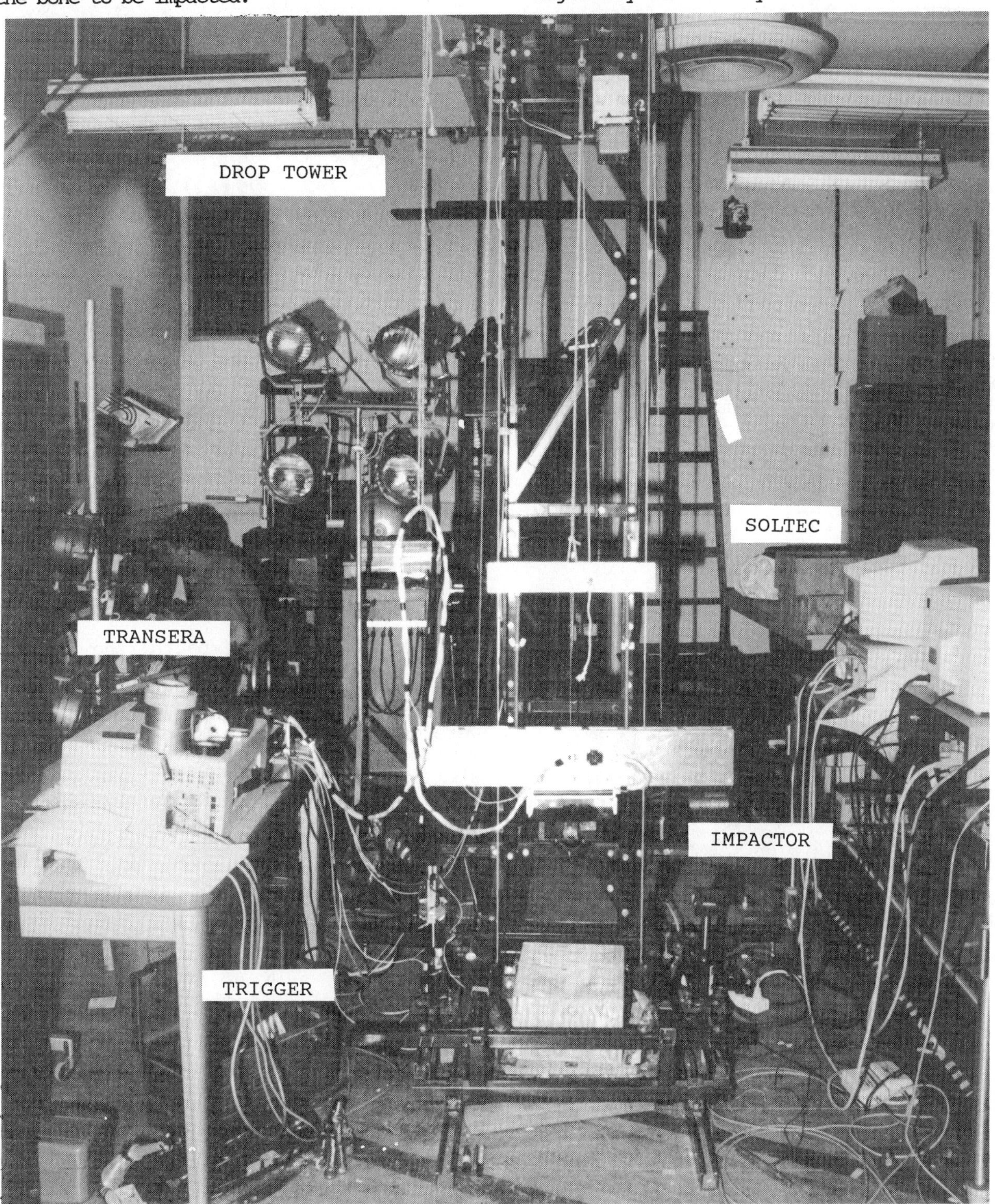

FIGURE 3
Cadaver Head Test Setup

on the right side of Figure 3.

SPECIMEN PREPARATION. For each of the fifteen unembalmed cadaver heads, (eleven female ages 59 to 90 and four male ages 39 to 84) a detailed set of anthropomorphic measurements was taken, and a full facial plaster casting was made. The specimens were then mounted facing upwards in individualized molded-plaster foundations, with the Frankfort plane vertical and the frontal plane laterally level.

TEST SEQUENCE. Two impacts per head were conducted to obtain the maximum amount of data per speciman: one to the midface region, either maxilla or zygoma, followed by an impact to the frontal region. It was felt that an impact to the midface region would not affect significantly the structural integrity or the force and strength characteristics of the frontal region. While nothing has been found in the literature to substantiate this, the geometry of the skull is such that minimal support of the frontal bone is derived from the midface structure.

Drop heights for the 14.5 kg impactor ranged from 460 to 915 millimeters for the frontal bone, and 305 to 610 millimeters for the maxillary and zygomatic regions. The impacted areas of the skull are shown in Figures 4a and 4b.

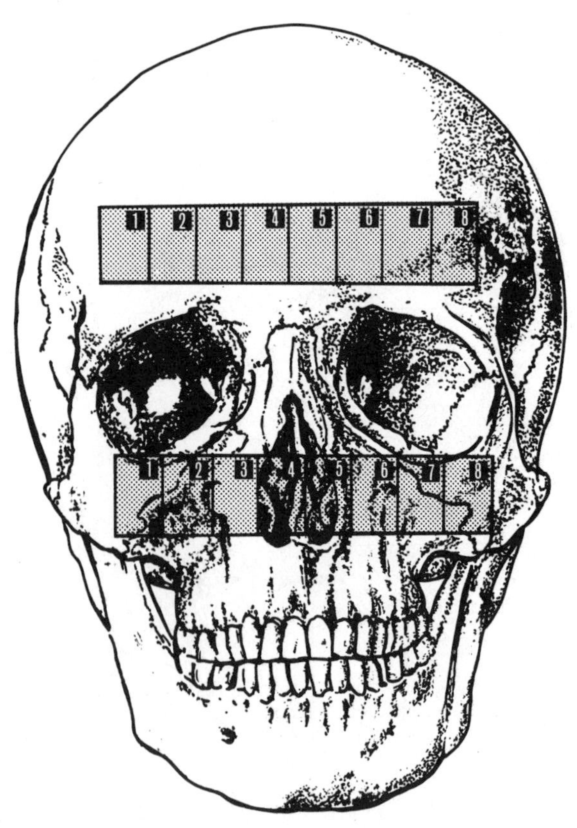

FIGURE 4(a)
Frontal/Zygoma Impact Location
(Numbers identify the different force tranducers on the impactor.)

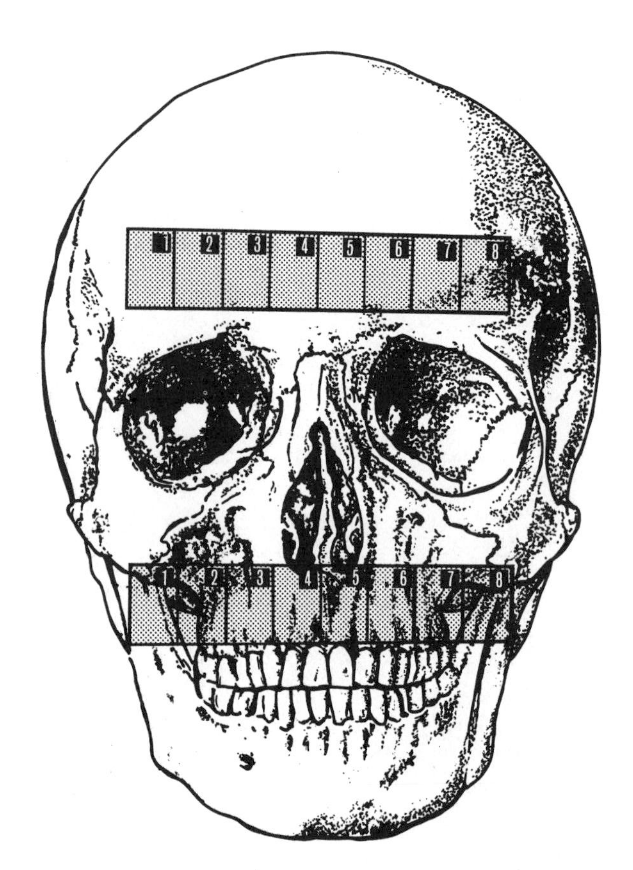

FIGURE 4(b)
Frontal/Maxilla Impact Location
(Numbers identify the different force tranducers on the impactor.)

For frontal bone impacts, the impactor centerline was set to strike the specimen approximately twenty millimeters above the supraorbital ridge. Impacts to the zygoma were centered on a line approximately ten millimeters below the suborbital ridge, which also impacted the nasal bone. Impacts to the maxilla were centered approximately ten millimeters below the tip of the nose.

HYBRID III TESTS

INSTRUMENTATION. Instrumentation used for the facial impact tests on the Hybrid III dummy was the same as that used for the cadaver head impacts as described above (see Figure 5). The only significant difference being that for the Hybrid III facial impacts no acoustic emission instrumentation was used.

HYBRID III HEAD PREPARATION. A standard, unmodified Hybrid III ATD skull was used with the rubber skin in place. The skull cap was removed and the skull with the facial skin in place was then mounted to a steel support plate.

FIGURE 5
Hybrid III Test Setup

TEST SEQUENCE. Impacts to the Hybrid III face were conducted at locations that corresponded to the impacts in the cadaver tests, i.e. the frontal, zygomatic, and maxillary regions. Drop heights used were 915, 760, 610 and 460 millimeters for the frontal zone, and 760, 610 and 460 millimeters for the zygomatic and maxillary zones.

RESULTS

CADAVER TESTS

FRACTURE DETECTION. From the data collected during the cadaver facial impact test series, force/time, force/displacement, and acoustic emission/time curves were generated.

During the test series only the central eight force transducers of the impactor were contacted. To generate the force/time curve for any one impact, the output data from each of the eight transducers was summed and the resultant was plotted against time. Figure 6 shows a typical force/time trace. The specimen used to generate the data in Figure 6 was from a sixty-eight year old female impacted in the maxillary region. The force/time history indicates a relatively gradual increase in force, and a slope which is nearly linear for approximately four milliseconds, followed by a very abrupt discontinuity in slope. This discontinuity indicates the initiation of significant fracture. Similar histories for each of the

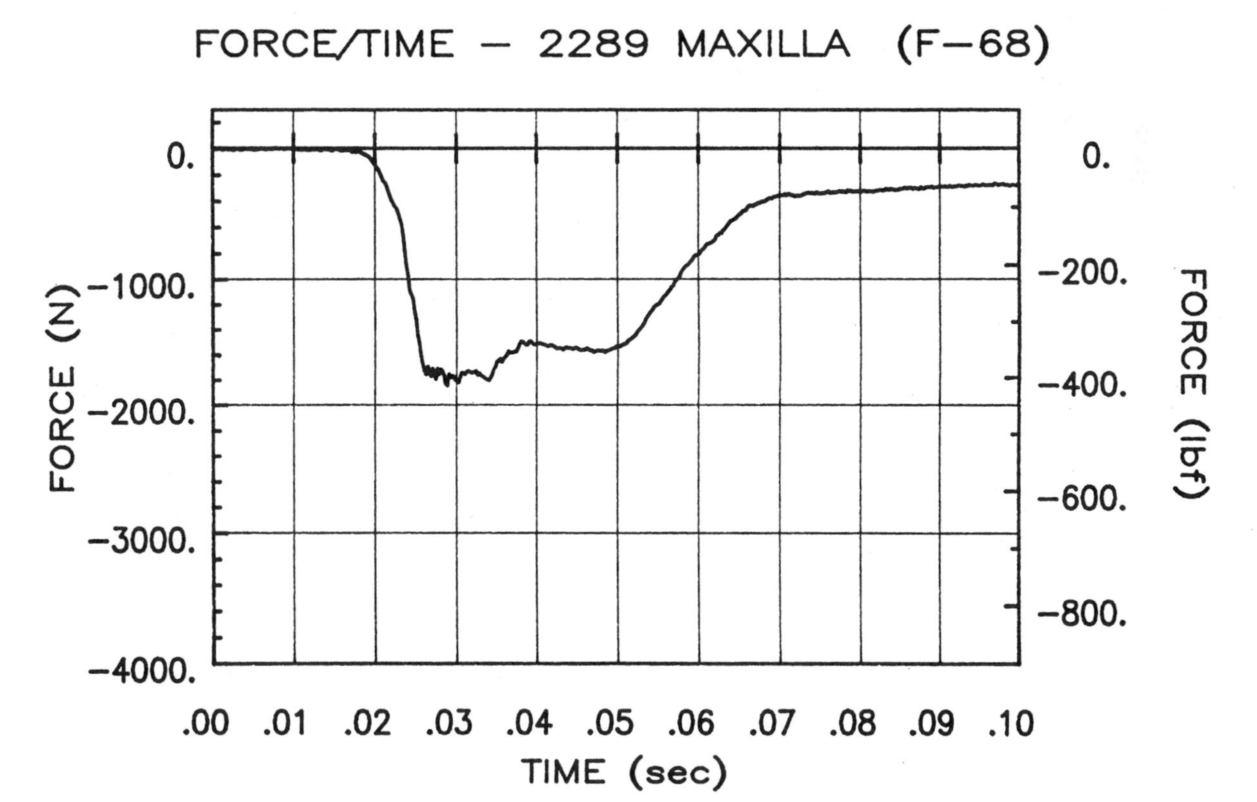

FIGURE 6
Force/Time History for Maxillary Impact

fifteen cadaver heads have been generated which showed similar discontinuities indicating fracture.

The force/time history for each of the impacts was cross plotted with its displacement/time history to generate force/displacement curves. The force/displacement curve generated with the data from Figure 6 is shown in Figure 7. This figure shows an increasing stiffness from zero to one cm, followed by a portion (from 1.0 to 1.5 cm) where the slope is approximately linear. At 1.5 cm, there is a very abrupt change in the slope of the curve. This change indicates that the nature of the impacted structure had been altered, probably because of a significant fracture of the bone in the impacted area.

If a homogeneous material were impacted without incurring fracture, Hooke's law would dictate that the maximum force would occur at the point of maximum penetration. While the human face is certainly not a Hookian structure, it is not unreasonable to expect a similar maximum force at the maximum point of penetration in the absence of bone fracture.

The force/displacement curve shown in Figure 7 clearly shows substantial impactor travel (over 3 cm) after the maximum force is reached. Coupled with the fact that the slope discontinuities on the force/time and force/displacement curves occurred at the same force levels, the sudden drop in stiffness supports the conclusion that fracture occurred at the discontinuity (at 26 milliseconds) shown in Figure 6. Another factor that supports this conclusion is the output from the acoustic emission sensors.

As was indicated previously, acoustic emission data are a measure of the fracture noise or energy associated with an event. The instrumentation used in this test series gave an indication of the frequency and intensity of the acoustic emissions associated with each impact. Figure 8 is a trace of the acoustic emissions for the maxillary impact depicted in Figures 6 and 7.

The solid line in Figure 8 marked with an "X" below it indicates a time of 20.56 msec. Prior to this point in time there were virtually no detectable acoustic emissions. After this point and up to the time marked by the dashed line with an "O" below it at 26.43 msec, some relatively low level emissions were detected. However, at about 26 msec a large burst of acoustic emissions is observed. These emissions occurred at essentially the same time as the discontinuity on the force/time curve in Figure 6, and was apparently caused by a major fracture of the maxilla. These two independent indicators of fracture are plotted on one graph in Figure 9. Comparing the two together yields an easily discernable point in time at which fracture occurred. Once the time at which fracture took place was determined, the fracture force level was readily obtained from the force/time curve.

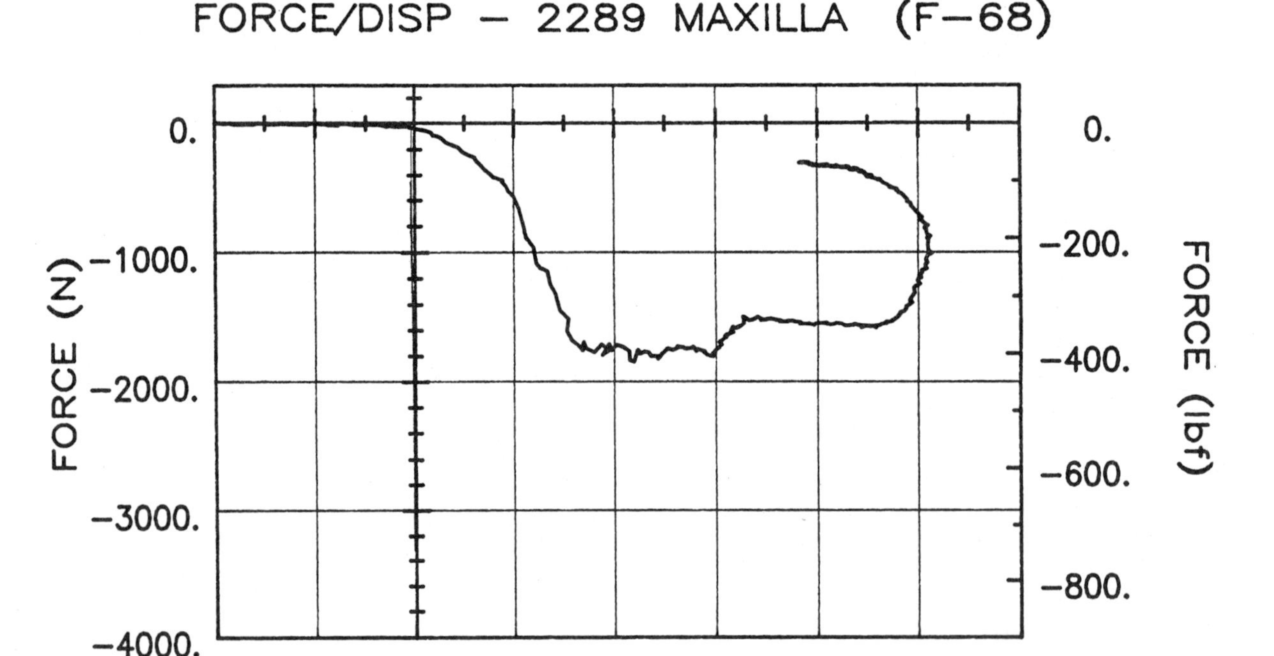

FIGURE 7
Force/Displacement History for Maxillary Impact

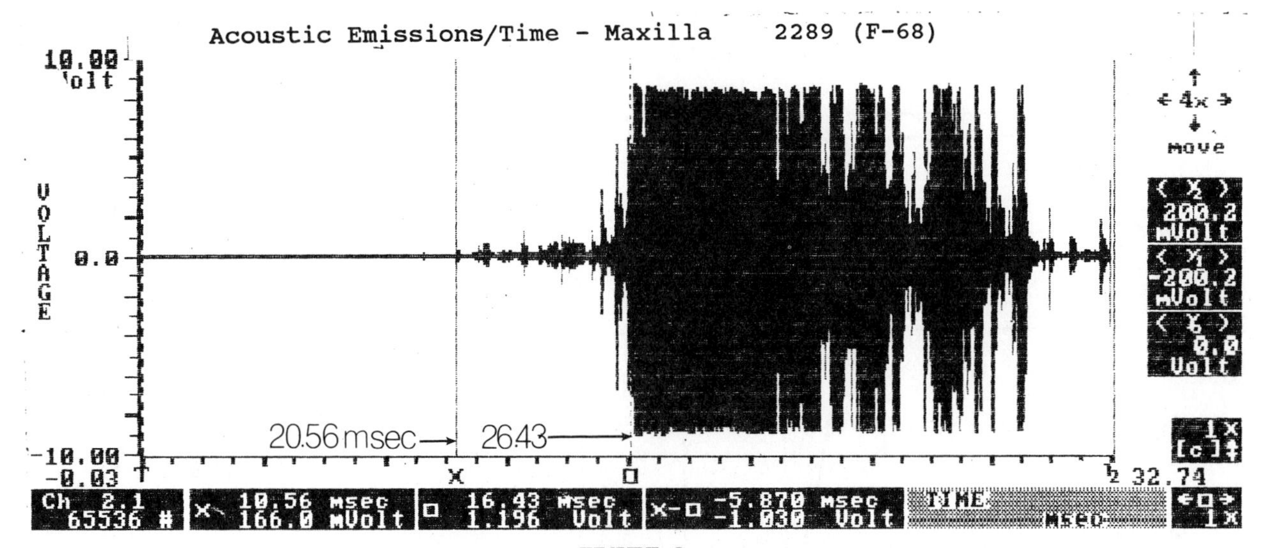

FIGURE 8
Acoustic Emissions/Time History

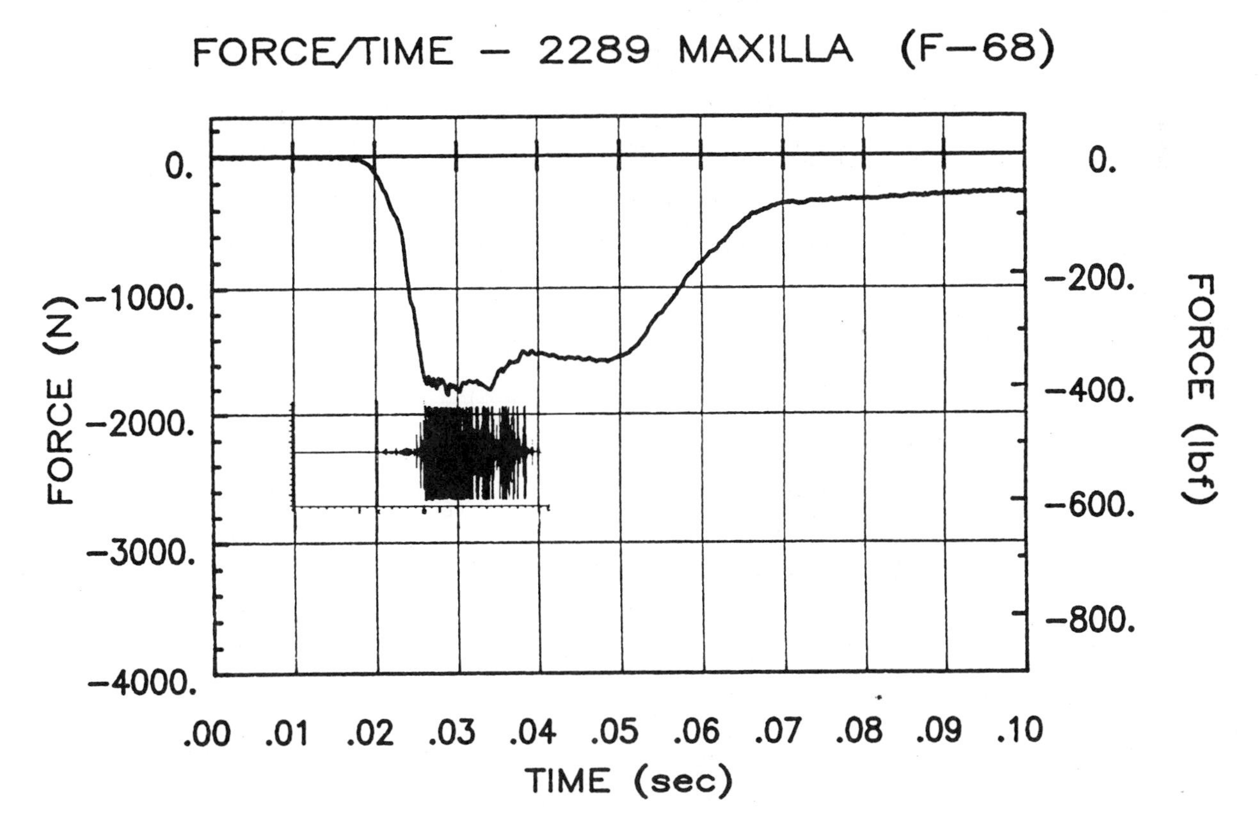

FIGURE 9
Comparison of Force/Time and Acoustic Emissions/Time Data

FRONTAL IMPACT. The force/ displacement curves for the frontal impacts are shown in Figure 10. These curves were clipped at the point where fracture occurred in order to make the figure less confusing. The curves are approximately linear from the point of contact until the point of fracture.

ZYGOMATIC IMPACT. Figure 11 shows force/displacement curves for each of the zygomatic impacts. A different trend is observed as compared to the frontal bone impacts. For all but one of the impacts a two-step linear curve is seen. Initially there is slope of about 10 newtons per millimeter. This is the area where the nasal bone is being loaded and fractured. This was determined by analyzing the force/ displacement curves for each of the eight force

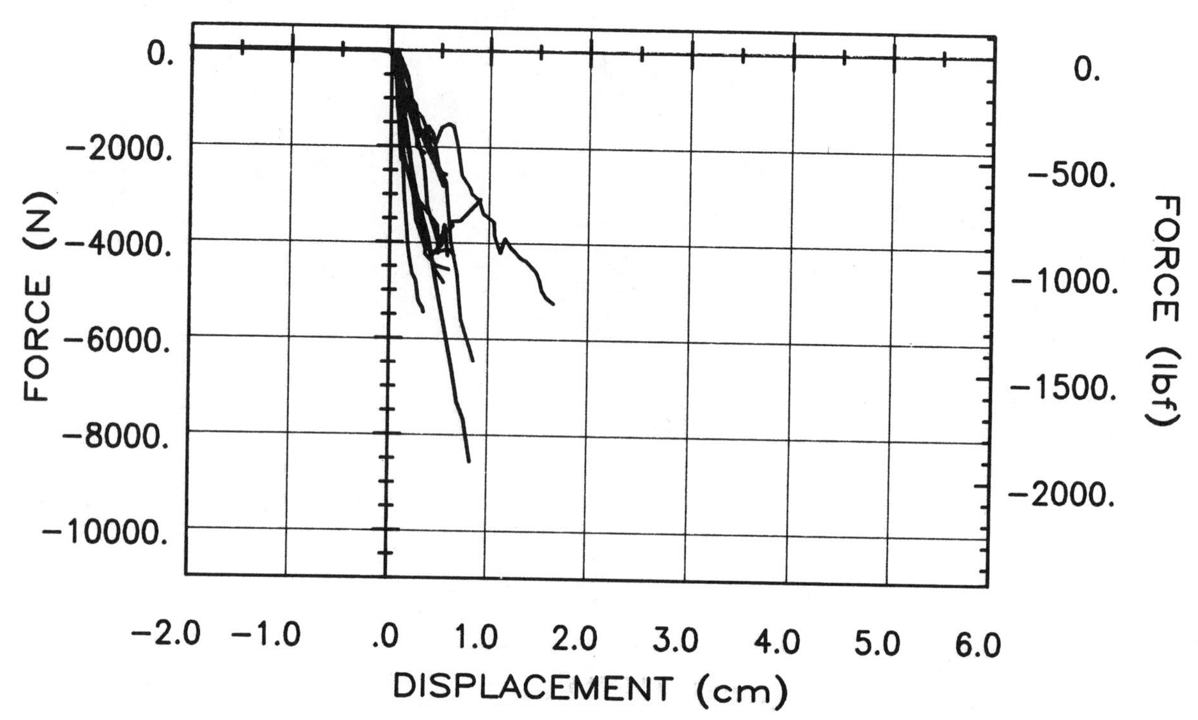

FIGURE 10
Force/Displacement Histories
For All Frontal Impacts

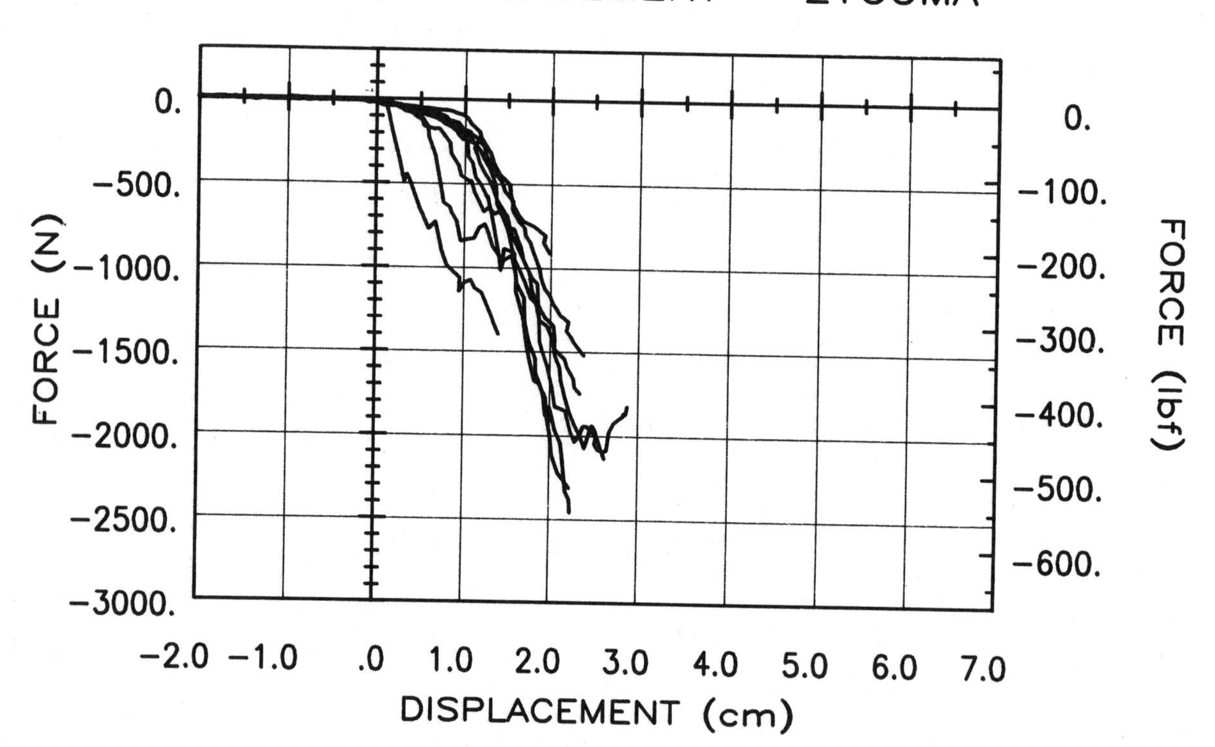

FIGURE 11
Force/Displacement Histories
For All Zygomatic Impacts

transducers on the impactor during a single impact. The impactor was centered on the head so that the center transducers contacted the nasal bone. The center transducers were contacted five to ten millimeters before any of the other sensors. After this initial contact, loading of the other more distal sensors indicated force readings as the impactor penetrated into the zygomatic region. At this point the slope on the force/displacement curve changes to yield the second, approximately linear segment with an average stiffness of 150 newtons per millimeter.

One force/displacement history deviated significantly from the description above. In this case, the force rose very quickly after initial contact with the impactor rather than rising slowly for the first five to ten millimeters. Upon examining the force/displacement histories of the individual force transducers for this impact, it was found that the center two sensors were the only ones loaded for the first 15 millimeters of impactor penetration. This indicated a nasal bone that was larger and significantly tougher than the rest of those in the test series.

<u>MAXILLARY IMPACT</u>. The force/displacement histories for the maxillary impacts are shown in Figure 12. Five of the six histories shown are very similar, and exhibit a characteristic stiffness about the same as that of the zygoma. There is a relatively flat portion for about five millimeters followed by a change in slope indicating increasing stiffness. As with the zygomatic impacts, the center force transducers were loaded first, followed by the outer ones. This was expected because of the curvature of the surface being impacted. The one curve that deviated substantially from the others was that of an 84 year old male. This specimen seemed to be much tougher than the rest. In the frontal bone impact of this specimen, only a hairline fracture was observed. Whereas, in all of the other specimens, very severe fractures occurred. It is felt that this variation in bone structure accounted for the shorter rise time and steeper slope for this impact.

<u>FRACTURE LEVELS AND STIFFNESS</u>. There is a fairly large spread in fracture forces and stiffness for each of the impact zones. The fracture-force level and stiffness for each of the impacts are listed in Table 1. Fracture force for frontal bone impacts ranged from 2200 N to 6500 N with stiffness ranging from 400 N/mm to 2200 N/mm. In the zygomatic region fracture forces varied from 900 to 2400 N, while stiffness ranged from 90 to 230 N/mm. Fracture forces of the maxilla varied from 1000 to 1800 N, with stiffness of 80 to 250 N/mm. While there is a large variation in the data, they are consistent with that seen in other facial impact research. Additionally, the fracture forces are consistent with those previously observed (2, 3, 10) even though the impact loading configuration, impactor shape and head support are significantly different.

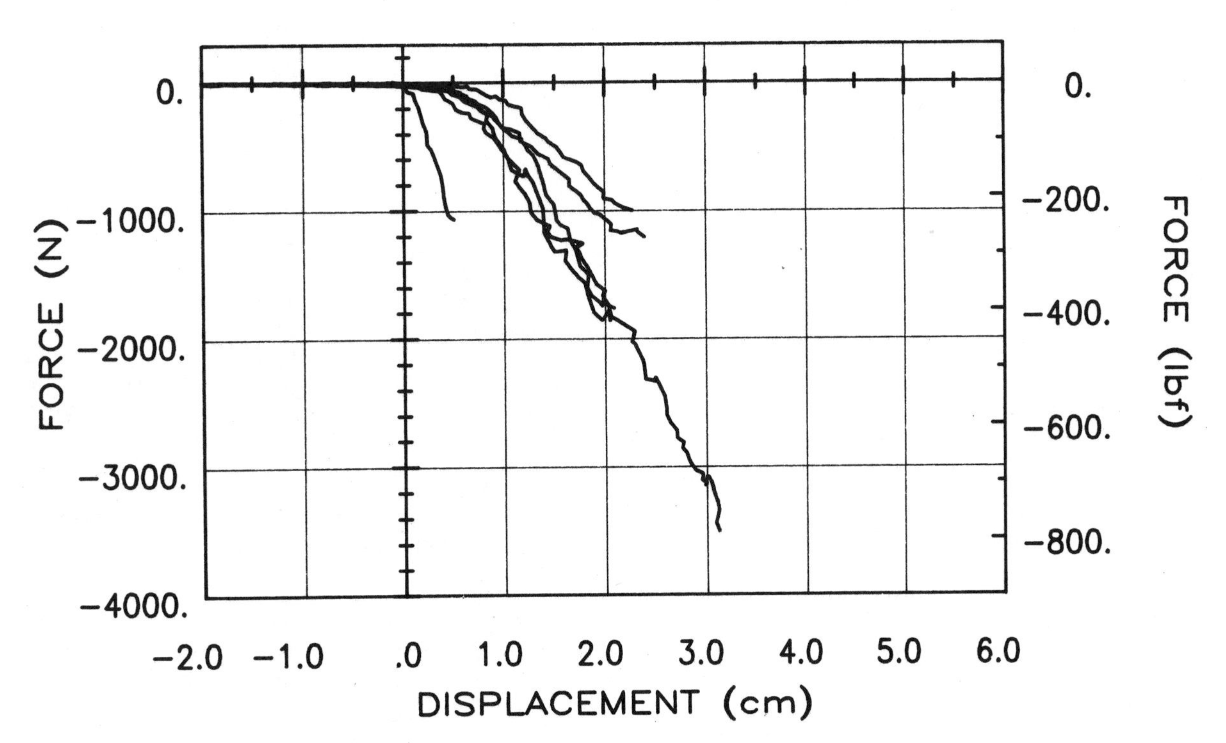

<u>FIGURE 12</u>
Force/Displacement Histories
For All Maxillary Impacts

Cadaver Fracture Force and Compliance Levels

IMPACT LOCATION	#	SEX/AGE	FRACTURE FORCE (N)	FRACTURE FORCE (LBS)	COMPLIANCE (N/mm)
FRONTAL	2308	F/68	5200	1170	400
	2215	F/78	2600	580	510
	2278	M/74	2800	630	550
	2185	M/39	2200	490	630
	2225	F/90	4000	1100	730
	2201	M/83	4800	1080	930
	2291	M/84	8600	1930	950
	2289	F/68	4300	970	970
	2245	F/75	4500	1010	980
	2151	F/79	4300	970	1090
	2297	F/82	6500	1460	1540
	2292	F/60	6000	1350	1670
	2230	F/74	5500	1240	2220
ZYGOMA	2215	F/78	900	202	90
	2230	F/74	1500	340	100
	2278	M/74	1700	380	110
	2245	F/75	1400	310	120
	2151	F/79	2100	470	150
	2308	F/68	1600	360	150
	2201	M/83	2300	520	230
	2188	F/81	2400	540	230
MAXILLA	2225	F/90	1000	220	80
	2213	F/72	1200	270	80
	2292	F/60	1200	270	130
	2185	M/39	1800	400	130
	2289	F/68	1800	400	180
	2291	M/84	1100	250	250

TABLE #1

There are essentially no entries in the literature with which to compare the stiffness data. However, the scatter in these results was similar to that observed in fracture-force measurements.

HYBRID III TESTS

IMPACT COMPARISON. As previously noted, the test setup and instrumentation for the Hybrid III tests were the same as that used for the cadaver head tests, with the exception of the head mount and the lack of acoustic emission sensors.

The force/time curves for the frontal, zygomatic and maxillary zones were very similar (see Figure 13). Each of the curves shown in Figure 13 were the result of a 460 millimeter drop. The maximum force reached for each region is also about the same, as is the shape of the curve. The force/displacement curves are also quite similar. However, there are some interesting differences that should be pointed out.

FRONTAL IMPACT. Figure 14 shows the force/displacement history for a 460 mm drop into the frontal region of the Hybrid III head. The curve shows a short 3 mm portion with one slope, and then a change in slope to a much steeper region from about 3 to 13 mm The first slope occurs while the rubber skin is being compressed. After this compression a steeper slope is encountered while deflection of the aluminum skull takes place.

ZYGOMATIC IMPACT. The zygomatic region exhibited similar behavior in the lower portion of the curve. However, the beginning of the curve is distinctly different. Figure 15 shows compression for nearly 20 mm at very low force levels. This corresponds to crushing of the hollow rubber nose structure molded into the skin. As soon as this structure is flattened, the behavior becomes very stiff, much like the frontal region.

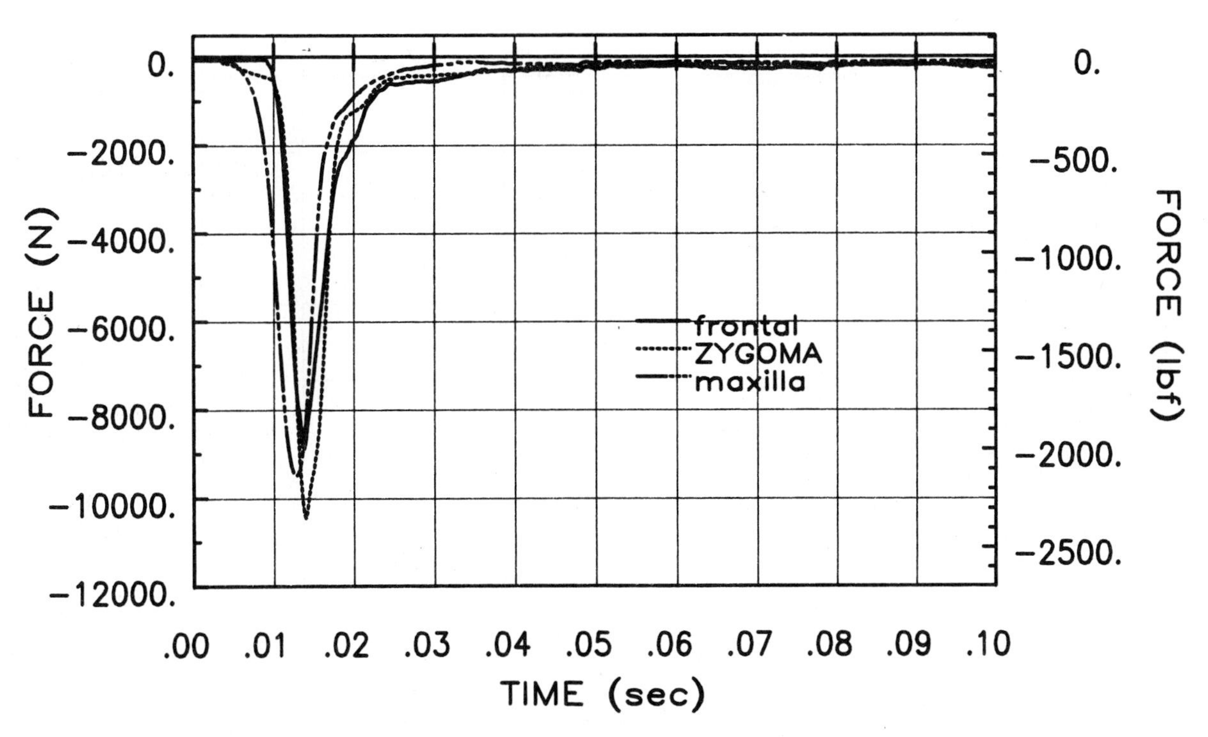

FIGURE 13
Force/Time Histories For Hybrid III Face

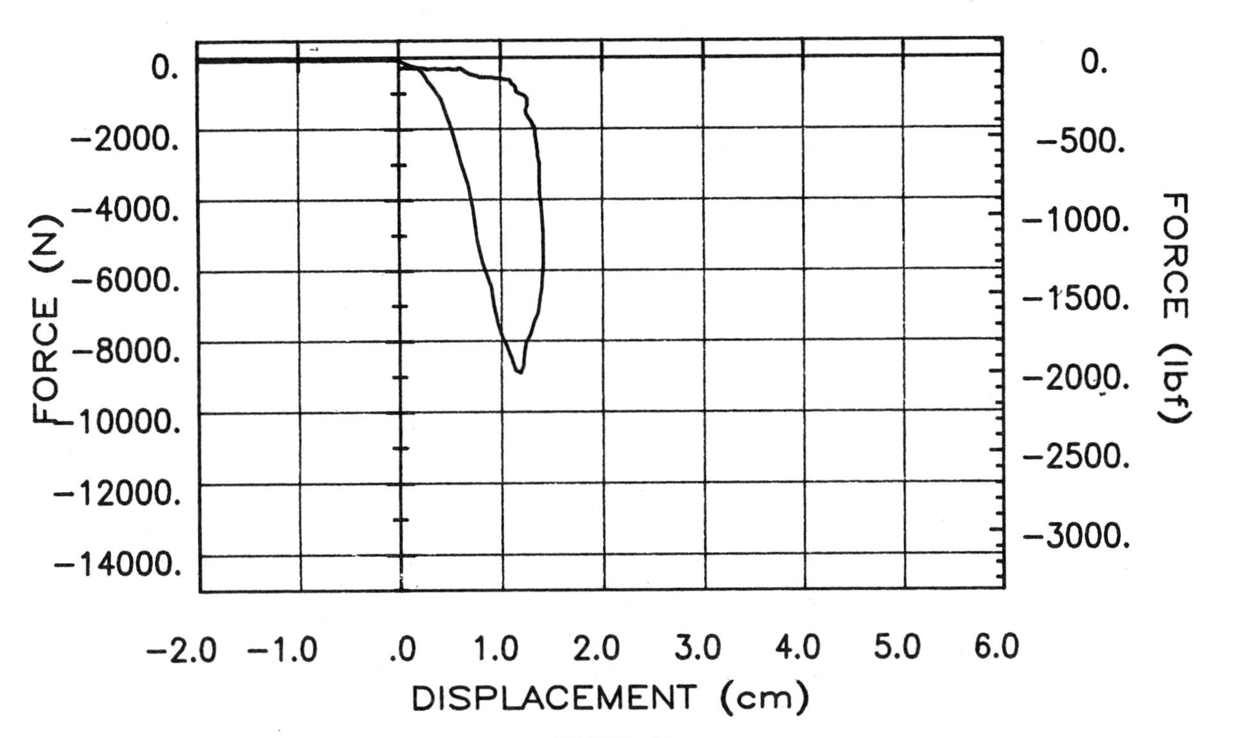

FIGURE 14
Force/Displacement History For
Hybrid III Frontal Impact

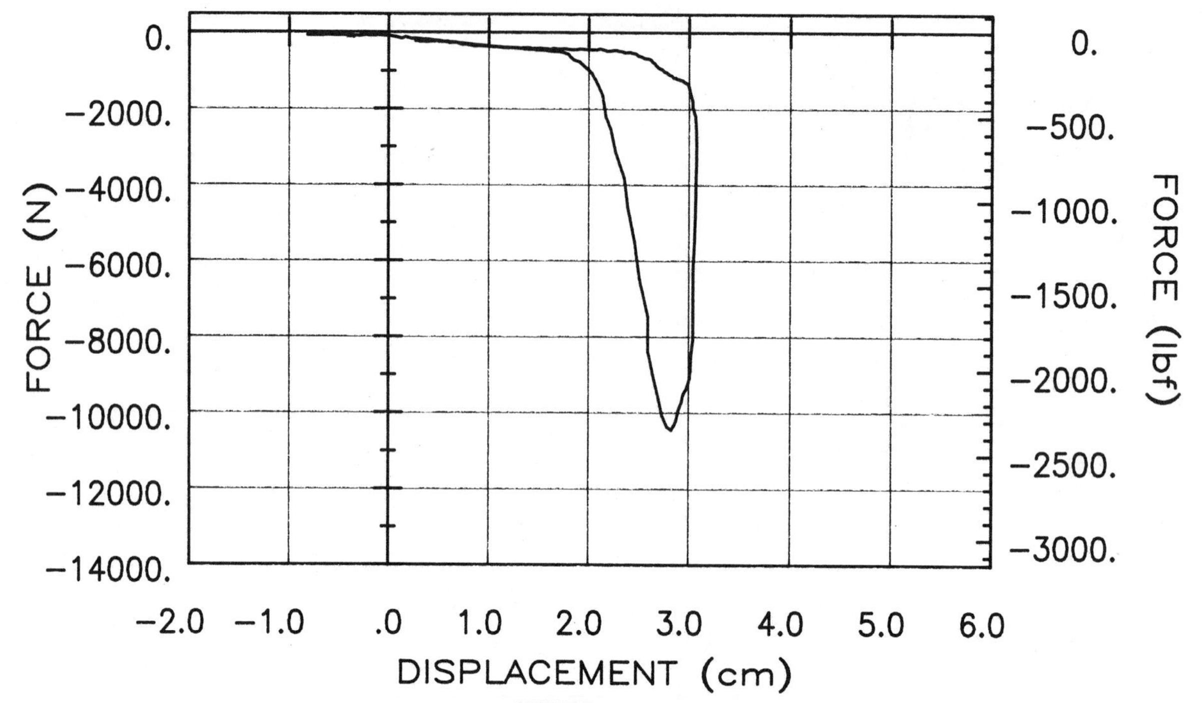

FIGURE 15
Force/Displacement History For
Hybrid III Zygomatic Impact

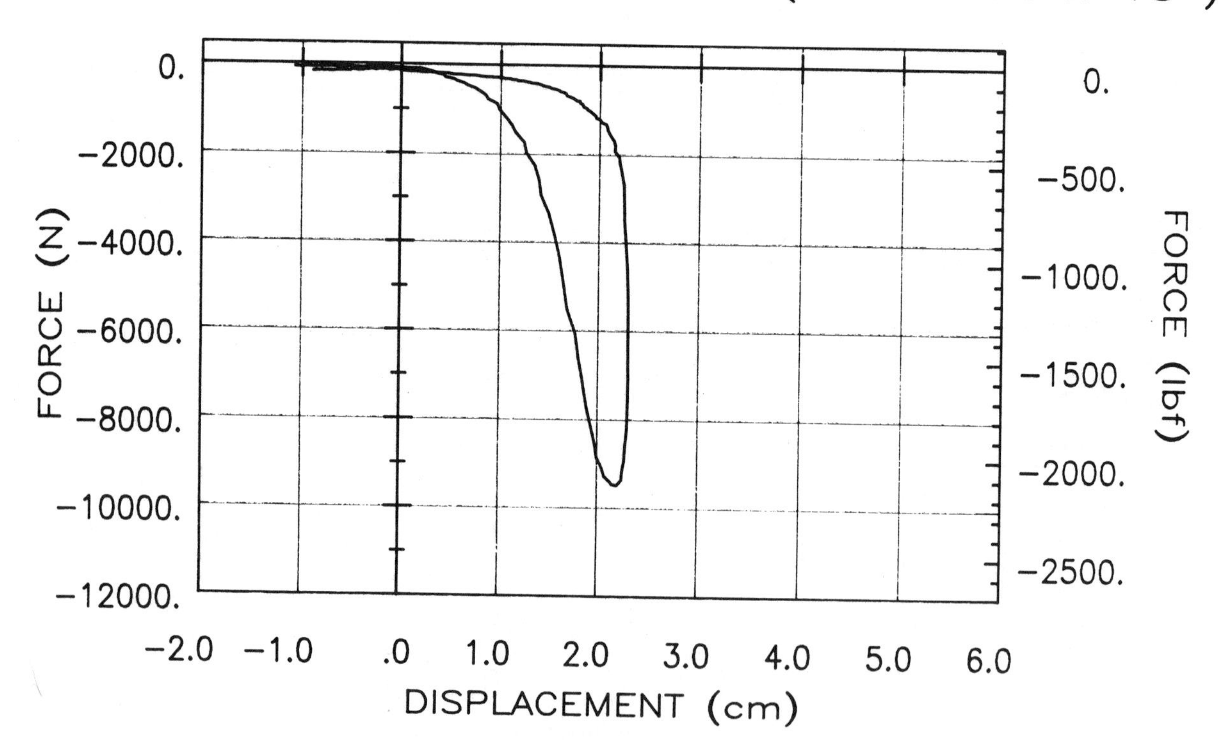

FIGURE 16
Force/Displacement History For
Hybrid III Maxillary Impact

MAXILLARY IMPACT. The stiffness of the maxillary zone, shown in Figure 16, ultimately becomes very stiff like the frontal and zygomatic regions, however there is a relatively gradual transition from the soft to stiff portions instead of the abrupt increase in stiffness observed for the frontal and the zygomatic zones. This is most likely due to the thickness of the rubber skin at the maxillary zone. The skin in both the frontal and zygomatic regions is 0.5 inches thick while the skin covering the maxillary zone is 0.9 inches. This somewhat explains the difference observed between the maxillary and the zygomatic and frontal regions.

DISCUSSION OF RESULTS

FRONTAL BONE. A comparison of the facial impacts of the human cadaver with those of the Hybrid III face is shown in Figure 17. This figure shows the force/displacement histories of frontal bone impacts on the cadavers with solid lines, and that of the Hybrid III with a diagonally striped line. Compression of the rubber skin and the subsequent deflection of the aluminum skull are very discernable in the dummy data. While the compliance of the dummy frontal zone does not fall exactly in the middle of the cadaver data, it does come close. The similarity of the dummy and cadaver data indicates reasonable biofidelity of the Hybrid III dummy face in the frontal region. However, there are substantial differences for the zygomatic and maxillary regions.

ZYGOMA. In Figure 18 the zygomatic impact of the Hybrid III dummy is again marked by a diagonally striped line, and the impacts into the cadavers are marked by solid lines with their average represented by the laterally striped line. The distinct areas of compression for the Hybrid III dummy face are observed, each with increasing slope or stiffness. The first is the crushing of the hollow rubber nose. The second is the compression of the rubber itself, and the third is the deformation of the aluminum skull. In comparison, the cadaver impacts exhibited only two distinct areas of compression. The first was the loading and fracture of the nasal bone, and the second was the loading of the zygomas. In comparing the Hybrid III and the cadavers, the observation is made that the Hybrid III nasal/zygomatic area is initially about equal to the cadavers in stiffness, but allows too much penetration relative to the cadaver data. The dummy then becomes much stiffer than all of the cadavers tested, having a stiffness of 1000 N/mm. This is about seven times greater than the average stiffness of 150 N/mm for the cadavers in this region.

MAXILLA. Force/displacement histories for each impact in the maxillary area from both the Hybrid III dummy and the cadaver

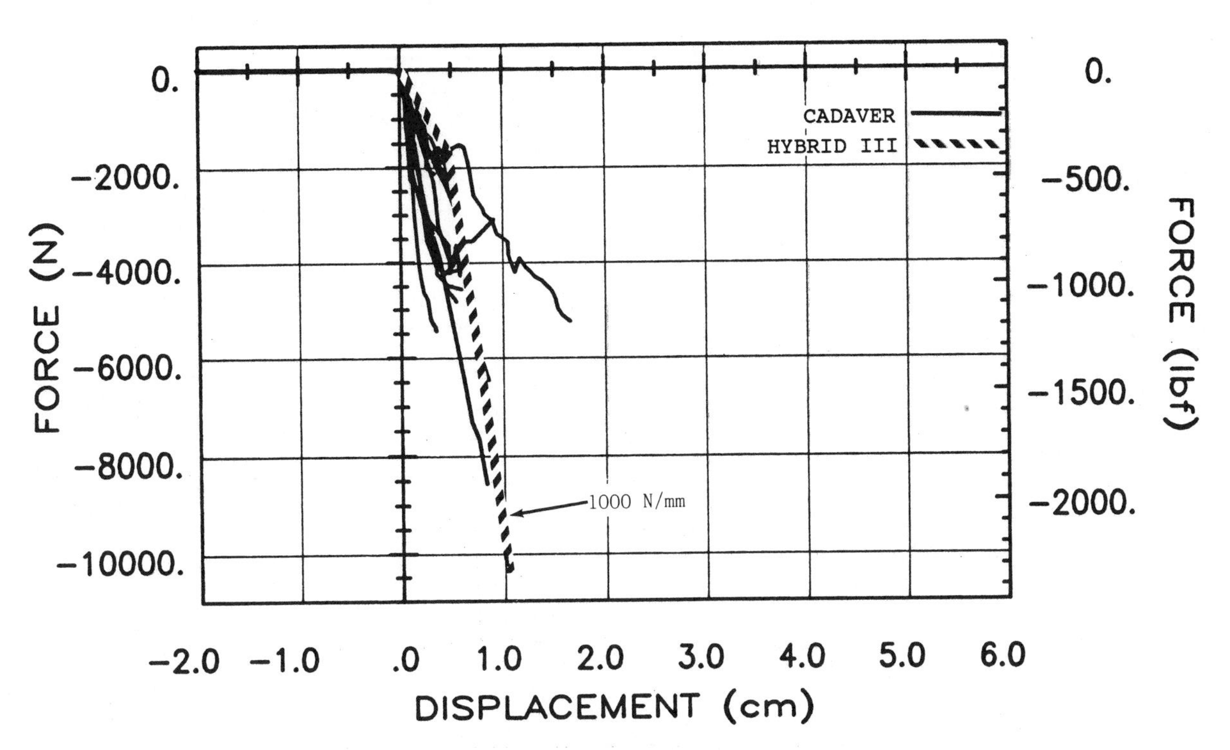

FIGURE 17
Comparison of Hybrid III and Cadaver
Frontal Bone Compliance

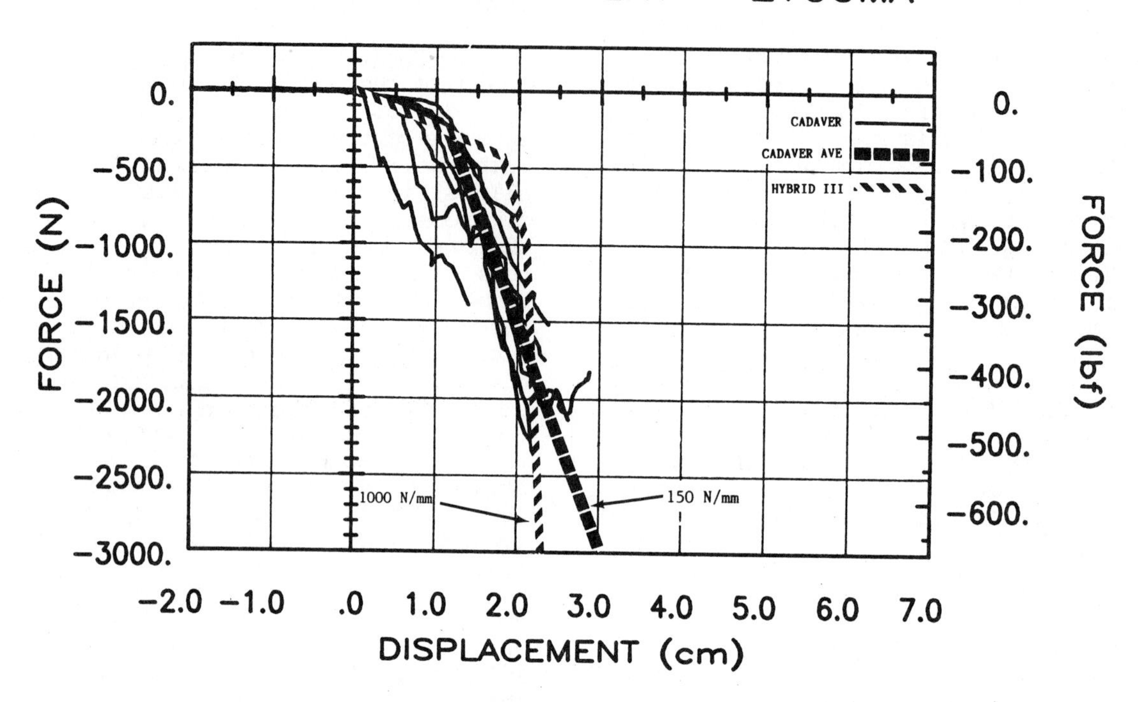

FIGURE 18
Comparison of Hybrid III and Cadaver
Zygomatic Compliance

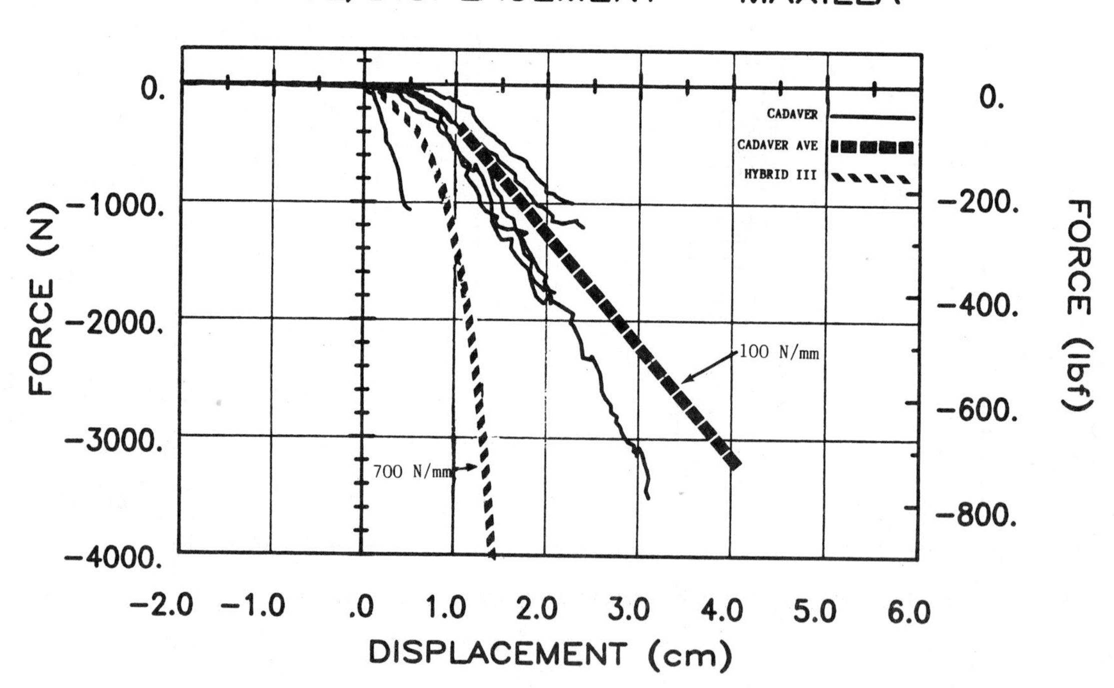

FIGURE 19
Comparison of Hybrid III and Cadaver
Maxillary Compliance

heads are plotted in Figure 19. The average for the cadaver impacts was plotted discounting the stiffest cadaver head for reasons described previously. As shown in the Figure 19, the Hybrid III dummy face is approximately six times stiffer than the cadavers in the maxillary region. Even when compared to only the very stiffest head, which seems to be out of line with the others, the Hybrid III maxillary region is still three times too stiff.

Two things become obvious in comparing the stiffness of the Hybrid III face with that of cadavers. First, the frontal region appears to have acceptable biofidelity as far as compliance is concerned. This should not be surprising, considering the fact that the Hybrid III head was designed based on cadaver frontal impact data (11). Secondly, the Hybrid III dummy head is many times stiffer in the midface region.

Even though there is a significant amount of scatter in the stiffness data for the zygomatic and maxillary regions of the human cadaver, the Hybrid III dummy face is at least three times stiffer than any cadaver head tested. This undoubtedly results in dummy-crash-test accelerations and forces that are artificially high, making correlation with biomedical research found in the literature nearly impossible. As observed in this test series, the midface impacts on the Hybrid III face resulted in forces of approximately 10,000 newtons, while the cadaver impacts averaged about 2000 newtons, and never exceed 3,500 newtons on any impact.

CONCLUSIONS

While recognizing the limitations of the small sample size employed in this research, the following conclusions and recommendations are indicated:

1) Average stiffness values for the human cadaver head in the frontal, zygomatic, and maxillary regions are approximately 1000, 150, and 120 newtons per millimeter respectively.

2) The Hybrid III face is several times stiffer in the midface region and should be redesigned if it is to yield accurate facial impact force and acceleration readings when used as an evaluation tool in automobile design.

3) Additional research is recommended to expand the sample base so that the quantitative conclusions listed can be refined.

REFERENCES

(1) J. J. Swearingen; "Tolerance of the Human Face to Crash Impact." Federal Aviation Agency, Office of Aviation Medicine, Civil Aeromedical Research Institute, Oklahoma City, OK., 1965.

(2) V. R. Hodgson; "Tolerance of the Facial Bones to Impact." Am J Anat 1967; 120:1,113.

(3) A. M. Nahum; J. D. Gatts; C. W. Gadd; J. Danforth; "Impact Tolerance of the Skull and Face," Proceedings of the Twelfth Stapp Car Crash Conference. New York, Society of Automotive Engineers, 1968.

(4) G. W. Nyquist; J.M. Cavanaugh; S.J. Goldberg; A.I. King; "Facial Impact Tolerance and Response," Proceedings of the Thirtieth Stapp Car Crash Conference. California, Society of Automotive Engineers, 1986.

(5) David C. Viano, John W. Melvin, Joseph D. McCleary, Richard G. Madeira, T. Rex Shee, and John D. Horsch; "Measurement of Head Dynamics and Facial Contact Forces in the Hybrid III Dummy," Proceedings of the Thirtieth Stapp Car Crash Conference. California, Society of Automotive Engineers, 1986.

(6) Charles Y. Warner, Milton G. Wille, Stefan Nilsson; "A Load Sensing Face For Automotive Crash Dummy Instrumentation," SAE #860197, International Congress and Exposition, Detroit, Michigan, February 24-28, 1986.

(7) J. M. Carlyle; "Imminent Fracture Detection in Graphite/Epoxy Using Acoustic Emission," Experimental Mechanics, 18, 1978.

(8) T. M. Wright; F. Vosburgh; A. H. Burstein; "Permanent Deformation of Compact Bone Monitored by Acoustic Emission," Journal of Biomechanics, 14, no.6, 1981.

(9) S. Hanagud; R. G. Clinton; J. P. Lopez; "Acoustic Emission in Bone Substance," Proceedings of Biomechanics Symposium of American Society of Mechanical Engineers (p. 74) ASME, New York, 1973.

(10) D. C. Schneider; A. M. Nahum; "Impact Studies of the Facial Bones and Skull," Proceedings of the Sixteenth Stapp Car Crash Conference. New York, Society of Automotive Engineers, 1973.

(11) R. P. Hubbard; D. G. McLeod; "Definition and Development of a Crash Dummy Head," Proceedings of the Eighteenth Stapp Car Crash Conference. Michigan, Society of Automotive Engineers, 1974.

(12) Carl E. Ellis; "Correlation of Fracture in Sheep Nasal Bones with Acoustic Emissions." Presented to the Department of Mechanical Engineering, Brigham Young University; Masters Thesis - August 1987.

Appendix #1

ACOUSTIC EMISSIONS

BONE AND FRACTURE. As indicated in the literature, acoustic emissions is a proven method of bone fracture detection. It is particularly useful because bone is a complex structure of materials with widely differing properties. Furthermore, initiation of bone fracture is difficult to detect otherwise. Acoustic emission measuring techniques involve the detection and recording of perhaps thousands or millions of minute microfractures with their attendant releases of strain energy.

ACOUSTIC EMISSIONS. The fibrous, cellular structure of bone produces a series of microfractures, each of which results in a pulse or stress waves because of the sudden release of strain energy associated with the microfracture. A major fracture will produce a succession of microfractures as each overstressed microneighbor of a previous microfracture takes its turn being stressed to fracture.

ACOUSTIC EMISSIONS MEASUREMENTS AND RECORDINGS. Acoustic emissions sensors are acoustically coupled to tissue in the vicinity of impact fracture. Means are provided to produce the desired acoustic coupling so that microfracture stress waves can be sensed by the acoustic emission detectors. Means must also be provided to record the acoustic emission data vs. time for correlation with other impact data such as force and displacement.

The traditional acoustic emissions recording systems are too slow for the impact speeds involved in this study. Furthermore, the form of their partially reduced data did not correlate well to the impact fracture process. A high speed (5 MHz sampling rate) data acquisition system was found to be suitable for recording acoustic emissions data.

Exemplary experiments were performed by some of the authors on sheep heads (12). These helped establish the desirable system characteristics for proper emissions detection and recording. Figures 8 and 9 show typical, graphical recordings of acoustic emissions with their corresponding impact-data curves.

Appendix #2

Complete Set of Force/Time Histories for Sensors 1-8

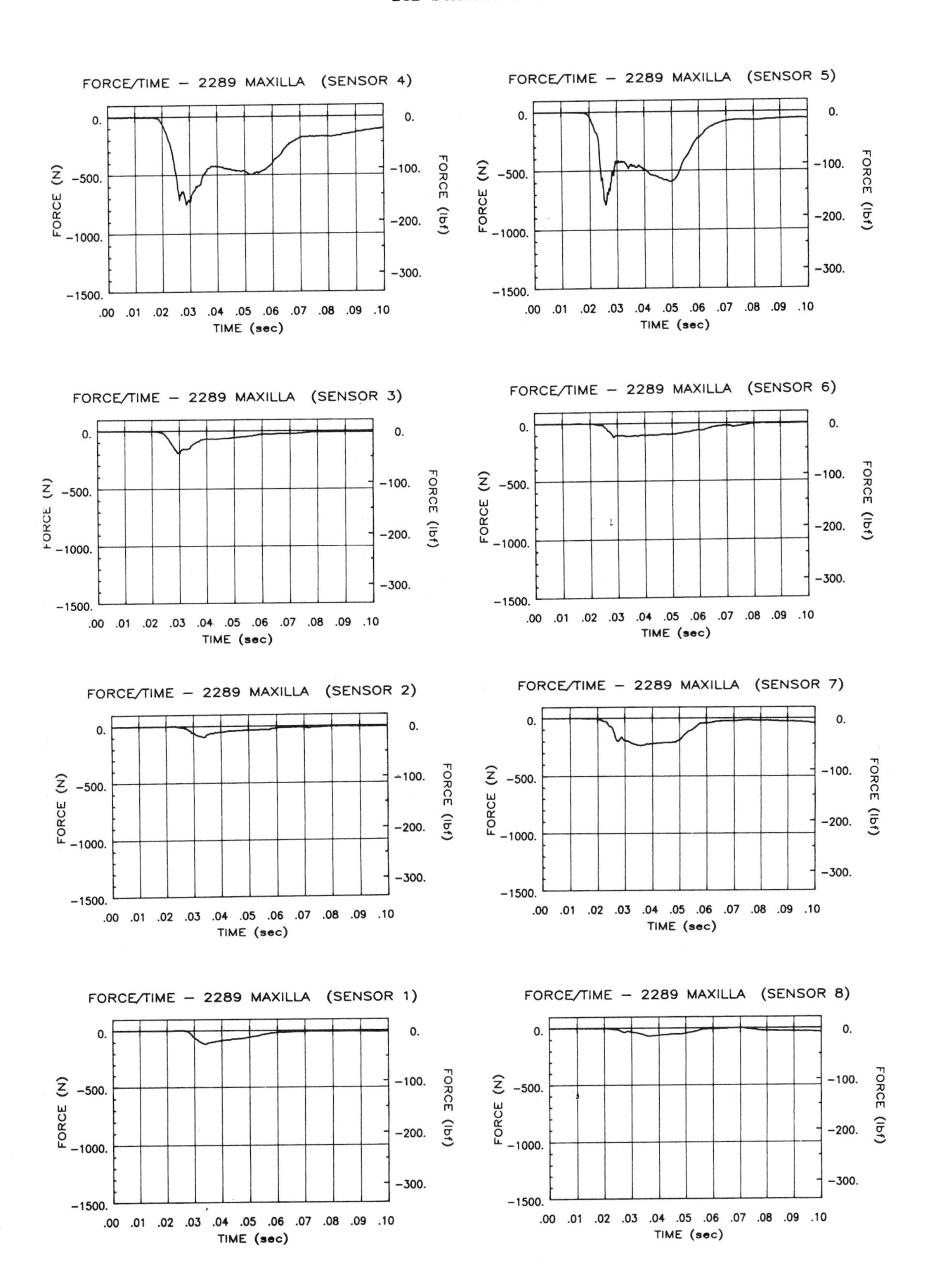

896072

Facial Injury Assessment Techniques

John W. Melvin and T. Rex Shee
General Motors Research Laboratories

Abstract

The interaction of the face with the steering system in crash testing is receiving increased attention as lap/shoulder belt restraint use increases and supplemental inflatable restraints for drivers are introduced. The impact response of the face of the test dummy is important for accurate assessment of both facial and brain injury potential. A number of facial injury assessment techniques have been proposed in recent years, including modifications to the Hybrid III face. However, none of the techniques were completely successful in producing biomechanically realistic forces and deflections when loaded by simulated steering wheel rims or hubs.

This paper reviews previously proposed facial injury assessment techniques with emphasis on the biomechanical realism of each of the procedures. In addition, biomechanical data on the response of the face to localized and distributed loads are analyzed to provide performance goals for a biomechanically realistic face. A new GM-Research modification to the Hybrid III dummy face is described, which produces biomechanically realistic frontal impact response and provides for contact force determination using conventional Hybrid III instrumentation. The modification retains the anthropometric and inertial properties and the forehead impact response of the standard Hybrid III head. It consists of a new facial skin molding overlying a deformable element that is replaced after each impact. Residual deflection of the element allows post-impact estimation of the loaded area. The magnitude and location of facial contact force is calculated through post-test processing of head accelerations and neck loads. Data demonstrating the utility of the prototype facial structure in steering system injury assessment tests are presented.

Introduction

The biomechanics of facial impact has been the subject of a number of research studies over the past twenty years, but the information developed in any one of the studies has never been sufficient to allow the development of a faceform for use in a crash test dummy. Many of the studies focused on isolated facial bone impact tolerance only and do not provide sufficient data to characterize response of the facial structure. The basic information needed for specifying the biomechanical characteristics of the face include the impact response forces due to both distributed and localized loading, the resulting deformations, and the tolerable loads for those conditions.

This study reviews the literature available on the biomechanics of facial impact response and anthropometric dummy faceforms for facial injury assessment. The general requirements for a biomechanically realistic faceform, developed from the literature, and the development of a simple structural modification to achieve this performance in the Hybrid III dummy are described.

Biomechanical Response Studies

The impact biomechanics literature contains a number of studies dealing with various individual aspects of facial impact injury. Only recently, however, have studies been performed to define the response of the face in terms of force-deflection or force-time behavior. The most extensive studies in terms of numbers of subjects and summary response data are those of Nyquist, et al. (1)* and Allsop, et al. (2). Both studies used a rigid cylindrical bar, transverse to the face to simulate steering wheel rim loading in a reproducible manner.

The Nyquist study used a 25 mm diameter bar, attached to either a 32 kg or 64 kg guided impactor mass, that impacted the faces of unembalmed cadavers across the nose and cheekbones (zygomas) as indicated in figure 1. Eleven subjects were used, and contact velocities ranged from 2.8 to 7.1 m/s. Contact force was determined by measuring the acceleration of the impactor and multiplying the data by the impactor mass. Deflection of the face was determined from high-speed movies. The data were analyzed to provide summary force-deflection curves and force-time histories for most tests. All tests produced fractures of the nasal bones and many produced more extensive fractures of the zygomas, the upper jaw bones (maxilla), and the bones behind the face. Our analysis of the impactor force-time

*Numbers in parentheses designate references at end of paper.

histories for those tests in which only the nasal bones were fractured (7 tests) produced an average peak force of 2.97 kN (range 2.2 to 4.15 kN) with an average rise time of 5.9 ms (range 5 to 8 ms). The rise time was defined for this analysis as the time from a significant increase in the slope of the impactor acceleration curve to the peak acceleration. The average contact velocity was 3.6 m/s (range 2.8 to 4.8 m/s).

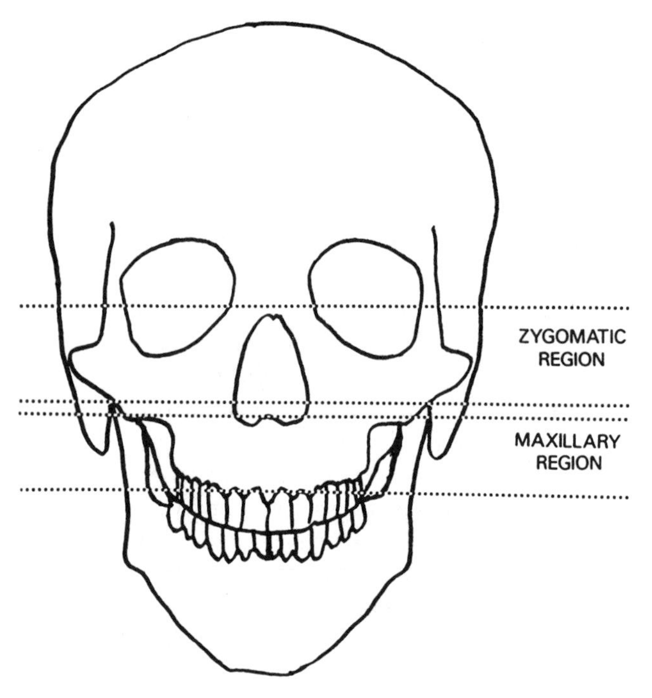

Figure 1. Impact locations for rigid bar tests with cadaver subjects.

The force-deflection data were quite varied, primarily due to large differences in the initial low-stiffness region of the concave-upward response curves. This region results from the deformation of the soft tissues of the face and the deflection of the nasal bone, factors that can vary greatly between subjects. Above loads of 0.25 kN, the response stiffens and the subjects exhibited force-deflection behaviors that were more consistent. A straight line approximation of this region above a load of 0.25 kN yields a slope of 2.0 kN/cm for the responses considered by Nyquist to represent the average response. The deflection at the 2.97 kN force level for this slope is approximately 25 mm.

The Allsop study simulated a rigid bar by a row of 20mm diameter semi-circular disks. Each disk was supported by an individual load cell which allowed the distribution of load across the face, as well as the total load, to be determined. The impactor mass was 14.5 kg and was mounted in a drop tower which guided the impactor onto the facial region of cadaver heads supported on individual molded-plaster foundations. Drop heights ranged from 305 to 610 mm (resulting in 2.45 to 3.46 m/s contact velocities) for impacts to the zygomatic and maxillary regions indicated in figure 1. Data were obtained with eight subjects for the zygoma region and six subjects for the maxilla region. A string potentiometer was used to record the impactor displacement which, because the heads were rigidly supported, was equivalent to the deflection of the impacted region of the face.

Allsop cross-plotted the total force versus the face deflection for all tests and then presented average response curves for the zygomatic and the maxillary regions. As with the Nyquist data, the zygomatic response above 0.25 kN could be characterized as a straight line, but with a slightly lower slope of 1.5 kN/cm. The average deflection at 0.25 kN was 1.2 cm and, at 3 kN the deflection was shown as 3 cm. The maxillary region was characterized by a straight line of 1.2 kN/cm slope from almost zero load at a deflection of 0.7 cm. Allsop did not present force-time histories for the tests, except for one maxilla test. Test data on the Hybrid III face were presented and showed it to be at least three times stiffer than the cadaver facial response.

Figure 2 summarizes the Nyquist data for force-time response. It defines a general response corridor for a biomechanically realistic facial structure for rigid bar impacts to the zygomatic region. The critical portion of the force-time response is the high slope region. The low initial slope of the response curve is due to soft tissue deformation at low loads and is not as important in determining the peak force response of the structure since very little impact energy is absorbed during that phase of the deformation. The presence of such soft material in a dummy face is important for maintaining proper facial anthropometry, however, to preserve proper spacing of facial contact points relative to the rest of the body surfaces.

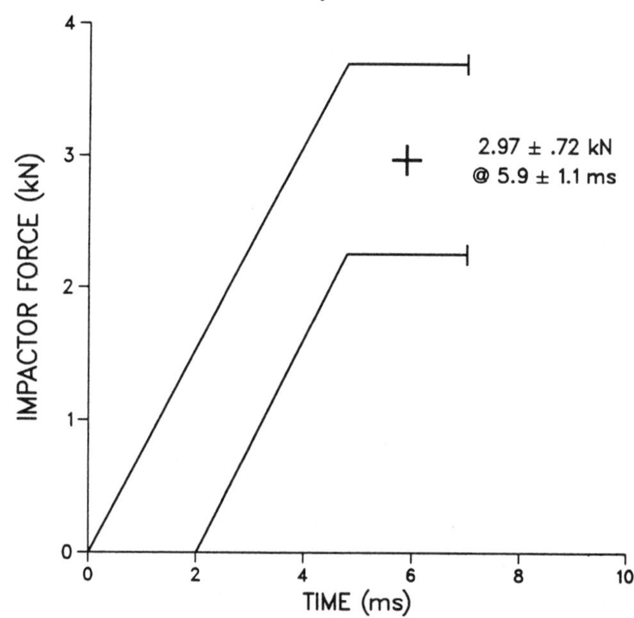

Figure 2. Summary force-time response corridor for rigid bar impacts to the zygomatic region of the face at 2.8–4.8 m/s.

Previous Development of Facial Injury Assessment Faceforms

There have been several impact headforms proposed and developed for facial injury assessment testing. This review of previous work will focus on only those devices that have the potential to be included in the head structure of a complete test dummy.

GMR frangible headform

McLeod and Gadd (3) reported on the development of a trauma indicating headform for assessment of localized and superficial head injury potential. An anatomically shaped skull was produced to achieve human-like fracture tolerances in the zygomatic and frontal regions. The design of the device was based on an actual skull with idealized geometric details. The skull material was a polyester resin reinforced by short glass fibers. Silicone rubber formulations were used to simulate skin and the underlying flesh. Limited test results were presented which showed that the fracture forces produced by the headform for concentrated rigid (6.5 cm^2) impacts to the zygomatic region were comparable to the cadaver data of Nahum, et al. (4) but the frontal bone forces were too high. No discussion was given as to the force-time response of the headform for either concentrated or blunt loading. The GMR Frangible Headform was a complex test device structure which attempted to achieve human-like biomechanical response through anatomical detail and matching of material properties. The device was never fully developed and was never used routinely in laboratory testing.

Association Peugeot-Renault faceform

Tarriere, et al. (5) reported on the development of a biomechanical facial structure for use on a modified Part 572 dummy head. The data base for the development consisted of four cadaver full-face impacts against a flat rigid surface, two with a rigid form-fitting mask and two without the mask. The modified dummy facial skeleton was simulated by a shaped piece of aluminum honeycomb supported by a flat plate which was welded in place of the facial region of the standard aluminum skull. A silicone rubber soft tissue simulation was used to cover the element and provide the original shape of the Part 572 facial exterior. The APR face performance was compared to that of a standard Part 572 face and a single test from the full-face cadaver impacts. The work demonstrated that the standard dummy facial structure can produce excessively high decelerations in a full-face flat rigid impact. The APR modification produced a close match to a first peak of the cadaver head acceleration-time history, but not a second, higher, peak. The cadaver test shown produced serious facial bone fractures including the frontal bone. The more significant bone fractures may have been associated with the second acceleration peak. The duration and thus area of the cadaver head acceleration-time curve was shown as much larger than the headform curves. Since the impact velocity was the same in all three tests, there may have been an error in the time scale of the cadaver test curve. In any event, the APR face response does not appear to provide adequate biomechanical response for the entire acceleration-time history. Accordingly, it may be useful in assessing facial injury potential at a threshold impact severity, but not at higher severity impacts. This limits the usefulness of the design in general impact testing where brain injury assessment as well as facial injury assessment is of concern.

Load-sensing and fracture-indicating devices

Recently, there have been a number of efforts to assess facial injury potential in steering systems either through measurment of facial loads in standard test dummies or by indicating the occurrence of facial structure by examining a fracture indicating feature incorporated in the dummy face or impactor surface. Warner, et al. (6) report on a piezoelectric load-sensing film matrix applied to the facial region of the Hybrid III skull. Grosch, et al. (7) used pressure-indicating paper film applied to face of a dummy to determine the pressure distribution from an impact. The pressure was integrated over the area to determine the load. Newman and Gallup (8) developed a frangible insert for a modified Hybrid III face. The insert was designed to fracture under impact severities associated with facial bone fracture in rigid impactor tests. Petty and Fenn (9) used a crushable aluminum honeycomb material on the flat front of a rigid impact mass. Injury potential was assessed by examining the impacted surface of the honeycomb for signs of permanent deformation.

The load-sensing devices using standard dummy faces have been shown by the APR study and the Allsop study to lack realistic biomechanical impact force response. Zuby (10) performed impact tests to evaluate the frangible facial insert and the aluminum honeycomb device. The honeycomb device was adapted to dummy testing by constructing an equivalent of its impactor structure which could be mounted to the Hybrid III neck. Tests were conducted with a rigid 25.4 mm diameter bar impact surface shape mounted on a 32 kg guided mass impactor. The test conditions were similar to those used by Nyquist, et al. (1). The results of the tests are summarized in figure 3 in terms of peak impactor acceleration as a function of test velocity. Figure 3 includes

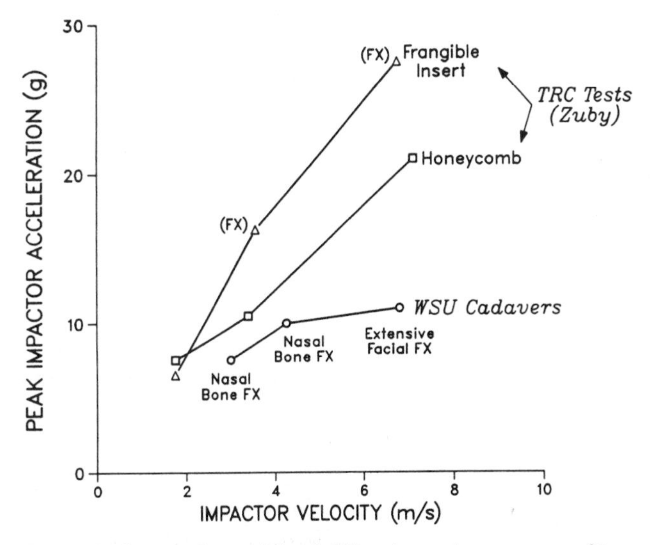

Figure 3. Comparison of the rigid bar impact responses of two fracture-indicating faceforms with the average responses of cadavers under similar test conditions.

the comparable data from the Nyquist study. Both faceforms produced unrealistically higher impactor accelerations and, therefore, contact forces at the higher impact severities.

Approach

The work reported here is part of a larger GMR program to develop improved head injury assessment techniques. The focus of the facial biomechanics part of the program is to obtain a basic understanding of facial injury mechanisms and impact response and to develop techniques for applying that knowledge to the assessment of facial injury potential and the generation of more realistic inputs to the head in crash testing. Application of biomechanical injury criteria in the laboratory requires a test device with realistic biomechanical response and a measurement method compatible with standard test laboratory instrumentation.

The specific goals of the facial injury assessment technique aspect of the program involved the following:

1. Develop a simple modification of the Hybrid III head to produce realistic force response during localized and distributed facial loading throughout the relevant range of impact severities.
2. Provide a technique capable of determining contact force magnitude, direction and location that is compatible with present testing procedures.
3. Combine the above into a comprehensive data analysis package for routine test laboratory use.

Additional biomechanical requirements included retaining the overall anthropometric and inertial properties and the forehead impact response of the standard Hybrid III head. Minimizing extra data channel requirements in the force measurement method was also emphasized. The force measurement technique and analysis package are summarized in this paper.

Development of a contact force measurement method

The achievement of a biomechanical face for the Hybrid III head, while allowing realistic head accelerations to be produced, does not allow the assessment of facial injury potential unless contact forces on the face can be determined. Adding facial load cells to the head instrumentation (6) is one direct way to provide a measurement capability. Such a method may not be desirable because of the extra complication and increased channel capacity necessary to accommodate them. Our goal was to develop a technique that could determine the magnitude and the location of the resultant facial contact force while adding little or no complication to the dummy and its associated data acquisition system. Since one of the overall goals of the program was to provide more complex and accurate assessment of brain injury potential, we had already chosen to measure angular acceleration as well as translational acceleration. This measurement allows the determination of the location of the resultant contact force.

The magnitude of the resultant contact force acting on the dummy head can be determined in terms of the x, y and z components of the force by applying Newton's Laws to the balance of inertial forces and neck reaction forces. As the face is initially loaded, the contact must be on the periphery of the dummy head. Since the shape of the head is known, it can be used in conjunction with the neck reaction moments and forces and the head angular accelerations to solve for the location of the resultant force by summing moments about the head center of gravity. The details of this procedure are given in the following section.

Derivation of equations

A free-body diagram of a Hybrid III dummy head is shown in figure 4, with a coordinate system fixed at its center of gravity. The forces acting on the head are in the impact force F, the neck load N, and the neck moment M. The impact force is to be determined and the reacting neck force and moment at the head/neck junction are measured by the standard Hybrid III load cell. These quantities are needed to solve the force and moment balance of the head.

Many frontal impact test configurations can be approximated as planar motion, particularly during the early head contact phase. The following derivation of the force and moment balance equations is given for mid-sagittal plane head motion for simplicity. Plane motion of a rigid body can be determined completely by three scalar equations, namely the force balance in two directions and the moment equation. The applied force components F_x and F_z are calculated from the force balance once the translational acceleration components are found.

$$F_x = ma_x - N_x$$
$$F_z = ma_z - N_z$$

Moment balance about the y-axis at the center of gravity (CG) of the head can be written as follows:

$$\Sigma M_{cg} = M_y + N_x z_n + N_z x_n + F_x r_z - F_z r_z = I_{cg}\alpha \quad (1)$$

where x_n and z_n are the known distances from the neck load cell to CG, r_x and r_z and the components of the position vector r of the contact point, which is located on the dummy head profile. I_{cg} is the mass amount of inertia of the dummy head about the CG, and α is the head angular acceleration calculated from the 2D accelerometry analysis (11).

Equation (1) can be rewritten as

$$-F_x r_z + F_z r_x = -I_{cg}\alpha + M_y + N_x z_n + N_z x_n = s_1 \quad (2)$$

Since all the quantities on the right-hand-side of the above equation are known, the sum s_1 can be calculated at each instant. The moment of inertia I_{cg} of the Hybrid III head is part of its design specifications. The unknowns in this equation are the contact point vector components r_x and r_z, and a second relation is needed to calculate them. At the initiation of contact, the impact force contact point must lie on the 2D head profile.

A digitized head profile was obtained from a plot of the head. The digitized data was curve-fitted by fifth-order polynomials. Two curves for $r_z > -3$ cm and $r_z \leq -3$ cm were selected for simplicty of the polynomials. Since we are interested in facial impacts with the car interior, only the facial area was considered, and the nose area was simplified for ease of calculation. The curve-fitted Hybrid III dummy head profile is shown in figure 4. Location of the contact point can be determined by simultaneously solving equations (2) and one of the two polynomial equations for r_x and r_z at each time instant. Two fifth-order polynomials were obtained after regrouping the terms.

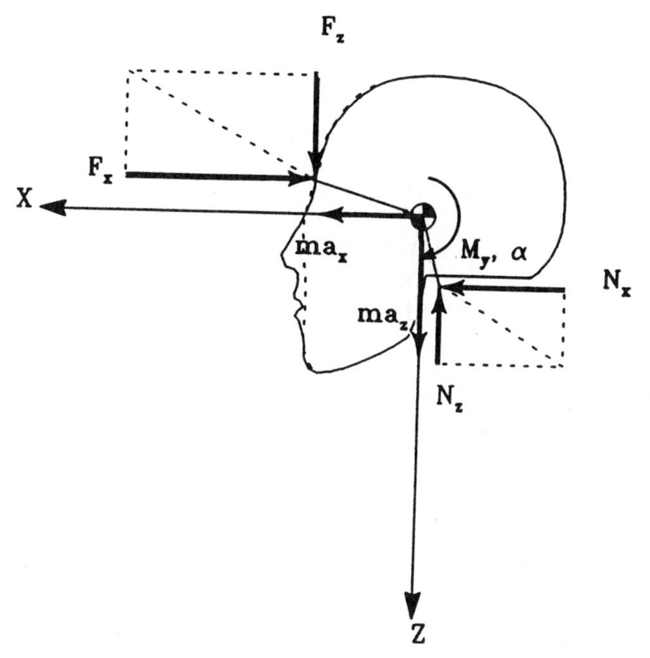

Figure 4. Free body diagram of the Hybrid III head. The dashed line represents the curve-fitted contour used to locate the contact point of the facial impact force F.

These equations were solved numerically and only real roots were considered. The format of the 2D motion analysis program was modified to accommodate the contact force components and the resultant and its location.

Development of a biomechanical face

The initial specifications for defining facial impact response consisted of a deflection requirement derived from localized loading data and a peak force requirement derived from distributed load data. The specification for the deflection of the facial structure was based on the average response from the Nyquist study which indicated that, for rigid 25.4 mm diameter cylindrical bar loading, the facial structure should allow approximately 25 mm of deflection, including soft tissue deformation, before bottoming out. The force level associated with the 25 mm deflection was about 3 kN. Many energy absorbing solid or dense foam materials exhibit bottoming out behavior in compression at strain levels of 50 to 70%. The Hybrid III face has about 12 mm of facial skin available to deform, so an additional 25 mm of space was estimated to be required to provide enough material to control the forces at the desired deflection level. Examination of the standard Hybrid III skull showed that there was sufficient room to provide such space through modified facial geometry without interfering with other features of the skull design and associated instrumentation.

Distributed load response data from Tarriere et al. (5) was used to establish initial response requirements in the form of impactor peak force as determined from three of the four APR tests and summarized in Melvin, et al. (12). The primary concern was to obtain a response which approximated the first average load peak of 7.5 kN, a unique characteristic of distributed facial loading. This load peak is related to the initial fracture of the facial skeleton and is indicative of a very stiff structure which yields or collapses. Continuing deformation produces a resumption of loading as the collapsed facial skeleton impinges upon its supporting skull bones, resulting in a second average load peak of 10.5 kN. This peak is also an important feature of the desired response, since one of the goals of a biomechanically realistic face is to provide realistic head accelerations, even when major facial bone fractures may be occurring.

Investigation of candidate materials for the modified face began by removing the face of a standard Hybrid III skull casting and welding in a flat aluminum plate with its front surface displaced 25 mm rearward from the original. This simplified headform was then used as a test platform to mount samples of the candidate materials for impact testing. The test consisted of impacting the face against a stationary 15.2 cm diameter rigid disc mounted on a biaxial load cell which was attached to a stationary support frame of a mini-sled. The modified head, on a standard Hybrid III neck, was attached to the moving carriage of the mini-sled. The carriage was guided by bearings and had a mass approximating that of the upper torso of the dummy. The mini-sled was pneumatically powered and could attain speeds up to 7 m/s at 690 kPa input air pressure. After the face struck the impactor surface/load cell, the sled carriage was arrested by a hydraulic cylinder.

The test velocity chosen for evaluating material performance was 6.7 m/s. Preliminary tests using a 15.2 cm flat disc impactor with cadavers at Wayne State University had indicated that 6.7 m/s was near the threshold for facial fracture under this type of loading. The tests also showed that, for a moving mass impactor of 13 kg, only the first peak in force occurred at that momentum level. The head moved away from the impactor with no further loading. Thus, the intent of the program, to reproduce the first portion of the force-time history became even more focussed. The data also indicated that for this test condition a first force peak lower than the 7.5 kN of the APR data was more representative of facial response when only the nasal bones were fractured. Preliminary data from five test subjects produced a mean peak force of 6.3 kN (1.9 kN SD) and a mean time at which the force peak occurred of 3.6 ms (0.9 ms SD). A summary response corridor based on the data is shown in figure 5.

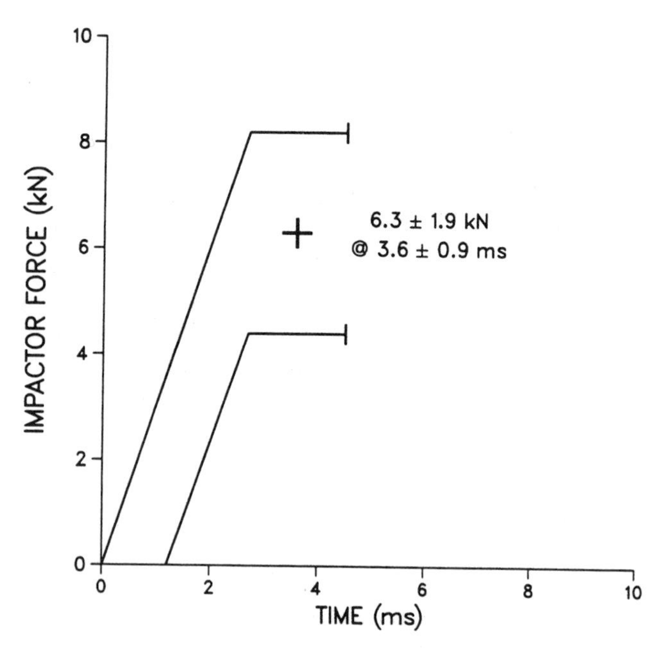

Figure 5. Preliminary force-time response corridor at 6.7/ms for full-face rigid impact to five cadaver subjects with only nasal bone fractures.

Replaceable face structures

The initial stiffness of the facial structure, coupled with the force limiting by the initiation of fractures associated with the first force peak is indicative of a facial structural material that exhibits a pronounced yield behavior in compression. Aluminum honeycomb is a classic approach to achieving this type of response and was used by Tarriere et al. (5) in developing a facial structure for the Part 572 dummy. Rigid polymer foams are another common class of materials that exhibit this type of behavior.

A series of tests with different rigid foams were conducted to determine the appropriate density range and composition. The results indicated that a foam composed of a rigid polymer with a density in the 50 to 60 kg/m^3 range was necessary to achieve both high initial stiffness and a suitable force limiting behavior/energy management behavior. The material chosen for further development was a high density (56 kg/m^3) extruded polystyrene foam (Dow HI 115 Styrofoam brand). It offered the advantage of a well-defined, commercially produced material that could easily be machined to the needs of the facial structure. The shape of the structure was also defined to allow a simplified facial geometry that would encompass the general configuration of the major facial bones in a planar form (figure 6). The projected area of the face was 58 sq. cm. The shape and area were determined from the facial anthropometry used by Hubbard and McLeod (13) to develop the Hybrid III skull.

The thickness of the foam insert, relative to the total 38 mm space available, was determined from rigid 25.4 mm diameter bar tests with the mini-sled. Various combinations of butyl rubber thickness and foam thickness were tested with a rigid bar in place of the flat disc. The first combination tested was 25.4 mm of rubber backed by 12.7 mm of the

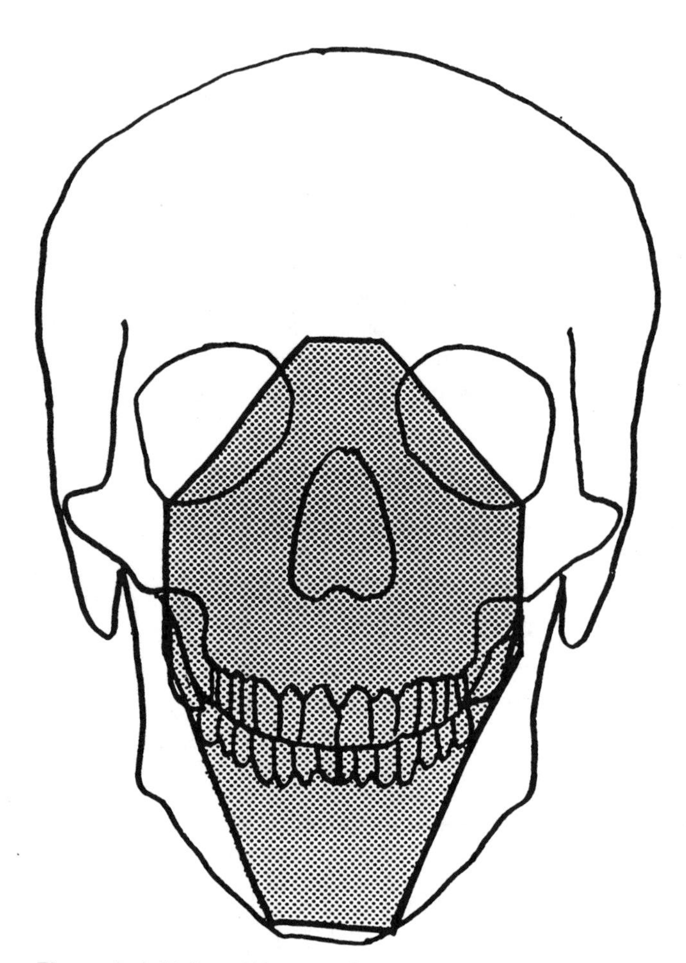

Figure 6. Initial prototype replaceable face geometry compared to the human facial skeleton.

HI 115 foam. This combination proved to be too stiff and the foam too strong to control the forces. Rather than try to obtain a lower density foam to reduce the crush force, an array of holes was drilled through the foam to reduce the effective loaded cross-sectional area while maintaining the overall geometry of the structure. This proved to be an effective method for adjusting the crush strength of the face and, potentially, could be used to account for any lot-to-lot variation in the material density. The area was reduced by about 25% and resulted in peak forces on the 3kN range. The tests also indicated that better response was achieved by placing half the rubber layer behind the foam.

Full face impacts were then conducted to evaluate the distributed load response of the structure. The higher energy of the full face test condition bottomed the 12.7 mm layer of foam. Accordingly, the foam thickness was increased to 19 mm and the rubber layer decreased to 19 mm. This combination proved to be able to control the force to the 5–6 kN level, but was not stiff enough during the initial loading phase. Placing 6.3 mm of rubber in front of the foam and the remaining 12.7 mm behind it increased the initial stiffness and resulted in the response shown in figure 7. Tests using a 25.4 mm diameter circular rigid bar were also conducted to determine whether the facial structure could meet the full face response requirement and still produce realistic forces under concentrated loading. The results, shown in figure 8, indicated satisfactory performance.

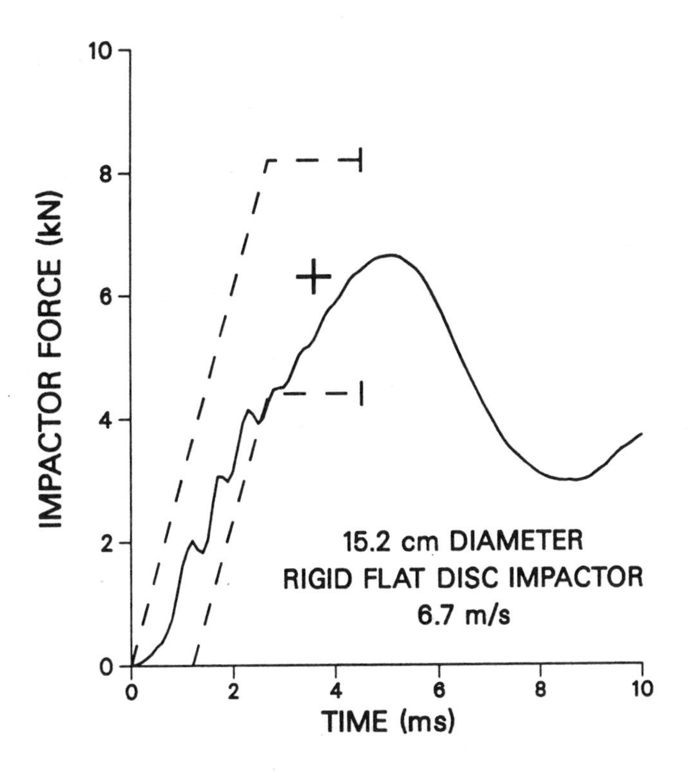

Figure 7. Full-face impact response of the prototype replaceable face structure compared to the response corridor at 6.7 m/s.

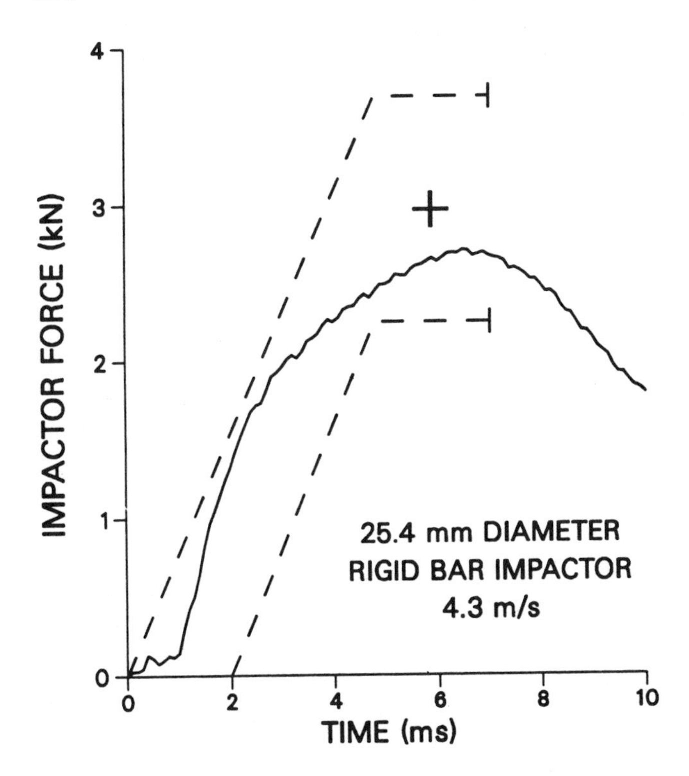

Figure 8. Rigid-bar 4.3 m/s impact response of the prototype replaceable face structure compared to the 2.8–4.8 m/s response corridor.

Evaluation of the replaceable face structure and contact force method

The modified head, including the facial structure and the vinyl head skin with the face portion cut away was ballasted to standard Hybrid III specifications. The result was termed the Prototype 1 deformable face. It was subjected to a variety of steering wheel impact tests in both the research setting and in the routine test laboratory. The major concern was to evaluate the performance of the facial structure in both wheel rim and hub impacts. The evaluation included observation of the deformation modes of the foam element, the apparent loaded area, and any tendency of the foam to fracture. Of secondary interest was the operational experience of having to replace the facial structure after every test and any problems associated with that process.

Experience was gained with steering wheel/face interactions using mini-sled tests, sled-mounted body buck tests, and full scale crash tests. The tests simulated both restrained and unrestrained drivers. The basic features of the Prototype I structure proved to be effective in use and, in particular the requirement to replace the deformable element after each test was not viewed as a detriment of the design. The consistency of the impact forces produced by structure was found to be excellent. In a mini-sled test series simulating an unrestrained driver striking the steering wheel rim the coefficient of variation was 2.7% for eleven different combinations of steering wheel design and test speed repeated twice for each combination. The contact force method was evaluted in a series of four mini-sled tests in which the zygomatic region of the face struck a rigid 25.4 mm diameter bar mounted on a load cell. The peak contact force values computed from the head acceleration/neck force method were within 2–3% of the load cell values. The computation method located the contact point typically within 5 mm.

The testing experience did suggest a number of modifications to the shape and structure of the foam element. Specifically, the mini-sled tests indicated occasions where the eyebrow area of the head became involved in rigid disc loading with resulting high loads. Ongoing testing of cadavers at WSU, in which similar loading of that region occurred, did not produce similarly high loads, although fractures of the forehead as well as the face occurred. Concern was also expressed over the ability of the Prototype 1 face to deal with oblique impacts to the cheekbone area. Finally, on occasion in steering rim impacts to the lower face, the foam was observed to have fractured transversely between the holes.

A Prototype 2 head and face structure was designed to reflect the experience gained with the first system. The foam element was redesigned to cover the frontal bone area above the eyes and side wings were added to provide oblique cheekbone average. As shown in figure 9, this was accomplished without changing the effective facial bone coverage provided by the Prototype 1 element. The hole pattern was changed to have more, slightly smaller holes arranged to minimize the occurrence of cracks connecting transversely between them. No holes were placed in the frontal sinus area above the eyes because there are no response data for this region. Future biomechanical studies will be needed to generate the information. The skull setback for the facial structure was extended up into the forehead as far as possible without changing the region of the forehead involved with the frontal impact response of the head. The resulting design is shown in figure 10.

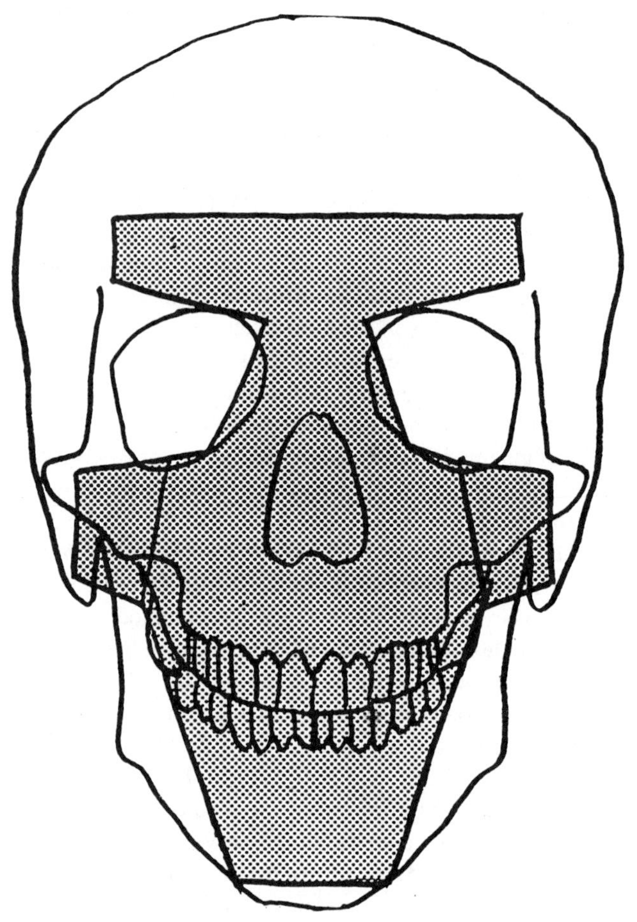

Figure 9. Prototype 2 replaceable face geometry compared to the human facial skeleton.

Figure 10. Prototype 2 Hybrid III head showing the deformable facial element in place. A featureless facial covering is shown.

Because of the flat support surface of the facial structure and the width required of the facial element to represent the facial bones, it was found necessary to increase the skull setback an additional 6.4 mm to preserve accurate Hybrid III facial anthropometry in the cheekbone areas on either side of the nasal area. The facial covering of the dummy will have to be constructed of a vinyl-coated soft foam to maintain facial dimensions. It will be backed by the required 6.3 mm of solid vinyl for proper impact response. It is possible to produce human-like nose and lip facial features by this technique. The durability of these features may be questionable and the nose and lips may be left out of the final design. Also, the presence of a mechanically durable nose and lips can result in reduced repeatability of the test device. Full-face rigid impact tests of a feature-less Prototype 2 face verified that the basic performance of the face was not changed by these modifications (figure 11).

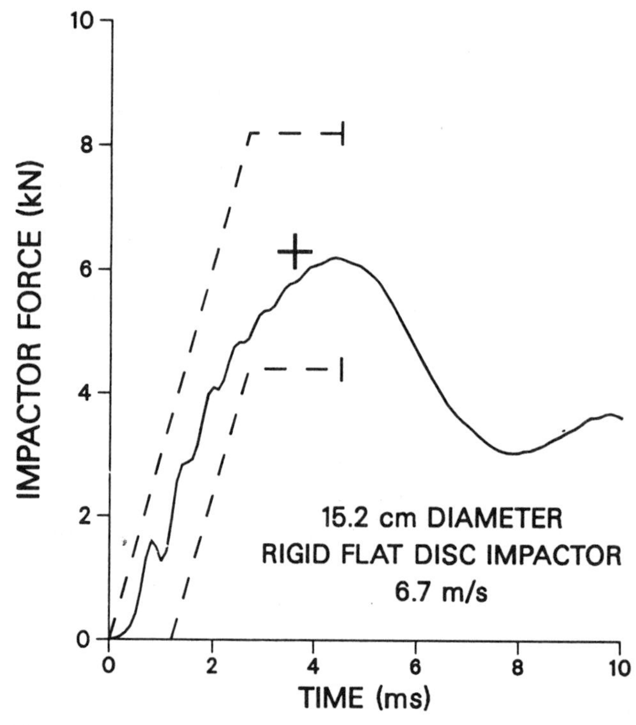

Figure 11. Full-face impact response of the Prototype 2 face structure compared to the response corridor at 6.7 m/s.

The Prototype 2 face was used in a series of sled tests to compare it with the standard Hybrid III in head impact injury assessment in high severity (67 kph) crash simulations. The dummies were restrained by a lap/shoulder belt restraint system which was adjusted to allow the face to strike the steering wheel hub as the torso was restrained by the shoulder belt. Identical steering systems were used in each test. The results of a pair of the tests, in terms of the resultant head accelerations, are shown in figure 12. The standard dummy head produced a 40 percent higher peak acceleration than the head with the deformable face. The 36 ms interval Head Injury Criterion (HIC) values for the two tests varied only by 6.1 percent, however, because the time duration of the acceleration in the test without the deformable face was greater. Although further testing over a range of impact conditions will be required, these preliminary data suggest that the use of the deformable face on the Hybrid III head in routine restraint system testing should allow improved facial injury assessment without significantly changing the HIC values associated with standard compliance certification.

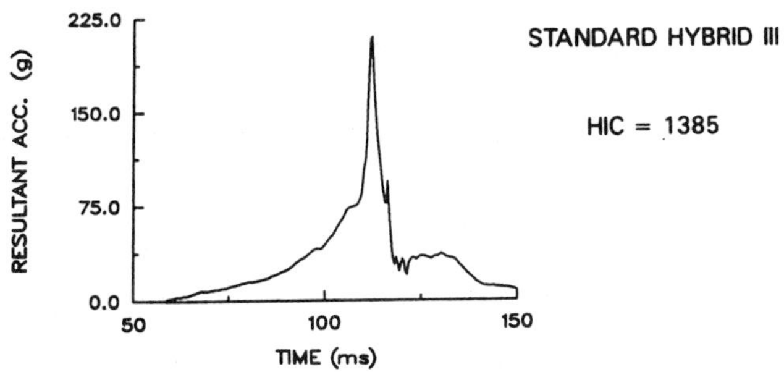

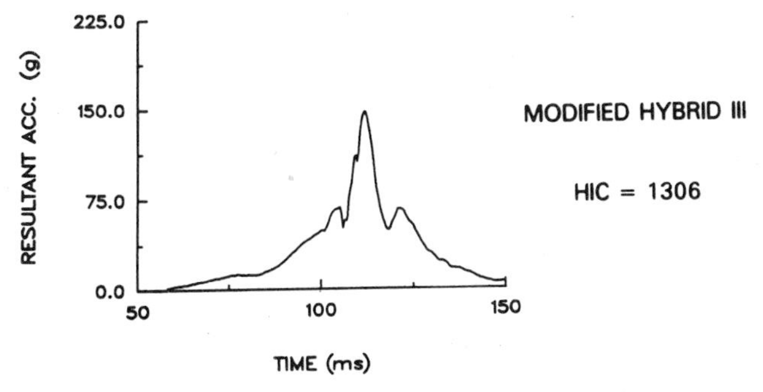

Figure 12. Comparison of the standard Hybrid III head acceleration with that of the Prototype 2 modified Hybrid III head when subjected to steering wheel hub impact. The tests were conducted on a sled with a complete dummy restrained by a lap/ shoulder belt, adjusted to produce face-to-hub contact, at 67 kph.

Conclusions

A facial structure which produces biomechanically realistic impact force response for localized and distributed loading has been developed for the Hybrid III dummy. The structure can be accommodated through simple modifications to the original Hybrid III head. The modifications do not interfere with the standard instrumentation, anthropometry or biomechanical forehead impact response of the Hybrid III and allow facial contact force to be determined using the standard instrumentation. The modified facial structure has been used in a variety of laboratory settings and the utility and efficacy of the design have been verified.

Sufficient data on human mid-face impact response to rigid-bar localized loading, analogous to steering rim loading, have been reported in the biomechanics literature to allow the definition of biomechanical response characteristics for such loading. Preliminary data for distributed facial loading are also available, but further work is needed to fully describe the response to distributed impact for the severity range of interest. This study has used the available data to generate suggested dummy faceform response corridors for both types of loading.

References

(1) Nyquist, G.W.; Cavanaugh, J.M.; Goldberg, S.J.; and King, A.I. (1986) Facial Impact Tolerance and Response. Proceedings of the 30th Stapp Car Crash Conference, pp. 379–400. Society of Automotive Engineers, Warrendale, Pa.

(2) Allsop, D.Y.; Warner, C.Y.; Wille, M.G.; Schneider, D.C.; and Nahum, A.M. (1988) Facial Impact Response—A Comparison of the Hybrid III Dummy and Human Cadaver. Proceedings of the 32nd Stapp Car Crash Conference, pp. 139–155. Society of Automotive Engineers, Warrendale, Pa.

(3) Nahum, A.M.; Gatts, J.D.; Gadd, C.W.; and Danforth, J. (1968) Impact Tolerance of the Skull and Face. Proceedings of the 12th Stapp Car Crash Conference, pp. 302–316. Society of Automotive Engineers, Warrendale, Pa.

(4) McLeod, D.G.; and Gadd, C.W. (1973) An Anatomical Skull For Impact Testing. Human Impact Response, pp. 153–177. Edited by King, W.F. and Mertz, H.J. Plenum Press, New York.

(5) Tarriere, C.; Leung, Y.C.; Fayon, A.; Got, C.; Patel, A. and Banzet, P. (1981) Field Facial Injuries and Study of Their Simulation with Dummy. Proceedings of the 25th Stapp Car Crash Conference, pp. 435–468. Society of Automotive Engineers, Warrendale, Pa.

(6) Warner, C.Y.; Wille, M.G.; Brown, S.R.; Nilsson, N.; Mellander, H.; and Koch, M. (1986) A Load Sensing Face Form for Automotive Collision Crash Dummy Instrumentation. SAE Paper No. 860197. Society of Automotive Engineers, Warrendale, Pa.

(7) Grosch, L.; Katz, E.; and Kassing, L. (1987) New Measurement Methods to Assess the Improved Injury Protection of Airbag Systems. SAE Paper No. 870333. Society of Automotive Engineers, Warrendale, Pa.

(8) Newman, J.A.; and Gallup, B.M. (1984) Biofidelity Improvements to the Hybrid III Headform. Proceedings of the 28th Stapp Car Crash Conference, pp. 87–99. Society of Automotive Engineers, Warrendale, Pa.

(9) Petty, S.P.F.; and Fenn, M.A. (1985) A Modified Steering Wheel to Reduce Facial Injuries and Associated Test Procedure. 10th International Technical Conference on Safety Vehicles, National Highway Traffic Safety Administration, Washington, D.C.

(10) Zuby, D. (1987) Steering Assembly Induced Facial Injury. ASME Paper No. 87–WA/SAF–3, The American Society of Mechanical Engineers, New York.

(11) Viano, D.C.; Melvin, J.W.; McCleary, J.D.; Madeira, R.G.; Shee, T.R.; and Horsch, J.D. (1986) Measurement of Head Dynamics and Facial Contact Forces in the Hybrid III Dummy. Proceedings of the 30th Stapp Car Crash Conference, pp. 269–289. Society of Automotive Engineers, Warrendale, Pa.

(12) Melvin, J.W.; King, A.I.; and Alem, N.M.; (1985) AATD System Technical Characteristics, Design Concepts, and Trauma Assessment Criteria. Task E–F Final Report, p. 28. National Highway Traffic Safety Administration Report Number DOT HS 807 224, Washington, DC.

(13) Hubbard, R.P.; and McLeod, D.G. (1974) Definition and Development of a Crash Dummy Head. Proceedings of the 18th Stapp Car Crash Conference, pp. 599–628. Society of Automotive Engineers, Warrendale, Pa.

Acknowledgements

The authors wish to recognize E.A. Jedrzejczak, W.C. Little, R.G. Madeira and J.D. McCleary for their technical assistance in this study.

912907

Force/Deflection and Fracture Characteristics of the Temporo-parietal Region of the Human Head

Douglas "L" Allsop, Thomas R. Perl and Charles Y. Warner
Collision Safety Engineering, Inc.

ABSTRACT

Impact tests were conducted on thirty-one unembalmed human cadaver heads. Impacts were delivered to the temporo-parietal region of fixed cadavers by two, different sized, flat-rigid impactors. Yield fracture force and stiffness data for this region of the head are presented. Impactor surfaces consisted of a 5 cm^2 circular plate and a 52 cm^2 rectangular plate. The average stiffness value observed using the circular impactor was 1800 N/mm, with an average bone-fracture-force level of 5000 N. Skull stiffness for the rectangular impactor was 4200 N/mm, and the average fracture-force level was 12,500 N.

INTRODUCTION

Efforts to quantify the impact response of the temporo-parietal region of the human skull were undertaken over a century ago.[1] Following that initial effort, little was recorded in the literature until the late sixties and early seventies when efforts by Melvin, Schneider, Nahum, Hodgson, and their colleagues, were published.[2-5] These studies concentrated on fracture-force levels, and did not address the dynamic deflection or stiffness of the skull during impact.

It is felt that documenting the human skull stiffness is a critical step in evaluating the biofidelity of current anthropomorphic test dummies, and is also necessary for future research and design work on dummy heads capable of measuring meaningful lateral impact loads. Due to Newton's second law (F=ma), and the inversely proportional relationship between deflection and acceleration during impact, unless the test device and the biological material have a similar stiffness, forces observed by the test device are not meaningful, because they cannot be directly compared to biomechanical data.

Therefore, objectives of research presented here were to: 1) Document the stiffness of the temporo-parietal region of the human skull during impact; and 2) Augment fracture-force data which has previously been gathered.

PROCEDURE/INSTRUMENTATION

THEORY - Many different approaches have been employed during the past few decades to determine the force at which skull bones fracture. Some approaches involved multiple impacts to a single specimen, with increased force until a fracture was detected. Other approaches bracketed the fracture force by impacting several specimens only once, but at various force levels, noting which levels produced fracture. This research used a new technique developed by Allsop et al.[6] It involves impacting each specimen with more than enough energy to cause fracture, recording the force-time history during impact, and identifying fracture force by the discontinuity observed on the force-time history. An example is shown in Figure 1 where fracture is observed at approximately 36 msec and the force is 16 kN. In contrast, note Figure 2 where a Hybrid III dummy head was impacted without causing fracture. The smooth nearly symmetric curve indicates no abrupt changes in material resistance, and therefore no fracture.

In this research, fracture force values determined from force-time traces were corroborated by acoustic emissions-time histories, as was previously done by Allsop.[ibid] A typical acoustic emissions (AE) trace is shown in Figure 3. Prior to fracture little or no acoustic energy is recorded, but even when a hairline fracture occurs, acoustic emissions are observed. The trace in the figure suggests fracture at 5.8 msec after the AE - time history began.

In Figure 4, force-time and AE - time histories for a single impact have been plotted on the same time scale. The discontinuity in the force-time history, and the onset of AE energy, clearly identify the force level at which fracture occurred.

Stiffness data were generated by measuring the impactor displacement during impact, and cross-plotting these data with force-time data. This eliminated the time axis, and left the desired stiffness characteristics in the form of a force-deflection curve, as shown in Figure 5.

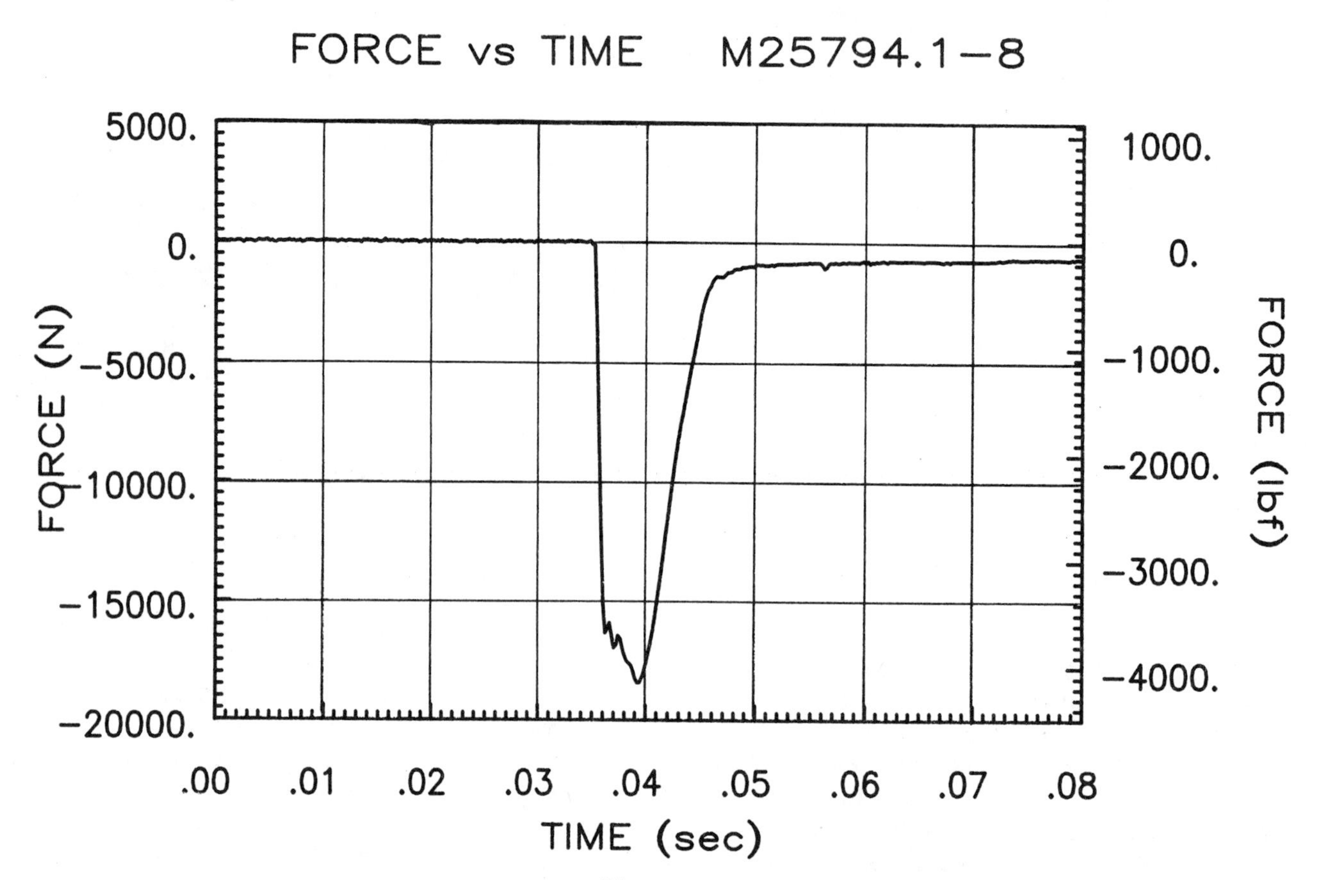

Figure 1
Summed Force vs. Time for Test Specimen M25794

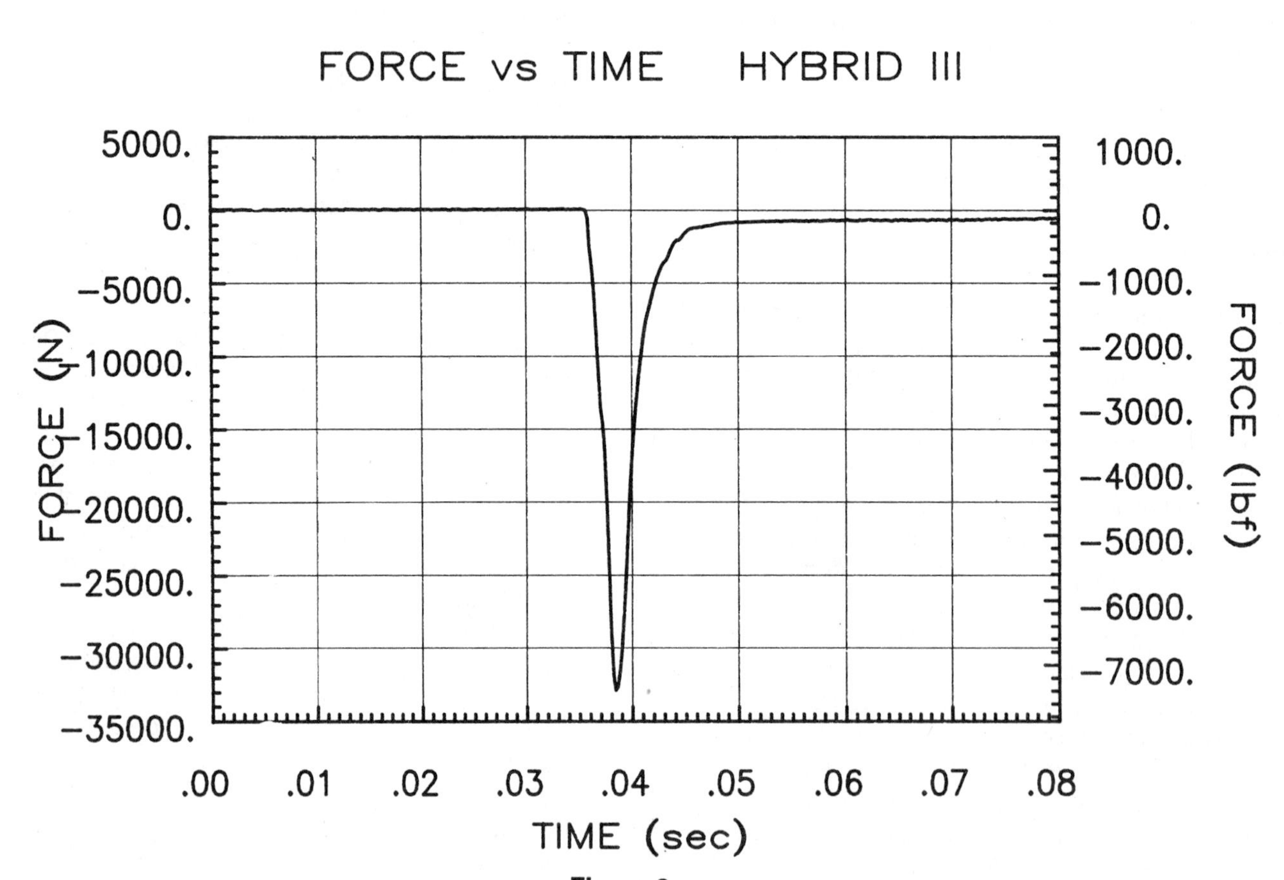

Figure 2
Summed Force vs. Time for Hybrid III Lateral Test Using Rectangular Plate Impactor

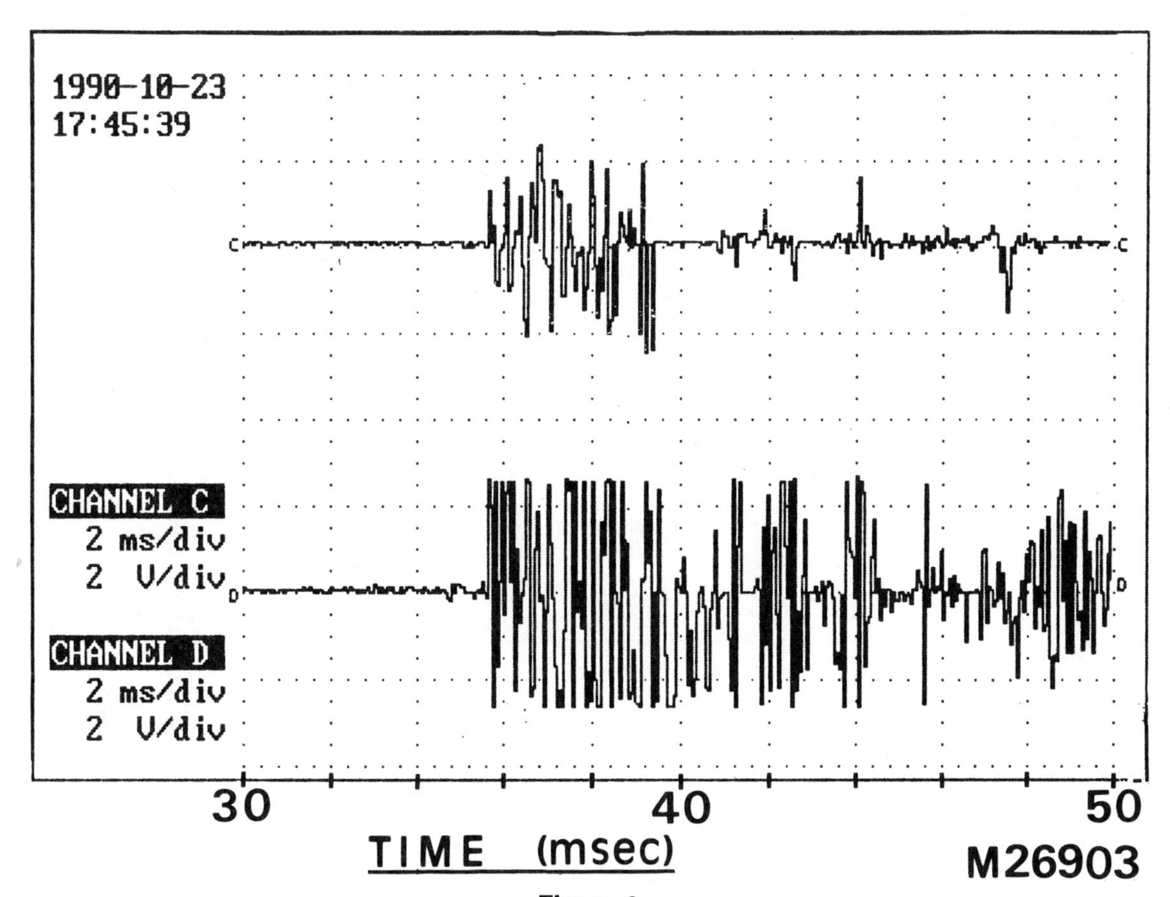

Figure 3
Acoustic Emissions v. Time Trace (Fracture at 35.8 msec After Trigger)

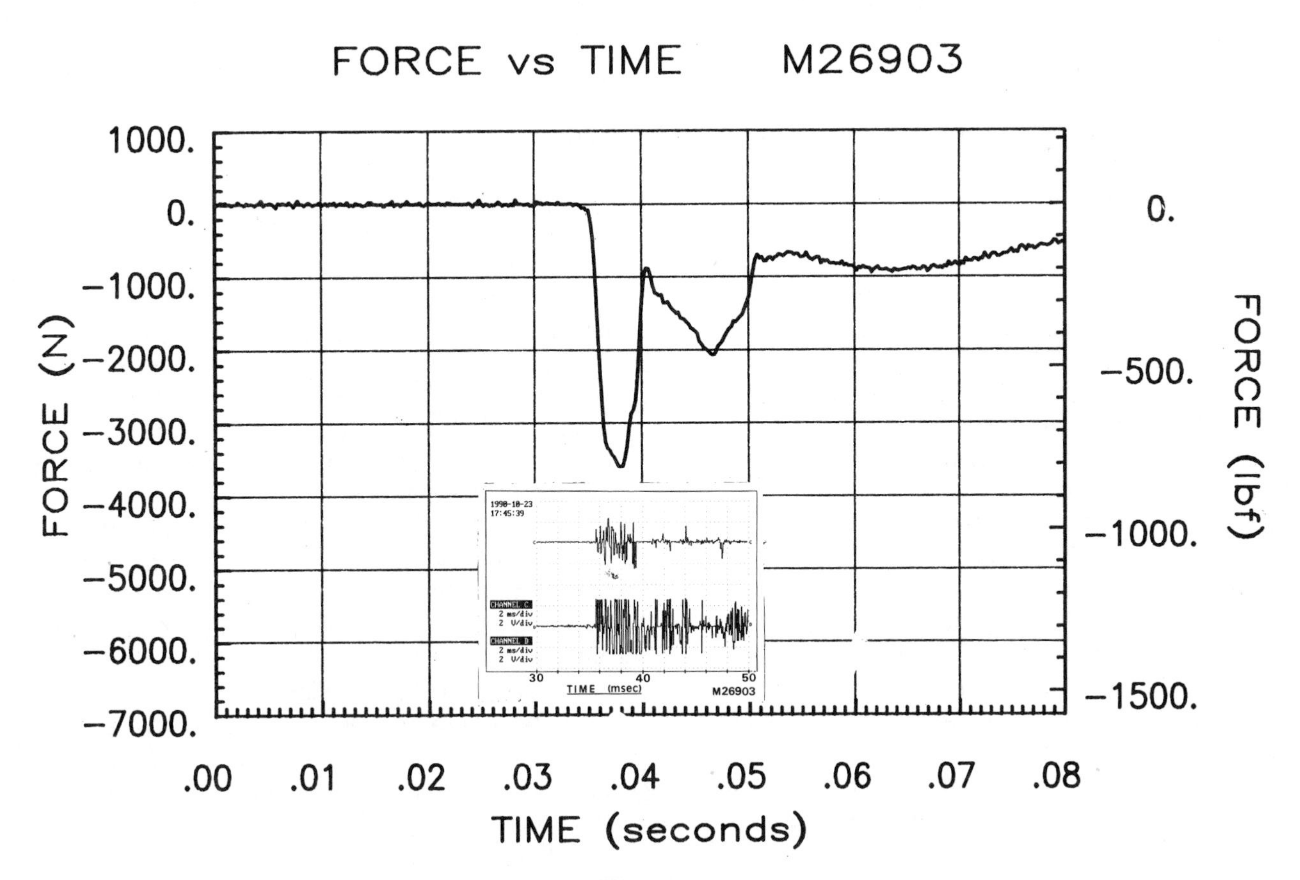

Figure 4
Force/Time and Acoustic Emissions/Time Curves

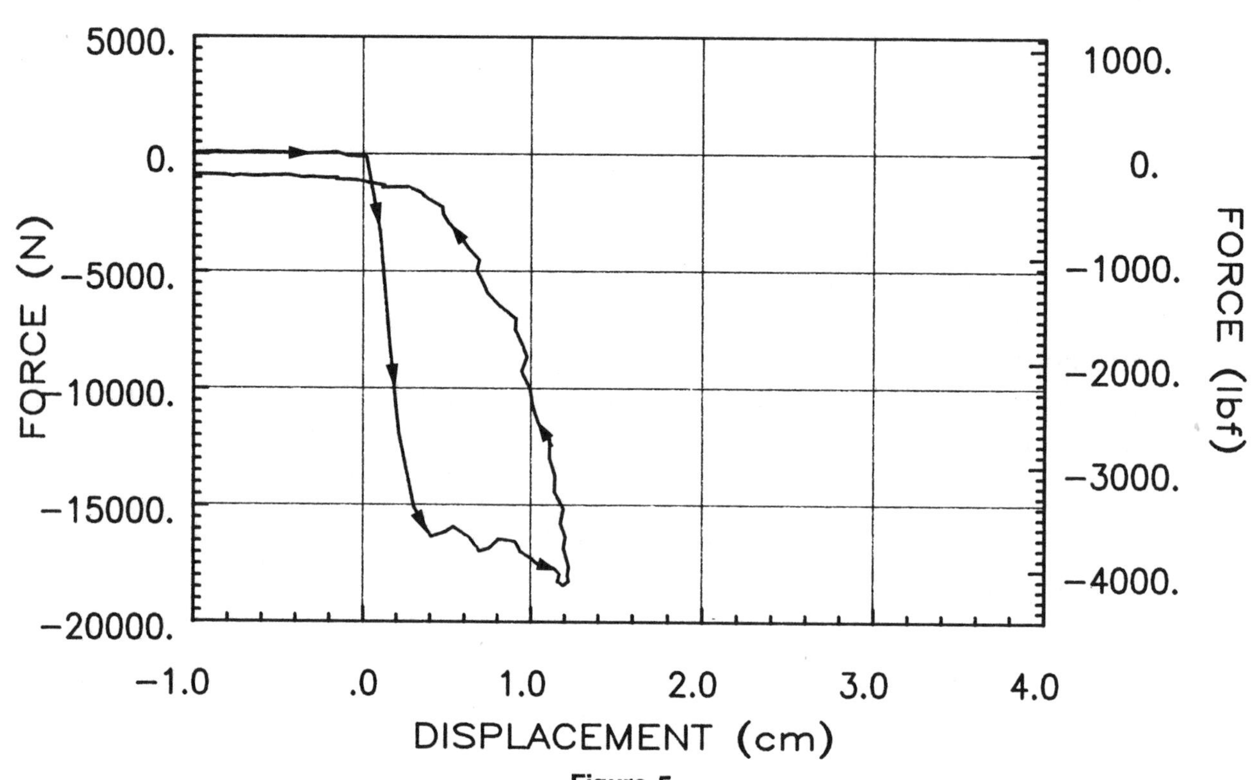

Figure 5
Force vs. Displacement Curves for a Circular Plate Impact

INSTRUMENTATION - Much of the instrumentation was the same as that used from the tests conducted by Allsop et al[ibid] with modifications for impacts to the side of the head. It consists of a drop tower, shown in Figure 6, which was unchanged with the exception of the actual impacting surface or impactor, and slightly modified AE recording equipment. For this series of tests two impactors were used. One was a flat rectangular plate approximately 5 cm by 10 cm which weighed 12 kg (Figure 7). This plate was actually a composite of eight smaller impactor inserts, each with dimensions of 5 by 1.1 cm and an edge radius of 5.1 mm as shown in Figure 8. The eight inserts connected to piezo-electric force transducers, so that a force-time curve was generated for each centimeter of length along the impactor surface. Therefore, eight force-time curves were generated for each rectangular plate impact, and were later summed to generate the total force-time curve.

The second impactor is seen in Figures 9 and 10. It is a flat circular plate with a contact surface diameter of 2.54 cm and an edge radius of 2 mm. This impactor weighs 10.6 kg and was connected to a Kistler 9251-A force transducer, which recorded data only in the vertical or "Z" direction.

Displacement or deflection was recorded by two string potentiometers attached to the impacting mass. The first was a Gynisco 100g string pot, and the second was a custom built continuous loop pot. The continuous loop pot was a precautionary measure employed to eliminate the possibility of potentiometer overshoot during the high decelerations encountered during impact.

All of the force and displacement instrumentation was digitally sampled by a TransEra MDAS 7000 data acquisition system at a rate of 5000 Hz.

As previously mentioned, AE - time histories were recorded during impact. This was accomplished by attaching conical-shaped wave guides to 6.4 mm diameter AE sensors (MAC-300L), making a small incision in the skin near the impact site, and wedging the wave guides into the flesh. The acoustic energy waves or acoustic emissions were transferred from the sensors to a digital oscilloscope board (SCC-1220) during impact, and recorded for later analysis. AE sampling rates were 1.0 MHz and 2.0 MHz for the rectangular and circular impactors respectively.

SPECIMEN PREPARATION - Specimens, which had been frozen, were allowed to thaw for at least 24 hours, but not more than 32 hours prior to impact. All specimens were rinsed, shaven, measured, and placed in plaster of paris with a rolled towel between their teeth to reduce erroneous acoustic emissions.

The depth of plaster of paris was approximately 40-50% of the head width. For rectangular plate impacts, heads were rotated 45 degrees from the horizontal plane as shown in Figure 11.

For circular plate impacts, specimens were oriented with the sagittal plane at a right angle to the impactor plane of travel. (See Figure 12)

Figure 6
Front View of Drop Tower

Figure 7
Rectangular Plate Impactor Head and Body

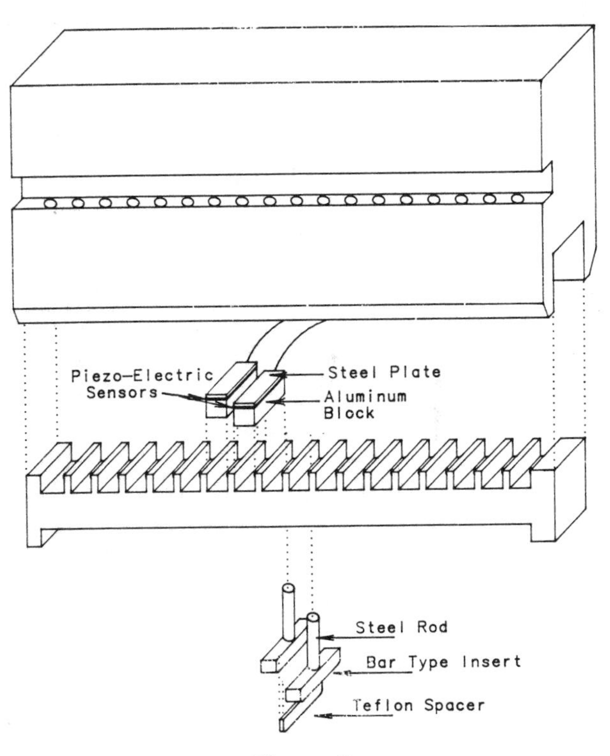

Figure 8
Rectangular Plate Impactor Head

TEST PROCEDURE - In the A-P direction, specimens were aligned, so that the impactor centerline would fall just slightly anterior to the external auditory meatus. The rectangular plate impact location is shown in Figure 13. The two impact locations employed with the circular impactor are illustrated in Figure 14.

The drop height for most of the rectangular plate impacts was 102 cm, with a nominal impact velocity of 4.3 m/s. Most of the circular plate impacts were conducted at a drop height of 38 cm, with a nominal impact velocity of 2.7 m/s.

RESULTS

FRACTURE FORCE - Analysis of the force-time and AE - time curves, following the procedure previously outlined, generated the fracture force levels listed in Tables 1 and 2. Table 1 lists results for the rectangular impactor, while Table 2 lists data for the circular impactor. The rectangular plate caused fracture at an average force level of 12,390 N with a standard deviation of 3654 N.

The circular plate fracture forces averaged to 4990 N for location one and 5400 N for location two. Standard deviations were 1801 N and 1984 N for locations one and two respectively.

Figure 9
Circular Plate Impactor Head and Body

Figure 10
Circular Plate Impactor Head and Body

It is interesting to note that if the exceptionally high fracture level from Specimen M26373 is left out of the data set, the average fracture value for impact sites one and two, are nearly identical at 4889 N and 4990 N.

STIFFNESS - The stiffness values for the rectangular and circular impactors (listed in Tables 3 and 4 respectively) were taken directly from the force-displacement curves generated from each impact. Force-displacement curves for each of the impacts are plotted in Figures 15 and 16.

To compute the stiffness value, the average slope of the force-displacement curve was evaluated between the force range of 4 to 12 kN for the rectangular impactor and 2 to 6 kN for the circular impactor.

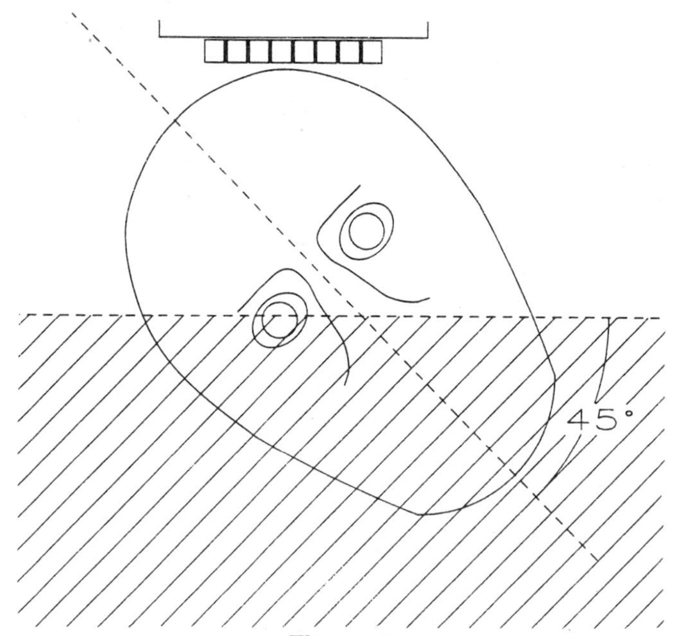

Figure 11
Specimen Orientation for Rectangular Plate Impacts
(Diagonal Lines Indicate Plaster of Paris)

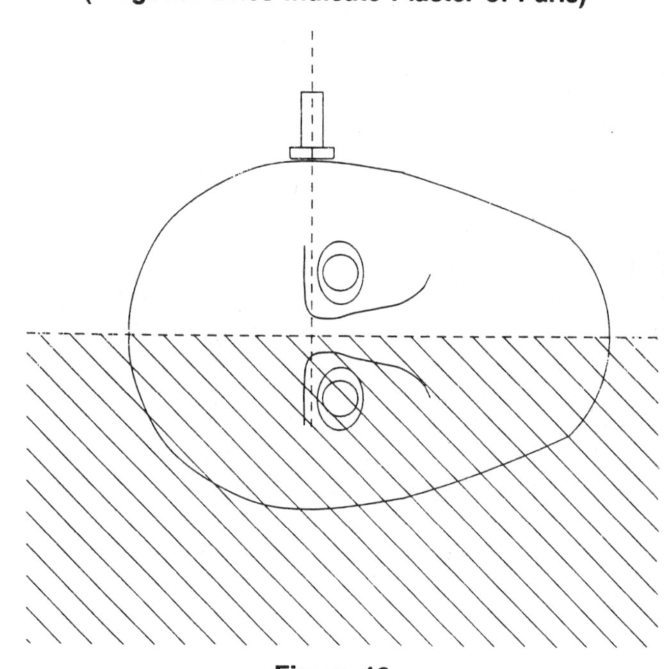

Figure 12
Specimen Orientation for Circular Plate Impacts
(Diagonal Lines Indicate Plaster of Paris)

DISCUSSION OF RESULTS

FRACTURE FORCE - A ratio of two and one-half to one was observed between the average fracture-force level of the rectangular and circular impactors. The impact angle and impact location for these impactors were slightly different, however, the difference in impactor area is likely the most significant factor in this variation. A four-to-one ratio between the high and low value in each set of

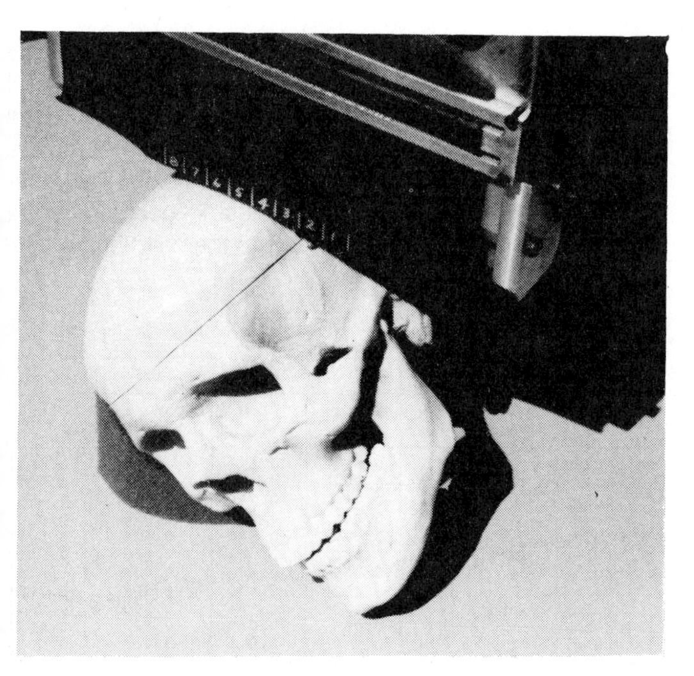

Figure 13
Impact Location for the Retangular Impactor

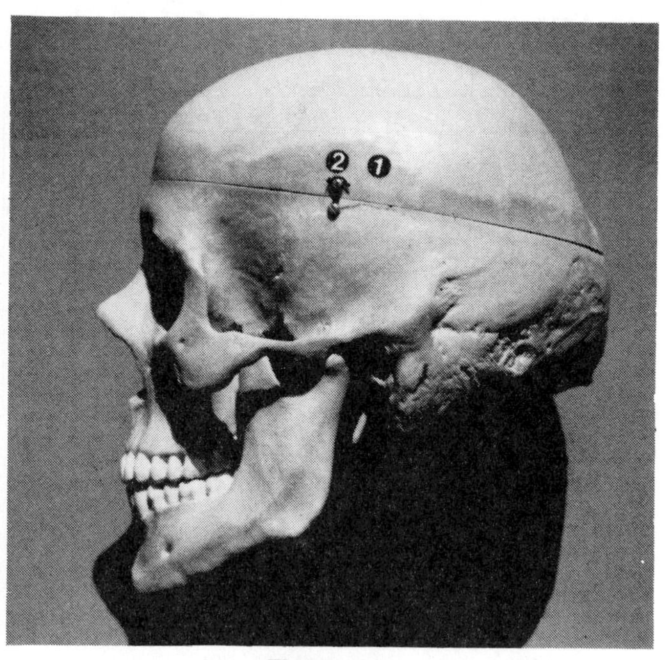

Figure 14
Impact Locations for the Circular Impactor

data also exists. This variability is not uncommon in biomechanical testing. The fracture-force range observed in previous research on the temporo-parietal region is bounded at 700 N on the low end, and 14,950 N on the high end. The averages for the previous studies range from 3120 N to 8500 N.[1-5] It appears that the variation observed in this study is similar to that observed by others. As expected, the higher forces were recorded by the larger area impactor, indicating a significant relationship between contact area and fracture force.

STIFFNESS - No other data have been found in the literature with which to compare the stiffness values

RECTANGULAR PLATE FRACTURE LEVELS

CADAVER #	AGE	SEX	FRACTURE FORCE (N)
F25746	72	F	5,800
F26378	57	F	7,000
M25744	44	M	10,000
F26107	75	F	11,500
F26341	70	F	12,500
M25725	68	M	13,000
F26379	31	F	13,000
F26115	63	F	14,000
M25794	19	M	16,000
F26352	90	F	16,500
M26338	61	M	17,000
Average: (approx. 12,500 N)			12,390
Standard Deviation:			3,654
High:			17,000
Low:			5,800

TABLE 1
Bone Fracture Levels for
Rectangular Plate Impacts

CIRCULAR PLATE FRACTURE LEVELS

CADAVER #	IMPACT LOCATION	AGE	SEX	FRACTURE (N)
F26588	1	90	F	2,500
M26903	2	37	M	3,100
F26342	1	73	F	3,200
F26374	1	82	F	3,800
F26380	2	60	F	4,000
M26922	2	72	M	4,000
F26377	1	72	F	4,000
F26371	2	46	F	4,700
M26329	1	62	M	4,800
F26361	2	75	F	4,800
M26383	2	92	M	5,000
F26925	2	67	F	5,100
F26587	1	62	F	5,200
F26384	1	89	F	5,300
F26573	2	63	F	6,000
M26350	1	75	M	6,400
F26354	1	66	F	7,000
M26372	2	80	M	7,300
M26368	1	81	M	7,700
M25373	2	73	M	10,000
Impact Locations #1 and #2				
Average: (approx. 5,200 N)				5,195
Standard Deviation:				1,801
High:				10,000
Low:				2,500
Impact Location #1 ONLY				
Average:				4,990
Standard Deviation:				1,678
High:				7,700
Low:				2,500
Impact Location #2 ONLY				
Average:				5,400
Standard Deviation:				1,984
High:				10,000
Low:				3,100

TABLE 2
Bone Fracture Levels for
Circular Plate Impacts

observed in this research. However, the stiffness of the temporo-parietal region with the circular impactor is similar to that observed previously for the frontal bone.[6] As with fracture force, stiffness appears to be affected significantly by the contact area of the impacting mass.

RECTANGULAR PLATE COMPLIANCE VALUES

CADAVER #	AGE	SEX	COMPLIANCE (N/mm)
F26378	57	F	1,600
F25746	72	F	1,620
M25744	44	M	3,000
F26115	63	F	3,270
M25725	68	M	4,090
M25794	19	M	4,500
F26352	90	F	4,860
F26341	70	F	5,140
F26379	31	F	5,630
F26107	75	F	5,710
M26338	61	M	6,430
Average: (approx. 4,200 N/mm)			4,168
Standard Deviation:			1,626
High:			6,430
Low:			1,600

TABLE 3
Rectangular Plate Compliance Values

CIRCULAR PLATE COMPLIANCE VALUES

CADAVER #	IMPACT LOCATION	AGE	SEX	COMPLIANCE (N/mm)
F26588	1	90	F	700
M26903	2	37	M	830
F26371	2	46	F	1,290
F26380	2	60	F	1,340
M26922	2	72	M	1,440
F26573	2	63	F	1,450
F26361	2	75	F	1,460
M26350	1	75	M	1,470
M26329	1	62	M	1,480
M26383	2	92	M	1,540
F26377	1	72	F	1,550
F26342	1	73	F	1,600
F26587	1	62	F	1,670
M26372	2	80	M	1,800
F26925	2	67	F	1,800
F26384	1	89	F	1,840
F26374	1	82	F	2,250
M26368	1	81	M	2,570
M26373	2	73	M	3,140
F26354	1	66	F	4,760
Impact Locations #1 and #2				
Average: (approx. 1,800 N/mm)				1,799
Standard Deviation:				881
High:				4,760
Low:				700
Impact Location #1 ONLY				
Average:				1,989
Standard Deviation:				1,093
High:				4,760
Low:				700
Impact Location #2 ONLY				
Average:				1,609
Standard Deviation:				604
High:				3,140
Low:				830

TABLE 4
Circular Plate Compliance Values

MINERAL CONTENT - It has been suggested that the strength of the skeletal structure is a function of mineral content. Particular interest has centered on calcium and magnesium. A sample coupon was extracted and measured for mineral content from each specimen. To determine the correlation between fracture force and mineral content, a correlation coefficient was computed for each test group and for various combined groups (see Tables 5 and 6). In each case, the correlation coefficients indicated no significant relationship between the mineral content and fracture force of subjects tested. This does not establish a lack of correlation between bone fracture strength and bone mineral content, but indicates that other effects (geometric, etc.) have a greater influence over the fracture strength for lateral impacts to the head of the type discussed in this paper.

			Correlation Coefficient			
Impactor Type	Impact Location	Sex	Linear	Log	Exponential	Power
Circular	1	M	0.617	0.621	0.667	0.671
Circular	2	M	-0.792	-0.794	-0.697	-0.699
Circular	1	F	0.750	0.736	0.667	0.655
Circular	2	F	-0.063	-0.062	-0.024	-0.024
Circular	Both	M	-0.570	-0.570	-0.528	-0.528
Circular	Both	F	0.551	0.559	0.555	0.561
Circular	1	Both	0.688	0.686	0.656	0.654
Circular	2	Both	-0.402	-0.420	-0.306	-0.321
Circular	Both	Both	0.170	0.184	0.250	0.263
Rectang.	1	M	-0.008	-0.019	-0.027	-0.038
Rectang.	1	F	0.002	0.011	-0.009	-0.003
Rectang.	1	Both	-0.019	-0.005	-0.030	-0.018

TABLE 5
Correlation Coefficients for Fracture Load vs. Calcium Mineral Content

			Correlation Coefficient			
Impactor Type	Impact Location	Sex	Linear	Log	Exponential	Power
Circular	1	M	0.894	0.894	0.922	0.922
Circular	2	M	-0.475	-0.466	-0.568	-0.559
Circular	1	F	-0.180	-0.184	-0.048	-0.050
Circular	2	F	0.413	0.402	0.398	0.386
Circular	Both	M	-0.359	-0.344	-0.439	-0.422
Circular	Both	F	0.131	0.118	0.231	0.223
Circular	1	Both	0.281	0.270	0.331	0.322
Circular	2	Both	-0.366	-0.362	-0.389	-0.383
Circular	Both	Both	-0.018	-0.006	0.029	0.044
Rectang.	1	M	0.788	0.792	0.822	0.827
Rectang.	1	F	-0.467	-0.460	-0.504	-0.496
Rectang.	1	Both	-0.131	-0.116	-0.204	-0.189

TABLE 6
Correlation Coefficients for Fracture Load vs. Magnesium Mineral Content

CONCLUSIONS

While recognizing the limitation of the relatively small sample size employed in this research, the following conclusions are made:

1) Fracture force for a 5 by 10 cm flat plate impacting the parietal region of the human cadaver head averaged 12.5 kN. The stiffness for the same region averaged 4200 N/mm.

2) Fracture force for a 2.54 cm diameter flat-circular plate impacting the temporo-parietal region of the human cadaver head averaged 5 kN. The stiffness for the same region averaged 1800 N/mm.

3) Fracture force and stiffness of the human cadaver head are significantly affected by impactor contact area, however, they do not seem to be significantly related to calcium nor magnesium content.

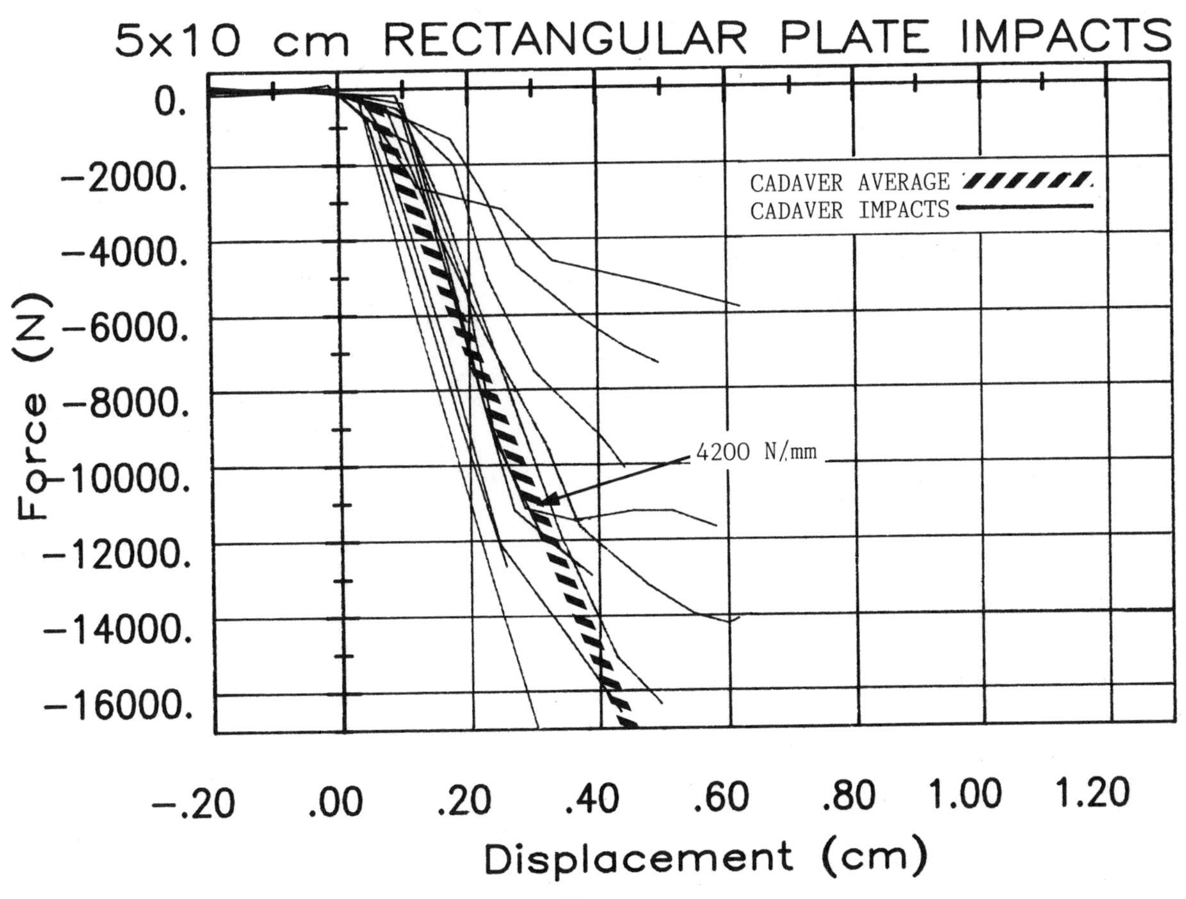

Figure 15
Force/Deflection Curves for the Rectangular Plate Impacts

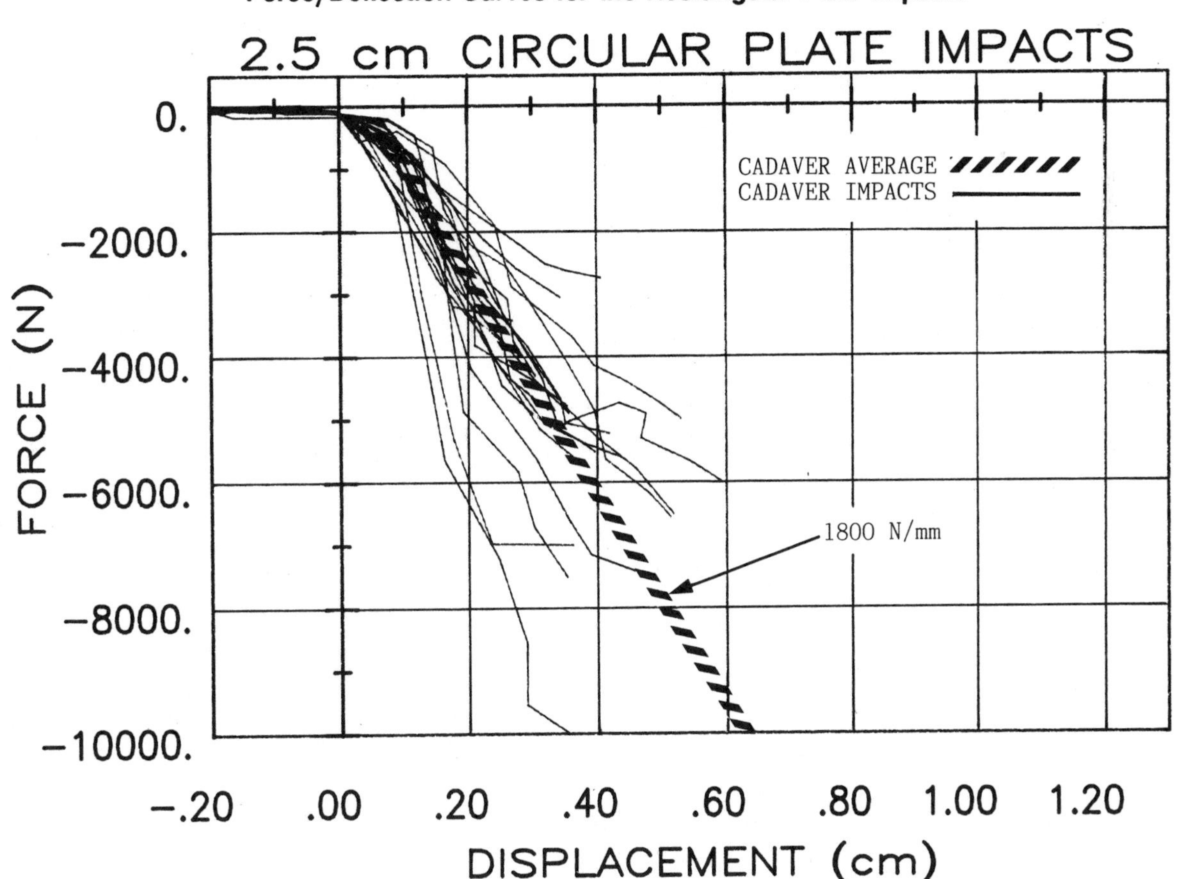

Figure 16
Force/Deflection Curves for the Circular Plate Impacts

ACKNOWLEDGEMENT

The authors wish to acknowledge the contributions of Scott McClellan, who conducted the experiments, and the staff of the Orthopaedics Research Laboratory, for their assistance in obtaining and processing test material. We also thank Sue LeBaron, Andy May and Gerald Carter for their efforts in the preparation of figures, tables, and typewritten drafts, and Dr. Milton Wille of BYU Mechanical Engineering Department, for his helpful counsel on many facets of this research.

REFERENCES

1. O. Messerer, "Elasticity and Strength of Human Bones." Stuttgart: Verlag der J.G. Cotta'schen Buchhandlung 1880.

2. A.M. Nahum, J.D. Gatts, C.W. Gadd and J. Danforth, "Impact Tolerance of the Skull and Face." SAE Paper No. 680785, 12th Stapp Car Crash Conference, Detroit, MI, October 22-23, 1968.

3. J.W. Melvin, P.M. Fuller, R.P. Daniel and G.M. Pavliscak, "Human Head and Knee Tolerance to Localized Impacts." SAE Paper No. 690477, presented at Chicago Mid-Year Meeting, May 19-23, 1969.

4. D.C. Schneider and A.M. Nahum, "Impact Studies of Facial Bones and Skull." SAE Paper No. 720965, 16th Stapp Car Crash Conference, Detroit, MI, November 8-10, 1972.

5. V.R. Hodgson and L.M. Thomas, "Breaking Strength of the Human Skull vs. Impact Surface Curvature." DOT Contract No. DOT-HS-146-2-230, Report No. DOT-HS-801-002, November 1973.

6. D.L. Allsop, C.Y. Warner, M.G. Wille, D.C. Schneider and A.M. Nahum, "Facial Impact Response - A Comparison of the Hybrid III Dummy and Human Cadaver." SAE Paper No. 881719, 32nd Stapp Car Crash Conference, Atlanta, Georgia, October 17-19, 1988.

Section 6: Neck Injury

710855

Strength and Response of the Human Neck*

H. J. Mertz
General Motors Corp.

L. M. Patrick
Wayne State University

SEVERAL RESEARCHERS, Mertz and Patrick (1)**, Mertz (2), Ewing, et al. (3, 4), and Tarriere and Sapin (5) have published papers dealing with the response and tolerance of the human neck in an impact environment. Ewing, et al. (3, 4) have presented two papers based on hyperflexion data obtained from human volunteer tests conducted at Wayne State University. In these tests, the subjects, restrained by a full harness, were exposed to increasing levels of sled decelerations which produced increasing severities of hyperflexion. Resultant accelerations measured by accelerometers secured to the occiput, mouth, and first thoracic vertebra, were given as functions of time as well as the angular velocity of the head. No attempt was made in these papers to analyze the acceleration data and determine the forces which acted on the head and produced these accelerations.

Tarriere and Sapin (5) presented hyperextension response data obtained on four human volunteers. These data consisted of the angular displacements of the head relative to the torso which were measured from high-speed movies of the experiments. These data were double-differentiated to obtain head acceleration values. No attempt was made to analyze the data for neck forces.

A detailed method for analyzing the forces and moments generated by the neck on the head during hyperextension was given by Mertz and Patrick (1). They used an instrumented human volunteer and cadavers as test subjects and showed that the magnitude of the torque developed by the neck on the head at the occipital condyles was an excellent indicator of neck trauma for hyperextension while the resultant shear and axial forces acting at the occipital condyles did not correlate with the degree of trauma. For a 50th percentile adult male, they recommended a noninjurious tolerance limit of 35 ft lb for the torque developed at the occipital condyles during hyperextension.

The objectives of this paper are to:

1. Describe the structure of the neck.

*Work sponsored in part by U.S. Army Natick Laboratories, Natick, Mass.

**Numbers in parentheses designate References at end of paper.

ABSTRACT

Human volunteers were subjected to static and dynamic environments which produced noninjurious neck responses for neck extension and flexion. Cadavers were used to extend this data into the injury region. Analysis of the data from volunteer and cadaver experiments indicates that equivalent moment at the occipital condyles is the critical injury parameter in extension and in flexion. Static voluntary levels of 17.5 ft lb in extension and 26 ft lb in flexion were attained. A maximum dynamic value of 35 ft lb in extension was reached without injury. In hyperflexion, the chin-chest reaction changes the loading condition at the occipital condyles which resulted in a maximum equivalent moment of 65 ft lb without injury. Noninjurious neck shear and axial forces of 190 lb and 250 lb are recommended based on the static strength data obtained on the volunteers. Neck response envelopes for performance of mechanical necks are given for the extension and flexion modes of the neck.

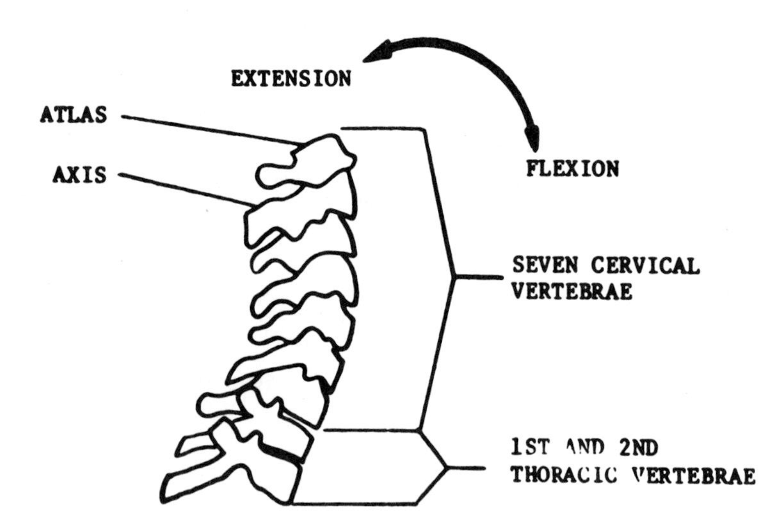

Fig. 1 - Bony structure of the neck showing the seven cervical vertebrae

2. Discuss data pertaining to the range of motion of the head relative to the torso in the sagittal plane.

3. Present data relative to the static strength of the neck in flexion and extension.

4. Present dynamic response and strength data for the human neck in flexion and extension.

5. Recommend noninjurious tolerance values for hyperextension and flexion of the neck.

STRUCTURE OF THE NECK

The bony structure of the neck consists of seven cervical vertebrae as shown in Fig. 1. No two cervical vertebrae of a given neck are identical, and very marked differences occur in the first and second vertebrae. The first cervical vertebra, the atlas, provides direct support for the head with the superior articular surface of the atlas bearing against the occipital condyles of the skull forming a synovial joint. The atlas does not have a body or a spinous process which are characteristic of the other cervical vertebrae.

The second cervical vertebra, the axis, forms a pivot around which the atlas, carrying the head, rotates from left to right. This pivot, the odontoid process, articulates through a synovial joint with the posterior portion of the anterior arch of the atlas and is prevented from moving posteriorly and compressing the spinal cord by the transverse ligament of the atlas.

The third through the seventh cervical vertebrae are similar in shape and function, except that the spinous process of the individual vertebra increases in length as one proceeds down the cervical spine. Each vertebra articulates with the adjacent ones through synovial joints. These joints are held in place by highly inextensible, fibrous ligaments. Separating the adjacent bodies of the vertebrae, but not including the union of the atlas and axis, are fibrocartilaginous intervertebral discs. The integrity of the cervical spine is maintained by numerous ligamentous connections. The anterior and posterior longitudinal ligaments are continuous over the length of the vertebral column. The anterior longitudinal ligament is attached to the base of the occipital bone and descends in front of the vertebral bodies to the sacrum. In between, it is firmly attached to the intervertebral discs and margins of the vertebrae. The posterior longitudinal ligament is attached superiorly to the membrana tectoria which is affixed to the occipital bone, extends posteriorly to the vertebral bodies to the sacrum, and is attached to the vertebral bodies and discs between.

Articulation of the neck is accomplished through muscle pairs which are attached to the skull, the individual vertebra, and the torso. These muscle pairs, which are symmetric with respect to the midsagittal plane, respond in various group actions to produce the desired movement of the head and neck. The muscle pairs which provide resistance to extension of the neck and rearward rotation of the head are the longus capitis, longus colli, rectus capitis anterior, scalenus anterior, the hyoids, sternothyroid, and sternocleidomastoid. Because of their attachment to the mastoid process which is slightly below and behind the occipital condyles, the sternocleidomastoids increase the extension of the neck when the head is rotated rearward and the neck is already extended. However, if the neck is flexed and head rotated forward, the action of this muscle pair flexes the head and resists extension.

The muscles which prevent flexion of the neck and forward rotation of the head are located posteriorly of the vertebral bodies and are the trapezius, levator scapulae, splenius capitis, longissimus capitis, splenius cervicis, semispinalis capitis, semispinalis cervicis, obliquus capitis inferior, obliquus capitis superior, rectus capitis posterior major, and rectus capitis posterior minor.

The relative position of attachment and cross-sectional size of the muscles which are inserted in the base of the skull are shown in Fig. 2. The total area of attachment of the postvertebral muscles is much greater than that of the prevertebral group. Also, the centroid of the area of the postvertebral muscles is further from the occipital condyles than is the centroid of the area of the prevertebral muscle. This implies that the head should be able to resist a larger applied flexing moment than extending moment. An additional resisting moment is developed in flexion when the chin contacts the torso. The contact force produces a moment about the occipital condyles which further limits flexion.

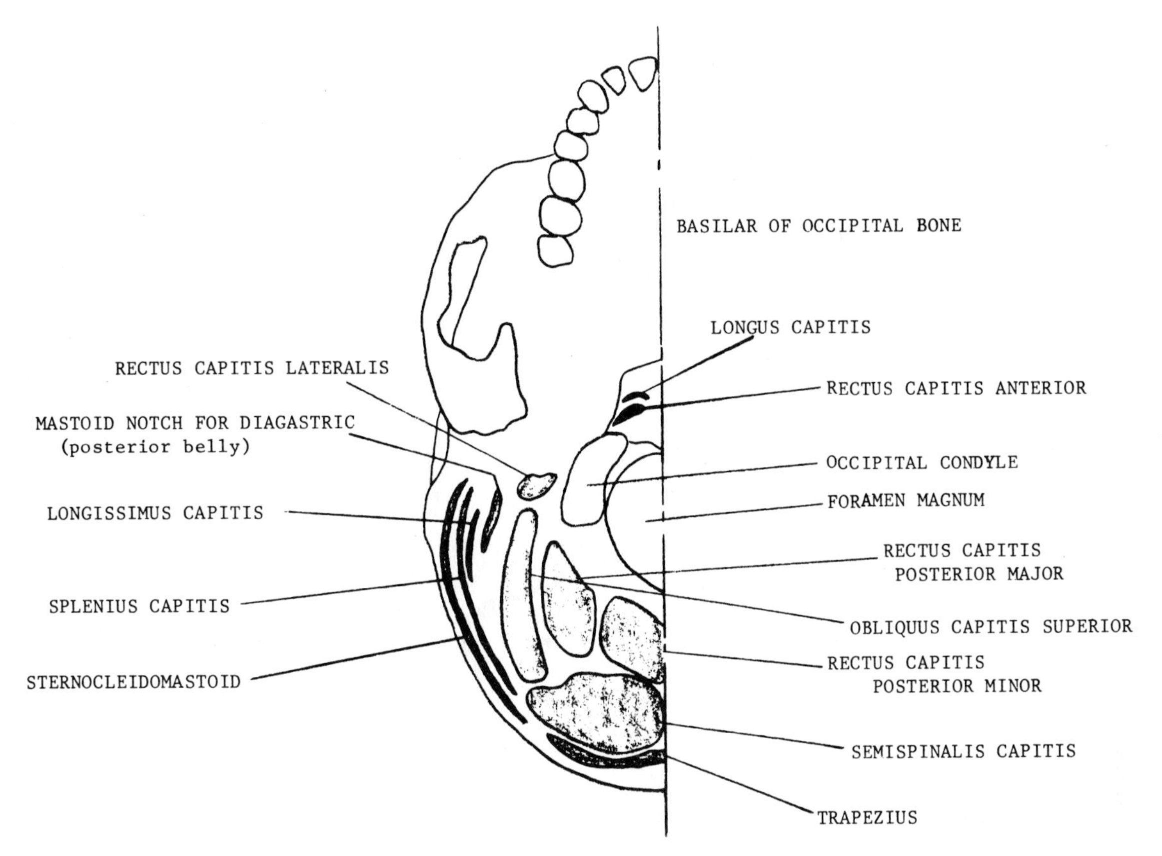

Fig. 2 - Relative position of attachment of the muscles which are inserted in the base of the skull

RANGE OF MOTIONS OF THE HEAD

The degree of articulation of the cervical spine varies from person to person. A person with a long, thin neck will have a greater range of head motion than a person with a short, thick neck. An athletic individual will have a greater range of head motion than his sedentary counterpart. Age and sex will also influence the range of neck articulation. Several investigators have studied the range of motion of the cervical spine for various classes of individuals.

Buck, et al. (6) measured the voluntary head motions of 100 individuals, 53 females and 47 males, whose ages ranged from 18-23 years. For the male subjects, the flexion range was 50-90 deg with a mean of 66 deg and a standard deviation of 8 deg. The extension range was 51-92 deg with a mean of 73 deg and a standard deviation of 9 deg. Combining these figures gives a total excursion range of 101-182 deg with a mean of 139 deg. The female subjects had a greater degree of flexibility in the cervical spine than the male subjects. The female means for flexion, extension, and total excursions were 69, 81, and 150 deg as compared to 66, 73, 139 deg for the male subjects. All of these measurements were taken with the subjects producing the head rotations with their neck muscles with no external forces being applied to the head.

Granville and Kreezer (7) made measurements on 10 male subjects between 20-40 years of age. The average height of the group was 5 ft 9 in, with a range of 5 ft 4 in-6 ft 1 in. The average weight was 142 lb with a range of 120-172 lb. Head movements were recorded for both voluntary and forced head-neck responses. The mean angulation for flexion was 59.8 deg with a standard deviation, σ of 11.7 deg for the voluntary response and 76.4 deg with a $\sigma = 9.2$ deg for the forced response. For extension, the mean voluntary response was 61.2 deg with a $\sigma = 26.8$ deg and the mean force response was 77.2 deg with a $\sigma = 25.1$ deg. Based on these means, average total excursions of 121 deg voluntary and 153.6 deg forced were computed. A questionable point of this data is the large spread of 76.4 deg in the forced angulation as compared to the voluntary 59.8 deg for flexion. In the flexed position, the chin is against the chest so the large spread between the forced and voluntary angulation is not expected.

Ferlic (8) made measurements of the motion of the cervical spines on subjects of various age groups. He did not classify his results in terms of flexion and extensions, but gave only total excursions all of which were voluntary and not forced. No attempt was made to immobilize the torso except that the subjects were instructed to move only the heads and necks. His results for various age groups which included both male and and female subjects were as follows: 15-24 years old, 139 ± 19 deg; 25-34 years old, 127 ± 22 deg; 35-44 years old, 120 ± 19 deg; 45-54 years old, 120 ± 15 deg; and 55-64 years old, 116 ± 22 deg. The average range of motion for all age groups was 127 ± 19.5 deg.

The volunteer, LMP, who was the subject of the dynamic hyperflexion and extension tests which are to be discussed in this paper, is quite representative of a 50th percentile adult male. He is 68 in tall and weighs 160 lb, and was 49 years old at the time of testing which is, according to the 1960 census,

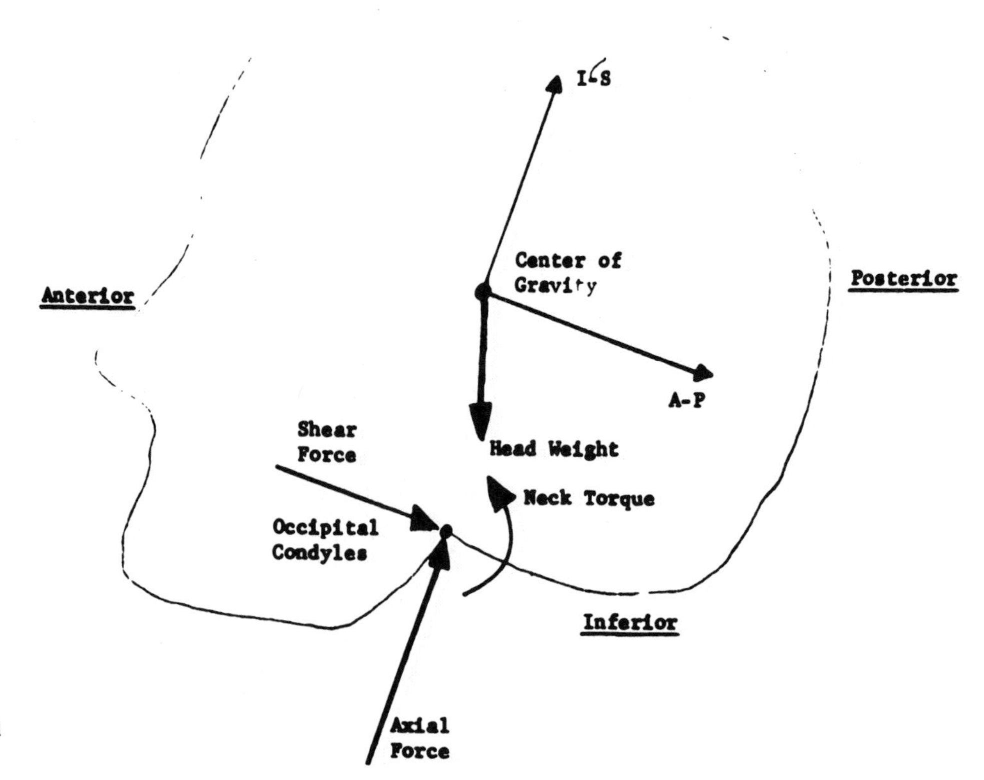

Fig. 3 - Free body diagram of the head for hyperextension

seven years older than the median age of 42 years for the age group of males over 18 years old. The volunteer LMP is employed in a sedentary position and does not have unusual muscle tone as a result of his occupation. His neck circumference is 15 in and length is approximately 5-1/2 in. Measurements were made of his normal range of motion of his cervical spine. The volunteer LMP was seated in a rigid chair with his torso inclined 15 deg rearward and his head in a normal upright position. The volunteer was capable of rotating his head relative to his torso 62 deg forward and 60 deg rearward using hand pressure on his head to achieve these extremes. The total excursion of 122 deg compares well with the average of 120 ± 15 deg given by Ferlic for the 45-54 year old group. Consequently, the neck articulation of the volunteer LMP will be used as a standard reference in the remainder of this paper.

METHOD OF ANALYZING NECK RESPONSE

Since the neck consists of seven cervical vertebrae which can move relative to each other, the neck cannot be considered a rigid body and, consequently, its motion cannot be analyzed by simple rigid body mechanics. To consider the motion of each vertebra requires the knowledge of the forces acting on each vertebra and resulting accelerations. In the case of human volunteer studies, these measurements would be impossible to make with any reasonable degree of accuracy.

An alternate approach is to analyze the kinematics and kinetics of the head, since in hyperextension and flexion the motion of the head is controlled by the forces generated by the neck. In the time domain of hyperextension and flexion response of the neck produced by torso acceleration, the head can be considered a rigid body and, consequently, Newton's Laws of rigid body mechanics can be used, as was noted by Mertz and Patrick (1).

For hyperextension resulting from torso acceleration, the forces acting on the head are produced entirely by the neck structure. For analytical purposes, these forces can be resolved into a force acting at the occipital condyles, the bony mating surface between the skull and the atlas, and a resultant torque. The only other external force acting on the head is the force of gravity. The resultant force is resolved into two components, an axial force directed along the axis of the vertebral column parallel to the odontoid process, and a mutually perpendicular shear force which produces a distributed bending moment along the cervical spine.

Applying Newton's Laws of rigid body motion to the head (Fig. 3), the following equations are obtained:

$$\overline{T}_O + \overline{r}_{G/O} \times \overline{W}_H = I_G\overline{\alpha} + m_H\overline{r}_{G/O} \times \overline{a}_G \qquad (1)$$

$$\overline{F}_O + \overline{W}_H = m_H\overline{a}_G \qquad (2)$$

The inertia properties of the head, I_G and m_H, and the geometric length $\overline{r}_{G/O}$ can be obtained or approximated by the methods described by Mertz (2). Experimentally, the acceleration of the head, $\overline{a}_G$ and $\overline{\alpha}$, during hyperextension can be measured using accelerometers affixed to the head. The corresponding torque, shear, and axial forces can be calculated

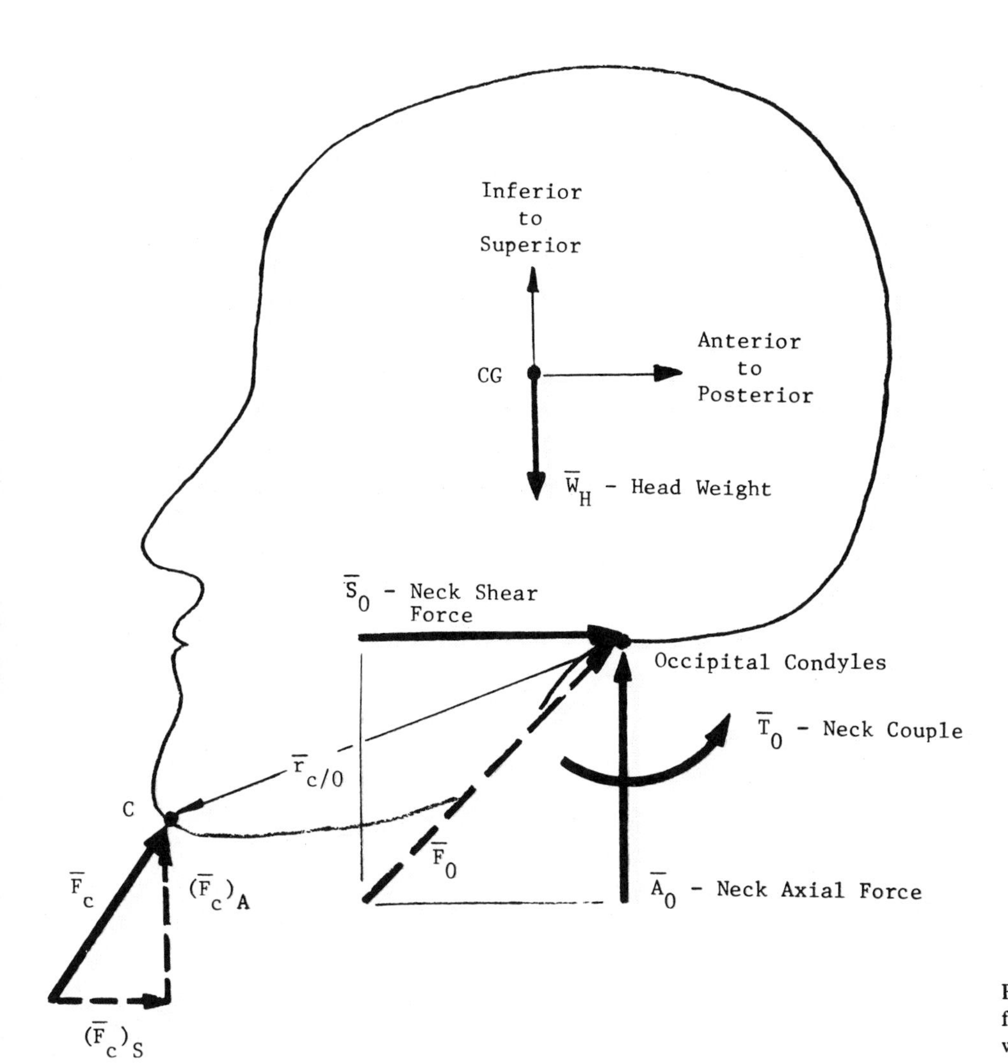

Fig. 4 - Free body diagram of the head for hyperflexion including chin contact with the chest

by Eqs. 1 and 2. Because of the number of force carrying elements, no attempt was made to distribute the loads throughout the neck structure.

Hyperflexion analysis is similar to that described for hyperextension except that the force developed by the chin when it contacts the chest must be taken into account (Fig. 4). The dynamic equations for hyperflexion are:

$$\overline{F}_O + \overline{F}_C + \overline{W}_H = m_H \overline{a}_G \qquad (3)$$

$$\overline{T}_O + \overline{r}_{G/O} \times \overline{W}_H + \overline{r}_{C/O} \times \overline{F}_C = I_G\,\overline{\alpha} + m_H \overline{r}_{G/O} \times \overline{a}_G \qquad (4)$$

Letting

$$\overline{T}_R = \overline{T}_O + \overline{r}_{C/O} \times \overline{F}_C$$

Eq. 4 becomes,

$$\overline{T}_R + \overline{r}_{G/O} \times \overline{W}_H = I_G \overline{\alpha} + m_H \overline{r}_{G/O} \times \overline{a}_G \qquad (5)$$

Again, knowing the inertial and geometrical properties of the head, the resultant forces and moments can be computed from Eqs. 3 and 5 by measuring the head accelerations. The resultant force, $\overline{F}_R$, acting on the head produced by the neck structure and chin contact force can be resolved into A-P and S-I components which are denoted as $\overline{F}_{A\text{-}P}$ and $\overline{F}_{S\text{-}I}$, respectively. The resulting torque, T_R, includes the moment of the chin contact force about the occipital condyles. This technique was used to analyze the static and dynamic hyperflexion data presented in this paper. For the static analysis, the acceleration terms in the equations are set equal to zero.

STATIC STRENGTH OF THE HUMAN NECK

The strength of the human neck under static loading systems has been studied by several investigators. Morehouse (9) reported on static strength tests performed on a 30-year old male, who was 5 ft 10 in tall, weighed 152 lb, and had a neck circumference of 13-1/2 in. This volunteer withstood A-P and P-A forces of 18 and 50 lb applied at a level slightly above the ear. Assuming a distance of 3 in from the line of action of the force to the occipital condyles, the torques developed by the neck structure to withstand extension and flexion of the neck were 4.4 and 12.5 ft lb, respectively.

Mertz (2) and Mertz and Patrick (1) describe static strength tests for determining values of maximum torque developed at

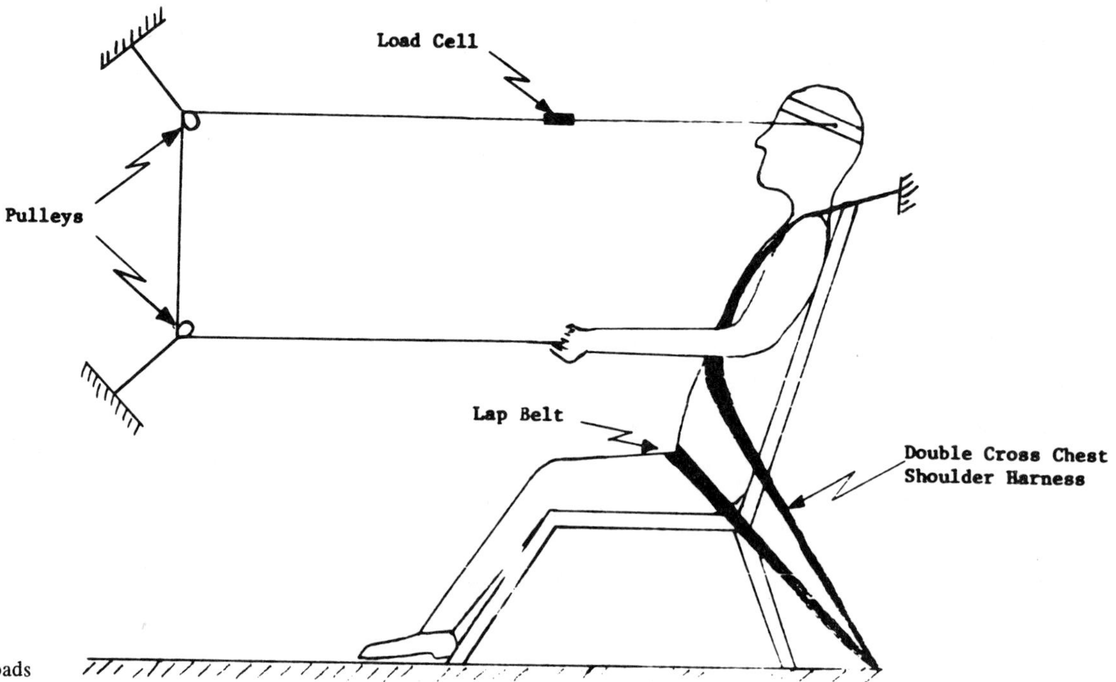

Fig. 5 - Test setup for applying static head loads

Table 1 - Summary of Voluntary Static Human Neck Torque Levels Developed at The Occipital Condyles for Various Neck Curvatures

	Neck Torque Developed at Occipital Condyles, ft-lb	
Neck Position	Resist Neck Flexion	Resist Neck Extension
Normal	23.5	10.5
Extended	25.0	17.5
Flexed	26.0	12.5

Notes:

1. Extension values obtained from paper by Mertz and Patrick (1).
2. Torque levels given to nearest 1/2 ft-lb.

the occipital condyles for resisting neck extension. They showed that the strength of the neck is dependent on its curvature and on the direction of the applied load. Mertz and Patrick combined their results with those of Carroll, et al. (10) and presented a summary of voluntary static human tolerance values for the neck based on the reactions developed at the occipital condyles. Part of these results are given in Table 1.

Additional tests have been conducted to determine the static strength of the human neck in resisting flexion. The test set-up and data analysis was similar to that employed by Mertz and Patrick (1). For each test, the volunteer was restrained in a rigid chair and applied a load to his head via a system of pulleys as shown in Fig. 5. A load cell was placed in series with the cord through which the load was applied. Films of the loading sequence, synchronized with load cell response, provided necessary information about the direction of the applied load from which the torque developed at the occipital condyles and the shear and axial forces generated by the neck structure were computed. Tests were conducted for three different initial curvatures of the neck: normal, extended, and flexed. For each of these initial positions, the volunteer applied a load which was directed anteriorly (P-A) relative to his head with the line of action of the force passing through the center of gravity of his head. With this loading condition, the neck structure generated forces which were similar to those required to prevent hyperflexion of the neck caused by sudden torso acceleration. The magnitude of the load was increased slowly until the volunteer could not maintain his head position or until he terminated the loading because of fear of injury and/or pain.

A total of 90 static neck strength tests were conducted on 10 volunteers. The maximum static reactions developed at the occipital condyles for the three neck positions of each of the ten volunteers are given in Table 2. The sign convention for these reactions is depicted on Fig. 4, the free-body diagram of the head. To resist the forward rotation of the head, a positive shear force and negative torque must be applied to the head at the occipital condyles. The axial force can be either positive or negative depending upon the position of the head and direction of the applied load. With the neck flexed, the maximum neck torques ranged from 14.8-25.9 ft lb. With the neck holding the head upright, the torque ranged from 16.1-23.4 ft lb. The listed values of the shear and axial forces are not maximum tolerable loads, since the resistive neck torque limited the application of higher loads.

Values for the maximum occipital torque for resisting neck

Table 2 - Maximum Static Neck Reactions Developed at The Occipital Condyles for Various Neck Positions, Load Applied Essentially in The P-A Direction

Volunteer	Maximum Occipital Torque, ft-lb			Maximum Shear Force, lb			Maximum Axial Force, lb		
	Extended	Normal	Flexed	Extended	Normal	Flexed	Extended	Normal	Flexed
LMP	-18.0	-23.4	-25.9	50.9	63.9	71.0	50.8	15.8	- 61.4
GDG	-16.8	-16.3	-15.5	39.8	41.9	73.9	31.5	19.3	-100.5
RAE	-16.9	-22.7	-14.8	33.4	57.3	45.1	50.5	16.1	- 17.2
DJV	-25.0	-23.1	-18.8	58.8	57.1	50.5	39.8	15.0	- 5.4
HA	-11.4	-20.5	-21.2	44.0	60.6	55.6	44.3	12.1	- 63.3
CJM	-13.1	-20.6	-24.1	45.0	49.5	57.2	40.2	11.4	- 44.9
KCH	-11.0	-16.1	-16.1	25.2	39.4	41.1	20.9	14.7	- 18.7
JVG	-10.9	-16.9	-22.5	33.6	48.0	65.0	86.1	16.8	- 74.9
FTD	-20.1	-21.6	-25.2	68.3	62.0	59.1	85.1	25.9	- 63.1
DAW	-12.6	-21.8	-17.6	36.2	43.6	40.7	59.6	13.5	- 11.4

Table 3 - Maximum Static Force Reactions Acting on The Head at The Occipital Condyles

Shear Force, lb		Axial Force, lb	
A-P	P-A	Tension	Compression
190	190	255	250

flexion are presented in Table 1 along with the values given by Mertz and Patrick for resisting neck extension. In its normal upright position, the neck is stronger in resisting flexion, 23.5 ft lb, than it is in resisting extension, 10.5 ft lb, because the major muscles of the neck are located posterior of the occipital condyles and can generate a larger resistive couple to limit neck flexion.

According to Mertz and Patrick (1), the maximum voluntary static neck reaction of one volunteer was 255 lb in tension and 250 lb in compression. For neck shear, the neck can withstand a force of 190 lb acting posteriorly (A-P) relative to the head. A lower bound for an anteriorly (P-A) directed shear force can be obtained by equating it to the A-P reaction of 190 lb because the neck structure is inherently stronger in this direction. For the neck to produce a P-A shear force on the head at the occipital condyles, the odontoid process bears against the bony anterior arch of the atlas, while to develop an A-P shear force the action tends to separate the joint. The values for the maximum tolerable static shear and axial forces are summarized in Table 3.

The static strength values listed in Tables 1 and 3 would apply to neck response characteristics in low g environments where the viscous resisting forces produced by the muscles are not a major portion of the resisting torque. In high g environments, the viscous contribution of the muscle reaction would be comparable to its static strength component, producing a higher resistive torque than that predicted by the static strength analysis. Consequently, the static strength values are considered as lower bounds for the neck strength in a dynamic environment.

Fig. 6 - Overall view of WHAM I with subject restraint in rigid chair

DYNAMIC NECK RESPONSE AND TOLERANCE LEVELS FOR HYPERFLEXION

DESCRIPTION OF EXPERIMENTS - The subjects used for the dynamic flexion tests were a human volunteer, LMP, and four human cadavers. The subjects were restrained in a rigid chair which was mounted on an impact sled, WHAM I, shown in Fig. 6. The sled travels on two horizontal rails and is accelerated pneumatically over a distance of 6 ft to the prescribed velocity. During the acceleration stroke, a headrest is used to maintain the head in an upright position. After reaching its prescribed velocity, the sled coasts for a short distance during which the sled velocity is measured with a magnetic pickup. The sled is stopped with a specially designed hydraulic cylinder which produces a repeatable deceleration pulse. The stroke of the stopping cylinder is continuously variable up to 22 in. Consequently, for a given sled velocity the magnitude and duration of the deceleration pulse is determined by the length of the stopping distance.

The seat, shown mounted on the sled in Fig. 6, was rigidly constructed using steel angles for the main structural components and plywood coverings for the seat back and bottom.

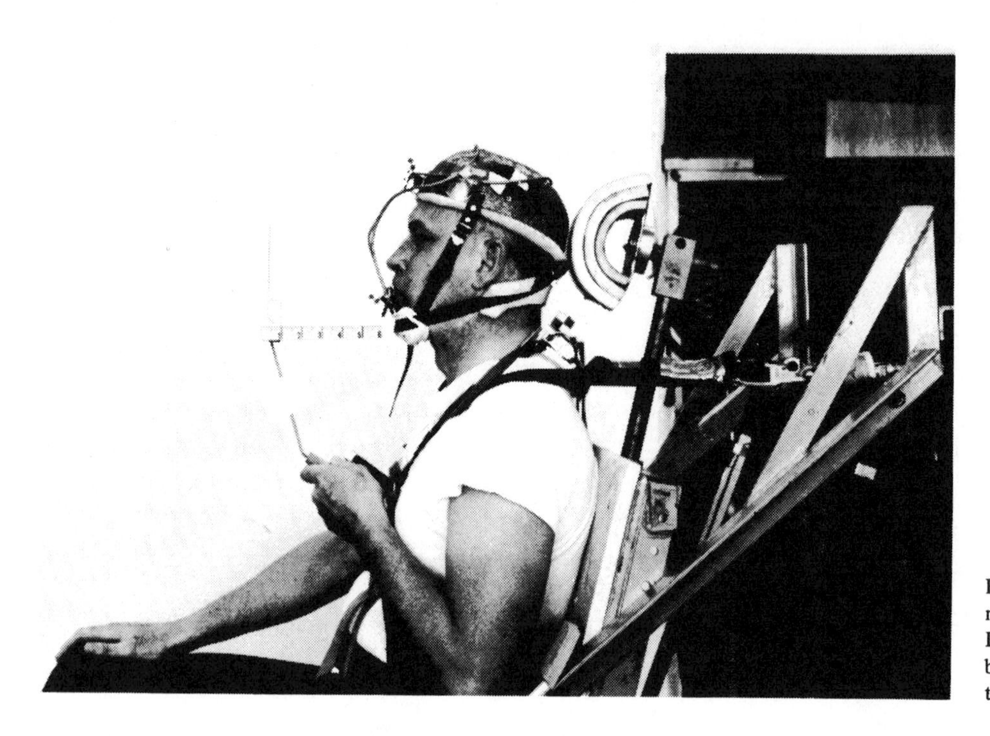

Fig. 7 - Close-up view of the head and neck transducer attached to volunteer LMP. Two accelerometers mounted to bite plate, front of helmet, and first thoracic vertebrae

These coverings were padded with a layer 5/8-in thick Rubatex to distribute the load to the subject's back and buttocks. A rigidly mounted headrest covered with three layers of Rubatex was used to maintain head position during sled acceleration.

The restraint system consisted of a lap belt and two individual shoulder harnesses which crisscrossed the chest at midsternum. Each belt was made of nominally 2-in wide, standard automotive webbing material fastened with a standard automotive seat belt buckle. Load cells were fastened to the ends of the belts to monitor the belt loads. The upper shoulder harness mounts were adjustable so that this portion of the harness could be kept horizontal, independent of the size of the subject. In addition to these restraints, the subject's feet were securely fastened to the foot support and his wrists were strapped to the armrests to prevent flailing of these appendages during the run.

The kinematics of the head of each subject were obtained from accelerometers which were attached to their heads (Fig. 7). For the volunteer, two uniaxial accelerometers were mounted on an acrylic bite plate which was molded to conform to his teeth. The accelerometers were oriented with their sensitive axes orthogonal and lying in the midsagittal plane of the head. For the cadaver tests, the accelerometers were attached to its mouth using dental acrylic, molded to the contour of the oral cavity and teeth as a mounting base. A second pair of accelerometers was attached to a lightweight fiber glass helmet which was securely fastened to the subject's head. The sensitive axes of these accelerometers were orthogonal in the midsagittal plane. From the outputs of these two pairs of accelerometers, the resultant acceleration of the center of mass of the head and the angular acceleration of the head were calculated as a function of time by the method outlined by Mertz (2). For the volunteer runs, a pair of accelerometers were mounted on the subject's back in the vicinity of the spinous process of the first thoracic vertebra, as shown in

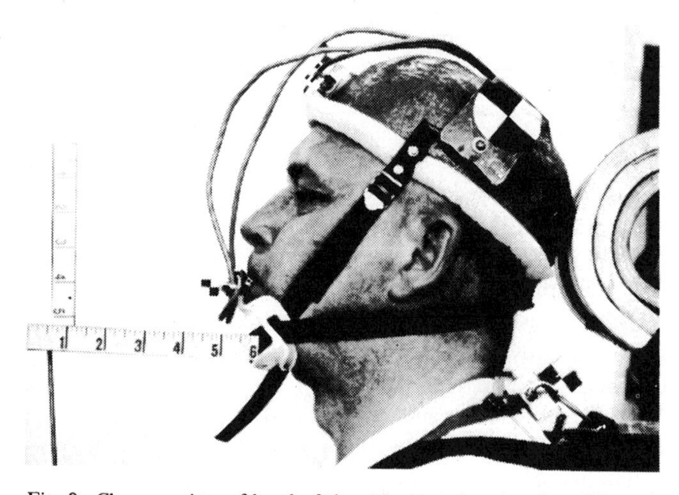

Fig. 8 - Close-up view of head of the volunteer showing the additional head weight located above the center of mass of the head

Fig. 7. The output of these accelerometers yields the translation acceleration of the first thoracic vertebra at the base of the neck.

The objective of the research project for which these hyperflexion tests were conducted was to determine the kinetic, kinematic, and physiological effects produced by varying the mass, center of gravity, and mass moment of inertia of the head by the addition of a helmet. Consequently, tests were conducted with lead weights attached to the fiber glass helmet. Four configurations shown in Figs. 7-10 were evaluated:

1. No additional weight except the lightweight fiber glass helmet.
2. Three pounds of additional weight located above the center of the mass of the head.
3. Three pounds of additional weight located approximately at the center of mass of the head.

Table 4 - Geometric and Inertial Properties of the Heads of the Subjects

Subject	Characteristic Lengths					Inertial Properties	
	Circum. in	Width in	A-P Length in	Radius of Gyration in	Vol., in^3	Wt., lb	Mass Moment of Inertia, lb-in-s-s
Volunteer-LMP	22.3	5.90	7.50	2.68	259	10.8	0.200
Cadaver–1404	22.3	6.19	7.88	2.80	244	10.1	0.205
Cadaver–1538	22.3	5.94	7.50	2.68	239	9.94	0.185
Cadaver–1548	22.6	6.56	7.75	2.75	249	10.35	0.203
Cadaver–1530	24.9	7.19	7.75	2.75	289	11.00	0.215

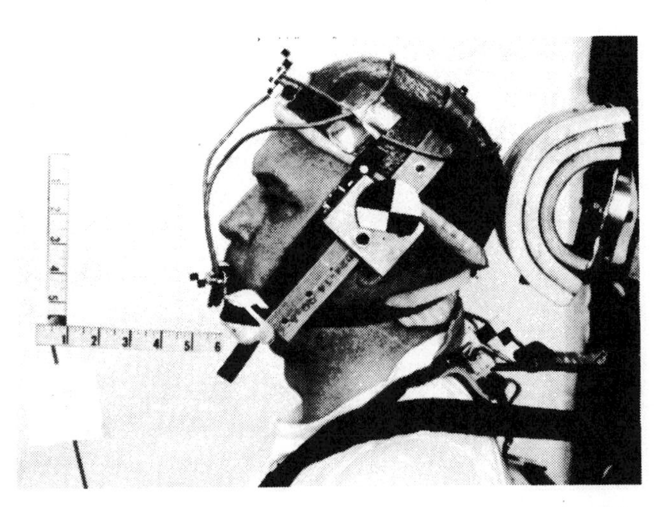

Fig. 9 - Close-up view of the head of the volunteer showing the additional head weight located approximately at the center of mass of the head

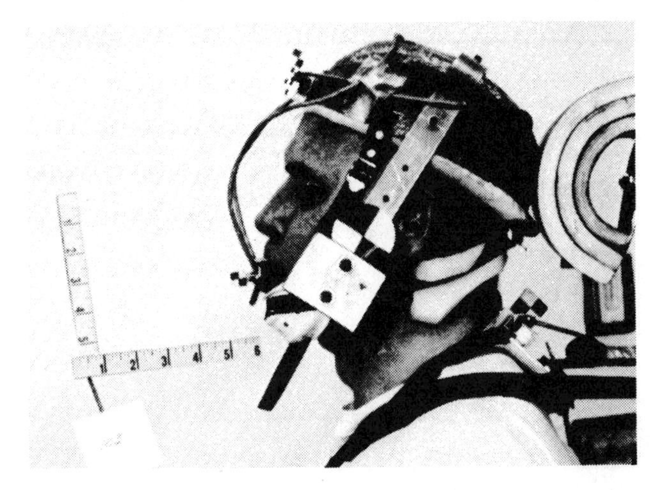

Fig. 10 - Close-up view of the head of the volunteer showing the additional head weight located below the center of mass of the head

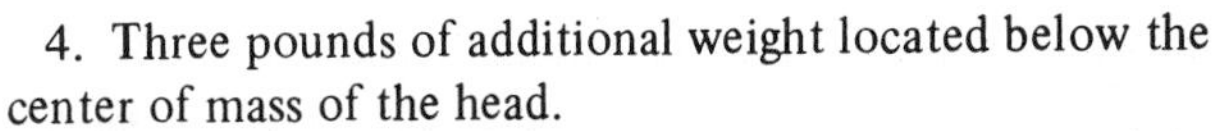

4. Three pounds of additional weight located below the center of mass of the head.

The geometric and inertial properties of the heads of the subjects are given in Table 4.

This paper will not address the question of varying the inertial properties of the head, but will utilize the data generated during these tests to describe response and tolerance limits for the neck in flexion.

The transducers used to monitor the various parameters were:

1. Four Statham strain gage accelerometers, Model A-52, to measure head acceleration.
2. Two Statham strain gage accelerometers, Model A-52, to measure the acceleration of the first thoracic vertebra.
3. Four strain gage load cells to measure the restraint harness loads.
4. A Statham strain gage accelerometer, Model A-6, to monitor sled deceleration.
5. A magnetic pickup which recorded the time required for the sled to traverse a known distance at a constant velocity to calculate sled velocity.

The outputs of all transducers were appropriately conditioned and recorded by a light-beam oscillograph.

Two high-speed, 16 mm cameras were used to photograph each run. A close-up view of head and neck motion was taken with a Photosonics rotating prism camera operating at 600 frames/s. An overall view of the subject's motion was obtained with a Milliken framing camera operating at 500 frames/s. A timing generator was used to mark the edges of the films and the oscillograph record and synchronization was obtained by switching the frequency of the timing generator from 1000 Hz to 100 Hz during each test.

VOLUNTEER EXPOSURES - The volunteer LMP was subjected to 46 sled runs of various degrees of severity for the four configurations of additional head weight. During these tests, the volunteer attempted to achieve two different degrees of initial muscle tenseness: relaxed and tense. For the relaxed condition, the volunteer relaxed all of his muscles insofar as he was able to do so while still maintaining an upright head position. For the runs with his muscles tensed, the volunteer tensed his muscles as completely as possible during the entire run.

With his muscles tensed, the volunteer was subjected to sled rides ranging from 1.9-6.8 g for the additional head weight configurations of Helmet Only, Weight Centered, and Weight Low (Table 5). With the weight placed above the center of mass of his head, the volunteer rode the sled at 9.6 g level—his most severe exposure. This run (Run 79) resulted in a pain in the neck and back which lasted for several days.

A summary of the impact conditions and restraint system

Table 5 - Volunteer Sled Runs Conducted at Various Levels of Sled Deceleration for Different Configurations of Additional Head Weights and for Different Degrees of Neck Muscle Tone

Additional Head Weight	Neck Muscle Tone	Sled Plateau Deceleration Level, g
Helmet only	Tensed	2.0, 2.1, 2.9, 3.3, 3.3, 3.7, 3.9, 4.4, 4.7, 5.3, 6.8
	Relaxed	2.0, 2.1, 2.9, 3.3, 3.8, 4.1, 4.2, 4.4
Weight high	Tensed	1.5, 2.2, 3.0, 3.1, 4.6, 5.1, 6.9, 8.0, 9.6
	Relaxed	None
Weight centered	Tensed	2.0, 2.9, 4.1, 4.1, 5.1, 6.2
	Relaxed	2.1, 2.8, 4.1
Weight low	Tensed	1.9, 2.9, 3.9, 4.1, 5.8, 6.6
	Relaxed	2.1, 2.7, 4.2

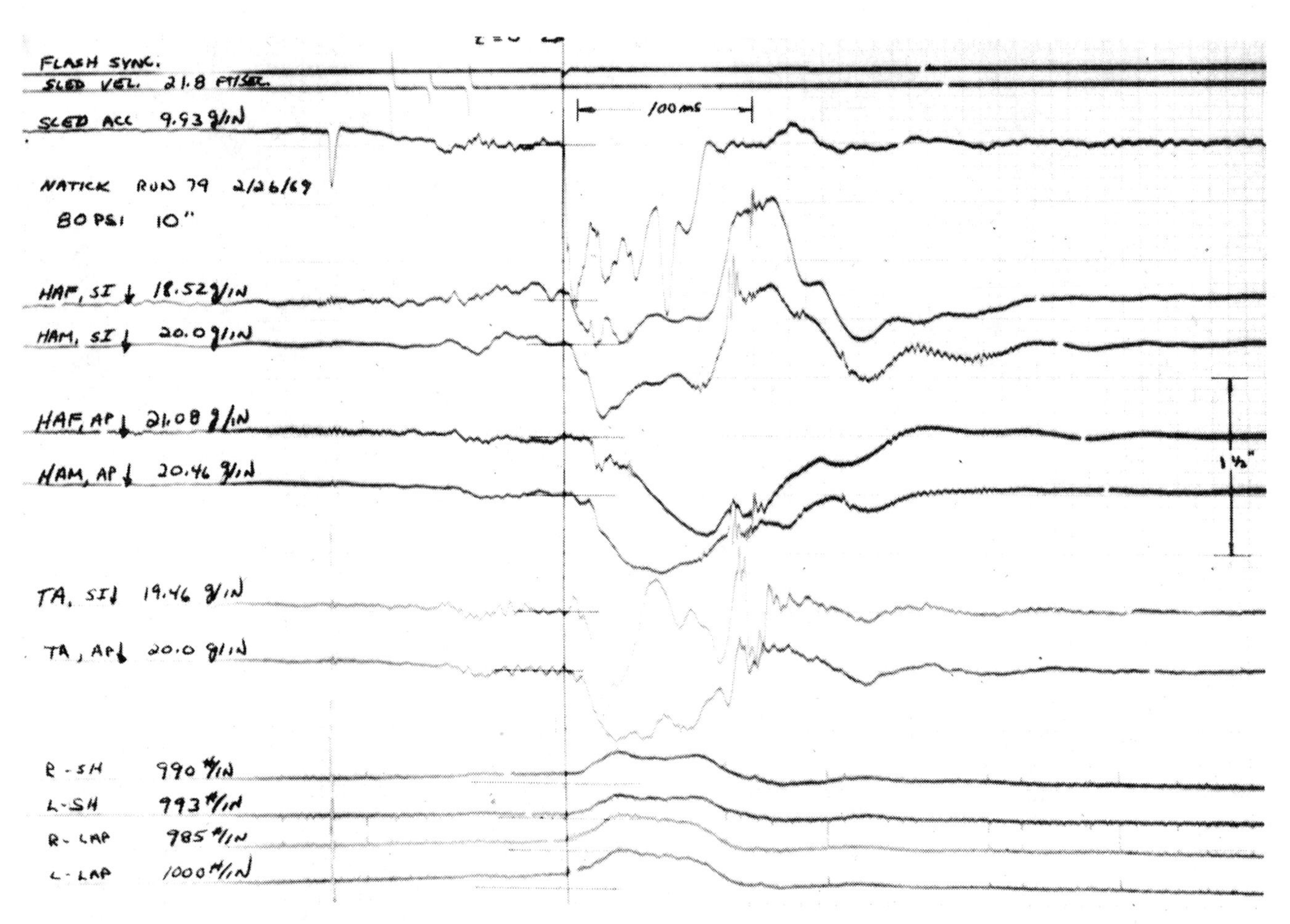

Fig. 11 - Oscillograph record of Run 79

load for all of the volunteer runs is given in Table 6. There does not appear to be a correlation between the magnitude of the onset of sled deceleration and the severity of the volunteer's exposure. Run 79, which was the most severe impact condition as noted by the pain experienced by the volunteer, had an onset of 6200 g/s with a plateau sled deceleration level of 9.6 g. Three other runs (Run 105, 108, and 109) had higher onset levels (6500 g/s, 6300 g/s, and 6400 g/s, respectively) without producing any adverse effects. The plateau sled deceleration level appears to be a good indicator of the volunteer's exposure since Run 79 had the highest level of all the volunteer runs. A more direct indicator of the severity of the exposure is the onset and magnitude of the shoulder harness loads. These two parameters give the "coupling" between the volunteer and the sled. The highest onsets for the shoulder harness loading (9700 and 8300 lb/s) and the largest restraint system loads (280 and 210 lb for the shoulder belts and 310 and 330 lb for the lap belt) occurred during Run 79.

An oscillograph record of Run 79 is shown in Fig. 11. The data read-out of the head acceleration traces of this record at

Table 6 - Summary of Impact Conditions and Restraint System Loads for Volunteer Runs

	Sled Kinematics					Restraint System								
			Deceleration Pulse			Onset		Maximum Shoulder Load*						
Run No.	Vel., fps	Stop. Dist. in	Onset g/s	Peak, g	Plat., g	R. Sh, lb/s	L. Sh, lb/s	R. Sh, lb	L. Sh, lb	R. Lap, lb	L. Lap, lb	Add.** Head Wt.	Neck† Mus-cles	Comments
22	10.9	15	–	2.9	2.0	–	–	20	30	30	40	W/O	T	
23	10.8	15	–	3.1	2.0	–	–	40	30	40	40	W/O	R	
24	16.1	15	–	5.3	3.7	–	–	60	60	60	60	W/O	T	
25	16.1	15	–	5.9	3.3	–	–	100	100	80	80	W/O	T	Subject involuntarily tensed neck muscles
26	15.9	15	–	5.7	3.8	–	–	80	80	80	80	W/O	R	
27	17.7	15	–	6.4	4.4	–	–	80	90	80	90	W/O	T	
28	17.7	15	–	7.2	4.4	–	–	120	130	110	120	W/O	R	
73	12.7	20	2450	4.2	1.5	400	800	60	80	90	100	H	T	
74	12.7	10	2400	5.7	3.1	1500	2100	100	100	110	110	H	T	
75	14.7	10	2400	7.0	4.6	4900	2800	120	110	170	140	H	T	
76	17.0	10	3600	8.8	5.1	4300	3900	130	150	200	200	H	T	
77	18.9	10	5600	10.2	6.9	6000	5500	190	190	260	230	H	T	
78	20.4	10	5500	12.2	8.0	9300	7500	250	220	310	310	H	T	Felt neck pain during run.
79	21.8	10	6200	14.0	9.6	9700	8300	280	210	310	330	H	T	Felt pain extending from neck into back. Does not desire to go higher
80	14.9	20	3000	5.6	2.2	1200	1200	90	80	100	100	H	T	
81	17.0	20	5000	6.9	3.0	1500	1700	90	110	110	110	H	T	
82	14.9	20	3500	5.8	2.1	800	1000	90	80	100	90	W/O	T	
83	14.9	10	2700	5.8	4.7	2300	2500	110	110	140	130	W/O	T	
84	17.0	20	4300	5.8	2.9	1000	1100	100	90	120	110	W/O	T	
85	14.9	20	3200	5.5	2.1	1000	1000	100	90	90	90	W/O	R	
86	14.7	10	3100	7.6	4.1	3400	2100	150	120	180	170	W/O	R	
87	17.0	20	4600	7.1	2.9	1700	1200	100	90	110	110	W/O	R	
88	14.9	20	3300	5.7	1.9	500	500	80	80	100	100	L	T	
89	14.9	10	3300	7.1	4.1	2900	2400	110	120	150	130	L	T	
90	17.0	20	3500	6.9	2.9	1000	1100	80	100	100	100	L	T	
91	14.8	20	3600	5.7	2.1	900	900	90	90	90	90	L	R	
92	14.8	10	3300	7.4	4.2	3400	3100	150	140	150	160	L	R	
93	17.0	20	3800	6.9	2.7	1800	1800	110	100	100	100	L	R	
94	14.9	20	3000	5.8	2.0	600	700	90	70	80	90	C	T	
95	14.9	10	3600	7.0	4.1	2700	1900	140	120	110	130	C	T	
96	16.8	20	4400	6.8	2.9	900	900	100	90	100	100	C	T	
97	14.7	20	3300	5.7	2.1	700	700	70	80	100	80	C	R	
98	14.7	10	3200	7.4	4.1	3500	2800	150	130	190	180	C	R	
99	16.8	20	4900	6.6	2.8	1500	1200	110	100	120	110	C	R	
100	17.0	10	2700	6.9	5.1	3200	2300	120	120	160	170	C	T	
101	18.9	10	4700	10.8	6.2	5000	4000	160	150	210	220	C	T	
102	20.3	20	6100	8.2	4.1	1800	1100	150	120	120	110	C	T	
103	17.0	10	4500	8.8	5.8	4300	3200	140	130	180	180	L	T	
104	18.9	10	5200	10.6	6.6	5100	3600	180	130	200	220	L	T	
105	20.4	20	6500	8.2	3.9	2100	1400	120	100	130	130	L	T	
106	16.8	10	3500	8.9	5.3	4000	3000	150	120	200	190	W/O	T	
107	18.7	10	5000	10.7	6.8	5300	4400	190	160	250	250	W/O	T	
108	20.6	20	6300	7.8	3.9	2500	1200	120	110	130	130	W/O	T	
109	18.9	20	6400	7.7	3.3	1400	1000	110	90	105	105	W/O	T	
110	18.9	20	4300	7.8	3.3	1100	1200	120	100	110	110	W/O	R	
111	20.8	20	6200	8.2	4.2	2400	1800	150	110	140	130	W/O	R	

Notes:

*Lap Loadings are combined loadings of the lower shoulder harness and lap belt corresponding to the maximum upper shoulder harness load.

**W/O - Helmet only; H - Added weight high; L - Added weight low; C - Added weight centered.

†T - Neck muscles tensed; R - Neck muscles relaxed.

Table 7 - Natick Project Oscillograph Read-Out of Run 79 Subject LMP

Time	Bite Plate Acc.		Forehead Acc.	
	AP	SI	AP	SI
ms	G	G	G	G
0.0	0.0	0.0	0.0	0.0
7.0	0.5	2.1	0.1	1.5
16.7	2.4	8.4	3.4	3.3
20.1	6.1	11.7	5.1	6.5
24.7	7.5	11.9	3.8	2.9
31.4	10.3	10.7	4.8	6.6
34.8	11.4	9.5	5.4	6.7
42.1	12.4	7.3	7.2	4.4
53.4	13.2	6.7	11.2	2.3
59.8	13.1	5.9	13.4	3.0
67.1	11.8	5.4	15.2	3.3
71.6	11.6	6.6	16.3	2.9
83.5	9.5	6.3	17.4	2.6
99.3	5.6	-10.3	12.5	-13.0
105.0	6.9	-7.7	14.0	-13.7
121.1	5.0	-9.7	10.2	-15.8
131.2	5.1	-5.1	6.9	-8.6
138.2	2.5	-5.0	5.7	-2.5
148.4	1.1	-2.6	4.0	-2.6
170.2	2.6	6.0	4.2	6.1
188.8	1.5	3.1	0.6	3.2
202.5	-0.1	0.8	-1.4	3.6
242.3	-0.7	2.2	-0.8	1.9
265.1	-1.0	0.1	-1.0	0.1
299.9	-0.7	0.8	0.3	-0.2
333.9	-0.7	-1.2	-0.5	-0.3

Table 8 - Natick Project Data Analysis of Run 79 Volunteer LMP

	Kinematics of the Head				Reactions at Occ. Cond.		
	Acc. of C.G.						
Time, ms	AP G	IS G	Angular Acc. RA/S/S	Rel. Angle, deg	Equiv. Moment, ft-lb	A-P Force, lb	Axial Force, lb
0.0	0.0	0.0	0.0	-13.0	0.7	-3.3	14.1
7.0	-0.2	-1.0	86.0	-13.0	3.8	-5.6	-0.1
16.7	0.0	-7.4	147.0	-10.0	5.5	-2.5	-92.6
20.1	3.2	-7.0	356.0	-8.0	-1.0	43.3	-87.2
24.7	1.4	-7.0	459.0	-6.0	8.4	18.5	-86.6
31.4	4.6	-3.1	573.0	-0.0	-1.4	65.6	-29.9
34.8	4.1	-2.6	573.0	3.0	0.2	59.4	-23.8
42.1	5.5	-2.1	451.0	10.0	-9.1	81.8	-16.6
53.4	8.4	-3.9	239.0	21.0	-26.5	126.3	-43.4
59.8	7.9	-5.8	97.0	30.0	-28.5	120.2	-71.4
67.1	7.6	-7.9	-78.0	38.0	-32.7	118.7	-103.0
71.6	9.1	-9.0	-125.0	43.0	-39.8	141.5	-119.2
83.5	8.9	-10.7	-301.0	58.0	-44.3	141.1	-145.6
99.3	9.5	1.0	-760.0	68.0	-60.8	150.5	21.7
105.0	11.3	-0.6	-729.0	70.0	-66.9	177.5	-1.6
121.1	7.9	1.1	-685.0	66.0	-52.6	127.1	22.9
131.2	6.3	1.8	-318.0	58.0	-36.2	102.6	34.8
138.2	0.6	-0.7	-317.0	52.0	-13.7	19.0	-0.8
148.4	1.5	-0.8	-250.0	42.0	-15.0	30.6	-0.8
170.2	-0.1	-5.6	98.0	20.0	2.4	3.2	-67.4
188.8	-0.6	-1.6	144.0	-1.0	6.3	-8.5	-8.5
202.5	-3.3	-0.5	129.0	-4.0	16.6	-49.5	7.6
242.3	-1.5	-1.6	72.0	-13.0	9.0	-26.1	-8.5
265.1	-1.4	-0.4	6.0	-15.0	6.4	-23.6	8.7
299.9	-0.9	-1.8	-43.0	-13.0	3.4	-17.0	-12.7
333.9	-0.7	0.4	-39.0	-13.0	2.7	-14.8	20.0

various time increments is given in Table 7. From this acceleration-time data and the movie data of head and torso position as a function of time, the acceleration of the center of gravity of the head, the angular acceleration of the head, the relative position of the head with respect to the torso, and the reactions of the neck and chin forces transferred to the occipital condyles were computed as described previously. A sample of the output from the computer program used to generate this data is given in Table 8 for Run 79.

The torque calculated at the occipital condyles as a function of the angular position of the volunteer's head relative to his torso for several sled deceleration levels with no additional head weight are shown in Figs. 12 and 13 for the conditions of muscles relaxed and tensed, respectively. There is a marked difference between these two families of response curves. For example, the neck response for the 3.3 g sled run with muscles relaxed is characterized by an initial overshoot of 10 ft lb followed by a plateau of 7 ft lb extending to 25 deg of relative head rotation with a maximum relative angular head position of 32.5 deg. On the other hand, the comparative 3.3 g run with muscles tensed, produced a response curve which rose to a peak of 9.5 ft lb followed by a decreasing torque-angle relation which extended to a maximum relative angular displacement of 8.5 deg, followed by a rapid return of the head toward its initial position.

The maximum responses which occurred during the various volunteer runs are presented in Tables 9-12 for the conditions of Helmet Only, Weight High, Weight Centered, and Weight Low, respectively. For each run, the tables give the plateau sled deceleration level; the maximum torque and the time at which it occurred with reference to the beginning of the sled deceleration pulse; the maximum A-P force acting on the head and its corresponding time to peak; the initial head position relative to a normal upright seating position (torso inclined 15 deg to vertical with A-P axis of the head horizontal); the maximum change in position of the head relative to its initial position and the time at which the maximum angulation occurs; and the degree of initial neck muscle tone.

The maximum neck response occurred during the sled sequence where the volunteer wore the additional head weight above the center of gravity of his head and tensed his neck muscles (Fig. 14). The curves represent the maximum neck response for a series of increasing sled deceleration levels and will be taken as the basis for the development of a maximum response envelope for the neck in flexion. It is evident that neck responses of lesser magnitude can be achieved by the same person, depending on his muscle tone as demonstrated by the curves in Figs. 12 and 13; however, specifying anything other than a maximum response condition would hinder the optimization of restraint system design. It should be emphasized that the torque calculated for the ordinate of these

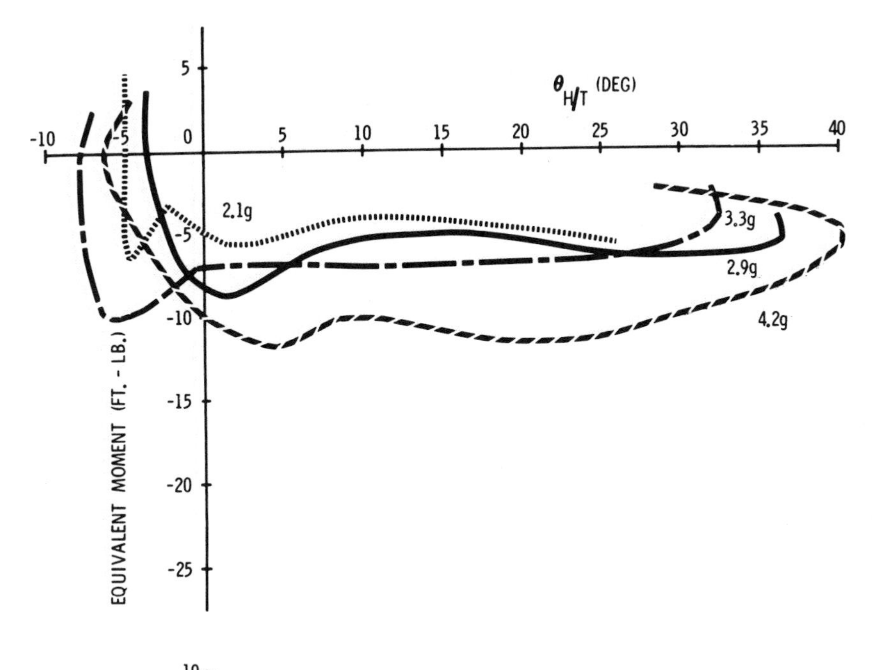

Fig. 12 - Equivalent moment about the occipital condyles as a function of the angular position of the head relative to the torso for various impact severities, volunteer LMP, neck muscles relaxed, no additional head weight

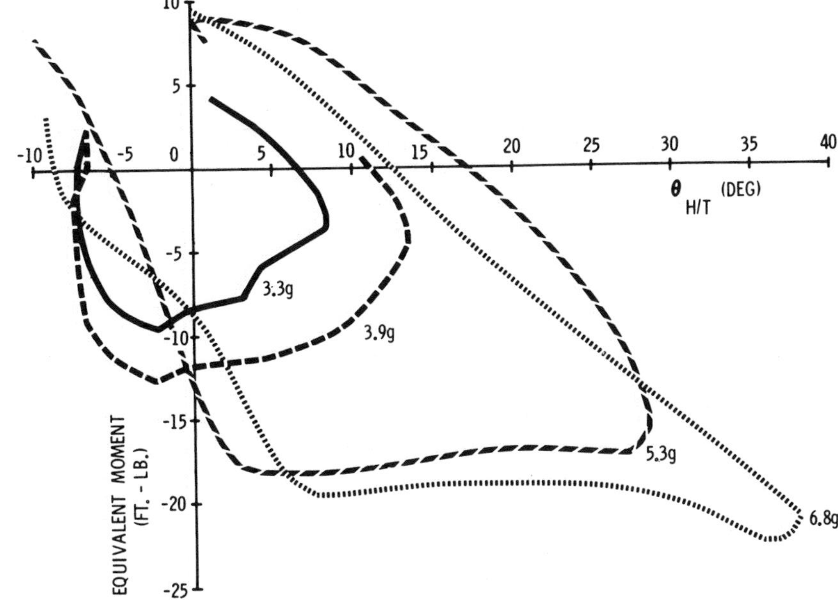

Fig. 13 - Equivalent moment about the occipital condyles as a function of the angular position of the head relative to the torso for various impact severities, volunteer LMP, neck muscles tensed, no additional head weight

curves includes the moment produced by the chin reaction on the chest as well as the moments produced by the muscles and ligaments of the neck structure.

The threshold of pain for the volunteer LMP occurred in Run 78 where the additional weight was positioned above the center of mass of the head. The torque level for this run was 43.5 ft lb with a maximum relative head position of 57 deg. His voluntary tolerance level was reached in Run 79 when he experienced a pain from the back of his neck, extending down into the middle of his back. This exposure produced a stiff neck which lasted several days. The maximum torque was 65 ft lb with a maximum relative angular displacement between his head and torso of 70 deg.

CADAVER EXPOSURES - The purpose of the cadaver runs was to obtain a comparison of the cadaver and volunteer responses to the same sled deceleration pulses for the various configurations of additional head weight and then to subject the cadaver to more severe conditions in order to extend the data into the injury region. A total of 132 runs were conducted using four human cadavers. These cadavers were identified as Cadaver 1404, Cadaver 1530, Cadaver 1538, and Cadaver 1548.

The transducers and instrumentation used for the cadaver runs were identical to those used for the volunteer runs, except that the thoracic acceleration was not recorded. X-rays were taken of the cadaver's neck to determine whether a particular sled ride caused any observable neck damage. The data obtained from the cadaver runs were analyzed in the same manner as the data obtained from the volunteer runs. A summary of the more germane findings follows.

Cadavers 1530, 1538, and 1548 were subjected to a sequence of high-severity sled rides at plateau accelerations up to

Table 9 - Various Maximum Responses for Volunteer LMP, Helmet Only

		Max. Moment		Max. A-P Force		Head Position			
						Initial	Max. Rotation		
Run No.	Plat. Sled, g	T_R, ft-lb	Time, ms	$F_{A\text{-}P}$, lb	Time, ms	$(\theta_{H/T})_0$, deg	$\Delta\theta_{H/T}$, deg	Time, ms	Initial Neck Muscle Tension
23	2.0	- 4.8	105	19	115	-13	26	204	Relaxed
85	2.1	- 6.3	54	25	125	- 4	36	305	Relaxed
87	2.9	- 8.6	92	36	92	- 4	40	250	Relaxed
110	3.3	- 9.8	75	39	75	- 4	36	246	Relaxed
26	3.8	- 7.5	75	33	95	-10	43	215	Relaxed
86	4.1	-15.3	96	63	107	- 6	42	147	Relaxed
111	4.2	-11.8	73	58	110	- 4	44	223	Relaxed
28	4.4	-13.0	99	50	99	-14	41	199	Relaxed
22	2.0	- 7.2	50	25	50	-14	1	69	Tensed
82	2.1	- 7.0	61	27	61	- 7	7	187	Tensed
84	2.9	- 9.0	70	36	82	- 6	6	227	Tensed
109	3.3	- 9.5	65	38	100	- 5	14	211	Tensed
25	3.3	-11.0	85	45	100	- 1	14	161	Tensed
24	3.7	-13.0	57	48	90	-19	11	118	Tensed
108	3.9	-12.2	80	56	100	- 6	19	213	Tensed
27	4.4	-13.5	87	60	87	-15	10	97	Tensed
83	4.7	-15.5	66	56	66	- 8	11	125	Tensed
106	5.3	-18.3	65	78	92	- 7	36	123	Tensed
107	6.8	-22.5	110	92	90	- 7	44	123	Tensed

Table 10 - Various Maximum Responses for Volunteer LMP, Additional Weight Above C. G. of Head

		Max. Moment		Max. A-P Force		Head Position			
						Initial	Max. Rotation		
Run No.	Plat. Sled, g	T_R, ft-lb	Time, ms	$F_{A\text{-}P}$, lb	Time, ms	$(\theta_{H/T})_0$, deg	$\Delta\theta_{H/T}$, deg	Time, ms	Initial Neck Muscle Tension
73	1.5	- 7.4	76	24	64	-12	4	92	Tensed
80	2.2	-10.9	185	36	175	-13	20	210	Tensed
81	3.0	-12.0	150	43	155	-11	25	202	Tensed
74	3.1	-17.0	94	59	94	-14	20	122	Tensed
75	4.6	-21.8	103	77	99	-13	33	138	Tensed
76	5.1	-28.7	87	104	87	-13	50	143	Tensed
77	6.9	-33.3	87	120	82	-13	59	130	Tensed
78	8.0	-43.5	80	150	80	-14	71	124	Tensed
79	9.6	-65.0	105	177	105	-13	83	105	Tensed

Notes:

1. The time listed represents the time interval from the initiation of sled deceleration to peak response.
2. No runs were conducted with the neck muscles initially relaxed.

14 g with the additional weight placed at the center of mass of the head. The corresponding neck response curves are shown in Figs. 15-17, respectively. These graphs are plots of the summation of the moments of the neck and chin forces with respect to the occipital condyles as a function of the angular position of the head with respect to the torso for various plateau sled deceleration levels. The response curves for the three cadavers are quite different. Cadaver 1530 had a very stiff neck and, consequently, for a given plateau deceleration level, his relative head angulation was quite small; it did not exceed 20 deg during the 12.0 g sled ride. The neck of this cadaver responded as a linear spring. The neck of Cadaver 1548 was not quite as stiff as Cadaver 1530 as noted by comparing their response curves. Cadaver 1538 had a loose neck with little neck resistance occurring until an appreciable relative head rotation was achieved. The main resistance developed by the neck of Cadaver 1538 was due to ligamentous straining caused by chin contact with the chest as noted by the rapid increase in slope, 12.5 ft lb/deg, at approximately 60 deg of relative head angulation.

Table 11 - Various Maximum Responses for Volunteer LMP, Additional Weight at CG of Head

		Max. Moment		Max. A-P Force		Head Position			
						Initial	Max. Rotation		
Run No.	Plat. Sled, g	T_R, ft-lb	Time, ms	$F_{A\text{-}P}$, lb	Time, ms	$(\theta_{H/T})_0$, deg	$\Delta\theta_{H/T}$, deg	Time, ms	Initial Neck Muscle Tension
97	2.1	- 9.0	64	33.0	100	- 5	-2/14	64/263	Relaxed
99	2.8	- 9.4	95	51.3	117	- 2	-6/27	39/276	Relaxed
98	4.1	-15.0	85	74.6	112	- 6	-3/47	36/164	Relaxed
94	2.0	- 8.7	58	32.9	122	- 6	-4	191	Tensed
96	2.9	- 9.7	63	39.4	63	- 4	-4/2	82/238	Tensed
95	4.1	-15.0	82	69.8	82	- 5	-4/18	43/144	Tensed
102	4.1	-12.2	74	65.9	94	-11	-6/22	40/212	Tensed
100	5.1	-20.7	74	95.6	97	-13	44.4	123	Tensed
101	6.2	-26.8	113	112.6	87	- 9	51.1	132	Tensed

Note:
1. The time listed represents the time interval from the initiation of sled deceleration to peak response.

Table 12 - Various Maximum Responses for Volunteer LMP Additional Weight Below CG of Head

		Max. Moment		Max. A-P Force		Head Position			
						Initial	Max. Rotation		
Run No.	Plat. Sled, g	T_R, ft-lb	Time, ms	$F_{A\text{-}P}$, lb	Time, ms	$(\theta_{H/T})_0$, deg	$\Delta\theta_{H/T}$, deg	Time, ms	Initial Neck Muscle Tension
91	2.1	- 5.7	100	28	135	- 4	- 2/38	41/282	Relaxed
93	2.7	- 7.8	100	44	100	- 6	- 6/33	63/273	Relaxed
92	4.2	-14.2	114	67	96	- 7	- 3/30	47/166	Relaxed
88	1.9	- 7.2	74	33	74	- 7	- 7/1	111/311	Tensed
90	2.9	-10.0	92	45	92	- 4	- 5/3	60/252	Tensed
105	3.9	-12.0	132	65	102	- 4	-13/11	47/238	Tensed
89	4.1	-14.0	95	72	95	- 8	- 4/20	34/146	Tensed
103	5.8	-16.4	103	92	94	- 5	- 8/38	27/155	Tensed
104	6.6	-23.0	98	124	85	-10	- 7/51	37/131	Tensed

Note:
1. The time listed represents the time interval from the initiation of sled deceleration to peak response.

Cadavers 1404, 1530, and 1548 were subjected to a sequence of high-severity sled runs with the additional weight placed above the center of mass of their heads. The corresponding neck response curves are given in Figs. 18-20. As before, the response curves for these cadavers are different due to the various degrees of neck stiffnesses. Cadaver 1404 had the loosest neck. The main resistance developed by the neck of this cadaver was due to the ligamentous straining caused by contact of the chin with the chest. The slope of the torque-head angle curve for Cadaver 1404 was 12.5 ft lb/deg which was the same as noted for Cadaver 1538 (Fig. 16). The main difference in the response curves for Cadaver 1404 (Fig. 18) and Cadaver 1538 (Fig. 16) is that Cadaver 1404 achieved a larger relative head angulation, 95 deg as compared to 73 deg. This difference is largely due to differences in neck geometry between the two cadavers.

The maximum responses of Cadavers 1404, 1538, 1548, and 1530 for their various sled rides are given in Tables 13-16 respectively. For each run, the maximum values of the summation of the moments about the occipital condyles and the summation of the A-P forces acting on the head are given. Also, the initial position of the head relative to its normal position with the subject seated (A-P axis of the head horizontal and torso inclined rearward at 15 deg to vertical) and the maximum change in the angular position of the head referenced to the torso are presented. The times which are listed represent the times required to reach the various peak responses referenced to the beginning of sled deceleration.

Maximum torque levels of 130 and 140 ft lb were achieved by Cadavers 1538 and 1404 without producing any damage to the cervical spine as noted by x-ray analysis. A maximum A-P force of 470 lb was developed on the head of Cadaver 1538 by

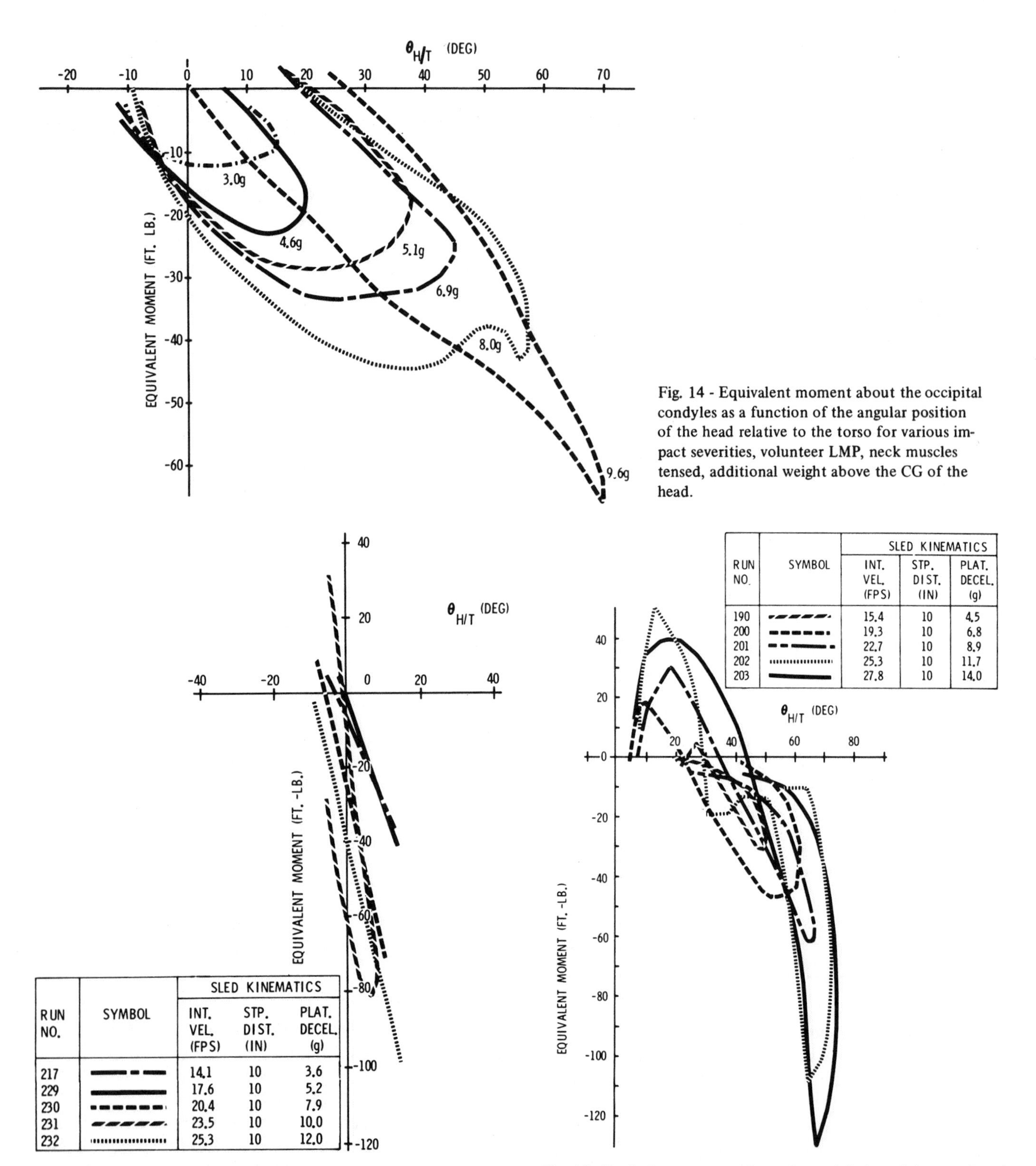

Fig. 15 - Equivalent moment about the occipital condyles as a function of the angular position of the head relative to the torso for various high-severity runs, Cadaver 1530, additional weight at the CG of the head

Fig. 16 - Equivalent moment about the occipital condyles as a function of the angular position of the head relative to the torso for various high-severity runs, Cadaver 1538, additional weight at the CG of the head

the neck structure and chin contact with the chest without producing any discernible neck trauma.

RESPONSE CHARACTERISTICS OF THE NECK IN FLEXION - As noted previously, the volunteer LMP is quite representative of a 50th percentile adult male from a weight, height, age, and occupation standpoint. Consequently, the neck response of the volunteer LMP is considered representative of a 50th percentile adult male.

With the added weight placed above the center of mass of his head, the volunteer was quite apprehensive about the possi-

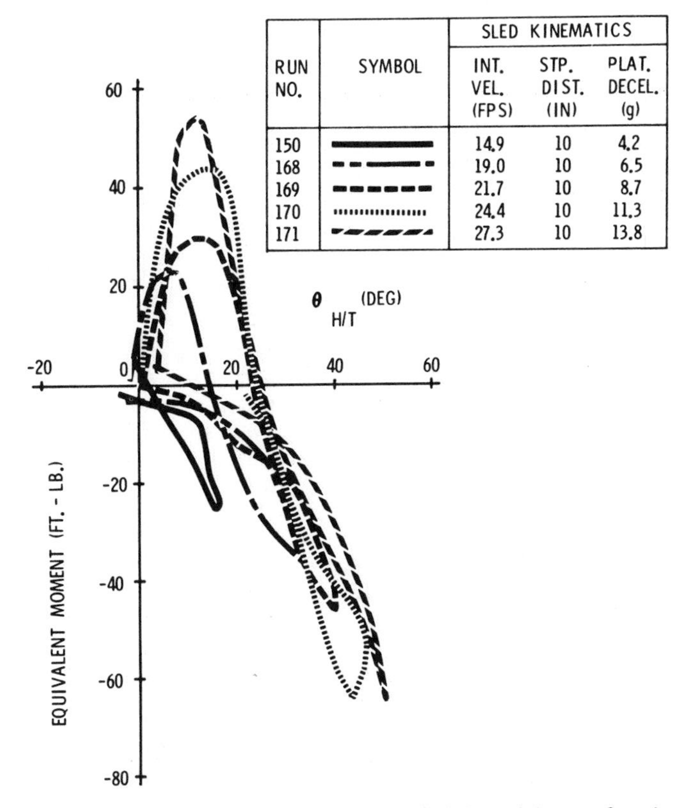

Fig. 17 - Equivalent moment about the occipital condyles as a function of the angular position of the head relative to the torso for various high-severity runs, Cadaver 1548, additional weight at the CG of the head

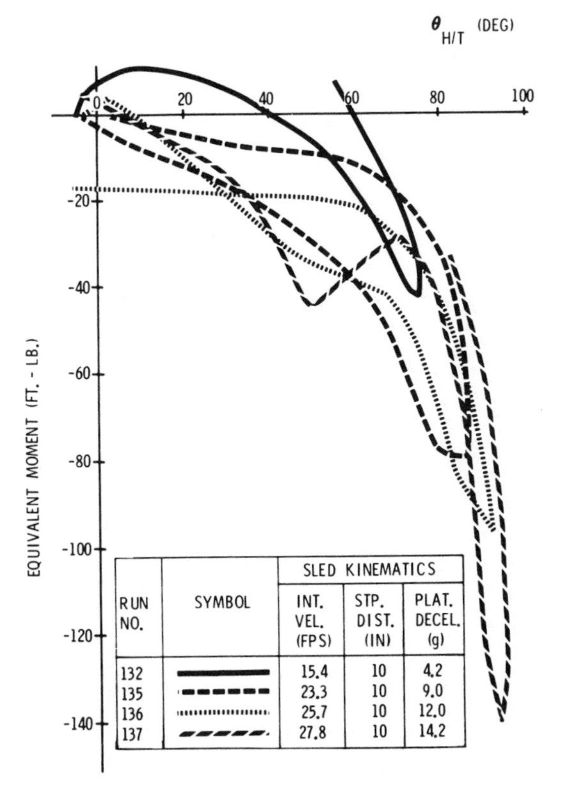

Fig. 18 - Equivalent moment about the occipital condyles as a function of the angular position of the head relative to the torso for various high-severity runs, Cadaver 1404, additional weight above the CG of the head

bility of incurring a neck injury and, therefore, exerted more effort to maintain control over his head motion than he did for any of the other configurations of additional head weight. Consequently, the family of torque-angle curves (Fig. 14), obtained with the added weight placed above the center of mass of the head with neck muscles tensed, will be taken as the basis for describing a response envelope for various degrees of neck flexion for the 50th percentile adult male.

This family of curves and a proposed response envelope are depicted in Fig. 21. The ordinate of the graph is the summation of the moments of the neck and chin forces with respect to the occipital condyles with angular position of the head relative to the torso as the abscissa. The zero angle is referenced to the position where the A-P axis of the head is horizontal and the torso is inclined rearward 15 deg to the vertical. The initial slope of the envelope curve is taken parallel to the initial slopes of the volunteer curves and is 3 ft lb/deg. This portion of the envelope is extended to the 45 ft lb level where the envelope assumes a constant torque level to 45 deg of relative head rotation which agrees with the trend of the volunteer data. From the point (45, -45) the envelope is extended on a straight line to the point (65, -65) which is the level of the maximum torque generated by the volunteer. The unloading portion of the envelope consists of two straight line segments which closely parallel the unloading portion of the volunteer curves. The bottom segment starts at the point (75, -65) and extends to (55, -20) from which the second segment starts and ends at (35, 0). The choice of the starting angle of 75 deg for the bottom segment of the unloading curve was arbitrarily taken as 10 deg greater than the end point of 65 deg for the bottom segment of the loading curve and places the maximum angle experienced by the volunteer midway between these two curves.

Ratios of the areas contained within the individual volunteer loading and unloading curves to the areas between the loading curves and the abscissa were computed and are 0.90, 0.83, 0.87, 0.80, 0.76, and 0.51 for the increasing degrees of severities as expressed by the maximum torque levels. This shows that as the rate of loading increases, the ratio of the energy dissipated by the neck muscles to that stored in the muscles decreases.

In order to extend the response envelope to higher torque levels, the cadaver response data were considered. The neck responses of cadavers used in the test program varied greatly as can be noted by comparing the curves in Figs. 15-20. The differences in response characteristics noted from the above comparisons are due primarily to differences in neck stiffnesses between the cadavers. Another interesting comparison is the change in neck stiffness which occurs after a number of tests have been conducted on the same cadaver, as shown in Figs. 22-25 for Cadavers 1404, 1530, 1538, and 1548, respectively. For each cadaver, there is a progressive loosening of the neck with the increase in the number and degree of severity of the runs conducted with the cadaver. Consequently, the neck response of the cadavers is constantly changing. The largest changes occurred for Cadavers 1538 and 1548, while the smallest change occurred for Cadaver 1404. Because of the consis-

Table 13 - Maximum Responses of Cadaver 1404 for Different Inertial Properties of the Head

		Max. Moment		Max. A-P Force		Head Position			
							Max. Rot.		
Run No.	Plat. Sled, g	T_R, ft-lb	Time, ms	$F_{A\text{-}P}$, lb	Time, ms	Initial $(\theta_{H/T})_0$, deg	$\Delta\theta_{H/T}$, deg	Time, ms	Weight Config.
133	2.2	- 14.7	173	58.0	170	-3	61	176	W/O
112	2.4	- 11.6	198	45.0	172	-5	46	198	W/O
113	2.4	- 12.2	170	50.0	170	-7	49	200	W/O
114	2.5	- 14.3	168	57.0	168	-6	50	188	W/O
128	2.9	- 22.0	175	86.5	157	-5	63	180	W/O
115	3.0	- 18.5	148	73.0	142	-3	57	142	W/O
122	3.0	- 19.5	167	71.0	167	-5	58	197	W/O
123	3.0	- 19.5	162	73.0	152	-5	60	182	W/O
116	3.4	- 23.8	142	92.0	132	-3	66	160	W/O
117	4.6	- 28.0	125	93.0	125	-5	71	150	W/O
118	2.0	- 16.8	175	80.0	170	-5	53	175	L
119	3.0	- 25.0	155	105.0	155	-7	58	155	L
120	3.5	- 28.4	138	125.0	128	-7	61	150	L
121	4.3	- 27.0	141	112.0	131	2	64	158	L
124	2.0	- 23.0	172	90.0	170	-2	63	172	C
125	3.0	- 32.0	153	117.0	153	-3	66	153	C
126	3.6	- 34.5	124	135.0	124	-5	62	124	C
127	4.3	- 43.0	135	147.0	135	-3	69	135	C
129	2.3	- 32.5	198	105.0	198	-4	68	198	H
130	3.0	- 41.0	174	135.0	174	-7	73	174	H
131	3.6	- 44.0	147	140.0	147	-8	74	155	H
134	3.7	- 45.0	153	144.0	153	-7	78	153	H
132	4.2	- 42.5	140	141.0	140	-5	75	140	H
135	9.0	- 80.0	100	264.0	100	-4	86	110	H
136	12.0	- 97.0	105	330.0	105	-5	92	105	H
137	14.2	-140.0	93	357.0	93	-7	95	100	H

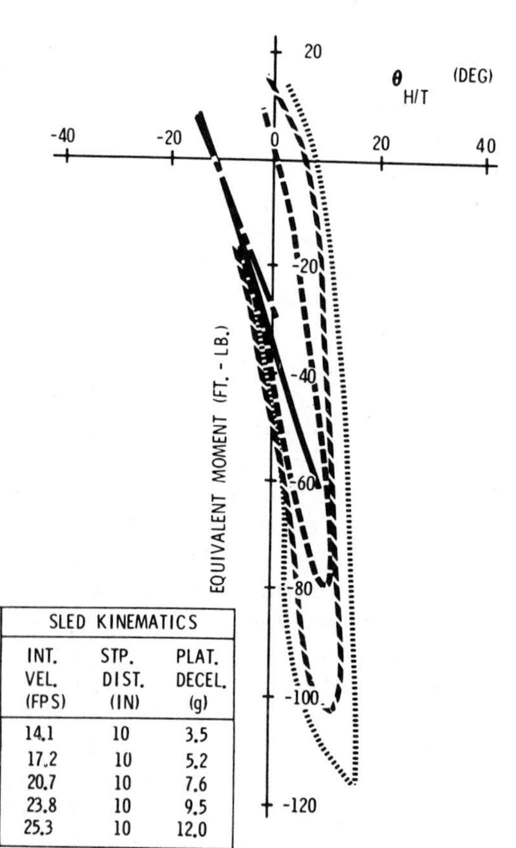

RUN NO.	SYMBOL	SLED KINEMATICS INT. VEL. (FPS)	STP. DIST. (IN)	PLAT. DECEL. (g)
222	— - —	14.1	10	3.5
239	———	17.2	10	5.2
240	·- - - - -·	20.7	10	7.6
241	▬ ▬ ▬ ▬	23.8	10	9.5
242	············	25.3	10	12.0

Fig. 19 - Equivalent moment about the occipital condyles as a function of the angular position of the head relative to the torso for various high-severity runs, Cadaver 1530, additional weight above the CG of the head

Table 14 - Maximum Responses of Cadaver 1538 for Different Inertial Properties of the Head

Run No.	Plat. Sled, g	Max. Moment: T_R, ft-lb	Max. Moment: Time, ms	Max. A-P Force: $F_{A\text{-}P}$, lb	Max. A-P Force: Time, ms	Head Position: Initial $(\theta_{H/T})_0$, deg	Head Position: Max. Rot. $\Delta\theta_{H/T}$, deg	Head Position: Max. Rot. Time, ms	Weight Config.
178	2.2	- 11.3	150	49.0	131	8	35	160	W/O
179	3.0	- 17.0	100	71.5	115	10	41	115	W/O
186	3.0	- 18.5	100	75.0	115	10	44	118	W/O
191	3.0	- 18.0	94	75.0	115	10	44	115	W/O
197	3.0	–	–	–	no film	–	–	–	W/O
198	3.0	- 16.5	130	70.0	123	10	49	142	W/O
204	3.0	- 28.5	154	100.0	149	−1	68	176	W/O
180	3.7	- 22.0	90	86.0	109	10	41	109	W/O
181	4.0	- 25.8	83	78.0	90	16	44	100	W/O
182	2.2	- 13.0	100	61.0	130	13	35	120	L
183	3.1	- 16.9	95	70.5	95	10	40	110	L
184	3.4	- 24.0	95	83.5	100	10	43	110	L
185	4.2	- 26.3	93	82.0	75	18	46	105	L
187	2.1	- 17.9	110	73.0	130	13	44	130	C
188	3.0	- 21.5	100	91.0	118	11	47	120	C
189	3.7	- 25.2	95	107.0	110	10	51	120	C
190	4.5	- 30.5	80	106.0	85	20	50	90	C
200	6.8	- 47.0	85	184.0	90	3	61	110	C
201	8.9	- 62.0	83	225.0	80	5	67	90	C
202	11.7	-109.0	73	360.0	73	7	72	90	C
203	14.0	-130.0	68	473.0	68	4	73	96	C
192	2.4	- 18.0	135	70.0	127	8	43	132	H
199	2.6	- 20.5	149	75.0	130	6	53	160	H
193	3.0	- 23.5	123	88.0	117	12	52	134	H
194	3.0	- 23.0	150	83.0	150	5	50	135	H
195	3.7	- 25.0	130	99.0	135	5	53	135	H
196	4.2	- 30.0	100	110.0	108	5	57	130	H

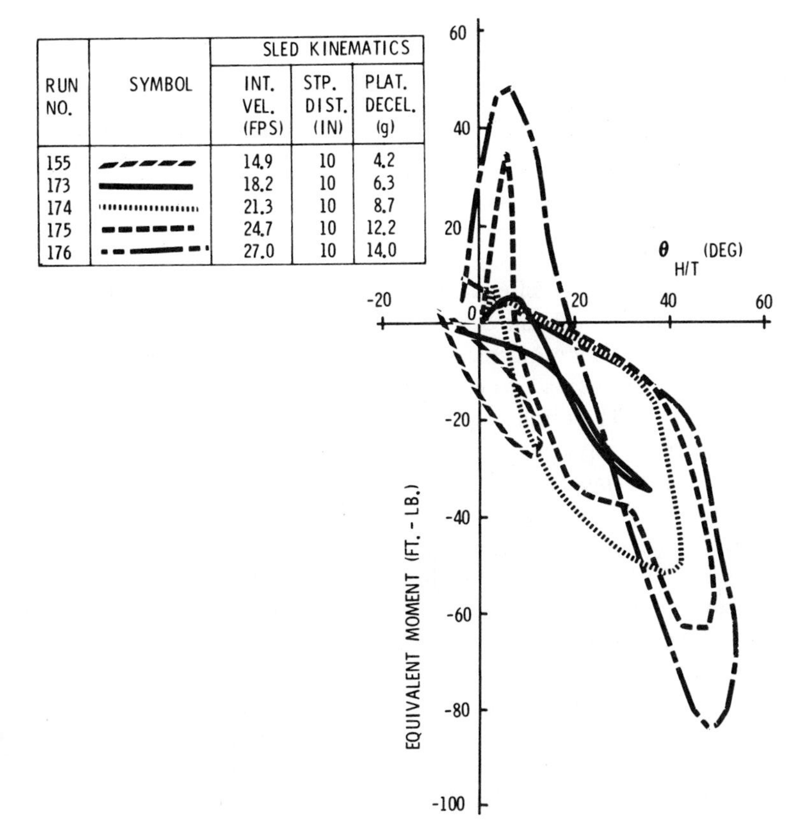

Fig. 20 - Equivalent moment about the occipital condyles as a function of the angular position of the head relative to the torso for various high-severity runs, Cadaver 1548, additional weight above the CG of the head

Table 15 - Maximum Response of Cadaver 1548 for Different Inertial Properties of the Head

		Max. Moment		Max. A-P Force		Head Position			
							Max. Rot.		
Run No.	Plat. Sled, g	T_R, ft-lb	Time, ms	$F_{A\text{-}P}$, lb	Time, ms	Initial $(\theta_{H/T})_0$, deg	$\Delta\theta_{H/T}$, deg	Time, ms	Weight Config.
138	2.3	-11.6	60	37.5	43	-0.2	4.7	70	W/O
139	2.8	-13.2	70	47.5	50	-0.2	5.8	80	W/O
156	3.0	-13.7	65	51.5	117	-2.2	9.5	95	W/O
162	3.0	-13.8	130	62.0	120	-2.8	17.4	140	W/O
177	3.0	-13.9	120	70.0	120	5.0	32.6	135	W/O
140	3.5	-15.5	75	62.0	110	0.0	7.7	75	W/O
141	4.1	-22.5	70	75.0	60	-0.7	11.5	80	W/O
157	6.3	-32.0	75	116.0	70	0.4	19.1	85	W/O
158	8.7	-39.5	65	163.0	70	-0.5	22.8	80	W/O
159	12.0	-51.7	70	204.0	70	-0.5	27.3	85	W/O
160	14.0	-45.0	55	223.0	55	0.2	31.7	67	W/O
142	2.0	-10.0	80	41.5	120	-0.9	4.9	90	L
143	3.0	-11.8	80	64.0	110	-0.5	8.7	90	L
144	3.6	-14.5	85	86.0	107	-0.6	10.0	85	L
145	4.1	-21.5	73	80.0	110	-2.6	13.5	80	L
163	6.3	-27.0	75	141.0	95	-1.7	24.8	85	L
164	9.3	-37.0	70	199.0	82	-1.1	29.3	100	L
165	11.7	-39.0	80	212.0	60	1.3	34.2	90	L
166	13.9	-52.0	80	222.0	70	-0.8	37.7	90	L
147	2.1	-11.1	120	56.5	132	-1.8	8.5	120	C
148	3.0	-13.5	80	70.0	120	-2.2	11.2	130	C
149	3.3	-16.5	80	90.0	110	-1.4	13.0	110	C
150	4.2	-25.0	80	106.0	90	-1.5	16.7	90	C
168	6.5	-36.0	95	174.0	95	-1.0	35.2	95	C
169	8.7	-46.2	90	223.0	80	1.7	40.2	100	C
170	11.3	-63.0	80	300.0	75	0.0	45.8	88	C
171	13.8	-64.0	85	280.0	70	2.8	50.0	85	C
152	2.0	-12.0	110	50.0	100	-8.2	4.0	130	H
153	2.9	-17.2	117	75.0	117	-8.6	6.5	117	H
154	3.5	-18.6	110	89.0	110	-9.0	9.4	120	H
155	4.2	-27.0	80	106.0	100	-9.0	12.7	110	H
173	6.3	-36.0	100	154.0	100	-0.3	34.7	100	H
174	8.7	-52.0	100	220.0	85	0.9	42.1	107	H
175	12.2	-64.0	90	245.0	80	-0.1	49.0	100	H
176	14.0	-85.0	80	320.0	80	-3.5	54.0	90	H

tency of the response curves, the data obtained from Cadaver 1404 was used to extend the volunteer response envelope shown in Fig. 21.

The slope of the torque-angle curve for Cadaver 1404 (Fig. 18), in the region of hyperflexion of the neck is 12.5 ft lb/deg and is due primarily to the moment produced by the chin reacting against the chest and the straining of the posterior neck ligaments. This loading rate was used to extend the voluntary response envelope from 65 ft lb to the level of 140 ft lb (Fig. 26). The unloading rate was taken parallel to the loading curve since the trend is for the loading and unloading curves to approach each other at the higher angles as indicated for Cadaver 1404 and the volunteer responses in Figs. 18 and 21.

The response envelope places the following conditions on the response of any mechanical simulations of the human neck:

1. The evaluation of the neck structure must be made with the neck mounted in a structure which includes a head and a chest so the chin can contact the torso.
2. The torque measured must include the moment produced by the contact force between the chin and chest, if any.
3. The torque-angle relationship must be determined dynamically and both the loading and unloading portions of the curve must lie within the response envelope of Fig. 26.
4. To insure adequate damping, the ratio of the area between the loading and unloading curves to the area between the loading curve and the abscissa should not be less than 0.5 for that portion of the response curve below the 45 ft lb moment level.

These conditions apply to the flexion response of the 50th percentile adult male and are necessary, but not sufficient, conditions to describe a unique head-neck response to flexion. The geometric and inertial properties of the head, neck, and chest must be defined as well as the linear displacement range

Table 16 - Maximum Response of Cadaver 1530 for Different Inertial Properties of the Head

		Max. Moment		Max. A-P Force		Head Position			
						Initial	Max. Rot.		
Run No.	Plat. Sled, g	T_R, ft-lb	Time, ms	$F_{A\text{-}P}$, lb	Times, ms	$(\theta_{H/T})_0$, deg	$\Delta\theta_{H/T}$, deg	Time, ms	Weight Config.
205	2.1	- 8.3	80	26.0	63	- 6.0	-1.0	80	W/O
223	2.9	- 12.2	60	38.0	70	- 6.0	-0.2	80	W/O
243	2.9	- 13.4	85	49.0	85	- 3.0	-5.2	85	W/O
206	3.0	- 12.4	73	27.0	73	- 8.0	-3.2	85	W/O
228	3.0	- 12.5	80	43.0	80	- 6.0	3.8	88	W/O
207	3.3	- 18.0	73	57.0	73	- 8.0	-1.0	73	W/O
208	3.8	- 30.0	70	90.0	70	- 8.0	3.0	80	W/O
234	5.7	- 43.0	70	148.0	70	- 4.0	13.4	70	W/O
235	7.3	- 52.0	65	188.0	65	- 3.0	11.9	65	W/O
236	10.0	- 62.0	65	248.0	65	- 3.0	13.0	65	W/O
237	12.0	- 74.0	63	310.0	63	- 4.0	11.2	53	W/O
209	2.1	- 5.8	90	34.0	130	- 7.0	-2.1	100	L
210	2.7	- 12.9	80	48.0	90	- 7.0	0.0	110	L
211	3.2	- 15.8	78	60.0	85	- 6.0	1.0	78	L
212	3.6	- 27.4	68	84.0	68	- 7.0	3.3	88	L
224	5.5	- 34.0	70	132.0	60	- 4.0	7.4	75	L
225	7.0	- 45.0	68	200.0	78	- 4.0	8.3	68	L
226	9.5	- 62.0	68	275.0	68	- 1.0	9.4	64	L
227	11.3	- 80.0	65	350.0	65	- 1.0	10.0	60	L
214	2.1	- 9.4	75	33.0	90	- 6.0	0.3	75	C
215	2.9	- 13.4	60	46.0	60	- 6.0	0.8	70	C
216	3.3	- 17.0	75	66.0	63	- 6.0	1.9	75	C
217	3.6	- 27.2	70	99.0	80	- 4.0	7.9	70	C
229	5.2	- 49.0	75	188.0	75	- 4.0	12.0	70	C
230	7.9	- 73.0	68	274.0	68	- 5.0	9.2	63	C
231	10.0	- 84.0	70	330.0	70	- 6.0	8.5	65	C
232	12.0	-110.0	58	375.0	65	- 3.0	13.8	65	C
219	2.2	- 12.0	65	35.0	110	-13.0	-4.4	110	H
220	2.9	- 16.6	75	49.0	115	-13.0	-3.5	110	H
221	3.4	- 20.0	78	63.0	110	-12.0	-2.2	100	H
222	3.5	- 28.0	70	92.0	80	-13.0	0.6	90	H
239	5.2	- 64.0	70	186.0	70	-11.0	7.6	70	H
240	7.6	- 80.0	70	248.0	70	- 8.0	11.0	67	H
241	9.5	-103.0	73	345.0	68	- 7.0	13.0	63	H
242	12.0	-116.0	58	390.0	63	- 6.0	16.1	58	H

of the center of gravity of the head taken relative to a fixed system on the torso.

TOLERANCE LEVELS FOR THE NECK IN FLEXION - Various tolerance levels for neck flexion are indicated in Fig. 26. The volunteer LMP withstood a static moment about his occipital condyles of 26 ft lb. This reaction was generated by the volunteer's neck structure without any contribution from the chin contacting the chest. No tests were conducted to establish a static moment tolerance level with the chin in contact with the chest.

A dynamic tolerance level for the moemnt about the occipital condyles for the initiation of pain occurred during Run 78 in which the maximum equivalent resisting moment was 44 ft lb. There was contact between the chin and the chest during the run.

The maximum dynamic moment generated by the volunteer LMP of 65 ft lb occurred during Run 79. This torque level produced sharp pain in the neck and upper back region with soreness persisting for several days. The 65 ft lb includes the moment produced by the chin reaction as well as the moments produced by the neck reactions. It is considered as noninjurious, but close to the injury threshold.

The maximum tolerable moment was not determined because the volunteer experienced no injuries of any consequence during any of his sled rides. The cadavers were exposed to a much more severe environment than the volunteer. Maximum moments of 85, 116, 130, and 140 ft lb were experienced by Cadavers 1548, 1530, 1538, and 1404, respectively. There was a progressive loosening of the neck structures of the cadavers with repeated exposures which is assumed to be due to the tearing of hardened connective tissue of the neck. Cadaver 1404 had the loosest neck structure and

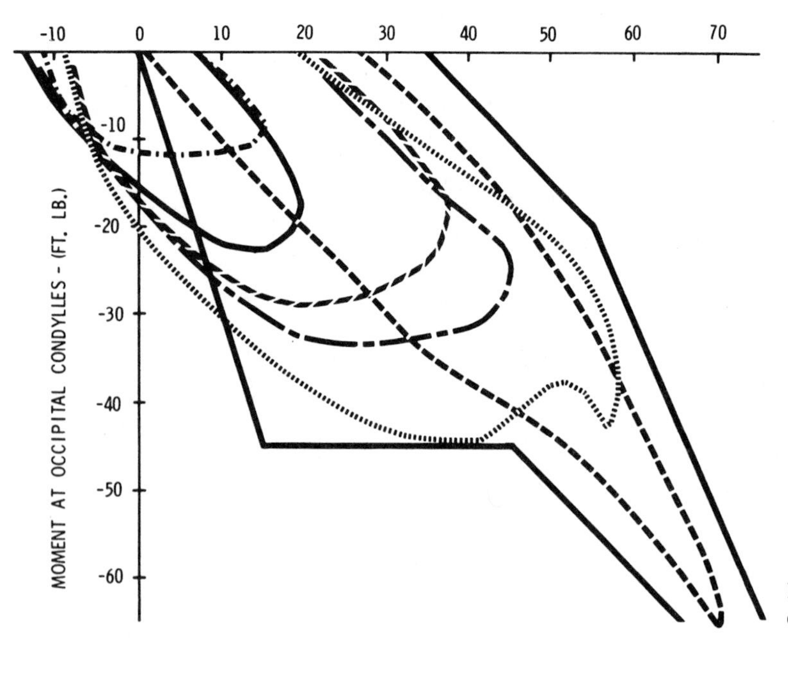

Fig. 21 - Comparison of the proposed flexion response envelope and the volunteer response curves

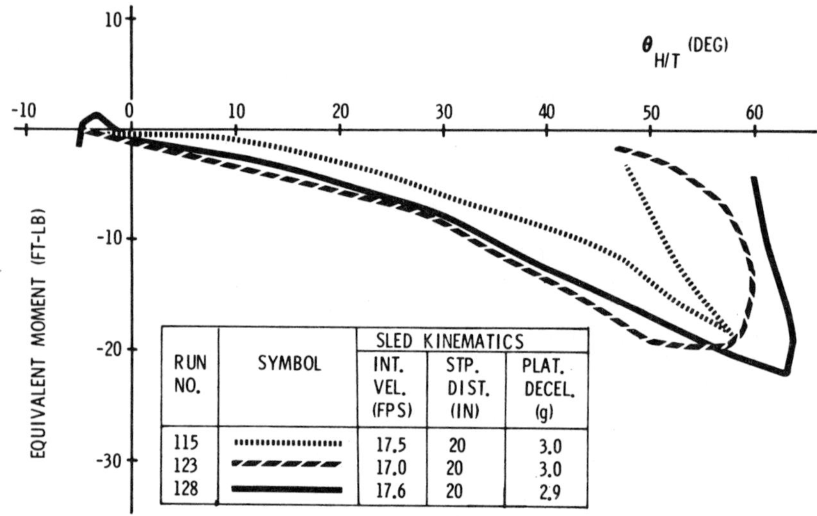

Fig. 22 - Comparison of neck response curves for identical 3 g plateau sled deceleration runs, Cadaver 1404, helmet only

showed little progressive loosening. None of these cadavers had any observable ligamentous, disc, or bone damage as noted from analysis of x-rays of their neck structures.

Based on this cadaver data, it would appear that the 50th percentile human could withstand equivalent moments of 140 ft lb without suffering ligamentous or bone damage. However, there is no guarantee that severe muscle injuries would not be produced at a lower value of equivalent moment. Consequently, the value of 140 ft lb should be used with discretion.

The static head rotational limit of 66 deg for the volunteer LMP compares favorably with his maximum dynamic head angle of 70 deg. The small difference between these two limits indicates that chin contact with the chest provides a stop for forward head rotation. Consequently, the position of the head relative to the torso is not a good physical measurement to be used in evaluating neck trauma. The equivalent moment about the occipital condyles is a better indicator.

The volunteer LMP withstood an A-P force of 190 lb with only his neck structure reacting. Dynamically, the volunteer LMP generated an A-P force of 177 lb, which includes the contribution of the chin contact force. None of these force levels produced any injury. The cadaver exposures can be used to obtain an approximation of a noninjurious level.

Cadavers 1548, 1404, 1530, and 1538 generated maximum A-P force levels of 320, 357, 390, and 473 lb, respectively. None of these force levels produced any observable neck damage. Consequently, an injury threshold of 450 lb is suggested for the A-P force acting on the head during hyperflexion when the chin is in contact with the chest.

DYNAMIC NECK RESPONSES AND TOLERANCE LEVELS FOR HYPEREXTENSION RESPONSE CHARACTERISTICS - Mertz and Patrick (1) presented data pertaining to the dynamic response and tolerance levels of the neck in hyperextension, including volunteer and cadaver data.

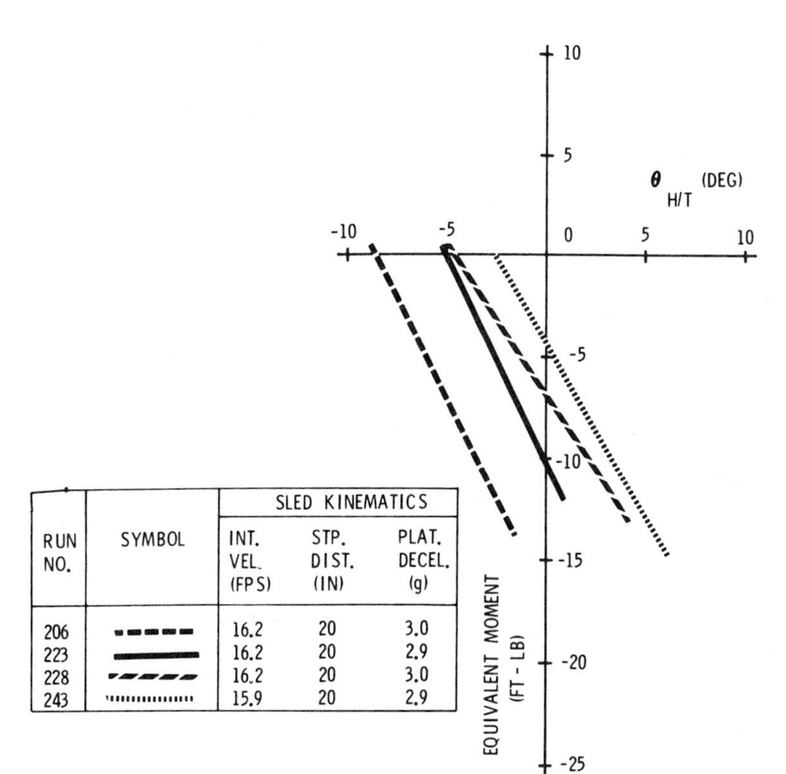

Fig. 23 - Comparison of neck response curves for identical 3 g plateau sled deceleration runs, Cadaver 1530, helmet only

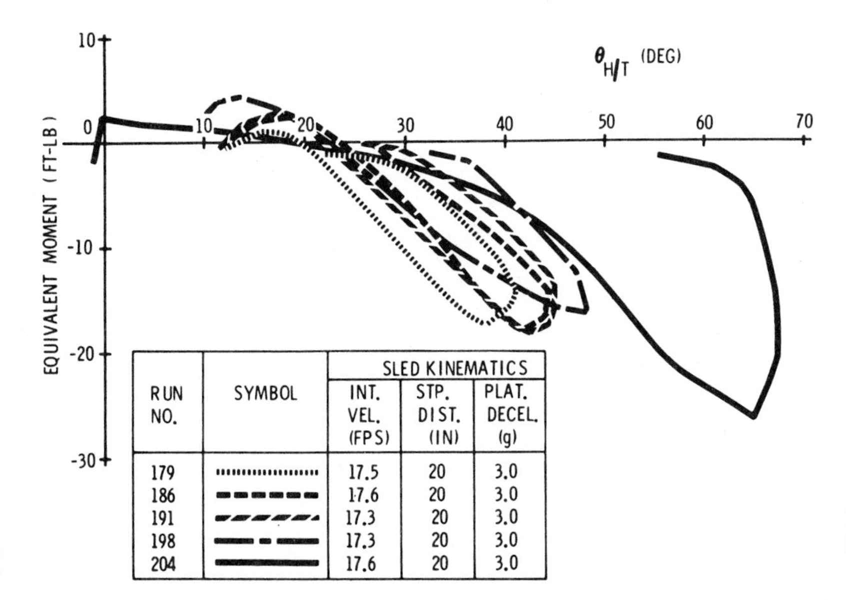

Fig. 24 - Comparison of neck response curves for identical 3 g plateau sled deceleration runs, Cadaver 1538, helmet only

The same volunteer was used in the earlier hyperextension experiments (1) and the hyperflexion experiments reported herein. The same data analysis techniques were also employed in the two programs. Therefore, the following development of hyperextension tolerance levels and a dynamic response envelope has a base consistent with the same development for hyperflexion.

The torque computed at the occipital condyles as a function of the position of the head relative to the torso is shown in Fig. 27 for the volunteer LMP and Cadavers 1035 and 1089. Only one curve is shown for the volunteer LMP because he did not wish to be exposed to higher severity sled runs for fear of neck injury. The response curves for the cadavers indicated a progressive loosening of their neck stıuctures with increased severity of the run.

An approximation of a hyperextension response envelope shown in Fig. 27 was made based on this volunteer and cadaver data, the static neck strength data for hyperextension (Table 1) and the dynamic neck strength data given by Carroll, et al. (10).

The initial slope of the response envelope was taken as 1 ft lb/deg which corresponds to the initial slope of the volunteer curve. This line is extended to a level of 22.5 ft lb. This level was chosen by noting that Carroll, et al. (10) gives a ratio of dynamic to static response of the neck in extension of 1.28 and that the static strength of the neck in resisting hyperex-

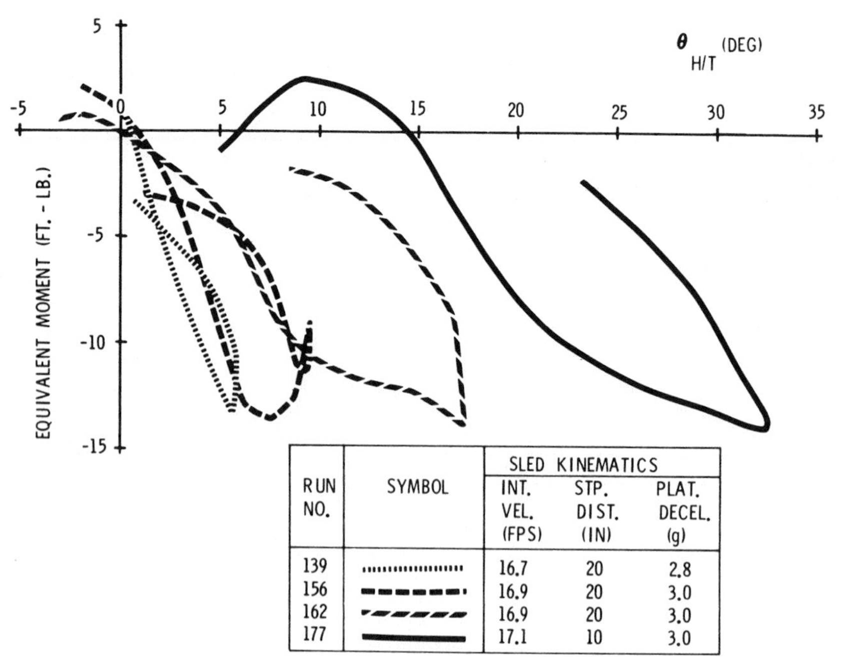

RUN NO.	SYMBOL	SLED KINEMATICS INT. VEL. (FPS)	STP. DIST. (IN)	PLAT. DECEL. (g)
139	··········	16.7	20	2.8
156	▬ ▬ ▬ ▬	16.9	20	3.0
162	▰▰▰▰	16.9	20	3.0
177	━━━━	17.1	10	3.0

Fig. 25 - Comparison of neck response curves for identical 3 g plateau sled deceleration runs, Cadaver 1548, helmet only

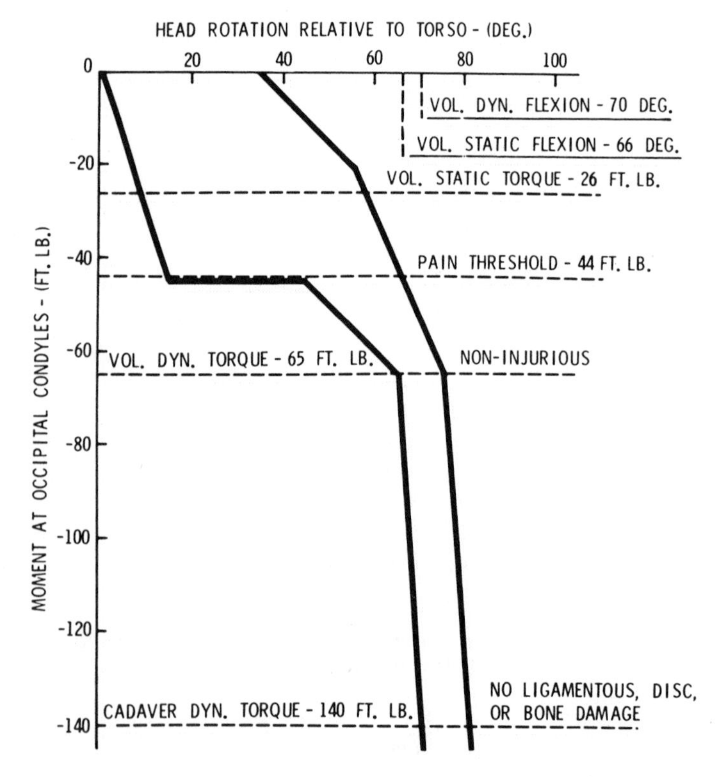

Fig. 26 - Head-neck response envelope for flexion and various tolerance levels

tension was 17.5 ft lb as indicated in Table 1. The product of these two factors gives a value of approximately 22.5 ft lb.

Similar to the flexion response curve, a constant torque level is assumed until the neck structure begins to "bottom out." For the volunteer LMP, this occurred at approximately 60 deg, as noted previously. The next portion of the response envelope was taken parallel to the maximum response curve for Cadaver 1089 and extended to the 35 ft lb level at approximately 80 deg. The torque level of 35 ft lb is twice the static strength level and is the noninjurious torque tolerance level recommended by Mertz and Patrick (1). The envelope is extended from this point by a line which is parallel to the maximum slope of the unloading portion of the response curve of Cadaver 1089. In this portion of the response envelope, the neck behaves as a linear spring and very little energy is dissipated. Consequently, the loading and unloading curves would be coincident.

The unloading portion of this envelope passes through a point 10 deg greater than the loading curve at the voluntary tolerance level of 35 ft lb and is taken parallel to the maximum unloading slope of the response curve for Cadaver 1089. Again, the 10 deg is arbitrary as it was in the flexion envelope.

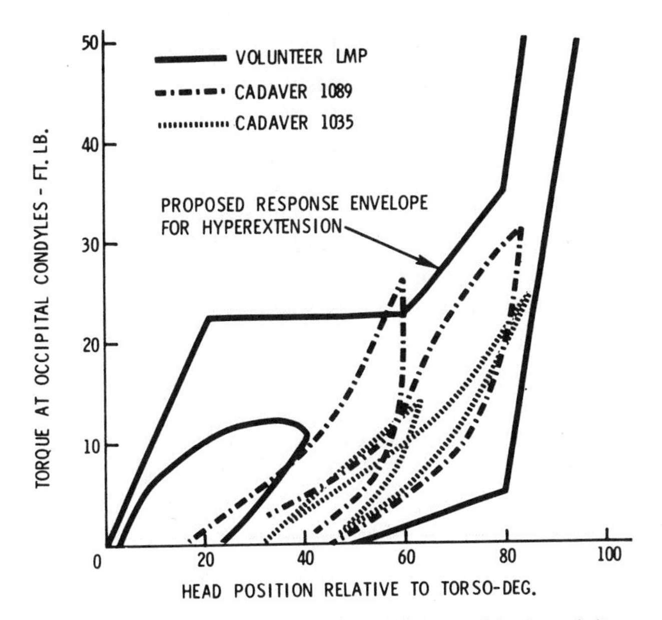

Fig. 27 - Torque about the occipital condyles as a function of the position of the head relative to the torso for the volunteer LMP and Cadavers 1035 and 1089

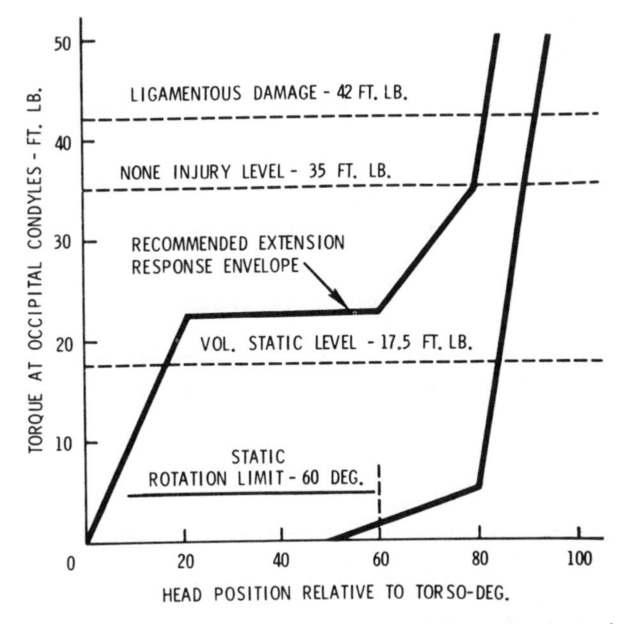

Fig. 28 - Head-neck response envelope for extension and various tolerance levels

This line is extended to the 5 ft lb and 80 deg point. At this point, the envelope is arbitrarily extended by a straight line to the abscissa at 50 deg. The envelope encloses the majority of the cadaver and volunteer response curves, except for the initial response curve of Cadaver 1089 in which the neck was quite stiff.

The logic employed in developing this response envelope is similar to that used to develop the response envelope for flexion. Until additional data is provided to indicate otherwise, this response envelope is recommended to describe the torque-angle response of the neck of a 50th percentile adult male in hyperextension.

The majority of the energy is dissipated by the neck muscles and occurs prior to bottoming out of the neck structure. There is no reason to believe that the ratio of the area between the loading curve and the abscissa will be any different for extension than flexion. Thus, the relationship developed for flexion (this ratio should not be less than 0.50) will be used for extension.

The response envelope shown in Fig. 27 places the following requirements on the response characteristics of any mechanical simulations of the human neck:

1. The evaluation of the neck structure must be made with an appropriate head structure.
2. The torque must be measured about the occipital condyles.
3. The torque-angle relationships must be determined dynamically and both the loading and unloading portions of the neck's response curve must lie within the response envelope of Fig. 27.
4. To insure adequate damping, the ratio of the area between the loading and unloading curves to the area between the loading curve and the abscissa should not be less than 0.5 for that portion of the response curve below the 22.5 ft lb moment level.

Again, these conditions apply to the extension response of the 50th percentile adult male and are necessary, but not sufficient conditions, to describe a unique head-neck response to extension. As mentioned in the discussion of the flexion response, the geometrical and inertial properties of the head and neck must be prescribed as well as the linear displacement range of the CG of the head measured relative to a prescribed system on the torso.

TOLERANCE LEVELS FOR EXTENSION - Some of the tolerance levels have been described in developing the response envelope for hyperextension. A static torque level of 17.5 ft lb was presented and a noninjurious dynamic torque level of 35 ft lb was recommended by Mertz and Patrick (1).

An injury tolerance level can also be obtained from the data presented by Mertz and Patrick. Cadaver 1035 suffered minor ligamentous damage between the third and fourth cervical vertebrae. The maximum neck torque experienced by Cadaver 1035 when this damage occurred was 24.6 ft lb. Since this cadaver was much smaller than the volunteer LMP, the torque value must be scaled in order to be applied to the volunteer LMP. Using the scaling technique presented by Mertz and Patrick (1), this torque would have been 42 ft lb for the volunteer LMP. Consequently, until data become available to indicate otherwise, an injury threshold for the torque developed at the occipital condyles of 45 ft lb is recommended for ligamentous damage of a 50th percentile adult male subjected to hyperextension.

The various injury thresholds for hyperextension are given for the response envelope shown in Fig. 28. It should be noted as it was for hyperflexion, that the measurement of the relative angle of the head to the torso is not an adequate physical measurement for describing hyperextension tolerance levels. The torque is a better indicator because, in the injury realm, a small error in angle measurement produces a large change in torque level.

CONCLUSIONS

The angle between the head and the torso is not a good physical measurement of trauma for neck extension or flexion because when the neck is hyperflexed or hyperextended, a small change in angle produces a large change in resisting torque. The best indicator for the degree of severity of neck flexion is the equivalent moment which consists of the moments of the neck and chin contact forces taken with respect to the occipital condyles. This is the same indicator used by Mertz and Patrick (1) for hyperextension. The resultant A-P and S-I forces acting on the head were well below tolerable levels and are not considered critical parameters.

Various tolerance levels proposed for neck flexion of a 50th percentile adult male are:

1. Equivalent moment about the occipital condyles of 44 ft lb for the initiation of pain.
2. Maximum voluntary equivalent moment about the occipital condyles of 65 ft lb based on volunteer tests.
3. Equivalent moment of 140 ft lb and A-P force level of 450 lb without producing ligamentous or bone damage based on cadaver responses with the chin in contact with chest. There is no guarantee that muscle injury will not occur at a lower level.

Various tolerance levels proposed for neck extension of a 50th percentile adult male are:

1. Noninjurious torque about the occipital condyles of 35 ft lb. Ligamentous damage at 42 ft lb.
2. Noninjurious shear force of 190 lb based on voluntary static strength tests. Noninjurious static strengths of the neck in tension and compression were 255 and 250 lb. These are considered lower bounds for the corresponding dynamic strengths.

There is no unique neck response curve for an individual because of the degrees of muscle tone that a person is capable of generating. The most repeatable neck response occurs when a person tenses his neck muscles. This condition produces the least amount of neck flexion for a given severity of exposure. Consequently, the muscle tensed response curves should be used for specifying performance of mechanical simulations of the human neck. This would optimize the protective systems for the head, neck, and torso.

Two necessary conditions required for the response of a mechanical simulation of the human neck of a 50th percentile adult male are:

1. The relationships between the equivalent moment and angular displacement of the head relative to the torso for loading and unloading must lie within the response envelopes shown in Figs. 26 and 28 for hyperflexion and extension. To evaluate the mechanical neck, it must be mounted between an appropriate dummy chest and head, and the testing must be done in a dynamic environment.
2. To insure adequate damping, the ratio of the area between the dynamic loading and unloading curves to the area between the loading curve and the abscissa must not be less than 0.5 for that portion of the particular response curve lying below constant plateaus (45 and 22.5 ft lb) of the flexion and extension response envelopes, respectively.

REFERENCES

1. H. J. Mertz and L. M. Patrick, "Investigation of the Kinematics and Kinetics of Whiplash." Proceedings of Eleventh Stapp Car Crash Conference, P-20, paper 670919. New York: Society of Automotive Engineers, Inc., 1967.

2. H. J. Mertz, "The Kinematics and Kinetics of Whiplash." Ph.D. Dissertation, Wayne State University, 1967.

3. C. L. Ewing, D. J. Thomas, G. W. Beeler, and L. M. Patrick, "Dynamic Response of the Head and Neck of the Living Human to $-G_x$ Impact Acceleration." Proceedings of Twelfth Stapp Car Crash Conference, P-26, paper 680792. New York: Society of Automotive Engineers, Inc., 1968.

4. C. L. Ewing, D. J. Thomas, L. M. Patrick, G. W. Beeler, and M. J. Smith, "Living Human Dynamic Response to $-G_x$ Impact Acceleration II—Accelerations Measured on the Head and Neck." Proceedings of Thirteenth Stapp Car Crash Conference, P-28, paper 690817. New York: Society of Automotive Engineers, Inc., 1969.

5. C. Tarriere and C. Sapin, "Biokinetic Study of the Head to Thorax Linkage." Proceedings of Thirteenth Stapp Car Crash Conference, P-28, paper 690815. New York: Society of Automotive Engineers, Inc., 1969.

6. C. A. Buck, F. B. Dameron, M. J. Dow, and H. V. Skowlund, "Study of Normal Range of Motion in the Neck Utilizing a Bubble Goniometer." Archives of Physical Medicine and Rehabilitation, Vol. 40 (September 1959).

7. A. D. Granville and G. Kreezer, "The Maximum Amplitude and Velocity of Joint Movements in Normal Male Human Adults." Human Biology, Vol. 9 (1937).

8. D. Ferlic, "The Range of Motion of the Normal Cervical Spine." Hopkins Hospital Bulletin 110, 1962.

9. L. E. Morehouse, "The Strength of a Man." Human Factors, Vol. 1 (April 1959).

10. D. F. Carroll, J. A. Collins, J. L. Haley, and J. W. Turnbow, "Crashworthiness Study for Passenger Seat Design—Analysis and Testing of Aircraft Seats." AVSER Report No. 67-4, May 1967.

730967

Response of Human Larynx to Blunt Loading

J. W. Melvin, R. G. Snyder, L. W. Travis and N. R. Olson
The University of Michigan

Abstract

Direct impact to the larynx is usually prevented in accidents by the protective nature of the chin. In some situations, the occupant motions leave the larynx unprotected and susceptible to impact by the steering wheel rim or instrument panel. As one of the unpaired vital organs of the body, there is no easy way to provide an alternative for its functions when the larynx is lost or damaged. Information available on the tolerance of the unembalmed human larynx to force is quite limited.

This paper describes a multidisciplinary study to determine the response of unembalmed human larynges to blunt mechanical loading and to interpret the response with respect to clinical data. Fresh intact larynges were obtained at autopsy and tested at either static or dynamic loading conditions utilizing special test fixtures in materials-testing machines. Load and deformation data were obtained up to levels sufficient to produce significant fractures in both the thyroid and cricoid cartilages. Additional information was obtained in the form of permanent dimensional changes through direct measurements and location of fracture sites by use of xeroradiography. Final evaluation of the damage was performed following dissection of the laryngeal structure. The results of the tests are analyzed and interpreted in relation to establishing tolerance criteria for laryngeal loading.

ALMOST ALL PATIENTS seen and treated for blunt trauma of the larynx are involved in motor vehicle collisions. A lesser number are seen from the consequences of other blunt trauma such as fist fights and sports accidents. As in most clinical situations associated with motor vehicle trauma, the patients are typically in young adulthood. Loading can be applied to the larynx and the surrounding neck structure of a vehicle occupant through contact with the steering wheel rim in the case of the driver, and through contact with the instrument panel (dashboard) in the case of the passenger. In the latter situation, the head of the occupant must be forced into extension by the dynamics of the collision in order to leave the front of the neck unprotected. One typical source of such motion occurs when a passenger strikes his head against the windshield with resulting rearward rotation of the head relative to the torso, as depicted in Fig. 1. The subsequent continued motion of the upper torso down onto the dashboard then produces impact loads on the exposed neck structure (1)*. This type of injury to the larynx has been called the "padded dash syndrome." Olsen (2) documents 12 such injuries verified by the recollection of the victim or other occupants of the car and by the use of accident investigations. The injuries to the larynx and associated structures resulted in chronic vocal or airway disability in 10 of the cases. Seven of the ten required tracheotomies, which indicates the life-threatening nature of these injuries. Although other injuries such as fractures of the extremities, cervical spine fractures, facial bone fractures, concussions, and various abrasions, contusions, and lacerations of the scalp, chin, and neck were sustained by the victims, personal interview and review of the records indicated that none of these associated injuries left chronic sequelae as significant as those due to the laryngeal injuries. Thus, although external injury to the larynx is not common, the long-term complications arising from laryngeal trauma require that consideration be given to the prevention of this injury wherever possible.

Consideration of the problem of laryngeal impact injury in the design of instrument panel structures requires a knowledge of the tolerance of the larynx to force. Past studies have produced limited data on the subject. Gadd, Culver, and Nahum (3) briefly reported the results of impacts to the necks of unembalmed human cadavers and impacts to dissected pig larynges (which are similar in structure to human larynges). Their findings indicate a force level of 90-100 lb (400-450 N) for marginal fracture of either the thyroid or cricoid cartilages in the human, while for the pig larynges the load range was found to be 60-65 lb (266-289 N). Miles, Olsen, and Rodriguez (4) reported on the production and treatment of laryngeal fractures by using a high-velocity animal stunner to deliver a blunt impact to the laryngeal region of anesthesized dogs.

Anatomical Considerations

The larynx is a tubular organ, the framework of which consists of cartilages and elastic membranes. It comprises the central portion of the airway, being immediately below the pharynx and above the trachea. The larynx is suspended

*Numbers in parentheses designate References at end of paper.

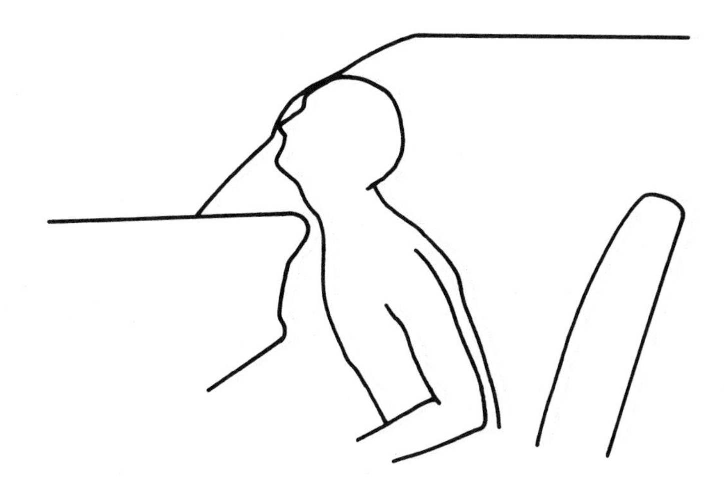

Fig. 1—Schematic representation of laryngeal impact in automobile frontal collision

by the thyrohyoid ligaments from above and is quite mobile laterally. The basic structure of the larynx is shown in a simplified manner in Fig. 2. The stiff cartilaginous structures of the larynx that serve to protect the airway are the tentlike thyroid cartilage and the ringlike cricoid cartilage. The thyroid cartilage is the largest element in the laryngeal skeleton and is composed of two broad laminae that meet and are joined anteriorly in the midline, forming an angle of 90 deg with each other in the male and 120 deg in the female. It is this forward joint of the laminae that produces the Adam's apple or laryngeal prominence.

The cricoid cartilage is the only structure to encircle the airway completely. As shown in Fig. 2A, the anterior portion of the cricoid is in the form of a narrow arch that broadens on the sides and forms a broad lamina posteriorly. The thyroid and cricoid cartilages are connected by various ligaments (Fig. 2) with the thyroid articulated on each side of the cricoid ring as shown. The arytenoid cartilages are a pair of structures interior to the larynx that articulate with the cricoid cartilage and serve as points of attachment for the vocal ligaments whose relations and state of tension are altered by the changes in position which these cartilages are constantly undergoing.

The larynx serves many important functions, including phonation, respiration, protection from aspiration, coughing, fixation of the thorax, emotional expression, alteration of blood flow, deglutition, and expectoration (5). As one of the unpaired vital organs of the body, there is no easy way to provide for these functions when the larynx is lost or damaged. Usually, a bypass procedure (tracheotomy) is needed or a different organ must be substituted (esophageal speech).

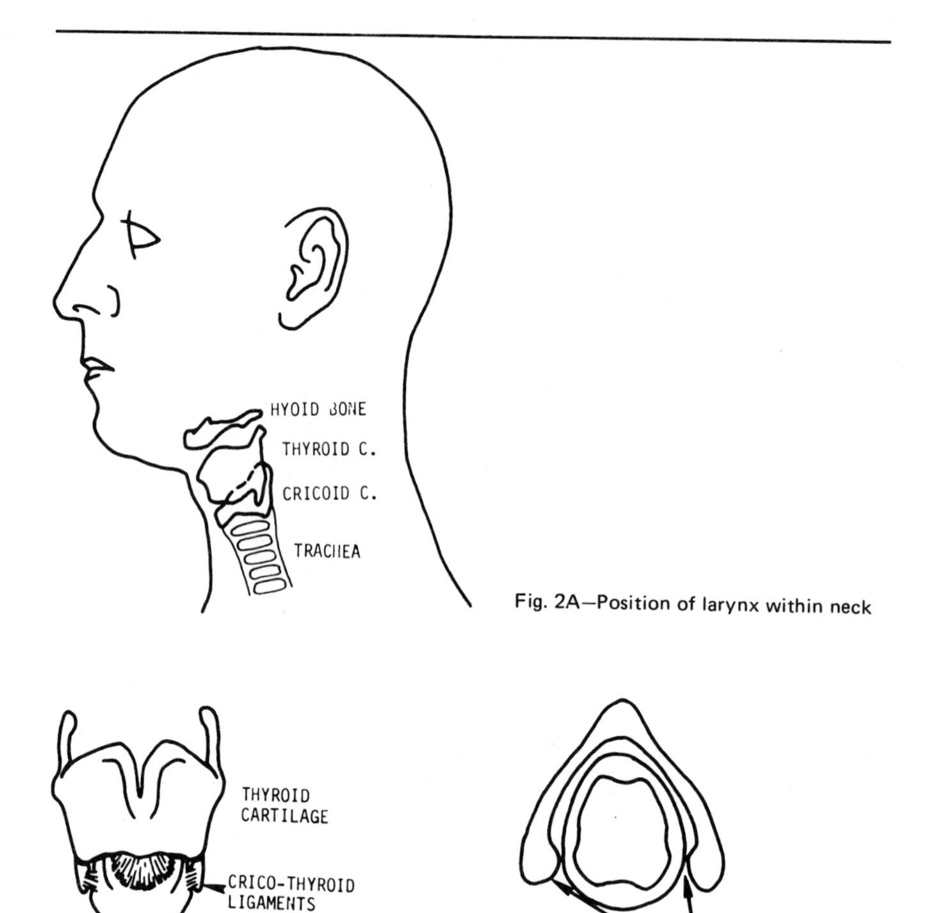

Fig. 2A—Position of larynx within neck

Fig. 2B—Frontal view of two main cartilage structures

Fig. 2C—View of larynx from below

Laryngeal Injury Mechanisms

Under normal conditions, a head-down position allows the larynx to be protected by the mandible, and the cricoid ring sinks low behind the sternal notch. In automobile collisions, as discussed in the introduction of this paper, normal compensatory reactions cannot be called upon to provide the necessary protection. If blunt loading is applied to the neck, the larynx may be crushed by impinging upon the vertebrae of the cervical spine. Such motion causes a spread-

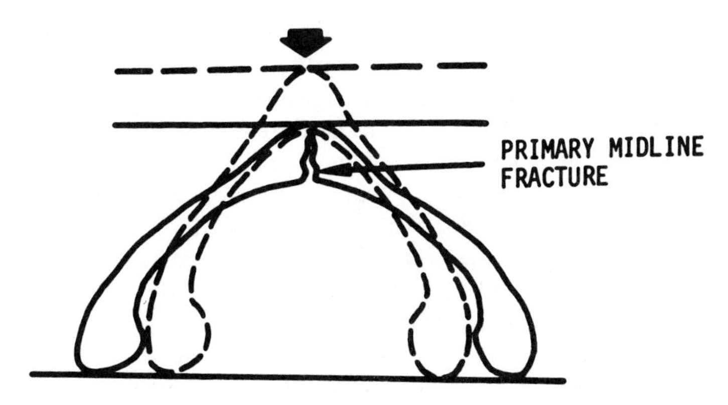

Fig. 3A—Thyroid cartilage fracture mode

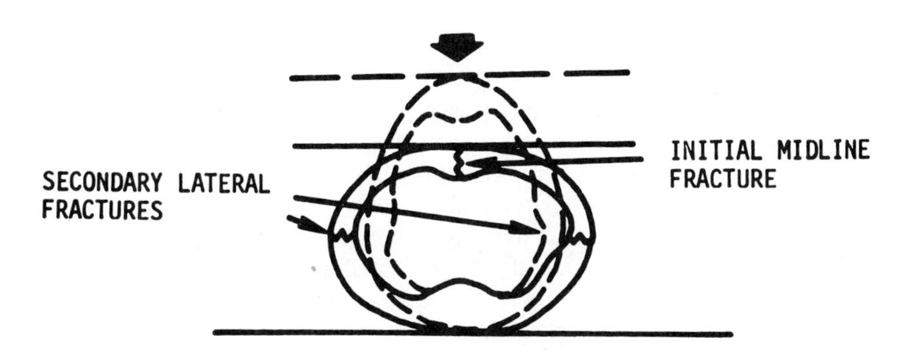

Fig. 3B—Cricoid cartilage fracture mode

ing of the thyroid lamina with subsequent fracture of the cartilage along the midline. With either injury, tears may occur in the area of the glottis and hematomas may form that threaten the airway. Injuries to the intrinsic muscles of the larynx may occur. The mucosa may become edematous and necrotic with raw glottic or ventricular edges. Arytenoid dislocation may occur anteriorly leaving the upper portion of the cricoid cartilage exposed.

Subglottic injuries may occur with blunt anterior trauma to the cricoid cartilage with possible cartilage fracture, hematoma, and tearing of mucosal surfaces—followed by fibrosis and perichondoitis. Tracheal avulsion, one of the most serious immediate injuries, may also occur. Injury to the recurrent laryngeal nerves, which pass near the cricothyroid articulations, may occur with resulting acute vocal cord abductor paralysis. While this paralysis may be temporary in most less serious injuries, it is quite often permanent in the tracheal avulsion injuries and other serious fractures. The unilateral vocal cord paralysis may be the only presenting symptom in a fracture near the cricothyroid articulation.

In the process of aging, the laryngeal cartilages ossify and thus become much more brittle. In the young, or in low-energy impacts, fracture may not occur; however, the laryngeal aperture may be decreased. In adults, fracture of these ossified cartilages occurs more frequently. The most common form of cartilage fracture is the midline fracture of the thyroid cartilage, as shown in Fig. 3A. Fractures of the thyroid lamina also occur at the inferior horns near the cricothyroid articulation. Midline fracture of the thyroid cartilage significantly reduces the load-carrying ability of the tentlike structure with resulting collapse of the airway with further deformation. The ringlike structure of the cricoid cartilage allows it to continue to carry load after a single anterior fracture, and its collapse requires additional fracture or excessive deformation at the sides of the cartilage, as shown in Fig. 3B.

Experimental Program

In order to obtain data on the forces necessary to produce significant injury to the human larynx, a cooperative program was carried out with the staff of the Highway Safety Research Institute (HSRI) Biomaterials Laboratory and members of the Department of Otorhinolaryngology, The University of Michigan Medical School. Fresh intact human larynges were obtained at autopsy from University, Veterans, and St. Joseph Mercy hospitals in Ann Arbor. A suitable thickness of soft tissue was left on the larynges, and they were frozen until needed for testing. Prior to testing, the specimens were thawed, the excess soft tissue was trimmed away, and the physical dimensions of the laryngeal structures were measured.

The test program was carried out in two phases. The first phase was to obtain data under quasi-static loading conditions in order to investigate the fracture levels in the thyroid and cricoid cartilages separately. The second phase consisted of high-velocity impact tests of a second group of larynges. Following each test, the specimen was examined by palpation for fractures. Selected specimens were subjected to xeroradiographic examination prior to final dissection of the laryngeal structures for a detailed analysis of the damages.

The static tests were performed in an Instron universal testing machine at crosshead rates of 1.2 and 12 in/min (0.05 and 0.5 cm/s). In all tests, a flat 1-1/2 in (3.81 cm) diameter loading platen was attached to the load cell and used to contact the anterior surface of the larynx. In initial tests, the posterior surface of the larynx was placed on a flat rectangular block whose width was sufficient to stabilize the specimen but was not wide enough to contact the posterior edges of the thyroid laminae. Different widths ranging from 1-2 in (2.5-5 cm) were used, depending upon the size of the specimen. Later tests

Fig. 4—Configuration for dynamic laryngeal compression tests

utilized a flat plate with the superior horns of the thyroid lamina trimmed off to stabilize the specimen. No significant differences were found in the loads necessary to achieve midline fractures of the thyroid cartilages by the two techniques, although the initial deflections were lower when the flat plate was used. One main advantage of the static test was that it was possible to stop the test as soon as an initial fracture was produced, examine the specimen for the source of the failure, and then continue the test to produce additional fractures. In many cases, an audible breaking sound was produced by the cartilage fracture; however, fracture was most reliably indicated by a sudden change in slope or a drop in the load-deflection curve as it was being recorded. The diameter of the loading platen was sufficient to contact both the laryngeal prominence and the cricoid ring area without repositioning the specimen. Loads in both the static and dynamic tests were transduced with a Kistler piezoelectric load cell and charge amplifier. Specimen deflection was determined by crosshead motion, and the resulting load-deflection curve was displayed and recorded on a Tektronix storage oscilloscope.

The dynamic loading tests were performed on the Plastechon high-speed universal testing machine at impact velocities of 7000 and 12000 in/min (300 and 500 cm/s). The stroke of the machine was limited mechanically to a predetermined platen separation to prevent complete destruction of the specimen. The test configuration is shown in Fig. 4 with the stroke-limiting collar on the right and the load cell and impacting arm on the left. The test procedure differed from the static tests in that the flat plate (magnesium in this case) was mounted to the load cell and the specimen placed on it as shown. The circular loading platen was mounted to the arm attached to the moving ram of the test machine. Ram displacement was transduced in these tests with a Physitech optical extensometer. The data-recording techniques were the same as in the static tests. The stroke was adjusted to produce approximately 60% compression of the specimen based upon its free height. The ram velocity was maintained constant during the stroke, with the exception of the arrest of the ram motion at the stroke limit. A total of 14 tests were performed dynamically with the majority of the tests at the 7000 in/min (300 cm/s) velocity.

Discussion of Test Results

A total of 24 unembalmed human larynges were tested, half of them under quasi-static conditions and the other half under dynamic impact conditions.

The results of the 12 static laryngeal compression tests are summarized in Table 1 in the form of the loads necessary to cause fracture of the thyroid cartilage and the cricoid cartilage, respectively. One characteristic of the laryngeal load-carrying structures that became evident in the static tests was the major effect of the depth of the thyroid cartilage with respect to the cricoid ring diameter. In those specimens with a marked laryngeal prominence, the initial loading occurred primarily in the thyroid cartilage, resulting in a rapid rise in load

Table 1—Static Laryngeal Fracture Loads

	Test Speed		Thyroid Fracture Load		Cricoid Fracture Load	
Specimen No.	in/min	cm/s	lb	N	lb	N
H-2	1.2	0.05	20	89	16	71
H-4	1.2	0.05	27	120	21	93
H-5	1.2	0.05	30	133	51	226
H-6	1.2	0.05	35	156	33	147
H-7	1.2	0.05	29	129	37	164
H-9	1.2	0.05	29	129	33	142
H-10	1.2	0.05	—	—	38	169
H-11	12	0.5	37	164	55	244
H-12	12	0.5	40	178	47	208
H-13	12	0.5	20	89	48	213
H-14	12	0.5	48	213	68	302

Table 2—Dynamic Laryngeal Fracture Loads

Specimen No.	Test Speed		Thyroid Fracture Load		Cricoid Fracture Load	
	in/min	cm/s	lb	N	lb	N
H-16	7000	300	85	377	50	222
H-19	7000	300	26	115	65	289
H-21	7000	300	27	120	57	254
H-22	7000	300	14	62	35	156
H-23	7000	300	41	182	52	231
H-25	7000	300	39	173	56	249
H-26	12000	500	46	204	68	302
H-28	12000	500	45	200	60	267

until fracture occurred. After the initial fracture, the stiffness of the thyroid structure was reduced, and subsequent deformation of the specimen continued at reduced load until cricoid cartilage began to carry the load. Thus, in such specimens, the thyroid cartilage always fractured first, followed later in the deformation process by cricoid fracture. In specimens where the depth of the thyroid cartilage was nearly the same as the cricoid ring diameter, the load tended to be shared by both structures. Such a case would typically be a female larynx. As noted in the previous discussion of laryngeal anatomy, the reduced prominence of the thyroid structure in the female is due to a more shallow angle between the lamina. This flatter configuration would produce a less stiff structure, which would also tend to allow load sharing to occur with the cricoid ring. The net result is a combined structure of generally lower initial stiffness than the larger laynges but with a more uniform load-carrying ability than many of the large larynges due to the load sharing between structures—that is, monotonic load-deflection curve in contrast to the multiple peaks of the large larynges. Occasionally, the load sharing worked to produce significantly higher total loads at a given deformation than in many of the large larynges.

All of the dynamic compression tests consisted of a single load application up to the maximum controlled deformation limit; thus, the interpretation of dynamic fracture loads for the cartilage structures was based upon the sequential behavior noted in the static tests. The values of the failure loads obtained for the cartilage structures are listed in Table 2 for those larynges that exhibited characteristic progressive structural failures. Additional analysis was performed on the dynamic load-deflection data by normalizing the specimen deflections with the initial specimen height to produce compressive strain data. A similar operation to reduce the load values to average stress values by dividing by the platen area was not felt to be warranted due to the stiff, irregular surfaces of the

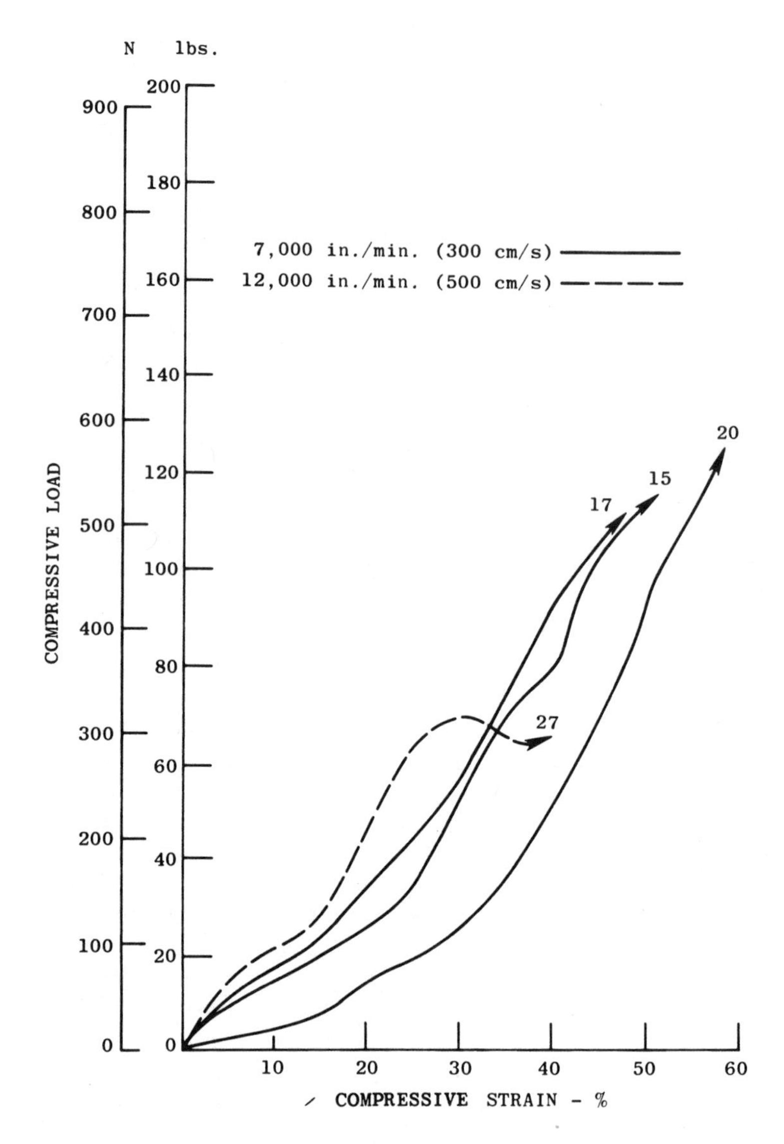

Fig. 5—Compressive load-strain behavior of small larynges

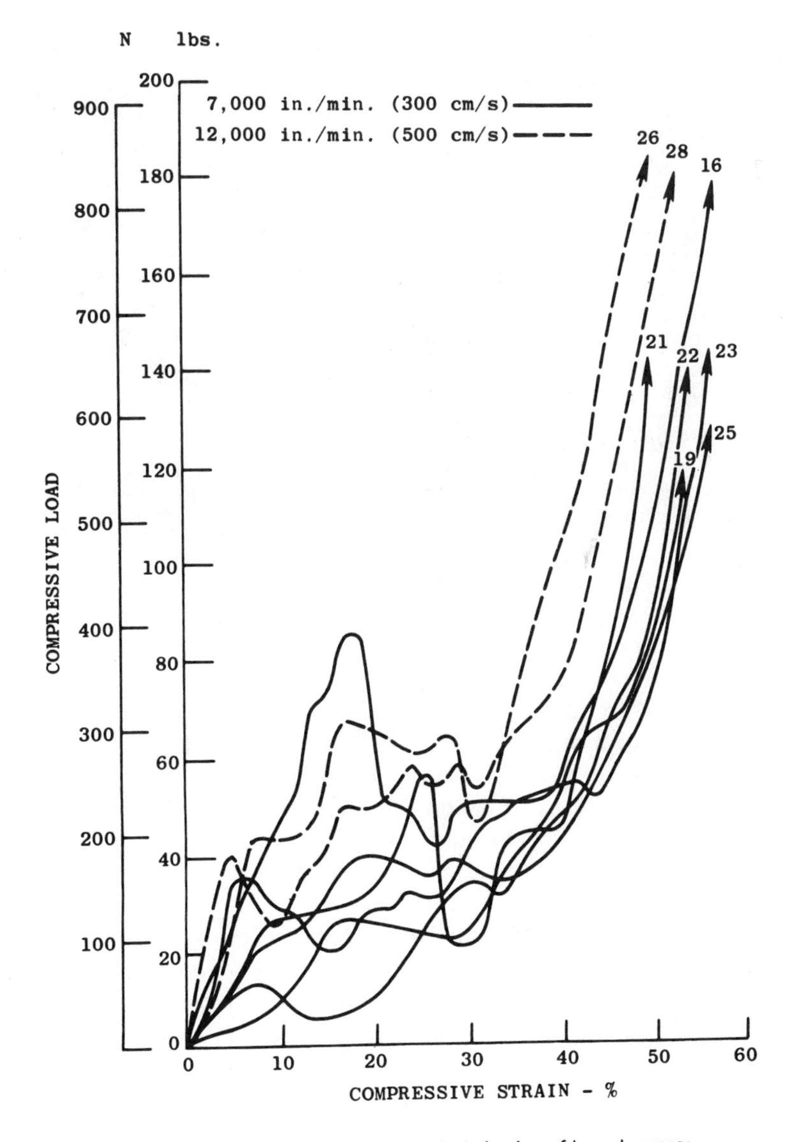

Fig. 6—Compressive load-strain behavior of large larynges

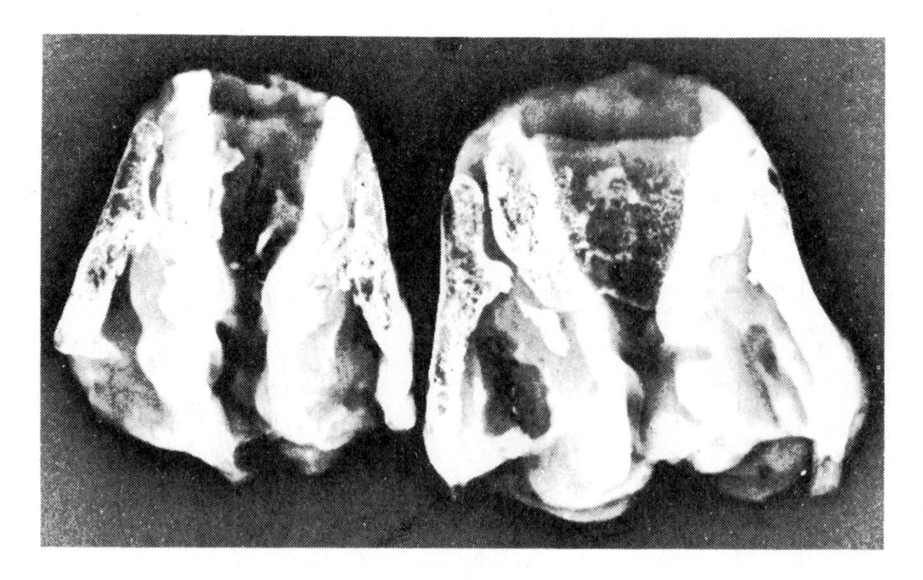

Fig. 7—Xeroradiographs of fractured larynges

cartilage structures. The resulting load-strain information is shown in Figs. 5 and 6. The curves in Fig. 5 represent the behavior of the smaller larynges, while the curves in Fig. 6 demonstrate the progressive failure behavior of the larger larynges, with the typical rapid rise in load followed by an oscillating load or load plateau region and ending in a rapidly rising curve, which is the onset of total collapse of the cricoid ring structure and subsequent bottoming out of the soft tissues of the airway.

Comparison of the data in Tables 1 and 2 indicates that dynamic fracture loads are generally higher than in the static tests—about 30% higher based upon mean values. In addition, the curves in Figs. 5 and 6 show that the loads at comparable strains are somewhat higher at 12000 in/min (500 cm/s) rates than at the 7000 in/min (300 cm/s) rates.

As discussed previously, fracture of the thyroid cartilage diminishes its load-carrying ability significantly due to its structural form, whereas a midline fracture of the cricoid ring may reduce its structural stiffness but does not destroy its capability to withstand load at continued deformation. Thus, in considering the relationship of structural fractures of the larynx to serious types of injuries of the airway, it is necessary to consider the point at which total structural collapse of the cartilaginous protective structures is eminent, for it is beyond that point that the soft tissues of the airway become severely damaged, as shown in Fig. 3B. The load at which such bottoming of tissues occurs is not as easily discerned as the abrupt load drops associated with prior fractures and in most cases manifests itself in the typical smooth, concave, upward load-strain relationship exhibited by soft tissues in compression. Examination of the load-strain curves in Fig. 4B indicates that the onset of this behavior occurs at strains ranging from 30-40%. Selecting the 50% strain level as a means of comparing the load levels in the individual larynges at a point beyond this onset strain level yields a mean value of 110 lb (490 N) with a range of 76–182 lb (337-810 N).

Examination of the larynges after testing revealed significant cricoid and thyroid fractures in all cases. In many cases, the central one-third of the cricoid ring was depressed. As a means of aiding the clinical determination of cartilage fractures, selected larynges were subjected to xeroradiographic examination prior to dissection. This technique allows greatly enhanced contrast to be obtained in the images of cartilagious tissues. Fig. 7 shows the xeroradiographs of two tested larynges with cricoid ring fractures.

Conclusions

Patients presenting to the otolaryngologist following acute trauma with pain in the larynx, hemoptysis, loss of laryngeal landmarks, subcutaneous emphysema, airway distress, or dysphonia are immediately suspected of having sustained a laryngeal fracture. Indirect laryngoscopy by mirror or direct operative laryngoscopic procedures are usually undertaken immediately following assessment of the patient's general status and evaluation of other injuries. Airway distress, or the presence of extensive mucosal lacerations, false passages, hematomas, severe edema, or severe crepitation of the cartilages, are indications for immediate tracheostomy, sometimes necessary as an emergency lifesaving procedure before laryngoscopy can be performed. Definitive surgical repair of the fragmented cartilages and mucosal lacerations can then proceed following stabilization of the patient's clinical status, often a period of days.

The majority of laryngeal injuries are classified as contusions. Standard x-ray techniques do not often reveal cartilaginous fracture lines, and in these less-serious injuries, tracheostomy is not usually necessary and surgical exploration is not performed. It is therefore quite probable that many less extensive cartilaginous fractures unaccompanied by airway distress or internal soft tissue disruption are undiagnosed.

It is felt that the injuries sustained by the specimens in these tests experimentally duplicate such a clinical entity. As the data in Tables 1 and 2 demonstrate, significant thyroid and cricoid cartilage fractures occur at mean force levels of 40–55 lb (178–244 N) without compromise of the airway or internal soft tissue disruption. Such an injury to the live larynx would probably be classified as a contusion and the patient treated with careful observation.

The data also emphasize the vulnerability of the cricoid cartilage, the only completely cartilaginous structure of the airway and its narrowest dimension. Significant fractures of this cartilage were also observed at the lower force levels—fractures that might well escape diagnosis by standard x-ray technique. In addition, continuity and integrity of this cartilaginous ring cannot be accurately assessed by palpation, in contrast to the easily palpable thyroid cartilage.

Finally, the data clearly show that imminent collapse of the vital laryngeal cartilages occurs at force levels below those reported previously; indeed, post-test dissection of these specimens suggests that such injuries may well be incompatible with acute survival. The low force levels at which laryngeal fractures can occur and the resulting serious nature of such injuries indicate the importance of the consideration of such factors in instrument panel design. The use of protruding hoods with narrow edges in regions of the panel where accidental occupant contact is likely can lead to laryngeal injuries, even though deformable padded materials are used.

Summary

The response of human larynges to static and dynamic compressive loading has been determined for 24 specimens. Significant fractures were obtained in all cases. The loads to cause fracture in both the thyroid cartilage and the cricoid cartilage were found to be higher under dynamic loading than under static loading. For the thyroid cartilage, the mean dynamic fracture load was 40.6 lb (180 N), and for the cricoid cartilage, the mean dynamic fracture load was 55.5 lb (248 N). At the 50% compressive strain level, all larynges had fractured and total collapse of the structure was eminent. The mean load at the 50% strain level was 110 lb (490 N) with a range of 76-182 lb (337-810 N).

References

1. R. M. Butler and R. H. Moses, "The Padded Dash Syndrome: Blunt Trauma to the Larynx and Trachea." The Laryngoscope, Vol. 78 (July 1968), pp. 1172-1182.

2. N. R. Olsen, "Dashboard Injuries of the Larynx." Proceedings of Fourteenth Annual Conference of American Association for Automotive Medicine, November 1970.

3. C. W. Gadd, C. C. Culver, and A. M. Nahum, "A Study of the Responses and Tolerances of the Neck." Paper 710856, Proceedings of Fifteenth Stapp Car Crash Conference, P-39. New York: Society of Automotive Engineers, Inc., 1971.

4. W. K. Miles, N. R. Olson, and A. Rodriguez, "Acute Treatment of Experimental Laryngeal Fractures." Annals of Otology, Rhinology and Laryngology, Vol. 80, No. 5 (October 1971), p. 710.

5. C. Jackson and C. L. Jackson, "Diseases of the Nose, Throat and Ear." Philadelphia: W. B. Saunders Co., Second Edition, 1959.

801300

Mechanisms of Cervical Spine Injury During Impact to the Protected Head

Voigt R. Hodgson and L. Murray Thomas
Wayne State University School of Medicine

ABSTRACT

Static and impact loading of the heads of embalmed cadavers wearing protective helmets have been conducted for the purpose of understanding the mechanics of fracture-dislocation injury to the cervical spine. Some of the cadavers were cut down on one side of the neck for high-speed photographic observation of the spine during impact. Others were instrumented with strain gages on the bodies and near the facets to assist in correlating spine movements and load configuration with strain distribution.

Results indicate that static loading can be a useful predictor of failure site under dynamic conditions. Those conditions which were found to be most influential on injury site and level of strain were: 1) The extent to which the head was gripped by the impact surface to allow or restrict motion at the atlanto-occipital junction; 2) Impact location; and, 3) Impact force alignment with the spine.

It was found that very little could be done with energy-absorbing material in the crown to reduce spine strain due to a crown impact. Also, the rear rim was not a 'guillotine' threat to fracture-dislocation from blows which cause hyperextension, and the higher cut rear rim recommended to reduce or eliminate this alleged hazard caused higher strain by virtue of allowing greater extension of the neck.

MOST OF THE RESEARCH on the mechanics of spinal injuries has been carried out on segments of the cervical spine. Notable in this respect has been the work of Roaf (1)* who used the basic spinal unit consisting of two intact vertebrae joined by an intervertebral disc, two posterior articulations and a number of ligaments. Roaf found the disc, joints and ligaments to be very resistant to compression, distraction, flexion and extension, but very vulnerable to rotation and horizontal shearing forces. In general he found that rotation forces produce dislocations, and compression forces produce fractures.

Bauze and Ardran (2) devised an experiment in which entire cervical spines, with basi-occiput attached, were subject to compressive loads with the lower part of the spine flexed and fixed, and the upper extended and free to move forward. They loaded the specimen with a combination of vertical compression, flexion, and horizontal shear forces. Bilateral dislocation of the facets was produced without fracture with only 1.42 kN. They found that the maximum load coincided with the rupture of the posterior ligaments (interspinous and capsular) and stripping of the anterior longitudinal ligament prior to dislocation. In contrast to this, Roaf (1) found that the compressive strength of end plates of the vertebral bodies was around 6.23 kN and that the intact disc was even stronger, failing around 7.12 kN. When Roaf loaded the spinal unit in slight flexion and applied rotational force to the posterior ligaments, joint capsules and posterior longitudinal ligaments tore in that order resulting in a typical dislocation. However, he was unable to succeed in producing pure hyperflexion (when disruption of posterior ligaments occur) injury of a normal intact spinal unit. Before the posterior ligaments ruptured, the vertebral body always became crushed. These experiments with segments of the cervical spine indicate that failure can be a complex process and the external load is not a good predictor of when failure will occur, what tissues will fail, and where the failure will occur.

The use of an entire human surrogate to study the response to axial compressive loads due to impacts on the crown were conducted by Mertz, et al. (3). A Hybrid III dummy was placed in an impact environment in which a football mechanical blocking and tackling

*Numbers in parentheses designate References at end of paper.

device had allegedly produced paralyzing neck injuries. Axial compressive loading was primarily produced by impacts from the resilient foam-padded steel cylinder weighing 245 N to the crown of the dummy during the impact intensities assumed to have produced the injuries. This group published two reference curves, one for football players with a maximum value of 6.67 Kn, and another with a peak value of 4.45 kN for the adult population. They cautioned that since injury can occur under a variety of loading conditions, being below the curve does not necessarily insure that neck injury will not occur when the dummy is placed in a particular loading environment.

Culver, et al. (4), studied direct impact to the crown of fresh cadavers in which the subjects were placed in a supine position and the cervical spine was aligned along the impactor axis. These investigators found that the predominantly spinous process fractures produced in their setup indicated a compressive arching which followed the normal lordotic curvature of the cervical spine and appeared to depend on the initial rotation of the head and axial alignment of the spine. They found that if the head rotated rearward or the head was placed above the axis of the spine the arching was increased. No dislocations, no anterior compressive fractures of the bodies of the vertebrae, nor any basal skull fractures were found. The data indicated that peak impact force of 5.7 kN is a level above which cervical spine fractures will begin to occur for an average cadaver under conditions of their experiment.

In addition to its inherent instability under the action of head impacts with a compressive component (crown impact), another primary reason for lack of understanding of cervical spine injury mechanisms due to crown impact has been the inability to visualize spine movements or quantify the effects of load variables. The present series was designed to enable visualization of parts of the spine by means of high-speed photography and measurement of vertebrae strain during impact to the helmeted head of embalmed cadavers.

METHODS

Shown in Figure 1 is the device used by Mertz, et al. (3), to impact the Hybrid III dummy. In the present series, it was used to propel any of three surfaces consisting of the soft foam-filled padded steel cylinder of Mertz, a padded knee from a Sierra 1050 dummy, or a load cell, padded or unpadded. When fitted with either the similar weight (245 N) padded cylinder or knee, the impact mass could be propelled by three combinations of tension springs to velocities of 3 m/s, 4.1 m/s, or 5.2 m/s, striking near the end of the allowable travel at which impact occurred. Using the load cell, the striking body weighed 445 N and moved at proportionately lower impact velocities.

The cadavers were strapped in a prone position to a 312 N aluminum pallet with provision for raising the chest and head to control the impact location. The pallet was placed on a roller-bearing conveyor such that the pallet was free to roll.

CADAVERS - Sixteen embalmed, male cadavers were used in these tests. Most of the cadavers were prepared by sectioning the left side of the neck to expose parts of either the spinous processes or the bodies and facet joints. Preliminary experiments on several of these cadavers in which the entire sides of the vertebrae were exposed, including the bodies, facets, and spinous processes, produced predominantly lower cervical interspinous ligament failures which are atypical of cervical spine injuries seen clinically. Four of the cadavers were fitted with strain gages, beginning with three on the anterior surfaces of the bodies of C3, C5, and C7 on the initial cadaver, to as many as twelve on both the anterior surfaces of the bodies of all except C1 and near the left facet joints of all the cervical vertebrae. Although it was necessary to remove part of the longitudinal ligament at each anterior body gage site, this did not appear to weaken the spine. The cadavers were fitted with a protective helmet of the resilient liner type for the purpose of distributing the impacts, especially with the rigid load cell, to prevent skull fracture, and minimize variables resulting from skin damage.

OTHER INSTRUMENTATION - In a few cases the helmet was cut away and a triaxial acclerometer was screwed to the cadaver skull for the purposes of measuring the resultant acceleration at the mounting location in the mid-sagittal plane. A LOCAM® and a HYCAM®, operating at 500 and 1000 fps, were used to record head and neck motion in selected cases. Velocity of the impact surfaces was measured by interception of two light beams within 25 mm of the impact site.

OTHER IMPACT SETUPS - Several tests used gravity propulsion of a cadaver strapped to a light pallet pivoted at the feet and allowed to free fall into impact of the extended head to a Sierra 1050 knee surface which was free to swing away. This produced an I-S component of force with resulting hyperextension of the neck. Facemasks were used on the helmet to investigate whether or not an impact on the mask could, by virtue of a rear rim 'guillotining' mechanism, produce fracture and/or dislocation of the cervical spine; and/or whether a high cut rear rim was beneficial from the standpoint of reducing this hazard (Fig. 2).

STATIC TESTS - For optimum control of tests and to understand the dynamics of the spine, static load deflection tests were conducted as shown in Figure 3. The cadavers were seated in a frame under a hydraulic press operated at a slow loading rate of 10 mm/s. Scissor jacks were used to position and brace the body. A loading fixture was clamped to

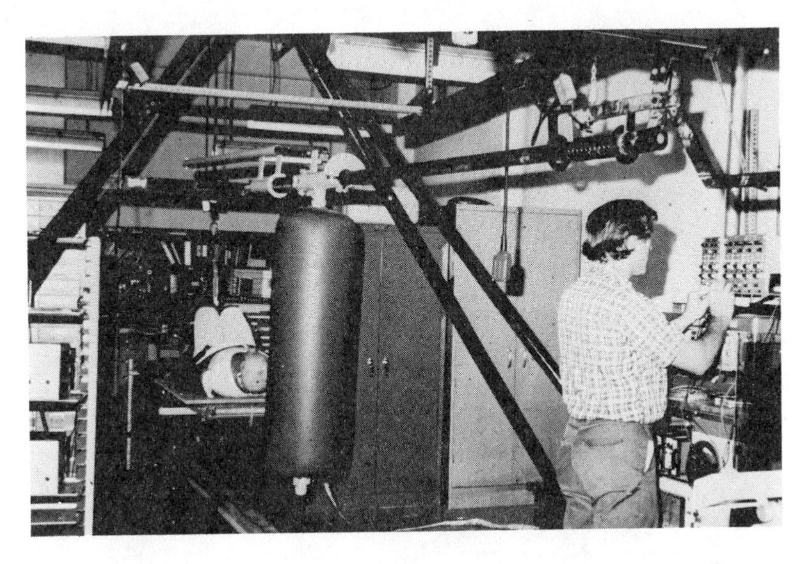

Fig. 1 - Springloaded propulsion system used to deliver head impacts

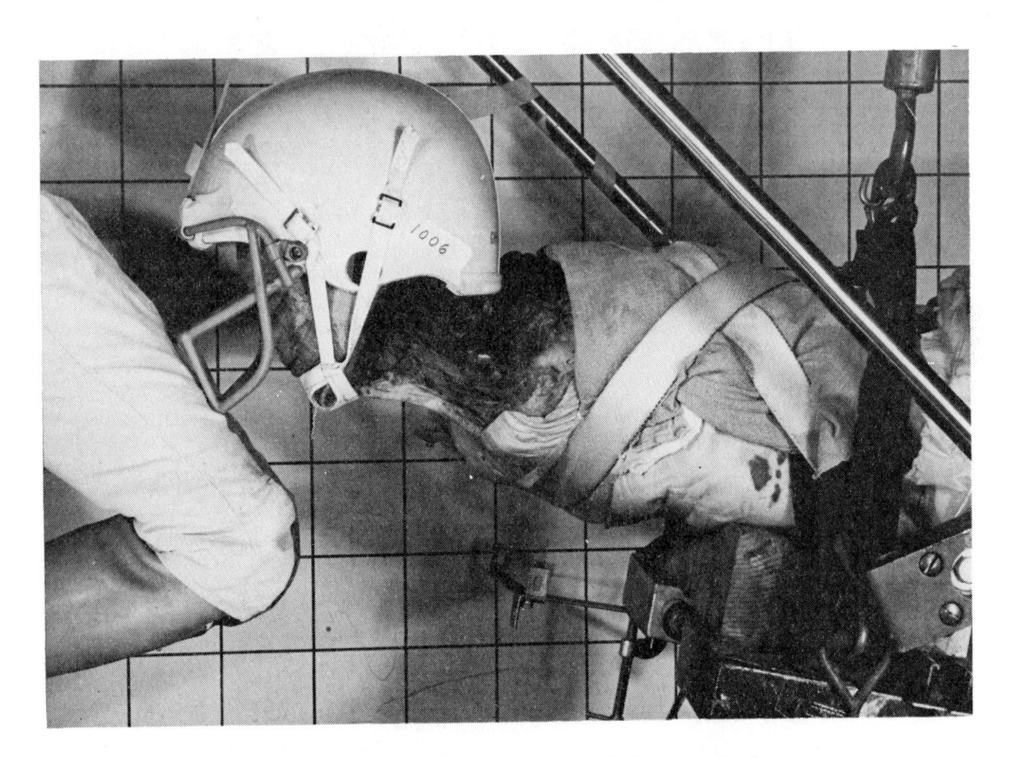

Fig. 2 - Position at impact of facemask drop test onto swing-away dummy knee to observe effect of rear rim on spine strain

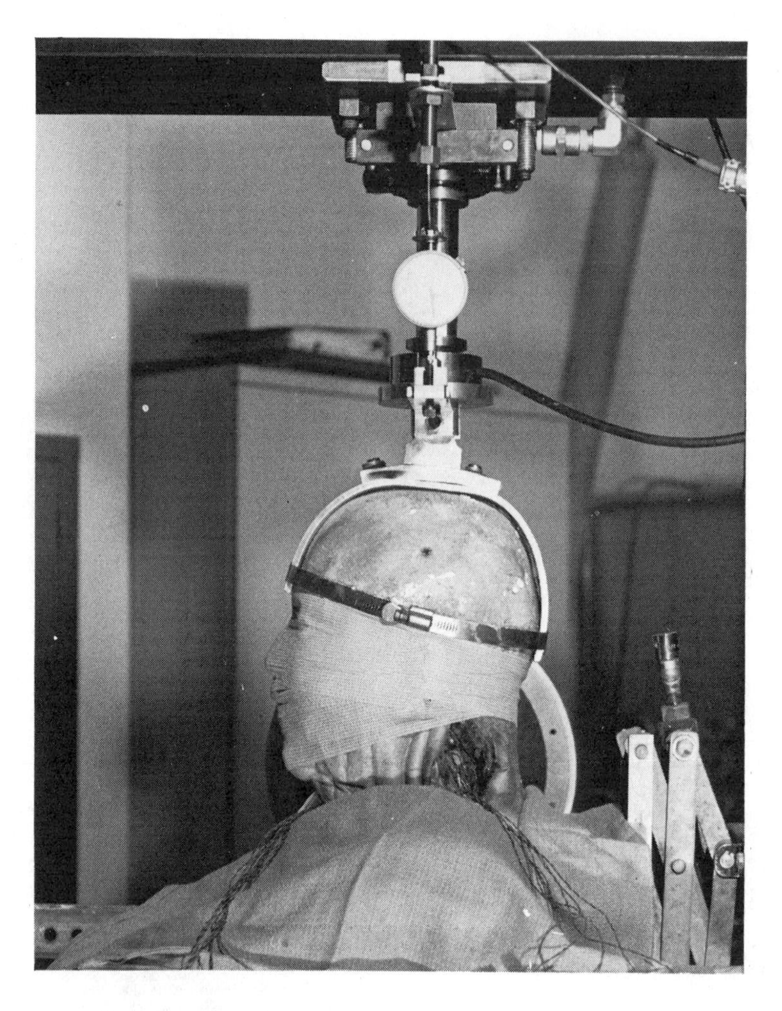

Fig. 3 - Static test setup for applying axial loads to head

the head for the purpose of loading at discrete points through a clevis which could be fixed to prevent head rotation, simulating the gripping action of a distributed dynamic impact, or a free clevis that simulated the action of a concentrated impact allowing nodding at the atlanto-occipital joint.

RESULTS

DISTRIBUTED CROWN IMPACT - Shown in Figure 4 is an excerpt from film taken at 500 fps of a distributed impact to the crown of the helmet. The gripping action of the bag on the helmet minimized rotation of the head on the neck at the atlanto-occipital joint even though the impact was anterior to the cervical axis. This resulted in a configuration of the spine similar to the so-called "ducking" shape described by Bauze and Ardran (2), typically sustained when diving into a shallow, sandy bottom pool. The modes of failure in this situation occur in the lower cervical with the possibility of crushing the anterior bodies of C5 or C6, a forward dislocation of C5 over C6, or a possible tear of interspinous ligaments in the region of C5-C7.

CONCENTRATED IMPACT - Concentrated impact against such as a padded knee of the 1050 Sierra dummy shown in Figure 5 anterior to the cervical axis, allows head nodding at the atlanto-occipital joint with resultant high cervical flexion. As the flexion increases the shear component increases with the likelihood of a bilateral dislocation (2) in the event of symmetrical loading, or what is more likely, a unilateral dislocation due to compression-flexion-rotation loading. Such a unilateral dislocation is shown in the sequence of frames in Figure 6 (left to right) which displays the dislocation of the C2 facet on C3, ending in locked facets. This dislocation appeared to be caused by compression-flexion loading with resultant p-a shear stress and was followed rather than preceded by rotation

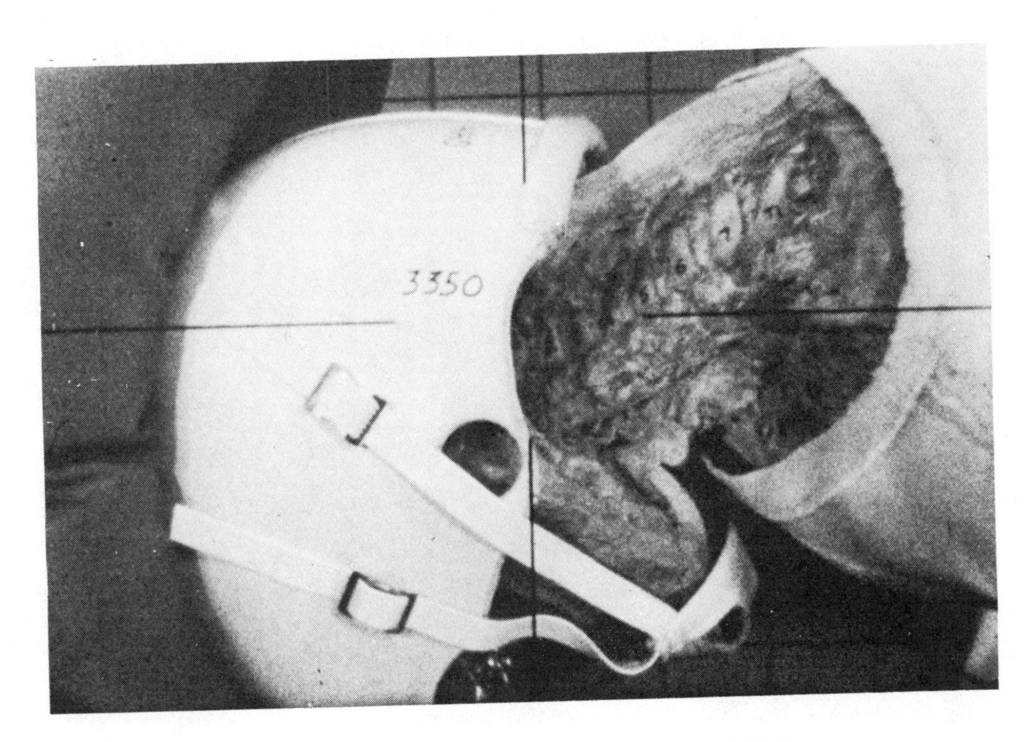

Fig. 4 - Excerpt from film taken at 500 fps showing distributed impact and resultant "ducking" shape of the cervical spine

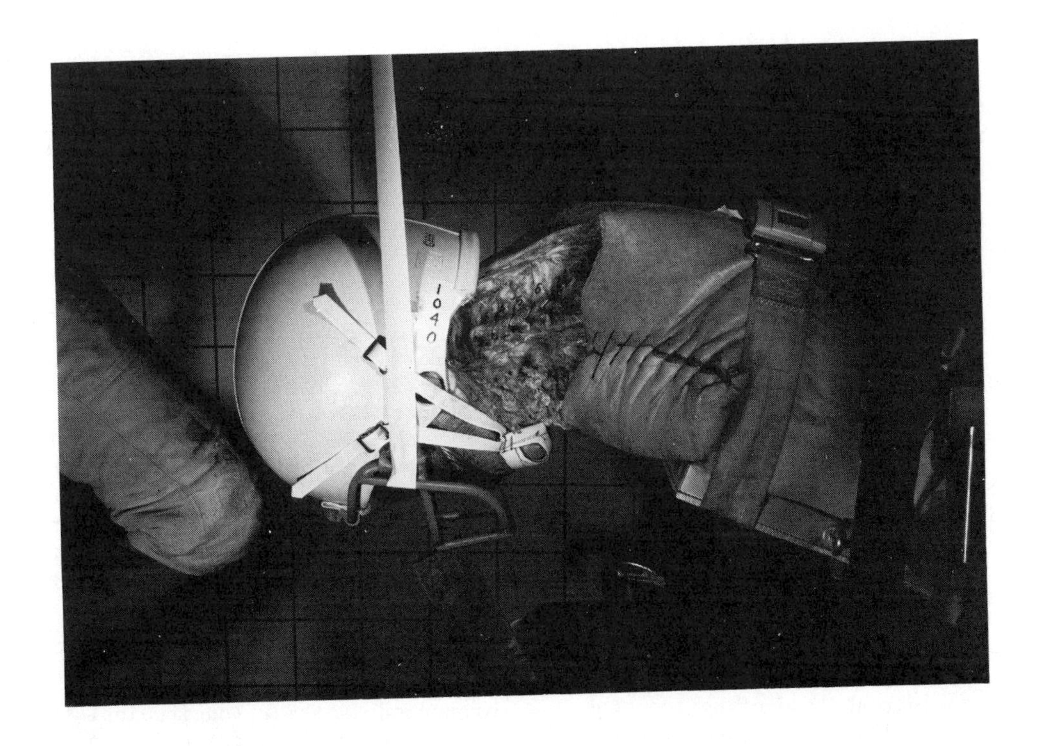

Fig. 5 - Concentrated load delivered by moving dummy knee to stationary cadaver anterior crown surface

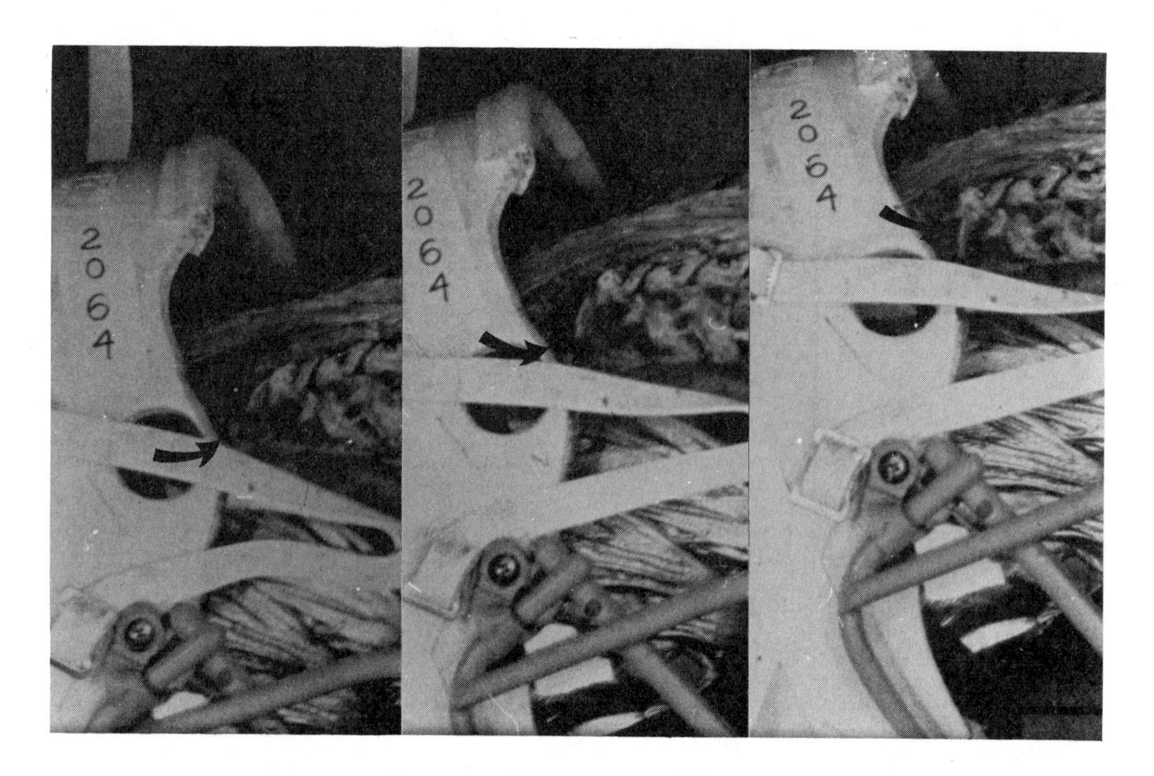

Fig. 6 - Excerpts from film taken at 500 fps showing the occurrence of a unilateral dislocation of C2 on C3 from a concentrated load anterior to the cervical axis

as predicted by Roaf's setup (1).

STATIC TESTS - After conducting numerous tests with distributed and concentrated loading it became apparent that many modes of failure could occur depending on the orientation of the neck prior to impact. As pointed out by Bauze and Ardran (2), if a person lands on his head with the whole neck in extension, fractured spinous processes with ruptured anterior longitudinal or anterior avulsion fracture may be expected. This was evidently the experience of the Culver, et al., study in which spinous process and intervertebral ligament failure occurred (4). Lateral flexion may produce rupture of the capsular ligaments around the middle of the neck on the convex side, as shown in the excerpt from high-speed film of C4-C5 separation in Figure 7 of the present series. This instability was afterward undetectable on x-ray as the joint returned to normal position. Experiences of these tests also indicated that if the striking surface is smooth and not centered on the smooth, hard helmet, either of which are free to move transverse to the initial line of motion, the result will be a harmless impact in which the head or striking body is deflected sideways, dissipating only a fracture of the kinetic energy of the moving body (see also reference 3).

For these reasons the static setup such as shown in Figure 3 was devised. Axial and anterior loading with the clevis pin locked and free were produced while measuring strain in the anterior surfaces of the body of C3, C5, C7 by means of foil strain gages. Also load deflection characteristics, primarily of the cervical spine, were obtained simultaneously as shown in Figure 8. For axial loading of the spine with the clevis fixed, the load is distributed predominantly as axial compression in the cervical spine bodies and facet joints as evident by the relatively straight lower plot. For anterior loading with free clevis, the curve labelled 2 was obtained and evidently results from stretching of the posterior longitudinal ligament and interspinous ligaments as the neck is flexed.

STRAIN DISTRIBUTED FROM STATIC LOADING - In Figure 9a is shown the shape of the cervical spine, diagramatically, under the condition of clevis free, axial loading and initial upper cervical spine extension. The strain distribution under these conditions corresponds to the prediction by Bauze and Ardran (2) from observations of their model, which suggested that if a person landed on his head with the neck in extension, fractured spinous processes and ruptured anterior longitudinal ligaments or anterior avulsion fractures could be expected. The spine shape and strain distribution also correspond to the cadaver damage experienced by Culver, et al. (4).

With similar loading of the initially straight spine, flexion occurred in the upper cervical spine which produced high compressive strain in C2. The model predicts the likelihood of a forward dislocation of C2 on C3

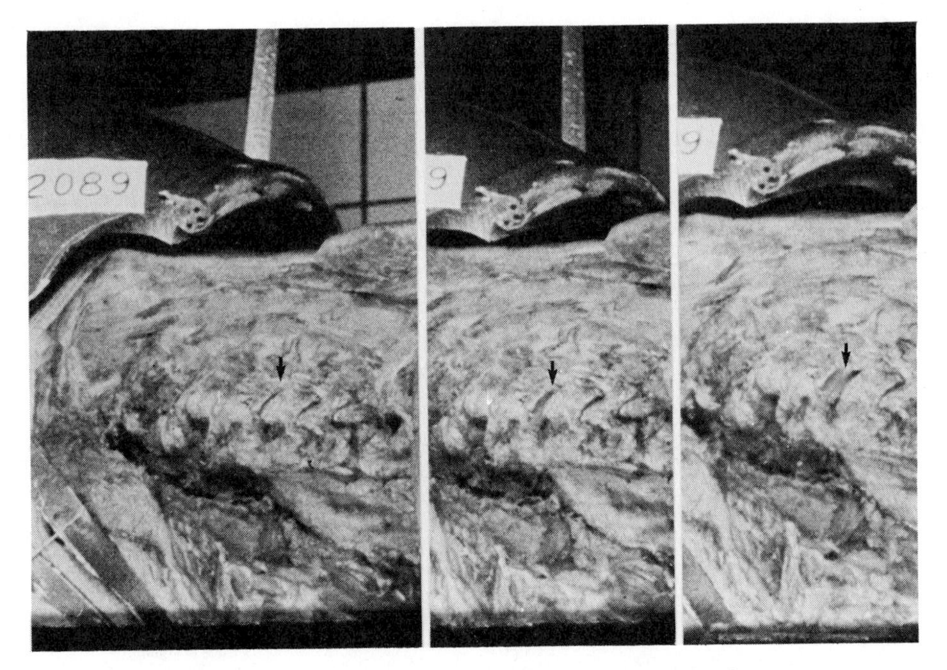

Fig. 7 - Excerpts from film taken at 500 fps showing the rupture and opening of articulating facets C3-C4 due to lateral bending caused by distributed crown loading

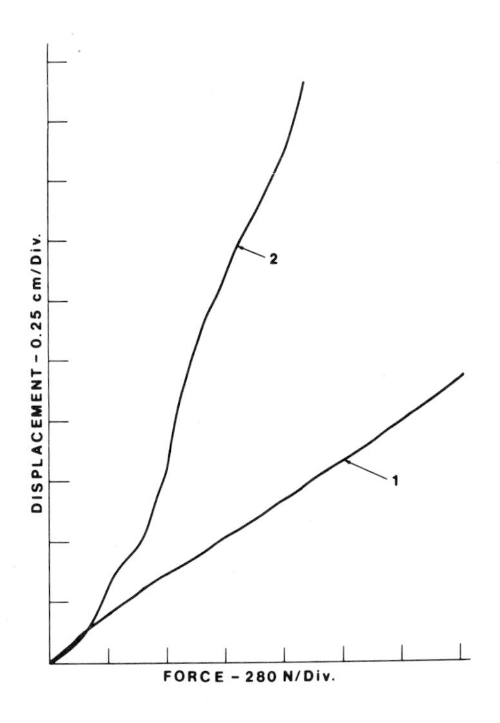

Fig. 8 - Static load deflection due to axial loading with no head rotation [(1) bone loading] and anterior to cervical axis loading with head free to rotate [(2) ligament loading]

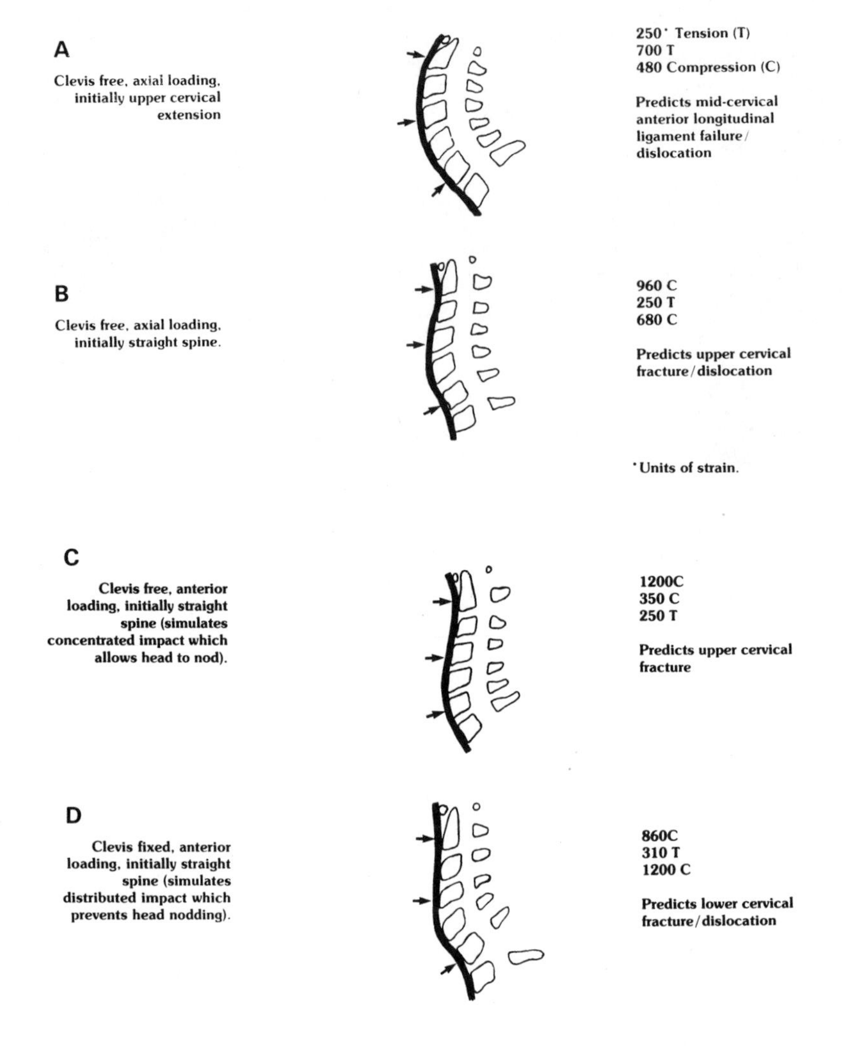

Fig. 9 A, B, C, and D - Diagrams of curvature and peak strain values for similar maximum load levels (~1.8 kN) under four different loading conditions

(Fig. 9b).

As the load is moved anteriorly with the clevis free, flexion in the upper cervical becomes more pronounced with the increased likelihood of either fracture or dislocation in the upper cervical spine at lower load levels (Fig. 9c).

With the clevis fixed, anterior loading, and with initially straight spine, the loading simulates an impact by a distributed surface anterior to the cervical spine which prevents rotation of atlanto-occipital joint (nodding). Moderate compressive loading is experienced in the upper cervical and high compression in C6 just below the inflection point at which the stress in anterior body surfaces changes from tension to compression. The model predicts lower cervical spine fracture and/or dislocation (Fig. 9d).

EFFECT OF HEAD-CERVICAL SPINE-BODY ALIGNMENT WITH LOAD LINE OF ACTION - In Figure 10 are shown three of the spine alignment configurations which were tested. In this series all impacts were centered on the crown and were produced by propelling the load cell into the stationary cadaver as shown in Figure 11. The tests began with the straight alignment of the lowest figure and progressively the cadaver chest and head were raised to the maximum height at which a crown impact could be produced, i.e., higher elevation would produce interference between the chin and the chest, preventing further flexion. The maximum anterior body strains of C3, C5, and C7 are shown on the figure, indicating a progressive severity as the upper arching and, therefore, initial cervical spine flexion becomes more pronounced. Because of the straight alignment, the lowest position involved a greater percentage of inertial resistance from body mass below C7, and the highest arched position the least amount, as indicated by the lower force recorded in the latter case. These comparative values of strain point out the difficulty in attempting to ascribe a tolerable force level for crown impact in three positions which would be described as axial loading. The top position which produced significantly higher strain at each of the three strain gage measuring locations, recorded the lowest force on the head.

EFFECT OF LOAD LINE OF ACTION RELATIVE TO THE CERVICAL AXIS - Shown in Figure 12 are the strains recorded for impacts at three locations on the head for a constant chest elevation. The least strain was recorded on the anterior bodies of the vertebrae with the load cell centered 76 mm above the front rim. Maximum levels of strain were recorded for the crown impact centered 191 mm above the front rim of the helmet. Contrasting with the data obtained in Figure 10, minimum strain on the anterior surfaces were recorded here for the minimum load on the head which was obtained for the blow centered near the front of the head. Maximum force and strain were

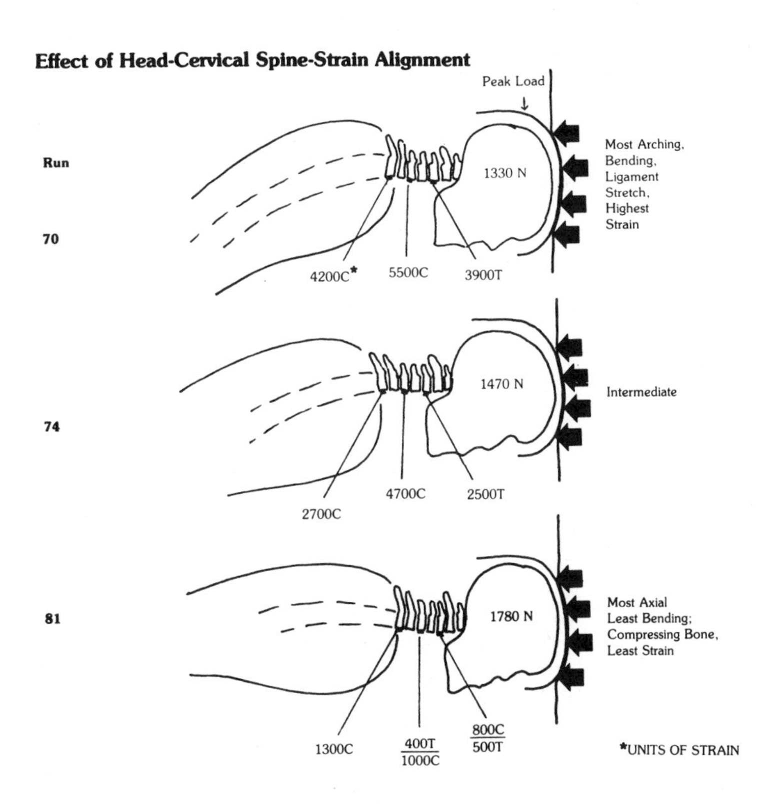

Fig. 10 - Effect of Head-Cervical Spine-Body alignment on anterior cervical body strain

recorded for the crown impact in which there was initially straightest alignment involving a greater amount of body mass below C7, whereas for the nearly frontal blow, the inertial reaction was provided primarily by the mass of the head and the neck. The intermediate position produced intermediate force and strain levels. It can be seen however, that there is not a proportionate increase of strain with load, but rather a precipitous increase in strain with small increase in load, again indicating the difficulties in establishing tolerable load levels for the neck due to head impacts.

EFFECT OF ENERGY ABSORBING MATERIAL IN THE CROWN - An attempt was made to determine if the strain in the cervical spine could be reduced by means of varying the energy absorbing material on the load cell which struck the crown of the cadaver oriented to produce cervical spine axis alignment as close as possible. Impacts were delivered against the unhelmeted head with no pad on the load cell, producing the load and strains at C3, C5, and C7 as shown for Run 61 in Table 1. A firm, resilient energy absorber, 76 mm thick, was applied to the load cell, with no helmet on the cadaver, producing the loads and strains for Run 68, resulting in slightly higher compressive strains at C5 and C7. When the helmet was placed on the cadaver, and using the same firm padding, strain levels were similar to the no pad, no helmet condition of Run 61, except being slightly lower compressive strains at C7. Essentially the same strains were produced with a soft foam pad of the same thickness and wearing a helmet as shown for Run 72. Apparently the only effect of inserting the padding was to slightly change the shape of the impulse by reducing the peak force and spreading out the time duration. It is assumed that practical amounts of padding did not significantly reduce strain because:

1. The loading is distributed over an area on the order of 26 cm^2, on the helmet surface, consequently, the padding is relatively stiff compared to the spine with which it is in series and does little to modify the loading of the spine.

2. Even when the helmet and energy absorbing material combination acts to alter the peak force, strain is not linearly related in the curved column which is trapped between the head and the body below C7. The critical load at which the cervical spine buckles further out of its initial alignment varies unpredictably due to the factors demonstrated previously in Figures 10 and 12.

3. Furthermore, the energy available in these tests or in a propelled body involved in a motor vehicle crash, is usually much greater than can be absorbed with practical amounts of deforming material or structures either worn on the head or mounted in a vehicle, to alleviate strain in the cervical spine due to a crown impact.

As shown for Run 102 in Table 1 the only protective procedure which produced significantly lower strain was by means of a silicon collar ('silly-putty') wrapped around the neck in a plastic bag. The collar had the effect of stiffening on impact to simulate the lateral stability supplied by tensed

Table 1

Crown 'Energy Absorption Inadequate' Theory:

EFFECT OF ENERGY ABSORBER (EA) MATERIAL ON CERVICAL SPINE STRAIN DUE TO CROWN IMPACTS BY 49 Kg MASS

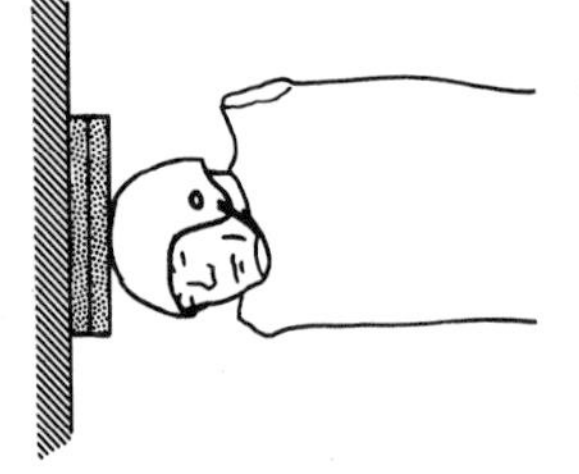

RUN	PAD TYPE	PAD THICKNESS mm	HELMET	PEAK FORCE kN	TENSION C_3	COMPRESSION C_5	C_7
61	no	—	no	2.0	3700	5200	4400
68	Firm	76	no	1.8	3700	5500	4700
69	Firm	76	yes	1.3	3700	5200	4000
72	Soft	76	yes	1.3	3800	5200	3840
'SILLY PUTTY' COLLAR							
102	NO	—	yes	2.0	700	1600	960

RESULTS:

Practical amounts of padding did not significantly reduce strain because:

1. Axial loading of a straight spine is usually distributed over a large head area ($\simeq$ 129 cm^2). Therefore padding is relatively stiff, doesn't dissipate much energy by deformation, and
2. In case of off-axis impact with a curved spine the padding is in series with a softer ligament stretching system over which it has little control.
3. Energy available is much greater than can be absorbed by practical amounts of padding worn on the helmet crown.
4. A 'silly putty' collar was very effective in reducing cervical spine strain. It is assumed to act by stiffening under impact to transfer loads from helmet to shoulder pads and provide a simulated tensed muscle lateral stability.

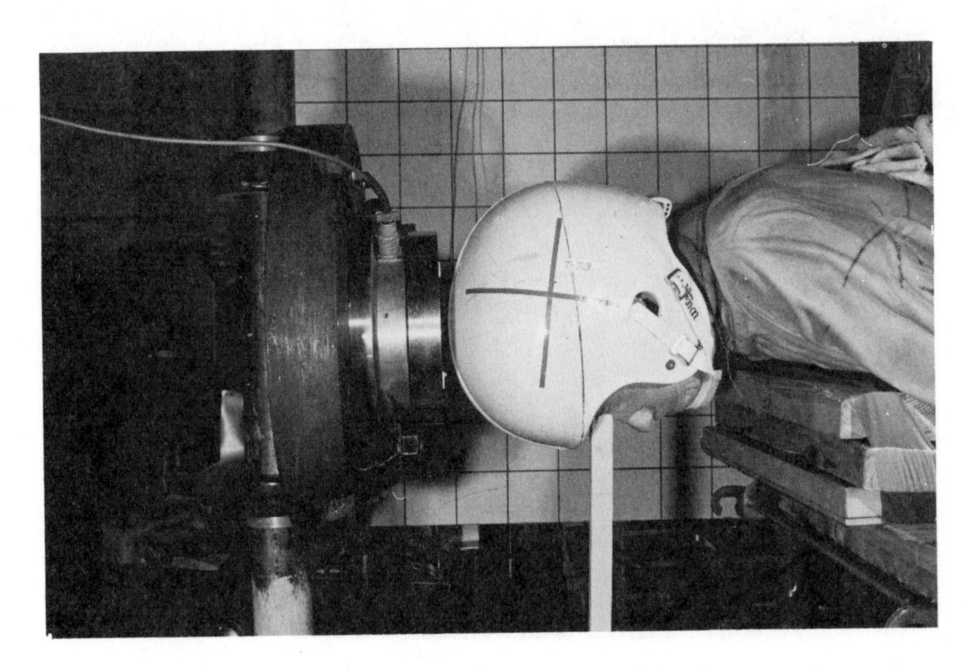

Fig. 11 - Load cell used to deliver padded or unpadded impacts to the cadaver crown

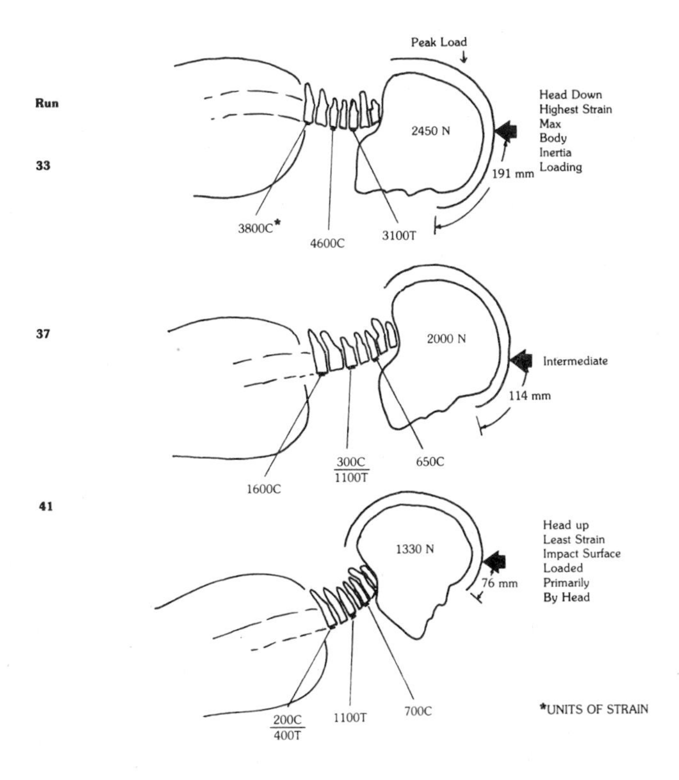

Fig. 12 - Effect of load line of action relative to the cervical spine axis on cervical anterior body strain

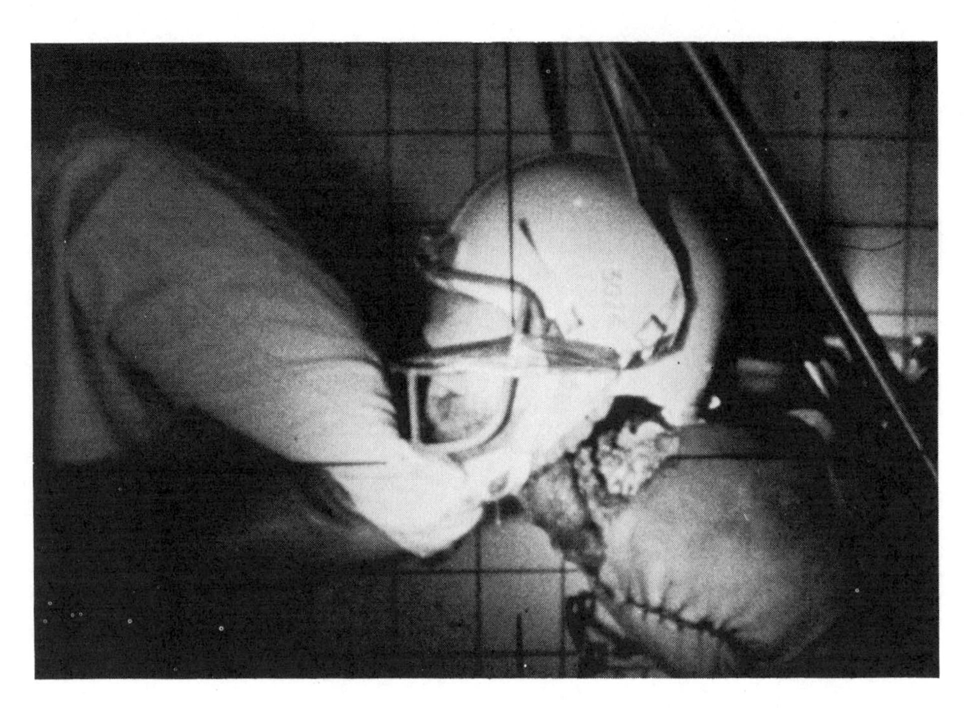

Fig. 13 - Rear rim bending flag stuck into T2 due to hyperextension from below under facemask

muscles, and also to transfer the load around the neck from the helmet into the shoulders.

EFFECT OF HELMET REAR RIM IN HYPER-EXTENSION INJURIES - There have been many medicolegal ramifications related to the design of the helmet rear rim (crash and football helmets) since Schneider (5) attributed three football neck injuries to a guillotine mechanism of the helmet. According to this theory, an impact to the facemask with an I-S component of force causes rotation of the helmet on the head, hyperextension of the neck and impingement of the helmet rear rim on the back of the neck with such force as to produce fracture and/or dislocation. Schneider recommended that the rear rim of helmets be cut higher to eliminate this hazard. To test the theory and the remedy, a cadaver was instrumented with seven strain gages on the anterior bodies and near the articulating facets from C2 through C7. The cadaver was strapped to a pallet free to pivot at the feet, and with head overhanging the raised end of the pallet. A Sierra 1050 leg, flexed at the knee, was mounted in the path of the fall of the head such that the facemask would strike the knee about 150 mm above the joint causing hyperextension of the neck of the cadaver as shown in Figure 2. The knee was free to move upon impact.

Preliminary tests on four cadavers with a standard helmet showed that the helmet contacted the rear of the body between C7 and T2 as shown in Figure 13, which is an excerpt taken at 500 fps at the instant of contact of the helmet with the body showing the flag, which is mounted in the spinous process of T2, being bent. Two drops each were conducted with a standard helmet and with a helmet cut high according to the recomendation. Maximum drop height of the head above the knee was 610 mm. The results of these tests for the maximum drop are shown in Table 2.

The high cut helmet had a padded rim to distribute the load. The standard helmet had a partial suspension (nape band) inside the rear rim to help maintain clearance between the rim and the neck in the event of hyperextension motions. High-speed motion pictures showed that both helmets contacted the lower neck. The cross on the helmets rotated through 41 degrees, from initial contact of the knee with the facemask until contact of the helmet rear rim with the neck for the high cut helmet, and 28 degrees for the standard helmet. At all but one (about equal) strain locations the strain was higher for the high cut helmet. These results would indicate that it is hyperextension of the neck, partially alleviated by the lower cut standard helmet shell interference, which is the determining factor in spinal strain distribution and not contact of the helmet with the neck. Furthermore, the remedy for this alleged mechanism appears to make it more likely that an injury from contact of the rear rim of the helmet with the neck will occur because not only does a higher rim allow more extension but tends to hit higher on the neck and at a greater angle.

CLOSING REMARKS

Although these tests have not been exhaustive of many modes of neck loading which can cause injury, the results indicate that the

Table 2

'GUILLOTINE' THEORY:

EFFECT OF KNEE TO FACEGUARD IMPACT
STANDARD VS HIGH CUT REAR SHELL RIM-610 mm DROP

	Strain¹ in/in							of Ext. Rotation-deg.
	Facets			Anterior Body				
HELMET	C2	C3	C5	C2	C3	C4	C5	
STANDARD SUSPENSION	-1030	550	-300	-480 150	1100	1140	270	28
HIGH CUT² (padded)	-1320	900	-320	-260 260	1380	1330	500	41
% CHANGE	28	64	7	—	24	16	85	46

¹ (-) indicates compression
²Done first

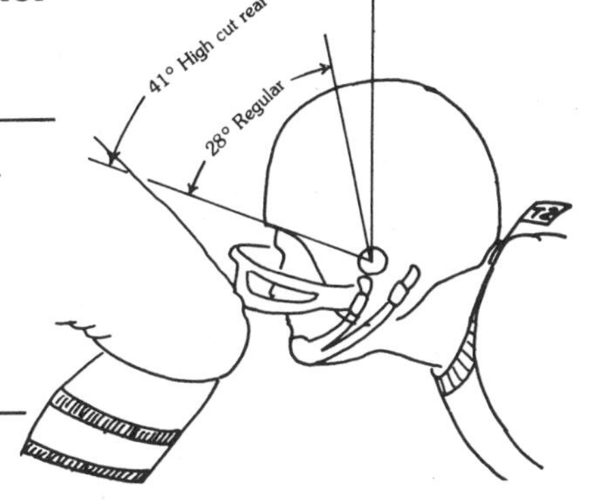

RESULTS:
No injury was produced
Rear rim impinged in C7-T2 area
Higher cutout caused more extension and larger strains

possibility of developing criteria such as tolerable neck loads for the general case of head impact seems unlikely. They suggest that the optimum method of head-face-neck protection for high risk of crash vehicles, and which may become more attractive to the general public with the transition to smaller vehicles, should include a helmet-facemask-neck collar-shoulder pad combination worn by passengers. The neck collar would serve the purpose of transferring loads around the neck in event of crown loading and would also help to limit cervical flexures in all directions.

CONCLUSIONS

1. Distributed crown impacts grip the head, tending to prevent the nodding motion at the atlanto-occipital joint, thereby forcing exaggeration of lower cervical flexion and thereby higher anterior body compression and ligament shear strain, with the likelihood of a fracture and/or dislocation most probable between C4-C7.

2. Concentrated crown impacts allow nodding at the atlanto-occipital junction with the result that dislocation failure in the upper cervical from compression-flexion-shear loading will be the threshold injury.

3. For crown impacts with a flat rigid impactor, the greatest risk of injury occurs when the body-cervical spine axis-head alignment is in a flexure posture prior to impact. The minimum risk occurs when the body-cervical spine and head are in alignment (near axial loading).

4. For impacts to the head with a flat rigid impactor, maximum risk of injury to the neck, holding the orientation of the spine below C7 constant, occurs for a crown impact; minimum risk for a forehead impact.

5. Firm or soft energy absorbing materials up to 76 mm did not reduce strain due to a crown impact with a flat rigid impactor.

6. The posterior rim guillotine injury mechanism theory was not validated by these experiments and the remedy of cutting the helmet higher appears to make the possibility of injury from this mechanism more likely to occur.

7. Crown loading was not found to correlate with cervical spine strain.

ACKNOWLEDGMENTS

This work was sponsored in part by the National Operating Committee on Standards for Athletic Equipment (NOCSAE), Detroit General Hospital Research Corporation, and the Office of Naval Research.

Eugene Dupuis conducted the surgery and instrumentation of the cadavers. Matthew Mason was responsible for the photography and Robert Neumann assisted in the experiments.

REFERENCES

1. R. Roaf, "A Study of the Mechanics of

Spinal Injury." The Journal of Bone and Joint Surgery, Volume 42B, Number 2, November 1960.

2. J.R. Bauze, and M.A. Ardran, "Experimental Production of Forward Dislocation in the Human Cervical Spine." The Journal of Bone and Joint Surgery, Volume 60B, Number 2, May 1978.

3. H.J. Mertz, V.R. Hodgson, L.M. Thomas, and G.W. Nyquist, "An Assessment of Compressive Neck Loads Under Injury-Producing Conditions." The Physician and Sportsmedicine, Volume 6, Number 11, November 1978.

4. R.H. Culver, M. Bender, and J.W. Melvin, "Mechanisms, Tolerance, and Responses Obtained Under Dynamic Superior-Inferior Head Impact." OSHA Final Report, May 1978.

5. R.C. Schneider, "Head and Neck Injuries in Football, Mechanisms, Treatment, and Prevention." Baltimore, MD: Williams and Wilkins, 1973.

811032

Experimental Studies of Brain and Neck Injury

A. Sances, Jr., J. Myklebust, J. F. Cusick, R. Weber, C. Houterman, S. J. Larson, P. Walsh, M. Chilbert and T. Prieto
Marquette University and VA Medical Center

Mark Zyvoloski
The Medical College of Wisconsin

C. Ewing and D. Thomas
Naval Biodynamics Laboratory

Bernard Saltzberg
Texas Research Institute of Mental Sciences

ABSTRACT

Static and dynamic axial tension loads were applied to the intact and isolated cervical column of the monkey and human cadaver. Radioactive microspheres were used to evaluate brain and spinal cord perfusion in the monkey. To determine neutal pathway damage, somatosensory evoked potentials were recorded with stimulation of the spinal cord, and in spinal cord with stimulation of sensorimotor cortex, and in spinal cord with stimulation of cauda equina. The evoked potential amplitude decreased prior to heart rate and blood pressure changes presumably due to brainstem distention. The preliminary studies show: 1) the brain and spinal cord were well perfused as measured with the microspheres when the evoked potentials decreased, 2) the cervical isolated cadaveric monkey spinal column ligaments failed statically at approximately 1/2 to 1/3 the force required for dynamic disruption, 3) In the intact monkey, the cervical ligaments failed statically at approximately 1/2 the dynamic failure force, 4) the isolated human cervical ligaments failed at loads approximately three times those observed in the isolated monkey cervical column.

AXIAL TENSION IS OFTEN APPLIED to the cervical spinal cord and brain during impact or inertial loading. These axial forces are routinely implicated as major factors of head and neck injury.

This study was therefore conducted to determine physiologic changes in brain and spinal cord with cervical axial tension. Tissue damage in the isolated monkey, and human cervical column and the intact monkey with axial tension forces applied to the cervical region was also studied. Previous studies have demonstrated that the evoked potential is useful for the evaluation of spinal and brain injury in patients with neurologic dysfunction (1-9)*. The mechanical strength of the isolated cervical column and physiologic response with static** axial tension forces in the monkey have been previously evaluated (10, 11, 12).

To further study the effects of axial tension, four separate studies were conducted. The first series was done in the isolated fresh cadaveric cervical spinal column of the monkey. These dynamic experiments, designed to evaluate tissue strength, were a follow-up to the axial tension static studies previously reported upon (10).

The second series of static studies was carried out in the living monkey to determine whether the spinal cord and brain evoked potential alterations are due to perfusion changes produced by the axial loading.

The third experiments investigated the physiologic effects of dynamic axial tension loads

*Numbers in parentheses designate References at end of paper.

**The term static represents quasi static applied loads, dynamic is considered 100 cm/sec or greater.

upon the intact monkey.

The final investigations measured the strength of fresh human cadaveric skull and cervical spinal columns and spinal cords with quasi static and dynamic axial tension loads.

METHODS

All animal studies were conducted with male rhesus (Macaca mulatta) monkeys.

ISOLATED CERVICAL MONKEY COLUMN METHODS - The first series of studies was conducted on two isolated monkey spinal columns including the base of the skull and the cervical and thoracic elements (Table 1). The two animals died prior to testing and were stored at 2°C. Animal #792 died two hours before and #793, 24 hours before testing. The isolated spinal columns were removed from the cooler and fixed with methylmethacrylate at the skull and at T1-T2 using previously reported techniques (10). Briefly, the skull was mounted in a swivel fixture. The thoracic region was fixed to the rigid frame.

The overlying tissues were carefully removed to preserve the ligaments and the column was kept moist upon removal from the refrigerator. Tests were conducted approximately one hour later when the tissues reached room temperature. Repeat runs in each spinal column were made. The remaining lower cervical elements were fixed with methylmethacrylate to the rigid frame. Dynamic axial tension loads at a rate up to 170 cm/sec were applied with an MTS Series 810 device using a 10 cm stroke cylinder (Fig. 3). The displacement between the occiput and T1 was measured with a Bourne 118 linear potentiometer fixed to the spinous process of T1 and the skull. The axial acceleration was measured with a single axis statham or 72-67 C-750 Endevco accelerometer mounted to the skull with methylmethacrylate. The piston displacement output was provided by the MTS device. The MTS was programmed for constant displacement up to 10 cm versus time. Following the second intact column test, the entire cervical spinal cord was removed and clamped into serrated jaws, attached to the cylinder piston with a swivel and the frame at the bottom. Approximately 4.0 cm of the central cervical cord was free. A dynamic load was applied to the spinal cord of animal #792 and a static load to animal #793. Failure was assumed to occur for the columns and spinal cords when the load dropped precipitously. The maximum force is given as the failure level. A Honeywell 1858 visicorder was used for all recording.

IN VIVO STATIC METHODS WITH MICROSPHERES - The second series of studies was conducted in four monkeys (Table 2). Methods for evoked potential studies including the electrodes, and implantation procedures are the same as those discussed in a previous publication (10). Briefly, bipolar electrodes were stereotaxically implanted in nucleus ventralis posterior lateralis of the thalamus (VPL). Three 2 mm diameter platinum electrodes 4 mm on center were placed bilaterally over sensorimotor cortices (SMC), over the cauda equina (CE), and the lower cervical dorsal columns of the spinal cord. In animal #787, a spinal cord (Cordis) stimulation electrode was passed epidurally to the lower cervical level for stimulation and for recording. During implantation, the animals were anesthetized with thiamylal sodium (10 mg/kg). A Clinical Technology 2000 evoked potential measurement system was used to retrieve the evoked responses. Afferent averaged evoked responses were recorded secondary to 0.2 ms duration, 4 Hz current application to the electrodes in cauda equina or cervical spinal cord. Efferent pathways were evaluated by passing the 4 Hz, 0.2 ms rectangular currents to the electrodes over the sensorimotor cortex with recordings from the electrodes placed over the thoracic or cervical dorsal column of the spinal cord.

Prior to force applications, the animals were given a single dose of sodium penthobarbital which lasted throughout the study. The

TABLE 1

DYNAMIC AXIAL TENSION STUDIES IN ISOLATED MONKEY CERVICAL COLUMNS AND SPINAL CORDS

ANIMAL #	Wt. (Kg)	RUN #	PREPARATION MOUNTING IN METHYLMETHACRYLATE	MACHINE LOADING RATE	LOCATION OF TISSUE FAILURE	MAX LOAD AT FAILURE N	lb.	POTENTIOMETER DISPLACEMENT AT FAILURE cm	in	MACHINE DISPLACEMENT AT FAILURE cm	in	AXIAL ACCELERATION (G)
# 792 Died 24 hours before test	9.0	1	Skull to (T1-T2)	170 cm/sec	Endplate C7-T1	1423	320	1.1	0.42			62
		2	Skull to (C6-C7)	152 cm/sec	Endplate C2-C3	580	130	1.8	0.7			55
		3	Cervical spinal cord (3.8 cm long) clamped in jaws	132 cm/sec	Midpoint	90	20			0.61	0.24	-
# 793 Died 2 hours before test	8.7	1	Skull to (T1-T2)	130 cm/sec	Endplate C7-T1	1289	290	1.1	0.42			58
		2	Skull to (C6-C7)	142 cm/sec	Endplate C5-C6	580	130	0.89	0.35			50
		3	Skull to (C3-C4)	152 cm/sec	Endplate C1-C2	1068	240	0.53	0.21			55
		4	Cervical spinal cord (4.2 cm long) clamped in jaws	0.2 cm/sec	Midpoint	36	8			0.61	0.24	-

TABLE 2
STATIC AXIAL TENSION IN VIVO MONKEY STUDIES

ANIMAL	WT KG	RESTRAINT*	LOCATION OF TISSUE FAILURE	MAXIMUM LOAD N	LB	CERVICAL COLUMN DISPLACEMENT (X-ray) (maximum) cm	in
#700	12.0	Belts	Complete C3-C4 disruption	1112	250	0.95	0.37
#731	11.5	Belts	Vertebral & carotid artery	888	200	-	-
#788	11.2	Belts	Slight blood atlanto occipital membrane	888	200	-	-
#787	10.1	Belts	None	888	200	2.0	0.79

*All heads fixed in Methylmethacrylate.

static forces were applied along the spinal axis of the monkey with an Instron device. The torso was restrained by means of a SAE J-386 seatbelt harness over each shoulder, and the head was fixed with methylmethacrylate molded into a block incorporating stereotaxic metal bars placed in the auditory meatus, fixed in a steel yoke (Fig. 1). In some studies, additional head fixation was provided by a 6.3 mm diameter stainless steel cable imbedded in the methylmethacrylate block. The cable was passed through the mouth and then immediately below the occipital protruberance and then beneath each zygomatic arch and brought up to form a loop above the vertex (Fig. 7-10). The cable loop was attached to a bar and swivel. The force was applied to the cervical column as the screw fixed to the surface of the support system was moved outward. An in series Dillon gauge measured the force which was increased in 112-224 Newton increments approximately once every 10 minutes. Periodic blood gases were taken. Blood pressure and heart rate were monitored with a catheter in the femoral artery. The rectal temperature was measured with a Yellow Springs thermistor. In two animals regional perfusion (14, 15, 16) was evaluated with radioactive microspheres injected prior to, during, and following force application. Following a thoracotomy, a catheter was inserted into the left atrium via the left atrial appendage for the microsphere injection. The thoracic cavity was closed and the animal was maintained on a Harvard apparatus respirator. Four different 15 micron diameter radioisotopes were used, ^{141}Ce, ^{51}Cr, ^{85}Sr, and ^{46}Sc. Scalars determined the amounts of radioactivity and the attendant perfusion of the samples of spinal cord, brainstem, and cerebral tissue. Samples were obtained and regional perfusion was calculated following the study. Lateral x-rays were taken of the animals at each force level. The distraction of cervi-

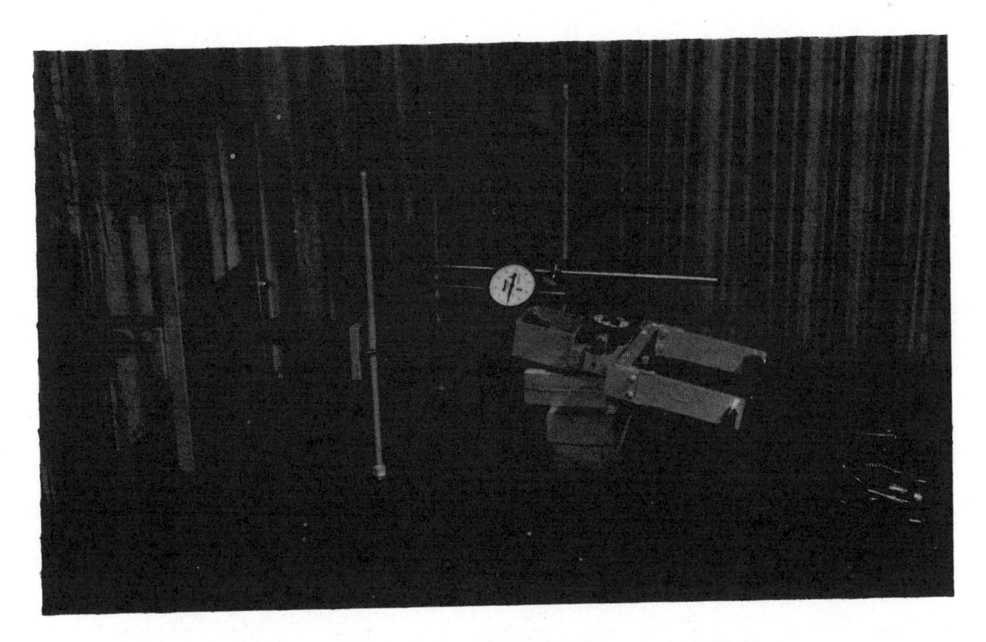

Fig. 1 - Screw at left is threaded out. The dial gauge measures displacement and Dillon force gauge the tension in series with the rigid table supports. The restraining seatbelts and yoke fixed to animals head at right.

cal column was determined from the spacing of vertebral bodies obtained from the x-rays (Table 3). The amplitude of the primary component of the evoked potential (10-30 ms) amplitude is plotted (Fig. 5).

INTACT DYNAMIC MONKEY METHODS - These studies were done in four living and two monkeys who expired 10 and 20 minutes prior to the test (Table 4). The dynamic force was applied with the 4480 Newton, 10 cm stroke cylinder controlled by the Series 810 materials test system at rates up to 150 cm per second. In animals #786 and #789, ear bars were inserted into the auditory meatus and 1 mm diameter nichrome wire was wrapped around the head. The entire system was fixed in methylmethacrylate. Because this system failed in animal #789, the technique was revised. A single 6.3 mm diameter stainless steel cable was passed through the oral cavity, immediately beneath one zygomatic process, around the occipital protruberance, then beneath the opposite zygomatic process through the mouth again and clamped at each zygomatic process; a

TABLE 3

Cervical distraction between vertebral bodies (cm) from serial lateral x-rays. (Figs. 7-10) Animal #787

LOAD N	O-C1	C1-C2	C2-C3	C3-C4	C4-C5	C5-C6	C6-C7	
0	0	1.0	1.4	1.2	1.1	1.1	1.1	
222	0.1	1.2	1.4	1.2	1.1	1.2	1.15	0.4
444	0.2	1.3	1.4	1.2	1.2	1.2	1.2	0.8
555	0.3	1.5	1.5	1.3	1.2	1.2	1.3	1.4
666	0.3	1.5	1.5	1.3	1.25	1.2	1.3	1.4
777	0.4	1.6	1.5	1.3	1.3	1.25	1.3	1.6
888	0.4	1.7	1.5	1.4	1.3	1.3	1.3	2.0

TABLE 4

DYNAMIC AXIAL TENSION STUDIES IN THE INTACT MONKEY WITH BELT RESTRAINTS (#786, #789, #790)

ANIMAL	WT KG	RESTRAINT*	RATE OF LOADING cm/sec	LOCATION OF TISSUE FAILURE	MAXIMUM LOAD		LINEAR POT DISPLACEMENT AT MAXIMUM LOAD		MAXIMUM ACCEL. AXIAL Gs
					N	LB	cm	in	
#786 Died 10 minutes prior to run	9.3	Belts	105	C1-C2, carotid, vertebral arteries	2668	600	1.76	0.69	-
#789 Living	11.1	Belts	110	Cap failed	1780	400	1.57	0.62	-
		Belts	110	Minimal damage at C1-C2	1770	397	2.13	0.84	-
#790 Died 20 minutes prior to run	11.2	Belts	100	C3-C4	2668	600	2.08	0.82	65
#785 Living	13.2	Yoke	100	Occiput-C1	2558	575	1.60	0.63	63
#794 Living	11.3	Yoke	100 Stroke increased increments in 1/2 cm	C3-C4 anterior and posterior long. ligaments & artial disc at C3-C4.	1690	380	1.25	0.43	-
#795 Living	12.1	Yoke	100 stroke increased increments in 1/2 cm	Post and lateral ligaments at C1-C2, vertebral artery and basilar artery	2669	600	1.7	0.67	-

*All heads fixed in methylmethacrylate

loop was left at the vertex for force application. The head was then blocked with methylmethacrylate. Animals #786, #789, and #790 were restrained with the SAE J-386 belt harness and seated in a G.M. child Love Seat (Fig. 2). For distraction of the cervical column, the previously described linear potentiometer was mounted on the spinous process at the thoracic level between Tl-T2 and the skull. In animals #785, #794, and #795 a metal yoke was bolted to the fixed force frame (Fig. 3). The piston was programmed to move 8 cm in animals #786, #789, #790, and #785. In animals #794 and #795, the piston stroke was increased from 1/2 cm to 4.5 cm in 1/2 cm displacements until failure occurred. A ycam camera took photos at 1000 frames per second. The previously described single axis Statham accelerometer or the triaxial Endevco accelerometer was mounted in the methylmethacrylate on the head along the spinal axis. The 1858 Honeywell visicorder recorded the transducer outputs.

Evoked potential electrodes were implanted bilaterally on sensorimotor cortices, and upon the lower cervical dorsal column, thoracic dorsal column, and in cauda equina. The implantation procedures and recording methods are the same as those described in the previous section. The amplitude of the early 10-30 ms latency components of the evoked potentials are reported upon. Failure was assumed to occur at the maximum force when the load dropped off sharply.

STATIC AND DYNAMIC HUMAN CADAVER METHODS - Fresh male human cadaveric spines (cervical to lumbar) including the head were removed. All specimens were determined to be normal from the medical history and x-ray examinations done prior to the tests. The supporting tissues were carefully removed to avoid damage to the ligaments. The spines were mounted in methylmethacrylate at the Tl-T2 level and the skull was incorporated in methylmethacrylate. The studies were conducted within 48 hours following death. Tissues were kept at 2°C until the study was conducted. Upon removal from the refrigerator, they were transported in ice and periodically moistened during the study. The quasi static axial studies were conducted with an Instron device. Dynamic axial tension studies were conducted with the previously described MTS 810 series system and the 10 cm stroke hydraulic cylinder. The intact unembalmed cadaver torso had the head mounted in methylmethacrylate. The shoulders were restrained with a yoke firmly fixed to the rigid force frame (Fig. 3). Axial tension loads at 125 cm/sec were applied. A linear potentiometer was fixed to the skull and the spinous process of Tl-T2. A Statham single axis and an Endevco triaxial accelerometer were mounted to the head. In HS-1 - HS-3, multiple studies were done (Table 5). The second and third studies followed within approximately 30 minutes after the specimens were remounted. In test #2 of HS-2, all posterior ligaments up to but excluding the posterior longitudinal at C4-C5 were cut. In test #3 of HS-2, all anterior ligaments up to and including the posterior longitudinal ligament were cut at C2-C3. Axial tension studies were conducted on HS-5 - HS-7 cervical spinal cords taken from fresh cadaveric specimens. The cord was clamped at each end with serrated jaws attached to the Instron or MTS system. The unclamped free cord section ranged from 3.8 to 7.6 cm. All records were made with the visicorder.

Fig. 2 - MTS 810 controller (rear) 10 cm stroke cylinder (upper) restraining belts and seat for monkey dynamic studies.

Fig. 3 - MTS 810 device and 10 cm stroke cylinder with adjustable frame and yoke at top for dynamic intact animal studies and human cadaver torso.

TABLE 5
IN VITRO HUMAN AXIAL TENSION STUDIES ON ISOLATED CERVICAL SPINAL COLUMNS AND CERVICAL SPINAL CORDS

Specimen	Weight (kg)	Age (yr)	Run #	Preparation Mounting	Rate of Loading	Location of Failure	Max Load at failure Newtons (lbs)	Machine Deflection at Failure cm (inches)
HS1 Cardiac Failure	77.1	50	1	Methylmethacrylate Isolated column C7, occiput	0.13 cm/min	C5-C6	1446 (325)	1.14 (0.450)
			2	C3-occiput	0.13 cm/min	C1-C2	1312 (295)	1.10 (0.44)
HS2 Renal Failure	52.2	36	1	Methymethacrylate Isolated column C7 and skull	0.13 cm/min	Circular skull fracture	1779 (400)	2.67 (1.05)
			2	C7 and C1 with (C4-C5 post-ablation)	0.13 cm/min	C4-C5	1289 (290)	1.59 (0.625)
			3	C4 and C1 with (C2-C3 ant. ablation)	0.13 cm/min	C2-C3	622 (140)	0.88 (0.345)
HS3 Respiratory Failure	70	61	1	Methylmethacrylate Isolated column T_1-T_2 occiput	0.13 cm/min	C6-C7	1940 (435)	3.0 (1.18)
			2	C5-occiput	100 cm/sec	None	2688 (600)	
HS4 Smoke Inhalation	67	67	1	Whole torso Yoke on shoulders	125 cm/s	Circular skull fracture	3780 (850)	2.03 (0.8)
HS5 Stroke Death	72.5	65	1	Clamped in jaws Isolated cerv. spinal cord (7.6 cm cord-free)	106 cm/s	Midpoint	278 (62.5)	0.76 (0.3)
HS6 Hepatic Failure	56.7	63	1	Clamped in jaws Isolated cerv. spinal cord (3.8 cm cord-free)	118 cm/s	Midpoint	389.2 (87.5)	1.0 (0.40)
HS7	58.9	76	1	Clamped in jaws Isolated cerv. spinal cord (3.8 cm cord-free)	1.6 cm/s	Midpoint	167 (37.5)	0.89 (0.35)

RESULTS

DYNAMIC MONKEY ISOLATED CERVICAL COLUMN STUDIES - The studies were conducted on two Macaca rhesus monkeys sacrificed with an overdose of barbiturate.

Animal #792 (Isolated Dynamic Stretch) - The spinal column and head were removed 24 hours prior to the test of this 9 kg animal (Table 1). The preparation was loaded at a rate of 110 cm/sec. The maximum axial acceleration of 62 G was recorded during the initial moment of the piston. The lateral and anterior posterior accelerations were less than 10 Gs. An end plate failure was observed at C7-T1 at 1423 Newtons. At this time the displacement between the occiput and T1 was 1.1 cm. No obvious damage could be observed at other levels; however, the remaining column was elongated 5-10% beyond its initial length. The posterior longitudinal ligament, ligament of flavum and posterior complex were transected (the piston was programmed to move up to 10 cm). However, the spinal cord was intact. The column failed at the C2-C3 end plate with 580 Newtons during the second run without damage at other levels. A mid cervical spinal cord failure was observed at 90 Newtons with a rate of 132 cm/sec. The spinal cord was flattened in the region of the jaws, but it did not slip.

Comment - The spinal columns were probably stressed during the first runs since the force levels were substantially less in the second tests. It was not possible to determine whether the ligaments internal to spinal cord at other levels were altered following the first test. The spinal cord elongated approximately 18% prior to failure.

Animal #793 (Isolated Dynamic Stretch) - The spinal column and skull of this 8.7 kg animal were removed and tested within two hours after death (Table 1). An end plate failure was observed at C7-T1 at 1289 Newtons on the first run with an elongation of 1.1 cm. For the second run, the column failed at C5-C6 at 580 Newtons. The column failed at the C1-C2 end plate with 1068 Newtons during the third run. The cervical cord which was loaded at a quasi static rate of 0.2 cm/sec failed at approximately 36 Newtons.

Comment - Because the tissues of #793 failed at a lower level than specimen #792, this suggests that the strength of the cervical elements of animal #792 probably did not deteriorate during the 24 hours after death. The cervical spinal cord of animal #793 failed at approximately half the level observed at the higher loading rate in animal #792.

Summary - The dynamic failure load of the first run isolated spinal columns failed with approximately -3 times the force level required for disruption previously reported in the static preparations (10). The spinal cord elongates 14-18% prior to failure as is anticipated from the literature (17). Repeated runs probably stressed the ligaments so that the second and third run failure values are probably lower than those which would have occurred in first runs.

IN VIVO STATIC MONKEY STUDIES WITH MICROSPHERES - The studies were conducted on four live animals. Monkeys #787 and #786 were injected with microspheres to determine brain and spinal cord perfusion.

Animal #700 (Static Stretch) - This 8 kg monkey was initially loaded at 222 Newtons. At 667 Newtons, the blood pressure decreased 20% and the early components of the cortical evoked potential, due to cauda equina stimulation, began to decrease in amplitude. However, the blood gases were within normal limits. The animal had breathing difficulties several minutes later, and was placed on a respirator. At 890 Newtons, the evoked potentials recorded at thoracic spinal cord with cortical stimulation began to decrease im amplitude along with the blood pressure. A 2 mm increase in spacing between the cervical vertebral bodies 3 and 4 was observed. At 1112 Newtons, the blood pressure and heart rate were markedly reduced, and an abrupt change in the cervical extension load was observed (Table 2). X-rays (Fig. 4) showed disruption at C3-C4. The core temperature and evoked potentials recorded from the thoracic cord with stimulation of the cauda equina were unaltered 20 minutes after application of 1112 Newtons, but decreased markedly thereafter. At autopsy, traces of blood were present at the base of the skull and from the occiput to C6 along the spinal cord. Complete disruption at the end plate and the posterior and anterior longitudinal ligaments at C3-C4 were observed; however, the spinal cord was intact.

Comment - The decrease in blood pressure at 667 Newtons was probably due to distention of the cervical cord or brainstem concommitant with a reduction in the somatosensory evoked cortical potential. The respiratory difficulties observed several minutes later were probably caused by brainstem distraction. The C3-C4 disruption probably occurred at 1112 Newtons when the load dropped off; however, the vital functions and evoked potential changes occurred earlier.

Animal #731 (Static Stretch) - In this 11.5 kg animal, the primary evoked potential amplitudes recorded at somatosensory cortex with stimulation of the dorsal columns of the thoracic spinal cord began to decrease at approximately 35 minutes with a force application of 667 Newtons. The evoked potentials from sensorimotor cortex to thoracic spinal cord followed this trend approximately four minutes later. A decrease in the evoked responses recorded at nucleus ventralis posterior lateralis of the thalamus followed several minutes thereafter. A reduction in blood pressure and heart rate was not observed until loads of 888 Newtons were applied (Table 2). At this level, the evoked potentials continued to deteriorate with the exception of the responses recorded at the thoracic dorsal column spinal cord with stimulation of cauda equina. This animal expired approximately 10 minutes following application of the 888 Newton load. At autopsy, the right carotid artery was ruptured and a vertebral artery disruption on the same side was observed. All ligaments were intact with no damage to the discs or vertebral bodies. Blood was present between the skull and C1 with hemor-

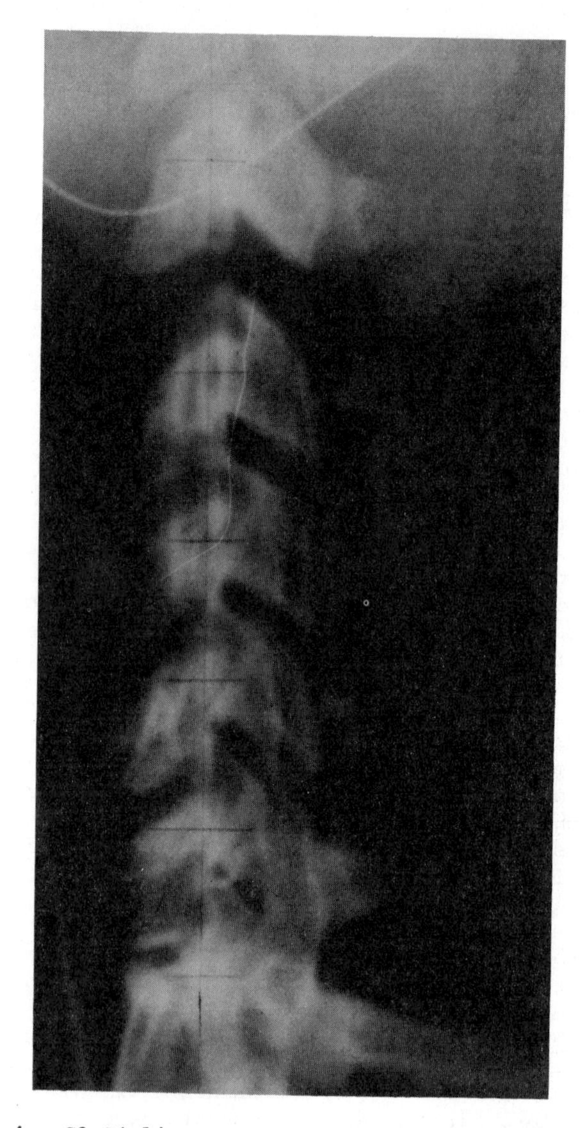

Fig. 4 - C3-C4 ligamentous disruption in Animal #700.

rhage in the region of the brainstem.

Comment - This animal apparently died from vascular trauma at the 888 Newton load level.

Animal #788 (Static Stretch, Microsphere Study) - In this 11.2 kg animal, the microspheres were injected immediately prior to the study, after 35 minutes at 666 Newtons and at 46 minutes at the same load level. The final injection was given at 60 minutes during application of 888 Newtons (Fig. 5). Samples were taken from different regions starting at the frontal cortex and going inferiorly for analysis. Six were taken in the brain, 3 from brainstem, 2 from cerebellum, 5 from cervical spinal cord, and 11 from the thoracic and lumbar spinal cord. At approximately 666 Newtons (and 32 minutes after the start) all primary evoked potential amplitudes including those from cauda equina to lower cervical spinal cord (CE-C6), from cauda equina to somatosensory cortex (CE-SMC) and somatosensory cortex to cervical spinal cord (SMC-C6) began to decrease markedly followed several minutes later by a slight decrease in the heart rate and in blood pressure. At autosy, all ligaments, discs, and vertebral bodies were intact. A slight disruption of atlanto-occipital junction in the posterior region was observed. A slight amount of epidural hemorrhage was observed from C2 through C4. However, the carotid and vertebral arteries were intact. At 46 minutes, the animal was slightly acidotic. This animal appeared to expire secondary to mechanical distraction of the brainstem. The microsphere studies indicate that the brain (B), brainstem (BS) and cervical spinal cord (C) and thoracic and lumbar (TL) spinal cord perfusion was within normal limits at 666 Newtons when the evoked potential amplitudes were reduced at least 50%. The perfusion was reduced 90% in the cervical cord (C) at the ^{85}Sr (3rd) radioisotope injection, was reduced 25% in the brain and brainstem, and 30% in the thoracic and lumbar spinal cord. The CE-SMC and SMC-C6 evoked potentials were obliterated at this time.

Comment - The 15 micron radioactive spheres lodge in the capillaries and reflect fiber ac-

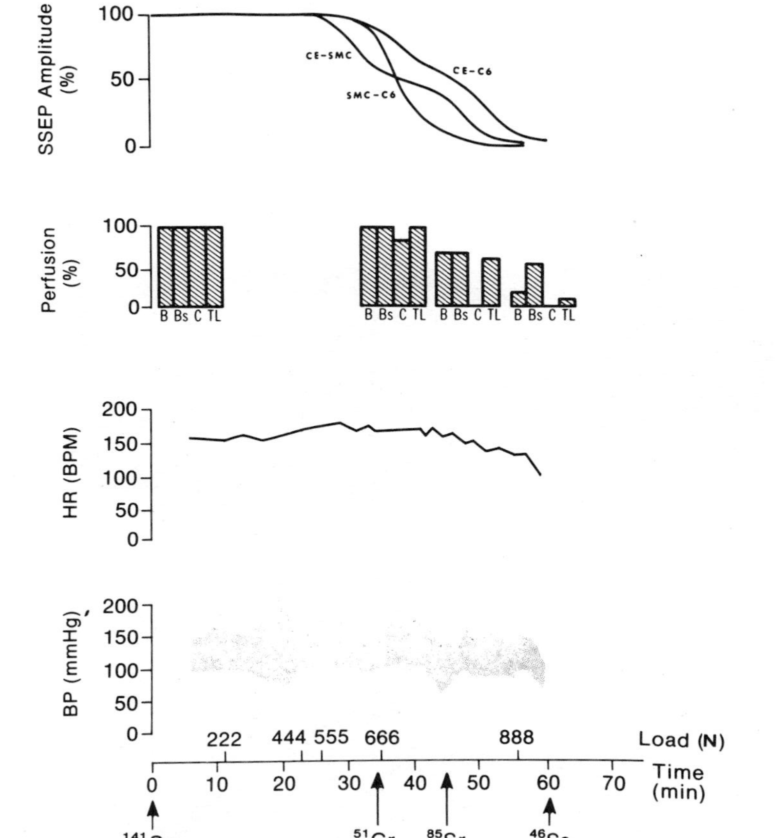

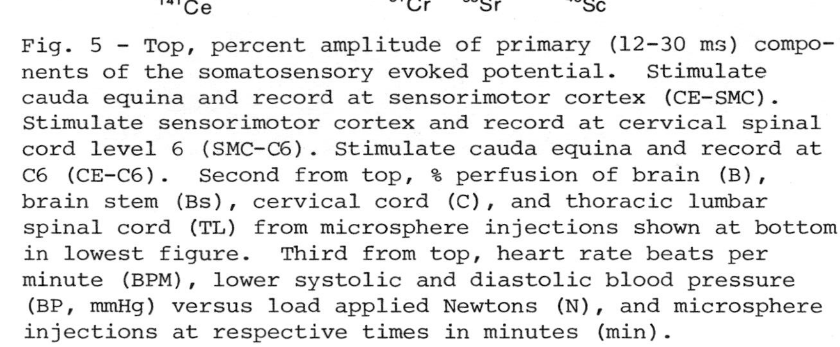

Fig. 5 - Top, percent amplitude of primary (12-30 ms) components of the somatosensory evoked potential. Stimulate cauda equina and record at sensorimotor cortex (CE-SMC). Stimulate sensorimotor cortex and record at cervical spinal cord level 6 (SMC-C6). Stimulate cauda equina and record at C6 (CE-C6). Second from top, % perfusion of brain (B), brain stem (Bs), cervical cord (C), and thoracic lumbar spinal cord (TL) from microsphere injections shown at bottom in lowest figure. Third from top, heart rate beats per minute (BPM), lower systolic and diastolic blood pressure (BP, mmHg) versus load applied Newtons (N), and microsphere injections at respective times in minutes (min).

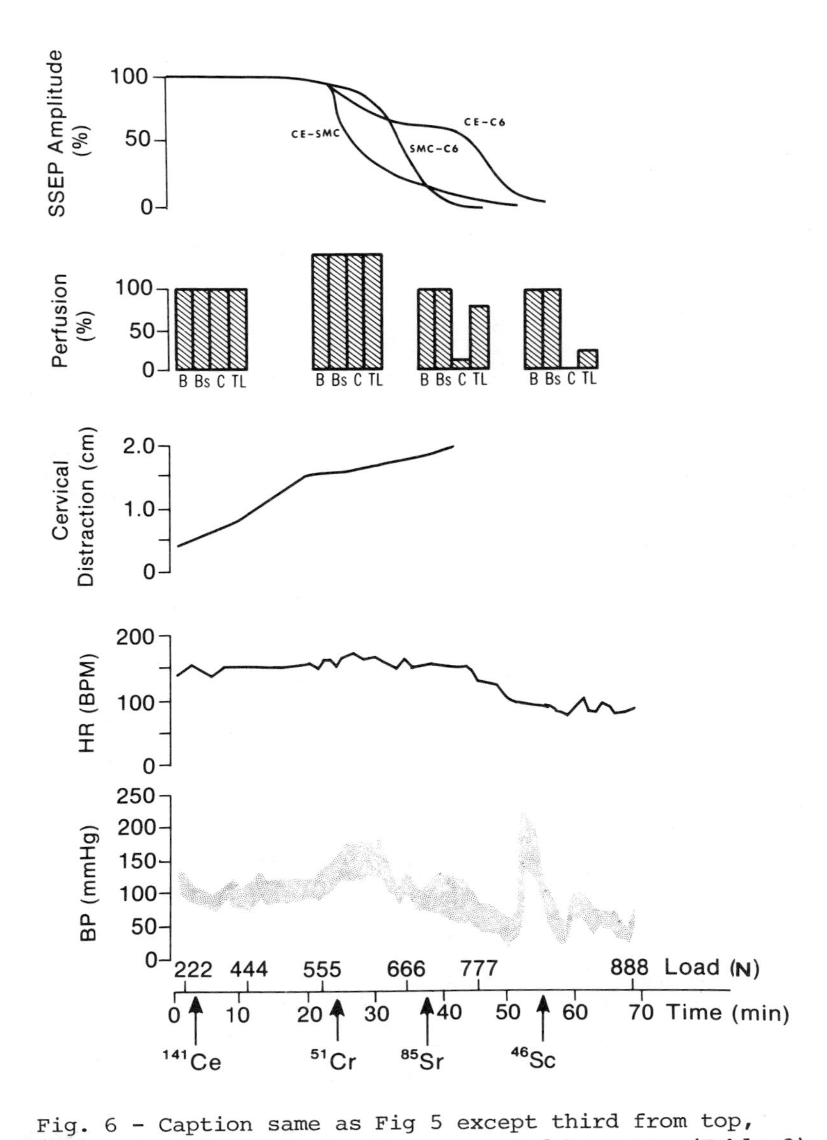

Fig. 6 - Caption same as Fig 5 except third from top, elongation of cervical column measured by x-ray (Table 3) versus applied load shown at bottom.

tivity (14, 15, 16). The alterations in evoked potentials preceded prefusion alterations in the spinal cord, brainstem, and cerebrum. The marked decrease in cervical perfusion during the third injection suggests that either the vessels in this area are mechanically occluded or the metabolism of the neural system is decreased and its perfusion requirements are markedly reduced. Nevertheless, this study demonstrates that the fibers are sensitive to mechanical distraction causing the evoked potentials to be reduced despite adequate perfusion.

Animal #787 (Static Stretch, Microsphere Study - This 10.1 kg monkey showed a cervical distraction of 2 cm at a load of 888 Newtons (Fig. 7, 8, 9, 10). The cervical column distended uniformly with load. The maximum distensions occurred between the occiput-Cl and Cl-C2 (Table 3). This animal was also given microspheres and maintained on a respirator throughout the experiment. The ^{141}Ce microspheres were injected 2 minutes into the study, the ^{51}Cr at 26 minutes during application of 555 Newtons, the ^{85}Sr at 39 minutes with application of 666 Newtons, and the ^{46}Sc at 55 minutes during application of 777 Newtons (Fig. 6). The evoked potentials recorded at sensorimotor cortex with stimulation of cauda equina (CE-SMC), those recorded at spinal cord with stimulation of sensorimotor cortex (SMC-C6), and those recorded at C5-C6 (with a monopolar 1/2 mm diameter electrode) with cauda equina stimulation (CE-C6) were reduced in amplitude from 20-50% at 555 Newtons prior to injection of the ^{51}Cr isotope. At this time, all regions of brain and spinal cord were well perfused. Ten minutes later at 777 Newtons, the heart rate and blood pressure began to deteriorate. The brain and brainstem perfusion were approximately normal; however, the cervical perfusion was 15% of control while the thoracic lumbar perfusion was reduced 15-20%. At autopsy no damage could be found. The microspheres study showed that the brain perfusion at cortex, hypocampus, thalamus, parietal lobe, midbrain and pons were not significantly altered prior to changes in the efferent and afferent evoked potentials. However, the perfusion of the spinal cord from C2 to midthoracic was essentially obliterated at 777 Newtons when ^{46}Sc was injected.

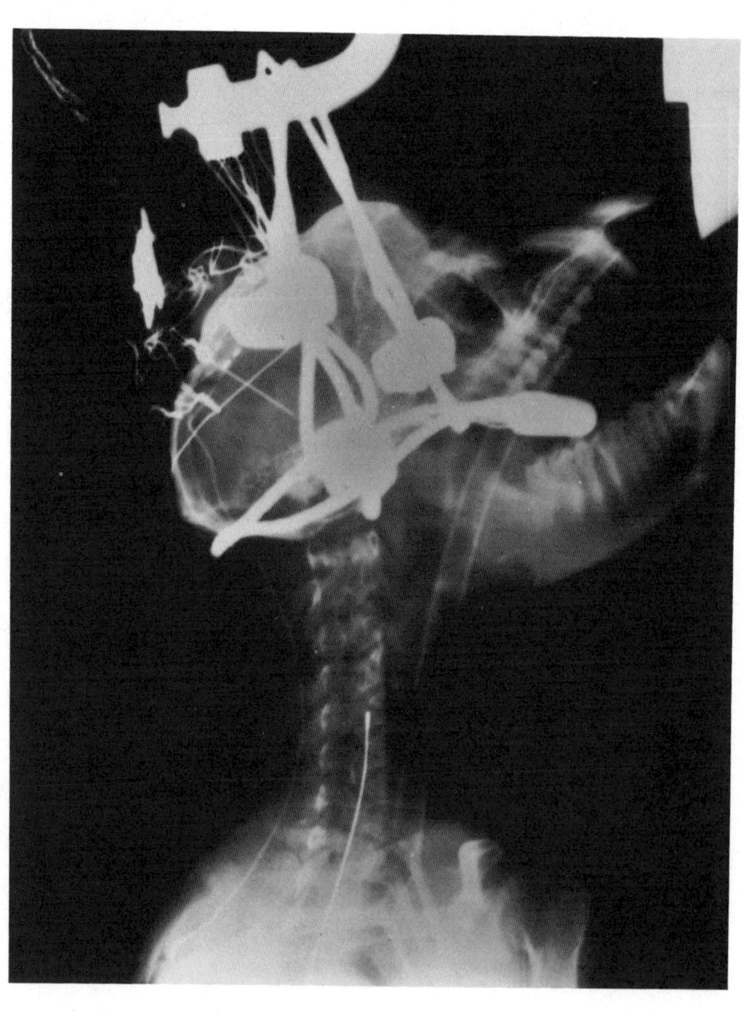

Fig. 7 - Lateral x-ray at 222 Newtons (Animal #787). The stainless steel restraining cable can be seen in upper region. The dorsal column (Cordis) stimulating electrode is shown at C5-C6.

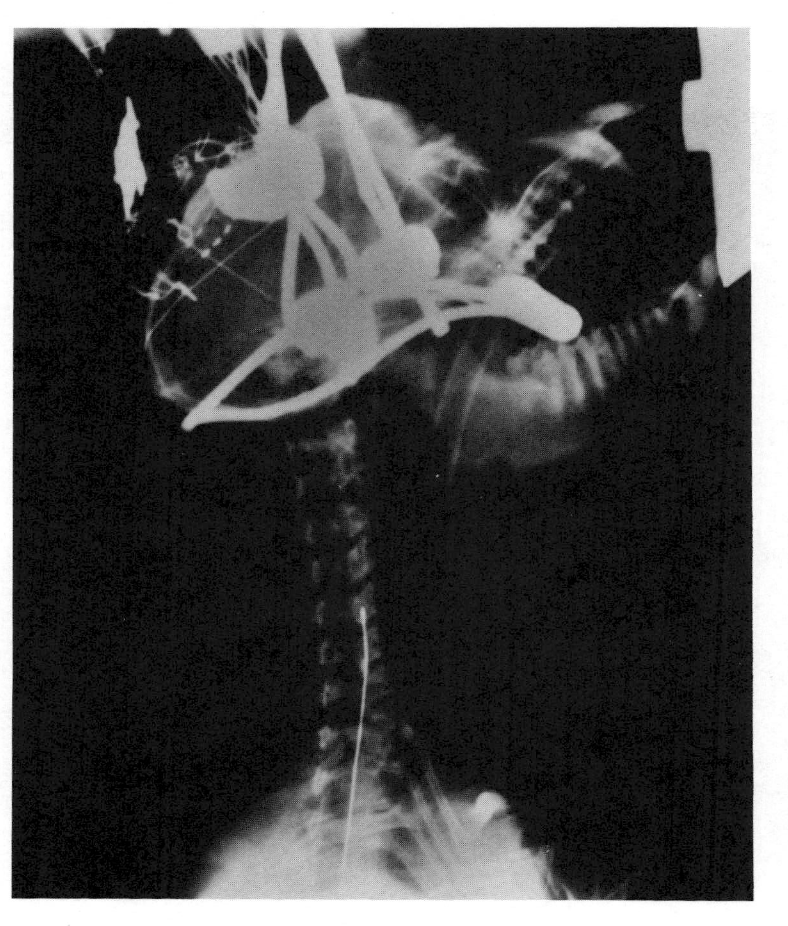

Fig. 8 - Lateral x-ray at 445 Newtons (Animal #787).

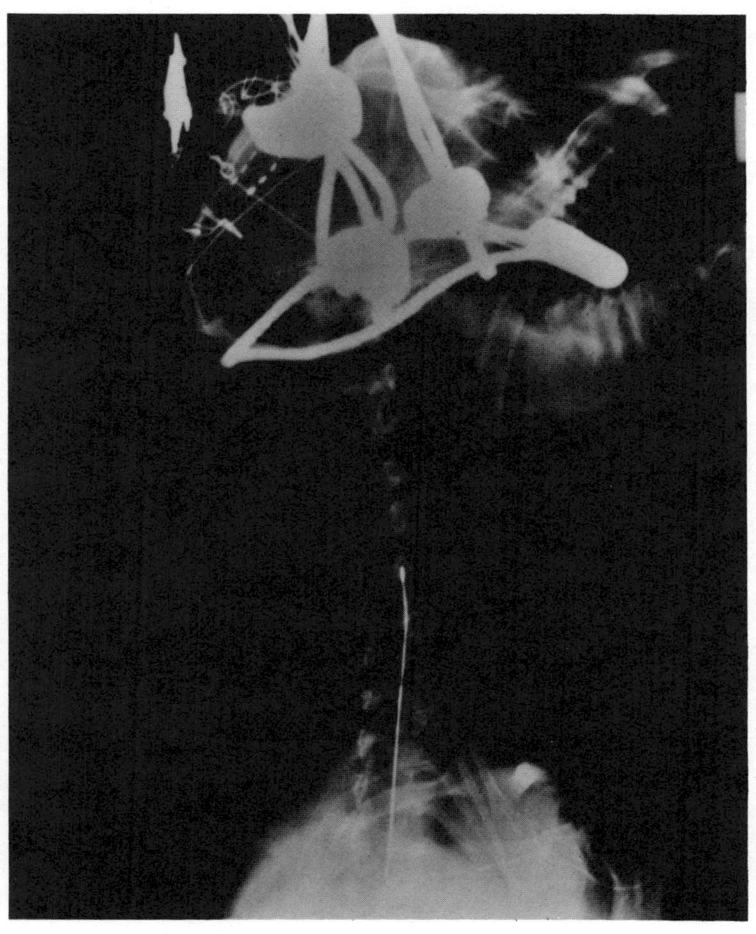

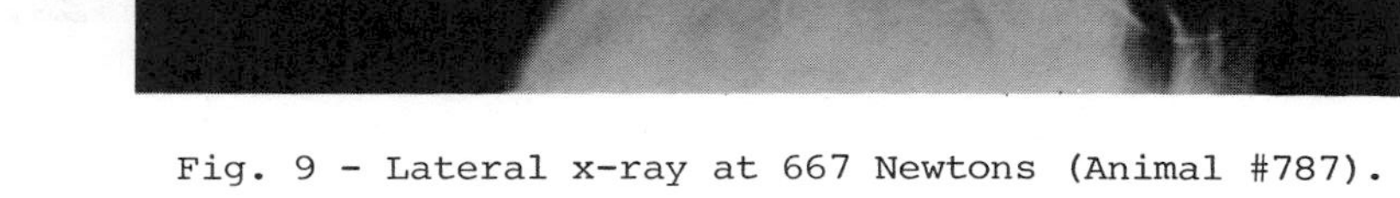

Fig. 9 - Lateral x-ray at 667 Newtons (Animal #787).

Fig. 10 - Lateral x-ray at 890 Newtons (Animal #787).

Comment - The perfusion studies in this animal show that the brain and brainstem are well perfused when the spinal cord to cortex, cortex to spinal cord and cauda equina to spinal cord evoked responses were depressed. Therefore, these changes are probably due to mechanical distention of the neural axis and not spasm or narrowing of the vessels.

Summary - Static studies indicate that the evoked potentials begin to decrease in animals prior to alterations in blood pressure and heart rate probably because of distention of the brainstem. Furthermore, the perfusion studies show that evoked potential amplitude decreases and heart rate and blood pressure alterations precede alteration in perfusion to the brain and brainstem of the animal. Therefore, in this study where the arteries and ligaments are intact, it can probably be assumed that the initial changes observed with axial tension are related to mechanical stretching of the neural fibers and are not due to inadequate perfusion of the nervous system. In those animals that died from vascular defects, the evoked potentials and heart rate deteriorated at approximately the same time, along with deterioration of blood gases and other metabolic indicators.

IN VIVO DYNAMIC MONKEY STUDIES - These studies were conducted on two recently dead and four living monkeys.

Animal #786 (Dynamic Stretch) - This 9.3 kg monkey died in hypotension 10 minutes prior to the test (Table 4). The animal sustained a C1-C2 column failure at 2668 Newtons at a loading rate of 105 cm/sec (Fig. 11). For these studies, the seatbelt restraint produced oscillations, measured by the load transducer, at approximately 200 Hz. However, minimal oscillations were observed in the linear plot measuring the displacement between T1-T2 and the skull. A review of the high speed films suggests that the failure occurred prior to the oscillations which commenced approximately 80 ms following load application. Studies conducted with a 9.3 kg dummy load produced similar oscillations. The peak loads during oscillation were approximately 25% greater than the maximum values observed at the assumed column failure levels. The oscillations were sustained for approximately 200 ms.

Two plateaus are observed in the linear pot deflection data (Fig. 11). One commences at approximately 20 ms and lasts until 35 ms; the

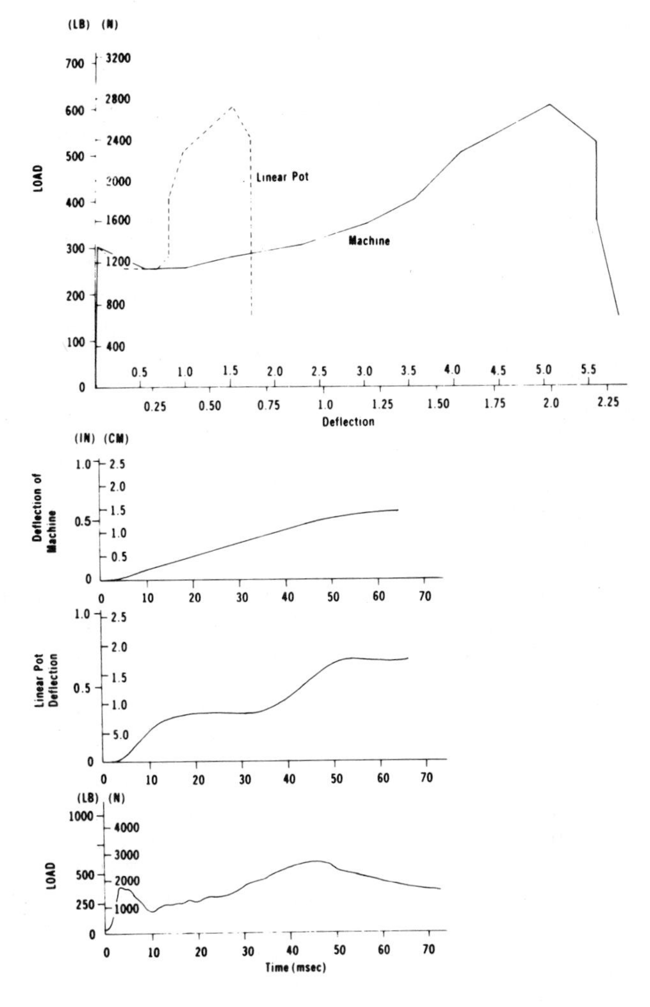

Fig. 11 - Top, machine and linear potentiometer (cervical elongation); 2nd, 3rd and 4th from top, machine deflection, linear potentiometer deflection and load versus time respectively (Monkey #786, Table 4)

second starts at approximately 55 ms. The initial sharp rise in the load was attendant with the peak accelerations. The machine displacement remained approximately linear throughout the tests. The hydraulic cylinder of the machine displaced approximately 8 cm. The seatbelt and seat deformed allowing approximately 4-5 cm of vertical movement. The cervical column displaced a total of approximately 1.76 cm (Fig. 11). The evoked potential record at the thoracic spinal electrode with cauda equina stimulation, remained for approximately 20 minutes following death. However, the evoked potential recorded at sensorimotor cortex with cauda equina stimulation and the evoked potential at thoracic spinal cord with stimulation of sensorimotor cortex were absent within 2 minutes following the maximum load application. The heart rate and blood pressure were absent within 3-4 minutes. At autopsy, a transection of all ligaments at C1-C2 was observed; however, the spinal cord was intact. The posterior and anterior ligaments and spinal column were intact from C2 to C7. The transverse ligament was torn and the dens was pulled out of C1. The carotid and the vertebral arteries were bilaterally disrupted.

Comment - The failure was vascular and ligamentus because of the continued movement of the piston up to 8 cm; however, substantial elongation was observed in the restraints. While the temporal events are difficult to correlate with the physiologic effects, the complex load displacement curve of the potentiometer suggests a multicompartmental load distribution which may be attributed to anterior spinal column failure followed closely by posterior ligament and supporting tissue load distribution.

Animal #789 (Dynamic Stretch) - This 11.1 kg animal was loaded at approximately 110 cm/sec (Table 4). During the first run, a failure in the methylmethacrylate cap occurred at approximately 1780 Newtons at a T1 to skull displacement of 1.57 cm. The displacement vs load characteristics suggested that failure of some of the elements of the spinal column probably occurred during the initial run. Following reapposition of the cap, a maximum load of 1770 Newtons was measured with a maximum T1 to skull displacement of 2.13 cm. The approximately 200 Hz oscillations were again observed in the load approximately 80 ms following application of the force. Because the cerebral electrodes were dislodged with the cap, evoked potentials were not available. The heart rate, blood pressure, and physiologic indicators were within normal limits following the first test, but deteriorated after the second run. The initial stiffness of the cervical column was decreased during the second run in contrast to the first force application. At autopsy, the ligamentum nuchae was intact along with the anterior and posterior neck muscles which showed no signs of hemorrhage. All ligaments were intact except those of the left articular capsule at C1-C2. Excessive rotation appeared to be present in the region of C1-C2 without substantial ligamentus damage.

Comment - The application of 1780 Newtons in the first run to this animal probably stressed some of the ligaments which reduced the failure force for the second run. Nevertheless, the animal maintained normal blood pressure and heart rate following the first 1780 Newton load application, but deteriorated, probably secondary to brainstem damage, following the second loading.

Animal #790 (Dynamic Stretch) - This animal expired 20 minutes prior to the run (Table 4). The 11.1 kg monkey was loaded at 100 cm/sec. The maximum laod was 2668 Newtons, with a T1-skull distention of 2.08 cm and a peak axial acceleration during the initial force application of 65 G. Five to ten G were measured in the lateral and anterior posterior directions. At autopsy, a C3-C4 disruption at the end plate was observed. The ligamentum flavum was intact and no disc or bony damage was observed. Traces of blood were present at cervical segments C3 to C7. Some stretching was observed in the C1 occipital, C1-C2 ligaments. However, the spinal cord was intact.

Comment - The force levels for failure in this dead animal were similar to those of #786 and #785, with the end plate disruption observed in monkey #700. The muscles probably did not contribute significantly to active force resistance.

Animal #785 (Dynamic Stretch) - This 13.2 kg monkey was restrained with a yoke over the shoulders which was fixed to the frame (Table 4). Maximum acceleration at the initial load application was 63 G with a loading rate of 100 cm/sec. Failure occurred between the skull and C1 at 2558 Newtons with a T1 to skull displacement of 1.6 cm (Fig. 12). The remainder of the cervical column appeared normal. The evoked potentials recorded at cortex with peripheral nerve stimulation of the spinal cord and those retrieved from the spinal cord with stimulation of the sensorimotor cortex were absent immediately following the force application. However, the responses recorded in the thoracic dorsal columns with stimulation of cauda equina were approximately normal for 20 minutes following the test. This animal expired approximately two minutes following the run. At autopsy, a complete separation was observed between the occiput and C1. The spinal cord was transected at the spino-medullary junction. There was elongation of the capsular and flaval ligaments between C1 and C2 with considerable stretching of the ligaments and looseness in this area. The 1.6 cm potentiometer displacement at failure more closely tracked the 2 cm movement of the piston (machine).

Comment - The total disruption observed in this study was probably because of the rigid fixation and the machine movement of approximately 8 cm.

Animal #794 (Incremented Dynamic Stretch, Table 4) - This 9.9 kg animal was placed in the force frame and loaded to maximum from 0.5 to 4.0 cm in 0.5 cm machine increments. Afferent and efferent evoked potentials were monitored.

The early components of the evoked response at the spinal cord with cortical stimulation was reduced 50% after the 1 cm machine displacement, with a corresponding cervical displacement (potentiometer reading) of 0.5 cm. The amplitude of the response at cortex with spinal stimulation was reduced 75% at a cervical displacement of 0.8 cm. All responses were obliterated immediately at a cervical displacement of 1.25 cm. The heart rate and blood pressure were within normal limits until the final run, but deteriorated within five minutes thereafter.

At autopsy, a disruption of the supra- and interspinous ligaments, the anterior and posterior longitudinal ligaments, and a partial disruption of the discs was observed at C3-C4. The occipito-atlanto ligament was stretched. The vertebral arteries were intact, but blood was found in the upper cervical spinal cord.

Comment - The changes in this animal were probably due to mechanical stretching of the cervical spinal cord and brainstem. The failure load in this monkey (1690 Newtons) was the lowest of the intact dynamic monkey runs. This might be due to weakening of the ligaments during the repeated load applications.

Animal #795 (Incremented Dynamic Stretch, Table 4) - This 9.0 kg monkey was placed in the force frame and restrained at the shoulders with the yoke. The piston was displacement limited and increased from 0.5 to 4.5 cm in 0.5 cm increments. The evoked potentials were observed at sensorimotor cortex with stimulation at cauda equina (CE-SMC) and at thoracic spinal cord with stimulation at sensorimotor cortex (SMC-T). No change was observed in the evoked potentials until the 2.5 cm run. At this time the cervical column was displaced 1.1 cm and the CE-SMC response was reduced 25%. When the cervical column was displaced 1.5 cm, the SMC-T response and the evoked potential CE-SMC response was reduced 50%. A failure was observed at a load of 2669 Newtons during the 4.5 cm run with a cervical displacement of 1.7 cm. All responses were eliminated within 2-3 minutes, closely followed by deterioration of the heart rate and blood pressure.

At autopsy, an avulsion of the posterior ligaments at C1-C2 was found. The right verte-

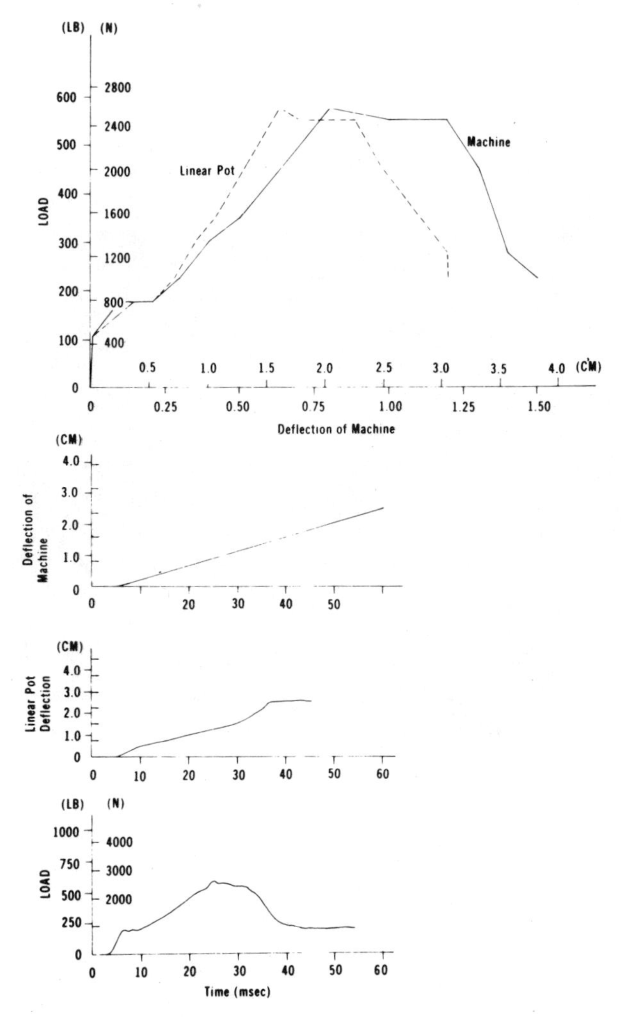

Fig. 12 - Top, machine and linear potentiometer (cervical elongation); 2nd, 3rd and 4th from top, machine deflection, linear potentiometer deflection and load versus time respectively (Monkey #785, Table 4)

bral artery was disrupted at this level. Blood was observed in the epidural space between C3 and upper midbrain levels.

Comment - The initial changes in the afferent and efferent responses were probably due to mechanical stretching of the cervical spinal cord. The final deterioration of the responses and the death of the animal were probably secondary to vascular disruption.

Summary - Studies #786, #789, #790 and #785 indicate that for the loading rates used, dynamic disruption of the cervical column probably occurs at approximately 2600 Newtons. With sufficient system stroke, total disruption occurs because the hydraulic piston stroke moves up to 8 cm. In these animals where seatbelts were used for fixation, the restraints often stretched sufficiently to prevent a complete transection. In contrast, with the yoke rigid fixation system, the disruption occurred with approximately 2 cm of piston displacement (#785). In all cases, either total or partial disruption of the cervical column occurred within approximately 2 cm distraction of the cervical column. The dynamic loads were 2-3 times those for static failure in the monkey, and the failure sites were in the upper cervical areas.

STATIC AND DYNAMIC HUMAN CADAVER STUDIES - These studies were conducted on fresh human cadaver isolated cervical spinal columns and cervical spinal cords. For comparison, studies were done on the intact human torso.

Case HS-1 (Isolated Column, Table 5) - This 50 year old isolated column was mounted on the Instron device and loaded at a constant rate of 0.13 cm per minute. In the first run, the specimen was mounted at C7 and the occiput. An end plate disruption was observed at C5-C6 with 1446 Newtons, and approximately 1.14 cm of distention (Fig. 13, 15, 15). Failure commenced in the anterior compartment (all anterior ligaments up to and including the posterior longitudinal) and proceeded posteriorly. The remaining tissue was remounted between the occiput and C3. With the second load application, an end plate failure was produced at C1-C2 which began anteriorly and occurred at 1312 Newtons.

Comment - The anterior to posterior failures

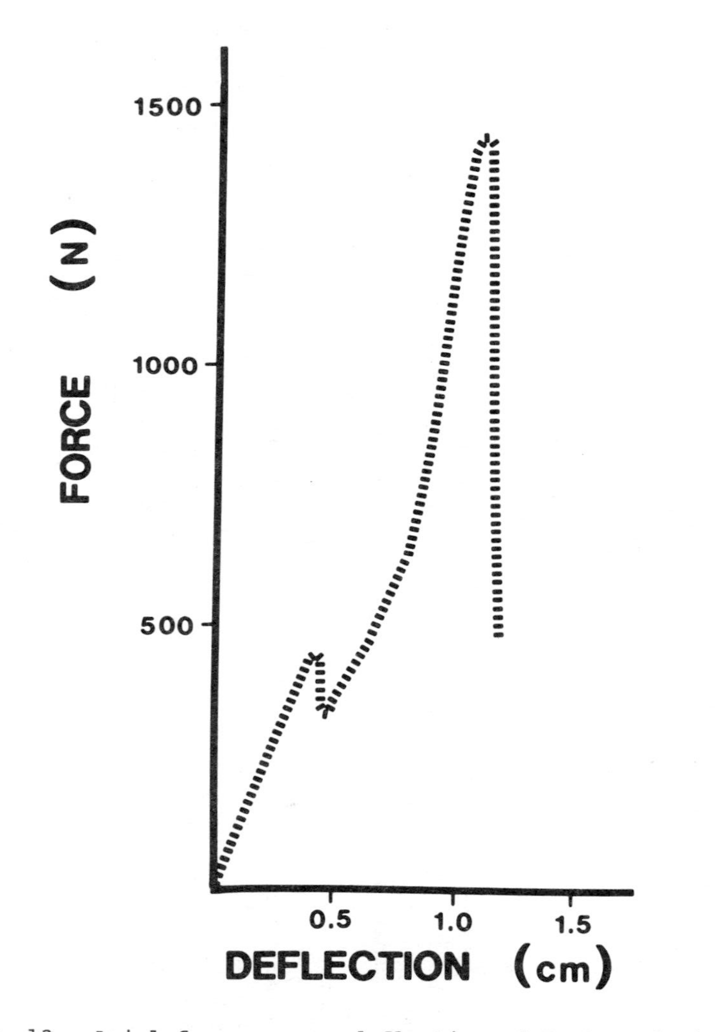

Fig. 13 - Axial force versus deflection of Instron device for a 50 year old male cervical column, with a C5-C6 disruption at the end plate (HS-1) (Table 5).

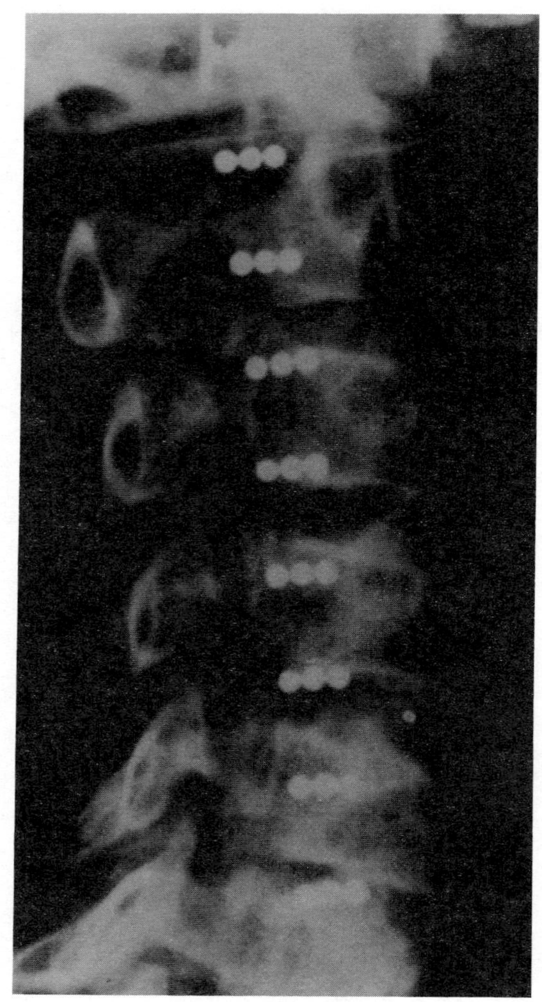

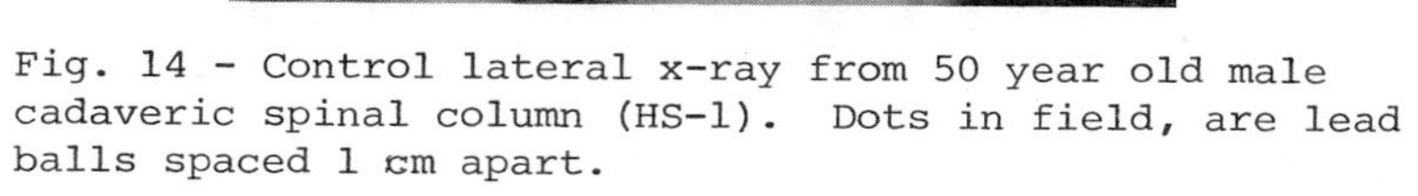

Fig. 14 - Control lateral x-ray from 50 year old male cadaveric spinal column (HS-1). Dots in field, are lead balls spaced 1 cm apart.

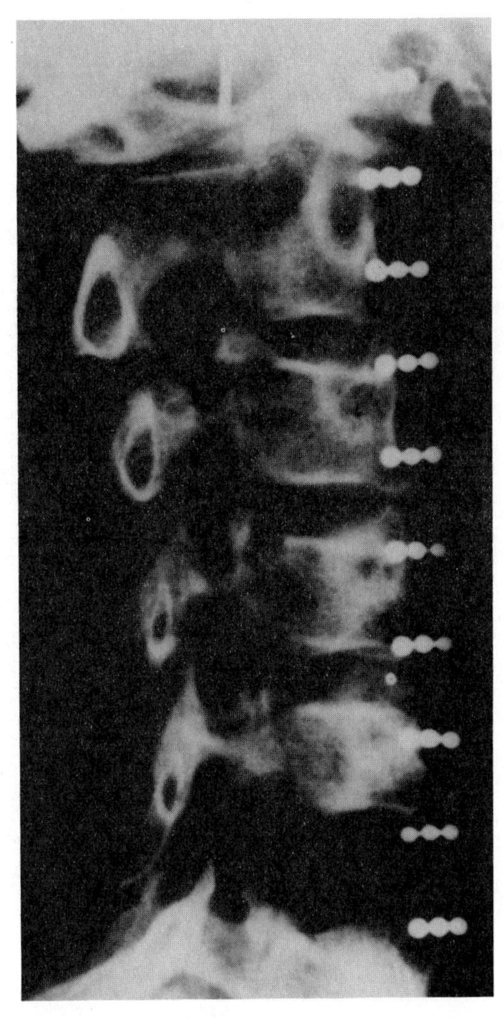

Fig 15 - Lateral view of the 50 year old male spinal column corresponding to Figure 14 (HS-1) with C5-6 endplate disruption.

were similar to those observed in the isolated cadaveric columns of the monkeys (10). The lower force value for failure between C1 and C2 is probably due to the stress imposed upon the column during the first run.

Case HS-2 (Isolated Column, Table 5) - This 36 year old 52 kg preparation sustained a circular fracture at the base of the skull at 1779 Newtons with a machine deflection of 2.67 cm at a loading rate of 0.13 cm per minute (Fig. 16, 17). For the second run, the preparation was mounted at C1 and C7 and the posterior ligaments were cut at C4-C5 up to but excluding the posterior longitudinal ligament. Failure occurred at the end plate between C4 and C5, at 1289 Newtons. For the third run the tissue was mounted at C4 and C1 with the anterior ligaments transected up to and including the posterior longitudinal ligament. Failure occurred at 622 Newtons at the end plate between C2-C3 (Fig. 18).

Comment - The basilar skull fracture of the first run shows that the strength of the cervical elements was in excess of 1779 Newtons. The posterior ligament and anterior ligament ablations demonstrate that the anterior ligaments are stronger than the posterior ligaments as found in the monkey (10). While repeated force application probably reduces the total strength of the column, this cervical section would probably have withstood an axial tension equal to 1911 Newtons, the sum of the failure loads of runs 2 and 3 (Fig. 18).

Case HS-3 (Isolated Column, Table 5) - This 61 year old preparation had a fracture of the C7 vertebra near the end plate at 1940 Newtons with 0.13 cm/min loading. For the second dynamic test, the machine malfunctioned and only provided a maximum of 2688 Newtons. This force was not sufficient to damage the spinal column.

Comment - The dynamic force applied to the isolated cervical column in the second test was at least 40% greater than that required statically, without failure.

Case HS-4 (Whole Torso, Table 5) - This 67 year old, 67 kg cadaver expired due to smoke inhalation. For this study, the entire torso, including the head and neck were mounted in the frame with a rigid yoke fixed on the shoulders (Fig. 3). At a loading rate of 125 cm/sec and a peak force of 3780 Newtons, a circular skull

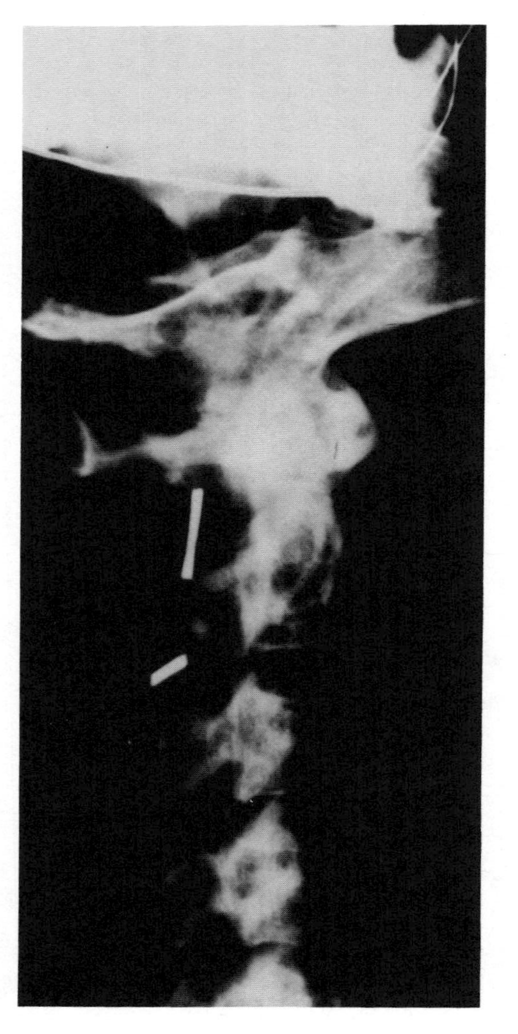
Fig. 16 - Lateral x-ray of the 36 year old male cadaver with basilar skull fracture (HS-2, Table 5).

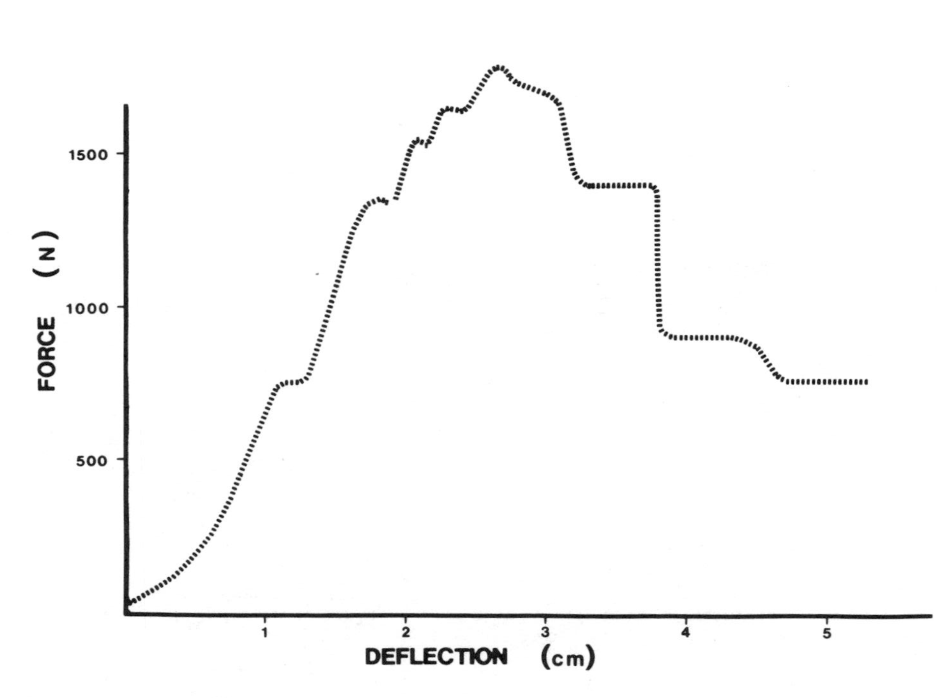

Fig. 17 - Axial force versus machine deflection for the 36 year old male with skull disruption at 1779 Newtons (HS-2).

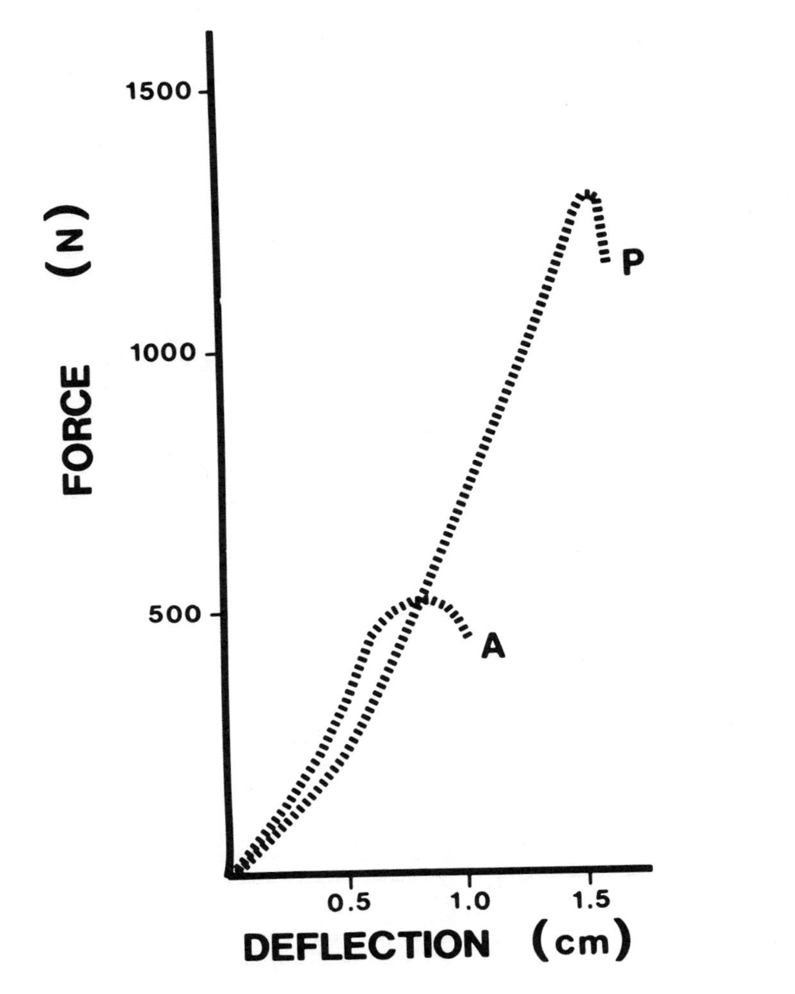

Fig. 18 - Force versus deflection following transection of the anterior ligaments (A) and transection of the posterior ligaments (P) for the spinal column of the 36 year old male (HS-2) (Table 5).

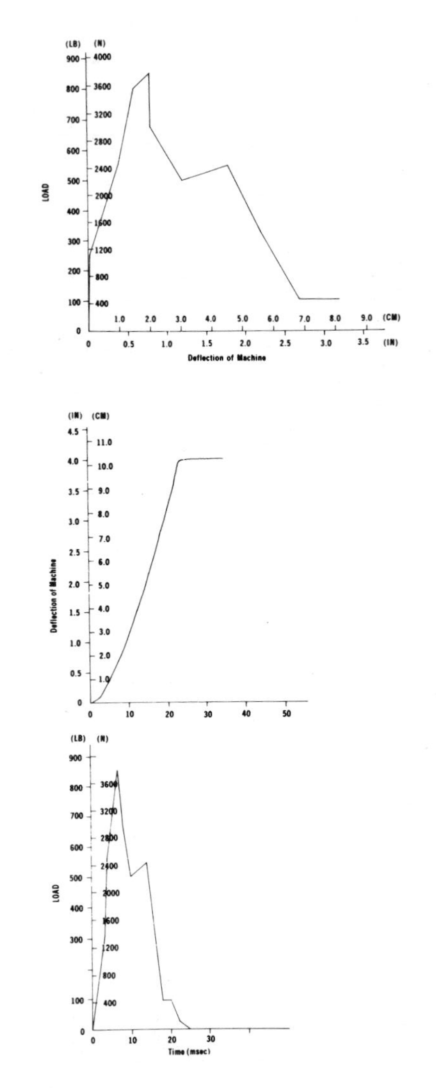

Fig. 19 - Machine deflection versus load, and machine deflection and load versus time (HS-4, Table 5).

fracture occurred at the base of the skull through the sellar region. However, the cervical ligaments did not fail (Fig. 19).

Comment - This failure was similar to that of run 1, preparation HS-1. The circular fracture observed for this preparation suggests that that dynamic force is approximately twice that required for a similar static skull fracture. The study suggests that this human cadaver neck could withstand at least 3780 Newtons without disruption.

Human Cervical Spinal Cords (HS-5, 6, 7), (Table 5) - The cervical cord was removed and mounted into the jaws of the apparatus. The center of the cord of HS-5 failed at a loading rate of 106 cm/sec between C3 and C4 at 278 Newtons. The free, unclamped region between the jaws was 7.6 cm long. For HS-6 the isolated human cervical spinal cord was transected at the midcervical level at 389 Newtons with a loading rate of 118 cm/sec. In HS-7, the human cervical cord had disruption at the midcervical section at 167 Newtons at a loading rate of 1.6 cm/sec (Fig. 20). None of the cords slipped in the jaws.

Comment - The dynamic forces were approximately twice those required for slower disruption. Failures occurred at the midsections with approximately 8-20% elongations.

Summary - The forces for the first run tissue failures in the isolated cervical columns ranged from 1440 to 1940 Newtons. The failures commenced in the anterior compartment and proceeded posteriorly. The ablation studies show that the anterior ligaments are stronger than the posterior ones. The sections are more resistant to damage with dynamic loading.

OVERALL DISCUSSION

Previous studies with statically applied forces between the head and shoulders of the Macaca rhesus monkey demonstrated a reduction in the amplitude of the efferent evoked potentials recorded at the thoracic level with application of currents to the sensorimotor cortex (10, 11, 12). The primary components of the afferent evoked potential recorded in sensorimotor cortex, nucleus ventralis posterior lateralis of the thalamus and the efferent evoked potentials record in spinal cord with stimulation of the sensorimotor cortex were reduced or obliterated with static force application between the shoulders and the head of the monkey. These evoked potential reductions always preceded or occurred with changes in heart rate and blood pressure (10). In this previous publication, the isolated monkey cervical column failed statically at approximately 400 Newtons. The isolated column studies in this study failed at substantially higher force levels with dynamic force application. The static monkey studies of this report showed reduction in the amplitude of the afferent and efferent evoked potentials which was probably related to mechanical distraction of the brainstem. In those animals where vertebral or carotid artery disruption occurred, the evoked potentials, heart rate and blood pressures deteriorated together. In contrast, in those animals where vascular disruptions were not observed, reduction in the evoked potentials preceded changes in heart rate and blood pressure. The static studies in the living animal also demonstrated that animals usually expired due to mechanical distraction of the

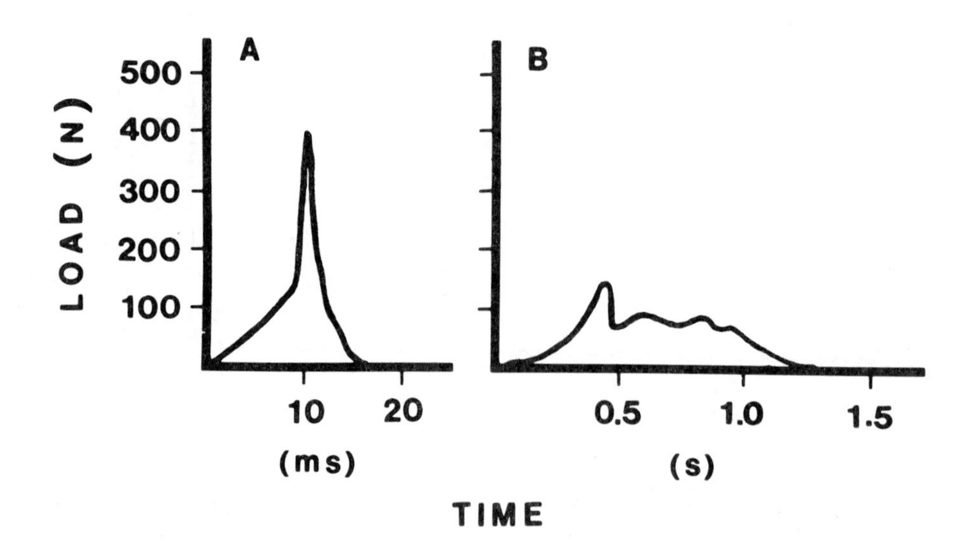

Fig. 20 - Force versus time curves for axial tension applied to cervical spinal cord at 118 cm/s (A) and 1.6 cm/s (B). B is from HS-6, and A is HS-7 (Table 5).

brainstem or vascular disruption with minimal cervical ligament destruction. However, in animal #700, an end plate disruption at C3-C4 was observed at a peak load of 1112 Newtons. The microsphere studies show that the cerebrum, brainstem and cervical and thoracic and lumbar spinal cords are well perfused during application of the axial distraction forces when the amplitude reduction in afferent and efferent evoked potentials occur. Furthermore, perfusion in the cervical spinal cord is reduced more than that in the brain or lower spinal cord elements with increased force application. The x-rays and distraction of the cervical column in the monkey with application of the static forces show that the forces straighten the column, distend it 2 cm (mainly in occiput-C2 region) and depress the shoulders downward similar to those demonstrated in the Padang tribes of Indonesia, who place rings about the neck to exaggerate the neck length (18). In other linear distraction studies in the monkey, an approximate 3 mm stretch in the cord was measured by means of marks placed directly on the dorsal columns during a 20 mm stretch measured between thoracic vertebral bodies. The amplitude of the afferent and efferent responses began to decrease with additional distraction. However, the somatosensory evoked potentials did not recover when distraction produced a more than 50% amplitude decrease in the early components (13). Breig (17) has shown that the 100 mm long freshly removed cervical human spinal cord elongates approximately 15% under its own weight. An additional 3.4 Newtons in the axial direction is then required for an additional 2 mm of extension. An additional load of approximately 9.8 Newtons beyond these levels produced no further elongation. The studies of Gray and Ritchie (19) suggest that an electrical conduction block occurs in the isolated nerve fiber when it is stretched. If the outer coverings of the nerves of the larger fibers are approximately the same as those of the smaller fibers, then the smaller ones are mechanically stronger per unit cross section than the larger axons. These findings are consistent with the studies conducted by Friede (20, 21). In his studies, cats that experienced an abrupt stretch of the vertebral column showed damage of the large fibers with preservation of the smaller diameter fibers.

For the dynamic experiments, approximately two to three times the statically applied force was required to produce total ligamentus disruption in the living intact monkey. It is difficult to explain why the upper cervical tissues were often damaged in the intact animals while the majority of failures in the isolated preparations occurred in the lower cervical regions.

Others have demonstrated that the strength of the ligaments increased with rate of loading (22, 23, 24). The muscles did not actively protect the neck in the dynamic runs since the cervical column failed with approximately the same load in the two monkeys that died minutes before the run. These studies suggest that the overlying tissue probably provides additional strength to the neck since the isolated columns failed dynamically at approximatly 1/2 the force applied to the intact monkey. For the second dynamic force application, the columns often failed at 1/2 the maximum values applied during the first run. While the initial dynamic accelerations were up to 65 Gs, the hydraulic cylinder moved at a constant velocity thereafter and during the failures. While the inertial forces might weaken the columns, their influences are probably minimal during the failure.

The studies in the isolated human cadaver columns show that with static loading the failures in the cervical area occur in the region of the end plate at 1446 to 1940 Newtons. A similar circular skull fracture without cervical column injury was observed dynamically at approximately twice the static applied loads in the intact human torso. Studies conducted by Patrick on himself demonstrated that the neck could withstand a hanging load of 1468 Newtons, equal to twice his body weight, without injury (25, 26). In control studies in 10 medical students from 23 to 34 years of age, traction forces of approximately 220 Newtons produced a separation of 3.5 mm posteriorly and 0.7 mm anteriorly (27). This would represent an average stiffness of 53 Newtons/mm. Similar stiffness factors were observed in our static studies.

For comparison with the axial tension studies of Table 5, two isolated fresh human cadaver cervical columns were mounted at T1, T2, and at the skull with methylmethacrylate. In one preparation with dynamic axial compression force applied at 120 cm/sec, a load of 4500 Newtons produced a C5 burst fracture with anterior subluxation of C5 on C6. The anterior and posterior longitudinal ligaments and discs were intact; however, the ligament of flavum and posterior ligament complex were disrupted. In the other cervical column, a C5 vertebral fracture without subluxation but with destruction of the anterior and posterior longitudinal ligaments occurred at 4410 Newtons at a 130 cm/sec loading rate. These levels are similar to that reported by Culver (28) in unembalmed cadavers placed in the supine position with the cervical spine along the axis of an impactor. Furthermore, these values are approximately two times greater than the static values observed by others (29, 30, 31).

SUMMARY

(1) The afferent and efferent evoked potentials decrease markedly in the monkey with static and dynamic axial tension loading prior to heart rate and blood pressure changes.

(2) Radioactive microsphere studies in the monkey suggest that perfusion is not altered in the brain and spinal cord when the evoked potentials decrease.

(3) The majority of the animal and human failures occurred at the end plate.

(4) The fresh isolated monkey cervical spinal column failed with axial tension between 325-534 Newtons statically (10) and 1289-1423 Newtons dynamically in the first runs.

(5) The anterior ligaments are approximately two times stronger than the posterior ligaments.

Failures routinely started in the anterior ligaments and proceeded posteriorly.

(6) The intact monkey spinal column failed at approximately 1100 Newtons statically and approximately 2600 Newtons dynamically.

(7) The isolated human cervical column failed in the first runs with static loading at 1446-1940 Newtons.

(8) A circular skull fracture was observed in the isolated human cadaver specimen at 1779 Newtons statically, and at 3780 Newtons dynamically in the intact torso; both without observable cervical ligamentus damage.

(9) The monkey and human cervical spinal cords were stronger dynamically.

ACKNOWLEDGMENT

This research was supported in part by the Office of Naval Research Contract N00014-77-C-0749.

REFERENCES

1. R. P. Greenberg, D. P. Becker, J. D. Miller and D.J. Mayer, "Evaluation of brain function in severe human head trauma with multimodality evoked potentials. Part II. Localization of brain dysfunction and correlation with posttraumatic neurological condition." J Neurosurg 47:163-177, 1973.

2. S. J. Larson, A. Sances, Jr., J. J. Ackmann and D. H. Reigel, "Non-invasive evaluation of head trauma patients." Surgery 74:34-40, 1973.

3. A. Sances, Jr., S. J. Larson, J. F. Cusick, J. Myklebust, C. L. Ewing, R. W. Jodat, J. J. Ackmann and P. R. Walsh, "Early somatosensory evoked potentials." Electroencephalogr Clin Neurophysiol 45(4):505-514, 1978.

4. S. J. Larson, R. A. Holst, D. C. Hemmy and A. Sances, Jr., "The lateral extracavitary approach to traumatic lesions of the thoracic and lumbar spine." J Neurosurg 45:628-637, 1976.

5. D. R. Giblin, "Somatosensory evoked potentials in healthy subjects and in patients with lesions of the nervous system." Ann NY Acad Sci 112:93-142, 1964.

6. S. J. Larson, A. Sances, Jr. and P. C. Christenson, "Evoked somatosensory potentials in man." Arch Neurol (Chic) 15:88-94, 1966.

7. J. F. Cusick, J. Myklebust, S. J. Larson and A. Sances, Jr., "Spinal evoked potentials in the primate: Neural substrate." J Neurosurg 49:551-557, 1978.

8. J. F. Cusick, J. Myklebust, S. J. Larson and A. Sances, Jr., "Spinal cord evaluation by cortical evoked responses." Arch Neurol 36(3):140-143, 1979.

9. A. Ommaya et al, "Pathologic biomechanics of central nervous system injury in head impact and whiplash trauma." Accident Pathology, Proc Int'l Conf on Accident Path, K. Brinkhous (ed), Government Printing Office, Washington, DC, pp. 160-181, 1970.

10. A. Sances, Jr., R. Weber, J. Myklebust, J. Cusick, S. J. Larson, P. R. Walsh, B. Saltzberg, D. Thomas, C. Ewing, T. Cristoffel and C. Houterman, "The evoked potential: an experimental method for biomechanical analysis of brain and spinal injury." Proc 24th Stapp Car Crash Conf, Society of Automotive Engineers, New York, pp. 63-100, 1980.

11. A. Sances, Jr., J. Myklebust, S. J. Larson, J. F. Cusick and R. Weber, "The evoked potential - a biomechanical tool." Chapter in Impact Injury of the Head and Spine, C. L. Ewing, D. J. Thomas, A. Sances, Jr., and S. J. Larson, (eds), Charles C. Thomas, Publ., Springfield, IL, (in press).

12. A. Sances, Jr., J. Myklebust, S. J. Larson, J. F. Cusick, R. C. Weber and P. R. Walsh, "Bioengineering analysis of head and spine injuries." CRC Crit Rev Bioeng 5(2):79-122, 1981.

13. S. J. Larson, P. R. Walsh, A. Sances, Jr., J. F. Cusick, D. C. Hemmy and H. Mahler, "Evoked potentials in experimental myelopathy. Spine 5(4):299-302, Jul/Aug 1980.

14. R. W. Hobson, II, C. B. Wright, M. J. Zinner and R. E. Lamoy, "Cerebral blood flow determinations by radioactive microspheres in the subhuman primate: Influence of unilateral internal carotid ligation, hypercapnic acidosis, and hypocapnic alkalosis." Surgery 80(2):224-230, Aug 1976.

15. A. Alm, "Radioactivity labelled microspheres in regional cerebral blood flow determinations. A study on monkeys with 15 and 35 μm spheres." Acta Physiol Scand 95:60-65, 1975.

16. A. M. Rudolph and M. A. Heyman, "The circulation of the fetus in utero: Methods for studying distribution of blood flow, cardiac output, and organ blood flow." Circ Res 21:163, 1967.

17. A. Breig, "Biomechanics of the Central Nervous System." Year Book Medical Publishing, Chicago, 1960.

18. R. Roaf, "Scoliosis." Edinburgh, Churchhill Livingston Ltd., 1966.

19. J. A. B. Gray and J. M. Ritchie, "Effects of stretch on single myelinated nerve fibers." J Physiol 124:84, 1954.

20. R. Friede, "Specific cord damage at the atlas level as a pathogenic mechanism in cerebral concussion." J Neuropathol Exp Neurol 19:266-279, 1960.

21. R. Friede, "The pathology and mechanics of experimental cerebral concussion." Wadd Technical Report 61-256, Air Research and Development Command, United States Air Force, Wright-Patterson Air Force Base, OH, 1961.

22. J. W. Fielding, A. H. Burnstein and V. H. Frankel, "The nuchal ligament." Spine 1 (1):3, 1976.

23. F. R. Noyes, J. L. DeLucas and P. J. Torvik, "Biomechanics of anterior cruciate ligament failure: an analysis of strain-rate sensitivity and mechanisms of failure in primates." J Bone Jt Surg 56A:236, 1974.

24. H. Tkaczuk, "Tensile properties of human lumbar longitudinal ligaments." Acta Orthop Scand Suppl 115, 1968.

25. H. J. Mertz, Jr. and L. M. Patrick, "Investigation of the kinematics and kinetics of whiplash." Proc 11th Stapp Car Crash Conf,

Society of Automotive Engineers, 175, 1967.
26. J. H. Mertz, Jr. and L. M. Patrick, "Strength and response of the human neck." Proc 15th Stapp Car Crash Conf, Society of Automotive Engineers, CA, 207, 1971.
27. S. C. Colachis, Jr. and B. R. Strohm, "Cervical traction: relationship of traction time to varied tractive force with constant angle of pull." Arch Phys Med Rehabil 46:815, 1965.
28. R. H. Culver, M. Bender and J. W. Melvin, "Mechanisms, tolerances and responses obtained under dynamics of superior inferior head impact." PB-299292, University of Michigan Highway Safety Research Institute, May 1978.
29. R. J. Bauze and G. M. Ardran, "Experimental production of forward dislocation in the human cervical spine." J Bone Jt Surg 60-B(2): 239, 1978.
30. B. R. Selecki and H. B. L. Williams, "Injuries to the cervical spine and cord in man, Australian Med Assoc, Mervyn Archdall Med Monograph #7, Australian Medical Publishers, South Wales, 1970.
31. H. Yamada, "Strength of Biological Materials." Robert E. Krieger, Huntington, NY, 1973.

821155

Injuries to the Cervical Spine Caused by a Distributed Frontal Load to the Chest

R. Cheng, K. H. Yang, R. S. Levine and A. I. King
Wayne State University

R. Morgan
NHTSA, U.S. Department of Transportation

ABSTRACT

Unembalmed cadavers were exposed to $-G_x$ acceleration while restrained by applying a frontal load to the chest. A pre-deployed non-venting production air cushion mounted on a non-collapsible horizontal steering column provided the distributed load. The sled deceleration pulse was determined from a series of Part 572 dummy runs in which the HIC, chest acceleration and knee loads were at but not in excess of the limits specified in the current FMVSS 208.

A total of six cadavers have been tested. In three of the runs, there were severe neck injuries of the type which have not been observed previously in belted tests. They include complete severance of the cord, complete avulsion of the odontoid process, atlanto-occipital separation with ring fracture. This study does not claim to establish the injury potential of air bags but uses the air bag to provide a uniform restraining load to the chest to investigate the mechanism of neck injuries. High neck loads were encountered in this mode of impact in which the head and neck kinematics were different from those of belted cadaveric subjects. Cadaveric and dummy neck loads were also different. The proposed neck fracture load is 6.2 kN (1,400 lb). This is the resultant neck load.

INTRODUCTION

Severe neck injuries are rarely reported in frontal automotive collisions, particularly for restrained occupants. In a carefully documented study by Patrick et al (1), only 3 cervical fractures were reported out of 128 cases of frontal collisions involving belted occupants of Volvo cars. Hartemann et al (2) compared injuries of belted and unbelted occupants in 200 pairs of matched field accidents. There was only one case of neck injury with a regional AIS of 5. Rattenbury et al (3) and Dalmotas (4) investigated biased samples of accidents involving serious injuries or those judged to have an AIS of 2 or greater. Neck injuries were also rare in these reports. Huelke et al (5) surveyed the literature and case histories from the files of the Highway Safety Research Institute of the University of Michigan. The objective was to locate cases in which cervical fractures and fracture dislocations occurred without head impact. For automotive occupants involved in frontal collisions, cervical injuries were rare but the serious ones that were reported involved the upper cervical area - hangman's fracture and atlanto-occipital separation.

Experimental studies on neck injuries due to frontal impacts have been conducted by several investigators. Schmidt et al (6) conducted 49 cadaveric impacts with a peak deceleration of 25 g. Thirty of these were restrained by a three-point belt and the remainder by a knee-bolster shoulder belt passive restraint system. Only minor neck injuries were observed in these specimens. Schmidt et al (7) extended their study to 103 restrained cadaveric subjects and conducted the tests at approximately 20 g with 3 different pulse durations between 63 and 90 ms. As far as cervical injuries were concerned, those that were most frequently observed were disc rupture and hemorrhage in the lower cervical spine. Fractures were rare. Of the five cadavers used by Cromack and Ziperman (8), 3 sustained cervical fractures during frontal sled impacts at a nominal peak deceleration of 21 g. All subjects were restrained by a three-point belt. Levine and Patrick (9) conducted a series of 9 belted cadaveric tests from 30 to 64 km/h (19 to 40 mph). The peak deceleration ranged from 7 - 16

g. Three cervical injuries were observed during runs made at 64 km/h. In another study by Levine et al (10), an additional 10 subjects were used. They were provided with knee braces to simulate quadriceps action and were prevented from submarining. The peak deceleration ranged from 13 to 18 g and the pre-impact speed was from 46 to 64 km/h (29 to 40 mph). There were 4 cervical injuries, 2 at the highest impact speed and 2 at lower speeds.

Walsh and Kelleher (11) conducted car-to-car crashes using cadavers and dummies which were restrained by either an air cushion or a three-point belt restraint system. Four of the eight cadavers were restrained by an air bag installed in 1973 Chevrolet Impala sedans. A pair of full-size Chevrolets was used in each test. At a closing speed of 97 km/h (60 mph) no neck injuries were found. In another study by Thomas and Jessop (12), fully restrained rhesus monkeys were subjected to high levels of $-G_x$ impacts (over 100 g), resulting in severe cervical injuries.

This paper reports on the results of a cadaveric study involving the use of a pre-deployed driver air bag system. A total of six subjects have been exposed to peak deceleration levels of 32 to 39 g. Severe neck injuries were found in 4 of 6 cadavers. It should be noted that although the air bag was a production item the air bag system was not representative of any known system in production. The experiment was originally designed to evolve a standard test methodology involving a driver air bag rather than the solicitation of human surrogate response to such a system.

METHODS

The experimental procedure can be divided into three phases. In the first phase, a suitable deceleration pulse was established to facilitate comparison of neck injuries with those sustained by other parts of the body. An instrumented Part 572 anthropomorphic test device (ATD) was used for the measurement of dummy parameters which could be related to current safety standards. The deceleration pulse was approximately triangular in shape. It had a nominal peak of 38 g with a duration of about 100 ms. This corresponded to a 48 km/h (30 mph) impact speed with a 0.47 m (18.5 in) stopping distance. The HIC, chest acceleration and knee loads were at but not in excess of the limits specified in the current FMVSS 208. Details of the experimental set-up are described below.

The second phase consisted of cadaveric* sled tests using the deceleration pulse obtained under Phase 1. There was a total of six runs involving six different unembalmed cadavers. The same experimental set-up was used for this series and the principal data were in the form of autopsy results and biomechanical response.

The third phase consisted of a series of runs using a Hybrid III dummy which was instrumented with a neck load cell capable of measuring axial and shear forces as well as bending moment in the mid-sagittal plane. The purpose of these runs is to compare dummy neck forces and moments with cadaveric head accelerations and to obtain a tolerance value for the neck under a combined axial and bending inertial load.

a) Sled Equipment

All of tests were run on the WHAM III deceleration sled. A schematic of the test set-up is shown in Figure 1. A stop-action photograph of a cadaveric impact using the test set-up is shown in Figure 2. The test subject was seated on a hard bench seat covered with a 1/4" thick sheet of teflon. The front edge of the seat was raised by setting the seat pan at an angle of 10 degrees with respect to the horizontal. The seat back was taken from a VW Rabbit bucket seat, remounted to pivot about hinges located in the rear of the seat frame which was built up with 2" square steel tubing. Two automotive type shock absorbers were installed in the back of the seat back to cushion the rebound of the cadaver. The foot rest was adjustable in location, height and angle to fit each cadaver individually. The distributed loading device used was a pre-deployed production air bag module mounted on a regular steering wheel. A standard steering column load cell was placed between the steering wheel and a horizontally positioned steel tube was used to simulate a non-energy absorbing steering column. The complete steering wheel and air bag unit was adjustable to position the pressurized air bag so that it just touched the cadaveric thorax, for a bag pressure of 8.6 kPa (1.25 psi). The height of the air bag module was also adjusted so that the height of the horizontal steering column was at the mid-sternum level for each run. The knees were set up to impact energy absorbing Hexcel blocks which were situated on adjustable knee restraint supports. The size of the pre-crushed blocks was 200 x 200 mm. Each block was 180 mm thick and was backed by a triaxial load cell which was fixed rigidly to the sled. The stiffness of the Hexcel was 317 kPa (46 psi). The initial spacing of 127 mm (5 in) between the knee and the Hexcel was based on Part 572 dummy runs performed in Phase 1. Peak knee loads were made to occur at about the same time as that of the steering column load. The height of these Hexcel blocks was adjusted so that the knee was slightly below the top of the block. The centers of the knees were 292 mm (11.5 in) apart.

*The protocol for use of cadaveric subjects in this research project was approved by the Human and Animal Investigation Committee of Wayne State University and is in compliance with NHTSA order 700-4.

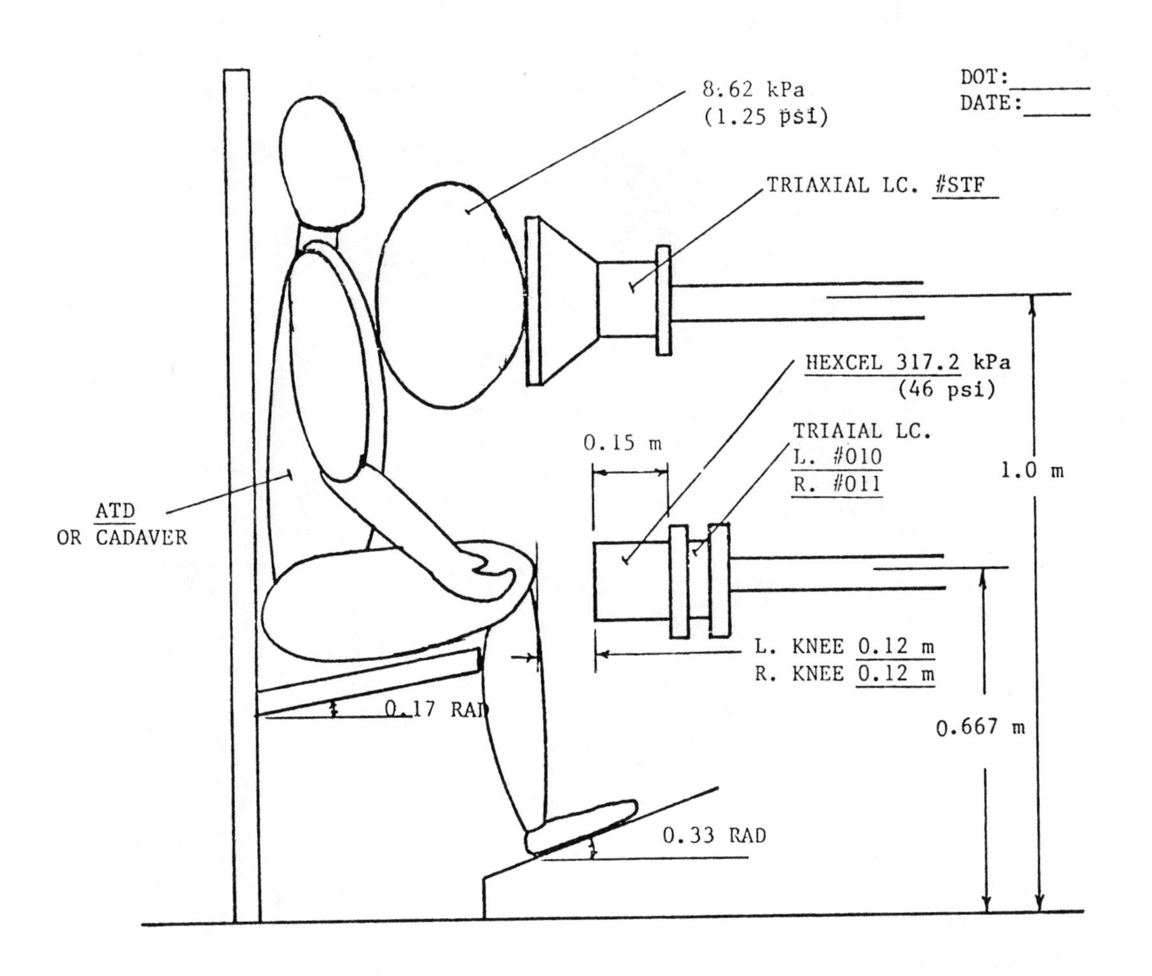

Figure 1. Schematic of Test Set-up

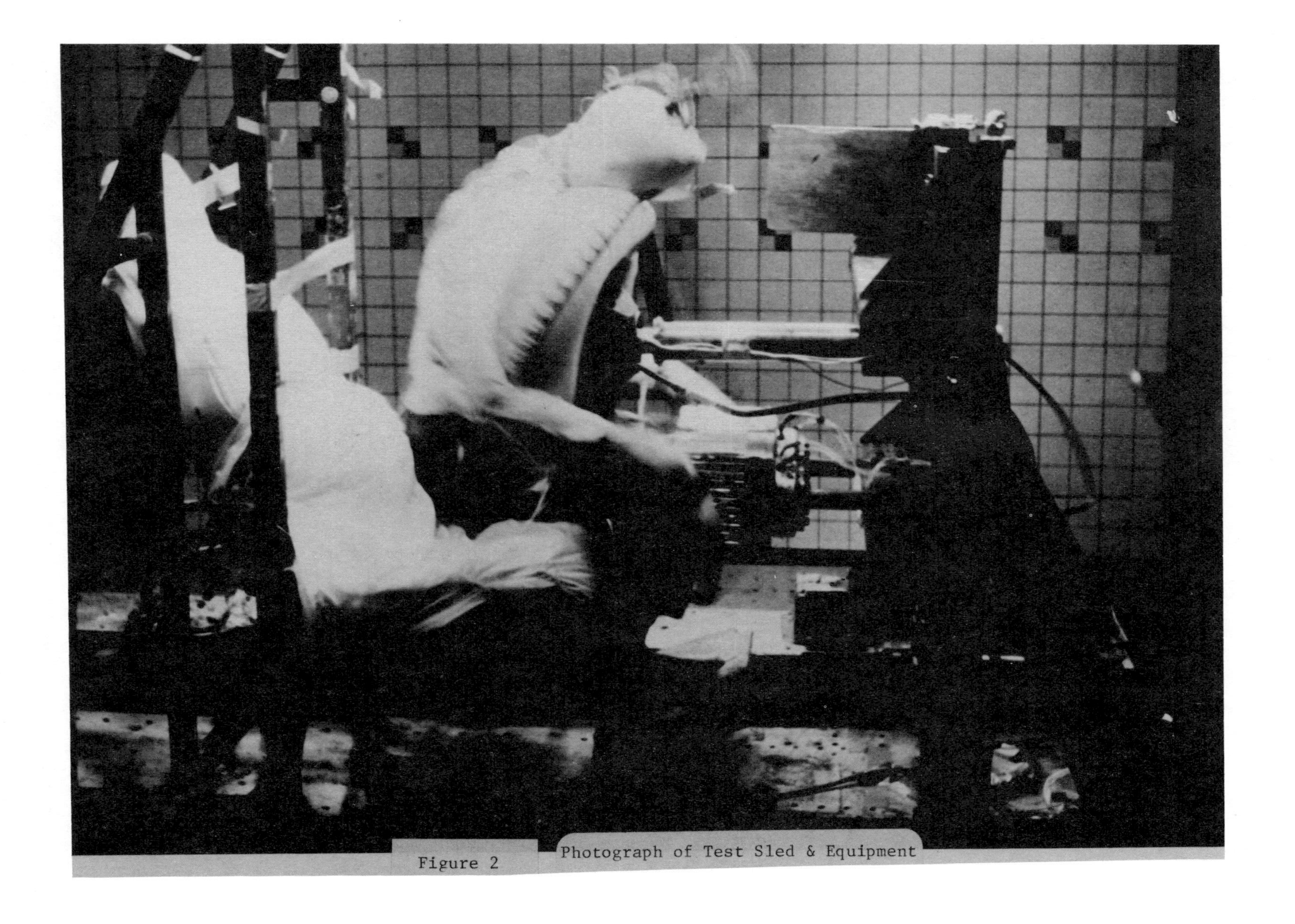

Figure 2 Photograph of Test Sled & Equipment

b) Instrumentation

Cadaveric instrumentaion included the following accelerometers:

Head	9	(3-2-2-2 configuration)
T1	3	
T4	3	
T12	3	
Pelvis	3	
U & L sternum	2	
R & L 4th rib	2	
R & L 8th rib	2	
Shoulder	3	

To render the cadaver more human-like, saline, colored with India ink, was fed into the aorta near the diaphragm through a Foley catheter. A vent hole was inserted in the brachial artery. Just prior to a run, the catheter balloon was pumped up to block the flow below it. Saline was allowed to flow until it came out of the vent tube. The vent was then clamped, thus ensuring that the system above the blockage, including the head, was filled with saline during the test. For the first four cadaveric runs, the lung was pressurized with air at about 3.4 kPa (0.5 psi) through a tube inserted into the trachea. Gauze and cotton were used to pack the space between the trachea and the tube to block air leakage. During these four runs, the lung was essentially a closed system. In the last two runs, however, a different approach was used. The lung was carefully drained of any fluid and then it was aerated repeatedly with room air prior to and just before the test. However, it was not pressurized.

RESULTS

A total of six cadaver runs were made using the test set up shown in Figures 1 and 2. Several Hybrid III dummy runs were also made mainly to obtain the neck loads. Figure 3 shows a typical sled deceleration pulse which was obtained by running a Part 572 dummy to assure that FMVSS 208 requirements were met. Figures 4 through 8 show the x-component* of acceleration of the head, T1, T4, T12 and pelvis for the six cadaver runs. Figures 9 through 13 are the z-component* of acceleration of the same points. The resultant steering column load for the six cadaver runs is shown in Figure 14 and the left and right knee loads in Figures 15 and 16. Figure 17 shows air bag pressure during impact. Figures 18, 19, 20, and 21 provide the Hybrid III neck responses of axial, shear and resultant load and the moment M_y.* A summary of cadaver and injury data for the six runs is provided in Table I.

*The x-axis is defined as being in the P-A direction, the y-axis in the R-L direction and the z-axis in the I-S direction. All accelerometer data are relative to body-fixed axes. Load cell data are relative to sled-fixed axes.

The data traces shown in Figures 4 - 21 have been shifted in time to align the largest peak. Although there is some loss of time phase relationship among different data channels, that relative to the sled pulse is still approximately correct. However, it should be noted that the peak steering column force (Figure 18) is actually coincident in time with the peak air bag pressure (Figure 21). The plots in these figures do not reflect this fact. Also, the knee loads tended to peak simultaneously at about 30 ms before the peak steering column load is reached.

In addition to transducer data, high speed film data were acquired, using 3 orthogonal cameras. The torso of the test subject was upright as it impacted the air bag. The head pitched forward and impacted the top rim of the air bag. For cadaveric subjects, the neck appeared to be in tension as the torso started to rebound. A rearward convexity of the thoraco-lumbar spine could be observed at the initiation of rebound. The kinematics of the Hybrid III dummy were different from those of the cadaver. The head appeared to be more rigidly attached to the torso which was also more rigid than that of the cadaver. Rebound of the head and torso occurred simultaneously. Four of the six cadaveric subjects sustained neck injuries, 3 of which were considered to be fatal. Two of these with severe neck injury also sustained multiple rib fractures. Rib fractures were also observed in one subject which did not have any neck injuries. The severe neck injuries were of two types, C1-C2 separation or atlanto-occipital separation due to ring fracture. There were also injuries to the thoracic vertebrae, as shown in Table. 1.

DISCUSSION

The many injury mechanisms of the neck have been described by Moffat et al (13) In the sagittal plane, the neck can simultaneously sustain axial and bending loads which combine to provide 4 possible mechanisms of injury. They are flexion with tension or compression and extension with tension and compression. In this study, the mechanism under investigation is combined axial tension and flexion. There does not appear to be any causal relationship between neck and rib fractures. The latter are well correlated with body mass since those with rib fractures had a mass of 72.5 kg or higher.

The discussion will be focussed on the observed cervical injuries which are uncommon and extremely severe. A series of Hybrid III runs were carried out in an attempt to correlate the observed injuries to measured neck loads in the dummy. Nyquist et al (14) proposed a neck tolerance value for belted occupants by attributing Hybrid III neck loads to field accident injury data reported by Patrick et al (1). The maximum AIS assigned to the 3 cervical fractures in the Volvo data was 3 and the average axial and shear forces were

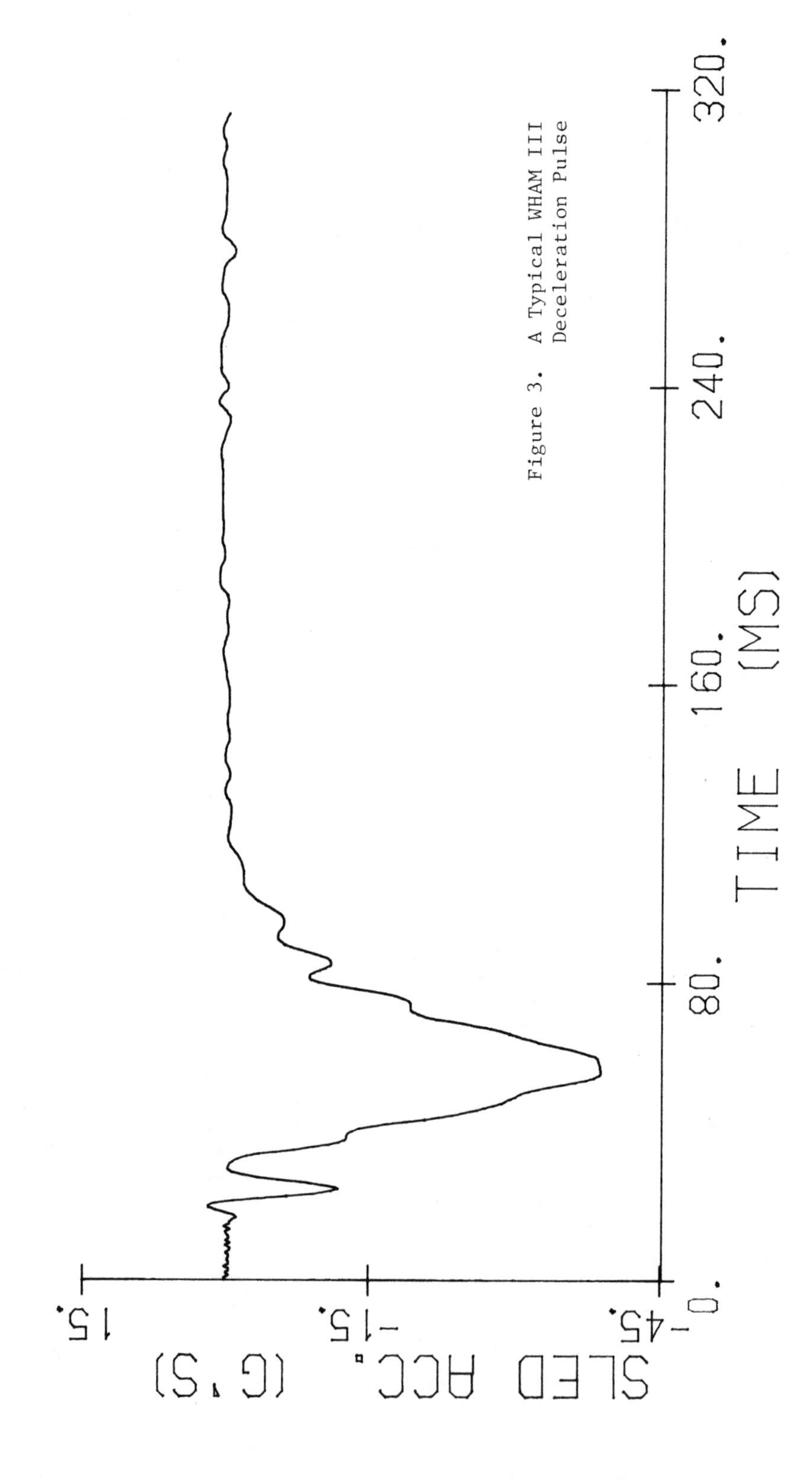

Figure 3. A Typical WHAM III Deceleration Pulse

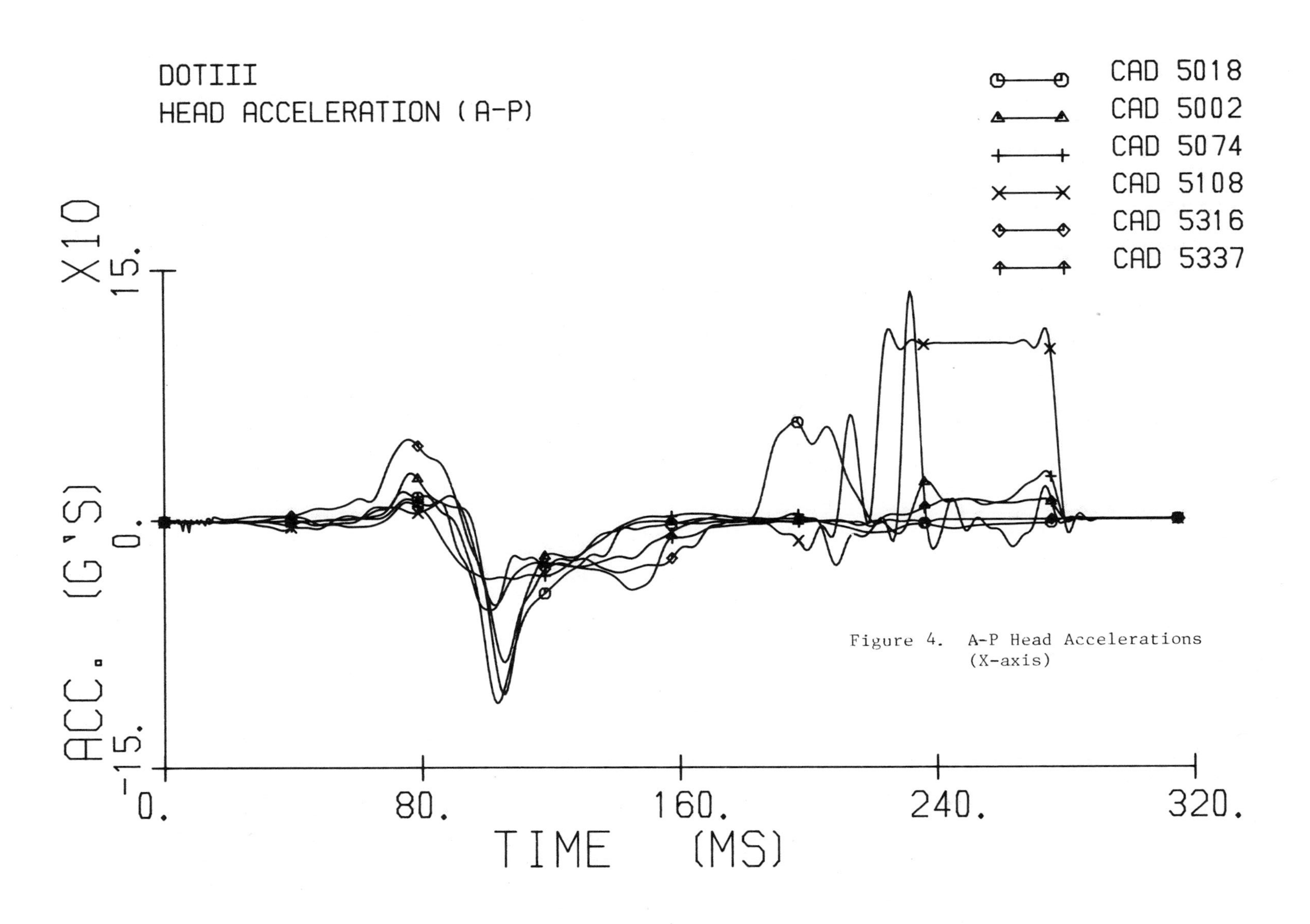

Figure 4. A-P Head Accelerations (X-axis)

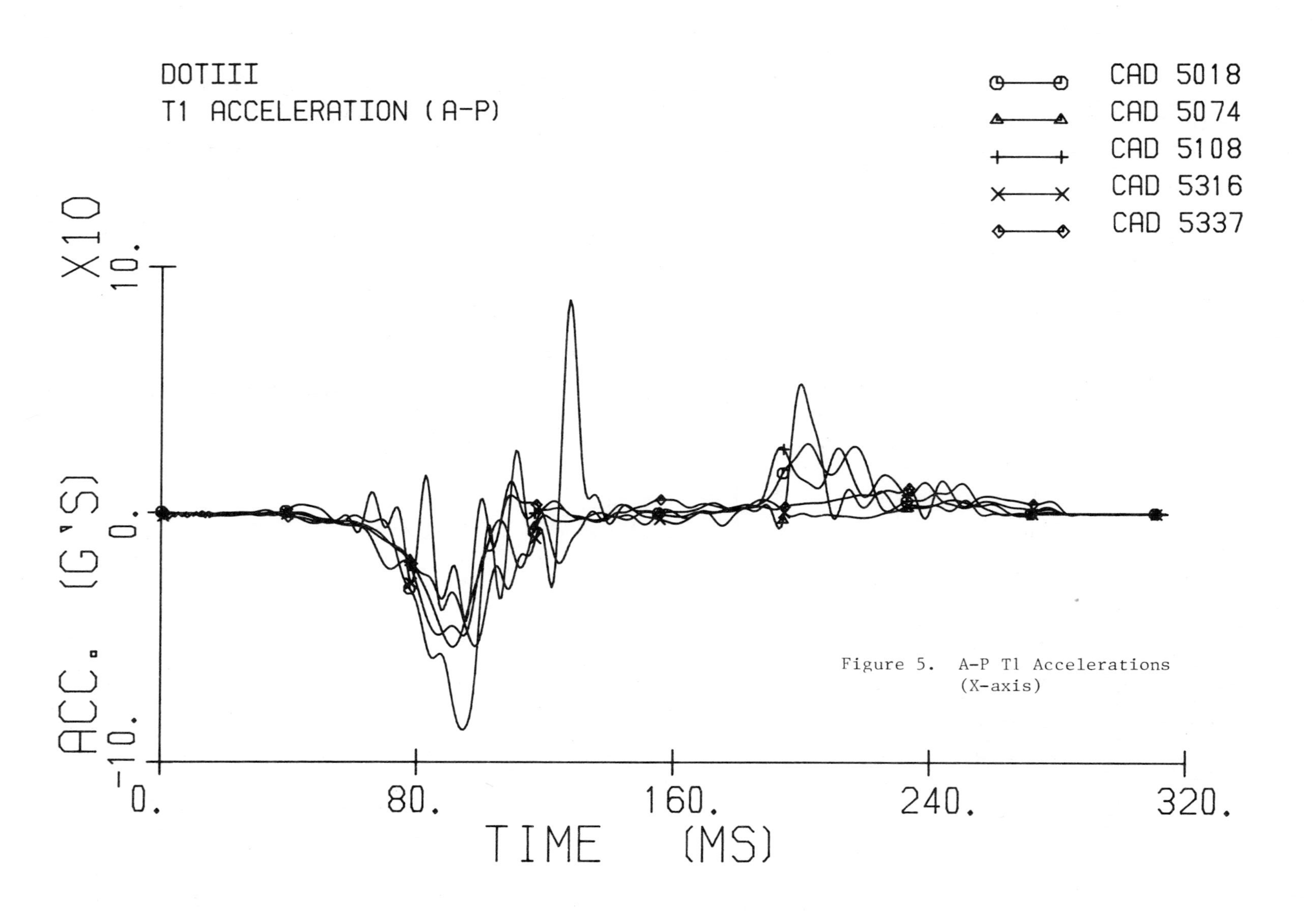

Figure 5. A-P T1 Accelerations
(X-axis)

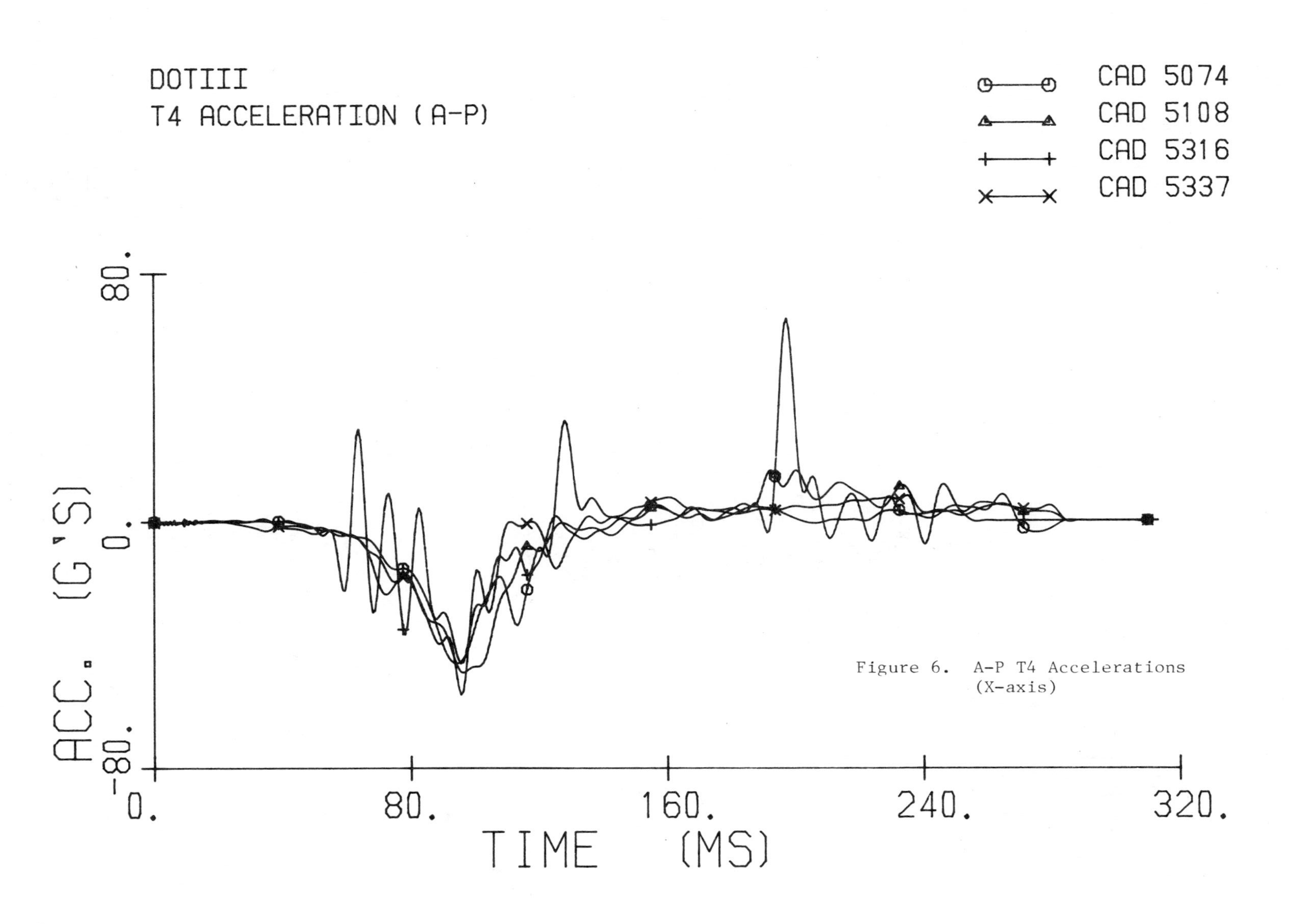

Figure 6. A-P T4 Accelerations (X-axis)

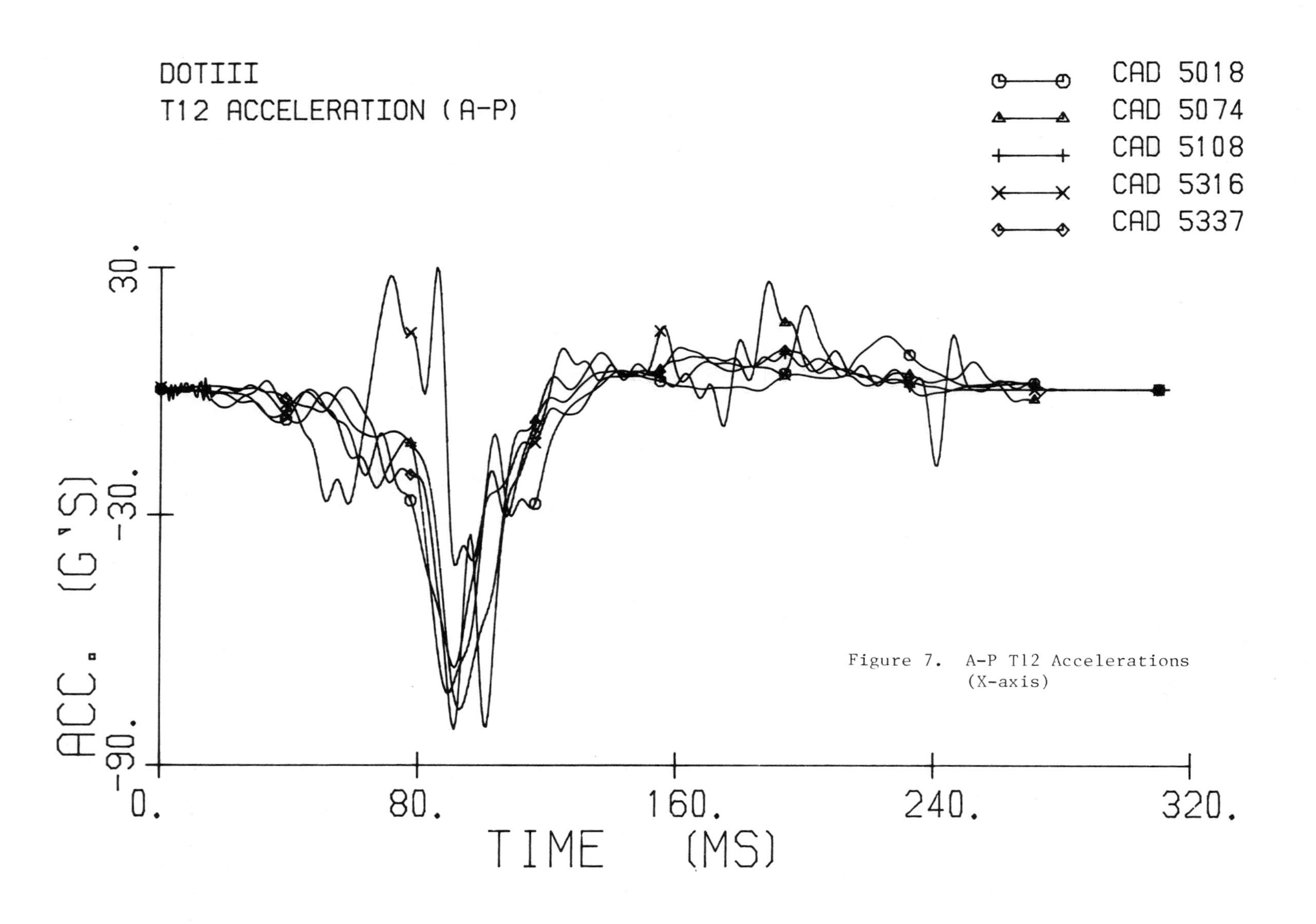

Figure 7. A-P T12 Accelerations (X-axis)

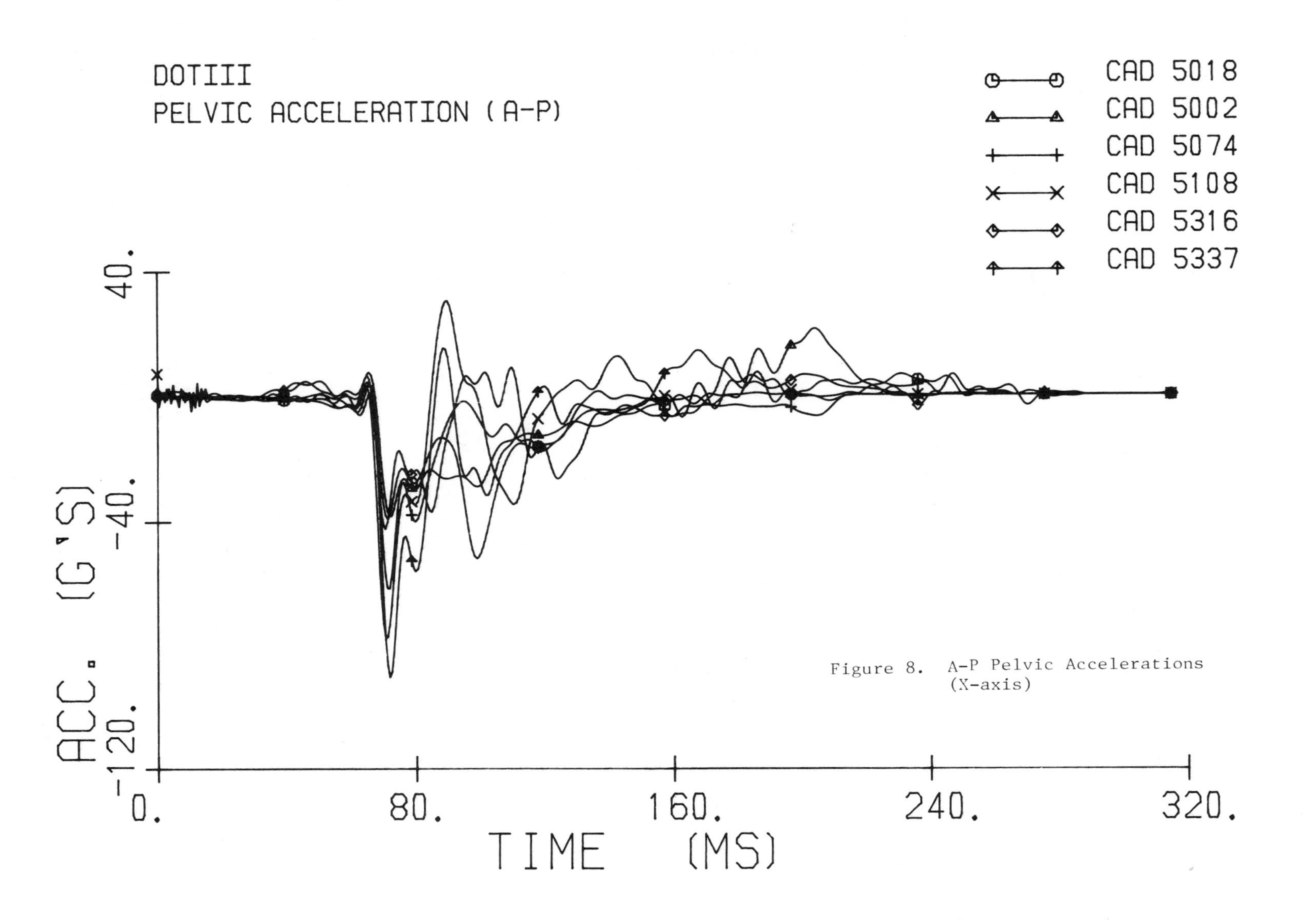

Figure 8. A-P Pelvic Accelerations (X-axis)

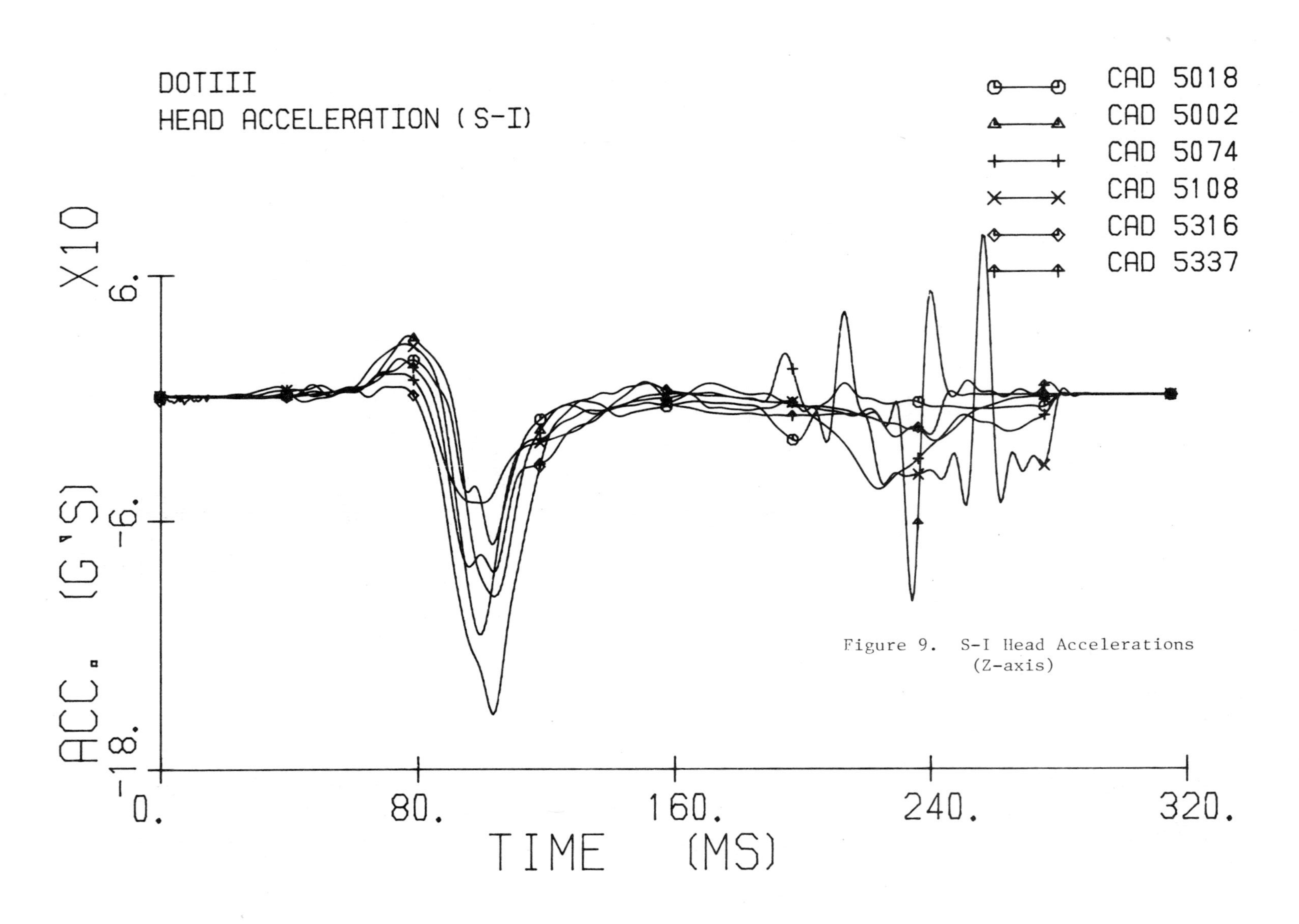

Figure 9. S-I Head Accelerations (Z-axis)

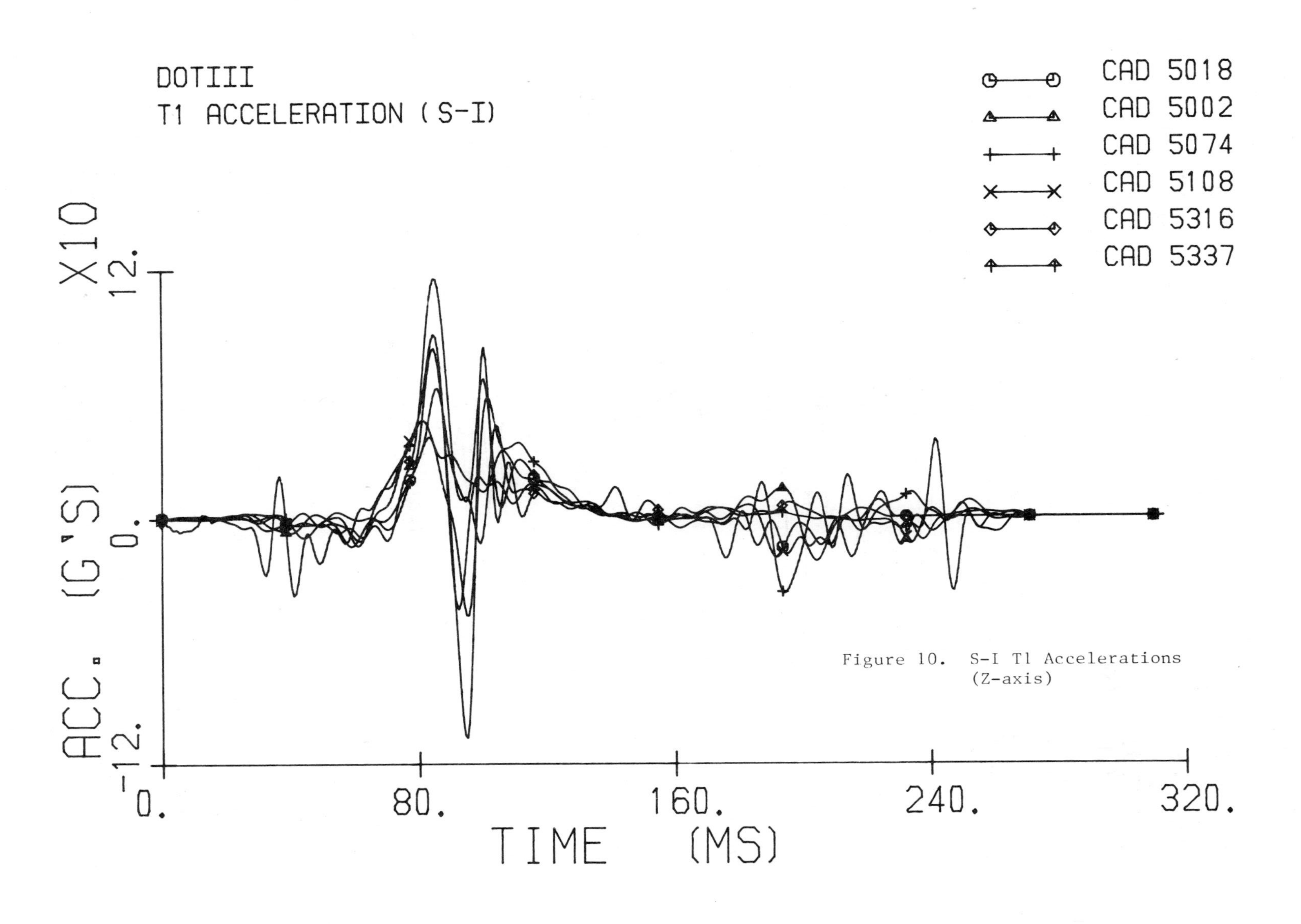

Figure 10. S-I Tl Accelerations (Z-axis)

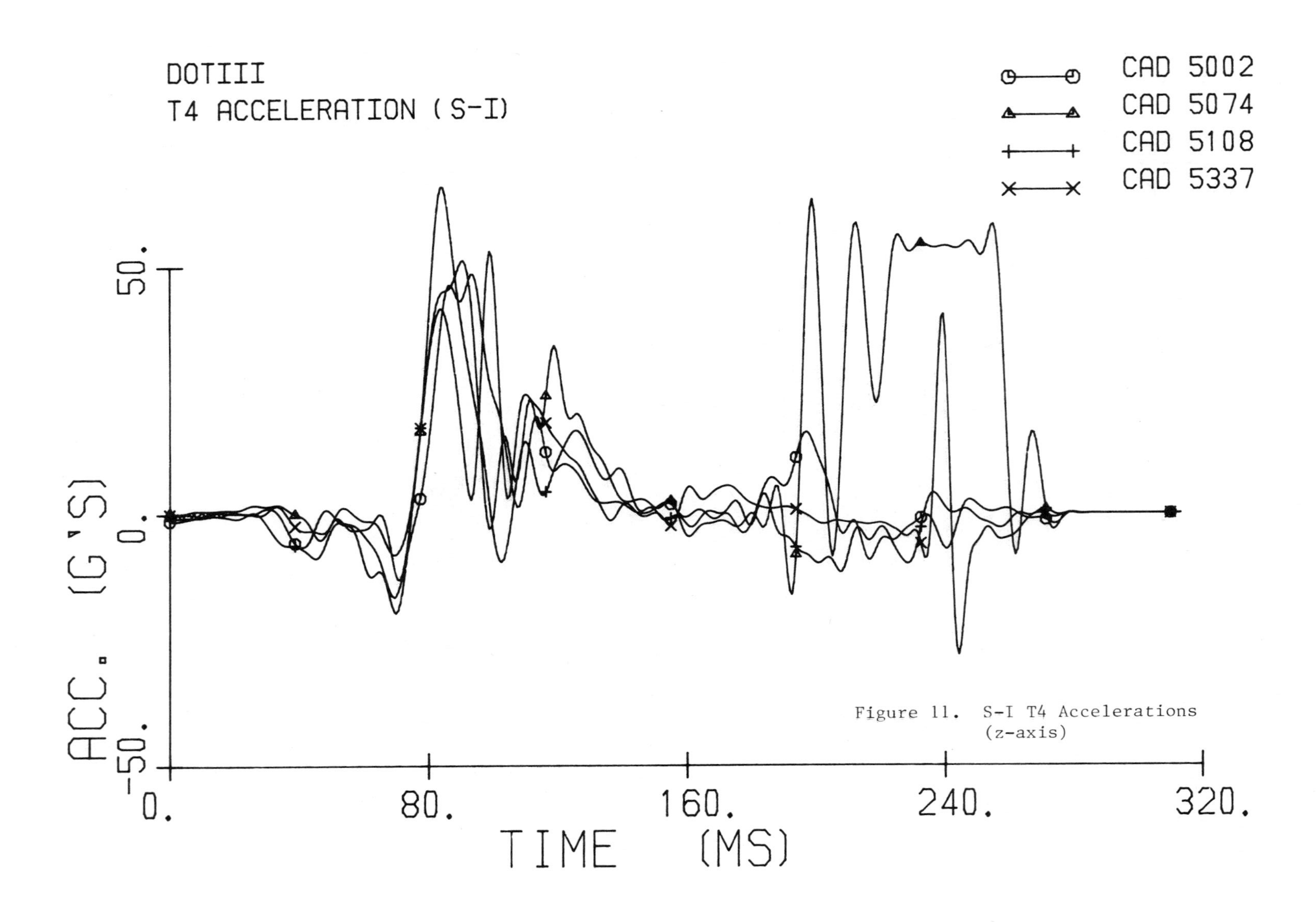

Figure 11. S-I T4 Accelerations (z-axis)

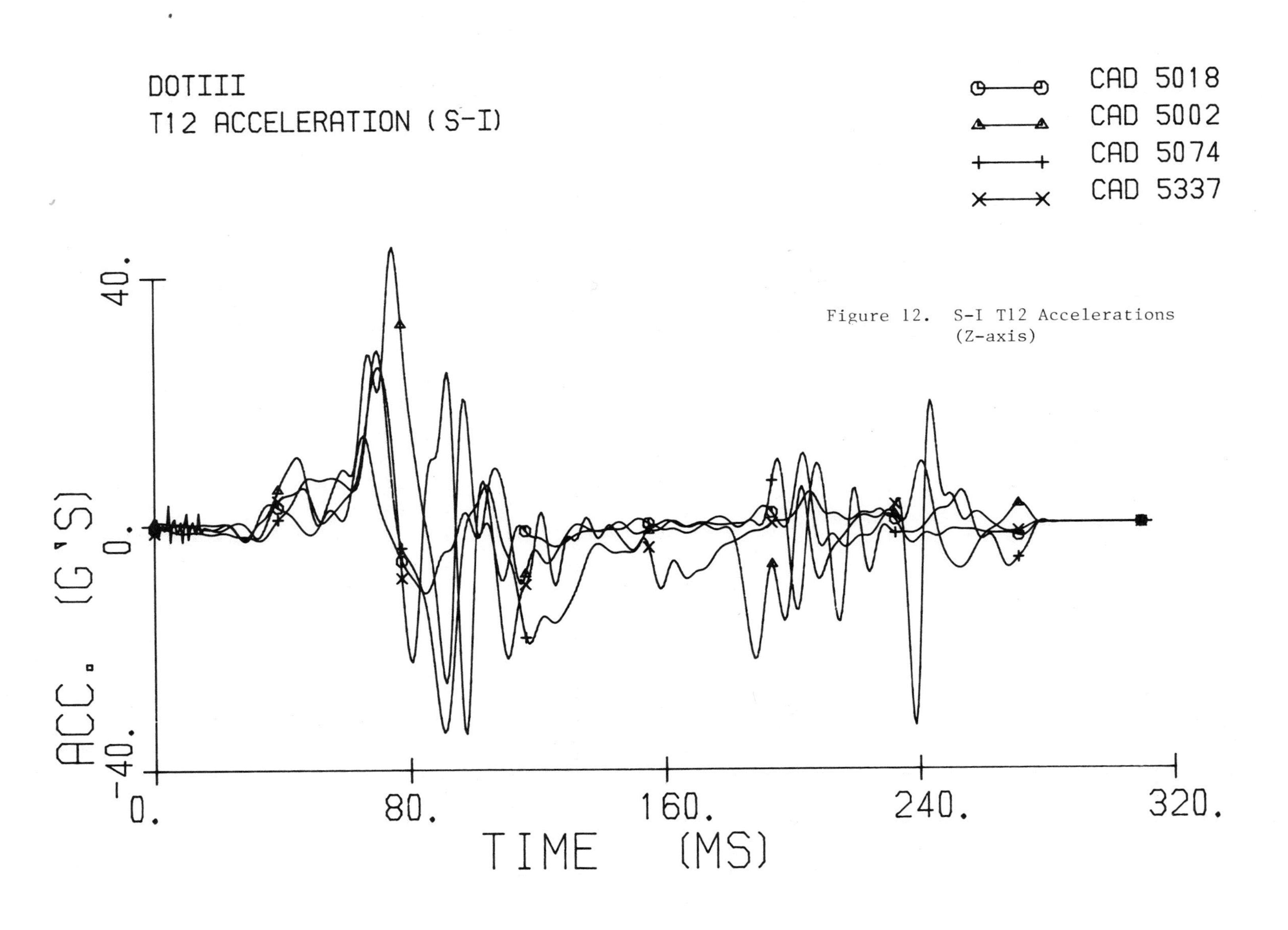

Figure 12. S-I T12 Accelerations (Z-axis)

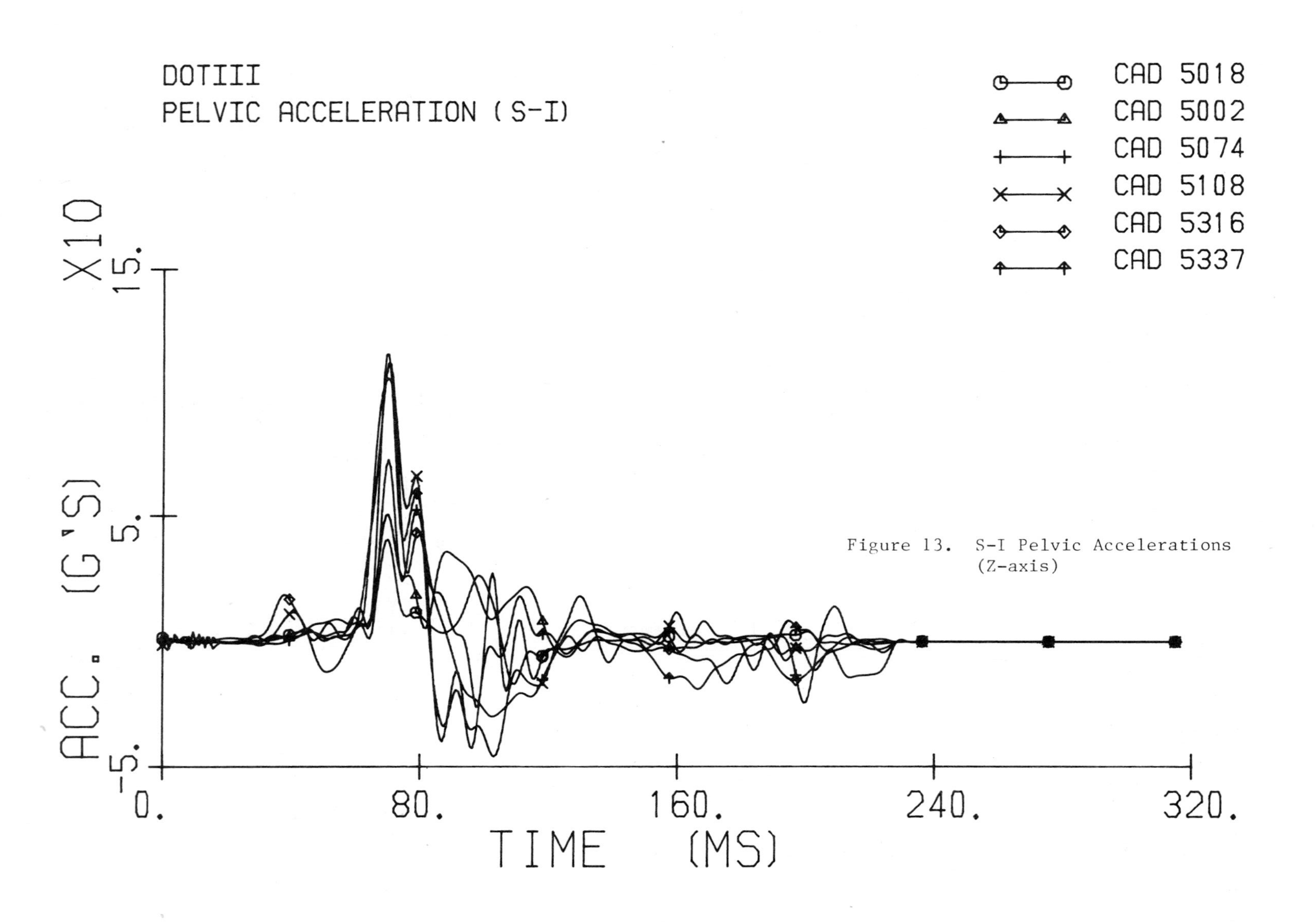

Figure 13. S-I Pelvic Accelerations (Z-axis)

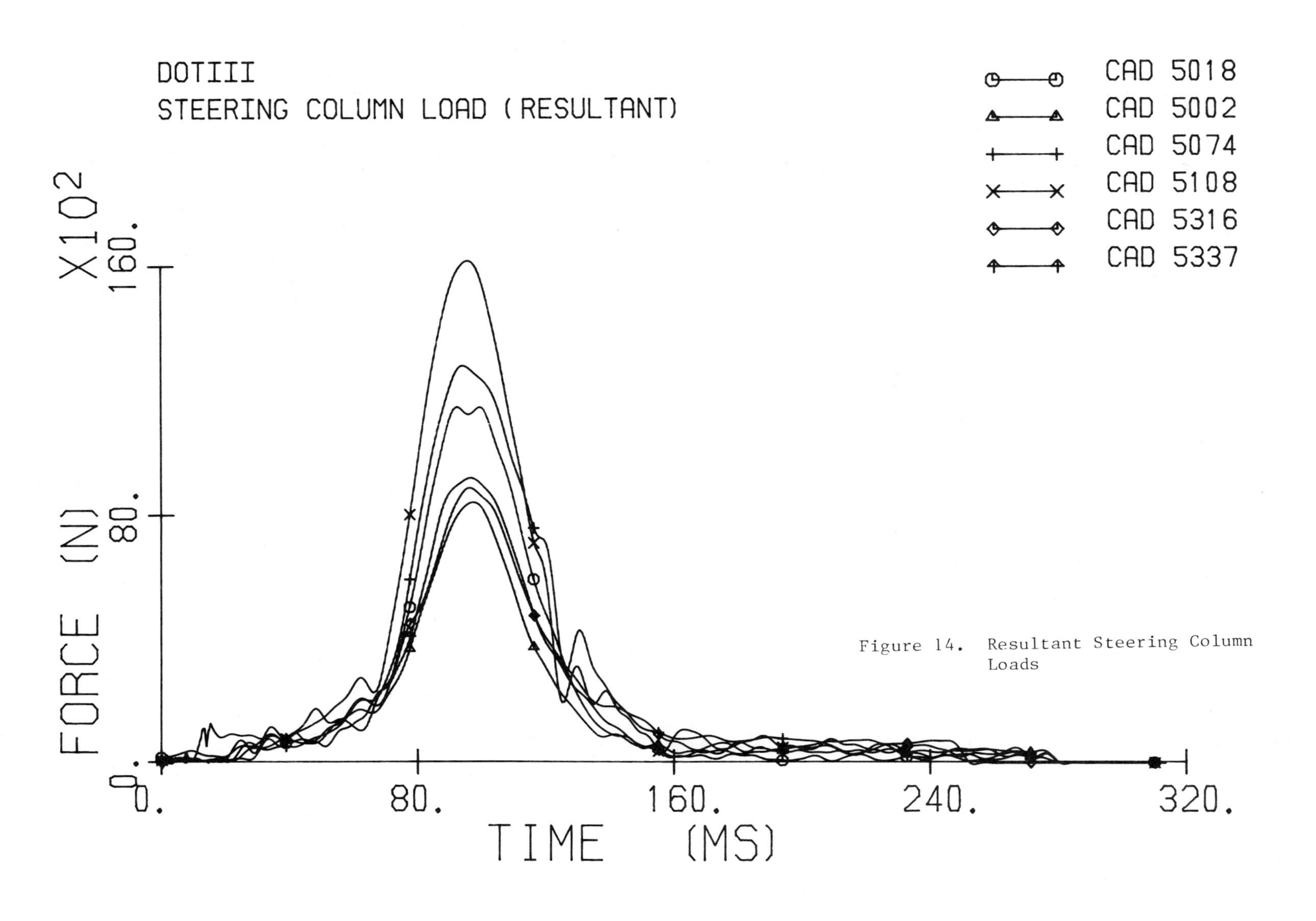

Figure 14. Resultant Steering Column Loads

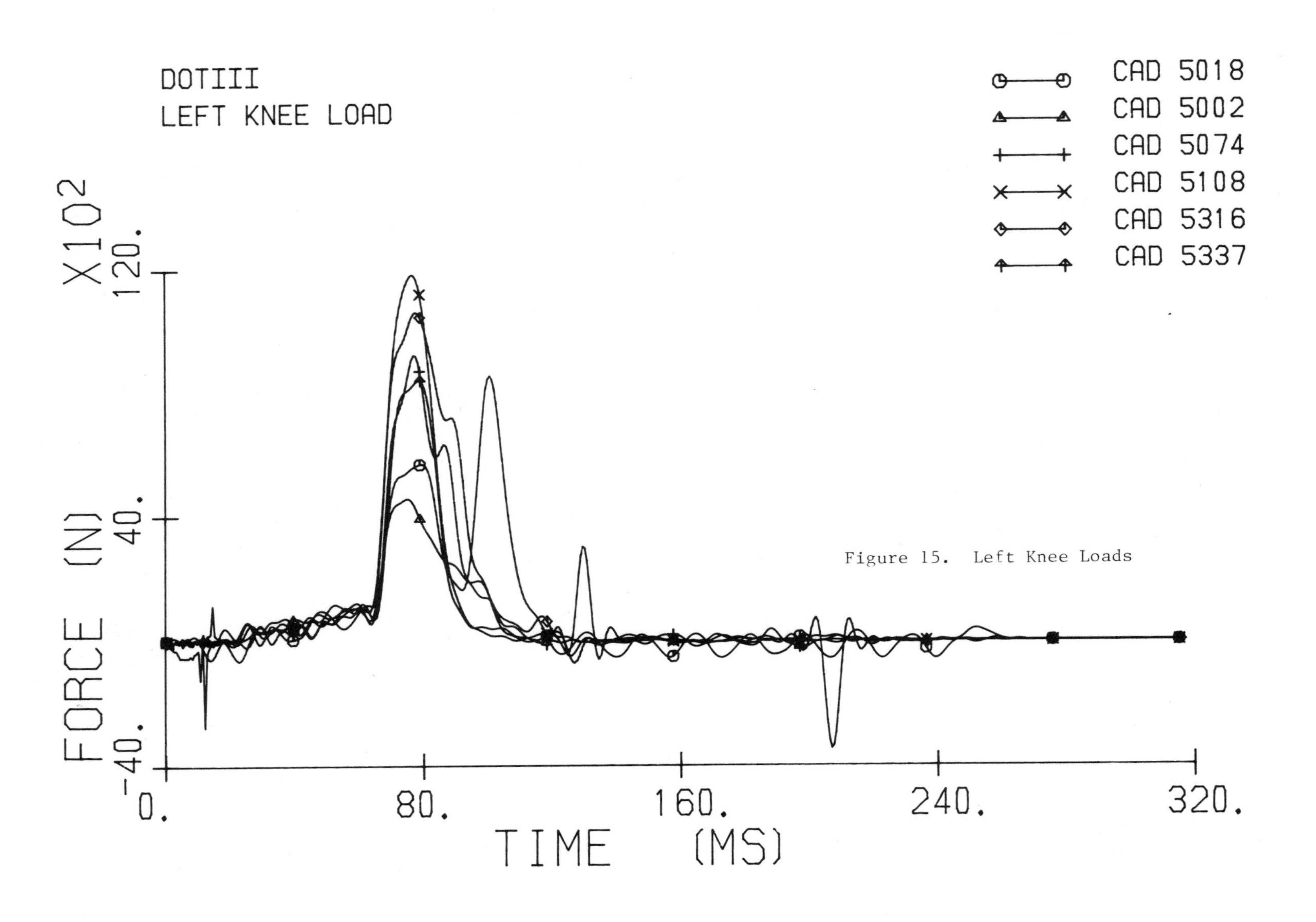

Figure 15. Left Knee Loads

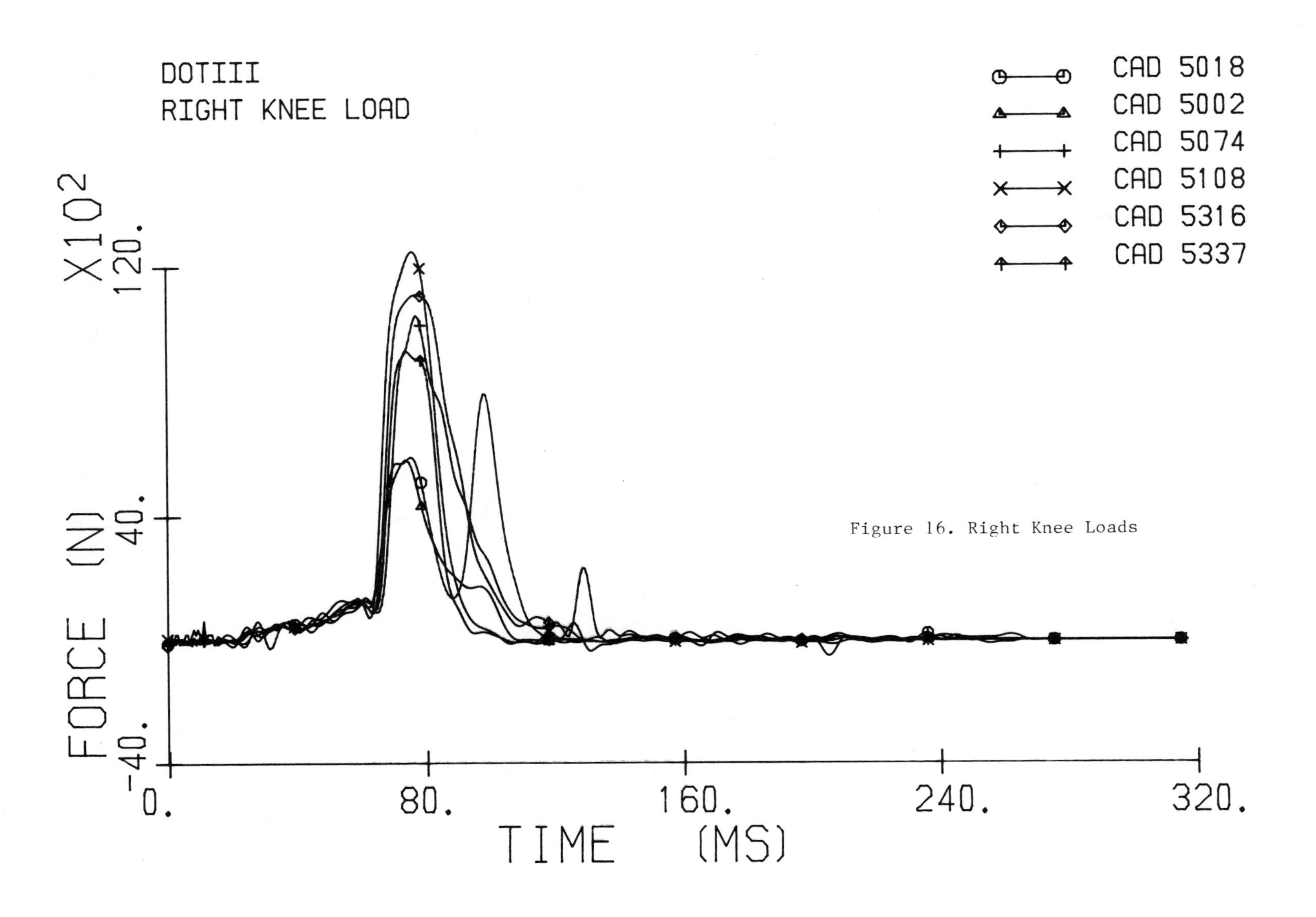

Figure 16. Right Knee Loads

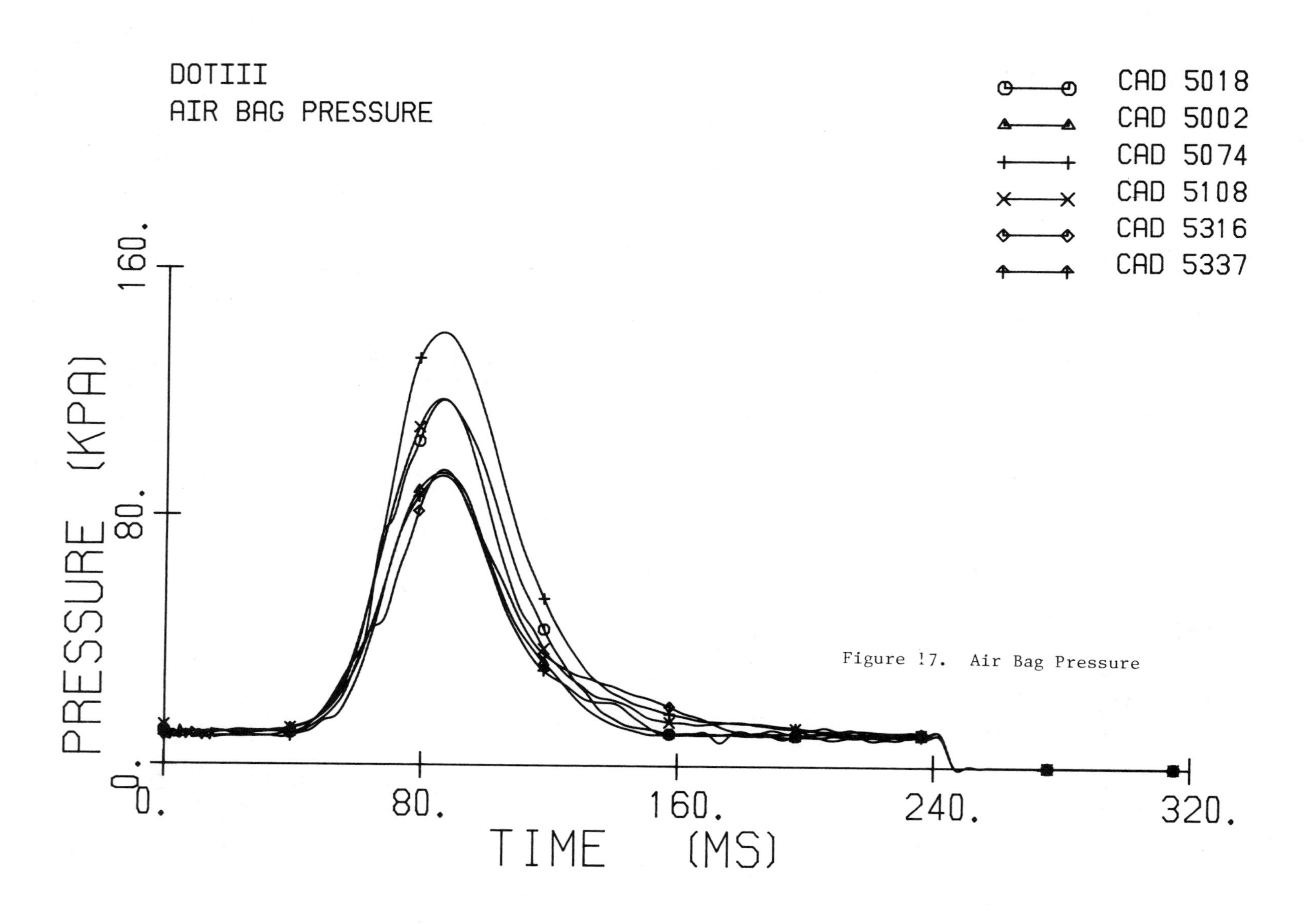

Figure 17. Air Bag Pressure

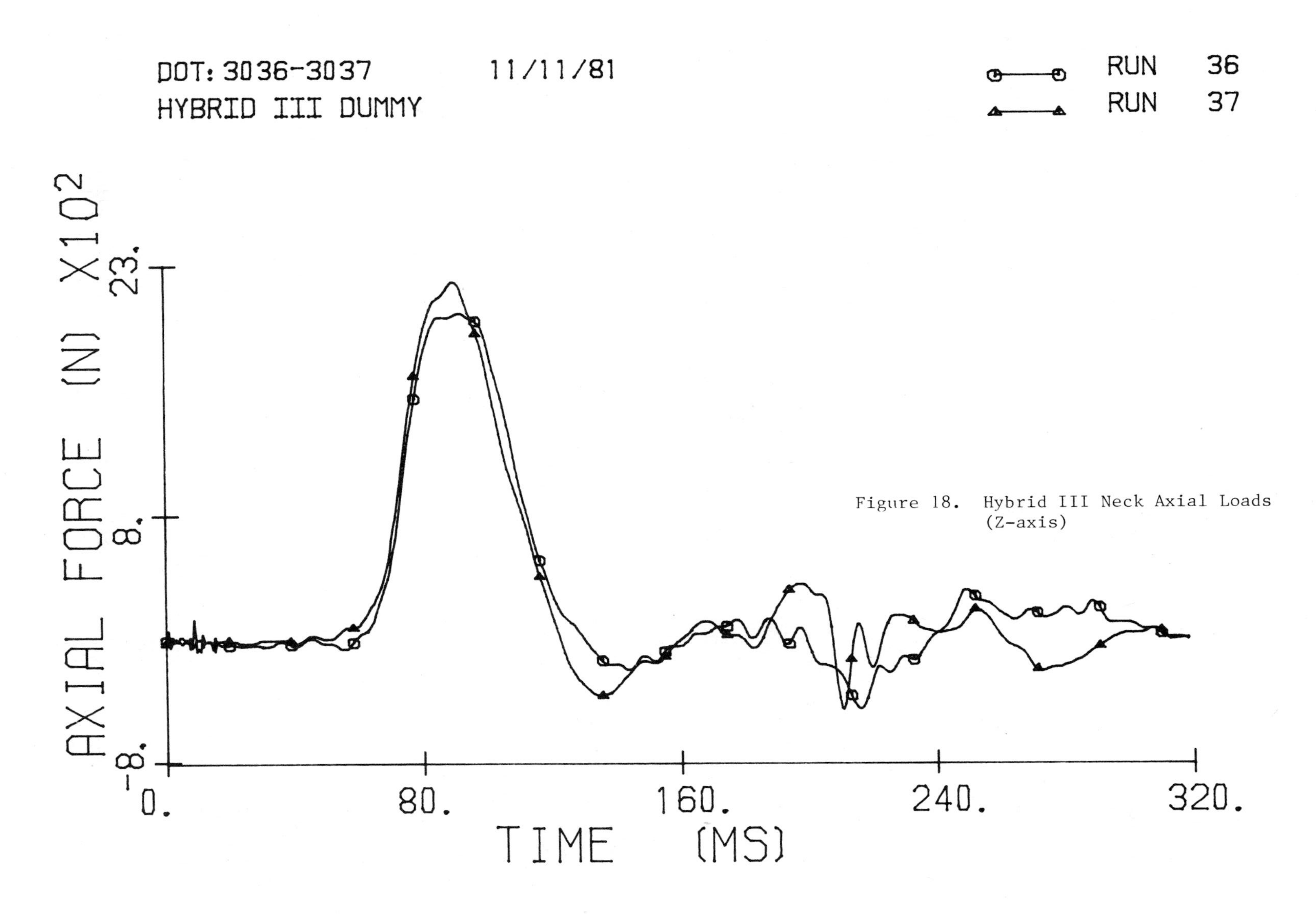

Figure 18. Hybrid III Neck Axial Loads (Z-axis)

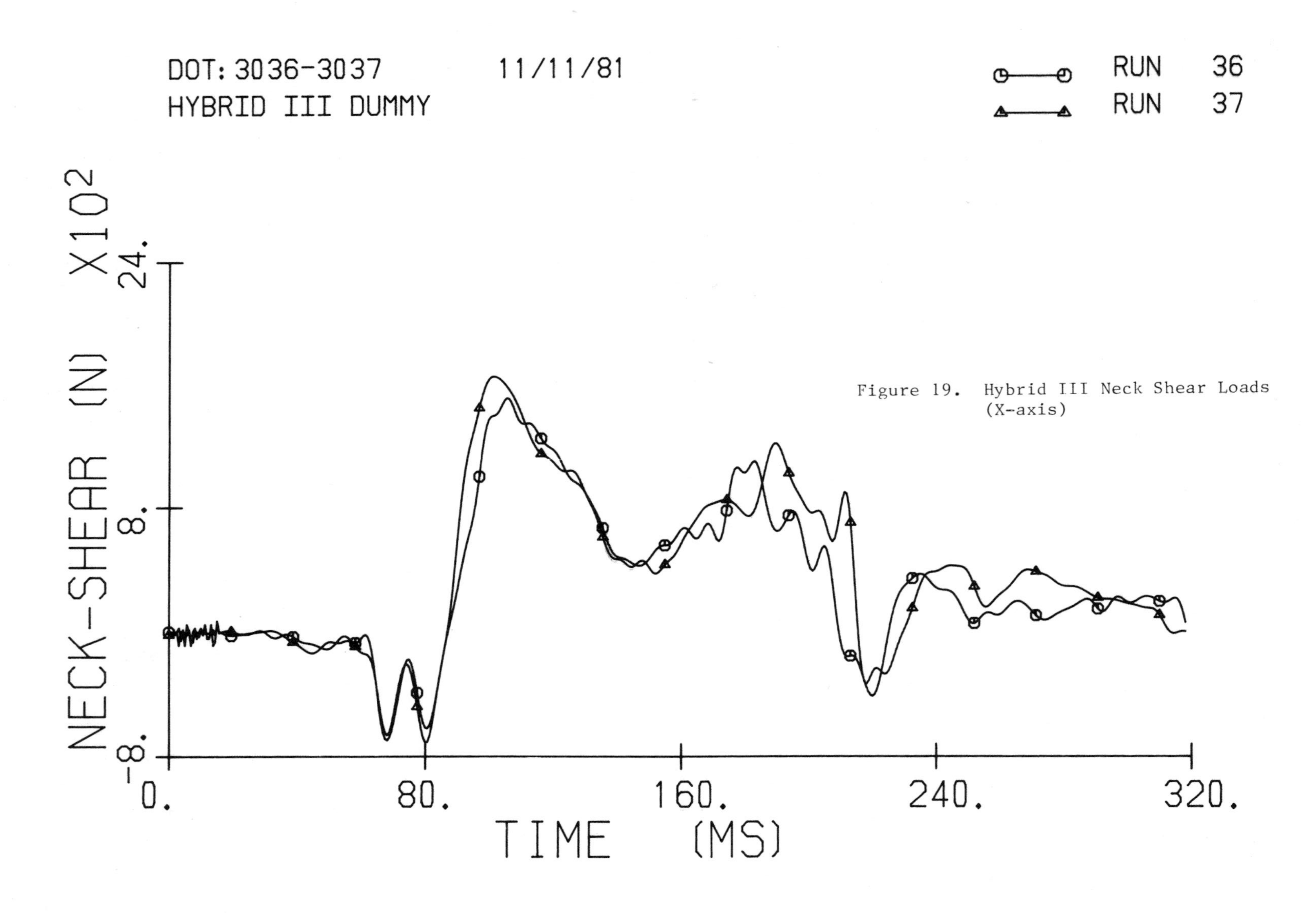

Figure 19. Hybrid III Neck Shear Loads (X-axis)

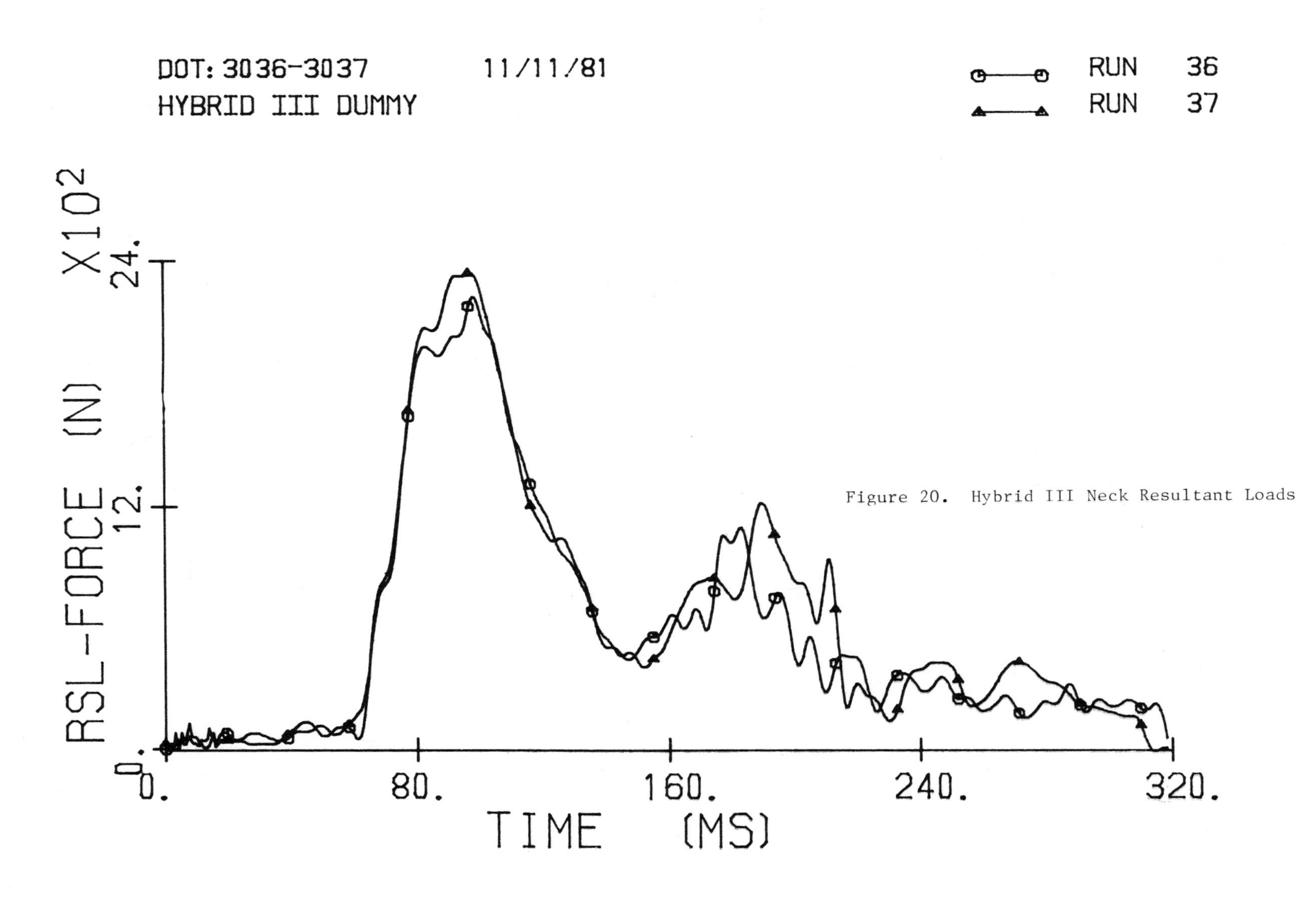

Figure 20. Hybrid III Neck Resultant Loads

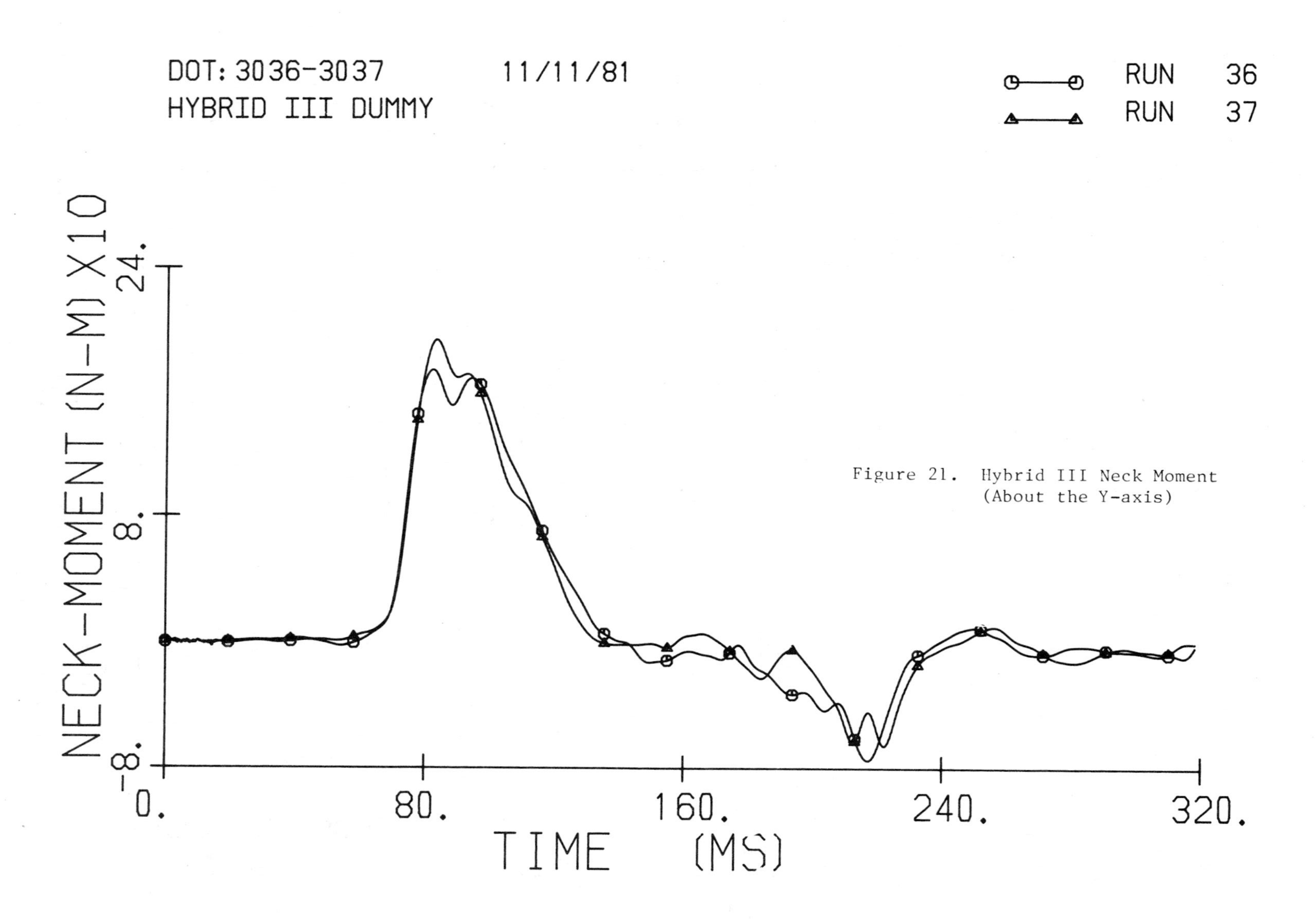

Figure 21. Hybrid III Neck Moment (About the Y-axis)

TABLE 1

Run No.	3022	3026	3032	3035	3038	3039
Date	1-13-81	2-27-81	3-20-81	5-1-81	3-26-82	4-29-82
Cad No.	5018	5002	5074	5108	5316	5337
Age	66	54	56	63	68	67
Sex	male	female	male	male	male	male
Body Mass (Kg)	72.5	50	96	72.5	83	60
Max sled g	32	37	38	36	37.5	39
Injury Rib fracture	L - 5, 6, 7, 8	None	L - 2, 4, 5, 6, 7, 8, 9 R - 2, 3, 4, 5, 6, 7, 8, 9	None	L - 2, 3, 4, 5, 6, 7, 8 R - 2, 7, 8	None
AIS	2	0	2	0	2	0
Spine/cord	C1-C2 fracture separation. Partial transection of cord	Ring fracture, Transection of cord, Post. ligament tear C1, C2	T4-T5 & T9-T10 fracture separation	C1, C2 posterior ligament tear	Ring fracture, Transection of cord, Fx of C7, T12	Fx of T12
AIS	6	6	5	2	6	2
Rib test (MPa)	173.7	156.5	182.0	115.8	NA	NA
E(GPa)	12.96	14.13	15.10	8.48	NA	NA
HIC	1487	1936	778	1670	4387	1252
Max steering col. load (kN)	12.0	8.5	12.0	16.0	9.0	9.4
Max air bag pres. (kPa)	117	94	139	117	94	93
Max lung pres. (kPa)	21.7 data saturated	54.3	2.3	152.0	29.3	NA
OAIS	6	6	5	2	6	2

NA = Not available

3.3 and 3.0 kN respectively. The input pulse for that study had a peak of 29 g and a duration of 150 ms. The peak loads measured in these air bag runs were 2.21 kN axial and 1.6 kN shear, as shown in Figures 18 and 19.

It is obvious that Hybrid III data have, in this case, become a source of contradiction and confusion rather than one of elucidation. As mentioned above, the clue to the puzzle was the difference in kinematic response of the cadaver and dummy. It was felt that the flexibility of the cadaveric torso permitted the air bag to start pushing it backward while the head has not completed its forward flexion swing. Thus, cadaveric neck loads had to be higher than those observed in the dummy in which, the head and torso moved more or less as a rigid body.

An indirect method of estimating cadaveric neck loads was developed, using the principles of a dynamic head model proposed by Mertz (15). The basic approach was to compute the acceleration of the head at its center of gravity (c.g.) in two dimensions and to estimate the air bag contact force against the face. Head acceleration and neck load data from the 2 Hybrid III runs were used to estimate the neck loads sustained by these cadaveric subjects. The procedure was as follows:

1. The airbag contact force was assumed to be in the body-fixed P-A or x-axis. Film data indicate that there is a good basis for making this assumption.

2. The axial neck load measured in the Hybrid III was compared with the z-axis acceleration of the head of the dummy to verify that the weight was 44.5N (10 lb). A comparison of the two sets of (axial force curves is shown in Figure 22).

3. It can be seen from Figure 23, that the air bag contact force, F_A, is the difference between the neck reaction F_{NX} and the inertial force F_{IX}. It is further assumed that

$$F_A = P_A \cdot A_F$$

where P_A = air bag pressure
and A_F = area of face in contact with the air bag

4. The objective was to determine A_F as a function of time which provided the best match between the measured neck shear and that computed from the equation

$$\begin{aligned} F_{NX} &= F_{IX} - F_A \\ &= F_{IX} - P_A A_F \end{aligned} \quad (1)$$

or $$A_F = (F_{IX} - F_{NX})/P_A$$

5. This area function, $A_F(t)$, was approximated by a bilinear curve shown in Figure 24. It had a peak of about 0.039 m^2 (60 sq in) and the duration of contact was 50 ms. In fact, the dummy A_F was of shorter duration (27 ms) but it was decided that a better match could be obtained by increasing the duration to 50 ms. Cadaveric contact was estimated to be longer in duration also.

6. The area function A_F was used to compute F_{NX} for the Hybrid III runs for which load cell (L.C.) data on F_{NX} were available. A comparison of the 2 sets of force curves is shown in Figure 25. The shear force computed from the assumed area function is designated as ACC.-AIR.

7. Head angular acceleration was computed for most of the cadaveric runs and was used to obtain the c.g. acceleration of the head. Average values for mass moments of inertia of the head were used. They were 0.027, 0.021 and 0.019 kg-m^2 about the x-(P-A), y-(R-L) and z-(I-S) axes respectively.

8. The contact force F_A was determined for each cadaveric run and the neck shear reaction was adjusted to account for this contact force. Actual air bag pressure data were used to compute F_A. The results are shown in Figures 26 - 33 for two of the 6 runs. Figure 26 shows the axial neck load in Cadaver 5316 which sustained a catastrophic atlanto-occipital separation. Its peak value is 6.52 kN (1,466 lb). The shear and resultant forces are shown in Figures 27 and 28. The peak shear force was estimated at 2.72 kN (611 lb) and the peak resultant force was 6.64 kN (1,710 lb). The peak moment shown in Figure 29 is 200 N.m if the second higher peak is ignored. These values are much higher than those measured in the Hybrid III. For Cadaver 5337, the peak axial, shear and resultant forces were 3.21 kN (724 lb), 1.82 kN (409 lb) and 3.22 kN (724 lb), as shown in Figures 30 - 32. The peak moment was 191 N.m, as shown in Figure 33. This cadaver did not sustain a neck injury. Data for all six runs are provided in Table 2. The resultant neck load for fracture at the base of the skull is conservatively estimated to be 6.2 kN (1,400 lb).

It is also interesting to compare neck loads and neck fracture data with the computed HIC. An approximate limiting value for HIC appears to be 1,500 s.

CONCLUSIONS

1. This study is not an investigation of potential injuries due to the use of an air bag

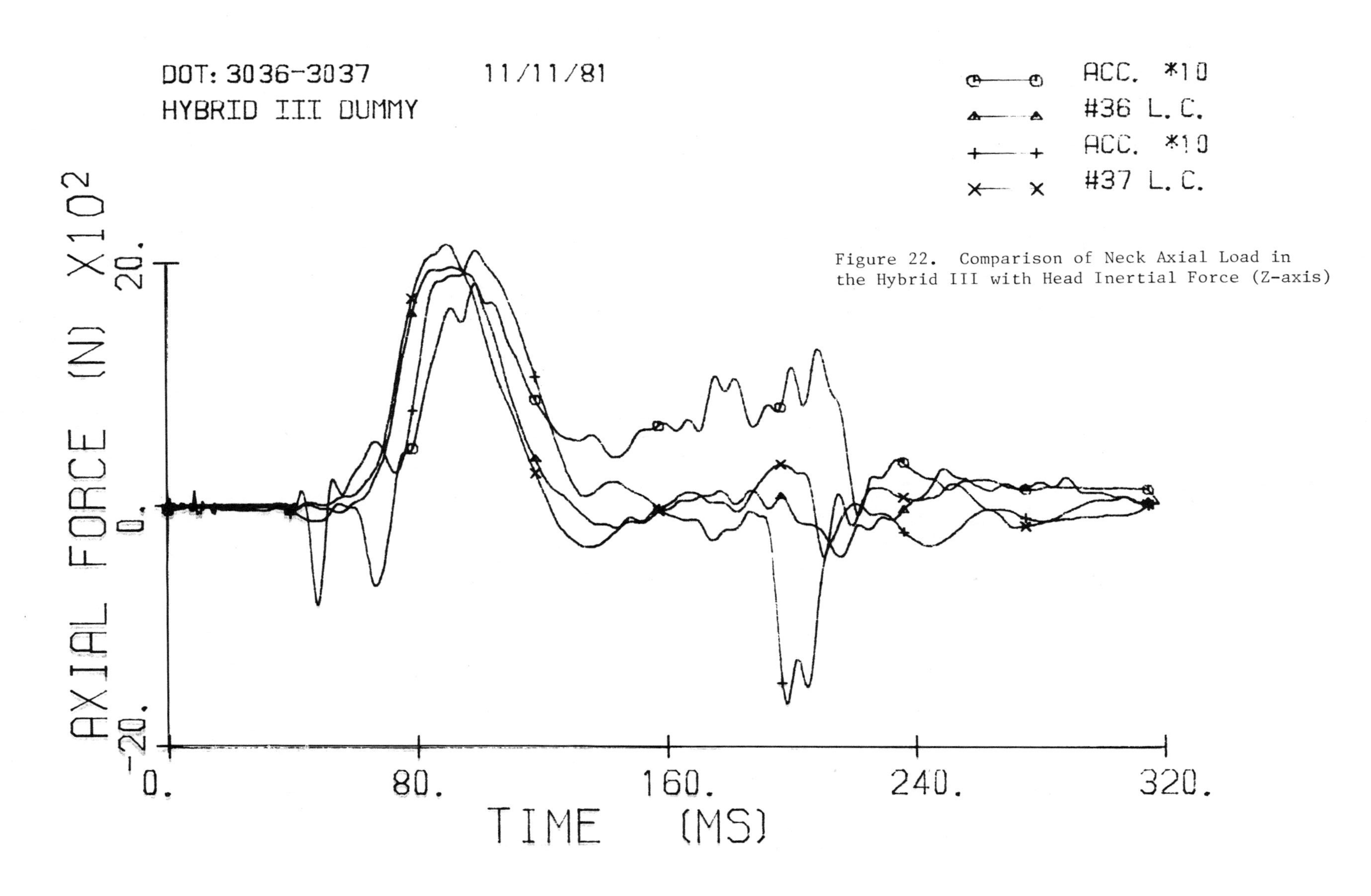

Figure 22. Comparison of Neck Axial Load in the Hybrid III with Head Inertial Force (Z-axis)

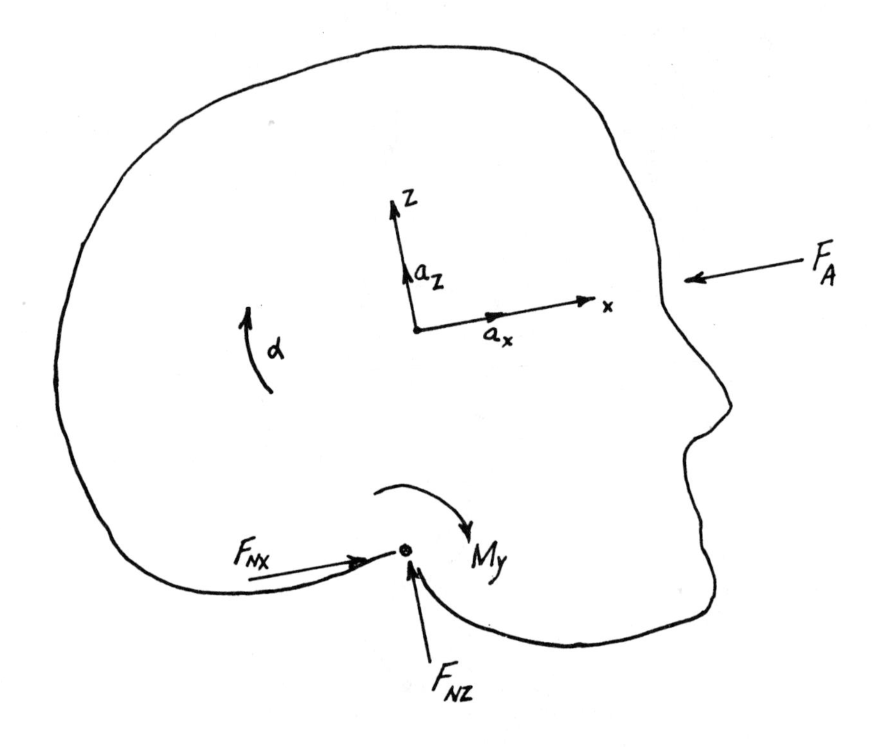

Figure 23. Free Body Diagram of the Head

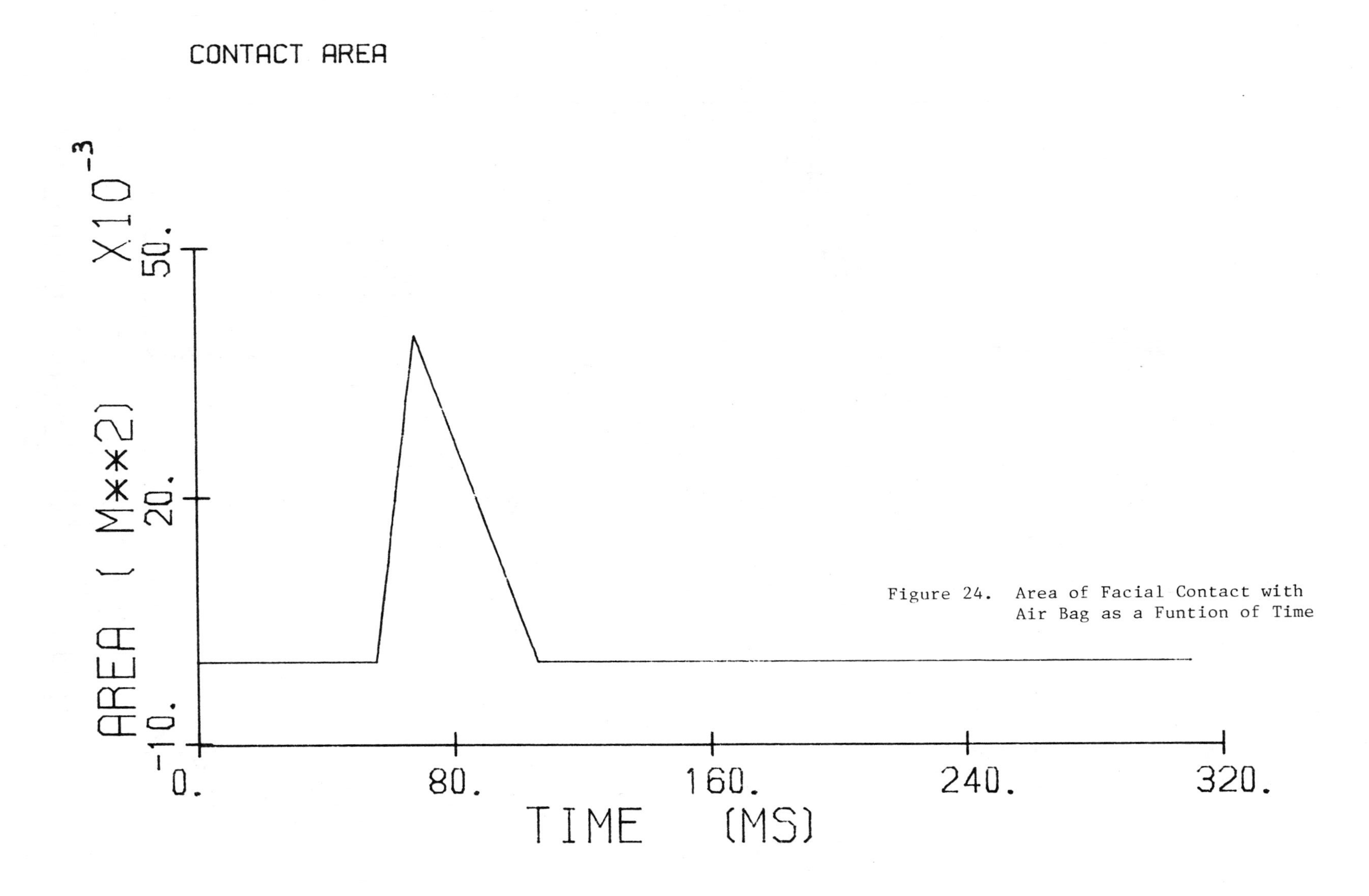

Figure 24. Area of Facial Contact with Air Bag as a Funtion of Time

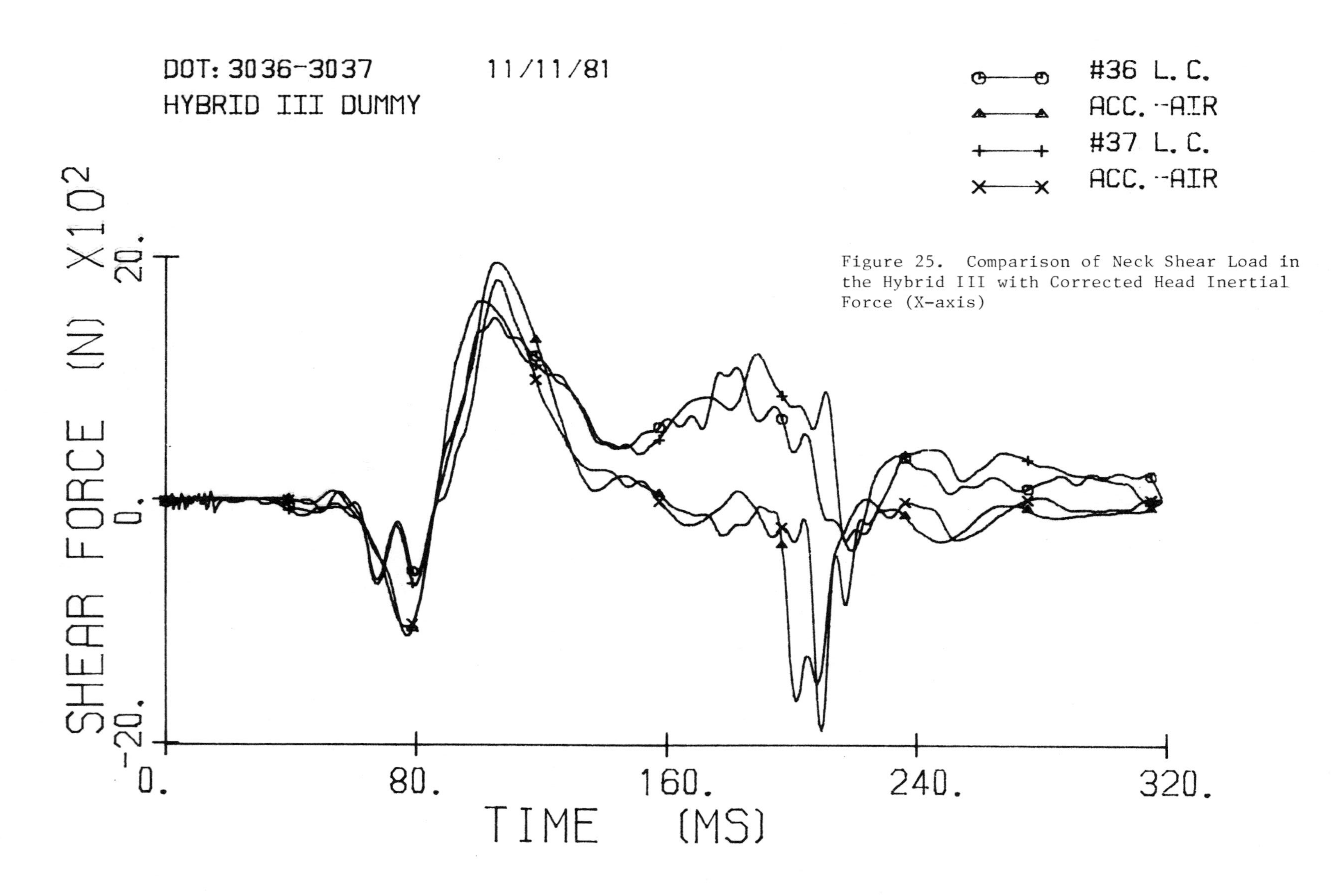

Figure 25. Comparison of Neck Shear Load in the Hybrid III with Corrected Head Inertial Force (X-axis)

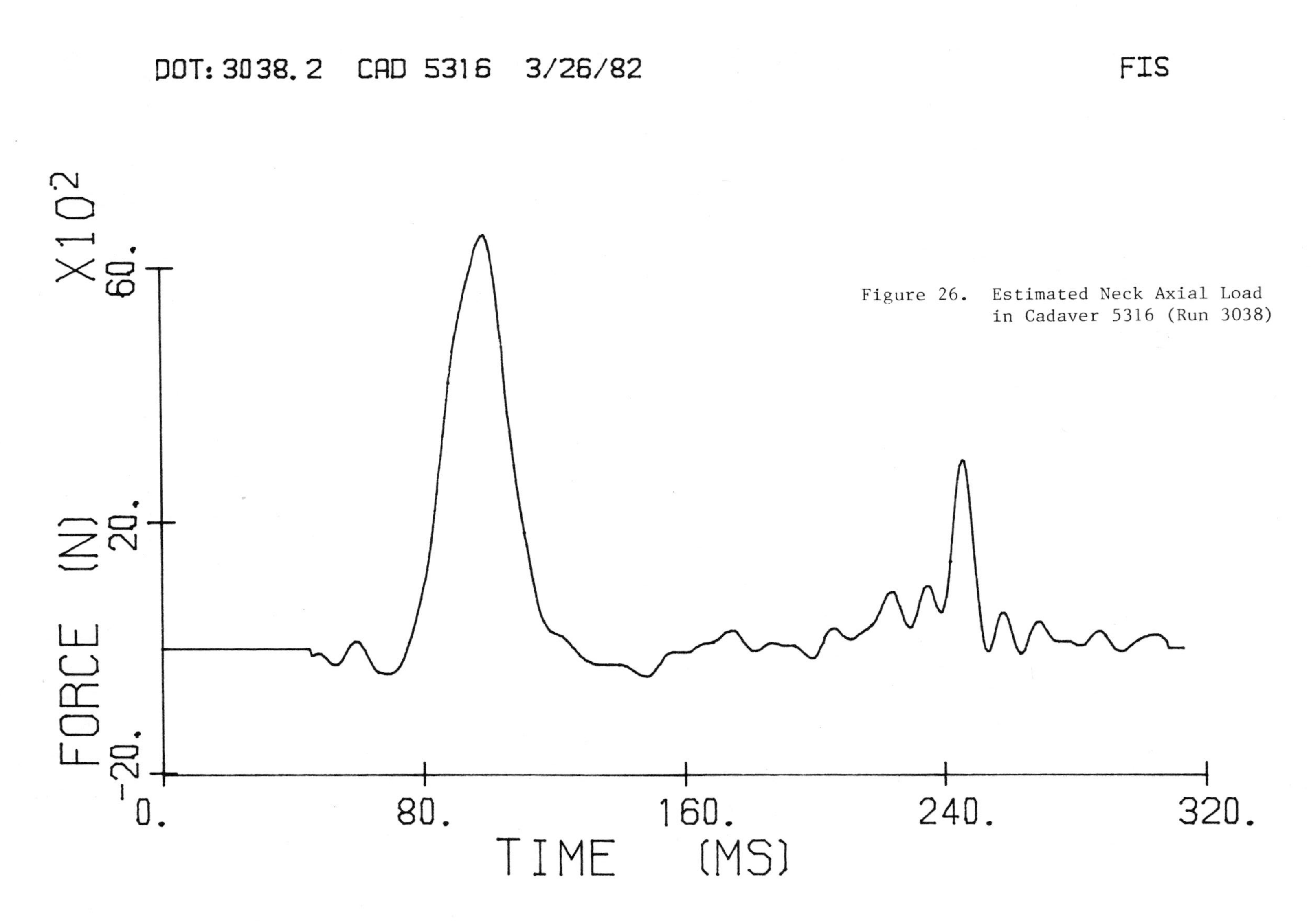

Figure 26. Estimated Neck Axial Load in Cadaver 5316 (Run 3038)

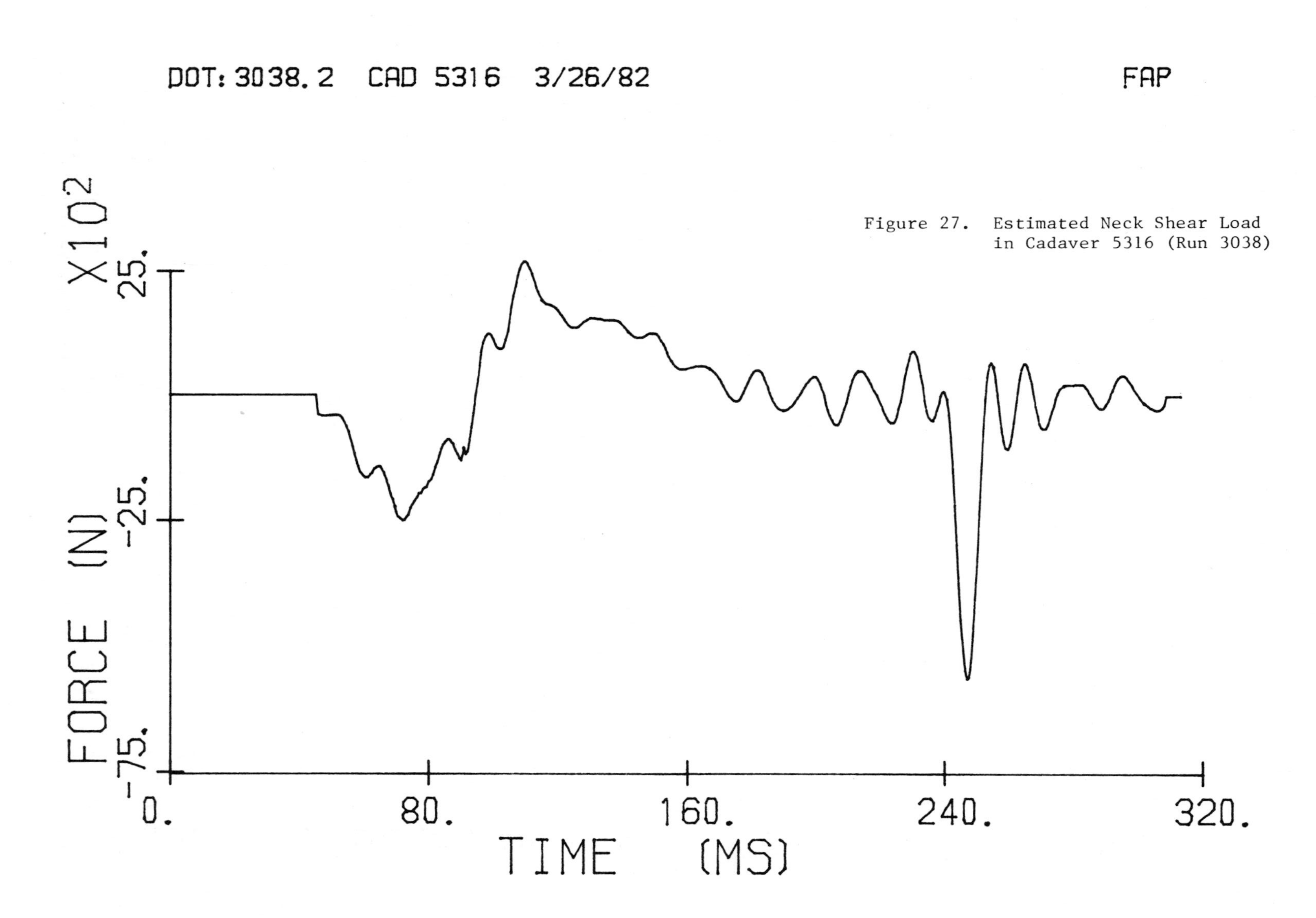

Figure 27. Estimated Neck Shear Load in Cadaver 5316 (Run 3038)

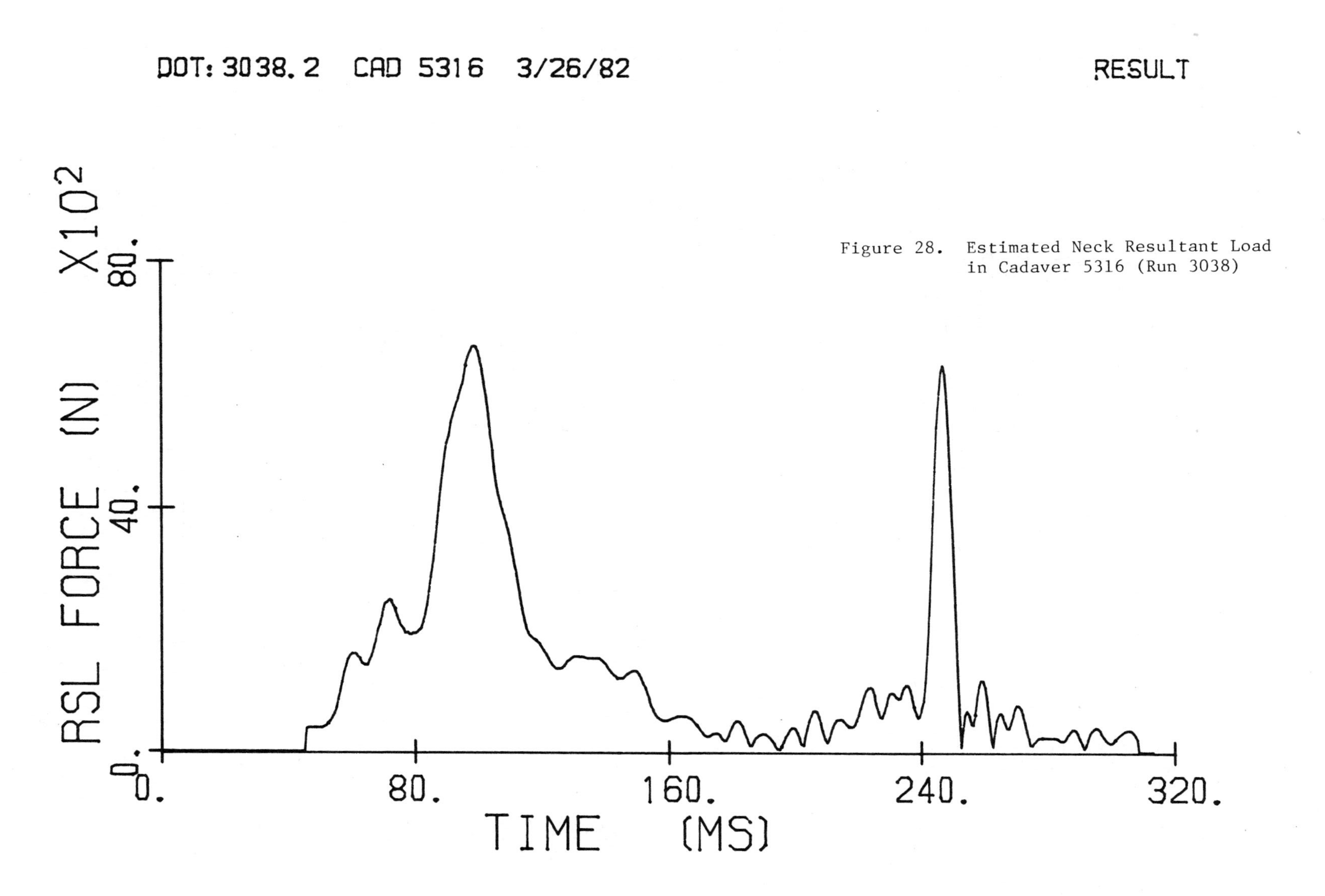

Figure 28. Estimated Neck Resultant Load in Cadaver 5316 (Run 3038)

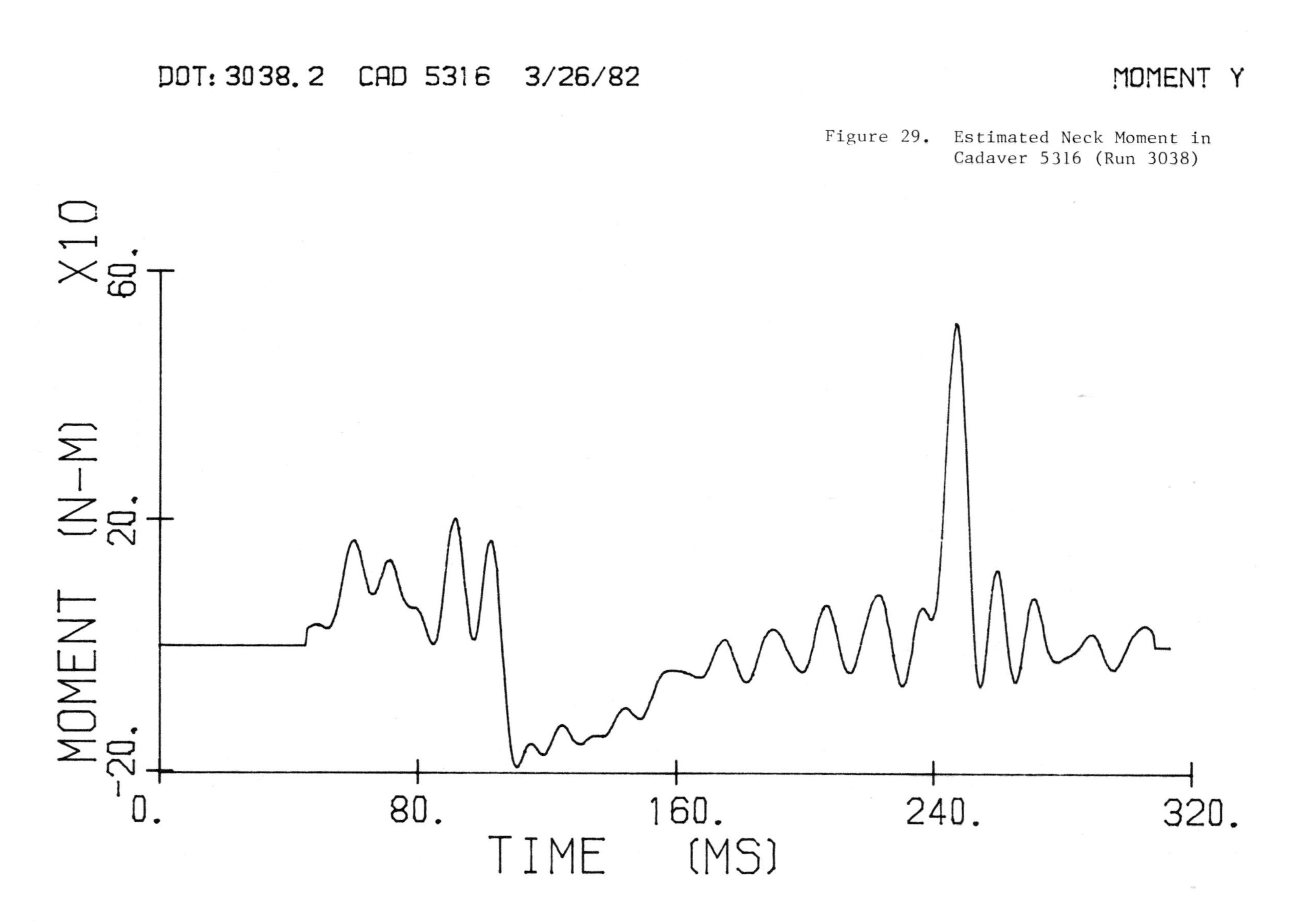

Figure 29. Estimated Neck Moment in Cadaver 5316 (Run 3038)

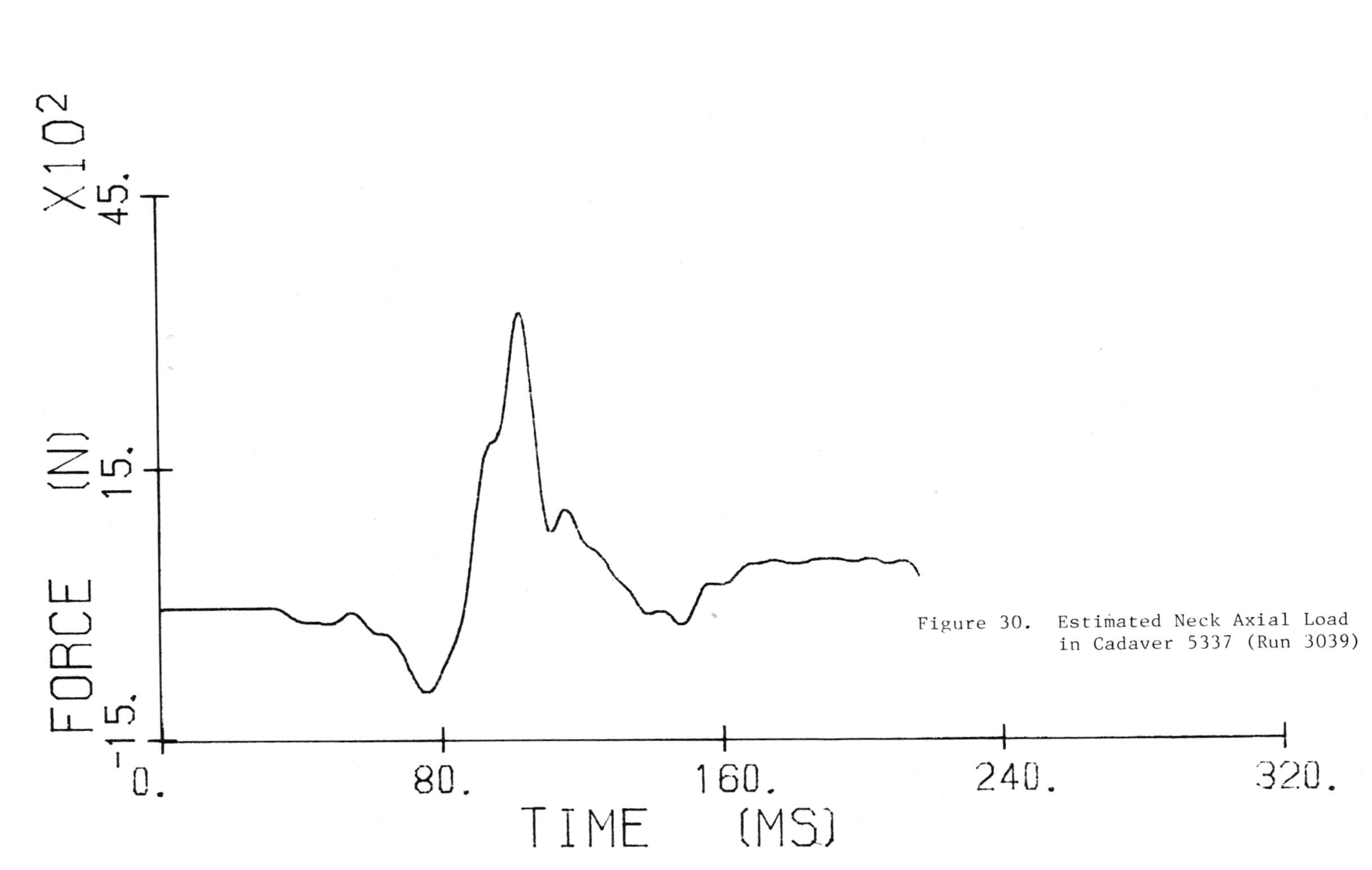

Figure 30. Estimated Neck Axial Load in Cadaver 5337 (Run 3039)

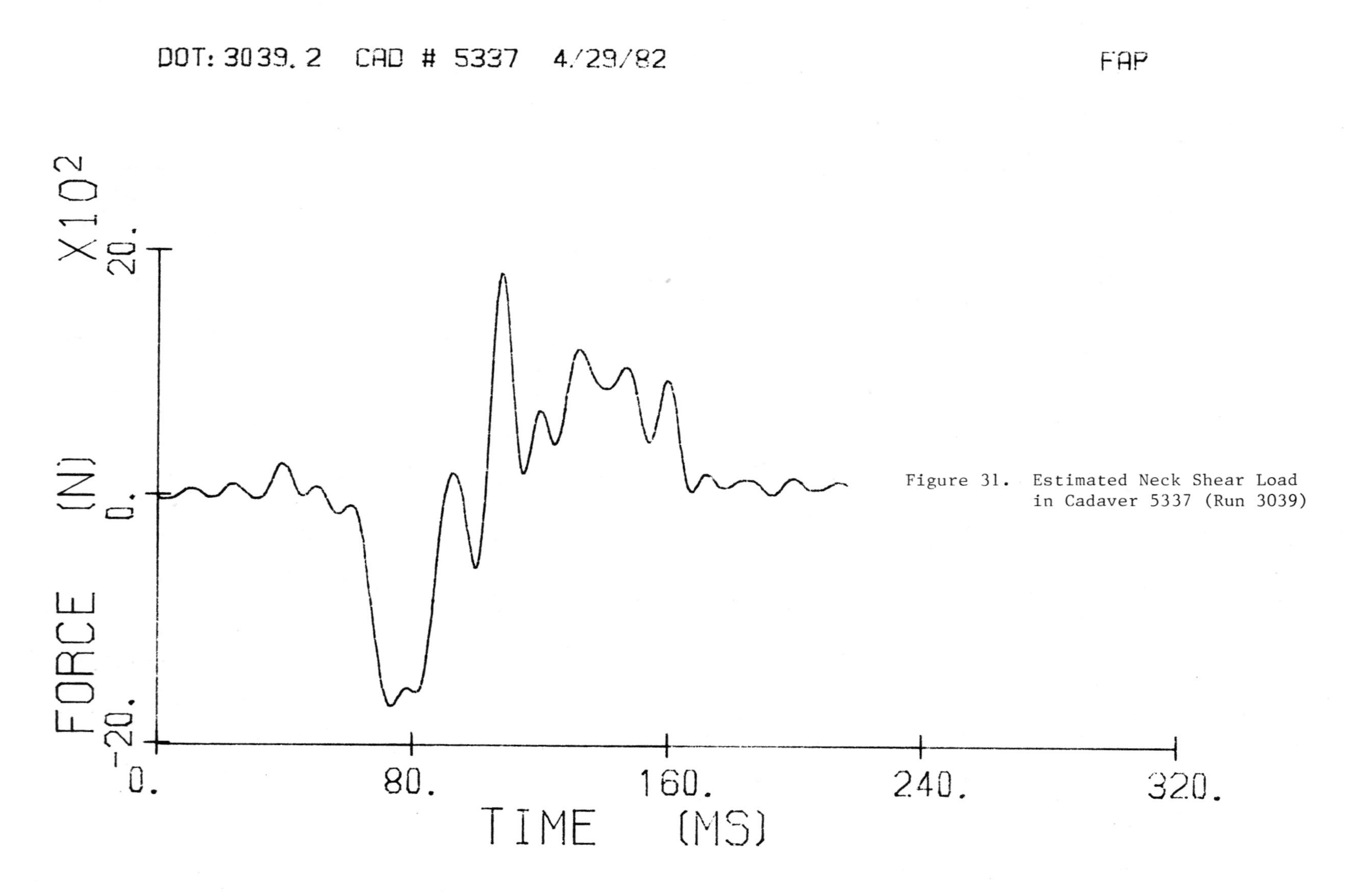

Figure 31. Estimated Neck Shear Load in Cadaver 5337 (Run 3039)

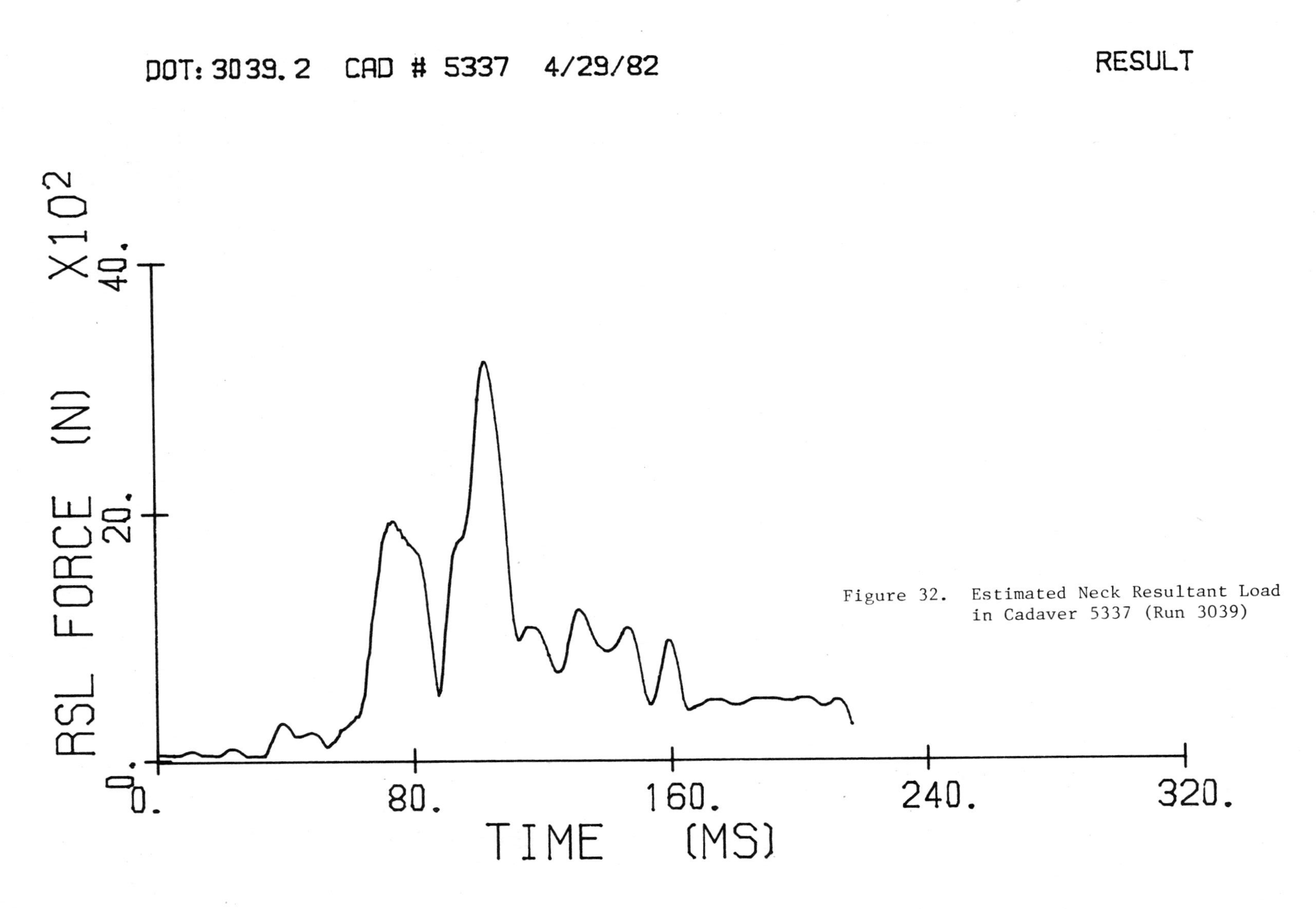

Figure 32. Estimated Neck Resultant Load in Cadaver 5337 (Run 3039)

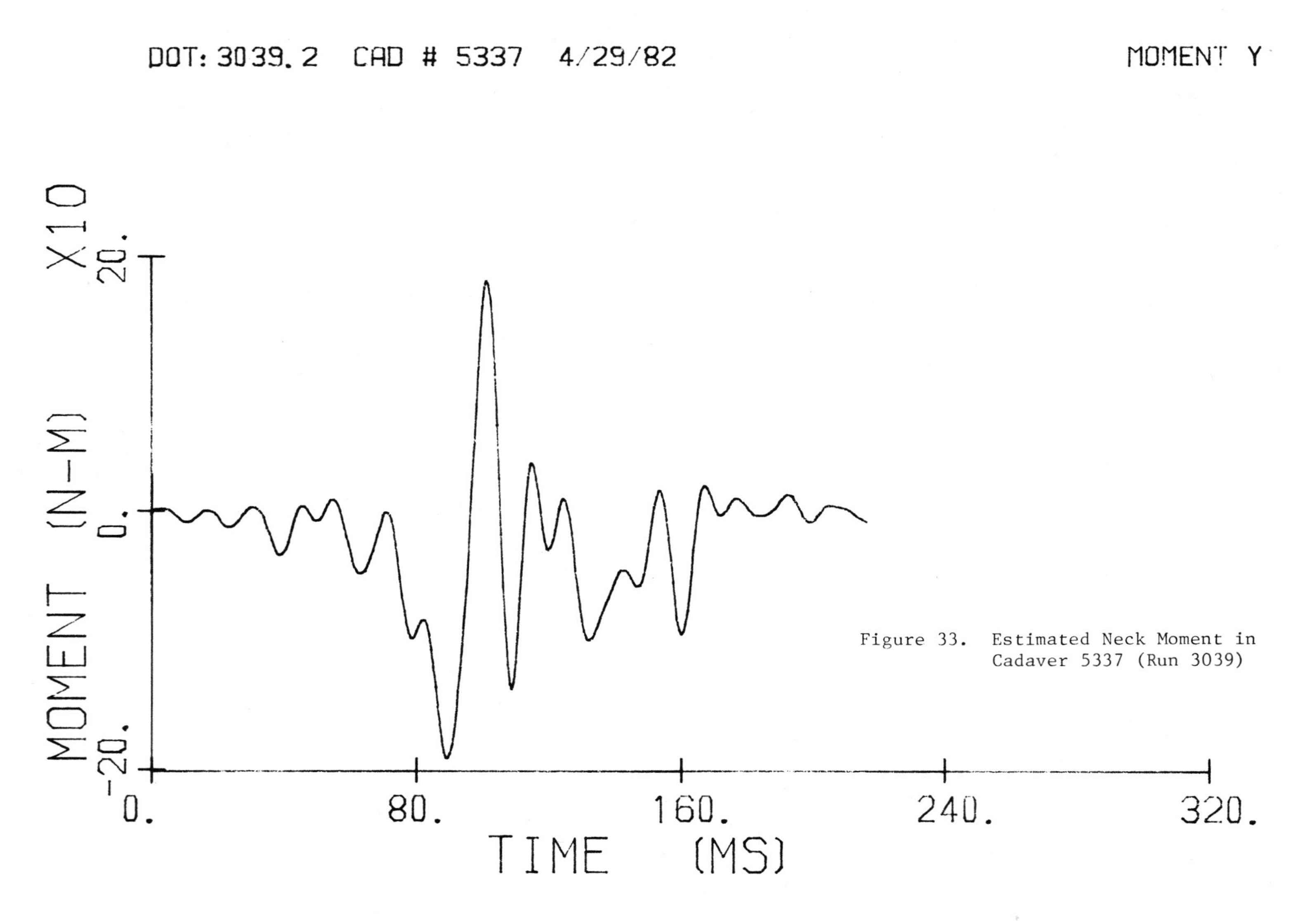

Figure 33. Estimated Neck Moment in Cadaver 5337 (Run 3039)

TABLE 2

ESTIMATED CADAVERIC NECK LOADS

Run No.	Cadaver No.	NECK LOADS kN (lb) Resultant	Axial	Shear	Moment N-m (ft-lb)	Neck Injury
3022*	5018	6.35 (1428)	3.49 (784)	5.34 (1200)	-	C1-C2 Separation
3026**	5002	7.22 (1623)	7.20 (1619)	1.48 (333)	329 (243)	Atl-occ. Separation
3032	5074	3.13 (704)	2.42 (544)	3.11 (649)	128 (94)	None
3035	5108	1.77 (398)	0.85 (191)	1.74 (391)	339 (250)	Post. lig tear at C1-C2
3038	5316	6.64 (1493) 6.30 (1416)+	6.52 (1466) 3.00 (674)+	2.72 (611) 5.66 (1272)+	200 (148) 517 (381)+	Atl-occ Separation
3039	5337	3.22 (724)	3.21 (722)	1.82 (409)	191 (141)	None
3037	HBIII	2.35 (528)	2.20 (415)	1.60 (360)	170 (125)	-

*Due to problems with head acceleration data, head angular acceleration and velocity could not be calculated for this run. The loads shown were computed directly from the measured acceleration.

**The loads for this run were computed after two accelerometer channels were interchanged to yield reasonable angular acceleration and velocity values.

+Loads computed from a second peak which may be an artifact.

restraint system.

2. Uniform chest and knee restraints which satisfy all three criteria in FMVSS 208 in a Part 572 ATD can cause severe neck injuries.

3. A conservative limit for neck fracture load is approximately 6.2 kN (1,400 lb).

4. In the combined axial tension and flexion mode, the critical parameter governing injury is axial tension. The role of shear and moment is unclear at this time.

5. Head and neck kinematics between belted cadavers and those restrained by this air bag system are quite different.

ACKNOWLEDGMENTS

This research was supported in part by DOT Contract DOT-HS-7-01792. All opinions given in this paper are those of the authors and are not necessarily representative of official DOT policy.

REFERENCES

1. Patrick, L. M., Bohlin, N., and Andersson, A.: Three Point Harness Accident and Laboratory Data Comparison. Proc. 18th Stapp Car Crash Conf., SAE Paper # 741181, pp. 201-282, 1974.

2. Hartemann, F., Thomas, C., Henry, C., Foret-Bruno, J. Y., Faveyon, G., Tarriere, C., Got, C. and Patel, A.: Belted or Not Belted: The Only Difference Between Two Matched Samples of 200 Car Occupants. Proc. 21st Stapp Car Crash Conf., pp. 97-150, 1977.

3. Rattenbury, S. J., Gloyns, P. F., Hayes, H. R. M. and Griffiths, D. K.: The Biomechanical Limits of Seat Belt Protection. Proc. 23rd AAAM Conf., pp. 162-176, 1979.

4. Dalmotas, D. J.: Mechanisms of Injury to Vehicle Occupants Restrained by Three-Point Seat Belts. Proc. 24th Stapp Car Crash Conf., pp. 441-476, 1980.

5. Huelke, D. F., Mendelsohn, R. A., States, J. D., and Melvin, J. W.: Cervical Fractures and Fracture-Dislocations Sustained Without Head Impact. J Trauma 18 (7), pp. 533-538, 1978.

6. Schmidt, G., Kallieris, D., Barz, J. and Mattern, R.: Results of 49 Cadaver Tests Simulating Frontal Collision of Front Seat Passengers. Proc. 18th Stapp Car Crash Conf., pp. 283-291, 1974.

7. Schmidt, G., Kallieris, D., Barz, J., Mattern, R. and Klaiber, J.: Neck and Thorax Tolerance Levels of Belt-Protected Occupants in Head-On Collisions. Proc. 19th Stapp Car Crash Conf., pp. 225-257, 1975.

8. Cromack, J. R. and Ziperman, H. H.: Three-Point Belt Induced Injuries: A Comparison Between Laboratory Surrogates and Real World Accident Victims. Proc. 19th Stapp Car Crash Conf., pp. 1-24, 1975.

9. Patrick, L. M. and Levine, R. S.: Injury to Unembalmed Belted Cadavers in Simulated Collisions. Proc. 19th Stapp Car Crash Conf., pp. 79-115, 1975.

10. Levine, R. S., Patrick, L. M., Begeman, P. C. and King, A. I.: Effect of Quadriceps Function on Submarining. Proc. 22nd AAAM Conf., pp. 319-329, 1978.

11. Walsh, M. J. and Kelleher, B. J.: Evaluation of Air Cushion and Belt Restraint Systems in Identical Crash Situations Using Dummies and Cadavers. Proc. 22nd Stapp Car Crash Conf., pp. 295-339, 1978.

12. Thomas, D.J. and Jessop, M.E., Exprerimental Head and Neck Injury, in Impact Injury of the Head and Spine, Ed. by Ewing, C. L. et al, Charles C. Thomas, Springfield, IL (In Press).

13. Moffatt, E. A., Siegel, A. W., Huelke, D. F. and Nahum, A. M.: The Biomechanics of Automotive Cervical Fractures. Proc. 22nd AAAM Conf., pp. 151-168, 1978.

14. Nyquist, G. W., Begeman, P. C., King, A. I. and Mertz, H. J.: Correlation of Field Injuries and GM Hybrid III Dummy Responses for Lap-Shoulder Belt Restraint. Journal of Biomechanical Engineering, Vol. 102, pp. 103-109, 1980.

15. Mertz, Jr., H. J.: The Kinematics and Kinetics of Whiplash. Ph.D. Dissertation, Wayne State University, Detroit, Michigan, 1967.

831616

Cervical Spine Injury Mechanisms

G. S. Nusholtz, D. E. Huelke, P. Lux and N. M. Alem
The University of Michigan

F. Montalvo
General Motors Corp.

ABSTRACT

A test series using eight unembalmed cadavers was conducted to investigate factors affecting the creation of cervical spine damage from impact to the crown of the head. The crown impact was accomplished by a free-fall drop of the test subject onto a load plate. The load plate striking surface was covered with padding to vary the contact force time characteristics. The orientations of the head, cervical spine, and torso were adjusted relative to a laboratory coordinate system to investigate the effects of head and spinal configuration on the damage patterns. Load and acceleration data are presented as a function of time and as a function of frequency in the form of mechanical impedance.

CERVICAL SPINE INJURY studies have been reported in both the clinical and the engineering literature (1-56). Clinical articles based on individual or a small number of accident injury histories have been well summarized in recently published texts on neck fractures (36-39) with approximately a dozen different types of neck fractures or fracture-dislocations having been described and classified.

The fracture or fracture-dislocation injuries of the cervical spine which have been observed most frequently in individual case histories have commonly been referred to as "flexion injuries" (51), because they were hypothesized to be caused by flexion motion combined with compression. These injuries involve fractures of the anterior aspect of the lower cervical vertebral bodies (possibly with dislocations of adjacent vertebral bodies, tearing of posterior ligaments, and/or fractures of the laminae or spines).

Recently Nusholtz et al. attempted to reproduce this "flexion-type" of damage pattern in unembalmed human cadavers under the conditions hypothesized to cause this type of injury in the live human (34). A pendulum impact device with a 56 kg free-moving mass was used to deliver cranial impacts to the cadaver subjects in the S-I direction. Some of the important conclusions from this preliminary study were:

1. The initial orientation of the spine relative to the impact axis was a critical factor influencing the type of kinematic response and damage produced.

2. Descriptive motion of the head relative to the torso was not a good indicator of the mechanism producing neck damage. "Flexion" damage was observed to occur with extension motion and "extension" damage was observed to occur with flexion motion.

3. Energy absorbing materials were effective methods of reducing peak impact force but did not necessarily reduce the amount of energy transferred to the head, neck, and torso or the damage produced.

Although in some tests flexion-compression damages were observed in the upper thoracic spine, only one of the twelve subjects sustained this type of damage in the cervical spine area. Usually extension-compression-type damage was observed.

More recently Alem et al. (35) attempted to obtain tolerance criteria for the cervical spine by superior-inferior crown impacts to unembalmed human cadavers using a 10 kg free-flying mass. Some important results from their study were:

1. Peak force was not found to be a reasonable predictor of cervical spine damage.

2. Forces as low as 3 kN produced cervical spine damage while forces up to 16 kN produced no cervical spine damage. Although tests were performed in which forces above 16 kN produced no cervical spine damage (peak forces as high as 35 kN), skull fractures, either basilar or local (under the impactor), were observed.

3. The impulse of the impact force and the maximum head velocity might be used to predict cervical spine damage.

These results are consistent with other research on cervical spine damage (Culver et al. (32), Hodgson and Thomas (33)).

This paper presents the results of a study investigating the effects of head/neck/torso configuration and impact conditions on the response and damage to unembalmed cadavers following crown impact. To address the importance of initial configuration in a three-dimensional sense, two test series were run. In the first series, (mid-sagittal series), the mid-sagittal plane of the head was aligned with the mid-sagittal plane of the thorax. The heads of the subjects in the second test series (non-mid-sagittal series) were rotated not only about the lateral axis, but also about the AP and SI axis; in addition, the thorax was turned about its SI axis. Impacts to the head in the superior-inferior direction were produced by a free-fall drop of the test subject onto a rigid structure to which padding had been added. Comparisons are made between the results from the free-fall drops and results from previous studies in which a free-flying mass delivered the impact (34,35).

CLINICAL STUDIES AND MECHANISM OF INJURY

There have been many clinical reports on the mechanism of the injury from actual cases wherein the authors postulated head/neck movements that caused the injury. These have been best summarized in the works by Braakman and Penning and by Kattan (36,37). Although there have been a dozen different types of fractures of the cervical spine previously described, the four most prominently discussed in the clinical literature have been flexion or extension, with either compression or tension. In almost all of the clinical literature it has been postulated that to get the classic flexion/compression fractures the head must have been bowed significantly forward with a marked downward-force causing the typical compression of the anterior cervical body, also causing, at times, associated injuries of the posterior cervical elements (36, 37, 52, 56). Flexion, with tension applied to the neck, would cause disruption of the posterior ligaments, separation of the articular facets, disc disruptions, dislocation, and possible tearing of the longitudinal ligaments. In general, the clinical literature had always described the need for the neck to be in a hyperflexed or hyperextended position, in association with the compression or tension. Injury mechanisms in sports have also been reported (38,39).

Recently, Torg et al. (39) indicated that many football injuries have been due to extreme axial loading on the straightened cervical spine and that the straight cervical spine, when axially loaded, acted like a segmented column. They pointed out that when the neck was in forward flexion, the cervical spine was in fact straightened.

METHODOLOGY

A detailed discussion of the techniques, procedures, and methodology for this test series was previously reported (57) and is briefly outlined below.

INITIAL CONDITIONS -- For all the tests, the subject was positioned upside down with a load platform centered beneath the crown of the head (Figure 1). The tests were divided into two series. The first being those experiments in which the initial conditions of the uninstrumented test subjects were constrained to rotations in the mid-sagittal plane (mid-sagittal series), the force being the only transducer time history obtained. In the second or non-mid-sagittal series of experiments, the head of the test subject was not only rotated about the lateral axis, but the A-P and I-S as well; also, the thorax was twisted about the I-S axis and the lateral axis. In this latter series of experiments, the head was instrumented with a nine accelerometer array and selected thoracic vertebra had triaxial accelerometer clusters affixed to them. The subjects' initial conditions were described using the method similar to that presented elsewhere (34) and outlined below.

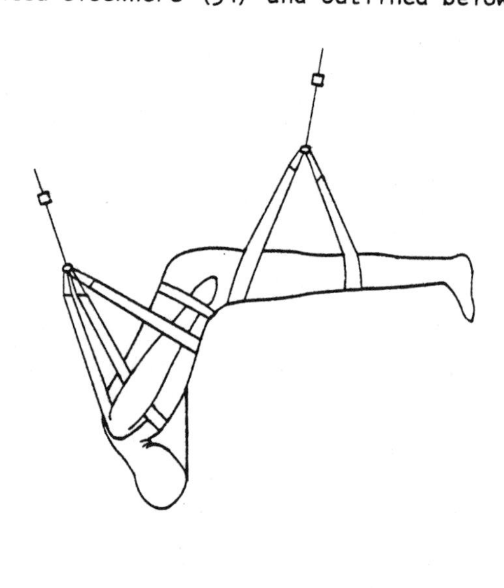

Fig. 1 - Restraint Configuration, Cadaveric Subject

Reference systems were defined on the head, neck, and thorax to document the initial position in a three-dimensional sense and to provide a method of comparison for the responses between subjects. The head, neck, and thorax reference systems are defined in Table 1 and shown in Figure 2. In addition the reference frame with which all initial angles of the head, thorax, and neck were measured is also shown in the table and figure.

TABLE 1: REFERENCE SYSTEM DESCRIPTIONS

Four reference systems were used to describe the initial configuration of each test subject. The conventional anatomical reference system for the head was used. However, the reference systems associated with the neck and thorax were not standard and have been defined for convenience of measurement with respect to the laboratory reference frame used here. All anatomical systems have been defined such that the 1 axis roughly corresponded to the medial-lateral direction, and the 3 axis roughly corresponded to the inferior-superior direction.

- W- The laboratory reference system defined as follows:
- Origin - Tip of the coccyx
- W_1- The projection onto the horizontal plane of the axis of the legs
- W_2- The vector to complete the orthogonal triad
- W_3- The normal to the horizontal plane in the vertical direction

- θ- The anatomical reference system for the head:
- Origin- The intersection of the mid-sagittal, coronal, and Frankfort planes
- θ_1- The posterior-anterior axis (the intersection of the mid-sagittal and Frankfort planes)
- θ_2- The medial-lateral axis (the intersection of the coronal and Frankfort planes)
- θ_3- The inferior-superior axis (the intersection of the mid-sagittal and coronal planes)

- α - The reference system associated with the neck:
- Origin- T1
- α_1- The cross product
- α_2- The medial-lateral axis running roughly between the acromion
- α_3- The inferior-superior axis of the neck

- γ - The reference system associated with the thoracic spine:
- Origin- T4
- γ_1- The cross product
- γ_2- The medial-lateral axis running roughly between the scapulae
- γ_3- The inferior-superior tangent to the spine at T4

- v - The vector originating approximately at T12 and describing the angle (Figure 2)

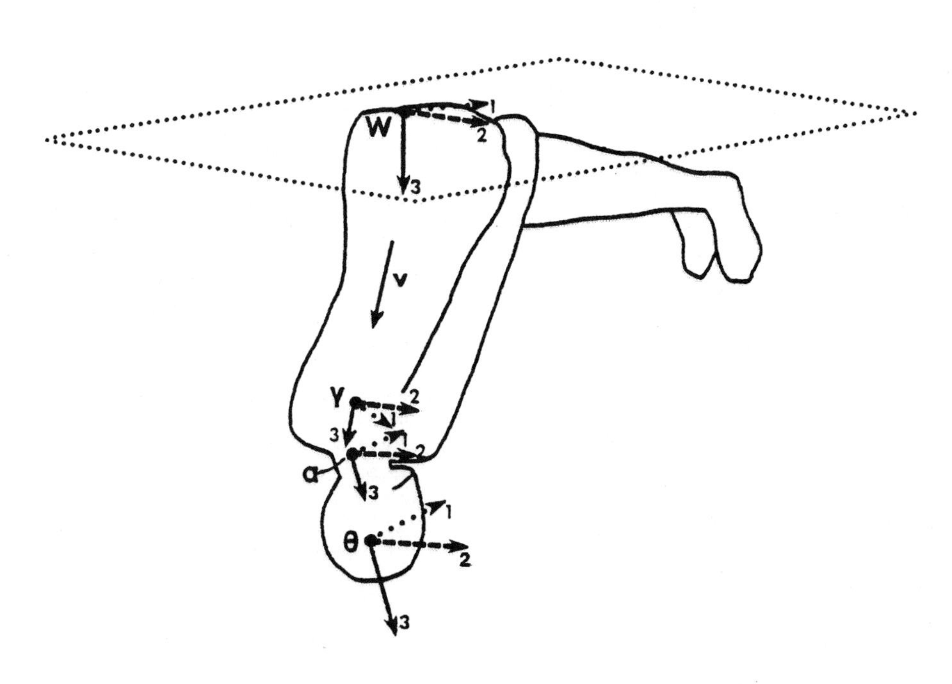

Figure 2. Three-Dimensional Initial Conditions

ACCELEROMETER INSTRUMENTATION -- The following is a brief description of the techniques used in mounting the accelerometer hardware.

Head Nine-Accelerometer Array - Several metal self-tapping screws were threaded through small pilot holes into the parietal bone of the skull. Feet were attached to the magnesium accelerometer mounting plate and were positioned near the screws on the exposed skull. To insure rigidity, plastic acrylic was molded around the screws, feet, and plate, such that the plate becomes rigidly affixed to the skull. Three triaxial clusters of accelerometers will later be attached to their positions on the plate.

Spine Triaxial Arrays - Incisions were made over the first, sixth, and twelfth thoracic vertebrae. The mounting platforms were screwed directly into the spinous processes of each vertebrae. Plastic acrylic was applied under and around the mounts to insure rigidity.

LOAD PLATE -- The apparatus to measure axial and shear forces in the first five test subjects consisted of a rectangular metal platform which rested on four piezoelectric force transducers. At the base of each transducer was a ball bearing which could rotate but was constrained from movement in the horizontal plane by a Plexiglas template (Figure 3). The output signals from the transducers underwent a series of processing steps prior to being recorded.

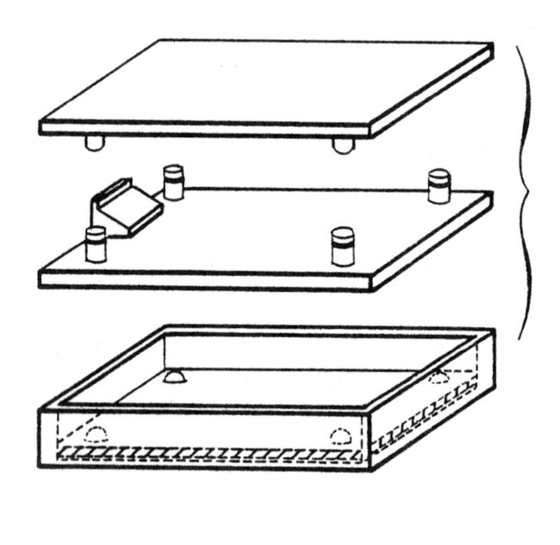

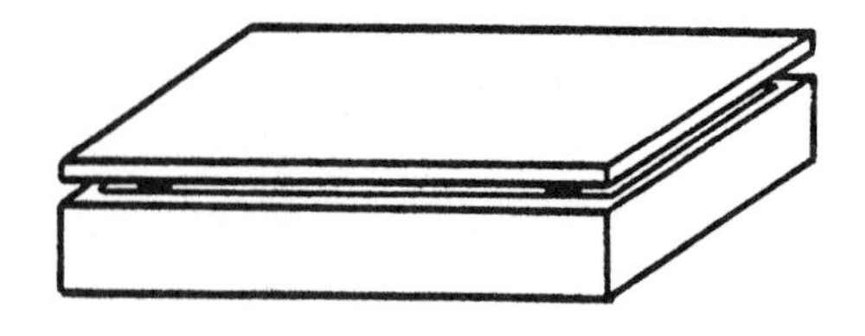

Fig 3 - Load Plate for Tests 82L484-82L494

The load plate system provided accurate axial and shear force readings for impacts of either relatively long duration and low frequency content (e.g., 100 to 300 ms), or for impacts which were associated with excessive motion of the subject on the surface of the plate.

However, when the impact was of a localized nature and the signal contained significant components of a high frequency content and short duration (10 to 30 ms), the load plate system failed to yield accurate results due to the fact that the natural frequency of the top plate was approximately the same as certain frequencies contained in the force signal, and oscillation of the plate might result. Although the effect of the plate oscillation in the force time history was detectable, it was not significant enough to effect the overall nature of the signal. For the non-mid-sagittal series (last three test subjects) the impactor surface was a 15 cm round rigid metal plane to which padding had been added. The impactor force transducer assembly consisted of a Kistler 904A piezoelectric load washer with a Kistler 804A piezoelectric accelerometer mounted internally for inertial compensation (Figure 4).

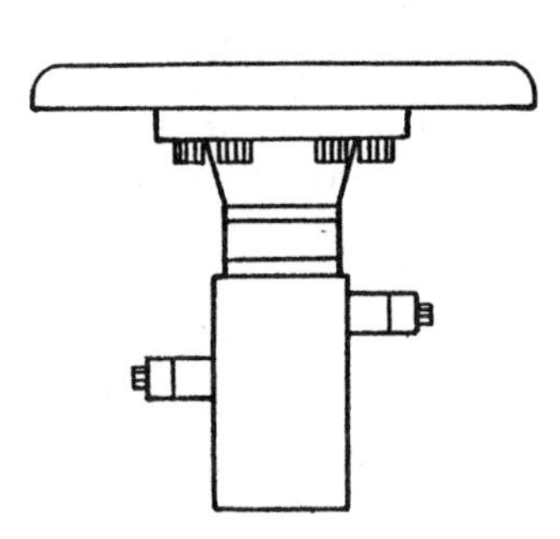

Fig 4 - Load Plate for Tests 83L495-83L501

OVERHEAD HOIST SYSTEM -- When positioned for the impact, the subject was suspended by two primary harness systems each of which was connected to the ceiling by a single rope passed through rope cutters (Figure 5). Initially, these two ropes were suspended from pulley systems attached to fixed points on the ceiling. When it had become necessary for the subject to be positioned with various head/neck/thorax angles as described above, an overhead hoist system was developed which would allow for increased relative motion between these two harness systems. It consisted of two power hoists which would move linearly in tracks oriented perpendicularly to the long axis of the subject. These tracks in turn moved linearly parallel to the long axis of the subject, such that each hoist could be positioned accurately anywhere within the horizontal plane (Figure 5).

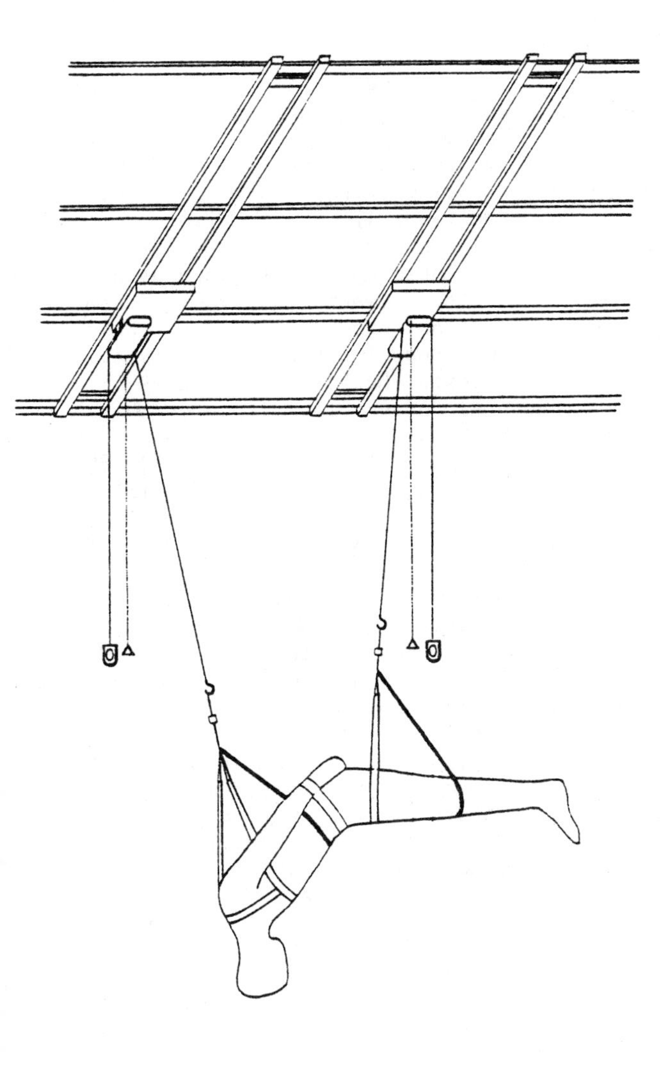

Fig. 5 - Overhead Hoist Positioning System

IMPACT TESTING -- Multiple non-damaging impact tests were performed on a select number of test subjects in the non-mid-sagittal series. To determine the level of energy input that could be applied without damage, an earlier series of tests using a pendulum impactor (with similar impact velocities and impactor padding) had been performed on three cadavers. Gross autopsies revealed no damage.

The protocol for the mid-sagittal series of tests was similar, but much less complicated than the non-mid-sagittal series. A brief outline of the protocol for the non-mid-sagittal series is given below.

Subjects were obtained from the University Anatomy Department and transported to UMTRI by Biomechanics Department personnel. Upon arrival they were weighed and pertinent information was logged in. Initial preparation was completed and pretest X-rays were taken of the head, thorax, pelvis, and femurs. If the skeletal structure was intact, the subject was taken to the anatomy lab for surgical preparation.

The subject was placed in a supine position while anthropometric measurements were made. Anatomical abnormalities, if any, were recorded and special note was taken if they would alter the usual instrumentation techniques. The subject was then placed in the prone position for installation of head and spinal accelerometer mounting hardware.

The subject was fitted with the belt harnesses and transducers were attached to the mounts. The subject was hoisted and positioned for a low-energy drop. After drop height and the initial angles of the head, neck, and thorax were measured and recorded, the test was run. This procedure was repeated three more times for the low-energy non-damaging drops.

Prior to the high-energy drop, the gain values of the amplifiers were changed, the subject positioned, drop height and initial angles recorded, and the test conducted.

When the testing sequence had been completed, instrumentation was removed and the autopsy was performed.

METHODS OF ANALYSIS

Both acceleration and photokinemetric data were obtained to document the motion of the head, neck, and thorax. The methods used to obtain these data are summarized below.

ACCELERATION TIME HISTORY -- The method used for documentation of three-dimensional motion from accelerometer data was based on a technique used to measure the general motion of a vehicle during a simulated crash.

For this application, three triaxial clusters of Endevco 2264-2000 accelerometers were affixed to a lightweight rigid magnesium plate which was then solidly attached to the skull. The nine acceleration signals obtained from the three triaxial clusters were used for computation of head motion using a least squares technique. This method took advantage of the redundancy of nine independent acceleration measurements to minimize the effect of experimental error to produce three angular accelerations and three estimates, in the least squares sense, of the true solution. The instrumentation frame and its origin were determined from the three triaxial accelerometer centers. Finally, the transformation matrix between the instrumentation frame and the anatomical frame was obtained through the use of X-rays.

In the case of a single triaxial accelerometer the complete three-dimensional motion description was impossible, but it was found that in many cases it was possible to find the most significant component of acceleration during impact, therefore the principal direction of motion could be obtained (34).

One method of determining the principal direction of motion and constructing the Principal Direction Triad was to determine the direction of the acceleration vector in the moving frame of the triaxial accelerometer cluster and then describe the transformation necessary to obtain a new moving frame that would have one of its axes in the principal direction. A single point in time at which the acceleration was a maximum was chosen to define the directional cosines for transforming from the triax frame to a new frame in such a way that the resultant acceleration vector (AR) and principal acceleration vector (A1) were co-directional. This then could be used to construct a new frame rigidly fixed to the triax but differing from the original one by an initial rotation. After completing the necessary transformation a comparison between the magnitude of the principal direction and the resultant acceleration was performed. In the case of the impacts presented here there was only a slight difference between the two quantities during the most significant part of the impact. However, for responses occurring after impact this was not always the case.

TRANSFER FUNCTION ANALYSIS -- With blunt impacts, the relationship between a transducer time history at a given point and the transducer time history at another given point of a system could be expressed in the frequency domain through the use of a transfer function. A fast Fourier transformation of simultaneously monitored transducer time histories could be used to obtain the frequency response functions of impact force and accelerations of remote points. Once obtained a transfer function of the form

$$Z(i\omega) = \omega \cdot \frac{\mathcal{F}(F(t))}{\mathcal{F}(A(t))}$$

could be calculated from the transformed quantities where is the given frequency, and $\mathcal{F}(F(t))$ and $\mathcal{F}(A(t))$ are the Fourier transforms of the impact forces and acceleration of the point of interest at the given frequency. This particular transfer function was closely related to a mechanical transfer impedance which was defined as the ratio between the simple harmonic driving force and the corresponding velocity of the point of interest. Mechanical transfer impedance (15) was a complex valued function which for the purpose of presentation will be described by its magnitude and its phase angle.

PHOTOKINEMETRICS -- Photographic documentation of the mid-sagittal series of tests consisted of a single lateral view by a high-speed Hycam movie camera operating at 1000 frames per second. Photographic documentation of the non-mid-sagittal series consisted of two orthogonal views with high speed movie cameras operating at 1000 frames per second.

RESULTS

The kinematic data, as well as injury response data, are summarized in Tables 2 through 5. The damage patterns observed from gross autopsy are described in Table 3, where neck damage was classified according to the type of motion traditionally hypothesized to have caused the injury (e.g., "extension-compression"). These categories were used only to differentiate between the types of damages observed, and do not necessarily correspond to the mechanism producing the injury listed. In addition, in tests 82L485 and 83L499 linear skull fractures under the impactor which extended into both parietal bones and the frontal bone were observed. Also, in tests 82L489 and 83L500 crushing fractures of the occipital condyles were observed.

Table 4 summarizes the individual test conditions, where the column "Initial Conditions" describes in a broad sense the pretest position of a subject's head relative to the thorax. As previously mentioned, in the mid-sagittal series, which was uninstrumented, the mid-sagittal plane of the head was constrained to lie parallel to the AP-SI plane of the thorax; however, in the second series, which was instrumented, the head did not lie in any specific orientation relative to the thorax. Table 5 contains the initial conditions of these tests as defined by Table 1 and Figure 2.

Some important kinematic quantities obtained from the instrumented tests by the UMTRI three-dimensional motion analysis computer program were presented in Table 6.

TABLE 2. ANTHROPOMETRIC DATA

Subject	Cadaver No.	Age	Ht (cm)	Wt (kg)
82L484	1	60	180	67
82L485	2	61	177	51
82L486	3	61	181	55
82L487-89	4	70	160	50
82L490-94	5	69	171	67
83L495-499	6	62	176	76
83L500	7	52	180	68
83L501	8	51	169	83

TABLE 3. DAMAGE SUMMARY

82L484
Extension-Compression Type:
- C4/C5 - Rupture of disc

Flexion-Compression Type:
- T3 - Fractured right superior articular process
- T2/T3 - Partially torn interspinous ligaments
- T3/T4 - Partially torn ligamentum flavum

82L485 (Extreme Osteoporosis)
- C1, C3, C4 - Fracture of laminae
- C2 - Fracture of dens
- C3 - Fracture of spinous process
- C4, C5 - Fracture of vertebral bodies
- C4/C5 - Ruptured disc
- T1, T3, T4 - Fracture of laminae
- T2 - Fractured body
- T3 - Anterior longitudinal ligament torn

82L486
Extension-Compression Type:
- C1 - Fracture of anterior ring at dens
- C4 - Fracture of vertebral body
- C3/C4 - Fractured spinous processes
- C2/C3 and C4/C5 - Ruptured discs
- C2/C3, C3/C4, C4/C5 - Anterior longitudinal ligament torn
- C3 - Posterior longitudinal ligament torn

82L489
Extension-Compression Type:
- C1 - Fractures of posterior arch with single fracture of right anterior arch
- C2 - Complete fracture of dens with separation of all anterior ligaments

Flexion-Compression Type:
- C5/C6 - Subluxation of C5 over C6 with disruption of all C5/C6 ligaments, capsules and discs
- T1/T2 - Compression of superior body T2; tear of posterior disc

82L494
Flexion-Compression Type:
- C6 - Fractures of right lamina and anterior-superior body

82L494 (cont.)
- C7 - Fracture of body and of right lateral facets at base
- C6/C7 - Anterior longitudinal ligament, ligamentum flava, and posterior aspect of dura torn

83L499
Extension-Compression Type:
- C5/C6 - Rupture of disc
- C6/C7 - Rupture of disc
- C5/C6 and C7/T1 - Anterior longitudinal ligament disrupted

Flexion-Compression Type:
- T1 - Compression of anterior superior body and associated compression of C7/T1 disc
- T2 - Fracture of anterior superior edge of vertebral body
- C7/T1 - Ligamentum flavum torn

83L500
Axial Loading
- C1 - Fracture of right transverse process

Extension-Compression Type:
- C4/C5 - Disc rupture
- C6/C7 - Disc rupture

Flexion-Compression Type:
- T3 - Fracture of left transverse process
- T3 - Chip fracture of anterior inferior body
- T4 - Fracture of left lateral vertebral body
- T4 - Compression of anterior superior body
- T3/T4 - Rupture of disc and partial tear of anterior longitudinal ligament
- T3/T4 - Articular facets very loose (possibly torn capsules) with tear of supra- and interspinous ligaments

83L501
Flexion-Compression Type:
- T1 - Compression of anterior superior body
- T2 - Compression of anterior superior body
- C7 - Fracture of right lamina
- C6/C7 - Nearly complete tear of posterior ligaments

TABLE 4. TEST SUMMARY

Test No.	Cadaver No.	Drop Height (m)	Peak Force (kN)	Contact Velocity (m/s)	Ensolite (cm)	Initial Conditions	HIC
82L484	1	1.0	6.7	4.4	.6	*constrained	
82L485	2	1.8	-	5.9	2.5	constrained	
82L486	3	1.5	5.9	5.4	2.5	constrained	
82L487	4	0.1	0.3	1.4	2.5	constrained	5
82L488	4	0.1	0.5	1.4	2.5	**unconstrained	7
82L489	4	1.5	5.4	5.4	2.5	unconstrained	2240
82L490	5	0.1	0.6	1.4	2.5	constrained	2
82L491	5	0.1	-	1.4	2.5	unconstrained	4
82L492	5	0.1	0.5	1.4	2.5	unconstrained	3
82L493	5	0.1	-	1.4	2.5	unconstrained	3
82L494	5	1.1	5.2	4.6	2.5	unconstrained	477
83L495	6	.1	.9	1.4	.6	unconstrained	14
83L496	6	.1	.9	1.4	.6	unconstrained	12
83L497	6	.1	1.0	1.4	.6	unconstrained	28
83L498	6	.1	.9	1.4	.6	unconstrained	18
83L499	6	.914	3.2	4.2	2.5	unconstrained	210
83L500	7	1.5	10.8	5.4	2.5	***constrained	
83L501	8	.8	5.6	4.0	2.5	unconstrained	540

*Constrained or initial conditions limited to mid-sagittal plane motion.
**Unconstrained or initial conditions not limited to mid-sagittal plane motion.
***Neck was wrapped with Dow Ethaform

TABLE 5. INITIAL CONDITIONS

Cadav.	Test	Location	*Axis 1	Axis 2	Axis 3
1	82L484	Head		30°	
		Neck		30°	
		Mid-spine		20°	
2	82L485	Head		5°	
		Neck		15°	
		Mid-Spine		27°	
3	82L486	Head		-5°	
		Neck		10°	
		Mid-Spine		30°	
4	82L487	Head	15°	5°	10°
		Neck	3°	0°	4°
		Thorax	0°	15°	7°
		Mid-Spine	0°	0°	17°
4	82L488	Head	15°	6°	12°
		Neck	4°	0°	6°
		Thorax	0°	15°	7°
		Mid-Spine	0°	0°	14°
4	82L489	Head	15°	6°	17°
		Neck	40°	0°	3°
		Thorax	0°	16°	8°
		Mid-Spine	0°	0°	15°
5	82L490	Head	3°	0°	0°
		Neck	0°	0°	0°
		Thorax	0°	15°	0°
		Mid-Spine	0°	0°	14°
5	82L491	Head	4°	19°	0°
		Neck	0°	0°	1°
		Thorax	0°	16°	0°
		Mid-Spine	0°	0°	12°
5	82L492	Head	0°	0°	10°
		Neck	0°	0°	2°
		Thorax	0°	16°	0°
		Mid-Spine	0°	0°	15°
5	82L493	Head	0°	3°	0°
		Neck	0°	2°	0°
		Thorax	0°	14°	10°
		Mid-Spine	0°	0°	13°
5	82L494	Head	15°	6°	15°
		Neck	40°	0°	0°
		Thorax	0°	24°	9°
		Mid-Spine	0°	0°	15°
6	83L495	Head	0°	0°	0°
		Neck	0°	0°	0°
		Thorax	0°	12°	15°
		Mid-Spine	0°	0°	12°
6	83L496	Head	0°	0°	20°
		Neck	0°	0°	0°
		Thorax	0°	0°	16°
		Mid-Spine	0°	0°	14°
6	83L497	Head	15°	0°	3°
		Neck	40°	0°	0°
		Thorax	0°	0°	16°
		Mid-Spine	0°	0°	13°
6	83L498	Head	0°	0°	0°
		Neck	0°	0°	0°
		Thorax	5°	0°	7°
		Mid-Spine	0°	0°	11°
6	83L499	Head	10°	3°	16°
		Neck	0°	0°	0°
		Thorax	5°	13°	7°
		Mid-Spine	0°	0°	14°
7	83L500	Head		0°	
		Neck		0°	
		Mid-Spine		22°	
8	83L501	Head	15°	0°	10°
		Neck	30°	3°	3°
		Mid-Spine	0°	0°	21°
		Thorax	0°	5°	0°

*These are the axes around which the angles are measured. See Table 1 and Figure 2.
(Axis 1 = P-A Axis 2 = R-L Axis 3 = I-S)

TABLE 6. SUMMARY OF KINEMATIC DATA FROM HEAD MOTION FOR INSTRUMENTED SUBJECTS

Test No.	Cadaver No.	Linear Vel. (m/s)	Linear Accel. (m/s²)	Angular Vel. (rad/s)	Angular Accel. (rad/s²)
82L487	4	1.6	157	5.9	375
82L488	4	2.3	182	5.0	384
82L489	4	8.5	3400	14	3800
82L490	5	1.3	91	5.1	354
82L491	5	1.6	148	1.9	184
82L492	5	1.5	119	1.5	124
82L493	5	1.7	101	2.8	192
82L494	5	6.0	2280	15	10000
83L495	6	1.4	320	8.4	960
83L496	6				
83L497	6	1.8	420	5.8	1200
83L498	6	1.6	344	6.5	880
83L499	6	4.3	1092	14.4	5440
83L500	7	N/A	N/A	N/A	N/A
83L501	8	4.2	1100	1.5	5500

DISCUSSION

In this study, kinematic and damage pattern data of the head and spine are gathered from a series of drop tests of cadaveric specimens upon a load cell platform. The impact results obtained by this method are in addition to the pendulum impact study of Nusholtz (34). The primary response characteristics investigated in the present study are the force time history of impact and the associated damage patterns, as well as the acceleration response of those subjects which had been instrumented. The results of the drop tests indicate that there are no gross differences in the kinematic response and damage patterns observed in this study (drop tests) versus the pendulum study. In addition, it has been observed that the involvement of the thoracic spine is an important factor in cervical spine impact response.

PREVIOUS IMPACT STUDIES -- Previously (34), it was reported that the initial head/neck/thorax position was a critical factor in the type of damage produced and the associated waveform of the force time history. The initial positioning of test subjects in the pendulum study included the alignment of the mid-sagittal plane of the head and neck with the mid-sagittal plane of the thorax, attempting to restrict the resulting motion in two directions. Positioning the head and neck of the test subject in the mid-sagittal plane, in addition to other initial configurations, resulted in damage patterns that are mainly of two types: extension-type damage is observed in the cervical spine and flexion-type damage is observed in the upper thoracic spine.

These force time histories are either unimodal or bimodal waveforms, ascribed to one maxima or two local maxima. The response of the test subject during the force time history interval for the unimodal impacts is attributed to inertial loading of the head during the initial part of the pulse with significant interactions of the cervical spine late in the event. The bimodal waveform is attributed to an initial inertial loading of the head (with some coupling of the cervical spine), followed by a second peak force, which is believed to be caused by an additional coupling between the head, neck and the whole body. Dynamically, the peak acceleration of the head precedes the first peak force, and becomes negative while the force remains positive, indicating that a greater force is exerted on the head from the neck and body in opposition to the action of the impactor piston of the pendulum. The mechanical impedance response of these subjects are non-linear and characteristic for the type of waveform of the force time history. Non-linearities in the response of the test subject is particularly apparent in the low frequency range of the mechanical impedance: a greater impedance is observed in frequencies below 50 Hertz for the bimodal force time histories than is seen in the unimodal tests.

MID-SAGITTAL PLANE MOTION FOR CURRENT AND PREVIOUS STUDIES -- An initial comparison of the force trace and damage response produced in the drop tests versus the pendulum supine cadaver subject was attempted in a series of uninstrumented test subjects. Tests 82L484 - 82L486 and 83L500 comprise this initial group, and for these tests the head was held with tape such that a slight amount of compression was maintained on the cervical spine. The head and thorax were aligned in the mid-sagittal plane with the thorax positioned at angles of 20°, 25°, 25°, and 17° relative to the vertical plane; the neck, 30°, 15°, 10° and 0°. The respective drop heights are 1.0 m, 1.8 m, 1.5 m, and 1.5 m.

The force time histories and the damage response of the four uninstrumented drop tests are similar, by gross comparison, to the pendulum results of (34). This comparison, however, is limited to the nature of the force time history waveform and the type of damage pattern, either extension or flexion, and not the degree of injury or the particulars of the damage patterns (such as fracture of the posterior processes or anterior wedge fractures of the vertebral bodies). Both cervical spine extension-type damages and thoracic spine flexion-type damages are observed in the four test subjects. One cervical spine flexion-type injury is observed in test 82L485; although this subject was extremely osteoporotic, sustaining many cervical fractures throughout the cervical and upper thoracic region that were not produced in those subjects with stronger skeletal structure. The motion of the head, as observed in the high-speed films, is similar to the parabolic motion (head forced into chest) previously observed in the pendulum impacts having the same initial conditions. The recorded forces are axial with negligible shear force between the head and load platform.

Three force time histories are obtained from the four uninstrumented tests: tests 82L484 and 82L486 are unimodal, and test 83L500 is bimodal (Figure 6).

Although in a gross sense the kinematic and damage response of the pendulum impact test and the free-fall drop test are similar; the following differences may occur in the force time history due to the different methods of testing. Although the unimodal and the first local maxima of the bimodal force time history may show no significant differences in the two test types, the second maxima of the bimodal test, however, can be different. For a given waveform the magnitudes of the peak force of the second maxima, relative to the first maxima, can be lower in the pendulum tests than in the drop tests. The duration of the waveform may also be different. Previously (34), it was suggested that the second local maxima of the force is due to in-

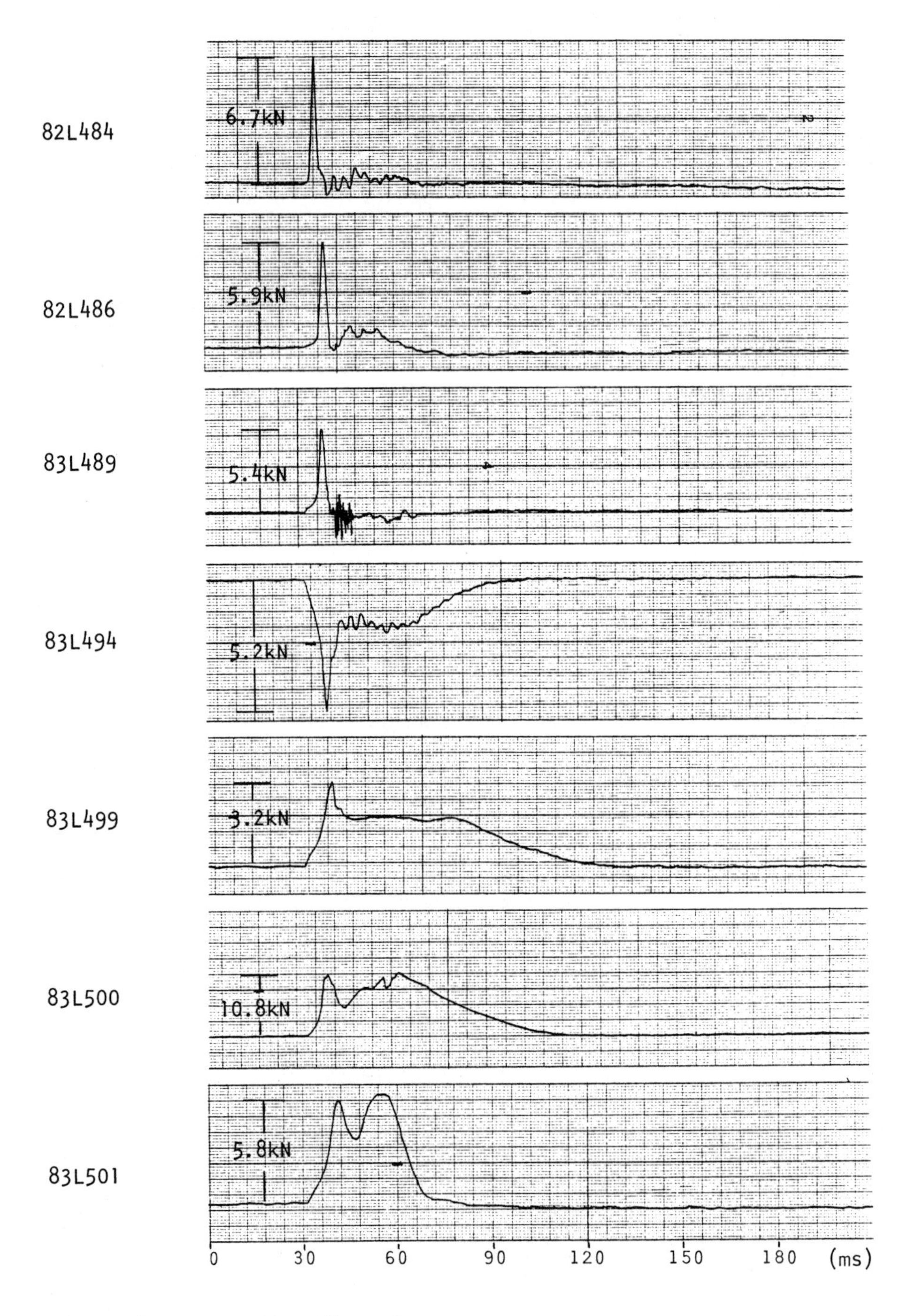

Figure 6. Force Time Histories

creased coupling of the whole body to the head and neck. If it is reasonable to assume a coupled whole body response occurs in the second maxima of the force trace, then the 18 cm stroke of the piston may significantly limit the duration of the force. (In some of the previous bimodal tests (34) the contact length is limited but it is uncertain as to the significance of the effect.) In addition, a 55 kg piston impacting a 70 kg test subject versus the free-fall body of a 70 kg subject upon the mass of the earth under constant gravitational acceleration may limit the magnitude of the second maxima for the pendulum tests. The short stroke of the pendulum piston combined with the smaller impact mass is therefore seen as a potential artifact limiting the impact response of the test subject for those tests that used the pendulum. This may have greater significance for different initial conditions (impact velocity, padding, or impact surface geometry) than those of this and the previous study (34).

It appears from these tests and those reported previously (34,35) that: 1) the possibility of obtaining cervical spine flexion-type damages is small when the mid-sagittal anatomical plane of the head is initially constrained in the AP-SI plane of the thorax (with resulting motion primarily in the mid-sagittal plane) and 2) that the initial conditions for a free-fall test which may result in cervical spine flexion-type damage should take into account the six degrees of freedom for each of the head, neck, and thorax, and the associated constraints. Allowing for six degrees of freedom, however, creates an extreme increase in the complexity of the test matrix. An attempt to reduce the number of test subjects required for such a complex test configuration was done in the following manner: a series of low energy (non-damaging) drops with varying head/neck/thorax angles were conducted for a select number of test subjects to obtain the response of these subjects for varying initial conditions. Given the various impact configurations possible in a low energy test, the response of these tests may be potentially useful in obtaining basic biomechanical information.

LOW-DROP TESTS -- In the mid-sagittal series of low-drop tests (82L487,82L488,82L490 to 82L493), the load platform was padded with 1" (2.54 cm) Ensolite (AL), analogous to the high-energy drop tests; the four subsequent tests, 82L495-82L498, had 1/4" (.6 cm) Ensolite (AH). Padding on the load platform is found to be a factor in the type of response elicited in the non-damaging drops. The force time histories as well as the acceleration response of the first six low-drop tests are significantly different from the high-energy impacts; a significant shear force, at times equal to or greater than the axial force (.3-.5 kN), in conjunction with an observed sliding of the head along the impact surface, results, in general, in a unimodal force trace five to ten times longer in duration than previous unimodal force trace time durations. It is observed in the last four tests that the Ensolite (AH) padding of .6 cm thickness resulted in bimodal force time histories similar to previously obtained bimodal waveforms with reduced magnitudes and somewhat longer contact times noted.

The mechanical impedance results of the last four low-energy drop tests have the following characteristics: the response of T6 and T12 vary to some extent, but, in general, have similar magnitudes of mechanical impedance (as obtained from the principle direction acceleration response) regardless of the initial positioning of the head, neck, and thorax. Figure 7 is a corridor for the mechanical impedance for these tests for T6 and T12. In addition, inspection of the individual impedance diagrams indicate that there is less variation in impedance response between T6 and T12 for a given test than the variation of either vertebrae from test to test, implying that the thoracic spine from T6 to T12 acts more like a coherent unit than individual segments in this instance; the impedance magnitude corridor of the T6 and T12 vertebrae in these tests indicates that the T6 and T12 section of the spine acts as an 8-10 kg mass in the frequency range of 20-200 Hz. The impedance response of the T1 vertebrae, however, varies to a greater degree for each of the four initial conditions (Figure 8), indicating that the response of Thoracic T1 is more strongly dependent on the initial position of the head, neck, and thorax than either T6 or T12 in a low-energy drop test.

HIGH-DROP TESTS -- When the initial positioning of the subject is not in the mid-sagittal plane, flexion-type damages are observed in the cervical spine in tests 82L489, 82L494, 83L499, and 83L501. This is consistent with other research which used animal models to study the cervical (53,54) and lumbar (55) spine. Other spinal damage patterns were observed in these tests that are not commonly reported in the clinical literature (36-39), namely, flexion-type damage in the thoracic spine (also, in some cases, compression fractures of the occipital condyles). These damages may be the result of the drop height and a subsequent overdriving of the system. In addition, normal changes in the vertebrae as a result of aging and/or postmortem differences between the cadaver test subject and that of the live human may be important contributing factors. However, it is also possible that injuries to the upper thoracic spine do occur with cervical flexion spine fractures, but are undetected.

Both unimodal and bimodal force time histories are observed: test 82L489 is unimodal, and tests 82L494, 83L499, and 83L501 are bimodal. The acceleration response and head motion observed in these four tests are similar to the pendulum impacts early in the impact event,

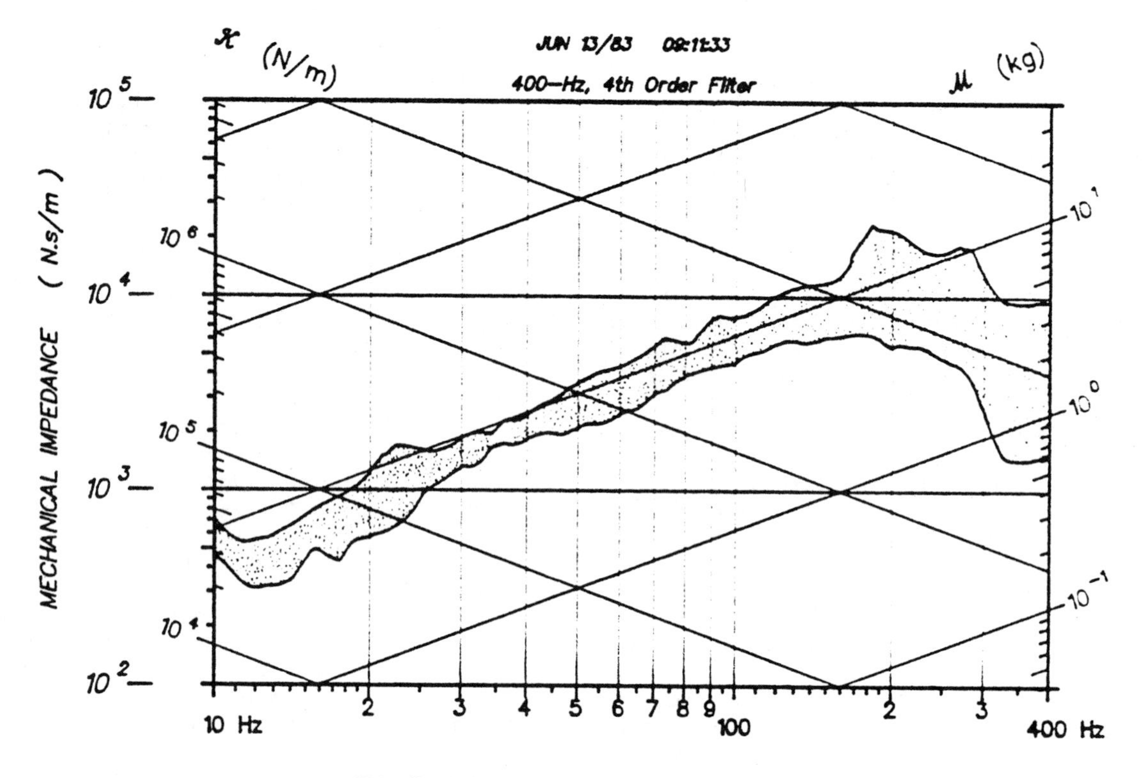

Z=F1/V1 for T6–T12 for 83L495 – 83L498

Figure 7

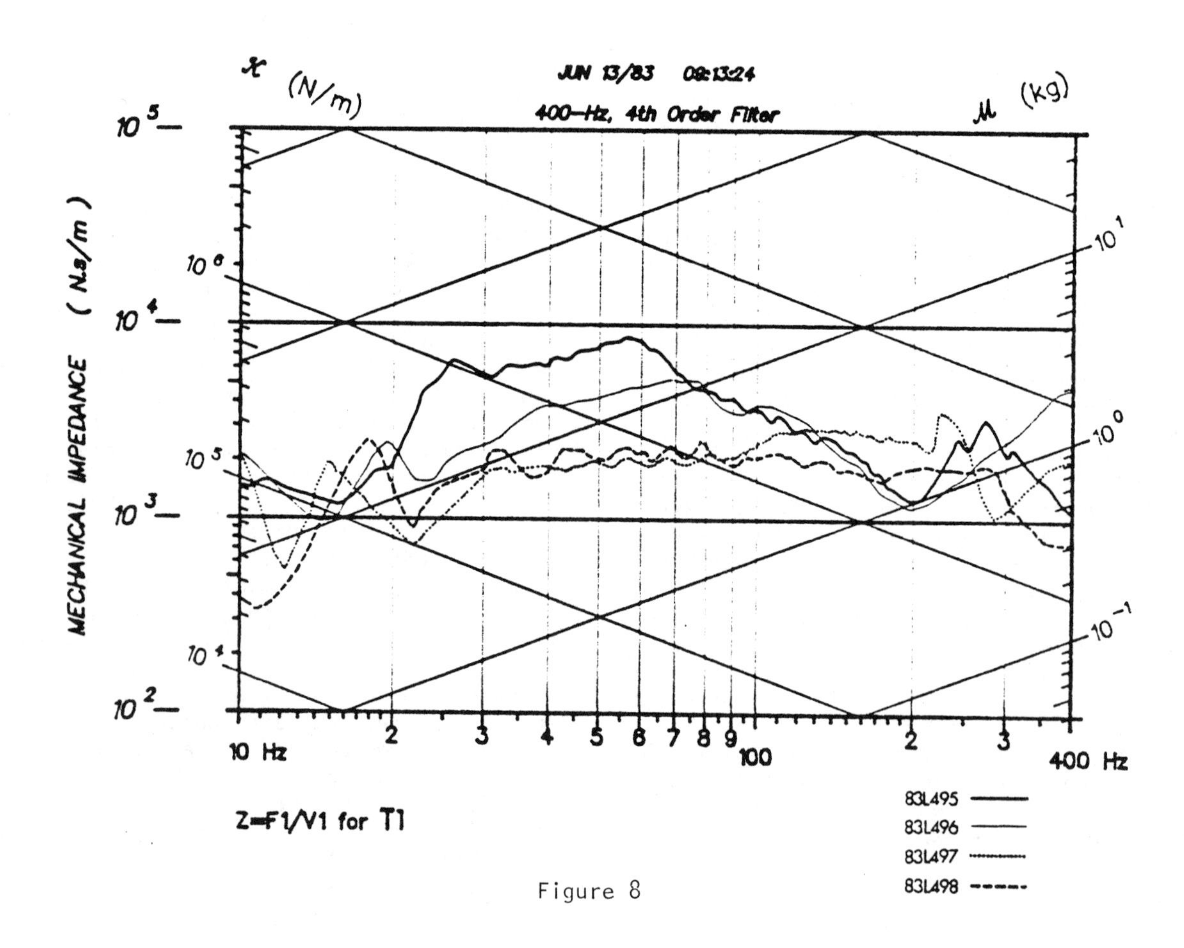

Figure 8

but differences are apparent later in time. Although the instrumentation procedures are different between the nine accelerometer array used in the drop tests and the triaxial cluster used in the pendulum study, a reasonable comparison is made between the tangential acceleration of the former and the principle direction acceleration of the latter. The impact response common to both test series is: a predominant head motion in the direction of impact in the initial part of the event, a peak tangential or resultant acceleration preceding the peak force, and a resultant negative acceleration while the force remains positive (Figure 9). In this study, however, those subjects not positioned in the mid-sagittal plane experience a significant rotation of the head about the P-A axis (Figure 10) as observed from the three-dimensional motion obtained from the nine accelerometer array and the high-speed films. In addition, the head did not move towards the chest to the same degree as previously observed in either the uninstrumented drop tests or other test series (34,35), but, instead, made contact with the shoulder. Therefore, compression with rotation in the P-A direction may be related to the flexion-type damage patterns produced in the cervical spine, but it is not possible to distinguish whether the flexion damage is a result of the observed head motion late in the impact event or a function of the initial conditions. In addition, the relative phasing between the tangential acceleration and the principal direction acceleration of the T6 spinal triax seems to be different, in general, from that of the previous study (34). This may be an artifact of the testing method (supine subject on sliding Styrofoam versus free-fall drop) or a function of the initial condition (mid-sagittal alignment, non-mid-sagittal alignment).

The mechanical impedance results of the thoracic vertebrae for tests 83L499 and 83L501 show that the response of thoracic T6 differs from T1 as much as it does from T12 (Figure 11, 12). Although mechanical impedance transfer functions can not be generated for tests 82L489 and 82L494 because of the nature of the load plate, direct transfer functions between the triaxes on the spine can be generated for those tests. The response of T6 again differs from T1 as much as it varies from T12. In the low-drop tests 83L495-82L498, T6 and T12 generate similar transfer functions, and, in comparison to the impedance response of 83L499 and 83L501, it seems that the transfer functions of the vertebrae vary with loading conditions and is indicative of a potential non-linear response of the thoracic spine. This may have importance in the consideration of the response of the thoracic vertebrae in relation to cervical spine damage patterns.

MECHANISM OF INJURY

Two types of mechanisms of injury were previously cited from the literature for cervical spine flexion-type injuries (36,37,39). In one case, injuries were sustained because the head was bowed significantly forward with a marked downward force, resulting in large bending moments at the base of the cervical spine. In the second case, when the cervical spine is straightened and large axial forces are applied to the top of the head, the spine acts as a segmented column that buckles under load. An important factor that may not be considered in either of these mechanisms of injury is the kinematic and injury response of the thoracic spine.

In this study and previous studies using unembalmed cadavers (34,35), large bending moments are not well correlated with flexion-type damage in the cervical spine. Primarily, it seems that when the head is forced down into the chest accompanied by compression, damage occurs in the upper thoracic spine. In addition, the complex response of the thoracic vertebrae under impact loading seems to indicate that a complete understanding of the mechanism of cervical spine injury must take into account the movement of the thoracic spine. Therefore, when the head/neck/thoracic movement is constrained to the mid-sagittal plane and the head is forced toward the chest beyond its normal range of motion, the thoracic spine is dynamically coupled such that the damage produced will most likely develop in the upper thoracic spine. Potentially, this area of the thoracic spine (T1-T4) is the area in which the largest stress concentrations occur during this type of impact event. Anatomically, the thoracic spine is easily distinguished from the cervical spine; but, kinematically, the coupling of response under impact conditions may have a more meaningful significance in regard to the mechanism of injury.

Consideration is given to the second mechanism of injury by looking at the impact response of the unimodal tests. Significant damage to the vertebrae and associated structures are observed in both the present study and previously (34) when the force time history is unimodal. The relationship of the force time history to the damage patterns and the time at which the damages occur is not well understood; however, for the unimodal impacts, one of the following three cases may be true: 1) Damage occurs solely during the impact event; 2) damage occurs as a result of the inertial loading of the head on the cervical spine after the most significant peak force; 3) damage is initiated during impact, compounded by the inertial loading of the head on the cervical spine after the impact event.

Results of this study and the pendulum tests indicate that the third possibility is most likely to occur. The cervical spine is

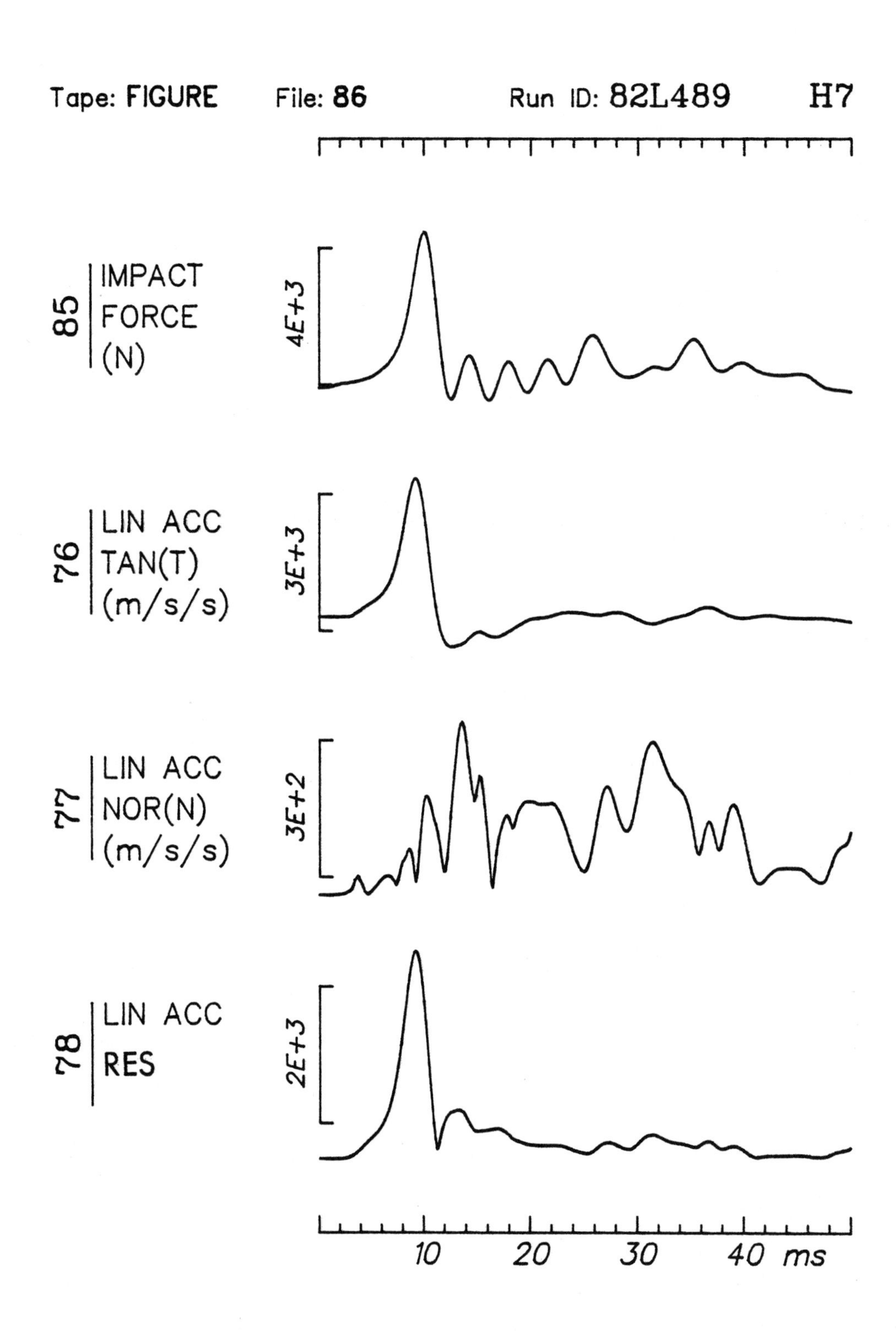
Tape: FIGURE
File: 86
Run ID: 82L489
H7
85
IMPACT FORCE (N)
4E+3
76
LIN ACC TAN(T) (m/s/s)
3E+3
77
LIN ACC NOR(N) (m/s/s)
3E+2
78
LIN ACC RES
2E+3
10
20
30
40 ms

Figure 9

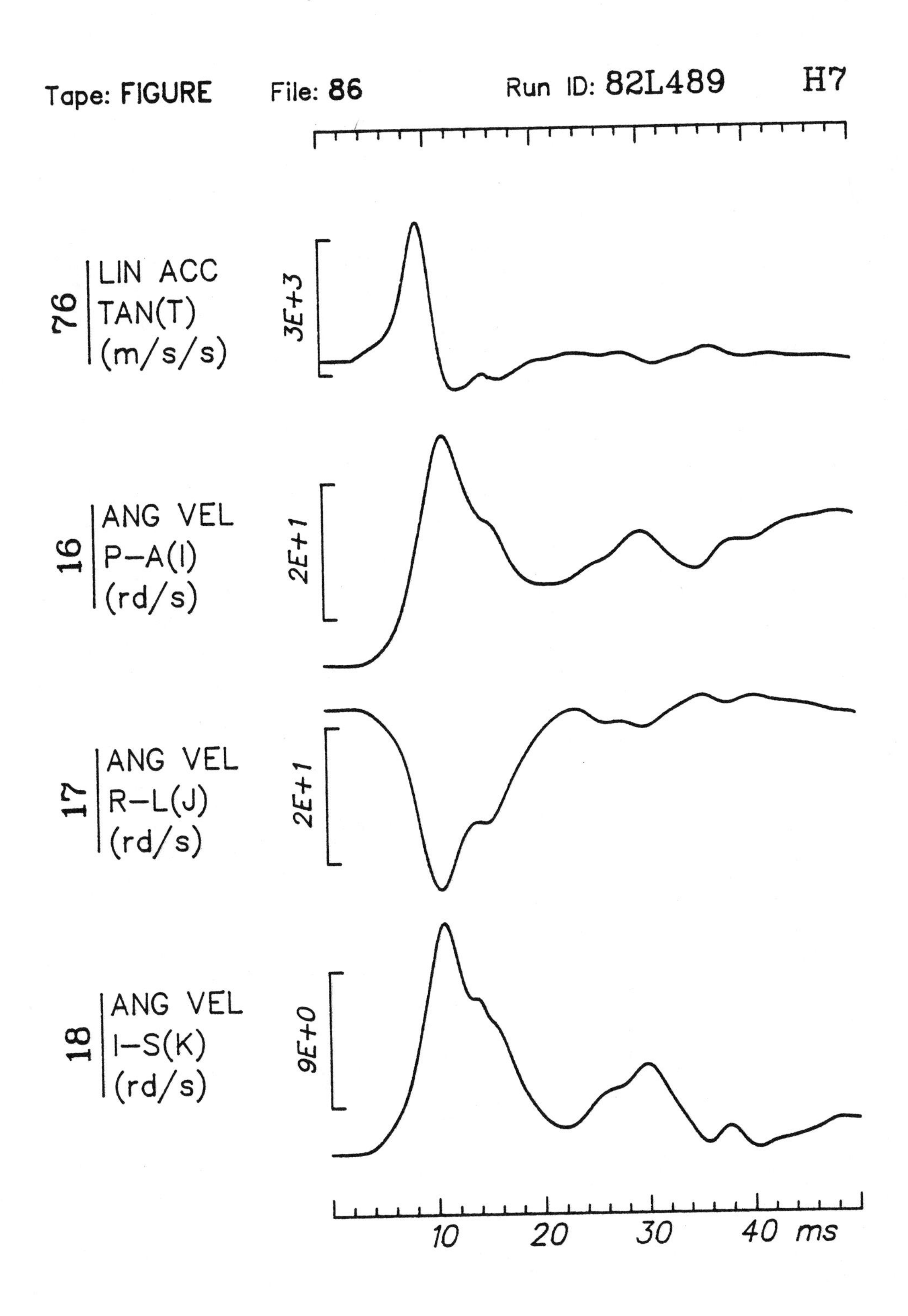
Tape: FIGURE
File: 86
Run ID: 82L489
H7
76
LIN ACC
TAN(T)
(m/s/s)
3E+3
16
ANG VEL
P-A(I)
(rd/s)
2E+1
17
ANG VEL
R-L(J)
(rd/s)
2E+1
18
ANG VEL
I-S(K)
(rd/s)
9E+0
10
20
30
40 ms

Figure 10

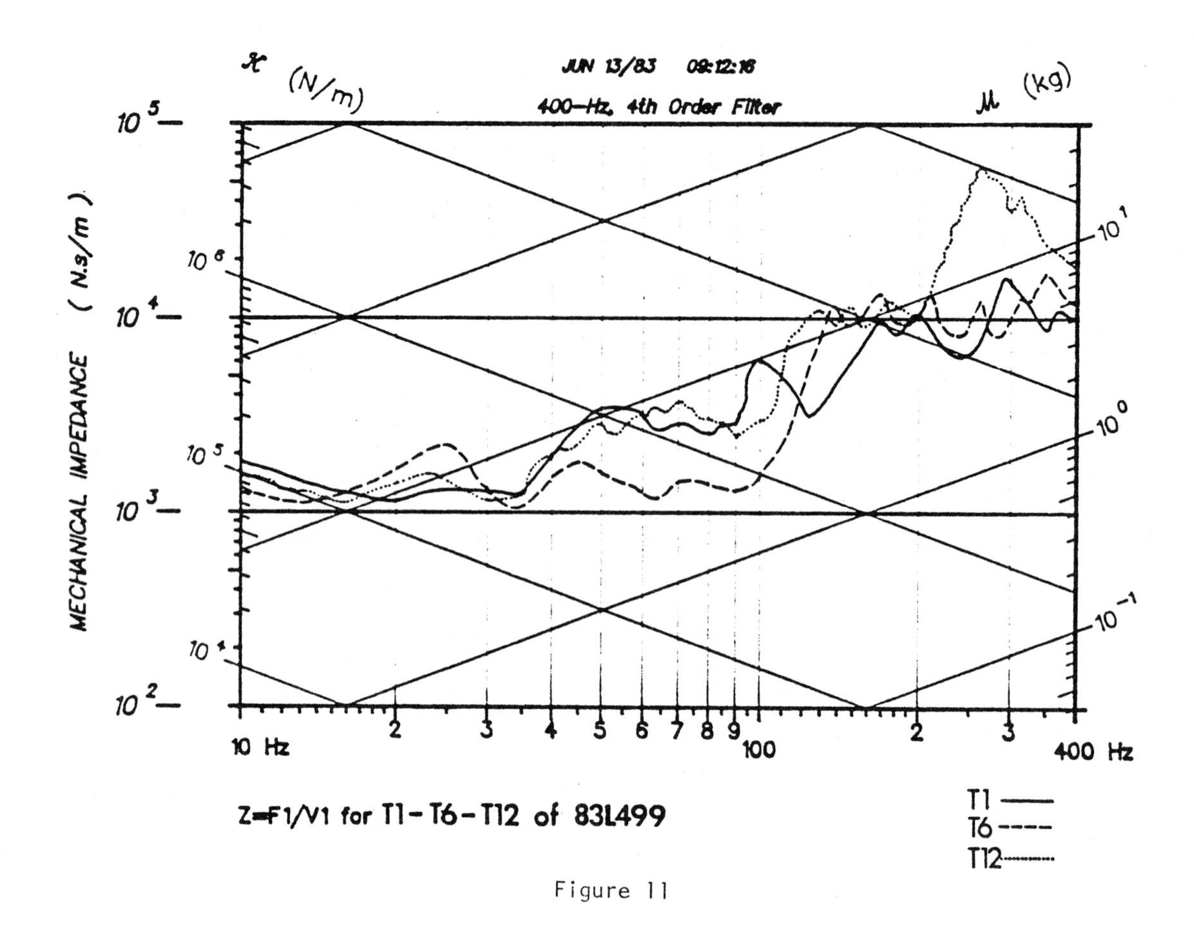

Figure 11

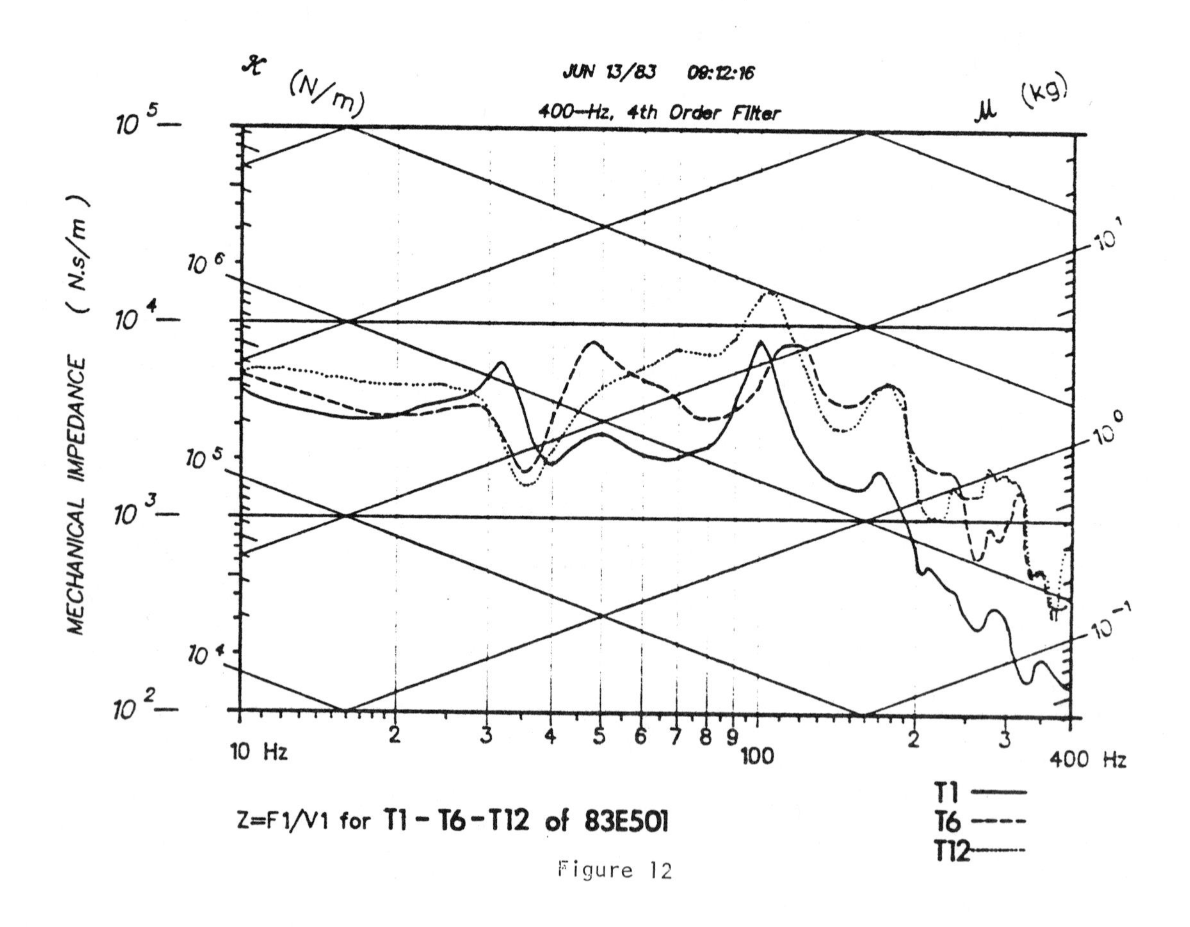

Figure 12

straighter initially for those tests in which unimodal force time histories are observed; the largest forces developed by the cervical spine on the head seem to occur during the interval from peak force to maximum negative acceleration; and, in addition, observations from the three-dimensional motion package show that the head moves less than 1 cm up to peak force and less than 3 cm at the time of maximum negative acceleration, implying that a large bending moment is not possible. It would therefore seem reasonable to assume that the neck is in maximum compression during the interval from peak force to maximum negative acceleration and that damage is initiated at this time in the impact event. For example, in test 83L500, the head, neck, and shoulders are wrapped with Dow Ethafoam to restrict the amount of bending and increase the amount of compression. This may have resulted in the observed increases in force when the subject is axially loaded, but even with this increase in force at the crown for a subject initially positioned in the mid-sagittal plane, flexion damage patterns are observed most often in the upper thoracic spine. Therefore, it seems that if compressive forces cause the spine to buckle, this is not in itself sufficient to cause damage to the cervical spine, and consideration should be given to the initial configuration of the head/neck/thorax of the test subject with respect to a three-dimensional motion as a strong contributing factor in the mechanism of injury.

CONCLUSIONS

This is a limited study of some of the important kinematic factors and cervical spine damage modes associated with crown impacts using unembalmed cadavers. More work is necessary before these factors can be generalized, however the following specific conclusions can be drawn.

1. When the head/neck is constrained to move in the mid-sagittal plane during crown impact, flexion-type cervical spine damage appears unlikely.

2. Flexion-type damage is produced when the head/neck/thorax is pre-positioned to produce non-mid-sagittal plane motion.

3. A two-dimensional description of the mechanism of injury that is restricted to the bending of the cervical spine in the mid-sagittal plane, i.e., hyperflexion, seems inadequate to predict the damage to the cervical spine in the cadaver model. However the two-dimensional description does seem appropriate to describe the damage pattern in the upper thoracic spine.

4. The response of the thoracic spine appears to be a critical factor influencing the type of response and damage patterns produced in the cervical spine.

5. In terms of damage response and force time history for subjects with similar initial conditions (impact velocity, padding, and contact surface geometry), free-fall tests do not seem to be significantly different from pendulum impacts in which a mass of 56 kg is used.

ACKNOWLEDGEMENTS

The authors wish to express their gratitude to the following people who contributed to this study: Jeff Pinsky, Marv Dunlap, Jean Brindamour, John W. Melvin, Carol Sobecki, and Gail Muscott.

REFERENCES

1. A. I. King. "Survey of the State of the Art of Human Biodynamics Response." Aircraft Crashworthiness. Charlottesville, VA.: University Press of Virginia, 1975, pp. 83-120.

2. R. G. Snyder. "State-of-the-Art - Human Impact Tolerance." 1970 International Automobile Safety Conference Compendium. Warrendale, PA.: Society of Automotive Engineers, 1970, pp. 712-782. SAE Paper No. 700398.

3 P. J. Van Eck, D. B. Chaffin, D. R. Foust, J. K. Baum, and R. G. Snyder. "A Bibliography of Whiplash and Cervical Kinematic Measurement." Ann Arbor: University of Michigan, Highway Safety Research Institute, November 1973. NTIS Order No. PB-225-203.

4. R. G. Snyder, D. B. Chaffin, L. W. Schneider, D. R. Foust, B. M. Bowman, T. A. Abdelnour, and J. K. Baum. "Basic Biomechanical Properties of the Human Neck Related to Lateral Hyperflexion Injury. Final Report." Ann Arbor: University of Michigan, Highway Safety Research Institute, 1975. NTIS Order No. PB-241-246.

5. J. W. Melvin. "Human Neck Injury Tolerances." The Human Neck--Anatomy, Injury Mechanisms and Biomechanics. Warrendale, Pa.: Society of Automotive Engineers, February 1979, pp. 45-46. SAE Paper No. 790136.

6. A. I. King. "Tolerance of the Neck to Indirect Impact. Technical Report." Detroit: Wayne State University, Bioengineering Center, 5 March 1979, 14 pp. Report No. N00014-75-C-1015/ Technical Report 9.

7. W. Goldsmith. "Some Aspects of Head and Neck Injury and Protection." Progress in Biomechanics. Alphen aan den Rijn, The Netherlands: Sijthoff and Noordhoff, 1979, pp. 211-245.

8. D. R. Foust, D. B. Chaffin, R. G. Snyder, and J. K. Baum. "Cervical Range of Motion and Dynamic Response and Strength of Cervical Muscles." 17th Stapp Car Crash Conference Proceedings. New York: Society of Automotive Engineers, 1973, pp. 285-308. SAE Paper No 730975.

9. R. Barnes. "Paraplegia in Cervical Spine Injuries." Journal of Bone and Joint Surgery, Vol. 30B, No. 2, May 1948, pp. 234-244.

10. S. Haughton. "On Hanging, Considered From a Mechanical and Physiological Point of

View." Philosophical Magazine, Series 4, Vol. 31-2. 1866, pp. 23-34.

11. R. J. Bauze and G. M. Ardran. "Experimental Production of Forward Dislocation in the Human Cervical Spine." Journal of Bone and Joint Surgery, Vol. 60B, May 1978, pp. 239-245.

12. T. R. Beatson. "Fracture and Dislocations of the Cervical Spine." Journal of Bone and Joint Surgery, Vol. 45B, February 1963, pp. 21-35.

13. B. R. Selecki. "The Effects of Rotation of the Atlas on the Axis: Experimental Work." Medical Journal of Australia, Vol. 1, 1969, p. 1012.

14. J. W. Fielding, G. V. B. Cochran, J. F. Loursing, and H. Mason. "Traumatic Lysis of the Transverse Ligament of the Atlas: A Clinical and Biomedical Study." Journal of Bone and Joint Surgery, Vol. 56A, December 1974, pp. 1683-1691.

15. K. F. Spence, S. Becker, and K. W. Sell. "Bursting Atlantal Fracture Associated with Rupture of the Transverse Ligament." Journal of Bone and Joint Surgery, Vol. 52A, April 1970, pp. 543-549.

16. M. M. Panjabi, A. A. White, and R. M. Johnson. "Cervical Spine Mechanics as a Function of Transection of Components." Journal of Biomechanics, Vol. 8, No. 5, September 1975, pp. 327-336.

17. W. Lang. "Mechanical and Physiological Response of the Human Cervical Vertebral Column to Severe Impacts Applied to the Torso." Symposium on Biodynamic Models and Their Applications. Wright-Patterson AFB, Ohio: Aerospace Medical Research Laboratory, December 1971, pp. 141-167.

18. J. R. Cromack and H. H. Zipperman. "Three-Point Belt Induced Injuries: A Comparison Between Laboratory Surrogates and Real World Accident Victims." 19th Stapp Car Crash Conference Proceedings. Warrendale, Pa.: Society of Automotive Engineers, 1975, pp. 1-24. SAE Paper No. 751141.

19. L. M. Patrick and R. S. Levine. "Injury to Unembalmed Belted Cadavers in Simulated Collisions." 19th Stapp Car Crash Conference Proceedings. Warrendale, Pa.: Society of Automotive Engineers, 1975, pp. 79-115. SAE Paper No. 751144.

20. A. S. Hu, S. P. Bean, and R. M. Zimmerman. "Response of Belted Dummy and Cadaver to Rear Impact." 21st Stapp Car Crash Conference Proceedings. Warrendale, Pa.: Society of Automotive Engineers, 1977, pp. 587-625. SAE Paper No. 770929.

21. D. Kallieris, B. Meister, and G. Schmidt. "Reactions of the Cervical Spine During Frontal Impacts of Belt Protected Cadavers." Biomechanics of Serious Trauma. 2nd International Conference Proceedings. Bron, France: IRCOBI, 1975, pp. 126-142.

22. H. J. Clemens and K. Burow. "Experimental Investigation on Injury Mechanisms of Cervical Spine at Frontal and Rear-Front Vehicle Impacts." 16th Stapp Car Crash Conference Proceedings. Warrendale, Pa.: Society of Automotive Engineers, 1976, pp. 76-104. SAE Paper N. 720960.

23. A. M. Jones, S. P. Bena, E. S. Sweeney. "Injuries to Cadavers Resulting from Experimental Rear Impact." Journal of Forensic Sciences, Vol. 23, No. 4, 1978.

24. B. C. Marar. "Hyperextension Injuries of the Cervical Spine." Journal of Bone and Joint Surgery, Vol. 56A, December 1974, pp. 1655-1662.

25. G. Schmidt, D. Kallieris, J. Barz, Mattern, and J. Klaiber. "Neck and Thorax Tolerance Levels of Belt-Protected Occupants in Head-On Collisions." 19th Stapp Car Crash Conference Proceedings. Warrendale, Pa.: Society of Automotive Engineers, 1975, pp. 225-257. SAE Paper No. 751149.

26. V. R. Hodgson, H. R. Lissner, and L. M. Patrick. "Response of the Seated Human Cadaver to Acceleration and Jerk With and Without Seat Cushions." Human Factors, Vol. 5, No. 5, October 1963, pp. 505-523.

27. A. I. King, A. P. Vulcan, and L. K. Cheng. "Effects of Bending on the Vertebral Column of the Seated Human During Caudocephalad Acceleration." 21st Conference on Engineering and Medicine in Biology. Proceedings. Vol. 10, 1968.

28. R. Cheng, K. H. Yang, R. S. Levine, A. I. King, and R. Morgan. "Injuries to the Cervical Spine Caused by a Distributed Frontal Load to the Chest." 26th Stapp Car Crash Conference Proceedings. Warrendale, Pa.: Society of Automotive Engineers, Inc., 1982, pp. 1-40. SAE Paper No. 821155.

29. C. L. Ewing, A. I. King, and P. Prasad. "Structural Considerations of the Human Vertebral Column Under $+G_z$ Impact Acceleration." Journal of Aircraft, Vol. 9, No. 1, January 1972, pp. 84-90.

30. P. Prasad, A. I. King, and C. L. Ewing. "The Role of the Articular Facets During $+G_z$ Acceleration." Journal of Applied Mechanisms, Vol. 41, 1974, pp. 321-326.

31. J. McElhaney, R. G. Snyder, J. D. States, and M. A. Gabrielson. "Biomechanical Analysis of Swimming Pool Neck Injuries." The Human Neck --Anatomy, Injury Mechanisms and Biomechanics. Warrendale, Pa.: Society of Automotive Engineers, February 1979, pp. 47-53. SAE Paper No. 790137.

32. R. H. Culver, M. Bender, J. W. Melvin. "Mechanisms, Tolerances, and Responses Obtained Under Dynamic Superior-Inferior Head Impact, A Pilot Study. Final Report." Ann Arbor: University of Michigan, HSRI, May 1978, 108 p. NTIS Order No. PB-299-292.

33. V. R. Hodgson and L. M. Thomas. "Mechanisms of Cervical Spine Injury During Impact to the Protected Head." 24th Stapp Car Crash Conference Proceedings. Warrendale, Pa.:

Society of Automotive Engineers, 1980, pp. 15-42. SAE Paper No. 801300.

34. G. S. Nusholtz, J. W. Melvin, D. F. Huelke, N. M. Alem, and J. G. Blank. "Response of the Cervical Spine to Superior-Inferior Head Impacts." 25th Stapp Car Crash Conference Proceedings. Warrendale, Pa.,: Society of Automotive Engineers, 1981, pp. 197-237. SAE Paper No. 811005.

35. N. M. Alem, G. S. Nusholtz, and J. W. Melvin. "Superior-Inferior Head Impact Tolerance Levels. Final Report." Ann Arbor: University of Michigan, Transportation Research Institute. November 1982. NTIS Order No. PB-83-144-501.

36. R. Braakman and L. Penning. "Injuries of the Cervical Spine." Amsterdam, The Netherlands: Excerpta Medica, 1971.

37. K. Kattan. "'Trauma' and 'No-Trauma' of the Cervical Spine." Springfield, Ill.: C. C. Thomas, 1975.

38. R. Schneider. "Head and Neck Injuries in Football-Mechanisms, Treatment, and Prevention." Baltimore, Md.: Williams and Wilkins Co., 1973.

39. J. Torg. "Athletic Injuries to the Head, Neck, and Face." Philadelphia, Pa.: Lea and Febiger, 1982.

40. J. Torg, L. Guedenfeld, A. Burstein, A. Spealman, C. Nichols III. "The National Football and Neck Injury Registry: Report and Conclusions 1978." American Medical Association Journal, Vol. 241, No. 14, 6 April 1979, pp. 1477-1479.

41. J. Torg, R. Truex, J. Marshall, V. Hodgson, T. Guedenfeld, A. Spealman, and C. Nichols." Spinal Injury at the Level of the Third and Fourth Cervical Vertebra from Football." Journal of Bone and Joint Surgery, Vol. 59A, 1977, pp. 1015-1019.

42. V. Frankel and A. Burstein. "Orthopaedics Biomechanics; The Application of Engineering to the Musculo-Skeletal System." Orthopaedic Biomechanics. Philadelphia, Pa.: Lea and Febiger, 1970.

43. H. J. Mertz and L. M. Patrick. "Strength and Response of a Human Neck." 15th Stapp Car Crash Conference Proceedings. New York: Society of Automotive Engineers, 1971, pp. 207-255. SAE Paper No. 710855.

44. C. W. Gadd. "A Study of Response and Tolerances of the Neck." 15th Stapp Car Crash Conference Proceedings." New York: Society of Automotive Engineers, 1972, pp. 256-268. SAE Paper No. 710856.

45. L. M. Patrick and C. C. Chou. "Response of the Human Neck in Flexion, Extension and Lateral Flexion. Final Report." Detroit: Wayne State University, Biomechanics Research Center, April 1976. Report No. VRI-7.3.

46. L. W. Schneider, D. R. Foust, B. M. Bowman, R. G. Snyder, D. B. Chaffin, T. A. Abdelnour, and J. K. Baum. "Biomechanical Properties of the Human Neck in Lateral Flexion." 19th Stapp Car Crash Conference Proceedings. Warrendale, Pa.: Society of Automotive Engineers, 1975, pp. 455-485. SAE Paper No. 751156.

47. D. T. Watts, E. S. Mendelson, H. N. Hunter, A. T. Kornfield, and J. R. Poppen. "Tolerance to Vertical Acceleration Required for Seat Ejection." Aviation Medicine, Vol. 18, 1947, p. 554.

48. C. L. Ewing, D. J. Thomas, G. W. Beeler, Jr., and L. M. Patrick. "Dynamic Response of the Head and Neck of the Living Human to $-G_x$ Impact Acceleration." 12th Stapp Car Crash Conference Proceedings. Warrendale, Pa.: Society of Automotive Engineers, 1968, pp. 424-439. SAE Paper No. 680792.

49. C. L. Ewing and D. J. Thomas. "Torque Versus Angular Displacement Response of Human Head to $-G_x$ Impact Acceleration." 17th Stapp Car Crash Conference Proceedings. New York: Society of Automotive Engineers, 1973, pp. 309-342. SAE Paper No. 730976.

50. C. L. Ewing, D. J. Thomas, L. M. Patrick, G. W. Beeler, and M. J. Smith. "Living Human Dynamic Response to $-G_x$ Impact Acceleration II--Accelerations Measured on the Head and Neck. 13th Stapp Car Crash Conference Proceedings. New York: Society of Automotive Engineers, 1969, pp. 400-415. SAE Paper No. 690817.

51. D. F. Huelke, E. A. Moffatt, R. A. Mendelsohn, and J. W. Melvin. "Cervical Fractures and Fracture Dislocations - An Overview." The Human Neck--Anatomy, Injury Mechanisms and Biomechanics. Warrendale, Pa.: Society of Automotive Engineers, February 1979, pp. 9-15. SAE Paper No. 790131.

52. J. L. Babcock. "Cervical Spine Injuries: Diagnosis and Classification." Archives of Surgery, Vol. III, No. 6, June 1976, pp. 646-651.

53. A. R. Taylor. "The Mechanism of Injury to the Spinal Cord in the Neck Without Damage to the Vertebral Column." Journal of Bone and Joint Surgery, Vol. 33B, No. 4, November 1951, pp. 543-547.

54. H. H. Gosch, E. Gooding, and R. C. Schneider. "An Experimental Study of Cervical Spine and Cord Injuries." Journal of Trauma, Vol. 12, No. 2, July 1972, pp. 570-575.

55. R. Roaf. "A Study of the Mechanics of Spinal Injuries." Journal of Bone and Joint Surgery, Vol. 42B, 1960, pp. 819-823.

56. J. H. Harris and J. Edeiken. "Acute Cervical Spine Trauma." Weekly Radiological Science Update, No. 17, 1976.

57. G. S. Nusholtz, M. A. Janicki, P. Lux, and D. F. Huelke. "Strength and Response of the Human Cadaver Cervical Spine Under Impact Loading. Interim Report." Ann Arbor: University of Michigan Transportation Research Institute, July 1982.

841667

Head and Neck Response to Axial Impacts

Nabih M. Alem, Guy S. Nusholtz and John W. Melvin
The University of Michigan

ABSTRACT

Two series of impacts to the head in the superior-inferior direction using 19 unembalmed cadavers are reported. The first series of five tests was aimed at generating kinematic and dynamic response to sub-injurious impacts for the purpose of defining the mechanical characteristics of the undamaged head-neck-spine system in the S-I direction. The second series of fourteen tests was intended to define injury tolerance levels for a selected subject configuration.

A 10-kg impactor was used to deliver the impact to the crown at a nominal velocity of 8 m/s for the first series, and between 7 and 11 m/s for the second series. Measurements made in the first series include the impact velocity, force, and energy, the head three-dimensional kinematics, forces and moments at the occipital condyles, and accelerations of the T1, T6, and T12 vertebrae. Impact impedance curves were also generated. Measurements made during the second series were the same as the first, except that no condyle reactions were calculated nor detailed autopsies were performed.

Peak impact force was not found to be a reliable predictor of cervical injury. Furthermore, since such injury often occurred without skull fractures indicating that the neck may be more vulnerable to S-I impacts than the head itself, the Head Injury Criterion (HIC) would not be useful to determine the effectiveness of head protective devices in protecting the neck. On the other hand, both the time integral of the force and the maximum head velocity correlated well with cervical spine damage and thus they may be useful in predicting neck injury.

EXPERIMENTAL RESEARCH ON AXIAL IMPACT to the head has been generally aimed at two goals. The first is to describe the kinematic and dynamic impact response of the head-neck-spine system for the purpose of designing humanlike mechanical surrogates, and the second is to understand the mechanisms of cervical spine injuries resulting from accidents in the automotive, industrial, and athletic environments, such as car rollovers, falling objects, and swimming pool diving accidents.

Culver et al. [1][1] conducted head axial impact studies on eleven unembalmed cadavers using a 10-kg padded impactor. Most of the fractures produced were of the posterior vertebral elements, with peak forces of 4.7 to 8.9 kN and impact energies of 260 to 645 J. Using a 56-kg impactor, Nusholtz et al. [2] conducted twelve cadaver tests to investigate the factors affecting cervical spine damage in axial impacts. In that study, study, the impact velocity was 4.6 to 5.6 m/s, and the generated forces were 1.8 to 11.1 kN. The reported damage to the cervical spine included fractures to the spinous and transverse processes, laminae, and vertebral bodies as well as ruptures of the discs and tears of the ligaments.

In a study of axial impact to the head when it is protected by a football helmet, Hodgson et al. [3] concluded that head freedom to rotate about the atlanto-occipital junction, location of impact, and alignment of impact force with the spine were influential factors on injury site and level. In other studies [4-7], the mechanical strength and properties of individual vertebrae and of segments of the spine and injury to the spine were investigated.

This paper summarizes the findings of two series of head axial impact studies that were carried out between 1979 and 1981 at the University of Michigan Transportation Research Institute. The objective of the first series of 5 cadaver tests was to generate kinematic and dynamic response data for sub-injurious impacts in order to describe the mechanical characteristics of the undamaged head-neck-spine system in the S-I direction and to design a helmet impact test device that is human-like in its S-I response [8]. The second series of 14 tests was conducted in order to determine head-neck

---

[1]Numbers in brackets designate references at the end of the paper.

tolerance levels that may be translated into helmet performance criteria [9].

DESCRIPTION OF EXPERIMENTS

SUBJECT SELECTION - Unembalmed cadavers provide reasonably good surrogates to live humans. Since their mechanical properties vary with age, size, time after death, integrity and condition of the skeletal structure, and medical history, every effort was made to keep the variability of these factors to a minimum by carefully selecting (or rejecting) an available subject. The subjects selected for testing in the two series were all male and are described in Table 1. Prior to each test, the subject was weighed and X-rayed to ensure the integrity of the head, cervical spine, and upper thorax. The head was instrumented with the UMTRI 9-accelerometer array, and triaxial accelerometer clusters were installed on T1, T12, and, in some tests, on T6 thoracic vertebrae. The surgical and instrumentation procedures used in these tests are described in detail in [2] and [10].

TABLE 1 - DESCRIPTION OF TEST SUBJECTS

Test No.	Subject			Head Dimensions (cm)			
	Age	Wt. (kg)	Ht. (cm)	Lth. (A-P)	Wth. (L-R)	Ht. (S-I)	Cir-cum.
79H201				19.8	17.7	24.2	57.8
79H202				19.7	14.7	22.6	56.5
79H203				18.5	15.0	23.3	54.7
79H204				19.5	15.4	22.8	55.8
79H205				21.2	15.4	23.7	59.2
81H401	59	74.9	170	20.1	13.5	23.8	56.7
81H402	41	52.3	171	20.7	14.3	24.1	56.9
81H403	64	44.7	172	18.4	14.7	25.0	54.0
81H404	65	62.9	184	20.4	15.5	18.9	75.2
81H405	49	85.8	182	19.7	16.0	23.7	57.5
81H406	72	64.7	166	19.4	15.5	21.4	56.5
81H407	63	55.9	169	17.9	14.5	23.4	53.0
81H408	63	69.9	175	20.3	15.0	23.6	57.0
81H410	66	81.5	177	20.0	15.6	23.1	58.1
81H411	63	50.5	171	20.4	15.9	23.7	54.0
81H412	66	54.5	178	20.3	15.0	22.9	58.5

TESTING PROCEDURES - The UMTRI air cannon was used to propel the impactor to the desired velocity. High speed movies were taken in a three-dimensional calibrated field, and, in the first series of tests, 35-mm high speed X-ray movies of the cervical spine were also taken during the impact. All tests were conducted with the subject placed either prone or supine on four layers of 10-cm seating foam atop an adjustable table. This soft cushion allowed for a relatively free motion of the cervical spine, with the head suspended with breakaway paper tapes, so that the initial and significant portion of the impact occurred when the head was aligned with the spine.

The initial position and alignment of the head and spinal column were adjusted to simulate the natural curvatures of the upper spine during normal seating or standing postures. In some tests, attempts were made to align the cervical and thoracic spines as closely as possible along the axis of impact. This was done to confirm the suspected effects of the initial alignment on the resulting injuries. In all cases, in-place lateral X-rays were used to document the initial angles, with respect to the horizontal, of the head anatomical posterior-anterior axis, and of the neck using the tangent to its mid-portion as a reference line. The convention used in defining these alignment angles is described in references [2] and [10], and the measured angles are given in Table 2, which also gives the test conditions.

TABLE 2 - TEST CONDITIONS

Test No.	Neck Angle	Head Angle	10-kg Impactor	
			Velocity (m/s)	Padding (cm)
H201	--	--	8.0	5.1
H202	--	--	8.0	5.1
H203	--	--	8.0	5.1
H204	--	--	8.0	5.1
H205	--	--	8.0	5.1
H401	30°	--	8.4	5.1
H402	20°	--	10.9	5.1
H403	25°	100°	10.9	5.1
H404	25°	95°	7.8	5.1
H405	5°	80°	7.7	5.1
H406	5°	80°	8.0	5.1
H407	5°	--	9.2	5.1
H408	10°	100°	9.7	5.1
H409	5°	--	10.4	0.0
H410	30°	--	9.0	5.1
H411	--	--	7.2	5.1
H412	10°	--	7.1	5.1
H413	0°	--	9.0	0.5
H414	--	--	6.9	0.5

The 10-kg impactor was propelled to a nominal velocity of 8 m/s for the first five tests. The impact velocity for the remainder of the tests was varied between 7 m/s and 11 m/s in an attempt to determine the tolerance level of the head-spine system. In most of the tests, the impactor was padded with 5-cm thick ensolite foam in order to control the duration of the impact force pulse and to minimize local skull fractures.

POST-TEST PROCEDURES - While no detailed autopsies were performed after the first five tests, the subjects were examined to confirm that no bone fractures were produced. A gross autopsy was performed on the head and neck of subjects in the second series. The scalp was reflected and the calvarium removed for examination. The epidural and subdural areas of the brain were inspected, the brain and dura were extracted, and the base of the skull was examined. All observed injuries were recorded and classified. Following the autopsy, the occipital condyles were marked with lead pellets, and a 3-D X-ray procedure was followed to determine the location of the two occipital condyles with respect to the anatomical reference frame. This information is necessary for computation of the reaction forces and moments at the condyles.

DATA ANALYSIS METHODS

IMPACT SEVERITY - This was characterized by the velocity of the impactor, the peak and duration of the contact force, the impulse, and the energy of the impactor at the instant of impact. The velocity was accurately measured from pulses generated magnetically by the moving impactor. All other parameters were calculated from the force-time history.

The contact-force duration was defined as the time interval between the beginning and end of the pulse. These two points were obtained by fitting straight lines to the rise and fall portions of the pulse and locating their intersections with the x-axis. The impulse was obtained by integrating the force time-history. Finally, the transferred energy was defined as half the ratio of the square of the impulse over the impactor mass.

HEAD KINEMATICS - The description of the impact response of the human head requires that the kinematic quantities measured experimentally be described in reference frames which vary from one instrumentation method to another. One method for comparing mechanical responses between subjects is to refer all results to a "standard" anatomical frame which may be easily identified. However, it may be impractical to require that transducers be aligned with this anatomical frame. An alternative is to mount transducers in an arbitrary and convenient reference frame and then describe the transformation necessary to convert the data from this instrumentation frame to a desired anatomical one.

A three-dimensional X-ray technique is used to accomplish this for head impacts. Four anatomical landmarks (two superior edges of the auditory meati and two infraorbital notches) are marked with four mutually distinguishable lead pellets. The nine-accelerometer plate is marked with lead pellets at the center of mass of each triaxial accelerometer cluster and also at the plate center of mass. The head containing this instrumentation is then radiographed in two orthogonal directions (the x-z and y-z planes). On each of the two radiographs the optical center and the laboratory vertical z-axis are simultaneously X-rayed. The subsequent computations reconstruct the laboratory coordinates of each of the lead targets. The Frankfort plane is determined and the anatomical reference frame is reconstructed from the four anatomical points. The instrumentation frame and its origin are determined from the three triaxial accelerometer centers. Finally, the transformation matrix between the instrumentation frame and the anatomical frame is obtained. The head kinematics are then computed from the 9 acceleration signals and expressed in the anatomical reference frame. The UMTRI 9-accelerometer method for computing the 3-D head motion has been fully documented in [11] and [12] and will not be described here.

HEAD DYNAMICS - In order to compute the reaction forces and moments at the occipital condyles, the location of the condyles, the head center of mass, and the head kinematics (acceleration of the anatomical center, angular acceleration and velocity vectors) must all be determined in the anatomical reference frame.

In addition, the mass and the moment of inertia matrix of the head must be estimated from anthropometric measurements. The location of the condyles was determined with a 3-D X-ray procedure similar to the one described above. The head center of mass was assumed to be in the midsagittal plane at 1.3 cm forward of and 2.1 cm above the anatomical origin [13]. Measurements of six cadaver heads by Chandler et al. [14] and a study by Lett et al. [15] were used as a basis for developing regression models to estimate the mass and moments of inertia matrix from 4 anthropometric measurements: head length, breadth, height, and circumference. The principal inertia ellipsoid was assumed to be pitched 49.6 degrees with respect to the anatomical forward axis.

The external forces applied to the head, assumed to balance the known inertial forces, include the six unknown reactions at the condyles (3 forces and 3 moments), as well as the contact forces applied to the top of the head. Since the head rotates under impact, the force signal is broken into 3 components along the anatomical moving reference frame. The 3 accelerations at the center of mass and the 6 unknown reactions are finally computed.

IMPACT IMPEDANCE - Classical impedance is defined as the ratio of magnitude of force over velocity, when the system has reached steady state under sinusoidal excitation. Impact impedance may be defined similarly except that those magnitudes are defined as the Fourier coefficients of the transient force and velocity signals. This is a useful characterization in impact testing of subjects, since it is impractical to repeat the application of impact until a steady state is reached without destroying the system being measured. Impact impedance curves were generated to describe the overall transfer function between the top of the head and the head anatomical center, T1, T6, and T12 thoracic vertebrae in the S-I direction.

RESULTS

The results of autopsies are summarized in Table 3. Since the impact level in the first five tests was intentionally designed to produce no injuries, none were found. However, no injuries were produced in tests H401, H405, H411, and H414 either. Skull fractures only were produced in tests H409, in which no padding was used, and in H413, in which the padding was only 0.5 cm thick.

The impact severity parameters associated with the force signals are given in Table 4. Head kinematic response is given in Table 5, while spinal acceleration response is given in Table 6. The computations of neck loads and moments at the condyles was performed only on the first five tests. These reactions are summarized in Table 7, and were based on the estimates of head inertial properties given in Table 8. Typical response time-histories are shown in Figure 1 for head linear and angular motion, Figure 2 for T1 and T12 accelerations, and in Figure 3 for the forces and moments of the condyles. Typical impact impedance curves are shown in Figures 4, 5, and 6 for the head, T1, and T12, respectively.

DISCUSSION OF RESULTS

SKULL INJURIES - In looking for impact conditions that produce skull injuries under the impactor, one finds that localized fractures can occur for impacts in the R-L and A-P direction [13,15] when there is no padding to distribute the impact force and reduce local stress. Similar injuries are produced in impacts in the S-I direction when no padding is used, as in test H409.

When a thin pad of 0.5-cm ensolite was added to the impactor surface while maintaining approximately the same velocity and producing approximately the same peak impact force, local skull fracture was avoided, but a basal skull fracture was produced in test H413. A similar type of injury was produced in test H200, in which the impact velocity was 12 m/s and impactor surface padding was 2.5-cm ensolite. Because test H200 was a trial run, it was not instrumented and thus no further analysis could be performed.

Although exact knowledge of the mechanisms of basal skull fracture cannot be directly observed, two mechanisms may be postulated on the basis of the system morphology. The first mechanism suggests that the neck reactive forces are transmitted to the relatively thin skull floor through the occipital condyles and the relatively strong ring opening of the foramen magnum, causing a basal skull ring fracture. The second mechanism is the initiation of a crack in the skull (due to excessive bending and stressing of the cranial shell), at a location removed from the skull base itself, that propagates toward the skull base and precipitates the basal skull fracture in question.

TABLE 3 - RESULTS OF AUTOPSIES

Test	Injuries
H402	Bilateral fracture T2 lamina at base of spinous process. Anterior body of T2 extremely compressed.
H403	Ruptured C2-C3 and C3-C4 discs, with anterior longitudinal ligament torn between C3-C4. Anterior-inferior chip fracture of C2 body. Vertical fracture of posterior C2 body. C3/C4 spinous process tip fractures. T1-T2 disc rupture with wedge fracture of T2 body. Rupture of posterior longitudinal T1-T2 ligament producing complete bilateral dislocation between T1 and T2. Fracture of T2 left transverse process. Partial separation of anterior longitudinal ligament at upper body of T2. All interspinal ligaments torn between T1-T2. Left first rib fractured adjacent to T1.
H404	Nearly complete tear of anterior longitudinal ligament at disc between C3-C4.
H406	Bilateral fracture of posterior C1 arch. Fracture of C2 dens. Fracture of spinous process of C3 and C4. Fracture of right lamina of C7. Fracture of anterior-superior T1 body.
H407	Rupture of anterior longitudinal ligament and disc between C5 and C6.
H408	Tear of anterior longitudinal ligament between C3 and C4. Tear of anterior longitudinal ligament between C4-C5 disc. Bilateral fracture of C1 posterior arch. Anterior inferior C2 body fracture extending through C2-C3 disc. Compression fracture of upper body of T2. Compression fracture of lower body of T3.
H409	Circular depressed fracture at apex of skull under impactor.
H410	Anterior longitudinal ligament torn, discs C3-C4 and C4-C5. Rupture of discs C3-C4 and C4-C5. Rupture of central portion of discs C5-C6 and C6-C7. Fracture of anterior of C4 body.
H412	Teardrop fracture of lip of C5 with complete anterior longitudinal ligament tear. Bilateral tears of anterior longitudinal ligament at C4. Rupture of disc at C4 and C5.
H413	Basilar skull fracture, from the foramen magnum area to the the bilateral areas.

In the first 5 tests (H201 through H205), attempts were made to position the subject such that the maximum possible force was transmitted to the spinal column through the head and neck. The test subject was placed in a supine position and the cervical spine was aligned along the line of action of the impact force. In the remaining tests (H401 through H414), an in-position X-ray was used to align as well as possible the "general spinal axis." The underlying assumption for this type of initial positioning is that, if the force level is high enough to exceed the strength of the skull floor near the foremen magnum, and the force is sufficiently distributed so as to avoid a local depressed fracture on the crown of the head, then direct loading of the condyles by the neck could cause the skull base to fracture. Despite these efforts, no skull fracture occurred. The injuries observed were fractures of the vertebral bodies and processes in the cervical and upper thoracic spine, and increased force only resulted in more severe spinal damage.

In the pilot S-I impact study [1] and in several tests in this study, attempts were made to orient the impact line of action along the spinal column. In tests H405 and H406, the impact velocity was nominally 8 m/s, the impactor surface padding was 5-cm ensolite, and the cervical spine was aligned approximately along the impactor axis. The peak forces in these tests were about 4 kN. While the padding and impactor velocity remained the same, tests H401 and H404 had a significantly different alignment of the cervical spine, but the peak forces produced remained at the same 4 kN level. This indicates that the angle formed between the axis of the cervical spine and the axis of the impactor for a supine test subject may not significantly affect peak force. However, the angle does profoundly affect both the impact force-time waveform and the head acceleration time history.

Although, in general, for a supine test subject the cervical spinal angle does not affect peak force, it has been suggested [2] that peak force may be increased by a combination of cervical and thoracic spinal angles which com-

TABLE 4 - IMPACT FORCE PARAMETERS

Test No.	Contact Force of 10-kg impactor			
	Peak (N)	Length (ms)	Impulse (N.s)	Energy (J)
H201	5100	15	--	--
H202	5210	12	--	--
H203	4400	14	--	--
H204	3930	17	--	--
H205	4800	16	--	--
H401	4200	15	34	61
H402	11000	9	49	122
H403	10500	6	40	82
H404	4000	14	36	65
H405	4100	15	35	61
H406	4000	22	48	115
H407	4500	13	40	82
H408	6000	16	49	118
H409	15000	3	36	66
H410	5200	20	42	88
H411	4100	19	35	61
H412	3000	20	35	61
H413	17000	3	26	33
H414	16000	3	24	29

TABLE 5 - HEAD KINEMATIC RESPONSE

Test No.	HIC	Peak Acceleration		Peak Velocity	
		Linear (G)	Angular (rad/s²)	Linear (m/s)	Angular (rad/s)
H201	325	67	4880	5.1	20
H202	286	63	5590	4.7	31
H203	333	80	2310	5.6	15
H204	249	72	2610	4.7	15
H205	167	64	2910	4.4	15
H401	--	130	7500	8.4	41
H403	1031	160	8100	8.1	41
H405	145	48	4000	3.7	39
H406	288	70	4200	5.8	29
H407	503	99	3690	6.9	25
H408	316	85	5080	5.9	28
H410	238	72	2200	5.0	16
H411	76	48	1150	3.5	7
H412	61	45	1400	3.5	12

NOTE: Tests M409 and M413 had skull fractures. No head 3-D analysis was done.

TABLE 6 - PEAK SPINAL RESPONSE

Test No.	Acceleration (G)			Velocity (m/s)		
	T1	T6	T12	T1	T6	T12
H201	68	--	31	2.1	--	2.7
H202	226	--	119	3.4	--	8.7
H203	60	--	111	2.4	--	2.1
H204	96	--	32	2.2	--	2.1
H205	78	--	30	2.7	--	2.1
H401	49	44	22	1.6	2.7	1.3
H403	130	91	27	2.4	3.0	1.6
H404	180	64	24	3.5	3.5	1.3
H405	46	41	12	1.7	2.0	1.0
H406	70	54	15	2.4	2.5	1.0
H408	59	88	20	2.2	1.7	1.1
H410	48	17	7	--	--	--
H411	12	15	7	--	--	--

TABLE 7 - NECK REACTIONS AT CONDYLES

Test No.	Peak Forces (N)			Peak Moments (N.m)		
	A-P	L-R	S-I	A-P	L-R	S-I
H201	1652	693	3491	61	91	43
H202	1268	432	2836	127	65	48
H203	1018	381	2945	75	85	18
H204	1343	491	2527	64	47	32
H205	1578	509	3163	170	107	68

TABLE 8 - ESTIMATES OF HEAD PARAMETERS

Test No.	Mass (kg)	Moments of Inertia, (kg.m²)		
		Ixx	Iyy	Izz
H201	4.23	0.0234	0.0228	0.0245
H202	3.85	0.0162	0.0188	0.0169
H203	3.32	0.0124	0.0145	0.0134
H204	3.64	0.0156	0.0174	0.0163
H205	4.64	0.0246	0.0274	0.0253
H401	3.91	0.0163	0.0208	0.0173
H402	3.96	0.0184	0.0224	0.0192
H403	3.11	0.0114	0.0142	0.0126
H404	4.05	0.0168	0.0168	0.0164
H405	4.14	0.0201	0.0216	0.0212
H406	3.85	0.0162	0.0172	0.0167
H407	2.81	0.0086	0.0110	0.0095
H408	3.99	0.0186	0.0215	0.0195
H410	4.32	0.0208	0.0226	0.0217
H411	3.11	0.0138	0.0158	0.0145
H412	4.44	0.0211	0.0237	0.0218

pensate the effects of the normal lordotic curvature of the cervical spine and the kyphotic curvature of the upper thoracic spine on the force load path.

The second mechanism of basal skull fracture, in which deformation in the skull causes stress distal to the point of impact, has been discussed by Gurdjian [16]. He postulates that, as long as the cranial skull remains intact under the impact, the shell undergoes an inbending under the impacted surface as well as an outbending away from the impacted region. It is at the outermost point of the bent cranial shell that a crack is initiated due to tension. The crack then propagates downward toward the foramen magnum around which the skull base is weakest, and a basal skull fracture is precipitated.

The tests in which basal skull fracture occurred were characterized by large forces of short duration. This type of force-time history is significantly different from the force-time history of the rest of the tests in this study. The implication is that, although skull deformation may be a necessary condition for basal skull fracture, it may not be the sole cause in S-I impacts.

CERVICAL INJURIES - In order for the forces to reach levels sufficient to cause the abovementioned skull deformation, padding on the impactor surface must be of sufficient depth to eliminate local skull fracture but not great enough to spread the transfer of energy from the 10-kg impactor to the skull over an extended time interval. When local or basal skull fractures occurred, neck injures were absent as in tests H200, H409, and H413.

From this observation it would seem that, if the available energy of the impactor is small enough not to overdrive the system, then, to produce damage to the neck, the skull must remain intact throughout the impact. The implication is that when the skull remains intact the initial curvature of the neck allows it to "buckle" under the load from the condyles, which results in extension or hyperextension motion of the neck.

Injuries to one or more cervical vertebrae occurred virtually in all the other tests where the neck "buckled" under the impact load. The common feature of these tests is that there was no attempt at aligning the neck with the spine and impactor axes. Instead, an initial curvature of the neck was allowed to simulate the natural attitude of normally standing or sitting persons.

Some observations could be made about the impact parameters and kinematic responses of these tests. In tests H402 and H403, fractures to the lower cervical vertebrae were caused. Both tests had peak forces of about 11 kN, an impactor velocity of 10.9 m/s, and a padding of 5 cm thickness. In the other tests (H404, H406, H407, H408, H410, and H412), the force level dropped to 4 to 5 kN. Yet injury was produced under a variety of impact conditions, including different paddings and different head/neck angles. Most of the injuries occurred in C3-C4, i.e., in the region of maximum neck bending with occasional damage done to the C1/C2 area or to the T1/T2 thoracic region.

Based on the tests in this study, it is clear that injuries to the cervical spine are occurring at impact force levels much lower than those required to produce skull fractures.

NON-INJURIOUS IMPACTS - Four impact tests conducted in this research project did not produce any damage to the neck or the head. These were tests H401, H405, H411, and H414. It might seem reasonable to draw a line between impact levels and kinematic parameters observed in these four tests and those observed in the remaining tests and call that line a threshold of tolerance. However, a closer look at these parameters reveals certain inconsistencies.

Based on tests H401, H405, and H411, a peak force level of about 4 kN seems at first glance to be just below the tolerance level of the neck. An exception is the level of 3 kN in test H412 where injury did occur. Another exception at the upper end of the force spectrum is test H414 where no injury occurred even for a force level of 16 kN, while identical test conditions and parameters produced a basal skull fracture in test H413.

IMPACT PARAMETERS AS INJURY PREDICTORS - Table 4 includes such impact force parameters as peak, duration and impulse, and the energy of impact that is absorbed by the head, as determined from conservation of momentum principles. By comparing these parameters to the results of autopsies, one should arrive at the critical value of some of the parameters previously mentioned below at which no injury should occur.

The data produced in this study indicates that none of the parameters chosen as predictors of injury consistently results in an infallible criterion. For example, based on three non-injurious impacts, it seems that a peak force of 4.2 kN is about the maximum that could be tolerated without injury. Two exceptions are test H414, where no injury occurred even when force reached a peak of 16 kN, and test H412, where injury was observed at a much lower peak force of 3 kN.

Because these tests had different force pulse durations, it was thought that a parameter that accounts for both duration and peak force may be more appropriate as an injury predictive measure. Thus the impulse defined as the integral of the force-time history, was calculated. Another related parameter that was computed was the impact energy transferred to the head. Of the two parameters, the impulse offers less inconsistency in predicting non-injury than energy.

RESPONSE PARAMETERS AS INJURY PREDICTORS - The two potential tolerance criteria discussed above (peak force and impulse) define tolerable levels of impact in the S-I direction. Tolerable levels of kinematic response of the head and neck, derived from kinematic measurements may also be defined.

Two other kinematic responses were considered as predictors of injury. The first is the peak head linear acceleration, and the other is the peak head linear velocity. These two parameters can easily be monitored in anthropomorphic devices. Of the two measures, the velocity offers a better predictive power than the acceleration, because it takes into account the duration of the impact. However, the number of tests available for determining a velocity threshold level is too small for that purpose.

CONCLUSIONS

1. Padding is essential in distributing the impact force and eliminating localized skull fracture.
2. Load-distributing materials are effective methods of reducing localized skull fractures, but they do not necessarily eliminate skull fractures in general.
3. The use of Head Injury Criterion (HIC) is not recommended for predicting the injury potential of S-I head impacts.
4. Of the response parameters that were examined, head velocity seems the best suitable indicator for injury. Of the response parameters that were not examined, the neck gross motions (either deflection or angle) may offer the greatest potential for accurate injury prediction.
5. Of the impact parameters that were examined, the integral of the force-time curve, i.e., the impulse, seems to be the most consistent injury indicator.
6. The size of the sample of tests available for determining S-I tolerance levels remains too small for accurate assessment. This sample must be enlarged by conducting more S-I head impacts, and by widening the scope of experimental documentation to focus on measurement of the neck motion.

ACKNOWLEDGMENTS

This research was sponsored by the National Institute for Occupational Safety and Health under contracts No. 210-78-0016 and No. 210-79-0028.

The protocol followed at the University of Michigan Transportation Research Institute for the use of cadavers in this study was approved by the Committee to Review Grants for Clinical Research and Investigations Involving Human Beings of the UM medical center and follows guidelines established by the U.S. Public Health Service and recommendations by the National Academy of Science/National Health Research Council.

REFERENCES

1. Culver, R.H.; Bender, M.; Melvin, J.S.: "Mechanisms, Tolerances, and Responses Obtained Under Dynamic Superior-Inferior Head Impact." Final Report UM-HSRI-78-21, May 1978. NTIS Order No. PB-299-292.
2. Nusholtz, G.S.; Melvin, J.W.; Huelke, D.F.; Alem, N.M.; Blank, J.G.: "Response of the Cervical Spine to Superior-Inferior Head Impact." Proc. 25th Stapp Car Crash Conf., SAE Paper No. 811005, 1981.
3. Hodgson, V.R.; Thomas, L.M.: "Mechanisms of Cervical Spine Injury During Impact to the Protected Head." Proc. 24th Stapp Car Crash Conf., SAE Paper No. 801300, 1980.
4. Roaf, R.: "A Study of the Mechanics of Spinal Injury." The Journal of Bone and Joint Surgery, Volume 42B, No. 2, November 1960.
5. Mertz, H.J.; Hodgson, V.R.; Thomas, L.M.; Nyquist, G.W.: "An Assessment of Compressive Neck Loads Under Injury-Producing Conditions." The Physician and Sportmedicine, November, 1978.
6. Sances, A.; Myklebust, J.; Houterman, C.; Weber, R.; Lepkowski, J.; Cusik, J.; Larson, S.; Ewing, C.; Thomas, D.; Weiss, M.; Berger, M.; Jessop, M.E.; Saltzberg, B.: "Head and Spine Injuries." AGARD Conf. Proc. No. 322 on Impact Injury Caused

by Linear Acceleration: Mechanism, Prevention and Cost, Koln, Germany, April 26-29, 1982.

7. McElhaney, J.H.; Paver, J.G.; McCrackin, H.J.; Maxwell, G.M.: "Cervical Spine Compression Responses." Proc. 27th Stapp Car Crash Conf., SAE Paper No. 831615, 1983.
8. Alem, N.M.: "Helmet Impact Test System Development." Final Report No. UM-HSRI-80-72, August 1980.
9. Alem, N.M.; Nusholtz, G.S.; Melvin, J.W.: "Superior-Inferior Head Impact Tolerance Levels." Final Report UMTRI-82-42, November 1982. NTIS Order No. PB-83-144-501.
10. Stalnaker, R.L.; Melvin, J.W.; Nusholtz, G.S.; Alem, N.M.; Benson, J.B.: "Head Impact Response." Proc. 21st Stapp Car Crash Conf., SAE Paper No. 770921, 1977.
11. Alem, N.M.; Benson, J.B.; Holstein, G.L.; Melvin, J.W.: "Whole Body Response Research Program - Methodology." Final Report No. UM-HSRI-77-39-2, April 1978.
12. Nusholtz, G.S.; Melvin, J.W.; Alem, N.M.: "Head Impact Response Comparison of Human Surrogates." Proc. 23rd Stapp Car Crash Conf., SAE Paper No. 791020, 1979.
13. Becker, E.B. et al.: "Measurement of Mass Distribution Parameters of Anatomical Segments."
14. Chandler, R.F. et al.: "Investigation of Inertial Properties of the Human Body."
15. Lett, D.G. et al.: "Estimating Moments of Inertia of the Head from Standard Anthropometric Data."
16. Gurdjian, E.S.; Gonzales, D.; Hodgson, V.R.; Thomas, L.M.; Greenberg, S.W.: "Comparisons of Research in Inanimate and Biologic Material: Artifacts and Pitfalls." Pages 234-253 in Impact Injury and Crash Protection, edited by Gurdjian et al., published by C.C. Thomas, 1970.

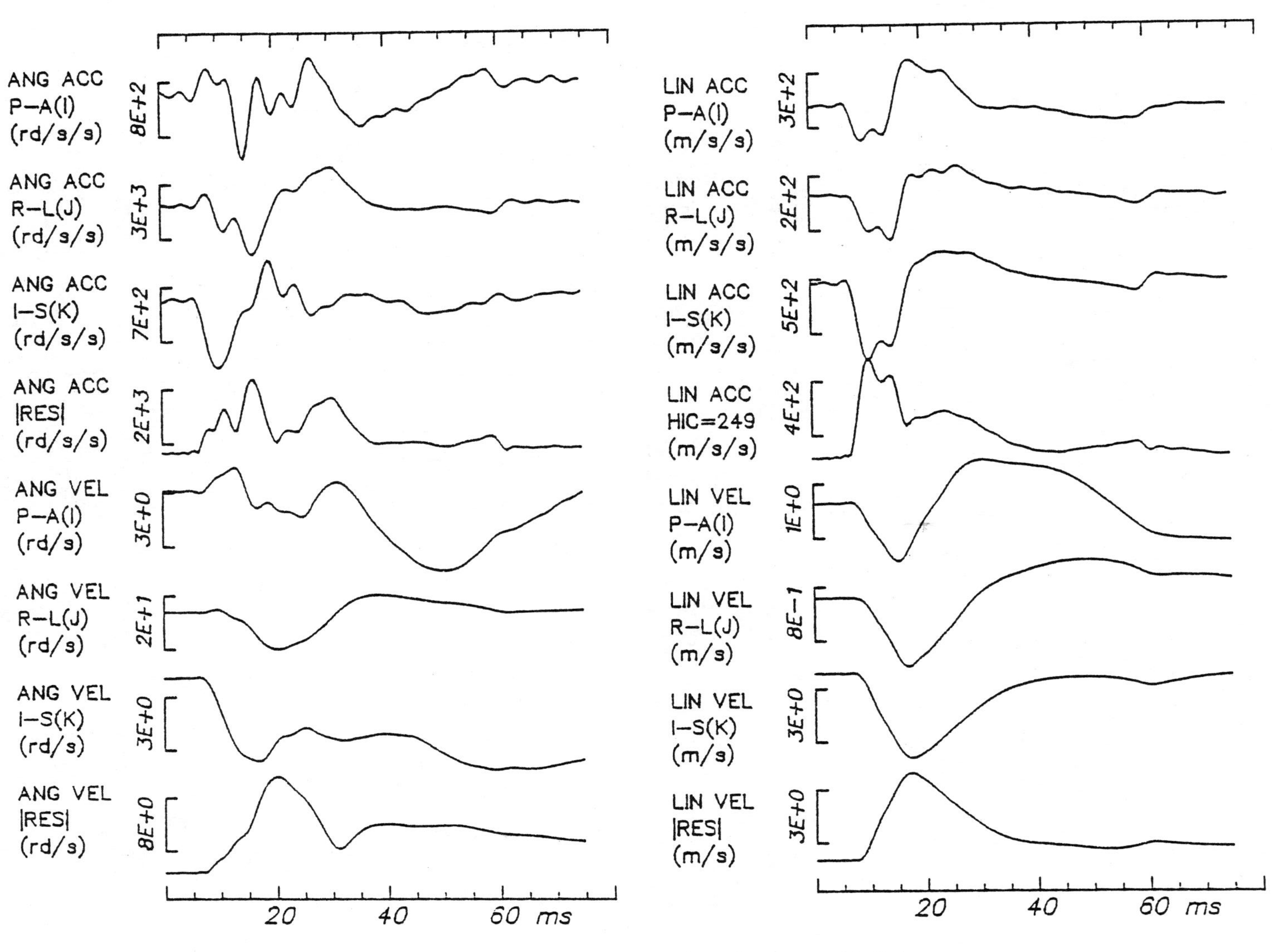

Fig. 1 - Head 3-D Motion for Test 79H204.

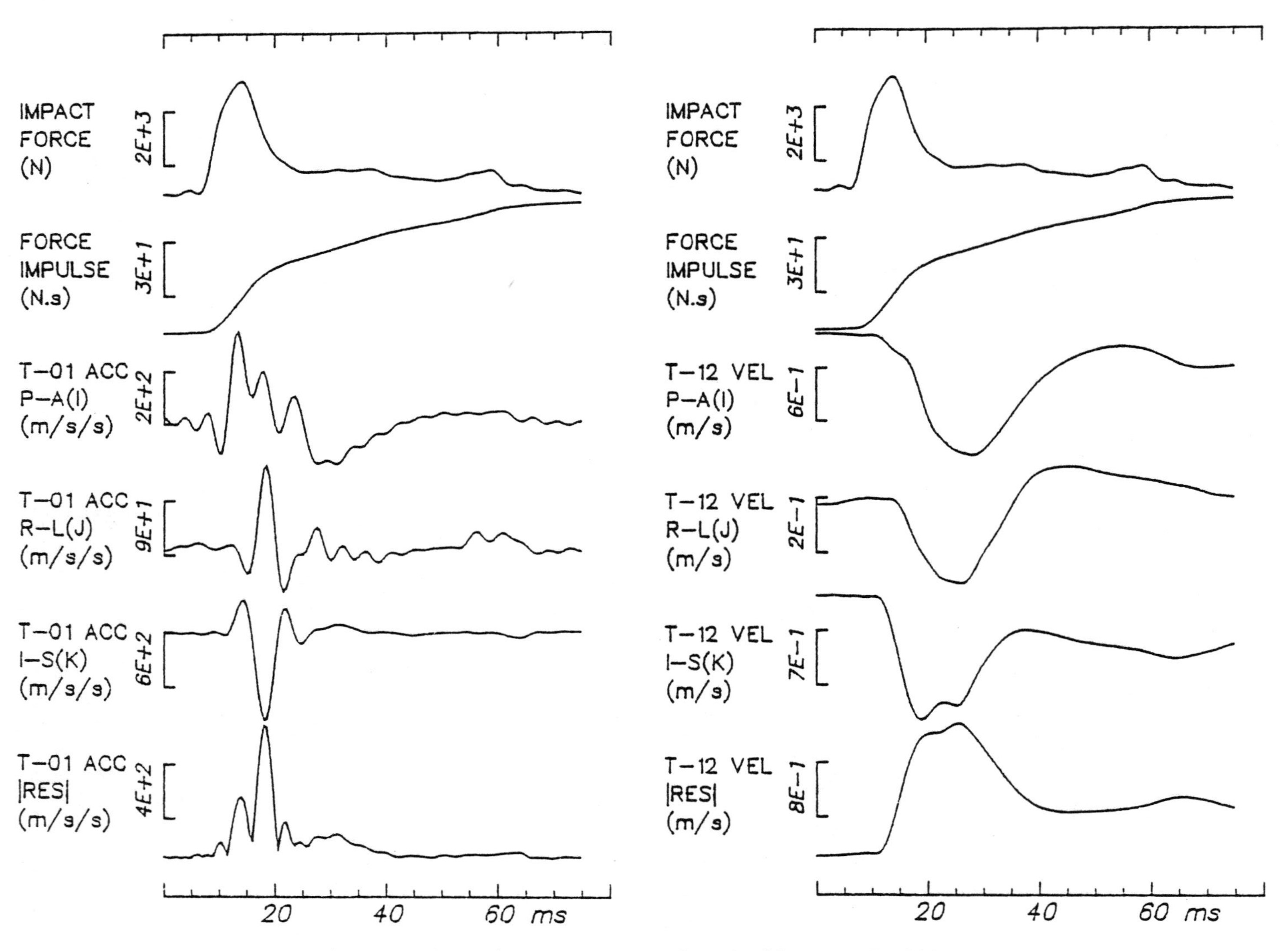

Fig. 2 - Thoracic Response at T1 and T12, Test 79H204.

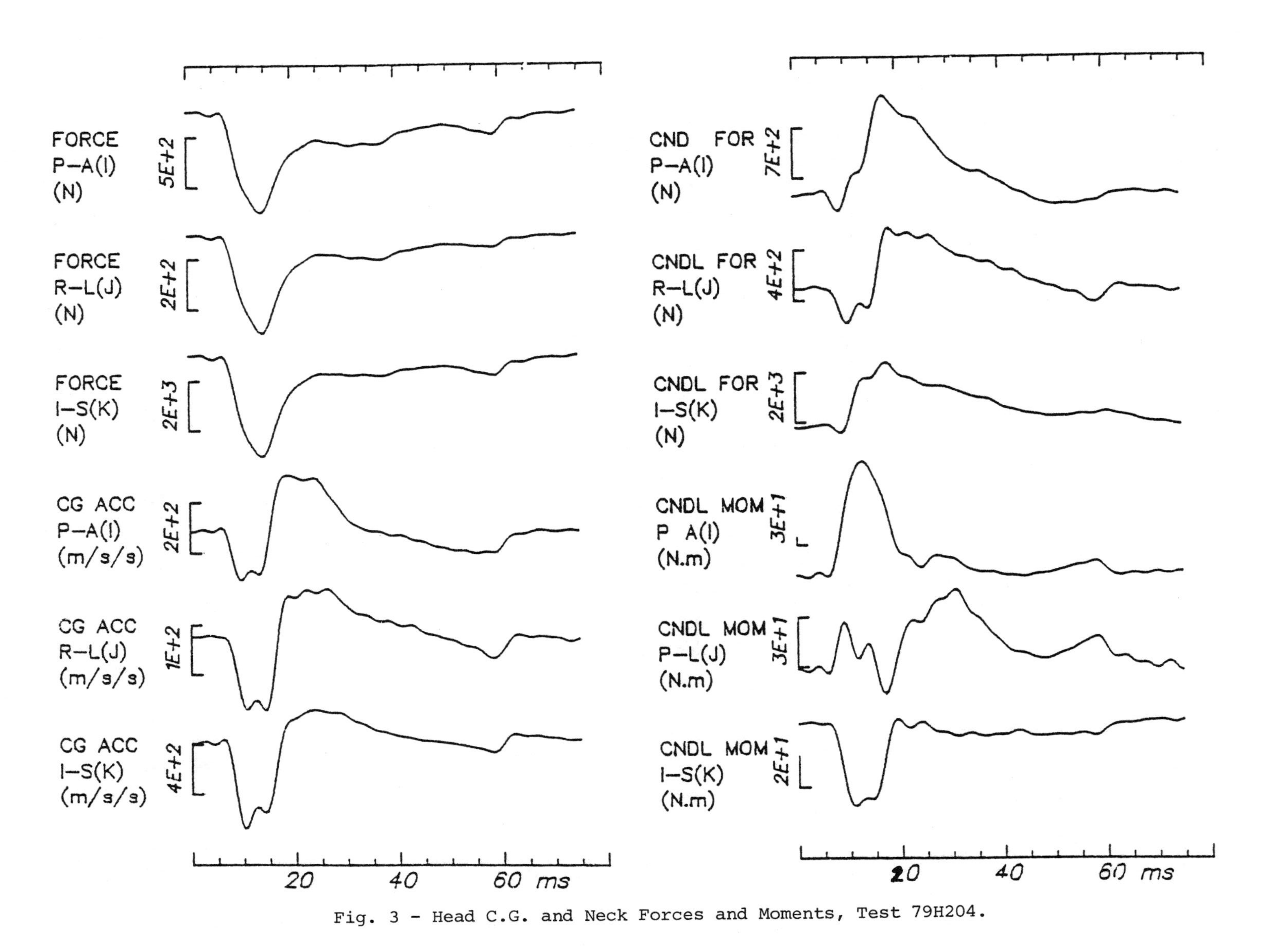

Fig. 3 - Head C.G. and Neck Forces and Moments, Test 79H204.

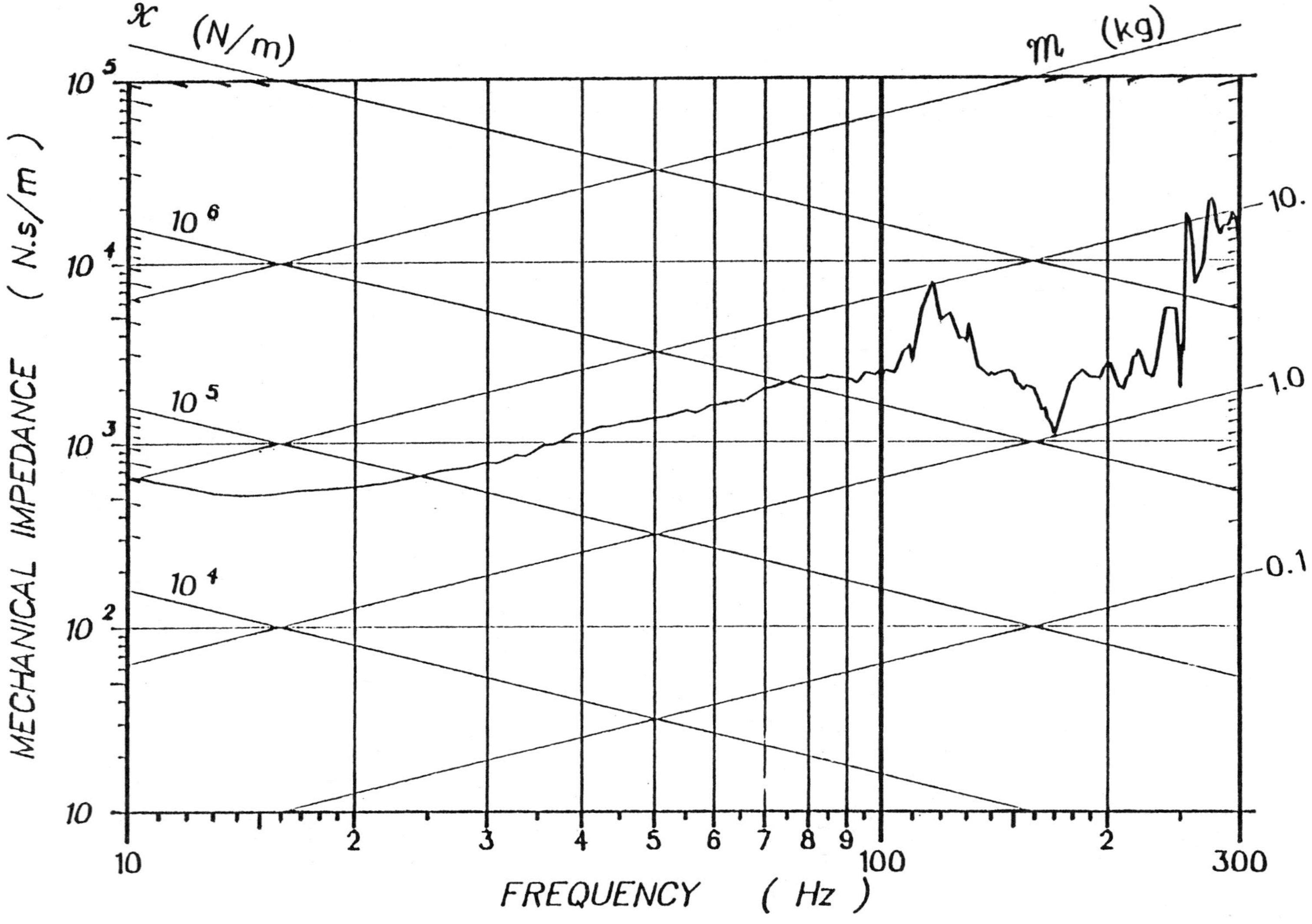

𝒳 (N/m)

𝔪 (kg)

10^6

10^5

10^4

10.

1.0

0.1

10^5

10^4

10^3

10^2

10

MECHANICAL IMPEDANCE (N.s/m)

10 2 3 4 5 6 7 8 9 100 2 300

FREQUENCY (Hz)

Fig. 4 - Head (I-S) Impedance, Test 79H204.

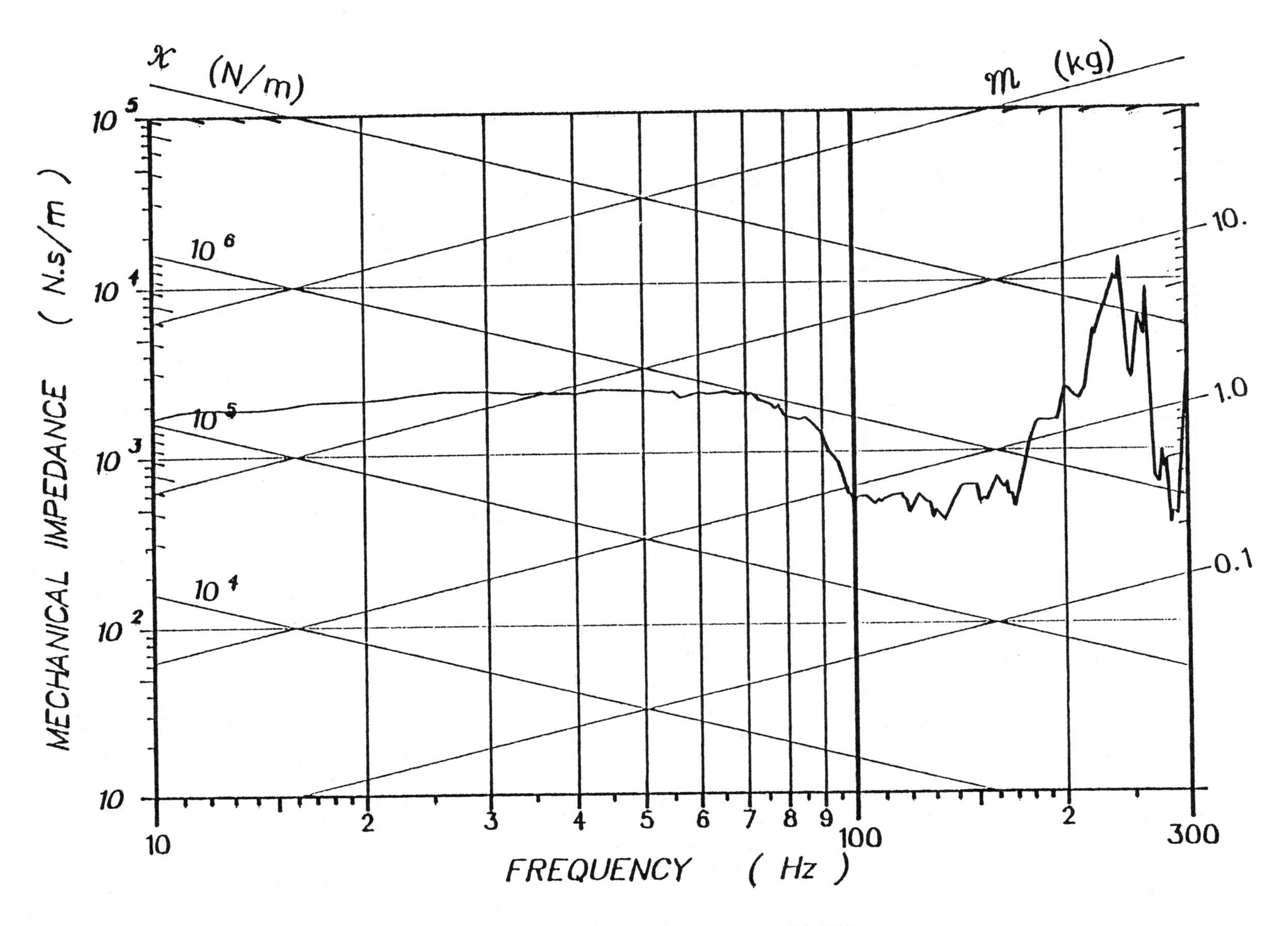

Fig. 5 - Tl (I-S) Impedance, Test 79H204.

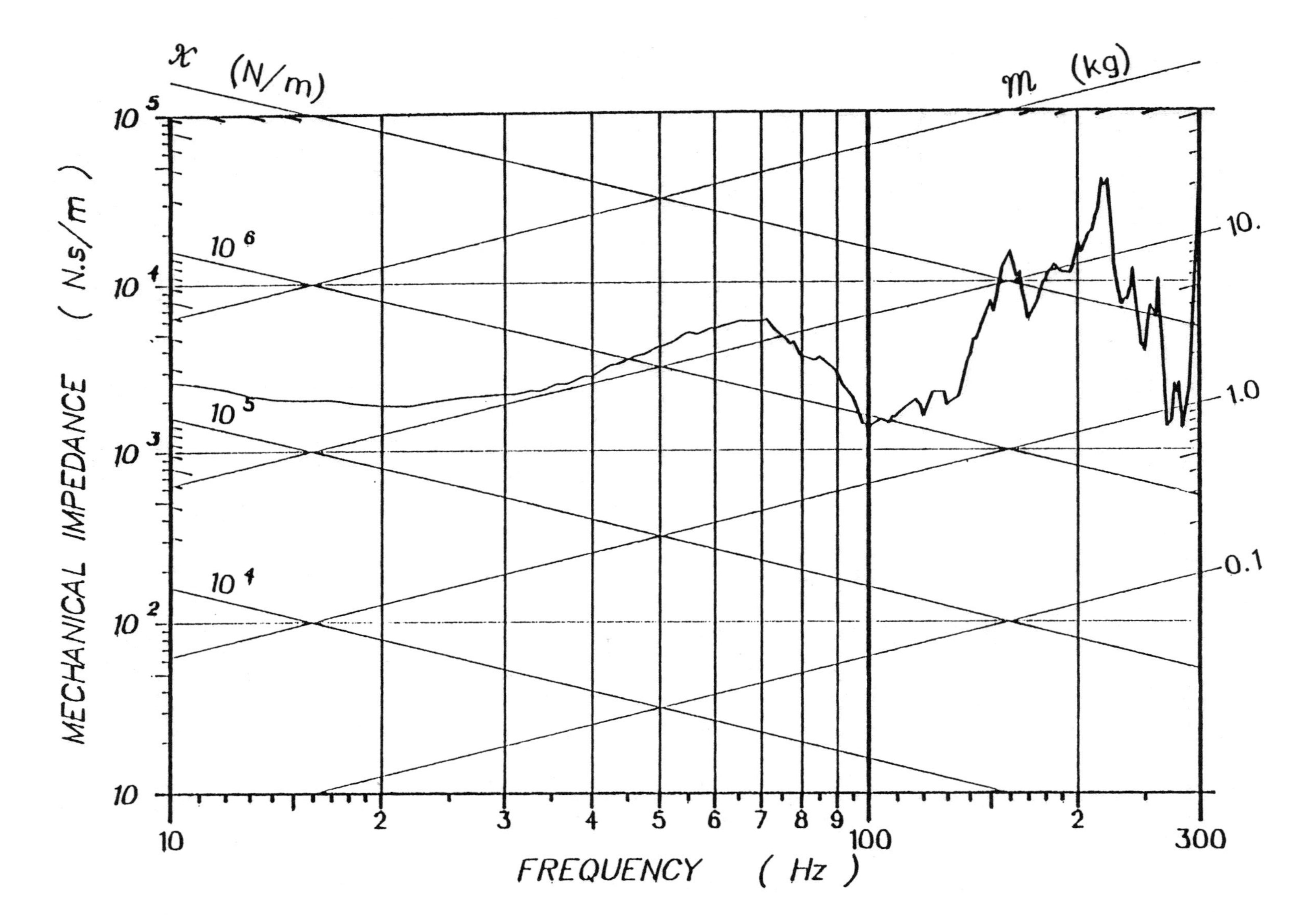

Fig. 6 - T12 Vertebra Impedance, Test 79H204.

872194

Comparison of Human Volunteer and Cadaver Head-Neck Response in Frontal Flexion

J. Wismans, M. Philippens and E. van Oorschot
TNO Road Vehicles Research Institute

D. Kallieris and R. Mattern
University of Heidelberg

ABSTRACT

At the 30th Stapp Conference an analysis was presented of human volunteer head-neck response in omni-directional impact tests. It was shown that the relative head motion can be described by a simple two-pivot analog system.

The present study extends this analysis to post-mortem human subject (PMHS) tests conducted at the University of Heidelberg. Two test series similar to the human volunteer frontal impacts tests were carried out. One having an impact severity identical to the most severe human volunteer tests. A second series with higher exposure levels are used to verify the proposed analog system for higher impact levels. Test results including neck injury data for five PMHS tests will be given with special attention to trajectories of the head center of gravity, head rotations and head accelerations.

It is concluded that the center of gravity trajectories for the PMHS and volunteer tests are similar for both impact levels. Head rotations, however, are larger for the PMHS than for the volunteer tests. The two-pivot linkage mechanism proposed for the volunteer head-neck motions also appears to be suitable to describe the PMHS head-neck response.

THE NAVAL BIODYNAMICS LABORATORY (NBDL) in New Orleans has conducted a large number of human volunteer tests to study omni-directional head-neck response in an impact situation. A detailed description of the NBDL instrumentation and test methods is provided by Ewing et al. (1-4)*. As part of our research program a large number of the NBDL tests conducted between 1981 and 1985 have been analyzed in order to develop performance requirements for a mechanical neck for crash dummies. Detailed results of these analyses have been presented in (5-8). It was shown that the volunteer relative head motion can be described quite well by a simple two-pivot analog system.

The purpose of the present study is to extend this analysis to post-mortem human subjects (PMHS) tests conducted at the University of Heidelberg. These subjects have been exposed to frontal impacts using a similar sled seat and restraint system as in the NBDL human volunteer tests. A detailed description of the test set-up and test results of the PMHS tests will be presented. Results will be compared with a representative set of nine tests out of the most severe frontal human volunteer tests conducted at NBDL. In these NBDL tests five different subjects were exposed to a 15 g sled acceleration. Table 1 summarizes these tests together with the important test conditions.

METHODS AND MATERIALS

TEST PROCEDURE - The post-mortem human subject (PHMS) tests were conducted on the decelerator of the Institute for Forensic Medicine of the University of Heidelberg. The experimental set-up was similar to the frontal NBDL volunteer tests (3). The subjects were placed on a 90 degree rigid seat and restrained by shoulder straps, a lap belt and inverted V-pelvic strap tied to the lap belt (Fig. 1). The arms were restrained by an additional belt at the mamillar level to prevent flailing.

Twelve tests have been conducted. Tabel 2 summarizes the most important test conditions. The first five tests are pre-test and can not be analyzed here due to limited visibility of the photographic targets. In two further tests (i.e. 8703 and 8705) no film-data are available. As a consequence five tests are suitable for a more detailed analysis. Two tests are of a similar impact severity as the NBDL volunteer tests (15 g sled acceleration), while three tests are more severe (23 g).

---

* Numbers in parentheses designate references at end of paper.

Table 1. Test characteristics of 9 selected human volunteer tests in frontal impacts.

Test no.	Subject no.	Imp. Vel. (m/s)	Peak Sled dec. (g)	Sex	Initial neck length[1]) (cm)	Anthropometric measurements at NBDL: Weight (kg)	Standing height (cm)	Sitting height (cm)	Head: Circumf. (cm)	Breadth (cm)	Length (cm)
3957	H00132	16.75	14.6	M	14.1	79.8	172.9	89.6	57.9	15.7	19.7
3959	H00127	16.84	14.8	M	16.2	62.1	172.3	89.8	54.2	14.9	18.5
3963	H00133	16.69	14.5	M	16.5	61.2	161.7	86.8	56.1	14.7	19.4
3965	H00135	16.67	14.6	M	15.0	68.9	171.6	90.7	53.5	14.6	17.9
3970	H00135	17.26	15.6	M	15.0	68.9	171.6	90.7	53.5	14.6	17.9
3982	H00132	17.47	15.6	M	14.2	79.8	172.9	89.6	57.9	15.7	19.7
3986	H00133	17.31	15.6	M	16.5	61.2	161.7	86.8	56.1	14.7	19.4
3987	H00131	16.76	14.5	M	15.6	67.6	167.0	90.0	57.5	15.4	19.6
3990	H00131	17.26	15.4	M	15.6	67.6	167.0	90.0	57.5	15.4	19.6

[1]) Defined as the average value of the initial distance between T1 and head anatomical origin in several tests.

Table 2. Test conditions, subject anthropometry and post test AIS values.

PMHS Test	Imp. Vel. (m/s)	Peak Sled dec. (g)	Sex	Initial neck length[1]) (cm)	Anthropometric measurements: Weight (kg)	Standing height (cm)	Sitting height (cm)	Head: Circumf. (cm)	Breadth (cm)	Length (cm)	Vol. (lit.)	AIS	Remarks
8615	16.4	15.5	M	-	80	176	96	61	17	24	-	0	pre-test
8616	16.4	15.0	M	-	66	170	92	60	17	19	4.4	1	pre-test
8618	16.4	15.1	F	-	58	175	84	55	15	18	4.1	1	pre-test
8620	16.4	16.4	F	-	78	165	91	55	15	18	4.6	2	pre-test
8621	16.4	14.8	F	-	72	166	91	55	15	18	4.8	0	pre-test
8622	16.4	16.0	M	16.7	80	182	95	57	15	19	4.6	1	-
8701	16.7	15.2	F	14.9	74	168	89	54	15	18	3.9	1	-
8703	16.7	15.2	M	-	66	154	85	55	15	19	4.0	1	no film
8705	16.7	15.2	M	-	71	175	92	59	17	19	4.0	1	no film
8706	16.4	23.0	M	16.0	72	172	91	59	16	19	5.2	1	-
8709	16.7	21.5	M	15.5	74	170	97	58	15	14	4.8	1	-
8710	16.7	23.3	F	13.6	53	175	90	53	14	18	3.3	2	-

[1]) Defined as the initial distance between T1 and head anatomical origin.

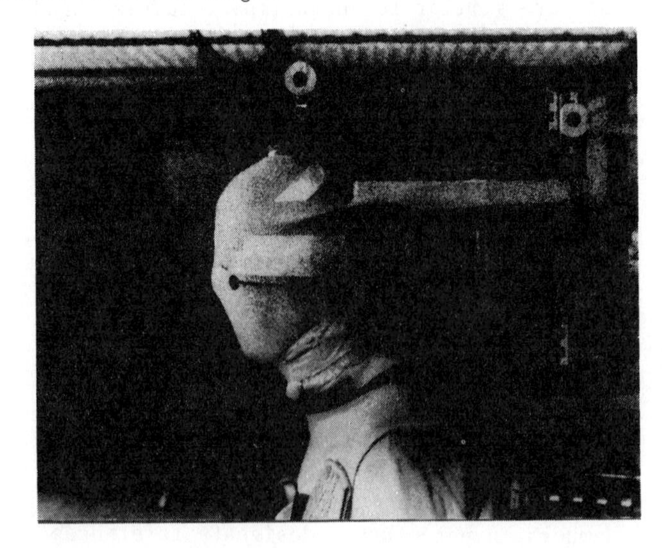

Fig. 1 Test set-up for PMHS tests (initial conditions).

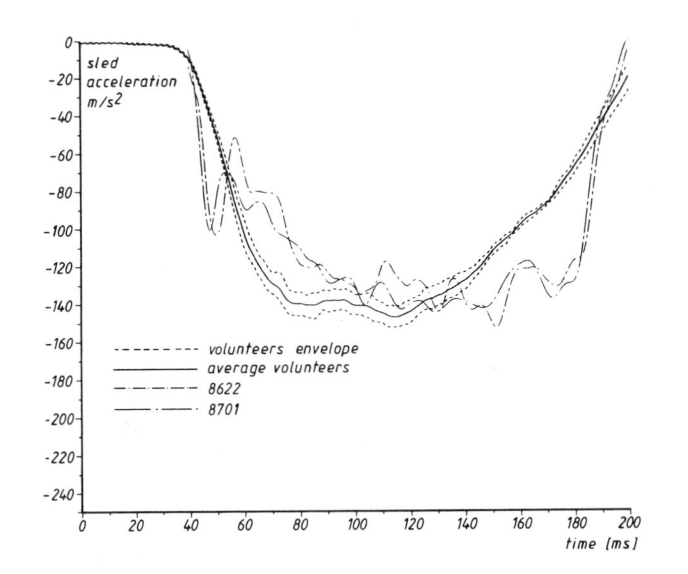

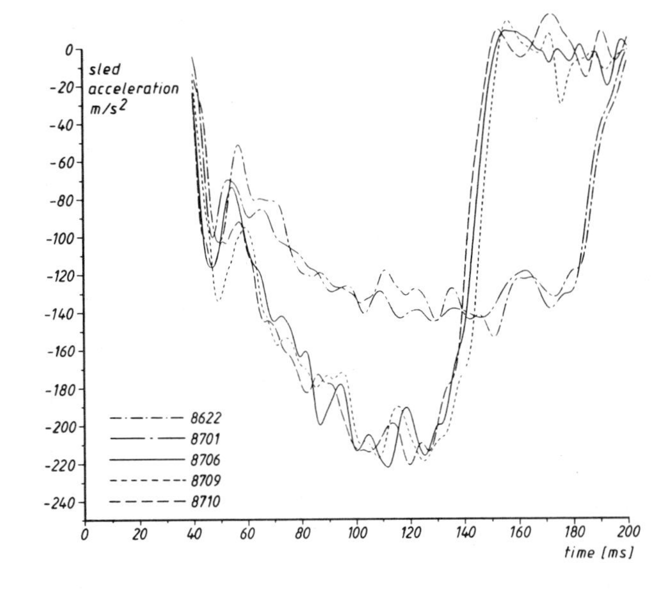

Fig. 2 Sled accelerations. Left: PMHS tests with volunteer envelope 15 g. Right: 15 g and 23 g PMHS tests.

INSTRUMENTATION - The positions of head and the first thoracic vertebral body (T1) were registered by a sled mounted high speed camera running at 1000 frames/second, assumed to be parallel to the plane of motion. Two photo targets are mounted to the head as well as T1. The positions of the head and T1 anatomical coordinate systems were related to these targets by two dimensional x-ray analysis. Sixteen channels of time-history data were recorded for each test. The instrumentation included a nine-accelerometer module as described by Padgaonkar (9) screwed to the top of the skull, a tri-axial accelerometer unit screwed at the clivus and a tri-axial accelerometer unit screwed to T1. All accelerometers used are ENDEVCO 2264-2000. Mass of the head instrumentation is about 0.165 kg. The signals were sampled at 10 kHz and filtered with a low pass 100 Hz 4th order Butterworth digital filter. The orientation of accelerometers in relation to the anatomical landmarks were obtained from lateral and anterior-posterior x-rays.

SLED ACCELERATION - Fig. 2 (left) shows the envelope of the NBDL sled acceleration-time histories in the nine volunteer tests together with the sled pulse in the two 15 g PMHS tests. It can be seen that the PMHS sled deceleration compares well with the volunteer tests. The velocity change of all PMHS tests is identical, which results in a shorter duration of the 23 g deceleration pulse as shown in Fig. 2 (right). The time bases of all PMHS tests has been shifted 40 ms in order to align the T1 acceleration time history of the volunteer and PMHS 15 g tests.

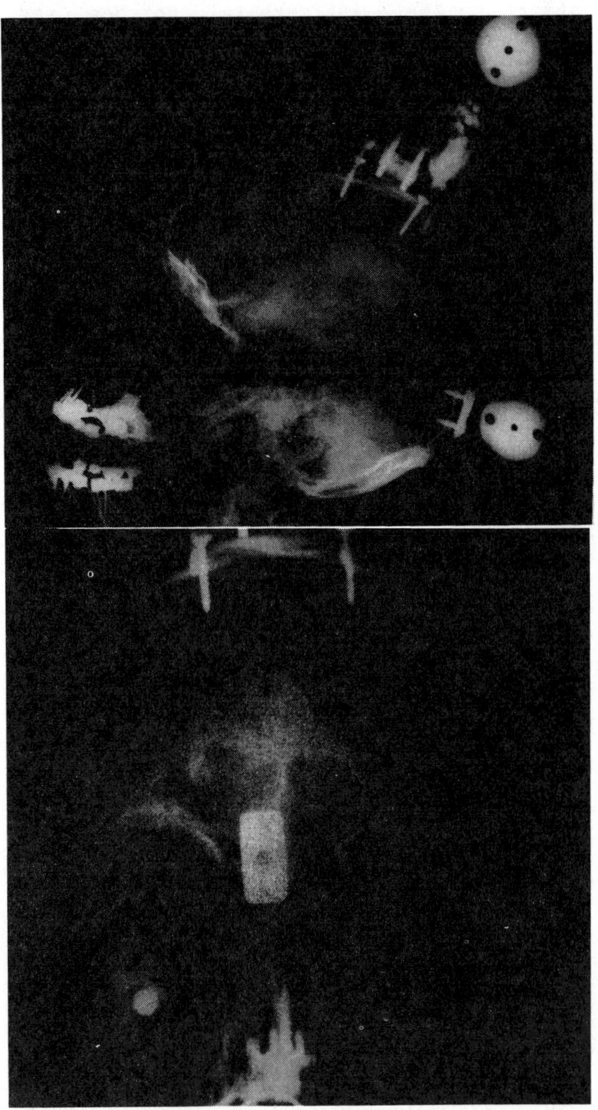

Fig. 3 Anterior-posterior and lateral x-rays with instrumentation and lead balls at the anatomical landmarks.

PRETEST PREPARATION OF POST-MORTEM HUMAN SUBJECTS - In an attempt to simulate the muscletone of living people, 100 ml 10% solution of formaldehyde was injected in the rear and side muscles of the neck of each post-mortem human subject PMHS. The time between the injection and the test amounted, in most cases, to more than 20 hours. By that it was attained that the contraction condition of the muscular system in the neck area was controlled in all tests. Fluctuations as they usually occur in the condition of rigor mortis, were reduced because of this measure. This means, that in the cases with a relaxed rigor mortis a contraction condition has been established which can be compared with a medium strong muscular tension in a living person. In a fully marked rigor mortis an additional contraction by means of formaldehyde injection could not be obtained. The fully marked rigor mortis corresponds to a very strong muscular tension in a living person.

The anatomical landmarks defining the Frankfurt plane, infraorbital notches and auditory meati, were marked with small lead balls. The position of these landmarks and positions of the optical targets and instrumentation were documented by anterior-posterior and lateral x-rays (Fig. 3).

POST-TEST PROCEDURE OF THE POST-MORTEM HUMAN SUBJECTS - A fully autopsy with a detailed investigation of the vertebral column has been performed after each test, as desribed by Mattern (10). The injury severity of observed lesions was scaled in accordance with AIS (11) (Table 2). Injuries not separately mentioned in the AIS vocabulary were scaled by analog application of the AIS criteria. Injuries in the PMHS were diagnosed as acute strain if hemorrhages occurred in the deep spinal column muscular system between the muscle bundles or in the region of the ligamental system, as well as in the vertebral joints and discs. Such cases were estimated with AIS 1. Macroscopically visible cuttings of above mentioned tissues did not occur in the investigated cases. Lacerations of ligaments, of intervertebral disc tissue and tear drop fractures of the vertebral bodies have been estimated with AIS 2.

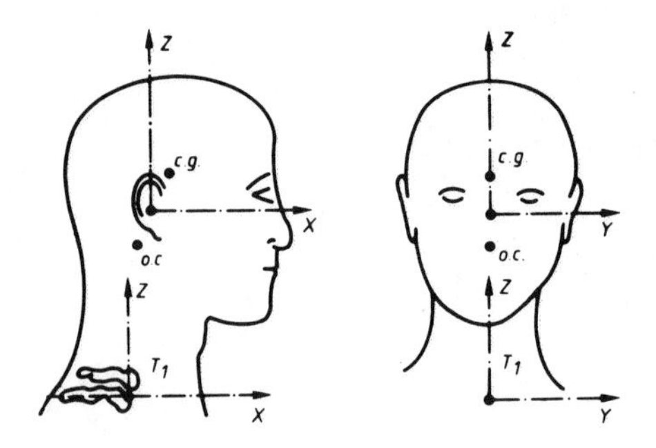

Fig. 4 Location of anatomical coordinate systems according to NBDL (o.c. = occipital condyles, c.g. = center of gravity).

COORDINATE SYSTEMS AND ANTROPOMETRY

COORDINATE SYSTEMS - Fig. 4 illustrates the location of the head and T1 anatomical coordinate system. The anatomical origin of the head is positioned at the midpoint of the connection of the auditory meati. The positive x-axis is defined as the line between this origin and the midpoint of the line connecting the infraorbital notches. The y-axis is perpendicular to the xz plane and positive toward the left ear. The plane defined by the x and y axis is approximately the Frankfurt plane. The origin of the spine (T1) anatomical coordinate system is at the anterior superior corner of the first thoracic vertebral body (T1). The orientation of the T1 coordinate system used for the analysis is taken parallel to the laboratory system.

HUMAN SUBJECT ANTHROPOMETRY - The most significant anthropometric data are summarized in Table 2. Definitions for these variables can be found in (3). These definitions are similar to the ones used by NBDL for the volunteers.

The volume of the PMHS head was measured by submersion (included in Table 2). The initial neck length is the distance between head and T1 anatomical origin as obtained from film-data at the start of the test.

TEST RESULTS

MEDICAL FINDINGS - No injuries occurred in three of the twelve conducted tests. In seven cases strains in the vertebral discs and small lacerations of the ligamenta flava and within the joints have been observed. In two tests, a severity of AIS 2 was determined: the first caused by a fracture in the upper front edge of the T2 vertebrae (Run no. 8620), the second caused by a laceration of the ligamentum flavum T2/T3 (Run no. 8710). Spinal column injuries for each test are reported in detail in Appendix A.

SPINE (T1) MOTION - Fig. 5 presents resultant T1 acceleration time-histories. In Fig. 5 (left) the two PMHS tests with a similar severity as the most severe human volunteer tests are presented together with the human volunteer envelope. It follows that the acceleration pulse is located near the lower boundary of the envelope and for a large portion within the envelope. This indicates that the input to the head-neck system is similar in the PMHS and the human volunteer tests. Fig. 5 (right) summarizes the T1 accelaration-time histories in all five PMHS tests. It can be seen that there is only a small difference in T1 accelerations between the 15 g and 23 g PMHS tests.

Table 3 summarizes maximum forward (horizontal) and vertical T1-origin displacements relative to the sled. The forward displacements appear to be of the same order of

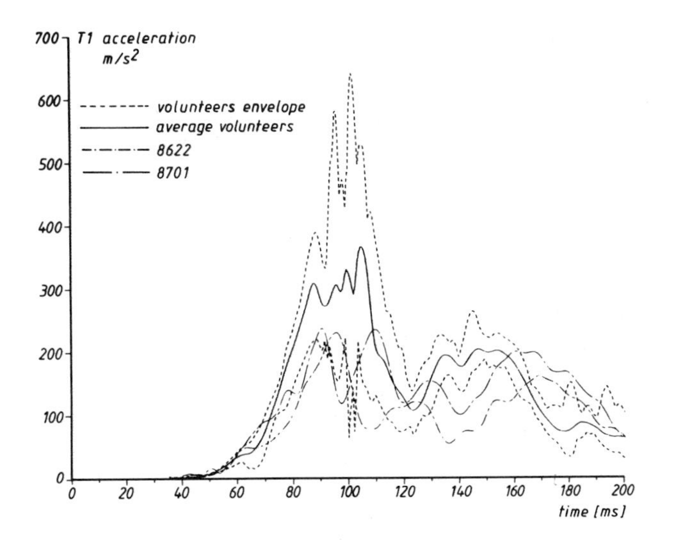

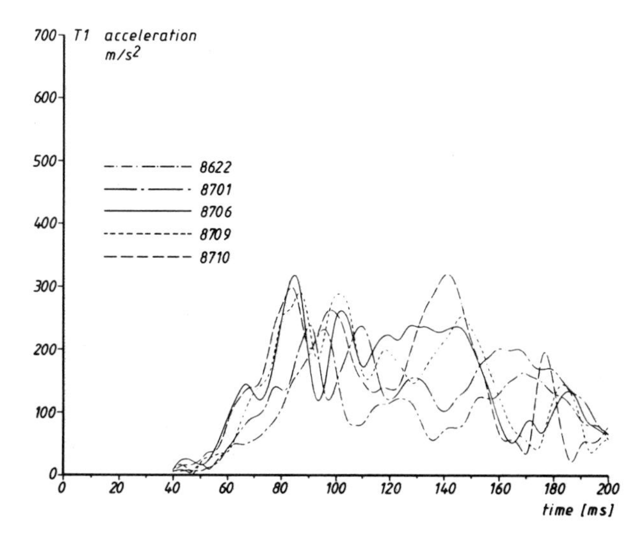

Fig. 5 Resultant T1 linear accelerations-time histories.

magnitude in the PMHS and the volunteer tests. The vertical displacements, however, show significant differences. In all PMHS tests a downward T1 displacement can be observed (up to 90 mm), while in the volunteer tests this downward motion is almost completely absent.

Table 3. T1 displacements and rotations.

Test number	Max. Displacement		Max. Rotation
	Horizontal (cm)	Vertical (cm)	(degrees)
Volunteer range	6.1/8.1	-0.1/-2.7	9.0/22.3
8622	5.2	-8.2	17.7
8701	7.9	-7.1	23.5
8706	6.8	-5.8	25.0
8709	10.8	-9.4	36.8
8710	13.2	-8.4	36.9

A comparison of T1 rotation-time histories in the human volunteer and PMHS tests is presented in Fig. 6. Note that in the volunteer tests initially a small backward rotation can be observed, which is completely absent in the PMHS tests. The peak T1 rotation appears to be slightly larger in the PMHS tests than in the volunteer tests. In the previous studies related to the human volunteer tests these T1 rotations have been neglected. The assumption of a non-rotating T1 will also be made for the PMHS tests in order to allow a direct comparison with the human volunteer tests. Note that in the severe PMHS tests (Fig. 6 top) larger T1 rotations can be observed than in the 15 g tests.

RELATIVE HEAD MOTIONS - The method used for the analysis of the relative head motions will be quite similar to the one used for the analysis of the volunteer tests. Head motions will be expressed relative to the T1 coordi-

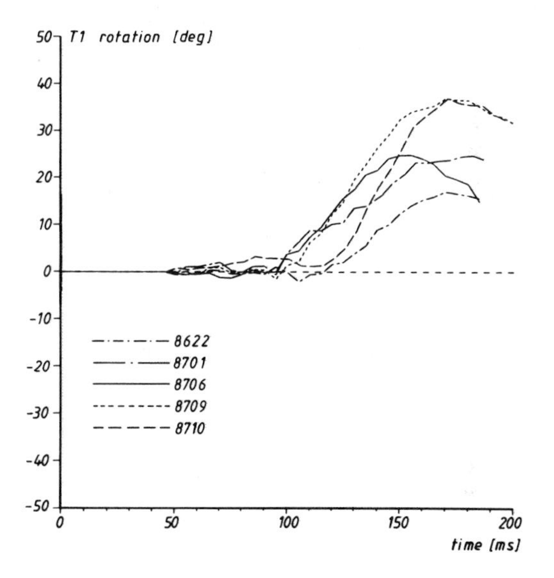

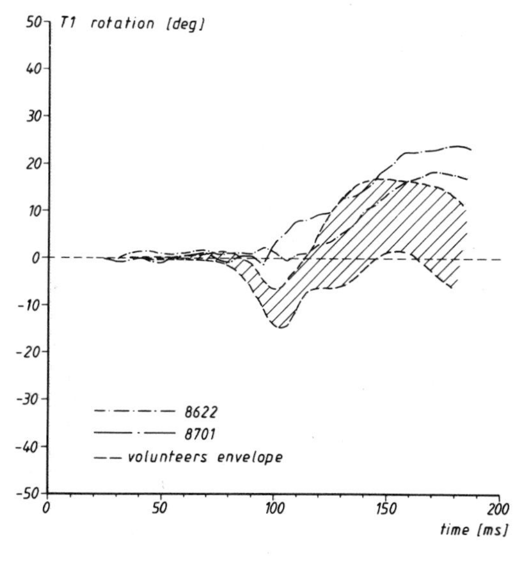

Fig. 6 T1 rotation-time histories.

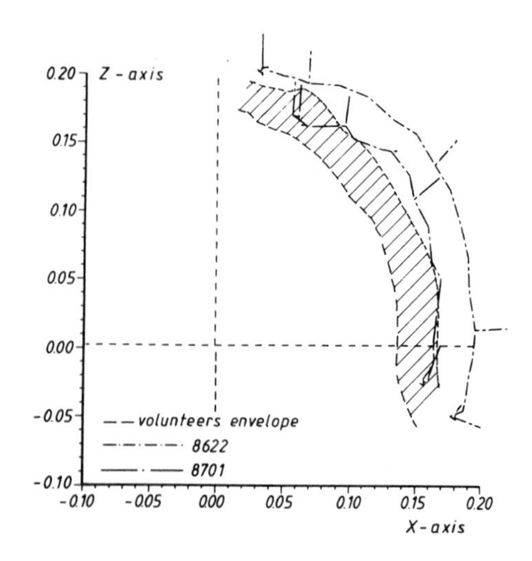

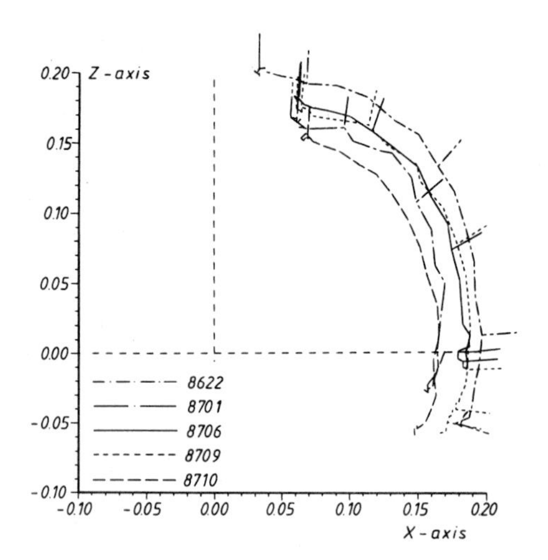

Fig. 7 Head c.g. trajectories relative to non-rotating T1 coordinate system.

nate system. Since T1 rotations will be neglected here, head motions will be presented with respect to a coordinate system which stays aligned with the laboratory system.

First an analysis will be made of head c.g. trajectories and head rotations. Fig. 7 shows the c.g. trajectories relative to T1. The left figure presents an envelope resulting from volunteer test together with results of the 15 g PMHS tests. Fig. 7 (right) compares the 15 g and 23 g PMHS tests. In a similar way in Fig. 8 head rotation-time histories are presented. In agreement with earlier volunteer test analysis, head rotational motion is defined here by the angle ϕ in the plane of impact between head and anatomical z-axis and

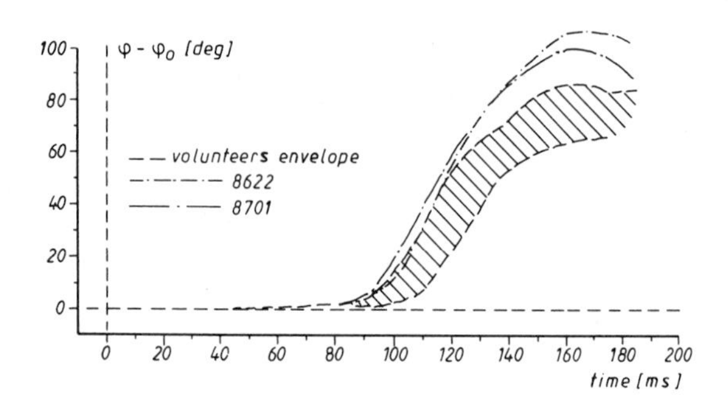

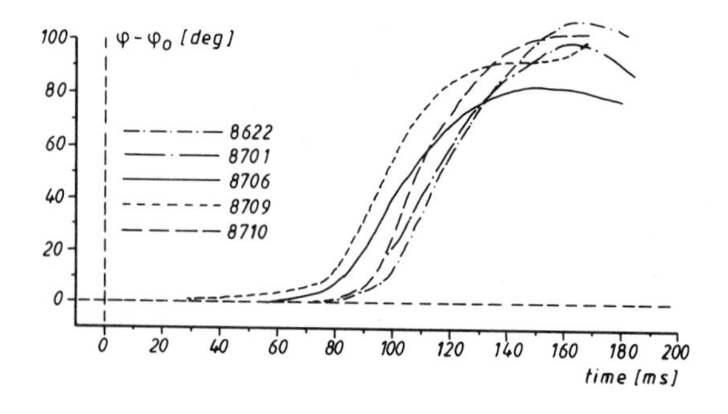

Fig. 8 Rotation of the head (ϕ-ϕ_0)

non-rotating T1 z-axis. Results presented in Fig. 8 relate to the angle ϕ-ϕ_0, where ϕ_0 is the initial head rotation angle (i.e. at time = zero). Peak c.g. translations and peak rotations are summarized in Table 4.

Table 4. Peak head rotations and c.g. displacement relative to T1.

Test number	c.g. displacement*		Rotation
	forward (cm)	downward (cm)	(ϕ-ϕ_0) (degrees)
Volunteer range	11.4/15.1	20.1/23.3	68.4/94.9
8622	16.3	25.1	108.8
8701	10.9	19.5	99.9
8706	12.5	19.5	84.4
8709	12.7	23.2	102.6
8710	9.6	21.2	102.6

* relative to initial c.g. position.

The following observations can be made:

- Except for test 8622, trajectories of the PMHS tests are within or close to the envelope defined by the volunteer tests. Also peak c.g. excursions in the PMHS tests appear to be close to the peak excursions observed in the volunteer tests.
- Peak c.g. excursions and head rotations in the severe PMHS tests do not differ significantly from the peak c.g. excursions in the 15 g PMHS tests.
- Except for test 8706, peak head rotations appear to be higher in the PMHS test than in the volunteer tests.
- In the severe PMHS tests an earlier rise of the head rotation-time histories can be observed compared to the moderate PMHS tests.

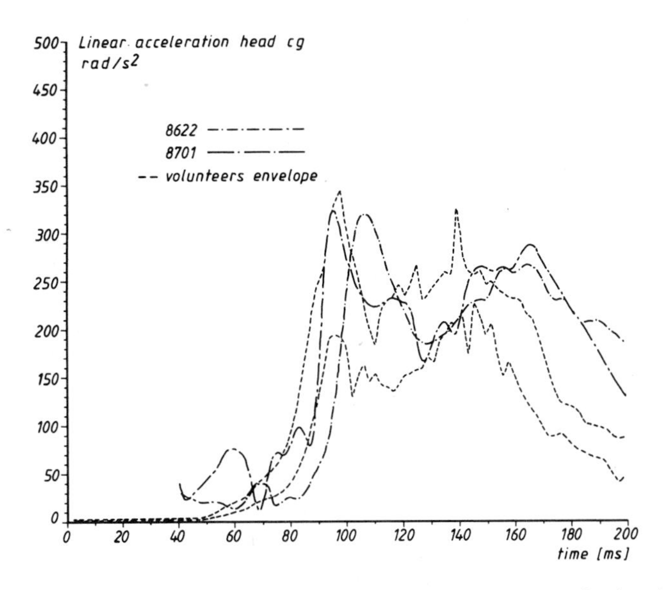

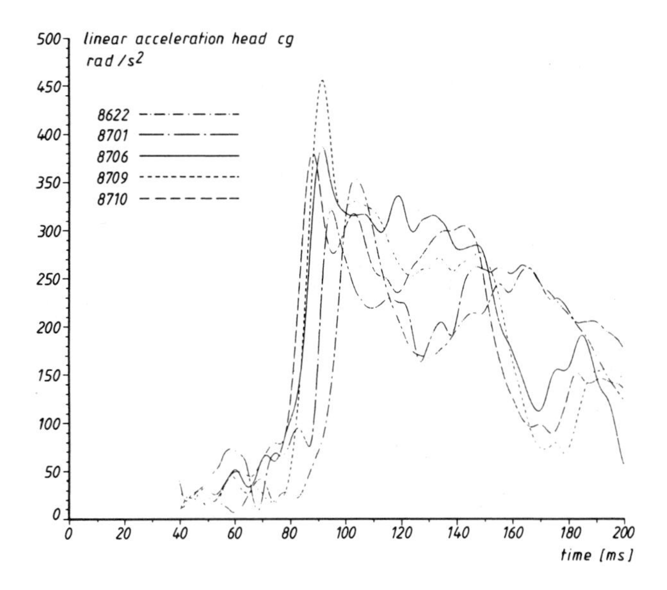

Fig. 9 Resultant linear acceleration of the head c.g.

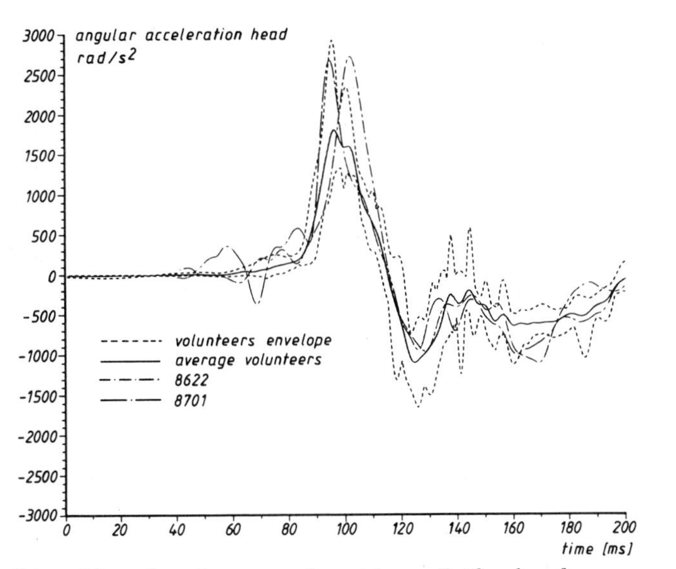

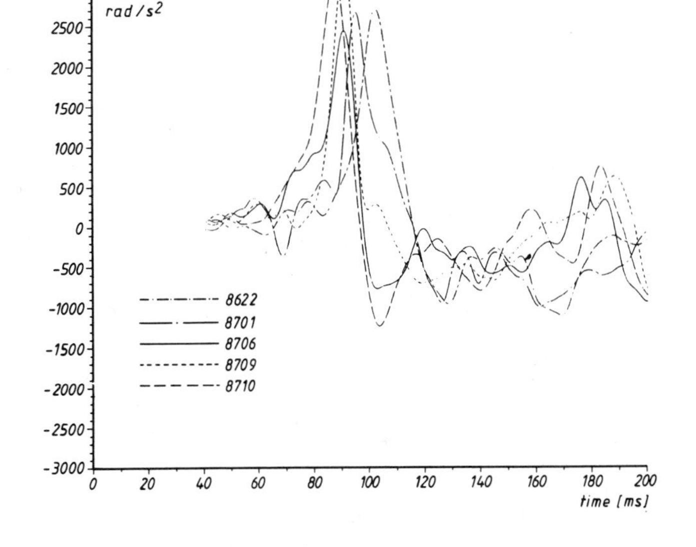

Fig. 10 Angular acceleration of the head.

LINEAR AND ANGULAR HEAD ACCELERATIONS - Fig. 9 presents head c.g. acceleration-time histories. Fig. 9 (left) presents results for two 15 g PMHS tests together with an envelope defined for the human volunteer tests. Fig. 9 (right) compares the c.g. accelerations in all PMHS tests. Fig. 10 presents in a similar way the head angular accelerations measured in the tests (i.e. the angular accelerations about the head local y-axis).

The linear head accelerations in the 15 g PMHS test are within the volunteer envelope for the first 170 ms and slightly higher for the remaining time. The linear head accelerations in the severe tests are higher than found in the 15 g tests except during the last 50 ms. This could be expected because the duration of the 23 g sled deceleration pulse is shorter.

The angular acceleration of the head for the 15 g PMHS tests appears to be well within the volunteers envelope. The first peaks are near the maximum values of the envelope. The 23 g tests show a peak of the same amplitude with a slightly smaller duration than the angular accelerations in the 15 g tests.

THE TWO-PIVOT LINKAGE MECHANISM - Results of the volunteer tests presented in previous studies (5-8) have been expressed in terms of geometrical properties and rotations of a two-pivot linkage mechanism. Fig. 11 illustrates this mechanism. This mechanism was found to be suitable for frontal, lateral as well as oblique impacts. The upper link represents the head, the middle link the neck and the lower link the torso. The upper pivot is located in the occipital condyles and the lower pivot in the center of the circular arc approximating the occipital condyle trajectories. This lower pivot is a pin joint i.e. a joint with one degree of freedom with the rotation axis perpendicular to the plane of impact. The rotation in this joint is denoted by θ and is defined as the angle between neck link and z-axis of the T1 coordinate system.

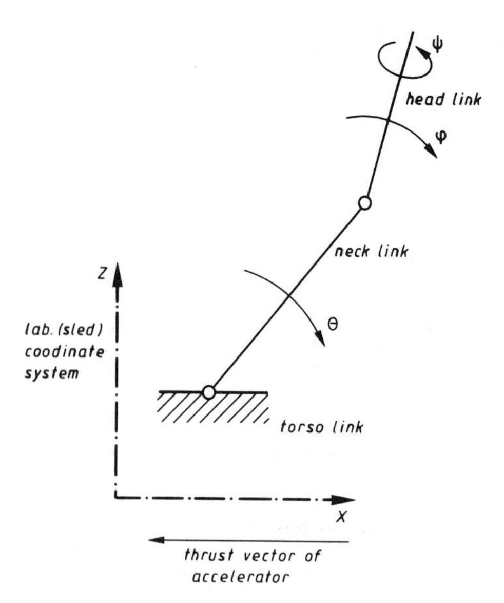

Fig. 11 Analog system for the description of the relative head motions.

The upper pivot is a joint with two degrees of freedom. The first degree of freedom allows the head link to rotate relative to the neck link in the plane of the impact. This rotation angle, denoted by ϕ, was defined in the preceding section. The second degree of freedom of this upper joint is the rotation Ψ of the head about the head anatomical z-axis indicating the head torsion or twist. In frontal impacts this twist motion can usually be neglected.

The geometrical properties of this mechanism (i.e. the neck link length and the lower pivot location relative to the torso) have been determined for the human volunteers using least squares estimation techniques (8). A neck link length of 0.129 m appeared to be a very realistic estimation for all volunteers tested and all impact directions. The same neck link length has also been applied to approximate the occipital condyle trajectories in the present PMHS tests. Table 5 shows the position of the lower pivot relative to the T1 origin, the maximal fitting error in the occipital condyle trajectory (devmax) and the residual standard deviation (sdw) corresponding to this neck link length. For reference purposes this table also includes the mean value resulting from the frontal volunteer tests.

Table 5. Lower pivot location relative to T1 origin and fitting accuracy (neck link length of 0.129 m).

	Xpivot (mm)	Zpivot (mm)	Max.fit. error (devmax) (mm)	Res.stand. deviation (sdw) (mm)
Volunt.*	-31.0	11.0	7.05	2.185
8622	-13.0	19.0	5.999	3.395
8701	-12.0	6.0	16.483	7.009
8706	- 1.0	9.0	7.440	3.737
8709	- 5.0	4.0	10.674	3.961
8710	-14.0	10.0	8.492	2.738

* average value for 46 volunteer tests (8).

The fitting error for most of the tests appears to be larger than the mean values for the fitting error in the volunteer tests. The coordinates of the lower pivot locations resulting from these numerical estimation techniques appeared to lie all within a distance of 2 cm from the T1 origin. Based on these estimations for the linkage geometrical properties the neck link rotation as function of time can be calculated for the PMHS tests.

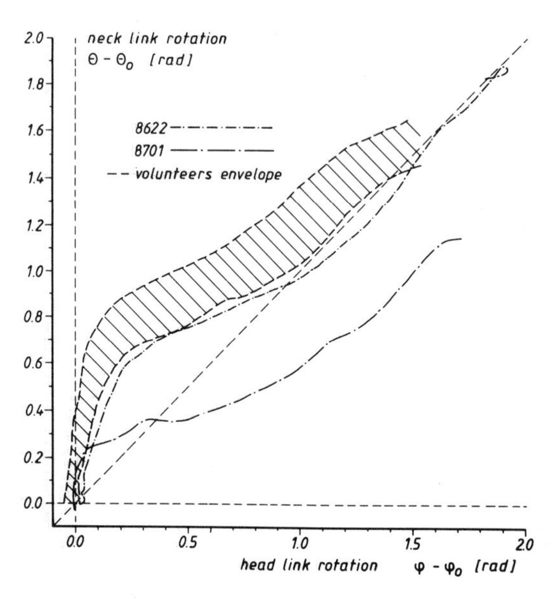

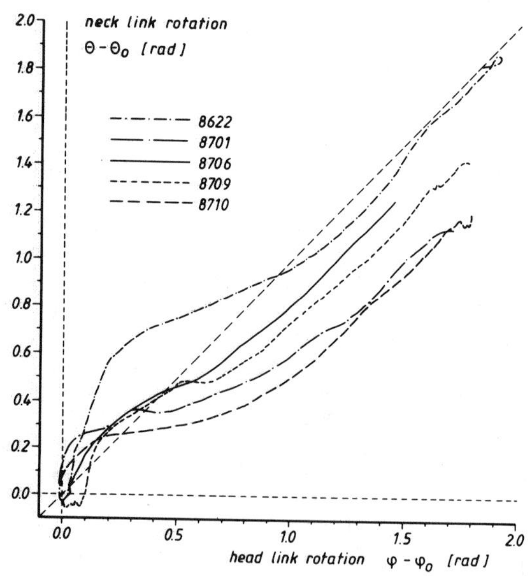

Fig. 12 Neck link rotation $\theta-\theta_0$ as function of head rotation $\phi-\phi_0$.

Fig. 12 presents the resulting neck link rotation $(\theta-\theta_0)$ as function of the head link flexion $(\phi-\phi_0)$ where θ_0 represents the initial rotation angle of the neck link. Fig. 12 (top) includes the envelope resulting from the volunteer tests.

One of the major findings in the volunteer tests was the translational nature of the initial head motion. The maximum relative rotation (i.e. $(\theta-\theta_0)-(\phi-\phi_0)$) between head and neck link appeared to be about 30 degrees in the frontal test (8). In the present PMHS test, however, this rotation is almost absent (except in test 8622). A second interesting finding from the volunteer tests was that the head flexion is smaller than the neck link rotation. In the PMHS test the opposite can be observed: Head flexions become larger than the neck link rotations in the second part of the motion. The most likely explanation for this response is the absence of muscle activity in the PMHS tests.

DISCUSSION

Twelve post-mortem human subject (PHMS) tests have been conducted at the decelerator of the University of Heidelberg. The set-up for these experiments was almost identical to the Naval Biodynamics Laboratory (NBDL) human volunteer tests, analyzed in an earlier phase of our research program.

All post-mortem human subjects were exposed to frontal impacts. In eight of the tests a similar sled pulse was applied as used for the most severe frontal human volunteer tests (15 g). In three tests a more severe pulse (23 g) was applied. Two of the 15 g tests and all three 23 g tests have been analyzed in detail in the present study. First the findings in the 15 g tests will be discussed here.

Due to the different type of sleds used at NBDL (HYGE) and the University of Heidelberg (decelerator), sled acceleration-time histories obtained in both laboratories show slight differences (Fig. 2, left). Peak sled accelerations and sled velocity change however are identical. The effect of the differences in pulse shape on subject response is small as illustrated by the T1 acceleration-time histories: resultant T1 accelerations appear to be located for a large portion within the envelope defined by the volunteer T1 accelerations (Fig. 5, left).

Horizontal T1 displacements, relative to the sled, appear to be of the same order of magnitude in the PMHS and the human volunteer tests. Vertical displacements, however, show a significant difference: in the PMHS tests a relatively large downward T1 motion (up to 9.5 cm) can be observed which is almost completely absent in the human volunteer tests. The absence of muscle activity in the post-mortem subjects seems to be the most likely explanation for this difference. In conjunction with this downward T1 motion a slightly larger T1 rotation can be observed in the PMHS tests. The human volunteer tests initially show a backward T1 rotation (about 10 degrees). This rotation appears to be absent in the PMHS tests. Sliding of the T1 mount relative to the vertebral body, due to skin compliance and interaction between instrumentation straps and restraint system, is probably responsible for this backward T1 rotation in the volunteer tests.

The two-pivot linkage analog system introduced in the previous studies to describe the characteristic head-neck motions in the human volunteer tests also appears to be adequate to describe the head-neck response in the PMHS tests. An identical neck link length (i.e. 0.129 m) as used for the human volunteer tests was selected. The fitting accuracy for the occipital condyle trajectories appeared to be slightly less than in case of the volunteer tests.

The most important finding of the present study is that the observed relative c.g. displacements in the PMHS tests are similar to the volunteer tests. Head rotations however appear to be larger in the PMHS tests. This response is clearly illustrated by the motions at the two-pivot analog system. Using this linkage concept head link rotation (i.e. head flexion) can be expressed as a function of neck link rotation (occipital condyle excursions Fig. 12). The upper pivot rotations in the PMHS tests appear to differ significantly from the human volunteer tests. The initial backward rotation in this pivot, which was about 30 degrees in the human volunteer tests appears to be much smaller in the PMHS tests. In the second phase of the volunteer head-neck motion the head flexion appears to lag behind the neck link rotation: head and neck link are almost "locked". This locking phenomenon however is absent in the PMHS tests: head link rotation becomes larger than the neck link rotation. This response which could be described by the term "overtipping" is most likely caused by the absence of muscle activity in the rear neck muscle group.

The head response as function of time in the PMHS and volunteer tests can be analyzed by a direct comparison of the head accelerations. For a large portion both the resultant linear c.g. and angular acceleration-time histories appear to fall within the envelopes defined for the human volunteer tests. In other words the observed differences in relative head motion between PMHS and human volunteer tests are not reflected in the head accelerations.

A major reason to conduct the PMHS tests is the head-neck response at higher impact levels. It follows that a more severe impact level (23 g) does not result in a significant increase of c.g. displacements and head rotations. In the NBDL volunteer tests the effect of impact severity (up to 15 g) on the head-neck response was studies extensively (8). In general larger head translations and rotations were obtained for higher impact levels. The absence of a significant increase in head excursions in the PMHS tests indicates that possibly the anatomical limits have been reached. A further increase might probably be realized by more excessive load conditions in conjunction with injuries and/or large flexions in the thoracic column.

It should be noted here that the relative head motions (i.e. head c.g. translations and head rotations) have been expressed relative to a non-rotating T1 coordinate system. In other words, a coordinate system which retains the same orientation as the laboratory coordinate system during the test. Consequently c.g. trajectories and head rotations presented here incorporate the influence of thoracic column flexibility which appears to become quite large in the 23 g PMHS tests. If relative head motions are expressed in a rotating T1 coordinate system in general smaller c.g. trajectories and head rotations will be observed. Separate performance requirements could be formulated on the basis of such an analysis. This type of requirements particularly would be beneficial for the evaluation of the performance of dummy designs in which realistic upper thoracic column flexibility has been introduced.

A full autopsy has been performed after each test. No injuries occurred in two of the 15 g and one of the 23 g tests. In seven tests strains in the vertebral discs and small lacerations were observed (AIS 1), mainly in the upper region. In two tests AIS 2 injuries were noted in the upper thoracic region (T2/T3), one test being a moderate and one test a severe impact.

Results obtained in the present study are mainly based on five PMHS tests. A more reliable validation of the results requires a greater number of tests to be analyzed. Calculation of the loads on the neck structure will be necessary to specify dynamic properties of the pivots in the two-pivot analog system. This will require a detailed analysis of the accuracy of the test methodology and will probably result in application of 3D x-ray and high speed 3D photogrammetric techniques. If results of load calculations become available an attempt could be made to correlate injuries to neck torque and shear and tension forces.

The PMHS tests conducted up to now do not effect significantly the human volunteer based performance requirements, except for the absence of the "locking mechanism", resulting in an increase in head flexion.

CONCLUSIONS

1. Twelve post-mortem human subject tests have been conducted. Five of these tests have been analyzed in detail.
2. The two-pivot analog system originally proposed to describe human volunteer head-neck motions appears to be adequate to characterize relative head-neck motion in the PMHS tests.
3. Head c.g. trajectories are of the same order of magnitude in human volunteer and post-mortem human subject test. Head rotations (flexion) however are higher in the post-mortem human subject tests.
4. Higher impact levels for the PMHS tests do not show a significant increase in relative head motions.
5. Injuries up to AIS 2 were observed.
6. Recommendations for future work include:
   - additional frontal flexion and hyperextension PMHS tests using 3D X-ray and 3D film techniques.
   - calculation of neck loads and correlation with neck injuries.
   - evaluation of existing and future dummy neck designs with respect to the findings of the PMHS tests.

ACKNOWLEDGEMENTS

This study has been supported by the Department of Transportation/National Highway Traffic Safety Administration. All opinions given in this paper are those of the authors and not necessarily those of DOT/NHTSA.

REFERENCES

1. Ewing, C.L., Thomas, D.J., Lustick, L., Williams, G.G., Muzzy III, W.H., Becker, E.B. and Jessop, M.E. (1978): "Dynamic Response of Human Primate Head and Neck to +Gy Impact Accelerations". Report DOT HS-803 058.

2. Ewing, C.L., Thomas, D.J. and Lustick, L. (1978): "Multiaxis Dynamic Response of the Human Head and Neck to Impact acceleration". Aerospace Medical Panel's Specialist's meeting. Paris, AGARD Conference Proceedings no 153. North Atlantic Treaty Organization. Advisory Group for Aerospace Research Development.

3. Ewing, C.L. and Thomas, D.J. (1973): "Human Head and Neck Response to Impact Acceleration". NAMRL Monograph 21. Naval Aerospace Medical Research Laboratory, Pensacola, Florida, 32512.

4. Ewing, C.L. and Thomas, D.J. (1973): "Torque versus Angular Displacement Response of Human Head to -Gx Impact Acceleration". In: Proceedings of the 17th Stapp Car Crash Conference.

5. Wismans, J. and Spenny, C.H. (1983): "Performance Requirements for Mechanical Necks in Lateral Flexion". In: Proceedings of the 27th Stapp Car Crash Conference. SAE Paper no. 831613.

6. Wismans, J. and Spenny, C.H. (1984): "Head-Neck Response in Frontal Flexion". In: Proceedings of the 28th Stapp Car Crash Conference. SAE Paper no. 841666.

7. Wismans, J. (1986): "Prelimanary Development Head-Neck Similator". Vol. 1: Analysis of Human Volunteer Tests. Report no. DOT HS 807034. Vehicle Research and Test Center, NHTSA.

8. Wismans, J., v. Oorschot, H. and Woltring, H.J. (1986): "Omni-Directional Human Head-Neck Response". In: P-189, 30th Stapp Car Crash Conference Proceedings, Paper 861983.

9. Padgaonkar, A.J., Krieger, K.W., King, A.I. (1976): "Measurements of Angular

Acceleration of a Rigid Body using Linear Accelerometers. Journal of Applied Mechanics.

10. Mattern, R. (1980): Wirbelsaulenverletzungen angegurter Fahrzeuginsassen bei frontal Kollisionen. Habilitationsschrift für das Fach Rechtsmedizin, Institut für Rechtsmedizin der Universtat Heidelberg, Heidelberg.

11. "The Abbreviated Injury Scale" 1980 Revision, American Association for Automotive Medicine, Morton Grove, IL 60053.

APPENDIX

Run No. 8615 DOT

Subject : Male, 50 years, body weight 80 kg, body length 176 cm
Cause of death : poisoning
Vertebral column condition : Small degenerative alterations in the cervical- and thoracic vertebral column area.
Medical findings : No injuries.

Run No. 8616 DOT

Subject : Male, 51 years, body weight 66 kg, body length 170 cm
Cause of death : poisoning acute
Vertebral column condition : Medium severe up to severe degenerative alterations in the cervical spine area; hardening of parts of the vertebral bodies near to upperplate at C6/C7.
Medical findings : Hemorrhage in the intervertebral disc C6/C7 (NCTJ1)

Run No. 8618 DOT

Subject : Female, 61 years, body weight 58 kg, body length 155 cm
Cause of death : suffocation
Vertebral column condition : Medium severe up to severe degenerative alterations in the cervical and thoracic vertebral column area. Discreet protrusion of intervertebral discs underneath the posterior longitudinal ligament in the level of C2/C3 and C3/C4 in the vertebral canal.
Medical findings : Small laceration of the ligamentum flavum Th5/Th6 (NPLJ1). Hemorrhage in the intervertebral discs Th2/Th3, Th3/Th4 dorsal (BCTJ1). Hemorrhage in the intervertebral discs Th1/Th2, Th3/Th4, Th4/Th5, Th5/Th6 ventral (BCTJ1).

Run No. 8620 DOT

Subject : Female, 51 years, body weight 78 kg, body length 165 cm
Cause of death : drown
Vertebral column condition : Moderate degree of degenerative alterations in the central thoracic spine in the sense of a reduction and exsiccation of the intervertebral discs and discreet intensity of Schmorl's node. The spongy bone appears alterated. In the cervical spine area occurred a partial fusion of the vertebral bodies 3 and 4.
Medical findings : Hemorrhage in the intervertebral disc C5/C6 dorsal (NPTJ1). Fracture of the upper front edge of the Th2 with extension to the center of the vertebral body (BAFV 2).

Run No. 8621 DOT

Subject : Female, 46 years, body weight 72 kg, body length 166 cm
Cause of death : poisoning acute
Vertebral column condition : Medium severe degenerations of intervertebral discs of the lower thoracic vertebral and upper lumbar vertebral column with protrusions underneath the posterior longitudinal ligament, reduction of the intervertebral discs and several Schmorl's nodes.
Medical findings : No injuries.

Run No. 8622 DOT

Subject : Male, 37 years, body weight 80 kg, body length 182 cm
Cause of death : poisoning acute
Vertebral column condition : Moderate up to medium severe degenerative alterations of the intervertebral discs in the lower thoracic column with exsiccation and reduction.
Medical findings : Hemorrhage in the intervertebral disc C3/C4 left dorsal (NPTJ1). Hemorrhage in the intervertebral discs C3/C4, C4/C5 ventral (NATJ1).

Run No. 8701 DOT

Subject : Female, 24 years, body weight 74 kg, body length 168 cm
Cause of death : poisoning acute
Vertebral column condition : No degenerative alterations.
Medical findings : Strain in the ligament apparatus between top of the clivus and top of C2 (NCTJ1).
Hemorrhage in the intervertebral disc C5/C6 (NPTJ1).

Run No. 8703 DOT

Subject : Male, 59 years, body weight 66 kg, body length 154 cm
Cause of death : poisoning acute
Vertebral column condition : Exsiccation of the intervertebral discs in the cervical spine region, main point C5/C6 with reduction in the dorsal segment; small protrusion of the intervertebral disc C6/C7 below the posterior longitudinal ligament.
Medical findings : Hemorrhage in the joint between base of the skull and C1 right (NRTJ1).

Run No. 8705 DOT

Subject : Male, 38 years, body weight 71 kg, body length 175 cm
Cause of death : poisoning acute
Vertebral column condition : No degenerative alterations.
Medical findings : Hemorrhage in the joint between C1 and C2 left (NLTJ1).

Run No. 8706 DOT

Subject : Male, 50 years, body weight 72 kg, body length 172 cm
Cause of death : poisoning acute
Vertebral column condition : Moderate severe degenerative alterations in the intervertebral discs of the thoracic vertebral column, increased thoracic kyphosis.
Medical findings : Hemorrhage in the intervertebral disc C4/C5 dorsal (NCTJ1). Hemorrhage in the joint between base of the skull and C1 left (NLTJ1). Hemorrhage in the muscles between base of the skull and arc of C1 (NLTM1).

Run No. 8709 DOT

Subject : Male, 27 years, body weight 74 kg, body length 170 cm
Cause of death : cardiac infarction
Vertebral column condition : No degenerative alterations.
Medical findings : No injuries

Run No. 8710 DOT

Subject : Female, 43 years, body weight 53 kg, body length 175 cm
Cause of death : suffocation
Vertebral column condition : Moderate severe degenerative alterations of all intervertebral discs of the thoracic vertebral column, increased thoracic kyphosis.
Medical findings : Laceration of the ligamentum flavum between Th2 en Th3 (BPLJ2). Fracture of the lower front edge of Th2 (BAFS2). Small hemorrhage in the intervertebral disc Th2/Th3 (BATJ1). Hemorrhage in the joint between C1 and C2 left (NLTJ1). Hemorrhage between the front area of axis and rear side of atlas (NPTJ1).

892436

Kinematic and Anatomical Analysis of the Human Cervical Spinal Column Under Axial Loading

Frank A. Pintar, Narayan Yoganandan, Anthony Sances, Jr., John Reinartz and Sanford J. Larson
Veterans Administration Medical Center

Gerald Harris
Marquette University

ABSTRACT

The patho-anatomic alterations due to vertical loading of the human cervical column were documented and correlated with biomechanical kinematic data. Seven fresh human cadaveric head-neck complexes were prepared, and six-axis load cells were placed at the proximal and distal ends of the specimens to document the gross biomechanical response. Retroreflective markers were placed on bony landmarks of vertebral bodies, articular facets, and spinous processes along the entire cervical column. Targets were also placed on the occiput and arch of C1. The localized movements of these markers were recorded using a video analyzer during the entire loading cycle. Pre-test two-dimensional, and three-dimensional computerized tomography (CT), and plane radiographs were taken. The specimens were loaded to failure using an electrohydraulic testing device at a rate of 2 mm/s. Preparations were deep frozen in the compressed state using dry ice and liquid nitrogen to preserve the deformations of the tissues as well as bony alterations. Two-dimensional CT's were done and sequential anatomic sections were taken using a cryomicrotome. These image data documented the soft and hard tissue strains. Biomechanical data (strength and localized kinematics) together with the image data correlated well with the pathology. Upper cervical injuries were observed under compression-extension modes while lower cervical injuries occurred under compression-flexion modes. Because the specimens were analyzed at failure, preserving tissue alterations, the pathology observed in the present study may provide an insight into the behavior of the cervical spine at the level of injury.

EPIDEMIOLOGICAL STUDIES INVOLVING motor vehicle accidents indicate that cervical injuries occur more often than injuries to other portions of the spinal column (35).* Furthermore, the majority of both complete and incomplete quadriplegic injuries were produced by flexion-compression loading with disruption of the vertebral bodies (35). The majority of previous studies in the cervical region has been directed to investigations of elements less than the total cervical column (2-4,12,13,16-19,22-27,29-31,34). More recently, studies have been conducted to determine mechanisms of injury to the total cervical column (1,5,6,9-11,14,15,25,32,33). The anatomical complexity of the cervical column permits motion in flexion, extension, rotation, and bending because of its material properties and architecture. Furthermore, pure motion in any direction is rarely encountered in the cervical column because of its articulation and change in the orientation of the facet joints. Therefore, the motions of the cervical column are coupled (7,8). While previous studies with segments of the cervical column are helpful, more recently, systems have become available to allow quantification of the movements of various components of the entire cervical spine with loading (28).

Because of the interest in the development of an improved anthropomorphic manikin head-neck system, a detailed analysis of the alterations of cervical column under compressive loading was undertaken in this study. Furthermore, the majority of our scientific knowledge regarding injury mechanisms of the total cervical column has been obtained from retrospective clinical or biomechanical studies designed to examine the tissues following load relaxation. This study, therefore, was conducted to examine the anatomical alterations of bony and soft tissues under failure loads. The corresponding movements of the bony elements at each level were recorded as a function of applied load, and the pre- and post-bony architecture was determined with conventional x-rays, two-dimensional (2-D) and three-dimensional (3-D) computerized tomography (CT).

MATERIALS AND METHODS

Seven fresh human cadavers were used. Table 1 includes details of the specimens used in the study. The specimens were evaluated radiographically and from their medical histories to preclude metastatic disease. The head and spine were isolated to avoid damage to the ligaments and other soft tissues. The head and neck was separated at the T2-T3 junction. Preparations were deep frozen at -40° Centigrade prior to testing. The specimens were radio-

*Numbers in parentheses designate references at end of paper.

graphed in the anteroposterior (AP) and lateral planes, and 2-D CT was done with the General Electric 9800 Scanner at overlapping 1.5 mm sections in the axial and sagittal planes. Three-dimensional CT reconstructions were obtained using the work station (Dimensional Medicine Inc., Minnetonka, MN).

Table 1: Details of Specimen Used

Specimen #	Age	Height	Weight
201	66	173	45
202	69	170	82
203	- -	173	54
204	72	170	75
205	65	178	88
206	66	157	45
207	59	173	68

One day prior to testing the specimen was removed from the freezer and thawed at room temperature and kept moist thereafter. The specimens were mounted in a specially designed test apparatus (Figure 1). The distal end of the specimen was rigidly mounted in a 75 mm diameter steel cylinder, and polymethyl-methacrylate (PMMA) was used for fixation. The cap of the head was removed with a bone saw through the frontal, temporal and occipital bones about 50 mm superior to the Frankfort plane. A 25 mm diameter aluminum rod with a 50 mm diameter spherical end was rigidly fixed in the skull with PMMA. The spherical end of the rod which protruded out of the skull was used for positioning and mounting in a custom designed socket. This socket was fixed to a six-axis Denton load cell (Denton, Rochester, NY). The load cell was fixed to a specially designed electrohydraulic Material Testing System (MTS, Minneapolis, MN) capable of applying 10,000 N of force at speeds up to 7.5 m/s. The distal end of the specimen was rigidly attached to a second identical six-axis load cell mounted on an x-y cross table. The orientation of the load cell was such that the z direction was superior-inferior and the x direction was AP, and the left to right or lateral measurements were in the y direction. The cervical spine was aligned to remove the lordosis of the column. This was achieved by preflexing the column approximately 10° to 20° prior to loading.

The tissues were prepared with retroreflective spherical pin targets 2 to 3 mm in diameter (Figure 2). These targets were securely inserted into the bony regions; at least two targets in each vertebral body, (sometimes four) one target in each facet column, and one target in the posterior aspect of the spinous process. Targets were also placed at the occiput, and the C1 arch. A total of 20 to 40 targets were used to obtain localized kinematic information of the various spinal elements along the entire cervical column during loading. Localized kinematic data was obtained with RS170 video signals and analyzed by a motion analyzer, (Motion Analysis System, CA) which allowed 2-D planar analysis in the AP (x) and the superior-inferior (z) directions. The precision/ reproducibility for 150 to 200 mm square field of view typical in the present study exceeds .08 mm and the system accuracy exceeds 0.1 mm (28).

The head-neck specimen was compressed at a quasi-static rate of 2 mm/s until a noticeable failure occurred. Failure was defined as a significant dip in the force time trace with a concomitant fracture. The deformation was maintained at this level, and a lateral x-ray was taken. A

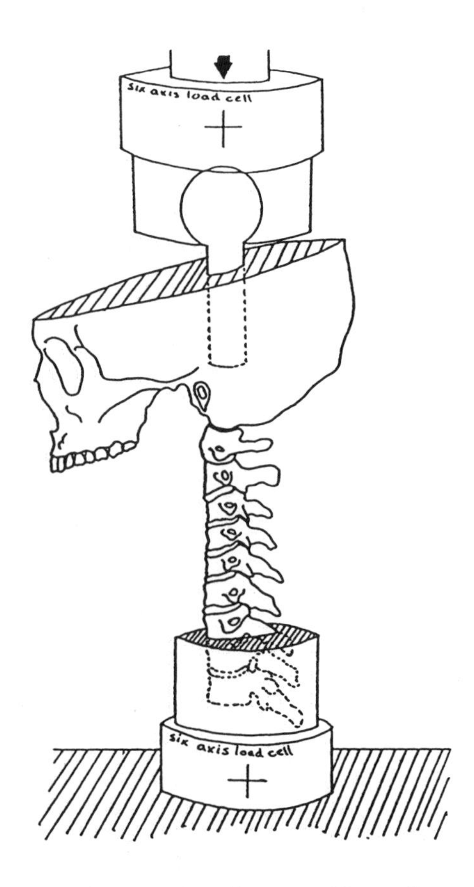

Figure 1: Schematic diagram illustrating the test preparation with six-axis load cells above and below the head-neck complex. Note the vertical orientation of the cervical spine.

styrofoam frame was constructed around the preparation to freeze the specimen in its compressed state. For this purpose dry ice and liquid nitrogen was poured around the specimen. Following this, 2-D and 3-D CT's were obtained. The entire cervical column was then sectioned with a heavy-duty cryomicrotome LKB (Broma, Sweden) capable of intervals down to 1 μm (21). Colored photographs of the tissue were taken after every 1 mm of tissue removal to determine the anatomical alterations of the soft and hard tissues.

Fourteen channels of biomechanical data were collected from the two six-axis load cells placed at the proximal and distal ends of the specimen, and from the load washer of the MTS device (Kistler Corp., Model 9352A). The axial compressive displacements were obtained from the built-in linear variable differential transformer (LVDT) attached in series with the piston. A modular data acquisition system (TransEra Corp. Provo, Utah, Model 7000) was used to gather the data as a function of time. A four-channel digital oscilloscope (Norland Corp., Madison, WI, Model 3001) provided a back up system. The generalized force histories (F_x, F_y, F_z), and moments (M_x, M_y, M_z) along with the localized two-dimensional kinematic data of the individual retroreflective targets were obtained. The pathoanatomical images were correlated to the biomechanical findings. Furthermore, the applied compressive force vs displacement response of the structure was obtained from the MTS piston records.

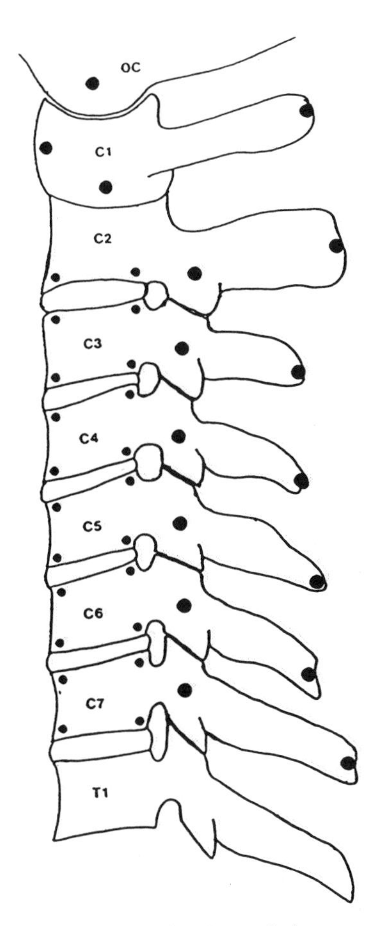

Figure 2: Lateral schematic view of the cervical column showing the placement of retroreflective targets. A maximum of six targets were used at each vertebral level of the column. Targets were rigidly fixed to the bony landmarks of the vertebrae.

RESULTS

Table 2 includes a summary of the biomechanical data and pathologic findings on each specimen. The specimens were subjected to compressive forces ranging from 1355 to 3613 N with corresponding displacements ranging from 9 to 37 mm. The axial compression force vs displacement response of all the specimens are indicated in Figure 3 . The peak represented by the point A in this figure is indicative of fracture and/or subluxation at any particular level in the preparation. For ex: In Figure 3 (for specimen #206), the decrease in loading occurring at the point A was associated with a posterior subluxation of C2 on C3 vertebra.

Figure 4 (a-g) illustrates the generalized force history data from all the specimens. The data includes the information from the proximal and distal load cells. As can be seen, there is a force balance in the three cartesian directions between the two sets of data. The maximum compressive forces recorded by the uni-axial load cell from the MTS piston was within ± 5% of the maximum compressive force (Fz) recorded by the load cell at the proximal end of the specimen (Figures 3, 4). In addition, the forces in the lateral-lateral (right to left or left to right) direction, F_y, were substantially smaller than the, AP and superior-inferior forces (F_x and F_z). The F_x shear forces were greater than the Fy forces but less than the compressive forces (F_z). The transverse moment (axial, M_z) were essentially insignificant compared to the flexion-extension moment (Sagital, M_y). Moments in the coronal plane (M_x) were also smaller than M_y. Therefore, with the axial compressive loading flexion or extension moments, often with shear, were induced on the column.

Table 3 (a-g) includes a summary of the kinematic data from all the specimens used in the study. The data indicates the compressions/distractions of the anterior and/or posterior vertebral body's discs and spinous processes at each level of the column. This data was generated by calculating the absolute distance between two adjacent retroreflective targets on every frame of the video image, and then recording the maximum change over the time sequence. For example, in specimen #206, the distraction of the C3-C4 disc (3.4 mm) shown in Table 3 (f) was obtained as the maximum change in distance (in this case elongation) between the C3 anterior-inferior target and the C4 anterior-superior target. In general, these kinematic data correlate well with the pathology (for ex: anterior longitudinal ligament tear is associated with distortion in the disc at that level).

A superposition of the pre-test and the post-test (with the specimen sustaining the maximum force) lateral projection obtained from radiographs is given in Figures 5a-5f. The occiput is compressed vertically in the superior-inferior direction by the piston of the testing device. The bottom vertebrae (T1 or T2) is rigidly fixed in PMMA. These diagrams correlate with the kinematic data and the pathlogy observed from the test. Figure 6 (a-b) illustrates the movements of the targets (in specimen 205 and 206) placed on the bony regions of the preparations. The right side of the figure corresponds to the targets placed in the vertebral bodies and the left side of the figure corresponds to those placed in the spinous processes. The middle trajectories are associated with the targets placed in the region of the facet. The closed circles indicate the initial points and the open circles the final point of each trajectory. The trajectories not terminated by an open circle are indicative of targets masked at sometime during the compression and, therefore, incomplete kinematic data was obtained. The trajectory information essentially demonstrates movements of the targets at each spinal level. The occiput compresses the preparation essentially vertically with some anterior motion secondary to the flexion-extension moment produced at the superior aspect of the preparation. The pathology observed from the CT and cryomicrotome sections are illustrated in Figures 7 (a-f). From these biomechanical data and pathological observations (Table 2), the mechanism of injury was axial compression in one (#203), compression-flexion in two (#201 and #205), and compression-extension in the remaining four specimens. All of the specimens experienced A-P shear forces, however, the lateral bending moments and lateral shear forces were small.

DISCUSSION

In this study, biomechanical strength, kinematic, and patho-anatomic (2-D CT, 3-D CT, and cryomicrotome) information were obtained from tests conducted on seven fresh human cadaveric head-neck

TABLE 2: Summary of Results

Test #	F_Z (N)	Δ (mm)	PATHOLOGIC FINDINGS: 3-D CT	2-D CT	Cryotome	MECHANISM OF INJURY
201	1355	9	Same	C2 Chip fx	Same	COMPRESSION-FLEXION
			Same	Flexion skull relative to C1 and C2	Same	
				alar ligament disrupted		
				Fx facet of C1	Same	
					Disruption of capsule C1-C2	
202	2232	30	Same	C2,C3 vert body fx	Same	COMPRESSION-EXTENSION
			Same	40% C7 ant comp fx	Same	
			Same	C2 on C3, C3 on C4 posterior disc location, facet fxs, and posterior complex disruption	Same	
					Posterior ligaments disrupted upper cervical spine	
203	1719	22	Same	Type III odontiod fx	Same	AXIAL COMPRESSION
204	2286	26	Sublux facet at C3-C4	C3 vertebral body displaced into canal; ant. long. lig and disc rupture at C3-C4; joint capsule disruption C3-C4 facet; C3 spinous process fx	Same	COMPRESSION-EXTENSION
205	3613	14	Minimal changes	Lig flavum tears at C6-C7, C7-T1	Same	COMPRESSION-FLEXION
				2mm anterior sublux C5 body on C6	Same	
206	2273	37	Same	Chip fx of ant inf body of C2	Same	COMPRESSION-EXTENSION
			Same	Complete posterior dislocation of C2 on C3, cord compromise	Same	
				Interspinous lig disruption at C7-T1 with fanning		
			Same	Mild comp fx C7	Same	
207	2597	34	Same	Chip fx of ant inf body of C2	Same	COMPRESSION-EXTENSION
			Same	Complete posterior dislocation of C2 on C3, cord compromise		
			Same	Complete ant dislocation of C5 on C6 with cord compromise	Same	
			Same	Compression fx of C7	Same	

Note: F_Z: SI load at failure
Δ: Deformation of the actuator at failure load
M_{yt}: Flexion-extension moment from the top load cell
M_{yb}: Flexion-extension moment from the bottom load cell
For M_{yt} and M_{yb} (+) values indicate flexion moments and (-) values indicate extension moments

complexes. Although many studies have been published in the literature on the biomechanical properties of the cadaveric human cervical spine, most of these focus on the physiologic motion at low level forces and on portions extracted from the neck. i.e., isolated functional units or segmental preparations (2-4,12,13,16-19,22,23,29-31,34). The failure strength and localized kinematic data were rarely reported. Because of the architecture and integrity of the human neck, it is essential to understand the behavior of the structure in its entire anatomical configuration.

Investigations have been conducted on the excised entire cadaveric cervical column (1,5,6,24-26). McElhaney et al studied the lateral and AP passive (without musculature) bending response of isolated ligamentous columns (9-11). Bending stiffness was measured using a five-axis load cell placed at the distal end of the specimen. Axial forces (F_Z) ranging from 108 to 2305 N with flexion-extension moments (M_y) ranging from 3.01 to 14.6 N.m produced wedged vertebral bodies and anterior or posterior ligament tears. In contrast, in another study, under compressive loading, with forces ranging from 960 to 5470 N McElhaney, et al reported fractured vertebral bodies in cervical column preparations (base of skull to C5, C6, C7 or T1) tested at velocities of 50 to 92 cm/s.

Hodgson and Thomas tested human cadavers with superior-inferior loading (5,6). In some cadavers, strain gauges were placed on the cervical vertebral bodies and articular facets. Soft tissues were removed from one side of the neck to examine the cervical spine kinematics. A stiffness of 1670 N/cm (maximum force = 1420 N, deflection = 12 mm) was reported under the axial compressive mode for a subject which underwent loading through a fixed swivel. This technique restrained atlanto-occipital movement. Lower cervical damage was observed. For anterior loading on the initially straight spine in which head rotation was permitted, an average stiffness of 1580 N/cm (force = 1420 N, deflection = 9 mm) was reported. Stretching of the posterior longitudinal ligament and interspinous ligaments were observed in these preparations.

Tests have also been conducted on intact human cadavers (14,15,24-26,32,33). Nusholtz et al studied the response to superior-inferior head impact using 12 fresh intact specimens. The impact tests were conducted with a free-falling moment-matched pendulum which struck a 56 kg impactor at velocities ranging from 4.6 to 5.6 m/s. The specimens were suitably oriented to study the effect of spinal configuration on the pathology induced due to impact. Seven out of 12 tests produced compression-extension injuries. One specimen did not have any injury. Of the remaining four specimens which resulted in a compression-flexion type injury, only one specimen demonstrated cervical vertebral body fracture.

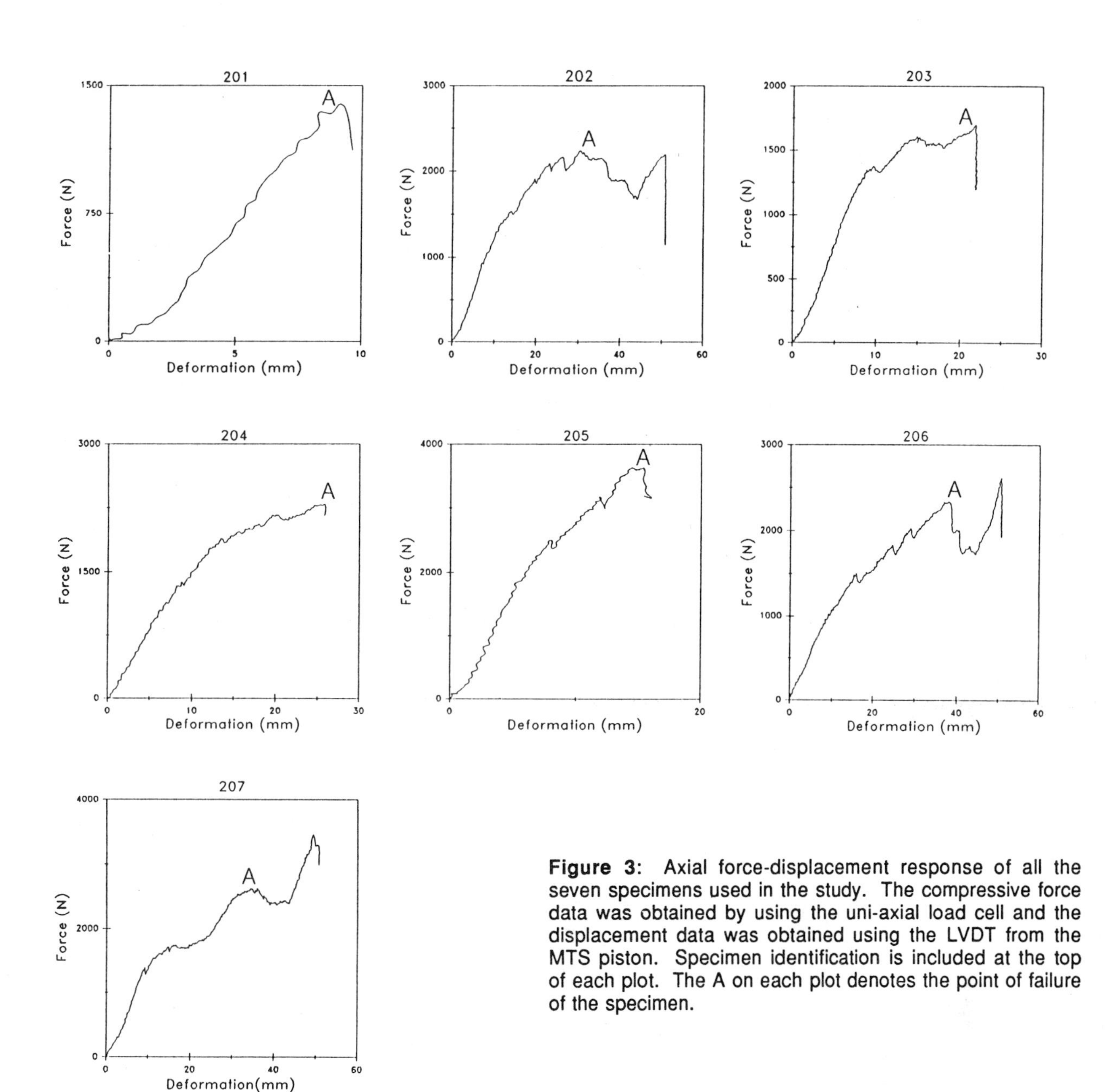

Figure 3: Axial force-displacement response of all the seven specimens used in the study. The compressive force data was obtained by using the uni-axial load cell and the displacement data was obtained using the LVDT from the MTS piston. Specimen identification is included at the top of each plot. The A on each plot denotes the point of failure of the specimen.

However, in all these four cases upper thoracic vertebral body fractures and soft tissue injuries of the cervical spine were observed. Impact forces ranged from 6000 to 11100 N. A force of 6000 N on the head produced vertebral body fractures at C5 and C6 in one specimen. Although higher impact forces at the head were recorded in the other three specimens, no cervical vertebral damage was present. This was, in part, due to the orientation of the specimen. A force of 11100 N did not result in bony trauma to the cervical spine in another specimen. In these experiments, the head angles ranged from 10° to 30°, neck angles from 5° to 25°. In contrast, in the present study, all the cervical columns were preflexed at approximately 10° to 20°.

In another study, Alem et al subjected 19 fresh human cadavers to superior-inferior impacts at velocities ranging from 6.9 to 10.9 m/s using a 10 kg force-flying mass (1). Impact forces ranged from 3000 N to 17000 N with peak accelerations varying from 45 to 160 G's. The study concluded that the peak impact force could not be used as a reliable predictor for cervical spine injury. In these studies using isolated cervical columns or intact human cadavers, the localized kinematic data and the detailed pathology at the point of failure were not obtained. However, the failure loads and deformation from the isolated cervical columns obtained in the present study compare favorably with the above cited isolated column studies (Table 4). Recent studies on isolated ligaments of cadaveric cervical spines, have shown that the tensile failure load, stiffness and energy absorbing capacity increase with increasing loading rates (34).

Recently, Yoganandan et al conducted a study to evaluate the mechanism of spinal injuries with vertical impact (32). Sixteen fresh intact human male cadavers were suspended head down and dropped vertically from a height of 0.9 to 1.5 m. In eight out of the sixteen specimens, the head were restrained to simulate the effect of muscle tone. The head-neck complexes of the specimens

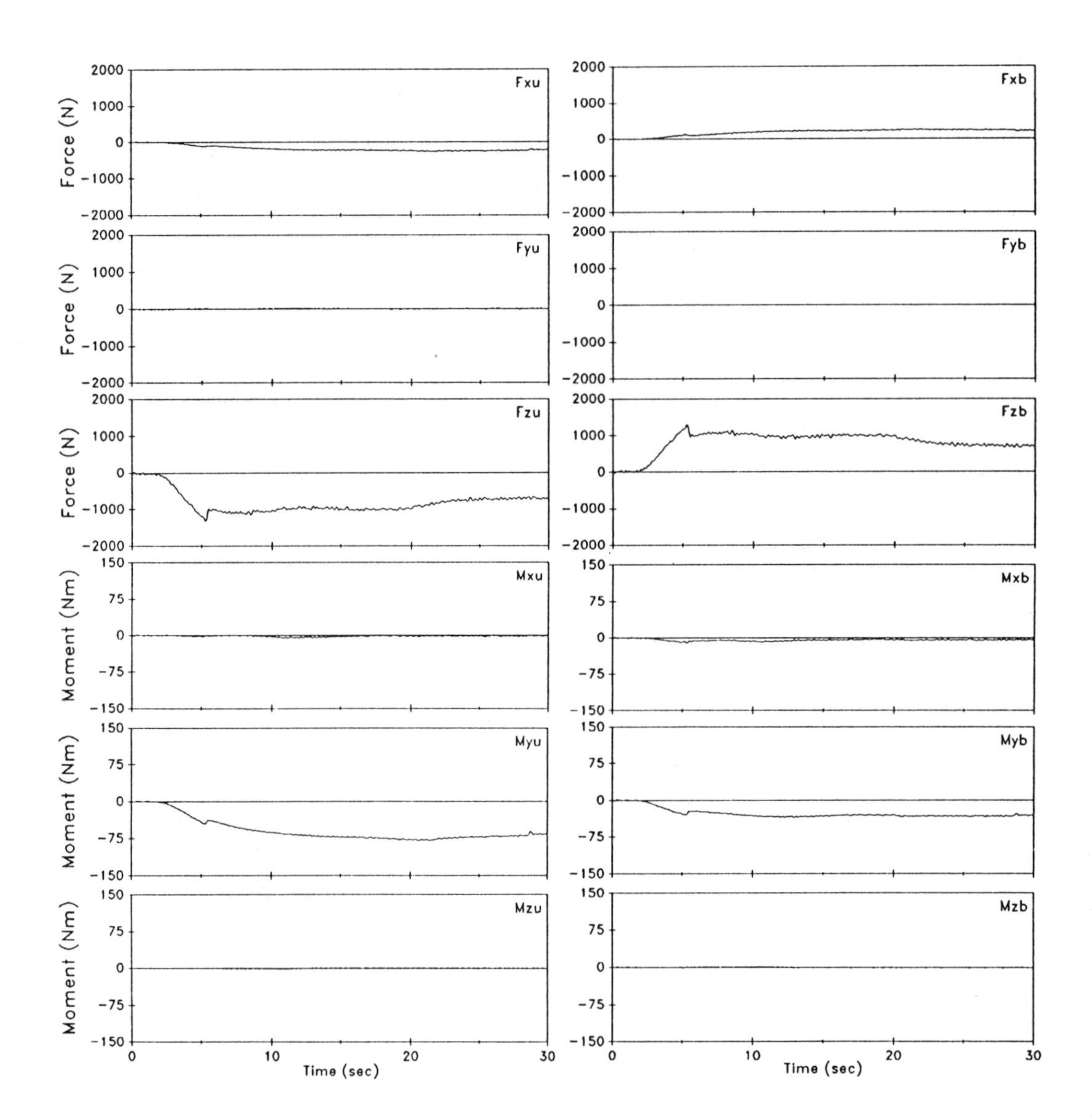

Figure 4a: Generalized force histories (Fx, Fy, Fz, Mx, My, Mz) from the proximal/upper load cell (subscript u) is shown on the left and from the distal/lower load cell (subscript b) is shown on the right. This data is for specimen #201.

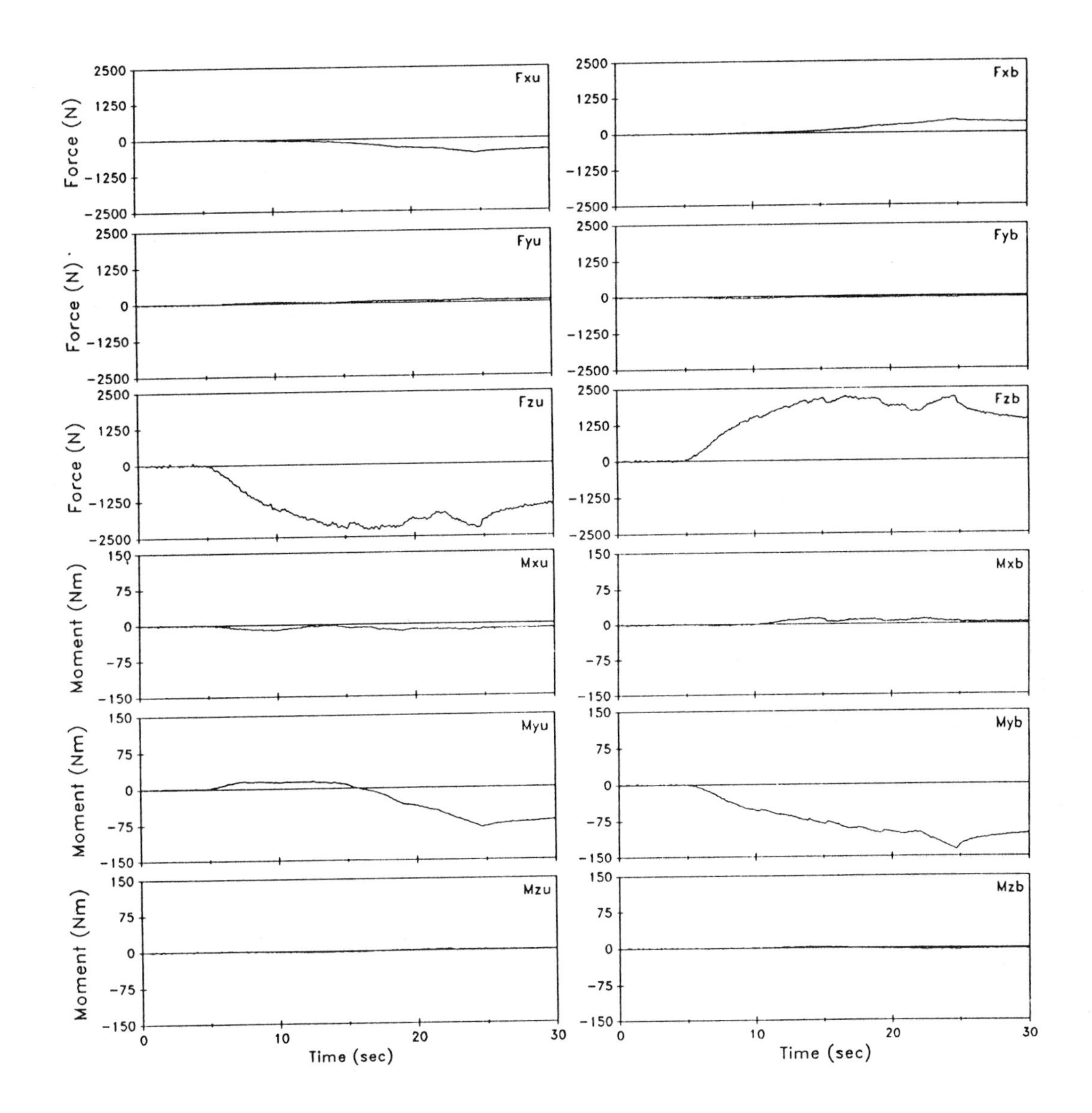

Figure 4b: Generalized force histories (Fx, Fy, Fz, Mx, My, Mz) from the proximal/upper load cell (subscript u) is shown on the left and from the distal/lower load cell (subscript b) is shown on the right. This data is for specimen #202.

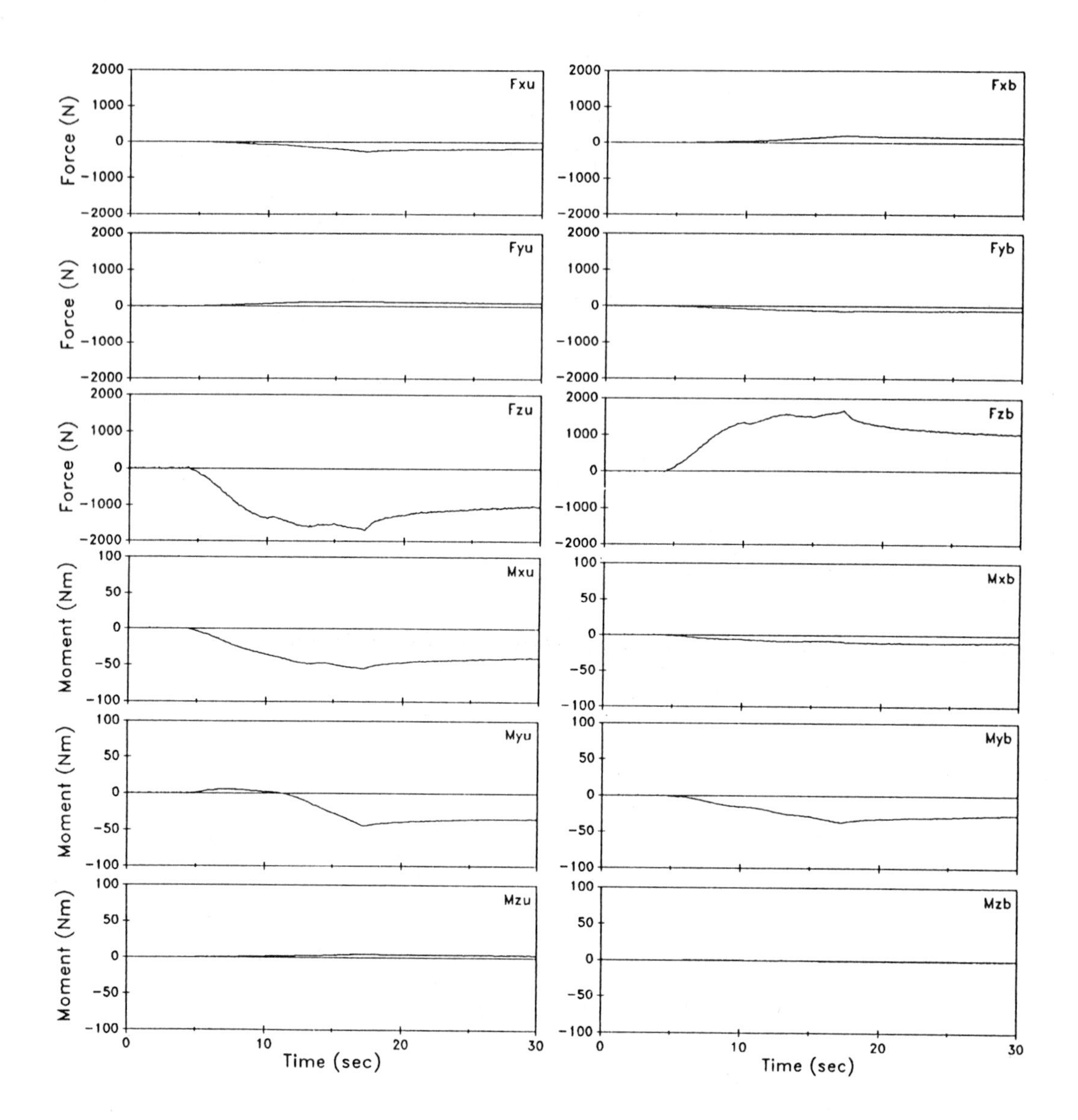

Figure 4c: Generalized force histories (Fx, Fy, Fz, Mx, My, Mz) from the proximal/upper load cell (subscript u) is shown on the left and from the distal/lower load cell (subscript b) is shown on the right. This data is for specimen #203.

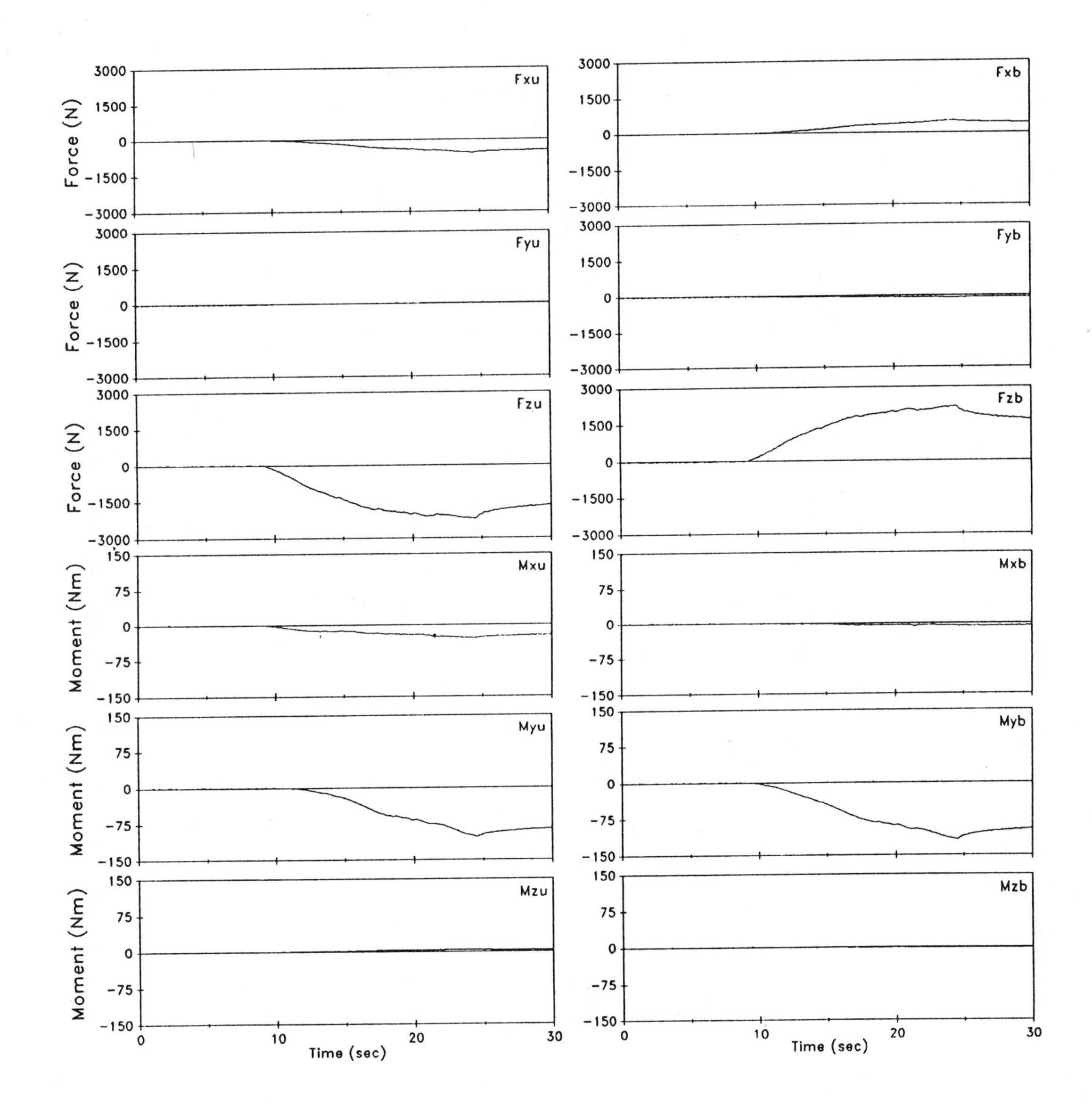

Figure 4d: Generalized force histories (Fx, Fy, Fz, Mx, My, Mz) from the proximal/upper load cell (subscript u) is shown on the left and from the distal/lower load cell (subscript b) is shown on the right. This data is for specimen #204.

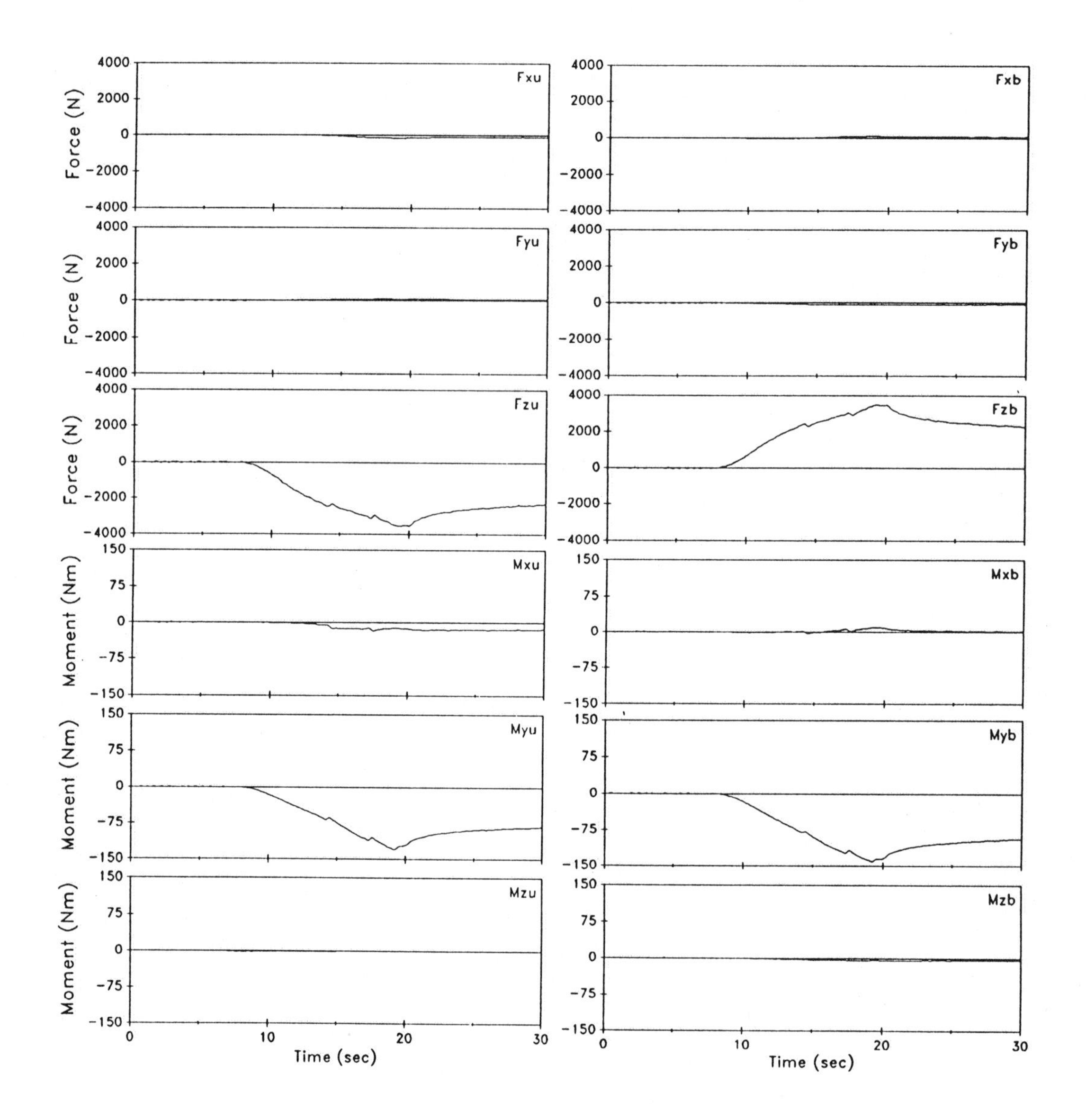

Figure 4e: Generalized force histories (Fx, Fy, Fz, Mx, My, Mz) from the proximal/upper load cell (subscript u) is shown on the left and from the distal/lower load cell (subscript b) is shown on the right. This data is for specimen #205.

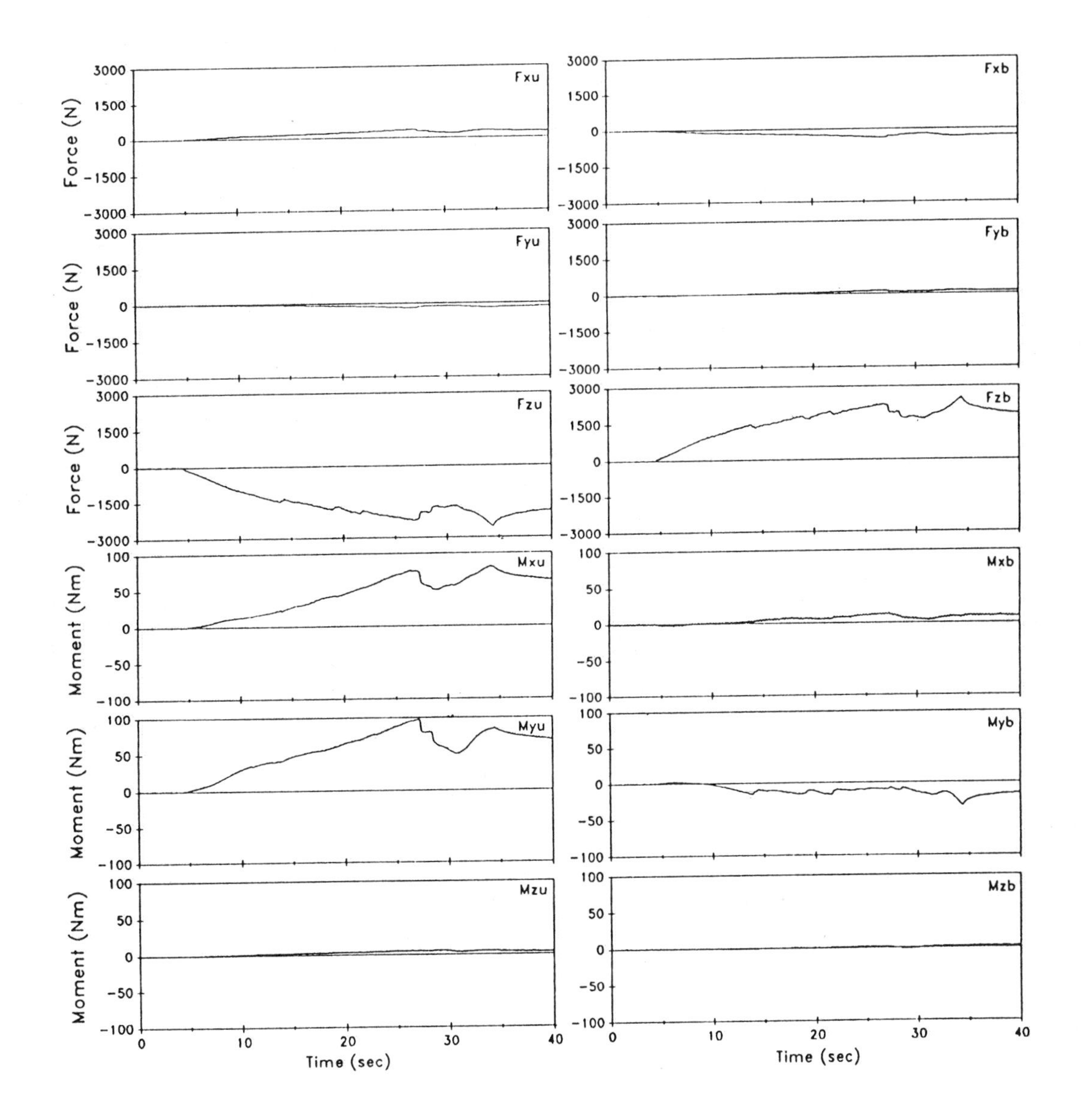

Figure 4f: Generalized force histories (Fx, Fy, Fz, Mx, My, Mz) from the proximal/upper load cell (subscript u) is shown on the left and from the distal/lower load cell (subscript b) is shown on the right. This data is for specimen #206.

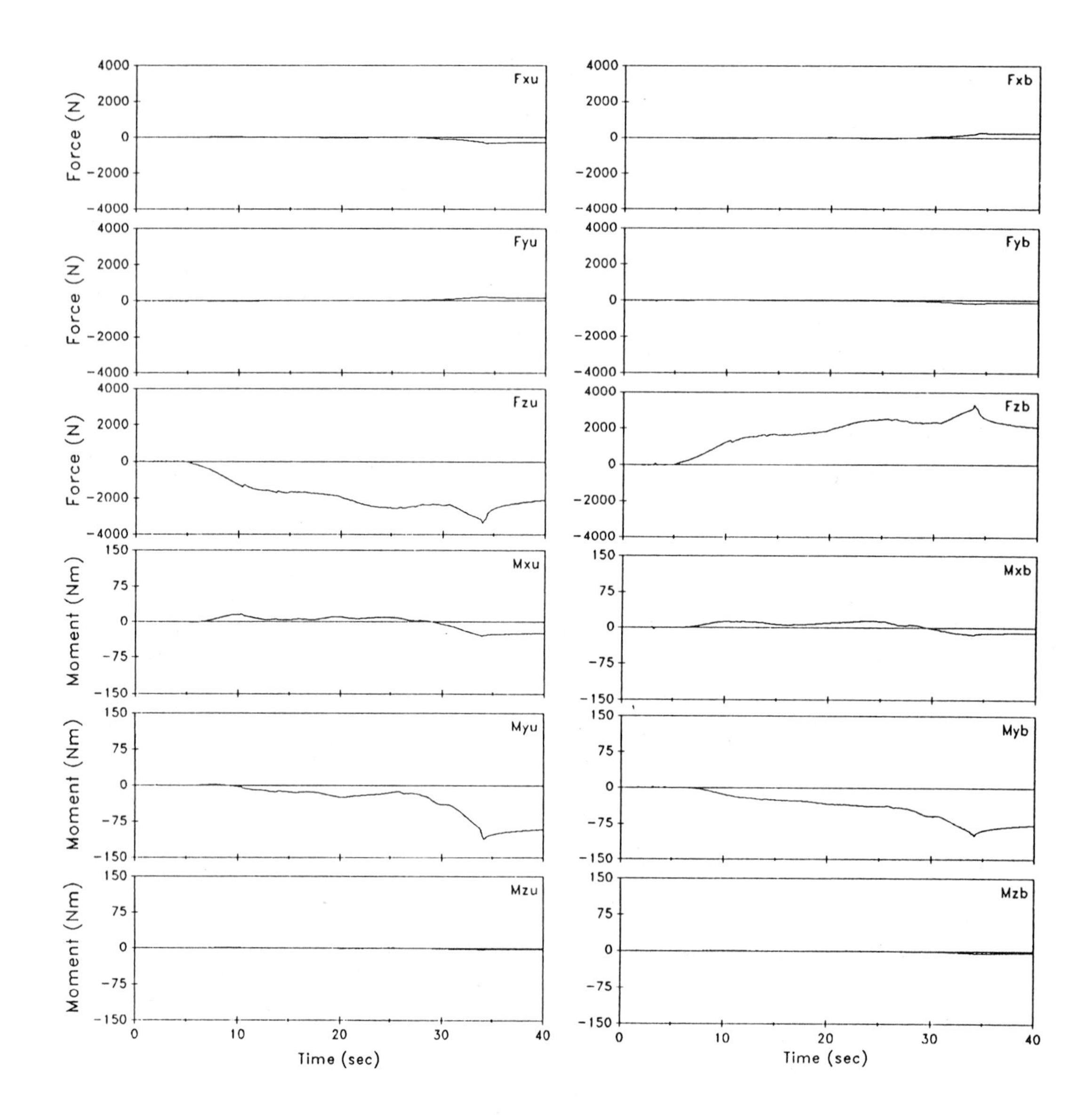

Figure 4g: Generalized force histories (Fx, Fy, Fz, Mx, My, Mz) from the proximal/upper load cell (subscript u) is shown on the left and from the distal/lower load cell (subscript b) is shown on the right. This data is for specimen #207.

Table 3a: Maximum Difference Movements (mm) of Targets on Vertebral Bodies and Spinous Processes (Specimen 201)

Element	Vertebral Body Targets	Spinous Process Targets
OC-C1	1.9	
C1-C2	9.4	-4.8
C2-C3	1.6	6.4
C3 body	2.6	
C3-C4	-1.3	2.9
C4 body	3.2	
C4-C5	-3.3	6.0
C5 body	1.3	
C5-C6	0.3	8.1
C6 body	-1.4	

Table 3b: Maximum Difference Movements (mm) of Targets on Vertebral Bodies and Spinous Processes (Specimen 202)

Element	Vertebral Body Targets	Spinous Process Targets
OC-C1		-12.6
C1-C2	7.5	12.0
C2-C3	3.8	4.1
C3 body	-6.3	
C3-C4	-2.0	6.8
C4 body	-0.3	
C4-C5	-1.8	12.3
C5 body	-1.8	
C5-C6	1.0*	2.5

*: Incomplete information: target was masked during loading

Table 3c: Maximum Difference Movements (mm) of Targets on Vertebral Bodies and Spinous Processes (Specimen 203)

Element	Vertebral Body Targets	Spinous Process Targets
C1-C2		4.6
C2-C3	0.7	1.5
C3 body	-0.6	
C3-C4	1.3	1.1
C4 body	0.7	
C4-C5	-0.8	4.2
C5 body	0.3	
C5-C6	-0.6	0.7*
C6 body	0.2*	

*: Incomplete information: target was masked during loading

Table 3d: Maximum Difference Movements (mm) of Targets on Vertebral Bodies and Spinous Processes (Specimen 204)

Element	Anterior Vertebral Body Targets	Posterior Vertebral Body Targets	Spinous Process Tartets
OC-C1			-16.3
C1-C2	1.2*		
C2-C3	-3.2	-1.7	8.5
C3 body	-0.3	-0.1	
C3-C4	-3.3*	-3.9	10.0
C4 body	-0.5	-0.2	
C4-C5	-0.9*	-1.9	8.4
C5 body	-0.5*	-1.3	
C5-C6	-0.2*	-0.7	-1.8
C6 body	0.1*	0.8*	
C6-C7	*	0.5*	*

*: Incomplete information: targets were masked during loading

Table 3e: Maximum Difference Movements (mm) of Targets on Vertebral Bodies and Spinous Processes (Specimen 205)

Element	Vertebral Body Targets	Spinous Process Targets
OC-C1		-6.3
C1-C2	-2.3*	-0.7
C2-C3	-1.1	16.5
C3 body	0.2	
C3-C4	-3.2	4.9
C4 body	-0.3	
C4-C5	-3.6	8.7
C5 body	-0.5	
C5-C6	1.3	7.8
C6 body	-0.6	
C6-C7	0.6	2.5
C7 body	0.2	
C7-T1	2.8	-1.3

*: Incomplete information: target was masked during loading

Table 3f: Maximum Difference Movements (mm) of Targets on Vertebral Bodies and Spinous Processes (Specimen 206)

Element	Anterior Vertebral Body Targets	Posterior Vertebral Body Targets	Spinous Process Targets
C2-C3 Disc	-0.2	-2.2	9.3
C3 body	0.2	0.6	
C3-C4 Disc	-3.4	-2.1	15.4
C4 Body	-0.6	0.4	
C4-C5 Disc	1.6	1.4	7.4
C5 Body	-1.5	-0.7	
C5-C6 Disc	-1.2	-1.3	9.4
C6 Body	0.6	0.5	
C6-C7 Disc	1.0	0.6	2.0

*: Incomplete information: targets were masked during loading

Table 3g: Maximum Difference Movements (mm) of Targets on Vertebral Bodies and Spinous Processes (Specimen 207)

Element	Anterior Vertebral Body Targets	Posterior Vertebral Body Targets	Spinous Process Targets
OC-C1			-1.1
C1-C2	-1.1		4.6
C2 Body	7.6		
C2-C3	-5.8	*	5.9
C3 Body	-0.3	-0.1	
C3-C4	-4.4	-3.4	8.7
C4 Body	-1.0	-0.3	
C4-C5	-2.8	-3.0	4.0
C5 Body	0.1	0.7	
C5-C6	6.1	2.8	-4.2

*: Incomplete information: targets are masked during loading

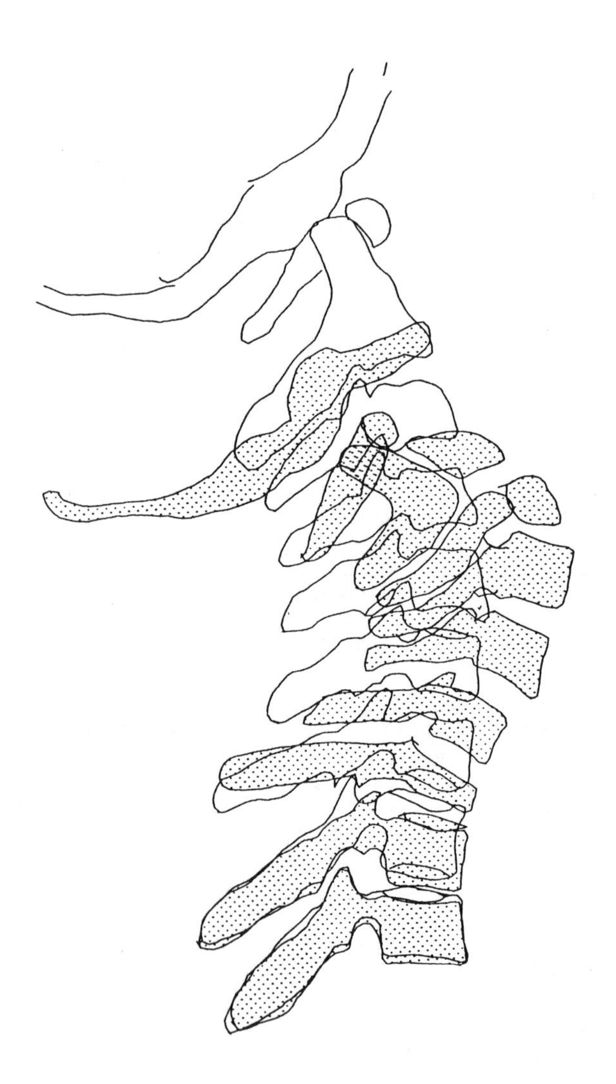

Figure 5a: A composite pre and post (shaded) lateral view of specimen #202. The images were reconstructed from x-rays of pre-test and post-test (in the compressed state). Note the deformation pattern in the 2-D plane.

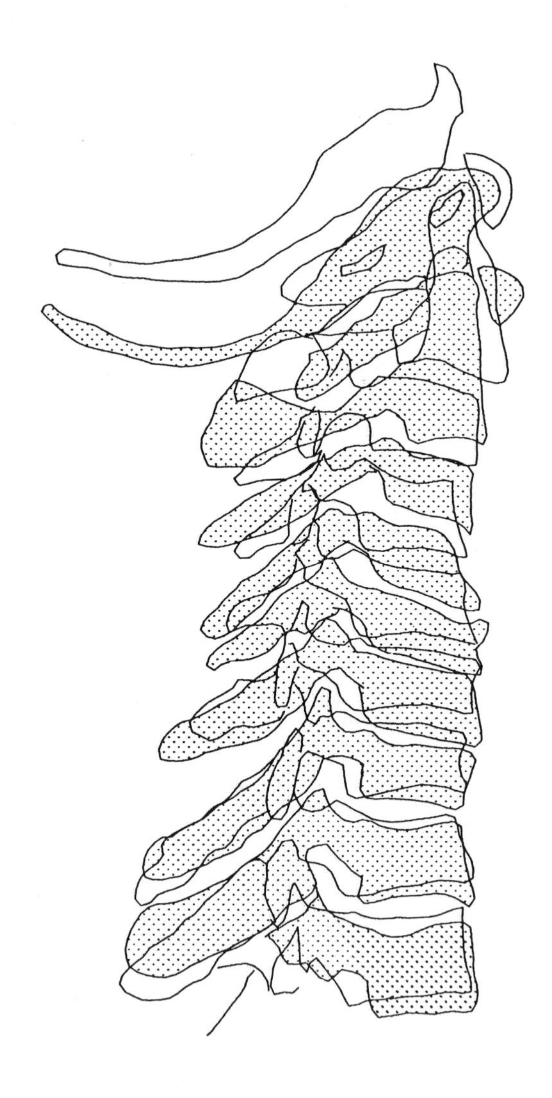

Figure 5b: A composite pre and post (shaded) lateral view of specimen #203. The images were reconstructed from x-rays of pre-test and post-test (in the compressed state). Note the minimal lateral movement of the spine indicating axial compressive mode of failure.

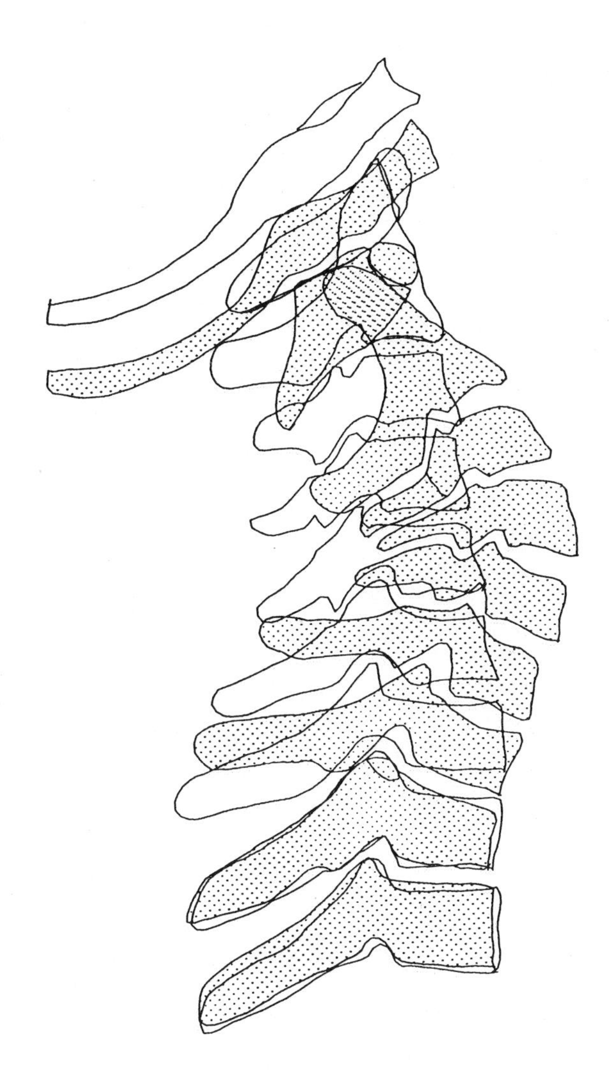

Figure 5c: A composite pre and post (shaded) lateral view of specimen #204. The images were reconstructed from x-rays of pre-test and post-test (in the compressed state). Note the compression-extension mode of injury.

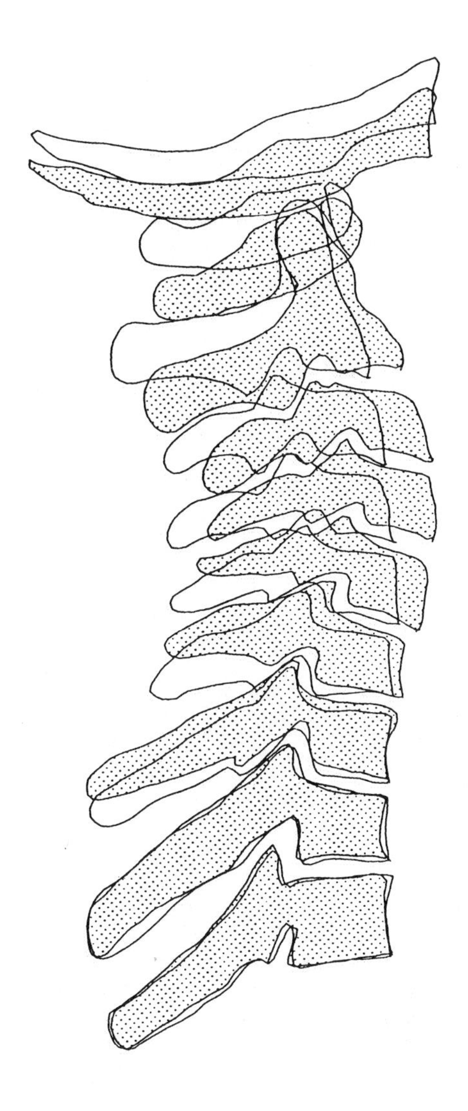

Figure 5d: A composite pre and post (shaded) lateral view of specimen #205. The images were reconstructed from x-rays of pre-test and post-test (in the compressed state). Some amount of anterior movement can be observed.

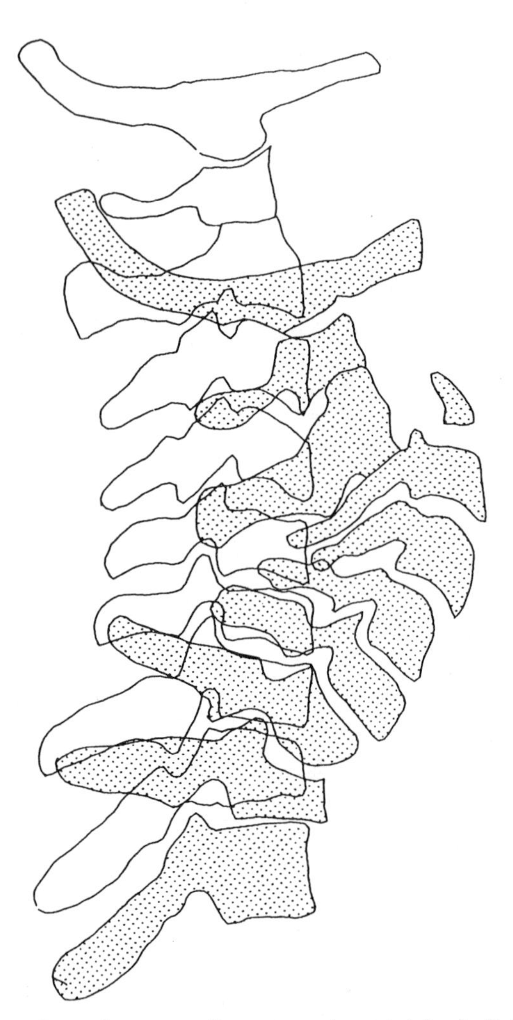

Figure 5e: A composite pre and post (shaded) lateral view of specimen #206. The images were reconstructed from the pre and post (in the compressed state) lateral radiographs. Notice that the occiput has essentially compressed the preparation in an axial mode. The chip fracture of the anterior body of C2 is observed. The cervical spine is translated anteriorly resulting in a compression-extension shear trauma.

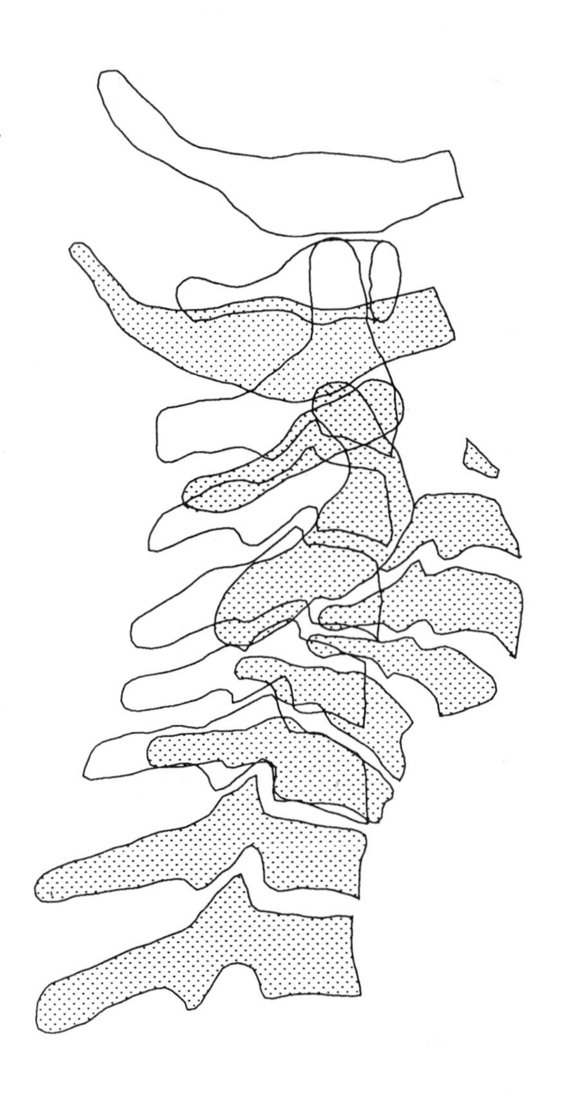

Figure 5f: A composite pre and post (shaded) lateral view of specimen #207. The images were reconstructed from the pre and post (in the compressed state) lateral radiographs. The chip fracture of the anterior body of C2 can be seen. The midcervical spine is translated anteriorly resulting in a compression-extension shear trauma.

Table 4: Vertical Failure Loads from Literature

Author/Year	Loading rate (mm/s)	Specimen	Failure load (N)	Failure deflection (mm)
Hodgson and Thomas (1981)	10.0	C spine	980-2450	9-25
McElhaney et al (1983)	1.27-640	Base of skull to C5,C6,C7 or T1	967-5470	25-45
McElhaney et al* (1988)	- -	Base of skull or C1 to T1	108-2305	- -
Present Study	2.0	C spine	1355-3612	9-37

*: These forces were accompanied by moments ranging from 3.01 to 14.6 N.m and shear forces of 0 to 35 N.

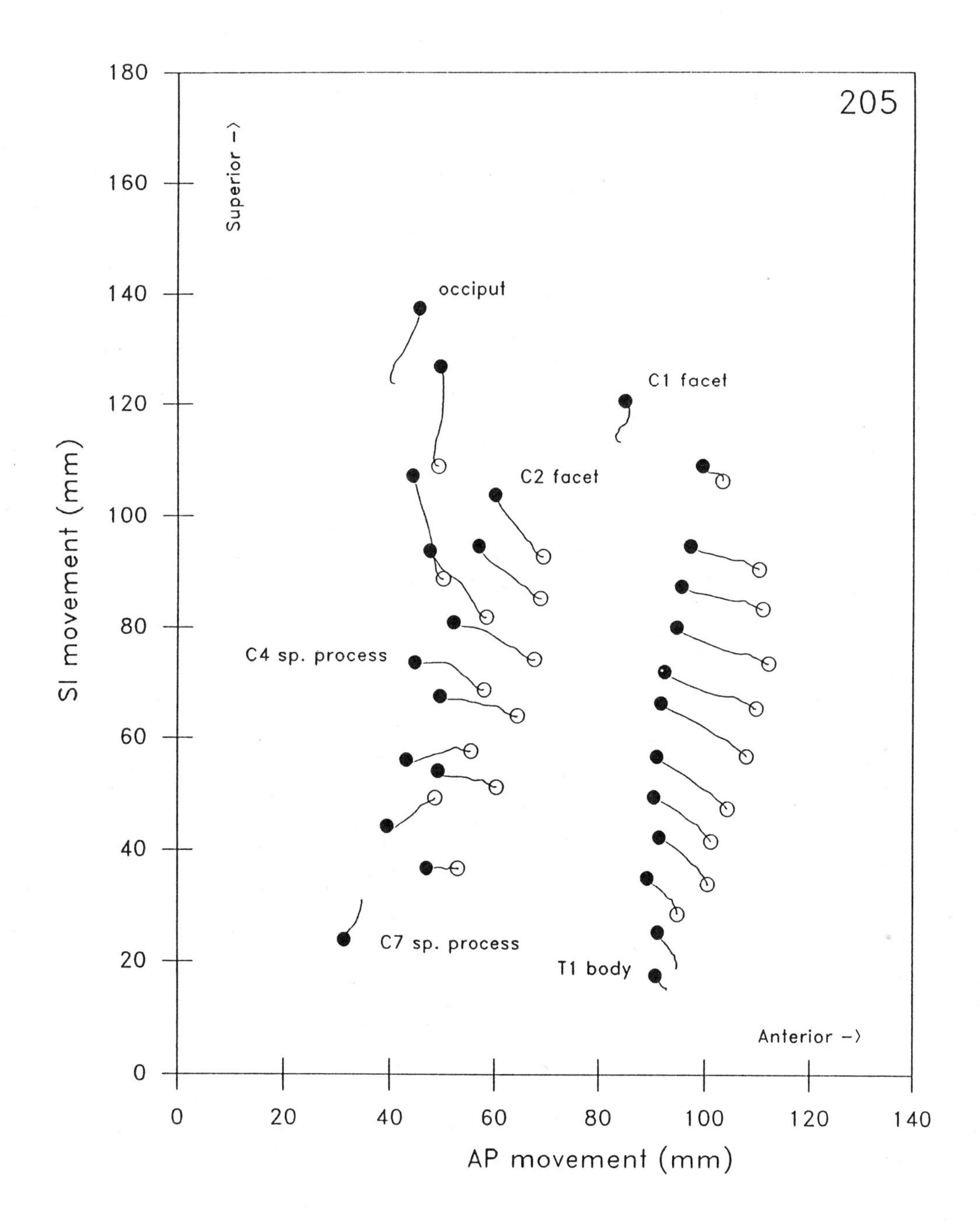

Figure 6a: Movements of the targets placed on the bony regions of specimen 205 as recorded by the anterolateral-view camera. The right side of the figure corresponds to the targets placed in the vertebral bodies and the left side of the figure corresponds to those placed in the posterior region, which is the spinous process. The middle trajectories are associated with the targets placed in the region of the facet. The closed circles indicate the initial points and the open circles the final point of each trajectory. Two vertebral body targets, 1 facet target, and 1 spinous process target were used at each level.

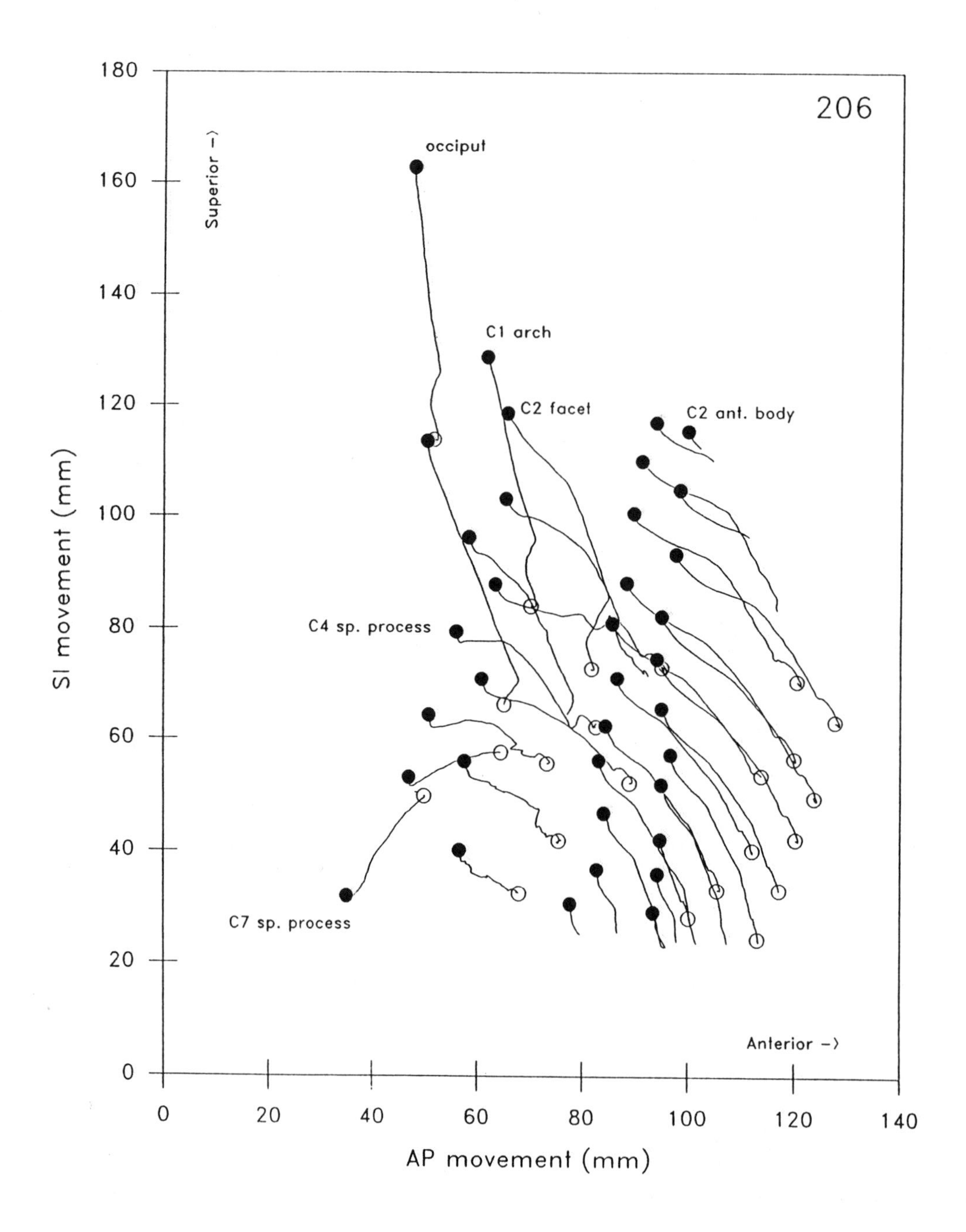

Figure 6b: Movements of the targets placed on the bony regions of specimen 206 as recorded by the anterolateral-view camera. The right side of the figure corresponds to the targets placed in the vertebral bodies and the left side of the figure corresponds to those placed in the posterior region, which is the spinous process. The middle trajectories are associated with the targets placed in the region of the facet. The closed circles indicate the initial points and the open circles the final point of each trajectory. Four vertebral body targets, 1 facet target, and 1 spinous process target were used at each level.

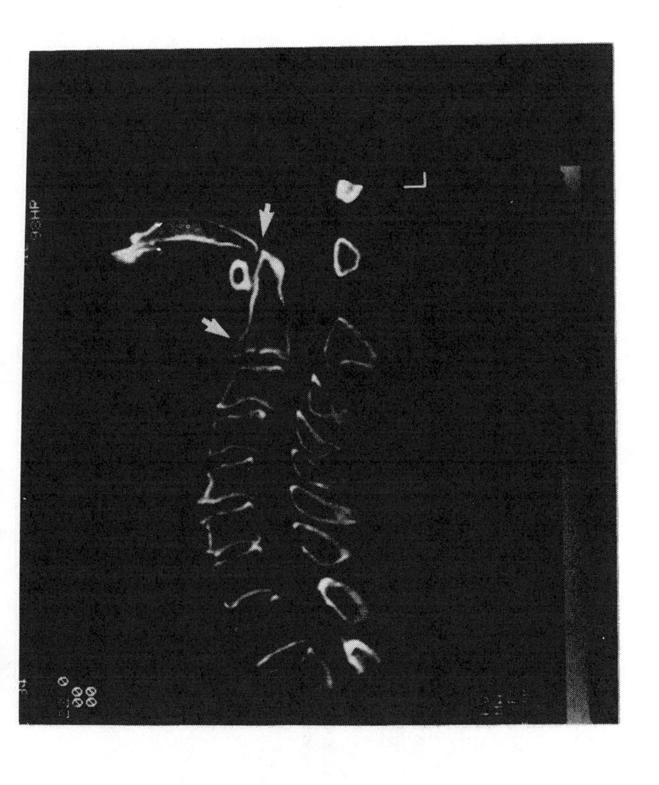

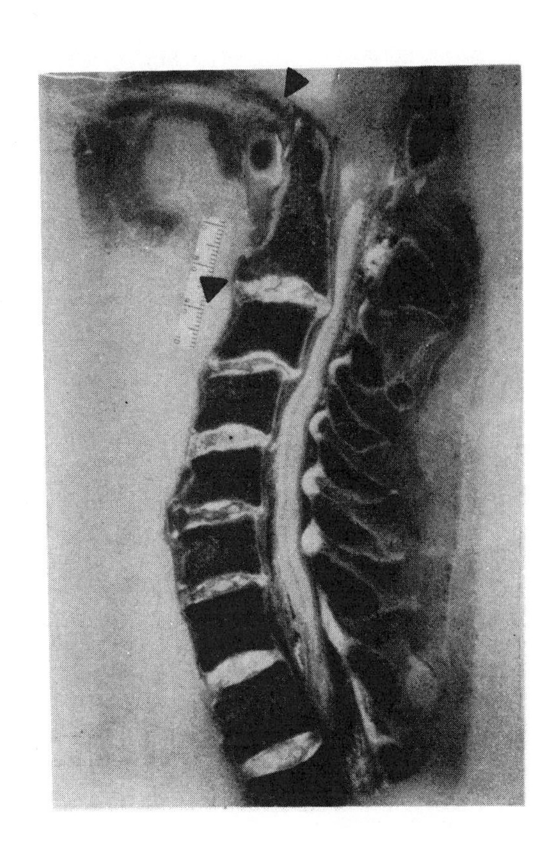

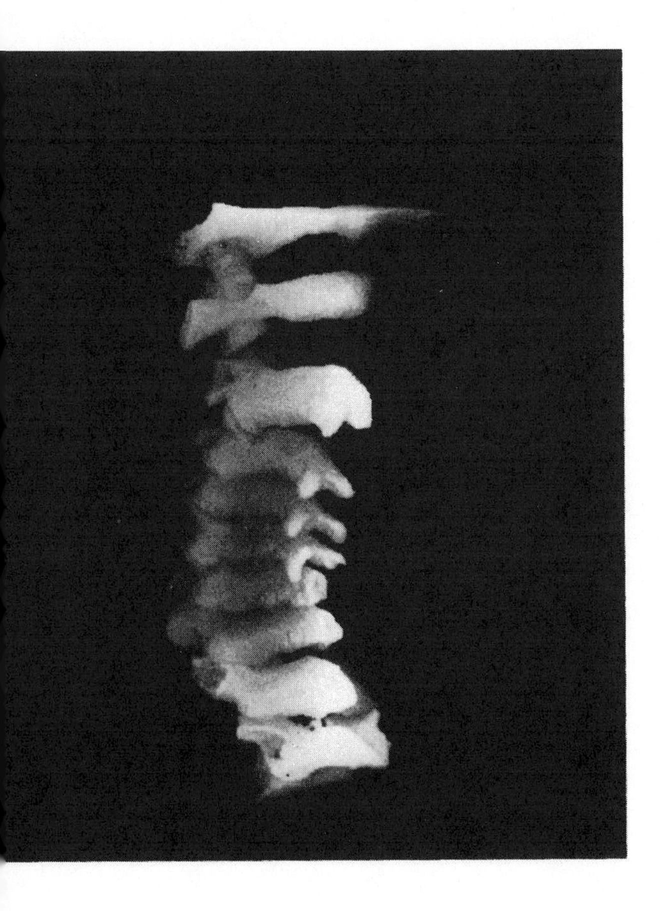

Figure 7a: Midsagittal 2-D CT (left top) and a corresponding cryomicrotome (right top) section for specimen #201. The corresponding 3-D CT reconstruction is shown (bottom). The specimen experienced a maximum force of 1355 N at a deformation of 9 mm. The anterior inferior aspect of C2 has a chip fracture seen on CT and cryotome images (arrow). The anterior region of the foramen magnum contacted the anterior tip of C2 with disruption of the alar ligament (arrow). There was also a disruption of the facet capsule at the C1-2 level, observable on the 3-D CT. This spine was injured in compression-flexion.

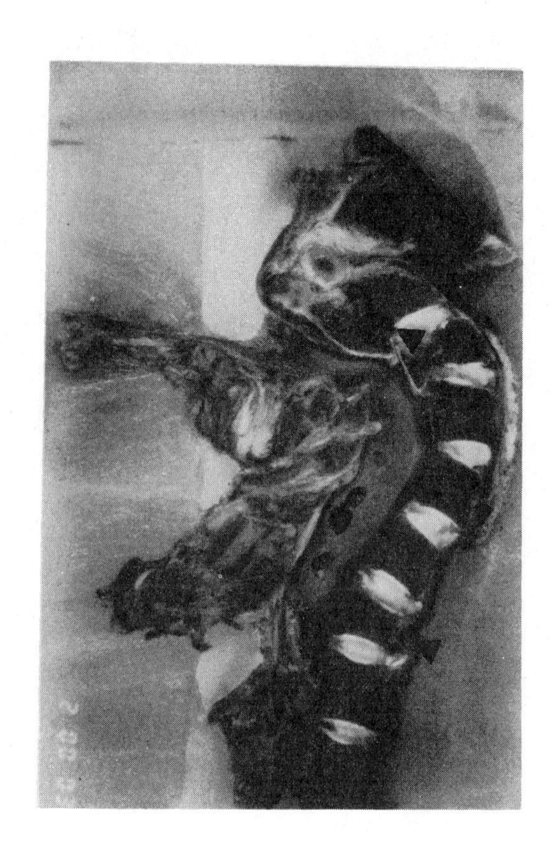

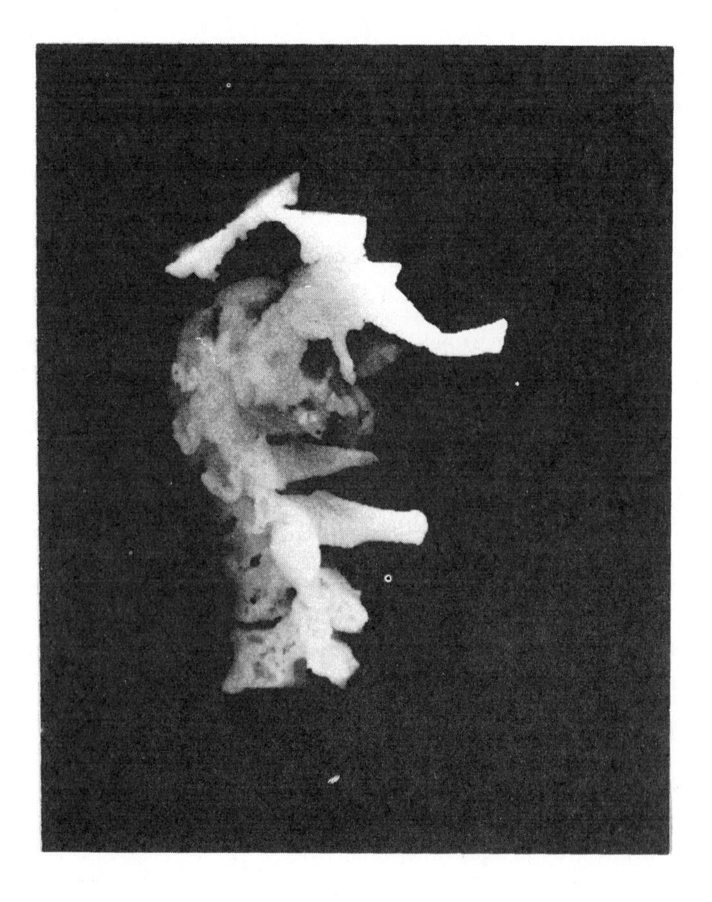

Figure 7b: Sagittal 2-D CT (left top) and the corresponding cryomicrotome image (right top) of specimen #202. C2 and C3 vertebral body fractures (arrows) on both the images as indicated. Notice the posterior longitudinal ligament tear at the C3-4 level, posterior dislocation of C2 and C3 on C4, and posterior ligament complex disruptions at the upper cervical region (on the cryomicrotome image). The mechanism of injury is compression-extension at the upper cervical level with the spinal cord compromised at this level. In the lower cervical spine, a wedge compression fracture (approximately 40%) is observed due to compression flexion. The 3-D CT reconstruction shown (bottom) illustrates fractures through the bodies of C2 and C3.

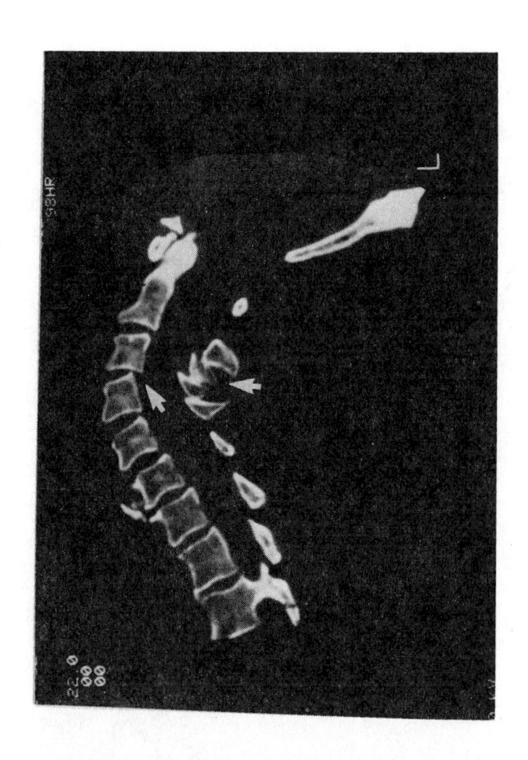

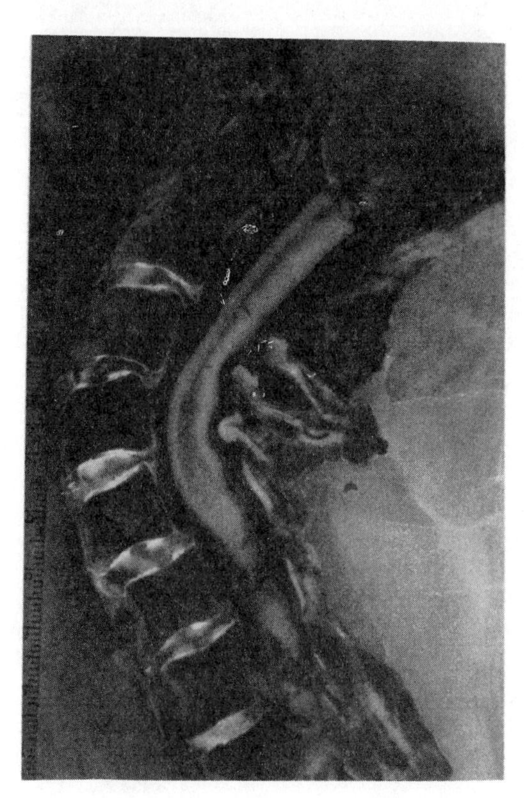

Figure 7c: 2-D CT (top) and the corresponding cryomicrtome anatomic section (bottom) of specimen #204. The C3 vertebral body is displaced into the spinal canal approximately 3 mm (arrow). There were anterior longitudinal ligament and disc ruptures at C3-4 level (arrow). The specimen also demonstrated C4-5 facet disruption and a C3 spinous process fracture (right arrow). The mechanism of injury is compression-extension.

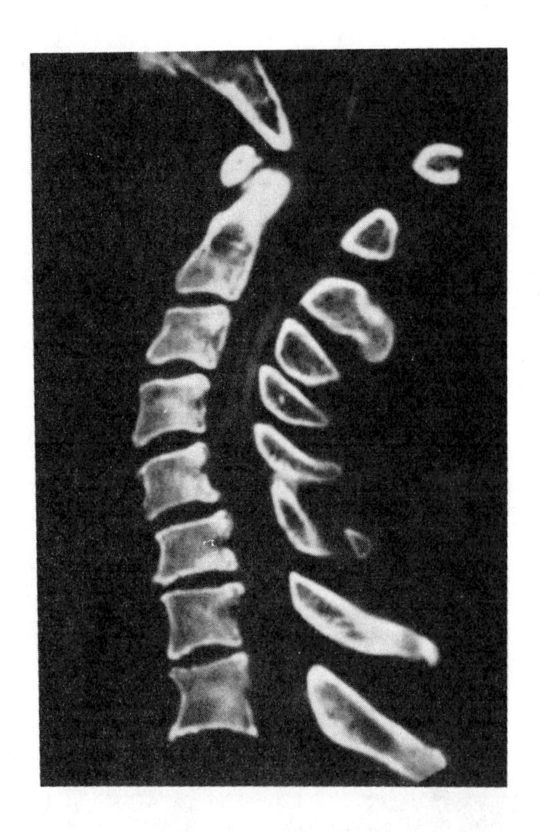

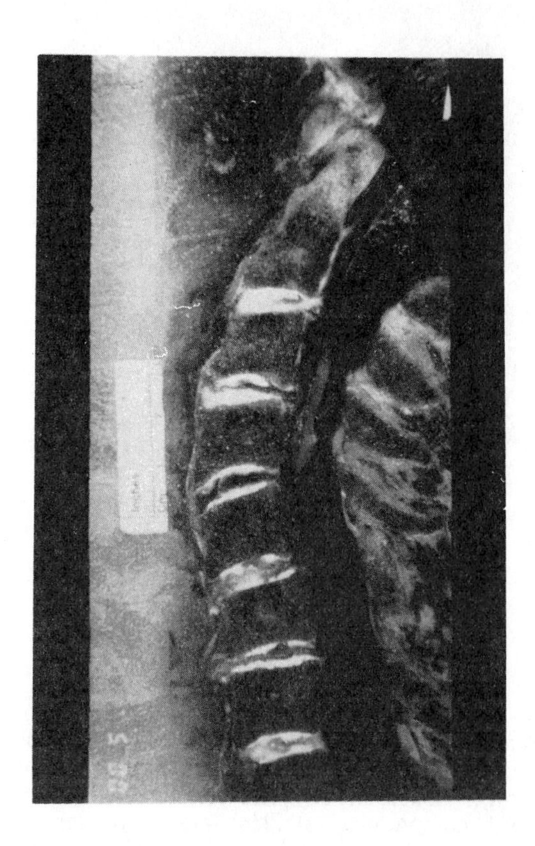

Figure 7d: A midsagittal 2-D CT (top) and cryomicrotome (bottom) image of specimen #205. Notice the ligamentum flavem tears at C6-7 and C7-T1).

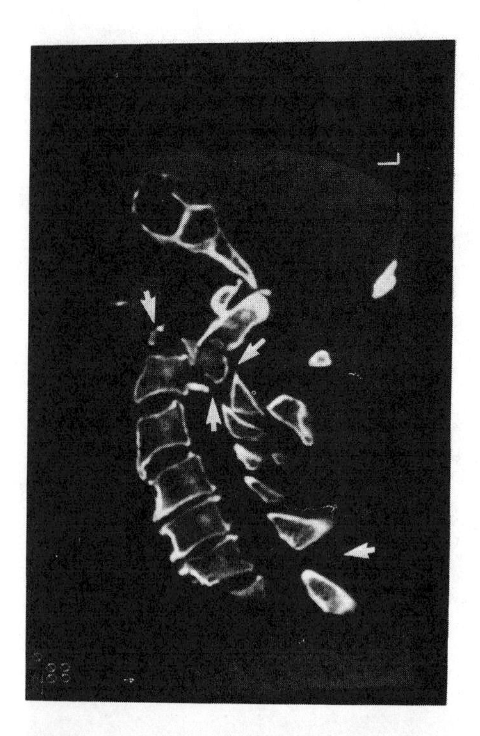

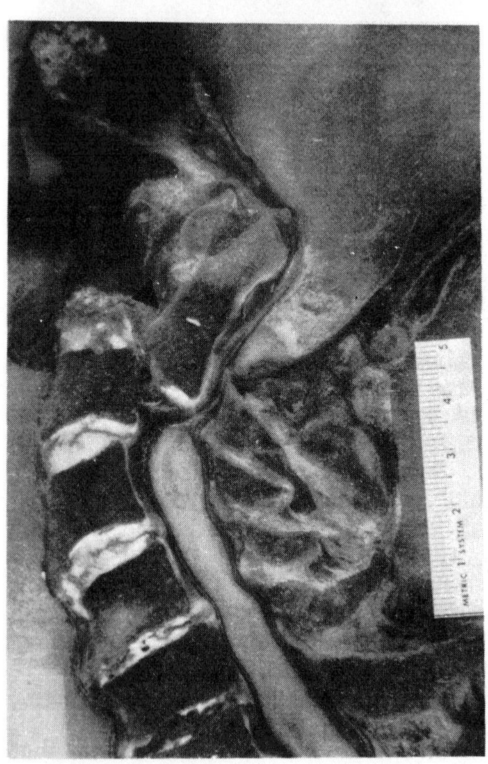

Figure 7e: Midsagittal images from 2-D CT (top) and cryomicrotome (bottom) of specimen #206. The displaced chip fracture of the anterior body of the axis is seen on the CT (arrow). Notice the complete dislocation of C2 on C3 and cord compromise at this level (arrows). The interspinous ligament at the C7-T1 level is disrupted. Fanning or spaying of the structure is observed. Compression-extension with shear is the mechanism of injury.

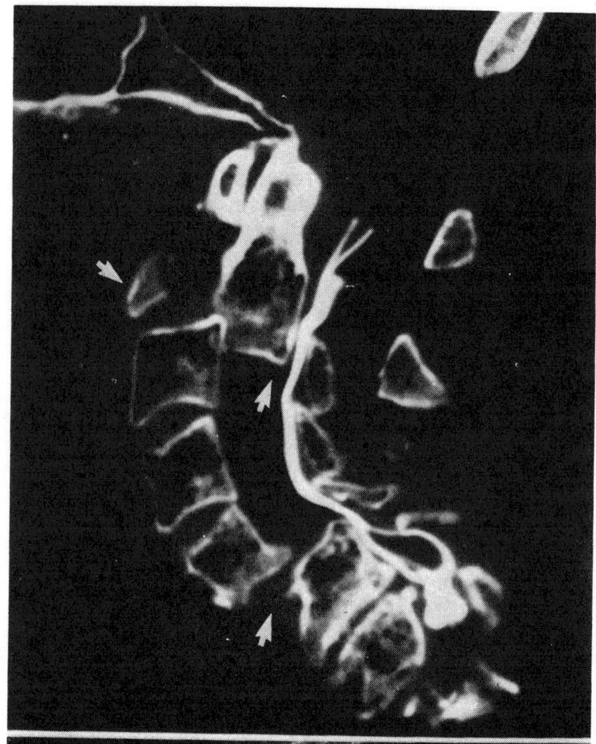

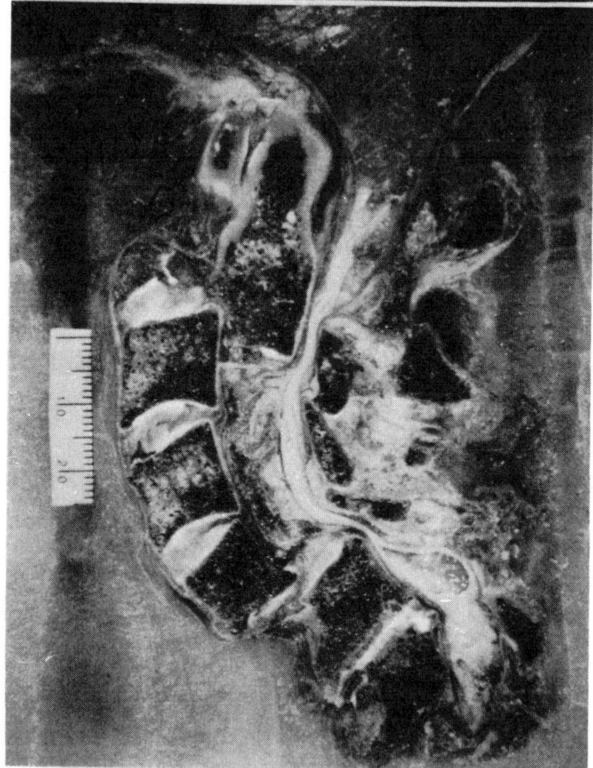

Figure 7f: Midsagittal images from 2-D CT (top) and cryomicrotome (bottom) of specimen #207. The displaced anterior chip fracture of C2 is seen on CT (arrow). Note the complete dislocation of C2 on C3 and anterior dislocation of C5 on C6 (arrows). Shear with compression-extension is the mechanism of injury.

were suitably oriented to achieve maximal axial loading of the cervical spine. Head impact forces ranged from 3000 N to 7100 N in the unrestrained and from 9800 N to 14700 N in the restrained specimens. There were more upper thoracic and cervical fractures in the restrained compared to the unrestrained case. Sixteen mm high-speed films taken at 1000 frames/s revealed that upper thoracic fractures occurred primarily due to the bending of the thoracic spine. Cervical vertebral body damage was observed most commonly when the cadavers remained in contact with the impacting surface without substantial rotation or rebound. In this study, although patho-anatomic correlations were obtained from CT and cryotome, the specimens were not frozen *in situ* at the point of failure.

This study emphasizes the soft and hard tissue alterations of the cervical spine at injury producing force levels. Because of preservation of the tissue alterations at failure, the patho-anatomic images reveal the non-relaxed injury state of the structure. Therefore, in field accident conditions, one would anticipate a different configuration and pathology for the cervical spine following trauma. Preliminary studies conducted in our laboratory have indicated that, the spine configuration secondary to axial loading are substantially different when relaxed in contrast to those shown in the present study. Consequently, retrospective studies which are used to determine the mechanisms of injury may not predict the exact position of the cervical column and its elements with respect to the spinal cord. It appears that substantial alterations and movements of the elements of the cervical column occur, and cannot be recorded in the absence of detailed analyses similar to those used in this study.

These studies indicate that substantial soft-tissue damage can occur with typical axial loading techniques which might not be evaluated with standard clinical methods. If predictable spinal alterations can be obtained using the methods of the present study, an improved extrapolation regarding cervical injuries is possible. Also, the deformations and alterations of the spinal column which occur can be used to advance an improved design for an anthropomorphic manikin.

ACKNOWLEDGMENT

This research was supported in part by PHS CDC Grant R49 CCR502508 and Veterans Administration Medical Research Funds.

REFERENCES

1. Alem NM, Nusholtz GS, Melvin JW: Head and neck response to axial impacts. Proc 28th Stapp Car Crash Conf, Society of Automotive Engineers, Warrendale, PA, 1984, pp 275-288.
2. Cusick JF, Yoganandan N, Pintar F, Myklebust J, Hussain H: Biomechanics of cervical spine facetectomy and fixation techniques. Spine 13(7):808-812, 1988.
3. Goel VK, Clark C, McGowan D, Goyal S: An *in vitro* study of the kinematics of the normal, injured and stabilized cervical spine. J Biomech 17:363-376, 1984.
4. Goel VK, Clark CR, Harris KG, Schulte KR: Kinematics of the cervical spine-effects of multiple total laminectomy and facet wiring. J Orthop Res 6:611-619, 1986.
5. Hodgson VR, Thomas LM: Mechanisms of cervical spine injury during impact to the protected head. Proc 24th Stapp Car Crash Conf, Society of Automotive Engineers, Warrendale, PA, 1980, pp 15-42.
6. Hodgson VR, Thomas LM: A model to study cervical spine injury mechanisms due to head impact. Institution of Mech Eng, London, 1980, pp 89-96.
7. Johnson RM, Hart DL, Simmons EF, Ramsby GR, Haven W, Southwick WO: Cervical orthoses. J Bone Joint Surg 59A:332-339, 1977.
8. Larson SJ: Evaluation and Treatment of Acute Cervical Spinal Cord Injury. In The Cervical Spine, 2nd Edition, HH Sherk, EJ Dunn, FJ Eismont, JW Fielding, DM, Long, K Ono, L Penning and R Raynor, eds, J.B. Lippincott Company, Publ, Philadelphia, PA, 1989, pp 496-503.
9. McElhaney JH, Roberts V, Paver J, Maxwell M: Etiology of trauma to the cervical spine. In Impact Injury of the Head and Spine, C.L. Ewing, D.J. Thomas, A. Sances, Jr, and S.J. Larson, eds, Charles C Thomas, Publ, Springfield, IL, 1983, pp 41-71.
10. McElhaney JH, Paver JG, McCrackin JH, Maxwell GM: Cervical spine compression responses. Proc 27th Stapp Car Crash Conf, Society Automotive Engineers, Warrendale, PA, 1983, pp 163-178.
11. McElhaney JH, Doherty BJ, Paver JG, Myers BS: Combined bending and axial loading responses of the human cervical spine. Proc 32nd Stapp Car Crash Conf, Society Automotive Engineers, Warrendale, PA, 1988, pp 21-28.
12. Moroney S, Schultz A, Miller J, Gunnar B, Andersson J: Load-displacement properties of lower cervical spine motion segments. J Biomech 21(9):767-779, 1988.
13. Myklebust JB, Pintar F, Yoganandan N, Cusick JF, Maiman D, Myers TJ, Sances A: Tensile strength of spinal ligaments. Spine 13(5):526-531, 1988.
14. Nusholtz GS, Melvin JW, Huelke DF, Alem NM, Blank JG: Response of the cervical spine to superior-inferior head impact. Proc 25th Stapp Car Crash Conf, Society of Automotive Engineers, Warrendale, PA, 1981, pp 197-237.
15. Nusholtz GS, Huelke DF, Lux P, Alem NM, Montalvo F: Cervical spine injury mechanisms. Proc 27th Stapp Car Crash Conf, Society of Automotive Engineers, Warrendale, PA, 1983, pp 179-198.
16. Panjabi MM, White AAIII: Basic biomechanics of the spine. Neurosurgery 7(1):76-93, 1980.
17. Panjabi MM, White AA, Johnson RM: Cervical spine mechanics as a function of transection of components. J Biomech 8:327-336, 1975.
18. Panjabi MM, White AA, Keller D, Southwick WO, friedlaender G: Stability of the cervical spine under tension. J Biomech 11:189-197, 1978.
19. Pintar FA, Myklebust JB, Yoganandan N, Maiman DJ, Sances A Jr: Biomechanics of human spinal ligaments. In Mechanisms of Head and Spine Trauma, A Sances Jr, DJ Thomas, CL Ewing, SJ Larson, F Unterharnscheidt, eds, Aloray Publisher, Goshen, New York, 1986, pp 505-527.

20. Prasad P, King AL, Ewing CL: The role of articular facets during +Gz acceleration Appl Mech, June 1974, pp 321-326
21. Rauschning W: Computed tomography and cryomicrotomy of lumbar spine specimens. A new technique for multiplanar anatomic correlation. Spine 8:170-180, 1983.
22. Raynor RR, Pugh J, Shapiro I: Cervical facetectomy and its effect on spine strength. J Neurosurg 63:278-282, 1985.
23. Raynor RB, Moskovich R, Zidel P, Pugh J: Alterations in primary and coupled neck motions after facetectomy. Neurosurgery 21(5):681-687, 1987.
24. Sances A Jr, Thomas DJ, Ewing CL, Larson SJ, Unterharnscheidt F, eds: Mechanisms of Head and Spine Trauma, Aloray Publisher, Goshen, NY, 1986, 746 pp.
25. Sances A, Jr, Myklebust JB, Weber R, et al: Head and spine injuries. Proc NATO AGARD Med Panel Specialists' Mtg, Koln, Germany, April 26-29, 1982, pp 13:1-34.
26. Sances A Jr, Myklebust JB, Maiman DJ, Larson SJ, Cusick JF, Jodat R: The biomechanics of spinal injuries. CRC Crit Rev Bioeng 11(1):1-76, 1984.
27. Sances A Jr, Yoganandan N, Myklebust JF: Biomechanics and accident investigation. In Hnadbook of Biomedical Engineering, J. Kline, ed., Academic Press, Orlando, FL, 1988, pp 525-562.
28. Walton JS: Accuracy and precision of a video-based motion analysis system. Proc 30th Intl Technical Symp Optical and Optoelectronic Applied Sciences and Engineering. Vol 693 "High-Speed Photography, Videography and Photonics IV", San Diego, CA, Aug 17-22, 1986.
29. White AA III, Panjabi MM. Clinical Biomechanics of the Spine, J.B. Lippincott, Philadelphia, 1978.
30. White AA III, Panjabi MM: The basic kinematics of the human spine: a review of past and current knowledge. Spine 3(1):12-20, 1978.
31. Yamada H: Strength of Biological Materials. Robert E. Krieger, Huntington, New York, 1973.
32. Yoganandan N, Sances A, Jr, Maiman DJ, Myklebust JB, Pech P, Larson SJ: Experimental spinal injuries with vertical impact. Spine 11(9):855-860, 1986.
33. Yoganandan N, Sances A Jr, Pintar F: Biomechanical evaluation of the axial compressive responses of the human cadaveric and manikin necks. J Biomech Eng (In Press).
34. Yoganandan N, Butler J, Pintar F, Reinartz J, Sances A Jr, Larson SJ: Dynamic response of human cervical spine ligaments. Spine (In Press).
35. Yoganandan N, Haffner M, Maiman DJ, Nichols H, Pintar FA, Jentzen J, Weinshel SS, Larson SJ, Sances A Jr: Epidemiology and injury biomechanics of motor vehicle related trauma to the human spine. Proc 33rd Stapp Car Crash Conf, Washington, D.C., Oct 5-7, 1989 (In Press).

912915

The Influence of End Condition on Human Cervical Spine Injury Mechanisms

Roger W. Nightingale, Barry S. Myers, James H. McElhaney and William J. Richardson
Duke University

Brain J. Doherty
Baylor University

ABSTRACT

The passive combined flexion and axial loading responses of the unembalmed human cervical spine were measured in a dynamic test environment. The influence of end condition (the degree of constraint imposed on the head by the contact surface) was varied to determine its effect on observed column stiffness and on failure modes of the cervical spine. Multi-axis load cells were used to completely describe the forces and moments developed in the specimen. Twenty three specimens were studied. The Hybrid III neckform performance was assessed to determine its suitability as a mechanical simulator of the neck during head impact. Changes in end condition produced significant changes in axial stiffness in both the Hybrid III neckform and the cadaver neck. The mode of injury also varied as a function of end condition in a repeatable fashion. Separation of injuries based upon imposed end condition identified groups with significantly different axial load to failure. These results also suggest that the risk of cervical injury may be strongly dependent on the degree of head constraint imposed by the contact surface, and that injury environments should be designed to minimize this constraint.

INTRODUCTION

THIS PAPER DESCRIBES the influence of imposed end condition on the structural and failure responses of the unembalmed human cadaver cervical spine in combined compression flexion loading. Implications of these results on compression based injury criteria and the design of safety equipment are also discussed.

Cervical injury remains an important social problem, especially in automotive and aircraft safety. A considerable portion of the literature has been devoted to the characterization of the responses of the cervical spine to loading in the sagittal plane (frontal impacts) and in the coronal plane (side impacts) [1-6]. The kinematics of the cervical spine have been studied in various ways, using static weight pulley, dynamic, and impact test systems [5,7,8]. However, because of the complexity of the spine, much remains unknown regarding its mechanisms of injury and methods for injury prevention.

Studies performed on the human cadaver have identified a large number of variables which influence neck injury. McElhaney *et al.* [9], noted that in fully constrained tests on isolated cervical spines, small changes in initial position of the head relative to the torso influenced injury mechanism. Nusholtz *et al.* [5], observed similar results in impacts of whole cadavers. Torg *et al.* [10], suggested that alignment of the cervical spine to remove the resting lordosis increased the ease with which the cervical spine was injured. Pintar *et al.* [1], also note the need for preflexion (i.e. removal of the resting lordosis) to create vertebral compression injuries. In contrast, Alem *et al.* [4], impacting whole cadavers noted that cervical injuries could be produced with the resting lordosis preserved.

The role of head constraint in cervical injury has also been investigated. Roaf [11], was unable to produce lower cervical ligamentous injury in unconstrained flexion. While he suggested that axial torsion was the mechanism of lower cervical ligamentous injury, subsequent work by Myers *et al.* [12], has mitigated the role of torsion in lower cervical injuries. Hodgson and Thomas [13] suggested that restriction of motion of the atlantoaxial joint greatly increased the risk of injury. Bauze and Ardran [14], produced bilateral facet dislocation in the cadaver by constraining rotation of the head, and inserting a peg in the neural foramen. Yoganandan *et al.* [15], noted that head constraint increased the measured axial load and the number of injuries in cadaver impacts. Doherty *et al.* [8], and McElhaney *et al.* [2], have shown that changes in the imposed end condition increased the observed axial and flexural stiffness of the cervical spine; and that these changes were significantly greater than those predicted by elementary beam theory. Liu and Dai [16], based on a theoretical analysis of a beam-column, suggest that the second stiffest axis may play a role in injury. Very few performance standards (restraint system or helmet) consider cervical spine protection and tolerance. The only ones we are aware of are Mertz and Patrick [17], and the industrial helmet standard ANSI Z89.

Mechanical analysis based upon the effective mass of the torso, and the energy absorption of the neck reveals that the cervical spine is capable of managing impacts equivalent to vertical drop of 0.50 meters when the neck is called upon to stop the torso [18]. Unfortunately, many impact situations have considerably larger impact energies. In these situations, the head–neck complex must either move out of the path of, or be at risk for injury from, the momentum of the impinging torso. Based on these considerations, the purpose of this paper is to examine the effect of imposed end condition, the degree of head constraint, on the observed axial stiffness of the cervical spine, and on the risk of cervical injury.

METHODS

SPECIMEN TYPES AND PROCUREMENT - Unembalmed human cervical spines were obtained shortly after death, sprayed with calcium buffered isotonic saline, sealed in plastic bags, frozen and stored at -20 degrees Celsius. Cervical spine specimens included the base of the skull through to the first thoracic vertebra (T1). All ligamentous structures were kept intact, with the exception of the ligamentum nuchae. Medical records of donors were examined to ensure that the specimens did not show evidence of serious degenerative disease, spinal disease, or other health related problems that would affect their structural responses. A total of 18 specimens were used in these experiments. Specimen ages varied from 40 to 75 years. Pretest A-P and lateral radiographs were performed prior to casting to rule out the possibility of pre-existing spinal pathology.

SPECIMEN PREPARATION - Prior to testing, each specimen was thawed at 20 degrees Celsius for 12 hours in a 100 percent relative humidity environment. The end vertebra were cleaned, dried and defatted for casting. Specimens were cast into aluminum cups with reinforced polyester resin so that the cup ends were parallel under no load. The cup centers were aligned along the center of the neural canal. Casting of the thoracic end of the specimen was performed such that the T1 vertebra was oriented at approximately 25 degrees to the horizontal plane, this preserved the resting lordosis of the spine. Screws were inserted into the T1 vertebral body and posterior elements of T1 so that the T1 vertebra only partially inserted into the resin of the lower cup. This ensured that the C7-T1 motion segment was clear of the cup surface. During casting, the aluminum cups were cooled in a flowing water bath to dissipate the heat of polymerization.

TEST INSTRUMENTATION - Tests were conducted with an MTS servo-controlled hydraulic testing machine composed of a load frame with linear actuator, a 25 gpm, 3000 psi hydraulic pump, two nitrogen filled accumulators and a linear feedback control system. Loads were measured at the caudal end of the specimen using a six axis array of strain gauge load cells composed of two GSE three-axes ATD neck load cells and a GSE torsion cell, arranged to quantify force and moment in three orthogonal axes. The caudal end of the specimen was fixed to the upper platen of the MTS. A system of linkages imposed a variety of different end conditions on the base of skull. Sagittal plane motion at the base of the skull was described using linear variable differential transformers (LVDT) and a rotational variable differential transformer (RVDT) as was required to fully describe the motion of the head relative to the fixed thoracic end for each end condition investigated. An MTS digital function generator and controller were used to apply waveforms to the linear actuator. In addition, *in situ* fluoroscopic images of the cadaver tests, and videotape images of the Hybrid III neckform [19] tests were recorded during testing.

A digital measurement and analysis system was developed utilizing a data logging computer to record the nine channels of transducer output. The multichannel microcomputer data acquisition system incorporated an RC-Electronics ISC-67 Computerscope for the digitization and storage of data. This system, which consists of a 16-channel A/D board, an external instrument interface box, and Scope Driver software, has a 1 MHz aggregate sampling rate, 12 bit resolution and writes data directly to a hard disk. Data analysis was performed on a Sun workstation.

EXPERIMENTAL METHODS - The thoracic cup was rigidly connected to the upper platen of the MTS. Three different end conditions were imposed on the head (base of skull) cup to represent the spectrum of head end conditions seen in the accident environment. The three end conditions investigated, and the motions they impose, are shown in Figure 1 and are denoted: unconstrained, rotation constraint, and full constraint. The reference axes used for these tests are spatially fixed to the centroid of the load cell, and are shown in Figure 2. The term 'axial' is used in this text to refer to loads and deflections in the 'z' (vertical) direction.

The first set of tests consisted of non-destructive testing of five specimens to determine the influence of end condition on axial stiffness. A preload equal to the weight of the head and neck was applied to each specimen and the amplifiers balanced so that the load cell showed no axial load, P, at the beginning of each test. A cyclic axial displacement using a 1 Hz haversine and the unconstrained end condition was applied for 50 cycles to exercise the specimen and place it in a mechanically stabilized (reproducible) state [9]. This was necessary so as to reduce the transient stiffening effects associated with prolonged (greater than 12 hours) inactivity [9]. A constant velocity axial displacement was then applied to the head over a 2.0 second interval for each end condition. The magnitude of the axial displacement was estimated to produce approximately 200 N of axial load. For comparison purposes axial stiffness was determined using a secant method defined by the axial deflection which occurred over an axial load range of from 0 to 200 N. Estimation of the C6-C7 motion segment flexion moment was accomplished using the initial position of the C6-C7 vertebral disk relative to the load cell and the following equation, derived from the free body diagram shown in Figure 2:

$$M_{6-7} = M_t + VB - PA \tag{1}$$

The second group of tests consisted of load to failure testing of 18 specimens. The specimens were divided amongst the unconstrained, rotational constraint and full constraint end condition. Data collected from McElhaney *et al.* [9], served to augment the full constraint end condition results. Each specimen was mechanically stabilized using the cyclic test described above, and the unconstrained end condition. To fail the specimen, a constant velocity axial displacement was performed using a 2.0 second ramp. The magnitude of the applied displacement was 9.0 cm for the unconstrained end condition, 4.0 cm for the rotational constraint end condition, and 2.0 cm for the full constraint end condition. The magnitude of the applied displacement was scaled based on the ratio specimen length against a mean specimen length of 18 cm, to give equal applied displacements as a percentage

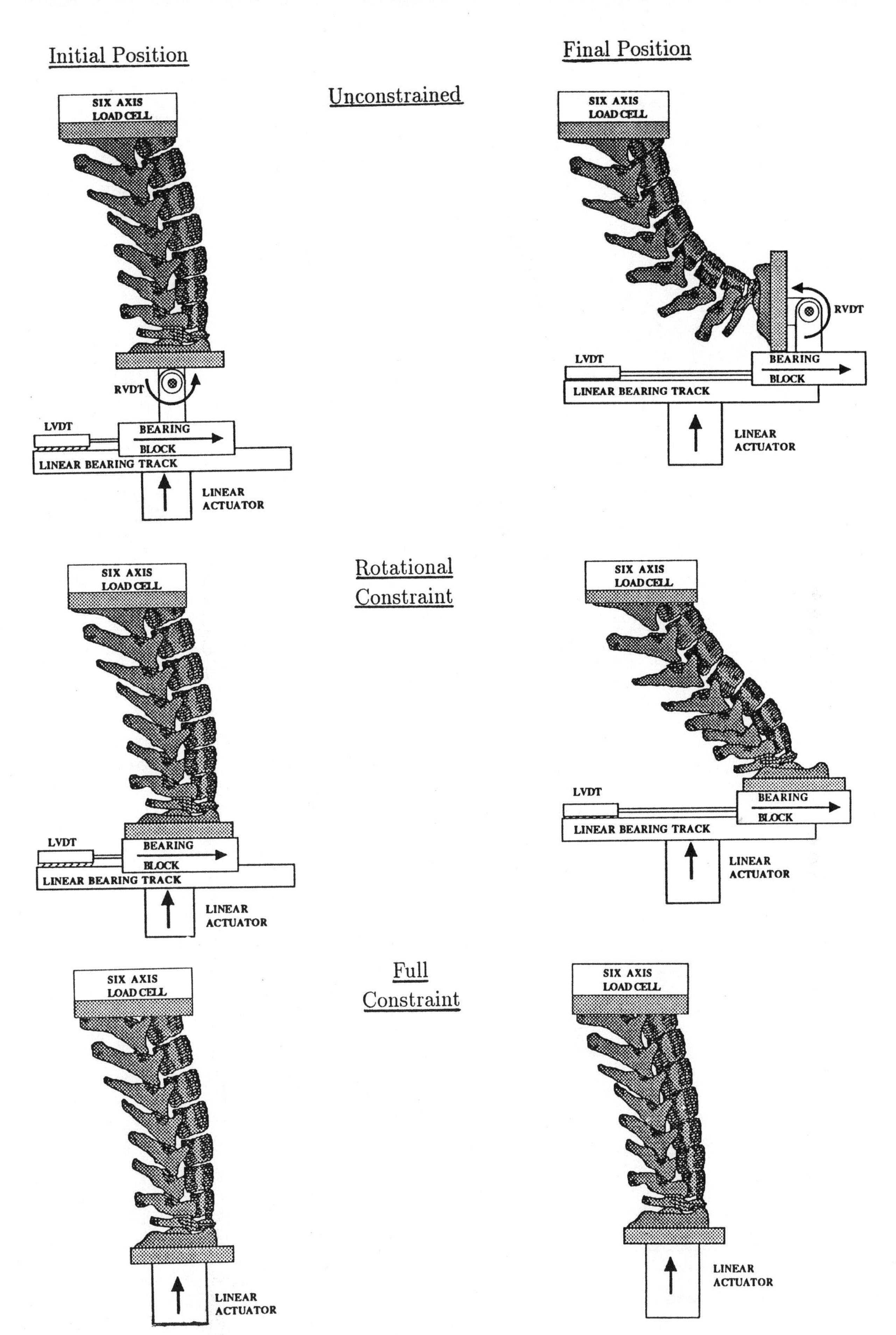

FIGURE 1. Schematics of the experimental apparatus showing the three end conditions used on the left, and the motions which result from the applied axial displacements on the right.

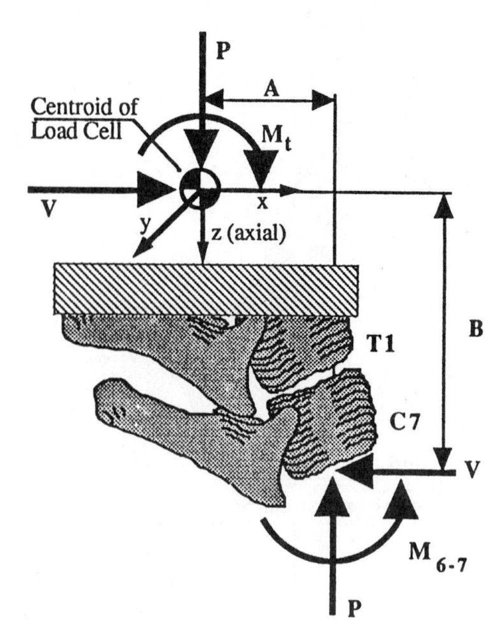

FIGURE 2. Free body diagram showing the method used for estimation of the C6–C7 bending moment from the load cell data, and the reference axes. P denotes axial compression force, V denotes a shearing force, and M denotes a flexion bending moment. A and B are the distances from the centroid of the load cell to the center of the C6-C7 intervertebral disk.

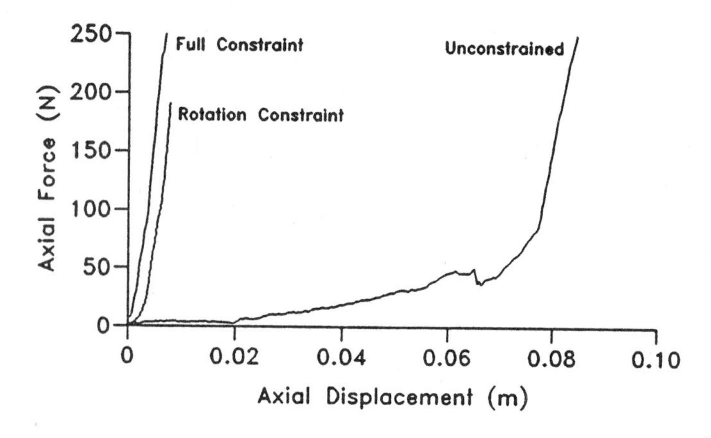

FIGURE 3. Influence of end condition on the axial load–deflection history for a typical specimen.

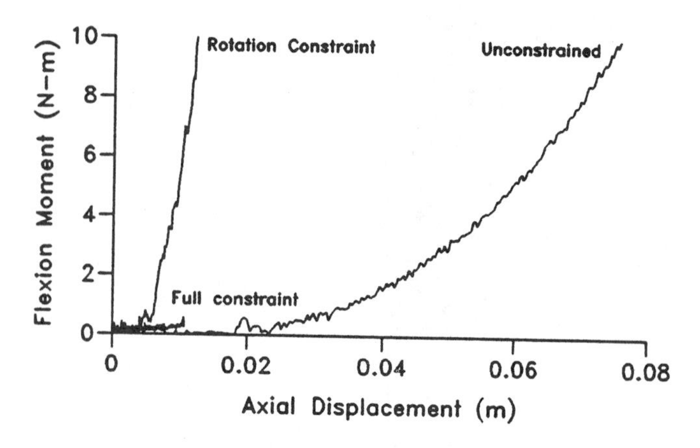

FIGURE 4. Influence of end condition on the flexion moment–axial deflection history for a typical specimen.

of specimen length for each specimen. Following loading, the specimen was removed from the actuator and failures were documented by magnetic resonance, computerized tomography, and physical dissection.

The test battery, including mechanical stabilization, and evaluation with each end condition was also performed on the Hybrid III neckform. The calculation of the flexion moment was performed at the base of the Hybrid III neckform.

TEST RESULTS

STIFFNESS TESTS – Qualitative analysis of fluoroscopic images revealed the expected kinematics of the spine associated with each end condition. Eccentricity, the effective bending moment arm for the resultant axial load, was defined as the horizontal distance from the center of the foramen magnum to the center of the C6-C7 vertebral disk. In the unconstrained group and the rotational constraint group, the applied axial displacement resulted in anterior translation of the head relative to the fixed thorax. The result was an increase in the eccentricity of the axial load. In the unconstrained group, the head was also free to rotate with applied axial deflection. In the full constraint group, no motion other than superior inferior translation was permitted. All specimens remained in the sagittal plane during axial loading, in the absence of significant constraint of out of plane motions.

Changes in axial stiffness were observed with the different end conditions. Increasing the degree of imposed constraint increased the observed axial stiffness in all five specimens. Figure 3 shows the influence of end condition on the axial load–deflection history for a typical specimen. Mean axial stiffness increased by a factor of 8.5 between unconstrained and rotational constraint end conditions, and by a factor of 12.2 between unconstrained and full constraint end conditions (Table 1).

The C6-C7 flexion moment resulting from the applied axial displacement was also affected by end condition. Unconstrained specimens had a significant "no moment" region with subsequent development of flexion moments at large displacements. Rotational constraint specimens developed large flexion moments at small axial displacements due to a moderate axial stiffness, large eccentricity, and no anterior to posterior shearing force. Full constraint specimens had small flexion moments despite a large axial stiffness, due to the development of an anterior to posterior shearing force which opposed flexion, and the small eccentricity of the resultant axial load (Figure 4). Figure 5 shows the change in eccentricity with applied axial displacement for each end condition for a typical cervical spine.

To summarize, full constraint produced large axial loads and comparatively small flexion moments when subjected to axial displacement. Rotational constraint produced moderate axial loads, and comparatively large flexion moments and the unconstrained specimens developed small axial loads and small flexion moments.

TABLE 1

THE EFFECT OF END CONDITION ON AXIAL STIFFNESS AT LOW LOADS

SPECIMEN #	UNCONSTRAINED (kN/m)	ROTATIONAL CONSTRAINT (kN/m)	FULL CONSTRAINT (kN/m)
1	4.46	54.6	95.5
2	3.93	22.4	24.3
3	2.21	27.2	26.1
4	4.21	25.6	41.0
5	1.97	12.2	18.2
Cadaver Average	3.36 ± 1.17	28.4 ± 15.8	41.0 ± 31.9
HYBRID III	169.2	336.4	416.5
HYBRID III to cadaver ratio	50.3	11.8	10.2

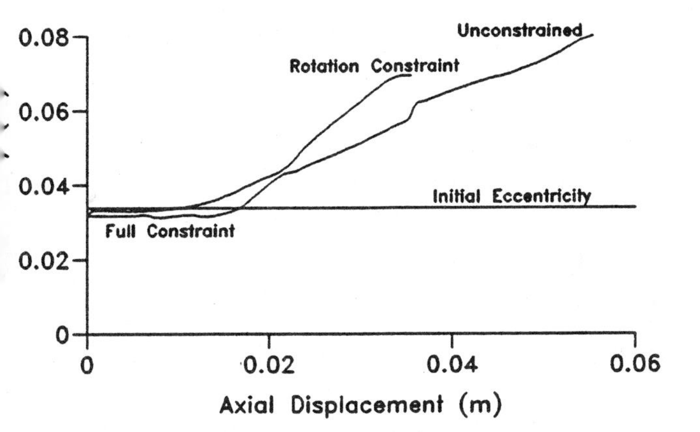

FIGURE 5. Change in eccentricity, the bending moment arm for the axial load, as a function of end condition for a typical cadaveric specimen. Note that in the full constraint end condition the eccentricity is constant.

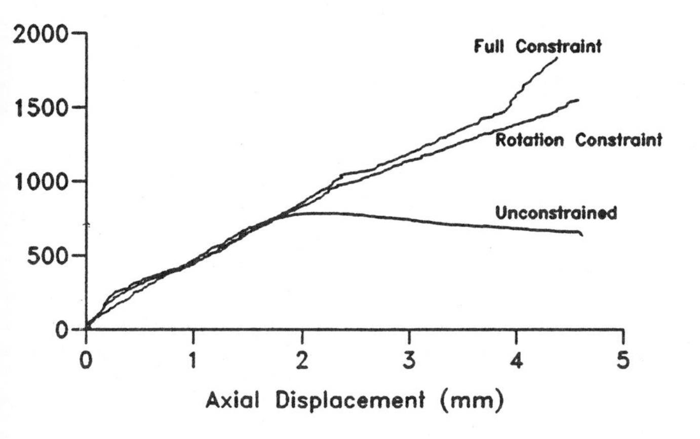

FIGURE 6. Influence of end condition on the axial load–deflection history for the Hybrid III neckform.

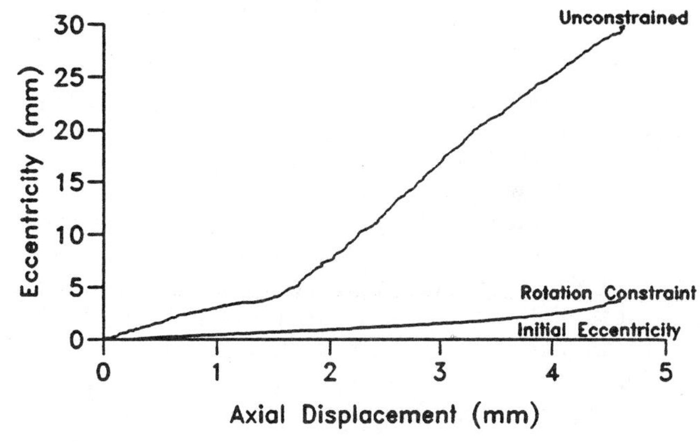

FIGURE 7. Change in eccentricity as a function of end condition for the Hybrid III neckform. The figure shows the absence of an initial eccentricity in all end conditions, and no eccentricity developing in the full constraint end condition.

The Hybrid III neckform was 10.2 to 50.3 fold stiffer than the human cadaver cervical spine depending on the imposed end condition (Table 1). The influence of end condition on the axial load–deflection history of the Hybrid III neckform is shown in Figure 6, on the eccentricity in Figure 7, and on the flexion moment in Figure 8. The unconstrained end condition developed larger flexion moments than the rotational constraint end condition, unlike the cadaver. This was due to the relatively small change in eccentricity with rotational constraint in the Hybrid III neckform as compared to the human cadaver (Figures 5 and 7).

FAILURE TESTS – End condition influenced the risk of injury, the failure mode, and the observed axial load to failure. In the full constraint group both upper and lower cervical injuries were observed. In this study only the lower cervical injuries have been included. Full constraint showed the largest axial stiffness, and the largest axial loads at failure. A mean peak axial load of 4810 ± 1290 N was measured at a mean axial displacement of 1.4 ± 0.4 cm. Compression and wedge compression injuries were produced. No dislocations or significant ligamentous injuries were observed in any of the specimens in this group.

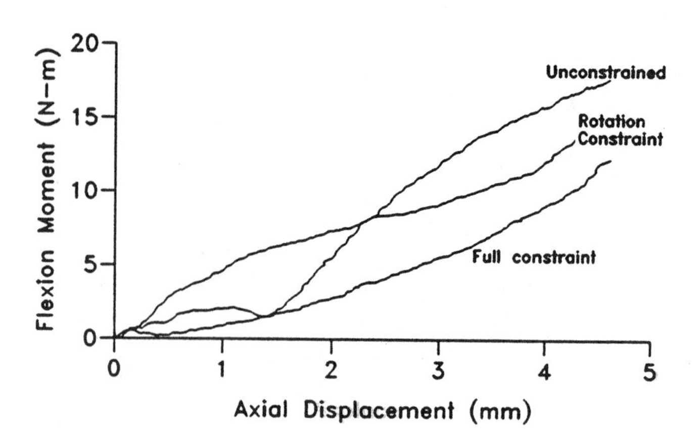

FIGURE 8. Change in flexion moment for the Hybrid III neckform as a function of end condition. Note the abrupt increase in flexion moment in the unconstrained end condition at approximately 1.5 mm of axial displacement indicative of buckling.

In the rotational constraint group, bilateral facet dislocation of the lower cervical spine was observed with posterior ligamentous and intervertebral disk disruption in each of the six specimens. No evidence of vertebral body compression injury was observed by dissection or imaging. Clear evidence of catastrophic failure was observed on the axial load–deflection data. Mean peak axial load to bilateral facet dislocation was 1720 ± 1230 N at a mean axial displacement of 2.9 ± 0.9 cm. Bending moments at failure were large due to the eccentric axial load and the absence of an anterior to posterior shearing force.

In the unconstrained group, the applied axial displacement was sufficiently large to place the chin (if it were present) through the sternum, and anterior portions of the thoracic vertebral bodies. These unconstrained specimens showed a large initial low stiffness region, and correspondingly small axial loads. Mean peak axial load was 289 ± 81.4 N at a mean axial displacement of 8.6 ± 1.8 cm and a total of 96 ± 7.3 degrees of rotation of the head relative to T1. Despite this applied axial displacement, no evidence of significant structural injury was observed by plain X rays, CT, MRI or physical dissection in all six specimens studied.

Evidence suggestive of failure (decreasing axial force with increasing deflection) was occasionally seen on the axial force–deflection history at low loads in the unconstrained end condition. Analysis of axial force–deflection, eccentricity–deflection, and moment–deflection data showed evidence of buckling of the cervical spine from a compression mode to a flexion mode when this decrease occurred (Figure 9). The Hybrid III neckform also buckled from a compression mode to a flexion mode with the unconstrained end condition, though at a considerably higher load (780 N). This is seen in Figures 6, 7 and 8 at an axial displacement of 1 to 2 mm, and was observed on the videotape. Figure 6 shows the decrease in load with increasing deflection. Figure 7 shows an abrupt change in the rate of development of eccentricity with axial displacement and Figure 8 shows an increase in the rate of flexion moment development with axial displacement.

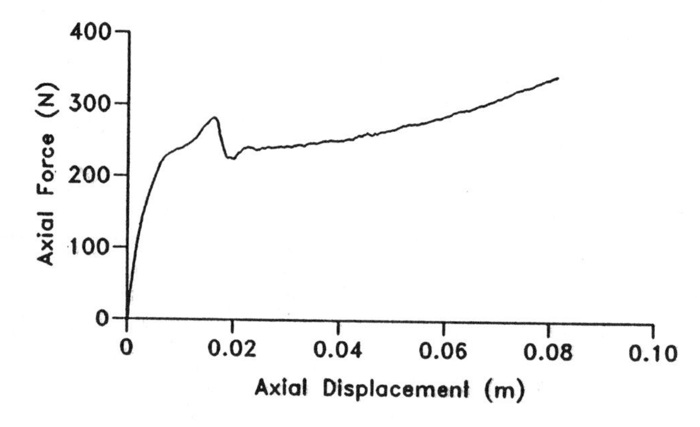

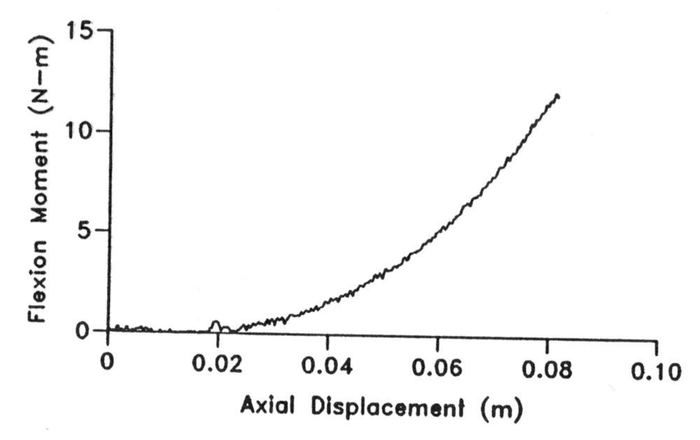

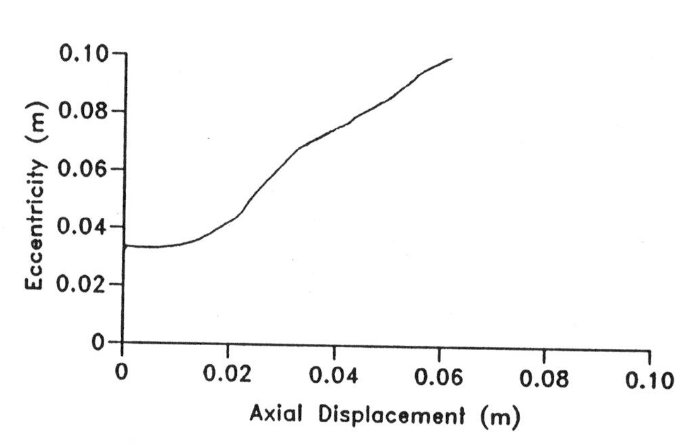

FIGURE 9. Three figures from a cadaveric specimen failure test with an unconstrained end condition showing evidence of low load buckling from a compression mode to a flexion mode. The figures show an abrupt decrease in the rate of development of axial load, and an increase in the rate of development of flexion moment and eccentricity occurring at approximately 0.016 m of axial displacement, indicative of buckling.

The device was undamaged by the loading, and subsequently carried larger loads at larger displacements. In addition, the event was reproducible, suggesting that it represented a buckle with post-buckling stability in the flexion mode.

Table 2 summarizes the injuries produced, and lists the measured peak axial loads, the axial displacements and strain energies to failure for all specimens studied. Figure 10 compares the load–deflection characteristics for three typical specimens. Using a student t-test, the rotational constraint group was found to have a significantly lower axial load to failure than the full constraint group ($p < 0.01$). The peak loads of the unconstrained group were also significantly lower than the rotational constraint group ($p < 0.01$).

TABLE 2

THE EFFECT OF END CONDITION ON FAILURE LOAD AND MECHANISM

SPECIMEN #	END CONDITION	PEAK AXIAL LOAD (N)	AXIAL DEFLECTION (cm)	ENERGY AT FAILURE (N-m)	INJURY
A	unconstrained	365	5.8	#4.2	None
B	unconstrained	367	10.2	#5.0	None
C	unconstrained	240	10.8	#6.1	None
D	unconstrained	343	8.2	#21.0	None
E	unconstrained	169	8.6	#9.9	None
F	unconstrained	250	8.2	#17.8	None
AVERAGE	unconstrained	289 ± 81.4	8.6 ± 1.8	#11.5 ± 6.5	
G	rotation constraint	3590	3.6	14.6	*BFD C5-C6
H	rotation constraint	2950	3.3	46.9	*BFD C6-C7
I	rotation constraint	1130	2.9	20.7	*BFD C7-T1
J	rotation constraint	1110	2.3	11.7	*BFD C7-T1
K	rotation constraint	600	1.5	4.8	*BFD C7-T1
L	rotation constraint	930	3.9	11.9	*BFD C7-T1
AVERAGE	rotation constraint	1720 ± 1234	2.9± 0.9	26.8 ± 23.7	
M	full constraint	5340	2.5	85.5	C2 comp. fx.
N	full constraint	4060	1.2	21.3	C3 comp. fx.
O	full constraint	6840	1.2	42.0	C4 & C5 wedge comp. fx.
P	full constraint	4700	1.2	32.9	C4 & C5 comp. fx.
Q	full constraint	3000	1.7	26.1	C3 & C6 comp. fx.
R	full constraint	4940	1.1	21.5	C4 wedge comp. fx.
AVERAGE	full constraint	4810 ± 1286	1.4 ± 0.4	32.9 ± 12.8	

*BFD = bilateral facet dislocation.
#Values denote stored strain energy.

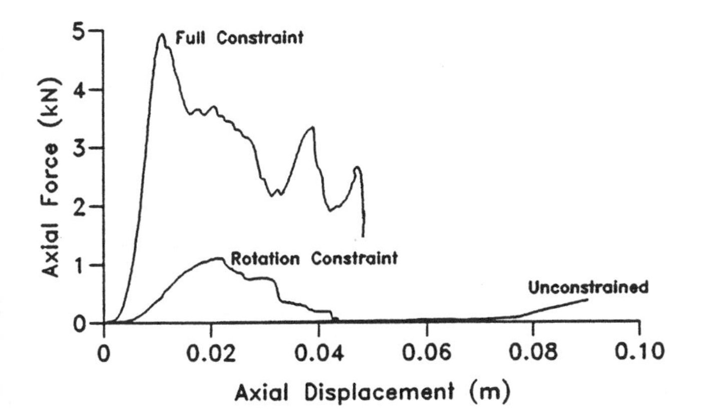

FIGURE 10. Comparison of typical axial load–deflection histories for three different cadaveric specimens loaded to failure with three different end conditions.

DISCUSSION

It is apparent from previous investigations that the neck is unable to stop the moving torso for all but the slowest velocities. Accordingly, in the injury environment, the head–neck complex must either move out of the way of the impinging torso or be at risk for injury.

Numerous investigations have been performed to understand the mechanisms of injury of the cervical spine. Based on these experiments a number of variables have been identified which predispose the spine to injury [5,9]. These include the initial position of the neck, the presence of a preflexed spine and the degree of the imposed head constraint. The purpose of these experiments was to create a set of end conditions which reasonably recreated the spectrum of end conditions imposed in the injury environment. The end conditions studied account for the degree of constraint imposed by the contact surface, and the trajectory of the thorax relative to the head.

Imposition of different end conditions resulted in large changes in the observed axial stiffness of the cervical spine within a given specimen. This was consistent with our previous work [2,8,9]. End condition also influenced the risk for injury. In the absence of significant constraint, the stiffness of the structure was low, the spine was able to flex out of the way of the torso and no injuries were observed. Increasing the constraint resulted in increased axial stiffness of the structure, but, the stiffened structure was neither able to manage the equivalent energy of a moving torso nor get out of the way of the torso.

This data suggests that increasing the constraint on the head increases the likelihood of neck injury. It also suggests that employment of deformable contact surfaces which do not manage the impact energy to reduce the force of impact, but, impose a significant head constraint, may be placing the neck at risk for injury. These results should be tempered however, realizing that the experiments discussed here were conducted at a low rate relative to real accidents. While the lower rate allowed for greater control of the end conditions, it neglects the potential contributions of inertial loading to the injury modes. Future work at higher rates of loading is therefore recommended to validate this hypothesis.

Each imposed end condition resulted in a repeatable injury mechanism. That is, rotational constraint, which might occur during impact of the head into a compliant surface with the torso moving posteriorly, resulted in bilateral facet dislocation in 6 of 6 specimens studied. Full constraint, which might occur when the head pockets in the contact surface and the trajectory of the torso is collinear with the axis of the neck, resulted in a compression type of injury in 6 of 6 specimens. Unconstrained, or a free end condition, resulted in no injury in 6 of 6 specimens. Of note, it was not necessary to preflex the spine to produce these injuries.

Buckling was observed in the unconstrained end condition in both the human cadaver, and the Hybrid III neckform at comparatively low axial loads. Analysis of the data, and the *in situ* video images showed that this represented first mode buckling from compression to flexion. The absence of injury in the specimens and the repeatabliity of the event in the Hybrid III neckform demonstrate that this event represents buckling with post-buckling stability. In the cadaver the axial load required to produce this event was of the same order as the weight of the head, suggesting that the *in vivo* cervical spine may be in a stable post-buckled condition. More importantly, the event was a first mode buckle with a free end condition; beam theory would suggest therefore that higher load, higher mode, potentially injurious buckling conditions may exist. The contributions of these modes, if they exist, to injury mechanics represents a topic of further investigation.

Understanding the basis for the structural failure of the neck remains a problem for biomechanical engineering. Much has been learned about injury by experiments performed in the cadaver. These tests are however, difficult to perform, show a large interspecimen variation, and are limited in the number which can be performed on a given specimen. The presence of nonlinear viscous effects, and buckling modes serve only to increase the complexity of the analysis. While the Hybrid III neckform was able to recreate some of these effects, most notably, the buckling of the unconstrained specimen, its was stiffer than the cadaver in the axial direction in all three end conditions. For the unconstrained, and rotational constraint end conditions, these differences are to be expected however, as the reported axial stiffnesses include contributions of both compression and flexion. The Hybrid III neckform correctly includes the stiffening effects of the active musculature in flexion, while the cadaver lacks these effects. In the full constraint end condition, which represents a nearly pure compression mode of loading, the contributions of the musculature are comparatively small, and the cadaver response approaches the volunteer response [20]. This suggests that the Hybrid III neckform is stiffer along its axis than the cervical spine. This result is consistent with the values reported by Yoganandan *et al.*[3]. Given the importance of compressive loading to cervical injury, particularly in non-automotive sources of cervical injury (i.e. sporting events and dives), the availability of a high biofidelity head–neck for compression would be of great benefit to the design and evaluation of safety equipment.

CONCLUSIONS

1. The imposed end condition reliably alters the failure mechanism of the cervical spine. This preliminary data suggests that safety equipment, and injury environments should be designed to minimize the degree of imposed constraint on the head. In particular, systems which tend to "pocket" without providing adequate energy management may provide an enhanced injury potential. As discussed, further work is necessary to better validate this conclusion.
2. The axial load to failure for lower cervical bilateral dislocation is significantly lower than the axial load to failure for vertebral compression type fractures.
3. First mode buckling can occur at very low loads in unconstrained specimens. The resulting structure is however, stable, and uninjured. The existence of other buckling modes and their contribution to cervical injury mechanics is unknown. Interpretation of failure data as defined by decreases in the axial load with increasing axial deflection are therefore difficult, given the frequent absence of significant mechanical injury, and the presence of buckling effects.

ACKNOWLEDGMENT

This study was supported by the Department of Health and Human Services, Centers for Disease Control Grant R49/CCR402396-02, and the Virginia Flowers Baker Chair.

BIBLIOGRAPHY

1. Pintar, F., Sances, A. Jr., Yoganandan, N., Reinartz, J., Maiman, D.J., Suh, J.K., Unger, G., Cusick, J.F., and Larson, S.J., "Biodynamics of the Total Human Cervical Spine," *The 32nd Stapp Car Crash Conference*, 55-72, 1990.
2. McElhaney, J.H., Doherty, B.J., Paver, J.G., Myers, B.S., and Gray, L., "Combined Bending and Axial Loading Responses of the Human Cervical Spine," *The 32nd Stapp Car Crash Conference*, 21-28, 1988.
3. Yoganandan, Sances, A. Jr., N., and Pintar, F., "Biomechanical Evaluation of the Axial Compressive Responses of the Human Cadaveric and Manikin Necks," *Journal of Biomechanical Engineering*, **111**:250–255, 1989.

4. Alem, N.M., Nusholtz, G.S., and Melvin, J.W., "Head and Neck Response to Axial Impacts," *The 28th Stapp Car Crash Conference*, 275-287, 1984.

5. Nusholtz, Huelke, D.E., Lux, P., Alem, N.M., and Montalvo, F., "Cervical Spine Injury Mechanisms," *The 27th Stapp Car Crash Conference*, 179-197, 1983.

6. Morgan, R.M., Marcus, J.H., and Eppinger, R.H., "Side Impact – The Biofidelity of NHTSA's Proposed ATD and Efficacy of TTI," *The 25th Stapp Car Crash Conference*, 301, 1981.

7. White A.A., and Panjabi, M.M., *Biomechanics of the Spine,* Lippincott Publishers, Philadelphia, 1978.

8. Doherty, B.J., McElhaney, J.H., and Myers, B.S., 'The Effect of End Conditions on the Responses of the Cervical Spine to Complex Loading," *Journal of Biomechanics*, submitted for review, 1991.

9. McElhaney, J.H., Paver, J.G., McCrackin, H.J., and Maxwell, G.M., "Cervical Spine Compression Responses", *The 27th Stapp Car Crash Conference*, 163-177, 1983.

10. Torg, J.S., *Athletic Injuries to the Head, Neck and Face.* Lea and Febiger, Philadelphia, 1982.

11. Roaf, R., "A Study of the Mechanics of Spinal Injury," *Journal of Bone and Joint Surgery*, **42B**:810-823, 1960.

12. Myers, B.S., McElhaney, J.H., Doherty, B.J., Paver, J.G., and Gray, L., "The Role of Torsion in Cervical Spinal Injury" *Spine*, **16(8)**:870-874, 1991.

13. Hodgson, V.R., and Thomas, L.M., "Mechanisms of Cervical Spine Injury During Impact to the Protected Head," *Society of Automotive Engineers Transactions*, Paper #801300, 17-42, 1980.

14. Bauze, R.J., and Ardran, G.M. "Experimental Production of Forward Dislocation of the Human Cervical Spine." *The Journal of Bone and Joint Surgery*, **60B**:239-245, 1978.

15. Yoganandan, N., Sances, A. Jr., Maiman, D.J., Myklebust, J.B., Pech, P., and Larson, S.J., "Experimental Spinal Injuries with Vertical Impact," *Spine*, **11(9)**:855-860, 1986.

16. Liu, Y.K., and Dai, Q.G., " The Second Stiffest Axis of A Beam Column: Implications for Cervical Spine Trauma." *Journal of Biomechanical Engineering*, **111**:122-127, 1989.

17. Mertz, H.J., and Patrick, L.M., "Strength and Response of the Human Neck," *The 15th Stapp Car Crash Conference*, SAE Paper No.710855, 1971.

18. McElhaney, J.H., Snyder, R.G., States, J.D., and Gabrielsen, M.A. "Biomechanical Analysis of Swimming Pool Injuries." *Society of Automotive Engineers*, paper no. 790137, 47-53, 1979.

19. Foster, J.K., Kortge, J.O., and Wolanin, M.J., "Hybrid III – A Biomechanically-Based Crash Test Dummy," *The 21st Stapp Car Crash Conference*, 975-1014, 1977.

20. Wismans, J., Philippens, M., van Oorschot, E., Kallieris, D., and Mattern, R. "Comparison of Human Volunteer and Cadaver Head-Neck Response in Frontal Flexion." *The 31st Stapp Car Crash Conference, 1-11, 1987.*

Section 7:
Head-Neck Structural Properties and Modeling

730975

Cervical Range of Motion and Dynamic Response and Strength of Cervical Muscles

David R. Foust, Don B. Chaffin, Richard G. Snyder and Janet K. Baum
University of Michigan

ABSTRACT

Basic physical characteristics of the neck have been defined which have application to the design of biomechanical models, anthropometric dummies, and occupant crash protection devices. The study was performed using a group of 180 volunteers chosen on the basis of sex, age (18-74 years), and stature. Measurements from each subject included anthropometry, cervical range-of-motion (observed with both x-rays and photographs), the dynamic response of the cervical flexor and extensor muscles to a controlled jerk, and the maximum voluntary strength of the cervical muscles. Data are presented in tabular and graphic form for total range-of-motion, cervical muscle reflex time, decelerations of the head, muscle activation time, and cervical muscle strength.

The range-of-motion of females was found to average 1-12 deg greater than that of males, depending upon age, and a definite degradation in range-of-motion was observed with increasing age. Average neck muscle reflex times range from 56-92 ms for flexors and 54-87 ms for extensors, with males being generally slower-reacting. This finding means that the cervical muscles generally cannot be activated in sufficient time to mitigate the hyperextension effects of a surprise rear-end collision. Reflex time becomes significantly longer after middle age. On the average, the stretch reflex was elicited with head accelerations at the c.g. of approximately 0.25 g. The time required to stop the head after initial detection of the muscle reflex was slightly longer for neck extensors. Strength tests revealed that males are on the average stronger than females in both flexor and extensor strengths; that males and females exhibit different aging characteristics, and that a slight stature effect was present in the younger and shorter portions of the population.

Age and sex were found to be important factors in cervical flexibility and response characteristics, and they should be included whenever accurate representation of neck parameters is desired.

THE NATURE OF cervical hyperflexion-hyperextension ("whiplash") injuries and their importance in the medical and highway safety fields is well documented. Vehicular accident statistics consistently list rear-end collisions as accounting for about 25% of both accidents and total injuries (1)*, and a recent literature survey resulting in more than 2300 references produced over 1350 references relating to the clinical aspects of cervical injuries (2). Many biomechanical models of the automobile passenger have also been developed in the past several years, with some of them incorporating highly sophisticated treatments of head-neck dynamics.

However, despite the mass of clinical references and increasing numbers of experimental papers in the literature, there has been a dearth of information on the basic response characteristics of the human neck structure and musculature. In particular, the influence of such physical parameters as sex, age, and body size on these characteristics has not been established. The study being reported herein was undertaken to define the

NOTE: This research was sponsored by The Insurance Institute for Highway Safety, Washington, D.C.

*Numbers in parentheses designate References at end of paper.

Table 1 - Comparison of Population Measures–50th Percentile Measurements

Subjects	Weight, kg (lb)		Stature, cm (in)		Erect Sitting Height, cm (in)	
	Study Population	U.S. Population	Study Population	U.S. Population	Study Population	U.S. Population
Females, 18-24 N = 31	58.3 (129)	56.2 (124)	162.7 (64.1)	162.3 (63.9)	85.7 (33.7)	85.6 (33.7)
Females, 35-44 N = 31	59.1 (130)	61.2 (135)	161.2 (63.5)	161.0 (63.4)	85.3 (33.6)	85.6 (33.7)
Females, 62-74 N = 31	64.9 (143)	64.9 (143)	158.3 (62.3)	156.5 (61.6)	82.6 (32.5)	81.8 (32.2)
Males, 18-24 N = 30	71.3 (157)	70.3 (155)	174.9 (68.9)	174.2 (68.6)	91.1 (35.9)	91.2 (35.9)
Males, 35-44 N=30	83.2 (183)	76.7 (169)	173.9 (68.5)	174.2 (68.6)	90.5 (35.6)	91.4 (35.9)
Males, 62-74 N = 27	71.3 (157)	72.1 (159)	170.2 (67.0)	169.7 (66.8)	88.1 (34.7)	88.4 (34.8)

range-of-motion, muscle reflex time, and muscle strength of the neck for a group of subjects representative of the United States adult population in the parameters of sex, age, and stature.

Of the three measurements of interest, only cervical range-of-motion has received significant attention. Ferlic (3) measured cervical range of motion on 105 men and 84 women in an age range of 15-74 years (though most subjects were younger than age 44). Delahaye (4), Defibaugh (5), Leighton (6), and Sigerseth (7) have all conducted cervical range-of-motion studies with young males and male college athletes, while Buck (8) and Bennett (9) have reported their measurements of college-age females. However, direct use of these results in cervical injury studies is difficult because of varying measurement techniques, lack of adequate landmark standards, and lack of comparability to the automobile seating position.

Information on cervical muscle characteristics has been almost totally lacking. Reflex times of the neck muscles have not been incorporated, not even in those biomechanical models that consider the musculature separately. (Instead, the muscles have been assumed to be tensed at the beginning of the simulated impact.) Neck strength data from volunteer subjects have been reported by Mertz and Patrick (10) in the United States and by Marotzky (11) in Germany.

In order to produce results directly applicable to the understanding of cervical injury susceptibility in the automotive safety context, it was concluded that it would be necessary to measure each of the characteristics of interest from the same group of people.

METHODOLOGY

SUBJECT ACQUISITION - The study population was chosen based upon the three primary variables of sex, age, and body stature. Sex was chosen as a primary variable because of indications that females more often incur whiplash injury than males (12). Since it is generally believed that the aging process adversely affects both joint range-of-motion and muscle reflexes, age was considered an important variable. Stature was included as the third primary variable on the supposition that neck characteristics could be affected by a person's overall height, sitting height, and neck length.

The final statistical design chosen was 2 by 3 by 3 factorial with 10 subjects per cell, for a total of 180 subjects. The three age groups selected were young adult (ages 18-24), early middle-age adult (ages 35-44), and elderly (ages 62-74). The three stature groups were selected to represent the 1-20 percentile, 40-60 percentile, and 80-99 percentile of the population within each age group. The selection of specific age and stature groups was based upon the latest available comprehensive study of the United States adult population (13).

Subjects were recruited from the local population in and around Ann Arbor, Michigan. Each subject's health was screened by means of a medical questionnaire (and later by x-rays), and each subject underwent a series of 48 anthropometric measures. A comparison of the final study population with the original design is presented in Table 1 for the key anthropometric measures of body weight, stature, and erect sitting height. Good agreement with the population data was achieved throughout, except in the weight category for males aged 35-44. Many of these subjects, especially in the short stature group, tended to be overweight.

CERVICAL MOBILITY - The study of cervical range-of-motion was carried out in two phases. First, three lateral x-rays of the head, neck, and upper torso were taken, using a range-of-motion sequence consisting of neutral, maximum voluntary flexion, and maximum voluntary extension positions, in that order. Second, the subject was photographed in the same three positions, using two orthogonally placed cameras. One x-ray sequence and three photographic sequences were taken for each subject, to be analyzed for repeatability of range-of-motion. The subject was seated in a simulated automobile seating position. The following position definitions were used:

1. Neutral: "normal, relaxed sitting position, looking straight ahead."

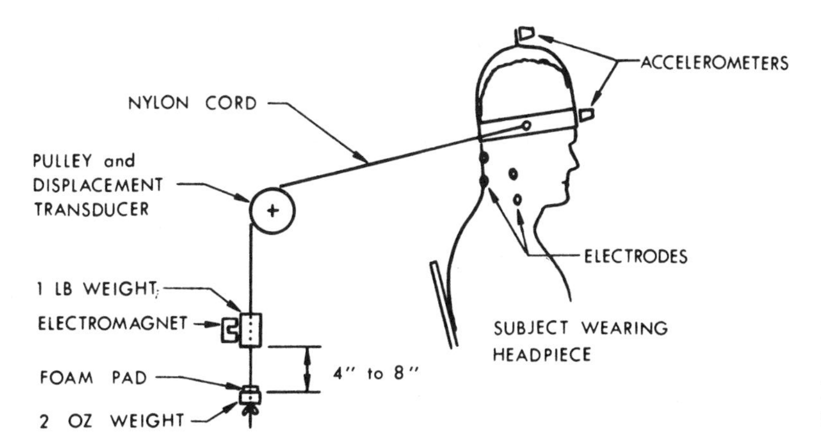

Fig. 1 - Method of applying controlled force for reflex time experiment

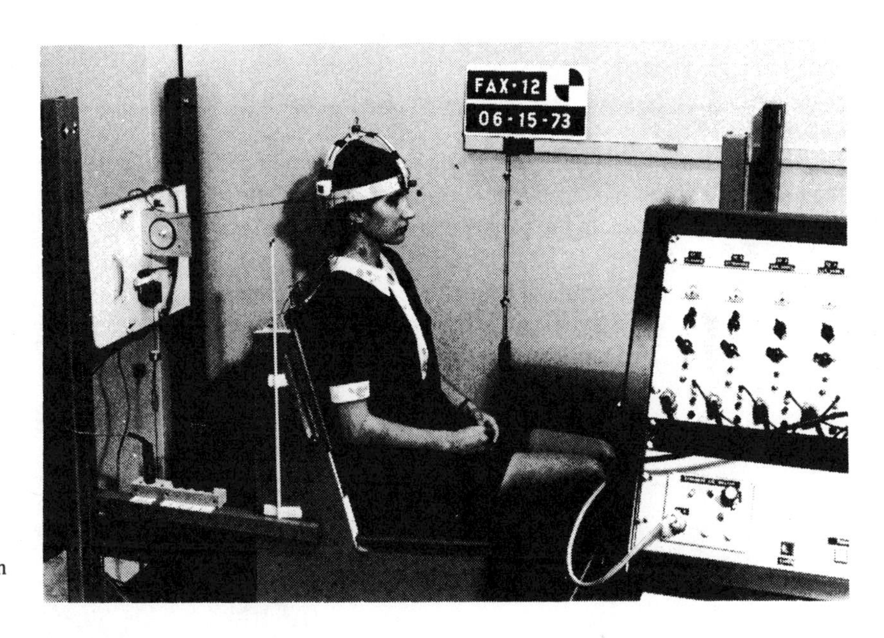

Fig. 2 - Subject seated in test fixture in preparation for test of reflex time of neck flexor muscles

2. Maximum voluntary flexion: "without moving shoulders or upper torso, thrust chin straight ahead and then tuck chin under as far as possible."

3. Maximum voluntary extension: "without moving shoulders or upper torso, and with jaw completely relaxed, allow head to rotate back as far as possible."

The flexion position definition was chosen to simulate impact accelerations in which the passenger is using an upper torso restraint, as demonstrated by Ewing and Thomas (14). Extension with jaw relaxed was intended to simulate a rear-end collision with complete surprise and no head restraint.

Range-of-motion was determined by measuring the angular change between landmarks on the body and a vertical source exposed in the background. High-contrast markers at nasion (nasal root depression) and tragion (top of cartilaginous notch on ear) and a plumb line were used for photographic analysis. A reference line established tangent to the base of the skull together with an x-ray opaque pendulum provided data from the x-rays.

CERVICAL MUSCLE STRETCH REFLEXES - Any influence of the neck muscles in reducing whiplash injury would be dependent upon two aspects–how fast the muscles can respond to a stimulus and how strongly they can respond. To prevent unnecessary hazards for the test population, these two responses were separated and measured independently at safe stimulus levels.

The stretch reflex times of the cervical flexor and cervical extensor muscle groups were measured, using electromyography (EMG) and a controlled dynamic "jerk" of the head. Prior to testing, surface electrodes were attached to the skin over the sternomastoid (flexor) and splenius and semispinalis capitis (extensor) muscles. Electrodes were positioned according to the recommendations of Davis (15), with modifications as necessary for subject size. The subject was then seated in a simulated car seat, and a headpiece weighing 225 g was fitted tightly around the head. A cord attached to the headpiece was then passed over a pulley and through a 455 g weight held by an electromagnet. A 57 g "pretensioning" weight was adjusted so that it would catch the 455 g weight after a fall of 10-20 cm (depending upon the subject). The test setup for a stretch reflex test of the neck flexor muscles is shown diagrammatically in Fig. 1. The same arrangement is illustrated in Fig. 2 to show a test subject in place and the relationship of the test operator's console to the subject. (Note that to measure the stretch reflex time of the extensor muscles, the mounting board for

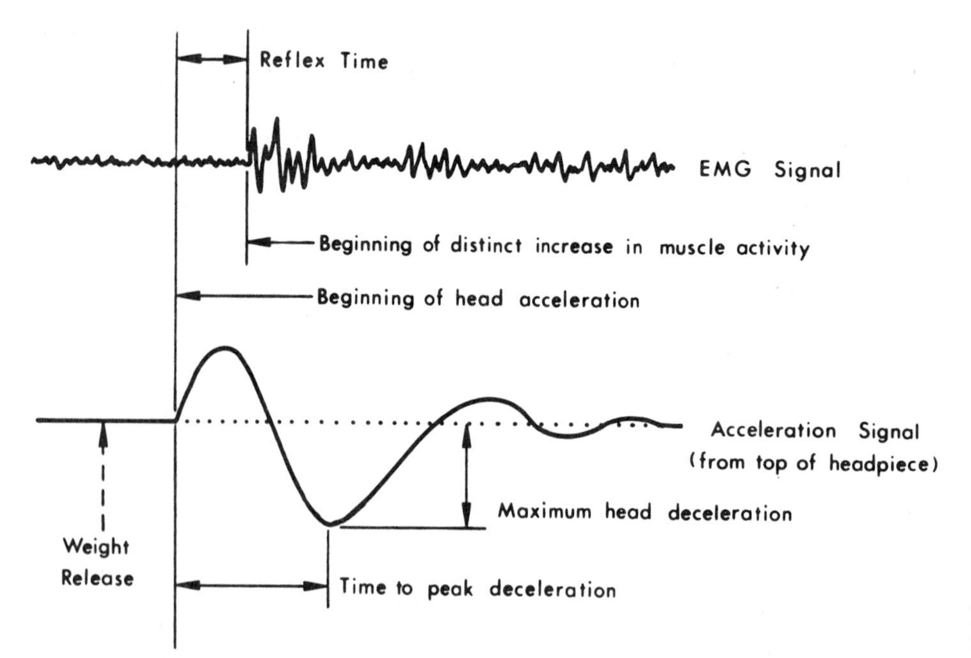

Fig. 3 - Strip-chart analysis of typical neck muscle reflex test

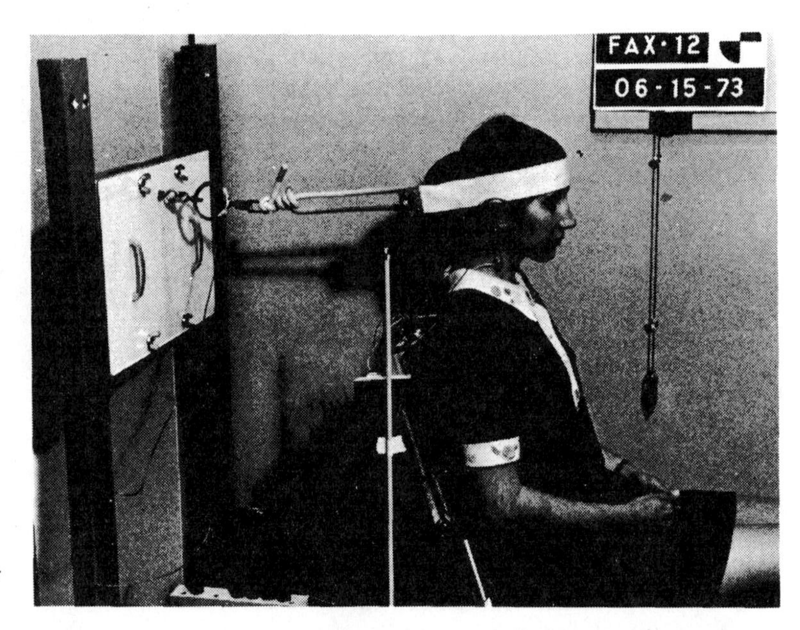

Fig. 4 - Subject positioned for strength test of neck flexor muscles

the pulley and electromagnet was moved to the upright guides in front of the subject. For those tests, a mask attached to the mounting board was used to block the subject's view of the weight.)

Testing was conducted in the following manner. The subject was encouraged to relax the neck muscles. When the test operator observed a relaxed EMG muscle signal with an oscilloscope, a switch was operated which momentarily interrupted the electrical power to the electromagnet, allowing the weight to drop and the head to be pulled slightly backward (or forward). Two Bruel & Kjaer type 4333 piezoelectric accelerometers, one mounted at the forehead and one at the top of the headpiece, measured head acceleration, a potentiometer mounted on the pulley axis directly measured head displacement, and the electrodes measured the EMG response of both flexor and extensor muscles. All data were recorded on an Ampex PR500 seven-channel recorder-reproducer. In addition, muscle response and head acceleration were monitored with a two-channel Brush strip-chart recorder. Three repetitions of each test were conducted.

Initial data reduction involved obtaining stretch reflex time, peak head deceleration, and time of peak deceleration from the strip-chart record. Stretch reflex time was defined as the time difference between onset of head acceleration and onset of increased muscular activity. The three measurements from a typical strip-chart record are illustrated in Fig. 3.

CERVICAL MUSCLE STRENGTH - To determine the second major aspect of cervical muscle response, isometric muscle strength tests were conducted with both the flexor and extensor muscles.

Cervical muscle strength was measured by having the subject exert force against a stainless-steel force ring which had been instrumented with four strain gages wired in a Wheat-

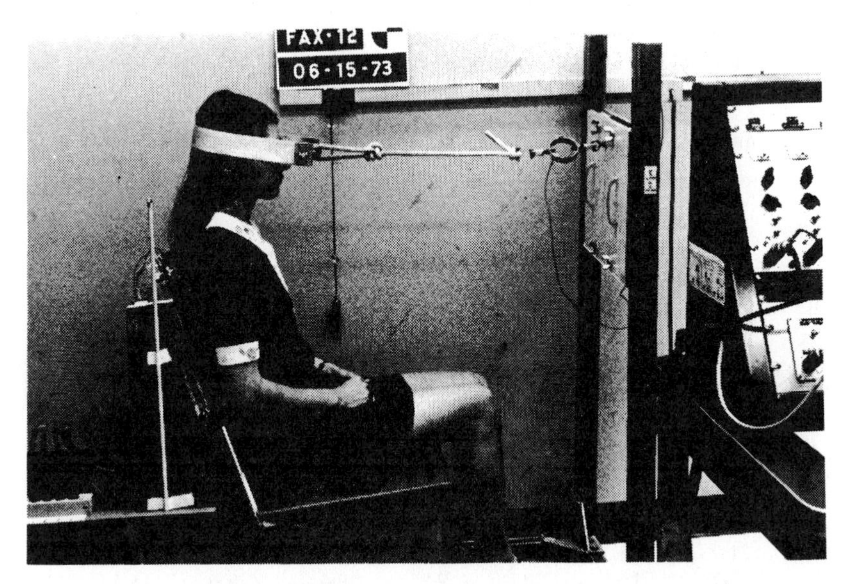

Fig. 5 - Subject positioned for strength test of neck extensor muscles

stone bridge circuit. The arrangement for measuring flexor muscle force is shown in Fig. 4. With the subject in a normal relaxed sitting position, a 4.5 cm inelastic headband was adjusted to fit around the forehead above the eyebrows (the adjusting cord was also inelastic woven dacron). The subject was asked to pull forward against the headband as hard as he or she was "voluntarily able." Three repetitions of the maximum voluntary strength test were performed, each test lasting 5 s. The subject was given a brief rest period between tests to preclude fatiguing the muscles. After completion of the flexor muscle tests, the testing apparatus was moved to the front of the subject, and the tests were repeated to measure the strength of the neck extensor muscles. This arrangement is illustrated in Fig. 5.

For each strength test, three channels of information were recorded on magnetic tape: neck flexor EMG, neck extensor EMG, and the strength signal from the force ring. A strip-chart record was also made of each strength test and was analyzed to provide the data for this report.

EXPERIMENTAL RESULTS

The large amount of data collected during this study has produced many results not heretofore available in the literature. Details of these results are of potential value to designers of biomechanical models and anthropometric dummies, and to other researchers in the field of human tolerance. For this reason, the authors are of the opinion that detailed tabular data are of more value than graphs of overall averages. Therefore, the test results are presented in terms of detailed tables, giving the summaries for each of the 18 subject data cells. Where graphic presentations help to illustrate significant trends, these are also used.

RANGE-OF-MOTION TESTS - The results of the x-ray and photographic sequences to determine total voluntary range-of-motion are presented in Table 2. The reader should note the comparison between the x-ray and photographic sequences, since this is an indication of voluntary repeatability. On the whole, subjects tended to assume the same position when given a subjective command to flex or extend "as far as possible," and the ranges-of-motion thus recorded reflect this repeatability. There are two cells, however (and several individual subjects in other cells) in which rather large differences in range-of-motion are noted between the x-ray and the first photograph. This difference might be attributable to the short time lapse during which the subject moved from the x-ray laboratory to the photographic laboratory, and it may also be related to increased difficulty in preventing motion of the upper torso in the x-ray laboratory. Also of interest is a slight tendency for the range-of-motion to increase during the three repetitions of the photographic sequence. This may be an indication that joint range-of-motion can be made to increase with repetition of maximum-effort exertions. Finally, while the mean values indicate trends, the very large standard deviations associated with these means emphasize the large variability among subjects that are otherwise classified by their similarities.

Analysis of the effects of age, sex, and stature on cervical range-of-motion is presented, in highly simplified form, in Fig. 6. The data points in Fig. 6 were obtained by averaging the x-ray and photographic results reported in Table 2 for each of the subject categories. Sex differences in range-of-motion are not substantial (within stature groups), but females tend to have slightly larger ranges-of-motion throughout life. Age has a pronounced effect upon cervical range-of-motion. However, aging effects in women are somewhat different from those of men; women tend to lose mobility gradually throughout their lives, but men deteriorate more rapidly between youth and middle age than they do thereafter. In youth and in middle age, the stature of a person seems to directly affect cervical range-of-motion, with taller people having a greater range. These effects largely disappear later in life. Range-of-motion of elderly females is similar regardless of stature, and the elderly males present mixed results. The percentage loss of range-of-motion tends to be greater for tall people than for short (as illustrated by the slopes of the lines in Fig. 6).

Table 2 - Range of Motion Analysis (Total Range of Motion, deg)

Age	1-20th Percentile				40-60th Percentile				80-99th Percentile			
	X-ray	Photo I	Photo II	Photo III	X-ray	Photo I	Photo II	Photo III	X-ray	Photo I	Photo II	Photo III
Female												
18-24	N = 11	10	10	10	14	14	12	11	11	11	10	10
	$\bar{x}$ = 128.9	126.0	129.2	131.5	139.9	134.0	139.1	135.8	148.9	143.3	147.7	146.7
	S.D. = 18.8	13.9	11.9	14.6	23.5	17.1	18.8	17.5	16.8	14.3	14.0	14.5
35-44	10	10	10	10	10	9	10	10	12	12	12	12
	118.5	116.9	119.5	117.2	121.7	121.7	120.5	123.3	120.2	124.2	130.1	129.8
	12.2	18.7	16.4	15.2	10.6	18.0	18.2	19.6	16.6	15.8	16.3	14.6
62-74	10	10	10	10	10	9	10	9	12	12	12	12
	95.1	96.1	103.3	106.0	97.1	102.0	101.1	104.0	96.8	96.8	96.8	97.7
	14.6	13.3	16.7	14.5	18.3	15.8	20.3	17.4	14.8	16.0	19.3	21.6
Male												
18-24	10	10	10	10	11	11	10	10	13	13	10	12
	135.2	130.2	131.2	133.3	143.6	131.2	134.8	139.2	146.3	139.9	138.2	139.9
	16.0	18.6	16.7	15.5	17.0	9.9	9.4	9.0	15.3	13.1	8.3	13.1
35-44	10	10	10	10	10	10	10	10	10	10	10	10
	101.9	103.5	103.3	105.1	106.4	106.9	107.9	109.0	115.7	115.8	116.7	117.9
	14.7	15.4	13.0	12.8	17.9	17.3	20.0	18.0	21.4	22.2	26.0	24.9
62-74	6	6	6	5	10	11	11	11	11	11	11	11
	99.9	92.5	92.1	100.6	87.9	86.0	83.8	82.6	102.7	102.2	104.7	102.8
	14.7	11.8	12.0	9.5	17.6	19.2	20.0	21.5	15.1	16.8	14.0	16.7

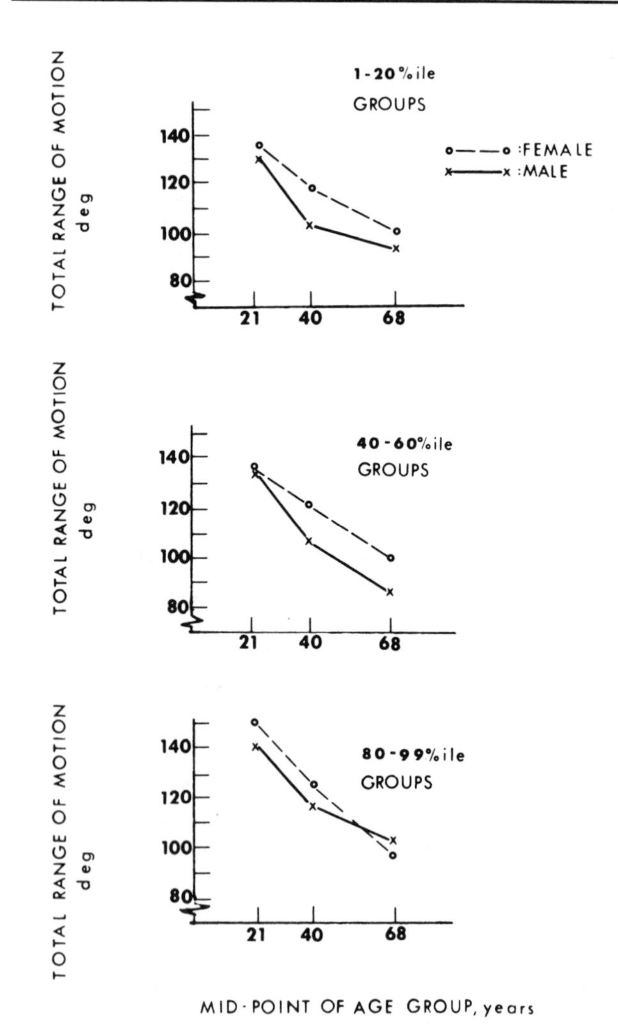

Fig. 6 - Effects of sex and aging on cervical range-of-motion

Table 3 - Comparison of Average Ranges-of-Motion, deg

	18-24 Years	35-44 Years	62-74 Years
Female	137.9	122.0	99.0
Male	137.6	109.1	94.4

If stature differences are averaged for each age and sex group, the results shown in Table 3 are obtained.

The data in Table 3 indicate that females exhibit greater average range-of-motion at all ages. However, it is in middle age where women maintain the largest difference over men. Both men and women can expect approximately 40% decrease in range-of-motion over their adult lifespan. However, men lose about 25% by middle age, whereas women decrease only about 13% in the same time period.

CERVICAL MUSCLE REFLEXES - As described in the methodology section, experiments were conducted to determine the stretch reflex time of both the neck flexor and extensor muscle groups. As the data were analyzed, it became apparent that the muscles responded in one of three typical ways. These typical results are illustrated in Fig. 7. A type 1 response was characterized by a low-intensity muscle reaction of rather long duration and by low-g, long-duration stopping forces. Type 2 responses were typified by more intense, short-duration muscle reflexes and a short, relatively high-g stopping force. A type 3 response was a hybrid of the others typified by two distinct bursts of muscle activity, each with an associated peak of stopping force. The bursts of muscle activity did not always differ in intensity, but they were distinct. (While these differences in response have been noted, no at-

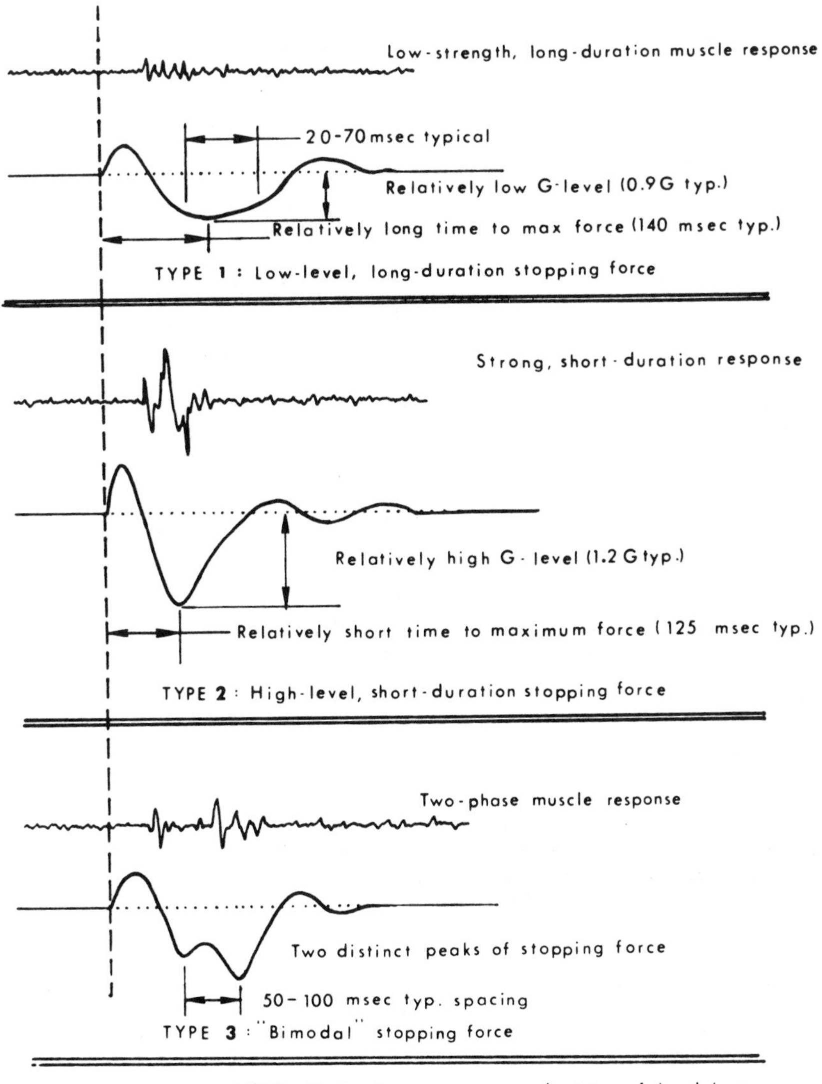

Fig. 7 - Typical results of reflex tests

tempt was made to segregate the reflex test results by response type for the data presented herein.)

Stretch reflex times recorded from the neck flexors (sternomastoid) muscles are presented in Table 4. Although the difference is typically only a few milliseconds, there is a consistent sex effect in evidence, with men having slower reflex times on the average than women in every comparable subject category. Within sex and stature groups, there is a general degradation of reflex time between middle and elderly age groups, but no specific trend toward slower reflexes between young and middle age groups. There is no consistency at all attributable to stature effects; in some groups, reflexes slow with increasing stature, while in others the middle or short statures have the slower reflexes. Standard deviations of the results tend to become greater with age in both sexes. This increase in variation may be partly due to the type of response and partly due to aging effects (some people age faster than others).

Neck extensor (semispinalis and splenius capitis) muscle reflex tests (Table 5) gave results generally similar to those of the flexor muscles, but not as pronounced. Sexual differences are still evident in many cells, but generally men are only slightly slower to react than women. There is, however, a pronounced degradation of reflex time between the middle and elderly age groups—a definite aging effect. Differences due to stature are even less pronounced than those observed for the flexors, and standard deviations within subject groups are generally smaller.

If stature effects are lumped to produce average reflex times for each sex and age group, the sex and age effects become more pronounced, and results of flexor and extensor tests may be compared (Tables 4 and 5, rightmost columns). It may be noted again that, for each muscle group, reflexes of males average slower than those of females and aging has a definite slowing effect on reflexes past middle age. It is also interesting to note that the extensor muscles exhibit faster reflexes than flexor muscles—in some cases substantially faster.

DECELERATION AND TIME ASSOCIATED WITH STOPPING HEAD - The deceleration of the head as it is brought to a stop by muscle reaction is of interest since it is thought to be indicative of the muscle force generated. This decelera-

Table 4 - Average Reflex Time of Neck Flexor Muscles (Weight Dropped Behind Head), ms

Age	1-20th Percentile	40-60th Percentile	80-99th Percentile	Average for Age Group
Female				
18-24	N = 10	10	10	62
	$\overline{x}$ = 59	61	67	
	S.D. = 8	8	11	
35-44	10	10	11	62
	56	66	64	
	12	15	11	
62-74	10	10	11	74
	74	78	72	
	20	18	15	
Male				
18-24	10	10	10	68
	65	65	74	
	12	10	13	
35-44	9	10	10	77
	82	75	74	
	12	15	14	
62-74	6	11	10	87
	79	92	89	
	15	11	22	

Table 5 - Average Reflex Time of Neck Extensor Muscles (Weight Dropped in Front of Head), ms

Age	1-20th Percentile	40-60th Percentile	80-99th Percentile	Average for Age Group
Female				
18-24	N = 10	10	11	57
	$\overline{x}$ = 54	57	59	
	S.D. = 7	7	9	
35-44	10	10	11	59
	55	62	61	
	10	12	7	
62-74	10	10	11	73
	72	73	75	
	10	11	11	
Male				
18-24	9	9	9	59
	54	65	58	
	6	14	5	
35-44	9	10	10	62
	61	61	65	
	6	6	10	
62-74	6	11	10	77
	67	76	87	
	4	9	16	

Table 6 - Average Peak Deceleration: Neck Flexor Muscle Tests, g as Measured at Top of Headpiece

Age	1-20th Percentile	40-60th Percentile	80-99th Percentile
Female			
18-24	N = 10	9	10
	$\overline{x}$ = 1.10	0.97	0.82
	S.D. = 0.25	0.26	0.19
35-44	10	10	11
	0.91	0.98	0.99
	0.14	0.12	0.23
62-74	10	9	11
	1.11	0.94	0.99
	0.13	0.18	0.18
Male			
18-24	10	10	10
	0.98	0.97	0.81
	0.13	0.23	0.31
35-44	9	9	10
	0.99	1.07	0.88
	0.15	0.17	0.11
62-74	6	11	10
	0.94	0.95	0.92
	0.18	0.13	0.15

Table 7 - Average Time to Peak Deceleration: Neck Flexor Muscle Tests, ms

Age	1-20th Percentile	40-60th Percentile	80-99th Percentile
Female			
18-24	N = 10	9	10
	$\overline{x}$ = 115	122	124
	S.D. = 8	8	14
35-44	10	10	10
	125	124	121
	11	10	13
62-74	10	10	11
	143	140	141
	17	14	15
Male			
18-24	10	10	10
	122	127	140
	20	10	22
35-44	10	10	10
	136	136	136
	20	14	17
62-74	6	11	10
	142	141	151
	11	14	27

tion is also believed to be directly related to the length of time required to generate sufficient muscle tension to resist the suspended weight and stop the head. Tables 6 and 7 present the peak deceleration and time to peak deceleration for the flexor muscle reflex tests. In general, the less the deceleration, the longer the time required to reach it. With respect to the three primary test variables, the results are similar to the reflex time results, although less pronounced. There is still an age effect, with males generally producing lower stopping forces but taking longer to stop the head. Aging has no definitive effect on force levels, but there is an age effect found in the time required for elderly persons to stop the head. There is something of a stature effect in young men and women (force decreasing, time increasing), but inconsistency prevails in the other age groups.

Similar deceleration and time data for the neck extensor

muscles are presented in Tables 8 and 9. Again, lower decelerations often produce longer stopping times, although the trend is not as consistent as noted with the flexors. A definite sex effect is noted in Table 8, with males requiring, on the average, less force to limit forward head motion. However, this effect is not directly related to stopping time, since Table 9 shows only a slight tendency for males to stop more slowly than females. Age effects are minimal with respect to force, but elderly persons stopped their heads more slowly as a rule than their younger counterparts.

Comparing neck flexor and extensor test results, it is notable that they are very similar on the average. The overall average deceleration for both flexors and extensors was 0.96 g as measured at the top of the headpiece. (This figure was developed for each muscle group by adding the mean of each cell and dividing by 18 cells.) The overall average time to peak deceleration was 132 ms for flexors and 134 ms for extensors. The corresponding figures for head acceleration due to loading by the weight were 0.77 g at 38 ms (identical results for both flexors and extensors). Since the test procedures followed the guideline of dropping the weight the minimum distance necessary to achieve a definitive reflex, this indicates that the acceleration levels required to elicit the involuntary stretch reflex of the neck muscles are approximately the same for either flexors or extensors.

MUSCLE ACTIVATION TIME - Given that the stretch reflex has occurred, the time in which the cervical muscles become fully activated could be an important factor in determining whether those muscles can influence susceptibility to injury. As illustrated in Fig. 3, reflex time is described in relation to the beginning of muscle activity, and the time to peak deceleration is indicative of the total time necessary for the neck muscles to react and stop the head from moving. The difference between the two times is a new quantity representative of the time required for the muscle to generate enough tension to stop the head. This average muscle activation time for both flexors and extensors is presented in Table 10. Comparison of the two muscle groups reveals that (except for tall, elderly persons) the extensor muscles develop tension consistently more slowly than the flexor muscles. On a cell-by-cell basis, sex, age, and stature effects are largely inconsistent.

CERVICAL MUSCLE STRENGTH - Coupled with the dynamic response time of the neck musculature is the maximum amount of strength that can be generated to prevent the head from being thrown forward or back. Tables 11 and 12 present the summarized data for the neck flexors and extensors, respectively. These tables are included to provide the reader with the precise means and standard deviations obtained. However, comparisons and contrasts will be based on Fig. 8, which is a series of plots of the mean values from Tables 11 and 12.

Sex, age, and some stature effects are evident in the flexor muscle test results (Fig. 8, top). The average muscle strength of males is greater than that of females in every age and stature group. Sex also seems to influence the effects of age. Females tend to decrease gradually in neck strength, but only slightly throughout their lives, while males are often stronger at middle age than they were when young. There are some notable stature effects (or noneffects). Taller men tend to be stronger in youth, but the advantage of size is less by middle age, and all male stature groups decline in strength at the same rate into old age. Stature effects are less pronounced in women. The shorter stature group in both sexes is slightly weaker throughout life.

Table 8 - Average Peak Deceleration: Neck Extensor Muscle Tests, g as Measured at Top of Headpiece

Age		1-20th Percentile	40-60th Percentile	80-99th Percentile
Female				
18-24	N =	10	10	11
	$\overline{x}$ =	1.05	0.92	0.82
	S.D. =	0.25	0.27	0.17
35-44		10	10	10
		1.07	1.12	1.06
		0.20	0.20	0.15
62-74		9	10	11
		1.14	0.98	0.93
		0.20	0.16	0.17
Male				
18-24		8	10	10
		1.02	0.83	0.80
		0.09	0.15	0.22
35-44		9	10	10
		0.92	0.99	0.92
		0.20	0.16	0.20
62-74		6	11	10
		0.86	0.94	0.84
		0.14	0.19	0.18

Table 9 - Average Time to Peak Deceleration: Neck Extensor Muscle Tests, ms

Age		1-20th Percentile	40-60th Percentile	80-99th Percentile
Female				
18-24	N =	10	10	9
	$\overline{x}$ =	125	134	133
	S.D. =	3	15	12
35-44		10	10	10
		126	134	124
		9	7	9
62-74		10	10	11
		142	142	140
		10	6	9
Male				
18-24		8	9	9
		128	128	133
		14	17	16
35-44		9	10	10
		130	137	129
		15	10	10
62-74		6	11	10
		137	141	147
		9	15	13

Table 10 - Average Muscle Activation Time
(Average Time to Peak Deceleration Minus Average Reflex Time), ms

Age	1-20th Percentile Flexors	1-20th Percentile Extensors	40-60th Percentile Flexors	40-60th Percentile Extensors	80-99th Percentile Flexors	80-99th Percentile Extensors
Female						
18-24	56	71	61	77	57	74
35-44	69	71	58	72	57	63
62-74	69	70	62	69	69	65
Male						
18-24	57	74	62	63	66	75
35-44	54	69	61	76	62	64
62-74	63	70	49	65	62	60

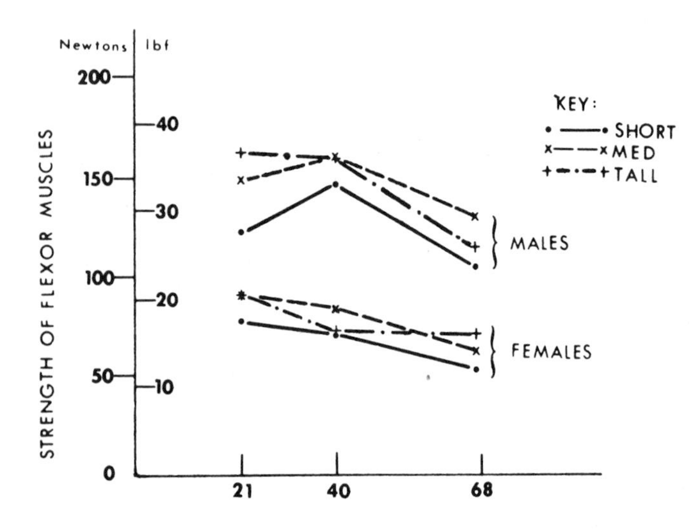

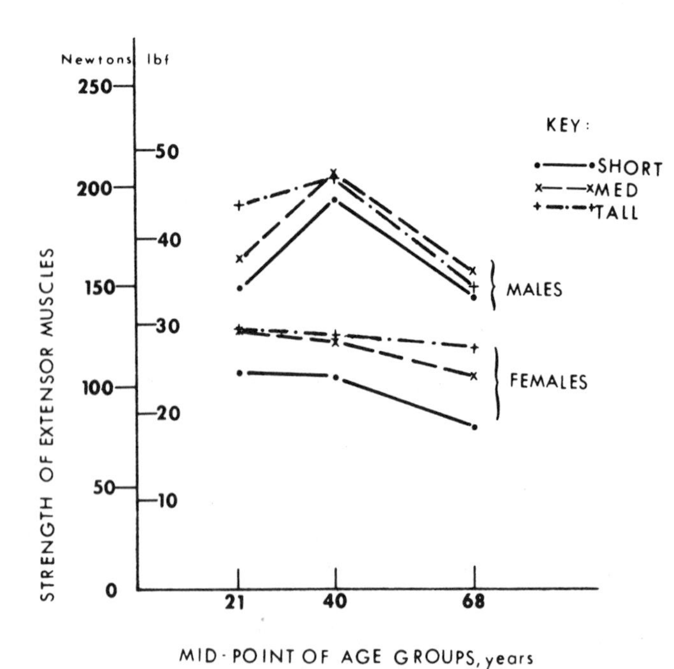

Fig. 8 - Mean strength of cervical muscles

Table 11 - Strength of Neck Flexor Muscles, N (lb=ft)

Age		1-20th Percentile	40-60th Percentile	80-99th Percentile
Female				
18-24	N =	10	10	11
	$\overline{x}$ =	77 (17.5)	91 (20.5)	91 (20.6)
	S.D. =	13 (2.9)	22 (4.9)	29 (6.7)
35-44		10	10	11
		70 (15.6)	83 (18.8)	72 (16.1)
		18 (4.0)	25 (5.5)	16 (3.5)
62-74		10	10	11
		52 (11.7)	62 (13.8)	70 (15.6)
		13 (2.9)	16 (3.6)	31 (7.1)
Male				
18-24		10	10	10
		123 (27.5)	148 (33.4)	162 (36.3)
		41 (9.2)	33 (7.5)	52 (11.7)
35-44		10	10	10
		147 (33.1)	160 (35.9)	158 (35.5)
		47 (10.6)	30 (6.9)	38 (8.6)
62-74		6	11	10
		104 (23.3)	128 (28.8)	113 (25.3)
		26 (5.9)	42 (9.5)	20 (4.3)

Table 12 - Strength of Neck Extensor Muscles, N (lbf)

Age		1-20th Percentile	40-60th Percentile	80-99th Percentile
Female				
18-24	N =	10	10	11
	$\overline{x}$ =	107 (24.1)	127 (28.7)	127 (28.7)
	S.D. =	33 (7.5)	27 (6.2)	36 (8.2)
35-44		10	10	11
		105 (23.5)	124 (27.7)	126 (28.2)
		29 (6.6)	26 (6.0)	28 (6.3)
62-74		10	10	11
		80 (17.9)	105 (23.5)	119 (26.7)
		24 (5.2)	28 (6.3)	46 (10.3)
Male				
18-24		10	10	10
		149 (33.6)	163 (36.6)	191 (43.0)
		20 (4.4)	52 (11.6)	38 (8.5)
35-44		10	10	10
		193 (43.5)	206 (46.3)	203 (45.6)
		39 (8.8)	47 (10.5)	45 (10.0)
62-74		6	11	10
		143 (32.2)	156 (35.1)	149 (33.5)
		40 (9.1)	47 (10.0)	22 (4.8)

Strikingly similar results were obtained from the extensor muscle tests (Fig. 8, bottom). Males average stronger than females in every age and stature group. Once again, females show a gradual and slight aging effect throughout life, while males increase in strength through middle age before their muscle strength decreases with increasing years. Finally, stature effects are also similar to those observed from the flexors, with taller young men being stronger, but losing most of this advantage by the time they reach middle age. Short persons are again weaker on the average throughout life.

A comparison of flexor and extensor test results reveals that the extensor muscles consistently average stronger than the flexor muscles. This is true of every category, and the strength difference ranges from 15 N (medium-size young men) to 54 N (tall, middle-age women), with an overall average difference of 37 N (8.3 lbf).

DISCUSSION

It is believed that the subject population was representative of those people whose state of health and neck characteristics could be called "normal" for their age. Normal arthritic degeneration with age was defined by the radiologist consultant to the study. Approximately one-third of those potential elderly subjects x-rayed were not allowed to complete the study—that is, to perform the reflex and strength test—primarily because of degenerative arthritis. Only 5.5% of other subjects were unacceptable, most often because of unusual spine configuration. Only x-rays and photographs of the subjects who completed all tests are included in the range-of-motion data presented herein. Inclusion of data from the entire population range ("abnormal" as well as "normal") would have resulted in somewhat smaller ranges-of-motion in the elderly groups. However, these data would unduly bias the test results, since comparable neck muscle data were not obtained from the same subjects. (It is recognized that these persons are part of the population using vehicles, and that they might have relatively different susceptibilities to hyperflexion/hyperextension injuries.)

The range-of-motion results obtained are in some disagreement with results obtained by Ferlic (3), specifically as regards young females and elderly males and females. However, Ferlic's sample size was extremely small in the elderly group (three men and three women), and he reports no stature trend, which was found to be significant among the young subjects. It is likely that the results of the present study are more representative of the effective range-of-motion of the seated automobile occupant.

From the results of the stretch reflex tests, it appears overall stature is less influential than originally hypothesized. If indeed any external size characteristic is a determinate factor in muscle dynamic response, perhaps sitting height or neck length or even body weight may be more correlated. A covariate analysis of these anthropometric measures is planned as a future exercise to help pinpoint influential body size characteristics.

As pointed out in the discussion of Table 10, the extensor muscles, even though they had a faster stretch reflex, take longer to activate than do the flexors. A possible physiological explanation for this lies in the anatomy of the muscles. The primary neck flexor is the sternomastoid, a single muscle, whereas the neck extensors are the semispinalis capitis, the splenius capitis, and various occipito-spinal and interspinal muscles—all of which are relatively short and are layered successively deeper in the neck. It is probable that the neural control system takes longer to activate the complicated system of extensor muscles sufficiently to stop forward head motion than it does to activate the single muscle necessary to stop rearward motion.

The rather remarkable finding that the same average level of head acceleration elicited stretch reflexes in both neck flexor and extensor muscles leads to the conclusion that a force of approximately one-quarter g is sufficient to produce a stretch relfex in the neck musculature. The acceleration data indicate that, for a given test, the accelerometer at the top of the headpiece registers about three times greater acceleration than does the accelerometer mounted at the forehead. Since the accelerometer at the forehead approximates the location of the c.g. of the head for a sagittal plane test, the forces measured by that accelerometer are representative of the anterior-posterior forces through the c.g.

The consistently higher strength of the neck extensor muscles is probably related to mechanical advantage. The extensors, which are located well posterior to the cervical spine, can exert greater torque than the flexors, which are attached to the skull very near the superior portion of the cervical spine.

The question of whether tensed neck muscles are strong enough to overcome hyperextension forces has recently been addressed by Robbins (16), using a biomechanical model with the stature, range-of-motion, and strength data from the present study. He found that, with maximum voluntary strength, all categories of occupants were able to influence head/neck dynamic response, but not all categories could prevent hyperextension. Since he used average strength values, it seems likely that many individuals of both sexes would not be strong enough to resist hyperextension forces in certain crash situations.

A matter of interest is the relationship between the maximum voluntary muscle strength as measured in these tests and the absolute muscle strength that might be generated by the same muscles in a panic situation. Chaffin and Baker (17) cite studies that indicate demonstrated maximum strength is always something less than absolute physiological capacity. This would seem especially true in the case of voluntary neck strength testing, since it is unlikely that test volunteers would want to induce neck muscle strain. It is the authors' opinion that the voluntary strength results represent about 80% of the maximum available physiological strength capacity. For biomechanical models in which maximum physiological strength is desired, a correction factor based on this percentage would seem reasonable as a first estimate.

CONCLUSIONS

The body of data presented herein establishes a description of important static and dynamic characteristics of the neck for a group that is representative of the entire driving population. The effects of sex, aging, and stature have been quantified in a manner useful to designers of biomechanical models, anthropometric dummies, and occupant crash protection systems, and to researchers in the field of human tolerance.

Biomechanical models and anthropometric devices that are intended to simulate neck responses must have at least sex and age factors incorporated into them. For example, on the average, females reflex about 11% faster than males, but are only 60% as strong. Average female range-of-motion is 12% greater than that of males during middle age. Over the adult lifespan, cervical range-of-motion is reduced by an average of 40%, cervical muscle reflexes slow by 23%, and voluntary strength capability diminishes by 25%. It is also important to consider the direction of impact. Reflex times of the neck extensor muscles average 11% faster than those of the flexor muscles, and the extensors are nearly 35% stronger than the flexors. (The reader should note that the above figures were developed from the entire data base. Specific values for the various age, sex, and stature categories are contained in the body of the text.)

The feedback control concept of Houk and Henneman (17) has direct application to biomechanical modeling of the neck, especially where the crash victim is assumed to be unaware of the crash. Delay times for both muscle stretch reflex and muscle activation must be representative of physiological reality.

Finally, given that a person is surprised by a rear-end collision, it is important to know if the neck flexor muscles can be activated in time to mitigate the forces that are trying to hyperextend the neck. If the crash pulse is similar to the 30 mph final velocity differential (closing rate of 50-60 mph) pulse cited by Robbins, et al (16), the muscles will probably have little or no effect, depending upon the sex and age of the crash victim. This pulse produces peak forces of 15 g at approximately 60 ms that decay linearly to 0 g at 192 ms. Assuming the head begins accelerating at 0 time, at 60 ms the young (and possibly middle-aged) woman is just starting to activate her muscles due to the stretch reflex. The elderly woman and most men have not yet had sufficient time to reflex. Assuming the muscles require another 60 ms to activate completely (Table 10), the worst of the crash may be over before the muscles can become fully effective. Therefore, in the situation of complete surprise, it is doubtful that the cervical muscles can be of much help in a high-speed rear-end collision. The implications of these results are important to designers of occupant protection devices such as seatbacks and head restraints. Since persons involved in a crash may neither react fast enough nor be strong enough to protect themselves, these devices must be designed to accommodate the physiological limitations of the occupant and provide effective protection.

ACKNOWLEDGMENT

This research involved developing and administering five diverse tests to each of 180 volunteer subjects, and it could not have been accomplished without the efforts of a large number of talented people. The authors are especially grateful to Brian O'Neill, who suggested the statistical design and monitored the project for the Insurance Institute for Highway Safety, to Dr. Laurence Rosenstein, the IIHS monitor in the early phases of the study, and to Dr. Harold Portnoy, neurosurgeon at Pontiac, Mich., General Hospital for the advice and consultation he generously provided. The authors are also deeply indebted to Ann Desautels, who expertly obtained x-rays and anthropometry data and assisted in all phases of data-taking, to Leonard Bryant and Paul Katz for reduction of x-rays and photographic data, and to Gary Gaul, whose conscientious efforts in recruiting elderly subjects and analyzing strip chart records provided for the timely completion of the testing and reporting phases of the study. Finally, the authors wish to express their appreciation to the volunteer subjects recruited from the general populace of Ann Arbor. Their willingness to participate insured the success of the study.

REFERENCES

1. National Safety Council, "Accident Facts." 1972 Edition, Chicago.

2. P. Van Eck, D. B. Chaffin, D. R. Foust, J. Baum, and R. G. Snyder, "A Bibliography of Whiplash and Cervical Kinematic Measurement." Highway Safety Research Institute, The University of Michigan (2326 references).

3. D. Ferlic, "The Range of Motion of the 'Normal' Cervical Spine." Johns Hopkins Hosp. Bulletin, Vol. 110 (1962), pp. 59-65.

4. R. P. Delahaye, et al., "Dynamic Radiology of the Cervical Spine of Flying Military Personnel; the Special Case of Jet Pilots." Rev. Corps Sante Armes, Vol. 9 (1968), pp. 593-614.

5. J. J. Defibaugh, "Measurement of Head Motion. Part I: A Review of Methods of Measuring Joint Motion. Part II; An Experimental Study of Head Motion in Adult Males." Physical Therapy, Vol. 44 (1964), pp. 157-168.

6. J. R. Leighton, "Flexibility Characteristics of Four Specialized Skill Groups of College Athletes." Arch. Phys. Med., Vol. 38 (1957), pp. 24-28.

7. P. O. Sigerseth and C. C. Haliski, "The Flexibility of Football Players." Res. Quart., Vol. 21 (1950), pp. 394-398.

8. C. A. Buck, F. B. Dameron, M. J. Dow, and H. V. Skowlund, "Study of Normal Range of Motion in the Neck Utilizing a Bubble Goniometer." Arch. Phys. Med., Vol. 40 (1959), pp. 390-392.

9. J. G. Bennett, L. E. Bergmanis, J. K. Carpenter, and H. V. Skowlund, "Range of Motion of the Neck." Jrl. Amer. Phys. Ther. Assn., Vol. 43 (1963), pp. 45-47.

10. H. J. Mertz, Jr., and L. M. Patrick, "Investigation of the Kinematics and Kinetics of Whiplash." SAE Transactions, Vol. 76, paper 670919.

11. H. J. Marotzky, "Isometric Energy Measurements on the Head-Neck System in Younger and Older People." Arch. Orthop. Unfallchir., Vol. 74 (1972), pp. 42-62.

12. B. O'Neill, W. Haddon, Jr., A. B. Kelley, and W. Sorenson, "Automobile Head Restraints—Frequency of Neck Injury Claims in Relation to the Presence of Head Restraints." Amer. Jrl. Pub. Health, Vol. 62, March 1972, pp. 399-406.

13. "Weight, Height and Selected Body Dimensions of Adults: United States, 1960-62." Public Health Service, 1963.

14. C. L. Ewing and D. J. Thomas, "Human Head and Neck Response to Impact Acceleration." NAMRL Monograph 21 (1972), p. 84.
15. J. F. Davis, "Manual of Surface Electromyography." WADC Technical Report 59-184 (1959), pp. 18-23.
16. D. H. Robbins, R. G. Snyder, D. B. Chaffin and D. R. Foust, "A Mathematical Study of the Effect of Neck Physical Parameters on Injury Susceptibility." Paper 740274 presented at SAE Automotive Engineering Congress, Detroit, February 1974.
17. D. B. Chaffin and W. H. Baker, "A Biomechanical Model for Analysis of Symmetric Sagittal Plane Lifting." AIIE Transactions (March 1970), pp. 16-27.
18. J. Houk and E. Henneman, "Feedback Control of Skeletal Muscles." Brain Research, Vol. 5 (1967), pp. 433-451.

751156

Biomechanical Properties of the Human Neck in Lateral Flexion

Lawrence W. Schneider, David R. Foust, Bruce M. Bowman, Richard G. Snyder, Don B. Chaffin, Thomas A. Abdelnour and Janet Baum
University of Michigan

THE TERM "WHIPLASH" has commonly been applied to encompass the complex interactions which occur when the occupant of a vehicle is struck from the rear, resulting in cervical hyperextension. The susceptibility of individuals to injury resulting from this "whiplash" is dependent on many factors, including the occupant's prior awareness of impact and muscle reflex time, neck muscle strength capability, and the normal head and neck range of motion of the individual. In a previous study by Foust, et al. (1)*, variations of these basic physical properties of the human neck with age, sex and stature of subjects were studied in sagittal plane bending. It is also the case, however, that individuals incur "whiplash" injuries from forces resulting in lateral flexion of the neck such as would occur on side impact or rear impact with the head turned. The present study on properties of the human neck in lateral flexion is a continuation and extension of the sagittal motion study. The influence of age, sex, and stature on neck lateral flexor muscle strength and reflex times, and on head/neck range of motion have been studied and results presented in a form which will be useful to designers of mathematical models and crash test dummies.

* Numbers in parentheses designate References at end of paper.

METHODOLOGY

SUBJECT ACQUISITION AND EXPERIMENTAL DESIGN - The results of the sagittal plane study indicated that the neck characteristics of interest were relatively insensitive to stature. As a consequence, the basic experimental design for the lateral motion study used only sex and age as primary variables while considering stature as a secondary variable. The final statistical design chosen for the subject pool was 2 X 3 factorial, with categorization by sex and by the same three age groups as were designated in the sagittal plane study (18-24 years, 35-44, and 62-74). Each of the six subject categories was intended to contain 16

ABSTRACT

Properties of the human neck which may influence a person's susceptibility to "whiplash" injury during lateral impact have been studied in 96 normal subjects. Subjects were chosen on the basis of age, sex, and stature and data were grouped into six primary categories based on sex (F, M) and age (18-24, 35-44, 62-74). The data include: measures of head, neck and body anthropometry in standing and simulated automotive seating positions, three-dimensional range of motion of the head and neck, head/neck response to low-level acceleration, and both stretch reflex time and voluntary isometric muscle force in the lateral direction. Reflex times are found to vary from about 30 to 70 ms with young and middle aged persons having faster times than older persons, and females having faster times than males. Muscle strength decreases with age and males are, on the average, stronger than females. Range of motion was determined by using three-dimensional photogrammetry to compute Euler angles relative to the Frankfort plane. Ranges of motion are reported for 6 planar movements (flexion, extension, left and right rotation, left and right lateral bend) and 3 combination movements (left rotation plus flexion, left rotation plus left lateral bend, and right rotation plus extension) chosen to simulate typical lateral impact conditions. For the young subjects, ranges of motion for males and females are similar. In all cases the range of motion decreases with age, with the rate of decrease for males being greater than that for females. Data are presented in a format applicable to biomechanical modeling of the seated human occupant.

NOTE: This research was sponsored by the Insurance Institute for Highway Safety, Washington, D. C.

I. GENERAL BODY MEASUREMENTS

1. Weight
2. Stature
3. Erect Sitting Height

II. HEAD MEASUREMENTS

4. Head Circumference
5. Bennett Ellipse Circumference
6. Bitragion Diameter
7. Head Breadth
8. Head Length
9. Sagittal Arc Length
10. Coronal Arc Length
11. Bitragion - Glabella Arc Length
12. Bitragion - Menton Arc Length
13. Bitragion - Inion Arc Length
14. Facial Height

III. NECK MEASUREMENTS

15. Lateral Neck Breadth
16. Anterior - Posterior Neck Breadth
17. Superior Neck Circumference
18. Inferior Neck Circumference
19. Posterior Neck Length

IV. MEASUREMENTS TO DETERMINE SIZE AND LOCATION OF MAJOR BODY MASSES

20. Biacromial Breadth
21. Shoulder Breadth (Bideltoid)
22. Chest Height
23. Chest Breadth
24. Chest Circumference
25. Waist Height
26. Waist Breadth
27. Waist Circumference
28. Hip Height
29. Hip Breadth (Standing Erect)
30. Hip Circumference
31. Acromion - Radiale Length
32. Upper Arm Circumference (at Axilla)
33. Upper Arm Circumference (above Elbow)
34. Biceps Flexed Circumference
35. Radiale - Stylion Length
36. Forearm Circumference
37. Wrist Circumference
38. Hand Length
39. Trochanter - Femoral Condyle Length
40. Upper Thigh Circumference
41. Lower Thigh Circumference
42. Fibula Length
43. Fibula Height
44. Calf Circumference
45. Ankle Circumference
46. Foot Length
47. Ball of Foot Breadth

V. MEASUREMENTS RELATED TO SOMATOTYPES

48. Humeral Biepicondylar Diameter
49. Femoral Biepicondylar Diameter
50. Triceps Skinfold
51. Subscapular Skinfold
52. Suprailiac Skinfold

VI. BODY ELEMENT LOCATIONS FOR THE SEATED OCCUPANT

53. Normal Sitting Height (Relative to Seat Reference Point)
54. Tragion Height (Rel SRP)
55. Tragion Depth (Rel SRP)
56. Glabella Height (Rel SRP)
57. Glabella Depth (Rel SRP)
58. Eye Ellipse Point Height (Rel SRP)
59. Eye Ellipse Point Depth (Rel SRP)
60. Eye Ellipse Point Width (Rel Glabella)
61. Cervicale Height (Rel SRP)
62. Cervicale Depth (Rel SRP)
63. Suprasternale Height (Rel SRP)
64. Suprasternale Depth (Rel SRP)
65. Shoulder Height (Rel SRP)
66. Shoulder Depth (Rel SRP)
67. Shoulder Breadth
68. Anterior Superior Iliac Spine Height (Rel SRP)
69. Anterior Superior Iliac Spine Depth (Rel SRP)
70. Bispinous Breadth
71. Trochanterion Height (Rel SRP)
72. Trochanterion Depth (Rel SRP)
73. Bitrochanterion Diameter
74. Hip Breadth (Seated)
75. Infraorbitale Height (Rel Tragion)
76. Infraorbitale Depth (Rel Tragion)
77. Tragion Height (Rel Cervicale)
78. Tragion Depth (Rel Cervicale)
79. Glabella Height (Rel Tragion)
80. Glabella Depth (Rel Tragion)
81. Eye Ellipse Point Height (Rel Tragion)
82. Eye Ellipse Point Depth (Rel Tragion)
83. Ectocanthus Height (Rel Tragion)
84. Ectocanthus Depth (Rel Tragion)

Fig. 1 - Anthropometric measurements

subjects, for a total of 96 in this study. Stature differences were accounted for within each category by selecting three subjects each of short and tall stature and the remaining ten of average stature. ("Short" was defined as the 1-20th percentiles of the appropriate sex and age category, according to the U.S. Public Health Survey, 1962 (2), "average" was selected to include statures in the 40-60th percentiles, and "tall" included the 80-99th percentiles.) This method of stature selection was intended to bias the results toward the 50th percentile groups without ignoring the extremes of the population. Thus, the subject sample population was selected to be representative of the entire U.S. adult population with respect to the major variables.

Subjects were recruited from the local population in and around Ann Arbor, Michigan, and 70% of these subjects tested in the lateral study were previously studied in the sagittal plane study. Each subject's health was screened by means of a medical questionnaire and x-rays of the cervical spine.

ANTHROPOMETRY - Anthropometric data were collected with the objectives of: 1. describing the subjects sufficiently to allow comparison with other study results, 2. determining the location of major body masses for use in mathematical modeling, and 3. locating important landmarks to describe the seated occupant. Toward these goals the 84 measurements listed in Fig. 1 were taken by traditional and photogrammetric methods. Of these 84 measurements, 60 were taken by traditional means in order to describe the general body, head, and neck characteristics and to locate and determine the size of major body masses for use in mathematical modeling. The remaining twenty-four measurements were obtained by computerized analysis of photographs using the technique of three-dimensional orthogonal photogrammetry. Group VI in Fig. 1 contains the 24 measures obtained from photogrammetry as well as traditional measures taken to describe the seated occupant in the relaxed seated position. For these measures, the subject was positioned on an unpadded simulated auto seat (seat pan angle 6 deg below horizontal, seat back 103 deg to seat pan) with the buttocks firmly against the seat back, and the mid-sagittal plane of the subject in the middle of the seat.

VOLUNTARY RANGE OF MOTION - With the realization that pure lateral bending is difficult to achieve voluntarily and unlikely to occur during an accident situation, and with the recognition of a need for more complete descriptions of the range of cervical motion on a general population, the technique of three dimensional orthogonal photogrammetry was adapted to the problem of determining the voluntary head and neck range of motion of subjects for selected head movements.

The basic technique to determine the true position of a point in three-dimensional space by photogrammetry requires only that the point be visible in the film plane of two cameras whose axes are oriented perpendicular to each other and intersect at a common point. For this study, a three camera system was used to insure that each point of interest could be seen by at least two cameras (Fig. 2). The three camera axes form the X-, Y-, and Z-axes of an inertial reference frame with origin at the "true origin." Since the true origin lay somewhere inside the subject during a test, a "visible origin" was rigidly attached to the test fixture at a known distance from the true origin so that it could be photographed by all three cameras. This point was used during data analysis to translate coordinate points into the proper inertial reference frame.

The laboratory arrangement is shown in Fig. 3. Camera mounts for the X and Y cameras were bolted to the floor, as was the test fixture with the

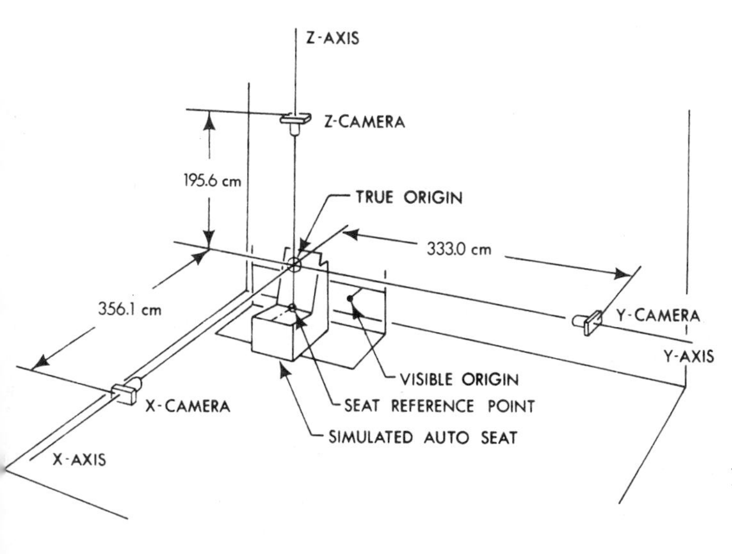

Fig. 2 - Laboratory arrangement for photogrammetric analysis of seated anthropometry and range of motion

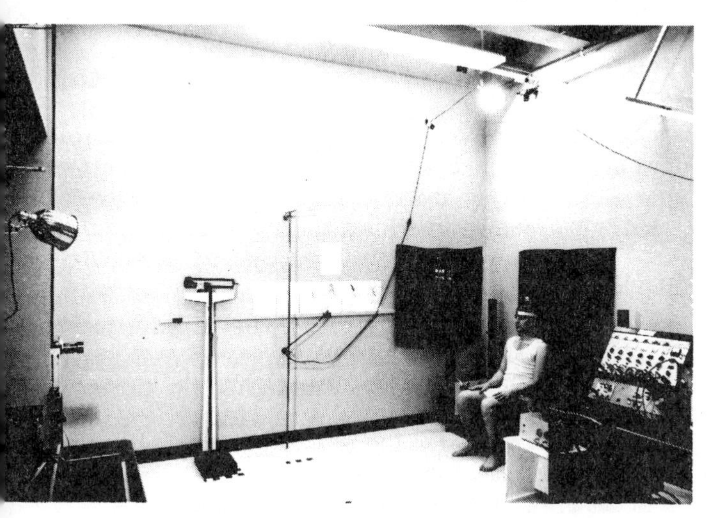

Fig. 3 - Laboratory setup for photogrammetry testing of subject. X and Z cameras are shown, Y camera is off the picture to the right

subject seat. The Z camera, equipped with a motor drive unit to advance the film, was mounted on a slide track attached to the ceiling and wall and was held in alignment by a magnet. The track mounting of this camera allowed it to be brought to a convenient level for film changing and returned to precisely the proper location for testing. The subject, when seated in the test seat, faced the X camera. The Y camera photographed the subject's left side and the Z camera photographed from above. A remote switch released the shutters of all three cameras simultaneously, thus assuring a consistent subject position for analysis.

The subject was prepared for the photogrammetry sequence in the following manner. High-contrast markers were placed at tragion, nasion, and infraorbitale landmarks on the head and at cervicale, suprasternale, and the acromions on the torso. The markers were used to calculate various anthropometric dimensions and to assess the

1. Frankfort Plane (anthropometric neutral)
2. Normal (neutral)
3. Extension
4. Flexion
5. Right rotation
6. Left rotation
7. Right lateral bend
8. Left lateral bend
9. Left rotation plus bend toward left
10. Left rotation plus bend toward rear
11. Right rotation plus bend toward left

Fig. 4 - Sequence of range-of-motion positions

amount of torso movement. A photogrammetry headpiece, consisting of an orthogonal coordinate axis system fixed to a modified liner of a welder's hood, was then fitted snugly to the subject's head. Preparations were completed by positioning the subject in the center of the test seat.

Upon completion of the traditional seated anthropometry, the subject's head was placed in the Frankfort plane neutral position, the sitting height measured, and the photogrammetry sequence begun. Subjects were instructed to keep their shoulders and torso from moving while turning the head and neck slowly and as far as possible in the requested directions. After completing a movement the subject returned to the neutral position, at which time the next movement was described. If a movement was performed incorrectly or poorly (for example, the shoulders moved while performing a movement), the subject was asked to return to the neutral position and begin again. When the subject reached the limits of range of motion at each position (indicated by no further movement) the shutter release switch was depressed, resulting in simultaneous recording of the position by the three cameras.

In all, a total of 11 subject positions were photographed in this manner as listed in Fig. 4. The first position is the Frankfort Plane in which the subject is placed for seated anthropometry and for referencing the range of motion angles. The second position is the subject's normal relaxed sitting position. The next 6 positions are "pure" planar movements in each of the three anatomical planes: rotation, flexion/extension, and lateral bend. Positions 9, 10, and 11 are combinations of two planar movements performed in sequence, with rotation always being performed first. These three movements were selected as having some practical relevance to: 1. left-side impact with occupant's head turned toward the left; 2. rear impact with occupant's head turned toward the left; and 3. left-side impact with occupant's head turned toward the right. Fig. 5 illustrates a subject performing each of these combination movements.

When computing or using Euler angles to describe the movement of a body in space, two factors must be known: 1. the order in which the angles are taken, and 2. the axes about which they are de-

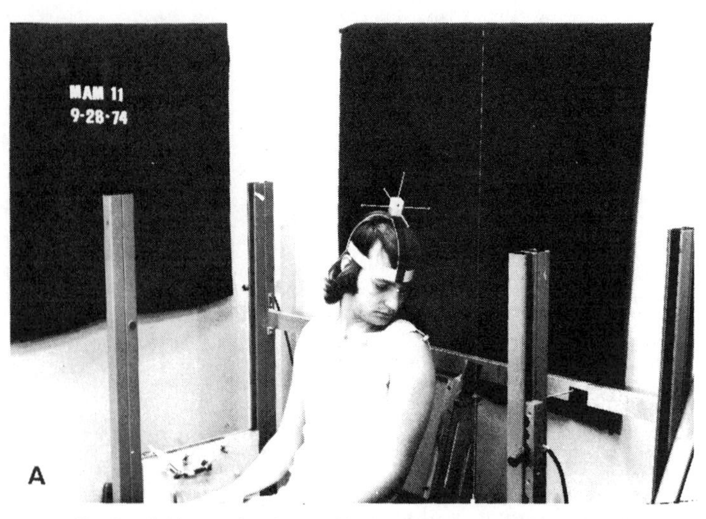

Fig. 5A - Subject performing combination movements of left rotation plus flexion

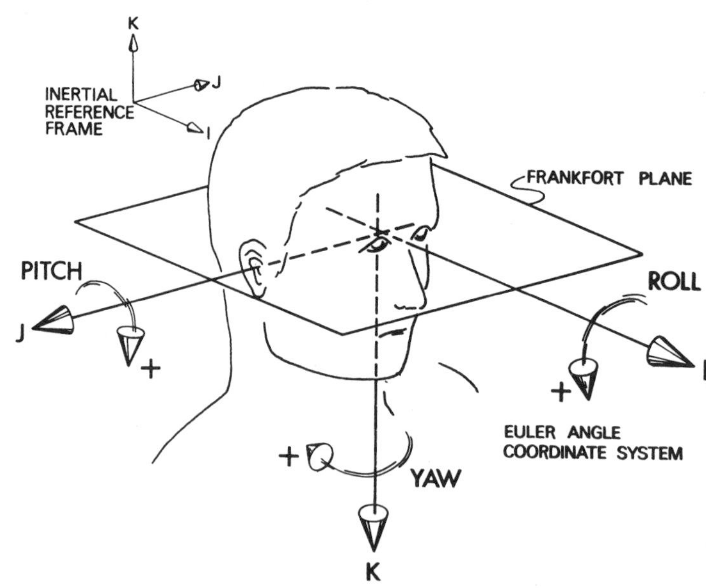

Fig. 6 - Anatomical axes in the head related to the Frankfort Plane used for Euler angle definitions

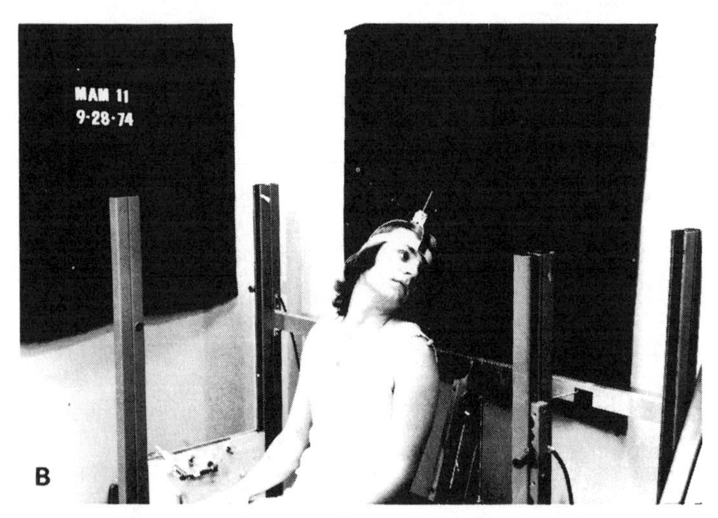

Fig. 5B - Subject performing combination movements of left rotation plus bend toward rear

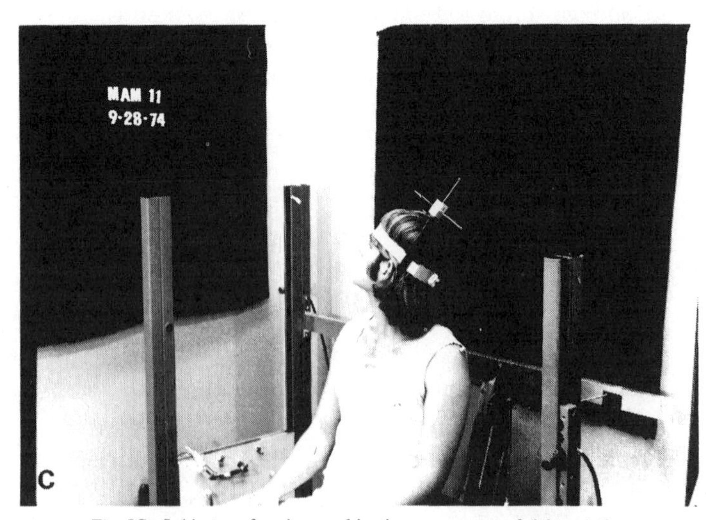

Fig. 5C - Subject performing combination movements of right rotation plus extension

fined. In this study the order of Euler angles is yaw (rotation), pitch (flexion/extension), and roll (lateral bend). The axes about which these angles have meaning are those anatomical axes in the head related to the Frankfort Plane, as illustrated in Fig. 6. That is, yaw is rotation about an axis perpendicular to the Frankfort Plane, pitch is rotation about an axis parallel to a line through left and right tragion, and roll is rotation about an axis parallel to the line formed by the intersection of the Frankfort and mid-sagittal planes. Using this convention and the convention that the positive Z axis is down, positive X is out of the subject's face, and positive Y is out of the subject's right ear, Euler angles which describe the orientation of the head in each of the above positions relative to the Frankfort position were computed. This involved determining the vectors of the headpiece coordinate system axes in each position from the positions in space of the end points as determined by orthogonal photogrammetry, and taking appropriate dot products and trigonometric relationships of these vectors to determine Euler angles. Since it is extremely difficult to align the headpiece coordinate system perfectly with the anatomical axes in the head about which the Euler angles are defined, a correction transformation was needed. The details of these computations as well as a review of basic three-dimensional photogrammetry techniques are given in the lateral neck motion study Final Report to the Insurance Institute for Highway Safety (3).

Photogrammetric data was put into digital form with a "Summographic Tablet" digitizer using a projector to enlarge the film frame. Digitized points were coded on paper tapes and processed by a minicomputer to produce range of motion and seated anthropometry results.

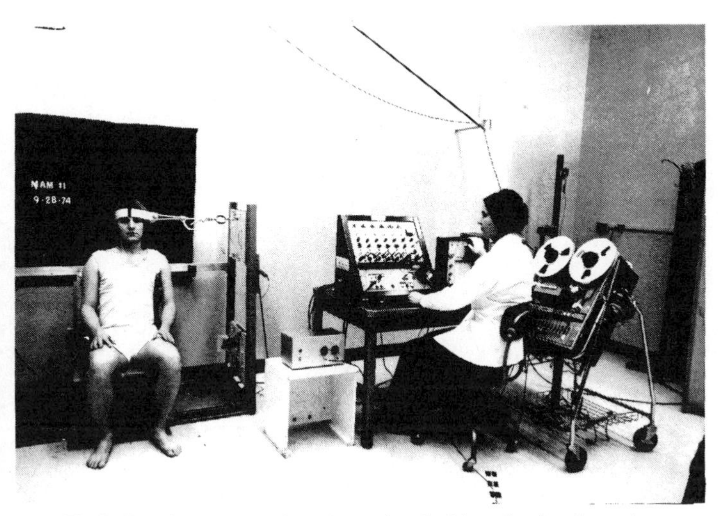

Fig. 7 - Experimental setup for active testing of subjects showing electronic equipment, subject in test seat prepared for strength testing, and research assistant

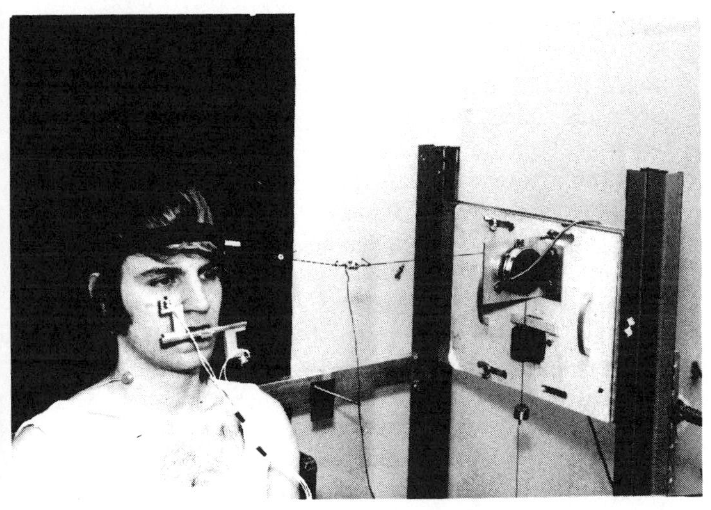

Fig. 8 - Closeup of subject ready for reflex test showing surface electrodes on skin over neck muscle, accelerometers mounted on bite bar, strain ring in weight drop line, and weight release mechanism: note that the shoulder is against a brace to prevent torso movement

REFLEX AND STRENGTH TESTS - The extent to which the neck muscles may reduce head movement and mitigate whiplash injury is a question of considerable importance to understanding injury mechanisms and modeling the response of the head and neck during impact. Two factors addressed in this study are thought to have a direct influence on head and neck response during impact. One is the force or strength which the neck muscles can exert to restrain the movement of the head. The second, which is primarily important during a surprise impact, is the reaction time of the neck muscles, or the time it takes the muscles to achieve their maximum force capability. This time can be divided into two parts: reflex time (the time difference between start of head movement and start of muscle activity; and muscle contraction time (the time from beginning of muscle activity to maximum contraction force). The active test portion of this study was designed to measure both the muscle strength and reflex times of the sample population to determine how these factors vary with age and sex, and to assess how they might affect response to impact. Two tests were conducted to acquire these data. The "reflex test" recorded the subjects' head acceleration and muscle activity in response to a known and sudden force applied to the head in the lateral direction. The "strength test" measured the maximum voluntary isometric pull that the subjects could exert by using their neck muscles and pulling in the lateral direction.

Fig. 7 shows the experimental setup for active testing of subjects (subject performing strength testing), including the amplifying, recording, and monitoring equipment. Complete control of the experiment was provided by the control console shown in front of the researcher. All signals passed through 6 channels on this unit and could be further amplified or filtered prior to recording on magnetic tape.

Fig. 8 shows a subject seated in preparation for reflex testing. A pulsed force was applied near the center of gravity of the head by means of a band placed tightly around the subject's head just over the ears. Attached to this band was a nylon cord which was passed over a pulley, threaded through a 455 g weight, and connected to a 56 g pretensioning weight at the other end. Prior to the test, a larger weight was held in place by an electromagnet. Force was applied to the head by allowing the 455 g weight to drop measured distances of 10, 15, or 20 cm (whichever was the minimum necessary to invoke the stretch reflex). A small strain ring in the nylon cord measured the tension or force applied to the head by the weight drop.

Muscle activity of the sternomastoid muscle group on the subject's right side was measured during the reflex test by bipolar surface electrodes placed over the belly of the muscle. A ground electrode was placed over C7. EMG activity was monitored on an oscilloscope during testing so that the weight could be dropped when the subject was relaxed.

Head response was detected and measured by four accelerometers contained in two biaxial units made by Entran Devices, Inc. These accelerometer units were fastened to a mounting bar and fixed to the subject's head by means of a bite plate which the subject held in his mouth. The accelerometer axes were oriented to allow for complete analysis of the head motion assuming planar movements of the head during the test. For each test, one EMG channel, four acceleration channels, and one force channel were recorded on magnetic tape.

Prior to reflex testing, each subject was given instructions to face straight ahead, close the

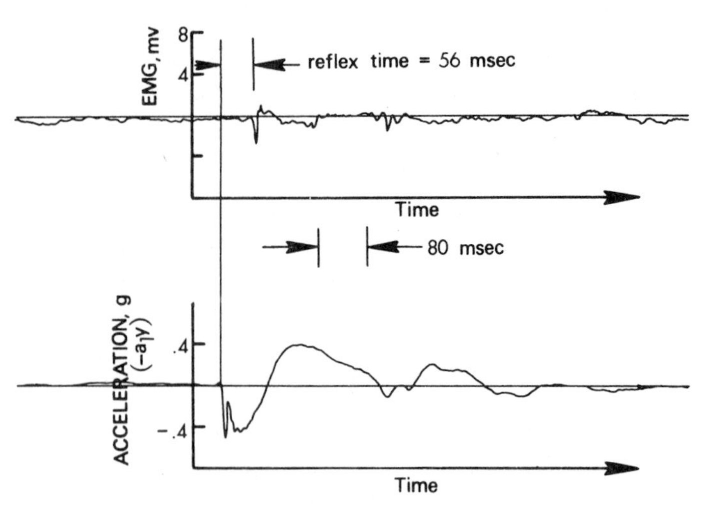

Fig. 9 - Typical EMG signal (top) and acceleration signal (bottom) in response to reflex test. The time from the start of head acceleration to the start of the increased EMG activity is the reflex time

eyes, and relax, but to attempt to keep the head upright when the jerk was felt. A second researcher observed the subject during the tests to ensure that the head did not rotate. If necessary, appropriate adjustments were made in the headband position to achieve planar motion. A series of several tests was run on each subject to obtain an average measure of reflex time.

Reflex time is defined as the time from onset of head acceleration to the time at which EMG signals show an increase in muscle activity. Fig. 9 illustrates a typical acceleration and EMG response to a weight drop of 10 cm. The onset of head motion is clearly indicated by the rapid rise in the acceleration signal. Similarly, the increase in EMG activity is noted by an initial spike of activity, followed by further increased activity. The time between these two points is the reflex time.

Isometric strength testing was measured as shown in Fig. 7. A head strap was placed around the head above the ears and isometric tension was measured by the large strain ring near the restraining fixture. Subjects were tested from both sides, although EMG activity was recorded from only the right sternomastoid muscles. After an initial training pull to check equipment and have the subject "get the feel," testing was begun. Subjects were asked to build to a maximum force in 1 or 2 s and to hold this level for a count of 4 s. Three trials were run from each side with at least one minute of rest between trials. For each trial, the EMG channel and the amplified strain ring output were recorded on magnetic tape. During the tests a second researcher observed the subject to ensure that only the neck muscles were being used (that is, the subject did not lift off the seat) and that the effort was in the lateral direction. A side brace was available for the subject to react against with the upper torso.

Table 1 - Comparison of Anthropometry Results

Subjects		Weight (kg) Lat.	Weight (kg) Sag.	Weight (kg) PHS	Stature (cm) Lat.	Stature (cm) Sag.	Stature (cm) PHS	Erect Sitting Ht (cm) Lat.	Erect Sitting Ht (cm) Sag.	Erect Sitting Ht (cm) PHS
Females										
18-24	N=	16	31	534	16	31	534	16	31	534
	$\bar{x}$=	57.3	58.6	57.7	162.3	163.1	162.1	86.2	85.8	85.3
35-44	N=	16	30	784	16	30	784	16	30	784
	$\bar{x}$=	59.3	59.5	64.5	160.5	161.4	161.3	85.4	85.4	85.6
62-74	N=	16	31	299	16	31	299	16	31	299
	$\bar{x}$=	66.0	65.2	65.5	156.3	158.5	156.2	82.0	82.7	81.5
Males										
18-24	N=	17	30	411	17	30	411	17	30	411
	$\bar{x}$=	71.6	71.4	71.8	175.0	174.9	174.5	91.5	91.1	90.9
35-44	N=	15	30	703	15	30	703	15	30	703
	$\bar{x}$=	78.5	83.5	77.3	173.7	173.9	174.0	89.7	90.5	91.2
62-74	N=	16	27	265	16	27	265	16	27	265
	$\bar{x}$=	70.7	72.9	71.8	169.8	171.3	169.9	88.8	88.8	88.1

RESULTS

Space limitations in this paper do not permit listing of all the tabulated data from this study which may be useful to other researchers. The following review of the results is intended to point out the general findings. For a more detailed presentation, the reader is referred to Appendixes B and C of the lateral neck motion study Final Report to the Insurance Institute for Highway Safety (3).

SUBJECT POOL AND ANTHROPOMETRY - The final subject pool of 96 persons matched closely the original subject pool design with 4 of the 6 subject groups containing 16 subjects including 3 short, 10 medium, and 3 tall in stature. In two of the categories (62-74 year old females and 35-44 year old males), it was not possible to locate enough subjects of average stature and additional short and tall subjects were used to balance these groups.

A common method used to check the assumption of similar subject pools involves comparing these measurements often used to describe a population: weight, stature, and erect sitting height. Table 1 shows a comparison of the lateral motion study results with those of two previous studies--the sagittal plane study and the U.S. Public Health Survey report cited previously. For the lateral motion study, it was intended that the subject pool: 1. be representative, in age and stature distribution, of the adult population of the United States, and 2. duplicate, insofar as possible with 96 subjects, the dimensional characteristics of the 180 subjects of the sagittal plane study. Table 1 shows that, in most categories, stature and erect sitting height matched very closely. Although subjects were not selected on the basis of weight, weights also matched well.

Twenty-eight of the measurements relate to the location of major body masses. These measurements describe the torso in three segments and the upper and lower extremities. The data are too voluminous to include herein, and the reader is referred to the statistical tabulations for all

Table 2 - Selected Results – Location of Upper Torso Mass

Subjects		Chest Dimensions (cm) Ht.	Br.	Cir.
Females				
18-24	$\bar{x}$ =	121.8	26.0	83.3
	S.D.=	3.7	1.9	4.0
35-44	$\bar{x}$ =	122.8	27.6	85.8
	S.D.=	4.5	2.1	5.6
62-74	$\bar{x}$ =	117.9	28.8	90.7
	S.D.=	5.1	2.8	6.7
Males				
18-24	$\bar{x}$ =	131.3	31.4	94.9
	S.D.=	3.8	2.6	5.3
35-44	$\bar{x}$ =	131.3	32.0	101.0
	S.D.=	5.6	1.6	4.8
62-74	$\bar{x}$ =	127.8	31.2	96.1
	S.D.=	4.1	2.7	4.9

subject categories which may be found in Appendix B of the Final Report (3). Table 2, which lists the upper torso measurements, is an example of the results. The upper torso is represented by chest height, breadth, and circumference. The chest height dimensions correlate closely to stature, averaging 75% of stature in each subject group. Upper torso mass (as represented by chest breadth and circumference) correlates well with weight, and remains relatively constant throughout life for males, but tends to increase with age in females. On the average, males are larger than females in upper torso dimensions.

Thirty-two measurements were used to describe the seated occupant with reference to a three-dimensional coordinate system. Twenty-one measures were taken to locate certain landmarks on the head, upper torso, and pelvis; eleven were calculated to locate landmarks with respect to other landmarks, particularly on the head. An example of the use of these measures is shown in Fig. 10. Descriptive statistics for these measures are also reported in Appendix B of the Final Report, together with statistics for 24 others which specify head and upper torso locations with respect to the Seat Reference Point (SRP) only and permit calculations relating one landmark to another.

That the anthropometric data obtained in this study present a logical set of dimensions directly applicable to three-dimensional modeling of the human body in the seated position is shown by the illustration in Figure 11 which was constructed from the data for one subject on isometric (3-D) graph paper.

RANGE OF MOTION - For a complete list of Euler angle statistics for the head positions and subject groups used in this study, the reader is again referred to Appendix C of the Final Report (3). Table 3 and Fig. 5. 12 and 13 summarize these data for the planar ranges of motion in the sagittal, rotational, and lateral planes for various subject groupings. For each plane, the ranges of motion for males and females are nearly the same but decrease with age. For all age groups, the largest range of motion is in rotation while the smallest is in lateral bend. The rate of decrease with age appears to be similar for each. As the figures show, however, males tend to show a slightly sharper decrease with age than females. Table 4 shows the percentage decrease from young to elderly age groups, for range of motion in the three planes, for males and females separately.

Table 5 shows the average Euler angles at each of the test positions for all subjects, females, and males, respectively. Again it is seen that there is little overall difference in results between males and females. Several other observations which can be made by inspection of these tables and similar data for each subject group are:

1. The amount of flexion possible after rotation is about half that possible from the Frankfort Plane position.

2. The amount of lateral bend possible after full rotation is nearly the same as that possible from the Frankfort Plane position.

3. The amount of extension possible after full rotation is about one-third that possible from the Frankfort Plane position, is relatively independent of age, and is usually accompanied by some right lateral bend.

4. Full right rotation is usually accompanied by a slight right lateral bend, while full left rotation is usually accompanied by a slight left lateral bend.

5. The average subject neutral position is nearly identical to that of the Frankfort Plane position.

6. The amount of extension possible from the Frankfort Plane position is usually greater than the amount of flexion, especially in the 18- 24- year-old groups.

7. The amount of left lateral bend is usually greater than the amount of right lateral bend, although it is accompanied by a greater amount of head rotation.

REFLEX TIME - Table 6 and Fig. 14 present the results of the reflex time analysis for the right sternomastoid muscle group resulting from lateral head jerks to the left. For females, the reflex time is nearly the same for the young and middle age groups (45.1 ms and 43.6 ms, respectively), but shows a significant increase to 53 ms for elderly females. For the males, the reflex time shows a steady increase with age and for each group is

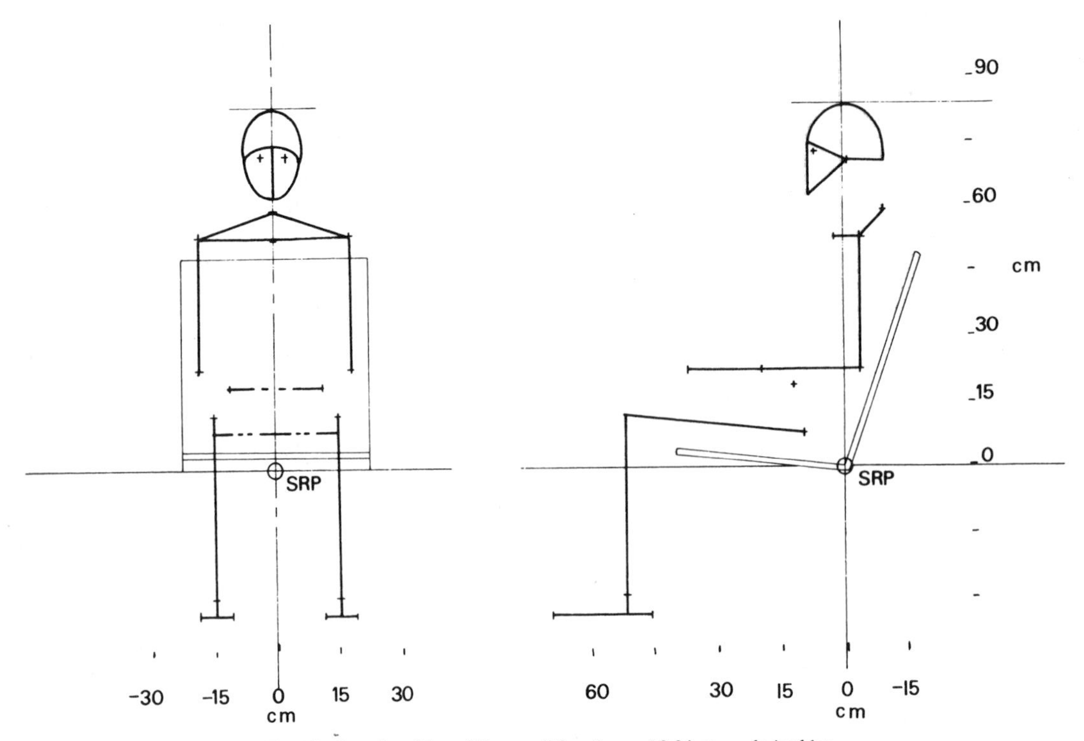

Fig. 10 - Seated position of "average" female, age 18-24 years, derived by combining seated anthropometry with anthropometry from photogrammetry

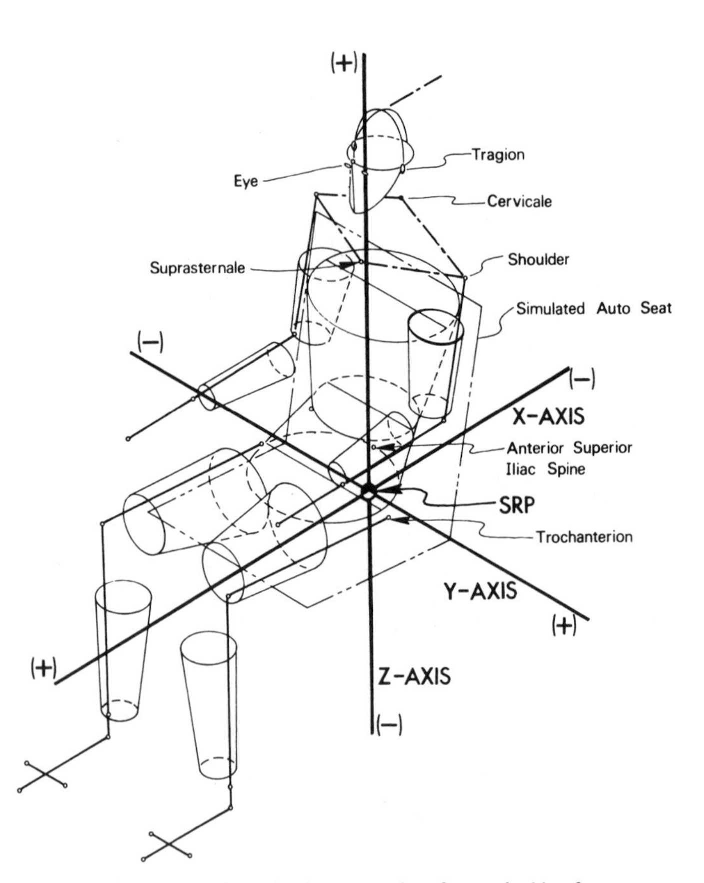

Fig. 11 - Three-dimensional representation of a seated subject from anthropometry data

Table 3 - Total Planar Range of Motion by Subject Group

Subjects		Planar Range of Motion (deg) Sagittal	Rotation	Lateral
Females				
18-24	$\bar{x}$=	124.1	150.6	86.0
35-44	$\bar{x}$=	104.6	143.6	73.9
62-74	$\bar{x}$=	84.3	123.6	56.3
Males				
18-24	$\bar{x}$=	129.0	149.5	86.3
35-44	$\bar{x}$=	102.7	137.1	73.0
62-74	$\bar{x}$=	76.6	113.9	48.0
All Females	$\bar{x}$=	104.2	139.3	72.0
All Males	$\bar{x}$=	103.3	133.7	69.8
All Subjects	$\bar{x}$=	103.7	136.5	71.0

greater than the average reflex time for the respective female group. This difference in reflex times between the sexes is most pronounced for the middle age groups, where the average for females was 43.6 ms compared to that for the males of 52.8 ms. The average reflex time for all males was 53.3 ms, compared to 47.1 ms for all females, and the overall average for all subjects was 50.2 ms.

STRENGTH - Table 7 and Fig. 15 show the average results by subject group of the strength

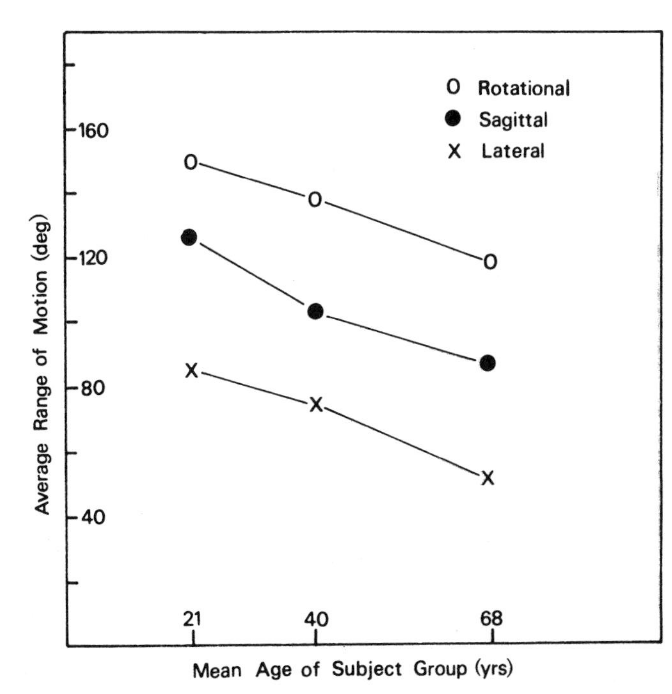

Fig. 12 - Average range of motion versus mean age of subject group for males and females combined

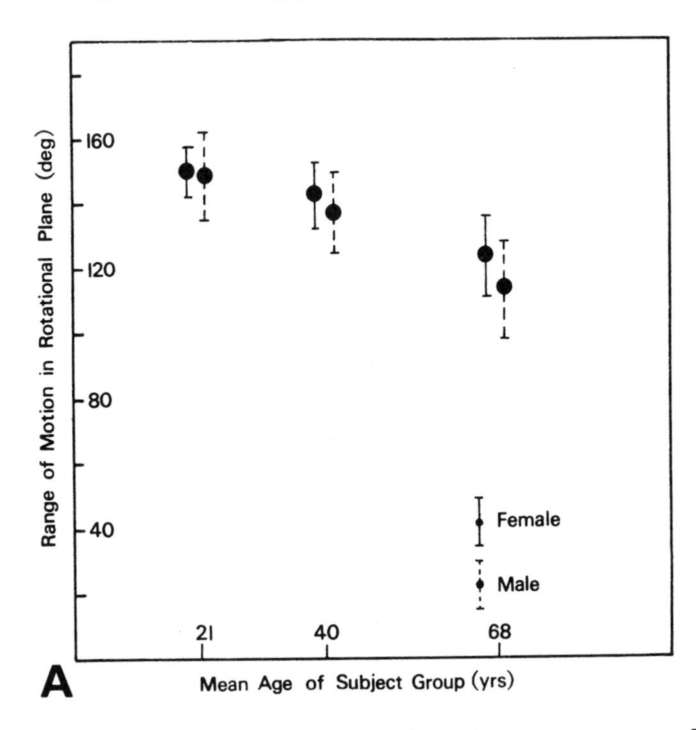

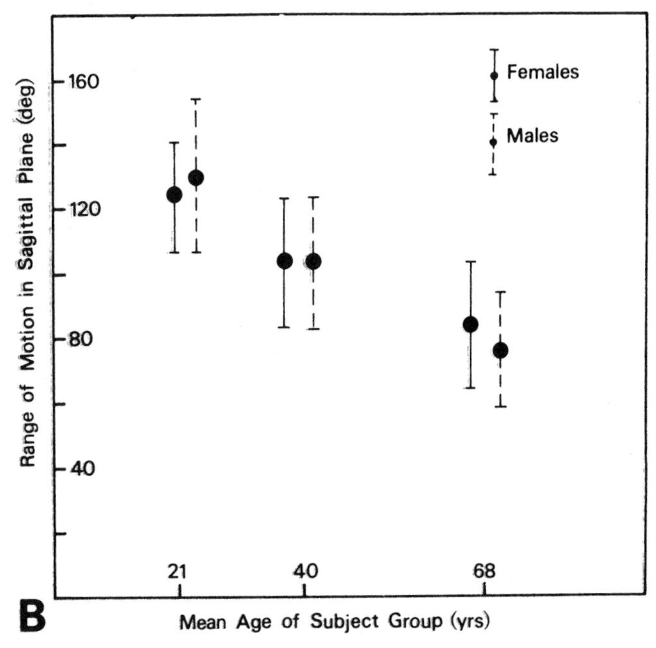

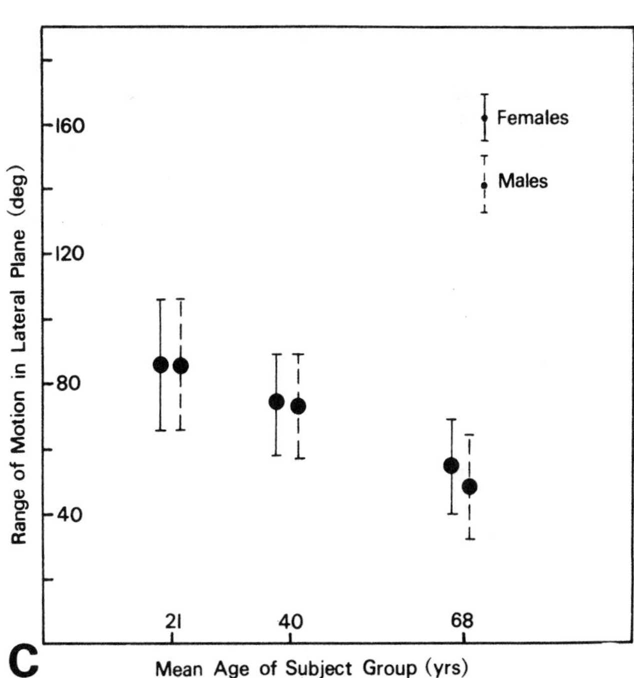

Figs. 13A-C - Range of motion in the sagittal, rotational, and lateral planes versus mean age of subject group; brackets indicate the standard deviation of the sample

Table 4 - Percentage Decrease in Total Range of Motion between Young and Elderly Age Groups

	Sagittal	Rotational	Lateral
emales	32.1	17.9	34.5
ales	40.6	23.8	44.4

Table 5 - Euler Angles for Test Positions Relative to Frankfort Plane

	All Subjects (deg)			Females (deg)			Males (deg)		
Position	Yaw	Pitch	Roll	Yaw	Pitch	Roll	Yaw	Pitch	Roll
Neutral	- .4	- 1.3	- 1.1	.7	- .5	- 1.4	- .2	- 2.2	- .8
Extension	.6	54.2	- .8	.7	53.6	- 1.4	1.1	54.7	- .5
Flexion	1.9	-49.5	- 4.5	1.1	-50.5	- 5.0	2.8	-48.6	- 4.0
Rt Rotation	67.5	- 1.1	4.3	68.5	- .2	5.4	66.4	- 2.1	3.4
Lt Rotation	-69.0	- 2.1	- 9.8	-70.8	- 1.9	-12.7	-67.3	- 2.3	- 7.0
Rt Lat Bend	2.9	1.8	32.5	1.7	2.5	31.7	4.1	1.2	33.3
Lt Lat Bend	- 6.1	2.2	-38.3	- 5.0	2.2	-40.3	- 6.2	2.2	-36.3
Lt Rot + Flexion	-67.2	-24.6	-14.1	-67.9	-25.3	-16.5	-66.5	-24.0	-11.8
Lt Rot + Lt Lat Bend	-68.1	7.9	-37.8	-69.7	5.9	-40.3	-66.5	9.8	-35.3
Rt Rot + Extension	64.5	18.4	7.0	66.4	19.0	9.2	62.6	17.9	4.8

Table 6 - Reflex Times of Right Sternomastoid Muscle Group in Lateral Bend

Subjects	Muscle Reflex Time (ms) N	$\bar{x}$	S.D.
Females			
18-24	16	45.1	10.0
35-44	16	43.6	12.7
62-74	15	53.0	11.2
Males			
18-24	17	48.9	6.3
35-44	14	52.8	9.1
62-74	16	58.3	14.9
All Females	47	47.1	11.8
All Males	47	53.3	11.2
All Subjects	94	50.2	11.9

testing in lateral flexion. Testing was performed on both right and left lateral flexors and it is seen that the average results from both sides are in excellent agreement. There is only a small change in strength from the young to middle age group (males showing an increase and females a decrease) but there is a considerable decrease in strength for both males and females from the middle age to the elderly subject groups. There is also a considerable difference in strength between males and females, males being 1 1/2 times stronger on the average.

DISCUSSION

RANGE OF MOTION - By using the technique of three-dimensional photogrammetry to compute Euler angles for range of motion, not only the planar ranges of motion but the deviations from these planar movements are obtained. In the present study, for example, it was found that most subjects performed some head rotation while attempting to perform a lateral bend and that this rotation was usually to the right for right lateral bend and to the left for left lateral bend. For some individuals, especially in the elderly groups, it was even difficult to perform a pure flexion or extension without some substantial amount of head rotation (even when the subject was asked to repeat the movement). This was apparent in the yaw angle reported for these subjects during flexion or extension.

In addition, this technique provides the ability to describe the total range of motion for the combination movements consisting first of rotation and then bending in some other direction. The three compound movements analyzed in this study are but a start in the task of describing the complete range

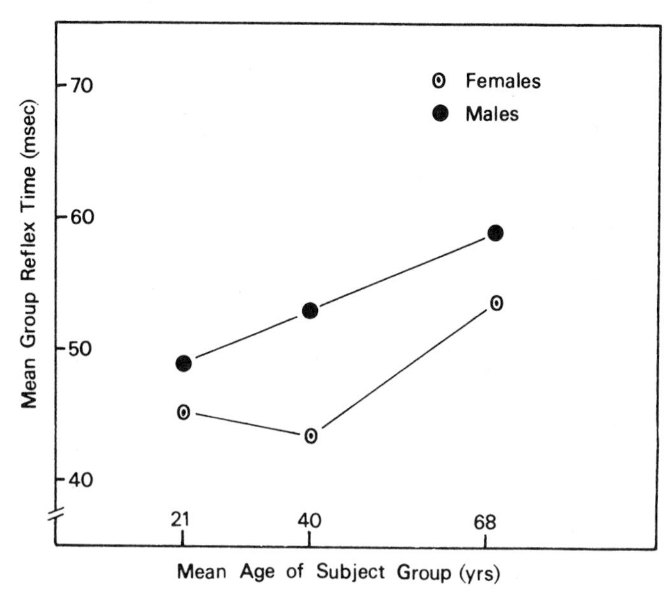

Fig. 14 - Mean reflex time versus mean age of subject group

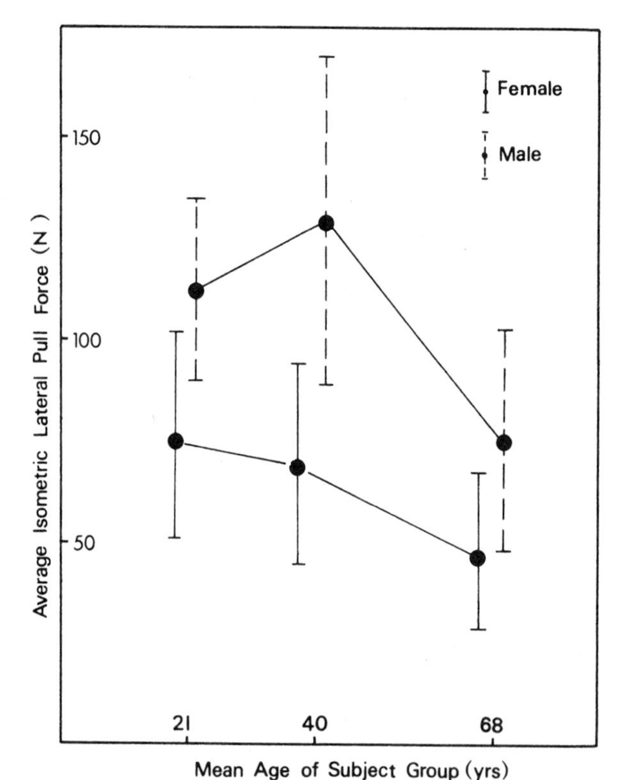

Fig. 15 - Average isometric pull force versus mean age of subject group; brackets indicate the standard deviation of the sample

of motion of the head and neck. For these and other complex movements, it must be noted that the values of the Euler angles are based on an assumption about the order in which they occurred. Thus, during left rotation plus bending toward the left shoulder, it is assumed that the order of the movements is yaw, pitch, and roll. This, of course, is a limitation to the use of Euler angles, since the second movement (bend toward the left) may be a rotation about some axis other than the pitch or roll axes (that is, pitch and roll may occur simultaneously), while in fact the values of the Euler angles assume that pitch occurred prior to roll. The computation of Euler angles from orthogonal photogrammetry provides one means of describing head movement in space. However, the necessity to specify an order to the Euler angles is a limitation to the technique, since it is very unlikely that the movements will be purely planar in a real situation.

The results for planar ranges-of-motion in the study are in general agreement with those found by Ferlic (4), although some differences exist--especially for the sagittal planes and for elderly groups. Comparisons are difficult in some cases, however, due to the small sample size of some groups in Ferlic's data. In general, where differences exist, the results of this study show smaller ranges-of-motion than those found by Ferlic. Two possible explanations should be mentioned.

The first is a consequence of the fact that the "stops" for range of motion are not solid limits, and therefore, subject motivation is an important factor in determining how far the subject will voluntarily move. A subject, when highly motivated, will try harder and achieve a greater range of motion than when poorly motivated. While subjects in this study were generally considered to be well motivated and were requested to move as far as they were able, this motivation factor was difficult to measure or to change. In particular, for extension, the fear of possible neck strain or injury may reduce the subject's voluntary effort.

The second factor which must be considered is the amount of torso movement. As previously mentioned, subjects were asked to perform all head and neck movements without moving the shoulders and torso. Subjects were watched carefully and when any torso movement did occur the subject was asked to repeat the movement. Thus torso movement was minimized in the present study and this could also account for the lower values of range of motion. Furthermore, it is possible that the restriction of subject torso movement caused subjects to be more cautious and less able to push against the "stops," thereby resulting in smaller ranges of motion.

REFLEX TIMES - The reflex times determined in this study are the times measured from onset of head movement to the beginning of muscle activity. Such times are an important consideration in modeling the human response to impact where the occupant was unaware of the impending impact through visual or other means. The reflex time alone, however, is not the time required for the muscle to exert a maximum force to resist head movement caused by the impact. To it must be added a muscle contraction time (on the order of 100 ms). The sum of these two times might then be called a reaction time, or, the time from onset of head movement to the time at which the muscle can provide a maximum restraining effect on limiting head movement. The effect of variations of reflex time (25 ms for the faster subject to 80 ms for the slowest) on the head response during impact may be studied by mathematical modeling techniques.

It is interesting that the reflex times for the sternomastoid muscle groups were significantly smaller for lateral bend (50.2 ms overall average) than for sagittal bending (extension) in the previous study (71.7 ms overall average). The reason for this difference may be a difference in rate of muscle stretch for the head jerks in the two directions. If, for example, the sternomastoid muscle is oriented such that it is stretched at a greater rate (that is, it receives a greater percentage of stretch in the same period of time) during lateral bend than during extension, then the initiation of impulses from a sufficient number of muscle spindles to cause a "measurable" EMG signal will occur sooner for lateral bend than for extension.

While the muscle reaction times for lateral bend may be too long to prevent neck injury during complete surprise impacts, it is interesting that the reflex times for lateral bend are shorter than for extension, and the range of motion is also less. In fact, the average lateral reflex time is 70% of the average extension reflex time (50.2 ms/71.7 ms), and the average lateral range of motion is 66% of the average extension range of motion (35.6°/54.2°). Thus, there may be some functional significance to these shorter lateral bend reflex times.

MUSCLE STRENGTH - The values of neck strength presented in Table 7 are the force values or tensions developed in a line attached about the head when the subject exerted a maximum pull with the lateral flexors. From these force values it is possible to calculate approximations of the actual muscle tension developed in the neck muscles. In a very simplified way, Fig. 16 represents a free body diagram of the head and neck. F_y is the force developed by the subject pulling on the cord and T_s is the tension in the sternomastoid muscle groups. There are the primary muscles which attach between the head and torso. The tension, T_s, can therefore be estimated by summing moments about the occipital condyles:

$$\Sigma M_{condyles} = 0 = T_s l_1 - F_y l_2$$

$$T_s = \frac{F_y \cdot l_2}{l_1}$$

Table 7 - Isometric Pull Force Using Neck Lateral Flexor Muscles

Subjects		Right	Left	Avg.
		Muscle Group Force [N (lbf)]		
Females				
18-24	N =	16	16	16
	$\bar{x}$ =	81.4 (18.3)	85.9 (19.3)	83.6 (18.8)
	S.D.=	28.0 (6.3)	32.0 (7.2)	28.9 (6.5)
35-44	N =	16	16	16
	$\bar{x}$ =	74.7 (16.8)	78.3 (17.6)	77.0 (17.3)
	S.D.=	25.4 (5.7)	32.9 (7.4)	28.5 (6.4)
62-74	N =	16	15	16
	$\bar{x}$ =	53.4 (12.0)	48.9 (11.0)	52.5 (11.8)
	S.D.=	21.8 (4.9)	18.7 (4.2)	20.5 (4.6)
Males				
18-24	N =	17	17	17
	$\bar{x}$ =	120.6 (27.1)	128.6 (28.9)	128.6 (28.9)
	S.D.=	22.7 (5.1)	29.4 (6.6)	24.9 (5.6)
35-44	N =	14	14	14
	$\bar{x}$ =	142.3 (32.0)	144.1 (32.4)	142.8 (32.1)
	S.D.=	44.5 (10.0)	47.6 (10.7)	45.4 (10.2)
62-74	N =	15	15	15
	$\bar{x}$ =	81.8 (18.4)	85.4 (19.2)	84.1 (18.9)
	S.D.=	32.0 (7.2)	31.1 (7.0)	30.7 (6.9)
All Subjects	N =	94	93	94
	$\bar{x}$ =	91.6 (20.6)	95.2 (21.4)	93.4 (21.0)
	S.D.=	41.4 (9.3)	44.9 (10.1)	42.3 (9.5)

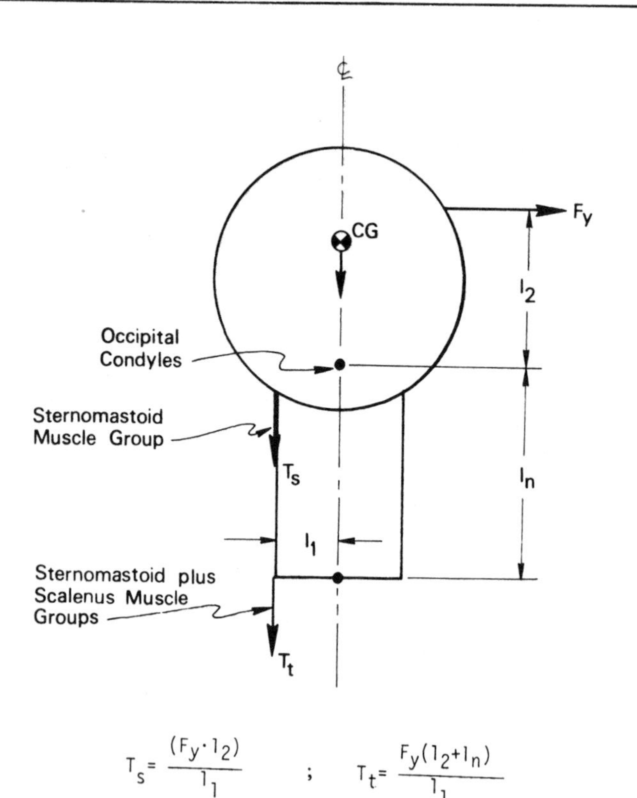

$$T_s = \frac{(F_y \cdot l_2)}{l_1} \quad ; \quad T_t = \frac{F_y(l_2 + l_n)}{l_1}$$

Fig. 16 - Simplified free-body diagram of head and neck showing approximate relations between measured force, F_y, and muscle tensions in sternomastoid (T_s) and scalenus plus sternomastoid (T_t) muscle groups

From anatomical observations (Grant, 1962), it can be shown that one centimeter is a reasonable estimate for the distance between the outside of the neck and the line of action of the sternomastoid muscle group. Accordingly, l_1 is equal to one-half the measured neck breadth minus one centimeter. The distance l_2 is determined by measuring the distance from the occipital condyles to tragion from x-rays and adding this to the distance from tragion to F_y, which was measured during testing.

The scalenus muscles are also involved in keeping the head and neck erect but attach from the torso along the entire length of the cervical spine. Therefore, to include their input into the model, moments can be summed about a point at the base of the neck, say $C_7 - T_1$. In so doing, the tension developed in all the neck muscles, sternomastoid plus scalenus, is estimated by:

$$T_t = \frac{F_y (l_2 + l_n)}{l_1}$$

In this way muscle tension may be computed from the measured pull force for use in computer modeling of crash impact. It must be considered, however, that while subjects were asked to pull with a maximum effort, the resulting tensions are probably somewhat less than would be developed in a "panic" or emergency situation. In any model therefore the values determined from these data should be considered as no more than 80% (5) of the maximum available physiological strength capacity.

CONCLUSION

The body of data described in this paper and documented more thoroughly in report UM-HSRI-BI-75-4 to IIHS provides information on properties of the human neck which will be useful in improving and extending representations of the human neck in either mathematical models or anthropometric test dummies for studies involving the general vehicle occupant population. By correlating these results and computer simulations which use these data with actual statistics of injuries, further insight into the mechanisms involved in injury and methods of prevention may be found.

In general it has been found that elderly persons have a 20 - 45% decrease in range of motion over young persons and this decrease is most severe in the lateral and sagittal planes and for males. Reflex times in lateral head jerks range from about 30 to 70 ms with elderly persons having longer times and females being slightly faster than males. These differences, however, may not be significant in preventing or causing neck injuries on complete surprise impact as the total time to maximum muscle force (including 100 ms contraction time) is on the order of 130-170 ms and is probably too long to prevent injury in a high-speed collision.

With regard to neck muscle strength, males show 1 1/2 to 2 times greater strength than females and for both males and females the strengths decrease with age. From the weakest group (elderly females) to the strongest (middle aged males) group there is a nearly three-fold increase in muscle strength capability. Such differences may be a major factor in spinal injury susceptibility suring impact, especially during conditions where the occupant has some prior warning to allow for pre-tensing of the neck muscles and where the impact acceleration is not extreme.

In conjunction with this study, these data were used in a series of computer simulations of crash impact using the MVMA-Two Dimensional Crash Victim Simulation Model (Version 3) (6). Variations in the responses were observed for variations in the above parameters for the different subject groups. These findings are presented and discussed in the following paper (7), and illustrate the manner in which these data may be used.

ACKNOWLEDGEMENTS

This multidisciplinary study could not have been accomplished without the support of the Insurance Institute for Highway Safety, Washington, D. C. and the contributions of many talented individuals. The authors are particularly grateful to Brian O'Neill, Vice President of Research, IIHS, who provided statistical design guidance and monitored the study. Others who contributed to this study include Ann Russ Desautels, who was responsible for taking and processing x-rays, and who took the anthropometry and assisted in all phases of data collection; Robert Eckstein and Gary Gaul, who recruited and scheduled subjects and assisted with subject testing; Paul Katz, who assisted in reducing the photographic data; Carl Christiansen, who prepared the illustrations and drawings; Walter Low, who wrote the photogrammetry data analysis computer programs; and Dr. Harold Portnoy, neurosurgeon at Pontiac General Hospital, who gave advice and consultation.

Additional assistance was provided by Dr. D. Hurley Robbins, Head, Biomathematics Department, HSRI, who advised on the mathematical modeling design and anthropometry measures useful to modeling; Dr. John Melvin, Head, Biomechanics Department, HSRI, who generously provided advice on instrumentation and data reduction problems; and Dr. Herbert Reynolds, physical anthropologist, Biomedical Department, HSRI, who suggested seated orthogonal photogrammetry measures and provided consultation on anthropometric measurements and techniques. The authors are grateful to the volunteer subjects from the Ann Arbor area, without whose willing participation this study could not have been conducted.

REFERENCES

1. D. R. Foust, et al., "Cervical Range of Motion and Dynamic Response and Strength of Cervical Muscles," Proceedings, 17th Stapp Car Crash Conference, SAE paper 730975, (1973).

2. "Weight, Height, and Selected Body Dimensions of Adults: United States, 1960-62," Public Health Service, 1963.

3. R. G. Snyder, et al., "Basic Biomechanical Properties of the Human Neck Related to Lateral Hyperflexion-Injury," Final Report UM-HSRI-75-4, March 1975.

4. D. Ferlic, "The Range of Motion of the 'Normal' Cervical Spine," John Hopkins Hosp. Bull., Vol. 110, (1962), pp. 59-65.

5. D. B. Chaffin and W. H. Baker, "A Biomechanical Model for Analysis of Symmetric Sagittal Plane Lifting," AIIE Transactions (March 1970), pp. 16-27.

6. B. M. Bowman, et al., "MVMA Two-Dimensional Crash Victim Simulation, Version 3," NTIS Nos. PB 235753, PB 236907, PB 236908. Final Report UM-HSRI-BI-74-1, Highway Safety Research Institute, Univ. of Mich., Ann Arbor (1974).

7. B. M. Bowman, L. W. Schneider, and D. R. Foust, "Simulated Occupant Response to Side-Impact Collisions," Proceedings, 19th Stapp Car Crash Conference, Nov. 1975.

770922

Intracranial Pressure Dynamics During Head Impact

Alan M. Nahum and Randall Smith
University of California

C. Ward
Naval Construction Battalion Center

Abstract

Two series of cadaver head impact experiments are presented. Series I consists of individual experiments and Series II multiple sequential impacts on a single specimen. Measured intracranial pressures were correlated with other impact parameters. The use of a linear finite element model is also presented

Supported by National Institute for Occupational Safety and Health, Grant # RO 1 OH 00404

THE DYNAMIC RESPONSE of the human brain to head impact has not yet been established. As a result, the hypothetical injury mechanisms which have been advanced over the years can not be substantiated. Factual data is required before conclusions can be made associating brain trauma and head impact response. The objective of this study is to provide such data for a frontal impact by measuring response quantities and observing the brain damage produced. The study is a continuation of an ongoing program to investigate the unembalmed cadaver head as a model for closed head injury. This series adds a new parameter to that of the previously reported study (1)*, that of intracranial pressure.

In a prior study, Roberts, et al., (2) reported a series of experiments using an unrestrained human skull from which the external soft tissues and brain were removed, leaving in some instances the internal membranes and sealing the foramen magnum. Linear pressure gradients were produced in the contained fluid by a frontal impact, which compares with the results reported herein. Further correlation would require a time reference frame which is not noted in (2). Lindgren (3) also measured pressures in a fluid filled skull. However, he supported the skull at the contrecoup location, restricting skull motion. Pressure correlation between (3) and the current experiments is not feasible because of this imposed kinematic constraint. No additional human intracranial pressure data is available. Head impact experiments where the pressure in fresh human brain material is measured, have not previously been reported.

Two types of experimental series are reported here. Series I consists of individual experiments on different specimens. Series II reports a series of multiple impacts on a single specimen. The latter series was performed in order to control some of the variables that might be introduced by differing anatomic specimens. A final section demonstrates the potential use of some of the data in a linear finite element model of the brain.

EXPERIMENTAL METHODOLOGY

A detailed description of specimen preparation, impact instrumentation, and data reduction techniques can be found in the authors' prior publication (1). Seated, stationary cadaver subjects were impacted by a rigid mass traveling at a constant velocity. The blow was delivered to the frontal bone in the mid-sagittal plane in an anterior-posterior direction. The skull was rotated forward so that the Frankfort anatomical plane was inclined 45° to the hori-

---

*Numbers in parentheses designate References at end of paper.

zontal. Various padding materials were interposed between the skull and impactor to vary the duration of the applied load. The input force and the biaxial acceleration-time histories of the skull were recorded during the impact event. Static fluid pressurization of the cranial vascular network and cerebral spinal fluid space to _in vivo_ pressure levels at impact was also performed. Following the impact exposure the brain was perfused with a 10 per cent formalin and carbon particle solution. Injury to the contents of the cranium, as evidenced by extravasation of the carbon particles into the brain tissue, was then assessed by pathologic examination.

In order to obtain greater pathologic information concerning the area surrounding the lateral ventricles of the brain, the ventriculostomy technique employed to provide input and monitoring sites for cerebral spinal fluid simulation used in the prior study was eliminated. An alternate method of pressurization by entering the dura over the superior surface of the brain for addition and removal of saline via 8 French catheters was employed in the present series.

In addition to the dynamic measurements of input force and head acceleration, a series of intracranial pressure-time histories were recorded. Endevco model 8510 piezo resistive pressure transducers (resonant frequency:180KHz) were used to monitor the dynamic intracranial pressure during the impact event. A 5 mm. diameter hole was made in the skull and the bone thickness measured. A stainless steel nipple was inserted a distance equal to the bone thickness and the pressure transducer threaded into the nipple such that the diaphragm of the transducer was flush with the inner surface of the skull to prevent bruising of the brain due to protrusion of the transducer into the cranium. In all but one (Expt. 36; occipital pressure #2) of the pressure transducer placements, the dura was opened at the insertion site to allow subdural pressure measurement. Transducers were placed in the frontal bone adjacent to the impact contact area, immediately posterior and superior to the coronal and squamosal sutures respectively in the parietal bone, and inferior to the lambdoidal suture in the occipital bone. Additionally, transducers were placed in the occipital bone at the posterior fossa. A second type of dynamic pressure measurement was obtained by insertion of a Kulite model MCP-808-9R (resonant frequency: 150 KHz) catheter tip pressure transducer in the internal carotid artery to the level of the carotid siphon.

Due to the limited number of transducers available for a given experiment and the desire to acquire information at various anatomic sites, pressure measurements were not duplicated at all locations for each experiment. Certain measurements were specifically paired to examine questions of pressure pulse symmetry and recording accuracy. Bilateral occipital (Expts. 37,38) pressures were monitored to gain information on pulse symmetry. Transducers were also placed adjacent to each other in the posterior fossa region (Expt. 46-52) and occipital (Expt.36) to determine if the measurement technique would yield similar results in essentially the same anatomical area of the skull.

EXPERIMENTAL RESULTS

SERIES I - Series I consisted of 8 individual impact experiments (Expts. 36-44, 54) on individual cadavers essentially continuing a prior impact series with the addition of intracranial pressure measurements. The cranial anthropometric measurements are described in Table I. Impactor mass and velocity combinations ranging from 5.23 to 23.09 kg. and 8.41 to 12.95 m/s, respectively, were selected for the series A detailed listing of these conditions is shown in Table II. The primary head impact response data and peak intracranial pressures recorded are tabulated in Table III. Peak input forces varied from 5.20 to 14.84 N x 10^3 with resulting peak head accelerations of 1.52 to 3.90 m/s^2 x 10^3. High positive peak pressures were recorded beneath the impact site in the frontal region. The pressure magnitudes decreased and eventually became negative as the area opposite the blow was approached. The greatest negative pressures were generated at the posterior fossa which, due to the inclination of the skull, was the point opposite the impact site. A representative data sample from Series I of the input forcing function, resulting head acceleration, and intracranial pressures developed is shown in Fig.1. The results of coding the individual findings according to an injury scale (Table IV) specifically designed for these experiments are shown in Table V. Also contained in this table are the impact severity ratings (GSI, HIC and associated computational period) for each experiment. Head rotation in the midsagittal plane versus time was monitored for all experiments with high speed motion picture photography. Dynamic pressure changes within the skull had ceased before significant rotation of the skull had occurred. Shown in Fig.2 is a crosshatched band containing rotation-time data for all impact experiments. Also plotted is the frontal pressure recorded in Experiment 37 to demonstrate the phasing relationship of intracranial pressures and head rotation. As depicted, the frontal pressure event was concluded before head rotation of more than a few degrees had occurred.

Table I - Cranial Anthropometry

Experiment	Spec/Sex	Age (yrs)	A	B	C	D	E	F	G	H
			cm							
36	101 F	87	15.2	18.7	17.8	12.7	22.4	55.0	33.0	36.5
37	108 M	42	14.5	16.7	19.2	13.6	22.3	56.0	37.2	35.5
38	109 M	68	15.7	19.1	19.0	15.4	25.6	58.0	34.5	37.0
41	111 F	61	13.6	17.5	17.9	11.8	20.0	52.8	34.0	34.0
42	112 F	65	14.0	16.6	18.0	10.6	21.5	51.8	33.0	32.0
43	115 F	47	15.0	18.0	18.3	13.3	23.8	54.0	31.5	36.0
44	117 F	83	15.0	18.5	19.0	11.3	23.6	55.5	36.5	34.0
46-52	120 M	60	14.7	17.7	18.1	12.4	20.0	54.0	36.0	35.5
54	122 F	50	15.0	18.2	18.3	11.8	21.8	55.5	37.5	33.6

A. Head breadth (maximum above ears)
B. Head length (inion to glabella)
C. Head length (ophistocranon to glabella
D. Head height (tragion to top of head)
E. Head height (gnathion to vertex)
F. Head circumference (across forehead, over ears, maximum)
G. Head midsagittal arc length (inion to glabella)
H. Head coronal arc length (tragion to tragion)

Table II - Impact Conditions

Experiment	36	37	38	41	42	43	44	54
Impactor Mass (kg)	5.36	5.59	5.32	23.09	5.23	5.23	5.23	5.23
Impactor Velocity (m/s)	8.75	9.94	9.60	12.83	12.95	12.95	4.36	8.41

Table III - Series I

Experiment	Peak Input Force (N x 10^3)	Peak Head Acceleration ($m/s^2 \times 10^3$)	Frontal	Parietal	Occipital #1	Occipital #2	Posterior Fossa	Carotid Siphon
			Pressures (mm Hg)					
36	7.78	2.30	1022	594	-205	-78*	-480	-
37	7.90	2.00	1059	552	-341	-363	-452	-
38	10.80	2.42	1041	494	-205	-	-485	-
41	14.84	3.90	3207	1414	857	-	-426	352
42	5.20	1.59	-	70	-	-	-329	550
43	10.59	2.23	2031	1664	482	-	-136	811
44	6.53	1.52	764	150	109	-	- 19	860
54	10.84	2.34	2059	1354	248	-	-483	356

* epidural

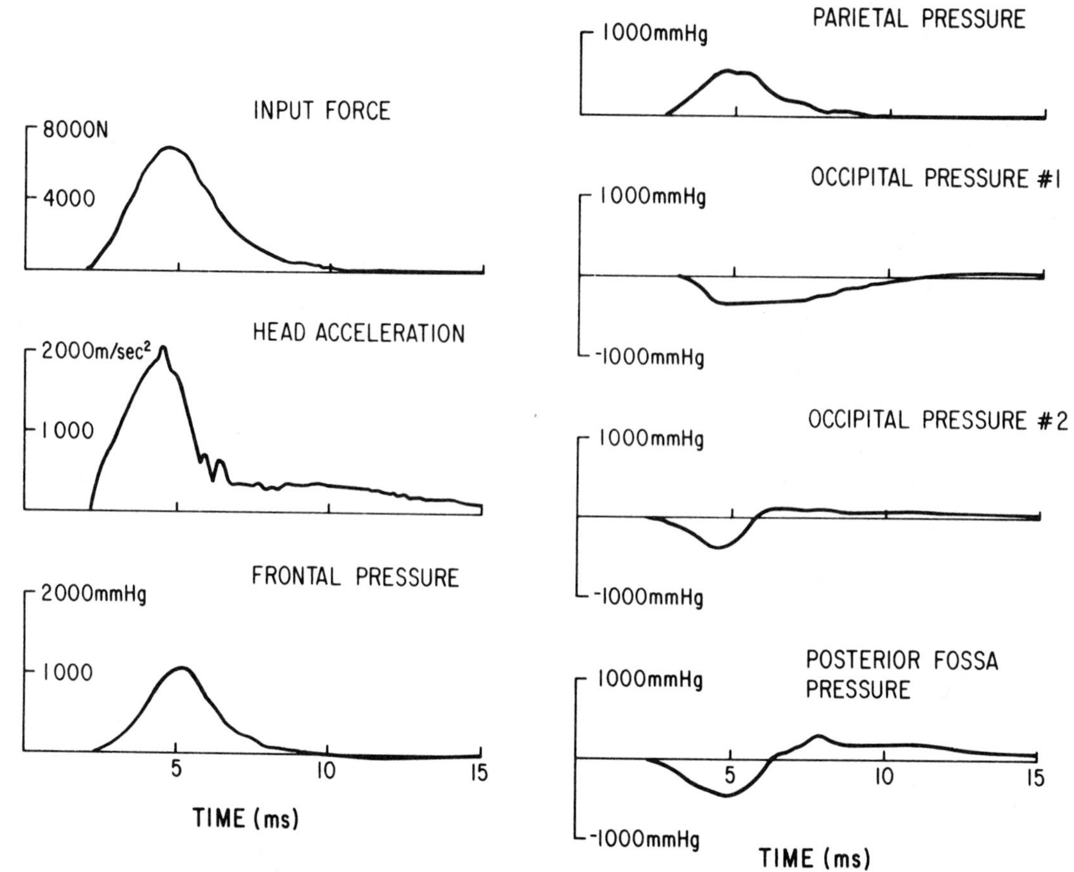

Fig. 1 - Sample data records for experiment 37

Table IV - Injury Coding Scale

GRADE 0 -		No discernable extravascular extravasation ("hemorrhage") of india ink marker
GRADE 1 -	(A)	Subarachnoid "hemorrhage" over less than one half of brain surface.
and/or -	(B)	Petechial "hemorrhage" ("contusion") confined to one lobe.
GRADE 2 -	(A)	Subarachnoid 'hemorrhage' over most of the brain surface.
and/or -	(B)	"Contusion" in two lobes.
and/or -	(C)	Small "contusion" of brain stem - not in central area.
GRADE 3 -	(A)	"Contusion" in three or more lobes.
and/or -	(B)	Contusion of central brain stem.
GRADE 4 -	(A)	Laceration of brain.
and/or -	(B)	Disruption of cerebral arteries at base of brain.

Table V - Results of Coding

Experiment	Peak Input Force (N x 10^3)	Peak Head Acceleration (m/s^2 x 10^3)	GSI	HIC	t_1 (ms)	t_2 (ms)	Injury Code
36	7.78	2.30	1068	923	2.4	4.4	1
37	7.90	2.00	861	744	3.1	5.5	0
38	10.80	2.42	1153	980	1.1	2.9	2
41	14.84	3.90	4756	3765	5.9	8.6	2
42	5.20	1.59	842	703	2.0	7.1	0
43	10.59	2.23	1008	804	3.7	6.1	0
44	6.53	1.52	675	551	2.8	7.0	2
54	10.84	2.34	1061	820	2.8	4.5	0

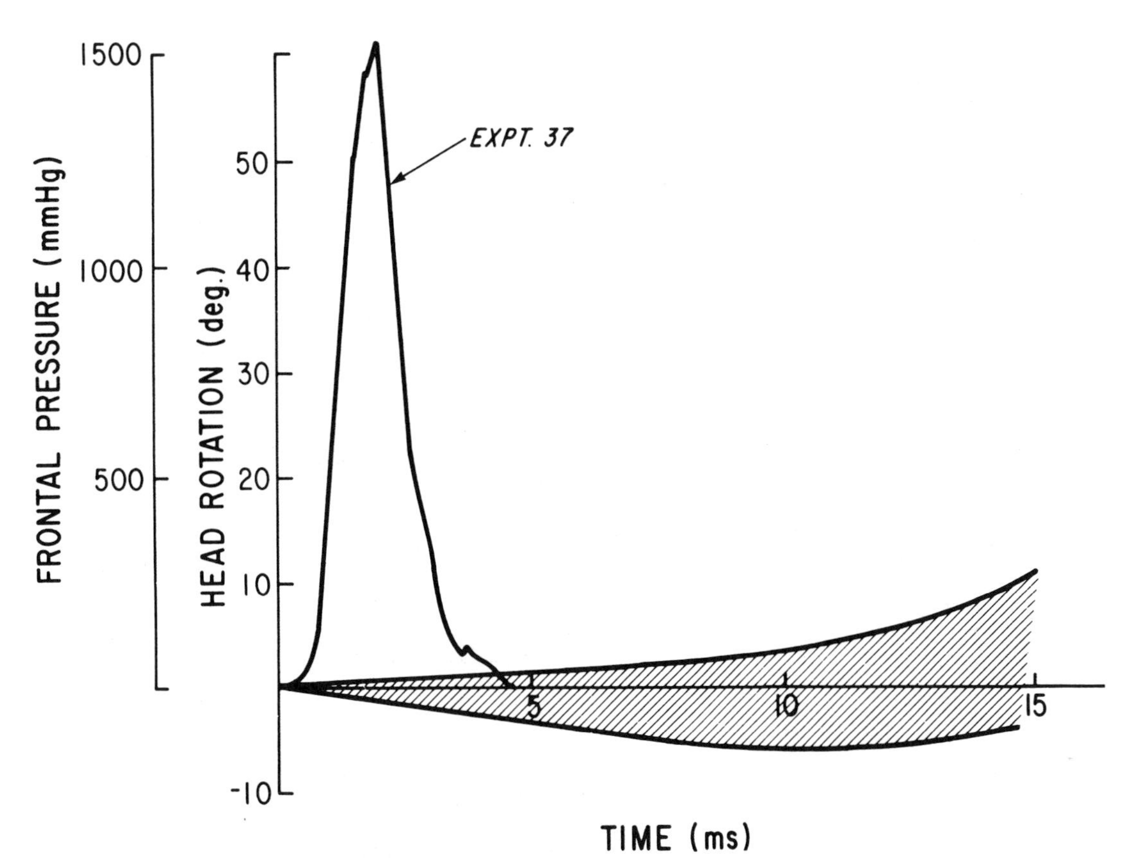

Fig. 2 - Head rotation related to the frontal pressure impact event

Table VI - Impact Conditions

Experiment	46	47	48	49	50	51	52
Impactor Mass (kg)	5.23	5.23	5.23	5.23	5.23	5.23	5.23
Impactor Velocity (m/s)	4.42	4.42	4.21	8.35	8.35	8.29	8.69

SERIES II - A second set of impact experiments (Expts. 46-52) was conducted sequentially on a single specimen with the impact conditions as described in Table VI. The input parameters were varied by maintaining a constant mass and altering the impactor velocity as well as the impact interface material. Peak head acceleration and intracranial pressure response values are contained in Table VII. These are displayed graphically in Figs. 3-6.

When peak pressures for individual locations were plotted against head acceleration (Fig.7) for the 5 sequential experiments a linear relationship could be demonstrated as shown by the plotted linear regression lines with associated correlation coefficients ranging from 0.89 to 0.99. Impact exposure severity ratings for Series II tests are given in Table VIII. The relationship of average frontal pressure to average head acceleration was also found to be linear (Fig.8). These linear relationships are apparent even though the impact events varied significantly in magnitude and time duration. Because peak pressure is linear with respect to peak head acceleration, it is necessarily non-linear with respect to the GSI and HIC. In these indices the 2.5 power of the acceleration term produces an exponential curve as shown in Fig.9.

These results are very encouraging since they show a linear relationship between the response of the brain and skull. This should greatly simplify the future study of brain response during impact. Once the relationship is established, skull acceleration (an easily measured quantity) can be used as a predictor of intracranial response and possible injury.

COMPUTER SIMULATION

Using a three dimensional linear finite element model of the brain, the impacts were simulated and the measured and computed intracranial pressures compared. The brain tissue and contained fluids are simulated with eight node brick elements; the partitioning folds of dura, the falx and tentorium, are simulated with four node membrane elements. A total of 406 degrees of freedom are utilized. An early version of the model was presented at the 19th Stapp Car Crash Conference (4). Since that time the model has been extensively revised (5,6). In the current model the external mesh (Fig.10) is defined on the skull and the internal elements are generated, using a Laplace generation program.

The selection of biological material properties for mathematical idealization is always a problem. Young's modulii of $.667 \times 10^6$ dyn/cm^2 and $.315 \times 10^9$ dyn/cm^2 for the

Table VII - Series II

Experiment	Peak Input Force (N x 10^3)	Peak Head Acceleration (m/s x 10^3)	Pressures: Frontal (mm Hg)	Parietal	Occipital	Posterior Fossa #1	Posterior Fossa #2	Carotid Siphon (mm Hg)
46	1.50	0.31	174	127	24	-44	-38	158
47	1.22	0.29	194	122	42	-24	-23	148
48	4.27	1.28	929	397	64	-282	-229	362
49	9.35	3.42	1969	1261	311	-405	-433	517
50	6.28	1.49	1173	628	122	-252	-233	1344
51	16.60	5.39	3785	2371	995	-609	-620	2619
52	13.34	4.29	3506	1883	533	-652	-662	2222

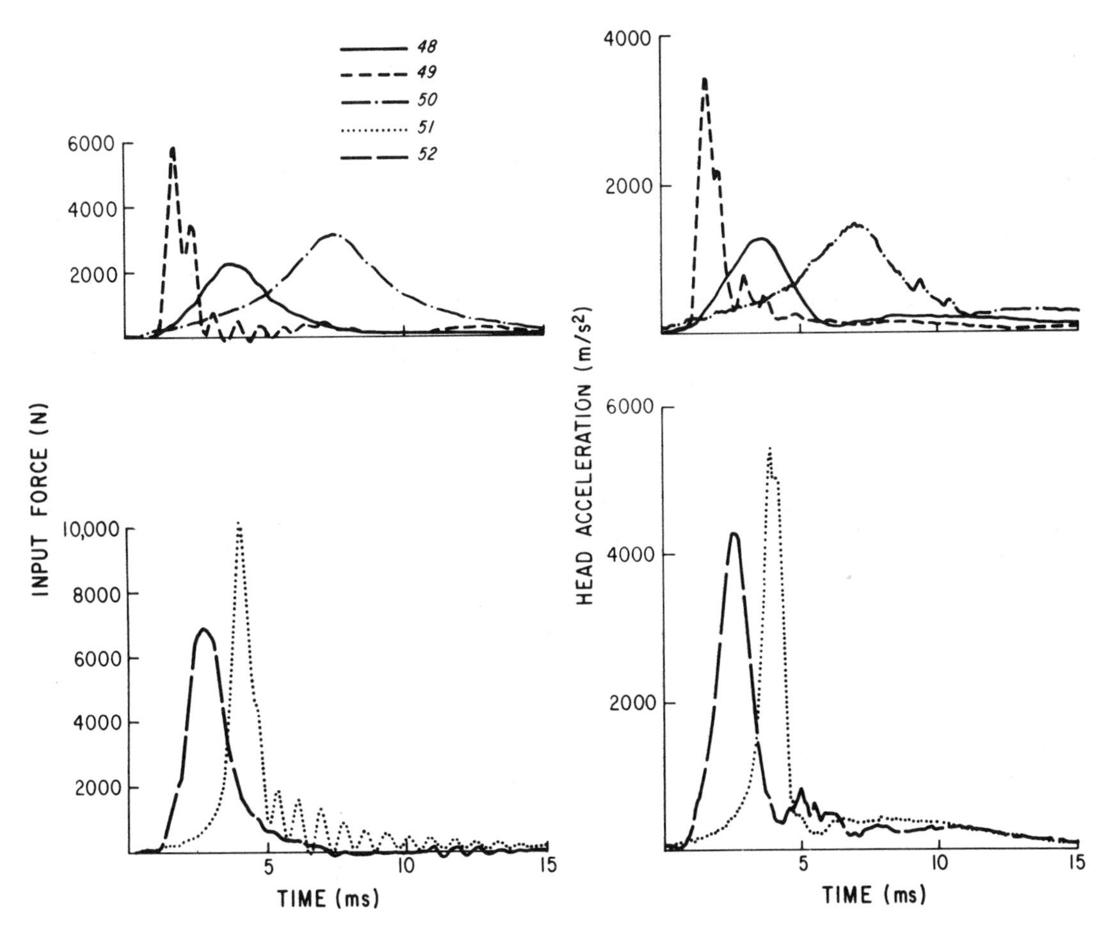

Fig. 3 - Combined data records for Series II multiple impacts

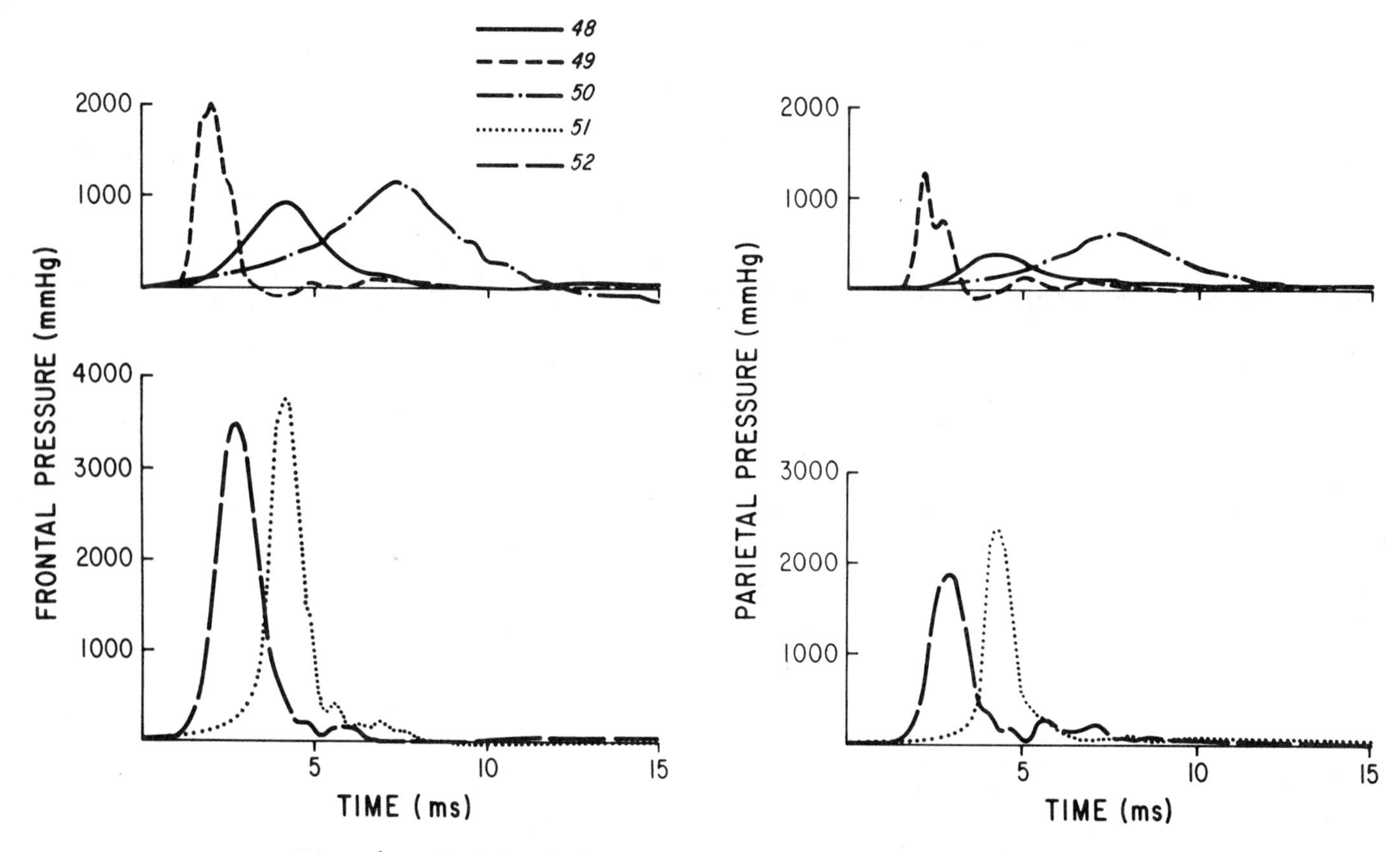

Fig. 4 - Combined data records for Series II multiple impacts

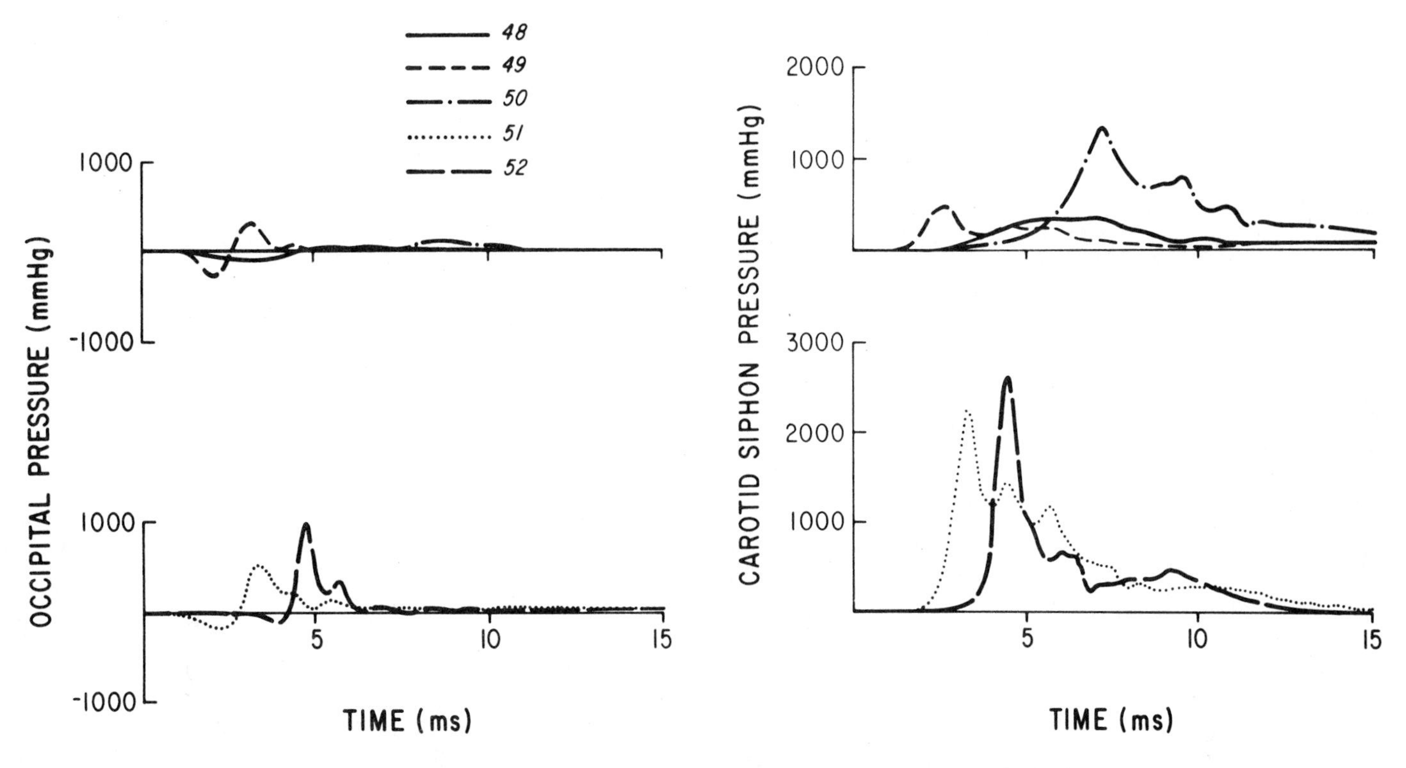

Fig. 5 - Combined data records for Series II multiple impacts

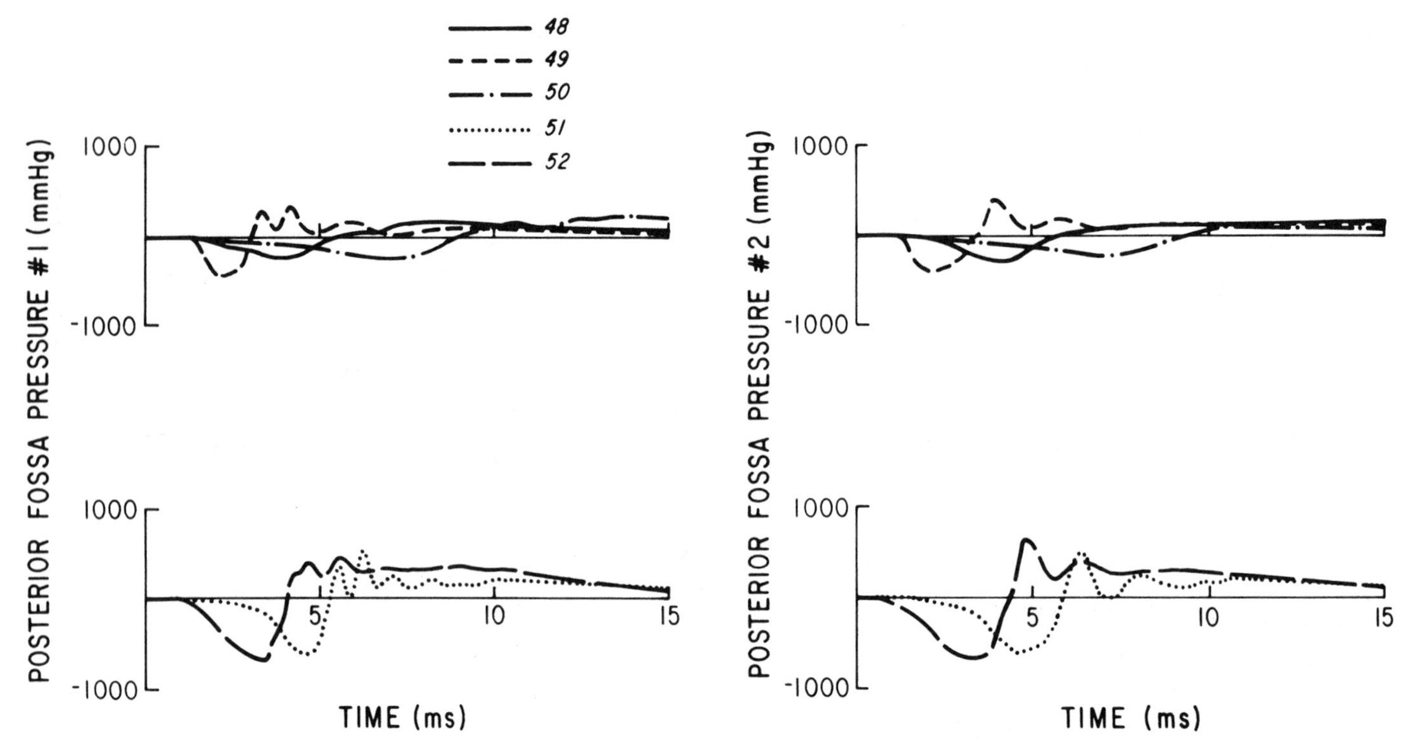

Fig. 6 - Combined data records for Series II multiple impacts

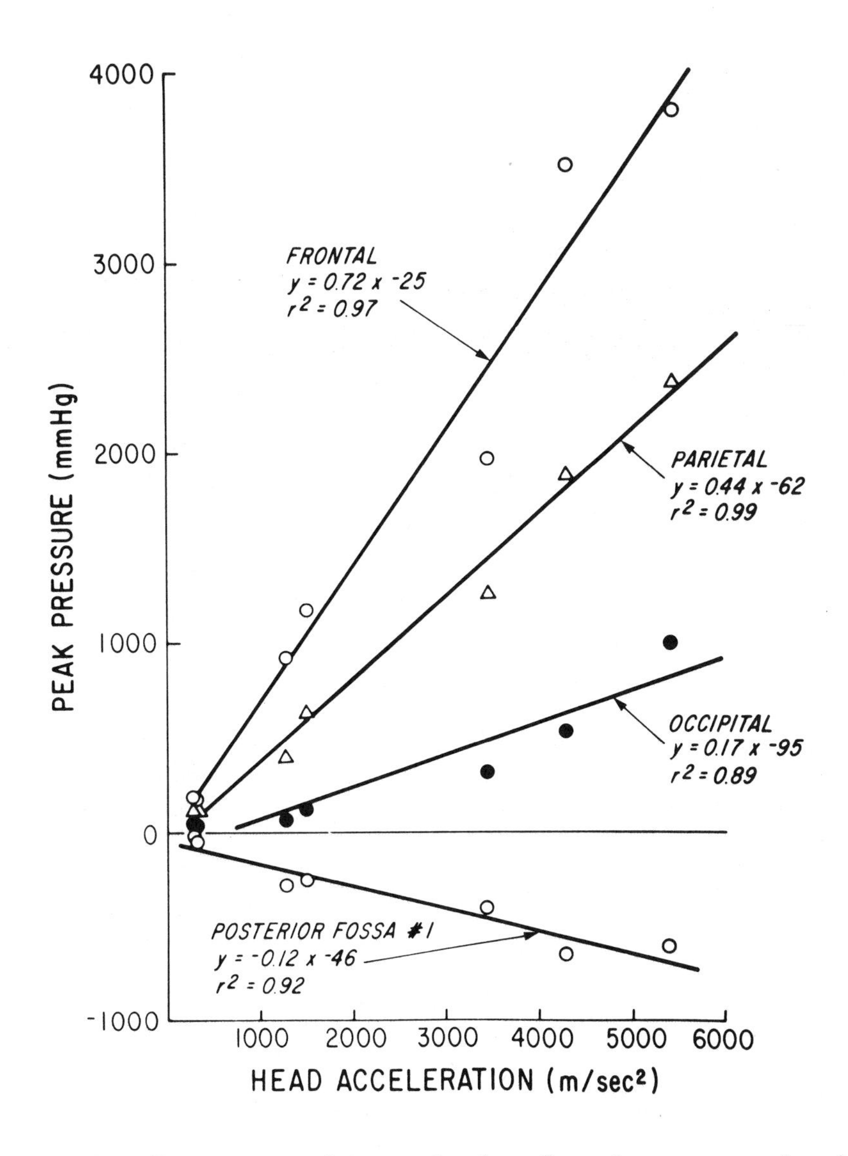

Fig. 7 - Regression analysis of peak pressure-head acceleration relationship

Table VIII - Impact Severity Ratings

Experiment	Peak Input Force ($N \times 10^3$)	Peak Head Acceleration ($m/s^2 \times 10^3$)	GSI	HIC	t_1 (ms)	t_2 (ms)
46	1.50	0.31	36	32	2.7	13.0
47	1.22	0.29	24	21	4.5	12.7
48	4.27	1.28	342	297	2.2	4.8
49	9.35	3.42	1153	1008	1.3	2.2
50	6.28	1.49	675	539	4.4	9.5
51	16.60	5.39	4394	3895	3.6	4.4
52	13.34	4.29	3687	3182	1.9	3.2

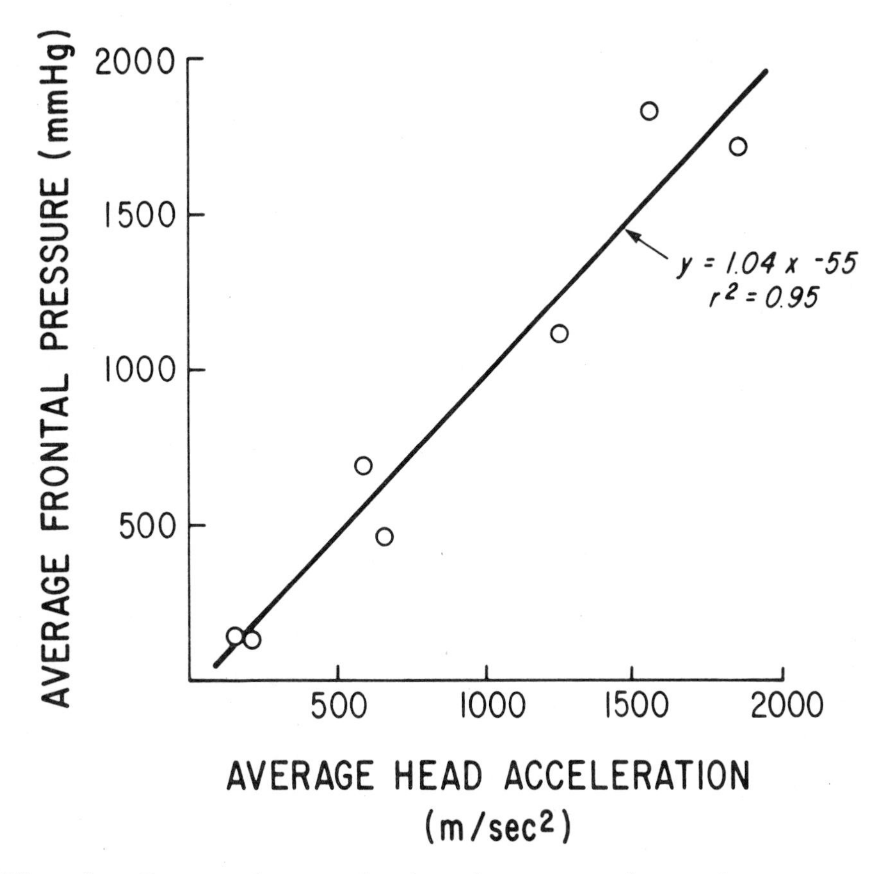

Fig. 8 - Regression analysis of average frontal pressure and average head acceleration relationship

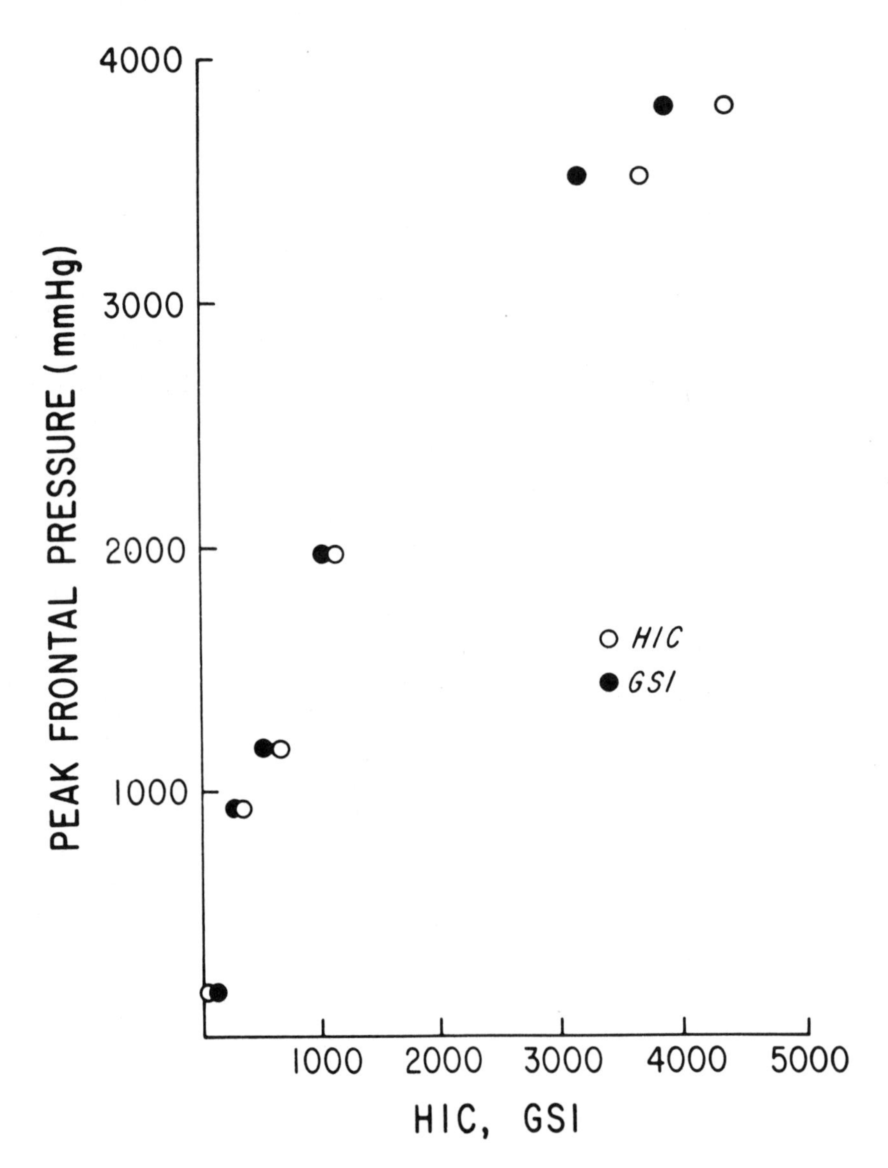

Fig. 9 - Relationship of peak frontal pressure and severity indices

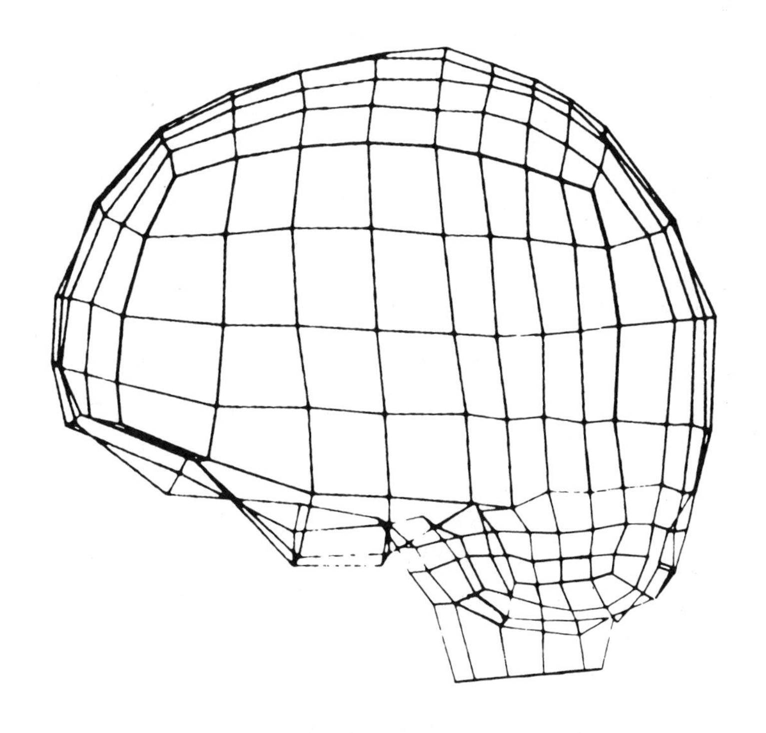

FINITE ELEMENT HUMAN BRAIN MODEL

Fig. 10 - External mesh of finite element human brain model

soft tissue and dura were obtained from specimen tests reported in (7). A tissue density nearly that of water, 1.04 gm/cm^3 was used (8). To determine the effective compressibility (the effective bulk modulus) a material sensitivity study was performed. Varying the compressibility of the intracranial material, the impacts were simulated. Bulk modulii which resulted in good pressure correlation were selected. As stated in (4), an effective bulk modulus is required to account for the pressure release mechanisms provided by the vascular and cerebrospinal fluid (CSF) systems. Therefore, in those regions near the large arteries and near the foramen magnum, a bulk modulus of 2.2×10^6 dyn/cm^3 was used. A value of 4.5×10^7 dyn/cm^2 was selected for the remainder of the brain material. (When a larger value is used, simulating near incompressibility, a variation in rise-time develops between the measured and computed pressures.) The equations of motion are solved on a digital computer using a special modified version of a Structural Analysis program (SAP V). The brain is treated separately by imposing appropriate inertial loads at each node (4). Displacements are measured relative to a skull fixed anatomical coordinate system. Using this technique inaccuracies due to large rotations and displacements are avoided and attention can be focused on the brain itself. A direct integration method is used to calculate the responses.

The intracranial pressures (stresses) are primarily a function of the brain's unique system response characteristics. In the test simulations, the response is excited purely by head motion. (Pressure waves which could be generated by skull deformation are not included.) All the impact tests were simulated and the measured and computed pressures compared at each transducer location. Because the point at which the pressure is computed (element center) and the transducers are not coincident, some variation is expected. Pressure comparisons for Expt. 37 are shown in Fig.11.

To evaluate correlation, the pressure pulse magnitudes, shapes, durations and rise times are compared. Correlation is always good in the frontal regions. That is, matching of the pressure pulse magnitude, duration and shape is reasonable. Correlation in the parietal region is usually somewhat better than shown in Fig.11 for Expt.37. Generally there is better magnitude and pulse shape correlation. Opposite the impact in the posterior fossa and occipital regions the correlation between measured and computed negative pressure magnitude is good for Expt.37 (Fig.11). The traces indicate that a pressure release occurs in the occipital region after 5 msec which is not simulated. This phenomenon is unique to Expt.37, and the cause has not been identified. Often the magnitude of the recorded negative pressure is

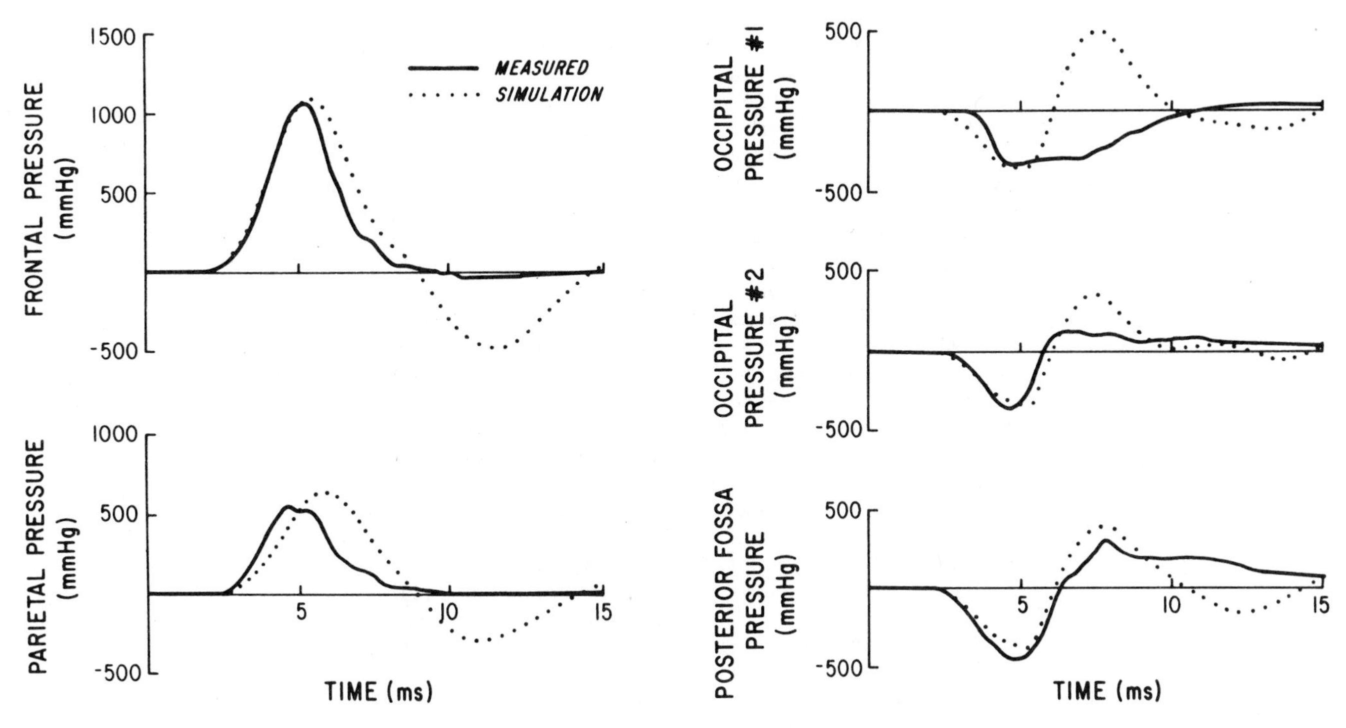

Fig. 11 - Comparisons of measured and simulated pressures for experiment 37

less than the computed stress. But the negative pulse duration usually compares and there is good initial rise time correlation (posterior fossa pressure, Fig.11). Negative pressure magnitude variations may be due to skull response characteristics which were not included in these simulations.

Similar pressure correlation was obtained in live animal test simulations. Using a monkey brain model, Highway Safety Research Institute (HSRI) live animal occipital head impact tests were simulated. Measured and computed pressures were compared (6). The same pattern evolved, good agreement throughout except opposite the impact site, where the magnitude of measured negative pressure was lower than the computed stress. However, it should be noted that the response of the smaller brain is markedly different from that of the human (5). The monkey brain has a different system response because of its smaller size, this results in a faster pressure rise time and a sensitivity to high frequency head acceleration. The response is also affected by the geometric shape; in the small primate brain, stress concentration develops in the frontal lobe due to recesses in the frontal bone. For these reasons it is not possible to correlate the animal response to that of the human cadaver. However, it is encouraging that the same pressure correlation pattern is obtained in both live animal and cadaver tests. The two types of tests can be inter-related with the aid of mathematical models, even though direct comparison is impossible. But even using models, interspecie scaling of brain response is inconceivable. These studies show the brain to be a very complex dynamic system, and as such, it has a complex response. A single response characteristic cannot be isolated and scaled.

The comparison of computed and measured brain response reveals some interesting facts.

1. The brain response is highly damped. A 20% damping factor was added to the model to improve correlation.
2. An effective bulk modulus should be used which provides some compressibility. A nearly incompressible value degrades the correlation.
3. High short duration positive and negative pressure in the cerebrum and cerebellum are produced by inertial loading of the brain.
4. The correlation between computed stress and intracranial pressure during impact is the same for both pressurized cadavers and live animals, indicating that the cadaver pressures are characteristic of the in vivo state.

The measured intracranial pressures have had a significant impact on the evolvement of the brain model. Much has been gained by combining experimental and analytical brain research. The experimental program provides factual benchmarks for the models and the model reveals information at locations where measurement is difficult if not impossible to obtain.

CONCLUSIONS

The human cadaver head appears to be an excellent model for obtaining certain types of impact related information. Post mortem brain condition permits vascular perfusion, whole head pressurization and the determination of vascular trauma following impact. In addition, reproducible intracranial pressure measurements can be obtained which bear a linear relationship to head acceleration and are positive at the point of impact and negative at the postulated location of contrecoup head injury. These pressures may be an additional index of potential brain injury. If one combines the previous impact data (1) with the limited number of experiments in Series I there appears to be a reasonable correlation to date between injury severity as determined by pathologic examination and injury severity indices. The relationship of intracranial pressure and head acceleration support the use of a linear finite element model to describe the response of the skull to blunt impact.

REFERENCES

1. A.M. Nahum and R.W. Smith, "An Experimental Model for Closed Head Impact Injury". Proc. 20th Stapp Car Crash Conf. 1976. (SAE 760825)

2. V.L. Roberts, V. Hodgson, L.M. Thomas, "Fluid Pressure Gradients Caused by Impact to the Human Skull." ASME paper (66-HUF-1), Biomechanics Monograph, 1967

3. S.O. Lindgren, "Experimental Studies of Mechanical Effects in Head Injury". Acta Chirurgica Scandinavia, Supplementum 360, Stockholm, 1966

4. C.C. Ward and R.B. Thompson, "The Development of a Detailed Finite Element Brain Model. Proc. 19th Stapp Car Crash Conf., 1975

5. C.C. Ward, P.E. Nikravesch and R.B. Thompson, "Biodynamic Finite Element Models Used in Brain Injury Research" Presented at Symposium on Biodynamic Models and their Applications, Dayton, Ohio, February, 1977. To be published in the Journal of Aviation, Space and Environmental Medicine.

6. C.C. Ward, "An Analytical Brain Model for Head Impact", Proc.IRCOBI International Conference on Biomechanics of Impact Trauma. Berlin, Germany, 1977

7. "Determination of the Physical Properties of Tissues of the Human Head - Final Report", Biomechanics Laboratories of Department of Theoretical and Applied Mechanics College of Engineering, West Virginia University. Contract # Ph-43-67-1137, May, 1970

8. T.W. Barber, J.A. Brockway and L.S. Higgins, "The Density of Tissue in and about the Head". Acta Neurol. Scandinav. 46: 85-92, 1970

811006

A Study of Impacts to the Lateral Protected and Unprotected Head

A. Nahum, C. Ward, D. Schneider, F. Raasch and S. Adams
University of California

ABSTRACT

In an earlier series of papers, we reported on the results of a study of frontal impacts to cadaver heads which were protected and unprotected. New data is presented to quantify head response to laterally directed impacts. The temporo-parietal area of the head in stationary unembalmed cadaver subjects was struck by a rigid impactor mass. Both helmeted and unhelmeted exposures were compared. Data collected included impactor energy, intracranial pressures at selected locations, and composite analyses from a nine accelerometer array system. In addition the data was entered into a finite element head model. Stresses and strains were predicted for various intracranial locations. The applicability of head injury indices in understanding lateral head impact are commented on.

THIS PAPER REPRESENTS a continuation of a comprehensive study of the human cadaver as a model for head impact; the study provides input data to a mathematical model which is then used in predictive capacity both to fill in missing data areas and to provide verification of the type of results which might be expected from the experiments conducted. Special attention has been paid to the protective characteristics of helmets for the varying impact exposures.

METHODOLOGY

A series of sequential impacts was conducted on the lateral aspect of a single unembalmed cadaver head. Paired tests were conducted in protected and unprotected modes. Impactor terminal velocity was measured by a magnetic probe. Intracranial pressure was regulated by both intravascular and intracranial (subdural) catheters connected to saline reservoirs. Just prior to impact the intravascular pressure was adjusted to approximately 100 mm Hg via a catheter inserted in the common carotid artery. Intracranial pressure was monitored by a water manometer and was adjusted to 0 mm Hg prior to impact.

Intracranial pressure was measured dynamically at the time of head impact by piezo resistive pressure transducers (Endevco model 8510, resonant frequency 180 KHz) which were inserted into the subdural space at designated locations. Head acceleration was measured by nine Endevco piezo resistive accelerometers (model 2264-2000; resonant frequency: 27 KHz) mounted on an Endevco triaxial bracket (model 21419).

The cadaver specimen was seated in an upright position. The Frankfort plane was maintained in a horizontal position by resting the mandible on a styrofoam support block prior to impact. The skull was impacted laterally in the area of the parietal bone by a 12.38 kg pneumatically actuated piston. The impactor surface was a circular disc, 12.5 cm in diameter.

In the unprotected tests the impactor surface was covered by a composite of two padding materials; 1.5 cm Ensolite and 1.5 cm open cell polystyrene. In the protected tests, the head was fitted with a Bell model R-T helmet.* This helmet utilizes an expanded polystyrene bead liner material.

The bracket was rigidly attached to the head opposite the side of impact by drilling a clearance hole through the skull and securing the apex of the bracket

---

*Helmets supplied courtesy of Bell Helmets, Inc.

by means of an expanding collet at the end of a threaded shaft. Two of the three accelerometer mount legs were also attached to the skull using threaded studs which terminated in drilled and tapped holes. Dental acrylic applied at each of the three attachment points acted to distribute the transmitted loads to a larger surface area of the skull. A portion of the helmet shell opposite the side of impact was removed to accomodate the accelerometer mount. Following the experiment the head was affixed in a measuring jig to establish the coordinate of the accelerometer bracket legs relative to the origin of the anatomic coordinate system. Three components of linear and rotational acceleration at the point of attachment of the mounting bracket were calculated from the nine accelerometer array output using the analysis reported by Padgaonkar et al. (1)*. A coordinate transformation was then implemented to represent these data at the origin of the anatomic axes. The sign convention followed for the anatomic axes was positive x, y and z directions being posterior-anterior, laterally right to left, and inferior-superior respectively. All acceleration data discussed subsequently represents values at the origin of the anatomic axes.

RESULTS

A tabulation of impact conditions, resulting linear head accelerations, severity indices, and intracranial pressures for the matched series of protected and unprotected head impacts is contained in Table I. Pressures listed are those magnitudes obtained from the initial pressure peak following impact. (Also comparable measurements from an unhelmeted and helmeted test, 92 and 89 respectively, are shown in Figure 1.) As can be seen from the table entries, it was not feasible to make intracranial pressure measurements at each of the anatomic sites for all experiments. An additional experiment (#84) was performed on a separate cadaver specimen for purposes of computer simulation. The reader should be cautioned against casual comparison of data from experiment 84 with the remaining tests due to differing skull masses and geometry of two separate test specimens. Although these data are included in Table I a discussion of the results is contained in the Computer Simulation section.

Several observations can be made regarding the energy absorption and cranial response attenuation

---

*Numbers in parentheses designate References at end of paper.

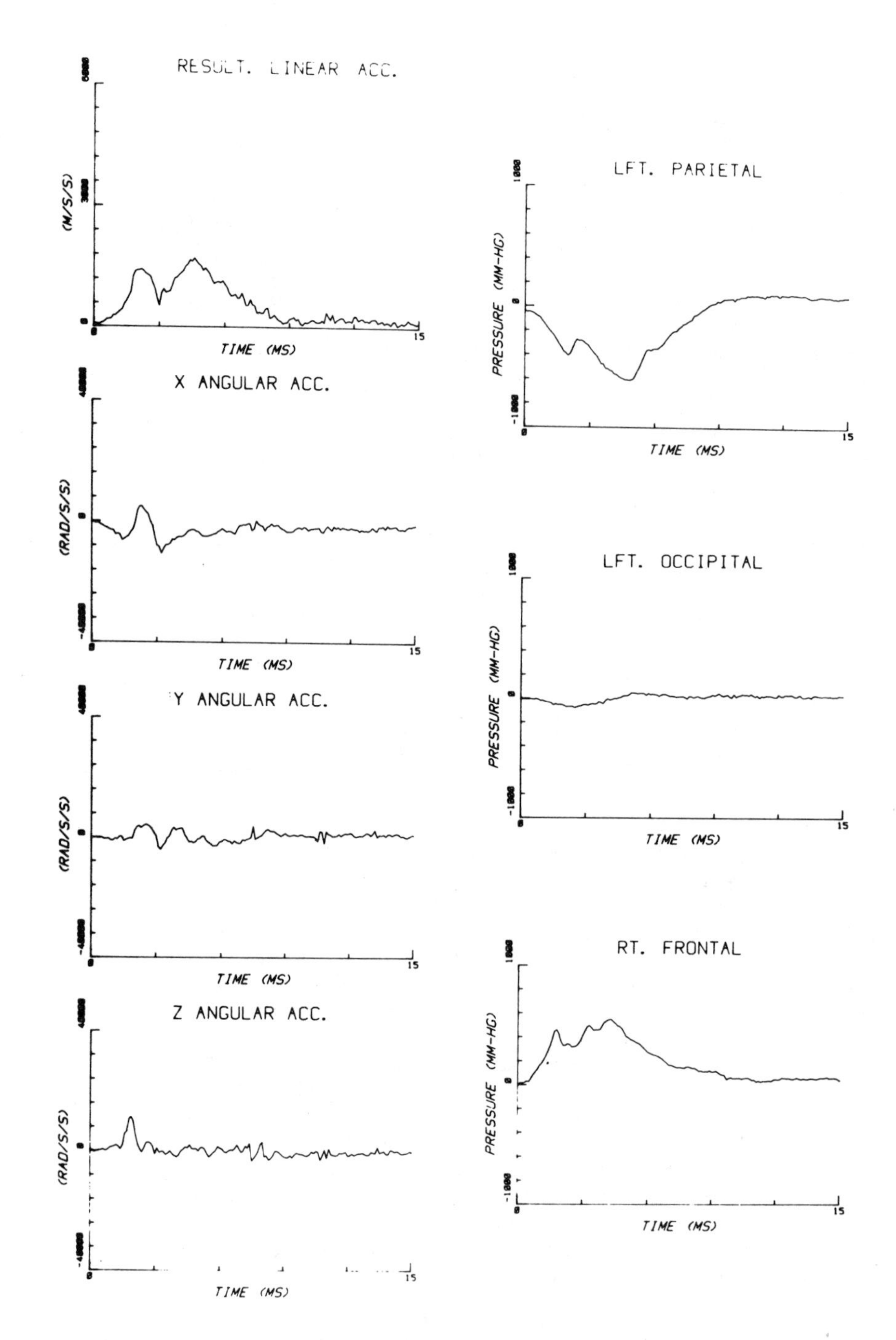

Fig. 1A - Experiment 89 data

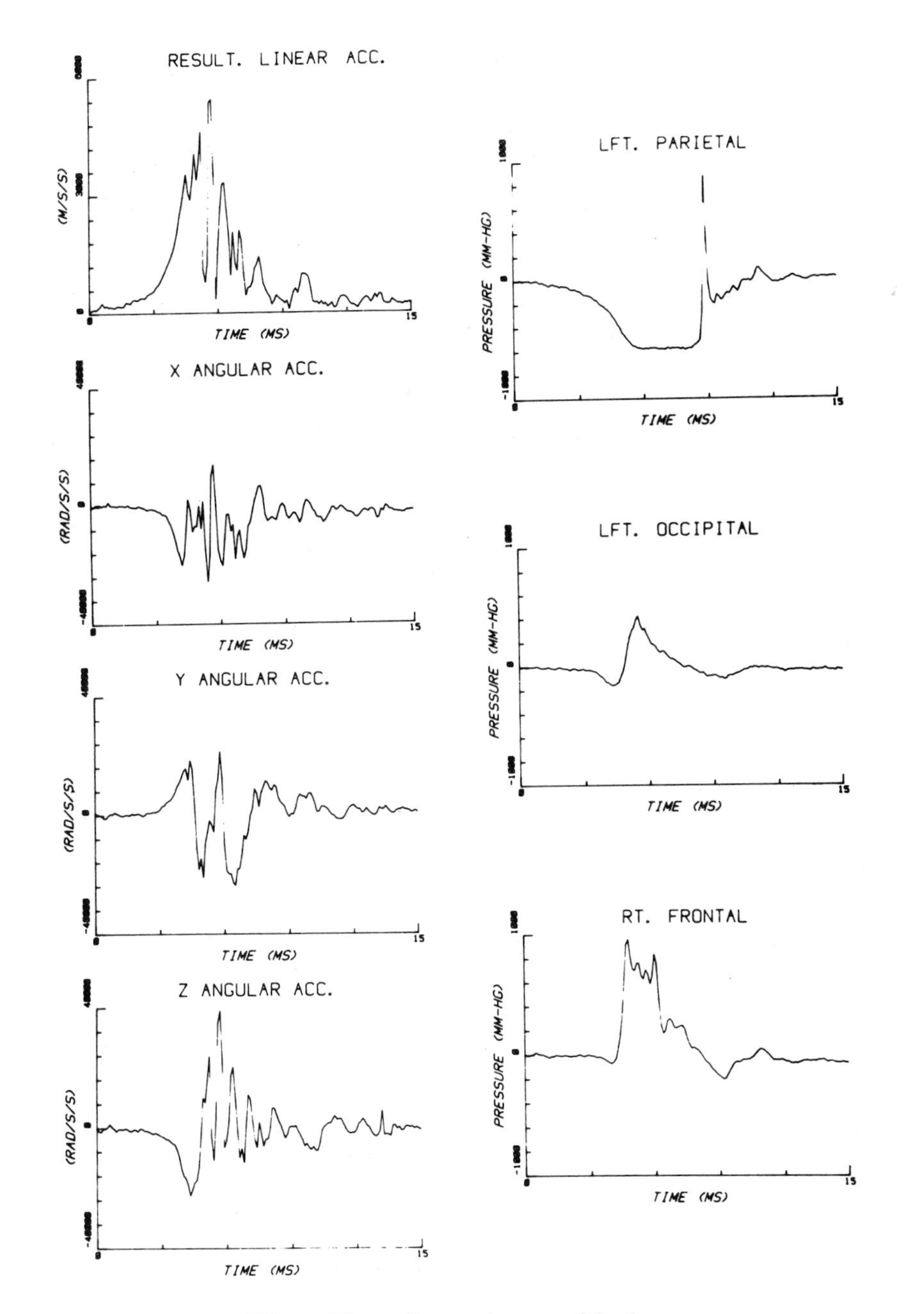

Fig. 1B - Experiment 92 data

Table I - Summary of Experimental Data

Experiment	Condition	Impactor Velocity m/s	Kinetic Energy joules	Maximum Linear Head Acceleration $m/s^2 x10^{-3}$	HIC	GSI	Pressure - mm Hg Right Frontal	Left Frontal	Left Parietal	Left Occipital
84	U*	6.03	225	1.47	411	488	-	-260	-350	8**
85	H*	5.08	159	1.25	564	711	348	-114	-597	-163
87	U	5.08	159	2.33	923	1152	815	-	-687	- 6
88	U	8.02	399	7.69	5337	7434	1276	-	-874	-141
89	H	7.25	323	1.71	810	945	552	-127	-606	- 78
90	H	8.46	440	2.83	1690	2001	662	-173	-726	- 79
91	H	8.80	476	3.08	2022	2517	863	-	-744	- 65
92	U	8.32	425	5.44	3234	5620	957	-	-683	-159
93	U	9.31	533	6.89	2583	5142	1432	-	-1012	-237

* U: No helmet H: Helmeted ** Epidural measurement

characteristics of the helmet. If peak linear head acceleration is plotted versus impactor kinetic energy (Figure 2), it can be seen that for equivalent impactor energies the linear head acceleration is reduced for the protected head. Tests 85 and 87 as well as tests 90 and 92 are comparable in impacted kinetic energy but in both pairs the head acceleration of the helmeted specimen was approximately one-half that of the unprotected skull.

In a similar fashion, a plot of severity index values, both HIC and Gadd Severity Index, versus impactor kinetic energy (Figure 3) shows an attenuation response. Although the helmeted response is lower for all energies employed, the difference increases at higher levels of impactor energies.

In all tests the maximum right frontal intracranial pressure which was measured adjacent to the impact site was positive while pressures on the contralateral side (left frontal, parietal and occipital) were negative. Figure 4 demonstrates the attenuation of maximum frontal pressure when a helmet is employed. In this regard experiments 87 and 91 are noteworthy. Although the impactor kinetic energy for test 91 was approximately three times higher than test 87, the resulting frontal pressure was comparable. However, this relationship is not demonstrated on the contralateral side due to the lower bound of a negative one atmosphere pressure. This bound is reflected in the pressure pulse shape which shows a squaring off at approximately minus one atmosphere. This phenomenon can be observed in the left parietal pressure-time history shown in Figure 1B. Therefore, the more severe impact conditions produced similar peak negative pressure responses for both the helmeted and unhelmeted tests.

A linear regression analysis of the severity indices versus right frontal pressure (Figure 5) indicates that the Gadd Severity Index correlates somewhat better than the Head Injury Criterion for the range of pulse durations studied. The correlation coefficients were 0.85 and 0.76 respectively.

Figure 6 demonstrates the dependence of maximum frontal pressure on peak linear head acceleration. This dependence which has been previously demonstrated (2) is not affected by the use of energy absorbing protective devices. However, as noted earlier in this section, the magnitude of the linear head acceleration and consequently intracranial pressure is reduced when a helmet is used for equivalent energy blows.

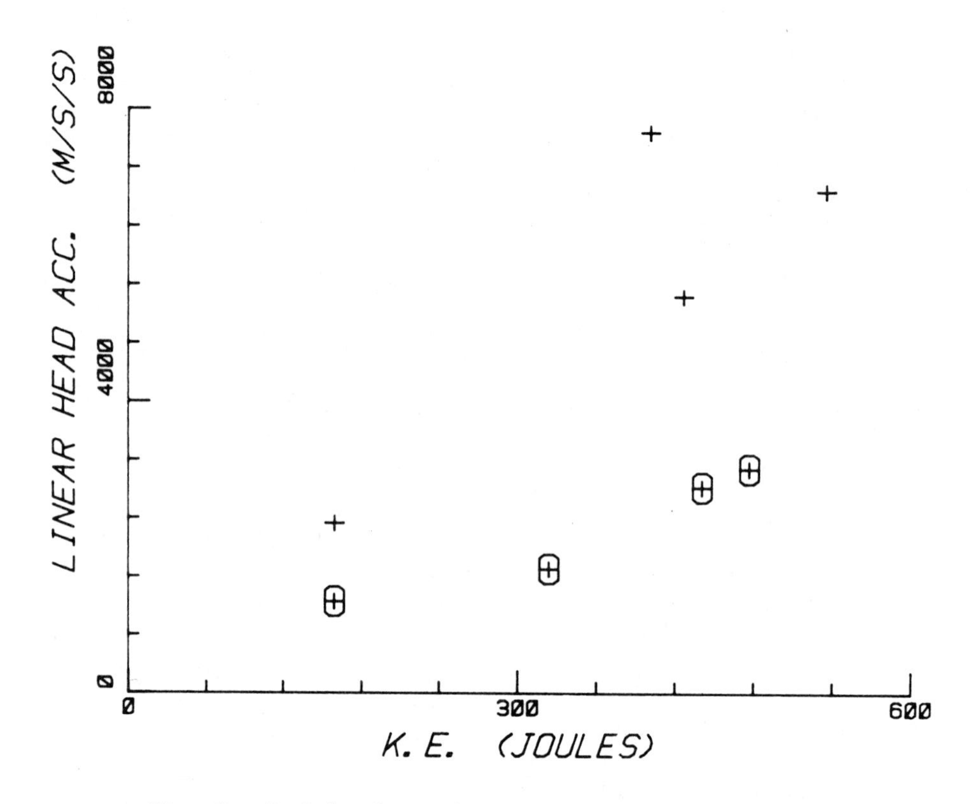

Fig. 2 - Peak head acceleration versus impactor kinetic energy (circled data points indicate helmeted test)

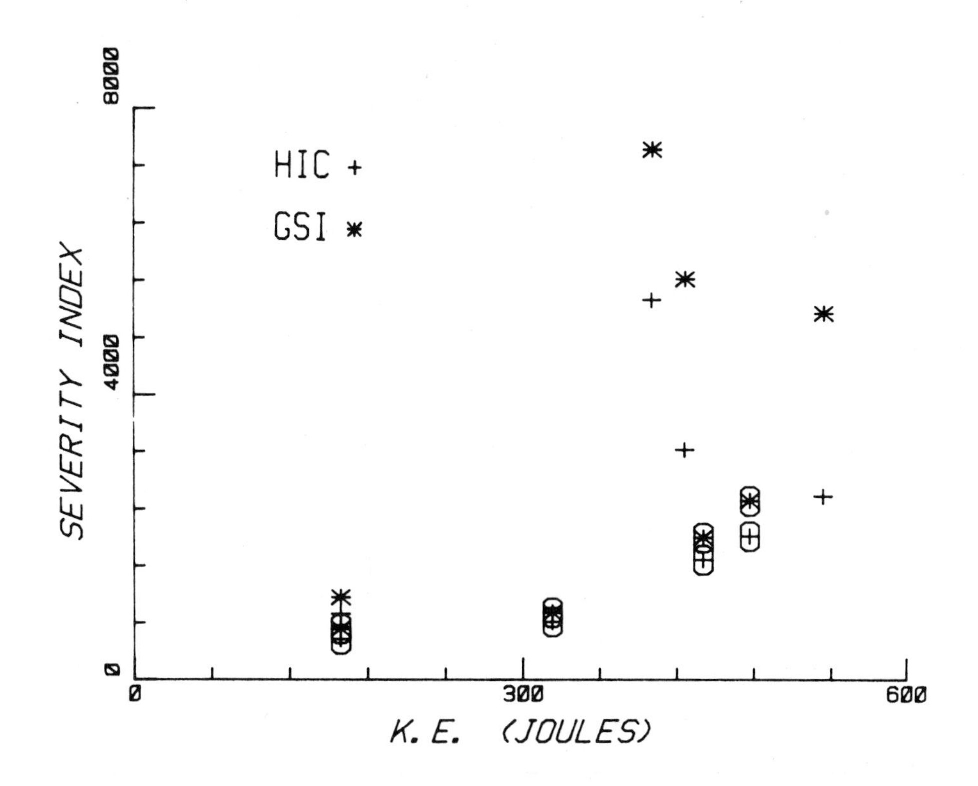

Fig. 3 - Severity index versus impactor kinetic energy

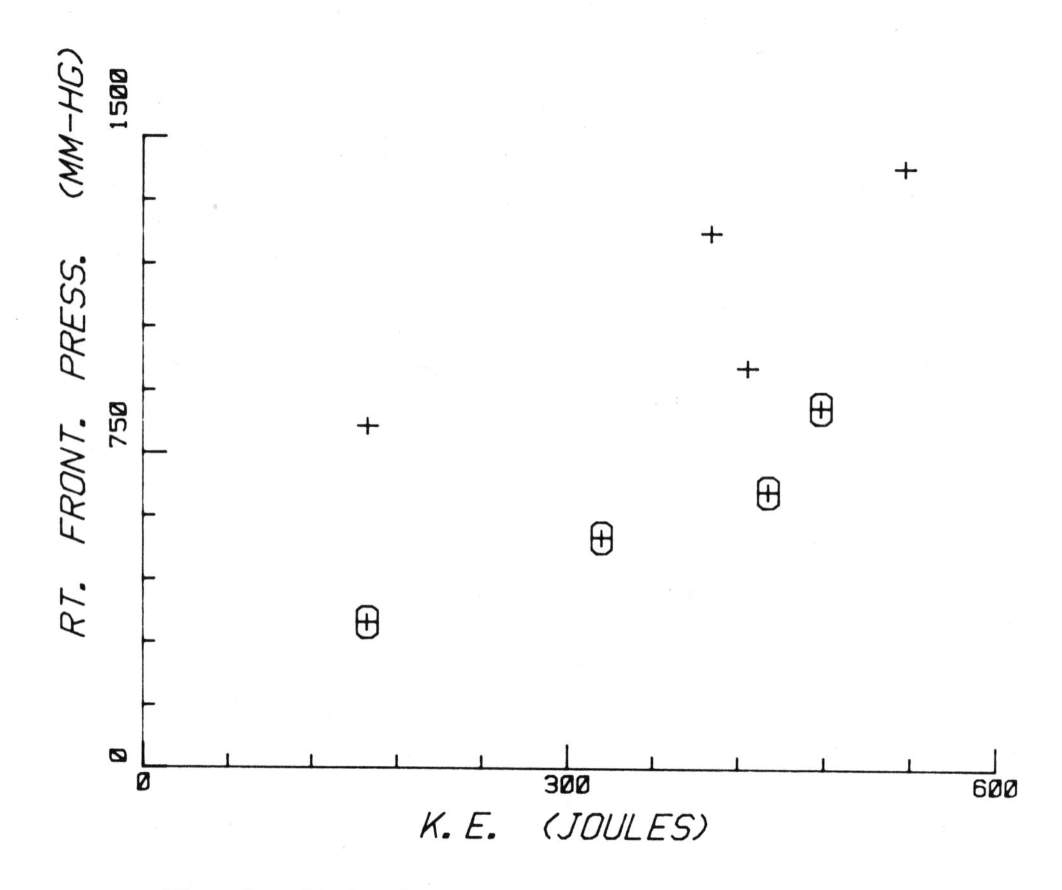

Fig. 4 - Right frontal pressure versus impactor kinetic energy

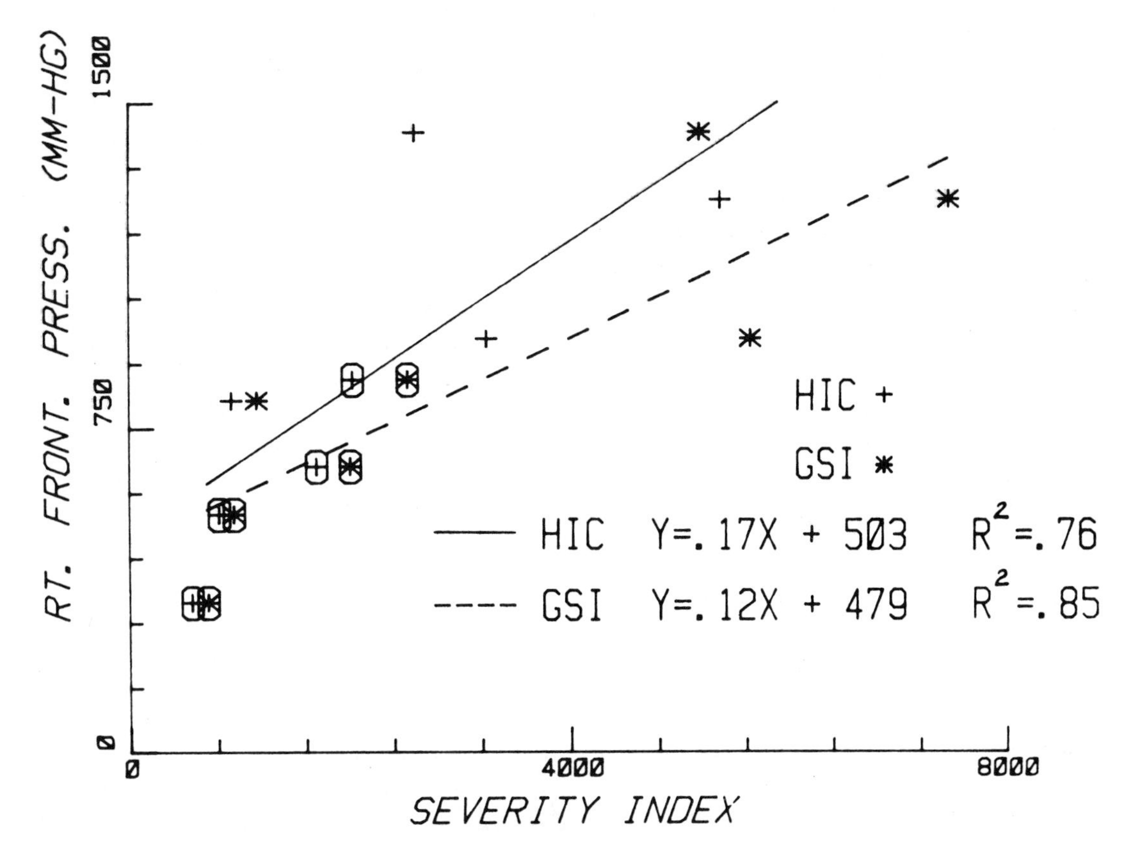

Fig. 5 - Right frontal pressure versus severity index

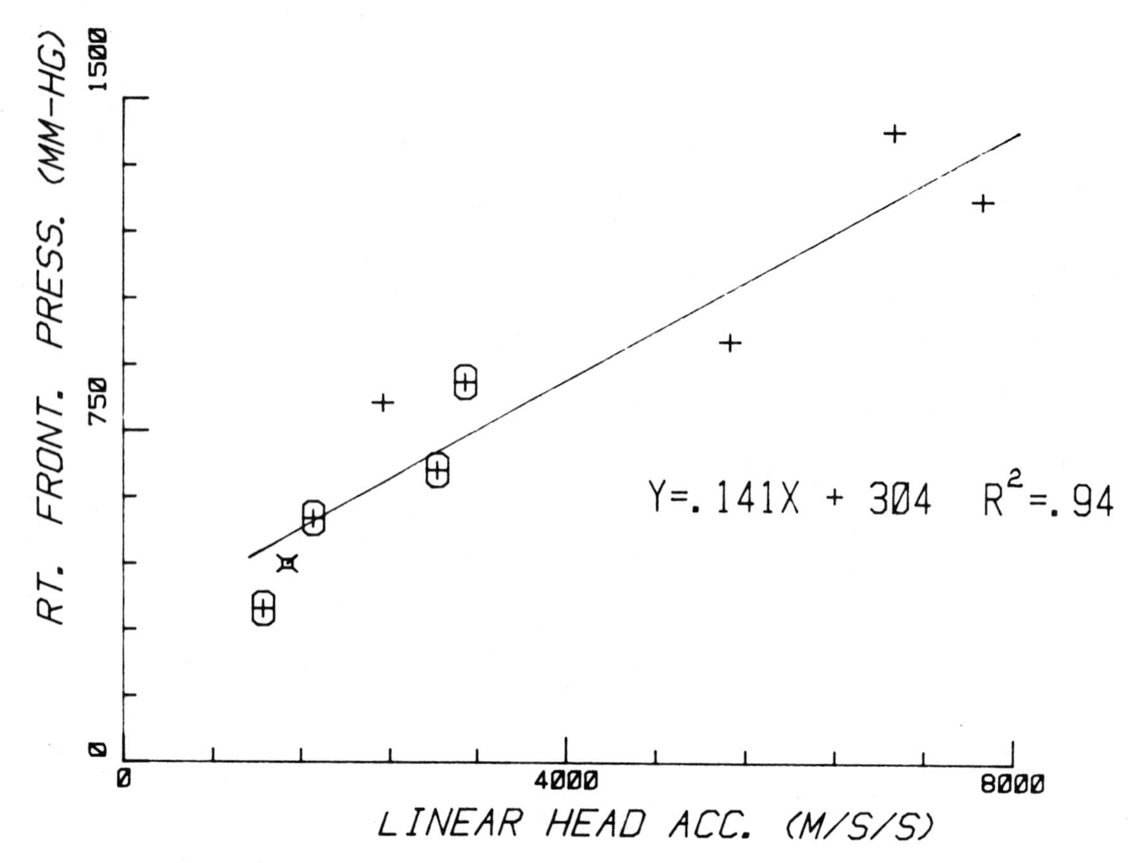

Fig. 6 - Right frontal pressure versus peak head acceleration

The calculated rotational head accelerations about each of the anatomic axes are listed in Table II. As can be seen in Figure 1A, the direction of the rotational acceleration changes during the impulse for each of the anatomic axes. The magnitude listed in Table II is the largest value computed for a given test regardless of direction. In general this was preceded by a pulse of the opposite sign. The direction of this initial pulse is a function of the spatial relationship between the impact location and the center of gravity of the head. If the blow was directed superior to the C.G. a positive rotational motion was produced while an impact point inferior to the C.G. would result in rotational displacements of the opposite sign. In this test series the impact site was along an axis passing nearly through the C.G. Therefore, a small change in the S-I position of the point of contact would result in cranial rotational motions of opposite directions. Similar considerations can be applied to motions about the Y and Z anatomic axes. Again, the use of a helmet reduced the magnitude of rotational acceleration produced in similar severity blows (Figure 7). Here the absolute value of rotational acceleration is plotted as a function of impactor kinetic energy.

As in most impact experiments involving a complex system the input-response relationships are not without individual variation (i.e. comparison of severity indices for tests 88 and 92). Carefully controlled impact studies utilizing anthropomorphic test dummies as subjects have demonstrated the difficulty in establishing repeatability and reproducibility in test results (2,3,4). Due the relative complexity of the human cadaver as a mechanical system when compared to ATD's, this problem is exacerbated. Therefore, these data collectively should be viewed as indicators of response trends.

COMPUTER SIMULATION

Finite Element Model. The impacts were simulated on the computer using a three dimensional finite element model of the human brain, Figure 8. This model was described in a previous Stapp Car Crash Conference (5). Brain tissue and contained fluids are modelled with eight node brick elements, and the partitioning folds of dura, the falx and tentorium are modelled with four node membrane elements. A special procedure was developed to treat the nearly incompressible brain material which avoids inaccuracies in the stiffness terms as Poisson's ratio approaches 0.5 (6). The face,

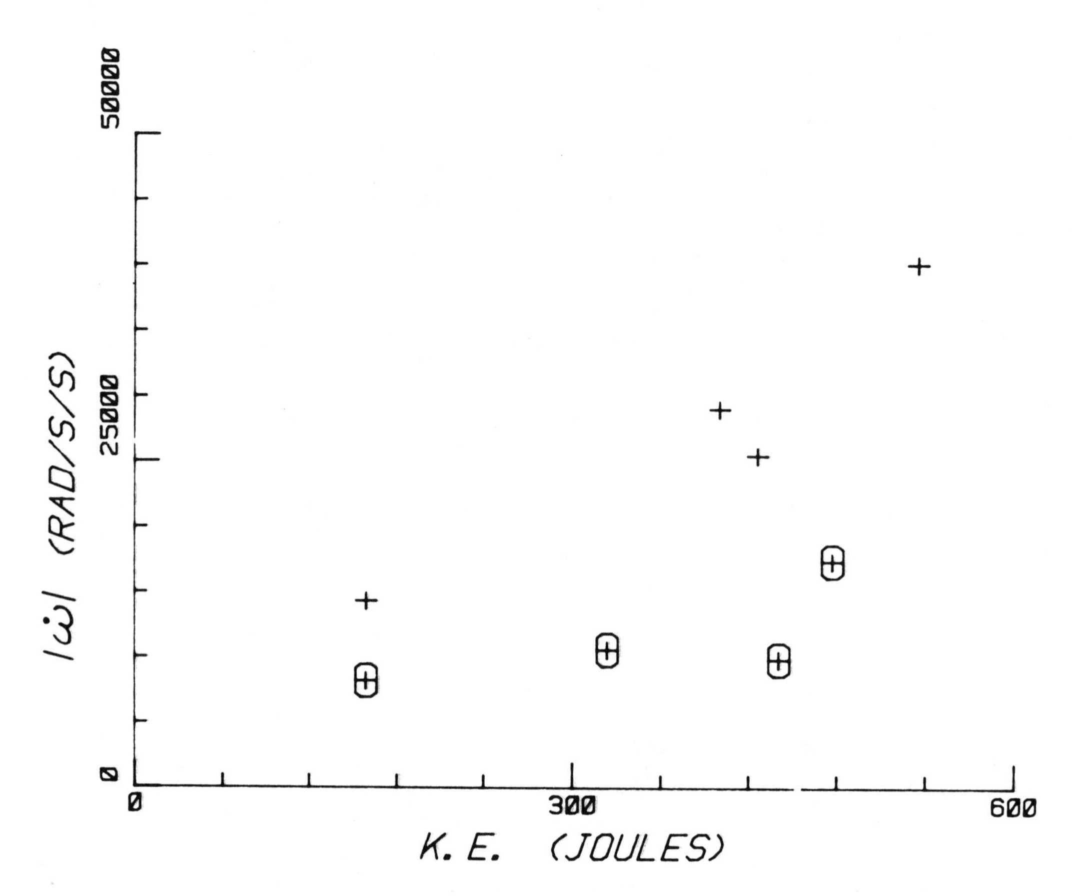

Fig. 7 - Impactor kinetic energy versus rotational acceleration

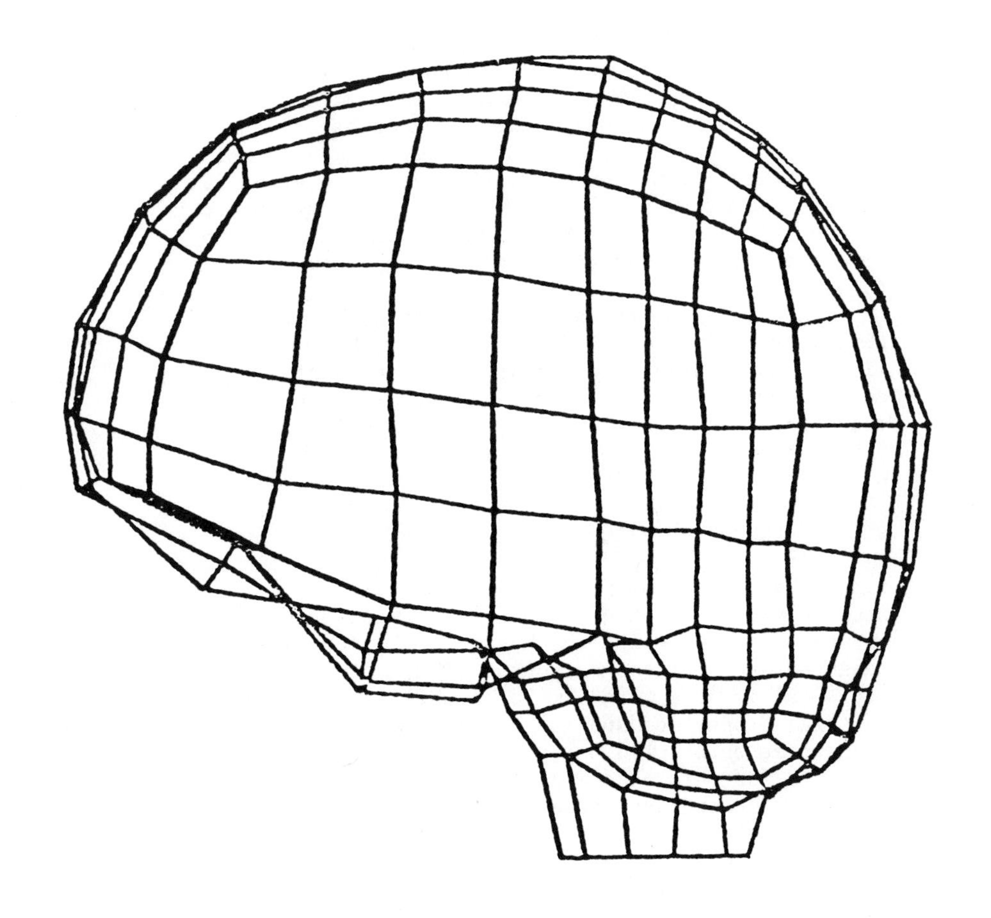

Fig. 8 - Finite element brain model

neck, scalp and skull are not included. Only the internal shape of the skull is modelled to contain the brain.

Solution. Mathematically the model is forced to move in space as the head did in the experiment. Measured head angular and linear acceleration are the input forcing functions. A general purpose finite element computer program (EASE 2) assembles and solves the equations of motion for each node, a total of 791 equations. Stresses, strains and displacements throughout the brain model are calculated versus time. The stresses are compared to the measured pressures for validation.

Material Properties. Properties of biological tissue are difficult to define because they are nonlinear, nonelastic, and strain rate dependent. The central nervous system is especially different to characterize because it is composed of soft tissue, vasculature and fluid. Furthermore mass is not constant, there is flow through the arteries, veins and foramen magnum. The constants selected must be considered effective values with restrictions on their use.

Since effective constant for the high strain rate impact event had not been defined, in 1978 parametric studies were conducted to determine which material constant would provide the best correlation between measured and computed response. Thirty frontal and five occipital impacts were simulated using a wide range of material properties and the results reported (7). The properties which produced the best correlation were selected for this side impact model (Table III). If values other than those shown in Table III were used the calculated response would lead or lag the measured trace.

The parametric study showed that brain response is most sensitive to changes in Poisson's ratio (ν). The appropriate value for ν depends on the duration of the acceleration pulse. The shorter the acceleration, the higher the value for ν. (Refer to Table III) These different ν values are needed to simulate the pressure release mechanisms or volume elastance. During short-duration spike-shaped acceleration pulses, these mechanisms have less time to act.

Simulation Results. (These calculations are based on data obtained from experiment 84, a single impact to an unhelmeted head.) Three angular and three linear acceleration traces are input to the model, these are the accelerations at the anatomical axes origin. Displacements of the brain relative to the

Table II - Maximum angular acceleration

Experiment	w_x	$\dot{w}_y$	$\dot{w}_z$
85H	-82	82	-54
87	-143	104	150
88	291	233	561
89H	-106	42	113
90H	- 99	97	112
91H	-174	-151	136
92	-255	-237	389
93	-402	-232	-490

Table III - Material Properties

Parameter	Brain	Membrane
Youngs Modulus (Pa)	66.7×10^4	315×10^5
Poisson's Ratio		
short pulse < 3msec	.499	.45
moderate pulse 3 to 8	.499 to .490	.45
long pulse > 8 msec	.490	.45

skull, and brain stresses and strains were computed. Analyses of these computed results reveals the following facts about brain response in a side impact.

Brain Pressure Gradients. High nearly hydrostatic stresses occur near and opposite the impact. Compression stresses (positive pressures) develop as the brain compresses against the struck side of the skull and tension stresses (negative pressures) develop as the brain is pulled along the opposite side. The result is a pressure gradient. In the earlier midsagittal frontal impacts the anterior-posterior gradient was easily identified (8), but in a side impact where there is acceleration in all three directions with varying pulse shapes the resulting gradient is less distinct. The side impact pressure response can be considered a superposition of three gradients, one from each of the three acceleration components.

In this series the stresses are highest on the lateral surfaces of the parietal lobes. Compression is greatest in the right posterior region and tension is greatest in the left anterior region. (Refer to Table IV) The direction of maximum pressure variation deviates from the impact direction depending on the magnitude of the x acceleration. Pressures are lowest in the center of the brain, the occipital lobe and the cerebellum.

Measured pressures and computed stresses in the element closest to the measurement site for test 84 are presented in Figures 9, 10 and 11. This was a single unhelmeted impact test performed according to the previously described methodology. Peak resultant acceleration was 1470 m/s/s, and the maximum angular acceleration about the x, y and z were -12199, 3680, 3500 (radians/sec/sec) respectively. The HIC value was 411. Agreement between the measured and computed brain response is good but not as good as in the earlier occipital and frontal impacts (8). In this three-dimensional head motion, pressures vary significantly from one element to another and correlation is difficult. Stresses are calculated at the element centroids which do not exactly coincide with the pressure transducer locations. Also energy dissipation due to localized tissue failure is not included in the model.

The measured response is lower and more damped. This corresponds with the findings in the previous side impact study where it was hypothesized that the subarachnoid tissue failure affected response (5). Inaccuracies in the calculation of angular acceleration due to vibration errors in nine acceleration traces was

Table IV - Maximum Stresses and Strains

Location	Stress in KPa	Strains cm/cm xx	yz	xz
Frontal lobe				
right	- 40.0	-.008	-.009	.002
left	+102.0	.009	.012	-.005
Occipital lobe				
right	- 70.0	-.007	-.024	-.021
left	+ 28.3	-.015	-.012	.004
Parietal lobe				
right	- 86.1	.006	-.027	-.015
left	+100.0	.012	-.016	-.008
Cerebellum				
right	- 46.0	.006	-.050	+.009
left	+ 38.6	-.020	-.030	-.010
Upper Brain Stem	- 17.8	.012	.018	.009
Lower Brain Stem	+ 23.1	.009	.021	.008

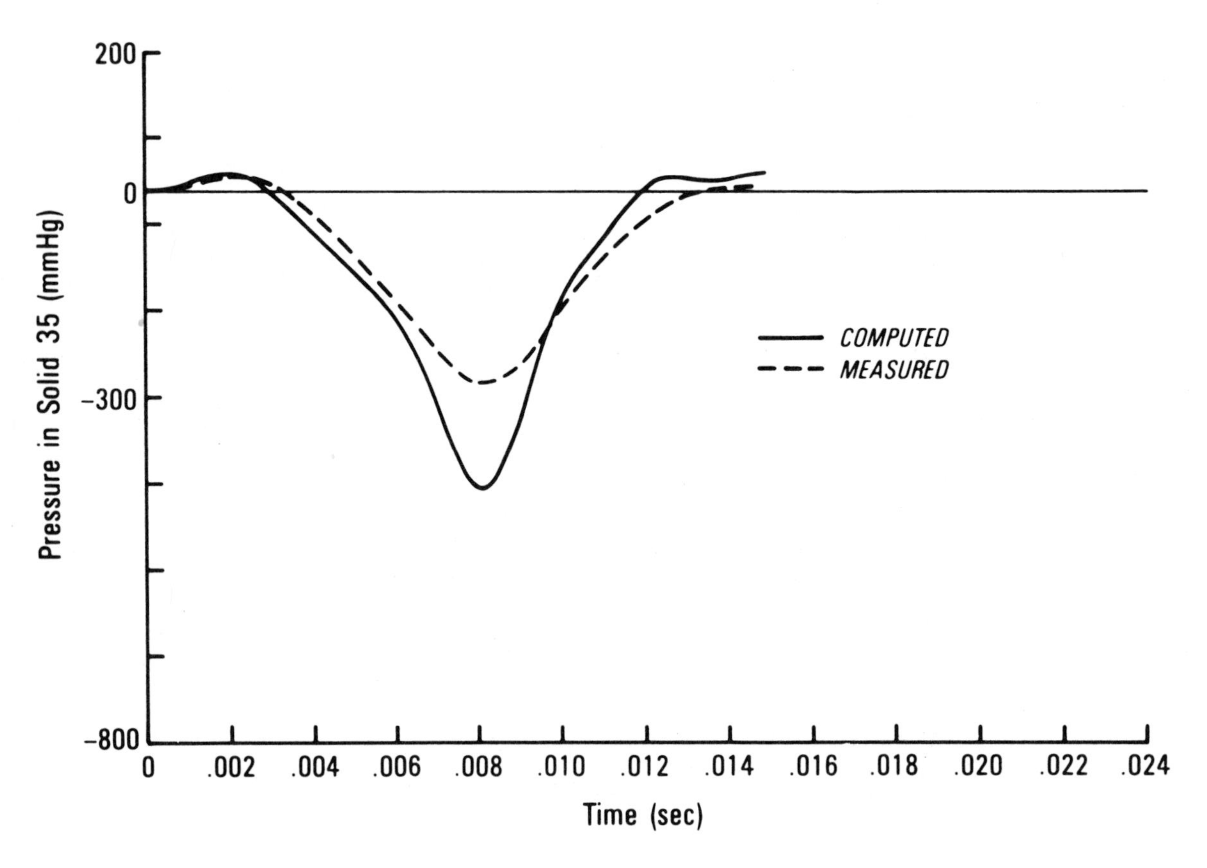

Fig. 9 - Measured and computed left frontal pressure

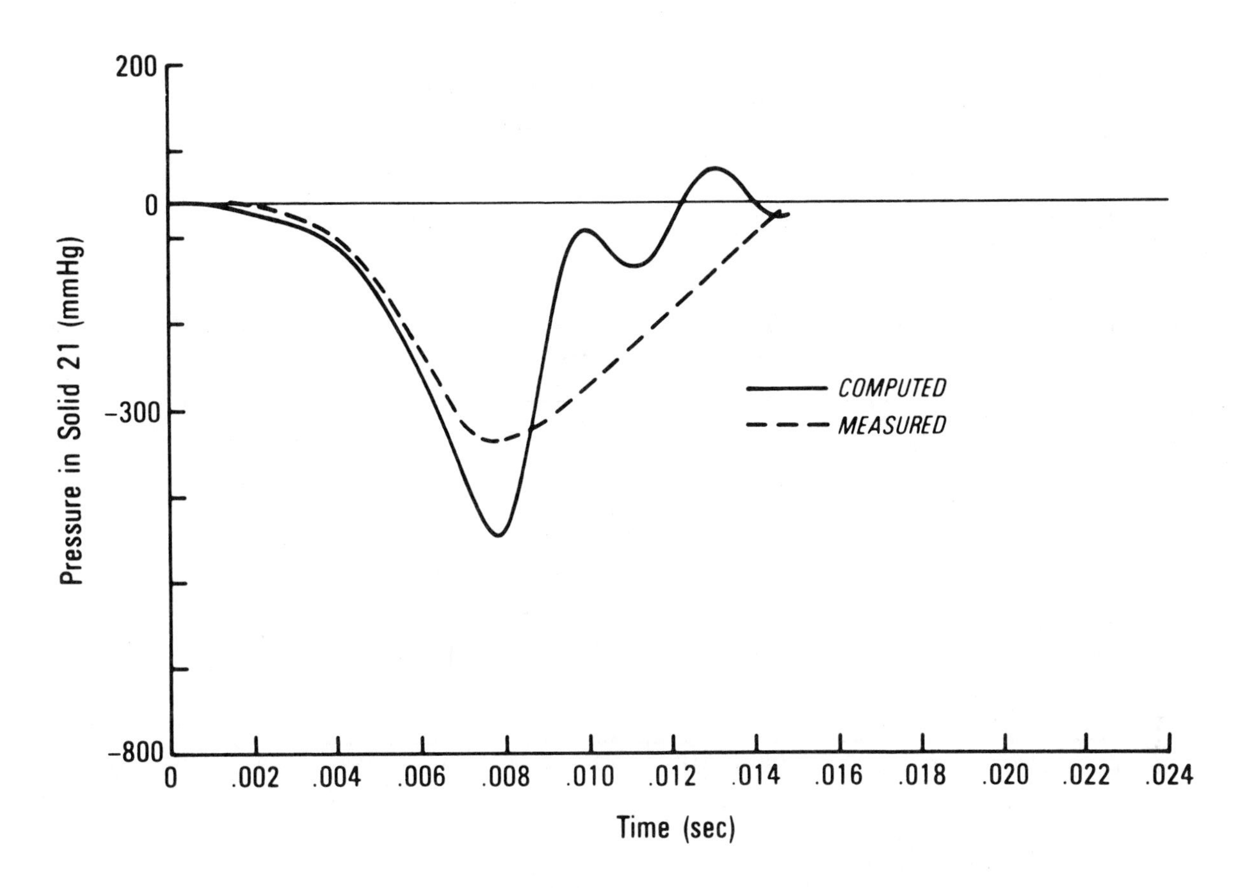

Fig. 10 - Measured and computed left parietal pressure

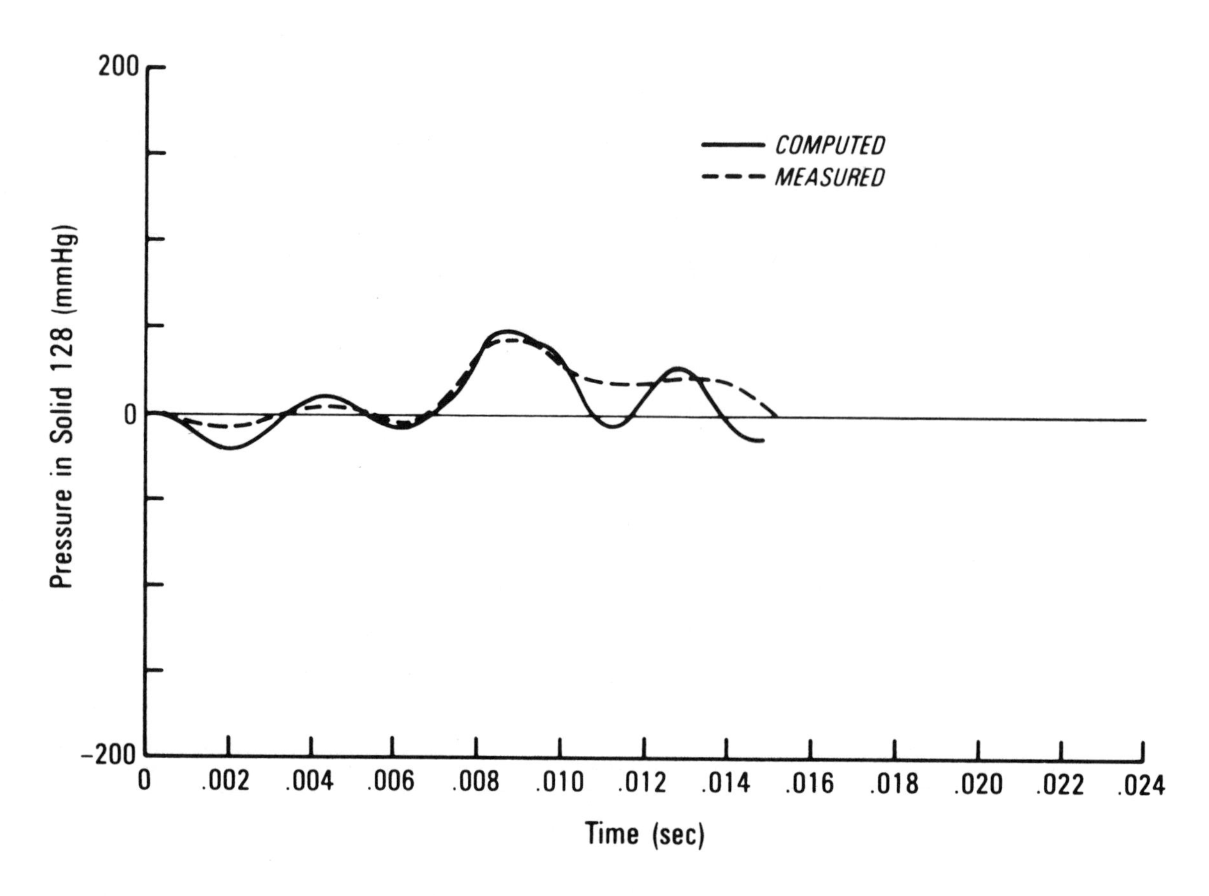

Fig. 11 - Measured and computed left occipital pressure

a problem in the earlier study. Although the accuracy has been improved, noise in the signals could still be contributing to the higher calculated head accelerations and as a result higher calculated pressures. These differences between the measured and computed values are still being investigated.

Brain Tissue Strains. Rotation and translation of the head produces a complex strain pattern in the brain. The largest strains are the shear strains and they occur along the brain-skull interface and in the brain stem and cerebellum. The internal folds of dura control the relative motion of the brain and determine the distribution of shear. The right and left cerebrum, and cerebellum tend to rotate individually producing a complex interaction with the brain stem. At the same time the lower brain stem is strained by motion through the foramen magnum. The motions which produce these shear strains in test 84 are shown in Figures 12, 13 and 14. A horizontal plane, Figure 12, shows the rotational motion of the upper cerebrum. A vertical plane, Figure 13, shows the interconnection and partitioning effect of the falx and tentorium. The individual rotations of the cerebellum, and right and left cerebrum are easily seen. Rotation in the anterior-posterior direction is shown in the sagittal plane, Figure 14. Maximum shear strains according to brain region are presented in Table IV.

These individual rotations of the cerebellum and cerebrum are different from those reported earlier by Holbourn (9). Holbourn used a water flask analogy to describe the rotation of the brain inside the skull. He neglected the effect of the internal folds of dura. Although these individual rotations have not been reported previously in the literature they are in agreement with experimental research being conducted by Aldman (10).

Varying the magnitude of the rotation acceleration has shown that the shear strains along the brain-skull interface are related to head orientation, and are reduced when head rotation is eliminated. Shear strains in the interior brain are less affected by head rotation, and at times the translational and rotational induced strains act to cancel one another.

CONCLUSIONS

1. Lateral impacts to the protected and unprotected head of equivalent impactor energy produced lower values of linear head acceleration, rotational

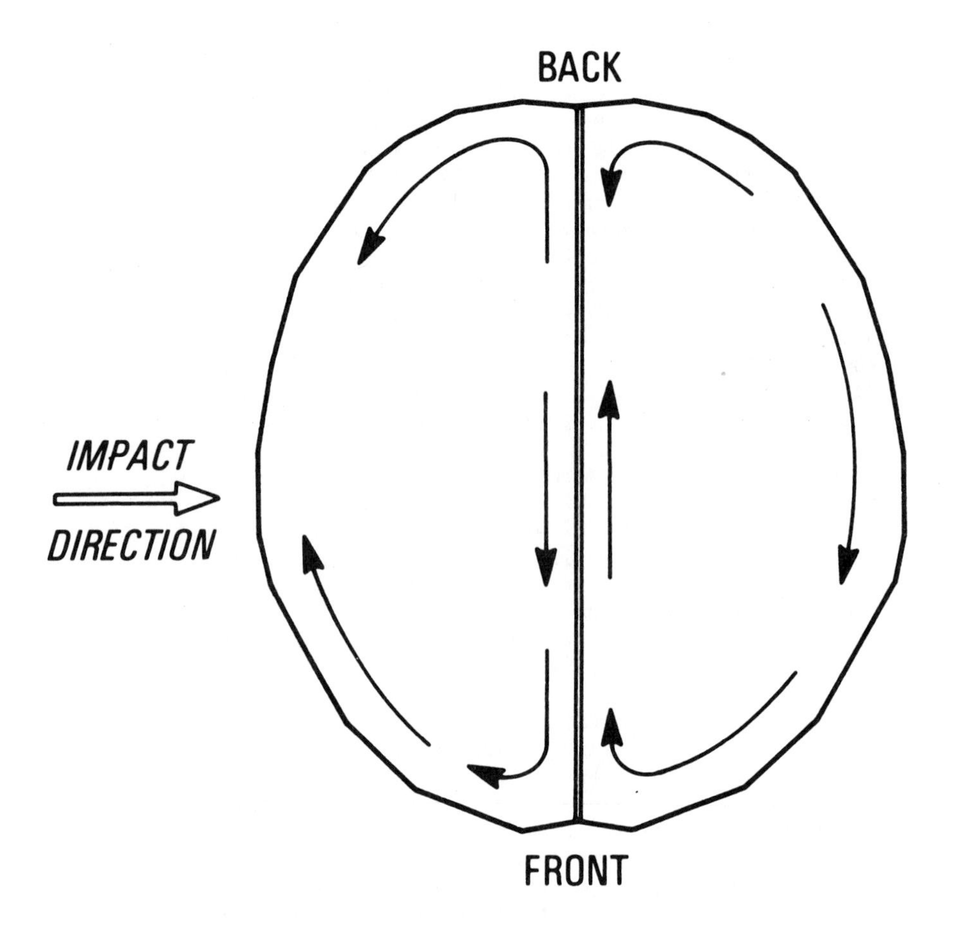

Fig. 12 - Motion of cerebrum in horizontal plane

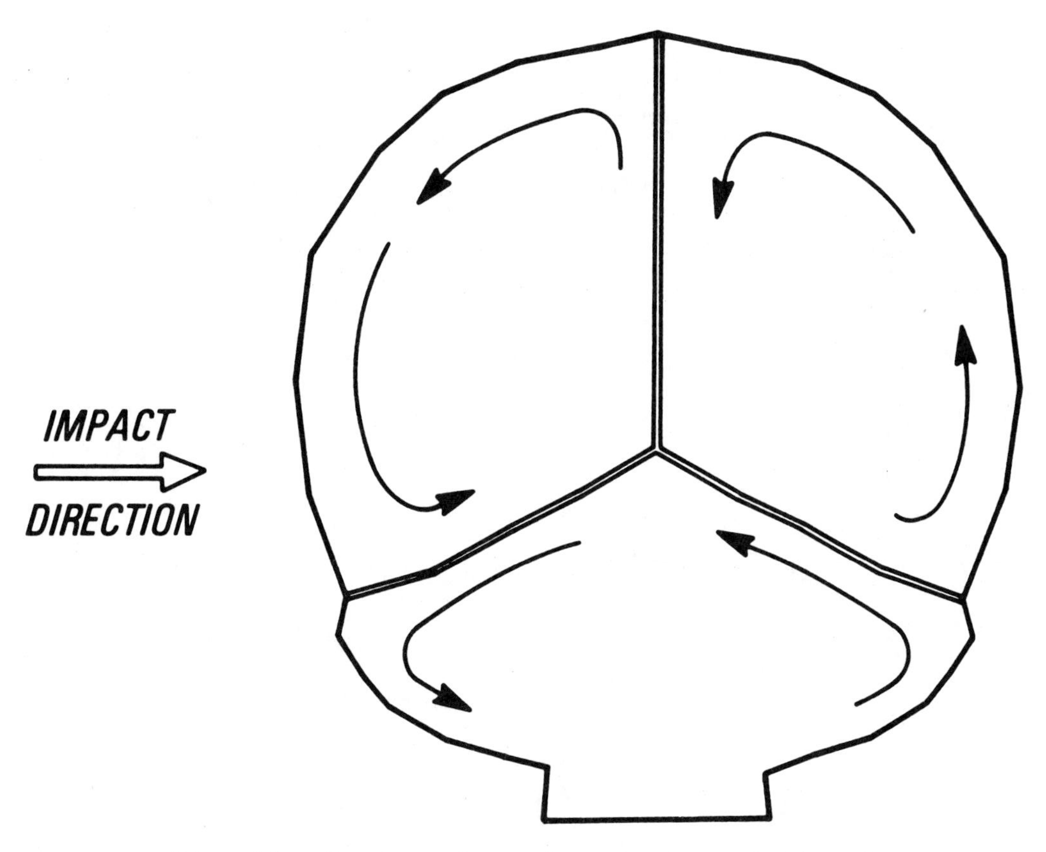

Fig. 13 - Motion of cerebrum and cerebellum in vertical xz plane

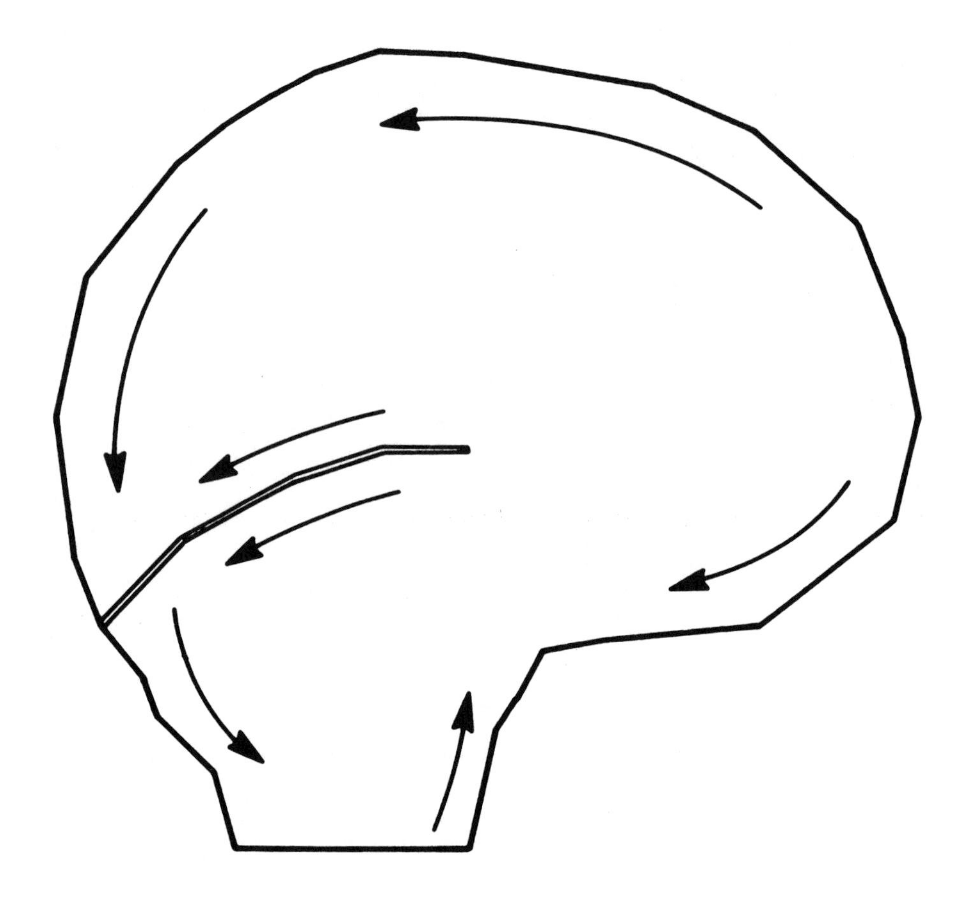

Fig. 14 - Motion of brain in sagittal plane

acceleration, severity indices, and intracranial pressure adjacent to the blow site when a helmet was used.

2. As found in prior studies, maximum frontal pressure is linearly dependent on peak linear head acceleration.

3. Lateral impact to the head produces a pressure gradient across the skull similar to that developed in anterior-posterior midsagittal impacts. However, the gradient is not as distinct in the lateral blow due to the larger magnitudes of acceleration normal to the impact direction.

4. The largest calculated displacements produce shear strains along the brain-skull interface, and in the brain stem and cerebellum.

5. Strain calculations indicate that the left and right cerebrum and the cerebellum rotate individually during impact while the lower brain stem is displaced relative to the foramen magnum.

6. Shear strain computations at the brain-skull interface were found to be dependent on the magnitude of skull rotational accelerations.

REFERENCES

1. A.J. Padgaonkar, K.W. Kreiger and A.I. King. "Measurement of Angular Acceleration of a Rigid Body Using Linear Accelerometers." J. of Applied Mechanics 42:552-556, 1975.

2. R.F. Chandler and R.A. Christian. "Comparative Evaluation of Dummy Performance Under - G_x Impact." Proceedings of 13th Stapp Car Crash Conference, paper 690798, 1969.

3. J.F. Sprouffske, W.H. Muzzy, E.M. Trout, T. D. Clarke and C.D. Gragg. "The Effect on Data Reproducibility of Dummy Modification." Proceedings of 15th Stapp Car Crash Conference, paper 710847, 1971.

4. C.K. Kroell, D.C. Schneider and A.M. Nahum. "Comparative Knee Impact Response of Part 572 Dummy and Cadaver Subjects." Proceedings of the 20th Stapp Car Crash Conference, paper 760817, 1976.

5. A.M. Nahum, C. Ward, F. Raasch, S. Adams and D. Schneider. "Experimental Studies of Side Impact to the Human Head." Proceedings of the 24th Stapp Car Crash Conference, Troy, Michigan. Pp. 43-62, October 1980.

6. D. Malkus and T. Hughes. "Mixed Finite Element Methods Reduced and Selective Integration

Techniques: A Unification of Concepts." J. of Computer Methods in Applied Mechanics and Engineering 15:63-81, 1978.

7. P.E. Nikravesch and C.C. Ward. "Evaluation of Composite Material Properties for a Human Brain Model--A Parametric Study." Navy TM No. 51-79-31.

8. A. Nahum, C.C. Ward, R. Smith and F. Raasch. "Intracranial Pressure Relationships in the Protected and Unprotected Head." Proceedings of 23rd Stapp Car Crash Conference. Pp. 613-636, 1979.

9. A.H. Halbourn. "Mechanics of Head Injuries." The Lancet Pp. 438-441, October 9, 1943.

10. Bertil Aldman. National Highway Traffic Safety Administration Consensus Workshop on Head and Neck Injury Criteria. March 26-27, 1981.

This paper supported by National Institute of Occupational Safety and Health Grant #R01 PH00404.

821150

Critical Issues in Finite Element Modeling of Head Impact

Tawfik B. Khalil and David C. Viano
General Motors Research Laboratories

ABSTRACT

Current finite element models of head impact involve a geometrically simplified fluid-filled shell composed of homogeneous, linear and (visco)elastic materials as the primary surrogate of the human skull and brain. The numerical procedure, which solves the mechanical response to impact, requires and presumes continuity of stress and displacement between elements, a defined boundary condition simulating the neck attachment and a known forcing function.

Our critical review of the models discussed, primarily, the technical aspects of the approximations made to simulate the head and the limitations of the proposed analytical tools in predicting the response of biological tissue. The following critical features were identified as major factors which compromised the accuracy and objectivity of the models:

- The brain was approximated by a fluid contained in an elastic or rigid shell with no provision for relative motion between the shell and fluid. The approximation is inadequate in view of clinical and experimental evidence asserting that brain injury may be strongly influenced or caused by skull deformation and relative brain motion.
- The computed shell strains were greater by as much as a factor of ten in a numerical comparison with experimental data. Moreover, the fluid pressure was dissimilar in phase and amplitude, with peak pressures representing about 25% of the experimental value.
- The model objectivity was compromised by using different fluid compressibilities which varied by a factor of 200 to match experimental data. No scientific justification could be found for the wide variation in fluid property.
- The first three resonant frequencies were about one-tenth of the values reported for the human head. The dynamic dissimilarity was probably due to an insufficient number of elements in the model and produced an incorrect impact response.
- The acceleration input was not sufficiently representative of a typical head impact. Similarily, the boundary condition representing the neck attachment was either excluded or unrealistically simulated inspite of its strong influence on the response.
- The proposed brain tolerance criteria, based on fluid pressures, indicated potential errors in space and time increments of the finite element solution.

Although significant progress has been made in the development and refinement of finite element models of head impact, the available computer programs represent mechanical simulations which are inadequate to establish tolerance criteria or predict head injury. In fact, the models are only a qualitative simulation of a simplified head surrogate, and as such, constitute merely the first approximation of one aspect of a highly complex biological problem.

HEAD INJURY INVOLVES ANATOMICAL alteration or physiological dysfunction of the scalp, skull, brain, or enclosed blood vessels, and usually excludes injury to facial tissues and sensory organs. Generally, injury is grouped by scalp damage, skull fracture, and brain contusion. For obvious reasons, brain injury has received the most attention; and yet, it remains the most difficult to understand. Several review papers and conferences with participants from medical and physical sciences have been devoted to the subject of head injury. The object has been, and remains, to develop a cause-effect or dose-response relationship between the mechanical load and subsequent injury.

The head mechanically responds to a direct collision with a moving or stationary object, or to indirect inertial effects transmitted through the cervical spine. The experimental simulation of impact responses has been addressed by use of anesthetized animals, human cadavers, inanimate replicas and mathematical representations to understand head responses. Each has advantages and drawbacks. Although the long range intent of the experimental work has been to understand injury mechanisms and establish tolerance criteria, all studies were challenged by the complexity of the mechanical response of the head models. Many problems have been identified in developing an analytical model; and today, we are only at the leading edge of a realistic approach.

The first mathematical model of the head is believed to have been contributed by Anzelius (1)* who investigated the response of a spherical liquid mass, simulating the brain tissue to an abrupt change in velocity. Because of the rigid boundary condition, the analysis produced a compressive wave at the frontal plane near the impact and simultaneously a reflective tensile wave at the back. The superposition of the two waves produced a node of zero pressure at the center. The pressure gradient, which included tensile pressure at the back or contrecoup location was compared with the cavitation hypothesis frequently cited in the literature for material failure at a point opposite the impact. On the other hand, the clinical experience indicated that the genesis of the so-called coup-contrecoup brain injury probably involves tissue compression and shear effects during the initial and rebound motion of the brain relative to the skull.

Following Anzelius's analysis and Goldsmith's suggestion (2) to model the head by an elastic fluid-filled spherical shell, a number of other simulations have been proposed. The models involve both spherical and oval shells, approximating the skull, with elastic or viscoelastic materials (3,4). Closed-form solutions using truncated series expansions were used. The models offered a two dimensional solution and assumed axial symmetry of the response. A pressure gradient was also reported along the impact axis of the shell, however the node was shifted from the center towards the point opposite to impact because of shell deformations. In addition, shear strains were predicted in the viscoelastic fluid (4).

Realizing the geometric and material complexities of the head and the limitations of closed form solutions to deal with such structures, finite element methods were used as a potentially promising technique to model head impact. At the beginning, the models were limited to two dimensional axisymmetric (5) or plane strain simulations of a midsagittal section of the head (6). Later, three dimensional models were developed and analyzed (7,8,9).

All mathematical head models are based on established principles of mechanics and provide the deformation response of a structure to impact. Mechanical parameters such as stress and strain are computed throughout the model. Strains have been used to study the potential for shell fracture. Pressure and shear strain in the fluid have been used to assess possible material damage by cavitation and tearing mechanisms (10,11). In order to use the model for material failure prediction, criteria were established in terms of mechanical variables. Although this was partially attempted for cranial bone, because of its resemblance to engineering materials (12), similar attempts for brain tissue have been largely unsuccessful due to the inadequacy of material property data obtained from postmortem tissue (13,14).

The purpose of this paper is to review the proposed finite element head models with particular emphasis on model objectivity and accuracy from the applied mechanics point of view. The strengths and limitations of the analytical approach need to be understood prior to any effort to assess the severity of head impact responses or the potential for head injury with finite element models. First of all, in order to satisfy the minimum technical criteria, a mathematical model of a physical system must simulate the important mechanical features in four areas: 1 - geometry, 2 - boundary/interface conditions, 3 - material properties, and 4 - frequency response. In addition, proper use of the finite element technique must be ascertained to insure accuracy of the analysis. In the following, model developments and responses will be reviewed and analyzed with reference to the previously cited features.

OVERVIEW OF MODELS: GEOMETRY

The head, unlike any engineering structure, has an extremely complex geometric configuration. Even though the calvarium resembles a shell it consists of eight layered bones of irregular thickness joined by sutures. The base of the skull is a thick irregular plate with several small foramina or openings and a large central foramen magnum. The skull is covered by the scalp, a loosely coupled layer of soft-tissue. The interior of the skull is occupied by the brain which is neither fluid nor solid, but rather a gel-like medium partitioned into many recesses by the folds of a system of soft, elastic covers. There is a complex network of fragile blood vessels inside the

---

* Numbers in parentheses designate references at end of paper.

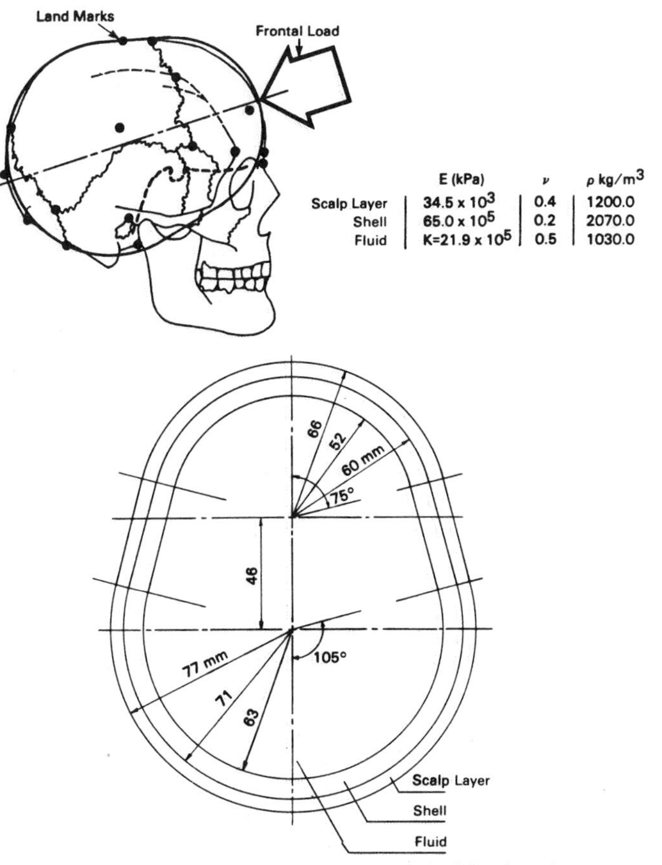

Figure 1. Schematic of human skull showing locations of main land marks and an oval model of a 50th percentile skull enclosing a majority of the land marks.

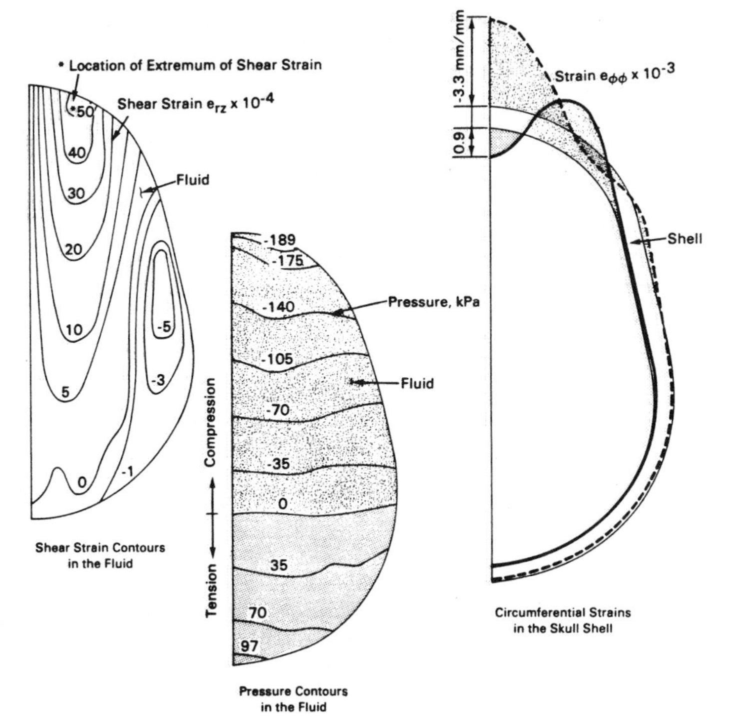

Figure 2. Response parameters of the nominal oval head model at 2 ms when subjected to frontal sine squared impact of duration=4 ms and an amplitude=4450 N.

brain tissue which communicate with the bridging veins from the dural venous sinuses to the brain cortex. Furthermore, the brain floats in pressurized cerebrospinal fluid within the dura. Most models simplify the head as a multilayered two or three dimensional structure, however, all models thus far, portray only the highly simplified general features of the multicompartmental system of the head.

In a first approximation, Khalil and Hubbard (5) simulated the skull geometry by a closed oval shell closely approximating the major landmarks of the skull (Fig. 1). An elastic layer encased the oval shell to simulate the scalp. The intracranial contents were approximated by an inviscid fluid. The response of this model and two companion models, simulating the skull by a single layer and multilayered spherical shells, respectively, were analyzed for a direct axisymmetric impact of the freely suspended structure. For impact durations of 2 ms to 12 ms, the model exhibited a vibrational response superimposed on rigid body motion. Tensile shell strains were developed on the inner surface and compression on the outer along a polar cap of approximately 60° half angle from the impact axis. A pressure gradient developed along the impact axis with compression near the impact and tension opposite. Modeling the intracranial contents by a viscoelastic material (15) showed shear strain maxima in the vicinity of the impact load. Spatial pressure and shear stress contours in the fluid, and strain distributions on the inner and outer surface of the shell (Fig. 2) demonstrate the general characteristics of the mechanical response of the fluid-filled shell at peak force.

Chan (16) analyzed the axisymmetric impact response of a spherical and ellipsoidal shell model of the head. The intracranial contents were simulated by a material supporting shear deformation and pressure. In order to accommodate more realistic geometry and maintain the simplicity of the two dimensional model, Shugar and Katona (6) analyzed a two dimensional plane strain midsagittal section of the human head (Fig. 3). The model included a layered shell encapsulating an elastic or viscoelastic material. The model was constrained against rigid body motion by fixed boundary nodes at the base of the shell. The anteroposterior impact response of the model revealed a pressure gradient along the impact direction with compression near the impact site and tension at the point opposite. However, the pressure gradient was quadratic and not linear as was noted in the axisymmetric models. Also, the magnitude of the tensile component was higher than the compressive. The different behavior was probably due to the opening at the base of the shell and the combined rotational and translational response associated with the

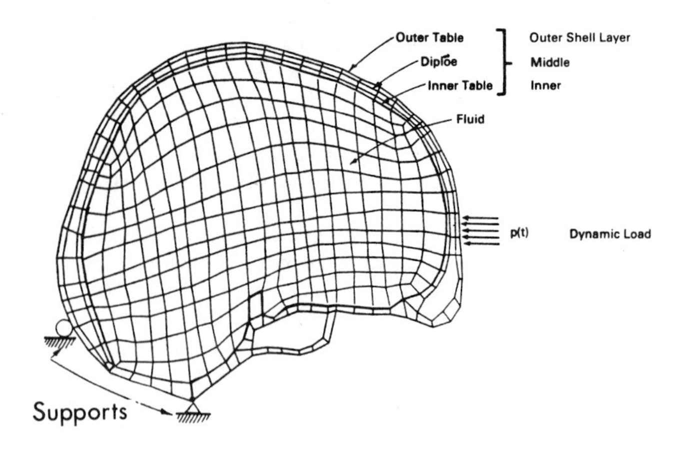

Figure 3. Plane strain model of the human head simulating the layered shell and interior fluid (Ref. 8).

prescribed boundary condition. Shell deformation and strain were in qualitative agreement with the axisymmetric spherical shell model response (5).

Ward (7) developed a three-dimensional model of the brain including representations for the dura folds, ventricles and brain stem. In the original model linear spring elements were used to simulate the tethering of the brain to the skull. This boundary condition was later abandoned and the model geometry was revised (17). The model was initially compared with experimental data on static brain-stem deflections obtained by x-rays of a human cadaver (7). The influence of the falx cerebri and tentorium on model responses was addressed by comparing brain stem deflections and resonances of a model with and without the membranes. Although it is not clear, why the authors selected uniform traction at the brain stem to compare model and experimental responses, inclusion of the membranes produced minimal effects.

After the dynamic response of the model was compared against experimental data, Ward modified the topology of the original human brain model and developed two additional models of Rhesus monkey and baboon (18). The models were subjected to an acceleration pulse at the boundary nodes and the solution was obtained in a relative coordinate system. However, the coordinate transformation did not allow the linear program to include large displacements and rotations (8) as the authors suggest. The animal models were used to simulate the responses obtained from head impacts on anesthetized animal and the cadaver model was used to simulate responses from experiments with a pressurized postmortem vascular system. Limited experimental observations from cadaver tests suggested small relative motion between the pressurized brain and skull during impact. Ward, therefore, modified the models for no slip at the boundary. Unfortunately, this assumption appears to be refuted by clinical observations of experimental neurosurgeons (Appendix A) and critical reviewers of the published data (19,20). The "no slip" condition remains a critical deficiency of all subsequent Ward's models. Although the models produced a pressure gradient during impact, they over predicted the magnitude of the negative pressure by 3-4 times at the contrecoup site. When the animal model was subjected to an acceleration pulse similar to that observed in experiments, the location, magnitude and duration of the maximum pressure in the fluid were different from the human model, possibly due to the high-frequency content associated with more elements used in the small animal head model. The finding led the authors to conclude that interspecies scaling of responses from the models was inconceivable.

All Ward's previous models and studies assumed midsagittal symmetry. In an attempt to simulate the side impact of a cadaver head (21), disagreement between measured and computed pressures were noted. The authors attributed the discrepancy to rotational accelerations and velocities which were not measured in the experiments and, therefore, not included in the analytical input. Although a better representation of the midsagittal sinus was suggested as a possible correction for differences between the model and experiment, further technical support for the argument is yet to be shown.

Shugar (8) developed and extensively analyzed a three dimensional model of the human head. The model approximated the skull by a layered structure and the brain by an inviscid fluid. A low modulus layer was introduced between the brain and skull to approximate the compliance of the subarachnoid space. The model included loads directly on the surface nodes, and initially, responses were compared against deflections and strains from static data obtained from an evacuated Rhesus monkey skull. Comparison of deflections showed that the model was stiffer than the evacuated skull. The author attributed this behavior to the eight-node brick element of the finite element analysis. Although an incompatible nine-node element was proposed to improve the bending stiffness of the element, no verification was presented. Further, comparison of static surface strains at different locations on the shell showed a generally poor representation where strains were both overpredicted and underpredicted. The author attributed the differences to comparing mid-element strains with surface measurements.

Shugar's model overpredicted dynamic strains by as much as three to ten times those observed in experiments. This contradictory shell response indicated the extreme complexity of skull geometry which include significant variations in bone thickness and the inability of the model to accurately model this sensitive mechanical feature. Comparison of fluid pressures showed disagreement in both phase and amplitude of the response. Attempts to obtain a more realistic pressure agreement, including adding representations for the foramen magnum, compressible subarachnoid space layer and facial mass, and adjusting the solution technique by reducing the time step, did not improve the situation.

Although agreement could not be reached, even in the original shell strains, the response of Shugar's human head model was computed for frontal and occipital impacts. Strain distributions in the shell were localized near the impact and the shell opening at the base. The latter was due to a stress concentration at the opening. The finding indicated that material failure would initiate near the impact load and at the base of the shell. This feature was consistent with experimental results on skull fracture reported by Gurdjian (22). The model predicted compressive pressure in the fluid near the impact and tensile pressure at the opposite location.

Hosey and Liu (9,23) developed a three-dimensional model of the head and neck. The model approximated a layered skull, subarachnoid space, dura, cervical vertebra and disks. The model simulated direct impact on the shell. Because of linearity of the solution and in view of Shugar's work (8), the response was limited to small rotations.
The occipital impact response of the model showed a pressure gradient with compression and tension regions in the fluid along the impact direction, as seen in other models. The model, so far, has not been rigorously tested or compared against experimental data.

One question should be addressed before the proposed three dimensional models are further pursued: what are the practical differences between a two and three dimensional model? Alternatively stated, what benefits may be gained by undertaking the costly task of developing, running and analyzing the responses in a three-dimensional situation? Obviously, the effects of inertia could be more accurately represented and the impact vector can be applied in any direction. However, it must be emphasized, that Nahum et al.'s, (21) attempt to simulate side impact responses using Ward's model (17) was unsuccessful, and indicated that three-dimensional models are probably not yet ready for study of the effects of side or oblique loading of the shell. Even the simulations of a midsagittal impact have not been sufficiently investigated. Table 1 presents a qualitative comparison between two and three dimensional finite element models. Clearly the two-dimensional models can provide nearly equivalent information by less complex computation and simpler output in many important responses. Pressure gradients, shear strains and their maxima near the foramen magnum, and shell strains and their distribution near the impact load are qualitatively similar for the two and three dimensional models (8).

In addition, since two-dimensional and three-dimensional models use nearly the same number of elements, it is obvious that the refined two-dimensional mesh would provide a better approximation of the response. The two-dimensional models use twenty-one elements to represent the midsagittal length of the fluid between the frontal and occipital shell elements. By contrast, the three dimensional models use only six to eight elements to simulate the same length. Without question, the smaller the element size, the higher its frequency content and, consequently, its ability to accurately simulate the dynamic response of the structure. Accordingly, two dimensional models, either axisymmetric or plane strain, may produce more satisfactory and more relevant information on mechanical responses.

The extensive studies of Ward and Shugar demonstrated that model responses do not faithfully replicate experimental results from biological specimens. In addition, since the reported parameters are limited to material deformation, stress and strain and in the absence of possible mechanisms of functional brain tissue damage, the prediction of brain trauma or impact tolerance remains far beyond the scope of any present finite element model and our current understanding of head injury. However, the models provide a valuable methodology to assess the relative impact severity and the efficacy of protective systems.

BOUNDARY CONDITIONS

The impact response of finite element models depends significantly on the boundary conditions and forces on the structure, including the spatial and temporal distributions. Additionally, in layered models proper interface conditions must be specified where displacement continuity or relative motion is anticipated.

NECK SIMULATION BOUNDARY CONDITION - Shugar (8) studied the influence of the boundary conditions approximating head/neck attachment during head impact. Strains in a hollow shell differed by more than a factor of three depending on whether the shell was fixed or hinged at the base. In addition, acceleration and intracranial pressure from experiments showed a sign reversal which was similar to model responses when a fixed

TABLE 1

COMPARISON BETWEEN FINITE ELEMENT MODELS

Author	Geometry	No. of Components	Material Characterization	Load Function	Comparisons	Boundary Condition	Significant Finding
Khalil and Hubbard (1977)	2D axisymmetric spherical and oval fluid-filled shells	3-simulating scalp, skull and brain including layered skull of human	linear elastic and viscoelastic	sine-squared in time, sine in space	skull fracture loads compared with human cadaver data	free	localized shell strains near impact site. Linear pressure gradient in fluid with compression near impact and tension opposite. Maximum shear in fluid near impact location.
Shugar and Katona (1975)	2D plane strain midsagittal section representing shell and fluid	2-simulating layered skull and brain of human	linear elastic and viscoelastic	sine-squared in time uniform in space	no validation	shell constrained against rotation and translation	localized shell strains. Quadratic pressure gradient with contrecoup pressure about twice the coup pressure. No shear strain results in the fluid.
Shugar (1977)	3D-symmetric about mid-sagittal plane shell and fluid	3-simulating layered skull-subarachnoid space and brain of human and rhesus monkey	linear elastic	either as experimentally measured or 1/2 sine in time and uniform in space	validated against empty rhesus skull for strain and in-vivo for pressure	fixed, hinged, sliding or free shell	localized shell strains near impact site and opening at base. Nearly linear pressure gradient in the fluid.
Ward (1974-1980)	3D fluid model.	5-simulating cerebrum, cerebellum, brain stem, ventricles and dura folds of human rhesus and baboon	linear elastic	acceleration pulse from experiments. Spatial and temporal distributions are not published.	brain displacement and pressures from human cadavers and in-vivo primates	spring elements were used between fluid and shell. Later, rigid and fixed boundary.	pressure gradient in the fluid. Due to frequency response difference between human and animal model no extrapolation is conceivable
Hosey and Liu (1980)	3D-symmetric about mid-sagittal plane	6-simulating layered skull, dura, CSF, brain, spinal cord and cervical column of human	linear elastic	1/2 sine in time and uniform in space	no validation	fixed at the base of C7	pressure gradient in the fluid. Small pressure difference (<10 kPa for 6000 N force) produced across the fluid membrane.

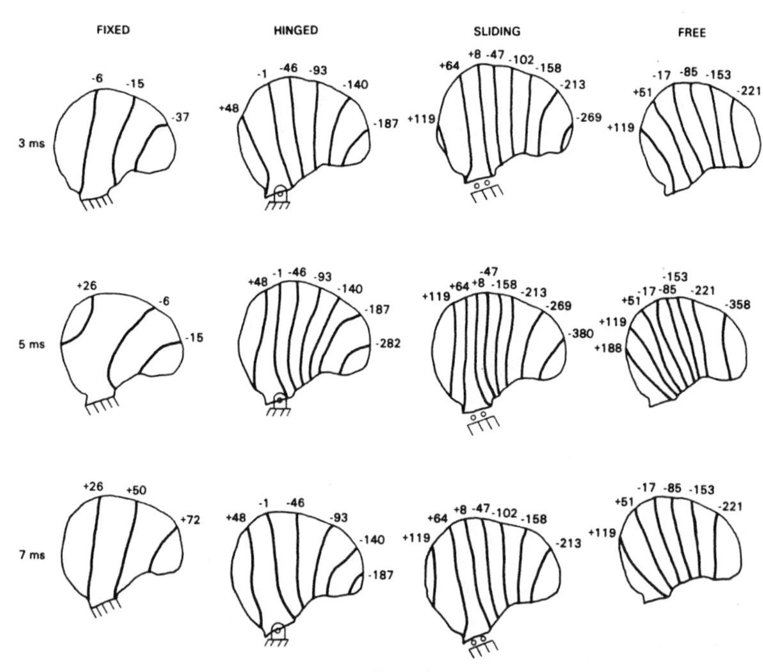

Figure 4. Fluid pressures (kPa) from a half sine load with 4450 N amplitude and 10 ms duration for various boundary conditions, (ref. 8).

boundary condition was used at the base of the shell. In order to study the effects of constraint at the base of the shell, Shugar compared the response of the human head model for four boundary conditions: fixed, hinged, sliding on a plane and free.

The analysis in terms of shell displacement, fluid pressure and shear strain showed that shell motion was strongly affected by, and different for, all boundary conditions. The fluid pressure showed sign reversal (Fig. 4) for the fixed shell. The hinged and sliding shells exhibited pressure contours that followed the rise and decay of the impact pulse. However, pressure contours in the free shell were more vertical and reflected the dominance of sliding motion. The maximum tensile pressure occurred in the free shell and the maximum compression in the sliding shell. Limited experimental studies show a sign reversal in pressure similar to the model response with fixed boundary conditions. The most reasonable boundary condition at the base of the shell probably depends on the impact condition and obviously needs further investigation.

The model of Hosey and Liu included a representation of the cervical spine fixed at the approximate level of the seventh vertebra. In one application of the model, the fluid pressure (23) attained a peak at 0.4 ms prior to the peak of the 4 ms half-sine impact. This behavior is inconsistent with the mechanics of the solution used. In fact one would expect a delay in the response due to the compliance at the base of the shell. Also, the full benefit of a compliant element at the base of the shell may not be realized in a linear model that allows only small rotation.

CONTACT/IMPACT FORCE - During the collision of the head with an object, a contact force is developed and removed in a short duration. Measuring the distribution of the contact force has not been experimentally achieved, so, most analytical efforts involve parametric study of the influence of contact area and impact duration. Using simple head models, Khalil and Hubbard (5) showed that the impact area significantly influences shell strain. The impact duration did not affect the responses, which followed the rise and decay of the load pulse.

The input to Ward's models is an acceleration pulse derived from experiments and represents the response of a rigid structure. The nodes to which this input is applied and the actual input profile to each node have not been reported. It seems likely that simple half-sine input accelerations in one direction have primarily been used as input functions to the rigidly contained fluid model. However, it may not be presumed that the boundary nodes are subject to the same acceleration pulse, if rotation of the model is to be included in the driving function. Obviously, the input, also, excludes the effects of skull deformation from the direct impact, which are generally considered important in the actual head impact process (see Appendix A).

INTERFACE BOUNDARY CONDITIONS - Proper interface conditions between dissimilar layers in the model must be specified for the various structural components, particularly between the shell and fluid. It is also important to permit relative motion. Ward (7), initially, used linear springs between the fluid and shell. Later, the spring elements were deleted and the boundary nodes of the fluid were assumed rigidly fixed, even though there is substantial evidence of relative motion between the brain and skull and between various brain compartments (24). Shugar (8) introduced a soft layer between the fluid and shell having a bulk modulus 1/100th to 1/1000th that of the fluid. Although this layer approximated the subarachnoid space, it introduced a significant impedance mismatch between the layers. The layer slightly reduced intracranial pressure, but did not improve the differences with the experimental data.

Interface boundary conditions also have a dramatic effect on model response and must be properly evaluated prior to use in models. To demonstrate this point, we analyzed the impact response of a simple plane-strain model consisting of an outer shell and an interior inviscid fluid, Fig. (5), for two interface boundary conditions: a) allowing no relative motion between fluid and shell and b) allowing sliding between the fluid and shell. The model was loaded by a uniform spatial pressure of 4 ms duration during which the load varied according to a sine-squared function. The peak force applied was 7 kN at 2 ms. Figure 5 shows that the pressure distribution in the fluid differs by about a factor of 8 for the two prescribed interface conditions.

MATERIAL PROPERTIES

The material properties, used in the models, significantly influence the responses and representativeness of the approximation. Medical and engineering scientists recognize that biological tissues are inhomogeneous, anisotropic, and nonlinear. Yet, approximations have been made for the sake of model development. All finite element models are based on small deformations of elastic or viscoelastic materials. In addition, linearity, homogeneity and isotropy are always assumed. Accordingly, the accuracy of model responses is directly limited by the simplicity of the material representation assumed in the mechanical theory.

SHELL PROPERTIES - Skull bone is a composite material consisting of a compact inner and outer layer encasing a soft spongy layer. Selected shell properties (Table 2) vary by as much as a factor of four in the available models due to variability in published experimental data (12,25,26). Khalil and Hubbard (5) noted that skull strains and fluid pressure slightly depended on the flexural stiffness of the shell when the membrane stiffness was held constant. By contrast, variations in the shell elastic modulus produced a significant inverse effect in shell strains and a slight effect on fluid pressures. To study this point, we exercised the oval head model shown in Fig. 1 with nominal elastic shell properties and with variations of ±50% in the skull representation. Table 3 shows that the maximum shell strain under the impact varied linearly in inverse proportion to the elastic property. Compressive strain developed at the outer surface and tensile strain at the inner surface. However, variations in fluid pressure were small, not exceeding 5%.

Also, we ran the oval head model for three shell thicknesses, representing the nominal and ±50% variations, to study the thickness influence on strain and pressure. Table 4 show that reducing the thickness by 50% increased the maximum bending strain from $\pm.45\times10^{-3}$ mm/mm to $\pm1.02\times10^{-3}$ mm/mm (127% higher) and the membrane strain from -0.25×10^{-3} mm/mm to -0.88×10^{-3} mm/mm (252% higher). On the other hand, increasing shell thickness by 50% reduced the bending strain to $\pm0.365\times10^{-3}$ mm/mm (20% lower) and the membrane strain to -0.16×10^{-3} mm/mm (36% lower). Obviously, the dependence of strain on skull thickness is not in conformity to what one expects from simple beam theory. This can be explained by the geometric complexity of the oval shell in comparison with a simple beam. The fluid pressure is also affected by variations in shell thickness. The decrease in shell thickness increased the pressure at the coup and contrecoup locations by about 9%. In contrast, the increase in shell thickness decreased the coup pressure by 34% and the contrecoup pressure by 19%. Although a rational and uniform procedure for selecting shell properties is needed, some investigators have used a range in properties to bound the expected response, such as Shugar who used two values for the elastic moduli (Table 2).

FLUID PROPERTIES - Elastic and viscoelastic properties of postmortem brain tissue were determined in cadaver experiments (27,28), and found to behave like a fluid. Accordingly, many models (Table 5) assumed that brain tissue can be simulated by a fluid having the compressibility and density of water. Other investigators (29) argued that in the simulation the fluid should be treated as a compressible material with a lower value

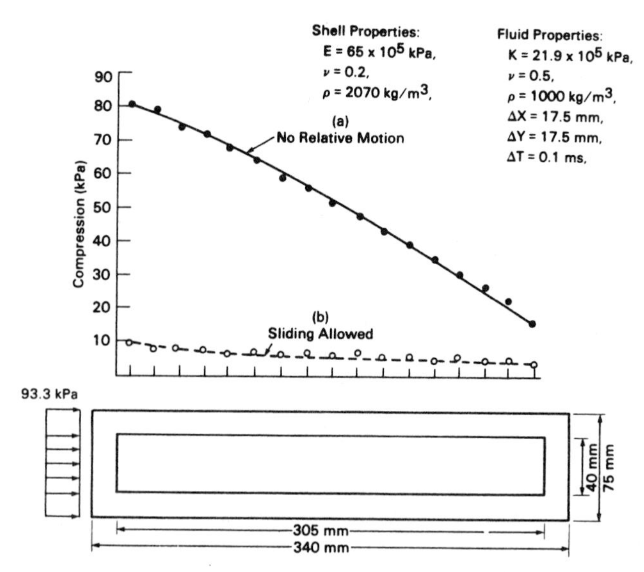

Figure 5. Spatial pressure distributions in the fluid at 2 ms for 2 interface conditions: (a) No relative motion was allowed and (b) Sliding motion was permitted.

for fluid compressibility than water. Ward (29) adjusted the fluid property by varying Poisson's ratio (ν) from 0.48 to 0.4999 depending on impact duration and type of specie modeled without sound theoretical and experimental verification. It is obvious that the fluid property was varied to match model predictions with experimental data (30).

We have investigated variations in fluid pressure in our oval model when Poisson's ratio (ν) was varied from 0.45 to 0.49999. Young's modulus of the fluid was selected as 66.7 kPa, similar to that used by Ward (7). A split energy element was used allowing for one point integration for the λ term and a 2 point integration for the μ term (31). This procedure approximated the response of compressible and nearly incompressible materials. For a variation in Poisson's ratio from 0.45 to 0.499 we noted that the compressive coup pressure increased from 7 kPa to 50 kPa and the tensile contrecoup pressure increased from 9 kPa to 47 kPa (Fig. 6). However, increasing Poisson's ratio to 0.4999 reduced, instead of increased, the coup and contrecoup pressures to 43 and 37 kPa, respectively. A further increase in Poisson's ratio to 0.49999 again increased the coup pressure to 46 kPa and decreased the contrecoup pressure to 33 kPa. The results showed a monotone increase in coup and contrecoup pressures with increasing Poisson's ratio to 0.499. However, this trend was interrupted for Poisson's ratio larger than 0.499. The pressure not only declined, it actually exhibited fluctuations (Fig. 6).

TABLE 2

SHELL BONE PROPERTIES

	Compact Bone			Spongy Bone			
	E kPa x 10^6	ν	ρ kg/m^3 x 10^3	E kPa x 10^6	ν	ρ kg/m^3 x 10^3	Comments
Nickell and Marcel (1974)	4.14	-	2.1	1.04	-	2.1	ν is not provided and one value for ρ is provided, presumably for both layers.
Khalil and Hubbard (1977)	17.94	0.35	3.0	0.73	.05	1.7	
Shugar (1977)	5.66 12.28	0.22	1.4	-	-	-	A low and a high value for E were used. middle layer properties are 0.1 of outer layers.
Hosey and Liu (1980)	4.46	0.21	1.4	0.00	0.00	0.0	No data is given for the middle layer of the shell.

TABLE 3

Variations in Maximum Shell Strain and Fluid Pressure of Oval Head Model at Maximum Load of 1000 N with Shell Elastic Modulus

Shell E (x10^5kPa)	Outer Surface Strain (mm/mmx10^{-3})	Inner Surface Strain (mm/mmx10^{-3})	Coup Pressure (kPa)	Contrecoup Pressure (kPa)
32.5	-1.40	+0.40	-39.4	20.4
65.0	-0.70	+0.20	-38.7	21.5
97.5	-0.46	+0.13	-37.3	22.3

TABLE 4

Variations in Maximum Shell Strain and Fluid Pressure of Oval Head Model at Maximum Load of 1000 N with Shell Thickness

Shell t (mm)	Outer Surface Strain (mm/mmx10^{-3})	Inner Surface Strain (mm/mmx10^{-3})	Coup Pressure (kPa)	Contrecoup Pressure (kPa)
4	-1.90	0.14	-41.9	23.5
8	-0.70	0.20	-38.7	21.5
12	-0.53	0.20	-25.5	17.5

TABLE 5

FLUID PROPERTIES

	Bulk Modulus, K kPa x 10^5	Elastic Modulus, E kPa	Poisson's Ratio, ν	Density ρ kg/m^{3}x10^3	Comments
Khalil and Hubbard (1977)	21.9	-	-	1.03	K replaced E and ν
Shugar (1977)	21.9	-	-	1.04	K replaces E and ν
Ward (1980)		66.7-667	0.48-0.4999	1.04	ν depends on pulse duration and type of model
Hosey and Liu (1980)		66.7	0.49- 0.49999462	1.04	

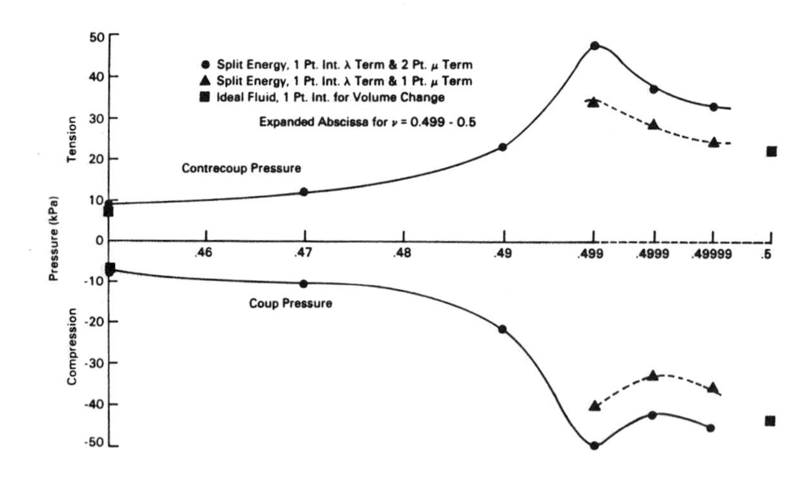

Figure 6. Variations in coup and contrecoup pressure with poisson's ratio ν.

The unexpected fluctuation in pressure was studied by varying Poisson's ratio (ν) from 0.4999 to 0.49999. We repeated the solution using a one point integration for both λ and μ terms for 3 values of ν=0.499, 0.4999 and 0.49999, and noted a reduction in the magnitude of the pressure (Fig. 6). However, the same fluctuation remained unchanged. Figure 6 also shows the coup and contrecoup pressures for ν=0.45 and E=66.7 kPa, and for ν=0.5 and K=21.9x10^5 kPa. In these two runs, an ideal fluid element was used with one point integration to account only for volume changes. In all these runs, shell strain did not vary with the change in Poisson's ratio. The results clearly demonstrated two importants points: a) variations in Poisson's ratio significantly effect fluid pressure, and b) using the split energy element for ν>0.499 produces an unexpected fluctuation in pressure. The reason for this behavior is not known and requires further investigation.

Properties of the membranes approximating the dura, falx, and tentorium were selected by Ward and Hosey from the literature. However, a close inspection of their properties (9) reveals an impedance mismatch (ρc) between the layers of nearly five orders of magnitude. This may contribute to the numerical inaccuracy in the solution.

DYNAMIC CHARACTERISTICS

When the head impacts another object, dynamic loads develop on the skull. The duration, area and amplitude of the contact force depend on the geometry, stiffness and relative velocity of the colliding structures. Whether the impact is direct or indirect, the head experiences transient motion, including translation, rotation, and structural vibration. The majority of head impacts have a duration of 4-25 ms, which is at least an order of magnitude larger than the wave transit time of approximately 130 μs (2) along the midsagittal plane. Accordingly, an inertial response and structural vibration will be encountered.

A finite element model must have representative resonances and mode shapes to have a reasonable dynamic response. Khalil et al. (32) identified the resonances and mode shapes of the dry human skull for frequencies of 10 Hz-5 kHz, and compared the results with empty shell models. In a subsequent study Khalil and Viano (33) investigated the adequacy of the available spherical shell models to approximate the dynamic response of the skull. The spherical shell was shown to be nearly twice as stiff as the dry skull. Consequently for the same load the shell would develop about half the deformation and strain as observed in the skull. Since the most appropriate criteria for failure of the shell material are based on the maximum tensile or compressive strain, the shell model would logically predict higher loads for failure of the shell than observed in experiments. This, indeed was the case as shown in Fig. 7. Although the elastic modulus of the shell could be reduced to bring both the resonances and the failure load of the shell into closer agreement with experimental data, it is clear that this is only a crude adjustment of the overly simplistic model. It, however, provides a more realistic response and does not contribute to other differences. Such a selection procedure for material properties may be appropriate if simplified models are to be further pursued, and made more realistic and relevant to the head impact response issue.

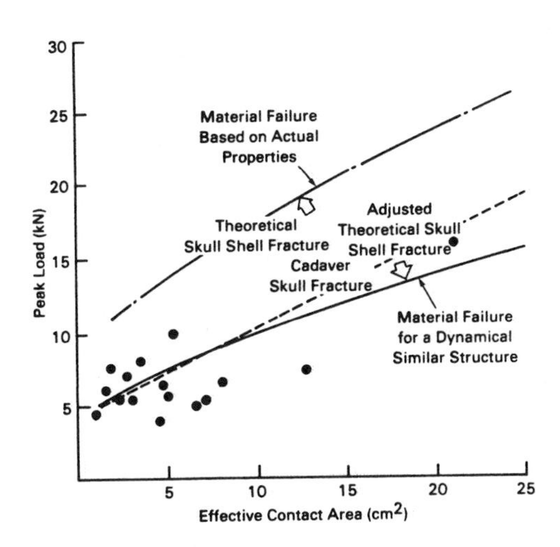

Figure 7. Material failure or shell fracture load vs. contact area. -·-·- Theoretical curve (Ref. 5). • Cadaver data and --- linear regression of cadaver data (Ref. 30). ——— Adjusted theoretical skull fracture curve obtained by imposing dynamic similarity between model and skull.

The currently available three dimensional finite element models do not have an acceptable frequency response. Nickell and Marcal (34) determined skull resonances for boundary support at the front, back and base. Four resonant frequences and mode shapes were identified. For frontal support the resonances were 86, 231, 458 and 864 Hz. For the back support the resonances did not differ by more than 20% from those with frontal support. When the skull was supported at the base the resonances were 194, 465, 635 and 1327 Hz. Nickell and Marcal's fundamental resonance for frontal support is about 1/10th the experimental value found by Khalil et al. (32).

Shugar (8) determined the four lowest resonances of the human shell model at: 43, 51, 55 and 57 Hz. The resonances are significantly lower than the frequencies at 313, 600 and 800 Hz reported by Gurdjian from cadaver tests (35). Ward (7) determined the resonances of the fluid model with and without the partitioning membranes. The first four resonances with the falx cerebri and tentorium were 23, 29, 32 and 33 Hz. Removal of the membranes slightly reduced the resonances. Since experimental data on postmortem brain resonances are not known, we compared Ward's resonances with those of a spherical liquid mass enclosed in a rigid shell. Using Engin's non-dimensional resonances (36) for a fluid sphere of 136 mm diameter and fluid properties as reported by Ward (7) (E = 66.7 kPa, and $\nu = 0.48 - 0.499$), the first four resonances were 128, 188, 250 and 314 Hz for $\nu = 0.48$, and 557, 820, 1089 and 1368 Hz for ν to 0.4999. Obviously the resonances presented by Ward (7) are much lower than those predicted for an ideal spherical fluid mass. Hosey and Liu (9,23) have not presented resonance data on their model. However, resonance information is necessary to assess the accuracy of the model, because of the large differences in the material properties and acoustic impedances of the adjacent components. Knowledge of model resonances, also, provides a basis for selecting the appropriate integration time-step.

SPACE-TIME DISCRETIZATION IN FINITE ELEMENT MODELS

The impact response of a fluid-filled shell approximating the head can be solved as a dynamic boundary value problem (BVP) of elastic or viscoelastic media. The governing equations are derived by a balance of momentum in the deformable media. A set of linear partial differential equations in space and time link boundary forces to internal deformations. The deformation function $\underline{u}\ (\underline{x},t)$ is determined at all points in the medium. Following this, computation of strain and internal stress is a straightforward application of the constitutive property of the material.

Closed form solutions of such BVP are possible for a small class of problems with simple geometry. However, the analysis of shells with complex geometry and multiple layers virtually relies on approximate numerical solutions using the computer. The finite element method is a commonly used technique for structural analysis of dynamic BVP. Formulation, convergence and accuracy of finite element techniques are well documented (37). Two approximations have been introduced in the formulation of dynamic finite elements. First, the equations of motion are derived from a virtual work or a variational principle, and are discretized in space by approximating the displacements in the element by a set of interpolation functions. The result is a set of second-order integral differential equations for viscoelastic materials:

$$\underline{M}\ \ddot{\underline{u}} + \int_{-\alpha}^{t} \underline{K}(t-s)\ \frac{\partial \underline{u}}{\partial s}\ ds = \underline{P}(t)\ . \qquad (1)$$

For an elastic material the system of equations reduces to a set of second-order, ordinary differential equations:

$$\underline{M}\ \ddot{\underline{u}} + \underline{K}\ \underline{u} = \underline{P}\ (t)\ , \qquad (2)$$

where $\underline{M}$ is the mass matrix, $\underline{K}$ is the stiffness matrix, $\underline{u}$ is the nodal displacements, and $\underline{P}$ is the load vector. The superimposed dots denote differentiation with respect to time.

The second approximation consists of discretizing equations (1) and (2) to a set of algebraic equations that can be solved numerically. Numerical solutions involving a transient response are usually accomplished by a step-forward integration technique (38) known as the Newmark-β method. This technique is used in the general purpose finite element program developed by Taylor (31). Khalil (5) and Shugar (8) used this program in their modeling efforts. A modified version of the step-forward technique was developed by Wilson (39) for the program SAP. Ward used this numerical solution technique. An analysis of these and other numerical techniques is provided by Bathe (39).

Newmark-β method includes two parameters γ and β controlling the numerical stability and damping, respectively. $\gamma = 1/2$ implies no numerical damping. For $\beta = 0$, a conditionally stable explicit solution requires that the time step $\Delta t \leqslant \Delta t_{cr}$, $\Delta t_{cr} = \frac{\Delta x}{c}$ with Δx the minimum distance in an element and c the fastest wave speed in the material. The selection of these parameters is recommended for wave solutions when impact duration is nearly the same magnitude as the wave transit time in the medium. For longer impact durations mainly producing vibrations an unconditionally stable implicit solution with $\beta = 1/4$ is preferred for computational economy. In this case larger time steps can be selected than those determined by the wave response, and there is no rigid rule for selecting Δt. However, there are some guidelines: Δt must be selected so that there are an adequate number of points to represent the rise-time and pulse-shape of the dynamic load. Alternatively stated, Δt must be selected to accurately simulate the frequency characteristics of the load. Bathe's analysis suggests that $\Delta t \leqslant \Delta t_{cr}$, $\Delta t_{cr} = T/100$ with T the largest period of oscillation of the structure. Obviously, the analyst must insure that the selected time-step provides an accurate solution. Figure (8) demonstrates, according to Shugar's analysis (8), the effect of different time-steps on the propagation of a triangular pulse in a bar. Selection of $\Delta t \leqslant \Delta t_{cr}$ produces a reasonable solution. However, when $\Delta t > \Delta t_{cr}$ pulse attenuation and dispersion result, so, we recommend at least ten time steps to represent a simple sine-squared pulse.

Selection of the spatial step Δx is also crucial to an accurate simulation of the impact response. Δx determines the frequency content of the structure and must be selected sufficiently small to capture the dominant frequencies of the input pulse. If a one dimensional bar is subjected to a uniform sine-squared pulse of duration T varying between 1 unit and 32 units. The maximum stress (Fig. 9) differs from unity for pulse durations less than 12. Such attenuation is

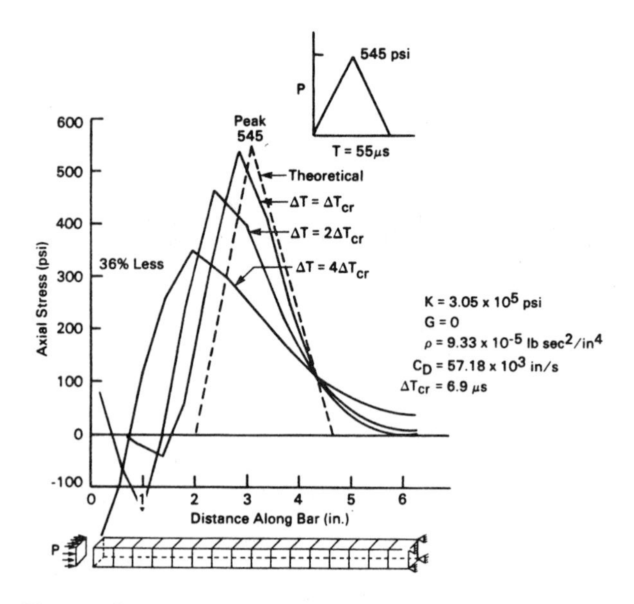

Figure 8. Propagation of a triangular pulse in a bar to demonstrate the attenuation produced by selecting $\Delta T > \Delta T_{cr}$ (Ref.8).

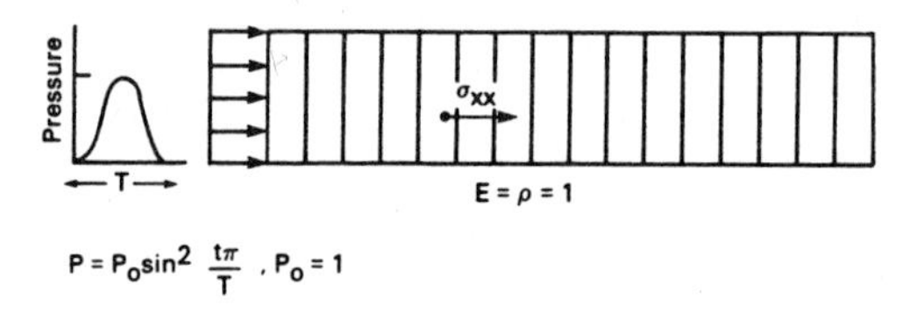

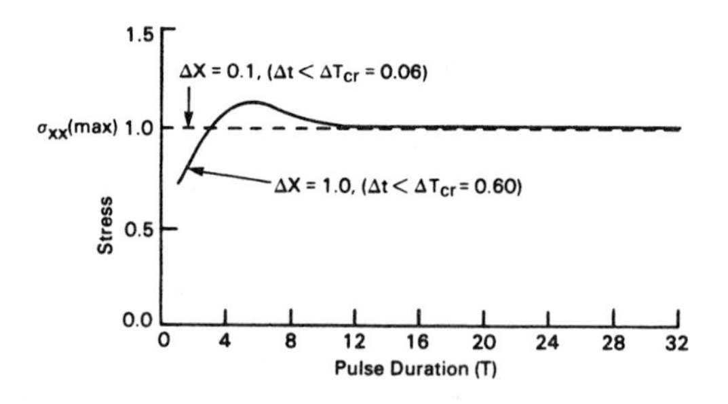

Figure 9. Influence of Δx on peak stress for various pulse durations (1-32 units). 8-10 elements are needed to simulate the pulse wave length.

a numerical inaccuracy due to improper selection of space and time steps and is not associated with a vibrational response of the bar. The fundamental frequency of longitudinal motion has a frequency of 1/32 Hz, and corresponds to a pulse duration of 32 for which there is a unit response. Theoretically a pulse travelling in an elastic bar will not attenuate. Since a reduction of Δx to 0.1 provides an accurate response, about 10 elements are needed to adequately represent the sine-squared load pulse.

Ward (40) developed a set of curves of constant maximum pressure in the fluid as the amplitude and duration of the acceleration input varied for a triangular or trapezoidal wave. The acceleration functions resulted in a minimum response for an impact of approximately 3 ms duration. This indicated that the fluid should have a resonance of about 330 Hz, which was 10 times greater than the published resonance for the model. The anomalous behavior may thus have been an artifact from an improper selection of the space-time discretization steps. Khalil and Hubbard's (5) theoretical analyses of the pulse duration dependancy of the maximum tensile strain in the shell or pressure in the fluid did not exhibit this phenomenon. Furthermore, Hickling and Wenner (4) showed that the magnitude of the negative pressure in the fluid in a spherical shell actually increased with impact duration of 2 to 20 ms due to excitation of resonances. However, the introduction of material damping in the fluid produced peak pressures that partially decayed with an increase in impact duration.

There is an impedance mismatch (ρc) as great as 10^5 times (9) in many fluid-filled shell models with multiple layers. Selection of space-time steps must account for the potential artifacts introduced by impedance differences. Otherwise the impact pulse will be erroneously attenuated as it interacts across the various layers.

CONCLUDING REMARKS

For practical purposes, current finite element models approximate the head as a fluid-filled shell. The dynamic solution is based on structural mechanics principles and is expressed in terms of mechanical quantities such as deformation, stress and strain in the structure. Due to obvious limitations, the models are mainly intended to provide the first approximation of the dynamic response to mechanical impact.

Our review analyzed the underlying mechanical features of the models with emphasis on the geometry, material properties, boundary/interface conditions and loading functions assumed in the solution, and we pointed out critical features in the analysis procedure. All models predict strain in the shell, except Ward's model, and pressure and shear strain in the fluid simulating brain tissue. Qualitative similarity was noted between the two- and three-dimensional model responses (Table 1). However the accuracy of the model responses is not known because of a lack of consistent experimental data on material properties and impact responses. Furthermore, most of available experimental results are not repeatable. Shugar concluded: "...although finite element analysis may provide a qualitative model, it is not possible to achieve a quantitative model at this time" (8). Obviously, the full potential of the three-dimensional models has not been realized. For practical purposes, two-dimensional axisymmetric and plane strain models provide a more refined mesh size with the same relevant information as obtained with the more complex three dimensional models.

Since brain injury is frequently associated with skull deformation, as noted by Gurdjian (22) and Ommaya (41), any model, such as proposed by Ward of an isolated fluid with rigid boundary constraint, must be inadequate to study brain response to mechanical impact. In addition, the inertial loading used in Ward's models is not clearly defined. Shugar's model is more suitable, since it includes a flexible shell and fluid. However, it does not allow relative motion between the fluid and the shell, which is critical for an accurate simulation of head impact. The model also needs to be modified for a more realistic frequency response and stiffness of the shell. It seems that all models should allow relative motion between the fluid and shell, and include a compliant outer layer. Gurdjian (22) estimated that the scalp significantly dampens an impact by increasing the load duration and area. Obviously, the contact area has a significant effect on shell strain (5) which is strongly affected by the presence or absence of a compliant layer. Hosey's model is extremely complex and has not been run sufficiently to judge its merits, or the validity of the empirical functions used to define the flexibility of the boundary condition at the base of the shell approximating the neck.

Boundary and interface conditions, particularly supporting the base of the shell and at the interface between the shell and fluid core, significantly affect model responses. Also, the temporal and spatial distribution of the loading function is important and needs evaluation for typical head impact conditions.

Even though it is generally accepted that biological tissues are inelastic, nonhomogeneous and anisotropic, all models approximate the tissues by elastic layers, homogeneous and isotropic materials, and visco(elastic) fluids. Of greater concern is the fact that available material properties, representing compressibility of the fluid core, vary by as much as a thousand. Clearly, consistency in material properties among the models must be established from sound scientific data, which is thus far lacking. Arbitrary variations of properties to curve-fit model responses into closer agreement with experimental data is obviously inappropriate, and only leads to an empirical description of one single experiment. There is also a need for a reasonable frequency response of the shell model. Models with a low frequency response may have too coarse a mesh which will attenuate the frequencies of impact and produce an incorrect dynamic response. If there is also an improper space-time discretization, the accuracy of the numerical solution will be questionable.

The currently available three-dimensional finite element models include approximations that render their results inadequate to represent the mechanical response of impact. It is therefore not surprising that their premature use for establishing tolerance criteria to crash injury has produced a significant skepticism in the scientific community, particularly when the models were used in the today still unjustifiable attempts to predict or even describe such a complex event as brain trauma.

REFERENCES

1. A. Anzelius, "The Effect of an Impact on a Spherical Liquid Mass." Acta Pathol. Microbiol. Scand., Supp. 48, 1943, pp. 153-159.

2. W. Goldsmith, "The Physical Processes Producing Head Injury," in W. F. Caveness and A. E. Walker (eds.), Head Injury Conference Proceedings, J. B. Lippincott, Philadelphia, 1966, pp. 350-382.

3. A. E. Engin, "The Axisymmetric Response of a Fluid-Filled Spherical Shell to a Local Radial Impulse. A Model for Head Injury," J. Biomech, Vol. 2, 1969, pp. 324-341.

4. R. Hickling and M. L. Wenner, "Mathematical Model of a Head Subjected to an Axisymmetric Impact," J. Biomech., Vol. 6, 1973, pp. 115-132.

5. T. B. Khalil and R. P. Hubbard, "Parametric Study of Head Response by Finite Element Modeling," J. Biomech., Vol. 10, 1977, pp. 119-132.

6. T. A. Shugar and M. G. Katona, "Development of Finite Element Head Injury Model," J. ASCE., Vol. 101, No. E173, 1975, pp. 223-239.

7. C. C. Ward and R. B. Thompson, "The Development of a Detailed Finite Element Brain Model," in Proceedings of 19th Stapp Car Crash Conference, San Diego, Calif., Oct 1975, pp. 641-674.

8. T. A. Shugar, "A Finite Element Head Injury Model," Report No. DOT HS 289-3-550-TA, Vol. I, 1977.

9. R. R. Hosey and Y. K. Liu, "A Homeomorphic Finite Element Model of Impact Head and Neck Injury," in International Conference Proceedings of Finite Elements in Biomechanics, Vol. 2, B. R. Simon (ed.), Tucson, Arizona, 1980, pp. 851-871.

10. F. Unterharnscheidt and K. Sellier, "Mechanics and Pathomorphology of Closed Brain Injuries," in W. F. Caveness and A. E. Walker (eds), Head Injury Conference Proceedings, J. B. Lippincott, Philadelphia, 1966, pp. 321-341.

11. A. H. S. Holbourn, "Mechanics of Head Injuries," Lancet, Vol. 2, 1943, pp. 438-441.

12. J. L. Wood, "Dynamic Response of Human Cranial Bone," J. Biomech., Vol. 4, 1971, pp. 1-12.

13. A. M. Nahum and W. R. Smith, "An Experimental Model for Closed Head Impact Injury," in Proceedings of 20th Stapp Car Crash Conference, Dearborn, Mich, 1976, pp. 783-813.

14. A. M. Nahum, R. W. Smith and C. C. Ward, "Intracranial Pressure Dynamics During Head Impact," in Proceedings of 21st Stapp Car Crash Conference, New Orleans, LA, 1977, pp. 337-365.

15. T. B. Khalil and D. C. Viano, "Impact Response of a Viscoelastic Head Model," in Proceedings of 30th ACEMB, Los Angeles, Calif., 1977, pp. 234-235.

16. H. S. Chan, "Mathematical Model for Closed Head Impact," in Proceedings of 18th Stapp Car Crash Conference, Ann Arbor, Mich., 1974, pp. 557-378.

17. C. C. Ward, "Analytical Brain Models for Head Impact," in Proceedings of Int. Conf. on Impact Trauma, IRCOBI, Berlin, Germany, 1977, pp. 389-398.

18. C. C. Ward, P. E. Nikravesh and R. B. Thompson, "Biodynamic Finite Element Models used in Brain Injury Research," J. of Aviation, Space and Environmental Medicine, 1978, pp. 136-142.

19. E. S. Gurdjian, V. R. Hodgson, L. M. Thomas, and L. M. Patrick, "The Significance of Relative Movements of Scalp, Skull and Intracranial Contents During Impact Injury of the Head," J. Neurosurg., Vol. 29, 1968, pp. 70-72.

20. A. K. Ommaya, J. W. Boretos and E. E. Beile, "The Lexan Calvarium: An Improved Method for Direct Observation of the Brain," J. Neurosurg., Vol. 30, 1961, pp. 25-29.

21. A. Nahum, C. Ward, E. Raasch, S. Adams and D. Schneider, "Experimental Studies of Side Impact to the Human Head," in Proceedings of 24th Stapp Car Crash Conference, 1980, pp. 345-362.

22. E. S. Gurdjian, "Deformation Studies of the Skull: Mechanisms of Skull Fracture," in Impact Head Injury, Charles C. Thomas, Publisher, 1975, pp. 113-139.

23. R. R. Hosey and Y. K. Liu, "A Homeomorphic Finite Element Model of the Human Head and Neck," in Finite Elements in Biomechanics, Wiley & Sons, Publisher, 1981, in print.

24. S. A. Shatsky, W. A. Alter III, D. E. Evans, V. Armbrustmacher and G. Clark, "Traumatic distortions of the primate head and chest: Correlation of biomechanical radiological and pathological data", in 18th Stapp, SAE Paper 741186, 1974, pp. 351-381.

25. J. H. McElhaney, et al., "Mechanical Properties of Cranial Bone," J. of Biomechanics, Vol. 3, 1970, pp. 495-511.

26. R. P. Hubbard, "Flexure of Layered Cranial Bone," J. of Biomechanics, Vol. 4, 1971, pp. 251-263.

27. J. H. McElhaney, et al., "Dynamic Mechanical Properties of Scalp and Brain," in Proceedings of the 6th Rocky Mountain Bio Engineering Symposium, Laramie, WY, 1969.

28. L. Z. Shuck, "Determination of the Dynamic Shear Modulus of Human Brain Tissue," Ph.D Dissertation, West Virginia University, 1970.

29. C. C. Ward, "Finite Element Modeling of the Head and Neck," in NAMRL Head and Neck Injury Workshop, Miami, Florida, 1980.

30. W. Goldsmith, "Current Controversies to the Stipulation of Head Injury Criteria," Letter to the editor, J. Biomechanics, V. 14, No. 12, 1981, pp. 883-884.

31. R. L. Taylor and J. L. Sackman, "Contact-Impact Problems," Structures and Materials Research Department of Civil Engineering, Report No. SESM 78-4, 1978.

32. T. B. Khalil, D. C. Viano and D. L. Smith, "Experimental Analysis of the Vibrational Characteristics of the Human Skull," J. of Sound and Vibration, Vol. 63 (1), 1979, pp. 351-376.

33. T. B. Khalil and D. C. Viano, "Comparison of Human Skull and Spherical Shell Vibrations, Implication to Head Injury Modeling," Accepted for publication in J. of Sound and Vibrations, 1981.

34. R. E. Nickell and P. V. Marcal, "In-Vacuo Modal Dynamic Response of the Human Skull," Transactions of ASME, J. of Eng. for Industry, 1974, pp. 490-494.

35. E. S. Gurdjian, V. R. Hodgson and L. M. Thomas, "Studies on Mechanical Impedance of the Human Skull: Preliminary Report," J. of Biomechanics, V. 3, 1970, pp. 239-247.

36. A. E. Engin, "The Axisymmetric Response of a Fluid-Filled Spherical Shell," Ph.D. Dissertation, University of Michigan, Ann Arbor, Michigan, 1968.

37. O. C. Zienkiewicz, The Finite Element Method, Third Edition, McGraw-Hill Book Company, (UK) Limited, 1977.

38. N. M. Newmark, "A Method of Computation for Structural Dynamics," J. of the Eng. Mech. Div., ASCE, EM3, 1959, pp. 67-95.

39. K. J. Bathe and E. L. Wilson, Numerical Methods in Finite Element Analysis, Prentice-Hall, Inc. Englewood Cliffs, New Jersey, 1976.

40. C. Ward, M. Chan and A. Nahum, "Intracranial Pressure - A Brain Injury Criterion," in Proceedings of 24th Stapp Car Crash Conference, 1980, pp. 163-185.

41. A. K. Ommaya, et al., "Comparative Tolerances for Cerebral Concussion by Head Impact and Whiplash Injury in Primates," International Automobile Safety Conference Compendium, 1970, Detroit, Mich., pp. 808-818.

42. E. S. Gurdjian, Detroit, Michigan, Personal Communication, Jan. 18, 1982.

43. A. K. Ommaya, Washington, DC, Personal Communication, Jan. 21, 1982.

44. E. S. Gurdjian, "Relative Movement of Scalp, Skull and Intracranial Contents at the Time of Impact," in Impact Head Injury, Charles C. Thomas publisher, 1975, pp. 162-180.

45. A. M. Nahum and R. W. Smith, "An experimental model for closed head impact injury," in proceedings of 20th Stapp Car Crash Conference, 1976, pp. 785-814.

APPENDIX A

ON RELATIVE MOTION BETWEEN THE BRAIN AND SKULL - Clinical experiences assert that relative motion between the brain and skull occurs under mechanical impact conditions (42,43). Gurdjian points out that relative motion was theoretically assumed as early as 1518 (44).

High speed photography at 7000 frames/s has been used to monitor the response of a midsagittal section of a Rhesus monkey to mechanical impact (44). The specimen was prepared by cutting a frozen animal head. A glass plate was used to seal the midsection and saline was introduced to support the brain in the cavity. Cavitation of the brain opposite the impact location was interpreted as a result of a combination of negative (tensile) pressure at this location and relative motion between the brain and skull.

The general opinion of the clinical neurosurgeons indicates (42,43) that brain contusion following head impact may develop due to local skull deformations, pressure gradients within the brain, and relative motions between the skull and brain. Brain movement may also result from the pressure gradients in addition to the relative displacement which takes place as the brain motion lags behind the skull during impact. The majority of brain injuries occur due to stresses resulting from the relative motion. Clinical observations involving tearing of olfactory nerve fibers, rupture of connecting veins and subdural hemmorhage is further evidence of relative motion. Not only relative motion between the brain and skull is conceivable but also motion between various brain compartments likely contributes to many and different lesions of the brain.

Experimental evidence (43), also, points out that brain motion exists within the skull. Experiments on primates with lexan calvarium and others using isodense pellets indicate that brain motion of few millimeters is possible. Critical reviews (43) of the experiments quoted by Ward (17), where little or no relative motion was observed, indicted that the window used to monitor the displacement was small, and consequently, brain tissue possibly extruded through and prevented relative movement. In addition, the postmortem pressurization of the arterial system (45) does not overcome the obvious deficiencies of the cadaver model. In fact, the nonuniform pressure in the intracranial contents may easily alter the mechanical response of the tissue and produce a completely unrealistic surrogate, because the physiological pressure gradients between the vascular network and the brain tissues or cerebrospinal fluid are different. Indirect evidence of relative brain motion was also observed in tests with primates subjected to linear and rotational motions. In these tests, brain concussion could have resulted from pure rotation of the skull as the brain was sheared from its attachments.

831615

Cervical Spine Compression Responses

James H. McElhaney, Jacqueline G. Paver and Hugh J. McCrackin
Duke University

G. Maxwell
National Institutes of Health

ABSTRACT

Time-varying compressive loading was applied to unembalmed human cervical spines using an MTS closed-loop hydraulic testing machine. Load programs included relaxation, cyclic loading, variable rate constant velocity loading (0.13-64 cm/sec), and constant velocity loading to failure. The failures produced were similar to those observed clinically. A generalized quasi-linear viscoelastic Maxwell-Weichert model incorporating a continuous relaxation spectrum was developed to predict the relaxation and constant velocity test responses. The fit was adequate considering the complexity of the structure involved.

WHILE INJURY STATISTICS generally attribute only 2 to 4% of serious trauma to the neck, any neck injury can have debilitating if not life-threatening consequences. The human neck not only contains vital neurologic, vascular and respiratory structures but it also provides for the support and motion of the head. From an anatomical, neurological and mechanical point of view, the neck is quite complex. Much research has been done to describe neck anatomy [17,24,37],* injury mechanisms [1,2,4,7,12,15,21,24-28,31,33,35,41], and the diagnosis [3,5,11,18,20], classification [3, 6,16,19,24,39] and treatment of neck injuries, but only a few serious efforts [5,7,32] have been made to quantitate the structural properties. Many authors have studied the viscoelastic behavior of the lumbar and thoracic discs [8-10,14,22,36,38,41] but rarely have they extended their work to include the cervical discs.

This paper summarizes both experimental and theoretical studies aimed at quantifying and predicting the responses of the cervical spine to compression loading. Time-dependent responses (i.e. force~time and deformation~time characteristics) of the cervical spine were measured and analyzed. Peak loads and deflections, strain energies and mechanisms have been summarized for all experimentally produced failures. A model that allows prediction of these responses has been proposed. Hopefully, the results presented in this paper will be useful in further mathematical modeling, in the design of protective devices, and in the development of anthropomorphic models.

RELATED LITERATURE

Probably the earliest empirical study was Messerer's (1880) work on the mechanical properties of the vertebrae. He reported compressive breaking loads ranging from 1.47-2.16 kN (330-486 lb) for the lower cervical spine. Later, Yamada (1970), in his extensive compilation on the strength of biological tissues, provided data on the static load~deflection properties of the vertebral bodies and discs. Roaf (1960) loaded single cervical spinal units in compression, extension, flexion, lateral flexion, horizontal shear and rotation. He found that the intact disc, which failed at approximately 7.12 kN (1600 lb), was more resistant to compression than wet vertebrae, which failed at approximately 6.23 kN (1400 lb). Bauze and Ardran (1978) loaded human cadaveric cervical spines in compression and reported flexion dislocations with loads of 135-145 kg [1.32-1.42 kN (298-320 lb)]. Fielding et al. (1974) conducted shearing studies of the atlas. In all cases, the 70-180 kp [0.69-1.77 kN (154-397 lb)] force required to fracture the odontoid process was greater than the 12-180 kp [0.12-1.77 kN (26-397 lb)] force required to tear the transverse ligament. Althoff et al. (1980) described

*References at end of paper.

experimental fractures of the odontoid process but did not report load or deformation data. Selecki and Williams (1970) conducted an extensive study of cadaveric cervical spines loaded with a manually-operated hydraulic jack. Unfortunately, they monitored the pressure in the hydraulic line and reported their results in terms of hydraulic pressure without indicating the ram piston diameter. They were able, however, to duplicate several types of clinically observed injuries. Panjabi et al. (1975) measured rotation and translation of the upper vertebra as a function of transection of the components in single units of the cervical spine. Liu and Krieger (1978) reported load~deflection responses from axial compression tests on single cervical spinal units. Sances et al. (1982) tested isolated cadaver cervical spines in compression, tension and shear. A quasi-static compression failure was observed at a load of 645 N (145 lb) and dynamic flexion/compression failures were reported at loads ranging from 1.78-4.45 KN (400-1000 lb). Except for studies by Fielding et al., Liu and Krieger and Sances et al., all of these tests were quasi-static and most researchers recorded only the maximum load.

Studies have also been done on impacts to intact cadavers which involved the neck. Most notable of these were the works of Hodgson et al. (1980), who measured strains on the anterior surfaces of the bodies of C2-C7 and near the left facet joints of all of the cervical vertebrae of embalmed cadavers during crown impacts, and Nusholtz et al. (1981), who studied neck motions and failure mechanisms on unembalmed cadavers due to crown impacts. They both reported significant influence of spinal configuration on the spinal response and damage.

INSTRUMENTATION

Tests were conducted with a Minneapolis Testing System (MTS) servo-controlled hydraulic testing machine which consisted of a rigid load frame, a 6000 lb ram, a temperature- and humidity-controlled environmental chamber, a 25 gpm 3000 psi hydraulic pump, two nitrogen-filled hydraulic accumulators and a displacement-controlled feedback system. Ram motion was monitored by an integral linear variable differential transformer. Load was measured by a strain gage load cell calibrated with proving rings certified by the National Bureau of Standards. Command voltages were provided by an Exact function generator. With the MTS system, we were able to apply displacements at ram speeds up to 127 cm/sec without overshoot.

Figure 1 illustrates the input displacement~time histories used in these studies. For the relaxation tests, the specimens were preloaded to 200 N and then subjected to ramp-and-hold command signals. For the cyclic modulus tests, sinusoidal waveforms were used. Upon completion of the cyclic modulus tests, a reference state was defined which corresponded to a 200 N preload. This preload defined a constant reference specimen length that became the initial condition for all subsequent constant velocity tests. Triangular waveforms were used for the constant velocity tests.

Specimen load~ and deformation~time histories were stored on a Tektronix 5223 Digitizing Oscilloscope and then recorded onto magnetic tape with a Tektronix 4052A Graphic Computer using the WP1310 Waveform Processing System Software.

High resolution x-ray images were obtained using a Hewlett-Packard Faxitron unit before and after testing.

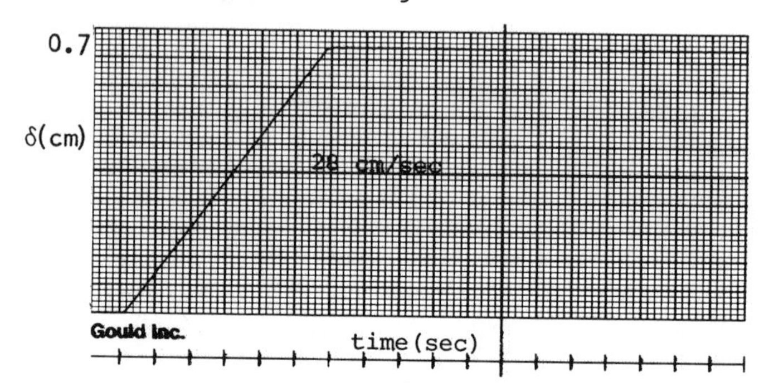

Relaxation Test -- Prescribed Deformation

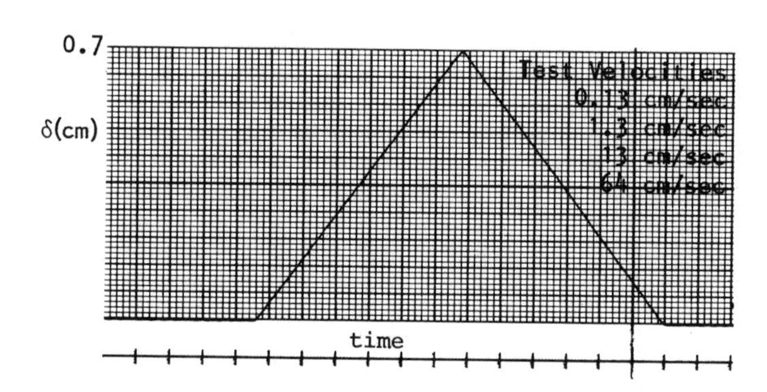

Constant Velocity Test

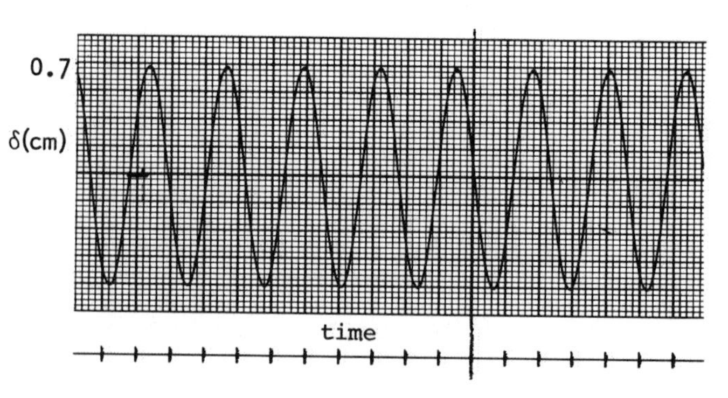

Cyclic Stiffness Test 20Hz

Figure 1, Test~Deformation Time Histories

PRELIMINARY RHESUS MONKEY TESTS

Preliminary tests were performed on eight rhesus (Macaca Mulatta) monkey cervical spines in order to develop the experimental protocol. The following questions were addressed:

(1) Do changes in the mechanical properties occur as time elapses after death?

(2) Can the properly stored cervical spine be reequilibrated after non-destructive testing and tested further? That is, can the cervical spine recover from tolerable load levels or are the effects of testing irreversible? If the results of the first test are, in fact, repeatable, what period of time is necessary for full recovery prior to subsequent testing?

(3) Does freezing significantly degrade the mechanical responses?

(4) Does the cervical spine exhibit a weak enough temperature dependence between room (25°C) and body (37°C) temperatures to justify room temperature testing?

All eight rhesus monkey necks were removed during nechropsy, sprayed with calcium-buffered saline, and kept sealed in waterproof plastic bags. Relaxation tests were performed on four of the specimens within 2, 24, 48 and 60 hours post-mortem. After testing, these four specimens were stored in the refrigerator, allowed to recover for varying times and then retested. Figure 2 shows typical results. The four remaining specimens were frozen post-mortem, stored at -20°C for two months, and thawed in the refrigerator for four days prior to relaxation testing.

These preliminary rhesus monkey tests demonstrated that:

(1) Properly stored specimens tested within 60 hours post-mortem did not exhibit detectable changes in their relaxation properties. A specimen was considered to be properly stored if it was moistened with calcium-buffered saline and then kept sealed in a waterproof plastic bag throughout the experiment. This procedure was selected in order to reduce the possibility of dehydration or chemical changes which could shift the osmotic gradient. Total immersion of the specimen in saline was deemed undesirable because of the possibility of disc swelling due to fluid imbibition. Anything that could change the fluid balance of the disc and/or ligaments could change the stiffness and, perhaps, even the failure characteristics since disc and ligament stiffness influence the strain distributions.

(2) The preliminary test results have shown that even loads of tolerable levels alter the mechanical behavior of the neck. This observation may be related to the capacity of the intervertebral discs to imbibe and release fluid. A detailed discussion of the osmotic action of the disc is beyond the scope of this paper. When externally loaded, the disc exhibits a tendency to lose fluid and, when the external loads are removed, the disc exhibits a tendency to absorb fluid. It is hypothesized that the reequilibration process involves the osmotic uptake of fluid into the discs and that the reequilibrated state is the end state characterized by an osmotic balance. Figure 2 shows that a 24-hour recovery period was required between the initial test and a subsequent test in order to achieve full reequilibration and test reproducibility. During this period, the specimen was properly stored and refrigerated.

(3) For periods up to two months, freezing had no observable effect on the relaxation responses. Panjabi et al. (1975), Hirsh and Galante (1967), and Casper (1980) also reported no degradation in mechanical properties due to storage by deep freezing and thawing prior to experimentation.

(4) Tests performed at room temperature were comparable to body temperature tests. This conclusion agrees with Casper's (1980) observations for the intervertebral disc.

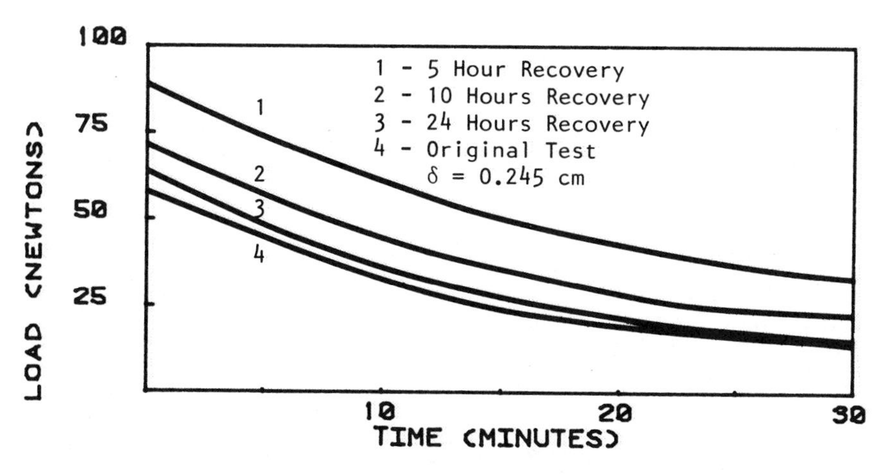

FIGURE 2, RHESUS MONKEY CERVICAL SPINE RELAXATION TESTS

EXPERIMENTAL PROTOCOL FOR HUMAN NECK TESTS

Fourteen intact unembalmed cervical spines were obtained at autopsy from cadavers. The donors, who ranged in age from 42 to 73 years, showed no evidence of cervical spine problems in their hospital records. All specimens included the base of the skull, approximately two centimeters around the foramen, at the proximal end and C5, C6, C7 or T1 at the distal end. All ligamentous structures were kept intact except the ligamentum nuchae where it attached to the base of the skull.

The specimens were sprayed with calcium-buffered saline, sealed in plastic bags until dissection and either tested on the day of removal or frozen and stored at -20°C. At the time of testing, the specimens were thawed to room temperature and allowed to fully equilibrate with their respective fluid environments. Using polyester casting resin, the ends of the specimens were cast in aluminum caps so that the caps were approximately perpendicular to the axes of the end vertebrae. During casting, the aluminum caps were cooled in a flowing water bath to minimize degradation due to the heat of polymerization. Figure 3 illustrates the lordotic curve configuration of the specimens after casting. Next, the initial values of α and L_0 were determined. α was measured with an adjustable protractor. L_0 was measured with vernier calipers. A moment M_0, measured with a spring scale operating on a known moment arm, was then applied to make the end caps parallel and the cervical vertebrae approximately vertically aligned. This moment varied from 5 to 30 N-m. The specimen was placed in the test fixture.

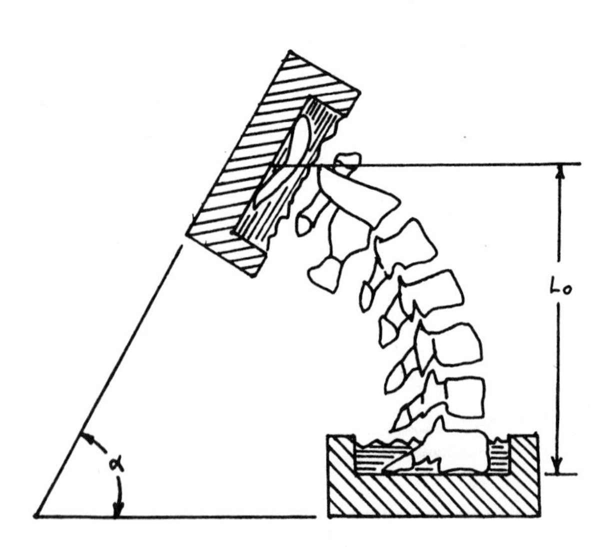

FIGURE 3. SPECIMEN WITH END CAPS

The test fixture (Figure 4), which consisted of a dovetail slide driven by a precision micrometer lead screw, allowed movement of the distal end of the specimen with respect to the proximal end. A displacement h in the anterior or posterior direction was applied with the slide and the lead screw in order to obtain the desired degree of flexion or extension. All specimens were x-rayed before testing in order to document the initial configuration. Finally, the test fixture was installed in the testing machine.

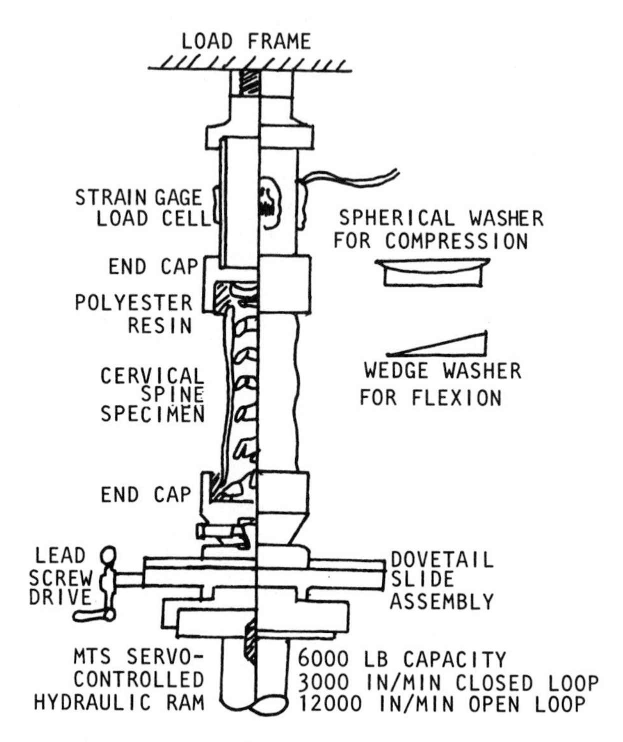

Figure 4, Test Fixture

The following tests were performed at room temperature:

(1) fully equilibrated relaxation test, δ=0.7 cm
(2) cyclic modulus test, δ=0.7 cm, 20 Hz, 150 cycles
(3) mechanically stabilized relaxation test, δ=0.7 cm
(4) constant velocity tests, δ=0.7 cm, ram speed = 0.13, 1.3, 13, and 64 cm/sec
(5) constant velocity load-to-failure test, ram speed ≈ 64 cm/sec.

With the computerized data collection system, tests #2 - #5 were completed in less than one hour.

After testing, the specimens were x-rayed in order to document the final configuration. Next, they were dissected. Failed ligaments and bones were noted and photographed. Critical dimensions were measured and recorded in Table 1.

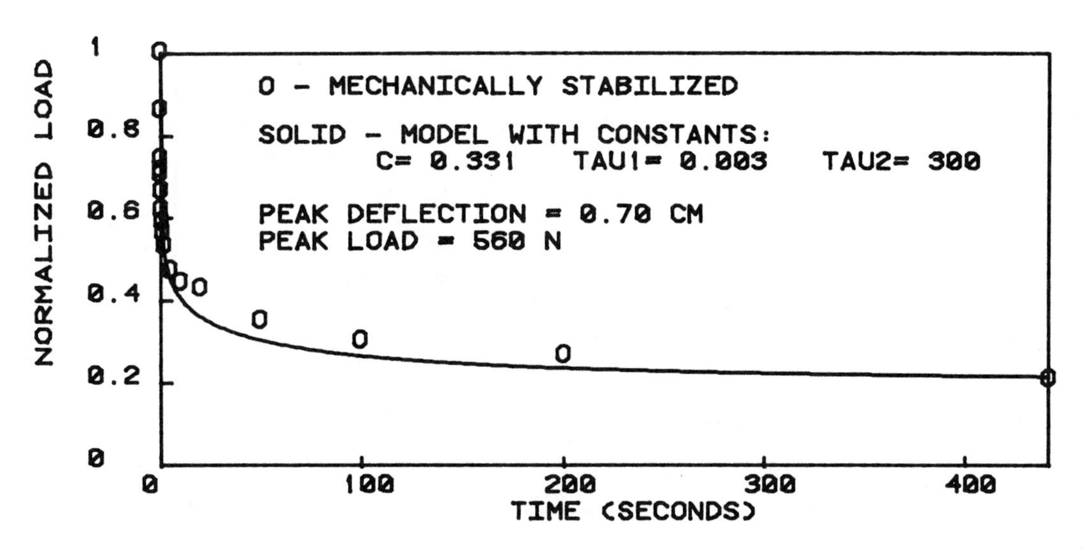

FIGURE 5, TYPICAL RELAXATION TEST FOR HUMAN CERVICAL SPINE A80-384

RELAXATION TESTS

The relaxation tests were performed by applying a ramp displacement of 0.7 cm in 25 msec followed by a constant displacement of 0.7 cm for 5 min. The load~time histories were monitored and recorded. Figure 5 shows a typical relaxation test for a human cervical spine.

A variable rate of load relaxation was demonstrated. Initially, for constant deformations, the load decay was extremely rapid. Thereafter, the load decayed at a much slower rate. This observation renders a standard lumped parameter viscoelastic model with a single dominant long-term time constant a poor predictor of neck behavior. Instead, a generalized Maxwell-Weichert model is proposed since it incorporates an ensemble of decay mechanisms and associated time constants.

CYCLIC MODULUS TESTS

By definition, the cyclic modulus G_c is given by:

$$G_c = \frac{P_c}{L_c}$$

where P_c is the load amplitude and L_c is the deformation amplitude. A sinusoidally-varying compressive displacement of 0.7 cm peak-to-peak amplitude at 20 Hz was applied for 150 cycles to the fully-reequilibrated specimen. The load~time history was monitored and recorded. Figure 6 shows a typical plot of G_c vs. deformation cycles.

Preconditioning behavior was demonstrated. When a specimen was subjected to a repeated deformation history about a fixed length, there was a decrease in the cyclic modulus G_c as the number of deformation cycles

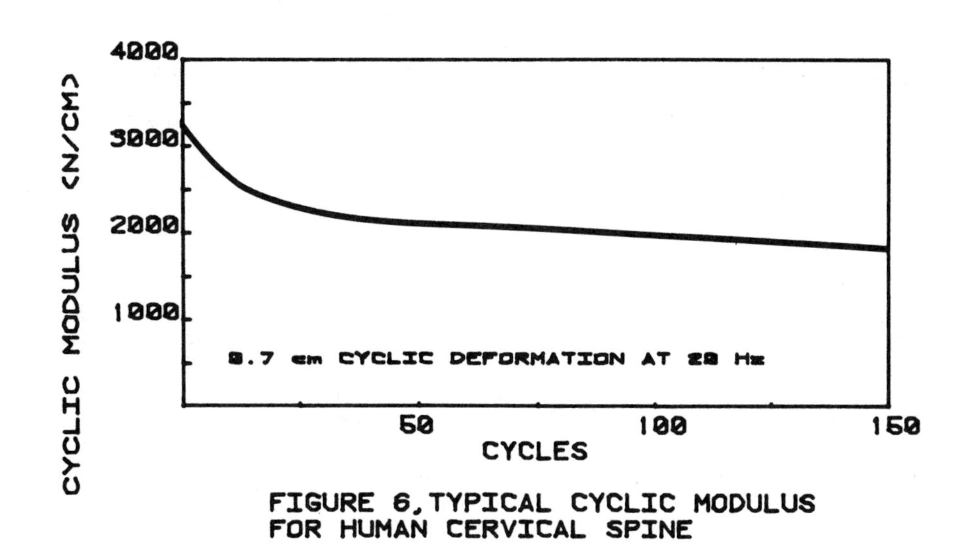

FIGURE 6, TYPICAL CYCLIC MODULUS FOR HUMAN CERVICAL SPINE

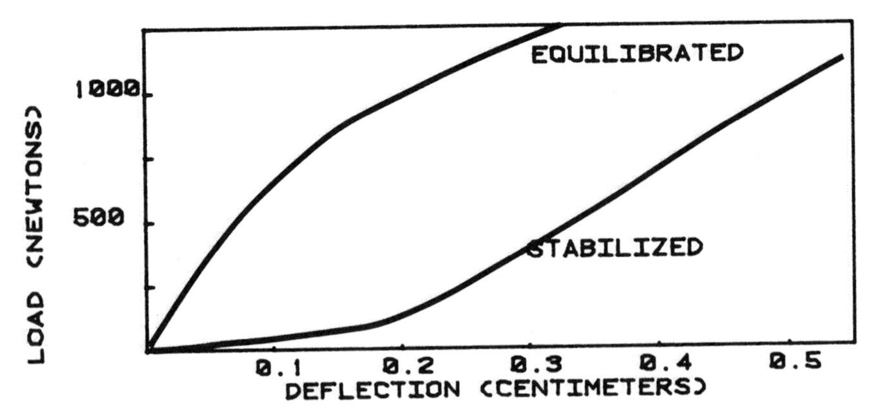

FIGURE 7. TYPICAL ELASTIC RESPONSE FOR EQUILIBRATED AND MECHANICALLY STABILIZED HUMAN CERVICAL SPINE

increased. The initial cycle was representative of the elastic response of the fully reequilibrated cervical spine. Eventually, a steady-state was reached, which we defined as the mechanically stabilized state, where G_C approached a constant value and the load-deflection response was repeatable. The cyclic modulus of the mechanically stabilized state ranged from 45% to 55% of the modulus of the reequilibrated state.

Figure 7 illustates, for a typical relaxation test, the difference between the fully equilibrated elastic response and the mechanically stabilized elastic response.

These cyclic modulus tests demonstrated the influence of the previous load history and the osmotic state of the cervical intervertebral discs on the mechanical response.

VARIABLE RATE CONSTANT VELOCITY TESTS

Figure 8 shows typical results for a human cervical spine in a mechanically stabilized state. The deformation rate was varied by a factor of 500. The stiffness ranged from 1285 to 2250 N/cm, less than a two times increase. The indicated points are measured data.

Deformation rate sensitivity is common in viscous, viscoelastic and plastic materials. For example, the ultimate strength and stiffness of compact bone increases with increasing strain rate [23]. Cancellous bone is sensitive to strain rate to a lesser degree [40]. The intervertebral disc is also sensitive to strain rate to a lesser degree [10].

The human cervical spine exhibits a dependence on deformation rate. This experimental result is consistent with a generalized quasi-linear viscoelastic Maxwell-Weichert model which incorporates a continuous spectrum of relaxation mechanisms and predicts a more distributed deformation rate sensitivity than the standard lumped parameter viscoelastic models.

CONSTANT VELOCITY LOAD-TO-FAILURE TESTS

The last test performed on each mechanically stabilized specimen was the constant velocity load-to-failure test. Ram velocity was nominally 64 cm/sec. By moving the base of the specimen one centimeter in the anterior or posterior direction via the slide and lead screw, the classical extension, compression and flexion injuries were produced.

Table 1 summarizes the type of failure, the maximum load and deflection, and the strain energy or area under the loading portion of the load-deflection curve. Figures 9 through 19 show representative curves.

The following four failure mechanisms were observed as the specimens buckled:

EXTENSION/COMPRESSION - As the body, discs and facet joints resisted the load, the posterior elements were compressed and, as failure of the disc and end plates occurred, the cervical spine extended in a forward buckling mode. Specimen A80-339 failed in this way with rupture of the anterior longitudinal ligament and distraction of the anterior section of the disc between C4 and C5. This occurred with a one centimeter posterior eccentricity.

JEFFERSON FRACTURES - In the clinical literature [18], the common etiology of a fracture of the atlas is a direct blow to the top of the head. In these tests, the experimentally produced atlas fractures, which were usually bilateral and symmetrical, involved the anterior and posterior arches. This was probably due to the compressive force driving the articular condyles outward and bending the arches.

A fourth-order polynomial was used to fit the high rate initial loading curve of the relaxation test. The polynomial used was

$$F^{e} = 26 + 2600\delta - 5360\delta^{2} + 18900\delta^{3} - 14700\delta^{4}, \ \delta < 0.76 \text{ cm}$$

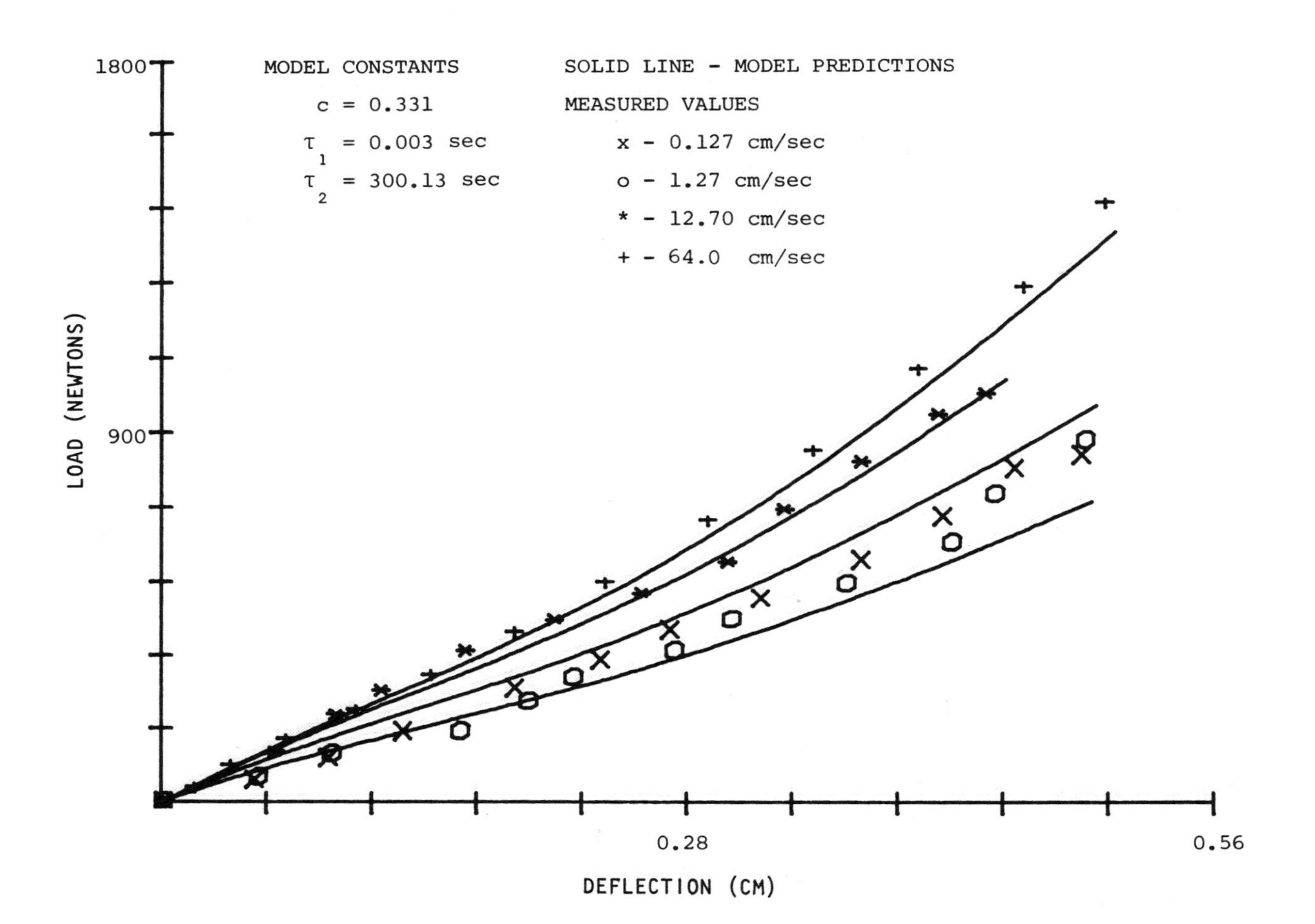

Figure 8, Strain Rate Sensitivity of Human Cervical Spine A80-384

BURST FRACTURES - Comminuted vertical fractures through the vertebral body produced fragmentation of the centrum into a number of large pieces. There were no obvious areas of compressed cancellous bone. Analysis of x-rays taken before and after each test indicated that the specimens that burst were slightly flexed to straight while the specimens that sustained the Jefferson fractures were slightly extended to straight. The burst fractures required larger forces and strain energies than the Jefferson fractures. The load-deflection diagram exhibited a characteristic M-shape or twin peak. Specimen A80-384 showed multiple spikes in the first peak which may be related to the multiple fracturing process.

ANTERIOR WEDGING - The addition of a small flexing moment arm (h < 1 cm) using the test fixture resulted in compression and fracture of the anterior section of the vertebral body. The addition of a slightly larger moment arm (h = 1 cm) produced buckling rearward. Pieces of the cortical shell were displaced in a random pattern. End plate failure occurred and the intervertebral disc was disrupted. However, the amount of displacement applied to the specimen did not result in large anterior dislocation or rupture of the anterior longitudinal ligament. By careful alignment and adjustment of the slide-positioning device, we were able to produce fractures similar to those observed clinically. But, after fourteen tests, we had the distinct impression that one or two centimeters forward or backward, right or left, made a tremendous difference in the outcome. Perhaps, this is the reason there is such a wide range of responses to cervical spine compression in the relevent literature [1,2,4,15,21,24-28,31,33,35,41].

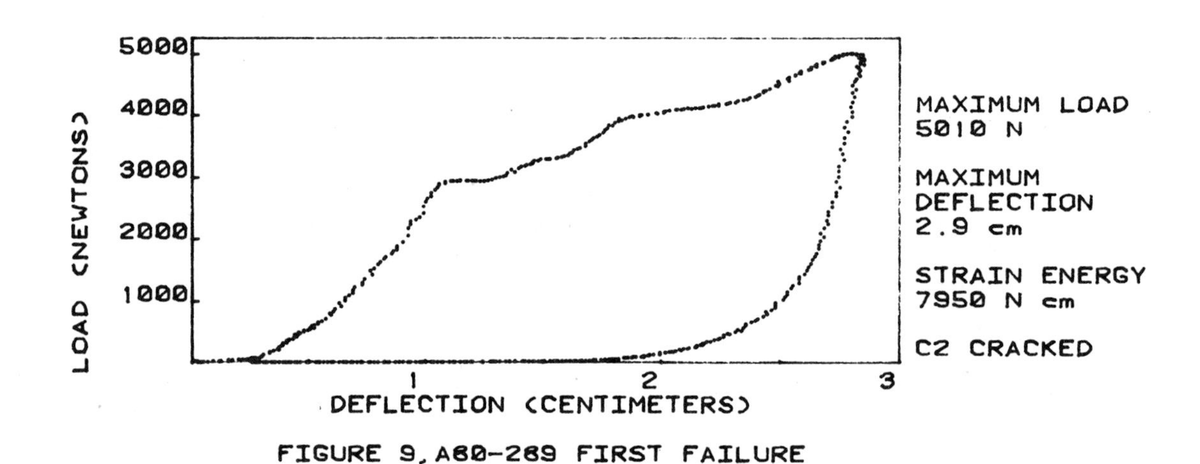

FIGURE 9. A80-289 FIRST FAILURE

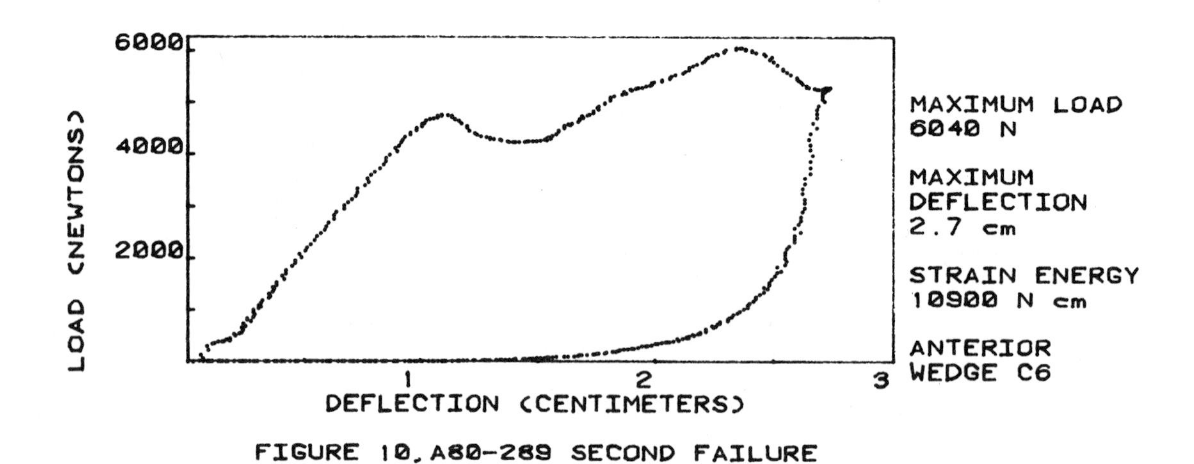

FIGURE 10. A80-289 SECOND FAILURE

TABLE 1

Specimen No.	Age (years) Sex	Description	L_o (cm)	α_o (°)	Ram Velocity (cm/sec)	Failure Mode	C5 Area (cm^2)	Max. Load (N)	Max. Deflection (cm)	Strain Energy (N-cm)
A79-409	58M	B.O.S.* to T2	23	65	50	Jefferson Fr.	5.71	3560	3.0	7470
A79-415	37M	B.O.S. to T1	21	57	50	Compression C5	5.98	5340	3.0	12800
A79-419	49F	B.O.S. to T2	22	63	50	Compression C4&C5	4.29	4860	3.0	10300
A79-423	52M	B.O.S. to T1	19	60	50	Jefferson Fr.	6.17	4190	3.0	7920
A79-431	46M	B.O.S. to T1	20	59	50	Anterior Wedge C5	6.30	4720	3.0	9340
A80-289	70M	B.O.S. to C7	14	60	54	C2 Cracked	5.43	5010	2.9	7950
Retest		C3 to C7	9.1	10	57	Anterior Wedge C6		6040	2.7	10900
A80-339	62F	B.O.S. to T1	21	63	84	Extension Failure	3.51	1930	4.0	4480
A80-352	62M	B.O.S. to C6	9.2	65	55	Jefferson Fr.	6.58	3120	3.0	5740
A80-357	46F	B.O.S. to C6	9.5	55	56	Jefferson Fr.	3.71	960	2.9	1800
A80-364	41M	B.O.S. to C6	12	55	45	C1&C2 Fractured	5.62	5270	2.5	8550
A80-368	77M	B.O.S. to C6 C3,4,5 Bodies Fused	11	45	57	C1 Fractured	5.77	3650	2.7	6350
A80-384	64F	B.O.S. to C7	16	47	92	C2 Fractured	4.38	4060	4.5	12300
Retest		C3 to C7	10	30	77	Burst C4 and Anterior Wedge C4&C5		6840	3.5	15500
A83-26	44M	C2 to T2	13	60	88	Burst Fracture C3,C4&C5	5.45	5470	4.4	15600
A83-42	63F	B.O.S. to C6	11	45	87	Burst Fracture C3&C6	3.28	3000	2.8	5550

*B.O.S. = Base of Skull

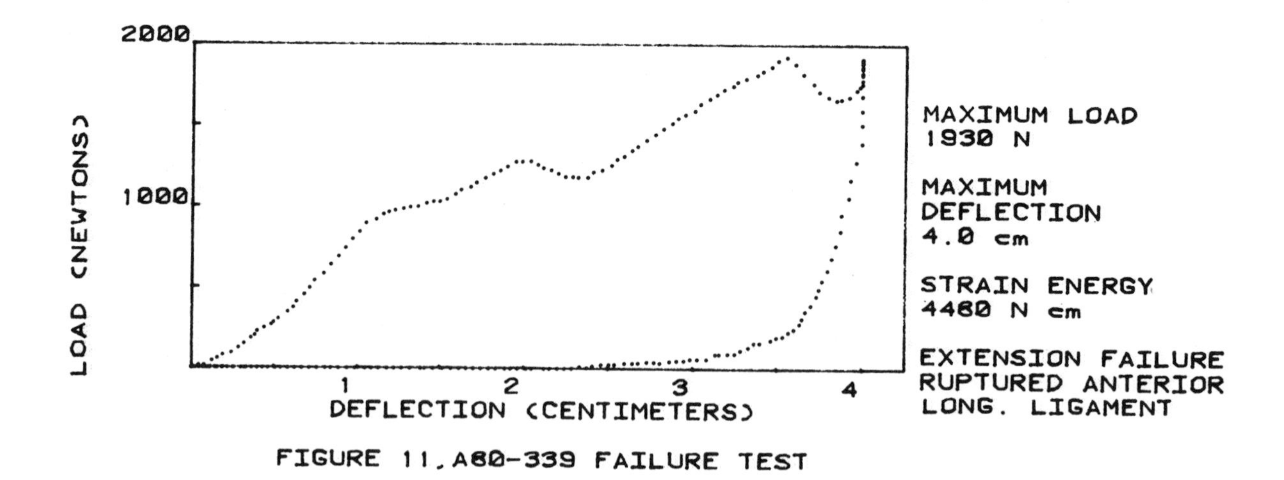

FIGURE 11. A80-339 FAILURE TEST

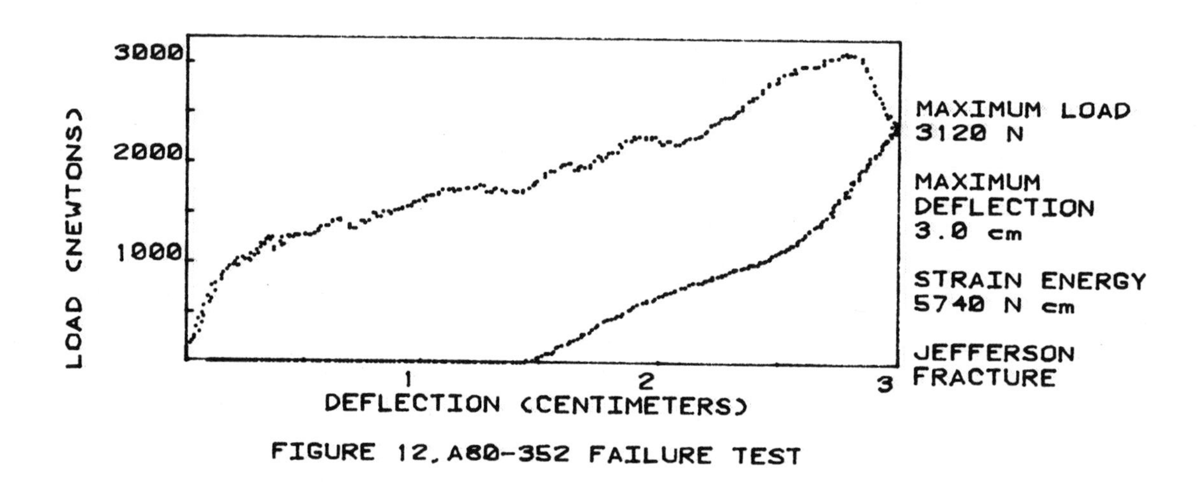

FIGURE 12. A80-352 FAILURE TEST

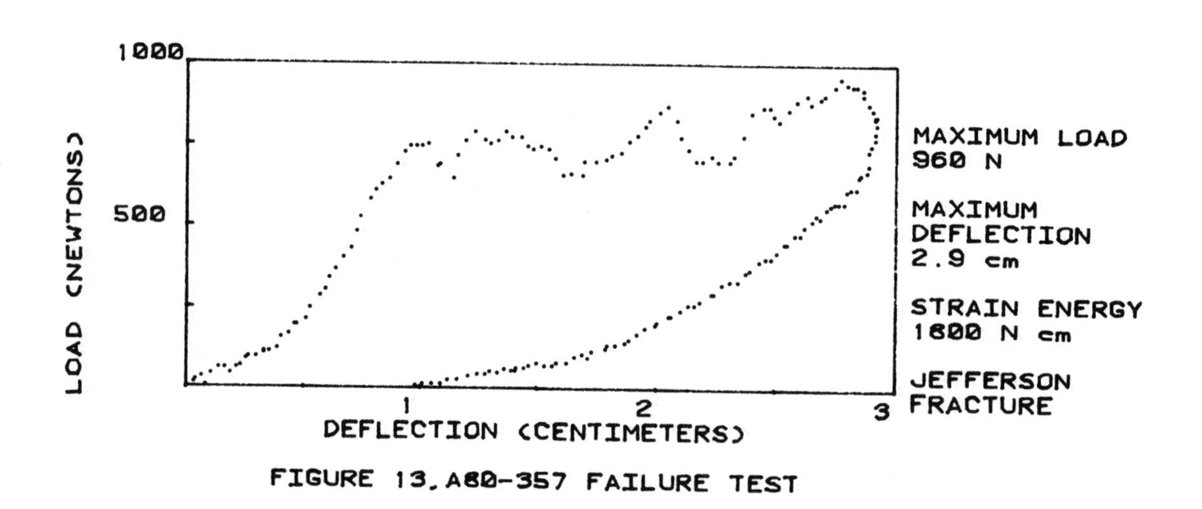

FIGURE 13. A80-357 FAILURE TEST

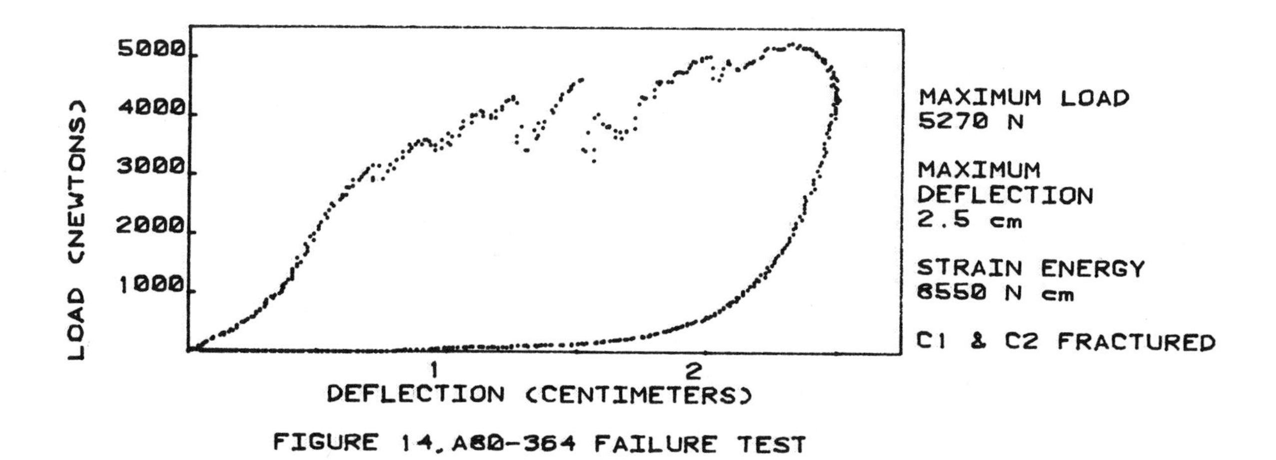

FIGURE 14. A60-364 FAILURE TEST

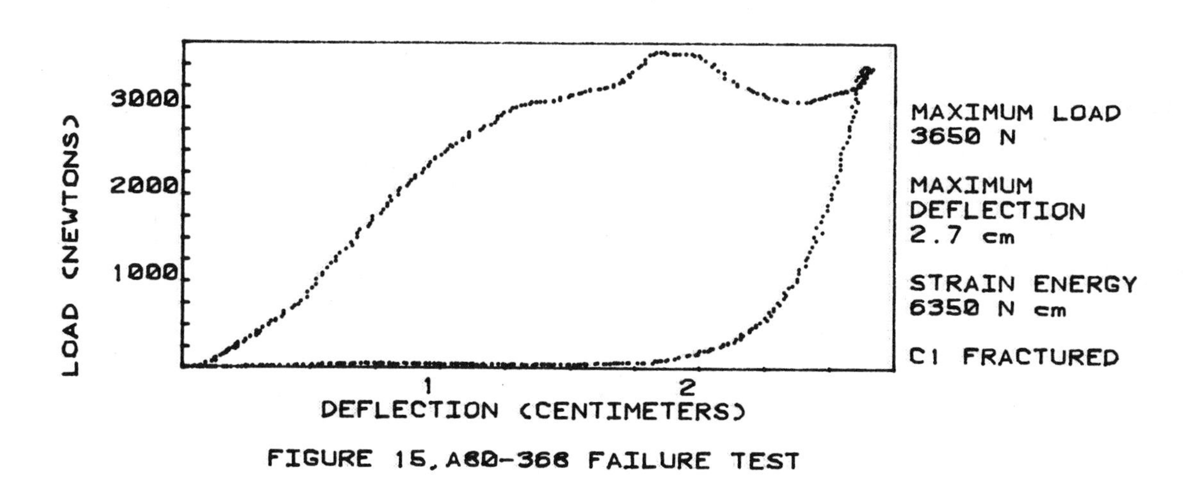

FIGURE 15. A60-366 FAILURE TEST

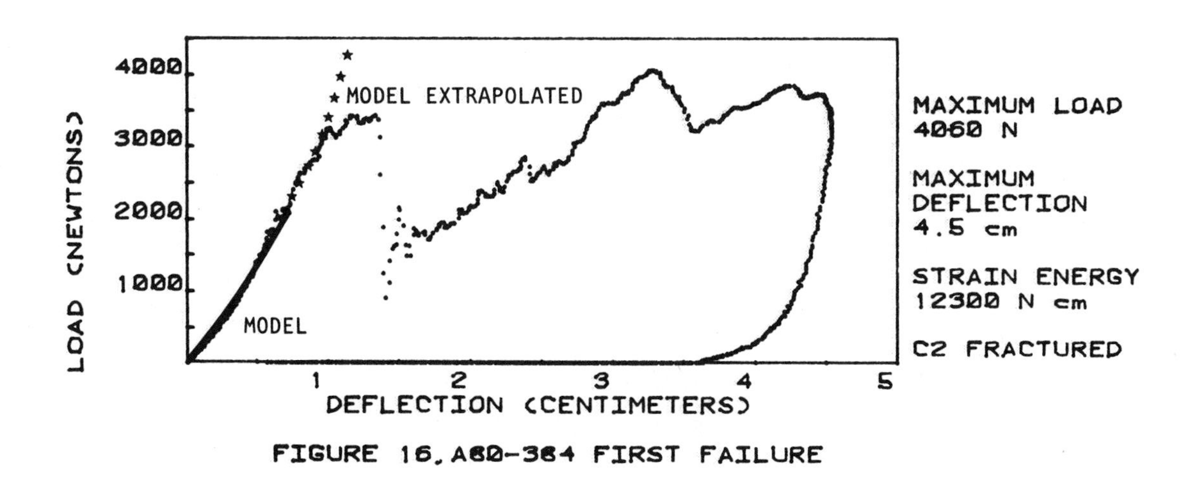

FIGURE 16. A60-364 FIRST FAILURE

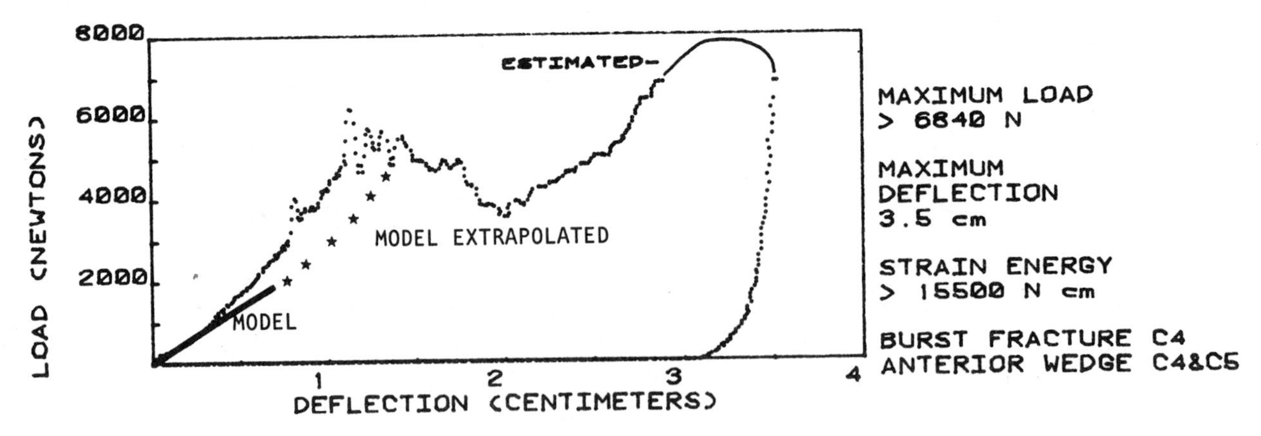

FIGURE 17, A80-384 SECOND FAILURE

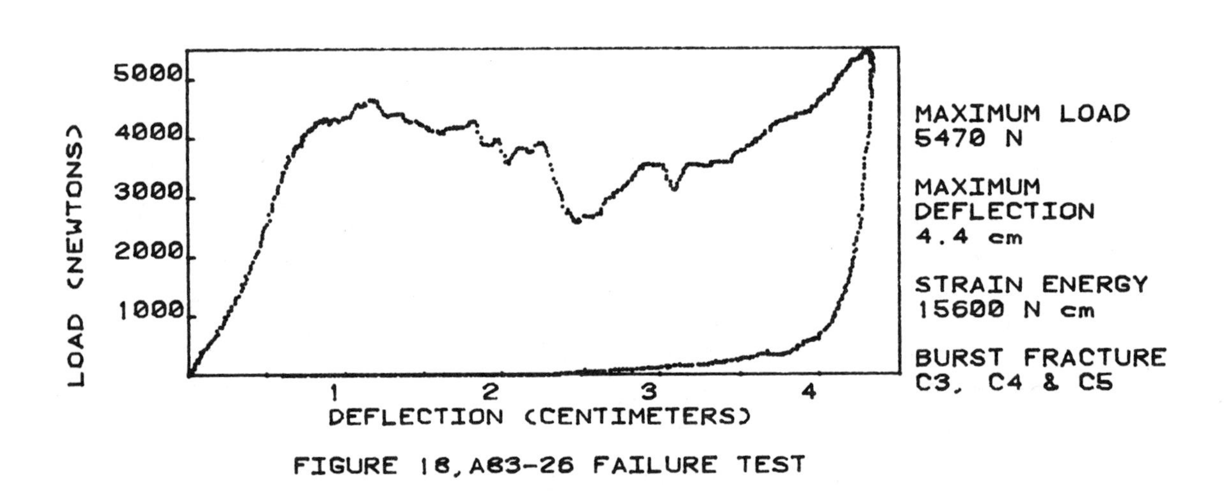

FIGURE 18, A83-26 FAILURE TEST

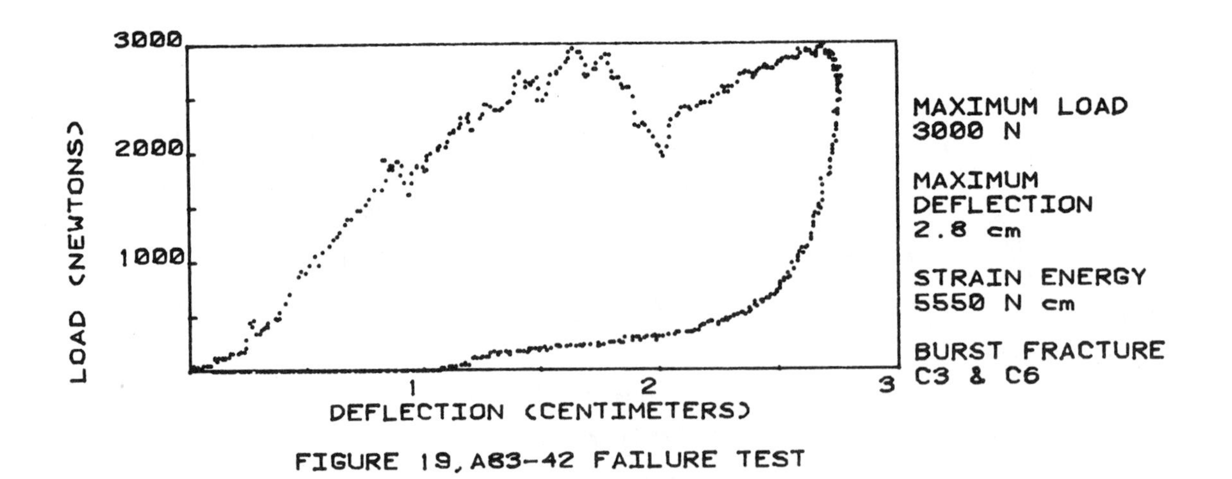

FIGURE 19, A83-42 FAILURE TEST

MODEL

We have attempted to develop a model that would allow the prediction of the compressive load-deformation responses of the human neck for arbitrary deformation-time histories. A variety of standard linear viscoelastic models were tried and proven inadequate. These included the Maxwell, Kelvin, three-parameter solid and four-parameter solid models. Both the presence of rapid initial load decay for fixed deformations and the fact that the hysteresis loop and the load- deformation response fail to exhibit the strong strain rate dependence predicted by standard lumped parameter viscoelastic models suggest that the behavior is governed by a broad distribution of relaxation times. The model developed is based on the quasi-linear viscoelastic constitutive law hypothesized by Fung (1972). Determination of model constants procedes in the manner used by Pinto and Patitucci (1980) for cardiac muscle, by Casper and McElhaney (1980) for the intervertebral disc, and by Sauren and Rousseau (1983).

In a mechanical sense, we may view this continuous spectrum as arising from a generalized Maxwell-Weichert model. The reduced relaxation function, $Y_r(t)$, may be written as

$$Y_r(t) = \frac{E_\infty + \int_0^\infty E(\tau)e^{-t/\tau}\,d\tau}{E_\infty + \int_0^\infty E(\tau)d\tau} \qquad (1)$$

Defining $H(\tau)$, the relaxation time distribution function, as

$$H(\tau) = \tau E(\tau) \qquad (2)$$

and substituting into equation (1), we find, after some rearrangement:

$$Y_r(t) = \frac{1 + \frac{1}{E_\infty}\int_0^\infty H(\tau)e^{-t/\tau}\,d(\ln\tau)}{1 + \frac{1}{E_\infty}\int_0^\infty H(\tau)d(\ln\tau)} \qquad (3)$$

$H(\tau)$ may be approximated from experimental data as the negative slope of the relaxation modulus vs. logarithmic time plot. This slope is roughly equal to a constant, C, over a major portion of the time domain. Therefore,

$$H(\tau) = -C \quad ; \tau_1 < \tau < \tau_2 \qquad (4)$$

$$H(\tau) = 0 \quad ; \tau < \tau_1,\ \tau > \tau_2$$

and the relaxation spectrum becomes:

$$E(\tau) = -C/\tau \quad ; \tau_1 < \tau < \tau_2 \qquad (5)$$

$$E(\tau) = 0 \quad ; \tau < \tau_1,\ \tau > \tau_2$$

Defining C* as

$$C^* = -C/E_\infty \qquad (6)$$

and substituting equations (5) and (6) into equation (3), we find:

$$Y_r(t) = \frac{1 + C^*\left(E_1(t/\tau_2) - E_1(t/\tau_1)\right)}{1 + C^*\ln(\tau_2/\tau_1)} \qquad (7)$$

where $E_1(t/\tau)$ is the exponential integral function.

The relaxation spectrum approximation can be incorporated in the hereditary integral representation to allow the prediction of load-deformation behavior at various deformation rates. Employing $Y_r(t)$ in the quasi-linear viscoelastic representation, we find:

$$F(\delta,t) = \int_0^t Y_r\,(t-\tau)\,\frac{dF^e(\delta(\tau))}{d\delta}\,\frac{d\delta(\tau)}{d\tau}\,d\tau \qquad (8)$$

where $F(\delta,t)$ is the force as a function of deformation and time, $dF^e/d\delta$ is the slope of the 'elastic' load-deformation curve, and $d\delta(\tau)/d\tau$ is the change in deformation with time.

In the integral representation, $Y_r(t-\tau)$ is obtained from a relaxation test on a mechanically stabilized neck and $d\delta(\tau)/d\tau$ is the deformation rate. For short test times (less than 0.025 sec for full displacement), the load-deformation curves for the neck undergo rapid compaction. Instantaneous deformation of the neck is impossible to achieve in a physical sense. Therefore, the elastic load-deformation curve is estimated from the 0.025 sec-full stroke load-deformation curve.

For integral calculations, the pseudo-elastic load-deflection curves were computer-fitted to a power series in δ. Little additional information was gained by extending the power series beyond a 4th-order expansion.

MODEL PREDICTIONS

The loading responses of the mechanically stabilized constant velocity tests were predicted using the model constants calculated from the corresponding mechanically stabilized relaxation test data. The theoretical results, plotted as solid lines, are compared to the experimental results, indicated as points, in Figures 8, 14, and 16. The measured constant velocity responses are predicted with reasonable accuracy within the load range of the relaxation tests upon which they were based. However, the prediction of failure loads is beyond the scope of this model. They must, therefore, be determined empirically. We are currently addressing the question of the validity of extrapolating this model beyond the load range of the corresponding relaxation test and for predicting unloading responses. This work is ongoing.

SUMMARY

In the engineering disciplines, a designer starts with a basic building material and shapes it into a structure with specified load and deformation responses. These load and deformation responses are defined as the structural properties. The structural properties are determined by the size, shape, configuration and material of which a structure is composed. In contrast, the material properties are independent of the structure or shape of the material under consideration. Since the human body exists, it exhibits load and deformation responses which determine its injury potential in traumatic environments. Knowledge of the properties of the material of which the human body is composed is useful in so far as it leads to a better understanding of these structural properties.

The structural properties of the cervical spine have been investigated, with particular emphasis on the quantification and prediction of the time-dependent responses to dynamic compression loading.

A key aspect of this research has been the development of an experimental protocol that produces accurate and repeatable test results and is biomechanically significant. The rhesus monkey tests established the validity of delayed post-mortem testing, specimen freezing, and room temperature testing. Results of these tests and the cyclic modulus tests on the human neck have led to procedures for proper specimen storage and the definitions of reequilibrated and mechanically stabilized states. These procedures and definitions are significant in that they have ensured test accuracy and reproducibility.

Relaxation tests were performed on equilibrated, reequilibrated, and mechanically stabilized specimens. An initially rapid and subsequently slow load decay pattern was observed. The dynamic variable rate constant velocity tests were performed on mechanically stabilized specimens. Results indicated a demonstrable deformation rate dependence. Dynamic load-to-failure tests were also performed on mechanically stabilized specimens. A discussion of failure modes, a summary data table, and load-deflection curves indicated that the failures produced were similar to those observed clinically. It was found that small eccentricities ($\pm$1cm) in the load axis could change the buckling mode from posterior to anterior.

A mathematical model with constants established from a relaxation test was developed. This model predicts with reasonable accuracy the specimen behavior in constant velocity tests at different rates within the load range of instantaneous elastic response data. Extrapolation beyond this range is probably not justified. We are also currently exploring the model's predictive ability for variable rate loading and unloading tests.

REFERENCES

1. Abel, M.S.: Experimental Studies and Cervical Spine Surveys. Occult Traumatic Lesions of the Cervical Vertebrae, W.H. Green Inc., St. Louis, 1971.

2. Althoff, B.; Goldie, I.F.; Romanus, B.: Experimental Fractures of the Odontoid Process. Transactions of the 26th Annual Meeting of the Orthopaedic Research Society, 1980.

3. Babcock, J.L.: Cervical Spine Injuries. ARCHIVES SURGERY 111:646-651, June 1976.

4. Bauze, R.J.; Ardran, G.M.: Experimental Production of Forward Dislocation in the Human Cervical Spine. J. BONE & JOINT SURGERY 60B(2):239-245, May 1978.

5. Beatson, T.R.: Fractures and Dislocations of the Cervical Spine. J. BONE & JOINT SURGERY 45B(1):21-35, February 1963.

6. Braakman, R.; Penning, L.: Injuries of the Cervical Spine, Excerpta Medica, Amsterdam, 1971.

7. Brieg, A.: Adverse Mechanical Tension in the Central Nervous System, John Wiley & Sons, New York, 1978.

8. Brown, T.; Hansen, R.J.; Yorra, A.J.: Some Mechanical Tests on the Lumbosacral Spine with Particular Reference to the Intervertebral Disc. J. BONE & JOINT SURGERY 39A:1135-1164, 1957.

9. Casper, R.; McElhaney, J.H.: Relaxation Response of the Intervertebral Disc. Proceedings of the Southeastern Conference on Theoretical and Applied Mechanics, 1980.

10. Casper, R.A.: Viscoelastic Behavior of the Human Intervertebral Disc (Ph.D. Dissertation). Duke University, Durham, N.C., 1980.

11. Dimnet, J.; Pasquet, A.; Krag, M.H.; Panjabi, M.M.: Cervical Spine Motion in the Sagittal Plane--Kinematic and Geometric Parameters. J. BIOMECHANICS 15(12):959-969, 1982.

12. Fielding, J.; Cochran, G.; Lawsing, J.; Hohl, M.: Tears of the Transverse Ligament of the Atlas. J. BONE & JOINT SURGERY 56A(8): 1683-1691, December 1974.

13. Fung, Y.C.: Stress-Strain-History Relations of Soft Tissues in Simple Elongation. Biomechanics--Its Foundations and Objectives, Y.C. Fung, N. Perrone, M. Anliker (Eds.), Prentice-Hall Inc., Englewood Cliffs, 1972.

14. Hirsch, C.; Galante, J.: Laboratory Conditions for Tensile Tests in Annulus Fibrosus from Human Intervertebral Discs. ACTA ORTHOPAEDICA SCANDINAVICA 38:148, 1967.

15. Hodgson, V.R.; Thomas, L.M.: Mechanisms of Cervical Spine Injury During Impact to the Protected Head. Proceedings of the 24th Stapp Car Crash Conference, SAE PAPER #801300, 1980.

16. Holdsworth, F.W.: Fractures, Dislocations, and Fracture-Dislocations of the Spine. J. BONE & JOINT SURGERY 52A:1534-1551, 1970.

17. Huelke, D.F.: Anatomy of the Human Cervical Spine and Associated Structures. SAE PAPER #790130, 1980.

18. Jefferson, G.: Fracture of the Atlas Vertebrae. BRITISH J. SURGERY 7(27):407-422, 1920.

19. Kazarian, L.: Classification of Simple Spinal Column Injuries. Impact Injury of the Head and Spine, C.L. Ewing, D.J. Thomas, A. Sances Jr., S.J. Larson (Eds.), C.C. Thomas Publishers, Springfield, 1983.

20. Lee, C.; Kim, K.S.; Rogers, L.F.: Triangular Cervical Vertebral Body Fractures--Diagnostic Significance. AMERICAN J. RADIOLOGY 138:1123-1132, June 1982.

21. Liu, Y.K.; Krieger, K.W.: Quasistatic and High-Strain Rate Material Properties of Young Cervical Spines in Axial Loading and Bending. Digest of the 1st International Conference on Mechanics in Medicine and Biology, 1978.

22. Markolf, K.L.: Deformation of the Thoracolumbar Intervertebral Joints in Response to External Loads. J. BONE & JOINT SURGERY 54A(3):511-533, April 1972.

23. McElhaney, J.H.: Dynamic Response of Bone and Muscle Tissue. J. APPLIED PHYSIOLOGY 21:1231-1236, 1966.

24. McElhaney, J.H.; Roberts, V.L.; Hilyard, J.F.: Handbook of Human Tolerance, Japan Automobile Research Institute Inc., Tokyo, 1976.

25. McElhaney, J.H.; Roberts, V.L.; Maxwell, G.M.; Paver, J.G.: Etiology of Trauma to the Cervical Spine. Impact Injury of the Head and Spine, C.L. Ewing, D.J. Thomas, A. Sances, Jr., S.J. Larson (Eds.), C.C. Thomas Publishers, Springfield, 1983.

26. Messerer, O.: Uber Elasticitat and Festigkeit der Meuschlichen Knochen, J.G. Cottaschen Buchhandling, Stuttgart, 1880.

27. Mourodian, W.H.; Fietti, V.G.; Cochran, G.V.B.; Fielding, J.W.; Young, J.: Fractures of the Odontoid--A Laboratory and Clinical Study of Mechanisms. ORTHOPEDIC CLINICS NORTH AMERICA 9(4):985-1001, October 1978.

28. Nusholtz, G.S.; Melvin, J.W.; Huelke, D.F.; Alem, N.M.; Blank, J.G.: Response of the Cervical Spine to Superior-Inferior Head Impact. Proceedings of the 25th Stapp Car Crash Conference, SAE PAPER #81005, 1981.

29. Panjabi, M.M.; White III, A.A.; Johnson, R.M.: Cervical Spine Mechanics as a Function of Transection of Components. J. BIOMECHANICS 8(5):327-336, September 1975.

30. Pinto, J.G.; Patitucci, P.J.: Visco-Elasticity of Passive Cardiac Muscle. ASME TRANSACTIONS, J. BIOMECHANICAL ENGINEERING 102:57-60, February 1980.

31. Roaf, R.: A Study of the Mechanics of Spinal Injuries. J. BONE & JOINT SURGERY 42B(2):810-823, November 1960.

32. Rogers, W.A.: Fractures and Dislocation of the Cervical Spine--An End Result Study. J. BONE & JOINT SURGERY 39A:341-376, 1957.

33. Sances Jr., A.; Myklebust, J.; Houterman, C.; Weber, R.; Lepkowski, J.; Cusick, J.; Larson, S.; Ewing, C.; Thomas, D.; Weiss, M.; Berger, M.; Jessop, M.E.; Saltzberg, B.: Head and Spine Injuries. AGARD Conference Proceedings on Impact Injury Caused by Linear Acceleration--Mechanism, Prevention, and Cost, 1982.

34. Sauren, A.A.H.J.; Rousseau, E.P.M.: A Concise Sensitivity Analysis of the Quasi-Linear Viscoelastic Model Proposed by Fung. ASME TRANSACTIONS, J. BIOMECHANICAL ENGINEERING 105:92-95, 1983.

35. Selecki, B.R.; Williams, H.B.L.: Experimental Study of Mechanisms of Injury. Injuries to the Cervical Spine and Cord in Man, Australian Medical Publishing Co. Ltd., Australia, 1970.

36. Sonnerup, L.: A Semi-Experimental Stress Analysis of the Human Intervertebral Disc in Compression. EXPERIMENTAL MECHANICS 12:142-147, 1972.

37. Veleanu, C.: Vertebral Structural Peculiarities with a Role in the Cervical Spine Mechanics. FOLIA MORPHOLOGICA 19(4):388-393, 1971.

38. Virgin, W.J.: Experimental Investigations Into the Physical Properties of the Intervertebral Disc. J. BONE & JOINT SURGERY 33B:607-611, 1951.

39. Whitley, J.E.; Forsyth, H.F.: The Classification of Cervical Spine Injuries. AMERICAN J. ROENTGENOLOGY 83:633-644, 1960.

40. Wood, J.L.: Dynamic Response of Human Cranial Bone. J. BIOMECHANICS 4(1):1-12, 1971.

41. Yamada, H.: Strength of Biological Materials, F.G. Evans (Ed.), Williams & Wilkins Co., Baltimore, 1970.

872200

The Temporal and Spatial Deformation Response of a Brain Model in Inertial Loading

Lawrence E. Thibault and Thomas A. Gennarelli
University of Pennsylvania

Susan Sheps Margulies
Mayo Clinic and Foundation

ABSTRACT

Physical models of the skull-brain system have been subjected to controlled inertial loading experiments in which the deformation response of the surrogate brain was measured. The propose of this report is to present the results of these studies. Two types of models are examined herein; an idealized right circular cylinderical geometry and a baboon skull, sectioned in a midcoronal plane. The surrogate brain, consisting of an optically transparent silicone-gel, contains a painted grid of orthogonal lines with approximately 5mm spacing. The experimental data are presented in the form of nodal displacements and associated strains with one millisecond temporal resolution. The loading conditions are described by the rigid body accelerations of the skull or cylinder models. In each case the motion of the model is a noncentroidal rotation. The experimental results permit one to investigate the relations between the deformation and the acceleration magnitude and temporal characteristics. The model was primarily developed in order to estimate the strains experienced by the various tissue components within the brain, but it may also prove to be useful in addressing issues of scaling and in experimental validation of analytical or numerical simulations.

INTRODUCTION

Experimental models for cerebral concussion (1,2), diffuse axonal injury with prolonged coma (3), and acute subdural hematoma (4) in the primate have been developed in our laboratory. These lesions represent discrete failure of neural and vascular tissue elements respectively, and their topographic distribution is shown to be identical when comparing the primate model to pathophysiological observations in man (5).

The experimental technique employed to produce these various models of brain injury is an impulsive, noncentroidal rotation of the head in various directions and with careful control of both the magnitude and time history of the inertial load. As a direct consequence of this inertial loading, deformation of the brain occurs thereby producing critical levels of strain and stress which cause the neural and neurovascular tissue elements to respond pathologically from a structural and/or functional point of view.

Since it is not yet possible to visualize the dynamic deformation within the living brain while it undergoes these loading conditons, a series of physical models of the head have been constructed which enable one to approximate the deformation response. These surrogate skull-brain systems can then be subjected to identical kinematic constrains and loading conditions as those employed in the primate experiments and, therefore, serve to build a bridge which will enable one to correlate the approximate deformations with the pathophysiological consequences.

It is reasonable to begin with relatively simple model systems since it is not known at this time how sensitive the deformation response of the model may be to a number of complexities. Among these are the shape, compartmentalization of the structure by various membrains, precise boundary conditions, constitutive properties of the materials and numerous other considerations.

Presented herein are the results of experiments which utilize a simple, right-circular cylinder and a baboon skull sectioned in a mid-coronal plane.

METHODOLOGY

The experimental system used to provide the loading conditions for these physical model studies is identical to that used for the animal model. A detailed description of this system can be found elsewhere (6,7). Briefly, the device depicted in figure 1A is a six inch HYGER^{R} linear actuator as manufactured by Bendix. However, the unit is customized to include a kinematic linkage which converts the linear translation of the thrust column into a 65 degree angular motion of a test fixture. Previously, the primate head was attached through a helmet system to this test fixture. Previously, the primate head was attached through a helmet system to this test fixture for the animal experiments. In this instance the physical models are fastened on the outboard of the test fixture for purposes of photography. The kinematics of the physical model and animal model studies are virtually identical.

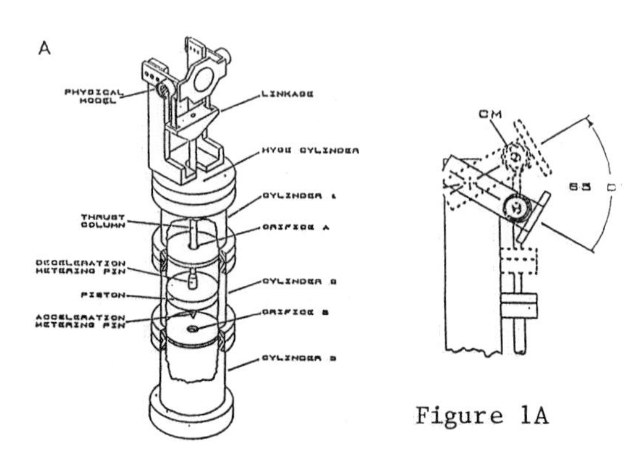

Figure 1A

The HYGE[R] device is further customized internally by redesign of the acceleration and deceleration metering pins. This modification permits waveshaping of the load-time history; a feature which we consider critical with regard to our ability to produce descrete forms of brain injury in the animal model.

Design and fabrication of the physical models is discussed in detail by Margulies (8). For purposes of these studies two models were constructed and are depicted in figure 1B. The shell of both models is a aluminum cylindrical can with one open end, machined to dimension from solid stock. In the case of the skull model an anterior half of a baboon skull, sectioned in a mid-coronal plane, is fixed within the aluminum shell using epoxy (Castolite Resin). The interior cavities now formed are the simple cylinder and the actual skull geometry. These surfaces are painted white for photographic contrast prior to pouring the surrogate brain material into place. This material is an optically transparent gel (Silicone Gel-System, **Dow Corning**) and consists of a catalyst and polymer which when mixed in a 1 = 1 ratio will cure in air to a visc-elastic solid with mechanical properties similar to those of brain. ($G \doteq 2$psi and $\mu \doteq 400$ poise) The material can be case in layers and adhere to itself at the layer interface. This property permits one to paint an enamel, orthogonal grid on a cured surface of interest and subsequently cause that surface to become embedded within the model by casting additional layers above the grid surface. The surrogate brain material naturally adheres to the surrounding model surfaces forming a non-slip boundary condition with respect to the cylinder wall and inner surface of the skull. A transparent plastic window is placed over the open end of the model and is spaced from the free surface of the gel. This space is then water-filled maintaining a free-slip condition and model, is then sealed.

The models are then fastened to the kinematic linkage and loaded in the desired configuration. Rigid body acceleration of the aluminum shell is measured with prizoelectric accelerometers (Endevco) attached to the kinematic linkage.

High speed photography of the model grid deformations is performed at 4000 frames per seconds using a HYCAM[R] camera (Redlake Industries). Analysis of the grid deformation is accomplished by first printing 8" x 10" black and white photographs of individual frames of the high speed film. These prints are then placed on a bit-pad and the position of each node point in the frame is digitized and stored on a microprocessor for further analysis and computational purposes. The bit-pad provides a spatial resolution of .001 inch and nodal points, defined as the intersection of the orthogonal grid, are measured within .010 inch.

RESULTS

When individual frames of high speed film are digitized and stored, the computer can then reconstruct the boundary of the model and the grid as shown in figure 2. To the left are the reconstructed underformed images of the cylinder and skull models and to the right are selected frames of the grid deformations. It is interesting to note the topographic distribution of these deformations as depicted in the figures. If one assumes that an infinitely-long, rigid cylinder, is filled with a visco-elastic material and rotated in a centroidal manner then the equation of motion for the system is given as:

B

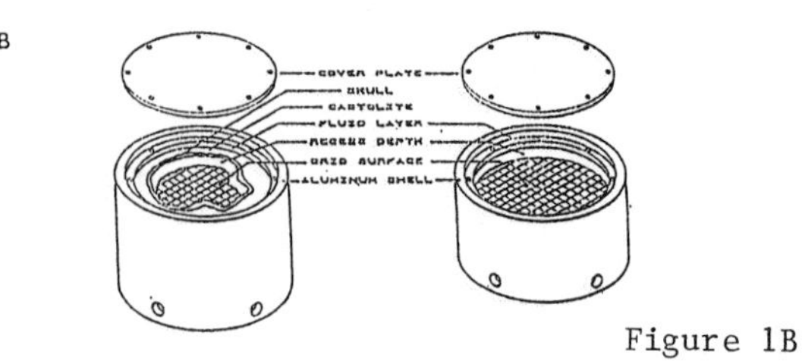

Figure 1B

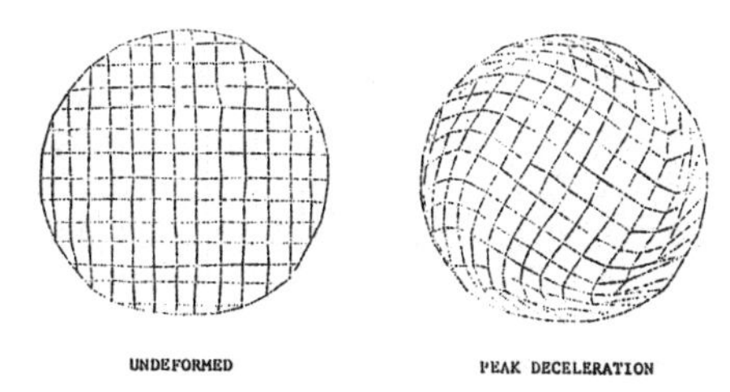

UNDEFORMED PEAK DECELERATION

Figure 2

$$\frac{\partial^2 u}{\partial t^2} = \left[\nu \frac{\partial}{\partial t} + c^2\right]\left(\frac{\partial^2 u}{\partial R^2} + \frac{1}{R}\frac{\partial u}{\partial R} - \frac{u}{R^2}\right)$$

u = displacement in the θ- direction
R = radius
t = time
ν = Kinematic viscosity
c = Shear wave speed $\sqrt{G/\rho}$
G = shear modulus
ρ = density

Dimensional analysis of the governing equation yields the following result:

$$L \doteq \sqrt{\nu t + c^2 t^2}$$

or,

$$L = \sqrt{\nu t + (G/\rho)^2 t^2}$$

where L represents a characteristic length associated with the depth of penetration of a shear wave. In this case the disturbance originates at the interface between the container (cylinder or skull) and the surrogate brain material and propagates inward. In the reconstructed frames in figure 2 the deformation layer is readily visible and is confined to a region which is approximately one grid element in thickness. The remainder of the field within the surrogate brain appears as a core undergoing an approximate rigid-body rotation.

The apparent asymmethry of the deformation in the case of the cylinder is attributed to the noncentroidal nature of the rotation, while in the case of the skull model, both the kinematics and the nonlinearity of the skull geometry play a role.

Figure 3 and 4 show selected frames from the cylinder and skull model experiments which span approximately 25 milliseconds in time. Between the underformed first frame and the next frame shown approximately 5 millisecond intervals. Technical difficulties precluded analysis of the early model deformations, however, peak deformations associated with the accerlation phase of the load-time history are presented.

The results of four experiments, performed with the cylinder model, will be presented and discussed herein. In each case the cylinder

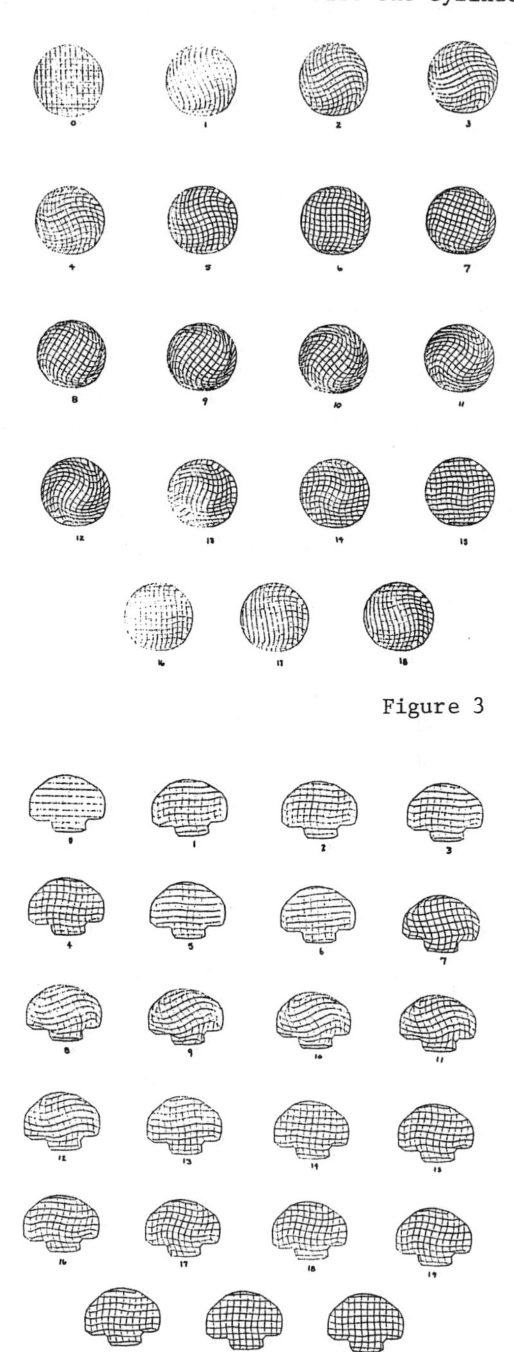

Figure 3

Figure 4

(internal diameter = 8cm) was subjected to a noncentroidal rotational acceleration where the distance from the center of rotation to the center of mass of the model was constant at 7.3 cm. The loading conditions are expressed in figure 5 for the four experiments, and one intended to depict the differences in the shapes of these acceleration traces. The first two acceleration time histories of the rigid body rotation are obtained from what we will refer to as metering pin #1. This internal metering pin configuration produces a relatively symmetrical and biphasic inertial load. The third and fourth acceleration traces are associated with metering pin #2 and possess an asymmetrical load-time history with a large predominate deceleration phase. The magnitudes of the inertial load are controlled by the pneumatic pressure which is used to set the HYGE[R] machine. Set pressures of 100 and 200 psi where with each metering pin producing the following results

PIN	PRESSURE	$\ddot{\theta}_{PEAK}$ (RAD/SEC^2)
1	100	2.19 X 10^4
1	200	4.60 X 10^4
2	100	11.40 X 10^4
2	200	17.10 X 10^4

Peaks angular accelerations occurred during the deceleration phase in all cases, however, the **pin** #1 configuration produced values of peak acceleration which were similar

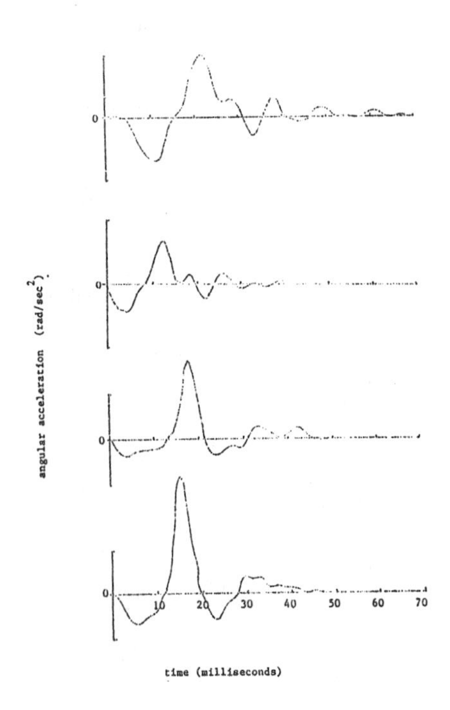

Figure 5

Figure 6 shows the calculated value of shear strain in radians as a function of time. The experiment was fun at a set pressure of 200 psi with metering pin #2 and the strains are calculated for three locations within the model, expressed as nondimensional radial positions r/R =.9, .6 and .3 where R is the cylinder **Radius**. As **can** be seen, the peak values of strain occur in close temporal proximity to the associated peak values of acceleration and deceleration. Further the strains are attenuated toward the central regions of the model and the peak values are shifted in **time** as shear **waves propagate** inward. Figure 7 shows the spatial distribution of shear strain from the boundary to the center of the model for four points in time and for the same experiment described in Figure 6. Note again that the peak shear strain occurs close to the boundary and corresponds temporally to the peak deceleration phase of the load-time history.

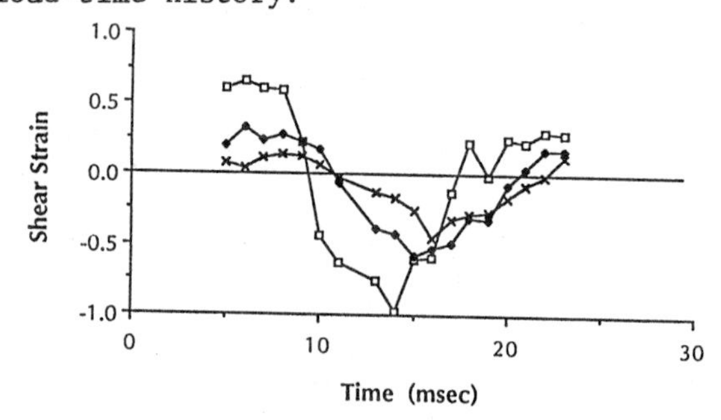

Figure 6
Shear Strain vs. Time Metering Pin #2 - 200 psi

□ - r/R = 0.9
◆ - r/R = 0.6
X - r/R = 0.3

Figure 8 serves as a comparision of the results of two experiments (100 psi, 200 psi) where the peak angular accelerations during the deceleration pahse where 11.4 X 10^4 and 17.1 X 10^4 rad/sec^2 respectively. Note that the peak value for the shear strain at a

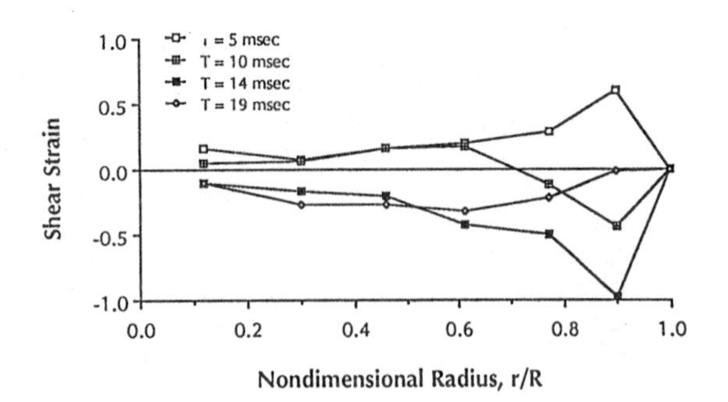

Figure 7
Shear Strain vs r/R at 200 psi with pin #2

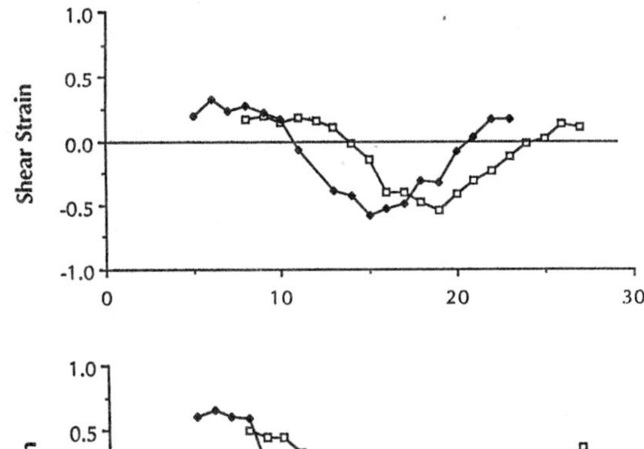

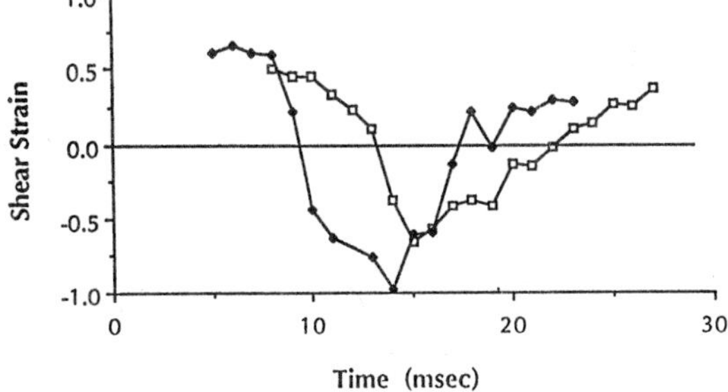

Figure 8
Shear Strain vs Time Pin #2
Top - r/R = 0.6
Bottom - r/R = 0.9

□ = 100 psi
◆ = 200 psi

nondimensional radius of 0.9 depends upon the peak deceleration magnitude, but at a nondimensional radius of 0.6 this relations no longer applies. Again there is a temporal shift in the occurrence of peak shear strain, presumably as a function of the waveshape of the acceleration-time history. **Figures** 9 and 10 present the results of the cylinder experiments using pin #1 with the more symmetrical inertial loading condition and longer pulse widths. The calculated strain reflect this symmetry in both magnitude and temporal distribution. That is the shear strains are of comparable magnitude during biphasic load-time history. The strains are also noted to be more uniformly distributed in the space as the thickness of the deformation annulus grows with longer time.

Figure 11 demonstrates that the magnitude of the shear strain is independent of the magnitude of the acceleration for the two cases shown (100 psi, 200 psi pin #1) where the peak values of the angular acceleration are 2.19×10^4 and 4.60×10^4 rad/sec^2 resepctively. We believe that this observation needs to be studied in a much more extensive manner. With longer pulse widths and a relatively symmetrical acceleration-time history a deformation pattern is developed which tends to depend more upon the change in angular velocity than the angular acceleration. For the two cases shown here the changes in angular velocity are similar compared to the magnitudes of the acceleration.

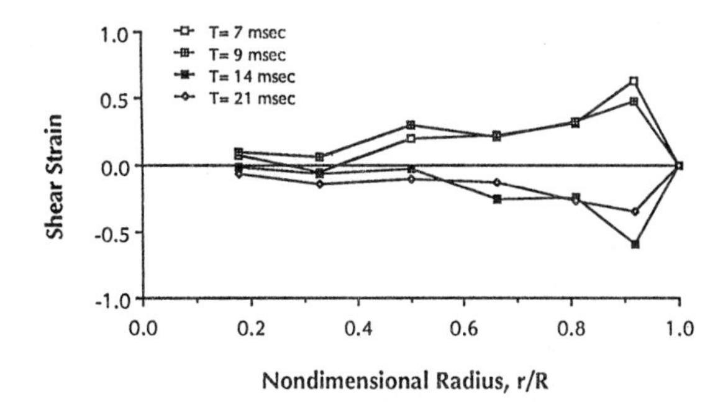

Figure 9
Pin #1 Shear Strain vs r/R
200 psi

DISCUSSION

The physical model experiments described herein are designed to provide approximations for the deformations **occurring** within the brain as the head experiences inertial loading. Estimates of the strain for example **enable** one to investigate the response of isolated tissue elements (single axons or small blood vessels) to similar mechanical stimulation in an effort to establish tissue failure criteria. However, the physical model experiments can also be used to compliment analytical, numerical and animal model studies. Digitized nodal displacements can serve as an experimental data set for use in the validation of mathematical models, and the topographic distribution of the deformations validation of the mathematical models, and the topographic distribution of the deformations can help shade light on the results obtained from the animal model pathophysiology.

The results presented in this report suggest that the sensitivity of the model to one single parameter such as peak angular acceleration, independent of the shape of the acceleration-time history, is suspect. The change in angular velocity and perhaps **the total displacement may be important parameters in describing the deformation response of the surrogate brain depending upon the temporal nature of the loading condition.**

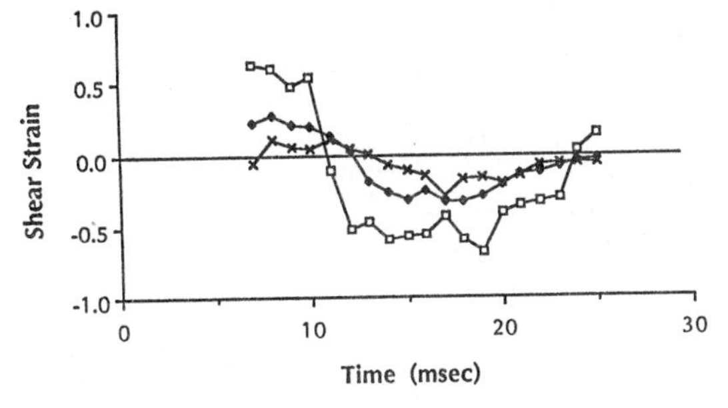

Figure 10
Pin #1 Shear Strain vs Time 200 psi
□ r/R = 0.9
◆ r/R = 0.6
✕ r/R = 0.3

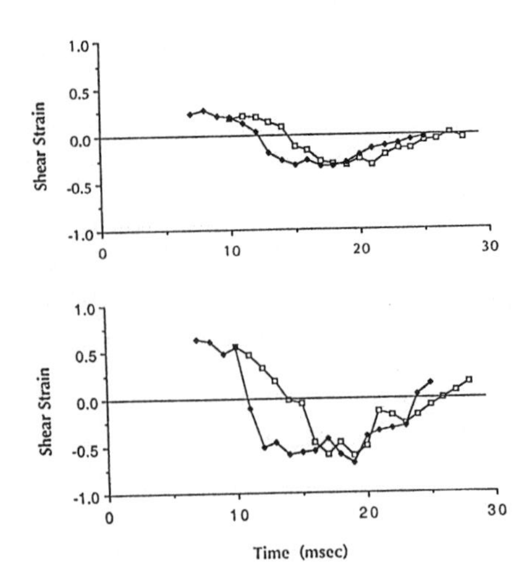

Figure 11
Shear Strain vs. Time Pin #1
Top - r/R = 0.6
Bottom - r/R 0.9

□ - 100 psi
◆ - 200 psi

BIBLIOGRAPHY

1. Abel, J.M., Gennarelli, T.A. and Seqawa, H. Incidence and Severity of Cerebral Concussion in the Rhesus Monkey Following Sagittal Plane Angular Acceleration. 22nd Stapp Car Crash Conf., Proc., SAE, New York pp. 33-53 , 1978
2. Gennarelli, T.A., Ommaya, A.K. and Thibault L.E. Comparison of Linear and Rotational Acceleration in Experimental Cerebral Concussion. 15th Stapp Car Crash Conf., Proc. SAE, New York, pp. 797-803, 1971
3. Gennarelli, T.A., Thibault, L.E. Adams, J.H., Graham, D.I., Thompson, C.J. and Marcincin, R.P. Diffuse Axonal Injury and Traumatic Coma in the Primate. Ann Neurol. 12:564-574, 1982.
4. Gennarelli, T.A. and Thibault, L.E. Biomechanics of Acute Subdural Hematoma J. Trauma 22:680-686, 1982.
5. Adams, J.H., The Neuropathology of Head Injuries, In: Viken, P.J., Bruyn, G.W. (eds) Handbook of Clinical Neurology, Vol. 23: Injuries of the Brain and Skull, Part I. New York, American Elsevier, 1975, p. 35.
6. Margulies, S.S., Thibault, L.E. and Gennarelli, T.A. A Study of Scaling and Head Injury Criteria Using Physical Model Experiments. Proc. IRCOBI pp. 223-234, 1985.
7. Thibault, L.E. and Gennarelli, T.A. Biomechanics of Diffuse Brain Injuries. Proc. 10th Experimental Safety Vehicle Conf. Oxford, England, 1985.
8. Margulies, S.S. Biomechanics of Traumatic Coma in the Primate, Ph.D. Dissertation, University of Pennsylvania, Department of Bioengineering, Jan. 1987.

881709

Combined Bending and Axial Loading Responses of the Human Cervical Spine

James H. McElhaney, Brain J. Doherty, Jacqueline G. Paver and Barry S. Myers
Duke University

Linda Gray
Duke University Medical Center

ABSTRACT

The lateral, anterior and posterior passive bending responses of the human cervical spine were investigated using unembalmed cervical spinal elements obtained from cadavers. Bending stiffness was measured in six modes ranging from tension-extension through compression-flexion. Viscoelastic responses studied included relaxation, cyclic conditioning and constant velocity deformation. A five-axis load cell was used to measure the applied forces. Results include moment-angle curves, relaxation moduli and the effect of cyclic conditioning on bending stiffness. The Hybrid III ATD neckform was also tested and its responses are compared with the human. It was observed that the Hybrid III neckform was more rate sensitive than the human, that mechanical conditioning changed the stiffness of the human specimens significantly, and that changing the end condition from pinned-pinned to fixed-pinned increased the stiffness by a large factor.

THIS PAPER DESCRIBES the results of a study of the bending responses of unembalmed human cervical spine segments. Considerable work has been done on the structural properties of the cervical spine and neck injury mechanisms. However, the cervical spine is an extremely complex structure, and many questions about its structural responses remain unanswered. Cervical injuries are of major interest because they occur with some frequency in automotive, recreational and sports activities, because they can involve the spinal cord with catastrophic and irreversible consequences, and because they usually involve the younger age groups. An extensive review of the literature was presented by Sances (1).

Many of the studies of the structural properties of the spine have involved compression. Perhaps the earliest such study was Messerer's work on the mechanical properties of the vertebrae (2). He reported compression breaking loads ranging from 1.47-2.16 kN for the lower cervical spine. Bauze and Ardran loaded human cadaveric cervical spines in compression and reported forward dislocations with loads of 1.32-1.42 kN (3). However, their experiments were designed to force the dislocations to occur at a given vertebral level. Sances tested isolated cadaver cervical spines in compression, tension and shear (4). A quasi-static compression failure was observed at a load of 0.645 kN, and dynamic compression-flexion failures were reported at loads ranging from 1.78-4.45 kN. McElhaney *et al.* applied time-varying compressive loading to unembalmed human cervical spines (5,6). Failures were produced which are similar to those observed clinically with maximum loads ranging from 1.93-6.84 kN. In addition, it was found that small eccentricities in the load axis could change the buckling mode from posterior to anterior. Panjabi *et al.* measured rotation and translation of the upper vertebra as a function of transection of the components in single units of the cervical spine (7). Selecki and Williams conducted a study of cadaveric cervical spines loaded with a manually operated hydraulic jack (8). They were able to duplicate several types of clinically observed injuries, but reported loads in terms of the hydraulic pressure. Nusholtz *et al.* studied neck motions and failure mechanisms on unembalmed cadavers due to crown impacts; failure loads ranged from 3.2 to 10.8 kN (9). They reported that spinal response and damage were significantly influenced by the initial configuration of the spine.

Very few tests have been conducted on longer spinal segments. Edwards *et al.* tested lumbar spine motion units in combined loading (10). They found that stiffness of the motion unit was nonlinear and increased with increasing load. Markolf and Steidel tested human cadaveric thoracolumbar spine motion units in flexion, extension, lateral bending, torsion, and tension (11). They conducted free-vibration tests, and reported stiffness and damping values for the various test modes and vertebral levels. Panjabi *et al.* measured the three-dimensional stiffness matrix for all levels of the thoracic spine by measuring all components of deflection of spinal units for various loading modes (12). Roaf loaded single cervical spinal units in compression, extension, flexion, horizontal shear, and rotation (torsion) (13). He found that the intact disc, which failed at approximately 7.14 kN, was more resistant to compression than wet vertebrae which failed at approximately 6.23 kN. It is his contention that ligamentous rupture cannot be caused by hyperflexion or hyperextension, but only by rotation and/or shear forces. Tencer *et al.* performed static tests on individual lumbar spinal units (14). They presented load-deflection data for all loading modes. Hodgson measured the strain at selected locations of the cervical vertebrae of cadavers under several head impact modes (15). He concluded that the effects of off-axis, torsional and transverse shear are important variables and influence the axial response. Seemann compared the dynamic responses of the human and Hybrid III neck (16). He concluded that there was a good match with some bending modes but a poor one in others.

A major problem with tests on spinal elements has been the proper measurement of the forces and moments applied to the specimen. The experiments reported here used a five-axis load cell in an attempt to better understand the reasons for the wide range of compressive failure loads and failure mechanisms reported in the literature.

METHODS

SPECIMEN TYPES AND PROCUREMENT - Unembalmed human cervical spines were obtained shortly after death, sprayed with calcium buffered, isotonic saline, sealed in plastic bags, frozen and stored at -20°C. Cervical spine specimens generally included the base of the skull, approximately two centimeters around the foramen, or C1 at the superior end and C5, C6, C7, or T1 at the inferior end. The associated ligamentous structures were kept intact. X-rays were taken and reviewed to assess specimen integrity. Medical records of donors were examined to ensure that the specimens were normal for their age group and did not show evidence of serious degeneration, spinal disease, or other health-related problems that would affect their structural responses.

SPECIMEN PREPARATION - Prior to testing, each specimen was thawed at 20°C for 12 hours. The pre-test specimen preparation was performed in an environmental chamber, which was designed to prevent specimen dehydration and deterioration. A variable flow humidifier pumped water vapor into the chamber to create a 100% humidity environment. The end vertebrae were cleaned, dried, and defatted for casting. The specimen was mounted in aluminum cups with a pin inserted into the spinal canal in order to provide a reference bending axis. Using polyester resin, the ends of the specimens were cast in the cups so that the cups were approximately perpendicular to the axes of the end vertebrae (17). During casting, the aluminum cups were cooled in a flowing water bath to minimize degradation due to the heat of polymerization.

TEST INSTRUMENTATION - A Minneapolis Test Systems (MTS) servo-controlled hydraulic testing machine was used to conduct the various viscoelastic tests. An eight-channel transducing system was used to measure the axial, lateral, and anterior forces, the flexion-extension and lateral bending moments, the linear motion of the ram, and the angular motion of the specimen ends. Loads and moments were measured with a five-axis load cell assembly, which was constructed using two GSE three-axis ATD neck load cells. The motion of the specimen ends was measured with an internal coaxial linear variable differential transformer (LVDT) and two external rotational variable differential transformers (RVDT). These transducers provided data to establish the motion of the two specimen ends from direct measurements of the total bending angle and calculations of the specimen length change. The internal LVDT was used to monitor the ram motion and hence the displacement of the clevis end of the lower transfer bar. One external RVDT was used in the pinned-pinned and fixed-pinned tests to track the rotation of the specimen end of the lower transfer bar relative to the ram; the second external RVDT was used in the pinned-pinned tests to track the rotation of the specimen end of the upper transfer bar. Figure 1 is a schematic diagram of the test apparatus.

A digital measurement and analysis system was developed utilizing a data logging computer. The multichannel microcomputer-based data acquisition system incorporated an RC Electronics ISC-16 Computerscope for the digitization and storage of data. This system, which consists of a 16-channel A/D board, external instrument interface box, and Scope Driver software, has a 1 MHz aggregate sampling rate capability with 12 bit resolution and writes data directly to a hard disk. In addition, during the failure tests, flouroscopic images were recorded on videotape.

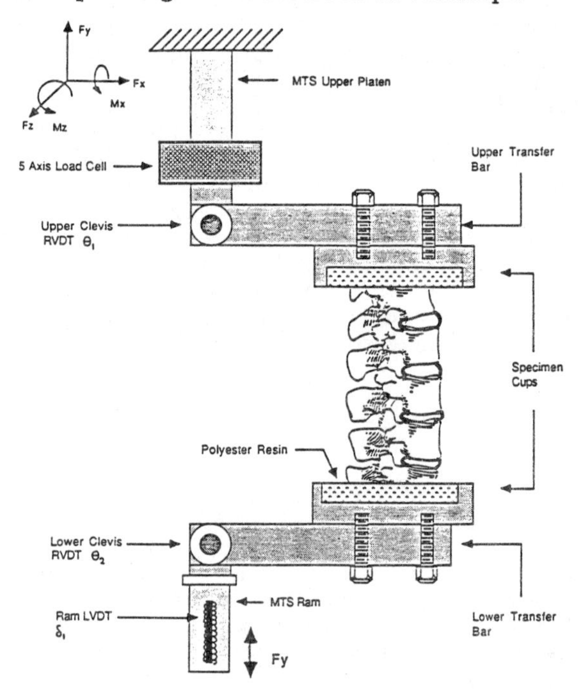

Figure 1: Pinned-Pinned Test Configuration

THE COMBINED AXIAL LOADING - BENDING TEST APPARATUS - A specially designed test jig was developed to place the specimen in a state of eccentric axial loading. This resulted in a combined axial load and bending moment applied at the ends of the specimen. The apparatus provided adjustable moment arms and accommodated the following six test modes: compression-flexion (CF), tension-flexion (TF), compression-extension (CE), tension-extension (TE), compression-lateral bending (CL), and tension-lateral bending (TL). Two test configurations were utilized: (1) pinned-pinned end conditions (PP), and (2) fixed-pinned end conditions (FP).

For the pinned-pinned end conditions, the upper transfer bar was attached via a clevis to the load cell assembly, which was rigidly mounted to the upper platen of the MTS. The lower transfer bar was attached via a clevis to the ram of the MTS. The centerline of the specimen was parallel to, but not coincident with, the line of action of the MTS ram. The clevis end of the upper transfer bar was constrained from translation. The two external RVDTs were mounted on the test apparatus in order to measure the angular displacement of each transfer arm. In this configuration, the specimen was mounted with the superior end attached to the upper transfer bar and the inferior end attached to the lower transfer bar.

For the fixed-pinned end conditions, the upper clevis and corresponding RVDT were removed. In this configuration, the specimen was mounted with the superior end attached to the pivoting lower transfer bar and the inferior end fixed to the load cell assembly, which was rigidly mounted to the upper platen of the MTS.

A free body diagram of the test configuration is presented in Figure 2. The reference center line of the specimen is the central axis of the spinal foramen. The moment at the center of the specimen is

$$M_A = P_y a - P_x b,$$

and the moment measured by the load cell is

$$M_O = P_x B.$$

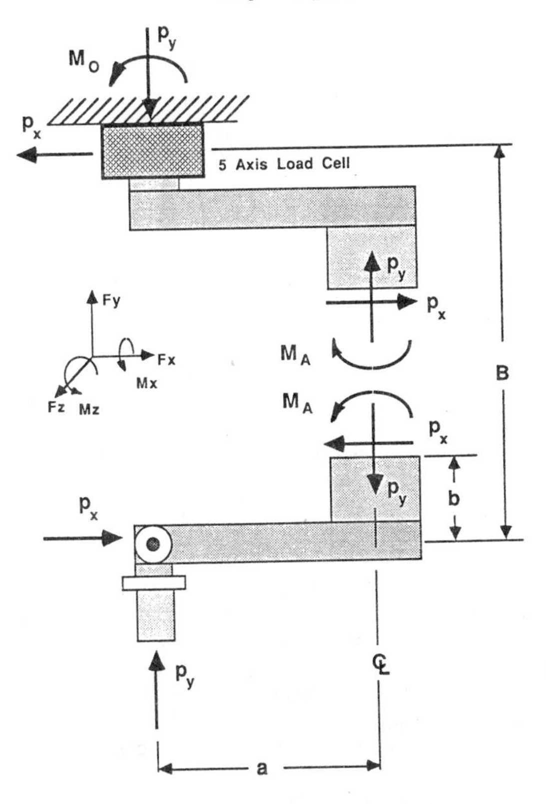

Figure 2: Freebody Diagram for the Fixed-Pinned Test Configuration

The moment induced by the shear force P_x was significant in the fixed-pinned configuration but was negligible in the pinned-pinned configuration. The apparatus had minimal overshoot and vibration below test frequencies of 5 Hz. Inertial forces begin to predominate above 10 Hz, and this is the current system's upper frequency range.

In this paper, test rates will be described in Hertz. The test period is the reciprocal of the frequency, and the time to peak load is one-half of the test period. The deformation rate is the maximum deformation in angular or linear units multiplied by twice the test frequency.

VISCOELASTIC TEST BATTERY – Five types of tests were performed in the following order:

1. Fully equilibrated relaxation test (the specimen was wet with isotonic saline for at least 24 hours pre-test.)
2. Cyclic test (nominally 40 cycles at 1.0 Hz until the waveform repeats.)
3. Mechanically stabilized or preconditioned relaxation test (the cyclic modulus reached an asymptotic plateau.)
4. Constant velocity tests (0.01 Hz, 0.1 Hz, 1.0 Hz, 5 Hz in the mechanically stabilized state.)
5. Constant velocity load to failure test (in the mechanically stabilized state.)

This battery of eight tests if performed in the six test modes results in forty-eight tests per specimen. If the end conditions were also varied (3 additional modes) there are one-hundred and forty-four combinations. Concern for specimen degradation resulted in approximately twenty-four tests per specimen actually being performed.

TEST RESULTS

RELAXATION TESTS – Pre-cyclic and post-cyclic relaxation tests were conducted to measure the viscoelastic responses of the spines in the initial equilibrated state and in the mechanically stabilized state. The relaxation tests were performed using ramp-and-hold command signals with 0.2 second rise times. The deflection was held constant for 125-150 seconds. Figure 3 is a typical example of the pre- (R1) and post-cyclic (R2) reduced (normalized by peak load) relaxation responses for spines in the equilibrated and mechanically stabilized states, respectively, tested in the pinned-pinned and fixed-pinned configuration. The response of the Hybrid III neck is shown for comparison.

The relaxation response of each spine reached a peak load for a finite loading rate, decreased monotonically, and approached asymptotically a non-zero equilibrium value during the test. A variable rate of moment relaxation was demonstrated. For constant deformations, the load decay was initially extremely rapid, decaying at a much slower rate thereafter. In all cases, the spines exhibited less load relaxation after cyclic preconditioning. The best least-squares fit, for all tests, of the reduced relaxation response was a linear regression in the natural log of time. This linear behavior in log time has also been observed for other biological tissues by Fung (18). These curves can be represented as a sum of exponentials, but this was not done since the exponents and amplitudes are quite sensitive to small changes in the data. Tables 2 and 3 show the mean and standard deviation (σ) of the instantaneous stiffness (K_O) and the asymptotic stiffness

TABLE 1. FAILURE TEST RESULTS.

SPECIMEN NUMBER	AGE/ SEX	VERTEBRAL LEVELS	MAXIMUM MOMENT (N-m)	MAXIMUM AXIAL FORCE (N)	MAXIMUM A-P SHEAR (N)	ANGLE AT MAX.MOMENT (deg)	FAILURE CLASSIFICATION
1C P-P	52/M	$C_1 - T_1$	14.6	192	0	54	C_4-C_5, C_5-C_6 ligamentum nuchae, ligamentum flavum, and post. long. ligament torn
2C P-P	64/F	$C_1 - T_1$	8.75	214	0	57	C_6-C_7 ligamentum nuchae and R capsular ligament torn
3C P-P	N/A	$C_1 - T_1$	3.01	108	0	31	wedging of C_4-C_5 bodies, C_5-C_6 ligamentum nuchae disrupted
4C P-P	69/M	$C_1 - T_1$	3.40	338	11.7	40	wedging and broadening of C_4-C_5 and C_5-C_6 bodies, tear of C_3-C_4 disc
5C P-F	77/M	$C_1 - T_1$					this specimen was not loaded to failure
6C P-F	76/M	$BOS - T_1$	6.7	1513	23.0	15	C_4-C_5 ant. disc disrupted, C_2-C_3, C_3-C_4, C_4-C_5 L capsular ligaments partially disrupted
7C P-F	86/M	$BOS - T_1$	10.2	2305	35	22	C_4-C_5, C_5-C_6, C_6-C_7 shortened discs and wedged bodies, disrupted C_7-T_1 disc, ligamentum nuchae and ligamentum flavum stretched

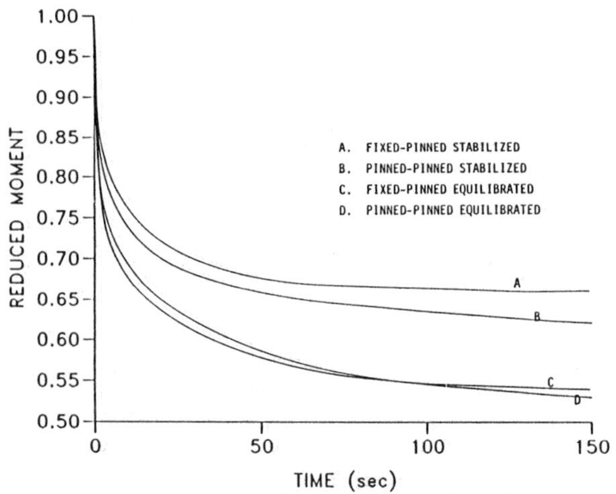

Figure 3: Typical Bending Moment Relaxation for Human Cervical Spines

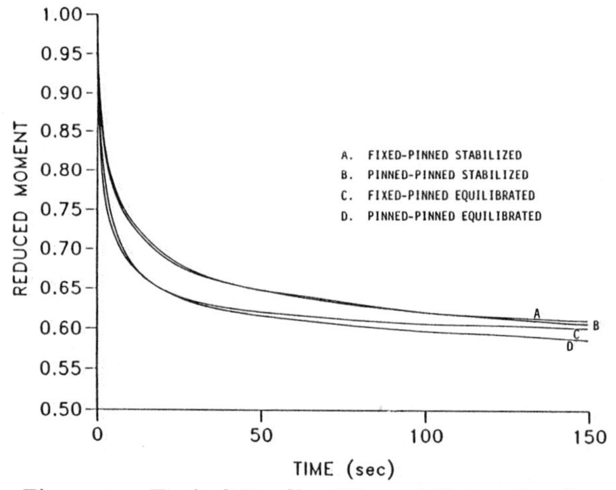

Figure 4: Typical Bending Moment Relaxation for the Hybrid III Neckform

(K_∞) for the various test modes of the human cervical spine and the Hybrid III neckform.

The ratio of K_O/K_∞ is a measure of the time dependent or viscous component. For the fully equilibrated state the average value of K_O/K_∞ for all tests was 1.838 for Hybrid III and 2.099 for the human cervical. For the mechanically stabilized state the average value was 1.658 for the Hybrid III and 1.507 for the human cervical. The ratio of the K_O's for the equilibrated and the stabilized state is a measure of the nonlinear response probably due to the release of fluid from the various tissues. This ratio averaged for all tests was 1.623. For the Hybrid III, this ratio was very close to 1.

CYCLIC TESTS – Each fully equilibrated specimen was subjected to a sinusoidally varying displacement for approximately 50 cycles. The cyclic loading established the mechanically stabilized or preconditioned state (6). When a specimen was subjected to a repeated deformation history about a fixed length, the cyclic modulus magnitude decreased as the number of deformation cycles increased. In these tests, the cyclic modulus is the peak moment divided by the peak angular deformation (N-m/Radian). The initial cycle was representative of the response of the fully equilibrated spine. Eventually, after about 30 cycles, a steady-state (i.e, the mechanically stabilized state) was reached. The achievement of the mechanically stabilized state

Table 2. RELAXATION STIFFNESS (N-m/rad) FULLY EQUILIBRATED.

MODES	HUMAN												HYBRID III			
	FIXED-PINNED						PINNED-PINNED						FIXED-PINNED		PINNED-PINNED	
	K_0			K_∞			K_0			K_∞			K_0	K_∞	K_0	K_∞
	Mean	σ	N	Mean	σ	N	Mean	σ	N	Mean	σ	N				
CF	59.2	13.7	3	34.5	13.5	3							680.0	280.3	132.9	75.6
TF	29.3		1	17.8		1	14.8		1	8.9		1	449.7	357.3	268.4	119.5
CE							3.9	0.9	2	2.1	0.0	3	683.9	347.1	120.2	65.3
TE							13.2		1	7.4		1	238.9	300.1	135.7	67.3
CL	8.2	1.7	3	2.0	0.2	3	3.3	0.6	2	1.1	0.3	3	931.7	347.1	120.2	65.3
TL	177.3		1	107.9		1	16.2		1	10.1		1	302.2	227.9	195.6	81.7

TABLE 3. RELAXATION STIFFNESS (N-m/rad) MECHANICALLY STABILIZED.

MODES	HUMAN												HYBRID III			
	FIXED-PINNED						PINNED-PINNED						FIXED-PINNED		PINNED-PINNED	
	K_0			K_∞			K_0			K_∞			K_0	K_∞	K_0	K_∞
	Mean	σ	N	Mean	σ	N	Mean	σ	N	Mean	σ	N				
CF	36.9	6.9	3	31.8	4.7	3							656.5	282.4	134.0	87.4
TF							10.5		1	7.0		1	524.2	349.0	157.2	89.2
CE							2.4	0.4	3	1.8	0.3	3	751.8	390.2	122.6	70.1
TE	67.5		1	58.0		1	7.1		1	4.5		1	271.3	313.2	114.4	64.9
CL	6.0	1.7	3	3.3	0.4	3	2.0	0.5	3	1.0	0.2	2	805.5	435.5	172.0	101.5
TL							9.3		1	6.4		1	317.6	242.5	182.7	116.1

σ = Standard Deviation; N = Number of Tests.

was evidenced by a constant cyclic modulus magnitude and a repeatable load-deflection response. Hopefully, after cyclic conditioning, the responses from one loading mode to another can be modeled using viscoelastic parameters which do not vary with time. The specimen could be returned to the fully equilibrated state by keeping it wet with isotonic saline in the refrigerator for 24 hours. Fung, who observed this phenomenon in other soft biological tissues, referred to this state as a preconditioned state (18). Figure 5 shows a typical family of moment conditioning cycles, while Table 3 provides the cyclic modulus for the various test modes. The instantaneous or fully equilibrated cyclic modulus K_0 (N-m/Radian) was computed as the ratio of the maximum moment and the bending angle. The mechanically stabilized or preconditioned cyclic modulus K_∞ was the above ratio after 40 cycles.

CONSTANT VELOCITY TESTS – Constant velocity tests were conducted on the mechanically stabilized spines using triangle wave deformations at frequencies of 0.01, 0.1, 1.0, 5 Hz, and, for some specimens, 10 Hz. The maximum ram displacement used for these test was the same as the maximum ram displacement used in the relaxation and cyclic tests. Thus, the deformation rate was varied by a factor of 500-1000.

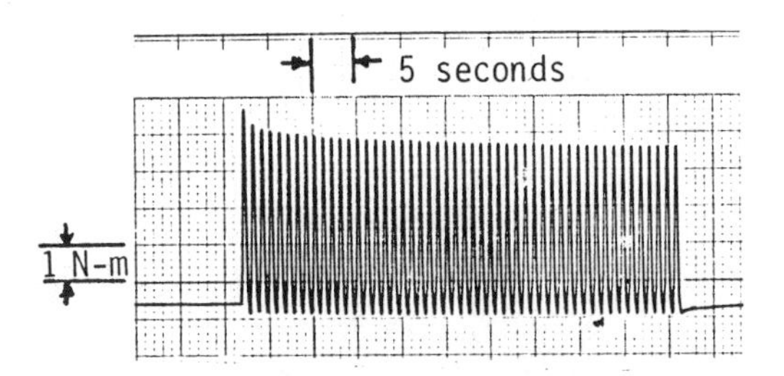

Figure 5: Typical Cyclic Test Envelope

Typical constant velocity moment-angle curves are presented for human and Hybrid III cervical spines in the pinned-pinned and fixed-pinned test configuration in Figures 6 and 7. All of the curves exhibit a hardening response (increasing stiffness) and hysteresis. The human and Hybrid III responses are fundamentally different. The Hybrid III shows the classic linear viscoelastic response of increasing stiffness with displacement rate while the human shows little change in stiffness or hysteresis over the rate range tested. Since these features of hysteresis, relaxation, and stiffness are not very sensitive to the rate of strain, simple linear viscoelastic models would not be appropriate pre-

TABLE 4. CYCLIC STIFFNESS (N-m/rad).

MODES	HUMAN 1st CYCLE FIXED-PINNED			HUMAN 1st CYCLE PINNED-PINNED			HUMAN 40th CYCLE FIXED-PINNED			HUMAN 40th CYCLE PINNED-PINNED			HYBRID III FIXED-PINNED	HYBRID III PINNED-PINNED
	Mean	σ	N	Mean	σ	N	Mean	σ	N	Mean	σ	N	Mean	Mean
CF	206.3		1	4.9	1.4	3	105.0		1	4.3	0.9	3	567.3	153.2
TF	27.6		1	16.1		1	28.1		1	14.0		1	1216.4	185.2
CE	2.0		1	1.8		1	2.7		1	2.4		1	868.5	173.9
TE													710.1	137.5
CL	7.8	1.9	3	2.4	0.3	3	8.0	3.9	3	2.3	0.3	3	1054.2	209.5
TL	202.0		1	13.2		1	142.3		1	9.8		1	957.1	232.5

TABLE 5. CONSTANT VELOCITY STIFFNESS (N-m/rad).

MODES	HUMAN FIXED-PINNED			HUMAN PINNED-PINNED			HYBRID III FIXED-PINNED	HYBRID III PINNED-PINNED
	Mean	σ	N	Mean	σ	N	Mean	Mean
CF	29.9	2.6	10	8.1	0.7	5	589.1	150.8
TF	41.8	5.6	5	14.8	1.3	5	608.4	199.0
CE				2.8	0.6	9	795.7	122.5
TE	309.0	26.9	5	10.3	1.2	11	232.1	138.8
CL	8.7	0.6	10	3.1	1.0	17	898.9	190.9
TL	254.1	34.6	5	13.0	1.9	5	442.0	226.1

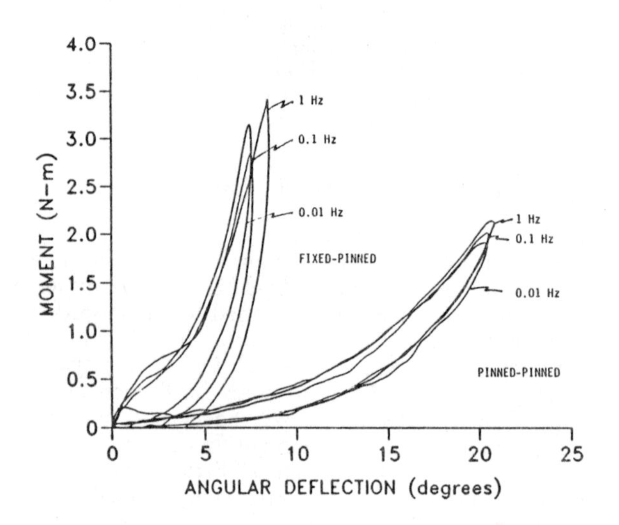

Figure 6: Typical Constant Velocity Profile for Human Cervical Spine

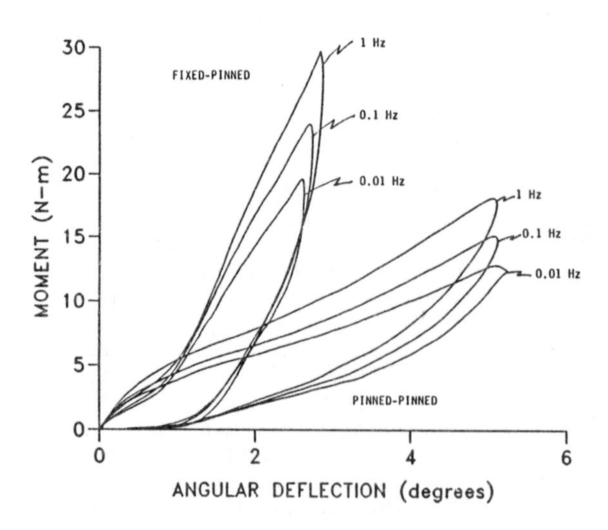

Figure 7: Typical Constant Velocity Profile for Hybrid III Neckform

dictors of the time dependent human spinal bending responses; and the more complex Maxwell-Weichert quasi-linear model is required (6).

Table 5 shows the stiffness averaged over four rates and all specimens. Three distinct tests of the Hybrid III were performed so that each value represents the mean of 12 tests. Several observations are apparent from this data. First, there are significant differences between the bending stiffnesses of the cadaver cervical spine and the Hybrid III. Second, there are significant differences in the bending stiffness of the cadaver cervical spine in the different modes. Tension-extension was the largest with a stiffness of 125 N-m/Radian, fixed-pinned and 15 N-m/Radian, pinned-pinned. Compression-lateral was the smallest with a stiffness of 10 N-m/Radian, fixed-pinned and 2.6 N-m/Radian pinned-pinned.

Figure 8 shows a typical response pattern for the human cervical spine to the various combined bending and axial loading modes. Figure 9 shows a typical response pattern for the Hybrid III.

FAILURE TESTS – After the battery of viscoelastic tests was accomplished, a constant velocity failure test at 0.1 Hz was performed. This rate was used so that flouroscopic images of the specimen motion could be obtained. All failure tests were in the compression-flexion mode (CF). After the tests the specimens were examined with magnetic resonance imaging (MRI) and computerized tomographic radiography (CT), then dissected. Table 1 provides the maximum moment axial force and shear force applied to the specimen and the bending angle at which these peaks occurred. The first four tests (1C, 2C, 3C, 4C) were performed in the pinned-pinned mode and the remainder (6C, 7C) were tested in the fixed-pinned mode. In the pinned-pinned configuration the specimens were very flexible and were able to bend through on average of 45 degrees without an unstable dislocation. These specimens contained C_1 through T_1 and seven intact intervertebral structures. This is approximately 6.4 degrees per vertebral level. The shear forces were very small. The axial forces were low enough that the major stresses were due to the bending moment. The primary failure mechanism was disruption of the interspinous ligaments (ligamentum nuchae), the ligamentum flavum and capsular ligaments. There was also minor anterior wedging of the middle vertebral bodies and discs. In the pinned-pinned configuration the moment is maximum in the middle of the specimen. This may be the reason that the most frequent spinal cord injury level observed clinically is C_4 - C_5 and C_5 - C_6 (5).

In the fixed-pinned configuration much larger axial forces are required to produce the same bending moment because the shear force produces a counteracting moment. This is reflected in the failure mechanisms by superimposing compressively induced failures (wedging of bodies and discs) to the posterior tensile failures due to bending.

Figure 10 shows a composite of the moment-angle diagrams for the failure tests. The maximum moment ranged from 3.01 to 14.6 N-m. This large range is probably due to the variation in the size of the specimens. Specimen 1C and 7C had much larger vertebrae than the others as demonstrated by the CT scans.

DISCUSSION – This study demonstrated the complex, time-dependent responses of the human cervical spine and the Hybrid III neckform in combined axial and bending deformations. In all test

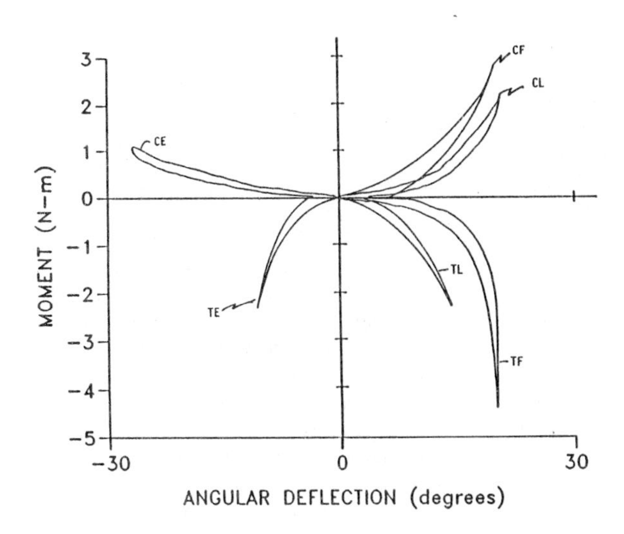

Figure 8: Typical Bending Responses of Human Cervical Spine

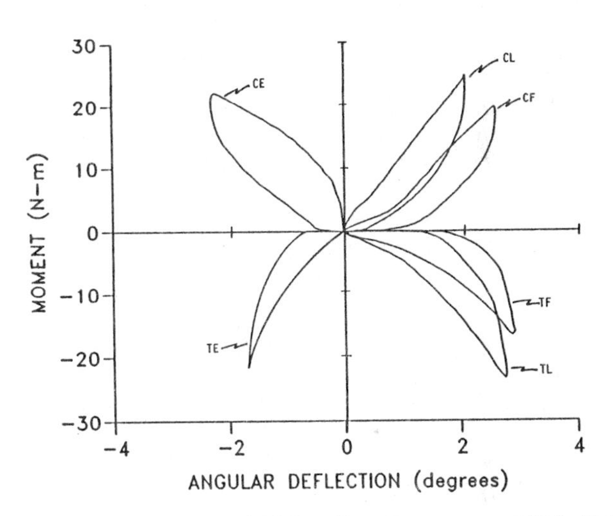

Figure 9: Typical Bending Responses of Hybrid III Neckform

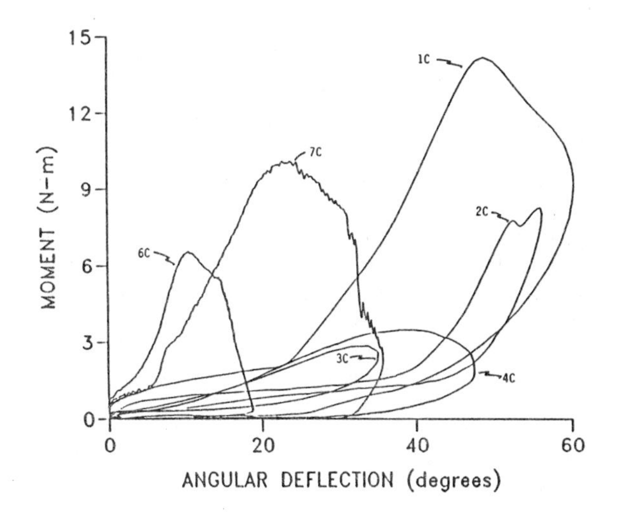

Figure 10: Failure Curves

modes (tension-extension, tension-flexion, tension-lateral bending, compression-extension, compression-flexion, compression-lateral bending) there was a large difference between the responses of spines in the fully equilibrated and mechanically stabilized states. In all test modes, the time-dependent responses included a significant viscoelastic exponential relaxation. The hysteresis and stiffness of the human specimens was only weakly dependent on strain rate.

There was a significant difference between the stiffness of the cadaver cervical spines and the Hybrid III. This was expected since the performance requirements of the Hybrid III were based on human volunteer data, and it is considered to represent a tensed human neck while the cadaver spines have no musculature present (19). The Hybrid III responses were the typical linear viscoelastic type. That is, a linear differential equation would provide an adequate model. The behavior of the human cervical spine was more complex, however, and requires a quasi-linear model (6).

The bending stiffness of the cervical spine was significantly influenced by the direction of the bending moment, the types of end restraint, the magnitude of the deformation, and the previous deformation history. After approximately thirty deformation cycles a mechanically stabilized state was attained that provided repeatable load-deformation responses. The tensile modes were consistently stiffer than the compressive modes. This may be due to a shift in the neutral axis toward the tensile side which pre-tensions slack ligaments and reduces the eccentricity.

Simple beam theory predicts doubling of the bending stiffness when comparing pinned-pinned and fixed-pinned ends. These tests showed an increase in stiffness of approximately eight times. The test apparatus used in these tests (and by most other researchers) constrained the pinned end to move in a straight line. This produced a shearing force which, acting over a relatively long moment arm, stiffened the specimen. This shearing force not only changes the moment acting on the specimen but also influences the failure mode. Several researchers have tested cervical specimens without well controlled and monitored end conditions. Most other works report only the axial load. These experiments indicate that when the loading is eccentric (as it almost always is), the primary deformation mode is bending; and the moment applied to the specimen is strongly influenced by shear forces and the magnitude of the eccentricity. The axial load is therefore a poor indicator of the type and magnitude of failure stresses.

ACKNOWLEDGEMENTS

This work was supported by the Department of Health and Human Services, Center for Disease Control grant R49/CCR402396-02, and by the Whitaker Foundation.

BIBLIOGRAPHY

1. Sances, A, Weber, R.C., Larson, S.J., Cusick, J.S., Myklebust, J.B., and Wash, P.R., "Bioengineering Analysis of Head and Spine Injuries," *CRC Critical Reviews in Bioengineering*, February, 1981.

2. Messerer, O., *Uber Elasticitat and Festigkeit der Meuschlichen Knochen*, J. G. Cottaschen Buchhandling, Stuttgart, 1880.

3. Bauze, R.J. and Ardran, M.J., "Experimental Production of Forward Dislocation in the Human Cervical Spine". *Journal of Bone and Joint Surgery*, 60B(2):239, 1978.

4. Sances Jr., A., Myklebust, J., Houterman, C., Webber, R., Lepkowski, J., Cusick, J., Larson, S., Ewing, C., Thomas, D., Weiss, M., Berger, M., Jessop, M.E., Saltzberg, B., "Head and Spine Injuries." *AGARD Conference Proceedings on Impact Injury Caused by Linear Acceleration — Mechanism, Prevention, and Cost*, 1982.

5. McElhaney, J.H., Roberts, V.L., Paver, J.G. and Maxwell, G.M., "Etiology of Trauma to the Cervical Spine," in *Impact Injury to the Head and Spine*, C.C. Thomas, 1982.

6. McElhaney, J.H., Paver, J.G., McCrackin, H.J., and Maxwell, G.M., "Cervical Spine Compression Responses," SAE Paper No. 831615, 1983.

7. Panjabi, M.M., White, A.A., and Johnson, R.M., "Cervical Spine Mechanics as a Function of Transection of Components," *Journal of Biomechanics*, 8(5):327, 1975.

8. Selecki, B.R., and Williams, H.B.L., "Injuries to the Cervical Spine and Cord in Man," Australian Med. Assoc., Mervyn Archdall Med. Monograph #7, Australian Medical Publishers, South Wales, 1970.

9. Nusholtz, G.S., Melvin, J.W., Huelke, D.F., Alem, N.M., and Blank, J.G., "Response of the Cervical Spine to Superior-Inferior Head Impact," *Proc. of the 25th Stapp Car Crash Conference*, SAE Paper No. 811005, pp. 197-237, 1981.

10. Edwards, W.T., Hayes, W.C., Posner, I., White, A.A. III, and Mann, R.W., "Variation of Lumbar Spine Stiffness With Load," *Journal of Biomechanical Engineering*, 109:35, 1987.

11. Markolf, K.L., and Steidel, R.S., Jr., "The Dynamic Characteristics of the Human Intervertebral Joint," ASME Paper No. 70-WA/BHF-6, 1970.

12. Panjabi, M.M., Brand, R.A., Jr., and White, A.A., "Three Dimensional Flexibility and Stiffness Properties of the Human Thoracic Spine," *Journal of Biomechanics*, 9:185, 1976.

13. Roaf, R., "A Study of the Mechanics of Spinal Injury," *Journal of Bone and Joint Surgery*, 42B:810, 1960.

14. Tencer, A.F., Ahmed, A.M., and Burke, D.L., "The Role of Secondary Variables in the Measurement of the Mechanical Properties of the Lumbar Intervertebral Joint," *Journal of Biomechanical Engineering*, 103:129, 1981.

15. Hodgson, V.R., and Thomas, L.M., "The Biomechanics of Neck Injury From Direct Impact to the Head and Neck," Head and Neck Injury Criteria, U.S. Government Printing Office, 1981.

16. Seemann, M.R., Muzzy, W.H. and Lustick, L.S., "Comparison of Human and Hybrid III Head and Neck Response," *Proc. of the Thirtieth Stapp Car Crash Conference*, 1986.

17. Hirsch, C., "Method of Stabilizing Autopsy Specimens in Biomechanical Experiments," *Acta Orthopaedica Scandinavia*, 34(4):374, 1964.

18. Fung, Y.C., "Stress-Strain History Relations of Soft Tissue in Simple Elongations," *Biomechanics — Its Foundation and Objectives*, Prentice-Hall, Inc., Englewood Cliffs, 1972.

19. Mertz, H.J., Neathery, R.F., and Culver, C.C., "Performance Requirements and Characteristics of Mechanical Necks, in *Human Impact Response*, Plenum Press, 1973.

Appendix 1: Bibliography

BIBLIOGRAPHY

Appendix 1

This Appendix references additional literature available on the biomechanics of impact injury and injury tolerances of the head-neck complex. Due to space constrainsts, the valuable technical papers included in this bibliography were not reprinted in this book. Instead, abstracts of each paper have been included.

Most of these papers have been published by SAE and are available in original or photocopy form. For ordering information, contact SAE's Customer Sales and Satisfaction Department, SAE International, 400 Commonwealth Drive, Warrendale, PA 15096-0001, USA, Telephone: 412/776-4970, Fax: 412/776-0790. Copies of papers referenced but not published by SAE must be obtained directly from the publisher listed.

Careme, L. M. M., "Biomechanical Tolerance Limits of the Cranio-Cervical Junction in Side Impacts," SAE Technical Paper 890383.

The high incidence of injuries in the upper cervical region and cranio-cervical junction point out the importance of continued research into safety measurements to protect this vital area of the human body. Therefore, we collected information on biomechanical tolerance limits of the occipito-atlanto-axial complex from postmortem radiography and laboratory biomechanical studies.

The forces responsible for the injuries seen in the CO-C1-C2 segment are generally complex. In more than 1/4th of the cases, the precise mechanism could not be determined. Many were quite likely due to lateral forces or a combination of several factors.

In the currently available European Sid Impact Dummy (EUROSID) the dummy neck is composed of a central section made of rubber that links two interfaces, each consisting of two metal disks with rubber elements and a spinal joint inside. With consideration to injury tolerance, instrumentation and data processing, additional measurement channels at the head-neck interface should be included in the design of further experimental developments.

Chamouard, F.; Trosseille, X.; Pincemaille, Y.; Tarrière, C., "Methodological Aspects of an Experimental Research on Cerebral Tolerance on the Basis of Boxers Training Fights," SAE Technical Paper 872195.

In order to obtain data about human head tolerance, the APR Lab. of Biomechanics has developed a specific methodology for volunteer boxers.

These ones are used because they expose themselves, in their normal body activities, to direct head impacts similar in nature to those experienced by vehicle occupants under crash conditions. This paper describes the specific experimental technique that permits association of the severity of the blows, measured in terms of physical parameters, to corresponding physiological effects, measured in medical terms.
Chou, Clifford C.; Nyquist, Gerald W., "Analytical Studies of the Head Injury Criterion (HIC)," SAE Technical Paper 740082.

The Head Injury Criterion (HIC) in FMVSS 208 for evaluating the potential head injury requires maximization of a mathematical expression, involving the time-average acceleration, by varying the limits (t1, t2) of the time interval over which the average is calculated.

This paper describes the HIC behavior through the analysis of a function of two independent variables t1 and t2. The analysis is carried out for any arbitrary acceleration profile a(t). It is found that maximization requires that a(t1) =

a(t2). Also, for the unique values of t1 and t2 that maximize HIC, the average acceleration between t1 and t2 is 5/3 times the acceleration at t1 and t2. Illustrative examples are provided by applying this condition to simple pulses. Numerical results are presented in tables and graphs. Clarke, Thomas D.; Gragg, C. D.; Sprouffske, James F.; Trout, Edwin M.; Zimmerman, Roger M.; Muzzy, William H., "Human Head Linear and Angular Accelerations During Impact," SAE Technical Paper 710857.

Head linear and angular accelerations of humans were investigated during exposure to abrupt linear deceleration (-Gx). The 14 subjects were restrained with three different restraints: lap belt only, Air Force shoulder harness, and air bag plus lap belt. Peak sled decelerations ranged from 7.7-10.3 g.

The results indicated that peak head angular and linear resultant accelerations were elevated with the air bag in contrast to the Air Force shoulder harness or lap belt only restraints. However, the peak angular and linear accelerations may have less traumatic consequences than the degree of head-neck hyperextension.

Clemens, H. J.; Burow, K., "Experimental Investigation on Injury Mechanisms of Cervical Spine at Frontal and Rear-Front Vehicle Impacts," SAE Technical Paper 720960.

This paper discusses the results of simulated head-on and rear-front vehicle crashes employing 53 human torsos. Measurements of deceleration of the head were taken, and the resulting injuries were noted. The most common and serious injury was to the cervical spine at the sixth vertebra. It is suggested that vehicle restraint systems be developed to avoid such injury, such as safety belts that would limit anteflexion of the head, airbags for head-on crash protection, and seat backs with integrated headrests to support the head at the c.g.

Culver, Clyde C.; Haut, Roger C.; Nusholtz, Guy S., "Significance of Head-To-Knee Impacts - A Comparison of Dummy and Cadaver Responses," SAE Technical Paper 841662.

Head-to-knee interaction of the right front passenger dummy can occur in some 30-35 HPH crash barrier tests. The biofidelity and significance of these interactions as related to predicting human response was addressed in this study. In a series of laboratory experiments an instrumented headform was dropped on the dummy knee to simulate the barrier interactions. These test results were then related to the human by dropping the same headform on the cadaver leg.

The instrumented headform was dropped from three heights to impact the Part 572 dummy knee at three velocities. Two impact sites and two impact angles were used. These test parameters bracketed the barrier conditions. Measurements from headform accelerometers permitted calculation of HIC value for comparison to barrier values. Comparable experiments were subsequently performed with three unembalmed cadaver subjects using the same headform and test procedures.

Impacts on the 90 degree flexed dummy knee at 6 and 7 m/s were found to correspond to the HIC vs impact velocity "window" of barrier test results (400-1400 HIC and 5.8-7.2 m/s). Identical tests on the knees of three unembalmed cadaver subjects produced HIC values averaging 45% of the dummy values.

Head-to-knee impact tests on cadaver subjects indicate a lack of biofidelity in the dummy lower leg structure for this type of impact. Assuming that head-to-knee interaction actually occurs in field accidents, values of HIC measured from dummy crash simulation experiments are likely to overestimate the severity of the interaction. Enouen, Susan Willke, "The Development of Experimental Head Impact Procedures for Simulating Pedestrian Head Injury," SAE Technical Paper 861888.

This paper describes the procedures and methodologies used to reconstruct cadaver test head impacts and pedestrian accident cases, respectively. Head impact dent depth and its relationship to impact energy was found to be useful in determining the effective head mass upon impact in the cadaver test reconstructions. Good correlation between impactor response and the cadaver head accelerations was achieved. Good correlation was also achieved between pedestrian accident case injuries and the reconstruction results using three different head injury criterion, HIC, Mean Strain Criterion (MSC), and Normalized G's.

Ewing, C. L.; Thomas, D. J.; Lustick, L.; Muzzy, W. H. III; Willems, G. C.; Majewski, P., "Effect of Initial Position on the Human Head and Neck Response to +Y Impact Acceleration," SAE Technical Paper 780888.

The response of the human head and neck to impact acceleration has been previously reported for the sDX (chest to back) and +Y (right to left) directions. Wide ranges of sled peak acceleration, rate of onset of acceleration and duration of acceleration have been investigated and reported. A major mechanical effect on the dynamic response due to initial position for the sDX direction has been reported. The purpose of this study is to report the initial position effect on the human head and neck response for +Y direction experiments.

Four initial positions of the head relative to the first thoracic vertebral body (T1) have been investigated over a range of sled acceleration peaks from 2 to 7G. The data from six young adult male volunteers representative of a wide range of anthropometry will be presented. There are 18 experiments for each volunteer for a total of 108 experiments.

The effect of initial position on the resulting head angular and linear acceleration, velocity and displacement will be presented. A comparison of the initial position effect for -X and +Y direction experiments will be made. The ability to model the initial position effect will be discussed.

Ewing, Channing L.; Thomas, Daniel J.; Beeler, G. W.; Patrick, L. M.; Gillis, David B., "Dynamic Response of Head and Neck of Living Human to Gravity Acceleration," SAE Technical Paper 680792.

An acceleration sled carrying living human subjects was used to measure the dynamic response of the head and neck to -Gx impact acceleration. Seated volunteers with complete pelvic and upper torso restraint were subjected to increasing impact accelerations beginning at 2.7 g and increasing in 1 g increments. The volunteers were selected to encompass the 5th to 95th percentile distribution of sitting height according to a selected reference.

Precision inertial transducers were used to determine the linear and angular acceleration of the head and the first thoracic vertebra. The inertial system consisted of a biaxial accelerometer and rate gyroscope on a bite-plate, a biaxial accelerometer over the bregma, and a biaxial accelerometer and rate gyroscope over the spinous process of the first thoracic vertebra. The transducers on the bite-plate and over the bregma were rigidly connected to one another.

Precision high-speed cameras were mounted on the sled to determine the displacements of the head and neck of the subject. One camera was mounted posteriorly and one mounted laterally with respect to the subject.

Data handling was effected by recording the transducer outputs on analog tape, converting the analog tape to digital form, and calculating the pertinent trajectories by digital computer. The distance of test markings from a point of reference was measured from each photographic exposure and committed to punch cards in digital form. The inertial and photographic systems were completely redundant. The time reference axes of both systems were interlocked. The results obtained from both systems were compared for validation.

Preliminary results are presented.

Ewing, C. L.; Thomas, D. J.; Lustick, L.; Becker, E.; Willems, G.; Muzzy, W. H. III, "The Effect of the Initial Position of the Head and Neck on the Dynamic Response of the Human Head and Neck to -Gx Impact Acceleration," SAE Technical Paper 751157.

In preparation of an analog of human head and neck, the reports by R. G. Snyder and others were noted which stated that initial position of the head and neck had a definite effect upon resulting response. An investigation was undertaken to attempt to quantitate this effect, as a part of a much larger study underway for several years.

Thirteen human volunteer subjects ranging from the 5th to the 97th percentile in sitting height were exposed to -Gx impact acceleration at peak sled accelerations of 6G and 10G. Two angles of the neck relative to chair and two angles of the head relative to the neck for a total of four conditions were tested for each subject for the 2 peak acceleration levels giving a total of 104 experiments. Instrumentation consisted of 6 accelerometers and two-axis rate gyro at the posterior spinous process of the first thoracic vertebral body, 6 accelerometers at the mouth, and a two-axis rate gyro at the top of the head. Three-dimensional photography from two orthogonally mounted onboard cameras was used also.

The input data at T1 along with the differential effects of initial head position relative to T1 on the linear acceleration at the origin of the head anatomical coordinate system and on the angular acceleration and angular velocity of the head will be presented along with the implications for modeling the response and a statistical comparison.

Foust, David R.; Bowan, Bruce M.; Snyder, Richard G., "Study of Human Impact Tolerance Using Investigations and Simulations of Free-Falls," SAE Technical Paper 770915.

A study of free-fall accidents and resulting injuries was conducted to determine how useful these types of data could be in establishing human injury tolerance limits. "Tolerance" was examined primarily for children and at two levels-reversible injury and threat to survival. The specific objectives were to investigate specific free-falls in sufficient depth to permit biomedical or mathematical reconstruction of the fall, simulate selected free-falls to estimate impact response, and compare predicted responses with observed injuries as a means of estimating human tolerance levels. From more than 2100 reported free-falls, 110 were investigated on-site. Seven head-first and three feet-first falls were then simulated using the MMMA 2-D Crash Victim Simulator. Newspaper reports of free-falls showed that males fell six times as often as females and most often while at work. Children fell from windows and balconies more often than from any other hazard. Case investigations showed that children were generally injured less severely than adults under similar fall circumstances, and tended to land on their heads a greater proportion of the time. The investigations also indicated that for impacts onto rigid surfaces: both children and adults may be expected to suffer skull fractures in head-first falls of greater than ten feet; adults are likely to incur lumbar spine fracture if they land in a sitting position after falling 10 feet; and adults will probably sustain pelvic fracture in feet-first falls of more than 30 feet. The head-first fall simulations predicted skull acceleration responses that were higher than previously considered survivable, although five of the six child subjects whose falls were simulated did survive. Analysis of simulation results indicated that, for some children under 8 years old, peak accelerations of as high as 350 G, with a rate of onset of up to 200,000 G/sec and a duration of up to 3 msec (approximate HIC = 1700-2800), would produce only moderate, reversible injury (as defined by AIS 2). For this same age group, the survival limit for head impacts-the level at which most people would sustain AIS 5 or greater injuries-was estimated to be as high as 600 G peak acceleration and 300,000 G/sec for a duration of up to 3 msec (approximate HIC = up to 11,000).

Gadd, Charles W.; Culver, C. C.; Nahum, Alan M., "Study of Responses and Tolerances of the Neck", SAE Technical Paper 710856.

The principal objectives of this study were first to obtain experimental curves of angulation versus moment of resistance of the human neck in hyperextension and lateral flexion, and second to determine angular limits short of significant injury observable in the unembalmed subjects employed in the study. The first of the tests were of the "static" type with load applied over a period of approximately 1 s.

To determine the applicability of the data to dynamic conditions, tests were also made of the dissected neck at angulation velocities comparable with those of typical accidental injury. Overall resisting moment and injury threshold were similar under the dynamic loading, but somewhat greater moment of resistance was noted during the (earlier) portion of the loading cycle when angular velocity was greatest.

It is believed the data obtained together with muscular restraint data of other investigators who have used volunteers should be of value in the selection of neck characteristics for anthropometric test devices.

Tolerances of the unembalmed laryngeal cartilages to direct frontal impact in situ were also obtained.

Greenberg, S. W.; Gonzalez, D.; Gurdjian, E. S.; Thomas, L. M., "Changes In Physical Properties of Bone Between the In vivo, Freshly Dead, and Embalmed Conditions," SAE Technical Paper 680783.

A knowledge of the state of the physical properties of cadaver material is important if such material is utilized for impact studies. Experiments were designed to evaluate changes in elastic and strength properties of bone in the experimental animal in the course of its transition from live to recent death to embalmed conditions.

Results indicate less than 5% change from the live to the fresh condition. In progressing to the embalmed wet condition variable degrees of stiffening are observed averaging around 8%. Drying of embalmed bone further increases stiffness about 24% and remoistening reconstitutes some of the initial flexibility.

Hodgson, V. R.; Patrick, L. M., "Dynamic Response of the Human Cadaver Head Compared to a Simple Mathematical Model," SAE Technical Paper 680784.

It is shown that the response of the occiput of a cadaver to sinusoidal vibration input to the frontal bone corresponds closely to that of a simple damped spring-mass system having a natural frequency equal to the first mode of the skull, 0.17 damping factor. The first and third bending mode of the skull occurred near 300 and 900 Hz for both the cadaver preparation with silicon gel filled cranial cavity and the live human head. A second mode was found near 600 Hz in the live human. Head acceleration levels at which opposite pole pressure reached near -1 atm were 170 g and 500-600 g in the human cadaver and live monkey head, respectively, which values are roughly inversely proportional to major intracranial diameters. A method is derived for comparing the impact response of a simple system to a general shaped pulse to that of the cadaver head. Under certain impact conditions it is found that a simple model responds to cadaver force-time input within 5% of cadaver occiput response over a broad range of pulse durations and acceleration levels.

Hodgson, V. R.; Thomas, L. M.; Brinn, J., "Concussion Levels Determined by HPR Windshield Impacts," SAE Technical Paper 730970.

The purpose of this study was to increase our understanding of the head impact level that will produce concussion in humans. The technique employed was that of accident restaging.

The investigation reported here was composed of three parts:

1. The Cornell accident records were reexamined to establish the frequency of brain concussion as a function of windshield damage.

2. Tests were conducted with instrumented cadavers to determine the head accelerations achieved when the appropriate windshield damage levels were obtained.

3. Head injury indexes were calculated from the measured accelerations, and their predictions were compared to the Cornell field data.

The present reexamination of the Cornell accident data found that the percentage of victims who received a concussion involving known unconsciousness reduces to, at most, 11% for the case of radial crack with bulge. The percentage obtained for radial crack - no bulge was, at most, 2.8%. Thus, the former condition is approximately four times as likely to produce a concussion as the latter condition.

A number of head injury indexes were calculated from the cadaver head accelerations obtained in the current study. It would be expected that these should be somewhat greater for the bulged windshield condition than for the nonbulged condition. This was not found to be the case. The calculated criteria were either approximately equal or else greater for the no-bulge condition than the bulged condition. The reason for this disparity is not known.

Most injury indexes treated the accelerations produced by the cadaver/windshield impacts in a fundamentally different way than accelerations produced by equivalent dummy impacts. For the cadaver data, the indexes focused on the spike caused by the glass breakage, whereas for dummy data, the indexes focused on the broad pulse resulting from the interlayer bulging. This difference arose from the fact that the cadavers produced lower accelerations during the bulge event than did dummies.

Hodgson, V. R.; Thomas, L. M.; Prasad, P., "Testing the Validity and Limitations of the Severity Index," SAE Technical Paper 700901.

The head acceleration pulses obtained from monkey concussion, cadaver skull fracture ($t = 0.002$ sec) and football helmet experiments ($0.006 < t < 0.011$ sec) have been subjected to injury hazard assessment by the Severity Index method. Although not directly applicable, the method correlates well with degree of monkey concussion. The range of Severity Indices for acceleration pulses obtained during impact to nine cadavers, all of which produced a linear fracture, was 540-1760 (1000 is danger to life) with a median value of 910. The helmet experiments showed good correlation between the Severity Index and the Wayne State University tolerance curve. These helmet tests also showed that a kinematics chart with curves of velocity change, stopping distance, average head acceleration, and time, with superimposed Wayne State tolerance curve can be useful in injury assessment.

The Vienna mathematical model of Slattenschek and Tauffkirchen proposed as an alternative method of hazard assessment to the Severity Index was tested with the football helmet data. Displacement response of the model increases relative to severity of impact and is in good agreement with predictions by the authors.

It is shown that the Severity Index method does not apply to a metal head form.

Hubbard, Robert P.; McLeod, Donald G., "Definition and Development of a Crash Dummy Head," SAE Technical Paper 741193.

Crash test dummies serve as human surrogates in automotive crash simulations, and accelerations monitored in the heads of these dummies are used for assessment of human head injury hazard. For these acceleration response of the human head must be a part of dummy head design. This paper describes the conception, design and development of a crash test dummy head. Geometric, inertial, and performance requirements based on biomechanical information are presented and discussed. The head design concept is compatible with current head injury assessment procedures, and the configuration is based on the GM Research skull and head geometry models. The manufacture and development are described, and the test procedures and results are presented and discussed with reference to the biomechanical and functional requirements. The resulting dummy head is shown to comply with these requirements.13 Huelke, D. F.; Mendelsohn, R. A.; States, John D.; Melvin, J. W., "Cervical Fractures and Fracture-Dislocations Sustained Without Head Impact," SAE Technical Paper 790132.

Because of its flexibility and structure, the cervical spine is disposed to various mechanisms of injury: although not so common as injuries caused by head impacts, cervical fractures and/ or fracture-dislocations have been reported without direct impact to the head. Some cervical injuries reported have been sustained by wearers of lap and shoulder belts in auto accidents; however, we do not consider belt use a potential hazard because ample evidence has accrued in the medical and engineering literature to document general injury and fatality reduction by use of seatbelts. We believe that in many instances occupants would be more seriously injured or killed were belts not worn.

The present paper reviews reports of cervical injuries without head impact found in the literature and case histories of such injuries from the Highway Safety Research Institute of The University of Michigan, as well as experimental studies in animals, cadavers, and volunteer subjects.

Huelke, D. F.; Moffatt, E. A.; Mendelsohn, R. A.; Melvin, J. W., "Cervical Fractures and Fracture Dislocations - An Overview," SAE Technical Paper 790131.

In that the neck has a wide range of movements-flexion, extension, lateral bending and rotation, there is a large variety of types of neck fractures and fracture-dislocations.

This paper describes these various fractures and dislocations emphasizing the mechanisms as determined from clinical experience and potential, neurological damage.

Fractures and fracture-dislocations with and without spinal cord involvement have been extensively described in the medical literature. This paper will give a brief overview of some of the types of fractures, as well as the mechanisms involved in these injuries. For more detailed descriptions, the reader is encouraged to review the articles in the list of suggested readings found in this symposium proceedings.

Huelke, Donald F.; Smock, William S.; Peter M.; Nichols, George R., "Basilar Skull Fractures Produced by Facial Impacts Case Histories and a Review of the Literature," SAE Technical Paper 881711.

Basilar skull fractures have previously been described as the result of cranial vault impacts. Such fractures resulting from impacts to the chin, face and/or orbital areas (non-cranial vault areas) have not been adequately documented. These types of fractures and the motor vehicle incidents in which the injuries were sustained will be described according to the type of crash and vehicles involved. Data on the tolerance levels of the basilar region of the skull are sparce at best. Case histories of non-cranial vault impacts, presented in a variety of crash types, document that facial impacts can be the sole cause of basilar skull fractures. An historical review of the basilar skull fracture literature is presented.

Kallieris, D.; Mattern, R.; Miltner, E.; Schmidt, Gg.; Stein, K., "Considerations for a Neck Injury Criterion," SAE Technical Paper 912916.

A kinematic analysis of the head-neck unit has been conducted in 37 simulated traffic accidents in order to investigate correlations between neck response and injuries. Belted fresh human cadavers in the age range 18 to 74 years have been used as front and rear-seat passengers.

The analysed data included 23 frontal collisions, impact velocity 30 km/h, 50 km/h and 60 km/h, barrier impact and 14 90°-car to car lateral collisions with near-side passengers (6 cases) as well as far- side rear-seat passengers with an inboard upper anchoring point for the shoulder belt (8 cases).

The head bending angle depended on the type of the collision. At the frontal collision, the mean head bending maxima amounted 79°, the evaluated mean angular velocity maxima and angular acceleration maxima corresponded to 41 rad/s and 2208 rad/s_2, the mean maximum velocity in trajectory of the head was 10 m/s, the mean maximum acceleration along the path amounted 23 g.

With the same collision velocity of 50 km/h in the 90°-car to car lateral collision higher mean values of head bending angle maxima were observed in the near-side front passengers (62°) than in the far-side rear-seat passengers (57°). The evaluated angular velocity maxima and angular acceleration maxima amounted in the mean for the near-side passengers 43 rad/s and 2887 rad/s,, the mean maximum velocity in trajectory of the head amounted 8 m/s, the mean maximum acceleration along the path was 33 g; for the far-side passengers the mean angular velocity maxima of 26 rad/s were evaluated, for the angular acceleration it was 1345 rad/s_2, for the head velocity in trajectory 10 m/s and for the acceleration along the path 18 g.

Injuries of AIS 1 (hemorrhage = strain) already occurred at an angular acceleration of 560 rad/s_2

or a head acceleration of 13 g along the path, whereas the cervical spine may remain uninjured at an angular acceleration of still 2470 rad/s2 or head acceleration of 23 g along the path. On the other hand, an AIS 3 was observed at an angular acceleration of 1320 rad/s2 or a head acceleration of 22 g along the path, and an AIS 4 of 4910 rad/s_2 or a head acceleration of 39 g along the path.

Lee, Maw-Chang; Melvin, John W.; Ueno, Kazunari, "Finite Element Analysis of Traumatic Subdural Hematoma," SAE Technical Paper 872201.

A two-dimensional finite element model of the head of a rhesus monkey was built to simulate the head acceleration experiments done by Gennarelli and his colleagues. The purposes of the study were to better understand the mechanisms of traumatic subdural hematoma and to estimate its threshold of occurrence.

Lenox, John B.; Stalnaker, Richard L.; White, Curtis D.; Moore, Gary T.; Anderson, Orin M.; Schleicher, Ralph R.; Peel, H. Herbert; Martin, Susan S.; Driscoll, Gerald D.; Huntington, Howard W.; Carey, K. Dee; Haffner, Mark P.; Ommaya, Ayubk, "Development of Neck Injury Tolerance Criteria in Human Surrogates - Static Tensile Loading in the Baboon Neck - Preliminary Observations," SAE Technical Paper 826039.

Currently available biomechanical data are insufficient for determining the neck loads required to produce cervical injuries in automotive accidents. The purpose of this study was to initiate a series of experiments required for identifying injury mechanisms and to develop criteria suitable for representing the spectrum of significant functional and structural neck injuries that could be sustained by car crash occupants. The first test conducted was a static neck tens ion test performed on an anesthetized, 12. 2-kg baboon (Papio hamadryas). A state of muscle paralysis was maintained throughout neck loading to mimic the flaccid muscle tone present in fresh human cadavers. Tensile neck loading was increased at a rate of 20.2 N every five minutes until definitive structural failure (atlas-occipital subluxation) occurred at the 1170 N load level. Neurophysiological testing in the anesthetized animal demonstrated that cervical spinal cord function was seriously impaired at the 595 N load level, i.e., at 51 percent of the structural failure load. The potential implications of this finding are discussed, relative to the understanding of injury mechanisms. In addition, a strategy is presented for deriving functional neural tissue impairment thresholds, from fresh human cadaver data, that may apply in real accident victims.

Lestrelin, D.; Tarrière, C.; Walfisch, G.; Fayon, A.; Got, C.; Patel, A.; Hureau, J., "Proper Use of HIC Under Different Typical Collision Environments," SAE Technical Paper 826044.

Authors present a synthesis of data concerning the relationship between HIC values and head injury levels, obtained from various methods and under several particular test conditions. Then, frontal collisions are discussed. Two cases are separately considered; when head impacts occur, and without head or neck impacts. Accidentological data allow one to conclude that no criterion is presently needed when no head impact is recorded. On the contrary, when head impacts are undergone, biomechanical data and analyses of kinematics are in favour of HIC 1500 as far as 3-point belt is used.

The other classical types of accidents are mostly relevant to the case of direct head impacts; HIC 1500 criterion may apply provided that a sufficient pressure distribution exists - besides, the protection of the face is a particular problem.

The head injury criterion (HIC) is an already old concept which was often criticized for both its mathematical formula and its relationships with any injury level. However, in 1982, no alternative candidate criterion appears which was sufficiently known and directly available from measurements on present dummies.

This paper aims in particular to determine if meeting the requirements of a HIC value in a dummy test can provide a better safety to real accident victims, by protecting them from rather severe injuries when they undergo similar impact conditions. Various accident circumstances will be examined.

Melvin, J. W., "Human Neck Injury Tolerance", SAE Technical aper 790136.

The specification of information and specific data on the biomechanical quantities which describe the injury processes produced in the human neck are, at best, minimal. This paper discusses the problems associated with the topic and lists the quantitative information that does exist on the human neck injury tolerance.

Melvin, John W.; Fuller, Peter M.; Daniel, Roger P.; Pavliscak, George M., "Human Head and Knee Tolerance to Localized Impacts," SAE Technical Paper 690477.

The results of recent dynamic load measurements on human skull and patella bone, conducted with less-than-one-square-inch penetrators, are discussed in relation to previously reported skull impact data from larger contact areas. These medical data are compared to the dynamic response of a large variety of natural and synthetic plastic materials, for use in trauma-indicating head form and kneeform design. Several bodyform designs are proposed as research tools.

Mertz, H. J.; Weber, D. A., "Interpretations of the Impact Responses of a 3-Year-Old Child Dummy Relative to Child Injury Potential," SAE Technical Paper 826048.

An analysis is presented that was used to interpret the significance of response measurements made with a specially instrumented, 3-year-old child dummy that was used to evaluate child injury potential of the second-generation, passenger inflatable restraint system that was being developed by General Motors Corporation. Anesthetized animals and a specially instrumented child dummy, both 3-year-old child surrogates, were exposed to similar inflating-cushion, simulated collision environments. The exposure environments were chosen to produce a wide spectrum of animal injury types and severities, and a corresponding broad range of child dummy responses. For a given exposure environment, the animal injury severity ratings for the head, neck, thorax and abdomen are paired with dummy response values corresponding to these body regions. These data are used to develop relationships that can be used to predict the probability of an animal experiencing significant injuries to these body regions based on the child dummy response measurements. A rationale is developed for interpreting the predicted animal injury severities relative to child injury severities.

Mucciardi, Anthony N.; Sanders, John D.; Eppinger, Rolf H., "Prediction of Brain Injury Measures From Head Motion Parameters," SAE Technical Paper 770923.

An analysis of experimental head impact data was performed to demonstrate (1) that kinematic waveforms contain information relating to head and brain injuries, and (2) that analysis techniques exist which can properly exploit this information to create injury predictive functions.

An experimental data base consisting of 26 monkey head impacts was utilized. Translational and rotational acceleration time histories of the head were available. Parameters computed from these kinematic waveforms were the input variables to an analysis technique. The output, or modeled, variable was the experimentalist's evaluation of the severity of injuries.

The results of the analysis are presented and it is concluded that it is possible to accurately model head and brain injury assessments from strictly head motion parameters. Subsequent analysis of the derived predictive functions reveals interesting insight into the relationship between the kinematic parameters and head injury and suggests potential strategies for head injury mitigation.

Muzzy, William H.; Bittner, Alvah C. Jr.; Seemann, Marjorie R.; Willems, Gilbert C.; Lustick, Leonard S., "The Effect of Mass Distribution Parameters on Head/Neck Dynamic Response," SAE Technical Paper 861886.

The Naval Biodynamics Laboratory (NBDL) has

collected a database which describes human dynamic responses for -X acceleration exposures as a function of mass distribution variations of the head. Kinematic responses were measured on subjects with, (a) no mass addition; (b) with a helmet and weight-carrier mass addition; (c) and with the helmet and added weights symmetrically located with respect to the mid-sagittal plane of the head. The total mass addition to the head with the weights was approximately 30 percent. The helmet and weights were positioned with reference to the head anatomical coordinate system for each subject, with mass moments of inertia and variations in center of gravity then being determined. This paper compares responses both as a function of a mass distribution parameters and as a model to simulate the observed responses. The maximum resultant force at the occipital condyles was found to be proportional to the product of the nominal impact acceleration and the sum of the head and added mass.

Myers, Barry S.; McElhaney, James H.; Doherty, Brian J.; Paver, Jacqueline G.; Nightingale, Roger W.; Ladd, Thomas P.; Gray, Linda, "Responses of the Human Cervical Spin to Torsion," SAE Technical Paper 892437.

The passive torsional responses of the human cervical spine were investigated using unembalmed cervical spines in a dynamic test environment. Kinematic constraints were designed to simulate in vivo conditions. A physiologic axis of twist was determined based on a minimum energy hypothesis. Six-axis load cells completely described the resultant forces. Results include viscoelastic responses, moment- angle curves, and piece-wise linear stiffness. The Hybrid III ATD neckform was also tested, and its responses compared with the human. The Hybrid III neckform was stiffer than the human, was more rate sensitive than the human, and unlike the human, was relatively insensitive to the axis of twist. A rotational element to improve the biofidelity of the Hybrid III neckform in rotation was developed, and the results presented. In addition, this data was compared with volunteer sled tests to determine the contributions of the cervical musculature to the stabilization of the neck in rotation during lateral acceleration.

Nahum, Alan M. M.D.; Smith, Randall W. M.D., "An Experimental Model for Closed Head Impact Injury," SAE Technical Paper 760825.

A series of blunt head impacts has been performed on stationary unembalmed human cadavers. The specimens were prepared to simulate realistic fluid pressures within the cerebrospinal fluid space and cerebral blood vessels. Translational acceleration-time histories of the head were recorded by biaxial accelerometers attached to the skull. Peak resultant head accelerations in excess of 3,000 m/s_2 and pulse durations of 5 ms. or less were observed in a series of 10 experiments. The cerebral vascular system was perfused with a carbon particle tracer solution. Following impact, careful gross and microscopic pathologic studies of the cranial soft tissues were performed to assess vascular hemorrhage as represented by extravasation of tracer solution into the brain tissue. Data is presented describing the input forcing function, resultant head acceleration, and detailed necropsy findings. The results suggest that the use of this vascular trauma simulation technique may add additional information concerning acceptable acceleration-time limits for closed head injury.

Newman, James A., "Head Injury Criteria in Automotive Crash Testing," SAE Technical Paper 801317.

The origin and development of the Head Injury Criterion (HIC) has been reviewed. The concept of correlating internal head injury with measured kinematics has also been re-examined. It is concluded that since no general kinematic correlation can be developed, the HIC concept is invalid. Recent attempts to develop relationships between head injury AIS and HIC, are shown to substantiate this view. Alternatives to HIC are reviewed and other approaches to head injury assessment are proposed.

Nusholtz, Guy S.; Kaiker, Patricia S.; Muscott, Gail J.; Suggitt, Bryan R., "UMTRI Experimental Techniques in Head Injury Research," SAE Technical Paper 851244.

This paper discusses techniques developed and

used by the Biosciences Group at the University of Michigan Transportation Research Institute (UMTRI) for measuring three-dimensional head motion, skull bone strain, epidural pressure, and internal brain motion of repressurized cadavers and Rhesus monkeys during head impact. In the experimental design, a stationary test subject is struck by a guided moving impactor of 10 kg (monkeys) and 25 or 65 kg (cadavers). The impactor striking surface is fitted with padding to vary the contact force-time characteristics. The experimental technique uses a nine-accelerometer system rigidly affixed to the skull to measure head motion, transducers placed at specific points below the skull to record epidural pressure, repressurization of both the vascular and cerebro-spinal systems, and high-speed cineradiography (at 1000 frames per second) of radiopaque targets. The results of several experimental tests imply intractable problems in determining potential injury criteria, such as the HIC, from accelerometer data.

Nusholtz, G. S.; Melvin, J. W.; Huelke, D. F.; Alem, N. M.; Blank, J. G., "Response of the Cervical Spine to Superior-Inferior Head Impact," SAE Technical Paper 811005.

A test series using 12 unembalmed cadavers was conducted to investigate factors affecting the creation of cervical spine damage due to impact to the top of the head. The test subjects were instrumented to measure head, T8 thoracic spine, and sternum acceleration responses. Photographic targets on the head and torso allowed analysis of impact motions from high-speed movies.

The stationary test subject was struck by a guided, moving impactor mass of 56 Kg at 4.6-5.6 m/s. The impactor striking surface consisted of a biaxial load cell with padding to vary the contact force-time characteristics of the head/impactor. The orientation of the head, cervical spine, and torso was adjusted relative to the impactor axis to investigate the effect of spinal configurations on the damage patterns. Load and acceleration data are presented as functions of time and as functions of frequency in the form of mechanical impedance.

Damage to the cervical spine was produced in all but one test, including fractures of the spinous processes, laminae, transverse processes, and the bodies of the vertebrae as well as ruptured discs and torn ligaments. Both anterior and posterior damage was produced and the sites of the damage ranged from C2 to T4. The peak forces produced during the impacts ranged from 1.8 kN to 11.1 kN.

The limited response data of this pilot study do not allow any specific conclusions with regard to cervical spine tolerance levels. However, it does attest to the influence of spinal configuration and impact conditions on both response and damage of the spine due to crown impact.

Pintar, Frank A.; Sances, Anthony Jr.; Yoganandan, Narayan; Reinartz, John; Maiman, Dennis J.; Suh, Jung Keun; Unger, George; Cusick, Joseph F.; Larson, Sanford J., "Biodynamics of the Total Human Cadaveric Cervical Spine," SAE Technical Paper 902309.

Spinal trauma produced from motor vehicle accidents, diving accidents, or falls occur at high rates of loading. This study was undertaken to reproduce clinically relevant cervical spine injuries under controlled conditions. Six isolated head - T2 human cadaveric preparations were tested using an electrohydraulic piston actuator at loading rates from 295 to 813 cm/sec. The hybrid III head-neck was tested similarly at rates from 401 to 683 cm/sec. The input forces for specimen tests were of higher magnitude and shorter duration than the distally measured forces. In contrast, the hybrid III head-neck revealed similar magnitude and duration force traces from input to output. The specimen preparations were analyzed kinematically at 1200 frames/sec with 20 to 30 retroref lective targets fixed to each level of the cervical spine. With this technique it is possible to temporally follow cervical damage as a function of applied force. Wedge, burst, and anterior compression fractures were documented using CT and cryosectioning techniques. This pathology can be evaluated quantitatively using the target analysis. For example, a burst fracture occurred suddenly and within 2.5 msec, whereas wedge fractures were progressive and took 4 to 5 msec to occur.

Portnoy, Harold D.; Benjamin, Durand; Brian, Michael; McCoy, Lowell E.; Pince, Bruce W.; Edgerton, Robert; Young, Joseph, "Intracranial Pressure and Head Acceleration During Whiplash," SAE Technical Paper 700900.

Baboons were whiplashed to measure intracranial pressure (ICP) in whiplash, to correlate ICP to linear and angular head acceleration, and to determine if head restraint configuration altered ICP. Clinical, biochemical, physiological, and pathological observations were made.

Intracranial pressure increased in whiplashed baboons, and is temporally related to translational and rotational head accelerations. With different head restraint configurations, the pressure increases are smallest when using a vertical restraint minimizing head movement, and are not reduced by partial restraints. There may be some correlation between EEG, clinical, and gross pathologic observations, and the use of the various restraints, but EKG, heart rate, and biochemical measurements were not significant.

Prasad, Priya; Daniel, Roger P., "A Biomechanical Analysis of Head, Neck, and Torso Injuries to Child Surrogates Due to Sudden Torso Acceleration," SAE Technical Paper 841656.

This paper reports on the injuries to the head, neck and thorax of fifteen child surrogates subjected to varying levels of sudden acceleration. Measured response data in the child surrogate tests and in matched tests with a three-year-old child test dummy are compared to the observed child surrogates injury levels to develop preliminary tolerance data for the child surrogate. The data are compared with already published data in the literature.

Prasad, P.; Mital, N.; King. A. I.; Patrick, L. M., "Dynamic Response of the Spine During +Gx Acceleration," SAE Technical Paper 751172.

A review of the existing mathematical models of a car occupant in a rear-end crash reveals that existing models inadequately describe the kinematics of the occupant and cannot demonstrate the injury mechanisms involved. Most models concentrate on head and neck motion and have neglected to study the interaction of the occupant with the seat back, seat cushion, and restraint systems. Major deficiencies are the inability to simulate the torso sliding up the seat back and the absence of the thoracic and lumbar spine as deformable, load transmitting members.

The paper shows the results of a 78 degree-of-freedom model of the spine, head, and pelvis which has already been validated in +Gz and -Gx acceleration directions. It considers automotive-type restraint systems, seat back, and seat cushions, and the torso is free to slide up the seat back. Results of simulation runs show a good correlation with the existing data and experimental cadaveric runs made with a high seat back (or head restraint) and low seat back. The effect of seat-back stiffness and the restraint system on the response of the head and spine is studied.

Sances, A. Jr.; Myklebust, J.; Cusick, J. F.; Weber, R.; Houterman, C.; Larson, S. J.; Walsh, P.; Chilbert, M.; Prieto, T.; Zyvoloski, Mark; Ewing, C.; Thomas, D., "Experimental Studies of Brain and Neck Injury," SAE Technical Paper 811032.

Static and dynamic axial tension loads were applied to the intact and isolated cervical column of the monkey and human cadaver. Radioactive microspheres were used to evaluate brain and spinal cord perfusion in the monkey. To determine neural pathway damage, somatosensory evoked potentials were recorded with stimulation of the spinal cord, and in spinal cord with stimulation of sensorimotor cortex, and in spinal cord with stimulation of cauda equina. The evoked potential amplitude decreased prior to heart rate and blood pressure changes presumably due to brainstem distention. The preliminary studies show, 1) the brain and spinal cord were well perfused as measured with the microspheres when the evoked potentials decreased, 2) The cervical isolated cadaveric monkey spinal column ligaments failed statically at approximately 1/2 to 1/3 the force required for dynamic disruption, 3)

In the intact monkey, the cervical ligaments failed statically at approximately 1/2 the dynamic failure force, 4) the isolated human cervical ligaments failed at loads approximately three times those observed in the isolated monkey cervical column.

Schmidt, G.; Kallieris, D.; Barz, J.; Mattern, R.; Klaiber, J., "Neck and Thorax Tolerance Levels of Belt-Protected Occupants In Head-On Collisions," SAE Technical Paper 751149.

Our tests with fresh human cadavers were continued (cf. Proceedings, 18th Stapp Car Crash Conference). Presented herewith are the results of 103 tests evaluated so far.

While the severity of injuries showed an increase with advancing age, it is not obviously dependent upon weight or sex. Under the conditions chosen by us, the 12 to 30-year age group reached the thorax tolerance level at an impact velocity of still below 50 km/h with a stopping distance of 40 cm, the 30 to 50-year age group of below 40 km/h, and the age group beyond 50 years of below 30 km/h.

A comparison of our results with volunteer tests and with evaluated real accidents as well as with similar cadavar tests is made. When introducing safety testing regulations for vehicles in the light of dummy tests, the broad spectrum of the respective age groups has to be considered. The thorax injuries may be slightly mitigated due to a lessening of the surface pressure. The injuries of the neck and thoracic vertebra, on the other hand are apparently not dependent on this factor and to a varying extent, are always occurring at an impact velocity of 50 km/h. Here too, the severity of the injuries increases according to the age.

Therefore, it was set great store by the recording of vertebral injuries. By employing careful dissection and preparatory-techniques, disc injuries, hemorrhages in the region of muscles and intervertebral joints were found in the majority of the cases, where as less frequently fractures of vertebral bodies, vertebral arcs and fractures of the processus spinosus and transversus, respectively were noticed. Highly dangerous vertebral injuries involving contusion or rupture of the cervical or thoracic spinal cord occurred only rarely. The Abbreviated ACIR scale proved to be sufficient for the graduation of the injuries. The grades ranging from 0 to 4 may well be compared with the AIS scale with grades from 0 to 9 (AIS 6 - 9 = fatal is equivalent to ACIR 4). States and Huelke suggest the simplification and reduction of the AIS scale. They recommend a condensation of figures 6 - 9, thereby approaching the ACIR scale. By measuring the bone hardness and bending strength of the ribs, the skeleton injuries observed in the tests were related to the age-dependent resistance capability.

In concluding, reports will be rendered on the photographic ebaluation of the phases of movement recorded laterally by a high-speed camera, and on the legal basis of cadaver tests.

Stalnaker, R. L.; Roberts, V. L.; McElhaney, J. H., "Side Impact Tolerance to Blunt Trauma," SAE Technical Paper 730979.

The object of this research program has been to extend the scope of earlier work to include long-duration head impacts and to develop new scaling relationships to allow extrapolation of impact data from infrahuman primates to living humans.

A series of living primate side impacts to the head and torso was conducted in parallel with a series of impacts to human cadavers. Dimensional analysis techniques were employed to estimate in vivo human tolerance to side injury.

The threshold of closed brain injury to humans was found to be 76 g for a pulse duration of 20 ms and an impact velocity of 43 ft/s (13.2 m/s). The maximum tolerable penetration to the chest was found to be 2.65 in (6.72 cm) for both the left and right sides.

Scaling of abdominal injuries to humans was accomplished by employing a factor that relates impact contact area, animal mass, impact force, and pulse duration to injury severity. The maximum tolerable contact pressure to the upper abdomen of a human was found to be 32 lbf/in_2 (220 kPa).

Thomas, L. M.; Sezgin, Y.; Hodgson, V. R.; Cheng, L. K.; Gurdjian, E. S., "Static Deformation and Volume Changes in the Human Skull," SAE Technical Paper 680782.

Three human male cadaver heads were statically loaded along anteroposterior, posterioanterior, side to side, and vertex to base lines of action, while simultaneously measuring skull deflections at four or five locations and intracranial volume changes. Volume changes due to loading along the long (A-P) axis were small and either increased or decreased, while loads transverse to the A-P axis decreased the volume. Transverse loads produced volume changes on the order of 10 times larger than those due to A-P forces. Two skulls loaded to fracture in the A-P direction, failed at 1150 and 2200 lb, respectively, into the right orbit. These magnitudes and linear fracture direction correspond to four fractures produced by impact to the frontal bone of intact cadavers in previous work.

Walker, Leon B. Jr.; Harris, Edward H.; Pontius, Uwe R., "Mass, Volume, Center of Mass, and Mass Moment of Inertia of Head and Head and Neck of Human Body," SAE Technical Paper 730985.

The mass, volume, center of mass, and mass moment of inertia of the head and the head and neck were determined for 20 human male cadavers. Anthropometric values and anatomic landmarks were obtained by external measurements and by use of x-ray procedures. The procedures used to determine the above measurements are described. Uniform planes for the separation of the head and neck from the torso and separation of the head from the neck were established and are described in detail.

The values of the physical properties of the head and neck and the head are tabulated and compared to data reported in previous studies.

Welbourne, E. R.; Ramet, M.; Zarebski, M., "A Comparison of Human Facial Fracture Tolerance With the Performance of a Surrogate Test Device," SAE Technical Paper 896071.

A series of tests was conducted to provide information on the fracture tolerance of the bones of the face under impact conditions simulating those occurring in motor vehicle collisions. The results of that series were compared with the results of tests conducted under similar conditions on a Hybrid III dummy equipped with a frangible face.

The faces of eight cadavers were impacted at the subnasal maxilla and at the nasion with a horizontal steel bar of 25 mm diameter. The energy levels used were intended to cause fracture in about half of the impacts at each site. Similar numbers of tests were performed at each of the two equivalent sites on the frangible face. The sequence of impact energies was intended to define the fracture threshold at each of the equivalent sites as precisely as possible. The data from both series of tests were analysed assuming that the probability of fracture could be described by a Weibull distribution.

The cadaver tests for the maxilla resulted in only one significant fracture; no inference could be made about the probability of fracture at a given impact energy level within the range of the data. For the nasion, the cadaver data implied the existence of a plausible distribution of fracture probability as a function of impact energy. Using maximum force as the variate instead of impact energy produced essentially similar results. The data from the frangible face showed well defined thresholds at both impact sites. Although the cadaver data for the maxilla were somewhat inconclusive, it seems probable that the frangible face provides a quite conservative indication of the strength of the human face at the maxilla. For the nasion, the fracture threshold of the frangible face coincides approximately with the 40th percentile of the distribution of the cadaver tolerance data.

Wismans, J.; Spenny, C. H., "Head-Neck Response in Frontal Flexion," SAE Technical Paper 841666.

At the 27th Stapp Conference an analysis was presented of human head-neck motion in lateral flexion. Based on this analysis performance requirements for mechanical necks in this type of impact were formulated.

This study extends the analysis to head-neck response in frontal flexion. Results will be presented of dynamical tests with human subjects conducted by the Naval Biodynamics Laboratory (NBDL) in New Orleans. Two of these subjects were also included in the analysis of lateral flexion. It will be shown that the mechanical (mathematical) system with two ball and socket joints describing the head-neck response in lateral flexion is also suitable for forward flexion. Geometrical parameters are identical for both impact directions in contrast to the dynamical properties which show significant differences.

By this mechanical analog the observed head-neck motions are completely defined. Results will be compared with earlier performance requirements proposed by Mertz et al.

Evaluations of the Part 572 and Hybrid III head-neck systems show that both necks are stiffer than any of the volunteers in the impact ranges tested.

Wismans, J.; Spenny, C. H., "Performance Requirements for Mechanical Necks in Lateral Flexion," SAE Technical Paper 831613.

Realistic simulation of the neck response in a dummy is of vital importance to obtain a human-like dynamical behavior of the head. Trajectories of the head and the nature of head contact with vehicle interior or exterior are critically dependent on the dummy's neck design.

Neck performance criteria in literature are limited to the neck response in forward flexion and extension. Recent research programs to develop dummies with omnidirectional biofidelity clearly show a need for additional requirements in lateral and oblique directions. In this study, dynamic lateral flexion tests with human volunteers conducted by the Naval Biodynamics Laboratory (NBDL) in New Orleans are analysed.

It follows that the observed head neck motions in this type of impact quite well can be represented by a system with three degrees of freedom: a head and neck rotation in the plane of impact and a head torsion about the head anatomical z-axis. Mechanical and geometrical properties of this system resulting from tests with 6 different subjects under various test conditions are presented.

Based on this analysis, a simple test set-up is proposed, derived from the most severe NBDL test conditions, for the evaluation of existing or future mechanical head-neck designs. Requirements for the head motion in this specific test set-up are presented.

Appendix 2: Related Reading

RELATED READING

Appendix 2

This Appendix is a collection of papers suggested for related reading by the individuals who assisted with the development of PT-43. Due to space constraints, abstracts for these papers were not included.

Most of these papers have been published by SAE and are available in original or photocopy form. For ordering information, contact SAE's Customer Sales and Satisfaction Department, SAE International, 400 Commonwealth Drive, Warrendale, PA 15096-0001, USA, Telephone: 412/776-4970, Fax: 412/776-0790. Copies of papers referenced but not published by SAE must be obtained directly from the publisher listed.

Advani, Sunder H.; Owings, Raymond P., "Evaluation of Head Injury Criteria", SAE Technical Paper 740083.

Aldman, B.; Ljung, C.; Thorngren, L., "Intercranial Deformation Patterns Due to Impulsive Loading - A Model Study", SAE Technical Paper 826037.

Chan, Han Sun, "Mathematical Model for Closed Head Impact", SAE Technical Paper 741191.

Clarke, Thomas D.; Smedley, David C.; Muzzy, William H.; Gragg, C. D.; Schmidt, Robert E.; Trout, Edwin M., "Impact Tolerance and Resulting Injury Patterns in the Baboon: Air Force Shoulder Harness-Lap Belt Restraint", SAE Technical Paper 720974.

Douglass, John M.; Nahum, Alan M.; Roberts, Sanford B., "Applications of Experimental Head Injury Research", SAE Technical Paper 680786.

Ewing, C. L.; Thomas, D. J.; Lustik, L.; Muzzy, W. H. III; Willems, G. C.; Majewski, P., "Dynamic Response of the Human Head and Neck to +Gy Impact Acceleration", SAE Technical Paper 770928.

Ewing, Channing L.; Thomas, Daniel J.; Patrick, L. M.; Beeler, G. W.; Smith, M. J., "Living Human Dynamic Response to -Gx Impact Acceleration - Accelerations Measured on Head and Neck", SAE Technical Paper 690817.

Ewing, C. L.; Thomas, D. J., "Torque Versus Angular Displacement Response of Human Head to -Gx Impact Acceleration", SAE Technical Paper 730976.

Ewing, C. L.; Thomas, D. J.; Lustick, L.; Muzzy, W. H. III; Willems, G.; Majewski, P. L., "The Effect of Duration, Rate of Onset and Peak Sled Acceleration on the Dynamic Response of the Human Head and Neck", SAE Technical Paper 760800.

Fan, William R. S., "The Role of Skull Fractures in Short Duration Head Impacts", SAE Technical Paper 870321.

Haut, Roger C.; Gadd, Charles W.; Madeira, Richard G., "Nonlinear Viscoelastic Model for Head Impact Injury Hazard", SAE Technical Paper 720963.

Hu, Anthony S.; Bean, Stewart P.; Zimmerman, Roger M., "Response of Belted Dummy and Cadaver to Rear Impact", SAE Technical Paper 770929.

Huelke, D. F., "Anatomy of the Human Cervical Spine and Associated Structures", SAE Technical Paper 790130.

Kallieris, Dimitrios; Schmidt, Georg, "Neck Response and Injury Assessment Using Cadavers and the US-SID for Far-Side Lateral Impacts of Rear Seat Occupants with Inboard-Anchored Shoulder Belts", SAE Technical Paper 902313.

MacLennon, Carol; Ommaya, Ayub K., "Head Injury and Outcome: A Critical Review of Methodology", SAE Technical Paper 860425.

Melvin, John W.; McElhaney, J. H.; Roberts, V. L., Development of a Mechanical Model of the Human Head - Determination of Tissue Properties and Synthetic Substitute Materials", SAE Technical Paper 700903.

Mertz, H. J.; Driscoll, G. D.; Lenox, J. B.; Nyquist, G. W.; Weber, D. A., "Responses of Animals Exposed to Deployment of Various Passenger Inflatable Restraint System Concepts for a Variety of Collision Severities and Animal Positions", SAE Technical Paper 826047.

Ommaya, A. K.; Hirsch, Arthur E.; Martinez, John L., "Role of "Whiplash" in Cerebral Concussion", SAE Technical Paper 660804.

Pintar, Frank; Sances, Anthony Jr.; Yoganandan, Narayan; Reinartz, John M.; Larson, Sanford J.; Kurakami, Chikaharu; Rauschning, W., "Injury Biomechanics of the Head-Neck Complex", SAE Technical Paper 896075.

Reynolds, Herbert M.; Clauser, Charles E.; McConville, John; Chandler, Richard; Young, J. W., "Mass Distribution Properties of the Male Cadaver", SAE Technical Paper 750424.

Sances, A. Jr.; Myklebust, J.; Kostreva, D.; Cusick, J. F.; Weber, R.; Houtewrman, C.; Larson, S. J.; Maiman, D.; Walsh, P.; Chilbert, M.; Unterharnscheidt, F.; Ewing, C., Thomas, D.; Siegesmund, K.; Ho, K.; Saltzberg, B., "Pathophysiology of Cervical Injuries", SAE Technical Paper 821153.

Sances, A. Jr.; Weber, R.; Myklebust, J.; Cusick, J.; Larson, S.; Walsh, P.; Christoffel, T.; Houterman, C.; Ewing, C.; Thomas, D.; Saltzberg, B., "The Evoked Potential: An Experimental Method for Biomechanical Analysis of Brain and Spinal Injury", SAE Technical Paper 801302.

Seemann, Marjorie R.; Muzzy,. William H.; Lustick, Leonard S., "Comparison of Human and Hybrid III Head and Neck Dynamic Response", SAE Technical Paper 861892.

Slattenschek, A.; Tauffkirchen, W., "Critical Evaluation of Assessment Methods for Head Impact Applied in Appraisal of Brain Injury Hazards, in Particular in Head Impact on Windshields", SAE Technical Paper 700426.

Smith, Randall W., "The Response of Unembalmed Cadaveric and Living Cerebral Vessels to Graded Injury - A Pilot Study", SAE Technical Paper 791021.

States, John D.; Balcerak, John C.; Williams, James S.; Morris, Alexander T.; Babcock, William; Polvino, Robert; Riger, Paul; Dawley, Raymond E., "Injury Frequency and Head Restraint Effectiveness in Rear-End Impact Accidents", SAE Technical Paper 720967.

Tong, P.; Eppinger, R.; Marcus, J.; Galbraith, C., "Finite Element Modeling of Head Injury Caused by Inertial Loading", SAE Technical Paper 896073.

Trosseille, X.; Chambouard, F.; Tarrière, C., "Reconsideration of the HIC, Taking Into Account the Skull Bone Condition Factor (SBCF) - Limit of Head Tolerance in Side Impacts", SAE Technical Paper 881710.

Versace, John, "A Review of the Severity Index", SAE Technical Paper 710881.

Viano, David C.; Gadd, Charles W., "Significance of Rate of Onset in Impact Injury Evaluation", SAE Technical Paper 751169.

Welbourne, E. R.; Ramet, M.; Zarebski, M., "A Comparison of Human Facial Fracture Tolerance with the Performance of a Surrogate Test Device", SAE Technical Paper 896071.

INDEX